Handbook of Conducting Polymers
Third Edition

CONJUGATED POLYMERS

THEORY, SYNTHESIS, PROPERTIES, AND CHARACTERIZATION

WITHDRAWN

Handbook of Conducting Polymers
Third Edition

CONJUGATED POLYMERS

THEORY, SYNTHESIS, PROPERTIES, AND CHARACTERIZATION

Edited by

Terje A. Skotheim and John R. Reynolds

CRC Press
Taylor & Francis Group
Boca Raton   London   New York

CRC Press is an imprint of the
Taylor & Francis Group, an informa business

Cover image by Ellen Skotheim: A collage, based on representations of different phases of intercalated, or doped, conjugated polymers.

CRC Press
Taylor & Francis Group
6000 Broken Sound Parkway NW, Suite 300
Boca Raton, FL 33487-2742

© 2007 by Taylor & Francis Group, LLC
CRC Press is an imprint of Taylor & Francis Group, an Informa business

No claim to original U.S. Government works
Printed in the United States of America on acid-free paper
10 9 8 7 6 5 4 3 2 1

International Standard Book Number-10: 1-4200-4358-7 (Hardcover)
International Standard Book Number-13: 978-1-4200-4358-7 (Hardcover)

Library of Congress Cataloging-in-Publication Data

Conjugated polymers : theory, synthesis, properties, and characterization / [edited by] Terje A.
    Skotheim, John R. Reynolds.
        p. cm.
    Rev. ed. of: Handbook of conducting polymers. 2nd ed., rev. and expanded. c1998.
    Includes bibliographical references and index.
    ISBN-13: 978-1-4200-4358-7
    ISBN-10: 1-4200-4358-7
    1. Conjugated polymers. 2. Conducting polymers. 3. Organic conductors. I. Skotheim, Terje A.,
1949- II. Reynolds, John R., 1956- III. Handbook of conducting polymers.

    QD382.C66H36 2006
    547'.70457--dc22   2006032904

Visit the Taylor & Francis Web site at
http://www.taylorandfrancis.com

and the CRC Press Web site at
http://www.crcpress.com

547.70457 SKO

Dedication

This book is dedicated to our spouses, Ellen Skotheim and Dianne Reynolds. Without their understanding and support, we would never have completed this project.

Preface to Third Edition

The field of conjugated, electrically conducting, and electroactive polymers continues to grow. Since the publication of the second edition of the *Handbook of Conducting Polymers* in 1998, we have witnessed broad advances with significant developments in both fundamental understanding and applications, some of which are already reaching the marketplace.

It was particularly rewarding to see that in 2000, the Nobel Prize in chemistry was awarded to Alan Heeger, Alan MacDiarmid, and Hideki Shirakawa, recognizing their pathbreaking discovery of high conductivity in polyacetylene in 1977. This capstone to the field was celebrated by all of us as the entire community has participated in turning their initial discovery into the important field that it now is, almost 30 years later. The vast portfolio of new polymer structures with unique and tailored properties and the wide range of applications being pursued are far beyond what we could have envisioned when the field was in its infancy.

It was developments in polymer synthesis that led to free-standing polyacetylene films and the discovery of conductivity in polymers. The synthesis of π-conjugated chains is central to the science and technology of conducting polymers and is featured in this edition. Examining the synthetic advances across the board, one is struck by refined and careful syntheses that have yielded polymers with well-controlled and well-understood structures. Among other things, it has led to materials that are highly processable using industrially relevant techniques. In aspects of processing, spin coating, layer-by-layer assembly, fiber spinning, and the application of printing technology have all had a big impact during the last 10 years.

Throughout the Handbook, we notice that structure–property relationships are now understood and have been developed for many of the polymers. These properties span the redox, interfacial, electrical, and optical phenomena that are unique to this class of materials.

During the last 10 years, we have witnessed fascinating developments of a wide range of commercial applications, in particular, in optoelectronic devices. Importantly, a number of polymers and compositions have been made available by the producers for product development. This has helped to drive the applications developments to marketable products.

While conductivity, nonlinear optics, and light emission continue to be important properties for investigation and have undergone significant developments as discussed throughout the Handbook, the advances in semiconducting electronics, memory materials, photovoltaics (solar cells), and applications directed to biomedicine are emerging as future growth areas.

As we have assembled this edition, it has become clear that the field has reached a new level of maturity. Nevertheless, with the vast repertoire of synthetic chemistry at our disposal to create new structures with new, and perhaps unpredictable properties, we can expect exciting discoveries to continue in this dynamic field.

Terje A. Skotheim
Tucson, Arizona

John R. Reynolds
Gainesville, Florida

Editors

Terje A. Skotheim is the founder and chief executive of Intex, a Tucson, Arizona technology company. Dr. Skotheim is an experienced developer of several technologies, a seasoned executive, and a successful founder of several startup companies in the United States, Norway, and Russia. His research interests over more than 25 years span several disciplines in materials science and applications, including electroactive and conjugated polymers, molecular electronic materials, solid-state ion conductors, new electronic nanoamorphous carbon- and diamond-like carbon materials, and thin-film and surface science. He has pursued a wide range of technology applications of advanced materials in OLEDs, biosensors, lithium batteries, photovoltaic cells, and MEMS devices. He has held research positions in France, Sweden, and Norway in addition to the United States and was head of the conducting polymer group at Brookhaven National Laboratory before launching his career as an entrepreneur.

Skotheim received his B.S. in physics from the Massachusetts Institute Technology and Ph.D. in physics from the University of California at Berkeley (1979). He is the editor/coeditor of the *Handbook of Conducting Polymers* (first and second editions, Marcel Dekker) and *Electroresponsive Molecular and Polymeric Systems* (Marcel Dekker), the author of more than 300 publications and more than 70 patents. He can be reached at terje.skotheim@intexworld.com.

John R. Reynolds is a professor of chemistry at the University of Florida with expertise in polymer chemistry. He serves as an associate director for the Center for Macromolecular Science and Engineering. His research interests have involved electrically conducting and electroactive-conjugated polymers for over 25 years with work focused on the development of new polymers by manipulating their fundamental organic structure in order to control their optoelectronic and redox properties. His group has been heavily involved in the areas developing new polyheterocycles, visible, and infrared (IR) light electrochromism, along with light emission from polymer and composite light-emitting diodes (LEDs) (both visible and near-IR) and light-emitting electrochemical cells (LECs). Further work is directed to using organic polymers and oligomers in photovoltaic cells.

Reynolds obtained his M.S. (1982) and Ph.D. (1984) degrees in polymer science and engineering from the University of Massachusetts. He has published more than 200 peer-reviewed scientific papers and served as coeditor of the *Handbook of Conducting Polymers* that was published in 1998. He can be reached by e-mail at reynolds@chem.ufl.edu or see http://www.chem.ufl.edu/~reynolds/.

Contributors

Kazuo Akagi
Department of Polymer Chemistry
Graduate School of Engineering
Kyoto University
Kyoto, Japan

A.N. Aleshin
School of Physics and Nano Systems Institute—
  National Core Research Center
Seoul National University
Seoul, Korea and A.F. Ioffe Physical—Technical
  Institute
Russian Academy of Sciences
St. Petersburg, Russia

P. Audebert
Laboratory Photophysique & Photochimie
  Supramoléculaires et Macromoléculaires
Ecole Normale Supérieure de Cachan
Cachan, France

David Beljonne
Laboratory for Chemistry of Novel Materials and
  Center for Research in Molecular Photonics
  and Electronics
University of Mons-Hainaut
Mons, Belgium

and

School of Chemistry and Biochemistry
  and Center for Organic Photonics and
  Electronics
Georgia Institute of Technology
Atlanta, Georgia

Philippe Blanchard
Groupe Systèmes Conjugués Linéaires
Laboratoire CIMMA, UMR CNRS 6200
Université d'Angers
Angers, France

Jean-Luc Brédas
School of Chemistry and Biochemistry
  and Center for Organic Photonics
  and Electronics
Georgia Institute of Technology
Atlanta, Georgia

and

  Laboratory for Chemistry of Novel Materials
  and Center for Research in Molecular
  Photonics and Electronics
University of Mons-Hainaut
Mons, Belgium

Thomas M. Brown
FREEnergy Laboratory
Department of Electronic Engineering
University of Rome–Tor Vergata
Rome, Italy

Uwe H.F. Bunz
School of Chemistry and Biochemistry
Georgia Institute of Technology
Atlanta, Georgia

Franco Cacialli
Department of Physics and Astronomy and
  London Centre for Nanotechnology
University College London
London, United Kingdom

Seung Hyun Cho
Polymer Technology Institute
Sungkyunkwan University
Kyunggi-do, Korea

and

Department of Organic Materials Engineering
Sungkyunkwan University
Kyunggi-do, Korea

Jérôme Cornil
Laboratory for Chemistry of Novel Materials
  and Center for Research in Molecular Photonics
  and Electronics
University of Mons-Hainaut
Belgium

and

School of Chemistry and Biochemistry and Center
  for Organic Photonics and
  Electronics
Georgia Institute of Technology
Atlanta, Georgia

Veaceslav Coropceanu
School of Chemistry and Biochemistry and Center
  for Organic Photonics and Electronics, Georgia
  Institute of Technology
Atlanta, Georgia

X. Crispin
Department of Science and Technology (ITN)
Linköping University
Norrköping, Sweden

Aubrey L. Dyer
The George and Josephine Butler Polymer
  Research Laboratories
Department of Chemistry and Center for
  Macromolecular Science and Engineering
University of Florida
Gainesville, Florida

Arthur J. Epstein
The Ohio State University
Columbus, Ohio

M. Fahlman
Department of Science and Technology (ITN)
Linköping University
Norrköping, Sweden

Pierre Frère
Groupe Systèmes Conjugués Linéaires
Laboratoire CIMMA, UMR CNRS 6200
Université d'Angers
Angers, France

R. Friedlein
Department of Physics (IFM)
Linköping University
Linköping, Sweden

Victor Geskin
Laboratory for Chemistry of Novel Materials and
  Center for Research in Molecular Photonics
  and Electronics
University of Mons-Hainaut
Mons, Belgium

G. Greczynski
Department of Physics (IFM)
Linköping University
Linköping, Sweden

Andrew C. Grimsdale
School of Chemistry
Bio21 Institute
University of Melbourne
Parkville, Australia

Andrew B. Holmes
School of Chemistry
Bio21 Institute
University of Melbourne
Parkville, Australia

Jiaxing Huang
Miller Institute for Basic Research
  in Science
University of California
Berkeley, California
and Department of Chemistry
University of California
Los Angeles, California

Malika Jeffries-El
Department of Chemistry
Iowa State University
Ames, Iowa

and

Department of Chemistry
Carnegie Mellon University
Pittsburgh, Pennsylvania

M.P. de Jong
Department of Physics (IFM)
Linköping University
Linköping, Sweden

Richard B. Kaner
Department of Chemistry and Biochemistry
 and California NanoSystems Institute
University of California
Los Angeles, California

Stephan Kirchmeyer
H.C. Starck GmbH & Co. KG
Leverkusen, Germany

O. Korovyanko
Chemistry Division
Argonne National Laboratory
Argonne, Illinois
and
Department of Physics
University of Utah
Salt Lake City, Utah

Roberto Lazzaroni
Laboratory for Chemistry of Novel Materials
 and Center for Research in Molecular Photonics
 and Electronics
University of Mons-Hainaut
Mons, Belgium
and
School of Chemistry and
 Biochemistry and Center for Organic Photonics
 and Electronics
Georgia Institute of Technology
Atlanta, Georgia

Philippe Leclère
Laboratory for Chemistry of Novel
 Materials and Center for Research in
 Molecular Photonics and Electronics
University of Mons-Hainaut
Mons, Belgium

Jun Young Lee
Polymer Technology Institute
Sungkyunkwan University
Kyunggi-do, Korea
and
Department of Organic Materials Engineering
Sungkyunkwan University
Kyunggi-do, Korea

Philippe Leriche
Groupe Systèmes Conjugués Linéaires
Laboratoire CIMMA, UMR CNRS 6200
Université d'Angers
Angers, France-

Emil Jachim Wolfgang List
Christian Doppler Laboratory Advanced
 Functional Materials
Institute of Solid State Physics
Graz University of Technology
Graz, Austria
and
Institute of Nanostructured Materials and
 Photonics
Weiz, Austria

Richard D. McCullough
Department of Chemistry
Carnegie Mellon University
Pittsburgh, Pennsylvania

Fabien Miomandre
Laboratory Photophysique & Photochimie
 Supramoléculaires et Macromoléculaires
Ecole Normale Supérieure de Cachan
Cachan, France

W. Osikowicz
Department of Physics (IFM)
Linköping University
Linköping, Sweden

Y.W. Park
School of Physics and Nano Systems Institute—
 National Core Research Center
Seoul National University
Seoul, Korea

Martin Pomerantz
Center for Advanced Polymer Research
Department of Chemistry
 and Biochemistry
The University of Texas
Arlington, Texas

Seth C. Rasmussen
Department of Chemistry and
 Molecular Biology
North Dakota State University
Fargo, North Dakota

Knud Reuter
H.C. Starck GmbH & Co. KG
Leverkusen, Germany

John R. Reynolds
The George and Josephine Butler Polymer
  Research Laboratories
Department of Chemistry and Center
  for Macromolecular Science and Engineering
University of Florida
Gainesville, Florida

Jean Roncali
Groupe Systèmes Conjugués Linéaires
Laboratoire CIMMA, UMR CNRS 6200
Université d'Angers
Angers, France

W.R. Salaneck
Department of Physics (IFM)
Linköping University
Linköping, Sweden

Kirk S. Schanze
Department of Chemistry
University of Florida
Gainesville, Florida

Ullrich Scherf
Bergische Universität Wuppertal
Macromolecular Chemistry
  Group and Institute for Polymer Technology
Wuppertal, Germany

Venkataramanan Seshadri
Department of Chemistry
  and the Polymer Program
University of Connecticut
Storrs, Connecticut

Demetrio A. da Silva Filho
School of Chemistry and Biochemistry and Center
  for Organic Photonics and Electronics
Georgia Institute of Technology
Atlanta, Georgia

Jill C. Simpson
Crosslink
St. Louis, Missouri

Ki Tae Song
Electronic Chemical Materials Division
Cheil Industries Inc.
Kyunggi-do, Korea

Gregory A. Sotzing
Department of Chemistry and the
  Polymer Program
University of Connecticut
Storrs, Connecticut

Sven Stafström
Department of Physics and
  Measurement Technology, IFM
Linköping University
Linköping, Sweden

Z. Valy Vardeny
Department of Physics
University of Utah
Salt Lake City, Utah

Francis J. Waller
Air Products and Chemicals Inc.
Allentown, Pennsylvania

Michael J. Winokur
Department of Physics
University of Wisconsin
Madison, Wisconsin

Xiaoyong Zhao
Department of Chemistry
University of Florida
Gainesville, Florida

Table of Contents

III: Properties and Characterization of Conjugated Polymers

I

Theory of Conjugated Polymers

1

On the Transport, Optical, and Self-Assembly Properties of π-Conjugated Materials: A Combined Theoretical/ Experimental Insight

David Beljonne,
Jérôme Cornil,
Veaceslav Coropceanu,
Demetrio A. da Silva Filho,
Victor Geskin,
Roberto Lazzaroni,
Philippe Leclère, and
Jean-Luc Brédas

1.1 Introduction

As the third edition of this Handbook exemplifies, remarkable progress has been made over the past few years in the field of organic electronics. In this chapter, we will review some of our recent efforts along three directions:

1. *The evaluation of the electronic couplings in organic semiconductors*: The electronic coupling is an important parameter that enters the description of charge transport in both the band regime and hopping regime. We will describe recent results on the rubrene single crystal.

Rubrene, a tetraphenyl derivative of tetracene, has recently attracted much attention since hole mobilities as high as 20 cm^2/V s at room temperature have been reported. In addition, the temperature dependent mobility around 300 K is indicative of band transport. These results are a priori surprising since the presence of the phenyl substituents attached to the side of the tetracene backbone is expected to lead to weak intermolecular interactions. We thus embarked on a theoretical investigation of the interchain transfer integrals for holes and electrons in rubrene and compared the results obtained for rubrene with those for pentacene and tetracene. It is found that the limited cofacial π-stack interactions that are present in rubrene actually result in very efficient electronic couplings, which are consistent with the setting up of a band regime at room temperature.

2. *The impact of charge injection on the optical properties*: Injection of charges into (disordered) conjugated polymers is believed to induce local distortions of the geometric structure. Direct experimental measurements of the geometric modifications occurring upon charge injection are scarce and subject to controversy. An alternative, indirect, way to assess these deformations is to monitor the resulting changes in optical absorption. Indeed, according to the seminal Su–Schrieffer–Heeger (SSH) model, addition of extra charges to the polymer chains is expected to lead to the formation of polarons, the appearance of localized electronic levels inside the otherwise forbidden bandgap, and the emergence of new optical transitions.

Here, the changes in geometric structure in the ionized state of isolated model oligomers have been reexamined by means of high-level ab initio quantum-chemical approaches, performed at both the Hartree–Fock (HF) and density functional theory (DFT) levels. While DFT is found to fully delocalize the charge and the resulting geometric distortions independently of the size of the system, HF leads to results that are consistent with the SSH picture, i.e., the formation of self-localized polarons. We show, however, that the mere emergence in the optical spectrum of conjugated systems of new, low-lying optical transitions upon charge injection is not necessarily indicative of polaron formation but is simply a signature of the molecular nature of these materials (though their energetic positions and relative intensities are quantitatively affected by the geometric and electronic relaxation taking place upon charge injection).

We have also investigated the changes in optical properties in the case in which, in the presence of increasing order, the excess charges would spread out over several neighboring chains. Optical signatures for such "delocalized polarons" are shifts in the monomer-based transition energies together with the appearance of a new charge-transfer (charge-resonance) transition in the far-infrared. Such changes have been observed when going from regio-random polythiophene derivatives to the corresponding regio-regular materials, which form ordered two-dimensional multilayer structures. Importantly, the more ordered polythiophenes also show the highest charge carrier mobilities reported so far for conjugated polymers. This again illustrates the critical impact of morphology on the electronic structure of conjugated polymers and the resulting charge-transport properties; therefore, efforts to control the morphology at the nanoscale form the basis for the last section of this chapter.

3. *Self-assembly of conjugated polymer chains (oligomers, homopolymers, and random or block copolymers)*: The solid state supramolecular organization of π-conjugated materials is described on the basis of a joint experimental and theoretical approach. This approach combines atomic force microscopy (AFM) measurements on thin polymer deposits, which reveal the typical microscopic morphologies, and molecular modeling, which allows one to derive the models for chain packing that are most likely to explain the AFM observations.

The conjugated systems considered here are based on fluorene and indenofluorene building blocks (substituted with alkyl or more bulky side groups to provide solubility; in block copolymers, the conjugated segments are attached to nonconjugated blocks such as polyethylene oxide). Films are prepared from solutions in good solvents in order to prevent aggregation processes in solution. Therefore, the morphology observed in the solid state is expected to result mostly from the intrinsic self-assembly of the chains, with little specific influence of the solvent. In such conditions, the vast majority of compounds shows deposits made of fibrillar objects, with typical width and height of a few tens of nanometers and a few nanometers, respectively. These results indicate that a single type of packing process, governed by the π-stacking of the conjugated chains, is at work. The prevalence of such a type of packing is supported by theoretical simulations. Molecular mechanics (MM) and molecular dynamics (MD) calculations show that the conjugated segments tend to form stable π-stacks. For the block copolymer molecules, assemblies can organize themselves in either a head-to-tail or head-to-head configuration. The former case appears to be most likely because it allows for significant coiling of the nonconjugated blocks while maintaining the conjugated blocks in a compact, regular assembly. Such supramolecular organization is likely responsible for the formation of the thin "elementary" ribbons, which can further assemble into larger bundles. We explore some other molecular architectures leading to the formation of untextured deposits. Finally, a clear correlation with the luminescence properties in the solid state is established.

1.2   Electronic Coupling in Organic Semiconductors

The charge-transport properties of conjugated materials critically depend on the packing of the molecules or chains and the degree of order in the solid state, as well as on the density of impurities and structural defects. As a result, the measured mobility values can vary largely as a function of sample quality [1]. In addition, in many instances, the mobility values are extracted from current or voltage characteristics (which depend on the nature of the device used) on the basis of analytical expressions often derived for inorganic semiconductors. This opens the way to variations and uncertainties in the determination of the actual mobility of the charge carriers in organic-based devices. Moreover, as recently shown by Blom and coworkers [2], the charge mobility can be a function not only of the electric field applied across the organic layer but also of the charge carrier density. Thus, the measured mobility values can vary with the experimental conditions under which the I/V curves are recorded.

   In the absence of chemical and physical defects, the transport mechanism in both conducting polymers and molecular single crystals results from a delicate interplay between electronic and electron-vibration (phonon) interactions [3]. The origin and physical consequences of such interactions can be understood by simply considering the tight-binding Hamiltonian for noninteracting electrons and phonons:

$$H = \sum_m \varepsilon_m a_m^+ a_m + \sum_{mn} t_{mn} a_m^+ a_n + \sum_Q \hbar\omega_Q (b_Q^+ b_Q + 1/2) \tag{1.1}$$

Here, a_m^+ and a_m are the creation and annihilation operators, respectively, for an electron on lattice site m (molecular unit or chain segment); b_Q^+ and b_Q are the creation and annihilation operators for a phonon with wavevector Q and frequency ω_Q; ε_m is the electron site energy and t_{mn} is the transfer (electronic coupling) integral, both of which depend on vibrational and phonon coordinates. The electron-vibration coupling arising from the modulations of the site energy is termed *local* coupling; it is the key interaction in the (Holstein) small radius polaron model [4,5]. The second source is related to the transfer integral t_{mn}, which is a function of the spacing and relative orientations of adjacent molecules (chain segments). The modulation of the transfer integrals by phonons (vibrations) is referred to as *nonlocal* coupling; it leads to Peierls-type models, such as the SSH Hamiltonian [6]. Although generalized Holstein–Peierls models are discussed in literature, it is generally believed that Peierls (SSH)-type mechanisms dominate the charge-transport and optical properties of conjugated polymers (see Section 1.3), while local coupling is more relevant for charge transport in molecular crystals as discussed in this section.

   In highly purified molecular single crystals such as pentacene, transport at low temperature can be described within a band picture, as shown by Karl and coworkers [7]. As a general rule of thumb, (effective) bandwidths of at least 0.1 eV are needed to stabilize a band regime [3], in which case the positive or negative charge carriers are fully delocalized; their mobilities are a function of the width and shape of the valence band (VB) or conduction band (CB), respectively, i.e., of the extent of electronic coupling between adjacent oligomers and chains. In pentacene, low-temperature charge carrier mobilities of up to 60 cm^2/V s have been reported [8]. When temperature increases, the mobility progressively decreases as a result of scattering processes due mainly to lattice phonons, as is the case in metallic conductors; transport can then be described on the basis of effective bandwidths that are renormalized and smaller than the bandwidths obtained for a rigid lattice. At the temperature around which localization energy becomes dominant vs. delocalization energy, transport crosses over to a thermally assisted polaron hopping regime in which localized charge carriers jump between adjacent chains.

At the microscopic level, polaron hopping can be viewed as a self-exchange electron-transfer reaction where a charge hops from an ionized oligomer or chain segment to an adjacent neutral unit. In the framework of semiclassical Marcus theory, the electron-transfer rate is written as

$$k_{ET} = \frac{4\pi^2}{h} t^2 \frac{1}{\sqrt{4\pi\lambda kT}} \exp\left[-\frac{(\lambda + \Delta G^\circ)^2}{4\lambda kT}\right] \tag{1.2}$$

The transfer rate depends on two important parameters:

1. The electronic coupling, which reflects the strength of the interactions between adjacent chains and needs to be maximized to increase the rate of electron transfer, i.e., of charge carrier hopping; in the present context, the electronic coupling is often assimilated to the transfer integral, t, between adjacent chains.
2. The reorganization energy, λ, which needs to be minimized; in an ultra-pure single crystal, a lower λ means a higher temperature for the crossover to a hopping regime. The reorganization energy is made up of two components: an inner contribution, which reflects the changes in the geometry of the molecules or chain segments when going from the charged state to the neutral state and vice versa; and and an outer contribution, which includes the changes in the polarization of the surrounding medium upon charge transfer.

It is useful to note that the reorganization energy, λ, is directly related to the (Holstein) polaron binding energy ($E_{pol} = \lambda/2$) [9,10]; in addition, for a self-exchange reaction, the driving force ΔG° is zero.

Since an in-depth discussion of the nature and magnitude of the reorganization energy has been presented in a recent review [9], we will focus in this chapter on the electronic coupling between adjacent π-conjugated chains and describe some recent developments.

The transfer integrals quantify the electronic coupling between two interacting oligomers or chain segments, M_a and M_b, and are defined by the matrix element $t = \langle \Psi_i | V | \Psi_f \rangle$, where the operator V describes the intermolecular interactions and Ψ_i and Ψ_f are the wavefunctions corresponding to the two charge-localized states $\{M_a^- - M_b\}$ and $\{M_a - M_b^-\}$ [or $\{M_a^+ - M_b\}$ and $\{M_a - M_b^+\}$], respectively. A number of computational techniques, based on ab initio or semiempirical methodologies, have been developed to estimate the electronic coupling t; they have recently been reviewed in several excellent publications [11–13]. The most simple and still reliable estimate of t is based on the application of Koopmans' theorem [11]. In this context, the absolute value of the transfer integral for electron [hole] transfer is approximated by the energy difference, $t = (\varepsilon_{L+1[H]} - \varepsilon_{L[H-1]})/2$, where $\varepsilon_{L[H]}$ and $\varepsilon_{L+1[H-1]}$ are the energies of the LUMO and LUMO + 1 [HOMO and HOMO − 1] orbitals taken from the closed-shell configuration of the neutral state of a dimer $\{M_a - M_b\}$. The sign of t can be obtained from the symmetry of the corresponding frontier orbitals of a dimer [14]; t is negative if the LUMO [HOMO] of the dimer $\{M_a - M_b\}$ is symmetric (i.e., represents a bonding combination of the monomer LUMOs [HOMOs]), and positive if the LUMO [HOMO] is asymmetric (antibonding combination of the monomer LUMO [HOMO]'s).

These considerations explain that many theoretical studies have made use of Koopmans' theorem to estimate electronic couplings [15–21]. However, caution is required when using Koopmans' theorem to estimate transfer integrals in asymmetric dimers. In such instances, part of the electronic splitting can simply arise from the different local environments experienced by the two interacting molecules, which create an offset between their HOMO and LUMO levels. In order to evaluate the actual couplings, this offset needs to be accounted for, either by performing calculations using molecular orbitals localized on the individual units as the basic set [22], by computing directly the coupling matrix element between the molecular orbitals of the isolated molecules, or by applying an electric field to induce resonance between the electronic levels, as done by Jortner and coworkers [19]. In this chapter, we will consider cases in which these site energy fluctuations are small and can thus be neglected.

The electronic splittings reported below have been calculated within Koopmans' theorem using the semiempirical HF intermediate neglect of differential overlap (INDO) method; interestingly, the INDO method provides transfer integrals of the same order of magnitude as those obtained with DFT-based approaches [23,24]. It is also interesting to note that when building (infinite) one-dimensional stacks of molecules, oligomers, or chains, the widths of the corresponding VBs and CBs are usually found to be nearly equal to four times their respective t integrals; when this is the case, it indicates that the tight-binding approximation is relevant (which is not surprising, since transfer integrals are short-ranged and hence interactions with non-nearest neighbors can often be neglected).

Before turning to actual packing structures, we recall that our simple analysis of perfectly cofacial configurations has shown that [3,25]:

1. By their very nature, cofacial configurations provide the largest electronic interactions (coupling) between adjacent chains.
2. As a qualitative rule, the lower the number of nodes in the wavefunction of the frontier level of an isolated chain, the larger the splitting of that level upon cofacial interaction.
3. In cofacial stacks of oligomers, the valence bandwidth is expected to be larger for small oligomers and the conduction bandwidth, larger for long oligomers; however, for any oligomer length, the valence bandwidth remains larger than the conduction bandwidth. The latter point is the basic reason why it was conventionally thought that organic materials displayed higher hole mobilities than electron mobilities. For actual packing structures, this belief has no reason to hold [3,4].

The high quality of rubrene crystals has allowed detailed measurements of the transport characteristics, including the recent observation of the Hall effect [26]. Charge transport in rubrene single crystals, while trap-limited at low temperature, appears to occur via delocalized states over the 150–300 K temperature range with an (anisotropic) hole mobility of up to 20 cm^2/V s at room temperature [27,28].

Due to the bulky phenyl substituents attached to the side of the tetracene backbone (see Figure 1.1), weak intermolecular interactions are a priori expected in rubrene. The large carrier mobilities that are observed are thus surprising from that standpoint. Therefore, in order to elucidate the origin of these mobilities and of their marked anisotropy along the crystallographic axes, we have performed quantum-chemical calculations of the electronic structure of rubrene [29]; we discuss the results of these calculations below and also make comparisons of tetracene and pentacene.

In the isolated (neutral) molecules, the shapes and energies of the HOMO and LUMO levels of rubrene and tetracene are very similar (see Figure 1.2). This reflects the large torsion angle between the phenyl rings and the tetracene backbone in rubrene (this angle is calculated in the isolated molecule and measured in the crystal [30] to be ∼85°). This near-perpendicular arrangement strongly reduces any mixing of the molecular orbitals between the backbone and the side groups.

The widths of the HOMO and LUMO bands, calculated at the INDO level, are plotted in Figure 1.3 as a function of $\cos(\pi/N+1)$, where N is the number of rubrene molecules in stacks built along the a-direction and d-direction (see Figure 1.4). Large transfer integrals are found along the a-direction and lead to bandwidths W of ∼340 and 160 meV for holes and electrons, respectively. Smaller interactions are calculated along the diagonal directions (d), while all other directions provide for negligible transfer integrals. Interestingly, the highest bandwidths in rubrene (along the a-direction) are comparable to the highest bandwidths calculated in pentacene (along the diag-

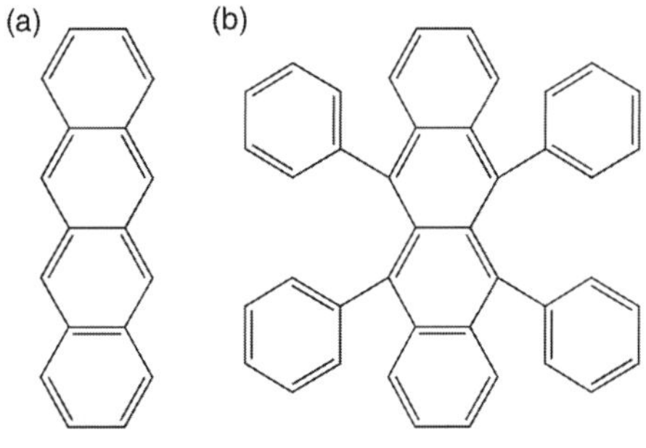

FIGURE 1.1 Chemical structures of (a) tetracene and (b) rubrene.

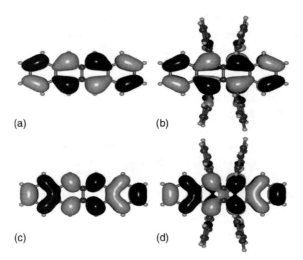

(a)   (b)

(c)   (d)

FIGURE 1.2 DFT-B3LYP/6-31G(d,p)-calculated HOMO and LUMO wavefunctions in the neutral ground state geometry: (a) tetracene HOMO (−4.87 eV); (b) rubrene HOMO (−4.69 eV); (c) tetracene LUMO (−2.09 eV); and (d) rubrene LUMO (−2.09 eV).

onal d-directions) [25,31]. The linear evolution of the calculated bandwidths (Figure 1.3) as a function of $\cos(\pi/N+1)$ implies that the intermolecular interactions in rubrene can be cast into a tight-binding formalism, with $W = 4t$, as was also observed for oligoacenes, sexithiophene, and bisdithienothiophene [31–33].

To understand the magnitude of the transfer integral values, we need to examine the packing of the molecules in the crystal structure. Both rubrene and pentacene (or tetracene) present a herringbone motif in the ab plane where the most significant electronic couplings are found. However, there are major differences that are triggered by the presence of the phenyl side groups in rubrene:

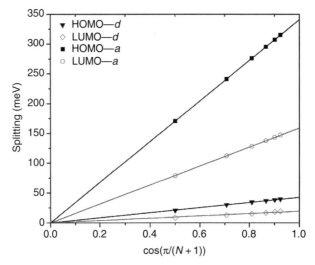

FIGURE 1.3 Evolution as a function of $\cos(\pi/N+1)$ of the INDO-calculated splitting formed by the HOMO and LUMO levels in rubrene stacks along the a-direction and d-direction. N is the number of interacting molecules. A linear fit was used to extrapolate the full bandwidth along each stack. The calculated valence (conduction) bandwidth is 341 meV (159 meV) along the a-direction and 43 meV (20 meV) along the d-direction. The bandwidths along other directions are vanishingly small and therefore not reported.

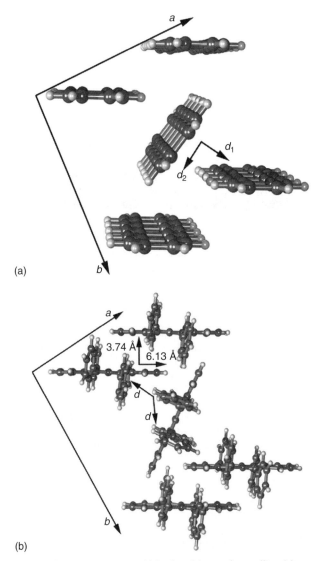

(a)

(b)

FIGURE 1.4 Illustration of the lattice parameters within the *ab* layer of crystalline (a) pentacene and (b) rubrene. The long-axis displacement and π-stacking distance in the *a*-direction are also indicated.

1. The long molecular axes all come out of the *ab* plane in pentacene (or tetracene), while in rubrene they are embedded in that plane (see Figure 1.4). As a consequence, the long molecular axes of adjacent molecules along the diagonal (herringbone) directions are parallel in pentacene, while they are almost perpendicular in rubrene. This explains the smaller transfer integrals along the diagonal directions in rubrene.

2. Along the crystal *a*-direction for which the highest bandwidths are calculated, the rubrene molecules are found to form a π-stack with a stacking distance of 3.74 Å; although this distance is larger than in a typical π-stack [34,35], a major feature is that there occurs no displacement along the short molecular axes (see Figure 1.4). In pentacene (or tetracene), the short-axis displacements are so large that adjacent molecules along *a* interact weakly in spite of a very short π-stack distance of 2.59 Å.

While there occurs no short-axis displacement along the *a*-direction in rubrene, the phenyl side groups promote a very large sliding, by 6.13 Å, of one molecule with respect to the next along the long molecular axis (see Figure 1.4). This leads to the appearance of the slipped-cofacial configuration illustrated in Figure 1.4. Such long-axis displacements are known to reduce the electronic coupling between adjacent molecules. As noted above, the coupling is maximum for the perfectly cofacial situation; it then oscillates between positive and negative values as a function of increasing displacement and eventually vanishes for large displacements [9,25].

The question that needs to be answered is how such a pronounced long-axis sliding can be consistent with the large bandwidths that are calculated along the *a*-direction of the rubrene crystal. To provide an answer, we examined at the INDO level the evolution of the HOMO and LUMO transfer integrals for a complex made of two tetracene molecules (we did not consider the phenyl side groups present in rubrene since the phenyls do not play any significant role in the HOMO or LUMO wavefunctions of rubrene (see Figure 1.2) and their presence would obviously lead to major steric interactions upon displacement). We started from a perfectly cofacial situation with the two molecules superimposed at a distance of 3.74 Å, as in the rubrene crystal structure, and then increasingly displaced the top molecule along the long molecular axis. The results are shown in Figure 1.5.

The HOMO and LUMO transfer integrals display a typical oscillating evolution upon displacement [3,4]. The most remarkable result is that the displacement of 6.13 Å observed in the rubrene crystal actually closely corresponds to extrema in the oscillations of both HOMO and LUMO transfer integrals. This result clarifies why, even at such a large long-axis displacement, the transfer integrals are still a significant fraction of the values found for the cofacial case.

Along the *d*-direction of rubrene, the calculated HOMO and LUMO bandwidths are much smaller than along the *a*-direction, in agreement with the experimental findings [36] that show a strong anisotropy of the mobility within the herringbone planes of rubrene single crystals. Two adjacent molecules in the *b*-direction are 14.4 Å apart (see Figure 1.4); for such a distance, there is no electronic overlap between the two oligomers. As a result, a charge carrier moving along the *b*-direction is expected to have to make its way via the intercalated molecules (in the *d*-direction illustrated in Figure 1.4); thus, the transfer integrals along the *d*-direction should be used to understand the transport in the *b*-direction. This suggestion has been confirmed by the results of the calculation of the three-dimensional electronic band structure of the rubrene crystal with the plane-wave (PW) DFT method [29].

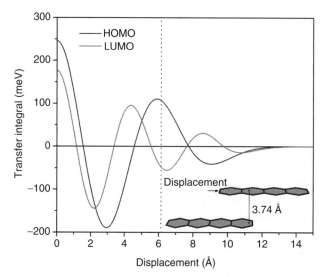

FIGURE 1.5 Evolution of the INDO HOMO and LUMO transfer integrals as a function of displacement, for a complex made of two tetracenes stacked along the rubrene *a*-direction with a π-stacking distance of 3.74 Å. The dotted line indicates the magnitude of the long-axis displacement (6.13 Å) found in the rubrene crystal. The molecular geometries were optimized at the DFT-B3LYP/6-31G(d,p) level.

The PW-DFT results are also consistent with a rubrene valence bandwidth on the order of a few tenths of an eV (ca. 0.4 eV). Thus, the INDO and DFT calculations concur to suggest that, in the absence of traps, a band regime should be present in rubrene single crystals at low temperature which could still be operative for holes at room temperature (as a consequence of larger electronic coupling for holes than for electrons); the latter is consistent with the decrease in mobility with temperature observed around room temperature [28,36].

1.3 Impact of Charge Injection on the Optical Properties of Conjugated Polymers

Charge injection into conducting polymers is readily achieved by (reversible) chemical or electrochemical redox transformations and leads to drastic changes in their electrical [37] and optical [38] properties. In particular, new optical transitions at lower energies than in the neutral polymer emerge upon doping. The SSH theory [6,39] has shaped the early understanding of these phenomena.

With appropriate parameterization, the SSH Hamiltonian depicts neutral conjugated polymers as semiconductors; the lowest optical transition is then interpreted as an electronic excitation from the (fully occupied) VB to the (unoccupied) CB. When a conjugated polymer is brought into a charged state (i.e., when it is "doped"), a few, typically two or three, new low-energy optical transitions are usually observed (see Figure 1.8). These subgap states are the result of local geometric distortions due to charge self-trapping (self-localization). In addition to the description of the characteristics of the ground state, the SSH Hamiltonian reproduces the appearance of several types of excitations [40]: polarons (singly charged species with spin $\frac{1}{2}$); bipolarons (doubly charged species that are spinless); and, in the case of degenerate ground state polymers such as trans-polyacetylene, solitons (either singly charged species without spin or neutral species with spin $\frac{1}{2}$). As a result of their *electron-phonon* origin, these excitations are associated with well-defined (self)-localized charges and/or geometric distortion distributions.

In fact, conjugated polymers often present a distribution of finite chain lengths. Furthermore, the conjugation lengths are rather short (typically 5–20 units), so that a molecular picture can also be applied. Within such a molecular approach, it is interesting to note that, even in the absence of any geometric and electronic relaxation upon ionization, new low-energy optical transitions would arise that would involve the discrete level affected by electron removal or addition (see bottom of Figure 1.6).

1.3.1 Methodological Aspects: Hartree–Fock vs. Density Functional Theory

In order to study theoretically the electronic excited states and, therefore, the optical properties of a charged conjugated chain, the molecular geometry is usually optimized beforehand. Thus, it is of interest to assess: (i) the characteristics of the ionized state geometry; and (ii) to what extent the calculated optical properties depend on the input molecular geometries. While the transport properties can be well described in a one-electron picture by considering the calculated splitting of the HOMO or LUMO levels (see above), excited states are generally not properly depicted at that level when considering simply the promotion of a single electron from an occupied molecular orbital to an unoccupied orbital. It is possible to go beyond the one-electron picture by using a configuration interaction (CI) technique; in this approach, excited states are described as a linear combination of electronic configurations where one or several electrons are excited from various occupied to unoccupied molecular orbitals. This is the underlying principle of the INDO/SCI calculations reported here, for which the molecular orbitals are initially calculated at the INDO level and singly excited configurations are involved in the CI expansion. The use of several electronic configurations is also exploited in time-dependent (TD) formalisms, as is the case for the TD-DFT results discussed later.

To elaborate point (i), we chose thiophene oligomers as model systems [41]. In the present context, it is convenient to characterize the geometry with a single relevant parameter, the inter-ring C–C bond length; this bond shortens in radical-cations in agreement with the valence-bond scheme shown in Figure 1.7. The results of ab initio and semiempirical HF spin-restricted open-shell self-consistent field

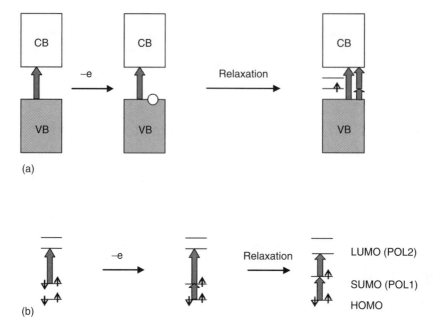

FIGURE 1.6 Schematic representation of the electronic structure of conjugated polymers according to (a) band theory and (b) a molecular picture.

(ROHF-SCF) calculations on oligothiophenes are in good mutual agreement. They indicate clear self-localization of spin and charge over five to six rings around the middle of the chain, with the accompanying quinoid geometric distortion being somewhat more confined. On the contrary, we find, in agreement with previous studies [42,43], that the spin-unrestricted HF (UHF) tends to equalize single and double bonds and to completely delocalize neutral radicals. This trend is completely reversed when, starting from the UHF reference, geometry optimization is performed (on an oligomer long enough to distinguish between polaron formation and charge delocalization) with a *correlated* ab initio unrestricted second-order Möller–Plesset (UMP2) method. A major result is that UMP2 significantly sharpens the localization of the polaron with respect to all HF methods.

The treatment of the spin characteristics in all these methods is not entirely satisfactory. The ROHF method insures a correct symmetry of the wavefunction but rules out spin polarization. On the other hand, the UHF open-shell solutions for conjugated systems are usually spin-contaminated [41,44]. In this context, DFT has emerged as an attractive option by providing for inclusion of both electron correlation and spin polarization effects with usually low spin contamination. However, DFT calculations rapidly became a subject of controversy when applied to conjugated polymers. When a pure DFT approach (e.g., the BLYP functional) is used in the case of a radical-cation, a complete charge/spin delocalization is predicted to occur over the whole chain (Figure 1.7, see also [45,46]), which is in marked contrast to the HF and MP2 results. A hybrid DFT approach (the BHandHLYP functional including as much as 50% of HF exchange) yields a relatively diffuse defect for $8T^+$, but with some degree of localization of the polaron toward the middle of the chain at this chain length [41]. It is interesting to note the absence of direct correlation between the degree of localization of geometry distortion and charge/spin distribution that can occur: while the UHF geometric distortion is extended (with intermediate bond lengths along the entire backbone), the UHF method yields localized charge and spin distributions (Figure 1.7).

In addition, there is a notable difference in the chain-length evolution of the HF- and BHandHLYP-optimized radical-cation structures [47]. With AM1 ROHF, self-localization occurs; this is first seen via an increased localization as the chain becomes longer (Figure 1.8a) and, after a certain chain-length threshold (decamer in this case), via symmetry breaking (the distortion is no longer forced to be in the

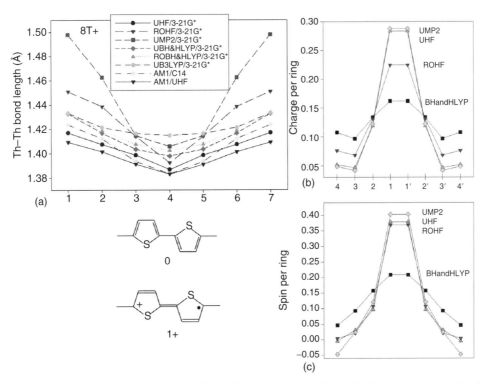

FIGURE 1.7 (a) Inter-ring C–C bond lengths, (b) Mulliken charge, and (c) spin distributions in the octathiophene (8T +) radical-cation as obtained by geometry optimization with different methods.

middle of the chain) and independence of the polaron segment geometry vs. the rest of the chain (Figure 1.8b). On the contrary, with hybrid DFT (Figure 1.8c), in spite of some degree of localization, the defect still spreads somewhat more along the chain as the oligomer becomes longer and no symmetry breaking is observed even for 13T +.

Similar geometric trends with HF and DFT methods were obtained by others and by our group [48,49] in the case of oligophenylenevinylene (OPV) radical-cations. Thus, it can be concluded that: (i) HF tends to localize the geometric distortion and the excess charge and spin; (ii) pure DFT with standard functionals leads to moderate changes that extend over the entire molecule; and (iii) the behavior is intermediate with hybrid DFT.

A brief discussion on the theoretical description of the structure of extended conjugated compounds is in order at this stage. As is well established for neutral conjugated oligomers, and polyenes in particular, taking electron correlation into account is crucial to obtain a correct degree of bond-length alternation (BLA). While an ab initio HF approach strongly overestimates BLA, MP2 correlated results are in good accord with experimental data. On the other hand, pure DFT functionals, even though they take electron correlation into account, tend to underestimate BLA strongly; this feature is equally true for the simplest LDA, the more precise GGA, and the more sophisticated meta-GGA families (see e.g., [50] for recent results, discussion, and references therein). The reason for the failure of common DFT functionals was shown [51] to be the self-interaction error (noncancellation of Coulomb and exchange terms for the same orbital), which is corrected only in rather exotic functionals. A practical, well-known way to improve on DFT results is to include a part of HF exchange in the so-called hybrid DFT. For polyenes in particular, a correct BLA is obtained with the popular B3LYP functional (which contains 20% of HF exchange). However, this is not a general rule; for polymethineimine, BLA is still underestimated with B3LYP [52]. Thus, a correct BLA can in principle be obtained in neutral conjugated oligomers within hybrid DFT, but the amount of required HF exchange is unknown beforehand.

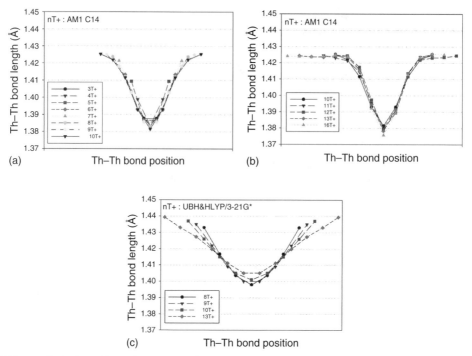

FIGURE 1.8   Evolution of inter-ring C–C bond lengths in thiophene oligomers as a function of chain length as optimized with AM1-CAS (a) before and (b) after symmetry breaking and (c) with BHandHLYP/3-21G*.

The situation is still more complicated for open-shell (radical) conjugated oligomers. In neutral polyene radicals, even the most precise and expensive ab initio correlated methods, such as CCSD(T), can give at best a qualitative prediction of the spin density distributions, and the behavior of common DFT schemes is also mediocre [42]. The problems with exchange and correlation are augmented in this case by the spin-restricted/unrestricted ansatz dilemma discussed above. An efficient practical solution is, somewhat unexpectedly, provided by correlated semiempirical methods with a simple Hamiltonian, such as PPP [53]. We therfore also often use semiempirical methods in our studies of conjugated oligomer radical-ions.

To get more insight into the effect of geometry on the optical properties, we now turn to an analysis of the first optically allowed transition in OPV + oligomers; this transition has a clear one-electron nature, the wavefunction of the associated excited state being dominated by the HOMO→SOMO (HOMO→POL1) excitation. Furthermore, this low-energy polaron transition has recently been singled out for the universal character of its energy evolution in different conjugated polymers [54].

Figure 1.9 illustrates the shape of the HOMO orbital in the neutral 6-ring OPV and that of the SOMO (POL1) in the unrelaxed (i.e., neutral) geometry and fully relaxed (i.e., singly charged) geometry, as provided by INDO calculations on the basis of AM1 geometries. The HOMO of the neutral molecule is concentrated near about the middle of the chain. Strong localization occurs for the SOMO (POL1) of the unrelaxed radical-cation, which exhibits a bonding–antibonding pattern very similar to the HOMO of the neutral molecule. Interestingly, the geometry relaxation characteristic of the radical-cation adds practically no modification to the shape of the SOMO (POL1) level (compare Figure 1.9b and Figure 1.9c).

It is usual in quantum chemistry to use different methods for geometry optimization and for the subsequent determination of the molecular properties. In addition to the different geometries used as input, we have therefore also applied different methods for the calculation of the optical transitions to the electronic excited states, namely, INDO-SCI and time-dependent DFT (TD-DFT). Interestingly, a Mulliken population analysis for the 7-ring OPV radical-cation, presented in Figure 1.10, demonstrates

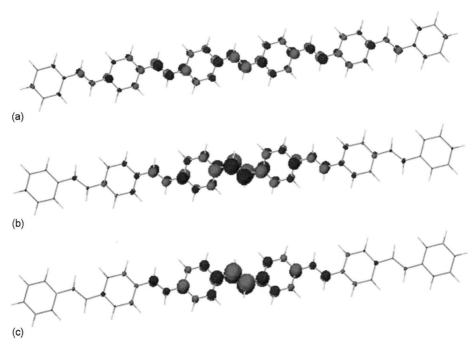

(a)

(b)

(c)

FIGURE 1.9 Linear combination of atomic orbitals (LCAO) pattern of (a) the HOMO level in the neutral 6-ring OPV; and the SOMO level of (b) unrelaxed and (c) fully relaxed 6-ring OPV +, as calculated at the INDO level on the basis of AM1-optimized geometries. The size and color of the circles reflect the amplitude and sign of the LCAO coefficients, respectively.

that, irrespective of geometry, the excess charge is systematically sharply localized when INDO calculations are performed and weakly localized for hybrid DFT results. Thus, it appears that it is primarily the choice of the method used for the electronic-structure calculations, rather than the molecular geometry, that drives the theoretical evolution of the optical properties.

To better illustrate these trends, the chain-length evolution of the lowest transition energy, as computed with different methods and different molecular geometries, is reported in Figure 1.11 for the OPV radical-cations. At the INDO-SCI level, the transition energies obtained on the basis of the AM1 geometries saturate quickly with increasing chain length and level off around OPV7-9. The same saturation is observed when the radical-cations are forced to adopt the AM1 geometry of the neutral OPV chains, except that the transition energies are red-shifted (by up to 0.4 eV for the longest oligomers). The transition is strongly optically allowed, as indicated by the increase in the oscillator strength from 0.47 to 1.31 throughout the series. A similar evolution is obtained for the transition energies computed at the INDO-SCI level on the basis of pure DFT geometries. In contrast, the transition energies calculated with the TD-DFT formalism for geometries obtained at the AM1 and pure DFT (VWN-BP) levels evolve in a very different way: in both cases, there is hardly any sign of saturation even for chain lengths as long as 8–9 repeat units.

The different curves in Figure 1.11 are crossing for oligomer lengths that typically correspond to the longest chains investigated at the experimental level; therfore, the assessment of which is the most reliable theoretical technique would actually require data collected for longer chains. Nevertheless, it is clearly seen at the long-chain limit that the calculated transition energies decrease in the following order: HF-CI//HF > HF-CI//DFT > TD-DFT//HF > TD-DFT//DFT (where the latter acronym represents the method used for geometry optimizations and the former that used for electronic-structure calculations). Combining DFT geometries with the TD-DFT formalism provides very low excitation energies in long chains. The rapid saturation of the INDO-SCI transition energies suggests that the structure of the frontier orbitals has already converged in short chains. The larger values calculated for the fully

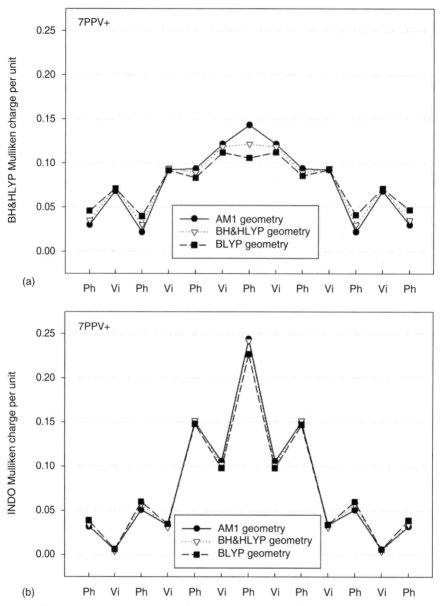

FIGURE 1.10   Charge distribution on the phenylene and vinylene units of the 7-ring OPV +, as obtained with the (a) hybrid DFT-BH and HLYP and (b) INDO methods for various starting geometries.

relaxed radical-cations compared to the unrelaxed systems point to a reinforcement of the electronic localization. The AM1//INDO-SCI values, which saturate around 0.8–0.9 eV, provide the best agreement with the experimental value initially reported for doped PPV (~1 eV) [38]; the other theoretical values saturate more slowly with chain length and are too low in energy or, even worse, do not seem to saturate at all. Recent studies based on vis-NIR spectroscopy of chemically doped PPV [55] and on photoinduced absorption in a series of phenylenevinylene oligomers [56] indicate a somewhat smaller first polaron transition energy of ~0.6 eV.

To provide a more general assessment of the TD-DFT results, it is worth noting that, while it usually gives accurate results for excitation energies, the method is known to fail for higher electronic states with doubly excited charge-transfer or Rydberg character, and for excitations in extended π-systems [57].

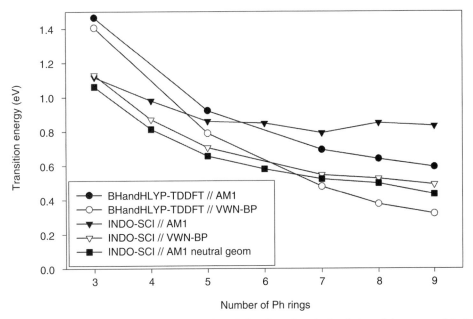

FIGURE 1.11 Evolution, as a function of the number of phenylene rings in the chains, of the energy of the lowest optically allowed vertical transition in OPV radical-cations, as obtained by different procedures (the first acronym represents the method used for the electronic-structure calculations and the second, that for the geometry optimizations: AM1 and INDO are semiempirical H–F methods, VWN-BP is a pure DFT functional, and BHandH-LYP is a hybrid DFT with 50% of H–F exchange).

In practice, it is not always easy to know how to classify a given excited state. In the case discussed above, the first excited state is a valence single excitation with no charge-transfer character. Although the polaron MOs are localized in the central segment of the chain, their extension increases with chain length, which can be related to DFT favoring delocalization. In turn, this can lead to a progressive underestimation of the excitation energy. The quality of the TD-DFT description is expected to become worse for longer oligomers.

The chain-length evolution of the lowest transition energy is calculated to depend primarily on the degree of *electronic* localization of the SOMO (POL1) level; the latter is to a large extent decoupled from the details of the actual geometry, making the calculated results mostly dependent on the nature of the theoretical method used to generate the electronic structure. In the range of oligomer sizes typically studied experimentally (up to 5–7 unit cells), the absolute values of the lowest transition energy calculated with different techniques on the basis of different geometries do not differ significantly due to the confinement of the charged species by the chain ends. Thus, these data cannot be exploited to assess the best theoretical approach. The chain-length evolution of the transition energies is, however, more sensitive to the choice of the theoretical approach when dealing with longer oligomers.

1.3.2 Optical Signature of Delocalized Interchain Polarons

The inherent disorder in organic conjugated polymers has long prevented the achievement of charge-carrier mobilities compared to those in molecular crystals (see above) or amorphous silicon. However, in the case of regioregular poly-3-hexylthiophene (RR-P3HT), self-organization of the polymer chains leads to a lamellar structure with two-dimensional sheets built by strongly interacting conjugated segments (with interchain distances on the order of 3.8 Å); as a result, the material displays room-temperature mobilities of up to 0.1 cm^2/V s [58,59]. (Films of RR-P3HT were successfully used in an FET device to drive a light-emitting diode (LED) based on a luminescent polymer, thereby providing the first demonstration of an all-organic display pixel [59].) The large mobilities reported in the case of

RR-P3HT suggest that the transport mechanisms might be different in such a highly ordered polymer and could involve charged species that are delocalized over several adjacent chains [58,59].

The nature of the charge carriers in RR-P3HT has been investigated experimentally using optical charge modulation spectroscopy (CMS) [60] and photoinduced absorption (PiA) [61]. These studies have pinpointed substantial differences between the optical signatures of polarons in the microcrystalline materials compared to chemically doped polythiophene in solution. Such differences were attributed to delocalization of the charges over several adjacent polythiophene chains, which is consistent with the high mobilities within the two-dimensional conjugated sheets [58,59].

To assess the validity of these propositions, we investigated the influence of interchain charge delocalization on the optical properties of singly charged conjugated systems. For the sake of simplicity (the electronic excited states of phenylene-based molecules are less subject to configuration mixing than in oligothiophenes), we focused our attention on a model phenylenevinylene oligomer, namely the 5-ring OPV, hereafter denoted 5PV. Ground state geometry optimizations of the isolated 5PV molecule in its neutral and singly charged states have been carried out at the Austin Model 1/Full Configuration Interaction (AM1/FCI) level (with four molecular orbitals included in the CI active space) [62]. Figure 1.12 shows the bond-length deformations associated with the formation of a positive molecular polaron on 5PV. Charge injection induces the appearance of a slight quinoid character on the phenylene

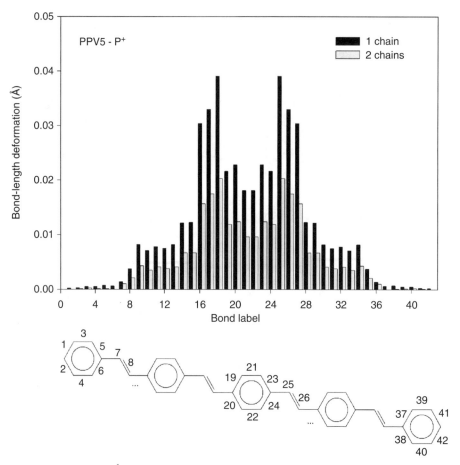

FIGURE 1.12 Differences (in Å) between the AM1/FCI calculated bond lengths in the neutral and singly charged states of 5PV, in the isolated molecule (black) and in the dimer (yellow). The structure and bond labeling of the 5-ring oligomer is shown at the bottom.

rings, together with a decrease in bond-length alternation in the vinylene linkages. These deformations are more pronounced around the center of the molecule and typically extend over three repeat units ($\sim$20 Å).

On the basis of the optimized geometry of the charged species, the one-electron energy levels were calculated using both the AM1 and INDO [63] semiempirical approaches. A simplified representation of the one-electron energy diagram typically obtained with these two techniques is given in Figure 1.13a and has been discussed above.

The AM1-optimized geometry of the charged species is then used as input for the computation of the excited states, which is performed at two different levels:

1. The AM1/FCI formalism, as adopted for geometry optimization, but considering a larger configuration interaction expansion (including up to 16 molecular orbitals) to insure convergence of the spectroscopic properties.
2. The INDO/SDCI method, in which all single excitations over the electronic π-levels and a limited number of double excitations (more than 16 MOs) were involved.

As seen from Figure 1.14, most of the optical absorption cross section of the charged species is shared between two excited states described as the C1 and C2 electronic transitions of Figure 1.13, namely the singly excited HOMO→P1 and P1→P2 configurations, respectively [63].

The influence of interchain interactions was assessed by considering a cofacial aggregate formed by two 5PV chains separated by 4 Å. This configuration is similar to that encountered in the microcrystalline domains of RR-P3HT [60]. When enforcing such a symmetric cofacial arrangement, the INDO and AM1 approaches lead to molecular orbitals completely delocalized over the two conjugated chains, which carry half a unitary charge each. To determine the ground state geometry of the two-chain system in its singly oxidized state, we applied the same AM1/FCI approach as in the single chain case (while doubling the active space to eight molecular orbitals for size consistency).

The lattice deformations in the aggregate, as provided by this approach, are compared to those calculated for the isolated molecule in Figure 1.12. Since the perturbation induced by charge injection in

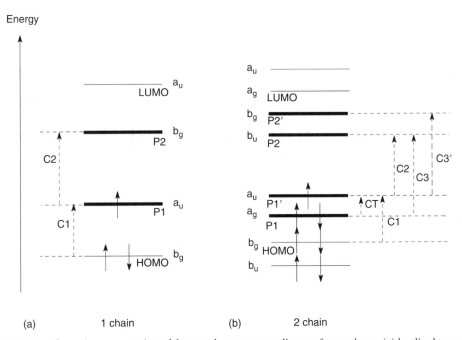

FIGURE 1.13 Schematic representation of the one-electron energy diagram for a polaron: (a) localized on a single conjugated chain; and (b) delocalized over two cofacial chains. The symmetry of the orbitals and the relevant electronic excitations are indicated.

the individual chains of the dimer is weaker than in the single-chain case, the bond-length modifications are less pronounced in the aggregate (especially around the center of the molecule) and extend in a symmetric way over the two conjugated chains.

The electronic structure and absorption spectrum of the cationic dimer can be interpreted in the framework of a two-site molecular (Holstein) polaron model (or equivalently, in terms of the Marcus–Hush model for self-exchange electron transfer). As mentioned in Section 1.2 and discussed in more detail elsewhere [9], the system properties depend on the interplay between the electronic coupling (transfer integral t between adjacent chains) and the reorganization energy λ (polaron binding energy, $E_{pol} = \lambda/2$). In the case when the electronic coupling $2t$ is smaller than the reorganization energy λ, the lower adiabatic potential surface exhibits two equivalent minima. Each of these minima corresponds to the situation in which the system is mainly localized on one of the valence structures $\{M_a^+ - M_b\}$ and $\{M_a - M_b^+\}$ (see Section 1.2), respectively. In terms of the Holstein model, this situation corresponds to the formation of a localized molecular polaron. The electronic spectrum is characterized in this case by the appearance in the visible or the near-infrared region of a charge-transfer band, with the band maximum being equal to the reorganization energy: $E_{CT} = \lambda$. In the case of strong electronic coupling ($2t > \lambda$), corresponding to the situation discussed in this section, the lower potential surface possesses only one minimum. In this case, the CT transition becomes a direct measure of the electronic coupling, since: $E_{CT} = 2t$. In a one-electron picture, this band is related to the resonance splitting of the oligomer SOMO levels (P1). The interchain coupling leads to a similar resonance splitting of all other monomer electronic levels (see Figure 1.13). As a result of C_{2h} symmetry of the cofacial dimer configuration considered here, all levels split into *gerade/ungerade* symmetry molecular orbital pairs. The appearance of these levels leads to new optical transitions in the dimer. As shown in Figure 1.13, the low-energy transitions involve the levels P1, P1′, P2, and P2′ referred to as polaronic levels in Section 1.2.

The absorption spectrum of the cationic dimer, as computed at the AM1/FCI level (with up to 26 molecular orbitals in the active space) and INDO/SDCI (with all possible single excitations within the π-manifold and double excitations over 22 π-orbitals), is shown in comparison to that of the one-chain polaron in Figure 1.14. The optical features discussed below are common to the descriptions provided by the AM1/FCI and INDO/SDCI formalisms.

The CT band (Figure 1.14) is predicted to be around 0.1 [0.25] eV at the AM1/FCI [INDO/SDCI] level. This result indicates that the interchain electronic coupling is about 400 [1000] cm^{-1}, i.e., comparable to the values derived for pentacene and rubrene (see Section 1.2) at Koopmans' theorem level. The CT transition is polarized along the interchain packing axis; all other absorption bands are mainly polarized along the chain axis. However, despite its polarization feature and its (commonly used) name, this band is not accompanied by a net charge transfer (in both ground and excited states, the charge is equally distributed over both oligomers). The intensity of this CT absorption band depends on the electronic coupling and is thus controlled by interchain separation and mutual orientation of the monomers.

The smaller geometric and electronic relaxation in the two-chain aggregate translate into shallower polaronic levels inside the gap and a consequent red-shift (by 0.1–0.2 eV) of the absorption band described by the CI electronic excitation. The nature of the latter transition remains essentially unaffected by interchain interactions. An additional striking feature in the optical spectrum of the singly charged dimer is the appearance of a new intense optical transition around 3.1 [3.2] eV (AM1/FCI [INDO/SDCI]). Single particle excitations between the P1 and P2 electronic levels produce two excited states with B_u symmetry (i.e., one-photon symmetry allowed with respect to the ground state), instead of one in the single-chain limit (note that C2 is asymmetrical in the dimer). These two B_u excited states result from the configuration mixing between the C3 and C3′ electronic transitions. Constructive combination of these optical excitations leads to the spectral feature around 3.1 eV (i.e., just below the optical gap of the neutral system (~3.2 eV)), which dominates the absorption spectrum of the two-chain aggregate (see Figure 1.14). Destructive coupling between the same configurations provides the weaker absorption line around 2.4 [2.1] eV (note that this excited state also involves significant contributions from doubly excited configurations).

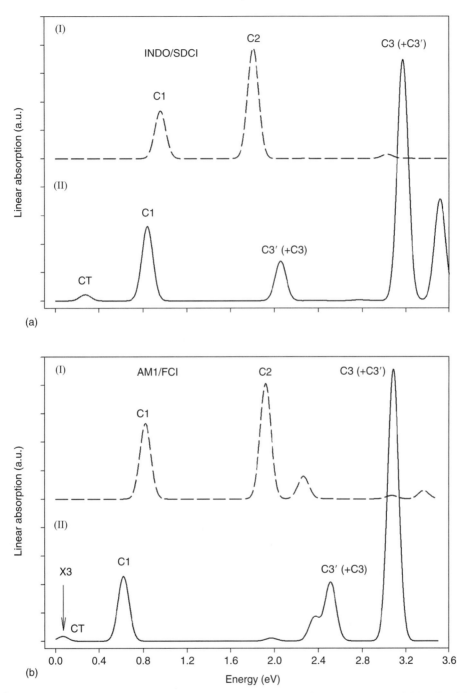

FIGURE 1.14   (a) Linear absorption spectra of the singly charged PPV5 (I) single molecule and (II) cofacial dimer, simulated on the basis of the INDO/SDCI description of the excited states. (b) Similar results from AM1/FCI calculations. Gaussian functions with 0.05 eV width at half maximum have been used to convolute the spectra.

The theoretical simulations we have reviewed here are consistent with the changes in optical absorption observed experimentally when the degree of interchain order is improved in alkyl-substituted polythiophene derivatives [60,61]. The CMS spectrum of RR-P3HT exhibits a pronounced charge-induced transition at 1.65 eV (C3) just below the bleaching signal of the neutral polymer absorption and a shoulder at 1.35 eV (C3′) [64]. In addition, a broad absorption signal (CT) is observed with a distinct shoulder around 0.35–0.45 eV (which we assign to C1). The CT band is characteristic of highly microcrystalline P3HT lamellae and appears to be more pronounced, the higher the regioregularity of the polymer [60]. Polymers with regioregularity lower than 70%–80% feature markedly different optical signatures compared with those of highly regioregular P3HTs. The dominant charge-induced transition is at 1.25 eV (C2), close to where the C3′ shoulder is observed in the highly regioregular polymers. At energies around 1.7 eV no clear transition is resolved. This spectrum is interpreted in terms of optical transitions involving charge carriers confined to short conjugated segments on individual, one-dimensional chains as a consequence of pronounced disorder.

1.4   Self-Assembly of Conjugated Polymers into Solid State Nanostructures

As emphasized throughout this chapter, molecular packing critically impacts on the electronic structure of conjugated oligomers and polymers and the resulting luminescence and charge-transport properties. Many important questions remain about the nature of the electronic states in conjugated polymers. In particular, the influence of interactions among chains on the electronic properties in the solid state is a matter of major interest [65–68]. Interchain interactions can have a major impact on the operation of devices, since they determine not only the charge-transport properties but also the spectrum and the efficiency of LEDs, or the charge-separation process in photodiodes. More generally, energy and charge transport in a molecular solid are critically dependent on the intermolecular interactions, and these interactions in turn are dependent on the molecular packing. Therefore, it is critical to investigate and control the effects of ordering and structure in conjugated polymers, from the nanoscale through the mesoscale (100–1000 nm) to the optical scale (>1000 nm). In the nanometer regime, the major goal is to synthesize stable materials with high luminescence efficiencies by control of the intra- and intermolecular structures [69], whereas at larger scales, well-defined structures can be generated by exploiting the self-assembling properties of conjugated chains.

Over the last decade, well-ordered thin films of conjugated polymers have been prepared by various techniques, e.g., stretch-drawing or the Langmuir–Blodgett technique, in which the conjugated backbones are preferentially oriented in one direction, along the substrate surface [70,71]. Oligomers can be processed by vacuum evaporation, yielding high-purity thin films that often exhibit a high degree of crystallinity and preferential orientation with respect to the substrate surface [72]. Such films display in-plane dichroism in photoluminescence, yielding polarized emission [73–75].

The general goal of those research efforts is to control the amount of π-aggregation of the conjugated polymers, which can lead to aggregate (even in solution) or excimer formation, and impact on the LED operation [76]. Approaches to avoid undesired aggregation of rigid polymer chains include the use of bulky side groups [77], end groups [78], or cross-linkable groups [79,80], and the synthesis of statistical copolymers [81,82]. π-Stacking can also be controlled via hydrogen bonding between side groups attached to the conjugated backbone [83–86].

Here, we describe our own approach to investigate and control the self-assembly on conjugated materials having different architectures, from the simplest (oligomers) to the more elaborate (e.g., block copolymers). For this purpose, we use a joint experimental/modeling methodology, combining scanning probe microscopy characterization of thin polymer deposits with molecular simulations of the polymer chain packing. The microscopic morphology on the nanometer to the micrometer scale, as determined by scanning probe techniques, is indeed a direct signature of the self-assembly processes taking place on the molecular scale. The structural details of the self-assembly can be addressed by molecular modeling. By analyzing the optical properties of these thin deposits, we can then establish the relationship between the molecular-scale packing, the mesoscale morphology in the solid state, and the macroscopic luminescence response.

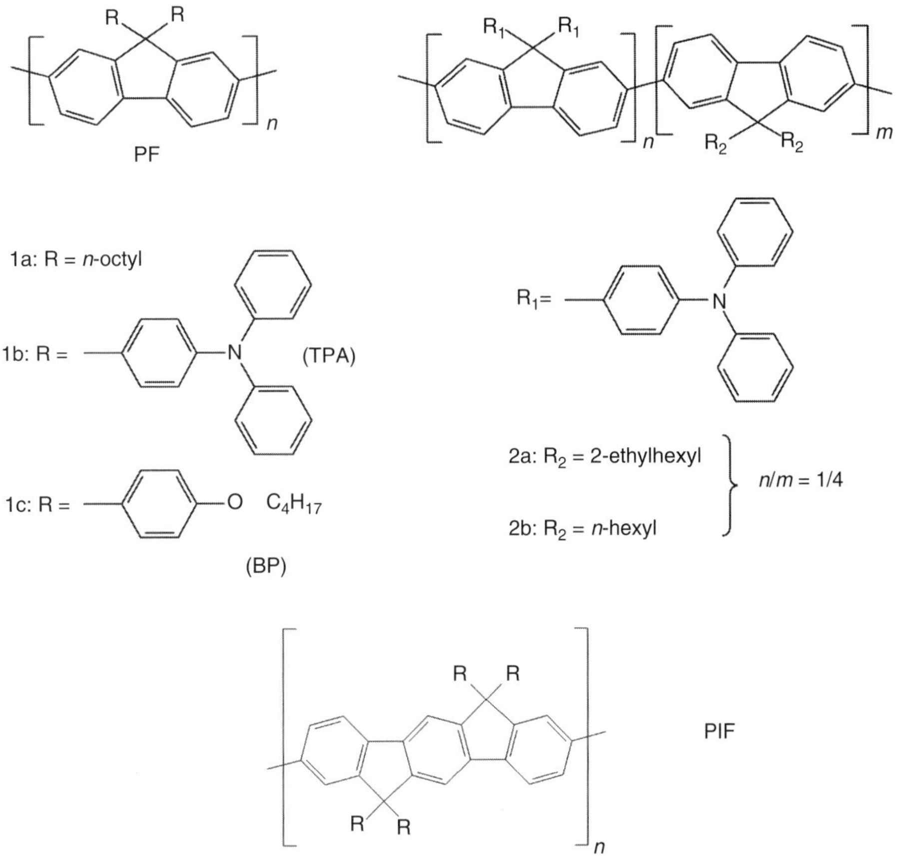

FIGURE 1.15 Chemical structure of poly(fluorene) (PF) and poly(indenofluorene) (PIF). R is either a linear or a branched alkyl chain or bulkier groups such as butoxyphenyl (BP) or triphenylamine (TPA). For PIF, R is either *n*-octyl or 2-ethylhexyl chain.

We illustrate this joint approach for a coherent series of compounds incorporating fluorene (F) and indenofluorene (IF) units (Figure 1.15). Polyfluorenes and poly(indenofluorenes) are good candidates as blue-emitters in polymeric light-emitting diodes (PLEDs) because of their very high photoluminescence quantum efficiencies compared with the other conjugated polymers (ϕ_{PL} approaching unity in solution and around 0.5 in the solid state), excellent solubility and thermal stability, and the ease in controlling their properties via facile substitution in the C9-position (the bridging site) [87–89]. Much effort has been devoted to tuning the optoelectronic properties of PFs through macromolecular engineering: e.g., specific substituents or end groups have been used to improve solubility [90], create cross-linkable polymers [80], and modulate the optical and charge injection properties [91–93]. Copolymers of fluorene with different conjugated moieties have been used to improve charge transport [94] or to tune the emission color [95]. Moreover, the liquid crystalline order of some PFs has opened the way to fabricate blue polarized electroluminescent devices [74].

1.4.1 A Joint Experimental/Modeling Methodology

In order to determine the microscopic morphology of materials, scanning probe microscopies are particularly interesting since they allow the visualization of nanoscale objects in direct space. This family of techniques is complementary to electron microscopies (because the measurements can be carried out on as-prepared samples, without special conditioning) and to scattering experiments (because they address the local scale rather than averaging over the whole sample). The surface morphology can be best observed when an AFM apparatus operates in tapping mode (TM), which greatly reduces tip-induced damage [96]. The ultimate resolution of TM-AFM depends on the sharpness of the probe; typically, it is on the order of a few nanometers.

Thin films of the conjugated compounds were prepared from solutions in good solvents, so that the final morphology reflects at best the "intrinsic" assembling behavior of the chains, rather than aggregation processes due to poor solvation. The amount of deposited material corresponds to a few nanomoles per square centimeter of the substrate. AFM measurements were performed on thin polymer deposits obtained by solvent casting on freshly cleaved mica or graphite (highly ordered pyrolitic graphite, HOPG) substrates. These surfaces are atomically flat, which precludes any influence of the substrate topography on the observed morphology. AFM images were recorded with a Nanoscope IIIa microscope operated in TM at room temperature in air, using microfabricated cantilevers (spring constant of 30/Nm).

AFM can provide an extremely detailed description of the microscopic morphology of the material under study. However, it usually does not allow to distinguish individual molecules or polymer chains. In order to understand the supramolecular organization leading to the observed morphology, models of the chain assembly can be proposed by using a theoretical approach based on MM and MD calculations. MM is a theoretical approach developed to provide information on the structure of large molecules such as proteins and polymers [97]. It basically uses a set of equations to reproduce the multidimensional surface of the energy of a molecule as a function of the nuclear positions, i.e., the "potential energy surface" [98]. This set of equations is referred to as the *force field*, and is based on analytical functions in which parameters are optimized to reproduce the properties of the molecules such as geometries, conformational energies, and heats of formation. The force field parameters are defined for each type of atom in a group of atoms and are obtained experimentally (mainly by spectroscopic methods) and/or using high-level quantum chemistry calculations. A typical expression of a force field relies on the decoupling of the different degrees of freedom: the energy of a molecule in one conformation is the sum of "valence energies" estimated for the given bond lengths, bending angles, torsion angles, etc. (and coupled terms in some force fields) and "nonbond" energies, such as van der Waals and electrostatic interactions. A force field is usually parameterized by the study of small, representative molecules. The assumption is made that these parameters can be transferred for larger, similar types of molecules. The accuracy in determining a stable structure is therefore dependent on the parameters and the analytical expressions of the force field. During the last 30 y, many types of increasingly reliable force fields have been developed. The first force fields were developed by Allinger et al. for the description of hydrocarbons, such as MM1 and related ones (MM2, MM3) (molecular mechanics programs 1–3, see, e.g., [99–101]). Later on, a number of force fields were developed. Without being exhaustive, we can cite the assisted model building with energy refinement (AMBER) [102–104] and chemistry Harvard macromolecular mechanics (CHARMm) [105] force fields for the description of biological molecules (proteins and nucleic acids), DREIDING [106], consistent force fields (CFF91 and CFF93) [107–109], and Tripos [110], for the study of small organic molecules and condensed-phase optimized molecular potentials for atomistic simulation studies (COMPASS) [111,112] and polymer-consistent force field (PCFF) [113–115] for polymers in the solid state.

For our study of poly(fluorene)- and poly(indenofluorene)-based compounds, one specific force field has been selected: PCFF, implemented in the Cerius [2] software by Accelrys. This choice is based on an extensive comparison between the geometrical results obtained with various force fields and the data obtained with high-level ab initio quantum-chemical methods (including correlation effects), and experimental data when available. Here, the most relevant geometrical factors are: the internal geometry of the fluorene units, the length alternation between carbon–carbon single and double bonds along the conjugated chain, the torsion potential around the single bond between the fluorene units, the torsion potential of alkyl and oligoethylene oxide groups, and the interchain separation in the solid state. The torsion behavior of alkanes and simple ethers has been described very accurately with correlated quantum-chemical techniques [116,117]. We find that PCFF accurately reproduces the positions of the energy minima (within a couple of degrees) and the height of the torsion barrier (within 1 kcal/mol) for these groups. The force field also provides reliable results for the internal structure of the conjugated units and the bond-length alternation, when compared to HF results. It correctly indicates that the adjacent fluorene units are expected to be tilted by about 40°, with a very low barrier (∼2 kcal/mol) to the fully planar structure. The distances between two polymer chains obtained from MM calculations are consistent with X-ray diffraction (XRD) studies of the crystalline phase in poly(fluorene) [89,118].

Such complex systems possess numerous local energy minima, and MM calculations tend to bring the system the closest possible to the initial geometrical configuration. It is thus essential to perform calculations starting from a set of "reasonable" but different configurations and to compare the relative stabilities of the final geometries. In order to reach stable final structures, it is also possible to combine MM calculations with other theoretical procedures, e.g., MD simulations, which account for the motion of atoms for a given time period at a given temperature. During a standard MD simulation, Newton's equations of motion are numerically integrated for the entire system of interest. This procedure allows the system to overcome local potential energy barriers and reach more stable energy minima. In the field of conjugated materials, only a few MM and/or MD studies have been reported on some homopolymers [119,120].

Once the typical microscopic morphology was established from TM-AFM measurements, MM or MD calculations were carried out on small clusters of the same polymers studied with AFM (considering monodisperse chains with a length corresponding to the molecular weight determined experimentally). These calculations allow us to propose models of solid state organization that are checked with regard to the observed microscopic morphology, with the goal of determining the arrangement of the polymer chains in thin solid films.

MD simulations are performed on the MM-optimized structures. They rely on the NVT ensemble, in which particle number, volume, and temperature are kept constant. The canonical Hoover heat bath coupling scheme [121] is used with a relaxation time (related to the coupling constant to the thermal bath) of 0.01 ps to maintain the temperature at 300 K. Verlet algorithms [122] update the atom coordinates and velocities with time steps of 0.001 ps. Run times are between 450 and 1050 ps.

1.4.2 Self-Assembly of Homopolymer Chains

As a starting point, we study the microscopic morphology of thin deposits of a low-molecular weight poly(fluorene) ($n = 8$) substituted with linear octyl substituents. Figure 1.16 illustrates the typical AFM results obtained for thin deposits on a mica and a graphite substrate. On both substrates, for deposits

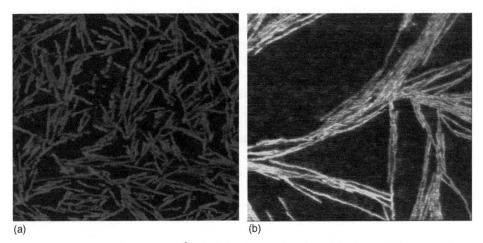

(a)   (b)

FIGURE 1.16 AFM images ($1.0 \times 1.0 \ \mu m^2$) of a $(F)_8$ oligomer deposit on (a) mica and (b) on graphite.

from good solvents like THF or toluene, the AFM images reveal the presence of "fibrillar" structures, i.e., one-dimensional objects with typical width of a few nanometers and length of several hundred nanometers to several micrometers, as shown in Figure 1.16. In these AFM images, the fibrils are seen as bright elongated objects, the substrate appearing dark. This fibrillar morphology appears throughout the sample and can be reproduced independently of the (good) solvent and the substrate used; it is thus the result of the intrinsic self-assembly properties of the chains. Here, we use AFM to determine the exact shape of the fibrils; this technique indeed provides three-dimensional topographic information, which is not the case in transmission electron microscopy (TEM). In thick deposits, the objects to be measured are often very close to each other and/or entangled, which makes the determination of their shapes rather delicate. In contrast, in the very thin deposits, individual fibrils are easily distinguished and their profile can be measured accurately. When analyzing the images, we have found that the width and height are rather constant along a given fibril, which suggests that they are made of regular arrangements of the polymer molecules. Wider fibrils most probably correspond to the aggregation of two (or more) narrow ones. Most importantly, the topographic section analysis of the fibrils indicates that their width is at least one order of magnitude larger than their height. In other words, these objects are not cylinders; they rather correspond to flat ribbons.

Based on the analysis of various similar systems including poly(paraphenylene) or poly(parapheny-leneethynylene) polymer chains [123], we can reasonably propose that the ribbon-like morphology observed in the fluorene-based compound is best understood if we consider that the molecules form long, one-dimensional π-stacks. While the formation of such stacks appears to be easy for PPP- and PPE-based systems (for which the alkyl substituents can remain in the plane of the conjugated backbone), the question can arise as to how packing takes place in fluorene-based polymers, where the alkyl groups are likely to lie out of the plane of the conjugated system, because they are attached to an sp^3-hybridized site.

We have modeled the assembly of fluorene oligomers with ten monomer units. As a first step, they are substituted with two methyl groups on position 9. Longer alkyl substituents (from hexyl upwards) are located in that position in the actual polymers to provide solubility. The use of methyl groups allows one to envisage many different arrangements in the aggregates (while keeping the computational effort at a reasonable level) and provides preliminary indications on the effects of steric hindrance. The next step in the modeling then consists in replacing the methyl groups by longer substituents.

MM calculations indicate that isolated oligofluorene molecules are nonplanar: the most stable configuration giving rise to linear chains is the anticonformation with alternating angles of $\pm 40°$ between the planes of adjacent monomer units (Figure 1.17a). The calculations indicate that the potential energy increase required to reach the antiplanar conformation is quite small at around 1.5 kcal/mol per monomer unit. When two methyl-substituted chains are brought into contact to form a π-stack, two geometric configurations are expected to yield minimal steric hindrance. In the first one, the monomer units are superimposed, parallel to each other, but they are flipped so that the methyl groups point in opposite directions (Figure 1.17b; this will be referred to as the "flipped" configuration). In the second one, the two chains are shifted by half a period relative to each other, so that the methyl groups in one chain are located in front of the interunit bond of the other chain (Figure 1.17c; this is the "shifted" configuration). Upon optimization, the fluorene units in the flipped system remain parallel to each other but slightly drift sideways, while in the shifted configuration, drift occurs along the chain axis. However, these effects are small, the main conclusion from the calculations being that the assemblies are stable: the equilibrium interchain distance is 0.37 nm and the binding energy (i.e., the energy gained by making the two molecules interact) is around 70 kcal/mol.

We now consider assemblies made up of four chains, either in the flipped or shifted configuration (Figure 1.18). The major differences (compared to the two-molecule assemblies) appear for the two central molecules, which experience more intermolecular contacts (since they have two neighbors), while the two outer molecules behave like those in the two-molecule systems described above. In the central part of the four-molecule aggregate, we find that the intermolecular distance tends to increase up to 0.45–0.50 nm and the chains tend to become more planar (the dihedral angle between adjacent

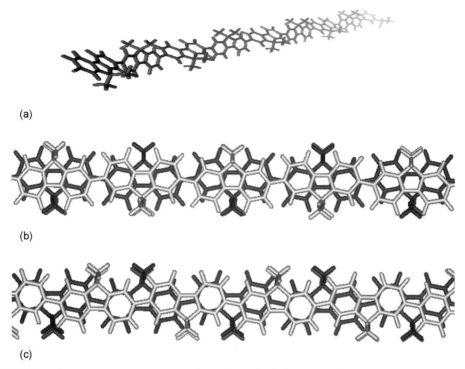

FIGURE 1.17 (a) Geometric structure of an isolated dimethyl-substituted poly(fluorene) chain; note the alternating angle of $\pm 40°$ between the planes of adjacent monomer units. Stacks of two dimethyl-substituted poly(fluorene) chains (b) in the flipped configuration and (c) in the shifted configuration.

monomer units decreases to $\pm 25°–30°$). These geometrical changes probably result from a compromise between the steric hindrance due to the methyl groups, which tends to push the molecules apart, and π–π interactions, which tend to make the conjugated backbone planar with the molecules close together. It is important to notice that the stability of those larger assemblies is similar to that of

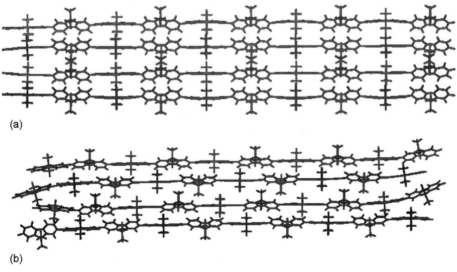

FIGURE 1.18 Stacks of four dimethyl-substituted poly(fluorene) chains (a) in the flipped configuration and (b) in the shifted configuration.

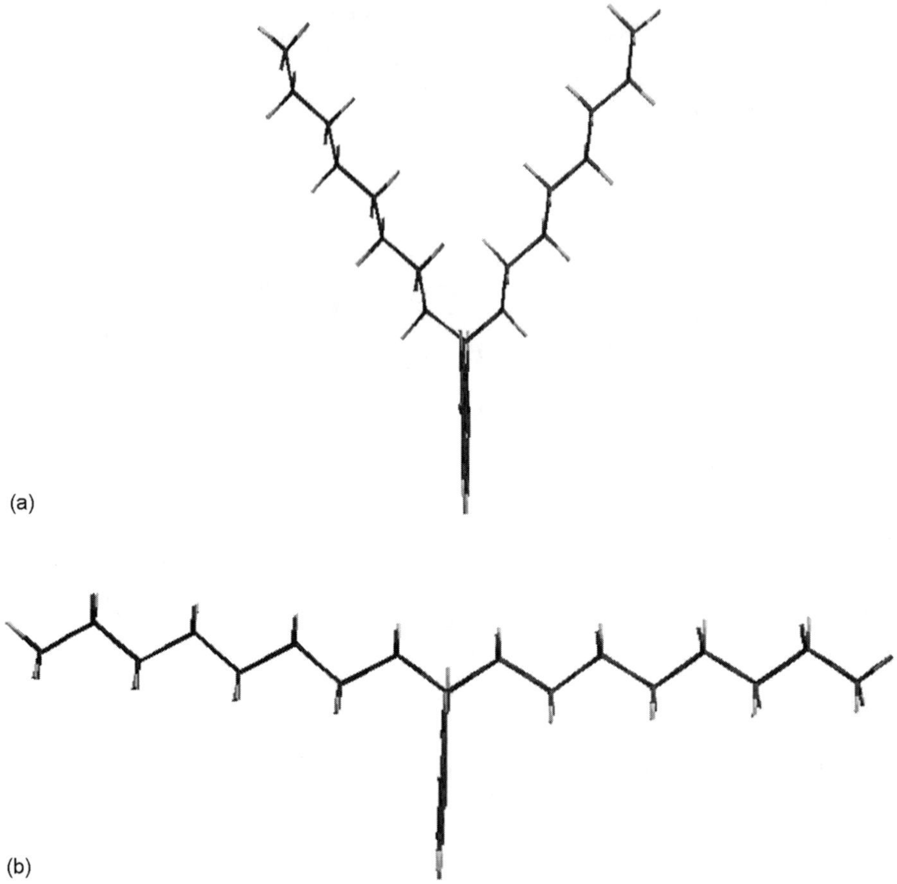

(a)

(b)

FIGURE 1.19 Geometric structure of dioctyl-substituted fluorene. The octyl groups are in the zig-zag planar conformation, with the second C–C bond either (a) roughly parallel (Y-shape) or (b) perpendicular (T-shape) to the fluorene plane.

the two-molecule systems described above. This confirms the possibility for the methyl-substituted poly(fluorene) chains to form π-stacks. As observed by AFM, for fibrils a few hundred nanometers long, considering that the equilibrium distance between chains is about 0.5 nm, the number of molecules in a given fibril is more than 1000.

The decrease in the torsion angle found between the fluorene units in the inner molecules of the assemblies prompted us to assess the packing of fully planar chains. We have found that both the flipped and shifted configurations of planar chains are stable, with the relative position of the molecules in the assembly very close to those found for the nonplanar systems (Figure 1.17b and Figure 1.17c). The stability of the assemblies of planar molecules is thought to originate from the improved intermolecular π–π interactions (because the conjugated systems are fully parallel). Therefore, from this point on, all assemblies have been built with planar poly(fluorene) backbones.

When considering longer alkyl substituents, the conformational behavior of the alkyl groups emerges as a new important factor in the modeling of chain packing. This aspect has been investigated for a series of systems with alkyl groups ranging from ethyl to *n*-octyl. We have considered that the alkyl groups remain in the zig-zag planar conformation, since it is the one most likely to allow for ordered packing of the chains. Globally, these studies indicate that the side groups can orient in two different ways with respect to the plane of the fluorene unit, as illustrated in Figure 1.19 for *n*-octyl moieties.

In the first conformation, the second C–C bond of the octyl unit is almost parallel to the fluorene plane; the dioctyl fluorene unit is then Y-shaped (Figure 1.19a). In the other conformation, the octyl groups are perpendicular to the fluorene plane; the monomer unit is then T-shaped (Figure 1.19b). The corresponding two-molecule assemblies are shown in Figure 1.20 for the flipped configuration. From the views along the axis of the conjugated backbone, it clearly appears that the space between the two poly(fluorene) chains is basically empty for the Y-shaped system (Figure 1.20b), which is expected to allow close packing of the polymer chains. In contrast, in the T-shaped aggregates, hydrogen atoms of the octyl groups are located within that space (Figure 1.20d).

Such a system is not stable; the calculations indicate that the octyl groups reorient toward a Y-shaped conformation. If the octyl groups are constrained into the T-shape conformation, the two molecules drift away from each other, to distances greater than 0.70 nm. Larger assemblies with Y-shaped monomer units are also stable, with equilibrium distances between the poly(fluorene) chains of about 0.50 nm. These results indicate that long alkyl side groups, provided they find the proper conformation,

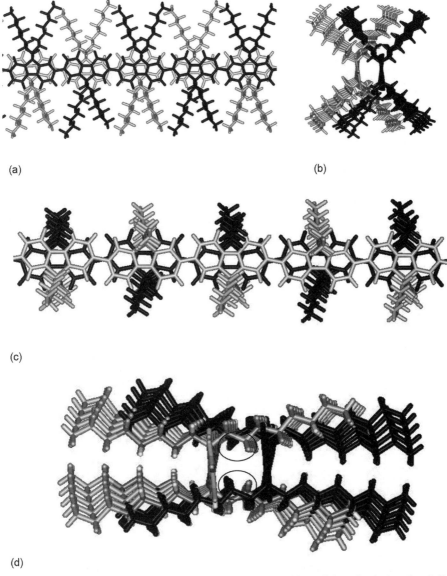

(a)   (b)

(c)

(d)

FIGURE 1.20 Top and side views of two-molecule "flipped" assemblies of dioctyl-substituted poly(fluorene) chains, with the octyl groups in (a,b) Y-shape conformation or (c,d) T-shape conformation.

do not prevent close packing of poly(fluorene) chains into π-stacks. This in turn enables the understanding of the fibrillar morphology observed for the thin deposits of the octyl-substituted poly(fluorene) homopolymer (see Figure 1.16).

1.4.3   Influence of the Side Groups

To describe the impact of substitution on the supramolecular organization, we investigate thin deposits of PFs with aryl-based substituents: poly(9,9-ditriphenylaminefluorene) (PTPAF) and poly(9,9-di-4-butoxyphenylfluorene) (PBPF) (1c). PTPAF (1b) is of particular interest, because triphenylamine has an ionization potential of 5.3 eV, i.e., between those of PDOF (around 5.7 eV) [124] and ITO (which acts as the anode in LEDs, work function $\phi = 5.0$ eV), which globally improves the hole injection in such devices. The microscopic morphology of thin deposits of PTPAF shows a uniform, smooth, very flat surface (the value of the roughness being 0.3 nm over 1 μm^2), without any specific structure (not shown here).

As described above, octyl-substitution of the polyfluorene chains allows for a close packing of the conjugated segments (π-stacking), while the steric hindrance due to the bulkier aryl-based substituents is expected to prevent this type of stacking. To confirm this hypothesis, we performed MM calculations on stacks of a few oligomer chains (from two to four oligomer stacks of hexamers to decamers). For stacks of two chains, the results show that the system strongly stabilizes when the distance between the PF backbones is increased from 0.4 to 0.7 nm. Despite an extensive conformational search, we have found no stable configuration in which the distance between two adjacent π-backbones comes down to about 0.5 nm, in contrast to the case of the alkyl-substituted polymer. Here, the chains tend to move away and drift sideways from each other due to the steric repulsion among the bulky phenyl substituents. This is illustrated in Figure 1.21a, which shows the most stable configuration obtained after minimization from the starting "shifted" configuration: the equilibrium distance between chains is found to be near 0.8 nm (PF backbones are shown perpendicular (left) and parallel (right) to the view).

The results for the "flipped" configuration are shown in Figure 1.21b: a nonparallel arrangement of the PF chains is found (see the arrows pointing at the two ends of one of the chains) and the equilibrium distance is between 0.75 and 0.80 nm. Thus, in the case of phenyl substituents, the PF chains cannot pack closely in a cofacial way. The calculated equilibrium distances between adjacent backbones are much larger than the values obtained in the case of di-octyl-substituted oligofluorenes; they are clearly

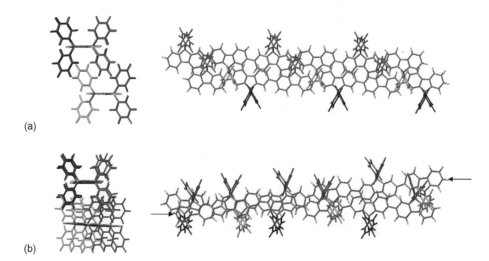

(a)

(b)

FIGURE 1.21   Molecular modeling of the assembly of phenyl-substituted PF chains. Optimal configurations obtained from (a) the "shifted" starting configuration and (b) the "flipped" starting configuration; views with PF chains perpendicular (left) and parallel (right).

inconsistent with the formation of regular, compact π-stacks, i.e., the interchain distance is too large for significant overlap of π-orbitals to occur. Again, this trend was confirmed in the calculations of the three-chain and four-chain clusters. Since the actual polymers have substituents that are phenyl-based and are even bulkier, it is expected that in the solid state they would form assemblies of molecules with no long-range organization and therefore nontextured microscopic morphologies.

The impact of the microscopic morphology on the optical properties clearly appears on the photoluminescence spectra. These spectra may contain the signature of the π−π interacting interchain species (optical species) with a well-defined vibronic structure, due to the diffusion of excitons to the lowest energy sites. Such spectra determine the applicability of the materials in PLEDs. Here we compare the photoluminescence spectra of poly(fluorenes) substituted by *n*-octyl 1a (PDOF; see Figure 1.15) and triphenylamine 1b (PTPAF; see Figure 1.15) groups, respectively. In solution (in chloroform, in which the compounds are molecularly dissolved), the two polymers produce essentially identical spectra with well-defined vibronic progression (Figure 1.22 top): the emission maxima are located at 419 nm (2.96 eV) and 442 nm (2.81 eV), and a band around 470 nm, which are assigned to intrachain singlet 0–0, 0–1, and 0–2 transitions, respectively. In contrast, the solid state fluorescence spectra of 1a and 1b are markedly different (Figure 1.22 bottom): PDOF 1a shows a vibronic progression with two well-defined maxima at 442 nm (2.81 eV) and 468 nm (2.65 eV), along with an additional broad band above 500 nm. The large red-shift (compared to the solution spectrum) and the broad band above 500 nm have been initially attributed to emission from excimers (interchain species, see Section 1.2.) [125]. Recently, the band in the 400–550 nm green region (often referred to as the *g-band*) has been shown to be due to emission from fluorenone-like defects due to oxidative (photo- or electro-) degradation of the polymer [126–128]. To solve this problem, an electron-transporting material acting as a buffer layer can be used between the emissive polymer and the Ca/Ag (Ba/Ag) cathode [129]. In this case, the buffer layer stabilizes the blue color and improves

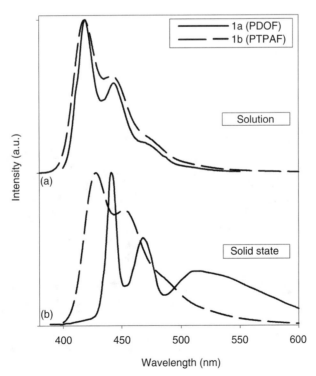

FIGURE 1.22 Normalized photoluminescence spectra of PDOF 1a (solid line) and PTPAF 1b (dashed line): (a) solution spectra (in chloroform); (b) solid state spectra.

the brightness of the device by blocking or limiting the diffusion of the calcium atoms into poly(fluorene), and thereby blocking the formation of fluorenone. In addition, the buffer layer prevents the formation of bipolaron states.

In contrast, the fluorescence of PTPAF 1b closely resembles its solution spectrum, with maxima at 428 nm (2.90 eV) and 452 nm (2.74 eV), and no peak in the low-energy region. The same is true for the solid state emission spectrum of PBPF (1c), i.e., another PF polymer with bulky side groups, which shows a peak at 457 nm (2.71 eV, pure blue emission), with no emission in the low-energy region.

In our opinion, the difference in the photoluminescence properties of PDOF on one side and PTPAF and PBPF on the other can be rationalized in the following way: since the PDOF chains are closely and regularly packed, electronic excitations can easily migrate (as we have shown) with quantum-chemical calculations [9] and reach fluorenone defects, from which red-shifted emission takes place. In contrast, the poor packing of PTPAF and PBPF chains most probably reduces the exciton migration range [9], thereby preventing the fluorenone units from being reached, and the emission then occurs from the poly(fluorene) segments. The weakness of the intermolecular interactions in the latter two polymers is also underlined by the similarity of the spectra in solution and in the solid state.

The conclusions drawn on the packing of PF chains can be extended to a parent polymer, poly (indenofluorene) (PIF; see Figure 1.15). Indeed, thin deposits of *n*-octyl-substituted PIF (OPIF) show fibrillar patterns very similar to those observed for the other polymers in the series (Figure 1.23a). This suggests that the packing of the PIF chains is similar to that proposed for the poly(fluorene) chains.

Interestingly, when the PIF backbone is substituted with 2-ethylhexyl groups (which possess the same number of carbon atoms as *n*-octyl substituents), the fibrillar morphology is completely suppressed; instead, thin deposits of 2-ethylhexyl-substituted PIF (EHPIF) are featureless (Figure 1.23b). On the basis of previous arguments, this suggests that the packing of EHPIF chains is markedly different from that of OPIF chains, despite the fact that the two polymers are isomers. This difference in microscopic morphology is thought to be due to the difference in sterical hindrance brought by the branched 2-ethylhexyl groups, compared to the linear *n*-octyl groups, which affects the way the polymer chains can pack. The geometric structure of the EH-substituted fluorene unit (Figure 1.24a) is indeed rather globular. As a consequence, when bringing two EH-substituted chains close together, strong steric hindrance sets in among the side groups of the two chains. The calculations indicate that the conjugated chains tend to move away from each other, to distances larger than 0.70 nm (whatever the conformation

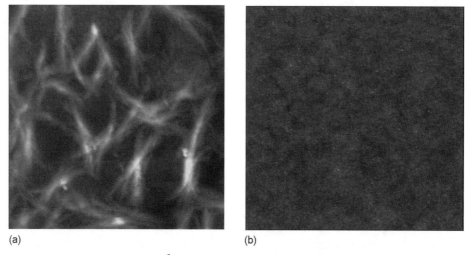

(a)   (b)

FIGURE 1.23   AFM images ($10 \times 10\ \mu m^2$) of thin films of (a) *n*-octyl-substituted and (b) 2-ethylhexyl-substituted poly(indenofluorene).

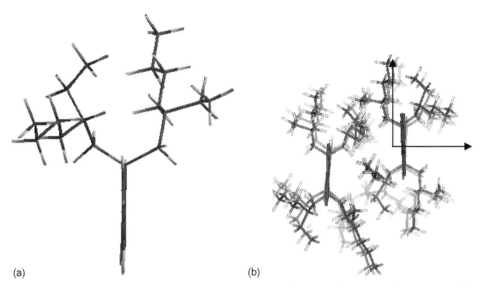

FIGURE 1.24 (a) Geometric structure of 2-ethylhexyl-substituted fluorene and (b) assembly of two 2-ethylhexyl-substituted poly(fluorene) chains, with the conjugated backbones oriented perpendicular to the sheet. The arrows indicate the direction of the drift of the chain upon geometry optimization.

of the EH groups) and the two backbones move sideways, in order to accommodate at best the branched alkyl groups. This is illustrated in Figure 1.24b, with the arrows indicating the direction of the drift of one chain relative to the other. Therefore, no π-stacking of the conjugated segments is expected to occur in such compounds, and the morphological signature of long-range π-stacking, i.e., fibrillar structures, is absent in the deposits observed with AFM for the parent EHPIF polymer.

The difference in the ability of the PIF chains to pack, due to the sterical hindrance induced by the alkyl groups, directly affects the luminescence properties in the solid state. Both OPIF and EHPIF are excellent blue emitters in solution (where polymer molecules are isolated). However, OPIF shows a broad, red-shifted emission in the solid state while the luminescence behavior of EHPIF is not significantly modified [130]. Again, the shift in the emission of OPIF is due to aggregation of chromophores in the solid state, excimer formation, and/or fluorenone-type emission. Conversely, the persistence of blue emission for EHPIF indicates that luminescence still originates from isolated chromophores. This observation is fully consistent with the models proposed for chain packing in *n*-octyl- and 2-ethylhexyl-substituted systems. Our simulations indeed show that *n*-octyl groups still allow the conjugated backbones to be close to each other ($\approx$0.50 nm) and oriented in such a way that intermolecular π–π interactions can take place. In contrast, the relative position of the conjugated systems in assemblies of 2-ethylhexyl-substituted chains is expected to greatly reduce any interchain π–π interaction [9], so that the luminescence is still that of isolated polymer molecules.

1.4.4 Self-Assembly of Copolymers

1.4.4.1 Block Copolymers

Using block copolymers is a powerful approach for the preparation of self-assembled materials [131–136]. Block copolymers in general are known to generate a variety of morphologies [137,138] due to the selective solvation in solution and the microphase separation in the solid state. Therefore, they provide an attractive, nonlithographic, alternative route to nanostructures. Among the different types of block copolymer architectures, coil–coil diblock copolymers have been studied most intensively and their phase diagram is now reasonably well understood. Replacing one (or more) block(s) of a coil–coil block copolymer by rigid segment(s) has a number of consequences. First, the self-assembly of such "rod–coil" block copolymers—so-called because at least one segment is rigid while the other(s) tend(s) to coil—is no longer solely determined by the microphase separation of the constituent blocks, but is also affected by the self-organization of the rod segment. The competition between these two processes can lead to morphologies that are distinctly different from those known for coil–coil block copolymers. A second important consequence of the rod–coil architecture is the stiffness asymmetry between the blocks, which leads to an increase in the Flory–Huggins χ-parameter and allows phase separation to take place at degrees of polymerization where coil–coil block copolymers are normally still miscible.

It is therefore not surprising that, in the field of conjugated polymers, controlling their morphology via the synthesis of block copolymers, including at least one conjugated sequence and one nonconjugated sequence, is attracting much interest. The demanding synthetic access to rod–coil or coil–rod–coil block copolymers has made these compounds available only recently [139–153]. The work of Stupp, François, et al. [83,84,143,144], Jenekhe [145–148], Hadziioannou [149–151], Swager [152], and Yu [153] and their coworkers can be emphasized. In all these cases, the properties rely on the capacity of the covalently linked rod–coil blocks to form phase-separated morphologies.

Figure 1.25 describes the fluorene-based block copolymers considered here as illustrating examples. Solvent-cast thin deposits of the block copolymers generally also show a fibrillar morphology. This is illustrated in Figure 1.26 for PF-*b*-PEO and PS-*b*-PF-*b*-PS block copolymers. The details in the shape of these elongated objects vary from one compound to the other: for instance, fibrils can be extremely long (up to 10 μm) and curved while in other cases they tend to be shorter and straighter. Nevertheless, from the AFM analysis of more than ten of these rod–coil block copolymers, it is clear that fibrils are the morphological signature of these compounds in the solid state, similar to the corresponding homopolymer. In other words, when the deposits are prepared by evaporation of a solution in a good solvent for both blocks, formation of fibrils appears to be favored over other morphologies, such as globular micelles or vesicles, or extended two-dimensional assemblies (lamellae).

It is important to recall that the appearance of the fibrillar morphology is not critically dependent on the film thickness; when very dilute solutions are used, discontinuous deposits are formed. In most cases, the fibrils are dispersed over the substrate, with narrow fibrils occasionally meeting to form broader features.

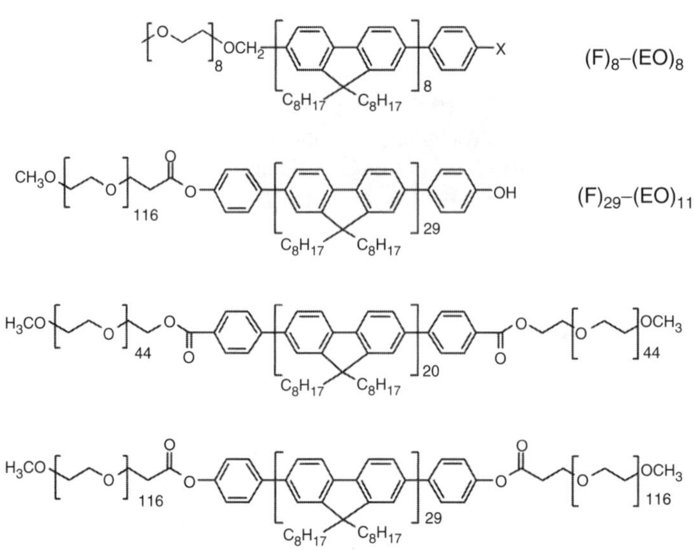

FIGURE 1.25 Chemical structures of the fluorene-based block copolymers.

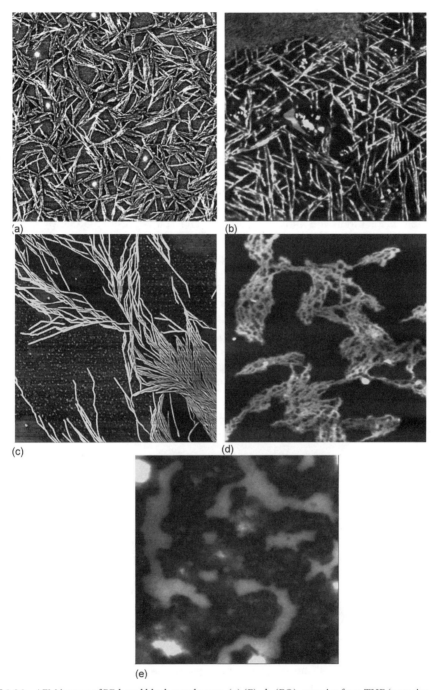

FIGURE 1.26 AFM images of PF-based block copolymers. (a) $(F)_8$-b-$(EO)_8$ on mica from THF (scan size $= 1$ μm); (b) $(F)_8$-b-$(EO)_8$ on graphite from THF (scan size $= 1$ μm); (c) $(EO)_{45}$-b-$(F)_{20}$-b-$(EO)_{45}$ on mica from toluene (scan size $= 2.5$ μm); (d) $(EO)_{116}$-b-$(F)_{29}$ on mica from toluene (scan size $= 1$ μm); (e) $(EO)_{116}$–$(F)_{29}$–$(EO)_{116}$ on mica from toluene (scan size $= 2$ μm).

At this stage, one can raise the question as to whether the fibrils observed with AFM are micelles pre-formed in solution that deposit on the substrate as the solvent evaporates, or assemblies forming upon adsorption of the polymer molecules on the substrate. As mentioned above, the preparation procedure used here (dilute solutions in good solvents for both blocks and slow solvent evaporation) was designed to prevent aggregation in solution. We thus believe that the observed morphologies are the result of molecular assembly at the substrate surface. This hypothesis is supported by the fact that on graphite substrates, the fibrils show preferential orientation along three directions at 120° from each other, as illustrated in Figure 1.26b for the case of PF-*b*-PEO. This is reminiscent of the crystallographic symmetry of the graphite surface, which indicates that molecule–surface interactions play a significant role during the deposition. A fibril growing at the surface, with new polymer molecules continuously reaching the substrate and interacting with it, is more likely to be influenced by the presence of the surface than a pre-formed fibril precipitating from the solution as the film dries. Preferential fibril orientation therefore suggests that molecular aggregation mostly takes place on the substrate surface.

The formation of quasi-one-dimensional objects such as fibrils indicates that the copolymer molecules assemble in a highly anisotropic way. Moreover, phase separation between the blocks most probably takes place within the fibrils (because rod and coil segments are known to be strongly immiscible). The final morphology must therefore be a compromise between the molecular organization typical of each subphase. This is a strong indication that the morphology found in the copolymers is governed by the assembly of the conjugated segment, which probably forms the central part of the fibril, with the nonconjugated sequences located at the periphery. Consistently, the luminescence spectrum of the $(F)_8$-b-$(EO)_8$ copolymer in the solid state shows red-shifted emission, typical of strongly interacting poly(fluorene) segments.

AFM investigations of thin films of $(EO)_{45}$–$(F)_{20}$–$(EO)_{45}$ deposited from toluene on mica show a morphology of large areas made of aggregated tortuous fibrils (Figure 1.26c). The measured height and width of those fibrils are 2.0 and 15–20 nm, respectively. The length of the fully extended molecule is estimated by molecular modeling to be nearly 50 nm (with the conjugated segment being 16.7 nm long). The presence of the fibrillar morphology suggests that π-stacking of the PF segments also takes place in this triblock copolymer. Possible configurations for the PEO chains could be: (i) in coiled conformation on the sides of the PF core, or (ii) extended over and/or (iii) under the ribbon formed by the PF chains.

For $(EO)_{116}$–$(F)_{29}$, i.e., a diblock copolymer with a longer nonconjugated chain but with an average volume ratio of PEO $f_{PEO} = 0.29$ close to that of $(EO)_{45}$–$(F)_{20}$–$(EO)_{45}$, AFM investigations show no formation of fibrils in the case of thin deposits from THF. When prepared from toluene, thin deposits of the same compound exhibit a microscopic morphology of aggregated fibrils, as depicted in Figure 1.26d. Again, this is most probably due to assembly of the PF blocks into long π-stacks. The height and the width of the fibrils are 2 and $\sim$20 nm, respectively. These dimensions suggest that the PEO chains are not in fully extended conformations along the PF chains forming the core of the fibrils, but probably lie under the ribbon, because of favorable interactions with the mica surface. Again, the height of the fibrils is consistent with such an organization. The $(EO)_{116}$–$(F)_{29}$–$(EO)_{116}$ compound does not form fibrils, but rather nontextured aggregates of various sizes (from 10 nm to few hundred nanometers), appearing in gray in the Figure 1.26e. Whatever the solvent, nonorganized aggregation overcomes the formation of nanoribbons; in this case, we propose that the hindrance and/or assembly of the PEO chains overcomes the π-stacking of the conjugated units. Although no fibrillar structure was observed in deposits of $(EO)_{116}$–$(F)_{29}$–$(EO)_{116}$, the solid state photoluminescence spectra indicate that poly(fluorene) blocks interact to a certain extent in those deposits, since a band in the 400–500 nm region is still observed, although with a lower intensity than that of $(F)_8$–$(EO)_8$.

The experimental results show that the block copolymers are able to form ribbon-like assemblies in the solid state, as a result of the assembly of the conjugated segments. In the corresponding simulations, we find that the conjugated segments tend to form stable π-stacks while the nonconjugated blocks tend to coil. The π-stacking of a large number of conjugated segments is expected to produce thin elongated objects (Figure 1.27). The calculations indicate that this process is energetically favored over the side-by-side assembly of the chains with interdigitation of the alkyl

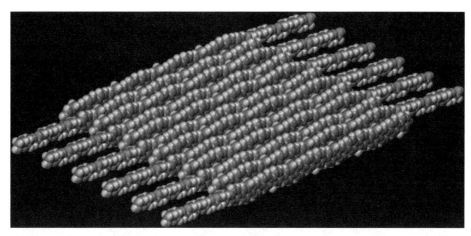

FIGURE 1.27   Molecular modeling of a $(F)_8$-b-$(EO)_8$ stack illustrating the most energetically favored (head-to-tail assembly) situation. The alkyl groups on the fluorene units are not represented.

substituents. This is related to the increased attractive van der Waals interactions due to the overlap of the π-systems. Straight fibrils are expected to form when the molecules pack in a head-to-tail fashion, i.e., with the nonconjugated blocks of neighboring chains located on opposite sides of the ribbon. This arrangement prevents steric hindrance betwen the nonconjugated segments while keeping compact π-stacking of the conjugated blocks at the center of the fibril. In contrast, head-to-head assembly could lead to curved objects, due to the steric hindrance between adjacent nonconjugated segments.

Compared to the other procedures for film preparation, the method we use (casting on a flat mica substrate and slow evaporation of the solvent) leads to individual fibrils where the conjugated segments are perpendicular to the nanoribbon axis. The Langmuir–Blodgett molecular processing of conjugated polymers into highly aligned films, as proposed by Swager et al. [154], leads to the formation of nematic liquid crystalline monolayer films that structurally evolve into fibril aggregates. There is a natural tendency of rigid rods to align perpendicular to the direction of compression and parallel to lines of flow between the water and the substrate induced by dipping [155].

As a technique complementary to AFM, near-field scanning optical microscopy (NSOM) studies have reported the nanoscale topographic and fluorescence features of poly(fluorene)s [156–158]. From the NSOM experiments, it is possible to quantify the film optical anisotropy on the local scale by measuring the polarization of the emitted light. The intensity of fluorescence is found to be the most when collected perpendicular to the fibril axis. Since the fluorescence is polarized along the conjugated backbone, this indicates that the ribbons are indeed composed of poly(fluorene) chains stacked orthogonal to the ribbon axis.

1.4.4.2 Statistical Copolymers

Another way to influence the organization of conjugated polymer chains is to consider random or alternated copolymers. In this context, we have investigated two PFs obtained by random copolymerization: triphenylaminefluorene copolymerized in a 1:4 ratio with *n*-ethylhexylfluorene 2a or with hexylfluorene 2b (Figure 1.15). Thin deposits of 2a reveal uniform, smooth featureless films, with a roughness of only 0.6 nm (Figure 1.28a). Such a low value indicates that these films are particularly flat; we did not observe any fibrils (or organized structures) for this system. Films of the copolymer with the same chemical structure as 2a but with a 1:1 ratio of the co-monomers also show a smooth microscopic morphology, without any specific structure. Molecular modeling calculations indicate that ethylhexyl-substituted PF chains cannot close-pack,

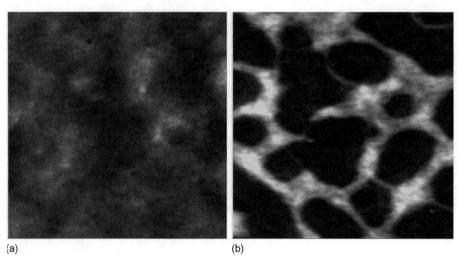

(a)   (b)

FIGURE 1.28   AFM images of thin deposits on mica of (a) 2a (2.0 × 2.0 μm^2) and (b) 2b (1.0 × 1.0 μm^2). The vertical gray scale is 5.0 and 7.0 nm, respectively.

because of steric hindrance among substituents; instead the conjugated backbones of adjacent chains tend to move away from each other, with distances between adjacent backbones larger than 0.70 nm. The fact that polymer 2a exhibits no fibrillar morphology is thus consistent with the fact that both homopolymers of triphenylaminefluorene and ethyhexylfluorene do not allow for a close packing of the chains.

The microscopic morphology of thin deposits of 2b is shown in Figure 1.28b. One can distinguish three levels of contrast, corresponding to the substrate (in dark) and to thin wires (in gray), which join brighter, flat areas. We find from image analysis that the wires have relatively constant widths, between 20 and 30 nm, and a constant height of about 2.0 nm, while the connected bright structures have varying heights (between 4.0 and 6.0 nm) and lateral dimensions (from few tens to few hundreds of nanometers). The large structures may originate from dewetting, while the constant dimensions of the wires and their aspect ratio are reminiscent of the fibrillar morphology. We believe that these objects are an indication that some long-range organization of the chains can occur. This is consistent with the fact that the chains are rich in *n*-hexylfluorene units (four times more abundant than triphenylamine-fluorene units). It is likely that relatively long sequences of *n*-hexylfluorene units are present along the chains; such sequences could pack locally with corresponding segments in neighboring chains, as occur in PDOF, thereby giving rise to organized structures (the wires). However, the presence of TPA-substituted units tends to disrupt close packing and long-range organization and leads to the prevalence of amorphous structures (the connected bright structures).

The emission spectrum of 2a in solution (in chloroform) shows two bands appearing at 417 and 439 nm (2.97 and 2.82 eV, respectively); the solid state spectrum is very similar to the bands slightly shifted to 422 and 446 nm (2.94 and 2.78 eV, respectively, see Figure 1.29). The absence of a significant red-shift in fluorescence is consistent with a nonorganized microscopic structure (smooth film), due to a molecular architecture with crowded substituents. Polymer 2b exhibits the same type of solid state spectrum as 2a (Figure 1.29) for the two major peaks; however, it also shows a long wavelength band extending into the green, which is not present for polymer 2a. Most likely, the difference in stability toward oxidation does not play a major role (since the only difference between the two systems is the structure of the alkyl substituents -hexyl vs. ethylhexyl-, with identical aryl content). Therefore, the long-wavelength peak is attributed to the possibility, within the wires, of facile migration of excitons to emissive defect sites arising from dense local packing and interactions among chains and/or fluorenone excimers.

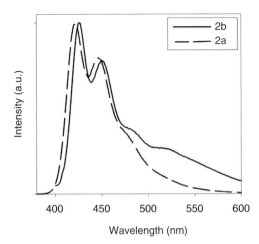

FIGURE 1.29 Normalized solid state photoluminescence spectra of 2a (dashed line) and 2b (solid line).

In random (PTOIF-TEHIF) copolymers, the ratio between the two monomers controls the microscopic morphologies of thin deposits: a 9:1 ratio (PTOIF-PTEHIF) leads to fibrillar structures while a 1:9 ratio leads to a nontextured, granular morphology. As a consequence of these different microscopic morphologies (and therefore intermolecular interactions), the luminescence properties strongly differ: the compound forming fibrillar structures shows a red-shifted "excimer-like" emission and a reduced solid state luminescence quantum yield compared to that forming untextured deposits.

1.5 Synopsis

In this chapter, by reviewing some of the recent theoretical and experimental work performed in the Georgia Tech/Mons-Hainaut group, we have described the progress in our knowledge of the relationships between the solid state organization of conjugated molecules and polymers and their electronic and optical properties. In particular, we have discussed:

1. The electronic interactions present in crystals of conjugated oligomers and the role of specific structural defects
2. The impact of charge injection on the optical properties of conjugated oligomers and polymers
3. The solid state organization of conjugated polymers

Fuelled by the prospect of an increasing number of applications, from television displays to solar cells, the field of organic electronics and photonics is more attractive than ever. There is no doubt that the fabrication of more efficient devices passes by a deeper understanding of the structure–property relationships reviewed in this chapter.

Acknowledgments

The work at Georgia Tech was partly supported by the National Science Foundation (through the STC Program under Award No. DMR 02-38307, the MRSEC Program under Award No. DMR-0212302, the CRIF Program under Award CHE-0443564, and Grant No. CHE-0342321) and the Office of Naval Research. The work in Mons was partly supported by the European Commission and Government of the Region of Wallonia (Phasing Out—Hainaut) and the Belgian National Fund for Scientific Research (FNRS/FRFC). The authors thank P. Brocorens, A. Calderone, C.D. Frisbie, E. Hennebicq, E.G. Kim, M. Surin, C. Risko, K. Schmidt, and E. Zojer for their collaboration in this work and many fruitful discussions. They also thank K. Müllen, A.C. Grimsdale, and their colleagues at the Max-Planck Institute for Polymer Research in Mainz, Germany, for providing the samples of conjugated materials investigated in Section 1.4 and recording their photoluminescence spectra. DB is "Maître de Recherche du Fonds National de la Recherche Scientifique (FNRS – Belgium)"; JC and PL are "Chercheurs Qualifiés du Fonds National de la Recherche Scientifique."

References

1. Fichou, D. 2000. Structural order in conjugated oligothiophenes and its implications on optoelectronic devices. *J Mater Chem* 10:571.
2. Pasveer, W.F., et al. 2005. Unified description of charge-carrier mobilities in disordered semiconducting polymers. *Phys Rev Lett* 94.

3. Duke, C.B., and L.B. Schein. 1980. Organic-solids: Is energy-band theory enough. *Phys Today* 33:42.

4. Holstein, T. 1959. Studies of polaron motion, 1. The molecular-crystal model. *Ann Phys* 8:325.

5. Holstein, T. 1959. Studies of polaron motion, 2. The small polaron. *Ann Physics* 8:343.

6. Heeger, A.J., et al. 1988. Solitons in conducting polymers. *Rev Mod Phys* 60:781.

7. Warta, W., R. Stehle, and N. Karl. 1985. Ultrapure, high mobility organic photoconductors. *Appl Phys A–Mater Sci Process* 36:163.

8. Jurchescu, O.D., J. Baas, and T.T.M. Palstra. 2004. Effect of impurities on the mobility of single crystal pentacene. *Appl Phys Lett* 84:3061.

9. Bredas, J.L., et al. 2004. Charge-transfer and energy-transfer processes in π-conjugated oligomers and polymers: A molecular picture. *Chem Rev* 104:4971.

10. Coropceanu, V., et al. 2003. The role of vibronic interactions on intramolecular and intermolecular electron transfer in p-conjugated oligomers. *Theor Chem Acc* 110:59.

11. Balzani, V. 2001. *Electron transfer in chemistry*. Weinheim/New York: Wiley-VCH.

12. Jortner, J., et al. 1999. *Electron transfer: From isolated molecules to biomolecules*. New York: John Wiley & Sons.

13. Newton, M.D. 1991. Quantum chemical probes of electron-transfer kinetics: The nature of donor–acceptor interactions. *Chem Rev* 91:767.

14. Kwon, O., et al. 2004. Characterization of the molecular parameters determining charge transport in anthradithiophene. *J Chem Phys* 120:8186.

15. Pietro, W.J., T.J. Marks, and M.A. Ratner. 1985. Resistivity mechanisms in phthalocyanine-based linear-chain and polymeric conductors: Variation of bandwidth with geometry. *J Am Chem Soc* 107:5387.

16. Wolfsberg, M., and L. Helmholz. 1952. The spectra and electronic structure of the tetrahedral ions Mno4-, Cro4-, and Clo4-. *J Chem Phys* 20:837.

17. Liang, C.X., and M.D. Newton. 1992. Ab initio studies of electron-transfer: Pathway analysis of effective transfer integrals. *J Phys Chem* 96:2855.

18. Paulson, B.P., et al. 1996. Investigation of through-bond coupling dependence on spacer structure. *J Am Chem Soc* 118:378.

19. Voityuk, A.A., et al. 2000. Energetics of hole transfer in DNA. *Chem Phys Lett* 324:430.

20. Grozema, F.C., et al. 2002. Intramolecular charge transport along isolated chains of conjugated polymers: Effect of torsional disorder and polymerization defects. *J Phys Chem B* 106:7791.

21. Palenberg, M.A., et al. 2000. Almost temperature independent charge carrier mobilities in liquid crystals. *J Chem Phys* 112:1541.

22. Senthilkumar, K., et al. 2003. Charge transport in columnar stacked triphenylenes: Effects of conformational fluctuations on charge transfer integrals and site energies. *J Chem Phys* 119:9809.

23. Mattheus, C.C. 2002. Polymorphism and electronic properties of pentacene. Ph.D. dissertation, University of Groningen, Groningen, The Netherlands.

24. Lemaur, V., et al. 2004. Charge transport properties in discotic liquid crystals: A quantum-chemical insight into structure–property relationships. *J Am Chem Soc* 126:3271.

25. Bredas, J.L., et al. 2002. Organic semiconductors: A theoretical characterization of the basic parameters governing charge transport. *Proc Natl Acad Sci USA* 99:5804.

26. Podzorov, V., et al. 2005. Hall effect in the accumulation layers on the surface of organic semiconductors. *Phys Rev Lett* 95:226601.

27. de Boer, R.W.I., et al. 2004. Organic single-crystal field-effect transistors. *Physica Status Solidi A* 201:1302.

28. Podzorov, V., et al. 2004. Intrinsic charge transport on the surface of organic semiconductors. *Phys Rev Lett* 93:086602.

29. da Silva, D.A., E.G. Kim, and J.L. Bredas. 2005. Transport properties in the rubrene crystal: Electronic coupling and vibrational reorganization energy. *Adv Mater* 17:1072.

30. Bulgarovskaya, I., et al. 1983. Growth, structure and optical properties of single crystals of rubrene. *Latvijas PSR Zinatnu Akademijas Vestis Fizikas un Tehnisko Zinatnu Serija* 53.

31. Cornil, J., J.P. Calbert, and J.L. Bredas. 2001. Electronic structure of the pentacene single crystal: Relation to transport properties. *J Am Chem Soc* 123:1250.

32. Cheng, Y.C., et al. 2003. Three-dimensional band structure and bandlike mobility in oligoacene single crystals: A theoretical investigation. *J Chem Phys* 118:3764.

33. Cornil, J., et al. 2000. Charge transport versus optical properties in semiconducting crystalline organic thin films. *Adv Mater* 12:978.

34. Janzen, D.E., et al. 2004. Preparation and characterization of π-stacking quinodimethane oligothiophenes: Predicting semiconductor behavior and bandwidths from crystal structures and molecular orbital calculations. *J Am Chem Soc* 126:15295.

35. Curtis, M.D., J. Cao, and J.W. Kampf. 2004. Solid-state packing of conjugated oligomers: From π-stacks to the herringbone structure. *J Am Chem Soc* 126:4318.

36. Sundar, V.C., et al. 2004. Elastomeric transistor stamps: Reversible probing of charge transport in organic crystals. *Science* 303:1644.

37. Shirakawa, H., et al. 1977. Synthesis of electrically conducting organic polymers: Halogen derivatives of polyacetylene, (Ch)X. *Chem Commun* 578.

38. Furukawa, Y. 1996. Electronic absorption and vibrational spectroscopies of conjugated conducting polymers. *J Phys Chem* 100:15644.

39. Bredas, J.L., and G.B. Street. 1985. Polarons, bipolarons, and solitons in conducting polymers. *Acc Chem Res* 18:309.

40. Fesser, K., A.R. Bishop, and D.K. Campbell. 1983. Optical-absorption from polarons in a model of polyacetylene. *Phys Rev B* 27:4804.

41. Geskin, V.M., A. Dkhissi, and J.L. Bredas. 2003. Oligothiophene radical cations: Polaron structure in hybrid DFT and MP2 calculations. *Int J Quantum Chem* 91:350.

42. Bally, T., D.A. Hrovat, and W.T. Borden. 2000. Attempts to model neutral solitons in polyacetylene by ab initio and density functional methods. The nature of the spin distribution in polyenyl radicals. *Phys Chem Chem Phys* 2:3363.

43. Rodriguez-Monge, L., and S. Larsson. 1995. Conductivity in polyacetylene, 1. Ab-initio calculation of charge localization, bond distances, and reorganization energy in model molecules. *J Chem Phys* 102:7106.

44. Geskin, V.M., J. Cornil, and J.L. Bredas. 2005. Comment on 'Polaron formation and symmetry breaking' by L Zuppiroli et al. *Chem Phys Lett* 403:228.

45. Brocks, G. 1999. Polarons and bipolarons in oligothiophenes: A first principles study. *Synthetic Met* 102:914.

46. Moro, G., et al. 2000. On the structure of polaronic defects in thiophene oligomers: A combined Hartree–Fock and density functional theory study. *Synthetic Met* 108:165.

47. Geskin Victor, M. unpublished.

48. Grozema, F.C., et al. 2002. Theoretical and experimental studies of the opto-electronic properties of positively charged oligo(phenylene vinylene)s: Effects of chain length and alkoxy substitution. *J Chem Phys* 117:11366.

49. Geskin Victor, M., et al. 2005. Impact of the computational method on the geometric and electronic properties of oligo(phenylene vinylene) radical cations. *J Phys Chem B* 109:20237.

50. Jacquemin, D., et al. 2005. Effect of recently-developed density functional approaches for the evaluation of the bond length alternation in polyacetylene. *Chem Phys Lett* 405:376.

51. Ciofini, I., C. Adamo, and H. Chermette. 2005. Effect of self-interaction error in the evaluation of the bond length alternation in trans-polyacetylene using density-functional theory. *J Chem Phys* 123.

52. Jacquemin, D., J.M. Andre, and E.A. Perpete. 2004. Geometry, dipole moment, polarizability and first hyperpolarizability of polymethineimine: An assessment of electron correlation contributions. *J Chem Phys* 121:4389.

53. Ma, H.B., et al. 2005. Spin distribution in neutral polyene radicals: Pariser-Parr-Pople model studied with the density matrix renormalization group method. *J Chem Phys* 122.

54. Wohlgenannt, M., X.M. Jiang, and Z.V. Vardeny. 2004. Confined and delocalized polarons in π-conjugated oligomers and polymers: A study of the effective conjugation length. *Phys Rev B* 69.

55. Fernandes, M.R., et al. 2005. Polaron and bipolaron transitions in doped poly(p-phenylene vinylene) films. *Thin Solid Films* 474:279.

56. Wohlgenannt, M., X.M. Jiang, and Z.V. Vardeny. 2004. Confined and delocalized polarons in p-conjugated oligomers and polymers: A study of the effective conjugation length. *Phys Rev B: Condensed Matter Mater Phys* 69:241204/1.

57. Dreuw, A., and M. Head-Gordon. 2005. Single-reference ab initio methods for the calculation of excited states of large molecules. *Chem Rev* 105:4009.

58. Bao, Z., A. Dodabalapur, and A.J. Lovinger. 1996. Soluble and processable regioregular poly(3-hexylthiophene) for thin film field-effect transistor applications with high mobility. *Appl Phys Lett* 69:4108.

59. Sirringhaus, H., N. Tessler, and R.H. Friend. 1998. Integrated optoelectronic devices based on conjugated polymers. *Science* 280:1741.

60. Sirringhaus, H., et al. 1999. Two-dimensional charge transport in self-organized, high-mobility conjugated polymers. *Nature* 401:685.

61. Osterbacka, R., et al. 2000. Two-dimensional electronic excitations in self-assembled conjugated polymer nanocrystals. *Science* 287:839.

62. AMPAC. Version 6.0, 1997. 7204 Mullen, Shawnee, KS 66216. In this approach, a subset of electronic configurations is selected using perturbation theory from the list generated by a full configuration interaction over a defined window of molecular orbitals (active space).

63. Cornil, J., and J.L. Bredas. 1995. Nature of the optical transitions in charged oligothiophenes. *Adv Mater* 7:295.

64. Beljonne, D., et al. 2001. Optical signature of delocalized polarons in conjugated polymers. *Adv Funct Mater* 11:229.

65. Cornil, J., et al. 1998. Influence of interchain interactions on the absorption and luminescence of conjugated oligomers and polymers: A quantum-chemical characterization. *J Am Chem Soc* 120:1289.

66. Tretiak, S., et al. 2000. Interchain electronic excitations in poly(phenylenevinylene) (PPV) aggregates. *J Phys Chem B* 104:7029.

67. Ruseckas, A., et al. 2001. Intra- and interchain luminescence in amorphous and semicrystalline films of phenyl-substituted polythiophene. *J Phys Chem B* 105:7624.

68. Cornil, J., et al. 2001. Interchain interactions in organic pi-conjugated materials: Impact on electronic structure, optical response, and charge transport. *Adv Mater* 13:1053.

69. Siddiqui, S., and F.C. Spano. 1999. H- and J-aggregates of conjugated polymers and oligomers: A theoretical investigation. *Chem Phys Lett* 308:99.

70. Bradley, D.D.C., et al. 1986. Infrared characterization of oriented poly(phenylene vinylene). *Polymer* 27:1709.

71. Cimrova, V., et al. 1996. Polarized light emission from LEDs prepared by the Langmuir–Blodgett technique. *Adv Mater* 8:146.

72. Prato, S., et al. 1999. Anisotropic ordered planar growth of alpha-sexithienyl thin films. *J Phys Chem B* 103:7788.

73. Herz, L.M., and R.T. Phillips. 2000. Effects of interchain interactions, polarization anisotropy, and photo-oxidation on the ultrafast photoluminescence decay from a polyfluorene. *Phys Rev B* 61:13691.

74. Grell, M., et al. 1999. Blue polarized electroluminescence from a liquid crystalline polyfluorene. *Adv Mater* 11:671.

75. Grell, M., et al. 2000. Intrachain ordered polyfluorene. *Synthetic Met* 111:579.

76. Nguyen, T.Q., V. Doan, and B.J. Schwartz. 1999. Conjugated polymer aggregates in solution: Control of interchain interactions. *J Chem Phys* 110:4068.

77. Klaerner, G., R.D. Miller, and C.J. Hawker. 1998. Dendrimers as end groups in rigid rod polymers based on di-n-hexylfluorene-2.7-diyl: Polymeric molecular dumbells. *Polym Preprints (Am Chem Soc, Div Polym Chem)* 39:1006.

78. Lupton, J.M., et al. 2001. Control of charge transport and intermolecular interaction in organic light-emitting diodes by dendrimer generation. *Adv Mater* 13:258.

79. Chen, J.P., et al. 1999. Efficient, blue light-emitting diodes using cross-linked layers of polymeric arylamine and fluorene. *Synthetic Met* 107:129.

80. Klarner, G., et al. 1999. Cross-linkable polymers based on dialkylfluorenes. *Chem Mater* 11:1800.

81. Klarner, G., et al. 1998. Colorfast blue-light-emitting random copolymers derived from di-*n*-hexylfluorene and anthracene. *Adv Mater* 10:993.

82. Kreger, M.A., et al. 2000. Femtosecond study of exciton dynamics in 9,9-di-*n*-hexylfluorene/anthracene random copolymers. *Phys Rev B* 61:8172.

83. Ikkala, O., et al. 1999. Self-organized liquid phase and co-crystallization of rod-like polymers hydrogen-bonded to amphiphilic molecules. *Adv Mater* 11:1206.

84. Hirschberg, J.H.K.K., et al. 2000. Helical self-assembled polymers from cooperative stacking of hydrogen-bonded pairs. *Nature* 407:167.

85. El-Ghayoury, A., et al. 2001. Supramolecular hydrogen-bonded oligo(p-phenylene vinylene) polymers. *Angewandte Chemie–International Edition* 40:3660.

86. Koren, A.B., M.D. Curtis, and J.W. Kampf. 2000. Crystal engineering of conjugated oligomers and the spectral signature of pi stacking in conjugated oligomers and polymers. *Chem Mater* 12:1519.

87. Neher, D. 2001. Polyfluorene homopolymers: Conjugated liquid-crystalline polymers for bright blue emission and polarized electroluminescence. *Macromol Rapid Commun* 22:1366.

88. Leclerc, M. 2001. Polyfluorenes: Twenty years of progress. *J Polym Sci Part A–Polym Chem* 39:2867.

89. Scherf, U., and E.J.W. List. 2002. Semiconducting polyfluorenes: Towards reliable structure–property relationships. *Adv Mater* 14:477.

90. Pei, Q.B., and Y. Yang. 1996. Efficient photoluminescence and electroluminescence from a soluble polyfluorene. *J Am Chem Soc* 118:7416.

91. Setayesh, S., et al. 2001. Polyfluorenes with polyphenylene dendron side chains: Toward non-aggregating, light-emitting polymers. *J Am Chem Soc* 123:946.

92. Miteva, T., et al. 2001. Improving the performance of polyfluorene-based organic light-emitting diodes via end-capping. *Adv Mater* 13:565.

93. Ego, C., et al. 2002. Triphenylamine-substituted polyfluorene: A stable blue-emitter with improved charge injection for light-emitting diodes. *Adv Mater* 14:809.

94. Redecker, M., et al. 1999. High mobility hole transport fluorene-triarylamine copolymers. *Adv Mater* 11:241.

95. Donat-Bouillud, A., et al. 2000. Light-emitting diodes from fluorene-based pi-conjugated polylmers. *Chem Mater* 12:1931.

96. Wiesendanger, R. 1994. *Scanning probe microscopy and spectroscopy: Methods and applications.* Cambridge, UK/New York: Cambridge University Press.

97. Burkert, U., and N.L. Allinger. 1982. *Molecular mechanics.* Washington, D.C.: American Chemical Society.

98. Cramer, C.J. 2002. *Essentials of computational chemistry: Theories and models.* West Sussex, UK/New York: John Wiley & Sons.

99. Allinger, N.L. 1977. Conformational-analysis, 130. Mm2: Hydrocarbon force-field utilizing V1 and V2 torsional terms. *J Am Chem Soc* 99:8127.

100. Allinger, N.L., Y.H. Yuh, and J.H. Lii. 1989. Molecular mechanics: The Mm3 force-field for hydrocarbons, 1. *J Am Chem Soc* 111:8551.

101. Allinger, N.L., F.B. Li, and L.Q. Yan. 1990. Molecular mechanics: The Mm3 force-field for alkenes. *J Comput Chem* 11:848.

102. Weiner, S.J., et al. 1984. A new force-field for molecular mechanical simulation of nucleic-acids and proteins. *J Am Chem Soc* 106:765.

103. Weiner, S.J., et al. 1986. An all atom force-field for simulations of proteins and nucleic-acids. *J Comput Chem* 7:230.

104. Cornell, W.D., et al. 1995. A 2nd generation force-field for the simulation of proteins, nucleic-acids, and organic-molecules. *J Am Chem Soc* 117:5179.

105. Brooks, B.R., et al. 1983. Charmm: A program for macromolecular energy, minimization, and dynamics calculations. *J Comput Chem* 4:187.

106. Mayo, S.L., B.D. Olafson, and W.A. Goddard. 1990. Dreiding: A generic force-field for molecular simulations. *J Phys Chem–US* 94:8897.

107. Hwang, M.J., T.P. Stockfisch, and A.T. Hagler. 1994. Derivation of Class-Ii force-fields, 2. Derivation and characterization of a Class-Ii force-field, Cff93, for the alkyl functional-group and alkane molecules. *J Am Chem Soc* 116:2515.

108. Maple, J.R., et al. 1994. Derivation of Class-Ii force-fields, 1. Methodology and quantum force-field for the alkyl functional-group and alkane molecules. *J Comput Chem* 15:162.

109. Peng, Z.W., et al. 1997. Derivation of class II force fields, 4. van der Waals parameters of alkali metal cations and halide anions. *J Phys Chem A* 101:7243.

110. Clark, M., R.D. Cramer, and N. Vanopdenbosch. 1989. Validation of the general-purpose Tripos 5.2 force-field. *J Comput Chem* 10:982.

111. Sun, H. 1998. COMPASS: An ab initio force-field optimized for condensed-phase applications: Overview with details on alkane and benzene compounds. *J Phys Chem B* 102:7338.

112. Bunte, S.W., and H. Sun. 2000. Molecular modeling of energetic materials: The parameterization and validation of nitrate esters in the COMPASS force field. *J Phys Chem B* 104:2477.

113. Sun, H., et al. 1995. Ab Initio calculations on small-molecule analogs of polycarbonates. *J Phys Chem–US* 99:5873.

114. Sun, H., et al. 1994. An ab initio Cff93 all-atom force field for polycarbonates. *J Am Chem Soc* 116:2978.

115. Sun, H. 1995. Ab initio calculations and force field development for computer simulation of polysilanes. *Macromolecules* 28:701.

116. Allinger, N.L., et al. 1997. The torsional conformations of butane: Definitive energetics from ab initio methods. *J Chem Phys* 106:5143.

117. Smith, G.D., R.L. Jaffe, and D.Y. Yoon. 1993. A force field for simulations of 1,2-dimethoxyethane and poly(oxyethylene) based upon ab initio electronic-structure calculations on model molecules. *J Phys Chem–US* 97:12752.

118. Chen, S.H., et al. 2005. Crystalline forms and emission behavior of poly(9,9-di-*n*-octyl-2,7-fluorene). *Macromolecules* 38:379.

119. Hu, D.H., et al. 2000. Collapse of stiff conjugated polymers with chemical defects into ordered, cylindrical conformations. *Nature* 405:1030.

120. Claes, L., J.P. François, and M.S. Deleuze. 2001. Molecular packing of oligomer chains of poly (p-phenylene vinylene). *Chem Physics Lett* 339:216.

121. Hoover, W.G. 1985. Canonical dynamics: Equilibrium phase-space distributions. *Phys Rev A* 31:1695.

122. Verlet, L. 1967. Computer experiments on classical fluids, I. Thermodynamical properties of Lennard–Jones molecules. *Phys Rev* 159:98.

123. Leclere, P., et al. 2003. Supramolecular organization in block copolymers containing a conjugated segment: A joint AFM/molecular modeling study. *Prog Polym Sci* 28:55.

124. Liao, L.S., et al. 2000. Electronic structure and energy band gap of poly(9,9-dioctylfluorene) investigated by photoelectron spectroscopy. *Appl Phys Lett* 76:3582.

125. Kreyenschmidt, M., et al. 1998. Thermally stable blue-light-emitting copolymers of poly(alkyl-fluorene). *Macromolecules* 31:1099.

126. Romaner, L., et al. 2003. The origin of green emission in polyfluorene-based conjugated polymers: On-chain defect fluorescence. *Adv Funct Mater* 13:597.

127. Zojer, E., et al. 2002. Green emission from poly(fluorene)s: The role of oxidation. *J Chem Phys* 117:6794.

128. List, E.J.W., et al. 2002. The effect of keto defect sites on the emission properties of polyfluorene-type materials. *Adv Mater* 14:374.

129. Gong, X., et al. 2003. Electrophosphorescence from a polymer guest–host system with an iridium complex as guest: Forster energy transfer and charge trapping. *Adv Funct Mater* 13:439.

130. Grimsdale, A.C., et al. 2002. Correlation between molecular structure, microscopic morphology, and optical properties of poly(tetraalkylindenofluorenes). *Adv Funct Mater* 12:729.

131. Klok, H.A., and S. Lecommandoux. 2001. Supramolecular materials via block copolymer self-assembly. *Adv Mater* 13:1217.

132. Radzilowski, L.H., and S.I. Stupp. 1994. Nanophase separation in monodisperse rod–coil diblock polymers. *Macromolecules* 27:7747.

133. Radzilowski, L.H., B.O. Carragher, and S.I. Stupp. 1997. Three-dimensional self-assembly of rod–coil copolymer nanostructures. *Macromolecules* 30:2110.

134. Sayar, M., et al. 2000. Competing interactions among supramolecular structures on surfaces. *Macromolecules* 33:7226.

135. Zubarev, E.R., et al. 2001. Self-assembly of dendron rod–coil molecules into nanoribbons. *J Am Chem Soc* 123:4105.

136. Ikkala, O., and G. ten Brinke. 2002. Functional materials based on self-assembly of polymeric supramolecules. *Science* 295:2407.

137. Bates, F.S., and G.H. Fredrickson. 1990. Block copolymer thermodynamics: Theory and experiment. *Annu Rev Phys Chem* 41:525.

138. Riess, G., and P. Bahadur. 1989. *Encyclopedia of polymer science and engineering*, 2nd edn. New York: John Wiley & Sons.

139. Mao, G., and C.K. Ober. 1997. Block copolymers containing liquid crystalline segments. *Acta Polymerica* 48:405.

140. Hempenius, M.A., et al. 1998. A polystyrene–oligothiophene–polystyrene triblock copolymer. *J Am Chem Soc* 120:2798.

141. Bazan, G.C., et al. 1996. Fluorescence quantum yield of poly(p-phenylenevinylene) prepared via the paracyclophene route: Effect of chain length and interchain contacts. *J Am Chem Soc* 118:2618.

142. François, B., et al. 1995. Block-copolymers with conjugated segments: Synthesis and structural characterization. *Synthetic Met* 69:463.

143. Widawski, G., M. Rawiso, and B. François. 1994. Self-organized honeycomb morphology of star-polymer polystyrene films. *Nature* 369:387.

144. Leclere, P., et al. 1998. Organized semiconducting nanostructures from conjugated block copolymer self-assembly. *Chem Mater* 10:4010.

145. Chen, X.L., and S.A. Jenekhe. 1996. Block conjugated copolymers: Toward quantum-well nanostructures for exploring spatial confinement effects on electronic, optoelectronic, and optical phenomena. *Macromolecules* 29:6189.

146. Jenekhe, S.A., and X.L. Chen. 1998. Self-assembled aggregates of rod–coil block copolymers and their solubilization and encapsulation of fullerenes. *Science* 279:1903.

147. Jenekhe, S.A., and X.L. Chen. 1999. Self-assembly of ordered microporous materials from rod–coil block copolymers. *Science* 283:372.

148. Jenekhe, S.A., and X.L. Chen. 2000. Supramolecular photophysics of self-assembled block copolymers containing luminescent conjugated polymers. *J Phys Chem B* 104:6332.

149. Stalmach, U., et al. 2000. Semiconducting diblock copolymers synthesized by means of controlled radical polymerization techniques. *J Am Chem Soc* 122:5464.

150. Huijs, F.M., et al. 2001. Formation of transparent conducting films based on core-shell latices: Influence of the polypyrrole shell thickness. *J Appl Polym Sci* 79:900.

151. Malliaras, G.G., et al. 1993. Tuning of the photoluminescence and electroluminescence in multi-block copolymers of poly[(silanylene)-thiophenes] via exciton confinement. *Adv Mater* 5:721.

152. Kim, J., and T.M. Swager. 2001. Control of conformational and interpolymer effects in conjugated polymers. *Nature* 411:1030.

153. Wang, H.B., et al. 2000. Syntheses of amphiphilic diblock copolymers containing a conjugated block and their self-assembling properties. *J Am Chem Soc* 122:6855.

154. Kim, J.S., S.K. McHugh, and T.M. Swager. 1999. Nanoscale fibrils and grids: Aggregated structures from rigid-rod conjugated polymers. *Macromolecules* 32:1500.

155. Ciferri, A. 1991. *Liquid crystallinity in polymers: Principles and fundamental Properties.* New York: VCH Publishers.

156. Teetsov, J., and M.A. Fox. 1999. Photophysical characterization of dilute solutions and ordered thin films of alkyl-substituted polyfluorenes. *J Mater Chem* 9:2117.

157. Teetsov, J.A., and D.A. Vanden Bout. 2000. Near-field scanning optical microscopy (NSOM) studies of nano-scale polymer ordering in thin films of poly(9,9-dialkylfluorene). *J Phys Chem B* 104:9378.

158. Teetsov, J.A., and D.A. Vanden Bout. 2001. Imaging molecular and nanoscale order in conjugated polymer thin films with near-field scanning optical microscopy. *J Am Chem Soc* 123:3605.

2

Theoretical Studies of Electron–Lattice Dynamics in Organic Systems

Sven Stafström

2.1 Introduction

As a result of the strong interaction between the electrons and the lattice in organic systems, such as conjugated polymers and molecular crystals, all dynamical processes in these materials involve both electronic and atomic motions. Detailed studies of these interactions are necessary to provide the most basic understanding of electronic as well as optoelectronic device applications based on organic materials. Such processes include charge transport and charge transfer, exciton formation, exciton dissociation, and energy transfer.

The laws of physics governing these dynamical processes are based on quantum mechanics for the electrons, electrodynamics for the photons, and quantum mechanics or Newton mechanics to describe the molecular mechanics part. In this review, we will describe a method based on solving the time-dependent Schrödinger equation together with the classical equation of motion for the lattice. This method was introduced by Hartman et al. [1] and later developed by Block and Streitwolf [2] and Johansson and Stafström [3]. We have applied this method to the studies of microscopic dynamical processes of charge transport along and between conjugated polymer chains [3,4] and in molecular crystals [5], thermalization of hot electrons in conjugated polymers [6], as well as to the studies of ultrafast optoelectronic processes such as exciton dissociation induced by an external electric field [7] or by a strong electron acceptor [8]. In all of these cases, we obtain a microscopic picture of the dynamics involving electrons and atoms. As exemplified above, this method is quite general in terms of the processes to be studied. To limit this presentation we will focus on charge transport. As a complement to the description of the electron–lattice dynamics approach, we will also discuss other transport theories that have been introduced to explain experimental data related to transport in organic systems. These methods are all based on hopping theories and as such restricted to low-mobility systems. As examples of systems belonging to this regime, we will discuss transport in disordered conjugated polymeric systems and weakly coupled molecular crystals. Molecular crystals such as pentacene and rubrene will also be discussed. We will particularly focus on the transition from thermally activated nonadiabatic hopping type of transport, which applied to the less-ordered organic materials, to the adiabatic drift process that occurs in the highly ordered high-mobility samples.

2.2   Experimental and Theoretical Background

Charge transport in synthesized organic systems has been a subject of intensive research over the last three decades. In the past, conjugated polymers and charge transfer complexes were the main focus of these studies. Today, a lot of attention is drawn toward high-ordered molecular crystals, particularly for their high mobility and potential use in molecular electronic devices [9]. Recent observations of mobilities of the order of 30 cm^2/V s in rubrene [10] and several reports of very high mobilities in pentacene [11–13] showed that it is possible to compete with inorganic semiconductors in this aspect. However, high-mobility state can easily be destroyed in the presence of disorder, which reduces the electronic coupling strength between neighboring molecules, induces localization of the charge carriers, and creates traps for the charge carriers. Thus, a high degree of structural order as well as low-impurity concentration are absolutely essential to achieve these high mobilities.

In addition to extrinsic effects, there are also intrinsic properties that are important in the context of transport. The electron–phonon coupling present in most low-dimensional carbon-based systems can give rise to self-localization of the charge carrier, i.e., polaron formation [14]. In a molecular crystal the charge carrier can be localized due to gain in energy caused by the (intramolecular) geometrical response to the presence of the carrier. This gain in energy by localizing the charge, E_b is competing with the delocalization energy that results from electronic intermolecular interactions, J [15]. Depending on the relative strength of these two interactions, we can describe the intrinsic transport in three different categories.

In the limit of weak intermolecular interactions ($J \ll E_l$), the so-called small Holstein polaron localized to a single molecule is stable [14]. As the electronic state is localized, transport occurs through hopping from one molecule to the next. This is a nonadiabatic process for which the carrier motion is slow compared with the molecular relaxation times and the mobility depends quadratically on the intermolecular interaction strength [16]. In this regime, the localization energy, E_b is approximately identical to the formation energy of the polaron and the barrier for polaron hopping is $E_l/2$ [17].

For stronger intermolecular interactions ($J \sim E_l$) the polaron is delocalized over several molecular units and the electron probability density can sample an even larger region in space. Under the influence of an external electric field the charge density associated with the polaron moves along the direction of the electric field and in an adiabatic way changes the molecular geometry as the carrier drifts along the system [18]. In this regime, in the absence of static disorder, the activation energy barrier is small and the mobility is weakly dependent on the intermolecular interaction strength. Finally, for $J \gg E_l$ the lattice relaxation can be neglected and transport is based on band theory and can be described by the relaxation time approximation or similar textbook theories [19,20].

The charge-carrier mobilities discussed above have different J dependences. There is also a difference in their temperature dependence. The nonadiabatic and the adiabatic processes are temperature-activated whereas in the band transport regime, the phonons generated act as scattering centers of the charge carriers. It is interesting to compare this behavior with the observed temperature dependence of the mobility for the highly purified pentacene [13] and rubrene crystals [10]. In both these systems, the low-temperature regime is temperature activated whereas above $T = 300$ K the mobility decreases with increasing temperature. The latter behavior is typical for a system exhibiting extended states whereas the behavior in the low-temperature regime occurs due to either residual defects or as a result of polaron formation [10,21].

The concept of polarons (and solitons) was further developed in connection with the observation of high electrical conductivity in doped conjugated polymers [22]. Su et al. [23] and Rice [24], in 1979, had shown that soliton excitations are possible in *trans*-polyacetylene (*t*PA) and that these defects could explain both magnetic and optical properties of doped systems. Exactly how these doping induced defects are involved in charge transport is, however, not yet settled. A vast amount of transport data [25] for lightly or moderately doped samples are in agreement with the hopping laws over a wide temperature range. This suggests that once the localized polaron, soliton, or bipolaron states are formed, conduction takes place predominantly through hopping process between these states [26]. In heavily doped samples, the metallic picture is likely to be more appropriate at least in the ordered regions and samples [25].

Studies on conjugated polymers have recently shifted to the undoped state, predominantly as the undoped state has shown to have very interesting luminescence [27] as well as photovoltaic [28] properties. In the undoped state the number of charge carriers is small and the interest from the point of view of transport is more focussed on the mobility of charge carriers. Similar to the situation in molecular crystals, the transport properties of conjugated polymers are also strongly dependent on structural order and the purity of the samples.

Charge transport in disordered polymers is regarded as hopping process between localized states. The confinement of the electronic states is, in this case, a combination of the effect of chain interruptions and the polaronic effect discussed earlier. Thus, to some extent this type of system is similar to the molecular systems discussed earlier, with the molecules replaced by the chain segments over which the electrons can delocalize. However, structural order as well as mobility are much lower in the polymer samples [29]. In disordered polymer system the overlap between the states localized to different chain segments is quite low [30], the separation between chain segments is of variable range, and the segments themselves are of different length and experience variations in the local field. Thus, for this type of systems the transport is regarded as a low-probability hopping process with mobilities in the order of 10^{-5} to 10^{-7} cm^2/V s.

The use of conjugated polymers in electronic devices, such as light emitting diodes and field effect transistors, is very much dependent on the mobility and there have been numerous attempts to improve the hopping rates, i.e., to process polymers with high degree of structural order and low concentration of impurities. Typical such polymers are poly(3-hexylthiophene) (P3HT) with mobility as high as 0.2 cm^2/V s when processed and characterized in inert atmosphere [31]. However, P3HT yielded poorer mobility (0.045 cm^2/V s) when the device was fabricated at ambient conditions [32]. Probably, the transport is still of hopping type even for these systems but with quite low barriers for the charge to hop between sites.

2.3 Transport

Hopping conduction is defined as the process in which charge carriers conduct the electric current by thermally activated tunneling from an occupied site to an empty site. More explicitly, the thermal energy that is required for the process to occur is gained from the phonon system. Hopping was suggested independently by Conwell [33], Mott [34], and Pines [35] to describe charge transport in inorganic semiconductors.

Shortly after the concept of hopping was introduced, Holstein [14] developed the concept of polaron motion. This theory was introduced to describe charge transport in molecular crystals. As a result of electron–lattice interaction, the surrounding lattice particles will be displaced to new equilibrium positions. The induced displacements provide a well for the electron. If the well is sufficiently deep, the electron will occupy a bound state unable to move unless accompanied by the well. The unit consisting of the electron together with its induced lattice deformation is termed "the polaron."

In the Holstein model the molecular crystal is described as a regular one-dimensional (1D) array of diatomic molecules. The Hamiltonian of the system is a sum of three terms H_L, H_{el}, and H_{int}. The lattice

Hamiltonian contains the intranuclear degree of freedom where u_n is the deviation of the nth intranuclear degree of freedom from its equilibrium value:

$$H_L = -\frac{1}{2M} \sum_n p_n^2 + \frac{1}{2} \sum_n M\omega_0^2 u_n^2 \tag{2.1}$$

The vibrational kinetic energy operator is shown here for completeness but is neglected in the derivation of the Holstein polaron (stationary solution).

The electron term is approximated as a tight-binding Hamiltonian:

$$H_{el} = -J \sum_n \left(\hat{c}_n^\dagger \hat{c}_{n+1} + \hat{c}_{n+1}^\dagger \hat{c}_n \right), \tag{2.2}$$

where J is the nearest neighbor electron interaction energy (introduced in Section 2.2).

On-site energies are the solutions to the Schrödinger equation for the diatomic molecule, for a given value of u_n. The energy of an electron localized to a particular site, n, is presumed to be proportional to the deformation parameter. The constant of proportionality is the electron–phonon coupling, A:

$$H_{int} = \sum_n \varepsilon_n \hat{c}_n^\dagger \hat{c}_n = \sum_n A u_n \hat{c}_n^\dagger \hat{c}_n \tag{2.3}$$

The results derived from the Holstein model are based on the solution to the time-dependent Schrödinger equation:

$$i\hbar |\dot{\Psi}(t)\rangle = (H_L + H_{el} + H_{int})|\Psi(t)\rangle. \tag{2.4}$$

The ground state stationary solutions for $\Psi(t)$ as well as the resulting displacement parameters can be derived analytically [14]. From this solution the localization energy, i.e., the energy gained by localizing the charge carrier (see also Section 2.2) can be expressed as

$$E_l = \frac{A^2}{2M\omega_0^2}. \tag{2.5}$$

The stabilization of the bound state is referred to as the polaron energy, E_P. In the nonadiabatic, or small polaron, limit, J is small compared to E_l and the polaron energy is identical to the localization energy:

$$E_P = \frac{A^2}{2M\omega_0^2} - 2J \approx \frac{A^2}{2M\omega_0^2}. \tag{2.6}$$

Holstein then proceeded to calculate transition probabilities between polaronic states localized at different sites (molecules). This treatment is essentially identical to that of Landau and Zener [15,36,37] and also has a lot of similarities with the more elaborate Marcus theory [16]. For small polarons at high temperatures the result is

$$W_T^{(Hol)}(n \to n \pm 1) = \frac{J^2}{\hbar} \left(\frac{\pi}{4kTE_a} \right) e^{-E_a/kT}. \tag{2.7}$$

Shortly after Holstein published his work on polarons in molecular crystals, Miller and Abrahams introduced a very useful description of hopping conduction in terms of a phonon-assisted electron tunneling process [38]. Miller–Abrahams theory does not include the polaronic effect. Nevertheless it has been used extensively in studies of transport in disorder conjugated system. The transition rate is a product of the tunneling probability and the phonon-assisted transition probability as follows:

$$W_T^{(M-A)}(n \rightarrow n \pm 1) = \begin{cases} \nu_0 \exp\left[-2\alpha_n R_{n,n+1}\right] \exp\left[-(\varepsilon_{n+1} - \varepsilon_n)/k_B T\right], & \varepsilon_{n+1} \geq \varepsilon_n, \\ \nu_0 \exp\left[-2\alpha_n R_{n,n+1}\right], & \varepsilon_n < \varepsilon_{n+1}, \end{cases} \quad (2.8)$$

where ν_0 is an intrinsic rate and $R_{n,n+1} \equiv |\mathbf{R}_{n+1} - \mathbf{R}_n|$ is the distance between sites n and $n+1$. The inverse localization length, α, determines the overlap of the wavefunctions associated with these sites and also the tunneling probability for a transition. The first exponential term in Equation 2.8 corresponds to the J^2 term in Equation 2.7, whereas the second term includes the activation energy, also in close correspondence to Equation 2.7.

To first order in the electric field, which is included in the on-site energies ε_n, the solution of the steady state and the current equations are equivalent to the solution of a linear resistance network [38].

The mobility, μ, can be determined from a (numerical) solution of the master equation representing hopping of charge carriers between lattice sites:

$$\sum_n \left[W_{n,n+1} p_n (1 - p_{n+1}) - W_{n+1,n} p_{n+1} (1 - p_n)\right] = 0, \quad (2.9)$$

where p_n is the probability that site n is occupied by a charge. The factors $(1 - p_n)$ account for the fact that, as a result of a large on-site Coulomb repulsion, only one carrier can occupy a site.

It should be noted that the small polaron model by Holstein model as well as the Miller–Abrahams model are based on the hopping concept that applies only to low-mobility systems. The J^2 factor in Equation 2.7 corresponds to the factor in Equation 2.8 describing the tunneling probability. Both these factors are assumed to be small and the resulting expressions for the transition rate were derived from perturbation theory. Holstein also derived a result for the large polaron for which the molecular units are strongly overlapping and the golden rule type of transition probability factor J^2 in Equation 2.7 is replaced by unity [14]. Transport is best described as a polaron drift process limited only by the activation energy barrier, i.e., the bare electronic tunneling barrier is zero in this case.

Even though the resulting expressions from the Holstein and the Miller–Abrahams theories are very similar, they are conceptually different in that the Holstein model includes the concept of localization energy in terms of the polaronic effect but no static disorder, and the Miller–Abrahams theory does not involve the localization energy. Instead, it is assumed that the localization occurs as a result of static disorder. In the Miller–Abrahams formula this is given an explicit form, which is related to the inverse localization length of the electronic wavefunction associated with donor and acceptor states.

The shift of interest from doped conjugated polymers of interest for conductivity, to pristine conjugated polymers for applications in OLEDs and other electronic or optoelectronic devices. In this case, disorder plays a dominating role over the polaronic effect of localizing charge-carrier wavefunctions. A certain amount of static disorder has to be accounted for in the studies of charge transport in organic systems. A part of this disorder is energetic, i.e., appears as fluctuations in the on-site energies. The transport properties depend strongly on the amount of this type of disorder as well as the spatial correlations in the values of the on-site energies between sites [39]. Successful comparisons with experimental data for disordered polymeric systems show, however, that a simple uncorrelated Gaussian disorder model [40] can account for the experimental determined electric field and temperature dependence of the mobility.

2.4 Methodology

To study the electron–lattice coupling dynamics of any molecular system without any constrains concerning the relation between the localization energy, E_b, and the electronic coupling, J, it is necessary to simultaneously solve the time-dependent Schrödinger equation

$$i\hbar \left|\dot{\Psi}(t)\right\rangle = \hat{H}_{el} |\Psi(t)\rangle, \quad (2.10)$$

and the lattice equation of motion

$$M\ddot{\mathbf{r}}_i = -\nabla_{\mathbf{r}_i} E_{\text{tot}}, \tag{2.11}$$

where the total energy of the system E_{tot} is obtained by taking the expectation value of the Hamiltonian H of the system, i.e., from the relationship $E_{\text{tot}} = \langle\, \Psi \mid \hat{H} \mid \Psi \,\rangle$. These calculations may readily be performed using state of the art numerical differential equation solvers, provided that $\mid \Psi \,\rangle$ is expanded as a linear combination of known basis functions. However, the evaluation and handling of the large number of two-electron integrals that would arise from an exact treatment of the electronic part of the Hamiltonian (H_{el}) even for small basis sets of moderately sized systems exclude ab initio treatments of most realistic systems. Instead, we are restricted to work with more cost-efficient approximative treatments of H. In the case of conjugated hydrocarbon system, this means invoking the $\sigma-\pi$ electron separability, i.e., to treat the π-electrons associated with the carbon $2p_z$-orbitals in an explicit fashion and to simply regard the σ-electrons residing in the bonding sp^2-hybrid orbitals as providing a background potential.

The Su–Schrieffer–Heeger (SSH) model [41] is the most commonly used theoretical model for describing carbon-based π-conjugated systems. Here, we have used an extended version of the SSH-model that includes a three-dimensional (3D) description of the system based on internal coordinates [42,43] (bond distances, bond angles, and dihedral angles). Since the changes in the geometrical variables following excitations or charge transport are small, these changes can be approximated with linear terms for π-electron hopping and classical harmonic terms for the potential energy of the σ-system.

The SSH Hamiltonian with an additional term describing the external electric field has the form:

$$H(t) = H_{\text{el}} + H_{\mathbf{E}} + H_{\text{latt}}, \tag{2.12}$$

where the electronic part is defined as

$$H_{\text{el}} = -\sum_{\langle nn'\rangle} \hat{c}_n^\dagger t_{nn'} \hat{c}_{n'}. \tag{2.13}$$

with intra- and intermolecular transfer integrals

$$t_{nn'} = \begin{cases} t_0 - \alpha(R_{nn'} - R_0) & \text{intramolecular,} \\ kS_{nn'}^{(0)} & \text{intermolecular,} \end{cases} \tag{2.14}$$

where α is the electron–phonon coupling constant. The intermolecular transfer integrals are assumed to be proportional to the overlap integrals, $S_{nn'}^{(0)}$, between $2p_z$ Slater-type $2p_z$ atomic orbitals on sites n and n', with the constant of proportionality given by k. These integrals are calculated analytically from the formulas of Mulliken et al. [44] using a method outlined by Hansson and Stafström [45].

All summations are over nearest neighboring carbon atoms, here denoted by $\langle nn'\rangle$, $t_{nn'}$ is the hopping integral between nearest neighbor carbon atom n and n', and t_0 is the hopping at a reference distance R_0. The actual interatomic distances are denoted $R_{nn'} \equiv \mid \mathbf{R}_n - \mathbf{R}_{n'} \mid$.

The external electric field, $\mathbf{E}$, is accounted for by adding the following term to the electronic part of the Hamiltonian:

$$H_{\mathbf{E}} = -e\sum_n \mathbf{R}_n \cdot \mathbf{E}(\hat{c}_n^\dagger \hat{c}_n - 1). \tag{2.15}$$

The field is constant both spatially and in time after a smooth turn on [3].

For the small anticipated lattice distortions [23], we adopt a classical description and expand the σ-bonding energy to second order around the ground state geometry obtained from geometry optimizations using ab initio techniques:

$$\hat{H}_{\text{latt}} = \frac{K_1}{2}\sum_{\langle nn'\rangle}(R_{nn'}-R_0)^2 + \frac{K_2}{2}\sum_n(\phi_n-\phi_0)^2 + \frac{K_3}{2}\sum_n(\theta_n-\theta_0)^2. \tag{2.16}$$

Here, K_1, K_2, and K_3 are harmonic spring constants accounting for σ-bond forces, ϕ_n denotes an angle in which atom n participates and θ_n is a dihedral angle to which atom n contributes. All summations are over unique distances and angles to avoid double counting of these energy contributions. The angles ϕ_0 and θ_0 are those of the "relaxed" ground state geometry.

The equations of motion for constituent atoms (see Equation 2.11) can be written more explicitly as

$$
\begin{aligned}
-M\ddot{q}_i^c(t) = &-\sum_{nn'}\left[t_0-\sum_j^{\text{bonds}}\alpha(R_{ij}-R_0)\delta_{in}\delta_{jn'}\right]\frac{\partial\rho_{nn'}}{\partial q_i^c}\\
&+\sum_{nn'}\sum_j^{\text{bonds}}2\alpha(\rho_{nn'}-\langle\rho\rangle)\delta_{in}\delta_{jn'}\frac{\partial R_{ij}}{\partial q_i^c}\\
&+\sum_j^{\text{bonds}}K_1(R_{ij}-R_0)\frac{\partial R_{ij}}{\partial q_i^c}\\
&+\sum_{j,k}^{\text{bond angles}}K_2(\theta_{ijk}-\theta_0)\frac{\partial\theta_{ijk}}{\partial q_i^c}\\
&+\sum_{j,k,l}^{\text{torsion angles}}K_3(\phi_{ijkl}-\phi_0)\frac{\partial\phi_{ijkl}}{\partial q_i^c}\\
&+\sum_n|e|E^c(t)(\rho_{nn}-1)\delta_{in},
\end{aligned}
\tag{2.17}
$$

where primed sums are only over the nearest neighbors, q_i^c with $c=\{x,y,z\}$ indicate the Cartesian coordinates of atom i and $\rho_{nn'}=\langle\hat{c}_n^\dagger\hat{c}_{n'}\rangle$ are density matrix elements.

With the electronic wavefunction expanded in the basic functions of the atomic sites, we can write the time-dependent Schrödinger equation as

$$i\hbar\dot{\psi}_{np}(t) = \sum_{n'}h_{nn'}(t)\psi_{n'p}(t), \tag{2.18}$$

where p is the molecular orbital index. In the nonadiabatic, or hopping, processes the time evolution of the time-dependent wavefunctions differ from the time-independent, instantaneous wavefunctions that are obtained from $\sum_{n'}h_{nn'}(t)\varphi_{n'p}(t)=\varphi_{np}(t)\varepsilon_p(t)$. The nonadiabatic behavior can, therefore, be analyzed by expanding the time-dependent wavefunctions in the basis of instantaneous eigenfunctions [2]

$$\psi_{np}(t) = \sum_{p'}\varphi_{np'}(t)\alpha_{p'p}(t). \tag{2.19}$$

The time-dependent occupation number of the eigenstates is

$$n_p(t) = \sum_{p'}f_{p'}|\alpha_{pp'}(t)|^2, \tag{2.20}$$

where $f_{p'}$ is the initial occupation number. The examples in Section 2.5 are simulating the dynamics of an electron polaron. Therefore, we set $f_{p'}$ to 2 for the highest occupied molecular orbital (HOMO) and 1 for the lowest unoccupied molecular orbital (LUMO) of the neutral system. In adiabatic dynamics the occupation numbers are unchanged as a function of time and the dynamics is restricted to a single Born–Oppenheimer potential energy surface. In the nonadiabatic processes, however, the occupation varies as a result of electronic excitation caused by electron–phonon interaction. In the following, we will use the time evolution of $n_p(t)$ as a signature of the type of transport process.

When optimizing the starting geometries of the systems we minimized the total energy with respect to the atomic positions in all three dimensions, with the constrain to keep the molecular size constant, i.e., $\sum_{\langle nn' \rangle} (r_{nn'} - a) = 0$. For the optimization, the resilient propagation (RPROP) method [46] was used.

The atomic positions and the charge densities were numerically integrated in time by solving the coupled differential equations (Equation 2.17 and Equation 2.18). This is done using an eighth-order Runge–Kutta integration with step-size control [47], which in practice means a time step of less than 10 as. The "global time step" is 1 fs.

2.5 Results And Discussion

2.5.1 Polaron Drift

Most of the discussions regarding transport in conjugated polymers today focus on the effects of disorder, i.e., extrinsic effects that cause localization and hopping conduction. A lot of effort is put into reducing the amount of disorder and to reach a state in which intrinsic polaronic effect is dominating the transport properties. One of the major issues of interest related to polaron transport has been the maximum polaron drift velocity that was shown to exceed sound velocity by approximately a factor of 4 [3,48]. The maximum velocity is reached at high-electric field strengths, 3.5×10^5 V/cm. At even higher field strengths the polaron becomes unstable and the electron is decoupled from the lattice deformation. In contrast to these reports, there are also studies that show saturation of the polaron velocity just below the sound velocity [48]. This result is obtained in an adiabatic simulation in which the electronic energy is treated within the Born–Oppenheimer approximation. In the present approach, we allow for nonadiabatic effects and present results from the simulations of field dependence of intrachain polaron dynamics. The nonadiabatic description of the system is of fundamental importance for these studies and here we emphasize the behavior of such effects on the electronic system during polaron drift.

To understand the transition to supersonic velocities we also make use of a description of the lattice in terms of both optical and acoustic order parameters. Here, we study a strictly 1D model that is a direct map of the *t*PA structure. However, as far as the properties of the drifting polaron are concerned, the model is more general and quantitatively describes these properties in other polymeric systems, e.g., poly(paraphenylene vinylene) or polythiophene. In the 1D model, the displacement of the lattice sites entering Equation 2.14 and Equation 2.16 is described by a single variable u_n. In the description of the geometry of conjugated polymers, *t*PA in particular, it is common to study a smoothed atomic displacement order parameter. This *optical* order parameter is given by

$$y_n = \frac{(-1)^n}{4} (2u_n - u_{n-1} - u_{n+1}).$$ (2.21)

However, this order of parameter hides the acoustic deformations (compressions and expansions) of the lattice. Sound waves are therefore nearly invisible when the geometry is described by y_n. To emphasize such deformations we introduce an *acoustic* order parameter given by

$$z_n = (u_{n+1} - u_n) + (u_{n+2} - u_{n+1}) = -u_n + u_{n+2}.$$ (2.22)

Both the optical and the acoustic order parameters are used in the analysis of time evolution of the polaron drift.

It is well known that electron (or hole) excitations are unstable with respect to the formation of an electron (or hole) polaron [49]. The ground state conformation of a *t*PA chain with one extra electron consists of a local deformation in the middle part of the chain [41,49]. We have used this type of conformation as the initial geometry in our simulations, but with the polaron shifted away from the center of the chain. This change is introduced to achieve a longer distance for the polaron to travel before it reaches the chain end.

The system we study is a *t*PA chain with 300 CH-units. The polaron center is initially located at 50 sites from one of the ends, which increases the energy with 1 meV compared to an optimized chain with the polaron in the middle of the chain. This additional energy will generate some lattice vibrations but will have very little effect on the dynamics of the polaron itself.

With an electric field applied to the polymer chain, the self-localized polaron starts to drift along the chain. After a short acceleration phase the velocity becomes constant and the excess energy is dissipated by creation of lattice vibrations. Since the description of the lattice is classical in our model, the energy is dissipated continuously.

Figure 2.1 shows the behavior of the polaron velocity as a function of the strength of the applied electric field. In the middle range of field strengths the (supersonic) velocity increases approximately linearly with the field. Above 3×10^5 V/cm, the polaron velocity becomes super linear and the polaron starts to break apart, i.e., the lattice displacements associated with the polaron can no longer follow the electronic motion and, therefore, decouple from the charge. This effect reduces the polaron effective mass that explains the increase in the velocity. Around 3.5×10^5 V/cm the polaron is no longer stable and the charge decouples completely from the lattice and becomes delocalized.

A very interesting feature is seen at low-field strengths. From the results displayed in Figure 2.1, we can see an approximately discontinuous change in velocity slightly below the sound velocity to a supersonic polaron velocity. This behavior is shown in detail in Figure 2.2. In our simulation the change occurs for a critical field strength, E_C between 1.35×10^4 and 1.40×10^4 V/cm. Figure 2.2 shows a nearly field independent polaron velocity below E_C indicating high-polaron mobility at low electric fields [50].

Figure 2.3 and Figure 2.4 show the optical and acoustical (grey scale) order parameters (see Equation 2.21 and Equation 2.22, respectively) of the polaron in the polymer chain as a function of time. These results were obtained from the numerical solution of Equation 2.17 and Equation 2.18. The black curve

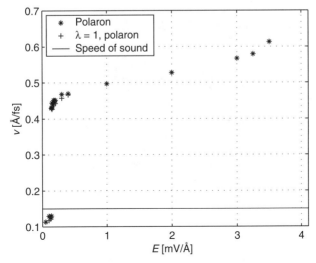

FIGURE 2.1   Polaron velocity as a function of the strength of the applied electric field (in units of mV/Å).

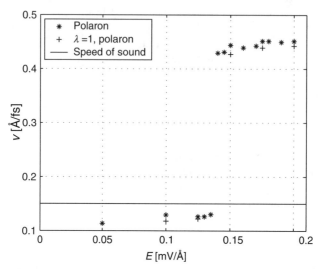

FIGURE 2.2   Polaron velocity as a function of the strength of the applied electric field (in units of mV/Å) around the critical values (part of Figure 2.1).

in Figure 2.4 shows the center of the polaron as obtained from Figure 2.3. The explicit numerical values for the acoustic order parameter are shown in Figure 2.5 for four different times. The rapidly oscillating feature seen in these graphs arises due to the deviations from the alternating bond length pattern of the pristine chain. These oscillations are cancelled by the prefactor in the expression for the optical order parameter and we obtain the typical signature of the polaron lattice deformation [41]. In all these figures the field strength is 1.35×10^4 V/cm that is just below the critical value to reach supersonic velocities. Figure 2.6 and Figure 2.7 show the acoustical order parameter as a function of time at 1.40×10^4 V/cm that is just above the critical field strength. We can now compare the polaron motion for the two cases shown in Figure 2.4 through Figure 2.7.

It is clear from the topmost graph in Figure 2.5 that the static polaron contains an acoustic deformation, more specifically, a contraction of the chain ($z_n < 0$). When an electric field lower than

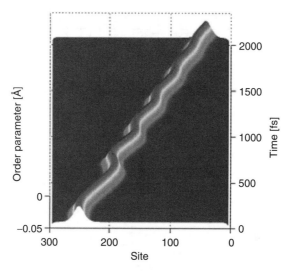

FIGURE 2.3   Polaron dynamics for an electric field strength just below the critical value ($E_0 = 1.35 \times 10^4$ V/cm). The order parameter is defined by Equation 2.21.

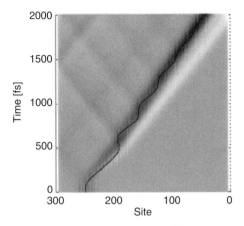

FIGURE 2.4 Same as Figure 2.3 but for the order parameter defined by Equation 2.22. The black line is the center of the polaron obtained from Figure 2.3.

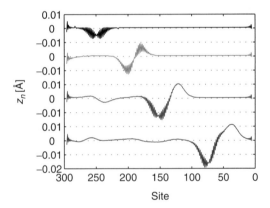

FIGURE 2.5 Order parameter taken from Figure 2.4 at different times (from the top) $t = 0$, 500, 1000, and 1700 fs.

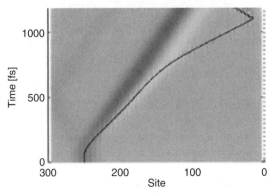

FIGURE 2.6 Dynamics of a polaron for an electric field strength just above the critical value ($E_0 = 1.40 \times 10^4$ V/cm) described by the order parameter defined by Equation 2.22. The black line is the center of the polaron.

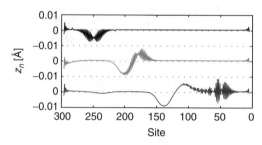

FIGURE 2.7   Order parameter taken from Figure 2.6 at different times (from the top) $t = 0$, 500, and 1000 fs.

E_C is applied, the polaron accelerates quickly to a velocity below the sound velocity, similar to the behavior reported by Arikabe et al. [48]. The acoustic deformation, which is a part of the optimized geometry of the static polaron, continues to travel along with the polaron. With the increasing time, the acoustic contraction wave that travels with the polaron becomes more compressed. At the same time, the acoustic wave develops and contains a region of expansion ($z_n > 0$) in addition to the initial contracted region ($z_n < 0$). These regions are seen in Figure 2.4 in light and dark, respectively. The front of the acoustic wave moves at the velocity of sound (0.15 fs/Å in tPA). At field strengths just below E_C this wave front that corresponds to a region of lattice expansion moves ahead of the polaron, which in this case has an oscillating velocity (see Figure 2.3). The polaron accelerates to a velocity just above the sound velocity and then slows down to zero (sometimes the polaron actually goes in the other direction). The reason for this is that the compressed part of the acoustic deformation associated with the polaron acts as a potential trap in which the electron exhibits an oscillating motion. This is a result of the fact that there is a net compression of the lattice around the ground state (static) polaron (as can be seen in the acoustic order parameter displayed in Figure 2.5). In contrast, the expanded region of the lattice deformation is energetically unfavorable for the polaron. This is the reason why the geometrical extension of the polaron is slightly smaller when this region has developed [48].

It is clear from Figure 2.4 that when the polaron is oscillating back and forth in the potential well it emits acoustic phonons in both directions. The phonons that go in the forward direction build up the sound wave that moves approximately at the same velocity (the sound velocity) as that of the polaron itself, whereas the phonons emitted with a velocity in opposite direction to that of the polaron move to the left in the figure.

We now turn to the supersonic polaron motion. As stated above, the existence of this type of motion has been debated in the literature [3,48]. Figure 2.6 and Figure 2.7 show in detail how the transition to supersonic velocities occurs. With electric field strength just above the critical value, the polaron first accelerates to a velocity slightly above the velocity of sound. After sometime the polaron has gained enough energy to overcome the potential barrier associated with the expanded region of the acoustic deformation and then accelerates to a velocity about three times the velocity of sound. The acoustic wave is now left behind the polaron (see Figure 2.7). The supersonic polaron emits optical phonons as seen from the rapidly oscillation trace that is left on the chain between the polaron and the acoustic wave. At higher field strength the polaron has enough energy to overcome the potential barrier associated with lattice expansion and accelerates more or less directly to its characteristic supersonic velocity.

The polaron moving at supersonic velocities has a distinct difference compared to the static polaron, i.e., the absence of an acoustic compression of the lattice around the localized charge. However, the deformation described by the optical order parameter is more or less the same for the two types of defects. Our simulations show clearly that these optical deformations can move along the polymer chain at supersonic velocities.

The change in the geometry associated with the polaron moving at supersonic velocities also affects the electronic structure of the system. The transition from below to above the sound velocity is clearly coupled to the phonons, i.e., the energy associated with the acoustic deformation of the static polaron is decoupled from the electron through a nonadiabatic process. This can be seen by analyzing the time-dependent occupation numbers introduced in Equation 2.20. In passing the sound velocity, the time-dependent wavefunction associated with the polaron is a linear combination of a few of the lowest π^* levels. This mixing of states gives the polaronic state the necessary probability density outside the acoustic potential well in order to decouple from this well. This effect also explains why earlier calculations of the polaron velocity based on an adiabatic theory [48] fail in describing this transition to supersonic velocities.

Also the steady state electronic structure of the system with the polaron moving at supersonic velocities is changed as compared to the static polaron. A small shift of about 40 meV of the two polaronic levels away from the band edges can be observed, i.e., these levels now appear slightly deeper into the bandgap. Also the wavefunction associated with this state is localized to the region of the optical deformation of the lattice. Thus, even though the lattice deformation differs between the static polaron and the polaron moving at supersonic velocities, the latter has, both from the point of view of lattice deformation and electronic structure, the signature of a localized state.

In summary, using the electron–lattice dynamics theory, we have been able to explore the details of the intrachain polaron motion. In particular, we have shown that the velocity of the polaron can exceed the sound velocity of the system. This is achieved by the decoupling of acoustic phonons from the polaron. With this knowledge about the intrachain behavior of the polaron dynamics, we will now go on to discuss the interchain transport of polarons.

2.5.2 Interchain Hopping

The next step of our study of polaron dynamics in organic systems includes interchain polaron transport in conjugated polymer systems. We study a system consisting of two *t*PA chains placed beside each other and overlapping by 50 sites. The first chain consists of 140 sites and contains the polaron at the start; the other chain is a perfectly dimerized 300 sites chains. The orthogonal hopping integral, i.e., the hopping between a site on one chain and a nearest neighbor site on an adjacent chain, was set to $t_\perp = 0.1$ eV. The diagonal hopping integral, i.e., the hopping between the next nearest neighbors on adjacent chains was set to $t_d = 0.05$ eV.

Simulations were carried out for a number of different field strengths ranging from 0.1×10^5 to 6.0×10^5 V/cm. Three regions are displayed in Figure 2.8 corresponding to markedly different behavior of the polaron dynamics. In all cases, the polaron starts to move on the short chain with constant speed. When it reaches the end of chain 1, i.e., the region where the chains interact, the polaron (both charge and distortion) gets stuck for fields lower than $E_0 = 0.8 \times 10^5$ V/cm, as shown in Figure 2.8a. For a slightly higher field strength, $E_0 = 0.9 \times 10^5$ V/cm, the polaron jumps over to the second chain but it has too low energy to continue moving with constant velocity, it oscillates at the beginning of the chain (this behavior is not displayed in Figure 2.8). At a field strength of 1.0×10^5 V/cm (see Figure 2.8b) the polaron remains at the first chain-end for about 100 fs. It oscillates back and forth due to the potential of the electric field and the energy minimum for the polaron at the middle of the chain. After this "waiting time" it has gained enough energy from the phonon system to jump over to chain 2 where it recollects and continues to move as a polaron first at a slightly reduced velocity as compared to the velocity on chain 1 but after some time it gains velocity and eventually behaves exactly as on the first chain.

Similar to the case of transition to supersonic velocities discussed above, this interchain transport process is also nonadiabatic. To pass through the barrier, the electron has to undergo an electronic transition from the polaron level localized to chain 1 to the π^*-level localized to chain 2. The energy needed for this transition, the activation energy, is taken from the phonon system. Actually, the process is very similar to that treated by the Holstein theory discussed in Section 2.2; the electronic coupling between the chains is weak enough to force the electronic states to localize on individual chains.

Our simulations show that interchain polaron transport is possible at sufficiently high field strengths. This behavior persists up to a field strength of about 3×10^5 V/cm but with shorter and shorter waiting times. Note that typical electric field strengths for polymer LEDs are about 1×10^5 V/cm. Also the

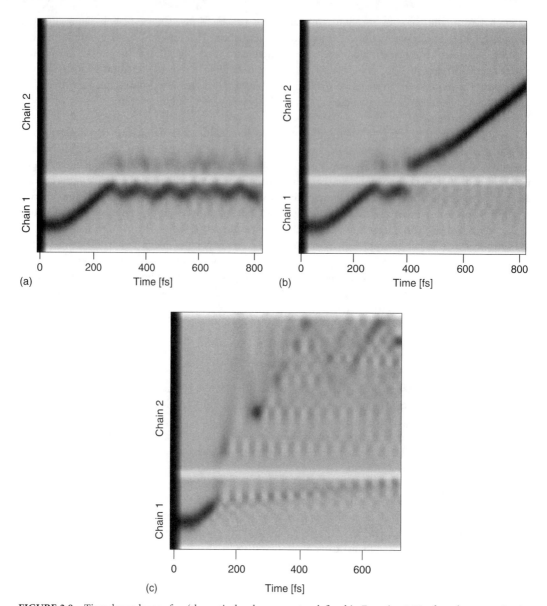

FIGURE 2.8   Time dependence of y_n (the optical order parameter defined in Equation 2.21 of a polaron moving in a two-chain system with different electric field strengths E_0).

interchain hopping strength and the overlap length play important roles in this context, as will be discussed further below.

Now we focus on the same system but for high fields, $E_0 \gtrsim 3.0 \times 10^5$ V/cm. We observe the situation pictured in Figure 2.8c, where $E_0 = 3.0 \times 10^5$ V/cm and the other parameters as above. We see that the polaron jumps over to chain 2, leaving a trace of phonons behind chain 1. However, the charge appears to be delocalized on chain 2, having difficulty to recollect. It is possible to see that some localized lattice distortion is traveling along the chain and scattered at the chain end but this can no longer be regarded as a polaron. In this case, the excess energy in the system creates too much disturbance, e.g., in the form of phonons for the polaron to be formed on the second chain. In a real physical system, e.g., an LED, excess energy will to some extent be transported away in the form of heat, which might lead to a stabilization of the polaron at slightly higher field strengths than 3×10^5 V/cm. For even higher fields, about 4×10^5 V/cm the polaron starts to dissociate already on the first chain as described for the single-chain system above.

We have also performed simulations of a three chain system, each chain 140 sites long. The two first chains are placed as described above, the third and second chains are spatially related to each other as the second chain to the first. The chain overlaps are again 50 sites. We study the system with the same intra- and interchain hopping as discussed above and for the electric field strength we choose to study the intermediate region, e.g., $E_0 = 2.0 \times 10^5$ V/cm. The polaron moves through the system first from chain 1 to chain 2 and then from chain 2 to chain 3. Both of these interchain transport processes are very similar, which show that the effects of the initial conditions, i.e., the appearance and start up process of the polaron on the first chain, are negligible. Therefore, we believe that the results of the simulations shown in Figure 2.8 correspond to a steady state situation of polaron transport in a system of coupled conjugated polymer chains.

To summarize this part, for systems consisting of one or several model conjugated polymer chains initially holding a polaron, all described using the SSH-model extended with an additional part for an external electric field, we have solved the time-dependent Schrödinger equation for the π-electrons and the equations of motion for the monomer displacements.

Basically, we identify three regions of the electric field strength. Region I, $E_0 \gtrsim 0.8 \times 10^5$ V/cm: the polaron moves with constant velocity along a chain, keeping its shape but with very little energy to jump over to another neighboring chain, it gets stuck at the end of the first chain. Region II, 1.0×10^5 V/cm $\gtrsim$ V/cm: $E_0 \gtrsim 3.0 \times 10^5$ V/cm: the polaron moves along a chain, jumps over to the neighboring chain, and continues moving in more or less the same way as on the first chain. Region III, $E_0 \gtrsim 3.0 \times 10^5$ V/cm: the polaron becomes totally delocalized, either before or after the chain jump. In this energy regime charge transport cannot be described by polaron motion.

As discussed above the coupling strength between the chains, i.e., the overlap length and the interchain hopping, is of course important for those electric field strengths that separate these regions. For lower interchain couplings the border between Region I and Region II is shifted toward higher electric fields. This leads to a narrowing of Region II and eventually, for very low couplings the electric field would have to be so high for interchain charge transport to occur that the polaron will no longer be stable, i.e., Region II vanishes. Note that Region III is more or less independent of the interchain coupling. The reason of our choice of interchain couplings is thus to put Region II in a regime of electric field strengths relevant to devices.

2.5.3 Molecular Crystals

As a last example of polaronic transport, we will now discuss polaron motion in a molecular crystal. As a model for this type of system we use pentacene and crystal structure data taken from the literature. In the dynamical simulations we used a 1D array of 30 pentacene molecules which results in a total of 660 molecular orbitals with the LUMO level being orbital number 331 (see Figure 2.11). The neutral ground state geometry of an isolated pentacene molecule is used as initial condition for each molecule in the system. We have worked out a suitable Hamiltonian parameter set capable of reproducing the molecular geometry and the polaron energies of a single pentacene molecule [5]. The mean molecular bondlength deviations lie within an error margin of less than 0.02 Å as compared to the ground state geometry obtained from first principles density functional calculations [5]. The localization energy (polaron binding energy) for the present set of parameters is $E_l = 97$ meV.

We simulate electron transport by introducing an additional electron into the LUMO of the system together with a momentary initial potential well of -0.05 eV at the first molecule in the array. Due to the presence of this potential trap the charge is forced to localize to this end of the system. At the first time step the potential well is removed and the accompanying lattice displacements start to develop.

In Figure 2.9a and Figure 2.9b are shown the time evolution of the total geometrical distortion of the molecules and the net charge on the molecules, respectively. The intermolecular interaction strength is in this case $J = 50$ meV and the field strength $E_0 = 0.5 \times 10^5$ V/cm. The consecutive curves are for molecule 1, 2,... along the array of 30 pentacene molecules. Since the initial state geometry is that of a noninteracting uncharged molecular system, turning on J will displace constituent atoms from their previous equilibrium positions and gradually shift the lattice geometry into that of a fully interacting molecular system. This is seen as a uniform shift in the level of reference. Superimposed thereon are the distortions induced by the propagating charge. Comparing Figure 2.9a with the time evolution of charge per molecule depicted in Figure 2.9b, we see that the large distortion of atomic positions is a result of the charge density initially localized to the first three molecules by the previously mentioned momentarily lowered on-site potential of molecule 1 at $t = 0$. As time goes on, the moving polaron finds its steady-state geometrical shape with the associated charge density. The geometrical and charge density waves are closely coupled indicating a polaron type of transport. Despite the rather weak interchain interaction strength the polaron is delocalized over $\sim$3 pentacene molecule. The total time during which the charge carrier affects a single pentacene molecule is above 50 fs whereas the time it takes for the charge maximum to move from one molecule to the next is $\sim$20 fs. The maximal net charge on a single molecule is 0.3 q_e. Based on these observations, the motion of the polaron through the sample is best described as a drift process rather than hopping from one molecule to another. Despite this feature, as will be shown below, the polaron dynamics obtained for $J = 50$ meV has a clear nonadiabatic behavior, i.e., the polaron transport involves a multiple of localized electronic states.

The time elapsed for the center of the polaron to move from one molecule to the next remains essentially constant after the effects of the initial conditions have decayed. The velocity of charge carrier is therefore constant and can be obtained from the time of flight of the polaron between a pair of molecules (here we obtain the polaron velocity as the time of flight from molecule 10 to 20 divided by the distance between this pair of molecules). The velocity as a function of intermolecular interaction

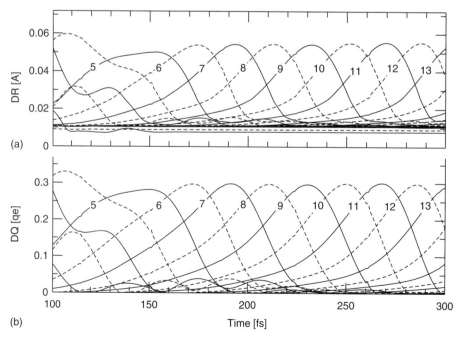

FIGURE 2.9 The time evolution of (a) total molecular bondlength deviation from the initial state geometry and (b) net charge per molecule. $J = 50$ meV and $E = 0.5 \times 10^5$ V/cm.

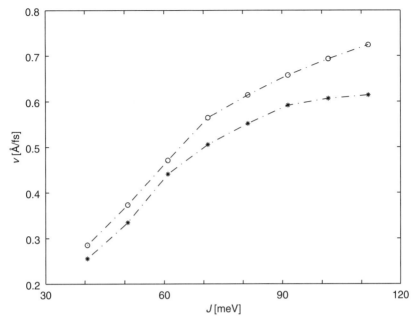

FIGURE 2.10 Charge-carrier velocity ν as a function of intermolecular transfer energy J for applied electric field strengths $E = 0.5 \times 10^5$ V/cm and $E = 1.5 \times 10^5$ V/cm.

strength in the range $0 < J < 120$ meV is shown in Figure 2.10 for two different electric field strengths 0.5×10^5 and 1.5×10^5 V/cm, respectively. For both these field strengths we found a threshold intermolecular integral strength below which the polaron remains stationary, and an upper limit to J above which the polaron continues to spread over more and more molecules as it moves through the system. In the latter case, the transport is best described in terms of band transport. Thus, from our simulations we can conclude that the perfect pentacene system exhibits band transport for intermolecular interaction strength above 120 meV. This is in agreement with the discussion above since for these values of J the interaction strength exceeds the localization energy ($E_l = 97$ meV). The experimentally observed decrease in mobility with increasing temperature indicates that it is possible to reach these high intermolecular interaction strengths in highly purified molecular crystals [13].

The abrupt trapping of the polaron for $J < 30$ meV is due to the fact that the stabilization of the local polaron level becomes much larger than the intermolecular interaction. In simulations presented here we have not included the thermal energy. Therefore, the energy barrier becomes impossible to pass and the polaron gets trapped. In a recent work [51] the method has been extended so that the atoms can be given an initial kinetic energy corresponding to a certain temperature. Preliminary simulations at 100 K show only minor effect on the polaron velocity except that we observe polaron transport at slightly lower values of J than if no thermal energy is added to the system. It should be noted, however, that so far these simulations do not include the effect of intermolecular phonons. Since such phonon modes are expected to change the intermolecular orientation it is highly likely that there will be a reduction of the polaron velocity for increasing temperature.

The regime of intermolecular interaction energies shown in Figure 2.10 most probably spans the value of J relevant for pentacene crystals [19]. The low-field mobility can be directly deduced from the calculated polaron velocity. These results set an upper limit on the mobility since the simulations are performed for a defect free system. Possibly, the mobility could change slightly if we would go from a 1D to 2D system. The fact that the polaron is rather delocalized even in the case of $J = 50$ meV indicates that the pure nonadiabatic (or hopping) treatment, such as that given by the Marcus theory [16] or the small polaron model [14] does not apply to a pentacene molecular crystal. Therefore, we do not expect a quadratic dependence of the velocity on J. From the results presented in Figure 2.10 this relation is more close to linear even though in the limit of low electric field strength we can expect a quadratic behavior for values of J below the range included in this study.

Despite the situation of a polaron drift rather than polaron hopping depicted in Figure 2.11, transport in the region of low J can be described as nonadiabatic, i.e., the dynamics requires multiple electronic states. We can obtain a very direct view of the nonadiabatic behavior of the polaron dynamics by studying the time evolution of the occupation number of the instantaneous eigenstates (see Equation 2.20 above). Figure 2.11a shows the occupation numbers for $J = 50$ meV. The occupancy, after the field has been turned on, is in the higher region of the unoccupied spectrum of the manifold of pentacene LUMO levels. Due to the weak intermolecular interaction relative to the localization energy, the electronic states are localized. The states that localize to the high (low) potential region of the system are Stark shifted upward (downward) in the spectrum. The electron is injected into the high potential energy region and will consequently occupy the higher lying states. Thus, the electron is initially in an excited state in the instantaneous picture. The motion toward the low potential region has a driving force in terms of the change in the occupancy of the instantaneous eigenstates toward lower energy as a function of time. This is seen from the Figure 2.11a. The process is driven by the external electric field and to some extent stimulated by the lattice vibrations, even though this a minor effect here since the

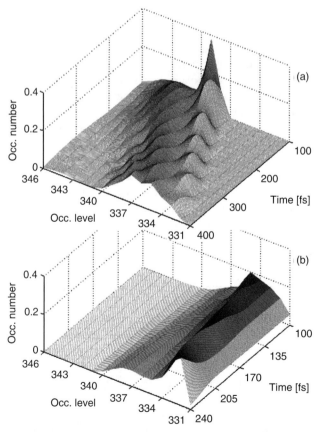

FIGURE 2.11 The time evolution of the occupation number at $E = 0.5 \times 10^5$ V/cm and (a) $J = 50$ meV and (b) $J = 100$ meV, respectively.

atoms are not given any initial kinetic energy. The lattice kinetic energy is instead generated as a result of the geometrical response of the molecule when the electron is entering.

Figure 2.11b shows the occupancy of the instantaneous eigenstates in the case of $J = 100$ meV. Clearly, the behavior is different from that shown in Figure 2.11a. The occupancy is essentially constant in time and only involves the LUMO and LUMO+1 orbitals. Still, however, we are dealing with a localized charge carrier. The difference as compared to the low J case is that all other states are delocalized and therefore do not exhibit such strong Stark shift as in the previous case. The only state which is really sensing the external field is the polaron state which itself drifts along the system toward the low potential energy end. Thus, the dynamics is in this case due to change in the spatial location of one and the same occupied state and not in change of the occupation of a manifold of states that are localized to different regions in space. Since the intermolecular interaction strength in this case is comparable or even larger than the localization energy, the activation barrier is very low in this case.

2.6   Conclusions

This study is aimed at presenting details of the microscopic picture of charge transport in molecular and polymeric solids. The results are not directly comparable to available experimental results such as charge-carrier mobility even though it is in principle possible to obtain the mobility by dividing the calculated velocity by the external electric field strength. Such a procedure corresponds to a time of flight measurement but over a very restricted distance or volume. Our simulations are performed on perfect systems in order to reveal the intrinsic features of polaronic motion. To include disorder, the simulations have to be performed on much larger systems in order to provide a statistical relevant distribution of traps and variations in the local electric field. In this case, for computational reasons, it is necessary to abandon the atomic description of the system and instead use polymer chain segments or molecules as the smallest unit. Based on the knowledge obtained from the type of studies presented here we now have the necessary information to take this step toward dynamical simulations of charge transport in organic materials. In combination with Monte Carlo simulations of charge transport based on the Miller and Abrahams or Marcus formulae this will provide the necessary simulation toolbox to relate measured mobility data to the microscopic morphology of the samples. This will be the next step in our effort to obtain a full understanding of charge transport inorganic materials.

Acknowledgments

First, I acknowledge the essential contribution from Å, Johansson who has written the first version of the computer code with which all simulations presented here are performed. Important contributions from A. Johansson, L. Gisslén, and Magnus Hultell-Andersson are also gratefully acknowledged. Financial support from the Swedish Research Council (VR) and the Center of Organic Electronics (COE), which is sponsored by the Swedish Foundation for Strategic Research (SSF) are also gratefully acknowledged. The author also wishes to thank the National Supercomputer Center (NSC) at Linköping University for providing computer facilities.

References

1. Hartman, M., M. Schreiber, H.W. Streitwolf, and S. Mukamel. 1996. *J Lum* 66–67:97.
2. Block, S., and H.W. Streitwolf. 1996. *J Phys Condens Matter* 8:889.
3. Johansson, Å., and S. Stafström. 2001. *Phys Rev Lett* 86:3602.
4. Johansson, Å., and S. Stafström. 2002. *Phys Rev B* 65:45207.
5. Hultell-Andersson, M., and S. Stafström, unpublished.
6. Johansson, A.A., and S. Stafström. 2004. *Phys Rev B* 69:235205.

7. Gisslén, L., Å. Johansson, and S. Stafström. 2004. *J Comp Phys* 121:1601.
8. Johansson, Å., and S. Stafström. 2003. *Phys Rev B* 68:35206.
9. Batail, P. (ed.). 2004. *Molecular conductors, Chemical Reviews*, vol. 104.
10. Podzorov, V., E. Menard, A. Borissov, V. Kiryukhin, J.A. Rogers, and M.E. Gershenson. 2004. *Phys Rev Lett* 93:086602.
11. Warta, W., L.B. Stehle, and N. Karl. 1985. *Appl Phys A* 36:163.
12. Klauk, H., M. Halik, U. Zschieschang, G. Schmid, W. Radlik, and W. Weber. 2002. *J Appl Phys* 92:5259.
13. Jurchescu, O.D., T. Baas, and T.T.M. Palstra. 2004. *Appl Phys Lett* 84:3061.
14. Holstein, T. 1959. *Ann Phys* 8:706.
15. May, V., and O. Kuhn. 2004. *Charge and energy transfer dynamics in molecular systems*. New York: Wiley-VCH, Chapter 6.
16. Marcus, R.A. 1993. *Rev Mod. Phys* 65:599.
17. Brédas, J.L., D. Beljonne, A. Coropceanu, and J. Cornil. 2004. *Chem Rev* 104:4971.
18. Holstein, T. 1959. *Ann Phys* 8:725.
19. Cheng, Y.C., R.J. Silbey, D.A. da Silva Filho, J.P. Calbert, J. Cornil, and J.L. Brédas. 2003. *J Comp Phys* 118:3764.
20. Ziman, J.M. 1982. *Electrons and phonons*. Oxford: Clarendon Press.
21. Marinov, O., M.J. Deen, and B. Iniguez. 2005. *IEE Proc Circ Dev Syst* 152:189.
22. Chiang, C.K., C.R. Fincher, Y.W. Park, A.J. Heeger, H. Shirakawa, E.J. Louis, S.C. Gau, and A.G. MacDiarmid. 1977. *Phys Rev Lett* 39:1098.
23. Su, W.P., J.R. Schrieffer, and A.J. Heeger. 1979. *Phys Rev Lett* 42:1698.a
24. Rice, M.J. 1979. *Phys Lett* 71A:152.
25. Kaiser, A.B. 2001. *Rep Prog Phys* 64:1.
26. Epstein, A.J., H. Rommelman, R. Bigelow, H.W. Gibson, D.M. Hoffmann, and D.B. Tanner. 1983. *Phys Rev Lett* 50:1866.
27. Burroughes, J., D. Bradley, A. Brown, R. Marks, K. MacKay, R. Friend, P.L. Burns, and A. Holmes. 1990. *Nature* 347:539.
28. Sariciftci, N.S., L. Smilowitz, A.J. Heeger, and F. Wudl. 1992. *Science* 258:1474.
29. Ong, B.S., Y. Wu, and P. Liu. 2005. *Proc IEEE* 93:1412.
30. Pasveer, W.F., J. Cottaar, C. Tanases, R. Coehoorn, P.A. Bobbert, P.W.M. Blom, D.M. de Leeuw, and M.A.J. Michels. 2005. *Phys Rev Lett* 94:206601.
31. Sirringhaus, H., P.J. Brown, R.H. Friend, M.M. Nielsen, K. Bechgaard, B.M.W. Langeveld-Voss, A.J.H. Spiering, R.A.J. Janssen, E.W. Meijer, P.T. Herwig, and D.M. de Leeuw. 1999. *Nature* 401:685.
32. Bao, Z., D.A., and A. Lovinger. 1996. *Appl Phys Lett* 69:4108.
33. Conwell, E. 1956. *Phys Rev* 103:51.
34. Mott, N.F. 1956. *Can J Phys* 34:1356.
35. Pines, D. 1956. *Can J Phys* 34:1367.
36. Landau, L.D. 1932. *Phys Z Sowjetunion* 2:46.
37. Zener, C. 1932. *Proc R Soc Lond Ser A* 137:696.
38. Miller, A., and E. Abrahams. 1960. *Phys Rev* 120:745.
39. Gartstein, Y.N., and E.M. Conwell. 1995. *Chem Phys Lett* 245:351.
40. Bässler, H. 1993. *Phys Status Solidi B* 175:15.
41. Su, W.P., and J.R. Schrieffer. 1980. *Proc Natl Acad Sci USA* 77:5626.
42. You, W., C. Wang, F. Zhang, and Z. Su. 1993. *Phys. Rev. B* 47:4765.
43. Bruening, J., and B. Friedman. 1997. *J Chem Phys* 106:963.
44. Mulliken, R.S., C.A. Reike, D. Orloff, and H. Orloff. 1949. *J Comp Phys* 17:1248.
45. Hansson, A., and S. Stafström. 2003. *Phys Rev B* 67:075406.
46. Riedmiller, M., and H. Braum. 1993. *Proceedings of the IEEE International Conference on Neural Networks*. San Francisco, CA, USA, pp. 586–591.

47. Brankin, R.W., and I. Gladwell. 1994. *Tech Rep.*
48. Arikabe, Y., M. Kuwabara, and Y. Ono. 1996. *J Phys Soc Jpn* 65:1317.
49. Brazovski, S.A., and N.N. Kirova. 1981. *Pris'ma Zh Eksp Teor Fiz* 33:6.
50. Wilson, E.G. 1983. *Solid State Phys* 16:6739.
51. Hultell-Andersson, M., and S. Stafström, *Chem. Phys. Lett,* in press.

II

Synthesis and Classes of Conjugated Polymers

3

Helical Polyacetylene Synthesized in Chiral Nematic Liquid Crystal

3.1 Introduction

Polyacetylene is the simplest linear conjugated macromolecule and a representative of conducting polymers [1]. Pristine polyacetylene is a typical semiconductor, but its electrical conductivity can be varied by over 14 orders of magnitude through doping [2]. The maximum conductivity reported to date is more than 10^5 S/cm [3], which is comparable to those of copper and gold. Strong interchain interaction gives rise to a fibrillar crystal consisting of rigidly π-stacked polymer chains [4]. This makes polyacetylene infusible and insoluble in any kind of solvent. Thus, the solid state structure and morphology of polyacetylene are determined during acetylene polymerization that is quite different from substituted polyacetylenes in terms of fusibility and solubility [5]. In addition, the fibril morphology of polyacetylene film is randomly oriented, as it is usually encountered with ordinary polymers, depressing an inherent one dimensionality of this polymer. Hence, several kinds of procedures and polymerization methods for macroscopic alignment of polymer have been developed to achieve higher electrical conductivity with anisotropic nature [3,6,7]. Introduction of a liquid crystal (LC) group into the side chain is an approach to align the polymer under an external force such as shear stress, rubbing, or a magnetic field [8–10]. The use of nematic LC as a solvent, on the other hand, gave us directly aligned polyacetylene with the aid of gravity flow or a magnetic force field [11,12].

It has been generally accepted that polyacetylene has a planar structure, irrespective of *cis* and *trans* forms, due to π-conjugation between the sp^2 hybridized carbon atoms in the polymer chain [1,4]. If it were possible to modify such a planar structure of polyacetylene into a helical one [13], one might expect novel magnetic and optical properties [14]. Here, we report polymerization of acetylene in asymmetric reaction field constructed with chiral nematic LCs, and show that polyacetylene films formed by helical chains and fibrils can be synthesized [15]. Polymerization mechanism giving helical structure from primary to higher order and hierarchical spiral morphology is discussed.

3.1.1 Chiral Dopants and Chiral Nematic Liquid Crystals

Chiral nematic LC used as an asymmetric solvent is prepared by adding a small amount of chiral compound, as a chiral dopant, into nematic LC (Figure 3.1). The formation of chiral nematic LC is recognized when a Schlieren texture characteristic of nematic LC changes into a striated Schlieren or a fingerprint texture in polarized optical microscope (POM). The distance between the striae corresponds to a half helical pitch of the chiral nematic LC. Note that as the degree of twist in the chiral nematic LC is larger, the helical pitch observed in POM is shorter. The helical pitch of the chiral nematic LC can be adjusted by two methods: changing the concentration or changing the twisting power of the chiral dopant [16]. However, the mesophase temperature region of the chiral nematic LC is affected by changing the concentration of the chiral dopant, i.e., it becomes narrow as the concentration increases, and finally the mesophase will be destroyed when the concentration is close to a critical value [17]. Consequently, owing to the limitation of the concentration method, an alternative approach of utilizing the chiral compound with large twisting power is adopted. Axially chiral binaphthyl derivatives are used as chiral dopants, since they have been reported to possess larger twisting powers than asymmetric carbon containing chiral compounds [15(c)].

(R)- and (S)-1,1′-bi-naphthyl-2,2′-di-[*para*-(*trans*-4-*n*-pentylcyclohexyl)phenoxy-1-hexyl]ether were synthesized through the Williamson etherification reactions of chiroptical (R)-(+)- and (S)-(−)-1,1′-bi-2-naphthols, respectively, with phenylcyclohexyl (PCH) derivatives. The products will hereafter be abbreviated as (R)- and (S)-PCH506-Binol (Scheme 3.1). The substituent is composed of PCH moiety, *n*-pentyl group (the number of carbon of 5), and hexamethylene chain linked with an ether-type oxygen atom, [-(CH$_2$)$_6$O-, 06], hence abbreviated as PCH506.

To prepare an induced chiral nematic LC, 5–14 wt% of (R)- or (S)-PCH506-Binol was added as a chiral dopant to the equimolar mixture of the nematic LCs 4-(*trans*-4-*n*-propylcyclohexyl)ethoxybenzene (PCH302) and 4-(*trans*-4-*n*-propylcyclohexyl)butoxybenzene (PCH304). The PCH506 substituent group in the (R)- and (S)-PCH506-Binol enhances the miscibility between the nematic LC mixture and the binaphthyl derivative used as the chiral dopant. Usage of similar substituent with a shorter methylene spacer, such as PCH503 or normal alkyl substituent, gave insufficient miscibility, yielding no chiral nematic phase.

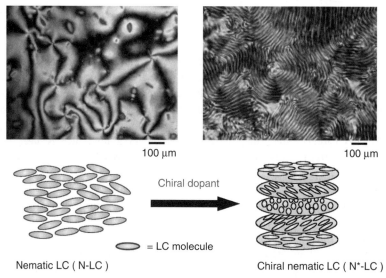

FIGURE 3.1 (**See color insert following page 14-20.**) Chiral nematic LC induced by the addition of chiral dopant into nematic LC. Schlieren texture (*left*) and fingerprint texture (*right*) are observed for nematic and chiral nematic LCs, respectively, in polarized optical microscope.

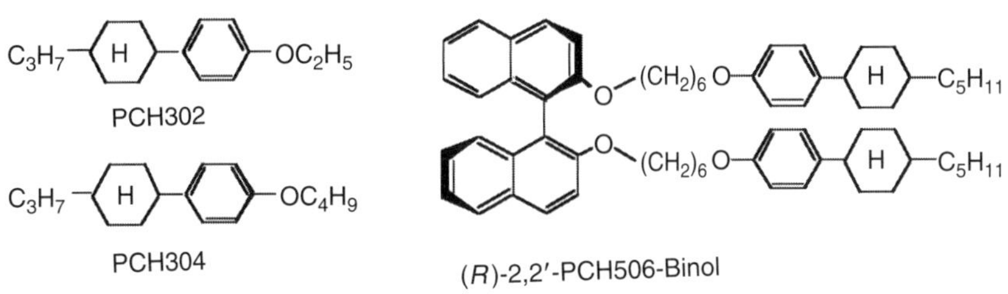

SCHEME 3.1 Two kinds of nematic liquid crystals and chiral dopants, (*R*)- and (*S*)-2,2′-PCH506-Binol.

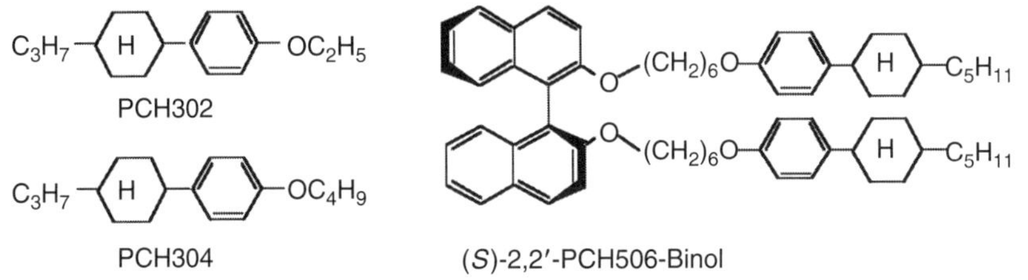

In polarizing optical micrographs of the mixture of PCH302, PCH304, and (*R*)-PCH506-Binol (abbreviated as *R*-1) and that of PCH302, PCH304, and (*S*)-PCH506-Binol (abbreviated as *S*-1), a striated Schlieren or fingerprinted texture characteristic of chiral nematic LC phases is observed (Figure 3.1). Cholesteryl oleyl carbonate is known to be a counterclockwise chiral nematic LC, and therefore it is available as a standard LC for miscibility test with the present LCs. The miscibility test method is based on the observation of the mixing area between the chiral nematic LC and the standard LC in POM, where the screw direction of the standard LC is known. If the screw direction of the chiral nematic LC is the same as that of the standard LC, the mixing area will be continuous. Otherwise, it will be discontinuous (shown as a Schlieren texture of the nematic LC). As shown in Figure 3.2, the mixture of (*R*-11)-chiral nematic LC and cholesteryl oleyl carbonate lost the striae characteristic of a chiral nematic phase in the POM, yielding features corresponding to an ordinary nematic phase. In contrast, the mixture of (*S*-1)-chiral nematic LC and cholesteryl oleyl carbonate showed no change in optical texture, keeping a chiral nematic phase. The results demonstrate that the screw directions of (*R*-1)- and (*S*-1)-chiral nematic LCs are opposite to and the same as that of cholesteryl oleyl carbonate, respectively, i.e., they are clockwise (right-handed screwed) and counterclockwise (left-handed screwed) chiral nematic LCs, respectively.

3.1.2 Acetylene Polymerization in Chiral Nematic Liquid Crystal

It should be noted that although each component (PCH302 or PCH304) shows an LC phase, the LC temperature region is very narrow, i.e., less than 1°C to 2°C. This is not suitable for acetylene polymerization in a nematic or chiral nematic LC reaction field, because the exothermal heat evoked during the acetylene polymerization would raise the temperature inside a Schlenk flask, and easily destroy the LC phase into an isotropic one. Hence, the LC mixture is prepared by mixing two equimolar LC components. In the LC mixture, the nematic–isotropic temperature, $T_{N–I}$, and the crystalline–nematic temperature, $T_{C–N}$, might be raised and lowered, respectively. In fact, the mixture exhibited the LC phase in the region from 20°C to 35°C. Subsequently, the change of $T_{N–I}$ upon an addition of

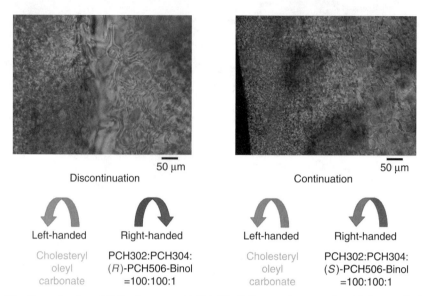

FIGURE 3.2 **(See color insert following page 14-20.)** Miscibility test between the chiral nematic LC induced by (*R*)- or (*S*)-PCH506-Binol and the standard LC, cholesteryl oleyl carbonate of counterclockwise (left-handed) screw direction.

Ti(O-*n*-Bu)$_4$–AlEt$_3$ catalyst was examined through DSC measurement. Taking account of the effect of supercooling for LCs, the catalyst solution consisting of the LC mixture and the chiral dopant was found available for room temperature polymerization ranging from 5°C to 25°C. This sufficiently wide temperature region enabled us to perform the acetylene polymerization in the N*-LC phase.

The Ziegler–Natta catalyst consisting of Ti(O-*n*-Bu)$_4$ and Et$_3$Al was prepared using (*R*)- or (*S*)-chiral nematic LC as a solvent (Figure 3.3). The concentration of Ti(O-*n*-Bu)$_4$ was 15 mmol/L, and the mole ratio of the cocatalyst/catalyst, [Et$_3$Al]/[Ti(O-*n*-Bu)$_4$], was 4:0. The catalyst solution was aged for 0.5 h at room temperature. During aging, the chiral nematic LC containing the catalyst showed no noticeable change in optical texture, but only a slight lowering of the transition temperature by 2°C to 5°C: The transition temperature between solid and chiral nematic phases was 16°C to 17°C, and that between chiral

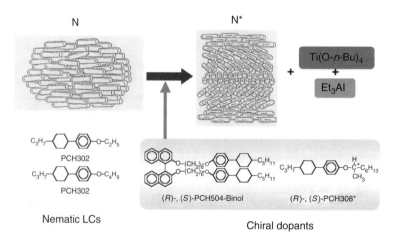

FIGURE 3.3 Construction of asymmetric reaction field for acetylene polymerization by dissolving Ziegler–Natta catalyst, Ti(O-*n*-Bu)$_4$–AlEt$_3$, into the chiral nematic LC. The chiral nematic LC includes an axially chiral binaphthyl derivative or an asymmetric center containing chiral compound.

nematic and isotropic ones was 30°C to 31°C. No solidification was observed at −7°C as a result of supercooling. Thus the (*R*)- and (*S*)-chiral nematic LCs are confirmed to be chemically stable to the catalyst. It is therefore allowed to employ these LCs as an asymmetric solvent for acetylene polymerization.

Acetylene gas of 6 to 9 grade (99.9999% purity) was used without further purification. The apparatus and procedure employed were similar to those in earlier studies [12(c–d)], except for the polymerization temperature. Here, the polymerization temperature was kept between 17°C and 18°C to maintain the chiral nematic phase, by circulating cooled ethanol through an outer flask enveloping the Schlenk flask. The initial acetylene pressure was 11.6 to 22.6 Torr and the polymerization time was 10 to 43 min. After polymerization, polyacetylene films were carefully stripped off from the container and washed with toluene several times under argon gas at room temperature. The films were dried through vacuum pumping on a Teflon sheet and stored in a freezer at −20°C.

3.1.3 Characterization of Helical Polyacetylene Film

Scanning electron microscope (SEM) images of polyacetylene films show that multidomains of spiral morphology are formed (Figure 3.4a), and each domain is composed of helical structure of bundle of fibrils with one-handed screwed direction (Figure 3.4b). The multidomain-type fibril morphology of polyacetylene seems to replicate the chiral nematic LC during interfacial acetylene polymerization.

Closer observation of SEM images indicates that helical polyacetylenes synthesized in the (*R*)- and (*S*)-chiral nematic LCs form screwed bundles of fibrils and screwed fibrils with counterclockwise and clockwise directions, respectively (Figure 3.5). This result implies that the screw direction of helical polyacetylene is controllable by choosing the helicity, i.e., optical configuration of the chiral dopant, so far as the chiral nematic LC induced by the chiral dopant is employed as an asymmetric polymerization solvent. In addition, it is clear that the screw directions of bundle and fibrils are opposite to those of the (*R*-1)- and (*S*-1)-chiral nematic LCs used as solvents. This is an unexpected and even surprising result, requiring a sound interpretation that is to be discussed later.

The helical pitch of the chiral nematic LC depends on the helical twisting power of the chiral dopant, as well as its concentration and optical purity. This means that the helical pitch of the polyacetylene chain can also be varied by changing helical twisting power of the chiral dopant. Another axially chiral dopant, (*R*)- or (*S*)-6,6′-PCH506-2,2′-Et-Binol (Scheme 3.2) [18] gave a shorter helical pitch of chiral nematic LC by 0.3 μm than the corresponding (*R*)-or (*S*)-PCH506-Binol. Acetylene polymerizations using these sorts of highly twisted chiral nematic LCs, designated (*R*-2)- and (*S*-2)-chiral nematic LCs, afforded clearer spiral morphologies consisting of helical bundles of fibrils (Figure 3.6). These bundles are aligned parallel to each other in the microscopic regime, and form spiral morphologies in the macroscopic regime. The hierarchical higher order structures observed in Figure 3.5 and Figure 3.6 resemble the helical self-assembled microstructure of biological molecules, such as lipids [19], but they, as well as those of Figure 3.4, are rarely formed in synthetic polymers. This indicates a validity of chiral nematic LC as a template polymerization medium for controlling a higher order structure of synthetic polymer.

The bundles of fibrils for helical polyacetylenes synthesized in the (*R*-2)- and (*S*-2)-chiral nematic LCs are screwed counterclockwise and clockwise, respectively: The screw directions of helical polyacetylene are opposite to those of the corresponding (*R*-2)- and (*S*-2)-chiral nematic LCs whose directions are confirmed to be clockwise and counterclockwise, respectively, through the miscibility test with cholesteryl oleyl carbonate. This is the same situation as that of the (*R*-1)- and (*S*-1)-chiral nematic LCs including (*R*)- and (*S*)-PCH506-Binol.

So far it was elucidated that the polyacetylene chains propagate along the director (an averaged direction for the LC molecules within a domain) of the chiral nematic LC. As the helical axis of polyacetylene is parallel to the polyacetylene chain and the director of the chiral nematic LC is perpendicular to the helical axis of chiral nematic LC, the helical axis of polyacetylene is perpendicular to that of chiral nematic LC. Considering these aspects, one can describe a plausible mechanism for interfacial acetylene polymerization in the chiral nematic LC, as shown in Figure 3.7. In the case of a right-handed chiral nematic LC, for instance, the polyacetylene chain would propagate with a

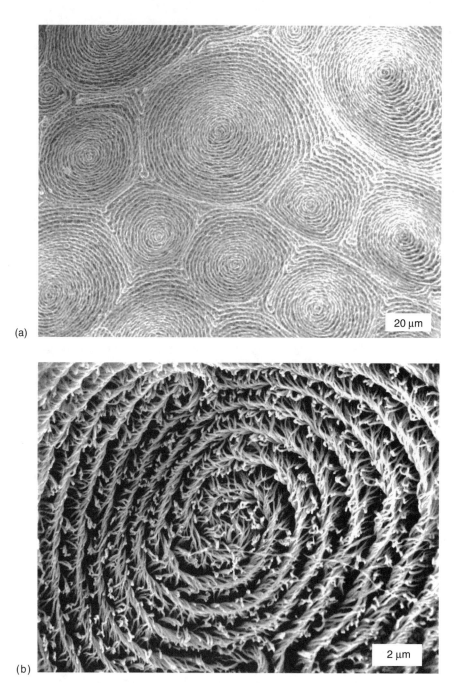

(a)

(b)

FIGURE 3.4 Hierarchical spiral morphology of helical polyacetylene film. (a) and (b) show scanning electron microscope (SEM) photographs of multidomain-type spiral morphology and helical bundles of fibrils in a domain, respectively.

left-handed manner, starting from the catalytic species, and not with a right-handed one. Because, the polyacetylene chains with the opposite screw direction to that of the chiral nematic LC could propagate along the LC molecules, but those with the same direction as that of the chiral nematic LC would

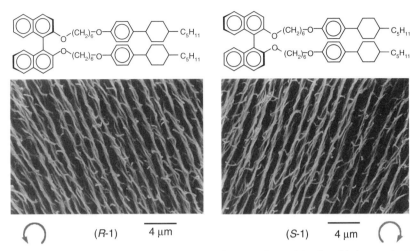

FIGURE 3.5 SEM photographs of helical polyacetylene films synthesized in the chiral nematic LCs including (*R*)- and (*S*)-PCH506-Binol.

encounter LC molecules, making the propagation stereospecifically impossible. Detailed mechanism of acetylene polymerization in chiral nematic LC has to be elucidated.

In circular dichroism (CD) spectra of the polyacetylene thin films synthesized under the (*R*-2)- and (*S*-2)-chiral nematic LCs, positive and negative Cotton effects are observed, respectively, in the region from 450 to 800 nm corresponding to $\pi \rightarrow \pi^*$ transition of polyacetylene chain (Figure 3.8), despite the absence of chiroptical substituent in side chains. This indicates that the polyacetylene chain itself is helically screwed. It is evident that the above Cotton effect is not due to the chiral dopant [(*R*)- or

(*R*)-6,6′-PCH506-2,2′-Et-Binol

(*S*)-6,6′-PCH506-2,2′-Et-Binol

SCHEME 3.2 Chiral dopants, (*R*)- and (*S*)-6,6′-PCH506-2,2′-Et-Binol.

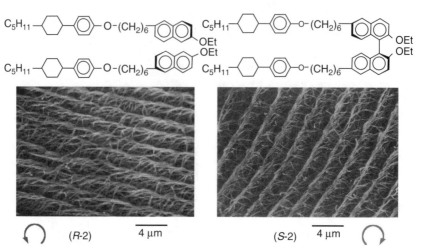

FIGURE 3.6 SEM photographs of helical polyacetylene films synthesized in the chiral nematic LCs including (*R*)- and (*S*)-6,6'-PCH506-2,2'-Et-Binol.

(*S*)-6,6'-PCH506-2,2'-Et-Binol], because the Cotton effect of the chiral dopant is only observed at shorter wavelengths such as 240 to ~340 nm.

From the above mentioned results, it can be remarked that counterclockwise and clockwise helical polyacetylene chains are formed in (*R*)- and (*S*)-chiral nematic LCs, respectively, and that these helical chains are bundled through van der Waals interactions to form helical fibrils with the opposite screw directions to those of the chiral nematic LCs. The bundles of fibrils further form the spiral morphology with various sizes of domains (Figure 3.9).

The dihedral angle between neighboring unit cells, (—CH=CH—), of the helical polyacetylene was estimated to be from 0.02° to 0.23°. Although such a very small dihedral angle may allow us to regard the present polyacetylene as an approximately planar structure, the polymer is rigorously screwed by one-handed direction with the nonzero dihedral angle. The present helical polyacetylene films have high

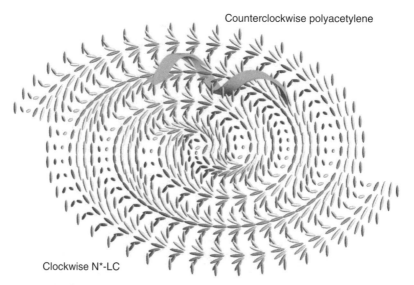

FIGURE 3.7 Schematic representation of plausible mechanism for acetylene polymerization in the chiral nematic LC. The helical polyacetylene with counterclockwise (left-handed) screw direction (*blue arrow*) grows up starting from catalytic species in the clockwise (right-handed) chiral nematic LC.

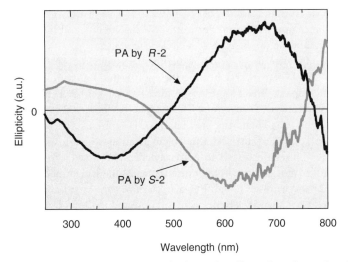

FIGURE 3.8 Circular dichroism (CD) spectra of helical polyacetylene films. The polyacetylene films synthesized in the (*R*-2)- and (*S*-2)-chiral nematic LCs including (*R*)- and (*S*)-6,6′-PCH506-2,2′-Et-Binol are designated as PA by *R*-2 and PA by *S*-2, respectively.

trans contents of 90%, and become highly conductive upon iodine doping. In fact, the electrical conductivities of the doped films are $1.5 \sim 1.8 \times 10^3$ S/cm at room temperature that are comparable to those of metals. The iodine-doped polyacetylene showed the same Cotton effect as that of nondoped polyacetylene, although the CD peak was slightly shifted to shorter wavelengths. This indicates that the helical structure is preserved even after iodine doping. Furthermore, CD and x-ray diffraction measurements showed that the helical structure was preserved after heating up to 150°C, which corresponds to the isomerization temperature from *cis* to *trans* form. The most stable structure of polyacetylene is the planar one. However, since the polyacetylene is actually insoluble and infusible, the helical structure

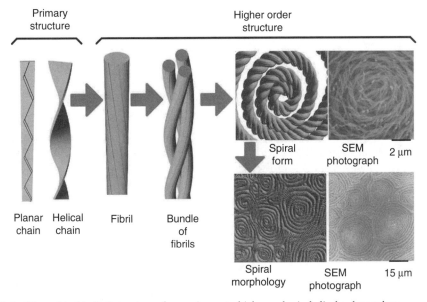

FIGURE 3.9 Hierarchical helical structures from primary to higher order in helical polyacetylene.

formed during the polymerization can be preserved even if it is washed by toluene or thermally heated below the isomerization temperature. In other words, the insolubility and infusibility of polyacetylene are indispensable for preserving the metastable helical structure.

It is worthwhile to emphasize the following experimental results: (i) Acetylene polymerization was carried out under nematic LC environment using the equiweighted mixture of PCH302 and PCH304, but without the chiral dopant. The polyacetylene film showed neither helical morphology in SEM photograph nor Cotton effect in CD spectrum: The morphology observed was composed of fibrils that were locally aligned owing to spontaneous orientation of LC solvent. This was also confirmed in the previous works [12(b–e)]. (ii) Next, the acetylene polymerization was performed in toluene and a small amount (<10%) of the chiral dopant. The polyacetylene synthesized showed usually encountered randomly oriented fibrillar morphology, but not helical one. At the same time, the polyacetylene showed no Cotton effect in CD spectrum, although the characteristic very broad absorption band due to the $\pi \rightarrow \pi^*$ transitions in the conjugated polyene chain was observed in the region from 450 to 800 nm. (iii) The acetylene polymerization at 35°C $\sim$ 40°C, where the LC mixture including the chiral dopant was isotropic, produced the polyacetylene with no helical morphology. These results demonstrate that the chiral nematic LC environment is essential for the helical polyacetylene.

3.2 Summary

Helical polyacetylene was synthesized in asymmetric reaction field consisting of chiral nematic LC. The chiral nematic LC was prepared by adding a chiroptical binaphthyl derivative as a chiral dopant to a mixture of two nematic LCs. Acetylene polymerizations were carried out using the catalyst Ti(O-n-Bu)$_4$–Et$_3$Al dissolved in the chiral nematic LC solvent. The polyacetylene film consisted of clockwise or counterclockwise helical structure of fibrils in SEM. Cotton effect was observed in the region of $\pi \rightarrow \pi^*$ transition of the polyacetylene chain in CD spectrum. The high-electrical conductivities of 1.5 $\sim$ 1.8 $\times$ 10^3 S/cm after iodine doping and the chiral helicity of the present films should be available for novel electromagnetic and optical properties. Macroscopic alignment of helical polyacetylene has also been successfully carried out to prepare the samples that are feasible for examination of novel electromagnetic properties, which will be presented in the near feature.

Lastly, it is worth noting that by using the chiral nematic LC as an asymmetric polymerization solvent, the helix formation is possible not only for polyacetylene but also for π-conjugated polymers without chiroptical substituents in side chains. Recently, other kinds of spiral morphology containing conjugated polymers such as polybithiophene, polyethylenedioxythiophne derivatives, and phenylene–thiophene copolymers [20] were synthesized through chemical or electrochemical polymerization in the chiral nematic LC [21].

Acknowledgments

The author is grateful to Dr. Hideki Shirakawa (Professor Emeritus of the University of Tsukuba) for his valuable discussion and stimulation during the collaboration. This work was supported by a Grant-in-Aid for Science Research in a Priority Area, "Super-Hierarchical Structures" (No. 446) from the Ministry of Education, Culture, Sports, Science, and Technology, Japan.

References

1. (a) Skotheim, T.A. (ed.). 1986. *Handbook of conducting polymers.* New York: Marcel Dekker. (b) Nalwa, H.S. (ed.). 1997. *Handbook of organic conductive molecules and polymers.* New York: Wiley. (c) Akagi, K., and H. Shirakawa. 1998. *Electrical and optical polymer systems: Fundamentals, methods, and applications,* eds. D.L. Wise, G.E. Wnek, D.J. Trantolo, T.M. Cooper, and J.D. Gresser. New York: Marcel Dekker, Chapter 28, p. 983.

2. (a) Shirakawa, H., E. Louis, A.G. MacDiarmid, C.K. Chiang, and A.J. Heeger. 1977. *J Chem Soc Chem Commun* 578. (b) Chiang, C.K., C.R. Fincher, Y.W. Park, A.J. Heeger, H. Shirakawa, E.J. Louis, S.C. Gau, and A.G. MacDiarmid. 1977. *Phys Rev Lett* 39:1098.

3. (a) Naarrmann, H., and N. Theophilou. 1987. *Synth Met* 22:1. (b) Tsukamoto, J., A. Takahashi, and K. Kawasaki. 1990. *Jpn J Appl Phys* 29:125.

4. (a) Shirakawa, H., and S. Ikeda. 1971. *Polym J* 2:231. (b) Ito, T., H. Shirakawa, and S. Ikeda. 1974. *J Polym Sci Polym Chem Ed* 12:11. (c) Tanabe, Y., H. Kyotani, K. Akagi, and H. Shirakawa. 1995. *Macromolecules* 28:4173.

5. Shirakawa, H., T. Masuda, and T. Takeda. 1994. *The chemistry of triple-bonded functional groups*, Vol. 2, Suppl. C2, ed. S. Patai. Chichester, UK: Wiley, Chapter 17.

6. (a) Akagi, K., M. Suezaki, H. Shirakawa, H. Kyotani, M. Shimomura, and Y. Tanabe. 1989. *Synth Met* 28:D1. (b) Cao, Y., P. Smith, and A.J. Heeger. 1991. *Polymer* 32:1210.

7. (a) Shibahara, S., M. Yamane, K. Ishikawa, and H. Takezoe. 1998. *Macromolecules* 31:3756. (b) Scherman, O.A., and R.H. Grubbs. 2001. *Synth Met* 124:431. (c) Scherman, O.A., I.M. Rutenberg, and R.H. Grubbs. 2003. *J Am Chem Soc* 125:8515. (d) Schuehler, D.E., J.E. Williams, and M.B. Sponsler. 2004. *Macromolecules* 37:6255. (e) Gu, H., R. Zheng, X. Zhang, and B. Xu. 2004. *Adv Mater* 16:1356.

8. (a) Oh, S.Y., K. Akagi, H. Shirakawa, and K. Araya. 1993. *Macromolecules* 26:6203. (b) Oh, S.Y., R. Ezaki, K. Akagi, and H. Shirakawa. 1993. *J Polym Sci A* 31:2977. (c) Akagi, K., and H. Shirakawa. 1996. *Macromol Symp* 104:137. (d) Kuroda, H., H. Goto, K. Akagi, and A. Kawaguchi. 2002. *Macromolecules* 35:1307. (d) Goto, H., X. Dai, T. Ueoka, and K. Akagi. 2004. *Macromolecules* 37:4783.

9. (a) Jin, S.H., S.J. Choi, W. Ahn, H.N. Cho, and S.K. Choi. 1993. *Macromolecules* 26:1487. (b) Choi, S.K., Y.S. Gal, S.H. Jin, and H.K. Kim. 2000. *Chem Rev* 100:1645. (c) Tang, B.Z., X. Kong, X. Wan, H. Peng, W.Y. Lam, X.D. Feng, and H.S. Kwok. 1998. *Macromolecules* 31:2419. (d) Kong, X., J.W.Y. Lam, and B.Z. Tang. 1999. *Macromolecules* 32:1722. (e) Geng, J., X. Zhao, E. Zhou, G. Li, J.W.Y. Lam, and B.Z. Tang. 2003. *Polymer* 44:8095.

10. (a) Koltzenburg, S., D. Wolff, F. Stelzer, J. Springer, and O. Nuyken. 1998. *Macromolecules* 31:9166. (b) Ting, C.H., J.T. Chen, and C.S. Hsu. 2002. *Macromolecules* 35:1180. (c) Schenning, A.P.H.J., M. Fransen, and E.W. Meijer. 2002. *Macromol Rapid Commun* 23:265. (d) Stagnaro, P., L. Conzatti, G. Costa, B. Gallot, and B. Valenti. 2003. *Polymer* 44:4443.

11. (a) Araya, K., A. Mukoh, T. Narahara, and H. Shirakawa. 1984. *Chem Lett* 1141. (b) Akagi, K., H. Shirakawa, K. Araya, A. Mukoh, and T. Narahara. 1987. *Polym J* 19:185. (c) Akagi, K., S. Katayama, H. Shirakawa, K. Araya, A. Mukoh, and T. Narahara. 1987. *Synth Met* 17:241. (d) Akagi, K., S. Katayama, M. Ito, H. Shirakawa, and K. Araya. 1989. *Synth Met* 28:D51. (e) Sinclair, M., D. Moses, K. Akagi, and A.J. Heeger. 1988. *Phys Rev B Condens Matter* 38:10724.

12. (a) Montaner, A., M. Rolland, J.L. Sauvajol, M. Galtier, R. Almairac, and J.L. Ribet. 1988. *Polymer* 29:1101. (b) Coustel, N., N. Foxonet, J.L. Ribet, P. Bernier, and J.E. Fischer. 1991. *Macromolecules* 24:5867.

13. Bozovic, I. 1987. *Mod Phys Lett B* 1:81.

14. (a) Suh, D.S., T.J. Kim, A.N. Aleshin, Y.W. Park, G. Piao, K. Akagi, H. Shirakawa, J.S. Qualls, S.Y. Han, and J.S. Brooks. 2001. *J Chem Phys* 114:7222. (b) Aleshin, A.N., H.J. Lee, Y.W. Park, and K. Akagi. 2004. *Phys Rev Lett* 93:196601. (c) Lee, H.J., Z.X. Jin, A.N. Aleshin, J.Y. Lee, M.J. Goh, K. Akagi, Y.S. Kim, D.W. Kim, and Y.W. Park. 2004. *J Am Chem Soc* 126:16722.

15. (a) Akagi, K., G. Piao, S. Kaneko, K. Sakamaki, H. Shirakawa, and M. Kyotani. 1998. *Science* 282:1683. (b) Akagi, K., G. Piao, S. Kaneko, I. Higuchi, H. Shirakawa, and M. Kyotani. 1999. *Synth Met* 102:1406. (c) Akagi, K., I. Higuchi, G. Piao, H. Shirakawa, and M. Kyotani. 1999. *Mol Cryst Liq Cryst* 332:463. (d) Akagi, K., S. Guo, T. Mori, M. Goh, G. Piao, and M. Kyotani. 2005. *J Am Chem Soc* 127:14647.

16. Gottarelli, G., P. Mariani, G.P. Spada, B. Samori, A. Forni, G. Solladie, and M. Hibert. 1983. *Tetrahedron* 39:1337.

17. (a) Semenkova, G.P., L.A. Kutulya, N.I. Shkol'nikova, and T.V. Khandrimailova. 2001. *Kristallografiya* 46:128. (b) Guan, L., and Y. Zhao. 2001. *J Mater Chem* 11:1339. (c) Hatoh, H. 1994. *Mol Cryst Liq Cryst Sci Technol Sect A* 250:1. (d) Lee, H., and M.M. Labes. 1982. *Mol Cryst Liq Cryst* 84:137.

18. (a) Kanazawa, K., I. Higuchi, and K. Akagi. 2001. *Mol Cryst Liq Cryst* 364:825. (b) Goh, M.J., M. Kyotani, and K. Akagi. 2006. *Current Appl Phys.* 6:948.

19. Schnur, J.M. 1993. *Science* 262:1669 and references therein.

20. (a) Osaka, I., A. Nakamura, Y. Inoue, and K. Akagi. 2002 *Trans Mat Res Soc Jpn* 27:567. (b) Yorozuya, S., I. Osaka, A. Nakamura, Y. Inoue, and K. Akagi. 2003. *Synth Met* 135–136:93. (c) Oh-e, M., H. Yokoyama, S. Yorozuya, K. Akagi, M.A. Belkin, and Y.R. Shen. 2004. *Phys Rev Lett* 93:267402. (d) Goto, H., and K. Akagi. 2005. *Angew Chem Int Ed* 44:4322.

21. (a) Goto, H., and K. Akagi. 2004. *Macromol Rapid Commun* 25:1482. (b) Goto, H., and Akagi, K. 2005. *Macromolecules* 38:1091. (c) Goto, H., N. Nomura, and K. Akagi. 2005. *J Polym Sci A* 43:4298.

4

Synthesis and Properties of Poly(arylene vinylene)s

Andrew C. Grimsdale and
Andrew B. Holmes

4.1 Introduction

This chapter discusses the synthesis and properties of poly(arylene vinylene)s (PAVs). These polymers are well known for their luminescent properties. They were the first class of conjugated polymers in which electroluminescence (EL) was demonstrated [1], the first to be used in a commercial polymer light-emitting diode (LED) [2], and they still form the most widely studied class of electroluminescent polymers [3]. They also offer much scope for synthetic effort to optimize their properties, so that their ongoing appeal to materials scientists is self-evident. Early history of the development of PAVs was thoroughly reviewed in the previous edition [4]; by providing a general overview of this field we aim to concentrate here on the more significant recent advances.

4.2 Routes to PAVs

Synthetic routes to PAVs fall into two classes: direct and precursor routes. The former are used only for the preparation of soluble materials, whereas the latter can be used for making both soluble and insoluble polymers. While soluble materials are more easily processed, there are advantages to using films of insoluble materials, as they are more robust in the presence of solvents. This is especially true if one is making multilayer devices, where one does not want the deposition of a later layer to damage or even remove an earlier one. As a result, and for the historical reason that most initial work on PAVs was based on the insoluble parent compound poly(p-phenylene vinylene) (PPV), the dominant routes for the synthesis of PAVs involve polymerizations of quinodimethanes to form soluble precursor polymers that are then converted, usually thermally, to the final polymers.

4.2.1 Precursor Routes to PAVs

As shown in Scheme 4.1 for the preparation of PPV (**1**), these quinodimethane routes involve treatment of a suitably functionalized *para*-xylene derivative (**2**) that is treated with just under one equivalent of base to form a quinodimethane, which then polymerizes to form the precursor polymer (**3**). This is then transformed into PPV by heating under reduced pressure. Three main quinodimethane precursor routes have been developed. The Wessling–Zimmerman method [5,6] uses sulfonium salts as leaving groups. The main disadvantage of this route is that the sulfides used to make the monomers, and which are produced during the reactions, are often toxic and have offensive odors. A variation developed by the group of Vanderzande [7,8] uses a sulfinyl leaving group. This method reportedly produces better quality PPV than the Wessling route [9], but the monomer synthesis is longer and less efficient, though it has been improved considerably since the initial work [10]. Finally the Gilch route [11] uses halide leaving groups. This method is now the most widely used method for making soluble substituted PAVs. For those polymers one can either isolate the halide precursor and thermally convert it to the final conjugated material or, by using an excess of base, prepare the PAV in one step in solution. The latter is only recommended if the product is highly soluble, as even partial conjugation tends to dramatically lower the solubility (and so the potential molar mass) of the polymer. Variations on these routes using other leaving groups, e.g., xanthates [12], have also been reported, but are not in general use.

The mechanism of these reactions has been the subject of much controversy [13]. The intermediacy of the quinodimethane **2** was readily established by spectroscopic means, but the question whether the polymerizations proceeded through a radical or an anionic mechanism was more difficult to answer. The radical nature of the Wessling process was finally established by Lahti et al. [14], who showed that radical trapping reagents suppressed the reaction. Similar methods confirmed the radical nature of the sulfinyl precursor route [15], but a nonradical mechanism for the Gilch reaction was still being proposed as late as 2002 [16]. This was supported by the ability to control the molecular weight of soluble PAVs by adding reagents such as poly(ethylene glycol) [17] that enhances anionic polymerization, of 4-methoxyphenol [18], a known anionic initiator, and of benzyl bromide chain stoppers [19]. However, recent studies by the groups of Vanderzande [20] and Rehahn [21] have established that the Gilch reaction also proceeds primarily through a radical pathway, with anionic polymerization being responsible at most only for the formation of low-molar mass material. The initiation step is proposed to be the coupling of two quinodimethanes to form a diradical (Figure 4.1) [21].

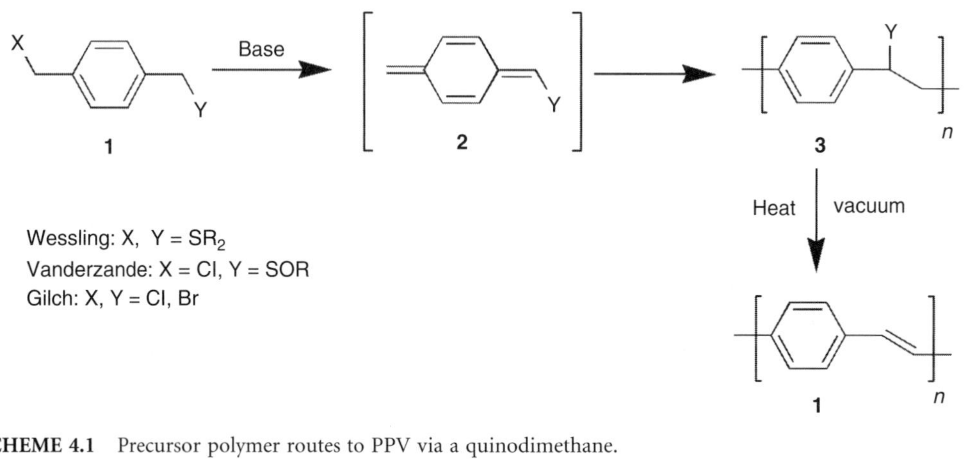

Wessling: X, Y = SR$_2$
Vanderzande: X = Cl, Y = SOR
Gilch: X, Y = Cl, Br

SCHEME 4.1   Precursor polymer routes to PPV via a quinodimethane.

FIGURE 4.1   Initiation step for quinodimethane polymerizations.

As discussed in the previous edition, not all aryl bis-sulfonium salts will undergo efficient polymerization by the Wessling method, due to either the loss of aromaticity in forming the quinodimethane (e.g., 2,6-naphthalene) being too great, or the intermediate being too stable (e.g., 9,10-anthracene). It has been established that the polymerizability of a bis-sulfonium salt (and presumably of the other monomers) is determined by the enthalpy of formation of quinodimethane and so the accessibility of a given PAV by these methods can now be predicted [22].

A key issue in the synthesis of any conjugated polymer is the minimization of defects. It is now well established that defects have a large, usually negative, impact upon the optical and electronic properties of conjugated polymers, as they can act as charge traps that reduce charge mobility and conductivity, sites for nonradiative decay of excited states that reduce luminescence efficiencies, or low band-gap emissive sites that affect emission color—this is a particular problem for blue-emitting polymers. One of the most important pieces of work on PAVs, in recent years, has been the investigation by a group at COVION into the formation of defects in the Gilch synthesis of soluble PAVs. They showed that precursor polymer **4** is formed by the coupling of quinodimethane **5** in the head-to-tail fashion, whereas coupling in head-to-head and tail-to-tail fashions produces diarylethyne (tolane, **6**) and bisbenzyl defects (**7**), respectively (Scheme 4.2) [23–25]. As discussed below they were then able to use this mechanism of defect formation to develop new materials with much lower defect levels and hence much better performance in LEDs [26].

Other important defects associated with PAVs are carbonyl groups formed by oxidation of the benzylic positions during synthesis of the precursor, or more especially during its conversion [27]. Here the use of an inert atmosphere is found to suppress the formation of the defects. Finally there is the question of *cis* vs. *trans* double bonds. The precursor methods described here produce largely *trans*-vinylenes, with the amount of *cis*-vinylenes being dependent upon the leaving group. For example, it is reported that using a xanthate leaving group in a modified Gilch synthesis of PPV results in a much higher level of *cis*-bonds in the product, which is accordingly amorphous rather than semicrystalline as

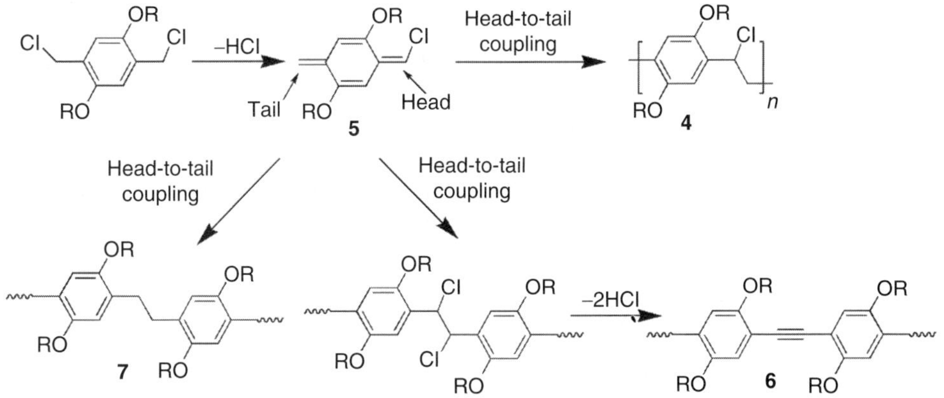

SCHEME 4.2   Formation of defects in PAVs during Gilch synthesis.

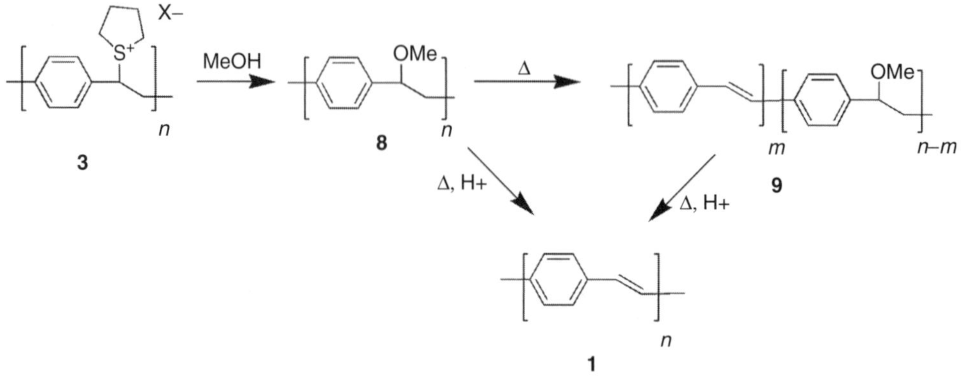

SCHEME 4.3   Methoxy-precursors to PPV.

seen for PPV made by the Wessling route [12]. While this is undesirable for applications in which high charge carrier mobility and conductivity are needed, e.g., transistors, it can be advantageous in other applications such as LEDs as it reduces exciton migration to defect sites that might quench emission, and so enhances the luminescence. Burn has shown that a soluble PPV derivative had higher EL efficiency when made from a xanthate precursor than by a halo-precursor, thus demonstrating the importance of the choice of leaving groups in optimizing materials properties [28]. The presence of *cis*-vinylene groups has been reported in only the low molar mass fractions of MEH-PPV, the best known soluble PAV, made by the Gilch route [29]. It is suggested that they hinder chain growth so that the higher mass fractions are all-*trans*.

A few other precursor methods have been used to make PAVs of which the best are chemical vapor deposition [30] or metathesis of cyclophanes [31], but these are of limited utility due to the difficulties in making such monomers.

A final feature of these precursor methods is that it is possible to chemically modify the precursor before the conversion step. This may be done to stabilize the precursor, e.g., replacement of sulfonium salts in **3** with methoxy groups produces a more stable precursor polymer **8** (Scheme 4.3), which can be stored for some time before final conversion without decomposition [32]. However, complete conversion of **8** requires the presence of acid that would dope the product, while conversion of this polymer under the normal conditions for preparing PPV produces copolymers **9**, which are not completely conjugated. As a result, by controlling the amount of displacement of the sulfides in **3** with methoxy or acetate groups, the degree of conjugation and thus the optoelectronic properties of the final product can be controlled [33,34]. This concept has been extended to other PAVs using methoxy [35], acetate [36,37], xanthate [38], or thiophenoxy [39] substituents to replace the sulfoniums, and has been used in a process for producing patterned emissive PAV films [40].

The conjugation length in PAVs made by the Gilch route can also be controlled by treating the chloro-precursor **10** with alcohols (Scheme 4.4). Thermal conversion of the resulting materials **11** eliminates the chlorines but not the alcohols. Methanol has been used as alcohol, but better efficiencies have been obtained using 2-dimethylaminoethanol [41]. Another route that has been examined to control the conjugation length in PAVs is the partial reduction of chloro-precursor polymers **10** with trialkyltinhydride to the copolymers **13** before thermal conversion to **14** under reduced pressure (Scheme 4.4) [42]. The degree of conjugation can be controlled very closely by varying the amount of reducing agent used.

4.2.2   Direct Routes to PPVs

Scheme 4.5 shows the most important direct routes for preparing soluble PAVs: Heck coupling, and Wittig and Horner polycondensations. The two biggest advantages of these routes are that they avoid the formation of tolane and other defects seen in the precursor routes, and also that they allow the synthesis

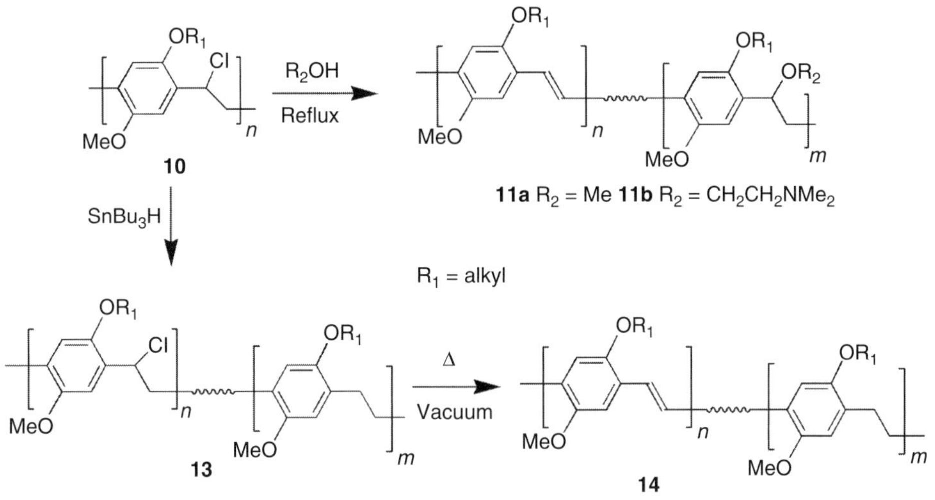

SCHEME 4.4 Reactions of chloro-precursors as a way to control conjugation lengths.

of alternating copolymers with two different arylene moieties, whereas the methods described above can be used only for making homopolymers or random copolymers. A disadvantage is that the molecular weights obtained are generally significantly lower than those obtained by quinodimethane routes. The first synthesis of PPV and its derivatives by Heck coupling involved reacting dihalobenzenes with gaseous ethylene [43], but it was soon found to be more synthetically convenient to use a divinylbenzene as the alkene component [44,45]. This latter method has since been used by many groups to make PAVs [46–51]. Homocoupling of a 2-alkoxy-4-bromostyrene has been used to make a regioregular PPV [52], as yet there is no evidence that such regioregularity produces any advantage in the resulting polymer. The Horner route has been used to make homopolymers such as MEH-PPV [53–55], but Horner and the Wittig methods have been used to prepare alternating copolymers [56–66]. The molecular masses from the Wittig methods tend to be slightly lower than the Horner and the latter produces much lower levels of *cis*-bonds [60,63,65,66]. The *cis*-bonds can be converted to *trans*-bonds by using iodine but this lowers the overall yield as the material has to be rigorously dedoped afterward, so that the Horner method is generally to be preferred.

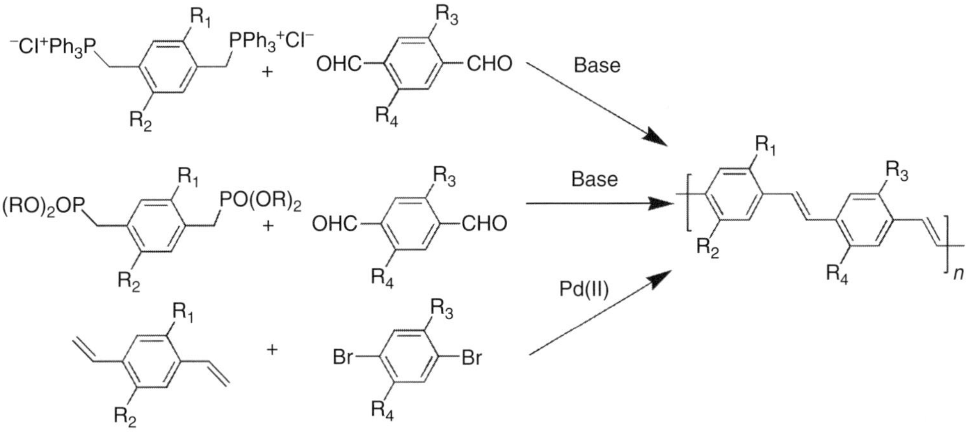

SCHEME 4.5 Wittig, Horner, and Heck routes to PAVs.

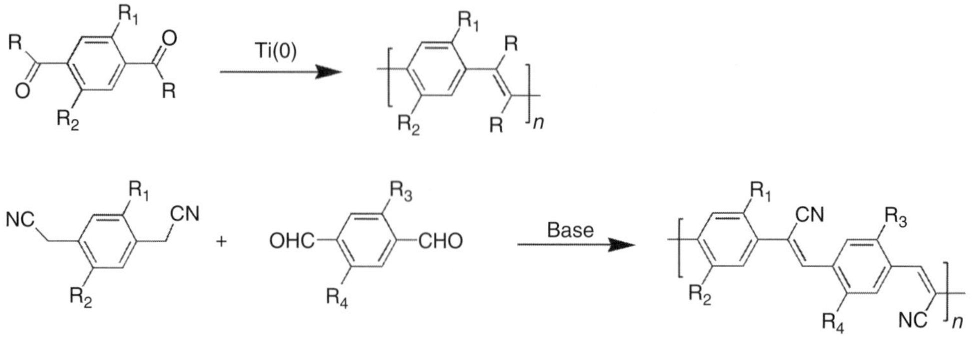

SCHEME 4.6   McMurray and Knoevenagel routes to PAVs.

Two other direct routes that are of some importance for making PAVs are McMurray coupling [67–69] and Knoevenagel polycondensation [70–72] (Scheme 4.6). The advantages of these are that one can easily make polymers with substituents on the vinylene, but the McMurray route produces a large amount of *cis*-bonds in some cases [73] and the Knoevenagel is only useful for making polymers with strongly electron-withdrawing groups such as nitriles on the vinylene. Other coupling reactions such as Stille coupling of dihaloarenes with vinylbisstannanes [74–76] or Suzuki coupling of arene bisboronates and vinyl dihalides [77] have also occasionally been used to make PAVs. A cascade Suzuki–Heck process in which dihaloarenes are coupled using a vinyltrifluoroborate has recently been reported [78]. Electropolymerization of bis(dihalomethyl)xylenes has been used to make PAVs directly, but the resulting materials are generally rather poorly defined, and contaminated with halide ions and electrolyte residues [79,80]. Metathesis of dialkyl- or dialkoxy-substituted divinylbenzenes has also been used to make soluble PAVs directly but the products are largely oligomers [81–83]. Even if the molar masses could be improved significantly, the difficulty of making the monomers would probably severely restrict the utility of this method.

4.2.3   Postsynthetic Modification of PPVs

The postsynthetic modification of PAVs is seldom of any preparative value as the reactions are hard to control. An exception is the oxidation of the PPV derivatives **15** and **16** to produce the phenathrene-containing polymers **17** and **18** (Scheme 4.7) [84]. It is interesting to note the difference in regiochemistry between the two reactions.

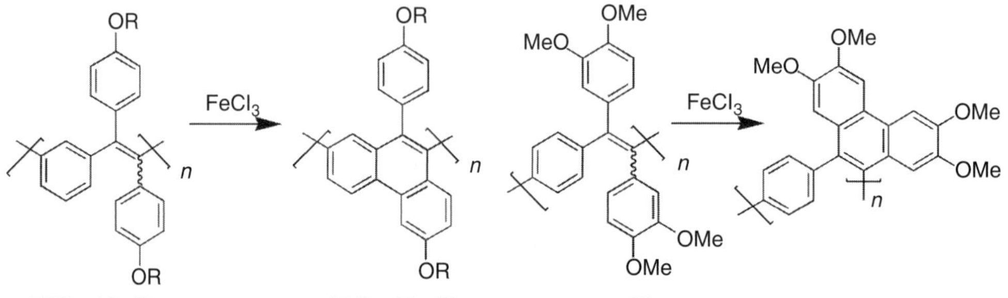

SCHEME 4.7   Oxidation of phenyl-substituted PPVs to form phenanthrene-based polymers.

4.3 Properties of PAVs

4.3.1 Electrical Properties of PAVs

As discussed in the previous edition [4], undoped PPV displays very low conductivity (10^{-13} S/cm) [85]. Even when doped it is only moderately conductive with values ranging from 10^{-3} S/cm (iodine doping) [86] to 100 S/cm (H_2SO_4 doping) [87] and the stability of the more conductive highly doped PPV is questionable. Combined with the low-charge carrier mobility values ($<10^{-5}$ cm^2/Vs) seen in transistors and diodes [88,89], it is clear that the electrical properties of PPV, which ironically were the main motivation for its investigation, are not good enough for it to be useful as a conductive material in any currently conceivable commercial device. Similar results have been reported for PPV derivatives. Attention has thus turned to the optical properties of PPV and its derivatives and in particular their luminescence.

4.3.2 Optical Properties of PAVs

PPV is a yellow–green emitter ($\lambda_{max} = 520$, 551 nm). Alkoxy groups redshift the emission so that monomethoxy derivative **19** (Figure 4.2) emits in the yellow ($\lambda_{max} = 550$ nm) [90], and the 2,5-dimethoxy polymer **20** shows orange–red emission ($\lambda_{max} = 603$, 650 nm) [91]. Longer side chains as in MEH-PPV (**21**) [92] are required to obtain solubility. They also lower the glass transition temperature and enhance the luminescence efficiency of PAVs. Alkyl or silyl groups by contrast do not redshift the emission so that green emission is obtained from polymers such as BuEH-PPV (**22**) [93] ($\lambda_{max} = 524$, 554 nm) or DMOS-PPV (**23**) [94] ($\lambda_{max} = 520$ nm). Both these last two polymers have notably high solid state ($>60\%$) PL efficiencies (cf. 22% for PPV) which makes them particularly attractive as emissive materials in LEDs.

The position of the substituents can be as important as their nature. For example, the 2,5-dialkoxy-PPVs **20** and **21** display orange–red emission, whereas the 2,3-dialkoxy-PPVs **24–25** are green emitters ($\lambda_{max} = 520$ and 505 nm, respectively) due to steric repulsions between the substituents that twist the backbone so as to reduce the degree of conjugation and thus blueshift the emission [95,96]. Conjugation along the backbone can also be interrupted by introducing *meta*- or *ortho*-phenylene units. For example, the emission maxima of the polymers **26–28** (Figure 4.3) occur at 550, 490, and 500 nm, respectively [59], whereas the blueshift emission from polymer **27** reflects a disruption in the through conjugation due to the *meta*-linkages that in the emission from polymer **28** arises from a twisting

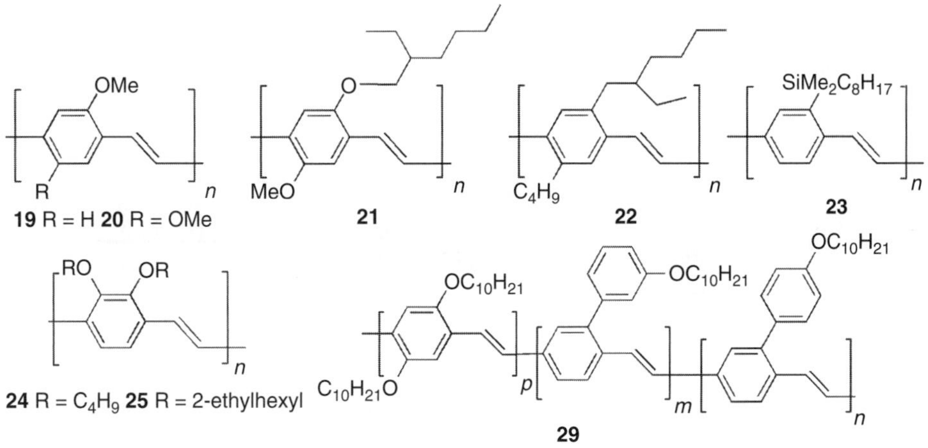

FIGURE 4.2 Some representative PPV derivatives.

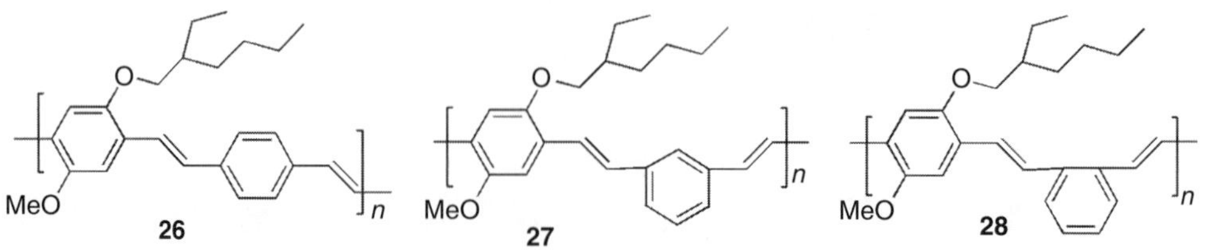

FIGURE 4.3   PAVs with *para-*, *meta-*, and *ortho-*linkages.

of the polymer backbone induced by the steric repulsion between the adjacent groups on the *ortho-*substituted phenylenes.

An obvious way to tune the emission color of PAVs is by copolymerization of different units, e.g., the emission of the copolymers **29** (Figure 4.2) can be tuned between green ($\lambda_{max} = 515$ nm) and orange ($\lambda_{max} = 567$ nm) by varying the amount of the dialkoxyphenylene units [26,97]. Incorporating units other than phenylenes into PAVs is another simple way to tune the emission color. PPV is an alternating copolymer of polyphenylene and polyacetylene. Since polyphenylenes are well known as blue-emitting polymers, increasing the proportion of phenylene units should blueshift the emission. Thus poly(bi-phenylene vinylene) (**30**) [98,99] and poly(fluorenylene vinylene) (**31**) [100–102] are blue–green emitters ($\lambda_{max} = 467, 497$ nm) (Figure 4.4). Further blueshifts have been seen from polymers with more phenylene units so that the poly(pentaphenylene vinylene) **32** is a blue emitter ($\lambda_{max} = 446$ nm) [103,104]. Polycyclic arylenes can have varying effects, e.g., poly(1,4-naphthalene vinylene) **33** is orange–red emitting ($\lambda_{max} = 605$ nm) [105,106], while poly(3,6-phenanthrene vinylene) **34** displays green fluorescence ($\lambda_{max} = 515$ nm) [31].

By contrast electron-rich heteroaromatic units such as 2,5-pyridines or 2,5-thiophenes provide a way to redshift the emission of PAVs (Figure 4.5). The lower symmetry of the pyridine than the phenylene ring means that poly(pyridine vinylene) can be produced as a random polymer or in two regioregular forms—head-to-tail (**35**) and head-to-head (**36**) [74]. The EL emission maxima of these appear at 575, 584, and 605 nm, respectively. The thiophene-containing copolymer **37** has even more redshifted emission ($\lambda_{max} = 620$ nm) [107,108]. The most redshifted emission yet to be reported from PAV is near-infrared emission ($\lambda_{max} = 800$ nm) from the polymers **38** ($\lambda_{max} = 740$ nm) [109] and **39** ($\lambda_{max} = 800$ nm) [110,111]. A wide range of other heteroaromatic units have been incorporated into PAVs with emission colors ranging from green to red.

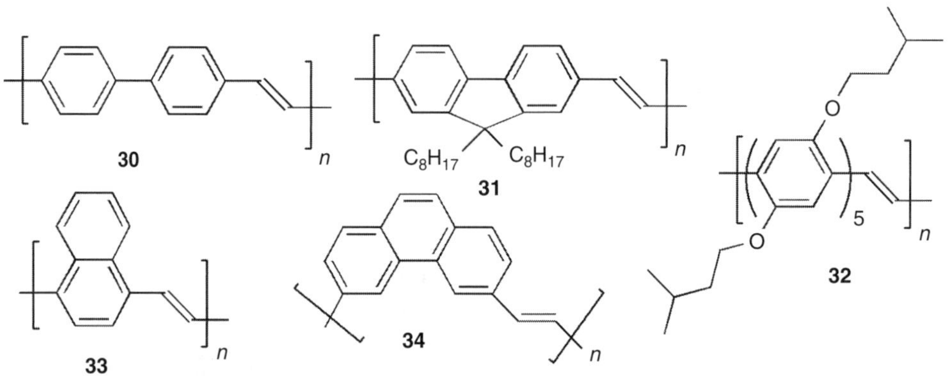

FIGURE 4.4   PAVs with arylene units other than phenylene.

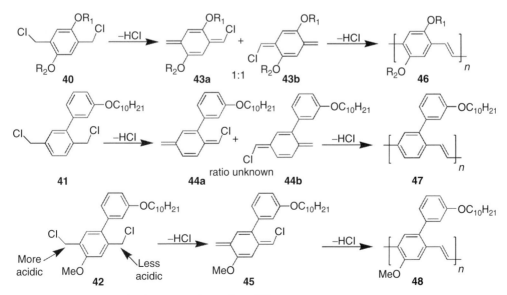

FIGURE 4.5 PAVs with heteroaromatic units.

4.3.3 Optimization of Properties of PAVs for LEDs

As previously mentioned the COVION group has utilized their understanding of the mechanism of defect formation in the Gilch reaction to make relatively defect-free polymers that show particularly good performance in LEDs (high efficiency and long lifetimes) [26].

Each of the three unsymmetrical monomers **40–42** can form two isomeric quinodimethanes on reaction with base (Scheme 4.8). There is no electronic reason why one of these forms should be favored for **40** and so the two intermediates **43a** and **43b** are probably formed in equal amount. However, the steric and electronic effects of the chlorine atoms would strongly favor head-to-tail coupling of either of these so the level of defects in the polymers **46** is typically only 3%. While there is probably some electronic effect affecting the ratio of the intermediates **44a–b** derived from **41**, it is not so strong that only one form is produced and steric repulsion between the phenyl groups means that the coupling of **44a** and **44b** can only proceed in a head-to-head fashion, so that the resulting polymer **47** has a high-defect level. The electronic effects of the methoxy group in **42** mean that the quinodimethane **45** is

SCHEME 4.8 Mechanism for formation of PAVs from different monomers.

strongly favored and steric repulsion hinders its head-to-head coupling, so that the defect levels are very low (<0.5%) in the final polymer **48**. Unfortunately this polymer tends to form gels, so as to obtain more processable materials they made copolymers of **47** and **48**. Such materials have now been optimized and comprise the active materials in the first commercial polymer LED devices.

A second approach toward optimizing the performance of PPVs in LEDs has been the improving of charge injection into the emissive polymers. Apart from device engineering approaches such as use of charge-transporting layers [3], this can be achieved by attaching charge-accepting groups onto the polymers. Since the limiting factor for efficiency of LEDs using PPV has been shown to be due to the injection of electron as opposed to holes [3], as shown by the much greater EL efficiency when using calcium than aluminum cathodes, the main effort has been into the attachment of electron-accepting groups to PAVs (Figure 4.6) either on the phenylene, e.g., in the fluoro-PPV (**49**) [112,113], or on the vinylene as in CN-PPV (**50**) [72]. In both cases the polymers display much higher electron affinities than PPV, and hence much greater efficiencies in LEDs using air-stable electrodes such as aluminum. In particular, double-layer devices utilizing **37** as electron-accepting emissive layer with PPV as a hole-transporting layer reached efficiencies of up to 4%, although their lifetime was limited [114,115]. The highest efficiency reported for an LED using a PPV-derivative is 21.1 cd/A using the polymer **51** bearing oxadiazole units [116]. The key to obtaining such high efficiencies seems to be achieving balanced injection of holes and electrons.

Polarized emission is desirable for certain applications, e.g., as backlights for liquid crystalline displays [117]. This is most easily obtained by alignment of the emissive layer. Here the ability of the polymer to form an ordered liquid crystalline phase is advantageous [118]. A nematic phase has been found for the dinonyloxy-PPV **52** (Figure 4.7) [119]. Cooling of the melted material to room temperature produced an oriented film with a lower energy gap (2.08 eV, 597 nm) than for an unoriented film (2.21 eV, 562 nm) formed by spin-casting of a chloroform solution of the polymer. Rubbing of a film of this material is reported to induce molecular orientation giving rise to polarized EL [120]. As discussed in the last edition [4], thermotropic phases have been seen for other PAV copolymers [121], including partially eliminated precursor polymers [122]. Some attempts have been made to induce liquid crystallinity and thus polarized emission by attaching mesogens as side chains onto PAVs but with mixed success [123,124]. Ordered films of PPV have been obtained by various methods, e.g., by conversion of Langmuir–Blodgett films of the sulfonium precursor **3** [125–127]. It has been reported that highly ordered PPV can be obtained by partial conversion of **3** with base followed by methanolysis to give **9**, and then conversion of **9** to **1** [128,129]. Chiral polymers can display circularly polarized emission. To date, the only PAVs for which this has been observed are the copolymers **53** bearing chiral substituents [130]. The best results were obtained for a copolymer with an *m:n* ratio of 8:1. A chiral PAV **54** has also been made containing chiral binaphthyl units in the main chain [131–133], yet the EL properties have not been reported.

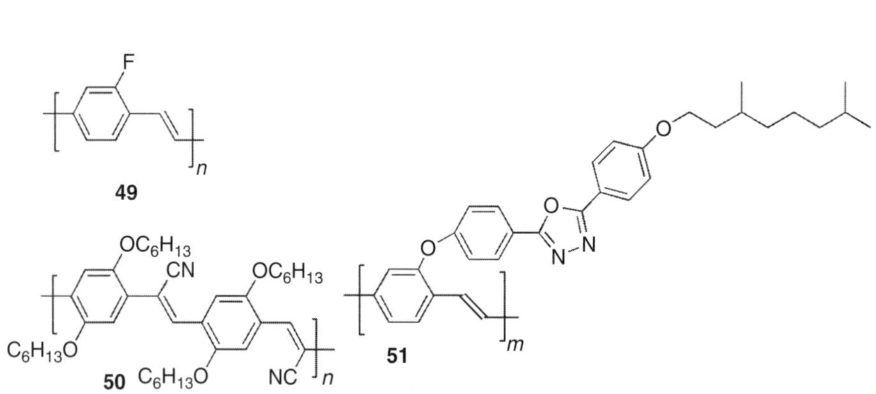

FIGURE 4.6   PPVs with improved electron affinities.

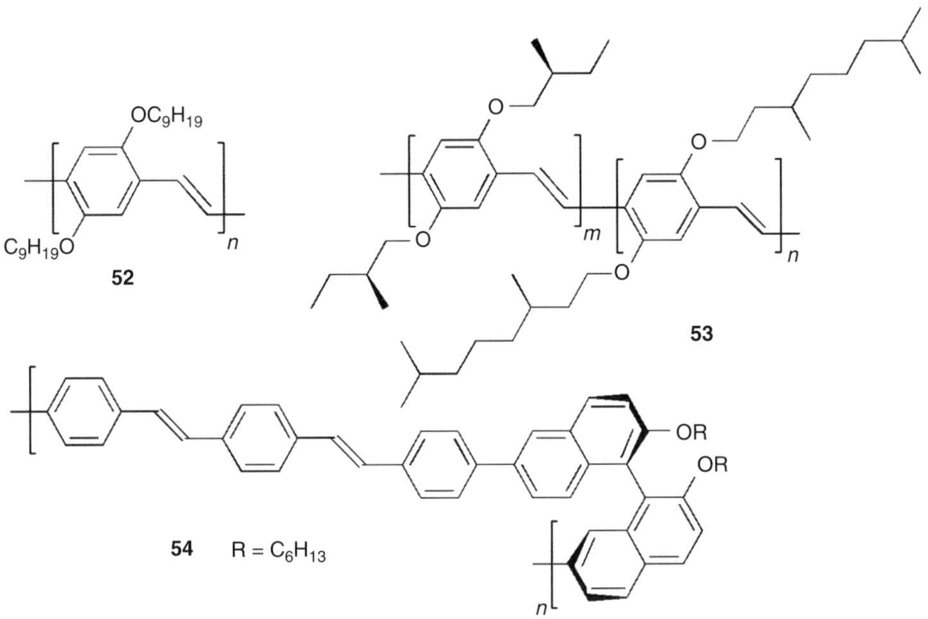

FIGURE 4.7 PAVs displaying polarized or circularly polarized EL.

Further enhancement in the device performance of PAVs is still required before more LEDs based on them can enter the commercial marketplace; it is clear that they should be able to establish a useful niche in the world of display technology. To what extent will depend not only upon the future development of OLED technology but also on that of competing technologies such as liquid-crystal displays. But the clearly established ability to make readily and inexpensively PAVs with emission colors ranging from blue to the infrared, plus the rapid advances in efficiency and lifetime for PAV-based LEDs suggest that they have a bright future.

4.4 Applications of PAVs Other Than OLEDs

While LEDs are currently the only application of PAVs to have reached the stage of commercialization, there remains considerable interest in developing other electronic devices utilizing them. As mentioned earlier the electrical properties of PAVs do not look sufficiently promising to permit commercially viable devices to be constructed, the use of PAVs in lasers and photovoltaic devices offers real promise for the future.

4.4.1 Lasing from PAVs

The attainment of electrically pumped solid state lasing from organic materials is one of the current "Holy Grails" of organic electronics research. While this goal remains elusive, optically pumped lasing of conjugated polymers including PAVs is now well known and has been widely reviewed [134–138]. The first report of lasing from a conjugated polymer was from a solution of MEH-PPV (**21**) in 1992 [139]. In 1996, solid state lasing was reported from a blend of MEH-PPV and titania nanoparticles in a polystyrene matrix [140], and shortly thereafter from a film of PPV (**1**) in a microcavity device [141]. Since then many other papers have appeared dealing with lasing from PPV [142–145] or from other PAVs [55,146–164]. One polymer that is seen as being particularly promising as a lasing material is the dialkyl-PPV **22** [148,153,155]. At present the future of lasing from conjugated polymers is unclear, but it remains one of the most exciting areas of organic materials research.

4.4.2   PAVs for Photovoltaic Applications

In view of the increasingly urgent need to develop new, renewable energy sources, organic-based solar cells have become an important area of research [165,166]. PAVs have become one of the most widely studied classes of organic polymers for such applications. PPV and its derivatives are photoconductive, e.g., irradiation of PPV films sandwiched between indium tin oxide (ITO) and aluminum electrodes produces a photocurrent with an efficiency of ~0.1% [167,168]. A PEDOT:PSS electrode can be used in place of ITO [169]. Creating an interpenetrating interface between the two polymer layers enhances the efficiency. Efficiency is enhanced by adding a second layer of an electron-accepting material. Thus, bilayer structures have been made using MEH-PPV and C_{60} [170,171], or the substituted fullerene PCBM (Figure 4.8, **56**) with efficiencies of up to 0.46% [172]. A bilayer device with an efficiency of 0.9% has recently been reported using aluminum tris(8-hydroxyquinolate) as the electron-accepting layer [173]. A bilayer device has even been made using layers of carbon nanotubes prepared by carbonization of PPV, and of PPV [174]. However, the blending of the PAV with an electron-accepting material has proved to be a better way to obtain high efficiency. It was discovered independently at Santa Barbara [175] and at Osaka [176,177] that the photoconductivity of soluble PPV derivatives was markedly enhanced by the addition of C_{60} (**55**), which acts as a highly efficient electron acceptor. For example, the photovoltaic efficiency of a 1:1 blend of MEH-PPV and C_{60} is two orders of magnitude higher than for neat MEH-PPV [170,178–181]. Solar cells have been made using other dialkoxy-PPV derivatives blended with C_{60} [182], but PCBM has now generally replaced the parent fullerene as the electron-accepting component in blends with PPV [183] or other PAVs [184–193]. Considerable progress has now been made toward making organic solar cells based on blends of PAVs with PCBM commercially viable, with efficiencies of up to 3% having been reported from blends of the polymer **57** with PCBM [192], and large-scale flexible devices having been fabricated by doctor-blading techniques [194], but reported lifetimes are still short [195]. A key parameter for obtaining high efficiency from such blends is the phase separation of the two components, which can be influenced by factors such as the choice of solvent, the solvent evaporation time, surface interactions, and annealing [166].

One potential way to avoid the problem of phase separation from polymer-fullerene blends is to prepare dyad materials with fullerenes attached to one end of a conjugated donor molecule, e.g., the phenylene vinylene oligomer **58** (Figure 4.9) [196] (the use of such dyad materials in solar cells has recently been reviewed [197]). Two advantages of such materials in solar cells are the minimizing of bad electrode contacts and the greater ability to determine structure–activity relationships. This approach has been extended to block copolymers **59** in which a fullerene-bearing polystyrene block is grafted onto a PAV [198]. One can also make polymers, e.g., polymer **60** bearing pendant fullerenes [199].

A number of materials other than fullerenes have been blended with PAVs to improve their photo-conductivity. These include viologens [200], fluorescent dyes [201–203], carbon nanotubes [204], and

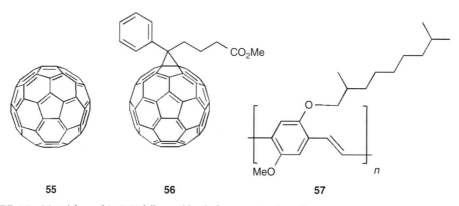

FIGURE 4.8   Materials used in PAV-fullerene blends for organic solar cells.

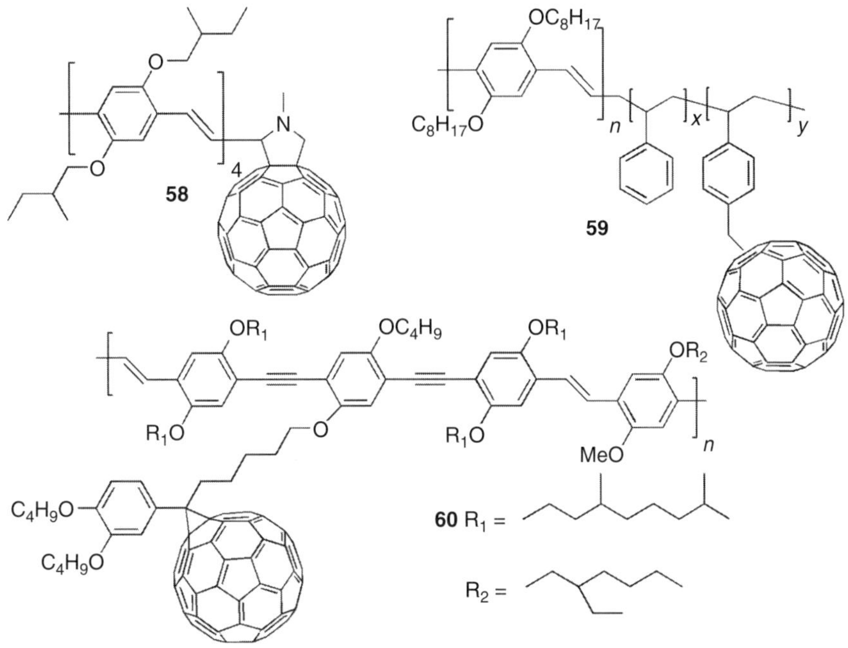

FIGURE 4.9 PAVs with fullerenes attached.

nanoparticles of cadmium selenide [205], or zinc oxide [206]. In an intermediate approach between blending and bilayer approaches, nanostructured titanium dioxide has been used as a charge-accepting layer in hybrid inorganic and organic solar cells using PAVs [207–209]. Such cells bear an obvious resemblance to those developed by Grätzel [210].

Finally an electron-accepting polymer can be blended with a PAV to enhance its photoresponse. Here cyano-substituted PPVs such as **50** have been used successfully. Thus, in 1995, it was reported by the groups of Friend and Heeger that blends of MEH-PPV with CN-PPV (**50**) [211], or with MEH-CN-PPV (Figure 4.10, **61**) [212] produce photocurrents with efficiencies up to 0.9%. However, despite further research efforts [213–217], the efficiency of devices using blends of PAVs with cyano-containing PPV

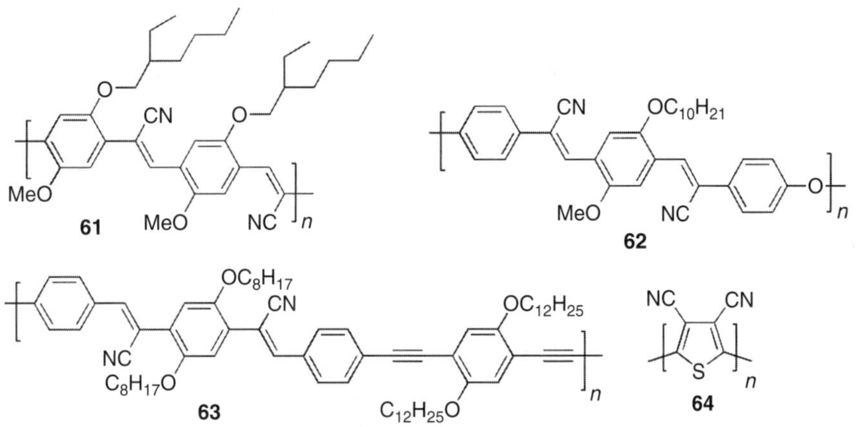

FIGURE 4.10 Electron-accepting polymers for use in PAV-based polymer–polymer bulk heterojunction solar cells.

derivatives such as **50** and **61–63** has not risen above 1%. CN-PPV has also been used as the electron-accepting component in a blend with other donors such as polythiophenes [218]. Another electron-accepting polymer that has been used with MEH-PPV is poly(3,4-dicyanothiophene) (**64**) but the efficiency was only 0.1% [219].

4.5   Conclusions

PAVs have emerged as one of the most important classes of conjugated polymers. Ironically, while their electrical properties, which were the original motivation for their study, have proved disappointing, their electro-optical properties are such that they comprise the active material in the first commercially available polymer-based LEDs, and they have been used in some of the best performing polymer-based solar cells yet tested. While considerable development remains before polymer-based electronic devices can represent more than a tiny share of the electronics market, there is every reason to believe that when they do, PAVs will form a significant fraction of the materials used.

References

1. Burroughes, J.H., et al. 1990. Light-emitting diodes based on conjugated polymers. *Nature* 347:539.
2. Available at http://www.theclockmag.com/oleds_2002/oleds_2002.htm.
3. Kraft, A., A.C. Grimsdale, and A.B. Holmes. 1998. Electroluminescent conjugated polymers—Seeing polymers in a new light. *Angew Chem Intl Ed* 37:403.
4. Moratti, S.C. 1997. The chemistry and uses of polyphenylenevinylenes. *The handbook of conducting polymers.* 2nd ed., eds. R.L. Elsenbaumer, and J.R. Reynolds. New York: Marcel Dekker, Chapter 13.
5. Wessling, R.A., and R.G. Zimmerman. 1968. US Patent 3,401,152.
6. Wessling, R.A. 1985. The polymerization of xylylidene bisalkylsulfonium salts. *J Polym Sci Polym Symp* 72:55.
7. Gelan, J., D. Vanderzande, and F. Louwet. 1995. European Patent Appl. EP 644,217, 1995.
8. Louwet, F., et al. 1995. A new synthetic route to a soluble high molecular weight precursor for poly(*p*-phenylene vinylene). *Macromolecules* 28:1330.
9. de Kok, M.M., et al. 1999. ESR spectroscopy of the elimination of sulfinyl precursor polymers towards PPV. *Synth Met* 102:949.
10. van Breemen, A.J.J.M., et al. 1999. Highly selective route for producing unsymmetrically substituted monomers towards synthesis of conjugated polymers derived from poly(*p*-phenylene vinylene). *J Org Chem* 64:3106.
11. Gilch, H.G., and W.L. Wheelwright. 1966. Polymerization of α-halogenated *p*-xylenes with base. *J Polym Sci A-1* 4:1337.
12. Son, S., et al. 1995. Luminescence enhancement by the introduction of disorder into poly(*p*-phenylene vinylene). *Science* 269:376.
13. Denton, F.R., and P.M. Lahti. 1998. Synthesis and properties of poly(phenylene vinylene)s and poly(arylene vinylene)s. *Plast Eng* 48:61.
14. Denton, F.R., P.M. Lahti, and F.E. Karasz. 1992. The effect of radical trapping upon formation of poly(α-tetrahydrothiophenio paraxylene) polyelectrolytes by the Wessling soluble precursor method. *J Polym Sci A* 30:2223.
15. Issaris, A., D. Vanderzande, and J. Gelan. 1997. Polymerization of a *p*-quinodimethane derivative to a precursor of poly(*p*-phenylene vinylene): Indications for a free radical mechanism. *Polymer* 38:2571.
16. Cho, B.R. 2002. Precursor polymers to PPV and its heteroaromatic derivatives: Polymerization mechanism. *Prog Polym Sci* 27:307.
17. Yin, C., and C.-Z. Yang. 2001. Mechanism analysis of the PEG-participated Gilch synthesis for soluble poly(*p*-phenylene vinylene) derivatives. *J Appl Polym Sci* 82:263.
18. Neef, C.J., and J.P. Ferraris. 2000. MEH-PPV: Improved synthetic procedure and molecular weight control. *Macromolecules* 33:2311.

19. Hsieh, B.R., et al. 1997. General methodology towards soluble poly(*p*-phenylene vinylene) derivatives. *Macromolecules* 30:8094.

20. Hontis, L., et al. 2001. The Gilch polymerization towards OC1C10-PPV; indications for a radical mechanism. *Polymer* 42:5793.

21. Wiesecke, J., and M. Rehahn. 2004. *Polym Prepr* 45(1):174.

22. Garay, R.O., et al. 2000. Synthesis of conjugated polymers. Polymerizability studies of bis-sulfonium salts. *Des Monomers Polym* 3:231.

23. Becker, H., et al. 1999. New insights into the microstructure of Gilch-polymerized PPVs. *Macromolecules* 32:4925.

24. Becker, H., et al. 2000. Development of high performance PPVs: Implications of the polymer-microstructure. *Synth Met* 111–112:145.

25. Becker, H., et al. 2001. Advances in polymers for PLEDs: From a polymerization mechanism to industrial manufacturing. *Synth Met* 122:105.

26. Becker, H., et al. 2000. Soluble PPVs with enhanced performance: A mechanistic approach. *Adv Mater* 12:42.

27. Papadimitrakopoulos, F., et al. 1994. The role of carbonyl groups in the photoluminescence of poly(*p*-phenylene vinylene). *Chem Mater* 6:1563.

28. Lo, S.C., et al. 2001. Control of polymer-electrode interactions: The effect of leaving group on the optical properties and device characteristics of EHPPV. *J Mater Chem* 11:2228.

29. Fan, Y.-L., and K.-F. Lin. 2005. Dependence of the luminescent properties and chain length of poly[2-methoxy-5-(2′-ethylhexyloxy)]-1,4-phenylene vinylene on the formation of *cis*-vinylene bonds during Gilch polymerization. *J Polym Sci A* 43:2520.

30. Vaeth, K.M., and K.F. Jensen. 1997. Chemical vapor deposition of thin polymer films used in polymer-based light emitting diodes. *Adv Mater* 9:490.

31. Miao, Y.-J., and G.C. Bazan. 1994. Paracyclophane route to poly(*p*-phenylenevinylene). *J Am Chem Soc* 116:9379.

32. Halliday, D.A., et al. 1993. Determination of the average molecular weight of poly(*p*-phenylene vinylene). *Synth Met* 55:902.

33. Burn, P.L., et al. 1992. Synthesis of a segmented conjugated polymer giving a blue-shifted electroluminescence and improved efficiency. *J Chem Soc Chem Commun* 32.

34. Carter, J.C., et al. 1997. Operating stability of light-emitting polymer diodes based on poly (*p*-phenylene vinylene). *Appl Phys Lett* 71:34.

35. Shim, H.-K., et al. 1997. Light-emitting properties of mono-substituted PPV derivatives. *Macromol Symp* 118:473.

36. Gowri, R., et al. 1998. Synthesis of novel poly[(2,5-dimethoxy-*p*-phenylene)vinylene] precursors having two eliminatable groups: An approach for the control of conjugation length. *Macromolecules* 31:1819.

37. Padmanaban, G., and S. Ramakrishnan. 2000. Conjugation length control in soluble MEH-PPV: Synthesis, optical properties and energy transfer. *J Am Chem Soc* 122:2244.

38. Padmanaban, G., and S. Ramakrishnan. 2001. An improved method for the control of conjugation length in MEH-PPV via a xanthate precursor route. *Synth Met* 119:533.

39. Hwang, D.-H., et al. 2001. Band gap tuning of PPV derivatives by thiophenoxy precursor polymer. *Synth Met* 119:393.

40. Burn, P.L., et al. 1992. Chemical tuning of electroluminescent copolymers to improve emission efficiencies and allow patterning. *Nature* 356:47.

41. Schoo, H.F.M., et al. 1997. Organic polymer LEDs with mobile and immobile ions. *Macromol Symp* 125:165.

42. Webster, G.R., and P.L. Burn. 2004. Controlling the conjugation length in poly[5-*n*-butyl-2-(2-ethylhexyl)-1,4-phenylenevinylene]: Exploring the scope of hydrogen radical substitution of leaving groups on precursor polymers. *Synth Met* 145:159.

43. Greiner, A., and W. Heitz. 1988. New synthetic approach to poly(1,4-phenylene vinylene) and its derivatives by palladium catalyzed arylation of ethylene. *Macromol Chem Rapid Commun* 9:581.

44. Heitz, W., et al. 1988. Synthesis of monomers and polymers by the Heck reaction. *Makromol Chem* 189:119.

45. Heitz, W. 1995. Metal-catalyzed polycondensation reactions. *Pure Appl Chem* 67:1951.

46. Bao, Z., et al. 1993. Conjugated liquid-crystalline polymers—Soluble and fusible poly(phenylene-vinylene) by the Heck coupling reaction. *Macromolecules* 26:5281.

47. Pan, M., Z. Bao, and L. Yu. 1995. Regiospecific, functionalized poly(phenylenevinylene) using the Heck coupling reaction. *Macromolecules* 28:5151.

48. Hilberer, A., et al. 1955. Synthesis and characterization of a new efficient blue-light-emitting copolymer. *Macromolecules* 28:4525.

49. Hilberer, A., et al. 1997. Poly(phenylenevinylene)-type conjugated alternating copolymers: Synthesis and optical properties in solution. *Macromol Chem Phys* 198:2211.

50. Cho, H.N., et al. 1999. Statistical copolymers for blue-light-emitting diodes. *Macromolecules* 32:1476.

51. Zhu, Z., and T.M. Swager. 2001. Conjugated polymers containing 2,3-dialkoxybenzene and iptycene building blocks. *Org Lett* 3:3471.

52. Liu, Y., P.M. Lahti, and F. La. 1998. Synthesis of a regiospecific, soluble poly(2-alkoxy-1,4-phenylenevinylene). *Polymer* 39:5241.

53. Pfeiffer, S., and H.-H. Hörhold. 1999. Synthesis of soluble MEH-PPV and MEH-PPB by Horner condensation polymerization. *Synth Met* 101:109.

54. Pfeiffer, S., and H.-H. Hörhold. 1999. Investigation of poly(arylenevinylene)s. Part 41. Synthesis of soluble dialkoxy-substituted poly(phenylene alkenylidenes) by applying the Horner-reaction for condensation polymerization. *Macromol Chem Phys* 200:1870.

55. Hörhold, H.-H., et al. 2001. MEH-PPV and dialkoxy phenylene vinylene copolymers. Synthesis and lasing characterization. *Synth Met* 119:199.

56. Doi, S., et al. 1997. Polymer light-emitting diodes utilizing poly(phenylene vinylene) derivatives. *Synth Met* 85:1281.

57. Ohnishi, T., et al. 1997. Polymer light emitting diodes using arylene vinylene copolymers. *ACS Symp Ser* 672:345.

58. Cho, H.N., et al. 1997. Blue and green emission from new soluble alternating copolymers. *Adv Mater* 9:326.

59. Ahn, T., et al. 1999. Blue electroluminescent polymers: Control of conjugation length by kink linkages abd substituents in the poly(*p*-phenylenevinylene)-related copolymers. *Macromolecules* 32:3279.

60. Davey, A., et al. 1999. Synthesis and optical properties of phenylene–vinylene copolymers. *Synth Met* 103:2478.

61. Peng, Z., J. Zhang, and B. Xu. 1999. New poly(*p*-phenylene vinylene) derivatives exhibiting high photoluminescence quantum efficiencies. *Macromolecules* 32:5162.

62. Peng, Z., et al. 2000. Towards highly photoluminescent and bipolar charge-transporting conjugated polymers. *Macromol Symp* 154:245.

63. Drury, A., et al. 2001. Systematic trends in the synthesis of (*meta*-phenylene vinylene) copolymers. *Synth Met* 119:151.

64. Menon, A., et al. 2002. Polydispersity effects on conjugated polymer light-emitting diodes. *Chem Mater* 14:3668.

65. Liao, L., et al. 2002. Effect of iodine-catalyzed isomerisation on the optical properties of poly[(1,3-phenylenevinylene)-*alt*-(2,5-hexyloxy-1,4-phenylenevinylene)]s. *Macromolecules* 35:6055.

66. Drury, A., et al. 2003. Investigation of different synthetic routes to and structure–property relationships of poly(*m*-phenylenevinylene-*co*-2,5-dioctyloxy-*p*-phenylenevinylene). *J Mater Chem* 13:485.

67. Feast, W.J., et al. 1985. Optical absorption and luminescence in poly(4,4′-diphenylenediphenyl-vinylene). *Synth Met* 10:181.

68. Feast, W.J., and I.S. Millichamp. 1983. The synthesis of poly(4,4′-diphenylene diphnylvinylene) via condensation polymerization of 4,4′-dibenzoylbiphenyl. *Polym Commun* 24:102.

69. Rehahn, M., and A.-D. Schlüter. 1990. Soluble poly(p-phenylenevinylene)s from 2,5-dihexylter-ephthalaldehyde using the improved McMurray reagent. *Makromol Chem Rapid Commun* 11:375.

70. Funke, V.W., and E.C. Schütze. 1963. Polykondensationreaktionen mit Xylylendicyaniden. *Makromol Chem* 74:71.

71. Hörhold, H.-H. 1972. *Z Chem* 12:41.

72. Greenham, N.C., et al. 1993. Efficient light-emitting diodes based on polymers with high electron affinity. *Nature* 365:628.

73. Cacialli, F., et al. 1999. Synthesis and properties of poly(arylene vinylene)s with controlled structures. *Opt Mater* 12:315.

74. Marsella, M.J., D.-K. Fu, and T.M. Swager. 1995. Synthesis of regioregular poly(methylpyridinium vinylene). An iso electronic analog to poly(phenylenevinylene). *Adv Mater* 7:145.

75. Bao, Z., W.K. Chan, and L. Yu. 1995. Exploration of the Stille coupling reaction for the synthesis of functional polymers. *J Am Chem Soc* 117:12426.

76. Babudri, F., et al. 1996. Synthesis, characterization, and properties of a soluble polymer with a polyphenylenevinylene structure. *Macromol Rapid Commun* 17:905.

77. Lopez, L.C., P. Strohriegl, and T. Stübinger. 2002. Synthesis of poly(fluorenevinylene-co-phenyle-nevinylene) by Suzuki coupling. *Macromol Chem Phys* 203:1926.

78. Grisorio, R., et al. 2005. A novel synthetic protocol for poly(fluorenylenevinylene)s: A cascade Suzuki–Heck reaction. *Tetr Lett* 46:2555.

79. Utley, J.H.P., and J. Gruber. 2002. Electrochemical synthesis of poly(p-xylylenes) (PPXs) and poly(p-phenylenevinylene s) (PPVs) and the study of xylylene intermediates: An underrated approach. *J Mater Chem* 12:1613.

80. Yoshikawa, E.K.C., et al. 2003. Synthesis and characterization of poly(decyloxyphenylene vinylene). *Synth Met* 135–136:3.

81. Thorn-Csányi, E., and P. Kraxner. 1997. All *trans*-oligomers of dialkyl-1,4-phenylenevinylenes—Metathesis preparation and characterization. *Macromol Chem Phys* 198:3827.

82. Reetz, R., et al. 2001. Substituted PPV oligomers by metathesis. *Synth Met* 119:539.

83. Oakley, G.W., and K.B. Wagener. 2005. Solid-state olefin metathesis: ADMET of rigid-rod poly-mers. *Macromol Chem Phys* 206:15.

84. Hörhold, H.-H., et al. 1995. A novel approach to light-emitting polyarylenes: Cyclisation of poly(arylene vinylene)s. *Synth Met* 69:525.

85. Gmeiner, J., et al. 1993. Synthesis, electrical conductivity and electroluminescence of poly (p-phenylene vinylene) prepared by the precursor route. *Acta Polymer* 44:201.

86. Murase, I., et al. 1987. Highly conducting poly(phenylene vinylene) derivatives via soluble precur-sor process. *Synth Met* 17:639.

87. Gagnon, D.R., et al. 1987. Synthesis, doping, and electrical conductivity of high molecular weight poly(p-phenylene vinylene). *Polymer* 28:567.

88. Parker, I.D., et al. 1993. Fabrication of a novel electro-optical intensity modulator from the conjugated polymer, poly(2,5-dimethoxy-p-phenylene vinylene). *Appl Phys Lett* 62:1519.

89. Ohmori, Y., et al. 1992. Fabrication and characterization of Schottky gated field-effect transistors utilizing poly(1,4-naphthalene vinylene) and poly(p-phenylene vinylene). *Jpn J Appl Phys* 31:L646.

90. Zyung, T., et al. 1995. Electroluminescence from poly(p-phenylenevinylene) with monoalkoxy substituent on the aromatic ring. *Synth Met* 71:2167.

91. Woo, H.S., et al. 1992. Photoinduced absorption and photoluminescence in poly(2,5-dimethoxy-p-phenylene vinylene). *Phys Rev B* 46:7379.

92. Braun, D., and A.J. Heeger. 1991. Visible light emission from semiconducting polymer diodes. *Appl Phys Lett* 58:1982.

93. Andersson, M.R., G. Yu, and A.J. Heeger. 1997. PL and EL of films from soluble PPV polymers. *Synth Met* 85:1275.

94. Hwang, D.-H., et al. 1996. Green light-emitting diodes from poly(2-dimethyloctylsilyl-1,4-phenylenevinylene). *J Chem Soc Chem Commun* 2241.

95. Martin, R.E., et al. 2000. Efficient blue–green emitting poly(1,4-phenylene vinylene) copolymers. *Chem Commun* 291.

96. Martin, R.E., et al. 2001. Versatile synthesis of conjugated aromatic homo- and copolymers. *Synth Met* 122:1.

97. Spreitzer, H., et al. 1998. Soluble phenyl-substituted PPVs. New materials for highly efficient polymer LEDs. *Adv Mater* 10:1340.

98. Van Der Borght, M., D. Vanderzande, and J. Gelan. 1998. Synthesis of high molecular weight poly(4,4'-biphenylene vinylene) and poly(2,6-naphthalene vinylene) via a non-ionic precursor route. *Polymer* 39:4171.

99. Van Der Borght, M., P. Adriaensens, and D. Vanderzande. 2000. The synthesis of poly(4,4'-biphenylene vinylene) and poly(2,6-naphthalene vinylene) via a radical chain polymerisation. *Polymer* 41:2743.

100. Cho, H.N., et al. 1997. Control of band gaps of conjugated polymers by copolymerization. *Synth Met* 91:293.

101. Nomura, K., et al. 2001. Synthesis of poly(9,9-dioctylfluorenyl vinylene) by metathesis. *J Polym Sci A* 39:2463.

102. Jin, S.-H., et al. 2002. Poly(fluorene vinylene) by Gilch for light emitting diode applications. *Macromolecules* 35:7532.

103. Wegner, G., et al. 1996. Materials engineered for polarized light emitting diodes. *Mat Res Soc Symp Proc* 413:23.

104. Remmers, M., et al. 1996. The optical, electronic, and electroluminescent properties of novel poly(*p*-phenylene) related polymers. *Macromolecules* 29:7432.

105. Antoun, S., et al. 1986. Preparation and electrical conductivity of poly(1,4-naphthalene vinylene). *J Polym Sci C* 24:503.

106. Stenger-Smith, J.D., et al. 1990. Preparation, spectroscopic and cyclic voltammetric studies of poly(1,4-naphthalene vinylene) prepared from a cycloalkylene sulfonium precursor polymer. *Polymer* 31:1632.

107. Hwang, D.-H., et al. 1998. Synthesis of PPV-PTV alternating copolymer and EL devices using the polymer. *Bull Kor Chem Soc* 19:332.

108. Hwang, D.-H., et al. 1999. Two-color emission from PPV-PTV copolymer and Alq$_3$ heterostructure EL device. *Synth Met* 102:1218.

109. Baigent, D.R., et al. 1995. Surface-emitting polymer light emitting diodes. *Synth Met* 71:2177.

110. Moratti, S.C., et al. 1994. Light-emitting polymer LEDs. *Proc SPIE* 2144:108.

111. Moratti, S.C., et al. 1995. High-electron affinity polymers for LEDs. *Synth Met* 71:2117.

112. Kang, I.-N., H.-K. Shim, and T. Zyung. 1997. Yellow-light emitting fluorine-substituted PPV. *Chem Mater* 9:746.

113. Gurge, R.M., et al. 1997. Light-emitting properties of fluorine-substituted poly(1,4-phenylene vinylenes). *Macromolecules* 30:8286.

114. Greenham, N.C., et al. 1994. Cyano-derivatives of poly(*p*-phenylene vinylene) for use in thin-film light-emitting iodes. *Mat Res Soc Symp Proc* 328:351.

115. Moratti, S.C., et al. 1994. Molecularly engineered polymer LEDs. *Mat Res Soc Symp Proc* 326:371.

116. Jin, S.-H., et al. 2004. High efficiency poly(*p*-phenylenevinylene)-based copolymers containing an oxadiazole pendant group for light-emitting diodes. *J Am Chem Soc* 126:2474

117. Grell, M., and D.D.C. Bradley. 1999. Polarized luminescence from oriented molecular materials. *Adv Mater* 11:895.

118. O'Neill, M., and S.M. Kelly. 2003. Liquid crystals for charge transport, luminescence and photonics. *Adv Mater* 15:1135.

119. Hamaguchi, M., and K. Yoshino. 1994. Lyotropic behavior of poly(2,5-dinonyloxy-*p*-phenylenevinylene). *Jpn J Appl Phys* 33:L1478.

120. Hamaguchi, M., and K. Yoshino. 1995. Rubbing-induced molecular orientation and polarized electroluminescence in conjugated polymer. *Jpn J Appl Phys* 34:L712.

121. Yu, L., and Z. Bao. 1994. Conjugated polymers exhibiting liquid crystallinity. *Adv Mater* 6:156.

122. Han, C.-C., and R.L. Elsenbaumer. 1990. Conveniently processable form of electrically conductive poly(dibutoxyphenylene vinylene). *Mol Cryst Liq Cryst* 189:183.

123. Onoda, M., et al. 2000. An electroluminescent diode using liquid-crystalline conducting polymer. *Thin Solid Films* 363:9.

124. Park, J.H., et al. 2001. PL and EL of LC aromatic conjugated polymers. *Synth Met* 119:633.

125. Nishikata, Y., M. Kakimoto, and Y. Imai. 1989. Preparation and properties of poly(*p*-phenylene vinylene) Langmuir–Blodgett film. *Thin Solid Films* 179:191.

126. Wu, A., et al. 1994. Fabrication of polymeric light emitting diodes based on poly(*p*-phenylene vinylene) LB films. *Chem Lett* 23:2319.

127. Kim, J.H., et al. 1995. Preparation and characterization of poly(*p*-phenylene vinylene) Langmuir–Blodgett film formed via precursor method. *Synth Met* 71:2023.

128. Halliday, D.A., et al. 1993. Large changes in optical response through chemical preordering of poly(*p*-phenylenevinylene). *Adv Mater* 5:40.

129. Halliday, D.A., et al. 1993. Extended π-conjugation in poly(*p*-phenylenevinylene) from a chemically modified precursor polymer. *Synth Met* 55:954.

130. Peeters, E., et al. 1997. Circularly polarized electroluminescence from a polymer light-emitting diode. *J Am Chem Soc* 119:9909.

131. Hu, Q.-S., et al. 1996. Conjugated polymers with main chain chirality. 1. Synthesis of an optically active poly(arylenevinylene). *Macromolecules* 29:1082.

132. Hu, Q.-S., D. Vitharana, and L. Pu. 1996. Synthesis of a 1,1′-binaphthyl based main chain chiral conjugated polymer. *Mat Res Soc Symp Proc* 413:621.

133. Pu, L. 1998. Binaphthyl dimers, oligomers and polymers. *Chem Rev* 98:2405.

134. Hide, F., et al. 1997. New developments in photonic applications of conjugated polymers. *Acc Chem Res* 30:430.

135. Lemmer, U. 1998. Stimulated emission and lasing in conjugated polymers. *Polym Adv Technol* 9:476.

136. Tessler, N. 1999. Lasers based on semiconducting organic materials. *Adv Mater* 11:363.

137. Tessler, N., et al. 2000. Properties of light-emitting organic materials in context of lasers. *Synth Met* 115:57.

138. McGehee, M.D., and A.J. Heeger. 2000. Semiconducting conjugated polymers as materials for solid-state lasers. *Adv Mater* 12:1655.

139. Moses, D. 1992. Lasing from MEH-PPV solution. *Appl Phys Lett* 60:3215.

140. Hide, F., et al. 1996. Semiconducting polymers: A new class of solid-state laser materials. *Science* 273:1833.

141. Tessler, N., G.J. Denton, and R.H. Friend. 1996. Lasing from conjugated-polymer microcavities. *Nature* 382:695.

142. Friend, R.H., et al. 1997. Electronic excitations in luminescent conjugated polymers. *Synth Met* 84:463.

143. Tessler, N., G.J. Denton, and R.H. Friend. 1997. Lasing characteristics of PPV microcavitites. *Synth Met* 84:475.

144. Friend, R.H., et al. 1997. Electronic processes of conjugated polymers in semiconductor devices. *Solid State Commun* 102:249.

145. Denton, G.J., et al. 1997. Stimulated emission, lasing, line-narrowing in conjugated polymers. *Proc SPIE* 3145:24.

146. Holzer, W., et al. 1996. Laser action in poly(*m*-phenylenevinylene-*co*-2,5-dioctyloxy-*p*-phenylenevinylene). *Adv Mater* 8:975.

147. Hide, F., et al. 1997. Highly photoluminescent conjugated polymers: Stimulated emission and device applications. *Proc SPIE* 3145:36.

148. Díaz-García, M.A., et al. 1997. Plastic lasers. *Appl Phys Lett* 70:3191.

149. Díaz-García, M.A., et al. 1997. Plastic lasers. *Synth Met* 84:455.

150. Berggren, M., et al. 1997. Solid-state droplet laser made from an organic blend with a conjugated polymer emitter. *Adv Mater* 9:968.

151. Hide, F. 1997. Light emission from semiconducting polymers: LEDs, lasers and white light. *Proc SPIE* 3148:22.

152. Eradat-Oskouvei, N., et al. 1997. Laser properties of luminescent conducting polymers in open resonators. *Proc SPIE* 3148:352.

153. Schwartz, B.J., et al. 1997. Stimulated emission and lasing in solid films of conjugated polymers. *Phil Trans R Soc London A* 355:775.

154. Kawabe, Y., et al. 1998. Whispering gallery mode micro-ring laser using a conjugated polymer. *Appl Phys Lett* 72:141.

155. McGehee, M.D., et al. 1998. Semiconducting polymer distributed feedback lasers. *Appl Phys Lett* 72:1536.

156. Frolov, S.V., et al. 1998. Cylindrical microlasers and light-emitting devices from conducting polymers. *Appl Phys Lett* 72:2811.

157. Chinn, D., et al. 1999. Synthesis and processing of conducting polymer microlasers and microsize light emitting diodes. *Synth Met* 102:930.

158. Fujii, A., et al. 1999. Polymer electroluminescent diodes with microcylinrical geometry. *Synth Met* 102:1010.

159. Park, J.Y., et al. 1999. ASE from MEH-PPV film in cylindrical capillary structure. *Synth Met* 106:35.

160. Turnbull, G.A., et al. 2001. Tuneable distributed feedback lasing in MEH-PPV films. *Synth Met* 121:1757.

161. Sheridan, A.K., et al. 2001. Tunability of ASE in thin organic films. *Synth Met* 121:1759.

162. Kranzelbinder, C., et al. 2002. Optically written solid-state lasers with broadly tunable mode emission based on improved poly(2,5-dialkoxy-phenylene-vinylene). *Appl Phys Lett* 80:716.

163. Manaa, H., et al. 2003. Light amplification and lasing in a (*meta*-phenylene vinylene) copolymer. *J Appl Phys* 93:1871.

164. Karastatiris, P., et al. 2004. Luminescent poly(phenylene vinylene) derivatives with *m*-terphenyl or 2,6-diphenylpyridine kinked segments along the main chain: Synthesis, characterization, and stimulated emission. *J Polym Sci A* 42:2214.

165. Brabec, C.J., N.S. Sariciftci, and J.C. Hummelen. 2001. Plastic solar cells. *Adv Funct Mater* 11:15.

166. Hoppe, H., and N.S. Sariciftci. 2004. Organic solar cells: An overview. *J Mater Res* 19:1924.

167. Karg, S., et al. 1993. Characterization of light emitting diodes and solar cells based on pol-phenylene-vinylene. *Synth Met* 57:4186.

168. Reiss, W., et al. 1994. Electroluminescence and photovoltaic effect in PPV Schottky diodes. *J Lumin* 60–61:906.

169. Arias, A.C., et al. 1999. Doped conducting-polymer–semiconducting-polymer interfaces: Their use in organic photovoltaic devices. *Phys Rev B* 60:1854.

170. Hayashi, Y., et al. 2005. Influence of structure and C_{60} composition on blends and bilayers of organic donor–acceptor polymer/C_{60} photovoltaic devices. *Jpn J Appl Phys* 44:1296.

171. Umeda, T., et al. 2005. Fabrication of interpenetrating semilayered structure of conducting polymer and fullerene by solvent corrosion method and its photovoltaic properties. *Jpn J Appl Phys* 44:4155.

172. Zhang, F., et al. 2002. Polymer photovoltaic cells with conducting polymer anodes. *Adv Mater* 14:662.

173. Lin, P., et al. 2005. Photovoltaic character of organic EL devices MEH-PPV/Alq$_3$. *Guangpuxue Yu Guangpu Fenxi* 25:23; *Chem Abstr* 143:88243.

174. Kim, K., et al. 2005. Photoconductivity of single-bilayer nanotubes consisting of PPV and carbon-ized-PPV layers. *Adv Mater* 17:464.

175. Lee, C.H., et al. 1993. Sensitization of the photoconductivity of conducting polymers by C_{60}: Photoinduced electron transfer. *Phys Rev B* 48:15425.

176. Yoshino, K., et al. 1993. Marked enhancement of photoconductivity and quenching of luminescence in poly(2,5-dialkoxy-*p*-phenylene vinylene) upon C_{60} doping. *Jpn J Appl Phys* 32:L357.

177. Morita, S., et al. 1993. Doping effect of buckminsterfullerene in poly(2,5-dialkoxy-*p*-phenylene vinylene). *J Appl Phys* 74:2860.

178. Yang, C.Y., et al. 1997. Nanostructured polymer blends: Novel materials with enhanced optical and electronic properties. *Synth Met* 84:895.

179. Liu, J., Y. Shi, and Y. Yang. 2001. Solvation-induced morphology effects on the performance of polymer-based photovoltaic devices. *Adv Funct Mater* 11:420.

180. Damodare, L., T. Soga, and T. Mieno. 2003. Studies on dark and photoconductivities of poly [2-methoxy,5-(2′ethylhexyloxy)-*p*-phenylene vinylene]:C_{60} thin films. *Jpn J Appl Phys* 42:2498.

181. Tada, K., and M. Onoda. 2005. Nanostructured conjugated polymer films for electroluminescent and photovoltaic applications. *Thin Solid Films* 477:187.

182. Padinger, F., et al. 1999. CW-photocurrent measurements of conjugated polymers and fullerenes blended into a conventional polymer matrix. *Synth Met* 101:1285.

183. Brabec, C.J., et al. 2001. Photoactive blends of PPV with methanofullerenes from a novel precursor. *J Phys Chem B* 105:1528.

184. Brabec, C.J., et al. 1999. Realization of large area flexible fullerene-conjugated polymer photocells: A route to plastic solar cells. *Synth Met* 102:861.

185. Gebeyehu, D. 2001. The interplay of efficiency and morphology in photovoltaic devices based on interpenetrating networks of conjugated polymers with fullerenes. *Synth Met* 118:1.

186. Brabec, C.J., et al. 2001. Organic photovoltaic devices produced from conjugated polymer/methanofullerene bulk heterojunctions. *Synth Met* 121:1517.

187. Manca, J.V., et al. 2003. State-of-the-art MDMO-PPV:PCBM bulk heterojunction organic solar cell: Materials, nanomorphology, and electro-optical properties. *Proc SPIE* 4801:15.

188. Zhou, Q., et al. 2003. Efficient polymer photovoltaic devices based on blend of MEH-PPV and C_{60} derivatives. *Synth Met* 135–136:825.

189. Mozer, A.J., et al. 2004. Novel regiospecific MDMO-PPV copolymer with improved charge transport for bulk heterojunction solar cells. *J Phys Chem B* 108:5235.

190. Hoppe, H., et al. 2004. Photovoltaic action of conjugated polymer/fullerene bulk heterojunction solar cells using novel PPE-PPV copolymers. *J Mater Chem* 14:3462.

191. Lu, S., et al. 2005. Novel alternating dioctyloxyphenylene vinylene-benzothiadiazole copolymer: Synthesis and photovoltaic performance. *Macromol Chem Phys* 206:664.

192. Al-Ibrahim, M., et al. 2005. Phenylene-ethynylene/phenylene vinylene hybrid polymers: Optical and electrochemical characterization, comparison with poly[2-methoxy-5-(3′,7′-dimethyloctyloxy)-1,4-phenylene vinylene] and its application in flexible polmer solar cells. *Thin Solid Films* 474:201.

193. Huang, H., et al. 2005. Properties of an alternating copolymer and its applications in LEDs and solar cells. *Thin Solid Films* 477:7.

194. Padinger, F., et al. 2000. Fabrication of large area photovoltaic devices containing various blends of polymer and fullerene derivatives by using the doctor blade technique. *Opto-Electron Rev* 8:280.

195. Sahin, Y., et al. 2005. Development of air-stable polymer solar cells using an inverted gold on top anode structure. *Thin Solid Films* 476:340.

196. Peeters, A., et al. 2000. Synthesis, photophysical properties, and photovoltaic devices of oligo (*p*-phenylene vinylene)-fullerene dyads. *J Phys Chem B* 104:10174.

197. Nierengarten, J.-F. 2004. Fullerene-(π-conjugated oligomers) dyads as active photovoltaic materials. *Solar Energy Mater Solar Cells* 83:187.

198. Stalmach, U., et al. 2000. Semiconducting diblock copolymers synthesized by means of controlled radical polymerization techniques. *J Am Chem Soc* 122:5464.

199. Ramos, A.M., et al. 2001. Photoinduced electron transfer and photovoltaic devices of a conjugated polymer with pendant fullerenes. *J Am Chem Soc* 123:6714.

200. Park, J.Y., et al. 1998. Doping effect of viologen on photoconductive device made of poly(*p*-phenylenevinylene). *Appl Phys Lett* 72:2871.

201. Angadi, M.A., D. Gosztola, and M.R. Wasielewski. 1998. Characterization of photovoltaic cells using poly(phenylenevinylene) doped with perylenediimide electron acceptors. *J Appl Phys* 83:6187.

202. Takahashi, K., et al. 2004. Performance enhancement by blending an electron acceptor in TiO$_2$/polyphenylenevinylene/Au solid-state solar cells. *Chem Lett* 33:1042.

203. Tanigaki, N., et al. 2005. Dye doping of poly(*p*-phenylenevinylene)s by vapor transportation for photovoltaic application. *Jpn J Appl Phys* 44:630.

204. Itoh, E., I. Suzuki, and K. Miyairi. 2005. Field emission from carbon-nanotube-dispersed conducting polymer thin film and its application to photovoltaic devices. *Jpn J Appl Phys* 44:636.

205. Greenham, N.C., X. Peng, and A. Alivisatos. 1997. A CdSe nanocrystal/MEH-PPV polymer composite photovoltaic. *AIP Conf Proc* 404:295.

206. Beek, W.J.E., et al. 2005. Hybrid zinc oxide conjugated polymer bulk heterojunction solar cells. *J Phys Chem B* 109:9505.

207. Breeze, A.J., et al. 2001. Charge transport in TiO$_2$/MEH-PPV polymer photovoltaics. *Phys Rev B* 64:125205.

208. Sirimanne, P.M., et al. 2003. An approach for utilization of organic polymer as a sensitizer in solid-state cells. *Solar Energy Mater Solar Cells* 77:15.

209. Ravirajan, P., et al. 2005. The effect of polymer optoelectronic properties on the performance of multilayer hybrid polymer/TiO$_2$ solar cells. *Adv Funct Mater* 15:609.

210. Hagfeldt, A., and M. Grätzel. 2000. Molecular photovoltaics. *Acc Chem Res* 33:269.

211. Halls, J.J.M., et al. 1995. Efficient photodiodes from interpenetrating polymer networks. *Nature* 376:498.

212. Yu, G., and A.J. Heeger. 1995. Charge separation and photovoltaic conversion in polymer composites with internal donor/acceptor heterojunctions. *J Appl Phys* 78:4510.

213. Gao, J., G. Yu, and A.J. Heeger. 1998. Polymer *p–i–n* junction photovoltaic cells. *Adv Mater* 10:692.

214. Veenstra, S.C., et al. 2004. Photovoltaic properties of a conjugated polymer blend of MDMO-PPV and PCNEPV. *Chem Mater* 16:2503.

215. Breeze, A.J., et al. 2004. Improving power efficiencies in polymer–polymer blend photovoltaics. *Solar Energy Mater Solar Cells* 83:263.

216. Egbe, D.A.M., et al. 2004. Synthesis, characterization, and photophysical, electrochemical, electroluminescent, and photovoltaic properties of yne-containing CN-PPVs. *Macromolecules* 37:8863.

217. Loos, J., et al. 2005. Morphology determination of functional poly[2-methoxy-5-(3,7-dimethyloctyloxy)-1,4-phenylenevinylene]/poly[oxa-1,1-phenylene-1,2-(1-cyanovinylene)-2-methoxy-5-(3,7-dimethyloctyloxy)-1,4-phenylene-1,2-(2-cyanovinylene)-1,4-phenylene] blends as used for all-polymer solar cells. *J Appl Polym Sci* 97:1001.

218. Tada, K., et al. 1997. Donor polymer (PAT6)–acceptor polymer (CNPPV) fractal polymer networks. *Synth Met* 85:1305.

219. Greenwald, Y., et al. 1998. Polymer–polymer rectifying heterojunction based on poly(dicyanothiophene) and MEH-PPV. *J Polym Sci A Polym Chem* 36:3115.

5

Blue-Emitting Poly(*para*-Phenylene)-Type Polymers

Emil Jachim Wolfgang List
and Ullrich Scherf

5.1 Introduction

In the last four decades, organic materials in general and conducting and electroactive-conjugated polymers in particular have been identified as a fascinating class of novel conductors and semiconductors that have the electrical and optical properties of metals and semiconductors and, in addition, have the processing advantages and mechanical properties of molecular materials [1–3]. As demonstrated over the last decade, conjugated polymers, e.g., poly(*para*-phenylene) (PPP), poly(*para*-phenylene vinylene)s, and polythiophenes, can be utilized as the active medium in organic light-emitting diodes (LEDs) [4], light-emitting electrochemical cells [5], solar cells [6], photo detectors [7], lasers [8], field-effect transistors [9], and all polymer-integrated electronic circuits [10]. It has been recently shown that printing of conjugated materials may also provide a new route to moderate-cost fabrication of

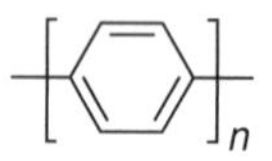

SCHEME 5.1   Structure of poly(*para*-phenylene) PPP.

integrated circuits [11–14]. Furthermore, the trend toward nanoscale electronics raises a quest for novel materials and concepts that allow to control the deposition of the active material on the nano- to mesoscopic scale. In contrast to inorganic semiconductors, organic semiconducting molecules, being small molecules or conjugated polymers, are inherently nanosized building blocks showing already various quantum effects in their original form. They allow for a defined structuring on the nanometer scale by means of both a *top-down* and a *bottom-up* concept, which is a conceivable benefit of this class of materials for future developments (Scheme 5.1).

The broad range of investigations of conjugated polymers over the last decade has already resulted in fundamental insight into the nature of electronic excitations, such as singlet excitons (SEs), triplet excitons (TEs), polarons, or bipolarons, as well as into the basic principles of operation of various devices fabricated from such materials. Moreover, it could be shown over the last 15 years that the electronic properties of conjugated polymers extremely sensitively depend not only on the principle molecular and supramolecular structure, but are also strongly determined by the presence of structural defects or impurities even at very low concentrations [15]. In particular, chemical defects formed during chemical synthesis or device operation have been identified to play an important role in the final performance of polymer light-emitting devices (PLEDs) [16–20] and field-effect transistors [21].

The first investigations on organic semiconductors have started in the 1950s on highly purified anthracene, resulting in the first organic light-emitting device in 1964 [2]. In 1987, the first thin film organic LED, based on small molecules, was realized at Kodak [22], followed by the first conjugated polymer light-emitting device in 1990 [4,23]. In particular, the advantages of organic molecule-based displays that have the higher brightness, the low power consumption, the ease and low cost of production, the 160° viewing angle, and the ultrathin and possible rollable design make this technology a real alternative to the now common LCD technology.

As part of the development of display applications based on conjugated polymers, significant attention has recently been directed to attaining stable blue electroluminescence (EL) from poly(*para*-phenylene)-type (PPP-type) polymers [24–26]. The generation of blue light is of crucial importance for red–green–blue (RGB) full color light-emitting devices. Moreover, it can be transformed into red and green light by internal [27] or external color conversion [28]. Among polyphenylene-based materials, highly emissive polyfluorenes (PFs) have received particular attention during the last decade as a promising class of conjugated polymers, which can be used as blue light-emitting materials in PLEDs. The first report on the synthesis of soluble poly(9,9-dialkylfluorene)s from the corresponding fluorene monomers with 9,9-alkyl chains up to C_{12} using $FeCl_3$ as the oxidative coupling agent was given by Yoshino and coworkers in 1989 [29,30], and first blue EL from such poly(9,9-dialkylfluorene) was shown by the same group 2 y later [31]. After these initial experiments, a broad variety of efficient synthesis of PFs has been developed, especially by Suzuki- and Yamamoto-type aryl–aryl couplings. The excellent optical and electronic properties of 9,9-disubstituted PFs, polyindenofluorenes (PIF) [32], and fluorene-based copolymers [33,34] have brought this class of materials into the focus of scientific and industrial interest. The photophysics of PF-type polymers has been intensively studied and impressive improvements concerning color purity and device stability of PF-based devices have been made. Still, the strict requirements for commercialization demanding tens of thousands of hours of operation are a hard to reach goal for blue-emitting PLEDs (Scheme 5.2).

Since PFs, PIFs, PPP-type ladder polymers (LPPPs), and other bridged PPPs are a very promising class of blue emitters of PLEDs [17,35,36], the synthesis, characterization, and application of such aromatic polymers have been very extensively worked on in the last decade. It seems impossible to review the enormous amount of scientific publications in this field. This review can only give a short overview that

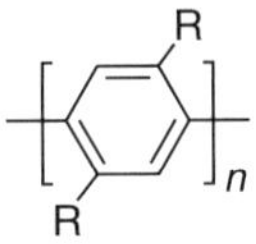

SCHEME 5.2   Structure of poly(2,5-dialkyl-1,4-phenylene).

summarizes some of the most attractive outcomes and trends of the last few years. Moreover, this review only touches the related topic of aromatic copolymers containing bridged oligophenyl segments as one structural component (especially fluorene-based statistical, alternating, and block copolymers). Such copolymers have been successfully applied in semiconducting polymer-based sensors [37], as semiconductors for organic polymer-based field-effect transistors [14,38] and as components in polymer-based "bulk heterojunction solar cells" [39,40].

The tasks of ongoing research will be addressed in the following paragraphs:

- Synthetic approaches to PPP-type polymers
- Photophysical processes in PPP-type polymers with an emphasis on excited state decay processes and the role of on-chain defects
- Optical and electrical properties of (*para*-phenylene)-type polymers with an increasing degree of aryl–aryl bridging
- Defect formation in PPP-type light-emitting polymers—low-energy electro- and photoluminescence bands and their origin in carbonyl-type defects
- An outlook toward novel synthetic strategies for PPP-type polymers with minimum amount of chemical and electronic defects

5.2 Synthetic Approaches to Poly(*para*-Phenylene)-Type Polymers

Unsubstituted PPP is characterized by its total insolubility in common organic solvents. Until today, no applicable route toward high-quality, defect-free PPP films is known. The approaches that have been investigated in the past include the use of soluble, nonconjugated precursor polymers [41,42] and the electrochemical polymerization of benzene [43].

The solubilization of PPP-type polymers was possible by the introduction of alkyl, alkoxy, ester, keto, or ionic side groups in a "hairy rod" approach initially developed by Schlüter and Wegner and coworkers [44–64]. Moreover, important progress has been made in the development of effective and selective aryl–aryl coupling methods toward structurally defined, soluble PPP derivatives.

The bundle of available synthetic methods today includes the Suzuki cross coupling of arylboronic acids or esters with chloro-, bromo-, iodo-, or tosylaryls (see Scheme 5.3), Yamamoto homocoupling of chloro-, bromo-, or iodoaryls, and Stille cross coupling of trialkylstannylaryls with haloaryls.

The "hairy rod" approach toward solubilized polymers of the PPP-type and the application of novel, sophisticated aryl–aryl coupling methods initiated an enormous progress in the field and today allow for the design of nearly defect-free soluble PPP derivatives of high molecular weight (up to 300,000 in the

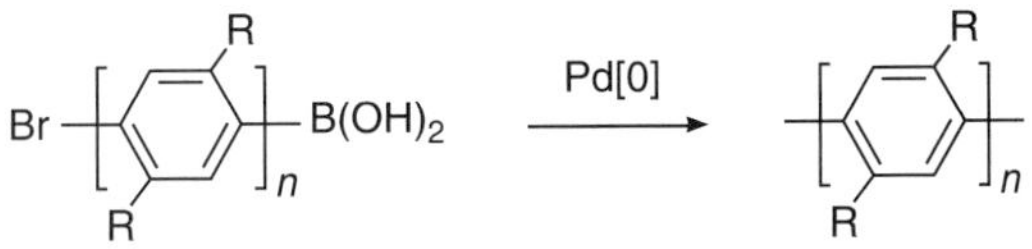

SCHEME 5.3   Initial synthesis of poly(2,5-dialkyl-1,4-phenylene) after Schlüter and Wegner [44] (for solubility reasons—R represents longer *n*-alkyls with *n*: 6–12).

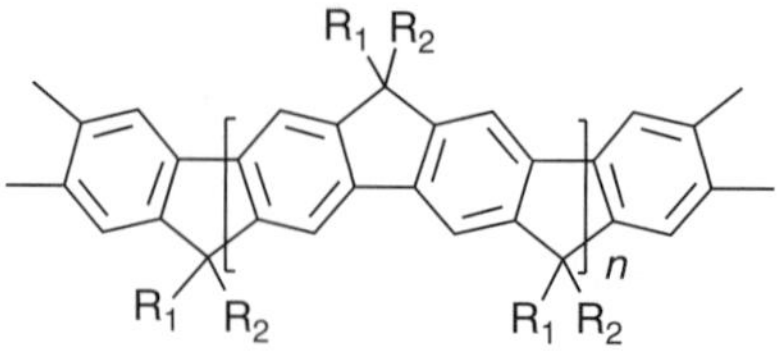

SCHEME 5.4 Structure of a *para*-phenylene ladder polymer (LPPP).

number average molecular weight M_n). However, the key shortcoming of such substituted PPPs is the distinct decrease of the main chain conjugative interaction that is caused by the steric bulkiness of the substituents. The effect is most pronounced for alkyls in the 2- and 5-positions of the 1,4-phenylene units (mutual distortion of neighboring aryl units of up to 80°) and somewhat lesser pronounced for alkoxy groups [63].

A strategy which overcomes this shortcoming was first proposed and developed by Scherf and Müllen, and includes the bridging of adjacent phenylenes by methylene bridges [64–66]. These bridges force neighboring aryl units into a coplanar arrangement and guarantee a full conjugative interaction of the aromatic moieties of the conjugated PPP main chain. A complete execution of this synthetic strategy leads to LPPPs (Scheme 5.4). The development of synthetic strategies toward well-defined ladder polymers has been a focus of intense research in the 1990s [67–69]. The "concerted" approaches, e.g., in a polycycloaddition after Diels-Alder as well as stepwise procedures, have been discussed. Müllen and Scherf developed a precursor strategy toward LPPPs (Scheme 5.5) including the generation of suitably substituted PPP precursors and their polymer-analogous cyclization to LPPPs. Surprisingly, the

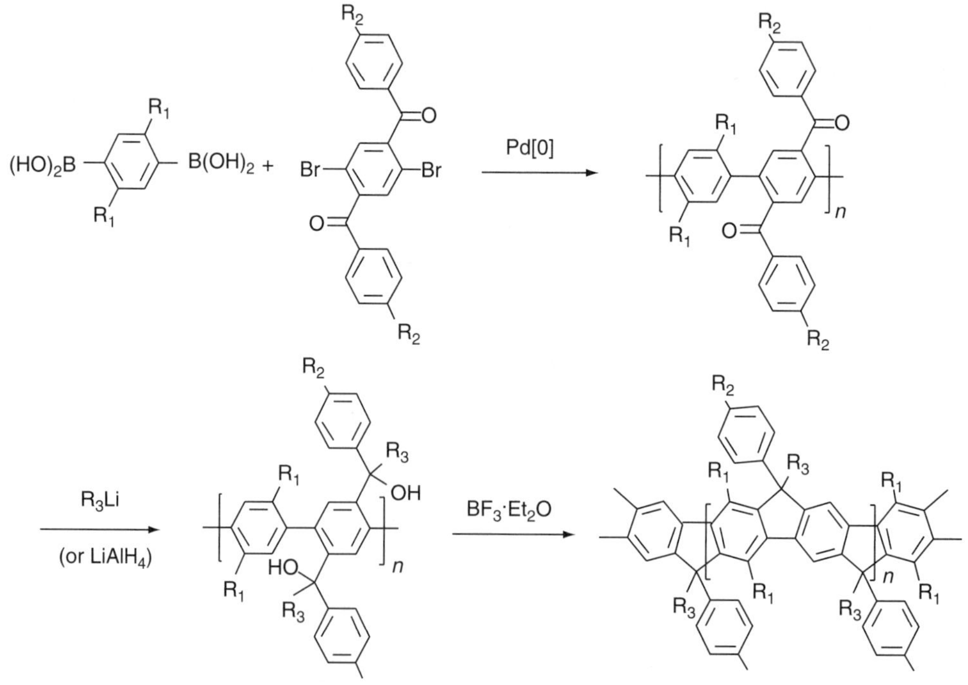

SCHEME 5.5 Synthesis of *para*-phenylene ladder polymers after Scherf and Müllen (MCP 1991) [64] (R_1 and R_2 are typically *n*-alkyls with C_6–C_{10}, R_3 is preferably a methyl group resulting in MeLPPP, R_3 = H leads to LPPP ladder polymers which are sensitive to oxidative degradation–formation of "keto" defects [16,17,70,71]).

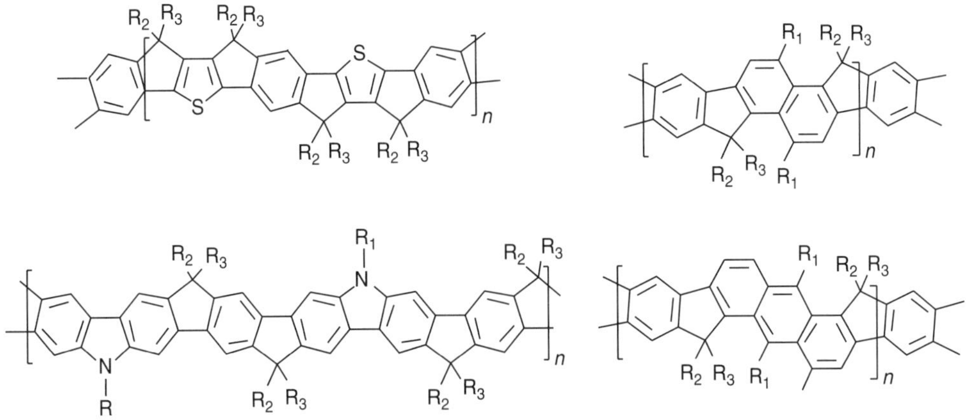

SCHEME 5.6 Structures of related polyarylene-type ladder polymers with main chain thiophene, carbazole, naphthalene, and anthracene units [32–35].

optimization of the substitution pattern of the single-stranded precursors allows for a regioselective and quantitative cyclization without the occurrence of unfavorable interchain cross-linking.

This approach leads to LPPPs with very unique optical and electronic properties, for more details see [67,68]. Following the general procedure outlined in Scheme 5.5, other aromatic building blocks (e.g., naphthalene, anthracene, thiophene, and carbazole) also have been incorporated into the main chain of such planarized arylene-type ladder polymers (Scheme 5.6).

An only partial execution of the aryl–aryl bridging approach at each second phenylene–phenylene linkage gives rise to PF. PFs have been first synthesized by Yoshino [29,30] by the oxidative coupling of 9,9-dialkylfluorene monomers with FeCl$_3$. However, these initial products were characterized by an ill-defined chemical structure since the coupling did not proceed regioselectively at the 2- and 7-positions of the fluorene core. Later on, more effective transition metal-catalyzed aryl–aryl coupling methods have been also applied for the PF synthesis.

A group at DOW Chemicals first developed and patented an approach based on the Suzuki-type cross coupling of 9,9-dialkylfluorene-2,7-diboronic esters and 9,9-dialkyl-2,7-dibromofluorenes toward high molecular weight, soluble poly(9,9-dialkylfluorene-2,7-diyl)s [72,73] (Scheme 5.7). Independently, Nothofer and Scherf developed an alternative approach based on the Yamamoto-type homocoupling of 9,9-dialkyl-2,7-dibromofluorene (Scheme 5.8) [74]. Both methods, if properly done, lead to the

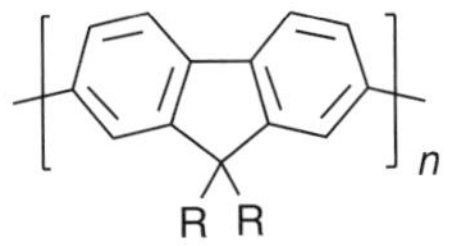

SCHEME 5.7 Structure of poly(9,9-dialkylfluorene-2,7-diyl) (PF).

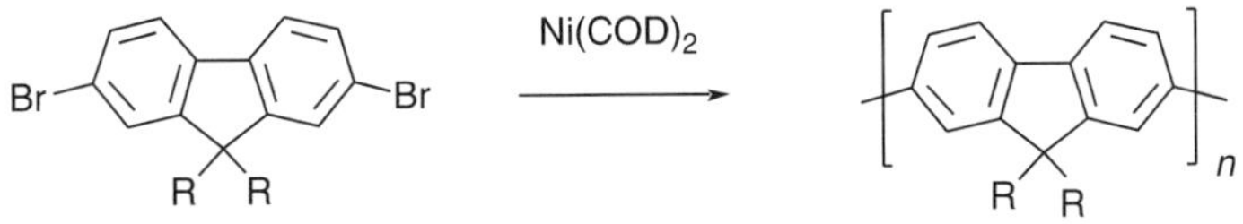

SCHEME 5.8 Synthesis of poly(9,9-dialkylfluorene-2,7-diyl) after Nothofer and Scherf [74]. Often used alkyls are *n*-octyl resulting in PFO (or PF8) and 2-ethylhexyl resulting in PF2/6.

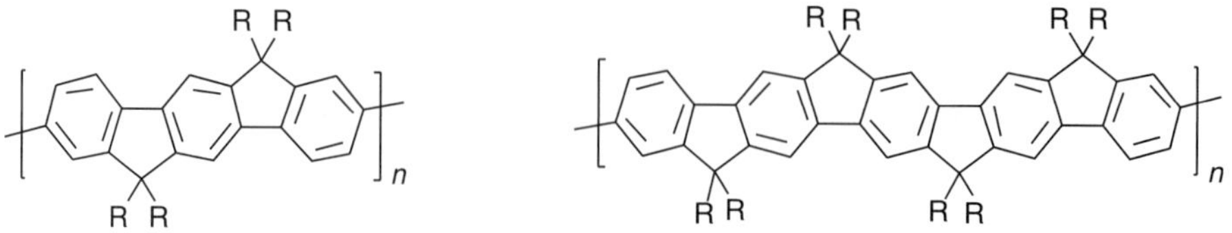

SCHEME 5.9   Structures of polyindenofluorene and a poly(*para*-phenylene) with bridged pentaphenyl units (the methylene bridges can carry different aryl and alkyl substituents, for simplicity all substituents are given as R).

highest-quality PFs with a number average molecular weight of up to 300,000, corresponding to a coupling of several hundreds of repeat units [17].

Especially poly(9,9-dialkylfluorene-2,7-diyl)s have been very broadly investigated for a potential application as blue emitters in polymer-based OLEDs [17,35,76]. PFs show a very rich variety of structures in the condensed state, including nematic LC mesophases and crystalline phases (e.g., the β-phase in PFO and a helical phase in PF2/6) [76,77]. The morphology and the underlying phase transitions of such PFs in the solid state are, until now, not fully understood. The substituents at the 9-position strongly influence the solid state packing behavior and have been varied in a broad range. Today many derivatives with linear and branched alkyls as well as aryls including dendritic oligophenyl and spirobifluorene-type side chains are known (R–R: alkylated biphenyl-2,2′-diyl) [46–48]. The fact that this nematic LC state can be quenched into a glass or crystallized [76,77] allows the fabrication of PLEDs that emit highly polarized light making them potential candidates for backlight illumination in LC displays [78].

In parallel, Müllen and Grimsdale have varied the pattern of bridged and nonbridged biphenyl units in PPP-type polymers by synthesizing PIFs (2 of 3 aryl–aryl linkages are bridged) [79,80] and recently PPPs containing ladder-type pentaphenyl segments (4 of 5 aryl–aryl linkages are bridged) as shown in Scheme 5.9 [81]. In that way, the authors filled the "gap" between PFs and LPPPs. Now, a complete series of PPP-type structures with increasing degree of bridging is available (for details see Chapter 7 on the properties of the polymers).

5.2.1   Poly(*para*-Phenylene)s with an Increasing Degree of Aryl–Aryl Bridging

After the extensive synthetic work during the last decade, now a complete series of PPPs with increasing degree of methylene bridging is available ranging from the nonbridged poly(2,5-dialkyl-1,4-phenylene)s by PFs, PIFs, and PPPs with bridged pentaphenyl units toward fully bridged LPPPs. Table 5.1 gives molecular weight as well as optical data (absorption and emission) of this series of aromatic, conjugated

TABLE 5.1   Poly(*para*-Phenylene)s with Different Degree of Aryl–Aryl Bridging

Polymer (Oligomer)	$\lambda_{max,abs}$ (nm)	$\lambda_{max,em}$ (nm)	M_n/M_w	PD
Poly(2,5-dihexyl-1,4-phenylene)[a]	+	—	16,000/73,000	4.5
Poly(2,5-dibutoxy-1,4-phenylene) [55]	336	404	6,600 (M_n from VPO)	—
Poly[9,9-*bis*(2-ethylhexyl)fluorene-2,7-diyl]	383	413 (film: 416)	83,500/163,000	1.95
Polyindenofluorene [80]	417	428 (film: 434)	66,400/257,000	3.87
Poly(*para*-phenylene) with bridged pentaphenyl units [81]	434	445 (film: 445)	136,000/298,000	2.19
Ladder-type 11-mer [82]	442	452	4116.73 (calc. M)	—
Para-phenylene ladder polymer [67]	453	455 (film: 456)	22,500/47,000	2.09

Note: PD: polydispersity, + no significant absorption detectable at $\lambda > 300$ nm.
[a]Witteler, H. Ph.D. thesis, Johannes-Gutenberg-Universität Mainz, 1993.

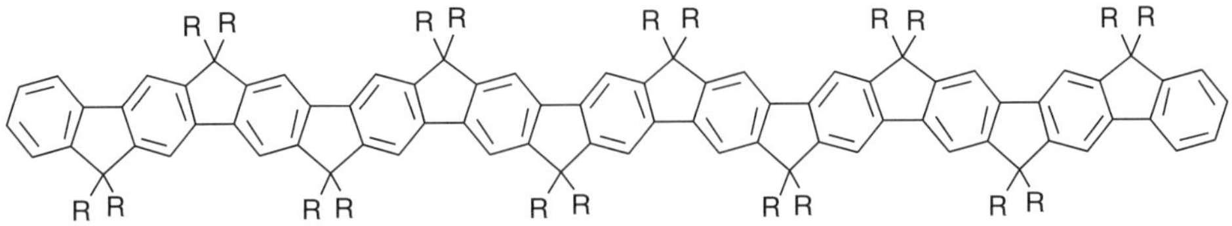

SCHEME 5.10 Structure of a ladder-type *para*-phenylene 11-mer (the 10 methylene bridges carry different aryl and alkyl substituents, for simplicity all substituents are given as –R).

polymers. For comparison, the data of a very recently synthesized and published ladder-type *para*-phenylene 11-mer (by Müllen and Grimsdale) are also included. The structure of the 11-mer is shown in Scheme 5.10 [82].

All partially bridged PPPs have been synthesized by the Yamamoto-type homocoupling protocol using preformed building blocks (e.g., dialkylfluorene, tetraalkylindenofluorene, and octasubstituted pentaphenyl blocks). The fully bridged ladder polymer MeLPPP has been synthesized by the already outlined cyclization protocol of suitably substituted, single-stranded PPP precursors (Scheme 5.2).

The data of Table 5.1 reflect the increasing planarization of the π-electron system with increasing degree of bridging. Since poly(2,5-dialkyl-1,4-phenylene)s display only a negligible degree of conjugative interaction in the main chain, the corresponding dialkoxy derivatives have been included as the first line of Table 5.1 [63]. The ongoing planarization is accompanied by an increasing conjugative interaction of the aromatic subunits as seen for the absorption as well as for the emission data later in this chapter. In the ladder-type 11-mer (data given for comparison), the electronic properties of MeLPPP are mostly reached, especially in emission. This finding illustrates the occurrence of a relatively short length of the effectively conjugated segments in ladder-type poly- and oligophenylenes of only 10–12 phenyls as proposed already 10 y ago in a study of a series of shorter LPPP oligomers up to the 7-mer [83].

Nevertheless, despite the fact that the optical and electronic properties already converge for relatively short chain lengths of 10–12 phenyls, there are a couple of simple reasons to favor polymers over defined oligomers such as the ladder-type phenylene 11-mer in "practical" applications: (1) in most cases, the synthetic effort toward the defined oligomers is enormous (extensive multistep procedures); and (2) only polymers with a certain molecular weight (favorable >10,000–20,000) allow the processing into high-quality, flexible thin films by solution techniques as favored for the production of "polymer electronic" devices.

5.2.2 Bridged Poly(*para*-Phenylene)s with On-Chain Phosphorescent Metal Complexes

As discussed later, fluorenone units can act as effective traps for excitations in PPP-type polymers. Also other chromophores can be incorporated within the conjugated main chain as such excitation energy traps. Phosphorescent chromophores are especially interesting since they allow for a capture of both singlet and triplet excitations. This is attractive for electroluminescent devices since both singlet and triplet excitations are produced in LEDs during the recombination of positively and negatively charged polarons.

Two examples of such PPPs with on-chain phosphorescent units are presented here. In the first example phosphorescent palladium centers are incorporated into the LPPP backbone by accident, and in the second example, phosphorescent platinum–salen (Pt–salen) complexes are incorporated by design into the PF-type copolymers.

In testing the flexibility of the synthetic approach toward LPPP-type polymers, ladder polymers with two aryl substituents at the methylene bridge (Scheme 5.4: R_1 and R_2:—aryl) were also produced. The resulting polymers showed the expected absorption and emission properties both in solution and the solid state [15]. Surprisingly, the EL spectrum displayed an intense low-energy emission feature peaking at 600/650 nm. In accordance with earlier investigations of the phosphorescence bands in LPPP-type polymers by Bässler and Romanovski and coworkers [84] this EL component was assigned to the occurrence of an "intrinsic" electrophosphorescence in this LPPP derivative. An increased amount of

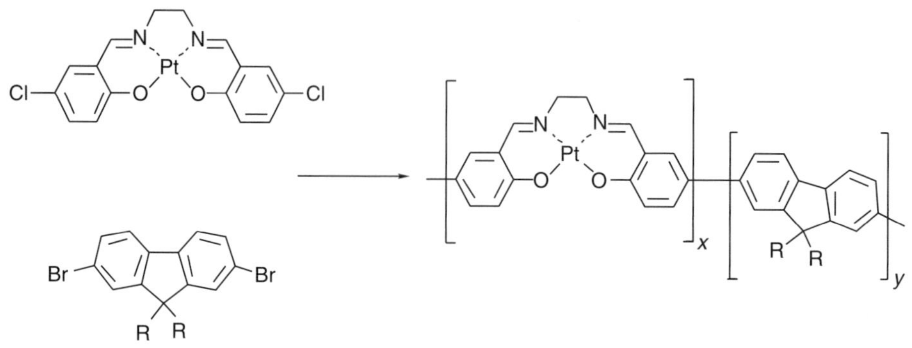

SCHEME 5.11   Synthesis of (9,9-dialkylfluorene-2,7-diyl)-based copolymers with on-chain Pt–salen phosphors.

80–200 ppm covalently bound palladium was found in this polymer. The palladium centers may be incorporated by a transmetallation-type side reaction during the polymer-analogous addition of phenyl lithium to the polyketone precursors. The on-chain palladium centers allow for a radiative deactivation of triplet excitations. The relatively long lifetime of such triplets guarantees an efficient transfer of such triplets to the small amount of palladium centers in the solid state (~1 Pd per 10 polymer chains). The system allows for an investigation of both singlet and triplet emissions without external dopant and has been used to investigate the possibility of spin randomization in polaron pairs (PP). No spin randomization under the experimental conditions could be detected [85]. This finding is in some contrast to the recently proposed deviations from the theoretical singlet–triplet ratio of 1:3 as derived from spin-statistical predictions determined from magnetic resonance measurements [86].

In the first example, the Pd centers have been incorporated by accident. However, there is much interest in a more controlled way of incorporating phosphorescent metal complexes. Therefore, the design of PF-type polymers with on-chain Pt–salen phosphors has been investigated [87].

The copolymers have been synthesized in a statistical copolymer approach (Scheme 5.11). The resulting copolymers show weak emission bands related to the phosphor in solid state PL measurements. However, the EL spectrum is dominated by the emission of the Pt–salen phosphors; the emission of the PF chromophore is negligible. This may be due to an additional pinning of negative charges at the phosphorescent metal complexes. The pure copolymer showed the occurrence of aggregate-related EL bands. A dilution of the phosphors in a PF2/6 matrix resulted in a suppression of these aggregation-related bands and allowed for a distinct increase of the OLED device efficiency up to 6 Cd/A.

5.3 Photophysics of Conjugated Polymers: Excited State Decay Processes and the Role of On-Chain Defects

The performance of PLEDs is, to a large extent, governed by the ongoing photophysical processes related to the chemical structure of the active molecule and different processes taking place in the solid state. In this section, the most important processes governing the fate of a SE after electrical or optical excitation are listed and briefly discussed for different PPP-type polymers. With respect to the focus of this contribution, however, mainly influences determined by the interplay of the excited state with chemical and structural defects as well as the mutual interaction of excited states will be investigated.

As depicted in Figure 5.1 the SE is the primary photoexcitation in conjugated polymers. In the EL process, the SE are, next to TEs, formed from injected charges with a ratio close to 1:3 as recently found [85]. In contrast to inorganic semiconductors, where one typically finds binding energies in the range from 5 to 30 meV (Wannier–Mott excitons), binding energies in the order of 0.2–0.5 eV (Frenkel excitons) have been reported for conjugated polymers [88].

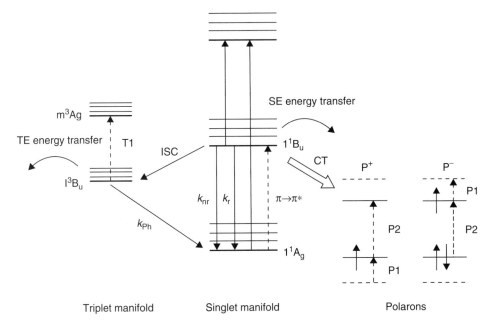

FIGURE 5.1 Jablonski diagram illustrating the excitation and recombination processes as found in organic semi-conductors. The 1^1A_g and the 1^3B_u states are the ground state and lowest TE states, respectively. The 1^1B_u is the lowest allowed excited SE state. k_r and k_{nr} are the radiative and nonradiative rate constants for the 1^1B_u SE decay. T1 denotes the first allowed triplet absorption from the 1^3B_u to the lowest excited m^3A_g state as observed in photoinduced absorption (PA), while k_{PH} is the radiative phosphorescent decay of triplets to the ground state. P$^+$ and P$^-$ represent polarons created upon charge transfer (CT), while P1 and P2 are the optical transitions within the polaronic states.

Following photoexcitation or recombination of an appropriate polaron pair, the SE can undergo various processes during its lifetime, which is typically $\sim$1 ns [88].* While it is well established that most of the processes listed below take place after the rapid thermalization (Kasha's rule) to the lowest $1B_u$ vibrational state on the conjugation segment the SE has been created on, there is still a debate [89] on the exact timescale of the thermalization process.[†] However, a list of the most important processes is given in the following:

1. Radiative recombination with typical recombination rates k_r of 10^9 s^{-1}.
2. Nonradiative recombination k_{nr} due to vibronic internal conversion (IC) at chemical defects, impurities, and structural defects.
3. Intersystem crossing (ISC) and formation of TE from the lowest $1B_u$ state to the lowest state in the triplet manifold, the 1^3B_u state. ISC rates k_{ISC} determined for conjugated polymers are found in the range of 10^7–10^8 s^{-1} [2].
4. Charge transfer (CT) of the relaxing (hot state)[‡] or thermalized excited state to an adjacent conjugation segment or conjugation segment of an adjacent chain [90–93].[§]
5. Intra- and interchain energy transfer from the lowest $1B_u$ to another conjugated segment or hetero-molecule [94]. (If this process is happening in multiple steps, it is denoted as excitation energy migration.)**

---

*Please note that SEs can also be created by a secondary process such as triplet–triplet annihilation, which is, however, an efficient process only at rather high triplet densities.

[†]For further details on this subject and on related ultrafast phenomena in conjugated polymers see Ref. [89].

[‡]The exact nature of this process is still under debate. For details see [89,90,92].

[§]Note that the CT process can also lead to the formation of geminate PP, which do not dissociate into charge carriers, but reform a SE giving rise to delayed fluorescence as discussed in detail in [93].

**Recent quantum chemical calculations have revealed that the on-chain migration in conjugated polymers is almost two orders of magnitude lower than the interchain migration. For details see [94].

6. Energy transfer onto intermediate metastable states such as polarons or TEs and energy transfer as well as exciton localization onto on-chain heteromonomer units or defects.

5.3.1 Recombination Dynamics in the Solid State

In the situation where the polymer chains are isolated in a solid state matrix or dissolved in a solvent the SE dynamics are governed by

$$\frac{dn_{SE}}{dt} = g_0 - (k_r + k_{nr})n_{SE} \qquad (5.1)$$

where g_0 is the generation rate of the SE (n_{SE}) from absorbed photons, k_r is the radiative decay, and k_{nr} is the nonradiative decay. As a consequence of the closer packing of the polymer chains and distinct solid state aspects arising from the interaction of π-electronic systems in films of conjugated polymers, one generally finds a clearly distinct behavior of the recombination and interaction dynamics of excited states. In most of the known polymer systems this leads to the observation of (a) a lowered photoluminescence quantum yield (η_{PL}) of the materials, (b) a more pronounced spectral relaxation, and (c) a broadening and bathochromic shift of the emission and absorption spectra.

5.3.2 Dynamics of Quenching Processes in the Solid State

The relaxation (migration in the film) of SEs makes it inevitable that the SE interacts with SE, other quasiparticles such as TE and polarons, and chemical and structural defects. Yet, as it was demonstrated by Rothberg as early as 1994 for different polymers, photooxidation of the material strongly quenches the PL. This is caused by the irreversible destruction of the polymer backbone as well as by the induced defects, such as carbonyl groups, acting as quenching sites for SEs consequently leading to the observed decrease of the photoluminescence [95–97].

In conjugated polymers a similar nonradiative quenching process as observed for molecular crystals [2] happens if a SE encounters a trapped or free polaron—acting as a charged defect—during the migration process:

$$P^{+/-} + S_1^* = S_0 + phonons + P^{+/-} \qquad (5.2)$$

a photoexcited intrachain SE (S_1^*) is quen*ched by polarons (P^+ or P^-) where S_0 is the singlet ground state. It should be noted, however, that there is a strong evidence that the quenching of SE by polarons does not result in dissociation of the SE, but rather in absorption of its energy by the polaron similar to the one found in a donor–acceptor system with the polaron being a metastable acceptor [2]. In such a scenario, the energy of the SE is absorbed by the polaron by a simple absorption process or transferred by an excitation energy transfer process due to the overlap of the emission with the subgap absorption.

Furthermore, the essentially same nonradiative quenching process happens if the SE encounters a TE. Following Agranovich's theory [98] SE–TE annihilation can be described by a nonradiative energy transfer, similar to a Förster-type transfer, of an excited SE (S_1^*) to a TE (T_1):

$$T_1 + S_1^* = S_0 + T_n + phonons \qquad (5.3)$$

where S_0 is the singlet ground state and T_n an excited state of the TE manifold [99]. Note that a TE generated from an excited SE will rapidly relax back to the T_1 state before it decays through one of the channels allowed for TEs.

For the quenching "dynamics" of SE at polarons and TE there is a fundamental difference to the quenching at chemical defects such as carbonyl groups: the density of chemical defects is independent of the excitation density in a chemically stable material. On the contrary, the steady state polaron and TE density increases with increasing excitation density as polarons and TE are created from SEs in photoexcitation or from injected charges. Hence the rate equation for the recombination dynamics of the SEs given for isolated molecules has to be modified and nonradiative bimolecular quenching processes with polarons and TE have to be included:

$$\frac{dn_{SE}}{dt} = N_0 - (k_r + k_{nr} + \gamma_{TE-SE} n_{TE} + \gamma_p n_p) n_{SE} \tag{5.4}$$

Here N_0 is the generation rate of the SEs, n_{SE} is their density, n_{TE} and n_p are the total density of polarons and TEs, and k_r and k_{nr} are the radiative and nonradiative decay constants, respectively. k_{nr} includes nonradiative recombination by production of phonons (IC) as well as the creation of polarons by dissociation. γ_{TE-SE} is the TE–SE annihilation rate constant and γ_P the polaron–SE annihilation rate constant, which are a measure for the interaction of either of the two quasiparticles.

The importance of these two nonradiative decay channels has been observed in devices at high current densities and under high excitation densities in photo pumping. In most PPP-type polymers, which possess more than 1×10^{17} cm^{-3} trap sites to stabilize polarons [113], the nonradiative recombination process becomes dominant as soon as $k_r \approx \gamma_p n_p$. In a similar manner SE–TE annihilation is a significant loss mechanism. However, SE–Polaron annihilation as well as the SE–TE annihilation are "competing" processes. Given the fact that only the order of magnitude for both processes is known up to now, no definite verdict on which of the processes becomes significant first can be given. This issue is subject to further studies. It has to be noted that the previous observations of nonlinear relaxation processes at higher excitation densities in both polymers and oligomers [100] were assigned to the nonlinear SE–SE annihilation mechanism. The detailed analysis shows, however, that the onset of the SE–TE and SE–Polaron annihilations should occur at similar excitation densities so that probably all three processes must be considered in future analyses of the dynamics and interaction of excited states in conjugated polymers. Moreover, as polaron or TE densities as high as 10^{17} cm^{-3} can be easily achieved upon carrier injection, their interaction with SE is also of great relevance for practical organic polymer light-emitting and laser applications as it has also been shown for small molecule-based devices [101].

5.3.3 Summary of Quenching Processes in the Solid State

For an assembly of conjugated polymer segments, the different mutual interactions of excited states are to a large extent assisted by different types of energy transfer processes of SE and TE as shown for PF segments in Figure 5.2. An important characteristic of both conjugated polymers and small-molecule films is energetic and morphological* disorder [102]. Although polymer chains may be quite long, typically the conjugation is interrupted by structural and chemical defects. Hence conjugated polymers can be considered as an assembly of individual conjugated segments as it has been beautifully illustrated by Lupton recently [82]. The local variations in the overlap of π-electrons give rise to the energetic disorder implying both an inhomogeneous broadening of the absorption spectrum [103] and a relatively broad "density-of-states" like energy distribution for neutral and charged excitations.

The width of this energetic distribution to a large degree determines the charge transport characteristics of the material as well as the migration of the charge-neutral states. The tail states of this distribution can, in principle, act as shallow trapping states for charge carriers and charge-neutral states as for the former observed at low temperatures in photoinduced absorption and thermally stimulated luminescence and current experiments [113]. According to the seminal work by Bässler et al. [104], energy (or charge) transport occurs through a sequence of incoherent hopping steps between localized chromophores. Therefore, in the relatively broad "density-of-states" like energy distribution, a strong migration of the charge-neutral states toward lower energetic sites can be expected prior to radiative or nonradiative

---

*A comprehensive survey of the structural properties of various conjugated polymers can be found in Ref. [102].

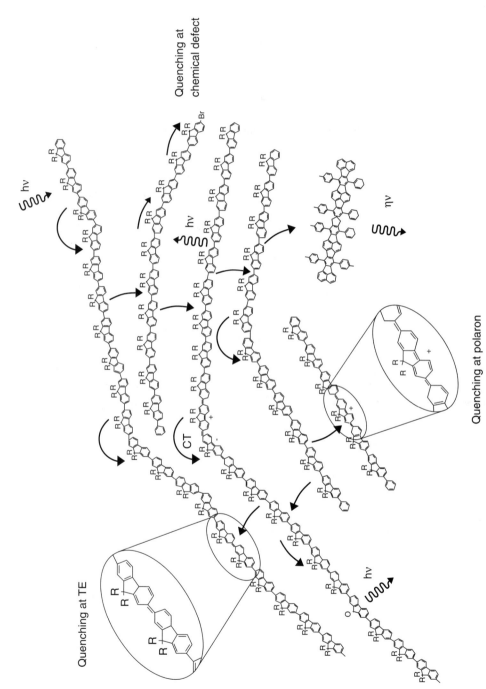

Quenching at chemical defect

Quenching at TE

Quenching at polaron

FIGURE 5.2 Schematics of migration-assisted processes in conjugated polymers. For explanation see text.

recombination or the quenching at a defect or other excited state. The excitation energy migration can be described as a combination of repeated steps of intrachain and interchain energy transfer steps which can be of *Förster-* [99] or *Dexter*-type [105]. Yet recent quantum chemical calculations and correlated experiments by Zojer and coworkers [106] have revealed that the Förster approximation can only be used to a limited extent.

As a consequence of the ongoing energy migration of SEs one can identify a number of processes to be rather strongly enhanced in the solid state.

5.3.3.1 Magnification of "Action-Radius" of Excitons in the Solid State

As depicted in Figure 5.2, a SE, after excitation in the solid sate, can undergo several hopping steps within different conjugation segments before it will recombine at the end of its natural lifetime (~1 ns). Each of these individual steps is established by the donor to acceptor transfer mechanism. In fact, in the solid state at RT one finds SE to randomly walk around within a radius of up to 15–20 nm after creation, which has been experimentally verified for example in m-LPPPs [107]. Yet, the entire migration process is strongly reduced with decreasing temperature, showing an Arrhenius-type behavior [108]—a radius of 6–7 nm at 20 K has been found in m-LPPPs. To the first approximation the temperature dependence can be modeled by means of the phenomenological approach developed by Miller and Abrahams [109] (where a penalty of the form $\exp(-\Delta E/kT)$ is applied for uphill hopping rates while downhill processes are assumed to be *T*-independent). More recently Wiesenhofer et al. [110] showed that the excitation diffusion in rigid-rod ladder-type polymers results from the interplay between the amount of inhomogeneous broadening and homogeneous broadening, and is determined by the coupling of the electronic excitations to low-frequency intramolecular vibrational modes.

5.3.3.2 Pronounced Spectral Relaxation in the Solid State due to Quasi-3D Migration

As a consequence of the migration, the SE also undergoes spectral relaxation. The SE is more likely to be transferred to a conjugation segment lower in energy than to a segment equal in energy as for a transfer between segments of different energy one finds a larger spectral overlap than for segments of similar energy.

5.3.3.3 Efficient Nonradiative Quenching at Defects, Polarons, and TE

A more important issue than the observation of spectral relaxation is the strong enhancement of quenching at structural and chemical defects and at metastable states such as polarons and TE. Note that these quenching processes are very likely to be established by an energy transfer from one conjugation segment to a segment sustaining a polaron or a TE acting as acceptor due to its given "photoinduced" absorption (see red arrows in Figure 5.2).

5.3.3.4 Enhancement of Overall Energy Transfer Efficiency in Polymer Blend Systems and Trapping at On-Chain Heteromolecules (e.g. Fluorenone Molecules in Polyfluorenes)

Due to the magnified SE action radius of up to 15–20 nm in the solid state [94], the SE in the solid state shows an overall enhanced and in nature different energy transfer behavior to heteromolecules in a blend system [27] as well as an enhanced trapping at on-chain heteromonomer units (e.g., fluorenone units in PFs) [16,17] compared to single isolated molecules in solution as discussed below in more detail [16,17].

5.4 Optical and Electrical Properties of (*para*-Phenylene)-Type Polymers with an Increasing Degree of Aryl–Aryl Bridging

After covering the interplay of the excited state with chemical and structural defects, we will now consider how changes in the chemical structure affect the properties of conjugated polymers, in particular the focus will be put on PPP-type structures with increasing degree of bridging. In the following, the most important issues related to both the chemical structure and the solid state properties

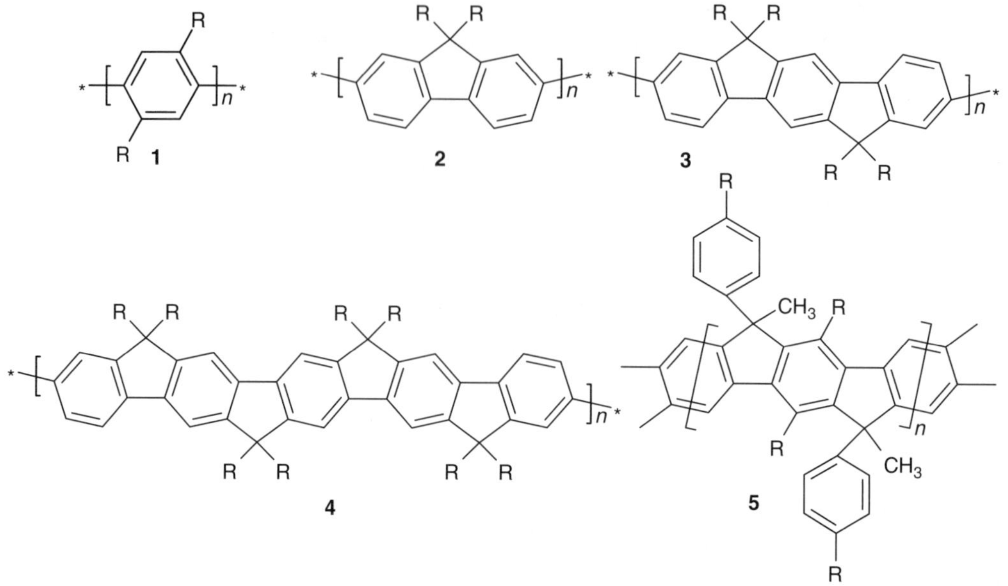

FIGURE 5.3 Polyphenylene-based conjugated polymers: **1** poly(*para*-phenylene), **2** polyfluorene, **3** polyindeno-fluorenes, **4** poly(*para*-phenylene) with bridged pentaphenyl units, and **5** fully planarized ladder-type poly(*para*-phenylene) (the methylene bridges can carry different aryl and alkyl substituents, for simplicity all substituents are given as –R).

of the bulk polymer shall be discussed. Polymer degradation, charge carrier injection, and interface-related issues are discussed in Chapter 6.

5.4.1 Electronic Properties of Poly(*para*-Phenylene)-Type Polymers

As discussed above a complete series of PPP-type structures with increasing degree of bridging is now available. All these polymers show efficient UV blue or blue–green emission in solution with an optical band-gab becoming continuously smaller with increasing chain rigidity (Figure 5.3).

5.4.1.1 Poly(*para*-Phenylene)-Type Polymers

Over the last 15 y, a vast number of unbridged PPPs have been optically characterized and tested for their applicability in different light-emitting devices such as PLEDs [24,25] and LECs [111]. The photoluminescence spectra of PPP-type polymers in films typically show a single broad unstructured emission maximum at $\sim$3.3 eV and an absorption spectrum that also shows just one unstructured peak around 4.1 eV. Typically, one finds a relatively large bathochromic shift between the film spectra and the spectra obtained for the dissolved polymer that can be as large as 150 meV as a consequence of the reduced interring twist in the solid state. In the photoinduced absorption spectrum one also finds one broad characteristic band at 1.6 eV, which has been assigned to an absorption in the triplet manifold—i.e., a transition from the 1^3B_u to a higher lying m^3A_g triplet state. The explicit signature of a well-resolved polaronic absorption, which would be located at $\sim$2.5 eV, has not been reported. The quantum yield of PPP derivatives has been reported to be as high as 85% in solution [112] and 35% in solid films [112] (Figure 5.4).

Blue light-emitting PLEDs fabricated from PPPs typically show EL emission spectra that are similar to the photoluminescence as long as the polymer is not degraded. As discussed in detail below, the emission maximum of PPP-type PLEDs is shifted from 3.2 to 2.6 eV if polymer degradation takes place. This polymer served as one of the first blue-emitting polymers in the PLED community and produced devices of good quality. However, it is not in the focus of blue-emitting materials anymore, which is due to the fact that the polymer is emitting most of its light in the UV region, where the eye is not

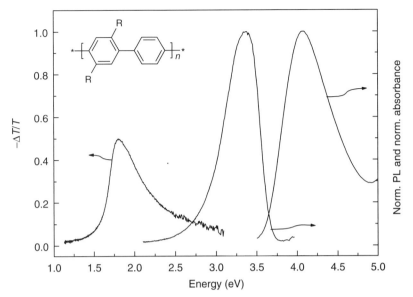

FIGURE 5.4 Absorption, photoluminescence emission, and photoinduced absorption spectrum of a typical poly(*para*-phenylene)-type polymer film. The inset shows a poly(2,5-disubstituted-1,4-phenylene).

sensitive. Moreover, there is an unsolved problem of fast polymer degradation under operation leading to device and color instability.

5.4.1.2 Polyfluorene-Type Polymers

Among the polyphenylene-based materials highly emissive PFs have received particular attention due to their extraordinary properties in device applications as originally demonstrated by Bradley [74,76–78]. In addition to the wide variety of synthetic routes allowing for the synthesis of differently substituted homopolymers as well as copolymers, PFs display attractive optical and electronic properties which make them ideal model systems for studying relevant structure to property relations in conjugated polymers as shown by Cadby et al. [77,116]. For this reason different properties which are related to the molecular structure as well as to intrachain order effects of the polymer shall be discussed in the following.

As depicted in Figure 5.5 PFs in film display an unstructured, long absorption maximum centered at ~3.3 eV. The photoluminescence emission spectrum of PFs shows a vibronic fine structure with an energetic spacing of ~180 meV (stretching vibration of the $C=C–C=C$ structure of the polymer backbone) with the π^*–π transition at ~2.9 eV yielding a deep blue emission. In dilute solution the spectra are very similar to that of the solid state and only a small bathochromic shift of ~20 meV is typically observed for both absorption and emission.

The photoinduced absorption shows one dominant band peaking at 1.54 eV. This band is assigned to a transition from the 1^3B_u to a higher lying m^3A_g triplet state [113]. The energetic spacing between the ground state 1^1A_g and the lowest TE 1^3B_u state was found to be 2.1 eV. This transition was also observed for phosphorescent decay of triplets yielding 2.1 eV photons. Typically, PF films do not exhibit a significant polaronic absorption band in steady state PA as a consequence of the rather low density of the traps and the low energetic disorder of the bulk polymer. The low energetic disorder is also reflected in the high mobility of charge carriers in this class of polymers and the observation of a nondispersive transport of the former [114,115]. Furthermore, the PA band has a remarkably low overlap with the PL emission. These results make PF a promising candidate for organic solid state lasers as such a low overlap is a required feature for this application. It is worth noting that PFs exhibit exceptionally high solid state quantum yields, which have been found to be as high as 50%. Typical radiative decay times of isolated molecules are in the order of 1 ns.

As mentioned earlier, the structure of the alkyl side chains in 9-position of the fluorene moieties strongly influences the solid state packing of the polymers. *Octyl–alkyl* derivatives display a unique packing behavior in condensed phase upon thermal treatment or in mixtures of solvents and non-solvents with increasing nonsolvent content [17]. They form a β-phase, in which the initially distorted backbone is flattened into a planarized conformation. The distinct redshift of the absorption spectra as well as the observation of a vibronic progression in the absorption spectrum in the β-phase can be explained with the planarized conformation improving the intrachain π-conjugation. The driving force for this unique packing behavior (β-phase formation) should, on the other hand, be the solid state packing of the *octyl–alkyl* side chains, described as "mechanical stress during β-phase formation" in Ref. [116]. The packing is suppressed in the case of branched alkyl chain-substituted PF derivatives (especially with 2-ethylhexyl, 3,7-dimethyloctyl, or most favorable 3,7,11-trimethyldodecyl side chains). The same behavior was described for dendron-substituted PF derivatives, at the same time independently by Müllen and coworkers [117]. A detailed description of the electronic properties of PF in the β-phase can be found in Refs. [17,76,118]. One of the main reasons that has so far prevented a breakthrough of PF-type polymers as the material for the realization of blue light-emitting devices is their poor long-term stability. Upon photooxidative or thermal degradation and during device operation, the formation of low-energy emission bands around 530 nm (~2.3 eV) is frequently observed in PF-type polymers. The origin and nature of this emission is discussed in detail below.

5.4.1.3 Polyindenofluorenes

By introducing a monomer unit where two aryl–aryl linkages are bridged the PIFs are established in which 2 of 3 aryl–aryl linkages are bridged [32,79,80]. The more planarized conformation leads to the observation of a redshifted emission better optimized to the eye's sensitivity. In addition, one finds a partially structured absorption spectrum as compared with PFs or PPPs. In particular, it has been shown that this polymer exhibits improved stability if aryl substituents are introduced instead of the alkyl chains [80].

The absorption maximum is centered at ~3.25 eV and shows a shoulder at 3.65 eV as shown in Figure 5.6. The photoluminescence emission spectrum shows a π*–π transition at ~2.85 eV and a

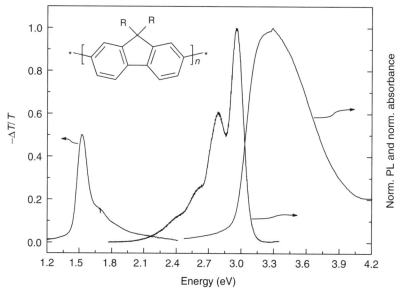

FIGURE 5.5 Absorption, photoluminescence emission, and photoinduced absorption spectrum of a typical polyfluorene-type polymer film. The inset shows the actual chemical structure with R typically being a branched or linear alkyl chain.

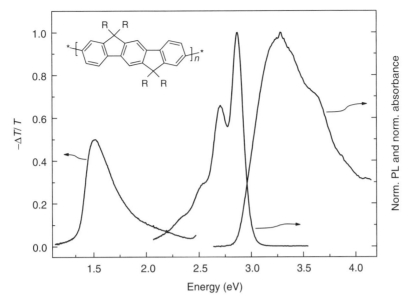

FIGURE 5.6 Absorption, photoluminescence emission, and photoinduced absorption spectrum of a typical polyindenofluorene-type polymer film. The inset shows the actual chemical structure with R being a linear alkyl chain (from Setayesh, S., Marsitzky, D., and Müllen, K., *Macromolecules*, 33, 2016, 2000) or an aryl-substituted group (from Jacob, J., Sax, S., Piok, T., List, E.J.W., Grimsdale, A.C., and Müllen, K., *J. Am. Chem. Soc.*, 126, 6987, 2004).

vibronic fine structure with an energetic spacing of ~180 meV. The absorbance and photoluminescence emission spectra for polymer films are very similar to the spectra observed in solution, however, exhibiting a bathochromic shift of ~60 meV. The 1^3B_u to a higher lying m^3A_g triplet state is located at 1.55 eV [113]. No explicit polaronic absorption feature is observed. Based on PIFs with aryl substituents on the methine bridges, efficient stable blue-emitting PLEDs have been fabricated [80].

5.4.1.4 Ladder-Type Polymer with Pentaphenyl Segments

In 2004, Müllen and Grimsdale and coworkers have successfully bridged the gap between the fully planar LPPP and the PIFs by introducing a ladder-type polymer with pentaphenyl segments (4 of 5 aryl–aryl linkages are bridged) [81].

As depicted in Figure 5.7, the absorption spectrum shows the typical structure of a ladder-type polymer. The spectrum exhibits its π–π* transition at 2.87 eV with a shoulder at 3.06 eV. The photoluminescence emission shows the first maximum at 2.8 eV and two vibronic replicas at 2.63 and 2.45 eV. The polymer shows a low stokes shift and a rather steep onset of emission and absorption. The bathochromic shift of the solid state spectra compared with the solution spectra is negligible. The transition from 1^3B_u to a higher lying m^3A_g triplet state is located at 1.34 eV and well-resolved polaronic absorption features are observed at 1.77 and 2 eV. The exceptionally low absolute value of the polaronic as well as the triplet absorption features in this polymer, which is attributed to the fact that the polymer possesses low energetic disorder in the bulk material makes this polymer an attractive candidate for PLEDs and solid state laser applications. Furthermore, similar to the PIFs Müllen and Grimsdale recently demonstrated that the polymer bearing aryl groups at all bridgehead positions show remarkably color-stable blue emission in PLEDs [119].

5.4.1.5 Methyl-Substituted Ladder-Type Poly(*para*-Phenylene)

As a consequence of its extraordinary properties MeLPPP and its derivatives are one of the most thoroughly studied conjugated polymers. In particular, the defined chemical structure and the rigidity of the backbone, giving rise to well-defined features in absorption and emission, make it an ideal candidate for spectroscopic studies. For this reason almost all available spectroscopic and electrical

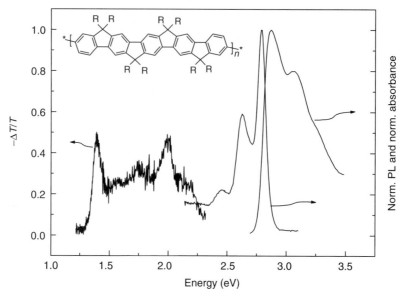

FIGURE 5.7 Absorption, photoluminescence emission, and photoinduced absorption spectrum of a polymer film of ladder-type polymer with pentaphenyl segments. The inset shows the actual chemical structure.

characterization methods have been applied to this polymer, yielding a quite complete picture of the photophysics of this polymer [93,113,120–122].

The photoluminescence spectrum of MeLPPP as depicted in Figure 5.8 is characterized by a steep onset at 2.69 eV and by the well-resolved vibrationally split maxima which are homologous to the excitation spectrum. The dominant photoluminescence maximum is only very slightly Stokes-shifted by ~35 meV, which is due to the ladder-type structure of the polymer hindering geometrical relaxations. The steep onset of the absorption spectrum reflects the high intrachain order of MeLPPP in the film.

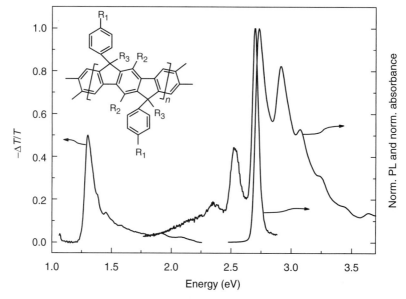

FIGURE 5.8 Absorption, photoluminescence emission, and photoinduced absorption spectrum of a fully planarized ladder-type poly(*para*-phenylene) polymer film. The inset shows the actual chemical structure.

The photoluminescence emission and absorption spectra of MeLPPP films and solutions are essentially identical [123,124] and the EL emission spectrum matches the photoluminescence spectrum. The photoluminescence quantum efficiency (η_{PL}) was found to be as high as 30% for pure MeLPPP [125]. The photoinduced absorption spectrum of MeLPPP exhibits two distinct features, the triplet–triplet transition from 1^3B_u to m^3A_g at 1.3 eV and the polaron band at ~1.9 eV. The ~2.1 eV band is a vibronic replica of the ~1.9 eV band. These assignments were deduced from a comparison of the PA with doping- and charge-induced absorption spectra, which only yielded the 1.9 eV bands [126]. Due to the high localization of the TE wavefunction, the triplet band only shows a vibronic progression of 80 meV [127].

The increasing degree of planarization is accompanied by an increasing conjugative interaction of the aromatic subunits as seen by a lowering of the optical bandgap in the absorption as well as for the emission spectra. A comparison of the shape of the absorption and emission spectra shows that for PPP and MeLPPP the absorption and emission are a mirror image of each other while for PF and PIF this symmetry is severely broken. As recently demonstrated by Heimel [128], the violation of the mirror image symmetry in the optical spectra of PPP-type model molecules has its origin in the shape and nature of the torsional potentials of the different molecules and is not so much related to a migration processes of the exciton as explained in detail in this paper.

5.4.1.6 CIE-Position and Shape of Emission Spectrum

The ideal blue light-emitting conjugated polymer should emit light only in the blue visible range with an emission spectrum stretching from 435 to 480 nm without any tail beyond 480 nm. This way the emission yields an almost pure blue color with CIE color values of $x < 0.20$ and $y < 0.15$ according to the CIE standard 1930. Utilizing PPP-type polymers, the task of optimizing the optical bandgap to the eye's sensitivity curve can be established by introducing a number of planarizing bridges to the PPP polymer backbone improving the π-electron overlap as a consequence of the suppressed twisting of the phenyl rings as can be seen in Figure 5.9.

The emission spectrum of PPP does not overlap with the sensitivity of the human eye so that it may only be used as the matrix for other emitting dyes. PF shows a low overlap so that it is already better

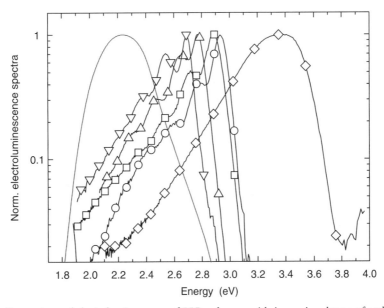

FIGURE 5.9 Comparison of electroluminescence of PPP polymers with increasing degree of aryl–aryl bridging with the eye-sensitize curve (full line with no symbols). Poly(*para*-phenylene) (diamonds), polyfluorene (squares), polyindenofluorenes (circles), poly(*para*-phenylene) with bridged pentaphenyl units (triangles up), and fully planarized ladder-type poly(*para*-phenylene) (triangles down).

suited for the human eye. For PIF the overlap becomes already significant. An analysis shows a level almost twice as high as for the peak of the emission of PF. The fact that both polymers give rise to a very similar color impression to the human eye is also well documented in the typical CIE color coordinates, where for PF one typically finds $x = 0.17$ and $y = 0.14$, for PIF $x = 0.18$ and $y = 0.13$. On the other hand the emission spectrum of MeLPPP gives rise to a blue–green emission with CIE coordinates at $x = 0.20$ and $y = 0.27$, which is already too far in the green spectral region to be utilized in an RGB display. As mentioned above, the ladder-type polymer with pentaphenyl segments offers the right overlap at PI-electrons giving rise to bright blue emission optimal for the human eye with a spectrum having CIE coordinates at $x = 0.17$ and $y = 0.09$. From an "energetic" point of view the ladder-type polymer with pentaphenyl segments provides an optimal solution. However, in terms of an industrial applicability these benefits have to hold up against the increased cost for the advanced synthesis.

5.5 Defect Formation in PPP-Type Light-Emitting Polymers—Low-Energy Electro-and Photoluminescence Bands and Their Origin in Carbonyl-Type Defects

As documented in numerous studies on device degradation behavior both the organic layers [129,130] and the low work function metal electrodes [131–134] were found to be prone to degradation upon exposure to oxygen or water, leading to the introduction of chemical defects into the conjugated backbone or metal electrode decomposition and ultimately delamination.

In addition to studies of overall device stability issues (e.g., current density, brightness or luminance, dark spot formation, and electrode delamination), the necessity for a high degree of color stability has come to the fore as a crucial parameter of device reliability. This is especially true for blue light-emitting devices. To date, the realization of stable blue PLEDs remains hindered by several problems. Specifically, the family of blue-emitting PPP-type polymers, especially PFs, is prone to oxidation resulting in a green emission band located at 2.2–2.3 eV as a consequence of a bulk polymer defect.

The origin of these low-energy photoluminescence and EL peaks was firstly attributed to an aggregate or excimer emission [77,135–138]. This primary assignment led to various synthetic efforts, such as copolymer generation, attachment of dendronic substituents [117,139,140], endcapping [136], and spiro-linking [141] in order to modify and stabilize the bulk emission properties. Frequently, the observed improvements of the bulk PL properties have then been related to the variations of the chemical structure of single polymer chains.

However, recently the low-energy emission bands in PF-type polymers were identified as the emission from an exciton and/or charge trapping on-chain emissive keto defect [16–19,142,143]. It was shown that by incorporating keto defects into the polymer backbone the emissive characteristics of oxidatively degraded PFs, especially the appearance of the 530 nm band (~2.3 eV), can be faithfully reproduced [20]. The absence of any concentration dependence of the 530 nm bands' intensity in solution as well as the appearance of a vibronic structure in the 530 nm band at low temperatures support the picture of an emissive on-chain defect as the origin of this particular spectral feature [144]. Efficient energy migration to these defects and the localization of the emissive state as illustrated earlier in this chapter, which is strongly confined to the fluorenone unit, lead to a significant modification of the optical properties of PF films already at low defect concentrations [18]. Despite the fact that there is rather strong experimental evidence for a scenario, where the keto groups in the PF emit the unwanted low-energy emission at 2.3 eV, there is still an ongoing discussion on this topic in literature [145–147]. However, Lupton and coworkers published a paper recently where they showed on-chain fluorenone defect emission from single PF molecules in the absence of intermolecular interactions using single molecule spectroscopy, which does not leave much room for an alternative, but the here given explanation [148].

Most wide bandgap polymers utilized in PLEDs require low work function cathode metals (e.g., Ca or Mg/Al) to provide proper injection of electrons. As shown recently, chemical reactions of these

PPP-linked polymers lead to defects on the polymer–metal interface that also shift the emission color under device operation [111,149,150].

5.5.1 Ketonic Defects Related to Polymer Synthesis

In contrast to the on-chain phosphorescent palladium centers, which are accidentally incorporated into the polymer backbone by a transmetallation-type side reaction during the synthesis, the formation of keto-type defects in PF derivates is found to be an oxidative degradation process taking place already during synthesis as a consequence of the chemical structure of the monomer units.

This formation of keto-type defects during the synthesis of the polymer can be best illustrated by a comparison of two derivatives of PF: (a) the 9-monoalkylated (MA-PF) and (b) the 9,9-dialkylated derivative (DA-PF). From the investigations it was found that MA-PF degrades much more rapidly than DA-PF. That can be attributed to the reduced chemical stability of MA-PF due to the presence of the relatively weakly bound, acidic methylene bridge hydrogen. Oxidative transformation of the 9-alkyl-fluorene moiety leads to the 9-fluorenone defect sites as depicted in Figure 5.10. Based on infrared (IR)-spectroscopic data it is proposed that the (photo)oxidative formation of fluorenone defects is responsible for the occurrence of the low-energy PL contributions. Such oxidative degradation processes are long known for nonconjugated polymers as "aliphatic CH (photo)oxidation" and represent a well-established pathway of polymer degradation. The CH-active functions present in the MA-PFs facilitate this type of degradation process. The formation of keto defects already happens during the work-up of the coupling mixtures under ambient conditions. It has been proposed that already trace amounts of only monoalkylated monomers lead to the formation of fluorenone defects in the corresponding polymer. The highly active Ni[0]-species used in the reductive coupling of the dibromo monomers reduce a certain amount of the 9-monoalkylated fluorene building blocks (I) to (aromatic) fluorenyl anions (II) under formation of hydrogen. These anions can form hydroperoxide anions (III) with atmospheric oxygen during the work-up of the reaction mixture. The hydroperoxide anions then undergo a final rearrangement to fluorenone moieties (IV).

As depicted in Figure 5.10b the incorporation of the 9-fluorenone defect sites in MA-PF dramatically changes the emission properties in photoluminescence of the polymer backbone as compared with the photoluminescence of defect-free pristine DA-PF.

In dilute solution MA-PF exhibits a broad absorption peak at 3.3 eV, the PL emission spectrum shows maxima at 2.95 and 2.8 eV as well as the low-energy band peaking at 2.3 eV. In the solid state, the

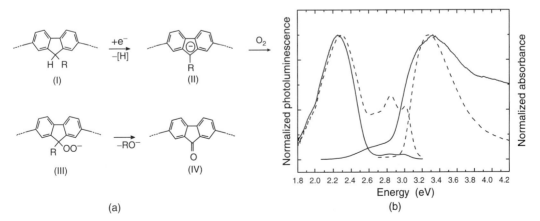

(a)   (b)

FIGURE 5.10 (a) Proposed mechanism for the generation of keto defect sites in MA-PF (for explanations see text. (b) Absorption and photoluminescence emission of MA-PF in dilute solution (dotted line) and in the solid state (full line). (Modified from Scherf, U. and List, E.J.W., *Adv. Mater.*, 14, 477, 2002.)

absorption of MA-PF becomes much broader and an additional weak contribution centered at $\sim$2.8 eV occurs. The solid state PL spectrum of MA-PF is dominated by the low-energy emission peak at $\sim$2.28 eV. In the IR absorption spectrum of MA-PF one observes an additional IR band at $\sim$1721 cm^{-1} already in pristine samples, while no such signal is detectable in pristine DA-PF. Since this band is an IR fingerprint of a carbonyl-stretching mode ($>$C$=$O) of the fluorenone building block [151] this demonstrates the clear correlation between the presence of 9-fluorenone defects and the presence of the low-energy emission band at 2.3 eV [152]. The formation of keto-type defects is, however, not limited to the PF-type polymers. Further evidence for this process also taking place in LPPPs is found when the methyl group in MeLPPP is replaced by hydrogen [71]. There the polymer with the hydrogen displays a broad emission band between 2.0 and 2.5 eV, that can be attributed to on-chain keto-type emission. From this point of view the defect formation mechanism presented for PF has to be extended to all bridged PPP-type polymers.

5.5.2 Oxidative Degradation of PF-Type Polymers

Regarding the stability of PPPs and PFs of different chemical structures, significant differences are observed with respect to the formation of ketonic defects as well as their impact on the spectral degradation. This has already been discussed above where mono- and dialkylated PFs are compared. This section will give a brief example of the degradation of three different DA-PFs with different substituents at the 9-position of the fluorene units, respectively, and a certain amount of spiro-links between the polymer chains. It will show that the obtained results can be reasonably explained within the theory laid out above for the degradation taking place during synthesis. For this review, these three different polymers have been chosen to compare the degradation taking place in the PF and elucidate the influence of spiro(fluorene) branches and triphenylamine endcappers for the first time under identical conditions, which has not been reported before.

The investigated polymers are poly(9,9 dialkylfluorene) with two 3,7,11-trimethyldodecyl side chains (PF 111/12) ($M_n = 80,000$ g/mol and $M_w = 197,400$ g/mol), a slightly branched PF derivative with ethylhexyl side chains containing 0.2 mol% spiro(fluorene) branches (sp-PF-1) (0.2 mol% branching agent, $M_n = 14,000$ g/mol, $M_w = 390,000$ g/mol, $D = 2.8$), and a similar branched PF derivative with triphenylamine endcappers (sp-PF-2) (1 mol% branching agent, 3 mol% endcapper, $M_n = 75,500$ g/mol, $M_w = 325,300$ g/mol, $D = 4.3$). Figure 5.11 shows the chemical structure and the PL and absorption spectra, while Figure 5.12 depicts the evolution of the solid state PL spectra of the three different PFs upon consecutive heating in air for 1 h at the specified temperatures, respectively. The inserts display the relevant region of the corresponding IR absorption spectra. While the blue fluorene emission at $\sim$2.95 and $\sim$2.8 eV initially increases, which is attributed to the removal of residual solvent and ordering effects, respectively, it decreases significantly (most strongly for PF 111/12) after the samples are heated to higher temperatures (166°C and 200°C, respectively). At the same time the broad emission band around 2.3 eV evolves. Although a relative increase of this band firstly becomes visible in sp-PF-2 and sp-PF-1 already after heating to 133°C, its intensity relative to the fluorene emission grows much faster in PF 111/12 upon heating at higher temperature. After the full stress cycle (1 h at 66°C, 100°C, 133°C, 166°C, and 200°C, respectively) peak height ratios between the degradation-induced peak and the first vibronic $\pi^*-\pi$ transition of 15.8:1 for PF 111/12, 5.4:1 for sp-PF-1, and 2.7:1 for sp-PF-2 are obtained.

For all three polymers the appearance of the green emission at 2.3 eV is accompanied by the appearance of new IR features that are related to keto defect sites as discussed above. The bandwidth of the induced IR features is larger for the PF 111/12 than for the other polymers. This is an indication that in this polymer also a considerable number of defects other than the keto at the 9-position of the fluorene unit are created. The discussed IR bands clearly appear in the PF 111/12 film after heating to temperatures $\geq$166°C for 1 h. Further heating to 200°C only slightly increases the peak height of the 1718 cm^{-1} band in this polymer while sp-PF-1 and sp-PF-2 require prolonged heating (6 h) at 200°C until the new IR features can be detected (Figure 5.13).

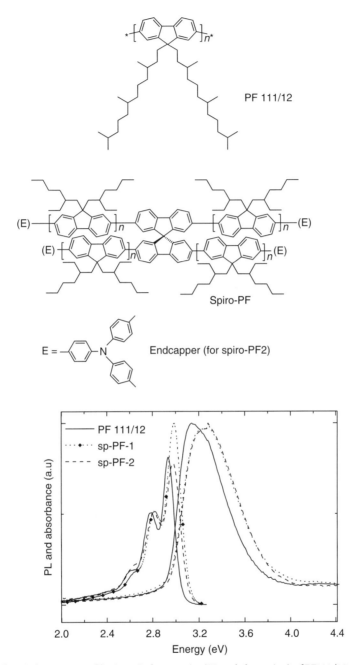

FIGURE 5.11 Chemical structure and basic optical properties (PL and absorption) of PF111/12, spiro-PF (sp-PF-1), and endcapped spiro-PF (sp-PF-2).

The overall quantum yield of the $\pi^*-\pi$ emission is finally reduced in all polymers. The highest reduction of the quantum yield is found for PF 111/12 and the lowest for sp-PF-2. In all polymers, an absolute increase of the keto emission, not only a relative increase with respect to the PF emission, is observed for moderate heating times (not more than 1 h at 200°C). For the spiro-type polymers this true increase is observed for excitation within the PF backbone absorption peak (370 nm), while it was only found for direct excitation of the keto defects with excitation energies lower than the optical bandgap of

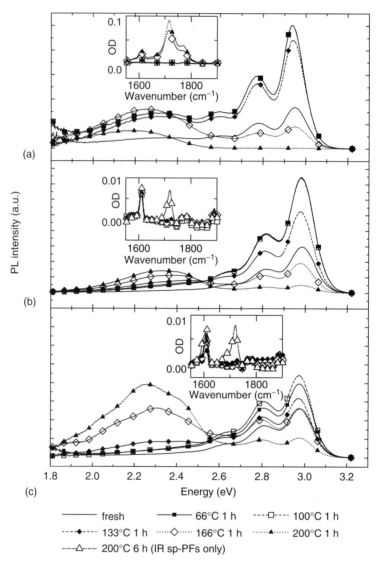

FIGURE 5.12   PL spectra of (a) PF111/12, (b) spiro-PF (sp-PF-1), and (c) endcapped spiro-PF (sp-PF-2) films in a thermal degradation experiment in air, the inserts show the corresponding IR absorption spectra (optical density).

the pristine material in PF 111/12. These observations are explained within the picture of excitation energy migration from the fluorene units to the emissive fluorenone sites. As the degraded PF 111/12 contains a considerably higher amount of nonemissive defects than the degraded spiro-PFs which is proven by the broader IR absorption peak at 1718 cm^{-1} and the lower overall quantum yield, a larger part of the SEs recombines nonradiatively before reaching an emissive fluorenone unit, thus affecting the lower fluorenone concentration. Direct excitation reduces the effective length of the migration path of the excitons and therefore the directly excited PL gives a better measure for the amount of emissive defects created.

The better spectral stability of sp-PF-2 as compared with sp-PF-1 may be attributed to a less effective exciton migration to the fluorenone defects. As shown in the inserts of Figure 5.12b and Figure 5.12c the amount of defects created in the two spiro-type materials is similar, if not even more defects are created in sp-PF-2. Nevertheless, sp-PF-1 shows an almost twofold relative increase in keto emission after

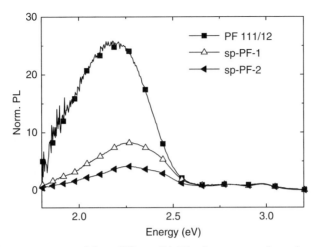

FIGURE 5.13 Solid state PL spectra of three different DA-PFs after common thermal treatment at 210°C. Data normalized to first vibronic progression of the PF backbone emission around 2.8 eV.

heating. Considering the higher amount of branching agent (1 mol% in sp-PF-2, compared with 0.2 mol% in sp-PF-1) and the additional presence of endcapping molecules lower order in sp-PF-2 and a related aggravated Förster-type excitation energy migration process in sp-PF-2 could be one of the reasons for the higher spectral stability of this material. It should be mentioned that this is in contrast to what has been observed before in literature [153], where a clear improvement in stability was observed for PF with triphenylamine endcappers as compared with regular uncapped PF.

Note that the results presented in this chapter, in particular the different stability with respect to the formation of keto defects, might not only be influenced by the particular chemical structure of the polymers, but also by the particular chemical purity of the used polymers.

5.5.3 Spectral Degradation of PPP-Based Light-Emitting Devices

Although improvements concerning color purity and device stability of PF-based PLEDs have recently been made the strict requirements for commercialization, demanding tens of thousands of hours of operation, are still an unreached goal for blue light-emitting PLEDs. While the EL spectra of unstressed PF-based light-emitting devices show very similar bands as observed in photoluminescence, prolonged operation of the devices frequently leads to the appearance of low-energy emission bands, i.e., to a change of the emission color from the desired blue to a greenish as shown in Figure 5.14.

A common mode of device degradation in PF-based PLEDs is the formation of keto defects during device operation and the related change of the emission spectrum with a broad peak emerging around 2.3 eV. The following discussion will, however, not only focus on bulk polymer defects, but also on the observation of additional distinct spectral features in EL of PLEDs as well as in photoluminescence of Ca/Al electrode covered films fabricated from PF derivatives. The emission of the new bands, peaking around 2.45 resp. 2.6 eV (480 and 506 nm, respectively), which has been observed but for long not *identified* in literature [32,137] as a distinct feature, is energetically located between the emission bands of pristine PF at 2.9 eV (416 nm) and the broad keto-defect related emission band at 2.3 eV (530 nm). It only emerges upon coverage of the polymer with metals as Al or Ca, e.g., and is attributed to the formation of chemical defects in the polymer in a chemical degradation reaction caused by the low work function metal electrodes.

5.5.3.1 Bulk Effects Emerging under Device Operation

As shown in Figure 5.15 in the case of PF-type polymers containing keto defects (e.g., MA-PF), a low-energy EL emission appears as a broad feature peaking at around 2.3 eV which is assigned to bulk keto

FIGURE 5.14 EL of PF-based PLEDs; left: pristine emission, right: emission after electrical stress (standard device operation).

defects as discussed above. The same emission may also be observed with a variety of presumably more stable DA-PFs where the intensity is influenced by a variety of device- and fabrication-related parameters Typically, the intensity of the defect emission increases with the ongoing operation time. Especially for the polymers containing an intrinsic amount of keto defects (MA-PF, Fluorene–fluorenone copolymers), the relative spectral contribution of the keto emission in EL is considerably stronger than in PL. This was initially attributed to the fact that the keto defects act as a trap for charge carriers, thus leading to the increased exciton localization at the defect sites. On the other hand, ongoing investigations, including device operation at very low temperatures have shown that the PL emission at 2.3 eV vanishes when the

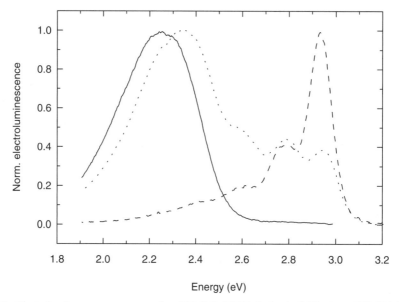

FIGURE 5.15 Electroluminescence spectrum of an ITO/MA-PF/Al device (solid line), an ITO/DA-PF/Al device (dashed line), and the latter device (dotted line) after 60 min continuous operation under air (electrooxidative degeneration). (Modified from List, E.J.W., Güntner, R., Scandiucci de Freitas, P., and Scherf, U., *Adv. Mater.*, 14, 374, 2002.)

temperature is reduced [154]. This behavior is not straightforwardly explained when charge trapping is held responsible for the enhanced defect emission in PF-PLEDs compared with PL results. In fact this explanation would now require the assumption that charge trapping on the keto sites occurs over an activation barrier that can only be overcome by thermal activation.

5.5.3.2 Degradation of the Polymer at the Cathode Interface

In contrast to the anode of the PLED, where a high work function electrode such as O_2–plasma-treated ITO or poly(3,4-ethylenedioxithiophene) doped with poly(styrenesulfonate) (PEDOT/PSS) is necessary to match the HOMO level of the polymer, the realization of a good electron-injecting contact in wide bandgap blue-emitting PPP-type polymers requires the use of low work function materials. Such materials as Ca or alkali metals have their Fermi levels close to the LUMO level and, therefore, no or only minor energy barriers have to be overcome when electrons are injected. Electrode–polymer interfaces of PLEDs made of PPP-type emitters and low work function metals have been thoroughly investigated by photoelectron spectroscopy (XPS and UPS) [155–157]. For PFO, a significant change of the electronic structure of the polymer with a broadening of the HOMO and LUMO as well as bipolaron states at the interface with Ca or alkali metal electrodes have been detected [149,156,157]. The thickness of the calcium layer and the calcium concentration within the interface region, as investigated by time of flight secondary ion mass spectroscopy (TOF-SIMS), strongly influence the EL efficiency through Ca-induced quenching of excitons [157]. Regarding the influence of oxygen at the Ca–PFO interface, small amounts of oxygen are described to reduce the number of gap states while a high amount of oxygen alters the original LUMO position due to changes of the chemical structure of the polymer.

As mentioned above, additional low-energy spectral features in PLEDs are often observed upon device operation that are energetically located between the "regular" blue emission bands at 2.9 eV and the broad keto-related emission band at 2.3 eV, i.e., between 2.45 and 2.6 eV. These emission features appear most strongly for devices with Ca electrodes. In the following it will be shown that the origin of the spectral features is located close to the cathode of the device and most probably related to a chemical degradation reaction caused by the low work function metal electrodes.

Figure 5.16a shows the EL spectra of ITO/PEDOT:PSS/sp-PF-1/Ca/Al PLEDs over time and for different treatments. The emission spectrum of the pristine device based on sp-PF-1 shows the typical PF emission with its maximum at 2.95 eV (419 nm) and the corresponding vibronic replica around 2.79 eV (444 nm). Within 30 s of operation in argon, the blue EL spectrum turns greenish white and additional peaks at 2.6 and 2.45 eV emerge. Further degradation of the devices by heating them in air (or taking less care to exclude oxygen and humidity, respectively in the preparation process) leads to the appearance of the already identified emission of bulk keto defects at 2.3 eV as depicted in Figure 5.16a. This behavior, i.e., the formation of the two bands at 2.45 and 2.6 eV has been observed for a variety of device configurations of PLEDs comprising Ca/Al or simple Al electrodes, and different PF-type materials such as PF 111/12 and sp-PF-2.

Devices where the top electrodes were prepared in the same evaporation step show very similar degradation behavior regarding the shape and magnitude of the 2.45–2.6 eV emission independent of the particular PF-type polymer used as the active material. As depicted in Figure 5.16b, the feature at 2.45–2.6 eV can be reversibly reduced as the driving voltage under pulsed operation is increased. The spectral change of the EL with driving voltage is explained by a shift of the recombination zone into a device region with different optical properties, i.e., with a reduced number of emissive defects [158] as sketched in Figure 5.17. Self-absorption-induced spectral shifts can be excluded since the higher energy PF emission at 2.95 eV, which has much stronger overlap with the PF absorption, is not affected by the bias change. This strongly indicates that the 2.45–2.6 eV emission bands stem from an emissive defect located near the cathode interface of the device. To test this hypothesis, thin PF films of approximately 20 and 4 nm, respectively, have been analyzed in Figure 5.16 before and after the evaporation of the Ca/Al electrode, which is shown in Figure 5.16c. Such thin films were chosen to increase the relative spectral contribution of the interface-associated emissive features compared with the bulk PF emission

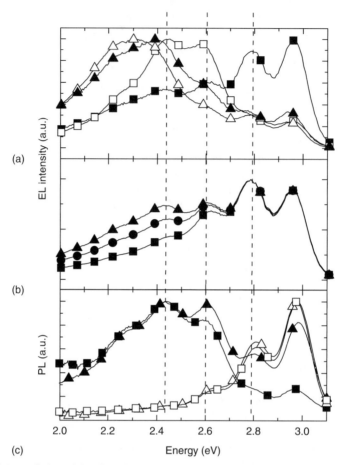

FIGURE 5.16 (a) Degradation of the electroluminescence spectrum of a sp-PF-1 PLED. Filled squares: pristine device emission; open squares: after operation in argon atmosphere; filled triangles: after operation in argon and storage in air (45 min at 150°C); open triangles: device prepared without proper exclusion of oxygen. (b) Voltage dependence of the electroluminescence: triangles: 5.3 V; circles: 6 V pulsed (100 kHz, 50% duty cycle); squares: 50 V pulsed (100 kHz, 5% duty cycle). (c) PL measurements of very thin (triangles: approximately 20 nm; squares: approximately 4 nm) films before (open symbols) and after (filled symbols) deposition of Ca/Al electrode.

as the 2.45–2.6 eV emission could not be observed in the photoluminescence of affected devices where the film thickness was in the order of 100 nm. Clearly after application of the electrode, the 4 nm-film displays an almost twofold increase of the relative contribution from the emission feature at 2.45–2.6 eV as compared with the 20 nm film which is consistent with the assignment that the novel emission band is indeed caused by the defects at the metal–polymer interface. A comparison clearly shows that the emission of bulk keto-related defects at 2.3 eV exhibits a different spectral signature than the interface-associated defects at 2.45–2.6 eV.

To clarify the nature of the novel 2.45–2.6 eV band, several electrode configurations such as evaporated Al or Ca/Al and sputtered Au layers on PF films as well as thin PF films on metal substrates (Al resp. Au) have been investigated [150]. The novel PL emission band was most strongly observed for Ca evaporated onto PF and only moderately for Al onto PF. No effect was observed for Au on PF indicating that the reduction potential of the metal is related to the observation. Except for a reduced PL intensity (PL quenching at the metal interface) the PF films on metal substrates showed nearly identical PL spectra as films on glass substrates. Obviously, not simply the presence of the metal, but the metal deposition process and resulting degradation reactions are responsible for the spectral changes.

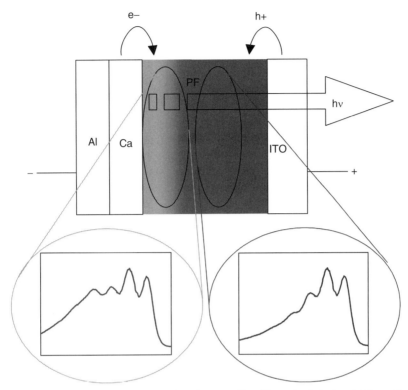

FIGURE 5.17   Schematic EL emission from a PF-based PLED affected by cathode-induced defects. Green emission at low driving voltage stems from a recombination zone close to the defects; blue emission is obtained from a recombination zone further in the bulk, when higher driving voltage is used.

In conclusion, the presented results clearly demonstrate that emissive defects located at the polymer–cathode interface are created in PF-based PLEDs during electrode deposition. The interface defects which emit at 2.45–2.6 eV are related to the degradation of the polymer in interplay of residual atmospheric species, most likely oxygen or water, and the low work function electrode. According to their spectral signature the interface defects are not directly related to the well-described keto-type bulk defects. Thus, having identified two independent defects, aside from the use of keto-free, structurally pure polymers proper interface control, optimized electrode deposition parameters and the use of suited electron transport, hole blocking, or protection layers may in future help to improve the spectral stability and enhance the lifetime of blue light-emitting PF-based PLEDs.

5.6 Outlook

After 15 y, the complete series of PPP-type structures with increasing degree of bridging is now available. All these polymers show efficient UV blue or blue–green emission in solution with an optical bandgap energy becoming continuously smaller with increasing degree of planarization, however, most of these polymers suffer from one or the other chemical instability.

The occurrence of low-energy PL bands of "unknown" origin in solid state PF samples is known for a while now and based on spectroscopic data it was proposed that the (photo)oxidative formation of fluorenone defects is responsible for the occurrence of the low-energy PL contributions. Such oxidative degradation processes are long known for nonconjugated polymers as "aliphatic CH (photo)oxidation" and represent a well-established pathway of polymer degradation. The CH-active functions present in the MA-PFs facilitate this type of degradation process. In accordance with this assumption, Meijer and

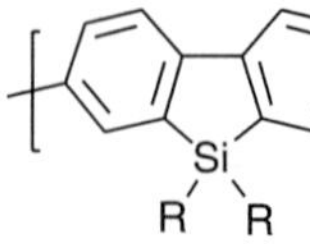

SCHEME 5.12   Structure of >SiR$_2$-bridged polyfluorene analogs.

coworkers published a sophisticated procedure for the purification of the 9,9-dialkylfluorene monomers toward defect-free poly(9,9-dialkylfluorene-2,7-diyl)s with a reduced low-energy emission feature in the solid state [159]. Devices made from this PFs do not exhibit the undesired green emission even after extensive electrical stress. An alternative approach has been suggested by Holmes. Silylene bridges have been very recently used to prepare the corresponding >SiR$_2$-bridged monomers and polymers as novel PF-analogs as depicted in Scheme 5.12 [160] with improved thermal and spectral stability.

As an alternative to stabilize PPP-type polymers with aryl–aryl methylene bridges the use of spiro-type PFs has attracted attention as an efficient way toward stable blue emitters [161]. Also, PF derivatives with bulky side chains, aggravating the exciton migration to the defect sites, have been reported to significantly improve the color stability [139,140]. There it was shown that shielding a PF backbone with Müllen-type dendronic side chains drastically reduces the interchain interactions and thus exciton migration to defect sites or emission from ketonic defect sites, while retaining good charge transport properties. Most recently Müllen et al. [119] also demonstrated that shielding the bridgeheads by full substitution with aryl groups shows remarkably improved resistance to oxidative degradation at the 9-position, which confirms that improving the chemical stability of the bridgeheads in bridged PPP-type polymers is one of the key steps toward stable blue-emitting PPP-type polymers.

Acknowledgments

The presented experiments have been sponsored and supported by the Christian Doppler Forschungs-gesellschaft, Sonderforschungsbereich "Elektroaktive Stoffe," the Fonds zur Förderung der wissenschaf-tlichen Forschung—Austria (P 12806-PHY), the Austrian Nanoinitiative (IOSOTEC-SENSPYS N702), SONY International (Europe), Stuttgart, Germany, StiftungVolkswagenwerk, Hannover, Germany, and the Max-Planck-Gesellschaft, München, Germany. CDL-AFM is an important part of the long-term research strategies of AT & S. We would like to thank the team of coworkers and collaborators who have been involved in the interdisciplinary research projects on semiconducting PFs, especially Stefan Gamerith, Josemon Jacobs, Lorenz Romaner, Martin Gaal, Heinz-Georg Nothofer, Roland Güntner, Michael Forster, Patricia Scandiucci de Freitas, Günther Lieser, Akio Yasuda, Gabi Nelles, Dieter Neher, Tzenka Miteva, Dessislava Sainova, Alexander F. Pogantsch, Franz P. Wenzl, Andrew C. Grimsdale, Egbert Zojer, Günther Leising, and Jean Luc Brédas. We also wish to thank Klaus Müllen, Mainz for their cooperation as well as continuous and generous support of our investigations.

References

1. Chiang, C.K., C.R. Fincher, Y.W. Park, A.J. Heeger, H. Shirakawa, E.J. Louis, S.C. Gau, and A.G. MacDiarmid. 1977. *Phys Rev Lett* 39:1098.
2. Pope, M. and C.E. Swenberg. 1998. *Electronic processes in organic crystals.* New York: Oxford University Press.
3. Bredas, J.L. and R. Silbey, eds. 1991. *Conjugated polymers.* Dordrecht, The Netherlands: Kluwer.
4. (a) Friend, R.H., R.W. Gymer, A.B. Holmes, J.H. Burroughes, R.N. Marks, C. Taliani, D.D.C. Bradley, D.A. Dos Santos, J.L. Brédas, M. Lögdlund, and W.R. Salaneck. 2000. *Nature* 397:121. (b) Burroughes, J.H., D.D.C. Bradley, A.R. Brown, R.N. Marks, K. Mackay, R.H. Friend, P. L Burn, and A.B. Holmes. 1990. *Nature* 347:539.

5. (a) Pei, Q., G. Yu, C. Zhang, Y. Yang, and A.J. Heeger. 1995. *Science* 269:1086. (b) Holzer, L., F. Wenzl, S. Tasch, G. Leising, B. Winkler, L. Dai, and A. Mau. 1997. *Appl Phys Lett* 75:2014.

6. (a) Sariciftci, N.S., L. Smilowitz, A.J. Heeger, and F. Wudl. 1992. *Science* 258:1474. (b) Granstrom, M., K. Petritsch, A.C. Arias, A. Lux, M.R. Andersson, and R.H. Friend. 1998. *Nature* 395:257.

7. Yu, G., J. Gao, J.C. Hummelen, F. Wudl, and A. Heeger. 1995. *Science* 270:1789.

8. Stagira, S., M. Zavelani-Rossi, M. Nisoli, S. DeSilvestri, G. Lanzani, C. Zenz, P. Mataloni, and G. Leising. 1998. *Appl Phys Lett* 73:2860.

9. Garnier, F., R. Hajlaoui, A. Yassar, and P. Srivastava. 1994. *Science* 265:1684.

10. Ho, P.K.H., D.S. Thomas, R.H. Friend, and N. Tessler. 1999. *Science* 285:233.

11. Yang, Y., S.C. Chang, J. Bharathan, and J. Liu. 2000. *J Mater Sci: Mater Electron* 11:89.

12. Sturm, J.C., F. Pschenitzka, T.R. Hebner, M.H. Lu, and S. Troian. 1999. *Proc SPIE* 3797:266.

13. Hebner, T.R. and J.C. Sturm. 1998. *Appl Phys Lett* 73:1775.

14. Sirringhaus, H., T. Kawase, R.H. Friend, T. Shimoda, M. Inbasekaran, W. Wu, and E.P. Woo. 2000. *Science* 290:2123.

15. Lupton, J.M., A. Pogantsch, T. Piok, E.J.W. List, S. Patil, and U. Scherf. 2002. *Phys Rev Lett* 89:167401.

16. List, E.J.W., R. Güntner, P. Scandiucci de Freitas, and U. Scherf. 2002. *Adv Mater* 14:374.

17. Scherf, U. and E.J.W. List. 2002. *Adv Mater* 14:477.

18. Zojer, E., A. Pogantsch, E. Hennebicq, D. Beljonne, J.L. Brédas, P. Scandiucci de Freitas, U. Scherf, and E.J.W. List. 2002. *J Chem Phys* 117:6794.

19. Lupton, J.M., M.R. Craig, and E.W. Meijer. 2002. *Appl Phys Lett* 80:4489.

20. Scandiucci de Freitas, P., U. Scherf, M. Collon, and E.J.W. List. 2002. *e-polymers* no. 0009.

21. Chua, L.-L., J. Zaumseil, J.-F. Chang, E.C.-W. Ou, P.K.-H. Ho, H. Sirringhaus, and R.H. Friend. 2005. *Nature* 434:194.

22. Tang, C.W. and S.A. Van Slyke. 1987. *Appl Phys Lett* 51:913.

23. Braun, D. and A.J. Heeger. 1991. *Appl Phys Lett* 58:1982.

24. Grem, G., G. Leditzky, B. Ullrich, and G. Leising. 1992. *Synth Met* 51:383.

25. Grem, G., G. Leditzky, B. Ulrich, and G. Leising. 1992. *Adv Mater* 4:15.

26. Grüner, J., P.J. Hamer, R.H. Friend, H.J. Huber, U. Scherf, and A.B. Holmes. 1994. *Adv Mater* 6:748.

27. Tasch, S., E.J.W. List, C. Hochfilzer, G. Leising, P. Schlichting, U. Rohr, Y. Geerts, U. Scherf, and K. Müllen. 1997. *Phys Rev B* 56:4479.

28. Tasch, S., C. Brandstätter, F. Meghdadi, G. Leising, L. Athouel, and G. Froyer. 1997. *Adv Mater* 9:33.

29. Fukuda, M., K. Sawada, and K. Yoshino. 1989. *Jpn J Appl Phys* 28:L1433.

30. Fukuda, M., K. Sawada, and K.J. Yoshino. 1993. *Polym Sci A* 31:2465.

31. Ohmori, Y., M. Uchida, K. Muro, and K. Yoshino. 1991. *Jpn J Appl Phys* 11B:L1941.

32. Grimsdale, A.C., P. Leclère, R. Lazzaroni, J.D. Mackenzie, C. Murphy, S. Setayesh, C. Silva, R.H. Friend, and K. Müllen. 2002. *Adv Funct Mater* 12:729.

33. Campbell, A.J., D.D.C. Bradley, H. Antoniadis, M. Inbasekaran, W.W. Wu, and E.P. Wu. 2000. *Appl Phys Lett* 76:1734.

34. Campbell, A.J., D.D.C. Bradley, and H. Antoniadis. 2001. *J Appl Phys* 89:3343.

35. Neher, D. 2001. *Macromol Rapid Commun* 22:1365.

36. Gamerith, S., Ch. Gadermaier, U. Scherf, E.J.W. List. 2005. *Physics of organic semiconductors*, ed. H. Bütting. Weinheim: Wiley-VCH, p. 153.

37. Gaylord, B.S., A.J. Heeger, and G.C. Bazan. *Proc Natl Acad Sci USA* 99:10954.

38. Dimitrakopoulos, C.D. and P.R.L. Malenfant. 2002. *Adv Mater* 14:99.

39. Ramsdale, C.M., J.A. Barker, A.C. Arias, J.D. McKenzie, R.H. Friend, and N.C. Greenham. 2002. *J Appl Phys* 92:4266.

40. Kietzke, T., D. Neher, M. Kumke, R. Montenegro, K. Landfester, and U. Scherf, 2004. *Macromolecules* 37:4882.

41. Ballard, D.G.H., A. Courtis, I.M. Shirley, and S.C. Taylor. 1983. *J Chem Soc Chem Commun* 17:954.

42. Ballard, D.G.H., A. Curtis, I.M. Shirley, and S.C. Taylor. 1987. *Macromolecules* 21:1787.

43. Tabata, M., M. Satoh, K. Kaneto, and K. Yoshino. 1986. *J Phys C* 19:L101.
44. Rehahn, M., A.-D. Schlüter, G. Wegner, and W.J. Feast. 1989. *Polymer* 30:1054.
45. Rehahn, M., A.-D. Schlüter, G. Wegner, and W.J. Feast. 1989. *Polymer* 30:1060.
46. Chaturvedi, V., S. Tanaka, and K. Kaeriyama. 1993. *Macromolecules* 26:2607.
47. Kanbara, T., N. Saito, T. Yamamoto, and K. Kubota. 1991. *Macromolecules* 24:5883.
48. Yamamoto, T., A. Morita, Y. Miyazaki, T. Maruyama, H. Wakayama, Z. Zhou, Y Nakumura, T. Kanbara, S. Sasaki, and K. Kubota. 1992. *Macromolecules* 25:1214.
49. Yamamoto, T., Y. Hayashi, and A. Yamamoto. 1978. *Bull Chem Soc Jpn* 51:2091.
50. Noll, A., N. Siegfield, and W. Heitz. 1990. *Makromol Chem Rapid Commun* 11: 485.
51. Ueda, M. and F. Ichikawa. 1990. *Macromolecules* 23:926.
52. Ueda, M., Y. Miyaji, and T. Ito. 1991. *Macromolecules* 24:2694.
53. Colon, I. and G.T. Kwiatkowski. 1990. *J Polym Sci A* 28:367.
54. Percec, V., S. Okita, and R. Weiss. 1992. *Macromolecules* 25:1816.
55. Vahlenkamp, T. and G. Wegner. 1994. *Macromol Chem Phys* 195:1933.
56. Wallow, T.I. and B.M.J. Novak. 1991. *J Am Chem Soc* 113:741.
57. Child, A.D. and J.R. Reynolds. 1994. *Macromolecules* 27:1975.
58. Rau, U.I. and M. Rehahn. 1994. *Acta Polymer* 45:3.
59. Rulkens, R., M. Schultze, and G. Wegner. 1994. *Macromol Rapid Commun* 15:669.
60. Huber, J. and U. Scherf. 1994. *Macromol Chem Phys* 15:897.
61. Fiesel, R., J. Huber, U. Apel, V. Enkelmann, R. Hentschke, U. Scherf, and K. Cabrera. 1997. *Macromol Chem Phys* 198:2623.
62. Phillips, R.W., V.V. Sheares, E.T. Samulski, and J.M. DeSimone. 1994. *Macromolecules* 27:2354.
63. Park, L.R., L.R. Dodd, K. Levon, and T.K. Kwei. 1996. *Macromolecules* 29:7149.
64. Scherf, U. and K. Müllen. 1991. *Makromol Chem Rapid Commun* 12:489.
65. Scherf, U., A. Bohnen, and K. Müllen. 1992. *Makromol Chem* 193:1127.
66. Scherf, U. 1999. *J Mater Chem* 9:1853.
67. Scherf, U. 1997. *Handbook of conducting polymers*, 2nd ed., eds. T. Skotheim, J.R. Reynolds. New York: Marcel Dekker, chap. 14, p. 363.
68. Scherf, U. 1999. *Topics in current chemistry: Carbon rich compounds II*, Vol. 201, ed. A. de Meijere. Berlin/Heidelberg: Springer, p. 163.
69. Scherf, U., and K. Müllen. 1999. *Semiconducting polymers*, eds. G. Hadziioannou, and P.F. van Hutten. Heidelberg: Wiley-VCH, chap. 2, p.37.
70. Cimrová, V., U. Scherf, and D. Neher. 1996. *Appl Phys Lett* 69:608.
71. Romaner, L., G. Heimel, H. Wiesenhofer, P. Scandiucci de Freitas, U. Scherf, J.L. Brédas, E. Zojer, and E.J.W. List. 2004. *Chem Mater* 16:4667.
72. Inbasekaran, M., W. Wu, P. Woo. 1998. US Patent 5, 777, 070.
73. Bernius, M.T., M. Inbasekaran, J. OBrien, and W. Wu. 2000. *Adv Mater* 12:1737.
74. Grell, M., W. Knoll, D. Lupo, A. Meisel, T. Miteva, D. Neher, H.-G. Nothofer, U. Scherf, and A. Yasuda. 1999. *Adv Mater* 11:671.
75. Leclerc, M. 2001. *J Polym Sci A Polym Chem* 39:2867.
76. Grell, M., D.D.C. Bradley, M. Inbasekaran, and E.P. Woo. 1997. *Adv Mater* 9:798.
77. Grell, M., D.D.C. Bradley, G. Ungar, J.Hill, and K.S. Whitehead. 1999. *Macromolecules* 32:5810.
78. Whitehead, K.S., M. Grell, D.D.C. Bradley, M. Jandke, and P. Strohriegl. 2000. *Appl Phys Lett* 76:2946.
79. Setayesh, S., D. Marsitzky, and K. Müllen. 2000. *Macromolecules* 33:2016.
80. Jacob, J., J. Zhang, A.C. Grimsdale, K. Müllen, M. Gaal, and E.J.W. List. 2003. *Macromolecules* 36:8240.
81. Jacob, J., S. Sax, T. Piok, E.J.W. List, A.C. Grimsdale, and K. Müllen. 2004. *J Am Chem Soc* 126:6987.
82. Schindler, F., J.M. Lupton, J.Feldmann, U. Scherf, J. Jacob, A. Grimsdale, and K. Müllen. 2005. *Angew Chem* 117:1544.

83. Grimme, J., M. Kreyenschmidt, F. Uckert, K. Müllen, and U. Scherf. 1995. *Adv Mater* 7:292.
84. Romanovski, Yu. V., A. Gerhardt, B. Schweitzer, U. Scherf, R.I. Personov, and H. Bässler. 2000. *Phys Rev Lett* 84:1027.
85. Walter, M.J., M. Reufer, P.G. Lagoudakis, U. Scherf, J.M. Lupton, and J. Feldmann. 2005. *Nature Mater* 4:340.
86. Wohlgenannt, M., K. Tandon, S. Mazumdar, S. Ramesesha, and Z.V. Vardeny. 2001. *Nature* 409:494.
87. Galbrecht, F., X.H. Yang, B.S. Nehls, D. Neher, T. Farrell, and U. Scherf. 2005. *Chem Commun* 18:2378.
88. Skotheim, T.A., R.L. Elsenbaumer, and J.R. Reynolds, eds. 1998. *Handbook of conducting polymers.* New York: Marcel Dekker, chaps. 1, 20, and 29.
89. Gadermaier, C. and G. Lanzani. 2002. *J Phys C* 14:9785.
90. Arkhipov, V.I., E.V. Emelianova, and H. Bassler. 1998. *Chem Phys Lett* 296:452; Schweitzer, B., V.I. Arkhipov, and H. Bassler. 1999. *Chem Phys Lett* 304:365.
91. Wohlgenannt, M., W. Graupner, G. Leising, and Z. Vardeny. 1999. *Phys Rev Lett* 82:3344.
92. Gadermaier, C., et al. 2002. *Phys Rev Lett* 89:117402.
93. Hertel, D., S. Setayesh, H.G. Nothofer, U. Scherf, K. Müllen, and H. Bässler. 2001. *Adv Mater* 13:65.
94. D. Beljonne, D., G. Pourtois, C. Silva, E. Hennebicq, L.M. Herz, R.H. Friend, G.D. Scholes, S. Setayesh, K. Müllen, and J.L. Bredas. 2002. *Proc. Natl. Acad. Sci. USA* 99: 10982.
95. Yan, M., L.J. Rothberg, F. Papadimitrakopoulos, M.E. Galvin, and T.M. Miller. 1994. *Phys Rev Lett* 72:1104.
96. Antoniadis, H., L.J. Rothberg, F. Papadimitrakopoulos, M. Yan, M.E. Galvin, and M.A. Abkowitz. 1994. *Phys Rev B* 50:14911.
97. Graupner, W., M. Sacher, M. Graupner, C. Zenz, G. Grampp, A. Hermetter, and G. Leising. 1998. *Mat Res Soc Symp Proc* 488:789.
98. Agranovich, V.M. 1968. *Theory of excitons.* Moscow: Nauka.
99. Förster, Th. 1948. *Ann Phys* 2:55.
100. Kranzelbinder, G., H.J. Byrne, S. Hallstein, S. Roth, G. Leising, and U. Scherf. 1997. *Phys Rev B* 56:11632; Maniloff, E.S., V.I. Klimov, and D.W. McBranch. 1997. *Phys Rev B* 56:1867; Denton, G.J., N. Tessler, N.T. Harrison, and R.H. Friend. 1997. *Phys Rev Lett* 78:733.
101. Kozlov, V.G., P.E. Burrows, G. Parthasarathy, and S.R. Forrest. 1999. *Appl Phys Lett* 74:1057.
102. Winokur, M.J. 1997. *Handbook of conducting polymers*, eds., T. Skotheim, R.L. Elsenbaumer, and J.R. Reynolds. New York: Marcel Dekker, p.707.
103. den Hartog, F.T.H., C. van Papendrecht, R.J. Silbey, and S. Völker. 1999. *J Chem Phys* 110:1010.
104. Bässler, H. and B. Schweitzer. 1999. *Acc Chem Res* 32:173.
105. Dexter, D.L. 1963. *J Chem Phys* 21:863.
106. Wiesenhofer, H., D. Beljonne, G.D. Scholes, E. Hennebicq, J.-L. Brédas, and E. Zojer. 2005. *Adv Funct Mat* 15:155.
107. Haugeneder, A., M. Neges, C. Kallinger, W. Spirkl, U. Lemmer, J. Feldmann, U. Scherf, E. Harth, A. Gügel, and K. Müllen. 1999. *Phys Rev B* 59:15346.
108. List, E.J.W., C. Creely, G. Leising, N. Schulte, A.-D. Schlüter, U. Scherf, K. Müllen, and W. Graupner. 2000. *Chem Phys Lett* 325:132.
109. Miller, A. and E. Abrahams. 1960. *Phys Rev* 120:745.
110. Wiesenhofer, H., E. Zojer, E.J.W. List, U. Scherf, J.-L. Brédas, and D. Beljonne. 2006. *Adv Mater* 18:310.
111. Mauthner, G., M. Collon, E.J.W. List, F.P. Wenzl, M. Bouguettaya, and J.R. Reynolds. 2005. *J Appl Phys* 97:635081.
112. Yang, Y., Q. Pei, and A. Heeger. 1996. *Appl Phys Lett* 79:934.
113. List, E.J.W., J. Partee, J.Shinar, U. Scherf, K. Müllen, W. Graupner, K. Petritsch, E. Zojer, and G. Leising. 2000. *Phys Rev B* 61:10807.
114. Redecker, M., D.D.C. Bradley, M. Inbasekaran, W.W. Wu, and E.P. Woo. 1999. *Adv Mater* 11:241.
115. Redecker, M., D.D.C. Bradley, M. Inbasekaran, and E.P. Woo. 1998. *Appl Phys Lett* 73:1565.

116. Cadby, A.J., P.A. Lane, H. Mellor, S.J. Martin, M. Grell, C. Giebler, and D.D.C. Bradley. 2000. *Phys Rev B* 62:15604.

117. Setayesh, S., A.C. Grimsdale, T. Weil, V. Enkelmann, K. Müllen, F. Meghdadi, E.J.W. List, and G. Leising. 2001. *J Am Chem Soc* 123:946.

118. Chunwaschirasiri, W., B. Tanto, D.L. Huber, and M.J. Winokur. 2005. *Phys Rev Lett* 94:107402.

119. Jacob, J., A.C. Grimsdale, K. Müllen, S. Sax, M. Gaal, and E.J.W. List. 2005. *Macromolecules* 38:9933.

120. Hertel, D., H. Bässler, U. Scherf, and H.H. Hörhold. 1999. *J Chem Phys* 110:9214.

121. Monkman, A.P., H.D. Burrows, M. da G. Miguel, I. Hamblett, and S. Navaratnam. 1999 *Chem Phys Lett* 307:303; Monkman, A.P., H.D. Burrows, I. Hamblett, S. Navaratnam, U. Scherf, and C. Schmitt. 2000. *Chem Phys Lett* 327:111.

122. Graupner, W., et al. 1998. *Phys Rev Lett* 81:3259; Wohlgenannt, M., et al. 82:3344.

123. Tasch, S., A. Niko, G. Leising, and U. Scherf. 1996. *Appl Phys Lett* 68:1090.

124. Tasch, S., G. Kranzelbinder, G. Leising, and U. Scherf. 1997. *Phys Rev B* 55:5079.

125. Leising, G., O. Ekström, W. Graupner, F. Meghdadi, M. Moser, G. Kranzelbinder, T. Jost, S. Tasch, B. Winkler, L. Athouel, G. Froyer, U. Scherf, K. Müllen, G. Lanzani, M. Nisoli, and S. De Silvestri. 1996. *SPIE Proc* 2852:189.

126. Graupner, W., J. Partee, J.Shinar, G. Leising, and U. Scherf. 1996. *Phys Rev Lett* 77:2033.

127. List, E.J.W., C.H. Kim, A.K. Naik, J. Shinar, G. Leising, and W. Graupner. 2001. *Phys Rev B* 64:155204.

128. Heimel, G., M. Daghofer, J. Gierschner, E.J.W. List, A.C. Grimsdale, K. Müllen, D. Beljonne, J.-L. Bredas, and E. Zojer. 2005. *J Chem Phys* 122:54501.

129. Scott, J.C., J.H. Kaufman, P.J. Brock, R. DiPietro, J. Salem, and J.A. Goitia. 1996. *J Appl Phys* 79:2745.

130. Aziz, H., Z. Popovic, S. Xie, A.-M. Hor, N.-X. Hu, C. Tripp, and G. Xu. 1998. *Appl Phys Lett* 72:756.

131. Burrows, P.E., V. Bulovic, S.R. Forrest, L.S. Sapochak, D.M. McCarty, and M.E. Thompson. 1994. *Appl Phys Lett* 65:2922.

132. Savvate'ev, V.N., A.V. Yakimov, D. Davidov, R.M. Pogreb, R. Neumann, and Y. Avny. 1997. *Appl Phys Lett* 71:3344.

133. Ke, L., S.-J. Chua, K. Zhang, and P. Chen. 2002. *Appl Phys Lett* 80:171.

134. Kolosov, D., D.S. English, V. Bulovic, P.F. Barbara, S.R. Forrest, and M.E. Thompson. 2001. *J Appl Phys* 90:3242.

135. Lemmer, U., S. Heun, R.F. Mahrt, U. Scherf, M. Hopmeier, U. Siegner, E.O. Göbel, K. Müllen, and H. Bässler. 1995. *Chem Phys Lett* 240:373.

136. Lee, J.I., G. Klärner, and R.D. Miller. 1999. *Chem Mater* 11:1083.

137. Weinfurter, K.H., H. Fujikawa, S. Tokito, and Y. Taga. 2000. *Appl Phys Lett* 76:2502.

138. Bliznyuk, V.N., S.A. Carter, J.C. Scott, G. Klärner, and R.D. Miller. 1999. *Macromolecules* 32:361.

139. Lupton, J.M., P. Schouwink, P.E. Keivanidis, A.C. Grimsdale, and K. Müllen. 2003. *Adv Funct Mat* 13:154.

140. Pogantsch, F., P. Wenzl, E.J.W. List, G. Leising, A.C. Grimsdale, and K. Müllen. 2002. *Adv Mater* 14:1061.

141. Nakazawa, Y.K., S.A. Carter, H.-G. Nothofer, U. Scherf, V.Y. Lee, R.D. Miller, and J.C. Scott. 2002. *Appl Phys Lett* 80:3832.

142. Sainova, D., D. Neher, E. Dobruchowska, B. Luszczynska, I. Glowacki, J. Ulanski, H.-G. Nothofer, and U. Scherf. 2003. *Chem Phys Lett* 371:15.

143. Gaal, M., E.J.W. List, and U. Scherf. 2003. *Macromolecules* 36:4236.

144. Romaner, L., A. Pogantsch, P. Scandiucci de Freitas, U. Scherf, M. Gaal, E. Zojer, and E.J.W. List. 2003. *Adv Funct Mat* 13:597.

145. Sims, M., D.D.C. Bradley, M. Ariu, M. Koeberg, A. Asimakis, M. Grell, and D.G. Lidzey. 2004. *Adv Funct Mat* 14:765.

146. Lu, S., T.X. Liu, L. Ke, D.G. Ma, S.J. Chua, and W. Huang. 2005. *Macromolecules* 38:8494.

147. Chochos, C.L., J.K. Kallitsis, and V.G. Gregoriou. 2005. *J Phys Chem B* 109:8755.

148. Becker, K., J.M. Lupton, J. Feldmann, B.S. Nehls, F. Galbrecht, D.Q. Gao, and U. Scherf. 2006. *Adv Funct Mater.* DOI: 10.1002adfm.200500550.

149. Gong, X., P.K. Iyer, D. Moses, G.C. Bazan, A.J. Heeger, and S.S. Xiao. 2003. *Adv Funct Mater* 13:325.

150. Gamerith, S., H.-G. Nothofer, U. Scherf, and E.J.W. List. 2004. *Jpn J Appl Phys* 43:L891.

151. Silverstein, R.M., G.C. Bassler, T.C. Morrill. 1981. *Spectroscopic identification of organic compounds*, 4th ed. New York: Wiley.

152. Ilharco, A.R.G., J. Lopes da Silva, M. João Lemos, and L.F. Vieira Ferreira. 1997. *Langmuir* 13:3787.

153. Miteva, T., A. Meisel, W. Knoll, H.G. Nothofer, U. Scherf, D.C. Müller, K. Meerholz, A. Yasuda, and D. Neher. *Adv Mater* 2001. 13:565.

154. Michael Graf. 2006. Diploma thesis. TU Graz.

155. Liao, L.S., L.F. Cheng, M.K. Fung, C.S. Lee, M. Inbasekaran, E.P. Woo, and W.W. Wu. 2000. *Chem Phys Lett* 325:405.

156. Fung, M.K., S.L. Liai, S.N. Bao, C.S. Lee, S.T. Lee, W.W. Wu, M. Inbasekaren, and J.J. OBrian. 2002. *J Vac Sci Technol* 20:91.

157. Stoessel, M., G. Wittmann, J. Staudigel, F. Steuber, J. Blässing, W. Roth, H. Klausmann, W. Rogler, J. Simmerer, A. Winnacker, M. Inbasekaran, and E.P. Woo. 2000. *J Appl Phys* 87:4467.

158. Becker, H., S.E. Burns, and R.H. Friend. 1997. *Phys Rev B* 56:1853.

159. Craig, M.R., M.M. de Kok, J.W. Hofstraat, A.P.H.J. Shenning, and E.W. Meijer. 2003. *J Mater Chem* 13:286.

160. Chan, K.H., M. McKiernan, C. Towns, and A.B. Holmes. 2005. *J Am Chem Soc* 127:7662.

161. Müller, C.D., A. Falcou, N. Reckefuss, M. Rojahn, V. Wiederhirn, P. Rudati, H. Frohne, O. Nuyken, H. Becker, and K. Meerholz. 2003. *Nature* 421:829.

6

Poly(*para*phenylene-ethynylene)s and Poly-(aryleneethynylenes): Materials with a Bright Future

Uwe H.F. Bunz

6.1 Introduction

This contribution describes the chemistry and uses of the poly(aryleneethynylene)s (PAEs). The different types of syntheses for PAEs are described first and the most important classes of PAEs are highlighted. Their properties and uses are outlined. Polymer analogous reactions of poly(*para*phenyleneethynylene)s (PPEs), i.e., hydrogenation, complexation with suitable organotransition metal fragments, and side chain manipulation are discussed with regard to the synthetic processes and the properties of the formed novel polymeric materials.

PAEs are chromic and reveal fascinating spectroscopic properties upon exposure to external stimuli. The UV–vis and fluorescence properties of selected PAEs are treated with regard to inter- and intrachain processes and a model for the spectral changes in PPEs is developed. In the last part, the uses of PAEs as fluorescent sensors and in semiconductor devices are discussed.

6.1.1 General Considerations

Conjugated polymers have found widespread applications in semiconducting devices and fluorescent sensing schemes. They are now incorporated as emitter layers into commercially produced polymer light-emitting diodes (PLEDs). What makes conjugated polymers alluring?

- Their synthesis is a playground for novel molecular architectures and a proving ground for organic reactions.
- Their conformational and structural properties are unusual.
- Their photophysics in dilute solution and thin films is complex and challenging, yet is governed by few relatively simple rules.
- Their film forming and processing properties are often excellent.
- Their fluorescent and semiconducting properties in the solid state make them powerful active ingredients in photovoltaic and other devices.

The conjugated polymer that has "made it big" is poly(*para*phenylenevinylene) (PPV), in part due to historical reasons: the first organic PLED was fabricated by Sir Richard Friend in 1990 [1]. He chose PPV as an emitter due to its overall satisfactory charge injection, transport, and fluorescence properties. PPV is an excellent hole conductor. The PPEs are apparently close relatives of the PPVs in that the vinyl groups are replaced by alkynes [2]. While the change appears miniscule in terms of structure, the properties of the PPEs are surprisingly different from those of the PPVs. Alkyne instead of vinyl linkages heavily influence the electronics of the resulting polymers, and PPEs are not particularly good hole injectors any more but are better electron injectors. In addition, the overall electronic transitions of the PPEs are blueshifted, i.e., both the HOMO and the LUMO of the PPEs are stabilized, but the HOMO moreso than the LUMO. Another nontrivial consequence of the introduction of the alkyne units is the possibility of planarization of the PPE chains due to the facile and almost barrier-free (<1 kcal/mol) rotation around the carbon–carbon single bond that connects the arenes to the alkynes. PPEs may prefer either planarized or twisted conformations in solution and thin films. PPEs are often ordered, poly-crystalline materials in the solid state and show sharp X-ray diffraction peaks in the q range of 0.02–1/Å. Some specific PPEs show potential in semiconductor applications, but their real strengths seem to be in sensory schemes, due to their chemical and photostability coupled with intense fluorescence. The human eye can perceive the fluorescence of approximately 100 ng of any dialkyl-PPE in 1 mL of chloroform solution. Intense chromic transitions triggered by external stimuli are typical for both dialkyl and dialkoxy-PPEs, leading PPEs to emerge as both potent sensor and biosensor platforms.

6.1.2 Historical Background

The first synthesis of PPEs was published by Schulz and Giesa in 1990 [3]. They reacted dibromodi(alkoxy)benzenes with different di(ethynyl)di(alkoxy)arenes in the presence of a Pd catalyst in an amine (Heck–Cassar–Sonogashira–Hagihara reaction) [4]. They obtained low molecular weight materials that were brownish or greenish in appearance and showed a multimodal molecular weight distribution according to gel permeation chromatography (GPC). Lemoigne et al. [5] improved the reaction conditions by the addition of THF, to give PPEs of higher molecular weight. When examining their (published) spectra, the obtained materials showed significant diyne and other defects that suggest that a PPE with approximately 20–50 repeat units had formed. The presence of diyne defects is a necessary consequence of their reaction conditions. The introduction of 5 mol% of catalytically active Pd as $PdCl_2$ will lead to a minimum of 10 mol% of diyne defects in the main chain, due to oxidative coupling of the terminal alkynes in the activation of the catalysts. The problem is discussed in detail (vide infra). In 1995, the groups of Wrighton, Bunz, and West independently described the synthesis of PPEs and their interesting optical properties, concluding that aromatic diiodides when compared with aromatic dibromides are far superior in the synthesis of PPEs when utilizing Sonogashira-type couplings [6–10]. Wrighton described the solid state of PPEs and noted that they attain a highly lamellar ordering that is reminiscent of aromatic polyesters described by Duran and Wegner [6–8,11]. Wrighton utilized $(Ph_3P)_4Pd$ in the Sonogashira coupling to minimize diyne formation. The problem of the presence of diyne defects prompted Bunz, Müllen, and Weiss to investigate alkyne metathesis as a tool for the synthesis of PPEs [12]. A Schrock-type tungsten alkylidyne produced dihexyl-PPE with a degree of polymerization (P_n) of approximately 100 repeating units [13–15]. While the acyclic diyne metathesis (ADIMET) is a powerful tool to synthesize PPEs, the Schrock complex makes matters less convenient. This carbyne is not commercially available and is quite sensitive toward air and water. Bunz and Kloppenburg investigated a much simpler system based on 4-chlorophenol and molybdenum hexacarbonyl in toluene [16–18]. These simple catalysts are powerful in the production of defect-free PPEs. At present (2005), the attention has shifted from the development of new synthetic tools to the synthesis of more sophisticated macromolecular architectures containing PPEs and their use in device and sensory applications.

6.1.3 Scope and Limitations of the Review

A series of reviews has been published on PPEs and Table 6.1 gives pertinent references, while Table 6.2 gives references to somewhat related fields. This contribution to the *Handbook of Conducting Polymers* will not be a comprehensive coverage of PAEs. Instead, the author attempts to highlight and discuss important developments.

TABLE 6.1   Reviews Directly Dealing with Poly(aryleneethynylene)s

Author(s)	Title	Source	Citation
Bunz U.H.F.	Poly(aryleneethynylene)s: syntheses, properties, structures, and applications	*Chem. Rev.*	100, 1605, 2000
Bunz U.H.F.	Poly(*para* phenyleneethynylene)s by alkyne metathesis	*Acc. Chem. Res.*	34, 998, 2001
Giesa R.C.	Synthesis and properties of conjugated poly(aryleneethynylene)s	*J. Macromol. Sci. Rev. Macromol. Chem. Phys.*	C36, 631, 1996
Weder C. (Editor)	Polyaryleneethynylenes	*Adv. Polym. Sci.*	177, whole issue, 2005
Naso F.	Synthesis of conjugated oligomers and polymers: the organometallic way	*J. Mater. Chem.*	14, 11, 2004
Yamamoto T.	π-Conjugated polymers bearing electronic and optical functionalities: preparation by organometallic polycondensations, properties, and their applications	*Bull. Chem. Soc. Jpn.*	72, 621, 1999
Yamamoto T.	π-Conjugated polymers with electronic and optical functionalities: preparation by organometallic polycondensation, properties, and applications	*Macromol. Chem. Rapid Commun.*	23, 583, 2002
Yamamoto T.	Synthesis of π-conjugated polymers bearing electronic and optical functionalities by organometallic polycondensations. Chemical properties and applications of the π-conjugated polymers	*Synlett*	425, 2003

TABLE 6.2   Reviews of Interest That Have Appeared in Adjacent Areas

Author(s)	Title	Source	Citation
Tour J.M.	Conjugated macromolecules of precise length and constitution. Organic synthesis for the construction of nanoarchitectures	*Chem. Rev.*	96, 537, 1996
Moore J.S.	Shape-persistent molecular architectures of nanoscale dimension	*Acc. Chem. Res.*	30, 402, 1997
Hill D.J., et al.	A field guide to foldamers	*Chem. Rev.*	101, 3893, 2001
McQuade D.T., Pullen A.E., and Swager T.M.	Conjugated polymer-based chemical sensors	*Chem. Rev.*	100, 2537, 2000
Swager T.M.	The molecular wire approach to sensory signal amplification	*Accounts*	31, 201, 1998
Haley M.M.	It takes alkynes to make a world—new methods for dehydrobenzoannulene synthesis	*Synlett*	557, 1998
Haley M.M., Pak J.J., and Brand S.C.	Macrocyclic oligo(phenylacetylenes) and oligo(phenyldiacetylenes)	*Top. Curr. Chem.*	201, 81, 1999
Marsden J.A. and Haley M.M.	Metal-catalyzed cross-coupling reactions, 2nd ed.	deMeijere A. and Diederich F., eds.	Wiley-VCH, Weinheim 2004, pp. 317–394
Sonogashira K.	Development of Pd–Cu catalyzed cross-coupling of terminal acetylenes with sp(2)-carbon halides	*J. Organomet. Chem.*	653, 46, 2002
Negishi E. and Anastasia L.	Palladium-catalyzed alkynylation	*Chem. Rev.*	103, 1979, 2003

Specific examples will be showcased as they contribute to the general understanding and advancement of the field. The synthesis and evaluation of PAE model compounds and their use as molecular wires as investigated by Tour [19] are not covered in this review.

6.2   Syntheses

The classic synthesis of PAEs is the Pd-catalyzed coupling of dihaloarenes (Br and I) with diethynylarenes (Scheme 6.1). In this general approach, different Pd compounds can be utilized, ranging from Pd(OAc)$_2$/PPh$_3$ combinations to Pd(PPh$_3$)$_4$ with the most popular catalyst being (Ph$_3$P)$_2$PdCl$_2$. This catalyst is stable, can be stored without any problem, and is not prone to oxidation, which is a potential problem with Pd(PPh$_3$)$_4$. A cocatalyst, CuI, is always added, even though its effectiveness has been questioned for these couplings in small molecule synthesis [20]. It would be attractive to examine if the presence of CuI is necessary for the formation of PPEs and PAEs. Triethylamine (works better for aryl bromides) and piperidine or other secondary and tertiary amines such as Hünigs base are used as

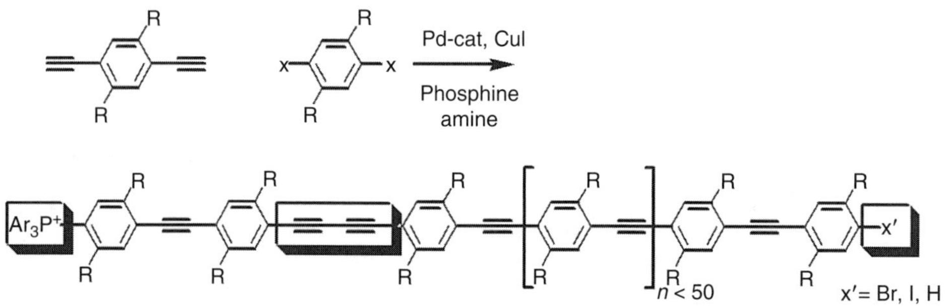

SCHEME 6.1   Synthesis of PPEs by the Pd-catalyzed reaction of aryldihalides with dialkynylarenes.

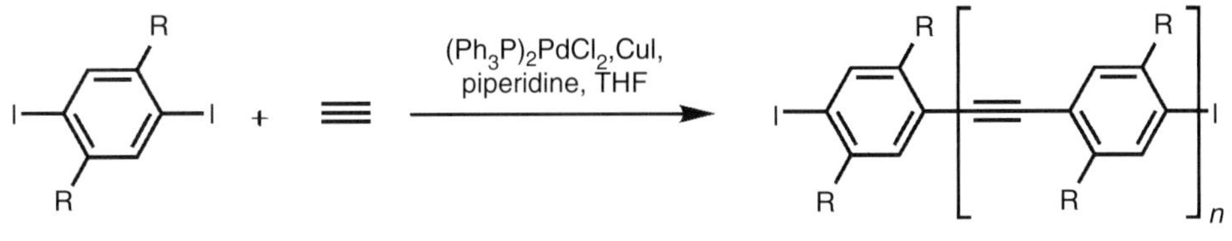

SCHEME 6.2   Synthesis of PPEs by the acetylene gas method.

reaction solvents. The choice of amine is important, but in a rather nonpredictable way. From the experience in the Bunz laboratories, the combination of diaryldiiodides with $(Ph_3P)_2PdCl_2$/CuI in piperidine seems to consistently give PAEs with good-to-excellent molecular weights. The addition of THF to these reaction mixtures as a cosolvent, described by Moroni et al. [5] and Thorand and Krause [21] helps to improve yield, degree of polymerization (P_n), and appearance of the PPEs. Other cosolvents that are sometimes beneficial are dichloromethane and toluene, perhaps because the solubility of the PAEs is better in these solvents than in the amine bases in which the small molecule couplings are performed. How much Pd catalyst is necessary to obtain PAEs of optimum quality? The typical recipes call for up to 5 mol% of catalyst. When utilizing acetylene gas as the alkyne, a much smaller amount of catalytic Pd (typically 0.1 mol%) is sufficient to effect the polycondensation reaction; generally Pd loadings of >0.5 mol% are probably unnecessary, unless for severely deactivated monomers [22]. An interesting side note is that even trace amounts of Pd compounds found in technical sodium carbonate catalyze the related Suzuki-type couplings [23].

The amount of reaction solvent should be kept to a minimum to maximize reaction rate and concentration of reactive end groups as a means of obtaining PAEs of sufficiently high molecular weight. These reactions are stirred for 48–72 h at temperatures ranging from 20°C to 50°C if diiodides are utilized. For dibromides, reaction around 100°C is necessary, due to their significantly decreased reactivity. There are now more active catalysts available [24]. It will be interesting to see if these catalysts work well in the formation of PPEs and PAEs from aromatic dibromides or even dichlorides.

An attractive way to PAEs is the Pd-catalyzed reaction of aromatic dihalides with acetylene gas (Scheme 6.2). This idea was first explored by Li to make *meta*-PPEs, but the product was brownish in appearance [25,26]. Largely defect-free PPEs look either colorless (*meta*-PPEs), lemon yellow (dialkyl-PPEs), or brilliantly deep yellow with a tinge of orange (dialkoxy-PPEs), suggesting that Li's *meta*-PPEs were defective in their nature. Acetylene gas is attractive, because (a) only one type of monomer has to be made; and (b) acetylene gas is inexpensive. Wilson et al. [22] investigated this approach further and found that brownish PPEs were formed when treating 2,5-diiodo-1,4-dialkylbenzenes with acetylene gas, 5 mol% of Pd catalyst, and CuI as cocatalyst (Scheme 6.2). However, when the catalyst amount was reduced to 0.1 mol% or less, the quality of the obtained PPEs was significantly improved and brilliantly yellow PPEs were produced (Scheme 6.2). This method is not restricted to dialkyl or dialkoxy-PPEs, but it is also useful for oligoethyleneglycol-substituted PPEs [27]. The degree of polymerization (P_n) in the acetylene gas reactions varies from 20 to >200 and the polydispersities (M_w/M_n and PDI) range from 2.5 to 5 with a monomodal molecular weight distribution.

6.2.1 Pd-Catalyzed Coupling: Mechanistic Cycle and the Formation of Side Products

What is the reaction mechanism of the Pd-catalyzed formation of PPEs and PAEs? While there has not been a full force mechanistic investigation of the Pd-catalyzed formation of PAEs, a working model is at hand (Scheme 6.3 and Scheme 6.4). If a Pd^{2+} compound is used in the reaction, it has to be reduced into the catalytically active zerovalent species. In step I, an alkyne is transformed into a cuprate by the cocatalyst CuI in the presence of an amine. Then L_2PdX_2 reacts under transmetallation with the cuprate to give a dialkynylated tetrahedral intermediate that reductively eliminates to furnish a diyne byproduct

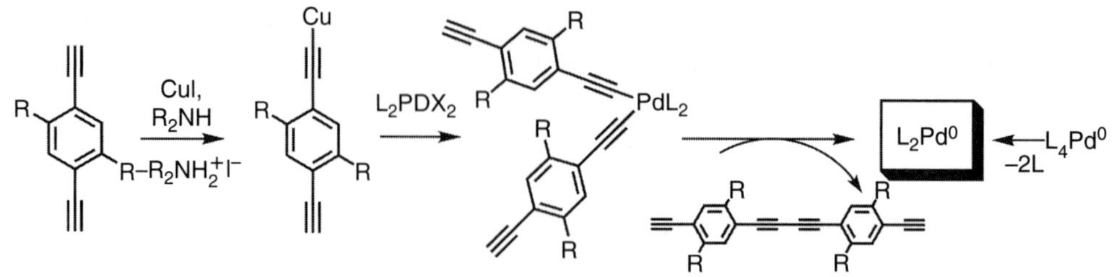

SCHEME 6.3 Catalyst activation. Formation of the active Pd^0 species from a Pd^{2+} species.

and the active catalyst, L_2Pd^0 (Scheme 6.3). In most cases $(Ph_3P)_2PdCl_2$ is utilized as catalyst precursor, but combinations of $Pd(OAc)_2$ and Ph_3P are also reported [5]. In some cases Ph_3P is added to stabilize the dicoordinate, reactive Pd catalyst. If a zerovalent Pd catalyst such as $(Ph_3P)_4Pd$ is chosen, then the catalytically active species forms simply by dissociative loss of two of the ligands [6–8].

The catalytic cycle (Scheme 6.4) starts with the oxidative addition of the aromatic diiodide to the zerovalent active Pd catalyst (I). The intermediate is a stable and isolable Pd^{2+} species. Müllen and Mangel have shown that such discrete species are active in the formation of endcapped PPEs [28]. In step II, transmetallation occurs to give a *cis*-configured, tetracoordinated Pd intermediate that is unstable toward elimination and furnishes (step III) the coupled product that carries an alkyne and an iodide functionality at its end: the Pd(0) catalyst is reformed in this step. In the final isolated PPEs, the end groups are mostly iodide functionalites and never alkynes according to NMR and capping reactions [6–8,28]. Repetition of the process forms the PPE in a polycondensation process. The Sonogashira polycondensation reactions obey only very roughly the Flory–Schulz law and can furnish PAEs with a polydispersity of 1.5–5.0 and a monomodal molecular weight distribution [29]. If PAEs with a polydispersity of <2 are reported, it is suggestive that partial precipitation of the PAE during the workup has occurred. The loss of the highest and the lowest molecular weight fractions is probable under such conditions. Flory–Schulz distribution of molecular weights of polymers with an

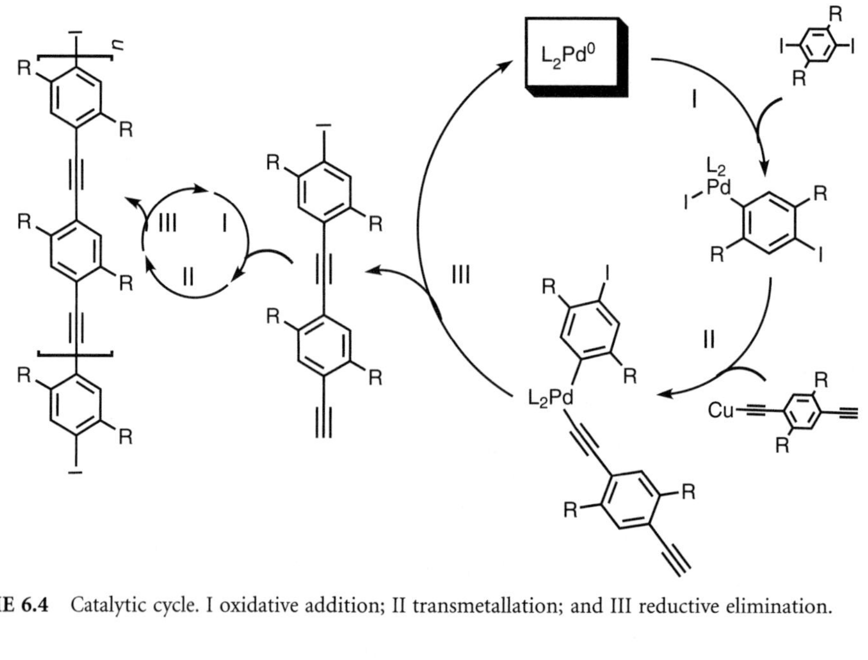

SCHEME 6.4 Catalytic cycle. I oxidative addition; II transmetallation; and III reductive elimination.

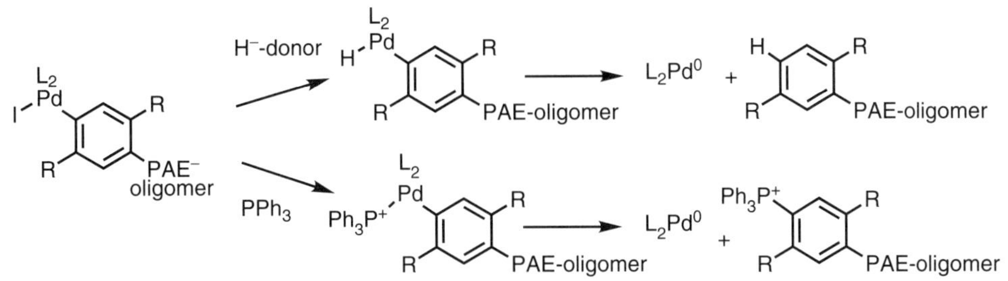

SCHEME 6.5 Side reactions in the formation of PPEs.

$M_w/M_n = \text{PDI} = 2$ is observed only if the reactivity of the monomer(s) and the formed oligomer(s) in the coupling (or condensation) reaction is identical. That is the case for acid-catalyzed polyester synthesis, but probably not for PPE formation, which could explain the deviating polydispersities.

The most common side reactions in the Sonogashira couplings are the formation of diyne defects, dehalogenation processes, and phosphonium salt formation (Scheme 6.4 and Scheme 6.5) [30–32]. With the exception of diyne formation, none of these processes has been observed *directly*, but there is precedence for these from small molecule and model studies. Diyne defects occur either if Pd^{2+} compound is used as catalyst precursor, or if traces of oxygen or other unknown oxidants are present in the reaction mixture. These oxidize Pd(0) into Pd(II) before oxidative addition of a haloarene occurs. As in the catalyst activation cycle, the double transmetallation with a copper acetylide of this (undesired) Pd(II) species will lead to a dialkynylated Pd complex. It undergoes reductive elimination to give the diyne, regenerating the catalytically active, zerovalent Pd species. The number of diyne defects can be significantly reduced by decreasing the catalyst loadings from 5 to <0.2 mol% and by strictly excluding oxygen. However, it does not seem to be possible to fully suppress the formation of diynes, even if the reaction is performed under an inert atmosphere. Water on the other hand seems to be tolerated.

Dehalogenation processes (Scheme 16) are the opposite of the diyne formation processes; here the intermediate of the oxidative addition is reduced by an unknown, hydride-donating species to give a Pd–arylhydride complex, which reductively eliminates into the dehalogenated product. A possible hydride source could be the amine solvent and the reduction could resemble that of the catalyst formation in the Heck reaction [30,31]. If the intermediate of the oxidative addition is not transmetallated by a copper acetylide, but attacked by triphenylphosphine, reductive elimination will lead to the corresponding phosphonium salts. Novak and associates have investigated this side reaction in the case of the Suzuki coupling for the synthesis of poly*para*phenylene derivatives. It is reasonable to assume that the same will happen in Sonogashira-type reactions [32].

Overall, the Pd-catalyzed couplings are a suitable way to make PPEs; these catalytic systems allow the introduction of almost any functional group into the PAEs. Disadvantage of the Pd-catalyzed couplings is sometimes the low degree of polymerization and the occurrence of defect structures that have to be minimized by the careful control of the reaction conditions.

6.2.2 Pd Nanoparticles in PAEs and Their Removal

An often unrecognized problem with the Pd-catalyzed synthesis of PPEs and other conjugated polymers is the presence of elemental Pd in the final product. One way to diminish the Pd content of these polymers is the use of low and ultralow amounts of Pd catalysts. However, even under these conditions there might be Pd residues in the polymers. The metal residues lead to an increase in electrical conductivity in PAEs, but inhibit electroluminescence according to Krebs [33]. To obtain strictly Pd-free PPEs, Krebs added the thiourea derivative **1** (Scheme 6.6) to a solution of conjugated polymers in chloroform; **1** sequesters Pd(0) as **2** and allows one to lower the residual Pd values to <0.1 ppm in the treated polymers. The complex **2** is soluble and can be washed out. In the Pd-free polymers, the

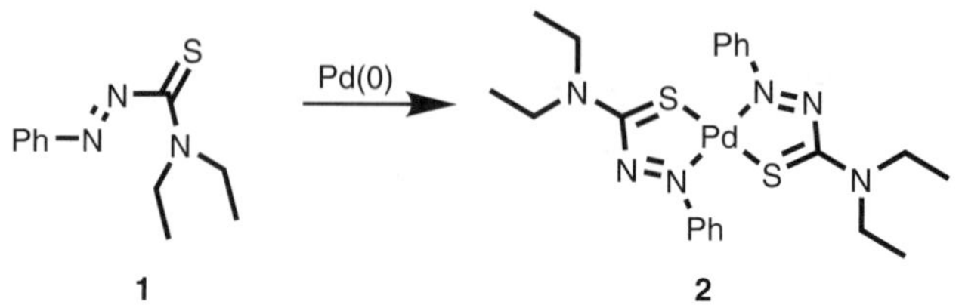

SCHEME 6.6   Removal of Pd residues from conjugated polymers.

conductivity resistance increased from 0.6 to 24 kΩ and device performance was improved. This process should help to obtain metal-free PAEs, PPEs, and other conjugated polymers.

6.2.3 Alkyne Metathesis

Alkyne metathesis (Scheme 6.7) is the reaction of two internal alkynes under permutation of their substituents [12–18,34–36].

The first catalyst to perform this reaction was reported by Mortreux in 1974. Initially, a mixture of molybdenum hexacarbonyl and resorcinol in toluene was used [17]. Shortly afterwards Vollhardt reported that alkyne metathesis can be initiated by the thermolytic cleavage of cyclobutadiene complexes [37]. While the use of $Mo(CO)_6$ and resorcinol was conceptually useful, the yields of the metathesis reactions were not very high, and the reaction conditions were quite drastic. Schrock developed tungsten carbyne complexes that performed alkyne metathesis at considerably lower temperatures. These catalysts are not easily synthesized, and are water and air sensitive, and not commercially available [13–15]. Bunz et al. prepared dihexyl-PPE **4a** by alkyne metathesis of dihexyl-dipropynylbenzene **3a** (Scheme 6.8) with a Schrock-type tungsten carbyne [12]. The PPE **4a** was bright yellow and fluorescent in the solid state with a P_n of approximately 90 repeat units according to GPC. This experiment demonstrated that alkyne metathesis was competitive to form PPE when compared with the Pd-catalyzed couplings, although P_n was not spectacularly high. It was noted that **4a** was not well soluble in organic solvents, suggesting that more powerfully solubilizing side chains might be advantageous. Attempts to generate poly(2,5-thio-phenyleneethynylene)s by this method failed. When the Schrock catalyst **5** was exposed to 3,4-dihexyl-2,5-dipropynylthiophene only small oligomers were formed.

What is the mechanism of this ADIMET? Schrock reported in depth mechanistic studies on alkyne metathesis utilizing **5** [13–15]. Scheme 6.9 shows the proposed mechanism. In the first step a tungsten carbyne reacts with the dipropynyldialkylbenzene to give an unstable metallacyclobutadiene that fragments under release of an alkyne. A new tungsten carbyne forms reacts with another dipropynylbenzene monomer to give a second metallacyclobutadiene. Upon fragmentation of the metallacyclobutadiene, a dimeric phenyleneethynylene is formed and a tungsten carbyne complex is released. Multiple repetition

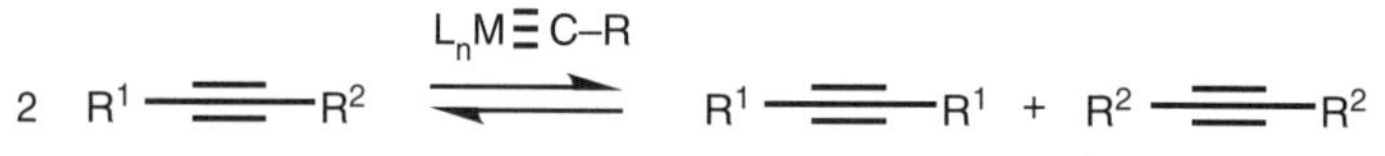

SCHEME 6.7   Alkyne metathesis.

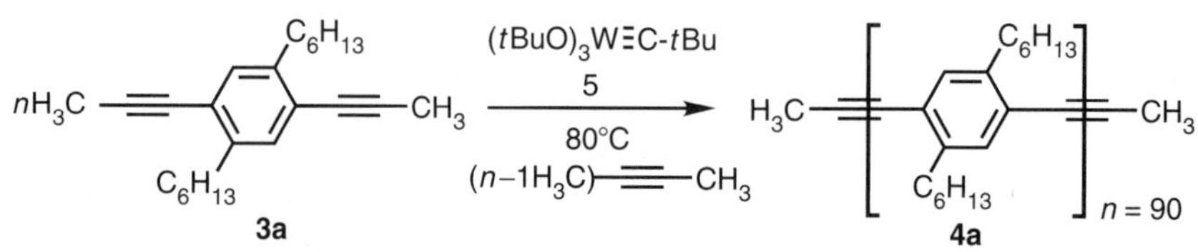

SCHEME 6.8   Synthesis of dihexyl-PPE by a Schrock-type catalyst.

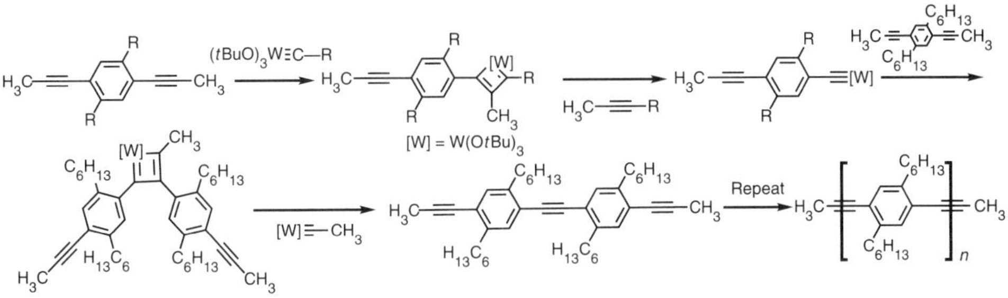

SCHEME 6.9   Mechanism of the formation of dialkyl-PPEs utilizing alkyne metathesis with a Schrock complex.

of this route furnishes PPE and butyne. The reaction is driven toward the PPE by removal of butyne from the equilibrium according to the mass law. The PPE is isolated by precipitation into methanol, redissolution in an organic solvent, and washing with acid and base to remove catalyst residues.

While the Schrock catalysts are active and give good results in ADIMET, it was of interest if simpler catalyst systems could do the same. A paper by Mori described that mixtures of $Mo(CO)_6$ and 4-chlorophenol were active at 105°C in the metathesis of propynylbenzene to give diphenylacetylene, albeit only in a 52% yield [18]. If the temperature is increased to 130–150°C, the yield of the metathesis reaction increases significantly even though the nature of the active catalyst is unknown. The formation of a mononuclear or oligonuclear molybdenum complex that is somewhat reminiscent of a Schrock-type carbyne is probable. This catalytically active species would carry aryloxy groups from the added phenol as spectator ligands. The mechanism would be similar to that shown in Scheme 6.9.

A series (Scheme 6.10) of dialkyldipropynylbenzenes was exposed to mixtures of $Mo(CO)_6$ and 4-chlorophenol in technical grade chlorobenzene, 1,2-dichlorobenzene, or 1,2,4-trichlorobenzene. Clean formation of PPEs was ensured if the formed butyne is swept out by a gentle stream of nitrogen or argon [16,38]. The polymers **4a–e** form in quantitative yields as yellow powders after workup and show high molecular weights with apparent $P_n's$ that can reach up to 10^3 repeat units (GPC). The use of dodecyl and ethylhexyl side chains is critical, because these enhance the solubility of the formed PPEs (**4b** and **c**). With hexyl side chains the degree of polymerization ($P_n = 100$) is not limited by the catalyst activity, but by the lack of solubility of **4a**.

It was discovered (Scheme 6.11) that in situ alkyne metathesis catalysts tolerate the presence of double bonds. Reaction of dipropynylstilbene **5** with the in situ catalyst system furnished the PPE–PPV-hybrid **6** [39]. However, instead of 4-chlorophenol the more active cocatalyst 4-trifluoromethylphenol has to be

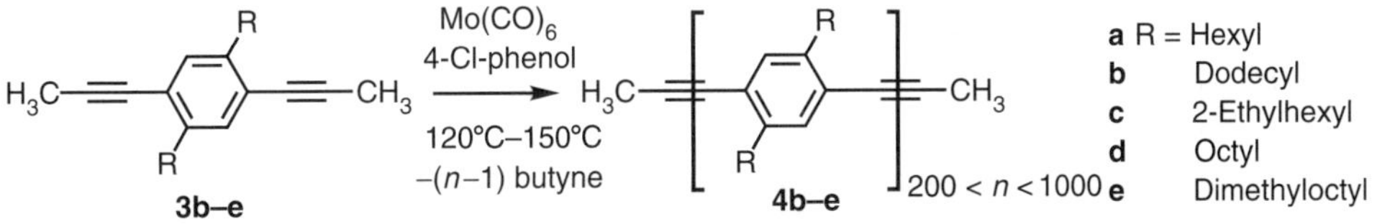

SCHEME 6.10   Synthesis of PPEs by alkyne metathesis utilizing in situ catalysts.

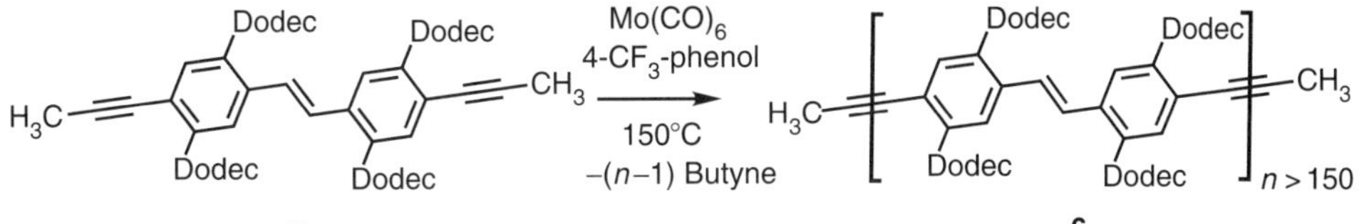

SCHEME 6.11   Synthesis of PPE–PPV hybrids by alkyne metathesis utilizing in situ catalysts.

utilized. Alkyne metathesis with simple in situ systems has been applied to a number of other PAEs (**7–10**) and works well as long as the monomers do not carry substituents that coordinate strongly to the catalytically active species [39–43]. For the synthesis of dialkoxy-PPEs either 2-fluorophenol has to be utilized as cocatalyst or the in situ catalyst has to be activated by treatment with 3-hexyne under heating [44,45]. The 2-fluorophenol protocol is simple and does not seem to have any disadvantages [44,45]. Mixtures of AlEt$_3$, phenols, and MoOCl$_2$ have been reinvestigated as catalysts, and seem to be active in ADIMET at temperatures as low as 100°C and furnish PPEs in high yields and good P_n after only 12 h reaction time [46].

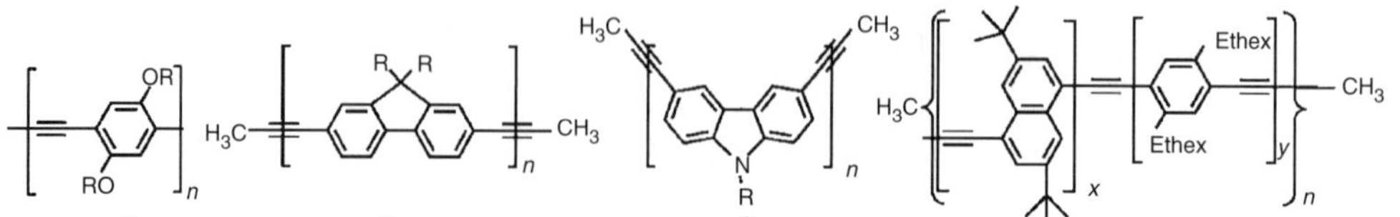

Dipropynylated thiophenes do not react at all under the in situ conditions, and only starting material is reisolated. If the temperatures are increased to above 170°C, a deep brown and insoluble material forms as a precipitate, suggesting decomposition of the monomers. Moore [47–49] has recently developed a new generation of ultra-active alkyne metathesis catalysts. He starts from a Cummins–Fürstner-type *tris*(arylamino)molybdenumcarbyne that is treated with 4-nitrophenol to give a Schrock–carbyne complex [50,51]. This complex produces PPEs at room temperature and metathesizes dipropynylthio-phenes **11** to give poly(2,5-thiophenyleneethynylene)s **12** (Scheme 6.12) in excellent yields and with a very good $P_n = 77$–128 (GPC). The same group reported the formation of cyclic aryleneethynylene macrocycles by precipitation-driven alkyne metathesis [52]. Overall, ADIMET is an excellent alternative to the classical Pd-catalyzed couplings to give PAEs of very high molecular weight and purity. The metathesis PAEs are not affected by butadiyne defect structures nor do they contain halogenated end groups. For electronic and semiconductor applications alkyne-metathesis PAEs will be the materials of choice.

6.2.4 Combinatorial Approach to PPEs

Lavastre et al. [53] explored the synthesis of PAEs in a combichem setup. The diynes **13–20** were coupled in THF utilizing a Sonogashira protocol to the dibromides **A–L** to give 96 different PAEs (Scheme 6.13). The authors reported that the polymers **J14, 16**, and **20** were fluorescent in thin film preparations, while **20H, J–L** were fluorescent in dilute solution. The results for the thin films are not unexpected; as it is

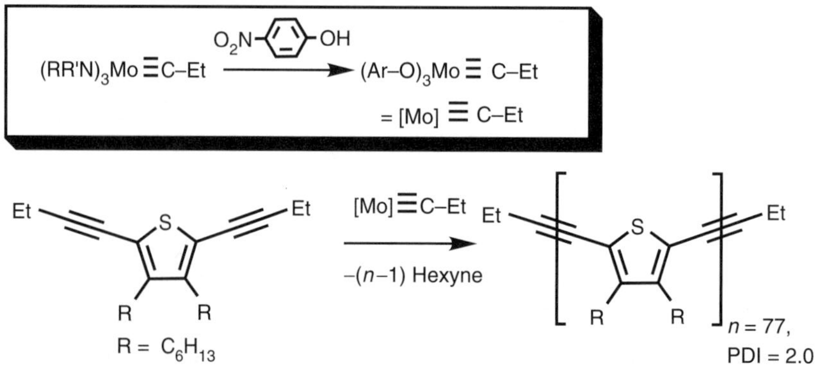

SCHEME 6.12 Synthesis of poly(2,5-thiophenyleneethynylene)s by alkyne metathesis utilizing Moore's catalyst system.

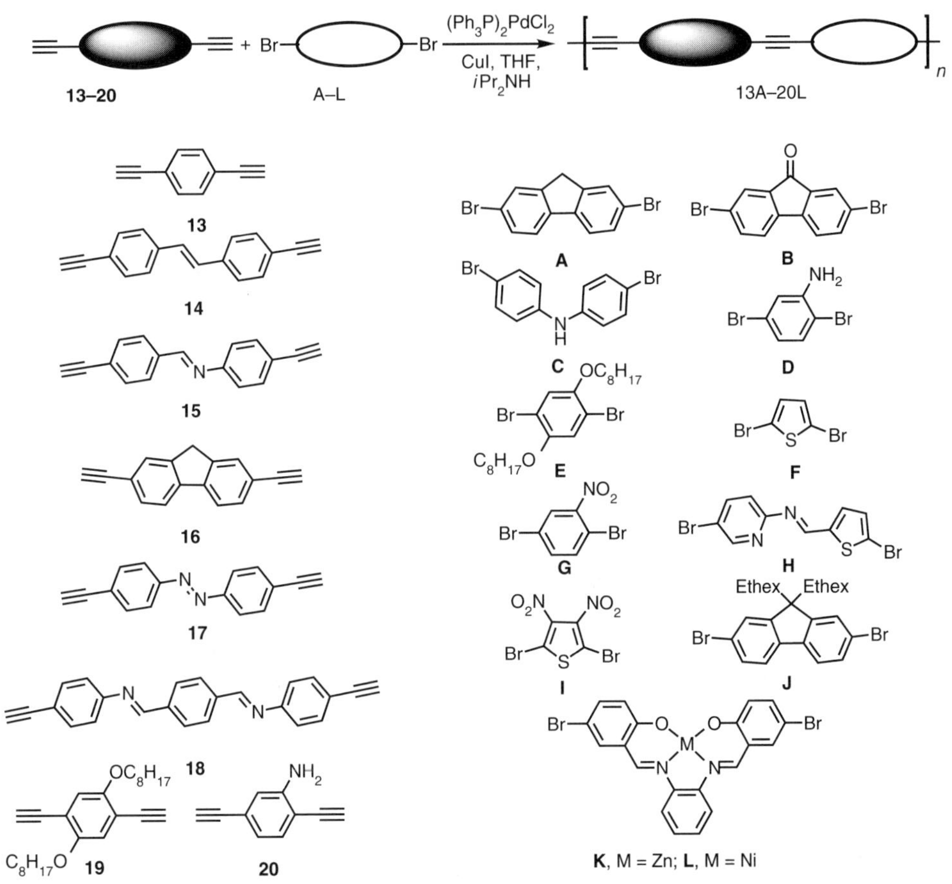

SCHEME 6.13 Synthesis of PAEs by a combinatorial approach.

known that dialkylfluorene-containing PAEs can show very high solid state quantum yields [41]. On the other hand, it was surprising that the salphene-containing PAEs **19K, L** were soluble and fluorescent. McLachlan reinvestigated Lavastre's claims and prepared **19K, L** independently [54]. He found that **19K, L** were highly insoluble and virtually not fluorescent both in solution and in the solid state. McLachlan has prepared heavily alkyl chain-substituted derivatives of the Lavastre polymers and showed that even these are quite insoluble and effectively nonfluorescent. McLachlan attributes the fluorescence of Lavastre's PAE **19K, L** to the presence of residual fluorescent dibromosalphene, and not to that of the polymer. While the concept of utilizing combichem to make and evaluate PAEs is clever, the problems involved are nontrivial. The use of (commercially available) aromatic dibromides instead of aromatic diiodides and the lack of powerful solubilizing groups make the results obtained by this specific approach somewhat problematic, but promising.

6.3 Important Classes of PAEs in Examples

6.3.1 Dialkyl and Dialkoxy-PPEs

The simplest PPEs are the dialkyl-PPEs **4** made by either alkyne metathesis of dialkyldipropynylbenzenes or by Pd-catalyzed couplings. The materials are thermally and photochemically very stable and form thermotropic liquid crystalline phases of the smectic variety [55]. Optical textures under crossed polarizers resemble those of nematic phases due to the homeotropic alignment of the chains on glass surfaces. In the solid state, dialkyl-PPEs **4** form lamellar structures [56]. Dialkyl-PPEs **4** form almost colorless, but blue fluorescent solutions in good solvents and yellow, less fluorescent solutions in poor solvents such as methanol [57]. Their spectroscopic data are discussed in Section 6.5.

Dialkoxy-PPEs are very similar to the dialkyl-PPEs, but considerably less stable. While dialkyl-PPEs can be melted without decomposition, dialkoxy-PPEs tend to degrade above 120°C and no thermotropic phases have been reported, however, nematic lyotropic preparations [58] were observed. The solid state ordering of dialkoxy-PPEs is similar to that of the dialkyl-PPEs [59]. Dialkoxy-PPEs are more electron rich and show a smaller bandgap than dialkyl-PPEs, i.e., their absorption and emission are redshifted with respect to that of the dialkyl-PPEs. Mixed systems have been reported by West, but seem to show properties that are more like those of dialkoxy-PPEs than those of the dialkyl-PPEs [10].

6.3.2   Grafted PPE Copolymers and PPE Block Copolymers

Several PPE-containing AB and ABA block copolymers have been made and investigated for their phase behavior, optical properties, and self-assembly [60–62]. Such AB block copolymers give well-developed morphologies, fibrous self-assembled nanotubes (Figure 6.1). Surprisingly, these materials do not form the interpenetrated networks observed by Hillmyer for polylactide/polystyrene (PS) blocks [63].

Okamoto, Swager, and others have made further PPE block types [64–66]. An ABA polyacrylamide–PPE–polyacrylamide triblock has been prepared by Kuroda and Swager [64] utilizing a di*tert* butylnitroxyl-substituted endcapper in the synthesis of PPEs. This hydroxylamine derivative was active in the radical polymerization of *N*-isopropylacrylamide and furnished the desired triblock **22**. However, **22** showed a multimodal molecular weight distribution and contained significant amount of the AB diblock; **22** was designed as a thermal precipitation assay, soluble in cold water but precipitating out at higher temperatures.

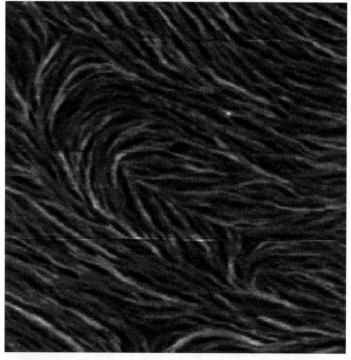

FIGURE 6.1   Self-assembled nanostructures of nanotubes from polymer **21**. The dimension of the picture is 5 × 5 μm. (From Leclere, P., et al., *Adv. Mater.*, 12, 1042, 2000. With permission.)

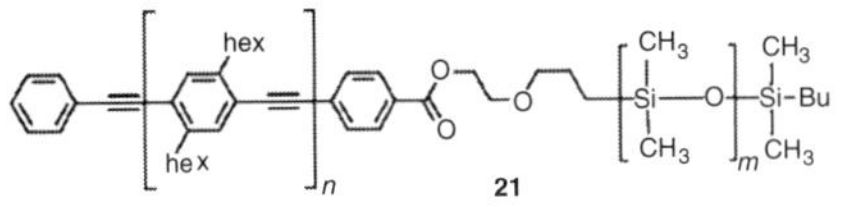

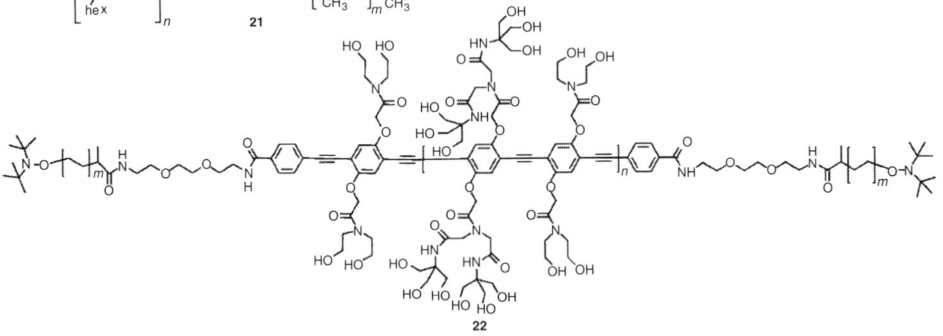

Polymeric substituents can be grafted onto PPEs (Scheme 6.14). The first example was reported by Wang et al. [67]. Starting from **23**, a tin-catalyzed reaction with caprolactone furnishes **24**, which was polymerized by the acetylene gas method to give **25** with a significant molecular weight. The grafts **25** were made to investigate the effect of the dilution of the (main chain: side chain) ratio on the structural and optical properties of PPEs. Thin films of **25** show a similar optical behavior as dialkyl-PPEs [57]. Upon annealing a significant blueshift (436–406 nm) of the absorption band took place. At the same time, a structural reorganization was observed. If one compares the powder diffraction pattern diagram of the pristine **25** to that of the annealed polymer, it is clear that the ordering of the side chains increases significantly. This behavior was interpreted such that the ordering of the side chains leads to a twist of the aryleneethynylene units in the solid state. Annealed samples of **25** show UV–vis spectra that are reminiscent of those of **4** obtained in dilute solution.

Breen et al. [68] (Scheme 6.15) utilized a postfunctionalization scheme to attach PS to the PPE backbone. The hydroxy-functionalized PPE **26** was treated with bromoacetylbromide and a subsequent ATRP of styrene onto **27** gave grafted copolymers **28** that could be doped into a PS-containing triblock

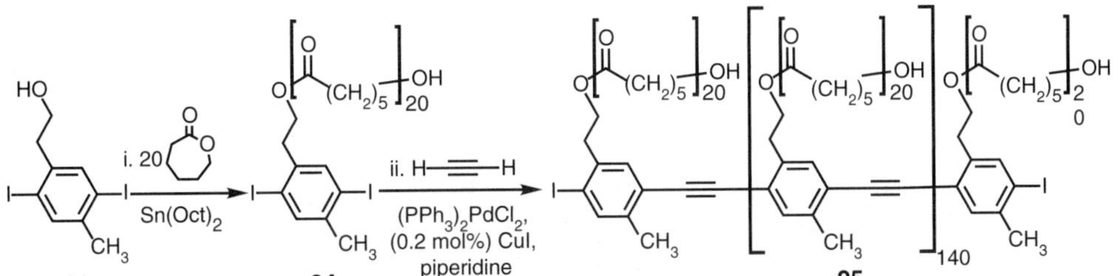

SCHEME 6.14 Synthesis of a grafted dialkyl-PPE.

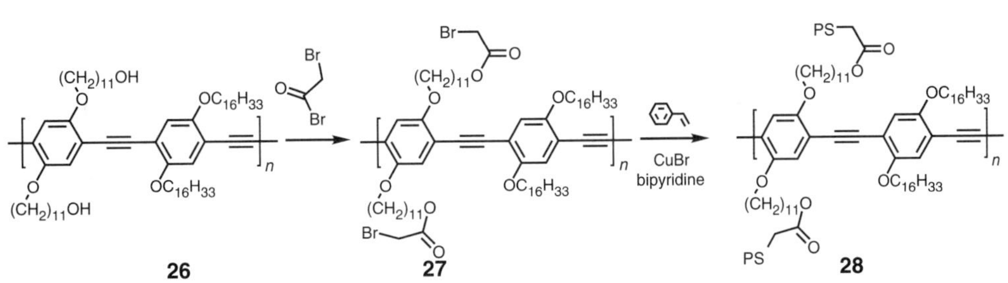

SCHEME 6.15 Synthesis of a grafted dialkoxy-PPE.

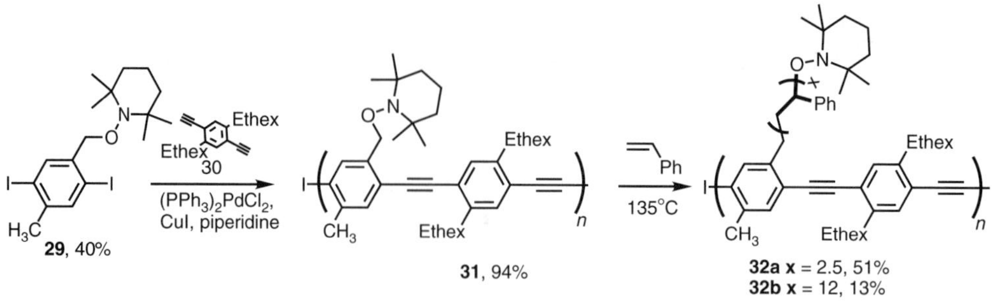

SCHEME 6.16 Synthesis of a grafted dialkyl-PPE.

copolymer with a cylindrical morphology. The PS graft **28** was roll cast into the triblock. The PPE backbone was sequestered into the cylindrical domains. Due to the roll cast orientation, the cylindrical phase was oriented in one direction and the PPE **28** shows a high degree of polarized emission. In further experiments the authors hydrolyzed the PS side chain of **28**, but did not find large differences in the molecular weights of **28** and its degrafted backbone.

Wang et al. [69] made a similar PS-substituted PPE utilizing nitroxyl-ATRP (Scheme 6.16). The polymer **31** was prepared by the condensation of **29** to **30**. The TEMPO functionality survived the conditions of the Sonogashira coupling. Addition of styrene to **31** at 135°C furnished **32** in excellent yields. The degree of polymerization could be adjusted by the amount of styrene that was added. According to GPC and viscosimetry, there was a significant difference in the molecular weights when compared **32** to the nongrafted PPE **31**, which is in variance with the results published by Breen et al. [68].

When chloromethylstyrene was utilized in the ATRP of **31** (Scheme 6.17), the polymer **33** was obtained. The benzylic chlorides do not interfere with the ATRP reaction, suggesting that the TEMPO group is a considerably more active initiator than the benzylic chloride [70,71]. Nucleophilic ring-opening polymerization with an oxazoline furnishes doubly grafted, water-soluble, fluorescent PPEs **34**.

6.3.3 Jacketed PPEs

Steric hindrance should prevent aggregation of PPEs. As a consequence, Aida prepared a series of dendronized PPEs by a Sonogashira reaction (Scheme 6.18) and investigated their emission quantum

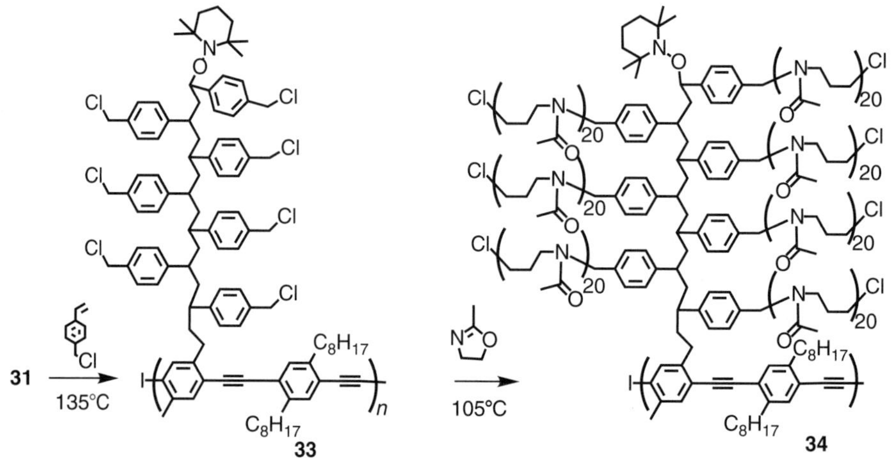

SCHEME 6.17 Synthesis of a doubly grafted dialkyl-PPE.

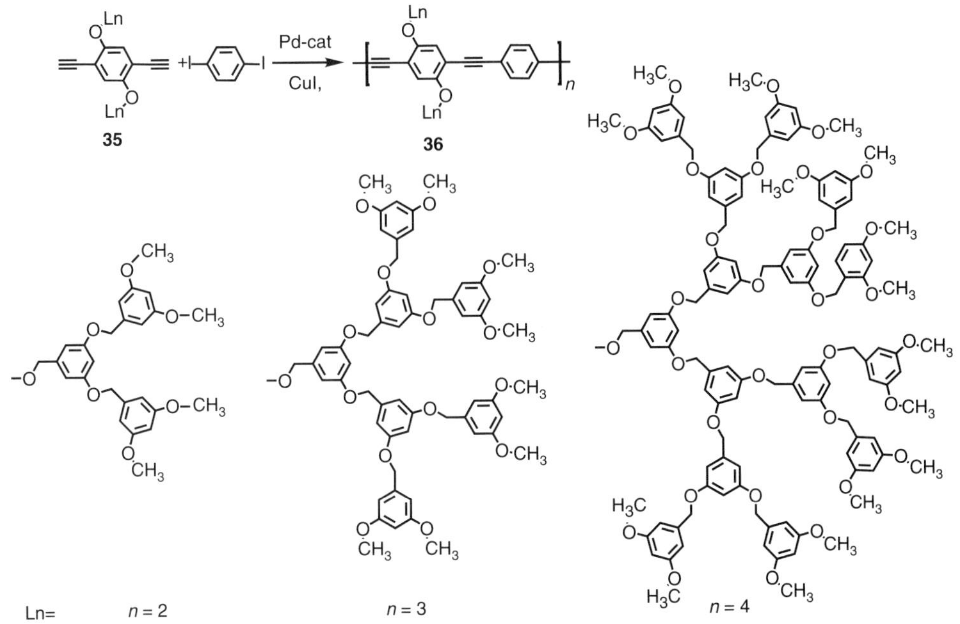

SCHEME 6.18   Synthesis of jacketed PPEs.

yield in dependence of their concentration [72–74]. He finds that the fluorescence quantum yield of **36L₄** to be almost independent of its concentration. This is not the case if the polymers are substituted with G-2 or G-3 dendrimers. The fluorescence quantum yield of the rods drops significantly at higher concentrations. It is believed that "...the large dendrimer framework in **36L₄** is likely to encapsulate the conjugated backbone as an 'envelope' and prevent the photoexcited state from collisional quenching...." These rods seem to harvest excitation energy if they are irradiated at 278 nm, because their emission at 454 nm is efficient, while no emission from the dendrimeric wedges is observed. These "dendrods" form unusual bagel-shaped nanostructures. When the dendrimeric wedges are decorated by ester functionalities, those PPEs displayed planarization of the backbone in solution.

The jacketed PPEs **38** (Scheme 6.19) are easily accessible and stable [75]. Their photophysical properties were evaluated with respect to the length of the chain that connects the conjugated

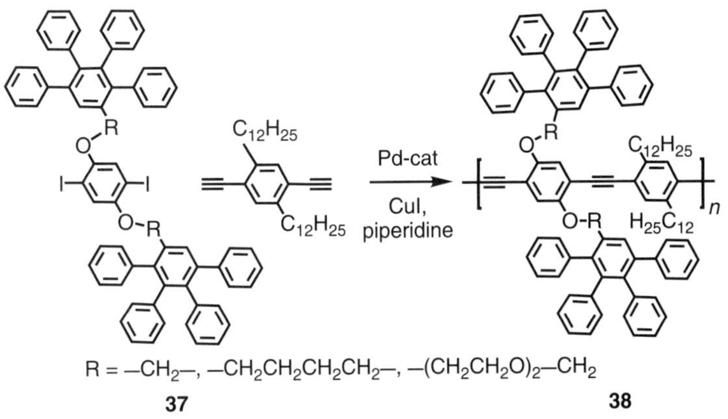

SCHEME 6.19   Synthesis of phenylene-dendron jacketed PPEs.

rod to the dendron wedge. The polymers **38** show a twist of the phenyleneethynylene unit in the solid state according to quantum chemical calculations and crystal structure analysis of a trimeric model compound.

6.3.4 Water-Soluble PPEs

Water-soluble PPEs **41** (Scheme 6.20) were investigated for applications in electroluminescent devices by Thünemann [76,77]. The structure and processability of this polyelectrolyte were manipulated by complexation with long-chain alkylammonium cations.

Schanze et al. [78,79] have imbued water solubility to PPEs by the attachment of sodium sulfonate side groups. The polymer **42** is highly aggregated in water, but dissolves unaggregated in methanol or other polar solvents. Fluorescence quenching with methyl viologen (paraquat, $K_{SV} \approx 10^7$) and its larger congeners is very effective due to the electrostatic interaction between quencher and polymer. Quenching is most effective in systems where the quencher leads to aggregation of the polymer system. The Stern–Volmer relationship was linear at very low quencher concentrations, but at higher quencher concentrations the graphs bent upward, suggesting that enhanced quenching occurs [80]. The quenching efficiency of these systems was defined by Schanze through the Stern–Volmer constant (K_{SV}) and the concentration of quencher necessary to decrease the fluorescence to 10% of its original value. The triscationic quencher **43** was particularly effective in quenching the fluorescence of **42**. It induced aggregation and served as an example of "self-amplified quenching." When a boronic acid containing quencher **44** was utilized, this system was sensitive in the detection of saccharides (see Section 9.2) [81]. The quencher–polymer complex (**42·44**) is non emissive and the addition of fructose, glucose, or galactose leads to a 50–100-fold increase in fluorescence. The complexation of **44** with saccharides leads to the formation of a noncharged complex (**42·45**) that binds less efficiently to the anionic PPE and is considerably less electron accepting than the starting viologen derivative and therefore does not quench PPE **42**'s fluorescence efficiently.

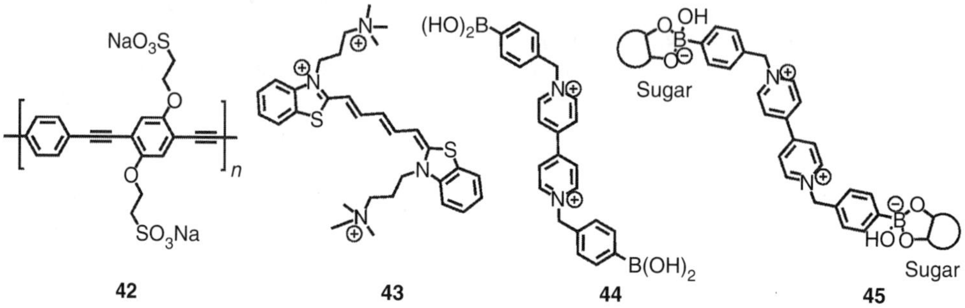

     42   **43**   **44**   **45**

A similar PPE (Scheme 6.21) containing phosphonate groups has been reported by the coupling of 1,4-diethynylbenzene to **46** [82]. The polymer **48** is soluble in organic solvents. Deprotection by boron tribromide in lutidine and hydrolysis of the mixed boronic and phosphonic acid anhydride in sodium

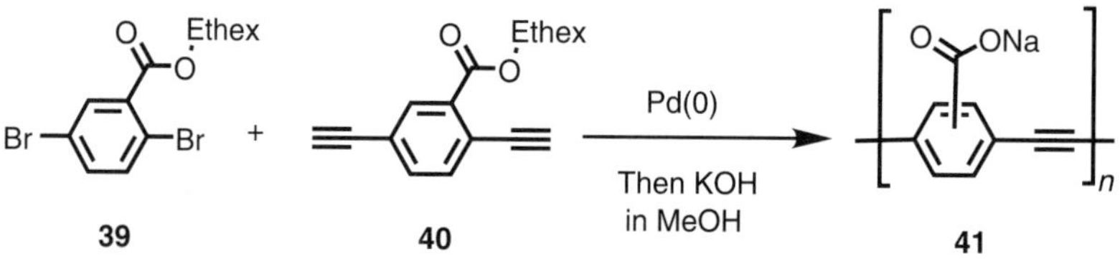

SCHEME 6.20 Synthesis of a carboxylate-substituted PPE.

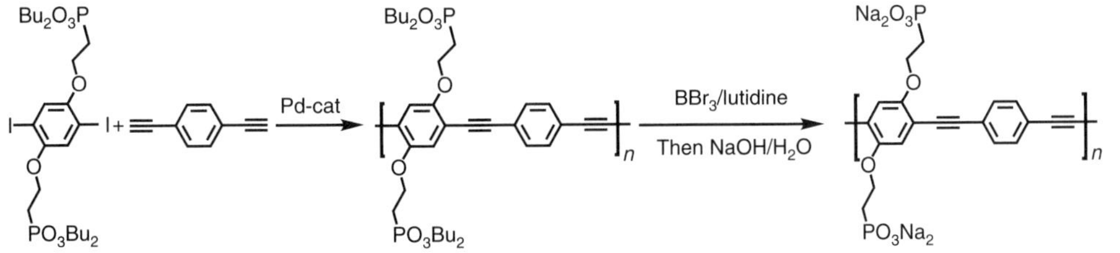

SCHEME 6.21   Synthesis of a phosphonate-substituted, water-soluble PPE.

hydroxide gives **49**. The phosphonate PPE **49** shows spectroscopic properties that are similar to those reported for **42**. Quenching with viologen salts shows K_{SV} constants in the same range as observed for **42**. The PPE **49** is amenable to layer-by-layer deposition. When treated with $ZrOCl_2$, a 10 bilayer containing, emissive preparation was utilized in an LED to emit light at 570 nm at a turn-on voltage of 6 V.

The water-soluble polymer **53** showed fluorescence, while **54** showed phosphorescence (Scheme 6.22) [83]. The luminescence quenching of **53** and **54** with viologen was investigated. The polymer **53** shows amplified—but static—quenching due to the short lifetime of the excitation (0.3 ns) while **54** shows dynamic, but not amplified, quenching if viologen types are added. The lifetime of **54**'s phosphorescence is 19 μs, i.e., almost 10^5 times higher than that of **53**, which is the reason for the dynamic quenching. The lack of amplified quenching in **54** was assigned to the spatial confinement of the exciton to 1.5 repeating units, while in **53** the singlet exciton is extended over 10–20 repeating units. Therefore, amplified quenching is the main mechanism of loss of fluorescence. Issues of energy transfer and details of typical quenching experiments are treated in Section 9.2 in more detail. Recently, nonionic fully water-soluble PPEs (Scheme 6.23) have been made by Sonogashira coupling of suitably minidendron-

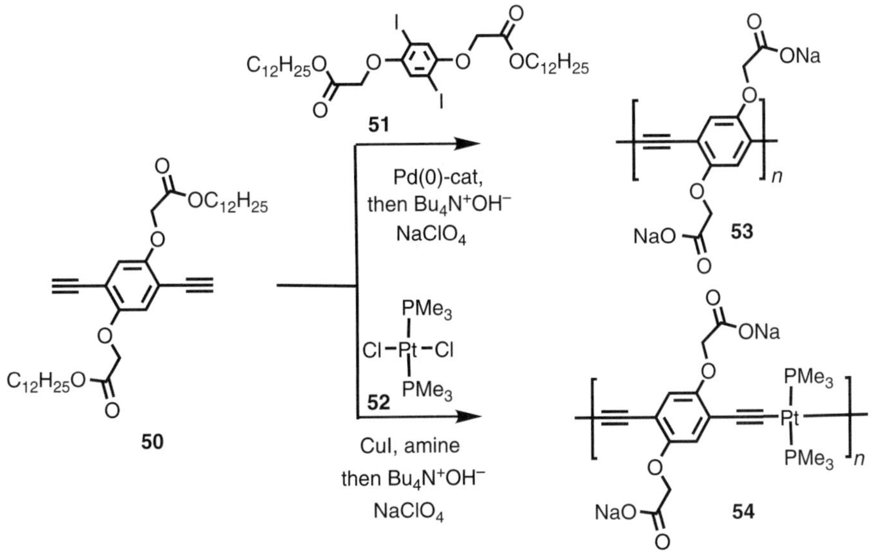

SCHEME 6.22   Synthesis of a carboxylate-substituted organometallic PPE and its organic congener.

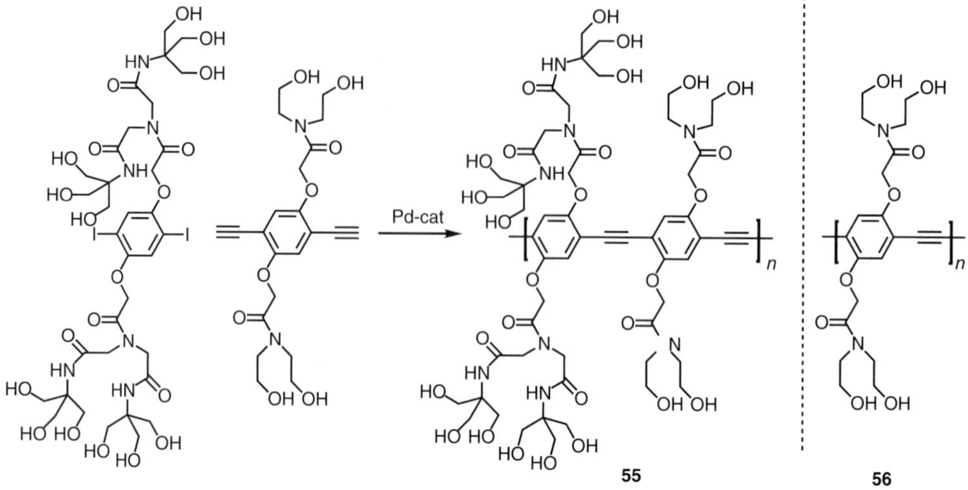

SCHEME 6.23   Water-soluble nonionic PPEs.

substituted monomers. These polymers do not aggregate in aqueous solution, but further properties were not evaluated [84,85].

6.3.5   *Meta*-PPEs

Meta-oligoPEs and *meta*-PPEs have been made and studied but are considerably less interesting as electronic materials than PPEs. Nonetheless, they were instrumental in investigating the folding of phenyleneethynylene chains into helical structures [9,86,87]. PPE types **57** in which 1,3- and 1,4-units are present have been reported [88–91]. These show increased flexibility and fluoresce, despite the *meta*-bridge, in the green. Their emissive properties are similar to those of the dialkyl-PPEs, suggesting that linear conjugation can be substituted by the suitable connection of low-bandgap oligomers utilizing cross-conjugated modules.

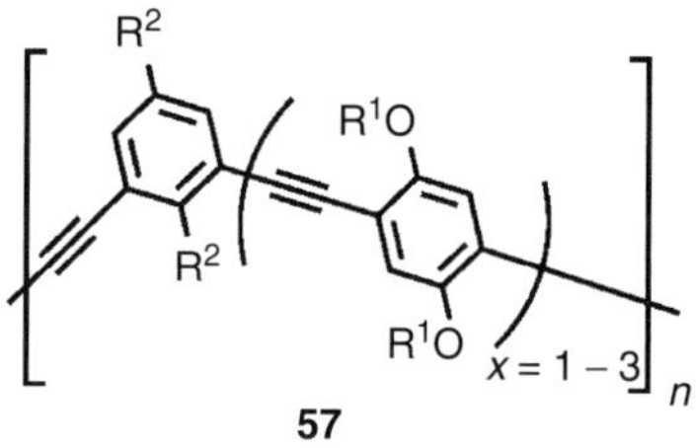

57

Tew and Arnt [92–94] have prepared amphiphilic *meta*-PPEs **58** and investigated their behavior at the air–water interface where they prefer an extended conformation. These PPEs are active in the lysis of phospholipid vesicles and might find application as bactericidal substances. While the derivative **58** with R = H forms clear solutions in water, the alkoxy-substituted congeners aggregate and precipitate out upon addition of water to their solution in DMSO. The structure of **58** in the solid state is assumed to be helical rather than extended. If acrylic ester substituents are placed on **59**, it is possible to capture the helical intermediates by a photochemical 2 + 2 cycloaddition (Figure 6.2) [95,96]. The presence of a fixated helical structure was evident, because addition of chloroform to the cross-linked form *did not*

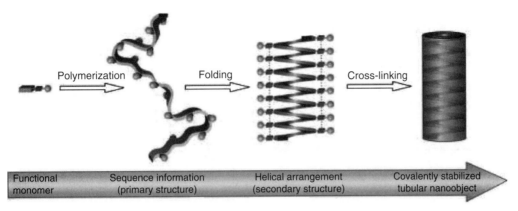

FIGURE 6.2 Synthesis of a cross-linked tubular PPE array by photochemical [2 +2] cycloaddition of **59**. (From Hecht, S. and Khan, A., *Angew. Chem.*, 42, 6021, 2003. With permission.)

result in spectral changes, strongly suggesting that unfolding or deaggregation is no longer possible. When mixed *meta–para*-PPEs **60** are utilized, their solvent-induced self-assembly is favored by enthalpy and disfavored by entropy [97]. Hydrophobic effects do not seem to play a large role in **60**. The emission of the extended form is structured and centered at 390 nm, while the emission of the folded form is broad, excimer-like, and has a maximum at 450 nm. In addition, the fluorescence of the helical form is more sensitive to the addition of external quenchers of the viologen type than that of the extended form (K_{SV} lin $= 2.6 \times 10^5$, K_{SV} hel $= 1.1 \times 10^6$).

R = H, OC$_6$H$_{13}$, OC$_{12}$H$_{25}$,

58 Tew　　　　　　　　　　**59** Hecht　　　　　　　　　**60** Schanze

6.3.6　Linear and Crossed PPE–PPV Hybrids

Linear PPE–PPV hybrids of the *meta* and the *para* varieties (**6**, **61–64**) have been produced by Sonogashira coupling and alkyne metathesis. These polymers show properties that are intermediate between those of the PPVs and the PPEs. Klemm et al. have shown that Horner reactions of suitable substrates lead to linear PPE–PPV hybrids, although the degree of polymerization is not particularly high [98]. The motivation behind these synthetic efforts is to combine the favorable hole injection properties of the PPVs with the blue emission, the stability, and the high fluorescence quantum yields of the PPEs. Several attempts have been made to obtain efficient LED action, but **62**, the best one, so far shows a luminance of 147 cd/m^2 at 602 nm, at 14 V, which is not very impressive compared with the numbers that are obtained for PPVs [99–102]. Cross-conjugated PPEs were first synthesized by Bunz et al. (**63**, **64**) and are useful as photodiode-active materials [102]. The dialkylamino substituents in

63 are so electron donating that photocurrents cannot be suppressed in a transistor arrangement even if a large cutoff current is applied [103]. PPEs such as **63** and **64** should be attractive in metal ion sensing, due to the binding capabilities of the attached substituents.

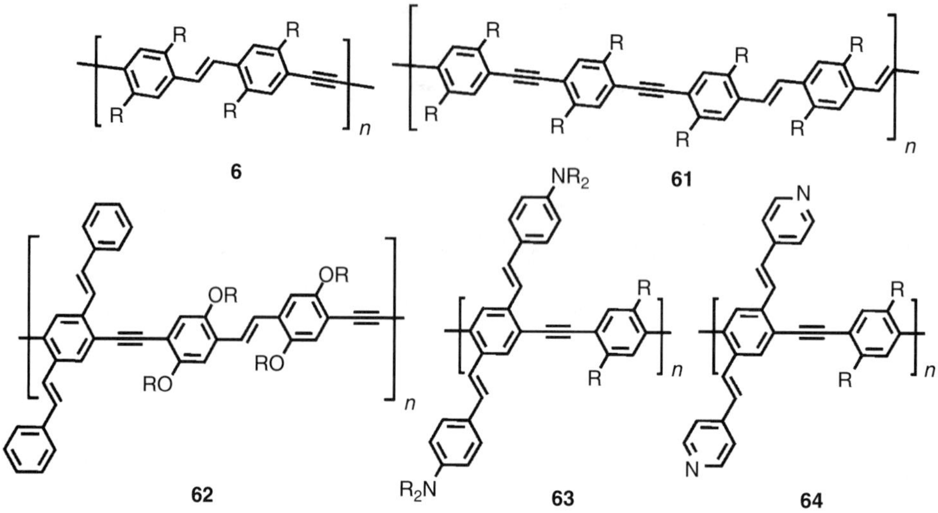

6.3.7 Organometallic PAEs

The first organometallic PPEs (**66**, **67**, Scheme 6.24) were obtained by the reaction of **65** with a series of 2,5-diiodo-1,4-dialkylbenzenes and 2,5-diiodo(3-alkyl)thiophenes by a Sonogashira protocol [104,105]. Later, elongated monomers were prepared by Steffen et al. [106] to give organometallic PAEs **69** by alkyne metathesis. All of the investigated materials are nonfluorescent but show distinct ordering on the nanoscale and are thermotropic or lyotropic liquid crystalline. While the benzene-containing polymers show a nematic ordering, the thiophenes induce a smectic ordering in **67** (Figure 6.3).

Yamamoto and Plenio introduced ferrocene units into PAEs and prepared redox-active PAEs **70** [107,109]. Yamamoto coupled 1,1′-diiodoferrocene to a diethynylarene by either a Sonogashira reaction or by utilizing alkynylgrignard reagents (Scheme 6.25). Plenio reported poly(1,3-ferrocenyleneethynylene)s **70d** (Scheme 6.26) by a clever metallation strategy. Polymers **70** are structurally novel but have not found any applications. They might be useful as active layers in transistor-types due to their ease of oxidation.

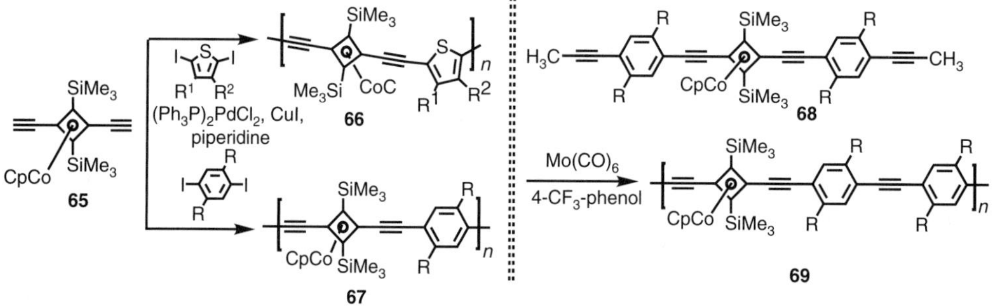

SCHEME 6.24 Synthesis of organometallic PAEs.

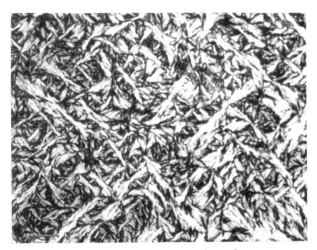

FIGURE 6.3　Polarizing micrograph of a texture of polymer **66**. $R^1 = R^2 = $ hexyl.

6.3.8　Thiophene-Containing PAEs

Alternating terpolymers [–≡–thienyl–≡–aryl–]$_n$– have been made by a Sonogashira coupling [110–113]. When compared with the dialkyl- and dialkoxy-PPEs, they show markedly redshifted absorption and emission. Poly[2,5-(3-hexyl)thienyleneethynylene] shows an absorption in solution at 438 nm, while its emission is centered at 510 nm. If in a terpolymer hexylthiophene and dialkoxybenzenes are interspersed by alkyne groups, the absorption shifts to 457 nm, while the emission is centered at 492 nm. For terpolymers in which benzene groups are utilized, absorption and emission are recorded at 418 and 460 nm, respectively. As a consequence, their spectral properties are intermediate to those of the PPEs and of the poly(thiophenyleneethynylene)s. The redshifted emission might make their use attractive in biological and sensory applications. As discussed, Moore has recently developed an alkyne metathesis strategy toward the synthesis of poly(thienyleneethynylene)s [49].

6.3.9　Nitrogen Heterocycles-Containing PAEs

Most nitrogen-containing heterocycles (e.g., pyridine, pyrimidine, pyrazine, quinoline, quinoxaline, or benzothiodiazole) are electron deficient. If these heterocycles are introduced into PAEs, polymers with

SCHEME 6.25　Synthesis of PAEs containing ferrocene units.

SCHEME 6.26　Synthesis of PAEs containing ferrocene units. PdCl$_2$(PPh$_3$)$_2$, CuI, solvent, iPr$_2$NH, and reflux.

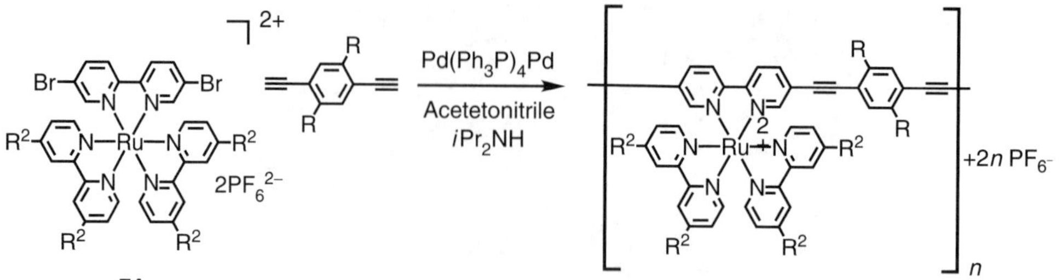

SCHEME 6.27   Synthesis of a bipyridine-containing PAE.

markedly different electronic structures result and due to their ability to coordinate metals, these PAEs could find use as sensory materials. If bipyridine units (Scheme 6.27, PAEs **71–73, 75**) are incorporated, metal coordination is strong for **71–75**. Their photophysics was investigated and is dominated by the organometallic fragments. Details of these systems were recently covered in a review and are not discussed further here [114–117].

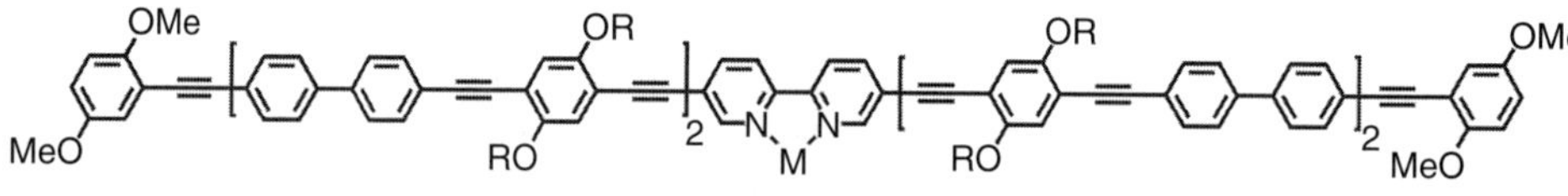

71: M=–; Re-4; Re(CO)$_3$CL

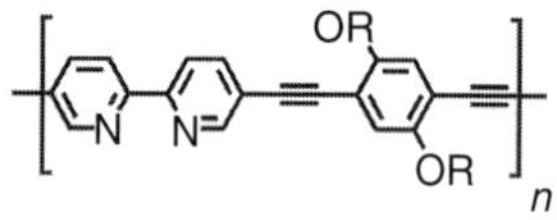

Dialkynyl monomers **76** and **77** undergo Sonogashira coupling to give quinoline-containing PAEs **79** (Scheme 6.28 and Scheme 6.29), stable with strong yellow to orange emission and high quantum yields in the solid state [118]. The direct coupling of diiodo- (**80**) or diethynylquinoline (**83**) to **30** or **82** under Sonogashira conditions formed PAEs **81** and **84** with a P_n of 95 and 48, respectively [119]. PAEs **81** and **84** are blue emissive in solution. Upon protonation, **81** gave yellow emission, while the emission of **84** was quenched. Exposure of **81** and **84** to metal cations led to either shift in the emission or to quenching. Particularly, Pd^{2+} was an excellent quencher of fluorescence for both **81** and **84**.

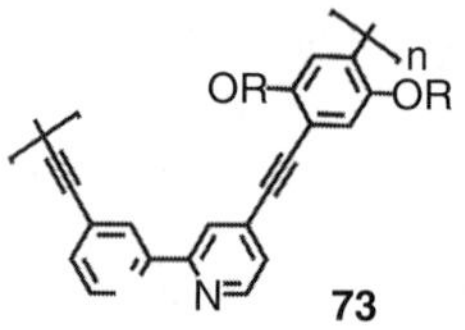

Similarly, a Pd-quenched (K_{SV} of 4.3×10^5), pyridyl-containing PAE **85** was synthesized [120]. The combination of flexible recognition elements and rigid rod energy transfer elements was hypothesized to be important for the efficiency of this sensing system. When shorter 1,4-phenyleneethynylene sequences were interspersed between the pyridine units, the quenching efficiency decreased.

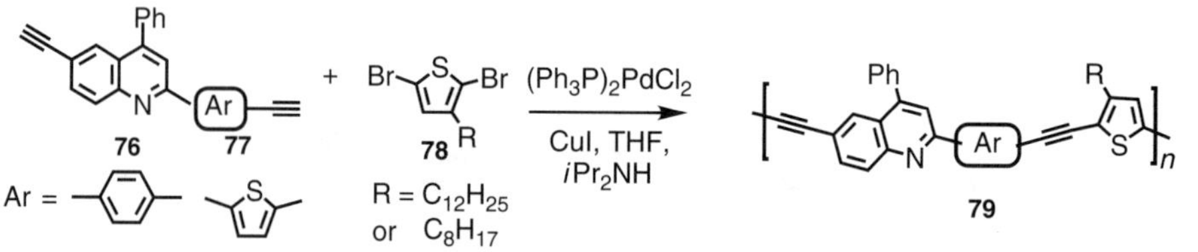

SCHEME 6.28   Synthesis of an arylquinoline-containing PAE.

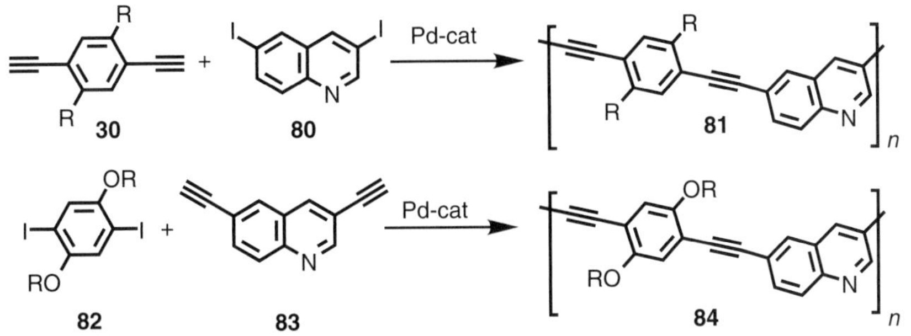

SCHEME 6.29 Synthesis of quinoline-containing PAEs.

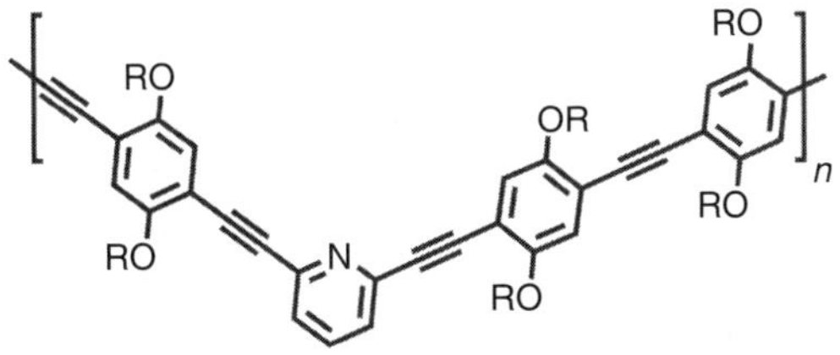

The quest for *n*-semiconducting and lower-bandgap PAEs led to **88** (Scheme 6.30) isolated as golden green lustrous solid. Its reduction potential is low (-1.42 V vs. Ag/Ag$^+$) and a bandgap of 1.9–2.0 eV results. For **88b**, the bandgap is somewhat lower than for **88a** [121–124]. A lamellar structural model has been proposed for **88b**, based on powder X-ray diffraction. Yamamoto reported the synthesis of **88b** by coupling dibromobenzothiadiazole to 1,4-diethynyl-2,4-bis(hexyloxy)benzene. Multiple attempts to repeat Yamamoto's experiment were not successful in the hands of the author of this contribution. Due to the very similar spectroscopic properties of **88a** and **88b** obtained by Bangcuyo and Bunz and by Yamamoto, it is however clear that Yamamoto et al. have made **88b**.

If the benzothiadiazole unit is exchanged for a quinoxaline ring (Scheme 6.31), an amorphous heterocyclic PAE **90** results that shows similar spectroscopic and electrochemical properties as **88**; **90**

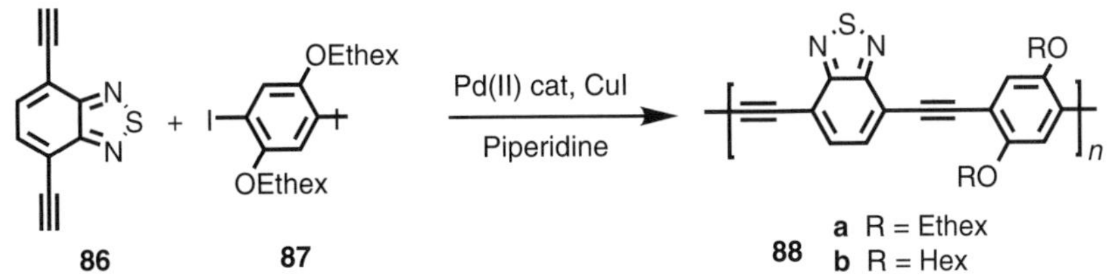

SCHEME 6.30 Synthesis of benzothiadiazol-containing PAEs.

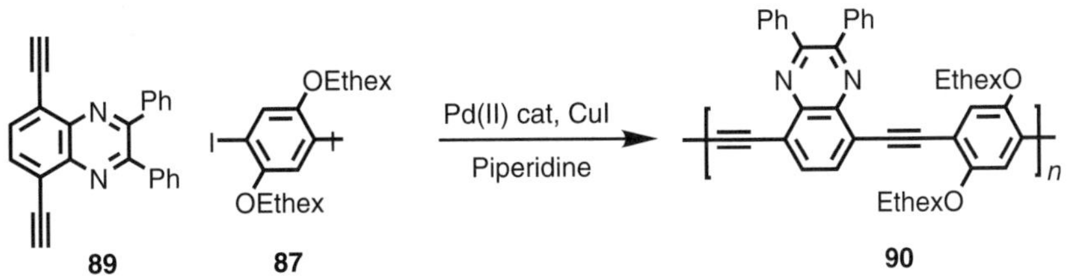

SCHEME 6.31 Synthesis of quinoxaline-containing PAEs.

is somewhat less easily (-1.54 V vs. Ag/Ag$^+$) reduced than **88** and **90**'s fluorescence is quenched by addition of trifluoroacetic acid or silver(I) salts [125].

6.4 Reactions of PPEs

PPEs should be amenable to postfunctionalization schemes. Side chain manipulation of PPEs has been discussed for the case of grafted polymers. Nuyken [126] has shown that a benzylic bromide-functionalized PPE is easily made by the polymer analogous conversion of a benzylic alcohol. This polymer is poised to react with any nucleophile to give side chain functionalized PPEs. Postfunctionalization schemes furnished biotinylated or mannosylated PPEs are useful in sensory schemes (vide infra) [27,127–131].

6.4.1 Reduction of Triple Bonds in PPEs

Catalytic hydrogenation of the triple bonds in dialkyl-PPEs **4** was effected by Wilkinson's catalyst at above 300°C and hydrogen pressures of 500 bar (Scheme 6.32) [132]. Only the triple bonds of **4** were reduced, but not the arene rings. Under less forcing conditions the triple bonds were only partially reduced. A material resulted in which triple, double, and single bonds would connect the phenylene units. The lack of reactivity of **4**'s triple bonds is surprising, but may be due to the steric hindrance that is exerted by the alkyl groups adjacent to the alkynes. To reduce dialkoxy-PPEs **7**, a milder method involving tosylhydrazine in the presence of tripropylamine is available (Scheme 6.32): diimine is added across the triple bond and forms **92** [133]. Attempts to reduce *dialkyl*-PPEs **4** by this method were not successful; **4** is inert under these conditions.

Successful reduction of the triple bonds of PPEs transforms a rigid rod, or more correctly, a wormlike semiflexible chain into a floppy coil (Figure 6.4). The determination of the molecular weight of a rigid rod polymer is not trivial, because GPC, typically with polystyrene (PS) as standard, overestimates the molecular weights. It is not clear, however, *how much* the molecular weight is overestimated. Molecular weight determination of PPEs by light scattering was reported by Cotts et al. [134] who demonstrated that PPEs are semiflexible with relatively short persistence lengths. With the pair of **4** and **91**, however, direct molecular weight comparison of the PPEs with their reduced congeners was possible by GPC. A scaling factor of 1.4 by which the PPE's molecular weight is overestimated, was obtained [135]. Consequently, GPC is a reasonable tool to obtain molecular weights for PAEs.

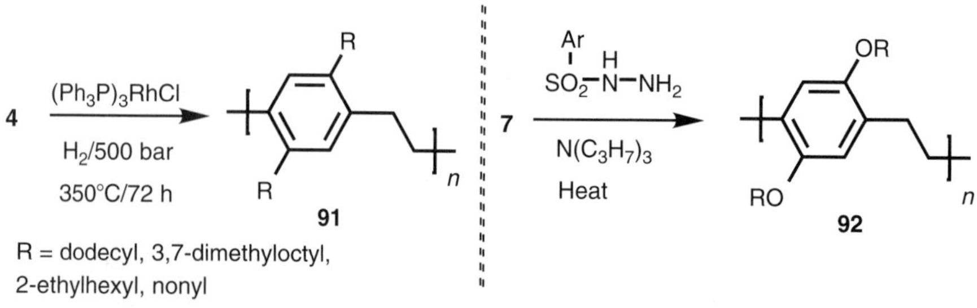

SCHEME 6.32 Hydrogenation of PPEs.

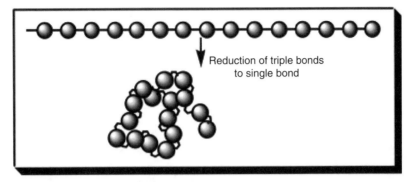

FIGURE 6.4 Transformation of a rigid rod into a flexible coil. (From Ricks, H.L., et al., *Macromolecules*, 36, 1424, 2003. With permission.)

6.4.2 Complexation of Triple Bonds in PPEs

The complexation of PAEs with organometallic fragments (Scheme 6.33) should lead to PAEs with greatly diversified properties. Scholz et al. [136] investigated the reaction of PPEs with dicobalt octacarbonyl: according to NMR and IR spectroscopies all of the triple bonds are occupied by dicobalt hexacarbonyl fragments. While hydrogenation and Diels-Alder type reactions of PAEs only occur under harsh conditions, the reaction with organometallic fragments is surprisingly smooth. When the dicobalt hexacarbonyl-coated PPE **93** was pyrolyzed at 500–600°C, microscopic spheres that formed contained only cobalt and carbon (Figure 6.5).

Neenan and Callstrom took phenylenebutadiynylene **94** and treated it with a Pt source to give the Pt(PPh$_3$)$_2$-functionalized polymer **95** (Scheme 6.34) [137]. Pyrolysis of **95** at temperatures above 600°C led to Pt nanoparticles embedded in a glassy carbon matrix, with attractive catalytic properties. The Pt-theme was revived by Weder who (Scheme 6.35) reacted **7** with [styrene$_2$Pt$_2$Cl$_4$] to give soluble **96** in which a significant portion of the triple bonds was complexed by the [styrene-Pt$_2$Cl$_4$] fragment [138]. The organometallic PPEs **96** showed somewhat blueshifted absorption and strongly quenched fluorescence. When thin films were cast from solutions of **96** in a halogenated solvent, the films (**97**) were insoluble and nonfluorescent. The insolubility is a consequence of cross-linking of two or more PPE chains by a Pt$_2$Cl$_4$ unit due to the decomplexation and evaporation of the formerly bound styrene ligand. Upon addition of an excess of styrene to **97** it was possible to dissolve the polymeric complex in organic solvents under reformation of **96**. This approach is important for reversibly cross-linking PPEs into insoluble semiconductor films.

SCHEME 6.33 Synthesis of the dicobalt hexacarbonyl-decorated PPE **93**.

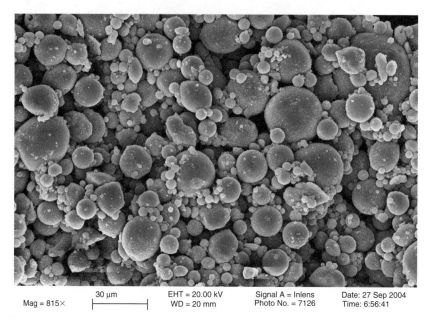

| | 30 μm | EHT = 20.00 kV | Signal A = Inlens | Date: 27 Sep 2004 |
| Mag = 815× | ⊢————⊣ | WD = 20 mm | Photo No. = 7126 | Time: 6:56:41 |

FIGURE 6.5   Carbon–cobalt spheres from the pyrolysis of dicobalt–hexacarbonyl-decorated PPEs. (From Scholz, S., et al., *Adv. Mater.*, 17. 1052, 2005, in press. With permission.)

94   **95**

SCHEME 6.34   Synthesis of Pt-complexed oligo(phenylenebutadiynylene)s.

96   **97**

SCHEME 6.35   Synthesis of reversible PPE networks utilizing a Pt$_2$Cl$_4$ cluster.

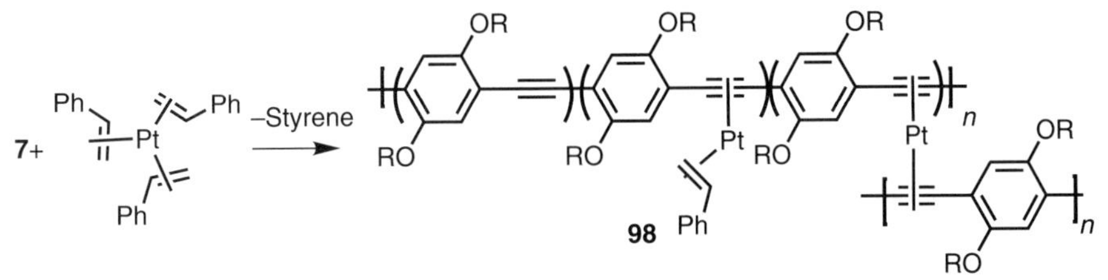

SCHEME 6.36 Synthesis of irreversible PPE networks as high charge carrier networks.

If [styrene$_3$Pt] is utilized as complexating agent instead of [styrene$_2$Pt$_2$Cl$_4$] (Scheme 6.36), irreversibly cross-linked networks of **98** are formed and addition of styrene does not lead to a reversible uptake of this ligand. Network **98** is of great interest because it shows substantial hole and electron mobilities (1.5×10^{-2} cm^2/V s) when investigated by time of flight (TOF) spectroscopy [139–141]. There is an increase of mobility by a factor of 10 when compared with the mobility of **7** in thin films. PPEs **7** as well as their organometallic complexes show a negative field dependency of their mobility. In TOF, a short intense pulse of light creates a layer of charge carriers in a thick polymer film. An applied electric field moves the charge carriers to the oppositely charged electrodes. The time dependence of the conductivity is recorded. From that measurement and the known film thickness, the mobility of electrons and holes in the sample is determined.

If the complexation of the PPE is carried out with [styrene$_3$Pt] in styrene as solvent, red, nonfluorescent PPE gels **98** with a significant mechanical resiliency are formed. The red color of these gels is due to a charge transfer band of **98** that appears at 650 nm. The concept of cross-linking PPEs into polymer networks allows the introduction of PPEs into mixed conducting polymer applications. Organometallic cross-linkers lead to a loss of fluorescent quantum yield, and as a consequence Weder et al. investigated the synthesis of cross-linked PPEs (Scheme 6.37) in the presence of water and sodium dodecyl sulfate under ultrasonication to obtain either micron (3–6 µm) or nanometer (100–400 nm) sized, highly fluorescent PPE spheres **99** that swell in toluene [142,143]. Light cross-linking does not change the optical properties of the PPEs much, however, if a large amount of 1,2,4-tribromobenzene cross-linker is added, a PPE network **99** with significantly redshifted emission is observed.

6.5 Spectroscopy, Aggregation, and Solid State Structures of PAEs

PAEs display extended π-systems that make them colored and fluorescent. Solutions of dialkoxy-PPEs **7** are yellow and turquoise fluorescent, while solutions of dialkyl-PPEs **4** in chloroform are almost

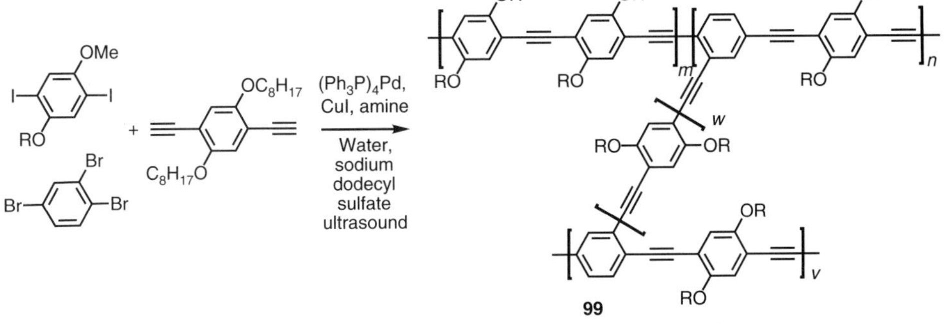

SCHEME 6.37 Synthesis of PPE networks by Sonogashira coupling.

colorless but intensely bluish-purple fluorescent [57]. In the solid state **4** are yellow while **7** are orange [57,87]. Depending upon sample treatment and history, such PPEs can either be very strongly or only slightly fluorescent.

6.5.1 UV–Vis Spectra

The difference between solutions of PPEs in good solvents and in thin films is dramatic. In dialkyl-PPEs **4** the color changes from almost colorless to deep yellow. Weder and Wrighton [6–8] were the first who recorded the differences of PPEs' spectra in solution and in the solid state. The thermotropic dialkyl-PPEs were later extensively investigated to unearth the origin of this effect [55–57,144]. There could be four explanations for the observed redshift in the UV–vis spectra of PPEs upon going from solution into the solid state [57,87]: (1) Formation of electronic aggregates, similar to J- or H-aggregates found in cyanine dyes. In this model the PPE chains show a strong electronic ground state interaction and this ground state interaction, i.e., aggregate formation, leads to the observed redshift [144,145]. (2) Planar-ization of the backbone of the PPEs. In this model the observed redshift would be purely a single-chain phenomenon due to the planarization of the backbone leading to increased conjugation. (3) Agglom-eration of the PPE chains occurs either upon going into the solid state or upon addition of a poor solvent to the solution of a PPE in a good solvent. The mechanical aggregation brings about the planarization of the PPE backbone and results in a concomitant redshift. While this would be a multichain process it would be reported by an optical response that only involves the π-system of a single chain. In this model there is no electronic interaction between the PPE backbones but chain aggregation. (4) A synthesis of explanations (1) and (3) would contend that there is a large conforma-tional effect to the observed spectral changes, but overlaid is a contribution of the chromophores interacting in the electronic ground state.

For different PPEs different explanations might be correct. In one case only planarization might play a role, while for another PPE the assumption of electronic ground state aggregates might be more reasonable. Bunz et al. have investigated the spectroscopic properties of dialkyl-PPEs, while the group of Swager has investigated donor–acceptor PPEs [55–57,87,144–149]. Conflicting interpretations were obtained. Bunz et al. favor the concept of planarization of the main chain evoked by mechanical aggregation of the PPEs as an overwhelming contributor for the spectral changes, while Swager et al. contend that electronic interchain interactions in the sense of ground state aggregates play a significant role in the changes of the spectral and chiroptical properties of PPEs.

What is the evidence for each point in case?

- *Dialkyl-PPEs*: in solution, all of the phenyleneethynylene units are rotationally disordered, because the barrier of rotation for diphenylacetylene has been measured and calculated to be below 1 kcal/mol [144]. A Boltzman-type distribution of the twist angles of neighboring phenyleneethy-nylene units in the macromolecular chains most likely results. In the solid state, powder diffrac-tion and electron diffraction of dialkyl-PPEs and single crystal X-ray diffraction of oligomeric model compounds are highly suggestive of an exclusively planar packing in the solid state [150]. Quantum chemical calculations utilizing the PM3 basis set on octameric model compounds **100P** and **100T** predict large differences in bandgaps of twisted vs. planarized phenyleneethynylenes [144]. Upon melting dialkyl-PPEs, their UV–vis spectra become superimposable to that of the PPEs in dilute solution, giving no sign of interchain interactions in the melt (Figure 6.6) [144]. *Ground state electronic interchain interactions are weak or nonexistent in dialkyl-PPEs.*

Twisted **100T**   Planar **100P**

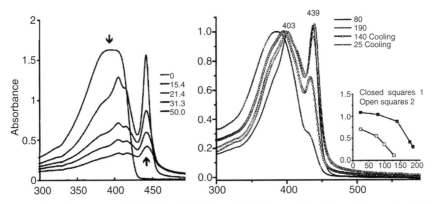

FIGURE 6.6 Solvatochromicity of bis(ethylhexyl)-PPE **4c** (left). Thermochromicity of **4c**. *x*-axis is given wavelength in nanometer.

- *Donor–Acceptor-PPEs*: Swager and Kim have performed Langmuir–Blodgett experiments on donor–acceptor-substituted PPEs (Figure 6.7). From the data it is credibly concluded that conformational effects are important and that ground state electronic interchain interactions must play a significant role [145–148]. When utilizing a pair of donor- and acceptor-substituted PPEs, electronic ground state aggregate formation will be favorable. The point is made by comparing **101** with **102** at the air–water interface. While **101** lays flat on the water surface, **102** is preferentially organized side ways at the interface by virtue of the suitably selected substituents. Due to this ordering, the authors contend, in **102** there will be interchain interaction while in **101** there will not be. According to the UV–vis spectra, the absorption of **101** is blueshifted as compared with that of **102**. Additionally, **101** experiences a redshift upon collapse of the LB layer. In the collapsed form the interchain interactions will be "turned on," explaining the observed redshift. In the case of **102**, collapse does not lead to a change in the interchain interaction and no change is visible in the UV–vis spectrum.

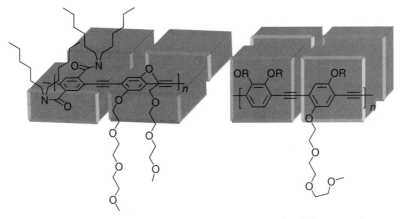

FIGURE 6.7 Intrachain vs. interchain electronic interactions in Langmuir–Blodgett experiments on the model polymers **101** (left) and **102** (right).

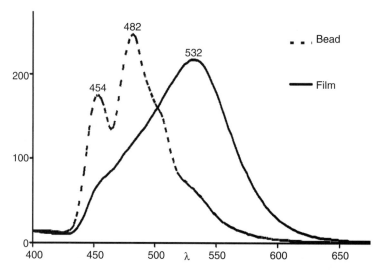

103   **104**

Tew demonstrated that, despite aggregation, spectral changes can be suppressed if large substituents are attached to the PPE's side chains [151]. A case in point was the investigation of **103** and **104** carrying either an amino group or a Boc-protected amino group. Both PPEs aggregate upon addition of methanol to their solution in chloroform, however, **103** did not show any changes in its UV–vis spectrum while **104** *did* show significant changes. This type of behavior may be attractive in biosensing applications, where one might not want the conjugated polymers' response in water to be determined by aggregation phenomena.

6.5.2   PPE's Properties in the Excited State

PPEs are highly emissive in solution and—depending upon sample history, thermal treatment, and side chains—can be very emissive as solids. In thin films, bis(ethylhexyl)-PPE **4c** (here as an example for all PPEs) shows either a broad emission at 532 nm or a structured emission with maxima at 454 and 482 nm (Figure 6.8). The structured band is mirror symmetrical to the absorption feature of the same polymer with a vanishing Stokes shift. The data suggest that the band at 532 nm stems from an excimer-like state, while the structured emission at 454 and 482 nm is assigned to planarized single chains or, potentially, aggregates.

104   **105**   **106, DMC**

FIGURE 6.8   Emission spectra of bis(ethylhexyl)-PPE **4c** in the solid state. Annealed samples or samples that are adsorbed on a polystyrene bead show emission at 454 nm, while pristine films show a broad emission at 532 nm.

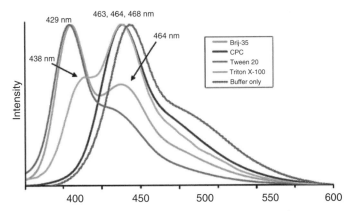

FIGURE 6.9 Emission spectra of **104** in different surfactant solutions (0.5 mg L^{-1} polymer in phosphate-buffered saline (PBS), 10% w/v surfactant, λ_{ex} 385 nm, 25°C): PBS, no surfactant; cetylpyridinium chloride (CPC); Triton X-100 (160 mM); Tween 20 (81 mM); and Brij 35 (82 mM). The emission spectra were normalized so that the emission maxima intensities of all the samples are identical. (From Lavigne, J.J., et al., *Macromolecules*, 36, 7409, 2003. With permission.)

Sugar-substituted PPEs **104** show a dramatic increase in fluorescence and deaggregation when exposed to surfactants [152,153]. Brij and Tween are the most effective surfactants in the series (Scheme 6.9) shifting the emission of **104** from 468 to 429 nm (Figure 6.9). The shift is attributed to the breakup of aggregates leading to single chain emission and was termed "surfactochromicity" [154–156]. The effect will be of importance when utilizing conjugated fluorescent polymers in bioanalytical and sensory applications to increase fluorescence quantum yield and detection limit.

The absorption and emission spectra of PPEs in dilute solution in good solvents are well known but give surprising insights with respect to PPE's excited state structure. In dilute solution, the UV–vis spectrum of a nonameric PPE-model **105** is broad and featureless [157,158]. The emission of **105** is much more narrow and a rather large Stokes shift is visible (Figure 6.10). Upon cooling of **105** in oligostyrene to 77 K, absorption and emission become mirror symmetric and the Stokes shift almost disappears. The emission broadens significantly, while absorption becomes more structured.

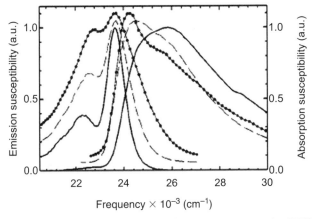

FIGURE 6.10 Absorption and emission susceptibilities of PPE nonamer **105** in CHCl$_3$ at 300 K (solid), in oligostyrene at 300 K (dashed), and in oligostyrene at 77 K (dotted). The spectra have been shifted vertically (+0.05, dashed; +0.1, dotted) and horizontally (+385 cm^{-1}, dotted) for clarity.

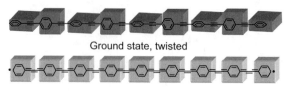

Ground state, twisted

Excited state, planar

FIGURE 6.11   Structure of the PPEs **4** and PPE model **105** in the ground and the excited states.

The spectra of **105** in oligostyrene at 77 K look as one would expect for a rigid rod, π-conjugated compound. The broad absorption spectrum and the narrow emission spectrum at room temperature and in chloroform are somewhat unusual and require explanation. The spectral properties of **105** can be explained if one considers that the rotational barrier of the phenyleneethynylene units, i.e., the aryl–alkynyl single bond, has a very shallow energy potential in the ground state, but a considerably steeper energy potential in the excited state. As a consequence, the excitation of the PPEs will occur from all possible rotamers, because all of them are populated in the ground state. In the excited state, immediate planarization will occur, and as a consequence, in the relaxed excited state only planarized molecules will be present. The molecular movement is much faster than the emissive lifetime (300 ps) and at room temperature *only* emission from the planar relaxed state is observed. This behavior is not too surprising, because the excited state of any PAE would be expected to have contributions from a butatrienic type, diradicaloid character, and butatrienes of course are planar (Figure 6.11). At 77 K, the rotation of **105** is frozen out in both ground and excited states, and planarization of the excited state will not occur. Broadening and blueshift of the emission occurs as a result.

6.6   PAEs as Sheet Polarizers and Energy-Transfer Materials

Weder has recently shown that dialkoxy-PPEs **7** carrying alternating ethylhexyl and octyl side chains can be dispersed in high-molecular weight polyethylene by codissolution (2%–25% PPE) in xylenes and subsequent spin casting. Absorption and emission spectra resemble those obtained for pristine films of pure **7**. These films can be drawn at 110°C to 80 times their original length. Films of a thickness of approximately 3 μm are obtained. In these drawn films, the emission is still redshifted from 474 (solution) to 496 nm, but is narrow. In addition, the films show a dramatic dichroic ratio emission$\|$/emission$\perp$ >70) parallel to the drawing direction. Typically state-of-the-art polarization ratios in conjugated polymers have not exceeded 10, rendering this result important. What is the structural model for the PPEs in polyethylene? In the drawn films the diffraction peaks due to the PPE moieties have disappeared and only the diffraction of polyethylene is recorded. This finding may be surprising at first, but single-crystal data collected on model compounds with long hydrocarbon substituents show that side chains dominate the packing of PPEs and the hydrocarbon tails form a polyethylene subcell in the crystalline state. This must be the reason for **7** orients perfectly into polyethylene matrices. The reported bathochromic shift of 20 nm (solution vs. solid state) can be explained by planarization of the PPE backbones in the matrix by the draw-induced order [57,159–161]. The high dichroic ratios in the thin PE/PPE film blends should have practical applications: **7** could prove useful as photoluminescent polarizers in LCD displays. Current LC displays are limited in efficiency and brightness because absorbing polarizers and color filters have to be used. If the highly polarized, PPE-containing polyethylene films are utilized instead, devices with substantially increased brightness, contrast, and efficiencies are obtained. If a sensitizer, **106**, is used in connection with **7** in polyethylene, the drawn ternary blends, in which **106** is oriented randomly, perform isotropic-to-polarized conversion of light efficiently. Energy transfer from excited but isotropic molecules **106** to the highly oriented PPE chains occurs.

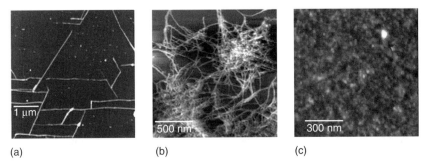

(a)   (b)   (c)

FIGURE 6.12   Nanostructures of dihexyl-PPE on mica, molecular weight dependence, tapping mode SFM in air. With the increasing length of PPE at low concentration there is a morphological transition of ribbons from ~9 nm wide → ~14 nm wide, → grains. SFM tapping mode images of PPE: (a) 7.9 nm, (b) 16.4 nm, and (c) 20.3 nm length. (SFM micrographs courtesy Prof. P. Samori.)

6.7   PPE Nanostructures

PPEs are "rigid rod" polymers. As such, one would expect self-assembly in the solid state. The driving force of the self-assembly would be van der Waals interactions of the side chains. Müllen, Samori, Rabe, et al. have shown that dialkyl-PPEs form nanowire and microwire-type arrangements when investigated by AFM and TEM [152–166]. Depending upon molecular weight, different morphologies form (Figure 6.12), which are potentially useful in molecular electronics and semiconductor applications.

Bunz et al. [56,167,168] found the same ordering in samples of dialkyl-PPEs. Figure 6.13 shows a lyotropic preparation of a sample of a trimethylhexyl-PPE with a spectacularly developed stranded structure. It seems that the PPEs almost "crystallize out" in ribbons or strands, approximately 30 nm wide and up to several micrometers long. These arrangements seem to preform in solution and form a fragile gel phase as Perahia et al. demonstrated by neutron diffraction and NMR experiments. These preformed flat aggregates seem to assemble on the surface and then form the nanoscopic ribbons. The force that governs the size and shape of these aggregates in solution is not clear.

These ribbons are also formed if thin spin cast films of dialkyl-PPEs are annealed. The seeding through solution-formed aggregates does not seem to be necessary. "Crystallization experiments" were conducted by annealing of a dialkyl PPE of the structure **108** containing both racemic and chiral side chains. The same type of 30-nm wide ribbons was found, but these ribbons showed an internal helical order according to dark field electron microscopy. They displayed very large chiroptical anisotropies, *g*-values, on the order of –0.38. Thin film preparations of chirally substituted PPEs that are spin cast and

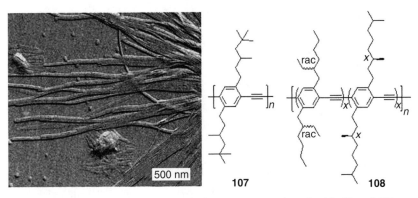

FIGURE 6.13   Transmission electron micrograph of a lyotropic preparation of a thin film of **107**.

do not show a nanoscopic order display g-values of only 0.005–0.01. The large g-values must therefore be a direct consequence of the increased supramolecular order in **108** [149,169]. Meskers et al. found also that large g-values are dependent upon the film thickness in chirally substituted polyfluorenes [170].

6.8   PPE Microstructures

While small nanostructures (e.g., ribbons, strands, and fibrous aggregates) are formed spontaneously upon annealing of thin films or slow evaporation of solutions of dialkyl-PPEs, their structuring in larger dimensions is not observed. Instead templating methods have to be employed. Martin [171] used nanoporous aluminum oxide membranes with discrete and well-defined pores, and infiltrated their channels with PAEs. Upon dissolution of the template by aqueous sodium hydroxide or acid solution, nanowires of the templated materials are isolated. Wilson et al. explored this have utilized this approach to generate 0.2 µm wide and up to 15 µm long nanotowers of different PAEs. The PAEs were dissolved in an organic solvent and the filter disk is soaked with the polymer solution. After drying, the alumina matrix is dissolved and the nanomaterials are examined by SEM. A typical example was obtained from **88a** (Figure 6.14) that gave well-developed, smooth nanotubes. If larger or different geometric shapes (e.g., spheres and hollow arrays of holes) are desired, templating with Whatman filter disks is not ideal and other templating methods have to be explored [172].

Breath figures are formed upon condensation of water vapor on a cold solid or liquid surface [173–176]. An example of this phenomenon is the fogging of mirrors and windows that was first examined by Lord Rayleigh. It has recently been utilized for the structuring of conjugated polymers into hexagonally ordered bubble arrays. Srinvasarao has examined this templating method and developed a mechanism for array formation (Figure 6.15). Moist air flows over a solution of a polymer (in this

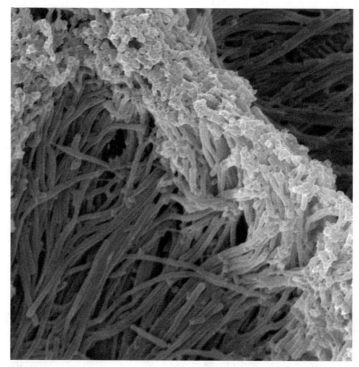

FIGURE 6.14   Benzothiadiazole-containing PAE **88a** forms templated nanotowers upon casting into 200 nm pore width Whatman aluminum disks. The towers are 180–200 nm thick and several micrometers long.

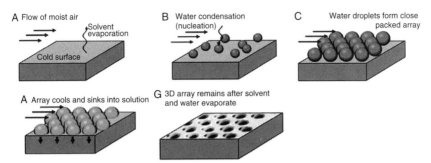

FIGURE 6.15 Array formation utilizing the breath-figure method. (Courtesy Prof. Srinivasarao.)

case didodecyl-PPE) in carbon disulfide or any other low boiling solvent. The solvent evaporates and the solvent reservoir cools down. Small water droplets start to condensate on the cold surface of the liquid. Upon further evaporation and condensation, these water droplets will grow larger, order themselves on the surface into a hexagonal array and start to sink into the carbon disulfide solution. Once all of the carbon disulfide has been evaporated, the water droplets have templated the polymeric bubble array. Upon drying, a 3D array of air bubbles is left as a fossil of the water droplet in the polymer (Figure 6.16). This method creates PPE-surrounded arrays of air bubbles that range in size from 700 nm to 7–8 µm. Almost all types of substituted PPEs give these patterns, as long as they are not too hydrophilic. In those cases misshapen and irregular arrays are formed [177].

There is great interest in small-scale containers with dimensions in the micrometer range and picoliter volume for bioanalytical assays [178]. A cross-linkable, azide-containing PPE (**111**) has been synthesized (Scheme 6.38) by the Sonogashira coupling of a **110** to 1,4-diethynyl-2,5-bis(ethylhexyl)benzene. Under the reaction conditions, the azide group does not undergo 1,3-dipolar cycloaddition, but furnishes the azide-substituted PPE **111** in high yield and with a good degree of polymerization [179,180]. Only upon heating to 300°C for 1 h do the azides react with the main chain triple bonds under cross-linking to give an insoluble array of picoliter beakers that are 5 µm in diameter, but are shrunken in the z-dimension from 5 to ∼1.5–2 µm, due to the cross-linking process. This process generates materials containing 40,000 of these picoliter beakers per square centimeter (Figure 6.17) [178].

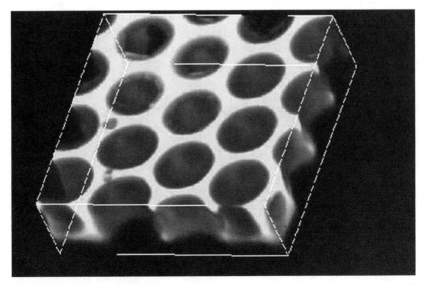

FIGURE 6.16 Confocal laser scanning light micrograph of a self-assembled honeycomb array formed from didodecyl-PPE **4c**. (Courtesy Prof. Srinivasarao.)

SCHEME 6.38   Synthesis of a cross-linkable, azide-containing PPE **111**.

Inorganic, ceramic nanostructures are generated from bubble arrays if organometallic PAEs are employed that do not melt. PAE **114** is generated when reacting (diethynylcyclobutadiene)cyclopentadienylcobalt with a suitable aromatic diiodide under Sonogashira conditions (Scheme 6.39) [104,181]. Polymer **114** formed bubble arrays when cast from carbon disulfide solution (1%–2% weight in carbon disulfide). Upon pyrolysis of these bubble arrays either under nitrogen or under air to >500°C, preserved inorganic ceramic or oxidic microstructures were obtained (Figure 6.18). In conclusion, the breath-figure method is a facile technique to generate microscopic structures of different dimensions in conjugated polymer films, particularly in PPEs. Applications of these bubble arrays in photovoltaics, photonic bandgap materials, and energy transfer materials are foreseeable.

6.9   Applications

6.9.1   Light-Emitting Diodes and Devices

PPVs have "made it big" as active materials in PLEDs. From the outset it was of interest if PPEs could likewise be useful [1]. Weder et al. investigated dialkoxy-PPEs **7** in different LED configurations and obtained devices with a peak brightness of 260 cd/m^2 and an onset at 5 V. The trick to improved efficiency is to make a device in which the PPE is mixed with a polymeric triphenyldiamine and a spiroquinoxaline as hole-injecting or hole-blocking layer and an aluminum cathode [182,183]. Bunz and Pschirer showed that PPEs could be used in blue LEDs if the nonaggregated polymer **115** with sterically obtrusive di(*tert*-butyl)naphthalene units was utilized as the emitting layer. Poly(ethylenedioxythiophene) was utilized as a hole injector in combination with aluminum cathodes [184]. Depending upon the amount of naphthalene monomer feed, the emission wavelength in these PLEDs could be engineered from 420 to 470 nm, and their maximum intensity was 100 cd/m^2 at 15 V.

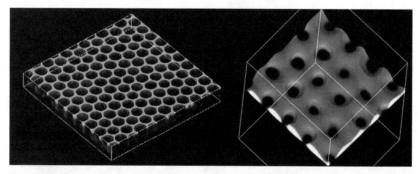

FIGURE 6.17   Laser scanning light micrograph of a self-assembled honeycomb array of azide-PPE **111**. The diameter of the holes is 5 μm (left). Same material after heating to 300°C for 1 h. Diameter of holes is 5 μm, note that the height is reduced (right).

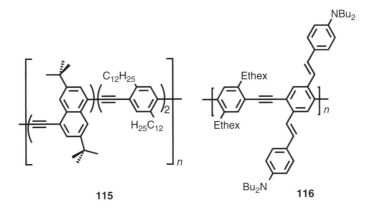

SCHEME 6.39   Synthesis of the organometallic polymer **114**.

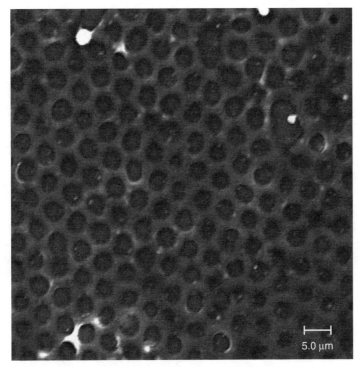

FIGURE 6.18   SEM micrograph of an inorganic bubble array formed by the thermolysis of polymer **114** under nitrogen at 500°C. (From Englert et al. *Chemistry Eur. J.* 11, 995, 2005. With permission.)

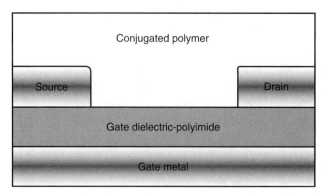

FIGURE 6.19   Photoresponsive thin film transistor based upon cross-conjugated PPE **116**.

While PPEs are not ideal candidates as active emitters for light-emitting diodes (LEDs), their cross-conjugated bisstyryl-functionalized derivatives **116** are excellent photodiodes that work well in a transistor configuration. Due to the hole-injecting nature of the dibutylamino groups, the source–drain current (Figure 6.19) increases by a factor of more than 6×10^3 upon illumination. The photocurrent cannot be pinched off even if a high positive gate bias is utilized. Further applications of PAEs in transistor and photovoltaic devices are possible, and only the surface has been scratched with respect to attractive applications of PAEs in electronic devices [103].

6.9.2   Sensory Applications of PPEs

Even simple dialkyl-PPEs show a marked response in their optical properties upon change of their environment: PPEs should be good candidates for sensory applications. Generally, sensing with fluorophores can work by different principles. The fluorescence could be turned off as a consequence of the sensing event (quenching and turn-off sensor), the fluorescence could be turned on as a consequence of the sensing event (turn-on sensor), or as a third possibility, the fluorescence wavelength could change as a consequence of the sensing event (ratiometric sensing).

As a quantitative measure of quenching, the Stern–Volmer equation is a useful tool [185–187].

$$(F_0/F_{[Q]}) = 1 + K_{SV}[Q] \quad \text{or} \quad K_{SV} = \{(F_0/F) - 1\}/[Q]$$

The concentration of quencher is $[Q]$, K_{SV} is the Stern–Volmer constant, F_0 is the fluorescence intensity measured without additional quencher, and $F_{[Q]}$ is the fluorescence intensity with quencher at a given concentration $[Q]$ [185–187]. The slope of the graph $(F_0/F_{[Q]})$ vs. $[Q]$ equals K_{SV}. The more sensitive a given system is to a specific quencher, the steeper the Stern–Volmer plot and the higher K_{SV}. Quenching processes can be generalized to have two mechanisms, i.e., they can be static or dynamic in nature [185]. In static quenching, the fluorophore and the quencher form a complex in the ground state. Upon irradiation, quenching of the excited fluorophore by the complexed quencher occurs. In dynamic quenching, the excited state of the fluorophore is quenched by collision with the quencher. Dynamic quenching is also termed collisional quenching. If either of those mechanisms is predominant in a given system, a linear Stern–Volmer relationship ensures. Unaggregated PPEs in solution show a short lifetime of fluorescence (around 0.3–0.5 ns) [2]. The fluorescence lifetime does not change upon addition of quencher [185]. It can therefore be assumed that static quenching is the prevalent mechanism operating in PPEs. In the case of static quenching, the Stern–Volmer constant is equal to the binding constant of the quencher to the fluorophore; K_{SV}, therefore conveniently delivers binding constants of complex formation [185,187].

In 1995, Zhou and Swager [185] elaborated on the idea of utilizing fluorescence quenching of PPEs as a concept to build more effective chemosensors. The model of analyte sensor system was a combination of a cyclophane with a highly acceptor-substituted arene, paraquat (methyl viologen, Figure 6.19). Upon binding of paraquat to the sensory cleft of the cyclophane, the fluorescence of the cyclophane-containing fluorophore is quenched. In the case of an isolated fluorescent chemosensor such as **117**, binding of 1/3 of the available binding sites would reduce the overall emission by 33%. In a second case, the sensor is wired in series (**118**), the quencher paraquat occupies a third of the receptor sites. However, complete shutdown of fluorescence results due to funneling of the exciton, i.e., the excited state energy, to the site of lowest energy, i.e., the complexed quencher.

Isolated fluorophore **117**   Fluorophores "wired in series" **118**   Paraquat, quencher **119**

When **118** of high molecular weight is compared with **117** in quenching experiments, its sensitivity toward paraquat is increased by a factor of approximately 65. This increase in sensitivity is attributed to the molecular wire effect, a direct consequence of the delocalization of the excited state over a large number of aryleneethynylene receptor sites (Figure 6.20). The enhancement factor was obtained by comparing the complex formation constants of both **117** as well as **118** with paraquat (**119**) by the Stern–Volmer relationship. This contribution established that conjugated polymers have advantages over small molecules in sensory applications that involve quenching processes, due to the increase in sensitivity brought about by the molecular wire effect.

If calix[4]arenes are attached to the PPEs, these polymers are useful in the sensing of *N*-methylquinolonium ion [188]. If the PPEs are appended with 15-crown-5, they can be utilized as sensors to detect potassium ions through ion-specific aggregation at the air–water interface. Redshift in absorption and emission was found in these systems [189]. A PPE was utilized to construct a fluoride dosimeter (Scheme 6.40) [190]. Upon exposure of **120** to fluoride anion, but not to hydroxide or any other anion, the TIPS group is deprotected and the free phenolate performs a lactonization reaction. The formed polymer has a redshifted emission, indicating the presence of fluoride. The change from blue to green emission, coupled with the increase of quantum yield can be observed with the naked eye and the assay is approximately 100 times as sensitive as the detection of fluoride with a similar small molecule sensor. The lactone acts as a trap for the excitonic energy of the excited polymer and leads to a ratiometric response of the system. This system is not a sensor but a dosimeter for the detection of fluoride, because the changes (lactonization) of the sensory appendix are not reversible.

Kim, Erdogan, and Bunz prepared a series of ethylene glycol and glucose-substituted PPEs (**104, 122, 123**) and found that the fluorescence of these polymers was quenched to a differing degree by mercury and lead salts [27,152]. In some cases the polymers are >1000 times as sensitive in the detection of metal salts than the monomeric model compounds **124** and **125**. The enhanced superquenching is probably not only due to the molecular wire effect (i.e., the delocalization of the excitons). The influence of multivalently binding side chains of PPEs to the metal cations must also play a role to explain these dramatic differences in sensitivity. Sugar-substituted PPEs also show great potential in the detection of

FIGURE 6.20 The molecular wire approach for the quenching of PPEs by paraquat.

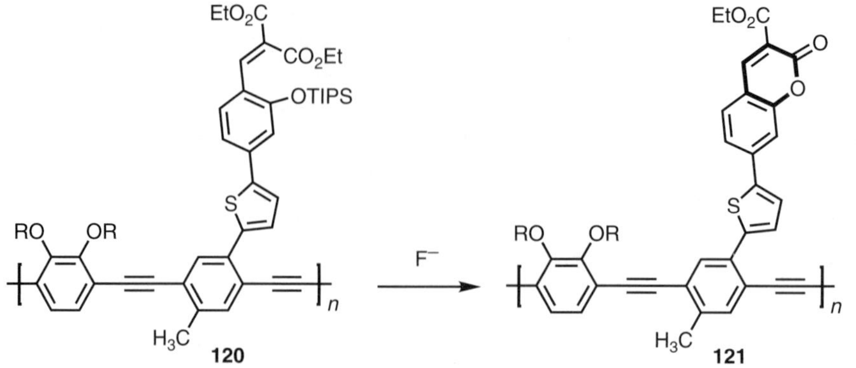

120 121

SCHEME 6.40 A PPE-based indicator for fluoride ions.

lectins, a class of proteins that mediate and regulate recognition events in cellular signaling [191–194]. A significant number of bacterial toxins (e.g., enterotoxin B and anthrax toxin) and plant toxins (ricin) have lectin-like substructures. In the early stages pathogen–cell interactions of lectin-like structures on the cell and pathogen surfaces occur. Lectins do not bind well to monomeric, free sugars such as glucose or mannose, perhaps due to their abundance in foodstuff. However, if a lectin is exposed to a sugar display that contains multiple copies of a sugar substituent, binding is efficient [193,194].

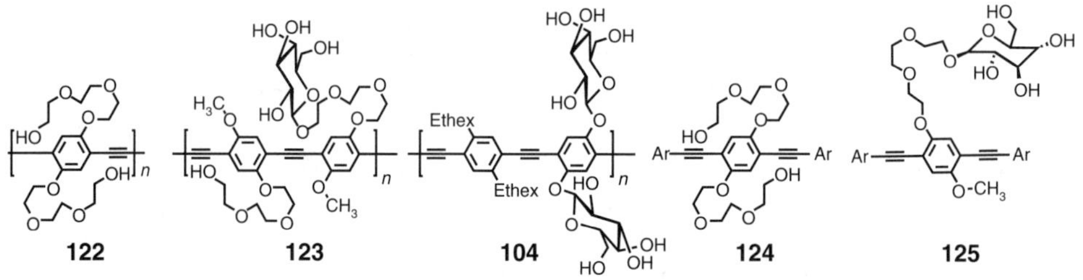

122 123 104 124 125

Independently Bunz et al. and Seeberger et al. have investigated the interaction of mannose-substituted PPEs **126**, **128**, and their model compound **127** with Concanavalin A. Quenching and formation of aggregates occurred [130,195]. It was shown that *Escherichia coli* bacteria can be stained by mannose-containing PPEs **128** and less than 10^5 *E. coli* bacteria can be detected by fluorescence microscopy of the now fluorescent cells [195]. These polymers inhibit ConA-induced agglutination of sheep blood at very low concentrations. The fluorescent assay is obtained in a short time and does not need growth of bacteria in liquid media or on a plate and is therefore a conceptually valuable low-tech alternative to the classic tools of microbiology.

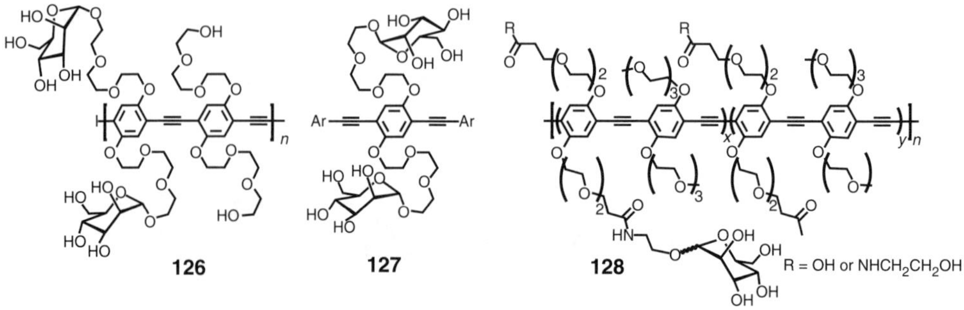

126 127 128

The fact that conjugated polymers are powerful tools in agglutination assays was first demonstrated by the exposure of biotinylated PPE **129** to avidin or streptavidin. Streptavidin binds to **129** and leads to agglutination and complete disappearance of the fluorescence in solution. Instead, a highly fluorescent pellet is formed. If **129** is treated with streptavidin-coated PS spheres, well-defined aggregates are formed. If similar biotinylated PPEs were treated with streptavidin, efficient energy transfer occurred when the streptavidin was labeled with Texas Red dye [127–129].

129   **130**

Biotinylated PPEs have been utilized as a fluorescent platform for the detection of DNA with single nucleotide mismatches in a PS microsphere-based sensing construct. In this approach a negatively charged, biotinylated PPE–neutravidin complex and the target DNA are adsorbed onto a positively charged PS microsphere. An unknown oligonucleotide is coadsorbed onto the positively charged and highly fluorescent beads. Upon binding to a shorter, quencher-substituted oligonucleotide with the complementary sequence, target oligonucleotides displace the quencher. Mismatched oligonucleotides do so to a lesser degree and the PPE's fluorescence, which is enhanced by the complexation to neutravidin, is turned on to a varying degree. The construct was utilized to detect the oligonucleotides that encode for the anthrax lethal factor, and while PPEs do not have any indigenous recognition function in this assay, they are the fluorescent part of the larger construct [127].

The PPE **130** forms 0.5 μm sized particles when dissolved in DMSO and precipitated into saline water [196]. These particles show emission at 450 nm that disappears if a quencher-substituted oligonucleotide is added. In aqueous solution, at neutral pH, a significant number of amino groups are protonated and the negatively charged, quencher-substituted oligonucleotide attaches itself electrostatically to the particles and efficiently quenches the PPE's fluorescence. There is no sequence specificity here, because the DNA is only used to impart multiple negative charges to the quencher. It is not clear how this concept will be further developed.

PPEs have recently been utilized in the dosimetric detection of proteases [197,198]. PPE **131** was substituted with an oligopeptide that carries a dinitroaniline quencher. The oligopeptide contains a motif that is susceptible toward protease action. This construct is nonfluorescent, because the dangling dinitroaniline units quench the PPE's fluorescence. Upon exposure to low loadings of trypsine, a nonselective hydrolytic enzyme, the fluorescence of the quencher-less PPE **132** is observed. The oligopeptide chain was selected because it could be a substrate for matrix metalloproteases (human collagenase III) associated with human cancer cells. The peptide does not have to be covalently attached to the PPE: Pinto and Schanze [198] were able to demonstrate that electrostatic interactions are likewise efficient in the detection of proteases by PPEs (Scheme 6.41).

Sensing in solution is important for most environmental and many biological applications, but it requires that the analyte is also in solution. If it is desired to detect a gaseous analyte, a different strategy has to be employed. An important case is the detection of land mines by Swager and Yang [199–202]. Buried land mines are difficult to locate if their position is not exactly recorded, but they exude dinitrotoluene (DNT). Significant concentrations of DNT, an electron-poor arene, are found over the location of buried land mines. A series of iptycene-containing PPEs **133–135** was prepared and their fluorescence was examined in thin films. Due to the presence of the bulky iptycene units, the solid state

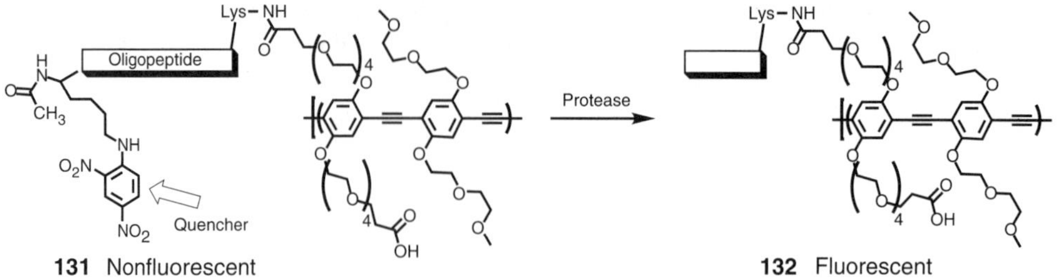

SCHEME 6.41   Sensing of proteases by a PPE-based turn-on assay.

fluorescence of these PPEs is efficient. Upon exposure of thin films of **133–135** to DNT, their fluorescence is dramatically quenched, due to the adsorption of the DNT into the thin films. The alkoxy-substituted **133** and **135** work better than **134**. The reason for the attenuated performance of **134** is the highly electron-accepting character of the amide groups. The authors hypothesize that the iptycene units render **133–135** porous. However, the crystal structure of a diethynyliptycene derivative showed the monomer's density to be $>1.2 \text{ g cm}^{-3}$, suggesting that the polymer is not particularly porous in the solid state. Perhaps surface-related effects might play a much larger role in the quenching experiments of **133–135** with DNT.

An exciplex band is observed in **135**'s solid state fluorescence spectrum. The exciplex band is significantly redshifted and has a considerably longer emission lifetime than the main fluorescence band. The authors contend that there is a significant ground state interaction between the chromophores of the main chain and the naphthalene units in the solid state. However, because the solution and the solid state UV–vis spectra are only very slightly different, this assertion is open to discussion. The polymers **133** and **135** are equal in their ability to sense acceptor-substituted arenes, and the use of the larger pentiptycene does not confer any special advantage to the sensing event. A working prototype of a DNT sensor has been made that detects land mines as efficiently and as precise as specially trained canines [203].

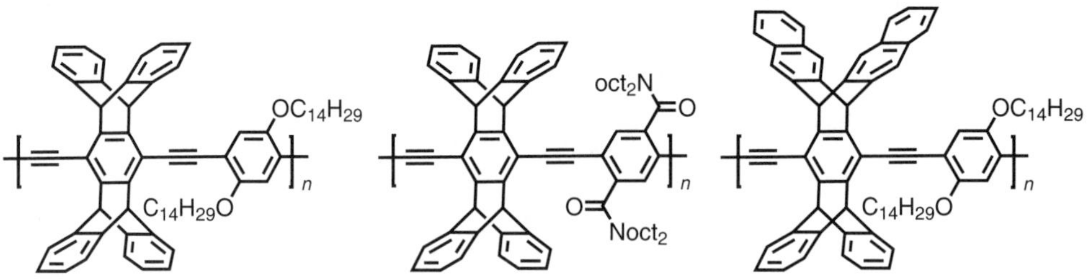

6.10   Conclusions

PPEs and PAEs have come a long way from the first synthesis in 1990 to sophisticated sensory and photonic materials; PAEs and PPEs are very different from PPVs—despite their obviously close structural relation. PPEs are oxidatively more stable than PPVs, which makes them less suitable for some organic semiconductor applications, but PPEs, high fluorescence coupled with their great environmental stability makes them attractive as parts for sensors that detect biogenic and toxigenic materials, transition and alkali metal ions, and vapors exuding from land mines.

The low energy of rotation around the C–C single bond that connects the alkynes with the benzene units leads to a twist between two adjacent benzene rings. In solution, the rotation generates a rich conformational behavior that has significant consequences for UV–vis and emission spectra and their aggregation behavior. Planarization leads to an increase in conjugation with concomitant redshift in absorption and emission, while twisting leads to blueshifted spectral features. PAEs and PPEs will be useful as superbly luminescent platforms for biologically and electronically active materials with a great potential in applications in the bioanalytical field as well as in materials science. The progress in this area is only limited by our creativity and insight.

Acknowledgments

I would like to thank the gifted and hardworking graduate and undergraduate students, and postdocs I have had in my group at the Max-Planck-Institut für Polymerforschung, the University of South Carolina, and the Georgia Institute of Technology. Funding for these endeavors was provided over the years by the Stiftung Volkswagenwerk, the Deutsche Forschungsgemeinschaft, the Petroleum Research Funds administered by the ACS, the National Science Foundation, the Dreyfus Foundation, the Department of Energy, and the National Institute of Health.

References

1. Friend, R.H., et al. 1990. Light emitting diodes based on conjugated polymers. *Nature* 347:539.
2. Bunz, U.H.F. 2000. Poly(aryleneethynylene)s: Syntheses, properties, structures, and applications. *Chem Rev* 100:1605.
3. Giesa, R. and R.C. Schulz. 1990. Soluble PPEs. *Makromol Chem Phys* 191:857.
4. Sonogashira, K. 2002. Development of Pd–Cu catalyzed cross-coupling of terminal acetylenes with sp^2-carbon halides. *J Organomet Chem* 653:46.
5. Moroni, M., J. Lemoigne, and S. Luzzati. 1994. Rigid-rod conjugated polymers for nonlinear optics. 1. Characterization and linear optical properties of PAE derivatives. *Macromolecules* 27:562.
6. Swager, T.M., C.J. Gil, and M.S. Wrighton. 1995. Fluorescence studies of PPEs—the effect of anthracene substitution. *J Phys Chem* 99:4886.
7. Ofer, D., T.M. Swager, and M.S. Wrighton. 1995. Solid-state ordering and potential dependence of conductivity in dialkoxy-PPEs. *Chem Mater* 7:418.
8. Weder, C. and M.S. Wrighton. 1996. Efficient solid-state photoluminescence in new poly(2,5-dialkoxy-*p*-phenyleneethynylene)s. *Macromolecules* 29:5157.
9. Mangel, T., et al. 1995. Synthesis and optical properties of some novel arylenealkynylene polymers. *Macromol Rapid Commun* 16:571.
10. Li, H., et al. 1998. Poly((2,5-dialkoxy-*p*-phenylene)ethynylene-*p*-phenyleneethynylene)s and their model compounds. *Macromolecules* 31:52.
11. Rodriguez-Prada, J.M., R. Duran, and G. Wegner. 1989. A comparative study of mesophase formation in rigid chain polyesters with flexible side chains. *Macromolecules* 22:2507.
12. Weiss, K., et al. 1997. Acyclic diyne metathesis (ADIMET), an efficient route to poly(phenylene)ethynylenes (PPEs) and nonconjugated polyalkynylenes of high molecular weight. *Angew Chem* 36:508.
13. Schrock, R.R., et al. 1982. Multiple metal carbon bonds. 31. Tungsten neopentylidyne complexes. *Organometallics* 1:1645.
14. Schrock, R.R. 2001. Transition metal–carbon multiple bonds. *J Chem Soc Dalton Trans* 2541.
15. Schrock, R.R. 2003. The discovery and development of high oxidation state alkylidyne complexes for alkyne metathesis, in *Handbook of metathesis*, ed., R.H. Grubbs. Weinheim: Wiley-VCH, chap. 1.11.
16. Kloppenburg, L., D. Song, and U.H.F. Bunz. 1998. Alkyne metathesis with simple catalyst systems: Poly(*p*-phenyleneethynylene)s. *J Am Chem Soc* 120:7973.
17. Mortreux, A. and M. Blanchard. 1974. Metathesis of alkynes by a molybdenum hexacarbonyl-resorcinol catalyst. *Chem Commun* 786.
18. Kaneta, N., et al. 1995. Novel synthesis of disubstituted alkynes using molybdenum-catalyzed cross-alkyne metathesis. *Chem Lett* 1055.

19. Tour, J.M. 1996. Conjugated macromolecules of precise length and constitution. Organic synthesis for the construction of nanoarchitecture. *Chem Rev* 96:537.

20. Gelman, D. and S.L. Buchwald. 2003. Efficient palladium-catalyzed coupling of aryl chlorides and tosylates with terminal alkynes: Use of a copper cocatalyst inhibits the reaction. *Angew Chem* 42:5993.

21. Thorand, S. and N. Krause. 1998. Improved procedures for the palladium-catalyzed coupling of terminal alkynes with aryl bromides (Sonogashira coupling). *J Org Chem* 63:8851.

22. Wilson, J.N., et al. 2002. Acetylene gas: A reagent in the synthesis of high molecular weight poly (*p*-phenyleneethynylene)s utilizing very low catalyst loadings. *Macromolecules* 35:3799.

23. Arvela, R.K., et al. 2005. A reassessment of the transition-metal free Suzuki-type coupling methodology. *J Org Chem* 70:161.

24. Hundertmark, T., et al. 2000. Pd(PhCN)$_2$Cl$_2$/P(*t*-Bu)$_3$: A versatile catalyst for Sonogashira reactions of aryl bromides at room temperature. *Org Lett* 2:1729.

25. Li, C.J., et al. 1997. Palladium catalysed polymerization of aryl diiodides with acetylene gas in aqueous medium: A novel synthesis of areneethynylene polymers and oligomers. *Chem Commun* 1569.

26. Li, C.J., et al. 1998. In aqua synthesis of a high molecular weight arylethynylene polymer with reversible hydrogel properties. *Chem Commun* 1352.

27. Kim, I.B., et al. 2004. Sugar-poly(*para*-phenyleneethynylene) conjugates as sensory materials: Efficient quenching by Hg^{2+} and Pb^{2+} ions. *Chem Eur J* 10:6247.

28. Francke, V., T. Mangel, and K. Müllen. 1998. Synthesis of alpha, omega-difunctionalized oligo- and poly(*p*-phenyleneethynylene)s. *Macromolecules* 31:2447.

29. Stevens, M.P. 1990. *Polymer chemistry*, 2nd ed., New York: Oxford University Press, chap. 2.

30. Beletskaya, I.P. and A.V. Cheprakov. 2000. The Heck reaction as a sharpening stone of palladium catalysis. *Chem Rev* 100:3009.

31. DeMeijere, A. and F.E. Meyer. 1994. Fine feathers make fine birds—the Heck reaction in modern garb. *Angew Chem* 33:2379.

32. Goodson, F.E., T.I. Wallow, and B.M. Novak. 1997. Mechanistic studies on the aryl–aryl interchange reaction of ArPdL2I (L = triarylphosphine) complexes. *J Am Chem Soc* 119:12441.

33. Nielsen, K.T., K. Bechgaard, and F.C. Krebs. 2005. Removal of palladium nanoparticles from polymer materials. *Macromolecules* 38:658.

34. Bunz, U.H.F. 2001. Poly(*p*-phenyleneethynylene)s by alkyne metathesis. *Acc Chem Res* 34:998.

35. Bunz, U.H.F. 2002. The ADIMET reaction: Synthesis and properties of poly(dialkyl*para*phenyleneethynylene)s, in *Modern arene chemistry*, ed., D. Astruc. Weinheim: Wiley-VCH, chap. 7.

36. Bunz, U.H.F. 2003. Acyclic diyne metathesis utilizing in situ transition metal catalysts: An efficient access to alkyne-bridged polymers, in *Handbook of metathesis*, ed., R.H. Grubbs. Weinheim: Wiley-VCH, chap. 3.10.

37. Fritch, J.R. and K.P.C. Vollhardt. 1979. Cyclobutadiene metal complexes as potential intermediates in alkyne metathesis—flash vacuum pyrolysis of substituted cyclobutadienecyclopentadienylcobalt complexes. *Angew Chem* 18:409.

38. Brizius, G., S. Kroth, and U.H.F. Bunz. 2002. Alkyne-bridged carbazole polymers by alkyne metathesis. *Macromolecules* 35:5317.

39. Brizius, G., et al. 2000. Alkyne metathesis with simple catalyst systems: Efficient synthesis of conjugated polymers containing vinyl groups in main or side chain. *J Am Chem Soc* 122:12435.

40. Brizius, G., et al. 2003. Conformational and electronic engineering of twisted diphenylacetylenes. *Org Lett* 5:3951.

41. Pschirer, N.G. and U.H.F. Bunz. 2000. Poly(fluorenyleneethynylene)s by alkyne metathesis: Optical properties and aggregation behavior. *Macromolecules* 33:3961.

42. Brizius, G. and U.H.F. Bunz. 2002. Increased activity of in situ catalysts for alkyne metathesis. *Org Lett* 4:2829.

43. Pschirer, N.G., et al. 2000. Novel liquid-crystalline PPE-naphthalene copolymers displaying blue solid-state fluorescence. *Chem Commun* 85.

44. Sashuk, V., J. Ignatowska, and K. Grela. 2004. A fine-tuned molybdenum hexacarbonyl/phenol initiator for alkyne metathesis. *J Org Chem* 69:7748.
45. Grela, K. and J. Ignatowska. 2002. An improved catalyst for ring-closing alkyne metathesis based on molybdenum hexacarbonyl/2-fluorophenol. *Org Lett* 4:3747.
46. Bly, R.K., K.M. Dyke, and U.H.F. Bunz. 2005. A study of molybdenum catalysts in the polymerization of 2,5-didodecyl-1,4-dipropynylbenzene. *J Organomet Chem* 690:825.
47. Zhang, W., S. Kraft, and J.S. Moore. 2003. A reductive recycle strategy for the facile synthesis of molybdenum (VI) alkylidyne catalysts for alkyne metathesis. *Chem Commun* 7:832.
48. Zhang, W., S. Kraft, and J.S. Moore. 2004. Highly active trialkoxymolybdenum (VI) alkylidyne catalysts synthesized by a reductive recycle strategy. *J Am Chem Soc* 126:329.
49. Zhang, W. and J.S. Moore. 2004. Synthesis of poly(2,5-thienyleneethynylene)s by alkyne metathesis. *Macromolecules* 37:3973.
50. Laplaza, C.E. and C.C. Cummins. 1995. Dintrogen cleavage by a 3-coordinate molybdenum complex. *Science* 268:861.
51. Fürstner, A. et al. 1999. Ring closing alkyne metathesis. Comparative investigation of two different catalyst systems and application to the stereoselective synthesis of olfactory lactones, azamacrolides, and the macrocyclic perimeter of the marine alkaloid nakadomarin A. *J Am Chem Soc* 121:11108.
52. Zhang, W. and J.S. Moore. 2004. Arylene ethynylene macrocycles prepared by precipitation-driven alkyne metathesis. *J Am Chem Soc* 126:12796.
53. Lavastre, O., et al. 2002. Discovery of new fluorescent materials from fast synthesis and screening of conjugated polymers. *J Am Chem Soc* 124:5278.
54. Leung, A.C.W., et al. 2003. Poly(salphenyleneethynylene)s: A new class of soluble, conjugated, metal-containing polymers. *Macromolecules* 36:5051.
55. Kloppenburg, L., et al. 1999. Poly(*p*-phenyleneethynylene)s are thermotropic liquid crystalline. *Macromolecules* 32:4460.
56. Bunz, U.H.F., et al. 1999. Solid-state structures of phenyleneethynylenes: Comparison of monomers and polymers. *Chem Mater* 11:1416.
57. Halkyard, C.E., et al. 1999. Evidence of aggregate formation for 2,5-dialkylpoly(*p*-phenyleneethynylenes) in solution and thin films. *Macromolecules* 31:8655.
58. Steiger, D., P. Smith, and C. Weder. 1997. Liquid crystalline, highly luminescent poly(2,5-dialkoxy-*p*-phenyleneethynylene). *Macromol Rapid Commun* 18:643.
59. Bunz, U.H.F., J.N. Wilson, and C. Bangcuyo. 2005. Chromicity in poly(aryleneethynylene)s. *ACS Symp Ser* 888:147.
60. Leclere, P., et al. 2000. Highly regular organization of conjugated polymer chains via block copolymer self-assembly. *Adv Mater* 12:1042.
61. Francke, V., et al. 1998. Synthesis and characterization of a poly(*para*phenyleneethynylene)-block-poly(ethylene oxide) rod-coil block copolymer. *Macromol Rapid Commun* 19:275.
62. Tsolakis, P.K., J.K. Kallitsis, and A. Godt. 2002. Synthesis of luminescent rod-coil block copolymers using atom transfer radical polymerization. *Macromolecules* 35:5758.
63. Rzayev, J. and M.A. Hillmyer. 2005. Nanoporous polystyrene containing hydrophilic pores from an ABC triblock copolymer precursor. *Macromolecules* 38:3.
64. Kuroda, K. and T.M. Swager. 2004. Fluorescent semiconducting polymer conjugates of poly (N-isopropylacrylamide) for thermal precipitation assays. *Macromolecules* 37:716.
65. Li, K. and Q. Wang. 2004. Synthesis and solution aggregation of polystyrene–oligo(*p*-phenyleneethynylene)–polystyrene triblock copolymer. *Macromolecules* 37:1172.
66. Huang, W.Y., et al. 2001. Organization and orientation of a triblock copolymer poly(ethylene glycol)-*b*-poly(*p*-phenyleneethynylene)-*b*-poly(ethylene glycol) and its blends in thin films. *Macromolecules* 34:7809.
67. Wang, Q.Y., et al. 2003. Grafted conjugated polymers: Synthesis and characterization of a polyester side chain substituted poly(*para*phenyleneethynylene). *Chem Commun* 1624.

68. Breen, C.A., et al. 2003. Polarized photoluminescence from poly(*p*-phenylene-ethynylene) via a block copolymer nanotemplate. *J Am Chem Soc* 125:9942.

69. Wang, Y.Q., et al. 2004. TEMPO-substituted PPEs: Polystyrene-PPE graft copolymers and double graft copolymers. *Macromolecules* 37:970.

70. Patten, T.E., et al. 1996. Polymers with very low polydispersities from atom transfer radical polymerization. *Science* 272:866.

71. Patten, T.E. and K. Matyjaszewski. 1999. Copper (I)-catalyzed atom transfer radical polymerization (ATRP). *Acc Chem Res* 32:895.

72. Sato, T., D.L. Jiang, and T. Aida. 1999. A blue-luminescent dendritic rod: Poly(phenyleneethynylene) within a light-harvesting dendritic envelope. *J Am Chem Soc* 121:10658.

73. Masuo, S., et al. 2003. Fluorescence spectroscopic properties and single aggregate structures of pi-conjugated wire-type dendrimers. *J Phys Chem B* 107:2471.

74. Li, W.S., D.L. Jiang, and T. Aida. 2004. Photoluminescence properties of discrete conjugated wires wrapped within dendrimeric envelopes: "Dendrimer effects" on π-electronic conjugation. *Angew Chem* 43:2943.

75. Englert, B.C., et al. 2004. Jacketed poly(*p*-phenyleneethynylene)s: Nonaggregating conjugated polymers as blue-emitting rods. *Macromolecules* 37:8212.

76. Häger, H. and W. Heitz. 1998. Synthesis of poly(phenyleneethynylene) without diine defects. *Macromol Chem Phys* 199:1821.

77. Thünemann, A.F. 1999. Nanostructured dihexadecyldimethylammonium-poly(1,4-phenylene-ethynylene-carboxylate): An ionic complex with blue electroluminescence. *Adv Mater* 11:127.

78. Tan, C.Y., M.R. Pinto, and K.S. Schanze. 2002. Photophysics, aggregation and amplified quenching of a water-soluble poly(phenyleneethynylene). *Chem Commun* 446.

79. Tan, C.Y., et al. 2004. Amplified quenching of a conjugated polyelectrolyte by cyanine dyes. *J Am Chem Soc* 126:13685.

80. Lakowicz, J.R. 1986. *Principles of fluorescence spectroscopy*. New York: Plenum Press.

81. DiCesare, N., et al. 2002. Saccharide detection based on the amplified fluorescence quenching of a water-soluble poly(phenyleneethynylene) by a boronic acid functionalized benzyl viologen derivative. *Langmuir* 18:7785.

82. Pinto, M.R., B.M. Kristal, and K.S. Schanze. 2003. A water-soluble poly(phenyleneethynylene) with pendant phosphonate groups. Synthesis, photophysics, and layer-by-layer self-assembled films. *Langmuir* 19:6523.

83. Haskins-Glusac, K., et al. 2004. Luminescence quenching of a phosphorescent conjugated polyelectrolyte. *J Am Chem Soc* 126:14964.

84. Kuroda, K. and T.M. Swager. 2003. Synthesis of a nonionic water soluble semiconductive polymer. *Chem Commun* 26.

85. Khan, A., S. Mueller, and S. Hecht. 2005. Practical synthesis of an amphiphilic, non-ionic poly (*para*-phenyleneethynylene) derivative with a remarkable quantum yield in water. *Chem Commun* 584.

86. Nelson, J.C., et al. 1997. Solvophobically driven folding of nonbiological oligomers. *Science* 277:1793.

87. Brunsveld, L., et al. 2001. Self-assembly of folded *m*-phenylene ethynylene oligomers into helical columns. *J Am Chem Soc* 123:7978.

88. Chu, Q. and Y. Pang. 2003. Molecular aggregation of poly[(1,3-phenyleneethynylene)-*alt*-oligo(2,5-dialkoxy-1,4-phenyleneethynylene)]: Effects of solvent, temperature, and polymer conformation. *Macromolecules* 36:4614.

89. Pang, Y., et al. 1998. A processible PPE with strong photoluminescence: Synthesis and characterization of poly[(*m*-phenyleneethynylene)-*alt*-(*p*-phenyleneethynylene)]. *Macromolecules* 31:6730.

90. Chu, Q. and Y. Pang. 2005. Aggregation and self-assembly of oligo(2,5-dialkoxy-1,4-phenyleneethynylene)]s: An improved probe to study inter- and intramolecular interaction. *Macromolecules* 38:517.

91. Chu, Q.H., et al. 2002. Synthesis, chain rigidity, and luminescent properties of poly[(1,3-phenyleneethynylene)-alt-*tris*(2,5-dialkoxy-1,4-phenyleneethynylene)]s. *Macromolecules* 35:7569.

92. Arnt, L. and G.N. Tew. 2004. Conformational changes of facially amphiphilic *meta*-PPEs in aqueous solution. *Macromolecules* 37:1283.

93. Arnt, L. and G.N. Tew. 2003. Cationic facially amphiphilic PPEs studied at the air–water interface. *Langmuir* 19:2404.

94. Arnt, L. and G.N. Tew. 2002. New poly(phenyleneethynylene)s with cationic, facially amphiphilic structures. *J Am Chem Soc* 124:7664.

95. Hecht, S. and A. Khan. 2003. Intramolecular cross-linking of helical folds: An approach to organic nanotubes. *Angew Chem* 42:6021.

96. Khan, A. and S. Hecht. 2004. Microwave-accelerated synthesis of lengthy and defect-free poly(*m*-phenyleneethynylene)s via AB′ and A(2) + BB′ polycondensation routes. *Chem Commun* 300.

97. Tan, C.Y., et al. 2004. Solvent-induced self-assembly of a meta-linked conjugated polyelectrolyte. Helix formation, guest intercalation, and amplified quenching. *Adv Mater* 16:1208.

98. Egbe, D.A.M., et al. 2004. Mixed alkyl- and alkoxy-substituted poly[(phenyleneethynylene)-alt-(phenylenevinylene)] hybrid polymers: Synthesis and photophysical properties. *Macromol Chem Phys* 205:2105.

99. Egbe, D.A.M., et al. 2003. Influence of the conjugation pattern on the photophysical properties of alkoxy-substituted PE/PV hybrid polymers. *Macromolecules* 36:9303.

100. Egbe, D.A.M., et al. 2003. Investigation of the photophysical and electrochemical properties of alkoxy-substituted arylene–ethynylene/arylene–vinylene hybrid polymers. *Macromolecules* 36:5459.

101. Egbe, D.A.M., et al. 2002. Side chain effects in hybrid PPV/PPE polymers. *Macromolecules* 35:3825.

102. Wilson, J.N., et al. 2002. Band gap engineering of poly(*p*-phenyleneethynylene)s: Cross-conjugated PPE–PPV hybrids. *Macromolecules* 35:8681.

103. Xu, Y.F., et al. 2004. Photoresponsivity of polymer thin-film transistors based on polyphenyleneethynylene derivative with improved hole injection. *Appl Phys Lett* 85:4219.

104. Altmann, M. and U.H.F. Bunz. 1995. Polymers with complexed cyclobutadiene units in the main chain—the first example of a thermotropic, liquid crystalline organometallic polymer. *Angew Chem* 34:569.

105. Altmann, M., et al. 1995. Synthesis of novel polymers containing cyclobutadiene, thiophene and alkyne units—polymeric organometallic mesogens. *Adv Mater* 7:726.

106. Steffen, W., et al. 2001. Conjugated organometallic polymers containing Vollhardt's cyclobutadiene complex: Aggregation and morphologies. *Chem Eur J* 7:117.

107. Yamamoto, T., et al. 1997. Poly(aryleneethynylene) type polymers containing a ferrocene unit in the pi-conjugated main chain. Preparation, optical properties, redox behavior, and Mössbauer spectroscopic analysis. *Macromolecules* 30:5390.

108. Plenio, H., J. Hermann, and A. Sehring. 2000. Optically and redox-active ferroceneacetylene polymers and oligomers. *Chem Eur J* 6:1820.

109. Plenio, H., J. Hermann, and J. Leukel. 1998. Synthesis of soluble 1,3-bridged ferrocene-acetylene polymers and the divergent-convergent synthesis of defined oligomers. *Eur J Inorg Chem* 2063.

110. Yamamoto, T., et al. 1998. Poly(aryleneethynylene) type polymers composed of *p*-phenylene and 2,5-thienylene units. Analysis of polymerization conditions and terminal group in relation to the mechanism of the polymerization and chemical and optical properties of the polymer. *Macromolecules* 31:7.

111. Yamamoto, T. 2003. Synthesis of pi-conjugated polymers bearing electronic and optical functionalities by organometallic polycondensations. Chemical properties and applications of the pi-conjugated polymers. *Synlett* 425.

112. Li, J. and Y. Pang. 1998. Regiocontrolled synthesis of poly[(*p*-phenyleneethynylene)-alt-(2,5-thienylene ethynylene)]s: Regioregularity effect on photoluminescence and solution properties. *Macromolecules* 31:5740.

113. Li, J. and Y. Pang. 1977. Regiocontrolled synthesis of poly((3-hexylthiopheneylene)ethynylenes): Their characterization and photoluminescent properties. *Macromolecules* 30:7487.
114. Pautzsch, T. and E. Klemm, 2002. Ruthenium-chelating poly(heteroaryleneethynylene)s: Synthesis and properties. *Macromolecules* 35:1569.
115. Yamamoto, T., et al. 1993. Preparation and optical properties of soluble pi-conjugated PAE-type polymers. *Chem Commun* 797.
116. Walters, K.A., et al. 2001. Photophysics of π-conjugated metal-organic oligomers: Aryleneethynylenes that contain the (bpy)Re(CO)$_3$Cl chromophore. *J Am Chem Soc* 123:8329.
117. Klemm, E. 2005. Organometallic PAEs, *Adv Polym Sci* 177, 53–90.
118. Jegou, G. and S.A. Jenekhe. 2001. Highly fluorescent poly(aryleneethynylene)s containing quinoline and 3-alkylthiophene units. *Macromolecules* 34:7926.
119. Bangcuyo, C.G., et al. 2002. Quinoline-containing, conjugated poly(aryleneethynylene)s: Novel metal and H$^+$-responsive materials. *Macromolecules* 35:1563.
120. Huang, H., et al. 2004. Design of a modular-based fluorescent conjugated polymer for selective sensing. *Angew Chem* 43:5635.
121. Bangcuyo, C.G., et al. 2001. Synthesis and characterization of a 2,1,3-benzothiadiazole-*b*-alkyne-*b*-1,4-bis(2-ethylhexyloxy)benzene terpolymer, a stable low-band-gap poly(heteroaryleneethynylene). *Macromolecules* 34:7592.
122. Morikita, T., I. Yamaguchi, and T. Yamamoto. 2001. New charge transfer-type *p*-conjugated poly(aryleneethynylene) containing benzo[2,1,3]thiadiazole as the electron-accepting unit. *Adv Mater* 13:1862.
123. Yamamoto, T., et al. 2000. Synthesis and chemical properties of pi-conjugated zinc porphyrin polymers with arylene and aryleneethynylene groups between zinc porphyrin units. *Macromolecules* 33:5988.
124. Yamamoto, T., Q. Fang, and T. Morikita. 2003. New soluble PAEs consisting of electron-accepting benzothiadiazole units and electron donating dialkoxybenzene units. Synthesis, molecular assembly, orientation on substrates and electrochemical and optical properties. *Macromolecules* 36:4262.
125. Bangcuyo, C.G., et al. 2003. Quinoxaline-based poly(aryleneethynylene)s. *Macromolecules* 36:546.
126. Wagner, M. and O. Nuyken. 2003. Benzyl bromide functionalized poly(phenyleneethynylene)s: A novel approach toward conjugated polymers with a well-defined chemical reactivity. *Macromolecules* 36:6716.
127. Kushon, S.A., et al. 2003. Detection of single nucleotide mismatches via fluorescent polymer superquenching. *Langmuir* 19:6456.
128. Wilson, J.N., et al. 2003. A biosensing model system: Selective interaction of biotinylated PPEs with streptavidin-coated polystyrene microspheres. *Chem Commun* 1626.
129. Zheng, J. and T.M. Swager. 2004. Biotinylated poly(*p*-phenyleneethynylene): Unexpected energy transfer results in the detection of biological analytes. *Chem Commun* 2798.
130. Disney, M.D., et al. 2004. Detection of bacteria with carbohydrate-functionalized fluorescent polymers. *Am Chem Soc* 126:13343.
131. Wagner, M. and O. Nuyken. 2004. Benzyl bromide functionalized poly(phenyleneethynylene)s as macroinitiators in the atom transfer radical polymerization of methyl methacrylate. *J Macromol Sci A* 41:637.
132. Marshall, A.R. and U.H.F. Bunz. 2001. Alkyne-bridged polymers as platform for novel macromolecular materials: Catalytic hydrogenation of poly[(*p*-dialkylphenylene)ethynylene]s. *Macromolecules* 34:4688.
133. Beck, J.B., et al. 2002. Facile reduction of poly(2,5-dialkoxy-*p*-phenyleneethynylene)s: An efficient route for the synthesis of poly(2,5-dialkoxy-*p*-xylylene)s. *Macromolecules* 35:590.
134. Cotts, P.M., T.M. Swager, and Q. Zhou. 1996. Equilibrium flexibility of a rigid linear conjugated polymer. *Macromolecules* 29:7323.

135. Ricks, H.L., et al. 2003. Rod vs coil: Molecular weight comparison of a poly(dialkyl-*p*-phenyle-neethynylene) with its reduced poly(2,5-dialkyl-*p*-xylylene). *Macromolecules* 36:1424.
136. Scholz, S., et al. 2005. Cobalt–carbon-spheres: Pyrolysis of dicobalthexacarbonyl-functionalized poly(*para*-phenyleneethynylene)s. *Adv Mater* 17:1052, in press.
137. Nicolas, L., et al. 1992. Nanoscale platinum (0) clusters in glassy carbon: Synthesis, characteriza-tion, and uncommon catalytic activity. *J Am Chem Soc* 114:769.
138. Huber, C., et al. 2001. Complexation of unsaturated C–C-bonds in *p*-conjugated polymers with transition metals. *J Am Chem Soc* 123:3857.
139. Kokil, A., et al. 2002. High charge carrier mobility in conjugated organometallic polymer networks. *J Am Chem Soc* 124:3857.
140. Kokil, A., et al. 2003. Charge carrier mobility in dialkoxy-PPEs. *Synth Met* 138:513.
141. Kokil, A., et al. 2003. Synthesis of π-conjugated organometallic polymer networks. *Macromol Chem Phys* 204:40.
142. Hittinger, E., A. Kokil, and C. Weder. 2004. Synthesis and characterization of cross-linked conju-gated polymer milli, micro and nanoparticles. *Angew Chem* 43:1808.
143. Hittinger, E., A. Kokil, and C. Weder. 2004. Synthesis and characterization of cross-linked PPEs. *Macromol Rapid Commun* 25:710.
144. Miteva, T., et al. 2000. Interplay of thermochromicity and liquid crystalline behavior in poly(*p*-phenyleneethynylene)s: π–π interactions or planarization of the conjugated backbone? *Macromol-ecules* 33:652.
145. Kim, J. and T.M. Swager. 2001. Control of conformational and interpolymer effects in conjugated polymers. *Nature* 411:1030.
146. Kim, J., et al. 2002. Structural control in thin layers of poly(*p*-phenyleneethynylene)s: Photophy-sical studies of Langmuir and Langmuir–Blodgett films. *J Am Chem Soc* 124:7710.
147. Deans, R., et al. 2000. A poly(*p*-phenyleneethynylene) with a highly emissive aggregated phase. *J Am Chem Soc* 122:8565.
148. McQuade, D.T., J. Kim, and T.M. Swager. 2000. Two-dimensional conjugated polymer assemblies: Interchain spacing for control of photophysics. *J Am Chem Soc* 122:5885.
149. Wilson, J.N., et al. 2002. Chiroptical properties of poly(*p*-phenyleneethynylene) copolymers in thin films: Large *g*-values. *J Am Chem Soc* 124:6830.
150. Samori, P., et al. 2003. Synthesis and solid state structures of functionalized phenyleneethynylene trimers in 2D and 3D. *Chem Mater* 15:1032.
151. Breitenkamp, R.B. and G.N. Tew. 2004. Aggregation of poly(*p*-phenyleneethynylene)s containing nonpolar and amine side chains. *Macromolecules* 37:1163.
152. Erdogan, B., J.N. Wilson, and U.H.F. Bunz. 2002. Synthesis and mesoseopic order of a sugar-coated poly(*p*-phenyleneethynylene). *Macromolecules* 35:7863.
153. Lavigne, J.J., et al. 2003. "Surfactochromic" conjugated polymers: Surfactant effects on sugar-substituted PPEs. *Macromolecules* 36:7409.
154. Chen, L.H., et al. 2000. Tuning the properties of conjugated polyelectrolytes through surfactant complexation. *J Am Chem Soc* 122:9302.
155. Stork, M., et al. 2002. Energy transfer in mixtures of water-soluble oligomers: Effect of charge, aggregation and surfactant complexation. *Adv Mater* 14:361.
156. Burrows, H.D., et al. 2005. Interactions between surfactants and {1,4-phenylene-[9,9-*bis*(4-phe-noxy-butylsulfonate)]fluorene-2,7-diyl}. *Colloids Surf A* 270:61.
157. Sluch, M.I., et al. 2001. Excited-state dynamics of oligo(*p*-phenyleneethynylene): Quadratic coup-ling and torsional motions. *J Am Chem Soc* 123:6447.
158. Beeby, A., et al. 2003. Studies of the S1, state in a prototypical molecular wire using picosecond time-resolved spectroscopies. *Chem Commun* 2406.
159. Weder, C., et al. 1998. Incorporation of photoluminescent polarizers into liquid crystal displays. *Science* 279:835.

160. Montali, A., et al. 1998. Polarizing energy transfer in photoluminescent materials for display applications. *Nature* 392:261.
161. Weder, C., et al. 1997. Highly polarized luminescence from oriented conjugated polymer/polyethylene blend films. *Adv Mater* 9:1035.
162. Samori, P., et al. 1999. Self-assembly of a conjugated polymer: From molecular rods to a nanoribbon architecture with molecular dimensions. *Chem Eur J* 5:2312.
163. Samori, P., et al. 2000. Macromolecular fractionation of rod-like polymers at atomically flat solid–liquid interfaces. *Adv Mater* 12:579.
164. Samori, P., et al. 1998. Poly-*para*-phenylene-ethynylene assemblies for a potential molecular-nanowire: An SFM study. *Opt Mater* 9:390.
165. Samori, P., et al. 1998. Growth of solution cast macromolecular pi-conjugated nanoribbons on mica. *Thin Solid Films* 336:13.
166. Samori, P., et al. 1999. Nanoribbons from conjugated macromolecules on amorphous substrates observed by SFM and TEM. *Nanotechnology* 10:77.
167. Perahia, D., R. Traiphol, and U.H.F. Bunz. 2002. From single molecules to aggregates to gels in dilute solution: Self-organization of nanoscale rodlike molecules. *J Chem Phys* 117:1827.
168. Perahia, D., R. Traiphol, and U.H.F. Bunz. 2001. From molecules to supramolecular structure: Self assembling of wirelike poly(*p*-phenyleneethynylene)s. *Macromolecules* 34:151.
169. Zahn, S. and T.M. Swager. 2002. Three-dimensional electronic delocalization in chiral conjugated polymers. *Angew Chem* 41:4226.
170. Craig, M.R., et al. 2003. The chiroptical properties of a thermally annealed film of chiral substituted polyfluorene depend on film thickness. *Adv Mater* 15:1435.
171. Martin, C.R. 1995. Template synthesis of electronically conductive polymer nanostructures. *Acc Chem Res* 28:61.
172. Wilson, J.N., et al. 2003. Nanostructuring of poly(aryleneethynylene)s: Formation of nanotowers, nanowires, and nanotubules by templated self-assembly. *Macromolecules* 36:1426.
173. Lord Rayleigh. 1911. Breath figures. *Nature* 86:416.
174. Lord Rayleigh. 1912. Breath figures. *Nature* 90:436.
175. Jenekhe, S.A. and X.L. Chen. 1999. Self-assembly of ordered microporous materials from rod-coil block copolymers. *Science* 283:372.
176. Srinivasarao, M., et al. 2001. Three-dimensionally ordered array of air bubbles in a polymer film. *Science* 292:79.
177. Song, L., et al. 2004. Facile microstructuring of organic semiconducting polymers by the breath figure method: Hexagonally ordered bubble arrays in rigid-rod polymers. *Adv Mater* 16:115.
178. Erdogan, B., et al. 2004. Permanent bubble arrays from a cross-linked poly(*para*-phenyleneethynylene): Picoliter holes without microfabrication. *J Am Chem Soc* 126:3678.
179. Huisgen, R., G. Szeimies, and L. Moebius. 1967. 1.3-Dipolare Cycloadditionen. 32. Kinetik der Additionen organischer Azide an CC-Mehrfachbindungen. *Chem Ber* 100:2494.
180. Huisgen, R., et al. 1965. 1.3-Dipolare Cycloadditionen. 23. Einige Beobachtungen zur Addition organischer Azide an CC-Dreifachbindungen. *Chem Ber* 98:4014.
181. Englert, B.C., et al. 2005. Templated ceramic microstructures by using the breath-figure method. *Chem Eur J* 11:995.
182. Montali, A., P. Smith, and C. Weder. 1998. PPE-based light-emitting devices. *Synth Met* 97:123.
183. Schmitz, C. et al. 2001. Polymeric light emitting diodes based on PPE, poly(triphenyldiamine), and spiroquinoxaline. *Adv Funct Mater* 11:41.
184. Pschirer, N.G., et al. 2001. Blue solid-state photoluminescence and electroluminescence from novel PPE copolymers. *Chem Mater* 13:2691.
185. Zhou, Q. and T.M. Swager. 1995. Fluorescent chemosensors based on energy migration in conjugated polymers: The molecular wire approach to increased sensitivity. *J Am Chem Soc* 117:12593.
186. Turro, N.J. 1978. *Modern molecular photochemistry*. Menlo Park: Benjamin Cummings.

187. Connors, K.A. 1987. *Binding constants: The measurement of molecular complex stability*. New York: Wiley-Interscience.
188. Wosnick, J.H. and T.M. Swager. 2004. Enhanced fluorescence quenching in receptor-containing conjugated polymers: A calix[4]arene-containing PPE. *Chem Commun* 2744.
189. Kim, J., et al. 2000. Ion-specific aggregation in conjugated polymers: Highly sensitive and selective fluorescent ion chemosensors. *Angew Chem* 39:3868.
190. Kim, T.H. and T.M. Swager. 2003. A fluorescent self-amplifying wavelength-responsive sensory polymer for fluoride ions. *Angew Chem* 115:4951.
191. Tanese, M.C., et al. 2004. Poly(phenyleneethynylene) polymers bearing glucose substituents as promising active layers in enantioselective chemoresistors. *Sens Actuators B Chem* 100:17.
192. Babudri, F., et al. 2003. Synthesis of poly(aryleneethynylene)s bearing glucose units as substituents. *Chem Commun* 130.
193. Mammen, M., S.K. Choi, and G.M. Whitesides. 1998. Polyvalent interactions in biological systems: Implications for design and use of multivalent ligands and inhibitors. *Angew Chem* 37:2755.
194. Lis, H. and N. Sharon. 1998. Lectins: Carbohydrate-specific proteins that mediate cellular recognition. *Chem Rev* 98:637.
195. Kim, I.-B., J.N. Wilson, and U.H.F. Bunz. 2005. Mannose-substituted PPEs detect lectins: A model for Ricin sensing. *Chem Commun* 1273.
196. Moon, J.H., et al. 2003. Capture and detection of a quencher labeled oligonucleotide by poly(-phenyleneethynylene) particles. *Chem Commun* 1:104.
197. Wosnic, J.H., C.M. Mello, and T.M. Swager. 2005. Synthesis and application of PPEs for bioconjugation: A conjugated polymer-based fluorogenic probe for proteases. *J Am Chem Soc* ASAP 127.
198. Pinto, M.R. and K.S. Schanze. 2004. Amplified fluorescence sensing of protease activity with conjugated polyelectrolytes. *Proc Natl Acad Sci USA* 101:7505.
199. Yang, J.-S. and T.M. Swager. 1998. Porous shape persistent fluorescent polymer films: An approach to TNT sensory materials. *J Am Chem Soc* 120:5321.
200. Yang, J.-S. and T.M. Swager. 1998. Fluorescent porous polymer films as TNT chemosensors: Electronic and structural effects. *J Am Chem Soc* 120:11864.
201. Williams, V.E. and T.M. Swager. 2000. Iptycene-containing poly(aryleneethynylene)s. *Macromolecules* 33:4069.
202. Zhu, Z. and T.M. Swager. 2001. Conjugated polymers containing 2,3-dialkoxybenzene and iptycene building blocks. *Org Lett* 3:3471.
203. Cumming, C.J., et al. 2001. Using novel fluorescent polymers as sensory materials for above-ground sensing of chemical signature compounds emanating from buried landmines. *IEEE Trans Geosci Remote Sens* 39:1119.

7

Polyaniline Nanofibers: Syntheses, Properties, and Applications

Jiaxing Huang and
Richard B. Kaner

7.1   Nanostructures of the Conducting Polymer Polyaniline: An Overview

7.1.1   Polyaniline

Since the discovery that conjugated polymers can be made to conduct electricity through doping [1], a tremendous amount of research has been carried out in the field of conducting polymers [2,3]. Polyaniline (Figure 7.1) is an excellent example of a conjugated polymer that can be tailored for specific applications through the doping process [4]. Since its conducting properties were rediscovered in the early 1980s, polyaniline has been studied for many other potential applications including lightweight battery electrodes [5], electromagnetic shielding devices [6,7], and anticorrosion coatings [8,9]. Bulk polyaniline is now commercially available from several sources [10]. Polyaniline is electrically conductive in its emeraldine oxidation state when doped with a salt that protonates the imine nitrogens on the polymer backbone. Dopants can be added in any desired quantity until all imine nitrogens (half of the total nitrogens) are doped, simply by controlling the pH of the dopant acid solution.

Polyaniline's conductivity increases with doping from the undoped insulating base form ($\sigma \leq 10^{-10}$ S/cm) to the fully doped, conducting acid form ($\sigma \geq 1$ S/cm) [4]. Doping and undoping processes are typically carried out chemically with common acids such as hydrochloric acid and bases such as ammonium hydroxide; electrochemical processes can also be used. In addition to affecting conductivity, doping and undoping processes can have dramatic effects on the morphology of the polymer films [11].

Nanostructures (nanorods, nanowires, nanofibers, and nanotubes) of conducting polymers have attracted a great deal of research interest with the expectation that the combination of organic conducting materials and nanostructures could yield new functional materials or enable new physio-chemical properties to be discovered. Here the term "nanostructure" refers to structures with at least one dimension smaller than 100 nm. Emphasis will be given to elongated nanostructures such as nanorods, nanowires, nanofibers, and nanotubes. This section overviews methods for making conducting polymer nanostructures using polyaniline as an example, followed by an outline to summarize the remaining sections in this chapter.

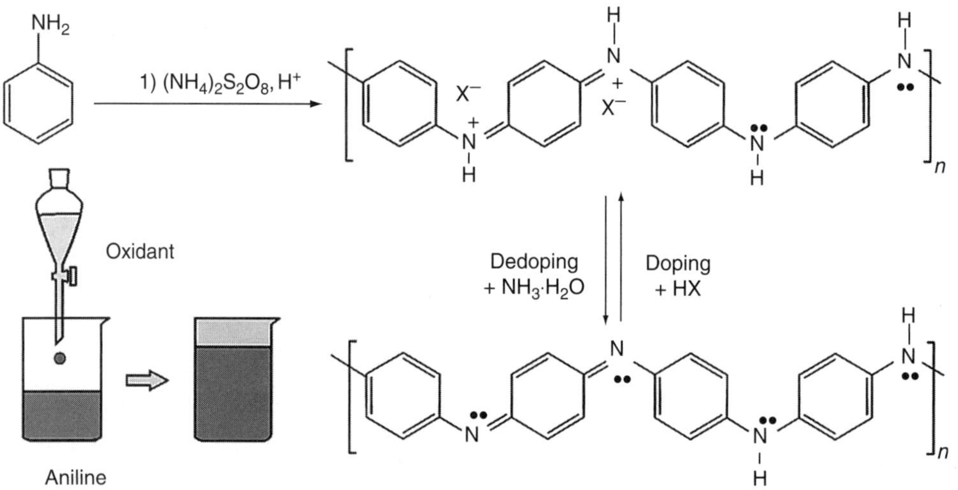

FIGURE 7.1   The oxidative polymerization of aniline in an acidic solution. The synthesized polyaniline forms in its doped emeraldine salt state that then can be dedoped by a base to its emeraldine base form. The *bottom left* scheme illustrates a typical reaction for making polyaniline.

7.1.2 Methods of Forming Polyaniline Nanostructures

7.1.2.1 Templated Synthesis of Polyaniline Nanostructures

The most straightforward route to nano-structures is to make them with a template of a desired size. This is especially applicable with polymeric materials due to their mechanical flexibility. Such a template can be a confined nanospace within which the polymer can be grown (Figure

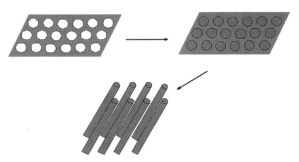

FIGURE 7.2   Schematic drawing illustrating templated synthesis of conducting polymer nanostructures.

7.2), or a nanosized entity allowing the polymer to encapsulate or wrap around it. For example, using a nanoporous membrane, conducting polymer nanofibers, nanowires, and nanotubes with diameters of tens of nanometers have been grown within the pores [12]. Polyaniline nanofilaments with diameters smaller than 10 nm have also been grown within zeolite channels [13]. Nanotubes of polyaniline have been obtained using sacrificial nanowires as templates [14,15]. However, to obtain free-standing poly-aniline nanostructures, the template has to be removed by postsynthetic physiochemical treatment. If the polyaniline nanostructures have to be recovered, they have already been subject to a round of physiochemical treatment before they are used. Therefore, their processing history becomes important. It is impractical to scale up such syntheses as the number of nanostructures created is determined by the number of templates.

7.1.2.2 Self-Assembly Route to Polyaniline Nanostructures

Using functional molecules as structural directors in the chemical polymerization bath can also produce polyaniline nanostructures. Such structural directors include surfactants [16–18], liquid crystals [19], polyelectrolytes (including DNA) [20,21], or complex bulky dopants [22–24]. It is believed that functional molecules can promote the formation of nanostructured soft condensed phase materials (e.g., micelles and emulsions) that can serve as "soft templates" for aniline polymerization (Figure 7.3). Polyelectrolytes such as polyacrylic acid, polystyrenesulfonic acid, and DNA can bind aniline monomer molecules, which can be polymerized in situ forming polyaniline nanowires along the polyelectrolyte molecules. Compared to templated syntheses, self-assembly routes are more scalable but they rely on the structural director molecules. It is also difficult to make nanostructures with small diameters (e.g., ≤ 50 nm). For example, in the dopant induced self-assembly route, very complex dopants with bulky side groups are needed to obtain nanotubes with diameters smaller than 100 nm, such as sulfonated naphthalene derivatives [23–25], fullerenes [26], or dendrimers [27,28].

7.1.2.3 Electrospinning Route to Polyaniline Nanostructures

Electrospinning [29] is a facile method to make almost any polymer into nano- and micro-fibers if the polymer can be solution-processed or melt-processed (Figure 7.4). However, due to the low solubility of polyaniline, it is very difficult to make polyaniline fibers thinner than 100 nm. To solve this problem, polyaniline is usually blended with a more soluble polymer to increase the weight percentage of polymer in solution. As a result, the obtained nanofibers are a polymer blend [29–33]; this significantly reduces the conductivity of the polymer. Additionally, when the size of the fiber decreases, phase separation may occur, yielding a mixture of nanofibers of polyaniline and other polymers [33].

7.1.2.4 Individual Polyaniline Nanofibers

The capability of making polyaniline nanowires and nanofibers individually at desired positions is the key to making single nanofiber devices. There are a couple of methods to make polyaniline nanofibers at

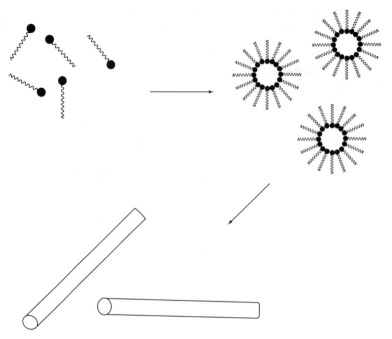

FIGURE 7.3 Schematic illustration showing how functional molecules can induce a self-assembly route to conducting polymer nanostructures.

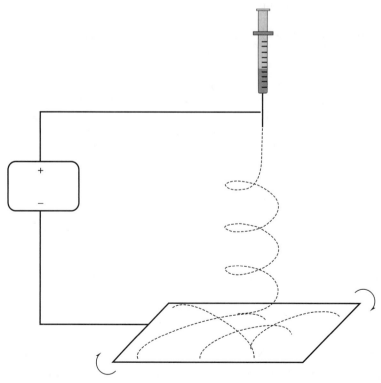

FIGURE 7.4 Schematic drawing illustrating the electrospinning process to make conducting polymer nanostructures. Typically, a potential of several thousand volts is applied to overcome the surface tension.

the single fiber level including templated polymerization in lithographically defined nanochannels [34,35], mechanical stretching [36], dip-pen writing [37], drawing [38], and single DNA strand templated polymerization [39].

7.1.3 Naturally Formed Polyaniline Nanofibers in Electrochemical Polymerization and Polymer Blends

It has been known since the early stage of conducting polymer research that polyaniline fibrils of ~100 nm in diameter can naturally form on the surface of an electrode [4,40–45] with a compact microsphriod underlayer. Some recent work demonstrates that pure polyaniline nanofibers can be obtained without the need for any template by controlling the polymerization rate [46–48]. Although this process is not readily scalable from a materials point of view, such work could be very important for making functional devices, since nanofiber-coated electrodes can be used as a platform to fabricate sensors and transistors. Interconnected network-like structures with polyaniline "nanolinkers" 10–50 nm wide have also been identified in polymer blends [49–51].

7.1.4 Outline of the Review

We have developed a conceptually new synthetic methodology that readily produces high-quality, small-diameter nanofibers in large quantities. In contrast to the previous work in which preparation conditions were designed to "shape" the polymer into nanostructures, our approach focuses on modifying the reaction pathway so that the intrinsic nanofibrillar morphology of polyaniline formation is favored and the formation of irregularly shaped agglomerates is suppressed (Section 7.2). The nanofibrillar morphology significantly improves the performance of polyaniline in many conventional applications involving polymer interactions with its environment. This leads to much faster and more responsive chemical sensors (Section 7.3), new inorganic or polyaniline nanocomposites, and ultrafast nonvolatile memory devices (Section 7.4). Additionally, the highly conjugated polymeric structure of polyaniline produces new nanoscale phenomena that are not accessible with current inorganic systems. As an example, the discovery of an enhanced photothermal effect that produces welding of the polyaniline nanofibers will be presented (Section 7.5). The nanoscale morphology of polyaniline derivatives and other conducting polymers is also discussed (Section 7.6). This chapter highlights our systematic body of research on syntheses, formation mechanism, and applications of polyaniline nanofibers, which are believed to open new research areas for both conducting polymers and nanomaterials.

7.2 Intrinsic Nanofibrillar Morphology of Polyaniline

7.2.1 Nanofibers in Conventional Polyaniline

The conventional chemical oxidative polymerization of aniline is carried out in an aqueous solution in which aniline is dissolved in a strong acidic solution (e.g., 1 M HCl) at ~0°C and the polymerization is initiated by adding an oxidant (e.g., ammonium peroxydisulfate) into the solution as shown in Figure 7.1 [4]. The traditional oxidative chemical polymerization route is known to yield granular polyaniline [3,52]. As discussed in Section 7.1, polyaniline nanostructures such as nanofibers, nanowires, nanorods, and nanotubes can be made by introducing "structural directors" into the chemical polymerization bath. By carefully examining the morphology of traditionally synthesized polyaniline powders, we have found that there is also a small portion of nanofibers in addition to granular particle agglomerations (Figure 7.5) [53]. These nanofibers are ~30 nm in diameter and they are much smaller than those formed by self-assembly methods. It is worth mentioning that unlike inorganic materials, many polymeric materials are known to have basic morphological units at the nanometer scale. For example, under proper synthetic conditions, fibers of 5–50 nm in diameter can be found in as-made polyacetylene films, the abundance of which is determined by

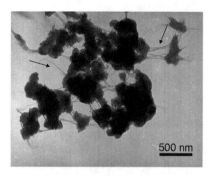 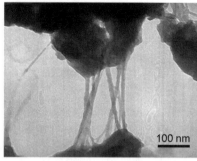

FIGURE 7.5 TEM images of polyaniline powders made by traditional chemical polymerization using 1.0 M HCl showing a small portion of nanofibers (*arrows*) in the sample. A magnified view of the nanofibers is shown on the right. (Reproduced from Huang, J.X. and Kaner, R.B., *J. Am. Chem. Soc.*, 126, 851–855, 2004. With permission.)

polymerization kinetics [54]. Therefore, nanofibers such as those shown in Figure 7.5 may also be an intrinsic morphological unit of polyaniline.

7.2.2 Morphological Evolution of Polyaniline during Chemical Polymerization

As small amount of nanofibers can be found even in conventional polyaniline, it is of great interest to look into the morphological evolution of polyaniline during traditional chemical oxidative polymerization. In a typical conventional one-phase polymerization, a solution of the oxidant ammonium peroxydisulfate dissolved in 1 M HCl is fed continuously into a solution of aniline dissolved in 1 M HCl using a syringe pump at a preset flow rate. This is analogous to the "drop-by-drop" titration used in traditional synthesis (Figure 7.6), but with a precisely controlled feeding rate. The reaction vessel is kept

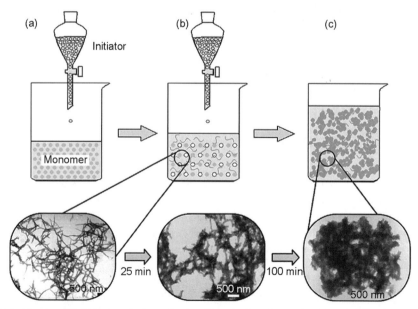

FIGURE 7.6 The morphological evolution of polyaniline during chemical polymerization (a) is explored by electron microscopy. It clearly shows that nanofibers (*green lines*) are produced in the early stages of polymerization (b) and then turn into large, irregularly shaped agglomerates (c) due to secondary growth. All the three TEM images have the same magnification.

in an ice bath at between −5°C and 0°C. Small amounts of product for transmission electron microscopy (TEM) studies are periodically extracted from the reaction bath as soon as the green color of polyaniline is visible. At this point, the samples are immediately diluted with distilled water and cast on TEM grids. The grids were placed on filter paper to absorb extra suspension and facilitate rapid drying, thereby quenching the polymerization.

Figure 7.6a shows the morphology of polyaniline formed (a) when polymerization is first initiated, (b) 25 min later, and (c) 100 min after initiation. Figure 7.6a clearly shows that polyaniline nanofibers form at an early stage in the polymerization process. These nanofibers have average diameter of 30–35 nm, which is consistent with those obtained using interfacial polymerization [53]. As more ammonium peroxydisulfate is fed into the reaction, the nanofibers become scaffolds for secondary growth of polyaniline (Figure 7.6b) and finally turn into irregularly shaped agglomerates containing nanofibers and particulates (Figure 7.6c). Therefore, it is not surprising to know that some nanofibers can escape the secondary growth and remain in the final product. The whole process is illustrated in Figure 7.6 [55].

7.2.3 Suppressing Secondary Growth to Make Pure Nanofibers

An important insight gained from the morphological evolution study is that if secondary growth of irregularly shaped polyaniline can be suppressed, the yield of nanofibers in the final product could be greatly increased. Pure polyaniline nanofibers may be obtained without the use of any template or structural-directing agents. Based on this idea, we have developed two facile approaches—interfacial polymerization and rapidly mixed reactions. Both methods can produce polyaniline nanofibers with nearly uniform diameters tunable between 30 and 120 nm with lengths varying from 500 nm to several microns. Gram scale products that contain almost exclusively nanofibers can be synthesized.

7.2.3.1 Interfacial Polymerization Route to Polyaniline Nanofibers

The interfacial polymerization is based on the well-known chemical oxidative polymerization of aniline in a strongly acidic environment, with ammonium peroxydisulfate as the oxidant [4]. Instead of using the traditional homogeneous aqueous solution of aniline, acid, and oxidant, the polymerization is performed in an immiscible organic–aqueous biphasic system, to separate the by-products (inorganic salts, oligomers, etc.) according to their solubility in the organic and aqueous phases. Interfacial polymerization does not depend on any specific template or dopant. High-quality polyaniline nanofibers are obtained even when common mineral acids, such as hydrochloric, sulfuric, or nitric acid, are used as dopants. The syntheses can be accomplished in one pot with a wide choice of solvent pairs, acid dopants, and reagent concentrations over a broad range of temperatures.

7.2.3.1.1 Experimental Procedure

Interfacial polymerization is performed in an aqueous–organic biphasic system (Figure 7.7) with aniline dissolved in an organic solvent and the oxidant, ammonium peroxydisulfate, dissolved in an aqueous acid solution. A great variety of organic solvents can be used, including benzene, hexane, toluene, carbon tetrachloride, chloroform, methylene chloride, diethyl ether, carbon disulfide, or *o*-dichlorobenzene.

FIGURE 7.7 Snapshots showing the interfacial polymerization route to polyaniline nanofibers. The reaction times are (a) 0, (b) 60, (c) 120, and (d) 180 s. The top layer is an aqueous solution of acid and oxidant, the bottom layer is aniline dissolved in an organic solvent. (e) A typical electron microscopy image of the nanofiber products.

The shape and size of the nanofibers do not appear to be affected by the solvent. Therefore, a less toxic organic solvent that is heavier than water, such as methylene chloride, is preferred due to safety considerations, because water can help to seal the organic vapor within the reaction vessel. The water layer is carefully spread onto an equal volume of the organic solvent, forming an aqueous–organic interface (Figure 7.7a). After a short induction period ranging from 30 s to several minutes depending on the acid used, green polyaniline appears at the interface (Figure 7.7b), migrating into the water phase (Figure 7.7c), and finally filling the entire water layer (Figure 7.7d). As the reaction proceeds, the color of the organic phase becomes darker, and finally stops changing, indicating the completion of the reaction. An overnight reaction time is generally sufficient. It is interesting to note that if the interfacial polymerization is performed with surfactants codissolved with aniline in the organic layer, then polyaniline forms only at the interface creating a free-standing film [17]. The morphology of the products can be imaged using electron microscopy. Highly uniform nanofibers are observed in samples prepared from dispersions, after dialysis or centrifugation under both TEM and scanning electron microscopy (SEM) (Figure 7.7e) [56]. The morphology of nanofibers is unaffected when the dopants are removed along with the base.

7.2.3.1.2   Effects of Dopants

Nanofibers appear to form irrespective of what acid dopant is used in the polymerization (Figure 7.8). For example, very uniform nanofibers are observed in SEM images after interfacial syntheses using the mineral acids: hydrochloric, sulfuric, nitric, and perchloric acid. Nanofibers are also obtained when interfacial polymerization is carried out using many other acids, including phosphoric, acetic, formic, tartaric, methylsulfonic, ethylsulfonic, camphorsulfonic, or 4-toluene sulfonic acid. The diameters of the resulting nanofibers are affected by the dopants used in the polymerization. The average diameter of nanofibers produced in HCl is ∼30 nm, those made in camphorsulfonic acid (CSA) approach 50 nm, and those synthesized in perchloric acid are centered at ∼120 nm [53]. Other acids, including sulfuric, nitric, and 4-toluene sulfonic acid, yield average diameters between 30 and 50 nm.

7.2.3.1.3   Effects of Synthetic Conditions

Interfacial synthesis of nanofibers appears to be insensitive to both the polymerization temperature and the monomer concentration. In the range between the freezing and the boiling points of the solvents used in the syntheses (typically 5°C–60°C), the nanofibers obtained look similar in both size and uniformity. However, when the reaction is carried out at −12°C with 5 M LiCl in the water phase, most of the products are particle agglomerations (Figure 7.9a). The monomer concentration in the organic phase can also be varied. For example, while keeping the aqueous phase unchanged, the relative volume of organic solvent can be decreased by increasing the concentration of aniline. When the aniline concentration varies from 0.032 to 1.6 M, no significant effect is found on the observed morphology. This is potentially a great advantage for scaling up reactions as less organic solvent is needed.

The quality and uniformity of the nanofibers seem to be affected only by the acid concentration of the aqueous phase. The lower the concentration of acid, the lower will be the fraction of nanofibers observed in the final product. For example, when HCl is used as the dopant, higher concentrations (0.5–2 M) are preferred because larger yields of high-quality nanofibers are obtained. As the concentration of HCl is lowered, the quantity of granular particles starts to increase and finally prevails over the nanofibers (Figure 7.9b). Similar results are obtained when the concentration of CSA is varied. When medium or weak acids are used, such as tartaric acid (pK_a = 3.03, Figure 7.8f) or pyrrolidone-5-carboxylic acid (pK_a = 3.32, Figure 7.8h), even at high concentrations (1 M), the products are still mixtures of particles and fibers. Therefore, high concentrations of strong acids are optimal for making polyaniline nanofibers. This is consistent with the traditional chemical polymerization that uses ≥1.0 M HCl. Therefore, forming an aqueous–organic interface is the only major difference in our synthesis of nanofibers vs. the classical method that produces polyaniline particles.

In most aqueous and organic systems, polyaniline is dispersed in the water phase since protonated polyaniline in the emeraldine salt form is hydrophilic. However, there are exceptions, e.g., when the

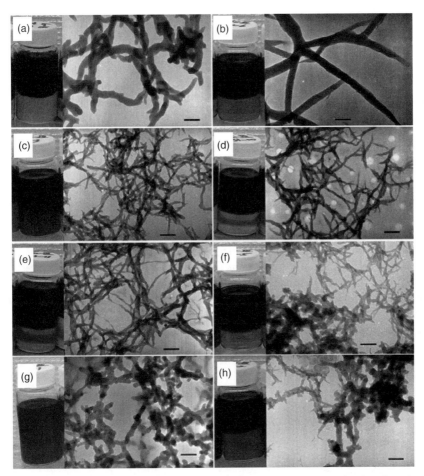

FIGURE 7.8 Typical TEM images showing nanofibers obtained through interfacial polymerization using (a) 4-toluenesulfonic, (b) perchloric, (c) formic, (d) methylsulfonic, (e) ethylsulfonic, (f) tartaric, (g) acetic, and (h) pyrrolidone-5-carboxylic acid. Scale bars = 200 nm. The *insets* are pictures of the reactions.

organic acid camphorsulfonic is used in the synthesis. CSA doped polyaniline nanofibers are found dispersed in the top water layer after reaction when carbon tetrachloride or carbon disulfide is used in the bottom organic layer. However, when chloroform or methylene chloride is used, polyaniline first diffuses into the aqueous phase, and then migrates into the bottom organic layer (Figure 7.9c, inset). This can be explained by the well-known counterion induced solubility of polyaniline in organic solvents [57,58]. Polyaniline collected in the organic layer shows no nanoscale features (Figure 7.9c), which differs greatly from the nanofibers collected from the water layer.

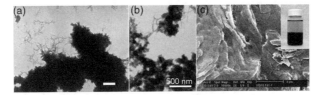

FIGURE 7.9 TEM and SEM images show the typical morphology of interfacially polymerized polyaniline when the reaction is carried out (a) at −12°C with 5 M LiCl in the water phase, scale bar = 1 μm, (b) with 0.001 M HCl, and (c) with 1.0 M of CSA in the aqueous phase when CHCl₃ is chosen as the organic phase. The *inset* shows that polyaniline is collected in the organic phase.

7.2.3.1.4 *Effects of Stirring*

In a stirred aqueous–organic biphasic system, emulsion droplets can form as templates for hollow spherical structures [59–61]. When the interfacial polymerization reaction of aniline is stirred (Figure 7.10a), polyaniline nanofibers are formed at the surface of the emulsion droplets. Therefore, cage-like nanofiber agglomerates are obtained, likely due to an interfacial templating effect (Figure 7.10b through Figure 7.10d). It is interesting to note that when CSA is used as the dopant in a stirred reaction, the cage is built from short polyaniline nanorods with diameters comparable to that of the nanofibers (Figure 7.10c and Figure 7.10d).

7.2.3.1.5 *Mechanism*

Interfacial polymerization represents one effective method to suppress secondary growth [53]. Since the monomer aniline and the initiator ammonium peroxydisulfate are separated by the boundary between the aqueous and the organic phase, polymerization occurs only at this interface where all the components needed for polymerization come together (Figure 7.7) [53,56,62]. Polyaniline then forms as nanofibers. Since these newly formed nanofibers are in the doped emeraldine salt form, they are hydrophilic and can rapidly move away from the interface and diffuse into the water layer (Figure 7.7c and Figure 7.7d). In this way, as the nanofibers form they are continuously withdrawn from the reaction front, thus avoiding secondary growth and allowing new nanofibers to grow at this interface. This explains why nanofibers are obtained no matter which solvent is used as the organic phase in interfacial polymerization. Hence, the interface between the immiscible aqueous–organic layers does not contribute directly to nanofiber formation; it simply separates nanofiber formation from secondary growth.

7.2.3.2 Rapidly Mixed Reactions

With the knowledge that the key to synthesizing polyaniline nanofibers is preventing secondary growth, we have now designed an even simpler method to make pure polyaniline nanofibers as illustrated in

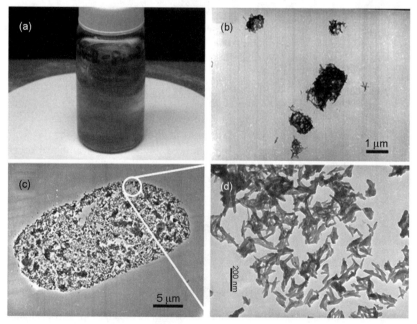

FIGURE 7.10   (a) A snapshot of a stirred interfacial polymerization in an H_2O/CH_2Cl_2 system. TEM images reveal that there are cage-like assemblies of polyaniline nanofibers in the product. (b) is HCl doped polyaniline nanofiber cages and (c)–(d) debris of a cage consisting of CSA doped nanofibers and rods.

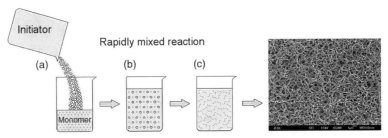

FIGURE 7.11 Schematic illustration showing a rapidly mixed reaction in which (a) the initiator and monomer solutions (in 1.0 M HCl) are rapidly mixed together all at once. Therefore (b and c), the initiator molecules are depleted during the formation of nanofibers, disabling further polymerization that leads to overgrowth. Pure nanofibers are obtained as shown in the SEM image.

Figure 7.11. The idea is that if all the reactants can be consumed during the formation of nanofibers, secondary growth will be greatly suppressed because no reactants will be available for further reaction. To achieve this goal, the initiator solution (ammonium peroxydisulfate in 1 M HCl) was added into the monomer solution (aniline in 1 M HCl) all at once (Figure 7.11a), rather than slowly feeding it by titration or syringe pumping. Using a magnetic stirrer or shaker, sufficient mixing can be achieved to evenly distribute the initiator and monomer molecules before polymerization (Figure 7.11b). As the polymerization begins, the initiator molecules induce the formation of nanofibers by rapidly polymerizing aniline monomers in their vicinity. Therefore, all the initiator molecules are consumed to form polyaniline nanofibers, suppressing the secondary growth of polyaniline (Figure 7.11b). The product, from a fast mixing reaction in an ice bath, is almost exclusively polyaniline nanofibers of uniform size as observed using SEM (Figure 7.11c, SEM). The nanofibers appear similar to those obtained by interfacial polymerization. In control experiments, the initiator solution was fed slowly into the aniline solution using a syringe pump (Figure 7.12). The product was a mixture of nanofibrillar agglomerates and irregularly shaped particulates (Figure 7.12, SEM). We have observed that the faster the feeding rate, the higher is the abundance of nanofibers in the product. Comparable results are obtained when the monomer solution is fed into the initiator solution.

Reactions that are rapidly mixed using other doping acids—including sulfuric, camphorsulfonic, and perchloric—also produce "pure" nanofibers with comparable shapes and sizes to those made by interfacial polymerization [55]. Figure 7.13 shows nanofibers obtained in a rapidly mixed reaction using H_2SO_4 and the agglomerates obtained in a reaction where the oxidant is fed slowly into the monomer solution. However, when the monomer solution (aniline + 1 M H_2SO_4) is fed into the initiator solution (ammonium peroxydisulfate + 1 M H_2SO_4), the morphological difference between the products from rapidly mixed to slowly mixed reactions is less pronounced than in Figure 7.13. It is worth noting that when 1 M of CSA or $HClO_4$ is used in the synthesis, a very high percent yield of nanofibers is obtained in the products from both the rapidly and the slowly mixed reactions (based on SEM studies). It seems that nanofibers of polyaniline readily form when dopants like 1 M CSA or $HClO_4$ are used, even in conventional syntheses.

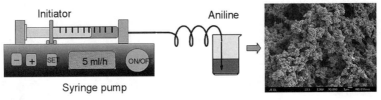

FIGURE 7.12 Schematic illustration showing a slowly mixed reaction in which the initiator solution (in 1.0 M HCl) is controllably injected into the monomer solution (in 1.0 M HCl) by a syringe pump. Irregularly shaped agglomerates are obtained as shown in the SEM image.

FIGURE 7.13   SEM images showing the morphology of polyaniline synthesized from (a) a rapidly mixed reaction and (b) a slowly mixed reaction using H_2SO_4. Good quality nanofibers are obtained in the rapidly mixed reaction, while irregularly shaped agglomerates form in the slowly mixed reaction.

The polymerization of aniline is exothermic and a rapidly mixed reaction would lead to an increase in the temperature of the solution. Therefore, it would be interesting to carry out fast mixing reactions at different temperatures. We have found that high-quality nanofibers can be obtained over a wide range of temperatures, such as in an ice bath ($\sim$0°C), near room temperature ($\sim$25°C), and in boiling water ($\sim$100°C) [197]. When the reactant concentrations are increased, the induction time of the reaction is reduced, but no apparent difference in the morphology of the product is observed.

7.2.4   Effect of Solvents on the Intrinsic Morphology of Polyaniline

The next question is why does polyaniline favor elongated, well-extended 1D nanofiber morphology in water? With the success of the rapidly mixed reaction route to making polyaniline nanofibrillar in water, the same reaction can be carried out in other solvents, which is difficult to do using interfacial polymerization. To test the effect of polarity of the solvent on polyaniline morphology, mixtures of isopropanol and water were used to continuously tune the polarity of the solvent. Nine reactions were carried out by decreasing the fraction of water by 10%, sequentially starting from pure water to 10% water in isopropanol. All other conditions are the same. The yields of the products based on the ratio between polyaniline emeraldine base to aniline are very consistent (19.9% $\pm$ 0.8%). Figure 7.14 shows five of the reactions carried out in different isopropanol–water mixtures. Although the final yields are about the same ($\sim$20%), their volumes are very different. This indicates a difference in the morphology and packing of the products. TEM images (Figure 7.15) of the product reveal the transition from nanofibers to agglomerates as the amount of the less polar solvent, isopropanol, increases. As nanofibers (Figure 7.15a) can pack loosely, they appear to have higher volume (Figure 7.14a). The agglomerates (Figure 7.15d) pack much more densely, therefore they appear to occupy less volume (Figure 7.14e).

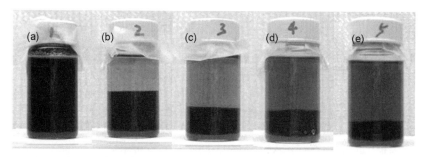

FIGURE 7.14   Images showing the volume occupied by polyaniline synthesized in a mixture of water and isopropanol. Vials (a) to (e) are **1** to **5** where the fraction of water decreases from 100% to 50%, respectively.

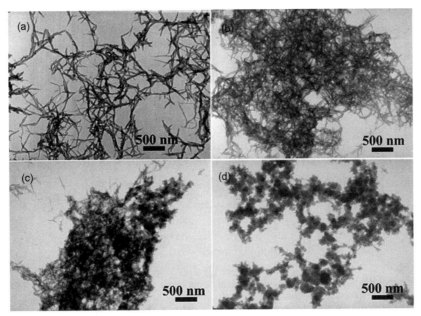

FIGURE 7.15 TEM images showing the progression from well-defined nanofibers to an entirely particulate polyaniline. (a)–(d) are the products from reaction **1**, **3**, **5**, and **9**, respectively.

7.2.5   Polyaniline Nanofibers Obtained by Solvent Exchange

The polymerization of aniline is also a precipitation reaction. In rapidly mixed reactions, the starting materials form a homogeneous single-phase solution. When aniline polymerizes, the monomer evolves into long macromolecular chains and the product precipitates out of solution. One can infer that there may be a critical size of polyaniline that the polymerizing solution can accommodate before precipitation begins. This critical size appears to be determined by the doping acid. To further understand the role of dopant, experiments were designed to precipitate polyaniline out of its solution in a good solvent by solvent exchange. In a typical experiment, a small amount of polyaniline solution in NMP is added into a 1 M of aqueous solution of the dopant (i.e., HCl or CSA). The polyaniline–NMP solution is prepared by dissolving 50 mg of conventional polyaniline powder in 5 mL of NMP. In this process, the particles in the powder disperse into the solvent, disentangling the polymer chains and simultaneously destroying any morphology the particles may possess. When the polyaniline–NMP solution is added into water, polyaniline precipitates. TEM images (Figure 7.16) of the precipitates reveal that there are

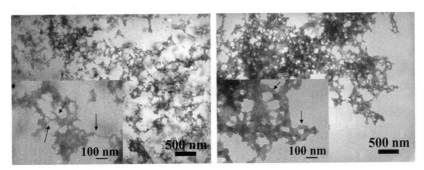

FIGURE 7.16   TEM images showing the morphology of polyaniline precipitates formed when 1 mL polyaniline/NMP solution is added into 20 mL of an aqueous solution of 1 M HCl (*left*) and CSA (*right*).

many nanometer-sized features corresponding to the diameter of polyaniline nanofibers synthesized using these acids. For example, 30 nm or even smaller filaments are observed in polyaniline re-precipitated in HCl. A significant amount of 50 nm-sized filaments are observed in CSA re-precipitated polyaniline, which resemble the 50 nm nanofibers obtained in "bottom–up" syntheses from aniline monomers. This suggests the possibility of constructing polyaniline nanostructures from both "top–down" (re-precipitation) and "bottom–up" approaches. Each polyaniline nanofiber is a collection, or an agglomeration of polyaniline macromolecular chains. It now appears possible to obtain polyaniline nanofibers starting from either the polymer chains or the monomers.

7.2.6 Processing and Characterization of Polyaniline Nanofibers

7.2.6.1 Purification of the Nanofibers and Their Dispersion

The product from either the interfacial polymerization or the rapidly mixed reactions can be purified through dialysis, centrifugation, or filtration. Dialysis yields the best water dispersions of nanofibers as no extra pressure is exerted on the product during the process. Dialysis has the least product loss as all the products are sealed in the dialysis tubing. One drawback is that dialysis can take more than a day, though only minimal effort is needed to change the dialyzing bath. Centrifugation can also produce aqueous nanofiber dispersions. An image of typical aqueous dispersions of doped and dedoped polyaniline nanofibers purified by either dialysis or centrifugation is given in Figure 7.17. Nanofibers dispersed in organic solvents can be made through solvent exchange by multiple centrifugation steps. However, it can be very tedious especially when the product does not settle down easily. Interfacially polymerized nanofibers in 1 M CSA water solutions can be transferred to chloroform or methylene chloride by agitated extraction, leading to organic dispersions of the nanofibers that could be very useful for spin-coating.

Filtration is the fastest purification method if a powder product is desired. For the powders obtained after filtration, SEM images reveal that they are actually agglomerations of nanofibers (Figure 7.18a). The nanofiber powders can be redispersed in water by sonication and they form better dispersion than conventional polyaniline powders (Figure 7.18b). The advantages and disadvantages of the three purification methods are listed in Table 7.1.

7.2.6.2 Optical Spectra of the Nanofibers

Figure 7.19 (left) presents the UV–vis spectra of CSA doped and dedoped nanofibers in water. The polyaniline–CSA complex forms a green suspension, after dialyzing against water. There are peaks centered ~340, 440, and 800 nm in the UV–vis spectrum, corresponding to doped polyaniline in its emeraldine salt form. The emeraldine oxidation state of polyaniline contains half imine and half amine nitrogens and can be represented by the formula $[-C_6H_4-N=C_6H_4=N-C_6H_4-NH-C_6H_4-NH-]_n$. The imine nitrogens can be completely protonated on exposure to strong acids, leading to doped polyaniline.

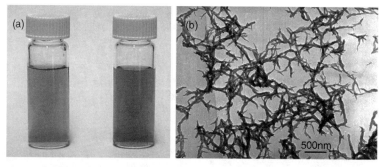

FIGURE 7.17 **(See color insert following page 14-20.)** Dialysis or centrifugation can create good polyaniline nanofiber aqueous dispersions. (a) HCl doped (*green, left*) and dedoped (*blue, right*) polyaniline nanofiber dispersions and (b) TEM image showing the morphology of the nanofibers.

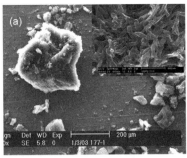

FIGURE 7.18 Filtration creates irregularly shaped micron scale polyaniline nanofiber powders as shown in (a) the SEM image. The *inset* shows a higher magnification view of a grain of the powder. (b) Aqueous dispersion of powders of polyaniline nanofibers (*left*) and conventional polyaniline (*right*). The concentration is 100 mg of polyaniline in 20 mL of water.

Since the polyaniline–CSA nanofibers have no absorption due to free CSA at ∼285 nm, CSA is most likely tightly incorporated as anions within the polyaniline backbone during the in situ polymerization of aniline in CSA solution. Dedoped polyaniline nanofibers (Figure 7.17a) can be obtained by dialyzing the pristine polyaniline–CSA complex against 0.1 M ammonium hydroxide, which produces the emeraldine base form of polyaniline.

Fourier transform infrared (FTIR) spectra (Figure 7.19, right) of the nanofibers are also consistent with the literature [63–65]. The absorption bands at ∼1590 and ∼1500 cm^{-1} correspond to quinoid and benzenoid rings, respectively. The relative intensities of these bands provide an index of polyaniline oxidation states. In Figure 7.19, right, the intensities of these two bands are about the same, indicating that there are approximately equal amounts of quinoid and benzenoid rings in the nanofibers, which is consistent with the emeraldine oxidation state of the nanofibers. Matching literature reports, a couple of other characteristics of doped polyaniline are also observed in Figure 7.19 (right). These include the disappearance of the weak band at ∼1380 cm^{-1} and the appearance of an intense broad band at ∼1150 cm^{-1}.

When an enantiomer of CSA is used in the polymerization, such as *R*-CSA, it is possible to create chiral polyaniline nanofibers. Figure 7.20 shows the circular dichroism (CD) spectrum of a water dispersion of as-prepared *R*-CSA doped polyaniline nanofibers. The positive peak at ∼450 nm is characteristic of chiral polyaniline [66–70], and is consistent with water's effect on the direction of the CD signals previously observed [71]. The peak at ∼290 nm is due to excess *R*-CSA in the dispersion. Recently, Wang et al. discovered that highly chiral polyaniline nanofibers can be produced by incremental addition of the oxidant, ammonium peroxydisulfate, into aniline solution with aniline oligomers and concentrated chiral dopants (>5 M *R*- or *S*-CSA) [72]. Chiral polyaniline nanofibers are very interesting for chiral recognition studies [68].

7.2.6.3 Molecular Weight, Conductivity, Surface Area, and Crystallinity

The molecular weight distribution of the polyaniline nanofibers is determined using gel permeation chromatography. Although a bimodal distribution of molecular weights is observed, similar to that of conventional polyaniline [73], the area under the high molecular weight peak (61.7%) is considerably larger than that under the low-molecular weight peak (38.3%). This contrasts with conventional

TABLE 7.1 Comparison of Purification Methods for Polyaniline Nanofibers

Purification	Product Form	Product Lost	Time	Effort
Dialysis	Dispersion (water only)	Lowest	Longest	Minimum
Centrifugation	Dispersion (in many solvents)	Highest	Medium	Maximum
Filtration	Powder	Medium	Shortest	Medium

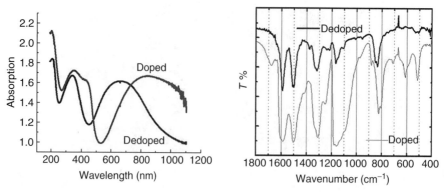

FIGURE 7.19   UV–vis (*left*) and FT-IR (*right*) spectra of HCl doped and dedoped polyaniline nanofibers.

chemically polymerized polyaniline, for which the low-molecular weight fraction dominates [73]. The electrical conductivity of a pressed pellet of polyaniline nanofibers made with 1.0 M HCl is ~0.5 S/cm. This is comparable to the conductivity reported for conventional polyaniline powders [74]. In fact, this is confirmed by a recent report at the International Conference on Synthetic Metals (ICSM), 2004 [75]. As a result, the electrical properties of the nanofiber powder sample are heavily dependent on the internanofiber transport behavior; single nanofiber level measurements are needed to find out the "true" conductivity of the nanofibers.

The surface area measured using nitrogen gas absorption (BET method, Table 7.2) of CSA doped nanofibers is 41.2 m^2/g. After dedoping with base, the surface area increases from 41.2 to 49.3 m^2/g. This indicates that the free volume of the nanofiber sample increases after the dopants are removed, consistent with observations of conventional polyaniline [11]. On the other hand, the surface area of the nanofibers increases as the average diameter decreases: the measured value of HCl dedoped nanofibers (average diameter = 30 nm) is 54.6 m^2/g, which is greater than that of CSA dedoped nanofibers (average diameter = 50 nm, 49.3 m^2/g) or $HClO_4$ dedoped nanofibers (average diameter = 120 nm, 37.2 m^2/g). The actual surface areas are higher than the value estimated based on a nanocylinder model, since the surface of the nanofibers is not smooth or completely dense.

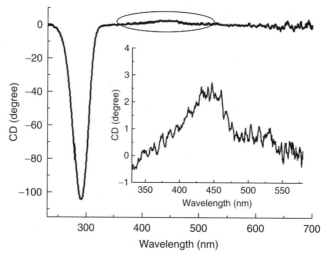

FIGURE 7.20   CD spectrum of interfacially polymerized polyaniline nanofibers using *R*-CSA as the dopant. The inset expands on the circled region from about 350–550 nm.

TABLE 7.2 Surface Areas of Polyaniline Nanofibers. The Estimated Values Are Based on a Cylindrical Model

Dopant	HCl	HCSA	HClO$_4$
Diameter (nm)	30	50	120
Estimation (m^2/g)	51.28	30.77	12.82
BET S_A (m^2/g)			
Doped	n/a[a]	41.21	34.24
Dedoped	54.66	49.31	37.21

[a]Surface areas of HCl doped nanofibers were not measured.

The nanofiber samples in either doped or dedoped forms do not appear to be more crystalline than conventional materials. For the three doped (HCl, CSA, and HClO$_4$) samples, there are peaks at ~9°, 15°, 19.5°, and 25.5° in their X-ray diffraction data (XRD) patterns. For the corresponded dedoped samples, only one peak is observed near 19°. This is consistent with XRD for conventional polyaniline [76]. In addition, high resolution TEM (HRTEM) images do not reveal any crystalline structures, either (Figure 7.21).

The synthetic methodology to polyaniline nanofibers presented here is conceptually different from the previous work and is much simpler and more effective. Following our methods, it is possible to make large-scale, high-quality nanofibers in any chemical laboratory, which has led to the discovery of many exciting properties and applications such as those described in the following sections.

7.3 Polyaniline Nanofiber-Based Vapor Sensors

7.3.1 Conducting Polymer–Vapor Interactions

For many conducting polymers including polyaniline, polypyrrole, and polythiophene, a rich chemistry of structural modifications has been developed making them potentially attractive materials for sensor applications [45,77–81]. Since, conjugated polymers respond to vapors at room temperature and can be deposited on a wide variety of substrates, they could be used in applications that prove to be difficult for traditional semiconductor gas sensors which often require high-temperature operation.

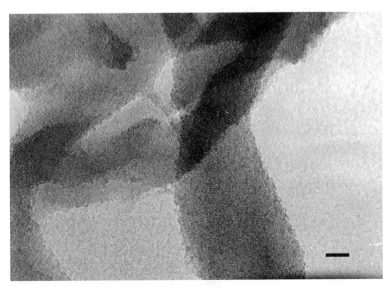

FIGURE 7.21 HRTEM images of polyaniline nanofibers. No ordered structure is evident (scale bars = 20 nm).

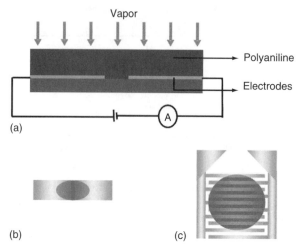

(a)

(b)   (c)

FIGURE 7.22 A schematic diagram of (a) a polyaniline chemiresistor consisting of a polyaniline film, electrodes, substrate, and electronic components needed to monitor current flowing through the resistor. (b) A polyaniline coated wide gap electrode and (c) interdigitated electrodes. Compared to gap electrodes, interdigitated electrodes offer better area efficiency as essentially all of the polymeric material contributes to the sensing process, while in gap electrodes only the material in the gap is useful. Interdigitated electrodes give a lower value of resistance as the distance between fingers is much smaller than that of the gap electrodes, therefore, they are preferred in doping experiments when the baseline resistance is high.

The performance of a polymeric sensor depends on the interaction between polymer and vapor, which must produce a measurable change in a physical property. For conducting polymers, conductivity sensors (chemiresistors; see Figure 7.22 for sensor structure) take advantage of the most outstanding property of these polymers, namely, their ability to undergo an insulator to metal transition. The conductivity of a conducting polymer depends on both its ability to transport charge carriers along the polymer backbone and for the carriers to hop between polymer chains (Figure 7.23) [82]. Therefore, any interaction with a conducting polymer that alters either of these processes can be used for sensor applications. In a real device, the contact between the sensing element and the electrodes could be another factor contributing to the sensing process [83].

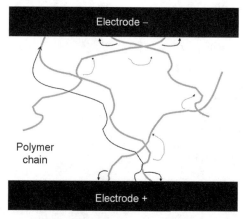

FIGURE 7.23 A schematic diagram illustrating the transport behavior of charge carriers in a conducting polymer film between two electrodes. The charge carriers can move along the polymer chains (*solid, long arrow*), hop between the polymer chains (*dotted, short arrows*), and hop between polymer chains and electrodes (*solid, short arrows*).

7.3.2 Enhanced Sensitivities of Polyaniline Nanofiber Sensors

The conjugated polymer polyaniline is a promising material for sensors [79] as its conductivity is highly sensitive to chemical vapors. Polyaniline sensor research has focused on chemically modifying the polymer structure to facilitate interactions between vapor molecules and polymer. However, little has been done so far to minimize the diffusional path length for vapor molecules; yet poor diffusion can readily outweigh any improvements made to the polymer chains because most of the material other than the limited number of surface sites is unavailable for interacting with the vapor, thus degrading sensitivity (Figure 7.24, left). This problem can be solved by using nanofibers of polyaniline as they have much greater exposed surface areas, as well as much greater penetration depths for gas molecules (Figure 7.24, right), leading to superior performance in both sensitivity and response time. For example, improvements in both sensitivity and time response of many orders of magnitude are now observed by this simple morphological change to the polymer [56,62]. In Section 7.3, HCl vapor is used as an example to illustrate the enhanced performance of polyaniline nanofiber sensors.

7.3.2.1 Response to HCl Vapor: Conventional Films vs. Nanofibers

Acidic vapors can be detected by polyaniline as they protonate the polymer and increase its conductivity [4]. The response given by conventional polyaniline (made by drop casting from a *n*-methylpyrrolidinone solution) and polyaniline nanofibers (made by drop casting from a water dispersion), both in the undoped emeraldine base form, on exposure to 100 ppm of HCl vapor is compared in Figure 7.25. As HCl vapor is highly acidic, it dopes polyaniline and increases its electrical conductivity. The nanofiber film (~2.5 μm thick) gives much greater response to HCl than conventional polyaniline film (~1 μm thick), although it is considerably thicker. This is due to the higher surface area of the nanofiber film that allows more interaction between vapor molecules and polyaniline. The BET surface area of 50 nm diameter polyaniline nanofibers is 49.3 m^2/g, while that of conventional films is limited to the area of the film surface [11].

Response time is a central issue to sensing applications. Data shown in Figure 7.25 were taken at relatively low flow rates in a large volume cell; hence the fastest data shown are flow rate limited. Data are plotted over many orders of magnitude on a log scale that makes the time responses appear "slow" even though they are not. Figure 7.26 shows the change in resistance of both the nanofiber and the polyaniline films plotted on a linear scale. The response time (90%) is measured to be ~2 s for the nanofiber film and ~30 s for the polyaniline film.

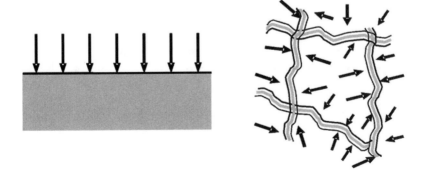

FIGURE 7.24 Schematic diagram of a conducting polymer thin film (*left*) and nanofibers (*right*) exposed to gas molecules (*arrows*). Compared to a thin film, nanofibers have a much higher fraction of exposed surface (*black*), and much shorter penetration depth (*gray*) for gas molecules. (Reproduced from Huang, J.X., Virji, S., Weiller, B.H., and Kaner, R.B., *Chem. Eur. J.*, 10, 1314–1319, 2004. With permission.)

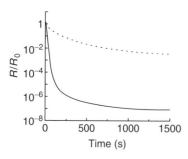

FIGURE 7.25 Resistance changes of a nanofiber emeraldine base thin film (*solid line*) and conventional film (*dotted line*) on exposure to 100 ppm HCl vapor in nitrogen. The resistance (R) is normalized to the initial value (R_0) prior to gas exposure. (Reproduced from Huang, J.X., Virji, S., Weiller, B.H., and Kaner, R.B., *J. Am. Chem. Soc.*, 125, 314–315, 2003. With permission.)

7.3.2.2 Other Chemical Vapors

A great variety of chemical vapors, including hydrochloric acid, ammonia, organic amines, hydrazine, chloroform, methanol, hydrogen sulfide, etc., have been tested and categorized according to the nature of their interactions with polyaniline. Five different mechanisms have been elucidated as follows:

1. Acid doping (e.g., HCl), resistance decreases.
2. Base dedoping (e.g., NH_3 and organic amines), resistance increases.
3. Reduction (e.g., N_2H_4), resistance increases.
4. Swelling (e.g., neutral volatile organic chemicals), resistance increases.
5. Polymer chain decoiling (e.g., CH_3OH), resistance decreases.

In all cases, the polyaniline nanofibers perform better than conventional thin films [84]. Their high surface area, porosity, and small diameters enhance diffusion of molecules and dopants into the nanofibers.

7.3.3   Thickness vs. Sensitivity

The sensitivity of conventional polyaniline films to vapors is strongly thickness-dependent. Generally thinner films result in better performance. As shown in Figure 7.27, when the thickness is decreased from 1.0 to 0.3 μm, the response time increases, and the magnitude of the response at a fixed time increases by more than five orders of magnitude. Since only the outermost surface is likely to interact with the vapor molecules, thicker films have more inactive material that does not contribute to the sensing process at short times. Therefore, the performance of conventional polyaniline thin film sensors is limited by the thickness of the underlying polyaniline film [56,62].

Figure 7.28 shows the first derivative plots constructed from Figure 7.27. The first derivative plot of ΔR vs. t tells how fast the resistance changes during the sensing process, offering a way to study the sensing kinetics.

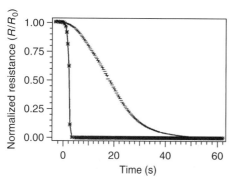

FIGURE 7.26 Response of 0.3 μm nanofiber (*solid line*) and conventional polyaniline (*dotted line*) thin films to 100 ppm HCl. The data are plotted on a linear scale instead of a log scale. The time response for the nanofiber data is ∼2 s but is instrument limited. (Reproduced from Virji, S., Huang, J.X., Kaner, R.B., and Weiller, B.H., *Nano. Lett.*, 4, 491–496, 2004. With permission.)

The plots in Figure 7.28 show that the speed of the resistance change increases rapidly at the beginning, reaches a maximum, then, decreases slowly, and finally stabilizes at a minimum value. For each of the three films, the fastest resistance changes occur at ∼40 s, and each response essentially stops changing after 450 s. This indicates that although the films have different thicknesses, their sensing kinetics are almost identical. The sensing processes are not affected by the thicknesses. Therefore, it is most likely that HCl vapor molecules are just saturating the three conventional film surfaces, rather than the entire films.

When nanofibers are used as the selective layer, the performance is essentially unaffected by the thickness, at least in the range of 0.2–2 μm (Figure 7.29). This is due to the porous nature of nanofiber films that allow vapor molecules to penetrate through the entire film and interact with all the fibers. Therefore, even in thicker films, all the fibers are able to contribute to the sensing process. The porous nature

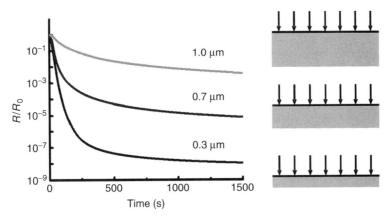

FIGURE 7.27 Resistance changes of conventional polyaniline films with different thicknesses upon exposure to 100 ppm of HCl vapor (*left*). The schematic diagram (*right*) shows the active polyaniline layer (*black*) interacting with vapor molecules (*arrows*) and the inactive supporting layer (*gray*). (Reproduced from Huang, J.X., Virji, S., Weiller, B.H., and Kaner, R.B., *Chem. Eur. J.*, 10, 1314–1319, 2004. With permission.)

of the nanofiber films can be observed directly using SEM. The thickness independence of sensitivity for nanofiber films eliminates the need to make very thin films to obtain good sensitivity.

It is also helpful to understand the thickness effect of the thin film sensors by looking at their surface areas. For a conventional polyaniline thin film, the area that is exposed to chemical vapors is its physical surface (Figure 7.30a). The surface area per unit mass, S_A, decreases as the thickness of the film increases.

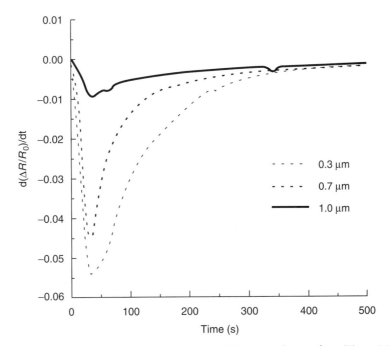

FIGURE 7.28 First derivative plot of the sensor's response to 100 ppm HCl vapor from (Figure 7.27). Such plots show how fast the resistances change during the sensing process. All three films, regardless of their thicknesses, show the fastest response at ~40 s and essentially stop responding after 450 s. This is consistent with the data presented in Figure 7.27.

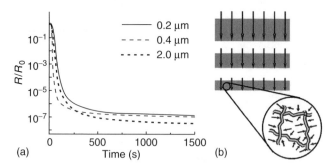

(a)   (b)

FIGURE 7.29 Response of polyaniline nanofiber sensors of different thicknesses upon exposure to 100 ppm of HCl vapor (*left*). The schematic diagram (*right*) illustrates that the vapor molecules (*arrows*) can interact rapidly with most of the polyaniline nanofibers (*gray*) since the films are porous. (Reproduced from Huang, J.X., Virji, S., Weiller, B.H., and Kaner, R.B., *Chem. Eur. J.*, 10, 1314–1319, 2004. With permission.)

However, if the film is composed of long nanofibers (Figure 7.30b), S_A is determined by the diameter of the nanofibers, rather than the film thickness. This may explain the thickness effects observed in the experiments. Since the surface area per unit mass of the nanofiber films is generally much greater than that of the conventional films (unless the thickness is reduced to less than the nanofiber diameter), a nanofiber sensor should have much better sensitivity than a conventional polyaniline sensor.

HCl doped polyaniline films can be used as base sensors. When these fully HCl doped polyaniline films are exposed to ammonia vapor, a drop in conductivity is observed. As expected, a nanofiber film outperforms a conventional polyaniline film. Again the nanofiber films are not dependent on thickness, whereas the sensitivity of conventional polyaniline films is completely dependent on thickness [62].

7.3.4 Reversibility of Ammonia Vapor Sensors

Figure 7.31a shows the response of a polyaniline sensor to 378 ppm ammonia vapor in a headspace system. The sensor response is clearly reversible even after being exposed for more than 1 h. The turn-on (Figure 7.31c) and turn-off (Figure 7.31d) of the sensor take place within ~10 s as revealed by the first derivative plot of the response (Figure 7.31b through Figure 7.31d). The reversibility of the ammonia sensor is observed directly by changes in the color of polyaniline nanofiber films on exposure to ammonia. A green polyaniline nanofiber film is exposed to ammonia vapor (~76%)

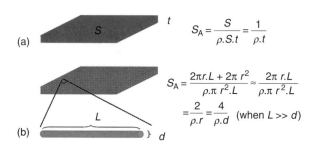

FIGURE 7.30 Schematic diagrams showing (a) a conventional film and (b) a nanofiber film along with the equations for their surface areas (S_A). For a convention thin film, S is the area of exposed surface, t is the thickness of the film, and ρ is the density of polyaniline. For a nanofiber film, r is the fiber radius, L is the length of the nanofiber, and d is the diameter.

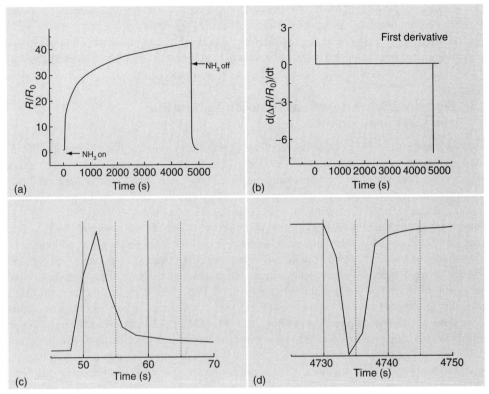

FIGURE 7.31　(a) A nanofiber sensor responding to 378 ppm of NH$_3$. (b) A first derivative plot of the sensor response, showing both a fast turn-on and turn-off. The details of the first derivative plot are shown in an expanded view in: (c) turn-on and (d) turn-off.

above concentrated ammonia hydroxide (14.8 M). The film turns from green to blue immediately but slowly changes back to green in ~1 min in air. It is known that ammonia dedopes polyaniline by competing with polyaniline for the acidic dopant. In the solution phase, aqueous ammonia can remove the acid dopant (HCl) from polyaniline by forming a water-soluble ammonium salt NH$_4$Cl. In the vapor phase, ammonia reacts with HCl to produce NH$_4$Cl in the solid state and the reaction is not reversible at room temperature.

$$HCl + NH_3 \rightarrow NH_4Cl$$

Then why are the polyaniline ammonia sensors reversible? In the presence of polyaniline, the reaction involves three reagents:

$$PANI \cdot HCl + NH_3\uparrow \rightleftharpoons NH_3 \cdot HCl + PANI$$

There are two weak bases, ammonia and polyaniline (emeraldine base) competing for the acid. When ammonia is present, the reaction moves to the right and polyaniline is dedoped. When the ammonia vapor is turned off (removed), the reaction may move back to the left and polyaniline becomes redoped, releasing ammonia. This process involves a reversible change in the mass of the film [85] in addition to the changes in color and resistance.

　　The nanofiber sensors have a considerably lower detection limit than conventional polyaniline. An HCl doped nanofiber film would respond clearly and reversibly to 0.9 ppm of ammonia vapor in air at room temperature. This is well below the human nose threshold for smelling ammonia (≥ 25 ppm). For conventional films, it is difficult to distinguish the response from the background noise. The detection of ammonia and amines has applications in sensing spoiled food products and in testing for diseases. For example, fish gives off amines as they spoil. Certain diseases, such as cholera and lung cancer, produce ammonia and organic amines as biomarkers [86].

7.3.5 Designing New Reactions between Polyaniline and Chemical Vapors

The underlying principle for polyaniline nanofiber chemiresistor vapor sensors is that the chemical–physical interaction between chemical vapors and polyaniline must produce a detectable change in the film's resistance. The sensor characteristics, such as sensitivity, time response, reversibility, and reusability, are then determined by such interactions. Therefore, by tailoring the interactions between polyaniline nanofibers and the target vapors, sensor characteristics can be improved. This can be accomplished through materials engineering approaches to modify the selective layer, including but not limited to modifying the polymer backbone, using functional dopants and adding catalysts. On the other hand, such interactions can be altered by converting hard-to-detect target vapor molecules into easily detected species. For example, adding functional "guests" into the polyaniline "matrix" could greatly broaden the scope of polyaniline sensors. Either the guest–matrix complex has new functionality introduced by the guests, or the guest–matrix interaction provides new mechanisms that are sensitive to new targets. The sensing capability of the polyaniline nanofibers can also be extended by introducing new chemical reactions that produce chemical species whose reaction products can be directly sensed by polyaniline nanofibers. For all these applications, using nanopolyaniline should be beneficial because the open structure of the nanofiber film would allow a better interaction between polyaniline and adducts. Here an example of enhancing the detection of weak acids by adding metal salts is given to demonstrate this idea [87].

Some acidic vapors such as H_2S are not strong enough to dope polyaniline in a reasonably short period. For example, solution doping experiments show that it takes several weeks for a saturated H_2S solution to dope polyaniline nanofibers. Therefore, H_2S detection is very difficult using neat polyaniline nanofibers. However, H_2S can react rapidly with many metal salts to form a metal sulfide precipitate and generate a strong acid as the by-product as shown in the following equation:

$$H_2S + PbX \rightarrow PbS\downarrow + HX$$

When a polyaniline nanofiber film is saturated with $Pb(NO_3)_2$ and then exposed to H_2S vapor, it becomes doped within a couple of seconds, indicated by a color change from blue to green. Therefore, the sensitivity of either neat polyaniline or nanofibers to H_2S vapor should be greatly enhanced. As shown in Figure 7.32, the sensitivity of neat polyaniline, even the nanofibers, is low. After the addition of $CuCl_2$, the sensitivity is significantly enhanced. Many other metal salts (e.g., Cu^{2+}, Bi^{3+}, and Ni^{2+}) with strong acid cations (e.g., Cl^-, Br^-, I, SO_4^{2-}, and ClO_4^-) should also work, as long as they can form a low K_{sp} metal sulfide compound with H_2S. We believe this idea is also applicable to enhance the sensitivity for detecting many other weak acid vapors.

7.4 Polyaniline Nanofiber-Based Composite Materials

There has been a great deal of interest in making polyaniline-based inorganic–organic nanocomposites because the impregnation of inorganic materials into a polyaniline matrix can introduce new properties to the polymer. This could create new functional materials combining the advantages of both the inorganic and the organic components. For example, metal-polyaniline [88–100] and inorganic-polyaniline [14,101–130] nanocomposites have been found to be useful for sensors, catalysts, photovoltaic devices, and magnetic materials. Polyaniline nanofiber-based nanocomposites could offer significant

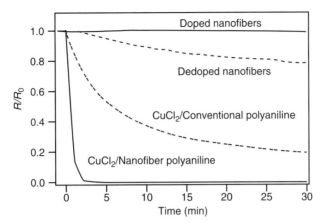

FIGURE 7.32 Resistance changes of polyaniline films upon exposure to 10 ppm hydrogen sulfide with 45% relative humidity.

improvements in applications such as sensors and catalysis compared to conventional polyaniline-based composites as the polymer matrix is already nanostructured. Therefore, the final nanocomposites should have both high-surface area and high-interfacial area between the inorganic and the organic components.

7.4.1 Metal-Polyaniline Nanocomposites Based on Polyaniline Nanofibers

Polyaniline is a conjugated polymer with multiple, switchable inherent oxidation states (Figure 7.33). Metal particles (e.g., Au [88,91,95–100,131], Ag [62,88,99], Pt [88,93,99,100,132,133], and Pd [88,90,99,100,133–136]) can be readily deposited on polyaniline by direct chemical or electro-chemical redox reactions between polyaniline and oxidative metal cations. However, with conventional

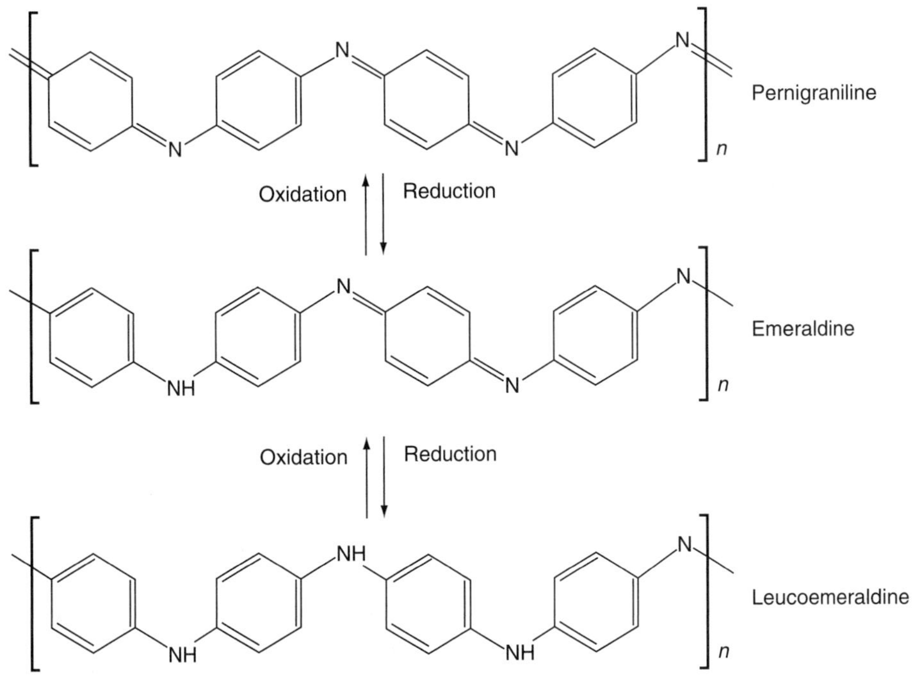

Pernigraniline

Oxidation ↑ Reduction

Emeraldine

Oxidation ↑ Reduction

Leucoemeraldine

FIGURE 7.33 Inherent oxidation states of polyaniline.

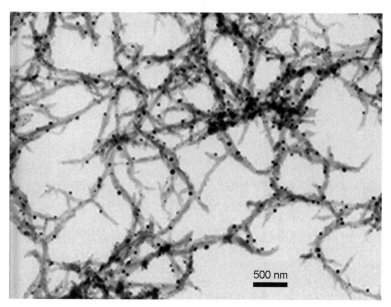

FIGURE 7.34 AgNO$_3$ treated polyaniline nanofibers ($\sim$25°C) showing a dot-ON-wire structure. (Reproduced from Huang, J.X., Virji, S., Weiller, B.H., and Kaner, R.B., *Chem. Eur. J.*, 10, 1314–1319, 2004. With permission.)

polyaniline controlling the size and distribution of the metal particles across the polyaniline matrix this process is very difficult, if not impossible.

With polyaniline nanofibers [53,55,56,62,137] the size of the metal particles and their distribution in the polymer matrix can be well controlled. The small diameters (<100 nm) of the nanofibers make it possible to form tiny metal particles due to the limited redox active sites in the nanofibers for reducing metal cations. The uniform diameters of the nanofibers allow relatively narrow size distribution of the metal nanoparticles. Treating the dedoped polyaniline nanofibers with 10 mM AgNO$_3$ solution at room temperature readily yields Ag nanoparticle decorated nanofibers in a dot-ON-fiber fashion (Figure 7.34). When the same reaction was carried out under reflux conditions, three morphologies: dot-ON-fiber, dot-IN-fiber, and silver shells on nanofibers were obtained (Figure 7.35). This can be explained by understanding the diffusion and reaction during the reaction (Figure 7.36). As the dot-ON-fiber structure is obtained at room temperature, at reflux temperature the rates of both reaction and diffusion are enhanced, forcing some Ag$^+$ into the nanofibers before the reaction is complete. Therefore, both dot-ON-fiber and dot-IN-fiber structures are obtained. For the silver nanoparticles outside the nano-fiber, they can merge into their neighboring particles and form a continuous shell on the nanofiber due to Ostwald ripening [138,139]. Under properly designed reaction conditions, it is possible to have either the diffusion or the reaction overwhelm the other, leading to pure dot-ON-fiber or dot-IN-fiber types of Ag (or Au, Pd, or Pt)-polyaniline composites.

7.4.2 Electrical Properties of Metal-Polyaniline Nanocomposite Materials

A new application for metal-polyaniline composites that we have demonstrated is electrical bistable memory devices (Figure 7.37) [140], in collaboration with Professor Yang Yang's group [141,142] in Material Science and Engineering at UCLA. The device employs Au-polyaniline nanocomposites as the active medium sandwiched between two aluminum electrodes (Figure 7.37). A very interesting bist-ability is observed in the *I–V* characteristics of the nanocomposite material (Figure 7.38). This electrical bistability indicates that the device using a nanocomposite of Au-polyaniline can be used for nonvolatile memory. Indeed, such polyaniline-based memory devices (PANI-MEM) [140] have been fabricated

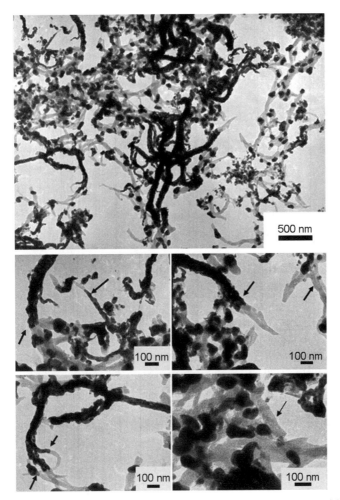

FIGURE 7.35 Polyaniline nanofibers refluxed in aqueous AgNO$_3$ solution show a mixture of dot-IN-wire and dot-ON-wire structures.

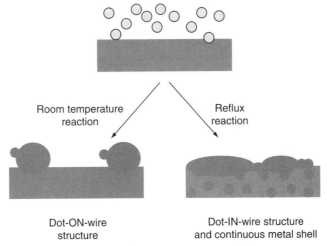

FIGURE 7.36 Schematic drawing illustrating the formation mechanism of the two types of Ag-polyaniline nano-composites.

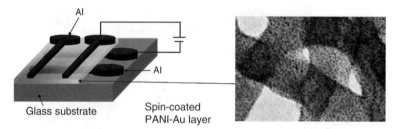

FIGURE 7.37   Schematic drawing showing the device structure of the polyaniline memory (PANI-MEM).

with nanosecond switching times, high on–off ratios, and low-manufacturing costs, making them very promising for faster and less expensive data storage media than what is currently available (e.g., flash memory).

7.4.3   Other Methods to Make Metal-Polyaniline Nanocomposites

The chemical polymerization of aniline involves a monomer, an acid, and an oxidant. $HAuCl_4$ is a strong acid and a strong oxidant as well as a precursor for gold nanoparticles. Therefore, polymerizing aniline using $HAuCl_4$ is possible. First, an attempt to make Au-polyaniline nanocomposites by interfacial polymerization was made using $HAuCl_4$ as the oxidant. The product stayed at the aqueous–organic interface instead of diffusing into the water layer. TEM studies revealed that the product was agglomerates of gold nanoparticles ($\sim$30 nm) with a polymer sheath. Second, a rapidly mixed reaction route as described in Section 7.2.3.2 was carried out using $HAuCl_4$. The product was found to be a random mixture of polyaniline nanofibers and gold nanoparticles (Figure 7.39). The result is consistent with the mechanistic study on the formation mechanism of polyaniline nanofibers discussed in Section 7.2.3.2.

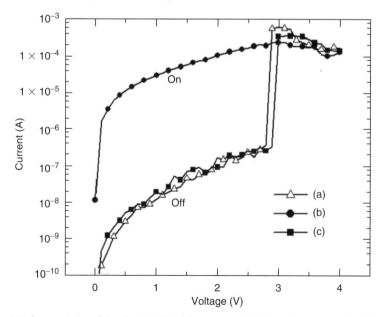

FIGURE 7.38   *I–V* characteristics of the electrical bistable memory device based on a Au-polyaniline nanocomposite. (Reproduced from Tseng, R.J., Huang, J.X., Ouyang, J., Kaner, R.B., and Yang. Y., *Nano. Lett.*, 5, 1077–1080, 2005. With permission.)

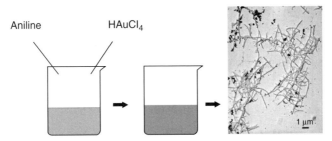

FIGURE 7.39   Rapidly mixed reaction route to polyaniline nanofibers using $HAuCl_4$ as both oxidant and acid.

When $AgNO_3$ was used in the rapidly mixed reactions, it took many days to observe the green color of doped polyaniline. The polymerization rate can be greatly enhanced by UV irradiation (Figure 7.40) and polyaniline nanofibers were obtained with highly dendritic silver microcrystals.

7.4.4   Growth of Inorganic Nanoparticles by In Situ Metathesis Reactions

Most of the current methods for preparing inorganic-polyaniline nanocomposites fall into the following categories: postsynthetic blending of polyaniline and inorganic materials [114,121,143,144], polymerization of aniline in the presence of inorganic components [110,112,115,116,127–129,145], and one-step reactions forming inorganic materials and polyaniline together [102,104,124,146,147]. Here, a new method is described to make inorganic-polyaniline nanocomposites based on in situ metathesis reactions on polyaniline nanofibers.

Many binary inorganic compounds can be prepared by simple metathesis reactions as given in Equation 7.1 through Equation 7.3:

$$AgNO_3 + HX \rightarrow AgX\downarrow + HNO_3 \quad (X = Cl,Br,I) \tag{7.1}$$

$$Pb(NO_3)_2 + H_2S \rightarrow PbS\downarrow + 2HNO_3 \tag{7.2}$$

$$BaCl_2 + H_2SO_4 \rightarrow BaSO_4\downarrow + 2HCl \tag{7.3}$$

Acids can act as dopants for polyaniline; therefore, if the nanofibers are first doped with these acids, and subsequently exposed to the metal ions, precipitates should be formed on the surface of the nanofibers, leading to inorganic-polyaniline nanofiber composites. This idea has been applied in Section 7.3.5 to improve the detection of H_2S. Another possibility is to use nanofibers as nucleation seeds to collect inorganic nanoparticles from supersaturated solutions [148].

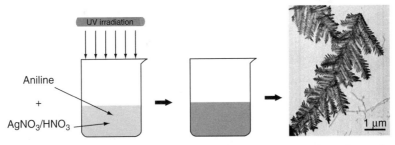

FIGURE 7.40   UV-irradiation assisted polymerization of aniline using $AgNO_3$ as the oxidant. Polyaniline nanofibers and silver dendrites are obtained.

7.5 Unusual Photothermal Effect of the Polyaniline Nanofibers: Flash Welding

The nanoscale structure of polyaniline nanofibers produces enhanced polymer functionalities; their polymeric nature also yields new nanoscale physicochemical phenomena that have not been observed in inorganic nanostructured materials. In this section, a flash welding technique is presented based on an enhanced photothermal effect discovered with the polyaniline nanofibers.

7.5.1 Photothermal Effect in Nanostructured Materials

The absorption of light by a material generates heat through nonradiative energy dissipation and exothermic photochemical reactions [149]. In nanostructured materials, the heat generated through photothermal processes will be confined within the individual nanostructures when heat transfer to neighboring nanostructures and the environment is slow. This enables unprecedented photothermal effects that cannot be observed in bulk materials, especially when a strong, pulsed light source is used [150,151]. Here, a photothermal phenomenon is demonstrated with conducting polymer nanofibers in which a camera flash causes instantaneous welding. Under flash irradiation, polyaniline nanofibers "melt" to form a smooth and continuous film from an originally random network of nanofibers. Flash welding technique is developed based on this photothermal effect to form asymmetric nanofiber films, to rapidly melt-blend polymer–polymer nanocomposites and to photopattern polymer nanofiber films. Polymer nanofiber films offer significant technological advantages over conventional polymeric materials in applications such as chemical sensors, tissue engineering, separation membranes, and composite materials, due to their high surface areas and enhanced physical properties. In many of these applications, forming connections between the nanofibers are essential to provide mechanical durability. Flash welding is a novel, pollution-free method for stitching polymers together.

7.5.2 Flash Welding of Polyaniline Nanofibers

7.5.2.1 Welding of Polyaniline Nanofibers by a Camera Flash

Polyaniline is a deeply colored conjugated polymer that has been extensively studied during the last 25 years for its electrical properties [152]. In powder form, undoped polyaniline is dark blue in color that changes to a deep green conducting form when doped by acids [4]. However, polyaniline has an extremely low luminescence efficiency, it converts most of the energy absorbed from light into heat [153–155]. We have now discovered that when exposed to a camera flash up close, i.e., within several centimeters of a sample, fine powders of polyaniline nanofibers respond audibly with distinct popping sounds and the concomitant formation of agglomerates within the exposed area. TEM images show that the nanofibers appear to have melted and merged together (Figure 7.41). This inspired us to explore welding and film formation using polyaniline nanofibers exposed to a camera flash [156].

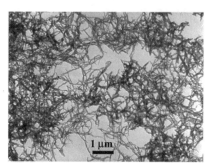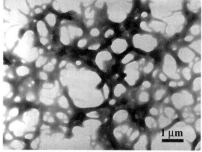

FIGURE 7.41 Typical TEM images showing polyaniline nanofibers before (*left*) and after (*right*) exposure to a camera flash. (Reproduced from Huang, J.X. and Kaner, R.B., *Nat. Mater.*, 3, 783–786, 2004.)

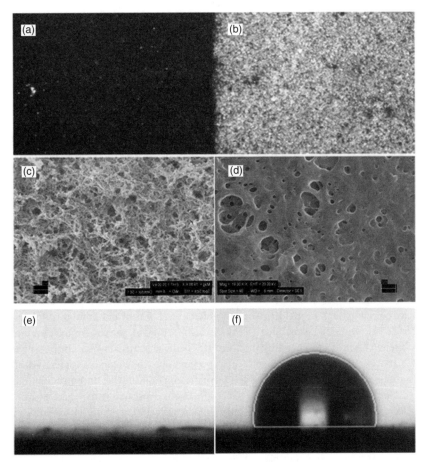

FIGURE 7.42   A polyaniline nanofiber film before (a, c) and after welding (b, d). (a) and (b) are optical microscope images (50×) showing the reflectance contrast of the unwelded and welded areas. (c) and (d) are SEM images (19,000×) of the nanofiber film before and after welding. (e) and (f) are camera images showing water contact angles on the nanofiber film before and after welding. (Adapted from Huang, J.X. and Kaner, R.B., *Nat. Mater.*, 3, 783–786, 2004.)

First, we found that a random network of nanofibers can be flash-welded to create a continuous film. Polyaniline nanofibers were cast from a water suspension on flat substrates, such as silicon wafers. On drying, a film containing a random matt of nanofibers was obtained (Figure 7.42a and Figure 7.42c). After exposure for milliseconds to a camera flash, the film became smooth and shiny (Figure 7.42b and Figure 7.42d). Change in surface roughness can be seen clearly with the naked eye. Optical microscope images in reflectance mode show distinct contrast in reflectivity between unwelded (Figure 7.42a) and welded (Figure 7.42b) areas of a nanofiber film. SEM images reveal that the nanofibers on the surface (Figure 7.42c) are welded together to create a continuous film (Figure 7.42d). The pinholes correspond to the free volume in the nanofiber film before welding. The difference in surface roughness before and after welding also alters the wettability of the film. For example, before flash welding, a porous undoped nanofiber film, although hydrophobic, absorbs water droplets like fibrous filter paper (Figure 7.42e). However, after flash welding, the surface repels water with a contact angle of 99.0° due to its bulk hydrophobic nature (Figure 7.42f).

7.5.2.2 Flash Welding Route to Asymmetrical Polyaniline Membranes

Flash-welded films can be readily removed from their substrates by water. They can also be peeled off by touching a corner to cellophane tape, leading to a free-standing nanofiber film (Figure 7.43, upper left).

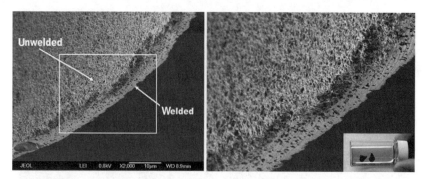

FIGURE 7.43 SEM images showing the cross section of an asymmetric free-standing nanofiber film produced by flash welding. The *inset* picture shows that the welded films become insoluble in *N,N*-dimethylformamide (DMF). The pale blue color is due to partial dissolution of uncross-linked polyaniline. (Adapted from Huang, J.X. and Kaner, R.B., *Nat. Mater.*, 3, 783–786, 2004.)

The nanofibers on the exposed side are welded together, while those on the unexposed side remain intact (Figure 7.43, lower right). The welding of the nanofibers gives the film enough mechanical strength to support itself without the substrate, forming an asymmetric film. There is also a dramatic change in solubility after the nanofibers are welded. For example, a cross-linked polyaniline nanofiber film is no longer soluble in *N,N*-dimethylformamide (DMF), which is a good solvent for polyaniline nanofibers (Figure 7.43, inset). This indicates that the nanofibers are chemically cross-linked. The free-standing nature of the welded nanofiber films enables their surface properties to be studied on each side. FTIR spectroscopy (Figure 7.44) studies in reflectance mode on both sides of the film indicate that the polyaniline nanofibers are chemically cross-linked upon flash irradiation [157–160]. This is consistent with the increase in mechanical strength and decrease in solubility after the nanofiber films are welded. Therefore, flash welding appears to be a very convenient method for making an asymmetric film. Asymmetric films are particularly useful in many applications including separation membranes [161,162], chemical sensors, and actuators [163–166]. Such films are usually made through multiple, relatively time-consuming steps [161–166]. Photothermally induced welding offers a rapid route to create free-standing asymmetric films.

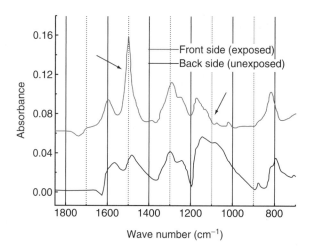

FIGURE 7.44 Attenuated total reflection infrared (ATR-IR) spectra of the front and back sides of a polyaniline nanofiber film after exposure to a camera flash. The dramatic changes in relative intensities of the peaks centered at 1500, 1150, and 1076 cm^{-1} indicate cross-linking of polyaniline as reported in the literature. (Adapted from Huang, J.X. and Kaner, R.B., *Nat. Mater.*, 3, 783–786, 2004.)

7.5.2.3 Flash Welding Route to Patterned Polyaniline Nanofiber Films

A great advantage of flash welding is its ability to selectively weld certain areas on a nanofiber film using a predesigned photomask. This enables the imprinting of a pattern defined by the mask into the film. A demonstration is shown in Figure 7.45, in which a well-defined pattern (Figure 7.45b) is developed on a nanofiber film that is identical to the mask used (Figure 7.45a). The areas exposed to the camera flash (the diamonds) are quite smooth and therefore appear bright gold due to the reflectance of incident light. The protected areas (the lines) remain as green nanofibers and appear darker as the rough surface scatters light and yields low reflectance. The high contrast between the resulting patterns (Figure 7.45c) is due to a change in reflectivity induced by flash welding. SEM studies confirm the morphological differences comparable to Figure 7.42c and Figure 7.42d. Since

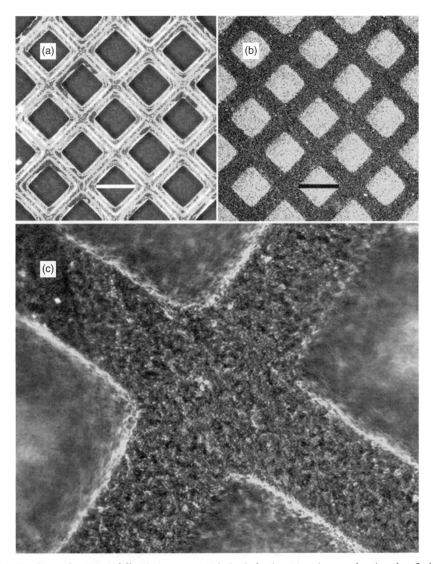

FIGURE 7.45 **(See color insert following page 14-20.)** Optical microscope images showing that flash welding through a copper grid (a) reproduces the grid pattern on a polyaniline nanofiber film (b). Scale bars = 100 μm. The high contrast of the pattern is revealed under a higher magnification (c). (Adapted from Huang, J.X. and Kaner, R.B., *Nat. Mater.*, 3, 783–786, 2004.)

many properties of polyaniline (i.e., conductivity, surface area, optical absorption, permeability, and thermal stability) change after cross-linking [167–171], flash welding through a mask represents a convenient way to fabricate polymer films into predesigned patterns, which could prove to be useful for organic micro and nano devices.

7.5.2.4 Flash Welding Route to Polymer–Polymer Composites

The heat generated during a camera flash can be employed to weld polyaniline to another polymer, offering a rapid and clean optical technique for polymer welding. This concept can be demonstrated by flash welding polyaniline nanofibers onto polystyrene spheres as follows: First, polyaniline nanofibers and polystyrene microspheres ($\sim$1 μm in diameter) were mixed in water. Then films of a random polyaniline–polystyrene mixture were made by drop casting. Such films look white due to strong light scattering (Figure 7.46e), since there is a significant amount of polystyrene beads on the surface as confirmed by an SEM image (Figure 7.46a). The blue hue is due to the underlying color of undoped polyaniline. After irradiation by a camera flash, the nanofibers and microspheres fuse together, leading to a dark blue colored film (Figure 7.46f). TEM studies (Figure 7.46c and Figure 7.46d) reveal that the polystyrene spheres are welded together with polyaniline nanofibers. Figure 7.46d is a typical TEM image of the resulting mixture of polyaniline and polystyrene after flash irradiation. All the polystyrene spheres in contact with polyaniline nanofibers are deformed after welding, while isolated spheres are intact. This supports the idea that only polyaniline nanofibers are responsible for the heat generated during welding. Polystyrene is visually clear and is also used to make sample cells for UV–vis spectroscopy, therefore it should have a negligible photothermal effect. Flash welding appears to be useful for making polymer–polymer blends and offers a novel method to embed guest polymer particles into a host matrix. Even polytetrafluoroethylene (Teflon®) particles can be welded using polyaniline nanofibers (Figure 7.47).

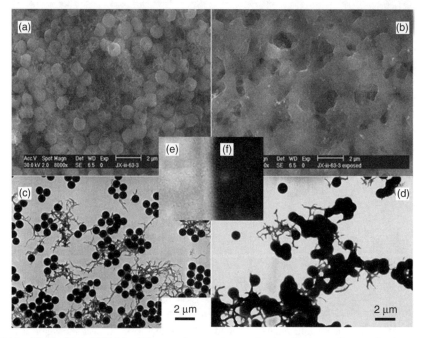

FIGURE 7.46   SEM (a, b) and TEM (c, d) images show a mixture of polyaniline nanofibers and polystyrene spheres (a, c) before and (b, d) after flash welding. The *inset* shows the visual color contrast between unexposed (e, *bluish white*) and exposed (f, *dark blue*) areas. (Adapted from Huang, J.X. and Kaner, R.B., *Nat. Mater.*, 3, 783–786, 2004.)

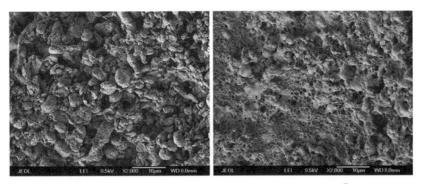

FIGURE 7.47   SEM images showing a mixture of polyaniline nanofibers and Teflon® particles before (*left*) and after (*right*) flash welding.

7.5.3   Flash Welding of Other Materials

One can speculate on the origin of the heat generated by a camera flash. A photographic flash produces a relatively high-intensity light within a short pulse (~1 ms) [150,151]. Although the energy is insufficient to melt the bulk polymer, local hot spots are likely. In single-walled carbon nanotubes and silicon nanowires, local hot spots of above 1500°C have been suggested upon flash irradiation [150,151,172,173]. In polyaniline nanofibers, hot spots are likely to form around the chromophores on the polyaniline chains due to highly efficient photothermal conversion of polyaniline [153–155], which may further initiate and propagate exothermic cross-linking reactions between polymer chains. Since polyaniline has a low thermal conductivity (~10^{-1} W/mK) [153,154], the heat from photothermal conversion and cross-linking reactions could accumulate within the nanosized domains leading to the welding of nanofibers. In contrast to carbon nanotubes and silicon nanowires, slight burning of the nanofibers only occurs when the flash is very close (within 0.5 cm) and is visible as smoke coming off from the powders. Melting of the polymer nanofibers may act as a benign way to drain the pulsed heat away avoiding complete structural breakdown or combustion [150,151,172,173], thus enabling many potentially useful applications.

Flash welding has the following advantages compared with other polymer welding techniques that use solvents, lasers, microwaves, or ultrasound:

1. It is a clean, chemical-free technique that does not involve any toxic solvents.
2. It can be performed across a wide range of scales from bulk materials to nanometer size objects.
3. It can selectively weld different areas of the same film, producing welded patterns using a photomask.
4. It is an extremely rapid process since the typical flash time is milliseconds.
5. It is an optical approach that can be easily interfaced with currently available optical techniques, opening up many new possibilities.
6. It can be performed anywhere without the need for expensive equipment.
7. It can be scaled to any area determined by the size of the flash used.

Flash welding is not limited to polyaniline nanofibers. For example, experiments involving nanostructures of polyaniline derivatives (Figure 7.48), polypyrroles, and polythiophenes indicate that flash welding is a general phenomenon. Flash welding should be a general phenomenon for nanomaterials with the following characteristics: high absorbance (i.e., deeply colored), high photothermal efficiency (i.e., low emission), low thermal conductivity, and a phase change before structural breakdown. It holds great promise for making asymmetric membranes, for photopatterning nanostructured films, and for welding polymer–polymer and polymer–inorganic nanocomposites.

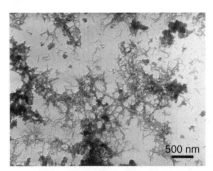

FIGURE 7.48   Typical TEM images showing polytoluidine nanofibers before (*left*) and after (*right*) exposure to a camera flash.

7.6   Outlook: Polyaniline Derivatives and Other Conducting Polymers

In Section 7.2, the relationship between synthetic conditions and the morphologies of polyaniline was examined. The following conclusions were reached: Nanofibers of polyaniline naturally form in water during chemical oxidative polymerization and can be made without the need for any template, special dopant, or seed. In organic solvents, the basic morphology of polyaniline is strongly affected by the solvent–polyaniline interactions. By tuning such interactions, polyaniline nanofibers, irregularly shaped agglomerates, and their mixtures can be obtained using the same chemical reaction.

Since many aniline derivatives (e.g., toluidine, anisidine, ethylaniline, and fluoroaniline) can be polymerized using oxidative polymerization reactions under conditions very similar to aniline polymerization [3,174–180], it would be very interesting to study the basic morphologies of substituted polyanilines using interfacial polymerization or rapidly mixed reactions. Substituted polyanilines are of great technical interest, because they can improve the properties of polyaniline [3]. For example, compared to pristine polyaniline improved processibility has been demonstrated for alkyl-substituted polyanilines [180–182]. Fluoro-substituted polyanilines are more stable against microbial and chemical degradation [174,176–178], therefore they are used in many biorelated applications such as bacterial-based fuel cells [183]. There have been few reports comparing the morphologies of chemically polymerized polyaniline derivatives. In this section, the effects of substituent groups on the basic morphology of polyaniline derivatives are investigated. It is found that not only the substituent groups but also their positions strongly affect the morphology. Other conducting polymers, such as polypyrroles and polythiophenes, are also synthesized using either interfacial polymerization or rapidly mixed reactions to obtain nanostructures.

7.6.1   *meta*- and *ortho*-Substituted Polyanilines

It has been noted that the polymer of *meta*- and *ortho*-substituted anilines is essentially identical (Figure 7.49) [176]. However, there are some differences between these two "identical" polymers, such as conductivity, solubility, and thermal stability [174,176,177,184,185]. With the success of the rapidly mixed reactions for making polyaniline nanofibers, we are able to reveal the basic morphologies of substituted polyanilines and compare the effect of substituent positions on the polymer morphology.

Figure 7.50 shows the *meta*- and *ortho*-polytoluidine products a week after polymerization. The reaction conditions were identical for the two reactions, but there is a big difference in the volume occupied by the polymer (Figure 7.50, insets). Based on the experience obtained with polyaniline synthesis (Figure 7.14) this indicates a big difference in the morphology of the products (see Section 7.2.4). For example, *m*-polytoluidine occupies much more volume than *o*-polytoluidine. Electron

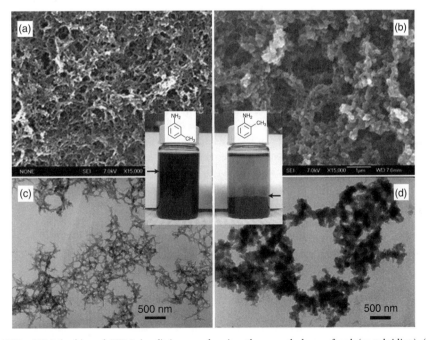

FIGURE 7.49 *meta-* and *ortho*-Substituted anilines and their polymers.

FIGURE 7.50 SEM (a, b) and TEM (c, d) images showing the morphology of poly(*m*-toluidine) (a, c) and poly(*o*-toluidine) (b, d) synthesized by rapidly mixed reactions using HCl as the dopant. The *insets* show the volume occupied by poly(*m*-toluidine) and poly(*o*-toluidine) as marked by the arrows.

FIGURE 7.51 Typical TEM images of poly(*m*-fluoroaniline) (*left*), poly(*o*-fluoroaniline) (*middle*), and poly (*p*- fluoroaniline) (*right*) produced by rapidly mixed reactions.

microscopy images (Figure 7.50) reveal that poly(*m*-toluidine) obtained using rapidly mixed reactions is composed of essentially pure nanofibers with average diameters ∼50 nm, while poly(*o*-toluidine) is composed of bigger, irregularly shaped particles with only negligible amounts of nanofibers. Compared to polyaniline nanofibers obtained under the same conditions, the poly(*m*-toluidine) nanofibers are shorter and thicker. The morphological difference between poly(*m*-toluidine) and poly(*o*-toluidine) has been discussed only previously using electrochemical polymerization [184]. It is interesting that the results obtained here are inverted with respect to electrochemically polymerized poly(*o*-toluidine), which appeared to be more fibrillar than poly(*m*-toluidine).

Figure 7.51 shows the effect of substituent position on the morphology of polyfluoroaniline. For poly(*m*-fluoroaniline), the dominating morphology is also fibrillar. However, compared to the nanofibers of polyaniline and poly(*m*-toluidine), the fibers of poly(*m*-fluoroaniline) are much thicker in diameter with a wider distribution (150–400 nm). For poly(*o*-fluoroaniline), the product is composed of big dendritic particles of 10 μm. For poly(*p*-fluoroaniline), the product seems to be a mixture of nanofibers with diameters ∼30 nm, thicker fibers with diameters ∼200 nm and big dendrites at the micron level, suggesting that the poly(*p*-fluoroaniline) obtained may be a mixture of multiple polymers: polyaniline, poly(*m*-fluoroaniline), and poly(*o*-fluoroaniline). It has been known that the fluorine substituent in the *para*-position may be displaced during the synthesis to allow for typical head-to-tail polymerization, giving polyaniline as the main product [174,176,177]. This is consistent with the abundance of nanofibers in the product, which are believed to be the basic morphological unit for polyaniline (see Section 7.2.2). The small fraction of thick fibers and large dendrites resembling poly(*m*-fluoroaniline) and poly(*o*-fluoroaniline) could be the by-products generated during the de-fluorinating process.

TEM studies on polyethylanilines show that poly(*m*-ethylaniline) also appears to be more fibrillar than the *ortho*-isomer. For anisidine, although no fibrillar structure is obtained for either *meta*- or *ortho*-polymer, the particles of the *meta*-isomer do look smaller than the *ortho*-isomer (Figure 7.52).

7.6.2 Effects of Solvent on the Basic Morphology of Poly(3,4-Ethylenedioxythiophene) and Polypyrrole

Poly(3,4-ethylenedioxythiophene) (PEDOT) is one of the most important conducting polymers industrially due to its superior electrical properties and high thermal stability [3,186–196]. PEDOT is also one of the few polythiophenes that can be synthesized through simple oxidative chemical polymerization without using any catalyst.

Since the EDOT monomer is not soluble in water, its reaction with the oxidant, $FeCl_3$ in water is inhomogeneous. As a result, in a rapidly mixed reaction, the polymerization occurs around EDOT droplets at the EDOT-water interface forming hollow capsule-like structures, such as the one shown in Figure 7.53.

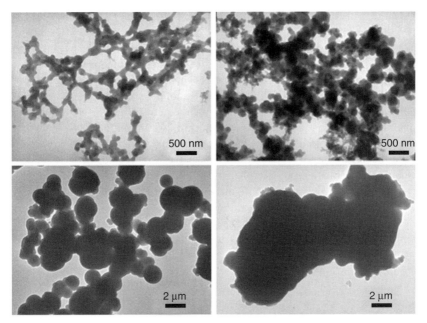

FIGURE 7.52   Typical TEM images showing the morphological difference between the *meta*-substituted (*left*) and *ortho*-substituted (*right*) poly(ethylaniline) (*top*) and poly(anisidine) (*bottom*).

In a mixture of water and methanol, a rapidly mixed reaction still produces mainly broken capsule-like structures. In isopropanol, both the monomer and the oxidant are soluble, therefore, no capsule-like structures are observed but rod-like nanostructures start to appear with large amounts of amorphous materials. Rapidly mixed reactions of pyrrole polymerization were carried out in water, methanol, and isopropanol. Large polypyrrole particles (micron size) were obtained in water, while smaller particles (~150 nm) were obtained in methanol. In isopropanol, the product disperses very well and agglomerates of tiny fibril structures (<30 nm) were found in the product. This indicates that further tuning of the solvent–polypyrrole–dopant interactions, reveal well-defined nanoscale morphological units for polypyrrole.

7.7 Conclusions

The synthetic methodology that has been developed is conceptually new from the previous approaches. It takes advantage of the nanofibrillar morphological unit of polyaniline itself and focuses on modifying the reaction pathway so that nanofiber formation is favored while their overgrowth, which would otherwise lead to irregularly shaped agglomerates, is suppressed. In summary, polyaniline preferentially forms as nanofibers in aqueous solution during chemical oxidative polymerization [55]. In slow feeding reactions, the nanofibers produced in the early stage of polymerization are subject to secondary growth,

FIGURE 7.53   TEM images showing the PEDOT product obtained by rapidly mixed reactions in water (*left*), a water–methanol mixture (*middle*), and isopropanol (*right*).

which leads to large agglomerates containing irregularly shaped particles and nanofibers. By preventing secondary growth, pure nanofibers can be obtained. In interfacial polymerization [53,56,62], secondary growth is suppressed when freshly formed nanofibers diffuse away from the reactive interface. In rapidly mixed reactions [55], secondary growth is limited by quickly consuming the reactants during the initial polymerization. Rapidly mixed reactions carried out in less polar solvents produce less perfect nanofibers, indicating that water is the best solvent for nanofiber synthesis.

The nanofibrillar morphology significantly improves the performance of polyaniline in many conventional applications involving polymer interactions with its environment. This leads to much faster and more responsive chemical sensors [56,62,84,87], new inorganic–polyaniline nanocomposites, and ultrafast nonvolatile memory devices [140]. Additionally, the highly conjugated polymeric structure of polyaniline should produce new nanoscale phenomena that are not accessible with current inorganic systems. As an example, an enhanced photothermal effect has been discovered that produces welding of the polyaniline nanofibers upon exposure to a camera flash [156].

The work presented here offers insights into the syntheses of conducting polymer nanostructures and clearly demonstrates the advantages of "nanostructures + conducting polymers." Thus, conducting polymers have an important role to play in the emerging field of nanoscience, which should lead to many exciting applications.

Acknowledgments

We thank Drs. S. Virji, B.H. Weiller (The Aerospace Corporation), and R.G. Blair for their collaborative work on the chemical sensors, and Professor Y. Yang and R. Tseng for their collaborative work on the memory devices. We thank the Microelectronics Advanced Research Corporation (MARCO) with its focus center on Functional Engineered NanoArchitectonics (FENA) and the National Science Foundation Grant DMR 0507294 for financial support.

References

1. Shirakawa, H., E.J. Louis, A.G. MacDiarmid, C.K. Chiang, and A.J. Heeger. 1977. Synthesis of electrically conducting organic polymers—Halogen derivatives of polyacetylene, $(CH)_x$. *Chem Commun* 16:578–580.
2. Skotheim, T.J., R.L. Elsenbaumer, and J.R. Reynolds. 1998. *Handbook of conducting polymers*, 2nd ed. New York: Marcel Dekker.
3. Chandrasekhar, P. 1999. *Conducting polymers, fundamentals and applications: A practical approach.* Boston, MA: Kluwer Academic Publishers.
4. Huang, W.S., B.D. Humphrey, and A.G. MacDiarmid. 1986. Polyaniline, a novel conducting polymer—Morphology and chemistry of its oxidation and reduction in aqueous-electrolytes. *J Chem Soc Faraday Trans I* 82:2385.
5. Desilvestro, J., W. Scheifele, and O. Haas. 1992. Insitu determination of gravimetric and volumetric charge-densities of battery electrodes—Polyaniline in aqueous and nonaqueous electrolytes. *J Electrochem Soc* 139 (10):2727–2736.
6. Joo, J., and A.J. Epstein. 1994. Electromagnetic-radiation shielding by intrinsically conducting polymers. *Appl Phys Lett* 65 (18):2278–2280.
7. Trivedi, D.C., and S.K. Dhawan. 1993. Shielding of electromagnetic-interference using polyaniline. *Synth Met* 59 (2):267–272.
8. Lu, W.K., R.L. Elsenbaumer, and B. Wessling. 1995. Corrosion protection of mild-steel by coatings containing polyaniline. *Synth Met* 71 (1–3):2163–2166.
9. Fahlman, M., S. Jasty, and A.J. Epstein. 1997. Corrosion protection of iron/steel by emeraldine base polyaniline: An x-ray photoelectron spectroscopy study. *Synth Met* 85 (1–3):1323–1326.
10. Conventional polyaniline can be purchased from Santa Fe Science and Technology (SFST), Inc., Santa Fe, NM 87507 USA, as well as Fisher Scientific and Sigma-Aldrich Co.

11. Anderson, M.R., B.R. Mattes, H. Reiss, and R.B. Kaner. 1991. Conjugated polymer films for gas separations. *Science* 252 (5011):1412–1415.

12. Martin, C.R. 1995. Template synthesis of electronically conductive polymer nanostructures. *Acc Chem Res* 28 (2):61–68.

13. Wu, C.G., and T. Bein. 1994. Conducting polyaniline filaments in a mesoporous channel host. *Science* 264 (5166):1757–1759.

14. Zheng, Z.X., Y.Y. Xi, P. Dong, H.G. Huang, J.Z. Zhou, L.L. Wu, and Z.H. Lin. 2002. The enhanced photoluminescence of zinc oxide and polyaniline coaxial nanowire arrays in anodic oxide aluminium membranes. *Physchemcomm* (9):63–65.

15. Dong, H., and W.E. Jones. 2002. A fiber templating approach to conducting polymer nanotubes. *Polym Mater Sci Eng* 87:273–274.

16. Li, G.C., and Z.K. Zhang. 2004. Synthesis of dendritic polyaniline nanofibers in a surfactant gel. *Macromolecules* 37 (8):2683–2685.

17. Michaelson, J.C., and A.J. McEvoy. 1994. Interfacial polymerization of aniline. *Chem Commun* 1:79–80.

18. Yu, L., J.I. Lee, K.W. Shin, C.E. Park, and R. Holze. 2003. Preparation of aqueous polyaniline dispersions by micellar-aided polymerization. *J Appl Polym Sci* 88 (6):1550–1555.

19. Huang, L.M., Z.B. Wang, H.T. Wang, X.L. Cheng, A. Mitra, and Y.X. Yan. 2002. Polyaniline nanowires by electropolymerization from liquid crystalline phases. *J Mater Chem* 12 (2): 388–391.

20. Liu, J.M., and S.C. Yang. 1991. Novel colloidal polyaniline fibrils made by template guided chemical polymerization. *Chem Commun* 21:1529–1531.

21. Shao, Y., Y.D. Jin, and S.J. Dong. 2002. DNA-templated assembly and electropolymerization of aniline on gold surface. *Electrochem Commun* 4 (10):773–779.

22. Wan, M.X. 2004. Conducting polymer nanotubes. In *Encyclopedia of nanoscience and nanotechnology*, ed. H.S. Nalwa. Los Angeles, CA: American Scientific Publishers, pp. 153–169.

23. Wei, Z.X., Z.M. Zhang, and M.X. Wan. 2002. Formation mechanism of self-assembled polyaniline micro/nanotubes. *Langmuir* 18 (3):917–921.

24. Wei, Z.X., and M.X. Wan. 2003. Synthesis and characterization of self-doped poly(aniline-co-aminonaphthalene sulfonic acid) nanotubes. *J Appl Polym Sci* 87 (8):1297–1301.

25. Kinlen, P.J., J. Liu, Y. Ding, C.R. Graham, and E.E. Remsen. 1998. Emulsion polymerization process for organically soluble and electrically conducting polyaniline. *Macromolecules* 31 (6): 1735–1744.

26. Langer, J.J., G. Framski, and R. Joachimiak. 2001. Polyaniline nano-wires and nano-networks. *Synth Met* 121 (1–3):1281–1282.

27. Qiu, H.J., M.X. Wan, B. Matthews, and L.M. Dai. 2001. Conducting polyaniline nanotubes by template-free polymerization. *Macromolecules* 34 (4):675–677.

28. Qiu, H.J., and M.X. Wan. 2001. Synthesis, characterization, and electrical properties of nanostructural polyaniline doped with novel sulfonic acids (4-{n-[4-(4-nitrophenylazo)phenyloxy]alkyl} aminobenzene sulfonic acid). *J Polym Sci A* 39 (20):3485–3497.

29. Reneker, D.H., and I. Chun. 1996. Nanometer diameter fibers of polymer, produced by electrospinning. *Nanotechnology* 7 (3):216–223.

30. MacDiarmid, A.G., W.E. Jones, I.D. Norris, J. Gao, A.T. Johnson, N.J. Pinto, J. Hone, B. Han, F.K. Ko, H. Okuzaki, and M. Llaguno. 2001. Electrostatically-generated nanofibers of electronic polymers. *Synth Met* 119 (1–3):27–30.

31. Desai, K., and C. Sung. 2002. Electrospinning nanofibers of PANI/PMMA blends. *Mater Res Soc Symp Proc* 736 (*Electronics on unconventional substrates—Electrotextiles and giant-area flexible circuits*):121–126.

32. Desai, K., and C. Sung. 2003. Phase characterization and morphology control of electrospun nanofibers of PANI/PMMA blends. *Mater Res Soc Symp Proc* 788 (*Continuous nanophase and nanostructured materials*):209–214.

33. Pinto, N.J., A.T. Johnson, Jr., A.G. MacDiarmid, G.H. Mueller, N. Theofylaktos, D.C. Robinson, and F.A. Miranda. 2003. Electrospun polyaniline/polyethylene oxide nanofiber field-effect transistor. *Appl Phys Lett* 83 (20):4244–4246.

34. Yun, M., N.V. Myung, R.P. Vasquez, J. Wang, and H. Monbouquette. 2003. Nanowire growth for sensor arrays. *Proc SPIE-ISOE* 5220 (*Nanofabrication technologies*):37–45.

35. Ramanathan, K., M.A. Bangar, M. Yun, W. Chen, A. Mulchandani, and N.V. Myung. 2004. Individually addressable conducting polymer nanowires array. *Nano Lett* 4 (7):1237–1239.

36. He, H.X., C.Z. Li, and N.J. Tao. 2001. Conductance of polymer nanowires fabricated by a combined electrodeposition and mechanical break junction method. *Appl Phys Lett* 78 (6):811–813.

37. Xu, P., and D.L. Kaplan. 2004. Nanoscale surface patterning of enzyme-catalyzed polymeric conducting wires. *Adv Mater* 16 (7):628–632.

38. Liu, H., J. Kameoka, D.A. Czaplewski, and H.G. Craighead. 2004. Polymeric nanowire chemical sensor. *Nano Lett* 4 (4):671–675.

39. Ma, Y., J. Zhang, G. Zhang, and H. He. 2004. Polyaniline nanowires on silicon surfaces fabricated with DNA templates. *J Am Chem Soc* 126 (22):7097–7101.

40. Anderson, R.E., D.D. Sawall, A.R. Hopkins, and R.M. Villahermosa. 2004. Electrochemical synthesis of polyaniline nanofibers. *Polymer* (Preprints) 45 (1):230–231.

41. Choi, S.J., and S.M. Park. 2002. Electrochemistry of conductive polymers—XXVI. Effects of electrolytes and growth methods on polyaniline morphology. *J Electrochem Soc* 149 (2):E26–E34.

42. He, H., Y. Zhao, B. Xu, and N. Tao. 2001. Electrochemical potential controlled electron transport in conducting polymer nanowires. *Proc Electrochem Soc* 19 (*Quantum confinement VI: Nanostructured materials and devices*):20–22.

43. Kan, J., R. Lu, and S. Zhang. 2004. Effect of ethanol on properties of electrochemically synthesized polyaniline. *Synth Met* 145 (1):37–42.

44. Zhou, H.H., S.Q. Jiao, J.H. Chen, W.Z. Wei, and Y.F. Kuang. 2004. Relationship between preparation conditions, morphology and electrochemical properties of polyaniline prepared by pulse galvanostatic method (PGM). *Thin Solid Films* 450 (2):233–239.

45. Janata, J., and M. Josowicz. 2003. Conducting polymers in electronic chemical sensors. *Nat Mater* 2:19.

46. Liu, J., Y.H. Lin, L. Liang, J.A. Voigt, D.L. Huber, Z.R. Tian, E. Coker, B. Mckenzie, and M.J. Mcdermott. 2003. Templateless assembly of molecularly aligned conductive polymer nanowires: A new approach for oriented nanostructures. *Chem Eur J* 9 (3):605–611.

47. Liang, L., J. Liu, C.F. Windisch, G.J. Exarhos, and Y.H. Lin. 2002. Direct assembly of large arrays of oriented conducting polymer nanowires. *Angew Chem Int Ed* 41 (19):3665–3668.

48. Choi, S.-J., and S.-M. Park. 2000. Electrochemical growth of nanosized conducting polymer wires on gold using molecular templates. *Adv Mater* 12 (20):1547–1549.

49. Yang, C.Y., Y. Cao, P. Smith, and A.J. Heeger. 1993. Morphology of conductive, solution-processed blends of polyaniline and poly(methyl methacrylate). *Synth Met* 53 (3):293–301.

50. Yang, C.Y., M. Reghu, A.J. Heeger, and Y. Cao. 1996. Thermal stability of polyaniline networks in conducting polymer blends. *Synth Met* 79 (1):27–32.

51. Yoon, C.O., M. Reghu, D. Moses, A.J. Heeger, and Y. Cao. 1994. Electrical-transport in conductive blends of polyaniline in poly(methyl methacrylate). *Synth Met* 63 (1):47–52.

52. Avlyanov, J.K., J.Y. Josefowicz, and A.G. Macdiarmid. 1995. Atomic-force microscopy surface-morphology studies of in-situ deposited polyaniline thin-films. *Synth Met* 73 (3):205–208.

53. Huang, J.X., and R.B. Kaner. 2004. A general chemical route to polyaniline nanofibers. *J Am Chem Soc* 126 (3):851–855.

54. Chien, J.C.W., Y. Yamashita, J.A. Hirsch, J.L. Fan, M.A. Schen, and F.E. Karasz. 1982. Resolution of controversy concerning the morphology of polyacetylene. *Nature* 299 (5884):608–611.

55. Huang, J.X., and R.B. Kaner. 2004. Nanofiber formation in the chemical polymerization of aniline: A mechanistic study. *Angew Chem Int Ed* 43 (43):5817–5821.

56. Huang, J.X., S. Virji, B.H. Weiller, and R.B. Kaner. 2003. Polyaniline nanofibers: Facile synthesis and chemical sensors. *J Am Chem Soc* 125 (2):314–315.

57. Cao, Y., P. Smith, and A.J. Heeger. 1993. Counterion-induced processibility of conducting polyaniline. *Synth Met* 57 (1):3514–3519.

58. Cao, Y., P. Smith, and A.J. Heeger. 1992. Counter-ion induced processibility of conducting polyaniline and of conducting polyblends of polyaniline in bulk polymers. *Synth Met* 48 (1):91–97.

59. Huang, J.X., Y. Xie, B. Li, Y. Liu, Y.T. Qian, and S.Y. Zhang. 2000. In-situ source–template-interface reaction route to semiconductor CdS submicrometer hollow spheres. *Adv Mater* 12 (11):808–811.

60. Wei, Z.X., and M.X. Wan. 2002. Hollow microspheres of polyaniline synthesized with an aniline emulsion template. *Adv Mater* 14 (18):1314–1317.

61. Zhang, L.J., and M.X. Wan. 2003. Self-assembly of polyaniline—From nanotubes to hollow microspheres. *Adv Funct Mater* 13 (10):815–820.

62. Huang, J.X., S. Virji, B.H. Weiller, and R.B. Kaner. 2004. Polyaniline nanofiber sensors. *Chem Eur J* 10 (6):1314.

63. Kang, E.T., K.G. Neoh, and K.L. Tan. 1998. Polyaniline: A polymer with many interesting intrinsic redox states. *Prog Polym Sci* 23 (2):277–324.

64. Tang, J.S., X.B. Jing, B.C. Wang, and F.S. Wang. 1988. Infrared-spectra of soluble polyaniline. *Synth Met* 24 (3):231–238.

65. Cao, Y., S.Z. Li, Z.J. Xue, and D. Guo. 1986. Spectroscopic and electrical characterization of some aniline oligomers and polyaniline. *Synth Met* 16 (3):305–315.

66. Ashraf, S.A., L.A.P. KaneMaguire, M.R. Majidi, S.G. Pyne, and G.G. Wallace. 1997. Influence of the chiral dopant anion on the generation of induced optical activity in polyanilines. *Polymer* 38 (11):2627–2631.

67. Guo, H.L., C.M. Knobler, and R.B. Kaner. 1999. A chiral recognition polymer based on polyaniline. *Synth Met* 101 (1–3):44–47.

68. Huang, J.X., V.M. Egan, H.L. Guo, J.Y. Yoon, A.L. Briseno, I.E. Rauda, R.L. Garrell, C.M. Knobler, F.M. Zhou, and R.B. Kaner. 2003. Enantioselective discrimination of D- and L-phenylalanine by chiral polyaniline thin films. *Adv Mater* 15 (14):1158–1161.

69. Majidi, M.R., L.A.P. Kane-Maguire, and G.G. Wallace. 1996. Facile synthesis of optically active polyaniline and polytoluidine. *Polymer* 37 (2):359–362.

70. Majidi, M.R., L.A.P. Kane-Maguire, and G.G. Wallace. 1995. Chemical generation of optically active polyaniline via the doping of emeraldine base with (+)- or (−)-camphorsulfonic acid. *Polymer* 36 (18):3597–3599.

71. Egan, V., R. Bernstein, T. Tran, L. Hohmann, and R. Kaner. 2001. Influence of water on the chirality of camphorsulfonic acid doped polyaniline. *Chem Commun* 9:801–802.

72. Li, W.G., and H.L. Wang. 2004. Oligomer-assisted synthesis of chiral polyaniline nanofibers. *J Am Chem Soc* 126 (8):2278–2279.

73. Tang, X., Y. Sun, and Y. Wei. 1988. Molecular-weight of chemically polymerized polyaniline. *Makromol Chem Rapid* 9 (12):829–834.

74. Chiang, J.C., and A.G. MacDiarmid. 1986. Polyaniline—Protonic acid doping of the emeraldine form to the metallic regime. *Synth Met* 13 (1–3):193–205.

75. King, R.C.Y., and F. Roussel. 2005. Morphological and electrical characteristics of polyaniline nanofibers. *Synth Met* 153 (1–3):337–340.

76. Pouget, J.P., M.E. Jozefowicz, A.J. Epstein, X. Tang, and A.G. Macdiarmid. 1991. X-ray structure of polyaniline. *Macromolecules* 24 (3):779–789.

77. Barisci, J.N., C. Conn, and G.G. Wallace. 1996. Conducting polymer sensors. *Trends Polym Sci* 4 (9):307–311.

78. Partridge, A.C., M.L. Jansen, and W.M. Arnold. 2000. Conducting polymer-based sensors. *Mat Sci Eng C Bio S* 12 (1–2):37–42.

79. Miasik, J.J., A. Hooper, and B.C. Tofield. 1986. Conducting polymer gas sensors. *J Chem Soc Faraday Trans I* 82:1117.

80. Wallace, G.G., M. Smyth, and H. Zhao. 1999. Conducting electroactive polymer-based biosensors. *Trend Anal Chem* 18 (4):245–251.

81. McQuade, D.T., A.E. Pullen, and T.M. Swager. 2000. Conjugated polymer-based chemical sensors. *Chem Rev* 100 (7):2537–2574.

82. MacDiarmid, A.G. 2001. Synthetic metals: A novel role for organic polymers. *Synth Met* 125 (1):11–22.

83. Chen, R.J., H.C. Choi, S. Bangsaruntip, E. Yenilmez, X.W. Tang, Q. Wang, Y.L. Chang, and H.J. Dai. 2004. An investigation of the mechanisms of electronic sensing of protein adsorption on carbon nanotube devices. *J Am Chem Soc* 126 (5):1563–1568.

84. Virji, S., J.X. Huang, R.B. Kaner, and B.H. Weiller. 2004. Polyaniline nanofiber gas sensors: Examination of response mechanisms. *Nano Lett* 4 (3):491–496.

85. Ding, B., J. Kima, Y. Miyazaki, and S. Shiratori. 2004. Electrospun nanofibrous membranes coated quartz crystal microbalance as gas sensor for NH_3 detection. *Sens Actuat B* 101:373–380.

86. Sotzing, G.A., J.N. Phend, R.H. Grubbs, and N.S. Lewis. 2000. Highly sensitive detection and discrimination of biogenic amines utilizing arrays of polyaniline/carbon black composite vapor detectors. *Chem Mater* 12 (3):593.

87. Virji, S., J.D. Fowler, C.O. Baker, J. Huang, R.B. Kaner, and B.H. Weiller. 2005. Polyaniline nanofiber composites with metal salts: Chemical sensors for hydrogen sulfide. *Small* 1 (6):624–627.

88. Antipov, A.A., G.B. Sukhorukov, Y.A. Fedutik, J. Hartmann, M. Giersig, and H. Mohwald. 2002. Fabrication of a novel type of metallized colloids and hollow capsules. *Langmuir* 18 (17):6687–6693.

89. Athawale, A.A., and S.V. Bhagwat. 2003. Synthesis and characterization of novel copper/polyaniline nanocomposite and application as a catalyst in the Wacker oxidation reaction. *J Appl Polym Sci* 89 (9):2412–2417.

90. Cioffi, N., L. Torsi, I. Losito, L. Sabbatini, P.G. Zambonin, and T. Bleve-Zacheo. 2001. Nanostructured palladium-polypyrrole composites electrosynthesised from organic solvents. *Electrochim Acta* 46 (26–27):4205–4211.

91. Dai, X.H., Y.W. Tan, and J. Xu. 2002. Formation of gold nanoparticles in the presence of *o*-anisidine and the dependence of the structure of poly(*o*-anisidine) on synthetic conditions. *Langmuir* 18 (23):9010–9016.

92. Henry, M.C., C.C. Hsueh, B.P. Timko, and M.S. Freund. 2001. Reaction of pyrrole and chlorauric acid—A new route to composite colloids. *J Electrochem Soc* 148 (11):D155–D162.

93. Mikhaylova, A.A., O.A. Khazova, and V.S. Bagotzky. 2000. Electrocatalytic and adsorption properties of platinum microparticles electrodeposited onto glassy carbon and into Nafion® films. *J Electroanal Chem* 480 (1–2):225–232.

94. Mascaro, L.H., D. Goncalves, and L.O.S. Bulhoes. 2004. Electrocatalytic properties and electrochemical stability of polyaniline and polyaniline modified with platinum nanoparticles in formaldehyde medium. *Thin Solid Films* 461 (2):243–249.

95. Sarma, T.K., and A. Chattopadhyay. 2004. Reversible encapsulation of nanometer-size polyaniline and polyaniline-Au-nanoparticle composite in starch. *Langmuir* 20 (11):4733–4737.

96. Sarma, T.K., D. Chowdhury, A. Paul, and A. Chattopadhyay. 2002. Synthesis of Au nanoparticle-conductive polyaniline composite using H_2O_2 as oxidising as well as reducing agent. *Chem Commun* 10:1048–1049.

97. Selvakannan, P.R., P.S. Kumar, A.S. More, R.D. Shingte, P.P. Wadgaonkar, and M. Sastry. 2004. One pot, spontaneous and simultaneous synthesis of gold nanoparticles in aqueous and nonpolar organic solvents using a diamine-containing oxyethylene linkage. *Langmuir* 20 (2):295–298.

98. Smith, J.A., M. Josowicz, and J. Janata. 2003. Polyaniline-gold nanocomposite system. *J Electrochem Soc* 150 (8):E384–E388.

99. Wang, J.G., K.G. Neoh, and E.T. Kang. 2001. Preparation of nanosized metallic particles in polyaniline. *J Coll Interf Sci* 239 (1):78–86.

100. Zhou, Y., H. Itoh, T. Uemura, K. Naka, and Y. Chujo. 2002. Synthesis of novel stable nanometer-sized metal (M = Pd, Au, Pt) colloids protected by a pi-conjugated polymer. *Langmuir* 18 (1):277–283.

101. Sawant, S.N., N. Bagkar, H. Subramanian, and J.V. Yakhmi. 2004. Polyaniline-Prussian blue hybrid: Synthesis and magnetic behaviour. *Philos Mag* 84 (20):2127–2138.

102. Schnitzler, D.C., M.S. Meruvia, I.A. Hummelgen, and A.J.G. Zarbin. 2003. Preparation and characterization of novel hybrid materials formed from (Ti,Sn)O-2 nanoparticles and polyaniline. *Chem Mater* 15 (24):4658–4665.

103. Schnitzler, D.C., and A.J.G. Zarbin. 2004. Organic/inorganic hybrid materials formed from TiO_2 nanoparticles and polyaniline. *J Braz Chem Soc* 15 (3):378–384.

104. Sui, X.M., Y. Chu, S.X. Xing, and C.Z. Liu. 2004. Synthesis of PANI/AgCl, PANI/$BaSO_4$ and PANI/TiO_2 nanocomposites in CTAB/hexanol/water reverse micelle. *Mater Lett* 58 (7–8):1255–1259.

105. Svechnikov, S.V., V.D. Pokhodenko, N.F. Guba, L.I. Fenenko, P.S. Smertenko, I.V. Prokopenko, L.N. Grebinskaya, P.M. Lytvyn, Y.P. Piryatinskii, O.P. Lytvyn, and G.P. Ol'khovik. 2004. Spectral characteristics and surface morphology of organic polymer films containing vanadium pentoxide nanoparticles. *Russ J Electrochem* 40 (3):259–266.

106. Li, Z.F., M.T. Swihart, and E. Ruckenstein. 2004. Luminescent silicon nanoparticles capped by conductive polyaniline through the self-assembly method. *Langmuir* 20 (5):1963–1971.

107. Machida, K., K. Furuuchi, M. Min, and K. Naoi. 2004. Mixed proton–electron conducting nanocomposite based on hydrous RuO_2 and polyaniline derivatives for supercapacitors. *Electrochemistry* 72 (6):402–404.

108. Khiew, P.S., N.M. Huang, S. Radiman, and M.S. Ahmad. 2004. Synthesis and characterization of conducting polyaniline-coated cadmium sulphide nanocomposites in reverse microemulsion. *Mater Lett* 58 (3–4):516–521.

109. Yang, Q.L., Z. Jin, Y.L. Song, M.X. Wan, J. Lei, W.G. Xu, and Q.S. Li. 2003. Electrical-magnetic properties of nanocomposites of conducting polyaniline and Y-Fe_2O_3 nanoparticles. *Chem J Chin U* 24 (12):2290–2292.

110. Zhang, Z.M., and M.X. Wan. 2003. Nanostructures of polyaniline composites containing nanomagnet. *Synth Met* 132 (2):205–212.

111. Deng, J.G., C.L. He, Y.X. Peng, J.H. Wang, X.P. Long, P. Li, and A.S.C. Chan. 2003. Magnetic and conductive Fe_3O_4-polyaniline nanoparticles with core-shell structure. *Synth Met* 139 (2):295–301.

112. Long, Y.Z., Z.J. Chen, Z.X. Liu, Z.M. Zhang, M.X. Wan, and N.L. Wang. 2003. Composites of nanotubular polyaniline containing Fe_3O_4 nanoparticles. *Chin Phys* 12 (4):433–437.

113. Tursun, A., and X.-G. Zhang. 2003. Effect of solvent on the solid-state polymerization reaction of polyaniline doped with $H_7PW_{12}O_{42}$. *Gongneng Gaofenzi Xuebao* 16 (4):451–455.

114. Wei, J.H., J.G. Guan, J. Shi, and R.Z. Yuan. 2003. The structure and electrorheological effect of PAn/$BaTiO_3$ nanocomposite. *Chin J Chem Phys* 16 (5):401–405.

115. Xia, H.S., and Q. Wang. 2003. Preparation of conductive polyaniline/nanosilica particle composites through ultrasonic irradiation. *J Appl Polym Sci* 87 (11):1811–1817.

116. Xia, H.S., and Q. Wang. 2002. Ultrasonic irradiation: A novel approach to prepare conductive polyaniline/nanocrystalline titanium oxide composites. *Chem Mater* 14 (5):2158–2165.

117. Yang, Q.L., Y.L. Song, M.X. Wan, L. Jiang, W.G. Xu, and Q.S. Li. 2002. Synthesis and characterization of the composite of conducting polyaniline with Fe_3O_4 magnetic nanoparticles. *Chem J Chin U* 23 (6):1105–1109.

118. Deng, J.G., Y.X. Peng, X.B. Ding, J.H. Wang, X.P. Long, P. Li, and A.S.C. Chan. 2002. Preparation and characterization of magnetic polyaniline microspheres. *Chin J Chem Phys* 15 (2):149–152.

119. Gan, L.M., L.H. Zhang, H.S.O. Chan, and C.H. Chew. 1995. Preparation of conducting polyaniline-coated barium-sulfate nanoparticles in inverse microemulsions. *Mater Chem Phys* 40 (2):94–98.

120. Gaponik, N.P., and D.V. Sviridov. 1997. Synthesis and characterization of PbS quantum dots embedded in the polyaniline film. *Ber Bunsen Phys Chem* 101 (11):1657–1659.

121. Godovsky, D.Y., A.E. Varfolomeev, D.F. Zaretsky, R.L.N. Chandrakanthi, A. Kundig, C. Weder, and W. Caseri. 2001. Preparation of nanocomposites of polyaniline and inorganic semiconductors. *J Mater Chem* 11 (10):2465–2469.

122. Gong, H., X.J. Cui, Z.W. Xie, S.G. Wang, and L.Y. Qu. 2002. The solid-state synthesis of poly-aniline/$H_4SiW_{12}O_{40}$ materials. *Synth Met* 129 (2):187–192.

123. Kryszewski, M., and J.K. Jeszka. 1998. Nanostructured conducting polymer composites—Super-paramagnetic particles in conducting polymers. *Synth Met* 94 (1):99–104.

124. Kumar, R.V., Y. Mastai, Y. Diamant, and A. Gedanken. 2001. Sonochemical synthesis of amorphous Cu and nanocrystalline Cu_2O embedded in a polyaniline matrix. *J Mater Chem* 11 (4):1209–1213.

125. Li, D.S., and G.X. Lu. 2002. Nanocomposite films of conductive polymer/semiconductor nano-particles (II)—Photoelectrochemical properties. *Chem J Chin U* 23 (4):685–689.

126. Li, J.P., and T.Z. Peng. 2002. Polyaniline/Prussian blue composite film electrochemical biosensors for cholesterol detection. *Chin J Chem* 20 (10):1038–1043.

127. Liu, S.Q., H.Z. Yu, H. Huang, and Y.Y. Xiong. 2002. Effect of TiO_2 nanoparticles on the properties of PANI. *Chem J Chin U* 23 (1):161–163.

128. Qu, L.Y., R.Q. Lu, J. Peng, Y.G. Chen, and Z.M. Dai. 1997. $H_3PW_{11}MoO_{40}$ center dot $^2H(2)O$ protonated polyaniline—Synthesis, characterization and catalytic conversion of isopropanol. *Synth Met* 84 (1–3):135–136.

129. Stejskal, J., O. Quadrat, I. Sapurina, J. Zemek, A. Drelinkiewicz, M. Hasik, I. Krivka, and J. Prokes. 2002. Polyaniline-coated silica gel. *Eur Polym J* 38 (4):631–637.

130. Vidya, V., N.P. Kumar, S.N. Narang, S. Major, S. Vitta, S.S. Talwar, P. Dubcek, H. Amenitsch, and S. Bernstorff. 2002. Molecular packing in CdS containing conducting polymer composite LB multi-layers. *Coll Surf A* 198:67–74.

131. Tian, S.J., J.Y. Liu, T. Zhu, and W. Knoll. 2003. Polyaniline doped with modified gold nanoparticles and its electrochemical properties in neutral aqueous solution. *Chem Commun* 21:2738–2739.

132. Kelaidopoulou, A., E. Abelidou, A. Papoutsis, E.K. Polychroniadis, and G. Kokkinidis. 1998. Electrooxidation of ethylene glycol on Pt-based catalysts dispersed in polyaniline. *J Appl Electro-chem* 28 (10):1101–1106.

133. Prasad, K.R. 2002. Electrooxidation of methanol on polyaniline without dispersed catalyst par-ticles. *J Power Sources* 103 (2):300–304.

134. Drelinkiewicza, A., A. Waksmundzka, W. Makowski, J.W. Sobczak, A. Krol, and A. Zieba. 2004. Acetophenone hydrogenation on polymer-palladium catalysts. The effect of polymer matrix. *Catal Lett* 94 (3–4):143–156.

135. Park, J.E., S.G. Park, A. Koukitu, O. Hatozaki, and N. Oyama. 2004. Electrochemical and chemical interactions between polyaniline and palladium nanoparticles. *Synth Met* 141 (3):265–269.

136. Park, J.E., S.G. Park, A. Koukitu, O. Hatozaki, and N. Oyama. 2004. Effect of adding Pd nano-particles to dimercaptan-polyaniline cathodes for lithium polymer battery. *Synth Met* 140 (2–3):121–126.

137. Hagiwara, T., M. Yamaura, and K. Iwata. 1988. Thermal-stability of polyaniline. *Synth Met* 25 (3):243–252.

138. Ostwald, W. 1897. Studien uber die Bildung und Umwandlung fester Korper. *Zeit Phys Chem* 22 (289).

139. Boistelle, R., and J.P. Astier. 1988. Crystallization mechanisms in solution. *J Cryst Growth* 90:14–30.

140. Tseng, R.J., J.X. Huang, J. Ouyang, R.B. Kaner, and Y. Yang. 2005. Polyaniline nanofiber/gold nanoparticle nonvolatile memory. *Nano Lett* 5 (6):1077–1080.

141. Ma, L.P., S. Pyo, J. Ouyang, Q.F. Xu, and Y. Yang. 2003. Nonvolatile electrical bistability of organic/metal-nanocluster/organic system. *Appl Phys Lett* 82 (9):1419–1421.

142. Ma, L.P., J. Liu, and Y. Yang. 2002. Organic electrical bistable devices and rewritable memory cells. *Appl Phys Lett* 80 (16):2997–2999.

143. Zimer, A.M., R. Bertholdo, M.T. Grassi, A.J.G. Zarbin, and L.H. Mascaro. 2003. Template carbon dispersed in polyaniline matrix electrodes: Evaluation and application as electrochemical sensors to low concentrations of Cu^{2+} and Pb^{2+}. *Electrochem Commun* 5 (12):983–988.

144. Chandrakanthi, R.L.N., and M.A. Careem. 2002. Preparation and characterization of CdS and Cu₂S nanoparticle/polyaniline composite films. *Thin Solid Films* 417 (1–2):51–56.
145. Zhang, L.J., and M.X. Wan. 2003. Polyaniline/TiO₂ composite nanotubes. *J Phys Chem B* 107 (28):6748–6753.
146. Wei, J.H., J.G. Guan, W.Y. Chen, and R.Z. Yuan. 2002. Preparation and characteristics of polyaniline-barium titanate nanocomposite particles. *Acta Phys Chim Sin* 18 (7):653–656.
147. Kumar, R.V., Y. Mastai, and A. Gedanken. 2000. Sonochemical synthesis and characterization of nanocrystalline paramelaconite in polyaniline matrix. *Chem Mater* 12 (12):3892–3895.
148. Kallitsis, J., E. Koumanakos, E. Dalas, S. Sakkopoulos, and P.G. Koutsoukos. 1989. The overgrowth of cadmium-sulfide on conducting polymers. *J Chem Soc Chem Comm* (16):1146–1147.
149. Rosencwaig, A. 1990. *Photoacoustics and photoacoustic spectroscopy.* New York: Wiley.
150. Ajayan, P.M., M. Terrones, A. de la Guardia, V. Huc, N. Grobert, B.Q. Wei, H. Lezec, G. Ramanath, and T.W. Ebbesen. 2002. Nanotubes in a flash—Ignition and reconstruction. *Science* 296 (5568):705.
151. Wang, N., B.D. Yao, Y.F. Chan, and X.Y. Zhang. 2003. Enhanced photothermal effect in Si nanowires. *Nano Lett* 3 (4):475–477.
152. MacDiarmid, A.G. 1997. Polyaniline and polypyrrole: Where are we headed? *Synth Met* 84 (1–3):27–34.
153. De Albuquerque, J.E., W.L.B. Melo, and R.M. Faria. 2000. Photopyroelectric spectroscopy of polyaniline films. *J Polym Sci B* 38 (10):1294–1300.
154. De Albuquerque, J.E., W.L.B. Melo, and R.M. Faria. 2003. Determination of physical parameters of conducting polymers by photothermal spectroscopies. *Rev Sci Instrum* 74 (1):306–308.
155. Toyoda, T., and H. Nakamura. 1995. Photoacoustic-spectroscopy of polyaniline films. *Jpn J Appl Phys Part 1* 34 (5B):2907–2910.
156. Huang, J.X., and R.B. Kaner. 2004. Flash welding of conducting polymer nanofibres. *Nat Mater* 3 (11):783–786.
157. Ding, L.L., X.W. Wang, and R.V. Gregory. 1999. Thermal properties of chemically synthesized polyaniline (EB) powder. *Synth Met* 104 (2):73–78.
158. Mathew, R., B.R. Mattes, and M.P. Espe. 2002. A solid state NMR characterization of cross-linked polyaniline powder. *Synth Met* 131 (1–3):141–147.
159. Pandey, S.S., M. Gerard, A.L. Sharma, and B.D. Malhotra. 2000. Thermal analysis of chemically synthesized polyemeraldine base. *J Appl Polym Sci* 75 (1):149–155.
160. Wei, Y., and K.F. Hsueh. 1989. Thermal-analysis of chemically synthesized polyaniline and effects of thermal aging on conductivity. *J Polym Sci A* 27 (13):4351–4363.
161. Huang, S.C., I.J. Ball, and R.B. Kaner. 1998. Polyaniline membranes for pervaporation of carboxylic acids and water. *Macromolecules* 31 (16):5456–5464.
162. Nunes, S.P., and K.-V. Peinemann. 2001. *Membrane technology in the chemical industry.* Weinheim: Wiley-VCH.
163. Gao, J.B., J.M. Sansinena, and H.L. Wang. 2003. Tunable polyaniline chemical actuators. *Chem Mater* 15 (12):2411–2418.
164. Gao, J.B., J.M. Sansinena, and H.L. Wang. 2003. Chemical vapor driven polyaniline sensor/actua-actuators. *Synth Met* 135 (1–3):809–810.
165. Sansinena, J.M., J.B. Gao, and H.L. Wang. 2003. High-performance, monolithic polyaniline electrochemical actuators. *Adv Funct Mater* 13 (9):703–709.
166. Wang, H.L., J.B. Gao, J.M. Sansinena, and P. McCarthy. 2002. Fabrication and characterization of polyaniline monolithic actuators based on a novel configuration: Integrally skinned asymmetric membrane. *Chem Mater* 14 (6):2546–2552.
167. Conklin, J.A., S.C. Huang, S.M. Huang, T.L. Wen, and R.B. Kaner. 1995. Thermal-properties of polyaniline and poly(aniline-co-O-ethylaniline). *Macromolecules* 28 (19):6522–6527.
168. Kieffel, Y., J.P. Travers, A. Ermolieff, and D. Rouchon. 2002. Thermal aging of undoped polyaniline: Effect of chemical degradation on electrical properties. *J Appl Polym Sci* 86 (2):395–404.

169. Liu, G., and M.S. Freund. 1997. New approach for the controlled cross-linking of polyaniline: Synthesis and characterization. *Macromolecules* 30 (19):5660–5665.

170. Rodrigues, P.C., G.P. de Souza, J.D.D. Neto, and L. Akcelrud. 2002. Thermal treatment and dynamic mechanical thermal properties of polyaniline. *Polymer* 43 (20):5493–5499.

171. Tan, H.H., K.G. Neoh, F.T. Liu, N. Kocherginsky, and E.T. Kang. 2001. Crosslinking and its effects on polyaniline films. *J Appl Polym Sci* 80 (1):1–9.

172. Braidy, N., G.A. Botton, and A. Adronov. 2002. Oxidation of Fe nanoparticles embedded in single-walled carbon nanotubes by exposure to a bright flash of white light. *Nano Lett* 2 (11):1277–1280.

173. Smits, J., B. Wincheski, M. Namkung, R. Crooks, and R. Louie. 2003. Response of Fe powder, purified and as-produced HiPco single-walled carbon nanotubes to flash exposure. *Mater Sci Eng A* 358 (1–2):384–389.

174. Cihaner, A., and A.M. Onal. 2001. Synthesis and characterization of fluorine-substituted polyanilines. *Eur Polym J* 37 (9):1767–1772.

175. Koval'chuk, E.P., N.V. Stratan, O.V. Reshetnyak, J. Blazejowski, and M.S. Whittingham. 2001. Synthesis and properties of the polyanisidines. *Solid State Ionics* 141:217–224.

176. Kwon, A.H., J.A. Conklin, M. Makhinson, and R.B. Kaner. 1997. Chemical synthesis and characterization of fluoro-substituted polyanilines. *Synth Met* 84 (1–3):95–96.

177. Sharma, A.L., M. Gerard, R. Singhal, B.D. Malhotra, and S. Annapoorni. 2001. Synthesis and characterization of fluoro-substituted polyaniline. *Appl Biochem Biotechnol* 96 (1–3):155–165.

178. Sharma, A.L., V. Saxena, S. Annapoorni, and B.D. Malhotra. 2001. Synthesis and characterization of a copolymer: Poly(aniline-co-fluoroaniline). *J Appl Polym Sci* 81 (6):1460–1466.

179. Wang, L.X., X.B. Jing, and F.S. Wang. 1991. The influence of protonic acids on the chemical polymerization of *ortho*-methylaniline. *Synth Met* 41 (1–2):745–748.

180. Yang, S.M., and J.H. Chiang. 1991. Morphological-study of alkylsubstituted polyaniline. *Synth Met* 41 (1–2):761–764.

181. Gruger, A., A. Novak, A. Regis, and P. Colomban. 1994. Infrared and Raman-study of polyaniline. 2. Influence of *ortho* substituents on hydrogen-bonding and UV/Vis near-IR electron charge-transfer. *J Mol Struct* 328:153–167.

182. Hugener, T.A., T.A.P. Seery, and A. Seghal. 1998. Dynamic light scattering studies of polyethylaniline solutions. *Abstr Pap Am Chem S* 215:U328–U328.

183. Niessen, J., U. Schroder, M. Rosenbaum, and F. Scholz. 2004. Fluorinated polyanilines as superior materials for electrocatalytic anodes in bacterial fuel cells. *Electrochem Commun* 6 (6):571–575.

184. Bedekar, A.G., S.F. Patil, and R.C. Patil. 1996. Comparative studies of electrochemically deposited poly(*o*-toluidine) and poly(*m*-toluidine) films. *Polym Int* 40 (3):201–205.

185. Mortimer, R.J. 1995. Spectroelectrochemistry of electrochromic poly(*o*-toluidine) and poly(*m*-toluidine) films. *J Mater Chem* 5 (7):969–973.

186. Caras-Quintero, D., and P. Bauerle. 2004. Synthesis of the first enantiomerically pure and chiral, disubstituted 3,4-ethylenedioxythiophenes (EDOTs) and corresponding stereo- and regioregular PEDOTs. *Chem Commun* 8:926–927.

187. Corradi, R., and S.P. Armes. 1997. Chemical synthesis of poly(3,4-ethylenedioxythiophene). *Synth Met* 84 (1–3):453–454.

188. Czardybon, A., and M. Lapkowski. 2001. Synthesis and electropolymerisation of 3,4-ethylenedioxythiophene functionalised with alkoxy groups. *Synth Met* 119 (1–3):161–162.

189. Dhanabalan, A., J.K.J. van Duren, P.A. van Hal, J.L.J. van Dongen, and R.A.J. Janssen. 2001. Synthesis and characterization of a low bandgap conjugated polymer for bulk heterojunction photovoltaic cells. *Adv Funct Mater* 11 (4):255–262.

190. Lefebvre, M., Z.G. Qi, D. Rana, and P.G. Pickup. 1999. Chemical synthesis, characterization, and electrochemical studies of poly(3,4-ethylenedioxythiophene)/poly(styrene-4-sulfonate) composites. *Chem Mater* 11 (2):262–268.

191. Louwet, F., L. Groenendaal, J. Dhaen, J. Manca, J. Van Luppen, E. Verdonck, and L. Leenders. 2003. PEDOT/PSS: Synthesis, characterization, properties and applications. *Synth Met* 135 (1–3): 115–117.

192. Pei, Q.B., G. Zuccarello, M. Ahlskog, and O. Inganas. 1994. Electrochromic and highly stable poly(3,4-ethylenedioxythiophene) switches between opaque blue-black and transparent sky blue. *Polymer* 35 (7):1347–1351.

193. Pickup, P.G., C.L. Kean, M.C. Lefebvre, G.C. Li, Z.Q. Qi, and J.N. Shan. 2000. Electronically conducting cation-exchange polymer powders: Synthesis, characterization and applications in PEM fuel cells and supercapacitors. *J New Mat Elect Syst* 3 (1):21–26.

194. Schwendeman, I., C.L. Gaupp, J.M. Hancock, L. Groenendaal, and J.R. Reynolds. 2003. Perfluoroalkanoate-substituted PEDOT for electrochromic device applications. *Adv Funct Mater* 13 (7):541–547.

195. Sotzing, G.A., J.R. Reynolds, and P.J. Steel. 1997. Poly(3,4-ethylenedioxythiophene) (PEDOT) prepared via electrochemical polymerization of EDOT, 2,2′-bis(3,4-ethylenedioxythiophene) (BiEDOT), and their TMS derivatives. *Adv Mater* 9 (10):795.

196. Syritski, V., K. Idla, and A. Opik. 2004. Synthesis and redox behavior of PEDOT/PSS and PPy/DBS structures. *Synth Met* 144 (3):235–239.

197. Huang, J.X., and R.B. Kaner. 2006. The intrinsic nanofibrillar morphology of polyaniline. *Chem Comm* (4):367–376.

8

Recent Advances
in Polypyrrole

Seung Hyun Cho,
Ki Tae Song, and
Jun Young Lee

8.1   Introduction

Most organic polymers are electrically insulating. Since the first discovery in 1977 that chemical treatment with iodine converts electrically insulating polyacetylene into a highly conducting material with electrical conductivity above 10^4 S/cm [1], many electrically conducting polymers have been reported. The electrically conducting polymers are able to transfer electrical charges to the same extent as an electrical conductor or a semiconductor. Due to their metal-like conductivity or semiconductivity and other fascinating properties, conducting polymers have played indispensable roles in specialized industrial applications in spite of their short history. However, the major aspect useful for most applications is not the metal-like electrical property itself, but the combination of electrical conductivity and polymeric properties such as flexibility, low density, and ease of structural modification that suffice for many commercial applications.

Among the conducting polymers, polypyrrole (PPy) has attracted great attention because of its high electrical conductivity and good environmental stability. PPy has been considered as the key material to many potential applications such as electronic devices, electrodes for rechargeable batteries and supercapacitors, solid electrolytes for capacitors, electromagnetic shielding materials, sensors, corrosion-protecting materials, actuators, electrochromic devices, or membranes. The heteroaromatic and extended π-conjugated backbone structure of PPy provide it with chemical stability and electrical conductivity, respectively. However, the π-conjugated backbone structure is not sufficient to produce appreciable conductivity on its own. Partial charge extraction from PPy chain is also required, which is achieved by a chemical or an electrochemical process referred to as doping. The conductivity of the neutral PPy are remarkably changed from an insulating regime to a metallic one by doping. This is a very worthwhile feature for applications in which the electrical conductivity of a material must be controlled.

It was reported that the electronic and band structures of PPy were changed with the doping level of the PPy chain [2,3]. Neutral PPy, with the benzenoid structure shown in Figure 8.1a, is categorized as an insulator and its proposed electronic energy diagram is shown in Figure 8.2a. The band gap of neutral PPy is reported to be 3.16 eV, which is too wide for electrons to transfer from the valence to the conduction band at room temperature. The PPy chain is, however, simultaneously doped during polymerization [4]. Counteranions in the reaction medium are incorporated into the growing PPy chains to maintain the electrical neutrality of the polymer system. Upon extraction of a negative charge from a neutral segment of a PPy chain by the doping process, a local deformation to the quinoid structure occurs (Figure 8.1b), since it is favored energetically. In combination with the quinoid structure, the positive charge and the unpaired spin are referred to as a polaron (Figure 8.1b). Referring to Figure 8.2b, the formation of a polaron induces two new intermediate states (bonding and antibonding) within the band gap while the unpaired electron occupies the bonding (low energy) state, thus giving the polaron a spin of 1/2.

As oxidation continues further, another electron has to be removed from a PPy chain that already contains a polaron, resulting in the formation of a bipolaron which is energetically preferred to the formation of two polarons. A bipolaron is known to extend over about four pyrrole rings (Figure 8.1c). The bipolaron states lie further from the band edges (Figure 8.2c). The lower energy state of the bipolaron is empty; thus the species has a zero spin. As the degree of oxidation increases, the bipolaronic

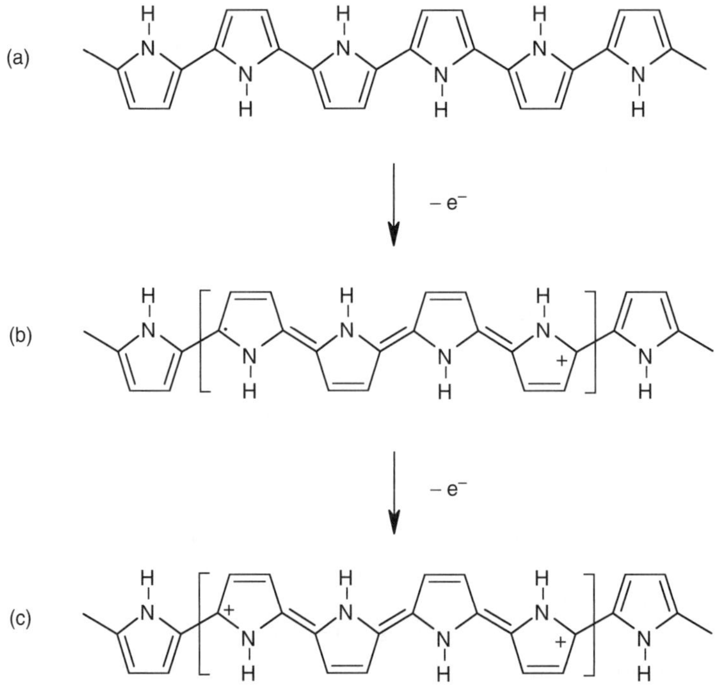

FIGURE 8.1 Electronic structures of (a) neutral PPy, (b) polaron in partially doped PPy and (c) bipolaron in fully doped PPy.

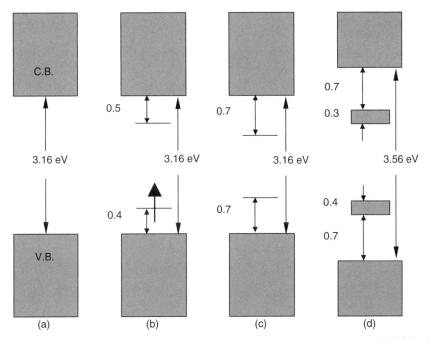

FIGURE 8.2 Electronic energy diagrams for (a) neutral PPy, (b) polaron, (c) bipolaron, and (d) fully doped PPy.

energy state overlaps, resulting in the formation of narrow intermediate band structures (Figure 8.2d). The energy diagram shown in Figure 8.2d corresponds to a doped state of about 33 mol%, which is close to the maximum value found in electrochemically oxidized PPy. The typical doping level of PPy is in the range of 20 to 40 mol%. At this doping level, bipolarons are predominant in PPy with few polarons, and thus the charge carriers in the conducting PPy have zero-spin number.

The electrical conductivity of a material is proportional to the product of the concentration of charge carriers and their mobility. Even though the concentration of charge carriers in fully doped conducting polymers, including PPy, is in the range of 10^{21} to $10^{23}/cm^3$, which is four to five orders of magnitudes higher than that of inorganic semiconductors, the electrical conductivities of most conducting polymers are merely in the same range as those of inorganic semiconductors. This indicates the relatively low mobility of carriers in conducting polymers, which results from the low degree of crystallinity and many other defects. Therefore, if higher electrical conductivities are to be achieved in conducting polymers, including PPy, the mobility of charge carriers must be improved.

The overall mobility of charge carriers in conducting polymers depends on two components: intrachain mobility, which corresponds to charge transfer along the polymer chain; and interchain mobility, which involves the hopping or tunneling of the charge from a chain to neighboring chains [5]. When an electric field is applied to conducting PPy, the charge carriers (polaron and bipolaron) necessarily begin to move along the PPy chain by the rearrangement of double and single bonds in the PPy backbone since the interchain charge transport requires considerably more energy than intrachain conduction [5]. When the charge carriers reach any defect point in the chain or the end of the PPy chain itself, the carriers have to hop to neighboring chains to deliver charges. Therefore, the total conductivity of conducting PPy is dominated by the combination of intrachain and interchain components.

The intrachain mobility is restricted by conformational and chemical defects in the PPy chain structure. Even though PPy chains are intrinsically planar and linear, many conformational and structural defects can be formed during polymerization (Figure 8.3) [5]. The defects break the planarity and linearity of the PPy chain and reduce the extent of π-orbital overlaps. The defects include conformational defects such as 2–2′ coupling with nonregular 180° rotation of the alternating pyrrole

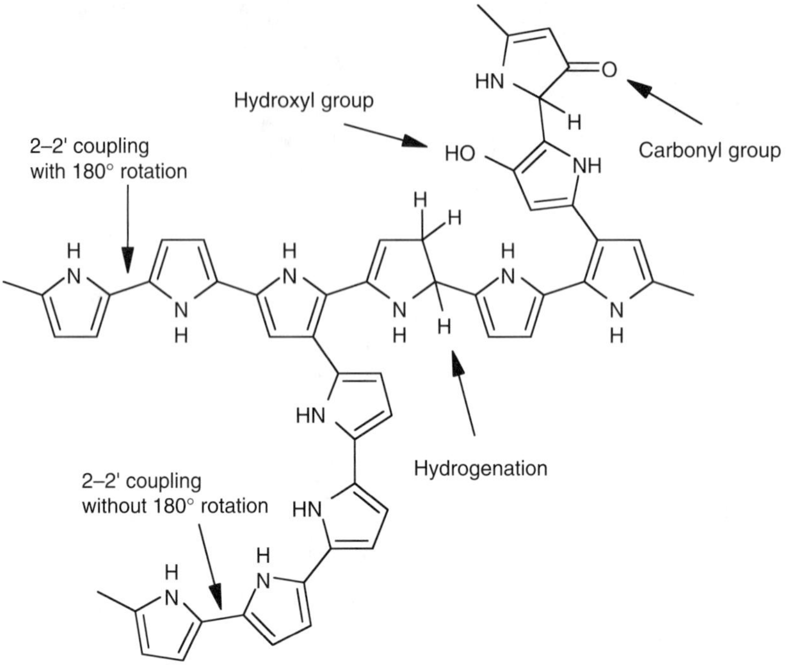

FIGURE 8.3 Chemical and conformational defects in PPy.

unit, chemical defects such as 2–3′ or 2–4′ coupling and nonaromatic bonding and formation of carbonyl or hydroxyl groups in PPy chains due to overoxidation (Figure 8.3).

Interchain mobility of carriers in conducting polymers including PPy is strongly dependent upon the structural order. A high degree of structural order is usually achieved in the crystalline region in a conducting polymer, which can be produced by mechanical stretching. Nogami et al. found a partial crystalline structure in the stretched films of PPy doped with hexafluorophosphate [6]. The PPy film showed an extremely high electrical conductivity of 2×10^3 S/cm in the direction parallel to the stretch direction. However, when PPy was synthesized using other dopants, no crystalline structure was found, which indicates that the nature of the counteranion considerably influences the structural arrangement of PPy chains.

8.2 Novel Synthesis of Polypyrrole

PPy may be prepared by either chemical or electrochemical oxidation of pyrrole monomer. The most widely accepted polymerization mechanism of PPy is the coupling between radical cations (Figure 8.4) [7]. In the initiation step, the oxidation of a pyrrole monomer yields a radical cation. Coupling of the two generated radical cations and deprotonation produces a bipyrrole, as confirmed by Andrieux et al. [8]. The bipyrrole is oxidized again and couples with another oxidized segment. In the propagation step, reoxidation, coupling, and deprotonation continue to form oligomers and finally PPy. The radical coupling between oligomeric pyrrole species is favored since the oxidation potential of oligomeric or polymeric pyrrole species is lower than that of the monomer [4]. Once the chain length of the oligomers exceeds the solubility limit of the solvent, precipitation of PPy occurs. The termination step has not been fully elucidated but is presumed to involve nucleophilic attack on the polymer chain [7]. On the other hand, an alternative mechanism has been proposed, in which a radical cation attacks a neutral pyrrole monomer as in common chain polymerization, but this explanation is not generally accepted [9,10].

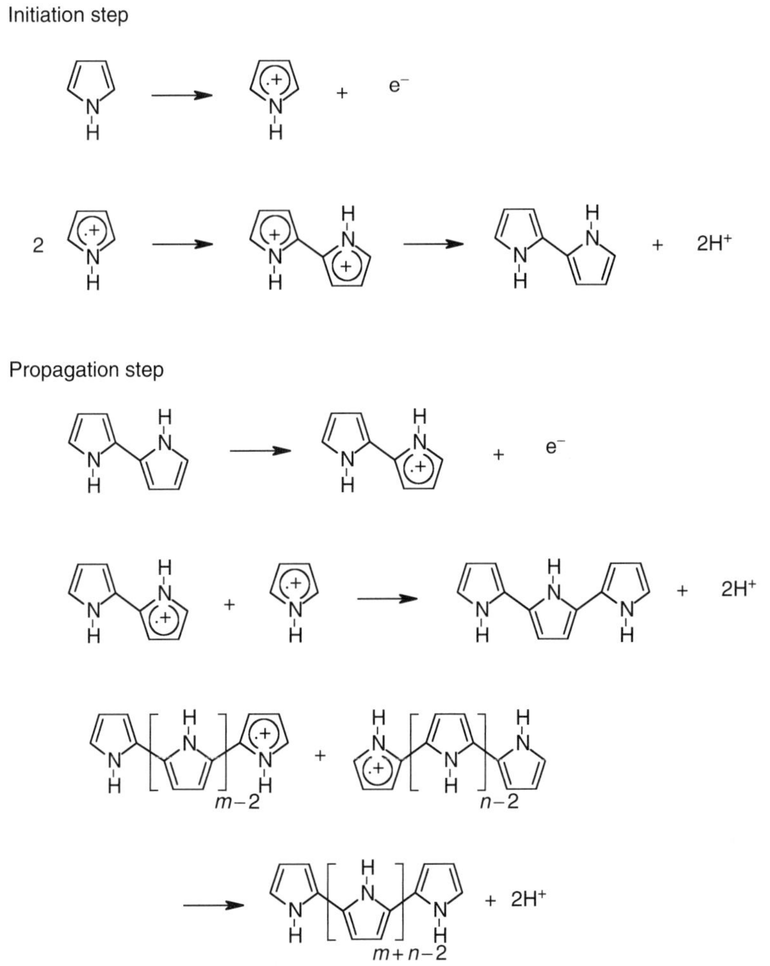

Initiation step

Propagation step

FIGURE 8.4 Polymerization mechanism of pyrrole through the coupling of two radical cations.

8.2.1 Electrochemical Polymerization

In electrochemical oxidation method, pyrrole and an electrolyte salt are dissolved in a suitable solvent and then the solution is subjected to oxidation, resulting in the growth of a conducting PPy film on the anodic working electrode. Diaz et al. [11] manufactured free-standing PPy films with excellent electrical and mechanical properties by the electrochemical method. Recently, electrochemically prepared PPy film doped with hexafluorophosphate exhibited considerably high conductivity of 2×10^3 S/cm at room temperature [6]. The electrochemical polymerization is a fast, easy, and clean method to obtain highly conductive PPy films. There are, however, some difficulties in correlating the properties of the material with the synthetic conditions and expanding the reaction system over a laboratory scale.

8.2.1.1 Electrochemical Polymerization on Oxidizable Metals

As potential materials for preventing common metals from corrosion, PPy and its derivatives have attracted much attention due to their advantageous properties such as ease of synthesis in aqueous solution and high stability in the oxidized state. In general, PPy films on inert anodes such as platinum, glassy carbon, stainless steel, and gold can be easily polymerized electrochemically in aqueous or organic media. However, in the case of electrochemical polymerization on the oxidizable metals, dissolution of metal occurs before electrochemical polymerization does, because the oxidation potentials of these metals are more negative than that of pyrrole. Therefore, much research has been focused on the conditions to passivate the metal without hindering electrochemical polymerization.

Smooth and uniform PPy coatings were obtained on low carbon steel by an aqueous electrochemical process using oxalic acid as the electrolyte at low current density [12–15], in which the steel was passivated by an iron oxalate (FeC_2O_4) interlayer formation as shown below:

$$Fe^{2+} + C_2O_4^{2-} + 2H_2O \rightarrow FeC_2O_4 \cdot H_2O$$

It was also reported that sodium salicylate produced strongly adherent PPy film on various oxidizable metals in aqueous media by one stop electrosynthesis [16]. In the process, sodium salicylate with the metal ion produces a thin protective layer, which slows dissolution of the metal without hindering pyrrole monomer oxidation.

Several parameters of electrochemical polymerization were found to affect the formation and properties of the resultant films. The thickness of the PPy coatings increased with the applied current density, while the induction time, polymerization potential, and the time needed for the dissolution of steel decreased with the current density. In acidic media, the induction (passivation) period was certainly observed before the electrochemical polymerization took place, while in alkaline media no such induction period was observed. Figure 8.5 shows the scanning electron microscope (SEM) micrographs of the interphases in the media at pH 1.4 and 6.0 [14]. As shown in Figure 8.5a, the interphase in the strongly acidic medium is composed of many crystals covering the steel substrate completely. However, no crystalline phases and only a very thin film on the steel surface is observed in the medium at pH 6.0 due to the formation of soluble iron oxides (Figure 8.5b). Uniformly smooth and strongly adherent PPy films were, therefore, obtained in low pH reaction media at low current density, while PPy films became brittle and poorly adherent as the current density and pH increased. It was also reported that treatment of iron and mild steel with 10% aqueous nitric acid prevented iron dissolution without hindering the oxidation of pyrrole, yielding thickness-controlled, strongly adherent PPy films [17].

It is even more difficult to polymerize PPy electrochemically on a zinc electrode than on iron because zinc is more electropositive than iron. It was reported that pretreatment of zinc with Na_2S resulted in homogeneous and adherent PPy films by two-step electrochemical polymerization [18]. In step I, very thin and mixed layers of ZnS and ZnO_xH_y were formed by the electrochemical treatment of Zn in 0.3 M Na_2S, whereas in step II, strongly adherent PPy films were produced by electrochemical polymerization. It was also reported that adherent PPy coating could be obtained on the zinc electrode by a one-step process when sodium sulfide was added to the electrolyte solution composed of pyrrole and oxalate [19].

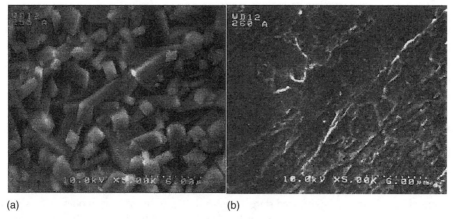

(a)   (b)

FIGURE 8.5 Scanning electron microscope (SEM) micrograph of (a) the interphase formed at 0.56 mA cm^{-2} in the medium of pH 1.4 and (b) the sample formed at 0.56 mA cm^{-2} in the medium of pH 6.0 for 30 min. (From Iroh, J.O., and Su, W., *J. Appl. Polym. Sci.*, 71, 2075, 1999. With permission.)

8.2.1.2 Electrochemical Polymerization on Modified Electrodes

Controlling various properties of the conducting polymers by highly ordered layers of organic molecules has been an attractive research area since the final characteristics of the polymer films can be controlled conveniently by modifying the electrode surface by the formation of spontaneously adsorbed monolayers [20–22]. Modifying polycrystalline Au surfaces with various kinds of ω-(N-pyrrole)alkanethiols produced interfaces influencing the morphology and properties of the resulting PPy films. When ω-(N-pyrrole)alkanethiols were used as the interfacial layer, the structure and hydrophobicity of the layer was found to be important in obstructing the electrode to ionic redox species, while the polarity and surface charge of the layer were even more important in preventing the polymer deposition [22]. PPy films formed on the ω-(N-pyrrole)alkanethiols/Au surfaces were found to be extremely smooth and strongly adherent compared with PPy films formed on pristine Au surfaces [20]. It was also reported that poly(3-ethylpyrrole) polymerized on the monolayer showed enhanced morphological dense packing, resulting in enhanced conductivity [22].

8.2.1.3 Electrochemical Polymerization of Substituted PPy

Even though PPy shows good conductivity and high stability in its oxidized form, its properties are degraded by the occurrence of α–β and β–β coupling during polymerization. Therefore, modified chemical structures with substituted 3 and 4 positions have been studied in recent years. Figure 8.6 shows the chemical structures of poly(3,4-dimethoxypyrrole) (PDMP) and poly(3,4-ethylenedioxypyrrole) (PEDP), which can be readily polymerized by electrochemical method [23]. PDMP exhibited maximum conductivity at 50% of the total charges and showed unusual solvatoconductive properties by the combination of redox type conduction and solvation.

Figure 8.7 shows the general synthesis scheme for 3,4-alkylenedioxypyrroles with especially low oxidation potential, which can be easily polymerized electrochemically to form poly(3,4-alkylenedioxypyrrole)s (PXDOPs) [24]. PXDOPs were electrochemically polymerized by multiple scan cyclic voltammetry in 0.1 M LiClO$_4$/propylene carbonate. The obtained PXDOP films exhibited outstanding electrochemical stability and color change from orange to blue with the doping state. The electrochromic property of the films made them an interesting material for display applications.

8.2.2 Chemical Polymerization

In the chemical oxidation method, an oxidizing agent such as lead dioxide, quinones, ferric chloride or persulfates is added to the pyrrole and a dopant dissolved in a suitable solvent, resulting in the precipitation of doped PPy powder. In general, the electrical conductivities of chemically prepared PPy are a little lower than those of PPy films prepared electrochemically. Nevertheless, the chemical oxidation method is suitable for commercial mass production of PPy and may produce processible PPy since the method has much greater feasibility to control the molecular weight and structural feature of the resulting polymer than the electrochemical oxidation method [25–27]. It is well known that various properties such as the electrical conductivity, stability, and morphology of synthesized PPy strongly depend on various reaction conditions such as the kinds and concentrations of oxidant and dopant, polymerization temperature and time, stoichiometry, and solvent.

8.2.2.1 Chemical Polymerization with Surfactants

To improve conductivity, stability or solubility in organic solvents, various kinds of surfactants have been used as additives in the chemical polymerization of pyrrole [28–32]. When PPy was prepared in aqueous solution containing

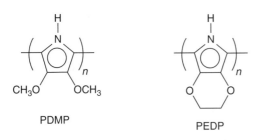

FIGURE 8.6 Structures of PDMP and PEDP. (From Zotti, G., Zecchin, S., and Schiavon, G., *Chem. Mater.*, 12, 2996, 2000. With permission.)

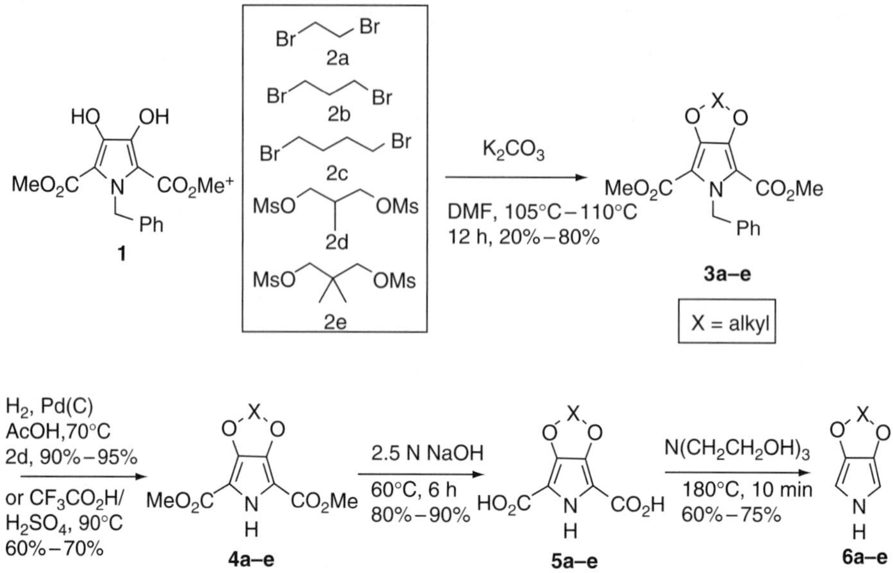

FIGURE 8.7 General reaction for the synthesis of 3,4-alkylenedioxypyrrole (XDOP) monomers. (From Schottland, P., Zong, K., and Gaupp, C.L., *Macromolecules*, 33, 7051–7061, 2000. With permission.)

$Fe_2(SO_4)_3$ as an oxidant and anionic surfactants such as sodium dodecylbenzenesulfone, sodium alkylnaphthalenesulfonate, or sodium alkylsulfonate, the resulting PPy exhibited enhanced conductivity and polymerization yield with an accelerated polymerization reaction rate [30].

When this solution contained *p*-nitrophenol as an additional additive, the resulting PPy possessed improved properties such as increased doping to a higher level with a more regular PPy backbone structure [31], which is caused by the strong interaction of the electron-withdrawing nitro group of *p*-nitrophenol with pyrrole monomer. The increased doping level seemed to result from the large-sized anions of the surfactants being effectively incorporated into PPy as the dopant. Fourier transform infrared (FT-IR) spectroscopy proved the incorporation of anionic surfactants into the PPy backbone [28,29] and SEM photographs revealed the morphological changes caused by the incorporation of the anionic surfactants (Figure 8.8) [29].

8.2.2.2 Chemical Polymerization with Sulfonic Acid Dopants

It was reported that the kind of sulfonic acid used as a dopant in chemical polymerization showed great effect on the properties of the resulting PPy such as morphology, electrical conductivity, and solubility [33]. PPy doped with camphor sulfonic acid (CSA), 5-butylnaphthalene sulfonic acid (BNSA), and *p*-methylbenzene sulfonic acid (MBSA) showed typical granular morphology, while PPy doped with β-naphthalene sulfonic acid (NSA) exhibited a fibrillar shape. The effects of different organic sulfonic acids as dopant on solubility and electrical properties are shown in Table 8.1 [33]. PPy prepared with dodecylbenzene sulfonic acid (DBSA) and NSAs possessed excellent solubility in *m*-cresol. It is believed that the solubility of PPy doped with DBSA is induced by good solvating ability of the sulfonic acid with a long alkyl group. Even though NSA does not possess a long alkyl group, the naphthalene ring may interact with the phenyl ring of *m*-cresol, thereby improving the solubility of PPy. Therefore, it can be expected that the solubility of doped PPy may be enhanced if a chemical group of sulfonic acid strongly interacts with the solvents.

8.2.2.3 In Situ Polymerization of PPy on Polymer Latex

To improve the poor processability of PPy, micrometer-sized PPy-coated polystyrene (PS) latexes were synthesized by in situ deposition process [34,35]. The schematic formation of PPy-coated latex particles

FIGURE 8.8 Scanning electron microscope (SEM) micrographs of PPy-sulfate (a and b) and PPy prepared in the presence of anionic surfactant, DBSNa (c and d). (From Omastová, M., Trchová, M., and Kovářová, J., *Synth. Met.*, 138, 447, 2005. With permission.)

is shown in Figure 8.9 [34]. Table 8.2 shows the effect of total surface area of latex on PPy loading, layer thickness, colloidal stability and electrical conductivity of the coated latex particles [34]. It was found that if the PPy layer was sufficiently thin, it could lie inside the steric stabilizer layer and the coated latexes could maintain reasonable stability. SEM study confirmed that the deposited PPy layer was very smooth and uniform at loadings lower than approximately 10% by mass. The "core-shell" morphology of the latex particles was confirmed by x-ray photoelectron spectroscopy, time-of-flight secondary ion mass spectroscopy, Raman and UV–Vis spectroscopy, and SEM [35].

8.2.2.4 In Situ Polymerization of PPy on Substrates

Highly transmissive, electroactive, and conductive poly[(3,4-alkylenedioxy)pyrrole-2,5-diyl] thin films were prepared by in situ chemical polymerization [36]. The process possesses a significant advantage as environmentally harmful organic solvents are not required since coatings can be obtained directly on the substrate. Many factors such as temperature, pH, nature of dopant, and oxidant were found to influence the film formation. PXDOPs were also synthesized from flexibly functionalized monomers, and showed highly stable aqueous compatibility and therefore unique potential for biomedical implications [37].

The effect of the substrate surface, where in situ PPy deposition process occurs by chemical oxidative polymerization, has also been investigated [38,39]. Deposition of continuous granular particles was observed on the hydrophobic surface, while spherical particles were deposited on the hydrophilic surface. The conductivity of the PPy deposited on the hydrophobic substrate was found to be $\sim10^4$ higher than that of PPy deposited on the hydrophobic surface.

8.2.2.5 Interfacial and Supercritical Fluid Polymerization

It was reported that thin PPy film could be prepared by chemical polymerization at the interface of chloroform solution of pyrrole and aqueous solution of $(NH_4)_2S_2O_8$ [40]. There are several conditions

TABLE 8.1 Solubility in *m*-Cresol and Electrical Properties of PPy In Situ Doped with Different Sulfonic Acids

Sulfonic Acid	Structure	Solubility[a] in *m*-Cresol	σ_{RT} (S/cm)	Charge Carrier
p-Methylbenzene sulfonic acid (MBSA)	 MBSA	×	16.0	Polaron and bipolaron
p-Hydroxybenzene sulfonic acid (HBSA)	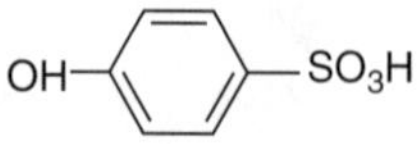 HBSA	△	11.0	Polaron and bipolaron
p-Dodecylbenzene sulfonic acid (DBSA)	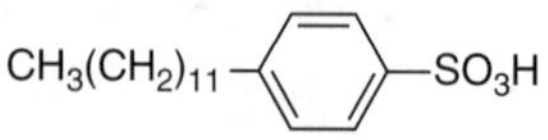 DBSA	⊕	2.0	Polaron and bipolaron
β-Naphthalene sulfonic acid (NSA)	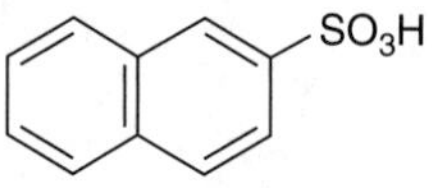 NSA	⊕	18.0	Polaron and bipolaron
5-*n*-Butylnaphthalene sulfonic acid (BNSA)	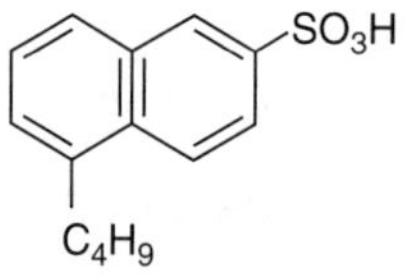 BNSA	⊕	0.5	Polaron and bipolaron
5-Sulfo-isophthalic acid (SIA)	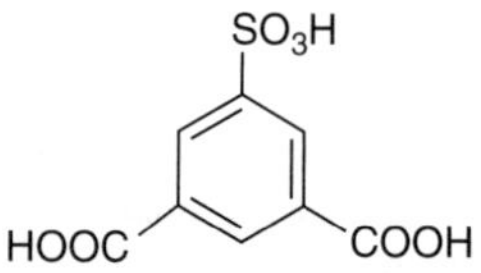 SIA	△	3.0	Polaron and bipolaron
8-Hydroxy-7-iodo-5-quinoline sulfonic acid (QSA)	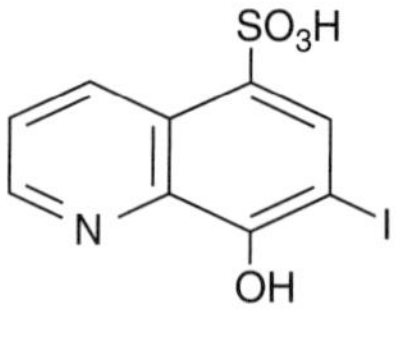 QSA	△	3.0	Bipolaron
Alizarin red acid (ARA)	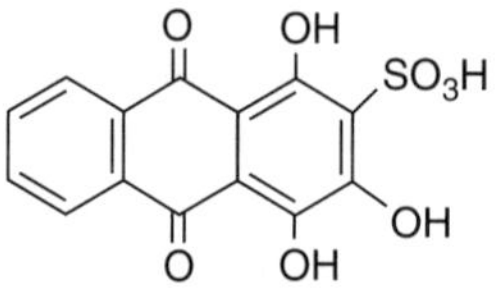 ARA	△	8	Bipolaron
Camphor sulfonic acid (CSA)	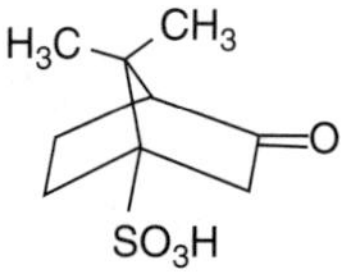 CSA	×	18	Polaron and bipolaron

[a]×, insoluble; ⊕, soluble; △, slightly soluble.

Source: Reprinted from Shen, Y., and Wan, M., *Synth. Met.*, 96, 128, 1998. With permission.

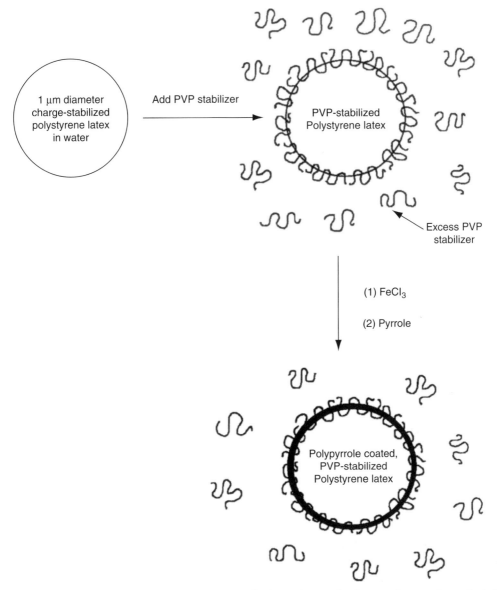

FIGURE 8.9 Schematic representation of the synthesis of polypyrrole-coated polystyrene latexes using a physically adsorbed poly(*N*-vinylpyrrolidone) stabilizer. (From Lascelles, S.F., and Armes, S.P., *J. Mater. Chem.*, 7, 1339, 1997. With permission.)

for the solvents to be used for interfacial polymerization of pyrrole. The oxidant and monomer should be soluble in different solvents for the reaction to occur only at the interface. To prevent PPy from depositing as powder or floating on the aqueous surface, the upper layer solvent should have lower density than PPy.

It was also reported that PPy could be polymerized using nontoxic supercritical fluids as solvents. In conventional chemical polymerization, incorporation of the oxidant into the polymerization process and/or washing to remove unwanted byproducts generates a large amount of environmentally hazardous solvent. In the process using supercritical fluid, which is nontoxic, nonflammable, and environmentally acceptable, PPy was polymerized within preformed polyurethane (PU) foam using supercritical carbon dioxide as solvent [41].

TABLE 8.2 Effect of Varying the Total Surface Area of Latex Available for the Deposition of Polypyrrole (PPy) on the Polypyrrole Loading and Layer Thickness, the Colloid Stability and the Electrical Conductivity of the Coated Latex Particles

Sample	Latex Surface Area/m^2	Theoretical PPy Loading (mass%)	Actual PPy Loading (mass%)[b]	Calculated PPy Layer Thickness/nm	Colloid Stability of PPy-Coated Latex[c]	σ^d/S cm^{-1}
1	0.4	52.7	51.1	164.3	Unstable	3
2[e]	1.2	25.6	25.1	59.7	Flocc'd	4
3	1.8	18.2	18.4	41.1	Flocc'd	6
4[e]	2.3	15.4	16.6	36.5	Flocc'd	3
5	3.6	10.0	9.9	20.5	Stable	2
6	5.4	6.9	6.1	12.3	Stable	2
7	7.1	5.3	5.6	11.2	Stable	2
8	7.1	4.7	4.6	10.3	Stable	0.8
9	10.0	3.4	3.5	7.8	Stable	0.2
10	15.0	2.3	2.1	4.6	Stable	6×10^{-2}
11	27.4	1.3	1.2	2.6	Stable	3×10^{-3}
12	39.4	0.9	1.0	2.2	Stable	$<10^{-6}$

[a]All polystyrene latexes were synthesized using PVP of $Mw = 44,000$ unless indicated otherwise.
[b]Determined by reduced nitrogen content relative to polypyrrole 'bulk powder' using CHN elemental microanalyses.
[c]Determined by DCP; flocc'd = flocculated.
[d]Determined by the four-point probe method on compressed pellets at room temperature.
[e]Stabilized using PVP of $Mw = 360,000$. The 1.57 μm diameter polystyrene latex was used for samples 1,3,5,6 and 7. The 1.64 μm diameter latex was used for samples 2 and 4 and the 1.80 μm diameter latex was used for samples 8–12.
Source: From Lascelles, S.F., and Armes, S.P., *J. Mater. Chem.*, 7, 1339, 1997. With permission.

8.2.3 Plasma Polymerization

Plasma polymerization has been recognized as an important process to obtain thin films of conductive polymer, which are formed by reactions in gas phase without any chemical oxidants [42]. The chemical structure of plasma-polymerized PPy is different from that obtained by conventional chemical or electrochemical polymerization processes because of fragment formation, trapped radicals, and a higher degree of branching and crosslinking [43]. Figure 8.10 represents one of the proposed PPy structures obtained by plasma polymerization, where some pyrrole rings are decomposed to form a three-dimensional (3D) network [44].

Plasma polymerization technique can also be used to obtain conducting copolymers. Aniline-pyrrole copolymer (PANI-PPy) thin films and the same copolymer with iodine added (PANI-PPy/I) were

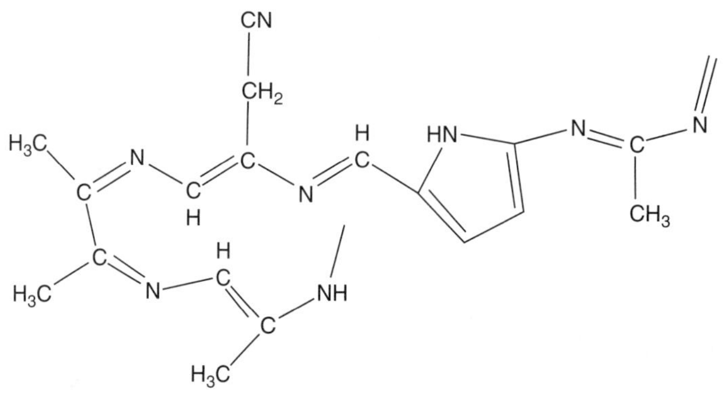

FIGURE 8.10 A structural model of PPy. (From Hosono, K., Matsubara, I., and Murayama, N., *Thin Solid Films*, 441, 72, 2003. With permission.)

synthesized by plasma process [45]. The growth rate of PANI-PPy/I was higher than that of the homopolymers, which suggests that the copolymer has the ability to include more iodine atoms than the homopolymers. Since plasma polymerization processes possess some critical limits such as high plasma energy, crosslinking, pinholes, or production of low molecular weight polymers, it has been difficult to deposit thin and continuous films of conducting polymers. Therefore, direct monomer injection method in the plasma bulk under a mild condition was proposed, by which continuous and homogeneous PPy film and pyrrole–ferrocene copolymer film were deposited on the substrate [46].

8.2.4 Copolymerization

Even though PPy may be used in many application fields, its poor processibility, mechanical, and physical properties have been a large obstacle. To improve the processibility, mechanical and other properties, various kinds of PPy copolymers have been polymerized with many conventional or conducting polymers. PPy was copolymerized electrochemically with polythiophene to improve its disadvantageous properties such as sensitivity to oxygen [47], leading to copolymer films with much less porosity than PPy film (Figure 8.11).

To improve the processibility of PPy, the sterically stabilized colloidal PPy particles have been synthesized using a tailor-made reactive copolymer [48]. Figure 8.12 represents the synthesis scheme of PPy colloids using statistical copolymer stabilizers containing reactive (bi)-thiophene groups. Figure 8.13 presents the chemical structures of the other comonomers for preparation of sterically stabilized PPy particles. PPy-silica microparticles functionalized by carboxylic acid were also synthesized using 3-substituted pyrrole comonomer (Figure 8.14) [49]. The silica sols act as high surface area colloidal substrates for PPy, leading to the production of stable colloidal dispersions of composite particles. It was reported that the copolymers of pyrrole and *m*-toluidine [50] or pyrrole and *o*-anisidine [51] possessed improved solubility.

Electroactive random copolymer was produced by the electrochemical copolymerization of pyrrole and aniline in acetonitrile in the presence of an organic acid and supporting electrolyte [52]. For the pyrrole and aniline system, changes in the order of coating and synthesis medium affected the structure and properties of the resultant samples, as shown in Table 8.3 [53]. It was also reported that there was a clear dependency of polymerization rate and yield, solubility, ability to form films, molecular weight, thermal stability, and conductivity of the copolymers on the pyrrole/ethylaniline comonomer ratio [54].

Pyrrole was copolymerized with siloxane since siloxane polymer has distinct and advantageous properties such as high temperature stability, high flexibility, and solubility in nonpolar or weakly polar solvents [55]. Figure 8.15 shows the synthesis procedure to prepare a block copolymer of pyrrole-functionalized polydimethylsiloxanes (PDMS-PPy). PDMS-PPy was polymerized by electrochemical polymerization of pyrrole on the electrode coated with the precursor PDMS-PPy.

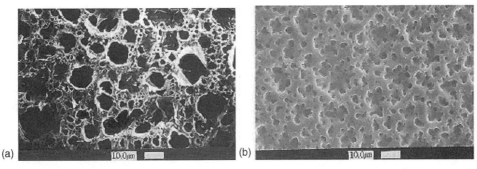

FIGURE 8.11 Scanning electron microscope (SEM) micrographs at the interface of the Pt/polymer; PF_6^- ions doped PPy porous polymer (a) and polymer composite of two monomers (b). (From Cha, S.K., *J. Polym. Sci. Part B Polym. Phys.*, 35,165, 1997. With permission.)

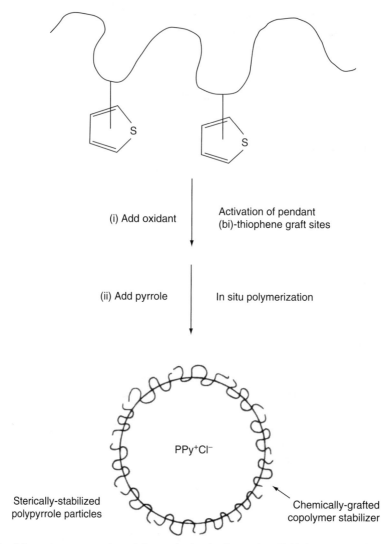

FIGURE 8.12 Schematic representation of the synthesis of polypyrrole colloids by aqueous dispersion polymerization using statistical copolymer stabilizers which contain reactive (bi)-thiophene groups. (From Simmons, M.R., Chaloner, P.A., and Armes, S.P., *Langmuir*, 14, 611, 1998. With permission.)

8.3   Soluble Polypyrrole

In the initial stages, conducting polymers were expected to possess properties of both common polymers (low density, flexibility, processibility, and low cost) and conductors or semiconductors. However, they tend to be insoluble and infusible since molecular interactions between conjugated polymers composed of ring-based backbones are relatively strong compared with the van der Waals force or hydrogen bonding between saturated polymers. The ionic interactions between doped regions on conducting polymer chains further increase the molecular interactions, making the conducting polymers even more intractable. As with most of the conducting polymers, PPy, polymerized either electrochemically or chemically, has been known to be insoluble and infusible due to the strong inter- or intramolecular interaction and crosslinking.

Research has been, therefore, focused on the development of new soluble or fusible PPy. A major breakthrough came with the modification of the chemical structure of PPy by adding solubilizing

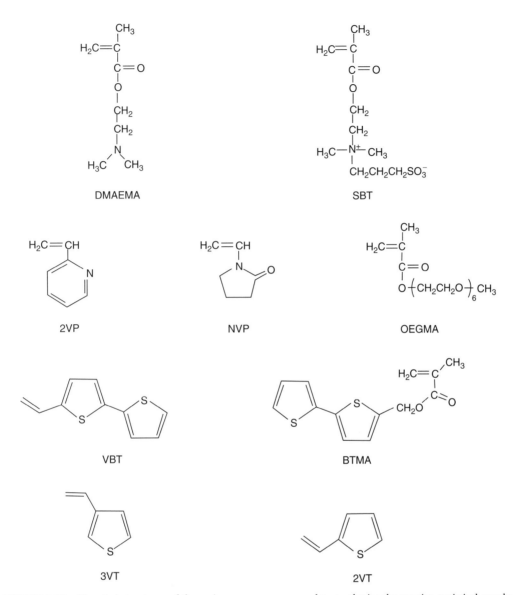

FIGURE 8.13 Chemical structures of the various comonomers used to synthesize the reactive statistical copolymers required for the preparation of sterically stabilized polypyrrole particles. (From Simmons, M.R., Chaloner, P.A., and Armes, S.P., *Langmuir*, 14, 611, 1998. With permission.)

substituents onto the pyrrole monomer unit. PPy derivatives containing alkyl or alkoxy substituents at the 1, 3, or 4 positions were synthesized to be soluble in chloroform, tetrahydrofuran and *o*-dichlorobenzene [56–58]. Since the bulky side groups reduce the molecular interactions between main chains and increase chain entropy, the substituted conducting polymers can be processed from the solution. However, as the addition of substituents induces a steric interference, the planarity of the polymer structure is destroyed to the detriment of the conductivity by the reduction of the degree of π-orbital overlap and obtainable conductivity. It was reported that the electrical conductivity of poly(3-alkylpyrrole) decreased with the size of the alkyl chain [58]. In the copolymerization of pyrrole and substituted pyrrole, a higher content of substituted pyrrole moiety also yielded a lower copolymer conductivity [59–61].

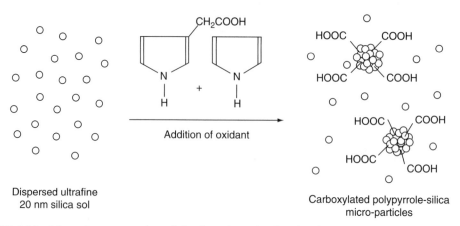

FIGURE 8.14 Schematic representation of the formation of carboxylated polypyrrole-silica microparticles by copolymerization of pyrrole-3-acetic acid with pyrrole using the hydrogen peroxide-based oxidant in the presence of an ultrafine silica sol. (From McCarthy, G.P., et al., *Langmuir*, 13, 3686, 1997. With permission.)

Great progress in the processing of conducting polymers has been achieved in one of the most well-known materials, polyaniline (PANi). PANi is the first conducting polymer found to be intrinsically soluble in organic solvents both in the insulating emeraldine base form and the conducting emeraldine salt form. It was reported that neutral PANi in the form of emeraldine base is soluble in 1-methyl-2-pyrrolidinone (NMP) [26]. PANi dissolved in NMP was processed in its blue emeraldine base form to make films, fibers, and blends, which had to be redoped to form the conducting emeraldine salt [26,62,63]. It was reported that PANi doped by CSA or DBSA was also soluble in various organic solvents in the conducting emeraldine salt form [64–66]. Since PANi is processed in the conducting form, no more doping processes are necessary for the final product. The solubility of PANi is distinguished from that of other conducting polymers by its simplicity. It is achieved without complicated processes such as the modification of the chemical structure of the monomer or careful design of the synthetic route.

Before the synthesis of soluble PPy in 1995, general applications using PPy were limited by its poor processibility in spite of some specific applications. However, a number of investigations have been

TABLE 8.3 Conductivity and Gouy Balance Measurements of Different Polymer Layers[a]

Polymer	Medium	Supporting Electrolyte	Conductivity ($\times 10^3$ S cm^{-1})	Gouy Balance Measurements
Pan	H_2O	0.2 M H_2SO_4	520	−36
PPy	Acetonitrile	0.5 M LiClO$_4$	130	−41
Pan/PPy	H_2O	0.2 M H_2SO_4		
	Acetonitrile	0.5 M LiClO$_4$	8.2	−35
PPy/PAn	Acetonitrile	0.5 M LiClO$_4$		
	H_2O	0.2 M H_2SO_4	44	−35
Pan	H_2O	0.2 M HclO$_4$	7.5	−34
PPy	H_2O	0.2 M HclO$_4$	0.36	−35
Pan/PPy	H_2O	0.2 M HclO$_4$		
	H_2O	0.2 M HclO$_4$	4.6	−33
PPy/PAn	H_2O	0.2 M HclO$_4$		
	H_2O	0.2 M HclO$_4$	7.9	−32
PPy/PPy	H_2O	0.2 M HclO$_4$		
	Acetonitrile	0.5 M LiClO$_4$	0.55	−37
PPy/PPy	Acetonitrile	0.5 M LiClO$_4$		
	H_2O	0.2 M HclO$_4$	0.038	−33

[a]Conductivities measured with four-probe method.

Source: Reprinted from Sari, B., and Talu, M., *Synth. Met.*, 94, 222, 1998. With permission.

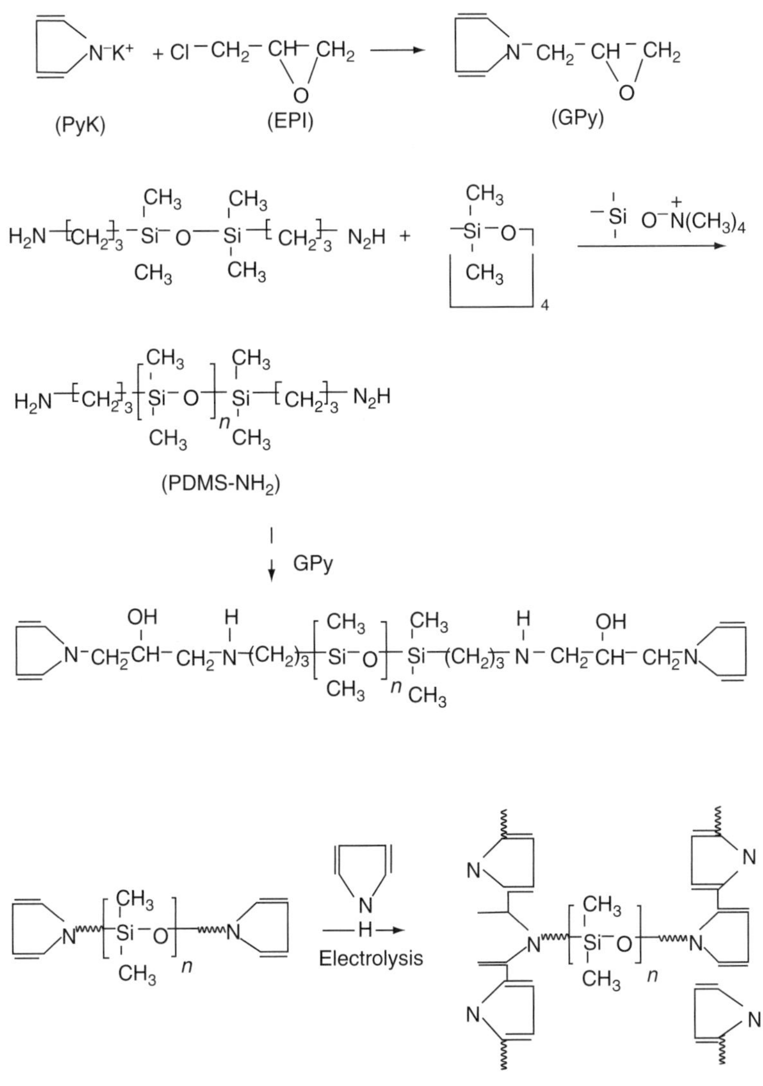

FIGURE 8.15 Synthesis scheme of block copolymer films of *N*-pyrrolyl-terminated polysiloxane and pyrrole. (From Kalaycioglu, E., et al., *Synth. Met.*, 97, 7, 1998. With permission.)

carried out on soluble PPy since Lee et al. [27] reported that soluble PPy in its doped state could be obtained by chemical polymerization in an aqueous medium with ammonium persulfate (APS) and DBSA as an oxidant and a dopant source, respectively. PPy doped with DBSA anion (PPy-DBSA) is readily dissolved in highly polar solvents such as *m*-cresol, NMP, and dimethyl sulfoxide (DMSO). PPy-DBSA can also be dissolved in weakly polar solvents including chloroform, chlorobenzene, tetrachloromethane, dichloromethane, xylene, and 1,1,2,2-tetrachloroethane with the addition of a certain amount of extra DBSA.

The solubility of PPy is similar to that of PANi in a broad sense as solubility is induced by the introduction of a bulky counter-anion. However, PPy is found to be soluble only when it is doped with DBSA at a reasonable doping level, while PANi is soluble when it is either undoped or doped. Since PPy is processed in the doped state, no postdoping process is required for the PPy films, coatings and blends prepared from the solution. The use of extra DBSA to dissolve PPy-DBSA in some solvents distinguishes the solubilization of PPy from that of PANi. The term "extra" means that the added DBSA is not identical to the dopant DBSA anion, but is rather used as a co-solvent.

8.3.1 Synthesis and General Characteristics of Soluble PPy

8.3.1.1 Synthesis Procedure for Soluble PPy

Soluble PPy can be polymerized by a relatively simple chemical oxidation procedure [27,67] beginning with the dissolution of 0.30 mol (20 g) of pyrrole and 0.15 mol (48 g) of DBSA in 400 mL of distilled water with vigorous stirring, forming a milky monomer emulsion. Various amounts of APS as an oxidant dissolved in distilled water were slowly added to the monomer solution. Polymerization was performed for various durations at various temperatures and then terminated by pouring excess methanol. The resultant PPy powder was filtered and washed sequentially with distilled water, methanol, and acetone several times, followed by drying in vacuum at room temperature for 12 h.

8.3.1.2 General Characteristics of Soluble PPy-DBSA

It was observed that PPy-DBSA was readily dissolved in *m*-cresol, NMP, and DMSO and that it could be dissolved in chloroform, chlorobenzene, tetrachloromethane, dichloromethane, xylene, and 1,1,2,2-tetrachloroethane with the addition of a certain amount of extra DBSA. The solubility of PPy-DBSA in both types of solvents was almost 100% up to a solution concentration of 7 wt%. The solution could be filtered through the Teflon membrane filter with a pore size of 0.5 μm, leaving almost no residue in the filter. As shown in Figure 8.16, PPy-DBSA thin films cast from *m*-cresol and chloroform solution exhibited an extraordinarily smooth surface, while PPy-DBSA prepared electrochemically showed fine protrusions [68]. The tensile strength and modulus of PPy-DBSA free-standing film cast from chloroform solution were measured to be 17 MPa and 1.9 GPa, respectively, which are in the same order of magnitudes as those of general polymers or PANi cast film [27]. All these properties indicate that PPy-DBSA can be dissolved in organic solvents to form a homogeneous PPy solution. The PPy cast film thickness can control the transmittance in the visible region and surface resistivity of the film (Figure 8.17) [68].

FT-Raman spectra of PPy films shown in Figure 8.18 confirm that soluble PPy possesses the same chemical structure as the electrochemically synthesized insoluble PPy in the doped state [27]. The electronic absorption band structure of the soluble PPy is also almost the same as that of the electrochemically polymerized PPy film (Figure 8.19), indicating the highly conducting, electronic structure of the soluble PPy [27]. Even though the chemical structures of PPy-DBSA prepared chemically and electrochemically are the same, the chemically synthesized soluble PPy becomes soluble in organic solvents, not only because the large dopant (DBSA⁻) reduces inter- and intramolecular interactions by placing itself between the polymer molecules, but also because the slow polymerization at low temperature retards the crosslinking reaction.

8.3.1.3 Synthesis Parameters to Determine the Solubility of PPy-DBSA

It was found that three parameters—oxidant concentration, polymerization temperature, and polymerization time—are the key factors to determine the electrical conductivity and solubility of the resulting PPy. The electrical conductivity of PPy increases with increasing oxidant concentration and with decreasing polymerization temperature (Figure 8.20) [67]. When PPy was polymerized with 0.2 mol ratio of APS/pyrrole at −2°C, the conductivity of the resulting PPy reached 10 S/cm, which is very close to that of PPy obtained electrochemically with DBSA anion as the dopant. The increased conductivity obtained by polymerization of PPy at lower temperature with higher APS/pyrrole mol ratio was caused by the increased number of radical cations of PPy monomer or oligomer resulting from the raised oxidant concentration, which increased the PPy chain length. In addition, polymerization temperature controls the reaction rate, which should be slowed down to obtain a molecular structure with high linearity. As can be seen in Figure 8.21, since the conductivity sharply increases with increasing intrinsic viscosity, which itself increases with increasing polymer molecular weight, it is believed that the molecular weight of PPy plays an important role in the electrical conductivity [67].

It was also reported that polymerization time significantly affected the electrical conductivity and solubility of the resulting PPy-DBSA, as summarized in Table 8.4 [68]. Doping level as well as the apparent yield, defined as the ratio of PPy-DBSA/monomer feed, increased with increasing polymerization time

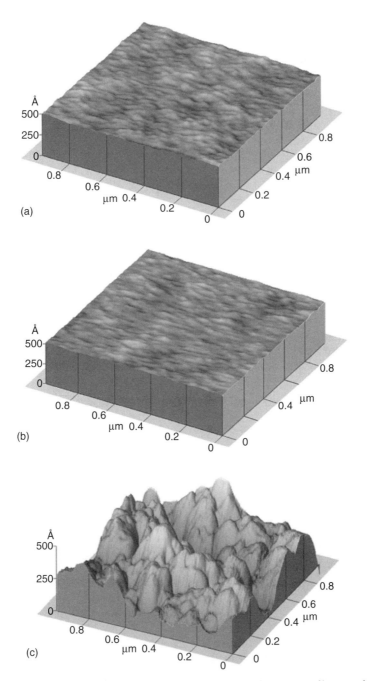

FIGURE 8.16 Atomic force microscopy (AFM) images of the surfaces of PPy-DBSA films cast from solutions in (a) NMP, (b) DBSA/chloroform, and (c) prepared electrochemically. (From Song, K.T., Synthesis of electrically conducting soluble polypyrrole and its characterization. Ph.D. thesis, 2000. With permission.)

over the range from 5 min to 4 h. However, further polymerization produced little increase in the doping level. Since the doping level increased with increasing polymerization time, the actual yield, which is defined as the ratio of pyrrole moiety only in PPy-DBSA/monomer feed, did not increase with extended polymerization time after 30 min. This implies that the polymer chain grows quite rapidly, while doping takes place relatively more slowly, possibly due to the slow diffusion of large-size dopant molecules. It was

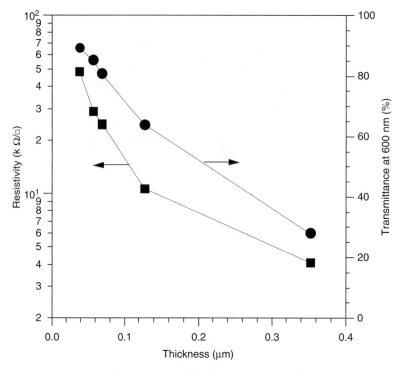

FIGURE 8.17 Transmittance and surface resistivity of PPy-DBSA spin-coated films as a function of film thickness. (From Song, K.T., Synthesis of electrically conducting soluble polypyrrole and its characterization. Ph.D. thesis, 2000. With permission.)

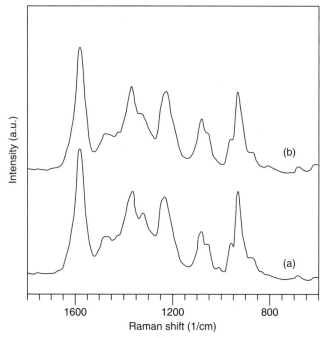

FIGURE 8.18 FT-Raman spectra of (a) chemically polymerized soluble and (b) electrochemically polymerized insoluble polypyrrole. (From Lee, J.Y., Kim, D.Y., and Kim, C.Y., *Synth. Met.*, 74, 103, 1995. With permission.)

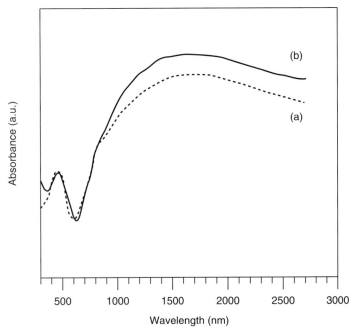

FIGURE 8.19 UV–Vis–NIR spectra of (a) chemically polymerized soluble and (b) electrochemically polymerized insoluble polypyrrole. (From Lee, J.Y., Kim, D.Y., and Kim, C.Y., *Synth. Met.*, 74, 103, 1995. With permission.)

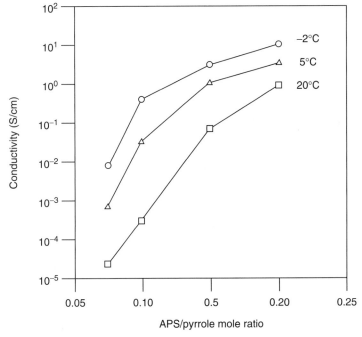

FIGURE 8.20 Conductivity of PPy synthesized with different concentrations of ammonium persulfate. (From Lee, J.Y., et al., *Synth. Met.*, 84, 137, 1997. With permission.)

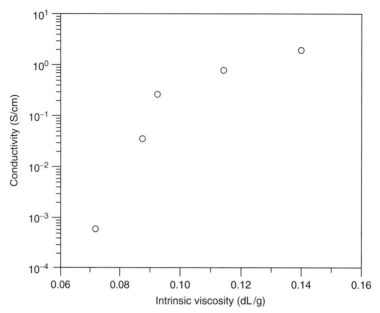

FIGURE 8.21 Conductivity as a function of intrinsic viscosity of the PPy solution in chloroform. (From Lee, J.Y., et al., *Synth. Met.*, 84, 137, 1997. With permission.)

also observed that increasing polymerization time up to 4 h led to the decrease in the conductivity of PPy powder from 5.8 to 2.5×10^{-1} S/cm and that PPy powder exhibited solubility only after the doping level exceeded about 20%. This indicates that the intermolecular interaction between PPy chains can be efficiently reduced only when PPy is doped with DBSA anions at a doping level above 20%. However, such a reduction of intermolecular interaction between PPy chains decreases the interchain conduction of charge carriers, resulting in the decreased bulk conductivity of PPy as suggested [69].

8.3.2 Solubility of PPy-DBSA

8.3.2.1 Solvents for PPy-DBSA

The solubility of PPy-DBSA powder in various solvents is listed in Table 8.5 [68]. It was observed that the solvent polarity index could be considered as an important parameter to determine the

TABLE 8.4 Properties of PPy-DBSA Powders as a Function of Polymerization Time

Polymerization Time	Apparent Yield (%)	True Yield (%)	Conductivity (S/cm)	Doping Level (mol%)	Solubility in NMP
5 min	8.4	5.2	5.8	16.9	X
10 min	9.4	6.2	5.0	14.8	X
15 min	11.6	7.1	3.5	16.9	X
30 min	12.6	7.2	1.3	18.8	X
1 h	14.6	8.6	6.2×10^{-1}	17.8	X
2 h	15.2	8.4	3.3×10^{-1}	19.5	O
4 h	14.6	7.8	2.5×10^{-1}	20.4	O
8 h	14.5	7.9	2.1×10^{-1}	19.9	O
24 h	15.2	7.9	2.0×10^{-1}	21.0	O
48 h	15.0	7.7	1.3×10^{-1}	21.3	O

Source: From Song, K.T., Synthesis of electrically conducting soluble polypyrrole and its characterization. Ph.D. thesis, 2000.

solvation of the pristine PPy-DBSA powder. PPy-DBSA powder was readily dissolved in polar solvents with high polarity index such as *N,N'*-dimethylformamide (DMF), NMP, DMSO, and *m*-cresol, resulting in a homogeneous solution, but only partially dissolved in solvents with intermediate polarity such as nitrobenzene, quinoline, and benzyl alcohol, and completely undissolved in nonpolar or weakly polar solvents characterized by low polarity index such as tetrachloromethane, xylene, chlorobenzene, dichloromethane, and chloroform. The good solubility of PPy-DBSA in polar solvents indicates a strong molecular interaction between PPy-DBSA and the polar solvents, whereas the insolubility in weakly polar solvents indicates lack of sufficient molecular interaction between PPy-DBSA and the solvents. Although there is minimum interaction between PPy-DBSA and weakly polar solvents, it is not strong enough to perturb the interaction between PPy-DBSA molecules. The critical polarity index that distinguishes the solubility of PPy-DBSA has been reported to range between 4.4 and 5.7.

Even though the weakly polar solvents are not able to dissolve the PPy-DBSA powder by themselves, the solvation power of the weakly polar solvents can be dramatically enhanced with the addition of a certain amount of extra DBSA, as shown in Table 8.5. It was also found that more extra DBSA should be added to the solvent in order to obtain a concentrated PPy-DBSA solution with homogeneity, which suggests that the molecular interaction between the extra DBSA and PPy-DBSA causes the solubilization of PPy-DBSA in weakly polar solvents. On the other hand, the addition of extra DBSA to the intermediately or highly polar solvents does not obviously promote the solvation powers of these solvents. The solvation power of NMP and DMSO is even decreased by the addition of extra DBSA, possibly because the extra DBSA perturbed the molecular interaction between PPy-DBSA and the polar solvent. PPy treated with 0. 1 N NaOH aqueous solution lost its solubility completely since some dopant anions were eliminated from the PPy-DBSA molecules.

It can, therefore, be concluded that the solubility of the pristine PPy-DBSA powder can be attributed to two important factors: the implantation of the bulky dopant anion that reduces the inter- or intrachain interactions in PPy and the carefully controlled chain length and structure that determines the solubility of PPy-DBSA. The electrochemical synthesis is also able to produce PPy-DBSA films with high conductivity. However, these films are totally insoluble in organic solvents, resulting from the high molecular weight of PPy and the crosslinks formed between PPy chains during rapid electrochemical polymerization. Therefore, in order to synthesize soluble PPy, the synthesized PPy needs bulky dopants that can reduce the inter- or intrachain interactions in PPy and the conditions should permit control of the chain length and structure.

TABLE 8.5 Solvent and Solubility of As-Polymerized PPy-DBSA Powder

Solvent	Polarity Index	Solubility of the PPy-DBSA Powder	
		Without Extra DBSA	With Extra DBSA
Tetrachloromethane	1.6	X	O
Xylene	2.5	X	O
Chlorobenzene	2.7	X	O
Dichloromethane	3.1	X	O
Chloroform	4.1	X	O
Nitrobenzene	4.4	△	△
Quinoline	5.0	O	O
Benzyl alcohol	5.7	△	△
Dimethyl formamide	6.4	O	O
N-methyl-2-pyrrolidinone	6.7	O	X
Dimethyl sulfoxide	7.2	O	X
m-Cresol	7.4	O	O

O: soluble; △: partially soluble; X: insoluble.
Source: From Song, K.T., Synthesis of electrically conducting soluble polypyrrole and its characterization. Ph.D. thesis, 2000.

8.3.2.2 Solvation Mechanism of PPy-DBSA

The polarity index of a solvent is a measure of the strength of a solvent or its ability to dissolve various polar solutes or to interact with them through hydrogen bonding or dipole interaction. In a chemical sense, PPy-DBSA is a sort of charge-transfer complex with oppositely charged ions, one of which is the positively charged PPy chain while the other is the dopant anion. Such an ionic characteristic produces the nonzero dipole moment around the doping spot on the PPy chain. Consequently, the PPy-DBSA molecules can be considered as a polar solute in a solution.

When PPy-DBSA powder is added to a polar solvent, Coulomb interactions between cationic PPy chains and anionic DBSA molecules may induce the close approach of the polar solvent molecules to the doping spot. New molecular interactions between solvent and PPy chain or solvent and dopant will be developed through hydrogen bondings or dipole interactions, which lead to the solvation of the PPy-DBSA. Since PPy-DBSA is dissolved in polar solvents at the price of partial loss in polymer–dopant interaction, the electrical properties of PPy-DBSA processed from the polar solvents worsened.

Intrinsically polar PPy-DBSA prevents itself from interacting with weakly polar solvents such as chloroform and PPy-DBSA, therefore, shows complete insolubility in the weakly polar solvents. The solvation of PPy-DBSA in a nonpolar solvent occurs with the addition of extra DBSA because the polarity difference between the PPy-DBSA and the solvent is adjusted by the extra DBSA. DBSA has a typical surfactant structure containing both a polar sulfonic group and a nonpolar alkyl group. Thus, DBSA molecule has good affinity with both polar and nonpolar molecules simultaneously. When a certain amount of PPy-DBSA powder is added to nonpolar solvents containing extra DBSA, the polar and nonpolar groups in the extra DBSA molecule interact with the doped PPy molecule and nonpolar solvent molecule, respectively. With the aid of the extra DBSA, the PPy-DBSA molecules are able to undergo second hand interactions with weakly polar solvent molecules, forming a homogeneous solution. As a result of the molecular interaction between the extra DBSA and the PPy-DBSA in a weakly polar solvent, a unique structure is evolved in the solution.

Figure 8.22 shows the change of hydrodynamic radius of PPy-DBSA in chloroform containing different amounts of extra DBSA [68]. The hydrodynamic radius of PPy-DBSA with the powder conductivity of 1 S/cm increased from 27 to 40 nm when the concentration of extra DBSA was increased from 0.02 to 0.5 g/dL. Further increase of the extra DBSA concentration beyond 1 g/dL resulted in the saturation of the hydrodynamic radius of PPy-DBSA at around 45 nm.

The hydrodynamic radius over 40 nm is too large to be considered as the radius of a single polymer molecule. Although some common polymers such as PS exhibit a hydrodynamic radius of about 40 nm, this is an extreme case that is observed only in a good solvent such as benzene with a molecular weight of several hundred thousand g/mol. The soluble PPy-DBSA is, however, hardly believed to have such a high molecular weight and exhibits a good interaction with solvent. Consequently, this unexpectedly large hydrodynamic volume indicates that the extra DBSA molecules form a micelle-like structure with PPy-DBSA molecules in which a small number of PPy-DBSA molecules are surrounded by a large number of extra DBSA molecules.

The saturation of the hydrodynamic radius beyond the extra DBSA concentration of 1 g/dL implies that there is an optimum radius of the micelle-like structure of PPy-DBSA, over which the structure becomes energetically unstable. The radius of the PPy-DBSA micelle-like structure increased with increased PPy-DBSA conductivity (Figure 8.23) [68]. PPy-DBSA is expected to have different conductivities depending on by the PPy chain length. As the chain length increases, the core volume of the PPy-DBSA micelle-like structure increases, thereby increasing the total volume of the micelle-like structure.

While collecting data from dynamic light scattering analysis of PPy-DBSA solutions, the obtained autocorrelation functions were exactly fitted with single exponential decay, as shown in the inset of Figure 8.22 [68]. This indicates that the micelle-like structure of PPy-DBSA has a very narrow size distribution close to mono-dispersity [71]. It was reported that PANi formed aggregates in NMP by the couplings between PANi molecules through hydrogen bonding. The formed PANi aggregates have a relatively broad size distribution and the autocorrelation function fitted with the Gaussian decay [72].

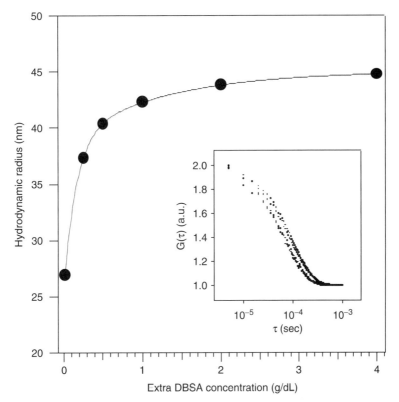

FIGURE 8.22 Hydrodynamic radius of PPy-DBSA in chloroform containing different amounts of extra DBSA at 25°C, where the concentration of PPy-DBSA is 0.01 g/dL. The inset shows autocorrelation functions obtained from the dynamic light scattering analysis. (From Song, K.T., Synthesis of electrically conducting soluble polypyrrole and its characterization. Ph.D. thesis, 2000. With permission.)

If the PPy-DBSA micelle-like structure were constructed by the aggregation of the multiple PPy-DBSA molecules, the obtained autocorrelation functions should be fitted by Gaussian decays as the aggregate of PANi does [72,73]. Therefore, it is expected that the micelle-like structure of PPy-DBSA is composed of a single chain of PPy-DBSA surrounded by many extra DBSA molecules.

The micelle-like structure of PPy-DBSA exists only in weakly polar solvents with extra DBSA in which the polymer–solvent interactions are relatively weak. The hydrodynamic volume of PPy-DBSA in polar solvents was too small to obtain an autocorrelation function on the available relaxation time scale. This failure was not improved by the addition of extra DBSA to the polar solvents because the molecular interaction between PPy-DBSA and the extra DBSA was limited by the polar solvents. On the other hand, this failure implies that PPy-DBSA exists as an isolated chain in polar solvents.

8.3.3 Effect of Solvent on Charge Transport of PPy-DBSA

Basically, the charge transport properties of a conducting polymer strongly depend on its electronic structure which is determined by the chemical and conformational structures of the polymer. Change of doping level or overoxidation of polymer backbone can be regarded as a change in the chemical structure of the conducting polymer, while chain extension with mechanical stretching is a change in conformational structure of the polymer.

8.3.3.1 Effect of Solvent on PPy Properties in Solution

In the case of soluble PPy, it has been reported that the solvent is a very important factor affecting the electronic structure of the polymer because the chemical or conformational structure of the PPy chain is

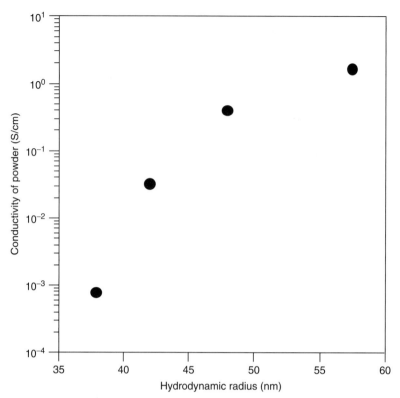

FIGURE 8.23 Electrical conductivity of PPy-DBSA as a function of hydrodynamic radius in chloroform containing 4 g/dL of extra DBSA at 25°C, where the concentration of PPy-DBSA is 0.01 g/dL. (From Song, K.T., Synthesis of electrically conducting soluble polypyrrole and its characterization. Ph.D. thesis, 2000. With permission.)

significantly influenced by the interaction of the solvent with the PPy molecule. PPy-DBSA is dissolved through two different solvation mechanisms depending on the polarity of the solvent as described in Section 8.3.2.2 PPy-DBSA dissolves in solvents with high polarity by partially sacrificing the polymer–dopant interaction, resulting in a less extended conformational structure. On the contrary, PPy-DBSA dissolves in weakly polar solvents containing extra DBSA by forming a micelle-like structure in the solution, yielding a more extended structure.

 Therefore, it is obvious that PPy-DBSA properties in solution are greatly affected by the solvent polarity. Figure 8.24 shows the UV–Vis–NIR spectra of PPy-DBSA solutions prepared with polar solvents such as *m*-cresol, DMSO, and NMP [68]. The three spectra resemble each other, and are composed of one major peak in the region near UV, two minor peaks in the visible region, and weak free-carrier absorption in the near IR region. The resemblance of the three spectra implies the similarity of the electronic structures of PPy-DBSA induced by the three polar solvents. The weak free-carrier absorption indicates that the PPy chain in these solvents has an electronic structure unfavorable for intrachain charge transfer. On the contrary, the PPy-DBSA solutions prepared using weakly polar solvents such as chloroform, chlorobenzene, and tetrachloromethane containing a certain amount of extra DBSA exhibit much stronger free-carrier absorption in the near IR region (Figure 8.25), implying that the PPy chain in the solution possesses an electronic structure favored for intrachain conduction [68]. The three spectra of PPy-DBSA in the weakly polar solvents also resemble each other very closely.

 The major peak near the UV region changes its position and shape according to the solvent polarity. The solutions prepared using polar solvents show a sharp major peak at 422 (*m*-cresol), 424 (DMSO),

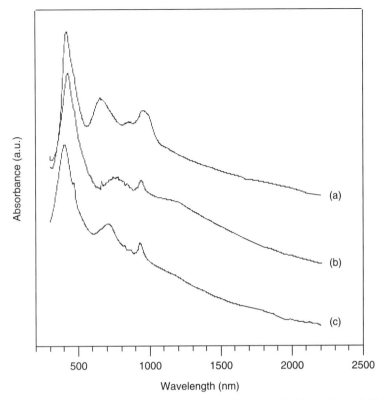

FIGURE 8.24 UV–Vis–NIR spectra of PPy-DBSA solutions in (a) *m*-cresol, (b) DMSO, and (c) NMP with the polymer concentration of 0.01 g/dL. (From Song, K.T., Synthesis of electrically conducting soluble polypyrrole and its characterization. Ph.D. thesis, 2000. With permission.)

and 406 (NMP) nm, which is associated with π–π* electronic transition in PPy. This indicates that some of the PPy segments in polar solvents exist in the neutral state. On the other hand, PPy-DBSA solutions prepared using weakly polar solvents with extra DBSA show a relatively broad major peak at the longer wavelength of 448 (chloroform), 440 (chlorobenzene), and 442 (tetrachloromethane) nm, which is associated with bipolaronic transition in PPy, indicating that the oxidized state is predominant in PPy segments. In the solution spectra in weakly polar solvents, π–π* transition is observed as just a small shoulder at 350 nm, confirming that the concentration of neutral species in PPy is very low.

The minor peaks observed at wavelengths between 600 and 1200 nm must be associated with some localized states formed only in solution state since they disappear in the spectra of solid PPy films. Minor peaks are predominant in the PPy-DBSA solution spectra in polar solvents compared to those in weakly polar solvents. Chloroform, which has the most polar character among the weakly polar solvents, revealed the best-developed minor peaks at these wavelengths. It is, therefore, concluded that these minor peaks originate from some localized states induced by the polar interactions between PPy-DBSA and the solvent.

Addition of extra DBSA to a polar solvent yields the obvious increase of free-carrier absorption in the PPy-DBSA solution spectrum (Figure 8.26) [68]. As the concentration of extra DBSA increased, the peak at 410 and 750 nm in the spectrum (a) was reduced dramatically and a new absorption shoulder was developed at around 1000 nm. The peak at 410 nm in the spectrum (a) was also shifted to the longer wavelength with the increasing extra DBSA concentration. The PPy-DBSA solution spectrum with 8 g/dL of extra DBSA in NMP is almost identical to the spectrum of a solution prepared with weakly polar solvents containing extra DBSA. These spectral changes induced by the addition of extra DBSA to

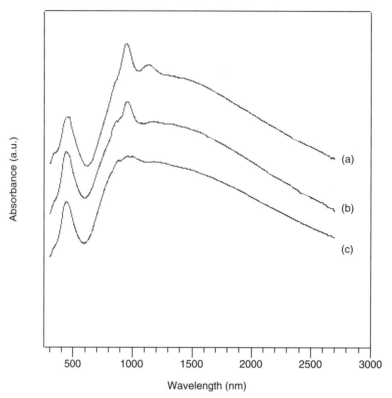

FIGURE 8.25   UV–Vis–NIR spectra of PPy-DBSA solutions in (a) chloroform, (b) chlorobenzene, and (c) tetra-chloromethane containing 0.02 g/dL of extra DBSA with the polymer concentration of 0.01 g/dL. (From Song, K.T., Synthesis of electrically conducting soluble polypyrrole and its characterization. Ph.D. thesis, 2000. With permission.)

a polar solvent indicate that there are some molecular interactions between PPy-DBSA and extra DBSA even in the polar solvent. The development of a new, free-carrier tail in the near IR region implies that the electronic structure of PPy is transformed to a more extended structure favored for the intrachain charge transport. In fact, the PPy-DBSA film cast from NMP solution containing extra DBSA showed increased electrical conductivity by one order of magnitude compared with the conductivity of the film obtained from extra DBSA-free solution. Nevertheless, molecular interaction between PPy-DBSA and extra DBSA is substantially obstructed by the polar solvent because of stronger interaction between PPy-DBSA and the solvent.

The UV–Vis–NIR spectra of a PPy-DBSA solution in chloroform changed little with increasing concentration of the extra DBSA (Figure 8.27), indicating that the increase of extra DBSA concentration had no influence upon the electronic structure of PPy-DBSA since extra DBSA restricts the position of PPy-DBSA at the core of the micelle-like structure [68]. Moreover, due to additional doping with extra DBSA at the core of the micelle-like structure, the PPy-DBSA chain requires more extended conformation, which is favorable for intrachain charge transport.

8.3.3.2   Effect of Solvent upon Film Properties

Figure 8.28 and Figure 8.29 show the UV–Vis–NIR spectra of PPy-DBSA films cast from the solutions prepared using polar and weakly polar solvents, respectively [68]. Each film spectrum was apparently similar to its solution spectrum shown in Figure 8.24 and Figure 8.25, implying a strong correlation between the solution properties and the final solid properties. The electronic structure of PPy-DBSA determined within the solution is preserved after extracting solvent from the solution. All the spectra of PPy-DBSA films showed much smoother absorption spectra than solution spectra. The minor peaks

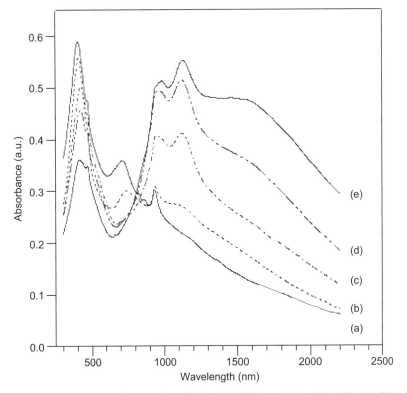

FIGURE 8.26 UV–Vis–NIR spectra of PPy-DBSA solutions in NMP containing (a) 0, (b) 0.02, (c) 0.1, (d) 2, and (e) 8 g/dL of extra DBSA, where the polymer concentration is 0.01 g/dL. (From Song, K.T., Synthesis of electrically conducting soluble polypyrrole and its characterization. Ph.D. thesis, 2000. With permission.)

were greatly reduced because in the film there were no localized states which existed in a solution induced by the polar interaction between PPy-DBSA and solvent.

Electrical conductivities and doping levels of the PPy-DBSA films cast from the solutions are listed in Table 8.6 [68]. The conductivities and doping levels of PPy-DBSA cast films can be obviously divided into two groups according to the solvent polarity. The PPy-DBSA films prepared with polar solvents have lower conductivity by two orders of magnitude than the films prepared using weakly polar solvents with extra DBSA. The PPy-DBSA films cast from weakly polar solvents show higher conductivity than the pristine PPy-DBSA powder.

The PPy-DBSA films prepared with polar solvents show a reduction in doping level because some labile dopants could be dissociated from PPy chain in the solution by formation of new strong interaction between the PPy chain and solvents perturbing the molecular interaction between PPy and dopant DBSA. However, a simple correlation between doping level and electrical conductivity such that a low doping level yields a low conductivity of the PPy-DBSA films is not acceptable since the conductivity of PPy-DBSA is inversely proportional to the doping level in the range of above 17 mol%. Instead, the conformational change of PPy chain induced by the reduced doping level is believed to be responsible for the reduction of the conductivity of the PPy-DBSA films prepared with polar solvents.

On the contrary, the PPy-DBSA films prepared using weakly polar solvents with extra DBSA exhibit slightly increased doping levels by 1 or 2 mol% which is considered within the error range. This unchanged doping level can be attributed to the micelle-like structure, through which the PPy-DBSA is dissolved in the weakly polar solvents. In the micelle-like structure in which the layer of extra DBSA interacts more strongly with PPy-DBSA in the core than the weakly polar solvents do, the PPy-DBSA will not show a decreased doping level. Moreover, an additional molecular interaction between extra DBSA and PPy-DBSA must yield an extended conformational structure of a PPy chain since the bending of a

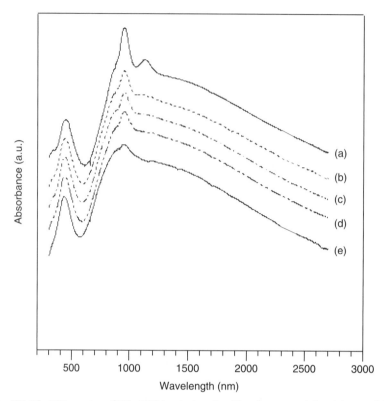

FIGURE 8.27 UV–Vis–NIR spectra of PPy-DBSA solutions in chloroform containing (a) 0.02, (b) 0.25, (c) 0.5, (d) 1, and (e) 8 g/dL of extra DBSA, where the polymer concentration is 0.01 g/dL. (From Song, K.T., Synthesis of electrically conducting soluble polypyrrole and its characterization. Ph.D. thesis, 2000. With permission.)

PPy chain is restricted by steric interference of extra DBSA. Thus an electronic structure favorable to the intrachain charge transport is achieved.

The difference in the chemical structure of PPy-DBSA films induced by different doping levels could be confirmed by FT-Raman and FT-IR spectroscopy. FT-Raman spectra of PPy-DBSA films prepared from solutions were similar to each other except for a small peak at 1290 cm^{-1} (Figure 8.30) [68]. The PPy-DBSA films prepared with polar solvents exhibited a small peak at 1290 cm^{-1} in Raman spectra, indicating that the PPy chain included some partially reduced structure presumed to be polaronic segments induced by the polar solvents. However, this peak was not observed from the spectra of PPy-DBSA films prepared using weakly polar solvents with extra DBSA. The relatively weak intensity of the peak at 1080 cm^{-1} in the spectra of the films from polar solvents corresponded with the lower doping levels of PPy-DBSA.

Compared with FT-Raman spectra, FT-IR spectra of the PPy-DBSA films were clearly divided into two groups according to the polarity of the solvent (Figure 8.31) [68]. This implies that it is a polaronic structure rather than a bipolaronic one that determines the chemical structure of PPy film according to the polarity of the solvents. The FT-IR spectra of PPy-DBSA films prepared using polar solvents exhibited a much stronger shoulder at 1290 cm^{-1} than those prepared using weakly polar solvents since the chemical species associated with the peak at 1290 cm^{-1} in a Raman spectrum is closely related with the polaronic structure in PPy. This indicates that the PPy-DBSA films include the more polaronic PPy segments when they are prepared using the more polar solvent.

The temperature dependencies of the conductivities of PPy-DBSA films cast from solutions are shown in Figure 8.32 [68]. The room temperature (300 K) conductivity of the films prepared using polar solvents were in the range from 2×10^{-2} to 1×10^{-1} S/cm, whereas those prepared using weakly polar

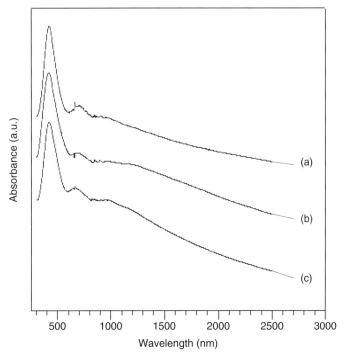

FIGURE 8.28 UV–Vis–NIR spectra of PPy-DBSA films cast from the solutions in (a) *m*-cresol, (b) DMSO, and (c) NMP. (From Song, K.T., Synthesis of electrically conducting soluble polypyrrole and its characterization. Ph.D. thesis, 2000. With permission.)

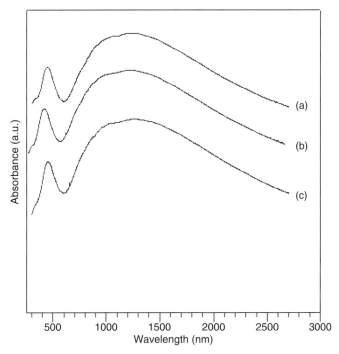

FIGURE 8.29 UV–Vis–NIR spectra of PPy-DBSA films cast from the solutions in (a) chloroform, (b) chloroben-zene, and (c) tetrachloromethane containing extra DBSA which is removed by washing after evaporation of the solvent. (From Song, K.T., Synthesis of electrically conducting soluble polypyrrole and its characterization. Ph.D. thesis, 2000. With permission.)

TABLE 8.6 Polarity Indices of the Solvents, Electrical Conductivities and Doping Levels of PPy Films Cast from the Solutions

Solvent	Polarity Index	Free-Standing Film Conductivity (S/cm)	Doping Level (mol%)
m-Cresol	7.4	3.1×10^{-2}	19
DMSO	7.2	1.6×10^{-2}	21
NMP	6.7	3.7×10^{-2}	16
Chloroform	4.1	1.7	24
Chlorobenzene	2.7	1.5	24
Tetrachloromethane	1.6	1.6	23
PPy-DBSA powder	—	1.0*	22

*Conductivity of PPy disk prepared by compression of the powder.

Source: From Song, K.T., Synthesis of electrically conducting soluble polypyrrole and its characterization. Ph.D. thesis, 2000.

solvents with extra DBSA were in the range from 3.5 to 5 S/cm. It is apparent that PPy-DBSA films with lower conductivity tend to have a greater temperature dependence. Nevertheless, all PPy-DBSA films cast from solutions showed considerably stronger temperature dependence of conductivity than electrochemically synthesized PPy-DBSA film.

The temperature dependency of conductivity of the PPy-DBSA films fits the following expression for variable-range hopping in disordered semiconductors:

$$\sigma(T) = \sigma_0 \exp[-(T_0/T)^m],$$

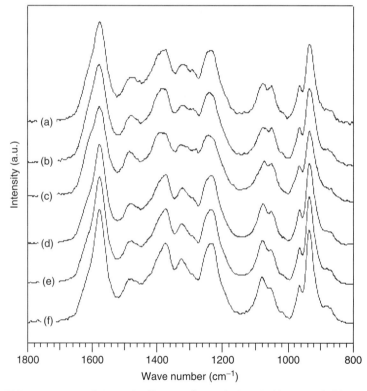

FIGURE 8.30 FT-Raman spectra of PPy-DBSA films cast from solutions in (a) *m*-cresol, (b) DMSO, (c) NMP, (d) chloroform, (e) chlorobenzene, and (f) tetrachloromethane, where solvents (d)–(f) contain extra DBSA. (From Song, K.T., Synthesis of electrically conducting soluble polypyrrole and its characterization. Ph.D. thesis, 2000. With permission.)

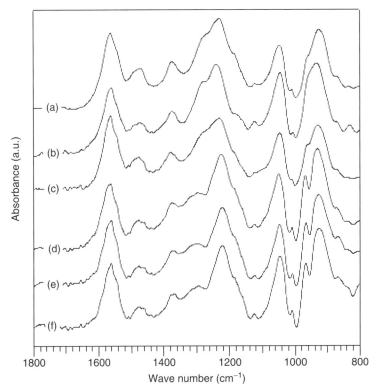

FIGURE 8.31 FT-IR spectra of PPy-DBSA films cast from solutions in (a) *m*-cresol, (b) DMSO, (c) NMP, (d) chloroform, (e) chlorobenzene, and (f) tetrachloromethane, where solvents (d)–(f) contain extra DBSA. (From Song, K.T., Synthesis of electrically conducting soluble polypyrrole and its characterization. Ph.D. thesis, 2000. With permission.)

where σ_0 is usually taken as a constant having a much weaker temperature dependence than the exponential term, T_0 is a parameter inversely dependent on the localization length of the localized electronic wavefunctions and on the density of states near the Fermi level and m is equal to $1/(d+1)$, where d is the dimensionality of the hopping [74–77].

The PPy-DBSA films prepared using weakly polar solvents with extra DBSA exhibited the temperature dependence of conductivity with $m = 1/2$ in the variable-range hopping equation (Figure 8.33) [68]. Quasi-one-dimensional (1D) variable-range hopping between the nearest neighboring chains was previously proposed to explain this charge transport property of PPy-DBSA [67]. However, the quasi-1D hypothesis was rejected due to the requirement that σ_0 in the equation should vary with temperature as $1/T$, which was not observed actually [78]. Therefore, the charge transport mechanism in the PPy-DBSA film prepared using weakly polar solvents is explained as the charging-energy limited tunneling model, where conduction is considered to proceed from tunneling between small conducting grains separated by insulating barriers [79,80]. The micelle-like structure of PPy-DBSA induced in weakly polar solvents containing extra DBSA has been found to be correlated with this transport model.

On the other hand, the PPy-DBSA films prepared using polar solvents exhibited the temperature dependence of conductivity following the variable-range hopping equation with $m = 1/3$ (Figure 8.34). This temperature dependence is very interesting because the value of $m = 1/3$ was not reported for the doped PPy. The charge transport with $m = 1/3$ could be explained in the framework of variable-range hopping theory as hopping in two dimensions [74]. Besides, some theoretical works predicted the same relation between σ and $T^{-1/3}$ for conducting polymers having short conjugation length [81].

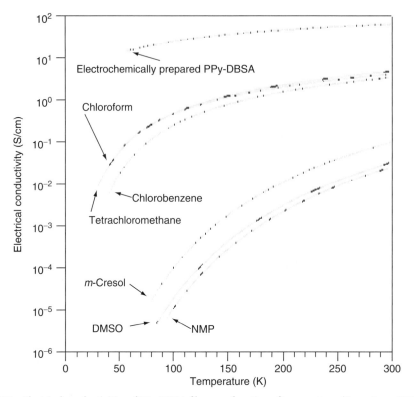

FIGURE 8.32   Electrical conductivities of PPy-DBSA films as a function of temperature. (From Song, K.T., Synthesis of electrically conducting soluble polypyrrole and its characterization. Ph.D. thesis, 2000. With permission.)

In the case of PPy-DBSA, the latter model would more reasonably explain the temperature dependence of conductivity with $m = 1/3$, because the partially dedoped structure with collapsed chain conformation of PPy-DBSA in polar solvents is considered to reduce the conjugation length of the PPy chain.

8.3.4   Solubility Changes with Crosslinking in PPy-DBSA

It was observed that the solubility of the as-polymerized PPy-DBSA powder gradually decreased as storage time elapsed. Some soluble PPy-DBSA powders lost their solubility completely after several months of storage time, being changed into fully insoluble materials. Generally, the pristine PPy-DBSA powder with higher conductivity tended to reveal a faster loss of solubility. Figure 8.35 shows the solubility change of a PPy-DBSA powder induced by isothermal heat treatment at 65°C for 5 days under inert or ambient atmosphere, indicating that the insolubilization of pristine PPy-DBSA powder was an irreversible process accelerated by both heating and oxygen [68]. It was also observed that the PPy-DBSA films cast from as-prepared solutions in weakly polar solvents could not be redissolved after the complete extraction of the casting solvent and extra DBSA.

   Time-dependent changes in PPy-DBSA solution properties were also investigated. In the case of PPy-DBSA solution prepared with weakly polar solvents containing extra DBSA, the solution was converted to a macroscopic irreversible gel during static storage. As shown in Figure 8.36, the gelation of the solution in chloroform was accelerated at an elevated temperature, which was identical to the process in which as-polymerized PPy-DBSA powders became insoluble after being stored for some period of time [68]. The gelation behavior was quantitatively studied in terms of the gelation rate defined as the reciprocal time for formation of macroscopic gel from a solution. The gelation rate of PPy-DBSA

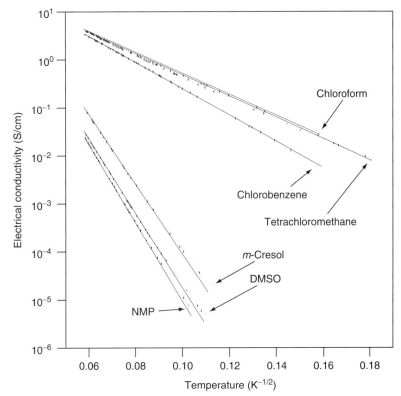

FIGURE 8.33 Electrical conductivities of PPy-DBSA films cast from solutions as a function of $T^{-1/2}$. (From Song, K.T., Synthesis of electrically conducting soluble polypyrrole and its characterization. Ph.D. thesis, 2000. With permission.)

solution increased with increasing solution concentration and PPy conductivity, but decreased with increasing concentration of extra DBSA in the solution (Figure 8.36). This indicates that the distance between neighboring PPy chains in the solution is the key factor to control gelation.

However, gelation of PPy-DBSA solutions in polar solvents rarely occurred. In a polar solvent, the polymer chains tend to be surrounded by solvent molecules with good affinity to PPy-DBSA since the polymer–solvent interactions are stronger than polymer–polymer interactions. This lack of polymer–polymer interactions might retard gelation. It took almost 10 months before partial gelation was observed for the solutions prepared with polar solvents, corresponding to a gelation rate of about 0.003 day^{-1}.

The physical property of the PPy-DBSA gel was similar to that of a pudding before the solvent was extracted from the gel since it was irreversibly destroyed into innumerable pieces by mechanical stirring. Moreover, the PPy-DBSA gel was never dissolved again in any solvent. After drying the solvent, the PPy-DBSA gel became a hard solid, but it never again underwent any degree of swelling in any of the solvents. These two features distinguish the PPy-DBSA gel from the PANi gel that is prepared from the emeraldine base solution in NMP [82,83]. The PANi gel can be deformed without permanent destruction and fragmented by the addition of a second solvent of appropriate character since the gelation of PANi results from interactions or physical crosslinks between the crystalline areas of the polymer [63]. This type of gelation normally results from conditions under which the polymer–polymer interactions are stronger than the polymer–solvent interaction. On the contrary, it is believed that PPy-DBSA gel is caused by irreversible chemical crosslinks between polymer chains. In the gelation system in which primary crosslinks take place, only a low degree of actual crosslinking of about 1% to 2% is required to form a 3D network or gel. This type of gel cannot be redissolved by stirring or addition of other solvents.

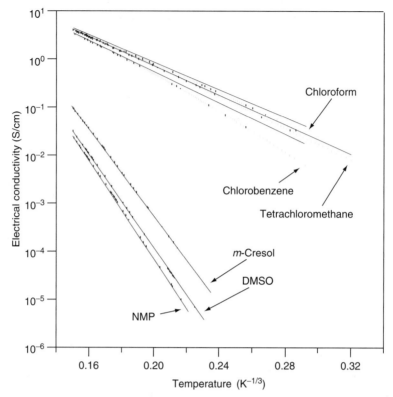

FIGURE 8.34 Electrical conductivities of PPy-DBSA films cast from solutions as a function of $T^{-1/3}$. (From Song, K.T., Synthesis of electrically conducting soluble polypyrrole and its characterization. Ph.D. thesis, 2000. With permission.)

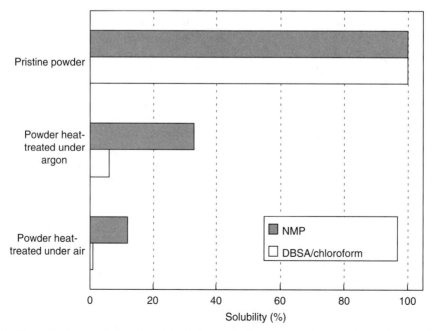

FIGURE 8.35 Solubility of PPy-DBSA powder before and after heat treatment under an inert and ambient atmosphere. (From Song, K.T., Synthesis of electrically conducting soluble polypyrrole and its characterization. Ph.D. thesis, 2000. With permission.)

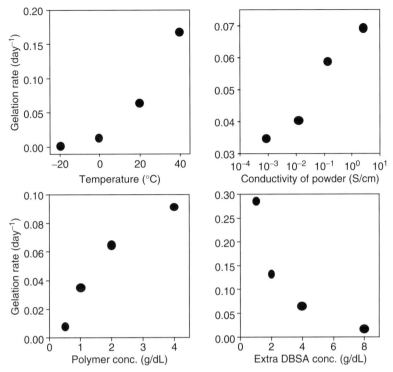

FIGURE 8.36 Gelation rate of PPy-DBSA solution prepared using chloroform containing extra DBSA. (From Song, K.T., Synthesis of electrically conducting soluble polypyrrole and its characterization. Ph.D. thesis, 2000. With permission.)

It is widely accepted that the bipolaronic structure is predominant in the fully oxidized PPy, while the polaronic structure is predominant in the partially reduced PPy. The polaronic structure can be formed by the partial oxidation of a neutral PPy or the injection of an electron into a bipolaronic structure. It is, however, obvious that the as-prepared soluble PPy cannot consist of only bipolarons, but some polarons must also exist. In the as-polymerized PPy-DBSA powder, bulky DBSA anions restrain effectively the molecular interactions between PPy chains. Therefore, even though there are some polaronic segments in the as-prepared PPy chain, they are maintained on the PPy chains without coupling to form crosslinks.

With the passage of time, PPy chains have opportunities to interact with neighboring PPy chains because of the molecular motion of PPy chains by changing their molecular conformation or chain movement to their neighboring chains. If the distance between two polarons in the adjacent chains is close enough for their coupling to take place, a primary valence bond is formed between the two PPy chains as the proposed mechanism shown in Figure 8.37 [68]. The coupling reaction results in the interchain crosslinks, making the polymer insoluble. Since the motion of PPy chains is a sort of self-diffusion process induced by thermal energy, the increase of temperature enhances the molecular motion to allow the formation of coupling, thereby accelerating the crosslinking reaction of the as-polymerized PPy-DBSA powder. Oxygen is thought to promote the crosslinking by introducing some reactive radical species into the PPy chains through their oxidation. This interchain crosslinking provides very reasonable explanations for various phenomena concerning the solubility change, the gelation of PPy-DBSA and the insoluble nature of PPy-DBSA cast film. Therefore, it is believed that the solubility of PPy-DBSA is not an inherent property simply induced by DBSA but is rather the result of the carefully controlled molecular structure.

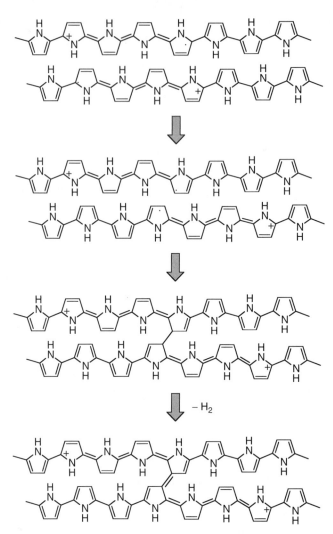

FIGURE 8.37   Schematic diagram of an interchain crosslink between neighboring PPy chains induced by radical coupling. (From Song, K.T., Synthesis of electrically conducting soluble polypyrrole and its characterization. Ph.D. thesis, 2000. With permission.)

8.3.5   Stability of Conductivity

The stability of conducting polymers is one of the most important properties that needs to be satisfied as many applications require a high level of thermal and environmental stability. Therefore, it is important to understand the factors that affect the stability of conducting polymers. Oxygen in ambient atmosphere is one of the most well-known factors affecting conductivity. Oxygen can easily react with the relatively high energy electrons in the valence band, forming structural defects on the conjugated system.

To investigate the long-term stability of the conductivity of PPy-DBSA, PPy films of about 0.1 μm thickness were prepared using chloroform with extra DBSA and the film resistivity was monitored with elapsed time. Thin PPy-DBSA films should be used to investigate the stability since the conductivity of thick films tends to be insensitive to their chemical degradation, which results from the slow diffusion of oxygen into the thick PPy-DBSA film.

Figure 8.38 shows the resistivity increase of thin PPy-DBSA films with the elapse of storage time at room temperature [68]. After 500 days, the resistivity of the PPy-DBSA film stored under ambient

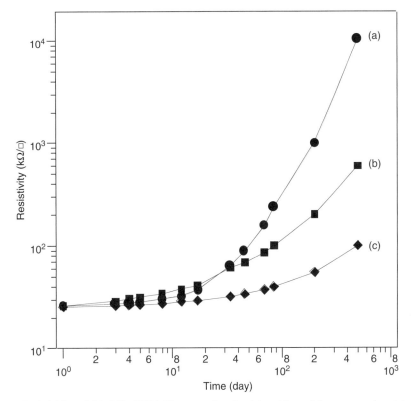

FIGURE 8.38 Resistivities of thin PPy-DBSA films stored under (a) ambient, (b) oxygen and moisture-free, and (c) oxygen, moisture and light-free conditions at room temperature. (From Song, K.T., Synthesis of electrically conducting soluble polypyrrole and its characterization. Ph.D. thesis, 2000. With permission.)

conditions, i.e., exposed to oxygen, moisture, and light, increased from 28 kΩ/$\square$ to 10 MΩ/$\square$, representing an increase of approximately two and a half orders of magnitude. The degradation of PPy-DBSA can be limited to some degree by the elimination of oxygen and moisture from the atmosphere which increases the resistivity by about one order of magnitude during the same storage duration. The most stable electrical property was achieved in the PPy-DBSA film stored under the condition free from oxygen, moisture, and light. The PPy-DBSA film stored under this inert condition showed only a small increase in the resistivity from 28 to 100 kΩ/$\square$ after 500 days, which indicates that light is another factor degrading PPy-DBSA.

Degradation of PPy-DBSA by oxygen occurred more seriously at elevated temperature. Figure 8.39 shows the UV–Vis–NIR spectra of the thin PPy-DBSA films during heat treatment at 200°C under an ambient atmosphere [68]. As heating time increased, the free-carrier absorption in the near IR region decreased significantly, which implies that the conductivity of PPy-DBSA films was seriously damaged. After 1 h, free-carrier absorption almost disappeared from the spectrum and the peak at 460 nm shifted to a shorter wavelength, indicating that the conjugation length of PPy had decreased. Figure 8.40 shows the UV–Vis–NIR spectra of the PPy-DBSA film during heat treatment at 200°C under an inert atmosphere [68]. The spectra of PPy-DBSA films exhibit little change with increased heating time, maintaining an almost constant level of free-carrier absorption. This indicates that PPy-DBSA is thermally stable in the absence of oxygen. With further increase in heating time, the peak at 460 nm becomes somewhat broader but still maintains its initial intensity and peak position.

The resistivity of the heat-treated, thin PPy-DBSA films measured as a function of heating time is shown in Figure 8.41 [68]. Under ambient conditions, the resistivity of PPy-DBSA film increased from

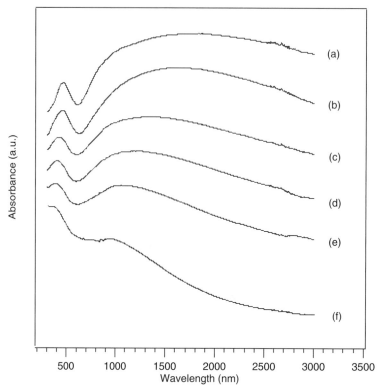

FIGURE 8.39 UV–Vis–NIR spectra of (a) a pristine PPy-DBSA film and of the same film after heat treatment at 200°C for (b) 2, (c) 10, (d) 20, (e) 30, and (f) 60 min under an ambient atmosphere. (From Song, K.T., Synthesis of electrically conducting soluble polypyrrole and its characterization. Ph.D. thesis, 2000. With permission.)

40 k$\Omega/\square$ to 10 G$\Omega/\square$ by over five orders of magnitude after heat treatment for 1 h. On the other hand, the heat treatment under inert condition increased resistivity by less than one order of magnitude after heat treatment for 1 h. The position of a peak in the UV–Vis region is closely correlated with the resistivity of PPy-DBSA films. Since the peak position in the UV–Vis region is associated with the conjugation length in a PPy backbone, the reduction in conjugation length of PPy chain was traced as the hypsochromic shift of this peak. Therefore, the reduction of conductivity upon heating under ambient conditions would be attributed to the decrease of the conjugation length of PPy induced by defects in the chemical structure of PPy through oxidation.

8.3.6   Application of Soluble PPy-DBSA

8.3.6.1   Semitransparent Bilayer Electrode

The role of surface morphology is very important for an electrode material. The advantage of the implantation of soluble PPy-DBSA in a device is the extremely smooth surface morphology of PPy-DBSA films. The PPy-DBSA film prepared by spin-casting of the PPy solution from both polar solvents and weakly polar solvents displayed a very smooth surface compared to the electrochemically prepared PPy-DBSA film (Figure 8.16) [68]. The transmittance and resistivity of PPy-DBSA coating on an electrode substrate can be controlled by the thickness of the coating, (Figure 8.17) [67]. Although the electrical conductivity of the soluble PPy-DBSA is not highly useful for an electrode material in electronic devices, its transparent property with a hole-injecting ability has been recognized to be suitable for various applications in the form of bilayer electrodes with other conducting materials.

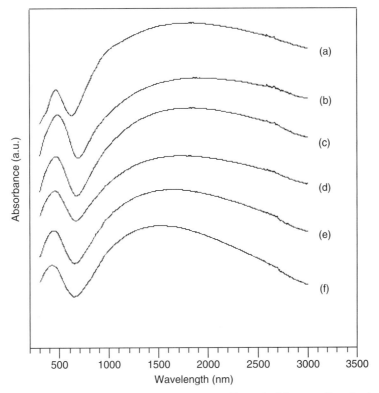

FIGURE 8.40   UV–Vis–NIR spectra of (a) a pristine PPy-DBSA film and of the same film after heat treatment at 200°C for (b) 2, (c) 10, (d) 20, (e) 30, and (f) 60 min under an inert atmosphere. (From Song, K.T., Synthesis of electrically conducting soluble polypyrrole and its characterization. Ph.D. thesis, 2000. With permission.)

8.3.6.2 Electrically Conducting Blend with Conventional Polymers

The solution processibility of PPy-DBSA enables the fabrication of conducting polymer blends with conventional polymers using solution blending method. PPy-DBSA/polymer blend solutions were prepared by adding PPy-DBSA to the solution of a thermoplastic polymer such as polystyrene (PS), polycarbonate (PC) and poly(methyl methacrylate) (PMMA) in chloroform containing extra DBSA under ultrasonification. Thin films of approximately 0.1 μm thickness were prepared by spin-casting the solutions onto glass substrates. The films were then washed with methanol to extract the extra DBSA and dried at 80°C under an ambient atmosphere. As shown in Figure 8.42, the surface morphology of PPy-DBSA/PMMA blend film is very uneven [68]. The other two thermoplastic polymers exhibit similar surface morphology. This uneven surface morphology indicates that the phase separation between PPy-DBSA and PMMA molecules occurred in the blend because of the lack of miscibility between the two polymer components. Precipitation of PPy was also observed even in the solution state after a sufficiently long time.

Therefore, to minimize the effect of phase separation, the as-prepared polyblend solution is poured into an excess amount of nonsolvent such as methanol to obtain a quenched bulk powder of polymer blend. The obtained polymer blends could be molded into the desired shape by hot-pressing at 180°C for 5 min. Figure 8.43 shows the electrical conductivity of the molded blend as a function of PPy-DBSA weight fraction [68]. The electrical conductivities of PPy-DBSA/PMMA blend exhibit a percolation threshold level at about 40%. Nevertheless, the PPy-DBSA/PMMA blends exhibit conductivities ranging from 10^{-5} to 10^{-4} S/cm with 2% PPy-DBSA, which satisfy the electrical conductivity required for static dissipative (10^{-5} to 10^{-3} S/cm) or antistatic applications (10^{-8} to 10^{-5} S/cm).

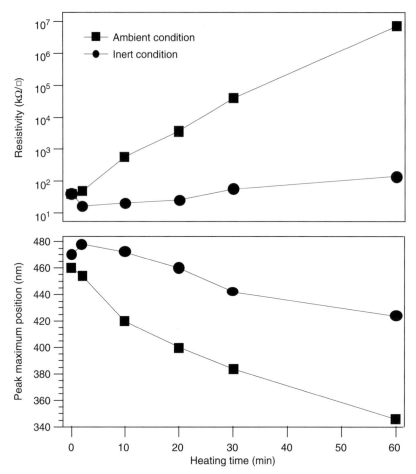

FIGURE 8.41 Change of the resistivities of PPy-DBSA films (upper) and the positions of absorption peak in the UV–Vis region (lower) during heat treatment at 200°C. (From Song, K.T., Synthesis of electrically conducting soluble polypyrrole and its characterization. Ph.D. thesis, 2000. With permission.)

8.3.6.3   Transparent Anode or Hole-Transport Layer in Polymer LED

It was reported that semi-transparent, PPy-DBSA film cast from the solution could be successfully used as the anode material in polymer light emitting diodes (LEDs) [84]. As a luminescent polymer, poly (2-methoxy-5-(2′-ethyl-hexyloxy)-1,4-phenylene-vinylene) (MEH-PPV) is used for the fabrication of polymer LEDs. As shown in Figure 8.44, a fully extracted and heat-treated device gave the best electroluminescent performance. The external quantum efficiency of the device was 0.3% which was increased to 0.5% by the optimization of MEH-PPV layer thickness. In addition, the hole-injection potential barrier between PPy and MEH-PPV as determined by Fowler–Nordheim analysis was $\Phi \neq 0.23$ eV, which is 0.1 eV larger than that of metallic PANI and MEH-PPV. It was also reported that PPy thin film could be used as the hole transport layer in organic LEDs [85]. As shown in Figure 8.45, the ITO/PECCP/Al device was greatly improved through the introduction of a PPy layer [85].

8.3.6.4   Reflective Bistable Cholesteric/Polymer Dispersion Display

Reflective cholesteric liquid crystal display has drawn much attention as an ideal flat portable panel display device due to its low power consumption. PPy film with a thickness of about 4 μm was spin-coated on a glass substrate. The PPy layer, opaque to light but very reflective, acted as a light-absorbing layer as well as the electrode in the device (Figure 8.46) [86]. The initial reflectance of the

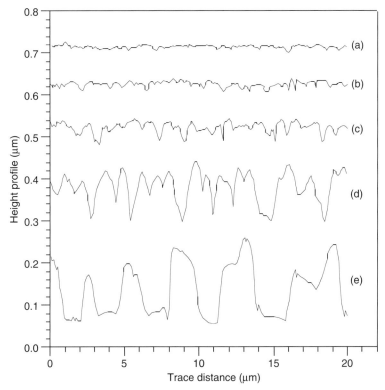

FIGURE 8.42 Surface morphology of PPy/polymer blend films cast from solutions of PPy-DBSA/PMMA in chloroform with the weight portion of (a) 2/98, (b) 5/95, (c) 10/90, (d) 20/80, and (e) 40/60. (From Song, K.T., Synthesis of electrically conducting soluble polypyrrole and its characterization. Ph.D. thesis, 2000. With permission.)

chiral nematic mixture (CNM) cell in the focal conic state remained the same until the applied voltage reached 85 V, after which it increased from 13% to 89% with switching to the planar state. On the other hand, the CNM cell initially in the planar state switched to the focal conic state in the voltage range of 55~90 V.

8.3.7 Other Soluble PPy

8.3.7.1 Soluble PPy with Various Functional Dopants

Several kinds of PPy soluble in organic solvents have been synthesized with various functional dopants [87–90]. PPy with high conductivity was synthesized by chemical polymerization with the dopant NSA [87], whose solubility in *m*-cresol and room temperature conductivity increased with increasing NSA concentration. High molecular weight soluble PPy has also been chemically synthesized by the incorporation of di(2-ethylhexyl) sulfosuccinate sodium salt (NaDEHS) as a dopant (Figure 8.47) [88–90]. The two different types of oxygen atoms in the two ester groups and the sulfonate group render the PPy soluble in various alcohols as well as polar solvents shown in Table 8.7 [89].

As shown in Figure 8.48, water soluble PPy, PPy(SO$_3$H)-DEHS, PPy(SO$_3$H)-BNS, and PPy(SO$_3$H)-DBSA bearing sulfonyl acid (SO$_3$H) were reported [91]. The polymer was formed by chemical modification with chlorosulfonic acid to give chlorosulfonyl group functionalized PPys, followed by hydrolysis with hot water. PPy(SO$_3$H)-BNS and PPy(SO$_3$H)-DBSA showed solubilities of 1.5 and 0.5 wt%, respectively, and PPy(SO$_3$H)-DEHS showed the highest solubility of 3.0 wt% in water.

Using tetraethylammonium tetrafluoroborate (TEABF$_4$) as a co-dopant in the PPy-DBSA system, PPy composites (PPy/DBSA/BF), which are soluble in *m*-cresol and NMP and partially soluble in chloroform,

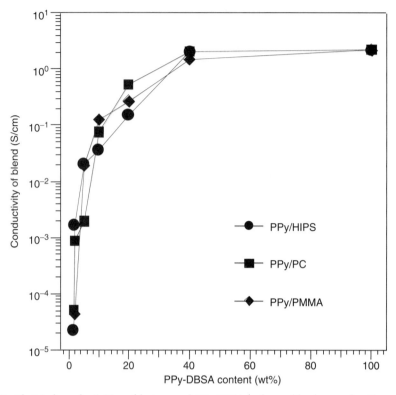

FIGURE 8.43 Electrical conductivities of hot-pressed PPy-DBSA/polymer blends as a function of PPy-DBSA content at room temperature. (From Song, K.T., Synthesis of electrically conducting soluble polypyrrole and its characterization. Ph.D. thesis, 2000. With permission.)

were chemically polymerized [92]. As shown in Table 8.8, the electrical conductivity of PPy/DBSA/BF was higher than that of PPy/DBSA due to the improved interchain interactions among PPy chains by small BF_4 anions. It was reported that the conductivity of PPy/DBS/BF is mainly influenced by the interchain interaction of polymer depending on the dopant size.

8.3.7.2 Soluble PPy with Polymeric Co-dopant

In an aqueous medium, using DBSA as a dopant, poly(2-acrylamido-2-methyl-1-propane sulfonic acid) (PAMPS) as a co-dopant, and APS as an oxidant, PPy/polymeric dopant composites are chemically synthesized [93]. Regardless of the dopant composition, the conductivities of PPy/DBSA/PAMPS composite powders are in the range of about 10^{-1} S/cm and those of cast films are about 2 S/cm as shown in Table 8.9. The mechanical properties of the composite films are greatly improved by PAMPS, as shown in Table 8.10. The high tensile strength value at break for PPy/DBSA/PAMPS implies that the composite has relatively high molecular weight and strong interactions between polymer molecules. Also, elongation at break for the composite with PAMPS is dramatically improved, possibly because the fracture in the rigid PPy chains is obstructed by the long chains of PAMPS.

Using DBSA as a primary dopant and a copolymer of N-isopropyl acrylamide (NiPAAm) and 2-acrylamido-2-methyl-1-propane sulfonic acid (AMPS) poly(NiPAAm-co-AMPS) as a co-dopant, electrically conducting PPy/DBSA/poly(NiPAAm-co-AMPS) composites are chemically synthesized in aqueous medium with APS as the oxidant (Figure 8.49) [94]. The PPy composites were readily dissolved in chloroform with the extra DBSA. The electrical conductivities of the PPy composite free-standing films are in the range of 10^{-2} to 5 S/cm and increase with increasing moisture concentration in air due

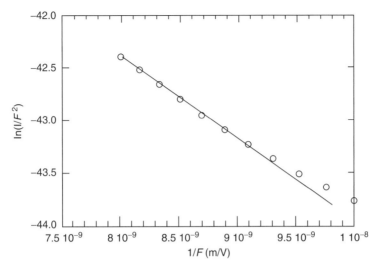

FIGURE 8.44 Fowler–Nordheim plot, $\ln(I/F^2)$ vs. $1/F$ of a "hole-only" device, ITO/PPy/MEH-PPV/Ag. (From Gao, J., et al., *Synth. Met.*, 82, 221, 1996. With permission.)

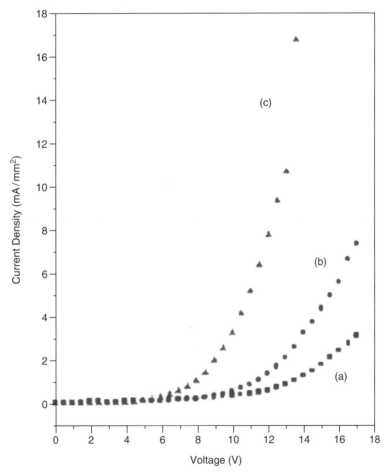

FIGURE 8.45 I–V characteristics of (a) ITO/PECCP/Al device, (b) ITO/PECCP/Al-Li device, and (c) ITO/PPy/PECCP/Al device. (From Park, J.W., et al., *Thin Solid Films*, 363, 259, 2000. With permission.)

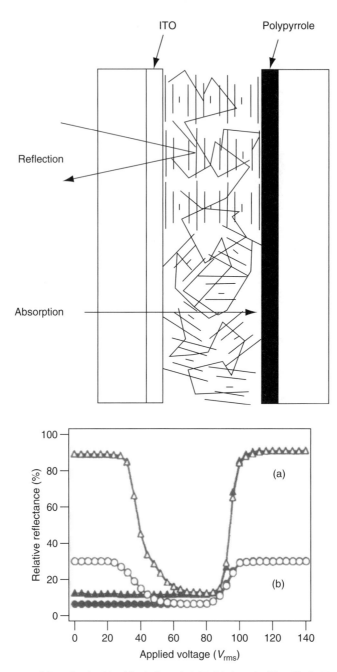

FIGURE 8.46 Structure of the reflective bistable cholesteric LCD fabricated with a black PPy film as the electrode and the light absorbing layer. Relative reflectance as a function of applied AC voltage for the (a) chiral nematic mixture (CNM) and (b) CPD cells initially in either the planar (open symbols) or focal conic (filled symbols) state. Reflectance at 510 and 560 nm for the CNM and CPD cells, respectively, was measured after 200 Hz-AC field was applied for 20 ms and then removed. (From Kim, Y.C., et al., *Mol. Cryst. Liq. Cryst.*, 327, 157, 1999. With permission.)

to moisture absorption by hydrophilic NiPAAm moiety in the composite. As shown in Figure 8.50, PPy composites show thermosensitive swelling behavior in water. Until about 35°C, the swelling ratio decreases considerably with increasing temperature due to the LCST behavior resulting in release of water by coagulation of the thermally sensitive NiPAAm moiety in the composite.

8.3.7.3 Water Soluble PPy

Water soluble, self-doped, PPy copolymer was also synthesized by introducing highly water-soluble, copolymer co-dopant, poly (2-acrylamido-2-methyl-1-propanesulfonic acid-*co*-N-(4-anilino-phenyl)-methacrylate), Poly(AMP-*co*-AMPA) (Figure 8.51) [95]. The effect of the weight ratio of poly(AMP-*co*-AMPA) to pyrrole on electrical conductivity is shown in Figure 8.52, in which the lower dopant/pyrrole ratio effectively shortens the conjugation length. At a low feed ratio, the conducting PPy segments are hardly formed, giving a maximum conductivity of 3.4 S/cm at 1:1 ratio.

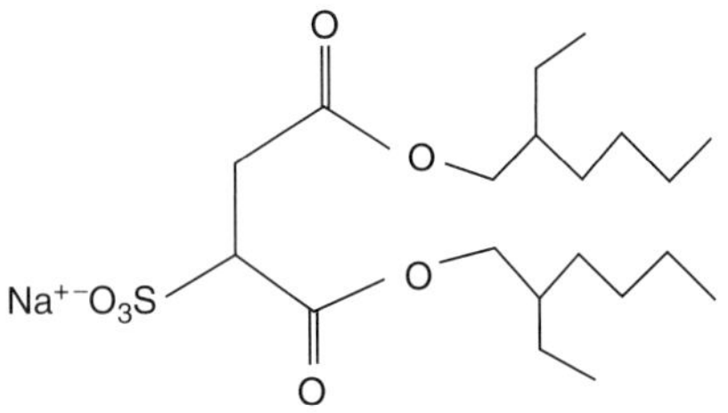

FIGURE 8.47 The structure of di(2-ethylhexyl) sulfo-succinate sodium salt (NaDEHS). (From Jang, K.S., et al., *Synth. Met.*, 119, 107, 2001. With permission.)

8.4 Polypyrrole with Nanostructure

Recently a number of research investigations have been carried out on the synthesis and characterization of nanoscale PPy materials for various applications because of their interesting properties such as high environmental stability, ion exchange capacity, electronic conductivity, and biocompatibility of PPy [96–104].

8.4.1 PPy Nanowires and Nanofibrils

PPy nanowires were electrochemically synthesized using a carbon electrode modified by cathodic electrografting of polyethylacrylate (PEA) as a template [96]. It was revealed that the nanowire grew from the electrode surface through a hole in the PEA coating. According to the suggested mechanism, oxidation of monomer would not occur on the surface of the PPy wire and there would be no inherent limit to the wire length [96].

Using porous aluminum oxide as a template, gold-capped, protein-modified PPy nanowires were electrochemically grown [97]. To investigate the effect of the electrochemical method, two different methods of constant potential referred to as Eco-nanowires, and potential cycling referred to as Ecy-nanowires were used to prepare the nanowires in aqueous phosphate-buffered saline solutions. During PPy polymerization, the proteins (Avidin and Streptavidin) were entrapped in the nanowires. As shown in Figure 8.53a and Figure 8.53b, the morphology and stiffness were greatly affected by the electrochemical

TABLE 8.7 Solubility of PPy-DEHS Powder and Conductivity (RT) of PPy-DEHS Free Standing Films Prepared from Various Alcohols

Solvent	Solubility (% wt/vol.)	Conductivity (S/cm)
Methyl alcohol	0.5 ~ 1	1.0×10^{-4}
Ethyl alcohol	1 ~ 2	5.5×10^{-4}
2-Propyl alcohol	2 ~ 3	4.3×10^{-3}
1-Butyl alcohol	2 ~ 3	9.2×10^{-3}
iso-Propyl alcohol	2 ~ 3	5.5×10^{-3}
iso-Butyl alcohol	2 ~ 3	7.1×10^{-3}
t-Butyl alcohol	2 ~ 3	8.2×10^{-3}
Benzyl alcohol	3 ~ 4	4.7×10^{-3}
2,2,2-Trifluoroethanol	2 ~ 3	4.5×10^{-4}
Oleyl alcohol	3 ~ 4	6.6×10^{-2}

Source: Reprinted from Jang, K.S., et al., *Synth. Met.*, 119, 108, 2001. With permission.

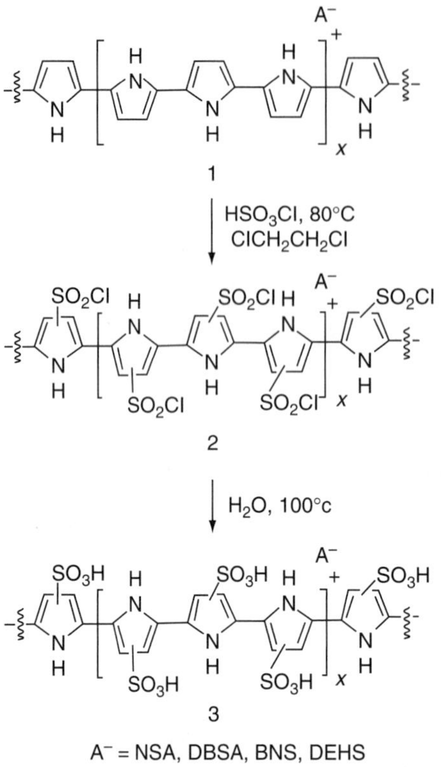

methods. The constant potential deposition method produced a larger cauliflower-like structure and straight nanowires with lengths even exceeding 10 μm, while the potential cycling method produced shiny black, smaller grains and flexible nanowires with spaghetti-like shapes. As shown in Figure 8.53c, the internal structure was porous compared to the external surface in the case of PPy prepared at constant potential [97].

A copolymer of pyrrole and thiophene nanofibrils was electrochemically polymerized within the pores of microporous, anodic, aluminum oxide template membranes [105]. The copolymer nucleated and grew on the pore wall of the membrane since the polymers were cationic and the membrane had anionic sites on the pore wall. The length, thickness, and diameter of the copolymer nanofibrils could be controlled and with higher applied potential, more thiophene units were incorporated into the copolymer nanofibrils [105]. Copolymer nanofibrils of pyrrole and aniline were also electrochemically polymerized within the pores of microporous, anodic, aluminum oxide template membranes [106]. Copolymer nanofibrils of PPy and poly(3-methylthiophene) prepared chemically in the microporous aluminum oxide template showed higher conductivity than the homopolymers did [107].

FIGURE 8.48 Synthesis scheme of water soluble, self-doped PPy. (From Jang, K.S., Lee, H., and Moon, B., *Synth. Met.*, 143, 289, 2004. With permission.)

8.4.2 PPy Nanotubes

To prepare nanostructures with monodispersed nanoscopic fibrils and tubules, the pores in the membranes have been used as templates in the "template-synthesis" method. With the pores of nanoporous PC membrane filters as templates, PPy nanotubules and nanofibrils were chemically synthesized with oxidizing agent, $FeCl_3$ [98,100]. Figure 8.54 shows transmission electron microscope (TEM) images of the PPy nanostructures after dissolution of the PC template membrane. The tubule wall thickness increased with increasing polymerization time, but the outer diameter did not change with polymer-

TABLE 8.8 Electrical Conductivity of Soluble PPy-DBS-BF at Room Temperature

Concentration (M)		
[DBS⁻]	[BF₄⁻]	Conductivity, σ (S/cm)
0.30	0.00	0.04
0.30	0.15	0.10
0.15	0.15	0.18
0.10	0.15	0.86
0.05	0.15	0.53

Source: From Lee, G.J., Lee, S.H., and Ahn, K.S., *J. Appl. Polym. Sci.*, 84, 2583, 2002. With permission.

TABLE 8.9 Electrical Conductivity and Solubility of PPy Composite

Feed Composition (DBSA/PAMPS)	Yield (%)	Powder Conductivity (S/cm)	Film Conductivity (S/cm)	Solubility in Chloroform
50/0	26.26	1.25×10^{-1}	2.56	Good
40/10	30.52	0.90×10^{-1}	1.55	Good
30/20	23.88	1.27×10^{-1}	1.87	Good
20/30	25.37	0.82×10^{-1}	1.93	Good
10/40	23.14	1.30×10^{-1}	N/A	Partially
0/50	29.48	4.03×10^{-4}	N/A	Partially

Source: Reprinted from Lee, Y.H., Lee, J.Y., and Lee, D.S., *Synth. Met.*, 114, 350, 2000. With permission.

ization time. Polarized infrared adsorption spectroscopy revealed that only the PPy layer deposited on the pore wall was oriented, implying that the template-synthesized solid fibrils were composed of a disordered, low conjugation length core with poor conductivity, which was surrounded by a thin, highly oriented, outer skin with high conductivity (Figure 8.55).

As shown in Figure 8.56, PPy nanotubules were also electrochemically synthesized using the pores of nanoporous PC track-etched membranes as templates [99,108]. Figure 8.57 shows a typical chronoamperogram obtained for the electropolymerization of PPy in pores of a PC track-etched membrane [108]. During the first stage, the polymer growth occurs in the pores and continues until the top surface of the membrane is filled (step II). Beyond this step, growth can continue in 3 dimensional and hemispherical shapes on the top of the tubules (step III). The electrochemical current increases during step III due to the increase of the effective electrode area. When the whole membrane surface is filled, the current reaches an asymptotic value (step IV). As shown in Figure 8.58, the thickness of the PPy tubules was found to depend on the pore diameter of the template membrane and the supporting electrolyte [108]. PPy, PEDOT, and PANi nanotubules and nanowires were also electrochemically synthesized using an Al_2O_3 nanoporous template as shown in the TEM and SEM images of Figure 8.59 [101].

8.4.3 PPy Nanocomposites

Due to its remarkable mechanical and electrical properties, carbon nanotubes (CNTs) have been investigated as a template to grow nanosized, ultrahigh surface/volume ratio catalysts, as a conductive filler in insulating polymer matrices and as a reinforcing material in structural materials. PPy is suitable for the manufacture of composites with CNTs since it can be polymerized in neutral aqueous solutions where the CNT suspensions are possible. A CNT-PPy composite with a high concentration of well-dispersed nanotubes wetted by a continuous polymer phase was electrochemically synthesized. The anionic CNTs were reported to act as a strong and conductive dopant and the growth of the polymer coating was accelerated with increasing CNT concentration [109]. No chemical reaction between the PPy and CNTs was observed by the structural characterization, indicating that the CNTs only played a role as a template for PPy polymerization [110].

TABLE 8.10 Mechanical Properties of PPy Films

Mechanical Properties	PPy-DBSA/PAMPS (20/30)	PPy-DBSA[a]	PPy-DS[b]
Tensile strength at break (MPa)	22.3	17.0	68.5
Elastic modulus	222	1945	
Elongation at break (%)	25.7	0.9	7.7
Toughness (MPa)	3.36	0.0765	

[a]PPy film cast from the chloroform solution.
[b]PPy film synthesized by an electrochemical method.
Source: Reprinted from Lee, Y.H., Lee, J.Y., and Lee, D.S., *Synth. Met.*, 114, 352, 2000. With permission.

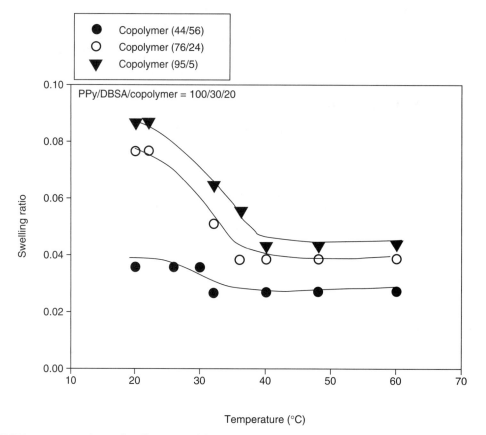

FIGURE 8.49 Schematic chemical structure of PPy composite doped with DBSA and copolymer. (From Han, J.S., Lee, J.Y., and Lee, D.S., *Synth. Met.*, 124, 301, 2001. With permission.)

FIGURE 8.50 Dependence of swelling ratio of the PPy composite on temperature. (From Han, J.S., Lee, J.Y., and Lee, D.S., *Synth. Met.*, 124, 301, 2001. With permission.)

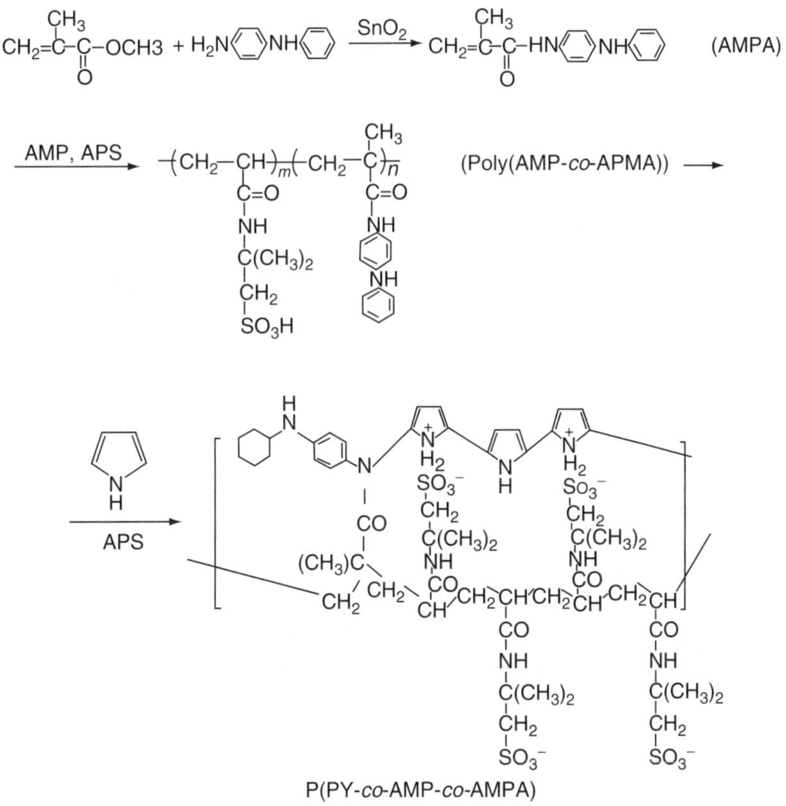

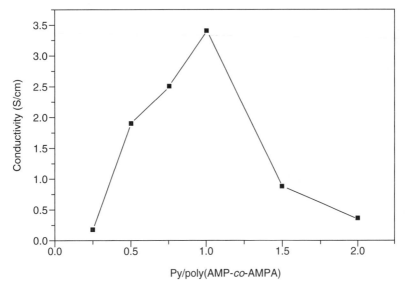

FIGURE 8.51 Synthesis scheme of water soluble, self-doped, PPy copolymer. (From Yin, W., and Ruckenstein, E., *J. Appl. Polym. Sci.*, 79, 86, 2001. With permission.)

FIGURE 8.52 Effect of PY/poly(AMP-*co*-AMPA) wt ratio on the conductivity of copolymer film. (From Yin, W., and Ruckenstein, E., *J. Appl. Polym. Sci.*, 79, 86, 2001. With permission.)

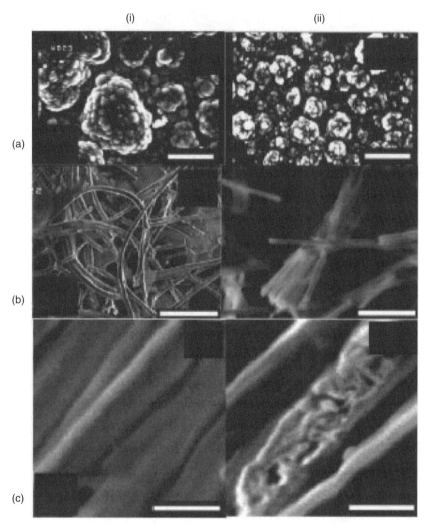

FIGURE 8.53 Field effect—scanning electron microscope (SEM) images of (a) PPy films synthesized by (i) constant potential (0.75 V) and (ii) potential cycling (−0.2 to 0.8 V), scale bar 3 μm. (b) Difference in stiffness of (i) Eco-nanowires and (ii) Ecy-nanowires, scale bar 4.3 μm. (c) Structure of Eco-nanowire: (i) external and (ii) internal, scale bar 300 nm. (From Hernandez, R.M., et al., *Chem. Mater.*, 16, 3431, 2004. With permission.)

A one dimensional, well-aligned nanocomposite of CNT with PPy was also synthesized electrochemically [111]. Figure 8.60 shows the SEM image of CNT with fairly uniform diameter and length. The PPy film coating was uniform along the entire length of the nanotubes and the thickness of the PPy film was easily controlled by the film-formation charge (Figure 8.61). When the catalytic cap was removed by acid pretreatment in HNO_3 aqueous solution, no PPy deposition occurred on the inside wall of the nanotubes as shown in Figure 8.61a and Figure 8.61d, because the electrolyte solution could not enter during PPy deposition due to the surface tension [111].

The composites of super paramagnetic materials and functional organic materials exhibit interesting properties such as useful optical or electrical properties as well as super-paramagnetic properties. PPy-based $\gamma\text{-Fe}_2\text{O}_3$ nanocomposites were prepared by two steps: presynthesis of hydrocarbon passivated $\gamma\text{-Fe}_2\text{O}_3$ nanocrystals from organic solution at high temperature and synthesis of $\gamma\text{-Fe}_2\text{O}_3/\text{PPy}$ nanocomposites by in situ polymerization in oil/water microemulsion. The blocking temperature increased with increasing content of the $\gamma\text{-Fe}_2\text{O}_3$ component and the resistivity was decreased with higher polymer concentration [112].

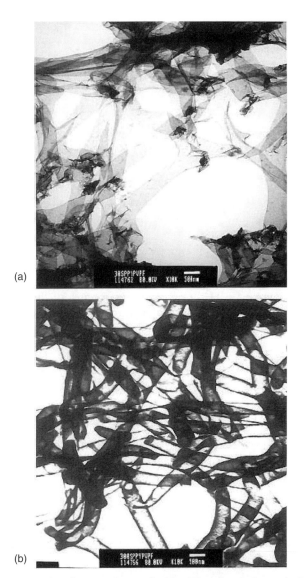

(a)

(b)

FIGURE 8.54 (a, top) Transmission electron micrographs of polypyrrole tubules synthesized in 400 nm polycarbonate membrane. Polymerization time = 30 s. (b, bottom) Transmission electron micrographs of polypyrrole tubules synthesized in 400 nm polycarbonate membrane. Polymerization time = 300 s. (From Menon, V.P., Lei, J., and Martin, C.R., *Chem. Mater.*, 8, 2382, 1996. With permission.)

Since hollow nanospheres can be applied to many fields, including drug delivery, enzyme and protein protection, dye encapsulation, removal of contaminated waste and nanostructured composites, they have recently drawn much attention to research. Figure 8.62 shows the procedure to prepare carbon nanocapsules from PPy nanocapsules, in which PPy hollow nanospheres are first fabricated using only PPy and subsequently carbonized to form carbon nanocapsules [113]. Figure 8.63a and Figure 8.63b show the soluble PPy nanoparticles prepared in decyltrimethylammonium bromide and linear PPy/crosslinked PPy core-shell nanoparticles with spherical particles, respectively. By etching the soluble PPy core, PPy hollow nanocapsules were obtained (Figure 8.63c), and carbon nanocapsule was obtained after carbonization of PPy nanocapsule (Figure 8.63d).

Apparent current density of the electrode can be improved by using high electrocatalytic activity, electrode material and by increasing the effective surface area. The SEM images of the naked and

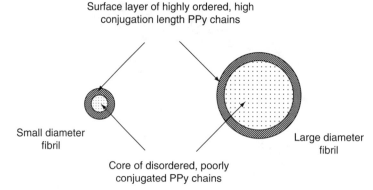

FIGURE 8.55 Proposed anatomy of template-synthesized PPy fibrils. (From Menon, V.P., Lei, J., and Martin, C.R., *Chem. Mater.*, 8, 2382, 1996. With permission.)

PPy-modified graphite electrode surface are shown in Figure 8.64 [114]. The electrode modification with PPy can increase the current density, especially at high electrolysis potentials, due to its high effective surface area (Figure 8.65) [114]. Since the surface area of the fibrils is larger than that of the cauliflower form, the current density of fibrillar PPy is higher. Further increase of the hydrogen evolution current density is obtained by redoping PPy fibrils with metal complex (Figure 8.66) [114]. Using naphthalene disulfonic acid disodium salt (NDSDNa) and benzenesulfonic acid sodium salt (BzSNa) as electrolytes, PPy is also electrochemically coated on the carbon fibers in aqueous medium [115].

8.4.4 PPy Fiber Spun by Electrospinning

PPy fiber was fabricated using electrospinning from PPy-DBSA solution [116]. Highly conducting, soluble PPy was first chemically polymerized using APS as the oxidant and DBSA as the dopant [27]. Electrospinning was carried out using PPy solution with the appropriate concentration for electrospinning. The electrospun, PPy fiber nonwoven web was finally washed with methanol to remove the remaining extra DBSA in the fiber, producing an electrically conducting PPy nonwoven web. The PPy fibers exhibited a circular cross section, an extraordinarily smooth surface, and a fiber diameter of about 3 μm (Figure 8.67). The electrical conductivity of the compressed PPy nonwoven web was about 0.5 S/cm, which is slightly higher than that of the powder or the cast films, possibly because of the molecular orientation induced during the electrospinning.

To improve the spinnability of the PPy spinning solution, poly(vinyl cinnamate), PVCN (PPy:PVCN = 0.8:0.2 by weight) was added to the PPy spinning solution. The morphologies of the PPy/PVCN blend fiber nonwoven web, and the fiber surface are shown in Figure 8.68. The PPy/PVCN blend solution produced a better quality of nonwoven web than the PPy solution did, because the high molecular weight of PVCN increased the intermolecular interaction of the polymer in the solution. The electrospun

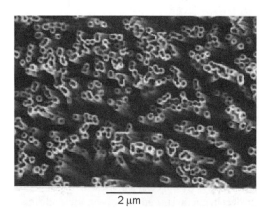

2 μm

FIGURE 8.56 Field effect—scanning electron microscope image of PPy/ClO₄ tubules obtained after dissolution of the PC membrane. (From Demoustier-Champagne, S., and Stavaux, P.Y., *Chem. Mater.*, 11, 829, 1999. With permission.)

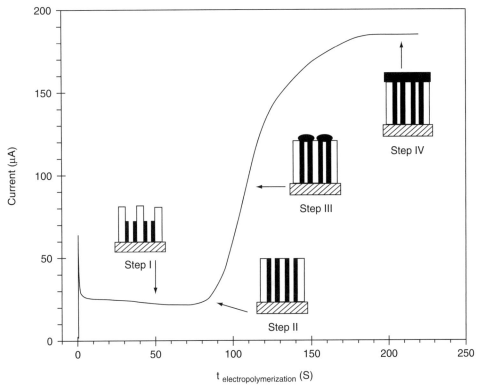

FIGURE 8.57 Typical chronoamperogram obtained for the electropolymerization of pyrrole in pores of a home-made PC track-etched membrane. The schematics display four different stages of the growth process: step I, PPy tubules are growing in the pores of the template membrane; step II, the pores are just completely filled (this point corresponds to $t_{filling}$); step III, PPy tips are appearing on the membrane surface; step IV, PPy growth appears over the whole membrane surface. (From Demoustier-Champagne, S., and Stavaux, P.Y., *Chem. Mater.*, 11, 829, 1999. With permission.)

PPy/PVCN fiber showed slightly smaller diameter with much better uniformity than PPy fiber, since the fibers were split much more uniformly during electrospinning. The surface of the PPy/PVCN fiber was also extraordinarily smooth and the electrical conductivity of the compressed PPy/PVCN blend fiber nonwoven web was about 0.3 S/cm.

8.5 Patterning of Polypyrrole

8.5.1 Patterning by Selective Deposition of PPy on Patterned Layer

The patterning of conducting polymer thin films with desired shapes has drawn technological and scientific interest. When electrochemical or chemical polymerization can be spatially controlled over the electrode surface, small features of electroactive and optically active polymers can be constructed.

8.5.1.1 Electrochemical Deposition of PPy

As shown in Figure 8.69, it was reported that PPy could be electrochemically deposited selectively on patterned self-assembled monolayers (SAM) because of the difference in electron transfer rate through the patterned SAM [117]. Figure 8.70 shows optical micrographs of polyaniline, poly(3-methylthiophene), and PPy polymer patterns with millimeter and micrometer resolutions. The SAM deposited on gold was patterned, as shown in Figure 8.69, where amines were attached by irradiating UV light to a

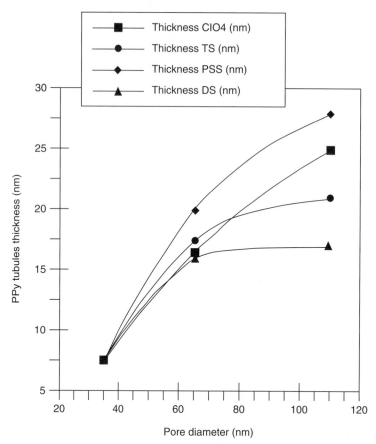

FIGURE 8.58 Dependence of the PPy thickness on pore diameter in terms of the electrolyte nature; [pyrrole] = 0.1 M, [electrolyte] = 0.1 M. (From Demoustier-Champagne, S., and Stavaux, P.Y., *Chem. Mater.*, 11, 829, 1999. With permission.)

monolayer of bis[11-(4-azidobenzoyl)oxy]-1-undecyl disulfide deposited on gold in the presence of various amines.

Microcontact printing is a soft lithographic and very simple technique that is useful in patterning SAMs. A patterned poly(dimethylsiloxanes) (PDMS) stamp, which is inked with an appropriate SAM solution, contacts the substrate surface to transfer SAM molecules to the substrate surface contacting the stamp. A copper pattern with feature sizes of 50–200 μm was first formed by electroless deposition on patterned *N*-[3(trimethoxysilyl)-propyl]diethylenetriamine (TMS) by microcontact printing. The PPy pattern was then formed by electrochemical deposition using the patterned copper as the working electrode (Figure 8.71) [118].

It was also reported that PPy pattern, formed by selective electrochemical deposition of PPy on patterned SAM, could be transferred to an insulating substrate [119]. The working electrode was patterned by microcontact printing using hexadecanethiol and was briefly immersed in 1 μM ethanolic solution of 2,2,2-trifluoroethanethiol to improve the contact adhesion of PPy after synthesis. By electrodeposition, a black pattern of PPy perchlorate appeared on the electrode and then the PPy pattern was peeled off from the gold electrode by pressing it against a piece of PDMS or Scotch double-sided tape using firm finger pressure. Figure 8.72 shows the patterns transferred to PDMS.

The two possible techniques for monolayer-guided electrodeposition of PPy, negative deposition, and positive deposition, are shown in Figure 8.73. The selective deposition and morphology of the PPy films were greatly affected by the compatibility between the chemical properties of conducting oligomer and the surface, the electric properties of the substrate surface, and the nature of the solvents [120].

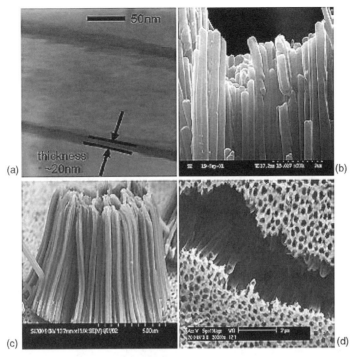

FIGURE 8.59 (a) Transmission electron microscope image of PPy-CSA nanotube; scanning electron microscope images of (b) PPy-CSA nanowire, (c) PEDOT-DBSA nanowire, and (d) PAN-HClO$_4$ nanotube. (From Joo, J., et al., *Synth. Met.*, 135, 7, 2003. With permission.)

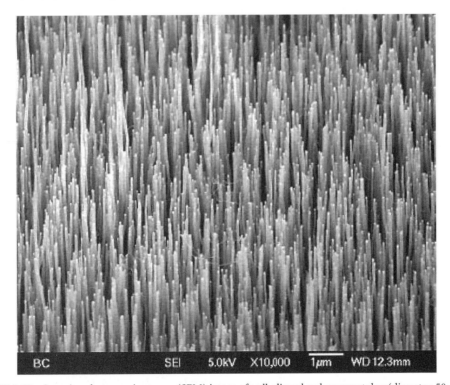

FIGURE 8.60 Scanning electron microscope (SEM) image of well-aligned carbon nanotubes (diameter, 50–70 nm; length, 3–4 μm) grown on a titanium substrate by plasma-enhanced chemical vapor deposition. (From Chen, J.H., et al., *Appl. Phys. A*, 73, 129, 2001. With permission.)

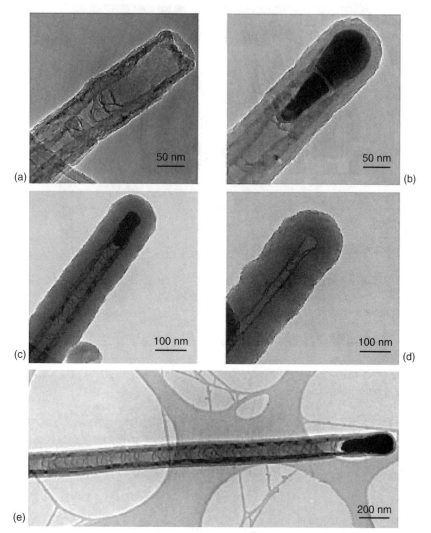

FIGURE 8.61 Transmission electron microscope (TEM) images of PPy/carbon nanotube nanoscale composites with different film thicknesses due to different PPy film-formation charges: (a) 86.1 mC/cm^2; (b) 207.9 mC/cm^2; (c) 681.9 mC/cm^2; (d) 1308.6 mC/cm^2; (e) Transmission electron microscope (TEM) image of a long PPy-coated carbon nanotube (681.9 mC/cm^2). (From Chen, J.H., et al., *Appl. Phys. A*, 73, 129, 2001. With permission.)

Octadecyltrichlorosilane-patterned Si surface in aqueous acetonitrile solution facilitated positive deposition to produce PPy film with a negative pattern, whereas negative deposition occurred in nonaqueous solution, resulting in a positive PPy pattern. The electrodeposition preferentially occurred on the exposed area of the gold substrate, even though the deposition on the octadecanthiol-covered area could not be prevented because of the hydrophobic–hydrophobic interaction.

8.5.1.2   Chemical Deposition of PPy

PPy pattern was chemically formed on a glass substrate by combining the selective deposition of PPy on a hydrophobic surface with the microcontact printing technique [121]. The computer-generated, hydrophobic pattern film was fabricated on hydrophilic glass substrates using microcontact printing. PPy thin film was deposited preferentially on the hydrophobic surface from dilute aqueous polymerization solution, while metals such as Ni were selectively deposited on the hydrophilic glass surface by an electroless process. Since the thickness and electronic properties of the PPy films formed on the

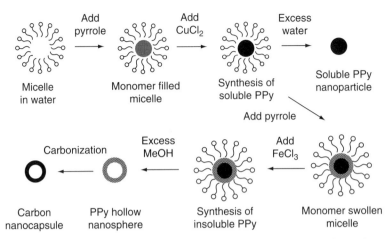

FIGURE 8.62 Schematic diagram of the fabrication of PPy and carbon hollow nanospheres. (From Jang, J.S., Li, X.L., and Oh, J.H., *Chem. Commun.*, 794, 2004. With permission.)

hydrophobic and hydrophilic glass surfaces were significantly different, this selectivity could be used for fabrication of a PPy pattern suitable for use as electrodes for many applications.

Even though conducting polymers have many advantages such as flexibility and relatively low cost, the complicated methods necessary to obtain customized conducting polymer patterns are obstacles in many applications. Especially in disposable electronics, the number of processing steps and chemicals should be minimized. It was reported that a novel method for the preparation of patterns, called "Line Patterning," could produce a PPy pattern with an acceptable resolution [122]. In line patterning, PPy

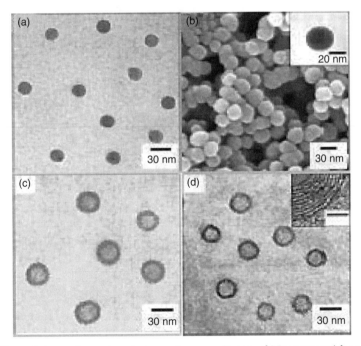

FIGURE 8.63 Transmission and scanning electron microscope images of PPy nanoparticles and hollow nanospheres: (a) soluble PPy nanoparticles fabricated using 0.45 M DeTAB; (b) linear PPy/crosslinked PPy core/shell nanoparticles; (c) PPy nanocapsules; (d) carbon nanocapsules (inset: a HRTEM image of the wall). (From Jang, J.S., Li, X.L., and Oh, J.H., *Chem. Commun.*, 794, 2004. With permission.)

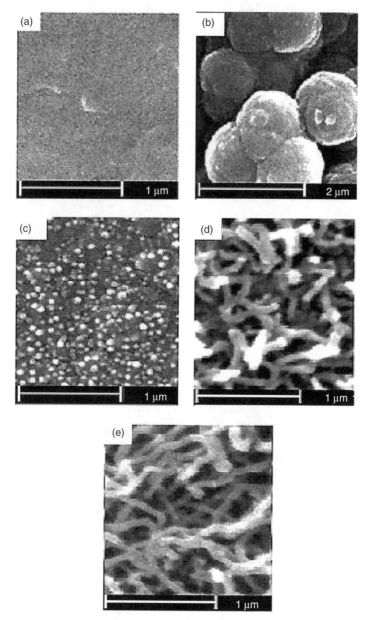

FIGURE 8.64 Scanning electron microscope pictures of naked and PPy-modified graphite electrode surface: (a) naked electrode, polished with 1200 emery paper; (b) PPy with "cauliflower" form electrogenerated in a solution without sodium carbonate by passing about 180 mC/cm^2 of charge; PPy fibrils electrogenerated in a solution with sodium carbonate by passing different amounts of charge; (c) 80 mC/cm^2; (d) 180 mC/cm^2; (e) 650 mC/cm^2. (From Mo, X., et al., *Synth. Met.*, 142, 217, 2004. With permission.)

was in situ deposited chemically on the electrode with a pattern printed by a commercial laser printer. Toluenesulfonate and chloride were used as the dopant anions in order to produce different surface resistivities on the printed and nonprinted areas of the pattern as shown in Table 8.11. PPy/toluenesulfonate was deposited preferentially on the printed lines, while PPy/chloride was deposited on both printed lines and bare transparency surface. After the electrode was sonicated in toluene, the printed lines were removed to leave PPy on the transparency, producing a PPy pattern.

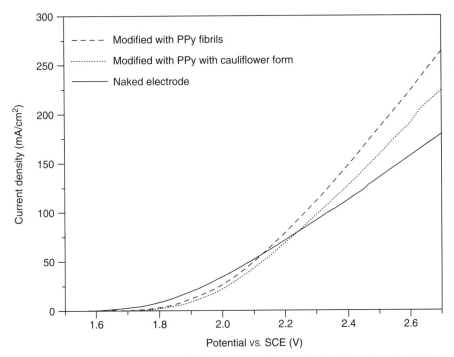

FIGURE 8.65 Comparison of current-potential relations on different cathodes in 30% KOH (w/w) solution at 25°C . (From Mo, X., et al., *Synth. Met.*, 142, 217, 2004. With permission.)

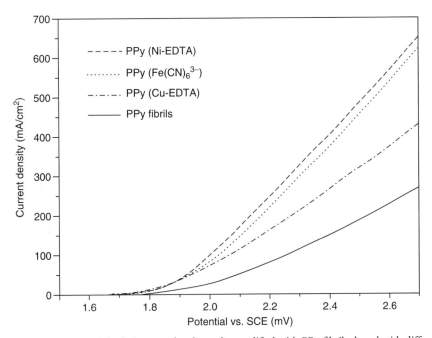

FIGURE 8.66 Current-potential relations on the electrodes modified with PPy fibrils doped with different metal complex ions in 30% KOH (w/w) solution at 25°C. (From Mo, X., et al., *Synth. Met.*, 142, 217, 2004. With permission.)

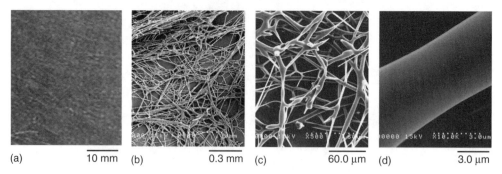

(a)   10 mm   (b)   0.3 mm   (c)   60.0 µm   (d)   3.0 µm

FIGURE 8.67 Morphologies of the electrospun PPy fiber and nonwoven web. (From Kang, T.S., et al., *Synth. Met.*, 151, 61, 2005. With permission.)

Using PPy film and PU elastic sheet, a stripe-patterned surface could be obtained (Figure 8.74) [123]. PPy was deposited on the mechanically drawn PU elastic sheet, after which the mechanical stress was released, leading to the stripe-patterned PPy film. The stripe-pattern could be obtained due to the inelasticity of PPy and the elasticity of PU. The width and depth of the stripe were controlled by the elongation ratio of PU sheet and the PPy film thickness. SEM images of the prepared PPy film in Figure 8.75 show a narrower width pattern due to the thinner PPy film. The cracks parallel to the drawing axis shown in Figure 8.75e were formed because the release of mechanical drawing resulted in the expansion of the sheet perpendicular to the drawing direction.

8.5.2 Patterning by Photolithography

Micro-patterns of the electrically conducting polymer thin films were fabricated using a simple photo-lithographic bleaching of the oxidant (Figure 8.76) [124,125]. The oxidant film was first formed on a glass or flexible and transparent polyester (PET) or PP film substrate by spin-coating the oxidant aqueous solution containing ferric p-toluene sulfonate (FTS) or ferric chloride as the oxidant and polyvinyl alcohol (PVA) as the matrix polymer. The oxidant film was exposed to UV light (365 nm) with an intensity of 10 mW/cm^2 on the surface of the film for 10 min through a photomask. Because Fe^{3+} in the oxidant is transformed into Fe^{2+} upon UV irradiation, the oxidant in the exposed area can no longer oxidize the monomer. The patterned oxidant film was then exposed to pyrrole vapor, resulting in polymerization of PPy only in the unirradiated region. Figure 8.77 shows optical photographs of PPy patterns obtained by the photolithographic method, in which patterns with a line width as low as 5 µm were successfully fabricated. Since the resistivity of the insulating region is about 10^{10}-fold higher than that of the PPy region, the intervening PVA region was able to work as an insulating barrier.

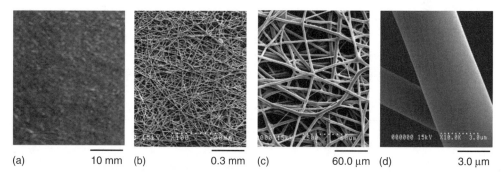

(a)   10 mm   (b)   0.3 mm   (c)   60.0 µm   (d)   3.0 µm

FIGURE 8.68 Morphologies of the electrospun PPy/PVCN blend fiber and nonwoven web. (From Kang, T.S., et al., *Synth. Met.*, 151, 61, 2005. With permission.)

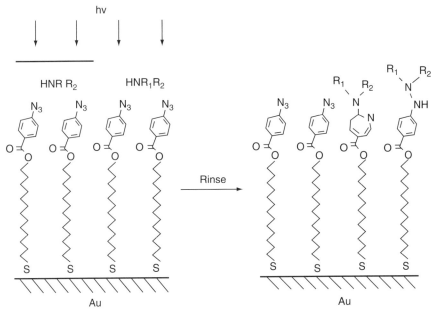

FIGURE 8.69 Photopatterning of SAMs on Au. (From Rozsnyai, L.F., and Wrighton, M.S., *Langmuir*, 11, 3913, 1995. With permission.)

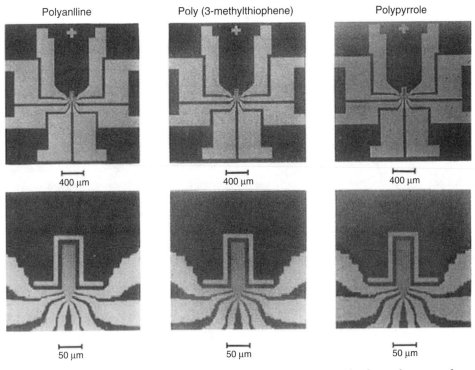

FIGURE 8.70 Optical micrographs of polyaniline, poly(3-methylthiophene), and polypyrrole patterns formed by oxidizing the monomers on patterned SAM substrates, as described in the text. The dark regions correspond to polymer deposited on nonirradiated self-assembled monolayers of I. The light regions contain little or no polymer and correspond to Au-I SAMs irradiated in the presence of $[CH_3(CH_2)_7]_2NH$. The microstructures in the bottom micrographs are 4 μm center-to-center. (From Rozsnyai, L.F., and Wrighton, M.S., *Langmuir*, 11, 3913, 1995. With permission.)

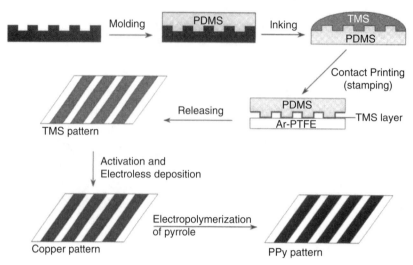

FIGURE 8.71 Schematic representation of the procedure to prepare the PPy patterns on the PTFE surfaces by using μCP. (From Prissanaroon, W., et al., *Thin Solid Films*, 477, 131, 2005. With permission.)

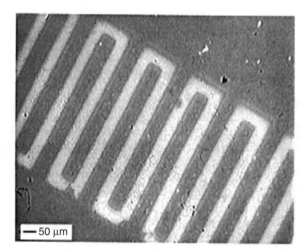

FIGURE 8.72 Patterns used to illustrate requirements for conducting polymer-based circuitry. Dots indicate placement of the probe wires for conductivity measurements. An optical micrograph of this interdigitated pattern after contact transfer to PDMS. Inherent limitations in self-assembled monolayers (SAM) are presumably responsible for the defects observed. (From Gorman, C.B., Biebuyck, H.A., and Whitesides, G.M., *Chem. Mater.*, 7, 526, 1995. With permission.)

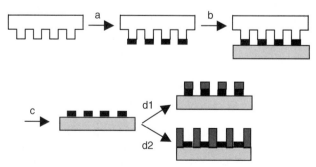

FIGURE 8.73 Summary of the procedures: (a) inking; (b) micro contact; (c) printing; (d1) negative electrodeposition; (d2) positive electrodeposition.

TABLE 8.11 In-Site Deposition of Polypyrrole on Laser-Printed Patterns on Overhead Transparency

	Surface Resistivity of Printed Lines/Areas (kΩ/□)	Surface Sensitivity of Bare Transparency (kΩ/□)
Polypyrrole/toluenesulfonate	12	>20,000 (out of measurement range)
Polypyrrole/chloride	18	80
Polypyrrole/chloride (after sonication)	>20,000 (out of measurement range)	80

Source: Reprinted from Hohnholz, D., and MacDiarmid, A.G., *Synth. Met.*, 121, 1328, 2001. With permission.

It was also reported that copper could be selectively formed on the pattern PPy electrode by electroless plating [124]. The PPy pattern, which was formed by the photolithographic patterning technique, was first immersed in acidic PdCl$_2$ solution, leading to selective adsorption of the palladium metal particles on the PPy pattern. The PPy pattern coated with palladium metal was then immersed in copper electroless plating solution, producing the selective copper-on-PPy pattern (Figure 8.78).

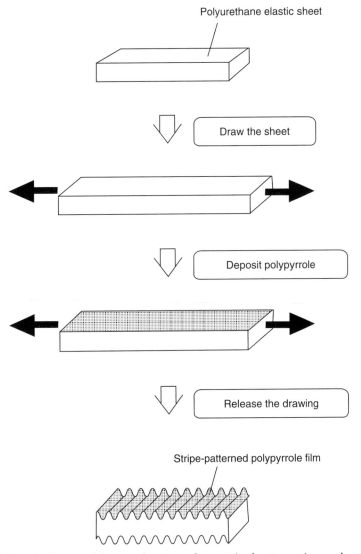

FIGURE 8.74 Schematic diagram of the procedure to produce a striped pattern using a polypyrrole film. (From Masashi, W., and Toshihiro, H., *J. Polym. Sci. Pol. Chem.*, 42, 2460, 2004. With permission.)

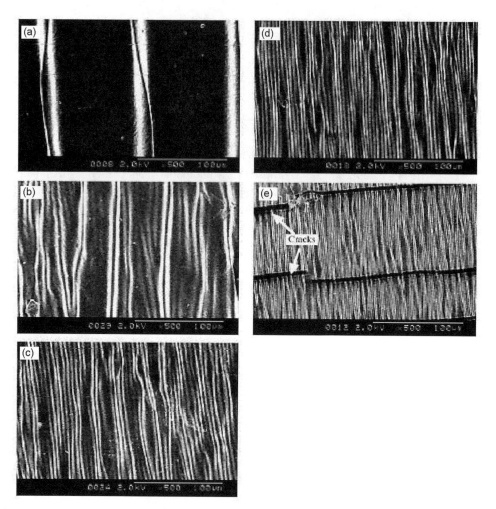

FIGURE 8.75 Scanning electron microscope (SEM) images of the surface of the polypyrrole films. These films were prepared by the electrochemical deposition for (a) 15, (b) 10, (c) 5, (d) 3, and (e) 1 min onto a drawn polyurethane sheet. The elongation ratio was 30%, and the drawing direction corresponded to the horizontal direction of these images. (From Masashi, W., and Toshihiro, H., *J. Polym. Sci. Pol. Chem.*, 42, 2460, 2004. With permission.)

8.6   Potential Applications of Polypyrrole

8.6.1   Sensors

Since protection of individuals and the environment has become more important, improved sensors to measure chemical and physical changes such as chemical species, temperature, pressure, or humidity are required. Much research has been carried out to apply conducting polymers as sensor materials.

8.6.1.1   Gas Sensor

Many studies have investigated the development of gas sensors containing PPy due to its gas sensing ability which is based on the dependence of conductivity on the kind or concentration of gases [126–128]. NO is an important bioregulatory molecule which plays a major role in many physiological and pathological processes. Sensitive, stable, and fast responding sensors for the detection of NO were obtained by electrochemical polymerization of pyrrole substituted Mn-*meso-p*-tetracarboxyphenylporphyrin (Mn(II)TCPPyP) [129]. In the case of PPy-PVA composite films, it was reported that sensitivity,

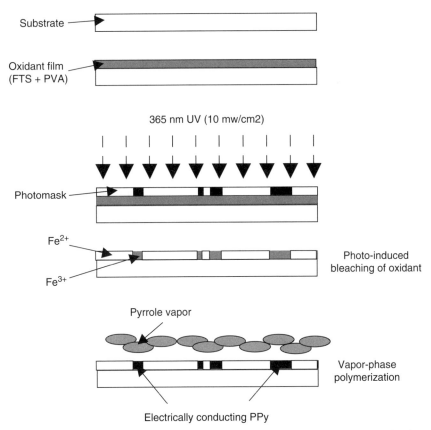

FIGURE 8.76 Scheme of photolithographic patterning of PPy film. (From Choi, M.S., et al., *Mol. Cryst. Liq. Cryst.*, 377, 181, 2002. With permission.)

response, and recovery time of PPy-PVA sensor to methanol depended on the electrical polymerization charge [127]. A higher sensitivity was obtained when the sensor was polymerized with a lower electrical charge. The PPy-PVA sensor prepared with 50 mC was about ninefold more sensitive to methanol than the PPy sensor [127].

Not only gas sensors but humidity sensors are also important in many fields including paper or electronic industries, domestic environment equipment such as air conditioner, and medical supplies

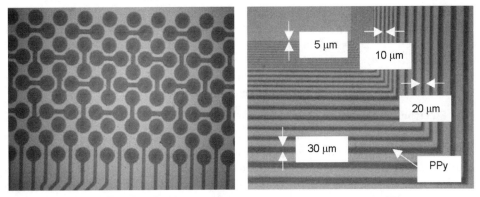

FIGURE 8.77 Optical microscopy photographs of PPy pattern obtained by the photolithographic patterning method. (From Choi, M.S., et al., *Mol. Cryst. Liq. Cryst.*, 377, 181, 2002. With permission.)

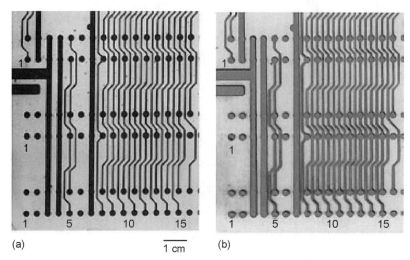

(a)   1 cm   (b)

FIGURE 8.78 Photograph of selective electroless Cu plating on a PPy pattern: (a) original PPy pattern; (b) after electroless Cu plating. (From Yasushi, O., Shoji, Y., and Masaki, K., *Synth. Met.*, 144, 265, 2004. With permission.)

such as respiratory equipment. Nanocomposites of PPy with iron oxide were prepared by sol–gel process to investigate the humidity and gas sensing properties [128]. The nanocomposites showed better sensitivity to humidity than α-Fe_2O_3 and the sensitivity increased with increasing PPy concentration.

8.6.1.2 Ion Sensor

PPy is known to exhibit anionic or cationic potentiometric responses depending on the doping ion. PPy films doped with mobile inorganic anions show anionic sensitivity, while PPy films doped with anions of low mobility such as sulfate, large organic anions, or polyanions show cationic sensitivity. This bifunctionality of ionic and redox sensitivity makes PPy applicable as an all-solid-state, ion-selective electrode. All-solid-state, potentiometric PPy sensors were developed for potassium and sodium sensing, which showed better response time, selectivity and most importantly long-term stability than the coated wire electrode [130].

An anion's lipophilicity plays an important role when the anion-selective electrodes involving ionophores with no functionalities are applied to the recognition of anions. The selectivity order for this recognition is known as the Hofmeister series:

$$\text{Large lipophilic anions} > ClO_4^- > SCN^- > I^- > NO_3^- > Br^- > Cl^- > H_2PO_4^-.$$

Therefore, to prepare a selective anion sensor for less lipophilic anions in the series, the selective interactions between the ionophores and the targeted anion should be considered. Truly selective nitrate electrodes were developed by the electrochemical deposition of PPy onto glassy carbon electrodes in the presence of $NaNO_3$ [131]. As seen in Figure 8.79, the selectivity pattern of $PPy(NO_3^-)$ electrodes exhibited a significant deviation from the Hofmeister series. The electrodes showed selectivity especially for nitrate and thiocyanate, while the commercially available nitrate-selective electrodes were affected by the more lipophilic anions including iodide and perchlorate.

8.6.1.3 Electronic Tongue

In the food and beverage industry, the use of artificial sensors to evaluate tastants is a very important tool to improve quality control since these compounds are present at high concentrations in foodstuffs. An artificial tongue must be able to distinguish the four basic tastes: sweet, sour, bitter, and salty. Since a single material cannot be sensitive to all four basic tastes, sensors are generally fabricated combining different components. The ultra thin, electronic tongue made of four different types of materials (Figure 8.80) was prepared by the Langmuir–Blodgett and self-assembly technique. PPy and self-assembled

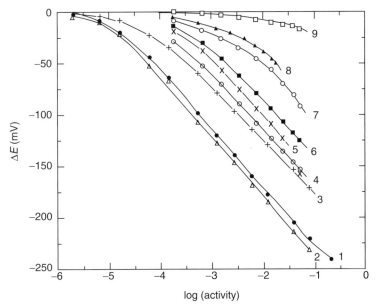

FIGURE 8.79 Selectivity pattern of PPy(NO_3^-) electrodes. The anions tested include (1) nitrate, (2) thiocyanate, (3) bromide, (4) iodide, (5) chloride, (6) perchlorate, (7) salicylate, (8) sulfate, and (9) phosphate ($n = 18$ for nitrate; $n = 3$ for all other ions). ΔE is the difference between the steady-state potential and the starting potential (i.e., potential of the cell before any addition of anions). Electrodes were conditioned in 1.00×10^{-2} M NaNO$_3$ for a minimum of 1 h between experiments. (From Hutchins, R.S., and Bachas, L.G., *Anal. Chem.*, 67, 1654, 1995. With permission.)

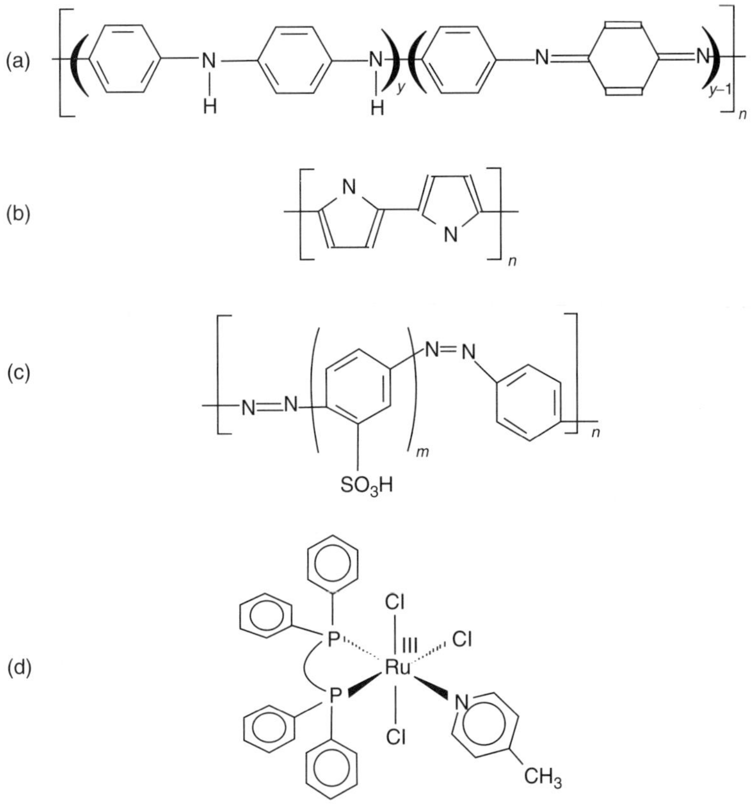

(a)

(b)

(c)

(d)

FIGURE 8.80 Structural formulas of the materials used. (From Riul, A., et al., *Langmuir*, 18, 239, 2002. With permission.)

polymer films were reported to detect salty and sour tastes better than Ru-Pic, while Ru-Pic and the conducting polymers were superior for detecting sweet substances [132]. In addition, an electronic tongue made of six different sensing units, bare interdigitated electrodes, interdigitated electrodes coated with stearic acid (SA), PANi oligomer (16-mer), PPy and mixtures of 16-mer/SA and PPy/SA, was prepared and investigated [133]. The sensor showed better sensitivity than the human tongue to detect concentrations as low as 5 mM.

8.6.2 Actuator

8.6.2.1 Electrochemical Actuator

Conducting polymers have attracted considerable attention because dimensional changes resulting from electrochemical doping and dedoping can be applied to produce polymeric actuators or artificial muscles [134–137]. The changes in dimensions of conducting polymers are related to the reversible reduction and oxidation processes in the solid state. The expansion of the polymer results from ionic and molecular influxes from the solution in order to maintain the electroneutrality of the polymer when a neutral conducting polymer film is oxidized either chemically or electrochemically. During the reduction process, the opposite phenomena are observed: by the injection of the electrons into the solid, positive charges are eliminated, while counterions and solvent molecules are expelled to the solution, thereby decreasing the volume of the polymer. These electromechanical actuators are termed artificial muscles when they can make angular movements of more than 360°, and are capable of dragging several hundred times their own weight.

Anode, cathode, and electrolyte are the three main elements in reversible electrochemical actuators and the conducting polymer actuators can be classified as extensional actuators and hydrostatic actuators. In order to perform mechanical work, linear or biaxial dimensional changes are used for the extensional actuators as illustrated in Figure 8.81 [134]. On the other hand, during the electrochemical redox processes, the overall volume change of the anode, cathode, and electrolyte is considered to perform the mechanical work in the hydrostatic actuators.

Using atomic force microscopy (AFM), the volume change of PPy doped with DBS was measured directly on thin films [137]. PPy-DBS was cycled electrochemically between oxidation and reduction states in 0.1 M NaDBS aqueous solution between 0 and -1 V at a rate of 5 mV/s (Figure 8.82). The large current peak during the first reduction in Figure 8.82a was associated with a large expansion. After the oxidation peak at -0.7 V ($t = 260$ s), the film contracted, but not to its original size, indicating the anisotropic nature of the expansion. For the third cycle shown in Figure 8.82b, the film is swollen to 1.3 μm in oxidized state and expands to 1.85 μm in the reduced state. Therefore, a 60% to 100% increase in volume from the as-grown film was observed during the first reduction, followed by a 30% to 40% volume change between oxidation and reduction state.

Generally, due to their intrinsic electrical resistance, PPy films show much lower force density than the theoretical value. This drawback was overcome by applying PPy as a coating on a metallized polymer substrate. It was reported that bundled PPy on Pt-coated polyester fiber had the maximum available force and stress among various types of films and fiber bundles as shown in Table 8.12 [135]. In order to improve the relatively low speed and low power/mass ratio of conducting polymer actuators, shaped voltage pulses were also applied to PPy linear and bilayer actuators with liquid or gel electrolytes [136]. Strain rates in PPy actuators increased with increasing applied potential and little or no degradation was observed following the application of potentials for relatively short periods of less than 1 s. For PPy films of 40 μm thickness, a strain rate of 3.2%/s and power/mass ratio of 39 W/kg were achieved, which is nearly equal to the power/mass ratio of mammalian skeletal muscle.

The volume changes of the polymer may result partly from the changes of the polymer backbone, due to the changes of bond lengths and conformation, and partly from osmotic expansion, which is the volume change associated with the solvent movement, of the polymer phase. The osmotic expansion of PPy-DBS was thermodynamically and experimentally described to investigate the osmotic effect on the

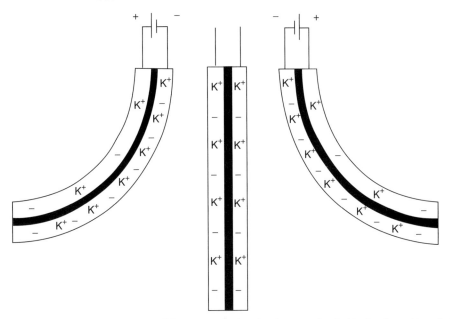

FIGURE 8.81 Schematic representation of three states during the electromechanical cycle of a rocking-chair-type, bimorph actuator. Both electrodes have the same concentration of dopant (K^+) when the cantilever is undistorted, and electrochemical transfer of dopant between electrodes causes bending either to the right or to the left. (From Baughman, R.H., *Synth. Met.*, 78, 339, 1996. With permission.)

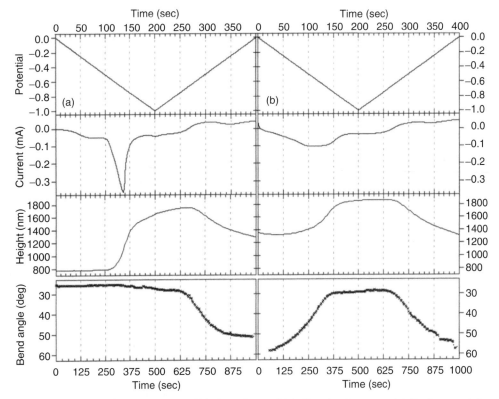

FIGURE 8.82 Cyclic voltammetry at 5 mV/s for a 3 μm thick film showing potential (top), current (upper middle), and film thickness (lower middle). Bilayer bending (bottom) at 2 mV/s is shown for comparison (the timescale is therefore 5/2 longer). (a) First reduction cycle. (b) Third reduction cycle. The currents have been corrected to reflect only that from the PPy by subtracting the current drawn by a bare gold reference surface. (From Smela, E., and Gadegaard, N., *Adv. Mater.*, 11, 953, 1999. With permission.)

TABLE 8.12   Summary of Stress Generated from Various Actuator Designs

Design	Stress Generated Based on Total Thickness (MPa)	Stress Generated Based on PPy Coating Thickness (MPa)	Force Generated per Unit Width (N/mm)	Comments
(1) Free-standing film	0.5	0.5	0.01	Limited by resistivity
(2) Pt-coated free-standing film	3.0	3.0	0.06	Very brittle films
(3) Laminated film	0.18	1.25	0.025	Stiffness of substrate must be overcome
(4) Axial force solid-state device	0.002	0.18	0.007	Nonoptimal design and low interface adhesion
(5) Fibre bundle	5.1	6.8	0.10	Could be improved by limiting fibre coagulation

Source: Reprinted from Hutchison, A.S., et al., *Synth. Met.*, 113, 127, 2000. With permission.

volume change [138]. Volume expansion decreased with increasing electrolyte concentration, indicating that the osmotic effect played an important role in the total expansion.

8.6.2.2   Chemomechanical Actuator

Without using ions, electric field or heat, electrochemically synthesized PPy films may exhibit fast and intensive bending in ambient air on the basis of a reversible van der Waals adsorption of polar, organic or water molecules in the vapor state [139–141]. As shown in Table 8.13, the chemomechanical response of the actuator strongly depended on the adsorbate [139]. It was reported that PPy film exhibited contraction when an electric field was applied in ambient air because of the contraction caused by desorption of water vapor and thermal expansion of polymer chains [142].

Figure 8.83 shows the schematic construction of a chemomechanical rotor, where anisotropic expansion causes rapid bending of the film due to the sorption of water vapor from one side of the film [140]. The rotation continued until the adsorbates were completely vaporized. This moving device may become a clean and silent power source for use as a molecular engine where the chemical free adsorption energy is transduced into the mechanical work.

8.6.2.3   Electroactive Polymer Actuators

A wrinkled PPy electrode was prepared as an electrode for a bending-electrostrictive, electroactive PU actuator [143]. Electroactive polymers are those that respond to electrical stimulation with changes in size or shape. For application as an electrode for electroactive polymer actuators, the conductivity should be maintained during the field-induced elongation of the electroactive polymer actuators. Since preparing electrodes from metal or a conducting polymer is not easy due to their inelastic properties, the wrinkled electrode, which could elongate easily by smoothing the wrinkles, was designed (Figure 8.84). The wrinkled PPy electrode was made by in situ deposition of PPy onto the mechanically drawn PU elastomer film. After the deposition, the film was released from the drawn state to form the wrinkled

TABLE 8.13   Chemomechanical Response of the PPy/ClO_4 Film to Various Adsorbates

Response	Adsorbate
Bending to the opposite side	Water, organic or inorganic aqueous solutions
Bending to the same side	Alcohols, ketones, aldehydes, amines, ethers, nitrils, monohaloakanes
No response	Alkanes, alkenes, aromatic and involatile compounds

Source: Okuzaki, H., and Kunugi, T., *J. Appl. Polym. Sci.*, 64, 383, 1997. With permission.

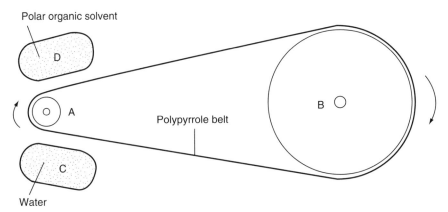

FIGURE 8.83 Schematic construction of a chemomechanical rotor (side view). Radius of pulley A: 1 mm; pulley B: 5 mm; wheel base: 20 mm. C: tissue containing water; D: tissue containing polar organic solvent. (From Okuzaki, H., and Kunugi, T., *J. Polym. Sci. Pol. Phys.*, 34, 1747, 1996. With permission.)

electrode. The wrinkles allowed the electrode to elongate without decreasing the conductivity and improved the bending-electrostrictive actuation of the PU film, especially when doped with sodium acetate as compared in Figure 8.85 [143].

8.6.3 Electrode for Secondary Battery, Solid Electrolytic Capacitor or Supercapacitor

8.6.3.1 Electrode for Secondary Battery

Due to its capacity to store recoverable charge, PPy has been recognized as a good candidate for the elements of advanced rechargeable batteries [144–150] which function by the interchange of ions between the electrodes and the electrolyte solution. It was reported that synthesis conditions greatly influenced the specific charge density of PPy [144]. PPy films with high specific charge density were obtained at low polymerization potential, short polymerization time, and temperature lower than 30°C. As shown in Figure 8.86, solvent was also an important factor to yield the PPy films with high specific charge density [144]. Some of the solvent parameters were analyzed by multiple regression analysis, suggesting that the best solvents with high charge storage ability were those having high dipole moments, low polarizability, and high capacity to donate electrons [145].

It was also reported that the Li/PPy batteries showed excellent performance with 3 V output voltage and 90% to 100% Coulombic efficiency when PMMA- and PAN-based gel electrolyte was used (Figure 8.87) [149]. Especially, a long charge/discharge cycle life of more than 8000 cycles was obtained by using a chemically more stable PMMA gel electrolyte [149]. PPy/graphite composites were prepared as negative electrodes in Li-ion cells by in situ polymerization of pyrrole onto commercial graphite

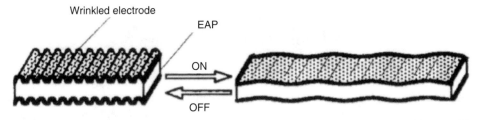

FIGURE 8.84 Concept of the wrinkled electrode for EAPs. The electrode can easily be elongated by smoothing the wrinkles. (From Watanabe, M., Shirai, H., and Hirai, T., *J. Appl. Phys.*, 92, 4631, 2002. With permsission.)

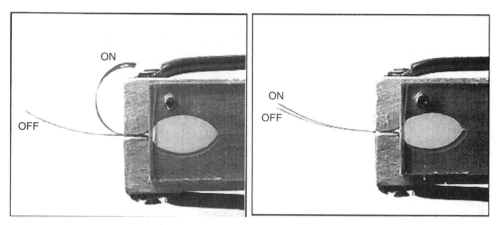

FIGURE 8.85 Video images of the bending actuation of the doped film with the wrinkled PPy electrode and the conventional gold electrode. The applied voltage was 1000 V (the electric field was 5 MV/m). (From Watanabe, M., Shirai, H., and Hirai, T., *J. Appl. Phys.*, 92, 4631, 2002. With permsission.)

[150]. Addition of PPy to the graphite decreased the charge-transfer resistance and overall polarization resistance, thereby exhibiting very good capability, conductivity efficiency, and cyclic behavior.

Since the viscosity and electrical conductivity of metal oxide powders such as $LiMn_2O_4$ or V_2O_5 are excessively low to make electrochemically active sheet or pellet, PPy has been used as both a binder and a conductor for metal oxide positive electrodes for rechargeable lithium batteries [146–148]. The composite pellets were prepared by chemical polymerization of pyrrole in the metal oxide-dispersed aqueous solution and by pressing the composite powder, in which PPy was found to work as a conducting matrix for the redox reaction. The metal oxide and conducting polymer composites exhibited good charge-

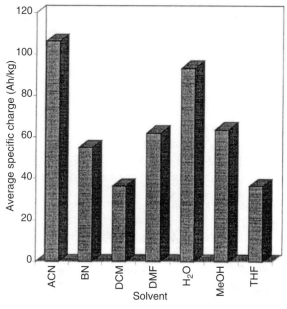

FIGURE 8.86 Average specific charges obtained in polypyrrole films using different conditions in acetonitrile (ACN), benzonitrile (BN), dichloromethane (DCM), *N,N'*-dimethylformamide (DMF), water (H_2O), methanol (MeOH), and tetrahydrofuran (THF). (From Otero, T.F., and Cantero, I., *J. Power Sources*, 81, 838, 1999. With permission.)

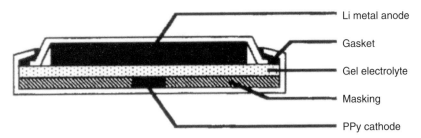

FIGURE 8.87 Schematic diagram of coin-type cell. (Osaka, T., et al., *J. Power Sources*, 68, 392, 1997. With permission.)

discharge performance when a conducting polymer had high electrical conductivity at potentials within which the redox reaction of the oxide occurred.

8.6.3.2 Electrode for Solid Electrolyte Capacitor

PPy has been used as a solid electrolyte for solid electrolyte capacitors because it possesses stability and higher conductivity, and can be easily prepared by chemical and electrochemical methods [151–158]. A solid state capacitor based on $PPy/Al_2O_3/Al$ was prepared by the constant current method and the factors affecting the film formation were investigated. The surface composition of Al foil pretreated with 0.1 M NaOH aqueous solution and anodized with 0.1 M DBSA aqueous solution was found to be AlO_2^- and Al_2O_3. The relationship between potential and the electrolysis time confirmed the presence of three stages as shown in Figure 8.88: the formation of Al_2O_3 and nucleation of PPy within the Al_2O_3 pores, the propagation of PPy on the Al_2O_3 barrier layer, and the overoxidation of PPy in the pores of Al_2O_3 [158]. When the preparation current density was greater than 0.9 mA/cm^2, the second stage became insignificant, resulting in very little PPy film formation on the Al_2O_3 barrier layer.

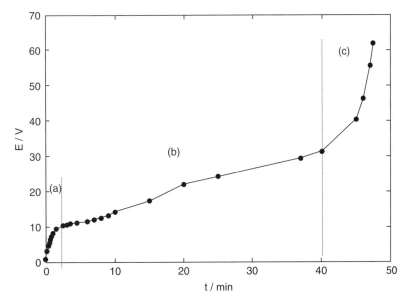

FIGURE 8.88 Effect of electrolysis time on potential for preparing $PPy/Al_2O_3/Al$. Area of Al foil = 1 × 1 cm, thickness of Al foil = 40 μm. Conditions for pre-treatment of Al foil: [NaOH] = 0.1 M, T = 30°C, t = 0.5 h. Conditions for preparing $PPy/Al_2O_3/Al/Al$: [pyrrole] = 0.1 M, [DBSA] = 0.1 M, initial pH = 1.3, T = 16°C. Counter electrode: 2 × 3 cm stainless-steel plate. Current density = 0.5 mA cm^{-2}. (a) First stage; (b) second stage, and (c) third stage. (From Tsai, M.L., Chen, P.J., and Do, J.S., *J. Power Sources*, 133, 302, 2004. With permission.)

8.6.3.3 Electrode for Supercapacitor

There has been growing interest in the field of supercapacitors due to their possible applications in medical devices, electrical vehicles, memory protection of computer electronics, and cellular communication devices. Their specific energies are generally greater than those of electrolytic capacitors and their specific power levels are higher than those of batteries. Supercapacitors can be divided into redox supercapacitors and electrical double layer capacitors (EDLCs). The former uses electroactive materials such as insertion-type compounds or conducting polymers as the electrode, while the latter uses carbon or other similar materials as the blocking electrode.

Advances in microtechnology and nanotechnology require micropower sources having higher power density with smaller dimensions. Microcapacitors using conducting polymer microelectrodes were fabricated by photolithography and electrochemical polymerization (Figure 8.89) [159]. By the photolithographic technique, the electron size, anode-cathode distance, and geometry to a sub-micron scale could be accurately controlled. Depending on the conducting polymers and electrolyte types, cell potentials between 0.6 and 1.4 V could be obtained while the charge of the conducting polymers and cell capacitances was easily controllable [159].

Multiwalled carbon nanotubes (MWNTs) have been investigated as an electrode material for supercapacitors [160–166]. When MWNT and single-walled nanotubes (SWNTs) were compared as active electrode materials for energy storage, MWNTs were reported to have twice the capacitance value of SWNTs due to their unique mesoporosity, which enables the accumulation of charges at the interface of the electrode and electrolyte [166]. To take advantage of the electronic and ionic conducting properties of PPy in the opened mesoporous network of nanotubes, PPy-modified CNTs were designed as an active electrode material for a supercapacitor. As shown in Figure 8.90, homogeneous coating was obtained by electrochemical polymerization of pyrrole on the nanotubular materials, leading to enhanced specific capacitance [165]. The electrochemically grown composite films of MWNTs and PPy were compared with the pure PPy films (Figure 8.91). The composite showed excellent charge storage and transfer capabilities, due to the high surface area, conductivity, and electrolytic accessibility of the nanoporous structure [164].

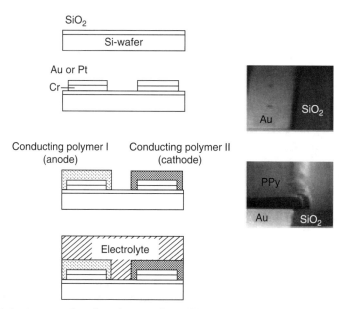

FIGURE 8.89 Fabrication procedure for microcapacitors. (From Sung, J.H., Kim, S.J., and Lee, K.H., *J. Power Sources*, 124, 343, 2003. With permission.)

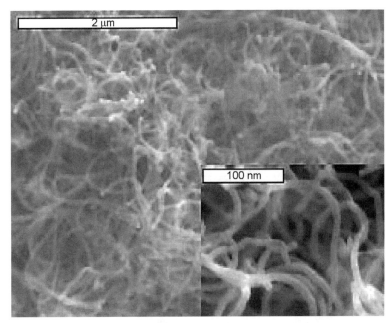

FIGURE 8.90 Scanning electron microscope (SEM) micrograph of a homogeneous population of carbon nanotubes A/CoNaY600 with electrodeposited polypyrrole. (From Jurewicz, K., et al., *Chem. Phys. Lett.*, 347, 36, 2001. With permission.)

8.6.4 Field Effect Transistor

Even though organic field effect transistors (FETs) have many advantageous properties such as light weight, flexibility as well as low temperature OTFTs process, and low preparation cost compared to conventional inorganic FETs, they do have some associated problems such as poor film forming ability of small organic molecules and current leakage due to the poor interfacial adhesion between the metal electrode and the organic active layer. Therefore, electrically conducting polymers have been considered as promising materials for both the electrodes and the active layers due to their high electrical conductivity, good film forming ability, flexibility, transparency, and environmental stability at room temperature as well as ease of synthesis [167–169].

The flexible, all-polymer FETs, in which the substrate, electrodes, active layer, and insulating layer are all composed of polymers, could be prepared by a simple, low-cost process, namely the simple photolithographic method at room temperature with high optical transmittance and mechanical flexibility [169]. The fabrication process was based on a simple photolithographic micro-patterning of the electrically conducting polymer [169,170], which produced the patterns with desired shape and good resolution (Figure 8.77). Using poly(3,4-ethylenedioxythiophene) or PPy as both electrodes and the active channel, with photocrosslinked polyvinyl cinnamate or epoxy/methacrylate polymers as the electric layer, all-polymer FETs were fabricated with the schematic structure shown in Figure 8.92. The p-type, all-polymer FETs were reported to work in a depletion mode with the application of a positive gate voltage [169].

Metal-oxide-semiconductor FETs (MOSFETs) were also prepared by electrochemical polymerization using PPy and N-alkyl substituted PPy films, including poly (N-methylpyrrole) (PNMePy) and poly (N-ethylpyrrole) (PNEtPy) as a p-type semiconductor and p-toluenesulfonic acid monohydrate as a supporting electrolyte [168]. Figure 8.93 represents the cross-sectional view of the fabricated MOSFET. The mobility of PPy and PNEtPy FETs was 1.7 cm^2 V^{-1} s^{-1}, which is close to the value of silicon inorganic transistors [168].

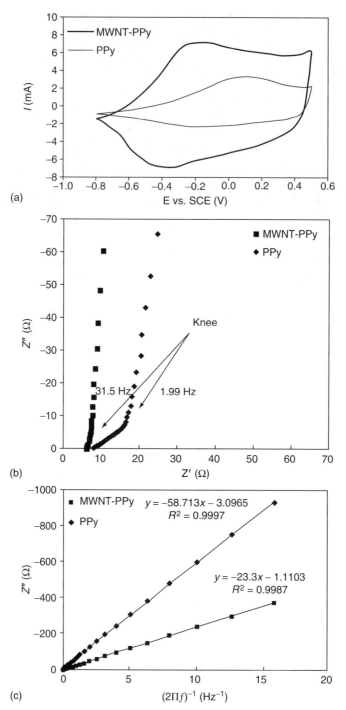

FIGURE 8.91 Comparison of MWNT-PPy composite films and pure PPy films prepared using similar conditions: (a) cyclic voltammograms (film-formation charge, 1.8 C cm^{-2}; scan rate, 50 mV s^{-1}; electrolyte, 0.5 M KCl); (b) complex plane impedance plots (film-formation charge, 1.0 C cm^{-2}; bias potential, 0.4 V vs. SCE; electrolyte, 0.5 M KCl); (c) relationship between the imaginary component of impedance and the inverse of frequency (film-formation charge, 1.0 C cm^{-2}; bias potential, 0.4 V vs. SCE; electrolyte, 0.5 M KCl). (From Hughes, M., et al., *Chem. Mater.*, 14, 1610, 2002. With permission.)

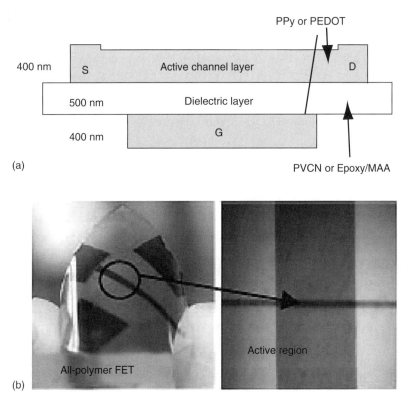

(a)

(b)

FIGURE 8.92 Schematic structure of the flexible all-polymer field effect transistor (FET) with optical transparency. (From Lee, M.S., et al., *Thin Solid Films*, 477, 169, 2005. With permission.)

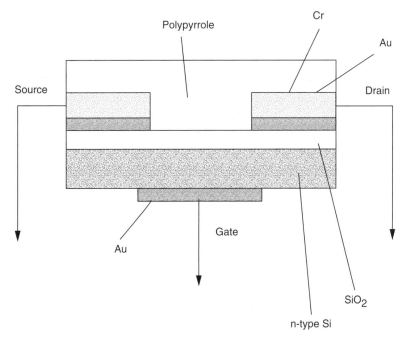

FIGURE 8.93 Cross-sectional view of fabricated MOSFET. (From Kou, C.T., and Liou, T.R., *Synth. Met.*, 82, 167, 1996. With permission.)

8.6.5   Electromagnetic Interference Absorbing Material

Electromagnetic interference (EMI) shielding by absorption rather than reflection is presently more important for many applications. Even though metals or metal-coated materials exhibit very high EMI shielding efficiency ranging from 40 to 100 dB, they cannot be used as an electromagnetic wave absorbent since their shallow skin depth makes them shield electromagnetic waves mainly through surface reflection. On the other hand, electrically conducting polymers are capable of not only reflecting but also absorbing the electromagnetic waves and therefore exhibit a significant advantage over the metallic shielding materials.

It was reported that coating PPy on textiles gave rise to electrically conducting textiles useful for many applications. PPy-coated textiles were produced by in situ chemical polymerization of pyrrole [171] or diffusion of pyrrole vapor in the presence of textile substrates [172]. It was also reported that PPy was polymerized chemically and electrochemically in sequence on a PET woven fabric, giving rise to a PET fabric/PPy composite with high electrical conductivity [173]. In the chemical polymerization, pyrrole, dissolved in an aqueous solution with polyvinyl alcohol as a surfactant, was first sprayed on the PET fabric and then oxidized by spraying an aqueous solution of an oxidant and a dopant. Electrochemical polymerization was carried out in an aqueous electrolyte solution by applying a constant current density to the PET fabric coated with chemically polymerized PPy and a stainless steel plate as the working and counter electrodes, respectively. The composite product shielded EMI by absorption as well as reflection, while the latter increased with increasing electrical conductivity. As shown in Figure 8.94, the specific volume resistivity of the composite was as low as about 0.2 Ω^- cm, producing an EMI shielding efficiency of about 36 dB over a wide frequency range in which 12% of the incident wave power was absorbed by the composite [173].

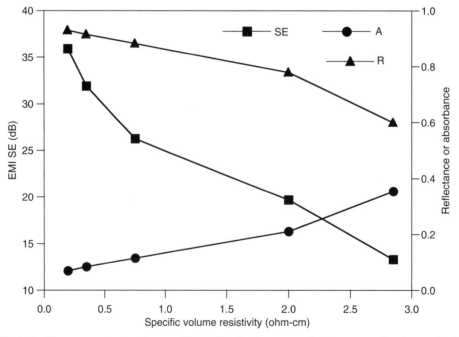

FIGURE 8.94   Electromagnetic interference (EMI) SE, absorbances, and reflectances of polyester fabric/PPy composites with various specific volume resistivities. (From Kim, M.S., et al., *Synth. Met.*, 126, 233, 2002. With permission.)

References

1. Shirakawa, H., et al. 1977. Synthesis of electrically conducting organic polymers: Halogen derivatives of polyacetylene (CH)n. *J Chem Soc Chem Commun* 578.
2. Brédas, J.L., et al. 1984. The role of mobile organic radicals and ions (solitons, polarons and bipolarons) in the transport properties of doped conjugated polymers. *Synth Met* 9:265.
3. Brédas, J.L., et al. 1984. Polarons and bipolarons in polypyrrole: Evolution of the band structure and optical spectrum upon doping. *Phys Rev B* 30:1023.
4. Diaz, A.F., et al. 1981. Electrooxidation of aromatic oligomers and conducting polymers. *Electroanal Chem* 121:355.
5. Mitchell, G.R., F.J. Davis, and C.H. Legge. 1988. The effect of dopant molecules on the molecular order of electrically-conducting films of polypyrrole. *Synth Met* 26:247.
6. Nogami, Y., J. Pouget, and T. Ishiguro. 1994. Structure of highly conducting PF_6^--doped polypyrrole. *Synth Met* 62:257.
7. Diaz, A.F., and J. Bargon. 1986. *Handbook of conducting polymers*, Vol. 1, ed. T.A. Skotheim. New York: Marcel Dekker, pp. 81–115.
8. Andrieux, C.P., et al. 1991. Identification of the first steps of the electrochemical polymerization of pyrroles by means of fast potential step techniques. *J Phys Chem* 95:10158.
9. Wei, Y., J. Tian, and D. Yang. 1991. A new method for polymerization of pyrrole and derivatives. *Makromol Chem Rapid Commun* 12:617.
10. Wei, Y., et al. 1991. Electrochemical polymerization of thiophenes in the presence of bithiophene or terthiophene: Kinetics and mechanism of the polymerization. *Chem Mater* 3:888.
11. Diaz, A.F., K.K. Kanazawa, and G.P. Gardini. 1979. Electrochemical polymerization of pyrrole. *J Chem Soc Chem Commun* 635.
12. Su, W., and J.O. Iroh. 1997. Formation of polypyrrole coatings onto low carbon steel by electrochemical process. *J Appl Polym Sci* 65:417.
13. Su, W., and J.O. Iroh. 1998. Effects of electrochemical process parameters on the synthesis and properties of polypyrrole coatings on steel. *Synth Met* 95:159.
14. Iroh, J.O., and W. Su. 1999. Characterization of the passive inorganic interphase and polypyrrole coatings formed on steel by the aqueous electrochemical process. *J Appl Polym Sci* 71:2075.
15. Iroh, J.O., and W. Su. 1999. Electropolymerization of pyrrole on steel substrate in the presence of oxalic acid and amines. *Electrochim Acta* 44:2173.
16. Petitjean, J., S. Aeiyach, and J.C. Lacroix. 1999. Ultra-fast electropolymerization of pyrrole in aqueous media on oxidizable metals in a one-step process. *J Electroanal Chem* 478 (1–2):92.
17. Ferreira, C.A., et al. 1996. Electrosynthesis of strongly adherent polypyrrole coatings on iron and mild steel in aqueous media. *Electrochim Acta* 41:1801.
18. Zaid, B., et al. 1998. A two-step electropolymerization of pyrrole on Zn in aqueous media. *Electrochim Acta* 43:2331.
19. Aeiyach, S., B. Zaid, and P.C. Lacaze. 1999. A one-step electrosynthesis of PPy films on zinc substrates by anodic polymerization of pyrrole in aqueous solution. *Electrochim Acta* 44:2889.
20. Willicut, R.J., and R.L. McCarley. 1995. Surface-confined monomers on electrode surfaces. 1. Electrochemical and microscopic characterization of ω-(N-pyrrolyl) alkanethiol self-assembled monolayers on Au. *Langmuir* 11:296.
21. Sayre, C.N., and D.M. Collard. 1995. Self-assembled monolayers of pyrrole-containing alkanethiols on gold. *Langmuir* 11:302.
22. Sayre, C.N., and D.M. Collard. 1997. Electrooxidative deposition of polypyrrole and polyaniline on self-assembled monolayer modified electrodes. *Langmuir* 13:714.
23. Zotti, G., S. Zecchin, and G. Schiavon. 2000. Conductive and magnetic properties of 3,4-dimethoxy- and 3,4-ethylenedioxy-capped polypyrrole and polythiophene. *Chem Mater* 12:2996.

24. Schottland, P., K. Zong, and C.L. Gaupp. 2000. Poly(3,4-alkylenedioxypyrrole)s: Highly stable electronically conducting and electrochromic polymers. *Macromolecules* 33:7051–7061.
25. Roncali, J., et al. 1987. Effects of steric factors on the electrosynthesis and properties of conducting poly(3-alkylthiophenes). *J Phys Chem* 91:6706.
26. Angelopoulos, M., et al. 1988. Polyaniline: Solutions, films and oxidation state. *Mol Cryst Liq Cryst* 160:151.
27. Lee, J.Y., D.Y. Kim, and C.Y. Kim. 1995. Synthesis of soluble polypyrrole of the doped state in organic solvents. *Synth Met* 74:103.
28. Stejskal, J., M. Omastová, and S. Fedorova. 2003. Polyaniline and polypyrrole prepared in the presence of surfactants: A comparative conductivity study. *Polymer* 44:1353.
29. Omastová, M., M. Trchová, and J. Kovářová. 2005. Synthesis and structural study of polypyrroles prepared in the presence of surfactants. *Synth Met* 138:447.
30. Kudoh, Y. 1996. Properties of polypyrrole prepared by chemical polymerization using aqueous solution containing $Fe_2(SO_4)_3$ and anionic surfactant. *Synth Met* 79:17.
31. Kudoh, Y., K. Akami, and Y. Matsuya. 1998. Properties of chemically prepared polypyrrole with an aqueous solution containing $Fe_2(SO_4)_3$, a sulfonic surfactant and a phenol derivative. *Synth Met* 95:191.
32. Omastová, M., J. Pionteck, and M. Trchová. 2003. Properties and morphology of polypyrrole containing a surfactant. *Synth Met* 135:447.
33. Shen, Y., and M. Wan. 1998. In situ doping polymerization of pyrrole with sulfonic acid as a dopant. *Synth Met* 96:127.
34. Lascelles, S.F., and S.P. Armes. 1997. Synthesis and characterization of micrometre-sized, polypyrrole-coated polystyrene latexes. *J Mater Chem* 7:1339.
35. Lascelles, S.F., et al. 1997. Surface characterization of micrometre-sized, polypyrrole-coated polystyrene latexes: Verification of a 'core-shell' morphology. *J Mater Chem* 7:1349.
36. Sonmez, G., P. Schottland, and K.K. Zong. 2001. Highly transmissive and conductive poly[(3,4-alkylenedioxy) pyrrole-2,5-diyl] (PXDOP) films prepared by air or transition metal catalyzed chemical oxidation. *J Mater Chem* 11:289.
37. Thomas, C.A., K. Zong, and P. Schottland. 2000. Poly(3,4-alkylenedioxypyrrole)s as highly stable aqueous-compatible conducting polymers with biomedical implications. *Adv Mater* 12:222.
38. Wang, P.C., Z. Huang, and A. MacDiarmid. 1999. Critical dependency of the conductivity of polypyrrole and polyaniline films on the hydrophobicity/hydrophilicity of the substrate surface. *Synth Met* 101:852.
39. Wang, P.C., and A.G. MacDiarmid. 2001. Dependency of properties of in situ deposited polypyrrole films on dopant anion and substrate surface. *Synth Met* 119:367.
40. Lu, Y., et al. 1998. Thin polypyrrole films prepared by chemical oxidative polymerization. *J Appl Polym Sci* 70:2169.
41. Fu, Y., et al. 1997. Synthesis of conductive polypyrrole/polyurethane foams via a supercritical fluid process. *Macromolecules* 30:7611.
42. John, R.K. and Kumar, D.S. 2002. Structural, electrical, and optical studies of plasma polymerized and iodine-doped polypyrrole. *J Appl Polym Sci* 83:1856.
43. Wang, J.G., K.G. Neoh, and E.T. Kang. 2004. Comparative study of chemically synthesized and plasma polymerized pyrrole and thiophene thin films. *Thin Solid Films* 446:205.
44. Hosono, K., I. Matsubara, and N. Murayama. 2003. Structure and properties of plasma polymerized and 4-ethylbenzenesulfonic acid-doped polypyrrole films. *Thin Solid Films* 441:72.
45. Morales, J., M.G. Olayo, and G.J. Cruz. 2002. Plasma polymerization of random polyaniline–polypyrrole–iodine copolymers. *J Appl Polym Sci* 85:52.
46. Nastase, F., D. Mihaiescu, and C. Nastase. 2005. Plasma polymerized ferrocene-pyrrole copolymer films. *Composites Part A Appl Sci Manuf* 36:503.
47. Cha, S.K. 1997. Electropolymerization rate of polythiophene/polypyrrole composite polymer with some dopant ions. *J Polym Sci Part B Polym Phys* 35:165.

48. Simmons, M.R., P.A. Chaloner, and S.P. Armes. 1998. Synthesis and characterization of colloidal polypyrrole particles using reactive polymeric stabilizers. *Langmuir* 14:611.
49. McCarthy, G.P., et al. 1997. Synthesis and characterization of carboxylic acid-functionalized polypyrrole-silica microparticles using a 3-substituted pyrrole comonomer. *Langmuir* 13: 3686.
50. Li, X.G., L.X. Wang, and M.R. Huang. 2001. Synthesis and characterization of pyrrole and *m*-toluidine copolymers. *Synth Met* 123:443.
51. Li, X.G., L.X., Wang, and M.R. Huang. 2001. Synthesis and characterization of pyrrole and anisidine copolymers. *Polym Plast Technol* 42:6095.
52. Fusalba, F., and D. Belanger. 1999. Electropolymerization of polypyrrole and polyaniline–polypyrrole from organic acidic medium. *J Phys Chem B* 103:9044.
53. Sari, B., and M. Talu. 1998. Electrochemical copolymerization of pyrrole and aniline. *Synth Met* 94:221.
54. Li, X.G., M.R. Huang, and M.F. Zhu. 2004. Synthesis and nitrosation of processible copolymer from pyrrole and ethylaniline. *Polym Plast Technol* 45:385.
55. Kalaycioglu, E., et al. 1998. Synthesis of conducting H-type polysiloxane–polypyrrole block copolymers. *Synth Met* 97:7.
56. Kanazawa, K.K., et al. 1981. Electrical properties of pyrrole and its copolymers. *Synth Met* 4:119.
57. Ezquerra, T., J. Rŭhe, and G. Wegner. 1988. Hopping conduction in 3,4-cycloalkylpolypyrrole perchlorates: A model study of conductivity in polymers. *Chem Phys Lett* 144:194.
58. Rŭhe, J., T. Ezquerra, and G. Wegner. 1989. New conducting polymers from 3-alkylpyrroles. *Synth Met* 28:C177.
59. Reynolds, J.R., P.A. Poropatic, and R.L. Toyooka. 1987. The physical and electrical properties of copolymers of polypyrrole. *Synth Met* 18:95.
60. Reynolds, J.R., P.A. Poropatic, and R.L. Toyooka. 1987. Electrochemical copolymerization of pyrrole with N-substituted pyrroles. Effect of composition on electrical conductivity. *Macromolecules* 20:958.
61. Nishizawa, M., et al. 1991. In situ characterization of copolymers of pyrrole and *N*-methylpyrrole at microarray electrodes. *Synth Met* 45:241.
62. Min, Y., A.G. MacDiarmid, and A.J. Epstein. 1993. The concept of "secondary doping " as applied to polyaniline. *Polym Prep* 35:231.
63. Hsu, C.H., J.D. Cohen, and R.F. Tietz. 1993. Polyaniline spinning solutions and fibers. *Synth Met* 59:37.
64. Cao, Y., P. Smith, and A.J. Heeger. 1992. Counter-ion induced processibility of conducting polyaniline and of conducting polyblends of polyaniline in bulk polymers. *Synth Met* 48:91.
65. Cao, Y., et al. 1992. Solution-cast films of polyaniline: Optical-quality transparent electrodes. *Appl Phys Lett* 60:2711.
66. Yang, C.Y., et al. 1993. Morphology of conductive, solution-processed blends of polyaniline and poly(methyl methacrylate). *Synth Met* 53:293–301.
67. Lee, J.Y., et al. 1997. Synthesis and characterization of soluble polypyrrole. *Synth Met* 84:137.
68. Song, K.T. 2000. Synthesis of electrically conducting soluble polypyrrole and its characterization. Ph.D. thesis.
69. Yamaura, T., T. Hagiwara, and K. Iwata. 1988. Enhancement of electrical conductivity of polypyrrole film by stretching: Counter ion effect. *Synth Met* 26:209.
70. Song, K.T., et al. 2000. Solvent effects on the characteristics of soluble polypyrrole. *Synth Met* 110:57.
71. Ford, N.C., Jr. 1985. *Dynamic light scattering: Applications of photon correlation spectroscopy*, ed. R. Pecora. New York: Plenum, pp. 7–58.
72. Seery, T.A.P., et al. 1997. Solution properties of polyaniline by light scattering measurements: Achieving spatial homogeneity. *Synth Met* 84:79.

73. Gettinger, C.L., et al. 1995. Solution characterization of surfactant solubilized polyaniline. *Synth Met* 74:81.

74. Mott, N.F., and E.A. Davis. 1979. *Electronic processes in non-crystalline materials*, 2nd ed. Oxford: Clarendon Press, p. 32.

75. Mott, N.F. 1993. *Conduction in non-crystalline materials*, 2nd ed. Oxford: Clarendon Press, p. 31.

76. Saunders, B.R., K.S. Murray, and R.J. Fleming. 1992. Physical properties of polypyrrole films containing sulfonated metallophthalocyanine anions. *Synth Met* 47:167.

77. Yoon, C.O., et al. 1994. Transport near the metal-insulator transition: Polypyrrole doped with PF_6. *J Phys Rev B* 49:10851.

78. Kemp, N.T., et al. 1999. Thermoelectric power and conductivity of different types of polypyrrole. *J Polym Sci Part B* 37:953.

79. Zupprioli, L., et al. 1994. Hopping in disordered conducting polymers. *Phys Rev B* 50:5196.

80. Sixou, B., N. Mermilliod, and J.P. Travers. 1996. Aging effects on the transport properties in conducting polymer polypyrrole. *Phys Rev B* 53:4509.

81. Baughman, R.H., and L.W. Shacklette. 1989. Conductivity as a function of conjugation length: Theory and experiment for conducting polymer complexes. *Phys Rev B* 39:5872.

82. Tzou, K.T., and R.V. Gregory. 1993. Mechanically strong, flexible highly conducting polyaniline structures formed from polyaniline gels. *Synth Met* 55–57:983.

83. Oka, O., S. Morita, and K. Yoshino. 1990. Gel characteristics of polyaniline and its anomalous doping effect. *Jpn J Appl Phys* 29:L679.

84. Gao, J., et al. 1996. Soluble polypyrrole as the transparent anode in polymer light-emitting diodes. *Synth Met* 82:221.

85. Park, J.W., et al. 2000. Characteristics of organic electroluminescent devices using polypyrrole conducting layer and undoped conjugated polymer layer. *Thin Solid Films* 363:259.

86. Kim, Y.C., et al. 1999. Reflective bistable cholesteric/polymer dispersion display with a black polypyrrole electrode cast from the solution. *Mol Cryst Liq Cryst* 327:157.

87. Shen, Y., and M. Wan. 1997. Soluble conductive polypyrrole synthesized by in situ doping with β-naphthalene sulphonic acid. *J Polym Sci A Polym Chem* 35:3689.

88. Oh, E.J., and K.S. Jang. 2001. Synthesis and characterization of high molecular weight, highly soluble polypyrrole in organic solvents. *Synth Met* 119:109.

89. Jang, K.S., et al. 2001. Synthesis and characterization of alcohol soluble polypyrrole. *Synth Met* 119:107.

90. Oh, E.J., K.S. Jang, and A.G. MacDiarmid. 2002. High molecular weight soluble polypyrrole. *Synth Met* 125:267.

91. Jang, K.S., H. Lee, and B. Moon. 2004. Synthesis and characterization of water soluble polypyrrole nanowires. *Synth Met* 143:289.

92. Lee, G.J., S.H. Lee, and K.S. Ahn. 2002. Synthesis and characterization of soluble polypyrrole with improved electrical conductivity. *J Appl Polym Sci* 84:2583.

93. Lee, Y.H., J.Y. Lee, and D.S. Lee. 2000. A novel conducting soluble polypyrrole composite with a polymeric co-dopant. *Synth Met* 114:347.

94. Han, J.S., J.Y. Lee, and D.S. Lee. 2001. A novel thermosensitive soluble polypyrrole composite. *Synth Met* 124:301.

95. Yin, W., and E. Ruckenstein. 2001. A water-soluble self-doped conducting polypyrrole-based copolymer. *J Appl Polym Sci* 79:86.

96. Jerome, C., and R. Jerome. 1998. Electrochemical synthesis of polypyrrole nanowires. *Angew Chem Int Ed* 37:2488.

97. Hernandez, R.M., et al. 2004. Template fabrication of protein-functionalized gold–polypyrrole–gold segmented nanowires. *Chem Mater* 16:3431.

98. Menon, V.P., J. Lei, and C.R. Martin. 1996. Investigation of molecular and supermolecular structure in template-synthesized polypyrrole tubules and fibrils. *Chem Mater* 8:2382.

99. Demoustier-Champagne, S., et al. 1998. Electrochemically synthesized polypyrrole nanotubules: Effects of different experimental conditions. *Eur Polym J* 34:1767.

100. Duchet, J., R. Legras, and S. Demoustier-Champagne. 1998. Chemical synthesis of polypyrrole: Structure–properties relationship. *Synth Met* 98:113.

101. Joo, J., et al. 2003. Conducting polymer nanotube and nanowire synthesized by using nanoporous template: Synthesis, characteristics, and applications. *Synth Met* 135:7.

102. Jerome, C., D. Labaye, and I. Bodart. 1999. Electrosynthesis of polyacrylic polypyrrole composite: Formation of polypyrrole wires. *Synth Met* 101:3.

103. Ge, D., et al. 2002. Electrochemical synthesis of polypyrrole nanowires on composite electrode. *Synth Met* 132:93.

104. Zhang, X., et al. 2004. Inorganic/organic mesostructure directed synthesis of wire/ribbon-like polypyrrole nanostructures. *Chem Commun* 1852.

105. Li, X., M. Lu, and H. Li. 2002. Electrochemical copolymerization of pyrrole and thiophene nanofibrils using template-synthesis method. *J Appl Polym Sci* 86:2403.

106. Li, X., X. Zhang, and H. Li. 2001. Preparation and characterization of pyrrole/aniline copolymer nanofibrils using the template-synthesis method. *J Appl Polym Sci* 81:3002.

107. Lu, M., X.H. Li, and H.L. Li. 2002. Synthesis and characterization of conducting copolymer nanofibrils of pyrrole and 3-methylthiophene using the template-synthesis method. *Mat Sci Eng A Struct* 334:291.

108. Demoustier-Champagne, S., and P.Y. Stavaux. 1999. Effect of electrolyte concentration and nature on the monophology and the electrical properties of electropolymerized polypyrrole nanotubules. *Chem Mater* 11:829.

109. Chen, G.Z., et al. 2000. Carbon nanotube and polypyrrole composites: Coating and doping. *Adv Mater* 12:522.

110. Fan, J., et al. 1999. Synthesis, characterizations, and physical properties of carbon nanotubes coated by conducting polypyrrole. *J Appl Polym Sci* 74:2605.

111. Chen, J.H., et al. 2001. Electrochemical synthesis of polypyrrole/carbon nanotube nanoscale composites using well-aligned carbon nanotube arrays. *Appl Phys A* 73:129.

112. Sunderland, K., et al. 2004. Synthesis of γ-Fe_2O_3/polylpyrrole nanocomposite materials. *Mater Lett* 58:3136.

113. Jang, J.S., X.L. Li, and J.H. Oh. 2004. Facile fabrication of polymer and carbon nanocapsules using polypyrrole core/shell nanomaterials. *Chem Commun* 794.

114. Mo, X., et al. 2004. The application of polypyrrole fibrils in hydrogen evolution reaction. *Synth Met* 142:217.

115. Iroh, J.O., and C. Williams. 1999. Formation of thermally stable polypyrrole–naphthalene/benzene sulfonate–carbon fiber composites by an electrochemical process. *Synth Met* 99:1

116. Kang, T.S., et al. 2005. Electrically conducting polypyrrole fibers spun by electrospinning. *Synth Met* 151:61.

117. Rozsnyai, L.F., and M.S. Wrighton. 1995. Selective deposition of conducting polymers via monolayer photopatterning. *Langmuir* 11:3913.

118. Prissanaroon, W., et al. 2005. Microcontact printing of copper and polypyrrole on fluoropolymers. *Thin Solid Films* 477:131.

119. Gorman, C.B., H.A. Biebuyck, and G.M. Whitesides. 1995. Fabrication of patterned, electrically conducting polypyrrole using a self-assembled monolayer: A route to all-organic circuits. *Chem Mater* 7:526.

120. Zhou, F., et al. 2004. Manipulation of the ultimate pattern of polypyrrole film on self-assembled monolayer patterned substrate by negative or positive electrodeposition. *Surf Sci* 561:1.

121. Huang, Z., et al. 1997. Selective deposition of polypyrrole, polyaniline and nickel on hydrophobic/hydrophilic patterned surfaces and applications. *Synth Met* 85:1375.

122. Hohnholz, D., and A.G. MacDiarmid. 2001. Line patterning of conducting polymers: New horizons for inexpensive, disposable electronic devices. *Synth Met* 121:1327.

123. Masashi, W., and H. Toshihiro. 2004. Polypyrrole film with striped pattern. *J Polym Sci Pol Chem* 42:2460.

124. Yasushi, O., Y. Shoji, and K. Masaki, 2004. Metal pattern formation by selective electroless metallization on polypyrrole films patterned by photochemical degradation of iron (β) chloride as oxidizing agent. *Synth Met* 144:265.

125. Choi, M.S., et al. 2002. Photolithographic patterning of electrically conducting polypyrrole film. *Mol Cryst Liq Cryst* 377:181.

126. Selampinar, F., et al. 1995. A conducting composite of polypyrrole α. As a gas sensor. *Synth Met* 68:109.

127. Lin, C.W., B.J. Hwang, and C.R. Lee. 1998. Methanol sensors based evanescent wave sensor to monitor the deposition rate of thin films. *Thin Solid Films* 325 (1–2):139.

128. Suri, K., et al. 2002. Gas and humidity sensors based on iron oxide–polypyrrole nanocomposites. *Sensor Actuat B Chem* 81:277.

129. Diab, N., and W. Schuhmann. 2001. Electropolymerized manganese porphyrin/polypyrrole films as catalytic surfaces for the oxidation of nitric oxide. *Electrochim Acta* 47 (1–2):265.

130. Michalska, A., A. Hulanicki, and A. Lewenstam. 1997. All-solid-state potentiometric sensors for potassium and sodium based on poly(pyrrole) solid contact. *Microchem J* 57:59.

131. Hutchins, R.S., and L.G. Bachas. 1995. Nitrate-selective electrode developed by electrochemically mediated imprinting/doping of polypyrrole. *Anal Chem* 67:1654.

132. Riul, A., et al. 2002. Artificial taste sensor: Efficient combination of sensors made from langmuir–blodgett films of conducting polymers and a ruthenium complex and self-assembled films of an azobenzene-containing polymer. *Langmuir* 18:239.

133. Riul, A., Jr., et al. 2003. An electronic tongue using polypyrrole and polyaniline. *Synth Met* 132:109.

134. Baughman, R.H. 1996. Conducting polymer artificial muscles. *Synth Met* 78:339.

135. Hutchison, A.S., et al. 2000. Development of polypyrrole-based electromechanical actuators. *Synth Met* 113:121.

136. Madden, J.D., et al. 2000. Fast contracting polypyrrole actuators. *Synth Met* 113:185.

137. Smela, E., and N. Gadegaard. 1999. Surprising volume change in PPy(DBS): An atomic force microscopy study. *Adv Mater* 11:953.

138. Bay, L., T. Jacobsen, and S. Skaarup. 2001. Mechanism of actuation in conducting polymers: Osmotic expansion. *J Phys Chem B* 105:8492.

139. Okuzaki, H., and T. Kunugi. 1997. Adsorption-induced chemomechanical behavior of polypyrrole films. *J Appl Polym Sci* 64:383.

140. Okuzaki, H., and T. Kunugi. 1996. Adsorption-induced bending of polypyrrole films and its application to a chemomechanical rotor. *J Polym Sci Pol Phys* 34:1747.

141. Okuzaki, H., T. Kuwabara, and T. Kunugi. 1997. A polypyrrole rotor driven by sorption of water vapour. *Polymer* 38:5491.

142. Okuzaki, H., and K. Funasaka. 2000. Electro-responsive polypyrrole film based on reversible sorption of water vapor. *Synth Met* 108:127.

143. Watanabe, M., H. Shirai, and T. Hirai. 2002. Wrinkled polypyrrole electrode for electroactive polymer actuators. *J Appl Phys* 92:4631.

144. Otero, T.F., and I. Cantero. 1999. Conducting polymers as positive electrodes in rechargeable lithium-ion batteries. *J Power Sources* 81:838.

145. Otero, T.F., I. Cantero, and H. Grande. 1999. Solvent effects on the charge storage ability in polypyrrole. *Electrochim Acta* 44:2053.

146. Kuwabata, S., S. Masui, and H. Yoneyama. 1999. Charge-discharge properties of composites of $LiMn_2O_4$ and polypyrrole as positive electrode materials for 4 V class of rechargeable Li batteries. *Electrochim Acta* 44:4593.

147. Kuwabata, S., et al. 2000. Charge-discharge properties of chemically prepared composites of V_2O_5 and polypyrrole as positive electrode materials in rechargeable Li batteries. *Electrochim Acta* 46:91.

148. Kuwabata, S., et al. 2000. Charge-discharge properties of composites of polypyrrole and vanadium oxide powder. *Macromol Symp* 156:213.
149. Osaka, T., et al. 1997. Performances of lithium/gel electrolyte/polypyrrole secondary batteries. *J Power Sources* 68:392.
150. Veeraraghavan, B., et al. 2002. Study of polypyrrole graphite composite as anode material for secondary lithium-ion batteries. *J Power Sources* 109:377.
151. Satoh, M., et al. 1995. Characterization of a tantalum capacitor fabricated with a conducting polypyrrole as a counter electrode. *Synth Met* 71:2259.
152. Larmat, F., J.R. Reynolds, and Y.J. Qiu. 1996. Polypyrrole as a solid electrolyte for tantalum capacitors. *Synth Met* 79:229.
153. Yamamoto, H., et al. 1999. Solid electrolytic capacitors using an aluminum alloy electrode and conducting polymers. *Synth Met* 104:38.
154. Krings, L.H.M., et al. 1993. Application of polypyrrole as counterelectrode in electrolytic capacitors. *Synth Met* 54:453.
155. Kudoh, Y., et al. 1991. An aluminum solid electrolytic capacitor with an electroconducting-polymer electrolyte. *Synth Met* 41–43:1133.
156. Kudoh, Y., et al. 1996. Covering anodized aluminum with electropolymerized polypyrrole via manganese oxide layer and application to solid electrolytic capacitor. *J Power Source* 60:157.
157. Satoh, M., et al. 1994. Highly conducting polypyrrole prepared from homogeneous mixtures of pyrrole/oxidizing agent and its applications to solid tantalum capacitors. *Synth Met* 65:39.
158. Tsai, M.L., P.J. Chen, and J.S. Do. 2004. Preparation and characterization of $Ppy/Al_2O_3/Al$ used as a solid-state capacitor. *J Power Sources* 133:302.
159. Sung, J.H., S.J. Kim, and K.H. Lee. 2003. Fabrication of microcapacitors using conducting polymer microelectrodes. *J Power Sources* 124:343.
160. Niu, C., et al. 1997. High power electrochemical capacitors based on carbon nanotube electrodes. *Appl Phys Lett* 70:1480.
161. Frackowiak, E.A., et al. 2000. Supercapacitor electrodes from multiwalled carbon nanotubes. *Appl Phys Lett* 77:2421
162. Frackowiak, E.B., and F. Guin. 2001. Carbon materials for the electrochemical storage of energy in capacitors. *Carbon* 39:937.
163. Frackowiak, E., et al. 1999. Electrochemical storage of lithium multiwalled carbon nanotubes. *Carbon* 37:61.
164. Hughes, M., et al. 2002. Electrochemical capacitance of a nanoporous composite of carbon nanotubes and polypyrrole. *Chem Mater* 14:1610.
165. Jurewicz, K., et al. 2001. Supercapacitors from nanotubes/polypyrrole composites. *Chem Phys Lett* 347:36.
166. Frackowiak, E., et al. 2001. Nanotubular materials for supercapacitors. *J Power Sources* 97:822.
167. Lee, M.S., S.B. Lee, and J.Y. Lee. 2003. All-polymer FET based on simple photolithographic micropatterning of electrically conducting polymer. *Mol Cryst Liq Cryst* 405:171.
168. Kou, C.T., and T.R. Liou. 1996. Characterization of metal-oxide-semiconductor field-effect transistor (MOSFET) for polypyrrole and poly(N-alkylpyrrole)s prepared by electrochemical synthesis. *Synth Met* 82:167.
169. Lee, M.S., et al. 2005. Flexible all-polymer field effect transistors with optical transparency using electrically conducting polymers. *Thin Solid Films* 477:169.
170. MacDiarmid, A.G. 2001. "Synthetic metals": A novel role for organic polymers. *Curr Appl Phys* 1:269.
171. Gregory, R.V., W.C. Kimbrell, and H.H. Kuhn. 1989. Application and processing conductive textiles. *Synth Met* 28:823.
172. Olmedo, L., P. Hourquebie, and F. Jousse. 1993. Microwave absorbing materials based on conducting polymers. *Adv Mater* 5:373.
173. Kim, M.S., et al. 2002. PET fabric/polypyrrole composite with high electrical conductivity for EMI shielding. *Synth Met* 126:233.

9

Regioregular Polythiophenes

Malika Jeffries-El and
Richard D. McCullough

9.1 Introduction

Over 30 years ago conjugated polymers were envisioned as futuristic new materials that would lead to the next generation of electronic and optical devices. The development of "plastic" electronic devices based on conjugated polymers has elevated the status of these materials from academic curiosity to the basis for a rapidly growing new electronics industry. Since polythiophenes (PTs) possess excellent thermal and environmental stability, solubility, processibility, and high conductivity when doped, they are considered an important class of conjugated polymers. To date PTs have been used in a variety of applications including: transistors [1], hole injection layer in polymer LEDs [2], electrical conductors [2], environmental sensors [2,3], and solar cells [4]. The primary [1] structure of the polymer backbone plays an essential role in determining the three-dimensional (3D) structure and morphology that critically determines the electrical and physical properties of conducting polymers. For this reason numerous research efforts have focused on tuning the structure and function of these materials by synthesis. For example, in the case of PT, the π-overlap along the backbone can be improved by using regiospecific synthetic protocols that eliminate structural defects. The ability of regioregular poly(3-alkylthiophenes) (rr-P3ATs) to form well-defined, organized 3D structures in the form of π-stacks leads to better materials and enhanced device performance in almost every category, ranging from electrical conductivity to stability [5]. It is apparent that both pioneering and future work in conjugated polymers strongly depend on synthetic chemists, who can create new polymers that can be fabricated into new devices whose physics and chemistry can be understood in detail.

Previously, we reviewed the synthesis and characterization of rr-P3ATs and its derivatives [5,6]. Since then the number of papers describing the synthesis, characterization, and applications of rr-P3ATs and its derivatives has doubled. In this chapter we will review some of the recent developments in this field. Owing to the enormous literature on this subject, some excellent work in this area will almost certainly have been inadvertently overlooked. However, it is our intention to review the pioneering work and to highlight the synthesis of new P3AT derivatives. In this chapter the use of P3ATs in devices will be discussed briefly, but will not be emphasized as it is the subject of another section of this volume.

9.2   The Evolution of Polythiophene

9.2.1   Polythiophene

The discovery of highly conducting polyacetylene, in 1977, by MacDiarmid, Shirakawa, and Heeger [7] prompted the synthesis of other polymers with conjugated π-systems. PT was first chemically prepared in the early 1980s by a metal-catalyzed polycondensation polymerization of the mono-Grignard of 2,5-dibromothiophene generated by treatment with magnesium metal [8,9]. This method is likely to give a 2,5-coupled PT; however, the structure was not determined due to the lack of solubility of the material. PT can also be synthesized by several other methods, such as metal-catalyzed polycondensation polymerization of 2,5-diiodothiophene [10], Wurtz coupling of 2,5-dilithiothiophene [11], and electrochemical polymerization. Wurtz coupling and electrochemical methods proceed by radical coupling mechanisms, increasing the possibility of 3,4-linkages in the PT structure and reducing the electrical performance. However, regardless of the method used the PT is intractable. Nevertheless, PT has an excellent thermal stability (42% weight loss at 900°C) and good conductivity (3.4×10^{-4} to 1.0×10^{-1} S/cm when doped with iodine). Despite the lack of processability, environmental stability, thermal stability, and high-electrical conductivity of the PT films still make it a highly desirable material [10].

9.2.2   Poly(3-alkylthiophenes)

Like most fully conjugated polymer systems PTs have poor solubility due to the strong π-stacking interactions between the aromatic rings [12]. In 1985, the first P3AT was prepared by Elsenbaumer through nickel-catalyzed cross-coupling, a similar fashion used for the synthesis of PT [13,14], with the hope of preparing a soluble and processable conducting PT. It had been demonstrated that the attachment of flexible side chains onto the backbone of an insoluble polymer can dramatically improve the solubility [12]. Soluble, film forming P3ATs were produced with regioregularity of around 50%–80% [15]. Molecular weights using this method were low ($M_w = 5000$; polydispersity index (PDI) $= 2$) for shorter side chains, such as butyl. Around the same time, Yoshino and coworkers prepared P3ATs on a large scale through oxidative polymerization using $FeCl_3$ [16–18]. Using this approach polymers with high-molecular weight ($M_n = 30,000$–$300,000$; PDI $= 1.3$–1.5) were obtained. P3ATs can also be synthesized by a variety of methods including the coupling of 5,5″-dilithiobithiophenes with $CuCl_2$ [11], a similar coupling of 5,5″-dilithiobithiophenes using $Fe(acac)_3$ [19], the Stille coupling of 2,5′-dibromobithiophenes with 2,5′-bis(trimethylstannyl)bithiophene using a catalytic amount of $PdCl_2(AsPh_3)_2$ [19], and the nickel(0) dehalogenative coupling [20]. P3AT with alkyl groups longer than butyl can readily be melt- or solution-processed into films, which, after oxidation, can exhibit reasonably high reported electrical conductivities of up to 40 ± 5 S/cm [13,21,22]. We have found that irregular P3ATs

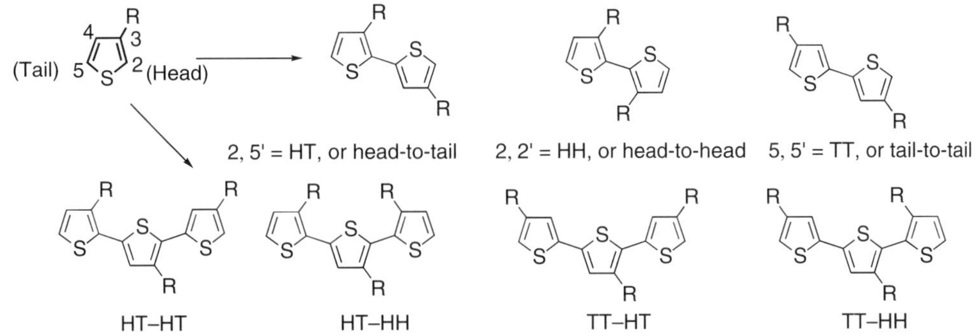

FIGURE 9.1 Classical synthetic methods lead to a number of regiochemical isomers.

have electrical conductivities from 0.1 to 10 S/cm, when doped with iodine. As an example, a 80% regioregular sample of poly(3-hexylthiophene) has a reported conductivity of 6 S/cm.

Although all of these methods reduce or eliminate 2,4-linkages, they do not solve the lack of regiochemical control over head-to-tail (HT) couplings between adjacent thiophene rings. Since 3-alkylthiophene is not a symmetric molecule, the fashion in which the rings are coupled together becomes an issue. If you consider the simple case of coupling, two thiophene rings are coupled between the 2- and 5-positions; it is apparent that there are three relative orientations available. The first of these is 2,5′ or HT coupling, the second is 2,2′ or head-to-head coupling (HH), and the third is 5,5′ or tail-to-tail coupling (TT). The situation becomes even more complex when you consider the coupling of three thiophene rings that leads to a mixture of four chemically distinct triad regioisomers (Figure 9.1) [23]. Polymers that are prepared without regiochemical control contain mixtures of the different coupling types. These structurally irregular polymers will be denoted as regioirregular or non-HT. Irregularly substituted PTs have structures where unfavorable HH couplings cause a sterically driven twist of thiophene rings, increasing the torsion angle between the rings, resulting in a loss of conjugation. This increase of the torsion angle between thiophene rings leads to greater bandgaps, with consequent destruction of high conductivity and other desirable properties (Figure 9.2) [24]. Polymers that are prepared with regiochemical control contain only HT coupling and are known as regioregular or HT coupled rr-PATs. Regioregularity provides planar polymer backbones that can self-assemble in three dimensions. This allows for efficient interchain and intrachain conductivity pathways, leading to highly conductive polymers (Figure 9.2).

9.3 Regioregular, HT Coupled Poly(3-alkylthiophenes)

9.3.1 Methods of Synthesis

P3ATs are generally synthesized by three methods commonly referred to as the McCullough, Rieke, and GRIM methods. The specific details of these methods are described below. These methods all produce comparable PATs, however, the Reike method offers the advantage of being tolerant to a variety of different functional groups, whereas McCullough and GRIM are limited to those that are tolerant to organolithium and organomagnesium reagents. The McCullough method has been modified with the use of organozinc reagents as discussed below. In general any transition metal catalyzed cross-coupling reaction can be adapted for regioregular synthesis. To date, there are also a few other limited examples that will be summarized at the end of this section.

The first synthesis of HT rr-P3AT (Scheme 9.1), known as the McCullough method was reported in 1992 (Scheme 9.1, Table 9.1, entry 1) [25,26]. The key to this method is the regiospecific generation of 2-bromo-5-bromomagnesio-3-alkylthiophene **3** from 2-bromo-3-alkylthiophene, which is then polymerized by Kumada and coworkers [27–29] cross-coupling methods using catalytic amounts of

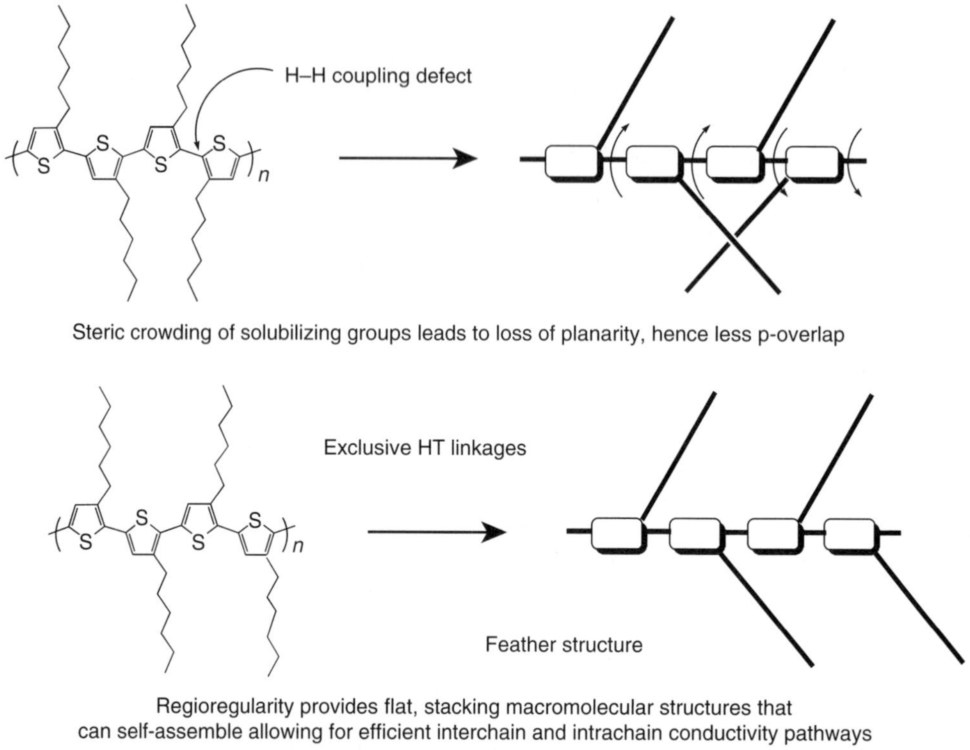

FIGURE 9.2 Regioirregular polythiophene in nonplanar (*top*). Regioregular polythiophene can be planar (*bottom*).

Ni(dppp)Cl$_2$. This method produces regioregular PTs **4** with 98%–100% HT–HT couplings in 44%–66% yields (chloroform soluble solid). Molecular weights are typically 20–40 K (PDI = 1.4) [26]. The generation of intermediate **3** is achieved by treating **1** with LDA at –78°C. The organolithium intermediate **2** is stable at this temperature and does not undergo metal halogen exchange. This intermediate is then treated with MgBr$_2$•Et$_2$O to form 2-bromo-5-bromomagnesio-3-alkylthiophene **3**. Quenching studies performed on the intermediates **2** and **3** indicate that 98%–99% of the desired monomer and less than 1%–2% of the 2,5-exchange product are produced [26]. The polymerization of these intermediates also occurs without any scrambling. This procedure was later modified by replacing

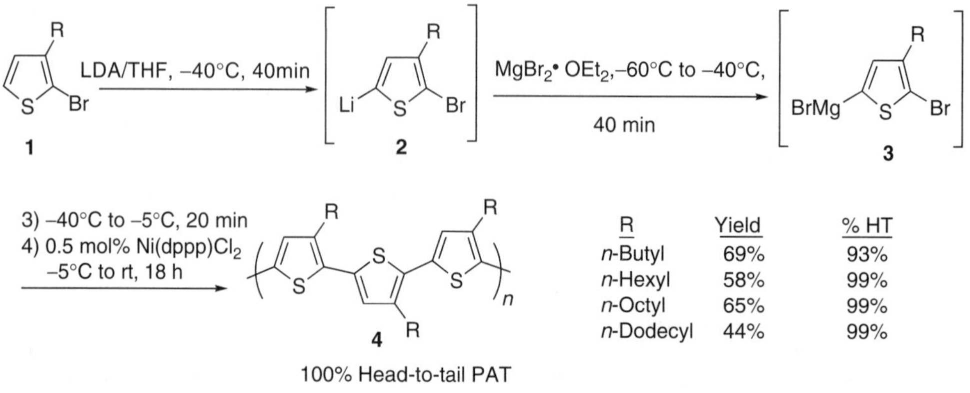

R	Yield	% HT
n-Butyl	69%	93%
n-Hexyl	58%	99%
n-Octyl	65%	99%
n-Dodecyl	44%	99%

100% Head-to-tail PAT

SCHEME 9.1 The McCullough method for the regioregular synthesis of poly(3-alkylthiophene) (P3AT) with 100% head-to-tail couplings.

TABLE 9.1 Summary of Side-Chain Functionalized PATs

Entry	Polymer	Method	σ (S/cm)	λ_{max} (nm)	M_n^c	PDI	Ref.
1	(a) R = C$_4$H$_9$ 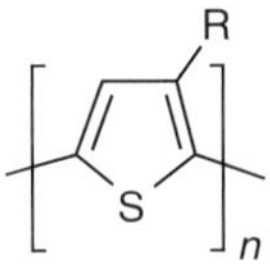	A	100	438[a]	—	—	26
		B	—	449[a]	—	—	31
	(b) R = C$_6$H$_{13}$	A	200	442[a]	10,000	1.60	26
		B	—	456[a]	25,500	1.48	31
		D	—	449[a]	16,118	1.22	44
	(c) R = C$_8$H$_{17}$	A	200	446[a]	24,424	1.98	26
		B	—	451[a]	34,580	1.13	31
		D	—	448[a]	27,000	1.50	48
	(d) R = C$_{10}$H$_{21}$	B	—	447[a]	30,480	1.39	31
	(e) R = C$_{12}$H$_{25}$	A	600	450[a]	11,600	—	25
		A	1,000	450[a]	12,500	—	25
	(f) R = C$_{14}$H$_{29}$	B	—	453[a]	34,650	1.42	59
	(g) R = C$_{18}$H$_{37}$	B	—	450[a]	30,230	1.46	31
	(h) R = C$_5$H$_{10}$OCH$_3$	C	—	—	21,300	—	31
	(i) R = C$_6$H$_{12}$OCH$_3$	C	—	451[b]	5,900	1.4	41
	(j) R = C$_{10}$H$_{20}$OCH$_3$	C	—	450[b]	9,900	1.5	41
	(k) R = (C$_6$H$_5$)C$_8$H$_{17}$	C	—	448[b]	9,450	1.6	41
		E	4	602[b]	23,000	—	49
2	(a) R = O(CH$_2$CH$_2$O)$_2$CH$_3$ 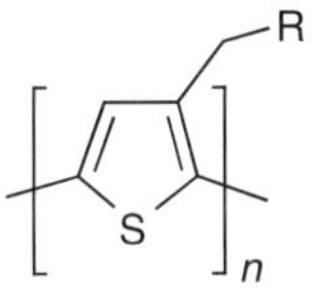	A	5,500	439[a]	71,000	1.6	65
		A	1,700	439[a]			65
	(b) R = OCH$_2$CH$_2$OCH$_3$	A	165		4,000	1.4	65
	(c) R = OCH$_3$	A	—				65
	(d) R = SCH$_3$	A	—				65
3	(a) R = (CH$_2$CH$_2$O)$_3$CH$_2$CH$_3$ 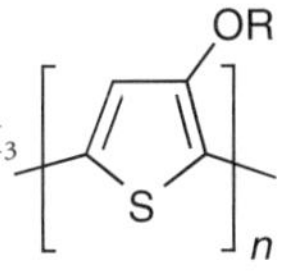	A, C	4	602[a]	18,000	1.6	68,69
	(b) R = CH$_2$O(CH$_2$CH$_2$O)$_2$CH$_3$	C	50	596	—	—	69
	(c) R = (CH$_2$CH$_2$O)$_2$CH$_3$	C	650	602	2,806	—	69,70
	(d) R = OC$_6$H$_{13}$	C	5	589	6,000	—	69,70
4	(a) R = C$_{12}$H$_{25}$ 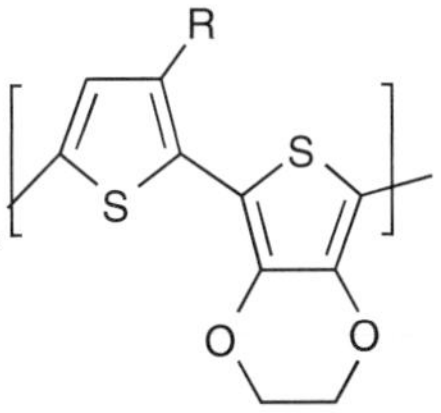	C	0.08	486	11,000	—	71
		C	1	533	13,400	1.50	70
	(b) R = O(CH$_2$CH$_2$O)$_2$CH$_3$	D	10^{-3}	—	1,690	1.1	68
5	(a) R = (CH$_2$CH$_2$O)$_m$CH$_3$ 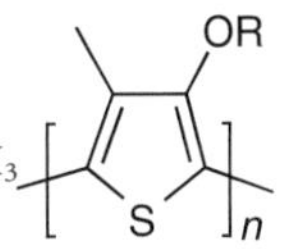	E	—	550	22,500	5.9	63
	(b) R = (CH$_2$CH$_2$O)$_7$CH$_3$	E	10^{-3}	550	—	—	64
	(c) R = CH$_2$CH(CH$_3$)(OC$_2$H$_5$)$_4$OCH$_3$	E	—	600	44,000	4.4	92
6	(a) R = C$_4$H$_9$ 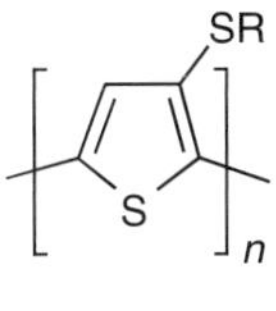	B	—	497	—	—	34
	(b) R = C$_6$H$_{13}$	B	500	503	6,090	1.15	34
	(c) R = C$_8$H$_{17}$	B	700	510	4,420	1.11	34
	(d) R = C$_{12}$H$_{25}$	B	600	505	4,030	1.03	34
7	(a) R = (CH$_2$)$_5$(CF$_2$)$_4$F 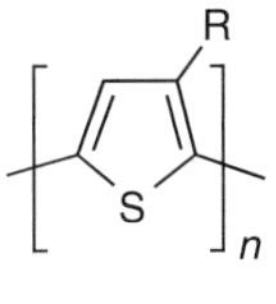	—	—	436			74
	(b) R = (CH$_2$)$_{11}$(CF$_2$)$_4$F	C	—	—	10,500	1.2	75
	(c) R = C$_8$F$_{17}$	A	—	326	14,000[d]	N/a	73

(continued)

TABLE 9.1 (continued)

Entry	Polymer	Method	σ (S/cm)	λ_{max} (nm)	M_n^c	PDI	Ref.
8	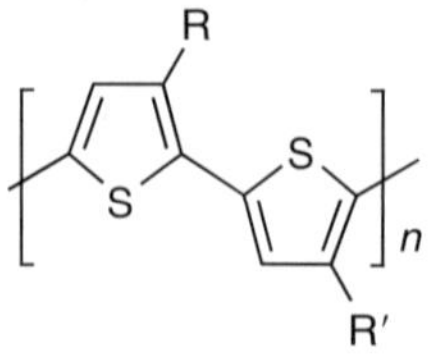 (a) R=(CH$_2$)$_5$(CF$_2$)$_4$F; R'=C$_9$H$_{19}$	C	—	445^a	12,700	1.29	74
	(b) R=C$_8$F$_{17}$; R'=C$_8$H$_{17}$	D	—	542	21,500	1.46	103
		F	—	Var	12,000	N/a	103
9	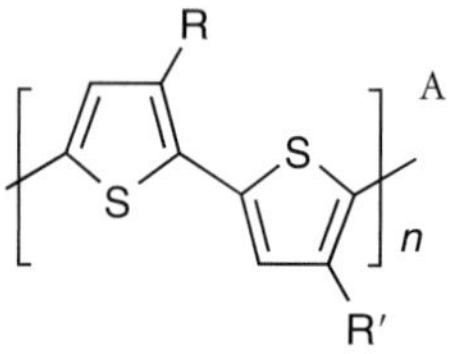 (a) R=C$_6$H$_{12}$PO$_3$(C$_2$H$_5$)$_2$; R'=C$_{12}$H$_{25}$	A	50	544^e	—	1.3	103
	(b) R=C$_6$H$_{12}$PO$_3$H$_2$; R'=C$_{12}$H$_{25}$						
10	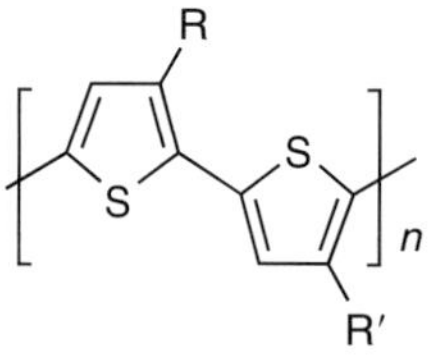 (a) R=(C$_2$H$_4$O)$_2$OCH$_3$; R'=C$_{12}$H$_{25}$	—	—	—	—	—	109
	(b) R=C$_2$H$_5$(C=O)C$_3$H$_7$; R'=C$_{12}$H$_{25}$						109
	(c) R=C$_2$H$_5$(C=O)C$_4$H$_9$; R'=C$_{12}$H$_{25}$						109
11	R= 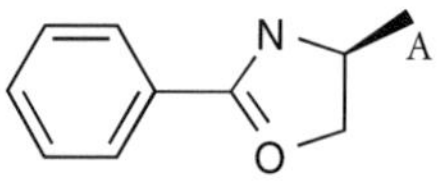 (structure)	A	—	448	5,300	1.3	88
12	(a) R=CH$_2$CH(CH$_3$)C$_2$H$_5$ (S) 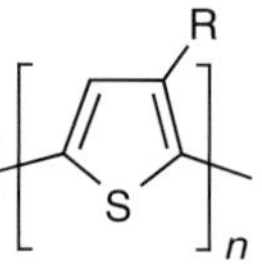	A	—	450^e	25,000	1.7	90
	(b) R=C$_2$H$_5$CH(CH$_3$)C$_2$H$_5$ (S)	A	—	—	6,200	1.7	89
	(c) R=C$_3$H$_6$OCH$_2$CH(CH$_3$)C$_2$H$_5$ (S)	A	—	444^f	16,900	1.4	89
	(d) R=C$_4$H$_8$OCH$_2$CH(CH$_3$)C$_2$H$_5$ (S)	A	—	451^f	21,300	1.5	89
13	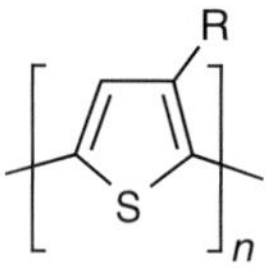 (a) R=C$_2$H$_4$OTHP	A	—	439^a	10,700	1.85	96
	(b) R=C$_6$H$_{12}$OTHP	A	—	454	10,670	1.55	95
	(c) R=C$_6$H$_{12}$OH	F	—	458	—	—	95
	(d) R=C$_{10}$H$_{20}$OSi(CH$_3$)$_3$	C	—	—	14,000	1.6	97
	(e) R=C$_{10}$H$_{20}$OH	F	—	451	11,000	1.6	97
14	(a) R=(CH$_2$)$_{11}$OTHP; R'=C$_6$H$_{13}$; y=1; x=9 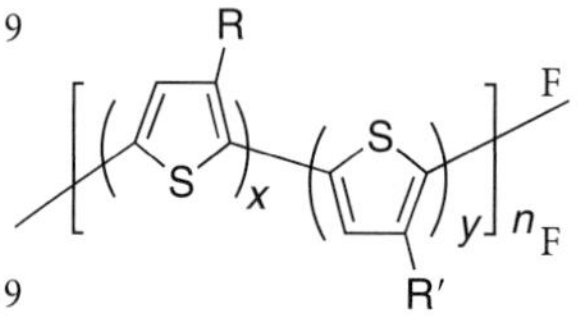	A	—	450^a	9,500	1.42	98
	(b) R=(CH$_2$)$_{11}$OH; R'=C$_6$H$_{13}$; y=1; x=9	F	—	450^a	11,500	1.65	98
	(c) R=(CH$_2$)$_{11}$N$_3$; R'=C$_6$H$_{13}$; y=1; x=9	F	—	450^a	12,900	1.80	98
	(d) R=C$_6$H$_{13}$; R'=(CH$_2$)$_2$OTHP; y=1; x=1	A	—	442^a	8,200	1.49	96
	(e) R=C$_6$H$_{13}$; R'=(CH$_2$)$_2$OH; y=1; x=1	F	—	—	—	—	96
15	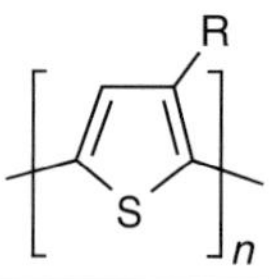 (a) R=(CH$_2$)$_6$PO$_3$(C$_2$H$_5$)$_2$	D	—	442^a	13,000	1.9	103
	(b) R=(CH$_2$)$_6$PO$_3$H$_2$	F	—	442^a	—	—	103

(continued)

TABLE 9.1 (continued)

Entry	Polymer	Method	σ (S/cm)	λ_{max} (nm)	M_n^c	PDI	Ref.
16	$n = 3$ 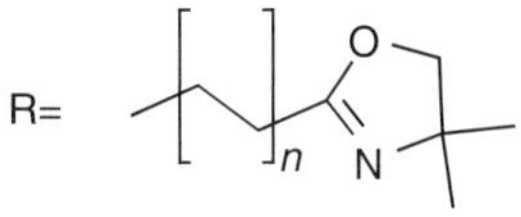	D	—	—	8,000	1.2	45
	$n = 6$	C	—	—	28,790	1.5	102
17	(a) R = C₃H₆COOH	D	—	—	—	—	45
	(b) R = C₃H₆COOC₄H₉	D	—	—	—	—	45
18	(a) R = CH₂NH(CH₃)	D	—	387	1,537	3.0	105
	(b) R = CH₂NH(C₁₂H₂₅)	D	—	416	8,769	1.83	
19	(a) R = (CH₂)₆Br	A	—	445	12,300	1.5	101
		C	—		26,430	1.6	102
	(b) R = (CH₂)₆N₃	F	—		25,290	1.6	102
	(c) R = (CH₂)₆NH₂	F	—		24,820	1.6	102
	(d) R = (CH₂)₆SH	F	—		20,938	1.6	102
	(e) R = (CH₂)₇COOH	F	—		—	—	—
20	R= 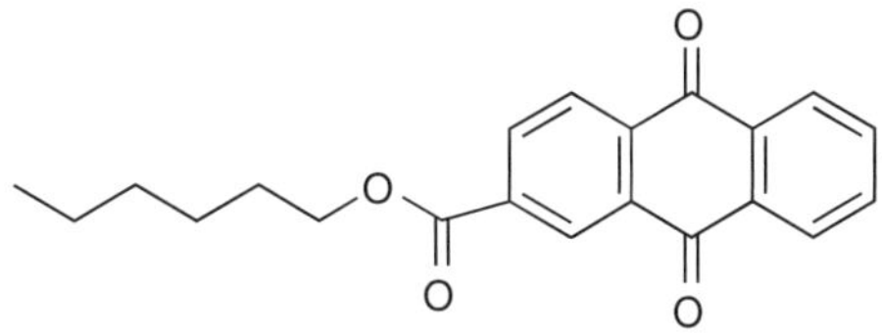	A	—	—	—	—	101

ᵃPolymer in chloroform.
ᵇPolymer in toluene.
ᶜMolecular weight determined by GPC.
ᵈMolecular weight determined by NMR.
ᵉThin film.
ᶠPolymer in dichloromethane.

MgBr·Et₂O with ZnCl₂, which has greater solubility at −78°C [30]. The resulting rr-PAT is precipitated in MeOH, washed and fractionated through sequential Soxhlet extractions (MeOH, hexanes, and chloroform) [26]. Polymers with average conductivities of 600 S/cm and maximum conductivities of 1000 S/cm have been prepared using this method [25].

Shortly after this initial report, the Rieke method for the synthesis of rr-PAT was reported [31–34]. The major innovation in this approach was the generation of an asymmetric organometallic intermediate by treating 2,5-dibromo-3-alkylthiophenes **5** with highly reactive "Rieke Zinc" (Zn*) [35,36] (Scheme 9.2). The metal reacts quantitatively to yield a mixture of the isomers, 2-bromo-3-alkyl-5-(bromozincio)thiophene **6** and 2-(bromozincio)-3-alkyl-5-bromothiophene **7**. The ratio of these

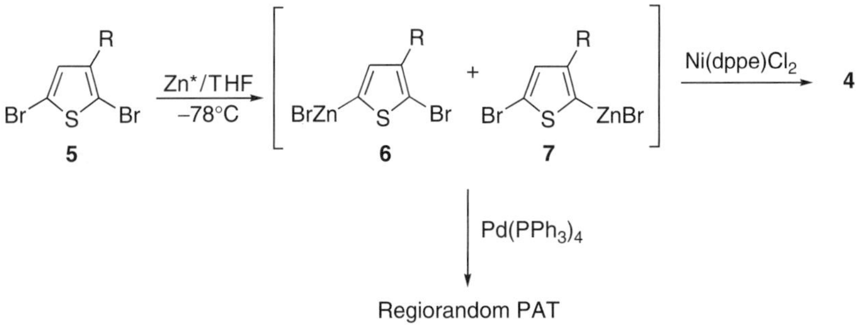

SCHEME 9.2 The Rieke method for the preparation of HT-PATs.

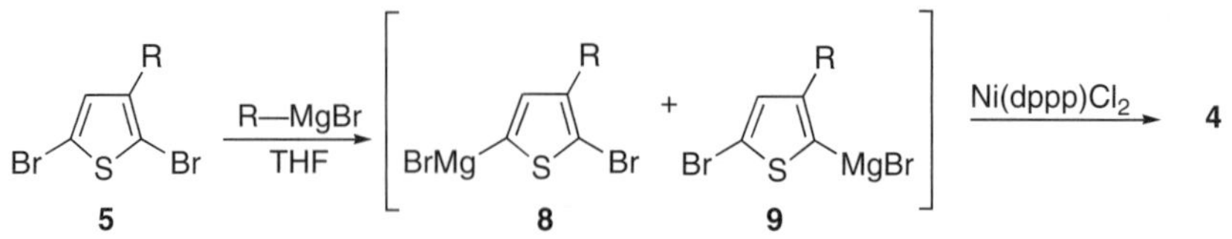

SCHEME 9.3 The Grignard metathesis method for the synthesis of regioregular HT-PATs.

isomers is dependent upon the reaction conditions and, to a lesser extent, the steric influence of the substituent. Although there is no risk of metal halogen exchange, cryogenic conditions must still be employed because the ratio of the isomers **6** and **7** produced is affected by temperature. The use of a nickel cross-coupling catalyst, Ni(dppe)Cl$_2$, yields a regioregular HT-PAT, whereas the use of a palladium cross-coupling catalyst, Pd(PPh$_3$)$_4$, yields a completely regiorandom polymer. The regiochemical control (or lack of) is due to the steric congestion (or lack of) at the reductive elimination step in the catalytic cycle. An alternate approach is the reaction of 2-bromo-3-alkyl-5-iodothiophene with Zn* that produces only one product, 2-bromo-3-alkyl-5-(iodozincio)thiophene. However, this species reacts in identical fashion, yielding regioregular HT-PAT **4** or regiorandom PAT, depending on the catalyst used [31]. After precipitation and Soxhlet extraction, the yield for these reactions is reported to be 75%. Molecular weights for polymers prepared using this method are $M_n = 24$–34 K (with PDI = 1.4). rr-PATs have improved conductivity and exhibit a smaller bandgap (1.7 eV) than regioirregular PATs (2.1 eV). McCullough et al. were the first to show that rr-PATs self-assemble into highly organized crystalline arrays by x-ray diffraction [37]. Cross-polarizing micrograph studies also show that the films of rr-PATs are self-organized and crystalline in contrast to the amorphous films obtained from regioirregular PATs.

In 1999 an economical new synthesis for rr-PATs, known as the Grignard metathesis (GRIM) method, was reported [38]. This method eliminates the need for both cryogenic temperatures and highly reactive metals; 2,5-dibromo-3-alkylthiophenes **5** are treated with 1 equivalent of any commercially available Grignard reagent to form a mixture of 2-bromo-5-bromomagnesio-3-alkylthiophene **8** and 2-bromomagnesio-5-bromo-3-alkylthiophene **9** in the ratio of 85:15 [39]. This ratio appears to be independent of the reaction time, temperature, and Grignard reagent used. Although two isomers are formed, as with the Rieke method, the use of a nickel cross-coupling catalyst, Ni(dppp)Cl$_2$, yields regioregular HT-PAT **4** with greater than 95% HT couplings (Scheme 9.3). This is due to a combination of kinetic and thermodynamic effects arising from the catalytic reaction. Typical molecular weights are in the range of $M_n = 20$–35 K with PDI = 1.2–1.4 [38]. Similarly, ethylmagnesium bromide preferentially reacts with 2,5-diiodo-3-alkyloxythiophenes to yield 2-iodo-5-iodomagnesio-3-alkyloxythiophenes (Table 9.1, entry 1h–j) [40]. Treatment with Ni(dppp)Cl$_2$ produces a regioregular polymer with 96%–100% HT couplings [41].

Other methods that have been applied to the synthesis of rr-PATs are Stille [42] and Suzuki [43] cross-coupling reactions. The Pd(0) catalyzed cross-coupling reaction of 3-hexyl-2-iodo-5-(tributylstannyl)thiophene **10** using a variety of solvents and catalyst was investigated (Scheme 9.4) [44,45]. The major advantage of this approach is the stability of the organometallic monomers. The degree of

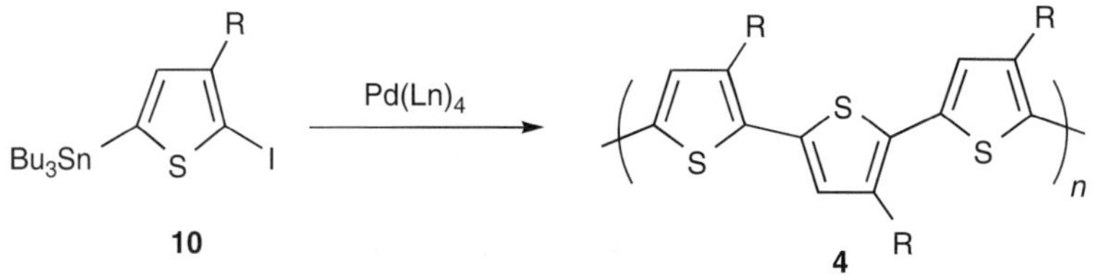

SCHEME 9.4 Synthesis of HT-PATs by Stille coupling reaction.

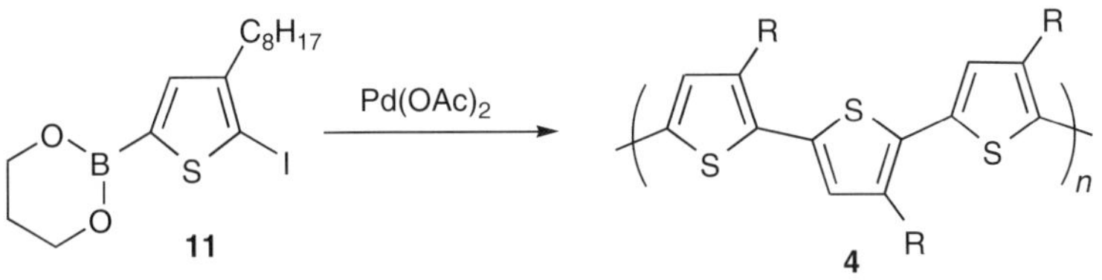

SCHEME 9.5   Synthesis of HT-PATs by Suzuki coupling reaction.

polymerizations varied depending on the reaction conditions used and ranged from $M_n = 7366$–11409 with PDIs in the range of 1.19–1.82, for unfractionated samples. When using this method THF was found to be the worst solvent and toluene worked the best. In all cases rr-PATs **4** with greater than 96% HT couplings were obtained. End group analysis indicates that polymers functionalized with both iodo and tributyltin end groups are obtained under certain conditions. This polymer has the potential for end group transformation through simple nucleophillic displacement of the bromide atom or coupling. The Stille coupling reaction has also been used to synthesize an rr-PAT functionalized with a masked carboxylic acid [45,46]. The polymerization of 2-bromo-3-(2,2-dimethyloxazoethyl)thiophene by the McCullough method using Ni(dppp)Cl₂ catalyst resulted in low-molecular weight polymer (1.5–3.0 K) and low yields (<5%) (entry 16, Table 9.1) [45]. This is presumably due to the coordination of the oxazoline-protecting group to the nickel catalyst hindering its reactivity. However, when the Stille approach is modified by the addition of CuO cocatalyst [47] higher molecular weight polymers ($M_n = 8$ K; PDI = 1.2) are obtained in high yield (>84%). The ability of this method to be used in the presence of functional groups opens the door for the synthesis of a variety of rr-PATs. The Suzuki coupling reaction has also been employed for the synthesis of rr-PATs (Scheme 9.5, entry 1 Table 9.1) [48]. Using 5-(2-iodo-3-octylthiophene) boronic acid **11** a polymer with 96%–97% HT coupling was obtained with $M_w \sim 27,000$ in 51% yield. One of the key factors in this approach was the choice of catalyst, with Pd(OAc)₂ giving the best results.

With few exceptions, the oxidative polymerization of 3-alkylthiophenes using FeCl₃ results in the formation of regioirregular polymers (Figure 9.3). One exception is the regioselective synthesis of poly(3-(4-octylphenyl)thiophene) by the slow addition of FeCl₃ to the reaction mixture (entry 1, Table 9.1). The reason reported for the success of this method is that slow addition keeps the Fe^{3+}/Fe^{2+} ratio low and thus lowers the oxidation potential providing a more controlled polymerization. The polymer obtained was found to have ~94% HT coupling by NMR. The conductivity of free films of the polymer was 4 S/cm [49]. This is an interesting method developed by Andersson that increases the regioregularity using specific conditions. The method gives between 70% and 74% HT couplings, when the thiophene contains typical alkyl groups such as hexyl, ethylhexyl, and cyclohexyl. However, when the 3-substituted side chain is dioctylphenyl, methoxyoctylphenyl, or octylphenyl, the method gives 90%–94% HT couplings. A second approach is to use monomers that preferentially coupled HT, such as 3-alkoxy-4-methylthiophenes (Figure 9.3). Experimental and theoretical studies of this monomer show that the reactivity of the monomer, i.e., the distribution of the unpaired electrons π-spin density on the different positions of the oxidized monomer significantly influences the regioregularity of the resultant

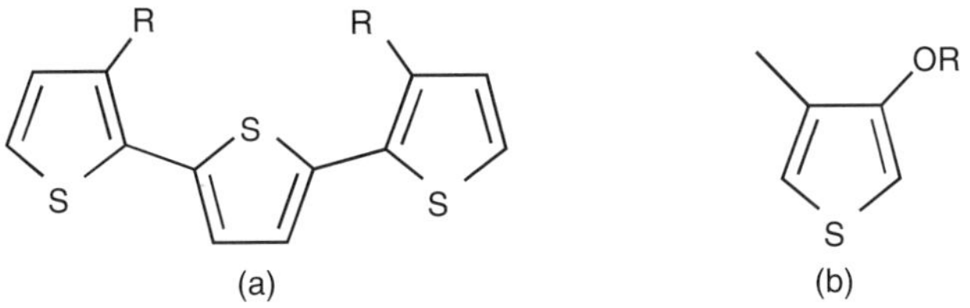

FIGURE 9.3   Examples of monomers that can be electrochemically polymerized to yield HT-PATs.

polymer [50]. Another way to avoid the need for a regiospecific synthesis is the use of symmetric monomers that are unable to produce regiorandom polymers, regardless of the polymerization method used. For example the use of symmetrically 3,3′- and 3′,4′-disubsubstituted 2,2′:5′,2″-terthiophenes (Figure 9.3a) have been used to yield regioregular PT through nonregiospecific polymerizations [51], although polymers of this type do not have substituents on every ring.

9.3.2 Mechanism of the Nickel-Catalyzed Cross-Coupling Reaction

A majority of the methods to synthesize regioregular PT are based on metal-catalyzed cross-couplings [52]. The generally accepted mechanism of these reactions is (1) oxidative addition of an organic halide with a metal-phosphine catalyst, (2) transmetallation between catalyst complex and reactive organometallic reagent (or disproportionation) to generate a diorganometallic complex, and (3) reductive elimination of the coupled product with regeneration of the metal-phosphine catalyst [28,53]. Numerous organometallic species such as organomagnesium, organozinc, organoboron, and organotin demonstrate sufficient efficiency to be used in cross-coupling reactions; further reports of regiospecific synthesis based on these types of monomers are increasing.

While the mechanism of metal-catalyzed cross-coupling reactions of small molecules is well understood, the exact mechanism as it applies to polymerizations has been reported only recently. Because the nickel-catalyzed dehalogenative polymerization is formally a polycondensation reaction, it is generally believed to proceed through a step-growth mechanism. Recently, it has been proposed that the nickel-promoted cross-coupling polymerization proceeds through a chain-growth mechanism rather than the generally accepted step-growth mechanism. These new insights into the mechanism are particularly exciting because the chain-growth mechanism allows for very narrow molecular weight distributions and defined molecular weights are to be made for the first time. In addition, the methods have produced a quasi-living conductive polymer synthesis. A living system will lead to new regioregular PTs that have never been made before. Yokozawa provided the evidence that the regioregular synthesis of poly(3-hexylthiophene) proceeds by the chain-growth polymerization in the case of the GRIM method (Figure 9.4). The conversion vs. M_n and conversion vs. M_w/M_n plots of the nickel-catalyzed polymerization of 2-bromo-5-chloromagnesio-3-hexylthiophene (obtained from the GRIM reaction of 2-bromo-3-hexyl-5-iodothiophene) were reported. These plots suggest that the monomer is polymerized in a chain-growth polymerization manner, in that the M_n values increased in proportion to the conversion of the monomer, indicating that an initiator species existed from which polymerization was propagated. At the same time, another study by the McCullough group showed that the mechanism

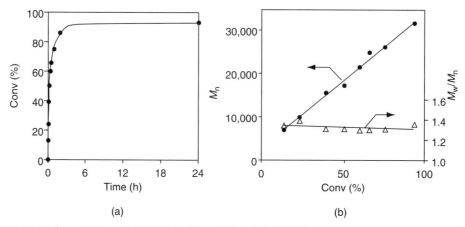

FIGURE 9.4 Polymerization of **2** with 0.4 mol% of Ni(dppp)Cl₂ in THF at room temperature ([**2**]0–0.12 M): (a) time conversion curve; (b) M_n and M_w/M_n values of HT-P3HT as a function of monomer conversion. (From Yokoyama, A., Miyakoshi, R., and Yokozawa, T., *Macromolecules*, 37, 1169, 2004.)

is chain growth and the degree of polymerization of P3AT increases with monomer conversion and can be predicted by the molar ratio of the monomer to the nickel initiator [54]. This was seen using two different monomers: 2-bromo-3-hexyl-5-chlorozincothiophene (obtained using the McCullough method) and 2-bromo-3-hexyl-5-bromomagnesiothiophene (obtained using the GRIM method) [54–56]. Based on these experimental results, the McCullough group proposed a nickel-initiated cross-coupling polymerization system that is essentially living (Scheme 9.6). This living system has produced PATs with low polydispersites (Figure 9.5 and Figure 9.6) and new block copolymers [57]. From the above data two possible explanations are possible. The first is that the oxidative addition of the polymer chain occurs selectively. This is because the oxidative addition of the monomer is kinetically slower, or thermodynamically less stable than the oxidative addition of the growing polymer chain, due to the decreased electron density of the thiophene ring on the polymer chain [58]. The second is the proposal by the McCullough group that the polymer chain and the metal catalyst exist as an associative pair [49] via the formation of a π-complex (Scheme 9.6, [**11•12**]), limiting polymerization to one end of the polymer chain. The mechanism for the regioregular polymerization of thiophene has been proposed by McCullough and is outlined in Scheme 9.6.

The effect of catalyst structure and reaction temperature on the regioregularity of PATs has also been investigated by Lucht (Table 9.1, entry 1g) [59]. Poly(3-octadecylthiophenes) (P3ODTs) were prepared by the GRIM method in the presence of palladium and nickel catalysts (Ni(dppe)Cl$_2$, Pd(dppe)Cl$_2$, Ni(PPh$_3$)$_4$, and Pd(PPh$_3$)$_4$). It was found that nickel catalysts provided P3ODTs with high regioregularity (>90% HT couplings), while palladium catalysts gave lower regioregularity. The effect of temperature on the regioregularity of P3ODT was also investigated. For all the catalysts studied it was determined that the regioregularity of the polymer increased as the reaction temperature was increased. Although a clear explanation of this effect was not presented the authors suggest that polymerizations with nickel catalysts occur primarily through chain-growth reactions, whereas polymerizations with

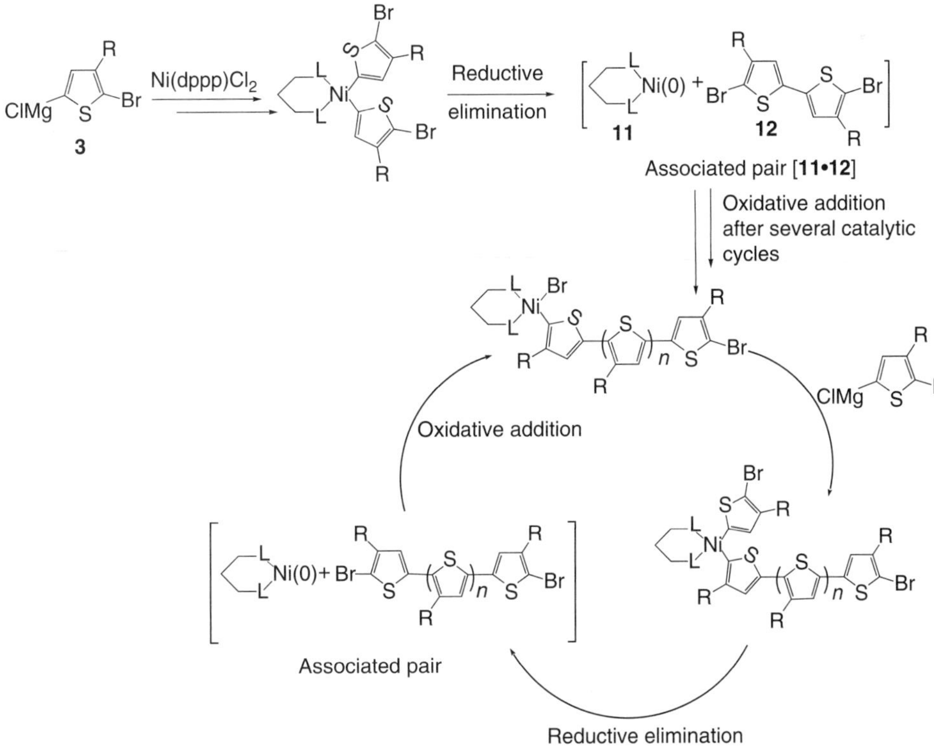

SCHEME 9.6 Proposed mechanism for the nickel initiated cross-coupling polymerization. (From Sheina, E.E., Liu, J., Laird, D.W., and McCullough, R.D., *Macromolecules*, 37, 3526, 2004.)

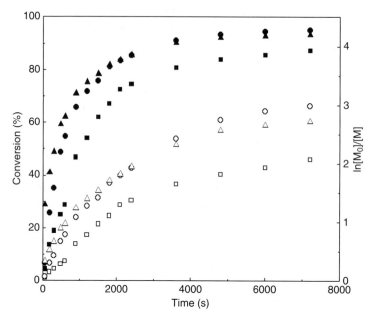

FIGURE 9.5 Conversion (*filled symbols*) and logarithm of monomer concentration (*open symbols*) vs. time plots for 2-bromo-3-hexylthiophene polymerization at different concentrations of Ni(dppp)Cl$_2$ initiator (23°C–25°C); $[M]_0 = 0.075$ mol/L: ($\blacksquare$, $\square$) $[M]_0$: [Ni(dppp)Cl$_2$] $= 136:1$; ($\bullet$, $\circ$) $[M]_0$: [Ni(dppp)Cl$_2$] $= 57:1$; ($\blacktriangle$, $\triangle$) $[M]_0$: [Ni(dppp)Cl$_2$] $= 49:1$. (From Sheina, E.E., Liu, J., Laird, D.W., and McCullough, R.D., *Macromolecules*, 37, 3526, 2004.)

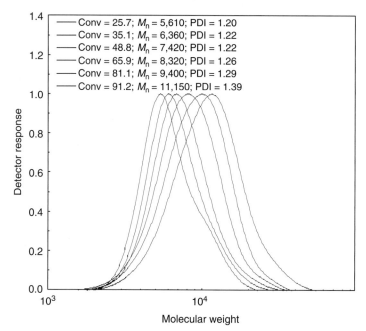

FIGURE 9.6 GPC traces for 2-bromo-3-hexylthiophene polymerization (23°C–25°C); $[M]_0 = 0.075$ mol/L; $[M]_0$: [Ni(dppp)Cl$_2$] $= 57:1$. (From Sheina, E.E., Liu, J., Laird, D.W., and McCullough, R.D., *Macromolecules*, 37, 3526, 2004.)

palladium catalysts have competing step-growth and chain-growth reactions. The temperature dependance of the regioregularity can be explained by the temperature dependance of the oxidative addition products and relative temperature independence of the transmetallation step of the catalytic cycle for the polymerization reaction. Both of which are critical factors for chain-growth polymerization. The ability to synthesize PTs with well-defined molecular weights and narrow PDIs is important in optimizing the electrical properties of conducting polymers. Further work is expected to lead to a truly living synthesis of conducting polymers.

9.4 Synthesis of Regioregular PATs with Functionalized Side Chains

Although the initial role of side chains on the PT backbone was to improve solubility it has been established that the introduction of various substituents along the backbone can not only enhance the processibility of these polymers, but can also alter their material properties, including their electrical properties and optical properties. PTs are known to exhibit a variety of sensory properties upon exposure to external stimuli such as affinity chromism, photochromism, piezochromism, solvatchromism, and thermochromism. These chromic effects are due to a planar and nonplanar transition of the conjugated backbone [60]. This first-order-like conformational transition (formation of twistons) is driven by side-chain disordering induced by the stimuli. The twisting of one repeat unit of this polymer provokes the twisting of a large number of adjacent thiophene units (a domino effect), which leads to an amplification of the resulting optical (i.e., absorption or emission band) signal. This has sparked an interest in the synthesis of a number of different derivatives of PT in an effort to create materials to fine tune the properties of the polymer or to utilize the polymer as a sensor. A large number of polymers will be described in this section, which will also be summarized in the form of a table. For cross-referencing of structures the number for each entry will be noted within the text.

9.4.1 Heteroatom Substituents on rr-PATs

Regioregular PTs containing etheric substituent are particularly interesting because they can enhance solubility or elicit a chemoselective response to external stimuli through binding phenomenon. Pioneering efforts by Garnier [61] on the development of P3ATs were absolutely critical to the development of these materials. A plethora of work has also been reviewed by Roncali [62], which is relevant to the field but will not be covered here. There have also been several reports of regioregular PT bearing oligo(oxyethylene) side chains. LeClerc has reported that poly(3-oligo(oxyethylene)-4-methylthiophene) was prepared by the iron trichloride polymerization of 3-oligo(oxyethylene)-4-methylthiophene, one of the few monomers that gives regioregular polymers using this method. Polymers were synthesized with side chains of varying length from 3 to 10 oxyethylene units with an average of 7 (Table 9.1, entry 5a) [63]. Later the synthesis of a regioregular PT bearing side chains with exactly six oxyethylene units was reported [64]. It was found that the hexa(oxyethylene) side chains induce a more planar conformation of the PT backbone than the random-length oxyethylene side chains. This polymer exhibited an ionochromic transistion in the presence of K^+. The polymer had a measurable ionic conductivity above 80°C and films had a doped conductivity of 3×10^{-3} S/cm (Table 9.1, entry 5b).

A series of P3ATs ($R = CH_2OCH_3$, CH_2SCH_3, $CH_2OCH_2CH_2OCH_3$, and $CH_2(OCH_2CH_2)_2OCH_3$) were synthesized by the McCullough method (Table 9.1, entries 2a–c). The short substituents were not adequate for improving the solubility of the growing polymer, therefore only low-molecular weight materials were produced [65]. The poly(3-(2,5,8-trioxanonyl)thiophene) was synthesized in high-molecular weight (71,000; PDI = 2) and was determined to have >99% HT coupling through ^{1}H NMR. The I_2 doped polymer possesses high-electrical conductivity, 500–1000 S/cm on an average, with one sample measured at 5500 S/cm. Poly(3-(2,5,8-trioxanonyl)thiophene) does exhibit ion-binding properties, and a mild 10–20 nm ionochromic response occurs upon exposure to Li^+, whereas Pb^{2+} and Hg^{2+} resulted in 190 nm blueshift in the absorption.

The introduction of an electron donating group, such as an alkoxy or alkyl thio group in the β-position of the thiophene unit, is expected to decrease the bandgap of the polymer by raising the energy of the highest occupied molecular orbital. As a result alkoxy-substituted PT possesses reduced bandgaps, low-oxidation potentials, and a highly stable conducting state [66]. One of the most widely studied alkoxyl-substituted PTs is poly(3,4-ethylenedioxythiophene) (PEDOT). There is a vast amount of research on this particular polymer and it has been reviewed, hence it will not be discussed in this chapter [67]. To date only limited examples of alkoxy rr-PATs have been synthesized. The synthesis of regioregular poly(3-(2-(2-methoxyethoxy)ethoxy)thiophene) (PMEET) by McCullough and GRIM method has been investigated (Table1, entries 3a–d) [68,69]. The polymer prepared using the McCullough method had λ_{max} of 602 nm in CHCl$_3$, whereas the polymer prepared using the GRIM method had λ_{max} of 587 nm. The conductivities of the polymer prepared by the McCullough method are also two to three times greater than those of the polymer prepared by the GRIM method, although specific values are not given. It is proposed that the difference in conductivity is due to the difference in the regioregularity of the polymers. PMEET prepared by the GRIM method has a measured conductivity of 650 S/cm, and an average conductivity of 200 S/cm when doped with iodine. The conductivity of the polymer remains high with values of 150 S/cm obtained 2 months later, 25–35 S/cm after 3 y, and in some cases the conductivity actually increased over time [69,70]. The synthesis of copolymers containing 3-alkoxy, 3-alkyl, and 3,4-ethylenedioxythiophene units has also been reported (Table 9.1, entries 4a–b). Poly (2,5-(3-dodecyl-3,4-ethylenedioxy)bithiophene) was synthesized by the GRIM method, to yield a polymer with $M_w = 11,000$ and a doped conductivity of 0.08 S/cm. Oligo(2,5-(3-(2-(2-methoxyethoxy)ethoxy)-3,4-ethylenedioxy)bithiophene) was obtained by the autopolymerization of the brominated bithiophene. The oligomers had $M_w = 2000$ and conductivities of 0.78 S/cm initially, which increased to 5 S/cm after the films were stored for 1 month [71].

Substitution of a sulfur atom directly onto the thiophene ring is also expected to lower the oxidation potential of the polymer. The synthesis of a series of regioregular P3ATs using Zn* has been reported (Table 9.1, entries 6a–d) [34,72]. These P3ATs were highly regioregular polymers that were completely soluble in carbon disulfide, and only slightly soluble in other common organic solvents, such as THF, chloroform, and xylenes. This poor solubility suggests that there is a stronger affinity between the polymer chains. As a result of the poor solubility the molecular weight of these polymers is low (~4400). The UV–vis spectra in chloroform for hexylthio-PAT, exhibited 3 peaks at 263, 324, and 513 nm, with a shoulder at 605 nm. The destabilization of the HOMO reduces the HOMO–LUMO gap and leads to a redshift of the absorbance of the polymer in solution. The p-doped poly(3-alkylthio)thiophene films are highly conductive in the range of 100 S/cm.

PTs bearing semifluoroalkyl side chains are materials with unique properties due to the hydrophobicity, rigidity, thermal stability, chemical, and oxidative resistance of the fluoroalkyl side chains [73]. Collard has shown that variations in the substitution of such polymers can be used to control the supermolecular packing of the polymer. With certain lengths of alkyl and perfluoroalkyl substituents, PATs can display liquid crystallinity and form highly ordered solid-state structures. The synthesis of regioregular PATs bearing fluoroalkyl chains and an alternating copolymer with nonyl thiophene was reported using the GRIM method [74,75]. In both cases alkyl spacers were used between the thiophene ring and the side chain, to insulate against the electronic effects of the fluoro substituents for ease of synthesis at 100°C. The UV–vis spectrum of a solution of the copolymer poly(3-nonylthiophene)-*alt*-(3-(6,6,7,7,8,8,9,9,9-nonafluorononyl)thiophene) in CHCl$_3$ has an absorbance maximum, λ_{max}, at 445 nm (Table 9.1, entry 8a). This is similar to that found for the homopolymers derived from the individual units: 3-nonylthiophene ($\lambda_{max} = 448$ nm) and 3-(6,6,7,7,8,8,9,9,9-nonafluorononyl)thiophene ($\lambda_{max} = 436$ nm) (Table 9.1, entry 7a). X-ray analysis of thin films of the polymers shows that the copolymer was found to form a self-assembled, well-ordered lamellar structure with an interlayer spacing of 44.8 Å, in contrast to the homopolymer that forms lamellar structures with 22.4 Å spacing. On the basis of this analysis, it was proposed that the alternating copolymer adopts a highly ordered bilayer lamellar structure in which the hydrocarbon chains pack with other hydrocarbon chains, while the fluorocarbon chains are segregated and packed among themselves. Thus the order is due to the immiscibility of the two types of chains, and their self-assembly that is enhanced by the regioregularity of the polymer. The liquid crystallinity of fluoroalkyl rr-PATs has also been examined. Using UV–vis, differential scanning calorimetry (DSC), the data obtained suggest that the semifluorinated rr-PATs form a weakly ordered thermotropic mesophase. Hence the semifluoroalkyl groups are not side-chain mesogens; they simply undergo conformational disordering prior to isotropization of the main chains with formation of a nematic phase [75].

The synthesis of a series of PATs with perfluoroalkyl chains directly attached to the thiophene ring has also been reported [73]. Using the McCullough method, poly(perfluorooctylthiophene) and poly (octylthiophene)-altwere(perfluorooctylthiophene) were prepared resulting in the formation of highly regioregular fluorescent polymers (Table entries, 7c and 8b). The redox properties of these polymers were examined and it was found that the alternating copolymer undergoes reduction at −2300 mV vs. SCE compared to −1180 mV for the perfluoroalkyl homopolymer. The alkyl-subtituted homopolymer cannot be reduced under normal conditions, thus the electron-withdrawing nature of the perfluoroalkyl chain stabilizes negative charge on the thiophene backbone. A higher density of perfluoroalkyl chains makes the reduced state more accessible resulting in a lower reduction potential [73]. The absorbance λ_{max} and the fluorescence λ of 384 and 547 nm, respectively, are between the values obtained for the corresponding alkyl homopolymer, poly(3-octylthiophene), and the fluorinated homopolymer, poly(3-perfluorooctylthiophene), at 441 and 570 nm and 326 and 506 nm, respectively. Hence variation of the ratio of the two monomers may serve as an excellent way to tune the electronic structure of the conjugated backbone and provide a polymer with enhanced fluorescence in comparison to the alkyl analog [73]. The solubility of the perfluoroalkyl polymer in supercritical CO_2 has also been studied [76]. It has been determined that the polymer begins to dissolve under 1500 psi of CO_2 at 50°C and the solubility continues to increase up to 2000 psi of CO_2. These studies are important for "green" processing of conducting polymers. Other fluoro-substituted PTs have been prepared such as the near perfluoro derivative of poly(3-dodecylthiophene) and copolymer of 3-[perfluoro(1,1,3,3,4,4,5,5,6,6,-decahydro-2-oxo)tetradecyl]thiophene and 3-(methoxyethoxyethoxymethyl)thiophene. However, these polymers are quite insoluble [77,78].

9.4.2 Chiral Substitutents on rr-PATs

Chiral conjugated polymers (CPs) with a well-defined structure are of interest, due to their potential for use in a variety of applications such as optoelectronic devices, enantioselective sensors, biosensors, catalysts, absorbants, and as artificial enzymes. Chiral PTs with an optically active substituent in the 3-position have been studied for these purposes. Typically chiral PTs exhibit optical activity in the π–π* transition region, as a result of the main-chain chirality created by the π-stacked self-assembly of the polymer chains, forming supramolecular aggregates in a poor solvent or at low temperature. These polymers show no activity in the UV–vis region in a good solvent or at high temperatures. The effect of regioregularity on the optical activity of chiral PTs can be evaluated by comparing regioirregular chiral PATs [79] with regioregular chiral PATs [80]. Regioirregular chiral PATs have been prepared by electro-chemically polymerizing $(S)(+)$- and $(R)(\pm)$-2-phenylbutyl ether of 3-propylthiophene. The resultant chiral PTs had reported specific rotations of $[\alpha]_{22} = 3000$ [80,81]. Using the McCullough method (entry 12, Table 9.1) [80,81], an optically active regioregular PT was synthesized that exhibited a specific rotation of $[\alpha]_{22} = 140,000$ for $\lambda = 513$ nm at the sodium D line and $[\alpha]_{22} = -9000$ for $\lambda = 589$ nm. This demonstrates the effect of regioregularity on the optical rotation of chiral PTs. Solvatochromic studies of the polymer ($M_n = 16,900$; PDI = 1.4) show that varying the solvent composition dramatically affects the shape and λ_{max} of the π–π* transition by altering the distribution of disordered and ordered, aggregated structures. This polymer also undergoes stereomutation in the solid state. The solid-state thermochromism in this polymer is typical for a PT, except that at the melting point of the polymer a complete loss of optical activity is observed. Even more interesting is the observation that when the polymer is cooled quickly from the disordered melt, by pouring the sample into a water bath at 0°C. The absorption spectra are unchanged, but a mirror image circular dichroism (CD) spectrum, relative to the original sample before melting, is found. Therefore the regioregular, chiral PT undergoes stereomutation that is believed to be driven by an aggregation effect. The effect is reversible, affording the opportunity to tune the chirality of the spectrum simply by controlling the cooling rate. Irregular chiral PTs do not show this effect. The same effects are also found in poly(3,4-di[(S)-2-methylbutoxy]thiophene), which has been studied in some detail by both CD and circular polarized luminescence [82,83]. Another chiral rr-PAT, namely poly(3-(S-3′,7′-dimethyloctyl)thiophene) [84], confirms that CD spectra can be used in chiral rr-PTs to probe aggregation states [81]. This particular polymer can also act as a sensor, by exhibiting large conformational changes induced by minute solvent variation.

The optical behavior of rr-PAT bearing a chiral phenyl oxazoline group has been reported (Table 9.1, entry 11) [85]. The polymer was prepared in two different ways: the McCullough method and via a modified Stille method using CuO cocatalyst. It was found that in good solvent the polymer prepared using the McCullough method did not show any induced circular dichroism (ICD). However, the polymer prepared by the Stille method did show ICD in the same solvent. It was believed that the ICD in the Stille sample was induced by the presence of residual Cu(II) from the polymerization coordinating with the oxazoline group. This was supported by the disappearance of the ICD peaks when the Cu(II) was removed by treatment with EDTA. Also when the McCullough method sample was treated with Cu(OTf), it exhibited the sample split-type ICD in the π–π* region. The supramolecular chirality of the polymer could also be reversibly controlled by the addition or removal of an electron [86]. The phenyloxazoline group can be easily converted into a chiral aminoacid bearing polymer by acid hydrolysis and further derivatized to achiral carboxylic acid groups through saponification. The aminoacid polymer that was soluble in water, DMSO, and methanol showed ICDs in these solvents (first Cotton ($[\theta]$) = 6×10^3 (607 nm, methanol) and 1×10^3 (474 nm, H_2O/DMSO (19/1)). Completely hydrolyzed polymer was insoluble, but the partially hydrolyzed polymer was soluble in DMSO and was found to respond to the chirality of amines such as (S)-2-amino-1-butanol, exhibiting an ICD depending on the absolute configuration of the amine [85]. A detailed study of the ICD of the phenyloxazoline polymer in a variety of solvents was performed [87,88]. It was found that the polymer exhibited an ICD in chloroform upon the addition of poor solvents, such as methanol and acetone, with similar cotton effect patterns (+1st, −2nd). Whereas with other solvents (acetonitrile and nitromethane) the cotton effect patterns were reversed. The ICD of a regiorandom PAT was also compared and found to be very weak even in the presence of a large excess of methanol, indicating that regioregularity is an important factor for inducing a chiral supramolecular aggregate. A similar solvent dependent inversion of ICD was observed in a study of rr-PATs bearing chiral alkyl and alkyloxy side chains [89]. The experiments demonstrate that the handedness of the chiral organization of the polymer main chain is very sensitive to the ordering conditions implying that the energy difference between the two diasteromeric forms is small.

The effect of temperature on the ordering of chiral polymers has also been examined. Poly-2-(S)-2-methylbutylthiophene (Table 9.1, entry 12) was prepared by the McCullough method, and the thermal and structural properties were studied [90]. This polymer was selected for study because the chiral center is close to the conjugated polymer backbone. X-ray diffraction studies show that the polymer has a high-density crystalline packing, similar to that seen in other rr-PATs. The polymer also shows strong thermochromism in solution and solid state. Increasing the temperature up to 100°C produces a redshift of the absorption band with the formation of a structured spectrum, due to an increase in the conjugation length and the 3D order of the side chains. The CD signals also increased as the temperature was increased with a maximum value obtained at 100°C. This is consistent with DSC measurements that show that the polymer becomes more crystalline at this temperature.

Cooperative effects of the monomer units along the polymer backbone may result in a nonlinear relation between the specific optical rotation and the enantiomeric excess (ee) of chiral units present in the polymer [91]. For stiff helical isocyanates, these cooperative effects have led to observations referred to as "majority-rules." The role of cooperative effects in chiral rr-PATs has also been examined using a series of PTs containing various ratios of chiral and achiral side chains. Data show that the cooperative interaction between the side chains affects the optical activity in a nonlinear fashion, while the majority-rules principle is applicable to chiral rr-PATs, the magnitude is less pronounced than with helical main-chain isocyanates [91].

Poly[3-[(2S)-2-methyl-3,6,9,12,15-pentaoxahexadecyloxy]-4-methylthiophene], a chiral PT bearing oligoethylene glycol side chains, was prepared by the FeCl$_3$ method [92]. This polymer is unique in addition to bearing enantiomerically pure side chains, it is also hydrophilic allowing for the polymer to be dispersed in water. Transmission electron microscopy (TEM) shows that the resulting dispersions form highly ordered hexagonal shaped aggregates with an average dimension of 100 nm. The aggregation process depends largely on the amount of THF used with the best results obtained at minimum THF content as observed by the strong bisignate Cotton effect in the CD spectrum. The dispersion was found to be stable even after removal of THF allowing for the generation of waterborne electroactive coatings. The synthesis and characterization of a chiral insulated molecular wire has recently been reported using a water-soluble rr-PAT and a polysacaccharide [93].

9.4.3 γ-Functionalized Side Chains on rr-PATs HT

Alkyl-substituted P3ATs are widely studied due to their ease of synthesis. Functionalization at the end of the side chains on regioregular PTs is desirable because it can be used to tune the properties of the polymers for various applications. The development of new synthetic techniques has allowed for the incorporation of a variety of substituents, including molecular recognition, cross-linking or grafting moieties, and electronic diversity units. The primary strategy has been the use of protecting groups at the gamma end of the side chain attached to the regioregular PT backbone, which allows for the incorporation of groups onto the polymer backbone that are usually reactive to the polymerization conditions.

Hydroxy-functionalized PATs are of particular interest as they can be used in hydrogen bond directed self-assembly or as cross-linkers [94]. The tetrahydropyranyl (THP) group has been widely used for the synthesis of such polymers as it is stable to the reaction conditions and readily removed. A highly regioregular poly[3-(6-tetrahydropyran-2-yloxyhexyl)-2,5-thiophene] (Table 9.1, entry 13b) was synthesized by the McCullough polymerization of 2-bromo-3-(6-tetrahydropryanyloxyhexyl)thiophene by Bolognesi. The removal of the pyranyl-function restores the hydroxy group yielding poly[3-(6-hydroxyhexyl)-2,5-thienylene] (Table 9.1, entry 13c), which is only soluble in 1-methyl-2-pyrrolidone. The polymer was characterized in its protected form, which was found to give Langmuir–Blodgett multilayered structures [95]. Similarly, Holdcroft has reported that 2-bromo-3-(2-(-tetrahydropyranyl-2-oxy)ethyl)-thiophene) was homopolymerized using the McCullough method and also copolymerized with 3-hexylthiophene to obtain functionalized HT P3AT homopolymer (Table 9.1, entry 13a) and random copolymers (Table 9.1, entries 14a–b) [96]. The polymers were comprised of various mixtures between 0% and 100% of 3-hexylthiophene and 3-(2-(2-tetrahydropyranyl-2-oxy)ethyl)thiophene. Thermolytic and catalytic removal of the bulky (THP) group from thin solid films of the polymers was studied, and appeared to occur in high yield. Deprotection can also be achieved catalytically in the presence of acids, at a significantly lower temperature to produce the hydroxy terminated polymers. The deprotected polymer is insoluble due to the short side chains and the influence of hydrogen bonding. The effect of the bulky protecting groups on the packing of the polymers was also reflected in the melting points of the polymer.

The synthesis of poly[3-(10-hydroxydecyl)-2,5-thiophene] (Table 9.1, entry 13d) by the GRIM method has been reported by Lanzi [97]. This report utilized a trimethylsilanyl group to protect the hydroxy group during polymerization (Table 9.1, entry 13e). This protecting group is easily removed after yielding the targeted polymer. The polymer contained 96% HT coupling and both the protected polymer and the hydroxy terminated polymer were soluble in a variety of common organic solvents. The chromic behavior of the hydroxy-functionalized polymer was investigated by UV–visible spectroscopy in different solvent and nonsolvent mixtures, and in the solid state by exposing the polymer adsorbed on hydroxylic matrixes to methanol vapors. In solution (DMPU) the polymer exhibits a maximum absorption λ_{max} at 442 nm. The addition of methanol results in more conjugated polymer conformation with λ_{max} of 556 nm at the highest methanol mole fraction. The solid films exhibited a redshift in absorbance spectra of 36 nm in cellulose and 50 nm in SiO$_2$. This effect is reversible with the initial λ_{max} values being restored upon removing methanol under vacuum.

Using the McCullough method, regioregular PT random copolymers bearing hexyl and 11-hydro-xyundecyl side chains have been prepared by Holmes (Table 9.1, entry 14c–e) [94,98]. The motivation of this work was to synthesize polymer bearing groups that could be used for cross-linking to develop materials for use in the fabrication of multilayer devices. The hydroxy groups were also protected as THP ethers, and subsequently transformed into azide groups. Heating these azide-functionalized copolymers under vacuum leads to azide decomposition, nitrene formation, and cross-linking. The polymers studied contained various ratios of the monomers ranging from 19:1 to 2:1 hexyl:undecylhydroxy. The resultant polymer films showed decreased solubility and a shift in the absorption spectrum to shorter wavelengths dependent on the azide content of the polymer. This color change is due to the conformational change in the PT backbone, which has been partially fixed by cross-linking at high temperature as seen by the modified thermochromism and photoluminescence behavior of the cross-linked polymer films.

A different approach toward the synthesis of side-chain functionalization is postpolymerization functionalizaion. In this approach a reactive group that is stable to the polymerization conditions is incorporated onto the polymer backbone. Once the polymer is formed the functionality of the polymer is increased by subsequent reactions. This provides a convenient route to PTs with useful functional groups that are not tolerant to the standard polymerization conditions. One of the most versatile examples of such a polymer is poly (3-(6-bromohexyl)thiophene) (Table 9.1, entry 19a). This polymer was first synthesized and developed by Iraqi and was also made by Pomerantz and several groups to make a plethora of interesting regioregular PTs [99–101]. Using the McCullough method poly(3-(6-bromohexyl)thiophene) was obtained ($M_n = 12,300$; PDI = 1.5) and then reacted with 2-carboxyanthraquinone (Anth), to give a highly redox active regioregular PT with four redox couples (Table 9.1, entry 20) [101]. ^{1}H NMR studies show that ~87% of the side chains are functionalized. Cyclic voltammetric studies of anthraquinone polymer-coated electrodes show that the observed response is coverage dependent, and thin films display four redox couples due to the $Anth^{0/-/2-}$ processes and the p- and n-doping of the conjugated thiophene backbone. Thick films are rectifying in the sense that reduction of the Anth groups is inhibited on the negative sweep. Spectroelectrochemical studies confirm the nature of the anodic p-doping process (the film turns red to nearly colorless) and show the characteristic changes on reduction (red to black). Poly(3-(6-bromohexyl)thiophene) has also been prepared by the GRIM method [102]. This resulted in the formation of a polymer of high-molecular weight (26,430; PDI = 1.6) in high yields (60%). The objective of this work is to synthesize poly(3-(6-bromohexyl)thiophene) and further functionalize the polymer (Table, entries 19a–e). When poly(3-(6-bromohexyl)thiophene) is reacted with sodium azide in DMF at reflux, the azide derivative is obtained. Subsequently, this polymer can be reduced with LiAlH$_4$ to form an amine-substituted polymer with quantitative conversion. The resulting amine-substituted polymer is soluble in aqueous acidic solvents. When poly(3-(6-bromohexyl)thiophene) is reacted with potassium thioacetate a thiol-substituted polymer is formed. When poly[3-(6-bromoalkyl)thiophene] is reacted with excess lithiated 2,4,4-trimethyloxazoline at low temperature the oxazoline polymer is formed. Hydrolysis in 3 M hydrochloric acid yields the octanoic acid polymer. The polymer is soluble in DMF and aqueous basic solvents. All conversions were near quantitative. This simple method provides access to a large number of chain end-functionalized derivatives.

9.4.3.1 Chromic rr-PATs

Like the etheric substituents, a variety of the side-chain functionalized polymers also exhibit chromic responses in the presence of external stimuli. Two examples are phosphonic acid-substituted regioregular P3AT homopolymers and an alternating copolymer containing a thiophene with a phosphonic acid and a dodecylthiophene (Table 9.1, entries 9c–d, 15a–b). These derivatives were synthesized [103] by the Stille polymerization method, which was found to be superior to the other methods that produce regioregular PTs. The GRIM method was found not to be compatible with the phosphonic ester moiety and yielded only short chain oligomers. The molecular weights obtained by GPC for the masked homopolymer and the copolymer were M_n of 13,000 (PDI = 1.9) and M_n of 21,500 g/mol (PDI = 1.46), respectively. ^{1}H NMR characterization showed a highly regioregular backbone as indicated by the single aromatic peak corresponding to a regioregular polymer. The UV–vis analysis of the homopolymer had λ_{max} of 442 nm in chloroform and λ_{max} of 524 nm for the thin film. The copolymer exhibited a blueshift in the spectrum relative to polyalkylthiophenes as shown by a thin film that exhibited a λ_{max} of 542 nm, while a CH_2Cl_2 solution had a λ_{max} of 443 nm. Hydrolysis of phosphonic esters with bromotrimethylsilane and water produced phosphonic acid terminated polymers. In both cases, the neutral polymer was insoluble in most solvents, but was very soluble in basic ammonium hydroxide and tetraalkylammonium hydroxide solutions. The polymers also exhibited sensitivity to counterion size. Treatment of the homopolymer with n-Bu$_4$NOH, n-Pr$_4$NOH, and n-Et$_4$NOH led to little solution order as indicated by the UV–vis spectrum (λ_{max} of 465 nm) of the solutions, while aqueous NH$_3$ solutions led to a λ_{max} of 510 nm, indicating that some induced self-assembly of the phosphonate PT polyelectrolyte did occur. In contrast, the amphiphilic PT showed a greater tendency toward hydrophobic and polyelectrolyte self-assembly. The polymer was soluble in all tetraalkylammonium hydoxide solutions (Bu$_4$N+, Pr$_4$N+, and Et$_4$N+). It was observed that the λ_{max} of the UV–vis spectrum in solution shifted according to the size of the counterion. A λ_{max} of 443 nm in chloroform was observed for the phophonic ester (λ_{max} was 542 nm for the thin film); the λ_{max} values obtained were 497 nm for the Bu$_4$N+ salt, 534 nm for the Pr$_4$N+ salt, and 532 nm for the Et$_4$N+ salt. The self-assembly of the amphiphilic phosphonate PT in solution was indicated by the large shift in the λ_{max} of solutions of the phosphonic ester when dissolved in alkyl ammonium salts.

Supramolecular self-assembly was observed with derivatives of poly-3-(2-(4,4-dimethyloxazolin-2-yl)-ethyl)thiophene (Table 9.1, entry 16a) [45,46]. This polymer was also synthesized using the Stille coupling reaction. Carboxylic acid groups, which are intolerant of the organometallic reagents used, were protected as oxazoline groups. Deprotection in aqueous HCl yields poly(thiophene-3-propionic acid) (Table 9.1, entry 17a) [45,46]. The carboxylic acid polymer is insoluble in all solvents, however, deprotonation produces a water-soluble salt, which displays marked ionochromic behavior and ion-induced self-assembly. Small bases drive an ordering of the conducting polymer structure to give highly conjugated purple solutions with λ_{max} of ~540 nm, while bulky bases give rise to disassembled, blueshifted solutions with λ_{max} of ~420 nm (schematically depicted in Figure 9.7). The protected

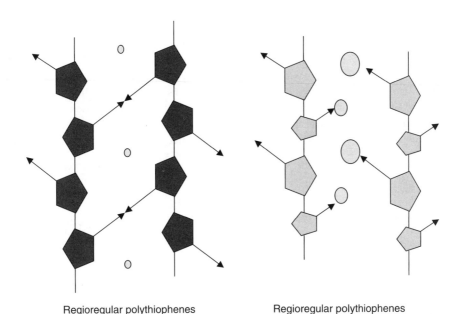

Regioregular polythiophenes
self-assembled by hydrogen bonding

Regioregular polythiophenes
disassembled by large cation binding

FIGURE 9.7 Carboxylic acid derivatives of regioregular polythiophene (Table 9.1, entry 17) can act as sensors of alkali metals in water through a self-assembly and disassembly mechanism.

polymer can also be transformed into its butyl ester, which is soluble in a variety of solvents allowing for characterization of the polymer. The shorter chained acid polymer poly(thiophene-3-propionic acid) was found to be sensitive to all alkali earth metals, whereas the longer chained acid poly(thiophene-3-octanoic acid) (Table 9.1, entry 19e) was only sensitive to ions larger than Et_4^+ [104].

Using a similar approach highly regioregular, HT coupled, 3-(amino-functionalized)-PTs were synthesized by the CuO cocatalyzed Stille coupling (Table 9.1, entries 19a–b) [105]. The polymer bearing a tetrahydropyran-substituted amine was insoluble, however upon hydrolysis, formed a methyl derivative that was soluble in aqueous HCl. The polymer had a λ_{max} of 387 nm and a molecular weight by GPC of ($M_n = 1537$; PDI = 3.0) corresponding to 4 repeat units. When the amine is functionalized with a long dodecyl chain the solubility is dramatically improved. The resulting polymer ($M_n = 8769$; PDI = 1.83) had a λ_{max} of 416 nm and was soluble in many organic solvents. Salts of methylamine are water soluble and become helically ordered upon addition of DNA [105–108].

9.4.3.2 Summary of Side-Chain Functionalized Regioregular Polythiophenes

Table 9.1 is a quick reference to the large number of side-chain functionalized PTs that have been synthesized to date. All of these have been covered throughout the chapter. The primary methods of synthesis used have been designated as Method A (McCullough), Method B (Rieke), Method C (GRIM), Method D (Stille), Method E (FeCl₃), and Method F (derivatization). Please refer to the previous section for the general schemes of these synthesizes.

9.4.4 End Group Functionalization of rr-PATs

End group functionalization of rr-PATs is expected to lead to a number of new uses for these polymers including end group driven self-assembly onto surfaces and into conducting polymer networks, and their use as building blocks for the synthesis of block copolymers [30]. While, a great deal of work has been reported for the modification and variation of the side chains of regioregular, HT coupled PTs (rr-PATs) [5,6], less attention has been given to the nature and control of the end groups of such polymers [30,44,110,111]. To date, procedures to functionalize end groups of conjugated polymers are limited in scope and number. Typical synthesis of PATs through Rieke, GRIM, or McCullough methods produces a polymer with two different end groups H/Br (13, Scheme 9.6) and H/H (trace amount of Br/Br are present in rr-PATs made using the McCullough method) [112], which cannot be separated from each other. Two approaches have been investigated to alter the end group composition: postpolymerization functionalization and in situ modification.

In one of the first examples, rr-PATs bearing a terminal hydroxy group were prepared by coupling thiophene bearing tetrahydropryanyl protected hydroxy groups onto the bromine end of an H/Br terminated poly(3-hexylthiophene) polymer 13 (Scheme 9.7) [30]. The reactions resulted in the formation of a polymer that has been functionalized ~90% with hydroxyl groups at one end. In addition, amino terminated poly(3-hexylthiophene) was prepared by the coupling of thiophene bearing

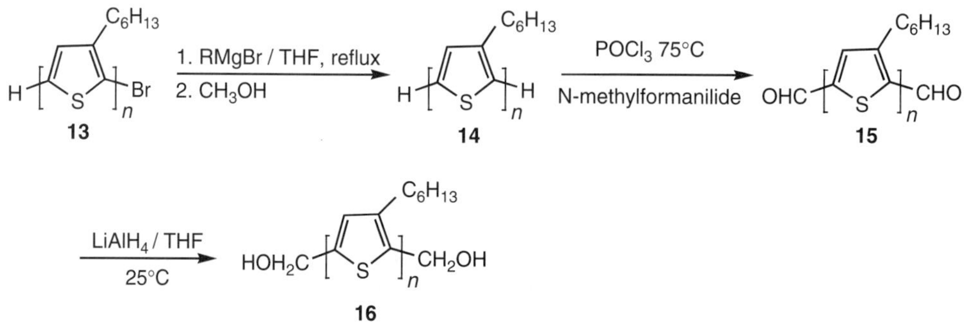

SCHEME 9.7   Synthesis of HT-poly(3-hexylthiophene) with –OH difunctionality at ends.

STABASE-protected amino groups to an H/Br terminated poly(3-hexylthiophene) polymer [30]. The amine group incorporation was slightly more difficult to monitor due to the loss of the protecting group during the work-up. McCullough and Liu describe the incorporation of aldehyde groups on both ends of rr-PATs by the Vilsmeier reaction of an H/H terminated polymer (Scheme 9.7). rr-PATs were prepared using the McCullough methods to produce a polymer bearing a distribution of end groups, H/H and H/Br. The end groups were then converted to H/H end terminated polymer (**14**) by treating the H/Br terminated polymer with excess *t*-butyl MgCl and aqueous work-up. After debromination, Vilsmeier conditions of POCl$_3$ and *N*-methylformanilide led to aldehyde groups onto both ends of the polymer **15**. To drive the reaction forward a large excess of reagents was used at 75°C. The aldehyde groups could be subsequently reduced to create an rr-PAT with hydroxymethyl groups on both ends. Hydroxyl terminated polymers (**16**) were used as building blocks for the synthesis of diblock and triblock copolymers [113]. The same method was employed by Liu and Frechet to synthesize amine and cyano terminated rr-PATs for use in hybrid solar cells [4].

The first attempt toward in situ end group modification was reported by Janssen and coworkers. Various rr-PATs were prepared using the McCullough method [110]. The standard work-up was altered by the addition of a second thienyl Grignard reagent, bearing a trimethylsilyl group and some fresh Ni(dppp)Cl$_2$ catalyst, prior to quenching with MeOH. MALDI-TOF analysis of the resultant polymer end groups shows that a mixture of different end products was obtained namely mono (trimethylsilyl thienyl) terminated PT, difunctionalized (trimethylsilyl thienyl) terminated PT, and H/H terminated PT. The molecular weights of the polymer were modest at M_n 4–5 K, with PDIs of 1.3–1.5. Iraqi and Barker also reported the presence of two different end groups, namely iodo and tributylstannyl, when 2-iodo-5-trimethylstannyl-3-alkylthiophene was polymerized using the Stille method [44].

After an extensive study of the mechanism for the polymerization of regioregular PTs, the McCullough group reported on a very versatile and simple method to do in situ end group functionalization of regioregular P3ATs using the GRIM method. This method eliminates the need to synthesize and isolate a polymer bearing predominately H/Br end groups. In the mechanism of polymerization (as discussed above), the nickel catalyst acts as an initiator and therefore at the end of the reaction, the regioregular PT will be still bound to the catalyst, along with a Br group and the diphospine ligand (**17**, Scheme 9.8). Therefore, it was postulated that the addition of a second Grignard reagent, prior to quenching the reaction, should lead to effectively terminating the reaction and "capping" the polymer (Scheme 9.8). This method has been demonstrated to be very successful for a variety of different types of Grignard reagents (i.e., aryl, alkyl, allyl, vinyl, etc.) (Table 9.2). Both monofunctionalized (**19**) and

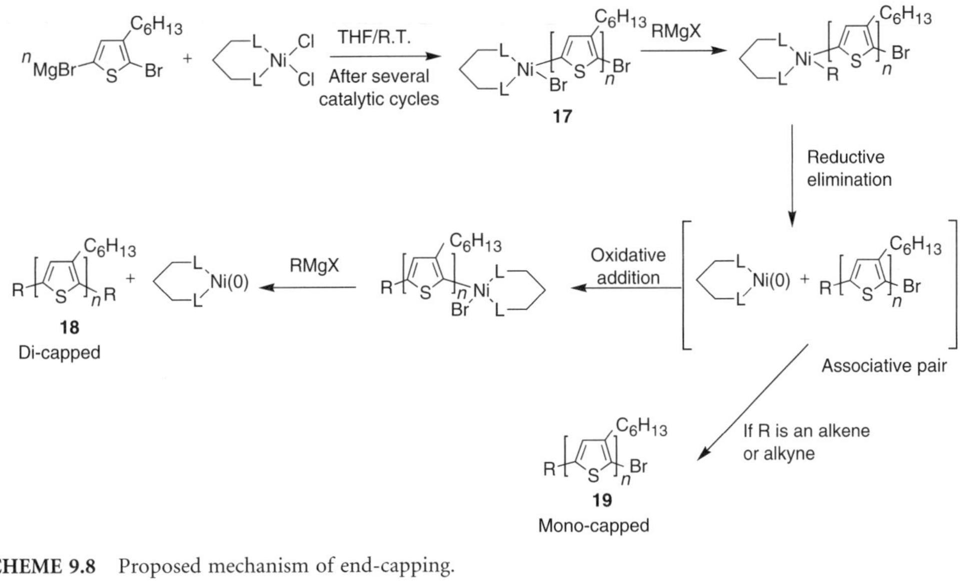

SCHEME 9.8 Proposed mechanism of end-capping.

TABLE 9.2　End Groups Synthesized and Resulting Polymers (Scheme 9.8)

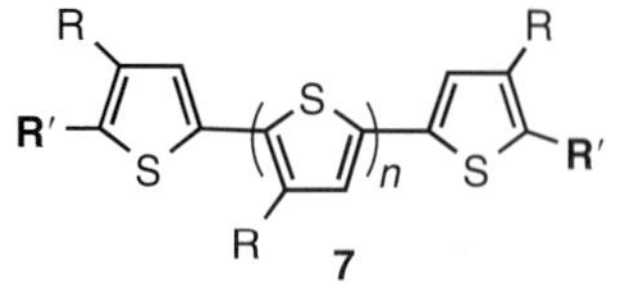

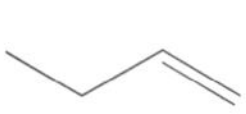

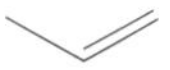

difunctionalized (**18**), end-capped regioregular PTs can be made. When the end group is allyl, ethynyl, and vinyl groups, the nickel catalyst is postulated to be bound to the end group through a nickel-π complex, shutting down the possibility of difunctionalization. When other groups, such as alkyl were used, the polymers further reacted to yield difunctionalized polymers (**18**) [114]. The end group composition of the polymers approaches 100% in most cases and percentages of end groups can be monitored using a combination of MALDI-TOF and ^{1}H NMR. By utilizing the proper protecting groups −OH, −CHO, and −NH$_2$ groups have been incorporated onto the polymer ends, as well [115]. The molecular weights of the polymers ranged from 6000 to 13,700: PDI 1.1–1.2. The main advantage of this method is that it allows for the in situ functionalization of regioregular PT, generating a variety of end-capped polymers in a simple, one-step procedure.

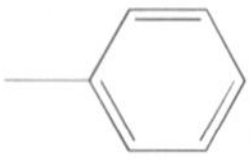

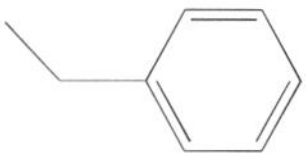

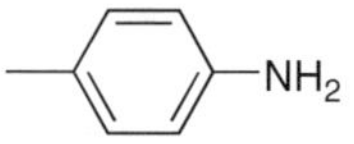

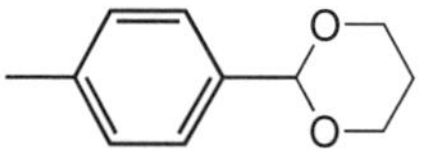

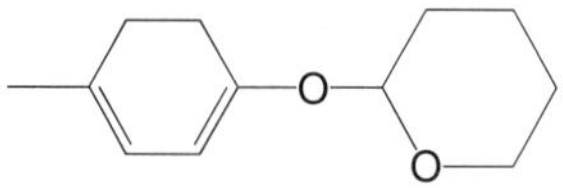

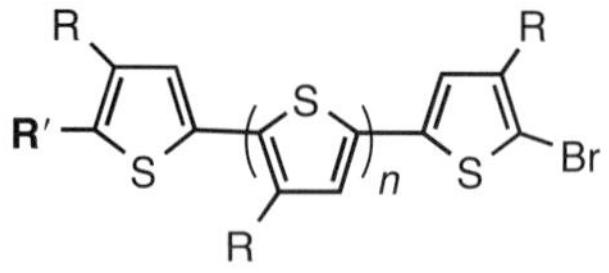

9.5　Self-Assembly of Regioregular Polythiophenes

Regioregular PTs have demonstrated great potentials such as active materials in various applications in microelectronics and electro-optics. A critical aspect of incorporating them into these applications is the reproducible formation of thin films with good morphologies, structures, and electrical properties on different surfaces. Several methods have been utilized for the ordering of electronic polymers but most of the methods use molecular self-assembly or directed self-assembly onto surfaces. Molecular self-assembly has been used as an effective method to direct the formation of supramolecular structures. Self-assembly is defined as the spontaneous organization of molecules or objects into stable, well-defined structure by noncovalent forces. This concept has been modeled after biological systems that display a diverse array of self-organized structures such as the folding of proteins and t-RNAs, the formation of the double helix of DNA, and the formation of cell membranes from phospholipids. All of these biological systems depend on hydrophobic interactions, hydrogen bonding, salt-pairing, and other molecular interactions to drive supramolecular ordering that leads to functional biological systems. Similarly, these techniques have been applied to the area of π-conjugated polymers to develop organized conducting films.

9.5.1 Supramolecular Ordering and Self-Assembly in rr-P3ATs

One of the most fascinating physical property differences between irregular PATs and regioregular HT-PATs is that supramolecular ordering occurs in regioregular HT-PATs and not in irregular PATs. Light scattering studies and UV–vis experiments clearly show that supramolecular association of HT-PATs occurs in solution [116]. X-ray studies, atom-force microscopy (AFM) studies, and TEM studies on solid thin films of rr-PATs show macromolecular self-assembly into polycrystalline, ordered domains, and nanowire morphologies of differing dimensions.

Aggregation or self-association of rr-PAT chains has been well established and it guides the ability of the conductive polymer structures to organize into thin film with optimized properties. In dilute solution, rr-PATs are less aggregated and can show little self-association between the polymer chains [117,118]. Even in the UV–vis spectra, aggregation is apparent. Dilute solutions exhibit standard UV–vis data with $\lambda_{max} \approx 450$ nm. However, concentrated solutions exhibit very broad UV–vis spectra with λ_{max} values ranging from 450 to 600 nm and are deep magenta in color. As the concentration of HT-PATs increases supramolecular association or aggregation drives the formation of colloids and eventually macroscopic, precipitated aggregates. Precipitation of irregular PATs occurs only with poor solvent-generating particles, which, by x-ray, are identical to disordered neutral films [117,118], In contrast, precipitation of regioregular HT-PATs occurs eventually even in good solvents due to highly crystalline nature of HT-PATs. It has been observed that clear solutions of HT-PATs that are allowed to sit overnight form precipitates [116]. Adding hexane to chloroform solutions can drive aggregation to a number of aggregated states, including a postulated helical solid proposed by Kiriy [119]. Solid aggregates can be (and should be) filtered out before film casting or physical studies proceed.

The solution properties of irregular PATs have been well studied by a number of research groups [118,120,121] and concepts such as thermochromism and solvatochromism have been studied. These studies have led to the observation of reversible thermochromism and solvatochromism in solutions of PATs. The presence of an isosbestic point indicates that two conformations are possible in solution, with the one possessing increased conjugation length relative to the other. Different forms are often attributed to either a conformationally disordered random coil structure or an ordered rod-like structure [118,120,122–124]. Solutions can be a mixture of the two phases (biphasic). The two forms can be distinguished by solution light scattering.

Static and dynamic light scattering studies, by Berry, have shown that solutions of regioregular HT-PDDT of different thermal histories and concentrations show strong intermolecular association between the regioregular PT chains [116]. This association (or aggregation) drives main-chain conformational order and a supramolecular association. The supramolecular structure may assume a variety of shapes dependent on concentration. One solution structure is a nematic liquid crystalline phase consisting of lamella of extended PT chains interspersed by lamella of alkyl chains [6,116]. The supramolecular aggregate structure is thought to have a disc-like shape and possesses long-range order. Supramolecular ordering increases as the concentration of the polymer increases. This behavior is consistent with the concepts embodied in the Flory phase diagram for semiflexible polymer chains. The theory predicts that with increasing concentration of semiflexible chains, these polymer solutions move from a disordered phase to a biphasic solution and then to an ordered phase. It is also found that at low temperature, the dodecyl side chains slowly self-order enhancing coplanarity of the thiophene rings along the polymer chain [116].

Another shape of the solution structure that can exist in HT-PDDT consists of needle-like aggregates of extended chains of PT surrounded by a sheath of ordered alkyl chains. The needle-like aggregate structure does not possess a great deal of two-dimensional (2D) order, in contrast to the lamellar form. It is very interesting to note that the supramolecular ordering is greatest in HT-PDDT. The aggregation affects are much smaller in HT-PHT and intermediate in HT-POT.

The solution and solid-state studies indicate that the ordered, supramolecular liquid crystalline structures are desirable precursors to self-assembled PAT films. The existence of two distinct aggregated structural phases of rr-PAT in solution creates a challenge for the preparation of thin films with reproducible properties. Films with a variety of morphologies, ranging from brittle, cracked films to fluffy films to dense, uniform films [125]. Films of rr-PATs can have microscopic cracks that cause physical discontinuity between regions of the film and low conductivities. In addition, less-dense morphologies have been observed and again show low conductivities, relative to dense, uniform films. As an example, rr-poly(3-dodecylthiophene) films drop cast from toluene can show a dense, uniform morphology and have an electrical conductivity of 1000 S/cm (iodine doped) [37].

The McCullough group was the first to show that regioregular rr-P3ATs have self-assembled structures by x-ray scattering studies [37,126]. X-ray data have shown strong intensity in three small-angle reflections, which correspond to a well-ordered lamellar structure with an interlayer spacing, in the case of rr-P3HT, of 16.0 ± 0.2 Å. A narrow (0.1 Å half-width), single wide-angle reflection is observed at a $90°$ angle of incidence that corresponds to 3.81 ± 0.02 Å—the π-stacking distance of the thiophene rings between two polymer chains [37]. These self-assembled structures are similar to the work presented by Winokur and coworkers on stretch-oriented PATs [127,128]. However, an important point is that in all other irregular non-rr-PAT samples examined previously, the wide-angle peak is quite broad [127–129]. Our observation of a very narrow width in the wide-angle region indicates that along the polymer chain, conformational order gives rise to a single stacking distance and not a distribution of stacking distances. The narrow widths of all these dominant x-ray features indicate self-oriented, highly ordered crystalline domains in PHT [37]. It appears, however, that the samples are disordered from domain to domain, which gives rise to the wide-angle ring found in the $90°$ picture. In striking contrast, irregular PHT prepared from $FeCl_3$ shows much weaker intensity reflections in the small-angle region, with a d-spacing of 17.3 ± 0.5 Å and a very broad amorphous halo centered at 3.8 Å. These results show an unprecedented structural order for regioregular PATs, indicating that the order is induced by the regiochemical purity afforded by the synthetic method used to prepare these samples.

Examples of morphological extremes were studied using x-rays by Winokur, revealing the existence of two distinct packing structures—the self-assembled PATs with a well-ordered lamellar structure with partially interdigitated side chains and a brittle, powdery structure, which is dominated by a packing pattern in which the side chains are more interdigitated than self-assembled structure [126]. The brittle, metastable structure has intrastack distance increases from 3.8 to 4.47 Å. (Interchain hole hopping is quite diminished with a stacking distance of 4.47 Å.)

The crystallization mechanism in regioregular P3ATs has also been examined by Malik and Nadi [130]. X-ray studies, coupled with thermal studies by Nandi, examine the relationship between alkyl side-chain length and degree of regioregularity in the cocrystallization of rr-P3HT and rr-P3OT blends [131]. Additional orientation can be seen in x-ray studies of P3HT and P3DDT upon "friction transfer" by Nagamatsu [132].

Recent works by Kline, McGehee, and Frechet as well as Kowalewski and McCullough have used AFM to show that P3ATs form a number of types of morphologies that depend on the molecular weight of the-samples [133–135]. AFM has shown that low-molecular weight rr-P3HT forms nanowire morphologies with samples of molecular weights around 3 K [133–135] and more of a disordered phase with better electronic connectivity between chains in samples of rr-P3HT of around 31 K. These results indicate that field-effect mobility of these materials can be further improved by increasing their molecular weight and through the choice of deposition methods favoring formation of ordered structures. Kowalewski and McCullough [136] have gone one step further, by demonstrating that the careful choice of processing conditions of rr-P3HTs with narrow polydispersities can lead to the formation of very well-defined nanofibrillar morphologies in which the width of the nanofibrils corresponds very closely to the weight average contour length of polymer chains. Moreover, they show that the charge carrier mobility in field-effect transistors (FETs) fabricated from these well-ordered polymers increases exponentially with the nanofibril width.

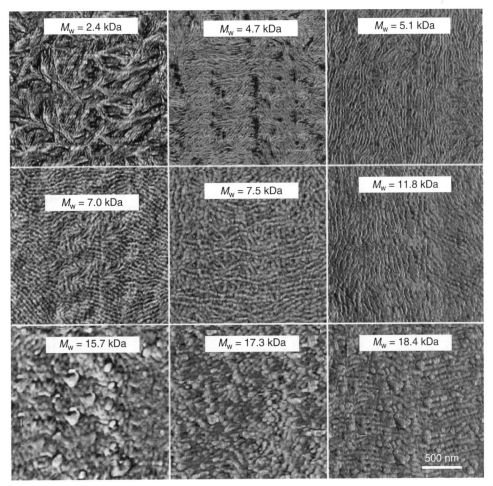

FIGURE 9.8 Tapping mode AFM images (phase contrast) of thin films of RR-P3HTs of various molecular weights in FET devices prepared by drop casting from toluene.

In all cases, phase contrast tapping mode atomic force microscopy (TMAFM) (Figure 9.8) images revealed the presence of densely packed, elongated features of highly uniform widths referred to as nanofibrils. The width of nanofibrils, w_{AFM} (determined by Fourier analysis of AFM images), initially increased linearly with M_w and then leveled-off (Figure 9.9c and Figure 9.9d). The weight average contour length showed a direct relationship with the width of the nanofibrils. This relationship is highly suggestive of a structure comprised of one-molecule wide, stacked sheets of rr-P3HT, with polymer backbones aligned perpendicular to the nanofibril axis. Such structure was consistent with grazing incidence x-ray diffraction (GIXRD) studies, which revealed the presence of familiar layered, π-stacked structures, with π-stacking plane perpendicular and periodic layering parallel to the film surface. Grazing incidence small-angle x-ray scattering (GISAXS) utilizing the synchrotron radiation source revealed nanofibril widths, which were similar to that measured by TMAFM (Figure 9.9b).

Most interestingly, as shown in Figure 9.9a (blue symbols) the log of charge carrier mobility μ in studied FET devices mapped directly onto the periodicities revealed by AFM/GISAXS observations, pointing to the exponential dependence of the form $\mu = \mu \exp(w_{AFM}/w_0)$, with $\mu_0 = 4.6 \times 10^{-7}$ cm^2/V s and $w_0 \approx 3$ nm (Figure 9.9c). Since, as shown above, the nanofibril widths w_{AFM} tend to directly reflect the contour length of polymer molecules; the observed exponential dependence of μ vs. w_{AFM} appears to directly point to the prominent role of extended, conjugated states in charge transport in

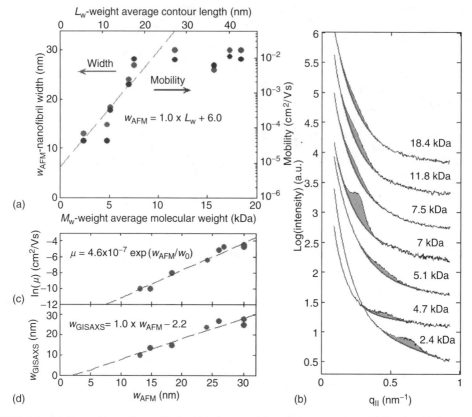

FIGURE 9.9 (a) Dependence of nanofibril width (w_{AFM}, *red*) and charge carrier mobility (μ, *blue*) on weight average molecular weight (M_w, *bottom axis*) and weight average contour length (L_w, *top axis*) of RR-P3HT. (b) GISAXS scattering profiles for thin films of RR-P3HTs with various molecular weights. (c) Exponential fit of μ vs. w_{AFM}. (d) Linear correlation between GISAXS periodicities and nanofibril widths (w_{GISAXS} vs. w_{AFM}).

semiconducting polymers. The authors suggest that the observed exponential dependance of mobility on nanofibril width could be influenced by the nature of the grain boundaries between adjacent fibrils.

9.5.2 Block Copolymers Containing Regioregular P3ATs

Block copolymers comprised of conjugated and nonconjugated segments are interesting materials that fall into the category of rod-coil polymers. Due to the immiscibility of the covalently connected segments, polymers of this type are expected to phase segregate into regimes resulting in the formation of nanoscale morphologies, such as lamella, spherical, cylindrical, and vesicular structures. In addition, tailoring the morphologies of such block copolymers may let us tune the macroscopic functional properties of their films, providing control over the supramolecular architecture. Moreover, incorporation with structural polymers may improve the mechanical and processing properties of conjugated polymers, such as rr-PATs. To date, the reports of di- and triblock copolymers containing π-conjugated segments are limited in scope and number, the majority of which contain only oligomeric conjugated units [137–145]. This is largely due to the difficulty in synthesizing suitably functionalized conjugated polymer blocks for incorporation into such polymers. The work by Hempenius and Meijer is a good example, where triblock copolymers of polystyrene–oligothiophene–polystyrene were made and apparent spherical micelles were formed [116] Recently, the synthesis of a number of well-defined di- and tri-block copolymers containing rr-PATs has been reported by our group [113]. These polymers

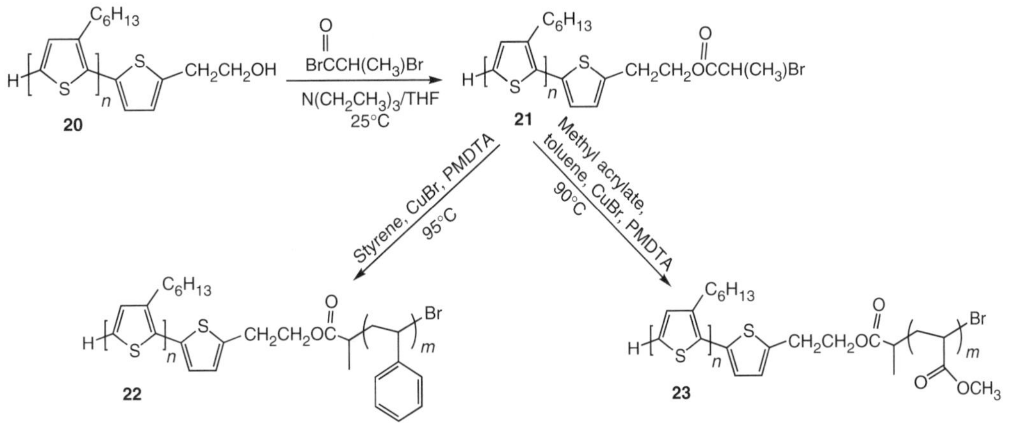

SCHEME 9.9 Synthesis of well–defined PHT–PS (**22**) and PHT–PMA (**23**) diblock copolymers.

were comprised of rr-PATs and methyl acrylate or styrene blocks. Regioregular poly(3-hexylthiophene) (rr-P3HT) with one hydroxyl end can be synthesized using previously reported end group modification methods [30,146,147]. An rr-P3HT with one hydroxyl end group **20** is subsequently converted to an ATRP macroinitiator **21** by reaction with 2-bromopropionyl bromide. Using this initiator, poly(3-hexyl thiophene)–polystyrene (PHT–PS) **22** and poly(3-hexyl thiophene)–poly(methyl acrylate) (PHT–PMA) **23** can be synthesized by ATRP (Scheme 9.9) [148]. Monitoring the feed ratio of the monomers provides full control over the percentage composition of the PS or PMA block [7]. The combination of the rr-P3HT macroinitiator **21** with a low PDI and the "living" nature of the ATRP method results in the formation of diblock copolymers of high-molecular weights and low polydispersities. In a similar fashion, rr-P3HT functionalized on both α and ω ends can be used for the synthesis of PS and PMA triblock copolymers [113]. Thin films of all of the polymers (PHT–PS) **22**, (PHT–PA) **23** were generated by the slow evaporation of toluene solutions to give magenta to purple films, which when doped with iodine had high-electrical conductivities. While 100% rr-P3HT has a conductivity of 110 S/cm, PHT–PS **22** has a conductivity of ~5 S/cm for a block copolymer containing 37% rr-P3HT. The conductivity drops down to 0.1 S/cm for samples containing ~22% of rr-P3HT or less. The PS–PHT–PS triblock copolymers have conductivities as high as 5 S/cm for a sample with 52% PHT.

AFM, TEM, and x-ray diffraction have been used to study the self-assembly and phase behaviors of the block copolymers. It can be seen that the films contain well-defined worm-like wires entangled with each other [113]. These interwoven wires are spaced laterally by ~30 nm that corresponds roughly to the fully extended rr-P3HT blocks of **22** (Figure 9.10). TEM images of a film also show the presence of "nanowire" structures (Figure 9.11). X-ray diffraction has indicated that the "nanowires" have layer rr-P3HT π-stacking structures in their cores. The "nanowires" in the diblock copolymers are believed to be the nanoscale domains of rr-P3HT aggregations.

9.5.3 Lithography and Microcontact Printing of rr-PATs

Regioregular polyalkylthiophenes have great potential in the fabrication of polymeric circuits due to their high conductivity and processibility. While currently, most of the focus have been on circuits as sensors, ultimately it is expected that radio frequency identification tags will be developed. As of now the critical aspect will be feature size and speed, however for the existing technology there are many potential applications. One critical aspect of fabricating polyalkylthiophene into devices is the controlled deposition onto different surfaces. Various techniques, developed by the Whitesides group, including photolithography and soft-lithography techniques have been investigated including micromolding in

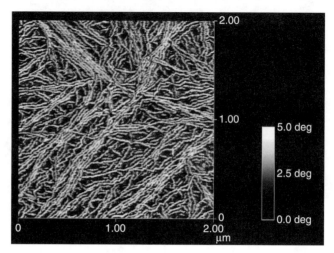

FIGURE 9.10   AFM phase image of a thin film drop cast from a toluene solution of PHT_{45}–PMA_{45} copolymer.

capillaries (MIMIC) [149], microtransfer molding (μTM) [150], and microcontact printing (μCP) [151]. HT rr-P3HT has been employed in making paper-like electronic displays using μCP [152]. In this study, a hexadecane thiol (HDT) self-assembled monolayer was patterned on a thin layer of gold-coated plastics by μCP. The removal of exposed gold with aqueous potassium ferrocyanide yielded the gold source and drain structures of FET. P3HT was then spin-coated and filled with these structures, producing a working FET [152]. The rr-P3HT transistors were used to drive microencapsulated electrophoretic inks to generate "electronic paper."

The soft-lithography of rr-PATs has also been reported. Poly(3-dodecyl-3′-hexylthiol-5,2′-bithiophene) was successfully synthesized in two steps from poly(3-dodecyl-3′-hexylbromo-5,2′-bithiophene), although the synthesis of poly(3-dodecyl-3′-hexylthiol-5,2′-bithiophene) is quite long. The strong

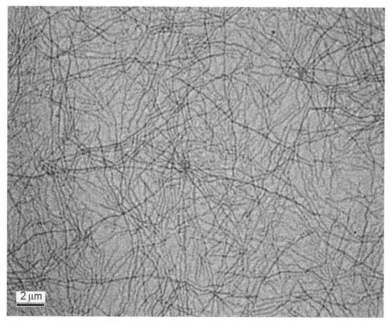

FIGURE 9.11   Representative TEM image (bright- field mode) of a thin film drop cast from the toluene.

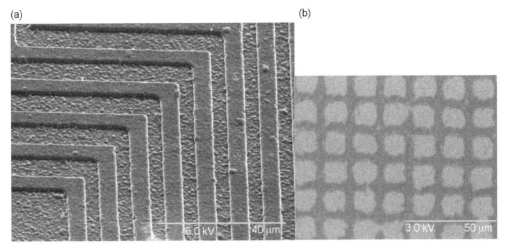

FIGURE 9.12 (a) SEM image of patterned poly(3-hexylthiophene) by capillary force lithography. (b) SEM image of a patterned gold surface of poly(3-dodecyl-3′-hexylthiol-5,2′-bithiophene) by microcontact printing. Square-shaped structures represent positive surface relief and are 13 mm on a side.

bonding of the thiol terminated side chains with the gold surface led to an energetically stable *anti* conformation of adjacent thiophene rings, creating a lamellar arrangement, resulting in the most compact packing [10]. Microstructures of poly(3-dodecyl-3′-hexylthiol-5,2′-bithiophene) were fabricated on a gold-coated glass slide through microcontact printing and microtransfer molding techniques (Figure 9.12). Patterns of poly(3-hexylthiophenes) were constructed from a patterned silicon wafer through capillary force lithography methodology (Figure 9.12).

9.5.4 Fabrication of rr-PATs Thin Films Using Langmuir–Blodgett

The use of the Langmuir–Blodgett (LB) technique has been investigated to form thin films that are conducting and completely self-assembled at the same time. This technique works with a specific type of molecule, such as amphiphilic hairy board polymers, nonamphiphilic hairy board polymers, and long-chain acids board polymers with polar substituents [153]. In the first example, a series of five amphiphilic, alternating copolymers of rr-PAT were synthesized and the LB films were studied by Bjornholm and McCullough (Figure 9.13) (Table 9.1, entries 8e–g) [109]. The primary PT was an alternating copolymer of 3-dodecylthiophene and 3-(2,5,8-trioxanonyl)thiophene (Table 9.1, entry 8e). The pressure–area isotherm reveals a collapse of ~29 Å^2 per polymer repeat units, to create the structure shown (Figure 9.13). The structure of the Langmuir monolayer was determined by diffraction and reflection of synchronotron x-rays. Two dominant peaks were found, one for the π-stacking at 3.84 Å and one for the disordered alkyl chains. Approximately, 15 stacked polymers gave rise to the coherent scattering. Monolayer films were transferred to give highly anisotropic ordered domains with a dichroic ratio of 4. The UV–vis stability of these films is outstanding. All attempts to bleach the films were unsuccessful. Monolayer films, upon doping with iodine, gave conductivities of 1–50 S/cm. Moreover, transfer of rr-PT monolayers to hydrophilic circuit patterns led to nanocircuits of PT. Collapse of monolayers on the LB trough led to the formation of the first PT nanowires. This method leads to the formation of large PT strands [109,154]. In contrast, the nonamphiphilic polymer poly(3-dode-cylthiophene) does not form a monolayer, emphasizing the importance of the amphiphilic nature of the polymer. PTs based on 3-dodecyl-3′-(3,6,9,12-tetraoxatridecanyl)-2′5-bithiophene can also be used to form Langmuir and Langmuir–Blodgett films. Mixtures of PT and poly(ethylene glycol) have also been employed [155]. Highly ordered Langmuir–Blodgett films of rr-PTs were found to be more ordered through the preparation of LB technique, relative to spin casting [156].

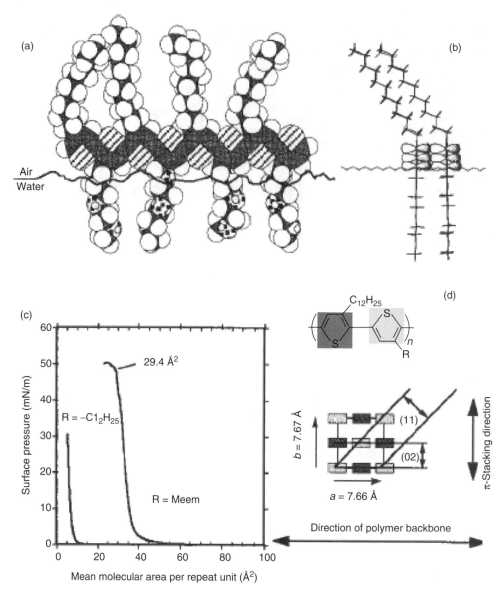

FIGURE 9.13 (a) Model of a segment of polymer **10** showing the void space between the alkyl chains present on the upper (hydrophobic) side of the polymer oriented at the air–water interface. (b) Side view of the π-stacked polymer indicating the presence of good electronic contacts from one polymer to the next. (c) Compression isotherm (T, 20°C) of amphiphilic and nonamphiphilic regioregular polythiophene. (d) Top view of unit cell obtained from x-ray diffraction from the self-assembled monolayer at the air–water interface. The π–π stacking distance is 3.84 Å, and the projected area taken up by one repeat unit is 3.84×7.66 Å $= 29.4$ Å^2. This area is close to the collapse area of the monolayer depicted on the isotherm (c), indicating full monolayer coverage of the water surface. (From Bjornholm, T., et al., *JACS*, 120, 7643, 2000.)

9.5.5   Layer-by-Layer Self-Assembly of rr-PATs

Electrostatic layer-by-layer self-assembly is a simple, versatile, and effective technique for the fabrication of ultrathin organic multilayer films [157,158]. This method involves sequential absorption of polyanions and polycations on a charged surface. During each dipping process, a molecule thin layer

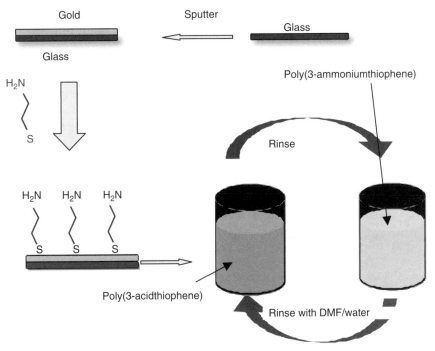

FIGURE 9.14 Layer-by-layer self-assembly of poly(3-octanoic acid thiophene) and poly(3-hexylammonium thiophene).

of the corresponding organic polyion is adsorbed on the substrate. This new layer saturates the substrate with opposite charges so that in the subsequent dipping process oppositely charged polyions can be adsorbed. Since electrostatic attraction between opposite charges is the driving force for the build-up of multilayers, the use of polyelectrolytes rather than small molecules is advantageous. Because good adhesion of a layer to the underlying substrate or film requires a certain number of electrostatic attraction bonds. Layer-by-layer self-assembly could provide control over the thickness of thin film and allow for fabrication of nanoscale devices. This method has been applied to the assembly of conjugated polymers [36,159] DNA [37], protein [38], and nanoparticles [38] in the formation of layer-by-layer self-assembled structures. Layer-by-layer rr-PAT self-assembly (Figure 9.14) was accomplished by a simple dipping procedure providing electrostatically self-assembled ultrathin films of carboxylate derivatives of rr-P3HT and ammonium derivatives of rr-P3HT. Repeating this simple procedure can produce multilayers of 40 bilayers or greater. The UV–vis absorption spectra generated during the fabrication of multilayer thin films of these PT derivatives indicated that the deposition process was completely reproducible from run-to-run. The linear relationship of the UV–vis absorption spectra demonstrated by poly(3-octanoic acid thiophene) and poly(hexylammonium thiophene) multilayer system (Figure 9.15 and Figure 9.16) clearly indicated that each layer deposited contributes an equal amount of material to the thin film. It was found that the absorption of poly(3-octanoic acid)thiophene–poly(3-(6-*N*,*N*-dimethylhexyl ammonium))thiophene system did not increase at the same rate as the more rapid absorption of poly(3-octanoic acid thiophene)–poly(hexylammonium thiophene) system. The increase of steric hindrance of the side chains of poly(3-(6-*N*,*N*-dimethylhexyl ammonium)thiophenes) leads to a decrease in conjugation length as the layers increase relative to the less sterically encumbered poly(hexylammonium thiophene). The conductivity found in the regioregular

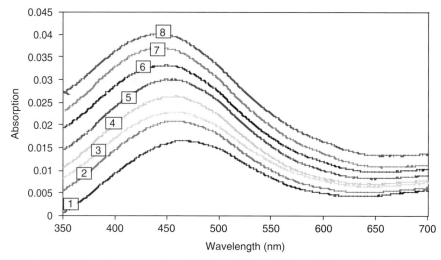

FIGURE 9.15   Optical absorption vs. number of bilayers (*indicated in boxes*) deposited for poly(3-octanoic acid thiophene) and poly(3-hexylammonium thiophene). (From Zhai, L., and McCullough, R.D., *Adv. Mat.*, 12, 901.)

PTs reported here is 0.04 S/cm, which was much higher than regiorandom PT multilayer systems (0.0006 S/cm) reported previously [160].

9.6   Electrical Properties of Regioregular Polythiophenes

9.6.1   Electrical Conductivity

The self-assembled structure leads to a large increase in electrical conductivity in HT-PATs relative to irregular PATs. While the measured conductivity of rr-PAT films cast from the same sample can differ markedly as a result of varying morphology from film to film, for the conductivities of rr-P3DDT (doped with I_2) the maximum reported values are 1000 S/cm [37]. Values for other rr-PATs synthesized

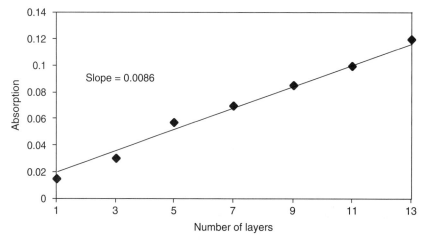

FIGURE 9.16   Optical absorption vs. number of layers deposited for poly(3-octanoic acid thiophene) and poly(3-hexylammonium thiophene). (From Zhai, L., and McCullough, R.D., *Adv. Mat.*, 12, 901.)

by McCullough et al. [25,65] exhibited maximum electrical conductivities of 200 S/cm (POT) and 150 S/cm (PHT) (Table 9.1) [37]. In contrast, PDDT from $FeCl_3$ generally gave conductivities of 0.1 ± 1 S/cm (58 ± 70% HT). Rieke and coworkers have reported that their HT-PATs have conductivities of 1000 S/cm [31], with an average conductivity for HT-PBT of 1350 S/cm, with a maximum conductivity of 2000 S/cm [32].

Careful conductivity studies for rr-P3HT, doped with iodine, give conductivities in the range of 50 to 100 S/cm. Regioregular PATs show conductivities as high as 650 S/cm, with typical conductivities of several hundreds, while PATs show conductivities as high as 700 S/cm.

9.6.2 Mobility

Optimization of conjugated polymers for use in transistors requires an understanding of the mechanism for charge transport. Typical mobilities for conjugated polymers are in the range of 10^{-5} to 10^{-7} cm^2/V s, with rr-PATs having one of the highest reported values at 0.01, due to the effective stacking of the conjugated backbone [161,162]. However, this value can vary by several orders of magnitude depending on the processing condition, including solvent used, sample history, casting conditions, and film-processing method. Hence a large amount of research has been done to determine the critical factors influencing mobility. It has been established that regioregularity strongly affects the mobility in PHT, with values varying from 2×10^{-5} to 0.05 cm^2/V s when regioregularity is increased from 70% to 90%. A similar study found that in addition to regioregularity, altering the doping level (0.01%–20%) can also enhance the mobility of rr-PATs [163].

The correlation between field-effect mobility and molecular weight has also been studied [135]. Data show that when molecular weight is increased from 3.2 to 36.5 K the mobility values increase from 1.7×10^{-6} to 9.4×10^{-3} cm^2/V s. On–off ratios varied from 10^3 to 10^5, but were not correlated to molecular weight. The polymers studied have been classified into three different sets based on the method used and the end group composition. The dependence of mobility on molecular weight was consistent regardless of which group the polymer was from indicating that charge trapping by polymer end groups did not affect the mobility. AFM studies show that low-molecular weight films have more defined grain boundaries than high-molecular weight films. Accordingly, it is believed that high-molecular weight films have higher mobilities because the regions are less defined, suggesting that the regions are connected, providing a pathway for charge transport between crystalline domains. A more detailed study of varying morphology at constant molecular weight shows that the mobility–molecular weight relationship is not due to in-plane π-stacking [134]. However, saturation of mobility is still an issue that has been recently revealed [132].

The correlation between structural order and mobility in rr-PATs has been probed by examining polymers with different substituents [164,165]. It was found that shorter side chains such as hexyl and octyl had higher mobilities (0.03–0.05 cm^2/V s) than longer one like dodecyl or bulky or carboxylic-substituted side chains ($<10^{-5}$–10^{-4} cm^2/V s). A chiral alkyl-substituted regioregular PT showed high crystallinity and had a greater π–π overlap distance between the PT backbones (4.3 vs. 3.8 Å for regioregular poly(3-hexylthiophene) (PHT)) due to the branched-side chain. The field-effect mobility of this polymer is reasonably high (10^{-3} cm^2/V s), but is still an order of magnitude lower than for rr-P3HT [164].

While it is clear that several factors (molecular weight, thermal processing, energy of the dielectric surface, and solvent evaporation rate) influence the carrier mobility of a polymer, the mechanism of charge transport is still not well understood. A study on structure and electronic conduction in rr-PAT thin film transistors (TFTs) found that structural modification through thermal processing can lead to differences in mobility by a factor of 25 [166]. The polymer used in this study poly[5,5′-bis(3-alkyl-2-thienyl)-2,2′-bithiophene] (PQT-12) has an intrinsic mobility of 1–4 cm^2/V s. The TFT showed thermally activated behavior below 200 K, such that the mobility decreases and the device turn-on is slow. A comparison of the experimental data with the theroretical models for transport suggests that a model based on the existence of a mobility edge and an exponential distribution of traps provides the best interpretation of the data.

In an effort to enhance field mobility surface-mediated 2D ordering has been employed. Using self-assembled monolayers (SAMs) functionalized with amino and methyl groups the rr-P3HT could be induced to adopt two different orientations, face-on or edge-on [167]. Different orientations were induced by interactions between the unshared electron pairs on the SAM end groups and the π–H interactions between the polymer backbones. The mobilities of these orientations differed by a factor of 4 with the edge-on one giving the best mobility (0.28 cm^2/V s). Similarly, modification of a SiO$_2$ gate dielectric surface with a silane SAM leads to a significant improvement in transistor performance. Mobilities of up to 0.18 cm^2/V s and a current on/off ratio of 10^7 were obtained for TFTs based on PQT-12, a 450-fold improvement over those using a nonmodified dielectric layer [168].

The nanostructure of rr-P3HT has also been fabricated into nanofibers using a nanostencil shadow mask [169]. The typical height of the nanofibers were 3–7 nm ($\sim$2–4 thiophene chains), typical widths were 15 nm ($\sim$40 repeat units), and typical lengths were 200 nm to 5 μm (500–1300 lamellar sheets). The mobility values were 0.02 cm^2/V s, with on/off ratios of 10^6 and current densities of $\sim$700 A/cm^2 per nanofiber. Subsequent studies on activation energies of the contacts and effect of temperature suggest that transport can be explained in terms of the multiple trap and release (MTR) or variable range hopping (VRH) formalism of trapping in a bimodal, exponential distribution of shallow and deep donor-like states.

9.7 Characterization of Regioregular Polythiophenes

9.7.1 Nuclear Magnetic Resonance Spectroscopy

As P3ATs and poly(3-substitutedthiophenes) are soluble in common organic solvents, ^{1}H and ^{13}C NMR can be used to determine their structure and regiochemistry [13,23,25,33,170–173]. For example, in a regioregular HT-PAT, there is only one aromatic proton signal in the ^{1}H NMR spectrum, due to the 4-proton on the aromatic thiophene ring, at $\delta = 6.98$, corresponding to only the HT–HT triad sequence. Proton NMR investigations of regioirregular, electrochemically synthesized PAT reveal that four singlets exist in the aromatic region that can be clearly attributed to the protons on the 4-position of the central thiophene ring in each configurational triad: HT–HT, TT–HT, HT–HH, and TT–HH. The synthesis of the four isomeric trimers by Barbarella led to the assignment of the relative chemical shift of each triad, with each trimer being shielded by $\sim$0.05 ppm relative to the polymer [171]. In this analysis the HT–HT ($\delta = 6.98$), TT–HT ($\delta = 7.00$), HT–HH ($\delta = 7.03$), and TT–HH ($\delta = 7.05$) couplings are readily distinguished by a 0.02 $\pm$ 0.03 ppm shift. These assignments are similar to that of Holdcroft and coworkers [15] but are different from those proposed by Sato and Morii [23]. The relative ratio of HT–HT to non-HT–HT can also be determined by an analysis of the protons that are on the α-carbon of the 3-substituent on thiophene [13]. Relative integration of the HT–HT peak relative to the other non-HT resonances can give the percentage of HT–HT couplings. From this NMR analysis, it has been determined that regioregular HT-PATs contain $\sim$100% HT–HT couplings, whereas samples from the FeCl$_3$ method contain 50%–70% HT–HT couplings [25,32,33,65]. The relative ratio of HT–HT coupling can also be determined by an analysis of the protons that are on the α-carbon of the 3-substituent on thiophene. For example, in poly(3-dodecylthiophene), the resonance for an HH coupling is observed at $\delta = 2.56$ ppm, and that of an HT coupling appears at $\delta = 2.79$ ppm [65]. The relative integration of the α-methylene protons of the 3-substituent reveals the ratio of HT–HT to non-HT–HT couplings in the polymer. The same information can be obtained from the β-methylene protons of the 3-substituent [31]. The ^{1}H NMR resonance for the HT coupled β-methylene proton appears at $\delta = 1.72$ ppm, and that of the HH coupled β-methylene proton appears at $\delta = 1.63$ ppm. Likewise, ^{13}C NMR can also be used to determine structural regularity. For rr-P3HTs, only four resonances are present in aromatic region corresponding to the four carbons of an HT coupled thiophene ring (128.5, 130.5, 134.0, and 140.0 ppm). The aromatic region of a regioirregular PHT shows the same four resonances as well as several other peaks due to the non-HT regioisomers (125.2, 126.6, 127.4, 128.3, 129.6, 134.9, 135.7, 140.3, 142.9, and 143.4 ppm).

More recently ^{1}H NMR has been used to provide information about the end group composition of rr-PATs. It has been earlier shown that the first step in the polymerization is the homocoupling of two

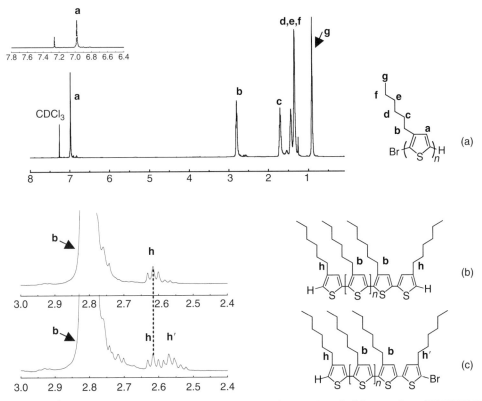

FIGURE 9.17 ^{1}H NMR (500 MHz) spectra of (a) HT-PHT H/Br terminated; (b) expansion of HT-PHT H/H terminated, and (c) expansion of HT-PHT H/Br terminated.

monomers through the organometallic portion of the molecule. As a result a TT dimer is formed and every polymer chain contains one bad coupling (TT–TT) at the end. The full 500 MHz ^{1}H NMR spectrum of an average molecular weight rr-P3HT is presented in Figure 9.17 [55]. The main absorption signals of rr-P3HT are assigned as shown. Two small triplets at $\delta \sim 2.6$ ppm of the same intensity for H/Br terminated rr-P3HT can be assigned to the methylene protons on the first-carbon substituent (**h** and **h′**) on the end units. Furthermore, the appearance of the two separate triplet signals at different resonance frequencies is due to different chemical environment around **h** and **h′** (Figure 9.17c). When the H/Br terminated polymer is converted to an H/H terminated rr-P3HT the spectra is different. The signal generated by the methylene protons **h′** is shifted downfield with the two groups (**h** and **h′**) resonating at the same frequency (Figure 9.17b). The intensity of the **h** peak is doubled in the absence of the bromine atom relative to the main peak (**b**) of first β-substituent methylene protons. These observations indicate that NMR analysis allows a relatively accurate determination of molecular weight from the integration of end group resonances relative to the bulk polymer. For instance, DP_n for the aforementioned polymer equals to the ratio of **b** to **h** and results in 50 monomer units corresponding to $M_n = 8300$. This is an excellent way to determine absolute molecular weights.

9.7.2 Matrix Assisted Laser Desorption Ionization-Time of Flight Mass Spectrometry (MALDI-TOF MS)

Matrix assisted laser desorption ionization-time of flight mass spectroscopy (MALDI-TOF MS) has proven to be a powerful tool for analysis of many biopolymers, such as proteins, oligonucleotides, and polysaccharides. Recently, it has been used to characterize many synthetic polymers [174]. Another advantage of MALDI-TOF MS for synthetic polymer analysis is that the *absolute* molecular masses can be determined as opposed to *relative* molecular weights by gel permeation chromatography (GPC), and this determination is independent of polymer structures. Determination of the molecular weights of PAT systems is almost invariably accomplished by GPC using polystyrene standards. GPC is based on correlating the hydrodynamic volume of the randomly coiled polymer chains with polymer molecular weight. Conjugated polymer systems such as PATs, however, adopt a more rod-like conformation in solution. Therefore, GPC tends to overestimate the molecular weights of such polymers [175]. Recently, the use of MALDI-TOF for the analysis of rr-PAT has been reported [112]. Several types of matrices, such as dihydroxybenzoic acid (DHB), dithranol, trans-3-indole acrylic acid (IAA), sinapinic acid, 9-nitroanthracene, and terthiophene, were explored, with and without cationization salts. Dithranol, 9-nitroanthracene, and terthiophene were found to assist the desorption of PAT. Terthiophene gave the best signal-to-noise ratios and required lower threshold laser powers to obtain good MALDI spectra, reducing the risk of fragmentation. Polymer samples of narrow polydispersity were obtained by fractionation of rr-P3HT using Soxhlet extraction with various solvents. Comparisons between the molecular weights calculated by MALDI and GPC of all polymer samples showed that GPC calculated molecular weights are a factor of 1.5–2.0 times higher than MALDI. The polydispersities calculated by MALDI were identical or slightly lower than those calculated by GPC.

MALDI-TOF can also be used to determine polymer end group composition. It was found that there are typically three types of end groups (H/H, H/Br, and Br/Br) depending on the synthetic method employed (Rieke, McCullough, or GRIM) [112]. MALDI-TOF has also been used to monitor transformations at the polymer end [30]. The end group composition can be calculated by $(166)n + 2(EG)$, $(166)n + 1 + EG$, or $(166)n + 80 + EG$, where EG is the molecular weight of the end group and n is the number of repeat units. One example is the Vilsmeier reaction of rr-P3HT bearing terminal hydrogen atoms. Using MALDI-TOF the reaction could be monitored allowing for complete conversion of the end groups (Figure 9.18). MALDI-TOF has also been used to verify the in situ incorporation of functional groups onto the polymer end [110,114,176].

9.7.3 UV–vis Absorption Properties

UV–vis spectroscopy has been established as a versatile and powerful tool for the characterization of conjugated polymers. It is a tool that is used to quantitatively determine the extent of π-orbital overlap in conjugated systems [120]. The extent of the conjugation is directly reflected in the maximum absorption for such systems. This is the manifestation of the π–π* transition (HOMO/LUMO or bandgap). Consequently, UV–vis can be used to determine information regarding the conformational state and structure of conjugated polymers [120].

UV–vis has been used in a qualitative comparison of regioregular (rr) vs. regioirregular (ir-PATs) with identical substituents [24,25,31,33,37,45,120,177]. The rr-PATs show a redshift in the λ_{max} compared to the ir-PATs (Table 9.1). This is indicative of a lower energy π–π* transition for the rr-PATs and, therefore, a longer conjugation length. The reason for the increase in orbital overlap is that rr-PATs establish a more planar conformation in solution that is more rodlike while ir-PATs are more coil like, twisting the orbitals away from interaction [117,118]. Studies show that different alkyl substituents on the thiophene backbone can drastically influence the absorption spectrum [3,5]. The longer the alkyl substituent, the longer will be the conjugation length; however, this may be a function of molecular weight. Most of the solution absorbtion maxima are around 450 nm. However, in some cases, longer alkyl chains can drive solution ordering, leading to an enhancement of higher ordered states. This is shown in UV–vis spectra as higher λ_{max} values (Table 9.1). Other methods of testing show that aggregation of the substituents occurs to a greater extent in PATs with longer alkyl substituents [116]. It has also been shown that in addition to regioregularity and substituent length, the reduction of the number of alkyl substituents can lead to increased conjugation length. By comparing poly(3-octylthiophene) to analogous but less substituted poly[4-octyl-2,2′-bi(thiophene)], both regioregular, the λ_{max} values in chloroform were 448 and 466 nm, respectively [178].

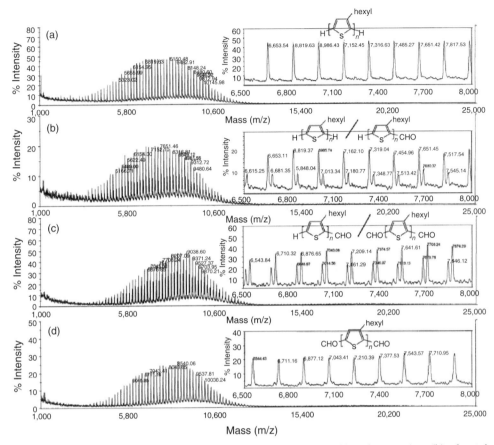

FIGURE 9.18 MALDI-TOF monitors the process of Vilsmeier reaction. (a) Before reaction, (b) after 1 h of reaction, (c) after 8 h of reaction, and (d) after 24 h of reaction. (From Liu, J., et al., *Macromolecules*, 27, 9882, 2002.)

Other studies into the effect of substituents on conjugation as monitored by UV–vis have shown interesting trends. For instance, the effect of placing an electron donating methoxy group on a thienylene–phenylene based polymer showed a shift in λ_{max} from 406 nm for the nonsubstituted polymer to 446 nm. However, the effects of electron-withdrawing groups on the same polymer (chloride, bromide, and cyano) had much less dramatic effects with λ_{max} values of 384, 382, and 384 nm, respectively [179]. The effect of molecular weight has been studied on thin films of regioregular poly(3-hexylthiophene) by UV–vis. Low-molecular weight polymer ($M_n = 2200$ g/mol; $M_w = 3146$ g/mol) exhibited a λ_{max} of 450 nm, while for high-molecular weight polymer ($M_n = 19{,}000$ g/mol; $M_w = 25{,}650$ g/mol) it was 555 nm. This clearly demonstrates the importance of chain length to conjugation length.

As temperature decreases, the aggregation of the polymers increases. This change in the conformational structure is manifested in a permutation of the absorption spectrum, hence thermochromism [1,11,26,27,92,122–124,180–182]. Thermochromism was clearly shown for poly(3-(2-methyl-1-butoxy)-4-methylthiophene) (PMBMT). A thin film of PMBMT, in the presence of acetone vapor, shows a λ_{max} of 502 nm with vibronic sidebands at 534 and 582 nm. Upon heating above the T_g (50°C), the spectrum blueshifts dramatically afford to a λ_{max} of 372 nm. This process was irreversible [183]. Thermochromism has also been shown in PATs with chiral substituents. Thin films of poly[3-(S)-2-methylbutylthiophene] show a redshift after heating, indicating a conformational transition that provides a structure with more efficient orbital overlap [90].

Since aggregation is greater in concentrated solutions, direct correlation of conjugation length to λ_{max} can be difficult. These concentrated solutions of PATs can form supramolecular aggregates that absorb over a broad area in the UV–vis region. This in turn can make determining the actual λ_{max} a challenge. Therefore, it could be said that λ_{max} may also be a function of concentration. Some drastic examples of concentrated solutions provide the polymer as colloidal suspensions of aggregates. As a result, it is typically necessary to prepare dilute solutions of PATs for study by UV–vis, while filtration can remove any residual suspensions.

Depending on solvent interactions with the polymer, changes in the spectrum are also observed. Solvatochromism is observed as blueshift or redshift in λ_{max} depending on how good or poor the extent of solvation. Good solvents, such as chloroform, allow the polymer to access a number of conformational states, leading to reduced conjugation relative to solutions with poor solvents that drive aggregational states. Kiriy et al. studied solvatochromism by way of solvent induced aggregation of regioregular poly(3-hexylthiophene) (PHT) and poly(3-octylthiophene) (POT) [119]. Using chloroform/hexane ratios, from chloroform only to 1/5, showed significant spectral changes from $\lambda_{max} = 448$ nm to 559 and 553 nm for POT and PHT, respectively. A well-defined isobestic point at 480 nm for PHT clearly indicates two distinct conformational states with the increase of hexane concentration in the solvent leading to a more ordered-π system.

Of notable significance to the discussion of UV–vis studies of PAT solvation is the work of Fraleoni-Morgera et al. [184]. They have found that the addition of a surfactant, Tween 80, can solubilize PHT in water. This process was studied using UV–vis by treating the polymer as its own optical probe. The spectrum of the aqueous polymer microemulsion shows a redshift with respect to the same polymer in THF, and vibronic structure similar to that of a drop cast film of the same. The results reveal that while the polymer is in solution, it is conformationally ordered due in part to the formation of aggregates and the interaction of the side chains with the solvent.

In addition to temperature and solvation effects, there are many other factors that may alter the UV–vis spectrum of conjugated polymers. Some examples of these are affinitychromism [19,65,185], electochromism [186], halochromism [187,189], and photochromism [189]. These effects can be deliberately employed in the design of novel devices such as optical sensors.

UV–vis has also been employed to monitor conformational changes in chiral formations, e.g., the formation of chiral supermolecular assemblies from PTs containing negatively charged substituents with positively charged peptides has been observed [190]. Yashima and coworkers compared chiral supramolecular assembly formations from regioregular vs. regioirregular PATs [88]. They found that the λ_{max} for a solution of rr-P3AT redshifted up to 200 nm upon addition of a poor (polar) solvent. A clear isobestic point was observed, indicating the conformational transformation to the desired assembly. On the contrary, the spectrum of the analogous regioirregular polymer showed little solvatochromism, indicating a lack of assembly.

Solid state, UV–vis measurements of rr-PATs vs. ir-PATs also reveal a wealth of conformational information. ir-PAT films absorb over broad regions of the UV–vis spectrum and offer very little fine structure. Alternatively, rr-PAT films are redshifted with respect to ir-PATs and do show fine structure. This additional fine spectral structure is usually manifested as shoulders [6,26] or even well-defined peaks [26]. In fact, thin films of HT-PDDT show at least three distinct peaks: $\lambda_{max} = 565$ nm, $\lambda = 530$, 620 nm. The intensity of these non-λ_{max} peaks is a function of the film thickness. It was observed that all three of these peaks are distinct conformational structures with different conjugation lengths. This gives the general trend that the higher the value of the wavelength, the higher will be the long-range order of the polymer backbone.

It was stated earlier that λ_{max} was to some extent a function of concentration. Thus it can be said that λ_{max} is a function of film thickness for solid-state spectroscopy. As films thicken, the morphological disorder of the polymer can increase. The evidence for these phenomena is studies on film thickness showing different spectra for thick vs. thin films. Roncali demonstrated that regioirregular poly(3-methylthiophene) (ir-PMT) exhibited thickness dependent spectra that correlated with the conjugation lengths and the conductivity of the films [191–193]. It was found that the λ_{max} for thick films (0.2 μm) was 510 nm, while for thin films (0.006 μm) it was 552 nm. By comparison, relatively thick films (1–3 μm) of rr-PAT show similar results to the thin films of ir-PMT. This indicates that the morphological order is the same for the two systems; consequently, order has been increased by control of regioregularity.

Thermochromic studies on thin films of rr-P3ATs and irregular P3ATs by Holdcroft show that the thermochromic behavior is controlled by the extent of regioregularity and side-chain length [194]. Holdcroft found that with an increase in regioregularity, the PTs are quite crystalline and as the regioregularity decreases the materials can be quasi-ordered to disordered. Both thermal and x-ray studies along with UV–vis studies were presented.

9.8   Conclusion

Regioregular PTs remain as one of the most promising conductive polymers for commercial applications and the unveiling of new chemistry and physics of organic electronic materials. New methods to prepare and new devices prepared from rr-PTs promise to usher in the next generation of electronic devices and new commercial products ranging from displays to components in a new age of printable electronics. The purity of structure and the amazing self-assembly properties make this one of the most exciting materials to study and to develop.

Acknowledgment

We thank the NSF CHE0107178 and CHE0415369 for support.

References

1. Torsi, L., A. Tafuri, N. Cioffi, M.C. Gallazzi, A. Sassella, L. Sabbatini, and P.G. Zambonin. 2003. Regioregular polythiophene field-effect transistors employed as chemical sensors. *Sens Actuat B Chem* B 93 (1–3):257–262.
2. Andersson, M.R., O. Thomas, W. Mammo, M. Svensson, M. Theander, and O. Inganas. 1999. Substituted polythiophenes designed for optoelectronic devices and conductors. *J Mater Chem* 9 (9):1933–1940.
3. McQuade, D.T., A.E. Pullen, and T.M. Swager. Chemical sensors—Conjugated polymer-based chemical sensors. 2000. *Chem Rev* 100 (7):2537–2574.
4. Liu, J., T. Tanaka, K. Sivula, A.P. Alivisatos, and J.M.J. Frechet. 2004. Employing end-functional polythiophene to control the morphology of nanocrystal-polymer composites in hybrid solar cells. *J Am Chem Soc* 126 (21):6550–6551.
5. McCullough, R.D. 1998. The chemistry of conducting polythiophenes. *Adv Mater* 10 (2):93–116.
6. McCullough, R.D. 1998. Regioregular, head-to-tail coupled poly(3-alkylthiophene) and its derivatives. In *Handbook of conducting polymers*, 2nd ed., eds. T.A. Skotheim, R.L. Elsenbaumer, J. R. Reynolds. NewYork: Marcell Dekker, pp. 225–258.
7. Chiang, C.K., C.R. Fincher, Jr., Y.W. Park, A.J. Heeger, H. Shirakawa, E.J. Louis, S.C. Gau, and A.G. MacDiarmid. 1977. Electrical conductivity in doped polyacetylene. *Phys Rev Lett* 39 (17):1098–1101.
8. Lin, J.W.P., and L.P. Dudek. 1980. Synthesis and properties of poly(2,5-thienylene). *J Polym Sci Polym Chem Ed* 18 (9):2869–2873.
9. Yamamoto, T., K. Sanechika, and A. Yamamoto. 1980. Preparation of thermostable and electric-conducting poly(2,5-thienylene). *J Polym Sci Polym Lett Ed* 18:9–12.
10. Kobayashi, M., J. Chen, T.C. Chung, F. Moraes, A.J. Heeger, and F. Wudl. 1984. Synthesis and properties of chemically coupled poly(thiophene). *Synth Met* 9 (1):77–86.
11. Berlin, A., G. Pagani, and F. Sannicolo. 1986. New synthetic routes to electroconductive polymers containing thiophene units. *J Chem Soc Chem Commun* (22):1663–1664.
12. Ballauff, M. 1989. Stiff-chain polymers-structure, phase behavior and properties. *Angew Chem Int Ed Eng* 28 (3):253–267.

13. Elsenbaumer, R.L., K.Y. Jen, and R. Oboodi. 1986. Processible and environmentally stable conducting polymers. *Synth Met* 15:169–174.

14. Jen, K.-Y., G.G. Miller, and R.L. Elsenbaumer. 1986. Highly conducting, soluble and environmentally-stable poly(3-alkylthiophenes). *J Chem Soc Chem Commun* (17):1346–1347.

15. Mao, H., B. Xu, and S. Holdcroft. 1993. Synthesis and structure–property relationships of regioirregular poly(3-hexylthiophenes). *Macromolecules* 26 (5):1163–1169.

16. Pomerantz, M., J.J. Tseng, H. Zhu, S.J. Sproull, J.R. Reynolds, R. Uitz, H.J. Arnott, and M.I. Haider. 1991. Processable polymers and copolymers of 3-alkylthiophenes and their blends. *Synth Met* 41 (3):825–830.

17. Sato, M., S. Tanaka, and K. Kaeriyama. 1986. *J Chem Soc Chem Commun* (11):873.

18. Sugimoto, R., S. Takeda, H.B. Gu, and K. Yoshino. 1986. *Chem Express* 1:635.

19. Marsella, M.J., P.J. Carroll, and T.M. Swager. 1995. Design of chemoresistive sensory materials: Polythiophene-based pseudopolyrotaxanes. *J Am Chem Soc* 117 (39):9832–9841.

20. Yamamoto, T., A. Morita, Y. Miyazaki, T. Maruyama, H. Wakayama, Z.H. Zhou, Y. Nakamura, T. Kanbara, S. Sasaki, and K. Kubota. 1992. Preparation of p-conjugated poly(thiophene-2,5-diyl), poly(p-phenylene), and related polymers using zerovalent nickel complexes. Linear structure and properties of the p-conjugated polymers. *Macromolecules* 25 (4):1214–1223.

21. Jen, K.Y., G.G. Miller, and R.L. Elsenbaumer. 1986. Highly conducting, soluble, and environmentally-stable poly(3-alkylthiophenes). *J Chem Soc Chem Commun* 17:1346–1347.

22. Hotta, S., S.D.D.V. Rughooputh, A.J. Heeger, and F. Wudl. 1987. Spectroscopic studies of soluble poly(3-alkylthienylenes). *Macromolecules* 20 (1):212–215.

23. Sato, M.-A., and H. Morii. 1991. Nuclear magnetic resonance studies on electrochemically prepared poly(3-dodecylthiophene). *Macromolecules* 24:1196–1200.

24. McCullough, R.D., S.P. Williams, M. Jayaraman, J. Reddinger, L. Miller, and S. Tristram-Nagle. 1994. Synthesis and physical properties of self-orienting head-to-tail polythiophenes. *Mater Res Soc SympProc* 328 (Electrical, Optical, and Magnetic Properties of Organic Solid State Materials):215–220.

25. McCullough, R.D., and R.D. Lowe. 1992. Enhanced electrical-conductivity in regioselectively synthesized poly(3-Alkylthiophenes). *J Chem Soc Chem Commun* 1:70–72.

26. McCullough, R.D., R.D. Lowe, M. Jayaraman, and D.L. Anderson. 1993. Design, synthesis, and control of conducting polymer architectures—Structurally homogeneous poly(3-alkylthiophenes). *J Org Chem* 58 (4):904–912.

27. Pham, C.V., H.B.J. Mark, and H. Zimmer. 1986. A convenient synthesis of 3-alkylthiophenes. *Synth Commun* 16 (6):689–696.

28. Tamao, K., S. Kodama, I. Nakajima, M. Kumada, A. Minato, and K. Suzuki. 1982. Nickel-phosphine complex-catalyzed grignard coupling—II. *Tetrahedron* 38:3347–3354.

29. Cunningham, D.D., L. Laguren-Davidson, H.B. Mark, Jr., C.V. Pham, and H. Zimmer. 1987. Synthesis of oligomeric 2,5-thienylenes. Their UV spectra and oxidation potentials. *J Chem Soc Chem Commun* 13:1021–1023.

30. Liu, J., and R.D. McCullough. 2002. End group modification of regioregular polythiophene through postpolymerization functionalization. *Macromolecules* 35 (27):9882–9889.

31. Chen, T.A., X. Wu, and R.D. Rieke. 1995. Regiocontrolled synthesis of poly(3-alkylthiophene) mediated by Rieke zinc: Their characterization and solid state properties. *J Am Chem Soc* 117:233–244.

32. Chen, T.-A., and R.D. Rieke. 1993. Polyalkylthiophenes with the smallest bandgap and the highest intrinsic conductivity. *Synth Met* 60 (2):175–177.

33. Chen, T.A., and R.D. Rieke. 1992. The first regioregular head-to-tail poly(3-hexylthiophene-2,5-diyl) and a regiorandom isopolymer: Nickel versus palladium catalysis of 2(5)-bromo-5(2)-(bromozincio)-3-hexylthiophene polymerization. *J Am Chem Soc* 114 (25):10087–10088.

34. Wu, X., T.A. Chen, and R.D. Rieke. 1996. A study of small band gap polymers: Head-to-tail regioregular poly[3-(alkylthio)thiophenes] prepared by regioselective synthesis using active zinc. *Macromolecules* 29 (24):7671–7677.

35. Rieke, R.D. 1977. Preparation of highly reactive metal powders and their use in organic and organometallic synthesis. *Acc Chem Res* 10:301–306.

36. Rieke, R.D., P.T.J. Li, T.P. Burns, and S.T. Uhm. 1981. Preparation of highly reactivity metal powder. A new procedure for the preparation of highly reactive zinc and magnesium powders. *J Org Chem* 46:4323–4324.

37. McCullough, R.D., S. Tristram-Nagle, S.P. Williams, R.D. Lowe, and M. Jayaraman. 1993. Self-orienting head-to-tail poly(3-alkylthiophenes): New insights on structure–property relationships in conducting polymers. *J Am Chem Soc* 115 (11):4910–4911.

38. Loewe, R.S., S.M. Khersonsky, and R.D. McCullough. 1999. A simple method to prepare head-to-tail coupled, regioregular poly(3-alkylthiophenes) using Grignard metathesis. *Adv Mater* 11 (3):250–258.

39. Loewe, R.S., P.C. Ewbank, J. Liu, L. Zhai, and R.D. McCullough. 2001. Regioregular, head-to-tail coupled poly(3-alkylthiophenes) made easy by the GRIM method: Investigation of the reaction and the origin of regioselectivity. *Macromolecules* 34 (13):4324–4333.

40. See table at the end of Section 9.4.

41. Bolognesi, A., W. Porzio, G. Bajo, G. Zannoni, and I. Fannig. 1999. Highly regioregular poly(3-alkylthiophenes). A new synthetic route and characterization of the resulting polymers. *Acta Polym* 50 (4):151–155.

42. Stille, J.K. 1986. The palladium-catalyzed cross-coupled reactions of organotin reagents with organic electrophiles. *Angew Chem Int Ed Eng* 25:508–524.

43. Suzuki, A. 1999. Recent advances in the cross-coupling reactions of organoboron derivatives with organic electrophiles, 1995–1998. *J Organomet Chem* 576 (1–2):147–168.

44. Iraqi, A., and G.W. Barker. 1998. Synthesis and characterization of telechelic regioregular head-to-tail poly(3-alkylthiophenes). *J Mater Chem* 8 (1):25–29.

45. McCullough, R.D., P.C. Ewbank, and R.S. Loewe. 1997. Self-assembly and disassembly of regioregular, water soluble polythiophenes: Chemoselective ionchromatic sensing in water. *J Am Chem Soc* 119 (3):633–634.

46. McCullough, R.D., and P.C. Ewbank. 1997. Self-assembly and chemical response of conducting polymer superstructures. *Synth Met* 84 (1–3):311–312.

47. Gronowitz, S., P. Bjork, J. Malm, and A.B. Hornfeldt. 1993. The effect of some additives on the Stille Pd0-catalyzed cross-coupling reaction. *J Organomet Chem* 460:127–129.

48. Guillerez, S., and G. Bidan. 1998. New convenient synthesis of highly regioregular poly(3-octylthiophene) based on the Suzuki coupling reaction. *Synth Met* 93 (2):123–126.

49. Andersson, M.R., D. Selse, M. Berggren, H. Jaervinen, T. Hjertberg, O. Inganaes, O. Wennerstroem, and J.E. Oesterholm. 1994. Regioselective polymerization of 3-(4-octylphenyl)thiophene with FeCl₃. *Macromolecules* 27 (22):6503–6506.

50. Frechette, M., M. Belletete, J.Y. Bergeron, G. Durocher, and M. Leclerc. 1997. Monomer reactivity vs. regioregularity in polythiophene derivatives: A joint synthetic and theoretical investigation. *Synth Met* 84 (1–3):223–224.

51. Andreani, F., L. Angiolini, D. Caretta, and E. Salatelli. 1998. Synthesis and polymerization of 3,3″-di[(S)-(+)-2-methylbutyl]- 2,2′:5′,2″-terthiophene: A new monomer precursor to chiral regioregular poly(thiophene). *J Mater Chem* 8 (5):1109–1111.

52. Meijere, A.D., and F.O. Diederich. 2004. *Metal-catalyzed cross-coupling reactions*, 2nd ed., eds., A. Meijere, and F. Diederich. Weinheim: Chichester.

53. Tamao, K., K. Sumitani, Y. Kiso, M. Zembayashi, A. Fujioka, S.I. Kodama, I. Nakajima, A. Minato, and M. Kumada. 1976. Nickel-phosphine complex-catalyzed Grignard coupling. I. Cross-coupling of alkyl, aryl, and alkenyl Grignard reagents with aryl and alkenyl halides: General scope and limitations. *Bull Chem Soc Jpn* 49 (7):1958–1969.

54. Sheina, E.E., J. Liu, M.C. Iovu, D.W. Laird, and R.D. McCullough. 2004. Chain growth mechanism for regioregular nickel-initiated cross-coupling polymerizations. *Macromolecules* 37 (10): 3526–2528.

55. Iovu, M.C., E.E. Sheina, R.R. Gil, and R.D. McCullough. 2005. Experimental evidence for the Quasi-"Living" nature of the Grignard metathesis (GRIM) for the synthesis of regioregular poly(3-alkylthiophenes). *Macromolecules* 38:8649–8656.

56. Iovu, M.C., E.E. Sheina, and R.D. McCullough. 2005. Grignard metathesis (GRIM) method for the synthesis of regioregular poly(3-alkylthiophenes) with well-defined molecular weights. *Polym Prepr (Am Chem Soc Div Polym Chem)* 46 (1):660–661.

57. Iovu, M.C., M. Jeffries-El, E.E. Sheina, J. Cooper, and R.D. McCullough. 2005. Synthesis of poly(3-hexylthiophene) di-block copolymers using atom transfer radical polymerization. *Polymer* 46 (19):8582–8586.

58. Yokozawa, T., and H. Shimura. 1999. Condensative chain polymerization. II. Preferential esterification of propagating end group in Pd-catalyzed CO-insertion polycondensation of 4-bromophenol derivatives. *J Polym Sci A* 37 (14):2607–2618.

59. Mao, Y., Y. Wang, and B.L. Lucht. 2004. Regiocontrolled synthesis of poly(3-alkylthiophene)s by Grignard metathesis. *J Polym Sci A* 42 (21):5538–5547.

60. Levesque, I., and M. Leclerc. 1995. Ionochromic effects in regioregular ether-substituted polythiophenes. *J Chem Soc Chem Commun* 22:2293–2294.

61. Garnier, F. 1989. Functionalized conducting polymers-towards intelligent materials. *Angew Chem Int Ed Eng* 101:529–533.

62. Roncali, J. 1997. Synthetic principles for bandgap control in linear p-conjugated systems. *Chem Rev* 97 (1):173–205.

63. Levesque, I., and M. Leclerc. 1996. Ionochromic and thermochromic phenomena in a regioregular polythiophene derivative bearing oligo(oxyethylene) side chains. *Chem Mater* 8 (12):2843–2849.

64. Levesque, I., P. Bazinet, and J. Roovers. 2000. Optical properties and dual electrical and ionic conductivity in poly(3-methylhexa(oxyethylene)oxy-4-methylthiophene). *Macromolecules* 33 (8):2952–2957.

65. McCullough, R.D., and S.P. Williams. 1993. Toward tuning electrical and optical properties in conjugated polymers using side-chains: Highly conductive head-to-tail, heteroatom functionalized polythiophenes. *J Am Chem Soc* 115 (24):11608–11609.

66. Faid, K., R. Cloutier, and M. Leclerc. 1993. Design of novel electroactive polybithiophene derivatives. *Macromolecules* 26 (10):2501–2507.

67. Groenendaal, L.B., F. Jonas, D. Freitag, H. Pielartzik, and J.R. Reynolds. 2000. Poly(3,4-ethylene-dioxythiophene) and its derivatives: Past, present, and future. *Adv Mater* 12 (7):481–494.

68. Sheina, E.E., and R.D. McCullough. 2003. Electrical and spectroscopic property differences in poly(3-alkoxythiophenes) synthesized by various methods. *Polym Prepr (Am Chem Soc Div Polym Chem)* 44 (2):885.

69. Sheina, E.E., S.M. Khersonsky, E.G. Jones, and R.D. McCullough. Highly conductive, regioregular alkoxy-functionalized polythiophenes: A new class of stable, low band gap materials. 2005. *Chem Mater* 17 (13):3317–3319.

70. Sheina, E.E., S.M. Khersonsky, E.G. Jones, and R.D. McCullough. 2003. Toward poly(3- and 3,4-alkoxythiophenes). *Polym Prepr (Am Chem Soc Div Polym Chem)* 44 (1):843–844.

71. Sheina, E.E., and R.D. McCullough. 2004. *Electrical properties of highly conductive EDOT based copolymers*, Abstracts of Papers, 228th ACS National Meeting, Philadelphia, August 22–26, 2004, POLY-151, 2004.

72. Wu, X., T.A. Chen, and R.D. Rieke. 1995. Synthesis of regioregular head-to-tail poly[3-(alkylthio)thiophenes]. A highly electroconductive polymer. *Macromolecules* 28 (6):2101–2102.

73. Li, L., and D.M. Collard. 2005. Tuning the electronic structure of conjugated polymers with fluoroalkyl substitution: Alternating alkyl/perfluoroalkyl-substituted polythiophene. *Macromolecules* 38 (2):372–378.

74. Hong, X.M., J.C. Tyson, and D.M. Collard. 2000. Controlling the macromolecular architecture of poly(3-alkylthiophene)s by alternating alkyl and fluoroalkyl substituents. *Macromolecules* 33 (10):3502–3504.

75. Hong, X.M., and D.M. Collard. 2000. Liquid crystalline regioregular semifluoroalkyl-substituted polythiophenes. *Macromolecules* 33 (19):6916–6917.

76. Li, L., K.E. Counts, S. Kurosawa, A.S. Teja, and D.M. Collard. 2004. Tuning the electronic structure and solubility of conjugated polymers with perfluoroalkyl substituents: Poly(3-perfluorooctylthiophene), the first supercritical-CO_2-soluble conjugated polymer. *Adv Mater* 16 (2):180–183.

77. Pilston, R.L. 2001. Synthesis and characterization of novel regioregular polythiophenes, random copolymers of thiophene and studies of the GRIM polymerization method for the synthesis of polythiophenes. Carnegie Mellon University.

78. Pilston, R.L., and R.D. McCullough. 2000. Toward highly fluorescent polythiophenes: Head-to-tail coupled copolymers of 3-(methoxyethoxyethoxymethyl)thiophene and 3-(perfluoroalkyl)thiophene. Thesis.

79. Lemaire, M., D. Delabouglise, R. Garreau, A. Guy, and J. Roncali. 1988. Enantioselective chiral poly(thiophenes). *J Chem Soc Chem Commun* 10:658–661.

80. Bouman, M.M., E.E. Havinga, R.A.J. Janssen, and E.W. Meijer. 1994. Chiroptical properties of regioregular chiral polythiophenes. *Mol Cryst Liq Cryst A* 256:439–448.

81. Bouman, M.M., and E.W. Meijer. 1995. Stereomutation in optically active regioregular polythiophenes. *Adv Mater* 7 (4):385–387.

82. Langeveld-Voss, B.M.W., M.M. Bouman, M.P.T. Christiaans, R.A.J. Janssen, and E.W. Meijer. 1996. Main-chain chirality of regioregular polythiophenes. *Polym Prepr (Am Chem Soc Div Polym Chem)* 37 (2):499–500.

83. Langeveld-Voss, B.M.W., E. Peeters, R.A.J. Janssen, and E.W. Meijer. 1997. Chiroptical properties of highly ordered di[(S)-2-methylbutoxy] substituted polythiophene and poly(p-phenylene vinylene). *Synth Met* 84 (1–3):611–614.

84. Bidan, G., S. Guillerez, and V. Sorokin. 1996. Chirality in regio-regular and soluble polythiophene. An internal probe of conformational changes induced by minute solvation variation. *Adv Mater* 8 (2):157–160.

85. Yashima, E., H. Goto, and Y. Okamoto. 1999. Metal-induced chirality induction and chiral recognition of optically active, regioregular polythiophenes. *Macromolecules* 32 (23):7942–7945.

86. Goto, H., and E. Yashima. 2002. Electron-induced switching of the supramolecular chirality of optically active polythiophene aggregates. *J Am Chem Soc* 124 (27):7943–7949.

87. Goto, H., E. Yashima, and Y. Okamoto. 2000. Unusual solvent effects on chiroptical properties of an optically active regioregular polythiophene in solution. *Chirality* 12 (5/6):396–399.

88. Goto, H., Y. Okamoto, and E. Yashima. 2002. Solvent-induced chiroptical changes in supramolecular assemblies of an optically active, regioregular polythiophene. *Macromolecules* 35 (12):4590–4601.

89. Langeveld-Voss, B.M.W., M.P.T. Christiaans, R.A.J. Janssen, and E.W. Meijer. Inversion of optical activity of chiral polythiophene aggregates by a change of solvent. 1998. *Macromolecules* 31 (19):6702–6704.

90. Catellani, M., S. Luzzati, F. Bertini, A. Bolognesi, F. Lebon, G. Longhi, S. Abbate, A. Famulari, and S.V. Meille. 2002. Solid-state optical and structural modifications induced by temperature in a chiral poly-3-alkylthiophene. *Chem Mater* 14 (11):4819–4826.

91. Langeveld-Voss, B.M.W., R.J.M. Waterval, R.A.J. Janssen, and E.W. Meijer. 1999. Principles of "majority rules" and "sergeants and soldiers" applied to the aggregation of optically active polythiophenes: Evidence for a multichain phenomenon. *Macromolecules* 32 (1):227–230.

92. Brustolin, F., F. Goldoni, E.W. Meijer, and N.A.J.M. Sommerdijk. 2002. Highly ordered structures of amphiphilic polythiophenes in aqueous media. *Macromolecules* 35 (3):1054–1059.

93. Li, C., M. Numata, A.H. Bae, K. Sakurai, and S. Shinkai. 2005. Self-assembly of supramolecular chiral insulated molecular wire. *J Am Chem Soc* 127 (13):4548–4549.

94. Murray, K.A., A.B. Holmes, S.C. Moratti, and R.H. Friend. 1996. The synthesis and crosslinking of substituted regioregular polythiophenes. *Synth Met* 76 (1–3):161–163.

95. Bolognesi, A., R. Mendichi, A. Schieroni, D. Villa, and O. Ahumada. 1997. Synthesis and characterization of poly[3-(6-tetrahydropyran-2-yloxyhexyl)-2,5-thienylene]. *Macromol Chem Phys* 198 (10):3277–3284.

96. Yu, J., and S. Holdcroft. 2000. Solid-state thermolytic and catalytic reactions in functionalized regioregular polythiophenes. *Macromolecules* 33 (14):5073–5079.

97. Lanzi, M., P. Costa-Bizzarri, C. Della-Casa, L. Paganin, and A. Fraleoni. 2002. Synthesis, characterization and optical properties of a regioregular and soluble poly[3-(10-hydroxydecyl)-2,5-thienylene]. *Polymer* 44 (3):535–545.

98. Murray, K.A., A.B. Holmes, S.C. Moratti, and G. Rumbles. 1999. Conformational changes in regioregular polythiophenes due to crosslinking. *J Mater Chem* 9 (9):2109–2116.

99. Pomerantz, M., and M.L. Liu. 1999. Synthesis and properties of poly[3-(w-bromoalkyl)thiophene]. *Synth Met* 101 (1–3):95.

100. Pomerantz, M. 1999. Poly(alkyl thiophene-3-carboxylates). Synthesis, properties and electroluminescence studies of polythiophenes containing a carbonyl group directly attached to the ring. *J Mater Chem* 9:2155–2263.

101. Iraqi, A., J.A. Crayston, and J.C. Walton. 1998. Covalent binding of redox active centers to preformed regioregular polythiophenes. *J Mater Chem* 8 (1):31–36.

102. Zhai, L., R.L. Pilston, K.L. Zaiger, K.K. Stokes, and R.D. McCullough. 2003. A simple method to generate side-chain derivatives of regioregular polythiophene via the GRIM metathesis and post-polymerization functionalization. *Macromolecules* 36 (1):61–64.

103. Stokes, K.K., K. Heuze, and R.D. McCullough. 2003. New phosphonic acid functionalized, regioregular polythiophenes. *Macromolecules* 36 (19):7114–7118.

104. Ewbank, P.C., R.S. Loewe, L. Zhai, J. Reddinger, G. Sauve, and R.D. McCullough. 2004. Regioregular poly(thiophene-3-alkanoic acid)s: Water soluble conducting polymers suitable for chromatic chemosensing in solution and solid state. *Tetrahedron* 60 (49):11269–11275.

105. Ewbank, P.C., G. Nuding, H. Suenaga, R.D. McCullough, and S. Shinkai. 2001. Amine functionalized polythiophenes: Synthesis and formation of chiral, ordered structures on DNA substrates. *Tetrahedron Lett* 42 (2):155–157.

106. Ho, H.A., and M. Leclerc. 2003. New colorimetric and fluorometric chemosensor based on a cationic polythiophene derivative for iodide-specific detection. *J Am Chem Soc* 125 (15):4412–4413.

107. Ho, H.A., M. Bèra-Abèrem, and M. Leclerc. 2005. Optical sensors based on hybrid DNA/conjuconjugated polymer complexes. *Chem Eur J* 11 (6):1718–1724.

108. Ho, H.A., M. Boissinot, M.G. Bergeron, G. Corbeil, K. Dore, D. Boudreau, and M. Leclerc. 2002. Colorimetric and fluorometric detection of nucleic acids using cationic polythiophene derivatives. *Angew Chem Int Ed Eng* 41 (Part 9):1548–1551.

109. Bjornholm, T., D.R. Greve, N. Reitzel, T. Hassenkam, K. Kjaer, P.B. Howes, N.B. Larsen, J. Bogerlund, M. Jayaraman, P.C. Ewbank, and R.D. McCullough. 1998. Self- assembly of regioregular, amphiphilic polythiophenes into highly ordered π-stacked conjugated polymer thin films and nanocircuits. *J Am Chem Soc* 120:7643–7644.

110. Langeveld-Voss, B.M.W., R.A.J. Janssen, A.J.H. Spiering, J.L.J. van Dongen, E.C. Vonk, and H.A. Claessens. 2000. End-group modification of regioregular poly(3-alkylthiophene)s. *Chem Commun* 2000 (1):81–82.

111. Jayakannan, M., J.L.J.V. Dongen, and R.A.J. Janssen. 2001. Mechanistic aspects of the Suzuki polycondensation of thiophenebisboronic derivatives and diiodobenzenes analyzed by MALDI-TOF mass spectrometry. *Macromolecules* 34 (16):5386–5393.

112. Liu, J., R.S. Loewe, and R.D. McCullough. 1999. Employing MALDI-MS on poly(alkylthiophenes): Analysis of molecular weights, molecular weight distributions, end-group structures, and end-group modifications. *Macromolecules* 32 (18):5777–5785.

113. Liu, J., E. Sheina, T. Kowalewski, and R.D. McCullough. 2002. Tuning the electrical conductivity and self-assembly of regioregular polythiophene by block copolymerization: Nanowire morphologies in new di- and triblock copolymers. *Angew Chem Int Ed Eng* 41 (2):329–332.

114. Jeffries-El, M., G. Sauve, and R.D. McCullough. 2004. In-situ end group functionalization of regioregular poly(3-alkylthiophene) using the Grignard metathesis polymerization method. *Adv Mater* 16 (12):1017–1019.

115. Jeffries-El, M., G. Sauve, and R.D. McCullough. 2005. Facile synthesis of end-functionalized regioregular poly(3-alkylthiophene)s via modified Grignard metathesis reaction. *Macromolecules* 38 (25):10346–10352.

116. Yue, S., G.C. Berry, and R.D. McCullough. 1996. Intermolecular association and supramolecular organization in dilute solution. 1. Regioregular poly(3-dodecylthiophene). *Macromolecules* 29 (3):933–939.

117. Inganaes, O. 1994. *Trends Polym Sci* 2:189–196.

118. Rughooputh, S.D.D.V., S. Hotta, A.J. Heeger, and F. Wudl. 1987. Chromism of soluble polythienylenes. *J Polym Sci B* 25 (5):1071–1078.

119. Kiriy, N., E. Jaehne, H.J. Adler, M. Schneider, A. Kiriy, G. Gorodyska, S. Minko, D. Jehnichen, P. Simon, A.A. Fokin, and M. Stamm. 2003. One-dimensional aggregation of regioregular polyalkylthiophenes. *Nano Lett* 3 (6):707–712.

120. Patil, A.O., A.J. Heeger, and F. Wudl. 1988. Optical properties of conducting polymers. *Chem Rev* 88:183–200.

121. Inganas, O., W.R. Salaneck, J.E. Osterholm, and J. Laakso. 1988. Thermochromic and solvatochromic effects in poly(3-hexylthiophene). *Synth Met* 22 (4):395–406.

122. Roux, C., and M. Leclerc. 1994. Structure–property relationships in thermochromic polythiophene derivatives. *Macromol Symp* 87 (Polymers: Progress in Chemistry and Physics):1–4.

123. Roux, C., and M. Leclerc. 1994. Thermochromic properties of polythiophene derivatives: Formation of localized and delocalized conformational defects. *Chem Mater* 6 (5):620–624.

124. Tashiro, K., K. Ono, Y. Minagawa, K. Kobayashi, T. Kawai, and K. Yoshino. 1991. Structural changes in the thermochromic solid-state phase transition on poly(3-alkylthiophene). *Synth Met* 41 (1–2):571–574.

125. McCullough, R.D., S.P. Williams, S. Tristram-Nagle, M. Jayaraman, P.C. Ewbank, and L. Miller. 1995. The first synthesis and new properties of regio-regular, head-to-tail coupled polythiophenes. *Synth Met* 69 (1–3):279–282.

126. Prosa, T.J., M.J. Winokur, and R.D. McCullough. 1996. Evidence of a novel side chain structure in regioregular poly(3-alkylthiophenes). *Macromolecules* 29 (10):3654–3656.

127. Prosa, T.J., M.J. Winokur, J. Moulton, P. Smith, and A.J. Heeger. 1992. X-ray structural studies of poly(3-alkylthiophenes): An example of an inverse comb. *Macromolecules* 25 (17):4364–4372.

128. Winokur, M.J., P. Wamsley, J. Moulton, P. Smith, and A.J. Heeger. 1991. Structural evolution in iodine-doped poly(3-alkylthiophenes). *Macromolecules* 24 (13):3812–3815.

129. Mardalen, J., E.J. Samuelsen, O.R. Gautun, and P.H. Carlsen. 1991. Molecular structure of stretch oriented poly(3-hexylthiophene) studied by an extended x-ray diffraction mapping. *Solid State Commun* 80 (9):687–689.

130. Malik, S. and A.K. Nandi, 2002. Crystallization mechanism of regioregular poly(3-alkyl thiophene)s. *J Polym Sci Part B Polym Phys* 40 (18):2073–2085.

131. Pal, S., and A.K. Nandi. 2003. Cocrystallization behavior of poly(3-alkylthiophenes): Influence of alkyl chain length and head to tail regioregularity. *Macromolecules* 36 (22):8426–8432.

132. Nagamatsu, S., W. Takashima, K. Kaneto, Y. Yoshida, N. Tanigaki, K. Yase, and K. Omote. 2003. Backbone arrangement in "friction-transferred" regioregular poly(3-alkylthiophene)s. *Macromolecules* 36 (14):5252–5257.

133. Goh, C., R.J. Kline, M.D. McGehee, E.N. Kadnikova, and J.M.J. Frechet. 2005. Molecular-weight-dependent mobilities in regioregular poly(3-hexyl-thiophene) diodes. *Appl Phys Lett* 86 (12):122110/1–122110/3.

134. Kline, R.J., M.D. McGehee, E.N. Kdnikova, J. Liu, J.M.J. Frechet, and M.F. Toney. 2005. Dependence of regioregular poly(3-hexylthiophene) film morphology and field-effect mobility on molecular weight. *Macromolecules* 38 (8):3312–3319.

135. Kline, R.J., M.D. McGehee, E.N. Kadnikova, J. Liu, and J.M.J. Frechet. 2003. Controlling the field-effect mobility of regioregular polythiophene by changing the molecular weight. *Adv Mater* 15 (18):1519–1522.

136. Zhang, R., B. Li, M.C. Iovu, M. Jeffries-El, G. Sauve, J. Cooper, S. Jia, S. Tristram-Nagle, D.M. Smilgies, D.N. Lambeth, R.D. McCullough, and T. Kowalewski. 2006. Nanostructure dependence of field-effect mobility in regioregular poly(3-hexylthiophene) thin film field effect transistors. *J Am Chem Soc* 128 (11):3480–3481.

137. Widawski, G., M. Rawiso, and B. Francois. 1994. Self-organized honeycomb morphology of star-polymer polystyrene films. *Nature* 369 (6479):387–389.

138. Hempenius, M.A., B.M.W. Langeveld-Voss, J.A.E.H. van Haare, R.A.J. Janssen, S.S. Sheiko, J.P. Spatz, M. Moeller, and E.W. Meijer. 1998. A polystyrene–oligothiophene–polystyrene triblock copolymer. *J Am Chem Soc* 120 (12):2798–2804.

139. Jenekhe, S.A., and X.L. Chen. 1998. Self-assembled aggregates of rod-coil block copolymers and their solubilization and encapsulation of fullerenes. *Science* 279 (5358):1903–1907.

140. Li, W., H. Wang, L. Yu, T.L. Morkved, and H.M. Jaeger. 1999. Syntheses of oligophenylenevinyl-lenes–polyisoprene diblock copolymers and their microphase separation. *Macromolecules* 32 (9):3034–3044.

141. Marsitzky, D., T. Brand, Y. Geerts, M. Klapper, and K. Muellen. 1998. Synthesis of rod-coil block copolymers via end-functionalized poly(p-phenylene)s. *Macromol Rapid Commun* 19 (7):385–389.

142. Messmore, B.W., J.F. Hulvat, E.D. Sone, and S.I. Stupp. 2004. Synthesis, self-assembly, and characterization of supramolecular polymers from electroactive dendron rodcoil molecules. *J Am Chem Soc* 126 (44):14452–14458.

143. Vriezema, D.M., J. Hoogboom, K. Velonia, K. Takazawa, P.C.M. Christianen, J.C. Maan, A.E. Rowan, and R.J.M. Nolte. 2003. Vesicles and polymerized vesicles from thiophene-containing rod-coil block copolymers. *Angew Chem Int Ed Eng* 42 (7):772–776.

144. Tsolakis, P.K., and J.K. Kallitsis. 2003. Synthesis and characterization of luminescent rod-coil block copolymers by atom transfer radical polymerization: Utilization of novel end-functionalized terfluorenes as macroinitiators. *Chem Eur J* 9 (4):936–943.

145. Hagberg, E.C., and V.V. Sheares. 2002. Poly(2,5-benzophenone) coil-rod-coil block copolymers and their phase separation behavior. *Polym Prepr (Am Chem Soc Div Polym Chem)* 43 (2):1055–1056.

146. Jeffries-El, M., M.C. Iovu, E. Sheina, G. Sauve, and R.D. McCullough. 2004. End functionalized poly(3-alkylthiophenes) as building blocks for the synthesis of diblock copolymers. *Polym Prepr (Am Chem Soc Div Polym Chem)* 45 (2):551–552.

147. Iovu, M.C., E.E. Sheina, G. Sauve, M. Jeffries-El, J. Cooper, and R.D. McCullough. 2004. Synthesis of poly(3-hexylthiophene) di-block copolymers using atom transfer radical polymerization. *Polym Prepr (Am Chem Soc Div Polym Chem)* 45 (1):278–279.

148. Matyjaszewski, K., and J. Xia. 2001. Atom transfer radical polymerization. *Chem Rev* 101 (9):2921–2990.

149. Kim, E., Y. Xia, and G.M. Whitesides. 1995. Polymer microstructures formed by molding in capillaries. *Nature* 376 (6541):581–584.

150. Zhao, X.M., Y. Xia, and G.M. Whitesides. 1996. Fabrication of three-dimensional microstructures. Microtransfer molding. *Adv Mater* 8 (10):837–840.

151. Xia, Y., J.A. Rogers, K.E. Paul, and G.M. Whitesides. 1999. Unconventional methods for fabricating and patterning nanostructures. *Chem Rev* 99 (7):1823–1848.

152. Rogers, J.A., Z. Bao, K. Baldwin, A. Dodabalapur, B. Crone, V.R. Raju, V. Kuck, H. Katz, K. Amundson, J. Ewing, and P. Drzaic. 2001. Paper-like electronic displays: Large-area rubber-stamped plastic sheets of electronics and microencapsulated electrophoretic inks. *Proc Natl Acad Sci USA* 98 (9):4835–4840.

153. Bjornholm, T., T. Hassenkam, and N. Reitzel. 1999. Supramolecular organization of highly conducting organic thin films by the Langmuir–Blodgett Technique. *J Mater Chem* 9 (9):1975–1990.

154. Reitzel, N., D.R. Greve, K. Kjaer, P.B. Hows, M. Jayaraman, S. Savoy, R.D. McCullough, J.T. McDevitt, and T. Bjornholm. 2000. Self-assembly of conjugated polymers at the air/water interface. Structure and properties of Langmuir and Langmuir–Blodgett films of amphiphilic regioregular polythiophenes. *J Am Chem Soc* 122 (24):5788–5800.

155. de Boer, B., P.F. Van Hutten, L. Ouali, V. Grayer, and G. Hadziioannou. 2002. Amphiphilic, regioregular polythiophenes. *Macromolecules* 35 (18):6883–6892.

156. Ochiai, K., M. Rikukawa, and K. Sanui. 1999. Novel highly ordered Langmuir–Blodgett films of regioregular poly(3-substituted thiophene). *Chem Commun* 10:867–868.

157. Decher. G., and Schlenoff, J.B., Multilayer Thin Films: Sequential Assembly of Nanocomposite Materials, Cambridge: Weinheim, pp. 1–46, 2003.

158. Decher, G., J. Maclennan, M. Straus, and U. Sohling. 1991. New amphiphilic terphenyl liquid crystals for the preparation of highly ordered ultrathin films. *Makromol Chem Macromol Symp* 46 (Eur. Conf. Organ. Org. Thin Films, 3rd, 1990):313–319.

159. Shiratori, S.S., and M.F. Rubner. pH-dependent thickness behavior of sequentially adsorbed layers of weak polyelectrolytes. 2000. *Macromolecules* 33 (11):4213–4219.

160. Lukkari, J., M. Salomaeki, A. Viinikanoja, T. Aeaeritalo, J. Paukkunen, N. Kocharova, and J. Kankare. 2001. Polyelectrolyte multilayers prepared from water-soluble poly(alkoxythiophene) derivatives. *J Am Chem Soc* 123 (25):6083–6091.

161. Bao, Z., A. Dodabalapur, and A.J. Lovinger. 1996. Soluble and processable regioregular poly(3-hexylthiophene) for thin film field-effect transistor applications with high mobility. *Appl Phys Lett* 69 (26):4108–4111.

162. Sirringhaus, H., P.J. Brown, R.H. Friend, M.M. Nielsen, K. Bechgaard, B.M.W. Langeveld-Voss, A.J.H. Spiering, R.A.J. Janssen, E.W. Meijer, P. Herwig, and D.M. De Leeuw. 1999. Two-dimensional charge transport in self-organized, high-mobility conjugated polymers. *Nature (Lond)* 401 (6754):685–688.

163. Jiang, X., Y. Harima, K. Yamashita, Y. Tada, J. Ohshita, and A. Kunai. 2003. A transport study of poly(3-hexylthiophene) films with different regioregularities. *Synth Met* 135–136: 351–352.

164. Bao, Z., and A.J. Lovinger. 1999. Soluble regioregular polythiophene derivatives as semiconducting materials for field-effect transistors. *Chem Mater* 11 (9):2607–2612.

165. Bao, Z., A.J. Lovinger, and O. Cherniavskaya. 2000. Material issues for construction of organic and polymeric driving circuits for display and electronic applications. *Macromol Symp* 154 (Polymers in Display Applications):199–207.

166. Salleo, A., T.W. Chen, A.R. Volkel, Y. Wu, P. Liu, B.S. Ong, and R.A. Street. 2004. Intrinsic hole mobility and trapping in a regioregular poly(thiophene). *Phys Rev B Condens Matter Mater Phys* 70 (11):115311/1–115311/10.

167. Kim, D.H., Y.D. Park, Y. Jang, H. Yang, Y.H. Kim, J.I. Han, D.G. Moon, S. Park, T. Chang, C. Chang, M. Joo, C.Y. Ryu, and K. Cho. 2005. Enhancement of field-effect mobility due to surface-mediated molecular ordering in regioregular polythiophene thin film transistors. *Adv Funct Mater* 15 (1):77–82.

168. Wu, Y., P. Liu, B.S. Ong, T. Srikumar, N. Zhao, G. Botton, and S. Zhu. 2005. Controlled orientation of liquid-crystalline polythiophene semiconductors for high-performance organic thin-film transistors. *Appl Phys Lett* 86 (14):142102/1–142102/3.

169. Merlo, J.A., and C.D. Frisbie. 2004. Field effect transport and trapping in regioregular polythiophene nanofibers. *J Phys Chem B* 108 (50):19169–19179.

170. McCullough, R.D., R.D. Lowe, M. Jayaraman, P.C. Ewbank, D.L. Anderson, and S. Tristramnagle. 1993. Synthesis and physical-properties of regiochemically well-defined, head-to-tail coupled poly(3-Alkylthiophenes). *Synth Met* 55 (2–3):1198–1203.

171. Barbarella, G., M. Zambianchi, A. Bongini, and L. Antolini. 1994. Conformational chirality of oligothiophenes in the solid state. X-ray structure of 3,4′,4″- trimethyl-2,2′:5′,2″-terthiophene. *Adv Mater* 6 (7/8):561–564.

172. Barbarella, G., A. Bongini, and M. Zambianchi. Regiochemistry and conformation of poly(3-hexylthiophene) via the synthesis and the spectroscopic characterization of the model configurational triads. 1994. *Macromolecules* 27 (11):3039–3045.

173. Maior, R.M.S., K. Hinkelmann, H. Eckert, and F. Wudl. 1990. Synthesis and characterization of two regiochemically defined poly(dialkylbithiophenes): A comparative study. *Macromolecules* 23 (5):1268–1279.

174. Nielen, M.W.F. 1999. MALDI time-of-flight mass spectrometry of synthetic polymers. *Mass Spectr Rev* 18 (5):309–344.

175. Holdcroft, S. 1991. Determination of molecular weights and Mark–Houwink constants for soluble electronically conducting polymers. *J Polym Sci B* 29 (13):1585–1588.

176. Miyakoshi, R., A. Yokoyama, and T. Yokozawa. 2005. Catalyst-transfer polycondensation mechanism of Ni-catalyzed chain-growth polymerization leading to well-defined poly(3-hexylthiophene). *J Am Chem Soc* 127 (49):17542–17547.

177. Mccullough, R.D., S. Tristramnagle, S.P. Williams, R.D. Lowe, and M. Jayaraman. 1993. Self-orienting head-to-tail poly(3-Alkylthiophenes)—New insights on structure–property relationships in conducting polymers. *J Am Chem Soc* 115 (11):4910–4911.

178. Lere-Porte, J.-P., J.J.E. Moreau, and C. Torreilles. 2001. Highly conjugated poly(thiophene)s—Synthesis of regioregular 3-alkylthiophene polymers and 3-alkylthiophene/thiophene copolymers. *Eur J Org Chem* (7):1249–1258.

179. Ng, S.C., J.M. Xu, and H.S.O. Chan. 2000. Synthesis and characterization of regioregular polymers containing substituted thienylene/bithienylene and phenylene repeating units. *Macromolecules* 33 (20):7349–7358.

180. Roux, C., K. Faid, and M. Leclerc. 1993. Thermochromic properties of polythiophenes: Cooperative effects. *Makromol Chem Rapid Commun* 14 (8):461–464.

181. Bertinelli, F., P. Costa-Bizzarri, C. Della-Casa, and M. Lanzi. 2001. Solvent and temperature effects on the chromic behaviour of poly[3-(10-hydroxydecyl)-2,5-thienylene]. *Synth Met* 122 (2):267–273.

182. Lu, H.-F., H.S.O. Chan, and S.C. Ng. 2003. Synthesis, characterization, and electronic and optical properties of donor-acceptor conjugated polymers based on alternating bis(3-alkylthiophene) and pyridine moieties. *Macromolecules* 36 (5):1543–1552.

183. Garreau, S., M. Leclerc, N. Errien, and G. Louarn. 2003. Planar-to-nonplanar conformational transition in thermochromic polythiophenes: A spectroscopic study. *Macromolecules* 36 (3):692–697.

184. Fraleoni-Morgera, A., S. Marazzita, D. Frascaro, and L. Setti. 2004. Influence of a non-ionic surfactant on the UV–vis absorption features of regioregular head-to-tail poly(3-hexylthiophene) in water-based dispersions. *Synth Met* 147 (1–3):149–154.

185. Bernier, S., S. Garreau, M. Bera-Aberem, C. Gravel, and M. Leclerc. A versatile approach to affinitychromic polythiophenes. 2000. *J Am Chem Soc* 124 (42):12463–12468.

186. Reeves, B.D., C.R.G. Grenier, A.A. Argun, A. Cirpan, T.D. McCarley, and J.R. Reynolds. 2004. Spray coatable electrochromic dioxythiophene polymers with high coloration efficiencies. *Macromolecules* 37 (20):7559–7569.

187. Viinikanoja, A., J. Lukkari, T. Aeaeritalo, T. Laiho, and J. Kankare. 2003. Phosphonic acid derivatized polythiophene: A building block for metal phosphonate and polyelectrolyte multilayers. *Langmuir* 19 (7):2768–2775.

188. Wang, F., and Y.H. Lai. 2003. Conducting azulene–thiophene copolymers with intact azulene units in the polymer backbone. *Macromolecules* 36 (3):536–538.

189. Zhao, X., X. Hu, C.Y. Yue, X. Xia, and L.H. Gan. 2002. Synthesis, characterization and dual photochroic properties of *azo*-substituted polythiophene derivatives. *Thin Solid Films* 417 (1–2):95–100.

190. Nilsson, K.P.R., J. Rydberg, L. Baltzer, and O. Inganaes. 2004. Twisting macromolecular chains: Self-assembly of a chiral supermolecule from nonchiral polythiophene polyanions and random-coil synthetic peptides. *Proc Natl Acad Sci USA* 101 (31):11197–11202.

191. Roncali, J. 1992. Conjugated poly(thiophenes): Synthesis, functionalization, and applications. *Chem Rev* 92 (4):711–738.
192. Yassar, A., J. Roncali, and F. Garnier. 1989. Conductivity and conjugation length in poly(3-methylthiophene) thin films. *Macromolecules* 22 (2):804–809.
193. Roncali, J., A. Yassar, and F. Garnier. 1988. Electrosynthesis of highly conducting poly(3-methylthiophene) thin films. *J Chem Soc Chem Commun* 9:581–582.
194. Yang, C., F.P. Orfino, and S. Holdcroft. 1996. A phenomenological model for predicting thermo-chromism of regioregular and nonregioregular poly(3-alkylthiophenes). *Macromolecules* 29 (20):6510–6517.

10

Poly(3,4-Ethylene-dioxythiophene)—Scientific Importance, Remarkable Properties, and Applications

Stephan Kirchmeyer,
Knud Reuter, and
Jill C. Simpson

10.1 Introduction

10.1.1 A Brief Overview and History of the Development of Conductive Polymers

The fundamental discovery of polymeric organic conductors by Shirakawa, MacDiarmid, and Heeger in 1977 [1,2] marked the beginning of an era of dramatic growth in a field that earned its pioneers the Nobel Prize in Chemistry in 2000 [3–5]. The number of publications and patents granted in the field, dubbed by Heeger as the "fourth generation of polymers," has grown steadily over the last 25 y, and within the last 10 y inherently conductive polymers (ICPs) have developed from laboratory curiosities into mature industrial products for real commercial applications.

Early versions of ICPs, mostly based on oxidatively doped polyacetylenes (PAcs), faced several intrinsic obstacles that prevented their industrial commercialization. The material degrades readily in air, and no known good methods exist for making easily processable PAc polymers. These obstacles led to the investigation of alternative polymer backbone structures in the search for stable, processable, high conductivity ICPs. Over the years, several promising polymer types have emerged as potentially useful alternatives to PAc for commercial applications, including polypyrroles, polyanilines, and polythiophenes. The goal of this chapter is to focus on one polythiophene, specifically poly-3,4-ethylenedioxythiophene, commonly known as PEDT or PEDOT, with particular attention paid to PEDT's practical and interdisciplinary aspects in commercial applications.

10.1.2   The Development of Poly(3,4-Ethylenedioxythiophene) or PEDT

Early progress in polythiophene chemistry was achieved by the synthesis of mono- and dialkoxy-substituted thiophene derivatives developed by Leclerc [6] and industrial scientists at Hoechst AG [7–9]. However, most polymers of mono- and dialkoxythiophenes exhibited low conductivity in the oxidized, doped state. A breakthrough in this area was the synthesis of polymers of the bicyclic 3,4-ethylenedioxythiophene (EDT or EDOT) and its derivatives—electrochemically polymerized by Heinze et al. and chemically polymerized by Jonas et al. of the Bayer Corporate Research Laboratories [10,11]. In contrast to the nonbicyclic polymers of mono- and dialkoxythiophenes, PEDT has a very stable and highly conductive cationic "doped" state. The low HOMO–LUMO bandgap of conductive PEDT allowed the formation of a tremendously stable, highly conductive ICP [12]. Technical use and commercialization quickly followed; today ICPs based on PEDT are commercially available in multiton quantities.

   PEDT, which is also known under the trade name Baytron®, plays a dominant role in antistatic and conductive coatings, electronic components, and displays. In particular, widespread applications have been developed using the conducting properties of both the PEDT complex with polystyrene sulfonic acid (PEDT:PSS, Baytron® P) and the in situ polymerized layers of the EDT (Baytron® M) monomer, hereinafter referred to as in situ PEDT. Antistatic coating applications for PEDT:PSS include, e.g., for photographic films [13], electronics packaging, cathode-ray tube (CRT) screens, and LCD polarizer films. Conductive films of PEDT:PSS are found in inorganic electroluminescent devices and all-organic field-effect transistors (FETs), and PEDT:PSS layers function as the material of choice for hole injection in polymeric organic light-emitting diodes (OLEDs) and polymer photovoltaic cells. In situ PEDT is also well established in industry and used as a polymeric cathode material for solid aluminum, tantalum, and niobium capacitors, and as a conductive template for copper-through-hole plating of printed wiring boards.

10.2   Basics of PEDT

10.2.1   In Situ PEDT

10.2.1.1   Oxidative Polymerization

The physical characteristics of an in situ PEDT film are affected by a variety of factors. Variables in the synthetic method used to manufacture PEDT will affect the resulting polymer morphology, crystallinity, doping level, conductivity, molecular weight, etc. In general, the more uniform or crystalline, a PEDT film is, the higher its conductivity. Although the overall conversion of EDT to in situ PEDT should roughly be parallel to the oxidative polymerization mechanism discussed for alkylthiophenes [14] in general, a detailed look reveals that the total reaction path is rather complex.

   The reaction of EDT with iron(III) tosylate is summarized in Figure 10.1. The overall polymerization reaction can be separated into two principal steps: (1) oxidative polymerization of the monomer to the neutral polythiophene, and (2) oxidative doping of the neutral polymer to the conductive polycation.

10.2.1.2   PEDT Polymerization Reaction Kinetics

Detailed studies of the reaction kinetics [15] carried out at Bayer revealed a complex reaction mechanism (Figure 10.2). Kinetic parameters obtained in the study explain the polymerization rate and stoichiometry of oxidized PEDT. The first step in the mechanism, EDT oxidation to a radical cation,

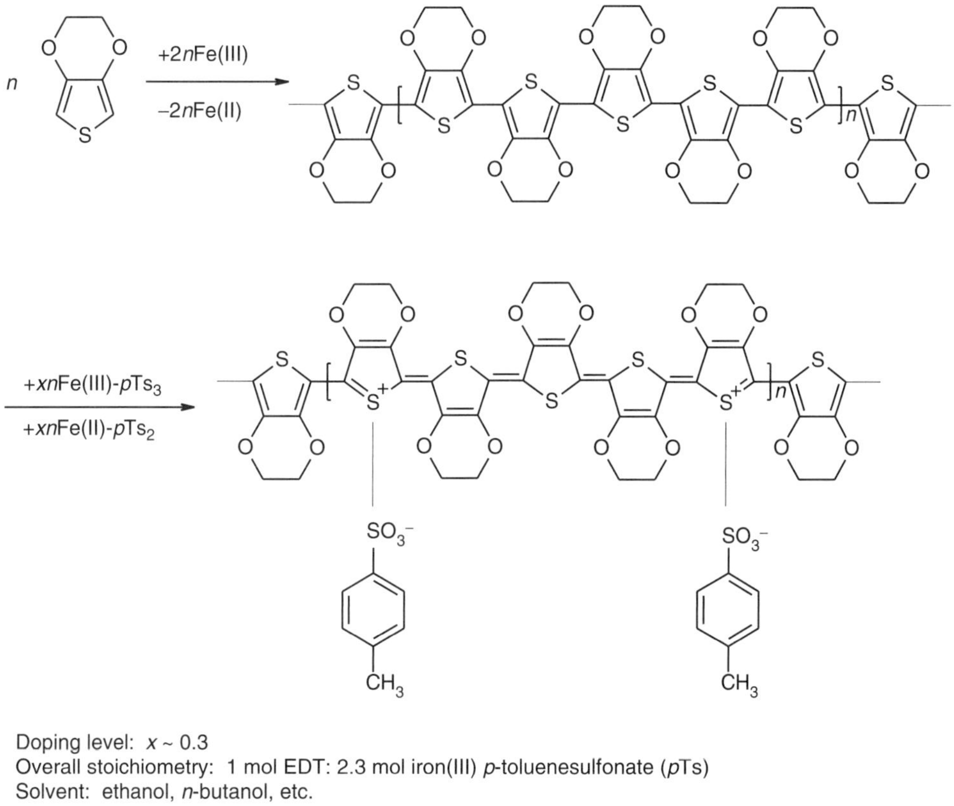

Doping level: $x \sim 0.3$
Overall stoichiometry: 1 mol EDT: 2.3 mol iron(III) p-toluenesulfonate (pTs)
Solvent: ethanol, n-butanol, etc.

FIGURE 10.1 Oxidation of ethylene-3,4-dioxythiophene with iron(III) toluenesulfonate.

is the rate-determining step (reaction rate constant $k_1 = 0.16$ $L^3/mol^3/h^1$). This is followed by dimerization of the free radical ($k_2 = 10^9$ L/mol/h). End group oxidation of EDT-oligomers is next, starting with dimer oxidation, which is faster than the monomer oxidation ($k_5 = 3000$ L/mol/h as proposed for all chain lengths) and which leads, after recombination of two radical cationic end groups, to higher oligomers with the same rate constant for recombination as for the monomer cations ($k_2 = 10^9$ L/mol/h). Finally, oligomers or polymers are doped by further oxidation. Figure 10.2 shows the paramagnetic polaron state as the first step of doping and as an intermediate to the highly conductive, diamagnetic bipolaron state (Figure 10.1). This bipolaron state is stabilized by the charge-balancing counterion—p-toluenesulfonate. A recent publication suggests a chain mechanism for the oxidative synthesis of aqueous PEDT microdispersions [16].

It is well known that acidification significantly enhances the reaction rate, but the detailed effects of protons on the reaction rates and mechanism have not yet been fully studied. It is known, however, that protic acids and a variety of Lewis acids can be used to catalyze an equilibrium reaction of EDT to the corresponding dimeric and trimeric compounds without further oxidation or reaction (Figure 10.3) [17].

Dimer yields of 25% to 50% by weight, along with about 5% to 15% of the stereoisomeric trimers, can be achieved by the addition of Lewis acids such as $AlCl_3$, $TiCl_4$, BF_3, $SnCl_4$, etc. to the reaction mixture. Strong protic acids such as trifluoroacetic or sulfuric acid are also effective for the formation of dimers and trimers.

The formed EDT dimer can then be converted to the oxidized form bis-(3,4-ethylenedioxythiophene) (BEDT or BEDOT) using quinoid oxidants such as chloranil or 2,3-dichloro-4,5-dicyano-1,4-benzoquinone [18].

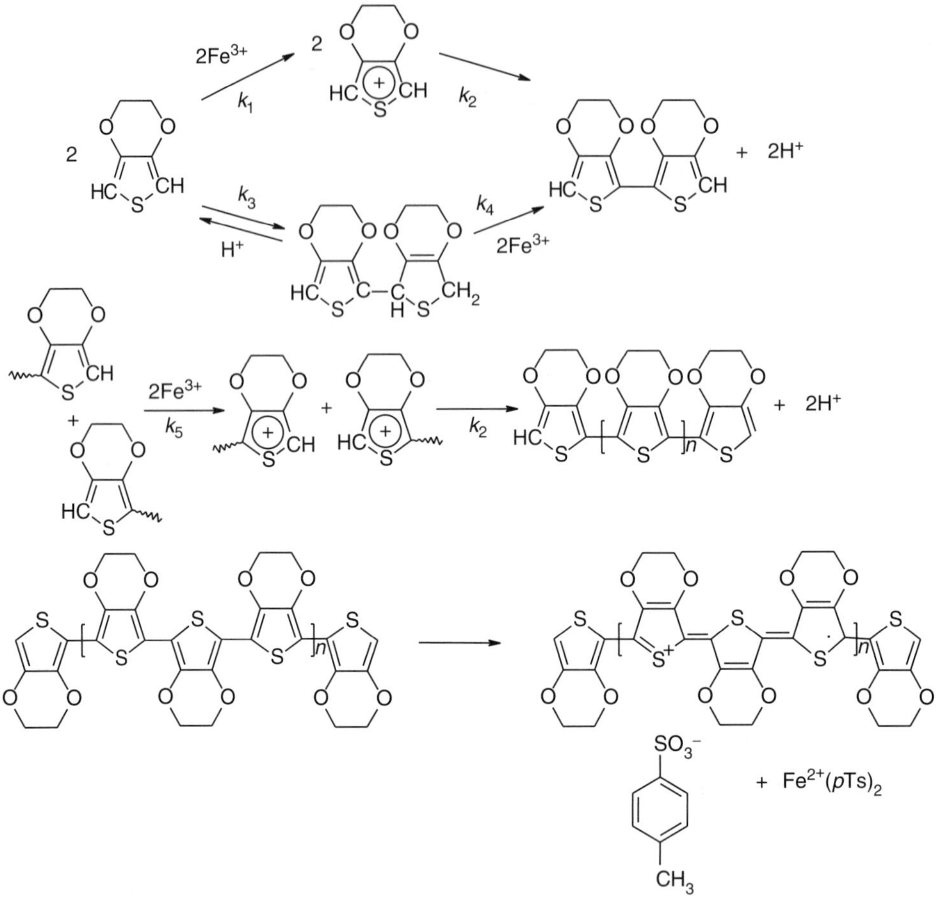

FIGURE 10.2 Proposed reaction mechanism for 3,4-ethylenedioxythiophene (EDT) oxidation to conductive PEDT-toluenesulfonate.

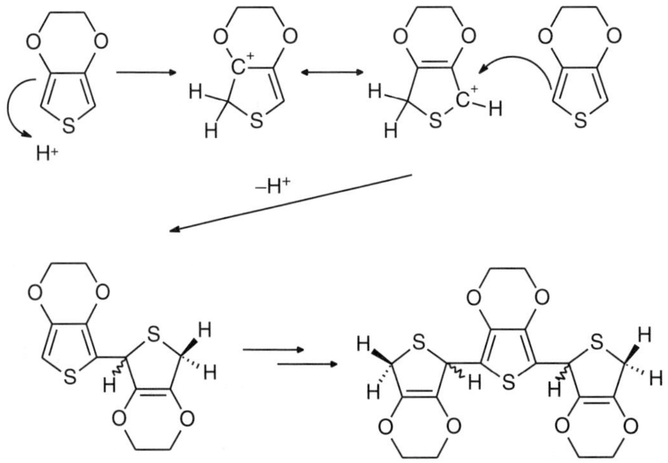

FIGURE 10.3 Reaction products of 3,4-ethylenedioxythiophene (EDT) di- and trimerization (only the *R*-isomer of the dimer and *R,S*-isomer of the trimer are shown).

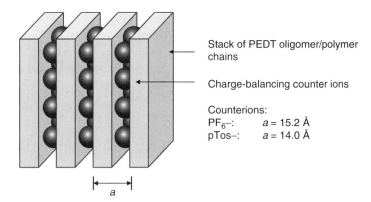

Stack of PEDT oligomer/polymer chains

Charge-balancing counter ions

Counterions:
PF_6^-: $a = 15.2$ Å
$pTos^-$: $a = 14.0$ Å

FIGURE 10.4   PEDT stacks: a = counterion-dependent layer distance between stacks.

10.2.1.3 Structure of In Situ PEDT Films

X-ray diffraction has been used to study the structure of in situ PEDT films. Specifically, PEDT polymeric salts of perchlorate [19], *p*-toluenesulfonate [20], and hexafluorophosphate [21] have been investigated. These studies reveal that PEDT chains are π-stacked with a characteristic repeat distance of 3.4 Å. The counterions of the PEDT salts are incorporated between the π-stacks leading to a structure resembling that is shown in Figure 10.4, in which the distance between the stacked layers is determined by the size of the counterion.

10.2.2 The PEDT:PSS-Complex

10.2.2.1 Template Polymerization with PSS

Since in situ PEDT polymers are quite insoluble in most commonly used solvents, in situ PEDT cannot be easily made into a processable, coatable solution. However, an industrially useful form of oxidized PEDT can be made by aqueous oxidative polymerization of the EDT monomer in the presence of a template polymer, usually polystyrene sulfonic acid (PSS or PSSA). PSS is a commercially available water-soluble polymer and can thus serve as a good dispersant for aqueous PEDT. Polymerization with the oxidant sodium peroxodisulfate yields a PEDT:PSS-complex in its conductive, cationic form (Figure 10.5).

The PSS in the complex has two functions. The first function is to serve as the charge-balancing counterion to the PEDT [11]. Without a PSS counterion in the system, the monomolecular thiolactone oxidation product 3,4-ethylenedioxy-2(5H)-thiophenone is formed instead of the desired PEDT:PSS polymer [22]. The second function of the PSS is to disperse the PEDT segments in the water. Although the resulting PEDT:PSS complex is not truly water soluble, the reaction forms a stable, easy-to-process, deep blue microdispersion of polymer gel particles.

Two key factors are important for understanding the nature of the PEDT:PSS-complex. First, the PEDT segments formed during polymerization are most likely oligomeric rather than polymeric. It has not been possible to directly observe high molecular weight PEDT polymers, and analyses of various PEDT-containing polymers through MALDI-TOF mass spectroscopy strongly support this assumption. Several measurements with PEDT:PSS or substituted PEDT derivatives, including neutral PEDT molecules which will be discussed later, indicate that the molecular weights of the individual PEDT molecules do not exceed 1000 to 2500 Da, or about 6 to 18 repeating units [23,24]. Second, the PEDT:PSS-complex has high stability. Ghosh and Inganäs demonstrated that the ionic species $PEDT^+$ and PSS^- could not be separated by standard capillary electrophoresis methods [25].

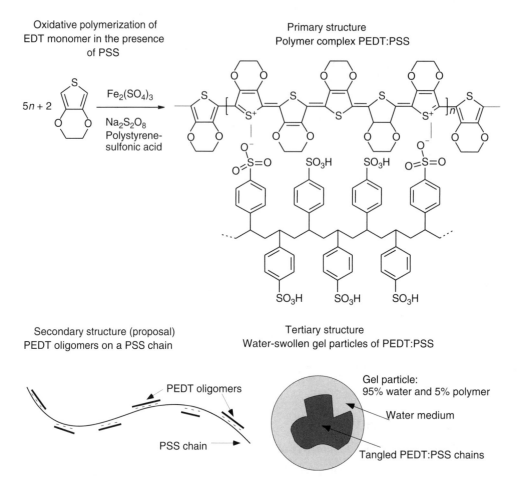

FIGURE 10.5   Synthesis, primary, secondary and tertiary structure of PEDT:PSS.

10.2.2.2   Structure of the PEDT:PSS Complex

The above evidence, taken together, indicates that it is therefore appropriate to draw a structural model for PEDT:PSS (Figure 10.5). In this model, oligomeric PEDT segments are tightly, electrostatically attached to PSS chains of much higher molecular weight. As in the PEDT stacks with monomeric counterions (Figure 10.4), high conductivity of PEDT:PSS can be attributed to stacked arrangements of the PEDT chains within a larger, tangled structure of loosely cross-linked, highly water-swollen PSS gel particles in films. These particles consist roughly of 90% to 95% water. The maximum solids content achievable, while maintaining a stable dispersion, depends on the PEDT:PSS ratio and increases with increasing PSS content. The PEDT:PSS gel particles have excellent film-forming properties and are easily processable into thin coatings on a variety of substrates. This processability led to the widespread availability of PEDT:PSS as a commercially useful material[26].

10.2.2.3   Properties of the PEDT:PSS Dispersion

Several typical properties of PEDT:PSS polymers that depend on the PEDT:PSS ratio are summarized in Table 10.1. To meet requirements for conductivity, antistatic grades of PEDT:PSS have relatively low PSS-contents, and therefore higher conductivity values. In contrast, PEDT:PSS grades designed for hole-injection in polymer OLEDs have larger PSS contents, smaller particles (Figure 10.6), and lower conductivities. Specifically, PEDT:PSS grades useful for passive matrix OLED displays have the lowest

TABLE 10.1 Typical PEDT:PSS Grades and Their Characteristics

PEDT:PSS Ratio	Solids Content, Approximate (%)	Conductivity, Approximate (S/cm)	Typical Application
1:2.5	1.3	10	Conductive coatings
1:2.5	1.3	1	Antistatics
1:6	1.5	10^{-3}	OLEDs
1:20	3	10^{-5}	Passive matrix displays

conductivity in order to prevent cross talk in multipixel devices. Increasing the PSS content logically reduces the electrical conductivity.

Particle size and film conductivity are also tightly linked: the smaller the particles, the lower the conductivity. This is demonstrated in Figure 10.6. The particle size can be varied by applying different high pressure shear rates to the PEDT:PSS dispersion during manufacture. Decreasing the median particle size not only decreases the conductivity, but also the viscosity.

10.2.2.4 Properties of Dry PEDT:PSS Films

Particle boundaries of the dried gel particles in a film contribute significantly to the overall resistivity of the film. The highest conductivities are achieved, therefore, when the particles are largest, which reduces the total number of particle boundaries. Alternatively, the PEDT:PSS film conductivity is enhanced when there is significant intermingling of individual gel particles, which reduces the effective number or "size" of the particle boundaries.

10.2.2.5 Neutral PEDT

For many years in the literature neutral, "undoped" PEDT with an intense blue color was postulated, and quite a few reports described unusual behavior for this "neutral" PEDT. For example, it is possible to reduce PEDT salts such as that of *p*-toluenesulfonate, tetrachloroferrate and PEDT:PSS to some extent [27] with strong reducing agents such as hydrazine, hydroxylamine, and others. Nevertheless, in all these

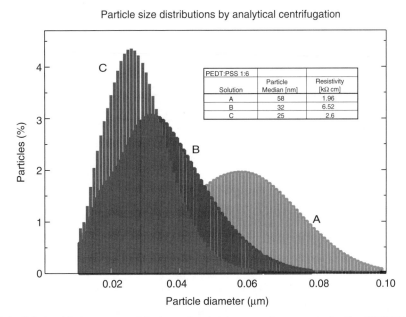

FIGURE 10.6 Relationship between particle size and resistivity, assuming a constant ratio of PEDT to PSS.

cases a truly neutral state of PEDT was never observed. Residual charged moieties—easily detectable by infrared (IR) spectroscopy—could not be removed completely [25].

Electrochemical reduction was also not successful at completely reducing the PEDT to its fully neutral form, although several attempts have been made. During electrochemical reduction the transparent, pale blue PEDT cations are reduced to a deep blue, reduced form [10,28]. Garreau et al. obtained a reduced PEDT that also exhibited the typical IR adsorption bands for residual charged PEDT structures at −1 V [29]. A more pronounced electrochemical reduction was achieved by the Inganäs group at −2.5 V. In this case, the results could be explained by an extensive reduction of PEDT, but certain characteristics of the material obtained led to the conclusion that complete reduction was improbable [30]. The residual conductivity of the reduced PEDT layers (5×10^{-4} S/cm) was several orders of magnitude higher than that theoretically expected for a fully neutral PEDT.

Additional attempts to directly synthesize undoped, neutral PEDT by organometallic means have also been made [25,31,32]. It is indeed possible to isolate fully neutral, unsubstituted PEDT by oxidation of the monomer EDT with iron(III) chloride while tightly controlling the stoichiometry of the oxidation during the course of the reaction, i.e., by maintaining far substoichiometric concentrations of $FeCl_3$ with respect to the doping reaction during the course of the synthesis. This allows isolation of the neutral PEDT from halogenated hydrocarbon solutions used as reaction media [33]. The PEDT obtained by this method is a reddish violet powder, soluble in chloroform, dichloromethane, or tetrahydrofuran and stable in air. Solutions are slowly oxidized by air, especially in the presence of acids, forming a blue precipitate. Gel permeation chromatography and MALDI-TOF mass spectra demonstrate the oligomeric to polymeric character of this neutral PEDT. Typical molecular weights (number average) of about 1000 to 1500 are found, which indicates PEDT chain lengths of between two and more than 20 EDT units.

10.2.2.6   Organosoluble PEDT Materials

Many attempts have been made to synthesize PEDT derivatives that are soluble in organic solvents. Some examples are EDT materials that incorporate long or medium chain alkyl [34,35], ether [36,37], or urethane [38] groups. Unfortunately, many of these organosoluble examples refer to either neutral, undoped PEDT, or to materials that are no more conductive than aqueous-based PEDT:PSS, and are therefore not generally useful for making highly conductive PEDT films.

10.3   Applications for PEDT

10.3.1   PEDT Coating Formulations

Applications using in situ PEDT must be distinguished from applications that utilize the PEDT:PSS complex. Since in situ PEDT films are formed by a chemical reaction, the process must be directly controlled during polymer formation. For example, kinetic parameters for in situ PEDT film formation can be controlled by addition of components, such as amines, that modify the reaction rate [39]. In general, in situ PEDT is used to obtain the highest conductivity. PEDT films with conductivities of 500 to 700 S/cm are easily attainable, and conductivities of 1000 S/cm and higher can be obtained with strict control of processing conditions. One caveat is that high conductivity in situ PEDT films are not generally highly transparent.

In contrast, the PEDT:PSS-complex is a "prefabricated" polymer, so film properties are instead adjusted by "formulation"—the addition of film-forming binders, surfactants, wetting agents, adhesion promoters, etc. The conductivity of unformulated PEDT:PSS layers is rather low. However, conductivity can usually be increased during formulation by as much as 1 to 2 orders of magnitude. This is clearly in contrast to what one might expect—that the addition of nonconductive components would lower the overall conductivity!

Common formulation components and their effects on PEDT:PSS films are:

1. *Low boiling solvents* like alcohols or low boiling ketones may be added to lower the surface tension of the PEDT:PSS dispersion, thereby increasing the wetting of substrates. These additives

are usually needed when plastic substrates such as polyester (PET), polycarbonate (PC), or polyethylene (PE) are to be coated with the aqueous PEDT:PSS dispersion.

2. *Surface active components* such as nonionic surfactants may be used for the same purpose as low boiling solvents. Because the solid contents of the aqueous PEDT:PSS dispersions are usually rather low (between 1.5% and 4%), and because nonvolatile additives will accumulate in the dried films, these additives are effective at extremely low loading levels. For example, a surfactant that has been added to the PEDT:PSS coating formulation at a 1% loading level will make up as much as 30% of the final, dried film.

3. *Polymeric binders* like waterborne PETs and polyurethanes are often used to improve the film's adhesive and mechanical properties. Especially, in cases where plastic substrates are coated by PEDT:PSS and subsequently mechanically treated after coating, such as by thermoforming, the PEDT:PSS needs a binder to maintain the overall conductivity. Choice of binder varies with the targeted properties of the final film. If, for example, an end-user desires a final film that is resistant to a particular solvent and well adherent to PC, then a binder designed for solvent-resistant coatings on PC should be chosen.

4. *Silanes and tetraalkylorthosilicates* are often used to increase the adhesion to the underlying substrate or to increase the hardness and wear resistance of the conducting film.

A very important class of additives for formulations is *high boiling solvents* and other *polar compounds*. Particularly useful are amides such as *N*-methylpyrrolidone and dimethylformamide, polyhydroxy compounds like ethylene glycol and sugar alcohols [40], and sulfoxides like dimethylsulfoxide. These solvents, often called "secondary dopants," [41] are used in small amounts to increase the conductivity of the final, dried film. The effects of these additives are independent of whether they remain in the film after drying or not. The mechanism of this conductivity enhancement has been discussed in depth [42,43]. The interpretation favored by these authors is that polar solvents at least partly dissolve the PEDT stacks in the PEDT:PSS complex, thereby creating an opportunity for a favorable morphological rearrangement and clustering of gel particles [44,45]. The rearrangement leads to a decreased resistance between dried gel particles, thus increasing the overall conductivity of the film.

10.3.2 Antistatic Coatings

Typical requirements for antistatic layers are:

1. A surface resistance of 10^5 to 10^9 Ω/sq
2. A transparent, practically colorless and haze-free appearance
3. Good adhesion to the substrate
4. Hardness of the coating

PEDT:PSS was first introduced in an industrial product as a roll-to-roll deposited antistatic layer onto biaxially oriented PET during photographic film production in order to avoid unwanted, stray electrostatic discharges within the photoactive layers during film processing. The use of PEDT:PSS as the conductive ingredient in the antistatic layer proved particularly advantageous because of its high conductivity, low-color, stability, processability, and moisture independent antistatic effect [46–48].

In a second commercial antistatic application, PEDT:PSS was used as the conductive ingredient in an outer surface antistatic layer on CRTs to avoid dust contamination during manufacture and use (Figure 10.7). In addition to the advantages described above, the PEDT:PSS layer was found to enhance optical contrast in the displays. PEDT:PSS antistatic layers for optical applications such as this require a low content of large particles, sufficient hardness, high adhesion on glass, and a surface resistance about 10^6 Ω/sq.

The other uses of PEDT:PSS for antistatic layers are numerous [43,44] and include antistatic gloves [49], carrier tapes, displays and video display panels [50–52], textiles [53], antistatic release films [54,55], protective films [56], recording tapes [57], and polarizers [58].

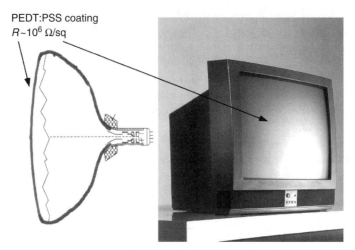

FIGURE 10.7   PEDT:PSS as an antistatic coating for cathode-ray tubes (CRTs) to avoid dust attraction.

10.3.3   Electrically Conducting Coatings

Conductive coatings are used to conduct a current used for the operation of a device, and in most cases the current density required for device operation is several orders of magnitude higher than that required for protection against antistatic discharges. Therefore, in these applications it is usually necessary to obtain layers with a maximum conductivity possible. In this regard, it is necessary to keep in mind that as the specific conductivity is increased, contact resistances can start to play a major role.

In many applications PEDT can function as a transparent conductor. For a transparent conductor the thickness is limited by the specific absorption of the conductor and the desired transparency of the conducting layer. As a first approximation the transparency is determined by calculating the absorption of PEDT in a given film, independent of whether this film was obtained from in situ generated PEDT, PEDT:PSS, or a formulation containing PEDT:PSS. In Figure 10.8, transmission is depicted vs. surface resistance for indium-tin oxide (ITO), a highly conductive PEDT:PSS grade, and

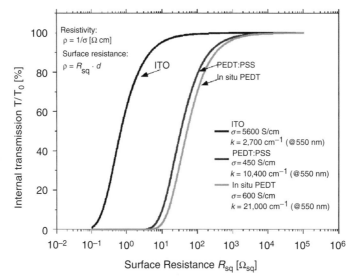

FIGURE 10.8   Relationship between conductivity and transparency of PEDT:PSS, compared to indium-tin oxide (ITO).

in situ PEDT layers. Surface resistance is adjusted by varying the layer thicknesses; consequently, a compromise must be made, near the upper part of the curves, between high transmission values and low surface resistances. The inset of Figure 10.8 indicates the specific conductivities and absorption coefficients (at 550 nm) of the three layers compared in this study.

10.3.4 PEDT:PSS as a Transparent Conductor in Electroluminescent Devices

Inorganic electroluminescent (EL) devices comprise a composite active layer of a zinc sulfide emitter and a dielectric such as barium titanate sandwiched between two conducting layers, one of which must be transparent. When an AC voltage of $\sim$100 V/400 Hz is applied, the zinc sulfide emits light, the hue of which can be tuned by the addition of appropriate doping agents. Figure 10.9 shows a schematic of the layered structure of typical EL device in which the transparent electrode is made with PEDT:PSS.

Due to the relatively high voltage used in typical EL devices, the specific resistance of the transparent conductor can be relatively low, at around 10^3 Ω/sq. Therefore, it is possible to replace the normally used ITO by a conducting polymer [59,60] such as PEDT:PSS. Even though the polymer has a lower specific conductance compared to ITO, all of the layers in the polymer-based device can be applied by printing techniques such as silkscreen printing. ITO, on the other hand, must usually be applied by costly sputtering deposition techniques.

In addition to lowered process costs, a technical advantage to the use of a PEDT:PSS-based printing paste for the transparent conductor is the flexibility of the contact layer. ITO is a brittle, inorganic material not ideally suited to destruction-free thermal deformation. In contrast, devices fabricated with transparent, conductive PEDT:PSS electrodes can be three-dimensionally [61] thermoformed after construction of the EL elements.

10.3.5 PEDT as Conducting Layer in Capacitors

One of the most widespread technical applications of PEDT is its use as a counterelectrode in solid aluminum, tantalum, and niobium electrolytic capacitors. As early as 1988, the application of PEDT as polymeric counter electrode in a tantalum capacitor was described in a patent application [62]. Since then, more than 100 patent applications have been filed which claim the use of PEDT in various capacitor configurations, and most major capacitor manufacturers today produce "polymer capacitors."

The motivations to introduce a polymeric counter electrode such as PEDT:PSS in solid metal capacitors are twofold. The first is to increase the conductivity of the cathode. A conducting polymer is roughly 1000 times more conductive than the traditionally used manganese dioxide cathode material, and penetrates more effectively into the porous metal oxide anode structures, creating more robust capacitor structures. The equivalent series resistance (ESR), an additive measure of the total resistance in a capacitor, is becoming more and more important to the industry as capacitors are used in higher and higher frequency applications. Currently, capacitor manufacturers strive to produce solid tantalum capacitors with mΩ-range ESR.

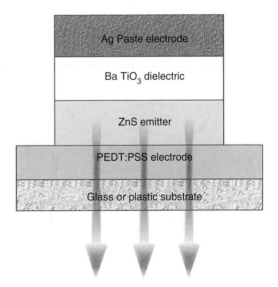

FIGURE 10.9 Schematic structure of an inorganic electroluminescent device with a transparent electrode made from a printing paste formulation of PEDT:PSS (screen printing technique).

The second motivation is to increase the safety of the capacitor during operation. Combining a metal like tantalum with manganese dioxide forms a strong redox couple separated by a dielectric layer of only a few microns in thickness. Even the smallest of defects in the dielectric layer can lead to leakage currents which cause increases in temperatures in the capacitor and ultimately ignition of the redox couple. When tantalum capacitors are pulse charged over their rated voltage, they tend to ignite, causing a small explosion.

In contrast, "self healing" properties have been attributed to PEDT:PSS cathodes in capacitors. As a potential mechanism, it has been postulated that high temperatures caused by leakage currents locally render PEDT:PSS nonconductive [63], thus preventing further leakage and capacitor failure.

The significance of contact resistivity is well demonstrated for capacitors. The increase in specific conductance in changing from manganese dioxide to a polymer (Figure 10.10) is only partly transferred to a decreased ESR due to contact resistances between polymer and outer graphite and silver layers in the devices.

The anode bodies of tantalum, aluminum, and most recently niobium capacitors are made of highly porous metals. These bodies are obtained either by sintering fine metal powders or electrochemical etching of thin foils. A thin dielectric layer is then electrochemically grown on the metal surface. Due to the porous structure of the anode bodies, the cathode polymer must be able to penetrate deep into the pores and coat all internal surfaces in order to utilize the full, potential capacitance of the anode.

Because of the need to deeply penetrate the porous anode structure, the PEDT counter electrode is preferentially formed by in situ PEDT with a chemical oxidant. The preferred oxidizing agent is often iron(III) toluenesulfonate. In most cases, monomer and oxidant are diluted in short-chain alcohols (e.g., ethanol or *n*-butanol) to obtain low viscosity impregnation solutions.

Impregnation is performed by dipping the anodes into the reactant solutions, using either a sequential dipping process in which the oxidant and monomer are deposited separately, or a one-pot dipping process in which the monomer and oxidant are deposited from the same solution. Although more uniform, and therefore more conductive, PEDT films can be deposited, a major obstacle to the one-pot method is limited pot life. Even at very low conversion rates, PEDT particles precipitate from the reaction solution and can clog the anode pores during impregnation. For this reason, most capacitor manufacturers have historically used a sequential dipping process to manufacture polymer capacitors (Figure 10.11).

However, the pot life can be enhanced by cooling the reaction mixture to +5°C, or more drastically by using iron(III) oxidant solutions that also contain polymerization retardant(s) [64,65]. The pot life of

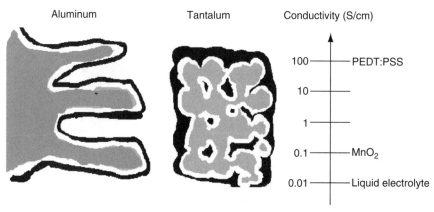

FIGURE 10.10   PEDT as cathode material (grey = metal; white = metal oxide dielectric; black = PEDT layer), boosting performance of solid electrolytic capacitors by better conductivity, lower equivalent series resistance (ESR), better impregnation, and avoidance of ignition.

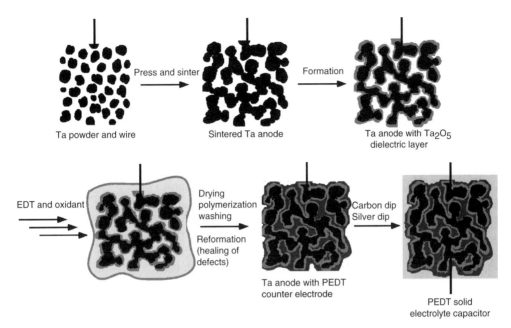

FIGURE 10.11   Process of forming the PEDT counter electrode in tantalum capacitors.

oxidant solutions that contain polymerization retardants can be increased from about 15 min to asmuch as 24 h. This extended pot life reduces obstacles to the use of a one-pot impregnation process in commercial polymer capacitor production [61], and many manufactures are now becoming more interested in one-pot processes. One-pot processes allow the manufacturer to have better control of the reaction stoichiometry and hence PEDT film quality. Not only does a one-pot process reduce EDT monomer yield loss, but it has also been shown that the capacitors made this way often exhibit better capacitance recovery than those made in sequential dipping processes [66].

Figure 10.12 shows the morphology of an impregnated tantalum capacitor with PEDT-layers of about 20 nm thickness.

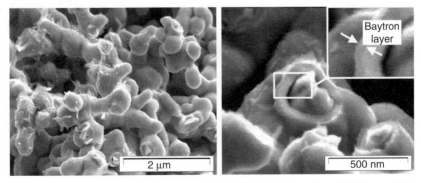

FIGURE 10.12   Microscopic photograph of PEDT-impregnated tantalum capacitor, PEDT layer thickness is about 20 nm (arrows).

10.3.6 Conducting Layers for Printed Wiring Board Manufacturing

In the mid 1990s Blasberg Oberflächentechnik GmbH, now Enthone GmbH, developed a process for the direct metallization of copper through-holes in printed wiring boards (PWB) and other multilayer structures called the Envision®* DMS-E process [67–69]. This process competes with the traditional chemical copper deposition procedures that employ strong complexing agents (like ethylenediamine-N,N,N', N'-tetraacetic acid) and formaldehyde. In Enthone's direct plating process a thin layer of PEDT conducting polymer is chemically deposited in the through-hole, and then this thin layer of PEDT is subsequently used as a conducting base for galvanostatic copper plating.

The EDT-based DMS-E process is a multistep process. The first step starts with the oxidation of the unstructured but predrilled copper substrate laminate. Potassium permanganate etches the organic substrate material while precipitating manganese dioxide on the oxidized surface. Following this surface activation, EDT is deposited on the manganese dioxide in the through-holes and polymerized. Usually, sulfonic acids are used as charge-balancing counter anions for the PEDT. Finally, copper is deposited electrochemically on the conductive PEDT layer, leading to fully plated through-holes (Figure 10.13).

10.3.7 PEDT Layers with "Electronic" Functions [70]

10.3.7.1 Poly(3,4-Ethylenedioxythiophene) as Hole Injection Layer in OLEDs

Since the pioneering work of Tang and VanSlyke [71], OLEDs have rapidly gained interest in the literature, industry, and media. The potential use of OLEDs in flat panel displays has encouraged

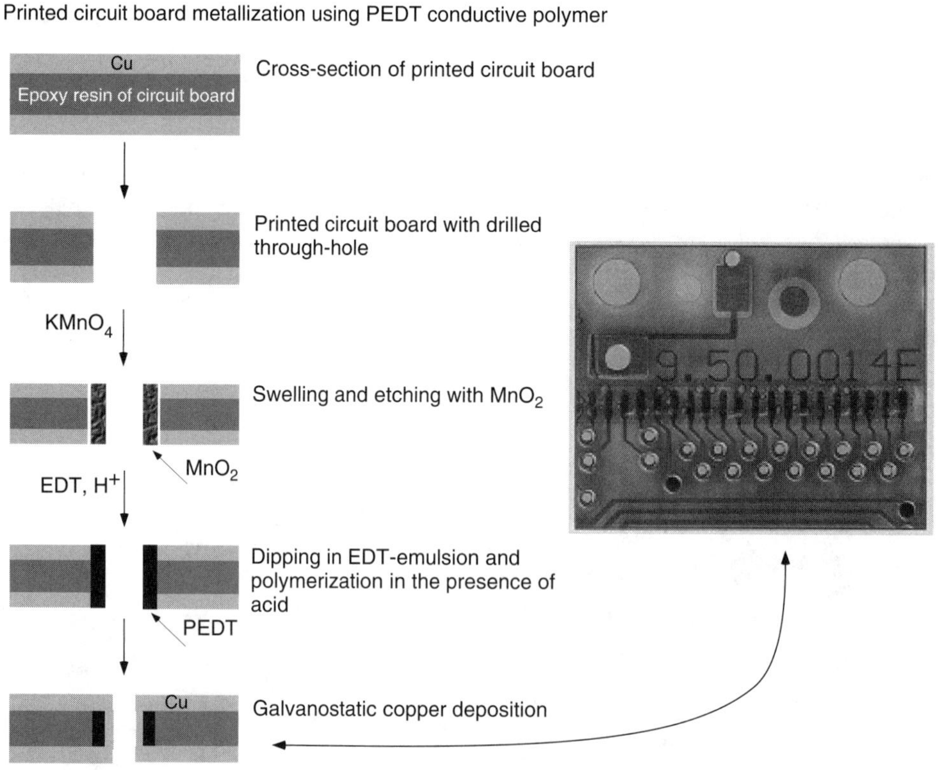

FIGURE 10.13   Scheme of printed circuit board metallization by the "DMS-E" process = direct metallization 3,4-ethylenedioxythiophene (EDT).

---

*A registered trademark of Enthone, Inc.

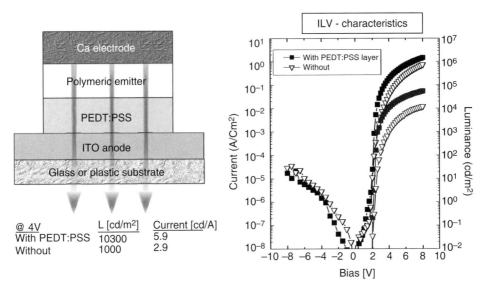

FIGURE 10.14 PEDT:PSS as hole injection layer for polymeric emitters: enhanced luminance, enhanced quantum-efficiency, and enhanced power-efficiency.

numerous scientific and industrial efforts to realize this new technology. The introduction of conjugated polymers as emitting materials by Friend and coworkers [72], which led to polymeric OLEDs or "PLEDs," was one of the milestones in the development of OLEDs.

The various functional layers in PLED devices are outlined in Figure 10.14. By applying a voltage across the device, electrons and holes are injected into the photoluminescent polymer, where they recombine to form excitons and emit light. Suitable, commonly used emissive polymers are conjugated materials such as poly(phenylenevinylene)s (PPVs) and polyfluorenes.

In 1992, Heeger et al. reported that an ICP such as polyaniline (Pani) could be used as a transparent anode material instead of ITO to make all-polymer, flexible OLEDs on plastic substrates [73,74]. Later work suggested that Pani could also serve as an interlayer between the ITO anode and the emissive polymer to improve hole injection in OLEDs, leading to devices with higher brightness and more efficiency [75–79]. PEDT was then introduced as a Pani alternative [80,81], and today is widely used commercially for hole injection in PLEDs. Figure 10.14 shows the *I–V* characteristics of an OLED device employing a PEDT:PSS hole injection layer. The shift to lower voltages indicates superior efficiency characteristics.

In addition to its role in improvement of hole injection, the PEDT:PSS can also effectively smoothen the normally rough ITO anode surface, thereby decreasing the occurrence of micro-shorts in OLED devices. The results depicted in Figure 10.15 [82] show the increased lifetime of an OLED device as well as the occurrence of fewer defects when a PEDT:PSS layer is employed. The lifetime improvement is attributed mainly to fewer micro-shorts, which manifest in devices as dark spots.

10.3.7.2 PEDT for Photovoltaics and Sensors

The fundamental physics of polymeric photovoltaic (PV) cells and sensors are related to that of polymeric LEDs. In polymeric LEDs, injected holes and electrons recombine in the active layer to generate photons. In PVs, charges are instead captured before recombination in the active layer to generate voltage. Currently a variety of device set-ups using various functional polymers for PVs are under development, among which one of the most promising and widely studied are titania-sensitized PVs [83]. A typical PV device (Figure 10.16) consists of a hole injection layer, often made of PEDT, and an active semiconducting layer sandwiched between two conductive electrodes, one of which must be transparent and is usually fabricated from ITO.

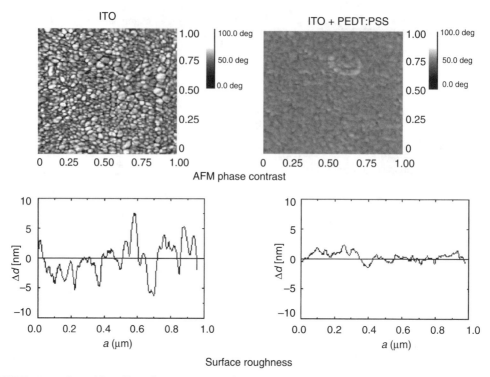

FIGURE 10.15   Smoothing effect of PEDT:PSS on indium-tin oxide (ITO).

While the task of photovoltaic cells is to generate energy, a subclass of these devices designed to simply detect light is photosensors. In this case, the conductivity requirement on the transparent conductor is lower. Hence, in addition to its electronic function as hole-injector, PEDT can be used in place of the transparent ITO electrode [84]. In this case, a highly conducting formulation of PEDT:PSS or in situ PEDT is used [85,86].

10.3.7.3 Conductive Layer in All-Organic Thin Film Transistors

Integrated circuits based on organic materials are another type of electronic device under development. Currently, target applications are low-end electronics [87] and drive electronics for flexible displays [88,89].

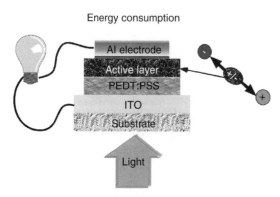

FIGURE 10.16   Photovoltaic element based on polymers, including PEDT:PSS.

Since organic thin film transistors (TFT) are targeted for inexpensive applications, low-cost fabrication methods with alternative materials are of interest. Fully patterned all-organic TFTs have been fabricated using PEDT:PSS by conventional structuring techniques [90], ink-jet processing [91–93], and other patterning techniques such as line patterning [94]. In most cases, gate, source, and drain contacts were prepared from PEDT:PSS (Figure 10.17), but some groups have also proposed using PEDT:PSS as the active, semiconducting, channel material [88,95].

A frequently used organic gate dielectric material is poly-4-vinyl phenol (PVP) cross-linked with melamine resin. In hybrid

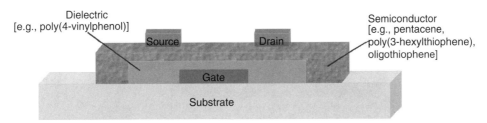

FIGURE 10.17 All-organic thin film transistor with PEDT:PSS source, drain and gate electrodes.

organic- and inorganic TFTs, thermally grown silicon dioxide is employed. Among the most widely employed semiconductors are pentacene and thiophene-based oligomers, which are generally vacuum-deposited, and poly-3-alkylthiophenes and poly(bithiophene-co-fluorene)s, which can be deposited by wet processing techniques such as printing or spin coating.

The current status of the research indicates that the performance of all-organic TFTs is approaching that of traditional, inorganic TFTs made from amorphous silicon, even though performance at higher temperatures and long-term stability is still a challenge to be addressed. Early results in organic TFT work indicate that the substitution of inorganic materials with organic ones for gate dielectrics and metal contacts does not necessarily result in reduced electrical performance. The largest reported charge carrier mobilities for organic TFTs with conducting polymer source and drain are about 0.05 cm²/V s for pentacene all-organic TFTs.

10.3.7.4 Utilizing the Redox Behavior of PEDT—Electrochromic Windows Based on PEDT:PSS

The perceived benefits of electrochromic glazing of architectural and automotive window-glass have been summarized in various reports. For the last two decades, the main effort in the development of electrochromic windows has focused on tungsten trioxide (WO_3) as the electrochromic material [96–98]. Still, commercialization of electrochromic windows struggles with high manufacturing costs, long-term optical stability, and high mechanical demands due to safety requirements. Nevertheless, introduction into the market has begun.

As demonstrated by the Inganäs group, the spectrum of PEDT:PSS, a cathodically coloring polymer, can be changed gradually by electrochemical means at different voltages [26]. PEDT:PSS therefore exhibits electrochromic properties that can be utilized in appropriate devices (Figure 10.18) when an electrical voltage is applied [26,99–101]. The electrochromic behavior is due to an electron transfer reaction that takes place during electrochemical oxidation and reduction of the polymer.

Starting from the oxidized state, PEDT:PSS accepts electrons from the cathode at a voltage of 2.5 V and is thereby switched to the reduced state. As a consequence of this electrochemical reduction step, the color turns from slightly blue and transparent to deep blue (Figure 10.19). Reverse biasing with a voltage of −2.5 V leads to the reverse reaction, where the polymer in the reduced state will be reoxidized on the other electrode. The original transparency, which depends on the layer thickness, is reversibly restored. Electrochromic layers of poly(3,4-dioxythiophene)s have been found to exhibit extremely high coloration efficiencies of up to 1400 cm²/C over narrow (∼100 nm) wavelength ranges, and to retain up to 60% of their optical response after 10,000 deep, double potential steps [102].

Optical spectra (UV/Vis/NIR) of electrochemically polymerized PEDT in both its reduced and fully oxidized state have been published as early as 1994 [10,103], demonstrating the nearly colorless, sky-blue conductive doped state. In addition, optical spectra of the PEDT:PSS complex have been measured in the same manner, exhibiting a λ_{max} value of about 600 nm for the reduced form [26]. The exact chemical nature of the deep blue, reduced state of PEDT and PEDT:PSS used in electrochromic devices has long been under discussion. As previously discussed in this review, fully undoped, reduced PEDT is not blue, but is reddish purple, and it cannot be readily obtained by electrochemical means [27,28]. Some recent

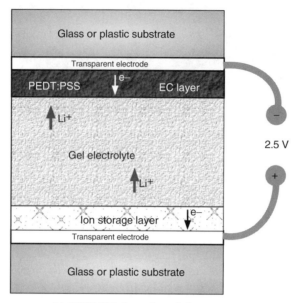

FIGURE 10.18   Device structure with PEDT:PSS electrochromic layer.

electron paramagnetic resonance investigations suggest a radical character of the blue electrochemically reduced PEDT [104,105], even after the most extensive dedoping achievable. On the contrary, a chemically synthesized neutral, completely undoped PEDT is clearly not paramagnetic and exhibits a λ_{max} value a bit below 500 nm [30], whether prepared by organometallic [30] or oxidative synthesis [31].

Reduction of PEDT during coloring requires the incorporation of cations, such as lithium ions, into the PEDT:PSS layer to compensate for the negatively charged sulfonate groups of the polyanionic PSS. The counterions migrate from an ion conductive electrolyte layer at certain speeds, limiting the overall switching speeds of the devices. The exact switching time from fully colored to nearly transparent and back is in the range of seconds when liquid electrolyte systems are used, and in the range of minutes when solid electrolyte systems are used. For safety reasons and to ensure the mechanical stability of the device stack, swollen polymeric electrolytes, called "gel electrolytes," are often preferred in commercial applications of electrochromic devices.

A gel electrolyte consists of an ion conducting polymer, such as polyoxyethylene or a polymer with polyoxyethylene segment, and an ion source. Hexafluorophosphate or tetrafluoroborate salts of lithium are commonly employed as the ion source. Other common components of the gel are liquid electrolytes like ethylene carbonate or propylene carbonate, both of which are used to increase the overall ion conductivity. Cross-linking agents, such as photo cross-linked acrylates or polyurethanes, can be incorporated to improve the mechanical stability of gel electrolytes.

FIGURE 10.19   Electrochromic glasses based on PEDT:PSS.

Also required in an electrochromic cell is an ion storage layer consisting of a redox active material. A variety of materials, including various metal oxides like titanium-doped ceria (CeO_2/TiO_2) have been proposed. While usually the ion storage layer alters its redox state without changing its absorption spectrum, other concepts employed to increase optical contrast in devices include the combination of PEDT:PSS with reversed switching metal oxides like tungsten trioxide (WO_3) [106] or other conducting polymers [107]. Unfortunately, these concepts often suffer from the fact that many ion storage layers differ in hue in comparison to the PEDT:PSS. As a consequence undesired color changes occur during the switching process.

An interesting approach to the fabrication of PEDT-based electrochromic cells is a setup that combines two PEDT:PSS layers [108]. This concept utilizes the nonlinear coloring behavior of PEDT:PSS, and the performance of the device is based on the fact that the sum of the absorption of both layers runs through a minimum. The drawback, however, to this dual-layer PEDT:PSS device is reduced optical contrast.

To improve performance, many PEDT derivatives used either alone or in combination have been proposed. While the electro-optical properties of WO_3 are fixed, the colors and hue of conducting polymers may be altered by modification of the monomers. For example, Reynolds has studied the electrochromic properties of a multitude of EDT derivatives [100,109], and recently reported an all-polymer electrochromic device based on two different PEDT derivatives [110]. Depending on their chemical structure, the various PEDT derivatives exhibit different colors upon switching from the oxidized to the reduced state. For example, poly(tetradecylethylenedioxythiophene) (C_{14}-EDT) is similar in switching to PEDT from transparent to blue, but it has an enhanced optical contrast.

10.4   Summary

PEDT has been widely studied over the last 20 y, and due to its commercial availability on a large scale, it has today found its way into myriad commercial applications. New applications for PEDT, whether utilizing the in situ polymerized form or the polymeric complex PEDT:PSS, continue to be developed.

Acknowledgments

This chapter has been published in a very similar form in the *Journal of Materials Chemistry*, 2005, 15:2077. The authors express their thanks to the Royal Society of Chemistry for the permission to reprint portions of the manuscript here.

The authors express their thanks to Dr. Klaus Wussow and Dr. Friedrich Jonas (Application Technology Baytron® PEDT:PSS), Dr. Andreas Elschner (Development Organic LED), Dr. Ursula Tracht (kinetic studies), Dr. Alexander Karbach (x-ray, AFM), Dr. Udo Merker (Development Capacitor Technology) and Dr. Helmut-Werner Heuer (electrochromic windows), and to Prof. Peter Bäuerle for helpful comments.

References

1. Chiang, C.K., C.R. Fincher, Y.W. Park, A.J. Heeger, H. Shirakawa, E.J. Louis, S.C. Gau, and A.G. MacDiarmid. 1977. *Phys Rev Lett* 39 (17):1098.
2. Shirakawa, H., E.J. Louis, A.G. MacDiarmid, C.K. Chiang, and A.J. Heeger. 1977. *J Chem Soc Chem Commun* 16:578.
3. (a) Shirakawa, H. 2001. *Angew Chem Int Ed* 40 (14):2575; (b) Shirakawa, H. 2002. *Synth Met* 125 (1):3.
4. (a) MacDiarmid, A. 2001. *Angew Chem Int Ed* 40 (14):2581; (b) MacDiarmid, A. 2002. *Synth Met* 125 (1):11.
5. (a) Heeger, A.J. 2001. *Angew Chem Int Ed* 40 (14):2591; (b) Heeger, A.J. 2002. *Synth Met* 125 (1):23.
6. Daoust, G., and M. Leclerc. 1991. *Macromolecules* 24 (2):455.
7. Feldhues, M., T. Mecklenburg, P. Wegener, and G. Kämpf. 1986. EP 257 573 (Hoechst AG), Prior: 26 August 1986.

8. Kämpf, G., and M. Feldhues. 1987. EP 292 905 (Hoechst AG), Prior: 26 May 1987.

9. Feldhues, M., G. Kämpf, H. Litterer, T. Mecklenburg, and P. Wegener. 1989. *Synth Met* 28 (1–2):C487.

10. Jonas, F., G. Heywang, W. Schmidtberg, J. Heinze, and M. Dietrich. 1988. EP 339 340 (Bayer AG), Prior: 22 April 1988.

11. Heywang, G., and F. Jonas. 1992. *Adv Mater* 4 (2):117.

12. Pei, Q., G. Zuccarello, M. Ahlskog, and O. Inganäs. 1994. *Polymer* 35 (7):1347.

13. Jonas, F., and W. Krafft. 1990. EP 440 957 (Bayer AG), Prior: 8 February 1990.

14. Fichou, D. 1999. *Handbook of oligo- and polythiophenes*, Weinheim, Germany: Wiley-VCH. ISBN: 3-527-29445-7.

15. (a) Tracht, U. Personal communication; (b) Kirchmeyer, S. 2002. *5th International Symposium on Functional π-Electron Systems*, Ulm, Germany.

16. Groenendaal, L., F. Louwet, P. Adriaensens, R. Carleer, D. Vanderzande, and J. Gelan. 2002. *Polym Mater Sci Eng* 86:52.

17. Reuter, K., V.A. Nikanorov, and V.M. Bazhenov. 2002. EP 1 375 560 (H.C. Starck GmbH), Prior: 28 June 2002.

18. Reuter, K. European patent filed (H.C. Starck GmbH).

19. Granström M., and O. Inganäs. 1995. *Polymer* 36 (15):2867.

20. Aasmundtveit, K.E., E.J. Samuelsen, L.A.A. Petterson, O. Inganäs, T. Johansson, and R. Feidenhans'l. 1999. *Synth Met* 101 (1–3):561.

21. Niu, L., C. Kvarnström, K. Fröberg, and A. Ivaska. 2001. *Synth Met* 122 (2):425.

22. Reuter, K. (unpublished).

23. Jonas, F., L. Groenendaal, and J. Pausch. (unpublished).

24. Reuter, K., A. Karbach, H. Ritter, and N. Wrubbel. 2003. EP 1 440 974 A2 (Bayer AG), Prior: 21 January 2003.

25. Ghosh, S., and O. Inganäs. 1999. *Synth Met* 101 (1–3):413.

26. Jonas, F., and G. Heywang. 1994. *Electrochim Acta* 39 (8/9):1345.

27. Yamamoto, T., and M. Abla. 1999. *Synth Met* 100 (2):237.

28. Gustafsson, J.C., B. Liedberg, and O. Inganäs. 1994. *Solid State Ion* 69:145.

29. Garreau, S., G. Louam, S. Lefrant, J.P. Buisson, and G. Froyer. 1999. *Synth Met* 101 (1– 3):312.

30. Johansson, T., L.A.A. Petterson, and O. Inganäs. 2002. *Synth Met* 129 (3):269.

31. Tran-Van, F., S. Garreau, G. Louarn, G. Froyer, and C. Chevrot. 2001. *Synth Met* 119 (1–3):381.

32. Tran-Van, F., S. Garreau, G. Louarn, G. Froyer, and C. Chevrot. 2001. *J Mater Chem* 11:1378.

33. Reuter, K., and S. Kirchmeyer. 2001. EP 1 327 645 (Bayer AG), Prior: 27 December 2001.

34. Kumar, A., and J.R. Reynolds. 1996. *Macromolecules* 29 (23):7629.

35. Shiraishi, K., T. Kanbara, T. Yamamoto, and L. Groenendaal. 2001. *Polymer* 42 (16):7229.

36. Schottland, P., O. Stephan, P.-Y. Le Gall, and C. Chevrot. 1998. *J Chim Phys* 95 (6):1258.

37. Luebben, S., B. Elliot, and C. Wilson. 2001. US Application 2003/0088032A1 and WO03/018648A1 (TDA Research), Prior: 31 August 2001.

38. Reuter, K. 2002. EP 1 352 918 (Bayer AG), Prior: 10 April 2002.

39. Staring, A.G.J., and D. Braun. 1994. WO 96 08047 (Philips Electronics N.V.), Prior: 6 September 1994.

40. Jonas, F., A. Karbach, B. Muys, E. van Thillo, R. Wehrmann, A. Elschner, and R. Dujardin. 1994. EP 686 662 (Bayer AG), Prior: 6 May 1994/3 March 1995.

41. Martin, B.D., N. Nikolov, S.K. Pollack, A. Saprigin, R. Shashidar, F. Zhang, and P.A. Heiney. 2004. *Synth Met* 142 (1–3):187.

42. Pettersson, L.A.A., S. Ghosh, and O. Inganäs. 2002. *Org Electron* 3 (3–4):143.

43. Jönsson, S.K.M., J. Birgerson, X. Crispin, G. Greczynski, W. Osikowicz, A.W. Denier van der Gon, W.R. Salaneck, and M. Fahlman. 2003. *Synth Met* 139 (1):1.

44. Ouyang, J., Q. Xu, C.-W. Chu, Y. Yang, G. Li, and J. Shinar. 2004. *Polymer* 45 (25):8443.

45. Timpanaro, S., M. Kemerink, F.J. Touwslager, M.M. De Kok, and S. Schrader. 2004. *Chem Phys Lett* 394 (4–6):339.

46. Jonas, F., W. Krafft, and B. Muys. 1995. *Macromol Symp* 100:169.

47. Jonas, F., and J.T. Morrison. 1997. *Synth Met* 85 (1–3):1397.

48. Jonas, F., and K. Lerch. 1997. *Kunststoffe* 87 (10):1401.

49. Thiess, A. 2001. *Deutsche Gebrauchsmusterschrift* (2001), DE 20021226 U1, Prior: 19 April 2001.

50. Hotta, O., S. Soga, and N. Sonoda. 1988. JP 02129284 (Matsushita Electric Ind. Co.), Prior: 8 November 1988.

51. Yoon, S.Y., C.H. Lee, and H.S. Son. 1998. KR 2000 009403 (Orion Electric Co.), Prior: 24 July 1998.

52. Tong, H., and J. Hu. 1996. CN 1 171 617 A (Zhonghua Kinescope Co.), Prior: 23 July 1996.

53. Hardtke, G., and H. Fuchs. 1999. TECHNOMER '99, 16th Fachtagung über Verarbeitung und Anwendung von Polymeren, Chemnitz, Germany. *CAN* 137:80211.

54. Kawashima, T., and T. Sekiya. 2002. JP 2002 241613 A2 (Sony Chemical Corp.), Prior: 28 August 2002.

55. Kawashima, T., and T. Sekiya. 2003. JP 2003 251756 A2 (Sony Chemical Corp.), Prior: 9 September 2003.

56. Morimoto, Y. 1998. JP 2000 026817 A2 (Teijin Ltd.), Prior: 14 July 1998.

57. Nittel, F., H. Randolph, and W. Himmelmann. 1993. EP 554 588 (Agfa AG), Prior: 23 December 1993.

58. Shinohara, H. 2002. JP 2003 246874 A2 (HS Planning Y.K.), Prior: 26 February 2002.

59. Andriessen, H. 2002. EP 1 231 251 A1 (Agfa-Gevaert), Prior: 14 August 2002.

60. Saito, A. 2000. JP 2002 124391 A2 (Seiko Precision Inc.), Prior: 8 August 2000.

61. Enz, E. 2001. WO 2003 037039 A1 (Lumitec AG), Prior: 24 October 2001.

62. Jonas, F., G. Heywang, and W. Schmidtberg. 1988. EP 340 512 (Bayer AG), Prior: 30 April 1988.

63. Prymak, J.D. 2001. Improvements with polymer cathodes in aluminum and tantalum capacitors. Vol. 2. Applied Power Electronics Conference and Exposition, 16th Annual IEEE Anaheim, p. 1210.

64. Wussow, K., U. Merker, et al. European patent filed (H.C. Starck GmbH).

65. Merker, U., K. Wussow, S. Kirchmeyer, C. Schnitter, and K. Lerch. 2003. *Proceedings of the 17th Passive Components Conference CARTS Europe*, Stuttgart, p. 79.

66. Merker, U., K. Reuter, K. Wussow, S. Kirchmeyer, and U. Tracht. 2002. *Proceedings of the 16th Passive Components Conference CARTS Europe*, Nice, p. 71.

67. Hupe, J., G.D. Wolf, and F. Jonas. 1995. *Galvanotechnik* 86 (10):3404.

68. Wolf, G.-D., F. Jonas, and R. Schomaecker. 1994. EP 707 440 (Bayer AG), Prior: 12 October 1994.

69. Kirchmeyer, S., and F. Jonas. 1999. WO 2000 045625 (Bayer AG), Prior: 27 January 1999.

70. Martin, H.-D. 2001. *Chem Lab Biotech* 52 (10):364.

71. Tang, C.W., and S.A. VanSlyke. 1987. *Appl Phys Lett* 51:913.

72. Burroughes, J.H., D.D.C. Bradley, A.R. Brown, R.N. Marks, K. Mackay, R.H. Friend, P.L. Burns, and A.B. Holmes. 1990. *Nature* 347 (6293):539.

73. Gustafsson, G., Y. Cao, G.M. Treacy, F. Klavetter, N. Colaneri, and A.J. Heeger. 1992. *Nature* 357:477.

74. Gustafsson, G., G.M. Treacy, Y. Cao, F. Klavetter, N. Colaneri, and A.J. Heeger. 1993. *Synth Met* 55–57:4123.

75. Heeger, A.J., I.D. Parker, and Y. Yang. 1994. *Synth Met* 67:23.

76. Yang, Y., and A.J. Heeger. 1994. *Appl Phys Lett* 64 (10):1245.

77. Yang, Y., and A.J. Heeger. 1994. *Mol Cryst Liq Cryst* 256:537.

78. Yang, Y., E. Westerweele, C. Zhang, P. Smith, and A.J. Heeger. 1995. *J Appl Phys* 77 (2):694.

79. Karg, S., J.C. Scott, J.R. Salem, and M. Angelopoulos. 1996. *Synth Met* 80:111.

80. Granström, M., M. Berggren, and O. Inganas. 1995. *Science* 267:5203.

81. Scott, J.C., S.A. Carter, S. Karg, and M. Angelopoulos. 1997. *Synth Met* 85:1197.

82. Elschner, A., F. Bruder, H.-W. Heuer, F. Jonas, A. Karbach, S. Kirchmeyer, S. Thurm, and R. Wehrmann. 2000. *Synth Met* 111–112:139.

83. Gazzotti, W.A., A.F. Nogueira, E.M. Girotto, L. Micaroni, M. Martini, S. das Neves, and M.-A. De Paoli. 2001. *Handbook of advanced electronic and photonic materials and devices*, Vol. 10, ed., H.S. Nalwa. San Diego, CA: Academic Press, pp. 53–58.

84. Zhang, F., M. Johansson, M.R. Andersson, J.C. Hummelen, and O. Inganäs. 2002. *Adv Mater* 14 (9):662.

85. Carter, S.A., M. Angelopoulos, S. Karg, P.J. Brock, and J.C. Scott. 1997. *Appl Phys Lett* 70 (16):2067.

86. Aemouts, T., P. Vanlaeke, W. Geens, J. Poortmans, P. Heremans, S. Borghs, R. Mertens, R. Andriessen, and L. Leenders. 2004. *Thin Solid Films* 451–452:22.

87. Wu, Y., B. Ong, P. Liu, Y. Li, and S. Gardner. 2004. Abstracts of Papers, 228th ACS National Meeting, Philadelphia, USA.

88. Andersson, P., D. Nilsson, P.-O. Svensson, M. Chen, A. Malmstrom, T. Remonen, T. Kugler, and M. Berggren. 2002. *Adv Mater* 14 (20):1460.

89. Gelinck, G.H., H.E.A. Huitema, E. van Veenendaal, E. Cantatore, L. Schrijnemakers, J.B.P.H. van der Putten, T.C.T. Geuns, M. Beenhakkers, J.B. Giesbers, B.-H. Huisman, E.J. Meijer, E.M. Benito, F.J. Touwslager, A.W. Marsman, B.J.E. Van Rens, and D.M. De Leeuw. 2004. *Nat Mater* 3:106.

90. Halik, M., H. Klauk, U. Zschieschang, T. Kriem, G. Schmid, W. Radlik, and K. Wussow. 2002. *Appl Phys Lett* 81 (2):289.

91. Kawase, T., H. Sirringhaus, R.H. Friend, and T. Shimoda. 2001. *Adv Mater* 13 (21):1601.

92. Sirringhaus, H., T. Kawase, R.H. Friend, T. Shimoda, M. Inbasekaran, W. Wu, and E.P. Woo. 2000. *Science* 290 (5499):2123.

93. Sirringhaus, H., R.H. Friend, and T. Kawase. 2001. WO 2001 047043 A1 (Plastic Logic Ltd.), Prior: 28 June 2001.

94. Okuzaki, H., M. Ishihara, and S. Ashizawa. 2003. *Synth Met* 137 (1–3):947.

95. Epstein, A.J., F.-C. Hsu, N.-R. Chiou, and V.N. Prigodin. 2003. *Synth Met* 137 (1–3):859.

96. Svensson, J.S.E.M., and C.G. Granqvist. 1985. *Thin Solid Films* 126 (1–2):31.

97. Batchelor, R.A., M.S. Burdis, and J.R. Siddle. 1996. *J Electrochem Soc* 143 (3):1050.

98. Green, M. 1999. *Ionics* 5 (3–4):161.

99. Pielartzik, H., H.-W. Heuer, R. Wehrmann, and T. Bieringer. 1999. *Kunststoffe* 89 (9):135.

100. Pielartzik, H., H.-W. Heuer, R. Wehrmann, and T. Bieringer. 1999. *Engin Plast* 89 (9):49 (Translation of Ref. [99]).

101. Heuer, H.-W., R. Wehrmann, and S. Kirchmeyer. 2002. *Adv Funct Mater* 12 (2):89.

102. Argun, A.A., A. Cirpan, and J. R. Reynolds. 2003. *Adv Mater* 15 (16):1338.

103. Dietrich, M., J. Heinze, G. Heywang, and F. Jonas. 1994. *J Electroanal Chem* 369 (1–2):87.

104. Zykwinska, A., W. Domagala, A. Czardybon, B. Pilawa, and M. Lapkowski. 2003. *Chem Phys* 292:31.

105. Zykwinska, A., W. Domagala, and M. Lapkowski. 2003. *Electrochem Commun* 5:603.

106. Heuer, H.-W., and R. Wehrmann. 1998. EP 961 159 (Bayer AG), Prior: 29 May 1998.

107. Sapp, S.A., G.A. Sotzing, J.R. Reddinger, and J.R. Reynolds. 1996. *Adv Mater* 8 (10):808.

108. Mecerreyes, D., R. Marcilla, E. Ochoteco, H. Grande, J.A. Pomposo, R. Vergaz, and J.M. Sanchez Pena. 2004. *Electrochim Acta* 49 (21):3555.

109. Kumar, A., and J.R. Reynolds. 1996. *Macromolecules* 29 (23):7629.

110. Sankaran, B., and J.R. Reynolds. 1997. *Macromolecules* 30 (9):2582.

11

Thienothiophenes: From Monomers to Polymers

Gregory A. Sotzing,
Venkataramanan Seshadri,
and Francis J. Waller

11.1 Introduction

Three decades ago, fused heterocycles evoked a new interest apart from being studied for their pharmacological activity and unique chemistry. The discovery of highly conductive polyacetylene [1] had led to the research on conjugated polyheteroaromatics such as polythiophenes [2], polypyrroles [3], and polyfurans [4] to name a few. In the quest for low bandgap conjugated polymers, Wudl et al. [5] had reported on low bandgap conjugated polymer using a fused thiophene, benzo[*b*]thiophene, or isothianaphthene. Bredas [6] indicated that the stabilization of the quinoidal form of poly(isothianaphthene) by aromatization of the fused benzene ring is a key factor responsible for reducing the bandgap. Following his work several fused systems have been synthesized and studied. This chapter will focus on fused thienothiophenes, their synthesis, characterization, and polymerization. Figure 11.1 shows the four isomers of thienothiophenes: thieno[2,3-*b*]thiophene (T23bT), thieno[3,2-*b*]thiophene (T32bT), thieno[3,4-*b*]thiophene (T34bT), and thieno[3,4-*c*]thiophene.

The first three of the four monomers are relatively stable for further research, and several papers have been reported on the synthesis of these thienothiophenes and their derivatives. The fourth isomer is an unstable nonclassical isomer and is the least studied, yet a few stable derivatives of this isomer have been prepared (Figure 11.2).

Reviews on polythiophenes, polysquarines, polybenzenes, etc. have appeared elsewhere, which also discuss the tunability of the bandgap of conjugated polymers [7]. Further sections of this chapter will focus on the synthesis and characterization of the underivatized thienothiophenes (a–c) and conjugated

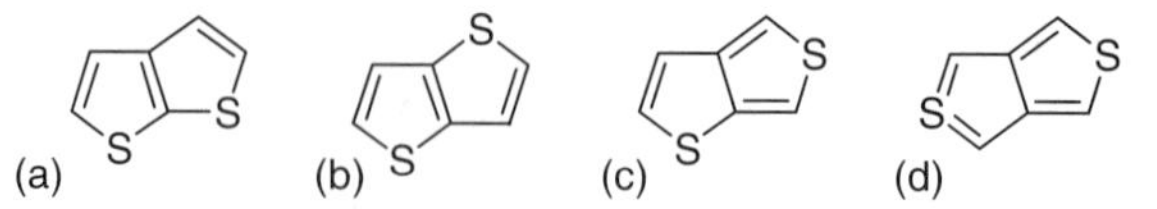

FIGURE 11.1   Four different thienothiophene isomers: (a) thieno[2,3-*b*]thiophene, (b) thieno[3,2-*b*]thiophene, (c) thieno[3,4-*b*]thiophene, and (d) thieno[3,4-*c*]thiophene.

polymers comprising these monomers. Recent spurt of interest in these monomers has been echoed by the reviews that have been published exclusively on the chemistry of thienothiophenes [8].

11.2   Synthesis and Characterization of Thienothiophenes

11.2.1   Thieno[2,3-*b*]thiophene

The earliest recorded T23bT derivative dates back to 1886 by Biedermann and Jacobson [9], although procedures for the synthesis of T23bT were not recorded. Gronowitz and Persson [10] first reported the synthesis of underivatized T23bT (b.p. 95°C/10 mm Hg, picrate m.p. 135.5°C–136.5°C) starting from 3-thiophenealdehyde ethylene acetal (Scheme 11.1).

Brandsma and De Jong have reported a one-pot synthesis of substituted as well as underivatized T23bT (Scheme 11.2) using derivatives of 1,3-pentadiynes in yields of roughly 50% [11]. This procedure relies on the reaction of the intermediate potassiated diyne through the allenic form. One of the major impurities found is the 2,5-disubstituted thiophene thiolate that is easily separated owing to its water solubility.

Earlier, cyclization of α-mercaptoacrylic acids in the presence of iodine was shown to yield fused thiophenes, which was utilized by Schneller and Petru [12] to synthesize thieno[2,3-*b*]thiophene carboxylic acid, which on decarboxylation yielded T23bT, an overall 60% yield (Scheme 11.3).

The α-mercaptoacrylic acid was reported to be obtained quantitatively from 3-formylthiophene. Similar attempts to ring close to prepare T32bT were not successful, possibly due to lowered electron density at the β-position. T23bT also been synthesized using the same synthetic strategy as shown in Scheme 11.4 using 2-bromo-3-trimethylsilylacetylenic thiophene instead.

^{1}H NMR [13]: Acetone (δ ppm vs. TMS (# H's; assignment)) 7.425 (2H, H-2), 7.232 (2H, H-3); long-range proton–proton couplings were studied and their calculated constants (Hz) are given as follows: J(2,2′) −1.20, J(2,3) −5.23, J(2,3′) −0.02, J(2′,3) −0.02, J(2′,3′) −5.23, J(3,3′) −0.18.

11.2.2   Thieno[3,2-*b*]thiophene

Several fused thienoheteroles have been reported to have been synthesized following the schematic as shown for T32bT (Scheme 11.4) [14]. Based on the positioning of bromide and acetylene, the closure to form the second ring yields [2,3-*b*], [3,2-*b*], or [3,4-*b*] isomers. Following the Teste and Lozac'h [15] method for preparing 3,6-dimethylthieno[3,2-*b*]thiophene, Nakayama and coworkers prepared 3,6-disubstituted T32bTs in moderate yields, ~33%, by one-pot synthesis by heating 2,5-dimethyl-3- hexyne-2,5-diol with elemental sulfur in the presence of *p*-toluenesulfonic acid (Scheme 11.5) [16]. They had also studied the

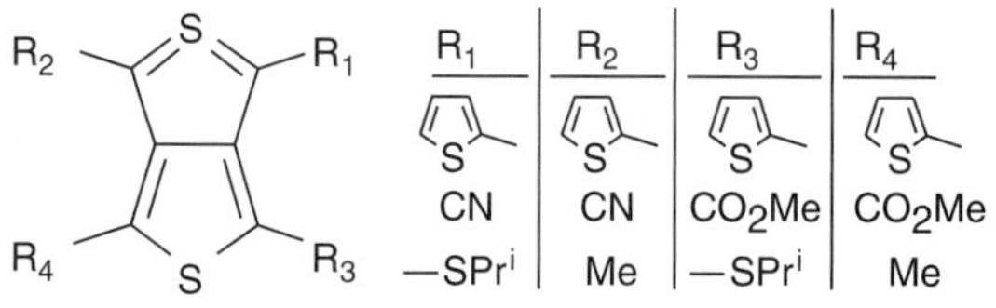

	R$_1$	R$_2$	R$_3$	R$_4$
	CN	CN	CO$_2$Me	CO$_2$Me
	−SPr^i	Me	−SPr^i	Me

FIGURE 11.2   Some derivatives of thieno[3,4-*c*]thiophene.

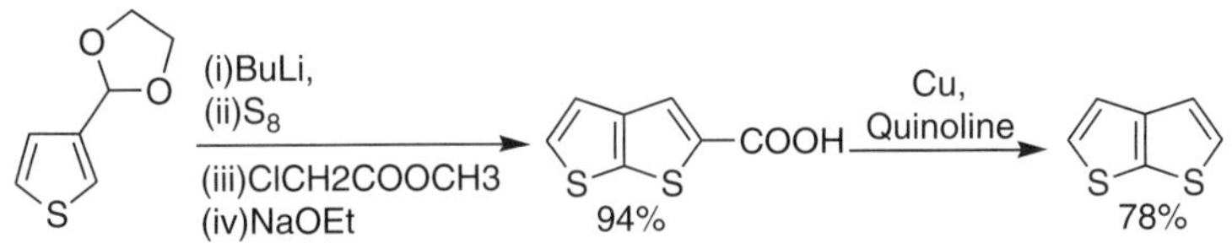

SCHEME 11.1 Synthesis of T23bT by Gronowitz et al.

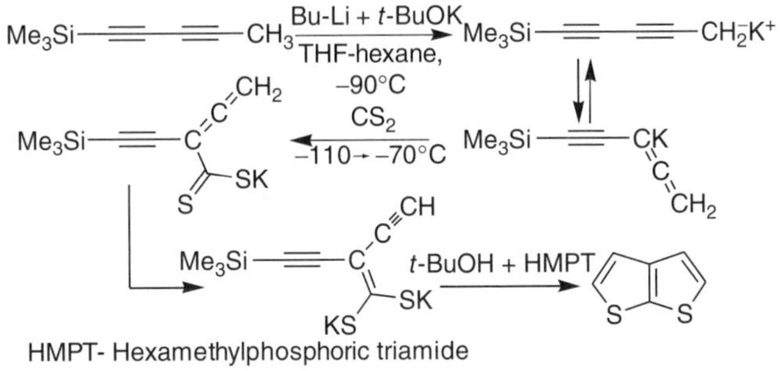

HMPT- Hexamethylphosphoric triamide

SCHEME 11.2 One-pot synthesis of T23bT reported by Brandsma.

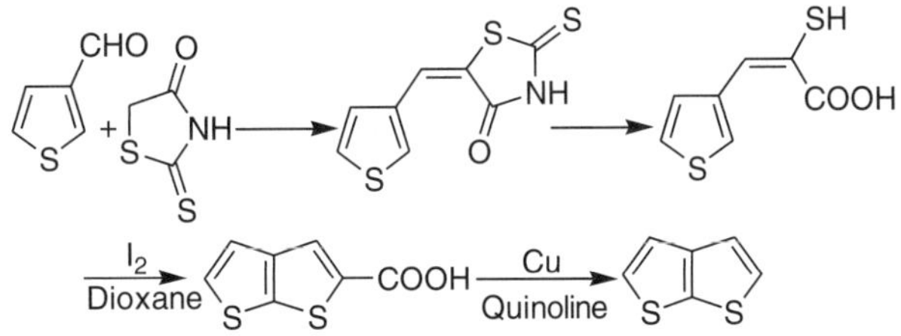

SCHEME 11.3 Schneller's synthesis of thieno[2,3-*b*]thiophene.

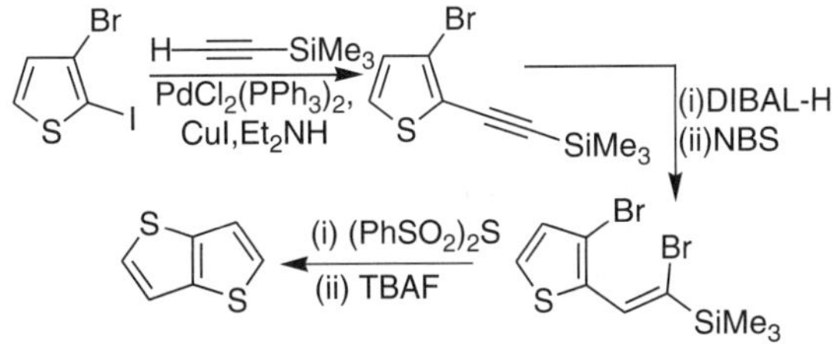

SCHEME 11.4 An example of the synthetic strategy applied toward making fused thienoheteroles (thieno[3,2-*b*]thiophene) (m.p. 55°C–56°C) [14].

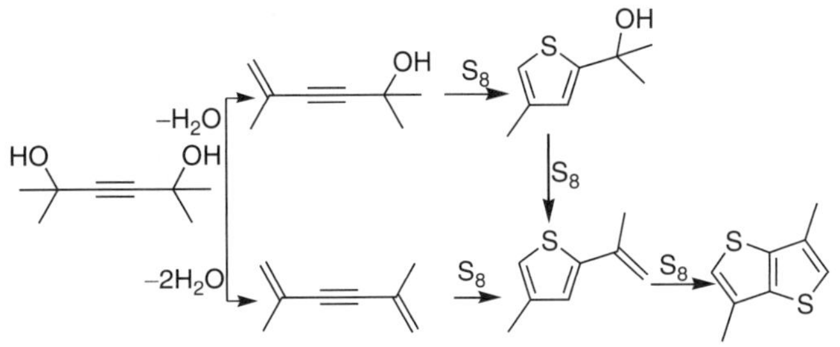

SCHEME 11.5 A simple one-step conversion of the inexpensive 2,5-dimethyl-3-hexyne-2,5-diol to 3,6-disubstituted thieno[3,2-*b*]thiophene.

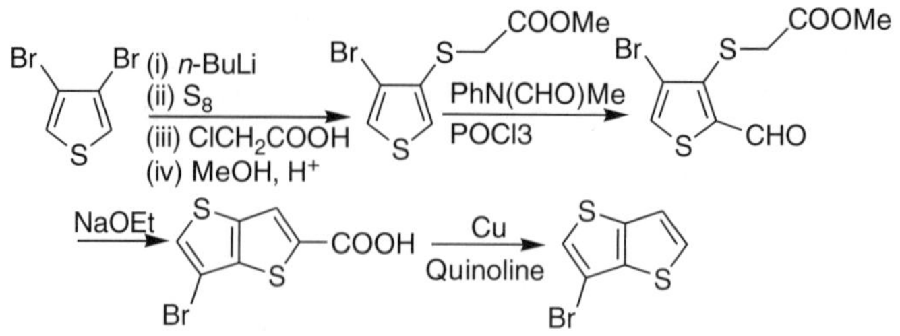

SCHEME 11.6 Synthesis of 3-bromothieno[3,2-*b*]thiophene.

reaction of elemental sulfur with the dehydrated diol as well to prove that indeed the reaction proceeds through these intermediates.

In another procedure reported by Kobayashi 3-bromothieno[3,2-*b*]thiophene was prepared starting from 3,4-dibromothiophene (Scheme 11.6). This was further utilized in making tetrathienoacene and pentathienoacene. Matzger and Zhang [17] have also synthesized and studied the absorption and emission spectra of thienoacenes; in a recent communication, they reported the synthesis of heptathienoacene [18]. Appending triisopropylsilyl groups at the end of thienoacenes imparted solubility to the longer α-oligothienoacenes; thereby improving the yields during the synthesis of the heptamer. Both penta and heptathienoacene were found to adopt a face-to-face π-stacking motif as opposed to a herringbone packing.

¹H-NMR [13]: Acetone (δ ppm vs. TMS (# H's; assignment)) 7.475 (2H, H-2), 7.295 (2H, H-3); long-range proton–proton couplings were studied and their calculated constants (Hz) are given as follows: $J(2,2')$ −1.55, $J(2,3)$ −5.25, $J(2,3')$ −0.20, $J(2',3)$ −0.20, $J(2',3')$ −5.25, $J(3,3')$ −0.75.

11.2.3 Thieno[3,4-*b*]thiophene

The third thienothiophene was the last to be synthesized, in 1967, by Wynberg [19] starting from dihydrothieno[3,4-*b*]thiophene sulfoxide (Scheme 11.7).

In 1991, Brandsma [20] reported the synthesis of T34bT by the conversion of 3,4-dibromothiophene to the monotrimethylsilyl acetylenic derivative, which was then ring closed to form T34bT after forming the thiolate on the 3-position (Scheme 11.8). We have reported in detail a slightly modified synthetic procedure through which we were able to obtain as high as 60%–70% yields in the final ring-closing step, whereas using the identical condition reported by Brandsma, we were able to obtain only a maximum of 8% yield [32].

Carbon peak assignments were not available in the earlier reports, and hence, a comprehensive study utilizing ¹H, ¹³C, correlation spectroscopy (COSY), heteronuclear correlation (HETCOR), and heteronuclear multibond correlation (HMBC) nuclear magnetic resonance was undertaken. ¹H NMR: CDCl₃ (δ ppm vs. TMS (multiplicity; # H's; coupling constant; assignment)) 7.36 (d; 1H; J 5.4 Hz, H-2),

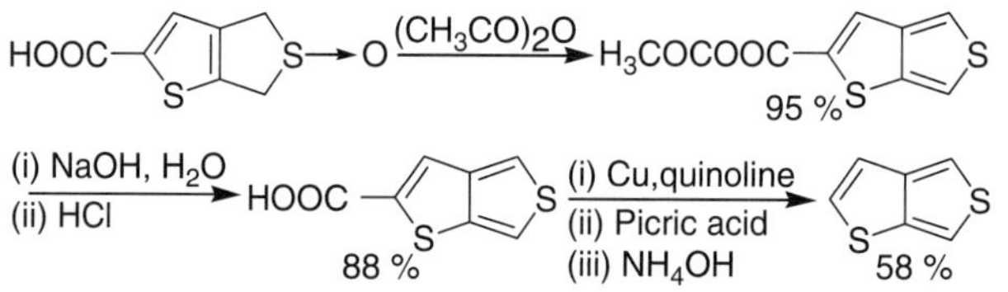

SCHEME 11.7 Wynberg's synthesis of thieno[3,4-*b*]thiophene (m.p. 7°C–7.5°C).

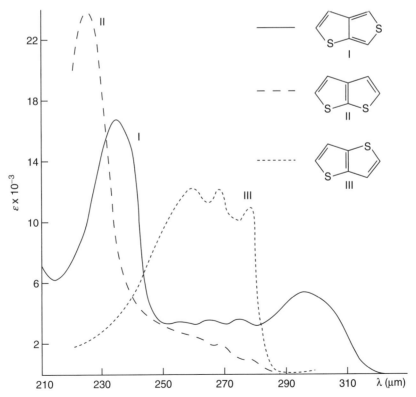

SCHEME 11.8 Synthesis of thieno[3,4-*b*]thiophene.

7.35 (d; 1H; *J* 3.1 Hz; H-6), 7.26 (dd; 1H; *J* 2.6 and 0.7 Hz; H-4), 6.94 (dd; 1H; *J* 5.6 and 0.5 Hz; H-3). ^{13}C NMR: CDCl$_3$ (δ ppm vs. TMS) 147.6 (C-8), 139.3 (C-7), 132.4 (C-2), 116.8 (C-3), 111.6 (C-6), 110.6 (C-4). Figure 11.3 shows the UV spectrum of the three thienothiophenes [20] (the solvent used for obtaining the spectra was not reported). The absorption maximum is redshifted in the following order T32bT > T34bT > T23bT.

Apart from these, Pomerantz and Ferraris have synthesized 2-substitued T34bT, which was further polymerized by the free 4- and 6-positions to obtain low bandgap conjugated polymers, the details of which will be given in the following sections. Detailed chemistry of thienothiophenes can be found in the reviews by Litvinov et al. [8].

FIGURE 11.3 UV spectra of the three isomeric thienothiophenes. (Reprinted from Wynberg, H. and Zwanenberg, D.J., *Tet. Lett.*, 9, 761, 1967. With permission.)

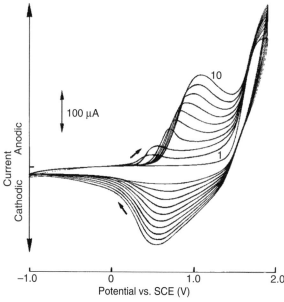

R	Mn
C_8H_{17}	15 k
$C_{10}H_{21}$	29 k
$C_{12}H_{25}$	26 k

FIGURE 11.4   Copolymers comprising of thieno[2,3-*b*]thiophene.

11.3   Polythienothiophenes

11.3.1   Poly(thieno[2,3-*b*]thiophene)

Oxidation of T23bT was reported to be 1.58 V (vs. SCE) which was lower than that for thiophene (2.06 V), but the polymer did not show any redox behavior on cycling in a monomer free electrolyte solution up to 1.6 V [21]. Lazzaroni et al. [22c] reported the polymer film to be green with no redox activity. Recently, Heeney et al. [22] reported copolymer comprising T23bT to disrupt the conjugation thereby reducing the ability of the polymer to oxidize in air or light giving a high on–off ratio of nearly 10^5 (Figure 11.4).

Due to the central cross-conjugated double bond a fully conjugated point T23bT is unable to form a fully conjugated polymer by 2,5-positions.

11.3.2   Poly(thieno[3,2-*b*]thiophene)

Historically, this monomer was reported within a year of the first report on poly(T23bT) by Danieli et al. [23]. Figure 11.5 shows the polymerization cyclovoltammogram from a 2 mM solution of T32bT in 0.1 M Bu4NClO4 /CH2Cl2 at a scan rate of 0.2 V/s. Other than the IR and mid-IR spectra, and conductivity no other characterization of the polymer is available. Electrochemical polymerization of T32bT and 3-methyl T32bT was also reported by Rutherford et al. [24]. Thin polymer films of T32bT were reported to be red in the neutral state and purple in the oxidized form. There is no further electrochemical characterization.

FIGURE 11.5   CV of T32bT polymerization on a Pt button electrode from a 2 mM solution of the monomer in 0.1 M Bu4NClO4/CH2Cl2 at a scan rate of 0.2 V/s. (Reprinted from Danieli, R., et al., *Synth. Met.*, 13, 325, 1986. With permission.)

Matzger and coworkers [25] reported the synthesis of alkyl substituted poly(T32bTs) and studied their optical properties. It was found that 3,6-dinonyl substitution leads to a drastic increase in bandgap due to the twisting of adjacent T32bT units away from planarity.

Poly(nonylthieno[3,2-*b*]thiophenes) were prepared by oxidative polymerization using FeCl$_3$ that is expected to give a regioirregular polymer as well as by Kumada and Stille coupling leading to a regioregular polymer. The UV–vis spectra of polymer solutions and films obtained from different alkyl substituted PT32bTs are shown in Figure 11.6. Lim et al. [26] have reported the synthesis of an alternating copolymer consisting of T32bT and fluorene, which was obtained by Suzuki coupling of 2,5-dibromoT32bT with 2,7-diboronic ester of 9,9′-dioctylfluorene. The bandgap was calculated to be 2.98 eV (electrochemical bandgap) with an oxidation onset at 0.99 V (vs. SCE). Furthermore, the photoluminescence and electroluminescence were also studied. This polymer was reported to emit a pure green light with an emission peak at 515 nm compared to *bis*-thienyl–fluorenyl copolymers (emission peak, 539 nm) in a device.

Recently, Turbiez et al. [27] reported a poly(3,4-ethylenedioxythiophene) (PEDOT) like polymer, poly(3,6-dimethoxythieno[3,2-*b*]thiophene), which exhibited a discernible vibronic fine structure with λ_{max} at 592 nm (Figure 11.7) and a band-edge gap of 1.65 eV. The electrochemically synthesized polymer was reported as pale blue in the oxidized form, while the spectrum of the oxidized polymer remained unchanged even after 1 week, indicative of a potential transparent semiconductor like PEDOT.

11.3.3 Poly(thieno[3,4-*b*]thiophene)

Poly(thieno[3,4-*b*]thiophene), like the monomers, was the last of the thienothiophenes to be synthesized. Hong and Marynick [28] had obtained the electronic structures of poly(thieno[3,4-*b*]thiophene) using modified extended-Huckel band calculations and predicted that PT34bT connected through 4- and 6-positions will have a bandgap of 1.5–1.6 eV (π–π^*, λ_{max}), comparable to that of polyacetylene. In fact, Pomerantz [29] and Ferraris [30] had synthesized PT34bTs with the 2-position blocked using decyl, dodecyl, or phenyl groups. Pomerantz had synthesized the alkyl substituted T34bT and polymerized using FeCl$_3$ in chloroform. Thin films of this polymer were found to exhibit a band-edge gap of 0.92 eV with a λ_{max} at 925 nm, while in solution (chloroform) the band-edge gap was found to be 0.98 eV with a λ_{max} at 739 nm (Figure 11.8).

2-Phenylthieno[3,4-*b*]thiophene was synthesized using Brandsma's procedure [21] except that phenylacetylene was used to place the phenyl group on the 2-position of T34bT. Polymer from this monomer was obtained electrochemically, and was found to exhibit a band-edge gap of 0.82 eV and a λ_{max} at 920–940 nm (Figure 11.8).

We reported the electrochemical polymerization of unsubstituted T34bT that yielded a polymer with a bandgap of 0.8 eV. The polymer was found to be transparent in the oxidized form even for a 0.8 μm thick film (Figure 11.9) [31]. The onset for polymer oxidation or p-doping was found to occur at approximately −0.6 V and the onset for reduction or n-doping was found at approximately −1.3 V (vs. nonaqueous Ag/Ag$^+$, calibrated to be 0.47 V vs. NHE). All T34bT polymers described so far have been polymerized either electrochemically or using an oxidant such as FeCl$_3$ and hence are not expected to be regioregular (T34bT does not possess C2 axis of symmetry). It would be of great interest to synthesize the regioregular version of PT34bT, owing to the fact that regioregular poly(3-alkylthiophenes) have been shown to have very high conductivities and an increase in the bandgap as a result of better packing of the polymers and reduced torsional angle between rings [32].

Unlike the 2-substituted T34bTs, the underivatized molecule has three open α-positions through which polymerization can proceed and yet not break the conjugation, thus acting as molecular interconnects [33]. To understand PT34bT with connectivites through the 2-position, we had blocked the 6-position using methanol as well as methoxymethyl groups and studied the possibility of electrochemically polymerizing the resultant monomer [34]. Recently, we reported the synthesis of the symmetrical dimer with 6–6′ connectivity and electrochemically polymerized the same [35]. As expected, the onset and peak for the oxidation of the monomer was reduced by ~0.4 and 0.66 V (vs. nonaqueous Ag/Ag$^+$, 0.45 V vs. NHE). The polymer had a broad oxidation potential

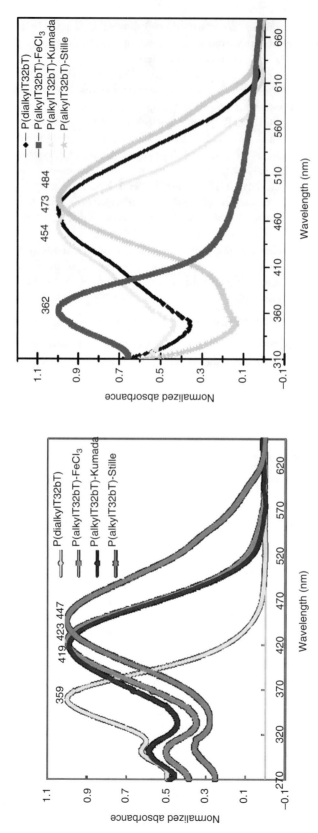

FIGURE 11.6   UV–vis spectra of poly(alkylT32bTs) in chloroform (*left*) and as thin films (*right*). (Reprinted from Zhang, X., Köhler, M., and Matzger, A.J., *Macromolecules*, 37, 6306, 2004. With permission.)

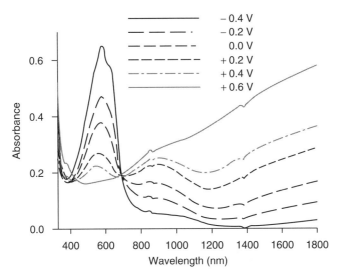

FIGURE 11.7 Electronic absorption spectra of PDMTT electrodeposited on ITO. (Reprinted from Turbiez, M., et al., *J. Chem. Soc. Chem. Commun.*, 1161, 2005. With permission.)

ranging approximately between -0.7 and 0.5 V (vs. nonaqueous Ag/Ag^+) in 0.1 M n-Bu$_4$N-ClO$_4$/nitrobenzene. This polymer has switched from deep blue (neutral state) to light blue (oxidized state). Figure 11.10 shows the absorption spectrum of the polymer in the neutral and oxidized states. The onset for the π–π^* transition was determined to be 0.90 eV with a peak at 890 nm (1.39 eV). Further studies are underway on the 2–2$'$ and 4–4$'$ dimers and polymerization of the same.

Copolymers containing T34bT and EDOT were synthesized electrochemically by simultaneously oxidizing both the monomers from a solution containing equimolar amounts of the monomer [36]. These copolymers were found to have bandgaps intermediate to that of either homopolymers. Figure 11.11 shows the electronic absorption spectra of a layered structure of the two homopolymers exhibiting two λ_{max} and the copolymers prepared using perchlorate and hexafluorophosphate anions as the dopant. The doping levels for the homopolymers were calculated using the measured chronocoulometric and chronogravimetric data simultaneously obtained using electrochemical quartz chemical microbalance (EQCM) studies. The T34bT containing homo- and copolymers exhibited nearly twice the doping levels (28%–38%) as that of PEDOT using the same anions. The ion-transport in and out of the polymer film was calculated to be highly anion dominant, approximately $\geq 80\%$. Furthermore, the copolymer obtained using hexafluorophosphate dopant was found to exhibit good stability to n-doping. The composition of the copolymers was found from the ratio of carbon to sulfur obtained using elemental analysis. Both the copolymers were found to be rich in T34bT, with copolymer prepared using n-Bu$_4$NClO$_4$ containing $\sim79.9\%$ T34bT and that prepared using n-Bu$_4$NPF$_6$ containing $\sim67.7\%$.

Introduction of vinylene spacers between aromatic units has been used to reduce the torsional angle thereby lowering the bandgap of the conjugated polymer [37]. Another advantage of vinylene and ethylenic bond between the rings is to impart restricted rotation of the rings due to the defined configuration of the double bond, i.e., *cis* or *trans*. Further lowering of the bandgap has been achieved by placing a cyano group on one of the vinyl carbons that helps in stabilizing the quinoidal state [38]. Based on this strategy, we synthesized two unsymmetrical T34bT monomers containing cyanovinylene spacers by a Knoevanagel condensation of 4- and 6-formylthieno[3,4-*b*]thiophene and thiophene-2-acetonitrile (Figure 11.12) [39]. These monomers were found to be very stable and the polymers from these monomers were obtained electrochemically. In spite of being nonequal monomeric structures the polymers obtained from them exhibited similar properties. The monomer oxidation potentials were similar and the onset for oxidation occurred at 0.7 V with a peak at 0.8 V (vs. nonaqueous Ag/Ag^+, calibrated to be 0.45 V vs. NHE).

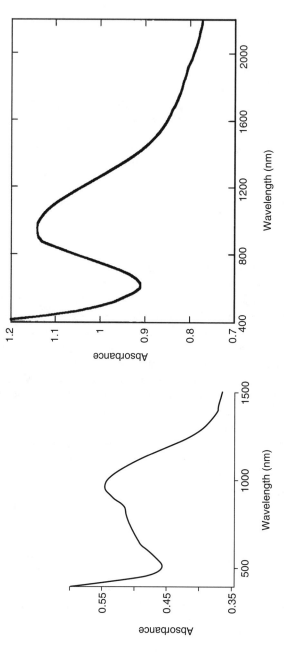

FIGURE 11.8 Electronic absorption spectra of poly(2-decylT34bT) (*left*) and poly(2-phenylT34bT) (*right*) as thin films. (Reprinted from Pomerantz, M., Gu, X., and Zhang, S.X., *Macromolecules*, 34, 1817, 2001 (*left*) and Neef, C.J., Brotherston, I.D., and Ferraris, J.P., *Chem. Mater.*, 11, 1957, 1999 (*right*). With permission.)

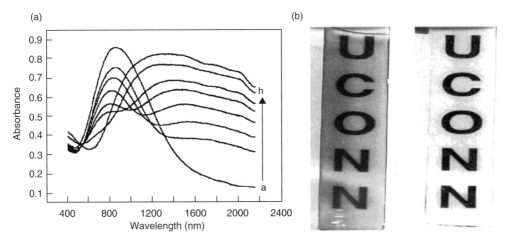

FIGURE 11.9 Vis–NIR spectrum of a 0.1 μm thick poly(thieno[3,4-*b*]thiophene) film on ITO-glass at (a) reduced to neutral state using hydrazine reduced, (b) −0.4, (c) −0.3, (d) −0.2, (e) −0.1, (f) 0.0, (g) 0.15, and (h) 0.4 V vs. Ag/Ag+ reference electrode (0.47 V vs. NHE) (left). Neutral and oxidized states of a 0.8 μm thick PT34bT film coated on ITO glass (right). (Reprinted from Sotzing, G.A., and Lee. K., *Macromolecules*, 35, 7281, 2002. With permission.)

The polymers were found to have a broad oxidation peak from −0.1 to 0.9 V (vs. Ag/Ag$^+$) in 0.1 M *n*-Bu$_4$NClO$_4$. The band-edge gaps for the polymers from 4- and 6-cyanovinylenes were found to be ~1.19 eV (1045 nm) and 1.20 eV (1035 nm) with a peak at ~1.72 eV (720 nm) and 1.85 eV (699 nm), respectively. These polymers also exhibited a very good n-doping behavior apart from being able to be p-doped, which was confirmed by the broad absorption spectra reaching to the IR for the n-doped polymer (Figure 11.13b). The inset in the figure shows a reversible change in the absorption at 669 nm indicative of the stability of the n-doped polymer.

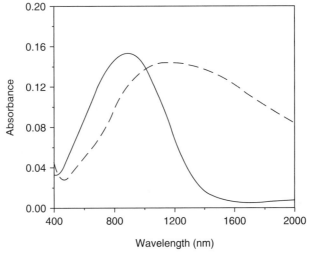

FIGURE 11.10 Vis–NIR spectrum of a poly(6,6′-*bis*T34bT) film on ITO glass substrate in oxidized state (*dashed line*) and neutral state (*solid line*). (Reprinted from Lee, B., Yavuz, M.S., and Sotzing, G.A., *Poly. Prepr.*, 46, 860, 2005. With permission.)

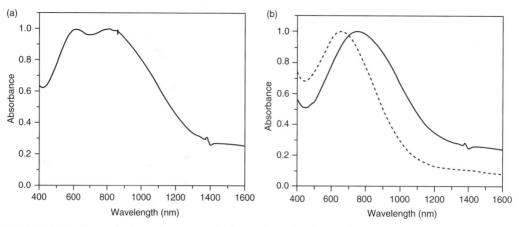

FIGURE 11.11   Electronic absorption spectra of (a) neutral PEDOT deposited over neutral PT34bT and (b) neutral copolymer containing T34bT and EDOT obtained from electrochemical polymerization in (—) n-Bu$_4$NClO$_4$ and (- - -) n-Bu$_4$NPF$_6$. (Reprinted from Seshadri, V., Wu, L., and Sotzing, G.A., *Langmuir*, 19, 9479. With permission.)

11.4   Processable Poly(thieno[3,4-*b*]thiophenes)

While several conjugated polyaromatics have been studied, their insolubility and intractability have always been an inherent problem associated with their utility. Some of the well-studied strategies for making conducting conjugated polymers processable include appending soluble groups to the rigid backbone such as alkyl chains ($\geq$C6) [40], alkyl sulfonates [41], and sulfonation of the ring itself [42]; counterion induced processability of conjugated polymers [43]; polymerization in the presence of a polyelectrolyte like polystyrenesulfonic acid [44]; postderivatization of polymer chain ends or repeat units [45]; and solid-swollen state oxidative cross-linking of pendant heterocycles on soluble polymers [46]. Polymerization in the presence of a polyelectrolyte offers a very easy and fast approach toward incorporating processability into these polymers. Typically, the monomer is oxidatively polymerized in an aqueous medium in the presence of polystyrenesulfonic acid using Iron(III) salts or persulfates. At the end of the polymerization, the salts from the oxidant are removed by passing through a column of acidic and basic ion-exchange resins to leave the pure dispersion of oxidized conjugated polymer–polystyrene sulfonate complex. Recently, Air Products and Chemicals, Inc. have announced research on aqueous dispersions based on poly(thieno[3,4-*b*]thiophene) for the hole injection layer in PLED [47].

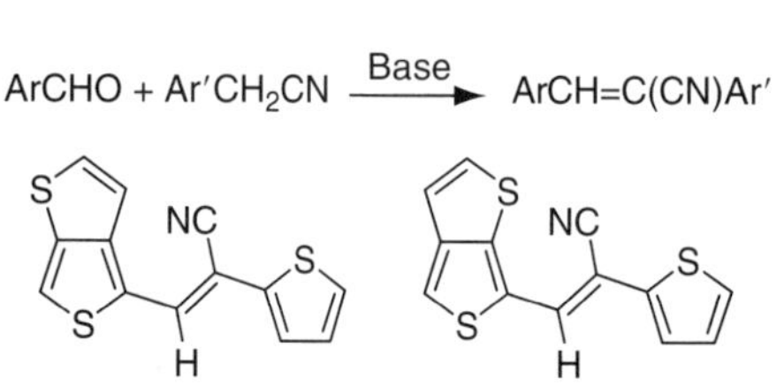

11.4.1   Colloidal Dispersions

Poly(thieno[3,4-*b*]thiophene)–polystyrene sulfonate dispersions were prepared similar to that as described for PEDOT–PSS [45]. Polymerization of T34bT was carried out in the presence of a variety of oxidants such as iron(III)sulfate, ammonium persulfate, hydrogen peroxide, mixtures of persulfate, or hydrogen peroxide with iron(III)sulfate [48]. It was observed that the absorption of the resultant polymers is controllably dependent on the type of oxidant used. PT34bT–PSS dispersions obtained using persulfate and peroxide exhibited a band-edge gap of 1.1 eV

ArCHO + Ar'CH$_2$CN $\xrightarrow{\text{Base}}$ ArCH=C(CN)Ar'

FIGURE 11.12   Knoevanagel synthesis of aryl-cyanovinylenes from 4- and 6-formylthieno[3,4-*b*]thiophene.

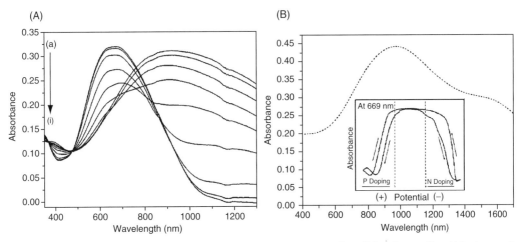

FIGURE 11.13 (A) In situ spectroelectrochemistry of poly((*E*)-3-(thieno[3,4-*b*]thiophen-6-yl)-2-(thiophen-2-yl) acrylonitrile) at (a) −0.15 V, (b) 0.25 V, (c) 0.35 V, (d) 0.45 V, (e) 0.55 V, (f) 0.65 V, (g) 0.75 V, (h) 0.85 V, (i) 0.95 V. (B) Vis–NIR spectrum of poly((*E*)-3-(thieno[3,4-*b*]thiophen-6-yl)-2-(thiophen-2-yl)acrylonitrile) in the n-doped form (−1.65 V, dashed line). *Inset* shows the change in absorbance at 669 nm (λ_{max} for the neutral poly(E)-3-(thieno[3,4-b]thiophen-6-yl)-2-(thiophen-2-yl) (acrylonitrile) as a function of scanning the potential from 0.8 to −1.5 V). (Reprinted from Seshadri, V. and Sotzing, G.A., *Chem. Mater.*, 16, 5644, 2004. With permission.)

and the λ_{max} was ∼830 and 840 nm, respectively; while dispersions obtained using iron(III) were found to have a λ_{max} at 910 nm. These dispersions were found to be very stable for prolonged periods of over a year. The conductivity of thin films of PT34bT–PSS varied depending on the oxidant system used; polymers obtained using persulfate showed 10^{-4} to 10^{-3} S/cm while that obtained from iron(III) showed 10^{-2} S/cm. Diluted PT34bT–PSS dispersions were a transmissive light green in the oxidized form and transmissive light blue in the neutral state.

Figure 11.14 depicts a hypothetical structure of the dispersion along with the dilute solution in the neutral and doped states. The neutral undoped state was obtained by adding a few drops of hydrazine solution to the dispersion.

11.4.2 Ring Sulfonated Poly(thieno[3,4-*b*]thiophene)

Ring sulfonation of the polymer backbone has been used to render the intractable polyaniline soluble in water, DMSO, DMF, etc. [43]. Toward this end it is necessary to have a few unsubstituted aromatic carbons in the conjugated polymer. Calculations on T34bT indicate that each of the α-carbons has different reactivities and hence PT34bT obtained by oxidative polymerization is expected to have few unreacted α-positions. Based on these assumptions T34bT was polymerized using FeCl₃ in chloroform and the precipitated polymer reduced to the neutral form by hydrazine treatment. The neutral polymer thus obtained was reacted with fuming sulfuric acid to obtain sulfonated poly(thieno[3,4-*b*]thiophene) (SPoT) [49].

To this date two different sulfonation levels have been achieved depending on the dedoping time with hydrazine. The longer time the polymer is stirred in hydrazine the lower the number of positive charges on the polymer backbone and hence more the percent of sulfonation. The sulfonation levels were calculated using both titration and from the carbon to sulfur ratio obtained from elemental analysis. The values obtained from the titration results indicate sulfonation levels of 65% and 56%. The as-prepared polymers exhibited a broad absorption throughout the visible and extending as high as 1400 nm. On reducing the polymer to the neutral form with hydrazine, the two polymers exhibited a single broad peak with λ_{max} at 764 (1.62 eV) and 620 nm (1.99 eV) with onsets at 1.05 and 1.18 eV for 56-SPoT and 65-SPoT, respectively. The polymer was rendered processable in aqueous and several organic media such as DMF, DMSO, NMP etc., as a result of sulfonation. Oxidation of the polymer film from the

(a)

(b)

FIGURE 11.14   (a) Preparation of PT34bT–PSS dispersion and (b) dilute aqueous solution of PT34bT in the oxidized and neutral states.

as-prepared form leads to further decrease in the absorption in the visible region indicating that this is in fact a partial doped state.

Furthermore, well-ordered films comprised of 20 bilayers of alternating SPoT and polyethyleneimine. Hydrochloride was obtained through layer-by-layer (LBL) technique. Figure 11.15 shows a thin LBL film of 56-SPoT coated onto ITO-glass and the spectra corresponding to each bilayer. The spectra of the LBL film remained unchanged even after several months of storage under ambient conditions, making it a potentially useful dye in solar cells. The λ_{max} of the LBL film prepared using SPoT–PEI was found to be ~165 nm blueshifted compared to the spectrum of neutral 56-SPoT providing another path to tune the optical properties of the material.

11.5   Conclusions

T23bT produces a nonelectroactive polymer whereas T32bT produces a high bandgap polymer. T32bT has shown the ability for lowering the bandgap through incorporation of electron donating groups. Underivatized poly(thieno[3,4-*b*]thiophene) is a very low bandgap conjugated polymer with the

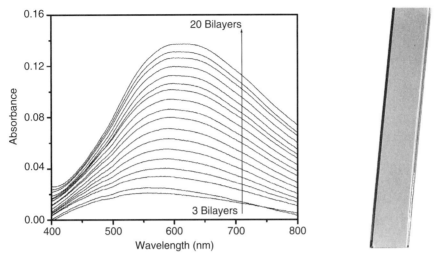

FIGURE 11.15 Optical spectra of LBL film of 56-SPoT–PEI as a function of bilayer numbers (*left*). A 16 nm thick 20 bilayer LBL film of 56-SPoT–PEI (*right*). (Reprinted from Lee, B., et al., *Adv. Mater.*, 17, 1792, 2005. With permission.)

capability of water dispersibility and processibility. Furthermore, through the process of direct sulfonation of the backbone, spectral properties can be tuned and the polymer is rendered processable using the LBL technique to generate organized thin layer films. With the properties of electrical conductivity and high visible transmissivity, this polymer is suitable for use in display applications particularly where band energy matching is required. Monomers like T34bT that produce low bandgap polymers are suitable for spectral tuning through the visible region by copolymerization with other conjugated units. For it is easy to take a low bandgap polymer and alter the structure through such a process to generate higher bandgap materials.

References

1. (a) Ito, T., H. Shirakawa, and S. Ikeda. 1974. Simultaneous polymerization and formation of polyacetylene film on the surface of a concentrated soluble Ziegler-type catalyst solution. *J Polym Sci Polym Chem* 12:1. (b) Chiang, C.K., et al. 1977. Electrical conductivity in doped polyacetylene. *Phys Rev Lett* 39:1098.
2. (a) Tourillon, G., and F. Garnier. 1982. New electrochemically generated organic conducting polymers. *J Electroanal Chem* 135:173. (b) Kobayashi, M., et al. 1984. Synthesis and properties of chemically coupled poly(thiophene). *Synth Met* 9:77.
3. (a) Dall'Olio, A., et al. 1968. Electron paramagnetic resonance and conductivity of an electrolytic oxypyrrole (pyrrole polymer) black. *CRS Acad Sci Ser C* 267:433. (b) Gardini, G.P. 1973. *Adv Heterocyc Chem* 15:67.
4. Zotti, G., et al. 1990. Electrochemical synthesis and characterization of polyconjugated polyfuran. *Synth Met* 36:337.
5. (a) Wudl, F., M. Kobayashi, and A.J. Heeger. 1984. Poly(isothinaphthene). *J Org Chem* 49:3382. (b) Kobayashi, M., et al. 1985. The electronic and electrochemical properties of poly(isothianaphthene). *J Chem Phys* 82:5717.
6. Bredas, J.L., A.J. Heeger, and F. Wudl. 1986. Towards organic polymers with very small intrinsic band gaps. I. Electronic structure of polyisothianaphthene and derivatives. *J Chem Phys* 85:4673.
7. (a) Roncali, J. 1997. Synthetic principles for bandgap control in linear π-conjugated systems. *Chem Rev* 97:173. (b) Patil, A.O., A.J. Heeger, and F. Wudl. 1988. Optical properties of conducting polymers. *Chem Rev* 88:183. (c) Yamamoto, T., and N. Hayashida. 1998. π-Conjugated polymers

bearing electronic and optical functionalities. Preparation, properties and their applications. *React Func Poly* 37:1. (d) Scherf, U., and K. Mullen. 1992. Design and Synthesis of extended π-systems: monomers, oligomers, polymers. *Synthesis* 23. (e) Ajayaghosh, A. 2003. Donor–acceptor type low band gap polymers: Polysquaraines and related systems. *Chem Soc Rev* 32:181.

8. (a) Litvinov, V.P. 2005. The latest achievements in thienothiophene chemistry. *Russ Chem Rev* 74:217. (b) Comel, A., G. Sommen, and G. Kirsch. 2004. Thienothiophenes: Synthesis and applications. *Mini-Rev Org Chem* 1:367. (c) Brandsma, L. 2001. Unsaturated carbanions, heterocumulenes and thiocarbonyl compounds—New routes to heterocycles. *Eur J Org Chem* 4569.

9. Biedermann, A., and P. Jacobson. 1886. ||||||||*Ber* 19:2444.

10. Gronowitz, S., and B. Persson. 1967. A convenient synthesis of thieno[2,3-*b*]thiophene. *Acta Chem Scand* 21:812.

11. De Jong, R.L.P., and L. Brandsma. 1991. A one-pot procedure for thieno[2,3-*b*]thiophene and some of its derivatives using derivatives of 1,3-pentadiyne and carbon disulfide as building units. *Synth Commun* 21:145.

12. Schneller, S.W., and J.D. Petru. Convenient preparation of thieno[2,3-*b*]thiophene. *Synth Commun* 4:29.

13. Bugge, A., B. Gestoblom, and O. Hartmann. 1970. Long-range proton–proton spin coupling constants in thienothiophenes. *Acta Chem Scand* 24:105.

14. Bugge, A. 1968. Metallation of thieno[2,3-*b*]thiophene and thieno[3,2-*b*]thiophene with butyl-lithium. *Acta Chem Scand* 22:63.

15. Teste, J., and N. Lozac'h. 1955. Sulfuration of organic compounds. IX. Sulfuration of alcohols and acetylenic glycols. *Bull Soc Chim Fr* 442.

16. Choi, K.S., et al. 1994. A one-pot synthesis of substituted thieno[3,2-*b*]thiophenens and selenolo[3,2-*b*]selenophenes. *Heterocycles* 38:143.

17. Zhang, X., and A.J. Matzger. 2003. Effect of ring fusion on the electronic absorption and emission properties of oligothiophenes. *J Org Chem* 68:9813.

18. Zhang, X., A.P. Côtè, and A.J. Matzger. 2005. Synthesis and structure of fused α-oligothiophenes with up to seven rings. *J Am Chem Soc* 127:10502.

19. Wynberg, H., and D.J. Zwanenberg. 1967. Thieno[3,4-*b*]thiophene. The third thiophthene. *Tet Lett* 9:761.

20. Brandsma, L., and H.D. Verkruijsse. 1990. An alternative synthesis of thieno[3,4-*b*]thiophene. *Synth Commun* 20:2275.

21. (a) Jow, T.R., et al. 1985. *J Electrochem Soc* 132:98C. (b) Jow, T.R., et al. 1986. Electrochemical studies of fused-thiophene systems. *Synth Met* 14:53. (c) Lazzaroni, R., et al. 1987. Electronic structure of conducting polymers from heteroaromatic bicyclic compounds. *Synth Met* 21:189.

22. Heeney, M., et al. 2005. Stable polythiophene semiconductors incorporating thieno[2,3-*b*]thiophene. *J Am Chem Soc* 127:1078.

23. Danieli, R., et al. 1986. Optical, electrical and electrochemical characterization of electrosynthesized polythieno(3,2-*b*)thiophene. *Synth Met* 13:325.

24. Rutherford, D.R., et al. 1992. Poly(2,5-ethynylenethiophenediylethynylenes), related heteroaromatic analogs, and poly(thieno[3,2-*b*]thiophenes): Synthesis and thermal and electrical properties. *Macromolecules* 25:2294.

25. Zhang, X., M. Köhler, and A.J. Matzger. 2004. Alkyl-substituted thieno[3,2-*b*]thiophene polymers and their dimeric subunits. *Macromolecules* 37:6306.

26. Lim, E., B.-J. Jung, and H.-K. Shim. 2003. Synthesis and characterization of a new light-emitting fluorene–thieno[3,2-*b*]thiophene-based conjugated copolymer. *Macromolecules* 36:4288.

27. Turbiez, M., et al. 2005. Poly(3,6-dimethoxy-thieno[3,2-*b*]thiophene): A possible alternative to poly(3,4-ethylenedioxythiophene) (PEDOT). *J Chem Soc Chem Commun* 1161.

28. Hong, S.Y., and D.S. Marynick. 1992. Understanding the conformational stability and electronic structures of modified polymers based on polythiophene. *Macromolecules* 25:4652.

29. (a) Pomerantz, M., and X. Gu. 1997. Poly(2-decylthieno[3,4-*b*]thiophene). A new soluble low-bandgap conducting polymer. *Synth Met* 84:243. (b) Pomerantz, M., X. Gu, and S.X. Zhang. Poly(2-decylthieno[3,4-*b*]thiophene-4,6-diyl). A new low band gap conducting polymer. *Macromolecules* 34:1817.

30. Neef, C.J., I.D. Brotherston, and J.P. Ferraris. 1999. Synthesis and electronic properties of poly(2-phenylthieno[3,4-*b*]thiophene): A new low band gap polymer. *Chem Mater* 11:1957.

31. (a) Lee, K., and G.A. Sotzing. 2001. Poly(thieno[3,4-*b*]thiophene). A new stable low band gap conducting polymer. *Macromolecules* 34:5746. (b) Sotzing, G.A., and K. Lee. 2002. Poly(thieno[3,4-*b*]thiophene): A p- and n-dopable polythiophene exhibiting high optical transparency in the semiconducting state. *Macromolecules* 35:7281.

32. (a) McCullough, R.D., et al. 1993. Self-orienting head-to-tail poly(3-alkylthiophenes): New insights on structure–property relationships in conducting polymers. *J Am Chem Soc* 115:4910. (b) McCullough, R.D., et al. 1993. Design, synthesis, and control of conducting polymer architectures: Structurally homogeneous poly(3-alkylthiophenes). *J Org Chem* 58:904.

33. Lee, K., L. Wu, and G.A. Sotzing. 2002. Thieno[3,4-*b*]thiophene as a novel low oxidation cross-linking agent. *Poly Mater Sci and Eng* 86:195.

34. Seshadri, V., et al. 2003. Conjugated polythiophene consisting of coupling through locked transoid conformation. *Poly Mater Sci and Eng* 88:292.

35. (a) Lee, B., M.S. Yavuz, and G.A. Sotzing. 2006. Poly(thieno[3,4-*b*]thiophene)s from three symmetrical thieno[3,4-*b*]thiophene dimmers. *Macromolecules* 39:3118. (b) Lee, B., M.S. Yavuz, and G.A. Sotzing. 2005. Conjugated polymers from thieno[3,4-*b*]thiophene dimers. *Poly Prepr* 46:860.

36. Seshadri, V., L. Wu, and G.A. Sotzing. 2003. Conjugated polymers via electrochemical polymerization of thieno[3,4-*b*]thiophene (T34bT) and 3,4-ethylenedioxythiophene (EDOT). *Langmuir* 19:9479.

37. (a) Jen, K-Y., et al. 1987. Highly-conducting, poly(2,5-thienylene vinylene) prepared via a soluble precursor polymer. *J Chem Soc Chem Commun* 309. (b) Barker, J. 1989. An electrochemical investigation of the doping processes in poly(thienylene-vinylene). *Synth Met* 32:43. (c) Sotzing, G.A., and J.R. Reynolds. 1995. Poly[trans-*bis*(3,4-ethylenedioxythiophene)vinylene]: A low band-gap polymer with rapid redox switching capabilities between conducting transmissive and insulating absorptive states. *J Chem Soc Chem Commun* 703.

38. (a) Ho, H.A., et al. 1995. Electrogenerated small bandgap π-conjugated polymers derived from substituted dithienylethylenes. *J Chem Soc Chem Commun* 2309. (b) Sotzing, G.A., C.A. Thomas, and J.R. Reynolds. 1998. Low band gap cyanovinylene polymers based on ethylenedioxythiophene. *Macromolecules* 31:3750.

39. Seshadri, V., and G.A. Sotzing. 2004. Polymerization of two unsymmetrical isomeric monomers based on thieno[3,4-*b*]thiophene containing cyanovinylene spacers. *Chem Mater* 16:5644.

40. (a) Elsenbaumer, R.L., K.Y. Jen, and R. Oboodi. 1986. Processible and environmentally stable conducting polymers. *Synth Met* 15:169. (b) Jen, K.Y., G.G. Miller, and R.L. Elsenbaumer. 1986. Highly conducting, soluble, and environmentally-stable poly(3-alkylthiophenes). *J Chem Soc Chem Commun* 1346. (c) McCullough, R.D., and R.D. Lowe. 1992. Enhanced electrical conductivity in regioselectively synthesized poly(3-alkylthiophenes). *J Chem Soc Chem Commun* 70. (d) Loewe, R.S., et al. 2001. Regioregular, head-to-tail coupled poly(3-alkylthiophenes) made easy by the grim method: Investigation of the reaction and the origin of regioselectivity. *Macromolecules* 34:4324.

41. Patil, A.O., et al. 1987. Water soluble conducting polymers. *J Am Chem Soc* 109:1858.

42. (a) Yue, J., et al. 1991. Effect of sulfonic acid group on polyaniline backbone. *J Am Chem Soc* 113:2665. (b) Wei, X.L., et al. 1996. Synthesis and physical properties of highly sulfonated polyaniline. *J Am Chem Soc* 113:2545.

43. Cao, Y., P. Smith, and A.J. Heeger. 1992. Counter-ion induced processibility of conducting polyaniline and of conducting polyblends of polyaniline in bulk polymers. *Synth Met* 48:91.

44. (a) Jonas, F., W. Krafft, and B. Muys. 1995. *Macromol Symp* 100:169. (b) Jonas, F., and J.T. Morrison. 1997. 3,4-polyethylenedioxythiophene (PEDT): Conductive coatings technical applications and properties. *Synth Met* 85:1397. (c) Chayer, M., K. Faied, and M. Leclerc. 1997. *Chem Mater* 9:2902.

45. (a) Li, Y., G. Vamvounis, and S. Holdcroft. 2002. Tuning optical properties and enhancing solid-state emission of poly(thiophene)s by molecular control: A postfunctionalization approach. *Macromolecules* 35:6900. (b) Zhai, L., et al. 2003. A simple method to generate side-chain derivatives of regioregular polythiophene via the GRIM metathesis and post-polymerization functionalization. *Macromolecules* 36:61.

46. (a) Jang, S-Y., M. Marquez, and G.A. Sotzing. 2004. Rapid direct nanowriting of conductive polymer via electrochemical oxidative nanolithography. *J Am Chem Soc* 126(31):9476. (b) Jang, S-Y., et al. 2005. Welded electrochromic conductive polymer nanofibers by electrostatic spinning. *Adv Mater* 17:2177. (c) Jang, S-Y., G.A. Sotzing, and M. Marquez. 2002. Intrinsically conducting polymer networks of poly(thiophene) via solid-state oxidative cross-linking of a poly(norbornylene) containing terthiophene moieties. *Macromolecules* 35:7293.

47. Reisch, M., and M. McCoy. 2004. *Chem Eng News* 82:44 p. 17.

48. (a) Lee, B., and G.A. Sotzing. 2002. Aqueous phase polymerization of thieno[3,4-*b*]thiophene. *Poly Prepr* 43(2):568. (b) Lee, B., V. Seshadri, and G.A. Sotzing. 2005. Water dispersible low band gap conductive polymer based on thieno[3,4-*b*]thiophene. *Synth Met* 152:177. (c) Lee, B., V. Seshadri, and G.A. Sotzing. 2005. Suspension polymerization of thieno[3,4-*b*]thiophene in water to produce low band gap conjugated polymers. *Poly Prepr* 46:1010. (d) Lee, B., V. Seshadri, and G.A. Sotzing. 2005. Poly(thieno[3,4-*b*]thiophene)–poly(styrene sulfonate): A low band gap, water dispersible conjugated polymer. *Langmuir* 21:10797.

49. Lee, B., et al. 2005. Ring-sulfonated poly(thienothiophene). *Adv Mater* 17:1792.

12

Low Bandgap Conducting Polymers

Seth C. Rasmussen
and Martin Pomerantz

12.1 Introduction

Electrically conducting, highly conjugated polymers continue to fascinate scientists and to be the subject of many publications. This chapter addresses the recent research activity in this field and will thus concentrate on publications since the second edition of this handbook was published in 1998. For the most part, Chapter 11 [1] covered the literature through 1995 and so we will, in this updated chapter, cover the literature from 1995 to 2005.

As with the previous Chapter 11, some introductory remarks are in order. Since conducting polymers are highly conjugated materials these tend to be highly colored. The bandgap (E_g), defined as the energy difference between the top of the valence band (HOMO for smaller molecules) and the bottom of the valence band (LUMO for smaller molecules), is of such an energy, e.g., about 2.1 eV for polythiophene, that the optical absorption is in the visible region of the spectrum (1 eV = 1240 nm; 2.1 eV = 590 nm).[*] Figure 12.1 shows the bandgap in the undoped conducting polymer. This material is generally a semiconductor and is nonconducting.

However, upon doping, usually carried out by oxidation (more common) or reduction, the polymer may be rendered electrically conductive. For example, upon oxidative doping, two new states are produced in the gap between the valence and conduction bands and this gives rise to two new lower energy, longer wavelength, and optical transitions. This often produces a more highly colored polymer that absorbs light at longer wavelengths, with the original absorption decreasing in intensity and the new

---

*We will define a low bandgap polymer using the same energy cut-off as used previously, namely $E_g < 1.5$ eV [1].

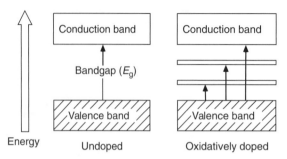

FIGURE 12.1   Optical transitions of doped and undoped conducting polymers.

absorptions increasing in intensity. These oxidized, positively charged sites, with electrons paired, are known as bipolarons.

In metals, which, of course, are inherently electrically conductive without doping, the valence and conduction bands form a single, overlapping band. This allows for easy electron mobility through the metal, hence the metallic conductivity. In doped conducting polymers, the conductivity is generally less than, and often considerably less than, that of a metal. There are a few isolated examples of conducting polymers where the conductivity approaches that of a metal like copper.

For many years scientists have tried to devise conductive organic polymers where the bandgap, between the valence and conduction bands, is zero or extremely small. The idea is that since electrical conductivity depends on electron mobility (in addition to applied voltage), zero or very low bandgaps would produce greater electron mobility and hence enhanced electrical conduction. Furthermore, as seen in Figure 12.1, as the bandgap gets smaller and smaller the optical absorptions are shifted to longer and longer wavelengths and at some point they will be in the infrared (IR) portion of the spectrum. This means that the polymer would no longer be colored and should, therefore, be transparent.

These objectives gave rise to a flurry of activity over the years and this is summarized, as indicated above, through 1995, in the second edition of this *Handbook* [1].

Clearly, an important aspect of the science and technology of conducting polymers would be to prepare and study low bandgap polymers and employ them in practical systems and devices where transparency is required and/or, where it might be possible to prepare highly conductive materials. Unfortunately, the very concept of a low bandgap polymer generally means an increase in the top of the conduction band (HOMO) and a decrease in the bottom of the valence band (LUMO). With an increase in the HOMO goes ease of oxidation (electron loss). Thus, most of the low bandgap polymers prepared up to the present are oxidatively unstable and cannot be used in devices under ambient conditions. A relatively low bandgap polymer, which will not be discussed in this chapter since its E_g is 1.6 eV, above the cutoff of 1.5 eV used here, namely poly(ethylenedioxythiophene) (PEDOT) (Figure 12.2) is an exception. It is rather transparent in its conducting, oxidized state and is quite stable toward ambient conditions [2]. Other low bandgap systems, such as poly(isothianaphthene) (PITN), synthesized over 20 y ago [3,4], are rather sensitive to atmospheric oxygen and will rapidly oxidize when exposed to the atmosphere.

Over the years several approaches have been developed to prepare lower bandgap polymers. One was to add additional π-conjugation to the polymer. One way to accomplish this was to fuse an additional

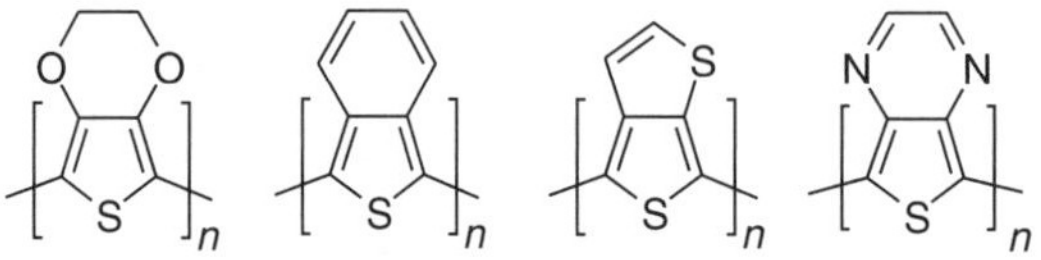

PEDOT   PITN   **1**   **2**

FIGURE 12.2   Polythiophene analogs with reduced bandgaps.

ring, such as benzo- or thieno- to produce a polymer such as PITN above. A second way to increase conjugation along the backbone of the chain is to eliminate steric effects that cause twisting. Thus, the bandgap could be lowered, e.g., by preparing regioregular head-to-tail 3-substituted polythiophenes [5]. Substituents on a conjugated polymer enhance the solubility of otherwise insoluble materials. Thus, when substituents are employed, it is important to arrange them in a regioregular way such that steric effects that would cause twisting are minimal or nonexistent. Another way to insure greater conjugation is to rigidify the individual monomer units by employing additional rings that serve to enforce planarity and this also minimizes or eliminates steric crowding [6]. Finally, substituents on individual monomers within the polymer backbone can serve to lower the bandgap. Thus, the use of alternating electron donor and acceptor units will also have a lowering effect on the bandgap. These concepts were discussed in Chapter 11 on low bandgap conducting polymers in the second edition [1].

In this chapter we will discuss progress in the preparation and study of low bandgap conducting polymers and, where appropriate, reference will be made to one of the various ways in which the bandgap was adjusted downward such that the molecule fits into the low bandgap polymer classification. There will also be compounds where it will be obvious which of these principles applies for the lowered E_g. For example, when a fused aromatic ring, such as a thiophene ring, is fused to the thiophene rings in a polythiophene as shown by structure **1** above, it should be anticipated that E_g would decrease as a result of the extra conjugation.

Initially, we wish to point out a number of review articles that appeared since the approximately 1995 cutoff of literature references in the chapter in the second edition [1]. We will not discuss the contents of each article but will refer to them as appropriate in the present chapter. There was a review in 1997 of synthetic principles involved in the control of the bandgap which discussed many of the same points as in the chapter on low bandgap, in the second edition of this handbook [6]. A review of PEDOT and derivatives appeared and because PEDOT has a bandgap of 1.6 eV, there are polymers containing ethylenedioxythiophene units in the backbone which are low bandgap ($E_g < 1.5$ eV) [2]. Several additional reviews of conducting polymers have appeared and as part of the reviews there is a brief discussion of low bandgap polymers. There is a review of polythiophene polymers [7], two general reviews [8,9], and a limited review on novel low bandgap conducting polymers [10].

12.2 Quantum Mechanical Calculations

A number of publications dealing with theoretical aspects of low bandgap conducting polymers have appeared since the second edition of this handbook. *Before this, an interesting paper appeared, in which the band structure of a variety of quasi-one-dimensional conjugated systems was examined, and in particular polymers with antiaromatic structural units were investigated [11]. In all cases where a calculational comparison could be made, the systems with antiaromatic units showed lower bandgaps than those with aromatic units. The calculations used an RHF variant of the MNDO method for geometries and the energy gaps were obtained using the Hubbard–Hückel approximation. A few examples of antiaromatic and corresponding aromatic polymers are shown in Figure 12.3. It should be noted that Dougherty had shown that poly(cyclobutadiene-1,3-diyl) had a low or zero bandgap (MNDO geometry with Hückel and valence effective Hamiltonian [VEH] band structure calculations). An interesting point that emerged from these calculations is that there is bond alternation in each of the cyclobutadienes as expected for an antiaromatic system. In a more recent paper, density functional theory (DFT) calculations showed two antiaromatic polymers (3 and 4) (Chart 12.1) that were inherent metals (zero bandgap) and three others (5–7) had low bandgaps, between 0.7 and 1.4 eV [12].*

Kertesz [13] briefly discussed design strategies, based upon calculations, for low bandgap polymers. In order to get reliable estimates of bandgaps he pointed out that it is important to start with accurate geometries because of electron–phonon coupling in these polymers. In certain systems the bandgap is a compromise between two competing minimum energy structures, aromatic and quinonoid. Further, some polymers that superficially appear to have low bandgaps based on their structures, such as some ladder polymers, in fact

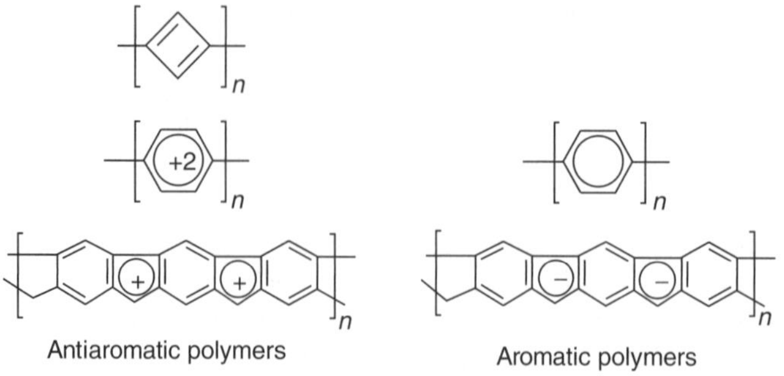

Antiaromatic polymers   Aromatic polymers

FIGURE 12.3   Examples of antiaromatic and aromatic polymers.

have large bandgaps, the result of substantial bond alternation as a consequence of π-electron delocalization. Others have very low bandgaps and the E_g values depend on how the rings are fused together.

Ratner and Marks employed a combination of DFT methods to derive ground-state electronic structures with ZINDO/CIS and time-dependent DFT (TDDFT) methods for optical properties in a multidimensional computational study of a wide variety of conjugated oligoheterocycles [14]. It was concluded that contrary to the conventional picture, midgap states lie much closer to doubly occupied valence states than to the unoccupied states. In addition, it was noted that simplistic structural metrics such as backbone bond length alternation and interannular dihedral angles are not particularly effective at predicting the overall electronic structure. This is due in part to the fact that very significant interaction effects exist due to the interplay of contributions from the heteroatom, substituents, and conjugation length. Thus, it was concluded that in order to design systems with reduced bandgaps, modeling potential individual systems is more reliable than attempting to sum individual effects.

PITN and poly(thieno[3,4-*b*]pyrazine) (**2**) have been recalculated using a variety of methods [15]. Oligomer geometries were optimized for octamers (AM1) and for tetramers (B3LYP/6-31G*) while excitation energies were calculated using ZINDO/(INDO/S), CIS, configuration interaction singles, and TDDFT theory. Bandgaps were extrapolated from the excitation energies. The benzo[*c*]thiophene oligomers were calculated to be nonplanar with interannular torsional angles of 40–45°, while the thieno[3,4-*b*]pyrazine oligomers were calculated to be planar, thus resulting in a lower predicted bandgap for **z** in comparison with PITN. The values calculated were dependent on the methodology used, but the best methods, ZINDO//DFT and TDDFT//DFT, gave values close to the experimental values; 1.13 and 1.47 eV, respectively for PITN, compared with the 1.0–1.2 eV experimental values [13,15] and 0.49 and 0.47 eV, respectively for **2**, compared with 0.9 eV experimental value for the dihexyl-derivative [16,17].

A number of papers have appeared describing calculations of polymers based on bridged bithiophenes and their carbon, silicon, or germanium analogs (Chart 12.2). In 1995, Toussaint and Brédas [18] reported band structure calculations on polymer **10** with the VEH technique. The calculated bandgap

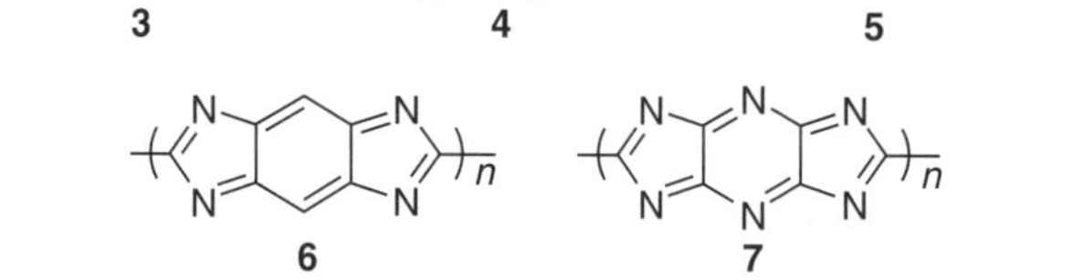

CHART 12.1   Calculated antiaromatic polymers.

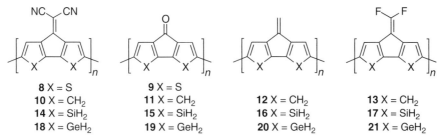

CHART 12.2   Polymers based on bridged bithiophenes and their analogs.

was determined to be 0.16 eV, which was 0.4 eV lower than the previously calculated value for polymer **8** using the same methods.

Bakhshi and coworkers [19–22] then reported studies of the electronic structures of polymers **8–21** in which the ab initio Hartree–Fock crystal orbital method was applied to optimized unit cell geometries obtained from the MNDO–AM1 method. It was pointed out that this method largely overestimated all E_g values, but correctly reproduced the trends observed in experimental values. For example, E_g values of 5.92 and 6.26 eV were determined for polymers **8** and **9**, respectively [19]. While these values are quite high, the differences between the two calculated values agreed well with the differences in the reported experimental values of 0.8 [23] and 1.2 eV [24]. For the most part, it was found that in each series the E_g values decreased with changing X in the following order; $S > GeH_2 > SiH_2 > CH_2$. Trends were also observed with changing the bridge, with the carbonyl unit giving the highest E_g values and the dicyanomethylene unit giving the lowest values. The lowest calculated E_g was 3.43 eV for polymer **18**, which the authors estimated to translate to an expected actual value of less than 1.0 eV [20]. The authors have also reported calculations of polymers utilizing fluorinated units for X (CF_2, SiF_2, and GeF_2), but the use of fluorine atoms in place of hydrogen seemed to have little effect [25,26].

Hong and coworkers [27–29] have reported electronic studies on polypentafulvalene (**22**) and a number of its functionalized analogs (Chart 12.3, **23–27**). The band calculations were performed at the AM1 level, which was implemented by MO-SOL, a solid state version of semiempirical methods that

CHART 12.3   Polypentafulvalene and various analogs.

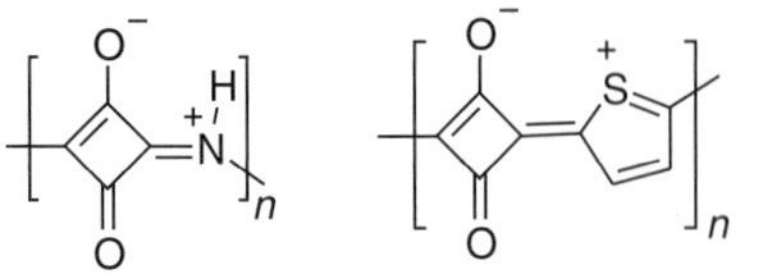

28   **29**

CHART 12.4   Squaraine-derived polymers.

adopts the Born-von Kármán periodic boundary condition and Bloch functions for crystal calculations. The E_g of polymer **22** was determined to be 1.13 eV, while the functionalized derivatives **23** and **24** were determined to fall in the range of 0.56–1.45 eV [27,28]. The lowest values were determined for derivatives **23** containing electron-donating groups (**23a–c**), while the isomeric polymers **24a–c** gave significantly higher values. The addition of electron-withdrawing groups did not show as clear a trend. The fused-ring polymers **25b** and **25c** exhibited extremely low values of only 0.3 eV [28,29], while the remaining polymers **25a**, **26**, and **27** gave values between 0.77 and 1.16 eV [29].

Lastly, two papers have reported electronic structure calculations on squaraine-based polymeric systems. Brocks [30] reported a study on polyaminosquaraine (Chart 12.4, **28**) that utilized ab initio calculations within the local density functional approximation (LDFA). The calculated E_g from these methods was 0.3 eV, to which a scaling factor was applied to give a final predicted E_g of 0.5 ± 0.1 eV. The addition of conjugated spacers to the parent structure resulted in an increase of the E_g value [31]. More recently, Dudis and coworkers [32] reported a study on poly(1,3-thienylsquaraine) (**29**). The methods employed utilized optimized geometries of multiple oligomers ($n = 1$–6) by AM1 calculations, followed by crystal orbital calculations for the periodic polymer chain using extended Huckel molecular orbital (EHMO) theory. The crystal orbital calculations gave an E_g value of 0.61 eV, in good agreement with the value reported for polymer **28**. The authors also point out that extrapolation of the HOMO–LUMO gap of the six calculated oligomers would result in an E_g below 0.3 eV.

12.3   Poly(isothianaphthene) and Related Fused-Ring Systems

As the first reported low bandgap polymer, PITN is still of considerable interest and a significant number of publications covering its preparation and derivatives have appeared since the second edition of this handbook. While the majority of these reports have dealt with various PITN derivatives, new methods for the preparation of the parent PITN have also been investigated. Before 1995, the preparation of PITN had focused primarily on the oxidative polymerization of either isothianaphthene or dihydroisothianaphthene. Starting in 1995, however, Vanderzande and coworkers reported a significantly different preparation method for PITN [33,34]. This method involved the reaction of phthalic anhydride (**30**) or phthalide (**31**) with P_4S_{10} at temperatures above 120°C (Scheme 12.1).

While these thermal polymerization reactions could be carried out by equimolar melts, it was found that such conditions were difficult to control on larger scales and the use of high boiling solvents gave more consistent results. The use of solvents such as mesitylene and 1,2-dichlorobenzene gave quantitative yields after 16 h, but the products isolated indicated significant amounts of impurities and defects. The use of the lower boiling xylene resulted in reduced conversion yields,

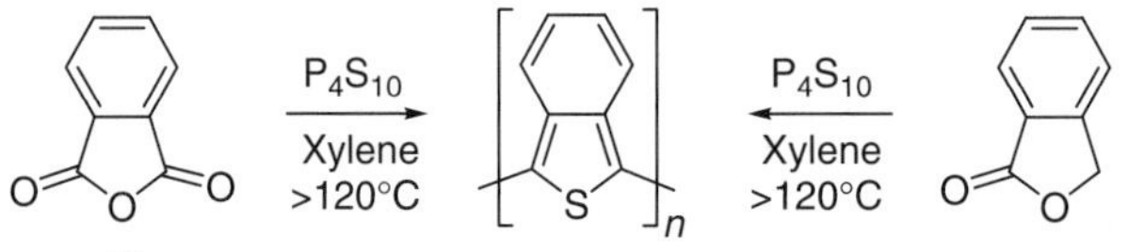

30   **PITN**   **31**

SCHEME 12.1   Thermal polymerization of PITN.

but allowed the isolation of better-defined products. The products produced in this manner exhibited IR and Raman spectra consistent with PITN, but elemental analysis did indicate higher than expected sulfur content and residual phosphorus contaminants. While E_g values were not reported, doped samples gave conductivities of 0.05–10 S cm^{-1} [33], similar to previously reported PITN values [4,35].

After identification of various intermediate compounds, a proposed mechanism of this process was reported in 1996 (Scheme 12.2) [34,36]. It was found that by limiting the reaction time to 3 h, only a trace of PITN was produced and the majority of the reaction mixture was composed of unreacted phthalic anhydride (**30**). However, the mixture was also found to contain a number of intermediates (Scheme 12.2, **32–35**) that could be identified by mass spectrometry. Intermediates **32** and **34** were also independently synthesized and successfully reacted with P$_4$S$_{10}$ to produce PITN in both shorter periods of time and with higher yields than by the initial anhydride **30**. The rate-determining step of the polymerization process was determined to be the initial conversion of carbonyl to thiocarbonyl through reaction of **30** with P$_4$S$_{10}$.

Although intermediate **36** was never detected in the mechanistic studies described above, it was thought to be a viable intermediate in the polymerization mechanism [34,37]. To further investigate this possible intermediate, 1,1,3,3-tetrachlorothiophthalan (**37**) was subjected to conditions that should result in the production of the intermediate **36** (Scheme 12.3). When allowed to react for extended periods of time (20 h), these conditions produced PITN as characterized by spectroscopy (IR, Raman, and solid state NMR), elemental analysis, and conductivity [37]. To further test that the applied conditions were generating the desired intermediate **36**, the reactions were also performed in the presence of an excess of thiobenzophenone as a trapping agent. The presence of the trapping agent resulted in no formation of PITN and successfully produced the trapped product 3-benzhydrylidene-3*H*-benzo[*c*]thiophene-1-thione (**38**) [37].

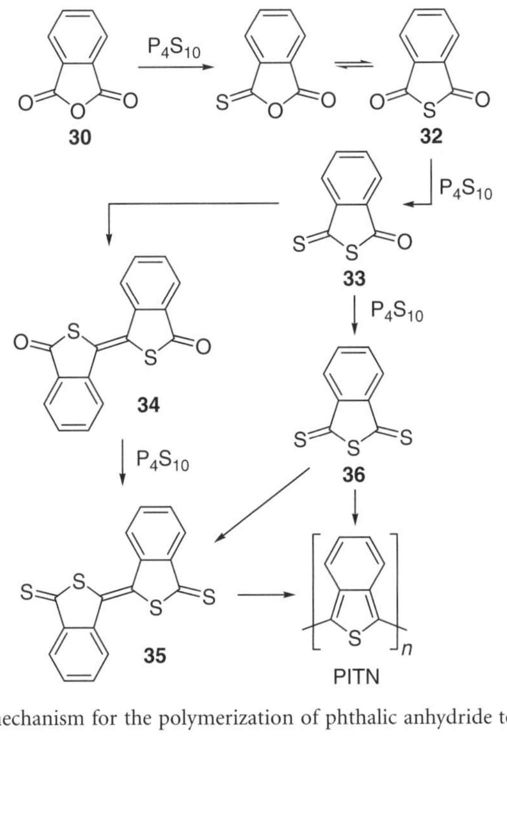

SCHEME 12.2 Proposed mechanism for the polymerization of phthalic anhydride to PITN.

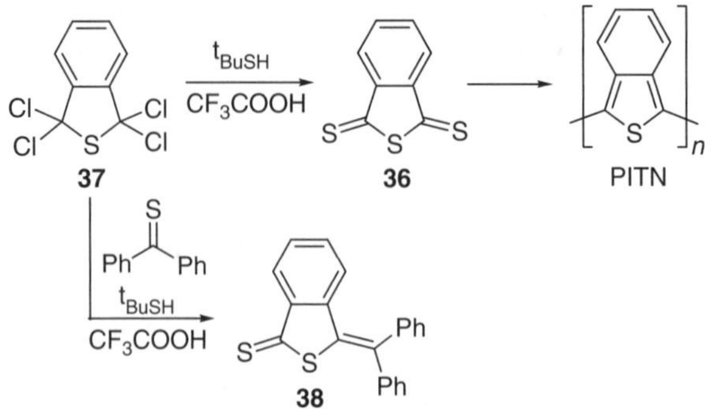

SCHEME 12.3   PITN from 1,1,3,3-tetrachlorothiophthalan (**15**).

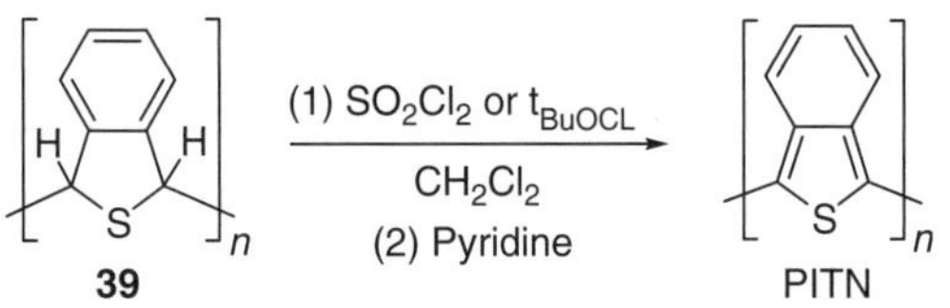

SCHEME 12.4   PITN from a soluble precursor polymer.

Also in 1995, Chen and Lee [38,39] reported the production of stable PITN solutions that could be cast to provide PITN films. These methods utilized the soluble precursor poly(1,3-dihydroisothianaphthene) (**39**) treated with a dehydrogenation agent (SO_2Cl_2 or ^tBuOCl) in order to provide a PITN solution (Scheme 12.4). It was reported that these solutions could be stabilized by the addition of pyridine, which would then allow its use for the casting of PITN films. The solutions prepared in this manner were proposed to be composed of aggregated PITN microgels with diameters below 0.25 μm [40]. The resulting cast films were reported to exhibit conductivities of 0.1 S cm^{-1} and high electrochromic contrast.

In a 2002 paper, Bowmaker and coworkers [41] extended the initial study of the oxygen-promoted polymerization of dihydroisothianaphthene (**40**) to PITN reported by Jen and Elsenbaumer [35] in 1986 (Scheme 12.5). As previously shown, this method could produce PITN as a blue powder. However, by treating solutions of **40** on various substrates, thin films of PITN could also be obtained which allowed characterization of the solid state optical properties. In the generation of the polymeric samples, this new study showed that light, in addition to oxygen, was required to accelerate the polymerization. In the absence of light, dibenzoyl peroxide could be used to successfully generate PITN. In the absence of additional doping, samples generated by either method gave conductivities of 10^{-4} S cm^{-1}, suggesting the as-prepared samples were partially doped. This was later verified by EPR studies of the as-prepared materials in comparison with dedoped samples [42]. Conductivities of iodine-doped samples were determined to be $1.5–1.6 \times 10^{-1}$ S cm^{-1}.

The optical spectra of the oxygen-polymerized samples exhibited a blueshift ($\lambda_{max} = 570$ nm) in comparison with electrochemically produced PITN, while the samples prepared using the dibenzoyl peroxide exhibited lower energy transitions at 670 nm [41]. It was concluded that the difference in the spectra was a result of different polymer chain lengths produced by the various polymerization methods, with the dibenzoyl peroxide method giving the greatest chain lengths.

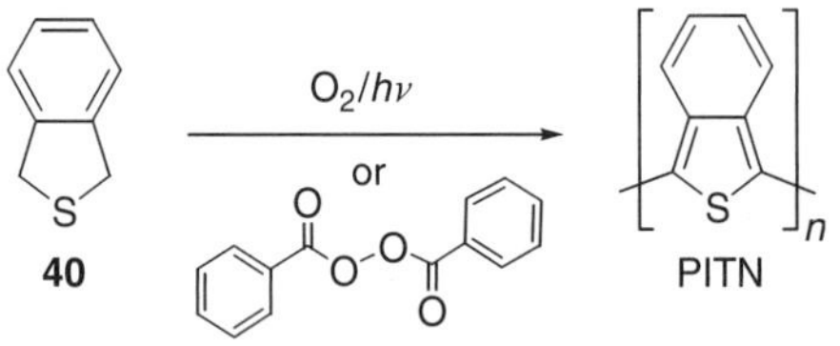

With the aim of tuning the properties of PITN, studies of a number of new substituted analogs have also been reported (Chart 12.5). This included two new alkyl derivatives containing simple methyl (**41a**) [43] or *tert*-butyl (**41b**) [44–47] functionalities. The methyl analog **41a** was reported to give E_g values of

SCHEME 12.5   Oxygen-promoted polymerization of dihydroisothianaphthene.

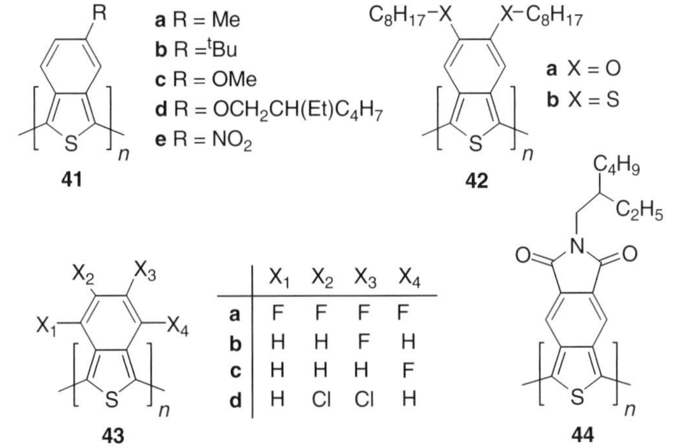

CHART 12.5 Functionalized PITN derivatives.

1.13 eV, consistent with the 1.0–1.3 eV values previously reported by Pomerantz and coworkers [17] for poly(5-decylisothianaphthene).

The *tert*-butyl analog **41b** was prepared independently by two groups with the first publication appearing in 1994 by Burbridge and coworkers [44]. In this initial report, polymer **41b** was prepared by oxidative polymerization of the functionalized dihydro precursor using either oxygen or $FeCl_3$. The resulting materials were reported to be soluble in common organic solvents, allowing the structure of **41b** to be confirmed by NMR spectroscopy. While GPC indicated low M_w samples by oxygen-promoted polymerization, the $FeCl_3$-polymerized material was reported to give an M_w of ~11,000. Solvent cast films of the high M_w samples exhibited maxima at ~750 nm with a corresponding E_g value of ~1.2 eV. The microscopic third-order nonlinearity of the materials as $CHCl_3$ solutions was also studied to give a γ value of 4×10^{-45} m^5 V^{-2} [44,45].

Following this, Vanderzande and coworkers [46] reported the preparation of **41b** through the reaction of a *tert*-butyl-functionalized phthalic anhydride with P_4S_{10}. The resulting material was reported to be only partially soluble in $CHCl_3$ and THF, with the soluble fractions giving M_w of 1500–3700. Solid state optical spectra exhibited maxima at 788 nm with an E_g of 1.42 eV, and the conductivity of the undoped material was determined to be as high as 4.0×10^{-4} S cm^{-1}. Solid state ^{13}C NMR data suggested a quinonoid structure for the material [47].

In a brief report, Hagan and coworkers [48] prepared three additional PITN derivatives, including two alkoxy-substituted analogs (**41c** and **41d**) and a NO_2 derivative (**41e**). These polymers were also prepared by the phthalic anhydride/P_4S_{10} method of Vanderzande and coworkers [33,34] and E_g values of 0.9, 0.75–1.0, and 1.0 eV were reported for **41c–e**, respectively. Little other characterization was given, but it was mentioned that **41d** was soluble in organic solvents.

More recently, Loi and coworkers reported the preparation of the dialkoxy and dialkylmercapto analogs, **42a** and **42b**, respectively [49,50]. The polymers were prepared by thermal polymerization of the difunctionalized dithiophthalide **46** (X = O, S), which was obtained from the phthalic thioanhydride intermediate **45** (X = O, S), as shown in Scheme 12.6. The resulting polymers (M_w = 6667; PDI 2.26) were reported to be highly soluble in $CHCl_3$ and other common organic solvents. Spectra of spin-cast films of **42b** exhibited a low energy transition at 710 nm with an onset corresponding to an E_g of ~1.16 eV. This value was in good agreement with the electrochemical measurements, which gave E_g values of 1.22 eV for **42a** and 1.16 eV for **42b**.

In an attempt to produce a PITN derivative less susceptible to atmospheric oxidation, the tetra-fluorinated analog **43a** was prepared in 1993 and was reported to be surprisingly soluble in organic solvents [51]. The E_g of **43a** was determined to be 2.1 eV, significantly higher than the corresponding

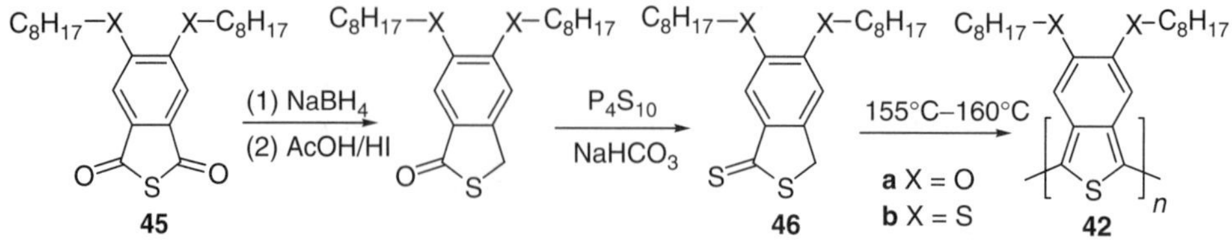

SCHEME 12.6 Thermal polymerization of PITN derivatives.

value of the parent PITN. The high E_g value was explained to be a consequence of the electronic effects of the fluorine atoms and increased steric hindrance. However, it must be noted that the E_g value was estimated from a solution spectrum, thus introducing solvent effects and limiting interchain coupling, both of which could also contribute to increases in E_g. NMR and Raman spectral studies were then undertaken to investigate the possibility that the increased sterics of the fluorine functionalities could result in a highly twisted aromatic structure [52,53]. While the Raman studies were inconclusive, the NMR results did suggest an aromatic structure for **43a**.

Higgins and coworkers [43,54] also reported several other halogen-substituted analogs (**43b–d**). It was found that halogen substitution at the 5- or 6-positions (**43b** and **43d**) had little effect on the oxidation potential, but did significantly lower the potential necessary to reduce the material. Surprisingly, this did not result in significant changes in the E_g values (0.95 eV for **43b**). As a possible explanation, it was suggested that the observed reduction may involve orbitals not involved in the π–π^* transition commonly associated with the bandgap transition [43]. For the 4-substituted analog **43c**, it was noted that the unusual color of the neutral form and the marked asymmetry of the redox processes suggest significant increase in the interannular torsional angle. It was explained that this would thus lead to an increase in E_g, although the actual E_g value for **43c** was not reported.

In an attempt to modify the HOMO and LUMO energies, Meng and Wudl [55] reported PITN derivative **44**, which contains an electron-withdrawing dicarboximide functionality. Polymer **44** was prepared by FeCl$_3$ polymerization of the corresponding benzo[c]thiophene-N-(2'-ethylhexyl)-4,5-dicarboximide monomer. Such methods resulted in a CHCl$_3$-soluble polymer of high M_w ($\sim$88,000). Visible spectra of spin-coated films exhibit maxima at 778 nm, which is blueshifted in comparison with the CHCl$_3$ solution spectra maxima at 832 nm. The absorption edge of the solid state spectra corresponds to an E_g of 1.24 eV, which is in good agreement with the E_g value of 1.29 eV determined from the electrochemical onset potentials. While the E_g is comparable with other PITN derivatives, the inclusion of the electron-withdrawing group did successfully shift the HOMO and LUMO energies. As a result, polymer **44** exhibits a higher oxidation potential than PITN, but is also easier to reduce and the shifts in redox potentials contribute to increased polymer stability.

a R = CH$_3$
b R = C$_2$H$_5$
c R = C$_6$H$_{13}$
d R = C$_8$H$_{17}$
e R = C$_{10}$H$_{21}$
f R = C$_{11}$H$_{23}$
g R = C$_{12}$H$_{25}$
h R = C$_{13}$H$_{27}$

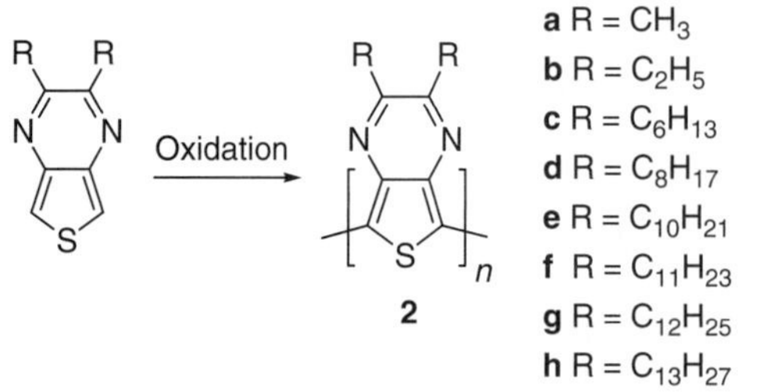

SCHEME 12.7 Oxidative polymerization of 2,3-dialkylthieno[3,4-b]pyrazines.

A number of studies on the related polymer, poly(thieno[3,4-b]pyrazine) (**2**), and its derivatives have also been reported since the second edition of this handbook. The first example of a poly(thieno[3,4-b] pyrazine) was reported by Pomerantz and coworkers [16,17] in 1992 with the FeCl$_3$ polymerization of 2,3-dihexylthieno[3,4-b]pyrazine to give **2c** (Scheme 12.7) exhibiting an E_g of 0.95 $\pm$ 0.10 eV.

Following this initial work, Kuzmany and coworkers published studies on a series of poly(2,3-dialkylthieno[3,4-b]pyrazine)s (**2a–c**, **2f**, and **2h**) in 1995 [56,57]. The polymers were prepared by FeCl$_3$ polymerization of the corresponding monomers in a manner similar to that reported by Pomerantz and coworkers [16,17]. Unfortunately, little general polymer characterization was reported, with the exception of some optical data for **2c** ($E_g \sim 0.9$ eV) [57]. The reports primarily focused on Raman spectral studies in an attempt to characterize the nature of the geometries of the ground-state polymers, as well as the corresponding doped materials. From this work, the authors concluded that the polymers studied were quinonoid in the ground state and aromatic upon doping [56,57].

The unfunctionalized parent polymer **2** was reportedly prepared from pyrazinedicarboxylic anhydride and P_4S_{10} in a manner analogous to the thermal polymerization of PITN [33,58]. The resulting material was reported to be completely insoluble, and as with the PITN analogs prepared in this fashion, elemental analysis indicated excess sulfur content and residual phosphorus contaminants. However, the sulfur content for the samples of polymer **2** were nearly twice the expected values and the phosphorus impurities were four times that found in the analogs of PITN samples. Pressed pellets doped with $NOBF_4$, I_2, or $FeCl_3$ gave conductivities of $5-7 \times 10^{-5}$ S cm^{-1} [33], significantly lower than that measured for **2c** [16,17]. Hagan and coworkers [48] later produced samples of **2** using the same methods and reported the E_g to be 1.0 eV.

More recently, Kenning and Rasmussen [59] reported the first electropolymerization of 2,3-dialkylthieno[3,4-*b*]pyrazines in 2003. In contrast to poly(thieno[3,4-*b*]pyrazine) samples produced by other methods [16,17,48,57], the electropolymerized materials exhibited E_g values of ~ 0.7 eV (Table 12.1). The reduced E_g values agree well with the originally calculated value of Marynick [60] and are close to the ~ 0.5 eV values calculated by Kwon and McKee [15].

The exact factors contributing to the reduction in E_g for the electropolymerized samples are still under investigation. However, it was reported that the basicity of the thieno[3,4-*b*]pyrazine nitrogen atoms ($pK_a \sim 1.66$) [61] could promote iron binding and analysis of samples prepared by $FeCl_3$ polymerization did indicate iron impurities, even after extensive purification [59]. Such iron impurities could also account for the paramagnetic nature of **2c** reported by Pomerantz and coworkers [16]. Another potentially contributing factor discovered during the electropolymerization studies is the rather strong susceptibility of poly(thieno[3,4-*b*]pyrazines) to overoxidation. Such difficulties with overoxidation have previously been reported for PITN at potentials above ~ 1.3 V [4], but for the poly(thieno[3,4-*b*]pyrazine)s this process begins at ~ 0.9 V. It was even found that attempts to electropolymerize at potentials above 1.4 V failed due to the strong competition with overoxidation [59]. Due to the high oxidizing potential of $FeCl_3$ (~ 2.1 V in $CHCl_3$), the use of $FeCl_3$ in oxidative polymerization would cause significant overoxidation, potentially leading to a reduction in conjugation. Lastly, while the electropolymerized samples were reported to be soluble in organic solvents (with the exception of **2a**), the preparation of larger sample sizes show that only a portion of these electropolymerized samples are completely soluble and it is likely that they contain higher M_w species than the fully soluble fractions isolated by the $FeCl_3$ polymerizations ($M_w = 3844$, $M_n = 3636$) [62].

TABLE 12.1 Solid State Absorption and Bandgaps
for Poly(dialkylthieno[3,4-*b*]pyrazine)s

Polymer[a]	Alkyl	λ_{max} (nm)	E_g (eV)[b]
2a	CH_3	1275	0.66
2c	C_6H_{13}	1365	0.69
2d	C_8H_{17}	1410	0.69
2e	$C_{10}H_{21}$	1400	0.67
2g	$C_{12}H_{25}$	1260	0.79

[a] *Source*: Kenning, D.D. and Rasmussen, S.C., A second look at polythieno [3,4-*b*]pyrazines: Chemical vs. electrochemical polymerization and its effect on band gap, *Macromolecules* 36, 6298, 2003.
[b] Optical bandgap.

12.4 Copolymers of Poly(isothianaphthene) and Related Fused-Ring Systems

In addition to the homopolymeric systems discussed above, a wide variety of mixed copolymeric systems incorporating isothianaphthene, thieno[3,4-*b*]pyrazines, and related materials have also been reported. The most common approach is the endcapping of a tailored fused-ring system (e.g., isothianaphthene) with external thiophene units to produce symmetrical mixed terthienyl precursors, which can in turn be polymerized by oxidative polymerization to produce the desired copolymeric material.

One of the first examples of this approach was reported in 1992 with the polymerization of the terthienyl **47a** that contains a central isothianaphthene unit [63–66]. More recently, these methods have been adapted to allow the production of functionalized derivatives **47b–g** (Scheme 12.8) [67–71]. The terthienyls **47b–g** were then polymerized by $FeCl_3$ polymerization to give the corresponding polymers **48b–g**. The related polymer **48h** was also prepared by similar methods. Although polymer **48b** was insoluble, polymers **48c–h** were soluble in $CHCl_3$, CH_2Cl_2, and THF and gave M_w of 5600–7600. The corresponding E_g values and conductivities are given in Table 12.2. As can be seen, polymers **48e–g** not only have the added benefit of solubility, but also give both lower E_g values and higher conductivities than the unfunctionalized polymer **48a**. Polymers **48d**, **48f**, and **48h** have found applications in photovoltaic devices [70,71].

Wudl and coworkers [72] reported a related analog utilizing a central benzo[*c*]thiophene-*N*-(2′-ethylhexyl)-4,5-dicarboximide in 2003. The terthienyl precursor **49** was prepared by Stille coupling (Scheme 12.9) to give a system that endcapped the central electron-poor unit with electron-rich EDOT donors. Terthienyl **49** was then electrochemically polymerized to give films of polymer **50**. The resulting polymer films exhibit a low energy transition at 780 nm, along with second higher energy absorption at 440 nm that was attributed to the dicarboximide group. The absorption onset of the low energy transition corresponds to an E_g of 1.10 eV, which was in excellent agreement with the difference of electrochemical onsets for p- and n-doping.

Wudl and coworkers also prepared the related alternating copolymer **51** by Stille coupling as shown in Scheme 12.10 [73,74]. The resulting material was blue in color and highly soluble in common organic solvents such as $CHCl_3$. Spin-cast films exhibited a λ_{max} at 807 nm with a tailed onset at 1240 nm ($E_g = 1.0$ eV). The difference of the redox onsets gave a very similar E_g value of 1.10 eV and both values are quite similar to the related polymer **50**.

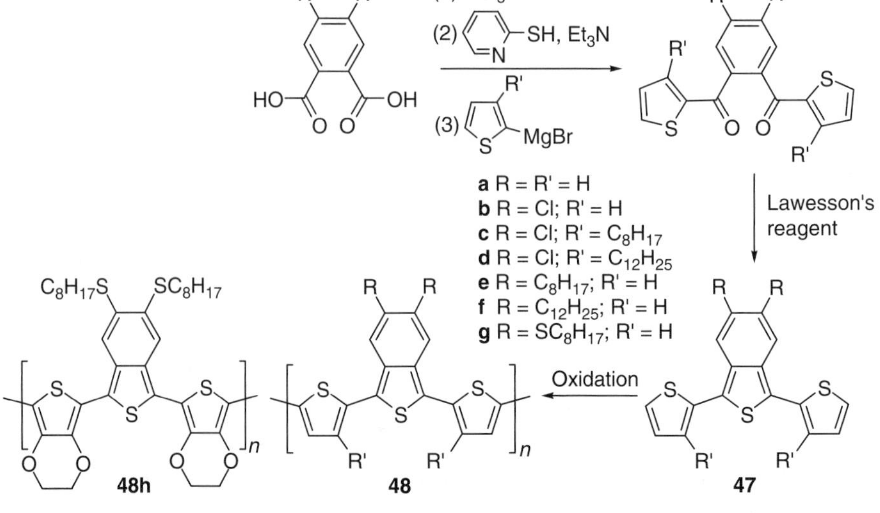
R R
(1) PCl5
(2) SH, Et3N
N
R'
HO OH
O O
(3) S MgBr
R R
R'
S
S
O O
R'
a R = R' = H
b R = Cl; R' = H
c R = Cl; R' = C8H17
d R = Cl; R' = C12H25
e R = C8H17; R' = H
f R = C12H25; R' = H
g R = SC8H17; R' = H
Lawesson's reagent
C8H17S SC8H17
R R
R R
Oxidation
48h
48
47

SCHEME 12.8 Synthesis of functionalized poly(dithienylisothianaphthene)s.

TABLE 12.2 Bandgaps and Conductivities of Various Poly(dithienylisothianaphthene)s

Polymer	E_g (eV)[a]	Conductivity[b] (S cm^{-1})	References
48a	1.58–1.77	5.0×10^{-3}	63
48c	1.76	1.2×10^{-2}	68, 69
48d	1.8	—	70, 71
48e	1.50	4.6	68, 69
48f	1.5	—	70, 71
48g	1.41	38.6	68, 69
48h	1.2	—	70, 71

[a] Optical bandgap.
[b] **48c**, **48e**, and **48g**: I$_2$ doped; **48a**: electrochemically doped.

Probably the most prolific class of such terthienyl polymer precursors utilize central thieno[3,4-*b*] pyrazine units (Scheme 12.11). Yamashita and coworkers [75] reported the first examples of this class in a 1994 communication, followed by a more in-depth study in 1996 [76]. The terthienyls **52a–e** were prepared in a similar manner to the synthesis of the parent thieno[3,4-*b*]pyrazines [61] and then polymerized by chemical or electrochemical oxidation (Scheme 12.11).

Electropolymerization produced polymers **53a–d** as dark blue-green to blue-black films. All films were completely insoluble, but showed relatively high stability under normal conditions. Although the longer side chains of polymers **53c** and **53d** did not produce soluble materials, their inclusion did lead to smoother polymer films than the powdery deposits of **53a** and **53b** [76]. Both the CV and optical spectra of **53a** gave an E_g of 1.0 eV [75,76], considerably lower than the analogous isothianaphthene-based polymer **48a** (1.58–1.77 eV) [63–66]. The incorporation of 2,3-dialkylthieno[3,4-*b*]pyrazine units in polymers **53b–d** led to a slight increase in E_g of less than 0.3 eV.

The addition of further alkyl functionalities in **52e** resulted in a failure to produce polymeric films of **53e** by electrochemical methods. However, polymer **53e** was successfully obtained using Cu(ClO$_4$)$_2$ as a chemical oxidant to give a material soluble in THF and CHCl$_3$ ($M_w = 8600$, $M_n = 5800$) [75]. While the additional side chains provided solubility, this also resulted in the highest determined E_g (1.5 eV). This is most likely due to some loss of planarity as a result of chain–chain interactions between the two exterior side chains of the terthienyl repeat unit.

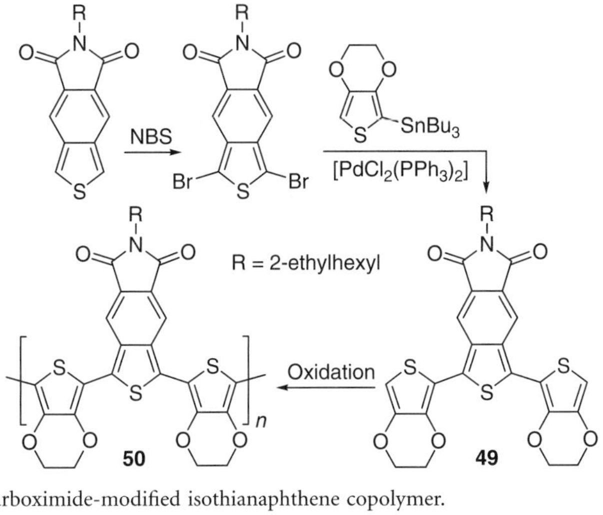

SCHEME 12.9 A dicarboximide-modified isothianaphthene copolymer.

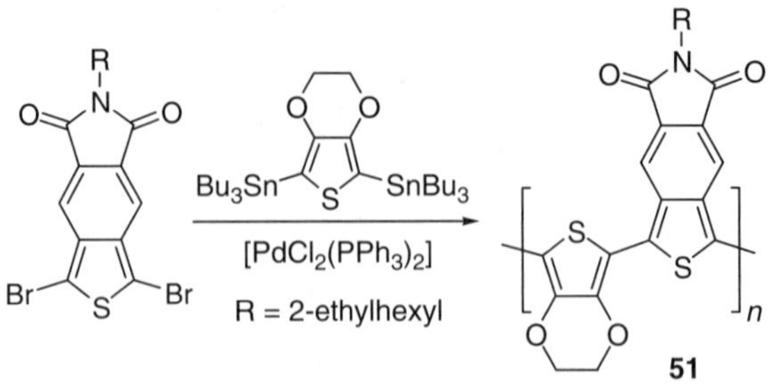

SCHEME 12.10   Alternating isothianaphthenedicarboximide–EDOT copolymer.

More recently, Wudl and coworkers reported the preparation of polymers **53f** and **53g** containing central 2,3-(3-thienyl)thieno[3,4-*b*]pyrazine units as new electrochromic materials [77–79]. While polymer **53f** was completely insoluble, the alkyl side chains of polymer **53g** provided solubility in chlorinated solvents. NMR spectra of **53g** are consistent with the structure given in Scheme 12.11, thus confirming that coupling of the two peripheral 3-thienyl functionalities at the 2,3-positions of the thieno[3,4-*b*]pyrazine ring does not occur under the polymerization conditions [77,78]. M_w studies of **53g** suggest the maximum degree of polymerization to be 14, corresponding to a backbone length of 42 thiophenes.

Both materials give green films, with **53f** exhibiting transitions at 370 and 725 nm separated by a valley at 550 nm [77,79]. The spectra of **53g** are fairly similar with bands at 410 and 670 nm, along with shoulders at 485 and 790 nm. From the spectral onsets, the E_g values were determined to be 1.1 eV for **53f** and 1.3 eV for **53g** [77,78]. The increased E_g resulting from the additional side chains of **53g** is consistent with the similar increase previously discussed for **53e**.

Thieno[3,4-*b*]pyrazine terthienyls utilizing external EDOT groups have also been reported by Zotti and coworkers [80]. Unlike the thieno[3,4-*b*]pyrazine-based terthienyls **52a–g** discussed above, these systems were prepared by Stille coupling from 5,7-dibromothieno[3,4-*b*]pyrazine in a manner analogous to the preparation of **49**. Electropolymerization of the resulting terthienyls produced insoluble polymer films of **54a** and **54b** (Chart 12.6). Optical spectra of **54a** exhibit a λ_{max} at 950 nm, with a reported E_g of

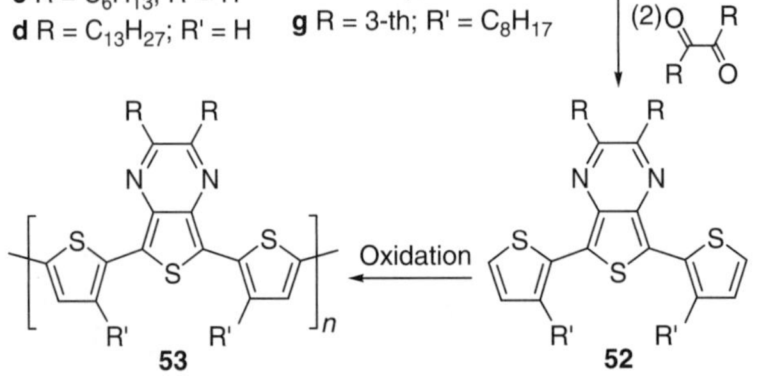

SCHEME 12.11   Synthesis of various poly(dithienylthieno[3,4-*b*]pyrazine)s.

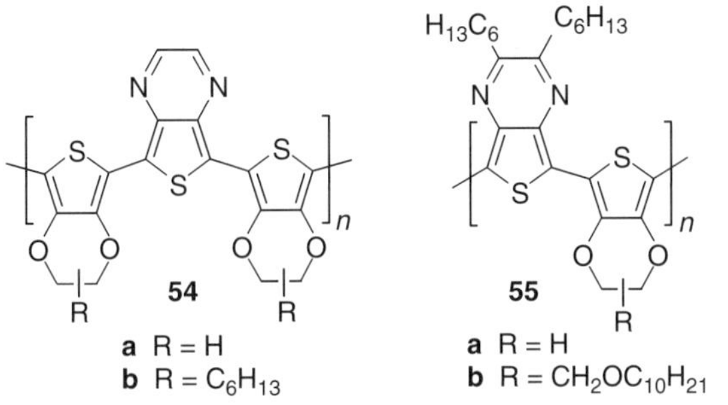

CHART 12.6 Thieno[3,4-*b*]pyrazine–EDOT copolymers.

1.3 eV [80]. However, it should be noted that this reported E_g value corresponds to the given peak maximum, whereas an E_g of ~0.86 eV may be estimated from the onset of the published spectra. In comparison with **53a**, this lower value is consistent with the expected reduction in E_g due to the addition of the alkoxy functionalities of the EDOT units. Electrochemically oxidized films gave p-doped conductivities of 0.5 and 15 S cm^{-1} for **54a** and **54b**, respectively, while reduced films gave lower n-doped values of 0.01 and 0.03 S cm^{-1} [80].

Roncali and coworkers [81–83] reported the related polymers **55a** and **55b** as shown in Chart 12.6. These polymers were prepared by oxidative polymerization of the bithienyl precursors, which were obtained in a manner similar to the thieno[3,4-*b*]pyrazine terthienyls shown in Scheme 12.10. While polymer **55b** exhibits properties consistent with analogs **54a–b** ($\lambda_{max} = 1070$ nm; $E_g = 0.83$ eV) [83], the E_g of **55a** is reported to be 0.36 eV [81,82]. This latter value is believed to be one of the lowest E_g values reported for a conjugated polymer. The significant difference between the **55a** and **55b** was attributed to weakening of the interchain interactions as a result of the sterics introduced by the EDOT side chain [83].

Tanaka and Yamashita [84,85] investigated the effects of extending the ring fusion of the thieno[3,4-*b*]pyrazine unit through the use of a central [1,2,5]thiadiazolo[3,4-*b*]thieno[3,4-*e*]pyrazine unit as shown in Scheme 12.12. Due to its high reactivity, attempts to isolate the unfunctionalized [1,2,5]thiadiazolo[3,4-*b*]thieno[3,4-*e*]pyrazine unit were unsuccessful, but its incorporation into the terthienyl precursor allowed its isolation and characterization [85]. The preparation of the monomeric terthienyl was accomplished by condensation of diaminoterthiophene (**56**) with dimethyl oxalodiimidate, followed by cyclization by SOCl$_2$. The central dihydropyrazine ring was then aromatized by oxidation with NiO$_2$ to produce the desired terthienyl **57**.

Electrochemical polymerization of **57** produced films of polymer **58** that exhibited an oxidation at 0.80 V and a reduction at -0.19 V. From the separation of the redox onsets, an E_g of 0.3 eV was determined. Electronic spectra of the films exhibit a low energy transition at 990 nm, with an onset

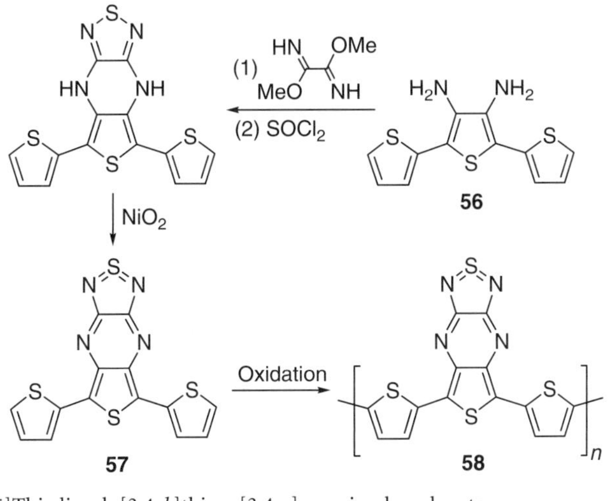

SCHEME 12.12 [1,2,5]Thiadiazolo[3,4-*b*]thieno[3,4-*e*]pyrazine-based systems.

below 0.5 eV [84,85]. Due to the conflicting absorption of the ITO electrode below 0.5 eV, determination of the exact onset was not possible [84]. The electrochemical and optical values both agree closely with the E_g of 0.47 eV recently predicted by theoretical calculations [86].

More recently, Wudl and coworkers have reported a terthienyl utilizing a central C_{60}-functionalized thieno[3,4-*b*]pyrazine (Scheme 12.13) [87]. Terthienyl **59** was electropolymerized to produce polymer **60** as greenish-brown thin films on either platinum or ITO. The CV of **60** was reported to exhibit three reversible and one quasireversible redox waves at −0.43, −0.93, −1.39, and −1.94 V (vs. Ag/Ag$^+$). An E_g of 0.215 eV was determined from the onsets of the p- and n-doping redox processes (−0.050 and −0.265 V, respectively). The p- and n-doping onset potentials were also determined by differential pulse voltammetry (DPV) to give an even lower E_g of 0.176 eV.

Spectroelectrochemistry of the neutral polymer deposited on ITO revealed a transition at 880 nm with a long tailing below 1400 nm. Ignoring the tailing, the extrapolated onset gave an E_g value of 0.88 eV. However, if the tailing were used for the onset, the E_g would fall below 0.3 eV and agree with the previous electrochemical data [87].

Inganas and coworkers have also produced a three-component copolymeric system utilizing thieno[3,4-*b*]pyrazine units for applications in solar cells [88]. Polymer **62** was prepared by Suzuki cross-coupling of the thieno[3,4-*b*]pyrazine-based terthienyl **61** with a fluorene unit as shown in Scheme 12.14. The resulting material was mildly soluble in CHCl$_3$ and the soluble fraction (5%) could be isolated by Soxhlet extraction and spin-cast to give deep green films of **62**. Visible spectra of the films exhibited a transition at 615 nm, with a reported onset at 850 nm ($E_g = 1.46$ eV).

A couple of attempts to produce random thieno[3,4-*b*]pyrazine-based copolymers by polymerization of monomer mixtures have also been reported (Chart 12.7). Copolymers **63** [89] and **64** [90] were prepared by FeCl$_3$ polymerization of either 1:1 or 1:3 mixtures of the respective monomers. For both systems, however, the resultant polymer properties are more consistent with poly(3-dodecylthiophene) and any contribution from the thieno[3,4-*b*]pyrazine units is quite minimal. Copolymers **65a** and **65b** were prepared by treating mixtures of phthalic anhydride and pyrazinedicarboxylic anhydride with P$_4$S$_{10}$ [48]. The resulting materials were reported to exhibit E_g values of 1.05 eV, but little other characterization was given.

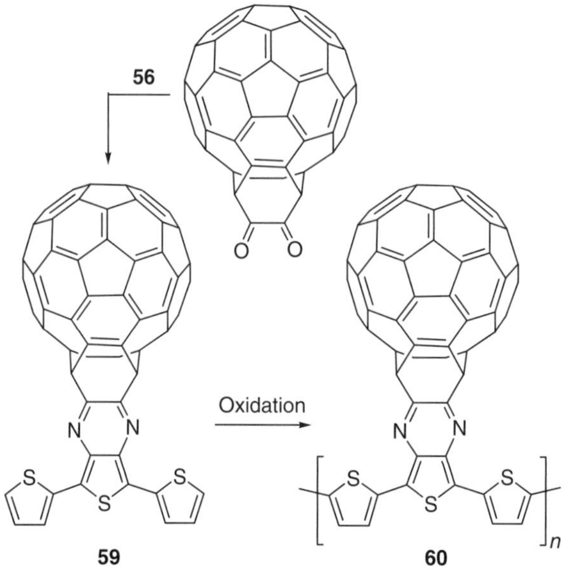

SCHEME 12.13 A C$_{60}$-functionalized thieno[3,4-*b*]pyrazine polymer.

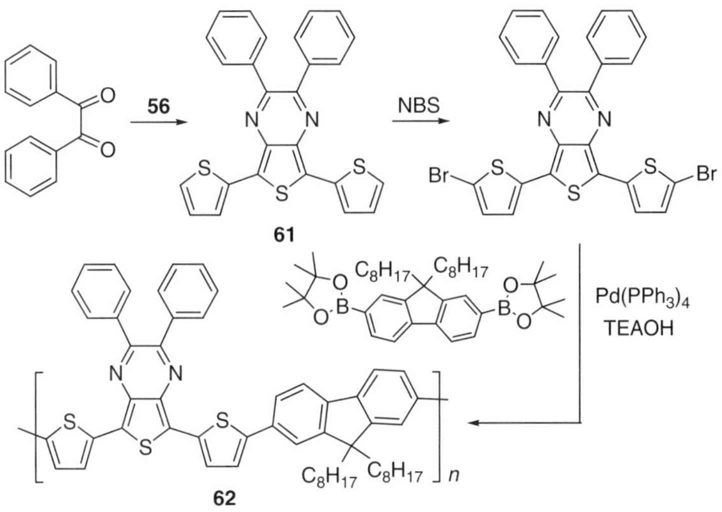

SCHEME 12.14 A three-component, thieno[3,4-*b*]pyrazine-based copolymer.

Due to their relative stability and ease of synthetic manipulation, polythiophene-based systems have been the focus of the majority of synthetic approaches to reducing bandgap. However, a few reports have utilized fused-ring units endcapped with pyrrole as precursors to low E_g materials (Chart 12.8). The first such examples were polymers **66** and **67** reported by Tanaka and Yamashita in 1995 [84]. X-ray structures of the trimeric precursor of **66a** show that the three-ring unit is highly planar with hydrogen bonding evident between the pyrrole N–H and pyrazine nitrogen atoms. The addition of the *N*-methyl in **66b**, however, results in dihedral angles of >40° as a consequence of fairly large steric interactions. This also results in a 118 nm difference in the λ_{max} of the two polymer precursors. Unfortunately, very little characterization of the resulting polymers was reported.

In the following years, Tanaka and Yamashita [85] reported the two additional polymers **68a** and **68b** as well. While the substituted **68b** suffers from the same steric problems as **66b** above, CVs of polymer **68a** were reported to exhibit overlapping onsets of p- and n-doping to give an electrochemical E_g of ~0 eV. If this is correct, this is significantly reduced in comparison with the analogous thiophene-capped system **58** ($E_g = 0.3$–0.5 eV) and illustrates the potential strength of this synthetic approach to reduce E_g. Theoretical studies have predicted an E_g of 0.33 eV for **68a**, which is less than the predicted value for **58** (0.47 eV), but not to the extent reported for the experimental results [86].

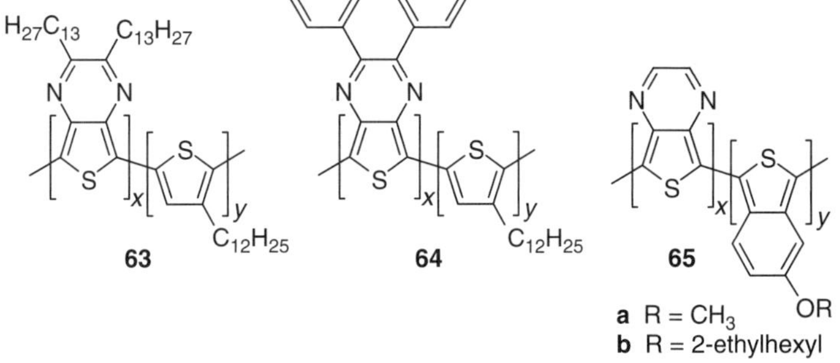

CHART 12.7 Random thieno[3,4-*b*]pyrazine copolymers.

66a   **66b**

67

a R = H
b R = CH$_3$

68

a R = H
b R = Boc

CHART 12.8   Polymers of pyrrole-capped oligomers.

12.5 Poly(thienothiophene)s and Related Systems

Much like PITN and its analogs, various fused thienothiophenes have also found interest as potential precursors for low bandgap systems (Chart 12.9). One of the early systems of promise were polymers of dithieno[3,4-*b*:3′,4′-*d*]thiophene (**69**), initially reported in 1988 by Taliani and coworkers [91–93] to have an E_g of 1.1 eV (for a relevant discussion of prior work see Ref. [1]). Since 1995, however, a number of new reports on poly(dithieno[3,4-*b*:3′,4′-*d*]thiophene) (**69**) have appeared.

Arbizzani and coworkers [94,95] reported electrochemical and spectroelectrochemical studies of **69** in order to probe the n-doping of the material. The spectroelectrochemistry revealed a transition at 610 nm and an E_g of 1.09 eV, in good agreement with previous studies [91–93]. From the studies, it was confirmed that reduction of the polymer resulted in n-doping and that the n-doping levels were less than, but still comparable, to the p-doping levels.

Neugebauer and coworkers [96–98] then reported studies on **69** using total reflection FTIR spectro-electrochemistry to characterize infrared active vibrational (IRAV) modes that were correlated with the generation of charge within the doped polymer. It was found that the IRAV bands arising upon doping had high intensity and broad shape, indicative of delocalized charge carriers. However, it was found that the results indicated differences between the delocalization upon p-doping in comparison with n-doping, with higher localization of the negative charge carriers. Additional papers also detailed Raman studies [99–101], from which it was determined that polymer **69** exhibits enhanced quinonoidal character in comparison with simpler polythiophenes.

The first polymerization of substituted analogs of **69** were reported by Inaoka and Collard [102] in 1997. The monomers were prepared as shown in Scheme 12.15 and then polymerized by both FeCl$_3$ oxidation and electrochemical methods [102,103]. Neutral films of the dimethyl-functionalized **72a** exhibit transitions at 500 and 950 nm, with an onset corresponding to an E_g of 0.9 eV. This agrees with the value of 0.7 eV determined from the onset of the oxidation and reduction peaks in the CV. Spectra of **72b** show similar results with transitions at 550 and 1100 nm and an E_g of 0.7 eV. The reduction in E_g when compared to the unfunctionalized parent **69** (1.1 eV) [91–93] is likely due to the fact that

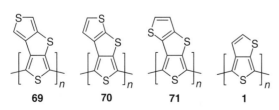

69   **70**   **71**   **1**

CHART 12.9   Isomeric poly(dithienothiophene)s.

the substitution blocks other potential coupling sites and thus eliminates the possibility of the alternate linkages within the polymer (Chart 12.10) that could contribute to increases in E_g [104,105]. The higher energy transitions in the spectra (500 and 550 nm, respectively) were proposed to be the result of overoxidation defects [103].

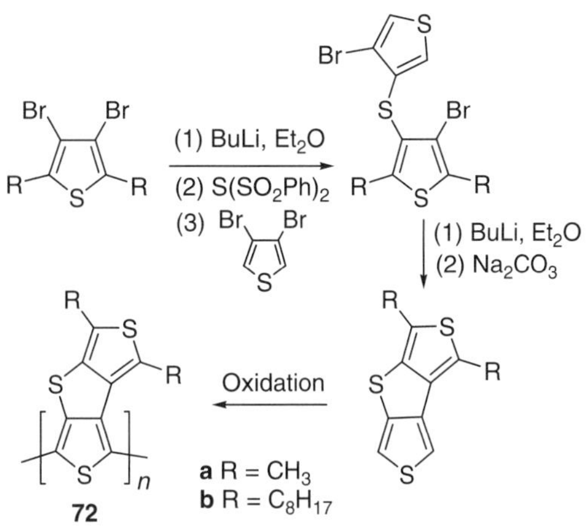

The remaining isomeric dithienothiophenes, poly(dithieno[3,4-*b*:3',2'-*d*]thiophene) (**70**) and poly(dithieno[3,4-*b*:2',3'-*d*]thiophene) (**71**), have also been the subject of a number of studies since their initial preparation in 1997 [106]. As with polymer **69**, polymers **70** and **71** were prepared by electropolymerization of the corresponding monomeric dithienothiophenes to give films, which undergo both n- and p-doping processes. Solid state spectra of **70** exhibited a transition at

SCHEME 12.15 Substituted poly(dithieno[3,4-*b*:3',4'-*d*]thiophene)s.

650 nm with an E_g of 1.21 eV, while **71** gave a transition at 760 nm and an E_g of 1.12 eV [95,106]. The E_g values determined from the CV onsets of oxidation and reduction showed a greater difference between the polymers with values of 1.26 eV for **70** and 0.61 eV for **71** [95,107]. As with polymer **69** above, the doping processes of polymers **70** and **71** have also been characterized by both optical and total reflection FTIR spectroelectrochemistry [95–98,108]. The results of these studies for **70** were fairly similar to those for **69** [95–98]. In the case of **71**, however, it was found that the delocalization for the charge carriers upon n- or p-doping was much more equivalent than in the other two isomeric systems [96–98,108].

In addition to the various dithienothiophene polymers, the related thieno[3,4-*b*]thiophene (**1**) and its derivatives have received considerable attention. The first reported example was the alkyl-functionalized derivative, poly(2-decylthieno[3,4-*b*]thiophene) (**73**) reported by Pomerantz and Gu [109]. Polymer **73** was prepared by the FeCl$_3$ oxidation of 2-decylthieno[3,4-*b*]thiophene, which was synthesized from 2-thiophenecarboxylic acid as illustrated in Scheme 12.16 [109,110]. Such methods produced a fairly high M_w material ($M_w = 90,000$, $M_n = 52,000$, PDI = 1.7) that was blue-green in color and soluble in a variety of organic solvents (CHCl$_3$, THF, and chlorobenzene). Solvent cast films exhibited a transition at 925 nm with an onset at 1350 nm (E_g of 0.92 eV). FeCl$_3$-doped films gave a conductivity of 3.1×10^{-3}. Films doped with either I$_2$ or NOBF$_4$ gave lower conductivities ($1.0–4.2 \times 10^{-6}$).

Ferraris and coworkers [111] then reported a new approach to thieno[3,4-*b*]thiophenes (Scheme 12.17) with the preparation of the phenyl analog **74** in 1999. Thin films of polymer **75** were then successfully produced by repetitive cyclic voltammetry of monomer **74**. The polymer CV exhibited broad waves for both the p- and n-doping cycles, from which an E_g of 0.85 eV was estimated based on the redox onsets. Spectra of films grown on ITO show a transition at 954 nm with an onset corresponding to an E_g of ~0.82 eV.

Lee and Sotzing [112] then used a similar approach for the synthesis of the unfunctionalized thieno[3,4-*b*]thiophene (**76**) and its corresponding polymer **1** in 2001 (Scheme 12.18). The electropolymerized films are deep blue in the neutral state, but colorless and transparent in the oxidized state. Films grown on ITO give optical spectra with a transition at 846 nm. The onset of this transition (1459 nm) corresponds to an E_g of 0.85 eV [112,113]. This agrees well with the E_g value of ~0.8 eV determined from the difference of the redox onsets in the polymer CV.

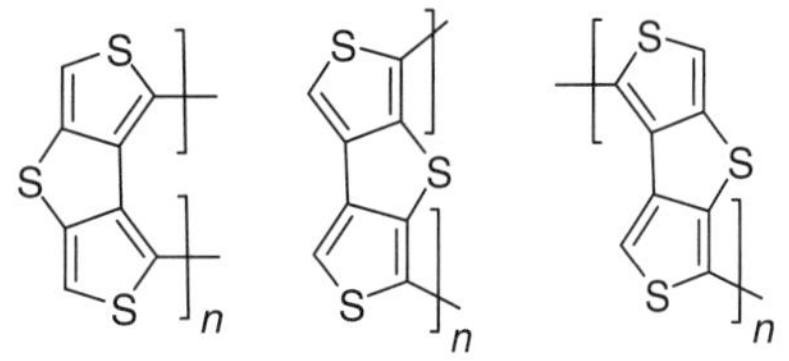

More recently, polymer **1** was obtained from electropolymerization of the neutral dimer 6,6-bis(thieno[3,4-*b*]thiophene) (**77**), which was prepared by CuCl$_2$ oxidation of the

CHART 12.10 Alternate polymerization forms of poly(dithieno[3,4-*b*:3',4'-*d*]thiophene)s.

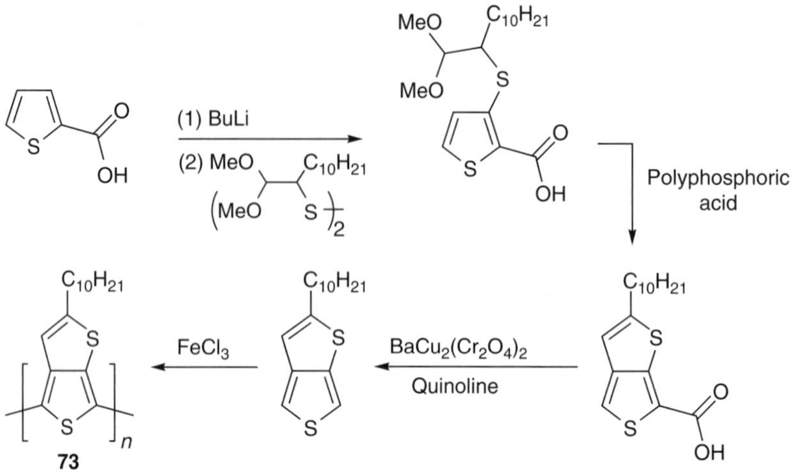

SCHEME 12.16   Synthesis of poly(2-decylthieno[3,4-*b*]thiophene).

monomeric anion as shown in Scheme 12.18 [114]. As expected, the oxidation potential of **77** was reduced in comparison with **76** (0.63 vs. 1.29 V) and could be easily electropolymerized by potential cycling. The resulting film exhibited a peak at 890 nm with an onset corresponding to an E_g of 0.90 eV, similar to films produced by electropolymerization of monomer **76**.

While polymer **1** is not soluble, aqueous dispersions of the polymer have been successfully produced by oxidative polymerization of **76** in aqueous solutions of poly(styrenesulfonic acid) (PSSA) using either $NH_4S_2O_8$ or ferric sulfate [115–117]. The resulting **1**-PSSA dispersions were green in color and could be cast as thin optically transparent coatings. After reduction with hydrazine, the neutral films from the persulfate polymerizations exhibited an E_g of ~1.1 eV with a corresponding λ_{max} of 810 nm [115,116]. The films produced from the ferric sulfate oxidation exhibited slightly lower E_g values (0.94–0.98 eV) [117], just slightly higher than the previous electropolymerized films.

An alternate method for the preparation of soluble forms of polymer **1** involved the chemical treatment of **1** with fuming H_2SO_4, which resulted in ~56%–65% of the repeat units undergoing ring sulfonation [118]. The resulting sulfonated products were water-soluble with onsets of 1.05–1.18 eV, depending on the extent of sulfonation. The M_n of the materials was determined to be 2.6–3.6 × 10^4 with PDIs of 1.3–1.7.

Unlike the previous 2-substituted thieno[3,4-*b*]thiophenes, monomer **76** has the potential to form multidimensional conjugated pathways. Sotzing and coworkers removed the possibility of 4,6-coupling by blocking the 6-position of **76** with a hydroxymethyl functionality (Scheme 12.19), thus allowing investigation of the alternate 2,4-conjugation pathway [119]. Electropolymerization of **78** resulted in polymer **79**, which exhibited a transition at 551 nm and an E_g of 1.55 eV. The higher E_g of **79** coupled with the fact that polymers **1**, **73**, and **75** all exhibit very similar E_g values suggest that the primary conjugation pathway is by the 4,6-couplings in the latter polymers.

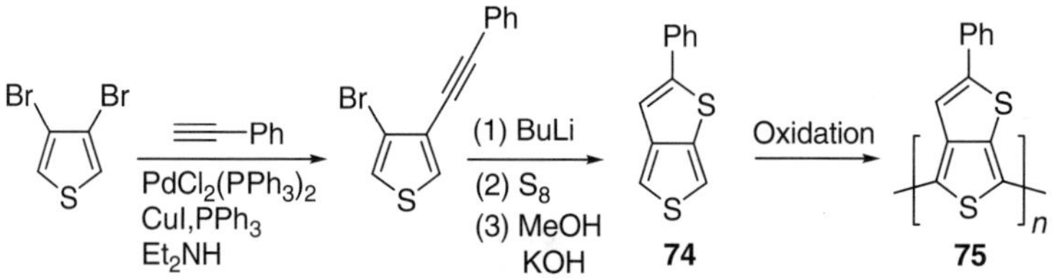

SCHEME 12.17   Synthesis of poly(2-phenylthieno[3,4-*b*]thiophene).

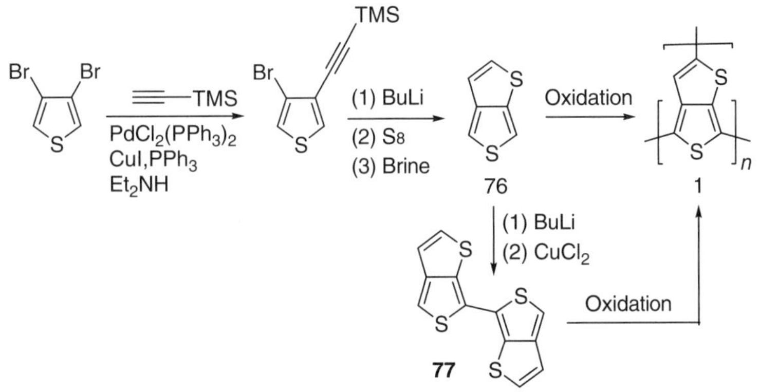

SCHEME 12.18 Synthesis of poly(thieno[3,4-*b*]thiophene).

Random copolymers of thieno[3,4-*b*]thiophene and EDOT have also been prepared by electropolymerization of 1:1 mixtures of the two monomeric species [120]. It was reported that the redox behavior of the resulting copolymeric films closely resembled that of polymer **1** and exhibited oxidation peak potentials of 1.14–1.18 V vs. Ag/Ag^+. Vis–NIR spectroscopy of the neutral films gave an E_g value of 1.06 eV when tetrabutylammonium perchlorate was used as the electrolyte. The use of the corresponding hexafluorophosphate salt gave a higher E_g of 1.65 eV. This difference was attributed to differences in the ratio of thieno[3,4-*b*]thiophene to EDOT in the resulting films. The ratios for the two materials were determined to be 79.9:20.1 ($TBAClO_4$) and 67.7:32.3 ($TBAPF_6$).

Finally, two related polymeric analogs of polymer **1** have been reported that incorporate either a fused furan (**80**) [121,122] or thiazole (**81**) [123] ring (Chart 12.11). Both polymers were prepared by electropolymerization of the corresponding monomer and the resulting materials exhibited E_g values of 1.03 and 1.3 eV for **80** and **81**, respectively. Polymer **80** was reported to be pale blue in the neutral form and transmissive light blue in the oxidized state. Polymer **81** was reported to exhibit p-doped conductivities of 1.0×10^{-4} (I_2 doped) and n-doped conductivities of 1.5×10^{-2} (NaC_6H_5 doped) [123].

12.6 Planarized Bridged Bithiophene Systems

Another approach to fused-ring polymeric systems is to insert a bridging unit between the adjacent thiophene rings of 2,2′-bithiophene. The resulting fused system is thus more planar than the parent bithiophene unit due to a lack of free rotation around the interannular bond resulting in a decrease of E_g. In addition, the use of electron-deficient bridging units in such systems can even further reduce the E_g to produce low E_g materials. Ferraris and Lambert [23,24] initially showed the success of this approach through polymers **8** and **9** (Chart 12.2), and a number of new papers on these systems have appeared since the second edition of this handbook.

Starting in 1997, Murphy and coworkers reported methodology for the electrodeposition of thin uniform films of polymers **8** and **9** for the formation of Schottky barrier diodes [124–127]. Spectro-electrochemical measurements indicated that while n-doping occurs, it does so to a lesser extent than the

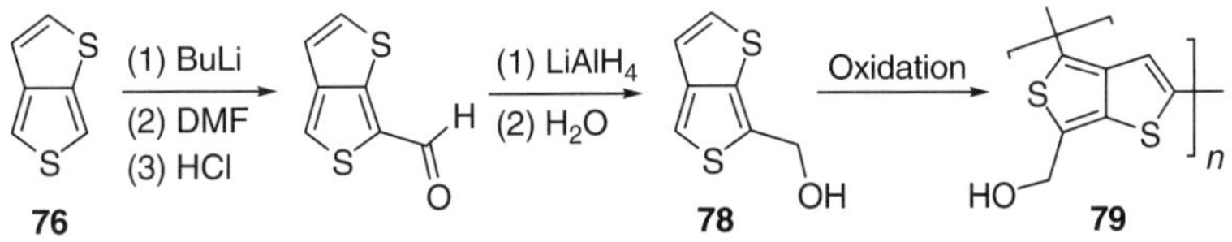

SCHEME 12.19 2,4-coupled poly(thieno[3,4-*b*]thiophene).

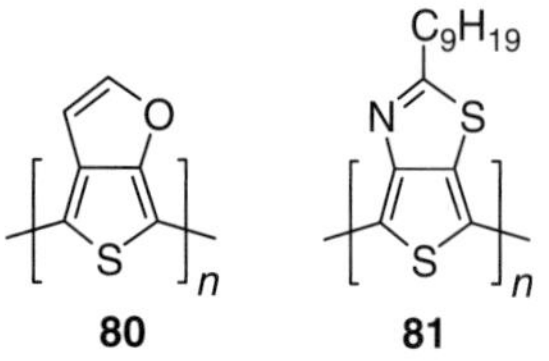

CHART 12.11 Poly(thieno [3,4-*b*]thiophene) analogs.

corresponding p-doping processes. The resulting diodes prepared displayed little or no rectification, but did exhibit characteristics of Schottky diodes.

Huang and Pickup [128] measured the in situ conductivity of polymer **8** to give a maximum p-doped conductivity of 0.59 S cm^{-1}. The n-doped conductivity was found to be 100 times smaller, giving a maximum value of 5.4×10^{-3} S cm^{-1}. From the electrochemical measurements, the E_g was estimated to be 0.75 eV. This value was in good agreement with the previously measured optical value of 0.8 eV [23]. In a more recent study by the same group, conductivity measurements gave slightly higher values of 1.1 and 1.2×10^{-2} S cm^{-1} for p-doping and n-doping, respectively [129]. In this same study it was shown that the n-doped material could also undergo attack by oxygen to produce a new modified polymer as shown in the simplistic Scheme 12.20 [129]. Not only did the resulting polymer **82** exhibit a reduced E_g of <0.5 eV, but also reduced p- and n-type conductivities. Raman spectroscopy suggests that substitution has occurred in the β-positions to give hydroxyl groups (**82b**), which are in equilibrium with the keto tautomer (**82a**).

Zotti and coworkers reported the new methylene-bridged polymer **84** in 1998 [130]. The monomer **83** was prepared as shown in Scheme 12.21 and then electropolymerized to produce films of the corresponding polymer **84**. The CV exhibited oxidations at −0.3 and 0.3 V and reduction at −2.2 V. In situ conductivity measurements gave a value of 20 S cm^{-1}. The visible spectrum of polymer **84** exhibited a transition at 670 nm. From this reported spectrum, the E_g of polymer **84** was estimated to be 1.46 eV.

Pickup and coworkers reported an interesting variant of the polymers by dimerizing cyclopenta[2,1-*b*;3,4-*b*]dithiophen-4-one to produce the corresponding monomer **85** (Scheme 12.22) [131]. Of particular interest was that since **85** contains four thiophene rings capable of polymerization, it should be possible to cross-link the conjugated system in a well-defined way, with conjugation in two dimensions. Polymer **86** was obtained by electropolymerization to give films exhibiting optical transitions at 410 and 800 nm, with extensive tailing to below 1600 nm. The onset on the low energy tailing corresponds to an E_g of <0.8 eV, which is one of the lowest values reported for such cyclopentadithiophene-based polymers. The large peak at 410 nm was reported to be due to the unpolymerized pendent unit and thus it is thought that **85** polymerizes primarily through only one of the two cyclopentadithiophene units as shown in Scheme 12.22.

A couple of studies of copolymers of bridged bithiophenes and EDOT have also been reported, starting with the random copolymer **87** reported by Huang and Pickup [132] in 1998 (Chart 12.12). Copolymer **87** was prepared by electropolymerization of 4:1 mixtures of 4-dicyanomethylene-4*H*-cyclopenta[2,1-*b*;3,4-*b*′]dithiophene and EDOT to give films with conductivities of 0.052−0.69 S cm^{-1}.

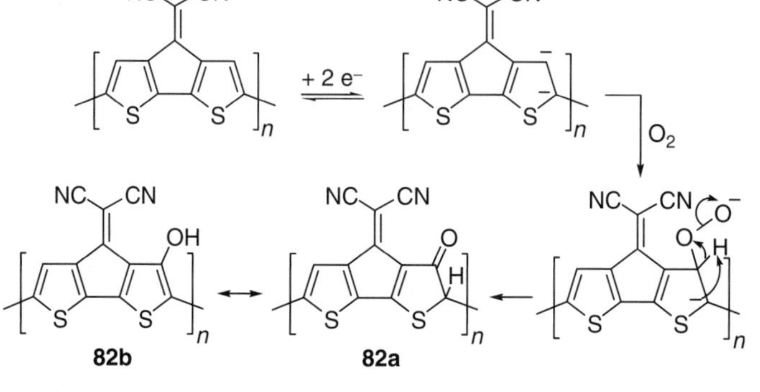

SCHEME 12.20   Postpolymerization O_2-modification of polymer **8**.

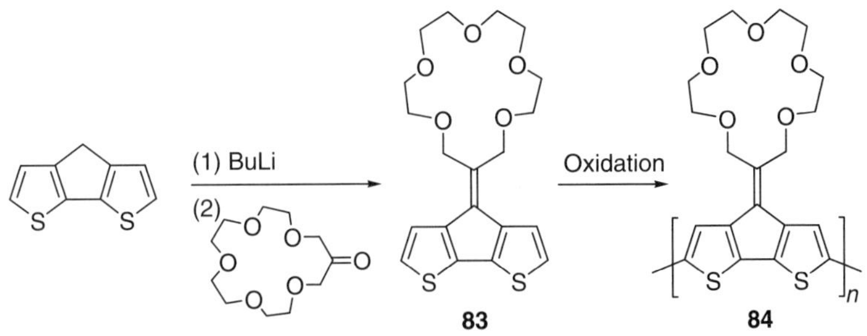

SCHEME 12.21 Synthesis of a crown ether-functionalized poly(cyclopentabithiophene).

From the difference in the p- and n-doping processes, E_g values for the copolymers were determined to be 0.19–0.33 eV.

More recently, Zotti and coworkers prepared copolymers **88–90** by the electropolymerization of oligomeric precursors [80]. While copolymers **89** and **90** exhibited E_g values of 1.55–1.65 eV, copolymer **88** exhibited a low energy transition at 1300 nm, with a reported E_g of 0.95 eV. A lower limit of the p-doped conductivity was determined to be >10 S cm^{-1}.

12.7 Donor–Acceptor Low Bandgap Polymers

Due to the success of the various systems prepared by thieno[3,4-*b*]pyrazine-based terthienyls, a variety of other systems have been developed utilizing a mixture of electron-rich and electron-poor subunits. A common approach is to employ a selection of other electron-deficient units as the central unit of trimeric precursors, while retaining the thiophene endgroups. Common electron-deficient units include either benzothiadiazole or benzobisthiadiazole, as first reported by Yamashita and coworkers in 1995 with the preparation of polymers **91a** and **92** (Chart 12.13) [133].

The trimeric precursors were prepared from the corresponding dibrominated central unit by Kumada coupling and then electrochemically polymerized. The E_g of polymer **91a** was reported to be 1.1–1.2 eV [76,133], while the absorption edge of neutral films of **93** corresponded to an E_g below 0.5 eV. The exact value for **93** could not be determined due to the overlapping absorption of the ITO electrode below 0.5 eV [76,133].

Akoudad and Roncali [82] then reported the preparation and study of polymer **92** in 1999. Thin films of **92** exhibited an absorption maximum at 665 nm and an E_g of 1.30 eV. Surprisingly, there is an increase of 0.1–0.2 eV in comparison with **91a**, potentially due to increased steric problems. Janssen and coworkers [134] later reported soluble derivatives of **91a** that utilized octyl side chains on the exterior thiophenes (**91b** and **91c**, Chart 12.13). While the addition of the side chains did result in soluble materials, the steric interactions introduced caused large increases in the E_g values to 2.01 eV and 1.96 eV for **91b** and **91c**, respectively.

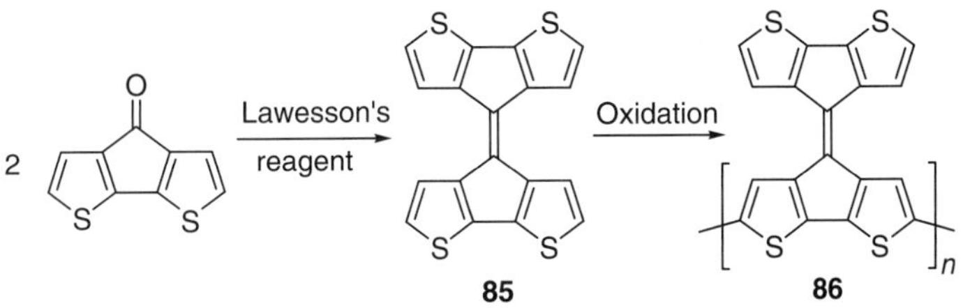

SCHEME 12.22 Synthesis of poly(dicyclopenta[2,1-*b*:3,4-*b'*]dithiophene).

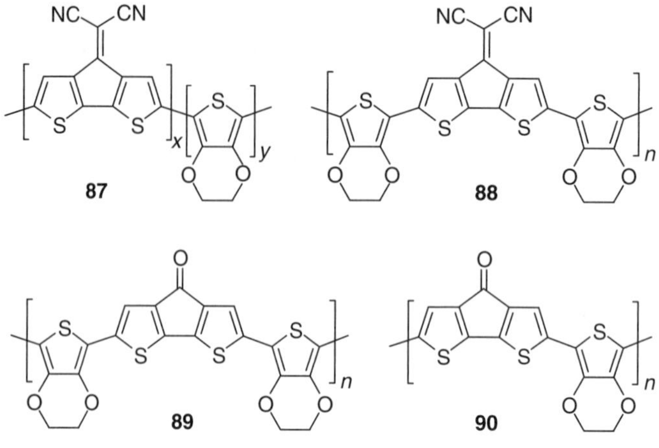

CHART 12.12   Copolymers of cyclopentadithiophenes and EDOT.

A similar approach using central quinoxaline units has also been investigated (Chart 12.14) [76,135,136]. Polymer **94a** exhibits an E_g of 1.4 eV, slightly higher than the analogous benzothiadiazole-based polymer **91a**, but consistent with the difference in electron-accepting nature of the central units. Surprisingly, the addition of hexyl side chains in **94b** did not result in soluble materials. No mention was given as to the effect of the longer side chains on the E_g [76].

More recently, Anzenbacher and coworkers [135,136] reported the related polymers **95a** and **95b** as anion-sensing materials. Although it may be expected that the external EDOT units would result in a reduction of the E_g as seen in previously discussed systems, the E_g values for both polymers were reported to be ~1.4 eV. It may be that the addition of pyrrole groups on the central quinoxaline balances the addition of the EDOT, thus resulting in no net electronic change and E_g values consistent with **94a**.

Reynolds and coworkers reported the related polymer **96** [137]. The polymer CV exhibited a broad oxidation at 0.5 V and a corresponding reduction centered at −1.4 V. Spectroelectrochemistry revealed a transition centered at 750 nm, the onset of which gave an E_g of 1.2 eV. This reduced value in comparison with polymer **94a** is similar to what was initially expected for polymer **95a**.

Yamashita and coworkers [76] have also reported polymers **97** and derivatives of **97** containing more extended quinoxaline systems. For polymer **97**, the fusion of an additional pyrazine ring resulted in a reduction of the E_g to 0.9 eV in comparison with the 1.4 eV of polymer **94a**. As might be expected, a greater reduction was seen in polymer **98a** to give an E_g of 0.7 eV.

A related material, polymer **99**, has found applications in photovoltaic devices [138–140]. This is a multicomponent alternating copolymeric system prepared in a similar manner to the previously discussed polymer **62** (Scheme 12.14). Spin-cast films of polymer **99** were green in color and exhibited

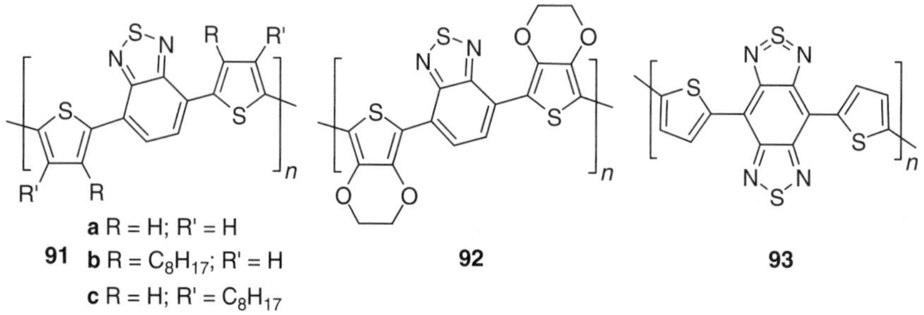

a R = H; R' = H
91   **b** R = C$_8$H$_{17}$; R' = H
c R = H; R' = C$_8$H$_{17}$

CHART 12.13   Benzothiadiazole-based polymers.

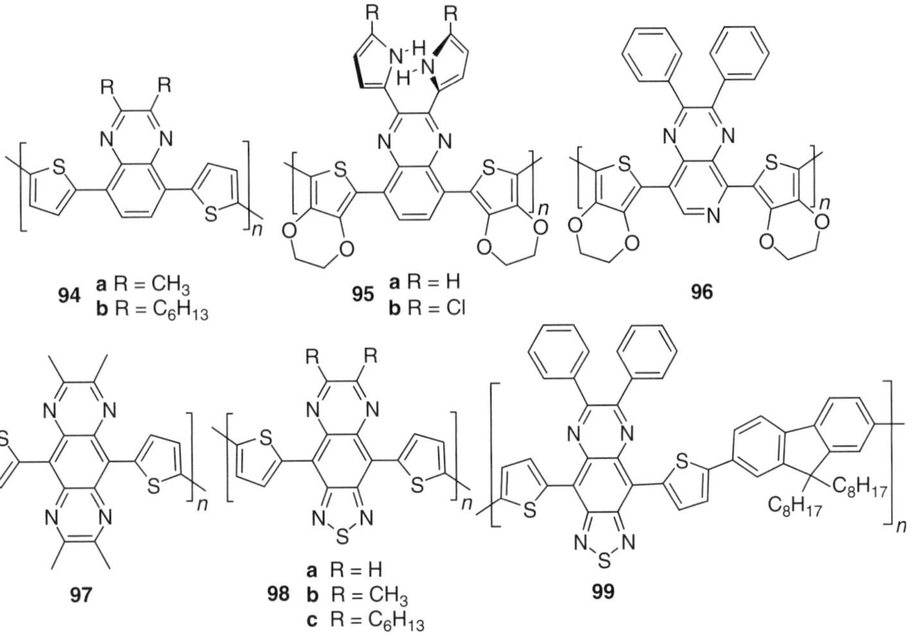

CHART 12.14 Quinoxaline-based polymers.

broad peaks at 400 and 780 nm. The onset of the low energy transition tails to ~1000 nm, corresponding to an E_g of 1.24 eV.

Ajayaghosh and coworkers [141–145] have also reported a series of low E_g polysquaraines prepared by the reaction of squaric acid with dialkoxydivinylbenzene-bridged bispyrroles (**100a–g**, Scheme 12.23). The optical data and conductivity are given in Table 12.3, and the E_g values are in the range of 0.79–1.02 eV. The use of short side chains for both functionalized units results in the lowest E_g values and as long as one of the two units uses methyl side chains, the E_g remains ~0.8 eV.

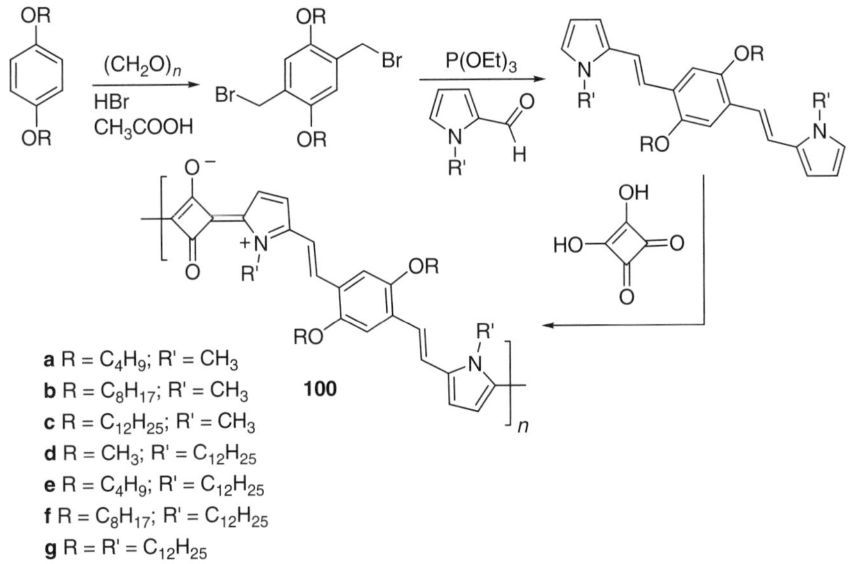

SCHEME 12.23 Squaraine-based polymeric systems.

TABLE 12.3   Solid State Absorption, Bandgaps, and Conductivities for Polysquaraines

Polymer[a]	R	R′	λ_{max} (nm)	E_g (eV)[b]	Conductivity (S cm^{-1})[c]
100a	C_4H_7	CH_3	880	0.79	5.20×10^{-2}
100b	C_8H_{17}	CH_3	875	0.81	2.98×10^{-2}
100c	$C_{12}H_{25}$	CH_3	860	0.82	6.17×10^{-2}
100d	CH_3	$C_{12}H_{25}$	870, 1028	0.83	1.10×10^{-2}
100e	C_4H_7	$C_{12}H_{25}$	890, 1020	0.99	2.80×10^{-2}
100f	C_8H_{17}	$C_{12}H_{25}$	905, 1025	1.02	5.10×10^{-2}
100g	$C_{12}H_{25}$	$C_{12}H_{25}$	881, 1021	1.02	1.10×10^{-2}

[a] *Source*: Eldo, J. and Ajayaghosh, A., New low band gap polymers: Control of optical and electronic properties in near infrared absorbing π-conjugated polysquaraines, *Chem. Mater.* 14, 410, 2002; Ajayaghosh, A., Intrinsically conducting low band gap polysquaraines via a–b type polycondensation, *Int. J. Plast. Technol.* 6, 117, 2003.
[b] Optical bandgap.
[c] I_2 doped.

12.8   Poly(dithienylethylene)s and Related Systems

Another successful approach to low E_g systems is the polymerization of diarylethylene precursors. Roncali and coworkers have reported investigations on the rigidification of the dithienylethylene unit through the incorporation of alkyl bridges between the thiophene and the central double bond. Synthesis of the bridged dithienylethylenes 101a–c was accomplished as illustrated in Scheme 12.24 and then electropolymerized to produce polymers 102a–c [146–148]. Solid state spectra of polymer 102a exhibit a peak at ~500 nm with a lower energy shoulder at ~760 nm, the onset of which corresponds to an E_g of ~1.40 eV. For the substituted systems 102b and 102c, the peak is slightly blueshifted and the intensity of the low energy shoulder decreased, possibly suggesting less highly conjugated material in the films. The absorption onsets, however, were still consistent with that of polymer 102a.

In attempts to further reduce the E_g of the poly(dithienylethylene) systems, polymer analogs were prepared containing either cyano functionalization at the olefinic linkage, EDOT units, or the combination of both (Chart 12.15). Polymers 103a, 103b, and 104a were initially reported by Roncali and coworkers in 1995 [149]. Spectra of electropolymerized films of 103a revealed an E_g of 0.5–0.6 eV, while polymers 103b and 104a gave the much higher values of 1.4–1.5 eV. All three materials underwent both p- and n-doping and conductivities of 7×10^{-2} and 4.8 S cm^{-1} were reported for 103a and 104a, respectively. More recently, Belanger and coworkers [150] have reported a study including all of the

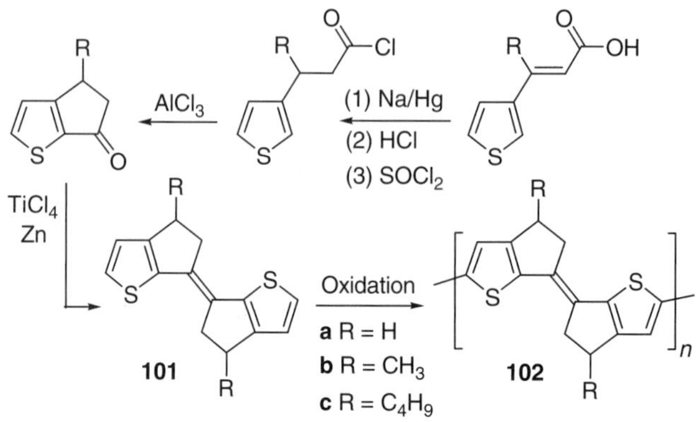

SCHEME 12.24   Synthesis of bridged poly(dithienylethylene)s.

previous materials reported by Roncali, as well as two new systems, **103c** and **104b**. The results found for polymers **103a**, **103b**, and **104a** agreed well with the previous study by Roncali, giving E_g values of 0.54, 1.55, and 1.4 eV, respectively. For the new isomeric **103c**, the E_g was found to be <1.5 eV, slightly less than the corresponding **103b**. The absorption maximum of **103c** was also redshifted in comparison with **103b** to 523 and 505 nm, respectively. It was thought that this difference may be due to reduced steric interactions or a difference in material M_w. Polymer **104b** exhibited an E_g very similar to **104a** ($\sim$1.4 eV). However, it was found that **104b** was less stable to potential cycling than the isomeric **104a**.

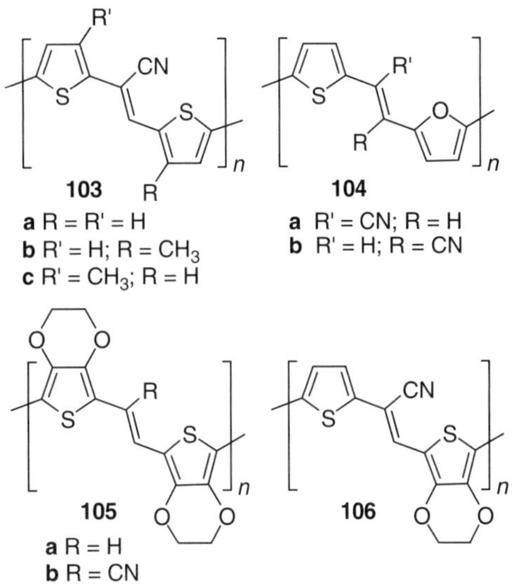

103 R	**104**
a R = R' = H	**a** R' = CN; R = H
b R' = H; R = CH$_3$	**b** R' = H; R = CN
c R' = CH$_3$; R = H	

105
a R = H
b R = CN

106

CHART 12.15 Cyano-functionalized poly(dithienylethylene)s.

The EDOT analog **105a** was first reported by Reynolds and coworkers in 1995 [151–153], followed soon after by the cyano derivatives **105b** and **106** [154,155]. Neutral films of polymer **105a** were reported to be deep purple, while oxidized films were sky blue. The neutral films exhibit a transition maximum at 600 nm, with an onset of 1.4 eV, $\sim$0.2–0.3 eV lower than that of the parent PEDOT. The addition of the cyano functionality further lowered the E_g and polymers **105b** and **106** were found to have values of 1.1 and 1.3 eV, respectively.

Gallazzi and coworkers then reported the difluorinated analog **107** (Chart 12.16) in 2002 [156]. The electropolymerized film was blue in the neutral form and transparent upon oxidation. From the absorption onset, the E_g of **107** was determined to be 1.42 eV.

More recently, Seshadri and Sotzing [157] have reported the two isomeric thieno[3,4-*b*]thiophene-containing analogs **108** and **109**. Neutral films of both polymers were reported to be deep blue, while becoming transmissive sky blue upon oxidation. From the onsets of p- and n-doping, E_g values of 1.0 and 1.2 eV were estimated for **108** and **109**, respectively. It was noted, however, that while both polymers exhibited similar electrochemical behavior, polymer **109** showed lower stability in comparison with **108**. Spectra of the films revealed a transition at 669 nm and E_g of 1.19 eV for **108**, while **109** gave a transition at 720 nm and E_g of 1.20 eV.

Hanack and coworkers [158] reported a study on **103a** and a new polymer **110** (Chart 12.17), which was produced from a larger analogous polymer precursor [158]. The properties of **103a** agreed well with the other studies, while polymer **110** exhibited almost identical electrochemical behavior. From the difference of the redox processes, an E_g value of 0.65 eV was determined for **110**, also very similar to the 0.60 eV value of **103a**. The primary difference between the two polymers was a reduced stability to n-doping for polymer **110**.

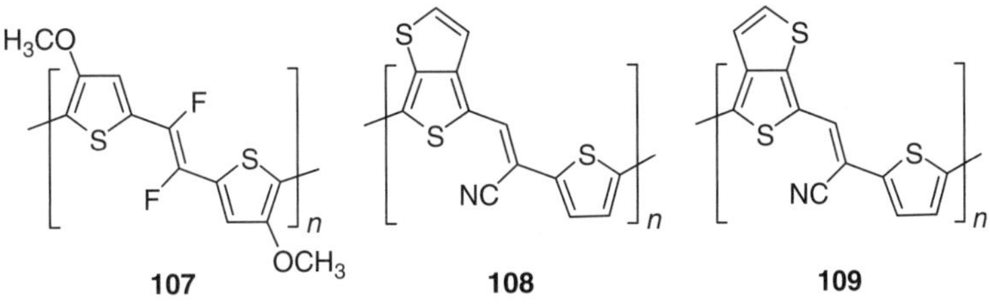

107	**108**	**109**

CHART 12.16 Additional functionalized poly(dithienylethylene)s.

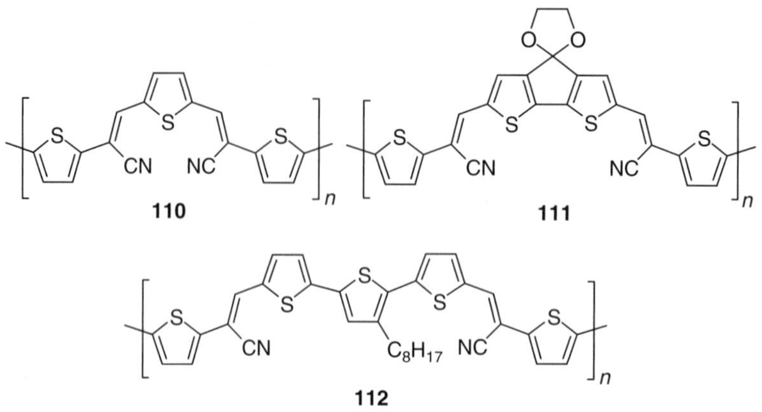

CHART 12.17   Polymers from larger dithienylethylene oligomers.

Extending the approach to the use of larger dithienylethylene polymer precursors, Roncali and coworkers reported the polymers **111** and **112** [159]. Both materials showed well-defined redox processes for both p- and n-doping. From the optical onsets, E_g values of 1.50 and 1.60 eV were determined for **111** and **112**, respectively. These values were considerably higher than **103a** or **110**, but are consistent with many of the other related systems discussed above.

While the majority of efforts have focused on the dithienylethylene approach, a couple of papers have appeared since the second addition of this handbook that looked at more conventional poly (thienylenevinylene) materials. Elsenbaumer and coworkers [160,161] reported a series of dialkoxy-functionalized poly(thienylenevinylene)s (**113a–d**). The materials were prepared by a soluble precursor polymer that could then be thermally converted to the desired materials as shown in Scheme 12.25. While polymers **113a** and **113c** were insoluble, the materials containing longer alkyl side chains were soluble in common organic solvents. The optical data and conductivity are given in Table 12.4, with E_g values of 1.0–1.2 eV.

Bazzi and Sleiman have also reported some interesting related polymers produced by ring opening metathesis polymerization (ROMP) as shown in Scheme 12.26 [162]. These methods allow the production of the homopolymers **114**, as well as the corresponding block copolymers **115a** and **115b**. While polymers **114a** and **114b** were only isolated as dark insoluble solids, films of polymer **114c** and of the copolymers **115a** and **115b** exhibited absorption onsets corresponding to E_g values of 0.85 eV.

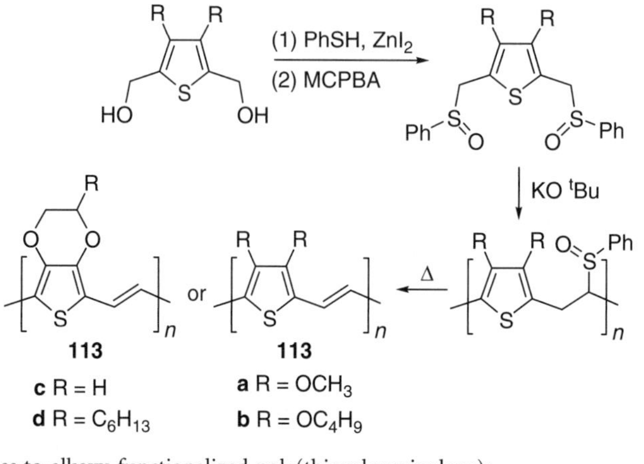

SCHEME 12.25   Routes to alkoxy-functionalized poly(thienylenevinylene)s.

TABLE 12.4 Solid State Absorption, Bandgaps, and Conductivities for Poly(thienylenevinylene)s

Polymer	λ_{max} (nm)	E_g (eV)	Conductivity (S cm^{-1})[b]	References
113a	680	—	25	160
113b	702	1.2	15	160
113c	680	1.0	2.0	160, 161
113d	680, 730[a]	<1.2	36 (10.6)[c]	160

[a] Broad absorption tailing to 2500 nm.
[b] FeCl$_3$ doped.
[c] I$_2$ doped.

12.9 Poly(thiophene-methine) and Related Polymers

Poly(thienylenemethine)s and various analogs have become of significant interest and a large number of papers in this area have appeared since the second edition of this handbook. In general, poly(thienyl-enemethine)s can be obtained by the condensation of an oligothiophene with an aldehyde to produce a nonconjugated precursor polymer (Scheme 12.27) [163]. Redox-based elimination of the bridge hydrogens then results in the formation of quinonoid domains to give the desired polymer. While this method was effective for bithiophenes and higher oligomers, it is more difficult to control the reaction using 3-alkylthiophenes and the resulting materials typically give higher E_g values. Polymers **116a–c**, **117a**, and **117b** were partially soluble in THF and DMF and could be processed into thin films by spin coating. Solid state visible spectra of **116a–c** exhibited a low energy transition between 618 and 675 nm, with E_g values of 1.14–1.45 eV. CVs of the films showed clear p- and n-doping processes, from which E_g values of 1.28–1.40 eV could be estimated. Iodine-doped films gave conductivities of 6×10^{-3}–0.16 S cm^{-1} and doping levels were found to be 28%–33% by weight. In all cases, **116c** gave the lowest E_g and the highest conductivity. While characterization focused on polymers **116a–c**, solution spectra of **117a** and **117b** in THF exhibited maxima at 687 and 667 nm, respectively.

More recently, Yang and coworkers reported the preparation of **117c** by the same methods [164]. Visible spectra exhibit a broad transition at ~550 nm, with an onset giving an E_g of 1.46 eV. However, there is also significant tailing out to ~1100 nm. Taking this as the absorption onset would give an E_g of 1.13 eV. From the polymer CV, an electrochemical E_g value of 0.90 eV was determined, which is in good agreement with the lower optical value.

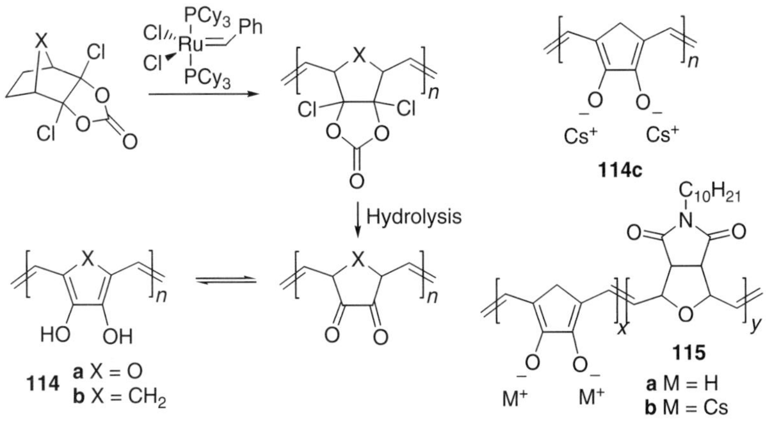

SCHEME 12.26 Low E_g poly(arylenevinylene)s by ROMP.

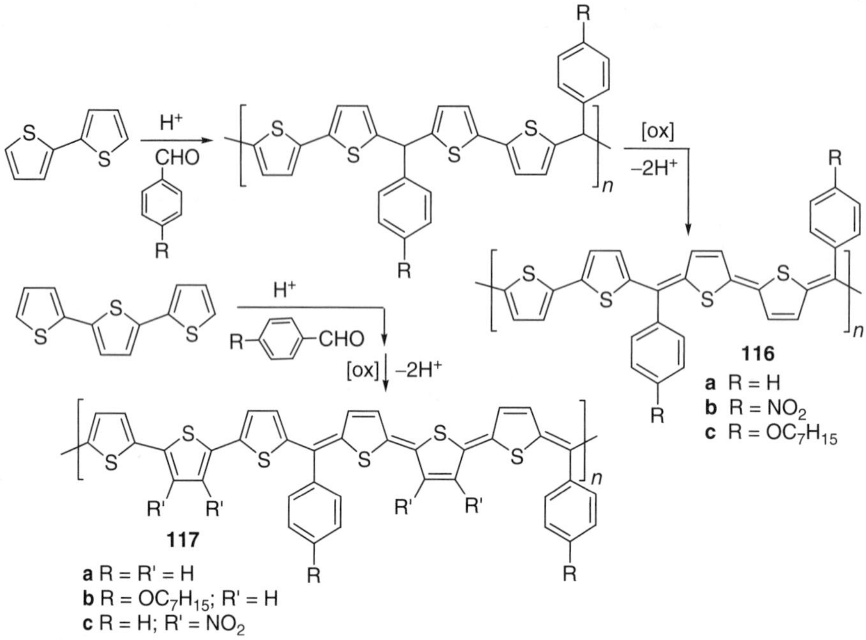

SCHEME 12.27   Synthesis of poly(thienylenemethine)s.

In a separate paper, Yang and coworkers [165] also investigated the use of oligomeric units containing various central aryl units endcapped with external thiophenes (Chart 12.18). Polymers **118–120** were all readily soluble in common organic solvents such as CHCl$_3$ and THF. Optical spectra gave low energy transitions in the range of 500–650 nm, with onsets corresponding to E_g values of 1.49, 1.40, and 1.24 eV for polymers **118**, **119**, and **120**, respectively. From the onset redox potentials, E_g values of 1.72, 1.41, and 0.93 eV were determined for **118**, **119**, and **120**, respectively.

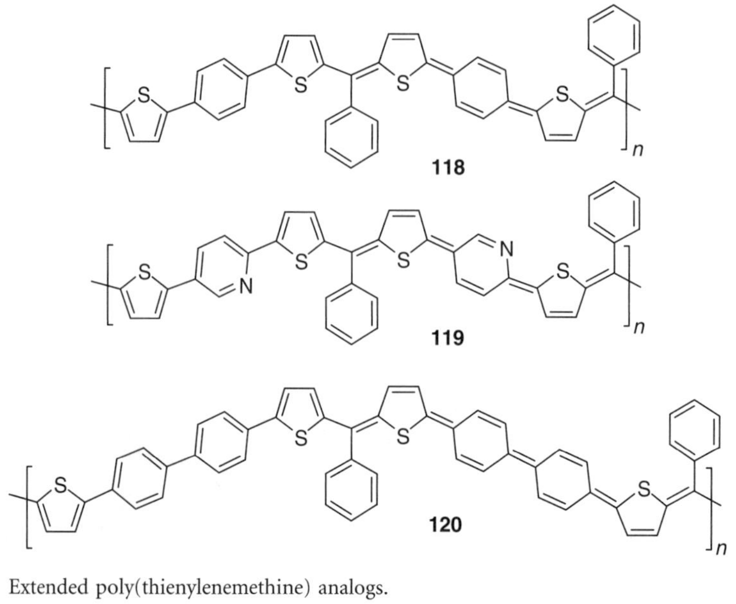

CHART 12.18   Extended poly(thienylenemethine) analogs.

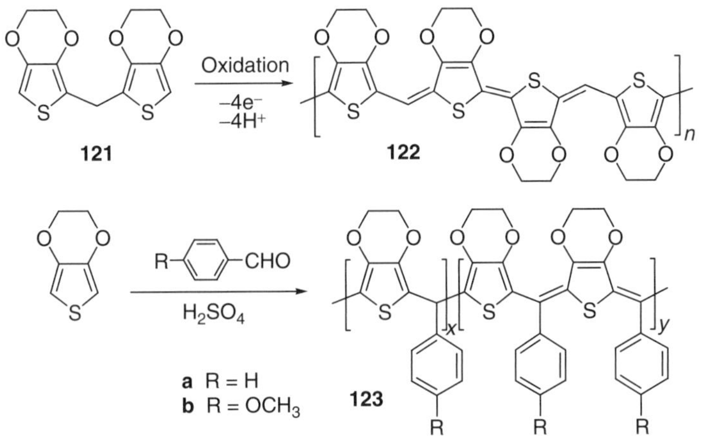

SCHEME 12.28 Synthesis of poly(3,4-ethylenedioxythienylenemethine)s.

The production of EDOT analogs was first reported by Zotti and coworkers in 2003 [166]. Polymer **122** (Scheme 12.28) was produced by the electropolymerization of bis[2-(3,4-ethylenedioxy)thienyl]-methane (**121**). The resulting insoluble material was brownish-gray in the neutral form and blue upon oxidation. The CV onsets were −0.95 V for oxidation and −1.35 V for reduction, corresponding to an E_g of 0.4 eV. Optical spectra exhibited absorption beyond 2000 nm, consistent with a very low E_g. In situ conductivity measurements gave a maximum conductivity of 10^{-3} S cm^{-1}.

In 2004, Chen and coworkers [167] prepared polymers **123a** and **123b** by chemical means as shown in Scheme 12.28. The polymers were soluble in THF and this allowed the preparation of solvent cast films. For polymer **123a**, solid state and solution UV–visible spectra gave almost identical results consisting of three transitions; one in the UV (255 nm), one in the visible (437 nm), and one in the NIR (1002 nm). Polymer **123b** gave very similar results in solution with a slight redshift in the solid state. From the onsets of the low energy transition, E_g values of 0.87 and 0.86 eV were determined for **123a** and **123b**, respectively. CVs of polymer **123b** exhibited peak potentials at 0.03 and −1.60 V and an E_g value of 1.01 eV was determined from the redox onsets. Iodine-doped samples gave conductivities of 10^{-6}–10^{-7} S cm^{-1}, considerably lower than the values measured for polymer **121**.

Beginning in 1999, Kiebooms and Wudl reported the production of isothianaphthene analogs as shown in Scheme 12.29 [168]. After the removal of lower M_w fractions by Soxhlet extraction with methanol and THF, polymer **124a** was isolated as a purple solid that was found to be only soluble in 1,1,2,2-tetrachloroethane. It was found that the initially isolated material was partially doped, but could be dedoped by conventional hydrazine treatment. From the absorption onset of solvent cast films, the E_g was determined to be 1.2–1.5 eV [168,169]. Characterization of the IRAV modes correlated with the generation of charge within the doped polymer revealed sharp bands of quite low intensity, indicative of strongly localized charge carriers [98,169].

More recently, polymers **124b–f** have been reported that contain liquid crystalline (LC) substituents [170]. Unlike **124a**, these polymers were soluble in CHCl$_3$ and THF and GPC characterization gave M_n of 5400–8100, with fairly narrow PDIs (1.2–1.4). While the polymers showed polymorphic texture, the texture exhibited was not satisfactory for a typical LC phase, potentially due to randomly oriented domains. All the pristine polymers were dark blue and E_g values of 1.2–1.3 eV were determined from the absorption onset. Iodine-doped films gave conductivities of 5.2×10^{-5}–3.2×10^{-4} S cm^{-1}. All properties seemed fairly independent of the LC side chain used.

A number of pyrrole-based systems have also been briefly reported [171–176]. These materials have been prepared through methods analogous to those outlined in Scheme 12.27, but in all cases either O$_2$ or 2,3-dicyano-5,6-dichloroquinone (DDQ) was used as a chemical oxidant for the final synthetic step and no dedoping workup was reported. As a result, it is unclear whether the reported materials were

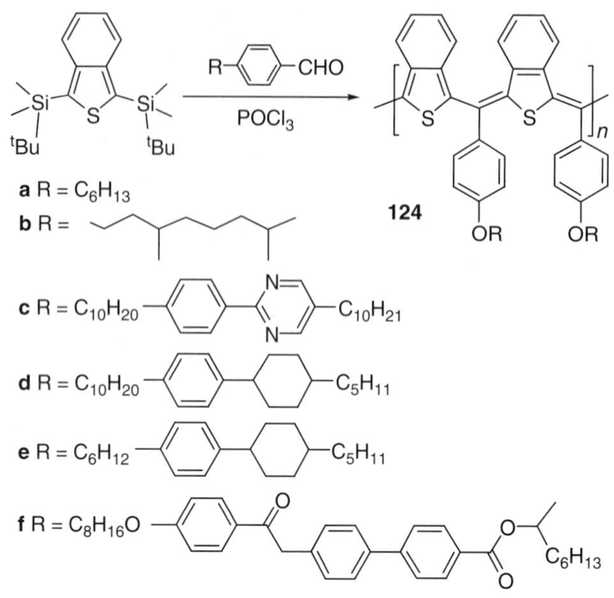

SCHEME 12.29   Synthesis of poly(isothianaphthenemethine)s.

neutral or doped to some extent. While all the reported materials are fairly similar, a variety of E_g values have been reported, from a low of below 0.5 eV [171,172] to a high of 1.14 eV [175]. These differences could also be indicative of a variety of doping levels. In some cases, no E_g values were reported [173,174].

12.10   Other Low Bandgap Conducting Polymers

In addition to the various EDOT-containing polymers discussed above, Reynolds and coworkers reported the silole-based system **126** as shown in Scheme 12.30 [177,178]. The precursor compound **125** was prepared by Pd-catalyzed cross-coupling and then electropolymerized to produce polymer **126**. The CV of the polymer exhibited an oxidation at −0.01 V and two reduction peaks at −0.19 and −0.35 V. Optical spectra of neutral films gave two absorptions at 706 and 767 nm, with an onset corresponding to an E_g of 1.3–1.4 eV.

Otsubo and coworkers reported the preparation of homo- and copolymers of two quite interesting isomeric fused-ring naphthodithiophenes (**127** and **128**) [179,180]. The naphthodithiophenes and their 2-thienyl derivatives **129** and **130** were prepared as shown in Scheme 12.31. The corresponding polymers **131–134** were then obtained by electropolymerization.

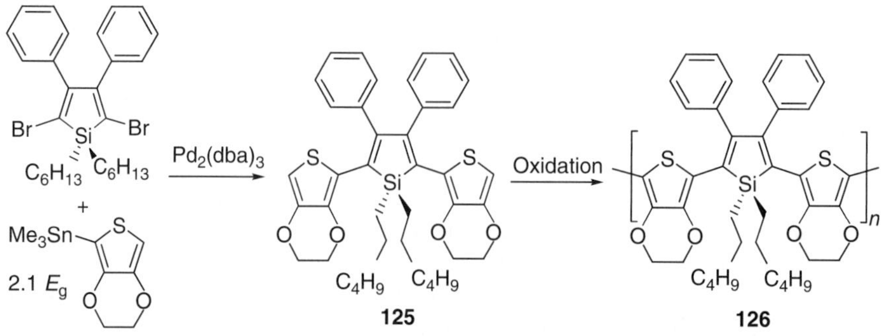

SCHEME 12.30   Synthesis of a silole–EDOT copolymer.

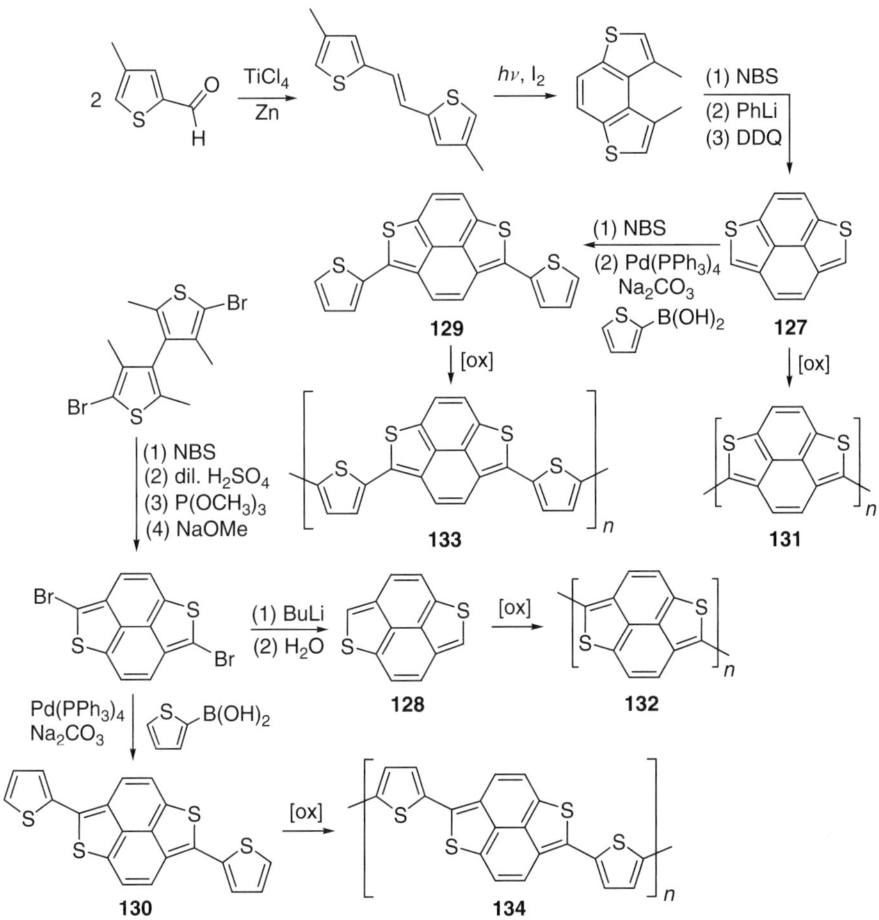

SCHEME 12.31 Synthesis of naphthodithiophene-based polymers.

The properties of the homopolymers **131** and **132** were almost identical. CVs of the electrodeposited films of polymer **132** showed an oxidation at 0.46 V and a reduction at -1.35 V, with a difference between the onsets of less than 1.0 V. Optical spectra of neutral films of **132** gave a transition at 900 nm, with an onset at 1500 nm ($E_g = 0.86$ eV). The E_g of **131** was determined to be 0.83 eV [179]. Polymer **134** exhibited two oxidations at 0.38 and 0.76 V and a fairly asymmetric reduction at -0.81 V. From the difference of the redox onsets, electrochemical E_g values of 1.19 and 1.39 eV were estimated for **134** and **133**, respectively. From the absorption onsets, optical E_g values of 1.32 and 1.49 eV were determined for **134** and **133**, respectively [180].

Two thienylethynylene polymers (Chart 12.19) have been reported, which were prepared by conventional alkyne-coupling methods [181]. The optical spectra of both polymers were quite broad with a maximum at 344 nm, with additional shoulders at 496 and 653 nm. From the onsets of both polymers, E_g values of 1.0 eV were determined. Iodine-doped samples exhibited a maximum conductivity of 10^{-6} S cm^{-1}.

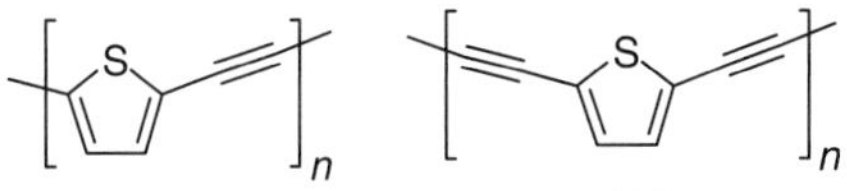

Kean and Pickup [182] reported the polymerization of the bis[1,2-di(2-thienyl)-1,2-ethenedithiolene]nickel complex **137** (Scheme 12.32) to produce a low E_g conjugated metallopolymer [182]. Complex **137** was electropolymerized to produce a metallopolymer that was

135   **136**

CHART 12.19 Thienylethynylene-based polymers.

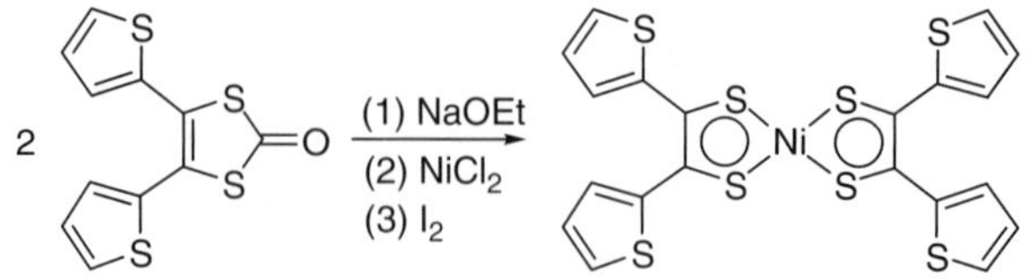

SCHEME 12.32 Synthesis of bis[1,2-di(2-thienyl)-1,2-ethenedithiolene]nickel.

expected to be highly cross-linked due to the multiple thienyl groups of the precursor **137**. The CV of the resulting metallopolymer film exhibited a broad wave at 0.5 V characteristic of an oligothiophene backbone. In addition, two Ni-dithiolene reductions were found at 0.19 and −0.49 V. From the difference of the redox onsets, an electrochemical E_g value of ∼0.35 eV was estimated. Optical spectra of the metallopolymer showed a low energy transition at 1193 nm, with a higher energy shoulder at 450 nm. From the absorption onset, the optical E_g value was determined to be <0.8 eV.

Lastly, a series of very interesting metallopolymers referred to as "porphyrin tapes" have been reported by Tsuda and Osuka [183]. These tape-shaped structures were prepared from the meso-like porphyrin arrays as shown in Scheme 12.33. The longest fused array, **138c**, exhibited a low energy transition at 3500 cm^{-1} (∼2860 nm). This transition tailed below 1500 cm^{-1} (∼6500 nm) would correspond to an E_g of <0.2 eV. The possibility that the electronic excitation energies of these materials are nearly as small as those of vibration stretching was confirmed by the observation of intense electronic bands in the IR spectrum.

12.11   Epilogue and Conclusions

Since the second edition of this handbook considerable progress has been made in the area of low bandgap conducting polymers. Much of it remains within the confines of basic fundamental science, but some of the molecules are being studied for more practical purposes. It was pointed out previously that

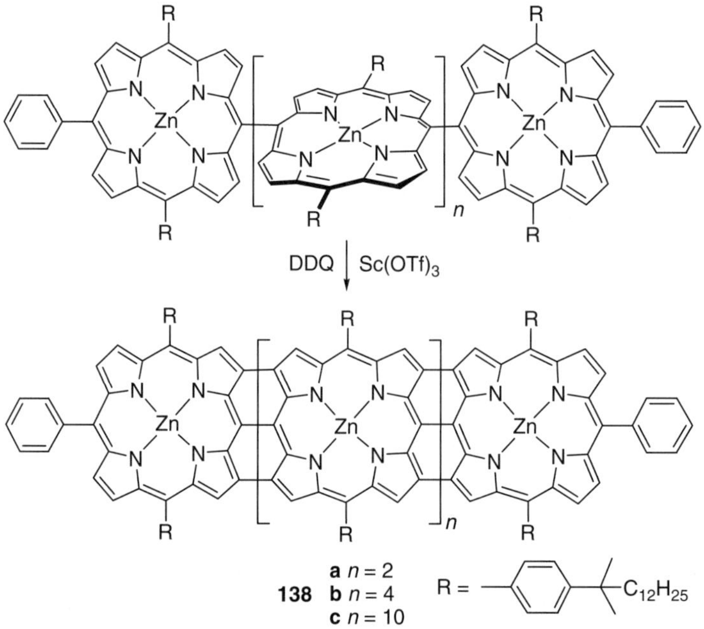

SCHEME 12.33 Synthesis of porphyrin tapes.

with low bandgap generally comes low oxidation potential and with that, relative environmental instability/reactivity. However, since the discovery of PEDOT, a relatively low bandgap polymer ($E_g \approx$ 1.6 eV) that is quite stable in its doped state, there has been considerable activity in examining not only this material, but also its lower bandgap copolymers and derivatives. Although there are molecules predicted by theory to have zero bandgaps, the values for known, synthesized molecules have not gotten down quite that low so the search for an organic polymer which is an inherent metal (where no doping is required for electrical conductivity) continues. Stable polymers that are relatively transparent and nearly colorless still remain elusive mainly because of the lack of environmental stability of the majority of these systems. As mentioned above, however, PEDOT, while not a true low bandgap polymer by our definition, is nevertheless an exception. We look forward to the day when there will be a report of an isolable (processable) zero bandgap conducting polymer.

Acknowledgments

We wish to thank the Robert A. Welch Foundation (MP, Grant Y-1407) and the National Science Foundation (SCR, Grant # CHE-0132886) for financial support of our work. This manuscript was partially written while M.P. was serving at the National Science Foundation.

Disclaimer

Any opinion, findings, and conclusions or recommendations expressed in this manuscript are those of the authors and do not necessarily reflect the views of the National Science Foundation.

References

1. Pomerantz, M. 1998. Low band gap conducting polymers. *Handbook of conducting polymers*, 2nd ed., eds. T.A. Skotheim, R.L. Elsenbaumer, and J.R. Reynolds. New York: Marcel Dekker, chap. 11, p. 277.
2. Groenendaal, L., et al. 2000. Poly(3,4-ethylenedioxythiophene) and its derivatives: Past, present, and future. *Adv Mater* 12:481.
3. Wudl, F., M. Kobayashi, and A.J. Heeger. 1984. Poly(isothianaphthene). *J Org Chem* 49:3382.
4. Kobayashi, M., et al. 1985. The electronic and electrochemical properties of poly(isothianaphthene). *J Chem Phys* 82:5717.
5. McCullough, R.D. and P.C. Ewbank. 1998. Regioregular, head-to-tail coupled poly(3-alkylthiophene) and its derivatives. *Handbook of conducting polymers,* 2nd ed., eds. T.A. Skotheim, R.L. Elsenbaumer, and J.R. Reynolds. New York: Marcel Dekker, chap. 9, p. 225.
6. Roncali, J. 1997. Synthetic principles for bandgap control in linear π-conjugated systems. *Chem Rev* 97:173.
7. Chan, H.S.O. and S. Ng. 1998. Synthesis, characterization and applications of thiophene-based functional polymers. *Prog Polym Sci* 23:1167.
8. Bakhshi, A.K. and P. Rattan. 1997. Electrically conducting polymers: An emerging technology. *Curr Sci* 73:648.
9. Bakhshi, A.K. and G. Bhalla. 2004. Electrically conducting polymers: Materials of the twenty-first century. *J Sci Ind Res* 63:715.
10. Bakhshi, A.K. and S. Kaul. 2001. Strategies for molecular designing of novel low-band-gap electrically conducting polymers. *Appl Biochem Biotechnol* 96:125.
11. Dietz, F., et al. 1994. Band structure of quasi-1-dimensional polycondensed π-systems, 3. *Macromol Theory Simul* 3:241.
12. Brocks, G. and E.E. Havinga. 2001. Small band gap polymers based upon anti-aromatic monomers. *Synth Met* 119:93.
13. Kertesz, M., C.H. Choi, and G. Sun. 1998. Design strategies for small bandgap conjugated polymers. *Polym Prepr (Am Chem Soc Div Polym Chem)* 39(1):76.

14. Hutchison, G.R., M.A. Ratner, and T.J. Marks. 2005. Electronic structure and band gaps in cationic heterocyclic oligomers, multidimensional analysis of the interplay of heteroatoms, substituents, molecular length, and charge on redox and transparency characteristics. *J Phys Chem B* 109:3126.

15. Kwon, O. and M.L. McKee. 2000. Theoretical calculations of band gaps in the aromatic structures of polythieno[3,4-*b*]benzene and polythieno[3,4-*b*]pyrazine. *J Phys Chem A* 104:7106.

16. Pomerantz, M., et al. 1992. Poly(2,3-dihexylthieno[3,4-*b*]pyrazine). A new processable low band-gap polyheterocycle. *J Chem Soc Chem Commun* 1672.

17. Pomerantz, M., et al. 1993. New processable low band-gap, conjugated polyheterocycles. *Synth Met* 55:960.

18. Toussaint, J.M. and J.L. Brédas. 1995. Geometric and electronic structure of polydicyanomethylene- cyclopenta-dicyclopentadiene, a conjugated polymer possessing a very small intrinsic band-gap. *Synth Met* 69:637.

19. Bakhshi, A.K., Deepika, and J. Ladik. 1997. Ab initio study of the electronic structures of polycyclopentadithiophene-4-one and polydicyanomethylene-cyclopentadithiophene: Two conjugated polymers with small band gaps. *Solid State Commun* 101:347.

20. Bakhshi, A.K., et al. 1996. Design of novel donor–acceptor polymers with low bandgaps. *Synth Met* 79:115.

21. Bakhshi, A.K. and P. Rattan. 1998. Ab initio study of novel low-band-gap donor–acceptor polymers based on poly(cyclopentadienylene). *J Chem Soc Faraday Trans* 94:2823.

22. Bakhshi, A.K. and G. Gandhi. 2004. Ab initio study of the electronic structures and conduction properties of some novel low band-gap donor–acceptor polymers. *Solid State Commun* 129:335.

23. Ferraris, J.P. and T.L. Lambert. 1991. Narrow bandgap polymers: Poly-4-dicyanomethylene-4H-cyclopenta[2,1-*b*;3,4-*b*]dithiophene (PCDM). *J Chem Soc Chem Commun* 1268.

24. Lambert, T.L. and J.P. Ferraris. 1991. Narrow band gap polymers: Polycyclopenta[2,1-*b*;3,4-*b*]dithiophen-4-one. *J Chem Soc Chem Commun* 752.

25. Bakhshi, A.K., et al. 1998. Theoretical design of donor–acceptor polymers with low bandgaps. *J Mol Struct* 427:211.

26. Bakhshi, A.K. and P. Bhargava. 2003. Ab initio study of the electronic structures and conduction properties of some donor–acceptor polymers and their copolymers. *J Chem Phys* 119:13159.

27. Hong, S.Y. and K.W. Lee. 2000. Small band-gap polymers: Quantum-chemical study of electronic structures of degenerate π-conjugated systems. *Chem Mater* 12:155.

28. Hong, S.Y. and S.C. Kim. 2003. Towards designing environmentally stable conjugated polymers with very small band-gaps. *Bull Korean Chem Soc* 24:1649.

29. Hong, S.Y. 2000. Zero band-gap polymers: Quantum-chemical study of electronic structures of degenerate π-conjugated systems. *Chem Mater* 12:495.

30. Brocks, G. 1995. Ab initio electronic structure of a small band gap polymer: Poly-aminosquaraine. *J Chem Phys* 102:2522.

31. Brocks, G. and A. Tol. 1996. Small band gap semiconducting polymers made from dye molecules: Polysquaraines. *J Phys Chem* 100:1838.

32. Duan, X., D.S. Dudis, and A.T. Yeates. 2001. Electronic structure and energy band gap studies for poly(1,3-thienyl-squaraine) and its dianion. *Synth Met* 116:285.

33. van Asselt, R., et al. 1995. New synthetic routes to poly(isothianaphthene). I. Reaction of phthalic anhydride and phthalide with phosphorus pentasulfide. *Synth Met* 74:65.

34. van Asselt, R., et al. 1996. New synthetic routes to poly(isothianaphthene). II. Mechanistic aspects of the reactions of phthalic anhydride and phthalide with phosphorus pentasulfide. *J Polym Sci Part A: Polym Chem* 34:1553.

35. Jen, K.-Y. and R. Elsenbaumer. 1986. Facile preparation of electrically conductive poly(isothianaphthene). *Synth Met* 16:379.

36. Paulussen, H., et al. 2001. New mechanistic aspects on the formation of poly(isothianaphthene) from P_4S_{10} and phthalic anhydride derivatives: Carbon–carbon bone formation and cleavage via a cyclic reaction mechanism. *Polymer* 41:3121.

37. Paulussen, H., et al. 1997. Synthesis of poly(isothianapththene) from 1,1,3,3-tetrachlorothiophthalan and tert-butylmercaptan: Mechanism and quantitative analysis by solid state NMR. *Polymer* 38:5221.

38. Chen, S.-A. and C.-C. Lee. 1995. Conversion of poly(1,3-dihydroisothianaphthene) into poly-isothianaphthene with the new dehydrogenation agent, tert-butyl hypochlorite. *Synth Met* 75:187.

39. Chen, S.-A. and C.-C. Lee. 1995. Processable low band gap π-conjugated polymer, poly(isothia-naphthene): Its synthesis and reaction mechanism. *Pure Appl Chem* 67:1983.

40. Chen, S.-A. and C.-C. Lee. 1996. Processable low band gap π-conjugated polymer, poly(isothia-naphthene). *Polymer* 37:519.

41. Chen, W.-T., et al. 2002. Modified synthetic routes to poly(isothianaphthene). *Synth Met* 128:215.

42. Chen, W.-T., G.A. Bowmaker, and R.P. Cooney. 2002. An electron paramagnetic resonance study of poly(isothianaphthene). *Phys Chem Chem Phys* 4:4218.

43. King, G. and S.J. Higgins. 1995. Synthesis and characterisation of novel substituted benzo[*c*]thio-phenes and polybenzo[c]thiophenes: Tuning the potentials for n- and p-doping in transparent conducting polymers. *J Mater Chem* 5:447.

44. Burbridge, S.J., et al. 1994. Nonlinear optical properties of a soluble form of polyisothianaphthene. *J Mod Opt* 41:1217.

45. Burbridge, S.J., et al. 1995. The third order nonlinear optical response of a soluble form of polyisothianaphthene. *Mol Cryst Liq Cryst Sci Technol Sect B Nonlin Opt* 10:139.

46. Paulussen, H., D. Vanderzande, and J. Gelan. 1997. The synthesis and characterization of soluble poly(isothianaphthene)-derivatives. *Synth Met* 84:415.

47. Ottenbourgs, B., et al. 1997. Characterization of poly(isothianaphthene) derivatives and analogs by using solid-state ^{13}C NMR. *Synth Met* 89:95.

48. Hagan, A.J., S.C. Moratti, and I.C. Sage. 2001. Synthesis of low band gap polymers: Studies in polyisothianaphthene. *Synth Met* 119:147.

49. Polec, I., et al. 2003. Convenient synthesis and polymerization of 5,6-disubstituted dithiophthalides toward soluble poly(isothianaphthene): An initial spectroscopic characterization of the resulting low-band-gap polymers. *J Polym Sci Part A Polym Chem* 41:1034.

50. Goris, L., et al. 2003. Poly(5,6-dithiooctylisothianaphthene), a new low band gap polymer: Spectroscopy and solar cell construction. *Synth Met* 138:249.

51. Swann, M.J., et al. 1993. Spectroscopic studies of the novel conducting polymer polytetrafluor-obenzo-c-thiophene. *Synth Met* 55:281.

52. Kiebooms, R., et al. 1996. Poly(tetrafluorobenzo[*c*]thiophene). Structure analysis of oligomers and model compound based on 1D and 2D NMR spectroscopy. *Macromolecules* 29:5981.

53. Kiebooms, R., et al. 1997. Spectroscopic analysis of poly(tetrafluoroisothianaphthene) and aro-matic model compounds. *Synth Met* 84:189.

54. Jones, C.L., S.J. Higgins, and P.A. Christensen. 2002. Some in situ reflectance fourier transform infrared studies of electrochemically prepared polybenzo[*c*]thiophene and poly-5-fluoroben-zo[*c*]thiophene films. *J Mater Chem* 12:758.

55. Meng, H. and F. Wudl. 2001. A robust low band gap processable n-type conducting polymer based on poly(isothianaphthene). *Macromolecules* 34:1810.

56. Kastner, J., et al. 1995. Raman spectra and ground state of the new low bandgap polymer poly(thienopyrazine). *Synth Met* 69:593.

57. Kastner, J., et al. 1995. Raman spectra of poly(2,3-R,R-thieno[3,4-*b*]pyrazine). A new low-band-gap polymer. *Macromolecules* 28:2922.

58. Huskic, M., D. Vanderzande, and J. Gelan. 1999. Synthesis of aza-analogues of poly(isothia-naphthene). *Synth Met* 99:143.

59. Kenning, D.D. and S.C. Rasmussen. 2003. A second look at polythieno[3,4-*b*]pyrazines: Chemical vs. electrochemical polymerization and its effect on band gap. *Macromolecules* 36:6298.

60. Nayak, K. and D.S. Marynick. 1990. The interplay between geometric and electronic struc-tures in polyisothianaphthene, polyisonaphthothiophene, polythieno(3,4-*b*)pyrazine, and poly-thieno(3,4-*b*)quinoxaline. *Macromolecules* 23:2237.

61. Kenning, D.D., et al. 2002. Thieno[3,4-*b*]pyrazines: Synthesis, structure, and reactivity. *J Org Chem* 67:9073.

62. Kenning, D.D. and S.C. Rasmussen. Unpublished results.

63. Lorcy, D. and M.P. Cava. 1992. Poly(isothianaphthene-bithiophene): A new regularly structured polythiophene analog. *Adv Mater* 4:562.

64. Bäuerle, P., et al. 1992. Synthesis and characterization of new annulated terheterocycles. *Adv Mater* 4:564.

65. Hoogmartens, I., et al. 1992. An investigation into the electronic structure of poly(isothianaphthene). *Synth Met* 51:219.

66. Musmanni, S. and J.P. Ferraris. 1993. Preparation and characterization of conducting polymers based on 1,3-di(2-thienyl)benzo[*c*]thiophene. *J Chem Soc Chem Commun* 172.

67. Staes, E., et al. 1999. Properties of a low band gap conducting polymer electrode used for amperometric detection in liquid chromatography. *Electroanalysis* 11:65.

68. Vangeneugden, D., et al. 1998. Formal copolymers based on 1,3-dithienylisothianaphthene derivatives: Promising materials for electronic devices. *Acta Polym* 49:687.

69. Vangeneugden, D.L., et al. 1999. A general synthetic route towards soluble poly(1,3-dithienylisothianaphthene) derivatives. *Synth Met* 101:120.

70. Shaheen, S.E., et al. 2001. Low band-gap polymeric photovoltaic devices. *Synth Met* 121:1583.

71. Gebeyehu, D., et al. 2002. Hybrid solar cells based on dye-sensitized nanoporous TiO_2 electrodes and conjugated polymers as hole transport materials. *Synth Met* 125:279.

72. Sonmez, G., H. Meng, and F. Wudl. 2003. Very stable low band gap polymer for charge storage purposes and near-infrared applications. *Chem Mater* 15:4923.

73. Meng, H., et al. 2003. An unusual electrochromic device based on a new low-bandgap conjugated polymer. *Adv Mater* 15:146.

74. Cravino, A., et al. 2003. Spectroscopic properties of PEDOTEHIITN, a novel soluble low band-gap conjugated polymer. *Synth Met* 137:1435.

75. Kitamura, C., S. Tanaka, and Y. Yamashita. 1994. Synthesis of new narrow bandgap polymers based on 5,7-di(2-thienyl)thieno[3,4-*b*]pyrazine and its derivatives. *J Chem Soc Chem Commun* 1585.

76. Kitamura, C., S. Tanaka, and Y. Yamashita. 1996. Design of narrow-bandgap polymers. Syntheses and properties of monomers and polymers containing aromatic-donor and o-quinoid-acceptor units. *Chem Mater* 8:570.

77. Sonmez, G., et al. 2004. A red, green, and blue (RGB) polymeric electrochromic device (PECD): The dawning of the PECD era. *Angew Chem Int Ed* 43:1498.

78. Sonmez, G., et al. 2005. A processable green polymeric electrochromic. *Macromolecules* 38:669.

79. Sonmez, G., et al. 2004. Red, green, and blue colors in polymeric electrochromics. *Adv Mater* 16:1905.

80. Berlin, A., et al. 2004. New low-gap polymers from 3,4-ethylenedioxythiophene-bis-substituted electron-poor thiophenes. The roles of thiophene, donor–acceptor alternation, and copolymerization in intrinsic conductivity. *Chem Mater* 16:3667.

81. Akoudad, S. and J. Roncali. 1998. Electrogenerated poly(thiophenes) with extremely narrow bandgap and high stability under n-doping cycling. *Chem Commun* 2081.

82. Akoudad, S. and J. Roncali. 1999. Electrogenerated poly(thiophenes) with extremely low bandgap. *Synth Met* 101:149.

83. Perepichka, I.F., E. Levillain, and J. Roncali. 2004. Effect of substitution of 3,4-ethylenedioxythiophene (EDOT) on the electronic properties of the derived electrogenerated low band gap conjugated polymers. *J Mater Chem* 14:1679.

84. Tanaka, S. and Y. Yamashita. 1995. Syntheses of narrow band gap heterocyclic copolymers of aromatic-donor and quinonoid-acceptor units. *Synth Met* 69:599.

85. Tanaka, S. and Y. Yamashita. 1997. A novel monomer candidate for intrinsically conductive organic polymers based on nonclassical thiophene. *Synth Met* 84:229.

86. Tachibana, M., et al. 2002. Small band-gap polymers involving tricyclic nonclassical thiophene as a building block. *J Phys Chem B* 106:3549.

87. Sonmez, G., et al. 2005. The unusual effect of bandgap lowering by C60 on a conjugated polymer. *Adv Mater* 17:897.

88. Zhang, F., et al. 2005. Polymer solar cells based on a low-bandgap fluorene copolymer and a fullerene derivative with photocurrent extended to 850 nm. *Adv Funct Mater* 15:745.

89. Krajcovic, J., et al. 1999. Synthesis of the copolymer of 2,3-di(1-tridecyl)thieno[3,4-*b*]pyrazine with 3-dodecylthiophene using the chemical oxidation with iron trichloride. *Synth Met* 105:79.

90. Cik, G., et al. 2001. Characterization and properties of the copolymer of dipyrido-[3,2-*a*;2′,3′-*c*]-thien-[3,4-*c*]azine with 3-dodecylthiophene. *Synth Met* 118:111.

91. Bolognesi, A., et al. 1988. Poly(dithieno[3,4-*b*:3′,4′-*d*]thiophene): A new transparent conducting polymer. *J Chem Soc Chem Commun* 246.

92. Taliani, C., et al. 1989. Optical properties of a low energy gap conduction polymer: Poly-dithieno[3,4-*b*:3′,4′-*d*]thiophene. *Synth Met* 28:C507.

93. Bolognesi, A., et al. 1989. Preparation and properties of a new conducting polyheterocycle: Polydithieno[3,4-*b*:3′,4′-*d*]thiophene (PDDT). *Synth Met* 28:C527.

94. Arbizzani, C., et al. 1995. N- and p-doped polydithieno[3,4-*b*:3′,4′-*d*]thiophene: A narrow band gap polymer for redox supercapacitors. *Electrochim Acta* 40:1871.

95. Arbizzani, C., et al. 1997. A spectroelectrochemical study of poly(dithienothiophenes). *J Electroanal Chem* 423:23.

96. Cravino, A., et al. 2000. Electrochemically- and photo-induced IR absorption of low band-gap polydithienothiophenes: A comparative study. *Mater Res Soc Symp Proc* 598:BB3.74.1.

97. Neugebauer, H., et al. 2003. Spectral signatures of positive and negative charged states in doped and photoexcited low band-gap polydithienothiophenes. *Synth Met* 139:747.

98. Neugebauer, H. 2004. Infrared signatures of postive and negative charge carriers in conjugated polymers with low band gaps. *J Electroanal Chem* 563:153.

99. Cravino, A., et al. 2002. Positive and negative charge carriers in doped or photoexcited polydithie-nothiophenes: A comparative study using Raman, infrared, and electron spin resonance spectros-copy. *J Phys Chem B* 106:3583.

100. Ehrenfreund, E., et al. 2004. Even parity states in small band gap π-conjugated polymers: Poly-dithienothiophenes. *Chem Phys Lett* 394:132.

101. Ehrenfreund, E., et al. 2005. Resonant Raman scattering dispersion in poly(dithieno[3,4-*b*:3′,4′-*d*]-thiophene): 2a$_g$ spectroscopy. *Synth Met* 150:251.

102. Inaoka, S. and D.M. Collard. 1997. Polymerization of substituted dithieno[3,4-*b*:3′,4′-*d*]thio-phenes. *Synth Met* 84:193.

103. Inaoka, S. and D.M. Collard. 1999. Synthesis, polymerization, and characterization of substituted dithieno[3,4-*b*:3′,4′-*d*]thiophenes. *J Mater Chem* 9:1719.

104. Quattrocchi, C., et al. 1993. Electronic structure of polythieno[3,4-*b*;3′,4′-*d*]thiophene, a small bandgap conjugated polymer. *Synth Met* 55–57:4399.

105. Quattrocchi, C., et al. 1993. Theoretical investigation of the structure and electronic properties of poly(dithieno[3,4-*b*:3′,4′-*d*]thiophene), a small-band-gap conjugated polymer. *Macromolecules* 26:1260.

106. Arbizzani, C., et al. 1997. Polydithienothiophenes: Two new conjugated materials with narrow band gap. *Synth Met* 84:249.

107. Catellani, M., et al. 1999. Electronic structure of polydithienothiophene materials. *Synth Met* 101:175.

108. Cravino, A., et al. 2001. Vibrational spectroscopy on pDTT3—a low band gap polymer based on dithienothiophene. *J Phys Chem B* 105:46.

109. Pomerantz, M. and X. Gu. 1997. Poly(2-decylthieno[3,4-*b*]thiophene). A new soluble low-bandgap conducting polymer. *Synth Met* 84:243.

110. Pomerantz, M., X. Gu, and S.X. Zhang. 2001. Poly(2-decylthieno[3,4-*b*]thiophene-4,6-diyl). A new low band gap conducting polymer. *Macromolecules* 34:1817.

111. Neef, C.J., I.D. Brotherston, and J.P. Ferraris. 1999. Synthesis and electronic properties of poly (2-phenylthieno[3,4-*b*]thiophene): A new low band gap polymer. *Chem Mater* 11:1957.

112. Lee, K. and G.A. Sotzing. 2001. Poly(thieno[3,4-*b*]thiophene). A new stable low band gap conducting polymer. *Macromolecules* 34:5746.

113. Sotzing, G.A. and K. Lee. 2002. Poly(thieno[3,4-*b*]thiophene): A p- and n-dopable polythiophene exhibiting high optical transparency in the semiconducting state. *Macromolecules* 35:7281.

114. Lee, B., M.S. Yavuz, and G.A. Sotzing. 2005. Conjugated polymers from thieno[3,4-*b*]thiophene dimers. *Polym Prepr (Am Chem Soc Div Polym Chem)* 46:860.

115. Lee, B., and G.A. Sotzing. 2002. Aqueous phase polymerization of thieno[3,4-*b*]thiophene. *Polym Prepr (Am Chem Soc Div Polym Chem)* 43:568.

116. Sotzing, G.A., et al. 2003. Optically transparent conductive polymers from thieno[3,4-*b*]thiophene. *Polym Mater Sci Eng* 88:268.

117. Lee, B., V. Seshadri, and G.A. Sotzing. 2005. Suspension polymerization of thieno[3,4-*b*]thiophene in water to produce low band gap conjugated polymers. *Polym Prepr (Am Chem Soc Div Polym Chem)* 46:1010.

118. Lee, B., et al. 2005. Ring-sulfonated poly(thienothiophene) *Adv Mater* 17:1792.

119. Seshadri, V., et al. 2003. Conjugated polythiophene consisting of coupling through locked transoid conformation. *Polym Mater Sci Eng* 88:292.

120. Seshadri, V., L. Wu, and G.A. Sotzing. 2003. Conjugated polymers via electrochemical polymerization of thieno[3,4-*b*]thiophene (T34bT) and 3,4-ethylenedioxythiophene (EDOT). *Langmuir* 19:9479.

121. Kumar, A. and G.A. Sotzing. 2005. Poly(thieno[3,4-*b*]furan), a new low band gap conjugated polymer. *Polym Prepr (Am Chem Soc Div Polym Chem)* 46:969.

122. Kumar, A., Z. Buyukmumcu, and G.A. Sotzing. 2006. Poly(thieno[3,4-*b*]furan). A new low band gap conjugated polymer. *Macromolecules* 39:2723.

123. Kim, I.T., S.W. Lee, and J.Y. Lee. 2003. A new low bandgap conducting polymer. *Polym Prepr (Am Chem Soc Div Polym Chem)* 44:1163.

124. Gunatunga, S.R., et al. 1997. Synthesis and characterization of low band gap polymers. *Synth Met* 84:973.

125. Beyer, R., et al. 1998. Spectroelectrochemical and electrical characterization of low bandgap polymers. *Synth Met* 92:25.

126. Kalaji, M., P.J. Murphy, and G.O. Williams. 1999. Spectroelectrochemistry of bridged dithienyl derived polymers. *Synth Met* 101:123.

127. Mills, C.A., et al. 1999. Investigations into a low band gap, semiconducting polymer. *Synth Met* 102:1000.

128. Huang, H. and P.G. Pickup. 1997. In situ conductivity of a low band-gap conducting polymer: Measurement of intrinsic conductivity. *Acta Polym* 48:455.

129. Huang, H. and P.G. Pickup. 1999. Oxygen-modified poly(4-dicyanomethylene-4*H*-cyclopenta[2,1-*b*;3,4-*b'*] dithiophene): A tunable low band gap polymer. *Chem Mater* 11:1541.

130. Sannicolo, F., et al. 1998. Highly ordered poly(cyclopentabithiophenes) functionalized with crown-ether moieties for lithium- and sodium-sensing electrodes. *Chem Mater* 10:2167.

131. Loganathan, K., et al. 2003. $\Delta^{4,4'}$-dicyclopenta[2,1-*b*:3,4-*b'*]dithiophene. A conjugated bridging unit for low band gap conducting polymers. *Chem Mater* 15:1918.

132. Huang, H. and P.G. Pickup. 1998. A donor–acceptor conducting copolymer with a very low band gap and high intrinsic conductivity. *Chem Mater* 10:2212.

133. Karikomi, M., et al. 1995. New narrow-bandgap polymer composed of benzobis(1,2,5-thiadiazole) and thiophenes. *J Am Chem Soc* 117:6791.

134. Jayakannan, M., P.A. Van Hal, and R.A.J. Janssen. 2002. Synthesis, optical, and electrochemical properties of novel copolymers on the basis of benzothiadiazole and electron-rich arene units. *J Polym Sci A Polym Chem* 40:2360.

135. Anzenbacher, P., et al. 2004. Materials chemistry approach to anion-sensor design. *Tetrahedron* 60:11163.

136. Dmitry, A. and P. Anzenbacher. 2004. Sensing of aqueous phosphates by polymers with dual modes of signal transduction. *J Am Chem Soc* 126:4752.

137. Du Bois, C.J., et al. 2001. Multi-colored electrochromic polymers based on BEDOT-pyridines. *Synth Met* 119:321.

138. Chen, M., et al. 2004. 1 Micron wavelength photo- and electroluminescence from a conjugated polymer. *Appl Phys Lett* 84:3570.

139. Chen, M.X., et al. 2004. Low band gap donor–acceptor–donor polymers for infra-red electroluminescence and transistors. *Synth Met* 146:233.

140. Wang, X., et al. 2004. Infrared photocurrent spectral response from plastic solar cell with low-band-gap polyfluorene and fullerene derivative. *Appl Phys Lett* 85:5081.

141. Ajayaghosh, A. and J. Eldo. 2001. A novel approach towards low optical band gap polysquaraines. *Org Lett* 3:2595.

142. Eldo, J. and A. Ajayaghosh. 2002. New low band gap polymers: Control of optical and electronic properties in near infrared absorbing π-conjugated polysquaraines. *Chem Mater* 14:410.

143. Ajayaghosh, A. 2003. Intrinsically conducting low band gap polysquaraines via a–b type polycondensation. *Int J Plast Technol* 6:117.

144. Ajayaghosh, A. 2003. Donor–acceptor type low band gap polymers: Polysquaraines and related systems. *Chem Soc Rev* 32:181.

145. Ajayaghosh, A. 2005. Chemistry of squaraine-derived materials: Near-IR dyes, low band gap systems and cation sensors. *Acc Chem Res* 38:499.

146. Roncali, J., et al. 1994. Control of the bandgap of conducting polymers by rigidification of the π-conjugated system. *J Chem Soc Chem Commun* 2249.

147. Verlhac, P., et al. 1998. Polydithiényléthylènes solubles dérivés de précurseurs a structure pontée. *J Chim Phys* 95:1274.

148. Blanchard, P., A. Riou, and J. Roncali. 1998. Electrosynthesis of a low band gap π-conjugated polymer from a multibridged dithienylethylene. *J Org Chem* 63:7107.

149. Ho, H.A., et al. 1995. Electrogenerated small bandgap π-conjugated polymers derived from substituted dithienylethylenes. *J Chem Soc Chem Commun* 2309.

150. Fusalba, F., et al. 2000. Poly(cyano-substituted diheteroareneethylene) as active electrode material for electrochemcial supercapacitors. *Chem Mater* 12:2581.

151. Sotzing, G.A. and J.R. Reynolds. 1995. Poly[*trans*-bis(3,4-ethylenedioxythiophene)vinylene]: A low band-gap polymer with rapid redox switching capabilities between conducting transmissive and insulating absorptive states. *J Chem Soc Chem Commun* 703.

152. Sotzing, G.A., J.R. Reynolds, and P.J. Steel. 1996. Electrochromic conducting polymers via electrochemical polymerization of bis(2-(3,4-ethylendioxy)thienyl) monomers. *Chem Mater* 8:882.

153. Sotzing, G.A., J.L. Reddinger, and J.R. Reynolds. 1997. Redox active electrochromic polymers from low oxidation monomers containing 3,4-ethylenedioxythiophene (EDOT). *Synth Met* 84:199.

154. Thomas, C.A. and J.R. Reynolds. 1999. Lowering the band gap of ethylenedioxythiophene polymers: Cyanovinylene-linked biheterocycles. *Semiconducting polymers: Applications, properties, and Synthesis*, eds. B.R. Hsieh and Y. Wei. Vol. 735. Washington DC: American Chemical Society, p. 367.

155. Sotzing, G.A., et al. 1998. Low band gap cyanovinylene polymers based on ethylenedioxythiophene. *Macromolecules* 31:3750.

156. Gallazzi, M.C., et al. 2002. A new low band gap dithienylene-fluorovinylene electrochromic polymer: An intriguing effect of silyl substitution on the polymerization reaction. *J Mater Chem* 12:2202.

157. Seshadri, V. and G.A. Sotzing. 2004. Polymerization of two unsymmetrical isomeric monomers based on thieno[3,4-*b*]thiophene containing cyanovinylene spacers. *Chem Mater* 16:5644.

158. Schlick, U., F. Teichert, and M. Hanack. 1998. Electrochemical and spectroelectrochemical investigations of small-bandgap π-conjugated polymers and their precursors. *Synth Met* 92:75.

159. Ho, H.A., et al. 1996. Thiophene-based conjugated oligomers and polymers with high electron affinity. *Adv Mater* 8:990.

160. Cheng, H. and R.L. Elsenbaumer. 1995. New precursors and polymerization route for the preparation of high molecular mass poly(3,4-dialkoxy-2,5-thienylenevinylene)s: Low band gap conductive polymers. *J Chem Soc Chem Commun* 1451.

161. Fu, Y., H. Cheng, and R.L. Elsenbaumer. 1997. Electron-rich thienylene-vinylene low bandgap polymers. *Chem Mater* 9:1720.

162. Bazzi, H.S. and H.F. Sleiman. 2002. Synthesis and self-assembly of conjugated polymer precursors containing dichlorocarbonate groups by living ring-opening metathesis polymerization. *Macromolecules* 35:624.

163. Chen, W.-C. and S.A. Jenekhe. 1995. Small-bandgap conducting polymers based on conjugated poly(heteroarylenemethines). 2. Synthesis, structure, and properties. *Macromolecules* 28:465.

164. Zhang, Q., Y. Li, and M. Yang. 2004. A novel low band gap polymer PDTNTBQ. *Synth Met* 146:69.

165. Zhang, Q., Y. Li, and M. Yang. 2004. Novel soluble low band gap polymers. *J Mater Sci* 39:6089.

166. Benincori, T., et al. 2003. An electrochemically prepared small-bandgap poly(biheteroarylidene-methine): Poly{bi[(3,4-ethlyenedioxy)thienylene]methine}. *Macromolecules* 36:5114.

167. Chen, W.-C., et al. 2004. Theoretical and experimental characterization of small band gap poly(3,4-ethylenedioxythiophene methine)s. *Macromolecules* 37:5959.

168. Kiebooms, R. and F. Wudl. 1999. Synthesis and characterisation of poly(isothianaphthene-methine). *Synth Met* 101:40.

169. Neugebauer, H., et al. 1999. Infrared spectroelectrochemical investigations on the doping of soluble poly(isothianaphthtene methine) (PIM). *J Chem Phys* 110:12108.

170. Kiebooms, R.H.L., H. Goto, and K. Akagi. 2001. Synthesis of a new class of low-band-gap polymers with liquid crystalline substituents. *Macromolecules* 34:7989.

171. Aota, H., et al. 1997. Syntheses and properties of π-conjugated polymers containing chromophore. 1. Water-soluble small-bandgap polymer backbone. *Chem Lett* 527.

172. Aota, H., et al. 1998. Syntheses and properties of π-conjugated polymers containing chromophore. 2. Introduction of porphyrin units into main chain of water-soluble small band gap polymer. *Chem Lett* 335.

173. Goto, H., K. Akagi, and H. Shirakawa. 1997. Syntheses and properties of small-bandgap liquid crystalline conjugated polymers. *Synth Met* 84:385.

174. Akagi, K., et al. 1998. Small bandgap type liquid crystalline conjugated polymers. *Mol Cryst Liq Cryst Sci Technol Sect A Mol Cryst Liq Cryst* 316:201.

175. Yan, W., C.-S., Hsu, and Y. Wei. 2002. Synthesis and characterization of small band-gap conjugated polymers—poly(pyrrolyl methines). *Chin Chem Lett* 13:988.

176. Yan, W., et al. 2001. Synthesis and characterization of poly(pyrrolyl methine). *Chin J Polym Sci* 19:499.

177. Lee, Y., et al. 2001. A new narrow band gap electroactive polymer: Poly[2,5-bis{2-(3,4-ethylene-dioxy)thienyl}silole]. *Chem Mater* 13:2234.

178. Lee, Y., et al. 2001. A new silole containing low band gap electroactive polymer. *Synth Met* 119:77.

179. Takimiya, K., et al. 2002. Synthesis, structures, and properties of two isomeric naphthodithiophenes and their methyl, methylthio, and 2-thienyl derivatives; application to conductive charge-transfer complexes and low-bandgap polymers. *Bull Chem Soc Jpn* 75:1795.

180. Casado, J., et al. 2004. A Raman and computational study of two dithienyl naphthodithiophenes: Synthesis and characterization of new polymers showing low band gap optical and electroactive features. *J Phys Chem B* 108:7611.

181. D'ilario, L., et al. 1995. Bis(ethynyl)-polymers. *J Mater Sci* 30:4273.

182. Kean, C.L. and P.G. Pickup. 2001. A low band gap conjugated metallopolymer with nickel *bis*(dithiolene) crosslinks. *Chem Commun* 815.

183. Tsuda, A. and A. Osuka. 2002. Discrete conjugated porphyrin tapes with an exceptionally small bandgap. *Adv Mater* 14:75.

13

Advanced Functional Polythiophenes Based on Tailored Precursors

Philippe Blanchard,
Philippe Leriche,
Pierre Frère, and
Jean Roncali

13.1 Introduction

Progress in material science generally results from a conjunction of advances in basic science and society demands. From the mid-1970s, the short history of conducting polymers provides a new illustration of this process. The discovery of the metallic conductivity of doped polyacetylene (PA) by MacDiarmid, Heeger, and Shirakawa in 1977 [1] has generated a considerable interest for the synthesis, electronic properties, and applications of conjugated polymers such as polyparaphenylene (PPP), polyparaphenylenevinylene (PPV), polyaniline (PANI), polypyrrole (PPy), and polythiophene (PT) [2–4]. In the 1980s, conducting polymers were considered bulk materials for applications resorting to the conductivity of the doped state such as antistatic coatings, energy storage, and electromagnetic interference shieldings.

The discovery of the electroluminescent properties of PPV by Friend and coworkers in 1990 [5] represents a turning point and the scientific and technological interest for conjugated polymers progressively left the conducting properties of the doped conducting state to focus on the semi-conducting properties of the neutral state. This renewed interest together with the concomitant development of well-defined conjugated oligomers has led to the emergence of the field of organic (opto)electronics with an increasing research effort invested in the development of organic light-emitting diodes (OLEDs), organic field-effect transistors (OFETs), and organic solar cells. During the last two decades, PT and its derivatives have progressively acquired a prominent position among conjugated polymers owing to a unique combination of high conductivity, environmental stability, and structural versatility allowing derivatization of the π-conjugated backbone in view of numerous technological applications.

The synthesis, the electronic properties and the applications of PT derivatives have already given rise to several reviews [3,6–11].

Functionalized conjugated polymers obtained by polymerization of thiophene monomers derivatized by covalently attached functional groups have been developed for almost two decades. However, while the field remains very active, it has been subjected to considerable changes in its objectives and methods during the last 10 years. Recent years have witnessed the use of new functionalized PT derivatives for advanced applications such as chemical, electrochemical, or bioelectrochemical sensors, and optoelectronic devices [12].

On the other hand, with the development of more elaborated systems capable to perform more complex tasks, the role of the precursor structure has progressively evolved from the initially viewed simple carrier of functional groups to become an important constitutive and active part of the whole system. Thus, by proper design of the precursor, it has been possible to improve the chemical or electrochemical polymerization process and to finely tune the electronic properties of conjugated polymers for specific applications. In this context, the structural control and manipulation of quantities such as absorption and emission spectra, oxidation and reduction potentials, or luminescence efficiency has become an important target. As most of these parameters depend on the width of the ionization potential, electron affinity, and bandgap of conjugated polymers, much work has been devoted to the analysis of structure–electronic properties relationships in functional π-conjugated systems to define synthetic principles for bandgap engineering [13].

The aim of this chapter is to emphasize the impact of bandgap engineering on the design and synthesis of functional π-conjugated PTs for advanced applications. After an introductory part summarizing the main structural approaches for bandgap control, a second part will focus on recent developments in gap engineering based on the use of chalcogeno-substituted thiophene precursors. The last part will cover advances in the synthesis of functional conjugated PTs derived from tailored precursors for applications in electrochemical and optical sensors, metal-containing PTs, electrochromic devices, and electronic devices such as OLEDs, OFETs, and organic solar cells. Special emphasis will be put on the relationships between the structure of the tailored conjugated PT backbone, its electronic properties, and each specific application.

13.2 Synthetic Principles for Bandgap Engineering

Various synthetic approaches for engineering the bandgap of conjugated polymers have already been reviewed [13] and consequently will be briefly summarized here. As shown below the magnitude of the bandgap depends on five parameters.

Because of their low dimensionality, linear π-conjugated polymers are subject to Peierls distorsion. Consequently, electron–phonon interactions and electron–electron correlation effects result in the localization of π-electrons along the conjugated backbone with the opening, at the Fermi level, of a bandgap, E_g, generally larger than 1.50 eV [14,15]. In polyaromatic systems such as PT, the ground state is nondegenerated and two different structures namely aromatic and quinoid are possible in contrast to PA [15]. Although the aromatic form is energetically more stable, theoretical work has demonstrated that the quinoid form should present a lower bandgap [16]. In both cases, the persistence of π-electron localization as well as single and double bond results in bond length alternation (BLA) that provides a major contribution $E^{\delta r}$ to the magnitude of the energy gap [16–19].

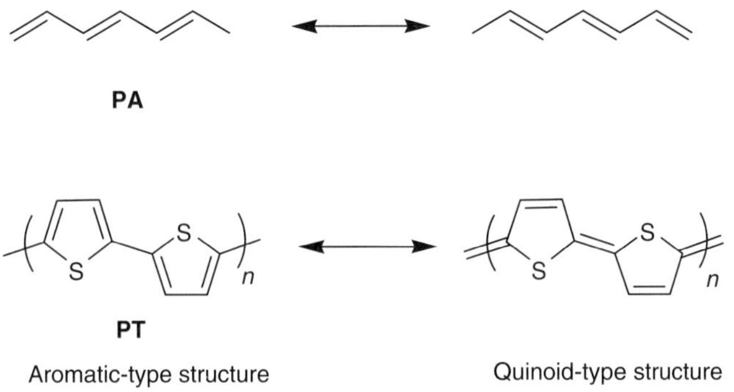

PA

PT

Aromatic-type structure   Quinoid-type structure

In polyaromatic systems, the single bonds between aromatic cycles allow for interannular rotations that may limit the orbital overlap of π-electrons along the conjugated backbone and hence contribute to E_g by a quantity E^θ. On the other hand, the aromatic resonance energy of the cycle (E^{Res}) tends to confine π-electrons within the rings and hence to limit their delocalization over the whole conjugated system. The lower bandgap (or more extended effective conjugation) of PT compared to PPP (2.0 vs. 3.4 eV) results from the combined effects of a more planar structure and lower E^{Res} (122 and 152 kJ/mol for thiophene and benzene, respectively). It must be mentioned that replacement of the sulfur atom of thiophene by other heteroatoms will also influence the value of E^{Res}. The introduction of groups with electron-withdrawing or electron-releasing inductive or mesomeric effects is the most direct way to modulate the HOMO and LUMO energy levels of a conjugated system and hence their difference. This structural factor is represented by the term E^{Sub}. Finally at the supramolecular level, the bandgap can be affected by intermolecular interactions denoted by the quantity E^{Int}.

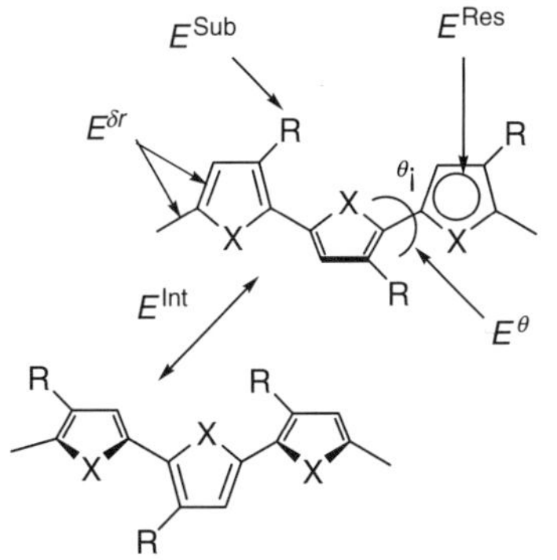

$$E_g = E^{\delta r} + E^\theta + E^{Res} + E^{Sub} + E^{Int}$$

X = CH=CH, S, SO$_2$, N-R, SiR$_2$, P-R

The increase of the quinoid character of the conjugated backbone to the detriment of its aromaticity has represented a main strategy for synthesizing small bandgap conjugated polymers. Poly(isothia-naphthene) (PITN) is the first prototype of low bandgap polymer based on this concept [20]. The presence of the benzene ring fused to the thiophene stabilizes some quinoid contributions to the ground-state geometry of PITN. Consequently, E_g decreases from 2.0 to 2.2 eV for PT to 1.0–1.1 eV for PITN. Following a related approach, thieno[3,4-b]pyrazine has also led to low bandgap PT derivatives such as **1** [21–23].

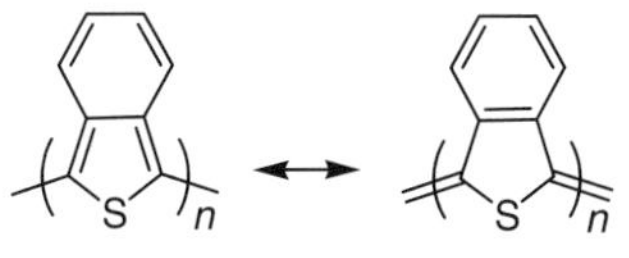

PITN (E_g= 1.0–1.1 eV) 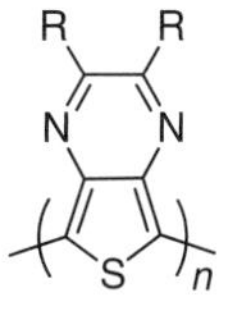   **1** (R = C$_6$H$_{13}$, E_g = 0.95 eV)

Low bandgap polymers have been synthesized from precursors based on fused thiophene rings such as thieno[3,4-*b*]thiophene. Substituted [24,25] and unsubstituted [26] precursors have led to polymers **2** with low bandgaps of 0.85–0.90 eV. These polymers are stable under redox switching between p- and n-doped states. Polymer **2** (R = H) is sky blue in the reduced form and transparent in the oxidized state [27].

Poly(heteroarylenemethines) **3** proposed by Jenekhe [28] represent an intermediate class between the aromatic and the quinoid forms. Bandgaps as low as 0.75 eV [28] were initially reported for these materials while their reexamination led to E_g values ~1.14 eV [29].

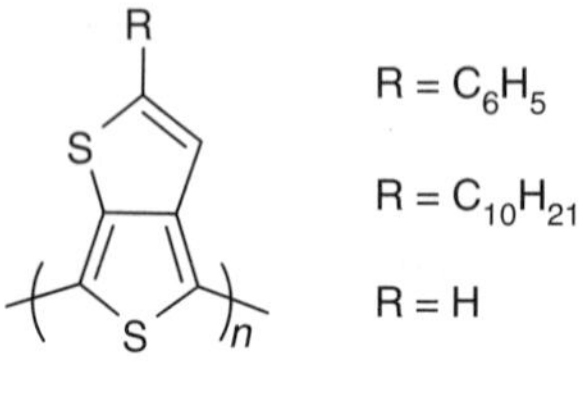

R = C_6H_5

R = $C_{10}H_{21}$

R = H

2 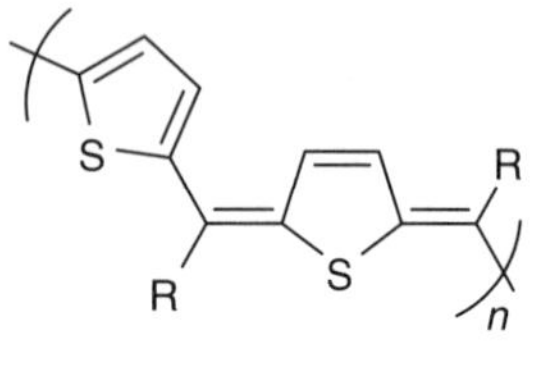   **3**

Since the pioneering work by Havinga and coworkers [30,31], one successful strategy to achieve narrow bandgap π-conjugated systems has involved the alternation of electron-rich and electron-deficient units along the same conjugated chain. Thus, Ferraris and coworkers reported the synthesis of a 2,2′-bithiophene bridged at the 3,3′-positions by an electron-withdrawing ketone group. Electropolymerization of this compound gave a polymer **4** with $E_g = 1.10$–1.20 eV [32]. The stronger electron-withdrawing effect of the dicyanomethylene groups of **5** reduces the LUMO level of the related monomer and leads to a further decrease of E_g to 0.80 eV [33]. Our group has reported the electrosynthesis of low bandgap polymers such as **6** ($E_g = 0.50$–0.60 eV) by the introduction of a cyanogroup on the double bond of *E*-1,2-(di-2-thienyl)ethylene [34] or other conjugated systems [35]. Recently, the use of stronger donor unit thieno[3,4-*b*]thiophene [36] has led to polymer **7** with a bandgap of 1.1–1.2 eV.

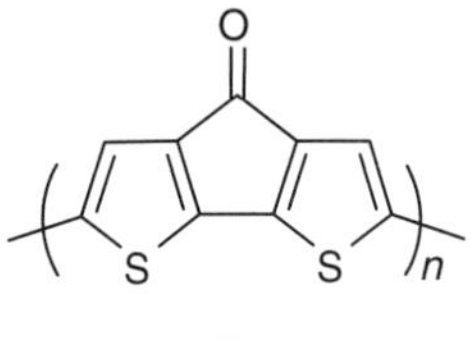

4 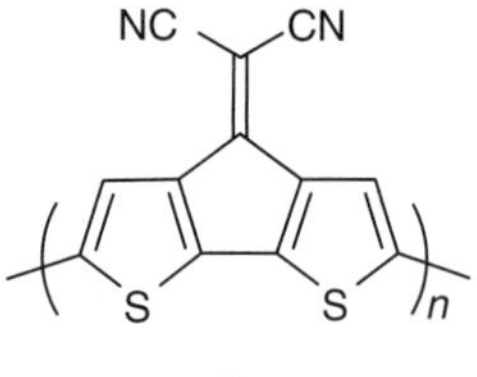   **5**

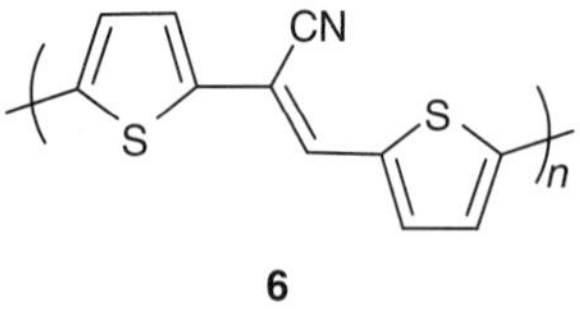

6 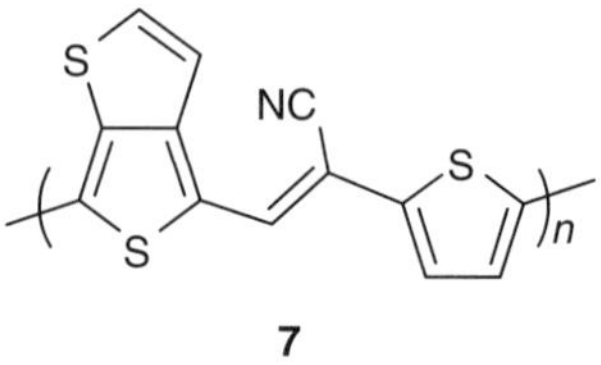   **7**

The insertion of ethylenic bonds between the thiophene rings represents another approach to lower E_g. The combined effects of reduced rotational freedom and lower overall aromaticity of the π-conjugated system result in a decrease of BLA and E_g from 2.20 eV for PT [8] to 1.74 eV for polythienylenevinylene PTV [37,38].

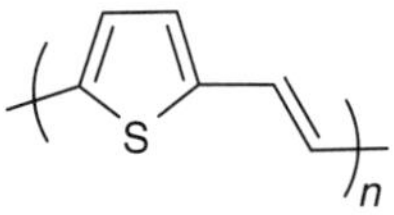

PTV

Enhanced planarity and reduced BLA have been achieved by covalent rigidification of the conjugated system. Thus, Scherf and Müllen have developed low bandgap ladder PPP systems [39]. Our group has extensively investigated the covalent rigidification of various classes of thiophene-based conjugated polymers [40–45] and oligomers [46,47]. For example, reduced bandgap polymers **8** and **9** were obtained by electropolymerization of bridged 2,2′-bithiophene [40] and dithienylethylene (DTE) [42] precursors, respectively. Bridged conjugated polymers based on the cyclopentadithiophene structure have been widely developed to extend π-conjugation [48].

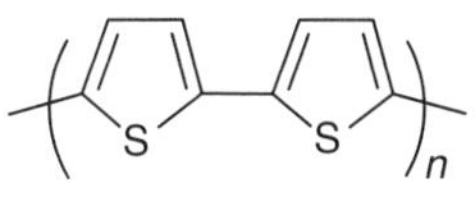

Poly(BT)

$E_g = 2.2$ eV

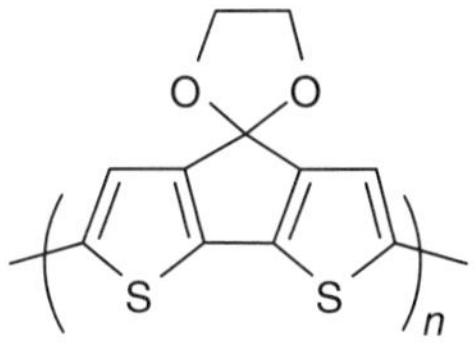

8

$E_g = 1.2$ eV

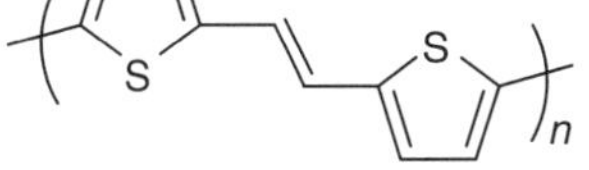

Poly(DTE)

$E_g = 1.7–1.8$ eV

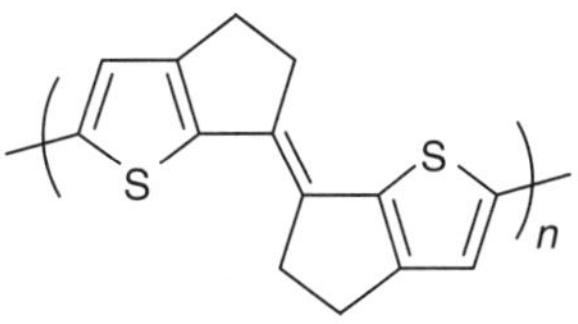

9

$E_g = 1.3–1.4$ eV

The replacement of some thiophene rings by other heteroaromatic rings, such as silole [49,50] or phosphole [51–53], represents an alternative solution to modulate the bandgap of PT. Thus, polymer **11** shows a relatively low E_g value of 1.60 eV due to the low aromatic character of phosphole. Additionally, optical and electrochemical properties depend on the substitution at the phosphorus atom [51–53]. Similarly, insertion of the nonaromatic thiophene-*S,S*-dioxide in conjugated systems, such as **12**, has been shown to produce an extension of the π-electron delocalization associated with the reduction of bandgap [54–57].

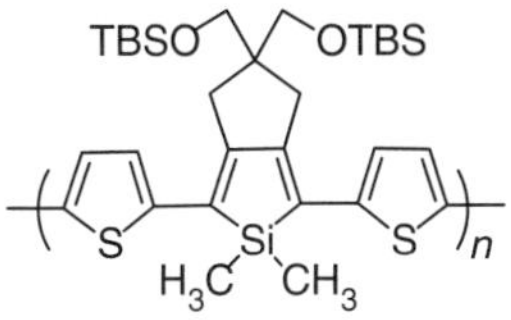

10

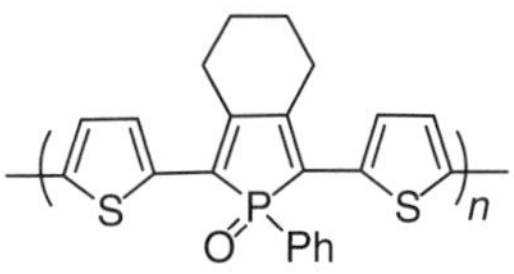

11

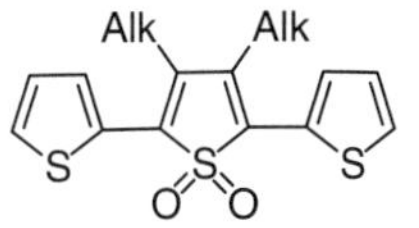

12

The increasing use of 3,4-ethylenedioxythiophene (EDOT) as a building block represents one of the most important step in the recent development of low bandgap PTs [58–60]. In this context, the following part will describe the synthesis of chalcogeno-substituted oligo- and polythiophenes and will discuss the impact of chalcogen substitution on the electrochemical, optical, and structural properties of the conjugated system.

13.3 Chalcogeno-Substituted Thiophene Derivatives

13.3.1 *o*-Alkylation of Thiophene

The attachment of alkoxy groups at the β-positions of thiophene rings leads to polymers with moderate bandgap, low oxidation potential, and stable oxidized conducting state. However, the nature and number of alkoxy groups grafted on the monomeric or oligomeric precursor can strongly influence the polymerization process and the properties of the resulting polymer.

Electropolymerization of 3-alkoxythiophenes **13** leads to poorly conjugated chains [61]. Chemical polymerization using ferric chloride as an oxidizing reagent leads to low-molecular weight polymer with a relatively irregular chemical structure. Regioregular poly(3-alkoxythiophene)s have been synthesized recently by a coupling reaction of 2-bromo-3-alkoxy-5-bromomagnesiothiophenes **14** [62,63]. Compared to the parent regiorandom polymers, regioregular PTs present lower oxidation potentials and redshifted λ_{max} indicative of longer effective conjugation.

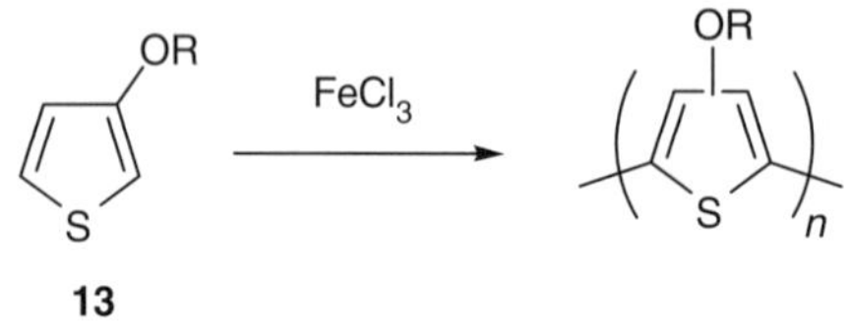

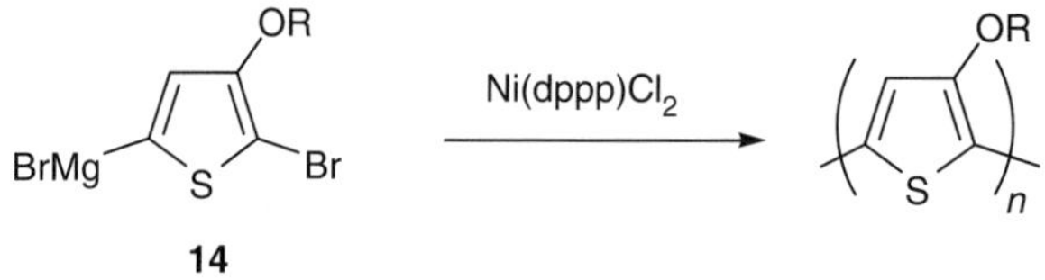

13

14

PTs β-substituted by alkoxy and aryolxy groups have been synthesized from symmetrical bithiophenes **15** and **16**. The position of the substituents proves to be crucial for the polymerization processes [64]. Bithiophene **15** substituted at the inner β-positions leads only to short-chain oligomers [64,65]. By contrast, substitution at the outer β-positions leads to a straightforward electropolymerization. As shown by electrochemical and theoretical studies, this latter mode of substitution contributes to enhance the density of unpaired electron at the coupling α-position [65].

As reported by Heinze et al. poly(4,4′-dimethoxy-2,2′-bithiophene) oxidizes at 0.20 V vs. Ag/AgCl, combines moderate bandgap (1.60 eV) and high stability in the conducting state [66]. Based on its transparency in the oxidized state, this polymer has been proposed for hole injection in OLEDs [67].

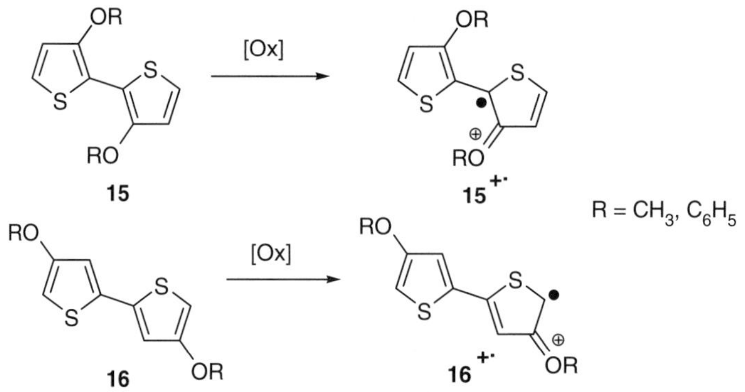

15   **15**$^{+\bullet}$ R = CH$_3$, C$_6$H$_5$

16   **16**$^{+\bullet}$

Terthiophenes with terminal alkoxy substituents such as **17**, **18**, and **19** have been also electropolymerized. It has been shown that for poly(**17**), the lengthening of the alkyl chain on the median thiophene leads to polymers with a lower bandgap [68]. For poly(**18**), electrooxidation of the lateral pentoxythiophene substituent has been proposed to enhance the conductivity by cross-linking of polymer chains [69]. Electropolymerization of **19** led to a polymer with a reduced bandgap (~1.20 eV) due to a donor–acceptor alternance in the conjugated chain [70]. Oligothiophenes **20** bearing alkylsulfanyl chains at the outer β-positions lead to polymers with a high degree of polymerization [71–73].

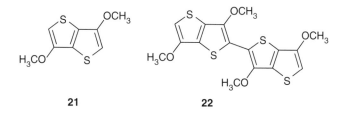

(structures 17, 18, 19, 20)

17

R₁ = CH₃, C₅H₁₁

R₂ = CH₃, C₆H₁₃, C₈H₁₇, C₁₂H₂₅

$R_1 = CH_3, C_5H_{11}$

$R_2 = CH_3, C_6H_{13}, C_8H_{17}, C_{12}H_{25}$

18

19

20   $n = 0, 1, 2$

We have recently shown that 3,6-dimethoxythieno[3,2-*b*]thiophene **21** leads to a polymer presenting low oxidation potential and moderate bandgap (1.7 eV) [74]. The advantage of thienothiophene unit compared to bithiophene one resides in the planar structure and absence of positional isomers. Furthermore, the crystallographic structure of the dimer **22** (Figure 13.1), shows a fully planar conjugated system stabilized by noncovalent intramolecular sulfur–oxygen interactions.

Disubstituted 3,4-alkoxythiophenes **23** allow to prepare regioregular polymers and devoid of parasitic α—β couplings that decreases the conjugation length. However, the effective conjugation of the polymers may be limited by steric hindrance. Thus, poly(3,4-dibutoxythiophene) obtained by chemical polymerization with FeCl₃ shows a higher oxidation potential and absorbs at shorter wavelengths ($\lambda_{max} = 480$ nm) than poly(3-butoxythiophene) prepared in the same conditions [75]. More recently, it has been shown that extended polymer chains can be obtained by electropolymerization of 3,4-dialkoxythiophenes in

21

22

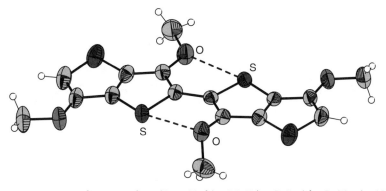

FIGURE 13.1 X-ray structure of compound **22**. (From Turbiez, M., Frère, P., Leriche, P., Mercier, N., and Roncali, J., *Chem. Commun.*, 1161–1163, 2005.) Intramolecular S–O interactions are represented by *dotted lines*.

acetonitrile in the presence of LiClO$_4$ [76]. The lengthening of the alkoxy chains produced a positive shift of the oxidation peak while the absorption maximum remains around 560 nm. A water soluble polymer based on 3,4-dialkoxythiophene **24** has been described and it was shown that the λ_{max} of 510 nm in THF shifts to 550 nm in water. This effect was attributed to an helical conformation which limits the rotational disorder of the conjugated chain [77].

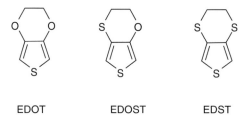

23   **24**

13.3.2 Poly(3,4-Ethylenedioxythiophene) and Sulfur Derivatives

Poly(3,4-ethylenedioxythiophene) (PEDOT) has acquired a prominent position among conducting polymers [58]. Compared to linear alkoxy groups, the ethylenedioxy bridge limits steric hindrance and hence strongly improves the electronic properties of the polymer. PEDOT presents a unique combination of moderate bandgap (1.65 eV), low oxidation potential (0.0 V vs. SCE) associated with a good stability, and optical transparency of the oxidized conducting state. Based on these properties, PEDOT has been widely used as hole injecting layer in OLEDs and solar cells and for various applications such as antistatic coating and electrode materials in supercapacitors [59].

The remarkable electronic properties of PEDOT are due to the synergy of electronic and geometric effects of oxygen atoms. In addition to electron donor effect of the ethylenedioxy group, our group has established that the oxygen atoms give rise to noncovalent intramolecular S–O interactions which result in a self-rigidification of the conjugated chain [60]. The crucial role of oxygen is clearly apparent when comparing the electronic properties of monomers and polymers containing oxygen or sulfur atoms at the 3- and 4-positions namely EDOT, thieno[3,4-*b*]-1,4-oxathiane (EDOST), and 3,4-ethylenedisulfanylthiophene (EDST).

EDOT   EDOST   EDST

Alkylsulfanyl groups are stronger donors than alkoxy ones, as shown by the decrease of the oxidation potential from 1.50 V for EDOT to 1.38 and 1.32 V (vs. Ag/AgCl) for EDOST and EDST, respectively [78]. However, this difference is not observed for polymers such as PEDOT, which has a lower oxidation potential and a smaller bandgap than poly(EDST) ($E_{ox} = 0.20$ vs. 0.90 V and $\lambda_{max} = 590$ vs. 448 nm); whereas poly(EDOST) represents the intermediate situation [78,79]. This result underlines the major impact of the sulfur atoms of the alkysulfanyl groups on the steric distorsion of the π-conjugated chain.

Comparison of the crystallographic structures and electronic properties of the dimers of EDOT (BEDOT) and EDST (BEDST) and the hybrid EDOT–EDST system [80,81] shows that BEDOT is perfectly planar and BEDST presents a median torsion angle of 45° (Figure 13.2). Surprisingly, the hybrid EDOT–EDST presents a quasiplanar conformation stabilized by one S–O interaction. The resulting polymer exhibits an optical bandgap close to that of the polymer obtained from BEDOT [80]. Recent works on terthiophene systems combining EDOT and EDST units support our conclusion [82].

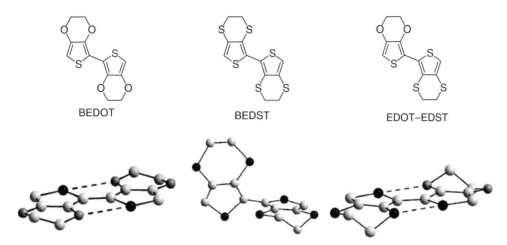

FIGURE 13.2 X-ray structures of the dimers BEDOT (From Raimundo, J.M., Blanchard, P., Frere, P., Mercier, N., Ledoux Rak, I., Hierle, R., and Roncali, J., *Tetrahedron Lett.*, 42, 1507–1510, 2001.), BEDST, and EDOT–EDST. (From Turbiez, M., Frere, P., Allain, M., Planas, N.G., and Roncali, J., *Macromolecules*, 38, 6806–6812, 2005.) The intramolecular interactions are represented by *dotted lines*.

The self-rigidification associated with noncovalent S–O intramolecular interactions between adjacent EDOT units has been observed for many EDOT-based π-conjugated systems [60].

Particularly, the absorption spectra of EDOT oligomers and hybrid EDOT-thiophene systems show that gradual replacement of thiophene rings by EDOT units produces the emergence of a vibronic fine structure [83–85] typical for rigid conjugated systems, while a concomitant decrease of the Stockes shift is observed in the emission spectrum (Figure 13.3). Such effects are consistent with a rigidification of the conjugated structure by the S–O interactions associated with the increase of the number of EDOT units.

The electron donor effect of the ethylenedioxy group increases the HOMO level of the monomer, thereby decreasing its oxidation potential [86]. Consequently, incorporation of EDOT in the structure of a precursor leads to a decrease of the potential required for electropolymerization thus allowing the derivatization of the precursor with low oxidation potential functional groups.

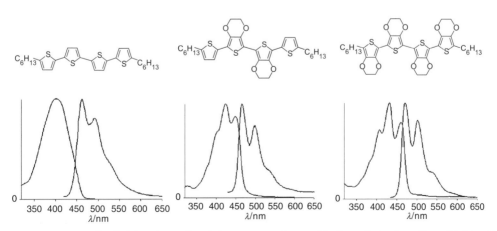

FIGURE 13.3 Normalized absorption and emission spectra of quaterthiophene derivatives. (From Turbiez, M., Frère, P., Allain, M., Videlot, C., Ackermann, J., and Roncali, J., *Chem. Eur. J.*, 11, 3742–3752, 2005.)

Furthermore, as shown in the following scheme, the increase of the spin density provides a high reactivity to the end α-position thus ensuring efficient electropolymerization of functionalized precursors at low potential and low monomer concentration.

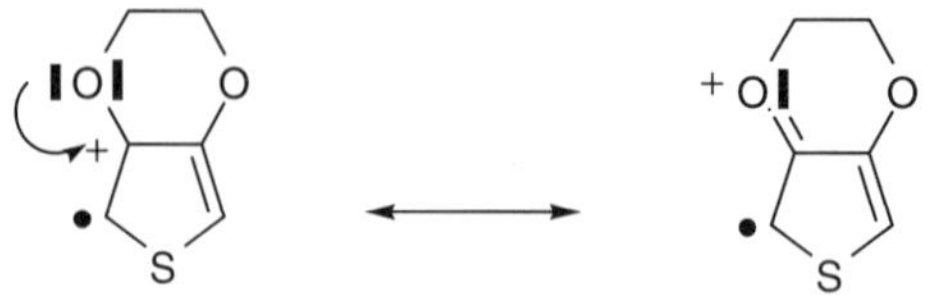

13.3.3 Reduced Bandgap Polymers Based on Chalcogen 3,4-Substituted Thiophenes

As previously indicated PEDOT exhibits a moderate bandgap of about 1.7 eV due to a combination of electronic and geometric effects. Based on the lower bandgap of PTV [37], the groups of Reynolds [87] and Elsenbaumer [88] have independently synthesized *trans*-bis(3,4-ethylenedioxythiophene)vinylene and have shown that the corresponding polymer **25** presents a lower bandgap (1.50 eV) than PEDOT. Replacement of EDOT by 3,4-dimethoxythiophene increases the bandgap to 2.0 eV probably due to steric interactions between consecutive thiophene rings [88]. Poly(3,4-ethylenedioxy-2,5-thienylene-vinylene) **26** obtained by thermolysis of films of a soluble precursor polymer presented an estimated bandgap of 1.0–1.1 eV [89]. Poly(3,4-dialkoxythienylenevinylene)s **27** with butoxy [90], (S)-2-methyl-butoxy and (S)-2-methylbutylsulfanyl [91] groups have been synthesized by McMurry reaction. The optical bandgap of polymer with butoxy groups was estimated to be 1.6 eV and the soluble polymers with chiral alkoxy or alkylsulfanyl substituents exhibited strong solvatochromism and thermochromism effects due to supramolecular aggregation.

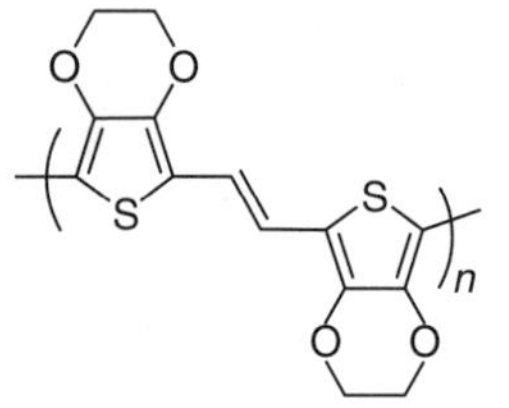

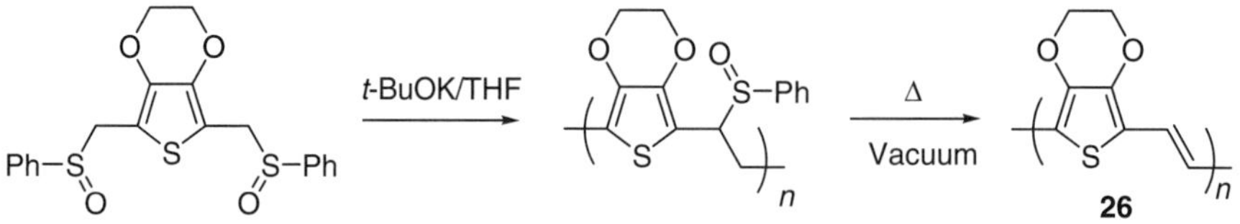

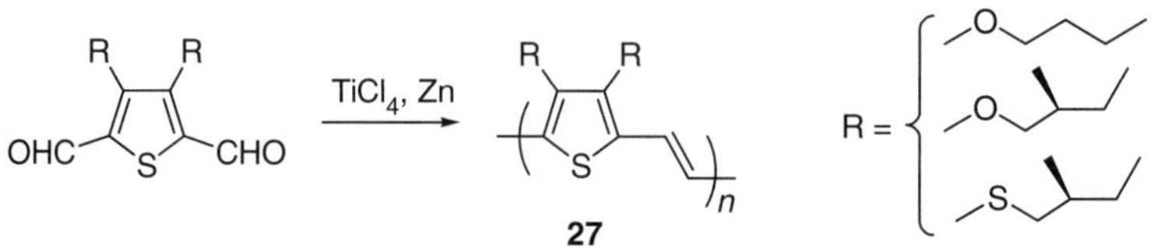

Alternation of electron-rich and poor units in conjugated systems is expected to reduce the bandgap due to the concomitant increase of HOMO level and decrease of LUMO level. Hence the use of stronger donor units, such as EDOT [92] and 3,4-ethylenedioxypyrrole (EDOP) [93], in PTV-like structure **28** and **29**, respectively, allowed a further decrease of the gap to 1.1–1.2 eV. Recently, following the same approach a soluble narrow bandgap polymer **30** ($E_g = 1.7$ eV) was described by Reynolds and coworkers by polycondensation of a dialdehyde derived from a 3,4-propylenedioxythiophene (ProDOT) with 1,4-dialkoxy-2,5-phenylene-diacetonitrile [94].

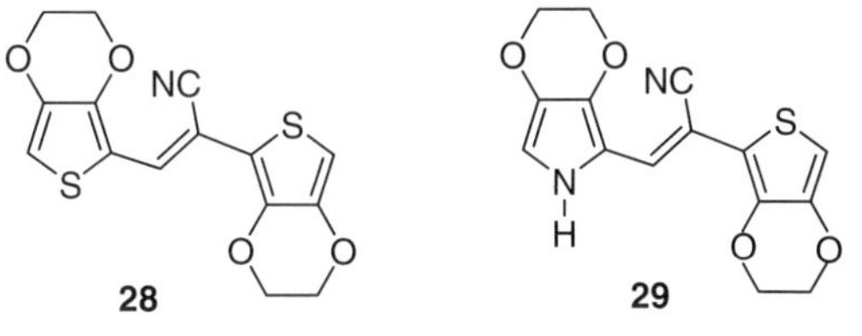

28   **29**

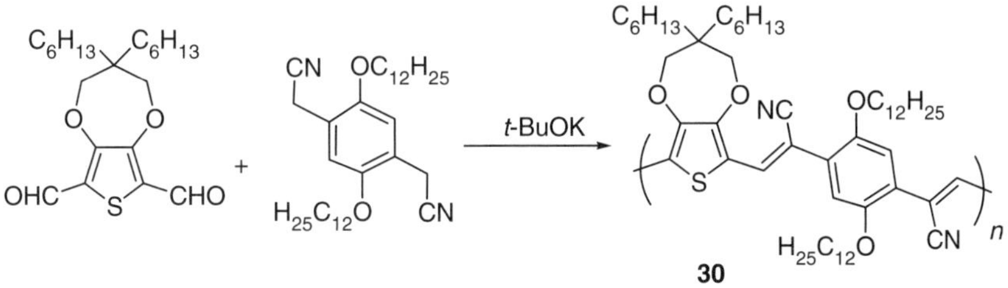

t-BuOK

30

The use of EDOT as basic units in poly(heteroarylenemethines) has been achieved only recently. Thus, electrochemical oxidation of bis[2-(3,4-ethylenedioxy)thienyl]methane **31** led to poly(bisEDOT-methine) **32**. The latter showed an electrochemical bandgap of 0.4 eV based on the onset of oxidation and reduction waves while the determination of the optical bandgap gave somewhat higher values [95]. Recently, the synthesis of poly(EDOT-methine) **33** was performed by the action of an excess of methanesulfonic acid on EDOT-carbaldehyde followed by the reduction of the resulting doped polymer. The related optical bandgap was estimated to be around 0.95 eV [96].

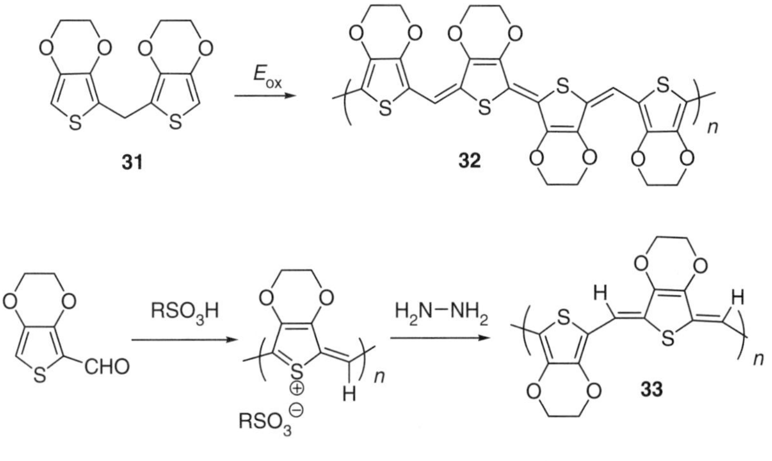

31 E_{ox} **32**

RSO$_3$H   H$_2$N—NH$_2$ **33**

Insertion of an acceptor group between two EDOT moieties [60] is the widely used approach. The width of the bandgap of the corresponding polymers depends on both the electronic and the geometric effects of the median acceptor unit. Thus, in spite of the strong acceptor character of *S,S*-dioxide-thiophene ring, polymers obtained from **34** [97] exhibited a relatively high bandgap of 1.9 eV due to the torsion of the conjugated systems as demonstrated by cristallographic and theoretical results [98]. Electropolymerization of 2,5-bis(3,4-ethylenedioxythiophene)-3,4-diphenylsilole **35** leads to a polymer with an optical bandgap of 1.30–1.40 eV [50], while the polymer based on **36** with a median benzo[*c*]thiophene-*N*-2″-ethylhexyl-4,5-dicarboximide unit shows an optical bandgap of 1.1 eV and can be p- and n-doped [99]. Recently, Berlin and coworkers have described the synthesis of derivatives **37** with a median 4-dicyanomethylenecyclopentadithiophene unit. The resulting polymer showed optical and electrochemical gaps of 0.95 and 0.80 eV, respectively [100]. Compound **38** with a dicyanomethylenefluorene has been described by Rault-Berthelot et al. An electrochemical bandgap of

0.38 eV was found but the optical bandgap was not reported [101]. Other systems containing a thieno[3,4-c]pyrazine unit **39** [100,102], **40** [100], and **41** [102] have been synthesized. The crystallographic structure of **41** showed that the conjugated system is rigidified by intramolecular S–O and S–N interactions [60,102]. The analysis of molecular structures of **39** and **41** is consistent with a partial quinoid structure of the conjugated systems [102]. The resulting electrogenerated polymers presented bandgaps in the range of 1.1–1.3 eV.

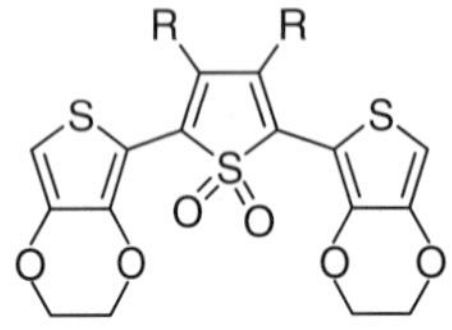

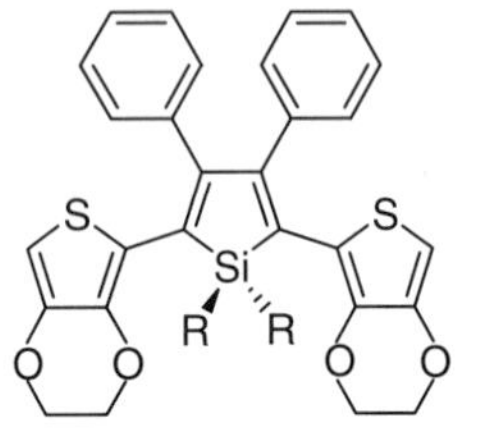

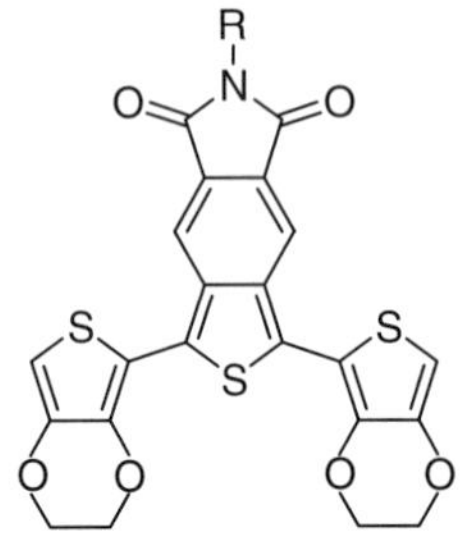

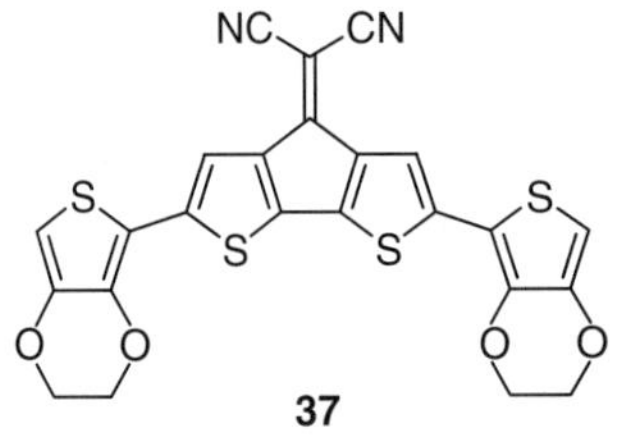

Proquinoid thieno[3,4-c]pyrazine unit has been associated with EDOT in a bithiophenic precursor **42** [103]. The resulting polymer presented a bandgap smaller than 0.40 eV and exhibited an exceptional stability under n-doping. Compared to the polymer obtained from tricyclic systems **39** and **41**, the decrease in the bandgap of poly(**42**) can be attributed to a regular alternation of donor and acceptor groups, which allows for an optimal gap reduction. Recently, polymer derived from the alkyl-substituted analog **43** has been synthesized, which showed a bandgap of 0.80 eV [104]. The 0.40 eV increase of the bandgap compared to poly(**42**) was attributed to a decrease of π-stacking interactions between the conjugated chains in the solid state due to the fixation of the solubilizing alkyl chain at an sp^3 carbon of the ethylenedioxy bridge.

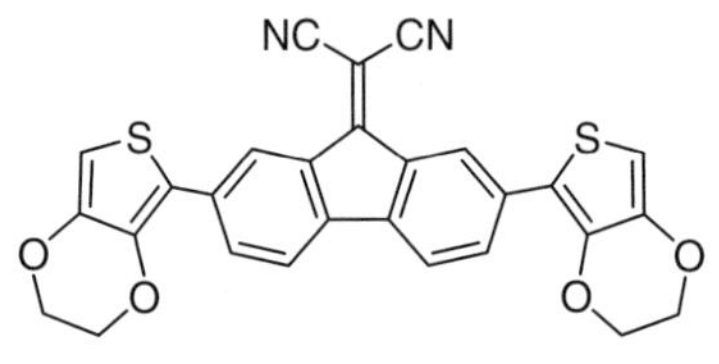

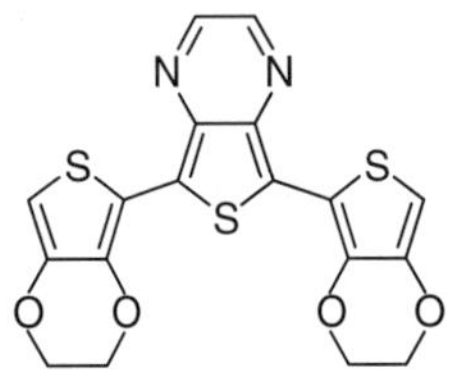

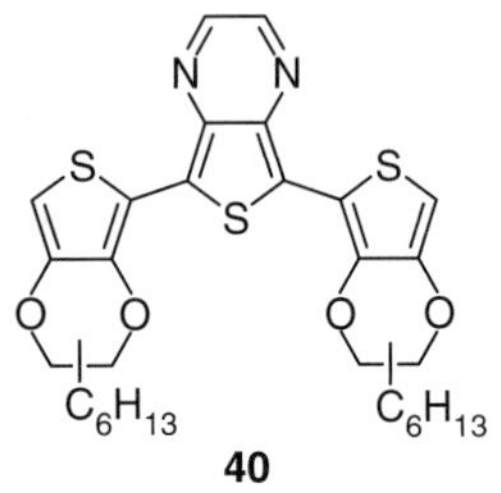

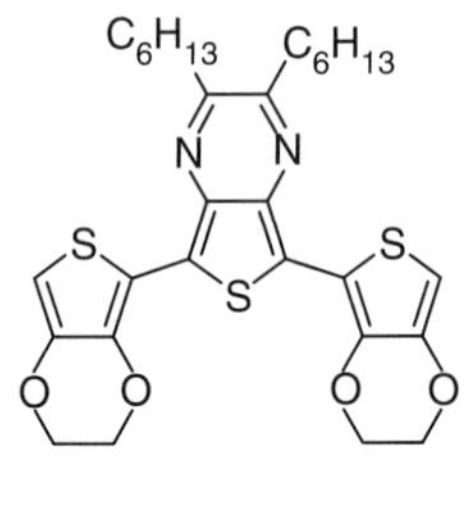

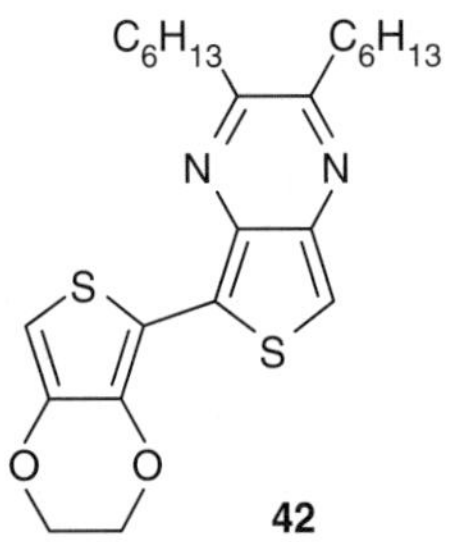

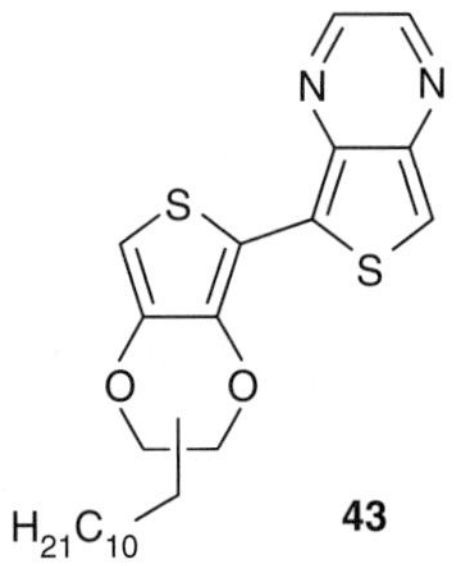

To summarize, the insertion of EDOT in precursors of polymer leads to a decrease of the polymerization potential and improves the efficiency of electropolymerization process due to the high reactivity of terminal EDOT radical cation. In addition to these electronic effects, the spontaneous planarization of some EDOT-based precursors by S–O or S–N intramolecular interactions contributes to reduce the bandgap of the polymers.

13.3.4 Structural Modifications of EDOT

The unique electronic properties of PEDOT combined with the high aptitude for polymerization of the EDOT monomer have given rise to the development of many EDOT-based functional conjugated systems by using either the EDOT building block or by modification of its structure.

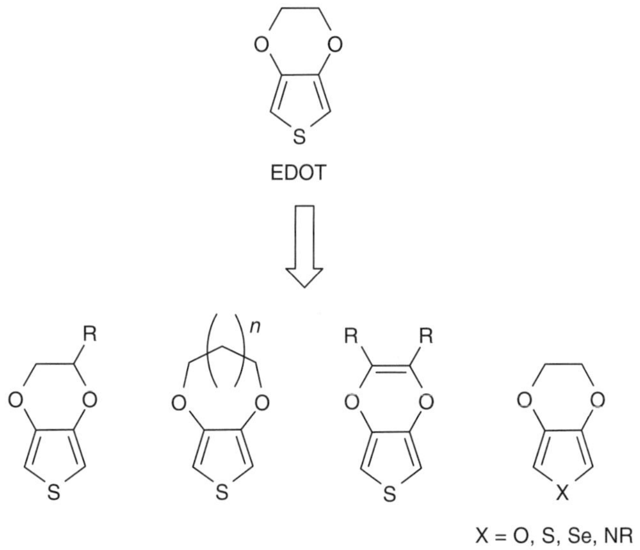

13.3.4.1 Modifications of the Ethylene Bridge

The EDOT system can be modified either by the substitution of ethylenedioxy bridge or by its replacement by another type of bridge. Both approaches involve a modification of the initial synthesis of EDOT [105]. Thus, alkylation of 3,4-dihydroxy-2,5-dicarboethoxythiophene [106] with 1-alkyl-1,2-dibromoethane or 2-alkyl-1,3-dibromobutane led to corresponding alkyl-EDOT and ProDOT derivatives [107].

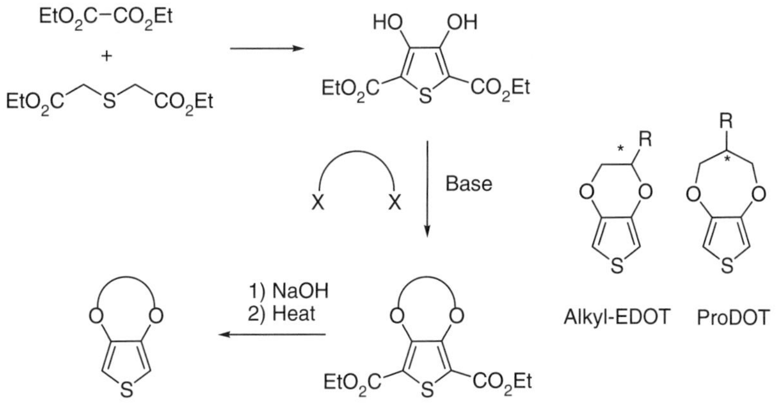

On this basis [108], many substituted EDOTs have been described and used as precursors for polymerization [107,109–116].

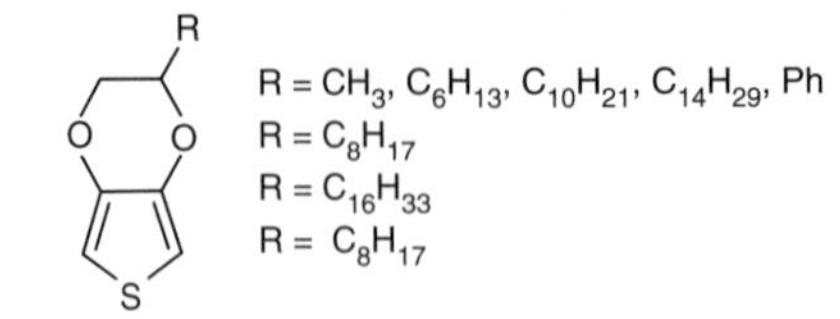
R = CH3, C6H13, C10H21, C14H29, Ph
R = C8H17
R = C16H33
R = C8H17

44

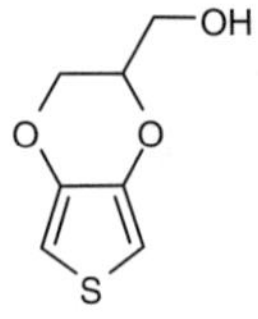

45

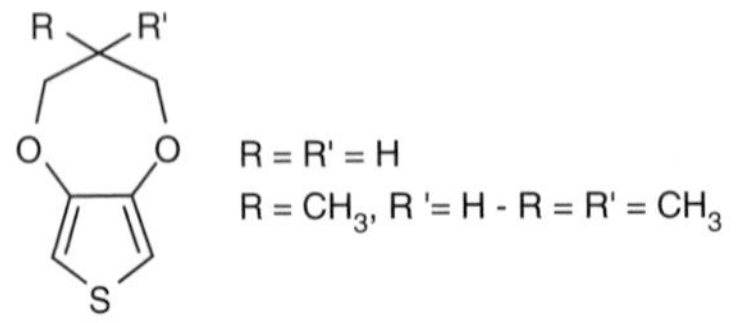
R = R' = H
R = CH3, R '= H - R = R' = CH3

46

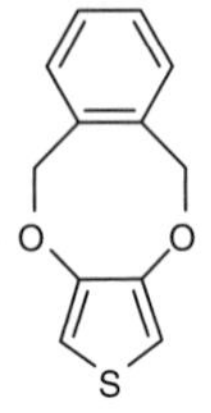

47

Despite its interest, the scope of this approach is limited by the low yield of the cross-etherification, which strongly decreases when the size of the substituent increases.

An alternative synthesis involves the transetherification of 3,4-dimethoxythiophene with diols. Initially obtained from 3,4-dibromothiophene, 3,4-dimethoxythiophene can be prepared on a large scale using the methodology recently described by Hellberg and coworkers [117]. The following scheme presents some examples of ProDOT analogs obtained by transetherification [118–124].

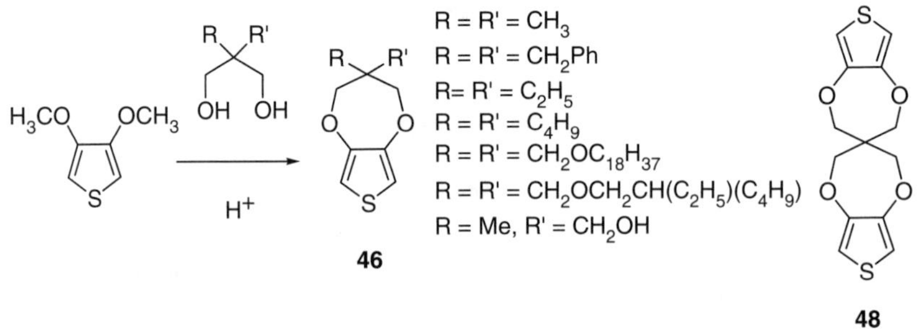
H3CO
OCH3
OH OH
H+
R = R' = CH3
R = R' = CH2Ph
R= R' = C2H5
R = R' = C4H9
R = R' = CH2OC18H37
R = R' = CH2OCH2CH(C2H5)(C4H9)
R = Me, R' = CH2OH
46
48

Applying this methodology to chiral glycols, Bäuerle and coworkers have prepared regio- and stereoregular PEDOTs **49** and demonstrated that the stereochemistry of the monomer influences the electronic properties of the polymer [125]. The synthesis of monomer **50** with an intramolecular disulfide bridge is usable for further functionalization or adsorption on gold.

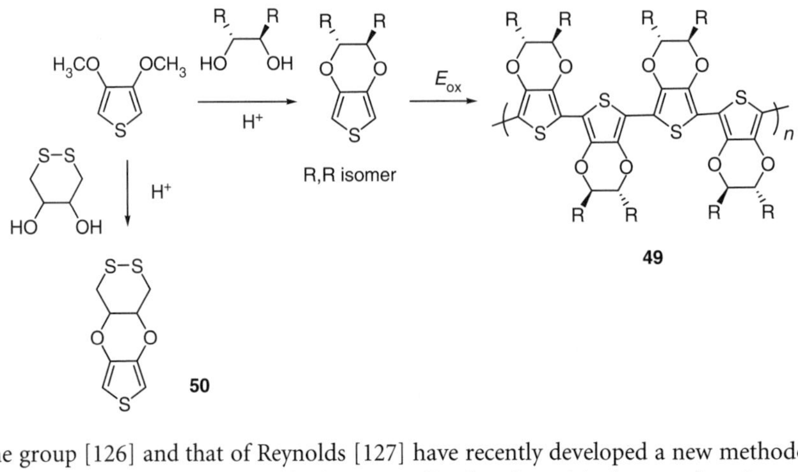
H3CO
OCH3
HO
OH
H+
R,R isomer
Eox
n
49
50

The same group [126] and that of Reynolds [127] have recently developed a new methodology based on a Mitsunobu reaction between 3,4-dihydroxy-2,5-dicarboethoxythiophene and various diols.

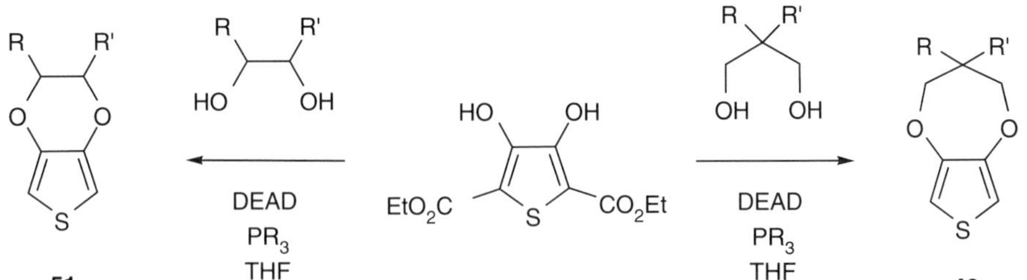

Introduction of a hydroxymethyl group at the ethylene bridge of EDOT has represented a practical approach for the functionalization of EDOT. Thus, reaction of 3,4-dihydroxy-2,5-dicarboethoxythiophene with epibromhydrine afforded hydroxymethylEDOT **45** and hydroxyProDOT **52** in a 3:1 ratio [113]. Chevrot et al. improved this synthesis and obtained pure hydroxymethylEDOT **45** after recrystallization [114]. An alternative synthesis starting from the 2,3-dibromopropan-1-ol and 2,5-dicarboethoxy-3,4-hydroxythiophene avoids the formation of **52** as side product [115].

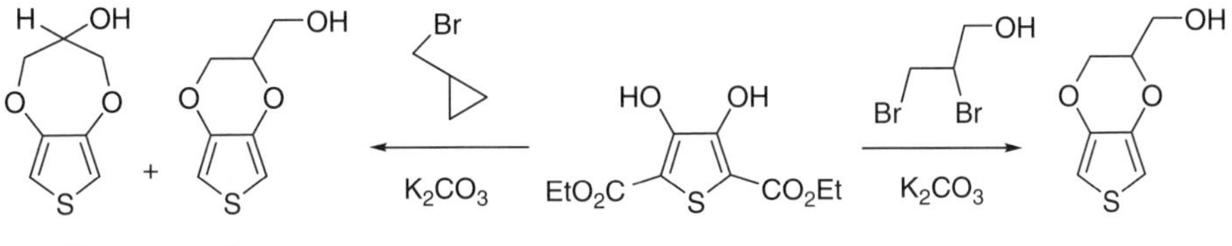

13.3.4.2 Phenyl-fused EDOT and 3,4-(Vinylenedioxy)thiophene

Substitution at the ethylenedioxy bridge of EDOT results in the creation of a stereogenic center that can give rise to many regio- and stereo-isomers in the resulting polymers. Furthermore, substitution at an sp^3 carbon does not allow the substituent to be coplanar with the conjugated system thus leading to weaker interchain interactions and hence to a loss of the charge-transport properties of the solid state material.

To circumvent this problem, new platforms derived from EDOT have been designed.

We have recently shown that replacement of the ethylene bridge by a phenyl group leads to a new system [phenyl-fused EDOT (PheDOT)] in which substitution can be achieved at some sp^2 carbons of the benzenic ring. Experimental and theoretical works have shown that it is difficult to polymerize PheDOT due to the high stability of its cation radical [128,129]. Further studies on its dimer and trimer have revealed that contrary to what is generally observed for conjugated oligomers, the aptitude for electropolymerization increases with chain length. As shown by theoretical analysis, this result is associated with relocalization of HOMO and SUMO on the conjugated thiophenic system for the PheDOT dimer and trimer [130].

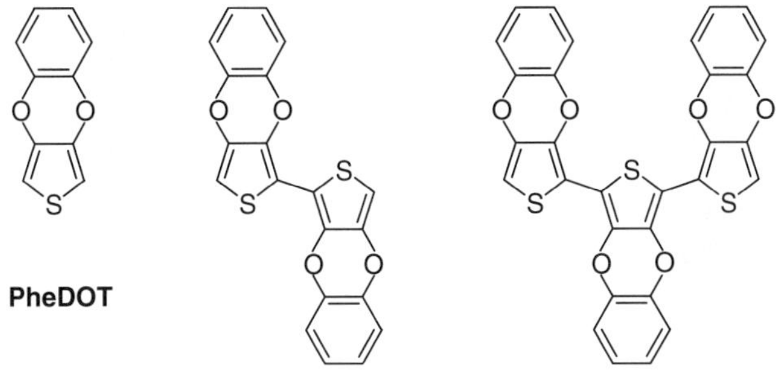

PheDOT

Recently, we have shown that 3,4-(vinylenedioxy)thiophene (VDOT), another EDOT analog bridged by sp^2 carbons, could be prepared from 3,4-dimethoxythiophene using final Grubbs reaction [131]. Similar to PheDOT, VDOT does not electropolymerize whereas its dimer BVDOT can be easily polymerized [132]. The crystal structure of BVDOT shows the presence of noncovalent S–O intramolecular interactions similar to

EDOT-based systems. These interactions and replacement of the sp^3 bridging carbons by an olefinic linkage lead to a perfectly coplanar structure (Figure 13.4). BVDOT and poly(BVDOT) show higher oxidation potential than BEDOT and poly(BEDOT) suggesting better environmental stability for the neutral form of the derived polymers and oligomers.

13.3.4.3   Other Heteroaromatic Analogs of EDOT

In addition to the modification of the ethylene bridge of EDOT, the replacement of the thiophene ring by other heteroaromatic rings such as pyrrole, furan, and selenophene has been reported.

EDOS   **EDOF**   **EDOP**

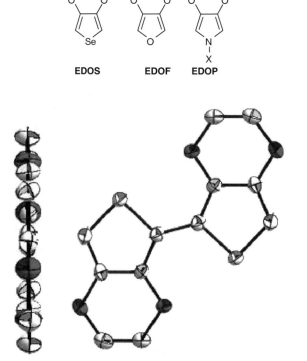

FIGURE 13.4 X-ray structure of BVDOT. (From Leriche, P., P. Blanchard, P. Frère, E. Levillain, G. Mabon, and J. Roncali. 2006. 3,4-Vinylenedioxythiophene (VDOT): A new building block for thiophene-(VDOT): A new building block for thiophene-based π-conjugated systems. *Chem. Commun.*, 275–277, 2006.)

Thus, Cava et al. have synthesized 3,4-ethylenedioxyselenophene (EDOS) [133]. Poly(alkylenedioxyfuran)s such as poly(EDOF) have been developed as photothermographic image-forming materials [134]. EDOP was first synthesized in 1995 by Merz et al. [135]. Recently, this monomer and its analogs, ProDOP and BuDOP, have been intensively investigated by Reynolds and coworkers and considered in relation with their potential applications in biosensors [136,137].

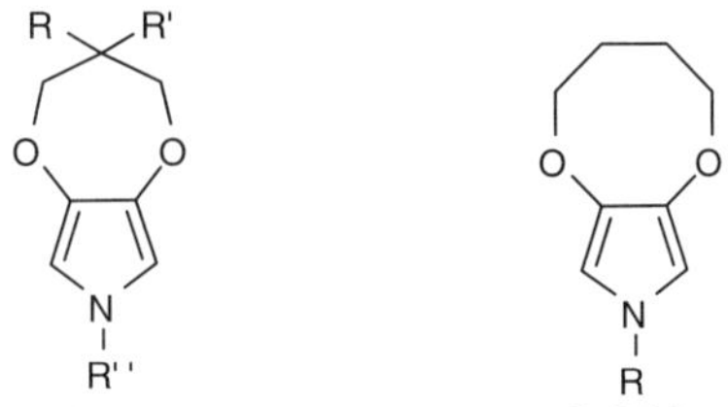

ProDOP   BuDOP

13.3.5 Functionalized EDOT Derivatives

Functionalized PT derivatives based on EDOT have been synthesized following two main approaches. The first one relies on the direct functionalization at the ethylene bridge to give monomeric precursors suitable for polymerization while in the second one, the functional group is grafted on one thiophene cycle incorporated in an EDOT-containing bithiophene or terthiophene structure.

13.3.5.1 Monomeric Precursors

Functionalization of EDOT can be easily accomplished by the use of the versatile hydroxymethyl-EDOT **45** whose hydroxyl group leads to specific properties and can be subjected to Williamson reaction.

First, the presence of the hydroxyl group in **45** allows to perform electropolymerization in water and to increase the electroactivity of the corresponding polymer in aqueous medium [115]. On this basis, poly(**45**) grown in the presence of nanopeptides has been used as biosensors [138], while a copolymer of **45** and EDOT has been used to bind glucose oxidase [139].

On the other hand, direct attachment of an alkyl sulfonate group to **45**, affording compound **53**, has been reported by Chevrot et al. [140]. A copolymer with EDOT was initially synthesized to produce a polymer with cation-exchange properties and was later applied to exchange radioactive ions such as Sr^{2+} [141]. Kumar et al. reported the electrochemical homopolymerization of another sulfonate containing EDOT derivative without supporting electrolyte [142]. Zotti et al. have also reported the electrosynthesis of poly(**53**) in the presence of strong acids [143]. Recently, Reynolds et al. have investigated the electrochromic properties of this material [144] and its efficiency as hole transporter in PLEDs [145].

Highly hydrophilic polymers derivatized with oligoethers chains have been prepared from **45**. These polymers show electroactivity in aqueous medium and present cation recognition properties. Moreover, they constitute a unique example of polymers with solid state ionochromic properties (see Section 13.4.1.1) [146]. The synthesis of **54** bearing perfluoro chains has also been reported starting from **45** and its related polymer has been investigated for electrochromic application [147].

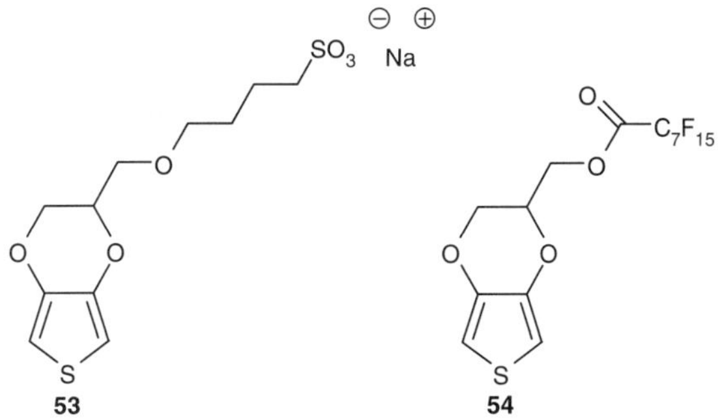

53   54

Cyanobiphenyl (**55**) [148] and redox pendant groups such as ferrocene (**56**) [149], viologen (see Section 13.4.3) [150], and Tetrathiafulvalene [151] (see Section 13.4.1.1) cores have been grafted on hydroxymethyl-EDOT **45**. A copolymer containing pendant ferrocene groups has been obtained by copolymerization of **56** and another EDOT derivative [149].

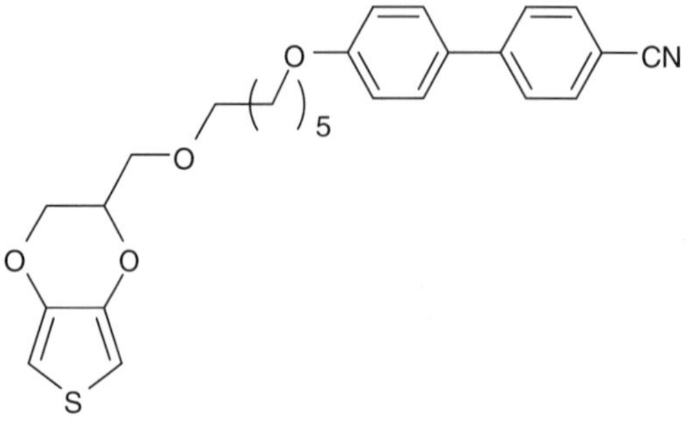

55 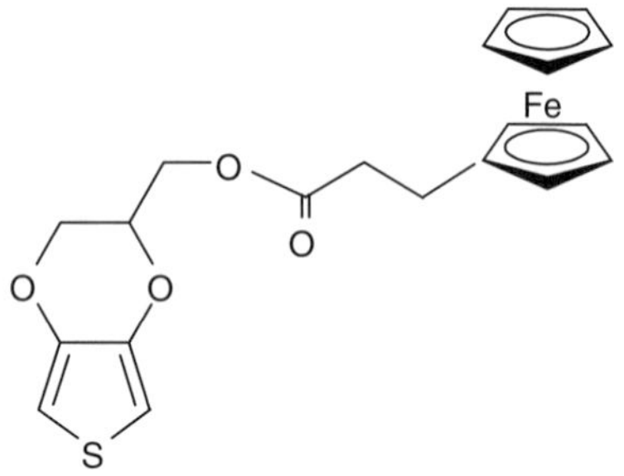   **56**

13.3.5.2 Bithiophenic Precursors

The use of bithiophenic precursors instead of thiophenic monomers leads to a decrease of the electropolymerization potential while maintaining a sufficient reactivity of the radical cation to produce an efficient polymerization [8,10]. Thus, functionalized PTs with groups sensitive to oxidation can be electrogenerated from tailored bithiophenic precursors. Additionally, the insertion of one EDOT unit in the precursor structure leads to a further decrease of the polymerization potential and enhances the efficiency of the electropolymerization process.

According to these principles, compound **57** was synthesized and electropolymerized in spite of the problems posed by the presence of an easily oxidized tetrathiafulvalene unit [152,153]. Other bithiophenic precursors **58** and **59** have been successfully electropolymerized [154], while polymers derived from precursors **60** [155,156] or **61** [157] have been prepared by chemical coupling.

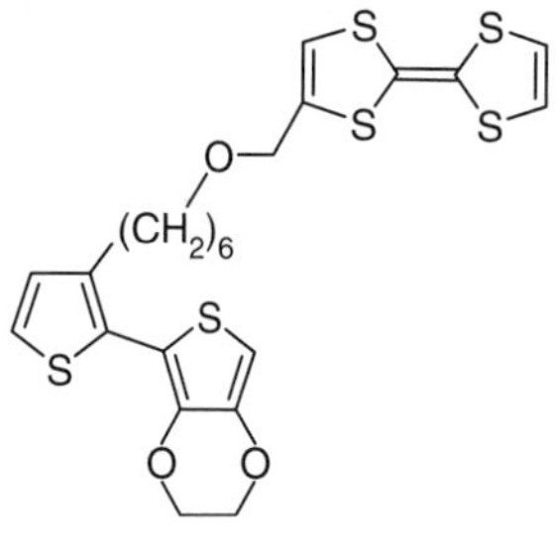

57 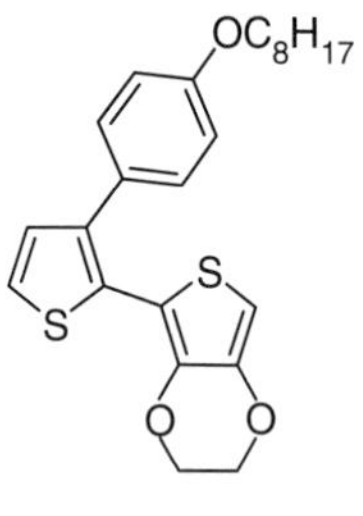   **58** 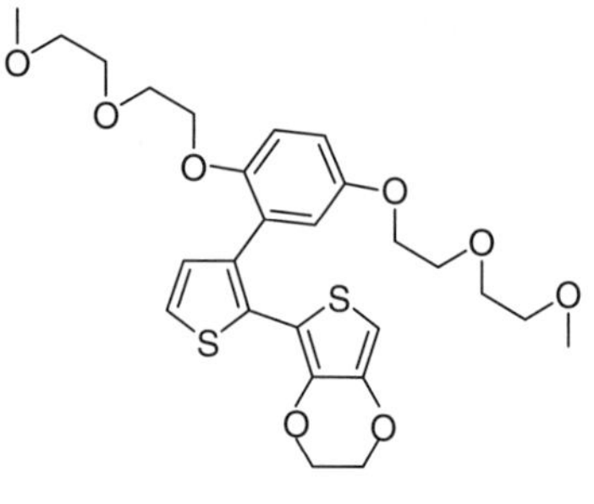   **59**

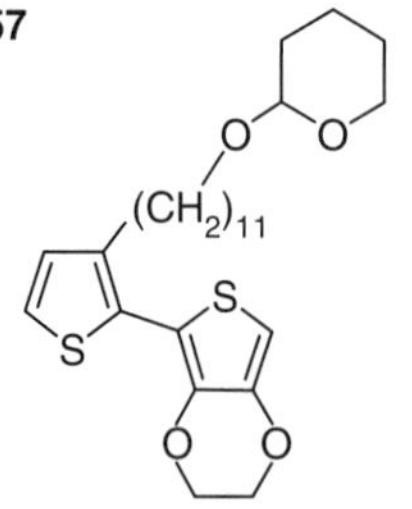

60 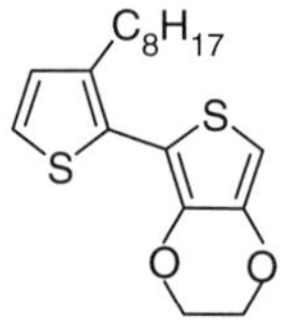   **61**

2-[(3,4-Ethylenedioxy)thienyl]-3-(2-cyanoethylsulfanyl)thiophene and its oxygen analog **62** (X = S, O) have been developed in our group as a new class of versatile precursors [158]. The protected thiolate functionality leads to a considerable simplification of the functionalization of PTs. Furthermore, the insertion of these chalcogen atoms at the 3-position of thiophene leads to a further decrease of the polymerization potential. Thiolate is easily generated by the elimination of 2-cyanoethyl protecting group in basic conditions and can then react with halogenoalkane derivatives [159]. Application of this synthetic methodology has led to the synthesis of compound **63** [160] as well as C_{60} or bypiridine functionalized PTs (*vide infra*) [158,161].

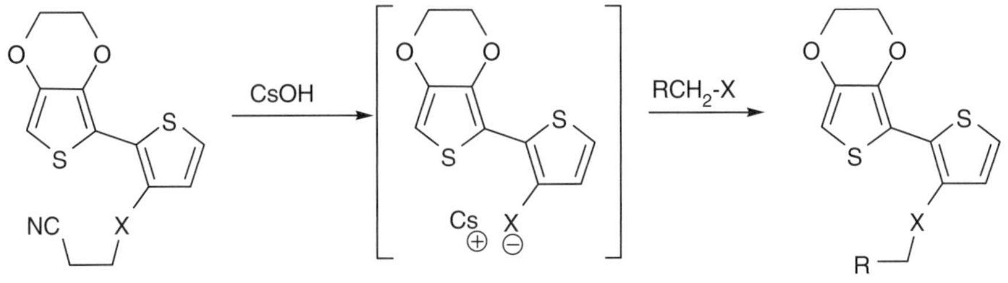

62 (X = O, S)

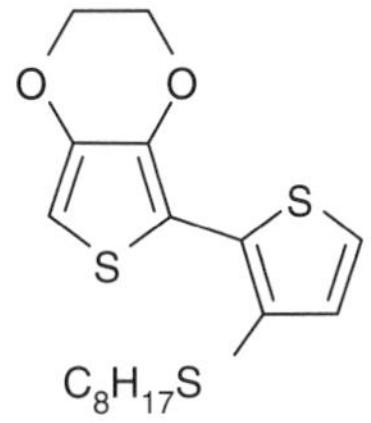

63

$C_8H_{17}S$

To summarize, EDOT has acquired a pivotal role among PT derivatives during the last decade. Thus, EDOT-based precursors lead to improved electropolymerization process at relatively low potential, allowing the grafting of functional groups sensitive to high oxidation potential. In addition, conjugated polymers based on EDOT usually exhibit better electronic delocalization associated to a combination of electron-donor mesomeric effect of oxygen atoms and enhanced planarization of the conjugated backbone. In this context, EDOT-based polymers have found various applications that will be described.

13.4 Applications of Functionalized Polythiophenes

Functionalized PTs have been investigated in various applications, e.g., their ability to detect, transduce, and amplify various physical or chemical informations into an electrical or an optical signal has led to the development of devices capable of detecting analytes or biomolecules in the field of environment, security, and biotechnology. Currently, functionalized PTs play a key role as active materials in the development of electrochromic devices and electronic devices, such as OLEDs, OFETs, and organic solar cells.

In addition to polymer processability and morphology, the development of such advanced technological applications relies on the specific electronic properties of π-conjugated skeleton and hence on the structural control and manipulation of quantities such as absorption and emission spectra, oxidation and reduction potentials, or luminescence efficiency. As most of these quantities depend on HOMO and LUMO energy level and on their difference, the control of these parameters has become a key issue for many technological applications of PTs.

In this context, this section deals with functionalized PTs developed for sensors, electrochromic devices, and plastic electronic devices as well as metal-containing PTs emphasizing on the relationships between bandgap control of the PT backbone and the corresponding application.

13.4.1 Polythiophenes for Electrochemical and Optical Sensors

Electropolymerization of a functionalized precursor represents a straightforward method for the realization of modified electrodes endowed with specific electrochemical or optical properties. Electrode materials based on electrogenerated functional conjugated polymers and their application as electrochemical sensors have been already reviewed [162–165]. On the other hand, the sensitivity of the optical properties of π-conjugated polymers to conformational changes has led to the realization of colorimetric sensors for the detection of various analytes extending from alkali metal ions to anions and biomolecules [165–168]. In general, the realization of sensors based on functional PTs relies on the fact that complexation at a side chain may lead to perturbation of the polymer conformation, which can be read by either electrochemical or optical methods.

13.4.1.1 Polythiophene-Based Cation-Sensors

PTs containing crown ethers or polyether chains have been largely investigated for metal-cation recognition. Poly[3-(3,6-dioxaheptyl)thiophene] (**64**) synthesized in 1989 is generally acknowledged as the first example of cation-sensitive PT [169–171]. Electrochemical investigations have shown that replacement of Bu_4N^+ by Li^+ in acetonitrile electrolytic media produces a 100–150 mV negative shift of the oxidation potential of the polymer [170,171] while the optical spectrum reveals a slight redshift of λ_{max} [172,173]. This result has been attributed to the complexation of Li^+ by the polyether chain, which contributes to planarize the conformation of the PT backbone. Furthermore, the hydrophilicity imparted by the polyether chain allows the polymer to be fully electroactive in aqueous media in contrast to poly(3-alkylthiophenes) [174].

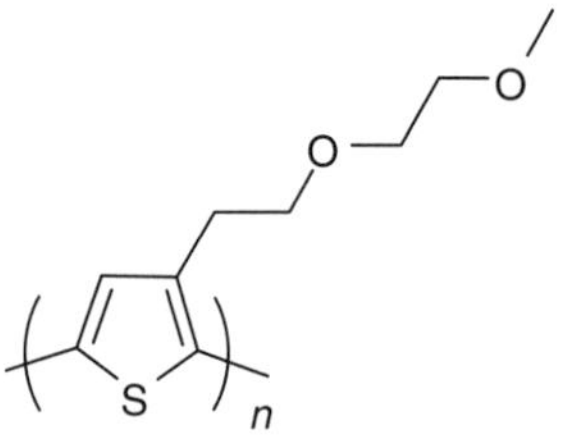

64

With the development of 3-(ω-bromoalkyl)thiophenes, several crown ether functionalized PTs have been synthesized by Bäuerle and coworkers. Electropolymerization was performed on mono-, bi-, and terthiophene monomers **65–67** substituted with pendant 12-crown-4 receptor tethered with alkyl chains [175–177]. While electropolymerization failed with **65**, compounds **66**, **67**, and **68** were easily electropolymerized and the chemosensing properties of the polymers were analyzed. Cyclic voltammetric analysis showed that addition of increasing amounts of Li^+, Na^+, or K^+ produces a positive shift of the oxidation potential of poly(**66**, $n = 5$), while this effect is less pronounced for poly(**67**). On the other hand, whereas the CV of poly(**68**, $n = 5$) is strongly affected by the presence of alkali ions, the lengthening of the alkyl spacer in poly(**68**, $n = 10$) produces a complete loss of ion sensitivity. Optical and spectroelectrochemical experiments revealed that the changes in electronic properties were due to hindered diffusion of the counteranions into the film during polymer oxidation [177].

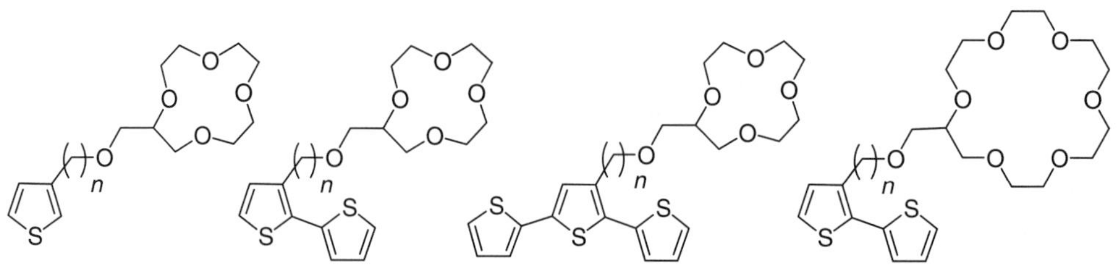

65 ($n = 5,10$)   **66** ($n = 5,10$)   **67** ($n = 5$)   **68** ($n = 5,10$)

Precursors **69**, **70**, and **71** where a 15-crown-5 unit is directly grafted onto the conjugated backbone have been synthesized by the same group [177,178]. Only **69** and **70** could be efficiently electropolymerized.

Whereas the CV of poly(**70**) was pratically insensitive to the presence of alkali cations, the CV of poly(**69**) shows a large positive shift of the oxidation peak in the presence of Na$^+$ in agreement with the size of the cavity of 15-crown-5-ether. Although this change in redox properties was ascribed to hinder diffusion of the counteranion, it is likely that the electron pairs of the oxygen atoms at the 3- and 4-positions of the thiophene ring participate to the complexation thereby decreasing the donating ability of oxygen toward the conjugated system. Recently, Berlin et al. have developed other terthiophenes with 18-crown-6 **72** and EDOT derivative **73** [179]. On the basis of electrochemical quartz crystal microbalance analysis, they found that the coordination constants of the crown ether ring toward alkali metal cation in the polymers are two orders of magnitude lower than those of typical 18-crown-6-ether molecules.

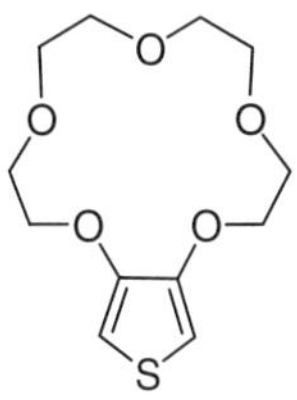

69 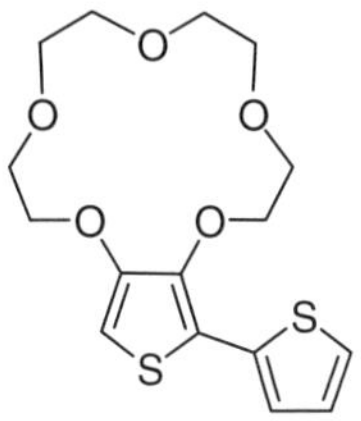   **70** 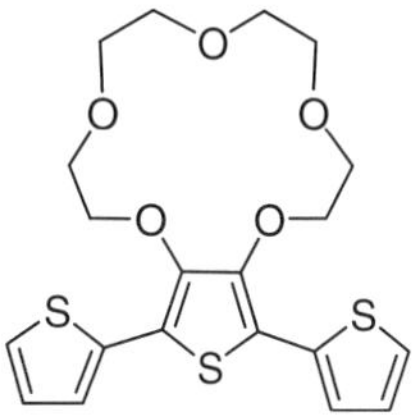   **71**

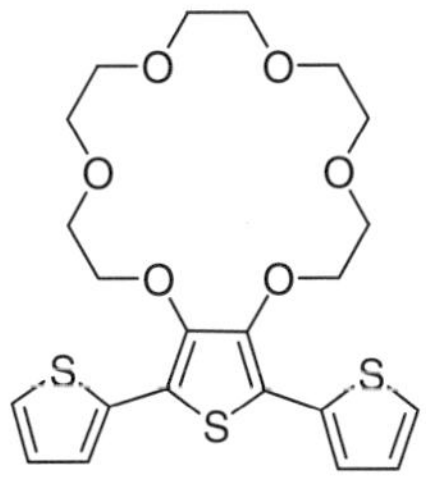

72 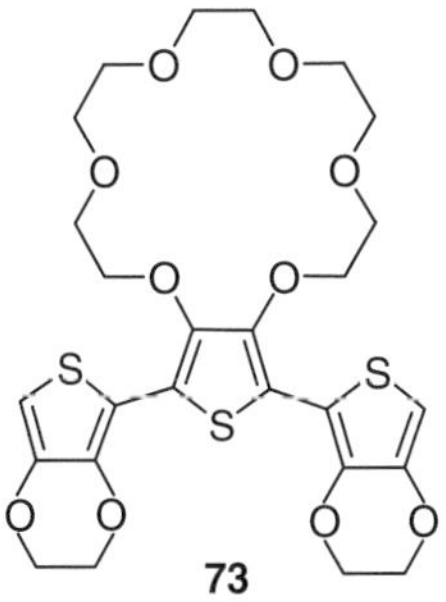   **73**

Another strategy proposed by Sannicolò et al. consists of the use of cyclopentabithiophene precursors with either a 16-crown-5-ether ring coplanar to the bithiophene moiety (**74**) or a 15-crown-5-ether perpendicular to it (**75**) [180]. This approach considerably reduces the steric interactions between the complexing site and the conjugated backbone. Consequently, the precursors were readily electropolymerized into extensively conjugated polymers as confirmed by optical data. The analysis of electrochemical properties in the presence of alkali cations showed that while poly(**75**) is completely insensitive to change of the cationic species, the redox potential of poly(**74**) undergoes a 350 mV positive shift in the presence of Na$^+$ [180].

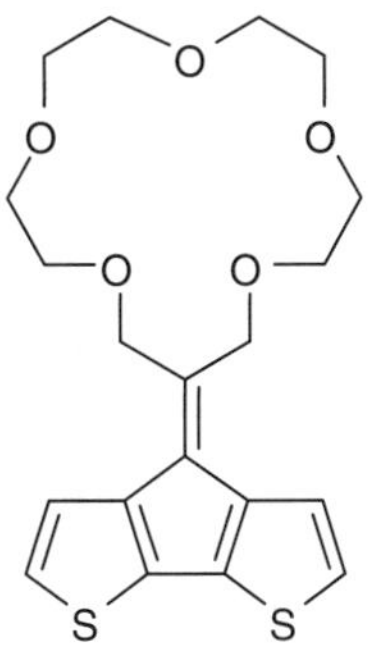

74 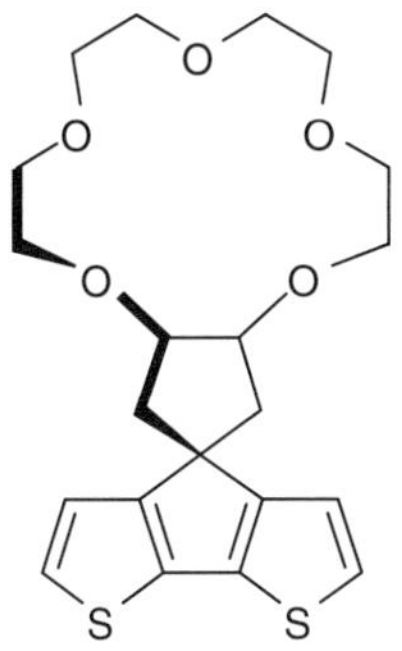   **75**

Electropolymerization of a precursor **76** involving two polymerizable groups linked by a flexible polyether chain was initially proposed by Roncali et al. as a possible route to conjugated polymers containing pseudocrown ether cavities [181]. Since then, analog precursors with polyether chains of various lengths have been studied [182,183].

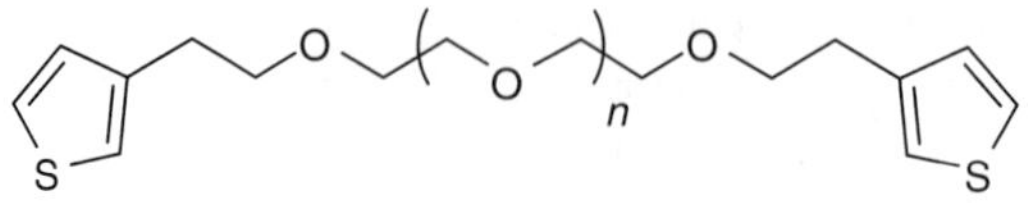

76 (*n* = 1–5)

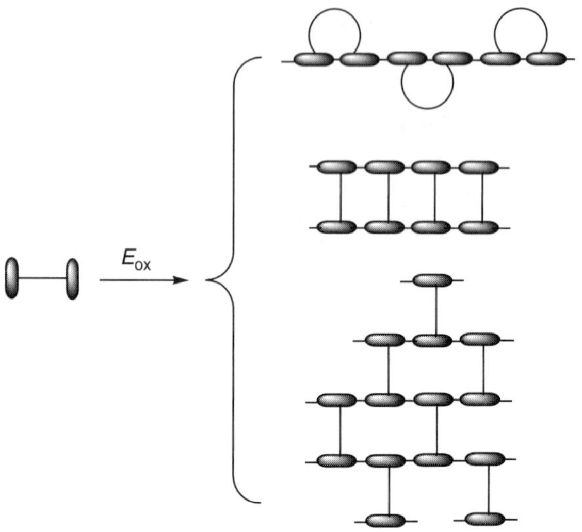

Another example **77** of this class of precursors with bithiophene groups has been recently described [184]. Electropolymerization in the presence of Li^+, Na^+, or Ba^{2+} shows that the nature of the cation strongly affects the polymerization process. The electrochemical and spectroelectrochemical properties of the resulting polymers show that the presence of Ba^{2+} in the electrosynthesis medium leads to 300 mV decrease of the oxidation potential of the polymer together with a significant redshift of the absorption maximum providing conclusive evidences for a metal template effect during electropolymerization [183].

The same approach has been extended to the electropolymerization of precursors **78** containing two EDOT units linked by long polyether chains [146]. The cyclic voltammograms of these polymers recorded in the presence of various metal cations show that depending on the length of the oligo-oxyethylene linker, doubly charge cations, such as Ba^{2+} and Sr^{2+}, produce positive shifts of the anodic peak potentials up to 400 mV. Furthermore, these highly hydrophilic polymers present the unique property to undergo immediate contraction of bandgap when immersed in water. Finally, a comparative analysis of the cyclability of the polymers showed that the polymers synthesized from the two-site precursors **78** are considerably more stable under long-term repetitive redox cycling than their analogs derived from the single-substituted EDOT monomer **79** [146].

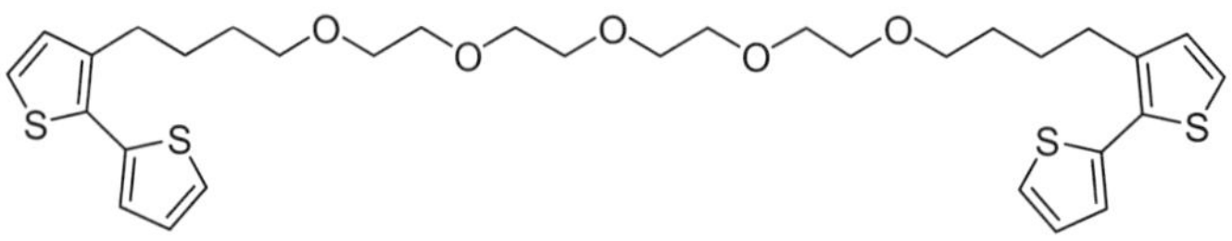

77

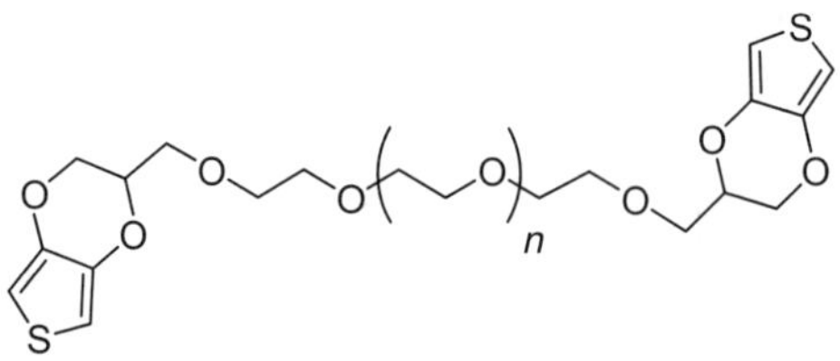

78 (*n* = 2, 3, 4)

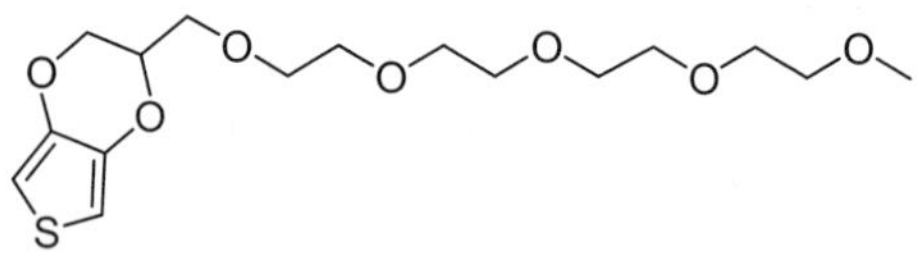

79

EDOT monomers bearing ω-iodo-alkyl and ω-iodo-polyether side chains (**80**) have been electro-polymerized into corresponding polymers. These polymers can be rapidly and quantitatively converted into functionalized polymers under mild conditions by postpolymerization reaction with functional blocks bearing a thiolate group, as demonstrated in the case of tetrathiafulvalene (**81**) [151]. Application of the same procedure led to a modified electrode containing a tetrathiafulvalene core substituted by two polyether chains (**82**). It was shown that the binding of Pb^{2+} by the polymer **82** could be electrochemically driven [185].

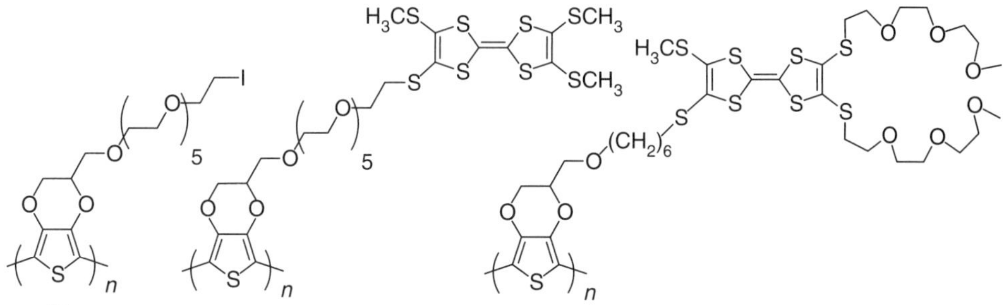

PTs functionalized with polyether chains have been also prepared chemically and the analysis of their optical properties in solution or in the solid state has evidenced ionochromic effects. Such effects have been observed for regioregular PTs **83** in the presence of Pb^{2+} or Hg^{2+} [186,187]. Leclerc and coworkers have used regioregular poly[3-oligo(oxyethylene)-4-methylthiophene] **84** for the determination of the concentration of alkali cations in methanol [188–190]. Addition of K^+ induced a conformational change in the PT backbone with a decrease of conjugation length as indicated by optical and electrochemical data. The optical response of polymers **85** bearing various pendant crown ethers was examined upon addition of alkali cations [191]. Contrary to what was expected in relation with the size of the cavity of each crown ether, the largest ionochromic effects were observed with Na^+ and K^+ for **85** ($m=1$) and **85** ($m=2$), respectively.

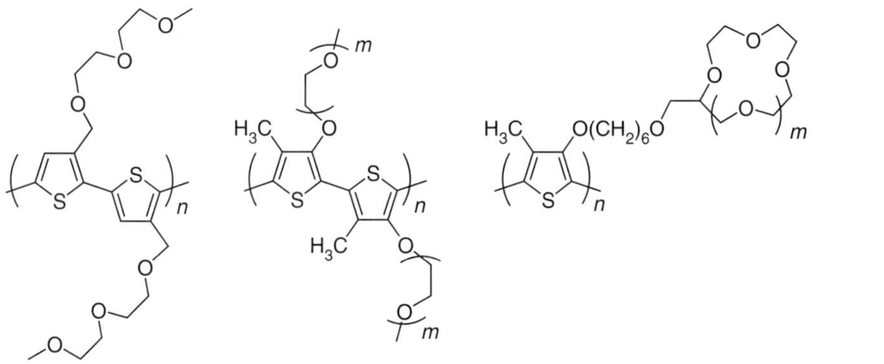

Marsella and Swager [192–196] have developed a series of precursors **86–88,** which have been chemically polymerized. Addition of Li^+, Na^+, or K^+ to poly(**86**) or poly(**87**) leads to a significant hypsochromic shift up to 90 nm of the absorption maximum. These large effects were attributed to a transition from a planar to a twisted conformation of the conjugated system. On the other hand, poly(**88**) exhibited little ionochromic activity.

Conformational changes induced by the complexation of alkali cations have been analyzed experimentally and theoretically in oligothiophenes **89** and **90** specifically tailored to serve as model compounds for molecular actuation [197].

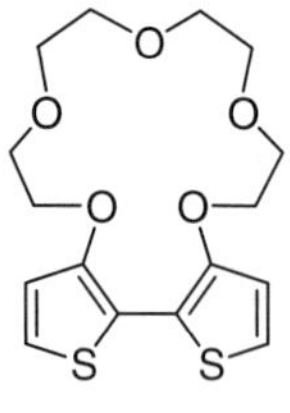

86 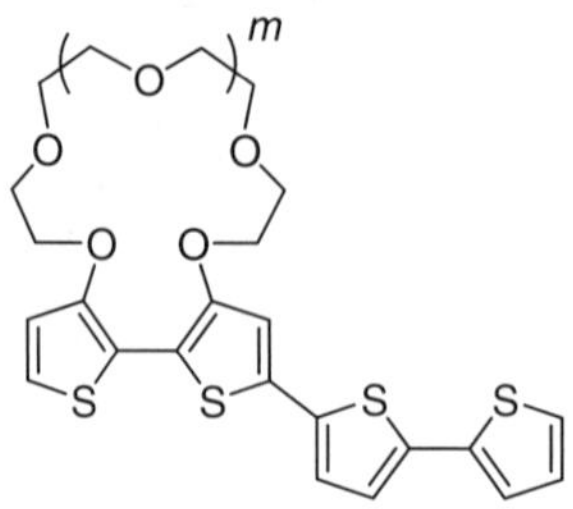   **87** (*m* = 1, 2) 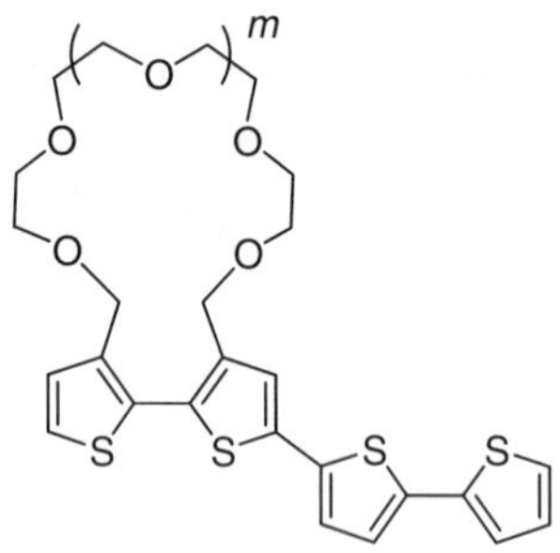   **88** (*m* = 1, 2)

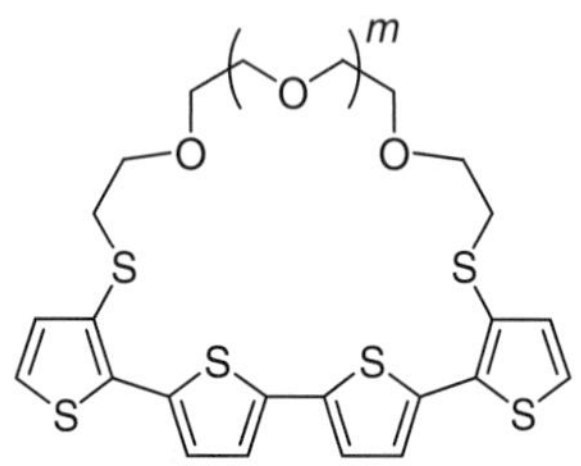

89 (*m* = 1, 2, 3) 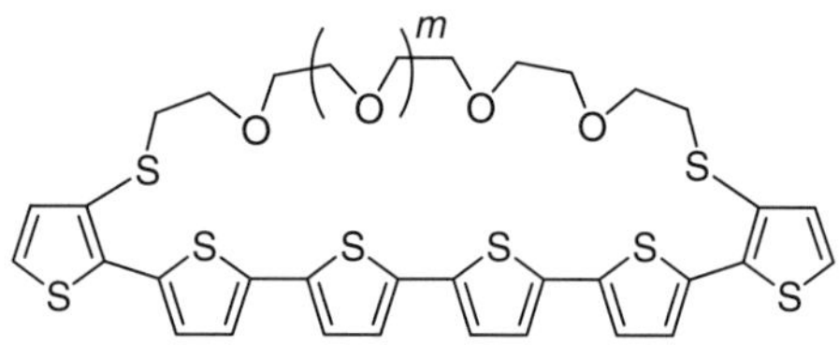   **90** (*m* = 1, 2)

Using postfunctionalization of a preformed polymer, Leclerc and coworkers have synthesized monomer **91** with a pendant chain terminated with a *N*-hydroxysuccinimide ester group known to react under mild conditions with amines to form the corresponding amides in high yields [198]. The solubility of electrogenerated poly(**91**) in CHCl$_3$, THF, and DMSO allows to cast thin polymer films with good mechanical properties. These films were then postfunctionalized in the solid state with substituted amines to give functional polymer such as poly(**92**) with a 15-crown-5. The UV–visible spectra recorded in the presence of various alkali cations revealed a redshift of λ_{max} from 435 to 589 nm in the presence of K$^+$. This phenomenon was ascribed to cation-complexation induced aggregation of the conjugated backbone.

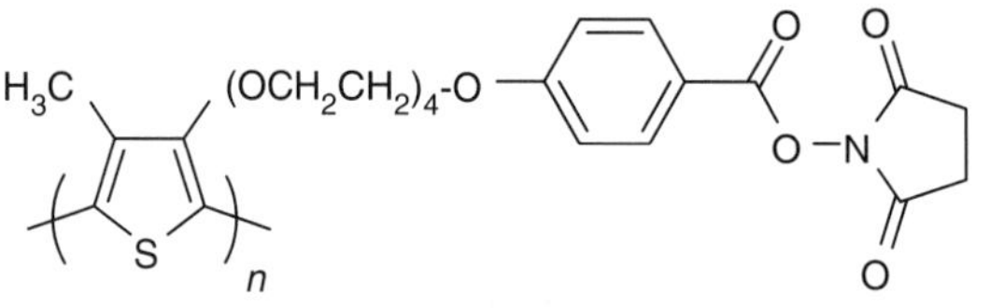

Poly(**91**)

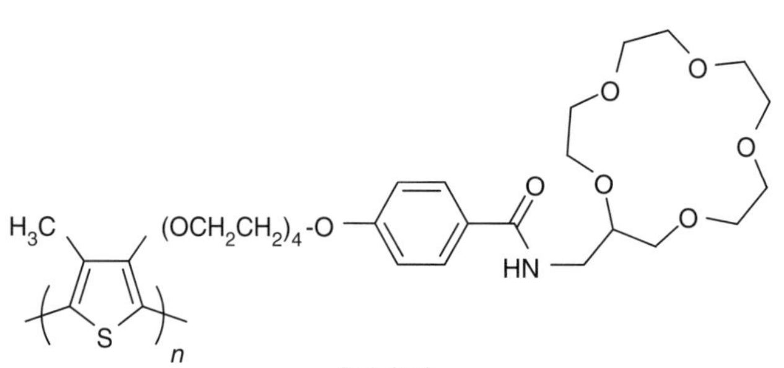

Poly(**92**)

PT **93** with pendant calix[4]arene linked through polyether chains has been synthesized by Swager and coworkers [199]. This polymer presents chemoselectivity toward Na$^+$ with a high binding constant. The cavity produced by the calix[4]arene and the alkylether chain fits well with the size of Na$^+$. The complexation of Na$^+$ leads to a 32 nm redshift of the absorption of the polymer associated to its planarization and to a positive shift of the oxidation potential. Other calixarene-PTs have been recently described. Among them, PT **94** and analog polymers were obtained by electropolymerization of a 1,3-bridged n-propoxy-calix[4]crown ethers linked to a 2,2′-bithiophene. After coating with a PVC membrane containing a lipophilic cation exchanger, the modified electrodes were tested for the potentiometric or amperometric recognition of alkali metal ions in aqueous solution [200]. A polymer derived from calixarene **95** was specifically designed for the recognition of Ca^{2+} ion. In situ conductivity measurements of films of electropolymerized poly(**95**) have demonstrated a selective Ca^{2+} ionoresistive response [201]. An original approach for the development of materials capable of transducing a conformational modification by an external stimulus into mechanical work was proposed by Swager who prepared proton-doped conducting polymers **96** based on a calix[4]arene scaffold with quaterthiophene segments [202].

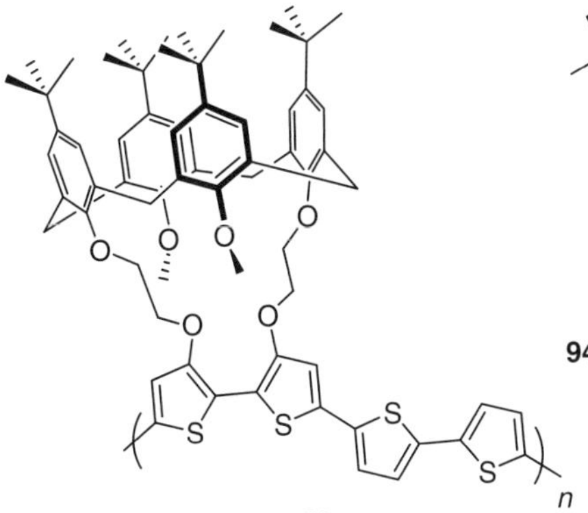

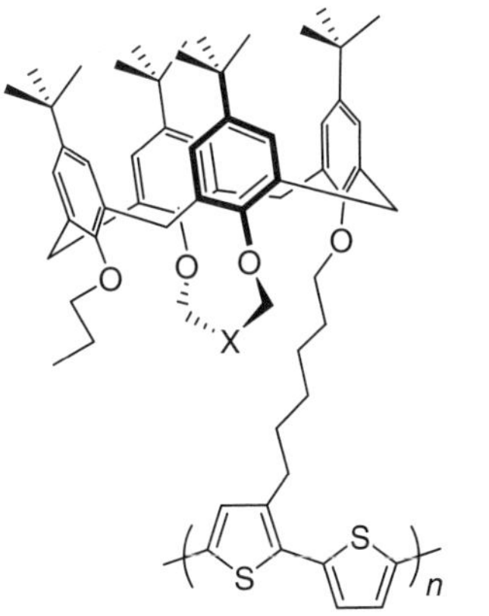

94 (X = (CH$_2$OCH$_2$)*m*, *m* = 2, 3)

93

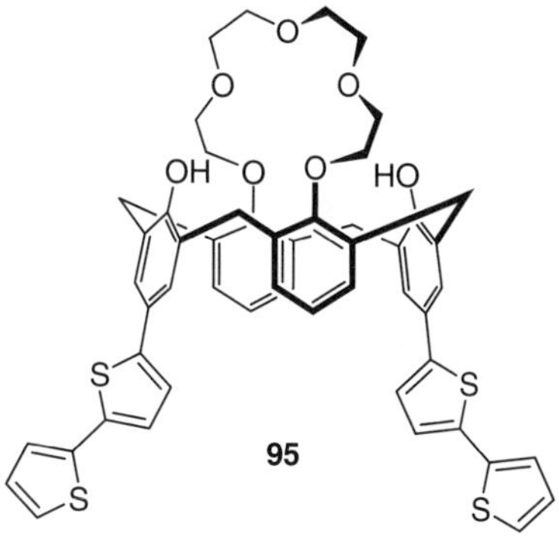

95

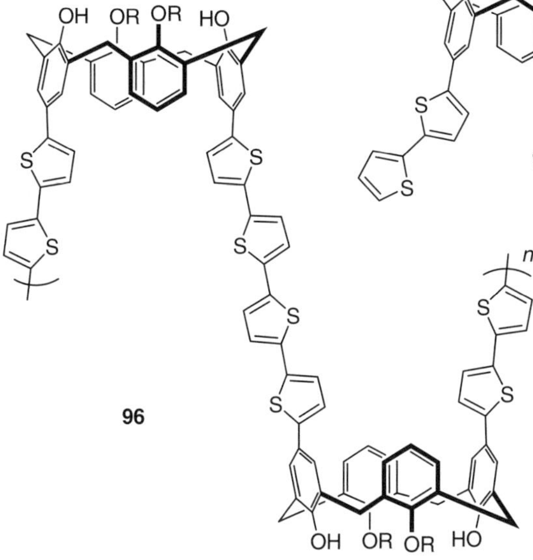

96

Electrogenerated PTs incorporating transition metal containing *N,N′*-ethylenebis(salicylidenimine) (salen) complexes have been developed by the Reynolds group [203,204]. A polyalkyl ether chain bridging the phenyl rings in **97** creates an additional site for chelation of a second metal ion. A positive shift of the oxidation potential of poly(**97**) was observed in the presence of Li^+, Na^+, Mg^{2+}, or Ba^{2+}. These electrodes were also sensitive to Lewis bases such as pyridine and triphenylphosphine.

A polyrotaxane sensitive to transition metal ion **98** was prepared by electropolymerization [205]. The precursor was obtained by complexation of a bipyridine sandwiched between two 2,2′-biEDOT moieties with Cu^+ or Zn^{2+} in the presence of a Sauvage's phenanthroline macrocycle incorporating a polyether chain. A reversible ionochromic process was observed. On removal of metal ions by ethylenediamine treatment, the color of the polymer film turns from blue-green to red while the reverse effect was observed upon reintroduction of the metal. Another salient feature is the fact that the matching of the polyrotaxane and $Cu^{1+/2+}$ couple redox potentials results in a $Cu^{1+/2+}$ contribution to conductivity.

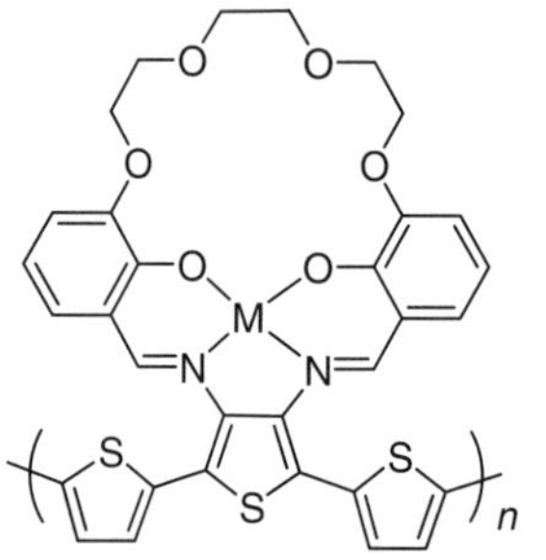

97 (M = Cu, Ni)    **98** (M = Zn^{2+}, Cu^+)

A porphyrin (**99**) linked at the 3-position of a thiophene ring by an alkyl spacer has been reported. An electrogenerated copolymer of **99** and 2,2′-bithiophene showed some ability to complex Ni^{2+} ions but the exact coordination environment of the metal ions in the polymer was not ascertained [206]. Coordination of protons and divalent transition metal ions (Fe^{2+}, Co^{2+}, Ni^{2+}, and Cu^{2+}) has been also reported in poly(**100**) where bipyridine units are inserted between two EDOT moieties in the polyconjugated chain. The spectral and redox changes accompanying coordination are used to produce ionochromic and potentiometric ion sensor electrodes [207]. Thiophene-based conjugated copolymers containing amino receptors have been also used for the detection of some metal cations [208]. Particularly, the fluorescence of **101** bearing a *N,N,N′*-trimethylethylenediamino receptor was increased by a factor of 2.7 in the presence of Hg^{2+} in aqueous solution. This "turn-on" sensing effect has been attributed to a photoinduced electron transfer in conjugated polymers, which is removed upon binding the analyte.

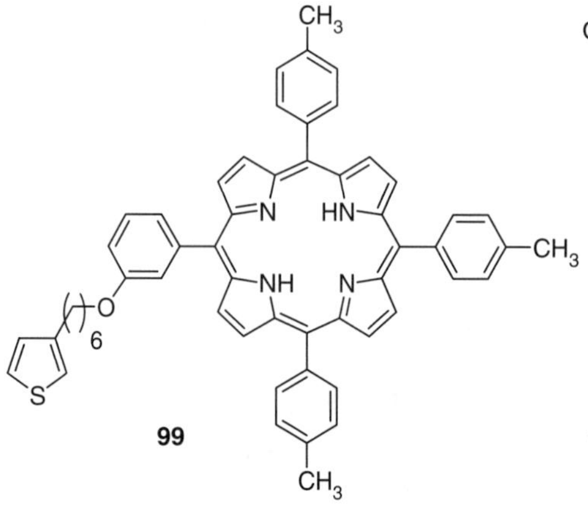

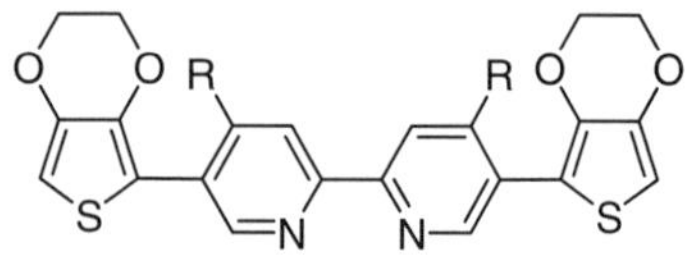

100 (R = H, CH_3, C_7H_{15})

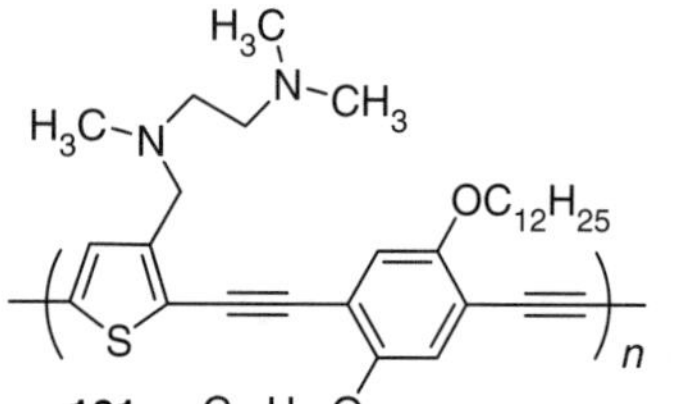

99

101 $C_{12}H_{25}O$

Swager and coworkers synthesized receptor polymers **102** and **103** [209,210]. The polymers were prepared by copolymerization of 2,5-dibromo-3-decylthiophene or 3,3′-bis(methoxyethoxy)-2,2′-bithiophene with the organozinc derivative of the macrocyclic 3,3′-dialkoxy-2,2′-bithiophene by palladium-catalyzed cross-coupling. Cyclic voltammetry and in situ conductivity measurements on polymer films have evidenced the complexation of paraquat derivatives. However, the complexation between polymer films and acceptors can lead to opposite shifts of oxidation potential of the polymer due to the interplay of donor–acceptor interaction and conformational changes of the polymer.

Polymer **102** was also used as a fluorescent chemosensory system in which the binding of paraquat leads to an attenuated emission intensity [211]. The same authors described related systems such as electrogenerated polymer **104**. Although, addition of paraquat **105** to the monomer of **104** resulted in a charge-transfer complex, the binding properties of **104** were not reported [212].

McCullough and coworkers have described chemoselective ammonium cation ionochromic sensing in water by regioregular, water-soluble PT functionalized with propionic acid **106**. When deprotonated the polymer becomes even more soluble in water and the ionochromic response of the related carboxylate polymer to alkylammonium counterions depends on the size of the counterion [213,214]. The λ_{max} can vary over a 130 nm range from purple, when small cations favor polymer chains aggregation, to yellow, when large cations completely disrupt the aggregated phase.

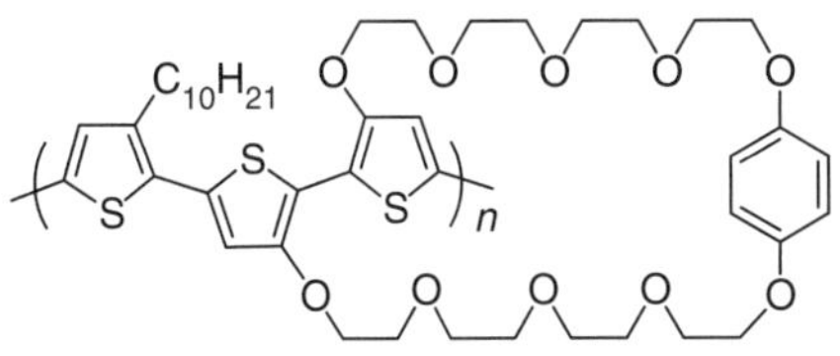

102 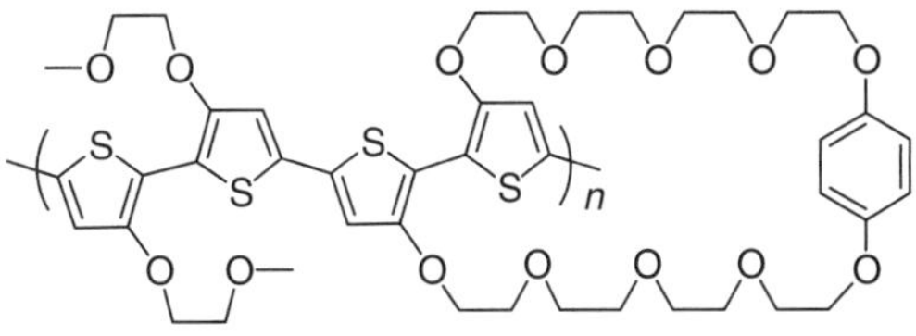   103

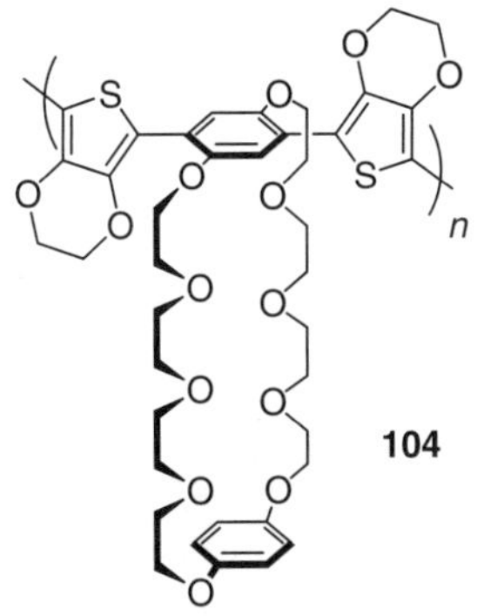

104 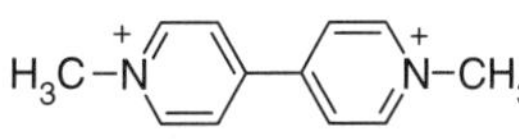   105 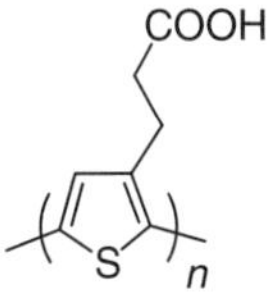   106

13.4.1.2 PT-Based Anion-Sensors

Receptors able to sense an anion recognition event through a macroscopic physical response have been previously reviewed [215,216]. However, only few examples of functionalized PTs for anion sensing have been reported.

First attempt to prepare functional PT with anion recognition properties was reported, in 2000, by Fabre and coworkers [217] who synthesized various boronic acid and boronate-substituted electropolymerizable precursors based on aniline, pyrrole, and thiophene. Boronic acid and boronate groups are known to strongly bind bases such as F⁻. However, while PPy (**107**) exhibited the most significant recognition properties toward F⁻ as deduced from electrochemical data [218], the results obtained for PT **108** were disappointing. Swager and coworkers synthesized a double-strapped porphyrin **109** for selective fluoride anion recognition taking advantage of the two small hydrogen-bonding cavities that are not able to bind larger anions [219]. Due to the presence of four dithienyl units, compound **109** was readily electropolymerized to give a cross-linked highly conducting film. Treatment of poly(**109**) with F$^-$ ion shifted the redox waves of the porphyrin to lower potentials that is consistent with the presence of an anion, which stabilizes proximate positive charges. Interestingly, the redox wave associated with the polymer was unaffected. On the other hand, the conductivity of the polymer was dramatically reduced upon addition of F$^-$.

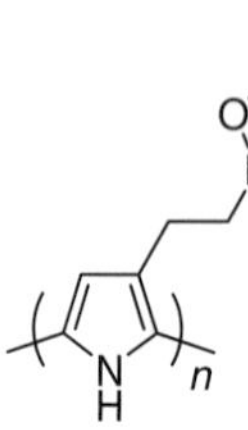

107 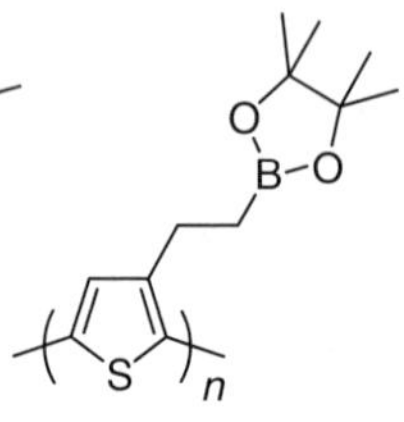   108

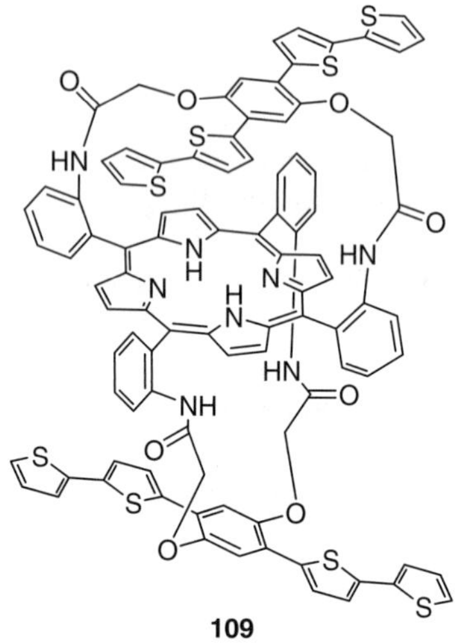

109

Leclerc and Ho have reported a water-soluble cationic PT derivative poly(**110**, $m = 1$) which can optically detect the presence of iodide over a wide range of other anions (F$^-$, Cl$^-$, Br$^-$, CO$_3^{2-}$, HCO$_3^-$, H$_2$PO$_4^-$, HPO$_4^{2-}$, CH$_3$COO$^-$, EDTA^{4-}, SO$_4^{2-}$, and (C$_6$H$_5$)$_4$B$^-$) [220]. Poly(**110**) consisted of a poly(3-alkoxy-4-methylthiophene) with a pendant imidazolium ring linked by an oxaethyl or oxapropyl chain. The detection principle is based on the different electrostatic interactions between the imidazolium ring and anions and the subsequent conformational change of the polymer backbone. These polymers are synthesized by oxidative polymerization of **110** with FeCl$_3$. The addition of NaI to a water solution of poly(**110**, $m = 1$) instantaneously produces a redshift of λ_{max} from 406 to 543 nm. In fact, the presence of I$^-$ promotes the aggregation and planarization of poly(**110**, $m = 1$). The addition of I$^-$ causes a quenching of the fluorescence of poly(**110**, $m = 1$) demonstrating that fluorometric detection of anion binding is also possible. While this attractive approach allows for a simple colorimetric and fluorimetric detection of iodide, further investigations are needed to understand the nature of the specific affinity of iodide toward imidazolium salt and the influence of the length of the side chain.

Aldakov and Anzenbacher have reported the preparation and electropolymerization of the tricyclic systems **111** [221,222] in which the dipyrrolylquinoxaline units are capable to bind anions by hydrogen bonding. Titration experiments with **111** dissolved in DMSO showed that addition of pyrophosphate and fluoride anions led to specific color changes indicative of anion binding [223–225]. The presence of these anions also led to significant modifications of the optical spectrum of films of poly(**111**). A dramatic increase of anion affinity was observed when the polymer was *p*-doped at a moderate positive potential as a result of stronger attractive electrostatic interactions between the polymer and the anion.

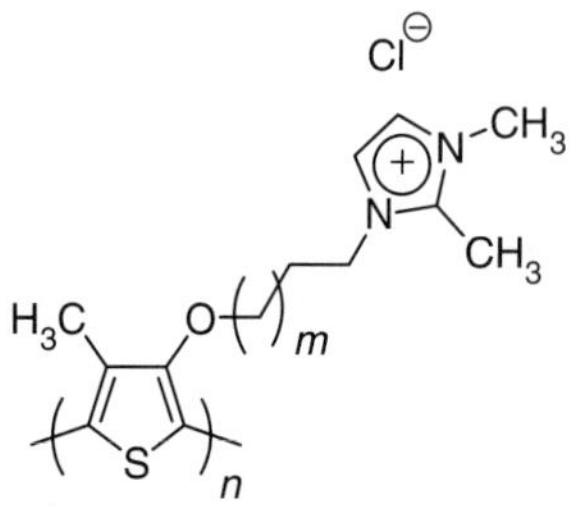

Poly(**110**) (*m* = 0, 1) 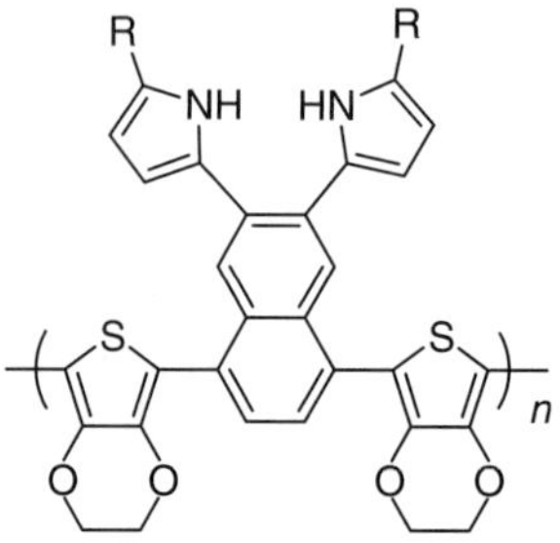   Poly(**111**) (R = H, Cl)

13.4.1.3 PT-Based Molecular and Biomolecular-Sensors

Conjugated polymers designed for the detection of molecules or biomolecules are subject to a growing interest [164,165,226]. PT derivatives have also been developed for molecular and biomolecular recognition. Selected examples are described in this section.

Few functionalized PTs capable of detecting small molecules have been reported. These molecules are essentially solvents, gaz (NO), nitrogen containing molecules, and polychlorinated phenols. For example, poly(cyclopenta[2,1-*b*:3,4-*b′*]bithiophene) **112** substituted at the 4-position by calix[4]arene group presents a strong affinity for toluene and acetone from the gas phase [227]. The conductivity of polymers incorporating tungsten-capped calixarenes **113** [228] was influenced by the presence of *p*-xylene making possible the use of these materials as sensors for aromatic analytes. The conductivity of parent conducting polymers of tungsten(VI)-oxo calixarene has also been shown to be influenced by substituted formamide [229].

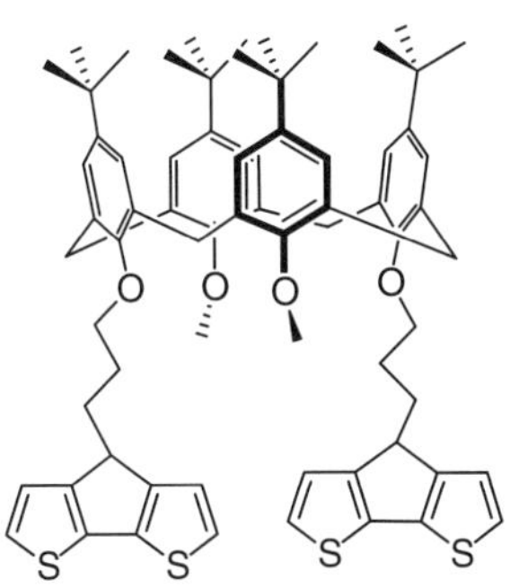

112

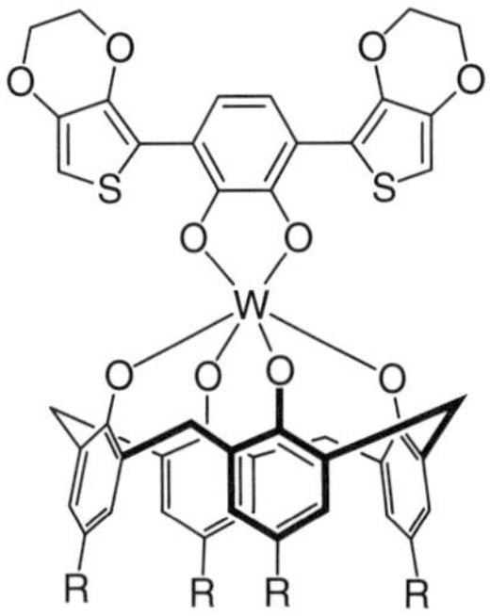

113 (R = H, *t*-Bu, Adamantyl)

The conductivity of polymer **114** constituted by a Co(II)-salen-3,4-ethylenedioxythiophene polymer backbone has been shown to be sensitive when exposed to Lewis bases, such as pyridine and 2,6-lutidine [230]. This polymer was recently used as reversible chemoresistive sensor for nitric oxide (NO) [231] as well as catalyst for the reduction of O_2 [232].

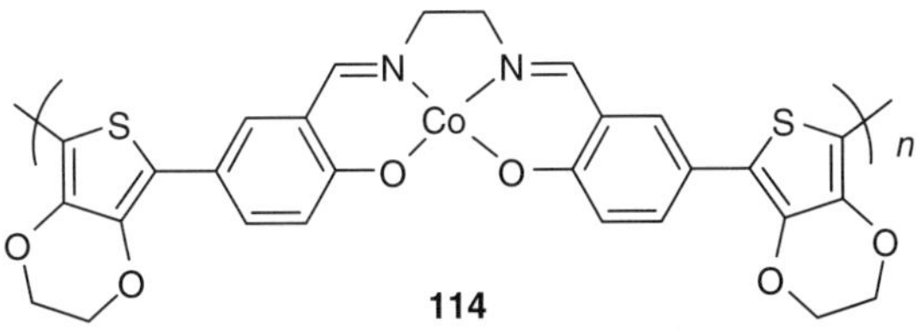

114

Synthesis of a series of bi- and terthiophenes (**115**) substituted with *meso*-tetraphenylporphyrin groups by an oxaalkyl chain has been described by Bäuerle and coworkers [233]. The corresponding electrogenerated polymers were considered for the detection of polychlorinated phenols in amperometric sensors.

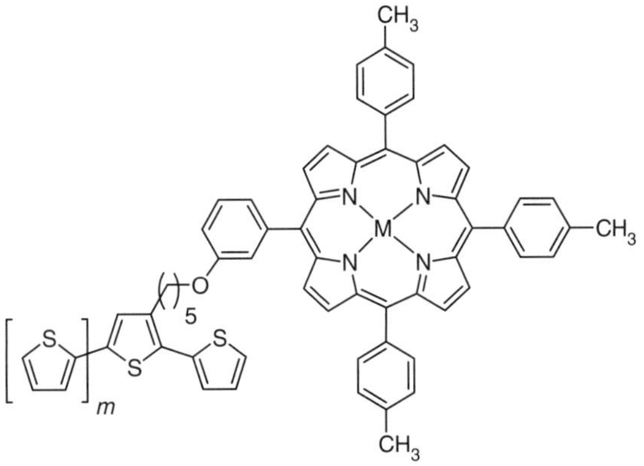

115 (*m* = 0, 1)

Jäkle and coworkers have analyzed the sensing properties of boron-modified PTs **116–118** toward pyridine or 2-picoline [234]. Polymerization was achieved under mild conditions by treatment of distannylated bithiophene with the corresponding bifunctional arylboron bromides at ambient temperature leading to polymers soluble in common organic solvents. Binding of pyridine was investigated by ^{11}B NMR, UV–vis, and fluorescence spectroscopy.

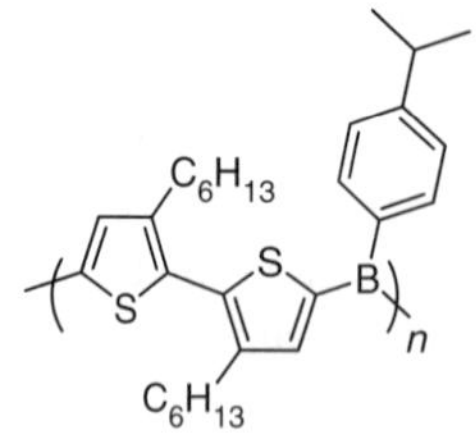

116 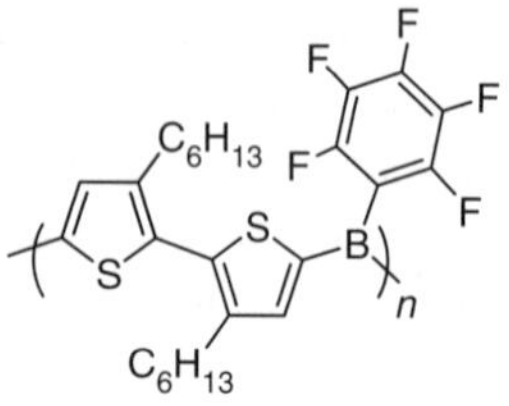   **117** 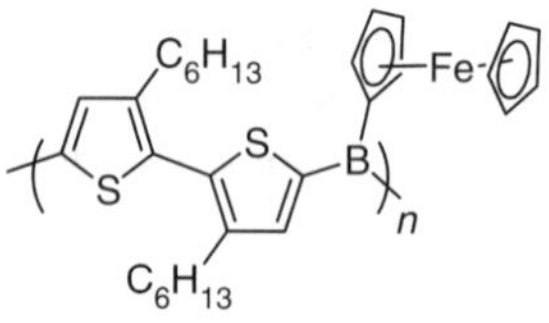   **118**

First, examples of fluorescence [235] and chemiluminescence-based biosensors [236], derived from PTs, were reported by Tripathy and coworkers. Later, the synthesis of **119** containing pendant biotin units was described. A water-soluble copolymer **120** with sulfonate and biotin in the side chains [237] exhibits a deep violet color ($\lambda_{max} = 550$ nm) which turns yellow ($\lambda_{max} = 400$ nm) on binding with avidin. An extension of this work based on the homopolymer **121** involved the preparation of monolayers of a biotinylated PT on an aminosilane-treated ITO surface by successive deposition of **121** and biocytin hydrazide in electrostatic interactions [238,239]. This ultrathin film modified electrode was shown to detect femtomoles of avidin in aqueous solution. Electrochemical and optical evidences for avidin binding were reported for a copolymer based on poly(terthiophene) **122** [240] and for the homopolymer **123** [198].

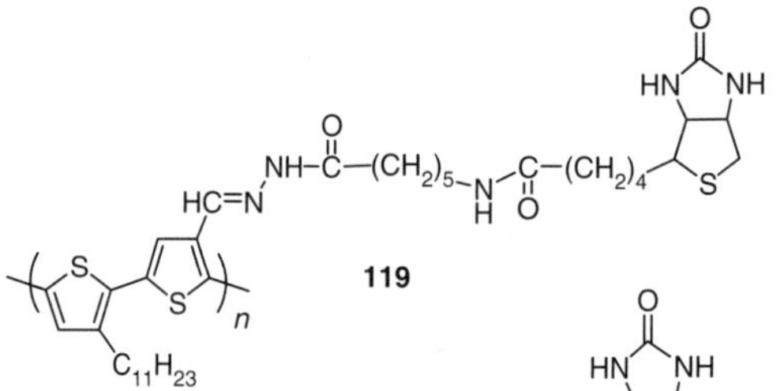

119

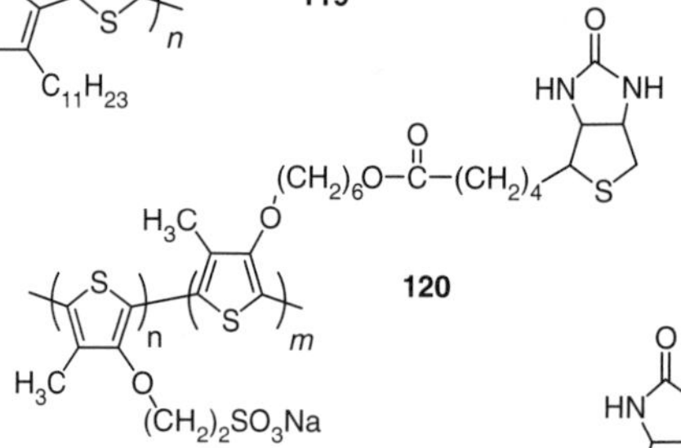

120

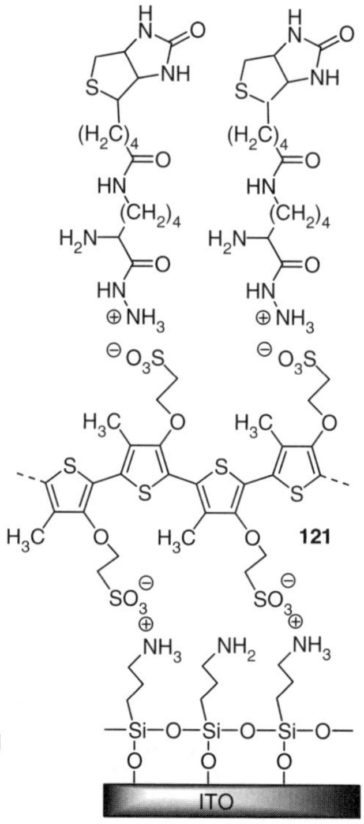

121

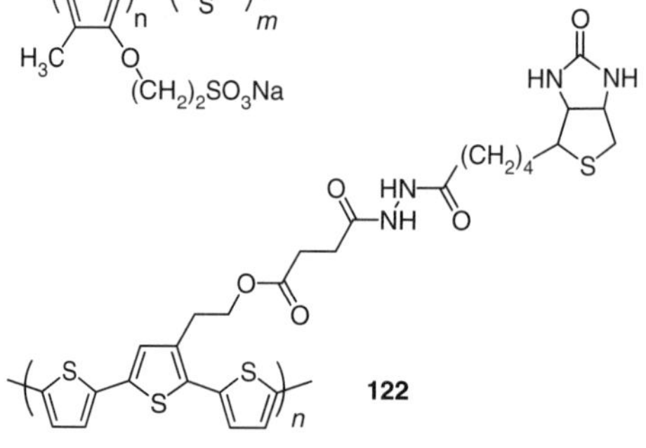

122

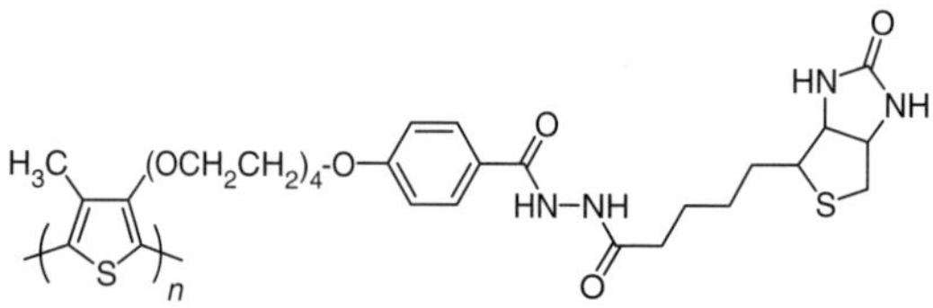

123

Bäuerle and coworkers have synthesized PTs derivatized with single nucleobases **124**, **125**, and **126** containing adenine, uracil, and pyrimidine [241–243]. The electrogenerated polymers were characterized by cyclic voltammetry in the presence of the complementary bases. Although, the nucleobases are electronically decoupled from the PT backbone, the stepwise addition of uracil **127** to adenine-poly(**124**) leads to the concomitant continuous positive shift of the oxidation potential of the polymer and the decrease in electroactivity.

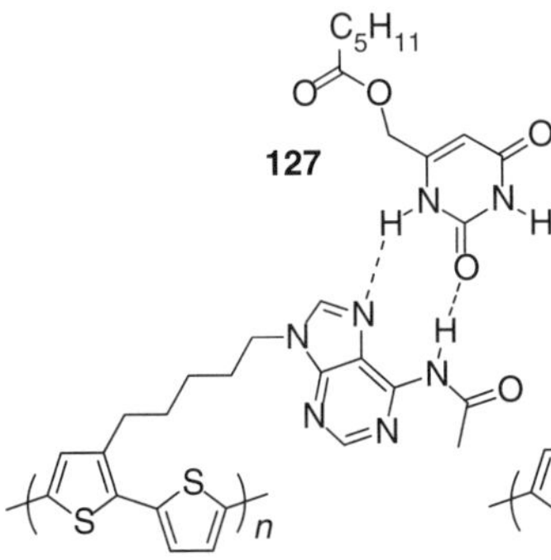

Poly(**124**) 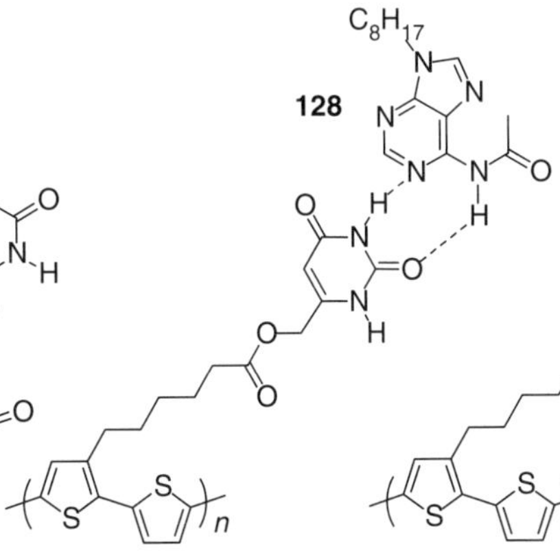   Poly(**125**) 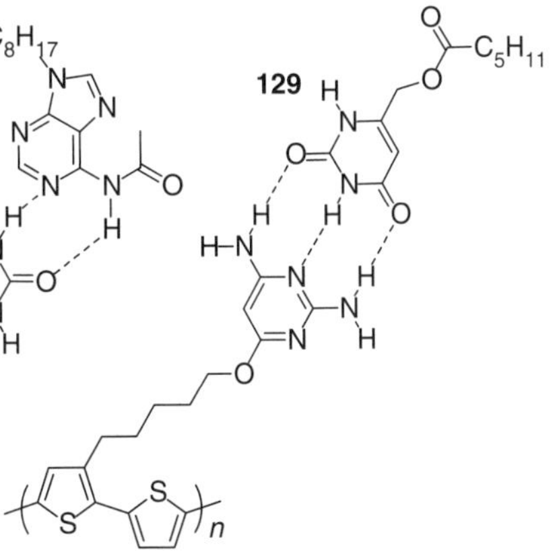   Poly(**126**)

Preformed polymers bearing a reactive or activated group that can react with a functional group of an enzyme have been reported. Schuhmann described the synthesis of polymers **130** and **131** derived from dithienylpyrrole containing a pendant amino group that can fix a glucose oxidase [244]. Bäuerle and coworkers have developed poly(bithiophene) **132** and the related poly(terthiophene) substituted with easily replaceable *N*-hydroxysuccinimide ester [245] for the immobilization of glucose oxidase [246].

An electrogenerated copolymer **133** of 3-methylthiophene and 3-thiopheneacetic acid has been developed for covalent attachment of lactate oxidase [247]. Polymer **134** with an activated ester was functionalized with an oligonucleotide probe for the electrochemical detection of DNA hybridization [248].

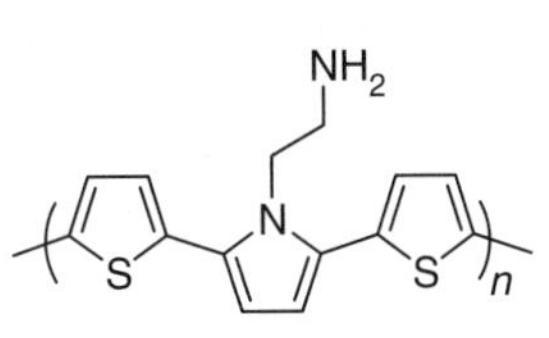

130 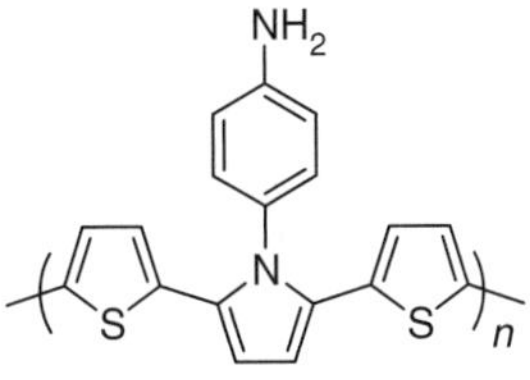   **131**

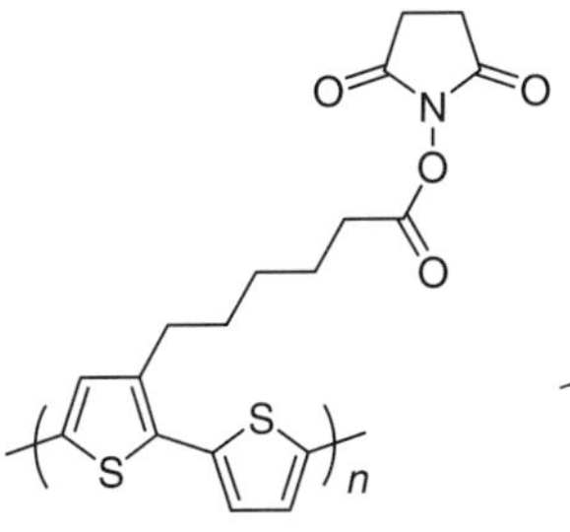

132 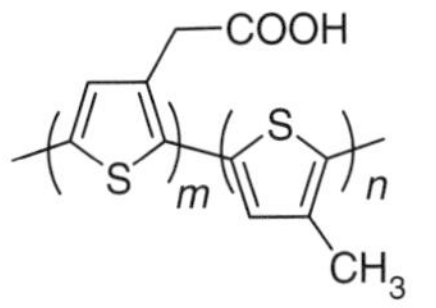   **133** 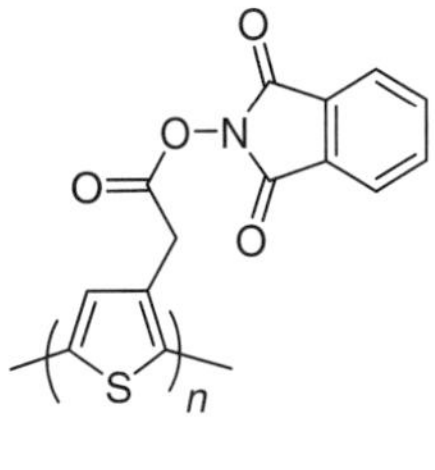   **134**

A copolymer **135** containing hydroxyl groups was electrochemically prepared and after covalent coupling with glucose oxidase, its ability to bind glucose was investigated [139]. A copolymer based on **136** and **78** ($n = 4$) was postfunctionalized, thanks to the activated-ester sites, with a series of ferrocene-oligonucleotides (ODNs). The electrochemical response of the resulting modified electrode was altered in the presence of ODN targets in agreement with hybridization [249].

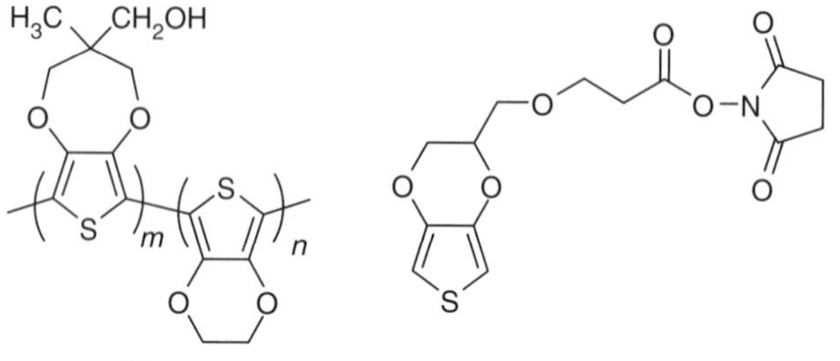

135   **136**

Shinkai and coworkers prepared a water-soluble cationic PT derivative **137** [250]. Addition of anion adenosine triphosphate (ATP) in an aqueous solution of **137** gives a dramatic color change with a redshift of λ_{max} from 400 to 538 nm. This effect was attributed to electrostatic and hydrophobic cooperative interactions inducing conformational changes and new modes of aggregation in the polymer. ATP also produces a quenching of fluorescence of **137**. A regioregular amine functionalized PT **138** forms a chiral self-assembly with DNA that further produces a low energy π–π^* absorption [251].

A solid state electrochemical approach based on a neutral peptide nucleic acid (PNA) fixed as a monolayer on gold has been developed by Leclerc. Polymer **139** does not bind the neutral PNA probes but interacts strongly with the negatively charged complementary ODNs bound to the PNA probes. In the presence of **139**, the complementary ODNs bound to neutral PNA move toward **139**, thus, modifying the electronic properties of the pendant ferrocenyl units of **139** [252].

The same group has also developed affinitychromic, water-soluble, cationic polymer **140** that forms electrostatic complexes with negatively charged ODN. The duplex formation with an ODN is accompanied by a color change from yellow to red. Subsequent addition of the complementary ODN leads to the formation of a yellow solution associated with the formation of a triplex between **140** and the two hybridized ODNs. Interestingly, when the complementary ODN is replaced by one- or two-mismatch complementary ODNs, the absorption and emission spectra exhibit distinct behaviors underlining the specifity of the hybridization process [253]. Furthermore, this system based on **140** is versatile enough to detect nucleic acids of various lengths with extremely high sensitivity and it has been applied for the detection of a segment from the RNA genome of the *Influenza* virus [254] and that of human α-thrombin [255]. A recent improvement of this system allowed the detection of nucleic acids at the zeptomole level using fluorescence amplification [256].

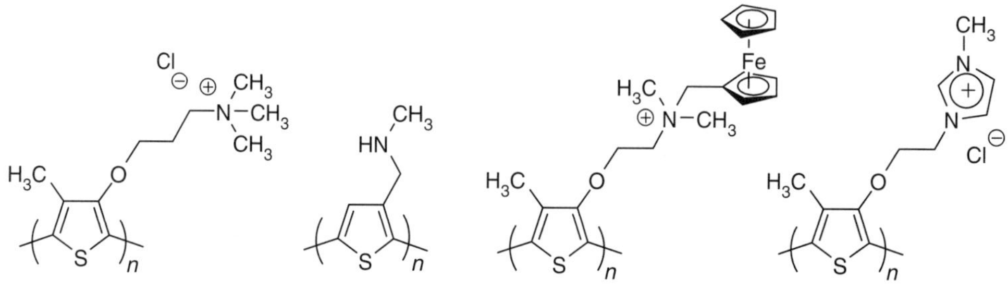

137   **138**   **139**   **140**

On the other hand, the use of poly(**141**) for the fluorimetric detection of conformational alterations in biomolecules and that of protein–protein interactions has prompted the group of Inganäs to develop other systems for recognition of polypeptides [257]. For example, a zwitterionic thiophene-based conjugated system poly(**142**) has been used at acidic pH as optical probe for the detection of amyloid fibril formation of bovine insuline [258]. Indeed, the conformational change from a highly compact native state of a protein to a fibrillar assembly is recognized as the cause of disease states, such as Alzheimer's disease and spongiform encephalopathies. Therefore, the detection of proteins in their fibrillar state is of importance. In this work, noncovalent interactions between the probe and the protein result in changes of the geometry and the electronic structure of the electrolyte chains of poly(**142**), which can be monitored with absorption and emission spectroscopy. Poly(**142**) has been prepared by oxidative polymerization of terthiophene **142** in the presence of $FeCl_3$. The same group has described a chiral recognition of a synthetic peptide by using two isomers of poly(**143**) with a cationic side chain [259]. In addition, it has been shown that depending on the structure of the peptide, poly(**143**) adapts its conformation which varies from well-defined structure to random-coil conformation as a function of pH.

Detection of the formation of amyloid fibrils in bovin insuline and chicken lysozyme was reported by using a conformational-sensitive optical method based on conformational changes of an anionic PT derivative poly(**144**) [260]. This conjugated polyelectrolyte forms a complex with the protein and, interestingly, the optical properties of the complex depend on the native state or the amyloid fibril conformation of the protein. Prior to this work, a chiral supermolecule based on poly(**144**) and a synthetic peptide resulting from electrostatic interaction between the positively charged lysine groups of the peptide and the negatively charged carboxylate groups of poly(**144**) have been synthesized. This supermolecule was shown to combine the three-dimensional ordered structure of the biomolecule and the electronic properties of the conjugated polymer [261].

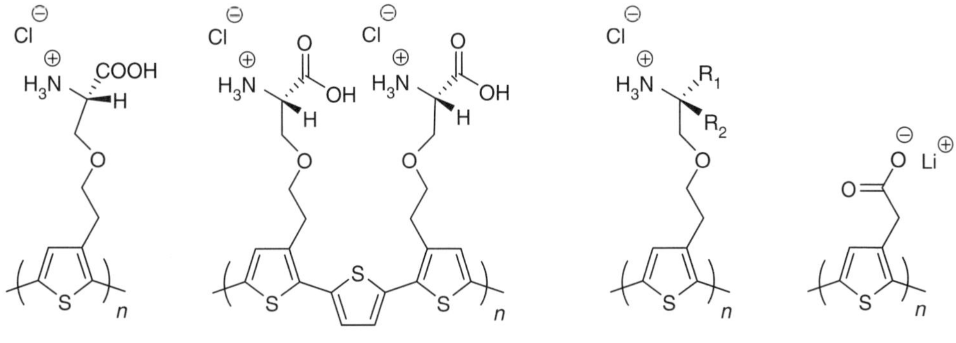

Poly(**141**)   Poly(**142**): n = 3, 4, 5   Poly(**143**)   Poly(**144**)
R_1 = COOMe, R_2 = H
R_1 = H, R_2 = COOMe

13.4.2 Metal-Containing Polythiophenes

Conjugated polymers containing transition metals combine the electronic properties of conjugated polymers and the redox and optical properties of metal complexes. From 1999 up to now, reviews have periodically surveyed this research field [162,262–272]. Particularly, conjugated metallo-PT derivatives have been investigated for the development of sensors for small molecules (*vide supra*) as well as modified electrodes for electrocatalysis and energy conversion [162,265,266,268,269,271]. From a fundamental point of view, the role of the redox metal complexes in the charge transport properties of conducting metallo-PTs has also been studied [268,271]. This section will be limited to the progress realized in the synthesis of PTs functionalized with bipyridine moiety and derivatives for metal complexation.

Examples of PTs functionalized with bipyridyl groups and the corresponding metal complexes were reported in 1989 by Parker et al., who described the electrodeposition of the Fe^{2+} complexes of thiophene monomers **145** and **146** bearing pendant bipyridine groups [273]. Recently, Wang and Keene have synthesized a series of thiophenic ligands **147** with a bipyridine group linked at the 3-position of thiophene through an alkyl spacer of variable length and analyzed the electropolymerization of ruthenium(II) complexes [274]. Crayston et al. have synthesized similar ligands **148** and **149** and investigated the electropolymerization of rhenium and ruthenium complexes [275]. Although the cyclic voltammogram of the obtained electrodeposited materials revealed in several cases the typical signature of the expected metal complexes, all these works have in common a weakly efficient electropolymerization process that leads to poorly defined and not very stable polymers.

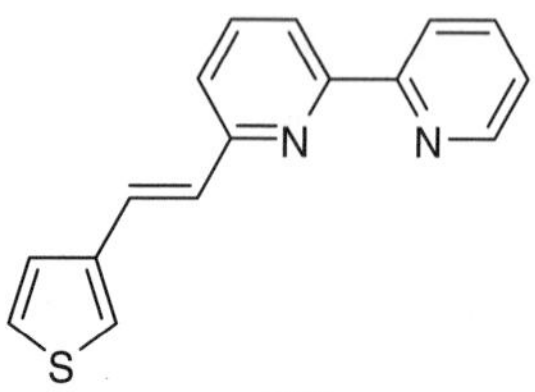

145 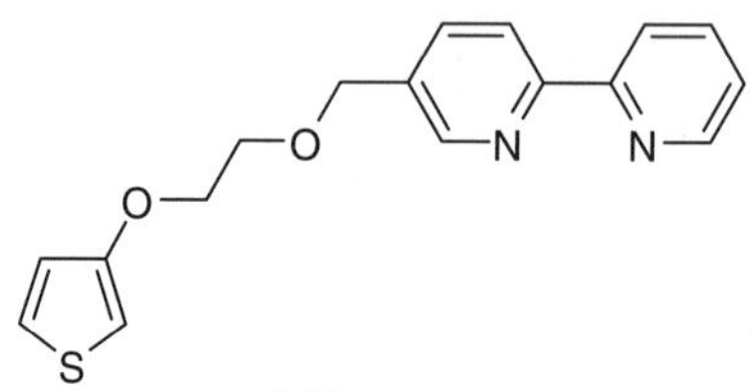   **146** 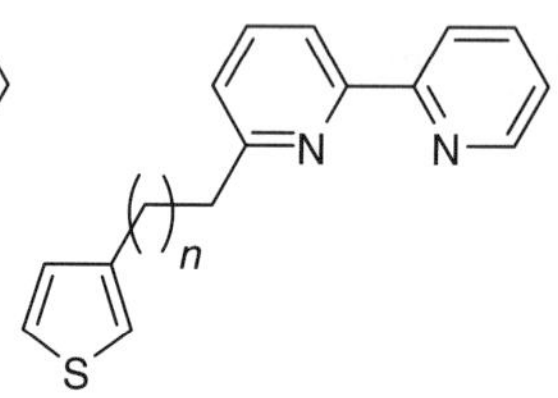   **147** (*n* = 1, 4, 6, 8, 10)

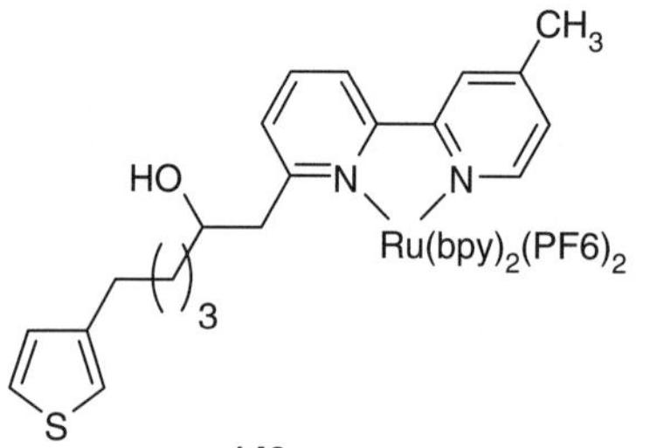

148 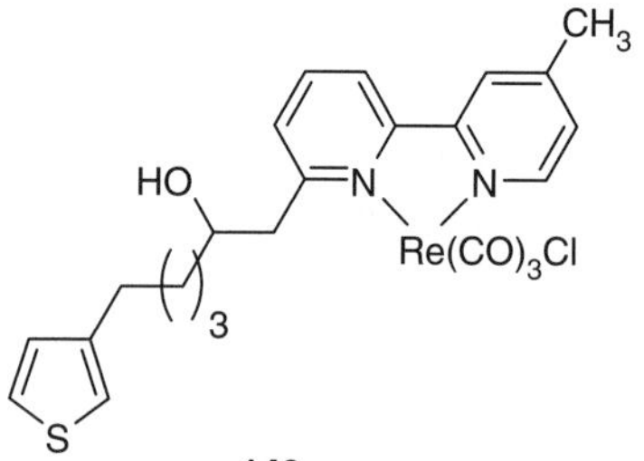   **149**

To solve these problems, we developed the synthesis of symmetrically disubstituted bipyridine ligands **150**, which possess two electropolymerizable bithiophenic groups fixed at an internal β-position of thiophene by an alkylsulfanyl or an alkoxy spacer [161]. The analysis of the electropolymerization of these compounds shows that the association of low oxidation potential polymerizable groups and two-site precursors allows us to synthesize stable functional polymers.

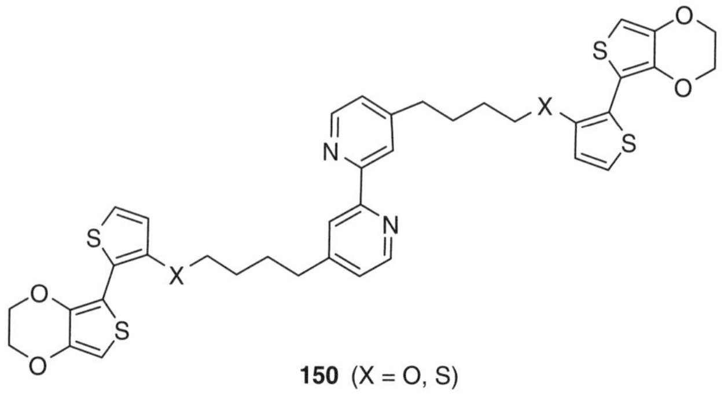

150 (X = O, S)

On this basis, ruthenium and iron complexes **151–153** with multiple polymerization sites (2, 4, and 6) have been synthesized and readily electropolymerized. The electrochemical behavior of the resulting metal-containing polymers displays the typical signature of both the π-conjugated PT backbone and the immobilized metal complex.

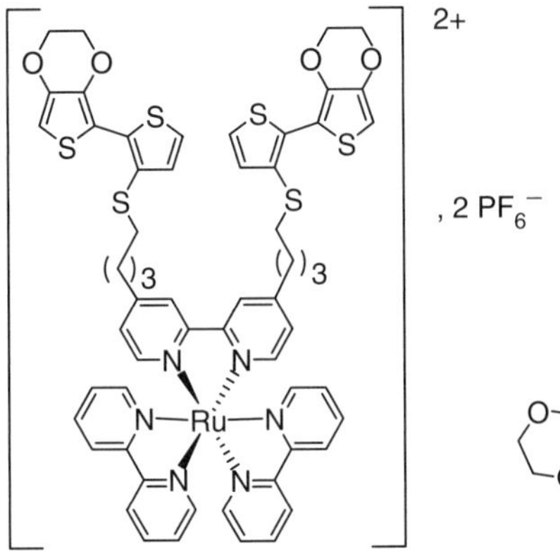

151

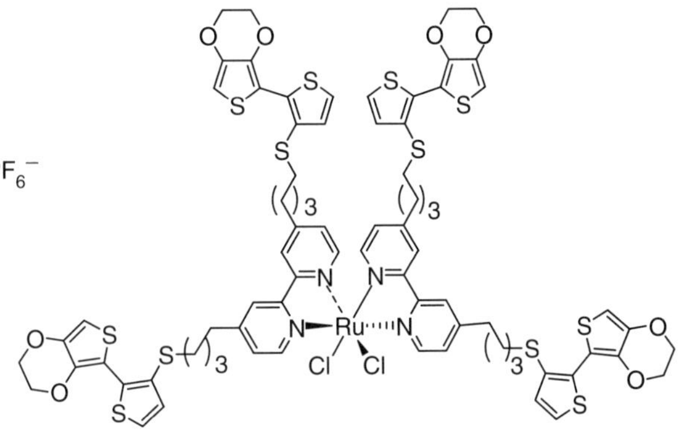

152

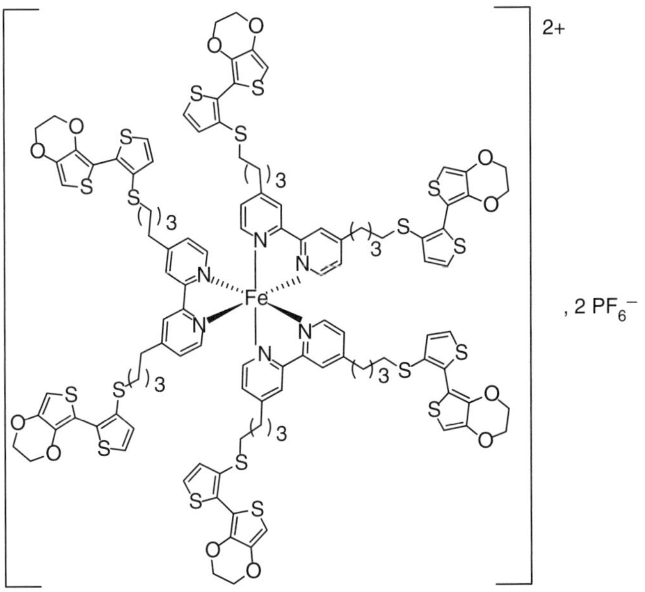

153

An alternative approach to the substitution at the 3-position of thiophene consists of the insertion of the bipyridine group directly into the π-conjugated system. Improved electronic communication between the metal center and the π-conjugated polymer was expected. Example of such PT derivative has concerned compounds **154** together with their Ru^{2+} complexes [276]. The electropolymerization of these ligands and the corresponding metal complexes occurred only for bithiophenic precursors. Pickup et al. have synthesized isomeric dithienyl-bipyridine structures **155** and **156**, but electropolymerization at relatively high potential was difficult and the electrochemical response of the polymer suggested a limited degree of polymerization [277]. Swager reported a series of polymeric systems based on polymerization of Ru^{2+} complexes incorporating bis(4,4'-dithienyl)bipyridine as ligand such as **157** [278]. Lemaire et al. prepared the chiral dithienyl-bipyridine **158** and rhodium, iridium, ruthenium, and cobalt complexes as catalysts for the reduction of acetophenone [279]. Whereas homoelectropolymerization of **158** failed, the compound was copolymerized with 3-methylthiophene.

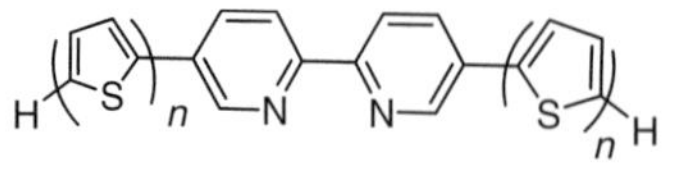

154 (*n* = 1, 2)

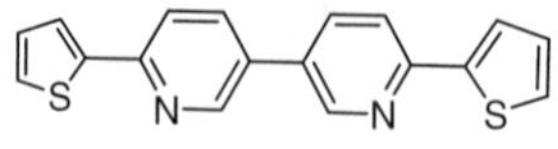

155

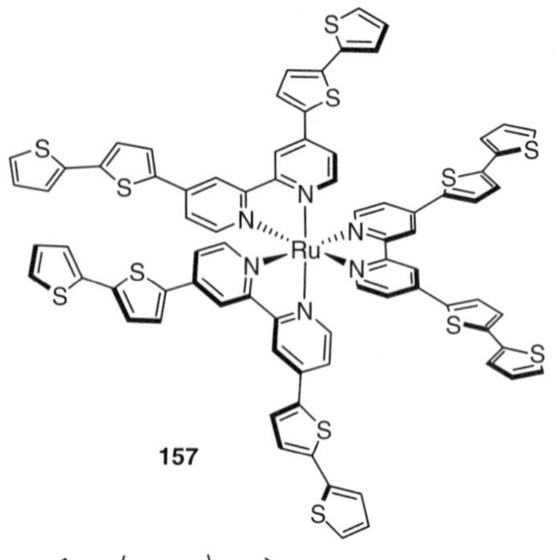

157

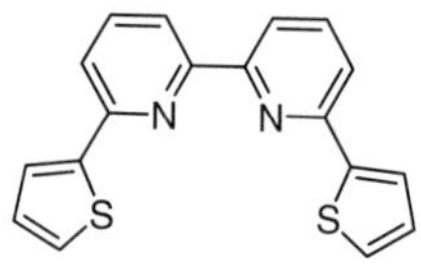

156

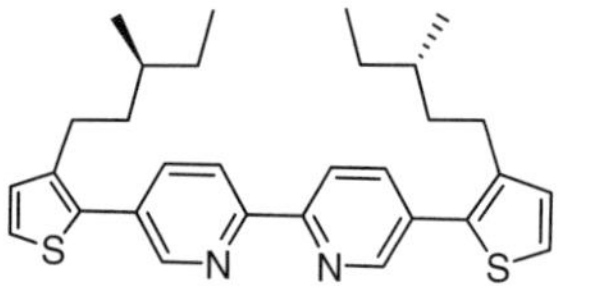

158

Bipyridine and the parent phenanthroline disubstituted with polymerizable thiophene moieties have been employed for the construction of conjugated polymers with metal-rotaxane. Swager and coworkers initially developed conjugated metallorotaxanes **159** based on 2,2′-bipyridine functionalized with thienyl groups in the 5,5′-positions and 1,10-phenanthroline **162** [205,280]. Studies on electropolymerized films showed that demetallation and remetallation processes are reversible leading to potential sensor applications. Independently, Bidan and Sauvage synthesized phenanthroline derivative **160** that was reacted with [Cu(I)–(MeCN)$_4$]BF$_4$ to form intertwined complex [281]. After electropolymerization, the CV of the resulting polyrotaxane showed the presence of Cu(I) ion. Addition of [Cu(I)–(MeCN)$_4$]BF$_4$ to a stoichiometric amount of **162** followed by the addition of **161** led to the formation of corresponding Cu(I) complexes that could be electropolymerized [282]. The presence of the CH$_3$ groups in *ortho*-positions of the nitrogen atoms of **161** enhances the stability of the polymer and makes its remetallation possible once the template Cu(I) center has been removed. The synthesis of a series of oligothiophene-[1,10]phenanthroline analogs of **161** by Negishi-type cross-coupling has been described but metal complexation has not been reported [283].

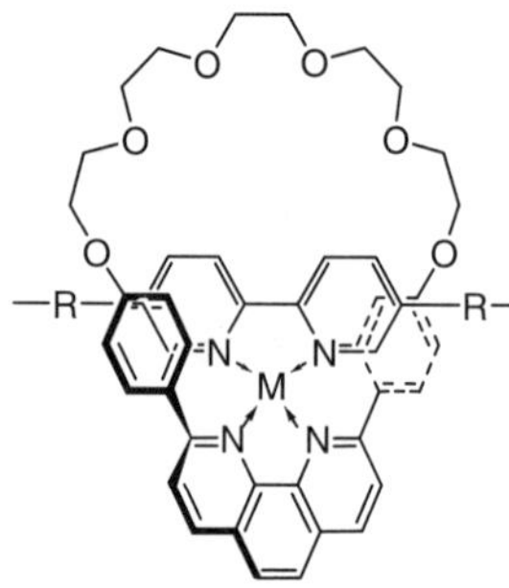

159 (M = Zn^{2+}, Cu$^+$)

R =

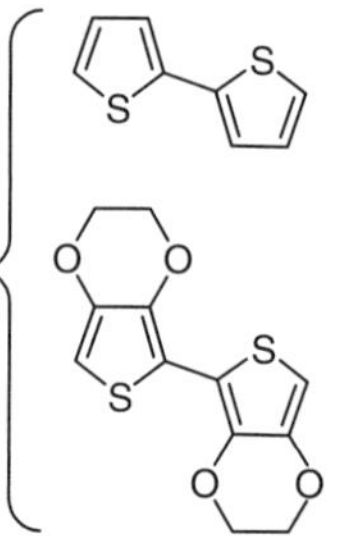

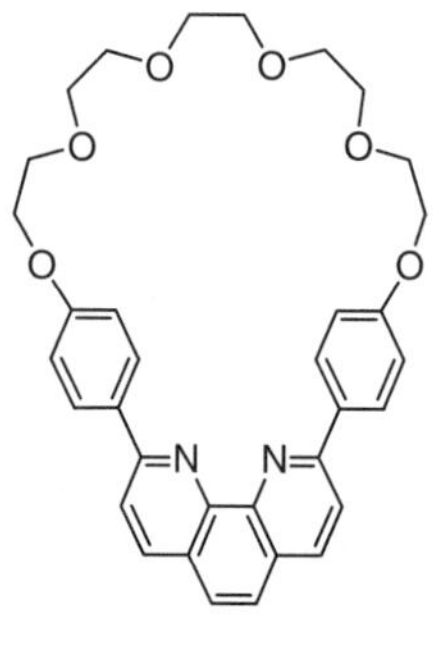

162

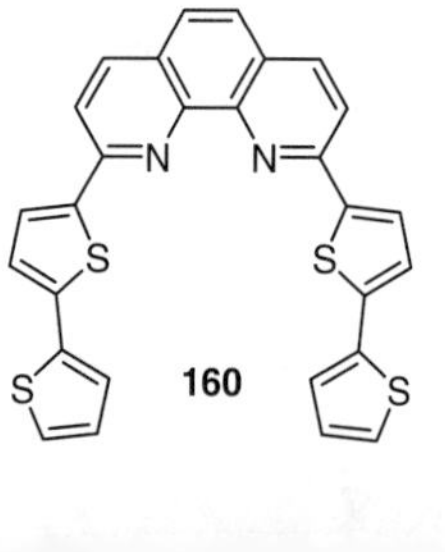

160

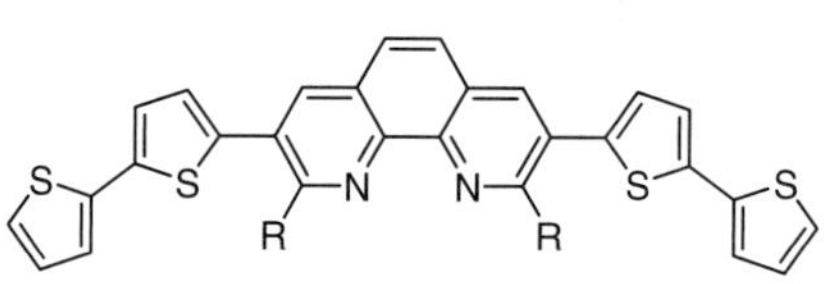

161 (R = H, CH$_3$)

More complex systems **163** and **164** have been synthesized by Swager et al. [284,285]. In the first one, the difference in oxidation potential of the two types of polymerizable groups, namely 2,2′-bithiophene and EDOT, makes possible a stepwise electropolymerization. The EDOT groups are polymerized first while polymerization of bithiophene occurs at higher potential. The resulting three-strand conducting metallorotaxane ladder polymer consists of two different conjugated polymers, one of them becoming conducting after selective electrochemical doping while the other can remain insulating [284]. Recently, polymers **164** with pendant rotaxanes were synthesized in two steps starting from the related metal-free rotaxane [285]. After oxidative polymerization of the latter in the presence of Cu(II)(OTf)$_2$ and subsequent treatment, the metal-free rotaxanated polymer was reacted with [Cu(I)–(MeCN)$_4$]PF$_6$ or Zn(ClO$_4$)$_2$ to give polymers **164**. In the metal-free rotaxanated polymer, the conjugated backbones are partially insulated by bulky rotaxane groups in agreement with low bulk-conductivity. However, insertion of Cu(I) in the rotaxane unit in polymer **164** introduces charge hoping pathways between the conjugated backbones and hence increases the conductivity.

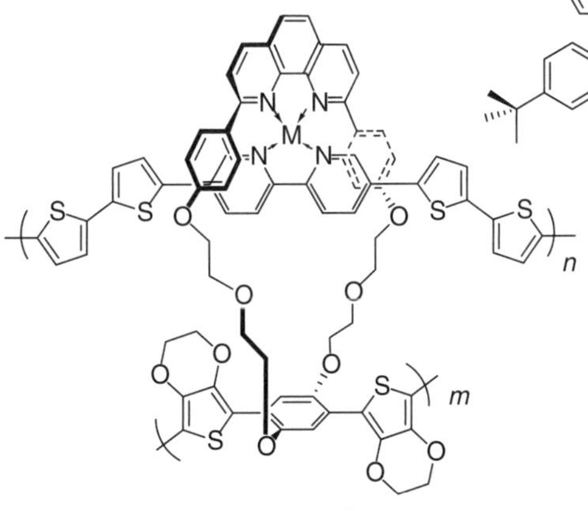

163 (M = Zn^{2+}, Cu$^+$) 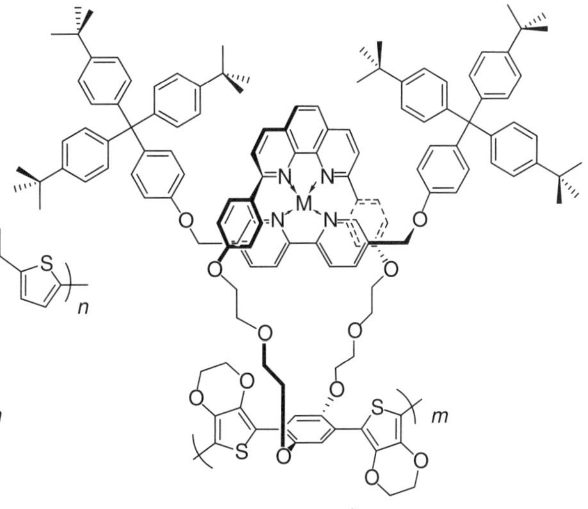 **164** (M = Zn^{2+}, Cu$^+$)

PT **165** [286,287] containing 2,2′-bipyridine has been also reported. The synthesis of polymers **165** containing Ru(II) or Os(II) was achieved by Stille copolymerization of the bistannyl derivative of the regioregular alkylated quaterthiophene and dibromobipyridine monomer. Low-molecular weight polymers were obtained. However, strong electronic coupling was observed between the metal-based chromophores and the π-conjugated segments [287].

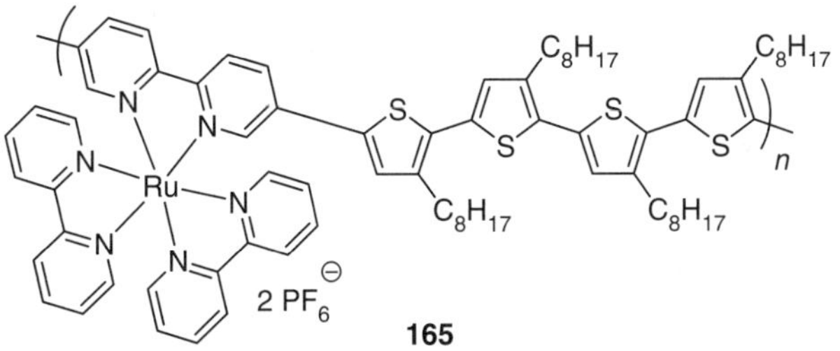

165

13.4.3 Polythiophenes for Electrochromic Devices

The use of PTs as electrochromic materials has been considered in the very early years of development of these polymers [288,289]. However, progress in this area has been hampered by the lack of stability of PTs in their oxidized and most transmissive form. The development of PEDOT, at the beginning of 1990s, has generated a renewal of interest in this field. In fact, PEDOT presents the advantage of a low oxidation potential combined with a stable doped conducting state, which is also the optically transparent state [107,108,290].

PEDOT is dark opaque blue in its reduced form and transmissive light blue in its oxidized form [107]. On this basis, efforts directed to the application of EDOT-based polymers in electrochromic devices have been focused on the improvement of the solubility, stability, switching time, and optical contrast [58,291,292].

Polymers with improved optical contrast have been obtained by introducing bulky groups on the EDOT structure to reach a more porous structure more appropriate to mass transport of ionic species. Another efficient approach has involved replacement of EDOT by ProDOT which leads to a less compact polymer [58].

Recently, it has been reported that polymer films of quality and performances comparable to those of electrogenerated films could be prepared from sprays prepared from poly(**46**) [118,121].

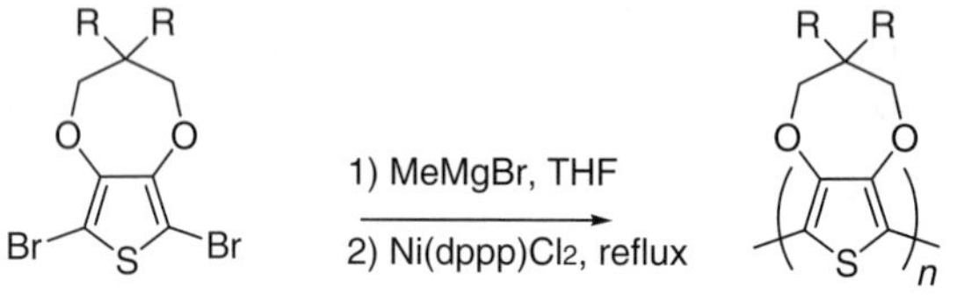

$R = C_4H_9, C_6H_{13}, CH_2CH(C_2H_5)C_4H_9, CH_2O(CH_2CH(C_2H_5)C_4H_9), C_{18}H_{37}$

Other recent examples **166–169** involving ProDOTs substituted by bulky groups have been obtained starting from a hydroxymethyl ProDOT derivative [123].

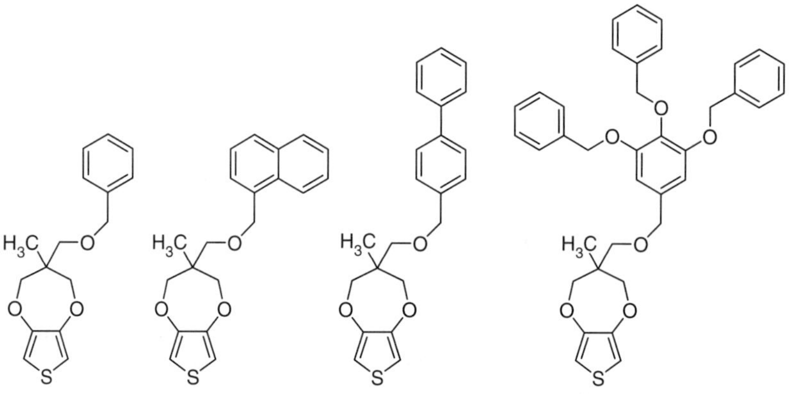

Examples of electrochromic polymers bearing perfluoro chains (poly(**54**)) [147], dendrons (poly(**170**)), or viologen groups, e.g., (poly(**171**)) [150] have been reported. In this later case, introduction of a second electrochromic viologen pendant group increases the contrast between the oxidized and the neutral states.

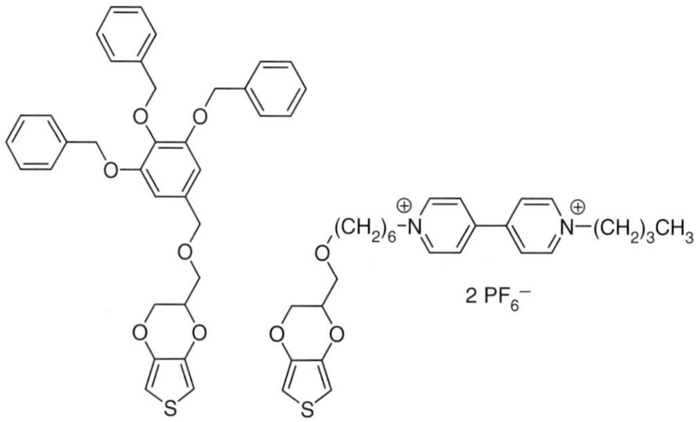

Electrochromic polymers have also been prepared by electropolymerization of a precursor involving a median core substituted by two EDOT units. Modification of the nature of the median core (vinyl, thienyl, phenyl, furan, etc.) allows for the tuning of the bandgap of the polymers and thus the color of the oxidized and reduced state [293].

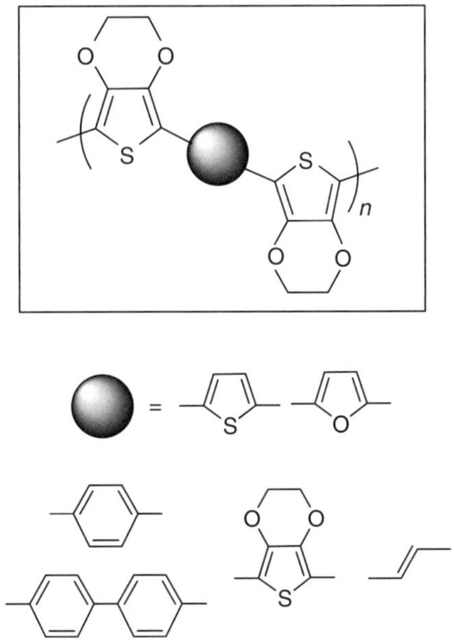

Based on this approach, polymers presenting successively different colors on oxidation have been described, e.g., polymer **172** [294,295]. Other electrochromes incorporating electroactive entities as pendant group, such as viologen in **173,** leads to the observation of two optical states on oxidation [296].

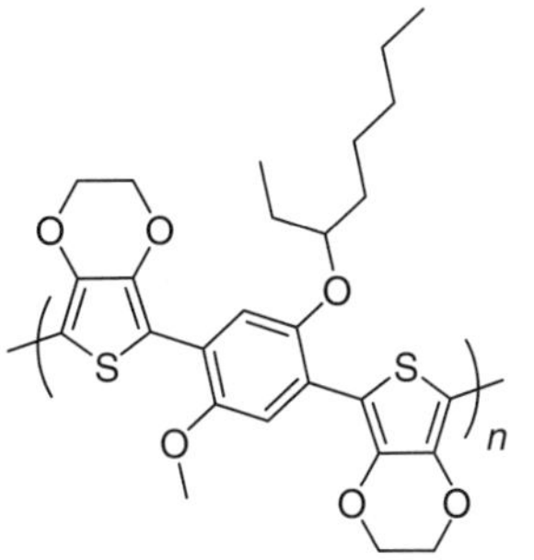

172

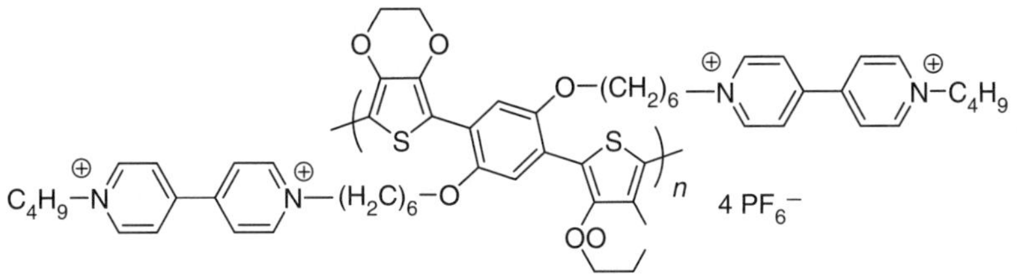

173

Similarly, introduction of carbazole in **174** also leads to a polymer presenting distinct redox processes and hence optical states [297,298]. Unlike electrochromic materials based on pure PEDOT, these compounds are colored in the oxidized state and colorless in the neutral one. The exploitation of the complementary properties of these two types of polymers has led to electrochromic devices exhibiting a large variety of colors [122].

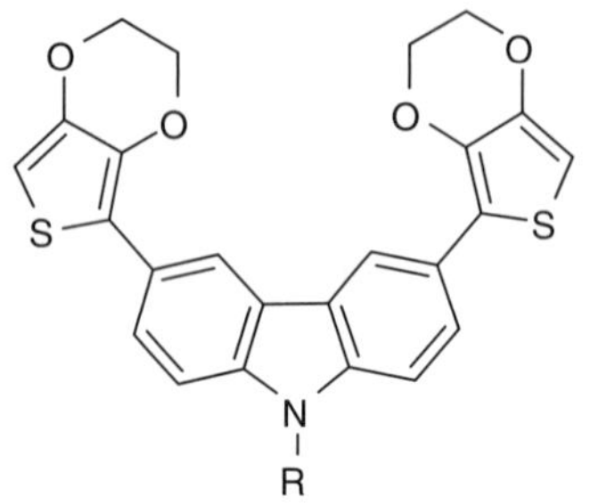

174

R = CH$_3$
R = –(CH$_2$)$_{19}$–CH$_3$
R = –(CH$_2$CH$_2$O)$_2$CH$_2$CH$_3$
R = –(CH$_2$)$_4$Th

Recently, new electrochromics having a distinct green color were obtained from a precursor containing a median thienopyrazine group disubstituted by two thienyl groups (**175**) [299]. This type of polymer can pave the way toward the development of red–green–blue electrochromic displays [300–303].

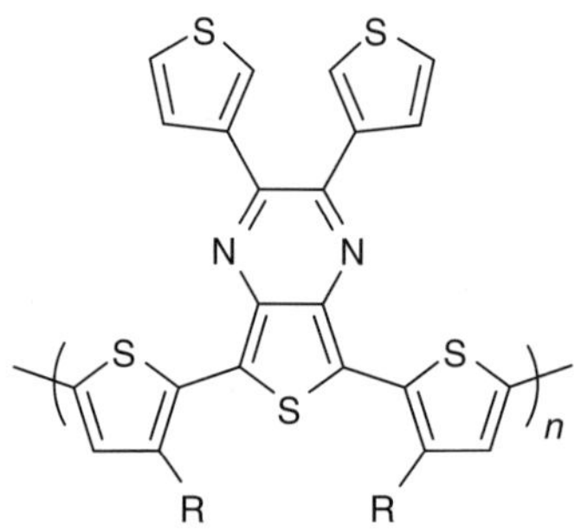

175 (R = H, C$_8$H$_{17}$)

13.4.4 Polythiophenes for Electronic Devices

Based on a combination of moderate bandgap, stability, and structural flexibility, PTs have progressively become widely investigated for the development of active materials for the realization of the various kinds of devices that constitute the emerging field of flexible or soft electronic namely OLEDs, OFETs, and organic photovoltaic devices.

13.4.4.1 PTs as Luminophores in OLEDs

Because of their tunable emission properties and environmental stability PT derivatives have been used as emitting material OLED [304]. However the device efficiency remains limited by both the low intrinsic emission quantum yield of PT and luminescence quenching by aggregation phenomena. As initially demonstrated by Inganäs and coworkers [305], the control of the torsion angle between adjacent thiophene rings by means of substituents of variable bulkyness allows to tune the emission wavelength of PTs over a wide spectral range.

A related approach involves the synthesis of PTs consisting of conjugated blocks connected by a single bond where head-to-head configuration interrupts the conjugation by torsion. On this basis, PTs **176–179** with emission maxima in the range 450 to 580 nm have been synthesized [306].

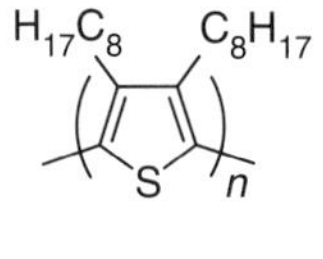

176 (λ_{em} = 450 nm)

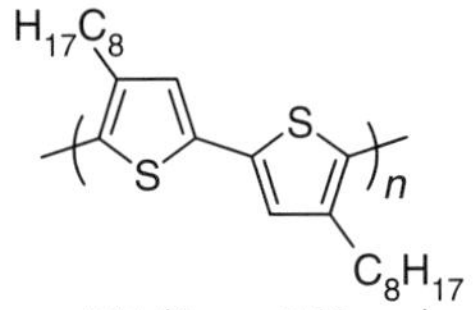

177 (λ_{em} = 540 nm)

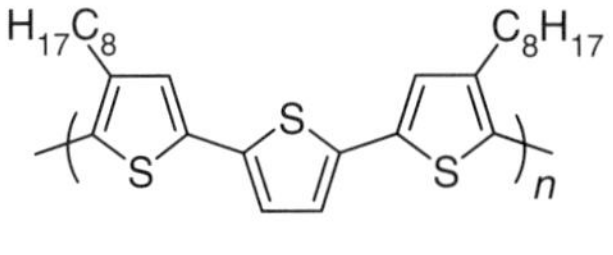

178 (λ_{em} = 560 nm)

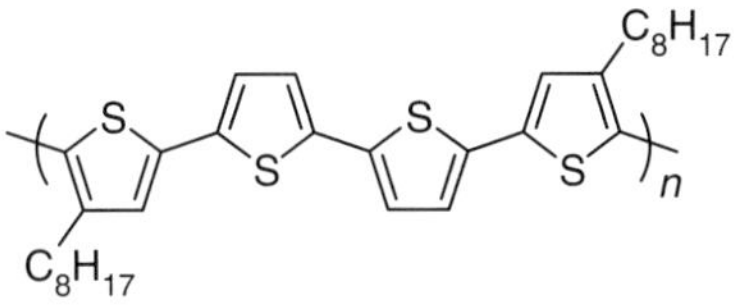

179 (λ_{em} = 580 nm)

Similarly, the emission properties of polymers **180** in which oligosilanylene blocks alternate with oligothiophenes depend mainly on the length of the oligothiophene block [307–309].

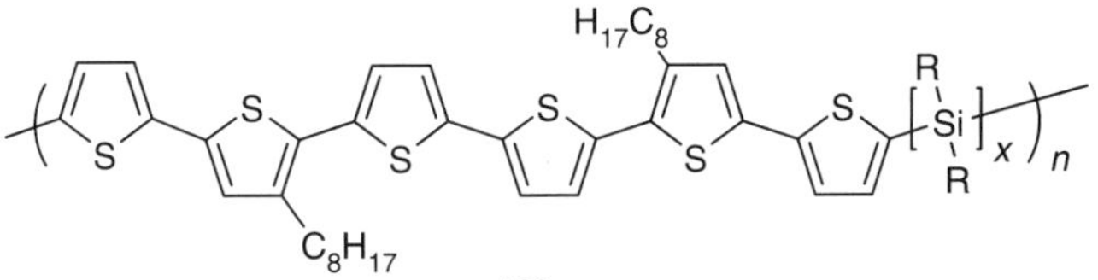

180

Block copolymers containing terthiophene units alternating with polyesters **181** and **182,** and poly-amide **183** spacers have been proposed for OLEDs fabrication [310,311]. These materials associate good film-forming properties to pure color of the emitted light [312]. Miller and coworkers have shown that quaterthiophene-based polyester **184** could be used to form a LED [313] as well as a photodiode [312].

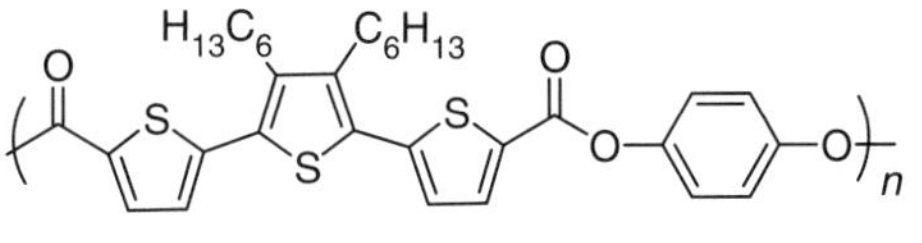

181

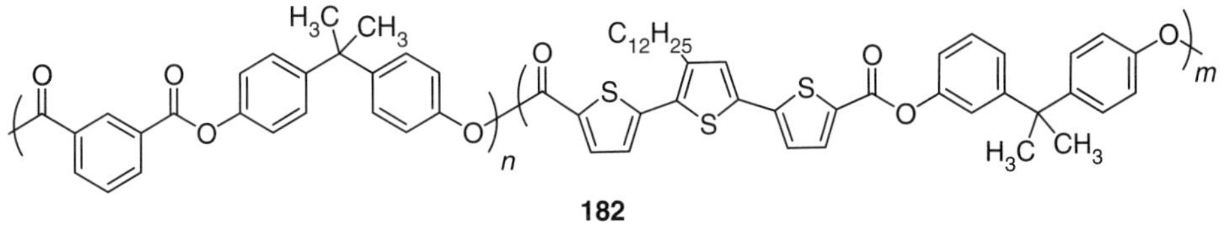

182

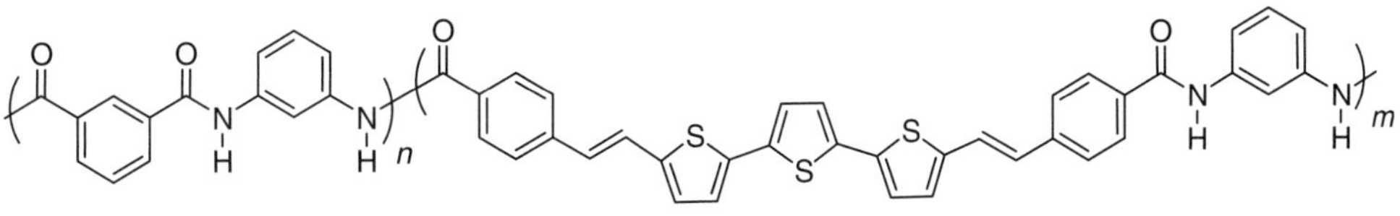

183

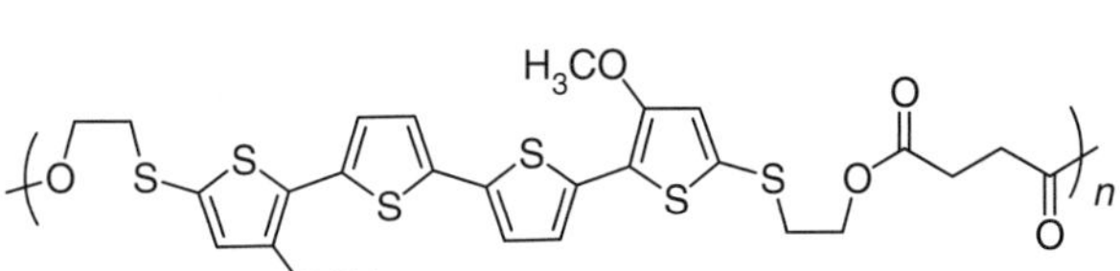

184

To improve the solubility and to facilitate processing of poly- and oligothiophenes, a large number of diblock or triblock copolymer based on polystyrene [314,315], polymethylmethacrylate [316], or poly-ethylene [317] have been reported. Interestingly, white photoluminescence was observed in thin films of **185** as a result of self-assembly of the side chains, however, no OLED was reported [317]. Dendrimers have been proposed as a possible alternative to limit aggregation and further self-quenching. Fréchet and coworkers have prepared hybrid dendrimers based on oligothiophene cores and poly(benzylether) dendrons [318,319]. A quinquethiophene **186** isolated by encapsulation with dendrons of different sizes containing peripheral triarylamines has been used in a single layer LED device [319].

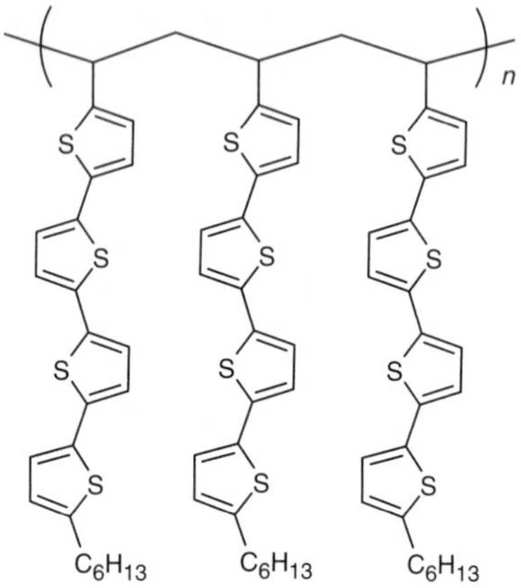

185

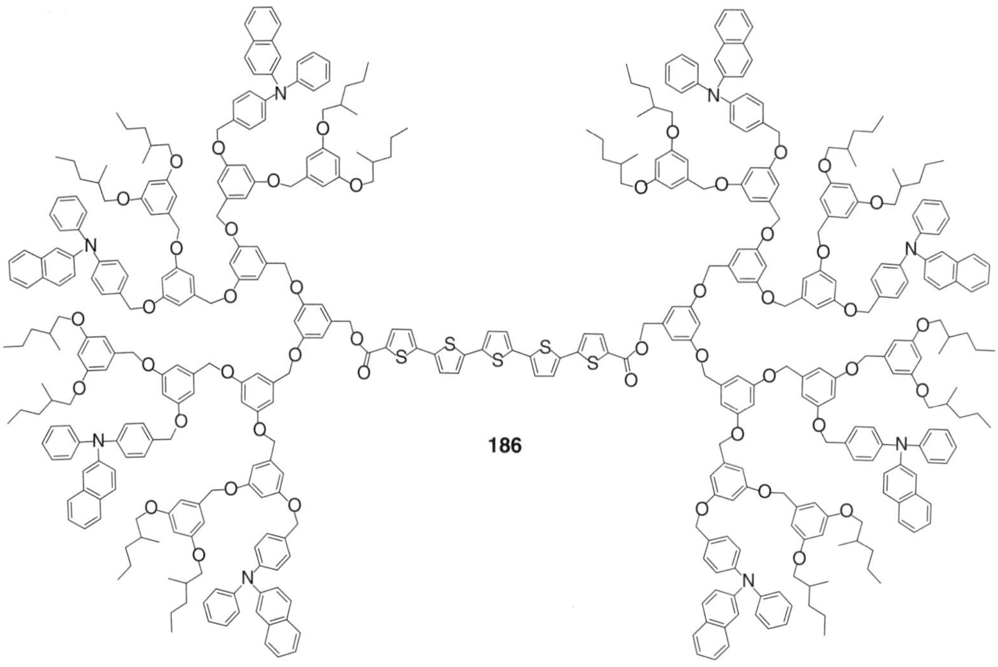

186

PTs substituted by electron-withdrawing groups have been scarcely used for OLEDs as they generally present low luminescence efficiency. A noticeable exception involves oligothiophenes containing thiophene-*S,S*-dioxide developed by the group of Barbarella [54–57]. These compounds that combine high electro-luminescence efficiency and electron-transport properties have been used to realize highly efficient OLED [320–323]. The incorporation of nonaromatic thiophene-*S,S*-dioxide moieties into oligothiophenes reduces the HOMO–LUMO gap and increases electron affinity. Moreover, the thiophene-*S,S*-dioxide prevents aggregation quenching in the solid state [324,325]. This approach has been recently extended to dibenzothiophene-*S,S*-dioxide-fluorene co-oligomers [326]. However, only few examples of PTs containing thiophene-*S,S*-dioxide moieties have been reported [327,328]. For example, polymers **187** and **188** were prepared by chemical and electrochemical oxidation of the corresponding monomers. However, a LED device has been reported only with **188** [328]. On the other hand, copolymers of fluorene [329–332] or carbazole [333] derivatives, and thiophene-*S,S*-dioxide units have been described and used as emitting material in LEDs [329–332].

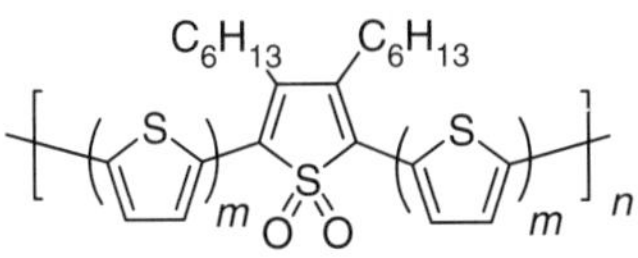

187 (*m* = 1–3) 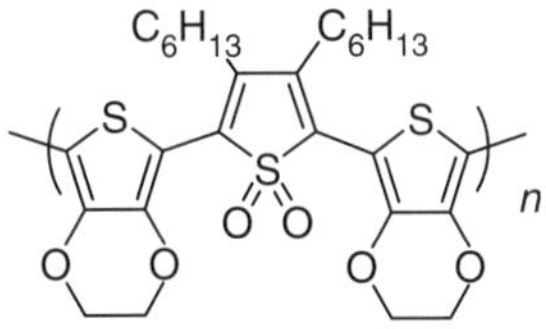   **188**

Conjugated polymers **189** and **190** containing a cyano group attached at a vinylene linkage have been synthesized by Knovenagel condensation [334,335]. These polymers that present reduced bandgaps of 1.75 and 1.56 eV, respectively, have been proposed for the fabrication of IR-emitting OLEDs [334].

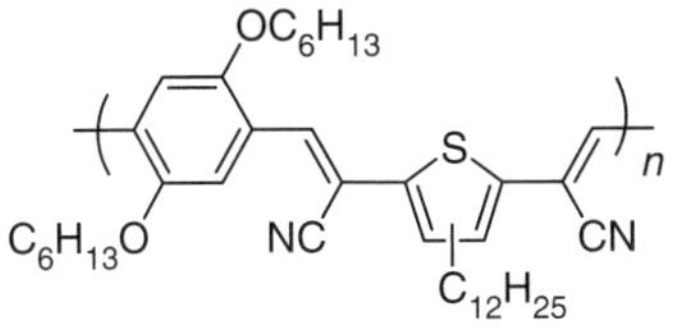

189 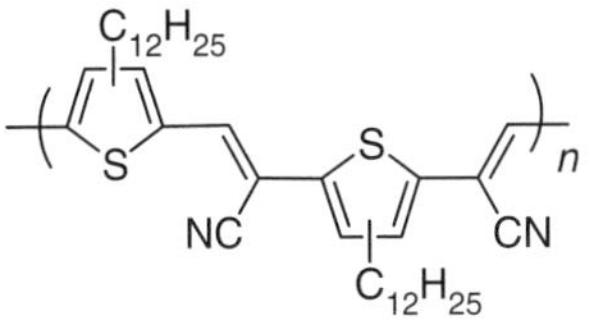   **190**

It has been shown that insertion of phenylene group in the PT backbone improves the PL efficiency of the polymer. A green OLED based on **191** gave an external EL quantum efficiency of 0.1% [336]. Thienylene–phenylene copolymers such as **192** and **193** with ethylene oxide side chains also exhibit electroluminescent properties [337,338]. In the case of the paracyclophane derivative **193**, a color variable light-emitting device generating two independent colors was demonstrated: red under forward bias and green under reverse bias [338].

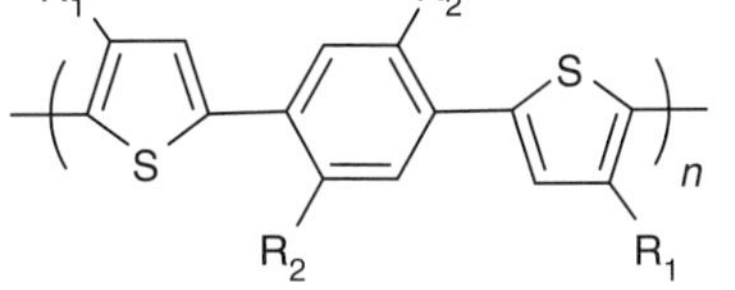

191

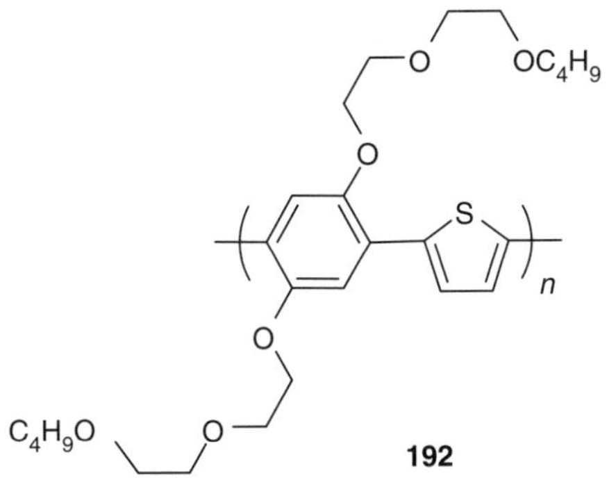

192 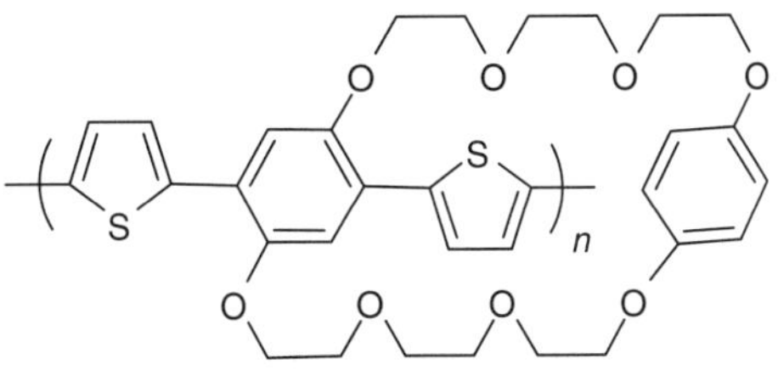   **193**

The high luminescence efficiency of fluorene has given rise to the synthesis of many hybrid conjugated systems, such as **194–198**, combining thiophene and fluorene [330,331,339–343]. In the case of **196**, a photoluminescence quantum efficiency as high as 32% was obtained [343]. A copolymer **199** based on fluorene and CN-substituted DTE has recently been described with the aim of achieving pure red-light emission [344]. A low bandgap polymer based on fluorene and thiophene **200** has recently been used in an OLED emitting at 1 µm [345].

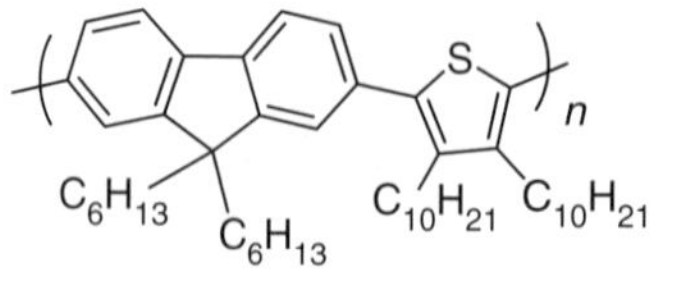

194

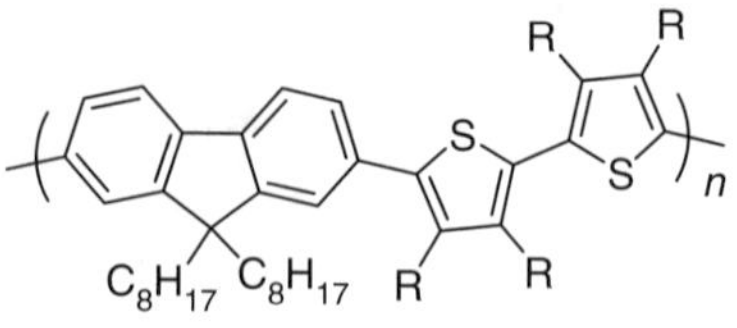

195: R = H, –OCH₂CH₂O–

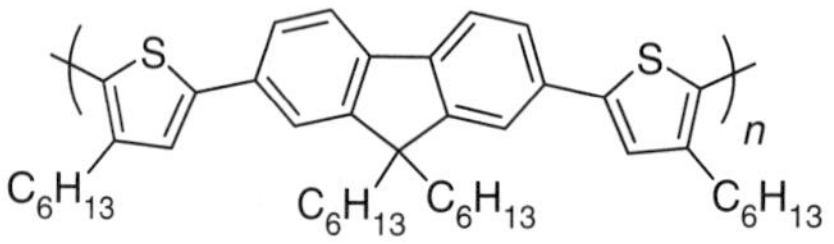

196

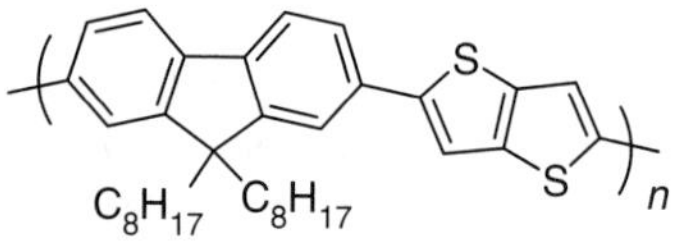

197

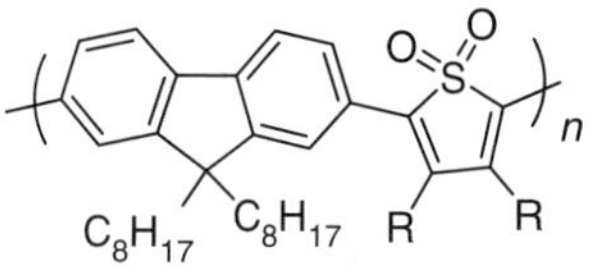

198: R = H, C₆H₁₃

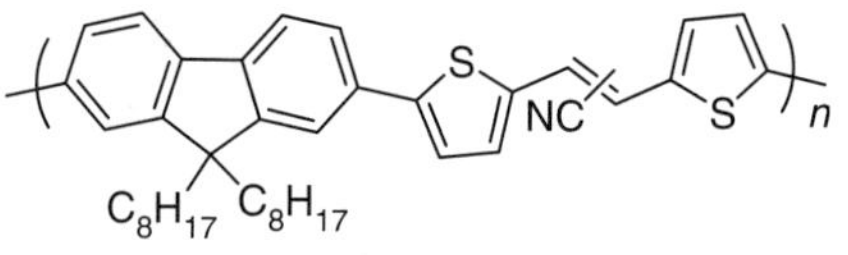

199

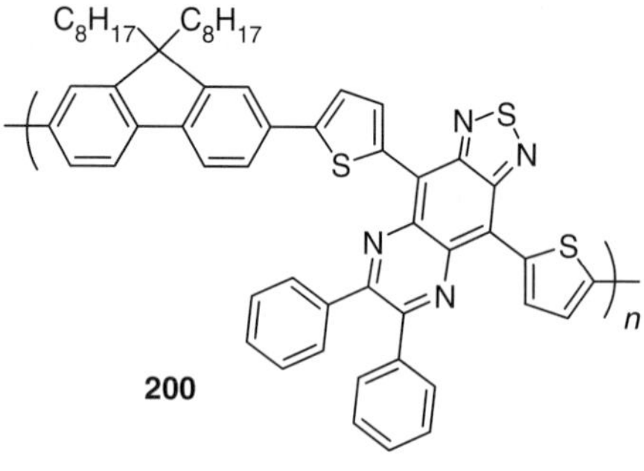

200

Various thiophene-based polymers have been used to improve hole injection in OLEDs such as PTV [346], electrogenerated poly(2,2′-bithiophene) [347], or a starburst-polymer poly(tris[4-(2-thienyl)phenyl]amine **201** [348]. In the last case, an OLED was also reported [349]. Polymer **202** combining the electron-poor phenylquinoline units with dialkylbithiophenes has been proposed as electron-transport material [350,351].

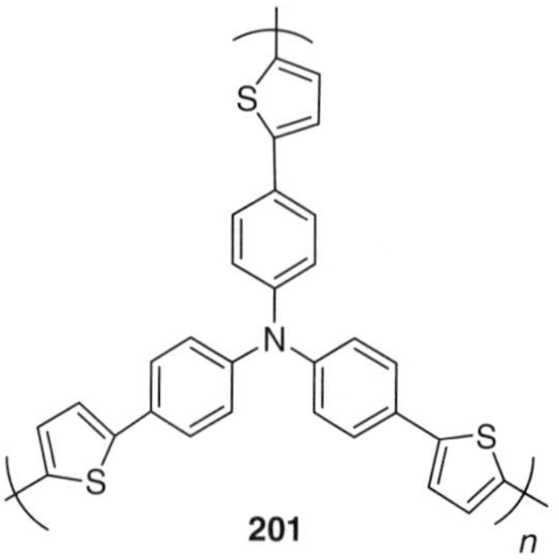

201

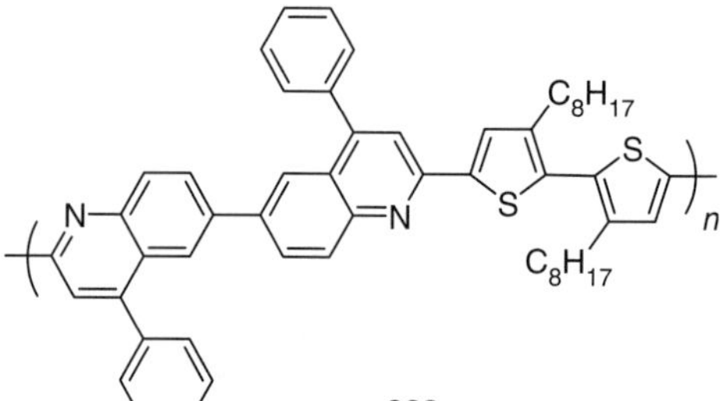

202

To summarize, PT derivatives can be used as emitting material, affording all colors of the visible spectrum, as well as hole transporting layer that is related to an intrinsic characteristic of PT or as electron-transporting material. OLEDs are also very promising for lighting applications [352]. In this context, the search for new electrophosphorescent material represents a real challenge. Recent synthesis of electrophosphorescent-substituted PT doped with iridium or platinum complexes opens new perspectives for the application of PT derivatives in light-emitting devices [353].

13.4.4.2 Thiophene-Based Conjugated Systems as Organic Semiconductors for Field-Effect Transistors

OFETs have attracted intense research interest for low cost, large-area electronic applications, such as active-matrix displays, electronic paper, flexible microelectronics, and chemical sensors [354–362].

The synthetic chemistry of thiophene-based organic semiconductors develops essentially along two main lines namely (1) processible materials that are mostly based on polymeric structures and (2) short-chain oligomers used either as single crystals or as polycrystalline films prepared by thermal evaporation under vacuum. Although this latter class of materials has given rise to much development during the last two decades, only polymeric materials will be considered here.

As the initial OFET prototype of Tsumura et al. (a hole mobility μ_h of 10^{-5} cm^2 V^{-1} s^{-1} was observed for PT) [363,364], the performance of OFETs based on PT derivatives has been improved significantly. Particularly, it has been shown that regioregular poly(3-hexylthiophene) (P3HT) shows much higher hole mobilities than regiorandom polymers and values up to 0.10–0.30 cm^2 V^{-1} s^{-1} have been claimed [365–367]. Thin films of P3HT can adopt a microcrystalline and anisotropic lamellar microstructure consisting of two-dimensional layers, with strong π–π interchain interactions [365]. Consequently, the mobility of P3HT depends not only on regioregularity [366,367] but also on the process of film deposition [365,368].

Ong et al. have synthesized solution-processable alkyl-substituted poly(ter- and quarterthiophenes) **203–205** [369–371] and realized OFETs with hole mobilities up to 0.14 cm^2 V^{-1} s^{-1} and on/off ratio over 10^7 in the case of **203** [369]. Similar results have been reported by others [372]. Another approach consists in limiting π-electron delocalization by incorporation of building block, such as thieno[2,3-*b*]thiophene that interrupts the conjugated pathway in the chain. The solution-processable polymers **206** leads to OFETs with hole mobility of 0.12–0.15 cm^2 V^{-1} s^{-1} [373].

Carboxylate-functionalized PTs such as **207** have been reported to show improved atmospheric stability [374] and leads to OFETs with hole mobility of ~0.06 cm^2 V^{-1} s^{-1} and on/off ratio over 10^5.

P3HT

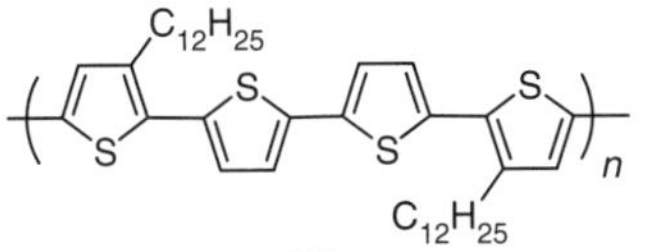

203

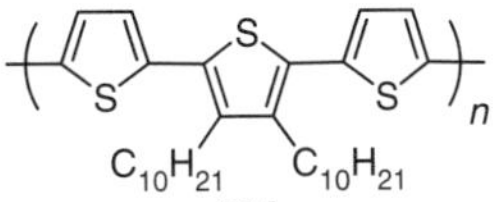

204

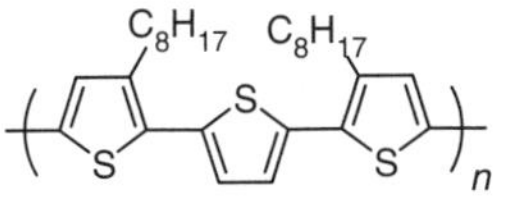

205

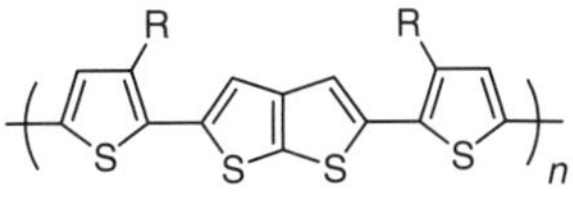

206 (R = C$_8$H$_{17}$, C$_{10}$H$_{21}$, C$_{12}$H$_{25}$)

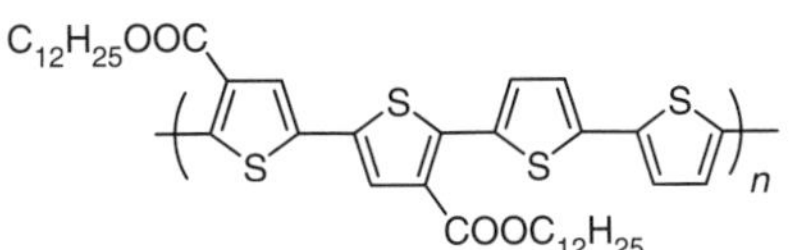

207

At the borderline between polymers and short-chain oligomers, dendritic thiophene derivatives have been synthesized by different groups [375–377] and proposed as solution-processing organic semiconductor. A hole mobility of 10^{-4} cm^2 V^{-1} s^{-1} has been claimed for an OFET based on **208** [376].

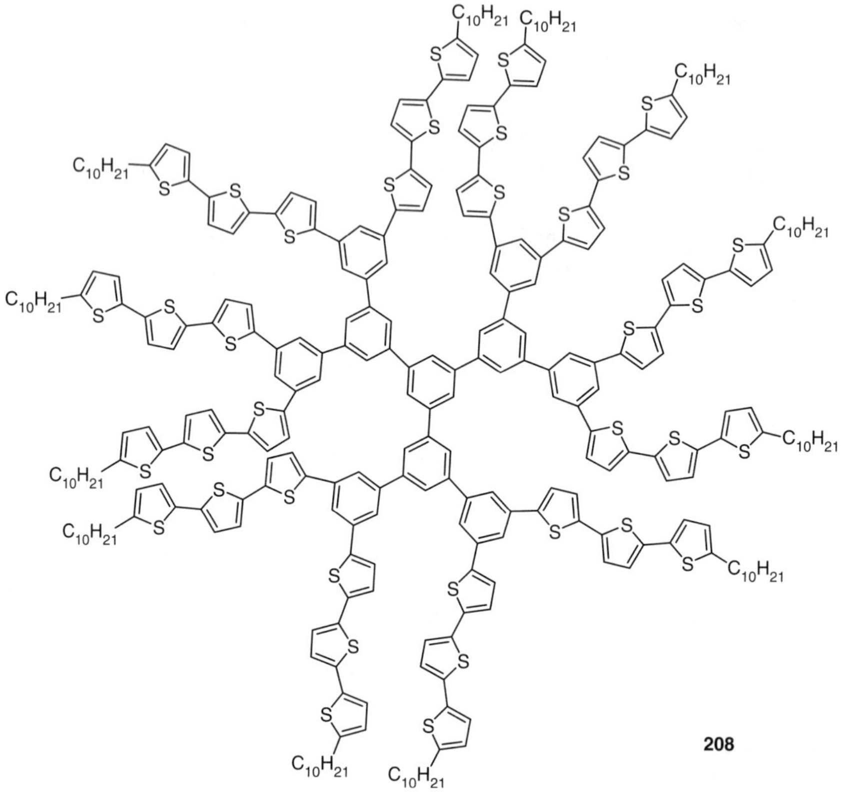

208

While a lot of work has been very recently devoted to oligothiophene-based n-type organic semiconductors [355,378–381], the number of n-type PT-based organic semiconductors remains very limited [103,382,383]. Polymer **209** and its model compound **210** [384] are solution-processable organic semiconductors that enable the realization of OTFTs by solution-casting. While compound **210** exhibits a high-electron mobility (μ_e) of 0.25 cm^2 V^{-1} s^{-1}, the mobility of polymer **209** remains weak ($\mu_e = 10^{-6}$ cm^2 V^{-1} s^{-1}).

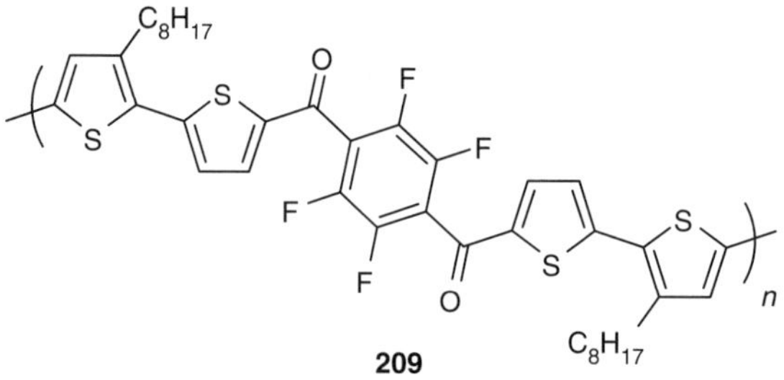

209

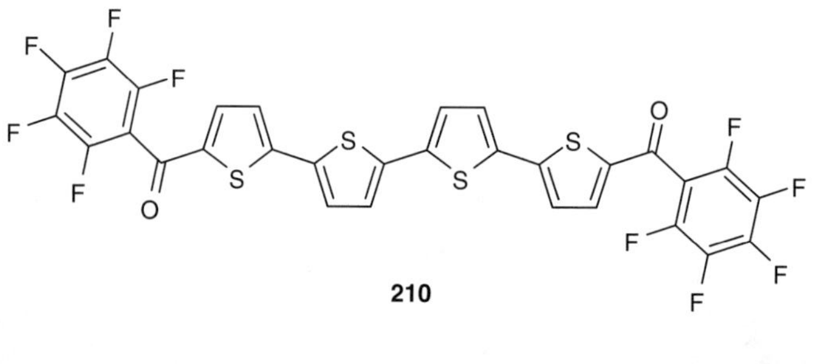

210

Other conjugated polymers based on thiophene have been investigated for transistor applications. For example, Sirringhaus and coworkers have reported the FET performance of a liquid crystal polymer consisting of an alternating copolymer of a bithiophene and a dioctylfluorene **211**. This one presents the advantage to be more resistant against chemical doping by atmospheric oxygen whereas after chain alignment in its liquid-crystalline phase, a mobility of 0.01–0.02 cm^2 V^{-1} s^{-1} has been obtained in the direction parallel to the alignment [385]. This polymer was also used in the fabrication of an all-polymer transistor circuits by inkjet printing [386]. According to Shim et al. [387], replacement of the relatively flexible bithiophene by a rigid fused thieno[3,2-*b*]thiophene in **212** leads to a threefold increase of mobility compared to **211**. A regioregular π-conjugated copolymer **213** alternating thiophene and electron-accepting 4-alkylthiazole leads to a hole mobility of 2.5×10^{-3} cm^2 V^{-1} s^{-1} [388].

Jenekhe et al. have synthesized a processable copolymer **214** of thiophene and phenoxazine [389], which presents a hole mobility of 6×10^{-4} cm^2 V^{-1} s^{-1}. PT derivatives incorporating fluorinated phenylene units in the main chain such as **215** have been investigated by Skabara and coworkers [390]. Although hole mobility (2×10^{-3} cm^2 V^{-1} s^{-1} for **215**) requires further improvement, they have shown that copolymers containing the tetrafluorophenyl unit are superior to the unsubstituted or difluorophenylene analogs.

While a hole mobility of 0.22 cm^2 V^{-1} s^{-1} has been claimed for PTV prepared by a soluble precursor route [391], this value has not been reproduced so far. Mobility of 1.5×10^{-3} cm^2 V^{-1} s^{-1} has been reported by the group of de Leeuw [392,393]. Using microwave conductivity measurements on PTV derivative **216**, Siebbeles et al. have suggested that this type of material could be used in principle for p- or n-type semiconductor [394]. It is worth noting that a hole mobility of 0.05 cm^{-2} V^{-1} s^{-1} has been obtained on OFETs based on a thienylenevinylene oligomer end-capped wit hexyl chains [395].

211   **212**

213   **214**

215   **216**

OFETs based on PT derivatives have been proposed for the realization of gas sensors [396]. Initial work dealt with the variation of field-effect mobility of P3HT upon exposure to NH_3 gas [397] or that of the source-drain current of OFETs prepared from various poly(3-alkylthiophene)s (PAT) on exposure to air, water, ethanol, and chloroform gases [398,399]. The Lucent group has developed field-effect sensors based on either various α,α′-disubstituted oligothiophenes or PAT, which are sufficiently sensitive (in the 1 ppm range) for use in electronic noses and respond to a range of analytes that includes alcohols, ketones, thiols, nitriles, esters, and ring compounds [359]. Soluble alkyl- and alkoxy-substituted regiochemically defined polyterthiophenes **217** and **218** have been used by Torsi et al. as active layers in alcohols sensing OFETs [400–402]. Comparison of these two polymers highlights the key role of the side chains in the device sensitivity.

PAT: *m* = 2,4,6,8,12   **217**   **218**

To summarize, almost 20 years after the first examples of PT-based OFETs [363,364,403], significant improvement of charge carrier mobility has been achieved from solution-processable polymers, thanks to the combination of technological optimization of the devices and progress in the design of materials combining better structural regularity, improved proccessability, and air-stability.

13.4.4.3   PTs as Active Materials for Organic Solar Cells

The use of PTs as active materials in solar cells has been considered already in 1982 in the Garnier's group. However, the poor efficiency of the initial prototype of Schottky diodes based on these materials [404] has led to a rapid decline of this field of research. The discovery of an ultrafast and efficient photoinduced electron transfer between conjugated polymers and fullerene C_{60} has represented an important step in the revival of organic photovoltaic devices [405]. This result rapidly followed by the first examples of bulk heterojunction solar cells based on interpenetrating networks of a donor conjugated polymer and an electron acceptor material such as a C_{60} derivative or a high electron affinity conjugated polymer has generated a strong renewal of interest for organic photovoltaic devices [406].

Regioregular P3HT is the most widely investigated PT for the fabrication of organic solar cells. While initial results were rather disappointing, technological optimization of solar cells based, in particular, on application of postproduction annealing, has led to continuous improvement of the performances and efficiency. Thus, Sariciftci and coworkers showed that optimization of the morphology of the P3HT/1-(3-methoxycarbonyl)propyl-1-phenyl[6,6]methanofullerene (PCBM) blend by postproduction thermal treatment allowed to increase the efficiency to 3.5% [407]. Recently, further optimization has led to efficiencies around 5% for bulk heterojunctions of regioregular P3HT and PCBM [408,409].

Regioregular poly[3-(4-octylphenyl)thiophene] (**219**) has been used with MEH-CN-PPV in two-layer polymer diodes which show an overall power conversion efficiency (PCE) of 1.9% [410]. Poly(3-(4′-(1″,4″,7″-trioxaocty)phenyl)thiophene) (**220**) mixed with C_{60}, was also considered for thin film organic PV devices [411].

219   **220**

Inganäs and coworkers have recently prepared bulk heterojunctions based on diversely substituted PTs and found a correlation between open circuit voltage (V_{oc}) of the cell and oxidation potential of the PT derivative [412].

Whereas tentatives to rationalize design rules for donors in bulk heterojunctions solar cells have been recently reported [413], the synthesis of PT derivatives for photovoltaic applications has been pursued following different approaches.

13.4.4.3.1 Low bandgap PTs

The efficiency of organic photovoltaic devices based on conjugated polymers is limited by the mismatch of their absorption spectrum and the solar spectrum. This problem has motivated the search for low bandgap PT derivatives. While polymers with very low bandgap are known, in the specific case of solar cells the problem is rendered more complicated by the fact that bandgap reduction should not have excessively detrimental effect on the open circuit voltage of the resulting solar cell [414].

Janssen and coworkers have synthesized solution-processable conjugated oligomeric materials (**221–222**) combining electron-rich 2,5-dithienyl-*N*-dodecylpyrrole and electron-deficient 2,1,3-benzothiadiazole units. Bandgap of 1.60 and 1.46 eV was measured for thin films of **221** and **222**, respectively. Bulk heterojunction realized with **221** and PCBM gave an efficiency of ∼0.34% [415].

Andersson and coworkers have prepared solar cells based on blends of poly(2,7-(9-(2′-ethylhexyl)-9-hexyl-fluorene)-*alt*-5,5-(4′,7′-di-2-thienyl-2′,1′,3′-benzothiadiazole) (**223**) and PCBM [416]. The polymer shows a λ_{max} (545 nm) with a broad optical absorption in the visible spectrum and an efficiency of 2.2% has been measured under simulated solar light. The same group has also reported the synthesis of low bandgap polymers **200** ($E_g = 1.25$ eV) and **224** ($E_g = 1.46$ eV) which have been blended with a soluble pyrazolino[70]fullerene and PCBM, respectively, to form bulk heterojunction solar cells of PCE of 0.7% [417] and 0.9% [418]. Incorporation of an electron-deficient silole moiety in a polyfluorene chain affords an alternating conjugated copolymer (**225**) with an optical bandgap of 2.08 eV. A solar cell based on a mixture 1:4 of **225** and PCBM exhibits 2.01% of PCE [419].

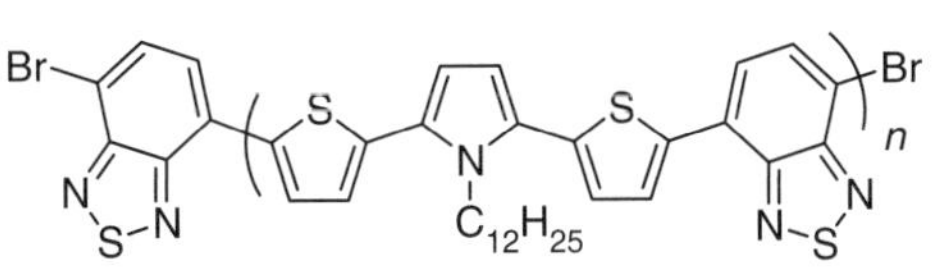

221 $n = 1–4$ ($E_g = 1.60$ eV)
222 $n = 3–8$ ($E_g = 1.46$ eV)

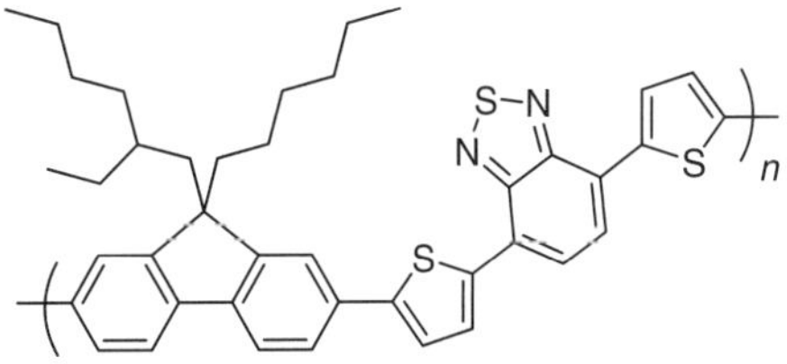

223 ($E_g ∼ 1.90$ eV)

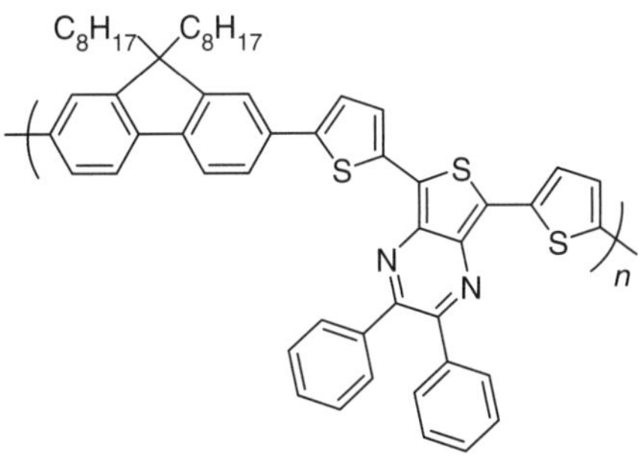

224 ($E_g = 1.46$ eV)

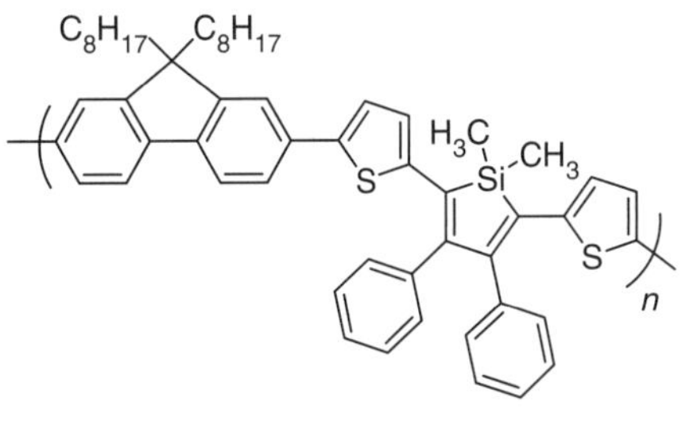

225 ($E_g = 2.08$ eV)

Several processable low bandgap PTs containing isothianaphthene units, such as **226**, **227**, and **228**, have been reported for their use in bulk heterojunction PV devices with PCBM, with efficiencies 0.008%, 0.24%, and 0.31%, respectively, under simulated solar illumination [420–422]. A solar cell made from a mixture of **229**/PCBM (1:1) led to an efficiency of 0.09% [423]. Polymer **230** presents a larger optical bandgap (2 eV) than that of **229** (1.3 eV) and a 1:1 mixture of **230** with PCBM leads to a PV cell of 0.024% of PCE under the same illumination conditions than for **229**.

PV devices constructed by using a blend of 80% of PCBM and 20% of narrow bandgap polymer **30** based on cyanovinylene and 3,4-propylenedioxythiophene-dihexyl ($E_g = 1.7$ eV) showed 0.1% of PCE. This value was increased to 1% by replacing **30** by a 50:50 blend of **30** and MEH-PPV whereas best values reported for MEH-PPV devices range from 1.2 to 2.9%. While no significant enhancement in device performance is observed compared to MEH-PPV, it has been shown that incorporation of **30** in MEH-PPV extends its light harvesting spectral range [94].

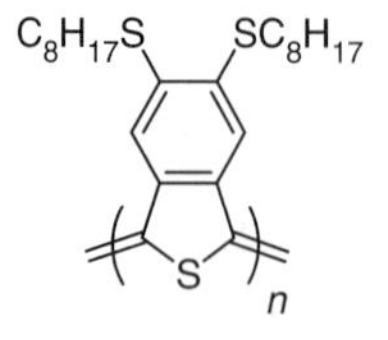

226 (E_g = 1.16 eV) 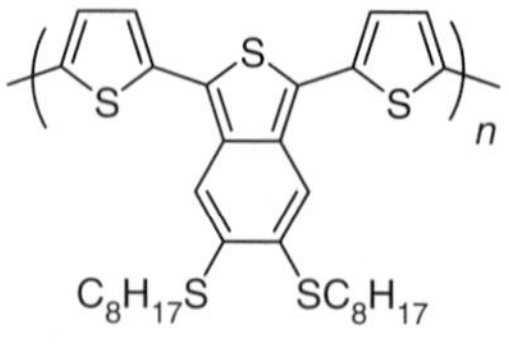   227 (E_g = 1.4 eV) 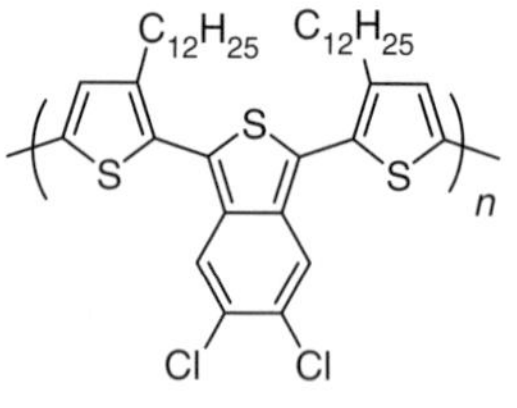   228 (E_g = 1.80–1.85 eV)

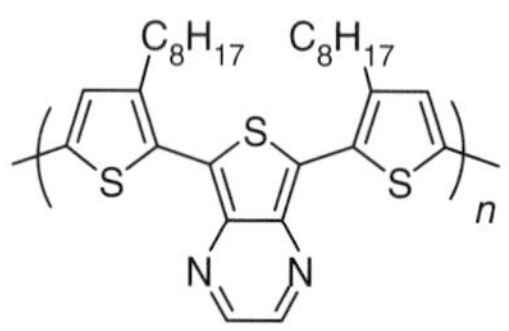

229 (E_g = 1.3 eV) 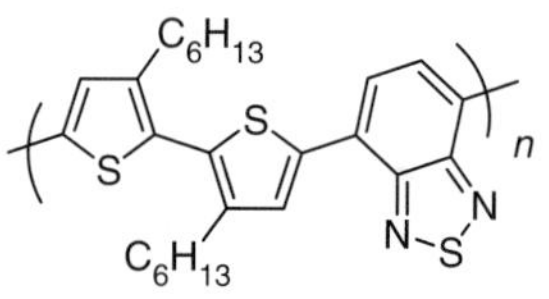   230 (E_g ~ 2 eV)

Because of their low bandgap ($E_g \sim$ 1.6–1.8 eV) and their high absorption in the visible spectrum, PTV derivatives appear as excellent candidates for bulk heterojunction solar cells. PV devices produced by thermal treatment of a blend of PCBM and a soluble sulfinyl precursor, which led to the in situ formation of the PTV derivatives, showed PCE of 0.18% and 0.12% for **231** and **232**, respectively [422]. Following a different strategy, Durstock and coworkers [424] have directly synthesized a soluble conjugated regioregular poly(3-dodecyl-2,5-thienylenevinylene) **233** according to the procedure of McCullough [425], for the realization of bulk heterojunction PV devices with PCBM. The best device showed a PCE of 0.24%. Recently, PCE of 0.13% has been reported for a solar cell based on a mixture of a soluble PTV **234** with fused-tetrathiafulvalene units and PCBM [426].

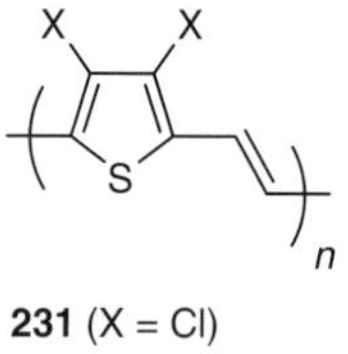

231 (X = Cl)
232 (X = Br)

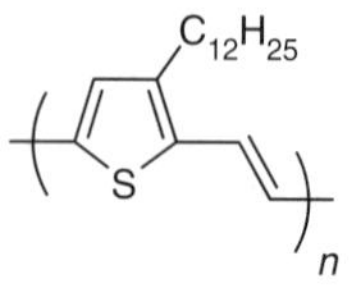

233

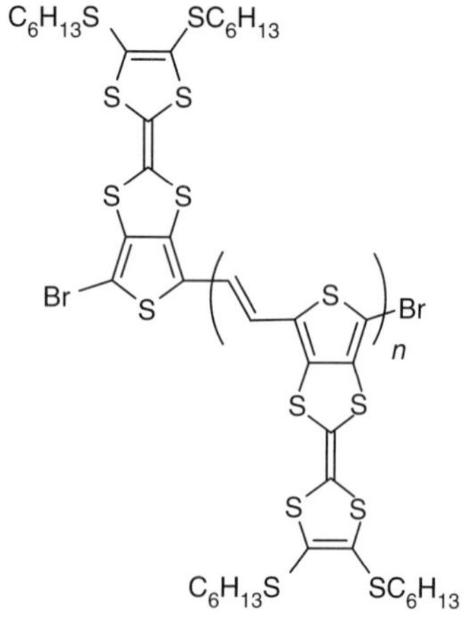

234

13.4.4.3.2 Fullerene-substituted PTs

As already discussed in the case of blends of P3HT and PCBM, the morphology of the film dramatically affects the device performance. Particularly, the limited miscibility of the two components in the biphasic system can lead to the formation of clusters of fullerene, which may increase intramolecular charge recombination and limit efficiency of electron transport [427,428]. An elegant solution to this phase segregation consists in the covalent fixation of the C_{60} group on the PT backbone [429,430].

Sannicolò et al. first reported the electropolymerization of a C_{60}-derivatized bridged bithiophenic precursor (**235**) [431]. Ferraris et al. have described the synthesis and the electropolymerization of a bithiophene with a C_{60} group attached at β-position by an alkyl spacer (**236**), whereas processable PT copolymers were postfunctionalized with C_{60} to give polymers [432]. Recently, PT with C_{60} groups connected to the π-conjugated polymer chain by a phenyl polyether linker has been prepared by electropolymerization of bithiophene **237** [427,433]. While the donor backbone and the acceptor moiety do not interact in the ground state, a photoinduced electron transfer between the PT chain and the attached C_{60} group has been demonstrated by ESR measurements.

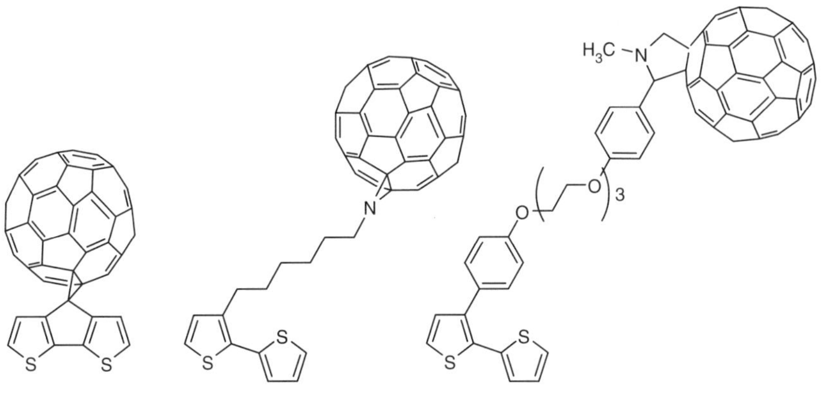

235   **236**   **237**

We synthesized a new series of C_{60}-derivatized PTs obtained by electropolymerization of bithiophenic precursors **238** and **239** in which one or two polymerizable groups are attached to C_{60} by alkyl spacers of variable length [158]. The precursor structure involves a 3,4-ethylenedioxythiophene associated with a 3-alkylsulfanylthiophene. In addition to the decrease of electropolymerization potential, the sulfide group represents a convenient way for the functionalization of thiophene by the facile deprotection of a thiolate function [159]. The analysis of the photoelectrochemical response of poly(**239**) ($n = 3$) and poly(**240**) on platinum microelectrodes polarized at -0.10 V and irradiated with intermittent white light shows that for both the initial peak and the stabilized current, the values of photocurrent are more than twice in the case of poly(**239**) ($n = 3$).

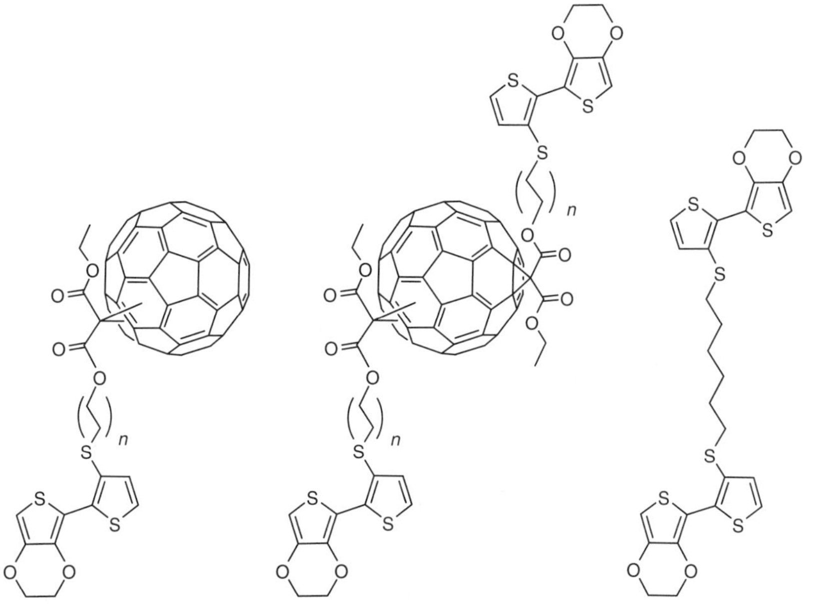

238 (n =1, 3)   **239** (n =1, 3)   **240**

Recently, two other C_{60} containing polymers have been synthesized by electropolymerization of terthiophene derivatives **241** and **242**, respectively, by Komatsu [434] and Wudl [435]. The efficient photoelectrochemical response observed for poly(**241**) and the low bandgap in the order of 0.7 eV for poly(**242**) suggest that these compounds could be useful for the realization of solar cells.

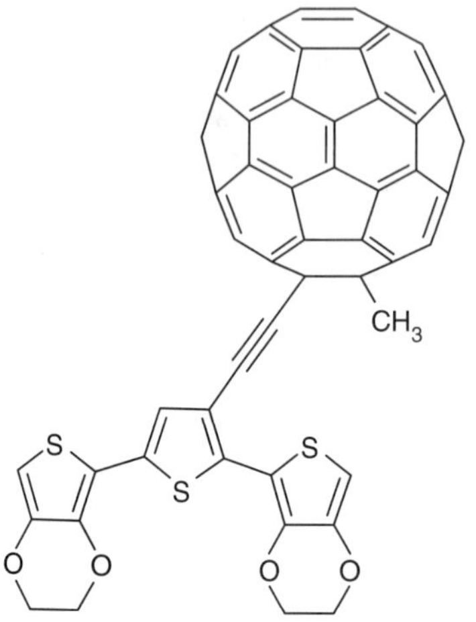

241 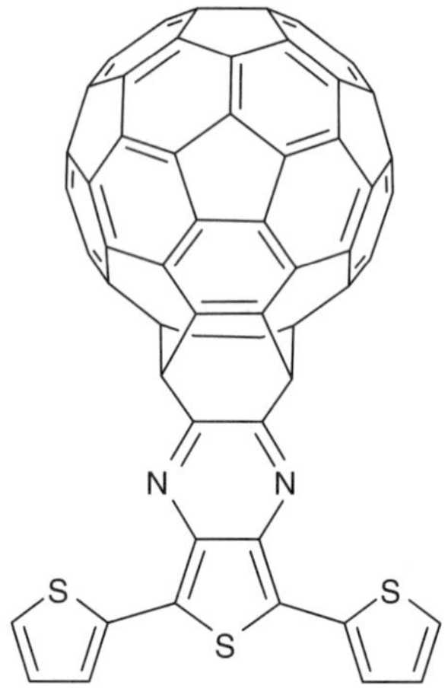   **242**

Zhang et al. [428] and Cravino et al. [436] synthesized [60] fullerene-thiophene polymers **243**. This soluble copolymer was used in solar cells that showed limited efficiency due to the low amount of fullerene incorporated in the copolymer that remained below the percolation threshold for efficient electron transport [436].

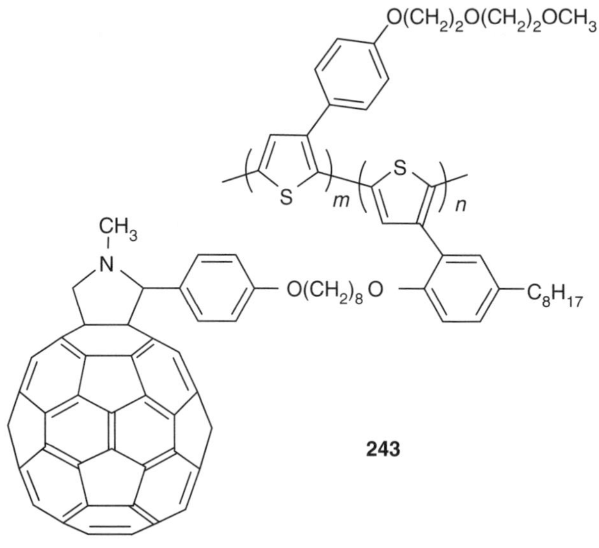

243

13.4.4.3.3   *Miscellaneous acceptor-substituted PTs*

PTs **244** [437], **245** [438], and **246** [439] with pendant anthraquinone or tetracyanoanthraquinodimethane moieties have been synthesized as possible active materials PV devices. The polymers were reported to be soluble and to exhibit a photoinduced electron transfer process in solution. Photoinduced electron transfer from PT to phthalocyanine units was observed in copolymer **247** [440].

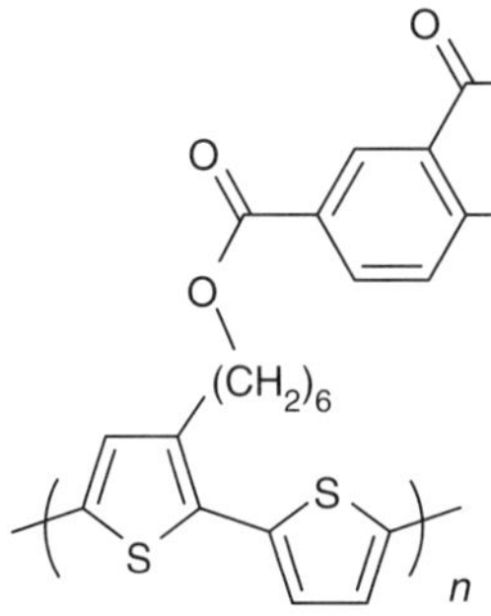

244

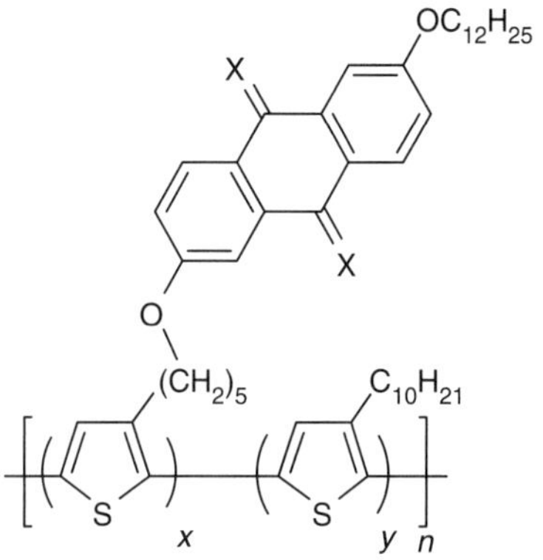

245 (X = O)
246 (X = C(CN)$_2$)

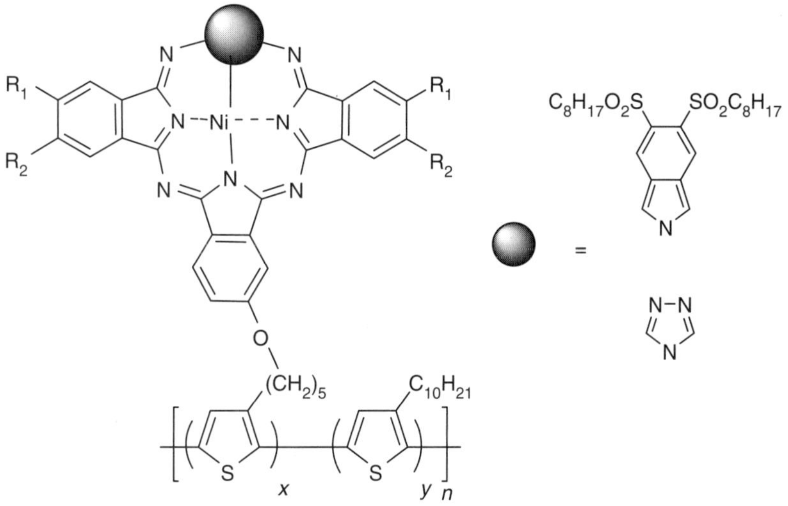

247

Recently, polymers prepared from thiophene-based conjugated systems **248** [441] or **249** [442] with covalently attached perylene bisimides have been reported. These polymers exhibit a broad absorption in the 400–800 nm range [442], which makes them potential candidates for PV application.

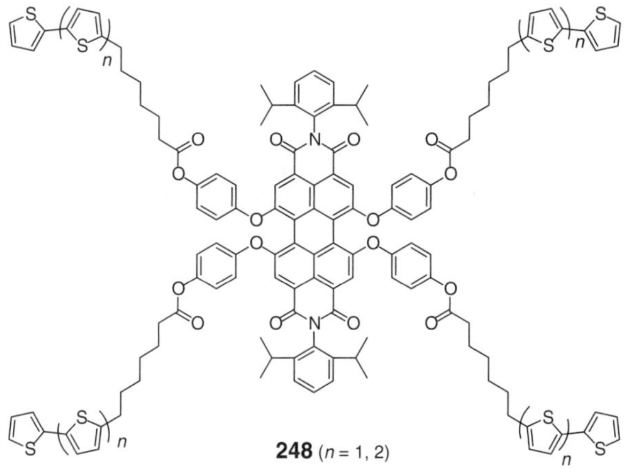

248 (n = 1, 2)

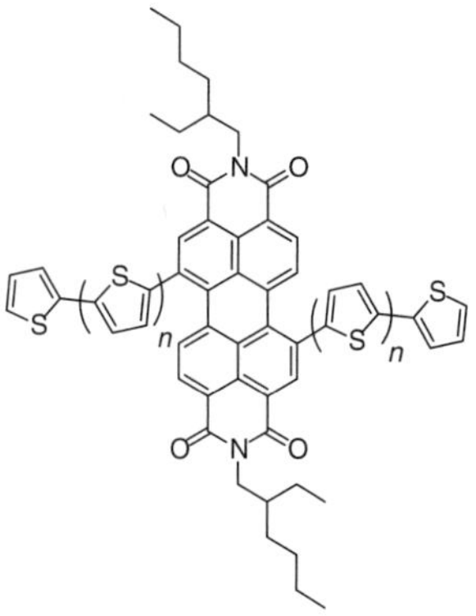

249 (n = 1, 2)

A polymer based on EDOT containing a perylenetetracarboxylic diimide unit has been prepared by electropolymerization of **250** [443]. The related absorption spectrum covers the visible range and extends up to 850 nm. Similarly, PV devices based on dyads **251** in which oligo(3-hexylthiophene)s are covalently linked to perylenemonoimide have been recently reported [444]. Preliminary results based on bulk heterojunction devices consisting of ITO/PEDOT-PSS/**251**:PCBM (1:4)/LiF/Al showed an open circuit voltage of 0.94 V and efficiencies of 0.33% (**251**, $n = 1$) and 0.48% (**251**, $n = 3$) under standard test conditions (AM 1.5G, 1000 W m^{-2}).

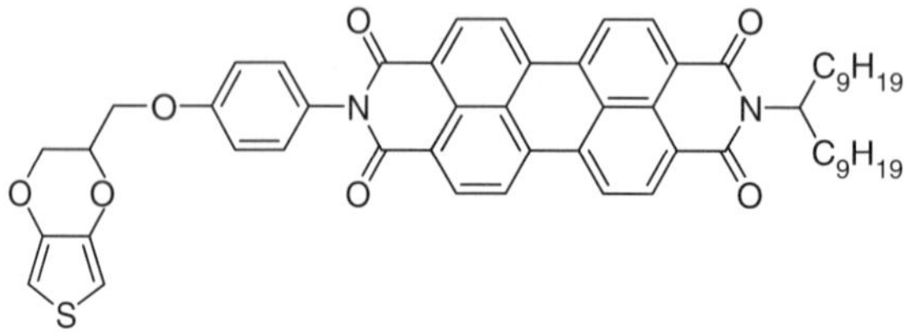

250

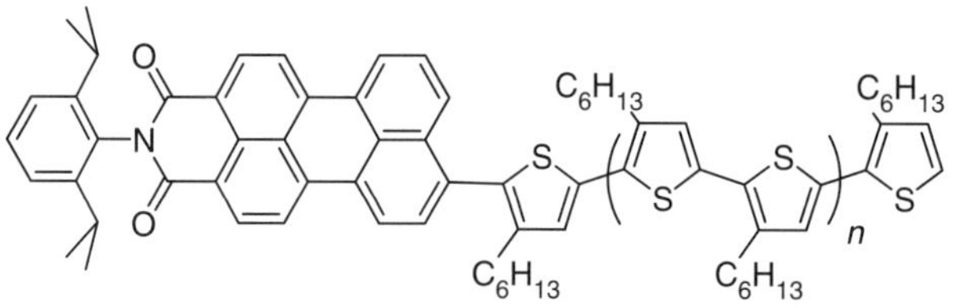

251 (n = 1, 3)

13.4.4.3.4 PT-inorganic hybrid solar cells

Current research focuses on the combination of PTs with inorganic materials of high electron affinity, such as nanostructured CdSe or TiO$_2$. Composite materials based on colloidal nanorods of CdSe and P3HT were first reported by Alivisatos to achieve a PCE of 1.7% [445]. Fréchet and coworkers have synthesized an end-functionalized PT **252** bearing an aminomethyl group that can effectively disperse CdSe nanocrystals to afford intimate nanocomposites [446]. Depending on the composition of the film, solar cells based on **252** can lead to PCE of 1.6% under solar illumination.

The use of PAT derivatives such as P3HT or poly(3-undecyl-2,2′-bithiophene) (**253**) as sensitizer and hole conductor in thin film and nanocrystalline TiO$_2$ has been explored for solar cell applications [447,448]. A novel poly(3-nonylthiophene)-TiO$_2$ hybrid material has been recently described in which the presence of the silane groups of the copolymer **254** prevents macroscopic aggregation of TiO$_2$ and improves the compatibility of the two components [449]. Processable PT **255** with thermocleavable side groups, for the purpose of anchoring and creating a better electronic contact between a conjugated polymer and TiO$_2$, has been recently described [447]. The facile thermal cleavage of esters derived from a carboxylic acid and a tertiary alcohol, at temperatures around 190–210°C, led to **256** that has been used with P3HT in a solar cell consisting of FTO/TiO$_2$/**256**/P3HT/Ag (FTO = fluorine-doped tin oxide). The presence of **256** in the cell leads to a 50% increase in PCE to reach a value of 1.10% under 39 mW/cm^2 514 nm illumination. This result stems from higher chromophore density and higher light absorption of the layer of **256** and probably from the chelation of –COOH groups to TiO$_2$, which contributes to improve forward interfacial electron transfer.

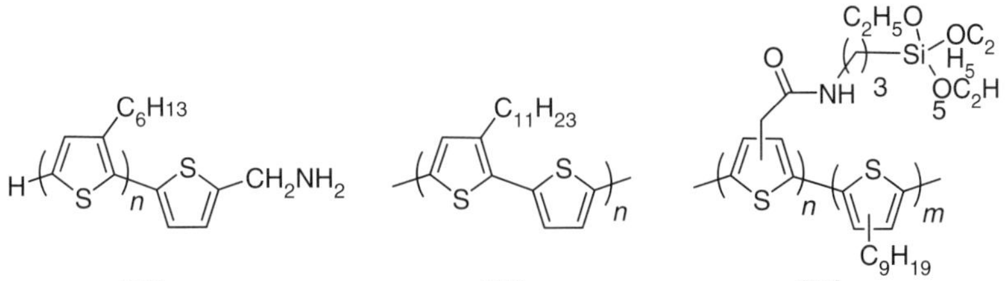

13.5 Conclusion

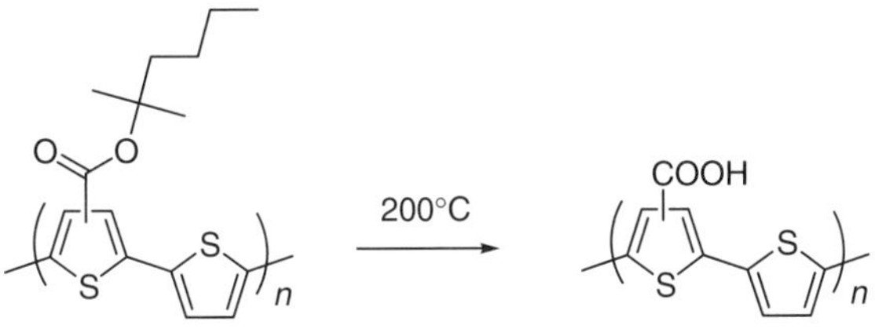

During the last two decades, π-conjugated molecules, oligomers, and polymers based on thiophenic building blocks have progressively acquired a prominent position among the vast family of conjugated structures.

In addition to the unique combination of stability and moderate bandgap of PT, this situation is largely due to the result of the structural versatility of thiophene-based systems structures, which has given rise to a considerable amount of synthetic work focused on the development of a growing number of applications.

The field of electrode materials has been one of the earliest areas of applications of PTs. Initially envisioned in the perspective of energy storage, the field has been progressively reoriented toward applications such as (bio)electrochemical sensors and electrochromic devices. Recently, the field of PT-based sensors has grown considerably and progressively evolved from purely electrochemical systems to optical sensors based on absorption or luminescence processes.

Soft or flexible (opto)electronics is one of the most important field of application of PTs with an enormous industrial potential for the future years. Oligomers and molecular materials have been widely used because of their well-defined chemical structure, relatively easy purification, and optimal molecular order especially for single crystals. While these materials have strongly contributed to the development of basic research, industrial production by means of printing technique imposes drastic prerequisites in terms of cost effectivness, processability, film-forming, and mechanical properties. It is likely that such a context will generate an increasing demand for polymeric materials combining processability, stability, and tailored mechanical properties. On the other hand, the development of active materials for soft electronic and optoelectronic devices, such as field-effect transistors, light-emitting devices, or solar cells, implies a continuation of effort of synthetic chemistry focused on the tuning of the electronic properties of π-conjugated systems. In this context, recent work has shown that dichalcogeno thiophenes and, in particular, EDOT represent major building blocks for the future chemistry of functional conjugated systems.

References

1. Shirakawa, H., E.J. Louis, A.G. MacDiarmid, C.K. Chiang, and A.J. Heeger. 1977. Synthesis of electrically conducting organic polymers: Halogen derivatives of polyacetylene, $(CH)_x$. *J Chem Soc Chem Commun* 578–580.

2. Skotheim, T.A., R.L. Elsenbaumer, and J.R. Reynolds (eds.). 1998. *Handbook of conducting polymers*, 2nd ed. New York: Marcel Dekker.
3. Kiebooms, R., R. Menon, and K. Lee. 2001. Synthesis, electrical, and optical properties of conjugated polymers, In *Handbook of advanced electronic and photonic materials and devices*, Vol. 8, ed. H.S. Nalwa. San Diego: Academic Press, pp. 1–102.
4. Pron, A., and P. Rannou. 2002. Processible conjugated polymers: From organic semiconductors to organic metals and superconductors. *Prog Polym Sci* 27:135–190.
5. Burroughes, J.H., D.D.C. Bradley, A.R. Brown, R.N. Marks, K. MacKay, R.H. Friend, P.L. Burn, and A.B. Holmes. 1990. Light-emitting diodes based on conjugated polymers. *Nature* 347:539–541.
6. Tourillon, G. 1986. Polythiophenes and its derivatives, In *Handbook of conducting polymers*, ed. T.A. Skotheim. New York: Marcel Dekker, p. 294.
7. Patil, A.O., A.J. Heeger, and F. Wudl. 1988. Optical properties of conducting polymers. *Chem Rev* 88:183–200.
8. Roncali, J. 1992. Conjugated poly(thiophenes): Synthesis, functionalization, and applications. *Chem Rev* 92:711–738.
9. Schopf, G., and G. Kossmehl. 1997. *Polythiophenes-electrically conductive polymers*. New York: Springer-Verlag.
10. Roncali, J. 1998. Advances in the molecular design of functional conjugated polymers, In *Handbook of conducting polymers*, eds. T.A. Skotheim, R.L. Elsenbaumer, and J.R. Reynolds. New York: Marcel Dekker, pp. 311–341.
11. Fichou, D. (ed.). 1999. *Handbook of oligo- and polythiophenes*. Weinhem: Wiley-VCH.
12. Barbarella, G., M. Melucci, and G. Sotgiu. 2005. The versatile thiophene: An overview of recent research on thiophene-based materials. *Adv Mater* 17:1581–1593.
13. Roncali, J. 1997. Synthetic principle for bandgap control in linear π-conjugated systems. *Chem Rev* 97:173–205.
14. Peierls, R.E. 1956. *Quantum theory of solids*. London: Oxford University Press.
15. Kertesz, M., C.H. Choi, and S. Yang. 2005. Conjugated polymers and aromaticity. *Chem Rev* 105:3448–3481.
16. Brédas, J.L. 1985. Relationship between band gap and bond length alternation in organic conjugated polymers. *J Chem Phys* 82:3808–3811.
17. Brédas, J.L., G.B. Street, B. Thémans, and J.M. André. 1985. Organic polymers based on aromatic rings (polyparaphenylene, polypyrrole, polythiophene): Evolution of the electronic properties as a function of the torsion angle between adjacent rings. *J Chem Phys* 83:1323–1329.
18. Wennerström, O. 1985. Qualitative evaluation of the band gap in polymers with extended π systems. *Macromolecules* 18:1977–1980.
19. Kertesz, M., and Y.-S. Lee. 1987. Energy gap and bond length alternation in heterosubstituted narrow gap semiconducting polymers. *J Phys Chem* 91:2690–2692.
20. Wudl, F., M. Kobayashi, and A.J. Heeger. 1984. Polyisothianaphthene. *J Org Chem* 49:3382–3384.
21. Pomerantz, M., B. Chaloner-ill, L.O. Harding, J.J. Tseng, and W.J. Pomerantz. 1992. Poly(2,3-dihexylthieno[3,4-b]pyrazine). A new processable low band gap polyheterocycle. *J Chem Soc Chem Commun* 1672–1673.
22. Kastner, J., H. Kuzmany, D. Vegh, M. Landl, L. Cuff, and M. Kertesz. 1995. Raman spectra of poly(2,3-*R,R*-thieno[3,4-b]pyrazine). A new low-band-gap polymer. *Macromolecules* 28:2922–2929.
23. Kitamura, C., S. Tanaka, and Y. Yamashita. 1994. Synthesis of new narrow bandgap polymers based on 5,7-di(2-thienyl)thieno[3,4-b]pyrazine and its derivatives. *J Chem Soc Chem Commun* 1585–1586.
24. Neef, C.J., I.D. Brotherston, and J.P. Ferraris. 1999. Synthesis and electronic properties of poly(2-phenylthieno[3,4-b]thiophene): A new low band gap polymer. *Chem Mater* 11:1957–1958.
25. Pomerantz, M., X.M. Gu, and S.X. Zhang. 2001. Poly(2-decylthieno[3,4-b]thiophene-4,6-diyl). A new low band gap conducting polymer. *Macromolecules* 34:1817–1822.
26. Lee, K., and G.A. Sotzing. 2001. Poly(thieno[3,4-b]thiophene). A new stable low band gap conducting polymer. *Macromolecules* 34:5746–5747.

27. Sotzing, G.A., and K.H. Lee. 2002. Poly(thieno[3,4-b]thiophene): A p- and n-dopable polythiophene exhibiting high optical transparency in the semiconducting state. *Macromolecules* 35:7281–7286.

28. Jenekhe, S.A. 1986. A class of narrow-band-gap semiconducting polymers. *Nature* 322:345–347.

29. Chen, W.C., and S.A. Jenekhe. 1995. Small-bandgap conducting polymers based on conjugated poly(heteroarylene methines). 2. Synthesis, structure, and properties. *Macromolecules* 28:465–480.

30. Havinga, E.E., W. ten Hoeve, and H. Wynberg. 1992. A new class of small band gap organic polymeric conductors. *Polym Bull* 29:119–126.

31. Havinga, E.E., W. ten Hoeve, and H. Wynberg. 1993. Alternate donor–acceptor small band-gap semiconducting polymers: Polysquaraines and polycroconaines. *Synth Met* 55–57:299–306.

32. Lambert, T.M., and J.P. Ferraris. 1991. Narrow band gap polymers: Polycyclopenta[2,1-b:3,4-b′]dithiophene-4-one. *J Chem Soc Chem Commun* 752–753.

33. Ferraris, J.P., and T.M. Lambert. 1991. Narrow bandgap polymers: Poly-4-dicyanomethylene-4H-cyclopenta[2,1-b:3,4-b′]dithiophene (PCDM). *J Chem Soc Chem Commun* 1268–1269.

34. Ho, H.A., H. Brisset, P. Frère, and J. Roncali. 1995. Electrogenerated small bandgap π-conjugated polymers derived from substituted dithienylethylenes. *J Chem Soc Chem Commun* 2309–2310.

35. Ho, H.A., H. Brisset, E.H. Elandaloussi, P. Frère, and J. Roncali. 1996. Thiophene-based conjugated oligomers and polymers with high electron affinity. *Adv Mater* 8:990–994.

36. Seshadri, V., and G.A. Sotzing. 2004. Polymerization of two unsymmetrical isomeric monomers based on thieno[3,4-b]thiophene containing cyanovinylene spacers. *Chem Mater* 16:5644–5649.

37. Jen, K.Y., M.R. Maxfield, L.W. Shacklette, and R.L. Elsenbaumer. 1987. Poly(thienylenevinylenes). *J Chem Soc Chem Commun* 309–310.

38. Yamada, S., S. Tokito, T. Tsutsui, and A. Saito. 1987. New conducting polymer film: Poly(2,5-thienylenevinylene) prepared via a soluble precursor polymer. *J Chem Soc Chem Commun* 1448–1449.

39. Scherf, U., and K. Müllen. 1991. Polyarylenes and poly(arylenevinylenes). A soluble ladder polymer via bridging of functionalized poly(p-phenylene)-precursors. *Makromol Chem Rapid Commun* 12:489–497.

40. Brisset, H., C. Thobie-Gautier, M. Jubault, A. Gorgues, and J. Roncali. 1994. Novel narrow bandgap polymers from sp^3 carbon-bridged bithienyls: Poly(4,4-ethylenedioxy-4H-cyclopenta[2,1-b:3,4-b′]dithiophene). *J Chem Soc Chem Commun* 1305–1306.

41. Roncali, J., and C. Thobie-Gautier. 1994. An efficient strategy towards small bandgap polymers: The rigidification of the π-conjugated system. *Adv Mater* 6:846–848.

42. Roncali, J., C. Thobie-Gautier, E.H. Elandaloussi, and P. Frère. 1994. Control of the bandgap of conducting polymers by rigidification of the π-conjugated system. *J Chem Soc Chem Commun* 2249–2250.

43. Brisset, H., P. Blanchard, B. Illien, A. Riou, and J. Roncali. 1997. Bandgap control through reduction of bond length alternation in briged poly(dithienylethylene)s. *Chem Commun* 569–570.

44. Blanchard, P., H. Brisset, B. Illien, A. Riou, and J. Roncali. 1997. Bridged dithienylethylenes as precursors of small bandgap electrogenerated conjugated polymers. *J Org Chem* 62:2401–2408.

45. Blanchard, P., A. Riou, and J. Roncali. 1998. Electrosynthesis of a low band gap π-conjugated polymer from a multibridged dithienylethylene. *J Org Chem* 63:7107–7110.

46. Blanchard, P., H. Brisset, A. Riou, R. Hierle, and J. Roncali. 1998. Bridged 1,6-dithienylhexa-1,3,5-trienes as highly photoluminescent and stable thiophene-based π-conjugated systems. *J Org Chem* 63:8310–8319.

47. Blanchard, P., P. Verlhac, L. Michaux, P. Frère, and J. Roncali. 2006. Fine tuning of the electronic properties of linear π-conjugated oligomers by covalent bridging. *Chem Eur J* 12:1244–1255.

48. Coppo, P., and M.L. Turner. 2005. Cyclopentadithiophene based electroactive materials. *J Mater Chem* 15:1123–1133.

49. Yamaguchi, S., and K. Tamao. 1998. Silole-containing σ- and π-conjugated compounds. *J Chem Soc Dalton* 3693–3702.

50. Lee, Y., S. Sadki, B. Tsuie, and J.R. Reynolds. 2001. A new narrow band gap electroactive polymer: Poly[2,5-bis{2-(3,4-ethylenedioxy)thienyl}silole]. *Chem Mater* 13:2234–2236.
51. Hay, C., C. Fischmeister, M. Hissler, L. Toupet, and R. Réau. 2000. Electropolymerization of π-conjugated oligomers containing phosphole cores and terminal thienyl moieties: Optical and electronic properties. *Angew Chem Int Ed Engl* 345:1812–1815.
52. Hay, C., M. Hissler, C. Fischmeister, J. Rault Berthelot, L. Toupet, L. Nyulaszi, and R. Réau. 2001. Phosphole-containing π-conjugated systems: From model molecules to polymer films on electrodes. *Chem Eur J* 7:4222–4236.
53. Fave, C., M. Hissler, T. Karpati, J. Rault Berthelot, V. Deborde, L. Toupet, L. Nyulaszi, and R. Réau. 2004. Connecting π-chromophores by σ-P-P-bonds: New types of assemblies exhibiting σ–π-conjugation. *J Am Chem Soc* 126:6058–6063.
54. Barbarella, G., O. Pudova, C. Arbizzani, M. Mastragostino, and A. Bongini. 1998. Oligothiophene-S,S-dioxides: A new class of thiophene-based materials. *J Org Chem* 63:1742–1745.
55. Barbarella, G., L. Favaretto, G. Sotgiu, M. Zambianchi, L. Antolini, O. Pudova, and A. Bongini. 1998. Oligothiophene S,S-dioxides. Synthesis and electronic properties in relation to the parent oligothiophenes. *J Org Chem* 63:5497–5506.
56. Barbarella, G., L. Favaretto, M. Zambianchi, O. Pudova, C. Arbizzani, A. Bongini, and M. Mastragostino. 1998. From easily oxidized to easily reduced thiophene-based materials. *Adv Mater* 10:551–554.
57. Barbarella, G., L. Favaretto, G. Sotgiu, M. Zambianchi, A. Bongini, C. Arbizzani, M. Mastragostino, M. Anni, G. Gigli, and R. Cingolani. 2000. Tuning solid-state photoluminescence frequencies and efficiencies of oligomers containing one central thiophene-S,S-dioxide unit. *J Am Chem Soc* 122:11971–11978.
58. Groenendaal, L.B., J. Friedrich, D. Freitag, H. Pielartzik, and J.R. Reynolds. 2000. Poly(3,4-ethylenedioxythiophene) and its derivatives: Past, present and future. *Adv Mater* 12 (7):481–494.
59. Kirchmeyer, S., and K. Reuter. 2005. Scientific importance, properties and growing applications of poly(3,4-ethylenedioxythiophene). *J Mater Chem* 15:2077–2088.
60. Roncali, J., P. Blanchard, and P. Frère. 2005. 3,4-Ethylenedioxythiophene (EDOT) as a versatile building block for avdanced functional π-conjugated systems. *J Mater Chem* 15:1589–1610.
61. Chang, A.-C., R.L. Blankespoor, and L.L. Miller. 1987. Characterization and spectroelectrochemical studies of soluble polymerized 3-methoxythiophene. *J Electroanal Chem* 236:239–252.
62. Sheina, E.E., S.M. Khersonsky, E.G. Jones, and R.D. McCullough. 2005. Highly conductive, regioregular alkoxy-functionalized polythiophenes: A new class of stable, low band gap materials. *Chem Mater* 17:3317–3319.
63. Koeckelberghs, G., M. Vangheluwe, C. Samyn, A. Persoons, and T. Verbiest. 2005. Regioregular poly(3-alkoxythiophene)s: Toward soluble, chiral conjugated polymers with a stable oxidized state. *Macromolecules* 38:5554–5559.
64. Tschuncky, P., and J. Heinze. 1993. Voltammetric studies on the electropolymerization mechanism of methoxythiophenes. *Synth Met* 55 (2–3):1603–1607.
65. Leriche, P., P. Frère, and J. Roncali. 2005. Structure–reactivity relationships in bithiophenic precursors based on the 3-phenoxythiophene building block. *J Mater Chem* 15:3473–3478.
66. Dietrich, M., and J. Heinze. 1991. Poly(4,4′-dimethoxybithiophene)—A new conducting polymer with extraordinary redox and optical properties. *Synth Met* 41 (1–2):505–506.
67. Gross, M., D.C. Muller, H.G. Nothofer, U. Scherf, D. Neher, C. Brauchle, and K. Meerholz. 2000. Improving the performance of doped π-conjugated polymers for use in organic light-emitting diodes. *Nature* 405:661–665.
68. Casalbore-Miceli, G., N. Carmaioni, M.C. Galazzi, L. Albertin, A.M. Fichera, A. Geri, and E.M. Girotto. 2002. Photoelectrical properties in poly(alky/alkoxy-terthiophenes): Dependance of the alkyl-chain length. *Synth Met* 125:307–311.
69. Zotti, G., R. Salmaso, M.C. Gallazzi, and R.A. Marin. 1997. In situ conductivity of a polythiophene from a branched alkoxy-substituted tetrathiophene. Enhancement of conductivity by conjugated cross-linking of polymers chains. *Chem Mater* 9:791–795.

70. Gallazi, M.C., F. Toscano, D. Paganuzzi, C. Bertarelli, A. Farina, and G. Zotti. 2001. Polythiophenes with unusual electrical and optical properties based on donor acceptor alternance strategy. *Macromol Chem Phys* 202:2074–2085.

71. Smie, A., A. Synowczyk, J. Heinze, R. Alle, P. Tschuncky, G. Gotz, and P. Bäuerle. 1998. Beta,beta-disubstituted oligothiophenes, a new oligomeric approach towards the synthesis of conducting polymers. *J Electroanal Chem* 452:87–95.

72. Iarossi, D., A. Mucci, L. Schenetti, R. Seeber, F. Goldoni, M. Affronte, and F. Nava. 1999. Polymerisation and characterisation of 4,4′-bis(alkylsulfanyl)-2,2′-bithiophenes. *Macromolecules* 32:1390–1397.

73. Bäuerle, P., G. Götz, A. Synowczyk, and J. Heinze. 1996. Synthesis and properties of a series of methylthio oligothiophenes. *Liebigs Ann* 279–284.

74. Turbiez, M., P. Frère, P. Leriche, N. Mercier, and J. Roncali. 2005. Poly(3,6-dimethoxy-thieno[3,2-b]thiophene): A possible alternative to poly(3,4-ethylenedioxythiophene) (PEDOT). *Chem Commun* 1161–1163.

75. Daoust, G., and M. Leclerc. 1991. Structure–property relationships in alkoxy-substituted polythiophenes. *Macromolecules* 24:455–459.

76. Szkurlat, A., B. Palys, J. Mieczkowski, and M. Skompska. 2003. Electrosynthesis and spectroelectrochemical characterization of poly(3,4-dimethoxy-thiophene), poly(3,4-dipropyloxythiophene) and poly(3,4-dioctyloxythiophene) films. *Electrochim Acta* 48:3665–3676.

77. Matthews, J.R., F. Goldoni, A.P.H.J. Schenning, and E.W. Meijer. 2005. Non-ionic polythiophenes: A non-aggregating folded structure in water. *Chem Commun* 5303–5505.

78. Blanchard, P., A. Cappon, E. Levillain, Y. Nicolas, P. Frere, and J. Roncali. 2002. Thieno[3,4-b]-1,4-oxathiane: An unsymmetrical sulfur analogue of 3,4-ethylenedioxythiophene (EDOT) as a building block for linear pi-conjugated systems. *Org Lett* 4:607–609.

79. Wang, C., J.L. Schindler, C.R. Kannewurf, and M.G. Kanatzidis. 1995. Poly(3,4-ethylenedithiathiophene). A new soluble conducting polythiophene derivative. *Chem Mater* 7:58–68.

80. Turbiez, M., P. Frere, M. Allain, N. Gallego Planas, and J. Roncali. 2005. Effect of structural factor on the electropolymerization of bithiophenic precursors containing a 3,4-ethylenedisulfanylthiophene unit. *Macromolecules* 38:6806–6812.

81. Raimundo, J.M., P. Blanchard, P. Frere, N. Mercier, I. Ledoux Rak, R. Hierle, and J. Roncali. 2001. Push–pull chromophores based on 2,2′-bi(3,4-ethylenedioxythiophene) (BEDOT) π-conjugating spacer. *Tetrahedron Lett* 42:1507–1510.

82. Spencer, H.J., P.J. Skabara, M. Giles, I. McCullough, S.J. Coles, and M.B. Hursthouse. 2005. The first direct experimental comparison between the hugely contrasting properties of PEDOT and the all sulfur analogues PEDTT by analogy with well-defined EDTT–EDOT copolymers. *J Mater Chem* 15:4783–4792.

83. Apperloo, J.J., L. Groenendaal, H. Verheyen, M. Jayakannan, R.A.J. Janssen, A. Dkhissi, D. Beljonne, R. Lazzaroni, and J.L. Brédas. 2002. Optical and redox properties of a series of 3,4-ethylenedioxythiophene oligomers. *Chem Eur J* 8:2384–2396.

84. Turbiez, M., P. Frère, and J. Roncali. 2003. Stable and soluble oligo-(3,4-ethylenedioxythiophene)s (EDOT) end-capped with alkyl chains. *J Org Chem* 68:5357–5360.

85. Turbiez, M., P. Frère, M. Allain, C. Videlot, J. Ackermann, and J. Roncali. 2005. Design of organic semiconductors: Tuning the electronic properties of pi-conjugated oligothiophenes with the 3,4-ethylenedioxythiophene (EDOT) building block. *Chem Eur J* 11:3742–3752.

86. Akoudad, S., and J. Roncali. 1998. Electrochemical synthesis of poly(3,4-ethylenedioxythiophene) from a dimer precursor. *Synth Met* 93:111–114.

87. Sotzing, G.A., and J.R. Reynolds. 1995. Poly[trans-bis(3,4-ethylenedioxythiophene)vinylene]: A low band gap polymer with rapid redox switching capabilities between conducting transmissive and insulating absorptive states. *J Chem Soc Chem Commun* 703–704.

88. Fu, Y., and R.L. Elsenbaumer. 1995. Structure property relationships in conjugated polymers: Oxidation potentials, reduction potentials, and band gaps for a series of regiospecifically

methoxysubstituted thiophene vinylene and poly(bithiophene vinylene). *Proc Am Chem Soc Div Polym Mater Sci Eng* 72:315–316.

89. Fu, Y., H. Cheng, and R.L. Elsenbaumer. 1997. Electron-rich thienylene–vinylene low band gap polymers. *Chem Mater* 9:1720–1724.

90. Iwatsuki, S., M. Kubo, and Y. Itoh. 1993. Poly(3,4-dibutoxythienylenevinylenes). *Chem Lett* 1085–1086.

91. Goldoni, F., R.A.J. Janssen, and E.W. Meijer. 1999. Synthesis and characterization of new copolymers of thiophene and vinylene: Poly(thienylenevinylene)s and poly(terthienylenevinylene)s with thioether side chains. *J Polym Sci A* 37:4629–4639.

92. Sotzing, G.A., C.A. Thomas, and J.R. Reynolds. 1998. Low band gap cyanovinylene polymers based on ethylenedioxythiophene. *Macromolecules* 31:3750–3752.

93. Thomas, C.A., K.W. Zong, K.A. Abboud, P.J. Steel, and J.R. Reynolds. 2004. Donor-mediated band gap reduction in a homologous series of conjugated polymers. *J Am Chem Soc* 126:16440–16450.

94. Thompson, B.C., Y.-G. Kim, and J.R. Reynolds. 2005. Spectral broadening in MEH-PPV: PCBM-based photovoltaic devices via blending with a narrow band gap cyanovinylene–dioxythiophene polymer. *Macromolecules* 38:5359–5362.

95. Benincori, T., S. Rizzo, F. Sannicolò, G. Schiavon, S. Zecchin, and G. Zotti. 2003. An electrochemically prepared small bandgap poly(biheteroarylidenemethine): Poly{bi[(3,4-ethylenedioxy)thienylene]methine}. *Macromolecules* 36:5114–5118.

96. Zaman, B.M., and D.F. Perepichka. 2005. A new simple synthesis of poly(thiophene-methine)s. *Chem Commun* 4187–4189.

97. Casado, J., G. Zotti, A. Berlin, V. Hernandez, R.P. Ortiz, and J.T.L. Navarrete. 2005. Combined theoretical and spectroscopic Raman study of 3,4-ethylenedioxy and *S,S*-dioxide substituted terthiophenes and their parent polymers. *J Mol Struct* 744:551–556.

98. Melucci, M., P. Frère, E. Levillain, G. Barbarella, and J. Roncali. Forthcoming.

99. Sonmez, G., H. Meng, and F. Wudl. 2003. Very stable low band gap polymer for charge storage purposes and near-infrared applications. *Chem Mater* 15:4923–4929.

100. Berlin, A., G. Zotti, S. Zecchin, G. Schiavon, B. Vercelli, and A. Zanelli. 2004. New low-gap polymers from 3,4-ethylenedioxythiophene-bis-substituted electron-poor thiophenes. The roles of thiophene, donor–acceptor alternation, and copolymerization in intrinsic conductivity. *Chem Mater* 16:3667–3676.

101. Rault Berthelot, J., E. Raoult, and F. Le Floch. 2003. Synthesis and anodic oxidation of a dimer EDOT-dicyanomethylenefluorene and a trimer EDOT-dicyanomethylenefluorene-EDOT: Towards mixed polymers with very low bandgap. *J Electroanal Chem* 546:29–34.

102. Casado, J., R.P. Ortiz, M.C.R. Delgado, V. Hernandez, J.T.L. Navarrete, J.M. Raimundo, P. Blanchard, M. Allain, and J. Roncali. 2005. Alternated quinoid/aromatic units in terthiophenes building blocks for electroactive narrow band gap polymers. Extended spectroscopic, solid state, electrochemical, and theoretical study. *J Phys Chem B* 109:16616–16627.

103. Akoudad, S., and J. Roncali. 1998. Electrogenerated poly(thiophenes) with extremely narrow bandgap and high stability under n-doping cycling. *Chem Commun* 2081–2082.

104. Perepichka, I.F., E. Levillain, and J. Roncali. 2004. Effect of substitution of 3,4-ethylenedioxythiophene (EDOT) on the electronic properties of the derived electrogenerated low band gap conjugated polymers. *J Mater Chem* 14:1679–1681.

105. Gogte, V.N., L.G. Shah, B.D. Tilak, K.N. Gadekar, and M.B. Sahasrabudhe. 1967. Synthesis of potential anticancer agents I: Synthesis of substituted thiophenes. *Tetrahedron* 23:2437–2441.

106. Fager, E.W. 1945. Some derivatives of 3,4-dioxythiophene. *J Am Chem Soc* 67:2217–2218.

107. Dietrich, M., J. Heinze, G. Heywang, and F. Jonas. 1994. Electrochemical and spectroscopic characterization of polyalkylenedioxythiophene. *J Electroanal Chem* 369:87–92.

108. Heywang, G., and F. Jonas. 1992. Poly(alkylenedioxythiophene)s-new, very stable conducting polymers. *Adv Mater* 4:116–118.

109. Kumar, A., D.M. Welsh, M.C. Morvant, F. Piroux, K.A. Abboud, and J.R. Reynolds. 1998. Conducting poly(3,4-alkylenedioxythiophene) derivatives as fast electrochromics with high-contrast ratio. *Chem Mater* 10:896–902.

110. Sankaran, B., and J.R. Reynolds. 1997. High-contrast electrochromic polymers from alkyl-derivatized poly(3,4-ethylenedioxythiophenes). *Macromolecules* 30:2582–2588.

111. Sapp, S.A., G.A. Sotzing, and J.R. Reynolds. 1998. High contrast ratio and fast-switching dual polymer electrochromic devices. *Chem Mater* 10:2101–2108.

112. Zotti, G., S. Zecchin, G. Schiavon, B. Vercelli, A. Berlin, and E. Dalcanale. 2003. Potential-driven conductivity of polypyrroles, poly-*N*-alkylpyrroles, and polythiophenes: Role of the pyrrole NH moiety in the doping-charge dependence of conductivity. *Chem Mater* 15:4642–4650.

113. Ng, S.C., H.S.O. Chan, and W.-L. Yu. 1997. Synthesis and characterization of electrically conducting copolymers of ethylenedioxythiophene and 1,3-propylenedioxythiophene with ϖ-functional substituents. *J Mater Sci Lett* 16:809–811.

114. Lima, A., P. Schottland, S. Sadki, and C. Chevrot. 1998. Electropolymerization of 3,4-ethylenedioxythiophene and 3,4-ethylenedioxythiophene methanol in the presence of dodecylbenzenesulfonate. *Synth Met* 93:33–41.

115. Akoudad, S., and J. Roncali. 2000. Modification of the electrochemical and electronic properties of electrogenerated poly(3,4-ethylenedioxythiophene) by hydroxymethyl and oligo(oxyethylene)substituents. *Electrochem Commun* 2 (1):72–76.

116. Welsh, D.M., A. Kumar, E.W. Meijer, and J.R. Reynolds. 1999. Enhanced contrast ratio and rapid switching in electrochromics based on poly(3,4-propylenedioxythiophene) derivatives. *Adv Mater* 11:1379–1382.

117. von Kieseritzky, F., F. Allared, E. Dahstedt, and J. Hellberg. 2004. Simple one-step synthesis of 3,4-dimethoxythiophene and its conversion into 3,4-ethylenedioxythiophene (EDOT). *Tetrahedron Lett* 45:6049–6050.

118. Reeves, B.D., C.R.G. Grenier, A.A. Argun, A. Cirpan, T.D. McCarley, and J.R. Reynolds. 2004. Spray coatable electrochromic dioxythiophene polymers with high coloration efficiencies. *Macromolecules* 37:7559–7569.

119. Krishnamoorthy, K., A.V. Ambade, M. Kanungo, A.Q. Contractor, and A. Kumar. 2001. Rational design of an electrochromic polymer with high contrast in the visible region: Dibenzyl substituted poly(3,4-propylenedioxythiophene). *J Mater Chem* 11:2909–2911.

120. Gaupp, C.L., D.M. Welsh, and J.R. Reynolds. 2002. Poly(ProDOT-Et-2): A high-contrast, high-coloration efficiency electrochromic polymer. *Macromol Rapid Commun* 23:885–889.

121. Welsh, D.M., L.J. Kloeppner, L. Madrigal, M.R. Pinto, B.C. Thompson, K.S. Schanze, K.A. Abboud, D. Powell, and J.R. Reynolds. 2002. Regiosymmetric dibutyl-substituted poly(3,4-propylenedioxythiophene)s as highly electron-rich electroactive and luminescent polymers. *Macromolecules* 35:6517–6525.

122. Cirpan, A., A.A. Argun, C.R.G. Grenier, B.D. Reeves, and J.R. Reynolds. 2003. Electrochromic devices based on soluble and processable dioxythiophene polymers. *J Mater Chem* 13:2422–2428.

123. Mishra, S.P., R. Sahoo, A.V. Ambade, A.Q. Contractor, and A. Kumar. 2004. Synthesis and characterization of functionalized 3,4-propylenedioxythiophene and its derivatives. *J Mater Chem* 14:1896–1900.

124. Reeves, B.D., B.C. Thompson, K.A. Abboud, B.E. Smart, and J.R. Reynolds. 2002. Dual cathodically and anodically coloring electrochromic polymer based on a spiro bipropylenedioxythiophene [poly(spiroBiProDOT)]. *Adv Mater* 14:717–719.

125. Caras-Quintero, D., and P. Bäuerle. 2004. Synthesis of the first enantiomerically pure and chiral, disubstituted 3,4-ethylenedioxythiophenes (EDOTs) and corresponding stereo- and regioregular PEDOTs. *Chem Commun* 926–927.

126. Caras-Quintero, D., and P. Bäuerle. 2002. Efficient synthesis of 3,4-ethylenedioxythiophenes (EDOT) by Mitsunobu reaction. *Chem Commun* 2690–2691.

127. Zong, K.W., L. Madrigal, L. Groenendaal, and J.R. Reynolds. 2002. 3,4-Alkylenedioxy ring formation via double Mitsunobu reactions: An efficient route for the synthesis of 3,4-ethylenedioxythiophene (EDOT) and 3,4-propylenedioxythiophene (ProDOT) derivatives as monomers for electron-rich conducting polymers. *Chem Commun* 2498–2499.

128. Roquet, S., P. Leriche, I. Perepichka, B. Jousselme, E. Levillain, P. Frère, and J. Roncali. 2004. 3,4-Phenylenedioxythiophene (PheDOT): A novel platform for the synthesis of planar substituted π-donor conjugated systems. *J Mater Chem* 14:1396–1400.

129. Storsberg, J., D. Schollmeyer, and H. Ritter. 2003. Route towards new heteroaromatic benzo[1,4]-dioxine derivatives. *Chem Lett* 32:140–141.

130. Perepichka, I.F., S. Roquet, P. Leriche, J.M. Raimundo, P. Frère, and J. Roncali. 2006. Electronic properties and reactivity of short-chain oligomers of 3,4-phenylenedioxythiophene (PheDOT). *Chem Eur J* 12:2960–2966.

131. Grubbs, R.H. 2004. Olefins metathesis. *Tetrahedron* 60:7117–7140.

132. Leriche, P., P. Blanchard, P. Frère, E. Levillain, G. Mabon, and J. Roncali 2006. 3,4-Vinylenedioxythiophene (VDOT): A new building block for thiophene-based π-conjugated systems. *Chem Commun* 275-277.

133. Aqad, E., M.V. Lakshmikantham, and M.P. Cava. 2001. Synthesis of 3,4-ethylenedioxyselenophene (EDOS): A novel building block for electron-rich π-conjugated polymers. *Org Lett* 3:4283–4285.

134. Takeshi, H. 2002. Method for producing 3,4-alkylenedioxyfuran, method for producing polymer thereof and photothermographic image-forming material using the same polymer. *Japanese Patent 2002–241383.*

135. Merz, A., R. Schropp, and E. Dötterl. 1995. 3,4-Dialkoxypyrroles and 2,3,7,8,12,13,17,18-octaalkyloxyporphyrins. *Synthesis* 7:795–800.

136. Thomas, C.A., K. Zong, P. Schottland, and J.R. Reynolds. 2000. Poly(3,4-alkylenedioxypyrrole)s as highly stable aqueous-compatible conducting polymers with biomedical implications. *Adv Mater* 12:222–225.

137. Schottland, P., K. Zong, C.L. Gaupp, B.C. Thompson, C.A. Thomas, I. Giurgiu, R. Hickman, K.A. Abboud, and J.R. Reynolds. 2000. Poly(3,4-alkylenedioxypyrrole)s: Highly stable electronically conducting and electrochromic polymers. *Macromolecules* 33:7051–7061.

138. Xiao, Y.H., X.Y. Cui, J.M. Hancock, M.B. Bouguettaya, J.R. Reynolds, and D.C. Martin. 2004. Electrochemical polymerization of poly(hydroxymethylated-3,4-ethylenedioxythiophene) (PEDOT-MeOH) on multichannel neural probes. *Sens Actuators B* 99:437–443.

139. Kros, A., and R.J.M. Nolte. 2002. Poly(3,4-ethylenedioxythiophene)-based copolymers for biosensor applications. *J Polym Sci A* 40:738–747.

140. Stéphan, O., P. Schottland, P.-Y. Le Gall, C. Chevrot, C. Mariet, and M. Carrier. 1998. Electrochemical behaviour of 3,4-ethylenedioxythiophene functionalized by a sulfonate group. Application to the preparation of poly(3,4-ethylenedioxythiophene) having permanent cation exchange properties. *J Electroanal Chem* 443:217–226.

141. Tran Van, F., M. Carrier, and C. Chevrot. 2004. Sulfonated polythiophene and poly(3,4-ethylenedioxythiophene) derivatives with cations exchange properties. *Synth Met* 142 (1–3):251–258.

142. Krishnamoorthy, K., M. Kanungo, A.V. Ambade, A.Q. Contractor, and A. Kumar. 2002. Electrochemically polymerized electroactive poly(3,4-ethylenedioxythiophene) containing covalently bound dopant ions: Poly{2-(3-sodiumsulfinopropyl)-2,3-dihydrothieno[3,4b][1,4]dioxin}. *Synth Met* 125:441.

143. Zotti, G., S. Zecchin, G. Schiavon, and L. Groenendaal. 2002. Electrochemical and chemical synthesis and characterization of sulfonated poly(3,4-ethylenedioxythiophene): A novel water-soluble and highly conductive conjugated oligomer. *Macromol Chem Phys* 203:1958–1964.

144. Cutler, C.A., M. Bouguettaya, and J.R. Reynolds. 2002. PEDOT polyelectrolyte based electrochromic films via electrostatic adsorption. *Adv Mater* 14:684–688.

145. Cutler, C.A., M. Bouguettaya, T.-S. Kang, and J.R. Reynolds. 2005. Alkoxysulfonate-functionalized PEDOT polyelectrolyte multilayer films: Electrochromic and hole transport materials. *Macromolecules* 38:3068–3074.

146. Perepichka, I.F., M. Besbes, E. Levillain, M. Salle, and J. Roncali. 2002. Hydrophilic oligo(oxyethylene)-derivatized poly(3,4-ethylenedioxythiophenes): Cation-responsive optoelectroelectrochemical properties and solid-state chromism. *Chem Mater* 14:449–457.

147. Schwendeman, I., C.L. Gaupp, J.M. Hancock, L. Groenendaal, and J.R. Reynolds. 2003. Perfluoroalkanoate-substituted PEDOT for electrochromic device applications. *Adv Funct Mater* 13 (7):541–547.

148. Krishnamoorthy, K., M. Kanungo, A.Q. Contractor, and A. Kumar. 2001. Electrochromic polymer based on a rigid cyanobiphenyl substituted 3,4-ethylenedioxythiophene. *Synth Met* 124:471–475.

149. Brisset, H., A.-E. Navarro, C. Moustrou, I.F. Perepichka, and J. Roncali. 2004. Electrogenerated conjugated polymers incorporating a ferrocene-derivatized-(3,4-ethylenedioxythiophene). *Electrochem Commun* 6:249–253.

150. Ko, H.C., M. Kang, B. Moon, and H. Lee. 2004. Enhancement of electrochromic contrast of poly(3,4-ethylenedioxythiophene) by incorporating a pendent viologen. *Adv Mater* 16 (19):1712–1716.

151. Besbes, M., G. Trippe, E. Levillain, M. Mazari, F. Le Derf, I.F. Perepichka, A. Derdour, A. Gorgues, M. Sallé, and J. Roncali. 2001. Rapid and efficient post-polymerization functionalization of poly(3,4-ethylenedioxythiophene) (PEDOT) derivatives on an electrode surface. *Adv Mater* 13:1249–1252.

152. Huchet, L., S. Akoudad, and J. Roncali. 1998. Electrosynthesis of highly electroactive tetrathiafulvalene-derivatized polythiophenes. *Adv Mater* 10 (7):541–545.

153. Huchet, L., S. Akoudad, E. Levillain, J. Roncali, A. Emge, and P. Bauerle. 1998. Spectroelectrochemistry of electrogenerated tetrathiafulvalene-derivatized poly(thiophenes): Toward a rational design of organic conductors with mixed conduction. *J Phys Chem B* 102:7776–7781.

154. Johansson, T., W. Mammo, M. Svensson, M.R. Andersson, and O. Inganäs. 2003. Electrochemical bandgaps of substituted polythiophenes. *J Mater Chem* 13:1316–1323.

155. Yu, J., and S. Holdcroft. 2001. Chemically amplified soft litography of a low band gap polymer. *Chem Commun* 1274–1275.

156. Yu, J., and S. Holdcroft. 2002. Synthesis, solid-phase reaction, and patterning of acid-labile 3,4-ethylenedioxythiophene-based conjugated polymers. *Chem Mater* 14:3705–3714.

157. Buvat, P., and P. Hourquebie. 1998. Synthesis and infrared properties of ethylenedioxythiophene and octylthiophene based copolymers. *J Chim Phys* 95:1180–1183.

158. Jousselme, B., P. Blanchard, E. Levillain, R. de Bettignies, and J. Roncali. 2003. Electrochemical synthesis of C_{60}-derivatized poly(thiophene)s from tailored precursors. *Macromolecules* 36:3020–3025.

159. Blanchard, P., B. Jousselme, P. Frère, and J. Roncali. 2002. 3- and 3,4-Bis(2-cyanoethylsulfanyl)thiophenes as building blocks for functionalized thiophene-based π-conjugated systems. *J Org Chem* 67:3961–3964.

160. Jousselme, B., P. Blanchard, and J. Roncali. 2004. Polymers based on bithiophene-type monomers, preparation process, and their applications. *French Patent FR2852320.*

161. Jousselme, B., P. Blanchard, M. Oçafrain, M. Allain, E. Levillain, and J. Roncali. 2004. Electrogenerated poly(thiophenes) derivatized by bipyridine ligands and metal complexes. *J Mater Chem* 14:421–427.

162. Roncali, J. 1999. Electrogenerated functional conjugated polymers as advanced electrode materials. *J Mater Chem* 9:1875–1893.

163. Goldenberg, L.M., M.R. Bryce, and M.C. Petty. 1999. Chemosensor devices: Voltammetric molecular recognition at solid interfaces. *J Mater Chem* 9:1957–1974.

164. Leclerc, M. 1999. Optical and electrochemical transducers based on functionalized conjugated polymers. *Adv Mater* 11:1491–1498.

165. McQuade, D.T., A.E. Pullen, and T.M. Swager. 2000. Conjugated polymer-based chemical sensors. *Chem Rev* 100:2537–2574.

166. Leclerc, M., and M. Faïd. 1997. Electrical and optical properties of processable polythiophene derivatives: Structure–property relathionships. *Adv Mater* 9:1087–1094.
167. Leclerc, M., and H.A. Ho. 2004. Affinitychromic polythiophenes: A novel bio-photonic tool for high-throughput screening and diagnostics. *Synlett* 380–387.
168. Ho, H.A., M. Béra-Abérel, and M. Leclerc. 2005. Optical sensors based on hybrid DNA/conjugated polymer complexes. *Chem Eur J* 11:1718–1724.
169. Lemaire, M., R. Garreau, J. Roncali, D. Delabouglise, H. Korri-Youssoufi, and F. Garnier. 1989. Design of poly(thiophene) containing oxyalkyl substituents. *New J Chem* 13:863–871.
170. Roncali, J., R. Garreau, D. Delabouglise, F. Garnier, and M. Lemaire. 1989. Modification of the structure and electrochemical properties of poly(thiophene) by ether groups. *J Chem Soc Chem Commun* 679–681.
171. Shi, L.H., F. Garnier, and J. Roncali. 1991. Electroactivity of poly(thiophenes) containing oxyalkyl substituents. *Synth Met* 41–43:547–550.
172. Roncali, J., L.H. Shi, and F. Garnier. 1991. Effects of environmental factors on the electroopical properties of conjugated polymers containing oligo(oxyethylene) substituents. *J Phys Chem* 95:8983–8989.
173. Shi, L.H., F. Garnier, and J. Roncali. 1991. Solid-state ionochromic and solvatochromic effects in substituted conjugated polymers. *Solid State Commun* 77:811–815.
174. Roncali, J., H.S. Li, R. Garreau, F. Garnier, and M. Lemaire. 1990. Tuning of the aqueous electroactivity of poly(3-alkylthiophenes) by ether groups. *Synth Met* 36:267–273.
175. Bäuerle, P., and S. Scheib. 1993. Molecular recognition of alkali-ions by crown-ether-functionalized poly(alkylthiophenes). *Adv Mater* 5:848–853.
176. Bäuerle, P., G. Götz, M. Hiller, S. Scheib, T. Fischer, U. Segelbacher, M. Bennati, A. Grupp, M. Mehring, M. Stoldt, C. Seidel, F. Geiger, H. Schweizer, E. Umbach, M. Schmelzer, S. Roth, H.J. Egelhaaf, D. Oelkrug, P. Emele, and H. Port. 1993. Design, synthesis and assembly of new thiophene-based molecular functional units with controlled properties. *Synth Met* 61:71–79.
177. Scheib, S., and P. Bäuerle. 1999. Synthesis and characterization of oligo- and crown ether-substituted polythiophenenes—A comparative study. *J Mater Chem* 9:2139–2150.
178. Bäuerle, P., and S. Scheib. 1995. Synthesis and characterization of thiophenes, oligothiophenes and polythiophenes with crown ether units in direct π-conjugation. *Acta Polym* 46:124–129.
179. Berlin, A., G. Zotti, S. Zecchin, and G. Schiavon. 2002. EQCM analysis of alkali metal ion coordination properties of a novel poly(thiophene)s 3,4-functionalized with crown-ether moieties. *Synth Met* 131:149–160.
180. Sannicolò, F., E. Brenna, T. Benincori, G. Zotti, S. Zecchin, G. Schiavon, and T. Pilati. 1998. Highly ordered poly(cyclopentabithiophenes) functionalized with crown-ether moities for lithium- and sodium-sensing electrodes. *Chem Mater* 10:2167–2176.
181. Roncali, J., R. Garreau, and M. Lemaire. 1990. Electrosynthesis of conducting poly-pseudo-crown ethers from substituted thiophenes. *J Electroanal Chem* 278:373–378.
182. Barker, J.M., J.D.E. Chaffin, J. Halfpenny, P.R. Huddleston, and P.F. Tseki. 1993. Novel thiophene-based macrocycles related to azacrown ethers. *J Chem Soc Chem Commun* 1733–1734.
183. Marrec, P., B. Fabre, and J. Simonet. 1997. Electrochemical and spectroscopic properties of new functionalized polythiophenes electroformed from the oxidation of dithienyls linked by long polyether spacers. *J Electroanal Chem* 437:245–253.
184. Blanchard, P., L. Huchet, E. Levillain, and J. Roncali. 2000. Cation template assisted electrosynthesis of a highly π-conjugated polythiophene containing oligooxyethylene segments. *Electrochem Commun* 2:1–5.
185. Lyskawa, J., F. Le Derf, E. Levillain, M. Mazari, M. Sallé, L. Dubois, P. Viel, C. Bureau, and S. Palacin. 2004. Univocal demonstration of the electrochemically mediated binding of Pb^{2+} by a modified surface incorporating a TTF-based redox-switchable ligand. *J Am Chem Soc* 126:12194–12195.

186. McCullough, R.D., and S.P. Williams. 1995. A dramatic conformational transformation of a regioregular polythiophene via a chemoselective, metal-ion assisted deconjugation. *Chem Mater* 7:2001–2003.

187. McCullough, R.D., and Williams, S.P. 1993. Toward tuning electrical and optical properties in conjugated polymers using side chains: Highly conductive head-to-tail heteroatom-functionalized polythiophenes. *J Am Chem Soc* 115:11608–11609.

188. Lévesque, I., and M. Leclerc. 1995. Ionochromic effects in regioregular ether-substituted polythiophenes. *J Chem Soc Chem Commun* 2293–2294.

189. Lévesque, I., and M. Leclerc. 1996. Ionochromic and thermochromic phenomena in a regioregular polythiophene derivative bearing oligo(oxyethylene) side chains. *Chem Mater* 8:2843–2849.

190. Lévesque, I., P. Bazinet, and J. Roovers. 2000. Optical properties and dual electrical and ionic conductivity in poly(3-methylhexa(oxyethylene)oxy-4-methylthiophene). *Macromolecules* 33:2952–2957.

191. Boldea, A., I. Lévesque, and M. Leclerc. 1999. Controlled ionochromism with polythiophenes bearing crown ether side chains. *J Mater Chem* 9:2133–2138.

192. Marsella, M.J., and T.M. Swager. 1993. Designing conducting polymer-based sensors: Selective ionochromic response in crown ether containing polythiophenes. *J Am Chem Soc* 115:12214–12215.

193. Swager, T.M., and M.J. Marsella. 1994. Molecular recognition and chemoresistive materials. *Adv Mater* 6:595–597.

194. Swager, T.M., M.J. Marsella, L.K. Bicknell, and Q. Zhou. 1994. New approaches to conducting polymer-based sensors. *Polym Prep* 35:206–207.

195. Marsella, M.J., and T.M. Swager. 1994. Ionochromic response in crown ether containing polythiophenes. *Polym Prep* 35:271–272.

196. Marsella, M.J., R.J. Newland, and T.M. Swager. 1995. The ionochromic response of crown ether and calix[4]arene substituted polythiophenes: Towards chemoresistive sensory materials. *Polym Prep* 36:594–595.

197. Jousselme, B., P. Blanchard, E. Levillain, J. Delaunay, M. Allain, P. Richomme, D. Rondeau, N. Gallego-Planas, and J. Roncali. 2003. Crown-annelated oligothiophenes as model compounds for molecular actuation. *J Am Chem Soc* 125:1363–1370.

198. Bernier, S., S. Garreau, M. Béra-Abérel, C. Gravel, and M. Leclerc. 2002. A versatile approach to affinitychromic polythiophenes. *J Am Chem Soc* 124:12463–12468.

199. Marsella, M.J., R.J. Newland, P.J. Caroll, and T.M. Swager. 1995. Ionoresistivity as highly sensitive sensory probe: Investigations of polythiophenes functionalized with calix[4]arene-based ion receptors. *J Am Chem Soc* 117:9842–9848.

200. Giannetto, M., G. Mori, A. Notti, S. Pappalardo, and M.F. Parisi. 2001. Calixarene–poly(dithiophene)-based chemically modified electrodes. *Chem Eur J* 7:3354–3362.

201. Yu, H.-H., A.E. Pullen, M.G. Büschel, and T.M. Swager. 2004. Charge-specific interactions in segmented conducting polymers: An approach to selective ionoresistive responses. *Angew Chem Int Ed Engl* 43:3700–3703.

202. Yu, H.-H., B. Xu, and T.M. Swager. 2003. A proton-doped calix[4]arene-based conducting polymer. *J Am Chem Soc* 125:1142–1143.

203. Reddinger, J.L., and J.R. Reynolds. 1998. A novel polymeric metallomacrocycle sensor capable of dual-ion cocomplexation. *Chem Mater* 10:3–5.

204. Reddinger, J.L., and J.R. Reynolds. 1998. Site specific electropolymerization to form transistion-metal-containing, electroactive polythiophenes. *Chem Mater* 10:1236–1243.

205. Zhu, S.S., and T.M. Swager. 1997. Conducting polymetallorotaxanes: Metal ion mediated enhancements in conductivity and charge localization. *J Am Chem Soc* 119:12568–12577.

206. Ballarin, B., S. Masiero, R. Seeber, and D. Tonelli. 1998. Modification of electrodes with porphyrin-functionalised conductive polymers. *J Electroanal Chem* 449:173–180.

207. Zotti, G., S. Zecchin, and G. Schiavon. 1999. Ionochromic and potentiometric properties of the novel polyconjugated polymer from anodic coupling of 5,5′-bis(3,4-ethylenedioxy)thien-2-yl)-2,2′-bipyridine. *Chem Mater* 11:3342–3351.

208. Fan, L.-J., Y. Zhang, and W.E. Jones, Jr. 2005. Design and synthesis of fluorene "turn-on" chemosensors based on photoinduced electron transfer in conjugated polymers. *Macromolecules* 38:2844–2849.

209. Marsella, M.J., P.J. Caroll, and T.M. Swager. 1994. Conducting pseudopolyrotaxanes: A chemoresistive response via molecular recognition. *J Am Chem Soc* 116:9347–9348.

210. Marsella, M.J., P.J. Caroll, and T.M. Swager. 1995. Design of chemoresistive sensory materials: Polythiophene-based pseudopolyrotaxanes. *J Am Chem Soc* 117:9832–9841.

211. Zhou, Q., and T.M. Swager. 1995. Fluorescent chemosensors based on energy migration in conjugated polymers: The molecular wire approach to increased sensitivity. *J Am Chem Soc* 117:12593–12602.

212. Simone, D.L., and T.M. Swager. 2000. A conducting poly(cyclophane) and its poly([2]-catenane). *J Am Chem Soc* 122:9300–9301.

213. McCullough, R.D., P.C. Ewbank, and R.S. Loewe. 1997. Self-assembly and disassembly of regioregular, water soluble polythiophenes: Chemoselective ionchromatic sensing in water. *J Am Chem Soc* 119:633–634.

214. McCullough, R.D., and P.C. Ewbank. 1997. Self-assembly and chemical response of conducting polymer superstructures. *Synth Met* 84:311–312.

215. Beer, P.D., and J. Cadman. 2000. Electrochemical and optical sensing of anions by transition metal based receptors. *Coord Chem Rev* 205:131–155.

216. Beer, P.D., and P.A. Gale. 2001. Anion recognition and sensing: The state of the art and future perspectives. *Angew Chem Int Ed Engl* 40:486–516.

217. Nicolas, M., B. Fabre, G. Marchand, and J. Simonet. 2000. New boronic-acid- and boronate-substituted aromatic compounds as precursors of fluoride-responsive conjugated polymer films. *Eur J Org Chem* 1703–1710.

218. Nicolas, M., B. Fabre, and J. Simonet. 1999. Boronate-functionalized polypyrrole as a new fluoride sensing material. *Chem Commun* 1881–1882.

219. Takeuchi, M., T. Shioya, and T.M. Swager. 2001. Allosteric fluoride anion recognition by a doubly strapped porphyrin. *Angew Chem Int Ed Engl* 40:3372–3375.

220. Ho, H.A., and M. Leclerc. 2003. New colorimetric and fluorometric chemosensor based on a cationic polythiophene derivative for iodide-specific detection. *J Am Chem Soc* 125:4412–4413.

221. Aldakov, D., and P. Anzenbacher, Jr. 2004. Sensing of aqueous phosphates by polymers with dual modes of signal transduction. *J Am Chem Soc* 126:4752–4753.

222. Anzenbacher, P., Jr., K. Jursikova, D. Aldakov, M. Marquez, and R. Pohl. 2004. Materials chemistry approach to anion-sensor design. *Tetrahedron* 60:11163–11168.

223. Aldakov, D., and P. Anzenbacher, Jr. 2003. Dipyrrolyl quinoxalines with extended chromophores are efficient fluorimetric sensors for pyrophosphate. *Chem Commun* 1394–1395.

224. Pohl, R., D. Aldakov, P. Kubat, K. Jursikova, M. Marquez, and P. Anzenbacher, Jr. 2004. Strategies toward improving the performance of fluorescence-based sensors for inorganic anions. *Chem Commun* 1282–1283.

225. Aldakov, D., M.A. Palacios, and P. Anzenbacher, Jr. 2005. Benzothiadiazoles and dipyrrolyl quinoxalines with extended conjugated chromophores-fluorophores and anion sensors. *Chem Mater* 17:5238–5241.

226. Englebienne, P. 1999. Synthetic materials capable of reporting biomolecular recognition events by chromic transition. *J Mater Chem* 9:1043–1054.

227. Rizzo, S., F. Sannicolò, T. Benincori, G. Schiavon, S. Zecchin, and G. Zotti. 2004. Calix[4]arene-functionalized poly-cyclopenta[2,1-*b*:3,4-*b*′]bithiophenes with good recognition ability and selectivity for small organic molecules for application in QCM-based sensors. *Italic NOT ALLOWEDJ Mater Chem* 14:1804–1811.

228. Vigalok, A., Z. Zhu, and T.M. Swager. 2001. Conducting polymers incorporating tungsten calix-arenes. *J Am Chem Soc* 123:7917–7918.

229. Vigalok, A., and T.M. Swager. 2002. Conducting polymers of tungsten(VI)-oxo calixarene: Inter-calation of neutral organic guests. *Adv Mater* 14:368–371.

230. Kingsborough, R.P., and T.M. Swager. 1998. Electroactivity enhancement by redox matching in cobalt salen-based conducting polymers. *Adv Mater* 10:1100–1103.

231. Shioya, T., and T.M. Swager. 2002. A reversible resistivity-based nitric oxide sensor. *Chem Commun* 1364–1365.

232. Kingsborough, R.P., and T.M. Swager. 2000. Electrocatalytic conducting polymers: Oxygen reduc-tion by a polythiophene–cobalt salen hybrid. *Chem Mater* 12:872–874.

233. Schäferling, M., and P. Bäuerle. 2004. Porphyrin-functionalized oligo- and polythiohenes. *J Mater Chem* 14:1132–1141.

234. Sundararaman, A., M. Victor, R. Varughese, and F. Jäkle. 2005. A family of main-chain polymeric Lewis acids: Synthesis and fluorescent sensing properties of boron-modified polythiophenes. *J Am Chem Soc* 127:13748–13749.

235. Samuelson, L.A., D.L. Kaplan, J.O. Lim, M. Kamath, K.A. Marx, and S.K. Tripathy. 1994. Molecular recognition between a biotinylated polythiophene copolymer and phycoerythrin utilizing the biotin–streptavidin interaction. *Thin Solid Films* 242:50–55.

236. Pande, R., S. Kamtekar, M.S. Ayyagari, M. Kamath, K.A. Marx, J. Kumar, S.K. Tripathy, and D.L. Kaplan. 1996. A biotinylated undecylthiophene copolymer bioconjugate for surface immo-bilization: Creating an alkaline phosphatase chemiluminescence-based biosensor. *Bioconjugate Chem* 7:159–164.

237. Faïd, K., and M. Leclerc. 1996. Functionalized regioregular polythiophenes: Towards the deve-lopment of biochromic sensors. *Chem Commun* 2761–2762.

238. Kumpumbu-Kalemba, L., and M. Leclerc. 2000. Electrochemical charcaterization of monolayers of biotinylated polythiophene: Towards the development of polymeric biosensors. *Chem Commun* 1847–1848.

239. Faïd, K., and M. Leclerc. 1998. Responsive supramolecular polythiophene assemblies. *J Am Chem Soc* 120:5274–5278.

240. Mouffouk, F., S.J. Brown, A.M. Demetriou, S.J. Higgins, R.J. Nichols, R.M.G. Rajapakse, and S. Reeman. 2005. Electrosynthesis and characterization of biotin-functionalized poly(terthiophene) copolymers, and their response to avidin. *J Mater Chem* 15:1186–1196.

241. Bäuerle, P., and A. Emge. 1998. Specific recognition of nucleobase-functionalized polythiophenes. *Adv Mater* 3:324–330.

242. Emge, A., and P. Bäuerle. 1997. Synthesis and molecular recognition properties of DNA- and RNA-base-functionalized oligo- and polythiophenes. *Synth Met* 84:213–214.

243. Emge, A., and P. Bäuerle. 1999. Molecular recognition properties of nucleobase-functionalized polythiophenes. *Synth Met* 102:1370–1373.

244. Rockel, H., J. Huber, R. Gleiter, and W. Schuhmann. 1994. Synthesis of functionalized poly(dithie-nylpyrrole) derivatives and their application in amperometric biosensors. *Adv Mater* 6:568–571.

245. Bäuerle, P., M. Hiller, S. Scheib, M. Sokolowski, and E. Umbach. 1996. Post-polymerization functionalization of conducting polymers: Novel poly(alkylthiophene)s substituted with easily replaceable activated ester groups. *Adv Mater* 8:214–218.

246. Hiller, M., C. Kranz, J. Huber, P. Bäuerle, and W. Schuhmann. 1996. Amperometric biosensors produced by immobilization of redox enzymes at polythiophene-modified electrodesurfaces. *Adv Mater* 8:219–222.

247. Welzel, H.-P., G. Kossmehl, G. Engelmann, B. Neumann, U. Wollenberger, F. Scheller, and W. Schröder. 1996. Reactive groups on polymer covered electrodes. Lactate-oxidase-biosensor based on electrodes modified by polythiophene. *Macromol Chem Phys* 197:3355–3363.

248. Cha, J., J.I. Han, Y. Choi, D.S. Yoon, K.W. Oh, and G. Lim. 2003. DNA hybridization electrochem-ical sensor using conducting polymer. *Biosens Bioelectron* 18:1241–1247.

249. Navarro, A.-E., F. Fages, C. Moustrou, H. Brisset, N. Spinelli, C. Chaix, and B. Mandrand. 2005. Characterization of PEDOT film functionalized with a series of automated synthesis ferrocenyl-containing oligonucleotides. *Tetrahedron* 61:3947–3952.

250. Li, C., M. Numata, M. Takeuchi, and S. Shinkai. 2005. A sensitive colorimetric and fluorescent probe based on a polythiophene derivative for the detection of ATP. *Angew Chem Int Ed Engl* 44:6371–6374.

251. Ewbank, P.C., G. Nuding, H. Suenaga, R.D. McCullough, and S. Shinkai. 2001. Amine functionalized polythiophenes: Synthesis and formation of chiral, ordered structures on DNA substrates. *Tetrahedron Lett* 42:155–157.

252. Le Floch, F., H.-A. Ho, P. Harding-Lepage, M. Bédard, R. Neagu-Plesu, and M. Leclerc. 2005. Ferrocene-functionalized cationic polythiophene for the label-free electrochemical detection of DNA. *Adv Mater* 17:1251–1254.

253. Ho, H.A., M. Boissinot, M.G. Bergeron, G. Corbeil, K. Doré, D. Boudreau, and M. Leclerc. 2002. Colorimetric and fluorometric detection of nucleic acids using cationic polythiophene derivatives. *Angew Chem Int Ed Engl* 41:1548–1551.

254. Doré, K., S. Dubus, H.A. Ho, I. Lévesque, M. Brunette, G. Corbeil, M. Boissinot, G. Boivin, M.G. Bergeron, D. Boudreau, and M. Leclerc. 2004. Fluorescent polymeric transducer for the rapid, simple, and specific detection of nucleic acids at the zeptomole level. *J Am Chem Soc* 126:4240–4244.

255. Ho, H.A., and M. Leclerc. 2004. Optical sensors based on hybrid aptamer/conjugated polymer complexes. *J Am Chem Soc* 126:1384–1387.

256. Ho, H.A., K. Doré, M. Boissinot, M.G. Bergeron, R.M. Tanguay, D. Boudreau, and M. Leclerc. 2005. Direct molecular detection of nucleic acids by fluorescence signal amplification. *J Am Chem Soc* 127:12673–12676.

257. Nilsson, K.P.R., and O. Inganäs. 2004. Optical emission of a conjugated polyelectrolyte: Calcium-induced conformational changes in calmodulin and calmodulin–calcineurin interactions. *Macromolecules* 37:9109–9113.

258. Herland, A., K.P.R. Nilsson, J.D.M. Olsson, P. Hammarström, P. Konradsson, and O. Inganäs. 2005. Synthesis of a regioregular zwitterionic conjugated oligoelectrolyte, usable as an optical probe for detection of amyloid fibril formation at acidic pH. *J Am Chem Soc* 127:2317–2323.

259. Nilsson, K.P.R., J.D.M. Olsson, F. Stabo-Eeg, M. Lindgren, P. Konradsson, and O. Inganäs. 2005. Chiral recognition of a synthetic peptide using enantiomeric conjugated polyelectrolytes and optical spectroscopy. *Macromolecules* 38:6813–6821.

260. Nilsson, K.P.R., A. Herland, P. Hammarström, and O. Inganäs. 2005. Conjugated polyelectrolytes: Conformation-sensitive optical probes for detection of amyloid fibril formation. *Biochemistry* 44:3718–3724.

261. Nilsson, K.P.R., J. Rydberg, L. Baltzer, and O. Inganäs. 2004. Twisting macromolecular chains: Self-assembly of a chiral supermolecule from non chiral polythiophene polyanions and random-coil synthetic peptides. *Proc Natl Acad Sci USA* 101:11197–11202.

262. Kingsborough, R.P., and T.M. Swager. 1999. Transition metals in polymeric π-conjugated organic frameworks. *Prog Inorg Chem* 48:123–231.

263. Nguyen, P., P. Gomez-Elipe, and I. Manners. 1999. Organometallic polymers with transistion metals in the main chain. *Chem Rev* 99:1515–1548.

264. Pickup, P.G. 1999. Conjugated metallopolymers. Redox polymers with interacting metal based redox sites. *J Mater Chem* 9:1641–1653.

265. Wolf, M.O., and Y. Zhu. 2000. Electropolymerization of oligothienyferrocene complexes. *Adv Mater* 12:599–601.

266. Wolf, M.O. 2001. Transition-metal-polythiophene hybrid materials. *Adv Mater* 13:545–553.

267. Schubert, U.S., and C. Eschbaumer. 2002. Macromolecules containing bipyridine and terpyridine metal complexes: Towards metallosupramolecular polymers. *Angew Chem Int Ed Engl* 41:2892–2926.

268. Stott, T.L., and M.O. Wolf. 2003. Electronic interactions in metallated polythiophenes: What can be learned from model complexes. *Coord Chem Rev* 246:89–101.

269. Moorlag, C., B.C. Sih, T.L. Stott, and M.O. Wolf. 2005. Metal-containing conjugated materials: Oligomers, polymers, and nanomaterials. *J Mater Chem* 15:2433–2436.

270. Sih, B.C., and M.O. Wolf. 2005. Metal nanoparticle-conjugated polymer nanocomposites. *Chem Commun* 3375–3384.

271. Holliday, B.J., and T.M. Swager. 2005. Conducting metallopolymers: The roles of molecular architecture and redox matching. *Chem Commun* 23–36.

272. Weder, C. 2005. Synthesis, processing and properties of conjugated polymer networks. *Chem Commun* 5378–5389.

273. Mirrazaei, R., D. Parker, and H.S. Munro. 1989. Stable redox and electronically conducting thin films containing $Fe(bpy)_3^{2+}$ bound to 3-substituted poly(thiophenes). *Synth Met* 30:265–269.

274. Wang, J., and F.R. Keene. 1996. Synthesis and electrochemical properties of poly[3-{ω-[4-(2,2′-bipyridyl)]alkyl}thiophenes], P{B(n)T}, and of poly[Ru(II){B(n)T}$_3^{2+}$]. *J Electroanal Chem* 405:71–83.

275. Crayston, J., A. Iraqi, J.J. Morrison, and J.C. Walton. 1997. Synthesis of thiophene substituted ruthenium and rhenium bipyridyl complexes. *Synth Met* 84:441–442.

276. Zhu, S.S., and T.M. Swager. 1996. Design of conducting redox polymers: A polythiophene-Ru(bipy)$_3^{n+}$ hybrid material. *Adv Mater* 8:497–500.

277. Jenkins, I.H., N.G. Rees, and P.G. Pickup. 1997. Conducting bipyridine–bithiophene copolymers. *Chem Mater* 9:1213–1216.

278. Zhu, S.S., R.P. Kingsborough, and T.M. Swager. 1999. Conducting redox polymers: Investigations of polythiophene-Ru(bpy)$_3^{n+}$ hybrid materials. *J Mater Chem* 9:2123–2131.

279. Papillon, J., E. Schulz, S. Gélinas, J. Lessars, and M. Lemaire. 1998. Towards the preparation of modified chiral electrodes for heterogeneous asymmetric catalysis: Synthesis and electrochemical properties of (S,S)-5,5′-bis-[3-(3-methyl-pentyl)-thiophen-2-yl]-[2,2′]-bipyridine. *Synth Met* 96:155–160.

280. Zhu, S.S., P.J. Caroll, and T.M. Swager. 1996. Conducting polymetallorotaxanes: A supramolecular approach to transition metal ion sensors. *J Am Chem Soc* 118:8713–8714.

281. Vidal, P.L., M. Billon, B. Divisia-Blohorn, G. Bidan, J.M. Kern, and J.P. Sauvage. 1998. Conjugated polyrotaxanes containing coordinating units: Reversible copper(I) metallation-dematallation using lithium as intermediate scaffolding. *Chem Commun* 629–630.

282. Sauvage, J.P., J.M. Kern, G. Bidan, B. Divisia–Blohorn, and P.L. Vidal. 2002. Conjugated polyrotaxanes: Improvement of the polymer properties by using sterically hindered coordinating units. *New J Chem* 26:1287–1290.

283. Ammann, M., and P. Bäuerle. 2005. Synthesis and electronic properties of a series of oligothiophene-[1,10]phenanthrolines. *Org Biomol Chem* 3:4143–4152.

284. Buey, J., and T.M. Swager. 2000. Three-strand conducting ladder polymers: Two-step electropolymerization of metallorotaxanes. *Angew Chem Int Ed Engl* 39:608–612.

285. Kwan, P.H., and T.M. Swager. 2005. Insulated conducting polymers: Manipulating charge transport using supramolecular complexes. *Chem Commun* 5211–5213.

286. Trouillet, L., A. De Nicola, and S. Guillerez. 2000. Synthesis and characterization of a new soluble, structurally well-defined conjugated polymer alternating regioregularity alkylated thiophene oligomer and 2,2′-bipyridine units: Metal-free form and Ru(II) complex. *Chem Mater* 12:1611–1621.

287. Walters, K.A., L. Trouillet, S. Guillerez, and K.S. Schanze. 2000. Photophysics and electron transfer in poly(3-octylthiophene) alternating with Ru(II)- and Os(II)-bipyrine complexes. *Inorg Chem* 39:5496–5509.

288. Kaneto, K., K. Yoshino, and Y. Inuishi. 1983. Characteristics of electro-optic device using conducting polymers, polythiophene and polypyrrole films. *Jpn J Appl Phys* 22:L412–L414.

289. Garnier, F., G. Tourillon, M. Gazard, and J.C. Dubois. 1983. Organic conducting polymers derived from substituted thiophenes as electrochromic material. *J Electroanal Chem* 148:299–303.

290. Jonas, F., and L. Schrader. 1991. Conductive modifications of polymers with polypyrroles and polythiophenes. *Synth Met* 41 (3):831–836.
291. Groenendaal, L., G. Zotti, P.H. Aubert, S.M. Waybright, and J.R. Reynolds. 2003. Electrochemistry of poly(3,4-alkylenedioxythiophene) derivatives. *Adv Mater* 15:855–879.
292. Heuer, H.W., R. Wehrmann, and S. Kirchmeyer. 2002. Electrochromic window based on conducting poly(3,4-ethylenedioxythiophene)-poly(styrene sulfonate). *Adv Funct Mater* 12:89–94.
293. Sotzing, G.A., J.R. Reynolds, and P.J. Steel. 1996. Electrochromic conducting polymer via electrochemical polymerization of bis(2-(3,4-ethylenedioxy)thienyl) monomers. *Chem Mater* 8:882.
294. Sonmez, G., H. Meng, Q. Zhang, and F. Wudl. 2003. A highly stable, new electrochromic polymer: Poly(1,4-bis(3′,4′-ethylenedioxy)thienyl)-2-methoxy-5,2″-ethylhexyloxybenzene. *Adv Funct Mater* 13 (9):726.
295. Sonmez, G., H. Meng, and F. Wudl. 2004. Organic polymeric electrochromic devices: Polychromism with very high coloration efficiency. *Chem Mater* 16 (4):574–580.
296. Ko, H.C., S. Kim, H. Lee, and B. Moon. 2005. Multicolored electrochromism of poly{1,4-bis[2-(3,4-ethylenedioxy)thienyl]benzene} derivative bearing viologen functional groups. *Adv Funct Mater* 15:905.
297. Reddinger, J.L., G.A. Sotzing, and J.R. Reynolds. 1996. Multicoloured electrochromic polymers derived from easily oxidized bis[2-(3,4-ethylenedioxy)thienyl]carbazoles. *Chem Commun* 1777.
298. Sotzing, G.A., J.L. Reddinger, A.R. Katrizky, J. Soloducho, R. Musgrave, and J.R. Reynolds. 1997. Multiply colored electrochromic carbazole-based polymers. *Chem Mater* 9:1578.
299. Sonmez, G., C.K.F. Shen, Y. Rubin, and F. Wudl. 2004. A red, green, and blue (RGB) polymeric electrochromic device (PECD): The dawning of the PECD era. *Angew Chem Int Ed Engl* 43:1498–1502.
300. Sonmez, G., H.B. Sonmez, C.K.F. Shen, and F. Wudl. 2004. Red, green and blue colors in polymeric electrochromics. *Adv Mater* 16:1905.
301. Sonmez, G., H.B. Sonmez, C.K.F. Shen, R.W. Jost, Y. Rubin, and F. Wudl. 2005. A processable green polymeric electrochromic. *Macromolecules* 38:669.
302. Sonmez, G., and F. Wudl. 2005. Completion of the three primary colours: The final step toward plastic displays. *J Mater Chem* 15:20.
303. Sonmez, G. 2005. Polymeric electrochromics. *Chem Commun* 5251.
304. Perepichka, I.F., D.F. Perepichka, H. Meng, and F. Wudl. 2005. Light-emitting polythiophenes. *Adv Mater* 17:2281–2305.
305. Andersson, M.R., M. Berggren, O. Inganäs, and G. Gustafsson. 1995. Electroluminescence from substituted poly(thiophenes): From blue to near-infrared. *Macromolecules* 28:7525–7529.
306. Gill, R.E., G.G. Malliaras, J. Wildeman, and G. Hadziioannou. 1994. Tuning of photo- and electroluminescence in alkylated polythiophenes with well-defined regioregularity. *Adv Mater* 6:132–135.
307. Herrema, J.K., P.F. Van Hutten, R.E. Gill, J. Wildeman, and G. Hadziioannou. 1995. Tuning of the luminescence in multiblock alternating copolymers. *Macromolecules* 28:8102–8116.
308. Malliaras, G.G., J.K. Herrema, J. Wildeman, R.H. Wieringa, R.E. Gill, S.S. Lampoura, and G. Hadziioannou. 1993. Tuning of the photoluminescence and electroluminescence in multiblock copolymers of poly[(silanylene)-thiophene]s via exciton confinement. *Adv Mater* 5:721–723.
309. Yoshino, K., A. Fujii, H. Nakayama, S. Lee, A. Naka, and M. Ishikawa. 1999. Optical properties of poly(disilanyleneoligothienylene)s and their doping characteristics. *J Apply Phys* 85:414–418.
310. Belletête, M., L. Mazerolle, N. Desrosiers, M. Leclerc, and G. Durocher. 1995. Spectroscopy and photophysics of some oligomers and polymers derived from thiophenes. *Macromolecules* 28:8587–8597.
311. Novikova, T.S., N.N. Barashkov, A. Yassar, M. Hmyene, and J.P. Ferraris. 1996. Aromatic copolyamides and copolyesters with vinylenarylene and terthiophene fragments in the polymer chain: Synthesis and photophysical properties. *Synth Met* 83:47–55.
312. Kunugi, Y., L.L. Miller, T. Maki, and A. Canavesi. 1997. Photodiodes utilizing polyesters that contain different colored oligothiophenes in the main chain. *Chem Mater* 9:1061–1062.

313. Hong, Y., L.L. Miller, D.D. Graf, K.R. Mann, and B. Zinger. 1996. Electroluminescence from a polyester containing oligothiophenes in the main chain, enhanced by a diimide electron transport agent. *Synth Met* 82:189–191.

314. Olinga, T., and B. François. 1991. Synthesis of soluble polystyrene-graft-polythiophene comblike copolymers: A new precursor for polythiophene film preparation. *Makromol Chem Rapid Commun* 12:575–582.

315. François, B., and T. Olinga. 1993. Polystyrene–polythiophene block copolymers (PS–PT) synthesis, characterization and doping. *Synth Met* 57:3489–3494.

316. Liu, J., E. Sheina, T. Kowalewski, and R.D. McCullough. 2002. Tuning the electrical conductivity and self-assembly of regioregular polythiophene by block copolymerization: Nanowire morphologies in new di- and triblock copolymers. *Angew Chem Int Ed Engl* 41:329–332.

317. Melucci, M., G. Barbarella, M. Zambianchi, M. Benzi, F. Biscarini, M. Cavallini, A. Bongini, S. Fabbroni, M. Mazzea, M. Anni, and G. Gigli. 2004. Poly(alpha-vinyl-omega-alkyloligothiophene) side-chain polymers. Synthesis, fluorescence, and morphology. *Macromolecules* 37:5692–5702.

318. Malenfant, P.R.L., L. Groenendaal, and J.M.J. Fréchet. 1998. Well-defined triblock hybrid dendrimers based on lengthy oligothiophene cores and poly(benzylether dendrons). *J Am Chem Soc* 120:10990–10991.

319. Furuta, P., J. Brooks, M.E. Thompson, and J.M.J. Fréchet. 2003. Simultaneous light emission from a mixture of dendrimer encapsulated chromophores: A model for single-layer multichromophoric organic light-emitting diodes. *J Am Chem Soc* 125:13165–13172.

320. Barbarella, G., L. Favaretto, G. Sotgiu, M. Zambianchi, V. Fattori, M. Cocchi, F. Cacialli, G. Gigli, and R. Cingolani. 1999. Modified oligothiophenes with high photo- and electroluminescence efficiencies. *Adv Mater* 11:1375–1379.

321. Gigli, G., G. Barbarella, L. Favaretto, F. Cacialli, and R. Cingolani. 1999. High-efficiency oligothiophene-based light-emitting diodes. *Appl Phys Lett* 75:439–441.

322. Mazzeo, M., V. Vitale, F. Della Sala, D. Pisignano, M. Anni, G. Barbarella, L. Favaretto, A. Zanelli, R. Cingolani, and G. Gigli. 2003. New branched thiophene-based oligomers for bright organic light-emitting devices. *Adv Mater* 15:2060–2063.

323. Barbarella, G., L. Favaretto, A. Zanelli, G. Gigli, M. Mazzeo, M. Anni, and A. Bongini. 2005. V-shaped thiophene-based oligomers with improved electroluminescence properties. *Adv Funct Mater* 15:664–670.

324. Antolini, L., E. Tedesco, G. Barbarella, L. Favaretto, G. Sotgiu, M. Zambianchi, D. Casarini, G. Gigli, and R. Cingolani. 2000. Molecular packing and photoluminescence efficiency in odd-embered oligothiophene *S,S*-dioxides. *J Am Chem Soc* 122:9006–9013.

325. Tedesco, E., F. Della Sala, L. Favaretto, G. Barbarella, D. Albesa-Jové, D. Pisignano, G. Gigli, R. Cingolani, and K.D.M. Harris. 2003. Solid-state supramolecular organization, established directly from powder diffraction data, and photoluminescence efficiency of rigid-core oligothiophene-*S,S*-dioxides. *J Am Chem Soc* 125:12277–12283.

326. Perepichka, I.I., I.F. Perepichka, M.R. Bryce, and L.-O. Palsson. 2005. Dibenzothiophene-*S,S*-dioxide-fluorene co-oligomers. Stable, highly-efficient blue emitters with improved electron affinity. *Chem Commun* 3397–3399.

327. Barbarella, G., L. Favaretto, G. Sotgiu, M. Zambianchi, C. Arbizzani, A. Bongini, and M. Mastragostino. 1999. Controlling the electronic properties of polythiophene through the insertion of nonaromatic thienyl *S,S*-dioxide units. *Chem Mater* 11:2533–2541.

328. Berlin, A., G. Zotti, S. Zecchin, G. Schiavon, M. Cocchi, D. Virgili, and C. Sabatini. 2003. 3,4-Ethylenedioxy-substituted bithiophene-alt-thiophene-*S,S*-dioxide regular copolymers. Synthesis and conductive, magnetic and luminescence properties. *J Mater Chem* 13:27–33.

329. Charas, A., J. Morgado, J.M.G. Martinho, L. Alcacer, and F. Cacialli. 2001. Electrochemical and luminescent properties of poly(fluorene) derivatives for optoelectronic applications. *Chem Commun* 1216–1217.

330. Charas, A., J. Morgado, J.M.G. Martinho, A. Fedorov, L. Alcacer, and F. Cacialli. 2002. Excitation energy transfer and spatial exciton confinement in polyfluorene blends for application in light-emtting diodes. *J Mater Chem* 12:3523–3527.

331. Pasini, M., S. Destri, W. Porzio, C. Botta, and U. Giovanella. 2003. Electroluminescent poly(fluorene-co-thiophene-*S,S*-dioxide): Synthesis, characterisation and structure–property relationships. *J Mater Chem* 13:807–813.

332. Beaupré, S., and M. Leclerc. 2002. Fluorene-based copolymers for red-light-emitting diodes. *Adv Funct Mater* 12:192–196.

333. Morin, J.-F., and M. Leclerc. 2002. 2,7-Carbazole-based conjugated polymers for blue, green, and red light emission. *Macromolecules* 35:8413–8417.

334. Baigent, D.R., P.J. Hamer, R.H. Friend, S.C. Moratti, and A.B. Holmes. 1995. Polymer electroluminescence in the near infra-red. *Synth Met* 71:2175–2176.

335. Moratti, S.C., R. Cervini, A.B. Holmes, D.R. Baigent, R.H. Friend, N.C. Greenham, J. Grüner, and P.J. Hamer. 1995. High electron affinity polymers for LEDs. *Synth Met* 71:2117–2120.

336. Pei, J., W.-L. Yu, W. Huang, and A.J. Heeger. 2000. A novel series of efficient thiophene-based light-emitting conjugated polymers and application in polymer light-emitting diodes. *Macromolecules* 33:2462–2471.

337. Bouachrine, M., J.-P. Lère-Porte, J.J.E. Moreau, F. Serein Spirau, R.A. Da Silva, K. Lmimouni, L. Ouchani, and C. Dufour. 2002. A thienylene–phenylene copolymer with di(ethylene oxide) side chains and its use in light emitting diodes. *Synth Met* 126:241–244.

338. Wang, Y.Z., D.D. Gebler, D.K. Fu, T.M. Swager, and A.J. Epstein. 1997. Color variable bipolar/ac light-emitting devices based on conjugated polymers. *Appl Phys Lett* 70:3212–3217.

339. Liu, C., W.-L. Yu, Y.-H. Lai, and W. Huang. 2001. Blue-light-emitting fluorene-based polymers with tunable electronic properties. *Chem Mater* 13:1984–1991.

340. Liu, B., W.-L. Yu, Y.-H. Lai, and W. Huang. 2000. Synthesis, characterization, and structure-relationship of novel fluorene-thiophene-based conjugated copolymers. *Macromolecules* 33:8945–8952.

341. Donat-Bouillud, A., I. Lévesque, Y. Tao, M. D'Ioro, P. Blondin, M. Ranger, J. Bouchard, and M. Leclerc. 2000. Light-emitting diodes from fluorene-based π-conjugated polymers. *Chem Mater* 12:1931–1936.

342. Lim, E., B.-J. Jung, and H.-K. Shim. 2003. Synthesis and characterization of a new light-emitting fluorene-thieno[3,2-b]thiophene-based conjugated copolymer. *Macromolecules* 36:4288–4293.

343. Pei, J., W.-L. Yu, W. Huang, and A.J. Heeger. 2000. The synthesis and characterization of an efficient green electroluminescent conjugated polymer: Poly[2,7-bis(4-hexylthienyl)-9,9-dihexyl-fluorene]. *Chem Commun* 1631–1632.

344. Lee, J., N.S. Cho, J. Lee, S.K. Lee, and H.-K. Shim. 2005. Emission color tuning of new fluorene-based alternating copolymers containing low band gap dyes. *Synth Met* 155:73–79.

345. Chen, M., E. Perzon, M.R. Andersson, S. Marcinkevicius, S.K.M. Jönsson, M. Fahlman, and M. Berggren. 2004. 1 micron Wavelength photo- and electroluminescence from a conjugated polymer. *Appl Phys Lett* 84:3570–3572.

346. Tak, Y.-H., S. Mang, A. Greiner, H. Bässler, S. Pfeiffer, and H.-H. Hörhold. 1997. Polythienylene-vinylene as promoter of hole injection from ITO into bilayer light emitting diodes. *Acta Polym* 48:450–454.

347. Zhang, F., A. Petr, U. Kirbach, and L. Dunsch. 2003. Improved hole injection and performance of multilayer OLED devices via electrochemically prepared-polythiophene layers. *J Mater Chem* 13:265–267.

348. Kunugi, Y., Y. Niwa, L. Zhu, Y. Harima, and K. Yamashita. 2001. Low operating voltage bilayer organic light-emitting diodes using electrochemically synthesized and p-doped starbust-polymer as hole transporting layer. *Chem Lett* 656–657.

349. Kunugi, Y., I. Tabakovic, A. Canavesi, and L.L. Miller. 1997. Light-emitting diodes based on linear and starburst electro-oligomerized thienyltriphenylamines. *Synth Met* 89:227–229.

350. Tonzola, C.J., M.M. Alam, and S.A. Jenekhe. 2002. New soluble n-type conjugated copolymer for light-emitting diodes. *Adv Mater* 14:1086–1090.

351. Tonzola, C.J., M.M. Alam, and S.A. Jenekhe. 2005. A new synthetic route to soluble polyquinolines with tunable photophysical, redox, and electroluminescent properties. *Macromolecules* 38:9539–9547.

352. Sheats, J.R. 2004. Manufacturing and commercialization issues in organic electronics. *J Mater Res* 19:1974–1989.

353. Wang, X., M.R. Andersson, M.E. Thompson, and O. Inganäs. 2004. Electrophosphorescence from substituted poly(thiophene) doped with iridium or platinum complex. *Thin Solid Films* 468:226–233.

354. Garnier, F. 1998. Thin-film transistors based on organic conjugated semiconductors. *Chem Phys* 227:253–262.

355. Dimitrakopoulos, C.D., and P.R.L. Malenfant. 2002. Organic thin film transistors for large area electronics. *Adv Mater* 14:99–117.

356. Horowitz, G. 1998. Organic field-effect transistors. *Adv Mater* 10:365–377.

357. Katz, H.E., and Z. Bao. 2000. The physical chemistry of organic field-effect transistors. *J Phys Chem B* 104:671–678.

358. Katz, H.E. 1997. Organic molecular solids as thin film transistor semiconductors. *J Mater Chem* 7:369–376.

359. Crone, B., A. Dodabalapur, A. Gelperin, L. Torsi, H.E. Katz, A.J. Lovinger, and Z. Bao. 2001. Electronic sensing of vapors with organic transistors. *Appl Phys Lett* 78:2229–2231.

360. Kraft, A. 2001. Organic field-effect transistors—The breakthrough at last. *Chem Phys Chem* 2:163–165.

361. Sun, Y., Y. Liu, and Z. Daoben. 2005. Advances in organic field-effect transistors. *J Mater Chem* 15:53–65.

362. Chabinye, M.L., and A. Salleo. 2004. Materials requirements and fabrication of active matrix arrays of organic thin-film transistors for displays. *Chem Mater* 16:4509–4521.

363. Tsumura, A., H. Koezuka, and T. Ando. 1986. Macromolecular electronic device: Field-effect transistor with a polythiophene thin film. *Appl Phys Lett* 49:1210–1212.

364. Koezuka, H., A. Tsumura, and T. Ando. 1987. Field-effect transistor with polythiophene thin film. *Synth Met* 18:699–704.

365. Sirringhaus, H., P.J. Brown, R.H. Friend, M.M. Nielsen, K. Bechgaard, B.M.W. Langeveld-Voss, A.J.H. Spiering, R.A.J. Janssen, E.W. Meijer, P. Herwig, and D.M. de Leeuw. 1999. Two-dimensional charge transport in self-organized, high-mobility conjugated polymers. *Nature* 401:685–688.

366. Bao, Z., A. Dodabalapur, and A.J. Lovinger. 1996. Soluble and processable regioregular poly(3-hexylthiophene) for thin film field-effect transistor applications with high mobility. *Appl Phys Lett* 69:4108–4110.

367. Sirringhaus, H., N. Tessler, and R.H. Friend. 1998. Integrated optoelectronic devices based on conjugated polymers. *Science* 280:1741–1744.

368. Chang, J.-F., B. Sun, D.W. Breiby, M.M. Nielsen, T.I. Sölling, M. Giles, I. McCulloch, and H. Sirringhaus. 2004. Enhanced mobility of poly(3-hexylthiophene) transistors by spin-coating from high-boiling-point solvents. *Chem Mater* 16:4772–4776.

369. Ong, B.S., Y. Wu, P. Liu, and S. Gardner. 2004. High-performance semiconducting polythiophenes for organic thin-film transistors. *J Am Chem Soc* 126:3378–3379.

370. Ong, B.S., Y. Wu, L. Jiang, P. Liu, and K. Murti. 2004. Polythiophene-based field-effect transistors with enhanced air stability. *Synth Met* 142:49–52.

371. Wu, Y., P. Liu, S. Gardner, and B.S. Ong. 2005. Poly(3,3″-dialkylterthiophene)s: Room-temperature, solution-processed, high-mobility semiconductors for organic thin-film transistors. *Chem Mater* 17:221–223.

372. Tierney, S., M. Heeney, and I. McCulloch. 2005. Microwave-assisted synthesis of polythiophenes via the Stille coupling. *Synth Met* 148:195–198.

373. Heeney, M., C. Bailey, K. Genevicius, M. Shkunov, D. Sparrowe, S. Tierney, and I. McCulloch. 2005. Stable polythiophene semiconductors incorporating thieno[2,3-b]thiophene. *J Am Chem Soc* 127:1078–1079.

374. Murphy, A.R., J. Liu, C. Luscombe, D. Kavulak, J.M.J. Fréchet, R.J. Kline, and M.D. McGehee. 2005. Synthesis, characterization, and field-effect transistor performance of carboxylate-functionalized polythiophenes with increased air stability. *Chem Mater* 17:4892–4899.

375. Xia, C., X. Fan, J. Locklin, R.C. Advincula, A. Gies, and W. Nonidez. 2004. Characterization, supramolecular assembly, and nanostructures of thiophene dendrimers. *J Am Chem Soc* 126:8735–8743.

376. Ponomarenko, S., and S. Kirchmeyer. 2004. Organic compounds having a core-shell structure. *European Patent 1398341.*

377. Mitchell, W.J., N. Kopidakis, G. Rumbles, D.S. Ginley, and S.E. Shaheen. 2005. The synthesis and properties of solution processable phenyl cored thiophene dendrimers. *J Mater Chem* 15:4518–4528.

378. Newman, C.R., C.D. Frisbie, D.A. da Silva Filho, J.-L. Brédas, P.C. Ewbank, and K.R. Mann. 2004. Introduction to organic thin film transistors and design of n-channel organic semiconductors. *Chem Mater* 16:4436–4451.

379. Ando, S., R. Murakami, J.-I. Nishida, H. Tada, Y. Inoue, S. Tokito, and Y. Yamashita. 2005. n-Type organic field-effect transistors with very high electron mobility based on thiazole oligomers with trifluoromethylphenyl groups. *J Am Chem Soc* 127:14996–14997.

380. Facchetti, A., M. Mushrush, M.-H. Yoon, M.H. Hutchison, M.A. Ratner, and T.J. Marks. 2004. Building blocks for n-type molecular and polymeric electronics. Perfluoroalkyl- versus alkyl-functionalized oligothiophenes(nT; $n = 2$–6). Systematics of thin film microstructure, semiconductor performance, and modeling of majority charge injection in field-effect transistors. *J Am Chem Soc* 126:13859–13874.

381. Facchetti, A., M.-H. Yoon, C.L. Stern, H.E. Katz, and T.J. Marks. 2003. Building blocks for n-type organic electronics: Regiochemically modulated inversion of majority carrier sign in perfluoroarene-modified polythiophene semiconductors. *Angew Chem Int Ed Engl* 42:3900–3903.

382. Meng, H., and F. Wudl. 2001. A robust low band gap processable n-type conducting polymer based on poly(isothianaphthene). *Macromolecules* 34:1810–1816.

383. Yu, W.-L., H. Meng, J. Pei, and W. Huang. 1998. Tuning redox behavior and emissive wavelength of conjugated polymers by p–n diblock structures. *J Am Chem Soc* 120:11808–11809.

384. Letizia, J.A., A. Fachetti, C.L. Stern, M.A. Ratner, and T.J. Marks. 2005. High electron mobility in solution-cast and vapor-deposited phenacyl-quaterthiophene-based field-effect transistors: Towards n-type polythiophenes. *J Am Chem Soc* 127:13476–13477.

385. Sirringhaus, H., R.J. Wilson, R.H. Friend, M. Inbasekaran, W. Wu, E.P. Woo, M. Grell, and D.C.C. Bradley. 2000. Mobility enhancement in conjugated polymer FETs through chain alignment in a liquid crystalline phase. *Appl Phys Lett* 77:406–408.

386. Sirringhaus, H., T. Kawase, R.H. Friend, T. Shimoda, M. Inbasekaran, W. Wu, and E.P. Woo. 2000. High-resolution inkjet printing of all-polymer transistor circuits. *Science* 290 (5499):2123–2126.

387. Lim, E., B.-J. Jung, J. Lee, H.-K. Shim, J.-I. Lee, Y.S. Yang, and L.-M. Do. 2005. Thin-film morphologies and solution-processable field-effect transistor behavior of a fluorene-thieno[3,2-b]thiophene-based conjugated copolymer. *Macromolecules* 38:4531–4535.

388. Yamamoto, T., H. Kotubo, M. Kobashi, and Y. Sakai. 2004. Alignment and field-effect transistor behavior of an alternative π-conjugated copolymer of thiophene and 4-alkylthiazole. *Chem Mater* 16:4616–4618.

389. Zhu, Y., A. Babel, and S.A. Jenekhe. 2005. Phenoxazine-based conjugated polymers: A new class of organic semiconductors for field-effect transistors. *Macromolecules* 38:7983–7991.

390. Crouch, D.J., P.J. Skabara, J.E. Lohr, J.J.W. McDouall, M. Heeney, I. McCulloch, D. Sparrowe, M. Shkunov, S.J. Coles, P.N. Horton, and M.B. Hursthouse. 2005. Thiophene and selenophene copolymers incorporating fluorinated phenylene units in the main chain: Synthesis, characterization, and application in organic field-effect transistors. *Chem Mater* 17:6567–6578.

391. Fujigami, H., A. Tsumura, and H. Koezuka. 1993. Polythienylenevinylene thin-film transistor with high carrier mobility. *Appl Phys Lett* 63:1372–1374.

392. Huitema, H.E.A., G.H. Gelinck, J.B.P.H. van der Putten, K.E. Kuijk, C.M. Hart, E. Cantatore, P.T. Herwig, A.J.J.M. van Breemen, and D.M. de Leeuw. 2001. Plastic transistors in active matrix displays. *Nature* 414:599.

393. Huitema, H.E.A., G.H. Gelinck, J.B.P.H. van der Putten, K.E. Kuijk, C.M. Hart, E. Cantatore, and D.M. de Leeuw. 2002. Active-matrix displays driven by solution processed polymeric transistors. *Adv Mater* 14:1201–1204.

394. Prins, P., L.P. Candeias, A.J.J.M. van Breemen, J. Sweelssen, P.T. Herwig, F.M. Schoo, and L.D.A. Siebbeles. 2005. Electron and hole dynamics on isolated chains of a solution-processable poly(thienylenevinylene) derivative in dilute solution. *Adv Mater* 17:718–723.

395. Videlot, C., J. Ackermann, P. Blanchard, J.M. Raimundo, P. Frère, M. Allain, R. de Bettignies, E. Levillain, and J. Roncali. 2003. Field-effect transistors based on oligothienylenevinylenes: From solution π-dimers to high mobility organic semiconductors. *Adv Mater* 15:306–310.

396. Torsi, L., N. Cioffi, C. Di Franco, L. Sabbatini, P.G. Zambonin, and T. Bleve-Zacheo. 2001. Organic thin film transistors: From active materials to novel applications. *Solid State Electro* 45:1479–1485.

397. Assadi, A., G. Gustafsson, M. Willander, C. Svensson, and O. Inganäs. 1990. Determination of field effect mobility of poly(3-hexylthiophene) upon exposure to NH_3 gas. *Synth Met* 37:123–130.

398. Ohmori, Y., H. Takahashi, K. Muro, M. Uchida, T. Kawai, and K. Yoshino. 1991. Gas-sensitive Schottky gated field effect transistor utilizing poly(3-alkylthiophene) films. *J Apply Phys* 30:L1247–L1249.

399. Ohomori, Y., K. Muro, and K. Yoshino. 1993. Gas-sensitive and temperature-dependent Schottky gated field effect transistors utilizing poly(3-alkylthiophene)s. *Synth Met* 57:4111–4116.

400. Torsi, L., A. Tafuri, N. Cioffi, M.C. Gallazzi, A. Sassella, L. Sabbatini, and P.G. Zambonin. 2003. Regioregular polythiophene field-effect transistors employed as chemical sensors. *Sens Actuators B* 93:257–262.

401. Torsi, L., M.C. Tanese, N. Cioffi, M.C. Gallazzi, L. Sabbatini, and P.G. Zambonin. 2004. Alkoxy-substituted polyterthiophene thin-film-transistors as alcohol sensors. *Sens Actuators B* 98:204–207.

402. Torsi, L., M.C. Tanese, N. Cioffi, M.C. Gallazzi, L. Sabbatini, P.G. Zambonin, G. Raos, S.V. Meille, and M.M. Giangregorio. 2003. Side-chain in chemically sensing conducting polymer field-effect transistors. *J Phys Chem B* 107:7589–7594.

403. Assadi, A., C. Svensson, M. Willander, and O. Inganas. 1988. Field-effect mobility of poly(3-hexylthiophene). *Appl Phys Lett* 53:195–197.

404. Glenis, S., G. Horowitz, G. Tourillon, and F. Garnier. 1984. Electrochemically grown polythiophene and poly(3-methylthiophene) organic photovoltaic cells. *Thin Solid Films* 111:93–103.

405. Sariciftci, N.S., L. Smilowitz, A.J. Heeger, and F. Wudl. 1992. Photoinduced electron transfer from a conducting polymer to buckminsterfullerene. *Science* 258:1474–1476.

406. Yu, G., J. Gao, J.C. Hummelen, F. Wudl, and A.J. Heeger. 1995. Polymer photovoltaic cells: Enhanced efficiencies via a network of internal donor–acceptor heterojunctions. *Science* 270:1789–1791.

407. Padinger, F., R.S. Rittberger, and N.S. Sariciftci. 2003. Effect of postproduction treatment on plastic solar cells. *Adv Funct Mater* 13:85–88.

408. Li, G., V. Shrotriya, J. Huang, Y. Yao, T. Moriarty, K. Emery, and Y. Yang. 2005. High-efficiency solution processable polymer photovoltaic cells by self-organisation of polymer blends. *Nature Mater* 4:864–868.

409. Ma, W., C. Yang, X. Gong, K. Lee, and A.J. Heeger. 2005. Thermally stable, efficient polymer solar cells with nanoscale control of the interpenetrating network morphology. *Adv Mater* 15:1617–1622.

410. Granström, M., K. Petritsch, A.C. Arias, A. Lux, M.R. Andersson, and R.H. Friend. 1998. Laminated fabrication of polymeric photovoltaic diodes. *Nature* 395:257–260.

411. Pettersson, L.A.A., S.R. Lucimara, and O. Inganäs. 1999. Modeling photocurrent action spectra of photovoltaic devices based on organic thin films. *J Apply Phys* 86:487–496.

412. Gadisa, A., M. Svensson, M.R. Andersson, and O. Inganäs. 2004. Correlation between oxidation potential and open-circuit voltage of composite solar cells based on blends of polythiophenes/fullerene derivative. *Appl Phys Lett* 84:1609–1611.

413. Scharber, M.C., D. Mühlbacher, M. Koppe, P. Denk, C. Waldauf, A.J. Heeger, and C.J. Brabec. 2006. Design for donors in bulk-heterojunction solar cells-towards 10% energy-conversion efficiency. *Adv Mater* 18:789–794.

414. Winder, C., and N.S. Sariciftci. 2004. Low bandgap polymers for photon harvesting in bulk heterojunction solar cells. *J Mater Chem* 14:1077–1086.

415. Dhanabalan, A., K.J. van Duren, P.A. van Hal, J.L.J. van Dongen, and R.A.J. Janssen. 2001. Synthesis and characterization of a low bandgap conjugated polymer for bulk heterojunction photovoltaic cells. *Adv Funct Mater* 11:255–262.

416. Svensson, M., F. Zhang, S.C. Veenstra, W.J.H. Verhees, J.C. Hummelen, J.M. Kroon, O. Inganäs, and M.R. Andersson. 2003. High-performance polymer solar cells of an alternating polyfluorene copolymer and a fullerene derivative. *Adv Mater* 15:988–991.

417. Wang, X., E. Perzon, F. Oswald, F. Langa, S. Admassie, M.R. Andersson, and O. Inganäs. 2005. Enhanced photocurrent spectral response in low-bandgap polyfluorene and C_{70}-derivative-based solar cells. *Adv Funct Mater* 15:1665–1670.

418. Zhang, F., E. Perzon, X. Wang, W. Mammo, M.R. Andersson, and O. Inganäs. 2005. Polymer solar cells based on a low-bandgap fluorene copolymer and a fullerene derivative with photocurrent extended to 850 nm. *Adv Funct Mater* 15:745–750.

419. Wang, F., J. Luo, K. Yang, J. Chen, F. Huang, and Y. Cao. 2005. Conjugated fluorene and silole copolymers: Synthesis, characterization, electronic transistion, light emission, photovoltaic cell, and field effect hole mobility. *Macromolecules* 38:2253–2260.

420. Vangeneugden, D.L., D. Vanderzande, J. Salbeck, P.A. Van Hal, R.A.J. Janssen, J.C. Hummelen, C.J. Brabec, S.E. Shaheen, and N.S. Sariciftci. 2001. Synthesis and characterization of poly(1,3-dithienylisothianaphthene) derivative for bulk heterojunction photovoltaic cells. *J Phys Chem B* 105:11106–11113.

421. Goris, L., M.A. Loi, A. Cravino, H. Neugebauer, N.S. Sariciftci, I. Polec, L. Lutsen, E. Andries, J. Manca, L. De Schepper, and D. Vanderzande. 2003. Poly(5,6-dithiooctylisothianaphthene), a new low band gap polymer: Spectroscopy and solar cell construction. *Synth Met* 138:249–253.

422. Henckens, A., M. Knipper, I. Polec, J. Manca, L. Lutsen, and D. Vanderzande. 2004. Poly(thienylene vinylene) derivatives as low band gap polymers for photovoltaic applications. *Thin Solid Films* 451–452:572–579.

423. Campos, L.M., A. Tontcheva, S. Günes, G. Sonmez, H. Neugebauer, N.S. Sariciftci, and F. Wudl. 2005. Extended phocurrent spectrum of a low band gap polymer in a bulk heterojunction solar cell. *Chem Mater* 17:4031–4033.

424. Smith, A.P., R.R. Smith, B.E. Taylor, and M.F. Durstock. 2004. An investigation of poly(thienylene vinylene) in organic photovoltaic devices. *Chem Mater* 16:4687–4692.

425. Loewe, R.S., and R.D. McCullough. 2000. Effects of structural regioregularity on the properties of poly(3-alkylthienylenevinylenes). *Chem Mater* 12:3214–3221.

426. Berridge, R., P.J. Skabara, C. Pozo-Gonzalo, A. Kanibolotsky, J. Lohr, W. McDouall, E.J.L. McInnes, J. Wolowska, C. Winder, N.S. Sariciftci, R.W. Harrington, and W. Clegg. 2006. Incorporation of fused tetrathiafulvalenes (TTFs) into polythiophene architectures: Varying the electroactive dominance of the TTF species in hybrid systems. *J Phys Chem B* 110:3140–3152.

427. Cravino, A., G. Zerza, M. Maggini, S. Bucella, M. Svensson, M.R. Andersson, H. Neugebauer, and N.S. Sariciftci. 2000. A novel polythiophene with pendant fullerenes: Toward donor/acceptor double-cable polymers. *Chem Commun* 2487–2488.

428. Zhang, F., M. Svensson, M.R. Andersson, M. Maggini, S. Bucella, E. Menna, and O. Inganäs. 2001. Soluble polythiophenes with pendant fullerene groups as double cable materials for photodiodes. *Adv Mater* 13:1871–1874.

429. Cravino, A., and N.S. Sariciftci. 2002. Double-cable polymers for fullerene based organic optoelectronic applications. *J Mater Chem* 12:1931–1943.

430. Cravino, A., and N.S. Sariciftci. 2003. Molecules as bipolar conductors. *Nat Mater* 6:360–361.

431. Benincori, T., E. Brenna, F. Sannicolò, L. Trimarco, G. Zotti, and P. Sozzani. 1996. The first "charm bracelet" conjugated polymer: An electroconducting polythiophene with covalently bound fullerene moieties. *Angew Chem Int Ed Engl* 35:648–651.

432. Ferraris, J.P., A. Yassar, D.C. Loveday, and M. Hmyene. 1998. Grafting of buckminsterfullerene onto polythiophene: Novel intramolecular donor–acceptor polymers. *Opt Mater* 9:34–42.

433. Cravino, A., G. Zerza, H. Neugebauer, M. Maggini, S. Bucella, E. Menna, M. Svensson, M.R. Andersson, C.J. Brabec, and N.S. Sariciftci. 2002. Electrochemical and photophysical properties of a novel polythiophene with pendant fulleropyrrolidine moieties: Toward "double cable" polymers for optoelectronic devices. *J Phys Chem B* 106:70–76.

434. Yamazaki, T., Y. Murata, K. Komatsu, K. Furukawa, M. Morita, N. Maruyama, T. Yamao, and S. Fujita. 2004. Synthesis and electrolytic polymerization of the ethylenedioxy-substituted terthiophene-fullerene dyad. *Org Lett* 6:4865–4868.

435. Sonmez, G., C.K.F. Shen, Y. Rubin, and F. Wudl. 2005. The unusual effect of bangap lowering by C_{60} on a conjugated polymer. *Adv Mater* 17:897–900.

436. Cravino, A., G. Zerza, M. Maggini, S. Bucella, M. Svensson, M.R. Andersson, H. Neugebauer, C.J. Brabec, and N.S. Sariciftci. 2003. A soluble donor–acceptor double-cable polymer: Polythiophene with pendant fullerenes. *Monatsh Chem* 134:519–527.

437. Huchet, L. 1998. Ph.D. thesis. University of Angers, France.

438. Catellani, M., S. Luzzati, N.-O. Lupsac, R. Mendichi, R. Consonni, A. Famulari, S.V. Meille, F. Giacalone, J.L. Segura, and N. Martin. 2004. Donor–acceptor polythiophene copolymers with tunable acceptor content for photoelectric conversion devices. *J Mater Chem* 14:67–74.

439. Giacalone, F., J.L. Segura, N. Martin, M. Catellani, S. Luzzati, and N.-O. Lupsac. 2003. Synthesis of soluble donor–acceptor double-cable polymers based on polythiophene and tetracyanoanthraquinodimethane (TCAQ). *Org Lett* 5:1669–1672.

440. Martinez-Diaz, M., S. Esperanza, A. de la Escosura, M. Catellani, S. Yunus, S. Luzzati, and T. Torres. 2003. New polythiophenes bearing electron-acceptor phthalocyanine chromophores. *Tetrahedron Lett* 44:8475–8478.

441. You, C.-C., C.R. Saha-Möller, and F. Würthner. 2004. Synthesis and electropolymerization of novel oligothiophene-functionalized perylene bisimides. *Chem Commun* 2030–2031.

442. Chen, S., Y. Liu, W. Qiu, X. Sun, Y. Ma, and D. Zhu. 2005. Oligothiophene-functionalized perylene bisimide system: Synthesis, characterization, and electrochemical polymerization properties. *Chem Mater* 17:2208–2215.

443. Segura, J.L., R. Gomez, E. Reinold, and P. Bäuerle. 2005. Synthesis and electropolymerization of a perylenebisimide-functionalized 3,4-ethylenedioxythiophene (EDOT) derivative. *Org Lett* 7:2345–2348.

444. Cremer, J., E. Mena-Osteritz, N.G. Pschierer, K. Müllen, and P. Bäuerle. 2005. Dye-functionalized head-to-tail coupled oligo(3-hexylthiophenes)-perylene-oligothiophene dyads for photovoltaic applications. *Org Biomol Chem* 3:985–995.

445. Huynh, W.U., J.J. Dittmer, and A.P. Alivisatos. 2002. Hybrid nanorod-polymer solar cells. *Science* 295:2425–2427.

446. Liu, J., T. Tanaka, K. Sivula, P.A. Alivisatos, and J.M.J. Fréchet. 2004. Employing end-functional polythiophene to control the morphology of nanocrystal-polymer composite in hybrid solar cells. *J Am Chem Soc* 126:6550–6551.

447. Liu, J., E.N. Kadnikova, Y. Liu, M.D. McGehee, and J.M.J. Fréchet. 2004. Polythiophene containing thermally removable solubilizing groups enhances the interface and the performance of polymer-titania hybrid solar cells. *J Am Chem Soc* 126:9486–9487.

448. Grant, C.D., A.M. Schwartzberg, G.P. Smestad, J. Kowalik, L.M. Tolbert, and J. Zhang. 2003. Optical and electrochemical characterization of poly(3-undecy-2,2′-bithiophene) in thin film solid state TiO_2 photovoltaic solar cells. *Synth Met* 132:197–204.

449. Wang, L., J.-S. Ji, Y.-J. Lin, and S.-P. Rwei. 2005. Novel poly(3-nonylthiophene)-TiO_2 hybrid materials for photovoltaic cells. *Synth Met* 155:677–680.

14

Structure–Property Relationships and Applications of Conjugated Polyelectrolytes

Kirk S. Schanze
and Xiaoyong Zhao

14.1 Introduction

π-Conjugated polymers such as poly(*para*-phenylene), poly(phenylene vinylene), poly(phenylene ethynylene), and polythiophene (PPP, PPV, PPE, and PTh, Scheme 14.1) have been of considerable interest for more than two decades, and during the past several years it has become increasingly evident that these materials will play an important role in new advanced optical, electronic, and optoelectronic devices. Delocalized π-electron energy levels and variations in extent of conjugation imbue conjugated polymers with a variety of useful properties including strong optical absorption and emission, photovoltaic response, reversible electrochemical switching and efficient charge, and exciton transport.

The processibility of conjugated polymers such as PPV is enhanced considerably by incorporation of solubilizing groups. For example, PPV is insoluble, however, substitution of alkyl or alkyloxy side chains on the phenylene rings renders the material soluble in common organic solvents [1]. The strategy of using alkyl- or alkyloxy-groups to render conjugated polymers soluble in and processible from organic solvents is now widely used and plays a key role in determining the electronic and optical properties of the materials after they are processed into active layers.

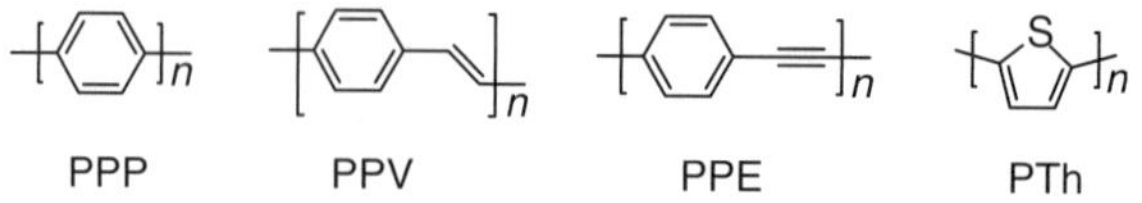

PPP   PPV   PPE   PTh

SCHEME 14.1

Over the past decade, there has been increasing interest in the properties of conjugated polymers that contain solubilizing side chains that feature charged ionic groups [2–11]. Examples of these materials are $PPV\text{-}SO_3^-$, $PPP\text{-}NEt_3^+$, $PPE\text{-}SO_3^-$, and $PT\text{-}CO_2^-$ (Scheme 14.2). These conjugated polyelectrolytes (CPEs) feature a variety of unique and useful materials properties. First, due to the charged and polar nature of the ionic side groups, CPEs are soluble in water and polar organic solvents such as alcohols. Second, CPEs interact strongly with other ionic species, such as metal ions, molecular ions, polyelectrolytes, proteins, and DNA. Third, due to their ionic and amphiphilic character, CPEs self-assemble into supramolecular assemblies such as colloids and polyelectrolyte layer-by-layer (LbL) films. These supramolecular assemblies are organized on the nanometer length scale. Moreover, the electronic and optical properties of these nanoassemblies can be controlled by synthetic chemistry at the molecular level [12].

In this chapter we will highlight some of the recent work carried out in the area of CPEs. The chapter is not designed to be an exhaustive review of the area—rather various areas of interest to the authors will be highlighted by taking examples from their own research, along with work from other research groups active in the field. The chapter will begin with a historical perspective describing work that was key in defining the field of CPEs as it stands today. This will be followed by sections covering properties and applications of CPEs. One important application that will not be covered in detail is the application of CPEs to optical biosensors. This area is covered in the chapter by Nilsson and Inganäs.

14.2 Historical Perspective

The first report of a CPE was published by Wudl, Heeger and coworkers in 1987 [13]. This group reported the synthesis and characterization of the conductivity of films of PThs substituted with ethylsulfonate and butylsulfonate side chains ($PT\text{-}PSO_3^-$ and $PT\text{-}BSO_3^-$, Scheme 14.3). The polymers

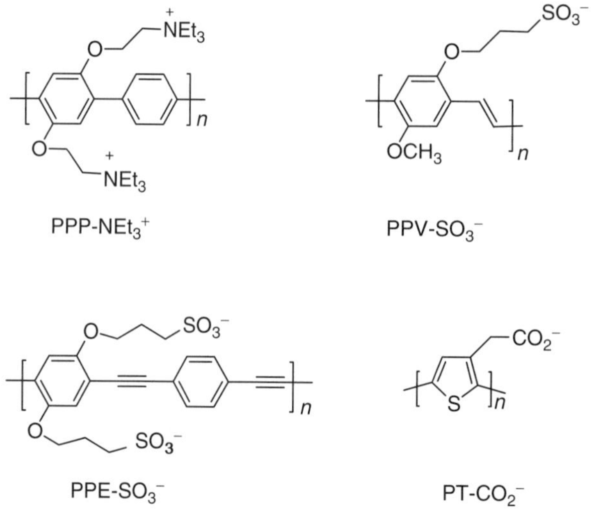

PPP-NEt$_3^+$ PPV-SO$_3^-$

PPE-SO$_3^-$ PT-CO$_2^-$

SCHEME 14.2

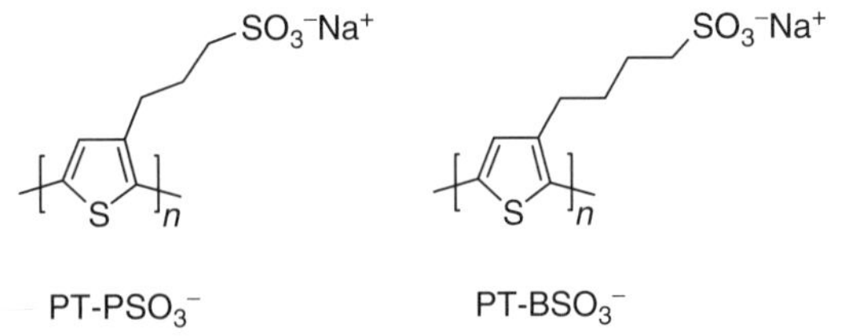

SCHEME 14.3

were prepared by electropolymerization of the corresponding methyl sulfonate esters, followed by conversion to the sodium salts by treatment with sodium iodide. The electropolymerized films of the polyelectrolytes were soluble in polar organic solvents as well as in water. The focus of this work was on the preparation of self-doped conducting polymers, where the pendant sulfonate groups could act as counterions to compensate the backbone charge in the oxidized forms of the polymers. This manuscript did not provide any data concerning the absorption or emission properties of the materials in solution or in the film state.

Several years later, Shi and Wudl reported a synthesis of the PPV derivative $PPV-SO_3^-$ (Scheme 14.2) that features a sulfonate-functionalized alkoxy side chain [14]. The material was synthesized by using the Wessling approach, and the sodium salt form of the material was water soluble. The study focused mainly on the chemical synthesis; however, it was noted that aqueous solutions of $PPV-SO_3^-$ exhibit an intense π, π^* absorption with $\lambda_{max} \approx 510$ nm. This absorption is very similar to that observed for organic soluble PPV derivatives. While this work did not provide much insight into the unique properties of conjugated polymers that are substituted with ionic solubilizing group, it is a key paper because the material that was synthesized by Shi and Wudl was used in several important later investigations which demonstrated the utility of CPEs in biosensor schemes based on fluorescence [6].

Another investigation which played an important role in the development and applications of CPEs was the work by Swager and coworkers which was the first to describe the concept of "amplified fluorescence quenching" in fluorescent conjugated polymers. Although this work was focused on neutral, organic soluble polymers (not CPEs), the concepts that were developed in the studies by the MIT group were important, because the fluorescence sensor concepts developed in this work have been applied extensively using CPEs. The concept first reported in 1995 centers on the use of a fluorescent conjugated polymer that is functionalized with receptor sites for a target analyte molecule. The target analyte molecule acts as a fluorescence quencher [15,16]. Because of the ability of the conjugated polymer backbone to serve as an efficient "conduit" for excitons (the *molecular wire effect*), a single analyte molecule that is bound to the receptor site can quench many repeat units in the polymer chain. Thus, the functionalized polymer is quenched much more efficiently compared to a solution that contains the same concentration of monomeric fluorophores (Figure 14.1). In one of the systems explored in this work cyclophane receptor sites were attached to each repeat unit in the conjugated polymer chain. The cyclophane units bind to the N,N'-dimethyl-4,4'-bipyridinium (MV^{2+}) ion, which quenches the fluorescence through a charge transfer mechanism. Stern–Volmer (SV) quenching experiments carried out on the polymer and a structurally analogous model compound at the same receptor unit concentration demonstrated that the polymer amplifies the quenching effect 60-fold. In both the polymer and the model compound the process is dominated by static quenching, which indicates that the active quenching unit is the receptor-bound MV^{2+}. Although this study did not directly lead to a sensor, the concept that was developed in the study pointed to a general method of using conjugated polymers as amplifying elements in fluorescence sensor schemes. Before this work was reported, most sensor schemes that used conjugated polymers as the active elements relied upon changes in conductivity or electrochemical response.

Isolated fluorescent chemosenors

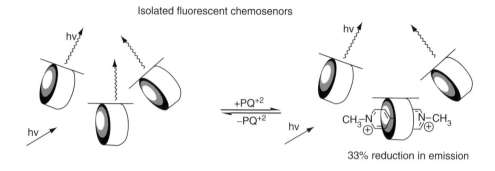

33% reduction in emission

Fluorescent chemosensors "wired in series"

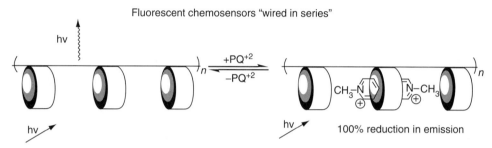

100% reduction in emission

FIGURE 14.1 Scheme illustrating the amplified quenching concept. Upper part of diagram shows that a single quencher (MV^{2+}, here called PQ^{2+}) quenches the fluorescence of a single chromophore in an ensemble. Lower part of the diagram shows how a single quencher can quench the fluorescence of an entire polymer chain. (Reprinted from Zhou, Q., T.M. Swager. *J. Am. Chem. Soc.*, 117, 12593–12602, 1995. With permission.)

In 1999 Whitten, McBranch and coworkers reported an investigation that set the stage for the recent considerable upsurge in interest in the properties and applications of CPEs [6]. Two key findings were reported in a single publication in *Proceedings of the National Academy of Sciences* (PNAS), which has been cited by a significant number of subsequent papers. First, the study demonstrated that the fluorescence of an aqueous solution of PPV-SO_3^- (Scheme 14.2) is quenched by the cationic quencher MV^{2+} with a SV constant of $\sim10^7$ M^{-1}. (This corresponds to $\sim$50% quenching at 100 mM [MV^{2+}]; see Section 14.6 for a discussion of SV quenching.) The authors attributed this effect, which they temed "superquenching," to binding of the cationic quencher to the anionic polyelectrolyte by ion pairing combined with extremely rapid diffusion of the exciton along the conjugated polymer backbone and between polyelectrolyte chains in aggregates. The superquenching effect is closely related to the amplified quenching effect reported several years earlier by Swager and coworkers [15,16], except that the degree of amplification achieved in the aqueous based PPV-SO_3^-/MV^{2+} system is several orders of magnitude larger than observed in the organic soluble system.

The second important result reported by Whitten and McBranch is the application of the super-quenching effect in the CPE quencher system to the development of a highly sensitive fluorescence-based sensor for the biotin–avidin interaction. Figure 14.2 illustrates the approach taken to this problem. Specifically, a quencher-tether-ligand (QTL) assembly was synthesized in which a viologen quencher is linked to a biotin unit by a flexible tether chain. The QTL assembly is then mixed with a solution of PPV-SO_3^-; because of the superquenching effect, only a small concentration of the biotin–viologen QTL is needed to quench a substantial fraction of the polymer's fluorescence. Then the sensor response is demonstrated when avidin is added to the PPV-SO_3^-/QTL mixture—small amounts of added avidin result in recovery of the PPV-SO_3^- fluorescence. The fluorescence recovery observed upon addition of avidin to the PPV-SO_3^-/QTL mixture was believed to arise because when the QTL conjugate is bound to

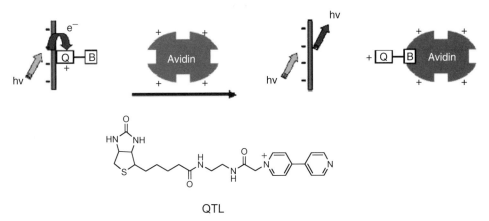

QTL

FIGURE 14.2 Mechanism for the quench/unquench response in the PPV-SO$_3^-$/QTL/avidin system. Addition of the QTL (labeled Q-B in the diagram) quenches the PPV-SO$_3^-$ fluorescence. Subsequent addition of avidin removes the QTL from the proximity of the polymer, causing the polymer's fluorescence to turn on. The chemical structure of the biotin–viologen QTL is shown at the bottom. (Reprinted from Dwight, S.J., B.S. Gaylord, J.W. Hong, and G.C. Bazan. *J. Am. Chem. Soc.*, 126, 16850–16859, 2004. With permission.)

the protein by the biotin ligand, the viologen moiety cannot approach the PPV-SO$_3^-$ closely enough to induce quenching.

While this mechanism seems reasonable, in a recent study Bazan and coworkers showed that addition of proteins that do not bind to biotin also induce an increase in the fluorescence of a PPV-SO$_3^-$/QTL mixture [17]. They argue that nonspecific interactions between proteins and the polymer change the polymer's conformation and state of aggregation, thereby influencing its intrinsic emission quantum yield and the quenching efficiency of the viologen-based QTL. Regardless of the fact that there may be some question as to the mechanism for the avidin-induced increase in fluorescence intensity in the PPV-SO$_3^-$/QTL system, the work by Whitten and McBranch remains significant because it demonstrated a general approach to using fluorescent conjugated polymers in biosensor applications, and this finding has set the stage for the recent surge of interest in CPEs.

14.3   Synthesis of Conjugated Polyelectrolytes

Due to the widespread interest in the properties and applications of CPEs, a variety of synthetic approaches have been developed for preparing of this family of materials. Because of the large body of work in this area, it is not possible to provide a comprehensive review in this chapter. Several reviews that are focused on the synthesis of conjugated polymers [1,12,18,19], especially on CPEs [20], are available in recent literature. Herein we present a few representative examples to illustrate methods that have been used to prepare CPEs.

Two general approaches have been used to prepare CPEs. The first approach (direct method) involves polymerization of monomers that already contain ionic groups to directly afford the CPE. The second approach (indirect method) involves polymerization of monomers in which the ionic groups are protected or masked; the ionic groups are produced in a subsequent reaction that is carried out on the polymer. Each of the two approaches has advantages and disadvantages. The direct method has the advantage that since the polymerization reaction produces the CPE, a postpolymerization functionalization reaction is not required, and consequently there is no danger that the polymer will not be completely functionalized. However, the direct approach has a major disadvantage in that it is very difficult to characterize the molecular weight of a CPE by using traditional methods such as gel permeation chromatography (GPC). In the indirect method, the polymer that is obtained is typically

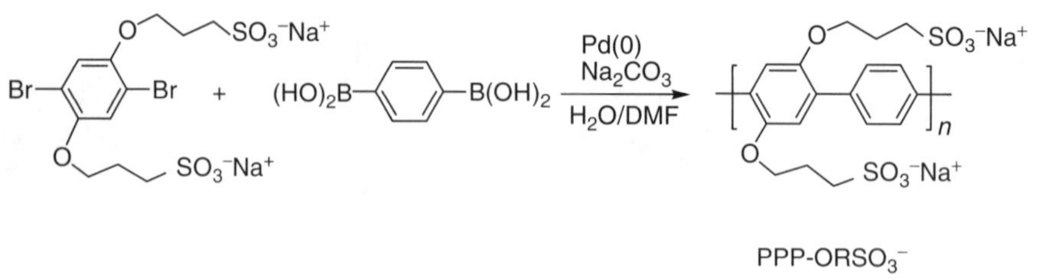

PPP-ORSO$_3^-$

SCHEME 14.4

soluble in organic solvents such as THF or CHCl$_3$, and consequently its molecular weight can be characterized by GPC. The disadvantage of the indirect approach is that the ionic groups are introduced in a postpolymerization reaction, and it is possible that complete conversion will not be achieved. This has the potential to afford a CPE in which there is some heterogeneity in the functional groups that are appended to the backbone.

Regardless of whether the direct or indirect approach is used to synthesize a particular CPE, most synthetic methods that have been used rely on step growth polymerization catalyzed by a transition metal catalyst. The reactions most frequently used are Suzuki coupling of bis(aryl bromides) with bis(aryl boronic acids) and Sonogashira coupling of bis(aryl iodides) with bis(terminal acetylenes). The first poly(p-phenylene)-based CPE was prepared by Reynolds and Child in a direct method which utilized a Suzuki coupling reaction. As shown in Scheme 14.4, these authors prepared the sulfonate-substituted PPP polymer (PPP-ORSO$_3^-$) by Suzuki coupling of an aryl dibromide with 1,4-benzene-bisboronic acid [21]. Subsequently this group optimized the Suzuki coupling reaction conditions, and they were able to obtain (PPP-ORSO$_3^-$) with a degree of polymerization (X_n) of 42 repeat units as assessed by NMR end-group analysis [2]. The sulfonated poly(phenylene ethynylene)-based CPE, PPE-SO$_3^-$, was also prepared by the direct method using a Sonogashira coupling reaction carried out in aqueous/dimethyl formamide (DMF) solution (Scheme 14.5). End group analysis of the PPE-SO$_3^-$ obtained from this synthesis indicates that the polymer has $X_n \approx 100$.

In 1999 Reynolds and coworkers reported a cationic PPP-type CPE that was prepared by an indirect approach [22]. PPP-NEt$_3^+$ was synthesized by quaternization of the neutral precursor PPP-NEt$_2$ by reaction of the latter with ethyl bromide in DMSO solution (Scheme 14.6). The neutral polymer was synthesized through Suzuki coupling reactions of boronate ester and bromobenzene monomers (Scheme 14.6). The advantage of this method is that it was possible to determine the molecular weight of the neutral precursor polymer, PPP-NEt$_2$ by GPC before quaternization reaction. The disadvantage is that the PPP-NEt$_2$ to PPP-NEt$_3^+$ quaternization reaction afforded only 90% conversion. A similar indirect method has been used by Bazan and coworkers to prepare cationic polyfluorene type polymers (Scheme 14.7) [23]. In this synthesis, a neutral precursor polyfluorene that contains dimethylamino-alkyl chains was prepared by a Suzuki coupling reaction. The neutral polymer was subsequently quaternized by reaction with excess methyl iodide.

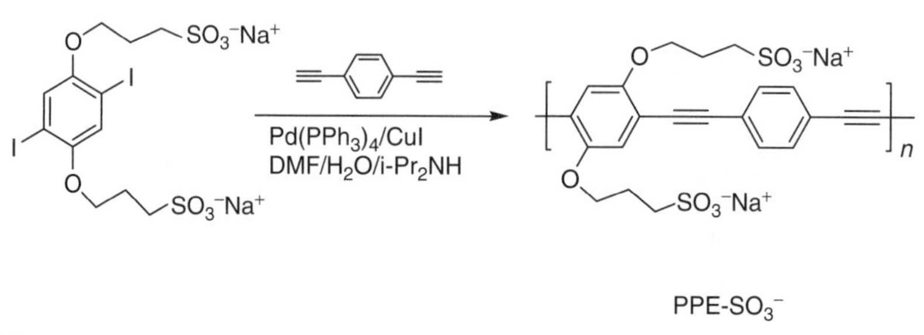

PPE-SO$_3^-$

SCHEME 14.5

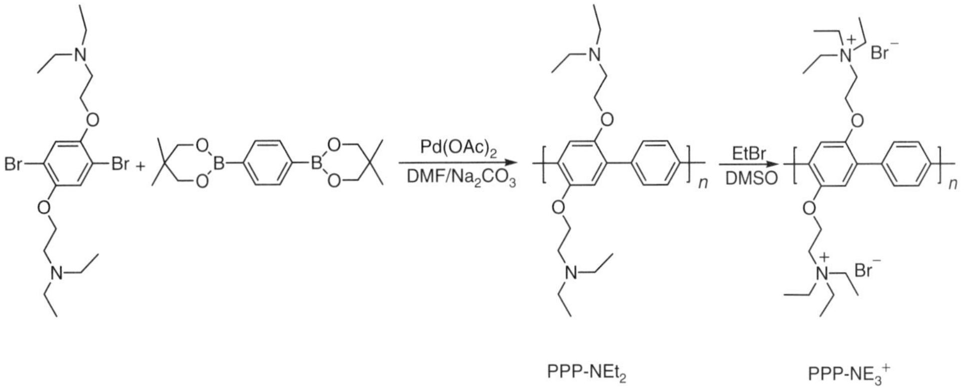

SCHEME 14.6

Esters have also been used extensively as protecting groups in the indirect method for the synthesis of CPEs. For example, carboxylic and phosphonic acid functionalized PPE-type polymers have been prepared by the corresponding neutral ester polymers (Scheme 14.8) [24]. These polymers are prepared by Sonogashira coupling reactions of the ester-functionalized monomers, followed by subsequent conversion of the esters to the acids through hydrolysis. The GPC analysis of the neutral ester-functionalized PPE-type polymers shows that molecular weights ranging from 10,000 to 100,000 are accessible with the Sonogashira coupling reaction.

14.4 Photophysical Properties of Selected Conjugated Polyelectrolytes—Effects of Polymer Aggregation

The optical properties of CPEs are determined by the chemical and electronic structure of the π-conjugated backbone. In general, the optical properties of a given CPE are similar to those of an organic soluble conjugated polymer that has the same backbone structure. However, important modifications in the properties of CPEs are often encountered due to the propensity of these materials to aggregate in solution. Examples of the absorption and emission properties of selected systems are described below.

PPEs are well investigated, both from the fundamental standpoint and with respect to applications in organic optoelectronic devices [18,19]. We have thoroughly investigated the photophysical properties of CPEs in which the backbone is a PPE-type structure. In this section we describe the photophysical properties of several of these PPE-type CPEs taken from our work; these systems provide representative examples of the interesting photophysical properties displayed by CPEs. Early optical studies of PPE-SO_3^- and related CPEs provided a clear indication that under specific solution conditions the polymers aggregate, and the aggregation process induces significant changes in the absorption and fluorescence of the polymers [9,24]. An example of these effects is provided by a study of the absorption and

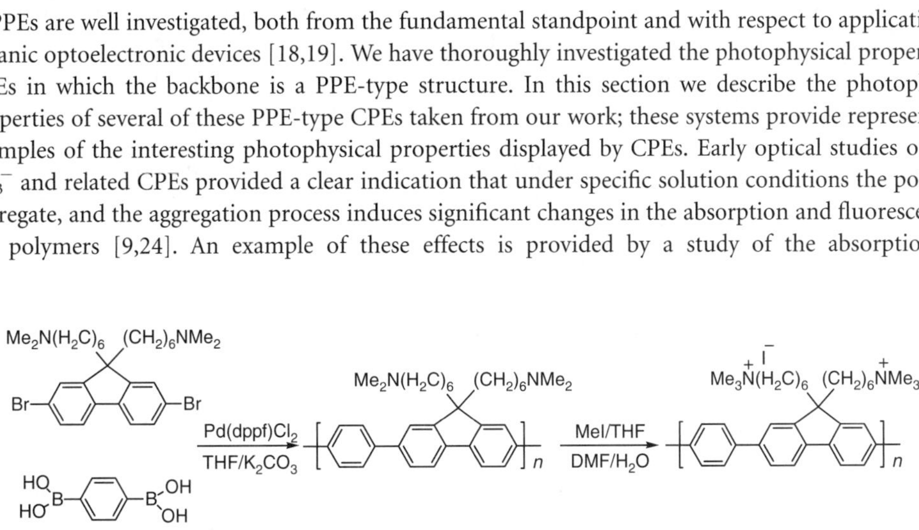

SCHEME 14.7

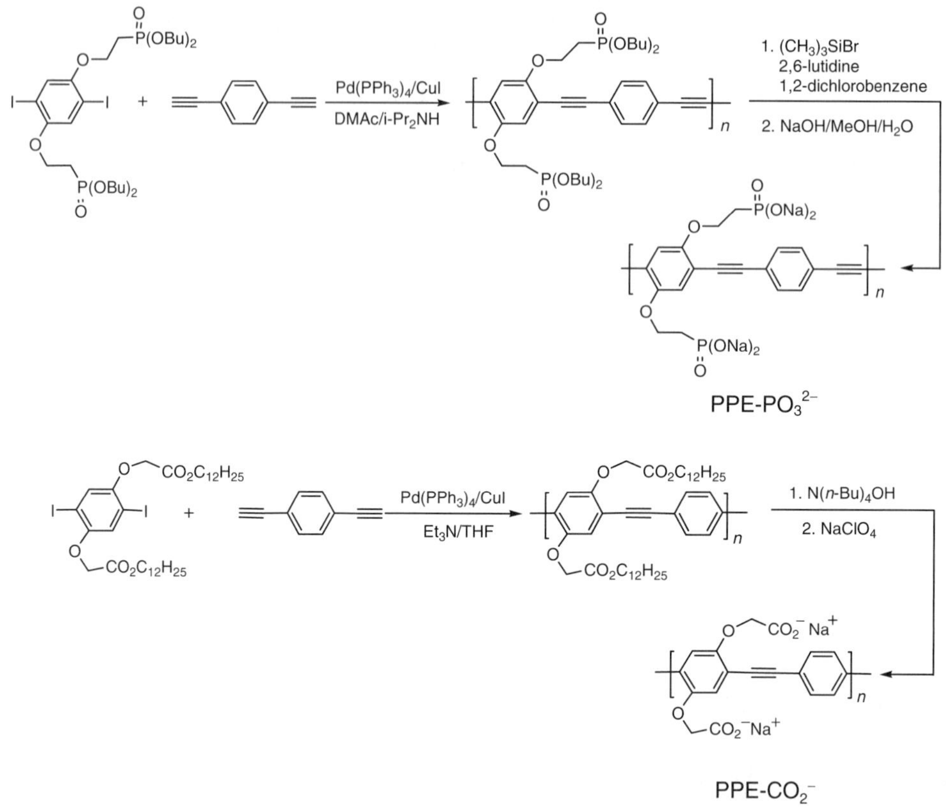

SCHEME 14.8

fluorescence of PPE-SO$_3^-$ in water, methanol, and a 1:1 water/methanol mixture. As the amount of water in the solvent increases, the absorption and fluorescence undergo a red-shift (Figure 14.3). The most pronounced changes are seen in the fluorescence. In methanol the fluorescence appears as a sharp, narrow band that has a very small Stokes shift relative to the absorption, whereas in water the fluorescence appears as a very broad and red-shifted band. The structured emission seen for PPE-SO$_3^-$ in methanol is similar to that reported for structurally related, neutral PPEs in dilute solutions of "good" organic solvents such as CHCl$_3$ or THF, where the polymer chains exist in an unaggregated "monomeric" state [16]. Thus, it is concluded that in methanol the PPE-SO$_3^-$ chains are well solvated and the material exists in a monomeric state. Consequently in this solvent the fluorescence is dominated by excitons that are confined to single chains. By contrast, in aqueous solution it is believed that PPE-SO$_3^-$ is strongly aggregated, and the fluorescence emission is dominated by excitons that are trapped in aggregate states arising from interactions between two or more PPE chains in the aggregate. It is important to note that the aggregate emission observed for PPE-SO$_3^-$ in water is very similar to that observed from excimers (excited π-dimers) of small aromatic fluorophores such as naphthalene or pyrene [25]. From this we conclude that a dominant structural feature in the aggregates must be π–π stacking between two or more polymer chains.

Studies of solvent effects on fluorescence of donor and/or acceptor substituted CPEs demonstrate that in these systems aggregation leads to complete or nearly complete quenching of the fluorescence. For example, in recent work we have shown that the anionic PPE-type CPEs EDOT-PPE-SO$_3^-$ and BDT-PPE-SO$_3^-$ display relatively strong fluorescence in methanol, DMSO, and DMF; however, in water the fluorescence is almost completely quenched. This effect is seen clearly in Figure 14.4 which illustrates the fluorescence of the two polymers in a series of methanol/water mixtures. This effect is believed to

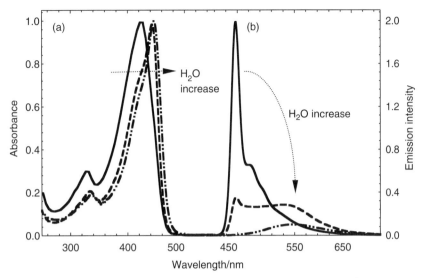

FIGURE 14.3 Absorption (a) and fluorescence (b) of PPE-SO$_3^-$ in methanol, methanol/water (50:50), and water. Arrows show direction of change with increasing water content.

arise because in these systems the presence of the donor (EDOT) or acceptor (BDT) units in the polymer backbone gives charge-transfer character to the aggregate excited state. In essence, the aggregate excited state takes on the character of an exciplex charge-transfer state [25], and consequently the radiative decay rate is strongly suppressed, leading to the low fluorescence quantum yield.

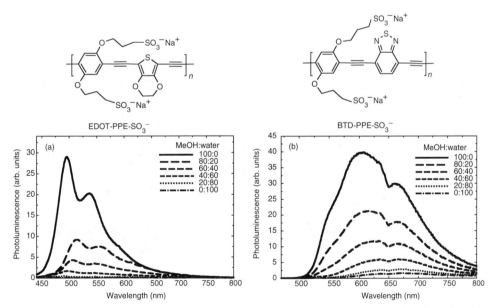

FIGURE 14.4 Fluorescence spectra of EDOT-PPE-SO$_3^-$ (a) and BDT-PPE-SO$_3^-$ (b) in methanol, methanol/water mixtures, and water. Fluorescence intensity decreases with increasing volume fraction of water.

14.5   Self-Assembly of Conjugated Polyelectrolytes: Aggregates, Helices, and Layer-by-Layer Films

As noted in the previous section, the photophysical properties of CPEs are dominated by the tendency of the materials to self-assemble in solution. This tendency of CPEs to undergo self-assembly in water is not surprising in view of the fact that typical CPEs have on average eight or more carbon atoms per ionic charge, and they contain relatively few heteroatoms (except those contained in the ionic group). Because the carbon-rich units are hydrophobic, whereas the ionic units are hydrophilic, the polymers are inherently amphiphilic.

The structure of self-assembled CPE structures have been explored in some detail using electron and scanning probe microscopy techniques (TEM and AFM, respectively). For example, Schnablegger, Antonietti and coworkers used TEM and AFM to explore the morphology of the carboxylate substituted polymer CPPE which was spread onto mica substrates [10]. As shown in the images in Figure 14.5, the polymer forms fibrillar structures. On the basis of analysis of the images, combined with solution phase light- and x-ray scattering data, these authors suggested a mechanism for formation of fibrils which involves sequential aggregation of individual polymer chains into "cylinders," which then assemble into the fibrils that can be seen in the microscopic images. Thünemann reported an investigation of the morphology of polyelectrolyte complexes formed by mixing CPPE (a polyanion) with the dihexadecy-lammonium cation [11,26]. Films cast from the polyelectrolyte complex featured an intense birefrin-gence indicating the presence of mesostructure in the material. Small angle x-ray scattering of the films revealed the presence of an intense scattering peak ($q = 0.3$ nm^{-1}) which corresponded to a *d*-spacing of ~30 nm. The scattering is attributed to the presence of a lamellar structure consisting of packed PPE chains separated by the interdigitated alkyl chains of the hexadecylammonium cations.

As noted in Section 14.4, aggregation of CPEs in solution gives rise to significant changes in the optical properties of the polymers. Although the evidence is indirect, it is believed that some informa-tion concerning the structure of the solution aggregates is provided by analysis of the optical spectra. For example, aggregation of PPE-SO$_3^-$ in water is accompanied by a red-shift and narrowing of the π, π^* absorption of the conjugated backbone (Figure 14.3). The absorption shift is believed to arise due to an increase in conjugation length in the polymer arising from planarization of the phenyl rings caused by packing of the chains in the aggregate [27–29]. This effect suggests that the aggregate structure consists

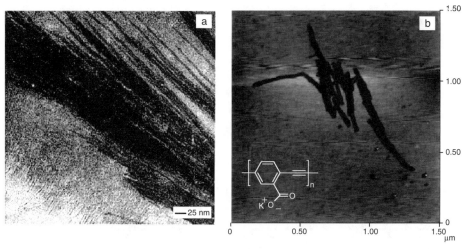

FIGURE 14.5   Transmission electron microscope (a) and atomic force microscope (b) images of aggregates of CPPE (structure shown in right panel) on mica substrates. (Reprinted from Schnablegger, H., M. Antonietti, C. Göltner, J. Hartmann, H. Cölfen, P. Samori, J.P. Rabe, H. Häger, and W. Heitz *J. Colloid. Interf. Sci.*, 212, 24–32, 1999. With permission.)

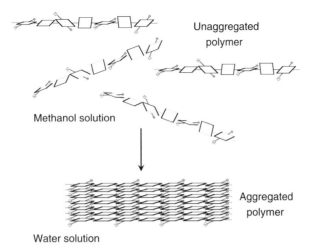

FIGURE 14.6   Effect of polymer aggregation on the structure of PPE-SO$_3^-$. (Reprinted from Tan, C., E. Atas, J.G. Müller, M.R. Pinto, V.D. Kleiman, and K.S. Schanze. *J. Am. Chem. Soc.*, 126, 13685–13694, 2004. With permission.)

of stacked PPE chains (Figure 14.6). The hypothesis regarding the relationship between the aggregate structure and the red-shift, and narrowing of the absorption seen for aggregated PPE-SO$_3^-$ is supported by the result of a very different experiment carried out by Swager and Kim [30]. These authors showed that compression of Langmuir films of facially amphiphilic PPEs induces a significant red-shift and narrowing in the π, π* absorption band, in a very similar pattern as observed for PPE-SO$_3^-$ when it aggregates in water (Figure 14.7). Swager and Kim attributed the spectral shift and narrowing to planarization of the π-electron system which occurs when the Langmuir film is compressed.

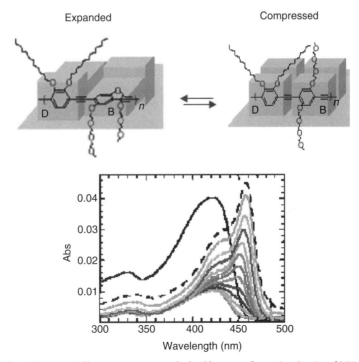

FIGURE 14.7   Effect of Langmuir film compression on the backbone conformation (top) and UV–visible absorption spectrum of a facially amphiphilic PPE-type polymer. Spectra increase red-shift and increase in intensity with surface pressure (compression). (Reprinted from Kim, J., and T.M. Swager. *Nature*, 411, 1030–1034. 2001. With permission.)

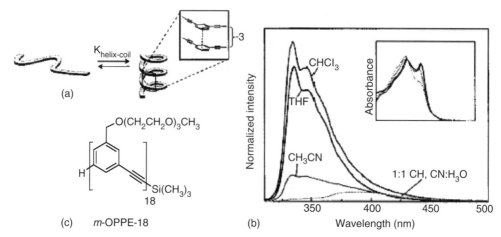

FIGURE 14.8   Coil helix transition for meta-linked PPE oligomer (a) and solvent dependence of absorption and fluorescence of *m*-OPPE-18 (b). Structure of *m*-OPPE-18 (c) is also shown. (Reprinted from Lahiri, S., J.L. Thompson, and J.S. Moore. *J. Am. Chem. Soc.*, 122, 11315–11319, 2000. With permission.)

The primary structure of a conjugated oligomer or polymer can program the material to self-assemble in solution into a helical secondary structure. For example, Moore and coworkers showed that meta-linked oligo(phenylene ethynylene)s (OPEs) spontaneously fold into a helical conformation [31,32]. Helical self-assembly is favored when the oligomers are exposed to "poor solvents"; folding is driven by solvophobic forces and favorable π–π interactions between adjacent aromatic residues in the helical conformation [32,33]. Studies of meta-linked OPEs in different solvents reveal that the random coil to helical conformational interconversion is accompanied by characteristic changes in the absorption and fluorescence spectra. Specifically, the fluorescence of *m*-OPPE-18 decreases significantly in intensity, red-shifts, and broadens when the solvent is changed from $CHCl_3$ to $CH_3CN:H_2O$ (Figure 14.8). These changes are believed to arise due to the strong π–π interactions present in the helical conformation. In the helical conformation the fluorescence is dominated by an excimer-like state arising from π-stacked chromophores in the helix. In addition to the changes in the fluorescence, the ratio of the absorption bands at 303 and 288 nm decreases significantly concomitant with helix formation. The latter effect is attributed to a change in the dominant backbone conformation from s-*trans* (transoid) to s-*cis* (cisoid) that accompanies helix formation.

On the basis of the investigations by Moore and coworkers, we speculated that CPEs that feature meta-links in the backbone might also form a helical secondary structure in water. In order to test this concept, we first prepared *m*-PPE-SO$_3^-$ (Figure 14.9) which features a meta-phenylene linkage in every other phenylene ethynylene repeat unit. Studies of *m*-PPE-SO$_3^-$ in various solvent mixtures show that the polymer features absorption spectroscopic characteristics that are remarkably similar to those of Moore's OPEs [34]. In particular, when the solvent is changed from methanol to water, the fluorescence of *m*-PPE-SO$_3^-$ decreases in intensity, red-shifts, and broadens (Figure 14.9). At the same time, the ratio of the absorption bands at 350 and 325 nm decreases significantly. These features are believed to arise due to a change in the polymer's secondary structure from random coil to helical when the solvent is changed from methanol to water.

Further evidence for the existence of a helical conformation for meta-linked CPEs comes from circular dichroism (CD) studies of *m*-PPE-CO$_2^-$ (Figure 14.10). The basis for this experiment lies in the fact that a helix is an inherently chiral structure. Because the helix is chiral, one would expect that if an enantiomeric excess (ee) of the right- or left-handed forms (the P and M forms) exists in solution, then the CD spectrum would feature a bisignate signal due to exciton coupling of the backbone chromophores which are in a chiral environment [35]. However, when a polymer folds into a helical structure, because the polymer itself is achiral, the ensemble of helically folded polymers exists as a

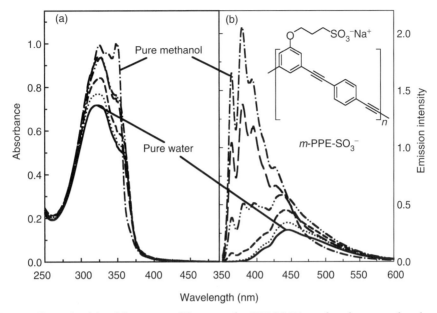

FIGURE 14.9 Absorption (a) and fluorescence (b) spectra of m-PPE-SO$_3^-$ in methanol, water, and methanol/water mixtures. Labels indicate spectra in pure water and methanol. Structure of m-PPE-SO$_3^-$ shown at upper right.

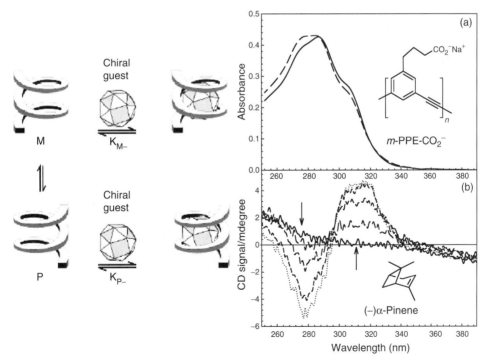

FIGURE 14.10 *Left:* Effect of a chiral guest molecule on the M/P equilibrium for a helical polymer. (Reprinted from Prince, R.B., S.A. Barnes, and J.S. Moore. *J. Am. Chem. Soc.*, 122, 2758–2762, 2000. With permission.) *Right:* Effect of added (−)-α-pinene on the absorption (a) and circular dichroism, (b) spectra of m-PPE-CO$_2^-$ (structure shown in panel (a). (−)-α-Pinene concentration ranges from 0 to 6 μM.

racemic mixture of the P and M forms, and consequently no CD is observed. However, Moore and coworkers showed that it is possible to perturb the P–M equilibrium for the helical structures (thereby producing an ee) by addition of an optically active, chiral guest to a solution of the helical polymer [36]. Addition of (–)-α-pinene to m-PPE-CO$_2^-$ in a CH$_3$CN/H$_2$O mixture leads to the appearance of a bisignate CD signal in the region of the spectrum corresponding to the π, π* absorption of the polymer backbone (Figure 14.10). The observation of this CD signal is clear evidence that m-PPE-CO$_2^-$ folds into a conformation that is helical, since this is the only way in which the polymer can exhibit a CD spectrum.

Another interesting aspect of the helical conformation of anionic meta-linked CPEs such as m-PPE-SO$_3^-$ and m-PPE-CO$_2^-$ is that these materials exhibit structural features that are analogous to those of DNA. In particular, double stranded DNA is a polyanion that exists in a helical conformation stabilized in part by π-stacking between the aromatic heterocyclic bases. By analogy, a meta-linked CPE such as m-PPE-SO$_3^-$ is a polyanion that can exist in a helical conformation stabilized by π-stacking between aromatic phenylene units. Given this similarity, it is not surprising that small molecules that interact with DNA, either by intercalation or by groove binding, interact in a similar manner with helically folded CPEs. A dramatic example of this effect is provided by the interaction of the cationic small molecule intercalator 9-aminomethylanthracene (9-AMA) and m-PPE-SO$_3^-$. Figure 14.11 compares the absorption spectrum of 9-AMA in water as a function of added double stranded DNA (a) and added m-PPE-SO$_3^-$ (b). In both cases, addition of the helical polyanion induces a decrease in the oscillator strength and red-shift of the 9-AMA absorption bands. This effect, which is referred to as hypochromism, is well known in DNA intercalator chemistry. It arises due to the effect of π-stacking interactions when the planar, aromatic chromophore binds to the polymer by slipping into the helix and π-stacking with heterocyclic DNA bases that are adjacent in the ladder-like structure. The remarkable fact is that m-PPE-SO$_3^-$ has almost the same effect on the 9-AMA absorption. We believe that this effect arises because the cationic, planar aromatic hydrocarbon binds to the helical polymer by intercalating in between π-stacked phenylene residues. This effect is further evidence that meta-linked CPEs fold into helical conformation in water, and provides a very interesting example of how synthetic polymers can take on secondary structures that mimic the structures adopted by biopolymers.

A final important motif for self-assembly of CPEs is by LbL deposition of polyelectrolyte multilayer films. The LbL multilayer film deposition method was first introduced by Decher [37,38], and since then it has been used to fabricate nanostructured films using a wide variety of synthetic and naturally occurring polyelectrolytes [39]. Deposition of LbL films involves a very simple sequence of alternately dipping a substrate into solutions that contain a cationic polyelectrolyte and an anionic polyelectrolyte. The sequential LbL deposition method leads to formation of polymer multilayer film structures. A single

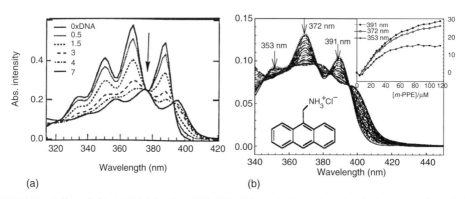

FIGURE 14.11 Effect of added DNA (a) and m-PPE-CO$_2^-$ (b) on the absorption of AMA (structure shown in "b"). Experiments carried out with AMA in water with 0–8 equivalents of the polymers added. Arrows indicate direction of change of spectra with increasing polymer concentration.

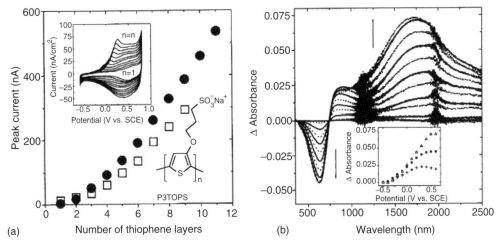

FIGURE 14.12 (a) Peak anodic current at gold electrode coated with poly(dimethyldiallylammonium) chloride/ P3TOPS LbL bilayer films as a function of number of bilayers. (b) Spectroelectrochemical response of a 5-bilayer film consisting of poly(dimethyldiallylammonium) chloride/P3TOPS on a tin oxide electrode. (Reprinted from Lukkari, J., M. Salomaki, A. Viinikanoja, T. Aaritalo, J. Paukkunen, N. Kocharova, and J. Kankare. *J. Am. Chem. Soc.*, 123, 6083–6091, 2001. With permission.)

"bilayer" consists of a polyanion/polycation pair having a thickness close to that expected for two molecular layers (i.e., the thickness of one bilayer is ~1 nm).

The LbL method has been used to deposit multilayer films of a variety of different CPEs. These films are of interest for a variety of applications, including electrochromics, light emitting diodes (LEDs), and solar photovoltaic cells. As an example of using this approach to construct active, ultrathin films, Lukkari and coworkers reported the fabrication of multilayer films consisting of bilayers of the anionic polythiophene-based CPE, P3TOPS alternating with the inert polycation poly(dimethyldiallylamine), PDDA [40]. Bilayers were constructed by alternately dipping a substrate into aqueous solutions that contained P3TOPS and PDDA. Multilayer buildup was monitored by using a quartz crystal micro-balance (QCM) and by cyclic voltammetry, where the anodic oxidation of P3TOPS was monitored. The anodic current response at a gold electrode coated with PDDA/P3TPOS bilayers increases linearly with an increase in the number of layers (Figure 14.12). The in-plane conductivity of a 5 bilayer film was ~10^{-5} S cm^{-2}. Subsequent to film deposition, the spectroelectrochemical response of the films was tested (Figure 14.12), and this experiment demonstrated that the P3TOPS deposited as an LbL film displayed a spectroelectrochemical response typical of polythiophene films.

14.6 Amplified Fluorescence Quenching

Much of the excitement regarding the properties and applications of CPEs is associated with the observation of the "amplified quenching effect," i.e., very efficient fluorescence quenching at low concentration of the quenching species. Before discussing this phenomenon, we provide a brief overview of fluorescence quenching and standard mechanisms used to explain the effect [41]. Fluorescence quenching can occur by two limiting mechanisms, dynamic, Equation 14.1a and static, Equation 14.1b. In the following equations, F* is an excited-state fluorophore, Q is a quencher, k_q is the bimolecular quenching rate constant and k_a is the association constant for formation of the ground state complex [F,Q].

$$F^* + Q \xrightarrow{K_q} F + Q \tag{14.1a}$$

$$F + Q \underset{}{\overset{K_a}{\rightleftharpoons}} [F,Q] \xrightarrow{h\nu} [F^*, Q] \xrightarrow{fast} F + Q \tag{14.1b}$$

Treatment of fluorescence intensity quenching data according to the SV equation yields [42]

$$\frac{I^o}{I} = 1 + K_{SV}[Q] \tag{14.2}$$

where I and I^o are the fluorescence intensity with and without Q, respectively, and K_{SV} is the SV quenching constant. In the limit where quenching is dominated by the dynamic pathway (Equation 14.1a), $K_{SV} = k_q \tau^o$, where τ^o is the fluorescence lifetime of F*, whereas in the limit where static quenching dominates, $K_{SV} = K_a$.

There are several ways to distinguish between dynamic and static quenching. First, note that k_q cannot exceed the diffusion rate constant ($\sim 10^{10}$ M^{-1}s^{-1}) when quenching is fully dynamic. Since $K_{SV} = k_q \tau^o$, for a fluorophore that has a 1 ns lifetime this places an upper limit on $K_{SV} \approx 10$ M^{-1} for dynamic quenching (i.e., 10^{10} M^{-1}s$^{-1} \times 10^{-9}$ s $= 10$ M^{-1}). Thus, static quenching is likely to be important if the experimentally observed K_{SV} is significantly greater than 10 M^{-1}. Another method for distinguishing when static quenching is important is to compare the ratio of emission lifetimes (τ^o/τ) and emission intensities (I^o/I) vs. quencher concentration. If the K_{SV} obtained from intensity quenching is greater than that from lifetime quenching, then static quenching is occurring.

When the quenching is dominated by either a purely static or dynamic pathway, the quenching behavior follows Equation 14.2 and consequently SV plots of I^o/I vs. [Q] are linear. However, in many situations (as shown below) the SV plots are curved upward (i.e., superlinear). Superlinear SV plots can arise from a variety of processes, including mixed static and dynamic quenching, variation in the association constant with quencher concentration, and chromophore (or polymer aggregation).

Whitten, McBranch, and coworkers first observed the amplified quenching effect in a CPE in the course of an investigation of the quenching of the fluorescence of PPV-SO$_3^-$ (Scheme 14.2) by the dicationic quencher N,N′-dimethyl-4,4′-bipyridinium (MV^{2+}) [6]. Specifically, addition of 100 nM of MV^{2+} to a solution of the polymer ($c \approx 10^{-5}$ M in polymer repeat units) affords very efficient quenching of the polymer's fluorescence coupled with a distinct redshift in the absorption spectrum (Figure 14.13). The SV plot for quenching of the PPV-SO$_3^-$ fluorescence by MV^{2+} is linear for very low concentrations of MV^{2+} (range 0–1 µM), with $K_{SV} \approx 2 \times 10^7$ M^{-1}. In view of the short lifetime of the unquenched PPV-SO$_3^-$ ($\tau \approx 200$ ps) the extremely large K_{SV} makes it clear that the predominant quenching mechanism involves a preformed complex between the PPV-SO$_3^-$ and MV^{2+}, i.e., static quenching, Equation 14.1b. In particular, the efficient quenching arises in part due to an ion pair complex formed between the cationic quencher and the anionic PPV-SO$_3^-$ chains. However, in addition to the importance of polymer quencher ion pair complex, it is evident that other processes must be active to give rise to the highly efficient quenching. Whitten and McBranch suggested that these other processes involve: (i) rapid and efficient diffusion of the fluorescent exciton within PPV-SO$_3^-$ chains; (ii) quencher-induced aggregation of the polymer chains, which enhances the efficiency of intrachain excimer formation and interchain hopping of excitons.

More detailed information concerning the mechanism for amplified quenching of PPV-SO$_3^-$ by cationic quenchers was provided in an investigation by Bazan and Heeger [43]. In this study, the effect of charge on the quenching of PPV-SO$_3^-$ was explored by using the series of bipyridinium salt quenchers DPy, MV$^+$, MV^{2+}, and MV^{4+} (Scheme 14.9). The structure of the actual polymer studied by Bazan and Heeger differs from PPV-SO$_3^-$ by having an additional methylene linker in the sulfonate side group, i.e., the chain is Ar-O-CH$_2$CH$_2$CH$_2$CH$_2$-SO$_3^-$. The charge on the quencher has a substantial influence on the overall quenching efficiency and the shape of the SV plots (Figure 14.14). The overall quenching efficiency increases substantially as the charge on the quencher increases. The K_{SV} values obtained from the linear region of the plots (at low quencher concentration, $c < 200$ nM)

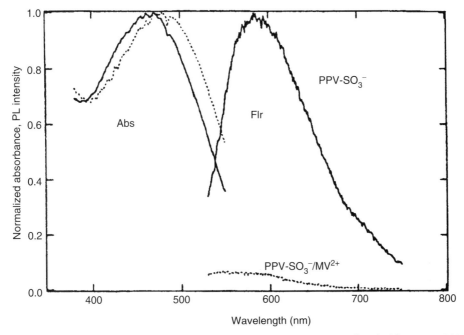

FIGURE 14.13   Absorption (*left*) and fluorescence (*right*) of PPV-SO$_3^-$ in water with and without 100 nM MV^{2+}. (Reprinted from Chen, L., D.W. McBranch, H.L. Wang, R. Helgeson, F. Wudl, and D.G. Whitten. *Proc. Natl. Acad. Sci. USA*, 96, 12287–12292, 1999. With permission.)

range from $\approx 10^5$ M^{-1} (DPy) to 2.2×10^7 M^{-1} (MV^{4+}). In addition to this effect, however, the SV plots for the charged quenchers become superlinear (upward curvature) at higher quencher concentrations. Moreover, the degree of superlinearity increases with charge on the quencher. The effect of increasing charge on the fluorescence quenching efficiency has its origins in two effects. First, with increasing charge, the stability of the PPV-SO$_3^-$/quencher ion pair complex increases. This is likely the origin of the increase in K_{SV} with quencher charge in the linear region of the SV plots. Second, as the charge on the quencher increases, the ability of the quencher to induce aggregation of multiple PPV-SO$_3^-$ chains increases. As suggested previously by Whitten and McBranch, aggregation of the polymer chains increases the probability for exciton migration to occur to a quencher "trap" site where quenching occurs.

In a recent series of investigations, our group, working in collaboration with Prof. V. Kleiman at the University of Florida, studied the mechanism and dynamics of quenching of PPE-SO$_3^-$ (Scheme 14.2) by a series of cyanine dyes [44]. These were used for several reasons. First, cyanines are cationic and

SCHEME 14.9

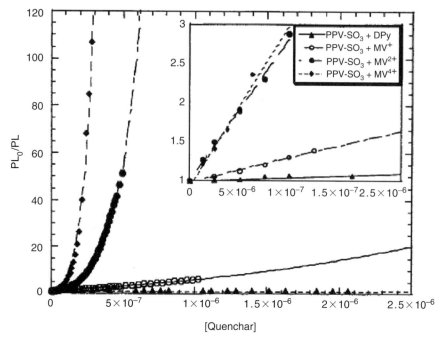

FIGURE 14.14   Stern-Volmer plots for quenching of PPV-SO$_3^-$ by viologen derivatives. In order of increasing slope: DPy, MV$^+$, MV^{2+}, and MV^{4+}. (Reprinted from Wang, D.L., J. Wang, D. Moses, G.C. Bazan, and A.J. Heeger. *Langmuir*, 17, 1262–1266, 2001. With permission.)

therefore interact with the anionic PPE-SO$_3^-$ chains by ion-pairing. Second, cyanines absorb and fluoresce strongly in the visible region, and the wavelength of the absorption and fluorescence can be systematically tuned by changing the length of the polymethine chain in the dye. Third, because the dyes absorb strongly in the visible, they quench the PPE-SO$_3^-$ by singlet–singlet energy transfer, and the dynamics of the energy transfer can be followed either by monitoring the decay of the polymer's fluorescence (the energy donor) or the *rise* of the dye's fluorescence (the energy acceptor).

Several key findings came from the work on the fluorescence quenching in the PPE-SO$_3^-$/cyanine dyes system. First, comparison of the SV quenching behavior of the dyes revealed that the quenching efficiency increases with dye size and dye charge. The latter characteristic is similar to the findings of Bazan and Heeger with the series of bipyridinium quenchers, and we interpret it in the same way, i.e., greater ionic charge on the quencher enhances its ability to induce aggregation of the CPE chains. This premise was supported by the clear observation of changes in the absorption of PPE-SO$_3^-$ concomitant with addition of small concentrations of a cyanine with a (3 +) charge. The absorption spectral changes were very similar to those observed when the polymer aggregates in water (Figure 14.3). The reason for more efficient quenching by larger dyes was ascribed to an increase in the PPE-SO$_3^-$/dye association constant caused by increased solvophobic interactions between the larger dyes and the methanol solvent.

The second key finding of this study was that the rate of PPE-SO$_3^-$ fluorescence quenching by a red emitting cyanine dye (HMIDC) is remarkably fast. Figure 14.15 shows the PPE-SO$_3^-$ fluorescence decays for a series of solutions that contain the polymer and HMIDC at concentrations ranging from 0 to 5 µM. The decay curves that are plotted in the main panel of the figure represent absolute fluorescence intensities, and therefore the amplitudes of the decays reflect the true relative instantaneous intensity of the time-resolved fluorescence signal from the different samples. Upon inspection of the data, two features are apparent. First, the initial amplitude of the PPE-SO$_3^-$ fluorescence decay is reduced by the addition of HMIDC. Second, with increasing HMIDC concentration, the fluorescence decays more rapidly, which can be better seen in the figure inset, where the same data is plotted in an intensity normalized format. Taken together, these observations indicate that the PPE-SO$_3^-$-to-HMIDC

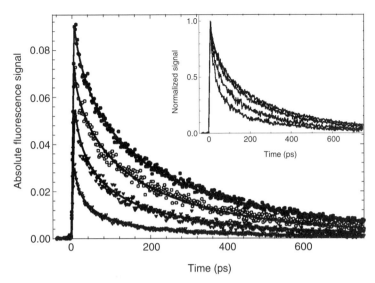

FIGURE 14.15 Fluorescence decays of PPE-SO$_3^-$ with added HMIDC cyanine dye. Main panel shows unnormalized decay traces, time zero intensity decreases with increasing [HMIDC]. Inset shows normalized decays, decay rate increases with increasing [HMIDC]. (Reprinted from Tan, C., E. Atas, J.G. Müller, M.R. Pinto, V.D. Kleiman, and K.S. Schanze. *J. Am. Chem. Soc.*, 126, 13685–13694, 2004. With permission.)

energytransfer takes place on two distinct timescales. A significant component of the transfer occurs by a pathway that is so fast that it cannot be resolved within the time-resolved fluorescence instrument response ($\approx$4 ps). This "prompt" quenching component accounts for more than 50% of the total fluorescence quenching that is observed. In addition to the prompt quenching component, the clearly noticeable increase in the rate of the fluorescence decay with increasing HMIDC concentration (inset of Figure 14.15) indicates that there is also a slow energy transfer process. The fluorescence quenching dynamics are interpreted by reference to the qualitative model (Figure 14.16). At relatively low quencher concentrations the PPE-SO$_3^-$ chains are relatively unaggregated and the dye is effective at quenching only a single chain (Figure 14.16a). The "prompt" quenching component arises due decay of an exciton that is produced (by the absorbed photon) on a chain in relatively close proximity to a complexed dye. The slower

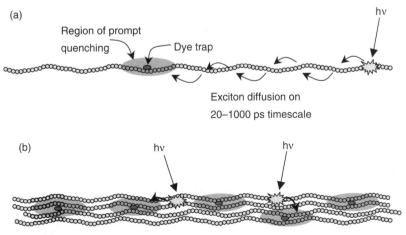

FIGURE 14.16 Model to explain exciton quenching dynamics in PPE-SO$_3^-$/cyanine dye systems. (a) Single (unaggregated) chains; (b) aggregated polymer. (Reprinted from Tan, C., E. Atas, J.G. Müller, M.R. Pinto, V.D. Kleiman, and K.S. Schanze. *J. Am. Chem. Soc.*, 126, 13685–13694, 2004. With permission.)

component arises due to decay of excitons that must first diffuse by hopping along the chain to a dye complex. As the concentration of dye quencher is increased, the polymer chains aggregate, and now a dye can effectively quench excitons that are produced on more than a single chain (Figure 14.16b). This leads to an overall increase in both the prompt and comparatively slower fluorescence quenching components.

Taken together the various studies that have been carried out to understand the mechanism of amplified quenching have demonstrated several characteristic features. First, the two key elements of the process are: (i) ion pairing between a CPE chain and an oppositely charged quencher ion; (ii) the delocalized nature of the exciton coupled with its ability to rapidly diffuse within a single CPE chain as well as between aggregated chains. Second, it is clear that the quenching is further augmented under conditions where the CPE chains are strongly aggregated. In general (although not clearly supported by data shown in this chapter), quenching is enhanced under conditions where the polymer chains are aggregated. Furthermore, as shown above, polyvalent quencher ions give enhanced quenching in part due to their ability to induce aggregation of the CPE chains.

14.7 Applications of Conjugated Polyelectrolytes

Much of the interest in the synthesis and properties of CPEs stems from their strong potential for use in applications ranging from optoelectronic devices to biological sensors. One key feature that distinguishes CPEs from nonionic conjugated polymers that are soluble in organic solvents is that the former can be dissolved in and processed from water. This feature allows the polymers to be used in an environment that is compatible with biological macromolecules and small molecules. A second unique feature with respect to application of CPEs is their propensity to undergo self-assembly driven by ionic and solvophobic interactions. This aspect of the materials is especially useful when considering their application in a thin film state.

In Section 14.7, we will review a number of investigations that have been carried out which have focused on the application of CPEs. As noted previously, the chapter by Nilsson and Inganäs provides a relatively thorough review of the applications of CPEs in biosensor schemes, and consequently we will not cover this aspect of CPE applications here. Rather, the following discussion will focus on the use of CPEs in optoelectronic devices as well as another novel bioapplication involving the use of cationic CPEs as antimicrobial agents.

14.7.1 Polymer Light Emitting Diodes

Rubner and coworkers devised an interesting method for the use of the LbL deposition technique for fabricating the active layer of a polymer light emitting diode (PLED) [45]. In their approach, they use the LbL method to deposit the sulfonium salt precursor to PPV (a cationic polyelectrolyte) along with a polyanion such as the sodium salt of polymethacrylic acid (PMA). Subsequent to LbL film deposition, the sulfonium salt is converted to PPV by heating to 210°C (Scheme 14.10). A typical PLED is fabricated by depositing 20 PMA/PPV bilayers onto an indium tin oxide (ITO) electrode, followed by deposition of aluminum as a low work function metal, i.e., ITO/(PMA/PPV)$_{20}$/Al. The devices exhibit good current rectification and light emission was observed from the PPV at applied voltages of 6 V and greater (Figure 14.17). The group also found that the light emission quantum efficiency was significantly improved if the structure of the films was slightly more complex. They first deposited a 5-bilayer film consisting of polystyrene sulfonate (PSS) along with the PPV sulfonium salt precursor, followed by 15 bilayers of PMA with the PPV precursor. After thermolysis, the active layer structure of the PLED was ITO/(PSS/PPV)$_5$(PMA/PPV)$_{15}$/Al. The device with the heterostructured active layer exhibited nearly a tenfold increase in light emission quantum efficiency, combined with a decreased light emission turn on voltage (turn on at 5 V). They attributed the improved performance that resulted from inclusion of the (PSS/PPV)$_5$ layer to the ability of this material to transport holes. The ability of (PSS/PPV) to act as

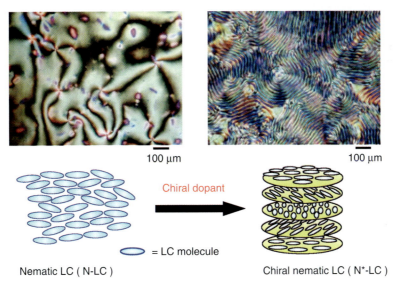

100 μm   100 μm

Chiral dopant

⬭ = LC molecule

Nematic LC (N-LC)   Chiral nematic LC (N*-LC)

FIGURE 3.1   Chiral nematic LC induced by the addition of chiral dopant into nematic LC. Schlieren texture (*left*) and fingerprint texture (*right*) are observed for nematic and chiral nematic LCs, respectively, in polarized optical microscope.

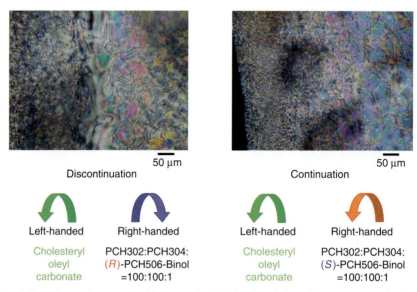

FIGURE 3.2   Miscibility test between the chiral nematic LC induced by (*R*)- or (*S*)-PCH506-Binol and the standard LC, cholesteryl oleyl carbonate of counterclockwise (left-handed) screw direction.

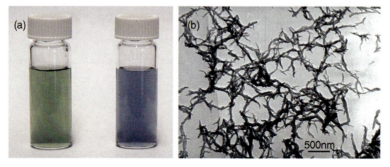

FIGURE 7.17   Dialysis or centrifugation can create good polyaniline nanofiber aqueous dispersions. (a) HCl doped (*green*, *left*) and dedoped (*blue*, *right*) polyaniline nanofiber dispersions and (b) TEM image showing the morphology of the nanofibers.

FIGURE 7.45 Optical microscope images showing that flash welding through a copper grid (a) reproduces the grid pattern on a polyaniline nanofiber film (b). Scale bars = 100 μm. The high contrast of the pattern is revealed under a higher magnification (c). (Adapted from Huang, J.X. and Kaner, R.B., *Nat. Mater.*, 3, 783–786, 2004.)

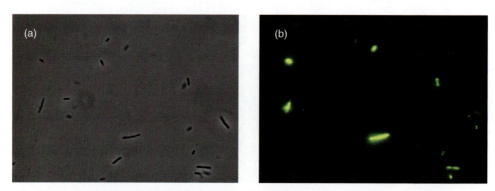

FIGURE 14.22 Phase contrast (a) and epifluorescence (b) microscope images of *E. coli* bacteria after exposure to a solution of PPE-OR8-NMe₃⁺. (Reprinted from Lu, L.D., F.H. Rininsland, S.K. Wittenburg, K.E. Achyuthan, D.W. McBranch, and D.G. Whitten. *Langmuir*, 21, 10154–10159, 2005. With permission.)

SCHEME 14.10

a hole transport material is believed to arise because the PSS (a strong polyacid) acts as a dopant to the PPV giving the $(PPV/PSS)_5$ layers a much higher conductivity compared to the $(PMA/PPV)_{15}$ layers.

Shortly after the work with the PPV precursor-based devices was reported, Rubner and Reynolds worked together to apply the LbL method to fabricate the active layer of PLEDs from CPEs [46]. In one report blue emitting PLEDs were fabricated by using the pair of charge-complementary PPP-based CPEs PPP-ORSO$_3^-$ and PPP-NMe$_3^+$ (Scheme 14.11). LEDs were fabricated on ITO substrates by depositing 30 (PPP-ORSO$_3^-$/PPP-NMe$_3^+$) bilayers, followed by thermal evaporation of an aluminum electrode. The 30 bilayer film was relatively thin at 21 nm, and the PLED operated with modest efficiency, exhibiting an external quantum efficiency for emission of blue light of 0.002%.

One reason that organic LEDs fabricated by thermal evaporation of small molecules operate more efficiently than polymer-based LEDs is that thermal evaporation allows the deposition of multilayered structures with different molecular composition. The ability to create heterostructured molecular thin films by thermal evaporation allows one to tailor the electronic structure of the active layer such that carriers are efficiently transported to and confined within the light emissive region. The ability to create heterostructures within PLED active layers is not feasible, because most polymers are soluble in the organic solvents used for spin-coating, and thus the solvent used for casting a second layer will dissolve the underlying film.

In a recent report, Bazan, Heeger and coworkers developed a novel use of a CPE that makes it possible to circumvent this problem and to fabricate multilayer heterostructures within the PLED active layer [47]. Thus, the alternating copolymer PFO-PBD-NMe$_3^+$ (Figure 14.18) which acts as an electron transport material was synthesized in a Suzuki coupling reaction. PFO-PBD-NMe$_3^+$ is soluble in water and methanol, but insoluble in typical organic solvents used to cast light emissive polymers such

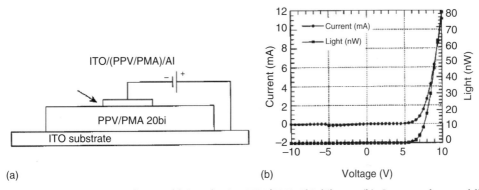

(a)   (b)

FIGURE 14.17 (a) Structure of PLED fabricated using PPV/PMA LbL bilayers. (b) Current–voltage and light output–voltage plots for ITO/(PPV/PMA)$_{20}$/Al PLED. (Reprinted from Onitsuka, O., A.C. Fou, M. Ferreira, B.R. Hsieh, and M.F. Rubner. *J. App. Phys.*, 80, 4067–4071.1996. With permission.)

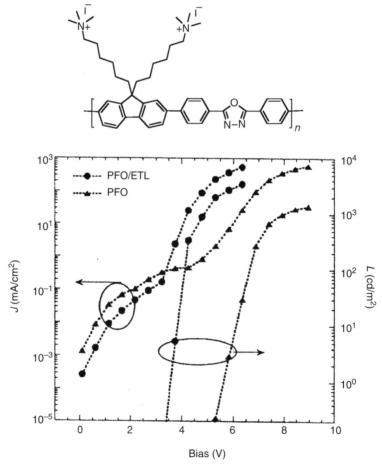

SCHEME 14.11

PPP-ORSO$_3^-$ PPE-NMe$_3^+$

as poly(octylfluorene) (PFO) and poly(2-meth-oxy-5-(2-ethylhexyloxy)-1,4-phenylene vinylene) (MEH-PPV). The group used spin-coating to construct PLED active layers producing devices with the configuration ITO/PEDOT-PSS/emis-emissive polymer/PFO-PBD-NMe$_3^+$/Ba/Al (where PEDOT is poly(3,4-ethylenedioxythio-phene) which serves as a hole transport layer). The emissive polymer (either PFO or MEH-PPV) was spin-coated from an organic solvent, and PFO-PBD-NMe$_3^+$ was spin-coated onto the emissive polymer from methanol. Because the emissive polymers are not soluble in methanol, the PFO-PBD-NMe$_3^+$ layer can be spin-coated onto the emissive polymer film without disturbing its structure. Comparison of the optoelectronic character-istics of devices constructed with or without the PFO-PBD-NMe$_3^+$ electron transport layer shows that inclusion of the PFO-PBD-NMe$_3^+$ layer improves device performance. In particular, PLEDs constructed

FIGURE 14.18 Current density–voltage and luminosity–voltage plots for PLEDs based on PFO with and without PFO-PBD-NMe$_3^+$ electron transport layer. (▲): Device configuration ITO/PEDOT/PFO/Ba/Al. (●): Device con-figuration ITO/PEDOT/PFO/PFO-PBD-NMe$_3^+$/Ba/Al. (Reprinted from Ma, W.L., P.K. Iyer, X. Gong, B. Liu, D. Moses, G.C. Bazan, and A.J. Heeger. *Adv. Mater.*, 17, 274, 2005. With permission.)

with PFO as the light emitting polymer along with the PFO-PBD-NMe$_3^+$ electron transporting layer turn on at a lower applied voltage and display a higher brightness and higher quantum efficiency compared to a similar device that lacks the PFO-PBD-NMe$_3^+$ layer (Figure 14.18).

14.7.2 Photovoltaic Devices

Since the early reports demonstrating the concept of organic photovoltaics based on donor–acceptor heterojunctions [48–50], there has been considerable interest in this area. Work in this field has been fueled by the promise for the development of low cost, high efficiency, and flexible devices for use in solar energy conversion. Like PLEDs, the efficiency of organic photovoltaic devices is strongly influenced by the structure of the active material. In particular, the key events of exciton diffusion, charge separation, and carrier transport must be optimized for efficient device operation, and these events are a strong function of the nanostructure of the active material. In view of this fact, several groups, including our own, have applied the LbL film deposition method to construct the active layer of organic photovoltaic devices.

The first report in this area was from Baur, Durstock and coworkers at Wright Patterson Air Force Base, in which PPV and C$_{60}$ were used as the donor (hole transport material) and acceptor (electron transport material), respectively [51,52]. The LbL method was used to deposit the sulfonium salt precursor to PPV along with an anionic derivative of C$_{60}$. In addition, layers that contained only C$_{60}$ were fabricated by LbL deposition of the anionic C$_{60}$ with a cationic C$_{60}$ derivative, whereas layers that contained PPV as the only electroactive material were constructed by LbL deposition of the cationic PPV precursor with polystyrene sulfonate (Scheme 14.12). As described in the preceding section, after LbL film deposition the PPV precursor was converted to PPV by thermal treatment. The effect of film nanostructure on the photovoltaic device efficiency, as assessed by the open-circuit voltage and short-circuit current produced when the devices were illuminated with simulated solar light. Optimal photovoltaic performance was observed for devices that contained carrier transport layers that contain C$_{60}$ or PPV separated by 5 or 10 intervening bilayers consisting of PPV and the anionic C$_{60}$ (Figure 14.19). Presumably photoinduced charge separation arises in the mixed layer,

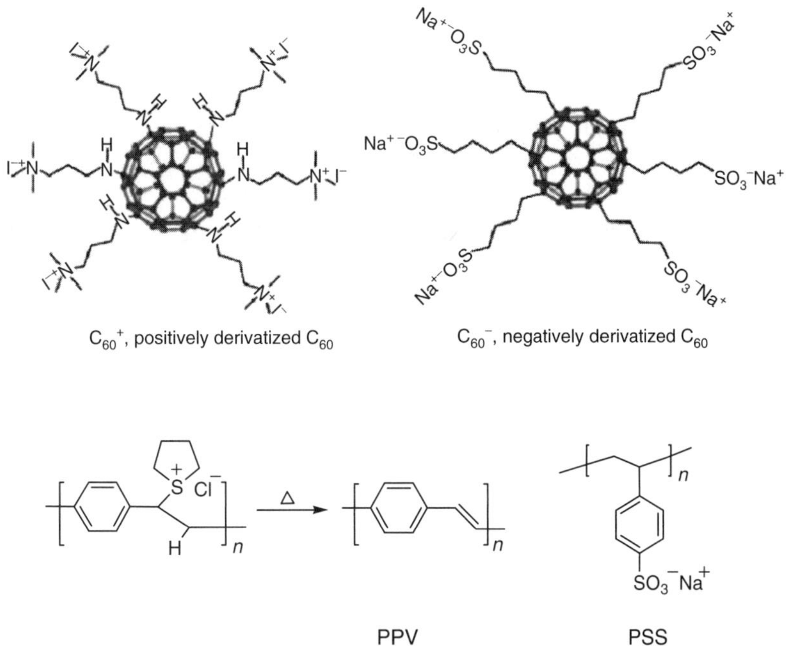

C$_{60}^+$, positively derivatized C$_{60}$ C$_{60}^-$, negatively derivatized C$_{60}$

PPV   PSS

SCHEME 14.12

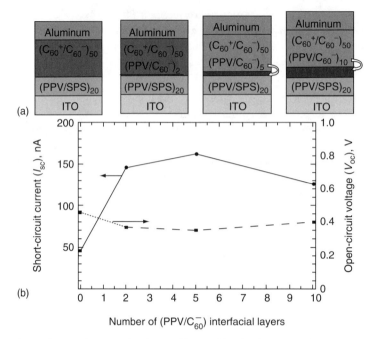

FIGURE 14.19   Photovoltaic performance for devices fabricated by the LbL approach using the materials described in the text. (a) The configuration of the heterostructured active layer; the variable is the number of (PPV/C_{60}^-) layers. (b) The short circuit photocurrent (•) and the open-circuit potential (■) as a function of the number of (PPV/C_{60}^-) layers. (Reprinted from Baur, J.W., M.F. Durstock, B.E. Taylor, R.J. Spry, S. Reulbach, and L.Y. Chiang. *Synth. Met.*, 121, 1547–1548, 2001. With permission.)

while the acceptor and donor only layers act to transport carriers to the electrodes. Despite the novelty of this work, the overall efficiency of these devices with respect to optical to electrical energy conversion was relatively low. The low efficiency presumably arises for a number of factors, including the relatively low optical cross section of the materials in the visible region, combined with poor carrier mobility in the LbL films which contain many ionic units that trap charges.

Working in collaboration with Reynolds, we have fabricated organic photovoltaic devices in which the active materials were assembled by using the LbL method [53]. In this work, the donor and hole transporting materials were the anionic CPEs PPE-SO_3^- and PPE-EDOT-SO_3^{3-}, whereas the acceptor and electron transport material was a cationic fullerene derivative, C_{60}–NH_3^+ (Scheme 14.13). The active layers were constructed atop an ITO substrate that was precoated with a PEDOT-PSS film (spin-coated). The PPE($-$)/C_{60}–NH_3^+ bilayers were deposited through the LbL method, and the effect of active layer thickness on device performance was explored. Figure 14.20 shows a schematic

PPE-SO_3^- EDOT-PPE-SO_3^- C_{60}–NH_3^-

SCHEME 14.13

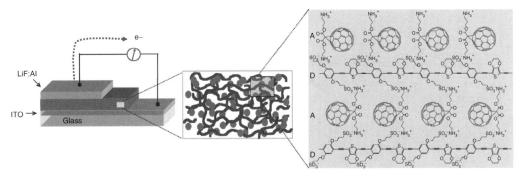

FIGURE 14.20   Structure of $(PPE(-)/C_{60}-NH_3^+)$ LbL active layers used in photovoltaic devices. (Reprinted from Mwaura, J.K., M.R. Pinto, D. Witker, N. Ananthakrishnan, K.S. Schanze, and J.R. Reynolds. *Langmuir*, 21, 10119–10126, 2005. With permission.)

representation of the possible structure of the LbL films. As suggested, it is likely that there is an interpenetration of the polymer chains within the multilayers leading to a relatively disordered structure. The disorder is favorable because it gives rise to intimate mixing of the donor and acceptor components within the film, and assuming that phase separation occurs within the layers, this leads to the possibility of the formation of a bulk heterojunction needed for efficient charge separation and transport through the active material.

Device studies show that the $PPE-SO_3^-/C_{60}-NH_3^+$ and $PPE-EDOT-SO_3^-/C_{60}-NH_3^+$ materials gave moderate quantum efficiencies for photocurrent generation (~5% at the absorption maximum), and the photocurrent action spectra mirrored the absorption spectra of the PPE polymers accurately. Interestingly, the photocurrent action spectrum of the $PPE-EDOT-SO_3^-$ based device was red-shifted relative to that of the $PPE-SO_3^-$ device, reflecting the lower bandgap of the PPE-EDOT backbone. The red-shift is advantageous for improving the energy conversion efficiency; this is because much of the energy in sunlight is in the red and near infrared regions of the spectrum. Unfortunately, the improvement in light harvesting efficiency for the $PPE-EDOT-SO_3^-/C_{60}-NH_3^+$ device does not translate into an overall improvement in energy conversion efficiency. This fact was borne out by device studies carried out under simulated one sun illumination conditions. Thus the current–voltage plot for the $PPE-SO_3^-/C_{60}-NH_3^+$ device showed a considerably higher short-circuit current and slightly higher open-circuit voltage compared to the $PPE-EDOT-SO_3^-/C_{60}-NH_3^+$ device (Figure 14.21). The poorer performance of the $PPE-EDOT-SO_3^-$ device was attributed to relatively rapid photobleaching of the PPE-EDOT film under one sun illumination. Despite this disappointment, the performance of the $PPE-SO_3^-/C_{60}-NH_3^+$ device, which gives an overall optical to electrical power conversion efficiency of 0.04%, is the best reported to date for a photovoltaic device fabricated by the LbL method.

14.7.3 Antimicrobial Applications

As a final example of the diverse applications being explored for CPEs, we describe some very recent studies that have demonstrated that under certain conditions cationic CPEs display antimicrobial activity. Tew and coworkers have sought to use abiotic polymers to mimic the antibacterial properties of "host defense peptides" which share the common structural feature of being "facially amphiphilic" [54]. The facially amphiphilic structure is mimicked by polyelectrolytes that can adopt a helical structure. Tew reasoned that meta-linked PPEs that are functionalized with cationic ammonium side groups would provide the motif necessary for effective antimicrobial action. Thus, they synthesized the series *m*-PPE-NH_3^+ with varying molecular weights and examined their activity against gram-negative (*Escherichia coli*) and gram-positive (*Bacillus subtilis*) bacteria. By using an optical density assay to quantify bacterial cell growth in the presence of varying concentrations of the polymers, they determined the minimum

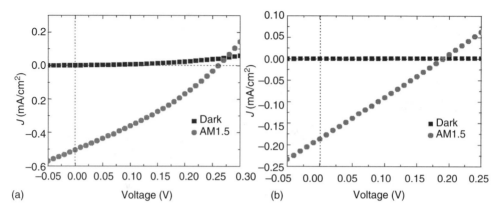

FIGURE 14.21 Current–voltage plots for photovoltaic devices that contain 60-bilayer films as the active layer. LbL films constructed with (a) PPE-SO$_3^-$/C$_{60}$–NH$_3^+$ and (b) PPE-EDOT-SO$_3^-$/C$_{60}$–NH$_3^+$. (Reprinted from Mwaura, J.K., M.R. Pinto, D. Witker, N. Ananthakrishnan, K.S. Schanze, and J.R. Reynolds. *Langmuir*, 21, 10119–10126, 2005. With permission.)

inhibitory concentration (MIC, concentration required to inhibit 90% of cell growth). This study showed that *m*-PPE-NH$_3^+$ polymers with molecular weights ranging from 1600 to 5500 g mol^{-1} (degree of polymerization = 5–20) exhibit moderate antimicrobial activity, with MIC values in the range of 25–50 μg mL^{-1}. Interestingly, the MIC increases with decreasing polymer chain length, and a monomer was also tested and it showed no measurable antimicrobial activity. This suggests that the antimicrobial activity is in part associated with the secondary structure adopted by the polymer chains in solution.

In a second investigation, Whitten and coworkers have explored the biocidal activity of another series of PPE-type CPEs against gram-negative bacteria and gram-positive spores [55]. This work focused on the properties of the linear chain, cationic polymer PPE-OR8-NMe$_3^+$ (Scheme 14.14). First, it was shown that exposure of *E. coli* bacteria or *B. Anthracis* spores to dilute solutions of PPE-OR8-NMe$_3^+$ affords modest reduction in survival of the colonies (30%–40% killing was typical). In a second set of investigations optical and fluorescence microscopy were used to show that exposure of bacteria or spores to solutions of the cationic CPE leads to relatively efficient coating of the microbes or spores by the CPE chains. This coating effect is seen clearly in the microscope images (Figure 14.22); the green yellow fluorescence typical of the PPE-type polymer is seen emanating from *E. coli* bacteria after they were exposed to a solution of the polymer. The authors suggest that the coating process involves electrostatic deposition of the cationic CPE chains onto the negatively charged surface of the bacteria. Further studies suggest that the bacteria and spore killing may occur in a photoinitiated process, possibly involving singlet oxygen (^{1}O$_2$) produced through sensitization by the polymer's excited state.

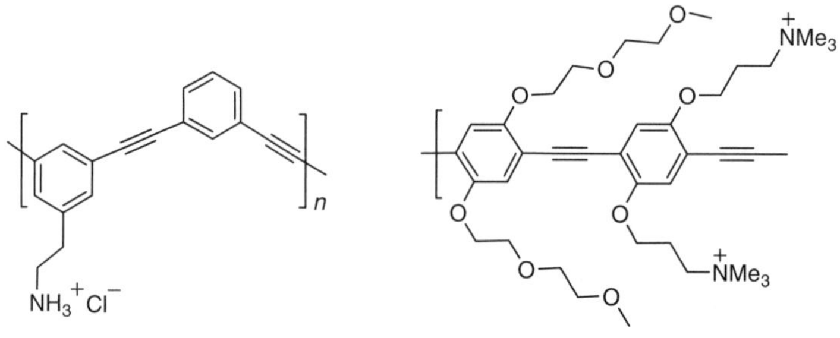

m-PPE-NH$_3^+$ PPE-OR8-NMe$_3^+$

SCHEME 14.14

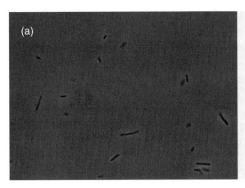

FIGURE 14.22 **(See color insert following page 14-20.)** Phase contrast (a) and epifluorescence (b) microscope images of *E. coli* bacteria after exposure to a solution of PPE-OR8-NMe$_3^+$. (Reprinted from Lu, L.D., F.H. Rininsland, S.K. Wittenburg, K.E. Achyuthan, D.W. McBranch, and D.G. Whitten. *Langmuir*, 21, 10154–10159, 2005. With permission.)

14.8   Summary, Conclusions, and Perspectives

This chapter provides a broad summary of recent research activity in the area of CPEs. It is evident that there is great enthusiasm in the conducting polymers community for CPEs. These materials provide a unique opportunity to explore the interactions between electro- and photoactive conjugated polymers with water-borne species such as metal ions and biological macromolecules, including proteins and DNA. Moreover, the ability to use the materials in LbL self-assembly affords researchers a tool for fabrication of heterostructured films of opto- and electronically active materials. This ability has not been easily possible with "conventional" conjugated polymers that are processed from organic solvents.

Future prospects for the application of CPEs seem bright. To date at least one small company has been formed with the objective of producing optical sensors that can be used in drug development assays and for detection of low levels of biological threat agents [56]. In addition, several chemical companies have begun to market water-soluble conjugated polymers, making them accessible for investigation by researchers who do not have synthetic capability in their own labs. With the increased interest in the field, and the broadening of groups working with the materials, it is very likely that the coming years will see vigorous development of the area with possible maturation leading to commercially viable products based on CPEs.

Acknowledgments

Work in the authors' laboratory in the area of CPEs is supported by the United States Department of Energy, Office of Basic Energy Sciences (DE-FG-02-96ER14617). KSS is also grateful to acknowledge long-term research collaborations with Profs. David G. Whitten and John R. Reynolds. In addition, Valeria Kleiman, Mauricio Pinto, Chunyan Tan, Benjamin Harrison, Jeremiah Mwaura, Hui Jiang, Evrim Atas, and Jürgen Müller made important contributions to the work described in the chapter.

References

1. Kraft, A., A.C. Grimsdale, and A.B. Holmes. 1998. *Angew Chem-Int Edit Engl* 37:402–428.
2. Kim, S., J. Jackiw, E. Robinson, K.S. Schanze, J.R. Reynolds, J. Baur, M.F. Rubner, and D. Boils. 1998. *Macromolecules* 31:964–974.
3. Balanda, P.B., M.B. Ramey, and J.R. Reynolds. 1999. *Macromolecules* 32:3970–3978.
4. Baur, J.W., M.F. Rubner, J.R. Reynolds, and S. Kim. 1999. *Langmuir* 15:6460–6469.
5. Harrison, B.S., M.B. Ramey, J.R. Reynolds, and K.S. Schanze. 2000. *J Am Chem Soc* 122:8561–8562.
6. Chen, L., D.W. McBranch, H.-L. Wang, R. Helgeson, F. Wudl, and D.G. Whitten. 1999. *Proc Natl Acad Sci USA* 96:12287–12292.
7. Chen, L.H., D. McBranch, R. Wang, and D. Whitten. 2000. *Chem Phys Lett* 330:27–33.

8. Chen, L.H., S. Xu, D. McBranch, and D. Whitten. 2000. *J Am Chem Soc* 122:9302–9303.
9. Tan, C., M.R. Pinto, and K.S. Schanze. 2002. *Chem Commun* 446–447.
10. Schnablegger, H., M. Antonietti, C. Göltner, J. Hartmann, H. Cölfen, P. Samori, J.P. Rabe, H. Häger, and W. Heitz. 1999. *J Colloid Interf Sci* 212:24–32.
11. Thünemann, A.F., and D. Ruppelt. 2000. *Langmuir* 16:3221–3226.
12. Reddinger, J.L., and J.R. Reynolds. 1999. *Adv Poly Sci* 145:57–122.
13. Patil, A.O., Y. Ikenoue, F. Wudl, and A.J. Heeger. 1987. *J Am Chem Soc* 109:1858–1859.
14. Shi, S., and F. Wudl. 1990. *Macromolecules* 23:2119–2124.
15. Zhou, Q., and T.M. Swager. 1995. *J Am Chem Soc* 117:7017–7018.
16. Zhou, Q., T.M. Swager. 1995. *J Am Chem Soc* 117:12593–12602.
17. Dwight, S.J., B.S. Gaylord, J.W. Hong, and G.C. Bazan. 2004. *J Am Chem Soc* 126:16850–16859.
18. Bunz, U.H.F. 2000. *Chem Rev* 100:1605–1644.
19. Bunz, U.H.F. 2005. *Adv Poly Sci* 177:1–52.
20. Pinto, M.R., and K.S. Schanze. 2002. *Synthesis* 1293–1309.
21. Child, A.D., and J.R. Reynolds. 1994. *Macromolecules* 27:1975–1977.
22. Balanda, P.B., M.B. Ramey, and J.R. Reynolds. 1999. *Macromolecules* 32:3970–3978.
23. Stork, M., B.S. Gaylord, A.J. Heeger, and G.C. Bazan. 2002. *Adv Mater* 14:361–366.
24. Pinto, M.R., B.M. Kristal, and K.S. Schanze. 2003. *Langmuir* 19:6523–6533.
25. Gordon, M., W.R. Ware. 1975. eds. *The Exciplex.* New York: Academic Press.
26. Thunemann, A.F. 1999. *Adv Mater* 11:127–130.
27. Halkyard, C.E., M.E. Rampey, L. Kloppenburg, S.L. Studer-Martinez, and U.H.F. Bunz. 1998. *Macromolecules* 31:8655–8659.
28. Miteva, T., L. Palmer, L. Kloppenburg, D. Neher, and U.H.F. Bunz. 2000. *Macromolecules* 33:652–654.
29. Bunz, U.H.F., J.M. Imhof, R.K. Bly, C.G. Bangcuyo, L. Rozanski, and D.A. Vanden Bout. 2005. *Macromolecules.* 38:5892–5896.
30. Kim, J., and T.M Swager. 2001. *Nature* 411:1030–1034.
31. Nelson, J.C., J.G. Saven, J.S. Moore, and P.G. Wolynes. 1997. *Science* 277:1793–1796.
32. Prince, R.B., J.G. Saven, P.G. Wolynes, and J.S. Moore. 1999. *J Am Chem Soc* 121:3114–3121.
33. Lahiri, S., J.L. Thompson, and J.S. Moore. 2000. *J Am Chem Soc* 122:11315–11319.
34. Tan, C., M.R. Pinto, M.E. Kose, I. Ghiviriga, and K.S. Schanze. 2004. *Adv Mater* 16:1208–1212.
35. Rodger, A., and B. Norden. 1997. *Circular Dichroism and Linear Dichroism.* Oxford: Oxford University Press.
36. Prince, R.B., S.A. Barnes, and J.S. Moore. 2000. *J Am Chem Soc* 122:2758–2762.
37. Decher, G., J.D. Hong, and J. Schmitt. 1992. *Thin Solid Films* 210/211:831–835.
38. Decher, G. 1997. *Science* 277:1232–1237.
39. Decher, G., and J. Schlenoff. 2002. eds. *Multilayer Thin Films. Sequential Assembly of Nanocomposite Materials.* Weinheim: Wiley-VCH.
40. Lukkari, J., M. Salomaki, A. Viinikanoja, T. Aaritalo, J. Paukkunen, N. Kocharova, and J. Kankare. 2001. *J Am Chem Soc* 123:6083–6091.
41. Lakowicz, J.R. 1999. *Principles of Fluorescence Spectroscopy;* 2nd ed. New York: Kluwer Academic/Plenum Publishers.
42. Turro, N.J. 1978. *Modern Molecular Photochemistry.* Menlo Park, CA: Benjamin/Cummings.
43. Wang, D.L., J. Wang, D. Moses, G.C. Bazan, and A.J. Heeger. 2001. *Langmuir* 17:1262–1266.
44. Tan, C., E. Atas, J.G. Müller, M.R. Pinto, V.D. Kleiman, and K.S. Schanze. 2004. *J Am Chem Soc* 126:13685–13694.
45. Onitsuka, O., A.C. Fou, M. Ferreira, B.R. Hsieh, and M.F. Rubner. 1996. *J App Phys* 80:4067–4071.
46. Baur, J.W., S. Kim, P.B. Balanda, J.R. Reynolds, and M.F. Rubner. 1998. *Adv Mater* 10:1452–1455.
47. Ma, W.L., P.K. Iyer, X. Gong, B. Liu, D. Moses, G.C. Bazan, and A.J. Heeger. 2005. *Adv Mater* 17:274.
48. Tang, C.W. 1986. *App Phys Lett* 48:183–185.
49. Yu, G., J. Gao, J.C. Hummelen, F. Wudl, and A.J. Heeger. 1995. *Science* 270:1789–1791.

50. Halls, J.J.M., C.A. Walsh, N.C. Greenham, E.A. Marseglia, R.H. Friend, S.C. Moratti, and A.B. Holmes. 1995. *Nature* 376:498–500.

51. Baur, J.W., M.F. Durstock, B.E. Taylor, R.J. Spry, S. Reulbach, and L.Y. Chiang. 2001. *Synth Met* 121:1547–1548.

52. Durstock, M.F., B. Taylor, R.J. Spry, L. Chiang, S. Reulbach, K. Heitfeld, and J.W. Baur. 2001. *Synth Met* 116:373–377.

53. Mwaura, J.K., M.R. Pinto, D. Witker, N. Ananthakrishnan, K.S. Schanze, and J.R. Reynolds. 2005. *Langmuir* 21:10119–10126.

54. Arnt, L., K. Nusslein, G.N. Tew. 2004. *J Poly Sci A Poly Chem* 42:3860–3864.

55. Lu, L.D., F.H. Rininsland, S.K. Wittenburg, K.E. Achyuthan, D.W. McBranch, and D.G. Whitten. 2005. *Langmuir* 21:10154–10159.

56. QTL Biosystems, LLC, Santa Fe, New Mexico, USA.

III

Properties and Characterization of Conjugated Polymers

15

Insulator–Metal Transition and Metallic State in Conducting Polymers

Arthur J. Epstein

15.1 Introduction

During the last 65 years, conventional insulating polymers have been increasingly used as substitutes for structural materials such as wood, ceramics, and metals because of their high strength, lightweight, ease of chemical modification and customization, and processability at low temperatures [1]. In 1977, the high electrical conductivity of an organic polymer, doped polyacetylene, was reported [2], spurring interest in "conducting polymers" [3–6] (for recent research activity, see, e.g., the conference proceedings listed in [5]). The frequency-dependent electrical transport in doped polyacetylene was investigated intensively to understand the charge transport mechanisms [6,7]. The common electronic feature of pristine (undoped) conducting polymers is the conjugated π-system that is formed by the overlap of carbon p_z orbitals forming alternating single and double bonds [4,7–9]. (In some systems, polyaniline (PAN), nitrogen p_z orbitals and benzene rings (C_6) are included in the conjugation path.) Figure 15.1a shows the chemical repeat units of the pristine forms of several families of conducting polymers, i.e., *trans-* and *cis*-polyacetylene [(CH)$_x$]; the leucoemeraldine base (LEB), emeraldine base (EB), and pernigraniline base (PNB) forms of PAN; polypyrrole (PPy), and polythiophene (PT). Figure 15.1b shows the doped form of PAN, emeraldine salt (shown here with camphor sulfonic acid as the counterion and one counterion for each two ring repeat, and self-doped sulfonated PAN with each $-SO_3^-$ counterion covalently bonded to a C_6 ring), PPy (shown here with one PF_6^- counterion for each three pyrrole ring repeat), and the PT derivative polyethylenedioxythiophene (PEDOT), with polystyrene sulfonic acid (PSSA) as the counterion). Figure 15.1c illustrates the protonic acid doping of semiconducting EB to form highly conducting emeraldine salt [10,11].

Due to the vulnerability of one-dimensional (1D) chains with partially occupied energy bands to Peierls distortions [12], and the vulnerability of finite chains with partially filled degenerate highest occupied energy level to Jahn–Teller distortions [13], charges introduced into the polymers and oligomers through doping are stored in novel states such as solitons, polarons, and bipolarons, which include a charge and a lattice distortion that surround it [7]. Despite the strong electron–phonon coupling in conducting polymers, the conductivities of the pristine polymers can be transformed from insulating to metallic through the process of doping, with the conductivity increasing as the doping level increases. Both *n*-type (electron-donating) and *p*-type (electron-accepting) dopants as well as protonic acid doping have been used to induce an insulator–metal transition (IMT) in electronic polymers [2–5,7–9,14].

An extraordinarily large range of conductivity is obtained upon doping [14,34]. Figure 15.2 shows that the conductivity of the undoped polymers can be increased by 10 orders or more through doping. For instance, the conductivity of polyacetylene processed by various routes [15–18,31,32, 35–38] when doped with iodine can be increased from $\sim10^{-10}$ [*cis*-(CH)$_x$]-10^{-5} [*trans*-(CH)$_r$] $(\Omega\cdot cm)^{-1}$ [$=S/cm$] to greater than 4×10^4 S/cm [3,17,39], comparable with the conductivity of good metals (e.g., $\sigma_{dc} \sim 4.8 \times 10^4$ S/cm for lead and 5.9×10^5 S/cm for copper at room temperature [40]). Recent advances in the processing of other conducting polymer systems have led to improvements in their σ_{dc} values to the range of $10^2–10^3$ S/cm [3,4,6,28,30,39,41–47]. With such high conductivity, conducting polymers become useful for applications as wires, in electromagnetic shielding [48,49] and for electrostatic dissipation [1,3,50], with the advantages that conducting polymers are lightweight and facilitate recycling, compared with metallized polymers [1]. Of growing interest is the use of a thin film of solution processable doped conducting polymers such as PEDOT and PAN to form "transparent" conducting layers in active devices including organic- and polymer-based light-emitting devices [51,52], photovoltaic devices [53], and field-effect transistors (FETs) [54–56].

Accompanied by the enhancement in σ_{dc}, many traditional signatures of an intrinsic metallic nature have become apparent, including negative dielectric constants [3,23,39,43,57–60], a Drude metallic response [23,29,39,44,45,47,60], temperature-independent Pauli susceptibility [14,39,61–68], and a linear dependence of thermoelectric power on temperature [69–73]. However, the conductivities of even the new highly conducting polymers, though comparable with those of traditional metals at room temperature, generally decrease as the temperature is lowered [3,39,41,42,45,73,74]. Some of the most highly conducting samples remain highly conducting even in the millikelvin range [41,42,44,45,75], demonstrating that a truly metallic state at low temperature has been attained.

These metallic features have been reported in conducting versions of both PAN and PPy. The ability to process PAN doped with camphorsulfonic acid (CSA) from solution [22,76] has resulted in freestanding films with high conductivity ($\sigma_{dc} \sim$ 100–400 S/cm) [44,77] that span the IMT even at low temperature [45,75]. Some samples of electrochemically prepared PPy doped with hexafluorophosphate (PF_6) are metallic to millikelvin temperatures [29,41,42]. However, when PPy is synthesized using different dopants or at high temperatures, the materials are more disordered and show insulating behavior [28–30]. Similar results are reported in the literature for conducting polyacetylene [41,42].

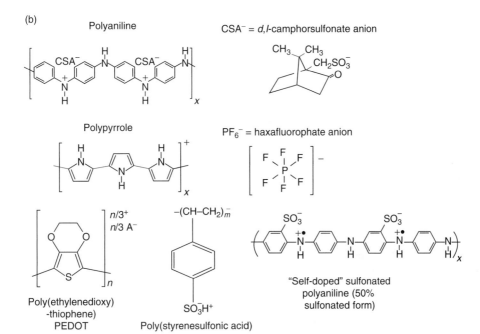

FIGURE 15.1 Schematic chemical structures of (a) undoped semiconducting polymers *trans*-polyacetylene, *cis*-polyacetylene, polythiophene, polypyrrole (PPy), and polyaniline (PAN) (leucoemeraldine base [$y = 0$], emeraldine base [$y = 0.5$], pernigraniline base [$y = 1$]), (b) doped form of polyaniline (emeraldine salt shown with camphor sulfonic acid as the counterion and one counterion for each two-ring repeat, "self-doped" sulfonated polyaniline shown with each $-SO_3^-$ counterion covalently bonded to alternating C_6 ring), polypyrrole (shown with one PF_6^- counterion for each three pyrrole ring repeat), and the polythiophene derivative polyethylenedioxythiophene (PEDOT), with polystyrene sulfonic acid as the counterion, and

(*continued*)

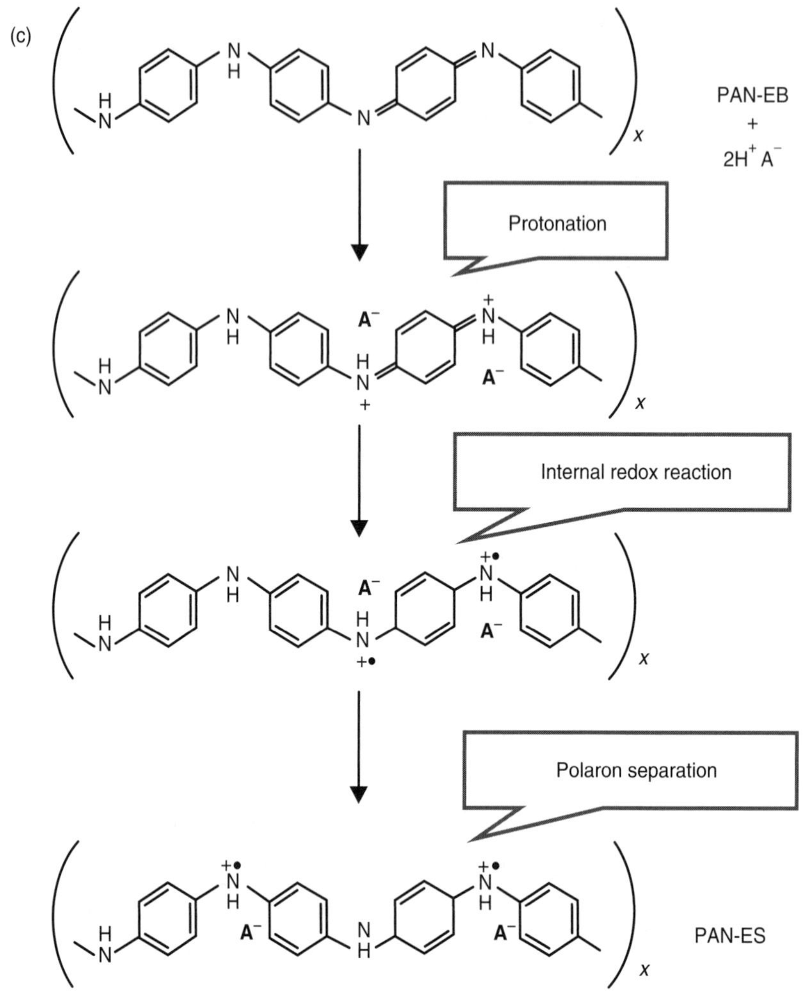

FIGURE 15.1 (**continued**) (c) protonic acid doping of polyaniline.

The transport properties of conducting polymers are highly dependent upon the structural disorder arising from sample quality, doping procedure, and aging [3,39]. For doped polyacetylene [41,42], PPy [41,42], and PAN [44,45], studies of the millikelvin resistivity ($\rho = 1/\sigma$) showed that the conductivity is thermally activated in strongly disordered films, the resistivity appears like that of a disordered metal ($\rho \propto \log T$) for samples of intermediate conductivity, and ρ has a very weak temperature dependence for the weakly disordered state. Careful studies of the microwave dielectric constant in PAN as the crystallinity of the films improves demonstrate that the carriers become more delocalized as the structural order is improved [23]. PAN-HCl samples have a range of structural order depending upon preparation methods including choice of solvent [78–81]. Variation of the solvent from which PAN-CSA is cast yields a change in the local order in the polymer [82–84] and a concomitant change in the crystallinity of the films, with the transport properties showing stronger localization as the crystallinity decreases [60–63]. Even the most metallic samples are not single crystals. PAN-CSA prepared by self-stabilized dispersion polymerization [85] has a similar variation of crystallinity with variation of the solvent [47]. Doped polyacetylene has a complex fibrillar morphology with ~80%–90% crystallinity, while doped PAN and doped PPy have ~50% crystallinity at best [82]. Recently, it has been shown that PAN [86] and PPy [87] may also be prepared in a fibrillar morphology depending upon the synthesis conditions.

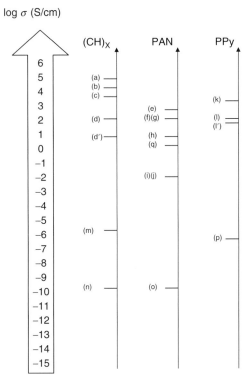

FIGURE 15.2 Overview of conductivity of conducting polymers at room temperature. (a) Stretched $[CH(I_3)]_x$. (From Tsukamoto, J., *Adv. Phys.*, 41, 509, 1992 and Tsukamoto, J., Takahashi, A., and Kawasaki, K., *Jpn. J. Appl. Phys.*, 29, 125, 1990.) (b) Stretched $[CH(I_3)]_x$. (From Naarmann, H. and Theophilou, N., *Synth. Met.*, 22, 1, 1987.) (c) $[CH(I_3)]_x$ (From Shirakawa, H., Zhang, Y.-X., Okuda, T., Sakamaki, K., and Akagi, K., *Synth. Met.*, 65, 93, 1994.) (d) $[CH(I_3)]_x$ (From Chiang, J.-C. and MacDiarmid, A.G., *Synth. Met.*, 13, 193, 1986.) (d′) $[CH(I_3)]_x$ (From Epstein, A.J., Rommelmann, H., Bigelow, R., Gibson, H.W., Hoffman, D.M., and Tanner, D.B., *Phys. Rev. Lett.*, 50, 1866, 1983.) (e) Stretched PAN-HCl. (From Adams, P.N., Laughlin, P., Monkman, A.P., and Bernhoeft, N., *Solid State Commun.*, 91, 895, 1994.) (f) PAN-CSA from *m*-cresol. (From Cao, Y., Smith, P., and Heeger, A.J., *Synth. Met.*, 48, 91, 1992.) (g) PAN-CSA from *m*-cresol. (From Joo, J., Oblakowski, Z., Du, G., Pouget, J.P., Oh, E.J., Weisinger, J.M., Min, Y., MacDiarmid, A.G., and Epstein, A.J., *Phys. Rev. B*, 69, 2977, 1994.) (h) PAN derivative: poly(*o*-toluidine) POT-CSA fiber from *m*-cresol. (From Wang, Y.Z., Joo, J., Hsu, C.-H., Pouget, J.P., and Epstein, A.J., *Phys. Rev. B*, 50, 16811, 1994.) (i) POT-HCl. (From Wang, Z.H., Javadi, H.H.S., Ray, A., MacDiarmid, A.G., and Epstein, A.J., *Phys. Rev. B*, 42, 5411, 1990.) (j) Sulfonated PAN. (From Yue, J. and Epstein, A.J., *J. Am. Chem. Soc.*, 112, 2800, 1990 and Yue, J., Wang, Z.H., Cromack, K.R., Epstein, A.J., and MacDiarmid, A.G., *J. Am. Chem. Soc.*, 113, 2655, 1991.) (k) Stretched PPy(PF$_6$). (From Yamaura, M., Hagiwara, T., and Iwata, K., *Synth. Met.*, 26, 209, 1988.) (l) PPy(PF$_6$) and (l′) PPy(TsO). (From Kohlman, R.S., Joo, J., Wang, Y.Z., Pouget, J.P., Kaneko, H., Ishiguro, T., and Epstein, A.J., *Phys. Rev. Lett.*, 74, 773, 1995 and Sato, K., Yamaura, M., Hagiwara, T., Murata, K., and Tokumoto, M., *Synth. Met.*, 40, 35, 1991.) (m) Undoped *trans*-(CH)$_x$. (From Epstein, A.J., Rommelmann, H., Abkowitz, M., and Gibson, H.W., *Phys. Rev. Lett.*, 47, 1549, 1981.) (n) Undoped *cis*-(CH)$_x$. (From Epstein, J., Rommelmann, H., and Gibson, H.W., *Phys. Rev. B*, 31, 2502, 1985.) (o) Undoped PAN (EB). (From Zuo, F., Angelopoulos, M., MacDiarmid, A.G., and Epstein, A.J., *Phys. Rev. B*, 39, 3570, 1989.) (p) Undoped PPy. (From Scott, J.C., Pfluger, P., Krounbi, M.T., and Street, G.B., *Phys. Rev. B*, 28, 2140, 1983.) (q) HClO$_4$ doped polyaniline nanofibers produced by dhilute polymerization. (From Chiou, N-R. and Epstein, A.J., *Adv. Mat.*, 17, 1679, 2005.). The conductivity reported for the undoped polymers should be considered an upper limit because of the possibility of impurities. (Based on Kohlman, R.S. and Epstein, A.J., *Handbook of conducting polymers*, 2nd ed., eds. Skotheim, T.A., Elsenbaumer, R.L., and Reynolds, J.R., Marcel Dekker, New York, 1988, chap. 3. Modified from Routledge/Taylor & Francis Group, LLC. With permission.)

The effect of the disorder and the one-dimensionality of the polymer on the nature of the metallic state and the IMT are still strongly debated. Disorder [88,89] and one-dimensionality [89] lead to localization of the electron wavefunctions. Even if the polymer chains are well-ordered, macroscopic transport is impossible unless charge carriers can hop or diffuse from one chain to another to avoid the chain breaks and defects [90–92]. Previous experimental work and theoretical calculations [23,25,43,93–95] have stressed the importance of interchain interaction as well as three-dimensionality of the electron states in highly conducting polymers to avoid 1D localization. There is evidence that the metallic states are three-dimensional (3D), though the transport properties are highly anisotropic [90,92,96–98]. The nature of the disorder is still an important question. Disorder can result in different properties depending on whether it is homogeneous or inhomogeneous.

The wavefunctions of the charge carriers may become localized to a few atomic sites if the disorder is strong enough. The behavior of a uniformly disordered material near an IMT has been discussed in the context of an Anderson transition [88,89,99,100]. For a 3D system, when the disorder potential becomes comparable with the electronic bandwidth, a mobility edge exists that separates localized from extended states [88,89]. If the Fermi level (E_F) lies in the region of extended states, σ_{dc} is finite [89] and the logarithmic derivative of the conductivity [$W \equiv d\ln\sigma_{dc}/d\ln T$] [101] has a positive slope at low temperature. When E_F lies in the range of localized states, the carriers have a hopping behavior (where $\sigma_{dc} \to 0$) [89] and the W function has a negative slope at low temperature [101]. Also, the dielectric function is positive owing to the polarization of the localized states. The strong disorder necessary to cause the localization of 3D electronic wavefunctions necessarily limits the mean free time (τ) to small values. As the IMT is crossed, the electronic localization length (L_{loc}) diverges and the system monotonically becomes more metallic, displaying higher σ_{dc} values with weaker temperature dependences. However, τ varies slowly with energy in crossing the IMT [99]. Therefore, τ remains short close to the IMT. In fact, the Ioffe–Regel condition requires that $k_F\lambda \sim 1$ for a material near an IMT [102], where k_F is the Fermi wavevector and λ is the mean free path. This implies that the scattering time $\tau \sim 10^{-15}$ s. For such short scattering times, the frequency-dependent conductivity, $\sigma(\omega)$, is monotonically suppressed at low frequency. Also, localization corrections to the metallic Drude response result in a positive dielectric function $\varepsilon(\omega)$ for short τ at low frequency, rather than the negative low frequency $\varepsilon(\omega)$ of normal metals [40,103].

In contrast, if the disorder is inhomogeneous with large variations on the length scale or larger than the characteristic electron L_{loc}, then the behavior characteristic for a composite system may be expected. The inhomogeneous disorder model [29,39,43–45] treats the metallic state of conducting polymers as a composite system comprising metallic ordered regions (with delocalized charge carriers) coupled by disordered quasi-1D regions through which hopping transport along and between chains occurs. Localization occurs in the disordered regions owing to the 1D electronic nature of the polymer chains in this region. When the polymer chains in the disordered region are tightly coiled, the in-chain L_{loc} is short and coupling between metallic regions is poor, so that free electrons are confined within the metallic regions [3,23,39,44,83,84,104,105]. The temperature-dependent transport is then dominated by hopping and phonon-induced delocalization in the disordered regions [43] or even tunneling between metallic islands [104,105], depending upon the morphology. When the polymer chains in the disordered regions are sufficiently straight (i.e., larger radii of curvature or longer persistence lengths), the in-chain L_{loc} is larger than the typical separation between metallic islands and carriers are able to diffuse macroscopically among the metallic regions [23,43–45]. In this case, a fraction of the carriers will percolate through the ordered paths. Just as in the Anderson transition, there is a crossover in slope for the W plot as percolation occurs, though the IMT is no longer necessarily a monotonic function of the room temperature σ_{dc} [44,105]. In this model, the magnitude of σ_{dc} depends on the number of well-coupled metallic regions across the sample. On the metallic side of the IMT, a fraction of the carriers will demonstrate free carrier response even at low temperature. Due to phonon-induced delocalization in the disordered regions, a fraction of carriers may appear to have percolated at room temperature even for samples on the insulating side of the IMT [44,105].

To determine the nature of the metallic state in conducting polymers, it is necessary to use a wide variety of probes. Direct current transport measurements provide insight into the insulating or metallic nature of electrons at the Fermi level. Measurement of high-frequency transport provides an important probe away from the Fermi energy to help discriminate between the homogeneously and inhomogeneously disordered metallic states [6,44]. For instance, $\varepsilon(\omega)$ and the microwave dielectric constant (ε_{MW}) can be used to determine the presence of Drude (free carrier) dispersion in the electrical response of the sample as well as the plasma frequency of the free carriers [3,23,29,39,44]. This free carrier behavior can be monitored as a function of processing and temperature. Also, $\varepsilon(\omega)$ and the optical frequency conductivity $\sigma(\omega)$ provide probes of the scattering times and mean free paths for the samples as the IMT is crossed.

Highly conducting polymers such as doped PAN and PPy in a metallic state have unusual frequency-dependent conductivity, including multiple zero crossings of the dielectric function. A low-frequency electromagnetic response in terms of a Drude metal is provided by an extremely small fraction of the total number of electrons ($\sim$0.1%), but with extremely high mobility or anomalously long scattering time $\sim 10^{-13}$ s. Prigodin and Epstein have shown that a network of metallic grains connected by resonance quantum tunneling has a Drude-type response for both the high- and low-frequency regimes and behaves as a dielectric at intermediate frequency in agreement with experimental observations. The metallic grains in polymers represent crystalline domains of well-packed chains with delocalized electrons embedded in the amorphous media of poor chain order. Intergrain resonance tunneling occurring through the strongly localized states in the amorphous media account for the anomalous frequency and temperature dependence of the conductivity and dielectric constant [106].

It was recently reported [56,107–112] that use of doped "metallic" polymers as the active channel in FET structures results in unexpected "normally on" transistor-like behavior. The analysis [111,112] of the ion flow into the active channel due to the gate voltage suggests that for the transistors studied, compensating $\sim$2 counterions per 100 dopant molecules suppresses conductance up to three or more orders of magnitude. It has been proposed [111,112] that removal of intermediate hopping sites through a small fraction of ion charge compensation causes carriers to hop over longer distances, thereby removing percolation paths and leading to a conductor–nonconductor transition.

In this chapter, coordinated studies of the temperature-dependent dc, optical (2 meV to 6 eV), and microwave (6.5 GHz) transport on two different polymer systems, PAN and PPy, are emphasized. From a comparison of this set of experiments with the reports in the literature for polyacetylene, it is determined that there is a universal type of behavior of the metallic state in conducting PAN, PPy, and $(CH)_x$. The metallic state is characterized by a small fraction of carriers that are delocalized down to low T with long mean free times ($\tau > \sim 10^{-13}$ s) and a large majority of carriers that are more strongly localized ($\tau_1 \sim 10^{15}$ s). These delocalized electrons result in a weak temperature dependence for σ_{dc} down to millikelvin temperatures. This metallic state is strongly dependent on structural order. More disordered polymers show strong localization and hopping transport (insulating behavior). It is concluded that although the IMT is due to structural disorder, it is not due to a conventional homogeneous Anderson transition. We propose that the IMT is instead better described by the inhomogeneous disorder model. This inhomogeneity is shown to account for the unexpected electric field effect in metallic polymers.

15.2 Models for Localization and Metallic Conductivity

It has been established experimentally that a finite density of states at the Fermi level [$N(E_F)$] can be induced by doping [3,14,20,39,62,63,64,65]. Even though there is a high density of conduction electrons at the Fermi level for the highly doped state, the carriers may be spatially localized so they cannot participate in transport except through hopping. The prime source of localization is structural disorder in the polymers [3,39,82]. X-ray diffraction studies of these systems show that they are generally of modest crystallinity, with regions of the material that are more ordered and other regions that are more disordered [82,113]. Also, the fibrillar nature of many of the conducting polymers may lead to localization by reducing the effective dimensionality of the electrons delocalized in a bundle of polymer chains and also may introduce an additional resistance in the connection between fibers [95]. In Section 15.2.1 through Section 15.2.3, the effect of inhomogeneous and homogeneous structural disorder on the resulting transport properties of the disordered metallic state is discussed.

15.2.1 Inhomogeneous Disorder-Induced Insulator–Metal Transition

Conducting polymers generally display a rich variety in their morphology [82], being partially crystalline and partially disordered as shown schematically in Figure 15.3. If the L_{loc} in the more disordered regions of the electrons is comparable with or smaller than the crystalline coherence lengths ($\sim$10 Å) in the polymer, then the disorder present in the conducting polymer is viewed as inhomogeneous [3,29,39,44,105]. The localization effects in the inhomogeneously disordered (partially crystalline) conducting polymers are proposed to originate from quasi-1D localization in the disordered regions [23,29,39,44] that surround the ordered regions. It is well known that for a 1D metallic chain the localization of charge carriers occurs for even weak disorder because of quantum interference of static backscattering [89], with L_{loc} increasing as the disorder decreases. For quasi-1D conducting polymers, the transfer integral perpendicular to the chain axis is crucial. As the polymer chains are finite, if the carrier cannot diffuse to another chain before being reflected back onto itself and thus experiencing quantum interference, then the polymer will not be a conductor. This leads to the condition $w_{\perp} > 1/2\tau$,

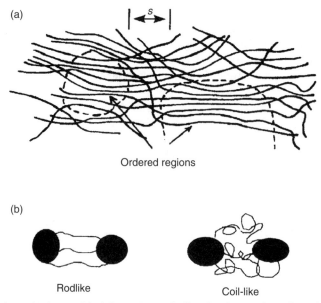

FIGURE 15.3   (a) Schematic picture of the inhomogeneously disordered state of metallic and insulating conducting polymers. (From Kohlman, R.S. and Epstein, A.J., *Handbook of conducting polymers*, 2nd ed., eds. Skotheim, T.A., Elsenbaumer, R.L., and Reynolds, J.R., Marcel Dekker, New York, 1988, chap. 3. Reprinted from Routledge/Taylor & Francis Group, LLC. With permission.) (b) Schematic picture of rodlike and coil-like morphology of disordered regions. (From Kohlman, R.S. and Epstein, A.J., *Handbook of conducting polymers*, 2nd ed., eds. Skotheim, T.A., Elsenbaumer, R.L., and Reynolds, J.R., Marcel Dekker, New York, 1988, chap. 3. Reprinted from Routledge/Taylor & Francis Group, LLC. With permission.)

(*continued*)

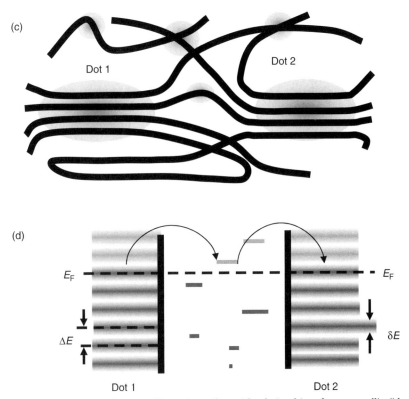

FIGURE 15.3 (continued) (c) Schematic illustration of spatial relationship of two metallic "dots" made of conducting polymers separated by insulating (disordered) regions. (From Prigodin, V.N. and Epstein, A.J., *Physica B*, 338, 310, 2003. From Elsevier. With permission.) (d) The electrical coupling between metallic grains is provided by resonance tunneling through localized states in the disordered region. (From Prigodin, V.N. and Epstein, A.J., *Physica B*, 338, 310, 2003. From Elsevier. With permission.)

where $w_\perp$ is the interchain transfer rate and τ is the mean free time, to avoid quasi-1D localization [90,114].

Prigodin and Efetov [95] studied the IMT of these quasi-1D conducting polymers using a random metallic network (RMN) model to represent weakly connected, fibrous bundles of metallic chains. In this zero temperature model, the phase transition from insulating to metallic behavior is a function of the cross section of electronic overlap between fibers (α) and $\rho = pL_{loc}$, the product of the localization radius (L_{loc}) and the concentration of cross-links between fibers (p). The metallic state can be induced by strengthening the interchain (or interfibril) interaction (increasing α), increasing the density of cross-links between fibers (increasing p), or increasing the L_{loc}. This model developed for contacts between fibers composed of parallel polymer chains can be generalized to the 3D delocalization transition that occurs in inhomogeneously disordered (partially crystalline) nonfibrillar polymers [3,39]: the metallic state is induced as the strength of connection between ordered or crystalline regions (α) is increased, as the density of interconnections between ordered or crystalline regions (p) increases, and as the L_{loc} within the disordered regions increases. L_{loc} depends on the morphology of the disordered region. Figure 15.3b schematically shows examples of rodlike and coil-like chain morphologies [76]. L_{loc} will be larger for the rodlike and smaller for the coil-like morphology. Within this model, conduction electrons are 3D delocalized in the "crystalline"-ordered regions (though the effects of paracrystalline disorder may limit delocalization within these regions [115]). In order to move between ordered regions, the conduction electrons must diffuse along electronically isolated quasi-1D chains through the disordered regions where the electrons readily become localized.

Because the IMT depends on the coupling between metallic regions, the inhomogeneous disorder model resembles a percolation transition. The main difference between percolating conducting polymers and more traditional percolating systems, such as silver particles in potassium chloride, is that the conducting polymer "crystals" do not have sharp boundaries. As a single polymer chain can be a part of both metallic crystalline regions and quasi-1D disordered regions, the percolating object is a fuzzy ellipsoid [3,44]. When L_{loc} is greater than the average distance between metallic regions, carriers may diffuse among the metallic regions and an IMT occurs. If L_{loc} is temperature-dependent due to phonon-assisted processes [43], it is possible to show the behavior of a percolated metallic system at high temperatures (L_{loc} greater than the average distance between metallic islands) and the behavior of an unpercolated insulating system at low temperatures (L_{loc} less than the average distance between metallic islands).

The dependence of the conductivity on transport through quasi-1D chains leads to some profound effects. For an ordered quasi-1D metal with its highly anisotropic electronic Fermi surface, σ in the chain direction is given by

$$\sigma_\| = \frac{e^2 na}{\pi\hbar} v_\mathrm{F}\tau = \frac{e^2 na}{\pi\hbar}\left(\frac{\lambda}{a}\right) \tag{15.1}$$

where n is the conduction electron density per unit volume, a is the interatomic distance in the chain direction, $v_\mathrm{F} = 2t_0 a/\hbar$, t_0 is the electron hopping element along the chain, and τ is the mean free time. Kivelson and Heeger [116] showed that only $2k_\mathrm{F}$ phonons can relax the momentum of the electrons for a quasi-1D chain from which the counterions are spatially removed and therefore effectively screened. Since the $2k_\mathrm{F}$ phonons have a relatively small thermal population (owing to their high energies, ~0.1 eV), the conductivity is predicted [116] to be quite high at room temperature ($\sim2 \times 10^6$ S/cm for polyacetylene) and increases exponentially upon cooling. A small interchain hopping integral $t_\perp \sim 0.1$ eV is all that is necessary to obtain 3D delocalization of carriers, and the resulting metallic state may take advantage of this large quasi-1D conductivity. The lack of effective scatterers in an ordered quasi-1D metal may lead to an anomalously long scattering time [23] compared with conventional 3D metals.

In a real system, disorder scattering must be taken into account as the dc conductivity is limited by the least conducting (most disordered) part of the conduction path. In disordered quasi-1D conducting polymers, static disorder scattering (with scattering time $\tau_{\mathrm{imp}}(2k_\mathrm{F})$) as well as $2k_\mathrm{F}$ phonons (with scattering time $\tau_{\mathrm{ph}}[2k_\mathrm{F},T]$) can relax momentum. The effect of phonon-induced delocalization (scattering time $\tau_{\mathrm{ph}}[0,T]$), which increases the conductivity with increasing temperature, must be taken into account. Within this model [43], the T-dependent dc conductivity is given by

$$\sigma_{\mathrm{dc}}(T) = \frac{\Omega_p^2}{4\pi}\tau_{\mathrm{tr}}(T)\frac{f(T)-1}{f(T)+1} \tag{15.2}$$

where $\Omega_p^2 (=4\pi ne^2/m^*)$ is the carrier plasma frequency, τ_{tr} is the effective transport scattering time determined by Matthieson's rule

$$\frac{1}{\tau_{\mathrm{tr}}(T)} = \frac{1}{\tau_{\mathrm{imp}}(2k_\mathrm{F})} + \frac{1}{\tau_{\mathrm{ph}}(2k_F,T)} \tag{15.3}$$

and

$$f^2(T) = 1 + 4\frac{\tau_{\mathrm{imp}}(2k_\mathrm{F})}{\tau_{\mathrm{ph}}(0,T)} \tag{15.4}$$

The behavior with temperature of the scattering times and the resultant $\sigma_{\mathrm{dc}}(T)$ [117] are shown schematically in Figure 15.4. In well-ordered materials, it is possible for $\tau_{\mathrm{imp}}(2k_\mathrm{F})$ to be greater than

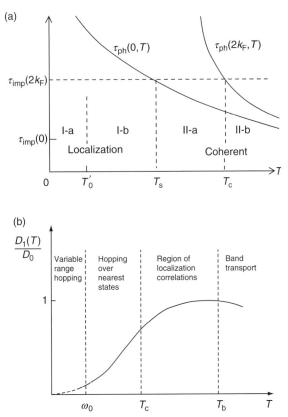

FIGURE 15.4 (a) Schematic picture of the temperature dependence of the various scattering times in the inhomogeneous disorder model and the respective localization domains. (b) The corresponding diffusion constant expected for the inhomogeneous quasi-one-dimensional system. (From Prigodin, V.N. and Efetov, K.B., *Synth. Met.*, 65, 195, 1994. Reprinted from Elsevier. With permission.)

$\tau_{\text{ph}}(2k_{\text{F}},T)$ near room temperature. Therefore, τ_{tr} is dominated by inelastic phonon scattering near room temperature. Since $\tau_{\text{ph}}(2k_{\text{F}},T)$ increases with decreasing temperature, σ_{dc} increases with decreasing T, similar to the behavior of metallic systems. When $\tau_{\text{imp}}(2k_{\text{F}}) < \tau_{\text{ph}}(2k_{\text{F}},T)$, localization effects begin to dominate (phonon-induced delocalization is important), resulting in a suppression in σ_{dc} at lower temperatures and a consequent maximum in the diffusion rate and the corresponding $\sigma_{\text{dc}}(T)$ (Figure 15.4). For weaker disorder (longer $\tau_{\text{imp}}(2k_{\text{F}})$), this maximum in σ_{dc} is shifted to lower temperature. At even lower temperature, if $\tau_{\text{imp}}(2k_{\text{F}}) > \tau_{\text{ph}}(0,T)$, there is a crossover to a more strongly localized hopping conductivity. This model accounts for localized behavior at low temperature despite conductivities at room temperature in excess of the Mott minimum conductivity. Only the most delocalized electrons on the percolating network contribute to the high dc conductivity. (The more localized electrons lead to a conductivity less than $\sim$10 S/cm [105].) The fraction of delocalized carriers can be determined from high frequency measurements.

For a truly metallic system, the dc conductivity will remain finite as $T \to 0$ [89] and the logarithmic derivative of the conductivity ($W = d\ln \sigma_{\text{dc}}(T)/d\ln T$) will show a positive slope at low T [101]. In contrast, for an unpercolated system, the dc conductivity will decrease rapidly at low T and the W plot will have a negative slope, characteristic of hopping systems. In the inhomogeneous disorder model, the W plot will show a crossover in the low T slope for the W plot from negative to positive, though the IMT is not necessarily a monotonic function of room temperature σ_{dc}, owing to its dependence on the density and strength of connections between metallic regions and the L_{loc} of electronic states in the disordered regions.

Effective medium theories characterize the frequency-dependent transport in systems with large-scale inhomogeneities such as metal particles dispersed in an insulating matrix [118,119]. An IMT in the effective medium model represents a percolation problem where a finite σ_{dc} as $T \to 0$ is not achieved until metallic grains in contact span the sample. To understand the frequency dependence of the macroscopic material, an effective medium is built up from a composite of volume fraction f of metallic grains and volume fraction $1 - f$ of insulator grains. The effective dielectric function $\varepsilon_{EMA}(\omega)$ and conductivity function $\sigma_{EMA}(\omega)$ are solved self-consistently.

The characteristic composite behavior of $\sigma_{EMA}(\omega)$ for a medium consisting of spherical particles with volume fractions f of Drude conductor and $1 - f$ of insulator is shown in Figure 15.5. For a volume fraction f less than the percolation value ($f = 1/3$ for spheres), $\sigma_{EMA}(\omega)$ is dominated by an impurity band of localized plasmon-like excitations. As the system approaches the percolation threshold, the localized peak $\sigma_{EMA}(\omega)$ shifts to lower frequency. Above the percolation threshold, a Drude peak corresponding to the carriers that have percolated through the composite structure occurs at low frequency. Only a fraction ($\sim(3f - 1)/2$ [119]) of the full conduction electron plasma frequency appears in the Drude peak, depending on the proximity to the percolation threshold. The same percolating free electron behavior is observable in the dielectric response $\varepsilon_{EMA}(\omega)$ for the system.

This characteristic behavior for traditional composites is expected to be modified for conducting polymer composites. As mentioned earlier, the geometry of the percolating objects resembles fuzzy ellipsoids with large aspect ratios rather than spheres. Also, phonon-assisted transport in the disordered regions of the conducting polymer (which are the insulating regions) may give the appearance of percolation in $\sigma(\omega)$ and $\varepsilon(\omega)$ at room temperature even though a particular system may not truly be metallic at low T [44,105]. For both $\sigma(\omega)$ and $\varepsilon(\omega)$, the distinct behavior of localized and delocalized

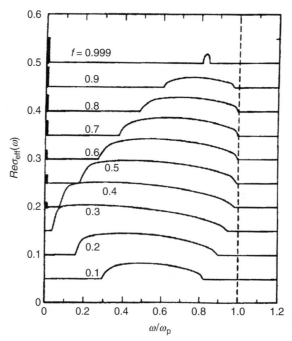

FIGURE 15.5 Schematic $\sigma(\omega)$ for an insulator–metal composite made up of volume fractions f of a Drude metal and $1 - f$ of an insulator, as calculated in the effective medium theory. The heavy line at $\omega = 0$ represents the Drude peak. The integrated strength of the delta function is proportional to the height of the delta function. The scattering time is chosen to be very long so that the width of the Drude peak is too narrow to be resolved in the plot, emphasizing the behavior of the localized modes. (From Bergman, D.J. and Stroud, D. *Solid state physics*, Vol. 46, eds. Ehrenreich, H. and Turnbull, D., Academic Press, New York, 1992, 148. Reprinted from Elsevier. With permission.)

charge carriers will be evident above percolation. Specifically, for percolated samples, $\sigma(\omega)$ will increase and $\varepsilon(\omega)$ will become increasingly negative with decreasing frequency below the unscreened free carrier plasma frequency Ω_p ($= \sqrt{4\pi\delta ne^2/m^*}$, where n is the full charge carrier density, δ is the fraction of the charge carrier density that is delocalized, e is the electronic charge, and m^* is the effective mass of the delocalized carriers). The response of localized carriers (the localized plasmon-like excitations) will occur for frequencies less than the "full" unscreened plasma frequency Ω_{p1} ($= \sqrt{4\pi ne^2/m_1^*}$, where m_1^* is the averaged effective mass of the carriers) [44,105]. The anomalously long scattering times possible for quasi-lD systems may lead to huge negative dielectric functions at low frequency.

15.2.2 Quantum Hopping in Metallic Polymers

The proposal of a new mechanism of transport in metallic polymers, quantum hopping [106], is based on puzzles, which are revealed in the frequency dependence of conductivity and dielectric constant of highly conducting polymers through the optical and low-frequency measurements of conductivity and dielectric constant (Figure 15.6). Experiments [45] (Figure 15.6) show that the high-frequency (≥ 0.1 eV) conductivity and dielectric constant generally following a Drude law with the number of electrons $\sim 10^{21}$ cm^{-3} correspond to the total density of conduction electrons and conventional scattering time

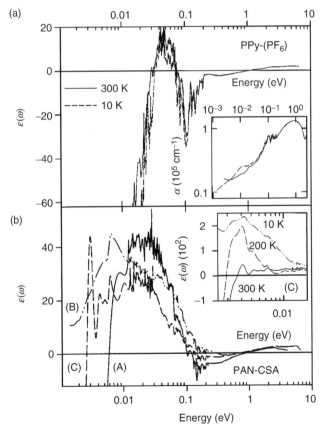

FIGURE 15.6 (a) Comparison of room temperature $\varepsilon(\omega)$ with 10 K $\varepsilon(\omega)$ for PPy (PF$_6$). Inset: comparison of the room temperature and 10 K absorption coefficient for PPy (PF$_6$). (b) Comparison of the room temperature $\varepsilon(\omega)$ for the PAN-CSA samples described in [45]. Inset: comparison of low temperature and room temperature $\varepsilon(\omega)$ for PAN-CSA Sample C which becomes insulating at low temperatures. (From Kohlman, R.S., Zibold, A., Tanner, D.B., Ihas, G.G., Min, Y.G., MacDiarmid, A.G., and Epstein, A.J., *Phys. Rev. Lett.*, 78, 3915, 1997. Reprinted from the American Physical Society. With permission.)

$\sim 10^{-15}$ s in both the metallic and dielectric phases. At decreasing frequency the polymers in the dielectric phase progressively display insulator properties and ε becomes positive for frequency ≤ 0.1 eV signaling that charge carriers are now localized. Microwave frequency (~ 6.6 GHz) ε experiments [3] yield $L_{loc} \sim 5$ nm, depending on sample, dopants, and preparation conditions.

A puzzling feature of the metallic phase in polymers is that ε is similar to that of dielectric samples with decreasing frequency, also changing sign from negative to positive at approximately the same frequency ~ 0.1 eV. However, for metallic samples ε changes again to negative at yet lower frequencies ≤ 0.01 eV indicating that free electronic motion is present [3,5]. The parameters of this low frequency coherent phase are quite anomalous. From the Drude model, the relaxation time is very long $\geq 10^{-13}$ s; also, the new plasma frequency below which ε is again negative is very small ~ 0.01 eV [3,5].

Recently, this second zero crossing of the dielectric constant at low frequency and the conclusion about a long relaxation time and a small plasma frequency were confirmed with radio frequency conductivity (see Figure 15.7) [60]. The results of [60] are important because they were obtained by direct measurement of conductivity. The early frequency dependence of conductivity was derived from the reflectance coefficient by using Kramers–Kronig procedure which requires some assumption on behavior of reflectance in the limit $\omega \to 0$.

These experimental findings for low-frequency electromagnetic response are in contrast with the expectations for the Anderson IMT [99] in which electronic behavior is controlled by disorder. In the dielectric phase, electrons are bound by fluctuations of the random potential. On the metallic side of the transition, free carriers have short scattering times. In the metallic phase near the transition ε is positive because the disorder causes dynamic polarization due to slowing diffusion due to localization effects. When approaching the IMT transition the localization effects increase and ε diverges (dielectric catastrophe [120]).

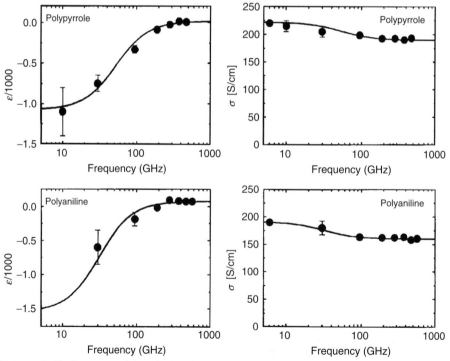

FIGURE 15.7 Radio frequency $\varepsilon(\omega)$ and $\sigma(\omega)$ for metallic doped polypyrrole (PPy) and polyaniline polymers. (From Martens, H.C.F., Reedijk, J.A., Brom, H.B., de Leuw, D.M., and Menon, R., *Phys. Rev. B*, 63, 073203, 2001. Reprinted from the American Physical Society. With permission.)

The small plasma frequency and very long τ of the metallic state in doped polymers can be explained [3,5,45,60,106] assuming that the conductivity is provided by a small fraction $\sim 0.1\%$ of the total carriers with long scattering time $> 10^{-13}$ s. However, it is difficult to reconcile this conclusion with the behavior for high frequencies which supports that the scattering time is the usual $\sim 10^{-15}$ s and all available electrons participate in conduction. To account for these anomalies the possible presence of a collective mode, as in a charge density wave conductor, or superconductor, was suggested [121].

An explanation of the low-frequency anomaly in doped polymers in the metallic phase, based on their chain morphology is very important. These materials are strongly inhomogeneous [82] with crystalline regions within which polymer chains are well ordered (Figure 15.3a). When the IMT is approached, delocalization first occurs inside these regions. Outside the crystalline regions, the chain order is poor and the electronic wavefunctions are strongly localized. Therefore, the crystalline domains can be considered as nanoscale metallic dots embedded in amorphous, poorly conducting medium [122]. The metallic grains remain always spatially separated by amorphous regions, and, therefore, direct tunneling between grains is exponentially suppressed. The intergrain tunneling is possible through intermediate localized states in the disordered portion with strong contribution from resonance states, whose energy is close to the Fermi level (Figure 15.3c and Figure 15.3d). The dynamics of resonance tunneling can account for the frequency-dependent anomalies in the conductivity and dielectric constant of the metallic phase of these doped polymers. Details of the analysis below can be found in [106].

We assume that grains have $N_\perp$ chains densely packed over $N_\parallel$ repeat units yielding $N_\perp \times N_\parallel$ unit cells in each grain. Neglecting intergrain coupling leaves the electronic levels inside grains quantized with mean level spacing $\Delta E = [\widetilde{N}(\varepsilon_F)(N_\perp \times N_\parallel)]^{-1}$, where $\widetilde{N}(\varepsilon_F)$ is the density of states per unit cell. For metallic doped PPy(PF$_6$) [82]: $N_\perp \approx 3 \times 8$, $N_\parallel \approx 7$, $\widetilde{N}(\varepsilon_F) = 0.8$ states/(eV $\times$ ring) and $\Delta E \approx 7.4$ meV. For doped metallic PAN(HCl) [78–82], $N_\perp \approx 9 \times 12$, $N_\parallel \approx 7$, $\widetilde{N}(\varepsilon_F) = 1.1$ states/(eV $\times$ 2 rings), and $\Delta E \approx 1$ meV.

We assume that the electronic states are delocalized inside grains and the electron's dynamics are diffusive with diffusion coefficient $D = v_F^2 \tau / 3$. This diffusive behavior is restricted to times smaller than the time, τ_T, for a charge carrier to cross the grain (Thouless time), $\tau_T = L_\parallel^2 / D$, where $L_\parallel \sim 5$ nm is taken for the size of a typical grain [79,82]. At frequency $\omega \tau_T \gg 1$, the system should show bulk metal behavior and $\sigma(\omega)$ is given by the Drude formula

$$\sigma(\omega) = [b/(4\pi)](\Omega_p^2 \tau)/(1 - iw\tau), \quad \Omega_p^2 = 4\pi e^2 n/m \tag{15.5}$$

where Ω_p is the unscreened plasma frequency and b is the degree of crystallinity (usually, $< \sim 50\%$ crystallinity). The Drude dielectric constant is

$$\varepsilon(\omega) = 1 - b\Omega_p^2 \tau^2/(1 + (\omega\tau)^2) \tag{15.6}$$

and is negative for $\omega \ll \Omega_p$; $\varepsilon(\omega) \sim -b(\Omega_p/\omega)^2$.

Optical data [44,45] show that at high frequencies $\gg 0.1$ eV, the system indeed is metal-like (Figure 15.6), which we attribute to the above inhomogeneous metallic island response (Equation 15.5 and Equation 15.6 (Figure 15.8). For PAN(CSA) metallic samples [44,45] the corresponding parameters are $\Omega_p \approx 2$ eV, $\tau \sim 10^{-15}$ s and τ_T is estimated to be $\sim 5 \times 10^{-14}$ s. The condition for applicability of the grain model, $\tau \ll \tau_T \ll 1/\Delta E \sim 10^{-13}$ s, is fulfilled.

The high-frequency Drude response (29,44,45) transforms with decreasing frequency into dielectric behavior at a frequency $1/\tau_T$. For $\omega \tau_T \ll 1$, electrons follow an external field and the conductivity is purely capacitive:

$$\sigma(\omega) = -i\omega e^2 bN(\varepsilon_F)L_\parallel^2 \tag{15.7}$$

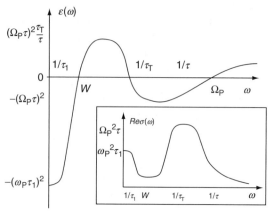

FIGURE 15.8 $\varepsilon(\omega)$ and $\sigma(\omega)$ (see inset) expected for intergrain resonance quantum tunneling in the chain-linked granular model in the metallic phase. (From Prigodin, V.N. and Epstein, A.J., *Physica B*, 338, 310, 2003. Reprinted from Elsevier. With permission.)

with positive dielectric constant given by the polarization of grains:

$$\varepsilon(\omega \leq 1/\tau_T) \sim b(\Omega_p L_\parallel / v_F)^2 \tag{15.8}$$

The behavior (Equation 15.7 and Equation 15.8) is in good agreement with indistinguishable experimental results for both dielectrics near the IMT and conductive phases at high and intermediate frequencies (0.01–0.1 eV) (Figure 15.6). Whether the lower frequency behavior is metallic or dielectric depends on the intergrain coupling.

Each grain is coupled to other grains by $2N_\perp$ independent chains through amorphous media. For simplicity, we assume the two nearest grains are electrically connected by $\sim N_\perp / z$ chains, where z is the number of nearest neighboring grains. In the metallic phase the intergrain coupling leads to broadening of quantized levels in the grains, $\delta E = 2N_\perp g \Delta E$, where g is the transmission coefficient between grains through a single chain (Figure 15.3d). The IMT occurs when $\delta E \sim \Delta E$ (Thouless criterion for the IMT [123]) and the critical chain-link coupling g_c satisfies

$$2N_\perp g_c = 1 \tag{15.9}$$

For PAN(HCl) and similar PAN(CSA) this yields $g_c \sim 10^{-2}$.

If $g < g_c$, the system is a dielectric and the behavior (Equation 15.8) is retained for all $\omega \tau_T \ll 1$. However, on the metallic side ($g > g_c$) electrons are delocalized over a network of grains and their low-frequency motion is a random hopping among the grains. The hopping between the grains is a quantum process and can be described by mean transition frequency W. Introducing the mean distance between the centers of neighboring grains, R ($b \sim (L_\parallel / R)^3$), the corresponding diffusion coefficient D_3 and the macroscopic conductivity of network are

$$D_3 = R^2 W, \quad \sigma(\omega - 0) = be^2 N(\varepsilon_F) D_3 \tag{15.10}$$

On approaching the IMT from the metallic side W tends to 0 as [124]: $W = [\Delta E/(2z)] \exp[-2\pi (g_c/(g - g_c))^{1/2}]$.

In the metallic phase the hopping frequency W is related to the above model parameters as

$$W = \delta E/(2z) = (N_\perp / z) g \Delta E \tag{15.11}$$

and the whole system can be represented as a network of random conductors. The nodes represent the grains where randomization of electronic motion occurs. Combining all together, the conductivity (Equation 15.10) also can be written in the simple form

$$\sigma(0) = (e^2 g)(N_\perp/z)(1/R) \tag{15.12}$$

Here, the first set of parentheses represent the conductance of a single intergrain chain-link, the second factor in parentheses is the number of chains connecting neighboring grains, and R in the last set of parentheses mimics the period of the grain network.

Thus, as described in [106], the problem of transport in the metallic phase far from the IMT is reduced to a study of the average transmission coefficient g in Equation 15.11 for a chain of finite length. For direct tunneling between grains, $g = \exp[-2L/\xi]$, where L is the length of chain connecting neighboring grains and ξ the L_{loc}. For 50% crystallinity, L is of the same order as a grain size $L_\parallel$, i.e., $L \sim 5$ nm. For weak localization, $\xi = 4l$, where $l = v_F \tau$ is the mean free path. For PAN(HCl) parameters assuming $v_F \sim 3 \times 10^7$ cm/s [121] and $\tau \sim 10^{-15}$ s, $\xi \sim 1.2$ nm. Therefore $g \sim 10^{-4}$ and direct tunneling is essentially suppressed.

However, the transmission coefficient is unity for resonance tunneling (Figure 15.3d). The probability of finding a resonance state is proportional to the width of the resonance level γ. For constant ξ the resonance state needs to be in the center of the chain, therefore, $\gamma \sim (1/\tau) \exp[-L/\xi]$. As a result, the average transmission coefficient is determined by the probability of finding the resonance state at the center, i.e., $\langle g \rangle = (\gamma\tau) \sim \exp[-L/\xi]$. For PAN(HCl), we have $\langle g \rangle \sim 10^{-2}$, which is close to the critical value for IMT, $g_c = 1/(2N_\perp) \sim 10^{-2}$. Thus, the probability of a resonance state is small $\sim 10^{-2}$, therefore resonance tunneling is not taken into account. However, as the number of interconnecting chains of a given grain with others is large ~ 100, resonance coupling occurs between grains.

A principal difference between direct and resonance tunneling is the time for tunneling. Direct tunneling that occurs in a conventional granular metal is an almost instantaneous process, i.e., its characteristic time is the scattering time τ. Resonance tunneling that is anticipated to be in the metallic polymers shows a delay determined by the level width γ. The frequency-dependent transmission coefficient $g = g(\omega)$ for resonance tunneling is given by a generalization of the Bright–Wigner formula [125]:

$$g(\omega) = [1 - i\omega/\gamma]^{-1} \tag{15.13}$$

Using Equation 15.10 through Equation 15.12, we find that in the region $\omega \ll W$, $\sigma(\omega)$ can be written in the standard Drude form

$$\sigma(\omega) = (1/(4\pi))\omega_p^2 \tau_1/(1 - i\omega\tau_1) \tag{15.14}$$

with ω_p being "the plasma frequency" determined by the frequency $\langle W \rangle$ of intergrain hops

$$(\omega_p/\Omega_p)^2 = (b\Delta E\tau)(\langle g \rangle)^2 (R/l)^2 \tag{15.15}$$

and the relaxation time $\tau_1 = (1/\gamma) = \tau\langle g \rangle \sim \tau \exp[L/\xi]$ determined by the Wigner transmission time.

The overall frequency dependence of dielectric function and conductivity in the metallic state of polymers dominated by quantum hopping as described in [106] are shown in Figure 15.8.

15.2.3 Anderson Disorder-Induced Insulator–Metal Transition

It is useful to describe the form of localization that occurs in a homogeneously disordered material in contrast to the models described in Section 15.2 and Section 15.3. The 3D models described below assume that the materials are isotropic; i.e., the materials should be electrically the same in all directions. In a perfect crystal with periodic potentials, the wavefunctions form Bloch waves that are delocalized over the entire solid [40]. In systems with disorder, impurities and defects introduce substantial scattering of the electron wavefunction, which may lead to localization. Anderson demonstrated [88] that electronic wavefunctions can be localized if the random component of the disorder potential is large enough compared with the electronic bandwidth (Figure 15.9). In this case, the localized wavefunctions have the form

$$\psi(r) \propto \exp(-r/L_{\text{loc}})Re(\psi_0) \qquad (15.16)$$

where L_{loc} is the localization length of the state. Later, Mott showed that band tail states are more easily localized than states in the center of the band [89]. Therefore, a critical energy called the mobility edge (E_c) exists that separates localized states in the band tails from extended states in the center of the band

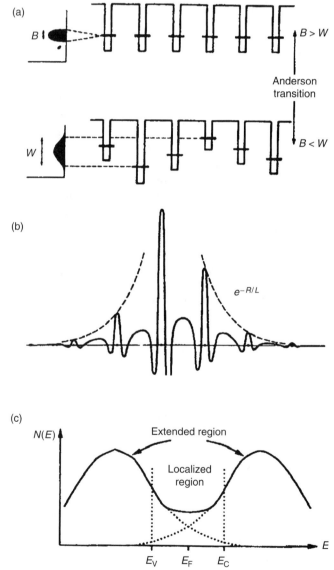

FIGURE 15.9 (a) The Anderson transition and (b) the form of the localized wavefunction in an Anderson metal–insulator transition. (c) The Fermi glass state where the Fermi level lies in the region of localized states. (From Kohlman, R.S. and Epstein, A.J., *Handbook of conducting polymers*, 2nd ed., eds. Skotheim, T.A., Elsenbaumer, R.L., and Reynolds, J.R., Marcel Dekker, New York, 1988, chap. 3. Reprinted from Routledge/Taylor & Francis Group, LLC. With permission.)

(Figure 15.9). The resulting electronic behavior of a material depends on where the Fermi energy (E_F) lies relative to E_c. If E_F lies in the range of extended states, then $\sigma_{dc}(T)$ is finite as $T \to 0$ [89] and the W plot has a positive slope at low temperature [101]. If the disorder potential is strong enough to cause E_F to be in the range of localized states, then the material will be nonmetallic, with $\sigma_{dc} \to 0$ as $T \to 0$ [89] and a negative slope for the low TW plot [101], even though there is a finite density of states at E_F.

When E_F approaches E_c on the insulating side of the IMT, the L_{loc} diverges as the electronic wavefunction becomes delocalized through the material. However, because of the strong disorder, the mean free path (λ) is still very short. In 1960, Ioffe and Regel [102] proposed that the lower limit for the metallic mean free path is the interatomic spacing, which occurs at and on the insulating side of the IMT. This condition led Mott to propose a minimum metallic conductivity ($\sigma_{min} \sim 0.03e^2/3\hbar a \sim 10^2$ S/cm in three dimensions, where a is the interatomic spacing). The Ioffe–Regel condition implies that $k_F\lambda \sim 1$ (k_F is the Fermi wavevector) when $\sigma_{dc} \sim \sigma_{min}$. If applied to conducting polymers, this leads to a very short mean free path and scattering time. For typical repeat distances along the polymer chain in doped PAN [82] (other doped conducting polymers have similar values), the Ioffe–Regel condition requires that

$$\lambda \sim 1/k_F \sim 1/[\pi/2(10 \text{ Å})] \sim 10 \text{ Å} \tag{15.17}$$

and

$$\tau = \lambda/v_F \sim (10 \text{ Å})/(10^6 \text{ m/s}) \sim 10^{-15} \text{ s} \tag{15.18}$$

Far into the metallic regime where the transport is due to diffusion of carriers in extended states, $k_F\lambda \gg 1$.

When the Fermi level lies in the localized region, the conductivity at zero temperature is zero even for a system with a finite density of states. The transport at higher temperature involves phonon-activated hopping between localized levels. The Mott variable range hopping (VRH) model is applicable to systems with strong disorder such that ΔV (disorder energy) $\gg B$ (bandwidth) [89]. The general form of the temperature-dependent conductivity of Mott's model is described as

$$\sigma = \sigma_0 \exp\left[-\left(\frac{T_0}{T}\right)^{1/(d+1)}\right] \tag{15.19}$$

where d is the dimensionality and, for 3D systems, $T_0 = c/k_B N(E_F)L^3$ (c is the proportionality constant, k_B the Boltzmann constant, and L the localization length). In Mott's model, electron correlations are neglected as for the classical Fermi liquid. Efros and Shklovski [126,127] pointed out that the interactions between localized electrons and holes play an important role in the hopping transport, especially at low temperature, changing the expected temperature dependence of the conductivity to

$$\sigma = \sigma_0 \exp\left[-\left(\frac{T_0'}{T}\right)^{1/2}\right] \tag{15.20}$$

where $T_0' = e^2/\varepsilon L$ (e is the electron charge and ε is the dielectric constant). For materials very close to the IMT, instead of an exponential temperature dependence, a conductivity with a power law temperature dependence is predicted [128]. A detailed discussion of the low-dimensional VRH conductivity for materials with various structural disorder is described in [129].

Typical behavior for the evolution of the temperature-dependent resistivity (ρ) as E_F crosses E_c is shown in Figure 15.10 for p-doped germanium as a function of the compensation (carrier density) [89]. For the material with the largest compensation (smallest carrier density), ρ has the largest magnitude and strongest temperature dependence. As the compensation is decreased (carrier density is increased),

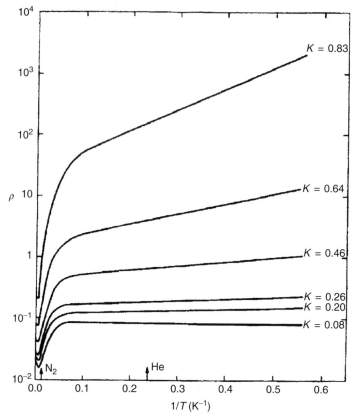

FIGURE 15.10   Temperature-dependent resistivity in an Anderson transition. Experimental temperature-dependent resistivity for *p*-type germanium with different levels of compensation (*K*) leading to a variation in the electron density. (From Kohlman, R.S. and Epstein, A.J., *Handbook of conducting polymers*, 2nd ed., eds. Skotheim, T.A., Elsenbaumer, R.L., and Reynolds, J.R., Marcel Dekker, New York, 1988, chap. 3. Reprinted from Routledge/Taylor & Francis Group, LLC. With permission.) The magnitude and temperature dependence change monotonically with increasing compensation (decreasing carrier density).

ρ decreases monotonically in magnitude and the temperature dependence decreases monotonically. For the most metallic sample, ρ shows a very weak *T* dependence at low temperature. This monotonic evolution for the magnitude and *T* dependence of σ $(= 1/\rho)$ is characteristic of an Anderson IMT [89].

At high frequencies, $\sigma(\omega)$ for ordered materials is given by the Drude formula,

$$\sigma_{\text{Drude}}(\omega) = \frac{\Omega p^2 \tau}{4\pi(1 + \omega^2\tau^2)} \tag{15.21}$$

where Ωp $(= \sqrt{4\pi ne^2/m^*})$ is the plasma frequency of the free electrons, *n* is the full carrier density, and m^* is the effective mass. This behavior follows from the Kubo formula when the mean free time (τ) is such that $\omega\tau \gg 1$ and breaks down when $\omega\tau \sim 1$ [89,130]. For 3D materials near an Anderson IMT, $\sigma(\omega)$ is suppressed at low frequencies relative to $\sigma_{\text{Drude}}(\omega)$ due to quantum interference of the electronic wavefunctions [131,132]. Localization corrections to σ_{Drude} calculated within scaling theories of the Anderson transition [89,99,100,132,133–135] yield the localization-modified Drude model:

$$\sigma_{\text{LMDM}}(\omega) = \sigma_{\text{Drude}}(\omega)\left[1 - \frac{C}{(k_{\text{F}}\lambda)^2}\left(1 - \frac{\lambda}{L(\omega)}\right)\right] \tag{15.22}$$

where C is an undetermined universal constant (~ 1), k_F is the Fermi wavevector, λ is the mean free path, and $L(\omega)$ is the distance a charge would diffuse during an oscillation of the electromagnetic wave. In this model, at zero frequency, the high dc conductivity of a sample near the IMT is due to localized charge carriers. Since in three dimensions,

$$L(\omega) = \sqrt{D/\omega} \qquad (15.23)$$

where $D = \lambda^2/3\tau$ is the diffusion constant, the localization-modified Drude model can be written as

$$\sigma_{\text{LMDM}}(\omega) = \sigma_{\text{Drude}}(\omega)\left[1 - \frac{C}{(k_F v_F \tau)^2} + \frac{C(3\omega)^{1/2}}{(k_F v_F)^2 \tau^{3/2}}\right] \qquad (15.24)$$

The real part of the dielectric function $\varepsilon_{\text{LMDM}}(\omega)$ corresponding to the localization-modified Drude model can be calculated using the Kramers–Kronig relations, giving

$$\varepsilon_{\text{LMDM}}(\omega) = \varepsilon_\infty + \frac{\Omega_p^2 \tau^2}{1 + \omega^2 \tau^2}\left[\frac{C}{(k_F v_F)^2 \tau^2}\left(\sqrt{\frac{3}{\omega\tau}} - (\sqrt{6} - 1)\right) - 1\right] \qquad (15.25)$$

where ε_∞ is the dielectric screening due to higher energy interband transitions. These expressions are not complete as they ignore cutoffs at low frequency when $L(\omega) \sim L_T$, where L_T is the Thouless length [99].

The behavior of both $\sigma_{\text{LMDM}}(\omega)$ and $\varepsilon_{\text{LMDM}}(\omega)$ for materials close to the IMT is determined in the Anderson model by the Ioffe–Regel condition, which requires $\tau \sim 10^{-15}$ s. $\sigma_{\text{LMDM}}(\omega)$ and $\varepsilon_{\text{LMDM}}(\omega)$ are shown as the mean free time τ ($\sim 10^{-15}$ s) is varied in Figure 15.11 (from [136]). $\sigma_{\text{LMDM}}(\omega)$ has Drude dispersion at high frequency, a maximum, and then monotonic suppression at low frequencies. As τ increases (reflecting a more ordered material), the Drude dispersion occurs over a wider frequency range and the maximum shifts to lower frequency. An important consequence of the localization-modified Drude model is that $\varepsilon_{\text{LMDM}}(\omega)$ becomes positive at very low frequencies, reflecting the short τ due to strong disorder scattering. This behavior should be contrasted with the negative value of $\varepsilon(\omega)$ for the Drude model at low frequency ($\omega \ll 1/\tau$),

$$\varepsilon_{\text{Drude}} \cong \varepsilon_\infty - \Omega_p^2 \tau^2 \qquad (15.26)$$

15.3 Experimental Techniques

15.3.1 Chemical Preparation

The final structural order and hence the electronic properties of conducting polymers are sensitive to preparation techniques. The high quality freestanding films of PAN EB that were doped with *d,l*-camphorsulfonic acid, for which data are presented here, were prepared using EB synthesized to provide different molecular weights [137]. Low molecular weight PAN, $\overline{M}_w < \sim 100{,}000$, was prepared by a well-known route [76,83,137,138–140]. PAN of high molecular weight, $\overline{M}_w > \sim 300{,}000$, was synthesized [137] by adding lithium chloride (LiCl) to the reaction vessel and lowering the temperature of the reaction. All molecular weights were determined by gel permeation chromatography using a 0.5% w/v LiCl/*N*-methyl pyrrolidone (NMP) solvent and a polystyrene standard.

To obtain fully doped PAN salt, the molar ratio of EB to the dopant (CSA) should be 1:2, assuming that all of the dopant ions successfully protonate an imine nitrogen. For example, 1.0 g (0.00276 mol) of EB would be mixed with 1.287 g (0.00552 mol) of HCSA. The EB and HCSA were mixed in two different ways. For some films, the EB and HCSA were mixed as powders using a mortar and pestle. This doped

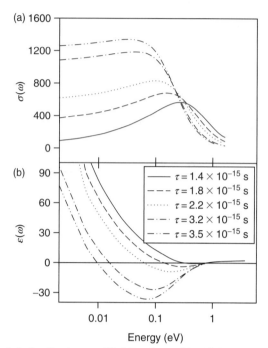

FIGURE 15.11 Behavior of the localization-modified Drude model with increasing mean free time (mean free path). (a) $\sigma(\omega)$ and (b) $\varepsilon(\omega)$. The parameters used were $\Omega_p{}^2 = 2eV^2$ and $C/(k_F v_F)^2 = 1.9 \times 10^{-30}\ s^2$. (From Kohlman, R.S. and Epstein, A.J., *Handbook of conducting polymers*, 2nd ed., eds. Skotheim, T.A., Elsenbaumer, R.L., and Reynolds, J.R., Marcel Dekker, New York, 1988, chap. 3. Reprinted from Routledge/Taylor & Francis Group, LLC. With permission.)

EB–CSA powder could then be dissolved in appropriate solvent mixtures. For the data reviewed here, these powders were dissolved in *m*-cresol, chloroform, or mixtures of these solvents. The resulting properties were shown to vary dramatically depending on which solvent was used through the effects of secondary doping [76,83,137]. The alternative way to obtain solutions of doped PAN-CSA is to separately add the EB powder and the HCSA into a volume of solvent and then mix the two solutions.

In addition, the concentration of PAN-CSA in the solvent was varied from 1 to 3 wt%, as the amount of solvent that must evaporate from the resulting film has an effect on the film's structural order. When the solution of PAN-CSA was prepared, it was stirred overnight with a magnetic stir bar and then sonicated for ∼2 h to completely dissolve all visible particles. The solution was then filtered and cast on a microscope slide upon which it dried under a hood. The resulting films could be peeled from the microscope slide to obtain freestanding films of thickness ∼40–100 μm. All PAN samples for which data are reported here were prepared in the laboratories of Alan G. MacDiarmid at the University of Pennsylvania. The details of the effects of different processing conditions on the freestanding films are given in [137]. For this study, samples of these films were chosen that illustrate the properties of PAN-CSA as the transport properties approach and cross the IMT. The samples for this study are listed in Table 15.1 along with their specific processing conditions.

The PPy films that were doped with PF_6 and *p*-toluenesulfonate (TsO) were obtained by anodic oxidation of pyrrole at −30°C under potentiostatic conditions. The electrolytic cell contained 0.06 mol/L of pyrrole and 0.06 mol/L of the appropriate salt in propylene carbonate (PC) containing 1 vol% of water. A glassy carbon plate and a platinum foil were used as the anode. The current was adjusted to ∼0.125 mA/cm². The reaction took place for ∼24 h under a stream of nitrogen gas. For $PPy(PF_6)$ and PPy(TsO), elemental analyses determined that there is approximately one unit of dopant per three pyrrole rings [28,30,41,42].

TABLE 15.1   Preparation Conditions for the PAN Film Samples Described in This Chapter[a,b]

Sample	EB $\overline{M}_w$	Mixed as	Solvent	wt%	σ_{dc}(300 K) (S/cm)
A	400,000	Powder	*m*-Cresol	2	240
B	300,000	Powder	*m*-Cresol	2	110
C	50,000	Solution	*m*-Cresol	6.8	400
D	400,000	Solution	*m*-Cresol	1	120
E	50,000	Powder	30% *m*-cresol	1	70
F	50,000	Powder	Chloroform	1	20
G	50,000	Powder	Chloroform	1	0.2

Source: From Kohlman, R.S. and Epstein, A.J. in *Handbook of conducting polymers*, 2nd ed., eds. Skotheim, T.A., Elsenbaumer, R.L., and Reynolds, J.R., Marcel Dekker, New York, 1988, chap. 3. Reprinted from Routledge/Taylor & Francis Group, LLC. With permission.)

[a]Information in Tables 15.1 through Table 15.6 was acquired from the same samples A through G.

[b]In the column describing the solvent used, when the solvent listed is a percentage of *m*-cresol, the balance of the solvent is chloroform. Also, samples F and G appear to have been exposed to *m*-cresol.

The PPy film doped with sulfonated polyhydroxyether (S-PHE) [141] was synthesized on rotating stainless steel electrodes in a solution of 0.05 mol/L S-PHE, 5 vol% pyrrole, and 1 vol% water in PC under galvanostatic conditions using a current density of 2 mA/cm^2. The film was peeled, washed with PC and ethanol, and then dried under vacuum at 50°C for 12 h. The molar ratio of sulfonation is defined as MR $= n/(n + m)$, where n is the number of S-PHE segments and m is the number of unsulfonated polyhydroxyether segments as shown in Figure 15.12. For samples with low molar ratios, a large fraction of the volume of the polymer is occupied by the PHE, a saturated polymer that does not transport electric charge effectively. Thus, a large volume of insulator is added to the film, drastically reducing the conduction electron density and the interaction between chains. Data presented here are from a study of a PPy(S-PHE) film with MR $= 0.125$.

In general, data presented here are for samples that are under vacuum or "pumped" conditions that substantially reduce or eliminate the effects of weakly bound moisture or solvents.

15.3.2   Transport Measurements

The dc conductivity (σ_{dc}) measurements were performed using a four-probe technique as described previously [90,114]. Four thin gold wires (0.05 mm thick and of 99.9% purity) were attached to the sample surface using a graphite paste (Acheson Electrodag 502) to improve the electrical contact. By placing the sample probe in a Janis DT dewar with helium (He) exchange gas, σ_{dc} (T) could be measured from 4.2 to 300 K. Measurements at millikelvin temperatures were carried out by placing the sample probe in thermal contact with the mixing chamber of a ^{3}He–^{4}He dilution refrigerator [75]. The milli-kelvin conductivity measurements were carried out using a low frequency (~19 Hz) alternating current and a lock-in amplifier. A magnetic field of 0–5 T was applied perpendicular to the plane of the samples.

Reflectance spectra from 2 meV to 6 eV were recorded as reported previously [29]. The high energy (0.5–6 eV) reflectance was recorded using a Perkin-Elmer λ-19 ultraviolet/visible/near-infrared (UV/Vis/NIR) spectrometer equipped with a Perkin-Elmer RSA-PE-90 reflectance accessory based on the Labsphere DRTA-9a integrating sphere. The low energy (2 meV–1.2 eV) reflectance measurements

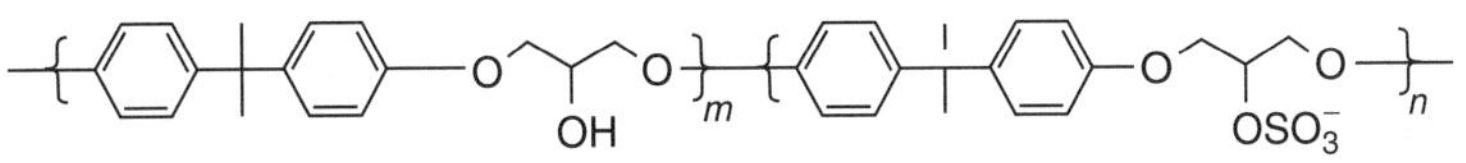

FIGURE 15.12   Chemical structure for sulfonated polyhydroxyethers. (From Chauvet, O., Paschen, S., Forro, L., Zuppiroli, L., Bujard, P., Kai, K. and Wernet, W., *Synth. Met.*, 63, 115, 1994. Reprinted from Elsevier. With permission.)

were made on a BOMEM DA-3 Fourier transform infrared (FTIR) spectrometer equipped with a homemade cryostat that could be lifted so that a reference mirror and sample could be placed alternately in the IR beam. A Michelsen interferometer (1–12 meV) was used to record temperature-dependent far-IR reflectance spectra [45]. The measured reflectance spectra from different spectrometers typically agreed to within 1% in the regions of overlap for samples with highly specular surfaces.

The samples were freestanding films with thickness $\sim$40–100 μm. This thickness is greater than the electromagnetic penetration depth [$\delta = c/\sqrt{2\pi\mu\omega\sigma(\omega)}$, where c is the speed of light and $\sigma(\omega)$ is the conductivity at frequency ω] when $\sigma > 10$ S/cm. Samples with or $\sigma_{dc} < 10$ S/cm were first checked to determine whether they had transmitted far-IR radiation. Therefore, the reflectance can be analyzed using the Fresnel reflection coefficients for semi-infinite media [103].

The optical conductivity $\sigma(\omega)$ and the real part of the dielectric function $\varepsilon(\omega)$ were calculated from the reflectance spectra using the Kramers–Kronig analysis and the Fresnel relations for semi-infinite media [103]. To calculate the optical functions using the Kramers–Kronig technique, reasonable extrapolations of the experimental reflectance data at low and high frequencies are necessary. In this study, the reflectance data were extrapolated at low energy using the Hagens–Rubens relation [103]. At high energies (6–12 eV), the reflectance spectra were extrapolated with a reflectance $R \propto \omega^{-2}$ (interband) followed by an $R \propto \omega^{-4}$ (free electron) at higher energies (>12 eV).

The microwave frequency conductivity and dielectric constant were measured using the cavity perturbation technique [90,114,142,143]. The resonant cavity used was cylindrical with a TM_{010} frequency of 6.5 GHz. The entire cavity is inserted into a dewar filled with He gas to provide a temperature range of 4.2–300 K. Alternatively, the microwave frequency conductivity and dielectric constant may be measured using a microwave impedance bridge [144].

To understand the nature of the spins in the polymers, electron paramagnetic resonance (EPR) experiments were carried out using a Bruker ESP 300 spectrometer equipped with a rectangular cavity that has a TE_{102} mode frequency of 9.5 GHz (X band). An ESR-900 continuous flow He cryostat from Oxford Instruments provided temperature control from 4 to 300 K.

15.4 Polyaniline

PAN doped with CSA has been shown to cross the IMT and shows metallic behavior at low temperature [44,45,74,75,77]. The proximity of the samples to the IMT can be controlled to some degree by the processing conditions [76,77,137,145]. By using the concept of secondary doping [76], the electrical behavior and structural order of PAN-CSA can be varied by varying the solvent from which the film is cast. The weight percent of the polymer in the solvent also can be varied to change the resulting electrical properties of PAN-CSA. Later in this section, the experimental results for the low- and high-frequency transports near the IMT are discussed. For simplicity, the real parts of the optical conductivity and dielectric function are referred to as $\sigma(\omega)$ and $\varepsilon(\omega)$, respectively.

15.4.1 X-Ray Diffraction

PAN forms a rich set of structures dependent upon the processing sequence and dopant [63,79–82,146–148]. Generally, PAN obtained from solution in the doped (conducting salt) form exhibits a local crystalline order of the emeraldine salt I (ES-I) type. In contrast, PAN obtained by doping powder or films cast as the base form from solution are of the ES-II type [79,82,148,149]. Both preparation methods lead to between a few percent and $\sim$50% crystallinity depending on the details of the processing route. In addition, there are significant differences in the type of local order that exists in the disordered regions between the crystalline ordered regions, varying from liquid benzene-like, to coil-like, to expanded coil-like, to more rodlike [60–63,106,107]. For many undoped and doped PANs, short-range local order in the disordered regions resembles that in the ordered regions [82,106].

Comparing the x-ray patterns for PAN-CSA cast from *m*-cresol with those of PAN-CSA cast from chloroform, there is a drastic difference in the structure. In PAN-CSA cast from *m*-cresol, the repeat distance along the chain direction is ~9.2 Å [82]. In the PAN-CSA cast from chloroform, the repeat distance along the chain increases to ~14.2 Å, possibly due to the periodicity of a snaking chain that has undergone a "rodlike to coil-like" transition [59]. There is evidence that the benzene rings of the PAN tend to align well with the plane of the freestanding film. In the most highly ordered, most conductive samples of PAN-CSA, a diffraction peak occurs at low scattering angles, suggesting a d spacing of ~21 Å [82,150], which may correspond to the distance between 2D stacks of PAN chains separated by CSA ions [82].

For each of these systems, the coherence length within the doped crystallographic regions generally is not more than 50–75 Å along the chain direction, with smaller values in the perpendicular direction. It has been proposed that these coherent crystalline regions form metallic islands and the disordered weak links between more ordered regions are areas where conduction electrons are subject to localization, as expected for charges moving through isolated 1D chains. That is, for each very highly conducting polymer system studied there are regions of 1D electronic character through which conduction electrons must pass [39].

It is pointed out that subtle disorder in the ordered crystalline regions can have drastic effects. For example, the percent crystallinity and the size of the crystalline regions of the HCl salt of poly(o-toluidine) (i.e., polymethylaniline) are similar to HCl-doped PAN, with the important difference of "paracrystallinity" introduced by two possible positions for the CH_3-substituted C_6 rings. That is, a rotation of the ring by 180° about its nitrogen–nitrogen axis changes the structure. The disorder introduced disrupts the interchain interaction, leading to relatively strong localization [25,115,148]. The disorder in HCl-doped poly(o-toluidine) is reduced (and the delocalization increased) through a processing-induced increase in structural order using fiber formation [24].

15.4.2   Magnetic Susceptibility

Magnetic susceptibility studies identify the charge storage mechanism at low doping levels as well as the density of states at the Fermi level and the density of localized "Curie" spins at higher dopant levels. The magnitude of χ_{Pauli} depends on the structural order and morphology of the polymers as these affect the uniformity of the doping.

For PAN, $N(E_F)$ is finite and has been shown to increase with the level of protonic acid doping and the volume fraction of crystalline material for both the ES-I and ES-II structures [14,782]. The $N(E_F)$ values differ for ES-I HCl and ES-II HCl, being 0.26 state/[eV·(C + N)], and 0.083 state/[eV·(C + N)], respectively. For highly conducting PAN-CSA (m-cresol) [56], $N(E_F)$ is ~0.07 state/[eV·(C + N)]. Differently prepared stretched PAN doped with HCl was reported to have a much higher $N(E_F)$, ~1.4 state/[eV·(C + N)] [21]. Some solutions of PAN-CSA have been reported to have a Pauli-like susceptibility [151]. It has been proposed that there is primarily spinless bipolaron formation in the disordered regions, especially for ES-II materials [78]. It has also been reported that oxygen and moisture can have a significant effect on both the Pauli and Curie susceptibility observed [143,152].

15.4.3   DC Conductivity

To establish that a material has undergone a transition into the metallic state, it is necessary to show that σ_{dc} is finite as $T \rightarrow 0$ [89] and that the logarithmic derivative of the temperature-dependent conductivity ($W = d\ln[\sigma(T)]/d\ln T$, a generalized activation energy) has a positive temperature coefficient at low temperature [101]. For a conductor very close to the IMT, called a critical sample, the resistivity follows a power law behavior with T [128]. The plot of log W vs. log T for a critical sample approaches $T = 0$ K at a constant value, providing a dividing line between the plot of log W vs. log T for insulating hopping behavior, which increases with decreasing T (i.e., the slope of log W vs. log T is equal to $-\gamma$ if $\sigma \propto \exp[T_0/T]^\gamma$), and the plot of log W vs. log T for metallic samples, which decreases with decreasing T.

The W plots for PAN-CSA samples A–G are shown in Figure 15.13. The lettering for each sample reflects the relative behavior of the W plot for the various samples. The W plot for sample A has a positive slope at low temperature, indicating that it is metallic at low temperature. Measurements of σ_{dc},

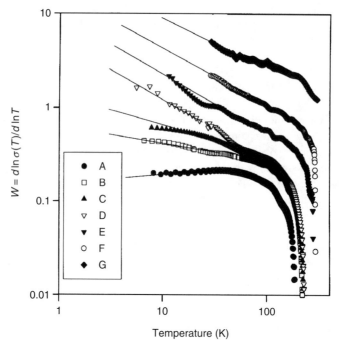

FIGURE 15.13 The reduced activation energy $W = d\ln[\sigma(T)]/d\ln T$ for selected PAN-CSA samples. (From Kohlman, R.S. and Epstein, A.J., *Handbook of conducting polymers*, 2nd ed., eds. Skotheim, T.A., Elsenbaumer, R.L., and Reynolds, J.R., Marcel Dekker, New York, 1988, chap. 3. Reprinted from Routledge/Taylor & Francis Group, LLC. With permission.)

for sample A at millikelvin temperature also show a large finite conductivity ($\sigma_{dc} > \sim$70 S/cm) down to $\sim$20 mK (discussed in greater detail in Section 15.4.4), confirming the presence of delocalized states at the Fermi level as $T \to 0$. Therefore, sample A has crossed the IMT and is metallic. However, the remaining samples show a negative slope at low temperature, varying from -0.2 to -0.7 (Table 15.2), which is characteristic of the hopping transport shown by samples that are insulating at low temperature (σ_{dc} $(T) = \sigma_0 \exp[-(T_0/T)^{\gamma}]$). As *W*, the reduced activation energy, increases in magnitude, the samples

TABLE 15.2 Comparison of Room Temperature DC Transport Properties and Far-Infrared Reflectance for PAN-CSA Films

Sample	σ_{dc} (300 K) (S/cm)	ρ_R[a]	γ[b]	T_M[c] (K)	Reflectance at 6 meV
A	230	1.7	0.08	188	0.8484
B	110	2.9	-0.2	218	0.7991
C	400	4.5	-0.3	224	0.9262
D	120	11	-0.6	240	0.8431
E	70	130	-0.5	288	0.7537
F	20	1.3×10^6	-0.7	>300	0.6915
G	0.7	—	-0.6	>300	0.3133

Source: From Kohlman, R.S. and Epstein, A.J. in *Handbook of conducting polymers*, 2nd ed., eds. Skotheim, T.A., Elsenbaumer, R.L., and Reynolds, J.R., Marcel Dekker, New York, 1988, chap. 3. Reprinted from Routledge/Taylor & Francis Group, LLC. With permission.)

[a]Resistivity ratio $\rho_R = \rho$ (4.2 K)/ρ (300 K).
[b]The parameter γ is the hopping exponent determined from the low-temperature *W* plot.
[c]T_M is the temperature at which the maximum occurs in σ_{dc}.

become more insulating and the exponent γ grows. For samples B and C, $\gamma \sim 0.25$, indicating 3D VRH at low temperature. For samples D–G, $\gamma \sim 0.5$–0.7, suggesting that the localization is now more quasi-1D. Therefore, as the materials become more insulating, the charge transport becomes more characteristic of hopping on isolated chains with reduced dimensionality (in the disordered regions).

Figure 15.14 shows $\sigma_{dc}(T)$ for each of these PAN-CSA samples. Table 15.2 gives σ_{dc} (300 K) and the resistivity ratio σ_{dc} (300 K)$/\sigma_{dc}$ (5 K) $\equiv \rho_R$ for each sample. A wide range is observed for ρ_R. For samples A–E, $\sigma_{dc}(T)$ has a positive temperature coefficient of resistivity near room temperature [23,39,77,145], a maximum, and then localization effects at lower temperatures. These experimental $\sigma_{dc}(T)$ values are consistent with the predictions of the inhomogeneous disorder model. The temperature at which the maximum in σ_{dc} occurs (T_M) is given for each sample in Table 15.2. The shift of the maximum in PAN-CSA to lower temperature as the materials become more metallic (as gauged by the W plot) is consistent with PAN-CSA sample A having the least disordered conduction paths. Assuming that the materials are homogeneous, sample A is the most ordered material.

Although PAN-CSA sample A shows intrinsic metallic behavior, it does not have the highest σ_{dc} (300 K) value compared with sample C. In fact, samples C and D each show stronger temperature dependence for σ_{dc} than materials with lower σ_{dc} (300 K) (Figure 15.14). This nonmonotonic behavior of the temperature dependence with increasing σ_{dc} (300 K) contrasts with the monotonic behavior shown in Figure 15.10 for p-doped germanium that undergoes an Anderson IMT. This nonmonotonic evolution of $\sigma_{dc}(T)$ is a strong argument against a homogeneous 3D Anderson IMT. Since PAN-CSA sample A is metallic, whereas sample C is insulating at low T, σ_{dc} (RT) $> \sigma_{min}$, the minimum metallic conductivity, for sample C. However, sample C becomes insulating at low T.

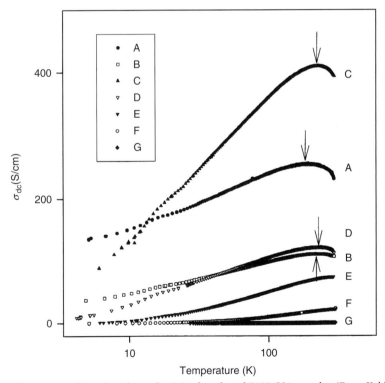

FIGURE 15.14 Temperature-dependent dc conductivity for selected PAN-CSA samples. (From Kohlman, R.S. and Epstein, A.J., *Handbook of conducting polymers*, 2nd ed., eds. Skotheim, T.A., Elsenbaumer, R.L., and Reynolds, J.R., Marcel Dekker, New York, 1988, chap. 3. Reprinted from Routledge/Taylor & Francis Group, LLC. With permission.)

15.4.4 Millikelvin Conductivity

The millikelvin σ_{dc} with no applied magnetic field is shown in Figure 15.15 along with the high temperature conductivity for samples A, B, and D. Sample A has a large finite $\sigma_{dc} > {\sim}70$ S/cm down to 18 mK. This indicates that there are extended states at E_F as $T \to 0$ for PAN-CSA sample A, i.e., PAN-CSA sample A is a metal at low temperature. The remaining samples show more complex behavior at low temperature, though they demonstrate hopping behavior at ${\sim}10$ K. Sample B shows a monotonic decrease in σ_{dc} down to 18 mK. Sample D shows a turnover in σ_{dc} below ${\sim}200$ mK. For sample D, σ_{dc} then increases and saturates at lowest temperatures. A similar increase in conductivity at low temperature has been reported for samples of PPy [28–30,153,154], though the turnover was observed at $T \sim 5$–20 K. In those materials, the increase at low temperature was ascribed to a tunneling mechanism [28–30,153]. This complex behavior implies that σ_{dc} does not go to zero as $T \to 0$ for samples B and D, even though σ_{dc} for these samples shows VRH behavior at higher temperatures. Therefore, σ_{dc} does not follow VRH behavior as $T \to 0$ as predicted for materials with homogeneous disorder and Anderson localization. It is conjectured that the dip in the conductivity may be due to reduced tunneling due to long-wavelength phonon scattering, which affects the PAN rings, reducing the overlap integrals and decreasing the wavefunction overlap. This saturation of σ_{dc} is not expected at low temperature for strongly disordered materials such as PAN-CSA samples F or G and PPy(S-PHE) due to larger separation between metallic regions and shorter L_{loc} for chains in the disordered regions.

The behavior of the millikelvin σ_{dc} when a magnetic field of 5 T is applied is shown in Figure 15.16. For sample A, there is a weak positive magnetoconductance ($\Delta\sigma/\sigma \sim 10^{-1}$). This same behavior was reported for doped polyacetylene [41,42,71] and PPy [75], which were highly conducting at millikelvin temperatures. For these samples, a positive magnetoconductance was attributed to magnetic field destruction of weak localization effects [41,42,71] (quantum interference of electronic wavefunctions leading to standing wave patterns or localization [99,100]). In such a model, the increase in σ_{dc} with increasing T may be seen as a result of activating parallel conduction paths through phonon scattering-induced destruction of weak localization in those parallel pathways. Positive magnetoconductance is also

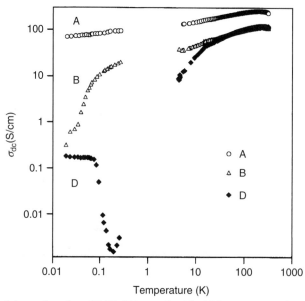

FIGURE 15.15   Millikelvin σ_{dc} for selected PAN-CSA samples. The high temperature (4–300 K) σ_{dc} data for each sample are included for comparison. (From Kohlman, R.S. and Epstein, A.J., *Handbook of conducting polymers*, 2nd ed., eds. Skotheim, T.A., Elsenbaumer, R.L., and Reynolds, J.R., Marcel Dekker, New York, 1988, chap. 3. Reprinted from Routledge/Taylor & Francis Group, LLC. With permission.)

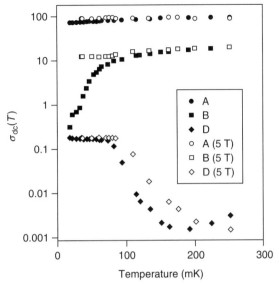

FIGURE 15.16 Dependence of the millikelvin transport of selected PAN-CSA samples on applied 5 T magnetic field. (From Kohlman, R.S. and Epstein, A.J., *Handbook of conducting polymers*, 2nd ed., eds. Skotheim, T.A., Elsenbaumer, R.L., and Reynolds, J.R., Marcel Dekker, New York, 1988, chap. 3. Reprinted from Routledge/Taylor & Francis Group, LLC. With permission.)

predicted for enhanced delocalization on percolating structures due to increased magnetic energy in the metallic regions [155]. Sample B also shows a positive magnetoconductance [$\Delta\sigma(H)/\sigma \sim 10^{1}$] that is larger than that of sample A at low temperatures ($\sim$30 mK) and reduced though positive [$\Delta\sigma(H)/\sigma \sim 10^{-1}$] at higher temperatures ($\sim$200 mK). The larger $\Delta\sigma(H)/\sigma$ for sample B compared with A may result from paths in series (bottlenecks) with the main conduction path in which weak localization is reduced, resulting in high σ_{dc}. The magnetic field shifts the minimum of the conductivity to higher temperature in sample D, which also shows positive magnetoconductance at the lowest temperatures.

Highly conducting samples of self-stabilized dispersion-polymerized PAN have temperature-independent conductivity between 5 and 70 K [47] though data were not reported to the millikelvin range to confirm the presence of a metallic state at lowest temperature. The temperature-independent conductivity here is similar to that of the highest conducting samples reported in Figure 15.15 and Figure 15.16 at lower temperatures.

Conductivity in regioregular PT FETs from room temperature to 4.2 K was reported as a function of gate voltage, V_{g}, and source-drain voltage, V_{sd} (Figure 15.17) [156]. It was proposed [156] that the gate voltage and source-drain voltage combine to induce an IMT at a carrier density of 5×10^{12} cm^{-2}. Alternatively, it has been argued [157] that there is no electric field-induced IMT associated with V_{sd}. Instead, it was argued that the crossover is from phonon-assisted hopping from the regime of phonon absorption (activation) to the regime of hopping with emission of phonons.

15.4.5 Reflectance

For free electron metals such as copper [103], silver [103], and aluminum [158], the reflectance approaches unity at low frequencies and remains high up to frequencies near the conduction electron plasma frequency (Ω_{p1}) for that system. Figure 15.18 shows the room temperature reflectance for samples A–G. At low energy, the reflectance approaches unity for PAN-CSA samples A–D. The value of the reflectance in the far-IR at 6 meV (50 cm^{-1}) scales with σ_{dc} for that sample (Table 15.2), providing an independent verification of the σ_{dc} (RT) for each sample. At higher energy, the reflectance for samples A–E decreases monotonically with increasing energy to a minimum near $\sim$2 eV, the

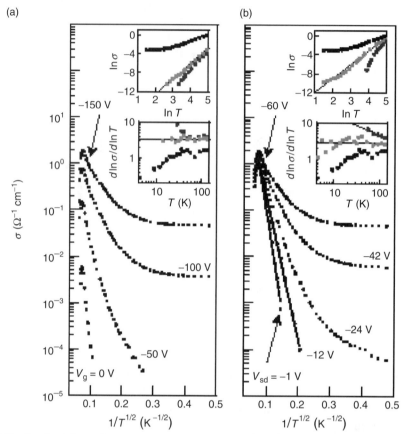

FIGURE 15.17 Conductivity vs. inverse square root temperature for a RR-P3HT FET with 0.07 cm^2/Vs (a) at $V_{sd} = -60$ V and varying V_g. Top inset (bottom to top) is a double logarithmic plot at $V_g = -40$ V, $V_g = -50$ V, and $V_g = -150$ V. Bottom inset shows the logarithmic derivative for these three gate voltages (top to bottom). The straight line is a power-law fit to the $V_g = -50$ V data. (b) At $V_g = -150$ V for varying V_{sd}. The straight lines are exponential fits to the data. Top inset is a double logarithmic plot at (bottom to top) $V_{sd} = -1$ V, $V_{sd} = -24$ V, and $V_{sd} = -60$ V. Bottom inset shows the logarithmic derivatives (top to bottom). The fit to the $V_{sd} = -1$ V data has slope $-1/2$ and the $V_{sd} = -24$ V data are fitted to a power law. (From Dhoot, A.S., Wang, G.M., Moses, D., and Heeger, A.J., *Phys. Rev. Lett.*, 96, 246403, 2006. Reprinted from the American Physical Society. With permission.)

conduction electron plasma frequency. Two distinct peaks are observed at higher energy near 2.8 and 6 eV, attributed to interband transitions within a polaron lattice [159,160].

For samples F and G, there is a strong peak at 1.5 eV, more pronounced in the lowest σ_{dc} sample, G. When the peak at 1.5 eV is strong, the reflectance in the far-IR is diminished. This indicates that there is a second "localization–delocalization" transition from isolated polarons localized to one or two C_6H_4NH repeat units to partially delocalized charges that are proposed [96] to extend over distances of ~10–100 Å.

The temperature-dependent reflectance data of PAN-CSA samples A and D in the far-IR are shown in Figure 15.19. With decreasing T, the reflectance at ~2 meV is decreased relative to the room temperature reflectance while the higher energy (>~4 meV) reflectance initially increases above the room temperature value. At the lowest T, the far-IR reflectance is suppressed over the whole energy range. The difference between the metallic (sample A) and insulating (sample D) materials is the magnitude by which the reflectance is suppressed. For sample A, the far-IR reflectance is suppressed by ~3% from its value at room temperature, whereas the suppression is ~10% for sample D. This suppression of the reflectance in the far-IR is consistent with the decrease in σ_{dc} at low T for each of

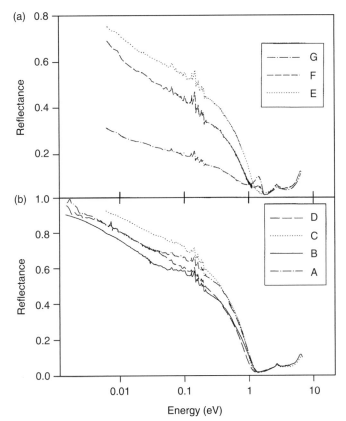

FIGURE 15.18 Room temperature reflectance for (a) PAN-CSA samples E–G and (b) PAN-CSA samples A–D. (From Kohlman, R.S. and Epstein, A.J., *Handbook of conducting polymers*, 2nd ed., eds. Skotheim, T.A., Elsenbaumer, R.L., and Reynolds, J.R., Marcel Dekker, New York, 1988, chap. 3. Reprinted from Routledge/Taylor & Francis Group, LLC. With permission.)

these materials and results from the localization (trapping) at low temperature of carriers that are delocalized at room temperature.

The IR active vibrations that appear at ∼0.1 eV in the reflectance data progressively weaken for highly conducting samples of progressively increasing conductivity (Figure 15.6, Figure 15.18, and Figure 15.20). This is as expected for increased screening. Samples of later reported highly conducting self-stabilized dispersion-polymerized PAN also show this reduction in amplitude of the IR active vibrations [47].

15.4.6 Optical Conductivity

Kramers–Kronig analysis of the reflectance data provides the optical conductivity function $\sigma(\omega)$. We can obtain information about the conduction electrons in PAN-CSA by comparing this function with the Drude expression for free electrons and the localization models. The room temperature $\sigma(\omega)$ is shown in Figure 15.20 for samples A–G. In sample G, there is a strong peak at ∼1.5 eV and a weak peak at ∼0.6 eV. For sample F, $\sigma(\omega)$ shows that the peak at 1.5 eV is reduced and the peak in the IR has grown and the maximum has shifted to lower energy. For the samples with higher σ_{dc}, the peak at 1.5 eV is absent, indicating that it is not an intrinsic transition in the highly conducting state. Comparison of $\sigma(\omega)$ for samples E–G shows that there is an isosbestic point at ∼1.3 eV, characteristic of composite behavior [44]. This directly supports the occurrence of composite behavior in bulk PAN.

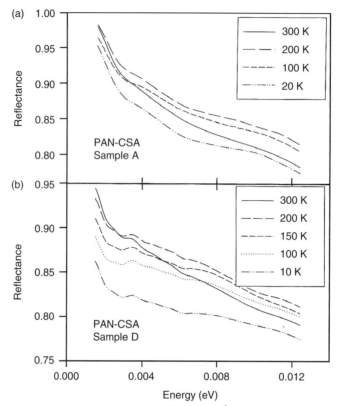

FIGURE 15.19   Temperature dependence of the far-IR (10–100 cm^{-1}) reflectance for (a) PAN-CSA sample A and (b) PAN-CSA sample D. (From Kohlman, R.S. and Epstein, A.J., *Handbook of conducting polymers*, 2nd ed., eds. Skotheim, T.A., Elsenbaumer, R.L., and Reynolds, J.R., Marcel Dekker, New York, 1988, chap. 3. Reprinted from Routledge/Taylor & Francis Group, LLC. With permission.)

For samples A–E, $\sigma(\omega)$ shows Drude dispersion with decreasing energy below ∼1.4 eV until reaching a maximum in the mid-IR. The oscillations of $\sigma(\omega)$ in the mid-IR correspond to the vibrational modes of doped PAN. $\sigma(\omega)$ is monotonically suppressed beneath the σ_{Drude} at lower energy except for sample C. For sample C, $\sigma(\omega)$ increases again with decreasing energy below 0.02 eV. Although the increase of $\sigma(\omega)$ in the far-IR is not as rapid, the behavior is qualitatively similar to what was reported for iodine- [97,98,161] and perchlorate- [162] doped $(CH)_x$ where free electrons are reported. This type of frequency behavior for $\sigma(\omega)$ is qualitatively similar to that of metallic particles in an insulating matrix after percolation of the metallic particles [118]. It is indicative of the small plasma frequency of conduction electrons that are delocalized through the material. Also, Figure 15.20 demonstrates that the far-IR $\sigma(\omega)$ scales with σ_{dc}, confirming the measured σ_{dc} values.

As σ_{dc} of the samples increases, the frequency at which the maximum in $\sigma(\omega)$ occurs (ω_{max}) shifts monotonically to lower energy (Table 15.3). This shift of the conductivity peak to low energies is the same behavior that the localization-modified Drude model displays as the mean free time (and therefore the mean free path) grows and the behavior that the inhomogeneous disorder model predicts as the samples near the percolation threshold. Since ω_{max} is lowest for PAN-CSA sample C, this indicates that the scattering time (mean free path) for sample C is the longest and therefore the disorder is weakest. Assuming a homogeneous material, sample C is inferred to be the most ordered sample. Recalling that the maximum in $\sigma_{dc}(T)$ implied that sample A is the most ordered, a contradiction is obtained from assuming that the materials are homogeneous. Within the inhomogeneous disorder model, there is no contradiction. This is because $\sigma_{dc}(T)$ reflects the behavior of percolated carriers in well-ordered

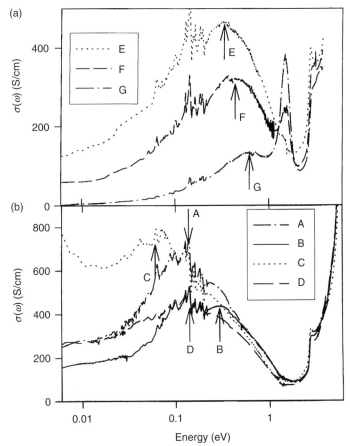

FIGURE 15.20 Room temperature optical conductivity $\sigma(\omega)$ for (a) PAN-CSA samples E–G and (b) PAN-CSA samples A–D. The shift of the maximum in $\sigma(\omega)$ indicates the growth of the electron mean free path. The arrows indicate the peak (maximum) in the intraband $\sigma(\omega)$. (From Kohlman, R.S. and Epstein, A.J., *Handbook of conducting polymers*, 2nd ed., eds. Skotheim, T.A., Elsenbaumer, R.L., and Reynolds, J.R., Marcel Dekker, New York, 1988, chap. 3. Reprinted from Routledge/Taylor & Francis Group, LLC. With permission.)

conduction paths while the high frequency $\sigma(\omega)$ is dominated by the contribution of localized carriers that contribute <10 S/cm to σ_{dc}, in disordered regions of the sample.

Drude conductivity fits to the energy range ~0.4–1.4 eV provide further information about the carriers. The plasma frequency (Ω_{p1}) and localized carrier scattering time (τ_1) for each of the samples (Table 15.3) fall into the range $\Omega_{p1} \sim 1.7$ eV and $\tau_1 \sim 10^{-15}$ s. The small scattering time $(\tau_1 \sim 10^{-15}$ s) is attributed to the disorder in the polymer. This scattering time is the magnitude predicted from the Ioffe–Regel condition for materials close to a 3D Anderson IMT. However, the Anderson IMT model is not appropriate for these materials. Analysis of the dielectric function $\varepsilon(\omega)$ (Section 15.4.7) demonstrates that $\sigma_{dc}(T)$ in conducting polymer systems is due to a small fraction of carriers with $\tau > \sim 10^{-13}$ s that percolate through the material. The increase in $\sigma(\omega)$ for sample C in the far-IR is evidence of the percolated free electrons.

The "average" effective mass (m^*) of the conduction electrons also can be calculated from knowledge of Ω_{p1} $(= \sqrt{4\pi n e^2 / m^*})$. From x-ray measurements [82,113], the volume of a two-PAN ring repeat with a CSA counterion is V ~ 760 Å^3 assuming a unit cell with dimensions 10.4 (the repeat length along the chain), 3.5, and 21 Å (which represents the repeat distance between alternating stacks of PAN and CSA). Since there is one doped charge added per two-ring repeat of PAN-CSA, $n = 1/760$ Å$^3 \sim 1.3 \times 10^{21}$ cm^{-3}. Therefore, $m^* \sim 0.6 m_e$ (m_e is the electron mass).

TABLE 15.3   Comparison of the DC and High-Frequency Conductivity Parameters for Selected PAN-CSA Films

Sample	σ_{dc} (300 K) (S/cm)	$\Omega_p{}^a$ (eV)	τ^a (10^{-15} s)	$\omega_{max}{}^b$ (eV)
A	240	1.7	1.1	0.09
B	110	1.7	0.9	0.3
C	400	1.6	1.3	0.06
D	120	1.5	1.0	0.14
E	70	2.0	0.7	0.33
F	20	1.8	0.6	0.37
G	0.7	1.3	0.4	0.7

Source: From Kohlman, R.S. and Epstein, A.J. in *Handbook of conducting polymers*, 2nd ed., eds. Skotheim, T.A., Elsenbaumer, R.L., and Reynolds, J.R., Marcel Dekker, New York, 1988, chap. 3. Reprinted from Routledge/Taylor & Francis Group, LLC. With permission.)

[a] The values for Ω_{p1} and τ_1 are obtained by fitting the Drude model to the optical conductivity near the plasma edge.

[b] ω_{max} represents the frequency where $\sigma(\omega)$ shows a maximum (ignoring phonon features).

Temperature-dependent far-IR $\sigma(\omega)$ values for samples A and D are shown in Figure 15.21. For both samples, the trend is clear. As the temperature is lowered, $\sigma(\omega)$ first increases and then suppresses. The resulting suppression for sample D is much stronger than for sample A, in agreement with the trend in σ_{dc}. The growth and then decrease of $\sigma(\omega)$ is consistent with the localization of charges at low

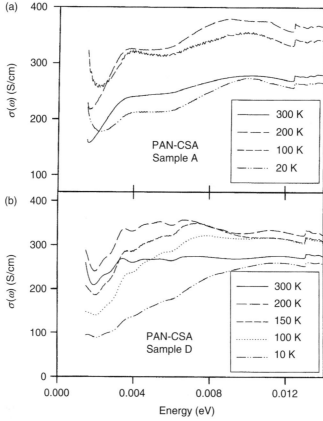

FIGURE 15.21   Temperature dependence of the far-IR optical conductivity in (a) PAN-CSA sample A and (b) PAN-CSA sample D. (From Kohlman, R.S. and Epstein, A.J., *Handbook of conducting polymers*, 2nd ed., eds. Skotheim, T.A., Elsenbaumer, R.L., and Reynolds, J.R., Marcel Dekker, New York, 1988, chap. 3. Reprinted from Routledge/Taylor & Francis Group, LLC. With permission.)

temperature that are delocalized at room temperature. Some of the free electrons contributing to the Drude plasma frequency are being frozen out at low (zero) frequency. Therefore, due to sum rules on $\sigma(\omega)$ [103], the transitions of the "frozen out" electrons must occur at higher frequency. From this point of view, sample D becomes insulating (hopping) at low temperature because the density of free electrons goes to zero at low temperatures. Sample A remains metallic at low temperatures because the free electron density is not sufficiently reduced.

15.4.7 Optical Dielectric Function

The dielectric function determined by Kramers–Kronig analysis provides information about whether the carriers are free or localized. The importance of localization is clear from $\sigma(\omega)$. As σ_{dc} increased, the mean free path also clearly increased. With increasing electron delocalization, a larger polarization of the electron gas is possible. The far-IR $\varepsilon(\omega)$ for samples E–G (Figure 15.22a) is positive in the far-IR, characteristic of localized carriers. The growth of the far-IR $\varepsilon(\omega)$ directly demonstrates the growing polarization of localized states with increasing σ_{dc}. For sample G, $\varepsilon(\omega)$ is small and positive in the far-IR with strong dispersion near the "localized polaron" peak at 1.5 eV. The growth of the average scattering time and mean free path is also made evident by the development of a zero crossing at ~1 eV. This zero crossing is termed ω_{p1} (the screened plasma frequency Ω_{p1}). When the mean free path grows, $\varepsilon(\omega)$ shows Drude dispersion to lower frequencies, so $\varepsilon(\omega)$ crosses zero at the screened plasma frequency of all the conduction electrons, even though localization corrections force $\varepsilon(\omega)$ positive at lower frequencies.

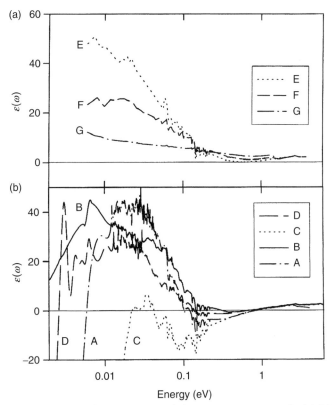

FIGURE 15.22 The real part of the dielectric response $\varepsilon(\omega)$ at room temperature for (a) PAN-CSA samples E–G and (b) PAN-CSA samples A–D. (From Kohlman, R.S. and Epstein, A.J., *Handbook of conducting polymers*, 2nd ed., eds. Skotheim, T.A., Elsenbaumer, R.L., and Reynolds, J.R., Marcel Dekker, New York, 1988, chap. 3. Reprinted from Routledge/Taylor & Francis Group, LLC. With permission.)

This behavior is reminiscent of the $\varepsilon(\omega)$ for the localization-modified Drude model of Section 15.2.2. The zero crossing of $\varepsilon(\omega)$ at ω_{p1} is therefore due to localized electrons.

However, the low frequency electrical response is not dominated by localized carriers for samples A–D near the IMT. Figure 15.22b shows $\varepsilon(\omega)$ for samples A–D. In addition to the dielectric response of localized electrons at ~1 eV, a third zero crossing of $\varepsilon(\omega)$ termed ω_p is evident in the far-IR. For energies lower than ω_p, $\varepsilon(\omega)$ remains negative as expected for free carriers. If the carriers are free electrons, then their frequency response is described by the Drude model. At sufficiently high frequencies ($\tau \gg 1/\omega$), the Drude dielectric function is given by

$$\varepsilon_{\text{Drude}}(\omega) = \varepsilon_b - \Omega_p^2/\omega^2 \tag{15.27}$$

where ε_b is the background dielectric response due to localized electrons and interband transitions and Ω_p is the plasma frequency for free carriers. Figure 15.23 shows $\varepsilon(\omega)$ plotted against $1/\omega^2$ for samples A, C, and D. (The plot for sample B is not shown because, though $\varepsilon(\omega)$ is clearly turning toward negative values, the zero crossing is not observed in the experimental frequency range. For this sample, we do not have low enough frequency data where $\varepsilon(\omega)$ is dominated only by free electrons.) The plots provide straight lines where the slope is Ω_p^2. The value of Ω_p for each of these samples is given in Table 15.4. The frequency response of $\varepsilon(\omega)$ for PAN-CSA samples A–D (i.e., the response of both localized carriers and delocalized carriers with long τ) is typical of percolating systems [118,119].

From Table 15.4, it is clear that ω_p (the plasma frequency of free electrons) scales with σ_{dc} for each sample. This supports the notion that the high dc conductivity and low frequency transport are controlled by the delocalized "free" carriers. Also, the scaling of Ω_p with σ_{dc} argues against the Anderson transition model for conducting polymers. Because sample C is insulating at low T within the Anderson model, the Fermi level for sample C lies in the region of localized (hopping) states. However, sample C has a larger free electron plasma frequency (Ω_p) at room temperature than sample A, which is metallic at low T and therefore has its Fermi level in the region of delocalized (free electron)

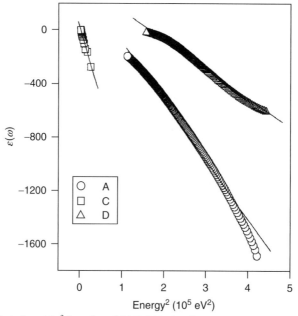

FIGURE 15.23 $\varepsilon(\omega)$ plotted vs. $1/\omega^2$ for selected PAN-CSA samples at room temperature. The linearity confirms that the low energy carriers are free (Drude) carriers. (From Kohlman, R.S. and Epstein, A.J., *Handbook of conducting polymers*, 2nd ed., eds. Skotheim, T.A., Elsenbaumer, R.L., and Reynolds, J.R., Marcel Dekker, New York, 1988, chap. 3. Reprinted from Routledge/Taylor & Francis Group, LLC. With permission.)

TABLE 15.4 Free Electron Plasma Frequency Ω_p and Scattering Time τ vs. Temperature for Selected PAN-CSA Film Samples

Sample	T (K)	σ_{dc} (300 K) (S/cm)	Ω_p (eV)	n_{free}/n_{cond}[a] (10^{-3})	τ[b] $(10^{-13}$ s)
A	300	230	0.07	1.2	2.3
	200	250	0.08	1.6	2.0
	100	240	0.06	0.9	2.9
	20	170	0.05	0.6	3.4
C	300	400	0.11	3.0	1.6
D	300	120	0.04	0.4	3.7
	200	120	0.03	0.2	6.6

Source: From Kohlman, R.S. and Epstein, A.J. in *Handbook of conducting polymers*, 2nd ed., eds. Skotheim, T.A., Elsenbaumer, R.L., and Reynolds, J.R., Marcel Dekker, New York, 1988, chap. 3. Reprinted from Routledge/Taylor & Francis Group, LLC. With permission.)

[a] Estimated fraction of the conduction electrons that are free.
[b] Estimated scattering time.

states. In addition, the scaling of Ω_p with σ_{dc} argues against the negative $\varepsilon(\omega)$ being due to the formation of an intrinsic many-body gap as proposed for PPy materials [121], especially since the negative far-IR $\varepsilon(\omega)$ is observed at room temperature in PAN-CSA samples.

The values obtained for Ω_p are very small compared with the full conduction electron density ($\Omega_{p1} \sim$ 2 eV), suggesting that only a small fraction (n_{free}/n_{cond}) of the conduction electrons are delocalized macroscopically. This small fraction of delocalized carriers is consistent with the small number of percolation paths that occur close to the percolation threshold in composite systems. The fraction of the carriers that are delocalized can be estimated by comparing the plasma frequency of free electrons (Ω_p) with the full conduction electron plasma frequency (Ω_{p1}),

$$\frac{n_{free}}{n_{cond}} = \frac{m^*_{free}}{m^*_{cond}}\left(\frac{\Omega_p}{\Omega_{p1}}\right)^2 \tag{15.28}$$

This ratio is estimated in Table 15.4 for samples A, C, and D, assuming that the effective masses (m_{free}^* and m_{cond}^*) are approximately the same. In each case,

$$n_{free}/n_{cond} \sim 10^{-3} \tag{15.29}$$

Even assuming a 10-fold increase in m_{free}^* as the free carriers may reside in a very narrow band (due to passage through the disordered regions), the fraction of delocalized carriers is only on the order of 1%. The fact that the current is carried by a small number of percolated paths is not surprising, as high conductivities (>10 S/cm) have been reported for PAN-CSA diluted to only 10% in insulating PMMA [163].

Since the plots of $\varepsilon(\omega)$ vs. $1/\omega^2$ in Figure 15.23 do not show a tendency to saturate down to 10 cm^{-1}, the mean free time for the free electrons (τ) can be estimated as

$$\tau \gg 1/(10\,\text{cm}^{-1}) \sim 5 \times 10^{-13}\text{s} \tag{15.30}$$

Such a huge mean free time in a disordered polymer is surprising, especially when compared with the typical room temperature scattering times for copper, $\tau \sim 10^{-13}$–10^{-14} s [40]. This small fraction is delocalized sufficiently that $k_F \lambda \gg 1$. Similar values for τ have been reported previously in doped polyacetylene [98]. The large scattering time likely results from the ramifications of an open Fermi surface as expected for an anisotropic metal [23,29,39,116], and that the usual scattering centers, being off the polymer chains, are likely screened, reducing their effectiveness. If the whole density of conduction electrons were able to diffuse with $\tau \sim 10^{-13}$ s, the dc conductivity would be $\sim 10^5$ S/cm. Using this estimate, it is suggested that substantial improvements in the electrical conductivity can still be obtained.

It is also possible to estimate τ within the Drude model where

$$\sigma_{dc} = \Omega_p^2 \tau / 4\pi \tag{15.31}$$

Since the free electron plasma frequency Ω_p scales with σ_{dc}, the scattering time is estimated as $\tau \sim 10^{-13}$ s for samples A, C, and D (Table 15.4), assuming that only the free electrons participate in the low frequency transport. This estimate of the free electron scattering time is in good agreement with the previous estimate. This large scattering time ($\tau \gtrsim 10^{-13}$ s) compares with $\tau \sim 10^{-14}$ s at room temperature in copper [164]. The presence of carriers with $\tau \sim 10^{-15}$ s indicates that the low frequency carriers are not subject to the Ioffe–Regel condition, which requires that $\tau \sim 10^{-15}$ s in conducting polymers. This constitutes further evidence against the applicability of the Anderson IMT model.

The temperature dependence of the far-IR $\varepsilon(\omega)$ explicitly demonstrates the essential difference between metallic and insulating samples. The temperature-dependent $\varepsilon(\omega)$ for sample A (Figure 15.6 and Figure 15.24) shows the crossover to negative values and the Drude dispersion in the far-IR. This free carrier dispersion is present in $\varepsilon(\omega)$ from 300 K down to ~ 20 K. The presence of free carriers down to low T is consistent with the presence of high σ_{dc} down to millikelvin T. In contrast, the temperature-dependent

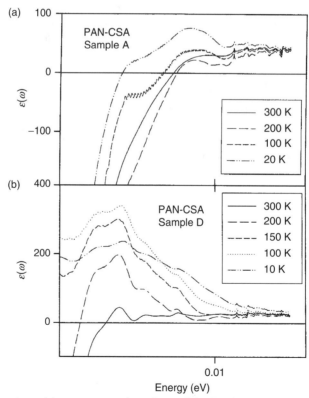

FIGURE 15.24 Comparison of the temperature dependence of the dielectric response $\varepsilon(\omega)$ for (a) metallic PAN-CSA sample A and (b) insulating PAN-CSA sample D. The value of the abscissa at the left-hand axis is 0.002 eV. (From Kohlman, R.S. and Epstein, A.J., *Handbook of conducting polymers*, 2nd ed., eds. Skotheim, T.A., Elsenbaumer, R.L., and Reynolds, J.R., Marcel Dekker, New York, 1988, chap. 3. Reprinted from Routledge/Taylor & Francis Group, LLC. With permission.)

$\varepsilon(\omega)$ for sample D is shown in Figure 15.24b. As the temperature is lowered to $\sim$10 K and σ_{dc} drops, $\varepsilon(\omega)$ crosses from negative to positive in the far-IR. Therefore, the free carriers become localized at low temperature in PAN-CSA sample D. It is proposed that the localization at low T is due to the ineffect-iveness of phonon-assisted delocalization in the disordered regions at low T [43]. The localization of the free carriers in PAN-CSA sample D at low T is consistent with the strong decrease of σ_{dc} at low T. Therefore, the difference between sample A, which remains metallic, and sample D, which becomes insulating at low T, is the presence of percolated free carriers down to low temperatures. The values for the low temperature Ω_p and τ (shown only for those temperatures where Drude dispersion is evident in $\varepsilon(\omega)$) determined from Drude fits to σ and ε [45] are shown in Table 15.4. The presence of free electrons at room temperature in insulating sample D indicates that percolation behavior can occur at high T even though at low T the composite is not percolated. This is proposed to reflect the importance of phonon-induced delocalization on the electronic L_{loc} in the disordered regions, which electronically couple crystalline metallic regions in inhomogeneous conducting polymers [29,39,44,45].

Figure 15.25a shows a plot of $\varepsilon(\omega)$ vs. $1/\omega^2$ at various temperatures for sample A. The slope varies with temperature, indicating that Ω_p is temperature-dependent. This is further evidence that the IMT is

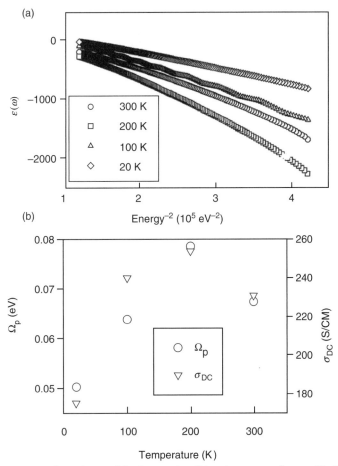

FIGURE 15.25 Temperature dependence of the free carrier dielectric response for metallic PAN-CSA sample A. (a) Far-IR $\varepsilon(\omega)$ plotted against $1/\omega^2$ as a function of temperature. (b) Comparison of the temperature dependence of the free electron plasma frequency Ω_p and σ_{dc}. (From Kohlman, R.S. and Epstein, A.J., *Handbook of conducting polymers*, 2nd ed., eds. Skotheim, T.A., Elsenbaumer, R.L., and Reynolds, J.R., Marcel Dekker, New York, 1988, chap. 3. Reprinted from Routledge/Taylor & Francis Group, LLC. With permission.)

due to a loss of free carrier density $n(T)$. Figure 15.25b shows directly that the free carrier plasma frequency Ω_p approximately scales with σ_{dc} at each temperature for metallic sample A. This behavior contrasts with the case in conventional metals where the temperature dependence of σ_{dc} is determined solely by the temperature dependence of the scattering time. Assuming that $\sigma_{dc}(T) = \Omega_p^2 \tau/4\pi$, the scattering time actually increases at low T (Table 15.4), as in conventional metals. Therefore, the temperature dependence of σ_{dc} in inhomogeneous metallic conducting polymers must be determined primarily by $n(T)$. Within the Anderson model, lowering the temperature would result in depopulation of more extended states above the mobility edge and therefore a decrease in the scattering time. The contradiction of the Anderson model prediction and the trend in the data argue further against the Anderson IMT. The increase in τ at low T may instead reflect the more robust percolation paths that remain at low T as the less robust ones freeze out.

Figure 15.26 shows the dielectric function parallel and perpendicular to the stretch direction for HCl-doped, stretched PAN obtained from polarized reflectance measurements [96]. There is a strong anisotropy evident in the electrical response. Parallel to the stretch direction, the high-energy ($\sim$1 eV) zero crossing of $\varepsilon(\omega)$ as well as strong dispersion in the IR is observed. Perpendicular to the stretch

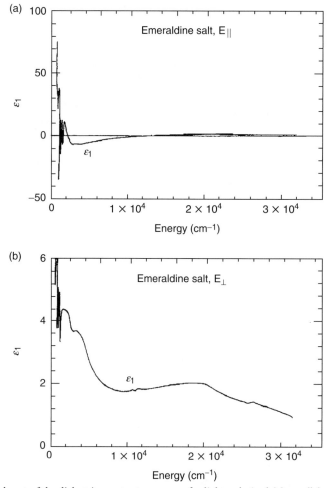

FIGURE 15.26   Real part of the dielectric constant vs. energy for light polarized (a) parallel and (b) perpendicular to the stretch direction for PAN doped with HCl. (From McCall, R.P., Scherr, E.M., MacDiarmid, A.G., and Epstein, A.J., *Phys. Rev. B*, 50, 5094, 1994. Reprinted from the American Physical Society. With permission.)

direction, $\varepsilon(\omega)$ is positive over the whole energy range, showing very weak dispersion. The anisotropy present along the chain direction must be accounted for within the model of the IMT. The homogeneous Anderson model that has been employed in the literature [121,133,159] assumes that the material is isotropic. The fact that the anisotropic nature of the polymer transport is not included in the framework of the homogeneous Anderson IMT model also rules out this model for the conducting polymer IMT.

In summary, $\varepsilon(\omega)$ is characteristic of strongly localized electrons ($\tau \sim 10^{-15}$ s) for samples E–G far from the IMT. For samples A–D near the IMT, a small fraction of the carriers become macroscopically delocalized ($\tau \geq \sim 10^{-13}$ s) while the majority of carriers are still strongly localized ($\tau \sim 10^{-15}$ s). The difference between metallic sample A and insulating sample D is that the free carriers freeze out in sample D at low *T*. This behavior is characteristic of percolating systems. Unlike the usual percolating systems, the percolation for doped polymers can be temperature-dependent. The dielectric response of oriented materials clearly points out the importance of dimensionality in the electronic behavior.

15.4.8 Discussion of Conductivity and Dielectric Functions

The dielectric function and optical conductivity provide insight into the nature of the disorder in the metallic state. In this section, $\varepsilon(\omega)$ and $\sigma(\omega)$ for the PAN-CSA samples are compared with the localization-modified Drude model for homogeneously disordered systems and a model for inhomogeneous disorder.

For the homogeneously disordered system, the mean free time is limited to a very short time owing to substantial disorder. For materials near the IMT, $k_F\lambda \sim 1$ [29,120,133,159]. Due to the limitation of λ, $\sigma(\omega)$ is suppressed and $\varepsilon(\omega)$ is driven positive at low energy. The frequency dependence of σ and ε in this model is described by the localization-modified Drude model [89,120,133,159], Section 15.2.2, with a short scattering time.

A typical fit of the localization-modified Drude model is shown in Figure 15.27 for sample E. To ensure that causality is satisfied, the parameters were chosen (Table 15.5) to describe both $\sigma(\omega)$ and $\varepsilon(\omega)$. The experimental $\sigma(\omega)$ and $\varepsilon(\omega)$ are well represented for PAN-CSA sample E. It is important, however, to determine whether the parameters obtained are reasonable. The values obtained for the localized carrier scattering time τ_1 are comparable with the values obtained from the Drude fits to $\sigma(\omega)$ (Table 15.3), confirming the short scattering time ($\tau_1 \sim 10^{-15}$ s). However, the full carrier plasma frequency Ω_{p1} is small compared with what was found for the Drude fits (Table 15.3), suggesting a sizable difference in oscillator strength between the two models. Integration of $\sigma(\omega)$ for the localization-modified Drude model shows that

$$8\int_0^\infty \sigma_{LMDM}(\omega)d\omega = \Omega_{p1}^2\left[1 + C/(k_Fv_F\tau_1)^2\right] \tag{15.32}$$

so that the oscillator strength is greater in the localization-modified Drude model than just Ω_{p1}^2 as in the unmodified Drude model [40,103]. For sample F, the full plasma frequency calculated using Equation 15.32 and the parameters from Table 15.5 is ~2.0 eV, reconciling the oscillator strength differences between the Drude model and localization-modified Drude model estimates of the oscillator strength.

The parameter $C/(k_Fv_F)^2$ can be used to estimate the Fermi velocity v_F, the mean free path ($v_F\tau$), and the 1D density of states ($N(E_F) = 2/\pi\hbar v_F$) [133]. To make the estimates, C is assumed to be of order unity [99,100,120,133]. Assuming that the sample is fully doped, k_F has the value $\pi/2c$ with $c \sim 10.2$ Å [82]. Using $\tau \sim 1.2 \times 10^{-15}$ s and $C/(k_Fv_F)^2 \sim 1.4 \times 10^{-30}$ s^2, $v_F \sim 5 \times 10^7$ cm/s and $\lambda = v_F\tau \sim 7$ Å. Using these parameters, $k_F\lambda \sim 1$. This value of $k_F\lambda$ is in accord with the Ioffe–Regel condition for conductors close to an IMT because the mean free path is of the same size as a nitrogen and benzene ring repeat in PAN (~5.2 Å). The prediction for the 1D density of states at the Fermi level, $N(E_F) \sim 2$ state/eV per formula unit (two nitrogens and two rings), is also in reasonably good agreement with the measured value of $N(E_F) \sim 1$ state/eV per formula unit [14,66]. The consistency of these estimates

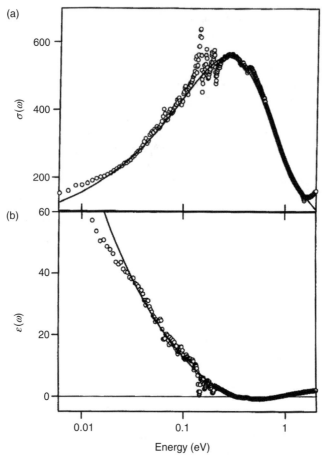

FIGURE 15.27 Typical comparison of the localization-modified Drude model fits to the experimental data. The data shown are (a) $\sigma(\omega)$ and (b) $\varepsilon(\omega)$ for PAN-CSA sample E. (From Kohlman, R.S. and Epstein, A.J., *Handbook of conducting polymers*, 2nd ed., eds. Skotheim, T.A., Elsenbaumer, R.L., and Reynolds, J.R., Marcel Dekker, New York, 1988, chap. 3. Reprinted from Routledge/Taylor & Francis Group, LLC. With permission.)

TABLE 15.5 Fit Parameters for PAN-CSA Using the Localization-Modified Drude Model

Sample	Ω_{p1} (eV)	τ_1 (10^{-15} s)	$C/(k_F v_F)^2$ (10^{-30} s^2)	ε_{inf}
A	1.2	2.0	3.1	2.8
B	1.3	1.4	1.6	2.6
C	1.3	2.4	1.6	3.6
D	1.2	1.6	1.4	2.8
E	1.4	1.2	1.4	2.5
F[a]	1.2	1.0	1.1	2.2
G[a]	0.9	0.7	0.6	1.8

Source: From Kohlman, R.S. and Epstein, A.J. in *Handbook of conducting polymers*, 2nd ed., eds. Skotheim, T.A., Elsenbaumer, R.L., and Reynolds, J.R., Marcel Dekker, New York, 1988, chap. 3. Reprinted from Routledge/Taylor & Francis Group, LLC. With permission.)

[a]An additional Lorentzian oscillator was included for samples F and G to model the 1.5 eV localized polaron peak.

obtained from the localization-modified Drude model indicates that the high frequency behavior of samples similar to E (dominated by localized excitations) is reasonably represented within a homogeneous Anderson transition picture.

The fits for $\sigma(\omega)$ and $\varepsilon(\omega)$ of PAN-CSA samples E–G using the parameters in Table 15.5 are shown in Figure 15.28 and Figure 15.29. For samples F and G, $\sigma(\omega)$ and $\varepsilon(\omega)$ are complicated by the presence of localized polarons as well. Since the localized polaron feature occurs in a narrow frequency window ($\sim$1.5 eV), it can be modeled by including a Lorentzian function. For sample F, the Lorentzian parameters were $\Omega_p = 0.8$ eV, $\gamma = 4.4 \times 10^{14}$ Hz, and $\omega_0 = 1.5$ eV, while $\Omega_p = 1.3$ eV, $\gamma = 5.4 \times 10^{14}$ Hz, and $\omega_0 = 1.5$ eV were used for PAN-CSA sample G.

One set of experimental trends is well represented by the localization-modified Drude model. τ_1 increases with increasing σ_{dc}, suggesting that the disorder decreases, consistent with the shift of ω_{max} in $\sigma(\omega)$. Samples A–D deviate in an important way from the predictions of the localization-modified Drude model in the far-IR (Figure 15.30 and Figure 15.31). Figure 15.30 shows that $\sigma(\omega)$ for PAN-CSA samples A–D can be reasonably modeled except for the upturn in $\sigma(\omega)$ for sample C at $\sim$0.02 eV. On the other hand, Figure 15.31 demonstrates that the localization-modified Drude model cannot account for the free carrier behavior in $\varepsilon(\omega)$ at low frequency. This is a consequence of the assumption within a homogeneous material that all of the carriers have short mean free times ($k_F v_F \tau_1 \sim 1$) limited by strong

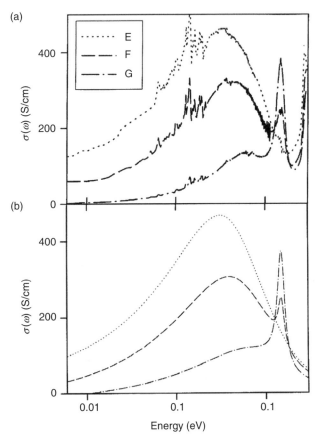

FIGURE 15.28 (a) Experimental $\sigma(\omega)$ compared with (b) localization-modified Drude model fits to $\sigma(\omega)$ for PAN-CSA samples E–G. (From Kohlman, R.S. and Epstein, A.J., *Handbook of conducting polymers*, 2nd ed., eds. Skotheim, T.A., Elsenbaumer, R.L., and Reynolds, J.R., Marcel Dekker, New York, 1988, chap. 3. Reprinted from Routledge/Taylor & Francis Group, LLC. With permission.)

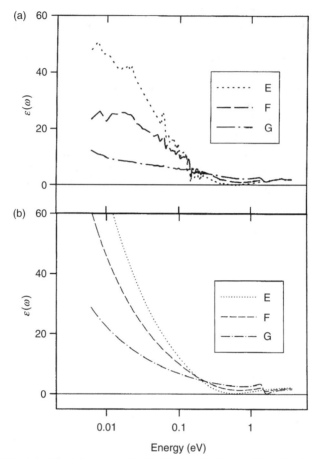

FIGURE 15.29 (a) Experimental $\varepsilon(\omega)$ compared with (b) localization-modified Drude model fits to $\varepsilon(\omega)$ for PAN-CSA samples E–G. (From Kohlman, R.S. and Epstein, A.J., *Handbook of conducting polymers*, 2nd ed., eds. Skotheim, T.A., Elsenbaumer, R.L., and Reynolds, J.R., Marcel Dekker, New York, 1988, chap. 3. Reprinted from Routledge/Taylor & Francis Group, LLC. With permission.)

disorder. It is proposed that this model explains well the behavior of only the relatively strongly localized electrons ($k_F\lambda \sim 1$) in conducting polymers with reasonable parameters.

The early studied PAN-CSA [133] and PPy doped with perchlorate [165] both had $\sigma_{dc} \sim 100$ S/cm, indicating that the carriers were reasonably localized. Therefore, the agreement of the optical properties with the localization-modified Drude model (with $\tau \sim 10^{-15}$ s) is expected. However, this model is unable to account for the free electron behavior observed in higher σ_{dc} samples because Drude dispersion requires that $k_F\lambda \gg 1$.

The necessity of treating the highly conducting polymers within a composite picture with inhomogeneous disorder is clear from an estimate of the mean free paths for the percolated electrons. Using the same Fermi velocity estimated from the localization-modified Drude model ($v_F \sim 5 \times 10^7$ cm/s), the mean free path for free electrons in PAN-CSA is estimated as $\lambda_{free} \sim 10^3$ Å, with $k_F\lambda_{free} \sim 10^2$ using τ determined from the IR measurements. (As noted earlier, band narrowing due to disorder will increase m^* and decrease estimates of λ_{free}.) This estimate of λ_{free} is consistent with previous estimates of the mean free time in doped polyacetylene [98,162] and the inelastic length obtained from magnetoresistance experiments [74,166]. Since $\lambda_{free} \gg \xi$, the crystalline coherence length ($\sim$50 Å), these free electrons are capable of diffusing large distances among ordered regions between scattering events.

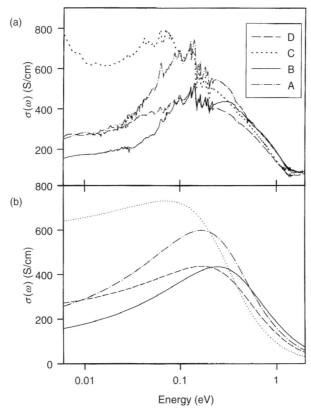

FIGURE 15.30 (a) Experimental $\sigma(\omega)$ compared with (b) localization-modified Drude model fits to $\sigma(\omega)$ for PAN-CSA samples A–D. (From Kohlman, R.S. and Epstein, A.J., *Handbook of conducting polymers*, 2nd ed., eds. Skotheim, T.A., Elsenbaumer, R.L., and Reynolds, J.R., Marcel Dekker, New York, 1988, chap. 3. Reprinted from Routledge/Taylor & Francis Group, LLC. With permission.)

In contrast to the macroscopically homogeneous Anderson transition, effective medium models allow for macroscopic inhomogeneity and composite behavior. Therefore, there can be percolated metallic regions of the sample where the mean free time is large. For composite systems, the disorder is macroscopic and there may be two very different phases. Most calculations within the effective medium theory are for composites of an insulator and a metal [118,119]. For the conducting PAN and PPy samples, the different parts of the composite are proposed to be the more ordered (metallic) and the disordered regions where localization becomes important. The percolating behavior of conducting polymers is expected to be modified from traditional insulator–metal composites as the ordered and disordered regions do not have sharp boundaries; a single polymer chain may be a part of both ordered and disordered regions. Also, phonon-assisted transport occurs in the disordered regions. Nevertheless, there are many qualitative similarities with the insulator–metal composite. Figure 15.5 shows the behavior of $\sigma(\omega)$ for a system of metal particles percolating in an insulating matrix. At low volume fractions before percolation, the Drude peak in $\sigma(\omega)$ is suppressed and shows up as a localized plasmon excitation at higher energy [118,119]. The maximum in this localized plasmon band in $\sigma(\omega)$ shifts to lower frequency as the materials near percolation. Above percolation, the Drude peak is observed at low frequency, though the conductivity is only a fraction of the conductivity of the bulk metal (alternatively, the Drude peak contains only a fraction of the density of carriers). The fraction depends on the proximity of the system to the percolation threshold. Although it was not calculated in [118], the appearance of a Drude peak in $\varepsilon(\omega)$ is required by causality when there is a Drude peak in $\sigma(\omega)$.

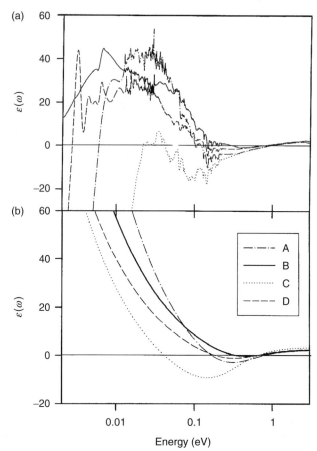

FIGURE 15.31   (a) Experimental $\varepsilon(\omega)$ compared with (b) localization-modified Drude model fits to $\varepsilon(\omega)$ for PAN-CSA samples A–D. (From Kohlman, R.S. and Epstein, A.J., *Handbook of conducting polymers*, 2nd ed., eds. Skotheim, T.A., Elsenbaumer, R.L., and Reynolds, J.R., Marcel Dekker, New York, 1988, chap. 3. Reprinted from Routledge/Taylor & Francis Group, LLC. With permission.)

For insulating PAN-CSA samples E–G, $\varepsilon(\omega)$ shows only localized excitations. The peak in $\sigma(\omega)$ shifts to lower frequency with increasing σ_{dc}, consistent with the behavior of a localized plasmon band in a composite system [161,162]. For metallic PAN-CSA samples A–D, $\varepsilon(\omega)$ has a large plasma frequency, ω_{p1}, for electrons with localized behavior and a small plasma frequency for free electrons. The fraction of the free carriers that percolate was estimated as $\sim 10^{-2}$–10^{-3}. This value may seem small for percolation, but the volume fraction required for percolation depends upon the aspect ratio of the percolating item [118,119]. Conducting polymers have already been shown to percolate in insulating polymers at volume-filling fractions below 1% [163,167]. The experimental $\sigma(\omega)$ for PAN-CSA sample C demonstrates an upturn at low energy (~ 0.02 eV) similar to the percolating Drude peak in Figure 15.5. Therefore, both the insulating and metallic samples are consistent with the expectations for a composite system.

The presence of two electron gases, one of which is strongly localized ($\tau \sim 10^{-15}$ s) and the other has percolated, showing Drude dispersion and a long mean free time ($\tau \sim 10^{-13}$ s), can be taken into account qualitatively by introducing a distribution function for scattering times into the localization-modified Drude model. This distribution function is strongly peaked with $k_F\lambda \sim 1$, but with a small tail with larger $k_F\lambda$. Such a distribution of scattering times assumes that there are some paths in the material that are more ordered so that τ is longer, taking into account the inhomogeneous nature of a polymer solid as determined from x-ray experiments [79,82]. For such a case, we assume for convenience that $\sigma_{inhomo}(\omega)$ and $\varepsilon_{inhomo}(\omega)$ are given by

$$\sigma_{\text{inhomo}}(\omega) = \int_0^\infty P(\tau)\sigma_{\text{LMDM}}(\omega,\tau)\,d\tau \tag{15.33}$$

and

$$\varepsilon_{\text{inhomo}}(\omega) = \int_0^\infty P(\tau)\varepsilon_{\text{LMDM}}(\omega,\tau)\,d\tau \tag{15.34}$$

where $P(\tau)$ is the distribution function for the scattering times.

For analytical simplicity, we consider the representative distribution function

$$P(\tau) = \frac{2\Delta}{\pi}\left[\frac{\tau^2}{(\tau^2 - \tau_0^2)^2 + \tau^2\Delta^2}\right] \tag{15.35}$$

where Δ is the width of the spread in scattering times and τ_0 is the average scattering time. The resulting functions were fit to $\varepsilon(\omega)$ and $\sigma(\omega)$ for sample C. The plots, shown in Figure 15.32, indicate that by including a distribution that allows for carriers with long scattering times, both the localized ($\tau \sim 10^{-15}$ s) and Drude carriers ($\tau \sim 10^{-13}$ s) can be roughly fit. The parameters used were $\Omega_p = 1.0$ eV, $\tau_0 = 2.7 \times 10^{-15}$ s, $C/(k_F v_F)^2 = 11 \times 10^{-30}$ s^2, and $\Delta = 1.3 \times 10^{-15}$ s. The behavior of these model functions as the width of the distribution function is varied is shown in Figure 15.33, which shows the crossover in

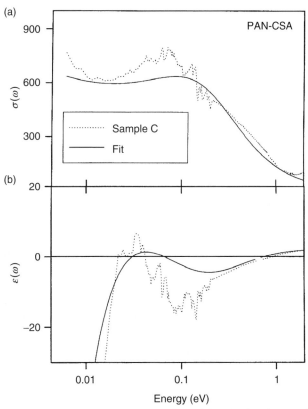

FIGURE 15.32 Typical fit of the localization-modified Drude model with a distribution of scattering times to the experimental data for PAN-CSA sample C. The plot shows (a) $\sigma(\omega)$ and (b) $\varepsilon(\omega)$. (From Kohlman, R.S. and Epstein, A.J., *Handbook of conducting polymers*, 2nd ed., eds. Skotheim, T.A., Elsenbaumer, R.L., and Reynolds, J.R., Marcel Dekker, New York, 1988, chap. 3. Reprinted from Routledge/Taylor & Francis Group, LLC. With permission.)

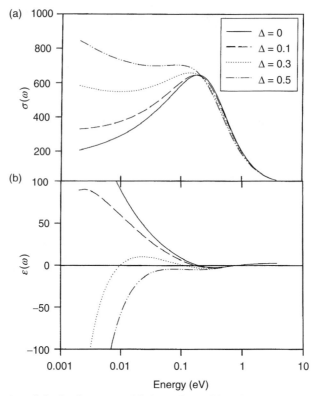

FIGURE 15.33 Behavior of the localization-modified Drude model with a spread in scattering times as the width of the distribution changes. Results are shown for (a) $\sigma(\omega)$ and (b) $\varepsilon(\omega)$. The parameters other than Δ are for PAN-CSA sample E in Table 15.3 and Table 15.5. The units for Δ are 10^{-15} s. (From Kohlman, R.S. and Epstein, A.J., *Handbook of conducting polymers*, 2nd ed., eds. Skotheim, T.A., Elsenbaumer, R.L., and Reynolds, J.R., Marcel Dekker, New York, 1988, chap. 3. Reprinted from Routledge/Taylor & Francis Group, LLC. With permission.)

behavior from insulating (with only localized carriers) to metallic (with some delocalized electrons). An increase in the width of spread of scattering times (Δ) is equivalent to having more carriers with long mean free times. In other words, an increase in Δ places the sample further above percolation so that there are more ordered paths where free electron diffusion is possible.

Qualitatively, $\sigma(\omega)$ and $\varepsilon(\omega)$ for PAN samples E–G show the behavior expected for the good conductor–poor conductor composites as well as for the 3D localization-modified Drude model with only short scattering times (corresponding to the Ioffe–Regel criterion). $\sigma(\omega)$ and $\varepsilon(\omega)$ of highly conducting PAN-CSA samples A–D show Drude behavior for a small fraction of the conduction electrons that essentially percolate through the film while the remaining conduction electrons are more localized, showing the behavior expected only within the composite picture. Therefore, $\sigma(\omega)$ and $\varepsilon(\omega)$ indicate that the metallic state in conducting PAN-CSA and PPy(PF$_6$) is inhomogeneous. This same experimental behavior has been reported for iodine- [97,98,161] and perchlorate-[162] doped polyacetylene and is shown for doped PPy in Section 15.5. Highly conducting samples of self-stabilized dispersion-polymerized PAN with temperature-independent conductivity between 5 and 70 K [47] have been reported to have Drude-like dielectric constant as a function of frequency. However, the reported frequency-dependent dielectric function data did not extend below ~0.1 eV, so comparison cannot be made to that of earlier reported highly conducting PAN-CSA dielectric data that show a complex $\varepsilon(\omega)$ at frequencies below 0.1 eV (see Figure 15.6, Figure 15.21, Figure 15.24, and Figure 15.34).

15.4.9 Microwave Frequency Dielectric Constant

The microwave frequency dielectric constant (ε_{MW}) is a key probe of the delocalization of charge carriers. For delocalized Drude electrons at frequencies less than their plasma frequency, the real part of the dielectric function $\varepsilon(\omega)$ is negative due to the inertia of the free electron in an alternating current field [40,103,130,164]. For a localized carrier, the charges can stay in phase with the field and $\varepsilon(\omega)$ is positive at low frequencies. Thus, the sign of the microwave dielectric constant serves as a sensitive probe of the presence of free electrons and provides independent verification of the IR results.

Figure 15.34 shows ε_{MW} for selected samples. For PAN-CSA samples E and F, ε_{MW} is positive from 300 K down to ~4 K, indicating the importance of localization in these materials at all temperatures. This also agrees well with the IR ε_{MW} for samples E and F, which were positive in the far-IR, characteristic of localized carriers. For samples A, C, and D, ε_{MW} is negative at room temperature, providing an independent confirmation of the free electron behavior observed in the IR. For sample A, ε_{MW} remains negative down to ~4 K, demonstrating independently of the IR data that the metallic state in conducting polymers is accompanied by free (Drude) carriers. Also, the trends observed in $\sigma_{dc}(T)$ for samples A and C are observed in $\varepsilon_{MW}(T)$. Near room temperature, ε_{MW} for sample C is metallic (negative), with an absolute value greater than that for sample A; however, at low T, sample A shows a weaker dependence on temperature. The strong temperature dependence of ε_{MW} at low T for sample C indicates that ε_{MW} will crossover to positive values at lower T, reflecting low T insulating behavior. For PAN-CSA sample D, the crossover of ε_{MW} from negative to positive values is directly observed at low T (~10 K). This provides a confirmation of the crossover observed for sample D from metallic (negative values) to insulating (positive values) in the T-dependent far-IR $\varepsilon(\omega)$ at ~150 K and reflects the progressively increasing localization of the charge carriers as the temperature is lowered. This changeover from delocalized (free) carriers to localized carriers at low temperature reflects the importance of phonon-induced delocalization in the disordered regions of the inhomogeneous metallic state [58,168,169]. For sample D, the crossover in ε_{MW} occurs at ~10 K. For other samples, this

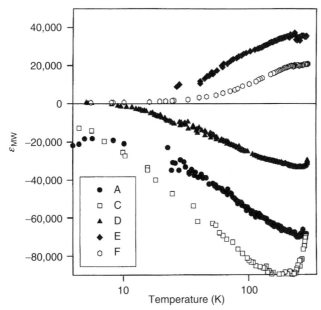

FIGURE 15.34 Temperature-dependent microwave (6.5 GHz) dielectric constant for selected PAN-CSA samples. ε_{MW} for sample C was reported in [19]. (From Kohlman, R.S. and Epstein, A.J., *Handbook of conducting polymers*, 2nd ed., eds. Skotheim, T.A., Elsenbaumer, R.L., and Reynolds, J.R., Marcel Dekker, New York, 1988, chap. 3. Reprinted from Routledge/Taylor & Francis Group, LLC. With permission.)

TABLE 15.6 Room Temperature Free Electron Plasma
Frequency ω_p and Scattering Time τ for Selected PAN-CSA
Samples Calculated from ε_{MW} and σ_{MW} Using the Drude
Model

Sample (300 K)	ω_p (eV)	T (10^{-11} s)
A	0.007	2.5
C	0.016	1.1
D	0.004	2.9

Source: From Kohlman, R.S. and Epstein, A.J. in *Handbook of
conducting polymers*, 2nd ed., eds. Skotheim, T.A., Elsenbaumer,
R.L., and Reynolds, J.R., Marcel Dekker, New York, 1988, chap.
3. Reprinted from Routledge/Taylor & Francis Group, LLC.
With permission.)

crossover has been observed at high temperatures ($\sim$20 K) [43], indicating the importance of the
specific material-processing conditions and the resulting composite network.

Using the Drude model at low frequency ($\omega\tau \ll$),

$$\sigma \sim \omega_p^2 \tau / 4\pi \tag{15.36}$$

and

$$\varepsilon \sim -\omega_p^2 \tau^2 \tag{15.37}$$

the plasma frequency Ω_p and scattering time (τ) can be estimated from $\varepsilon(\omega)$ and $\sigma(\omega)$ (not shown)
(Table 15.6). The predicted Ω_p for free electrons in samples A, C, and D is in the far-IR, in good
agreement with the observed zero crossings of $\varepsilon(\omega)$ in the far-IR, though the values are smaller than
those obtained from the Drude fits in the far-IR. The relative size of the plasma frequencies for the
different samples (ω_p [sample C] $> \omega_p$ [sample A] $> \omega_p$ [sample D]) is in good agreement with the IR
measurements. The scattering time predicted in each case is $\sim 10^{-11}$ s. The values are two orders of
magnitude larger than the τ predicted in Section 15.4.7.

The quantitative difference between the Ω_p and τ estimated from IR and microwave transport
measurements may be a reflection of the inhomogeneity of the percolating network [105]. From the
discussion of the optical conductivity and dielectric functions, a distribution of scattering times for the
conduction electrons is likely involved in the transport. Therefore, at the lowest frequencies, the carriers
that are the most delocalized (with the longest scattering times, $\tau \sim 10^{-11}$ s) may dominate the
transport. Ω_p appears smaller because a smaller fraction of the charge carriers have such a long mean
free time ($\tau \sim 10^{-11}$ s).

The trends observed in the IR are confirmed by the transport measurements at 6.5 GHz. The high σ_{dc}
observed as $T \to 0$ in metallic PAN-CSA is accompanied by the presence of free electrons down to low
temperature. In the samples where the free electron density "freezes out" at low temperature, σ_{dc} has
much stronger insulating temperature dependence.

15.4.10 Pseudoprotonic Acid Doping of Polyaniline

Similar to the doping of the EB form of PAN with protonic acids, the pseudodoping of EB with Li$^+$ ions
was reported [170]. This assignment of Li$^+$ doping is based upon the observation of defect states in the
bandgap and associated spins for EB processed with LiCl. However, the signatures of the doped phase do
not grow monotonically with increasing LiCl doping level, suggesting a unique equilibrium between the
Li$^+$ ion associating with the imine nitrogen of EB and possibly a LiCl cluster. Even at maximal doping
level, LiCl-doped EB is not a metal, but rather a strongly localized charged insulator.

15.4.11 Comparison of Doped Aniline Tetramers, Aniline Octamers, and Polyaniline

Transport and magnetic properties of a charge transfer complex of a phenyl-end-capped tetramer of aniline were reported [171]. This oligomer is an analog of the emeraldine salt form of PAN. Magnetic susceptibility, four-probe conductivity, and thermopower of BF_4^- aniline tetramer single crystals were reported together with ESR susceptibility, linewidth, and g-factor of the ClO_4^- aniline tetramer powder. Experimental results demonstrate that the bipolaron state is stabilized in the doped tetramer, in contrast to the polaron state being preferentially stabilized in the octamer and polymer (emeraldine salt) and in agreement with bipolaron stabilization in the protonated amorphous emeraldine and emeraldine salt in solution. The conductivity of the doped tetramer is thermally activated in contrast to that of the doped polymer. The cause of the stability of the bipolaron with respect to the polaron in the tetramer was suggested to be the reduced delocalization energy of these systems.

15.4.12 Chiral Metallic-Doped Polyaniline

Chiral metallic free carriers were prepared by doping PAN with chiral CSA and casting films from different solvents (*m*-cresol and chloroform) [172]. It was demonstrated that the chirality lies in the order metallic regions of the polymer films. Samples were studied by absorption, UV–Vis–IR circular dichroism, and charge transport (temperature-dependent dc and microwave frequency conductivity and dielectric constant) measurements. The metallic films (cast from *m*-cresol) show strong optical activity of the conducting electrons in the visible and IR regions, showing that a chiral metal can be prepared by the methods of organic chemistry. In contrast, the nonmetallic chloroform cast films show weaker optical activity of the polymer backbone and a completely different interaction with chiral CSA compared to *m*-cresol cast films. However, little difference in temperature-dependent dc conductivity is observed with PAN doped with chiral vs. achiral CSA. It was postulated that chiral regions of the polymer backbone segregate in the metallic regions of the polymer film upon evaporation of *m*-cresol. Secondary doping of chloroform cast films with the vapors of *m*-cresol yields an increase in electron delocalization but a decrease in optical activity of the polymer backbone, in accordance with a picture of inhomogeneous order in these materials.

15.5 Polypyrrole

The dc and high-frequency transport in PPy prepared with different dopants has also been systematically investigated. Introducing different dopants varies the disorder in the PPy system. For instance, PPy doped with PF_6 can be highly conducting down to millikelvin temperatures [41,42,75]. When the slightly larger dopant TsO is used, the conductivity is reported to be more strongly temperature-dependent [28,29], showing 3D VRH behavior. When doped not with an anion but with a polyanion (sulfated polyhydroxyether), the conductivity decreases drastically from ~ 10 S/cm at room temperature down to $\sim 10^{-10}$ S/cm at ~ 1.5 K [141], showing a stronger temperature dependence ($\sigma_{dc} \propto \exp[-(T_0/T)^{1/2}]$).

The key difference between PPy doped with PF_6 and TsO is the disorder introduced into the polymer during polymerization. When, for instance, PF_6 was chemically exchanged with TsO as the dopant for PPy in a film polymerized with PF_6, the magnitude and temperature dependence of the conductivity of the resulting film were nearly identical to those of the parent film, PPy(PF_6) [154]. The following set of experiments used the structural order of the doped PPy sample as a background to understand the difference between the metallic states in these systems.

15.5.1 X-Ray Diffraction

The x-ray diffractometer tracing for each of the films of this study is shown in Figure 15.35. X-ray studies of PPy(PF_6) indicate that these films are $\sim 50\%$ crystalline, with a structural coherence length of ~ 20 Å [82,173]. The reduction in structural disorder induced during synthesis compared with that of PPy

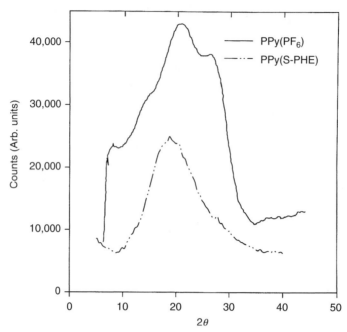

FIGURE 15.35   X-ray diffractometer tracings for PPy doped with PF_6 and S-PHE, showing the change in crystallinity of the PPy as the dopant is varied. (From Kohlman, R.S. and Epstein, A.J., *Handbook of conducting polymers*, 2nd ed., eds. Skotheim, T.A., Elsenbaumer, R.L., and Reynolds, J.R., Marcel Dekker, New York, 1988, chap. 3. Reprinted from Routledge/Taylor & Francis Group, LLC. With permission.)

materials studied earlier has been suggested to account for the improved metallic behavior of these new films [30,62,154,174,175]. The structure assigned to the crystalline portion of $PPy(PF_6)$ samples consists of two-dimensional (2D) stacks of PPy chains in the *bc* plane separated by an intervening layer of PF_6 counterions [82,173]. The PPy(TsO) for this study was ~25% crystalline, approximately half that of $PPy(PF_6)$, with a crystalline domain size that is decreased to ~15 Å [29] compared with ~20 Å. The bulky TsO dopant not only increases the separation between *bc* layers of PPy chains by 50%, but also likely enhances the conformational disorder of the PPy chain since intrachain x-ray reflections are not recorded for PPy(TsO). In contrast, the PPy(S-PHE) films show only a single broad peak reflecting disordered chains [136,141]. The fact that the maximum average *d* spacing for the disordered chains in PPy(S-PHE) is different from those of $PPy(PF_6)$ and PPy(TsO) may reflect the average stacking of the S-PHE polyanions, which account for the majority of the volume in this material. The differences in crystallinity allow us to probe the effects of structural disorder on the metallic state. For electrochemically prepared films up to 20% of the pyrrole rings are involved in cross-linking with neighboring chains [174,175].

15.5.2 Magnetic Susceptibility

The spin susceptibility was measured from room temperature down to 4 K for PPy doped with PF_6, TsO, and S-PHE. The temperature dependence provides insight into the nature of the spin states. A simple model for the total susceptibility is

$$\chi = \chi^{\text{Pauli}} + \chi^{\text{Curie}} \qquad (15.38)$$

where $\chi^{\text{Pauli}} [= 2\mu_B^2 N(E_F)]$ is temperature-independent and χ^{Curie} is proportional to T^{-1} [40]. Figure 15.36 shows χT vs. *T* for each of the samples. The solid lines are fits to $\chi T = \chi_p T + C$, where χ_p is the temperature-independent Pauli susceptibility and *C* is the Curie constant.

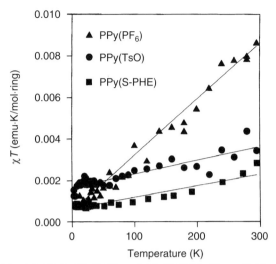

FIGURE 15.36 χT vs. T for PPy(PF$_6$), PPy(TsO), and PPy(S-PHE). (From Kohlman, R.S. and Epstein, A.J., *Handbook of conducting polymers*, 2nd ed., eds. Skotheim, T.A., Elsenbaumer, R.L., and Reynolds, J.R., Marcel Dekker, New York, 1988, chap. 3. Reprinted from Routledge/Taylor & Francis Group, LLC. With permission.)

The positive slopes in Figure 15.36 indicate a finite $N(E_F)$ for each of the materials. $N(E_F) \approx 0.80$ state/(eV · ring) for PPy(PF$_6$); $N(E_F) \approx 0.20$ state/(eV · ring) for PPy(TsO); and $N(E_F) \approx 0.17$ state/(eV · ring) for PPy(S-PHE). This value is comparable with the $N(E_F)$ measure for other highly conducting polymers [3]. For doped polyacetylene [39], $N(E_F) \approx 0.18$ state/(eV · C), and in PAN-CSA and PAN-AMPSA, $N(E_F) \approx 0.1$ state/[eV · (C + N)] [66,176]. The positive intercept corresponds to a finite number of uncorrelated Curie spins. PPy(TsO) ($C = 1.7 \times 10^{-3}$ emu·K/(mol·ring) $\sim$ 1 spin/200 PPy rings) and PPy(S-PHE) ($C = 6.3 \times 10^{-4}$ emu·K/(mol·ring) $\sim$ 1 spin/500 PPy rings) show a larger Curie component than PPy(PF$_6$) ($C = 3.6 \times 10^{-4}$ emu·K/(mol·ring) $\sim$ 1 spin/10^3 PPy rings). The fact that PPy(TsO) has a larger number of Curie spins than PPy(S-PHE) per chain may reflect that bipolarons are preferentially formed in the more disordered chains of PPy(S-PHE).

The larger $N(E_F)$ and smaller density of uncorrelated spins for PPy(PF$_6$) correlates with the larger fraction of the sample having 3D order [173], i.e., crystallinity stabilizes a metallic density of states. Similar to doped PAN, where the metallic Pauli susceptibility is associated with the 3D-ordered regions [14,63], a large finite $N(E_F)$ is present in PPy when there are large 3D-ordered (crystalline) regions. In more disordered materials, localized polarons or bipolarons predominate.

15.5.3 dc Conductivity

The reduced activation energy [101] ($W \equiv d \ln \sigma / d \ln T$) vs. T is shown in Figure 15.37 for PPy(PF$_6$), PPy(TsO), and PPy(S-PHE). For PPy(PF$_6$), the positive slope of W with increasing T implies that it is in the metallic regime. Millikelvin dc conductivity measurements on the same PPy(PF$_6$) materials show metallic behavior [41,42] and a negative magnetoresistance [75], confirming that PPy(PF$_6$) has crossed the IMT and is metallic at low T. In contrast, PPy(TsO) has a negative slope for its W plot, placing it in the localized regime. The slope of the line is $\sim$0.25, indicating 3D VRH. For PPy(S-PHE), the W plot also has a negative slope ($\sim$0.5) at low temperature, characteristic of quasi-1D VRH. Therefore, with increased disorder and chain isolation, doped PPy becomes more insulating, with charge transport that becomes more characteristic of reduced dimensionality hopping, just as for the PAN-CSA samples.

The same reliance on structural order is seen for σ_{dc} in Figure 15.38. For PPy(PF$_6$) σ_{dc} very slowly decreases as temperature decreases from room temperature ($\sigma_{dc} \sim$ 300 S/cm) to $\sim$20 K, and $d\sigma/dT < 0$ for $T < 20$ K. The increase in σ_{dc} with decreasing T at low temperature was reported previously

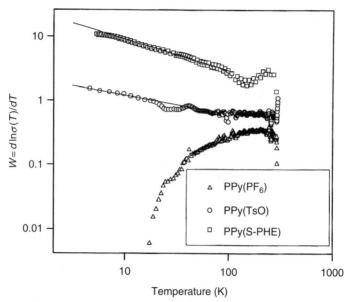

FIGURE 15.37   *W* plot for PPy doped with PF$_6$, TsO, and S-PHE. (From Kohlman, R.S. and Epstein, A.J., *Handbook of conducting polymers*, 2nd ed., eds. Skotheim, T.A., Elsenbaumer, R.L., and Reynolds, J.R., Marcel Dekker, New York, 1988, chap. 3. Reprinted from Routledge/Taylor & Francis Group, LLC. With permission.)

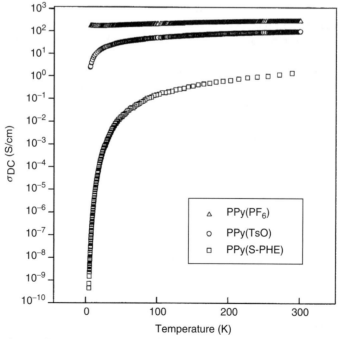

FIGURE 15.38 σ_{dc} for PPy doped with PF$_6$, TsO, and S-PHE. (From Kohlman, R.S. and Epstein, A.J., *Handbook of conducting polymers*, 2nd ed., eds. Skotheim, T.A., Elsenbaumer, R.L., and Reynolds, J.R., Marcel Dekker, New York, 1988, chap. 3. Reprinted from Routledge/Taylor & Francis Group, LLC. With permission.)

[28,30,41,42,152,153] and attributed to a tunneling mechanism [152]. The ratio of σ (RT)/σ (20 K) is <1.8, relatively small in comparison with that of other highly conducting polymers.

For more disordered PPy(TsO), $\sigma_{dc}(T)$ shows a stronger temperature dependence than PPy(PF$_6$), consistent with a 3D VRH model; $\sigma_{dc}(T) = \sigma_0 \exp[-(T_0 / T)^{1/4}]$, where $T_0 = 16/k_B N(E_F)L^3$ (k_B is the Boltzmann constant). Using the slope of ln σ vs. $T^{-1/4}$, $T_0 \approx 4100$ K, and $N(E_F) = 0.2$ state/(eV·ring) determined from the magnetic studies, the L_{loc} is estimated as $L \sim 30$ Å. This estimate of L_{loc} is comparable with the x-ray crystalline coherence length ξ.

For PPy(S-PHE), which has the least structural order of the doped PPy samples studied here, $\sigma_{dc}(T)$ shows the highest degree of localization, with a large resistivity ratio (ρ (4 K)/ρ (300 K) $\sim 10^{10}$). The temperature dependence is that of quasi-1D VRH.

15.5.4 Reflectance

Similar to the results for highly conducting PAN, the reflectance of PPy doped with PF$_6$ and TsO, shown in Figure 15.39, shows Drude-like behavior, increasing monotonically with decreasing energy from the conduction electron plasma edge at $\sim$2.1 eV for PPy(PF$_6$) and $\sim$1.9 eV for PPy(TsO) and approaching unity at low energy [29]. The oscillations in the mid-IR ($\sim$0.1 eV) are due to the phonons of doped PPy. The S-PHE-doped PPy samples have a more modest reflectance in the far-IR, only approaching 40% near 0.006 eV[136]. Unlike PAN, there is not a separate peak at higher energy ($\sim$1.5 eV) that develops with decreasing conductivity.

The reflectance for PPy(PF$_6$) measured at 10 K (2 meV–1.2 eV) is compared with its room temperature reflectance in Figure 15.40. There is very little difference at high frequencies (>0.1 eV). In the far-IR, the 10 K reflectance is higher than the room temperature reflectance down to $\sim$0.01 eV and then drops beneath the room temperature reflectance. The decrease of the reflectance measured in the far-IR is consistent with the drop in σ_{dc} as T decreases. The region where the 10 K reflectance rises above the room temperature reflectance is due to free carriers that are localized at low temperature, similar to what

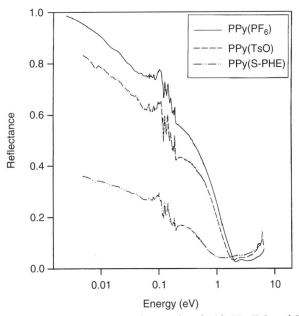

FIGURE 15.39 Room temperature reflection spectra for PPy doped with PF$_6$, TsO, and S-PHE. (From Kohlman, R.S. and Epstein, A.J., *Handbook of conducting polymers*, 2nd ed., eds. Skotheim, T.A., Elsenbaumer, R.L., and Reynolds, J.R., Marcel Dekker, New York, 1988, chap. 3. Reprinted from Routledge/Taylor & Francis Group, LLC. With permission.)

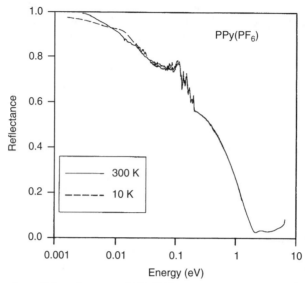

FIGURE 15.40   Comparison of the reflectance of PPy(PF$_6$) at room temperature and 10 K. Note the weak T dependence at high energy and the strong T dependence in the far-infrared. (From Kohlman, R.S. and Epstein, A.J., *Handbook of conducting polymers*, 2nd ed., eds. Skotheim, T.A., Elsenbaumer, R.L., and Reynolds, J.R., Marcel Dekker, New York, 1988, chap. 3. Reprinted from Routledge/Taylor & Francis Group, LLC. With permission.)

has been reported for polyacetylene [161] and PAN earlier in this chapter. This decrease in the reflectance in the far-IR is inconsistent with the opening of an intrinsic gap in the electronic spectrum of PPy(PF$_6$) as reported earlier [163].

15.5.5   Absorption Coefficient

The absorption coefficients (α) for PPy doped with PF$_6$, TsO, and S-PHE obtained by Kramers–Kronig analysis of the reflectance data are shown in Figure 15.41. For each of the samples, α is similar in character at low energy. There are peaks at $\sim$1.0 and 2.8 eV. The absorption at higher energy depends upon the counterion that was used. Aside from the presence of a peak for doped PAN at 1.5 eV (attributed to isolated polarons [76,177–179]), the absorption coefficients for PAN-CSA [76] and for PPy doped with PF$_6$, TsO, and S-PHE are very similar, implying that highly conducting PPy forms a polaron lattice like doped PAN instead of a bipolaron lattice [180]. In contrast, a peak at $\sim$1.5 eV was observed in lightly doped poorly conducting PPy [165].

The absorption coefficient for PPy(PF$_6$) at 10 K is compared with the room temperature α in Figure 15.42. There is very little change in the absorption coefficient at high energy as the temperature is lowered. In the far-IR, a peak at $\sim$0.01 eV is formed due to absorption by carriers that are delocalized at room temperature but localized at 10 K. This behavior is very similar to what has been reported for iodine-doped polyacetylene with decreasing temperature [161]. The lack of change of the absorption at high energy indicates that the localized electrons do not contribute significantly to the high σ_{dc}; the temperature dependence of their conductivity contribution is different from that of the highly conducting carriers at zero frequency.

15.5.6   Optical Conductivity

Similar to the results for PAN, $\sigma(\omega)$ for doped PPy (Figure 15.43) does not show Drude-like conductivity at low energies. Instead, $\sigma(\omega)$ shows Drude dispersion at high energy ($\sim$0.4–2 eV), a maximum at ω_{max}, and then suppression at low energy. The lack of structural order in PPy(S-PHE) results in a much

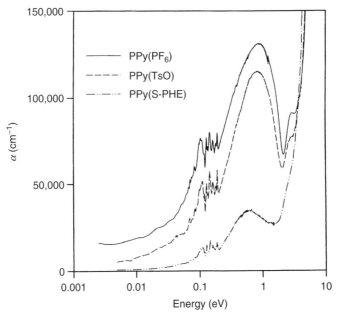

FIGURE 15.41 Room temperature absorption coefficient for PPy doped with PF_6, TsO, and S-PHE. (From Kohlman, R.S. and Epstein, A.J., *Handbook of conducting polymers*, 2nd ed., eds. Skotheim, T.A., Elsenbaumer, R.L., and Reynolds, J.R., Marcel Dekker, New York, 1988, chap. 3. Reprinted from Routledge/Taylor & Francis Group, LLC. With permission.)

greater suppression in $\sigma(\omega)$. The ratio $\sigma(6\text{ meV})/\sigma(\omega_{max})$ gives an indication of the localization. For $PPy(PF_6)$, this ratio is $\sim$0.9; for PPy(TsO), $\sim$0.3; and for the PPy doped with S-PHE, $\sim$0.05. The fact that $\omega_{max} \sim 0.3$ eV for $PPy(PF_6)$ is smaller than $\omega_{max} \sim 0.4$ eV for PPy(TsO) is consistent with the increased disorder in PPy(TsO), resulting in a smaller mean free path. The values obtained from a Drude fit to the high-energy region of each sample are given in Table 15.7.

The plasma frequency for conduction electrons is higher in PPy doped with PF_6, TsO, and S-PHE than in PAN-CSA. For $PPy(PF_6)$, a three-pyrrole-ring repeat and dopant has dimensions $a = 13.4$ Å, $b = 10.95$ Å, and $c = 3.4$ Å [173]. For PPy(TsO), the unit cell dimensions are the same except that $a = 18.0$ Å [29] as the TsO dopant is larger. In heavily doped $PPy(PF_6)$ and PPy(TsO), there is one dopant unit added per three rings [28–30,41,42]; therefore, the doping density for $PPy(PF_6)$ is $n \sim 2 \times 10^{21}$ cm^{-3} and for PPy(TsO), $n \sim 1.5 \times 10^{21}$ cm^{-3}. For PPy(S-PHE), the carrier density can be approximated from the density (1.3 g/cm^3) [141] as $n \sim 2.9 \times 10^{20}$ cm^{-3}. From these conduction electron densities and the $\Omega_{p1}\left(=\sqrt{4\pi ne^2/m*}\right)$ determined from Drude fits, $m^* \sim 0.5 m_e$, for $PPy(PF_6)$, $m^* \sim 0.5$ m_e, for PPy(TsO), and $m^* \sim 0.2$ m_e, for PPy(S-PHE). Therefore, the effective masses (m^*) for doped PPy are lower than for PAN-CSA, implying that the overlap integrals are larger.

The scattering times determined from Drude fits (from 1.4 to $\sim$2.4 eV) for the conduction electrons are small, $\tau_1 \sim 10^{-15}$ s. This small value is comparable to though smaller than what was found for τ_1, in PAN-CSA. The smaller values of τ for PPy compared with PAN may reflect greater disorder present in the disordered regions of PPy [82]. With increasing structural order within the PPy samples, τ increases, implying a larger mean free path for carriers in $PPy(PF_6)$, in agreement with the behavior of ω_{max}.

15.5.7 Optical Dielectric Function

The percolation behavior seen in $\varepsilon(\omega)$ for doped PAN is also observed for doped PPy. For metallic $PPy(PF_6)$, $\varepsilon(\omega)$ shows three zero crossings (Figure 15.44a). The dispersion of the localized electrons at high energy (>0.06 eV) and the Drude dispersion in the far-IR are both evident. For PPy(TsO), the

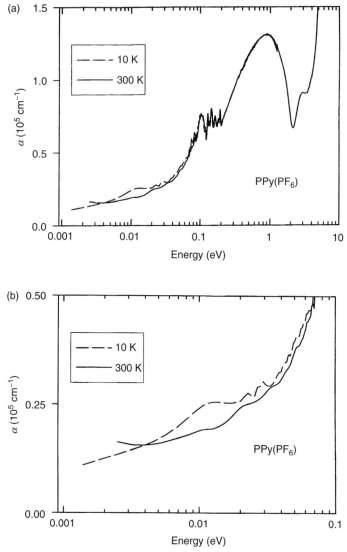

FIGURE 15.42 Comparison of the absorption coefficient of PPy(PF$_6$) at room temperature and 10 K (a) for the full optical range and (b) in the far-infrared. (From Kohlman, R.S. and Epstein, A.J., *Handbook of conducting polymers*, 2nd ed., eds. Skotheim, T.A., Elsenbaumer, R.L., and Reynolds, J.R., Marcel Dekker, New York, 1988, chap. 3. Reprinted from Routledge/Taylor & Francis Group, LLC. With permission.)

scattering time is reduced sufficiently by disorder that ε does not cross zero within the entire experimental frequency range. For PPy(S-PHE), the dispersion is very weak throughout the optical frequency range. The plot of $\varepsilon(\omega)$ vs. $1/\omega^2$ for PPy(PF$_6$) is shown in Figure 15.44b. The slope of the curve gives $\Omega_p \sim 0.17$ eV. Compared with the plasma frequency of the full conduction electron gas (~ 2.5 eV), only a small fraction of the conduction electrons are free. Estimating in the same manner as for PAN-CSA (Equation 15.28),

$$n_{\text{free}} \sim (0.17/2.5)^2 \sim 5 \times 10^{-3} \tag{15.39}$$

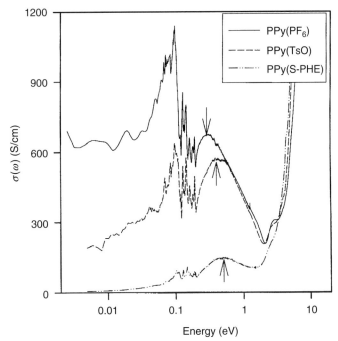

FIGURE 15.43 Room temperature optical conductivity for PPy doped with PF$_6$, TsO, and S-PHE. $\sigma(\omega)$ for disordered PPy(S-PHE) is suppressed strongly in the far-infrared. (From Kohlman, R.S. and Epstein, A.J., *Handbook of conducting polymers*, 2nd ed., eds. Skotheim, T.A., Elsenbaumer, R.L., and Reynolds, J.R., Marcel Dekker, New York, 1988, chap. 3. Reprinted from Routledge/Taylor & Francis Group, LLC. With permission.)

As discussed for PAN-CSA, the estimate of the fraction of delocalized electrons is dependent upon the ratio of m^* in the disordered regions to m^* in the ordered regions. Due to the linearity of $\varepsilon(\omega)$ vs. $1/\omega^2$ down to 20 cm^{-1}, $\tau > {\sim}1/20$ cm$^{-1} \sim 3 \times 10^{-13}$ s. Similar to the metallic state in PAN-CSA, a large majority of conduction electrons are localized ($\tau_1 \sim 10^{-15}$ s) and a small fraction of the conduction electrons are delocalized ($\tau \geq {\sim}10^{-13}$ s) in metallic PPy(PF$_6$). With this long τ, a mean free path comparable with that of delocalized carriers in PAN-CSA is estimated. For more disordered PPy(TsO) and PPy(S-PHE), the electrical response is determined by only localized electrons. Similar to the behavior of $\varepsilon(\omega)$ in PAN-CSA, $\varepsilon(\omega)$ is larger in the far-IR for more ordered, higher conductivity PPy(TsO) than for PPy(S-PHE).

TABLE 15.7 Parameters from Drude Fit to High Energy (0.4–2 eV) $\sigma(\omega)$ for PPy Samples with Different Dopants

Sample	Ω_{p1} (eV)	τ_1 (10^{-15} s)
PPy (PF$_6$)	2.5	0.53
PPy (TsO)	2.5	0.51
PPy (S-PHE)	1.5	0.43

Source: From Kohlman, R.S. and Epstein, A.J. in *Handbook of conducting polymers*, 2nd ed., eds. Skotheim, T.A., Elsenbaumer, R.L., and Reynolds, J.R., Marcel Dekker, New York, 1988, chap. 3. Reprinted from Routledge/Taylor & Francis Group, LLC. With permission.)

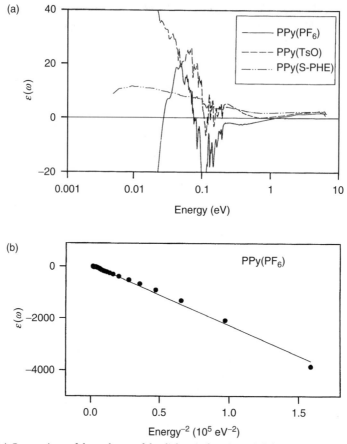

FIGURE 15.44   (a) Comparison of the real part of the dielectric function $\varepsilon(\omega)$ for PPy doped with PF_6, TsO, and S-PHE. (b) Far-infrared $\varepsilon(\omega)$ for $PPy(PF_6)$. (From Kohlman, R.S. and Epstein, A.J., *Handbook of conducting polymers*, 2nd ed., eds. Skotheim, T.A., Elsenbaumer, R.L., and Reynolds, J.R., Marcel Dekker, New York, 1988, chap. 3. Reprinted from Routledge/Taylor & Francis Group, LLC. With permission.)

For $PPy(PF_6)$, $\varepsilon(\omega)$ at room temperature is compared with $\varepsilon(\omega)$ at ~10 K as shown in Figure 15.45. Two important points are observed. At high energy, there is very little difference in the dispersion of the localized carriers. $\varepsilon(\omega)$ remains negative to slightly lower frequency, indicating that the scattering time changes only slightly as the temperature is lowered. This lack of a strong dependence of the scattering time on thermal processes indicates that the (static) disorder is the predominant scattering mechanism for the charge carriers. Similar results were reported for PAN doped with CSA[133]. The slight increase in the scattering time for localized electrons while σ_{dc} decreases implies that there is not a direct correlation of the localized carriers and the dc transport.

The temperature dependence of $\varepsilon(\omega)$ in the far-IR is shown in Figure 15.45b. In contrast with the weak temperature dependence at high energy, the Drude dispersion is more strongly affected. Down to 10 K, the dielectric function remains negative in the far-IR, indicating that free carriers are present down to low T in $PPy(PF_6)$, consistent with the high millikelvin σ_{dc} measured for this material [41,42] and also with the behavior of metallic PAN-CSA. At 10 K, the plasma frequency for free electrons (Ω_p) is 0.1 eV, decreased from 0.17 eV at room temperature. In comparison with σ_{dc}, which decreases by ~1.8, ω_p decreases by ~1.7. The scaling of Ω_p with σ_{dc} indicates that free carriers dominate the high dc conductivity for doped PPy and PAN. The free carriers that are frozen out at low temperature give rise to the additional absorption at ~10 meV seen in the absorption coefficient (Figure 15.41). As for

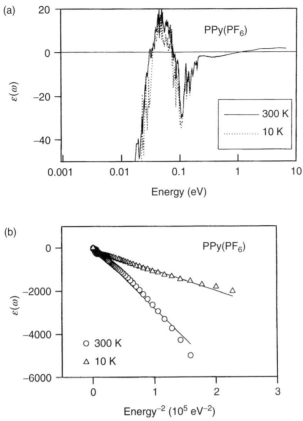

FIGURE 15.45   Comparison of the dielectric response of PPy(PF$_6$) at room temperature and 10 K. (a) $\varepsilon(\omega)$ in the range of localized carriers. (b) $\varepsilon(\omega)$ vs. $1/\omega^2$ in the far-infrared. (From Kohlman, R.S. and Epstein, A.J., *Handbook of conducting polymers*, 2nd ed., eds. Skotheim, T.A., Elsenbaumer, R.L., and Reynolds, J.R., Marcel Dekker, New York, 1988, chap. 3. Reprinted from Routledge/Taylor & Francis Group, LLC. With permission.)

PAN-CSA, the presence of a negative far-IR $\varepsilon(\omega)$ at room temperature excludes the possibility of an intrinsic many-body gap [121].

15.5.8   Microwave Transport

Microwave measurements for PPy samples provide independent confirmation of the results reported at IR frequencies. Figure 15.46 compares $\sigma_{dc}(T)$ with $\sigma_{MW}(T)$ for PPy(PF$_6$) [29], PPy(TsO) [29], and PPy(S-PHE) [136]. The absolute values and the temperature dependence of σ in the dc and microwave frequency ranges for PPy(PF$_6$) are nearly identical, in agreement with the Drude theory. The stronger temperature dependence of σ_{dc} for PPy(TsO) and PPy(S-PHE) in comparison with $\sigma_{MW}(T)$ is expected because the sample is in the localized regime [181].

Figure 15.47 shows $\varepsilon_{MW}(T)$ for PPy(PF$_6$) [29]. ε_{MW} at 265 K is huge and negative, $\sim -10^5$, corresponding to the Drude dielectric response at microwave frequencies. ε_{MW} remains huge and negative down to 4.2 K ($\sim -4 \times 10^4$). The absolute value of ε_{MW} has a very weak T dependence and a maximum at ~ 20 K. This behavior qualitatively agrees with $\sigma_{dc}(T)$ and $\sigma_{MW}(T)$. Using the Drude model in the low frequency limit ($\omega\tau \ll 1$), $\sigma_{MW} \simeq (\omega_p^2/4\pi)\tau$ and $\varepsilon_{MW} \simeq -\omega_p^2\tau^2$. Hence, ω_p and τ ($\simeq 4\pi\varepsilon_{MW}/\sigma_{MW}$) are ~ 0.007 eV and $\sim 3 \times 10^{-11}$ s, respectively. The prediction of a small plasma frequency of free electrons with a zero crossing in the far-IR agrees well with the IR measurements, providing an independent confirmation of the Drude carriers. Again, as for PAN-CSA, the plasma

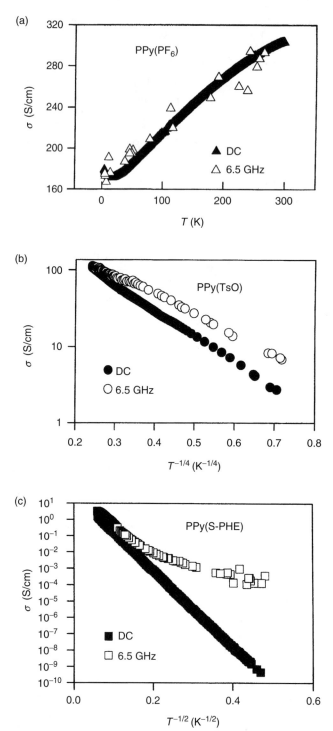

FIGURE 15.46   Comparison of the dc and microwave conductivity for (a) PPy(PF$_6$), (b) PPy(TsO), and (c) PPy(S-PHE). (From Kohlman, R.S. and Epstein, A.J., *Handbook of conducting polymers*, 2nd ed., eds. Skotheim, T.A., Elsenbaumer, R.L., and Reynolds, J.R., Marcel Dekker, New York, 1988, chap. 3. Reprinted from Routledge/Taylor & Francis Group, LLC. With permission.)

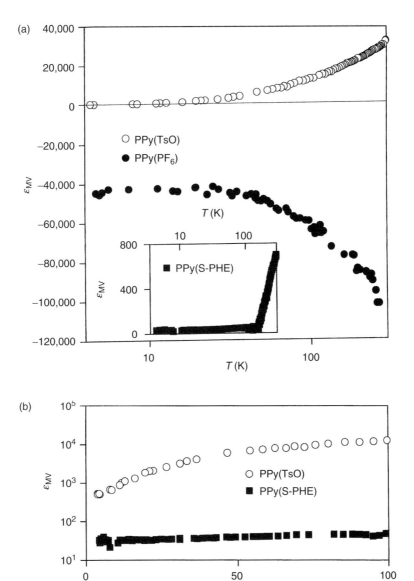

FIGURE 15.47 (a) Microwave (6.5 GHz) dielectric constant for metallic PPy(PF$_6$) and insulating PPy(TsO) and PPy(S-PHE). (b) Low T dielectric functions for PPy(TsO) and PPy(S-PHE) showing their saturation. (From Kohlman, R.S. and Epstein, A.J., *Handbook of conducting polymers*, 2nd ed., eds. Skotheim, T.A., Elsenbaumer, R.L., and Reynolds, J.R., Marcel Dekker, New York, 1988, chap. 3. Reprinted from Routledge/Taylor & Francis Group, LLC. With permission.)

frequency for free carriers is small and the scattering time is large compared with the IR estimates. Therefore, a distribution of scattering times is important for highly conducting PPy samples as well.

Although its room temperature σ_{MW} is one-third that of PPy(PF$_6$), ε_{MW} for more disordered PPy(TsO) is positive in the entire temperature range [29]. The $T \to 0$ L_{loc} can be estimated as ~25 Å using the metallic box model [90,92] and the ESR $N(E_F)$ in agreement with the crystalline coherence length ($\xi \sim 15$ Å). ε_{MW} linearly increases as temperature increases even though the L_{loc} is small, which implies that the charge can easily delocalize through the disordered regions, and the phase segregation between the metallic and disordered regions is weak in the PPy(TsO) sample.

For PPy(S-PHE) also, ε_{MW} is positive over the entire temperature range and smaller in absolute value than ε_{MW} for PPy(TsO), consistent with stronger localization of charges in the PPy(S-PHE) sample [136]. For PPy(S-PHE), ε_{MW} does not change much from its low temperature saturation value until the temperature increases above ~150 K, at which point it increases rapidly at higher temperatures. This indicates that there may be an energy barrier to hopping below 150 K. Using the metallic box model for this sample and the ESR $N(E_F)$, the $T \rightarrow 0$ L_{loc} is estimated as ~7 Å, comparable with two pyrrole rings and smaller than that of PPy(TsO), consistent again with stronger localization.

15.6 Polyacetylene

Similar behavior for highly conducting doped polyacetylene samples has been reported, including a finite $N(E_F)$ [3,17,39,61,65], σ_{dc} finite as $T \rightarrow 0$ [41,42], large and negative ε_{MW} [3,39,59]. There is a rapid increase in the negative value of ε in the far-IR (Figure 15.48 and Figure 15.49) [97,182] as expected for a Drude metallic response. Further, samples for which $\sigma_{dc} \rightarrow 0$ as $T \rightarrow 0$ generally have positive ε_{MW} with the magnitude related to sample conductivity [183] and a positive ε_{MW} in the IR [98,162]. Transport in the poorly conducting regime for doped polyacetylene has been thoroughly reviewed [6]. Models very similar to those applied to doped PAN and PPy can be applied to doped polyacetylene [3,39]. Stretching of polyacetylene leads to orientation of the polyacetylene chains. However, the induced order is not uniform, with the center portion having the highest orientation and largest negative microwave frequency dielectric constant with regions near the clamps having small negative or even positive dielectric constants [184].

15.7 Electric-Field-Controlled Conductance of Metallic Polymers in a Transistor Structure

Use of a doped metallic polymer as the active channel in a field effect transistor structure results in unexpected "normally on" transistor-like behavior. The importance was shown for ion migration

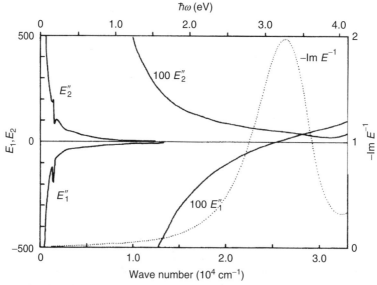

FIGURE 15.48 Real and imaginary parts of the dielectric functions (solid lines) and loss function Im$(-1/\varepsilon)$ (dotted line) from Kramers–Kronig analyses of the reflectance spectra for light polarized parallel to the chain direction for highly oriented AsF$_5$ doped *trans*-polyacetylene. (From Leising, G., *Phys. Rev. B*, 38, 10313, 1988. Reprinted from the American Physical Society. With permission.)

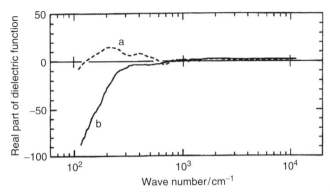

FIGURE 15.49 The frequency-dependent dielectric function for highly oriented polyacetylene doped with perchlorate from Kramers–Kronig analyses of the reflectance spectra. (a) Light polarized perpendicular to polyacetylene chain direction. (b) Light polarized parallel to polyacetylene chain directions. (From Miyamae, T., Shimizu, M., and Tanaka, J., *Bull. Chem. Soc. Jpn.*, 67, 2407, 1994. Reprinted from the Chemical Society of Japan. With permission.)

between the dielectric layer and the active channel for transistors based on "metallic" poly(3,4-ethylenedioxythiophene), doped with poly(styrene sulfonic acid) [(PEDOT:PSS)] [56, 107–112], conducting polypyrrole doped with Cl⁻, and conducting emeraldine salt coped with Cl⁻ [56, 187–189]

The analysis of the ion flow into the active channel due to the gate voltage suggests that for the transistors studied compensating ∼2 counterions per 100 dopant molecules suppresses PEDOT conductance up to three orders of magnitude [107–112]. It was determined from the temperature dependence of the conductivity of the active channel that the change in conductance of the active channel is throughout the channel in contrast to confinement of gate field induced charge carriers to the channel/dielectric interface in conventional transistors. The decrease of channel conductance reflects an increase of activation energy of carriers. It was proposed [112] that removal of intermediate hopping sites in the poorly conducting disordered regions of the polymers by a small fraction of ion charge compensation causes carriers to anisotropically hop over longer distances leading to a conductor–nonconductor transition.

15.8 Discussions and Conclusions

The experimental data for doped PAN, PPy, and polyacetylene show a wide range of electrical behavior, varying from localized charge carrier hopping behavior for all charge carriers to free Drude carrier diffusion for a small fraction of the carriers while the remaining carriers are localized. Regardless of the family of conducting polymers, the metallic state possesses certain universal properties, including a finite density of states at the Fermi level, a high σ_{dc} as $T \to 0$, and a dielectric function that becomes negative in the IR and remains negative to the microwave frequency range, as expected for Drude carriers. The IR-active vibrations weaken for the more highly conducting polymers due to increased screening. Conducting polymers exist that range from insulating to metallic and span the IMT at low temperature.

Through analyses of the high-frequency dielectric and conductivity functions, new insights have been obtained concerning the metallic state in conducting polymers. For highly conducting metallic-doped PAN and doped PPy, with decreasing energy, there are two negative-going zero crossings for the optical dielectric frequency. The dielectric frequency response for highly conducting doped polyacetylene is very similar, with a high energy zero crossing at ∼3 eV and a very rapid increase in the absolute value of the dielectric function in the far-IR. These two electrical responses correspond to the plasma response of delocalized (Drude) carriers at low frequencies (far-IR, ∼0.05 eV) and localized carriers at high frequencies (∼1–3 eV). The delocalized carriers have an unusually long scattering time, $\tau \sim 10^{-13}$ s that is attributed to the intrinsic anisotropy of the Fermi surface for quasi-1D conducting polymers and to the fact that most off-chain impurities are effectively screened. However, the fraction of the charge carrier density that is delocalized is estimated to be <1%. Analyses of the temperature dependence of the dc conductivity and dielectric function indicate that σ_{dc} is high at low T when $\varepsilon(\omega)$ has free carrier behavior at low T. For materials for which the free carrier behavior in $\varepsilon(\omega)$ freezes out at low T, σ_{dc} decreases rapidly at low T. This suggests that the high σ_{dc} in metallic conducting polymers is controlled by only the small fraction of delocalized carriers. This indicates that a large potential increase in σ_{dc} (up to $\sim 10^5$ S/cm) can be obtained if the entire charge carrier density has $\tau \sim 10^{-13}$ s. The conductivity for the metallic state involving all conduction electrons may be smaller than 10^5 S/cm as τ may be smaller than 10^{-13} s due to increased effects of 3D scattering due to easier motion between chains.

The zero crossing (ω_{p1}) at higher energy (1–3 eV) is attributed to the majority of carriers that are localized with short mean free times ($\tau \sim 10^{-15}$ s). For conducting polymers whose dielectric response is dominated by only localized carriers, $\varepsilon(\omega)$ is positive in the far-IR and σ_{dc}, has a strong temperature dependence, and becomes insulating at low T.

The IMT was explicitly controlled by disorder in PPy through structural studies and in PAN by varying the local chain conformation by casting PAN-CSA from different solvents. Although the IMT is controlled by disorder, it is not a homogeneous 3D Anderson transition. This is asserted because

1. PAN-CSA samples with conductivity (σ_{dc}) higher than the minimum metallic conductivity (σ_{min}) may become insulating at low temperature.
2. The evolution of $\sigma_{dc}(T)$ through the IMT is not monotonic.
3. Millikelvin σ_{dc} for selected insulating samples is not consistent with hopping transport.
4. The density of free electrons present in a sample scales with $\sigma_{dc}(T)$, so a sample that demonstrates metallic behavior at low temperature may have a smaller density of free electrons at room temperature than a sample that demonstrates insulating behavior at low temperature due to the nonmonotonic evolution of $\sigma_{dc}(T)$ through the IMT.
5. $\varepsilon(\omega)$ and $\sigma(\omega)$ for metallic samples are consistent with macroscopically inhomogeneous models, but not Anderson localization models.
6. A long scattering time ($\tau \geq \sim 10^{-13}$ s) inconsistent with the Ioffe–Regel condition for homogeneous 3D Anderson localization of carriers close to the IMT is determined.
7. Doped oriented films of PAN and polyacetylene demonstrate a known strong anisotropy along the chain direction, implying a strong influence of dimensionality on the entire class of polymers that is not accounted for in the Anderson IMT model.
8. Intergrain resonance quantum tunneling occurring through the strongly localized states in the amorphous media account for the anomalous frequency and temperature dependence of the conductivity and dielectric constant of highly conducting doped PAN and PPy samples with inhomogeneous disorder.
9. It has been shown that the large electric field effect is related to the inhomogeneous structure of the doped conducting polymers and the migration of a small fraction of mobile ions from the dielectric layer into the less ordered regions of the conducting polymer under application of gate voltage.

Therefore, it is suggested that the metallic state and IMT are not controlled by homogeneous disorder.

Instead, the metallic state and IMT are proposed to be controlled by inhomogeneous disorder, consistent with the percolation (composite) behavior evident in $\varepsilon(\omega)$ and $\sigma(\omega)$ at high frequencies, x-ray diffraction results, and the growth of a Pauli component of the magnetic susceptibility with increasing crystallinity. The composite nature results from the distinct transport among ordered metallic regions and the surrounding disordered regions where localization is strong. The IMT occurs when L_{loc} in the disordered regions exceeds the distance between ordered metallic regions so that carriers effectively percolate among metallic islands through the ordered paths. A small number of percolated paths could account for the small fraction of delocalized carriers. For metallic samples, a long mean free path up to $\sim 10^3$ Å (or perhaps longer if the microwave τ is used to estimate λ) is estimated (assuming that m^* is the same in the ordered and disordered regions), much larger than the crystalline domain sizes ($\sim 10^1$ Å). Such a long mean free path is consistent with percolation among many ordered regions.

For systems where the carriers have not percolated among the ordered regions, the transport is controlled by hopping and phonon-induced delocalization in the intermittent disordered regions. It is suggested that for materials with more rodlike chain conformation in the disordered regions, free carriers are present at room temperature that may freeze out at low T, causing a temperature-dependent IMT. For systems with more coil-like local order in the disordered regions, free carriers are not observed and transport is due to hopping among localized states, which contributes less than ~ 10 S/cm to σ_{dc}.

An explanation based on the chain morphology of polymers is very important for the low-frequency anomaly in the dielectric constant of doped polymers in the metallic phase. These materials are strongly inhomogeneous [82] with crystalline regions within which polymer chains are well ordered (Figure 15.3a). When the IMT is approached, delocalization first occurs inside these regions. Outside the crystalline regions the chain order is poor and the electronic wavefunctions are strongly localized. Therefore, the crystalline domains can be considered as nanoscale metallic dots embedded in amorphous poorly conducting medium [104,185,186]. The metallic grains remain always spatially separated by amorphous regions; therefore, direct tunneling between grains is exponentially suppressed. The inter-grain tunneling is possible through intermediate localized states in the disordered portion with strong contribution from resonance states whose energy is close to the Fermi level (Figure 15.3d). The dynamics of resonance tunneling can account for the frequency-dependent anomalies in the conductivity and dielectric constant of the metallic phase of these doped polymers. Details of the analysis below can be found in [106].

Acknowledgments

This work was partly supported by the Office of Naval Research, National Institute for Science and Technology under contract NIST ATP 1993-01-0149, and the National Science Foundation. It benefited immensely from the efforts of Randolph Kohlman who coauthored the chapter in the 2nd edition upon which the present chapter is based. I am deeply indebted to collaborators over many years, especially Alan G. MacDiarmid, Jean Paul Pouget, Vladimir Prigodin, Gary Ihas, Takaheko Ishiguro, Yongong Min, David Tanner, Libero Zuppiroli, and students and postdocs at The Ohio State University including Jinsoo Joo, Randolph Kohlman, Richard McCall, Yunzhang Wang, and Zhao-hui Wang. Without the cooperation of these scientists, this work would not have been done.

References

1. Mark, J.E. ed. 1996. *Handbook of the physical properties of polymers.* Woodbury, CT: AIP Press.
2. Chiang, C.K., C.R. Fincher, Jr., Y.W. Park, A.J. Heeger, H. Shirakawa, E.J. Louis, S.C. Gau, and A.G. MacDiarmid. 1977. *Phys Rev Lett* 39: 1098.
3. Epstein, A.J. 2006. Conducting polymers: electrical conductivity. In *Physical properties of polymers handbook*, ed. J.E. Mark. Berlin: Springer-Verlag, chap. 46.
4. (a) Skotheim T.A. ed. 1986. *Handbook of conducting polymers*, New York: Marcel Dekker; (b) Skotheim T.A., R.L. Elsenbaumer, and J.R. Reynolds eds. 1986. *Handbook of conducting polymers*, 2nd ed. New York: Marcel Dekker.
5. Proceedings of International Conferences on Science and Technology of Synthetic Metals: ICSM '06, Dublin, Ireland, July 2–7, 2006. *Synth Met* XXX (2007); ICSM '04, Wollongong, Australia, June 28–July 2, 2004. *Synth Met* 152–154 (2005); ICSM '02, Shanghai, Peoples Republic of China, June 29–July 5, 2002. *Synth Met* 133–137 (2003); ICSM '00, Gastein, Austria, July 15–21, 2000. *Synth Met* 119–121 (2001); ICSM '98, Montpellier, France, July 12–18, 1998. *Synth Met* 101–103 (1999); ICSM '96, Snowbird, Utah, July 28–Aug. 2, 1996. *Synth Met* (in press); ICSM '94, Seoul, Korea, July 21–29, 1994. *Synth Met* 69–71 (1995); ICSM '92, Goteborg, Sweden, Aug. 12–18,

1992. *Synth Met* 55–57 (1993); ICSM '90, Tubingen, FRG, Sept. 2–9, 1990. *Synth Met* 41–43 (1991); ICSM '88, Santa Fe, NM, June 26–July 2, 1988. *Synth Met* 27–29 (1988–1989); ICSM '86, Kyoto, Japan, June 1–6, 1986. *Synth Met* 19–21 (1987); ICSM '84, Abano Terme, Italy, June 17–22, 1984. *Mol Cryst, Liq Cryst* 117–121 (1985); ICSM '82, Les Arcs, France, December 11–15, 1982. *Journal de Physique Colloque (Paris)* 44-C3 (1983); ICSM '81, Boulder, Colorado, Aug. 9–14, 1981. *Mol Cryst, Liq Cryst* 77 (1981), 79,81,83,85,86 (1982); ICSM '80, Helsingor, Denmark, Aug. 10–15, 1980. *Chemica Scripta* 17 (1981); Dubrovnik, Yugoslavia, Sept. 4–8, 1978. Lecture Notes in Physics 96 (Springer-Verlag, Berlin) (1979); ICSM '77, New York, NY, June 13–16, 1977. *Ann NY Acad Sci* 313 (1978).

6. Epstein, A.J. 1986. AC conductivity of polyacetylene: distinguishing mechanisms of charge transport. In *Handbook of conducting polymers*, ed. T.A. Skotheim. New York: Marcel Dekker, p. 1041.

7. Heeger, A.J., S.A. Kivelson, J.R. Schrieffer, and W.P. Su. 1988. *Rev Mod Phys* 60: 781.

8. Baeriswyl, D., D.K. Campbell, and S. Mazumdar. 1992. *Conjugated conducting polymers*, ed. H.G. Keiss. Berlin: Springer-Verlag, p. 7.

9. Conwell, E.M. 1987. *IEEE Trans Electr Insul* EI-22: 591.

10. MacDiarmid, A.G. and A.J. Epstein. 1989. Faraday discs. *Chem Soc* 88: 317.

11. Ray, A., A.F. Richter, A.G. MacDiarmid, and A.J. Epstein. 1989. *Synth Met* 29: E151.

12. Peierls, R.E. 1955. Quantum theory of solids. Oxford: Clarendon Press, p. 108.

13. Jahn, H.A. and E. Teller. 1937. *Proc Roy Soc (London) Ser A – Math and Phys Sci* 161: 220.

14. Ginder, J.M., A.F. Richter, A.G. MacDiarmid, and A.J. Epstein. 1987. *Solid State Commun* 63: 97.

15. Tsukamoto, J. 1992. *Adv Phys* 41: 509.

16. Tsukamoto, J., A. Takahashi, and K. Kawasaki. 1990. *Jpn J Appl Phys* 29: 125.

17. Naarmann, H. and N. Theophilou. 1987. *Synth Met* 22: 1.

18. Shirakawa, H., Y.-X. Zhang, T. Okuda, K. Sakamaki, and K. Akagi. 1994. *Synth Met* 65: 93.

19. Chiang, J.-C. and A.G. MacDiarmid. 1986. *Synth Met* 13: 193.

20. Epstein, A.J., H. Rommelmann, R. Bigelow, H.W. Gibson, D.M. Hoffman, and D.B. Tanner. 1983. *Phys Rev Lett* 50: 1866.

21. Adams, P.N., P. Laughlin, A.P. Monkman, and N. Bernhoeft. 1994. *Solid State Commun* 91: 895. The value of conductivity reported is for samples kindly provided by Monkman and coworkers, and measured at The Ohio State University.

22. Cao, Y., P. Smith, and A.J. Heeger. 1992. *Synth Met* 48: 91.

23. Joo, J., Z. Oblakowski, G. Du, J.P. Pouget, E.J. Oh, J.M. Weisinger, Y. Min, A.G. MacDiarmid, and A.J. Epstein. 1994. *Phys Rev B* 69: 2977.

24. Wang, Y.Z., J. Joo, C.-H. Hsu, J.P. Pouget, and A.J. Epstein. 1994. *Phys Rev B* 50: 16811.

25. Wang, Z.H., H.H.S. Javadi, A. Ray, A.G. MacDiarmid, and A.J. Epstein. 1990. *Phys Rev B* 42: 5411.

26. Yue, J. and A.J. Epstein. 1990. *J Am Chem Soc* 112: 2800.

27. Yue, J., Z.H. Wang, K.R. Cromack, A.J. Epstein, and A.G. MacDiarmid. 1991. *J Am Chem Soc* 113: 2655.

28. Yamaura, M., T. Hagiwara, and K. Iwata. 1988. *Synth Met* 26: 209.

29. Kohlman, R.S., J. Joo, Y.Z. Wang, J.P. Pouget, H. Kaneko, T. Ishiguro, and A.J. Epstein. 1995. *Phys Rev Lett* 74: 773.

30. Sato, K., M. Yamaura, T. Hagiwara, K. Murata, and M. Tokumoto. 1991. *Synth Met* 40: 35.

31. Epstein, A.J., H. Rommelmann, M. Abkowitz, and H.W. Gibson. 1981. *Phys Rev Lett* 47: 1549.

32. Epstein, A.J., H. Rommelmann, and H.W. Gibson. 1985. *Phys Rev B* 31: 2502.

33. Zuo, F., M. Angelopoulos, A.G. MacDiarmid, and A.J. Epstein. 1989. *Phys Rev B* 39: 3570.

34. Scott, J.C., P. Pfluger, M.T. Krounbi, and G.B. Street. 1983. *Phys Rev B* 28: 2140.

35. Ito, T., H. Shirakawa, and S. Ikeda. 1974. *J Polym Sci Polym Chem Ed* 12: 11.

36. Ito, K., Y. Tanabe, K. Akagi, and H. Shirakawa. 1992. *Phys Rev B* 45: 1246.

37. Edwards, J.H. and W.J. Feast. 1980. *Polym Commun* 21: 595.

38. Chien, J.C.W. 1984. *Polyacetylene: chemistry, physics, and material science.* New York: Academic Press, p. 24.
39. Epstein, A.J., J. Joo, R.S. Kohlman, G. Du, A.G. MacDiarmid, E.J. Oh, Y. Min, J. Tsukamoto, H. Kaneko, and J.P. Pouget. 1994. *Synth Met* 65: 149.
40. Kittel, C. 1986. *Introduction to solid state physics.* New York: Wiley, p. 157.
41. Ishiguro, T., H. Kaneko, Y. Nogami, H. Nishiyama, J. Tsukamoto, A. Takahashi, M. Yamaura, and J. Sato. 1992. *Phys Rev Lett* 69: 660.
42. Kaneko, H., T. Ishiguro, J. Tsukamoto, and A. Takahashi. 1994. *Solid State Commun* 90: 83.
43. Joo, J., V.N. Prigodin, Y.G. Min, A.G. MacDiarmid, and A.J. Epstein. 1994. *Phys Rev B* 50: 12,226.
44. Kohlman, R.S., J. Joo, Y.G. Min, A.G. MacDiarmid, and A.J. Epstein. 1996. *Phys Rev Lett* 77: 2766.
45. Kohlman, R.S., A. Zibold, D.B. Tanner, G.G. Ihas, Y.G. Min, A.G. MacDiarmid, and A.J. Epstein. 1997. *Phys Rev Lett* 78: 3915.
46. Tzamalis, G., N.A. Zaidi, and A.P. Monkman. 2003. *Phys Rev B* 68: 245106.
47. Lee, K., S. Cho, S.H. Park, A.J. Heeger, C.-H. Lee, and S.-H. Lee. 2006. *Nature,* 441: 65.
48. Javadi, H.H.S., K.R. Cromack, A.G. MacDiarmid, and A.J. Epstein. 1989. *Phys Rev B* 39: 3579.
49. Joo, J. and A.J. Epstein. 1994. *Appl Phys Lett* 65: 2278.
50. Epstein, A.J. and A.G. MacDiarmid. 1991. Proceedings of the Europhysics Industrial Workshop on the Science and Applications of Conducting Polymers, Lofthus, Norway, 28–31 May 1990, 141–152, ed. W.R. Salaneck. IOP Publishing Ltd., Bristol, England.
51. Wang, Y.Z., D.D. Gebler, D.K. Fu, T.M. Swager, and A.J. Epstein. 1997. *Appl Phys Lett* 70: 3215.
52. Kim, J.S., M. Granström, R.H. Friend, N. Johansson, W.R. Salaneck, R. Daik, W.J. Feast, and F. Cacialli. 1998. *J Appl Phys* 84: 6859.
53. Kushto, G.P., W. Kim, and Z.H. Zakya. 2005. *Appl Phys Lett* 86: 093502.
54. Bäcklund, T.G., H.G.O. Sandberg, R. Österbacka, T. Stubb, T. Mäkelä, and S. Jussila. 2005. *Synth Met* 148: 87.
55. Lee, M.S., H.S. Kang, H.S. Kang, J. Joo, A.J. Epstein, and J.Y. Lee. 2005. *Thin Solid Films* 477: 169.
56. Epstein, A.J., F.-C. Hsu, N.-R. Chiou, and V.N. Prigodin. 2002. *Curr Appl Phys* 2: 339.
57. Joo, J., Z. Oblakowski, G. Du, J.P. Pouget, E.J. Oh, J.M. Wiesinger, Y. Min, A.G. MacDiarmid, and A.J. Epstein. 1994. *Phys Rev B, Rapid Commun* 49: 2977.
58. Joo, J., A.J. Epstein, V.N. Prigodin, Y. Min, and A.G. MacDiarmid. 1994. *Phys Rev B, Rapid Commun* 50: 12226.
59. Joo, J., G. Du, A.J. Epstein, V.N. Prigodin, and J. Tsukamoto. 1995. *Phys Rev B* 52, 8060.
60. Martens, H.C.F., J.A. Reedijk, H.B. Brom, D.M. de Leuw, and R. Menon. 2001. *Phys Rev B* 63: 073203.
61. Epstein, A.J., H. Rommelmann, M.A. Druy, A.J. Heeger, and A.G. MacDiarmid. 1981. *Solid State Commun* 38: 683.
62. Joo, J., J.K. Lee, J.S. Baeck, K.H. Kim, E.J. Oh, and A.J. Epstein. 2001. *Synth Met* 117: 45.
63. Jozefowicz, M.E., R. Laversanne, H.H.S. Javadi, A.J. Epstein, J.P. Pouget, X. Tang, and A.G. MacDiarmid. 1989. *Phys Rev B* 39: 12958.
64. Kahol, P.K., H. Guan, and B.J. McCormick. 1991. *Phys Rev B* 44: 10393.
65. Ikehata, S., J. Kaufer, T. Woerner, A. Pron, M.A. Druy, A. Sivak, A.J. Heeger, and A.G. MacDiarmid. 1980. *Phys Rev Lett* 45: 1123.
66. Saricifti, N.S., A.J. Heeger, and Y. Cao. 1994. *Phys Rev B* 49: 5988.
67. Mizoguchi, K., M. Nechtschein, J.-P. Travers, and C. Menardo. 1989. *Phys Rev Lett* 63: 66.
68. Nechtschein, M., F. Genoud, C. Menardo, K. Mizoguchi, J.-P. Travers, and B. Villeret. 1989. *Synth Met* 29: E211.
69. Park, Y.W. 1991. *Synth Met* 45: 173.
70. Park, Y.W., A.J. Heeger, M.A. Druy, and A.G. MacDiarmid. 1980. *J Chem Phys* 73: 946.
71. Javadi, H.H.S., A. Chakraborty, C. Li, N. Theophilou, D.B. Swanson, A.G. MacDiarmid, and A.J. Epstein. 1991. *Phys Rev B* 43: 2183.

72. Subramaniam, C.K., A.B. Kaiser, P.W. Gilberd, C.J. Liu, and B. Wessling. 1996. *Solid State Commun* 93: 235.

73. Joo, J., S.M. Long, J.P. Pouget, E.J. Oh, A.G. MacDiarmid, and A.J. Epstein. 1998. *Phys Rev B* 57: 9567.

74. Menon, R., C.O. Yoon, D. Moses, and A.J. Heeger. 1998. *Handbook of conducting polymers*, 2nd ed., eds. T. Skotheim, R. Elsenbaumer, and J. Reynolds. New York: Marcel Dekker, p. 27.

75. Clark, J.C., G.G. Ihas, A.J. Rafanello, M.W. Meisel, R. Menon, C.O. Yoon, Y. Cao, and A.J. Heeger. 1995. *Synth Met* 69: 215.

76. MacDiarmid, A.G. and A.J. Epstein. 1994. *Synth Met* 65: 103.

77. Menon, R., C.O. Yoon, D. Moses, A.J. Heeger, and Y. Cao. 1993. *Phys Rev B* 48: 17685.

78. Jozefowicz, M.E., A.J. Epstein, J.P. Pouget, J.G. Masters, A. Ray, Y. Sun, X. Tang, and A.G. MacDiarmid. 1991. *Synth Met* 41: 723.

79. Pouget, J.P., M.E. Jozefowicz, A.J. Epstein, X. Tang, and A.G. MacDiarmid. 1991. *Macromolecules* 24: 779.

80. Laridjani, M., J.P. Pouget, E.M. Scherr, A.G. MacDiarmid, M.E. Jozefowicz, and A.J. Epstein. 1992. *Macromolecules* 25: 4106.

81. Laridjani, M. and A.J. Epstein. 1999. *Euro Phys J B* 7: 585.

82. Pouget, J.P., Z. Oblakowski, Y. Nogami, P.A. Albouy, M. Laridjani, E.J. Oh, Y. Min, A.G. MacDiarmid, J. Tsukamuto, T. Ishiguro, and A.J. Epstein. 1994. *Synth Met* 65: 131.

83. MacDiarmid, A.G. and A.J. Epstein. 1994. *Synth Met* 65: 103.

84. Min, Y., A.G. MacDiarmid, and A.J. Epstein. 1994. *Polym Prep* 35: 231.

85. Lee, S.-H., D.-H. Lee, K. Lee, and C.-W. Lee. 2005. *Adv Funct Mater* 15: 1495.

86. Chiou, N.R. and A.J. Epstein. 2005. *Adv Mater* 17: 1679.

87. Huang, K., M. Wan, Y. Long, Z. Chen, and Y. Wei. 2005. *Synth Met* 155: 495.

88. Anderson, P.W. 1958. *Phys Rev* 109: 1492.

89. Mott, N.F. and E. Davis. 1979. *Electronic processes in non-crystalline materials*, Oxford: Clarendon Press, p. 6.

90. Wang, Z.H., C. Li, A.J. Epstein, E.M. Scherr, and A.G. MacDiarmid. 1991. *Phys Rev Lett* 66: 1745.

91. Ma, J., J.E. Fischer, E.M. Scherr, A.G. MacDiarmid, M.E. Jozefowicz, A.J. Epstein, C. Mathis, B. Francois, N. Coustel, and P. Bernier. 1991. *Phys Rev B* 44: 11609.

92. Wang, Z.H., E.M. Scherr, A.G. MacDiarmid, and A.J. Epstein. 1992. *Phys Rev B* 45: 4190.

93. Salkola, M.I. and S.A. Kivelson. 1994. *Phys Rev B* 50: 13962.

94. Kivelson, S.A. and M.I. Salkola. 1991. *Synth Met* 44: 281.

95. Prigodin, V.N. and K.B. Efetov. 1993. *Phys Rev Lett* 70: 2932.

96. McCall, R.P., E.M. Scherr, A.G. MacDiarmid, and A.J. Epstein. 1994. *Phys Rev B* 50: 5094.

97. Leising, G. 1988. *Phys Rev B* 38: 10313.

98. Tanaka, J., C. Tanaka, T. Miyamae, M. Shimizu, S. Hasegawa, K. Kamiya, and K. Seki. 1994. *Synth Met* 65: 173.

99. Lee, P. and T.V. Ramakrishnan. 1985. *Rev Mod Phys* 57: 287.

100. Fukuyama, H. 1985. *Electron–electron interactions in disordered systems*, eds. A.L. Efros and M. Pollak. New York: Elsevier, p. 155.

101. Zabrodskii, A.G. and K.N. Zeninova. 1984. *Zh Eksp Teor Fiz* 86: 727 [*Sov Phys JETP* 59: 425].

102. Ioffe, A.F. and A.R. Regel. 1960. *Prog Semicond* 4: 237.

103. Wooten, F. 1972. *Optical properties of solids*. New York: Academic Press, p. 173.

104. Zuo, F., M. Angelopoulos, A.G. MacDiarmid, and A.J. Epstein. 1987. *Phys Rev B* 36: 3475.

105. Kohlman, R.S., D.B. Tanner, G.G. Ihas, Y.G. Min, A.G. MacDiarmid, and A.J. Epstein. 1997. *Synth Met* 84: 709.

106. Prigodin, V.N. and A.J. Epstein. 2003. *Physica B* 338: 310.

107. MacDiarmid, A.G. 2001. *Angew Chem Int Ed Engl* 40: 2581.

108. Epstein, A.J., F.-C. Hsu, N.-R. Chiou, O. Waldmann, J.H. Park, Y. Kim, and V.N. Prigodim. 2003. *Proc SPIE Conf Org Field Effect Trans. II* (San Diego, CA, 3–4 August 2003), 5217: 141.

109. Lu, J., N.J. Pinto, and A.G. MacDiarmid. 2002. *J Appl Phys* 92: 6033.

110. Okuzaki, H., M. Ishihara, and S. Ashizawa. 2003. *Synth Met* 137: 947.

111. Prigodin, V.N., F.C. Hsu, Y.M. Kim, J.H. Park, O. Waldmann, and A.J. Epstein. 2005. *Synth Met* 153: 157.

112. Hsu, F.-C., V.N. Prigodin, and A.J. Epstein. 2006. *Phys Rev B* (in press).

113. Pouget, J.P., C.-H. Hsu, A.G. MacDiarmid, and A.J. Epstein. 1995. *Synth Met* 69: 119.

114. Wang, Z.H., E.M. Scherr, A.G. MacDiarmid, and A.J. Epstein. 1992. *Phys Rev B* 45: 4190.

115. Wang, Z.H., A. Ray, A.G. MacDiarmid, and A.J. Epstein. 1991. *Phys Rev B* 43: 4373.

116. Kivelson, S. and A.J. Heeger. 1988. *Synth Met* 22: 371.

117. Prigodin, V.N. and K.B. Efetov. 1994. *Synth Met* 65: 195.

118. Bergman, D.J. and D. Stroud. 1992. *Solid state physics*, Vol. 46, eds. H. Ehrenreich and D. Turnbull. New York: Academic Press, p. 148.

119. Carr, G.L., S. Perkowitz, and D.B. Tanner. 1985. *Infrared Millimeter Waves* 13: 171.

120. Mott, N.F. and M. Kaveh. 1985. *Adv Phys* 34: 329.

121. Lee, K., R. Menon, E.L. Yuh, N.S. Saricifti, and A.J. Heeger. 1995. *Synth Met* 68: 287.

122. Beau, B., J.P. Travers, and E. Banka. 1999. *Synth Met* 101: 772.

123. Thouless, D.J. 1974. *Phys Rep* 13: 3475.

124. Viehweger, O. and K.B. Efetov. 1992. *Phys Rev B* 45: 11546.

125. Bolton-Heaton, C.S., C.J. Lambert, V.I. Falko, V.N. Prigodin, and A.J. Epstein. 1999. *Phys Rev B* 60: 10569.

126. Efros, A.L. and B.I. Shklovski. 1975. *J Phys C* 8: L49.

127. Shklovski, B.I. and A.L. Efros. 1984. *Electronic properties of doped semiconductors*. Heidelberg: Springer-Verlag.

128. Larkin, A.I. and D.E. Khmelnitskii. 1982. *Zh Eskp Teor Fiz* 83: 1140 [*Sov Phys JETP* 56: 647].

129. Epstein, A.J., W.-P. Lee, and V.N. Prigodin. 2001. *Synth Met* 117: 93.

130. Drude, P. 1900. *Ann Phys* 1: 566; 3: 369.

131. Mott, N.F. 1990. *Metal–insulator transitions.* New York: Taylor & Francis.

132. Mott, N.F. and M. Kaveh. 1985. *Adv Phys* 34: 329.

133. Lee, K., A.J. Heeger, and Y. Cao. 1993. *Phys Rev B* 48: 14884.

134. Mott, N.F. 1986. *Localization and interaction in disordered metals and doped semiconductors*, ed. D.M. Finlayson, Proceedings of the 31st Scottish University Summer School in Physics.

135. Mott, N.F. 1990. *Localization, Inst Phys Conf Ser* 108, eds. K.A. Benedict and J.T. Chalker. Bristol: Institute of Physics (Paper presented at the Localization Conference held at the Imperial College, London).

136. Kohlman, R.S. and A.J. Epstein. 1998. *Handbook of conducting polymers*, 2nd ed., eds. T.A. Skotheim, R.L. Elsenbaumer, and J.R. Reynolds. New York: Marcel Dekker, chap. 3.

137. Min, Y.G. 1995. Determination of factors promoting increased conductivity in polyaniline, Ph.D. thesis, University of Pennsylvania.

138. MacDiarmid, A.G., J.M. Weisinger, and A.J. Epstein. 1993. *Bull Am Phys Soc* 38: 311.

139. MacDiarmid, A.G. and A.J. Epstein. 1993. Trans. 2nd Congresso Brazileiro de Polimeros, Sao Paulo, Brazil, Oct. 5–8, p. 544.

140. Min, Y., A.G. MacDiarmid, and A.J. Epstein. 1994. *Polym Prep* 35: 231.

141. Chauvet, O., S. Paschen, L. Forro, L. Zuppiroli, P. Bujard, K. Kai, and W. Wernet. 1994. *Synth Met* 63: 115.

142. Javadi, H.H.S., A. Chakraborty, C. Li, N. Theophilou, D.B. Swanson, A.G. MacDiarmid, and A.J. Epstein. 1991. *Phys Rev B* 43: 2183.

143. Javadi, H.H.S., R. Laversanne, A.J. Epstein, R.K. Kohli, E.M. Scherr, and A.G. MacDiarmid. 1989. *Synth Met* 29: E439.

144. Joo, J. and A.J. Epstein. 1994. *Rev Sci Instr* 65: 2653.

145. Menon, R., Y. Cao, D. Moses, and A.J. Heeger. 1993. *Phys Rev B* 47: 1758.

146. Fosong, W., T. Jinsong, W. Lixiang, Z. Hongfang, and M. Zhishen. 1988. *Mol Cryst, Liq Cryst* 160: 175.

147. Moon, Y.B., Y. Cao, P. Smith, and A.J. Heeger. 1989. *Polym Commun* 30: 196.
148. Jozefowicz, M.E., A.J. Epstein, J.P. Pouget, J.G. Masters, A. Ray, and A.G. MacDiarmid. 1991. *Macromolecules* 25: 5863.
149. Pouget, J.P., S.L. Zhao, Z.H. Wang, Z. Oblakowski, A.J. Epstein, S.K. Manohar, J.M. Wiesinger, A.G. MacDiarmid, and C.H. Hsu. 1993. *Synth Met* 55: 341.
150. Abell, L., S.J. Pomfret, E.R. Holland, P.N. Adams, and A.P. Monkman. 1996. Proceedings of the Society of Plastics Engineers Annual Technical Conference (ANTEC 1996), p. 1417.
151. Cao, Y. and A.J. Heeger. 1992. *Synth Met* 52: 193.
152. Kahol, K., A.J. Dyakonov, and B.J. McCormick. 1997. *Synth Met* 89:17.
153. Menon, R. and S.V. Subramanyam. 1989. *Solid State Commun* 72: 325.
154. Yamaura, K. Sato, T. Hagiwara, and K. Iwata. 1992. *Synth Met* 48: 337.
155. Movaghar, B. and S. Roth. 1994. *Synth Met* 63: 163.
156. Dhoot, A.S., G.M. Wang, D. Moses, and A.J. Heeger. 2006. *Phys Rev Lett* 96: 246403.
157. Priogodin, V.N. and A.J. Epstein (submitted for publication).
158. Ehrenreich, H. and H.R. Phillip. 1962. *Phys Rev* 128: 1622.
159. Lee, K., A.J. Heeger, and Y. Cao. 1995. *Synth Met* 72: 25.
160. Stafstrom, S., J.L. Brédas, A.J. Epstein, H.S. Woo, D.B. Tanner, W.S. Huang, and A.G. MacDiarmid. 1987. *Phys Rev Lett* 59: 1464.
161. Woo, H.S., D.B. Tanner, N. Theophilou, and A.G. MacDiarmid. 1991. *Synth Met* 41–43: 159.
162. Miyamae, T., M. Shimizu, and J. Tanaka. 1994. *Bull Chem Soc Jpn* 67: 40253.
163. Yoon, C.O., R. Menon, D. Moses, A.J. Heeger, and Y. Cao. 1994. *Synth Met* 63: 47.
164. Burns, G. 1985. *Solid state physics*. New York: Academic Press, p. 187.
165. Yakushi, K., L.J. Lauchlan, T.C. Clarke, and G.B. Street. 1983. *J Chem Phys* 79: 4774.
166. Menon, R., K. Vakiparta, C.O. Yoon, Y. Cao, D. Moses, and A.J. Heeger. 1994. *Synth Met* 65: 167.
167. Du, G., J. Avlyanov, C.Y. Wu, K.G. Reimer, A. Benatar, A.G. MacDiarmid, and A.J. Epstein. 1997. *Synth Met* 85: 1339.
168. Joo, J., S.M. Long, J.P. Pouget, E.J. Oh, A.G. MacDiarmid, and A.J. Epstein. 1998. *Phys Rev B* 57: 9567.
169. Joo, J., Y.C. Chung, H.G. Song, J.S. Baeck, W.P. Lee, A.J. Epstein, A.G. MacDiarmid, S.K. Jeong, and E.J. Oh. 1997. *Synth Met* 84: 7390.
170. Saprigin, A., R.S. Kohlman, S.M. Long, K.R. Brenneman, M. Angelopolous, Y.-H. Liao, W. Zheng, A.G. MacDiarmid, and A.J. Epstein. 1997. *Synth Met* 84: 767.
171. Javadi, H.H.S., S.P. Treat, J.M. Ginder, J.F. Wolf, and A.J. Epstein. 1990. *J Phys Chem Solids* 51: 107.
172. Tigelaar, D.M., W.P. Lee, K.A. Bates, A. Saprigin, V.N. Prigodin, X. Cao, L.A. Nafie, M.S. Platz, and A.J. Epstein. 2002. *Chem Mater* 14: 1430.
173. Nagomi, Y., J.P. Pouget, and T. Ishiguro. 1994. *Synth Met* 62: 257.
174. Joo, J., J.K. Lee, J.K. Hong, J.S. Baeck, W.P. Lee, A.J. Epstein, K.S. Jang, S.J. Suh, and E.J. Oh. 1998. *Macromolecules* 31: 479.
175. Joo, J., J.K. Lee, A.J. Epstein, K.S. Jang, and E.J. Oh. 2000. *Macromolecules* 33: 5131.
176. Lee, W.P., K.R. Brenneman, A.D. Gudmundsdottir, M.S. Platz, P.K. Kahol, A.P. Monkman, and A.J. Epstein. 1999. *Synth Met* 101: 819.
177. McCall, R.P., M.G. Roe, J.M. Ginder, T. Kusumoto, A.J. Epstein, G.E. Asturias, S.P. Ermer, A. Ray, and A.G. MacDiarmid. 1989. *Synth Met* 29: E433.
178. McCall, R.P., J.M. Ginder, M.G. Roe, G.E. Asturias, E.M. Scherr, A.G. MacDiarmid, and A.J. Epstein. 1989. *Phys Rev B* 39: 10174.
179. Ginder, J.M. and A.J. Epstein. 1990. *Phys Rev B* 41: 10674.
180. Brédas, J.L., J.C. Scott, K. Yakushi, and G.B. Street. 1984. *Phys Rev B* 30: 1023.
181. Long, A.R. 1982. *Adv Phys* 31: 553.
182. Miyamae, T., M. Shimizu, and J. Tanaka. 1994. *Bull Chem Soc Jpn* 67: 2407.
183. Joo, J., G. Du, A.J. Epstein, V.N. Prigodin, and J. Tsukamoto. 1995. *Phys Rev B* 52: 8060.
184. Joo, J., G. Du, J. Tsukamoto, and A.J. Epstein. 1997. *Synth Met* 88: 1.

185. Epstein, A.J., J.M. Ginder, F. Zuo, R.W. Bigelow, H.-S. Woo, D.B. Tanner, A.F. Richter, W.-S. Huang, and A.G. MacDiarmid. 1987. *Synth Met* 18: 303.
186. Bean, B. and J.P. Travers. 1999. *Synth Met* 65: 101.
187. Epstein, A.J., F.-C. Hsu, N.-R. Chiou, and V.N. Prigodin. 2003. *Synth Met* 137: 859.
188. Prigodin, V.N., F.C. Hsu, Y.M. Kim, J.H. Park, O. Waldmann, and A.J. Epstein. 2005. *Synth Met* 153: 157.

One-Dimensional Charge Transport in Conducting Polymer Nanofibers

A.N. Aleshin and Y.W. Park*

16.1   Introduction

The use of inherently nanostructured conjugated polymers as components of nanoelectronics may be the promising way of manufacturing high-density nanochips in the future [1,2]. Conjugated polymers can be synthesized in a precisely controlled way to have nanostructures in many forms down to molecular scale. The common feature of conjugated polymers is that their conductivity can be increased by many orders of magnitude upon doping [3–5]. The variety of structural, optical, and electrical properties of three-dimentional (3D) conjugated polymers have been studied thoroughly during the past decades [6]. The polymer nanostructures (wires and tubes) made of conjugated polymers such as polyacetylene (PA), polypyrrole (PPy), polyaniline (PANI), polythiophene (PT), poly(phenylenevinylene) (PPV) have attracted considerable attention (see Figure 16.1 for chemical structure of the most important conjugated polymers) [2,5]. Transport properties of low-dimensional systems such as

*Corresponding author:ywpark@phya.snu.ac.kr

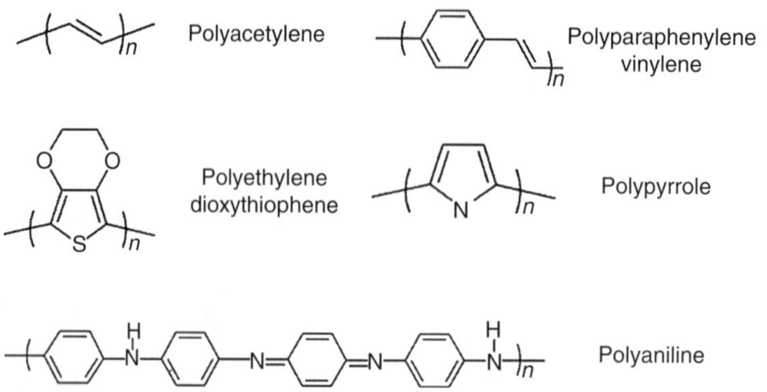

FIGURE 16.1   Chemical structure of some of the most important conjugated polymers.

polymer nanowires and nanotubes are a subject of interest in recent years [7]. This activity is based on previous studies of the charge transport in 3D conducting polymers made in the past [6]. Transport mechanism in 3D polymer films is still a topic of controversy because of strong influence of the effect of disorder [8,9]. On the one hand heavily doped conjugated polymers demonstrate properties that are characteristic for disordered metals and the metal–insulator transition (MIT) would be described by conventional 3D localization-interaction model for transport in disordered metals near the MIT [8]. According to another point of view, the transport properties are dominated by more macroscopic inhomogeneities and the MIT would be better described in terms of mixed metal semiconductor character with metallic islands surrounded by disordered barriers [9]. At the same time there are some transport features (especially at low temperatures), which cannot be explained either in localization-interaction or percolation models [10]. Moreover, recent results of magnetotransport studies in heavily doped oriented PA films at magnetic fields up to 30 T revealed strong evidences that even in 3D films charge transport is determined by nanojunctions between conducting nanofibers [11,12]. The observed anomalous linear behavior of magnetoresistance (MR) was attributed to weak localization arising from the inherently nonhomogeneous character of transport in polymers. It was concluded that current transport in a 3D PA film is mainly controlled by a network of point (interfibrillar) contacts between "bulk volumes" (nanofibers) with higher conductance [11,12]. These results have stressed the importance of nanotransport studies in single polymer fibers and tubes. One can note that the amount of experimental data on electronic properties of single polymer fibers and tubes is growing rapidly. At the same time there are only few reports related to the low temperature transport mechanism in such nanostructures. It is quite reasonable to suggest that in polymer fibers and tubes with high degree of crystallinity their inherent one-dimensional (1D) nature should clearly be manifested in transport properties.

The main goal of this work is to give a brief review of recent advances in synthesis and electrical transport characterization of polymer nanofibers in order to understand the intrinsic transport mechanism in these materials.

16.2   Advances in Synthesis of Polymer Nanofibers

Polymer nanofibers can be synthesized by several general methods. The most common techniques are template synthesis [13], chiral reaction [14], self-assembly [15], interfacial polymerization [16,17], and electrospinning technique [18].

The advantage of template synthesis is that the length and diameter of the polymer fibers and tubes can be controlled by the selected porous membrane which results in more regular nanostructures. The template method has been used to synthesize nanofibers and tubes of PPy [19,20], PANI, poly(3,4-ethylenedioxythiophene) (PEDOT) [21], and some other polymers. The general feature of conventional template method is that the membrane should be soluble to be removed after the synthesis in order to obtain single fibers or tubes. It restricts practical application of this method and gives rise to a search for other techniques. It was shown recently that PANI and PPy nanotubes can be synthesized by a self-assembly method without an external template [22–24]. One can obtain PANI nanotubes doped with camphor sulfonic acid (CSA) with 175 and 120 nm in outer and inner diameter respectively. Interfacial polymerization is another promising method to synthesize PANI nanofibers, where aniline is chemically and oxidatively polymerized to PANI at the interface of two immiscible liquids at ambient conditions without templates or functional dopants [16,17]. This method allows synthesis of the PANI fibers with uniform diameters between 30 and 120 nm using hydrochloric or perchloric acid; it has attracted significant attention in recent years [25,26]. At the same time fibers of PANI-CSA blended with polyethylene oxide (PANI-CSA/PEO) were fabricated by electrostatic spinning (electrospinning) technique [2,18]. It was shown that fiber diameters below 30 nm (near 5 nm) could be obtained for optimized electrospinning process parameters [27]. Nanowires with a diameter below 5 nm of PEDOT doped with polystyrenesulfonic acid (PPS), (PEDOT/PSS), promising polymer for organic devices, were prepared recently by molecular combining method (AFM lithography) [28].

Among the above-mentioned fibers and tubes, PA nanofibers are of particular interest because of its unique simple chemical structure. PA has the bandgap of $E_g \sim 1.5$ eV, and consists of linear chains of CH units with alternating single and double bonds (Figure 16.1). Networks of PA nanofibers can be prepared with diluted Ziegler–Natta catalyst exposed to acetylene [3]. Figure 16.2a presents the SEM image of such PA nanofiber ropes [29]. After disintegration of PA network by ultrasonic treatment, the nanofibers with diameter ranging from 100 nm down to 5 nm can be obtained. Due to increase of conductivity upon doping [3,4], PA fibers are promising for detail nanotransport studies. The structural and electronic properties of a doped PA network and single PA fibers were studied in detail in the past decade [30–33].

A new, advanced form of PA–helical PA was synthesized using chiral nematic liquid crystal (LC) as a solvent of Ziegler–Natta catalyst [14]. There are two types of helical films, R-helical PA (R-hel PA) and S-helical PA (S-hel PA). The direction of helicity of R-hel PA is counterclockwise and that of S-hel PA is clockwise in the form of entangled ropes as is evident from the SEM (Figure 16.2a) [14,34]. X-ray diffraction pattern of both R- and S-type helical PA films gave a broad reflection indicating that films are polycrystalline and corresponding to a spacing of 3.68 Å characteristic to *trans*-PA [35]. In order to extract a single helical PA fiber from the ropes a strong force, which overcomes the van der Waals force between the fibers, is required. To obtain a single fiber effectively we have used hexaethylene glycol mono-*n*-dodecyl ether ($C_{12}E_6$) as a nonionic surfactant in *N,N*-dimethylformamide (DMF) solution. The DMF solution enables avoidance of aging effect in aqueous solution while $C_{12}E_6$ works as a stabilizing agent. Well-dispersed helical PA fibers in $C_{12}E_6$/DMF solution showed a deep blue color, which is consistent with the transmission of light through *trans*-PA. No large or visible aggregation of dispersed S-hel PA fibers was observed in a proper concentration of $C_{12}E_6$ (58.2 g/L). Although the R-hel PA fibers are more tightly screwed compared to the S-hel PA fibers, the well-dispersed suspension of R-hel PA fibers can be acquired after increasing the concentration of $C_{12}E_6$ up to 163.2 g/L in DMF. The detailed procedure and proper concentrations are given [36]. As a result, the single fiber of the optimized suspension is well dispersed (Figure 16.3). We obtained helical PA single fibers with a cross-section, typically 40–65 nm (wide) and 130–300 nm (high). This is smaller than that observed previously from the SEM image of helical PA film [14,34]. The typical length of freshly prepared R- and S-hel PA fibers is of the order 10 μm—much longer than that of the traditional PA. It was reported that in helical PA film the screw direction of the fibers is the same as of the chiral nematic LC used as a solvent [34,35]. This fact is supported by our AFM results, which show the periodic modulation in height along the single fiber due to the chiral structure of helical PA fiber. As can be seen from Figure 16.2b and Figure 16.3, the main single fiber might consist of a few micro-fibers with smaller diameter [37]. For transport measurements, well-dispersed R-hel PA fibers were deposited on a Si substrate with a SiO_2 layer and gold or platinum electrodes thermally evaporated on the top, 2 μm apart(Figure 16.3). The fibers were doped with iodine from vapor phase up to the saturation level (1–2 wt%) [36].

The progress in synthesis enables us to obtain the variety of polymer nanofibers. The knowledge of charge transport mechanism in such nanostructures is of great interest for both basic and applied point of view aiming at the future nanoscale device making.

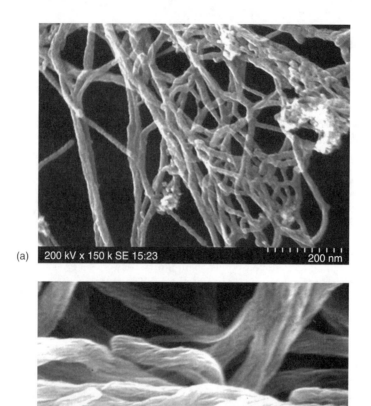

FIGURE 16.2 SEM images of: (a) polyacetylene nanofiber ropes. (From Park, J.H., Electronic and scanning tunneling spectroscopic studies of conducting polymer nanostructures: Polyacetylene nanofibers, PPV nanotubes and MEH-PPV nanowires, Ph.D. thesis, Seoul National University, Seoul, 2004.) (b) R-helical polyacetylene nanofiber ropes. (From Akagi, K., Unpublished data, 2004.)

16.3 Electronic Transport Properties of Polymer Nanofibers

As mentioned above there is no complete understanding of charge transport mechanism even in 3D conducting polymer films till now [8–10]. The intensive discussion on this subject is currently underway. At the same time the charge transport in such low-dimensional systems as polymer nanofibers is even less studied than that in 3D conducting polymer films. Generally speaking one can observe three different transport regimes in conducting polymers depending on the degree of disorder: metallic, critical, and insulating [8,9,38,39]. In the insulating regime of the MIT, conduction occurs by phonon-assisted tunneling between electronic localized states in the bandgap. The conductivity follows the exponential temperature dependence, $\sigma(T)$, characteristic of variable-range hopping (VRH) [40]:

$$\sigma(T) \propto \sigma_0 \exp[-(T_0/T)^p] \qquad (16.1)$$

where T_0 is the parameter and $p = 1/(d+1)$, d is the dimensionality of the system. Thus one can obtain $p = 0.25, 0.33, 0.5$ for 3D, 2D, and 1D VRH, respectively. For 3D disordered systems in the critical regime of the MIT the temperature dependence of the resistivity, $\rho(T)$, follows a universal power law [41]:

$$\rho(T) \approx (e^2 p_F/\hbar^2)(k_B T/E_F)^{-1/\eta}$$
$$\approx T^{-\gamma} \tag{16.2}$$

where p_F is the Fermi momentum, and $1 < \eta < 3$, i.e., $0.33 < \gamma < 1$. In heterogeneous systems the conduction is by electronic tunneling through nonconducting regions (barriers) separating metallic islands (rather than between localized states). If the metallic islands are large enough, the $\sigma(T)$ can be described by the fluctuation-induced tunneling (FIT) model [42] as

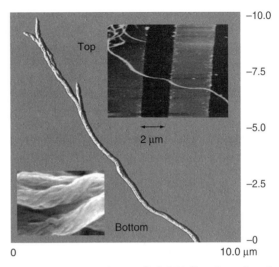

FIGURE 16.3 AFM image of R-hel PA fiber. *Insets* show the R-hel PA fiber on top of Pt electrodes (*top*) and SEM image of R-hel PA micro-fibers (*bottom*).

$$\sigma(T) = \sigma_t \exp\left[T_t/(T+T_S)\right] \tag{16.3}$$

where T_t represents the temperature at which the thermal voltage fluctuations across the tunneling junction become large enough to rise the energy of electronic states to the top of the barrier, the ratio T_t/T_S determines the tunneling in the absence of fluctuation. For small metallic islands, the FIT theory predicts $\sigma(T) \sim -T^{-0.5}$ similar to that for 1D VRH. The applicability of different transport models for 3D conducting polymer films has been analyzed thoroughly [38].

Recent advances in synthesis of polymer fibers and tubes helped in obtaining first results on electrical transport properties of fibers made of PA [30–33], PPy [19,20], PEDOT [21], PANI [18,23,24,27], and some others. In particular, the room temperature electrical conductivity, $\sigma_{300\ K}$, of iodine doped PA fiber network and PA single fiber was found to be $\sim$420 and $\sim$0.1 S/cm respectively [30,32]. The current–voltage (I–V) characteristics of doped single PA nanofiber show strong nonlinearities, which become stronger as temperature decreases from 300 K down to 1.8 K (Figure 16.4). The results for some single PA fibers revealed the absence of the I–V temperature dependence below 10–30 K with low ($\sim$0.1%) [30] or almost zero [33] MR in fields up to 7 T. Regarding other polymers, one can mention the I–V characteristic studies of PPy nanotube (diameter $\sim$120 nm), that allowed to estimate the $\sigma_{300\ K} \sim 1$ S/cm [19], which is consistent with $\sigma_{300\ K}$ values obtained for bulk PPy films [8]. PEDOT nanofibers (diameter $\sim$75–150 nm) synthesized by the template method were studied at 70–300 K [21]. It was found that for PEDOT fibers ($\sigma_{300\ K} \sim 37$ S/cm) the temperature dependence of the conductance, $G(T)$, is stronger than that for a PEDOT film, and this variation increases for a smaller fiber diameter. $G(T)$ of PEDOT nanofibers is close to the critical regime of the MIT and follows the power law described by Equation 16.2 similar to that for a PEDOT film [43]. PEDOT/PSS nanowires synthesized by AFM lithography revealed the $\sigma_{300\ K} \sim 0.07$–$0.15$ S/cm [28]—also of the same order of magnitude as for PEDOT/PSS films reported earlier [44]. PANI/PEO single fibers fabricated by electrospinning technique with diameters 70 and 20 nm revealed the conductivity of $\sim10^{-2}$ and $\sim10^{-3}$ S/cm, respectively [27]. Recently the $\sigma(T)$ of a single PANI-CSA nanotube with $\sigma_{300\ K} \sim 31.4$ S/cm (175 nm in outer diameter) synthesized by a self-assembled template-free method was studied by a standard four-terminal technique [23,24,45]. The typical $\sigma(T)/\sigma(300\ K)$ dependence of the PANI-CSA nanotube follows:

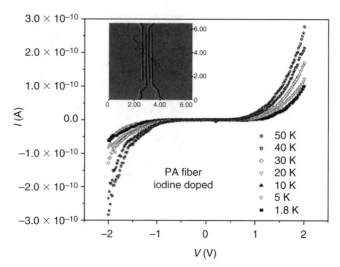

FIGURE 16.4 *I–V* characteristics of iodine doped PA nanofiber. *Inset* shows scanning force microscope image of PA nanofiber on top of Pt electrodes (with 100 nm separation). Typical diameter of PA nanofiber is ∼16–20 nm (From Park, J.G., et al. *Synth. Met.*, 119, 53, 2001 and Park, J.G., Electrical transport properties of conducting polymer nanostructures: Polyacetylene nanofiber, polypyrrole nanotube/nanowire, Ph.D. thesis, Seoul National University, Seoul, 2003.).

$\ln \sigma(T) \sim -T^{-0.25}$ with a crossover to $\ln \sigma(T) \sim -T^{-0.5}$ at low temperature [24] characteristic of a 3D VRH model [40,60]. This means the presumed low-dimensional PANI-CSA tubes behave as 3D systems that imply the high degree of disorder of these samples.

Among all above-mentioned polymers, the charge transport in doped PA nanofibers was studied more thoroughly [30–33]. One of the promising ways to increase the conductivity of PA nanofibers is the modification of polymer fiber itself during the synthesis with the following doping. As a result the polymer fiber can be twisted in different directions and thus its structural and electrical properties can be modified. This approach has been materialized in helical PA fibers synthesized by Akagi et al. [14] as described in Section 16.2. The unusual length of these fibers (up to 10 μm) and relatively high conductivity ($\sigma_{300\,K} \sim 1$ S/cm) enable the detail charge transport studies of helical PA films [46] and fibers [36,47,48] down to low temperatures (these results will be discussed in detail).

One can summarize that, despite the amount of experimental data on electronic properties of single polymer fibers and tubes is growing rapidly, the results on $\sigma(T)$ behavior are limited. In most cases, the $\sigma(T)$ has an activated nature and describes either 3D VRH [23,24] or FIT [30] models because of highly inhomogeneous or amorphous structure of polymer nanofibers under consideration. At the same time the results for different fibers are not uniform and there is no generally accepted model of the dominant transport mechanism in such systems till now. We suggest that in polymer fibers and tubes with higher degree of crystallinity, their inherent 1D nature might manifest itself more clearly in their transport properties.

16.4   1D Nature of Conjugated Polymers and Theories for Tunneling in 1D Conductors

16.4.1   1D Transport in Inorganic Nanowires—Luttinger Liquid

In this section we will discuss the transport features of polymer nanofibers and tubes characteristic of 1D systems. It is known that electron–electron interactions (EEIs) are strongly affecting the transport in 1D systems by leading to phases different from conventional Fermi liquid, namely the repulsive short-range

EEIs result in Luttinger liquids (LLs) [49], while long-range Coulomb interactions (LRCIs) cause a Wigner crystal (WC) [50]. The characteristic feature of 1D systems is the power-law behavior of the tunneling density of states near the Fermi level, which manifests itself in the power-law temperature dependencies of the conductance, $G(T)$, and the I–V characteristics, $I(V)$. For a clean LL a power-law variation of $G(T)$ is predicted at small biases (eV $\ll k_B T$), $G(T) \propto T^\alpha$, and a power-law variation of $I(V)$ at large biases (eV $\gg k_B T$), $I(V) \propto V^\beta$, where the exponents of the power laws depend on the number of 1D channels. The LL state survives for a few 1D chains coupled by Coulomb interactions (CIs) as well as the LL can be stabilized in the presence of impurities for more than two coupled 1D chains [51,52]. According to the LL theory, I–V curves taken at different temperatures should be fitted by the general equation [53]:

$$I = I_0 T^{\alpha+1} \sinh (eV/k_B T)|\Gamma(1 + \beta/2 + ieV/\pi k_B T)|^2 \qquad (16.4)$$

where Γ is the Gamma function, I_0 is a constant. The parameters α and β correspond to the exponents estimated from $G(T)$ and $I(V)$ plots. Equation 16.4 means that I–V curves should collapse into a single curve, if $I/T^{\alpha+1}$ are plotted as a function of $eV/k_B T$. These power-law variations have been found recently for inorganic 1D systems including single wall carbon nanotubes (SWNTs), multiwall carbon nanotubes (MWNTs) (α, $\beta \sim 0.36$) [54,55], doped semiconductor nanowires InSb ($\alpha \sim 2$–7, $\beta \sim 2$–6) [56], and NbSe$_3$ ($\alpha \sim 1$–3, $\beta \sim 1.7$–2.7) [57], and for fractional quantum Hall edge states in GaAs (α, $\beta \sim 1.4$–2.7) [58]. The LL, WC, or environmental Coulomb blockade (ECB) [55] models are currently debated in order to explain the above-mentioned power-law behavior in 1D systems. It is worthy to note that nanofibers that are made of conjugated polymers are another example of initially 1D systems [4]. Our recent results on transport in conventional and helical single PA fibers, PPy nanotubes revealed some features in both $G(T)$ and $I(V)$ behavior characteristic of 1D systems such as LL [47,48].

16.4.2 Recent Low Temperature Transport Experiments on Conducting Polymer Nanofibers

The charge transport in iodine doped R-hel PA fibers synthesized by K. Akagi [12] has been studied in the temperature range of 30–300 K [36,47]. The details of a single R-hel PA fiber dispersion are described elsewhere [36]. Figure 16.3 shows the AFM image of the R-hel PA fiber on top of Pt electrodes. The conductance of the sample was determined from the ohmic regime of the I–V characteristics at each temperature. The typical $G(T)$ for R-hel PA nanofibers as well as the results of our analysis of early data for iodine doped single PA nanofiber (diameter ~20 nm) and PPy nanotube (diameter ~15 nm) available [30] and [59] are presented in Figure 16.5 and Table 16.1. The conductance for all types of polymer fibers follows the power-law behavior $G(T) \propto T^\alpha$ starting from room temperature and down to ~30 K for the most conductive R-hel PA nanofiber (Figure 16.5). The power exponent increases from ~2.2 to ~7.2 as the fiber diameter or its cross-section (the amount of polymer chains) becomes smaller. The I–V characteristics of R-hel PA fiber at low temperature also follow the power law $I(V) \propto V^\beta$ (Figure 16.6). The same power-law variations for both $G(T)$ and $I(V)$ are found for all other PA fibers and PPy tubes and even for four crossed R-hel PA fibers (Figure 16.7) with a variety of power exponents $\alpha \sim 2.2$–7.2 and $\beta \sim 2$–5.7. The variation of α and β from one fiber to another may result from the nanofiber diameter scattering or doping level variation. We suggest that the power-law variations in $G(T)$ and $I(V)$ are a general feature of charge transport in the polycrystalline polymer nanofibers and nanotubes. Note that the similar power-law variations are found for inorganic 1D systems including SWNT and MWNT, InSb and NbSe$_3$ nanowires, and quantum Hall edge states in GaAs [54–58]. It is evident that polymer nanofibers and nanotubes differ from 1D carbon nanotubes and semiconductor nanowires. Each R-hel PA fiber consists of a number of oriented 1D chains, which are to some extent coupled and disordered as a result of doping. Therefore, polymer fibers are in fact quasi-1D and the confinement effects are expected to be not

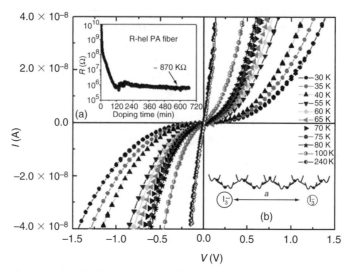

FIGURE 16.5   *I–V* characteristics of R-hel PA fiber at different temperatures. *Insets:* (a) Resistance vs. doping time for in situ doping with iodine for the same fiber. (b) PA chain with dopant (iodine) atoms. (From Lee, H.J., et al. *J. Am. Chem. Soc.*, 126, 16722, 2004.)

significant. However, the remarkable similarities observed in $G(T)$ and $I(V)$ behavior, in particular, with inorganic 1D nanowires, brought us to examine various theoretical models for tunneling in 1D systems.

16.4.3   Applicability of Different 1D Tunneling Models for Polymer Nanofibers

Polymer fibers and tubes under consideration are initially 1D conductors; thus one can expect that either short-range EEI or LRCI affect the transport in such quasi-1D systems. However, the applicability of some traditional models should be also considered first. Let's analyze the $G(T)$ curves in the framework of 1D VRH [40,60,61] and FIT [42] models. The VRH model implies $\sigma(T)$ in the form of Equation 16.1, i.e., $G(T) \propto G_0 \exp[-(T_0/T)^p]$. The standard derivation of the VRH transport for the d-dimensional hopping predicts power exponent p in Equation 16.1 in the form: $p = (\gamma+1)/(\gamma+d+1)$ [40,60,61]. In this expression γ is power exponent in a power-law density of states $g(\varkappa) \propto \varkappa^\gamma$ and d is the dimensionality of the system. The exponents p of VRH conductivity for 1D systems are strongly dependent on the shape of density of states $g(\varepsilon)$ near the Fermi level E_F that arises due to 3D CI. Thus, for $d=1$ one can obtain $p=0.5, 0.67$, and 0.75 by setting γ to 0, 1, and 2 respectively. However, even FIT with $p \sim 0.25$ for

TABLE 16.1   The Parameters of Polymer Fibers and Tubes

	Sample	Diameter or Cross-section, nm	$G_{300\,K}$ S	$\sigma_{300\,K}$ S/cm	α	β (at min T)	Refs
1	R-hel PA fiber	65 × 290	8.4×10^{-7}	1.13	2.2	—	47
					2.8	2.5 (30 K)	47
2	R-hel PA fiber	60 × 134	1.1×10^{-7}	0.85	5.5	4.8 (50 K)	47
3	R-hel PA fiber	47 × 312	2.1×10^{-9}	0.0036	7.2	5.7 (95 K)	47
4	Single PA fiber	20	7.3×10^{-9}	0.01	5.6	2.0 (94 K)	30
5	PPy tube	15	1.7×10^{-8}	0.83	5.0	2.1 (56 K)	59
6	PPy tube	50	2.8×10^{-8}	0.83	4.1	2.8 (50 K)	59
7	R-hel PA 4 fibers	—	2×10^{-8}		3.7	2.3 (90 K)	48

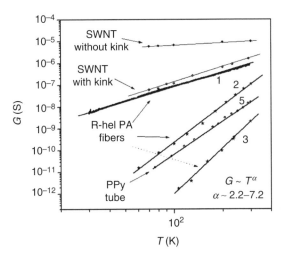

FIGURE 16.6 Conductance vs. T for different polymer fibers and tubes in comparison with $G(T)$ for SWNT with a kink. (From Yao, Z., et al. *Nature (London)*, 402, 273, 1999.)

3D VRH is too strong to explain the $G(T)$ for polymer fibers. Note that one may obtain the power law for $G(T)$ with $p \sim 0.25$ (3D VRH), if we would suggest that the preexponent $G_0(T)$ is temperature dependent [58]. This approach works in case of rather narrow interval for $G(T)$ and not valid for our polymer fibers where the power-law behavior is observed within several decades of $G(T)$. The latter pure FIT model reveals unreasonable fitting parameters to get the power law in $G(T)$ [59]. The tunneling transport mechanism based on FIT model has been recently considered for PA fibers [62,63]. The FIT model fits more metallic PA films data, i.e., it requires sizeable metallic islands between tunneling barriers. In our case the resistance of PA nanofiber is too high; this is not consistent with the FIT model. For a 3D disordered system in the critical regime of the MIT the $\rho(T)$ follows the universal power law, Equation 16.2, with $0.33 < \gamma < 1$ [41]. This model with $\gamma < 1$ can explain the transport in nonoriented 3D R-hel PA films at $T > 30$ K [46]. In the case of quasi-1D polymer nanofibers all exponents are above unity, and therefore this model should be ruled out. The power law $I(V) \propto V^\beta$ with $\beta \approx 2$ is known for the space charge limited current (SCLC) transport mechanism in semiconductors [64]. The values $\beta > 2$ observed for R-hel PA fibers and PPy tubes are inapplicable to the SCLC model. The analysis given above has motivated us to consider the applicability of other theories for tunneling in 1D conductors for polymer nanofibers.

As mentioned above, the power-law variations of $G(T)$ and $I(V)$ in inorganic 1D systems, such as SWNTs, MWNTs, doped semiconductor nanowires, are discussed in terms of LL, WC, or ECB theories

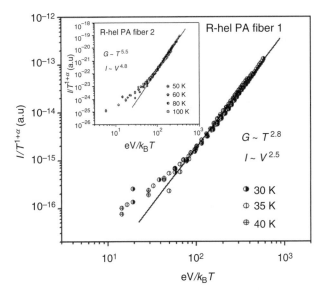

FIGURE 16.7 $I/T^{\alpha+1}$ vs. eV/k_BT for R-hel PA fibers 1 and 2 (*inset*) from Table 16.1 at different temperatures, where α is the exponent in $G(T) \propto T^\alpha$.

[54–58]. According to the LL or ECB models, all I–V curves taken at different temperatures should be fitted by the general Equation 16.4, i.e., they should collapse into a single plot $I/T^{\alpha+1}$ vs. eV/$k_B T$. All thus scaled I–V curves for an R-hel PA fibers collapse into a single curve at low temperatures by plotting $I/T^{\alpha+1}$ vs. eV/$k_B T$, where α is the power-law exponent estimated from the $G(T)$ curve for the same sample (Figure 16.6). The similar scaled behavior is found for all other R-hel PA fibers as well as for a single PA fiber and PPy tubes [47,48]. The exponents α and β (at the lowest possible temperatures) are presented in Table 16.1. One can again conclude that the power-law variations in $G(T)$ and $I(V)$ as well as the scaled I–V behavior are characteristic of such quasi-1D systems as polymer fibers and tubes. To a certain degree this behavior correlates with the LL theory predictions for transport in 1D systems. This theory implies that both tunneling along LLs through impurity barriers and tunneling between the chains with LLs provide the conduction of a set of coupled LLs. The system is characterized by measuring the LL interaction parameter g from the tunneling density of states [54]. LL theory predicts the conduction exponent α as $(1/g - 1)/2$ for impurity barriers and $(g + 1/g - 2)/4$ for interchain tunneling [65]. The relative contribution of these parallel conduction channels depends on the strength of impurities, temperature, and electric field. The latter equation for the interchain tunneling allows estimating $g \sim 0.08$ for the most conductive R-hel PA sample. This is much lower than $g \sim 0.2$ reported for tunneling into an SWNT from metal electrodes [53,54]. The low g value indicates that the strong repulsive EEI, characterized by $g \ll 1$, affects the transport in polymer fibers, whereas for the noninteracting electrons $g = 1$. It is worthy to note that the LL model for a single chain requires the correlation between the exponents $\beta = \alpha + 1$. However, as can be seen from Table 16.1, for all polymer fibers under consideration $\beta \neq \alpha + 1$, and, moreover, β is always less than α, which cannot be explained by LL or ECB theory [55]. The same disagreement with a pure LL model is found earlier for inorganic 1D nanowires [56,57] and can be associated with the fact that LL in the presence of a disorder may not exhibit the features typical for a clean case [66]. The interchain interactions may cause a renormalization of exponents at high temperatures whereas the interchain hopping destroys the LL state at low temperatures. In view of the interchain hopping to be one of the dominant transport mechanism in doped conjugated polymers at low temperatures [8,9], one can expect the manifestation of the LL state in doped polymer fibers at relatively high temperatures only. This correlates with the results obtained for R-hel PA nanofibers at $T > 30$ K [47].

On the other hand, the consideration of repulsive LRCI is believed to cause a WC, which is pinned by impurity [50]. According to the theory, WC occurs in solids with a low electronic density, i.e., with a large parameter $r_S = a/2a_B$ where a is the average distance between electrons and a_B is the effective Bohr radius. This corresponds to a negligible quantum fluctuation and can be rewritten as $r_S = E_C/E_F$ where the Coulomb energy E_C is larger than the kinetic energy E_F of the electrons. The WC transition has been shown to take place at $r_S \sim 36$ [67]. Such a pinned WC can adjust its phase in the presence of disorder to optimize the pinning energy gain, similar to classical CDW systems. Polymer nanofibers and tubes are quasi-1D conductors with a low electronic density, where WC may occur. Because the molecular structure of PA consists of alternating single and double bounds in the C–H chain, the Peierls distortion occurs. This creates the potential wells where the electrons can get trapped in and hence form the WC (Figure 16.8). At the modest doping level of the PA chain the distance between impurities (iodine) is of the order 10 nm, so r_S is large enough to get WC in the PA nanofiber. When the localization length is larger than the distance between impurities, the tunneling density of states follows the power law [50,68] with high values of the power exponents ~3–6 [69], similar to those found in $G(T)$ for polymer nanofibers. Some features characteristic of CDW behavior have recently been found in THz conductivity of doped PA films [70]. However, there are strong arguments against the applicability of a WC model to the polymer fibers. First, the impurities in doped conjugated polymers are sited outside of the polymer chains and, therefore, they only supply charge carriers without real pinning the chain. That results in the polymer chain either not pinned or pinned weakly by conjugated defects. Secondly, for a classical 1D WC pinned by impurities, one should expect the distinct VRH regime at low temperatures with an exponential $G(T)$ [71]. The absence of effective pinning prevents the observation of VRH in polymer fibers at least at $T > 30$ K that argues against the applicability of WC theory to polymer nanofiber results.

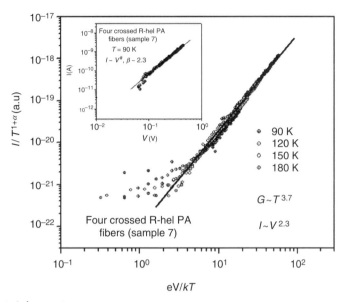

FIGURE 16.8 $I/T^{\alpha+1}$ vs. $eV/k_B T$ for R-hel PA sample 7 (four crossed R-hel PA fibers) with α 3.7 at different temperatures. The *insets* show *I–V* curves for the same samples at low temperature. (Reproduced from Aleshin, A.N., et al. *Microelectr. Eng.*, 81, 420, 2005. With permission.)

As compared with LL, WC is even strongly affected by the disorder and cannot survive in the doped polymer system.

Therefore, either pure VRH, LL, ECB or WC models cannot describe precisely the power-law variations observed in $G(T)$ and $I(V)$ for polymer nanofibers despite the fact that some relationship with LL and WC models is found. We suppose that the real transport mechanism in quasi-1D polymer fibers obeys either the superposition of the above-mentioned models or the single LL-like model valid for the different parts of the metallic polymer fiber separated by intramolecular junctions. By analogy with LL transport in bent metallic carbon nanotubes [72], we suggest that the conductance across the intramolecular junction is more temperature dependent than that of the two (or more) straight segments, but still obeys the power law $G(T) \propto T^{\alpha}$. In the case of SWNT for the end-to-end tunneling between two LLs separated with intramolecular junction (kink) the power exponent $\alpha_{end-end} = (1/g - 1)/2 \sim 1.8$, if the LL interaction parameter $g \sim 0.22$. Figure 16.5 demonstrates that for the SWNT with a single kink the exponent $\alpha = 2.2$ [72], which is close to the calculated value and surprisingly close to the α value for the most conductive R-hel PA fiber 1. It is evident from the AFM studies, that each R-hel PA fiber contains kinks at a distance between electrodes ~ 2 μm (Figure 16.3 and Figure 16.9). In our opinion each kink in the polymer fiber acts as the tunneling junction between the ends of two LLs. Therefore, we can approximate every fiber by a system of LL parts connected in series by intramolecular junctions with a power-law behavior of the density of states near the Fermi level in each part. This effect results in stronger $G(T)$ and $I(V)$ power-law variations with respect to a clean LL, and may cause the above-mentioned contradictions with a pure LL model. The similar strong power-law behavior of $G(T)$ and $I(V)$ observed for the four crossed R-hel PA fibers reminds the LL behavior in crossed metallic SWNT [73] and argues for our model. Recent structural studies of helical PA fibers by SEM reveal that each fiber is composed of helical micro-fibers that are small bundles of polymer chains (Figure 16.2b and inset to Figure 16.3) [37]. This is different from that in conventional, nonhelical PA (Figure 16.2a) and may affect the transport in more complicated way by leading to higher power exponent in $G(T)$ and $I(V)$. We suppose that in such micro-fiber system, long-range interactions between electrons in each micro-fiber may be screened through CI of these electrons with electrons of neighboring micro-fiber. This leads to a short-range intra fiber EEI which is the basic assumption of the LL theory.

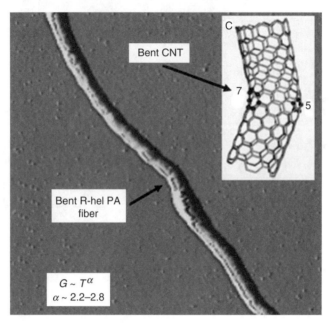

FIGURE 16.9   R-hel PA fiber and carbon nanotube with kinks.

In all other respects the above-described model of LLs separated with intramolecular junctions should be valid for the micro-fibers case. Therefore, we summarize that the power-law behavior observed in R-hel PA fibers for both $G(T)$ and $I(V)$ is characteristic of such 1D systems as a several LL parts connected in series. At the same time there is a discrepancy between our results for polymer fibers and theories for tunneling in truly 1D systems. Further efforts in this field are necessary to clarify the origin of the power laws in $G(T)$ and $I(V)$ observed in polymeric and inorganic 1D systems.

16.4.4 Non-Ohmic Transport in Lightly Doped Single Polymer Nanofibers at Low Temperatures

16.4.4.1 Transport Features in Lightly Doped PA Nanofibers

The above-discussed results were obtained for polymer fibers doped up to the saturation level (1–2 wt% in the case of iodine doped PA fiber). Let us consider the I–V characteristics of lightly doped single PA nanofiber to analyze the dominant transport mechanism in this case. Figure 16.10 shows the I–V characteristics of a PA nanofiber (diameter 40 nm) lightly doped with $FeCl_3$ for temperatures between 100 and 2 K [33,62]. These I–V characteristics were measured with the gap between electrodes ~270 nm, thus the maximum strength of the electric fields was of the order ~2×10^5 V/cm. The I–V characteristics of lightly doped single PA nanofiber show very strong nonlinearities without ohmic regime even at relatively high temperature (Figure 16.10). This makes impossible the determination and analysis of the conductance behavior from I–V characteristics, contrary to relatively highly doped PA fibers described in the previous chapter. For lightly doped PA fibers the current increases dramatically at high fields, with the increase occurring at lower voltages as temperature increase (negative temperature coefficient of the threshold electric field E_T). This behavior is rather similar to that reported in such quasi-1D polymer systems as polydiacetylene (PDA) single crystals and thin films [74,75], charge-density wave (CDW) and spin-density wave (SDW) systems [76,77].

The non-ohmic conductivity at high electric fields was discussed in crystalline and amorphous semiconductors and dielectrics in terms of such theoretical models as Pool-Frenkel effect [78], Shklovskii nonlinear hopping conduction in disordered solids [79,80], and tunneling in granular structures [81].

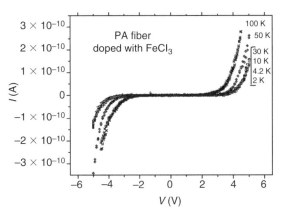

FIGURE 16.10 *I–V* characteristics of a PA nanofiber lightly doped with $FeCl_3$ (diameter 40 nm) for temperatures between 100 and 2 K. The gap between electrodes is 270 nm, thus maximum electric fields is ~2 × 10^5 V/cm. (Reproduced from Park, J.G., et al. *Synth. Met.*, 135–136, 299, 2003. With permission.)

Other approaches are related to electrical breakdown in dielectrics such as impact ionization and band-to-band (Zener) tunneling [82]. The former mechanisms imply temperature dependence of ohmic conductivity at low voltage, whereas the later models predict sharp increase of current without strong temperature dependence. The narrow (or absence) of ohmic part of *I–V* characteristics in the case of PA nanofibers, the nonlinear, weakly temperature dependent *I–V*s behavior contradicts the above-mentioned pure hopping models. It means that at low temperature we deal mainly with some tunneling mechanism in lightly doped PA nanofibers. At the first glance the weak temperature dependence of *I*(*V*), the negative temperature coefficient of E_T, and the absence of *I–V* saturation at high electric field (the signature of impact ionization) are in agreement with band-to-band (Zener) tunneling model. (Note that impact ionization is characterized by positive temperature coefficient of the breakdown voltage.) Zener tunneling is usually observed at $E \sim 10^5$ V/cm, where the current follows the expression [82]:

$$\sigma = j/E = \sigma_0 \exp(-E_0/E) \qquad (16.5)$$

The parameter E_0 is given by $E_0 = 4[(2m^*)^{1/2}(E_g)^{3/2}]/3e\hbar$, where E is the electric field, E_g is the bandgap, m* is the effective mass. According to Equation 16.5 the tunneling current should follow $\log(I/V^2) \sim V^{-1}$, in agreement with experimental data(Figure 16.11a). Equation 16.5 does not predict any temperature dependence of the tunneling current, which correlates with the results for PA nanofibers at $T < 30$ K, where the tunneling current is temperature independent. However, the temperature dependence do exists at $T > 30$ K, where the current increases as temperature increases with activation energy $\varepsilon_a \sim 5$–10 meV. The observed weak temperature dependence at $T > 30$ K can probably be related to the influence of an additional minor transport mechanism such as nonlinear hopping through some localized states within the PA bandgap introduced by random impurities. The probability of band-to-band tunneling through energy bandgap of the PA molecule, which is ~1.5 eV [4,5], seems to be low enough especially at electric fields lower than 10^5 V/cm. These facts stimulate us to look for another explanation for the low temperature *I–V*s behavior in the PA nanofibers taking into account the fact that this material exhibits both electronic and structural disorder.

By analogy with quasi-1D PDA films [74], one may suppose that the charge within the polymer nanofiber is not expected to obey the neutrality condition. Correspondingly, the charge transport can be supported due to a charge injection from the metallic banks. This injection is supported by tunneling of the electrons from the banks into the conductance band of the polymer (or holes into the valence band). Such a tunneling is possible as due to the presence of an electric field within the polymer the corresponding band edges have finite slopes and at some distance from the bank the band edge coincides with the Fermi level within the bank. One easily estimates the tunneling probability for such an injection as: $\sim \exp\{-4[(2m^*)^{1/2}(E_g/2)^{3/2}]/3eE\hbar\}$, where E_g is the polymer bandgap. One notes that this probability is much larger than the probability of the Zener tunneling as in the case of injection the electrons tunnel from the Fermi level of the metal and thus the effective gap is $E_g/2$ in contrast to the value E_g for the Zener tunneling. Let us discuss the nature of the temperature behavior. The simplest

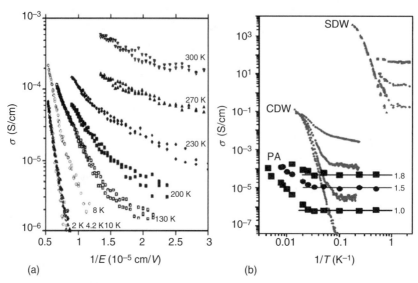

FIGURE 16.11   (a) Conductivity vs. $1/E$ at different T = 2–300 K. (From Kaiser, A.B., and Y.W. Park. *Curr. Appl. Phys.*, 2, 33, 2002.). (b) Conductivity vs. $1/T$ for an electric field strengths, E: 1.0, 1.5, and 1.8×10^5 V/cm for the same lightly FeCl$_3$ doped PA nanofiber. Data for the CDW material TaS$_3$ (from Zaitsev-Zotov, S.V. *Phys. Rev. Lett.*, 71, 605, 1993) at field strengths of 30, 40, 50, and 60 V/cm and the SDW material (TMTSF)$_2$PF$_6$ (from Mihaly, G., Y. Kim, and G. Gruner. *Phys. Rev. Lett.*, 67, 2713, 1991) at field strength of 0.1, 0.3, 0.5, and 1.0 V/cm are shown for comparison in (b). (Reproduced from Kaiser, A.B., and Y.W. Park. *Curr. Appl. Phys.*, 2, 33, 2002. With permission.)

assumption can be related to the fact that the disorder produces a modulation of the band edge with a scatter ∂E_g. If the spatial scale of the modulation is larger than the characteristic electron wavelength, one deals with a classical potential relief. Thus the value of $\sim\partial E_g$ plays a role of activation energy for the charge transport for such percolation problem. One notes that the applied electric field produces a sort of a "washboard" and, being strong enough, suppresses the localization introduced by the disorder potential. For the electric fields of the order of 10^4–10^5 V/cm, it would give the corresponding spatial scale for the disorder potential $\sim 10^{-7}$ cm which seems to be too small. Let us, however, assume that the low values of the carriers mobility (discussed in Section 16.4.4.2) are mainly due to the presence of tunneling barriers between the different "islands" within the polymer fiber; the latter can be related to the different polymer chains or different granules. One also expects that a disorder induces some scatter between the positions of the band edges within the different "islands" of the order of ∂E_g. In this case the observed activation energy $\varepsilon_a \sim$ 5–10 meV is related to the bandgap mismatch between the different "islands". The suppression of the localization at $T = 0$ K would correspond to electric fields of the order of $E \sim \partial E_g / e d_{tun}$, where d_{tun} is the thickness of the tunneling barrier which is indeed expected to be of the order of $\sim 10^{-7}$ cm.

The *I–Vs* behavior in lightly doped PA nanofibers has recently been discussed in the framework of a novel conduction mechanism arising from tunneling of a segment of the conjugated bond system in the presence of an electric field [62]. This approach is similar to the solitons pair creation mechanism in CDW materials that yields an expression of the form of Equation 16.5 [83]. The model implies that the tunneling process in lightly doped PA nanofiber would involve solitons rather than quasi-particle excitations. While neutral topological solitons in PA carry a spin, charged solitons are spinless [4] and, therefore, such transport mechanism would not be affected by a magnetic field in agreement with experimental results. As seen in Figure 16.11b, the strong similarities between $\sigma(T)$ behavior at strong electric fields in lightly doped PA fiber and that in quasi-1D conductors exhibiting CDWs—TaS$_3$ [76] and SDWs—(TMTSF)$_2$PF$_6$ [77] (as well as in PDA thin films [74]) are really existing [42]. However, the CDW-like transport implies the creation of such 1D system as WC, which is unlikely for doped conjugated polymers as shown in the previous chapter. It was shown that conjugated bond system in PA represents a "bond-order wave" [84], in which the charge density between the carbon positions is modulated. One can assume that the motion of charged (spinless) solitons in a system which consists of such segments is possible, but the exact attribution of $\sigma(T)$ behavior in lightly doped PA nanofibers to this transport mechanism still requires additional proofs from the low temperature MR and electron-spin resonance studies.

16.4.4.2 Coulomb Blockade Transport at Low Temperature in Polymer Nanofibers

In this content our recent results on the low temperature ($T < 30$ K) transport in iodine doped R-hel PA single fibers [85] as well as the analysis of the data for conventional lightly doped PA fibers[59,62] are of interest for better understanding of the dominant low temperature transport mechanism in such systems. Single R-hel PA fibers were synthesized and prepared similar to that in [47, 48], the details are described elsewhere [36]. Figure 16.12a presents the I–V characteristics at different temperatures for such an R-hel PA nanofiber shown in inset. The I–Vs at $T < 60$ K are symmetric and strongly nonlinear with no ohmic regime even at relatively high temperatures (Figure 16.12a). The temperature dependence of current, $I(T)$, has an activated nature with small activation energies, $\varepsilon_a \sim 2$–4 meV at high temperatures and low electric fields, while at lower temperatures and high electric fields the I–Vs are almost temperature independent (Figure 16.12b). This $I(V)$ behavior is characteristic of quasi-1D systems such as conventional PA nanofibers [33,62], PDA single crystals and films [74,75]. As can be seen from Figure 16.12a and Figure 16.13, as temperature goes below 30–40 K, the current starts to flow only above some voltage threshold, V_t, with negative temperature coefficient of V_t (V_t decreases as temperature increases, as shown in Figure 16.14). There is a pronounced power-law behavior in G(T) and $I(V)$ with a good I–Vs scaling as $I/T^{\alpha+1}$ vs. eV/k_BT at $T > 60$ K, where $\alpha \sim 4.4$ is the exponent in $G(T) \sim T^\alpha$, similar to our previous report on transport in R-hel PA nanofibers [47,48]. As temperature decreases, the I–Vs of R-hel PA nanofiber still follow the power law, but the $G(T)$ plot is not available due to presence of the V_t. As can be seen from Figure 16.13, the $I(V)$ at low temperature, $T < 30$–40 K, and high biases, $eV \gg k_B T$, follows another power-law scaling fitted by

$$I \sim (V - V_t)^\zeta \tag{16.6}$$

which is characteristic of Coulomb blockade (CB) dominated transport mechanism [86,87]. The nonzero V_t can be estimated from IdV/dI vs. V plots at each temperature. Figure 16.14 and inset to Figure 16.13 demonstrate that $V_t \sim 1.5$ V at $T = 1.45$ K and below this value at higher temperatures. The scaling exponents, ζ, obtained from I vs. $(V - V_t)/V_t$ log–log plots at high biases are found to be $\zeta \sim 1.78$–2.14 and weakly temperature dependent (Figure 16.13, inset to Figure 16.14). The same behavior has been found for several other iodine doped R-hel PA nanofibers. It is worthy to note that the I–V behavior found in doped polymer nanofibers at low temperature is similar to that observed earlier in 2D metal nanocrystal arrays [87–90] and 1D chains of graphitized carbon nanoparticles [91], where the CB is considered as the dominant transport mechanism. However, there are some features in the low temperature I–Vs of doped R-hel PA nanofibers. First, the temperature dependence of the V_t is not linear as in the case of nanocrystal/nanoparticle arrays [89,91], but at $T > 4$ K rather follows the power law $V_t \sim T^{-\gamma}$ with the power exponent $\gamma \sim 1$ (Figure 16.14). Secondly, there is an apparent transition in I vs. $(V - V_t)/V_t$ plots from the scaling exponent $\zeta \sim 1$ at low biases to $\zeta \sim 2$ at higher biases (Figure 16.13). The cusp point shifts to the lower values as temperature decreases, thus at the lowest temperature, $T \sim 1.45$ K, we observed the only slope with $\zeta \sim 1.8$. Figure 16.15a shows the differential conductance (dI/dV) derived from the I–V characteristics of the R-hel PA nanofiber at different temperatures. The dI/dV vs. V plots demonstrate a clear parabolic behavior of the conductance with a pronounced conductance oscillation whose magnitude greatly increases with decreasing temperature. The period of these oscillations at the lowest temperature $T = 1.45$ K was found to be ~ 200 mV.

The low temperature I–V characteristics behavior of doped R-hel PA nanofibers is similar to that observed in 2D metal nanocrystal arrays [87–90] as well as in 1D chains of graphitized carbon

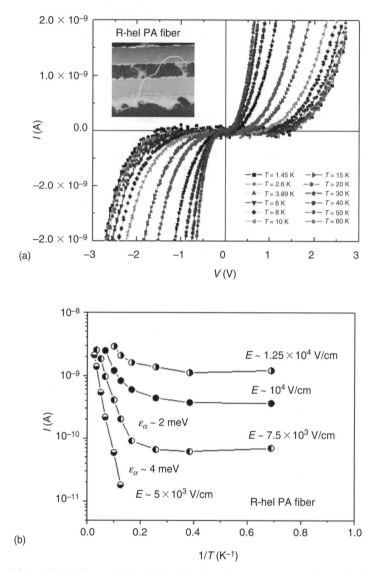

FIGURE 16.12 (a) Typical *I–V* characteristics of R-hel PA fiber at low temperatures. *Inset*: AFM image of R-hel PA fiber used in our study on top of Pt electrodes. (b) Temperature dependence of current at different electric fields for the same sample.

nanoparticles [91], where the CB is considered as the dominant transport mechanism. In the case of iodine doped R-hel PA nanofibers the localization of electron states by disorder at low temperature results in the appearance of the transport threshold voltage—V_t. The existence of V_t, its temperature dependence, the scaling exponents $\zeta \sim 1.78$–2.14 in Equation 16.6 are characteristic of the CB transport similar to that in quasi-1D arrays of metallic or carbon nanoparticles, separated by tunneling nanobarriers [89–91]. The general overall parabolic shape of the differential conductance and clear low temperature conductance oscillations indicate tunneling through the entire metal–insulator–metal junction for a structure with multiple tunnel barriers in the CB regime. The results of analysis of early data for conventional iodine doped PA fiber (diameter ~16 nm) available [33,62] revealed the qualitatively similar CB behavior of current at $eV \gg k_B T$ and $T < 50$ K with a weakly temperature dependent power exponents $\zeta \sim 1.9$–2.3 in Equation 16.6 (inset to Figure 16.14). Based on these experimental

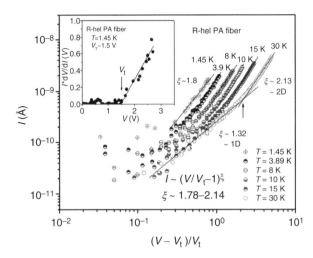

FIGURE 16.13 Current vs. $(V - V_t)/V_t$ at different temperatures. *Inset* shows the IdV/dI vs. V and the threshold voltage, $V_t \sim$ 1.5 V at $T = 1.45$ K for the same R-hel PA fiber.

results we suggest that at low temperature the inherent quasi-1D polymer fibers can, to first order, also be considered as an array of small conducting regions made of doped conjugated polymer separated by nanoscale barriers (Figure 16.15b), which is somewhat similar to the heterogeneous model of metallic islands of several nm size separated by barriers have been deduced for bulk polymers [38,39]. This suggestion is supported by early results of x-ray diffraction studies that revealed the polycrystalline structure of R-hel PA [35]. Recent SEM studies have demonstrated that each R-hel PA nanofiber is composed of smaller helical nanowires that are bundles of polymer chains [37]. This structure implies the superposition of several conducting channels inside the main R-hel PA fiber and may affect the transport by leading to some scattering in power exponents, threshold voltages, and conductance oscillation periods. The CB transport in such an array of tunnel junctions implies that the *I–V*s behavior depends on the size, dimension and connectivity of the arrays, the screening length, and the disorder [92]. In the case of doped polymer nanofiber, the size of an array of metallic regions separated by tunnel nanojunctions should be comparable to the size of polycrystalline grains and thus to the diameter of a smallest nanowire, approximately some tens of nanometers. On the other hand at the modest doping level of the PA chain the distance between impurities (iodine) is of the order 10 nm. Iodine atoms affect transport in the PA chain by the creation of conjugation defects, domain walls (soliton kinks), etc., which results in the creation of conductive islands of the size comparable to the distance between impurities. The CB transport in chains of graphitized carbon nanoparticles of diameter $\sim$30 nm[91] revealed the values of

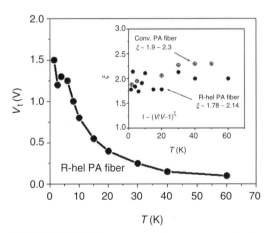

FIGURE 16.14 Temperature dependence of the threshold voltage. *Inset:* power exponent, ζ vs. T for the R-hel PA fiber and for conventional iodine doped PA fiber (diameter 16 nm).(From Park, J.G., et al. *Synth. Met.,* 135–136, 299, 2003.)

V_t and ζ similar to those in case of R-hel PA fibers. All the above arguments allow us to estimate the average size of conductive islands inside the R-hel PA nanofiber of the order $\sim$10 nm. The CB model predicts that the scaling exponent $\zeta \sim 1$ is characteristic of 1D system, while $\zeta \sim 2$ is typical for 2D tunneling [87]. From this point of view the existence of two slopes in *I* vs. $(V - V_t)/V_t$ plots at low temperature, namely $\zeta \sim 1.3$ at low biases and $\zeta \sim 1.8$–2.1 at high biases, can be attributed to the transition from 1D to 2D tunneling with increase in excitation bias [85].

In summary, the *I–V* characteristics of lightly doped quasi-1D PA nanofibers are non-ohmic and superlinear within the temperature interval 300–1.4 K. As temperature decreases the threshold voltage shifts to higher values with a weak influence of a magnetic field up to 7 T. We assume that at low temperatures such a doped polymer fiber can be considered as an array of small conducting regions separated by nanoscale

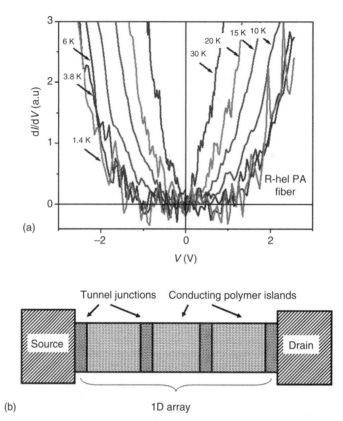

FIGURE 16.15 (a) Differential conductance vs. voltage at different temperatures for the same R-hel PA fiber. Note the clear conductance oscillations whose magnitude increases with decreasing temperature. (b) Schematic representation of the metal–insulator–metal tunneling junction structure of a conducting polymer nanofiber.

tunneling junctions where the CB tunneling is the dominant transport mechanism. We suppose that such a behavior is a general feature of the low temperature transport in other types of conducting polymer nanofibers, which is an essential issue for potential practical applications in nanoelectronics.

16.5   Device Applications of Conducting Polymer Nanofibers

Application of polymer nanofibers and nanotubes as elements of nanoelectronic devices is the current topic of research . For example, thin films made of the PANI nanofibers synthesized by interfacial polymerization have superior performance as gas sensors in both sensitivity and time response to vapors of acid (HCl) and base (NH$_3$) [16,17]. The conducting PPy, PANI and PEDOT nanowires ($\sim$100–200 nm diameter) synthesized by electrochemical polymerization method using Al$_2$O$_3$ nanoporous template have a potential to be used as nanotips in field emission displays [93,94]. It has been demonstrated that electrical properties of PA fibers (diameter $\sim$20 nm) can be efficiently studied using field-effect transistor (FET) geometry by changing their conductivity with application of gate voltage [32]. Figure 16.16a shows the schematic diagram of such FET based on PA single fiber. Similar FET technique was also used to study the electronic transport in doped PPy [19,20] and PANI/PEO nanofibers [27,95]. The *I–V* characteristics of polymer single fiber FETs are strongly nonlinear similar to that for *I–V*s of doped single PA fibers (Figure 16.16b). This is different from that of semiconductors [82] or organic FETs [96,97] that show the saturation behavior. At the same time this behavior is similar to that in SWNT

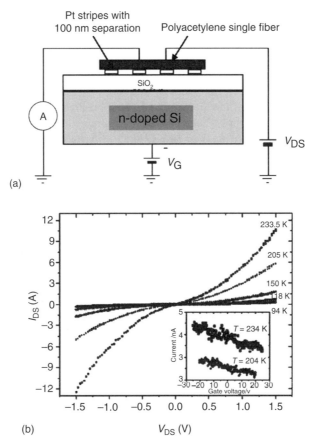

FIGURE 16.16 (a) Schematic diagram of PA single fiber FET. (b) Typical *I–V* characteristics of PA nanofiber FET ($D \sim 20$ nm) at different temperatures. *Inset* shows the transfer curves for the same fiber at $V_{DS} = 1.0$ V. (Reproduced from Park, J.G., et al. *Thin Solid Films*, 393, 161, 2001. With permission.)

[98], MEH-PPV thin strip FETs [32], DNA—templated SWNT FETs [99], and some other low-dimensional structures [1]. To evaluate the field-effect mobility (μ_{FET}) of polymer nanofiber one can consider the relatively low voltage region of the *I–V* characteristics in terms of the linear region in conventional MOSFETs. Then from the transconductance, g_m, the μ_{FET} can be estimated [82] as

$$g_m = \partial I_{DS}/\partial V_g|_{Vds=const} = -(Z/L)\mu_{FET}C_i V_{DS} \qquad (16.7)$$

where Z is the channel width, L is the channel length and C_i is the capacitance per unit area. For the PA single fiber FET the μ_{FET} at 233 K was found to be $\mu_{FET} \sim 1.5 \times 10^{-3}$ cm^2/Vs ($L \sim 100$ nm, $D \sim 4.4$ nm, $C_i \sim 12$ nF/cm^2) [32]. The observed nonlinear *I–V* characteristics of PA single nanofiber FET can be explained by either transport mechanisms discussed in previous sections or simply SCLC [63] or Schottky-barrier [82] models. The presence of traps, nonoptimal contact barriers as well as rather large channel length results in rather low μ_{FET} values in PA nanofiber FET structures. In order to increase the electrical stability and FET mobility performance the significant attention has recently been focused on PANI nanofibers. In particular, conducting polymer PANI nanofibers made by electrodeposition and self-assembly method were under intensive discussion [23,27,95]. Electrospinning is a well-established approach to polymer fiber fabrication with the possibility of large scale production of long polymer fibers for incorporation into smart textiles and wearable electronics [100]. It was shown that electrospinning is a promising method to obtain PANI-based nanocables [27,101]. In addition the FET behavior has been reported recently for PANI-CSA/PEO nanofibers made by electrospinning technique [95]. Saturation channel currents were observed at surprisingly low source–drain voltages −0.4 to 0.6 V, while the 1D charge density (at zero gate bias) was calculated to be ∼1 hole per 50 two-ring repeat units of PANI, consistent with the rather high channel conductivity (∼10^{-3} S/cm). It was found that for PANI-CSA/PEO nanofiber FET the hole μ_{FET} in the depletion regime is ∼1.4×10^{-4} cm²/Vs—lower than that for FET structures based on PA fibers. Reducing or eliminating the PEO content in the fiber is expected to enhance device parameters. The relatively low μ_{FET} values observed for FETs based on PA, PPy, PANI/PEO polymer fibers have motivated us to look for polymers of better quality as well as for the improved geometry of the nanofiber FETs. To make the polymer fiber FET more efficient one should decrease the amount of traps as well as the concentration of conjugation defects during the polymer synthesis. The higher degree of crystallinity of polymer fibers should provide the better device performance. Another problem for polymer nanofiber FETs is the search for the cost-effective way to position the fibers and tubes (from solution) in a given direction in a high-density electronic circuit. In parallel one should try to decrease the channel length down to several nanometers. The use of the polymer fibers with a diameter of the order ∼10 nm in ultimate double-gate FET structure (Figure 16.17a) in combination with high-k gate dielectric would provide the significant improvement of device characteristics. The proposed tubular structure (Figure 16.17b) is almost ideal in terms of electrostatic control of charge carriers in the channel if coupled with well-controlled high-k gate insulator and a metal gate with an appropriate work function. Such presumed FET structure may be realized by either carbon nanotube or polymer nanofiber [102]. The knowledge of the transport mechanism in such ideal nanowire FET structures should be the promising task for the future. We believe that by using conjugated polymer nanofibers, it might be possible to combine the insulating and semiconducting properties of such low-dimensional organic materials to achieve the best nanodevice characteristics.

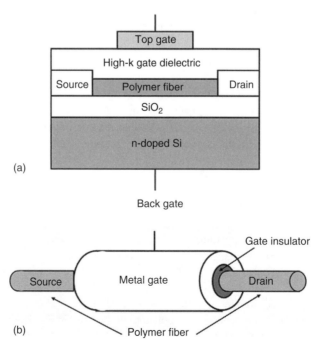

FIGURE 16.17   (a) Schematic of a double-gate FET based on polymer fiber. (b) The ideal nanofiber FET (not realized yet).

16.6 Conclusions

There has been significant progress in synthesis and electrical characterization of nanofibers and tubes made of conjugated polymers during the last decade. Our understanding of the electrical transport in polymer nanofibers and tubes has improved dramatically but is still the subject of intensive discussion. Recent advances in electrical properties and device performance characteristics have resulted from the introduction of novel, more stable polymeric materials and methods promising for applications in nanotechnology. We have considered the applicability of various theoretical models in order to under-stand the recent experimental results on transport in conducting polymer nanofibers and tubes. The power-law behavior observed recently in polymer fibers and tubes for both $G(T)$ and $I(V)$ is charac-teristic of such 1D systems as several LLs connected in series. This result indicates that the inherent 1D nature of conjugated polymers clear manifests itself in the transport properties of conducting polymer nanofibers and tubes. At low temperatures such a doped polymer fiber can be considered as an array of small conducting regions separated by nanoscale tunneling junctions where the CB tunneling is the dominant transport mechanism. However, some discrepancy between our results and theoretical models for tunneling in truly 1D systems still exists especially at low temperature. This makes for important further experimental and theoretical efforts to clarify the origin of the low temperature transport in such quasi-1D systems as conducting polymer nanofibers and tubes. We have suggested some promising nanoelectronic applications of polymer nanofibers and tubes as double-gate, nanofiber FETs in order to achieve the better device performance in the near future.

Acknowledgments

The authors are grateful to H.J. Lee, K. Akagi, and V.I. Kozub for their collaboration. This work was supported by the Nano Systems Institute–National Core Research Center (NSI-NCRC) program of KOSEF, Korea. Support from the Brain Pool Program of Korean Federation of Science and Technology Societies for A.N.A. is gratefully acknowledged. Partial support for Y.W.P. was provided by the Royal Swedish Academy of Science.

References

1. Duan, X., et al. 2003. Nanowire nanoelectronics assembled from the bottom-up. In *Molecular nanoelectronics*, eds. M.A. Reed, and T. Lee. California: ASP, Chapter 2, p. 199.
2. MacDiarmid, A.G. 2001. Nobel lecture: Synthetic metals: A novel role for organic polymers. *Rev Mod Phys* 73:701.
3. Chiang, C.K., et al. 1977. Electrical conductivity in doped polyacetylene. *Phys Rev Lett* 39:1098.
4. Heeger, A.J., et al. 1988. Solitons in conducting polymers. *Rev Mod Phys* 60:781.
5. Heeger, A.J. 2001. Nobel lecture: Semiconducting and metallic polymers: The fourth generation of polymeric materials. *Rev Mod Phys* 73:701.
6. For review, see: Skotheim, T.A., R.L. Elsenbaumer, and J.R. Reynolds, eds. 1996. *Handbook of conducting polymers*, 2nd ed. New York: Marcel Dekker.
7. Aleshin, A.N. 2004. Organic microelectronics based on polymer nanostructures. In *Future trends in microelectronics: The nano, the ultra, the giga, and the bio*. New York: Wiley, p. 253.
8. Menon, R., et al. 1996. Metal–insulator transition in doped conducting polymers. In *Handbook of conducting polymers*, 2nd ed., eds. T.A. Skotheim, R.L. Elsenbaumer, and J.R. Reynolds. New York: Marcel Dekker.
9. Kohlman, R.S., J. Joo, and A.J. Epstein. 1996. Conducting polymers: Electrical conductivity. In *Physical properties of polymers handbook*, ed. J. Mark. New York: American Institute of Physics.
10. Kozub, V.I., and A.N. Aleshin. 1999. Transport anomalies in highly doped conjugated polymers at low temperatures. *Phys Rev B* 59:11322.

11. Kozub, V.I., et al. 2002. Evidence of magnetoresistance for nanojunction-controlled transport in heavily doped polyacetylene. *Phys Rev B* 65:224204.
12. Suh, D.S., et al. 2002. Linear high-field magnetoconductivity of doped polyacetylene up to 30 Tesla. *Phys Rev B* 65:165210.
13. Martin, C.R. 1994. Nanomaterials: A membrane-based synthetic approach. *Science* 266:1961.
14. Akagi, K., et al. 1998. Helical polyacetylene synthesized with a chiral nematic reaction field. *Science* 282:1683.
15. Wan, M.X., et al. 2003. Studies on nanostructures of conducting polymers via self-assembly method. *Synth Met* 135:175.
16. Huang, J., et al. 2003. Polyaniline nanofibers: Facile synthesis and chemical sensors. *J Am Chem Soc* 125:314.
17. Huang, J., and R.B. Kaner. 2004. A general chemical route to polyaniline nanofibers. *J Am Chem Soc* 126:851.
18. Norris, I.D., et al. 2000. Electrostatic fabrication of ultrafine conducting fibers: Polyaniline/polyethylene oxide blends. *Synth Met* 114:109.
19. Park, J.G., et al. 2002. Electrical resistivity of polypyrrole nanotube measured by conductive scanning probe microscope: The role of contact force. *Appl Phys Lett* 81:4625.
20. Park, J.G., et al. 2003. Current–voltage characteristics of polypyrrole nanotube in both vertical and lateral electrodes configuration. *Thin Solid Films* 438–439:118.
21. Duvail, J.L., et al. 2002. Transport and vibrational properties of poly(3,4-ethylenedioxythiophene) nanofibers, 131:123.
22. Wan, M.X., et al. 2001. Template-free synthesized microtubules of conducting polymers. *Synth Met* 119:71.
23. Long, Y., et al. 2003. Electrical conductivity of an individual polyaniline nanotube synthesized by a self-assembly method. *Macromol Rapid Commun* 24:938.
24. Long, Y., et al. 2005. Electronic transport in single polyaniline and polypyrrole microtubes. *Phys Rev B* 71:165412.
25. Li, W., and H.-L. Wang. 2004. Oligomer-assisted synthesis of chiral polyaniline nanofibers. *J Am Chem Soc* 126:2278.
26. Zhang, X., et al. 2004. Nanofibers of polyaniline synthesized by interfacial polymerization. *Synth Met* 145:23.
27. Zhou, Y., et al. 2003. Fabrication and electrical characterization of polyaniline-based nanofibers with diameter below 30 nm. *Appl Phys Lett* 83:3800.
28. Samitsu, S., et al. 2005. Conductivity measurements of individual poly(3,4-ethylenedioxythiophene)/poly(styrenesulfonate) nanowires on nanoelectrodes using manipulation with an atomic force microscope. *Appl Phys Lett* 86:233103.
29. Park, J.H. 2004. Electronic and scanning tunneling spectroscopic studies of conducting polymer nanostructures: Polyacetylene nanofibers, PPV nanotubes and MEH-PPV nanowires. Ph.D. thesis, Seoul National University, Seoul.
30. Park, J.G., et al. 2001. Nanotransport in polyacetylene single fiber: Toward the intrinsic properties. *Synth Met* 119:53.
31. Park, J.G., et al. 2001. Gating effect in the I–V characteristics of iodine doped polyacetylene nanofibers. *Synth Met* 119:469.
32. Park, J.G., et al. 2001. Quantum transport in low-dimensional organic nanostructures. *Thin Solid Films* 393:161.
33. Park, J.G., et al. 2003. Tunneling conduction in polyacetylene nanofiber. *Synth Met* 135–136:299.
34. Akagi, K., et al. 1999. Helical polyacetylene synthesized under chiral nematic liquid crystals. *Synth Met* 102:1406.
35. Piao, G., et al. 2001. Synthesis of well-controlled helical polyacetylene films using chiral nematic liquid crystals. *Curr Appl Phys* 1:121.

36. Lee, H.J., et al. 2004. Dispersion and current–voltage characteristics of helical polyacetylene single fibers. *J Am Chem Soc* 126:16722.

37. Akagi, K. 2004. Unpublished data.

38. Kaiser, A.B. 2001. Electronic transport properties of conducting polymers and carbon nanotubes. *Rep Progr Phys* 64:1.

39. Kaiser, A.B. 2001. Systematic conductivity behavior in conducting polymers: Effects of heterogeneous disorder. *Adv Mater* 13:927.

40. Mott, N.F., and E.A. Davis. 1979. *Electronic processes in non-crystalline materials*, 2nd ed. Oxford: Clarendon Press.

41. Khmelnitskii, D.E., and A.I. Larkin. 1981. Mobility edges shift in external magnetic field. *Solid State Commun* 39:1069.

42. Sheng, P. 1980. Fluctuation-induced tunneling conduction in disordered materials. *Phys Rev B* 21:2180.

43. Aleshin, A., et al. 1997. Electronic transport in doped poly(3,4-ethylenedioxythiophene) near the metal–insulator transition. *Synth Met* 56:61.

44. Aleshin, A.N., S.R. Williams, and A.J. Heeger. 1998. Transport and magnetic properties of poly(3,4-ethylenedioxythiophene)/poly(styrenesulfonate) films. *Synth Met* 94:173.

45. Long, Y., et al. 2003. Electrical conductivity of a single conducting polyaniline nanotube. *Appl Phys Lett* 83:1863.

46. Suh, D.-S., et al. 2001. Helical polyacetylene heavily doped with iodine: Magnetotransport. *J Chem Phys* 114:7222.

47. Aleshin, A.N., et al. 2004. One-dimensional transport in polymer nanofibers. *Phys Rev Lett* 93:196601.

48. Aleshin, A.N., et al. 2005. One-dimensional transport in polymer nanowires. *Microelectr Eng* 81:420.

49. Voit, J. 1995. One-dimensional Fermi liquids. *Rep Prog Phys* 58:977.

50. Schulz, H.J. 1993. Wigner crystal in one dimension. *Phys Rev Lett* 71:1864.

51. Mukhopadhyay, R., C.L. Kane, and T.C. Lubensky. 2001. Sliding Luttinger liquid phases. *Phys Rev B* 64:045120.

52. Artemenko, S.N. 2004. Impurity-induced stabilization of Luttinger liquid in quasi-one-dimensional conductors. *JETP Lett* 79:277 [*Pis'ma Zh Eksp Teor Fiz* 79:335, 2004].

53. Balents, L. 1999. Orthogonality catastrophes in carbon nanotubes. *cond-mat*/9906032.

54. Bockrath, M., et al. 1999. Luttinger-liquid behaviour in carbon nanotubes. *Nature* (*London*) 397:598.

55. Bachtold, A., et al. 2001. Suppression of tunneling into multiwall carbon nanotubes. *Phys Rev Lett* 87:166801.

56. Zaitsev-Zotov, S.V., et al. 2000. Luttinger-liquid-like transport in long InSb nanowires. *J Phys C* 12:L303.

57. Slot, E., et al. 2004. One-dimensional conduction in charge-density wave nanowires. *Phys Rev Lett* 93:176602.

58. Chang, A.M., L.N. Pfeiffer, and K.W. West. 1996. Observation of chiral Luttinger behavior in electron tunneling into fractional quantum hall edges. *Phys Rev Lett* 77:2538.

59. Park, J.G. 2003. Electrical transport properties of conducting polymer nanostructures: Polyacetylene nanofiber, polypyrrole nanotube/nanowire. Ph.D. thesis, Seoul National University, Seoul.

60. Shklovskii, B.I., and A.L. Efros. 1984. *Electronic properties of doped semiconductors*. New York: Springer.

61. Fogler, M.M., S. Teber, and B.I. Shklovskii. 2004. Variable-range hopping in quasi-one-dimensional electron crystals. *Phys Rev B* 69:035413.

62. Kaiser, A.B., and Y.W. Park. 2002. Conduction mechanisms in polyacetylene nanofibers. *Curr Appl Phys* 2:33.

63. Kaiser, A.B., S.A. Rogers, and Y.W. Park. 2004. Charge transport in conducting polymers: Poly-acetylene nanofibers. *Mol Cryst Liq Cryst* 415:269.

64. Kao, K.C., and W. Hwang. 1981. *Electrical transport in solids.* Oxford: Pergamon Press.

65. Hausler, W., L. Kecke, and A.H. MacDonald. 2001. Strongly Correlated Electrons Mesoscopic Systems and Quantum Hall Effect. *cond-mat*/0108290.

66. Gornyi, I.V., A.D. Mirlin, and D.G. Polyakov. 2006. Dephasing and weak localization in disordered Luttinger liquid. *cond-mat*/0407305.

67. Tanatar, B., and D.M. Ceperley. 1989. Ground state of the two-dimensional electron gas. *Phys Rev B* 39:5005.

68. Maurey, H., and T. Giamarchi. 1995. Transport properties of a quantum wire in the presence of impurities and long-range Coulomb forces. *Phys Rev B* 51:10833.

69. Jeon, G.S., M.Y. Choi, and S.R. Yang. 1996. Eric, Coulomb gaps in one-dimensional spin-polarized electron systems. *Phys Rev B* 54, R8341.

70. Jeon, T.-I., et al. 2005. Electrical and optical properties of polyacetylene film in THz frequency range. *Curr Appl Phys* 5:289.

71. Shklovskii, B.I. 2004. Coulomb gap and variable-range hopping in a pinned Wigner crystal. *Phys Stat Sol(c)* 1:46.

72. Yao, Z., et al. 1999. Carbon nanotube intramolecular junctions. *Nature (London)* 402:273.

73. Gao, B., et al. 2004. Evidence for Luttinger-liquid behavior in crossed metallic single-wall nano-tubes. *Phys Rev Lett* 92:216804.

74. Aleshin, A.N., et al. 2004. Hopping conduction in polydiacetylene single crystals. *Phys Rev B* 69:214203.

75. Aleshin, A.N., et al. 2005. Nonohmic conduction in polydiacetylene thin films. *Curr Appl Phys* 5:85.

76. Zaitsev-Zotov, S.V. 1993. Classical-to-quantum crossover in charge-density wave creep at low temperatures. *Phys Rev Lett* 71:605.

77. Mihaly, G., Y. Kim, and G. Gruner. 1991. Crossover in low-temperature collective spin-density-wave transport. *Phys Rev Lett* 67:2713.

78. Simmons, J.G. 1967. Pool-Frenkel effect and Schottky effect in metal–insulator–metal systems. *Phys Rev Lett* 155:657.

79. Shklovskii, B.I. 1972. Percolation mechanism of electrical conduction in strong electric fields. *Fiz Tekh Poluprovodn* (Leningrad) 6:2335 [*Sov Phys Semicond* 6:1964, 1973].

80. Shklovskii, B.I. 1979. Hopping conduction in semiconductors subjected to a strong electric fields. *Fiz Tekh Poluprovodn* (Leningrad) 13:93 [*Sov Phys Semicond* 13:53].

81. Paschen, S., et al. 1995. Tunnel junctions in a polymer composite. *J Appl Phys* 78:3230.

82. Sze, S.M. 1985. *Semiconductor devices physics and technology.* New York: Wiley.

83. Maki, K. 1977. Creation of soliton pairs by electric fields in charge-density—wave condensates. *Phys Rev Lett* 39:46.

84. Roth, S., and H. Bleier. 1987. Solitons in polyacetylene. *Adv Phys* 36:385.

85. Aleshin, A.N., et al. Coulomb-blockade transport at low temperature in quasi-one dimentional polymer nanofibers. *cond-mat*/0506069.

86. Imry, Y. 2002. *Introduction to mesoscopic physics.* Oxford: Oxford University Press.

87. Middleton, A.A., and N.S. Wingreen. 1993. Collective transport in arrays of small metallic dots. *Phys Rev Lett* 71:3198.

88. Clarke, L., et al. 1997. Transport in gold cluster structures defined by electron-beam lithography. *Appl Phys Lett* 71:617.

89. Parthasarathy, R., et al. 2004. Percolation through networks of random thresholds: Finite tempera-ture electron tunneling in metallic nanocryctal arrays. *Phys Rev Lett* 92:076801.

90. Ancona, M.G., et al. 2001. Coulomb blockade in single-layer Au nanoclaster film. *Phys Rev B* 64:033408.

91. Bezryadin, A., R.M. Westervelt, and M. Tinkham. 1999. Self-assembled chains of graphitized carbon nanoparticles. *Appl Phys Lett* 74:2699.

92. Grabert, H., and M.H. Devoret. 1992. *Single charge tunneling*, eds. H. Grabert, and M.H. Devoret. NATO ASI Series. New York: Plenum, Chapter 1.

93. Kim, B.H., et al. 2003. Characteristics and field emission of conducting poly(3,4-ethylenedioxythiophene) nanowires. *Appl Phys Lett* 83:539.

94. Kim, B.H., et al. 2005. Synthesis, characteristics, and field emission of doped and de-doped polypyrrole, polyaniline, poly(3,4-ethylenedioxythiophene) nanotubes and nanowires. *Synth Met* 150:279.

95. Pinto, N.J., et al. 2003. Electrospun polyaniline/polyethylene oxide nanofiber field-effect transistor. *Appl Phys Lett* 83:4244.

96. Horowitz, J. 1998. Organic field-effect transistors. *Adv Mater* 10:365.

97. Brown, A.R., et al. 1997. Field-effect transistors made from solution-processed organic semiconductors. *Synth Met* 88:37.

98. Tans, S.J., A.R.M. Verschueren, and C. Dekker. 1998. Room-temperature transistor based on a single carbon nanotube. *Nature (London)* 393:49.

99. Keren, K., et al. 2003. DNA-templated carbon nanotube field-effect transistor. *Science* 302:1380.

100. Reneker, D.H., et al. 2000. Bending instability of electrically charged liquid jets of polymer solutions in electrospinning. *J Appl Phys* 87:4531.

101. MacDiarmid, A.G., et al. 2001. Electrostatically-generated nanofibers of electronic polymers. *Synth Met* 119:27.

102. *Semiconductor Industry Association ITRS 2001 roadmap.*

17

Structure Studies of π- and σ-Conjugated Polymers

17.1 Introduction

Over the last decade the science and applications of conducting or, more specifically, conjugated polymers (CPs) have continued to evolve and progress through coordinated research activities spanning a large number of disciplines. This broad effort has proved essential because these materials display a unique fusion of electronic properties, traits often exploited in traditional inorganic semiconductors and metals, with characteristics more typical of soft matter and conventional polymers [1]. The range of organic compounds that can be formed is immense and, not surprisingly, CPs exhibit a complex interplay between structure, charge transport, and photophysics that is intimately tied to physical and chemical characteristics starting at the molecular level. Although most CP applications are based on specific electronic or optical attributes, there is widespread recognition that structural order, both along the polymer chain and between chains, is ultimately responsible for defining their function in virtually all settings.

Broadly speaking, the actual resemblance of CPs to ordinary inorganic semiconductors and their electronic device applications may be superficial and fleeting. Most traditional semiconductors are based on well-defined structural repeat units with a high degree of three-dimensional (3D) order. These structures tend to be rigid and, at the molecular level, relatively isotropic so that once the band structure is specified much of the microscopic behavior is only of peripheral concern. As working device length scales have shrunk far more attention has been placed on interfacial phenomenon and the role of disorder, but there is still an overt reliance on models that assume a large degree of periodicity and structural rigidity at the molecular level.

In contrast, only a relatively small fraction of CPs manifest a high degree of structural order beyond that imposed by chemical synthesis. Polymers, by nature, are usually characterized by a twofold coordination geometry of the covalent bonds that constitute the polymer backbone. Torsional degrees of freedom are almost always present and in CPs, because of charge conjugation along the backbone, there is a high degree of torsional stiffness in comparison with most conventional polymers. Simply stated, CPs tend to be rigid rods. In directions orthogonal to the skeletal backbone the interactions are very much weaker. Depending upon the specific chemical architecture these interactions can be identified as ionic, hydrogen (H) bonding and/or van der Waals in nature. Competing interactions can give rise to enormous diversity in the resulting structures and phase behavior.

In terms of working materials in a real world setting there are two overwhelmingly important structural characteristics that impact device behavior. These key attributes are: (1) the overall 3D organization of the polymers themselves, and (2) the relative positioning of these chains with regard to the physical environment around them. These properties correspond to bulk structure and orientation. Specifying and exploiting these characteristics are challenging. One heavily represented subject is, e.g., supramolecular ordering [2]. In this case, molecular self-assembly is exploited for the design of large-scale structural entities. Somewhat less emphasis has been placed in deciphering the nature of intrinsic disorder and structural inhomogeneities. Elucidating deviations from the norm now constitute an increasingly larger part of CP structure studies and this topic, as already suggested, has a major impact on material properties. This facet is especially important with regard to these electroactive polymers and their applications.

The main goal of this chapter is to provide both a historical and up-to-date introductory survey of CP research with an emphasis not only on highlighting structure and structural phase behavior in the solid state, but also strives to put these results into a larger context. Thus, there are a number of topical short segments intended to give a brief synopsis of how structure at the molecular level influences other physical properties. Even the most crystalline of CPs include far more disorder and structural inhomogeneities than their crystallized oligomer counterparts. A number of structure-related summaries have appeared previously [3,4] and this work will tend to refer to these articles rather than revisit and reiterate all earlier issues in their entirety. The first part of the text is devoted to reviewing a number of prototypical CP families. Thereafter, the text will introduce a selected subset of functionalized CPs. Finally, in the last part, two important structure-related topics will be very briefly summarized. These are: (1) direct methods for assessing local structure, and (2) observations and control of chain orientation. As with all reviews, it is impossible to cover every aspect of CP structure in a wholly objective and balanced fashion. This chapter will only touch a selected subset of topics reflecting structural characteristics at or near molecular-level length scales.

17.2 Conducting Polymer Families and Basic Structural Characteristics

Over the last 25 years the conducting polymer field has existed and there has been a dramatic increase in the sheer number of compounds available [5,6]. Originally linear unsubstituted materials and various "doped" intercalation compounds were the central focus with the wistful idea that CPs might ultimately replace wires and interconnects [7]. Many of the originally synthesized materials exhibited appreciable long-range order and so deciphering the average structure of these compounds was a major undertaking. This information provided a much needed foundation for pursuing complementary experimental studies of optical properties and charge transport. Moreover, this information could be used for specifying a geometrical framework for theoretical calculations [8–12]. In this initial phase of research the importance of understanding the nature of the polymer chain bond alternation and interchain interactions was immediately recognized [13]. Better knowledge of these characteristics remains relevant in ongoing studies of new materials.

In regard to efforts directly aimed at relating the explicit physical structure to both charge motion and photophysics, the pendulum has shifted somewhat toward oligomers [14,15] and oligomer crystal measurements [16]. Single crystals clearly have a high degree of structural order and so many of the techniques developed for systems with periodicity are applicable. Moreover, ab initio quantum chemical computation tools can be used to assess the ground and excited states of various oligomers with rapidly improving accuracy [17,18]. Even for these materials (and a proper review is clearly a subject itself), because of the weakness of the intermolecular van der Waals forces, there can be appreciable disorder. The influence of this disorder is very pronounced and the manner in which this disorder is introduced into theoretical calculations is still being developed. For a herringbone (HB)-type packing (see Section 17.2.1) the interchain interactions, in terms of impact on band structure, can be very substantial and the effects of aggregation are almost always evident [19]. In materials with very low levels of disorder there can be a strong enhancement of the optical emission from the direct π–π^* transition (or, equivalently, the 0–0 transition or zero phonon line) and this is identified as superradiance [16,20]. CPs invariably have much higher levels of disorder. However, interchain interactions generally enhance nonradiative decay routes and often produce an overt reduction in the strength of the main π–π^* emission signatures, especially the direct 0–0 optical transition in terms of a Franck-Condon manifold of vibronic states. Current CP studies tend to be more qualitative. For example, Schwartz and Nguyen [21–23] have carefully prepared a series of functionalized CPs with systematically increasing levels of aggregation and this strongly affects the observed optical spectroscopy. Quantifying the structural characteristics with an explicit molecular-level model is difficult [24].

17.2.1 Linear Unsubstituted CPs

17.2.1.1 Polyacetylene

Polyacetylene (PA), the quintessential infinite polyene as sketched in Figure 17.1a, is still acknowledged as having a defining role in establishing the CP field. Of all events it was the unanticipated discovery of a synthetic route to PA in dense fibrillar film form by Shirakawa and Ikeda [25] and, somewhat later, "doping" by oxidizing or reducing agents (e.g., halogens or alkali metals) in conjunction with Heeger, MacDiarmid, and collaborators [26,27] that led to widespread excitement and recognition of the potential of polymer electronics [28,29]. Much of the early enthusiasm was tempered by the fact that PA itself was not only intractable and infusible, but also degraded in the presence of air. Innumerable papers with PA as the central subject have appeared and, although clearly no longer a prime candidate for most applications, PA studies continue to provide insight into the nature of charge transport. Recent studies of iodine-doped helical PA nanofibers [30,31] have, e.g., demonstrated that at these very small fiber dimensions, conventional models for electron transport cannot adequately describe the observed power law variations in the conductance with temperature.

However, as a prototypical doubly degenerate ground state material (labeled as A and B in Figure 17.1a) the chemical and structural simplicity of PA has provided a robust reference frame. As is now well established, the wavefunction overlap of the singly occupied p_z orbital at each and every carbon atom leads to π conjugation and, as a consequence of a Peierls instability [32], a pronounced double bond–single bond dimerization occurs. Pure PA exhibits a filled valence band and an empty conduction band and so is innately a semiconductor with a π–π^* transition that ranges from approximately 1.4 to 1.8 eV depending on the explicit chain morphology. There is a well-known conformational isomerism in which PA may be prepared in either of the "two forms," *cis* or *trans*. More accurately there are four distinct conformational isomers [33]. Of these the *trans* planar conformer has the smallest bandgap. Synthesis at reduced temperatures gives a preponderance of the *cis* isomer and interconversion [34–36] to *trans*-PA (or *t*-PA) is facilitated by thermal annealing or doping. The degenerate ground state of PA requires that

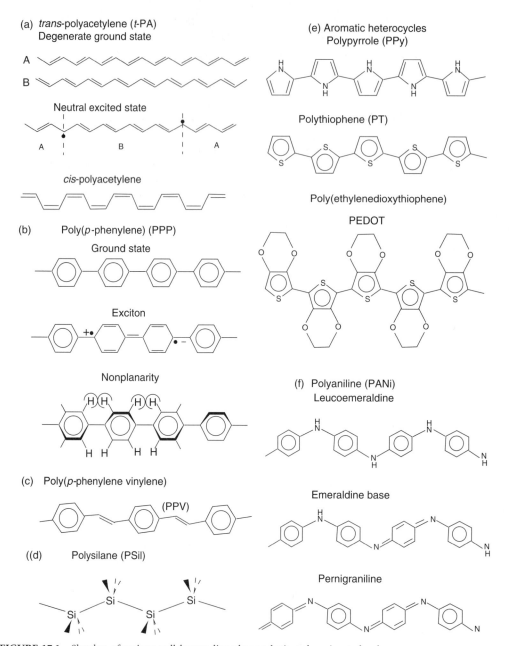

FIGURE 17.1   Sketches of various well-known linearly unsubstituted conjugated polymers.

the excited electronic states of *t*-PA are marked by a domain wall or change in the bond alternation [1,37] when passing from a type A dimerization to type B regions and then back to A. A more recent advance in PA synthesis and, by intended molecular engineering design, structure as well is the advent of chiral PA [38]. Development of these extended helices is completely consistent with a local conformational degree of freedom in regard to rotations about individual C–C single bonds along the polymer backbone.

PA in bulk has structural qualities typical of many of the linear unsubstituted CPs. Early papers by Baughman et al. [39], Akaishi et al. [40], and Fincher et al. [13] identified that PA readily forms small crystalline domains of approximately 200 Å in extent wherein there is a classic HB packing of the

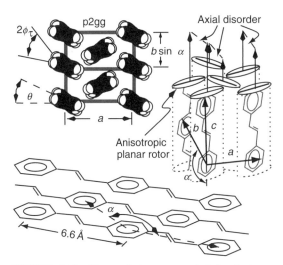

FIGURE 17.2 Views of the prototypical herringbone packing in poly(*p*-phenylene vinylene) and this motif is often seen in crystalline unsubstituted π-conjugated polymers. (Reprinted from Winokur, M.J. and Chunwachirasiri, W., *J. Polym. Sci.*, 41, 2630, 2003. With permission.)

individual chains to give a two polymer chain unit cell. This structural packing motif is common to many organic substances. Figure 17.2 displays this explicit packing for the polymer poly(*p*-phenylene vinylene) (PPV) (and this polymer will be discussed later) along with many of the molecular-level structural elements that need to be considered in real polymers. In terms of the single-chain morphology, there can be torsional deviations by the constituents comprising the chain and in this figure this degree of freedom is designated by ϕ_τ. Approximately, the equatorial projection of each individual chain can be interpreted as an anisotropic rotor. The orientation of the major axis with respect to a crystallographic unit cell direction (in this case a) is defined as a setting angle θ and this reflects hard-core packing and the true intermolecular nearest-neighbor spacings. There can also be shifts in the relative axial chain-to-chain positioning, along the *c*-axis, that act to lower the unit cell

symmetry. For PPV, this is reflected by the angle α in a monoclinic unit cell. For PA, the alignment of single and double bonds is an important issue [13,41]. Disorder between chains is evidenced and this takes many forms. In PPV, and many crystalline CPs as well, there are anisotropic axial chain-to-chain fluctuations that can be quantified [42] and produce characteristic streaking and broadening of the Bragg scattering peaks [43,44] in directions perpendicular to the meridional (i.e., along layer lines).

17.2.1.2 Poly(*p*-Phenylene Vinylene) and Poly(*p*-Phenylene)

Beyond PA, there is a litany of well-known CPs. Although synthetically more challenging to prepare in dense film form [45], a linear concatenated sequence of *para*-bonded phenylene rings can be fabricated [46] (as a partially crystalline powder [47]) and this polymer is generally referred to as poly(*p*-phenylene) (PPP). This also approximately doubles the energy gap between the ground and excited states from 1.4 eV in *t*-PA to ~2.8 eV and, equally important, PPP exhibits a nondegenerate ground state, also shown in Figure 17.1b, along with a sketch of an excited state. Obtaining quantitative results from first-principle calculations of the chain geometry remains as a challenging problem [48]. A very large majority of CPs have a nondegenerate ground state. In the excited state there is still pronounced change in the bond alternation but, in the case of PPP, there is also a changeover from a benzenoid (B) to a quinoid (Q) moiety. This bond alternation also includes a significant energy per unit length over the region that constitutes the excited state. In the case of neutral excitons, in regard to quantum mechanically dictated charge and spin degrees of freedom, there are opportunities for singlet and triplet states. The spatial extent of the bond alternation depends on the nature of the spin state.

In terms of single-chain morphology the impact of the phenylene ring architecture is also very significant. In *t*-PA, the shortest hydrogen–hydrogen (H–H) pair distance is on the order of 2.4 Å. At this spacing the direct H–H interactions are relatively weak. Thus, *t*-PA can adopt a planar or near-planar structure with low-energy torsional modes extending over many monomer units [49,50]. The nature of this static and dynamic structure has been the subject of numerous studies [51,52]. For PPP, the minimum H–H pair distance decreases to below 1.5 Å and, at that spacing, the hard-core repulsive interactions become significant. This α hydrogen repulsion leads to large torsional deviations between neighbor phenylene rings across the C–C single bonds. An earlier structural study of PPP [53] indicated that in the solid state there are alternating torsional variations from planarity on the order of ±20°.

Increasing nonplanarity of the phenylene rings increases the $\pi - \pi^*$ transition energy and reduces the effective conjugation length. The actual torsional potential well is very broad and at ambient temperature there is significant dynamical motion populating these torsional or librational modes. Because of the bond alternation associated with the excited state the nature of the torsional potential is dramatically altered and this impacts both the steady state geometry and thereby the occupancies of these librations. The influence on optical properties is very important and this issue is now being addressed at a rigorous quantitative level in oligomer studies. In *ter*-phenyl (i.e., three phenylene rings), e.g., ab initio calculations by Heimel et al. [54] have been able to quantitatively reproduce the large asymmetry between optical absorption and emission spectra of *ter*-phenyl in dilute solutions.

Intermediate to *t*-PA and PPP, is the well-studied polymer. PPV (shown in Figure 17.1c) shows PPV has the noteworthy distinction that it can be easily processed from a soluble precursor and then thermally converted to its final form [55,56]. (PA films can be prepared from a precursor polymer as well [5].) PPV is simply an AB-type block polymer of repeating phenylene and vinylene linkages and its properties are in many ways intermediate to *t*-PA and PPP. The bandgap is approximately 2.4 eV and the nearest-neighbor distance between vinyl and phenyl hydrogens is closer to 1.8 Å. This increase (over PPP) strongly modifies the nature of the phenylene ring torsion potential. In this case, the average phenylene ring deviation from planarity is approximately halved [57] to 10° and the barrier to librational motion is very much reduced. NMR studies [58–61] reveal both small- and large-scale torsional motions (i.e., librations and ring flips, respectively). These librational motions can be inferred by neutron diffraction of labeled samples [57]. In terms of interchain packing, PPV, as already noted, also adopts the HB-type motif with relatively large coherence lengths [62–65].

17.2.1.3 Polyheterocycles, Polypyrrole, and Polythiophene

Complementing CPs consisting solely of carbon and hydrogen are a large number of available polyheterocycles. Among these, polythiophene (PT) and polypyrrole (PPy) are probably the most well known (see Figure 17.1e) and, in terms of synthesis, both chemical and electrochemical approaches have been widely used. Both routes yield materials that are intractable and infusible but the influence of this synthesis, with respect to chemical variations at the molecular level, can be clearly discerned. Chemically synthesized PT powders exhibit the ubiquitous HB packing [66,67] while PPy is substantially more disordered [68]. Electrochemical synthesis generally yields dense films containing a very high proportion of amorphous material [69,70]. Many of these papers are only concerned with films that include dopant and not the neutral polymer. Cross-branching and other structural defects abound.

The vast majority of recent research has focused on functionalized PTs and, to a lesser extent, PPys. One of the simplest of CP "functionalizations" has been to directly alter the basic chemical structure of the monomer unit. Of great technical importance [71] has been the addition of an ethyl-dioxy moiety at the two hydrogen atom positions of the thiophene ring to obtain poly(ethyl-dioxythiophene) (PEDOT) (also shown in Figure 17.1e). A main advantage is that a resonance electron donation from both oxygens effectively lowers the observed bandgap to, in some cases, below 1.2 eV. In its doped form, very often with polystyrene sulfonate (PSS), this PSS–PEDOT complex is both conductive and, in thin-film form, partially transparent. This latter property derives from the reduced bandgap in the polymer itself, which then, after doping, produces a relatively smooth absorption band throughout the visible spectrum. PEDOTs can therefore be used in a myriad of thin-film applications requiring a transparent electrode. Perhaps, the best known use is that of a hole transport layer in polymer light-emitting diodes. Recent reports identify routes to PEDOT formulations with conductivities in excess of $1000 \; \Omega^{-1} \; cm^{-1}$ [72]. Many other pendant functional groups have also been attached [71,73], but these materials are not in widespread use and even less is currently known about their structure.

A few suggestive PEDOT structural studies [74,75] have been reported, but quantitative knowledge of the molecular packing in ordered regions is limited. An amorphous phase fraction, as shown in Figure 17.3, always dominates the reported scattering data and there are very few Bragg peaks from which this packing can be deduced although the data are sufficient to propose a basic model [76]. A major finding is in regard to the 3D orientation and both these papers identify a large scale of anisotropy

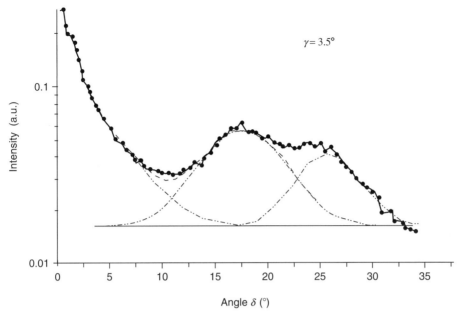

$\gamma = 3.5°$

FIGURE 17.3 A typical x-ray diffraction profile from a PEDOT thin film as recorded by Breiby et al. These data come from a thin-film sample and δ refers to the in-plane scattering angle (nominally 2θ, $\lambda = 1.550$ Å) and γ refers to the out of plane angle of the detector. (Reprinted from Breiby, D.W., Samuelsen, E.J., Groenendaal, L., and Struth, B., *J. Polym. Sci. Polym. Phys.*, 41, 945, 2003. With permission.)

in the PEDOT structure with respect to the film surface. Microscopic orientation of the polymer chains is one important aspect in interface engineering for CP-based devices requiring efficient charge injection. In one doping/dedoping experiment Aasmundtveit et al. [77] observed reversible changes in the widths of the peaks, but the peak positions remained essentially constant. Specifying the overall structural reorganization was difficult because, in this and similar cases, one does not necessarily know if the dopant itself is leaving and/or the counterion is coinserting. A few studies, primarily based on surface scanning probe measurements, indicate a lamellar-type arrangement of the PEDOT polymer and PSS dopant [78]. At comparable or slightly larger length scales there is evidence of a more granular-type morphology [79] with enormous variations in the transport properties. These inhomogeneities can also be seen in neutron reflection studies [80]. Again the majority of this work is mostly in regard to doped samples and not the neutral parent polymer.

17.2.1.4 Polyaniline

Polyaniline (PANi), sketched in Figure 17.1f, is one of the few linear unsubstituted CPs that is fusible and directly processable. The most common neutral form is referred to as emeraldine base (EB) and EB has a four monomer repeat characterized by four phenylene rings, three benzenoid (B), and one quinoid (Q). PANi exhibits an extremely facile chemistry [81] and can either be partially or fully oxidized or reduced to two additional forms each with specific material properties. These chemical structures are also depicted in Figure 17.1 but will not be discussed further except to note that there is one early structure study [82]. EB–PANi can also be doped with various acids to partially or completely protonate the imine nitrogens, thus giving emeraldine salt (ES), and this change is accompanied by increases in the measured conductivity more than 10 orders of magnitude to values near 1000 Ω^{-1} cm^{-1} [83]. Two special features are the presence of torsional degrees of freedom across the amine nitrogen–carbon bond in combination with opportunities for H-bonding. PANi is better described as a semirigid rod.

Both PANi EB and ES are extremely variable in terms of structure. By varying the processing conditions PANi, in the solid state, can be prepared in forms that range from essentially 100% amorphous [84] to those with moderate levels of crystallinity [85]. In the latter case, the coherence lengths are typically not more than 100 Å. Polymorphism extends to the crystalline phases and, depending on the specific processing history, a series of articles by Pouget and coworkers [85–88] described two very distinct structural forms of EB as EB-I and EB-II. Figure 17.4 shows the pseudoorthorhombic packing of EB-II alongside the proposed HCl-doped ES-I and ES-II crystal structures. Emeraldine samples of class I type are formed when protonated ES films or powders are prepared. EB-I is then obtained by dedoping and is essentially amorphous in this work. Class II samples are obtained when preparing films or powders from nonprotonated PANi solutions. Treatment of EB-II with aqueous HCl solutions yields ES-II.

Although the model EB-II structure has merit as a starting template, more recent reports give evidence of deviations and local distortions in the actual PANi unit cell. The likelihood of these effects was already recognized in the earlier papers. Over the approximately 15 years that have elapsed there has been steady improvement in the modeling of structure at the molecular level. A full-fledged Car-Parrinello molecular dynamics study by Cavazzoni et al. [89] of EB-II provides better agreement with experimental data after lowering the overall unit cell symmetry from orthorhombic to triclinic. Interchain packing interactions are also discerned. Variations in the single-chain morphology include "squeezing" the phenylene rings and, as a consequence, a reduction in the extent of ring torsions and nonplanarity. In another example [90], deprotonation of ES-I gave a partially crystalline EB-I material. Direct modeling of this EB-I scattering data, assuming a single average structure, required the presence of systematic defects in the local chain packing. These PANi chain-packing irregularities were intermediate between that of ES-I and EB-II.

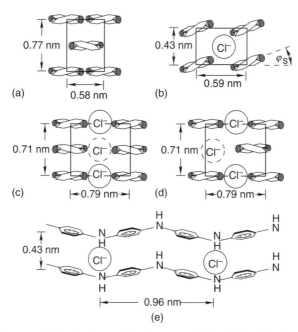

FIGURE 17.4 Proposed equatorial projections and lattice repeats (adapted from Jòzefowicz, M.E., Laversanne, R., Javadi, H.H.S., Epstein, A.J., Pouget, J.P., Tang, X., and MacDiarmid, A.G., *Phys. Rev. B*, 39, 12958, 1989) for various PANi forms including: (a) EB-II; (b) ES-I; (c) ES-II, space group *Pc 2a*; and (d) ES-II, P2₁22₁. In addition, (e) contains a *c*-axis projection of ES-I showing the relative position and spacing of the Cl⁻ anion. (Adapted from Pouget, J.P., Jòzefowicz, M.E., Epstein, A.J., Tang, X., and MacDiarmid, A.G., *Macromolecules*, 24, 779, 1991; and printed from Winokur, M.J. and Mattes, B.R., *Macromolecules*, 31, 8183, 1998. With permission.)

These structural variations strongly impact the observed charge transport [91] and much of this early work is reviewed by Epstein et al. [92]. A major conclusion was that the high degree of heterogeneity between ordered and disordered regions resulted in charge transport suggestive of a granular metal. The best ordered regions exhibited properties on the metallic side of metal–insulator transition. A number of parallel studies of PANi structure (in both polymers and oligomers) appear as well [93–97].

The H-bonding and local structure have been studied by Colomban and collaborators [98–101] in a series of reports. Inelastic neutron scattering experiments, in conjunction with infrared (IR) absorption and Raman spectroscopy studies, indicate that the properties of the doped state (whose larger length scale structural characteristics will be discussed later) do not radically vary between amorphous and crystalline phase fractions. Thus, the abrupt interface often depicted in PANi model sketches appears to be more gradual in nature. Moreover, there is a good evidence that proton conductivity and chain torsions, either by BQB segments or B alone, are strongly correlated with temperature-sensitive variations in the conductivity.

This propensity for H-bonding has also been studied by Zheng et al. [102] to better understand the solution characteristics prior to film formation. H-bonding in solution leads to aggregation and, as one may expect, the resulting film morphology is highly sensitive to the prior solution properties. Addition of LiCl salts was found to suppress H-bonding and reduce aggregation. Mattes et al. [103] obtained concentrated, stable (i.e., nongelling) PANi solutions by using organic H-bonding antagonists and subsequent processing led to completely amorphous PANi EB fibers and films [84]. These are presumably of EB-II type in terms of local structure.

More recent structural studies have also examined the influence of processing and aging on the physical properties in various doped PANi salts [90,104], although the importance of this on transport properties was an important component of earlier reports [105–109]. A study by Wolter et al. [110] can resolve a two-step process in which the more amorphous regions of HCl-doped ES are the first to undergo structural change followed by the crystalline ES domains. Although these aging studies clearly resolve correlated variations in the structural characteristics and significant drops in the measured conductivities, it is still not possible to identify the explicit molecular-level changes that are ultimately responsible for this degradation. (Some of these issues are revisited in the next section.)

Improving the overall extent of structural order to a level approximating that seen in crystalline PA or PPV films has had very limited success [111,112]. Notwithstanding this result, PANi may still be processed and oriented, even in a fully amorphous state [84], and then doped to conductivities in excess of 100 Ω^{-1} cm^{-1}. Even higher levels are seen in partially crystalline films [83]. Achieving substantially larger values probably will require improved interchain ordering, but there are many applications that are perfectly acceptable at this lower level.

17.2.1.5 Polysilane

In this survey of linear unsubstituted CPs some mention of polysilanes must be made. This polymer, sketched in Figure 17.1d, has an all silicon backbone and, to first order, many structural characteristics common to the aforementioned carbon-based polymers. However, there is only a single σ-bond between every Si atom and the onset of the σ−σ* transition, as measured by photoabsorption, is typically located in the UV portion of the spectrum. Thin films are perfectly transparent to the eye. When photoexcited, polysilanes exhibit high hole mobilities [113–115]. Thus, these σ-conjugated polysilanes are just as much a conducting polymer as their π-conjugated counterparts.

Although known to only a minority among the π-conjugated materials research community, the nature of σ-bonding is acknowledged to be less well understood than that of π-bonding and far more ubiquitous [116,117]. As is true in the π conjugates, there have been significant advances in obtaining quantitative agreement with experiment; this is especially true for short polysilane oligomers [118,119]. Particularly notable was the recognition of the importance of a Si through bond resonance integral (i.e., hyperconjugation) by Schepers and Michl [118]. Virtually, all of the torsion angle-dependent σ-conjugation can be attributed to this and studies [118,120] identify an approximate cos φ dependence on Si—Si torsion angles (as opposed to the nominal cos 2ϕ for linear π-conjugated systems [121]). These characteristics are analogously important for specifying the torsional energy potential seen in σ-bonded hydrocarbons [122]. Once again conjugation is seen to be greatest when there is main chain planarity and the backbone adopts an all-*anti* conformation (or in π-conjugation parlance an all-*trans* planar conformation). One major drawback, where device lifetime is an issue, is that photoexcitation in σ-conjugation leaves only a single electron behind to maintain the integrity of the chemical bond between two adjacent Si atoms. Photoscission may be a small statistical possibility during any single photoexcitation, but photodegradation is still quite prevalent and the measured rate is very much dependent on the explicit chain conformation [123,124].

17.2.2 Polymer Intercalation Compounds

17.2.2.1 Polyacetylene, PPV, and PPP Intercalation

Although there has been more emphasis of late in developing applications based on CP semiconductor properties, the electroactive nature of these materials also allows for doping or, more accurately, intercalation [125–127]. Intercalation is marked by the interdiffusion of a guest specie into a host matrix and, more often than not, a concomitant charge transfer between the guest and the host. In CPs the interactions in directions orthogonal to the CP backbone may include ionic, H-bonding and/or van der Waals contributions. The electronic states that form on the backbone are of intermediate extent and, because of the one-dimensional (1D) nature of the polymer, they are able to migrate along the backbone. This leads to the well-known and very useful property that intercalated CPs are inherently conductive. Conductivities [1] near $10^5 \ \Omega^{-1} \ cm^{-1}$ have been reported for iodine-doped PA although values near or slightly above $10^3 \ \Omega^{-1} \ cm^{-1}$ are more typically seen for acid-doped PANi, PEDOT, or halogen-doped alkyl-substituted PTs. There are necessarily anisotropic interactions at molecular length scales between the guest and host so that in order to accommodate the intercalate there must be a self-consistent structural reorganization by all the constituents. Guest/host concentrations can approach or exceed 1:1 in heavily doped CPs. (The terms dopant and intercalant will be used interchangeably although the latter is technically incorrect.) Large-scale incorporation of these dopants necessarily involves both translational and rotational degrees of freedom in the polymer host as well as the guest with, in many instances, an irreversible decline in the long-range order.

In terms of potential applications numerous examples have been identified. The first was obviously as a conductor for electrostatic dissipation in polymer composites or textiles. In some instances, the process of doping and dedoping alters the local interchain morphology so that the molecular pore size may be precisely tuned. In this case, it is possible to prepare high-performance membranes for gas or liquid separations [128–130]. The change in physical dimensions upon intercalation can be quite dramatic and so, in the context of biological mimicry, it is possible to use CPs as the active working element in artificial muscles or actuators [131,132]. Polymer intercalation also involves change in the chemical potential and, in the same guise, of the band structure. These properties have been exploited in the development of electrochromic devices and anticorrosion applications. In these two later cases, the nature of the molecular structure is very much of secondary importance except to note that structural order at the molecular level impacts the overall dopant ion diffusion. The more crystalline CPs generally have smaller diffusion constants and, by inference, slower response times. These issues are covered in a number of excellent review articles in this handbook and elsewhere [133–136].

In many of these just-mentioned CP applications, the polymers themselves have a limited structural order at the onset and so the actual mechanism of structural reorganization at the molecular level is very difficult to distinguish. Although technologically less important, studies of the most crystalline CP materials have provided a much clearer window into the inner workings of the structural evolution and phase behavior. In these polymers Rietveld analysis of the Bragg scattering peak positions and intensities can be employed to verify a particular structural model. Early studies [137–142], most notably by Shacklette, Baughman, Murthy, and coworkers, were instrumental in identifying the underlying structural evolution of intercalated PA and PPP. In classic graphite intercalation compounds [143], this process is conventionally referred to as staging and this term was similarly applied to the CP phases as well. Subsequent to these early papers were reports [144–153] by Bernier, Chien, Fischer, Pouget, Winokur, and many others in which both quasi-1D channel structures and quasi-2D layered structures could be resolved. For polymer intercalation compounds (PICs) with a high degree of structural symmetry (e.g., tetragonal or hexagonal), indexing of the Bragg peaks was relatively straightforward and overall guest/host packing motif could be readily identified. Many channel structures could be characterized as a quasi-1D column of intercalants surrounded by either three or four polymer chains depending on the size of the dopant ion. A selection of these is sketched in Figure 17.5. With increasing intercalant ion dimensions there is a general crossover from three- to fourfold to finally a layered structure [142].

There were however many instances in which the observed diffraction patterns had peaks that could not be easily indexed in terms of a simple high-symmetry phase. This was especially true of the larger dopants in which there were large-scale distortions and/or formation of a lamellar phase. Coexistence of multiple lower symmetry phases or invoking domain wall-type structures represents some of the hypothesized explanations [142,154]. In a lamellar-type structure, the layers of intercalate need not be in perfect registry with that of the polymer and so, in directions orthogonal to layer normal, there can be incommensurate lattice repeats between the intercalate and the polymer.

There are very few explicit examples in which this complicated phase behavior can be satisfactorily understood from a theoretical framework. In a notable sequence of papers [155–158], Choi, Mele, and Harris were able to adapt a well-known Hamiltonian, that of the generalized planar rotor model, to provide a clear theoretical basis for interpreting this complex phase behavior in instances where the overall structural relaxation of the host polymer matrix was relatively modest. In this model the perpendicular cross section of the polymer chains were assumed to represent an anisotropic rotor, as shown in Figure 17.6 inset, and the dopants were allocated their own lattice sites. Mean-field-type calculations could identify a regime in which a complex domain-walled structure could exist. Through Monte Carlo modeling methods, the use of this Hamiltonian was extended to address a larger range of dopant concentrations [159,160]. Figure 17.6 displays a generic phase diagram that underlies the structural evolution during alkali-metal intercalation of PPV. This phenomenological approach provides a good vantage point for elucidating the more complex channel structures associated with the most heavily doped phases. However, one sobering caveat is that, even in the restricted setting of a triangular lattice, the overall phase diagram is very sensitive to the interaction parameters. Polymer intercalation should (and does) reflect this complexity.

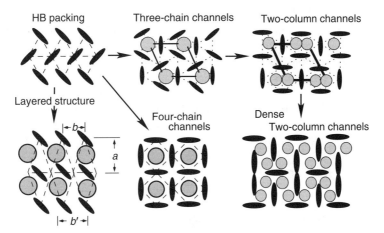

FIGURE 17.5 Models depicting the equatorial packing (i.e., perpendicular to the polymer chain axis) of various channel and layered polymer intercalation compounds.

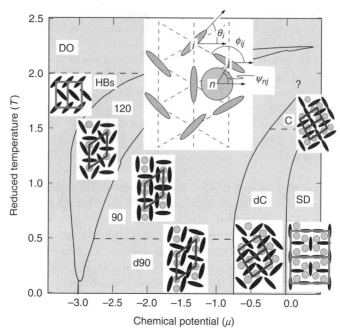

FIGURE 17.6   Extended phase diagram obtained by Monte Carlo calculations of the generalized anisotropic rotor model (GAPR) (see Refs. [157–159]) showing a series of structural phases analogous to those seen in alkali metal intercalation of *t*-PA and PPV as the chemical potential (μ) is varied. *Inset*: view of the rotors (i.e., polymer chain perpendicular cross sections) and an intercalate ion showing the explicit geometrical parameters (θ_i, ϕ_{ij}, and ψ_{nj}) used in the GAPR model calculation. (Adapted from Harris, A.B., *Phys. Rev. B*, 50, 12441, 1994; and Mao, G., Chen, D., Winokur, M.J., and Karasz, F.E., *Phys. Rev. Lett.*, 83, 622588, 1999. With permission.)

The structural attributes along the chain axis have also been examined. In addition to physical insertion of a guest specie, CP intercalation involves charge transfer from/to the CP backbone. Self-consistent with these are changes in the C−C bond angles and the dimerization. Reducing *n*-type dopants (e.g., by alkali metals) generally produce a systematic decrease in the lattice repeat along the chain axis whereas *p*-type oxidizing agents (e.g., iodine and SbCl$_5$) tend to increase the lattice constant [161,162].

17.2.2.2   Polyaniline Intercalation

Unfortunately there is much less structural information available in most other intercalated CPs [163,164] and, in these cases, structural modeling methods are often less quantitative and involve greater uncertainties. PANi is one of the few other intercalated CPs exhibiting crystalline phases sufficiently well developed to enable critical comparisons with structure factor calculations. The already cited 1991 paper by Pouget et al. [86] also identified the presence of structural polymorphism in HCl-doped PANi. Two basic structural motifs were observed and these were designated HCl ES-I and ES-II. This halogen anion is sufficiently small that these phases are, to first order, channel-like in appearance (as already sketched in Figure 17.4).

Once again there are a wide range of possible outcomes dependent on the choice of dopant and subsequent processing. Acid doping of PANi often employs aqueous solutions and, with this procedure, coinsertion of water molecules is of secondary concern. This coinsertion could be advantageous if the water acts as a plasticizer at the molecular level and facilitates structural reorganization. The hygroscopic nature of PANi ES is easily verified because, on exposure to water vapor, dried PANi ES powders exhibit significant changes in the x-ray scattering profiles [90] and many other physical properties as well [105]. The basic ES-II packing of Pouget et al. [86] again represents a good starting point, but results from

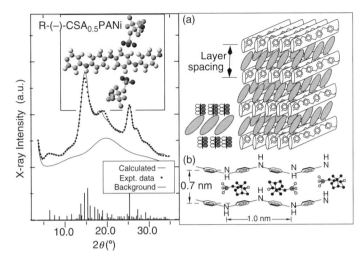

FIGURE 17.7 *Right*: nominal layer-type structure of R-(−)-CSA-doped PANi. *Left*: comparison of powder x-ray diffraction data with a calculated structure factor using the unit cell, $a = 14.3$ Å, $b = 7.1$ Å, $c = 20.4$ Å, $\alpha = 90°$, $\beta = 125°$, $\gamma = 95°$, and a CSA–PANi single-chain model as shown in the *inset*. Black vertical bars reflect the Bragg peak intensities. (Adapted from Winokur, M.J., Guo, H.L., and Kaner, R.B., *Synth. Met.*, 119, 403, 2001. With permission.)

modeling studies that attempt to accommodate the coinsertion of water have many characteristics more in common with layered structures. Layered structures with enhanced interchain overlap of the π-orbitals are beneficial to charge transport.

Larger molecular guests, e.g., camphorsulfonic acid (CSA), induce quantitatively different types of structures and opportunities for thermochromic behavior. CSA also has chiral enantiomers and so is optically active. This optical activity can appear in CSA–PANi salts or, after dedoping, in the neutral polymer itself [165–167]. In terms of molecular structure, both channel- and lamellar-type phases have been proposed in the case of CSA-doped PANi ES [168–170]. This molecular specie is quite bulky and so there are many possible ways to organize the CSA guest within the PANi host matrix. Moreover, the resulting unit cells are sufficiently large as to give numerous and overlapping Bragg reflections in any recorded diffraction pattern. Rigorous structure factor calculations have been attempted and, as shown in Figure 17.7, can yield good agreement with data, but the uniqueness of a particular model is always an important consideration [168,170]. Even larger protonating agents, e.g., dodecylbenzene sulfonic acid (DBSA) or other plasticizing dopants, yield structures [171–173] that begin to resemble those obtained by CPs having functionalizing side chains (and evidence of molecular self-assembly) and so these are further reviewed in Section 17.3.4.

17.2.3 Main Chain Conformation

In all previously mentioned CPs the importance of the chain torsion and conformational isomerism, in terms of the single-chain structure and its central role CP behavior, cannot be overemphasized. Dynamical motions are temperature dependent and, in response, the observed energy transport and optical emission features can change dramatically [174–176]. Static disorder is also very important. Optical probes of local structure, including hole burning [177,178] and single-chain spectroscopy [179–181], demonstrate that while there are broad distributions in the chain structure and environment (this leads to spectroscopic features labeled as inhomogeneous broadening) at the local level, there can be individual chain conformational sequences that function essentially as a single unique chromophore. There are qualitative similarities with biopolymers in which secondary or tertiary structural responses (i.e., folding or aggregation) can be even more important for function than that of the primary chemical sequencing or that of the mean structure over larger length scales. Ab initio calculations also clearly demonstrate the importance of chain morphology in the framework of large-scale periodicity [48,182] and in terms of local structure [54]. Indirect hard-core packing constraints can be used to alter these CP properties, but there are also extremely facile methods that exploit chemical bonding. Functionalization can be used to systematically alter the intrinsic single-chain morphology [183]. This facet of molecular design is one key component of supramolecular self-assembly.

Figure 17.8 displays just a few of the many ways in which the main chain morphology can be directly modified by forming bridges that span two or more molecular units. In the case of PPP, a planarizing carbon bridge across every other phenylene ring yields the polyfluorene polymer family. These materials constitute an important class of blue emitters. By bridging every second and third phenylene rings one can obtain a series of polyindofluorenes [184,185]. Both these families are representative of a generic grouping referred to as stepladder polymers. The bridging process can be extended to give the limiting case in which all phenylene rings are linked to give ladder PPP-type materials.

Heterocycles, e.g., PT, can also be chemically bridged [186,187]. Two examples of stepladders are depicted in Figure 17.8 and the degree of planarity may be systematically engineered by modifying the carbon bridge atom to an oxygen with its longer C−O bond length. The resulting material properties are quite distinct and this modification, in oligomers, allows for molecular engineering of the lifetimes of various photoexcited states [186]. This approach has been exploited by Swager and others to engineer families of ion or chemical-sensing polymers [188]. The actual number of polymers that only contain this type of chemical modification and nothing more (i.e., additional side-chain functionalization) are rare and, in terms of direct structure studies, there is relatively little information available.

A final limiting-case example is that of an extended array of fused phenylene rings to give the polymer polyacene that represents another well-known family of ladder-type materials. The oligomers (e.g.,

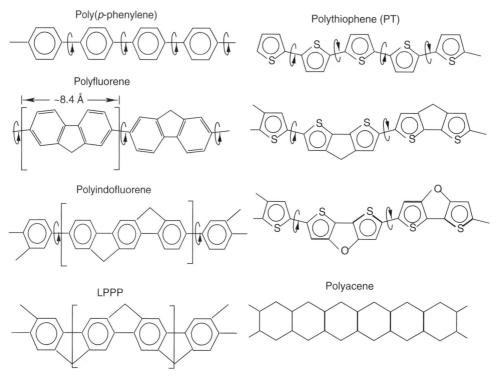

FIGURE 17.8   Sketches of various π-conjugated polymers having direct chemical modifications that alter the main chain morphology and torsion angles.

pentacene) are clearly important for molecular electronics but, for the moment, polyacenes exist mostly in the realm of the theoretical chemistry and physics [189].

17.3 Side-Chain Functionalization

In much of the CP field, the overall focus has shifted considerably and the most technologically important materials are now functionalized with an ever-expanding litany of possible variations on a theme. This transformation reflects the fact that most linear unsubstituted CPs are not easily processed using traditional polymer film-forming methods. At the very minimum, the addition of flexible side chains to the stiff CP backbone has been extremely effective for obtaining tractable and fusible materials. This approach is not restricted to just CPs, but has been used in conventional rigid rod-like polymers [190–193] as well. In many instances, this modification has created new materials exhibiting enhanced electronic properties [194]. Even more captivating is the appearance of wholly new qualities not present in the unsubstituted parent polymers. Examples of these include thermo- and solvatochromism [195], thermotropic and lyotropic liquid crystallinity, and structural self-assembly [2,196]. These materials are more complex physically and chemically and so correlating structure with key technological properties can be extremely challenging.

As a prelude to any explicit discussion of structure in the functionalized CPs it is worthwhile mentioning some general features of polymers that are comprised of subunits with differing chemical and physical properties. All of these materials include cooperative and competing interactions originating at molecular distances and the resulting structures and phase behavior are extraordinarily diverse [193,197,198]. In many instances, there is a well-orchestrated ordering of large-scale structures and well-defined interfaces that isolate and confine the various substituents. However, beyond this self-segregation there may be little structural ordering at molecular length scales. In many applications this factor is not of major concern, but for the CPs already mentioned, seemingly small variations in the average structure or in the overall distribution can have a major impact on both charge transport and optical properties.

Variations in the local environment necessarily influence chain morphology and thereby torsional degrees of freedom. Small stochastic twists between monomer units are specified as worm-like disorder and, on an individual basis, impact conjugation only minimally [121,174,199]. However, the integrated effect can be important. Large torsional variations are termed segmental disorder and tend to isolate chain sequences [177]. Most real materials appear to incorporate a combination of these two "limiting" cases [200]. Additional issues, such as variations in the local interchain packing, the presence of chemical defects, and/or changes in the bond alternation (because of excitation and charge transfer), must also be reconciled.

Many of these polymers have structural properties that are intimately related to the explicit forming conditions. Casting or precipitation from solution often results in nonequilibrium structures that do not represent anything resembling a thermodynamically stable state. Even if the environmental conditions are fixed, the properties can evolve and change over time. Subsequent processing will often initiate a series of irreversible structural changes that are likely to impact technologically important CP properties. There are kinetic factors as well and the energy landscape of accessible structures is very much influenced by the prior history. Many structured polymers exhibit phase diagrams with extensive hysteresis and populated by a myriad of monotropic (e.g., seen on heating or cooling) or enantiotropic structural phases. Modest cooling can easily overwhelm nucleation and growth rates leading to kinetically limited structures and the observation of buried metastable phases. These issues are extensively treated in reviews by Zhang and Keller [201,202]. Conformational entanglements, molecular weight, and polydispersity can also influence the particular outcome of a specific polymer and further complicate matters. Finally, the presence of impurities either accidentally (e.g., as residual solvent) or intentionally can significantly alter the phase behavior. Blomenhofer et al. [203] have recently documented the impact of additives in isotactic polypropylenes.

17.3.1 Polyacetylene, PPV, and PPP

Functionalizing pedant groups have been anchored to all of the aforementioned CP main chains. These include a vast number of alkanes, alkoxys, and ethers. These side chains can incorporate biphenyl (and related) liquid crystal mesogens, phenylene rings and/or with branches, chiral centers, and so on. The impact on the physical properties is substantial and, for the moment, it appears that current experiment and theory only reflect the tip of the iceberg. In an early paper, Gorman et al. [204] synthesized and characterized a series of sparsely monosubstituted PAs. These polymers were both soluble and dopable. Bulky substituents tended to increase the effective conjugation length, but all polymers were amorphous in the solid state. Addition of a phenylcyclohexyl mesogen to PA [205] has been instrumental for obtaining soluble polymers that exhibit thermotropic liquid crystal phases.

Moreover, in direct correspondence between structure, energy flow, and photophysics, pure PA exhibits minimal or no fluorescence, whereas functionalized acetylenes [206,207] or phenylacetylenes [208] can be highly emissive with some derivatives exhibiting photoluminescent yields that approach unity. Liquid crystallinity is often observed with the formation of various nematic or smectic phases. These can be easily inferred through polarized microscopy and the appearance of Schlieren and related mesomorphic textures. Functionalization with chiral-branched substituents [209,210] leads to optically active polymers.

Many of the functionalized PA-based polymers first synthesized included a large degree of chemical and stereoisomerism due to the nonspecificity of the synthetic approach. Recent reports have identified a range of catalysis agents that allow for partial or complete control of the chain structure so that properties such as tacticity, regioregularity (rR), and chirality are reliably obtained. Stereospecific polymerization of monosubstituted acetylene allows for appreciable control of conformational isomerism, shown as helicoidal-type structure in Figure 17.9, in work by Tabata et al. [211], and thus defines the single-chain morphology. The signatures of the hexagonal columnar phase are the sharp peaks at low angles whose positions approximate the ratio of $1:\sqrt{3}:2$. The column spacing grows proportionately as the side chain lengthens. There is also a distinct drop in the peak widths. The nature of the final structural form is very much a function of the physical or chemical processing. In some cases, exposure to toluene vapor facilitated conversion to a hexagonal columnar-type phase. There is sufficient information to specify the average chain position within the unit cell but, on the basis of just three reflections, one cannot deduce many finer substructural details. More than likely there are substantial torsional deviations from perfection along the backbone.

In a manner completely analogous to the aforementioned PA derivatives, both PPV and PPP have been functionalized. The presence of a phenylene ring gives additional opportunities with up to four possible substitution sites per ring. In the intermediate case, that of PPV, substitution at vinyl hydrogen positions is yet another option. There are not only multiple substitutional sites to choose from, but phenylene ring rotations about the 1,4 *para* positions will constitute conformational isomerism. This degree of freedom can, at times, frustrate order both locally and more globally. Many of these compounds exhibit only a few identifiable peaks in the respective diffraction patterns. Often an isolated low-angle peak is observed and this peak position varies inversely with the overall length of the functionalizing substituent. In terms of structure, this typically reflects a frozen nematic glass-like state in which there is local parallel orientation of the polymer chains at the molecular level [212,213]. In cases where there are a series of evenly spaced low-angle peaks, this indicates a lamellar-type phase. The rapid drop of peak intensity with scattering angle (usually given at 2θ) reflects the high level of disorder and the loss of phase coherence.

Poly(2-methoxy-5-(2′-ethylhexyloxy)-1,4-phenylenevinylene) (MEH-PPV) is probably one of the best known polymers representing this family. Ostensibly the use of two dissimilar side chains (and one that is bulky and branched) is intended to minimize crystallization and limit interchain cofacial contacts (i.e., molecular aggregation). As a prototypical red emitter it has formed the basis of numerous device studies. Emission signatures representative of single chain (i.e., singlet exciton) and multiple chains (i.e., molecular aggregates) are reported. These results are extremely sensitive to solvent, temperature,

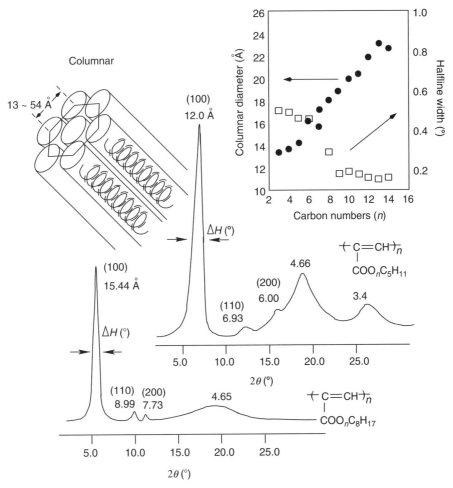

FIGURE 17.9 Two examples of powder x-ray diffraction profiles from a series of functionalized polyacetylene designed to give helicoidal-type structures. Upper *right*: plot of *d*-spacing and peak half-width vs. alkyl side chain length. Upper *left*: sketch of hexagonal columnar phase. (Adapted from Tabata, M., Sone, T., and Sadahiro, Y., *Macro. Chem. Phys.*, 200, 265, 1999. With permission.)

film-forming conditions and aging [214–216]. Physical gelation can occur in aged samples and this process itself impacts the observed spectroscopy. Morphological changes can be clearly seen in TEM studies [217]. Many aspects of MEH-PPV are still enigmatic.

Despite the bulky and asymmetric nature of the side chains, MEH-PPV diffraction studies of uniaxially oriented films are suggestive of a semicrystalline state albeit with a very short coherence length [215,218,219]. The optical spectroscopy absorption and emission features are typically very broad and do not narrow very much on cooling. This would indicate a high degree of local disorder in terms of the explicit chain conformational sequencing. Detailed structural models are sparse although there is one recent modeling approach by Claudio and Bittner [24,220] to specify the chain morphology and map between a chain conformational structure and the optical properties of MEH-PPV/mesoporous silica composite [221].

There is also one recent and somewhat unusual example in which a homologous family of dialkyl-substituted PPVs yields a modest level of structural order. Krebs and Jørgensen [222] prepared a large series of poly(2,5-dialkylphenylene vinylenes) (functionalized at every other phenylene ring) ranging from hexyl to dodecyl. In this case, a substantial number of Bragg reflections could be resolved and the

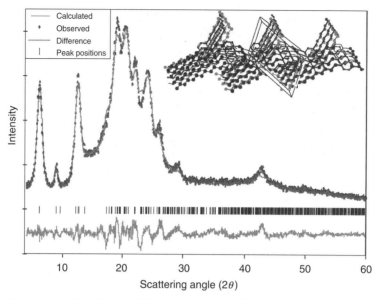

FIGURE 17.10 Comparison of powder x-ray diffraction data from a diakyl-functionalized PPV derivative (see text) with a profile obtained by Rietveld refinement using the unit cell, $a = 5.19$ Å, $b = 11.02$ Å, $c = 16.84$ Å, $\alpha = 63.8°$, $\beta = 96.7°$, and $\gamma = 97.6°$, as shown in the *inset*. (Reprinted from Krebs, F.C. and Jørgensen, M., *Macromolecules*, 36, 4374, 2003. With permission.)

Rietveld refinement of the dioctyl-substituted sample is shown in Figure 17.10 using a model in which the alkyl chains are oriented by twisting about the innermost methylene carbon to phenylene C–C bond so as to be nearly perpendicular to the plane of the attached phenylene ring. Strong preference for this orientation is found in a general search of the Cambridge Structural Database (CSD). The collected results of this search are reproduced in Figure 17.11. Interestingly, the alkoxyphenyl moiety has torsion angles lying near or in the plane of the benzene ring. It is important to reemphasize that this structural refinement reflects only the average chain positions and there is, as duly noted by the authors, substantial conformational disorder. This could be in the form of *gauche* isomers along the side chain or by rotations of the phenylene ring. Many of the ordered structures are metastable and permanently lost on thermal cycling. Studies of the (2,5-dialkoxy-*p*-PPVs) have also appeared and the scattering data from these CPs are more typical with only a single low-angle peak indicative of a nematic glass [212].

17.3.2 Polyalkylthiophenes

One of the first functionalized CPs was PT and, among the many examples, a very common substitution has been with alkyl substituents to yield the poly(3-alkylthiophene) (P3AT) family. These brush-like polymers exhibited a number of new physical properties including mesomorphic phase behavior [223], solvatochromism, and thermochromism [224]. Even substitution by a single methyl unit wholly disrupted the HB-type packing of the parent polymer [225]. At longer alkyl chain lengths one could distinguish a series of uniformly spaced low-angle XRD peaks [226] with *d*-spacings in direct proportion to the number of CH_2 units in the alkyl chain [223]. Thus, a simple lamellar-type phase was proposed [226], as shown in Figure 17.12, and thereafter heavily studied. The overall structure and structural phase behavior was addressed in parallel publications from a number of groups [223,227–237]. A very common postulated motif was a linearstack of the main chains approximately 3.8 Å apart and a *trans* planar sequence (i.e., a 2_1 helix) along the thiophene backbone with a 7.8 Å repeat. Thus, P3ATs are an example of a double inverse comb. Closely related motifs have also been proposed [238]. The actual nearest-neighbor spacing is somewhat less than 3.8 Å because there is a setting angle tilt ϕ_R (see Figure 17.12). This basic construction is widely seen in functionalized CPs and its signature hallmark is a series

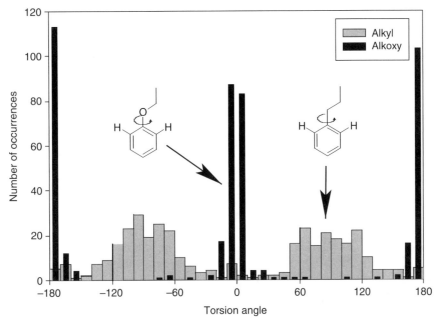

FIGURE 17.11 Histogram of the number of times a given indicated torsion angle appears in the Cambridge Structural Database for alkyl- or alkyloxy-aryl compounds. (Reprinted from Krebs, F.C. and Jørgensen, M., *Macromolecules*, 36, 4374, 2003. With permission.)

of evenly spaced sharp peaks at low to moderate 2θ scattering angles, a broad peak at 2θ angles of ~20° to 24° (assuming an x-ray wavelength of ~1.542 Å) and a single isolated peak situated between 23° and 24° [4].

The orientation of the side chains was still an unresolved issue and a number of groups forwarded possible packing schemes [229,239–241]. The overall level of structural order is generally modest and only at side chain lengths of decyl or longer was there evidence of a 3D packing clearly discerned [228]. Refinement of these x-ray data suggested a model in which the side chains were strongly twisted about the thiophene to CH$_2$ C—C linkage so that the side chains tipped out of the plane as defined by the adjoining thiophene ring. This polymorph is designated a "type I phase" and the tilted side chain construction is consistent with the alkylphenyl trends of the survey as depicted earlier in Figure 17.11. In this lamellar construction, the average alkyl–alkyl side chain spacing and overall areal density (~18 Å^2/chain) approximate the chain packing observed in polyethylene crystals [242].

In support of this twisted side chain conformation are more recent results from computational modeling studies (i.e., molecular mechanics, molecular dynamics, and Monte Carlo) by Corish, Morton-Blake, and coworkers [243–245]. These modeling papers also resolve correlated conformational changes in the side chain tilts and main chain interthiophene torsional angles that are both temperature and pressure dependent. Room temperature and ambient pressure are associated with mean intrachain thiophene torsional deviations of approximately 20° from planarity. There is also evidence of shifts in the axial positioning (i.e., along the chain or *c*-axis) from an eclipsed to a staggered construction. Direct comparisons to experiment are not immediately possible because the work of Corish and Morton-Blake is confined to just poly(3-butylthiophene) (to reduce computational overhead) while the majority of the experimental studies deal with hexyl and longer alkyl side chains. Poly(3-butylthiophene) samples are not as well ordered as P3ATs with longer alkyl moieties.

The modeling of Prosa et al. [228] only reflects the overall average structure of the PA3T chain. The persistence of diffuse scattering features, even in well-ordered films, suggests a high degree of residual disorder. The simulations of Corish and Morton-Blake [243,244] also result in pronounced disorder in

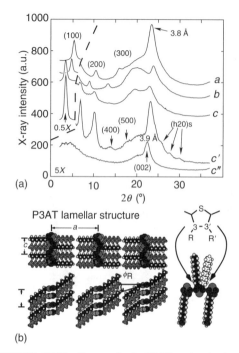

(a)

P3AT lamellar structure

(b)

FIGURE 17.12 *Bottom*: basic P3AT molecular-level packing motif that results in a lamellar phase. The chain axis repeat, *c*, is typically 7.8 Å and the intra-stack spacing for the most common type I crystalline phase is 3.8 Å (or *b*/2). *Right*: a trimer that shows two substitutional positions that lead to variations in the regioregularity. *Top*: typical diffraction profiles for unoriented (a) poly(3-hexylthiophene), (b) poly(3-octylthiophene), and (c) poly(3-dodecylthiophene) films in the type I phase. Also (c′) an equatorial (*hk*0) profile and (c″) a meridional (00*ℓ*) profile from a uniaxially stretched film of poly(3-dodecylthiophene). The arrows identify weak (*h*20) peaks associated with a full 3D ordering. (Adapted from Winokur, M.J. and Chunwachirasiri, W., *J. Polym. Sci.*, 41, 2630, 2003. With permission.)

terms of the main chain torsional deviations and in the side chain twists. Electron microscopy of whiskers formed during slow recrystallization by Ihn et al. [246] and recent scanning probe microscopy of ultrathin films [247,248] identifies an innate tendency for P3ATs to chain fold. This structured folding temporizes between the stiff nature of the backbone and chain collapse as the films precipitate [249].

These lamellar stacks constitute a linear aggregate and provide appreciable p_z overlap between adjacent chains within each layer. The net result is a strong enhancement of interchain transport. Measured mobilities can be quite high in P3AT field-effect transistors (FET). However, interchain interactions introduce an increase in nonluminescent decay paths and consequently photoluminescent or electroluminescent quantum yields are relatively low. Doping of P3ATs can proceed without fundamentally disrupting the intralayer stack construction [227,229,232, 250–252] and, as a result, these materials exhibit excellent conductivities. Extended review articles by Samuelson and Mårdalen [3] and Hotta [253] exclusively encompass the PT family of polymers and oligomers, respectively.

Perhaps, one of the more striking attributes intrinsic to P3ATs is the existence of a structural isomorphism referred to as rR in which the anchoring of the alkyl side chains can take one of two possible carbon sites, the C3 or C3′ positions. The degree of rR strongly varies with the chosen synthesis route [254–257]. P3ATs that include a high degree of rR can have noticeably higher phase transition temperatures, poorer solubilities, and better transport properties. The P3AT synthesis routes employed by McCullough [255] and Rieke and coworkers [256] have more than 95% of side chains anchored at the C3 position. Other types of regioregular dyad sequences have been prepared as well [258,259].

rR is just one of many important issues affecting P3AT properties. Not surprisingly the measured charge transport is very much a function of molecular weight. However, there is an anomalous observation of improved charge mobility as the molecular weight increases and the degree of structural order, in terms of Bragg scattering, diminishes [260,261]. Two distinct mechanisms have been forwarded as possible explanations. The first is that in high-molecular weight P3ATs the individual chains span across grain boundaries and regions of disorder to minimize the losses in interfacial regions. An alternative explanation is that the chain-packing structure systematically varies as the molecular weight drops and thus while the degree of structural order improves at lower molecular weight, the intrinsic transport properties worsen. A molecular modeling study by Curtis et al. [262] finds that the equilibrium structures of sequentially longer oligomer homologs do indeed vary with chain length, but, for the time being, this anomaly remains as an unresolved issue. One may expect that there are compounding contributions from both these proposed effects.

P3ATs also exhibit sensitivity to the explicit film-forming conditions and subsequent treatments. Bolognesi et al. [263,264] and Prosa et al. [265] observed a metastable structure designated as type II phase and these authors postulated somewhat different unit cell constructions. Prosa et al. [265] invoked a metastable interlayer interdigitation of the alkyl side chains. A follow-up article by Meille et al. [266] suggests that this type of packing may be inconsistent with general thermodynamic considerations.

Numerous follow-up studies have looked at the mesomorphic behavior of P3ATs. Differential scanning calorimetry (DSC) measurements resolved multiple endothermic peaks on heating [223,264,267–270] and this reflects the complexity of thermotropic transitions in functionalized polymers. For P3ATs there exists a highly anisotropic "melting" to distinct liquid crystal mesophases. Correlated with these transitions are very pronounced changes in the optical absorption and emission features. In particular, there is a large-scale hypsochromic shift in the observed spectral features at elevated temperatures with, in many cases, a clearly defined isosbestic point in UV–vis absorption measurements. In the longer alkyl P3ATs, two very different molecular-level transitions occur on heating and subsequently, in regard to cooling, there is a memory effect associated with the thermal history. This behavior results from sluggish kinetics and differing relative rates of nucleation and growth as the temperature varies [264,270]. An Avrami analysis by Causin et al. [271] suggests that crystallization occurs by heterogeneous nucleation followed by 1D growth. On heating, the lower temperature transition is associated primarily with a melting of the side chain endgroups while the second transition is dominated by an intralayer disordering of the thiophene chains [231,272,273]. These processes lead to systemic variations in the observed peak widths and line shapes.

At a minimum there are two structure-related factors acting at the molecular level that directly impact the nature of π-conjugation. The first is an increase in the intrathiophene ring torsion angles. The second is a change in the degree of aggregation as the intralayer chain spacing and relative axial positioning between neighboring chains varies. A full accounting of this exact mechanism is still an ongoing topic of current CP research. In terms of molecular dynamics calculations of local structure and thereafter, calculations of optical properties based on the average chain structure, one can reasonably conclude that thermally driven torsional deviations from the nominal starting planarity [243,244] average less than $25°$ and the corresponding changes in the π–π* gap energy are just too small to explain the large shifts observed experimentally [274]. The energy bandwidth in terms of interchain interactions is also relatively modest. Not only this issue is problematic, but, additionally, a number of recent reports have postulated the need for two distinct chromophore units in modeling of both the optical absorption and, of special concern, emission spectra in P3ATs [275]. Even in the presence of an admixture of two phases there is usually large-scale flow of charge excitations by interchain energy transfer and intrachain migration to low-energy chromophores. Generally, a small number of low-energy sites, presumably within a single ensemble of chromophores, will often dominate the steady state emission processes.

In the continuing quest to provide a direct framework for understanding the relationship between molecular-level structure and CP photophysics, a more subtle quantity that reflects structure on more intermediate length scales is the effective conjugation length per P3AT chain. Thermally derived increases in the torsion motions will produce a concomitant reduction in the effective conjugation length. Spano [276] has recently modified a formalism applied to HB and pinwheel-type aggregates [19], and applied this model to the case of linear aggregates. Based on an exciton model that directly accounts for spatially correlated site disorder it is possible to quantitatively match both the absorption and emission features without resorting to the somewhat cumbersome two-phase model. A few experimental groups are now reassessing this assertion of a two-emitter model as a general PAT feature [277].

Current P3AT structure studies encompass a myriad of specialized syntheses and settings that relate to specific applications. One interesting variant, clearly with a more fundamental relevance, is a class of inclusion compounds in which molecular structural self-assembly yields channel-type structures that can be occupied by linear arrays of individual polymers or oligomers [278,279]. In this way, the interchain distances and couplings can be systematically defined and then studied.

17.3.3   Other Heterocycles (Nonhydrogen Bonding)

Another recently synthesized class of materials are the poly(3,3'''-dialkyl-quaterthiophene)s (PQTs) in which only every third and forth thiophene rings are functionalized at alternating carbon 3 and 3' positions [280]. In this case, a reasonable balance has been struck between a presumed increase in the average main chain torsion angles (away from planarity) and reduced π-conjugation. This diminishes the ionization potential, thereby improving oxidative stability in juxtaposition with the need for solubility and structural self-assembly. Ong et al. [281] report charge mobilities approaching 0.2 V/cm^2 in thin-film transistors fabricated under ambient conditions. X-ray diffraction profiles show the classic lamellar construction and notably sharp equatorial reflections after thermal annealing. Thin films XRD spectra show little, if any, of the expected broad amorphous scattering halo near $2\theta = 24°$.

With a somewhat different purpose, Collard and colleagues [282–284] have prepared a series of alternating alkyl/fluoroalkyl- and alkyl/perfluoroalkyl-substituted PTs. The CF_2 unit is substantially bulkier than linear alkanes and, moreover, linear CF_2 sequences are helical. These polymers remain both mesomorphic and thermochromic and, in some cases, include a reasonably high level of structural order. In one formulation, that with $(CH_2)_5(CF_2)CF_3$ alternating with $(CH_2)_8CH_3$ units (Figure 17.13), up to six orders of (h00) reflections were observed. An interesting effect is the enhancement in the relative intensities of the even order (h00) from the variations in the electron density along the layer d-spacing. The differences in the chemical affinities between the fluorocarbon and hydrocarbon moieties indicate a self-assembled bilayer construction. Engineering the sequencing in different ways leads to nonaggregating samples that are highly fluorescent in the solid state.

By combining regiospecific synthesis with alternating hydrophilic and hydrophobic side chain moieties, Bjørnholm, McCullough, and coworkers [285,286] formed regioregular amphiphilic PATs that can spontaneously self-assemble at hydrophillic or hydrophobic interfaces to form oriented monolayers or bilayers. Finally, there are also efforts to induce chiral liquid crystalline phases. Goto et al. [287] have recently reported on a family of ferroelectric LC substituents. These materials also exhibit a rich array of

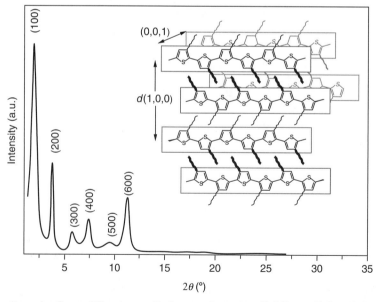

FIGURE 17.13   Example of x-ray diffraction profile from an alternating alkyl/fluoroalkyl- and alkyl/perfluoroalkyl-substituted polythiophenes. The specific side chains are $(CH_2)_5(CF_2)CF_3$ alternating with $(CH_2)_8CH_3$. Here, the polythiophene self-assembles into a bilayer repeat. (Adapted from Dong, X.M. and Collard, D.M., *Macromolecules*, 33, 6916, 2000. With permission.)

mesomorphic behavior with complex dynamics and highly textured polarizing optical micrographs albeit with limited structural perfection.

Although there are numerous other functionalized polyheterocycles (e.g., polyfurans [288] and PPys [289]), there are far fewer detailed structural studies. In many instances, there is little or no long-range order. For example, Orecchini et al. [290] have recently published a combined x-ray and neutron scattering study of poly(3-*n*-decylpyrrole) and extracted the interchain pair correlation function (see Section 17.4.1). They deduce, on a more local length scale, the presence of a stacked lamellar structure with tilted side chains containing a large fraction of *gauche* conformational isomers. The trend toward well-ordered lamellar phases can be "restored" by cofunctionalizing a polyelectrolyte surfactant at the hydrogen position in the NH moiety [291,292]. In this material the diffraction evidence points to a double layer stack alternating between layers of the alkyl chains and the polyelectrolyte.

17.3.3.1 Intercalation of Functionalized CPs

Intercalation of functionalized polyheterocylic CPs proceeds as readily as it does in the linear unsubstituted polymers although, the presence of the functionalizing side chains necessarily complicates matters. Many of the proposed structural models are formulated with minimal structural information from the scattering profiles. During P3ATs intercalation there can be minimal disruption of the lamellar stacks because the intercalate galleries open up within the alkyl chain portions of the polymer microstructure. In regioregular P3ATs, relatively high conductivities [194] have been reported (above $1000\,\Omega^{-1}\,cm^{-1}$) even though the actual fraction of material contributing to charge transport is relatively small. Both translational and rotational displacements may occur. In iodine-intercalated P3ATs, Tashiro et al. [293] presented one of the first reports suggesting the occurrence of small axial shifts parallel to the thiophene chain axis that would bring the side chains into registry and open up these intercalate channels. The transition itself was shown to be sensitive to the alkyl chain length and could be quasicontinuous or, alternatively, proceed by a two-phase coexistence between a lightly and a heavily doped phase with extremely large variations in the layer spacing [227]. Structure factor refinements by Prosa et al. [251] could quantify changes in the orientation of the side chains and that of the iodine intercalate as the structural evolution proceeded.

17.3.4 Polyanilines and Other Hydrogen Bonding Analogs

PANi, as already described, may be reversibly protonated with a variety of organic acids to a conductive state. Thus, there are special opportunities to tailor the functionalization and self-assembly through the dopant molecule itself rather than relying solely on covalently linked side chains. In principle, both avenues could be used simultaneously in a comprehensive and synergistic molecular design strategy but, for the moment, there is still much to be learned at a more basic level. Recent work has often been motivated by the widely heralded supposition that, starting at nanometer distances, with appropriate molecular-level engineering one can exploit competing and/or complementary interactions in order to arrive at a designated structure spanning many length scales [294,295]. This biologically inspired scenario has spawned intense activity encompassing a wide range of research areas. Recent reviews [296,297] by ten Brinke and Ikkala provide an excellent overview with the latter including an emphasis on current CP efforts and structural order at the larger length scales.

Within the CP field an oft-cited paper by Cao et al. [298] demonstrated the potential for using the amphiphilic acid dopant DBSA for simultaneously enhancing PANi processibility and improving electrical conductivity. These samples had transport properties that spanned the metal–insulator transition [299]. Included in a follow-up report [300] are diffraction data from oriented samples and these results indicate a lamellar-type phase. A staggered nesting of the PANi chains was proposed in a tentative model that more closely resembles EB-II rather than the stacked P3AT motif. The latter is probably more likely even though it has not been rigorously established as the correct one in this case. DBSA doping of PEDOT, which clearly does not hydrogen bond, appears to give the typical lamellar phase [301]. As a prelude to more recent efforts toward engineered self-assembly-type methods of controlling structure and charge transport, Zheng et al. [302,303] synthesized and studied a range of *n*-alkylated PANis. Although the level of order was relatively modest, this functionalization led to lamellar-type phases. These compounds were thermochromic and mesomorphic, and, at side chain lengths longer than *n*-dodecyl, they observed secondary crystallization of the alkyl side chains into a pseudohexagonal packed structure with a 4.1 Å *d*-spacing. These polymers could also be protonated to a conductive state, and in the case of long-chain dopants, there was evidence of a cooperative effect.

Transitioning to an approach that emphasizes supramolecular self-assembly was facilitated by a series of studies spearheaded first by Ikkala [304–306] and then by ten Brinke and Ikkala [307]. This work sought to identify and characterize the specific molecular associations and their respective strengths. Control of these factors is crucially important in order to prevent macroscopic phase separation and to tune the nature of the microscopic morphology. Interaction free-energies are associated with the surfactant polyelectrolyte and CP in terms of side chain stretching, interface formation, and polymer confinement. Longer surfactant species require stronger bonding interactions to the active CP site in order to sustain integrity of the complex mesomorphic structures during processing.

The strength of the intermolecular forces can be suitably enhanced by multiple pair interactions that mimic molecular recognition. PANi is just one of the many possible candidate CPs. Complexing PANi simultaneously with both alkyl (resorcinols) and 2-acrylamido-2-methyl-1-propanesulfinic acid (AMPSA) leads to cylindrically shaped supramolecules that pack hexagonally. Tiitu et al. [308] reported *d*-spacings, in shear-oriented films, typically on the order of 20 to 30 Å, but the extent of long-range order was modest with coherence lengths of approximately 60 Å.

Poly(2,5-pyridinediyl) (PPY) is yet another simple linear polymer [309] that includes all important H-bonding site necessary for protonation. One interesting set of mixtures, studied by Knaapila et al. [310] is CSA-doped PPY in combination with 5-pentyl-1,3-dihydroxybenzene (PRES), 4-hexyl-1,3-dihydroxy-benzene (HRES), octylphenol (OP), and/or 1-octyl-3,4,5-trihydroxybenzoate (OG). Here, protonation of the pyridyl nitrogen by CSA leads to conditions whereby the amphiphile will complex through H-bonding to the PPY(CSA) composite. The resulting cast films were mechanically sheared to further introduce structural anisotropy. Post processing of oriented films was achieved by, in this instance, thermally activated cleaving of the side chains under vacuum. A few formulations resulted in structural self-assembly exhibiting a very high degree of 3D texturing. These films also exhibited polarized luminescence. The very well-ordered lamellar structures formed is clearly demonstrated in the x-ray diffraction patterns of Figure 17.14. Coherence lengths approaching 80 nm are observed.

The phase behavior of these complexes, referred to as polyelectrolytic hairy-rod supramolecules, is also quite diverse and highly specific to the chemical formulation. Lyotropic and thermotropic phases consisting of isolated chains or multiple chain aggregates are also possible. Stepanyan et al. [311] have employed a mean-field theoretical approach and obtained phase diagrams containing microphase-separated structures as shown in Figure 17.15. Lamellar, square, elliptical oblique, and hexagonal motifs (and probably others as well) can all exist within a single extended phase diagram [312]. The kinetics can also be relatively sluggish and require considerable soak times during phase formation. In many ways there are qualitative parallels to the complex structural phase behavior seen earlier in alkali metal-intercalated PA and PPV.

As with all supramolecular assemblies, the degree of structural order over large length scales is immediately apparent in the diffraction profiles. At short distances there are methods to model the cooperative chemical association of the hydrogen bonds and phenyl–phenyl interactions [305]. More difficult to reckon is the heterogeneity and disorder at intermediate length scales.

17.3.5   Stepladders, Mostly Polyfluorenes

Within this somewhat generic classification of polymers, in-depth structural studies have so far focused mostly on the polyfluorenes (PFs, as sketched earlier in Figure 17.8). An interesting feature of the bridging carbon atom is that it provides two closely set substitutional sites for functionalization. This creates new

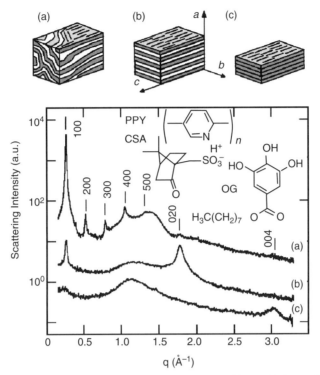

FIGURE 17.14 *Top*: sketches of lamellar phase in functionalized and self-assembled CSA-doped PPY complexed with OG (a) before mechanical shearing, (b) after mechanical shearing, and (c) after thermal cleaving. *Bottom*: x-ray scattering profiles postprocessed CSA-PPY films with the incident beam along the lattice vectors as indicated by the center top sketch. (Reprinted from Knaapila, M., et al., *Appl. Phys. Lett.*, 81, 1489, 2002. With permission.)

opportunities for even more complex architectures (e.g., spirofluorene). If two dissimilar side chains are chosen, then tacticity becomes a potentially important issue with formation of isotactic and syndiotactic dyads. With respect to the PF family as a whole, just two alkyl substituents have received a lion's share of recent attention and these are the linear *n*-octyl and branched 2-ethylhexyl groups [313–322]. The resulting disubstituted polymers are poly(di-9,9-*n*-octylfluorene) (or PF8) and poly(9,9-bis-2-ethylhexylfluorene) (or PF2/6). The latter PF2/6 polymer includes a chiral center at the ethyl to hexyl carbon and so left-handed (S) and right-handed (R) enantiomers can be specified. In a racemic mixture, four possible monomer enantiomers are formed (i.e., [S,S], [R,R], [S,R], and [R,S]).

Despite the seemingly rather modest differences in these two side chains, the impact on the structure and the resulting photophysics is striking and highlight both subtlety and complexity that functionalization can provide. For

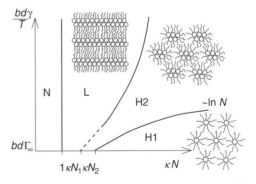

FIGURE 17.15 Theoretical phase diagram obtained by Stepanyan et al. that describes self-organization of hairy-rod polymers into various microphase-separated structures (N: nematic; L: lamellar; and H: hexagonal). Here, T is temperature, N_1 and N_2 represent various lengths of the side chains, and κ reflects the relative ratio of the volume of the side chain to that of the backbone on a per-segment basis. (Reprinted from Stepanyan, R., Subbotin, A., Knaapila, M., Ikkala, O., and ten Brinke, G., *Macromolecules*, 3760, 2003. With permission.)

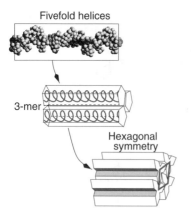

FIGURE 17.16 Molecular-level "supramolecular" self-assembly of the polyfluorene PF2/6 polymer from five-fold helices to a three-chain ensemble and, finally, into an ordered hexagonal array.

reasons that are now only partly understood, Lieser et al. [315] established that PF2/6 adopts a fivefold helical chain conformational sequencing along the polymer skeletal backbone (shown at top in Figure 17.16). Helical structures themselves are important and provide opportunities for, in the case of CPs, producing circularly polarized light emission [323] and chiral amplification [324]. These irregular fivefold helices [319] assemble into nested three-chain entities which themselves pack into hexagonal arrays. As cast films are initially in a state most closely resembling a frozen nematic glass. Thermal cycling through a thermotropic nematic liquid crystal (n-LC) mesophase induces large-scale ordering between chains. The extent over which periodicity extends is significant and coherence lengths in excess of 40 nm (or 30 unit cells) are reported for the hexagonal (Hex) phase. A notable property of PF2/6 and other polyfluorenes is the presence of an intrinsic molecular-level disorder with respect to the local, single-chain conformation sequencing [180,200,319] and the observed optical properties are entirely consistent with this claim. Regardless of the structural form, amorphous vs. ordered, or low molecular weight vs. high, the observed photoluminescence and photoabsorption spectra remain largely unchanged [319,322].

However, the overall phase behavior between the thermotropic n-LC and Hex phases is strongly molecular weight dependent [322,325]. As a consequence, only a small subset of PF2/6 thin films, those within a narrow range of molecular weights on the n-LC side of the n-LC to Hex phase boundary, are best suited for achieving a high level of orientation in thermally annealed thin films casted onto aligned polyimide substrates [322]. This issue is of major concern in applications where polarized emission is desired. The constancy of the main chain conformation throughout the n-LC to Hex phase change facilitates a more coarse-grained analysis of the phase behavior. Knaapila et al. [326] have recently shown that a mean-field theory approach can accurately model the temperature dependence of the n-LC to Hex transition vs. PF2/6 molecular weight. This phase diagram is shown in Figure 17.17.

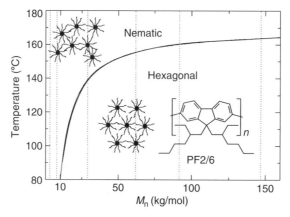

FIGURE 17.17 Phase diagram obtained from mean-field theory by Knaapila et al. for the molecular weight dependence of the thermal transition from a hexagonal ordered phase and a nematic mesophase in the PF2/6 polyfluorene polymer. (Adapted from Knaapila, M., et al., *Phys. Rev. E*, 71, 2005. With permission.)

The overt sensitivity to even small changes in the side chain architecture is also very notable. Replacing the 2-ethylhexyl moiety with *n*-octyl yields a polyfluorene polymer with a far more complex phase diagram and a completely different symmetry of the crystallographic-type phases in terms of the long-range order [314,316,320]. Grell et al. [314] documented the existence of at least two distinct ordered PFO polymorphs, now designated as α and α' phases, in addition to a nematic glass and high-temperature *n*-LC phase [320,321]. A series of oriented fiber 2D diffraction patterns as recorded by Grell et al. [314] are displayed in Figure 17.18. In the case of α phase, Chen et al. [320] have proposed an orthorhombic unit cell containing eight chains. Although a refinement of the equatorial reflections identifies average positions for the individual chains the reasons underlying this complex packing scheme are at present unclear.

PF8 is also widely known for the appearance of a very distinctive low-energy optical emission or absorption band designated as the "β phase" [327–330]. A pronounced narrowing in the spectral emission features of this β phase on cooling to widths below 20 meV is especially unusual. Far sharper line shapes [331] may be observed in molecular crystals or glasses and in isolated CPs [332,333], but this narrowing is rarely seen in dense CP phases due to heterogeneous broadening effects.

Although the β phase has often been interpreted as a distinct bulk structural form, a number of studies [181,200,334] suggest that this "phase" is more accurately depicted as isolated 1D threads or "impurities" of enhanced structural order that are embedded in a sea of more disordered conformational isomers which, as a whole, reflect the bulk structure. Becker and Lupton [181] have isolated and studied this β

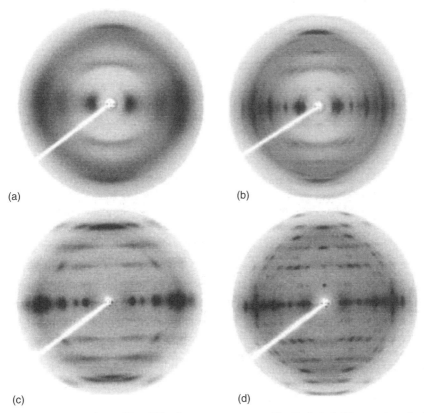

(a) (b)

(c) (d)

FIGURE 17.18 X-ray PF8 polymer fiber diffraction patterns, reported by Grell et al. (fiber *c*-axis vertical) after the following processes: (a) fiber drawn from liquid crystalline melt; (b) same fiber as (a), after thermalcrystallization (140°C for 1 h); (c) fiber exposed to saturated toluene vapor (40°C for 3 d); (d) same fiber as (c) after subsequent annealing (135°C for 1 h). (Reprinted from Grell, M., Bradley, D.D.C., Ungar, G., Hill, J., and Whitehead, K.S., *Macromolecules*, 32, 5810, 1999. With permission.)

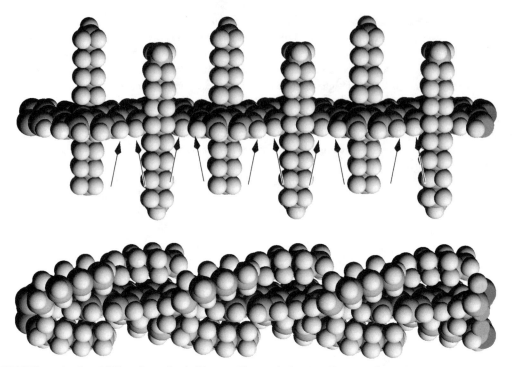

FIGURE 17.19   Space-filling views of poly(di-*n*-octylfluorene) showing, at top, a planar all-*trans* sequence of the side chains emphasizing the H–H hard-core repulsions along the polyfluorene backbone and, on the bottom, the proposed *trans-gauche-gauche* side chain conformational isomer [200] that stabilizes the β phase in a near-planar backbone conformation.

phase in single molecule spectroscopy measurements. Chunwachirasiri et al. [200] have recently proposed a single-chain model in which a single conformational isomer is stabilized through a local packing motif wherein the side chains self-organize parallel to the backbone through a *trans/gauche/gauche* sequence of the alkyl side chains. In contrast, a molecular di-*n*-octylfluorene derivative has been shown to crystallize into an all-*trans* side chain conformation [335]. Molecuar models of these two conformers are displayed in Figure 17.19. The main overriding theme in terms of a single-chain morphology is that direct H−H repulsions between monomer segments drive these polymers away from planarity, thereby reducing π-conjugation [175,176,220], but conformational isomerism and steric interactions of the side chains can orchestrate secondary interactions that, at times, take precedence over the primary H–H interactions. These side chain interactions are a general feature of functionalized polymers [336].

Another trend evident in these PF8 studies (and also in P3ATs [273]) is that enhancement in structural order over short distances may actually conflict with the development of order at longer length scales. However, interactions at the molecular level determine the chain morphology and these in turn establish the conditions under which ordering over larger distances will occur. Understanding these issues is still very much a work in progress and will impact the properties and performance of both conjugated and more traditional polymers.

17.3.6 Polysilanes

The study of functionalized polysilanes (this includes polygermanes as well [337–339]) in many ways parallels to that of the π-CPs. At a minimum, both polyfluorene and polysilane have two operationally identical substitutional sites for anchoring side chains. The commonly employed Wurtz coupling [123] presents very harsh synthesis conditions and, consequently, many polysilane studies pertain to alkyl, alkoxy, or arylalkyl substituents. Within this somewhat restricted setting there is an extensive literature addressing charge transport, spectroscopy, and structure [115,213,340,341]. Structural studies have demonstrated that alkyl side-chain substitution creates semicrystalline polymers with varied main chain conformations and a multitude of thermotropic mesophases [342–358]. A majority of these polymers are thermochromic. A widely occurring high-temperature mesophase is conventionally referred to as a hexagonal columnar mesophase. In this structure the individual chains form, on average, a cylindrical object and this close packs onto a hexagonal lattice. A typical powder diffraction profile is characterized by two or three sharp peaks at low angles with positions in the ratio of $1:\sqrt{3}:2$ and a broad halo at much higher angle indicative of short-range interchain packing disorder.

The shortest symmetric polysilanes (with a monomer unit specified as $-Si[(CH_2)_nCH_3]_2-$ and $n = 0$ to 2) typically adopt crystalline phases [345,346] dominated by a near-planar backbone conformations [359] with little or no thermochromism [360]. The monomer axial repeat distance averages about 2 Å and, with opportunities for two side chains per Si atom, the linear number density of side chains is high. In a perfect zigzag 2_1 helix the interchain near-neighbor spacing, excluding the second side chain on the same Si atom, is just above 4 Å. These polymers may be considered to be representative of dense bottlebrushes [361]. In di-*n*-butyl ($n = 3$) and di-*n*-pentyl ($n = 4$) substituted polysilanes the net influence of the side chain's bulkiness becomes more pronounced and there is substantial evidence for 7/3-type helices [344]. At di-*n*-hexyl or longer lengths, the polymer reverts back to a near-planar backbone conformation and the phase behavior becomes even more complex. More than five or more phases have been reported for poly(di-*n*-octylsilane) or poly(di-*n*-decylsilane) [357].

Many papers have observed crystallographic polymorphism and extremely slow-ordering kinetics [347,355,362,363]. In many instances, the number of individually resolved XRD peaks is substantially greater than that is typically seen in many other CP polymer scattering data sets. However, the specific details of the main chain and side chain structures are less well established and in-depth structural refinements are relatively few [213,348,350,351,353,354,356–358]. An example of this is shown in Figure 17.20 in which two different starting side orientations, a *syn-anti* and a *gauche–gauche* conformation in terms of the dihedral angles ϕ_1 and ϕ_2 depicted in Figure 17.21, are used in a structure factor refinement. The differences are relatively minimal until above 2θ angles above 24° at which point short-

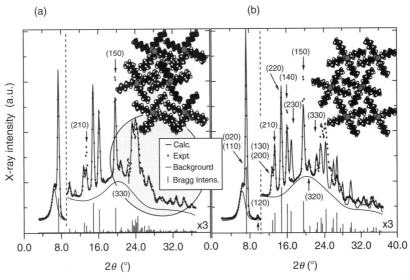

FIGURE 17.20 Comparisons of poly(di-*n*-hexylsilane) experimental x-ray diffraction powder profile (at room temperature) to those calculated in two Rietveld refinements using slightly different starting side chain conformational isomers (and short-range hard-core packing constraints). *Left*: (a) model based on a *cis-anti* starting conformation of the two Si–C torsion angles (per $Si[(CH_2)_5CH_3]_2$ monomer) and (b) a *gauche +, gauche –* starting conformation. See Figure 17.21 for the explicit dihedral angles ϕ_1 and ϕ_2. (Reprinted from Winokur, M.J. and West, R., *Macromolecules*, 36, 7338, 2003. With permission.)

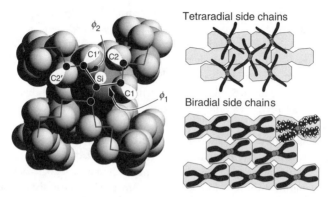

FIGURE 17.21   *Left*: cross-sectional sketch of poly(di-*n*-butylsilane) showing the two Si–C torsion angles (ϕ_1 and ϕ_2) involved in poly(di-*n*-alkylsilane) conformational isomerism. *Right*: schematic model depicting the crossover of side chain packing from splayed tetraradial to biradial construction as the alkyl side chain length increases. (Reprinted from Winokur, M.J. and Chunwachirasiri, W., *J. Polym. Sci.*, 41, 2630, 2003. With permission.)

range order becomes more important. At higher 2θ angles the first model is clearly superior. Small but significant backbone conformations away from true planarity are also necessary. Once again there is substantial long-range order, but the actual structural perfection along the polymer chain axis is found to be limited. Optical absorption and emission studies of these polymers in thin film always form almost yield moderately broad spectral absorption features and a significant redshift (i.e., a Stokes shift) in regard to the photoluminescence. Some derivatives exhibit a single monotonically varying peak in their absorption and emission spectra but, depending on the thermal history, multiple absorption and emission bands can be observed for ostensibly single heterogeneous phases [364].

There have been a modest number of more recent reports that emphasize structural modeling and structure factor refinement of both symmetric [357] and asymmetric polysilanes [356]. In regard to the di-*n*-alkyl chain moiety there is one significant trend, which has broader implications. Structure factor refinements indicate a distinctive crossover in the alkyl chain packing as the side chains lengthen [357]. In di-*n*-hexyl and shorter sequences, there is a tendency for the two immediately adjacent alkyl chains to adopt a splayed and open formation. In contrast, in polysilanes containing di-*n*-octyl and longer sequences there is an impetus for these two side chains to self-align and zip together. To describe these differing motifs the terms *tetraradial* and *biradial* have been coined [365] and examples are shown in Figure 17.21.

X-ray structural studies of polysilanes containing asymmetric and branched side chains also appear [353,366,367]. Tacticity issues are again present (refer Section 17.3.5). Branched side chains also provide opportunities for adding chiral centers and the effect of this addition has been extensively studied, primarily in solutions, by Fujiki, Saxena, and coworkers in numerous reports [368–373]. In particular, there is a very pronounced interaction between these polymers and a specific solvent so that one may observe switching of the chirality by varying the temperature or solvent. Cast films of these polymers can retain this optical activity in the solid state but the kinetics are, as one may expect, very much slower. In some instances, the propensity for forming well-ordered helices is very much enhanced and retained in the solid state [369]. In this case, modeling studies indicate the presence of a tacticity-insensitive nesting of the side chains along the polymer backbone [213].

In terms of a single-chain morphology recent molecular modeling and oligomer studies suggest the existence of at least three backbone conformations intermediate to the well-known *anti* (i.e., *trans*-planar, A or T_0) and *gauche* (G) arrangements [117,364,374,375] seen in alkanes. These have been specified by Fogarty et al. [376,377] as *transoid*, *ortho*, and *deviant*. The *transoid* conformer is reflective of a doubling in the degeneracy of the number of minima when moving away from a perfectly *anti* conformation. The existence of these multiple minima underlies the high degree of polymorphism in this family of polymers. Similar behavior likely extends to many π-CPs. A necessity prerequisite for deciphering these main chain conformations is simply a better understanding of the relationships between side chain structure and that of the backbone.

The impact of side chains for influencing the conformation energy landscapes [378] has long been known in the study of polypeptide chain folding and in more conventional polymers. Of primary importance are the 2D π–ψ maps and this surface reflects the torsional energetics of nearest-neighbor molecular base units along the skeletal backbone. This approach has been employed in recent studies of conformational structure of both polysilanes [379] and MEH-PPV [220]. However, the conformations adopted by the side chains will modify this surface. This is especially true of the dialkyl-substituted polysilanes although, in this case, both side chains must be self-consistently reconciled [358]. For linear alkanes this may not be as complicated as one may assume on first glance because there are, at the onset, only three low-energy alkyl chain conformational isomers (i.e., *trans*, *gauche*, and *gauche′*) that need to be systematically tested across any specific C–C pair of interest. In the polysilanes these conformational searches can identify a relatively modest number of low-energy conformation isomers with correlated backbone and side chain torsion angles and these "conformers" are in accord with the aforementioned oligomer studies [376,377].

17.3.7 Polydiacetylenes, Poly(*p*-Phenyleneethynylene)s and Copolymers

Polydiacetylenes (PDAs) and poly(*p*-phenyleneethynylene)s (PPEs), as shown in Figure 17.22, are examples of π-CPs incorporating acetylenic linkages and have many parallels with the CPs already

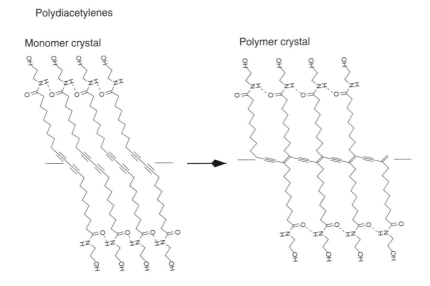

FIGURE 17.22 *Top*: a sketch of the topochemical transition from monomer to polymer observed in polydiacetylenes. *Bottom*: sketch of a PPE showing the rational degrees of freedom about the para linkages.

reviewed. In the simplest picture local π-conjugation always occurs individually with each of the neighboring two carbons immediately adjacent (to the C≡C linkage), but conjugation fully across the acetylene bridge requires that the two p_z-orbitals of these two adjoining carbons be coplanar. Without the steric hindrance of direct hardcore H–H interactions, torsional motions about the acetylene linkage are nominally less constrained; this is especially true in the case of PPE polymers. Despite over 30 years of study since the advent of single crystal PDA [380], a concise quantitative understanding of the structure–property relationships, with respect to the excited states and molecular ordering, remains as an unresolved issue although progress has been made [381,382].

Side chain functionalization was part and parcel to the earliest PDA investigations because of an unusual topochemical solid state polymerization that occurs in monomer crystals. Thus, the PDA family is renowned for forming nearly perfect single crystals. Once the monomer crystal is prepared, polymerization is initiated by heat, intense UV exposure, or ionizing radiation. Early x-ray studies by Baughman [383] or more recently through selected area electron diffraction (SAED) by Martin et al. [384] have demonstrated that polymerization generally proceeds by a solid solution of the polymer in the monomer crystal and occurs uniformly throughout. This polymerization process is sufficiently robust that monolayer or near-monolayer thin films of the monomer can be formed on a Langmuir trough, polymerized, and then transferred onto supporting substrates [385,386].

Once polymerized, these PDAs can be subsequently dissolved in organic solvent and cast into films that are considerably less well ordered. These irreversible changes in chain conformation modify the extent of π-conjugation and permanently alter the optical adsorption and emission properties. The optical properties are also sensitive to the specific length of flexible alkyl spacers within the side chains. In some derivatives, there can be a pronounced odd–even effect, based on CH_2 count, with the presence or absence of thermochromic transitions (especially in solution) [387]. Even when in single crystal form, the starting PDA structure can be metastable and, with mechanical treatment, the crystal will distort to a phase with differing optical properties [388]. In some instances, discrete color/structure changes may be correlated with reversible first-order phase transitions [389].

The existence of near-perfect single crystals has provided a model system to study intrinsic properties of π-conjugation. By carefully controlling the polymerization process, Schott and collaborators [332,333] obtained low concentrations of PDA polymer as dilute impurities in individual 3BCMU monomer crystals. The explicit 3BCMU monomer molecular formula is (=CR≡C=C≡CR) where R is $(CH_2)_3OCONHCH_2COOC_4H_9$. At least two structural polymorphs have been observed and these distinct states, seen in many BCMU PDAs, are typically designated as red and blue phases. The red form is very fluorescent while the blue phase is not. In these monomer crystals, heterogeneous broadening of the optical spectra is almost entirely absent and, in addition, the optical emission of the red form can be clearly discerned. At low temperatures the reduction in dynamical disorder was accompanied by a monotonic decrease in the emission line widths to below 100 μeV. Under these conditions, the classic quasi-1D $E^{-1/2}$ energy dependence in the density of states could be unambiguously observed.

There are other interesting settings to observe these distinctive PDA structural phases. Carpick et al. [386,390] fabricated PCDA (10,12-pentacosadiynoic acid) PDA thin-film samples that were initially in a blue phase and then, by applying shear stress by laterally sweeping a scanning probe tip, they could locally and microscopically, drive the transition (irreversibly) from the blue to a red form. Because of the large increase in photoluminescent quantum yield, after transforming to the red phase, this transition could be clearly discerned with near-field scanning optical microscopy. The explicit molecular-level response associated with this "blue to red" phase transformation has not, as yet, been elucidated (see the recent review by Carpick et al. [391]).

Synthesis of PPE polymers and PPE copolymers relies on more traditional methods and so they have structural properties more in common with many other CP families. Study of PPEs homopolymers and, more recently, a subset of copolymers has been very much intensified over the last few years. Much of this interest has been motivated by the development of new facile synthesis routes, as discussed by Bunz [392], and specialized side chain functionalizations in reports by Zhou and Swager [393,394], designed to exploit molecular recognition in conjunction with rotational degree of freedoms about the acetylene linkage for use as high-performance chemical sensors.

There are also many similarities to the previously described structure studies of the functionalized polyenes and polyphenylenes with the appearance of thermotropic LC mesophases and formation of lamellar phases containing substantial amounts of disorder [395–400]. In a few cases, constituting samples of relatively low molecular weight, the degree of structure order is moderately high [395]. Oriented fiber studies also appear. In these, a variety of suggestive side organizations have been postulated ranging from tilted to interdigitated [396,400]. Once again it is difficult to provide a quantitative framework whereby the side chain and backbone conformations are fully reconciled at the molecular level. For device applications where oriented thin films are desirable there have been a few reports demonstrating a high degree of chain alignment through friction transfer or epitaxy on oriented poly(tetrafluoroethylene) substrates [401]. SAED measurements of these samples can identify single crystal-like formations.

17.4 Special Topics

17.4.1 Pair Distribution Function Analysis

A number of complementary methods are used for probing structure of materials in environments ranging from bulk to thin film to solution. Bragg diffraction is especially useful in well-ordered materials because of the strong enhancement of scattering intensity at reciprocal lattice points intersecting the Ewald sphere [402]. Even in the most well-ordered CPs, the actual extent of Bragg diffraction is still limited in comparison with the total scattering. The number of readily identifiable peaks is often below 50 and, in rough correspondence to the limited number of chains comprising a crystalline domain, these peaks are relatively broad. The effect of this static and dynamic disorder, collectively represented by the Debye-Waller factor, causes the Bragg scattering to fall off exponentially with the scattering angle 2θ or wavevector q (where $\lambda q/4\pi = \sin\theta$ and λ is the wavelength of the scatterer) so that by wavevectors of only 3 to 4 Å^{-1} these Bragg peaks are no longer distinguishable from the diffuse scattering background (e.g., for x-rays, elastic and thermal diffuse as well as incoherent Compton scattering). Furthermore, a large proportion of CPs exists in a state that is better described as semicrystalline, liquid crystalline, or amorphous and exhibits little or no Bragg scattering at the onset. For all these reasons there is strong motivation for using a combination of probes, both direct and indirect (e.g., NMR, IR spectroscopy, Raman scattering, and EXAFS)

One established method for accessing structure at shorter length scales is pair distribution function (PDF) analysis. PDF analysis resolves structure at short to intermediate length scales with, in the case of polymers, tactic understanding that these data represent a superposition of all intrachain and interchain pair correlations. However, the anisotropy of the molecular architecture itself is a useful constraint. In these rigid rod polymers, the intrachain correlations over relatively modest distances, ~12 Å or less, are typically far narrower and well defined than those of the interchain pair correlations. An example of this is shown in Figure 17.23 in which a nominal single PANi chain conformation is used to represent the intrachain correlations and, by direct subtraction, resolve the interchain pair correlations. Much of the large-scale oscillatory behavior seen in the rescaled scattering data (i.e., minus the self-scattering), often written in a q-weighted form as $qH(q)$, derives from pair correlations of the more rigid near-neighbor bond lengths and angles. This is typically not of interest and thus the more salient intensity deviations are superimposed on this "background." This complication places a premium on datasets with good statistics and few systematic errors.

Although clearly less routine that x-ray diffraction, PDF analysis of conventional polymers has been used, perhaps most notably, in the study of polyolefins [403,404] and polyethylene oxides [405–407]. This PDF approach has also been utilized in structural analysis of PANis [408–413], functionalized polysilanes [356,414], and PPys [290]. An early PDF PANi EB study by Laridjani et al. [410] resolved differences in the local structure of type I and type II samples. Modeling is often used in conjunction

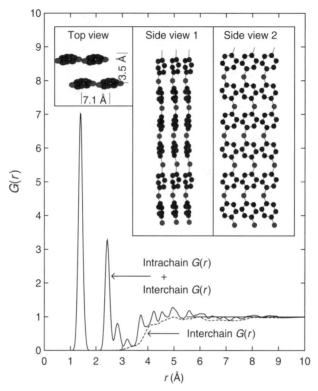

FIGURE 17.23   Example of RDF calculation using the PANi EB-II structure depicted in Figure 17.5 and explicitly showing the interchain and total (interchain + intrachain) $G(r)$'s when the level of intrachain disorder is much less than that for the interchain pair correlations. (Reprinted from Maron, J., Winokur, M.J., and Mattes, B.R., *Macromolecules*, 28, 4475, 1995. With permission.)

with experiment to provide more quantitative interpretation. An example of this is Winokur and Mattes [412] in a study of amorphous PANi. PDF analysis indicates an average local chain conformation in which one of every four phenylene rings, presumably the benzenoid unit in the BBBQ repeat sequence, is strongly twisted out of the plane specified by the remaining three quinoidal rings. Recent theoretical *ab initio* modeling results have reported qualitatively similar features [89]. In hydrohalogen acid-doped PANi samples, it is possible to deduce a probable position of the halogen dopant anions. These lie ~3 to 4 Å away from the PANi nitrogen atoms, thereby frustrating close packing of the PANi chains. After extracting the interchain correlation function, one can discern the subtle effects of cycling dopant in and out of the PANi matrix. The latter has implications for tuning the pore size in separation membranes. The work of Orecchini et al. [290], mentioned earlier, resolved both intrachain chain morphology and interchain packing characteristics.

   In polysilanes, the primary motivation has been to identify the chain conformational sequences and, to a lesser extent, orientation of the side chains. PDF analysis [356] of three atactic poly(methyl-*n*-alklysilanes) in bulk is indicative of a near-planar *transoid/deviant* sequencing of the skeletal backbone with relatively low concentrations of either *ortho*- or *gauche*-type conformations. Figure 17.24 shows data for atactic poly(methyl-*n*-propylsilane) and a single-chain model fit. The difference at lower q, ~<5 Å^{-1}, is due to scattering from interchain correlations.

17.4.2   Macroscopic Orientational Order

Although this section is too cursory, it is still worthwhile to mention just a few of the many efforts that have been made in this area. At a minimum, achieving orientational order within CP film and fibers

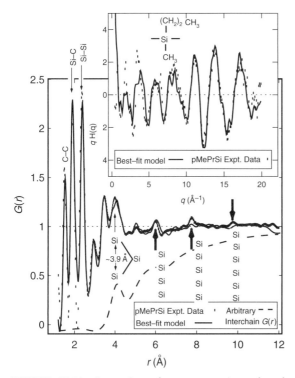

FIGURE 17.24 Comparisons between experimental and model-derived pair correlation function, $G(r)$, and, in the *inset*, q-weighted structure function for atactic poly(methyl-*n*-propylsilane). Arrows identify major silicon backbone pair correlations at intermediate distances. (Reprinted from Chunwachirasiri, W., Kanaglekar, I., Winokur, M.J., Koe, J.C., and West, R., *Macromolecules*, 34, 6719, 2001. With permission.)

facilitates structural analysis. It also has extremely important technological consequences for CP device performance.

Orientation of crystalline, semicrystalline, and liquid crystalline polymers is useful for identifying the origin of the various repeat distances (especially in the case of overlapping Bragg reflections) and identifying the nature of a structural phase[193]. SAED is especially well suited for this because of the small spot sizes. In this way, a single, ordered domain can be individually imaged and studied (Lieser et al. [315], Chen et al. [320], or Moggio et al. [401]).

Altering the orientational order within polymer films and fibers has long been used to exploit the inherent molecular-level anisotropy. Fully extended conventional polymers (e.g., gel-spun polyethylene [415]) have exceedingly high tensile strengths. The exact same arguments apply to CPs and, depending on application, there are notable improvements in the structural, optical, and/or charge transport properties. Large-scale chain orientation leads to greatly increased conductivities in the direction parallel to chain axis as well as providing a means for obtaining polarized absorption and emission. In an early study, Voss et al. [416] demonstrated that uniaxial alignment of MEH-PPV, by gel processing of a polyethylene blend, improves the microscopic structural order and thereby enables a rigorous analysis of the observed optical spectra.

The uniaxial stretching process typically used for aligning polymers has many molecular-level consequences. There can be stress-induced crystallization and also systematic variations in the local packing within the amorphous fractions [417]. The overall alignment of crystalline domains can exceed that of amorphous film fractions or, alternatively, the reverse can occur. The influence of stretching on PANi EB-II-type films and fibers has been studied by Fischer et al. [97]. In this case, stretching was found to induce nucleation and dramatically increase the fraction of crystallinity (~30 %) although the actual extent of crystallization was far less than that observed in the most crystalline PANi powders. Here, the orientational alignment of the crystalline domains was better than that of the residual amorphous film fractions. On the other hand, Wang et al. [103] observed no crystallization within stretched PANi fibers formed from solutions containing H-bonding inhibitors. In a more comprehensive combined optical, electrical, and structure study, of PANi films cast from N,N′-dimethylpropylene urea solution, Ou et al. [418] observed no stretching-induced crystallization although there were significant changes in the optical and dielectric properties.

Orientational ordering may be observed even without explicit mechanical deformation. Processing procedures as routine as simply casting a film or electrochemical synthesis at an interface can lead to pronounced anisotropy in the molecular-level order and orientation of the polymer chains [70]. This anisotropy, in combination with uniaxial stretching, often leads to a 3D texturing of the observed structure [65,419]. Interaction of CPs at interfaces is another important area of study. The structural anisotropy of CP and P3AT thin films [420,421] has been studied by a number of groups especially in studies by Samuelsen and coworkers [74,422–424]. In P3ATs, both the extent of rR and the film-forming conditions can play supporting roles. Aasmundtveit et al. [422] reported that for thick films the choice of solvent was a dominant factor and they observed P3AT *a*-axis (parallel to the layer spacing) orientation along the surface normal. Thin, spin-cast films were characterized by a bimodal distribution of crystallites, one with the layer spacing parallel to the surface normal and one perpendicular to the surface normal. These results were independent of either solvent or substrate. This anisotropy also strongly influences the optical properties. Zhokhavets et al. [425] have recently studied the optical anisotropy in poly(3-octylthiophene) as a function of film thickness. They have observed a strong exponential drop in the anisotropy as the film thickness increases from 40 to 160 nm.

The implications of orientational ordering are by now well known for affecting the performance characteristics of CP FET devices. The impact of bulk orientation is perhaps best demonstrated in the report by Sirringhaus et al. [426] in which small changes in the nature of the chain perfection (i.e., rR) in P3ATs altered the orientation of the lamellar unit cell with respect to the substrate and thereby greatly affected the mobility of these planar devices. The structural aspects of this result are shown in Figure 17.25. To date no complete explanation has been reported as to the underlying causes.

Obtaining full 3D control of orientation in CP thin films over large length scales remains a centrally important goal. Large-scale orientation of the chain axis has been typically achieved in thin films through friction transfer of an aligned substrate layer and then thermal annealing of the CP overlayer [427–429] or by direct application of the polymer itself. An example of the latter can be seen in Nagamatsu et al. [430]. These P3AT thin films (~100 nm or less) exhibited a very high degree of chain alignment with Herman's order parameters approaching unity. The x-ray diffraction profiles exhibited strong layer (*h*00) reflections in the direction perpendicular to mechanical alignment and, significantly, a notable absence of the amorphous halo that universally accompanies cast film scattering data.

Noncontacting methods for obtaining alignment have also been developed. A report by Fukuda et al. [431] convincingly demonstrates the potential for using photoalignment techniques. In the first step, an azobenzene-functionalized poly(vinylalcohol) was deposited onto a substrate by Langmuir–Blodgett deposition and, after a number of intervening steps, photoalignment was induced by exposure to linearly polarized blue light. Thereafter, thin (10 to 100 nm) films of poly(di-*n*-hexylsilane) were spin-casted on

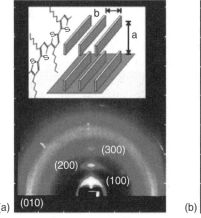 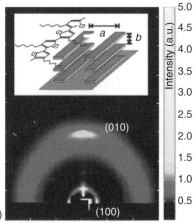

FIGURE 17.25 Grazing-incidence XRD patterns obtained by Sirringhaus et al. demonstrating two widely different orientations of ordered poly(3-*n*-hexylthiophene) domains with respect to the planar SiO₂/Si substrate. In (a) a 96% regioregular sample was spin-casted while for (b) an 81% regioregular sample was used. The vertical (horizontal) axes correspond to scattering normal (parallel) to the plane of the film. The *insets* show schematically the different orientations of the microcrystalline grains with respect to the substrate. (Reprinted from Sirringhaus, H., et al., *Nature*, 401, 685, 1999. With permission.)

top of this alignment layer. UV–vis absorption measurements verified the presence of orientational ordering.

As an alternate to friction transfer, one may also utilize classic polyimide substrate rubbing techniques to achieve alignment [318]. The thin-film studies mentioned so far have tended to focus on either uniaxial orientation or texturing about the surface normal. In a series of thin-film studies Knaapila et al. [318,432] have looked carefully at the orientation of the helical polyfluorene PF2/6 before and after thermal annealing on rubbed polyimide sublayers. A high degree of uniaxial orientation is observed and, as already noted earlier, chain alignment works best for lower molecular weight samples in the *n*-LC phase. Another striking attribute seen in these PF2/6 studies is that, in the case of the Hex phase, the orientational ordering extends to crystallographic directions corresponding to the equatorial (*hk*0) plane as shown in Figure 17.26. Thus, their grazing-incidence XRD patterns exhibit a full triaxial ordering with a complex distribution of the crystallites in terms of the anisotropy about the equatorial plane. The underlying molecular-level factors necessary for controlling and optimizing this 3D anisotropy have yet to be fully explored.

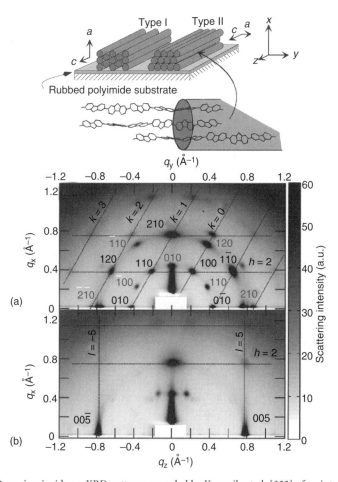

FIGURE 17.26 2D grazing-incidence XRD patterns recorded by Knaapila et al. [322] of an intermediate molecular weight PF2/6 polymer thin film thermally annealed on top of an oriented polyimide substrate. Patterns are: (a) the equatorial (*hk*0) and (b) the (*h*0ℓ) planes. The dotted lines highlight a few reflection planes of the type I crystal orientations. The black and white indices corresponds to the types I and II, respectively. These two orientations are shown schematically on top. (Reprinted from Knaapila, M., et al., *Macromolecules*, 38, 2744, 2005. With permission.)

17.5   Summary

This review has provided a broad overview of CP structure studies primarily with references to both historical developments and some of the more recent work in the field. Space and time constraints have limited the discussion almost entirely to the application of direct methods in bulk samples and, in terms of single-chain morphology and interchain packing, analysis of these structures through use of modeling studies. Gaining a fundamental understanding of these two molecular-level processes represents one key focus area in the now very diverse and expanding field of CP research. In a few instances, the relationship between molecular order and technologically important properties has been briefly summarized. Structure is a ubiquitous term and there are a number of glaring omissions. These include a vast majority of the reported work in the now burgeoning area of monolayer and interfacial studies through use of scanning probe measurements and other powerful surface-sensitive techniques. CP blends, block polymers, and composites are also important topical areas missing in this review. Most references to indirect probes of structure and structural behavior at short length scales are also notably absent. Raman scattering and IR spectroscopies probe dynamic characteristics and thereby provide yet another important window into the inner workings of structure–property relationships at the molecular level. NMR, both solution and solid state, is still another powerful tool for evincing CP properties over relatively short length scales. Although a modest number of reports have already appeared, far more could be done with this technique. Other direct methods, e.g., EXAFS, have been applied and published as well. This list of omissions itself is incomplete.

Another central issue that remains as a long-range goal is to provide a seamless framework for addressing CP structure starting at nanometer length scales and scaling up to more macroscopic entities. The application of computer-based modeling methods is another bright growth area and the capacity to replicate real physical systems is rapidly improving. A recent paper by Sumpter et al. [433] models chain folding of moderate length MEH-PPV oligomers and, through extended molecular dynamics calculations, can discern MEH-PPV morphological differences in response to good and bad solvents. CPs therefore represent a sort of proving ground for these studies because small, almost imperceptible differences in molecular forces, starting at the relatively weak and simple van der Waals interactions, can profoundly alter the electronic and optical properties. These latter attributes, of course, are now being actively studied from the perspective of comprehensive total energy calculations (and reflect much greater energy scales). Both static and dynamic properties play demonstratively important roles.

This paragraph brings this chapter to a somewhat open-ended but almost tidy closure, but reinforcing the original assertion that CPs are fundamentally different from conventional inorganic electronic materials for a myriad of reasons. In particular, there is a fascinating complexity in the manner in which extended organic molecules can associate and self-assemble so that, within a local region, order and disorder are heuristically arranged. Deciphering this fundamental underlying thread remains as a work in progress and future studies will continue to advance the conducting polymer field.

Acknowledgments

Much of the research conducted by the Wisconsin group is the result of major contributions by many students and collaborators for which I am indebted. I am especially grateful to T. Bjørholm, S. Guha, G. Heimel, D. Huber, M. Kertesz, M. Knaapila, J. Michl, F. Spano, F. Weinhold, and R. West for enlightening discussions. This work was partly supported by the US National Science Foundation under Grant DMR-0350383.

References

1. Heeger, A.J. 2001. Nobel lecture: Semiconducting and metallic polymers: The fourth generation of polymeric materials. *Rev Mod Phys* 73:681.
2. Hoeben, F.J.M., P. Jonkheijm, E.M. Meijer, and A.P.H. Schenning. 2005. About supramolecular assemblies of π-conjugated systems. *Chem Rev* 105:1491.

3. Samuelsen, E.J. and J. Mårdalen. 1997. Structure of polythiophenes. In *Handbook of organic conductive molecules and polymers*, Vol. 3, ed. H.S. Nalwa, New York: John Wiley & Sons, 87.

4. Winokur, M.J. 1998. Structural studies of conducting polymers. In *The handbook of conducting polymers*, 2nd ed., eds. T. Skotheim, R.L. Elsenbaumer, and J.R. Reynolds. New York: Marcel-Dekker, 123.

5. Feast, W.J., J. Tsibouklis, K.L. Pouwer, L. Groenendaal, and E.W. Meijer. 1996. Synthesis, processing and material properties of conjugated polymers. *Polymer* 37:5017.

6. Schwab, F.H., J.R. Smith, and J. Michl. 2005. Synthesis and properties of molecular rods. 2. Zig–Zag rods. *Chem Rev* 105:1197.

7. Jagur-Grodzinski, J. 2002. Electronically conductive polymers. *Polym Adv Technol* 13:615.

8. DaCosta, P.G., R.G. Dandrea, and E.M. Conwell. 1993. 1st-Principles calculation of the 3 dimensional band-structure of poly(phenylene vinylene). *Phys Rev B* 47:1800.

9. Filatov, I. and S. Larsson. 2002. Electronic structure and conduction mechanism of donor-bridge acceptor systems where PPV acts as a molecular wire. *J Chem Phys* 284:575.

10. Ruini, A., M.J. Caldas, G. Bussi, and E. Molinari. 2002. Solid state effects on exciton states and optical properties of PPV. *Phys Rev Lett* 88:206403.

11. Ferretti, A., A. Ruini, E. Molinari, and M.J. Caldas. 2003. Solid state effects on exciton states and optical properties of PPV. *Phys Rev Lett* 90:086401.

12. Capaz, R.B. and M.J. Caldas. 2003. Ab initio calculations of structural and dynamical properties of poly(*p*-phenylene) and poly(*p*-phenylene vinylene). *Phys Rev B* 67:205205.

13. Fincher, C.R., C.E. Chen, A.J. Heeger, A.G. MacDiarmid, and J.B. Hastings. 1982. Structural determination of the symmetry-breaking parameter in *trans*-(CH)*x*. *Phys Rev Lett* 48:100.

14. Gierschner, J. and D. Oelkrug. 2004. Optical properties of oligopheylenevinylenes. In *Encyclopedia of nanoscience and nanotechnology*, Vol. 8, ed. H.S. Nalwa, New York: American Science Publication, 219–238.

15. Gierschner, J., H.G. Mack, D. Oelkrug, and I. Waldner. 2004. Modeling of the optical properties of cofacial chromophore pairs: Stilbenophane. *J Phys Chem A* 116:257.

16. Lim, S.H., T.G. Bjorklund, F.C. Spano, and C.J. Bardeen. 2004. Exciton delocalization and superradiance in tetracene thin films and nanoaggregates. *Phys Rev Lett* 92:107402.

17. Beljonne, D., J. Cornil, R. Silbey, P. Millié, and J.L. Brédas. 2000. Interchain interactions in conjugated materials: The exciton model versus the supermolecular approach. *J Chem Phys* 112:4749.

18. Brédas, J.L., D. Beljonne, V. Coropceanu, and J. Cornil. 2004. Charge-transfer and energy-transfer processes in pi-conjugated oligomers and polymers: A molecular picture. *Chem Rev* 104:4971.

19. Spano, F.C. 2002. Absorption and emission in oligo-phenylene vinylene nanoaggregates: The role of disorder and structural defects. *J Chem Phys* 116:5877.

20. Meinardi, F., M. Cerminara, A. Sassella, R. Bonifacio, and R. Tubino. 2003. Superradiance in molecular H aggregates. *Phys Rev Lett* 91:247401.

21. Nguyen, T.Q., V. Doan, and B.J. Schwartz. 1999. Conjugated polymer aggregates in solution: Control of interchain interactions. *J Chem Phys* 110:4068.

22. Nguyen, T.Q., I.B. Martini, J. Liu, and B.J. Schwartz. 2000. Controlling interchain interactions in conjugated polymers: The effects of chain morphology on exciton–exciton annihilation and aggregation in MEH-PPV films. *J Phys Chem B* 104:237.

23. Nguyen, T.Q. and B.J. Schwartz. 2002. Ionomeric control of interchain interactions, morphology, and the electronic properties of conjugated polymer solutions and films *J Chem Phys* 116:8198.

24. Claudio, G.C. and E.R. Bittner. 2003. Excitation transfer in aggregated and linearly confined poly(*p*-phenylene vinylene) chains. *J Phys Chem A* 107:7092.

25. Shirakawa, H. and S. Ikeda. 1971. Infrared spectra of poly(acetylene). *Polym J* 2:231.

26. Shirakawa, H., E.J. Louis, A.J. MacDiarmid, C.K. Chiang, and A.J. Heeger. 1977. Synthesis of electrically conducting organic polymers—halogen derivatives of polyacetylene, (CH)*x*. *J Chem Soc Chem Commun* 16:578.

27. Chiang, C.K., C.R. Fincher, Y.W. Park, A.J. Heeger, H. Shirakawa, E.J. Louis, S. Gau, and A. Macdiarmid. 1977. Electrical-conductivity in doped polyacetylene. *Phys Rev Lett* 39:1098.

28. Heeger, A.J., J.R. Schriefer, and W.-P. Su. 1988. Solitons in conducting polymers. *Rev Mod Phys* 40:3439.

29. Tsukamoto, J. 1992. Recent advances in highly conductive polyacetylene. *Adv Phys* 41:509.

30. Park, J.G., G.T. Kim, V. Krstic, B. Kim, S.H. Lee, S. Roth, M. Burghard, and Y. Park. 2001. Nanotransport in polyacetylene single fiber: Toward the intrinsic properties. *Synth Met* 119 (1–3):53.

31. Aleshin, A.N., H.J. Lee, Y.W. Park, and K. Akagi. 2004. One-dimensional transport in polymer nanofibers. *Phys Rev Lett* 93:196601.

32. Peierls, R.E. 1939. *Proc R Soc A* 169:413.

33. Simionescu, C.I. and V. Percec. 1982. Progress in polyacetylene chemistry. *Prog Polym Sci* 8:133.

34. Ito, T., H. Shirakawa, and S. Ikeda. 1975. Thermal *cis–trans* isomerization and decomposition of polyacetylene. *J Polym Sci Polym Chem* 13:1943.

35. Shimamura, K., F.E. Karasz, J.A. Hirsch, and J.W. Chien. 1981. Crystal-structure of transpolyacetylene. *Makrom Chem Rapid Commun* 2:473.

36. Robin, P., J.P. Pouget, R. Comes, H.W. Gibson, and A.J. Epstein. 1983. Structural study of the *cis*-to-*trans* thermal-isomerization in polyacetylene. *Phys Rev B* 27:3938.

37. Su, W.P., J.R. Schrieffer, and A.J. Heeger. 1979. Solitons in polyacetylene. *Phys Rev Lett* 42:1698.

38. Akagi, K., G. Piao, S. Kaneko, K. Sakamaki, H. Shirakawa, and M. Kyotani. 1998. Helical polyacetylene synthesized with a chiral nematic reaction field. *Science* 283:1683.

39. Baughman, R.H., S.L. Hsu, G.P. Pez, and A.J. Signorelli. 1978. Structures of *cis*-polyacetylene and highly conducting derivatives. *J Chem Phys* 68:5405.

40. Akaishi, T., K. Miyasaka, K. Ishikawa, H. Shirakawa, and S. Ikeda. 1980. Crystallinity of bulk polyacetylene. *J Polym Sci B* 18:745.

41. Zhu, Q., J.E. Fischer, R. Zuzok, and S. Roth. 1992. Crystal-structure of polyacetylene revisited—an x-ray study. *Solid State Commun* 83:179.

42. Hosemann, R. and S.N. Bagchi. 1962. *Direct analysis of diffraction by matter*, Amsterdam: North-Holland.

43. Granier, T., E.L. Thomas, and F.E. Karasz. 1989. Paracrystalline structure of poly(*p*-phenylene vinylene). *J Polym Sci Polym Phys Ed* 27:469.

44. Zhang, X.B., G. VanTendeloo, J. VanLanduyt, D. VanDijck, J. Briers, Y. Bao, and H.J. Geise. 1996. An electron microscopic study of highly oriented undoped and FeCl$_3$-doped poly (*p*-phenylenevinylene). *Macromolecules* 29:1554.

45. Gin, D.L. and V.P. Conticello. 1996. Poly(*p*-phenylene): New directions in synthesis and application. *Trends Polym Sci* 4:217.

46. Froyer, G., J. Goblot, J.L. Guilbert, F. Maurice, and Y. Pelous. 1983. Poly(*para* phenylene): Some properties related to the synthesis method. *J Phys Colloq (France)* 44:745.

47. Boudet, A. and P. Pradere. 1984. Morphology and structure of poly(*p*-phenylene) as revealed by electron-microscopy. *Synth Met* 9:491.

48. Willnauer, C. and U. Birkenheuer. 2004. Quantum chemical ab initio calculations of correlation effects in complex polymers: Poly(*p*-phenylene). *J Chem Phys* 120:11910.

49. Sauvajol, J.L., D. Djarado, A.J. Dianoux, N. Theophilou, and J.E. Fischer. 1991. Polarized vibrational density of states of polyacetylene from incoherent inelastic neutron-scattering. *Phys Rev B* 43:14305.

50. Ma, J., et al. 1991. Intrachain dynamics and interchain structures of polymers—a comparison of polyacetylene, polyethylene, polyaniline and poly(*p*-phenylene vinylene). *Phys Rev B* 44:11609.

51. Dianoux, A.J., G.R. Kneller, J.L. Sauvajol, and J.C. Smith. 1993. The polarized density-of-states of crystalline polyacetylene-molecular-dynamics analysis and comparison with neutron-scattering results. *J Chem Phys* 99:5586.

52. Sauvajol, J.L., P. Papanek, J.E. Fischer, A.J. Dianoux, P.M. McNeillis, C. Mathis, and B. Francios. 1997. Dynamics of pristine and doped conjugated polymers: A combined inelastic neutron scattering and computer simulation analysis. *Synth Met* 84:941.
53. Sasaki, S., T. Yamamoto, T. Kanbara, A. Morita, and T. Yamamoto. 1992. Crystal-structure of poly(*para*-phenylene) prepared by organometallic technique. *J Polym Sci Polym Phys Ed* 30:293.
54. Heimel, G., et al. 2005. Breakdown of the mirror image symmetry in the optical absorption/emission spectra of oligo(*para*-phenylene)s. *J Chem Phys* 122:054501.
55. Kanbe, M. and M. Okawara. 1968. *J Polym Sci* A-1 6:1058.
56. Wessling, R.A. and R.G. Zimmerman. 1968. US Patent 3 401:152.
57. Mao, G., J.E. Fischer, F.E. Karasz, and M.J. Winokur. 1993. Non-planarity and ring torsion in poly(*p*-phenylene vinylene): A neutron diffraction study. *J Chem Phys* 98:712.
58. Simpson, J.H., D.M. Rice, F.E. Karasz, F.C. Rossito, and D. Lahti. 1993. A multitechnique investigation of sodium-doped poly(*p*-phenylene vinylene). *Polymer* 34:4595.
59. Simpson, J.H., D.M. Rice, and F.E. Karasz. 1991. Investigations of H_2SO_4-doped, ring-deuterated poly(*p*-phenylene vinylene) using solid-state H-2 quadrupole echo NMR-spectroscopy. *Polymer* 32:2340.
60. Simpson, J.H., D.M. Rice, and F.E. Karasz. 1992. Characterization of chain orientation in drawn poly(*p*-phenylene vinylene) by H-2 quadrupole echo. *Macromolecules* 25:2099.
61. de Azevedo, E.R., R.W.A. Franco, A. Marletta, R.M. Faria, and T.J. Bonagamba. 2003. Conformational dynamics of phenylene rings in poly(*p*-phenylene vinylene) as revealed by C-13 magic-angle-spinning exchange nuclear magnetic resonance experiments. *J Chem Phys* 119:2923.
62. Granier, T., E.L. Thomas, D.R. Gagnon, R.W. Lenz, and F.E. Karasz. 1986. Structure investigation of poly(*p*-phenylene vinylene). *J Polym Sci Polym Phys Ed* 24:2793.
63. Bradley, D.D.C., R.H. Friend, T. Hartmann, E.A. Marseglia, M.M. Sokolowski, and P.D. Townsend. 1987. Structural studies of oriented precursor route conjugated polymers. *Synth Met* 17:473.
64. Gagnon, D.R., F.E. Karasz, E.L. Thomas, and R.W. Lenz. 1987. Molecular-orientation and conductivity in highly drawn poly(*p*-phenylene vinylene). *Synth Met* 20:85.
65. Chen, D., M.J. Winokur, M.A. Masse, and F.E. Karasz. 1992. A structural study of poly(*p*-phenylene vinylene). *Polymer* 33:3116.
66. Mo, Z., K.B. Lee, Y.B. Moon, M. Kobayashi, A.J. Heeger, and F. Wudl. 1985. X-ray scattering from polythiophene-crystallinity and crystallographic structure. *Macromolecules* 18:1972.
67. Brüchner, S. and W. Porzio. 1988. The structure of neutral polythiophene—an application of the Rietveld method. *Makromol Chem* 189:961.
68. Chu, X., V. Chan, L.D. Schmidt, and W.H. Smyrl. 1995. Crystalline fibers in chemically polymerized ultrathin polypyrrole films. *J Appl Phys* 12:6658.
69. Geiss, R.H., G.B. Street, W. Volksen, and J. Economy. 1983. Polymer structure determination using electron-diffraction techniques. *IBM J Res Dev* 27:321.
70. Mitchell, G.R. and A. Geri. 1987. Molecular-organization of electrochemically prepared conducting polypyrrole films. *J Phys D* 20:1346.
71. Groenendaal, B.L., F. Jonas, D. Freitag, H. Pielartzik, and J.R. Reynolds. 2000. Poly(3,4-ethylene-dioxythiophene) and its derivatives: Past, present, and future. *Adv Mat* 12:481.
72. Winther-Jensen, B. and K. West. 2004. Vapor-phase polymerization of 3,4-ethylenedioxythiophene: A route to highly conducting polymer surface layers. *Macromolecules* 37:4538.
73. Mishra, S.P., R. Sahoo, A.V. Ambade, A.Q. Contractor, and A. Kumar. 2004. Synthesis and characterization of functionalized 3,4-propylenedioxythiophene and its derivatives. *J Mater Sci* 14:1896.
74. Aasmundtveit, K.E., E.J. Samuelsen, L.A.A. Pettersson, O. Inganäs, T. Johansson, and R. Feidenhans. 1999. Structure of thin films of poly(3,4-ethylenedioxythiophene). *Synth Met* 101:561.
75. Pettersson, L.A.A., T. Johansson, F. Carlsson, H. Arwin, and O. Inganäs. 1999. Anisotropic optical properties of doped poly(3,4-ethylenedioxythiophene). *Synth Met* 101:198.

76. Breiby, D.W., E.J. Samuelsen, L. Groenendaal, and B. Struth. 2003. Smectic structures in electrochemically prepared poly(3,4-ethylenedioxythiophene) films. *J Polym Sci Polym Phys* 41:945.

77. Aasmundtveit, K.E., E.J. Samuelsen, O. Inganäs, L.A.A. Pettersson, T. Johansson, and S. Ferrer. 2000. Structural aspects of electrochemical doping and dedoping of poly(3,4-ethylenedioxythiophene). *Synth Met* 113:93.

78. Ionescu-Zanetti, C., A. Mechler, S.A. Carter, and R. Lal. 2004. Semiconductive polymer blends: Correlating structure with transport properties at the nanoscale. *Adv Mater* 5:385.

79. Kemerink, M., S. Timpanaro, M.M. de Kok, E.A. Meulenkamp, and F.J. Touwslager. 2004. Three-dimensional inhomogeneities in PEDOT:PSS films. *J Phys Chem B* 108:18820.

80. Jukes, P.C., S.J. Martin, A.M. Higgins, M. Geoghegan, R.A.L. Jones, S. Langridge, A. Wehrum, and S. Kirchmeyer. 2004. Controlling the surface composition of poly(3,4-ethylene dioxythiophene)-poly(styrene sulfonate) blends by heat treatment. *Adv Mat* 16:807.

81. Genies, E.M., A. Boyle, M. Lapkowski, and C. Tsintavis. 1990. Polyaniline—a historical survey. *Synth Met* 36:139.

82. Jòzefowicz, M.E., A.J. Epstein, J.P. Pouget, J.G. Masters, A. Ray, Y. Sun, X. Tang, and A.G. MacDiarmid. 1991. X-ray structure of polyanilines. *Synth Met* 41–43:723.

83. Adams, P.N., P. Devasagayam, S.J. Pomfret, L. Abell, and A.P. Monkman. 1998. A new acid-processing route to polyaniline films which exhibit metallic conductivity and electrical transport strongly dependent upon intrachain molecular dynamics. *J Phys Cond Mat* 10:8293.

84. Wang, H.L., R.J. Romero, B.R. Mattes, Y.T. Zhu, and M.J. Winokur. 2000. Effect of processing conditions on the properties of high molecular weight conductive polyaniline fiber. *J Polym Sci Polym Phys* 38:194.

85. Jòzefowicz, M.E., R. Laversanne, H.H.S. Javadi, A.J. Epstein, J.P. Pouget, X. Tang, and A.G. MacDiarmid. 1989. Multiple lattice phases and polaron-lattice spinless-defect competition in polyaniline. *Phys Rev B* 39:12958.

86. Pouget, J.P., M.E. Jòzefowicz, A.J. Epstein, X. Tang, and A.G. MacDiarmid. 1991. X-ray structure of polyaniline. *Macromolecules* 24:779.

87. Laridjani, M., J.P. Pouget, E.M. Scherr, A.G. MacDiarmid, M. Jòzefowicz, and A.J. Epstein. 1992. Structural aspects of the polyaniline family of electronic polmers. *Synth Met* 51:95.

88. Pouget, J.P., M.E. Jòzefowicz, A.J. Epstein, J.G. Masters, A. Ray, and A.G. MacDiarmid. 1991. X-ray structure of the polyaniline derivative poly(ortho-toluidine)—the structural origin of charge localization. *Macromolecules* 24:5863.

89. Cavazzoni, C., R. Colle, R. Farchioni, and G. Grosso. 2004. Ab initio molecular dynamics study of the structure of emeraldine base polymers. *Phys Rev B* 69:115213.

90. Winokur, M.J. and B.R. Mattes. 1998. Structural studies of halogen acid doped polyaniline and the role of water hydration. *Macromolecules* 31:8183.

91. Wang, Z.H., C. Li, E.M. Scherr, A.G. MacDiarmid, and A.J. Epstein. 1991. 3 Dimensionality of metallic states in conducting polymers—polyaniline. *Phys Rev Lett* 66:1745.

92. Epstein, A.J., et al. 1994. Inhomogeneous disorder and the modified drude state of conducting polymers. *Synth Met* 65:149.

93. Shacklette, L.W., J.F. Wolf, S. Gould, and R.H. Baughman. 1988. Structure and properties of polyaniline as modeled by single-crystal oligomers. *J Chem Phys* 88:3955.

94. Baughman, R.H., J.F. Wolf, H. Eckhardt, and L.W. Shacklette. 1988. The structure of a novel polymeric metal—acceptor-doped polyaniline. *Synth Met* 25:121.

95. Moon, Y.B., Y. Cao, P. Smith, and A.J. Heeger. 1989. X-ray-scattering from crystalline polyaniline. *Polymer* 30:196.

96. Annis, B.K., J.S. Lin, E.M. Scherr, and A.G. MacDiarmid. 1989. Evidence for the development of one-dimensional array of crystallites in stretched polyaniline and the effect of Cl⁻ doping. *Macromolecules* 25:429.

97. Fischer, J.E., X. Tang, E.M. Scherr, V.B. Cajipe, and A.G. MacDiarmid. 1994. Polyaniline fibers and films: Stretch-induced orientation and crystallization, morphology and the nature of the amorphous phase. *Macromolecules* 27:5094.

98. Colomban, P., A. Gruger, A. Novak, and A. Régis. 1994. Infrared and Raman-study of polyaniline. 1. Hydrogen-bonding and electronic mobility in emeraldine salts. *J Mol Struct* 317:261.

99. Gruger, A., P. Colomban, A. Novak, and A. Régis. 1994. Infrared and Raman-study of polyaniline. 2. Influence of ortho substituents on hydrogen-bonding and UV/vis. *J Mol Struct* 328:153.

100. Colomban, P., S. Folch, and A. Gruger. 1999. Vibrational study of short-range order and structure of polyaniline bases and salts. *Macromolecules* 32:3080.

101. Khalki, A.E. and P. Colomban. 2002. Vibrational study of short-range order and structure of polyaniline bases and salts. *Macromolecules* 35:5203.

102. Zheng, W., M. Angelopoulos, A.J. Epstein, and A.G. MacDiarmid. 1997. Experimental evidence for hydrogen bonding in polyaniline mechanism of aggregate formation and dependency. *Macromolecules* 30:2953.

103. Mattes, B.R., H.L. Wang, D. Yang, Y. Zhu, W.R. Blumenthal, and M.F. Hundley. 1997. Formation of conductive polyaniline fibers derived from highly concentrated emeraldine base solution. *Synth Met* 84:45.

104. Łużny, W. and E. Bañka. 2000. Relations between the structure and electric conductivity of polyaniline protonated with camphorsulfonic acid. *Macromolecules* 33:425.

105. Angelopoulos, M., A. Ray, A.G. MacDiarmid, and A.J. Epstein. 1987. Polyaniline—processability from aqueous-solutions and effect of water-vapor on conductivity. *Synth Met* 21:21.

106. Boyle, A., J.F. Penneau, E. Geniès, and C. Riekel. 1992. The effect of heating on polyaniline powders studied by real-time synchrotron radiation diffraction, mass spectrometry and thermal analysis. *Polymer* 30:265.

107. Price, W.E. and G.G. Wallace. 1996. Effect of thermal treatment on the electroactivity of polyaniline. *Polymer* 37:917.

108. Pinto, N.J., P.D. Shah, P.K. Kahol, and B.J. McCormick. 1996. Conducting state of polyaniline films: Depenence on moisture. *Phys Rev B* 53:10690.

109. Kahol, P.K., A.J. Dyakonov, and B.J. McCormick. 1997. An electron-spin-resonance study of polymer interactions with moisture in polyaniline and its derivatives. *Synth Met* 89:1997.

110. Wolter, A., P. Rannou, J.P. Travers, B. Gilles, and D. Djurado. 1998. Model for aging in HCl-protonated polyaniline: Structure, conductivity, and composition studies. *Phys Rev B* 58:7637.

111. MacDiarmid, A.G., Y. Min, J.M. Wiesinger, E.J. Oh, E.M. Scherr, and A.J. Epstein. 1993. Towards optimization of electrical and mechanical-properties of polyaniline—is cross-linking between chains the key. *Synth Met* 55:753.

112. Joo, J., S.M. Long, J.P. Pouget, E.J. Oh, A.G. MacDiarmid, and A.J. Epstein. 1998. Charge transport of the mesoscopic metallic state in partially crystalline polyanilines. *Phys Rev B* 69:115213.

113. Miller, R.D. and J. Michl. 1989. Polysilane high polymers. *Chem Rev* 89:1359.

114. van der Laan, G.P., M.P. de Haas, A. Hummel, H. Frey, and M. Möller. 1996. Charge carrier mobilities in substituted polysilylenes: Influence of backbone conformation. *J Phys Chem* 100:5470.

115. Lacave-Goffin, B., L. Hevesi, S. Demoustier-Champange, and J. Devaux. 1999. Synthesis and properties of polysilanes: Versatile new organic materials. *ACH-Mod Chem* 136:214.

116. Dewar, M.J.S. 1984. Chemical implications of sigma-conjugation. *J Am Chem Soc* 106:669.

117. Michl, J. and R. West. 2000. Electronic structure and spectroscopy of polysilanes. In *Silicon-containing polymers: The science and technology of their synthesis and applications*, eds. J. Chojnowski, R.G. Jones, and W. Ando, Dordrecht, The Netherlands: Kluwer Academic, 499.

118. Schepers, T. and J. Michl. 2002. Optimized ladder C and ladder H models for sigma conjugation: Chain segmentation in polysilanes. *J Phys Org Chem* 15:490.

119. Rooklin, D.W., T. Schepers, M.K. Raymond-Johansson, and J. Michl. 2003. Time-dependent density functional theory treatment of the first UV absorption band in all-transoid permethyloligosilanes Si_nMe_{2n+2} ($n = 2$–8, 10). *Photochem Photobio Sci* 2:511.

120. Plitt, H.S., J.W. Downing, M.K. Raymond, V. Balaji, and J. Michl. 1994. Photophysics and potential-energy hypersurfaces of permethylated oligosilanes. *J Chem Soc Farad Trans* 90:1653.

121. Rossi, G., R.R. Chance, and R. Silbey. 1990. Conformational disorder in conjugated polymers. *J Chem Phys* 90:7594.

122. Pophristic, V. and L. Goodman. 2001. Hyperconjugation not steric repulsion leads to the staggered structure of ethane. *Nature* 411:565.

123. Trefonas, P., R. West, R.D. Miller, and D. Hofer. 1983. Organosilane high polymers-electronic-spectra and photodegradation. *J Polym Sci Polym Lett Ed* 21:819.

124. Karatsu, T., R.D. Miller, R. Sooriyakumaran, and J. Michl. 1989. Mechanism of the photochemical degradation of poly(di-*n*-alkylsilanes) in solution. *J Am Chem Soc* 111:1140.

125. Baughman, R.H., J.L. Brédas, R. Chance, R.L. Elsenbaumer, and L.W. Shacklette. 1982. Structural basis for semiconducting and metallic polymer/dopant systems. *Chem Rev* 82:209.

126. See *Chemical physics of intercalation*, Vol. I NATO ASI Series, eds. A.P. Legrand, and S. Flandrois, New York: Plenum. 1987.

127. See *Chemical physics of intercalation*, Vol. II NATO ASI Series, eds. P. Bernier, J.E. Fischer, S. Roth, and S.A. Solin, New York: Plenum. 1993.

128. Kaner, R.B., H. Reiss, B.R. Mattes, and M.R. Anderson. 1991. Conjugated polymer-films for gas separations. *Science* 252:1412.

129. Huang, S.C., I.J. Ball, and R.B. Kaner. 1998. Polyaniline membranes for pervaporation of carboxylic acids and water. *Macromolecules* 31:5456.

130. Pellegrino, J. 2003. The use of conducting polymers in membrane-based separations–A review and recent developments. In *Advanced membrane technology annals of the New York academy of sciences*, Vol. 984, New York: New York Academy of Sciences, 289.

131. Smela, E., O. Inganäs, and I. Lundstrom. 1995. Controlled folding of micrometer-size structures. *Science* 268:1735.

132. Smela, E. 1999. Microfabrication of polypyrrole microactuators and other conjugated polymer devices. *J Micromech Microeng* 9:1.

133. Sitaram, S.P., J.O. Stoffer, and T.J. O'Keefe. 1997. Application of conducting polymers in corrosion protection. *J Coating Tech* 69:65.

134. See *Electroactive Actuators as Artificial Muscles-Reality, Potential and Challenges*, ed. Y. Bar-Cohen, Bellingham, Washington: SPIE Press. 2001.

135. See *Conductive Electroactive Polymers: Intelligent Materials Systems*, Second Edition, eds. G.M. Spinks, L.A.P. Kane-Maguire, P.R. Teasdale, and G.G. Wallace, Boca Raton, FL: CRC Press. 2003.

136. Smela, E. 2003. Conjugated polymer actuators for biomedical applications. *Adv Mat* 15:481.

137. Baughman, R.H., L.W. Shacklette, N.S. Murthy, G.G. Miller, and R.L. Elsenbaumer. 1985. The evolution of structure during the alkali metal doping of polyacetylene and poly(*p*-phenylene). *Mol Cryst Liq Crys* 118:253.

138. Baughman, R.H., N.S. Murthy, and G.G. Miller. 1983. The structure of metallic complexes of polyacetylene with alkali metals. *J Chem Phys* 79:515.

139. Baughman, R.H., N.S. Murthy, G.G. Miller, and L.W. Shacklette. 1983. Staging in polyacetylene iodine conductors. *J Chem Phys* 79:1065.

140. Murthy, N.S., G.G. Miller, and R.H. Baughman. 1988. Structure of polyacetylene iodine complexes. *J Chem Phys* 89:2523.

141. Murthy, N.S., L.W. Shacklette, R.H. Baughman, H. Fark, and J. Fink. 1991. A hexagonal structure for alkali-metal doped poly(*p*-phenylene). *Solid State Commun* 78:691.

142. Murthy, N.S., L.W. Shacklette, and R.H. Baughman. 1989. Structure of lithium-doped polyacetylene. *Phys Rev B* 40:12550.

143. Dresselhaus, M. and G. Dresselhaus. 1981. *Adv Phys* 30:139.
144. Djurado, D., J. Ma, N. Theophilou, and J.E. Fischer. 1989. Structural characterization of a new polyacetylene—preferred orientation and coherence length. *Synth Met* 30:395.
145. Djurado, D., J.E. Fischer, P.A. Heiney, J. Ma, N. Coustel, and P. Bernier. 1990. Staging in potassium-doped polyacetylene: In situ x-ray diffraction. *Synth Met* 34:683.
146. Ma, J., D. Djurado, J.E. Fischer, and P.A. Heiney. 1990. Structure of cesium doped polyacetylene. *Phys Rev B* 41:2971.
147. Heiney, P.A., et al. 1991. Channel structures in alkali-metal-doped conjugated polymers—broken symmetry 2-dimensional intercalation superlattices. *Phys Rev B* 44:2507.
148. Winokur, M.J., J. Maron, Y. Cao, and A.J. Heeger. 1992. Disorder and staging in iodine-doped polyacetylene. *Phys Rev B* 45:9656.
149. Pouget, J.P., et al. 1985. Structural study of polyacetylene doped with tetrahedral anions. *Mol Cryst Liq Cryst* 117:75.
150. Pouget, J.P., P. Robin, R. Comes, H.W. Gibson, A.J. Epstein, and D. Billaud. 1984. Structural properties of a conducting polymer: Doped (CH)*x*. *Physica B & C*, 127B:158.
151. Robin, P., J.P. Pouget, R. Comes, H.W. Gibson, and A.J. Epstein. 1983. X-ray-diffraction studies of iodine-doped polyacetylene. *Polymer* 24:1558.
152. Flandrois, S., C. Hauw, and B. Francois. 1983. Analogies between graphite intercalation compounds and doped polyacetylene. *J Phys (Paris)* 44:C3.
153. Chien, J.W., F.E. Karasz, and K. Shimamura. 1988. Crystal-structure of pristine and iodine-doped *cis*-polyacetylene. *Macromolecules* 15:1012.
154. Winokur, M.J., Y.B. Moon, A.J. Heeger, J. Barker, D.C. Bott, and H. Shirakawa. 1987. X-ray scattering from sodium doped polyacetylene: Incommensurate–commensurate and order–disorder transformations. *Phys Rev Lett* 58:2329.
155. Choi, H.-Y., E.J. Mele, J. Ma, and J.E. Fischer. 1988. Staging in doped polymers. *Synth Met* 27:A75.
156. Choi, H.-Y., A.B. Harris, and E.J. Mele. 1989. Mean-field theory for interchain orientational ordering of conjugated polymers. *Phys Rev B* 40:3766.
157. Choi, H.-Y. and E.J. Mele. 1989. Doping-induced structural phase-transition in polyacetylene. *Phys Rev B* 40:3439.
158. Harris, A.B. 1994. Mean-field theory for alkali-metal-doped polyacetylene. *Phys Rev B* 50:12441.
159. Mao, G., D. Chen, M.J. Winokur, and F.E. Karasz. 1999. The generalized anisotropic planar rotor model and its application to polymer intercalation compounds. *Phys Rev Lett* 83:622588.
160. Mao, G. and M.J. Winokur. 1999. A Monte Carlo study of structure and intercalation in conducting polymers. *Synth Met* 124:101.
161. Hong, S.Y. and M. Kertesz. 1990. Dependence of young modulus of *trans*-polyacetylene upon charge-transfer. *Phys Rev Lett* 64:3031.
162. Sun, G.Y., J. Kurti, M. Kertesz, and R.H. Baughman. 2002. Dimensional changes as a function of charge injection for *trans*-polyacetylene: A density functional theory study. *J Chem Phys* 117:7691.
163. Nogami, Y., J.P. Pouget, and T. Ishiguro. 1994. Structure of highly conducting PF_6^--doped polypyrrole. *Synth Met* 62:257.
164. Graupner, W., L. Oniciu, M. Brie, and R. Turcu. 1996. Electrochemical and x-ray diffraction studies on polypyrrole films. *Mat Chem Phys* 46:55.
165. Majidi, M.R., L.A.P. Kane-Maguire, and G.G. Wallace. 1994. Enantioselective electropolymerization of aniline in the presence of (+)-camphorsulfonate or (−)-camphorsulfonate ion: A facile route to conducting polymers with preferred one-screw-sense helicity. *Polymer* 35:3113.
166. Norris, I.D., L.A.P. Kane-Maguire, and G.G. Wallace. 1998. Thermochromism in optically active polyaniline salts. *Macromolecules* 31:6529.
167. Majidi, M.R., L.A.P. Kane-Maguire, and G.G. Wallace. 1998. Electrochemical synthesis of optically active polyanilines. *Aust J Chem* 51:23.
168. Łużny, W., E.J. Samuelsen, D. Djurado, and Y.F. Nicolau. 1997. Polyaniline protonated with camphorsulfonic acid: Modeling of its crystalline structure. *Synth Met* 90:19.

169. Minto, C.D.G. and A.S. Vaughan. 1997. Orientation and conductivity in polyaniline. *Polymer* 38:2683.

170. Winokur, M.J., H.L. Guo, and R.B. Kaner. 2001. Structural study of chiral camphorsulfonic acid doped polyaniline. *Synth Met* 119:403.

171. Dufour, B., P. Rannou, P. Fedorko, D. Djurado, J.P. Travers, and A. Pron. 2001. Effect of plasticizing dopants on spectroscopic properties, supramolecular structure, and electrical transport in metallic polyaniline. *Chem Mat* 13:4032.

172. Jana, T., J. Chatterjee, and A.K. Nandi. 2002. Sulfonic acid doped thermoreversible polyaniline gels. 3. Structural investigations. *Langmuir* 18:5720.

173. Dufour, B., P. Rannou, D. Djurado, H. Janeczek, M. Zagorska, A. de Geyer, J.P. Travers, and A. Pron. 2003. Low T-g, stretchable polyaniline of metallic-type conductivity: Role of dopant engineering in the control of polymer supramolecular organization and in the tuning of its properties. *Chem Mat* 15:4032.

174. Yaliraki, S.N. and R.J. Silbey. 1996. Conformational disorder of conjugated polymers: Implications for optical properties. *J Chem Phys* 104:1245.

175. Blondin, P., J. Bouchard, S. Beaupré, M. Belletête, G. Durocher, and M. Leclerc. 2000. Molecular design and characterization of chromic polyfluorene derivatives. *Macromolecules* 33:5874.

176. Dufresne, G., J. Bouchard, M. Belletête, G. Durocher, and M. Leclerc. 2000. Thermochromic and solvatochromic conjugated polymers by design. *Macromolecules* 33:8252.

177. Tilgner, A., H.P. Trommsdorff, J.M. Zeigler, and R.M. Hochstrasser. 1992. Poly(di-normalhexyl-silane) in solid-solutions-experimental and theoretical-studies of electronic excitations of a disordered linear-chain. *J Chem Phys* 96:781.

178. Romanovskii, Y.V., H. Bässler, and U. Scherf. 2004. Relaxation processes in electronic states of conjugated polymers studied via spectral hole-burning at low temperature. *Chem Phys Lett* 383:89.

179. Müller, J.G., U. Lemmer, G. Raschke, M. Anni, U. Scherf, J.M. Lupton, and J. Feldmann. 2003. Linewidth-limited energy transfer in single conjugated polymer molecules. *Phys Rev Lett* 91:267403.

180. Schindler, F., J.M. Lupton, J. Feldmann, and U. Scherf. 2004. A universal picture of chromophores in π-conjugated polymers derived from single-molecule spectroscopy. *Proc Natl Acad Sci* 101:14695.

181. Becker, K. and J.M. Lupton. 2005. Dual species emission from single polyfluorene molecules: Signatures of stress-induced planarization of single polymer chains. *J Am Chem Soc* 127:7306.

182. Hutchinson, G.R., Y.J. Zhao, B. Delley, A.J. Freeman, M. Ratner, and T. Marks. 2003. Electronic structure of conducting polymers: Limitations of oligomer extrapolation approximations and effects of heteroatoms. *Phys Rev B* 68:035204.

183. Wang, J.F., J.K. Feng, A.M. Ren, X.D. Liu, Y.G. Ma, P. Lu, and H.X. Zhang. 2004. Theoretical studies of the absorption and emission properties of the fluorene-based conjugated polymers. *Macromolecules* 37:3451.

184. Setayesh, S., D. Marsitzky, and K. Müllen. 2000. Bridging the gap between polyfluorene and ladder-poly-*p*-phenylene: Synthesis and characterization of poly-2,8-indenofluorene. *Macromolecules* 33:2016.

185. Silva, C., et al. 2002. Exciton and polaron dynamics in a step-ladder polymeric semiconductor: The influence of interchain order. *J Phys Cond Mat* 14:9803.

186. Benincori, T., et al. 1998. Tuning of the excited-state lifetime by control of the structural relaxation in oligothiophenes. *Phys Rev B* 58:9082.

187. Oyaizu, K., T. Iwasaki, Y. Tsukahara, and E. Tsuchida. 2004. Linear ladder-type π-conjugated polymers composed of fused thiophene ring systems. *Macromolecules* 37:1257.

188. Swager, T.M. 1998. The molecular wire approach to sensory signal amplification. *Acc Chem Res* 31:201.

189. Petelenz, P. and M. Slawik. 1991. 2-Dimensional model of charge-transfer excitons in polyacene crystals. *Chem Phys* 157:169.

190. Wunderlich, B. and J. Grebowicz. 1984. Thermotropic mesophases and mesophase transitions of linear, flexible. *Adv Polym Sci* 60:1.

191. Plate, N.A., R.V. Talrose, and Y.S. Freidzon. 1987. Polymeric liquid-crystals-problems and trends. *Polym J* 19:135.

192. Brostow, W. 1990. Properties of polymer liquid crystals-choosing molecular structures and blending. *Polymer* 31:979.

193. Davidson, P. 1996. X-ray diffraction by liquid crystalline side-chain polymers. *Prog Polym Sci* 21:893.

194. McCullough, R.D. and P.C. Ewbank. 1998. In *The handbook of conducting polymers*, 2nd ed., eds. T. Skotheim, R. Elsenbaumer, and J. Reynolds, New York: Marcel-Dekker, 225.

195. Hotta, S., S.D.D.V. Rughooputh, A.J. Heeger, and F. Wudl. 1987. Thermochromic and solvatochromic effects in poly(3-hexylthiophene). *Macromolecules* 20:212.

196. Leclerc, P., A. Calderone, K. Mullen, J.L. Brédas, and R. Lazzaroni. 2002. Conjugated polymer chains self-assembly: A new method to generate (semi)-conducting nanowires? *Mat Sci Tech* 18:749.

197. Ge, J.J., A. Zhang, K.W. McCreight, R.M. Ho, S.Y. Wang, X. Jin, F.W. Harris, and S.Z.D. Cheng. 1999. Phase structure, transition behaviors, and surface alignment of polymers containing rigid-rodlike backbones with flexible side chains. 1. Monotropic phase behavior in a main-chain/side-chain liquid crystalline polyester. *Macromolecules* 30:6498.

198. Cheng, X.H., et al. 2003. Calamitic bolaamphiphiles with (semi)perfluorinated lateral chains: Polyphilic block molecules with new liquid crystalline phase structures. *J Am Chem Soc* 125:10977.

199. Beenken, W.J.D. and T. Pullerits. 2004. Spectroscopic units in conjugated polymers: A quantum chemically founded concept? *J Phys Chem B* 108:6164.

200. Chunwachirasiri, W., B. Tanto, D.L. Huber, and M.J. Winokur. 2005. Chain conformations and photoluminescence in poly(di-*n*-octylfluorene). *Phys Rev Lett* 94:107402.

201. Keller, A. and S.Z.D. Cheng. 1998. The role of metastability in polymer phase transitions. *Polymer* 39:4461.

202. Cheng, S.Z.D. and A. Keller. 1998. The role of metastable states in polymer phase transitions: Concepts, principles, and experimental observations. *Ann Rev Mat Sci* 28:533.

203. Blomenhofer, M., et al. 2005. "Designer" nucleating agents for polypropylene. *Macromolecules* 38:3688.

204. Gorman, C.B., E.J. Ginsburg, and R.H. Grubbs. 1993. Soluble, highly conjugated derivatives of polyacetylene from the ring-opening metathesis polymerization of monosubstituted cyclooctatetraenes—synthesis and the relationship between polymer structure and physical-properties. *J Am Chem Soc* 115:1397.

205. Oh, S.Y., R. Ezaki, K. Akagi, and H. Shirakawa. 1993. Polymerization of monosubstituted acetylenes with a liquid-crystalline moiety by Ziegler-Natta and metathesis catalysts. *J Polym Sci Polym Chem* 31:2977.

206. Tada, K., R. Hidayat, M. Hirohata, M. Teraguchi, T. Masuda, and K. Yoshino. 1996. Optical properties and blue and green electroluminescence in soluble disubstituted acetylene polymers. *Jpn J Appl Phys* 35:L1138.

207. Huang, Y.M., W.K. Ge, J.W.Y. Lam, and B.Z. Tang. 1999. Strong photoluminescence from monosubstituted polyacetylenes containing biphenylyl chromophores. *Appl Phys Lett* 75:4094.

208. Lam, J.W. and B.Z. Tang. 2003. Liquid crystalline and light emitting polyacetylenes. *J Polym Sci Polym Chem* 41:2067.

209. Ciardell, F., S. Lanzillo, and O. Pieroni. 1974. Optically active polymers of 1-alkynes. *Macromolecules* 7:174.

210. Mitsuyama, M. and K. Kondo. 2001. Induced chiral helical effect on the main chain of aliphatic polyacetylenes. *J Polym Sci Polym Chem* 39:913.

211. Tabata, M., T. Sone, and Y. Sadahiro. 1999. Precise synthesis of monosubstituted polyacetylenes using Rh complex catalysts. Control of solid structure and π-conjugation length. *Macro Chem Phys* 200:265.

212. Chen, S.A. and E.C. Chang. 1998. Structure and properties of cyano-substituted poly(2,5-dialkoxy-*p*-phenylene vinylene)s. *Macromolecules* 31:4899.

213. Winokur, M.J. and W. Chunwachirasiri. 2003. Nanoscale structure/property relationships in conjugated polymers: Implications for current and future device applications. *J Polym Sci* 41:2630.

214. Collison, C.J., L.J. Rothberg, V. Treemaneekarn, and Y. Li. 2001. Conformational effects on the photophysics of conjugated polymers: A two species model for MEH-PPV spectroscopy and dynamics. *Macromolecules* 34:2346.

215. Marseglia, E.A., B.A. Weir, S.M. Chang, and A.B. Holmes. 1999. Changes in structure and morphology in the conjugated polymer MEH-PPV. *Synth Met* 101:154.

216. Shi, Y., J. Liu, and Y. Yang. 2000. Device performance and polymer morphology in polymer light emitting diodes: The control of thin film morphology and device quantum effciency. *J Appl Phys* 87:4254.

217. Chen, S., A. Su, Y.F. Huang, C.H. Su, K.Y. Peng, and S.A. Chen. 2002. Supramolecular aggregation in bulk poly(2-methoxy-5-(2′-ethylhexyloxy)-1,4-phenylenevinylene). *Macromolecules* 35:4229.

218. Yang, C.Y., F. Hide, M.A. Diaz-Garcia, A.J. Heeger, and Y. Cao. 1998. Microstructure of thin films of photoluminescent semiconducting polymers. *Polymer* 39:2299.

219. Chen, S., A. Su, H.L. Chou, K.Y. Peng, and S.A. Chen. 2004. Phase behavior and molecular aggregation in bulk poly(2-methoxy-5-(2′-ethylhexyloxy)-1,4-phenylenevinylene). *Macromolecules* 37:167.

220. Claudio, G.C. and E.R. Bittner. 2001. Random growth statistics of long-chain single molecule poly-(*p*-phenylene vinylene). *J Chem Phys* 115:9585.

221. Nguyen, T.Q., J. Wu, V. Doan, B.J. Schwartz, and S.H. Tolbert. 2000. Control of energy transfer in oriented conjugated polymer-mesoporous silica composites. *Science* 288:652.

222. Krebs, F.C. and M. Jørgensen. 2003. High carrier mobility in a series of new semiconducting PPV-type polymers. *Macromolecules* 36:4374.

223. Chen, S.A. and J.M. Ni. 1992. Structure properties of conjugated conductive polymers: 1. Neutral poly(3-alkylthiophene)s. *Macromolecules* 25:6081.

224. Rughooputh, S.D.D.V., S. Hotta, A.J. Hegger, and F. Wudl. 1987. Chromism of soluble polythie-nylenes. *J Polym Sci Polym Phys Ed* 25:1071.

225. Garnier, F., G. Tourillon, J.Y. Barraud, and H. Dexpert. 1985. First evidence of crystalline structure in conducting polythiophene. *J Mat Sci* 20:2687.

226. Winokur, M.J., D. Spiegel, Y.H. Kim, S. Hotta, and A.J. Heeger. 1989. Structural and absorption studies of the thermochromic transition in poly(3-hexylthiophene). *Synth Met* 28:C419.

227. Winokur, M.J., P. Wamsley, J. Moulton, P. Smith, and A.J. Heeger. 1991. Structural evolution in iodine-doped poly(3-alkylthiophenes). *Macromolecules* 24:3812.

228. Prosa, T.J., M.J. Winokur, J. Moulton, P. Smith, and A.J. Heeger. X-ray studies of poly(3-alkylthio-phenes): An example of an inverse comb. *Macromolecules* 25:4364.

229. Kawai, T., M. Nakazono, R. Sugimoto, and K. Yoshino. 1992. Crystal structure of poly(3-alkythio-phene) and its doping effect. *J Phys Soc Jpn* 6:3400.

230. Tashiro, K., K. Ono, Y. Minagawa, K. Kobayashi, T. Kawai, and K. Yoshino. 1991. Structural-changes in the thermochromic solid-state phase-transition on poly(3-alkylthiophene). *Synth Met* 41:571.

231. Tashiro, K., K. Ono, Y. Minagawa, K. Kobayashi, T. Kawai, and K. Yoshino. 1991. Structure and thermochromic solid-state phase-transition of poly(3-alkylthiophene). *J Polym Sci Polym Phys* 29:1223.

232. Tashiro, K., Y. Minagawa, K. Kobayashi, S. Morita, T. Kawai, and K. Yoshino. 1992. Crystal structure change of poly(3-alkylthiophene) induced by iodine doping. *Polym Prepr Jpn* 41:4595.

233. Gustafsson, G., O. Inganäs, S. Stafström, H. Österholm, and J. Laakso. 1991. Stretch-oriented poly(3-alkylthiophenes). *Synth Met* 41:593.

234. Porzio, W., A. Bolognesi, S. Destri, M. Catellani, and B. Bajo. 1991. Molecular-organization in polyalkylthiophene films. *Synth Met* 41:537.

235. Bolognesi, A., M. Catellani, S. Destri, and W. Porzio. 1991. Evidence of chain planarity in PAT: Low-temperature structural analysis of poly(3-decylthiophene). *Makromol Chem Rapid Commun* 12:9.

236. Gustafsson, G., O. Inganäs, H. Österholm, and J. Laakso. 1991. X-ray diffraction and infrared spectroscopy studies of oriented poly(3-alkylthiophenes). *Polymer* 32:1574.

237. Mårdalen, J., E.J. Samuelsen, O.R. Gautun, and P.H. Carlsen. 1991. Molecular-structure of stretch oriented poly(3-hexylthiophene) studied by an extended X-ray diffraction. *Solid State Commun* 80:687.

238. Mårdalen, J., E.J. Samuelsen, O.R. Gautun, and P.H. Carlsen. 1991. Chain configuration of poly(3-hexylthiophene) as revealed by X-ray diffraction studies. *Solid State Commun* 77:337.

239. Łużny, W., Niżioł, S., Zagorska, M., and Proñ, A. 1993. X-ray-diffraction comparative-study of poly(3-decylthiophenes) and poly(4,4′-didecyl-2,2′-bithiophenes). *Synth Met* 55:359.

240. Tashiro, K., K. Kobayashi, K. Morita, T. Kawai, and K. Yoshino. 1995. An organized combination of X-ray-diffraction, infrared-spectroscopy, and computer-simulation in a study of crystal structural-change in thermochromic phase-transition of poly(3-alkyl thiophene)s. *Synth Met* 69:397.

241. Chen, S.A. and S.J. Lee. 1995. The importance of molecular-dynamics in the determination of crystalline-structure of poly(3-dodecylthiophene). *Polymer* 36:1719.

242. Busing, W. 1990. X-ray-diffraction study of disorder in allied Spectra-1000 polyethylene fibers. *Macromolecules* 23:4068.

243. Corish, J., D.E. Feeley, D.A. Morton-Blake, F. Beniere, and M. Marchetti. 1997. Atomistic investigation of thermochromism in a poly(3-alkylthiophene). *J Phys Chem B* 101:10075.

244. O'Dwyer, S., H.W. Xie, J. Corish, and D.A. Morton-Blake. 2001. An atomistic simulation of the effect of pressure on conductive polymers. *J Phys Cond Mat* 13:2395.

245. Xie, H.W., S. O'Dwyer, J. Corish, and D.A. Morton-Blake. 2001. The thermochromism of poly(3-alkylthiophene)s: The role of the side chains. *Synth Met* 122:287.

246. Ihn, K.J., J. Moulton, and P. Smith. 1993. Whiskers of poly(3-alkylthiophenes). *J Polym Sci Polym Phys* 31:735.

247. Mena-Osteritz, E., A. Meyer, B.M.W. Langeveld-Voss, R.A.J. Janssen, E.W. Meijer, and P. Bauerle. 2000. Two-dimensional crystals of poly(3-alkylthiophene)s: Direct visualization of polymer folds in submolecular resolution. *Angew Chem Int Ed* 39:2680.

248. Brun, M., R. Demadrille, P. Rannou, A. Pron, J.P. Travers, and B. Grévin. 2004. Multi-scale scanning tunneling microscopy study of self-assembly phenomena in two-dimensional polycrystals of π-conjugated polymers: The case of regioregular poly(dioctylbithiophenealt-fluore-none). *Adv Mat* 16:2087.

249. Kiriy, N., et al. 2003. One-dimensional aggregation of regioregular polyalkylthiophenes. *Nano Lett* 3:707.

250. Kawai, T., M. Nakazono, and K. Yoshino. 1992. Effects of doping on the crystal structions of poly(3-alkylthiophene)s. *J Mater Chem* 2:903.

251. Prosa, T.J., M.J. Winokur, J. Moulton, P. Smith, and A.J. Heeger. 1995. X-ray diffraction studies of the 3-dimensional structure within iodine-intercalated poly(3-octylthiophene). *Phys Rev B* 51:159.

252. Tashiro, K., K. Kobayashi, T. Kawai, and K. Yoshino. 1997. Crystal structural change in poly(3-alkyl thiophene)s induced by iodine doping as studied by an organized combination of X-ray diffraction, infrared/Raman spectroscopy and computer simulation techniques. *Polymer* 38:2867.

253. Hotta, S. 1997. In *Handbook of organic conductive molecules and polymers*, Vol. 2, ed. H.S. Nalwa, New York: John Wiley & Sons Ltd., 309.

254. McCullough, R.D. and R.D. Lowe. 1992. Enhanced electrical-conductivity in regioselectively synthesized poly(3-alkylthiophenes). *J Chem Soc Chem Commun* 1:72.

255. McCullough, R.D., S. Tristram-Nagle, S.P. Williams, R.D. Lowe, and M. Jayaraman. 1993. Self-orienting head-to-tail poly(3-alkylthiophenes)—new insights on structure–property relationships in conducting polymers. *J Am Chem Soc* 115:4910.

256. Chen, T.A., X. Wu, and R.D. Rieke. 1995. Regiocontrolled synthesis of poly(3-alkylthiophenes) mediated by rieke zinc—their characterization and solid-state properties. *J Am Chem Soc* 117:233.

257. McCullough, R.D. 1998. The chemistry of conducting polythiophenes. *Adv Mat* 10:93.

258. Maior, R.M.S., K. Hinkelmann, H. Eckert, and F. Wudl. 1990. Synthesis and characterization of two regiochemically defined poly(dialkylbithiophenes): A comparative study. *Macromolecules* 23:1268.

259. Aasmundtveit, K.E., E.J. Samuelsen, K. Hoffmann, E. Bakken, and P.H.J. Carlsen. 2000. Structural studies of polyalkylthiophenes with alternating side chain positioning. *Synth Met* 113:7.

260. Zen, A., et al. 2004. Effect of molecular weight and annealing of poly(3-hexylthiophene)s on the performance of organic field-effect transitors. *Adv Funct Mat* 14:757.

261. Kline, R.J., M.D. McGehee, E.N. Kadnikova, J.S. Liu, J.M.J. Frechet, and M.F. Toney. 2005. Dependence of regioregular poly(3-hexylthiophene) film morphology and field-effect mobility on molecular weight. *Macromolecules* 38:3312.

262. Curtis, M.D., J. Cao, and J.W. Kampf. 2004. Solid-state packing of conjugated oligomers: From pi-stacks to the herringbone structure. *J Am Chem Soc* 123:4328.

263. Bolognesi, A., W. Porzio, F. Provasoli, and T. Ezquerra. 1993. The thermal-behavior of low molecular-weight poly(3-decylthiophene). *Makromol Chem* 194:817.

264. Bolognesi, A., W. Porzio, G. Zjuo, and T. Ezquerra. 1996. The thermal behaviour of poly(3-octylthienylene) synthesized by an Ni-based catalyst: DSC, optical microscopy and XRD analyses. *Eur Polym J* 32:1097.

265. Prosa, T.J., R.D. McCullough, and M.J. Winokur. 1996. Evidence of a novel side chain structure in regioregular poly(3-alkylthiophenes). *Macromolecules* 29:3654.

266. Meille, S.V., V. Romita, T. Caronna, A.J. Lovinger, M. Catellani, and L. Belobrzeckaja. 1997. Influence of molecular weight and regioregularity on the polymorphic behavior of poly(3-decylthiophenes). *Macromolecules* 30:7898.

267. Liu, S.L. and T.S. Chung. 2000. Crystallization and melting behavior of regioregular poly(3-dodecythiophene). *Polymer* 41:2781.

268. Zhao, Y., G.X. Yuan, P. Roche, and M. Leclerc. 1995. A calorimetric study of the phase-transitions in poly(3-hexylthiophene). *Polymer* 36:2211.

269. Zhao, Y., D. Keroack, G.X. Yuan, A. Massicotte, R. Hanna, and M. Leclerc. 1997. Melting behavior of poly(3-alkylthiophene)s with long alkyl side-chains. *Macromol Chem Phys* 198:1035.

270. Park, K. and K. Levon. 1997. Order–disorder transition in the electroactive polymer poly(3-dodecylthiophene). *Macromolecules* 30:3175.

271. Causin, V., C. Marega, A. Marigo, L. Valentini, and J.M. Kenny. 2005. Crystallization and melting behavior of poly(3-butylthiophene), poly(3-octylthiophene), and poly(3-dodecylthiophene). *Macromolecules* 38:409.

272. Nakazono, M., T. Kawai, and K. Yoshino. 1994. Effects of heat treatment on properties of poly(3-alkylthiophene). *Chem. Mater* 6:864.

273. Prosa, T.J., J. Moulton, A.J. Heeger, and M.J. Winokur. 1999. Diffraction line-shape analysis of poly(3-dodecylthiophene): A study of layer disorder through the liquid crystalline polymer transition. *Macromolecules* 32:4000.

274. Elmaci, N. and E. Yurtsever. 2002. Thermochromism in oligothiophenes: The role of the internal rotation. *J Phys Chem* 106:11981.

275. Brown, P.J., D.S. Thomas, A. Köhler, J.S. Wilson, J.S. Kim, C.M. Ramsdale, H. Sirringhaus, and R.H. Friend. 2003. Effect of interchain interactions on the absorption and emission of poly(3-hexylthiophene). *Phys Rev B* 67:064203.

276. Spano, F.C. 2005. Modeling disorder in polymer aggregates: The optical spectroscopy of regioregular poly(3-hexylthiophene) thin films. *J Chem Phys* 122:234701.

277. Spano, F.C. 2005. Private communication.

278. Bongiovanni, G., C. Botta, J.L. Brédas, D.R. Ferro, A. Mura, A. Piaggi, and R. Tubino. 1997. Conformational analysis and optical characterization of oligothiophene inclusion compounds. *Chem Phys Lett* 278:146.

279. Gierschner, J., L. Lüer, D. Oelkrug, E. Musluoglu, B. Behnisch, and M. Hananck. 2000. Preparation and optical properties of oligophenylenevinylene/perydrotriphenylene inclusion compounds. *Adv Mater* 12:757.

280. Ong, B.S., Y.L. Wu, P. Liu, and S. Gardner. 2004. High-performance semiconducting polythiophenes for organic thin-film transistors. *J Am Chem Soc* 126:3378.

281. Wu, Y.O., P. Liu, B.S. Ong, T. Srikumar, N. Zhao, G. Botton, and S.P. Zhu. 2005. Controlled orientation of liquid-crystalline polythiophene semiconductors for high-performance organic thin-film transistors. *Appl Phys Lett* 86.

282. Dong, X.M., J.C. Tyson, and D.M. Collard. 2000. Controlling the macromolecular architecture of poly(3-alkylthiophene)s by alternating alkyl and fluoroalkyl substituents. *Macromolecules* 33:3502.

283. Dong, X.M. and D.M. Collard. 2000. Liquid crystalline regioregular semifluoroalkyl-substituted polythiophenes. *Macromolecules* 33:6916.

284. Ling, L. and D.M. Collard. 2005. Tuning the electronic structure of conjugated polymers with fluoroalkyl substitution: Alternating alkyl/perfluoroalkyl-substituted polythiophene. *Macromolecules* 38:372.

285. Greve, D.R., et al. 1999. Directed self-assembly of amphiphilic regioregular polythiophenes on the nanometer scale. *Synth Met* 1–3:1502.

286. Reitzel, N., et al. 2000. Self-assembly of conjugated polymer at the air/water interface. Structure and properties of Langmuir and Langmuir–Blodgett films and amphiphilic regioregular polythiophenes. *J Am Chem Soc* 122:5788.

287. Goto, H., X.M. Dai, H. Narihiro, and K. Akagi. 2005. Synthesis of polythiophene derivatives bearing ferroelectric liquid crystalline substituents. *Macromolecules* 37:2353.

288. Politis, J.K., J.C. Nemes, and M.D. Curtis. 2001. Synthesis and characterization of regiorandom and regioregular poly(3-octylfuran). *J Am Chem Soc* 123:2537.

289. Costantini, N. and J.M. Lupton. 2003. Infrared spectroscopic study of polaron formation in electrochemically synthesised poly(3-alkypyrroles). *Phys Chem Chem Phys* 5:749.

290. Orecchini, A., C. Petrillo, G. Ruggeri, and R. Cagnolati. 2003. Neutron and x-ray scattering study of fully deuterated poly(3-*n*-decylpyrrole). *J Chem Phys* 118:7690.

291. Collard, D.M. and M.S. Stoakes. 1991. A lamellar conjugated polymer by self-assembly of an electropolymerizable monomer. *J Am Chem Soc* 113:9414.

292. Collard, D.M. and M.S. Stoakes. 1994. Lamellar conjugated polymers by electrochemical polymerization of heteroarene-containing surfactants-potassium 3-(3-alkylpyrrol-1yl)propanesulfonates. *Chem Mat* 6:850.

293. Tashiro, K., Y. Minagawa, K. Kobayashi, S. Morita, T. Kawai, and K. Yoshino. 1994. Crystal structure change of poly(3-alkylthiophene) induced by iodine doping as revealed by X-ray diffraction and infrared/Raman spectroscopic measurements. *Jpn J Appl Phys* 33:L1023.

294. Muthukumar, M., C.K. Ober, and E.L. Thomas. 1997. Competing interactions and levels of ordering in self-organizing polymeric materials. *Science* 277:1225.

295. Thunemann, A.F. 2002. Polyelectrolyte-surfactant complexes (synthesis, structure and materials aspects). *Prog Polym Sci* 27:1473.

296. ten Brinke, G. and O. Ikkala. 1997. Ordered structures of molecular bottlebrushes. *Trends Polym Sci* 5:213.

297. Ikkala, O. and G. ten Brinke. 2004. Hierarchical self-assembly in polymeric complexes: Towards functional materials. *Chem Commun* 16:2131.

298. Cao, Y., P. Smith, and A.J. Heeger. 1992. Counter-ion induced processibility of conducting polyaniline. *Synth Met* 48:91.

299. Reghu, M., Y. Cao, D. Moses, and A.J. Heeger. 1993. Counterion-induced processibility of polyaniline—transport at metal-insulator boundary. *Phys Rev B* 47:1758.

300. Yang, C.Y., P. Smith, A.J. Heeger, Y. Cao, and J. Österholm. 1994. Electron-diffraction studies of the structure of polyaniline dodecylbenzenesulfonate. *Polymer* 35:1143.

301. Choi, J.W., M.G. Han, S.Y. Kim, S.G. Oh, and S.S. Im. 2004. Poly(3,4-ethylenedioxythiophene) nanoparticles prepared in aqueous DBSA solutions. *Synth Met* 141:293.

302. Zheng, W.Y., K. Levon, J. Laakso, and J.E. Österholm. 1994. Characterization and solid-state properties of processable *n*-alkylated polyanilines in the neutral state. *Macromolecules* 27:7755.

303. Zheng, W.Y., K. Levon, T. Taka, J. Laakso, and J.E. Österholm. 1996. Doping-induced layered structure in *n*-alkylated polyanilines. *Polym J* 28:412.

304. Ikkala, O.T., L. Pietilä, L. Ahjopalo, H. Österholm, and P.J. Passiniemi. 1995. On the molecular recognition and associations between electrically conduction polyaniline and solvents. *J Chem Phys* 103:9855.

305. Vikki, T., et al. 1996. Molecular recognition solvents for electrically conductive polyaniline. *Macromolecules* 29:2945.

306. Ikkala, O.T., L.O. Pietilac, P. Passiniemi, T. Vikki, H. Osterholm, L. Ahjopalo, and J.E. Osterholm. 1997. Processible polyaniline complexes due to molecular recognition: Supramolecular structures based on hydrogen bonding and phenyl stacking. *Synth Met* 84:55.

307. Ruokolainen, J., M. Torkkeli, R. Serimaa, S. Vahvaselka, M. Saariaho, G. ten Brinke, and O. Ikkala. 1996. Critical interaction strength for surfactant-induced mesomorphic structures in polymer-surfactant systems. *Macromolecules* 29:6621.

308. Tiitu, M., N. Volk, M. Torkkeli, R. Serimaa, G. ten Brinke, and O. Ikkala. 2004. Cylindrical self-assembly and flow alignment of comb-shaped supramolecules of electrically conducting polyaniline. *Macromolecules* 37:7364.

309. Samuelsen, E.J., A.P. Monkman, L.A.A. Pettersson, L.E. Horsburgh, K.E. Aasmundtveit, and S. Ferrer. 2001. The structure of polypyridine. *Synth Met* 124:393.

310. Knaapila, M., et al. 2002. Polarized luminescence from self-assembled, aligned, and cleaved supramolecules of highly ordered rodlike polymers. *Appl Phys Lett* 81:1489.

311. Stepanyan, R., A. Subbotin, M. Knaapila, O. Ikkala, and G. ten Brinke. 2003. Self-organization of Hairy-rod polymers. *Macromolecules* 3760.

312. Knaapila, M., et al. 2003. Structure and phase equilibria of polyelectrolytic hairy-rod supramolecules in the melt state. *J Phys Chem B* 107:14199.

313. Teetsov, J. and M.A. Fox. 1999. Photophysical characterization of dilute solutions and ordered thin films of alkyl-substituted polyfluorenes. *J Mater Chem* 9:2117.

314. Grell, M., D.D.C. Bradley, G. Ungar, J. Hill, and K.S. Whitehead. 1999. Interplay of physical structure and photophysics for a liquid crystalline polyfluorene. *Macromolecules* 32:5810.

315. Lieser, G., M. Oda, T. Miteva, A. Meisel, H.G. Nothofer, and U. Scherf. 2000. Ordering, graphoepitaxial orientation, and conformation of a polyfluorene derivative of the "hairy-rod" type on an oriented substrate of polyimide. *Macromolecules* 33:4490.

316. Grell, M., D.D.C. Bradley, M. Inbasekaran, G. Ungar, K.S. Whitehead, and E.P. Woo. 2000. Intrachain ordered polyfluorene. *Syth Met* 111:579.

317. Kawana, S., et al. 2002. X-ray diffraction study of the structure of thin polyfluorene films. *Polymer* 43:1907.

318. Knaapila, M., et al. 2003. X-ray diffraction studies of multiple orientation in poly(9,9-bis(2-ethylhexyl)fluorene-2,7-diyl) thin films. *J Phys Chem* 107:12425.

319. Tanto, B., S. Guha, C.M. Martin, U. Scherf, and M.J. Winokur. 2004. Structural and spectroscopic investigations of bulk poly[bis(2-ethyl)hexylfluorene]. *Macromolecules* 37:9438.

320. Chen, S.H., H.L. Chou, A.C. Su, and S.A. Chen. 2004. Molecular packing in crystalline poly(9,9-di-*n*-octyl-2,7-fluorene). *Macromolecules* 37:6833.

321. Chen, S.H., A.C. Su, C.H. Su, and S.A. Chen. 2004. Crystalline forms and emission behavior of poly(9,9-di-*n*-octyl-2,7-fluorene). *Macromolecules* 38:379.

322. Knaapila, M., et al. 2005. The influence of the molecular weight on the thermotropic alignment and self-organized structure formation of branched side chain hairy-rod polyfluorene in thin films. *Macromolecules* 38:2744.

323. Oda, M., H.G. Nothofer, U. Scherf, V. Sunjic, D. Richter, W. Regenstein, and D. Neher. 2002. Chiroptical properties of chiral substituted polyfluorenes. *Macromolecules* 35:6792.

324. Prins, L.J., P. Timmerman, and D.N. Reinhoudt. 2001. Amplification of chirality: The sergeants and soldiers principle applied to dynamic hydrogen-bonded assemblies. *J Am Chem Soc* 123:10153.

325. Knaapila, M., et al. 2004. Influence of molecular weight on self-organization, uniaxial alignment, and surface morphology of hairy-rodlike polyfluorene in thin films. *J Phys Chem B* 108:10711.

326. Knaapila, M., et al. 2005. Influence of molecular weight on the phase behavior and structure formation of branched side-chain hairy-rod polyfluorene in bulk phase. *Phys Rev E* 71.

327. Grell, M., D.D.C. Bradley, X. Long, T. Chamberlain, M. Inbasekaran, and E.P. Woo. 1998. Chain geometry, solution aggregation and enhanced dichroism in the liquid-crystalline conjugated polymer poly(9,9-dioctylfluorene). *Acta Polym* 49:439.

328. Cadby, A.J., et al. 2000. Film morphology and photophysics of polyfluorene. *Phys Rev B* 62:15604.

329. Ariu, M., et al. 2003. Exciton migration in beta-phase poly(9,9-dioctylfluorene). *Phys Rev B* 67:195333.

330. Winokur, M.J., J. Slinker, and D.L. Huber. 2003. Structure, photophysics and the order–disorder transition to the beta phase in poly(9,9-(di-*n,n*-octyl)fluorene). *Phys Rev B* 67:184106.

331. Port, H., H. Nissler, and R. Silbey. 1987. Optical-line broadening of triplet excitons in dibromo-naphthalene-isotopic impurity and phonon-scattering. *J Chem Phys* 87:1994.

332. Gulliet, T., J. Berrehar, R. Grousson, J. Kovensky, C. Lapersonne-Meyer, M. Schott, and V. Voliotis. 2001. Emission of a single conjugated polymer chain isolated in its single crystal monomer matrix. *Phys Rev Lett* 87:087401.

333. Dubin, F., J. Berrehar, R.T.G. Grousson, C. Lapersonne-Meyer, M. Schott, and V. Voliotis. 2002. Optical evidence of a purely one-dimensional exciton density of states in a single conjugated polymer chain. *Phys Rev B* 66:113202.

334. Ariu, M., D.G. Lidzey, M. Sims, A.J. Cadby, P.A. Lane, and D.D.C. Bradley. 2002. The effect of morphology on the temperature-dependent photoluminescence quantum effciency of the conjugated polymer poly(9,9-dioctylfluorene). *J Phys Cond Mat* 14:9975.

335. Leclerc, M., M. Ranger, and F. Belanger-Gariepy. 1998. 2,7-dibromo-9,9-dioctylfluorenechloroform (1/0.25). *Acta Cryst C* 54:799.

336. He, Z. and R.E. Prud'homme. 1999. Conformational and packing modeling of optically active polyesters. 2. Helical structure of an isotactic polyactone. *Macromolecules* 32:7655.

337. Katz, S.M., J.A. Reichl, and D.H. Berry. 1998. Catalytic synthesis of poly(arylmethylgermanes) by demethanative coupling: A mild route to sigma-conjugated polymers. *J Am Chem Soc* 38:9844.

338. Patnaik, S.S., A.J. Greso, and B.L. Farmer. 1992. The low-temperature structure of poly(di-*n*-hexyl germane). *Polymer* 33:5115.

339. Bukalov, S.S., L.A. Leites, I.A. Krylova, and M.P. Egorov. 2001. UV and Raman study of thermochromic phase transition in poly(di-*n*-hexylgermane). *J Organomet Chem* 636:164.

340. West, R. 1994. Organopolysilanes, in *Comprehensive Organometallic Chemisty*, Vol. 2, ed. A.G. Davies, Oxford: Pergamon Press, 77.

341. Ganicz, T. and W.A. Stanczyk. 2003. Organosilicon mesomorphic polymer systems. *Prog Polym Sci* 28:303.

342. Weber, P., D. Guillon, A. Skoulios, and R.D. Miller. 1989. Liquid-crystalline nature of poly(dinormal-hexylsilane). *J Physq* 50:793.

343. Lovinger, A.J., F.C. Schilling, F.A. Bovey, and J.M. Zeigler. 1986. Characterization of poly(di-*n*-hexylsilane) in the solid-state 1. X-ray and electron diffraction studies. *Macromolecules* 19:2657.

344. Schilling, F.C., A.J. Lovinger, J.M. Zeigler, D.D. Davis, and F.A. Bovey. 1989. Solid-state structures and thermochromism of poly(di-*n*-butylsilylene) and poly(di-*n*-pentylsilylene). *Macromolecules* 22:3055.

345. Lovinger, A.J., D.D. Davis, F.C. Schilling, F.A. Bovey, and J.M. Zeigler. 1989. Structures of poly(di-ethyl silylene) and poly(di-*n*-propyl silylene). *Polym Comm* 30:356.

346. Lovinger, A.J., D.D. Davis, F.C. Schilling, F.J. Padden, F.A. Bovey, and J.M. Zeigler. 1991. Solid-state structure and phase-transitions of poly(dimethylsilylene). *Macromolecules* 24:132.

347. Karikari, E.K., A. Greso, B.L. Farmer, R.D. Miller, and J.F. Rabolt. 1993. Studies of the conformation and packing of polysilanes. *Macromolecules* 26:3937.

348. Patnaik, S.S. and B.L. Farmer. 1992. X-ray structure determination of poly(di-*n*-hexyl silane). *Polymer* 33:4443.

349. Yuan, C.-H. and R. West. 1994. Side chain effect on the nature of thermochromism of polysilanes. *Macromolecules* 27:629.

350. Takeuchi, K. and S. Furukawa. 1993. The Crystal-structure of poly(di-*n*-hexylsilane) and its orientation in the film. *J Phys Cond Mat* 5:L601.

351. Furukawa, S., K. Takeuchi, and M. Shimana. 1994. Crystal-structures of poly(di-*n*-butylsilane) and poly(di-*n*-pentylsilane). *J Phys Cond Mat* 6:11007.

352. KariKari, E.K., B.L. Farmer, C.L. Hoffman, and J.F. Rabolt. 1994. Polymorphism of poly(di-*n*-pentylsilane). *Macromolecules* 27:7185.

353. Jambe, B., A. Jonas, and J. Devaux. 1997. Crystalline structure of poly(methyl-*n*-propylsilane). *J Polym Sci Polym Phys* 35:1533.

354. Furukawa, S. 1998. Structure and orientation control of organopolysilanes and their application to electronic devices. *Thin Solid Films* 331:222.

355. Kanai, T., H. Ishibashi, Y. Hayashi, T. Ogawa, S. Furukawa, R. West, T. Dohmaru, and K. Oka. 2000. A new cooling-rate dependent thermochromism of poly(dioctylsilane). *Chem Lett* 6:650.

356. Chunwachirasiri, W., I. Kanaglekar, M.J. Winokur, J.C. Koe, and R. West. 2001. Structure and chain conformation in poly(methyl-*n*-alkyl)silanes. *Macromolecules* 34:6719.

357. Chunwachirasiri, W., R. West, and M.J. Winokur. 2000. Polymorphism, structure and chromism in poly(di-*n*-octylsilane) and poly(di-*n*-decylsilane). *Macromolecules* 33:9720.

358. Winokur, M.J. and R. West. 2003. X-ray diffraction and molecular modeling studies of poly(di-*n*-alkylsilanes): The near planar type phases of poly(di-*n*-butylsilane) and poly(di-*n*-hexylsilane). *Macromolecules* 36:7338.

359. Patnaik, S.S. and B.L. Farmer. 1992. Energy calculations of the crystal-structure of poly(di-methyl silane). *Polymer* 33:5121.

360. Leites, L.A., T.S. Yadritseva, S.S. Bukalov, T.M. Frunze, B.A. Antipova, and V.V. Demet'ev. 1992. Synthesis, structure and conformations of side-chains of polydiethylsilane from the data of vibrational and electronic-spectra, *Vysokomol Soedin A+* 34:104.

361. Rathgeber, S., T. Pakula, A. Wilk, K. Matyjaszewski, and K.L. Beers. 2005. On the shape of bottle-brush macromolecules: Systematic variation of architectural parameters. *J Chem Phys* 122:124904.

362. Bukalov, S.S., L.A. Leites, R. West, and T. Asuke. 1996. A detailed UV and Raman study of poly(*n*-butyl-*n*-hexylsilylene) phase transitions. *Macromolecules* 29:907.

363. Mueller, C., H. Frey, and C. Schmidt. 1999. Phase behaviour of poly(di-*n*-decylsilane). *Monatsh Chem* 130:175.

364. West, R. 2003. A new theory for rotational isomeric states: Polysilanes lead the way. *J Organomet Chem* 685:6.

365. Abdallah, D.J., R.E. Bachman, J. Perlstein, and R.G. Weiss. 1999. Crystal structures of symmetrical tetra-*n*-alkyl ammonium and phosphonium halides. Dissection of competing interactions leading to biradial and tetraradial shapes. *J Phys Chem* 103:9269.

366. Klemann, B., R. West, and J.A. Koutsky. 1993. Structure and properties of poly(*n*-pentyl-*n*-alkylsilanes) 1. *Macromolecules* 26:1042.

367. Klemann, B., R. West, and J.A. Koutsky. 1996. Structure and properties of poly(*n*-pentyl-*n*-alkylsilanes) 2. *Macromolecules* 29:198.

368. Fujiki, M. 1994. Ideal exciton spectra in single- and double-screw-sense helical polysilanes. *J Am Chem Soc* 116:6017.

369. Fujiki, M. 1996. A correlation between global conformation of polysilane and UV absorption Characteristics. *J Am Chem Soc* 118:7424.

370. Fujiki, M. 2001. Optically active polysilylenes: State-of-the-art chiroptical polymers. *Macromol Rapid Commun* 22:539.

371. Fujiki, M. 2000. Helix magic: Thermo-driven chiroptical switching and screw-sense inversion of flexible rod helical polysilylenes. *J Am Chem Soc* 122:3336.

372. Koe, J.R., M. Fujiki, M. Motonaga, and H. Nakashima. 2000. Temperature-dependent helix–helix transition of an optically active poly(diarylsilylene). *Chem Commun* 5:389.

373. Saxena, A., G.Q. Guo, M. Fujiki, Y.G. Yang, A. Ohira, K. Okoshi, and M. Naito. 2004. Helical polymer command surface: Thermodriven chiroptical transfer and amplification in binary polysilane film system. *Macromolecules* 37:3081.

374. Neumann, F., H. Teramae, J.M. Downing, and J. Michl. 1998. Gauche, ortho, and anti conformations of saturated A(4)X(10) chains: When will all six conformers exist? *J Am Chem Soc* 120:573.

375. Albinsson, B., D. Antic, F. Neumann, and J. Michl. 1999. Conformers of *n*-Si5 Me12: A comparison of ab initio and molecular mechanics methods. *J Phys Chem* 103:2184.

376. Fogarty, H.A., C.-H. Ottoson, and J. Michl. 2000. Conformations of oligosilanes with ethyl and methyl substituents. *J Mol Struct* 556:105.

377. Fogarty, H.A., C.-H. Ottoson, and J. Michl. 2000. The five favored backbone conformations of n-Si4Et10. *J Mol Struct (Theochem)* 506:243.

378. Chakrabarti, P. and D. Pal. 2001. The interrelationships of side-chain and main-chain conformations in proteins. *Prog Biophys Mol Biol* 76:1.

379. Sasanuma, Y., H. Kato, and A. Kaito. 2003. Conformational analysis of poly(di-*n*-butylsilane), poly(di-*n*-hexylsilane), and poly(methyl-*n*-propylsilane) by a rotational isomeric state scheme with molecular dynamics simulations. *J Phys Chem B* 107:11852.

380. Wegner, G. 1971. Topochemical reactions of monomers with conjugated triple-bonds, 4. polymerization of bis-(para-toluene sulfonate) of 2,4-hexadiin-1,6-diol. *Makromol Chem* 145:85.

381. Lapersonne-Meyer, C. 2001. Excitons on a 1D periodic conjugated polymer chain: Two electronic structures of polydiacetylene chains. *Int J Mod Phys* 15:28.

382. Katagiri, H., Y. Shimoi, and S. Abe. 2004. A density functional study of backbone structures of polydiacetylene: Destabilization of butatriene structure. *Chem Phys* 306:191.

383. Baughman, R.H. 1972. Solid-state polymerization of diacetylenes. *J Appl Phys* 43:4362.

384. Liao, J. and D.C. Martin. 1993. Direct imaging of the diacetylene solid-state monomer–polymer phase transformation. *Science* 260:1489.

385. Day, D. and J.B. Lando. 1980. Morphology of crystalline diacetylene monolayers polymerized at the gas–water interface. *Macromolecules* 13:1478.

386. Carpick, R.W., D.Y. Sasaki, and A.R. Burns. 2000. First observation of mechanochromism at the nanometer scale. *Langmuir* 16:1270.

387. Hoofman, R.J.O., G.H. Gelinck, L.D.A. Siebbeles, M.P. de Haas, J.M. Warman, and D. Bloor. 2000. Influence of backbone conformation on the photoconductivity of polydiacetylene chains. *Macromolecules* 33:9289.

388. Young, R.J., R.T. Read, D.N. Batchelder, and D. Bloor. 1981. Structure and morphology of heavily deformed single crystals of a polydiacetylene. I. Electron microscopy and x-ray diffraction. *J Polym Sci Polym Phys* 19:293.

389. Lee, D.C., S.K. Sahoo, A.L. Cholli, and D.J. Sandman. 2002. Structural aspects of the thermochromic transition in urethane-substituted polydiacetylenes. *Macromolecules* 35:4347.

390. Carpick, R.W., T.M. Mayer, D.Y. Sasaki, and A.R. Burns. 2000. Spectroscopic ellipsometry and fluorescence study of thermochromism in an ultrathin poly(diacetylene) film: Reversibility and transition kinetics. *Langmuir* 16:4639.

391. Carpick, R.W., D.Y. Sasaki, M.S. Marcus, M.A. Eriksson, and A.R. Burns. 2004. Polydiacetylene films: A review of recent investigations into chromogenic transitions and nanomechanical properties. *J Phys Cond Mat* 16:R679.

392. Bunz, U.H.F. 2000. Poly(aryleneethynylene)s: Syntheses, properties, structures, and applications. *Chem Mater* 100:1605.

393. Zhou, Q. and T.M. Swager. 1995. Methodology for enhancing the sensitivity of fluorescent chemosensors—energy migration in conjugated polymers. *J Am Chem Soc* 117:7017.

394. Zhou, Q. and T.M. Swager. 1995. Fluorescent chemosensors based on energy migration in conjugated polymers: The molecular wire approach to increased sensitivity. *J Am Chem Soc* 117:12593.

395. Ofer, D., T.M. Swager, and M.S. Wrighton. 1995. Solid-state ordering and potential dependence of conductivity in poly(2,5-dialkoxy-*p*-phenyleneethynylene). *Chem Mater* 7:418.

396. Weder, C. and M.S. Wrighton. 1996. Effcient solid-state photoluminescence in new poly(2,5dialkoxy-*p*-phenyleneethynylene)s. *Macromolecules* 29:5157.

397. Bunz, U.H.F., V. Enkelmann, L. Kloppenburg, D. Jones, K.D. Shimizu, J.B. Claridge, H.C. zur Loye, and G. Lieser. 1999. Solid-state structures of phenyleneethynylenes: Comparisons of monomers and polymers. *Chem Mat* 11:1416.

398. Perahia, D., T. Traiphol, and U.H.F. Bunz. 2001. From molecules to supramolecular structure: Self assembling of wire-like poly(*p*-phenyleneethynylene)s. *Macromolecules* 34:151.

399. Egbe, D.A.M., B. Carbonnier, L.M. Ding, D. Muhlbacher, E. Birckner, T. Pakula, F.E. Karasz, and U.W. Grummt. 2004. Supramolecular ordering, thermal behavior, and photophysical, electrochemical, and electroluminescent properties of alkoxy-substituted yne-containing poly(phenylene-vinylene)s. *Macromolecules* 37:7451.

400. Carbonnier, B., T. Pakula, and D.A.M. Egbe. 2005. Self-organization of comb-like macromolecules comprised of four-fold alkoxy substituted (PPV-PPE) rigid backbone repeat units: Role of length variation of side chains attached on phenylene ring surrounded by vinylene moieties. *J Mat Chem* 15:880.

401. Moggio, I., J. Le Moigne, E. Arias-Marin, D. Issautier, A. Thierry, D. Comoretto, G. Dellepiane, and C. Cuniberti. 2001. Orientation of polydiacetylene and poly(*p*-phenyleneethynylene) films by epitaxy and rubbing. *Macromolecules* 34:7091.

402. Klug, H.P. and L.E. Alexander. 1974. *X-Ray Diffraction Procedures for Polycrystalline and Amorphous Materials*, New York: Wiley-Interscience.

403. Narten, A.H. 1989. Radial distribution of carbon atoms in crystalline and molten polyethylene. *J Chem Phys* 90:5857.

404. Narten, A.H. and A. Habenschuss. 1990. X-ray diffraction study of some liquid alkanes. *J Chem Phys* 92:5692.

405. Mao, G.M., M.L. Saboungi, D.L. Price, M.B. Armand, and W.S. Howells. 2000. Structure of liquid PEO-LiTFSI electrolyte. *Phys Rev Lett* 84:5536.

406. Annis, B.K., O. Borodin, G.D. Smith, C.J. Benemore, A.K. Soper, and J.D. Londono. 2001. The structure of a poly(ethylene oxide) melt from neutron scattering and molecular dynamics simulations. *J Chem Phys* 115:10998.

407. Carlsson, P., D. Andersson, J. Swenson, R.L. McGreevy, W. Howells, and L. Borjesson. 2004. Structural investigations of polymer electrolyte poly(propylene oxide)-LiClO4 using diffraction experiments. *J Chem Phys* 121:12026.

408. Annis, B.K., A.H. Narten, A.G. MacDiarmid, and A.F. Richter. 1988. A covalent bond to bromine in HBr-treated polyaniline from from X-ray diffraction. *Synth Met* 22:191.

409. Laridjani, M., J.P. Pouget, E.M. Scherr, A.G. MacDiarmid, M. Jòzefowicz, and A.J. Epstein. 1992. Amorphography—the relationship between amorphous and crystalline order 1. The memory effects in polyaniline. *Macromolecules* 25:4106.

410. Laridjani, M., J.P. Pouget, A.G. MacDiarmid, and A.J. Epstein. 1992. Structural study of amorphous polyaniline. *J Phys I*, 2:1003.

411. Maron, J., M.J. Winokur, and B.R. Mattes. 1995. Processing-induces changes in the local-structure of amorphous polyaniline by radial-distribution-function analysis of X-ray scattering data. *Macromolecules* 28:4475.

412. Winokur, M.J. and B.R. Mattes. 1996. Determination of the local molecular structure in amorphous polyaniline. *Phys Rev B* 54:R12637.

413. Winokur, M.J. and B.R. Mattes. 1997. Differential anomalous scattering studies of amorphous HBr-doped polyaniline. *Synth Met* 84:725.

414. Resel, R., G. Leising, F. Lunzer, and C. Marschner. 1998. Structure in amorphous polysilanes determined by diffuse X-ray scattering. *Polymer* 39:5257.

415. Barham, P.J. and A. Keller. 1985. High-strength polyethylene fibers from solution and gel spinning. *J Mat Sci* 20:2281.

416. Voss, K.F., et al. 1991. Enhanced order and electronic delocalization in conjugated polymers oriented by gel processing in polyethylene. *Phys Rev B* 43:5109.

417. Murthy, N.S., H. Minor, C. Bednarczyk, and S. Krimm. 1993. Structure of the amorphous phase in oriented polymers. *Macromolecules* 26:1712.

418. Ou, R.Q., R.A. Gerhardt, and R.J. Samuels. 2003. Structure-electrical property study of anisotropic polyaniline films. *J Polym Sci Polym Phys* 41:823.

419. Alexander, L.E. 1969. *X-Ray Diffraction Methods in Polymer Science*, New York: Wiley-Interscience.

420. Erb, T., S. Raleva, U. Zhokhavets, G. Gobsch, B. Stuhn, M. Spode, and O. Ambacher. 2004. Structural and optical properties of both pure poly(3-octylthiophene) (P3OT) and P3OT/fullerene films. *Thin Solid Films* 450:97.

421. Łużny, W., E.J. Samuelsen, and D.W. Breiby. 2001. Polyaniline thin films-structural anisotropy study by use of synchrotron radiation surface diffraction. *Synth Met* 119:203.

422. Aasmundtveit, K.E., et al. 2000. Structural anisotropy of poly(alkylthiophene) films. *Macromolecules* 33:3120.

423. Breiby, D.W., E.J. Samuelsen, O. Konovalov, and B. Struth. 2004. Ultrathin films of semiconducting polymers of water. *Langmuir* 20:4116.

424. Breiby, D.W. and E.J. Samuelsen. 2003. Quantification of preferential orientation in conjugated polymers using X-ray diffraction. *J Polym Sci Polym Phys* 41:2375.

425. Zhokhavets, U., G. Gobsch, H. Hoppe, and N.S. Sariciftci. 2005. A systematic study of the anisotropic optical properties of thin poly(3-octylthiophene)-films in dependence on growth parameters. *Thin Solid Films* 451:69.

426. Sirringhaus, H., et al. 1999. Two-dimensional charge transport in self-organized, high-mobility conjugated polymers. *Nature* 401:685.

427. Wittman, J.C. and P. Smith. 1992. Highly oriented thin-films of poly(tetrafluoroethylene) as a substrate for oriented growth of materials. *Nature* 352:414.

428. Misaki, M., Y. Ueda, S. Nagamatsu, Y. Yoshida, N. Tanigaki, and K. Yase. 2004. Formation of single-crystal-like poly(9,9-dioctylfluorene) thin film by the friction-transfer technique with subsequent thermal treatments. *Macromolecules* 37:6926.

429. Misaki, M., M. Chikamatsu, N. Tanigaki, M. Yamashita, Y. Ueda, and K. Yase. 2005. Polymer-supported anisotropic submicrometer-patterned electrodes for displays. *Adv Mat* 17:297.

430. Nagamatsu, S., W. Takashima, K. Kaneto, Y. Yoshida, N. Tanigaki, K. Yase, and K. Omote. 2003. Backbone arrangement in friction-transferred regioregular poly(3-alkylthiophene)s. *Macromolecules* 36:5252.

431. Fukuda, K., T. Seki, and K. Ichimura. 2002. Photoalignment of poly(di-*n*-hexylsilane) by azobenzene monolayer. 1. Preparative conditions of poly(di-*n*-hexylsilane) spin cast film. *Macromolecules* 35:2177.

432. Knaapila, M., R. Stepanyan, B.P. Lyons, M. Torkkeli, and A.P. Monkman. 2006. Towards general guidelines for aligned, nanoscale assemblies of Hairy-rod polyfluorenes. *Adv Funct Mat* 16:599.

433. Sumpter, B.G., P. Kumar, A. Mehta, M.D. Barnes, W.A. Shelton, and R.J. Harrison. 2005. Computational study of the structure, dynamics, and photophysical properties of conjugated polymers and oligomers under nanoscale confinement. *J Phys Chem B* 109:7671.

18

Electrochemistry of Conducting Polymers

P. Audebert and
Fabien Miomandre

This chapter is devoted to the presentation of the new trends in the electrochemistry of electronically conducting polymers (ECPs) that appeared during the past decade. It is divided into three major sections corresponding to the main uses of electrochemistry in the field of conducting polymers: electrosynthesis from specially designed monomers (Section 18.2), characterization of polymer films (Section 18.3), and investigation of related materials as modified electrodes for specific applications (Section 18.4).

These three areas will be detailed in the following review; however, it seemed necessary for us to introduce this chapter with a paragraph reminding the most striking features of electrosynthesis and redox behavior of ECPs, especially for readers who are not totally familiar with this field (a recent review of Heinze is also worth mentioning for further details on some aspects [1]).

18.1 Introduction

Historically, the story of most conducting polymers is tightly connected to electrochemistry, since most of the classical conducting polymers, like the ones based on heterocycles (polypyrrole, PPy [2], polythiophene, PTh [3], polyaniline, PANI [4]), were initially synthesized by electrochemical oxidation. The mechanism for electrochemical polymerization, initially suggested by Geniès et al. [5], was finally proved by Andrieux et al. about 10 years later [6], and is depicted in Scheme 18.1. It consists of an oxidation step generating a cation radical followed by coupling and deprotonation steps generating the dimer; this being more easily oxidized than the monomer, the same reaction suite occurs to produce longer species until precipitation takes place at the electrode/electrolyte interface.

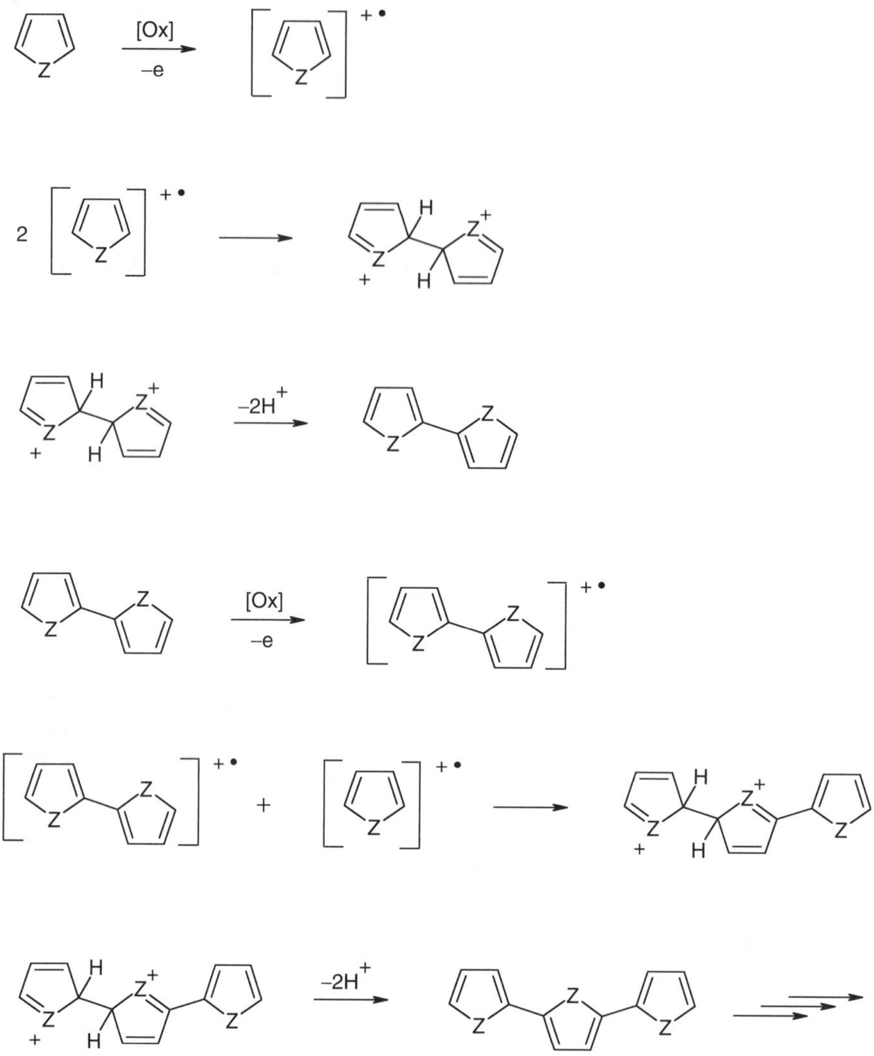

SCHEME 18.1   Mechanism of the first steps of the electrooxidation of electronically conducting polymers (ECPs).

Nevertheless the following steps of the mechanism are not clear, because on the one hand precipitation of the generated oligomers can occur, and on the other the spin and charge repartition change in the oligomers, and a cation-radical—neutral monomer coupling can be favored as the later steps arise, as Lacroix et al. have suggested on the basis of calculations [7].

The polymer growth can be monitored and followed by a cyclic voltammetry (CV), by scanning up to the oxidation potential of the monomer and then down to the reduction potential of the polymer formed. Therefore the most classical CV for an electropolymerization process is as featured in Figure 18.1, where the growth of the conducting polymer clearly appears as an increasing reversible peak on each incoming CV, at lower potential than the monomer one. The CV clearly depicts that the polymer film is formed upon electrodeposition of solution species and is conductive in one of its redox states,[1] generally the oxidized one. ECPs can also be electrogenerated potentiostatically by applying a

---

[1] If the polymer film were not conductive, the signal corresponding to redox switching of the polymer would stop to rise after a few cycles.

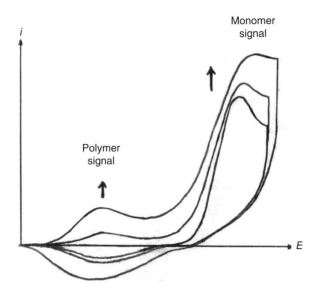

FIGURE 18.1 Classical feature for cyclic voltammetries (CVs) recorded upon electropolymerization process. The arrows indicate the current variation on each successive scan.

potential step and recording the current vs. time (chronoamperometry) or, albeit less frequently, galvanostatically by applying a current step and recording the potential vs. time (chronopotentiometry).

In addition to electrosynthesis, one should mention that later on many ECPs were prepared through chemical methods, mainly through organometallic coupling, applying Suzuki, Kumada, or Stille catalysts to polymerization reaction. This was mainly developed by McCullough [8] and proved to yield more regular and better-defined polymers. Interestingly, the pyrrole compounds, which are easier to synthesize if considering the electropolymerization potential, resisted to some extent to attempts of chemical routes because of the acidic hydrogen on the nitrogen atom.

Electrochemistry can also be used to monitor the redox state of the ECP generated at the electrode surface. Upon electrooxidation, ECP is generated as a film deposited on the electrode in a partially oxidized state, as shown in the followimg equation:

$$\text{NHMH} - n(2+\delta)ne^- + (n\delta)e^- \rightarrow \left[(M)_n^{n\delta+},(n\delta)X^-\right] + 2n\text{H}^+ \tag{18.1}$$

where M represents the repetition unit (HMH being the starting monomer) and X a counterion coming from the supporting electrolyte. The partial charge δ is called the doping level, reminding that oxidation (resp. reduction) is accompanied by insertion of compensating negative (positive) charges coming from the electrolyte; the doping level can be estimated from the coulombic charges (per area unit) passed respectively upon electrosynthesis Q_s and redox switching of the polymer Q_r (assuming an electropolymerization yield of 100%):

$$\delta \approx \frac{2Q_r}{Q_s - Q_r} \tag{18.2}$$

δ lies usually in the range [0.2–0.3] for heterocycle-based ECPs.

On the other hand, the polymer film thickness can be estimated through Q_s:

$$\ell \approx \frac{Q_s(M_{\text{Mono}} + \delta M_X)}{(2+\delta)F\rho} \tag{18.3}$$

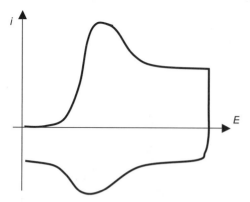

FIGURE 18.2 Typical cyclic voltammetry (CV) corresponding to redox switching of electronically conducting polymers (ECPs).

where M_j is the molar mass of species j and ρ is the polymer density. In the case of PPy [9], we can roughly estimate that a film thickness of 1 μm requires a charge of $\sim$400 mC cm^{-2}.

Typical CVs for redox switching of ECPs have the shape shown in Figure 18.2: broad peaks are usually obtained while changing the redox state. Reversibility is often not complete (there is a cathodic shift for the reversal peak potential) due to relaxation effects that generate hysteresis phenomena. At last, a plateau characteristic of capacitive behavior is frequently observed at high doping levels [10–12].

At low scan rate, peak currents vary linearly with the scan rate, which is characteristic of the behavior of a thin layer material deposited on the electrode. Conversely, at higher scan rates, linear variation with the square root of the scan rate is frequently observed, corresponding to diffusion limited currents, arising either from charge transfer process (hopping mechanism) or from an ionic contribution (diffusion of counterions). The response is also strongly dependent upon the nature and concentration of electrolyte (especially the anion [13]).

18.2   Electrochemistry of Monomers

18.2.1   Foreword

Despite its quasi-universal character, the research on the electrochemistry of new monomers as well as the electrosynthesis of ECPs has suffered some slowing down till the beginning of the 1990s, because of the apparition of several new synthetic pathways, e.g., following the pioneering work of McCullough [8] and Bauerle [14]. Since it was generally admitted that modern chemical routes yield much more regular polymers, with less defects and an overall much better quality and processibility, the electrochemical synthesis has backed to a point where it was somewhere necessary to get the conducting polymer on the electrode, to achieve another electrochemical process (e.g., electrocatalysis or sensing). Since conversely electrochemical applications of conducting polymers proportionally decreased compared to e.g., light involving devices (LEDs, photovoltaic cells, etc.), this tended also to restrict the amount of work in the field of electrosynthesis.

However, some scattered works have continued to refine and perfect existing electrochemical synthesis, playing e.g., on the temperature or changing solvents. Recently, some new impulse has been given by the appearance of the new electrolytic media, in particular ionic liquids, which have renewed considerably the field of organic synthesis and electrosynthesis.

On the other hand, a lot of work has been spent on the synthesis of new electropolymerizable monomers, because not all polymers are chemically accessible, and for getting some special structures the electropolymerization was again the privileged path.

Finally, a small community of people interested in fundamental electropolymerization processes has subsisted, and several works have appeared in the field of understanding fundamental electrode processes upon electropolymerization of ECPs. These works are usually of good quality, and allow one to have a new insight into the electropolymerization reactions, thus helping electrochemists to improve the preparation of their polymers.

18.2.2 Advances in the Electrochemical Synthesis of Classical ECPs

18.2.2.1 Electropolymerization into Ionic Liquids

The use of ionic liquids figures among the most noticeable works on the new electropolymerization conditions. Ionic liquids are new original organic salts that are liquid at ambient temperature [15]. They are much more conductive than ordinary electrolytes, and surprisingly only little more viscous for some of them: their viscosities range from roughly 10^{-2} to 10 Pa s, which is compare to aqueous (about 10^{-3} Pa s) or organic (5 to 9 10^{-3} Pa s) electrolytes at room temperature [16,17]. In addition, many ionic liquids also have the advantage that they can be obtained in a very dry state, making them suitable for applications in electrochemical systems from which moisture must be excluded over long periods. Furthermore, some ionic liquid relatives are plastic solids at room temperature, yet still maintaining reasonable conductivity while others can be transformed into soft, elastomeric solids at room temperature by the addition of small amounts (about 5%) of a suitable polymer [18]. The most common example of ionic liquid is [Bmim][BF_4], which stands for butylmethylimidazolium tetrafluoroborate. While the very stable alkylimidazolium cations, and especially among them the aforementioned butyl-methylimidazolium, generate the most important family of ionic liquids, the choice of the anion is relatively wide. The classical nonreactive tetrafluoroborate or bistrifluorosulfonimide (TFSI) anions are the most frequently chosen ones, both for their chemical inertia, as well as for their quasi-impossibility to be electrochemically oxidized or reduced.

The pioneering work was published by Mattes and Wallace [19], who in a single paper described the synthesis and the electrochemical cycling of several of the most classical ECPs in mainly [Bmim][BF_4]. Their main discovery is that the stability upon cycling considerably increases when compared to another solvent, especially in the case of PANI. The synthesis does not appear to proceed very differently, while on the other hand small differences can be found in the redox potentials. The same authors also report that the efficiency of electrochemical actuators using the same polymers was improved [20,21].

They reported the results of further investigations on the electropolymerization of pyrrole in various ionic liquids and found that, as a general rule, the films prepared into ionic liquids displayed a much larger apparent doping level for identical electrosynthesis conditions [22,23]. They also studied, in a recent work, the response of PANI films in two different ionic liquids and confirmed their unusual stability upon cycling in this medium. Figure 18.3 represents the differences between the cyclic voltammograms (CV) of three PPy films: one grown in a classical solvent, and two others grown in two different ionic liquids.

18.2.2.2 Electrochemically Nanostructured Conducting Polymers

Among other advances, the nanostructuration of electrochemically synthesized conducting polymers has raised a lot of interest. To achieve this, one of the most straightforward and common way is the use of a scanning electrochemical microscope (SECM). Since the pioneering work of Bard, several authors have refined the conditions, improving the resolution. Heinze et al. used a PMMA matrix to confine more cleanly the electrodeposition of PPy [24], and in a further work, they could deposit micropatterns of poly(dimethoxybithiophene) with the help of electrocatalysis by a ruthenium complex, obtaining well-defined plots [25,26].

The interest in nanopatterned electrodes continues to stimulate activity, especially with the desire to prepare one-dimensional (1D) or two-dimensional (2D) structures (Section 18.3.4), this being associated sometimes to applications in the field of sensors.

18.2.2.3 Solution Structuration

Among the efforts undertaken to realize original electrooxidative synthesis of ECPs, the attempts to structure the solution environment of the monomer should be recalled. An obvious way, which was

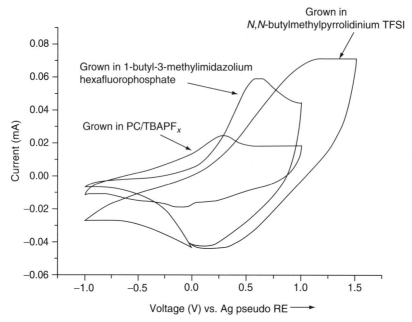

FIGURE 18.3 Comparison of cyclic voltammograms of polypyrrole (PPy) films in classical electrolyte and ionic liquids. Scan rate: 100 mV s^{-1}. (Reprinted from Pringle, J.M., J. Efthimiadis, P.C. Howlett, D.R. MacFarlane, A.B. Chaplin, S.B. Hall, D.L. Officer, G.G. Wallace, and M. Forsyth. *Polymer*, 45,1447–1453, 2004. With permission.)

introduced by Sadki et al. [27,28], is the emulsion polymerization. The quality of the films is not extremely different from the one for parent polymers in conventional electrolytes. However, it allows the use of water, which is a cheap, easy-to-handle, and environment-friendly solvent.

An original and worthwhile signaling approach was introduced by Chane-Ching and Lacroix [29], who tried to form inclusion complexes of polymerizable monomers like pyrrole and bithiophene [30] with cyclodextrines. The initial objective was to encapsulate the polymer chains, so that better organization of the film could be obtained, with the possible appearance of unusual properties like fluorescence. Actually, although the incorporation of the cyclodextrine in the films has been proven in the case of bithiophene, in the case of a two-head pyrrole monomer, the cyclodextrine seems to be kicked out after the electron transfer. In both cases the polymer is relatively close to the one obtained in organic medium, and displays an analogous electrochemical behavior. It remains that this approach, even if it did not yield fundamentally different polymer films, allows one to perform electropolymerization in aqueous medium and is thus an alternative to micellar solutions.

18.2.2.4 Conducting Polymers for Corrosion Protection

Recently more interest has come out in the synthesis of conducting polymers especially for corrosion protection. This mainly concerns the electropolymerization processes, because corrosion protection implies deposition on oxidizable metals and therefore explains the place for this topic here. One leading team in the field is Lacaze's, who investigated the electrodeposition of PPy on several nonnoble metals including zinc [31] and soft steel [32]. Their approach, also explored by other groups, involves adding to the polymerization feed a strong complexing agent of the electrode metal (oxalates or salicylates have been mainly used), which more or less passivates the electrode during the first steps of the electrodeposition; the complexing agent can be used as the only electrolyte counteranion in some cases. The team also studied PANI for corrosion protection [33]. Other groups have also examined the protection by PPy [34,35], as well as by PANI [36–41]. The Tallman's team also examined the case of aluminium in detail with some success, considering that this is an example of a very easily oxidizable metal [42–44].

A special mention should be awarded to Plieth [35], who prepared ultrathin coatings, and to Blackwood, who investigated multilayered films [45].

ECPs, in their oxidized form, protect metals because they bend the surface potential toward more positive potentials, making therefore the attack of oxidants more difficult and retarding the establishment of a local corrosion microcell; this phenomenon has been studied both from a general point of view [46] and at defects of the polymer coatings [47]. However, one major problem with ECPs is their sensitivity to chemical oxidative degradation [48]. Finally, the best solutions for corrosion protection might lie upon the assembly of different layers, especially if their protecting effects can be combined. Audebert et al. recently demonstrated that a modified PPy layer, electropolymerized according to Lacaze's conditions, but starting from a monomer blend, and covered by a special hybrid silica xerogel layer, can provide a much better protection than the simple independent addition of the two components [49]. This may open the way to further investigations in this field.

18.2.2.5 Miscellaneous

Many people have prepared and organized [50] layers of monomers on electrodes, typically using the well-known adsorption of thiols and thioethers on gold, and electrochemically grown polymers in the following steps. Not surprisingly little recent work can be found in this field, except in the emerging framework of nanoparticles [51]. For example, gold nanoparticles have been incorporated into multi-layered PANI films [52]. Also, some authors, e.g., E. Kim et al., have turned successfully to differently organized layers, such as alternate polyelectrolyte and ECP layers [53–55]. An interesting and recent example even describes CdSe nanoparticles functionalized with PANI in order to improve the optical sensing properties [56], and also CdTe nanoparticles functionalized with PPy [57]. However, no work seems to have really demonstrated the advantage of having a conjugated link with the inorganic substrate compared to previous works using a nonconjugated spacer.

Less effort was spent recently on perfecting the conditions of electrodeposition in conventional or related conditions, except when a subsequent application was aimed at, which is indicative of, either a rising lack of interest in electropolymerization conditions, or that this technique has reached its limits. However, De Paoli et al. have recently investigated the optimization of the deposition conditions of a terthiophene derivative on indium tin oxide (ITO), with the further purpose to make a LED device [58].

Finally, electrocatalysis has been used in electropolymerization processes. As an example of this, the cation radical of terthiophene has been used to mediate the electropolymerization of 3-methylthiophene [59], and a gold complex has also been used to mediate PPy formation [60]. Like analogous works in this direction [50] this did not seem to lead to major improvements in the film's characteristics nor properties (this latter aspect will be further detailed in Section 18.3.3).

18.2.3 New Monomers and Related Oligomers

A lot of work has been devoted to the elaboration of more and more purposely tailored monomers, followed by electropolymerization, in order to improve a given property of the resulting polymer, e.g., a low bandgap. Approaches have also been made to synthesize oligomers, in order to prepare a better-defined polymer with improved properties. However, little work has been published outside the three main families, namely thiophene (especially EDOT), pyrrole, and aniline derivatives. In this section, we will only recall which kind of approach in monomer tailoring has been followed, and focus on some examples where the electrochemical behavior of the monomers has been found different from what could be expected.

In the 1980s and early 1990s, many functionalized heterocyclic monomers have been produced, bearing the functional active group linked to the ring through an alkyl spacer, on the 3-position (in the case of thiophene or pyrrole) or directly on the nitrogen in the case of pyrrole. This approach, which initially followed the easiness of the monomer synthesis, suffers from a disadvantage that has been recognized: the weakness of the cooperativity between the polymer and the functional group. With the progress in the synthesis methods, a new approach has been introduced mainly by Reynolds [61–68] and

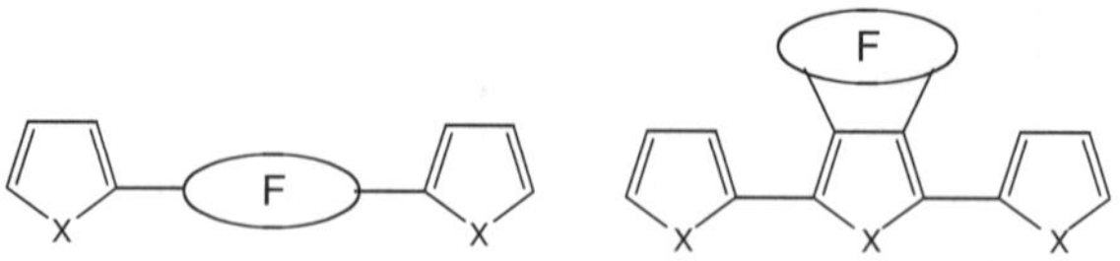

SCHEME 18.2   The two main ways to covalently link an active functional moiety F to an electropolymerizable monomer.

Zotti's groups [69–76], which consists in inserting a functional species between two polymerizable moieties, e.g., two thiophene or pyrrole rings (Scheme 18.2). This approach has spread, especially in France with the noticeable contribution of Roncali [77–82], Bidan [83], Chane-Ching [84–88], and the elegant polycatenanes developed by Divisia-Blohorn and Kern [89–92] on one part and Swager [93] on the other part. Pron [94,95], Audebert [96–102] and others [103–118] also published some original systems, including a nice example of a polyfullerene linked to a ferrocene [119] and an osmium complex included into PTh [120]. In this new approach, the functional system is either included into the polymer chain, or intimately linked to it, and the two subsystems are thus forced to cooperate together, mostly as represented in 1a and 1b of Scheme 18.2.

Usually electropolymerization of these monomers occurs without any problems, and the properties of the subsequent polymers are detailed in Section 18.3.3. However, in very few occasions, the classical electropolymerization reaction does not work. This has been especially demonstrated in the case of tetrazines [121], where probably the too strong electron attracting character of the tetrazine impedes the classical cation-radical coupling, favoring another type of decomposition reaction, e.g., a proton loss. However, while even EDOT functionalized tetrazine does not polymerize, the same tetrazine ring can be incorporated into an ECP with the help of a bithienyl functionalized moiety [97].

The classical approach introduced about 20 years ago by Deronzier, Bidan, and Audebert [122] which consisted of attaching a functional group by a nonconjugate spacer, still continues to attract interest, with for example the recent work of Vorotyntsev [123–125] and some others [62,76,87,111,126–132]. However, at noticeable exceptions of low bandgap polymers and some electrochromic displays of Reynolds, this approach has met a mitigated success till now, because in many cases, despite the remarkable efforts of synthetic chemists, little cooperativity has been found between the polymer and the introduced functionality. Chart 1 and Chart 2 below give a nonexhaustive overview of the most elegant new monomers published during the last 15 years.

18.2.4 Electropolymerization Reaction: Mechanism, Influence of Additives, Kinetic Considerations and Influence on the Related Polymers

About 15 years ago was published the unequivocal demonstration of the fact that the first steps of the polymerization of pyrrole and several pyrrole derivatives in dry acetonitrile were the coupling of cation-radicals [6,133]. Further studies by Hapiot and Audebert have shown that this could be extended to several types of thiophene [134] and pyrrole oligomers [98,135], from four to six units length. It has been assumed since then (and even before) that this was the same for all heterocycles, although the study of thiophene itself is impossible (the cation-radical is far too much unstable), as well as bithiophene and terthiophene, whose cation radicals are also too much reactive. This is also unfortunately the case of EDOT and its pyrrole analog EDOP, although the lifetime of the cation radical of EDOP could be recently estimated in particular conditions [136].

However, as a later review explains [137], it has come to light progressively that the mechanisms could differ, especially in the follow up after the initial coupling of two cation radicals, and more fundamental studies could still be undertaken in this field. While the following steps leading to the polymer formation are too remote to be precisely investigated by direct electrochemistry, it remained that both theoretical

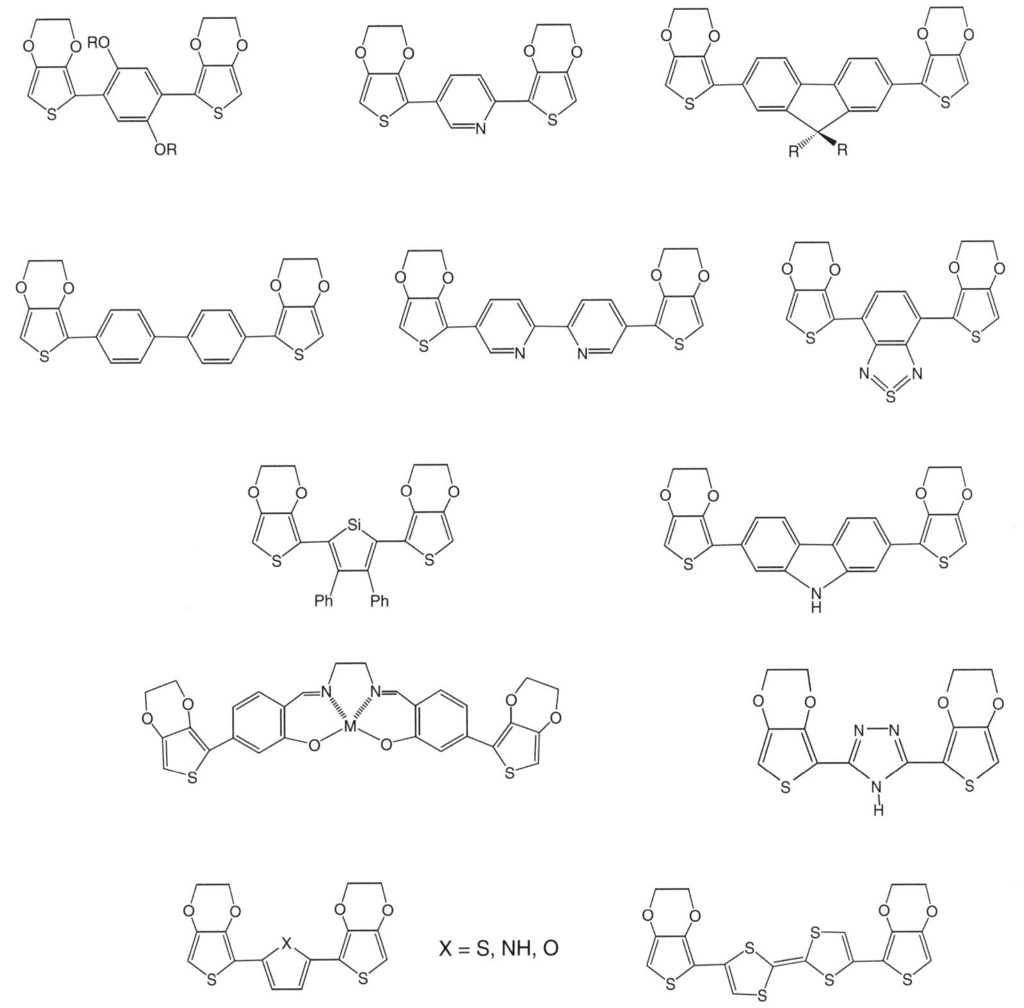

CHART 18.1 Ethylenedioxythiophene (EDOT)-based monomers.

considerations and analysis of the polymer structure in relation to the various synthetic conditions, could give new insights into the understanding of the electrochemical polymerization mechanism.

An interesting theoretical work by Lacroix et al. [7,138] confirms that the monomer-cation radical coupling is effectively the favored mechanism in many solvents, but also that the cation-radical coupling is less and less favored when the number of rings in the monomer increases (although it has been observed with oligothiophenes up to five units). However, the most important amount of work in the field is due to Heinze's group [139–141]. They published a series of articles analyzing finely the fate of the PPy in relation to the synthetic conditions, the polymerization potential, and especially the water content [5,141–144]. It had been known almost from the very beginning, thanks to Geniès et al., that water addition in acetontrile lead to dramatic changes in the CV of the subsequent polymer [5]. Heinze showed that in many cases the most important point was the deprotonation of the dimer produced by the initial coupling [142], and its afferent kinetics. They investigated not only the influence of water, but also of several bases in relation to their relative strength, and showed that optimum conditions were obtained with a nonnucleophilic mild base, which could deprotonate the dimer (and subsequent coupling issued oligomers) without affecting the monomer cation-radical. They showed in addition

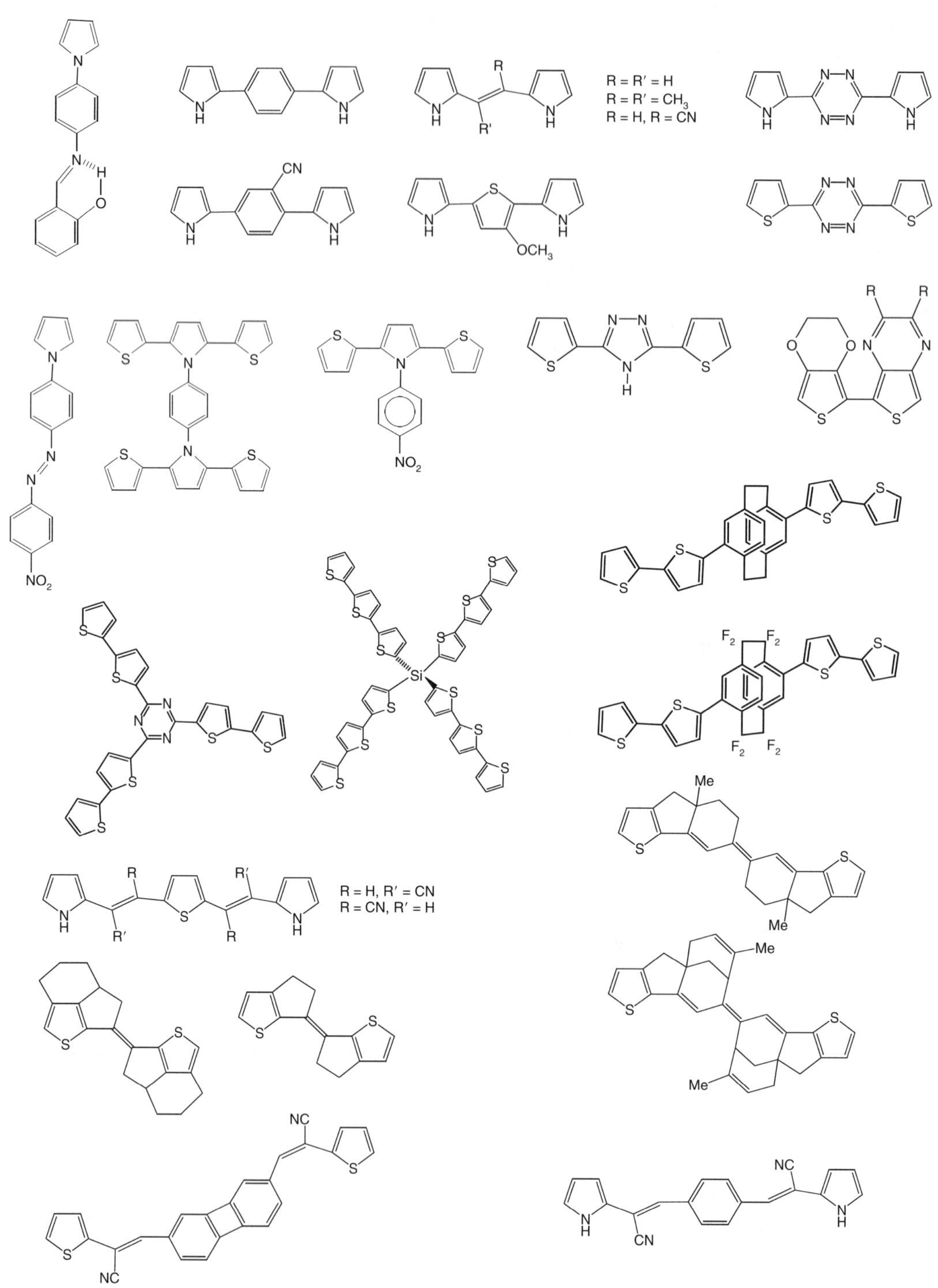

CHART 18.2   Thiophenes, pyrroles, and miscellaneous monomers.

that in different conditions, various forms of unsaturated PPy could be prepared, and they even named them and proposed structures for them [144].

Another noticeable improvement of the electropolymerization of conjugated heterocycles comes from the use of "super-acidic" conditions. Especially most of the mixtures used contain borontrifluoride etherate (BTFE) [145–147]. This research was triggered by the original report of Shi and Li, who showed that PPy with unusual mechanical strength could be obtained from BTFE solutions [148–149]. This field was developed a little later by Reynolds [150] who used mixtures of solvents (acetonitrile) and BTFE and Seeber [151] who used BTFE with concentrated sulphuric acid to polymerize high redox potentials monomers like 3-chlorothiophene. However, despite the apparent efficiency of the polymerization process in that case, no mechanistic studies have been performed so far.

Some works have been recently published on several monomers using various techniques: especially EQCM [152], AFM [153], and ellipsometry [154] have been used in parallel to electrochemistry [155–157] to analyze the growth of the polymer, but always without focusing on the first steps. In another approach, polymeric additives have been used in the polymerization feed to act as templates for the formation of the conducting polymer. A remarkable recent work shows that PPy nanotubes can be directly prepared from a polymer blend solution, without an external porous membrane (or clay) as formerly classically prepared [158].

18.3 Electrochemical Features of ECPs

18.3.1 Fundamental Processes upon Redox Switching of ECPs

18.3.1.1 General Features

A common feature among conjugated polymers is their ability to be reversibly switched between at least two redox states: one is generally a neutral form, and the other is a charged state (usually positive). However, contrary to redox polymers (like nonconjugated polymers bearing redox moieties, e.g., polyvinylferrocene), the redox sites of ECPs are not clearly defined and charges incorporated during redox switching are partially delocalized: indeed the polymer ability to bear deformations with rather low elastic energy cost makes it possible to create energy levels within the bandgap, thus stabilizing charge incorporation within the macromolecular backbone. Such charge species associated with local deformation of the polymer chains are called polarons (for single charge species with spin) or bipolarons (for spinless double charge species). Charge creation along the polymer backbone is possible only if it can be compensated by opposite charges generally coming from ionic species (called dopants). ECPs are thus both electronic and ionic conducting materials. Electronic conductivity is related to charge carriers mobility, both along a polymer chain and between adjacent chains: the first one is related to the actual conjugation length, which depends upon the orbital overlap between repeating units and the defects number; the second one is related to the orbital overlap between adjacent chains, that can occur either by stacking (formation of π-dimers) or simply by the presence of dopant ions, acting as bridges for electronic coupling. Usually electronic conductivity (in the redox state that allows charge delocalization) is much higher than ionic conductivity in ECPs, so that this latter is the rate-determining process.

From the electrochemical point of view, several common features can be evidenced from the typical CVs recorded when scanning between the redox states of ECPs (Figure 18.1): the signal is often very broad, not fully reversible, displays a kind of plateau at high potentials and major differences between the first and subsequent cycles (hysteresis phenomena). Extensive studies during the last two decades led to well-admitted interpretations of this behavior. First, oxidation–reduction processes occur with severe reorganization in the polymer structure, especially from the solvation aspect: the transition between a hydrophobic neutral form and a hydrophilic charged form induces swelling phenomena upon doping. When scanning backwards, relaxation processes take place leading to hysteretic behaviors. In addition, the polymeric nature of the material with usually high polydispersity values makes it likely that a distribution of oxidation–reduction potentials exist, thus broadening the CV signal relatively to individual redox species. At last, the high surface vs. volume ratio of ECPs generates a capacitive component of the current that superimposes to the faradaic one and leads to a characteristic plateau at high potentials.

Recent works aimed to investigate the electrochemical behavior of ECPs from the various points of view that have been described above: relaxation phenomena occurring during charge–discharge process, role of dopant, ionic and solvent exchanges accompanying redox switching, influence of the doping level on conducting properties, etc. These various points will be the topic of discussion in the following sections.

18.3.1.2 Relaxation Phenomena during Charge–Discharge Processes of ECPs

ECPs are easily and reversibly charged and discharged by varying the electrode potential either by scanning (CV) or by applying steps (chronoamperometry, CA). However, the mechanisms by which ECPs become oxidized or reduced are not straightforward to model, due to the complexity of the system: first, one has to deal with faradaic and capacitive components of the injected charge [159], which are always difficult to separate in the electrochemical response, as demonstrated in the mid-1980s by Feldberg [10]; then, the model must account for strong hysteresis phenomena that induce, e.g., large differences between oxidation and reduction peak potentials in CV. Rather than considering strong electrostatic interactions between charged polymer sites and dopants, this key point can be interpreted by postulating different kinds of redox sites within the polymer matrix, some exchanging directly with the electrode and others involving intermolecular bonds between adjacent chains [160], as evidenced in the case of conjugated oligomers by Heinze et al. [139]. In other words, polarons and bipolarons can be shared between adjacent chains through intermolecular interactions, which stabilize them. When the potential scan is reversed, one has to provide energy through overvoltage to destroy these dimeric charged species (sigma-dimers), thus leading to an electrochemical "irreversible" behavior.

Relaxation phenomena associated to redox switching of ECPs have been extensively studied [161–164], notably because of their implication in some important application fields like electrochromism [165]. In particular, the influence of various parameters like the counterion nature [166], the polymer thickness [167], the solvent properties [168], or the waiting potential [169] has been investigated.

Several models have been proposed in the literature to try to explain the observed electrochemical behavior:

— Otero et al. introduced the electrochemically stimulated conformation relaxation (ESCR) model [162,170] to account for the dependence of the electrochemical response of PPy or PANI on the cathodic potential limit of the scan and possible waiting times at this potential. This model is based on the idea of a more compact form of the polymer matrix as the potential is scanned further in the cathodic direction (or remains a longer time at negative values E_c). This compact structure hinders counterion exchange with the solution inducing relaxation processes upon reoxidation. The model proposes a dependence of the relaxation time τ according to the following equation [171]:

$$\tau = \tau_0 \exp\left[\frac{\Delta H^* + z_c(E_s - E_c) - z_r(E - E_0)}{RT}\right] \tag{18.4}$$

where E_s is the characteristic potential for closure of the polymer matrix and z_c is the charge required to compact one mol of polymeric segments. E_0 and z_r are the similar quantities for the reverse relaxation process and ΔH^* is the conformational energy consumed by one mol of polymeric segments in absence of applied electric field. It results from Equation 18.4 that relaxation requires higher anodic overvoltages, as much as the potential has been scanned further in the cathodic direction, explaining the hysteretic behavior. The parameters E_s, z_c, and ΔH^* have been determined for several ECPs by analyzing CA at various step potentials [172]. When the potential is scanned (or stepped) between more anodic potential than E_s, then oxidation is under counterion diffusion control and apparent diffusion coefficients can be deduced, that are dependent upon oxidation potentials, accounting for swelling of the polymeric structure during redox switching.

—Another approach that appears well-suited to understand the peculiarities of the electrochemical behavior of ECPs is based upon the dramatic change in conductivity that occurs during redox switching, which makes the situation very different from charging a redox polymer: this point can be accounted for by a two-phase model [173], with a moving boundary between the insulating and conductive parts of the polymer. Redox switching induces the propagation of a conductive front within the polymer film perpendicularly to the electrode: such a moving boundary delimiting conductive and insulating zones in the polymer matrix has been experimentally evidenced by Aoki using a diode-array detector [174]. Thus, the conversion process is characterized by a percolation threshold that can explain why the conversion rate passes through a maximum when the doping level increases [175]; Monte Carlo simulations have demonstrated that the percolation threshold corresponds approximately to the peak potential in CV [176]. Another point worth mentioning is the role played by migration phenomena in the ionic motion, when this latter becomes the rate-determining step of the conversion process [177]: as the conductive area increases, the potential drop across the film becomes restricted to a small area, inducing strong electric fields: migration-coupled diffusion of the dopant must be taken into account to simulate the redox behavior of the ECP [178].

18.3.1.3 Ion and Solvent Exchanges during Redox Switching of ECPs

Two major changes are associated with the redox switching process in ECPs: one is the charge of the polymer chains that induce ionic motion to maintain electroneutrality (and in the particular case of PANI, possibly proton exchange [179]); the other one is related to the hydrophobic/hydrophilic balance of the polymer matrix that makes the solvent play an important role in the doping process and often induce swelling phenomena [180]. It has been shown that counterion insertion during oxidation of PPy is accompanied by desolvation processes as evidenced by variation of the diffusion coefficient of the counterion [181]. These desolvation processes could be involved in the achievement of the quasi-metallic state of ECP [182].

Electrochemical impedance spectroscopy (EIS) is an efficient tool to analyze the major influence of the anion nature on several parameters characterizing the charge transfer across the ECP, among which is the fractal dimension of the polymer/electrolyte interface [183]. This fractal character can be evidenced by studying the electrochemical response of a redox probe (e.g., ferrocene or hexacyanoferrate): in the case of p-doped PEDOT, an anomalous diffusion has been clearly evidenced and fractal dimensions measured using rotating disk electrode (RDE) and EIS [184–185]. Such fractal dimensions can be useful to quantify the interface roughness and the values deduced from electrochemical measurements can be effectively compared with those obtained from small angle X-ray scattering (SAXS) techniques [186].

Solvent and ionic fluxes can also be analyzed by electrochemical quartz crystal microbalance (EQCM) experiments [187,188] and discriminated by recording simultaneously charge and mass changes during redox switching. Various ionic fluxes (for example, counterion and coion moving in opposite directions) contributing to doping can also be discriminated by EQCM [189] or by probing the diffusion layer by probe beam deflection technique (PBD) and analyzing the data with the convolution tool [190]. Coupling both EQCM and PBD allows one to discriminate all the contributions of the moving species, as demonstrated by Vieil and Hillman [191,192]. Typical current, mass, and laser beam deflection variations with applied potential can be seen in Figure 18.4, showing that oxidation of PPy is dominated by anionic doping (although the cationic coion plays a significant role at the early stage of doping) and mass loss due to desolvation of the polymer matrix.

The respective contributions of counterion and coion in the doping process depend on several factors, among which, apart from the nature of the species themselves, are the polymer thickness and the doping

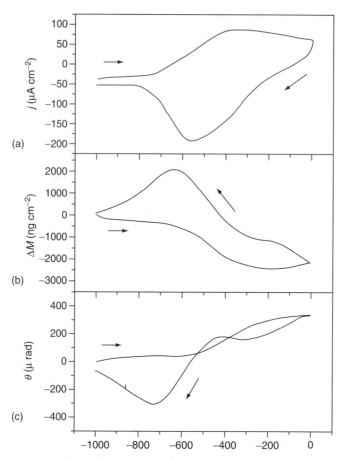

FIGURE 18.4   Current (a), mass (b), and probe beam deflection (c) variations with potential upon redox switching of a polypyrrole (PPy) film exposed to a 0.5 M salicylate sodium solution. Scan rate: 10 mV s^{-1}; Coverage ratio: 9.10^{-8} mol cm^{-2}. (Reprinted from Henderson, M.J., H. French, A.R. Hillman, and E. Vieil. *Electrochem. Solid. State. Lett.*, 2, 631–633, 1999. With permission.)

level [193]. An electrochemical study using rotating ring disk electrode (RRDE) has shown that halogenide anions, by remaining trapped within the film, induce a dramatically different behavior for PPy doping as compared to other anions (e.g., hexafluorophosphate) [194]. A model has been recently proposed to account for the respective contributions of both anion and cation in the p- and n-doping of PEDOT, according to EQCM results [195].

From a theoretical point of view, Vorotyntsev et al. proposed a model based upon the concept of free and bound dopant ions inside the film to account for the voltammetric and EQCM responses of PPy films upon charge–discharge process [196].

EQCM experiments can also be coupled to EIS to study the changes in the viscoelastic properties of the polymer matrix during redox switching [197]. Hillman et al. have shown that solvation changes induce large variations of storage and loss moduli of the polymer with applied potential and frequency. They proposed a scheme-of-squares to summary the behavior of a hexylthiophene film in propylene carbonate, reflecting that solvation changes are much slower than charge transfer processes (Scheme 18.3).

18.3.1.4   Influence of the Doping Level on Conducting Properties

This key point has been partly elucidated by using in situ conductivity measurements coupled with electrochemical techniques. Zotti demonstrated a few years ago that characteristic conductivity vs.

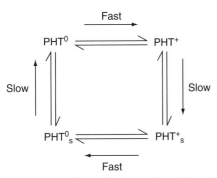

SCHEME 18.3 Scheme-of-squares for polyhexylthiophene, exhibiting two redox states, each of which existing in one solvation state. (Reprinted from Brown, M.J., A.R. Hillman, S.J. Martin, R.W. Cernosek, and H.L. Bandey. *J. Mater. Chem.*, 10, 115–126, 2000. With permission.)

potential curves exhibited a three-step variation [198]: a threshold followed by linear variation and a plateau (Figure 18.5). The same features can be observed on conductivity vs. charge (i.e., vs. doping level) curves. The threshold is related to percolation phenomenon, while the plateau indicates that a nearly constant density of states is reached at high doping levels.

Interestingly, the conductivity profile changes with the natures of dopant and ECP. Substituting perchlorate by tosylate as the dopant for PPy enhances the threshold doping level from 1% to 10%, as if coordinating anion like tosylate could pin polarons, hindering their propagation ability. A similar situation of trapped charged polarons was also evidenced in the case of PTh n-doping [199]. Another interpretation in terms of sigma-dimer formation by interaction be-tween charged polarons of adjacent polymer chains was suggested by Heinze et al., in the light of electrochemical studies of oligomer model compounds [200]. Conversely, Zotti invoked the formation of π-dimers to explain the different conducting behavior of PTh vs. PPy. Functionalization of poly (cyclopentadithiophenes) by crown ethers coplanar or perpendicular to the dithiophene plane allowed him to discriminate the role of π-dimers in the conductivity and to show that π-dimerized polarons were basically nonconductive carriers [198].

In some cases (for β-substituted PPy), a different shape for the conductivity vs. charge curve can be observed, with the conductivity reaching a maximum at doping levels around 0.25 [201]. An attempt to draw a general trend was proposed by Zotti and one can summarize these various behaviors along with the corresponding charge carriers in the Scheme 18.4.

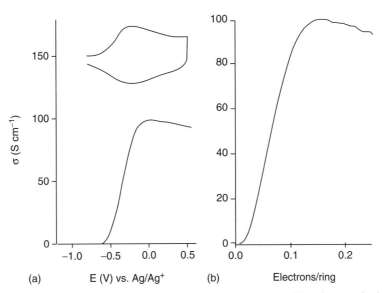

FIGURE 18.5 Conductivity vs. potential (a) and charge per ring (doping level) (b) in the case of polypyrrole (PPy) in acetonitrile. (Reprinted from Zotti, G. *Synth. Met.*, 97, 267–272, 1998. With permission.)

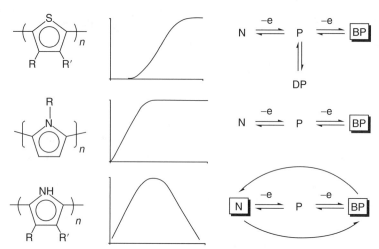

SCHEME 18.4 General features for the conductivity vs. doping level curves and the corresponding redox states involved (N = neutral state; P = polaronic state; BP = bipolaronic state; DP = dimerized polaronic state). Frames indicate the main carrier type. (Reprinted from Zotti, G., S. Zecchin, G. Schiavon, B. Vercelli, A. Berlin, E. Dalcanale, and L. Groenendaal. *Chem. Mater.*, 15, 4642–4650, 2003. With permission.)

Another interesting question arising from the behavior mentioned above is why the conductivity sometimes decreases at high doping levels, in contrast to the predictions of the bipolaron model. In addition to overoxidation phenomena leading to the polymer chemical degradation, an interpretation suggested by Zotti et al. was related to the stabilization of bipolaronic state through H-bonding, since this behavior is frequently observed in the case of unsubstituted nitrogen PPy [201]. A recent study on polydialkoxy-substituted thiophenes showing a similar behavior led to postulate that this feature is more generally inherent to charge carrier mobility in ECPs: this latter decreases as much as the number of uncharged repeat units decreases [202].

18.3.2 Electrochemistry of Low Band Gap ECPs

Such materials have known a large interest during the past decade due to promising applications in display [203] (LED) or charge storage devices [204]. Electrochemistry and especially CV is a very useful tool to define unambiguous bandgap values in ECPs, from the measurement of the difference between oxidation and reduction potentials.[2] This bandgap value can be compared with the one deduced from absorption wavelengths (electrochemical bandgaps are usually higher than optical ones [205]) and in many cases, spectroelectrochemical measurements allow one to fully characterize the investigated material. Electrochemistry is also used as a synthetic method to convert the monomer into the corresponding polymer (see Section 18.2), since most of the following examples of low bandgap ECPs have been electrosynthesized, either potentiostatically or by potential cycling.

Most of low bandgap ECPs are based upon thiophene moiety [206]. PTh has a bandgap value of ~2.3 eV [207], but this can be lowered either by fusing the thiophene ring with aromatic cycles, like in isothianaphthene (ITN) or thieno[3,4-*b*]thiophene [63], or by substituting the thiophene ring on the 3,4 positions by donor groups, like in ethylenedioxythiophene (EDOT). In the first case, bandgap lowering is due to the quinoid character of the thiophene rings in the main skeleton compelled by the aromatic character of the fused ring, while in the latter case, donor effect of the substituent induces a raise in the energy of the valence band edge of the polymer with little effect on the position of its conduction band.

---

[2]Very often, bandgap values are given according to the difference of potentials between the onset of oxidation and reduction, leading to lower values than the actual ones.

Thus, a very common moiety used to design low bandgap ECPs is ITN (Scheme 18.5), whose corresponding polymer exhibits a very low bandgap value of 1.1 eV, due to the existence of a very stabilized quinoid form [208].

Several authors have synthesized and studied new ECPs based upon ITN derivatives [209–212]. Among them, Wudl et al. [208] have shown that EHI-PITN (Scheme 18.6) exhibited an electrochemical behavior characteristic of low bandgap (1.3 eV) material with both well-defined *p*- and n-doping signals (Figure 18.6).

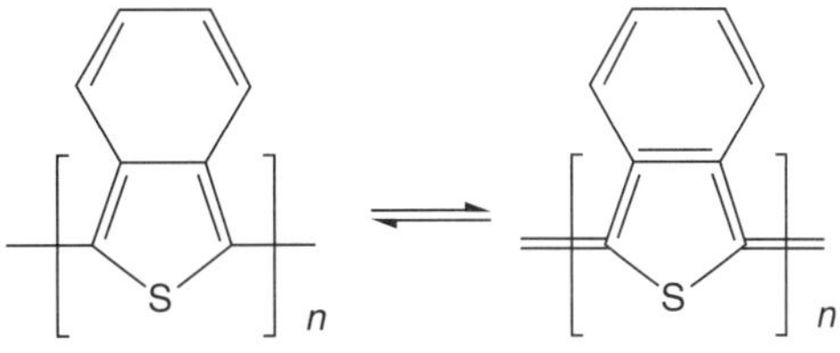

In the same family of compounds as polyisothianaphthene (PITN), two main examples of precursors for low bandgap ECPs can be given: first, poly(2-phenylthieno-3,4-*b*-thiophene) was electrosynthesized and characterized by CV, revealing a bandgap of 0.85 eV based on the onset of the oxidation and reduction signals [108]. Second, $\Delta^{4,4'}$-dicyclopenta[2,1-*b*:3,4-*b'*]dithiophene (inset in Figure 18.7) has been successfully electropolymerized by Pickup et al. [213]: comparison of the electrochemical and in situ conductivity responses upon potential scanning reveals a much better p-type conductivity than the n-type one, probably because smaller mobili-

SCHEME 18.5 Aromatic and quinoid forms of polyisothianaphthene (PITN).

EHI-PITN (R = 2-ethylhexyl)

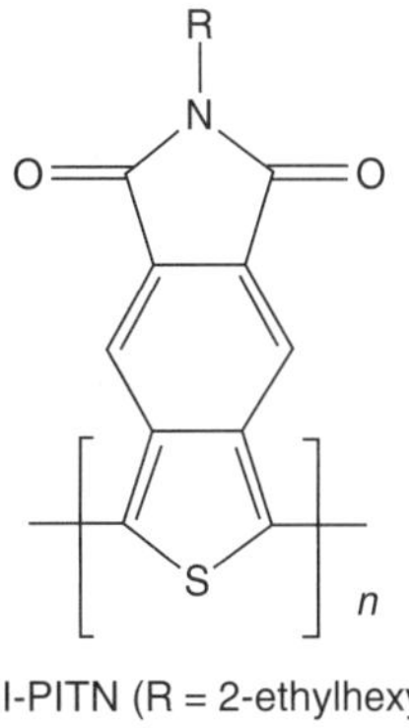

SCHEME 18.6 Formula of poly(benzo[c]thiophene-N-2-ethylhexy-4,5-dicarboxylic imide) (EHI-PITN).

ties for this latter carrier type are reached; this can be assigned to smaller bandwidth of the conduction band relatively to the valence bandwidth.

Another very common thiophene-based monomer used to design low bandgap ECPs is EDOT, whose corresponding polymer PEDOT has a bandgap of 1.2 eV [214]. Zotti et al. have shown that bis-(3,4-ethylenedioxythienyl)-methane could give rise to thick polymer films with an electrochemical bandgap (based upon standard redox potentials)[3] of 1.0 eV [215].

A widespread idea to decrease the bandgap has been to polymerize compounds made from alternate electron-rich and electron-poor moieties. The donor–acceptor concept [216] relies upon the fact that each moiety should contribute to lower the bandgap either by decreasing the ionization potential (for the electron-rich moiety) or by increasing the electronic affinity (for the electron-poor moiety). Although the limits of this concept have been clearly highlighted by molecular orbital calculations [207,217], a large number of ECPs have been designed according to this idea [218–220].

Among these examples, copolymers of thiophene and acrylonitrile moieties can be electropolymerized, giving ECPs with electrochemical bandgap as low as 0.6 eV [221]. The same features were also observed in the case of copolymers of thiophene and thieno[3,4-*b*]thiophene separated by a cyanovinylene moiety [222]. Various cyanoethylene thiophene polymers have been cycled between their n-doped and p-doped states to measure the stability and potential window for supercapacitor applications [223]. The best compound, poly-(*E*)-alpha-[(2-thienyl)methylene]-2-thiophenacetonitrile, displays unfortunately 60% of electrochemical charge loss after 2000 cycles, for a 2 V potential range.

Another example of the strategy of donor–acceptor copolymers uses monomers containing an electron-rich moiety like EDOT associated with an electron-poor one like 4H-cyclopenta-[1,2-*b*;3,4-*b'*] dithiophen-4-one [224] or 4-dicyanomethylene-4H-[1,2-*b*;3,4-*b'*]cyclopentadithiophene (CDM) [225]

[3]This corresponds to a bandgap value as low as 0.4 eV, based upon the onset of the oxidation and reduction signals.

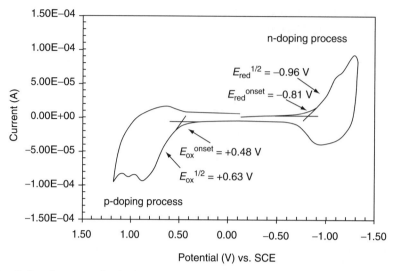

FIGURE 18.6   Cyclic voltammetry (CV) of EIH-PITN films on glassy carbon in acetonitrile showing n-doping and p-doping. (Reproduced from Meng, H., and F. Wudl. *Macromolecules*, 34:1810–1816, 2001. With permission.)

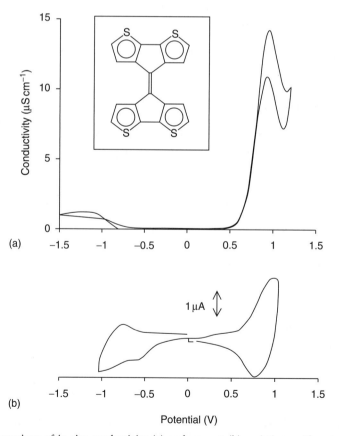

FIGURE 18.7   Comparison of in situ conductivity (a) and current (b) variations with potential upon p- and n-doping of the polymer derived from $\Delta^{4,4'}$-dicyclopenta[2,1-*b*:3,4-*b'*]dithiophene (see inset). (Reprinted from Loganathan, K., E.G. Cammisa, B.D. Myron, and P.G. Pickup. *Chem. Mater.*, 15, 1918–1923, 2003. With permission.)

(Scheme 18.7a): both have been shown to lead to ECPs with bandgaps lower than 1 eV [226]. Zotti et al. have also investigated the role played by copolymerization and donor–acceptor alternation in the conductivity and bandgap values [226]. The conclusion is that an increasing alternation between donor and acceptor moieties in the polymer chain leads to the localization of the n-doping charge carriers; in addition, bandgap values in the copolymers are limited by the bandgap of the individual homopolymers, but

SCHEME 18.7 Formulae of (a) 4H-cyclopenta-[1,2-*b*;3,4-*b′*]dithiophen-4-one and (b) 4-dicyanomethylene-4H-[1,2-*b*;3,4-*b′*]cyclopentadithiophene.

increasing the amount of the electron-rich compound (EDOT) leads to a decrease in the bandgap value because it makes the valence band edge lie at higher energies, while the conduction band edge remains nearly unaffected.

The electron-poor moiety associated to the electron-rich heterocycle can be varied from the examples given above. Audebert et al. proposed an original tetrazine moiety acting as an attracting conjugated spacer between bithiophene moieties and leading by electropolymerization to a reducible and oxidable polymer [227]. The incorporation of transition metal complex within the conjugated backbone has been also envisaged. For instance, Pickup et al. have incorporated nickel dithiolene into PTh by electropolymerization of bis[1,2-di(2-thienyl)-1,2-ethenedithiolene]nickel (Figure 18.8) [228]. The resulting polymer exhibits very low bandgap (0.35 eV based on the onset of oxidation and reduction processes) between oxidation of the PTh backbone and the two successive nickel centered reduction processes.

Apart from thiophene-based precursors, a few other monomers have been tested to design low bandgap ECPs, like the fluorene-based ones [229]. A copolymer of polyfluorenylidene, bearing alternatively a donor and an attracting group (Figure 18.9), exhibits reversible cycling with both p-doped and n-doped states, as shown by Rault-Berthelot et al. Interestingly, the electrochemical response depends on the scanning direction: when scanning toward anodic potential, one reaches the fully doped state where each fluorene-based moiety is oxidized; conversely when scanning toward cathodic potentials, only

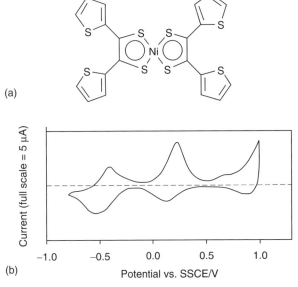

FIGURE 18.8 (a) Formula of bis[1,2-di(2-thienyl)-1,2-ethenedithiolene]nickel and (b) cyclic voltammetry (CV) of the resulting polymer showing p-doping of the polythiophene backbone and n-doping due to the nickel centred reduction. (Reprinted from Kean, C.L., and P.G. Pickup. *Chem. Commun.*, 815–816, 2001. With permission.)

the fluorene moiety bearing the dicyanomethylene group is reduced into an anion, so that when scanning backward, one passes through the fully undoped neutral form of the copolymer and then reaches a partially oxidized form, where only the dithiolate-substituted fluorene is in a cationic form (Figure 18.9). A gap value of 1.4 eV (between redox peaks) is obtained.

18.3.3   Electrochemical Features of Functionalized ECPs

When designing functional ECPs, one has often in mind to combine intrinsic properties of the conducting polymer backbone (electroactivity, conductivity, transducing properties, etc.) with other ones related to the additional moiety (ion sensing, nonlinear optics, electrocatalysis, etc.). This latter can be either organic (porphyrin [112], cyclodextrin [230], calixarenes [231], etc.), or (albeit less often) inorganic (metallic ion [232], metallic particles [233], etc.). Among the functional organic moieties, much of them behave as ligands that can undergo metallation, either before or after electropolymerization. Apart from this case, several examples exist in the literature of noncomplexing organic modifiers, such as NLO-active substituents (azo- or stilbene type) [234], electro- and photochromic substituents like anils [235], aromatic substituents like phenylene [236], paranitrophenyl [237], paracyclophane [238] or styryl [239], fullerene substituents [240], or a pendant dianiline group [241] (see Section 18.2 for a listing of most of the corresponding functionalizable monomers).

In the first part of this section, we focus on the role of electrochemical methods in the design of functional ECPs, then on the main features of these materials. In the last part, we turn on the main properties and the possible applications that can be expected from these properties.

18.3.3.1   Electrosynthesis

Several methods exist to incorporate the functional moiety in an ECP: prefunctionalization of the monomer [122,239,242,243], postfunctionalization of the electrosynthesized polymer [244], or simple doping of the ECP [230]. Each one has its own advantages and drawbacks: functionalization of the monomer is usually easier than one of the polymer but it usually makes electropolymerization more difficult due to steric hindrance and/or electronic (e.g., inductive) effects. Of course this can be overcome by copolymerization but the functional moiety is more diluted and sometimes less efficient. Conversely, postfunctionalization sometimes leads to an unhomogeneous distribution of the functional moiety in the material due to differences in the reactivity of surface and bulk functional groups.

The electropolymerization method of functional monomers has proved successful in many cases, in particular to incorporate various ligands and their metallic complexes, like salen [66,245], porphyrin [246], diphosphine [247], pyridine [71,80,248,249], crown ether [250,251], metallofullerene [252], tetraazacyclotetradecane [253], or Prussian blue type [254] in a conjugated organic material like PPy, PTh, or PANI. Thick films are likely to be obtained provided that the polymer is redox active at the deposition potential: due to this, Zotti et al. demonstrated that 5,5′-bis(3,4-(ethylenedioxy)thien-2-yl)-2,2′-bipyridine could be electropolymerized when complexed by iron and ruthenium, but not in the case of complexation by nickel or copper [71].

Usually the functional moiety is covalently linked to the conjugated backbone, but it can be sometimes added to ECP as a dopant when it is under an ionic form: e.g., sulfonated β-cyclodextrins have been successfully incorporated in one step as anionic dopants in PPy by electropolymerization [230]. In some cases, electrochemistry can be combined to chemical reactions to derive functional ECPs: functional PANI materials can be obtained from the reduction of the emeraldine form by alkylthiols [255] and the functionality degree on the PANI backbone can be monitored by successive oxidation–reduction cycles; various functionalized PANIs can also be synthesized through electrophilic substitution or nucleophilic addition reactions [256]. For example, Figure 18.10 shows the similar electrochemical behaviors of sulfonated PANI made from reduction of emeraldine by sulfite salt followed by reoxidation, compared to thin films deposited from solution of highly sulfonated PANIs.

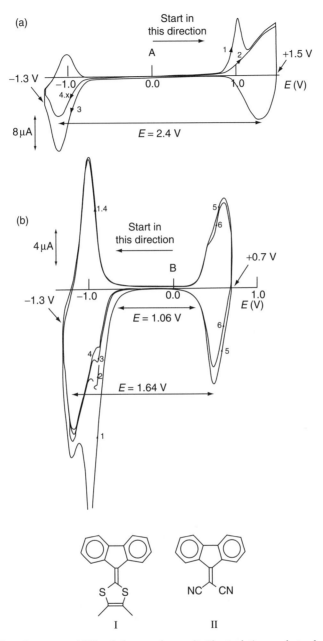

FIGURE 18.9 Cyclic voltammetry (CV) of the copolymer (I–II) at platinum electrode in dichloromethane: (a) scanning first toward anodic potentials; (b) scanning first toward cathodic potentials. Scan number is indicated on the curves. (Reprinted from Rault-Berthelot, J., and Raoult, E., *Synth. Met.*, 123, 101, 2001. With permission.)

18.3.3.2 Electrochemical Characterization

The electrochemical response of the functionalized ECP is usually very informative:

- It can give evidences of the actual incorporation of the functional moiety inside the polymer film, provided that this latter has redox properties. For instance, in the case of electropolymerized PThs functionalized by porphyrin rings, electrochemical and spectroelectrochemical experiments show charge delocalization on the whole structure, involving the porphyrin moiety through a hopping

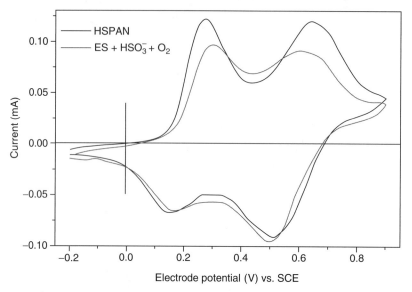

FIGURE 18.10 Cyclic voltammetry (CV) of thin films of highly sulfonated polyaniline deposited from solution (HSPAN) and similar material synthesized by reduction of emeraldine salt by sulphite ion followed by oxidation (ES + HSO_3^- + O_2). (Reprinted from Barbero, C., H.J. Salavagione, D.F. Acevedo, D.E. Grumelli, F. Garay, G.A. Planes, G.M. Morales, and M.C. Miras. *Electrochim. Acta.*, 49, 3671–3686, 2004. With permission.)

process [112]. On the other hand, PEDOT incorporating ferrocene covalently linked through ester function can be characterized by CV, showing that the electroactivity of ferrocene is actually retained in the polymer matrix [257].

- It can reveal possible modifications in the charge transfer ability of the ECP due to functionalization. For instance, in the case of PTh bearing nitrophenyl substituent, the electrochemical behavior of the ECP is strongly affected by the redox state of the nitro groups [237]. Conversely, there are some examples where the properties of the functional moiety can be modulated by the redox state of the conjugated main chain of the polymer: this effect was demonstrated in the case of conjugated phenyl linkers between PPy chains [258] or for transition metal complex linked to a PTh backbone [88]. Cooperative effects between the redox properties of a metallic complex and a conjugated polymer chain were also evidenced in the case of conjugated polyrotaxanes by Bidan's and Swager's groups [91–93].

In some cases, it can be useful to incorporate an additional redox probe like a ferrocene moiety into the main chain of an ECP. Wallace et al. have electrosynthesized copolymers of ferrocene- [259] and ferrocenophane-substituted [260] pyrroles with pyrrole itself and investigated the electrochemical properties of the related materials. When a complexing unit like bipyridyl or crown ether is directly connected to the ferrocene modified ECP, strong modifications of the redox properties can be expected in presence of a metallic cation, leading to recognition properties of Cu(I) [261] or Ba(II) [262]. Conversely, anion sensing materials based on ferrocenyl-substituted ECPs can also be designed, using ion pairing interactions induced modifications of the electrochemical signal [263]. In some cases, additional effects like dramatic conformational changes of the polymer structure increase the potential shift values up to several hundreds of mV, when the modified electrode is in presence of the ionic species to be detected [264].

18.3.3.3 Electrocatalytic Properties

Among the huge number of applications of functionalized ECPs, when focusing on electrochemical properties, one has to deal with electrocatalysis. Several reactions, notably of biological interest, are rather sluggish and require catalysis to operate in mild conditions: oxygen reduction, nitric oxide or

NADH oxidation, carbon dioxide reduction, etc. As electroactive, reversibly redox switching materials, ECPs are good candidates for electrocatalytic applications, as demonstrated in the case of oxygen reduction [265]: in that case, calculations show strong chemisorption ability of almost all classical ECPs toward oxygen. Such an electrocatalytic behavior was experimentally evidenced in the case of electropolymerized polybenzimidazole [266] toward several redox reactions.

The electrocatalytic behavior of ECPs toward the electropolymerization of other monomers has also been mentioned and recently investigated by Zotti et al. [59]: it has been demonstrated that electropolymerization of 3-methylthiophene could be catalyzed by previously electrodeposited polyterthiophene because of the formation of a charge transfer complex during the electropolymerization process (see also Section 18.2).

A lot of examples of electrocatalysis by functional ECPs concern oxygen reduction: among them several examples rely upon ECPs containing platinum particles [267–269]; one can also mention the use of PANI functionalized by dinuclear cobalt porphyrin [270] or composite layers of PANI and molybdenum doped ruthenium selenide [271], both appearing efficient to promote the four electron reduction of oxygen gas. In this case, the electrocatalytic mechanism relies upon multiple electron transfer between PANI and the porphyrin complex, as was demonstrated in the case of a self-doped co-PANI-cobalt porphyrin modified electrode [272], the ECP acting as an efficient redox mediator between the electrode and the complex. Analogous systems can also prove efficient to catalyse oxidation reactions: poly(chloroaniline) incorporating cobalt phtalocyanine shows electrocatalytic behavior for oxidation of 2-mercaptoethanol in acidic medium, because diffusion of the substrate through the polymer is improved when this latter is in its conducting state [273].

In the case of hydrogen oxidation, most of the systems are based upon metallic particles incorporated into an ECP; the electrocatalytic behavior has been investigated in relation to the distribution of metallic particles inside the conjugated matrix (which is itself dependent on the synthetic procedure) through analysis of Koutecky-Levich plots [274]. It was shown that real homogeneous distribution of colloidal platinum could only be achieved by incorporating the metal during the electropolymerization step (e.g., from $PtCl_6^{2-}$ acting as a dopant ion).

ECPs including coordination complexes are also able to show electrocatalytic properties: e.g., toward the oxidation of nitric oxide in the case of porphyrin functionalized polypyrroles containing various metallic centres [275], for the oxygen or hydrogen peroxide reduction in the case of cobalt-salen PEDOT [276] or iron-containing polysalen [245], or for oxidation of ascorbic acid in the case of osmium bipyridyl functionalized PPy [277].

ECPs functionalized by redox mediators like ferrocene [278] or ferricyanide [279] have also proved efficient to catalyze electrooxidation of biologically active compounds.

18.3.4 Controlled Structure of ECPs

18.3.4.1 Regioregular ECPs

There has been a lot of interest to get well-controlled structures of ECPs in the recent years. One method consists in synthesizing regioregular polymers, usually by electrochemical oxidation of short oligomers, themselves obtained from the corresponding monomers by organometallic coupling. Several groups have performed systematic studies comparing the properties of the regioregular polymers with those of the parent random polymer. In that framework, Swager et al. showed that regioregular alternate copolymers of aniline (actually electropolymerized from a methoxy derivatized dimer of aniline) exhibited different electrochemical behaviors and higher conductivity values than the corresponding random copolymer [280] (Figure 18.11). Regioregular polymers lead to more crystalline structures, which can explain the differences in conductivity.

The same kind of systematic study was also performed in the case of polyalkylterthiophene [281], with comparison of the p-doping features of the regioregular vs. random polymer. The electrochemical results were interpreted in terms of the influence of the substituant position on the packing ability of the polymer chains. Another research group used electrochemical techniques (chronocoulometry

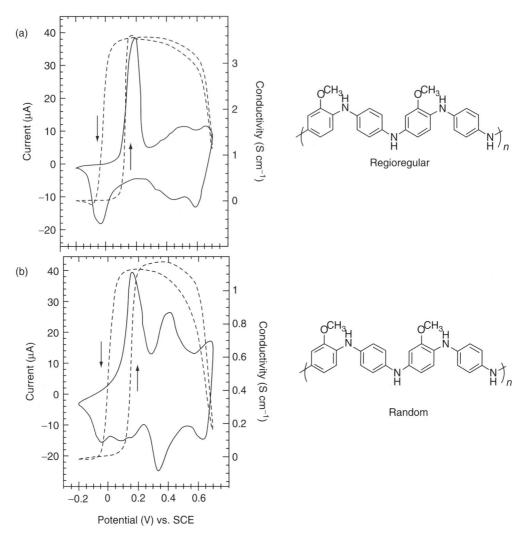

FIGURE 18.11 Comparison of cyclic voltammetry (full line) and in situ conductivities (dotted line) of (a) regioregular alternate copolymer of aniline derivative (b) with the random analogous copolymer. (Reprinted from Pullen, A.E., and T.M. Swager. *Macromolecules*, 34, 812–816, 2001. With permission.)

coupled to in situ conductivity measurement) to evaluate the carrier mobility in regioregular poly-octylthiophene (POT) [282–283]: the stacking ability of regioregular ECPs results in the formation of π-dimers, thus lowering the apparent carrier mobility at low doping levels. Regioregular polyalkylphe-nylene-alt-thienylene was also investigated by CV, showing both n-type and p-type doping [284].

Pron et al. proposed an alternate way to regioregular ECPs in order to obtain PANIs with high molecular order [285]: the use of plasticizers (e.g., octyl phtalate) results in polymers with high conductivity values (with a metallic behavior) and noticeably improved electromechanical properties.

18.3.4.2 Multidimensional ECPs

There are several ways to obtain multidimensional ECPs: one of them consists of the formation of ladder polymers, i.e., polymers formed by coupling conjugated monomers by two bonds rather than only one. The main example of such ladder polymer is the one derived from benzimidazobenzophenanthroline (BBL, Scheme 18.8) and we will focus on the electrochemical properties of this polymer, although other examples of ladder ECPs exist, among which are those based on fused thiophene ring monomers [286].

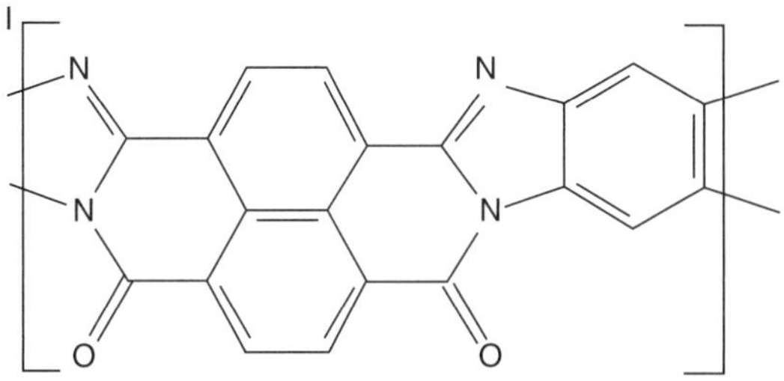

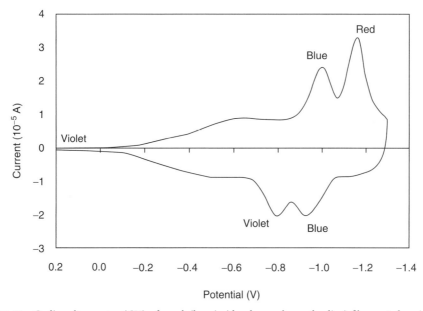

SCHEME 18.8 Formula of poly(benzimidazobenzophenanthroline).

Conjugated polymers derived from BBL present interesting features like n-type conductivity, photoluminescence and large nonlinear optical properties, which make them good candidates to be used in devices like light-emitting electrochemical cells [287]. They generally exhibit a four-step reduction with alternatively insulating and conductive states along the reduction process [288,289]. ESR experiments have allowed to assess that up to 2 electrons per ring were involved in the fully doped state resulting from the reduction process of the polymer [290]. The characteristic CV of a poly(BBL) film is shown in Figure 18.12, along with the various colors of the film depending on the redox state [291].

One can also obtain mutidimensional ECPs from actual multifunctional monomers, such as bispyrrolylazobenzene or bis(2,5-dithenylpyrrolyl)azobenzene [100]. The electropolymerization of monomers with two heterocyclic moieties (pyrrole or bisthienylpyrrole) connected through conjugated phenyl spacers gives rise to cross-linked structures [86], since it was demonstrated that both "heads" took part to the polymerization process [85]. Another example of cross-linked ECP can be electrochemically obtained from a chiral dicarbazole monomer [292].

Another interesting starting compound is the spirobifluorene, which contains two almost perpendicular electropolymerizable units [293], thus able to polymerize in two directions. Inserting a vinylogous tetrathiafulvalene (TTF) moiety into the spirobifluorenyl framework enables to design a new three-dimensional (3D) copolymer with the structure represented below [294] (Scheme 18.9); the copolymer exhibits both electroactivities due to oxidation of TTF and p-doping of polyfluorene.

Another method to obtain 3D ECPs consists in starting with a molecule designed with multiple electropolymerizable moieties connected to a central conjugated core, like bis-*N*-pyrrolyl phenylene

FIGURE 18.12 Cyclic voltammetry (CV) of a poly(benzimidazobenzophenanthroline) film coated on indium tin oxide (ITO) in acetonitrile. (Reprinted from Quinto, M., S.A. Jenekhe, and A.J. Bard. *Chem. Mater.*, 13, 2824–2832, 2001. With permission.)

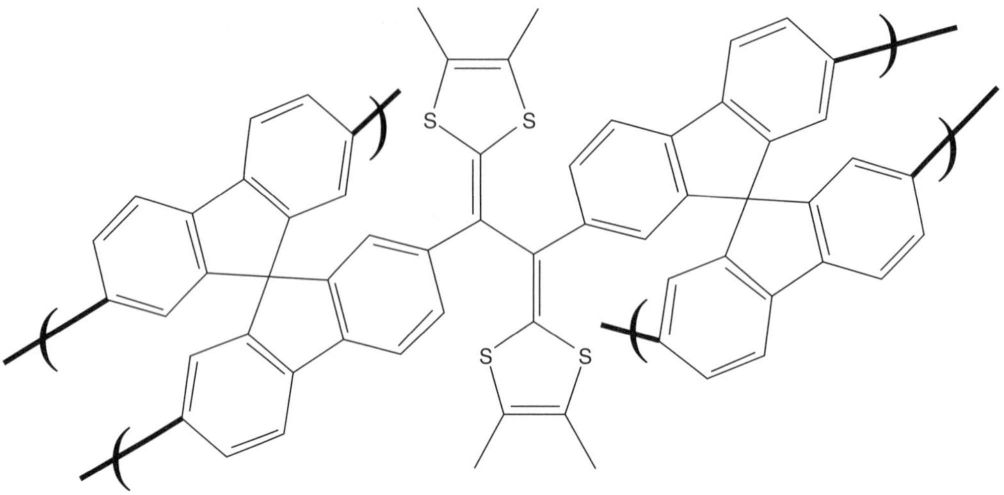

SCHEME 18.9   Structure of a three-dimensional copolymer of vinylogous tetrathiafulvalene (TTF) spirobifluorene. (Reprinted from Lorcy, D., L. Mattiello, C. Poriel, and J. Rault-Berthelot. *J. Electroanal. Chem.*, 530, 33–39, 2002. With permission.)

[295], bis-1,4(*N*-pyrrolylalkoxy)benzene [296], or thienylphenylamines [297]. In this case, an improvement in the electroactivity and conductivity has been obtained for the 3D network compared to the linear analogous ECP. 3D ECPs have also been obtained with electropolymerizing 2,2′-bithiophene with 1,3,5-(oligothienyl)benzene acting as cross-linking knots [298].

18.3.4.3   Superstructured ECPs

Superstructured assemblies of ECPs can easily be obtained through the template method. As electropolymerization of most conjugated monomers produces cationic species, the use of anionic template [299], such as DNA strands [300–301] and synthetic lipids [302–303], allows one to obtain well-defined morphologies of the corresponding polymer: e.g., helical superstructures of PEDOT have been obtained by electropolymerization in presence of lipid aggregate with amphiphilic character [303] (Figure 18.13).

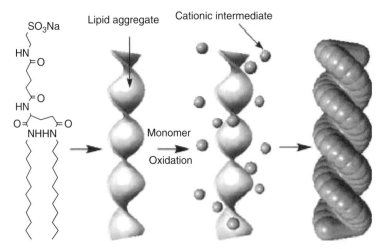

FIGURE 18.13   Obtention of helical superstructured PEDOT by electropolymerization in presence of amphiphilic lipid acting as a template. (Reprinted from Hatano, T., A.H. Bae, M. Takeuchi, N. Fujita, K. Kaneko, H. Ihara, M. Takafuji, and S. Shinkai. *Angew. Chem. Int. Ed.*, 43, 465–469, 2004. With permission.)

The use of plasmid DNA as a template during electropolymerization of pyrrole leads also to nanostructured rod-like, circular or supercoiled structures for the corresponding PPy [300].

Another interesting example of well-structured 3D ECPs is provided by Bartlett et al., who used PS latex beads as colloidal templates to electrosynthesize highly ordered macroporous PPy, PANI, and poly(bithiophene) [304].

Multilayers materials can be designed by the layer-by-layer (LbL) method using electrostatic interactions between an ECP (e.g., PANI) and a polyelectrolyte (e.g., poystyrenesulfonate) [305]. The growth can be followed by several methods, among which CV can be used to estimate the variation of diffusion coefficient for PANI redox switching as a function of the layer number. Several examples of mutilayers of this kind can be found in the literature [306–308].

Inorganic materials can also be used to generate well-structured ECPs, and electrochemical techniques can be used both to synthesize and to characterize the related material. One example of this is given by PANI electrosynthesized in porous aluminium oxide matrix, leading to highly ordered nanofibril arrays [309]. In the same idea, it is worth mentioning that the electrochemical behavior of PTh grown by a template electrosynthesis in a silica sol–gel matrix is rather different from the one of a classical PTh [310], reflecting the higher molecular order induced by the template method. Hybrid composites with better electrochemical performance can thus be obtained by electropolymerizing the organic part into the heterogeneous porosity of a sol–gel matrix [311,312].

At last, one can mention the possible design of ECP microcontainers by electropolymerization around the wall of gas bubble templates stabilized by surfactants [313]. The electrochemical potential at which hydrogen bubbles are generated directly influence the morphological parameters of the final material.

Apart from the template method, the use of chiral dopant like camphorsulfonic acid can also lead to superstructured materials by electropolymerization [314,315]. For example, aligned microribbons of ECP have been electropolymerized from alkylpyrrole derivatives in acetonitrile containing one enantiomer of camphorsulfonic acid [316].

18.4 Electrochemistry of Related Materials

It is somewhat challenging to write down this last part of the chapter, because up to now ECPs have been incorporated in so many devices and composites that the limit of such a scope is not easy to appreciate. Therefore this last part, and especially the very last one on biological materials will aim at giving an insight into the more recent and original features involving some electrochemistry than provide extensive data on all the work done in the field.

18.4.1 Electroactive Nanoobjects and Nanostructures Related with Conducting Polymers

Of course there are now many examples of nanostructuration of conducting polymers or related materials, as thoroughly quoted above (or in other chapters of this book). We would like to mainly limit ourselves to the ones where electrochemistry is used for the synthesis, or at least plays an important role in the subsequent applications of the materials. Among the most widespread nanocomposites are ECPs incorporating nanoparticles of metals [56,233,317] or oxides in view of catalysis [318], especially in fuel cells [319–322]. Oxide nanoparticles coated with ECPs have also been prepared [323], and form now an independent branch among the composites prepared from the growth of conducting polymers into sol–gel derived materials [312,320,324–326], and also polyoxometallates [327]. The very promising applications of anticorrosion coatings [40,312,328,329] play a large role in the triggering of this research field.

Microstructuration of ECPs by electropolymerization into channels or other polymer networks is still an active area of research [26,330–338], although this kind of work was initiated a long time ago. Nowadays some workers prefer to explore the nanostructuration of conducting polymers without the addition of template [339], or elegantly use epitaxy to promote the electrochemical building of polymer nanowires [340]. However, despite the continuing work in this area, little applications to electrocatalysis are still investigated [341–343], at the exception of methanol electrooxidation [342,344,345].

18.4.2 Polymer Blends, Interpenetrated Networks and Related Electromechanical Devices

Among materials derived from conducting polymers, blends with other polymers play a particularly important role. Of course here we will only focus on blends involving an electrochemical preparation, or further applications. Many composites of this type have been prepared, especially by the group of De Paoli, who recently reviewed the field [346]; other groups also prepared PPy blends with PVA [347], acrylonitrile [348], polyimide [349] or PANI with acrylic resin [350]. Despite the easy synthesis of most of these blends, probably the best conducting polymers of this type remain the PPy/Nafion composites, where it has been shown that the Nafion helps to nanostructure the PPy during the electrochemical growth; therefore unusually high conductivities (up to 500 S cm^{-1}) could be reached [351,352].

However, besides the conducting character, the main use of these composite systems is the realization of actuators. Since this will make the topic of a separate chapter we will only recall the work of Otero [162,170,353–356] and Inganas [357], and signal the emergence of interpenetrated polymer networks [358]. This concept consists in growing a conducting polymer into a nonconducting polymer network with mechanical appropriate properties, and be able in a following step to cross-link both polymers so that they become intrinsically linked together at a nanoscale level. This has been developed recently by Chevrot and has lead to actuators with an especially long working life [359,360].

18.4.3 Biological Applications and Bioelectrocatalysis

The implication of conducting polymers derived materials into biological applications and bioanalytical devices is an intensively growing field of research over the last 10 years. The systems studied can be divided into the analytical ones (sensing devices) and the electrocatalytic ones (test laboratory plants). The sensor field has been much more explored up to now and has been reviewed recently [361].

18.4.3.1 Sensing Devices Using Conducting Polymers and Enzymes

The main challenge in the field of electrochemical sensors and devices is the use of the conductivity of the polymer matrix to address electronically the active centre. For example, a major issue lies in the devices based on the biotine–avidine interaction. Of course it is possible to operate with a monolayer, but this diminishes the efficiency of the sensor. However, if the material between the sensing part and the electrode is a conducting polymer, the electron transfer occurs usually more efficiently and allows to operate with multilayered systems, as it has been demonstrated by Cosnier et al. [362–364] (Scheme 18.10).

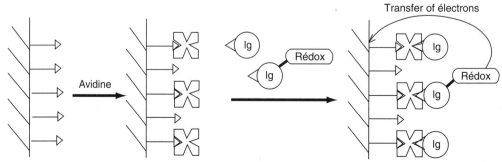

SCHEME 18.10 Bioelectrochemical sensor using the biotine–avidine recognition and an ECP matrix as a mediator for the electron transfer.

Many enzymatic displays based upon ECPs have been published in the last decade. The most widely studied system is for sure the glucose oxidase (GOD) one. This stems from the fact that this enzyme is naturally robust, cheap, and that glucose or more generally sugar analysis is still a field of active research. Several research teams have developed ECP-GOD devices, using PPy [365–370], PTh [371], PANI [368], or other derivatives [372,373]. An "enzyme" PPy transistor has even been developed [374].

Horseradish peroxydase, another very useful enzyme, is also more and more used in combination with conducting polymers [105,375]. Some other enzymes, like urease [376,377], uricase [378], or tyrosinase [379] have also been incorporated in ECPs. Interestingly, in the case of the incorporation of quinohemoprotein alcohol dehydrogenase into PPy, evidence of electron transfer through the film has been obtained [380]. Octylgallate in food has been analyzed using an electrode modified with a PPy-phtalocyanine polymer. Finally, an expanding field is the immobilization of DNA with the help of an electrodeposited conducting polymer. Despite this approach has mainly been applied to optical (fluorescence) sensing, electrochemical sensors have also been prepared [381–384]. Piro in particular has studied the entrapment of DNA and successfully investigated the hybridization of included DNA strands [385,386] by electrochemistry. Korri-Youssoufi has prepared very accurate DNA hybridization sensors with modified PPy, using either the polymer response [387], or the one of covalently grafted ferrocene inside the film [388]. A very recent work not only describes DNA immobilization, but also its use in the electrochemical detection of dopamine and epinephrine without ascorbic acid interference [389]. Finally, DNA biochips were currently developed by the Livache's group [390], although the use of electrochemistry is more limited in the case of this research.

Another critical issue is drug delivery. Conducting polymers can achieve drug delivery, by switching from a neutral to a cationc state. This has been investigated by Reynolds and Pernaut [391], as well as by other authors [392]. Immunosensors featuring antibodies included into PPy have also been described [393,394].

18.4.3.2　Electrooxidation of Biomolecules on ECP Modified Electrodes and Functional Systems

While ECPs are, at few exceptions [395], usually not used for reduction reactions, on the other hand a few compounds of biological interest have been oxidized on modified conducting polymers, among which, ascorbic acid [396] and of course glucose (see Section 18.4.3.1). A PPy-heparin composite has been applied to the separation of thrombin [397].

18.5　Conclusion

The electrochemistry of conducting polymers remains an active research field, although it does not have the impact it had at the beginning of the conducting polymer history. While electrochemical methods of synthesis are slightly regressing in contrast to chemical methods since organometallic coupling has become very popular, they do not decrease in absolute number, thus showing the vitality of the interest in the field, in particular because of their simplicity of handling to create well-defined structured polymers in template synthesis, for example. Since applications involving electrochemistry are still numerous, it remains a very valuable tool to investigate the properties of conducting polymers, often by coupling with other techniques like spectroscopy or microscopy. In addition, even if the fundamental processes appear now to be better understood and thus generate much less papers than in the 1980s or 1990s, original new synthetic methods are currently being developed, e.g., with the use of ionic liquids for electrochemical synthesis that appears very promising, especially to improve the mechanical properties of the polymers.

It should be also emphasized that the recent applications involving bioelectrochemistry increase rapidly and steadily, while those for corrosion protection, although still in the infancy, seem promised to a brilliant future, especially when the ECPs are associated to other materials like sol–gel coatings. This latter field should probably open on extremely promising applications, especially in the field of the protection of very oxidizable metals like aluminum.

Therefore we hope that the research field of electrochemistry of conducting polymers will continue to attract young scientist and specialist interest all over the world.

References

1. Heinze, J. 2000. *Organic electrochemistry*, eds. H. Lund, and O. Hammerich. New York: Marcel Dekker.
2. Diaz, A.F., K.K. Kanazawa, and G.P. Gardini. 1979. *J Chem Soc Chem Commun* 635.
3. Tourillon, G., and F. Garnier. 1982. *J Electroanal Chem* 135:173.
4. Diaz, A.F., and J.A. Logan. 1980. *J Electroanal Chem* 111:111.
5. Géniès, E.M., G. Bidan, and A.F. Diaz. 1983. *J Electroanal Chem* 149:101–113.
6. Andrieux, C.P., P. Audebert, P. Hapiot, and J.M. Savéant. 1991. *J Phys Chem* 95:10158.
7. Lacroix, J.C., F. Maurel, and P.C. Lacaze. 2001. *J Am Chem Soc* 123:1989–1996.
8. McCullough, R.D., R.D. Lowe, M. Jayaraman, and D.L. Anderson. *J Org Chem* 58:904–912.
9. Diaz, A.F., J.I. Castillo, J.A. Logan, and W.Y. Lee. 1981. *J Electroanal Chem* 129:115–132.
10. Feldberg, S.W. 1984. *J Am Chem Soc* 106:4671–4674.
11. Tanguy, J., and N. Mermillod. 1987. *Synth Met* 21:129–134.
12. Cai, Z., and C.R. Martin. 1991. *J Electroanal Chem* 300:35–50.
13. Ren, F., and P.G. Pickup. 1993. *J Phys Chem* 97:5356–5362.
14. Bauerle, P., F. Wurthner, G. Gotz, and F. Effenberger. 1993. *Synthesis* 1099–1103.
15. Welton, T. 1999. *Chem Rev* 99:2071.
16. Wasserscheid, P.W.K. 2000. *Angew Chem* 112:3926.
17. Sheldon, R. 2001. *Chem Commun* 2399.
18. Dupont, J., R.F. de Souza, and P.A.Z. Suarez. 2002. *Chem Rev* 102:3667.
19. Lu, W., A.G. Fadeev, B.H. Qi, E. Smela, B.R. Mattes, J. Ding, G.M. Spinks, J. Mazurkiewicz, D.Z. Zhou, G.G. Wallace, D.R. MacFarlane, S.A. Forsyth, and M. Forsyth. 2002. *Science* 297:983–987.
20. Lu, W., A.G. Fadeev, B. Qi, and B.R. Mattes. 2003. *Synth Met* 135:139–140.
21. Lu, W., and B.R. Mattes. 2003. *J Electrochem Soc* 150:E416–E422.
22. Pringle, J.M., M. Forsyth, D.R. MacFarlane, K. Wagner, S.B. Hall, and D.L. Officer. 2005. *Polymer* 46:2047–2058.
23. Pringle, J.M., J. Efthimiadis, P.C. Howlett, D.R. MacFarlane, A.B. Chaplin, S.B. Hall, D.L. Officer, G.G. Wallace, and M. Forsyth. 2004. *Polymer* 45:1447–1453.
24. Borgwarth, K., N. Rohde, C. Ricken, M.L. Hallensleben, D. Mandler, and J. Heinze. 1999. *Adv Mater* 11:1221–1226.
25. Marck, C., K. Borgwarth, and J. Heinze. 2001. *Chem Mater* 13:747–752.
26. Marck, C., K. Borgwarth, and J. Heinze. 2001. *Adv Mater* 13:47.
27. Lima, A., P. Shottland, S. Sadki, and C. Chevrot. 1998. *Synth Met* 93:33.
28. Sadki, S., and C. Chevrot. 2003. *Electrochim Acta* 48:733.
29. Chane-Ching, K.I., J.C. Lacroix, M. Jouini, and P.C. Lacaze. 1999. *J Mater Chem* 9:1379.
30. Lagrost, C., M. Jouini, J. Tanguy, S. Aeiyach, J.C. Lacroix, K.I. Chane-Ching, and P.C. Lacaze. 2001. *Electrochim Acta* 46:3985–3992.
31. Petitjean, J., S. Aeiyach, J.C. Lacroix, P.C. Lacaze, K.I. Chane-Ching, E. Cheauveau, and J.L. Camalet. 2000. *J Electroanal Chem* 481:3.
32. Petitjean, J., S. Aeiyach, J.C. Lacroix, and P.C. Lacaze. 1999. *J Electroanal Chem* 478:92–100.
33. Camalet, J.L., S. Aeiyach, J.C. Lacroix, and P.C. Lacaze. 1998. *J Electroanal Chem* 445:117.
34. Le, H.N.T., B. Garcia, C. Deslouis, and Q. Le Xuan. 2001. *Electrochim Acta* 46:4259–4272.
35. Rammelt, U., P.T. Nguyen, and W. Plieth. 2003. *Electrochim Acta* 48:1257–1262.
36. Ogurtsov, N.A., A.A. Pud, P. Kamarchik, and G.S. Shapoval. 2004. *Synth Met* 143:43–47.
37. Kilmartin, P.A., L. Trier, and G.A. Wright. 2002. *Synth Met* 131:99–109.
38. Nguyen, T.D., M.C. Pham, B. Piro, J. Aubard, H. Takenouti, and M. Keddam. 2004. *J Electrochem Soc* 151:B325–B330.

39. Tallman, D.E., Y. Pae, and G.P. Bierwagen. 1999. *Corrosion* 55:779–786.

40. Rout, T.K., G. Jha, A.K. Singh, N. Bandyopadhyay, and O.N. Mohanty. 2003. *Surf Coat Technol* 167:16–24.

41. Kraljic, M., Z. Mandic, and L. Duic. 2003. *Corrosion Sci* 45:181–198.

42. He, J., V.J. Gelling, D.E. Tallman, G.P. Bierwagen, and G.G. Wallace. 2000. *J Electrochem Soc* 147:3667–3672.

43. Tallman, D.E., Y. Pae, and G.P. Bierwagen. 2000. *Corrosion* 56:401–410.

44. He, J., D.E. Tallman, and G.P. Bierwagen. 2004. *J Electrochem Soc* 151:B644–B651.

45. Tan, C.K., and D.J. Blackwood. 2003. *Corrosion Sci* 45:545–557.

46. Nguyen, T.D., T.A. Nguyen, M.C. Pham, B. Piro, B. Normand, and H. Takenouti. 2004. *J Electroanal Chem* 572:225–234.

47. Nguyen, T.D., M. Keddam, and H. Takenouti. 2003. *Electrochem Solid State Lett* 2003, 6:B25–B28.

48. Refaey, S.A.M., G. Schwitzgebel, and O. Schneider. 1999. *Synth Met* 98:183–192.

49. Audebert, P., S. Roux, J. Pagetti, and M. Roche. 2001. *J Mater Chem* 11:3360.

50. Skotheim, T.A., R.L. Elsenbaumer, and J.R. Reynolds. 1998. *Handbook of conducting polymers*, 2nd ed. New York: Marcel Dekker.

51. Grose, J.E., A.N. Pasupathy, D.C. Ralph, B. Ulgut, and H.D. Abruna. 2005. *Phys Rev B* 71.

52. Tian, S.J., J.Y. Liu, T. Zhu, and W. Knoll. 2004. *Chem Mater* 16:4103–4108.

53. Kang, Y., E. Kim, and C. Lee. 2001. *Mol Cryst Liq Cryst* 371:87.

54. Kim, E., K.Y. Lee, M.-H. Lee, J.-S. Shin, and S. Bong Rhee. 1997. *Synth Met* 85:1367.

55. Kim, E., M. Lee, M.-H. Lee, and S. Bong Rhee. 1995. *Synth Met* 69:101.

56. Querner, C., P. Reiss, M. Zagorska, O. Renault, R. Payerne, F. Genoud, P. Rannou, and A. Pron. 2005. *J Mater Chem* 15:554–563.

57. Gaponik, N.P., D.V. Talapin, A.L. Rogach, and A. Eychmuller. 2000. *J Mater Chem* 10:2163–2166.

58. Girotto, E.M., N. Camaioni, G. Casalbore-Miceli, M.-A. De Paoli, A.M. Fichera, and M.C. Galazzi. 2001. *J Appl Electrochem* 31:335.

59. Zotti, G., S. Zecchin, G. Schiavon, B. Vercelli, and A. Berlin. 2005. *J Electroanal Chem* 575:169–175.

60. Liu, Y.C., and Y.S. Liang. 2003. *Electroanalysis* 15:200–207.

61. Groenendaal, L., G. Zotti, P.-H. Aubert, S.M. Waybright, and J.R. Reynolds. 2003. *Adv Mater* 15:855.

62. Reeves, B.D., C.R.G. Grenier, A.A. Argun, A. Cirpan, T.D. McCarley, and J.R. Reynolds. 2004. *Macromolecules* 37:7559–7569.

63. Sotzing, G.A., and K.H. Lee. 2002. *Macromolecules* 35:7281–7286.

64. Sotzing, G.A., J.L. Reddinger, A.R. Katritzky, J. Soloducho, R. Musgrave, and J.R. Reynolds. 1997. *Chem Mater* 9:1578–1587.

65. Sotzing, G.A., J.R. Reynolds, and P.J. Steel. 1997. *Adv Mater* 9:795.

66. Reddinger, J.L., and J.R. Reynolds. 1998. *Chem Mater* 10:1236–1243.

67. Tsuie, B., J.L. Reddinger, G.A. Sotzing, J. Soloducho, A.R. Katritzky, and J.R. Reynolds. 1999. *J Mater Chem* 9:2189–2200.

68. Larmat, F., J.R. Reynolds, B.A. Reinhardt, L.L. Brott, and S.J. Clarson. 1997. *J Polym Sci Part A Polym Chem* 35:3627–3636.

69. Zotti, G., R.A. Marin, and M.C. Gallazzi. 1997. *Chem Mater* 9:2945–2950.

70. Zotti, G., G. Schiavon, S. Zecchin, and A. Berlin. 1998. *Synth Met* 97:245–254.

71. Zotti, G., S. Zecchin, G. Schiavon, and A. Berlin. 2001. *J Electroanal Chem* 506:106–114.

72. Zotti, G., S. Zecchin, G. Schiavon, and A. Berlin. 2001. *Macromolecules* 34:3889–3895.

73. Zotti, G., S. Zecchin, G. Schiavon, A. Berlin, L. Huchet, and J. Roncali. 2001. *J Electroanal Chem* 504:64–70.

74. Zotti, G., S. Zecchin, G. Schiavon, A. Berlin, G. Pagani, and A. Canavesi. 1997. *Chem Mater* 9:2940–2944.

75. Zotti, G., S. Zecchin, G. Schiavon, B. Vercelli, and A. Berlin. 2005. *Electrochim Acta* 50:1469–1474.

76. Zotti, G., S. Zecchin, G. Schiavon, B. Vercelli, A. Berlin, and S. Grimoldi. 2004. *Macromol Chem Phys* 205:2026–2031.

77. Roncali, J., C., Thobie-Gautier, Brisset, H., J.-F., Favart, and A., Guy. 1995. *J Electroanal Chem* 381:257.

78. Roncali, J. 1999. *J Mater Chem* 9:1875–1893.

79. Blanchard, P., H. Brisset, A. Riou, R. Hierle, and J. Roncali. 1998. *J Org Chem* 63:8310–8319.

80. Jousselme, B., P. Blanchard, M. Ocafrain, M. Allain, E. Levillain, and J. Roncali. 2004. *J Mater Chem* 14:421–427.

81. Blanchard, P., H. Brisset, A. Riou, R. Hierle, and J. Roncali. 1998. *New J Chem* 547.

82. Blanchard, P., A. Riou, and J. Roncali. 1998. *J Org Chem* 63:7107.

83. Guillerez, S., M. Kalaji, F. Lafolet, and D.N. Tito. 2004. *J Electroanal Chem* 563:161–169.

84. Just, P.E., K.I. Chane-Ching, and P.C. Lacaze. 2002. *Tetrahedron* 58:3467–3472.

85. Just, P.E., K.I. Chane-Ching, J.C. Lacroix, and P.C. Lacaze. 1999. *J Electroanal Chem* 479:3–11.

86. Just, P.E., K.I. Chane-Ching, J.C. Lacroix, and P.C. Lacaze. 2001. *Electrochim Acta* 46:3279–3285.

87. Chane-Ching, K.I., J.C. Lacroix, R. Baudry, M. Jouini, S. Aeiyach, C. Lion, and P.C. Lacaze. 1998. *J Electroanal Chem* 453:139–149.

88. Mangeney, C., J.C. Lacroix, K.I. Chane-Ching, M. Jouini, F. Villain, S. Ammar, N. Jouini, and P.C. Lacaze. 2001. *Chem Eur J* 7:5029–5040.

89. Divisia-Blohorn, B., F. Gennoud, C. Borel, G. Bidan, J.M. Kern, and J.P. Sauvage. 2003. *J Phys Chem B* 107:5126.

90. Sauvage, J.P., J.M. Kern, G. Bidan, B. Divisia-Blohorn, and P.L. Vidal. 2002. *New J Chem* 26:1287–1290.

91. Vidal, P.L., B. Divisia-Blohorn, G. Bidan, J.L. Hazemann, J.M. Kern, and J.P. Sauvage. 2000. *Chem Eur J* 6:1663–1673.

92. Vidal, P.L., B. Divisia-Blohorn, G. Bidan, J.M. Kern, J.P. Sauvage, and J.L. Hazemann. 1999. *Inorg Chem* 38:4203–4210.

93. Zhu, S.S., and T.M. Swager. 1997. *J Am Chem Soc* 119:12568–12577.

94. Demadrille, R., B. Divisia-Blohorn, M. Zagorska, S. Quillard, S. Lefrant, and A. Pron. 2003. *New J Chem* 27:1479.

95. Demadrille, R., B. Divisia-Blohorn, M. Zagorska, S. Quillard, S. Lefrant, and A. Pron. 2005. *Electrochim Acta* 50.

96. Audebert, P., F. Miomandre, S.G. DiMagno, V.V. Smirnov, and P. Hapiot. 2000. *Chem Mater* 12:2025–2030.

97. Audebert, P., S. Sadki, F. Miomandre, and G. Clavier. 2004. *Electrochem Commun* 6:144.

98. Audebert, P., L. Guyard, M. N'Guyen Dinh An, P. Hapiot, C.P. Andrieux, E.W. Mejer, and L. Groenendaal. 1997. *Chem Mater* 9:723.

99. Aubert, P.H., A. Neudeck, L. Dunsch, P. Audebert, P. Capdevielle, and M. Maumy. 1999. *J Electroanal Chem* 470:77–88.

100. Audebert, P., S. Sadki, F. Miomandre, P. Hapiot, and K. Chane-Ching. 2003. *New J Chem* 27:798–804.

101. Audebert, P., L. Guyard, and F. Cherioux. 1998. *J Chem Soc Chem Commun* 2225.

102. Guyard, L., M. N'Guyen Dinh An, and P. Audebert. 2001. *Adv Mater* 13:133.

103. Lin, S.C., J.A. Chen, M.H. Liu, Y.O. Su, and M.K. Leung. 1998. *J Org Chem* 63:5059.

104. Divisia-Blohorn, B., F. Genoud, F. Salhi, and H. Muller. 2002. *J Phys Chem B* 106:6646–6651.

105. Mulchandani, A., and S.T. Pan. 1999. *Anal Biochem* 267:141–147.

106. Naudin, E., N. El Mehdi, C. Soucy, L. Breau, and D. Belanger. *Chem Mater* 13:634–642.

107. Naudin, E., H.A. Ho, M.A. Bonin, L. Breau, and D. Belanger. 2002. *Macromolecules* 35:4983–4987.

108. Neef, C.J., I.D. Brotherston, and J.P. Ferraris. 1999. *Chem Mater* 11:1957 + .

109. Neef, C.J., D.T. Glatzhofer, and K.M. Nicholas. 1997. *J Polym Sci Part A Polym Chem* 35:3365–3376.

110. Ranger, M., and M. Leclerc. 1999. *Macromolecules* 32:3306–3313.

111. Schaferling, M., and P. Bauerle. 2001. *Synth Met* 119:289–290.

112. Schaferling, M., and P. Bauerle. 2004. *J Mater Chem* 14:1132–1141.

113. Sezer, E., A.H.S. Sarac, and E.A. Parlak. 2003. *J Appl Electrochem* 33:1233–1237.

114. Tada, K., M. Onoda, and K. Yoshino. 1997. *J Phys D Appl Phys* 30:2063–2068.

115. Rault-Berthelot, J., and E. Raoult. 2000. *Adv Mater Opt Electron* 10:267–272.

116. Ribeiro, A.S., W.A. Gazotti, P.F. dos Santos, and M.A. De Paoli. 2004. *Synth Met* 145:43–49.

117. Jenekhe, S.A., L.D. Lu, and M.M. Alam. 2001. *Macromolecules* 34:7315–7324.

118. Ng, S.C., H.S.O. Chan, T.T. Ong, K. Kumura, Y. Mazaki, and K. Kobayashi. 1998. *Macromolecules* 31:1221–1228.

119. Plonska, M.E., A. de Bettencourt-Dias, A.L. Balch, and K. Winkler. 2003. *Chem Mater* 15:4122–4131.

120. MacLean, B.J., and P.G. Pickup. 1999. *Chem Commun* 2471–2472.

121. Audebert, P., F. Miomandre, S. Sadki, G. Clavier, M.-C. Vernières, M. Saoud, and P. Hapiot. 2004. *New J Chem* 28.

122. Deronzier, A., and J.-C. Moutet. 1989. *Acc Chem Res* 22:249.

123. Skompska, M., M.A. Vorotyntsev, J. Goux, C. Moise, O. Heinz, Y.S. Cohen, M.D. Levi, Y. Gofer, G. Salitra, and D. Aurbach. 2005. *Electrochim Acta* 50:1635–1641.

124. Vorotyntsev, M.A., M. Casalta, E. Pousson, L. Roullier, G. Boni, and C. Moise. 2001. *Electrochim Acta* 46:4017–4033.

125. Vorotyntsev, M.A., M. Skompska, E. Pousson, J. Goux, and C. Moise. 2003. *J Electroanal Chem* 552:307–317.

126. Tran-Van, F., T. Henri, and C. Chevrot. 2002. *Electrochim Acta* 47:2927.

127. Ng, S.C., P. Miao, and H.S.O. Chan. 1998. *Chem Commun* 153–154.

128. Marrec, P., C. Dano, N. Gueguen-Simonet, and J. Simonet. 1997. *Synth Met* 89:171–179.

129. Marrec, P., B. Fabre, and J. Simonet. 1997. *J Electroanal Chem* 437:245–253.

130. Marrec, P., and J. Simonet. 1998. *J Electroanal Chem* 459:35–44.

131. Sung, H.H., and H.C. Lin. 2004. *Macromolecules* 37:7945–7954.

132. Watson, K.J., P.S. Wolfe, S.T. Nguyen, J. Zhu, and C.A. Mirkin. 2000. *Macromolecules* 33:4628–4633.

133. Audebert, P., C.P. Andrieux, P. Hapiot, and J.M. Savéant. 1990. *J Am Chem Soc* 112:2439.

134. Audebert, P., and P. Hapiot. 1995. *Synth Met* 75:129.

135. Audebert, P., J.M. Catel, G. Le Coustumer, V. Duchenet, and P. Hapiot. 1998. *J Phys Chem B* 102:8661–8669.

136. Miomandre, F., P. Audebert, K. Zong, and J.R. Reynolds. 2003. *Langmuir* 19:8894.

137. Sadki, S., P. Schottland, N. Brodie, and G. Sabouraud. 2000. *Chem Soc Rev* 29:283.

138. Lacroix, J.C., R.J. Valente, F. Maurel, and P.C. Lacaze. 1998. *Chem Eur J* 4:1667–1677.

139. Heinze, J., H. John, M. Dietrich, and P. Tschuncky. 2001. *Synth Met* 119:49–52.

140. Heinze, J., P. Tschuncky, and A. Smie. 1998. *J Solid State Electrochem* 2:102–109.

141. Pagels, M., J. Heinze, B. Geschke, and V. Rang. 2001. *Electrochim Acta* 46:3943–3954.

142. Zhou, M., and J. Heinze. 1999. *J Phys Chem B* 103:8443–8450.

143. Zhou, M., and J. Heinze. 1999. *Electrochim Acta* 44:1733–1748.

144. Zhou, M., M. Pagels, B. Geschke, and J. Heinze. 2002. *J Phys Chem B* 106:10065–10073.

145. Huang, Z.M., L.T. Qu, G.Q. Shi, F. Chen, and X.Y. Hong. 2003. *J Electroanal Chem* 556:159–165.

146. Xu, J.K., G.Q. Shi, F.E. Chen, and X.Y. Hong. 2002. *Chin J Polym Sci* 20:425–430.

147. Xu, J.K., G.Q. Shi, F.G. Chen, F. Wang, J.X. Zhang, and X.Y. Hong. 2003. *J Appl Polym Sci* 87:502–509.

148. Shi, G., S. Jin, and C. Li. 1995. *Science* 267:991.

149. Shi, G.Q., C. Li, and Y.Q. Liang. 1999. *Adv Mater* 11:1145+.

150. Alkan, S., C.A. Cutler, and J.R. Reynolds. 2003. *Adv Funct Mater* 13:331.

151. Innocenti, M., F. Loglio, L. Pigani, R. Seeber, F. Terzi, and R. Udisti. 2005. *Electrochim Acta* 50:1497–1503.

152. Abrantes, L.M., C.M. Cordas, and E. Vieil. 2003. *Electrochim Acta* 47:1481–1487.

153. Cai, X.W., J.S. Gao, Z.X. Xie, Y. Xie, Z.Q. Tian, and B.W. Mao. 1998. *Langmuir* 14:2508–2514.

154. Abrantes, L.M., and J.P. Correia. 1999. *Electrochim Acta* 44:1901–1910.

155. Randriamahazaka, H., V. Noel, and C. Chevrot. 1999. *J Electroanal Chem* 472:103–111.

156. Otero, T.F., and I. Boyano. 2003. *J Phys Chem B* 107:6730–6738.

157. Otero, T.F., and I. Boyano. 2003. *Chemphyschem* 4:868–872.

158. Ge, D., J. Wang, S. Wang, J. Ma, and B. He. 2003. *J Mat Sci Lett* 22:839.

159. Barzukov, V.Z., V.G. Khomenko, S.V. Chivikov, I.V. Barzukov, and T.I. Motronyuk. 2001. *Electro-chim Acta* 46:4083–4094.

160. Vorotyntsev, M.A., and J. Heinze. 2001. *Electrochim Acta* 46:3309–3324.

161. Malinauskas, A., and R. Holze. 1997. *Berichte Der Bunsen Gesellschaft Phys Chem Chem Phys* 101:1851–1858.

162. Otero, T.F., H. Grande, and J. Rodriguez. 1997. *J Phys Chem B* 101:8525–8533.

163. Aoki, K., T. Edo, and J. Cao. 1998. *Electrochim Acta* 43:285–289.

164. Noel, V., H. Randriamahazaka, and C. Chevrot. 2003. *J Electroanal Chem* 542:33–38.

165. Malta, M., E.R. Gonzalez, and R.M. Torresi. 2002. *Polym* 43:5895–5901.

166. Okamoto, H., and T. Kotaka. 1999. *Polym* 40:407–417.

167. Presa, M.J.R., D. Posadas, and M.I. Florit. 2000. *J Electroanal Chem* 482:117–124.

168. Abou-Elenien, G.M., A.A. El-Maghraby, and G.M. El-Abdallah. 2004. *Synth Met* 146:109–119.

169. Mazeikiene, R., and A. Malinauskas. 2002. *Synth Met* 129:61–66.

170. Otero, T.F., H. Grande, and J. Rodriguez. 1997. *Synth Met* 85:1077–1078.

171. Grande, H., and T.F. Otero. 1999. *Electrochim Acta* 44:1893–1900.

172. Otero, T.F., and I. Boyano. 2003. *J Phys Chem B* 107:6730–6738.

173. Lapkowski, M., and A. Pron. 2000. *Synth Met* 110:79–83.

174. Tezuka, Y., K. Aoki, H. Yajima, and T. Ishii. 1997. *J Electroanal Chem* 425:167–172.

175. Aoki, K., and Y. Teragishi. 1998. *J Electroanal Chem* 441:25–31.

176. Malek, K. 2003. *Chem Phys Lett* 375:477–483.

177. Lacroix, J.C., K. Fraoua, and P.C. Lacaze. 1998. *J Electroanal Chem* 444:83–93.

178. Miomandre, F., M.N. Bussac, E. Vieil, and L. Zuppiroli. 2000. *Chem Phys* 255:291–300.

179. Ybarra, G., C. Moina, M.I. Florit, and D. Posadas. 2000. *Electrochem Solid State Lett* 3:330–332.

180. Lizarraga, L., E.M. Andrade, and F.V. Molina. 2004. *J Electroanal Chem* 561:127–135.

181. Cascales, J.J.L., and T.F. Otero. 2004. *J Chem Phys* 120:1951–1957.

182. Krivan, E., C. Visy, and J. Kankare. 2005. *Electrochim Acta* 50:1247–1254.

183. Refaey, S.A.M. 2004. *Synth Met* 140:87–94.

184. Randriamahazaka, H., V. Noel, and C. Chevrot. 2002. *J Electroanal Chem* 521:107–116.

185. Randriamahazaka, H., V. Noel, and C. Chevrot. 2003. *J Electroanal Chem* 556:35–42.

186. Neves, S., and C.P. Fonseca. 2001. *Electrochem Commun* 3:36–43.

187. Bauermann, L.P., and P.N. Bartlett. 2005. *Electrochim Acta* 50:1537–1546.

188. Weidlich, C., K.M. Mangold, and K. Jüttner. 2005. *Electrochim Acta* 50:1547–1552.

189. Bruckenstein, S., K. Brzezinska, and A.R. Hillman. 2000. *Phys Chem Chem Phys* 2:1221–1229.

190. Vieil, E., and C. Lopez. 1999. *J Electroanal Chem* 466:218–233.

191. Vilas-Boas, M., M.J. Henderson, C. Freire, A.R. Hillman, and E. Vieil. 2000. *Chem A Eur J* 6:1160–1167.

192. Henderson, M.J., H. French, A.R. Hillman, and E. Vieil. 1999. *Electrochem Solid State Lett* 2:631–633.

193. Inzelt, G., V. Kertesz, and A.S. Nybäck. 1999. *J Solid State Electrochem* 3:251–257.

194. Salzer, C.A., and C.M. Elliot. 2000. *Chem Mater* 12:2099–2105.

195. Niu, L., C. Kvarnström, and A. Ivaska. 2004. *J Electroanal Chem* 569:151–160.

196. Vorotyntsev, M.A., E. Vieil, and J. Heinze. 1998. *J Electroanal Chem* 450:121–141.

197. Brown, M.J., A.R. Hillman, S.J. Martin, R.W. Cernosek, and H.L. Bandey. 2000. *J Mater Chem* 10:115–126.

198. Zotti, G. 1998. *Synth Met* 97:267–272.

199. Levi, M.D., Y. Gofer, D. Aurbach, M. Lapkowski, E. Vieil, and J. Serose. 2000. *J Electrochem Soc* 147:1096–1104.

200. Meerholz, K., H. Gregorious, K. Müllen, and J. Heinze. 1994. *Adv Mater* 6:671.
201. Zotti, G., S. Zecchin, G. Schiavon, B. Vercelli, A. Berlin, E. Dalcanale, and L. Groenendaal. 2003. *Chem Mater* 15:4642–4650.
202. Skompska, M., K. Miecnikowski, R. Holze, and J. Heinze. 2005. *J Electroanal Chem* 577:9–17.
203. Yang, R.Q., R.Y. Tian, J.G. Yan, Y. Zhang, J. Yang, Q. Hou, W. Yang, C. Zhang, and Y. Cao. 2005. *Macromolecules* 38:244–253.
204. Sivakkumar, S.R., and R. Saraswathi. 2004. *J Power Sources* 137:322–328.
205. Johansson, T., W. Mammo, M. Svensson, M.R. Andersson, and O. Inganas. 2003. *J Mater Chem* 13:1316–1323.
206. Roncali, J. 1997. *Chem Rev* 97:173–205.
207. Salzner, U. 2002. *J Phys Chem B* 106:9214–9220.
208. Meng, H., and F. Wudl. 2001. *Macromolecules* 34:1810–1816.
209. Hagan, A.J., S.C. Moratti, and I.C. Sage. 2001. *Synth Met* 119:147–148.
210. Jones, C.L., S.J. Higgins, and P.A. Christensen. 2002. *J Mater Chem* 12:758–764.
211. Kisselev, R., and M. Thelakkat. 2004. *Macromolecules* 37:8951–8958.
212. Hung, T.T., and S.A. Chen. 1999. *Polymer* 40:3881–3884.
213. Loganathan, K., E.G. Cammisa, B.D. Myron, and P.G. Pickup. 2003. *Chem Mater* 15:1918–1923.
214. Huang, H., and P.G. Pickup. 1998. *Chem Mater* 10:2212–2216.
215. Benincori, T., S. Rizzo, F. Sannicolo, G. Schiavon, S. Zecchin, and G. Zotti. 2003. *Macromolecules* 36:5114–5118.
216. Havinga, E.E., W. Ten Hoeve, and H. Wynberg. 1993. *Synth Met* 55:299–306.
217. Salzner, U., and M.E. Kose. 2002. *J Phys Chem B* 106:9221–9226.
218. Du Bois, C.J., F. Larmat, D.J. Irvin, and J.R. Reynolds. 2001. *Synth Met* 119:321–322.
219. Jayakannan, M., P.A. Van Hal, and R.A.J. Janssen. 2002. *J Polym Sci Part a-Polym Chem* 40:2360–2372.
220. Jayakannan, M., P.A. Van Hal, and R.A.J. Janssen. 2002. *J Polym Sci Part a-Polym Chem* 40: 251–261.
221. Schlick, U., F. Teichert, and M. Hanack. 1998. *Synth Met* 92:75–85.
222. Seshadri, V., and G.A. Sotzing. 2004. *Chem Mater* 16:5644–5649.
223. Fusalba, F., H.A. Ho, L. Breau, and D. Belanger. 2000. *Chem Mater* 12:2581–2589.
224. Ferraris, J.P., and J. Lambert. 1991. *J Chem Soc Chem Commun* 11:752–754.
225. Ferraris, J.P., and J. Lambert. 1991. *J Chem Soc Chem Commun* 18:1268–1270.
226. Berlin, A., G. Zotti, S. Zecchin, G. Schiavon, B. Vercelli, and A. Zanelli. 2004. *Chem Mater* 16: 3667–3676.
227. Audebert, P., S. Sadki, F. Miomandre, and G. Clavier. 2004. *Electrochem Commun* 6:144–147.
228. Kean, C.L., and P.G. Pickup. 2001. *Chem Commun* 815–816.
229. Rault-Berthelot, J., and E. Raoult. 2001. *Synth Met* 123:101–105.
230. Temsamani, K.R., H.B. Mark, W. Kutner, and A.M. Stalcup. 2002. *J Solid State Electrochem* 6: 391–395.
231. Yu, H., B. Xu, and T.M. Swager. 2003. *J Am Chem Soc* 125:1142–1143.
232. Marawi, I., A. Khaskelis, A. Galal, J.F. Rubinson, R.P. Popat, F.J. Boerio, and H.B. Mark. 1997. *J Electroanal Chem* 434:61–68.
233. Hwang, B.J., R. Santhanam, and Y.L. Lin. 2003. *Electroanalysis* 15:1667–1676.
234. Ballarin, B., M. Facchini, M. Lanzi, L. Paganin, and C. Zanardi. 2003. *J Electroanal Chem* 553: 97–106.
235. Thompson, B.C., K.A. Abboud, J.R. Reynolds, K. Nakatani, and P. Audebert. 2005. *New J Chem* 29:1–7.
236. Ding, A.L., J. Pei, Y.H. Lai, and W. Huang. 2001. *J Mater Chem* 11:3082–3086.
237. Li, G.T., G. Kossmehl, H.P. Welzel, W. Plieth, and H.S. Zhu. 1998. *Macromol Chem Phys* 199: 2737–2746.
238. Guyard, L., and P. Audebert. 2001. *Electrochem Commun* 3:164–167.

239. Burrell, A.K., J. Chen, G.E. Collis, D.K. Grant, D.L. Officer, C.O. Too, and G.G. Wallace. 2003. *Synth Met* 135:97–98.

240. Ferraris, J.P., A. Yassar, D.C. Loveday, and M. Hmyene. 1998. *Opt Mater* 9:34–42.

241. Buga, K., K. Kepczynska, I. Kulszewicz-Bajer, M. Zagroska, R. Demadrille, A. Pron, S. Quillard, and S. Lefrant. 2004. *Macromolecules* 37:769–777.

242. Schweiger, L.F., K.S. Ryder, D.G. Morris, A. Glidle, and J.M. Cooper. 2000. *J Mater Chem* 10: 107–114.

243. Dai, J., J.L. Sellers, and R.E. Noftle. 2003. *Synth Met* 139:81–88.

244. Bello, A., M. Giannetto, and G. Mori. 2005. *J Electroanal Chem* 575:257–266.

245. Miomandre, F., P. Audebert, M. Maumy, and L. Uhl. 2001. *J Electroanal Chem* 516:66–72.

246. Ballarin, B., R. Seeber, L. Tassi, and D. Tonelli. 2000. *Synth Met* 114:279–285.

247. Higgins, S.J., H.L. Jones, M.K. McCart, and T.J. Pounds. 1997. *Chem Commun* 1907–1908.

248. Hjelm, J., R.W. Handel, A. Hagfeldt, E.C. Constable, C.E. Housecroft, and R.J. Forster. 2005. *Inorg Chem* 44:1073–1081.

249. Aranyos, V., A. Hagfeldt, H. Grennberg, and E. Figgemeier. 2004. *Polyhedron* 23:589–598.

250. Berlin, A., G. Zotti, S. Zecchin, and G. Schiavon. 2002. *Synth Met* 131:149–160.

251. Fabre, B., P. Marrec, and J. Simonet. 1998. *J Electrochem Soc* 145:4110–4119.

252. Fan, L.Z., S.F. Yang, and S.H. Yang. 2005. *Thin Solid Films* 483:95–101.

253. Higgins, S.J., T.J. Pounds, and P.A. Christensen. 2001. *J Mater Chem* 11:2253–2261.

254. Kulesza, P.J., K. Miecznikowski, M.A. Malik, M. Galkowski, M. Chojak, K. Caban, and A. Wieckowski. 2001. *Electrochim Acta* 46:4065–4073.

255. Han, C.C., W.D. Hseih, J.Y. Yeh, and S.P. Hong. 1999. *Chem Mater* 11:480–486.

256. Barbero, C., H.J. Salavagione, D.F. Acevedo, D.E. Grumelli, F. Garay, G.A. Planes, G.M. Morales, and M.C. Miras. 2004. *Electrochim Acta* 49:3671–3686.

257. Brisset, H., A.E. Navarro, C. Moustrou, I.F. Perepichka, and J. Roncali. 2004. *Electrochem Commun* 6:249–253.

258. Mangeney, C., P.E. Just, J.C. Lacroix, K.I.C. Ching, M. Jouini, S. Aeiyach, and P.C. Lacaze. 1999. *Synth Met* 102:1315–1316.

259. Jun, C., C.O. Too, G.G. Wallace, G.F. Swiegers, B.W. Skelton, and A.H. White. 2002. *Electrochim Acta* 47:4227–4238.

260. Chen, J., C.O. Too, G.G. Wallace, and G.F. Swiegers. 2004. *Electrochim Acta* 49:691–702.

261. Ion, A., I. Ion, J.C. Moutet, A. Pailleret, A. Popescu, E. Saint-Aman, E. Ungureanu, E. Siebert, and R. Ziessel. 1999. *Sens Actuat B Chem* 59:118–122.

262. Ion, A.C., J.C. Moutet, A. Pailleret, A. Popescu, E. St Aman, E. Siebert, and E.M. Ungureanu. 1999. *J Electroanal Chem* 464:24–30.

263. Reynes, O., J.C. Moutet, G. Royal, and E. Saint-Aman. 2004. *Electrochim Acta* 49:3727–3735.

264. Leclerc, M. 1999. *Adv Mater* 11:1491–1498.

265. Khomenko, V.G., V.Z. Barsukov, and A.S. Katashinskii. 2005. *Electrochim Acta* 50:1675–1683.

266. Taj, S., S. Sankarapapavinasam, and M.F. Ahmed. 2000. *J Appl Polym Sci* 77:112–115.

267. Kulesza, P.J., M. Chojak, K. Karnicka, K. Miecznikowski, B. Palys, A. Lewera, and A. Wieckowski. 2004. *Chem Mater* 16:4128–4134.

268. Coutanceau, C., M.J. Croissant, T. Napporn, and C. Lamy. 2000. *Electrochim Acta* 46:579–588.

269. Lai, E.K.W., P.D. Beattie, F.P. Orfino, E. Simon, and S. Holdcroft. 1999. *Electrochim Acta* 44:2559–2569.

270. Matsufuji, A., S. Nakazawa, and K. Yamamoto. 2001. *J Inorg Organomet Polym* 11:47–61.

271. Alonso-Vante, N., S. Cattarin, and M. Musiani. 2000. *J Electroanal Chem* 481:200–207.

272. Yamamoto, K., and D. Taneichi. 1999. *J Inorg Organomet Polym* 9:231–243.

273. Retamal, B.A., M.E. Vaschetto, and J.H. Zagal. 1997. *J Electroanal Chem* 431:1–5.

274. Bouzek, K., K.M. Mangold, and K. Juttner. 2000. *Electrochim Acta* 46:661–670.

275. Diab, N., J. Oni, A. Schulte, I. Radtke, A. Blochl, and W. Schuhmann. 2003. *Talanta* 61:43–51.

276. Kingsborough, R.P., and T.M. Swager. 2000. *Chem Mater* 12:872–874.

277. Foster, K., A. Allen, and T. McCormac. 2004. *J Electroanal Chem* 573:203–214.

278. Galal, A. 1998. *J Solid State Electrochem* 2:7–15.

279. Pournaghi-Azar, M.H., and R. Ojani. 2000. *J Solid State Electrochem* 4:75–79.

280. Pullen, A.E., and T.M. Swager. 2001. *Macromolecules* 34:812–816.

281. Dini, D., F. Decker, G. Zotti, G. Schiavon, S. Zecchin, F. Andreani, and E. Salatelli. 1999. *Chem Mater* 11:3484–3489.

282. Kunugi, Y., Y. Harima, K. Yamashita, N. Ohta, and S. Ito. 2000. *J Mater Chem* 10:2673–2677.

283. Kunugi, Y., Y. Harima, K. Yamashita, N. Ohta, and S. Ito. 2000. *Chem Lett* 260–261.

284. Ng, S.C., J.M. Xu, and H.S.O. Chan. 2000. *Macromolecules* 33:7349–7358.

285. Dufour, B., P. Rannou, P. Fedorko, D. Djurado, J.P. Travers, and A. Pron. 2001. *Chem Mater* 13:4032–4040.

286. Oyaizu, K., T. Iwasaki, Y. Tsukahara, and E. Tsuchida. 2004. *Macromolecules* 37:1257–1270.

287. Pachler, P., F.P. Wenzl, U. Scherf, and G. Leising. 2005. *J Phys Chem B* 109:6020–6024.

288. Yohannes, T., H. Neugebauer, S. Luzzati, M. Catellani, S.A. Jenekhe, and N.S. Sariciftci. 2000. *J Phys Chem B* 104:9430–9437.

289. Yohannes, T., H. Neugebauer, S.A. Jenekhe, and N.S. Sariciftci. 2001. *Synth Met* 116:241–245.

290. Zheng, T., F. Badrun, I.M. Brown, D.J. Leopold, and T.C. Sandreczki. 1999. *Synth Met* 107:39–45.

291. Quinto, M., S.A. Jenekhe, and A.J. Bard. 2001. *Chem Mater* 13:2824–2832.

292. Diamant, Y., E. Furmanovich, A. Landau, J.P. Lellouche, and A. Zaban. 2003. *Electrochim Acta* 48:507–512.

293. Rault-Berthelot, J., M.M. Granger, and L. Mattiello. 1998. *Synth Met* 97:211–215.

294. Lorcy, D., L. Mattiello, C. Poriel, and J. Rault-Berthelot. 2002. *J Electroanal Chem* 530:33–39.

295. Quinto, M., and A.J. Bard. 2001. *J Electroanal Chem* 498:67–74.

296. Ono, K., S. Yamada, M. Ohkita, K. Saito, S. Tanaka, and T. Hanaichi. 2003. *Chem Lett* 32:516–517.

297. Yamamoto, K., M. Higuchi, K. Uchida, and Y. Kojima. 2002. *Macromolecules* 35:5782–5788.

298. Cherioux, F., and L. Guyard. 2001. *Adv Funct Mater* 11:305–309.

299. Shinkai, S., M. Takeuchi, and A.H. Bae. 2005. *Supramol Chem* 17:181–186.

300. Bae, A.H., T. Hatano, M. Numata, M. Takeuchi, and S. Shinkai. 2004. *Chem Lett* 33:436–437.

301. Bae, A.H., T. Hatano, M. Numata, M. Takeuchi, and S. Shinkai. 2005. *Macromolecules* 38:1609–1615.

302. Hatano, T., A.H. Bae, M. Takeuchi, N. Fujita, K. Kaneko, H. Ihara, M. Takafuji, and S. Shinkai. 2004. *Chem Eur J* 10:5067–5075.

303. Hatano, T., A.H. Bae, M. Takeuchi, N. Fujita, K. Kaneko, H. Ihara, M. Takafuji, and S. Shinkai. 2004. *Angew Chem Int Ed* 43:465–469.

304. Bartlett, P.N., P.R. Birkin, M.A. Ghanem, and C.S. Toh. 2001. *J Mater Chem* 11:849–853.

305. Ram, M.K., M. Salerno, M. Adami, P. Faraci, and C. Nicolini. 1999. *Langmuir* 15:1252–1259.

306. Lukkari, J., M. Salomaki, A. Viinikanoja, T. Aaritalo, J. Paukkunen, N. Kocharova, and J. Kankare. 2001. *J Am Chem Soc* 123:6083–6091.

307. Cutler, C.A., M.B. Bouguettaya, and J.R. Reynolds. 2002. *Adv Mater* 14:684.

308. Cheung, J.H., W.B. Stockton, and M.F. Rubner. 1997. *Macromolecules* 30:2712.

309. Zhao, Y.C., M. Chen, T. Xu, and W.M. Liu. 2005. *Coll Surf A Phys Eng Aspects* 257–58:363–368.

310. Ballarin, B., M. Facchini, L. Dal Pozzo, and C. Martini. 2003. *Electrochem Commun* 5:625–631.

311. Neves, S., C.P. Fonseca, R.A. Zoppi, and S.I.C. de Torresi. 2001. *J Solid State Electrochem* 5:412–418.

312. Roux, S., P. Audebert, J. Pagetti, and M. Roche. 2002. *New J Chem* 26:298–304.

313. Bajpai, V., P.G. He, and L.M. Dai. 2004. *Adv Funct Mater* 14:145–151.

314. Yuan, J.Y., D.Q. Zhang, L.T. Qu, G.Q. Shi, and X.Y. Hong. 2004. *Polym International* 53:2125–2129.

315. Qu, L.T., G.Q. Shi, J.Y. Yuan, G.Y. Han, and F. Chen. 2004. *J Electroanal Chem* 561:149–156.

316. Han, G.Y., G.Q. Shi, J.Y. Yuan, and F. Chen. 2004. *J Mater Sci* 39:4451–4457.

317. Shimanouchi, T., S. Morita, H.S. Jung, Y. Sakurai, Y. Suzuki, and R. Kuboi. 2004. *Sens Mater* 16:255–265.

318. Guney, S., I. Becerik, and F. Kadirgan. 2004. *Bull Electrochem* 20:157–163.
319. Choi, J.H., K.W. Park, H.K. Lee, Y.M. Kim, J.S. Lee, and Y.E. Sung. 2003. *Electrochim Acta* 48:2781–2789.
320. Pokhodenko, V.D., V.A. Krylov, Y.I. Kurys, and O.Y. Posudievsky. 1999. *Phys Chem Chem Phys* 1:905–908.
321. Porter, T.L., D. Thompson, M. Bradley, M.P. Eastman, M.E. Hagerman, J.L. Attuso, A.E. Votava, and E.D. Bain. 1997. *J Vacuum Sci Technol A Vacuum Surf Films* 15:500–504.
322. Cioffi, N., L. Torsi, I. Losito, L. Sabbatini, P.G. Zambonin, and T. Bleve-Zacheo. 2001. *Electrochim Acta* 46:4205–4211.
323. Roux, S., G.J.A.A. Soler-Ilia, S. Demoustier-Champagne, P. Audebert, and C. Sanchez. 2003. *Adv Mater* 15:217.
324. Roux, S., P. Audebert, J. Pagetti, and M. Roche. 2003. *J Sol–Gel Sci Technol* 26:435–439.
325. Roux, S., P. Audebert, J. Pagetti, and M. Roche. 2000. *New J Chem* 24:877 and 885.
326. Nakayama, M., J. Yano, K. Nakaoka, and K. Ogura. 2003. *Synth Met* 138:419–422.
327. Lira-Cantu, M., and P. Gomez-Romero. 1998. *Chem Mater* 10:698–704.
328. Cossement, D., F. Plumier, J. Delhalle, L. Hevesi, and Z. Mekhalif. 2003. *Synth Met* 138:529–536.
329. Akundy, G.S., R. Rajagopalan, and J.O. Iroh. 2002. *J Appl Polym Sci* 83:1970–1977.
330. Schultze, J.W., T. Morgenstern, D. Schattka, and S. Winkels. 1999. *Electrochim Acta* 44:1847–1864.
331. Yang, J.Y., and D.C. Martin. 2004. *Sens Actuat B Chem* 101:133–142.
332. Yang, J.Y., and D.C. Martin. 2004. *Sens Actuat A Phys* 113:204–211.
333. Ackermann, J., C. Videlot, T.N. Nguyen, L. Wang, P.M. Sarro, D. Crawley, K. Nikolic, and M. Forshaw. 2003. *Appl Surf Sci* 212:411–416.
334. Tao, Y.T., K. Pandian, and W.C. Lee. 1998. *Langmuir* 14:6158–6166.
335. Allard, D., S. Allard, M. Brehmer, L. Conrad, R. Zentel, C. Stromberg, and J.W. Schultze. 2003. *Electrochim Acta* 48:3137–3146.
336. Mazur, M., M. Tagowska, B. Palys, and K. Jackowska. 2003. *Electrochem Commun* 5:403–407.
337. Subianto, S., G.D. Will, and S. Kokot. 2003. *J Polym Sci Part A Polym Chem* 41:1867–1869.
338. Li, X.H., M. Lu, and H.L. Li. 2002. *J Appl Polym Sci* 86:2403–2407.
339. Liu, J., Y.H. Lin, L. Liang, J.A. Voigt, D.L. Huber, Z.R. Tian, E. Coker, B. McKenzie, and M.J. McDermott. 2003. *Chem Eur J* 9:605–611.
340. Sakaguchi, H., H. Matsumura, and H. Gong. 2004. *Nat Mater* 3:551–557.
341. Rajesh, B., K.R. Thampi, J.M. Bonard, H.J. Mathieu, N. Xanthopoulos, and B. Viswanathan. 2005. *J Power Sources* 141:35–38.
342. Golabi, S.M., and A. Nozad. 2002. *J Electroanal Chem* 521:161–167.
343. Becerik, I., and F. Kadirgan. 1997. *J Electroanal Chem* 436:189–193.
344. Golabi, S.M., and A. Nozad. 2003. *Electroanalysis* 15:278–286.
345. Yang, H., T.H. Lu, K.H. Xue, S.G. Sun, G.Q. Lu, and S.P. Chen. 1997. *J Electrochem Soc* 144:2302–2307.
346. De Paoli, M.A., and W.A. Gazotti. 2002. *Macromol Symp* 189:83–103.
347. de Oliveira, H.P., M.V.B. dos Santos, C.G. dos Santos, and C.P. de Melo. 2003. *Mater Charact* 50:223–226.
348. Park, Y.H., S.H. Chang, J.Y. Lee, Y. Lee, and D.H. Baik. 2000. *Mol Cryst Liquid Cryst* 349:355–358.
349. Iroh, J.O., K. Levine, K. Shah, Y. Zhu, M. Donley, R. Mantz, J. Johnson, N.N. Voevodin, V.N. Balbyshev, and A.N. Khramov. 2004. *Surf Eng* 20:93–98.
350. de Souza, S., J.E.P. da Silva, S.I.C. de Torresi, M.L.A. Temperini, and R.M. Torresi. 2001. *Electrochem Solid State Lett* 4:B27–B30.
351. Audebert, P., P. Aldebert, and M. Pineri. 1989. *Synth Met* 32:1.
352. Girault, N., P. Aldebert, M. Pineri, and P. Audebert. 1990. *Synth Met* 38:277.
353. Otero, T.F., and M.J. Ariza. 2003. *J Phys Chem B* 107:13954–13961.
354. Otero, T.F., and M.T. Cortes. 2003. *Sens Actuat B Chem* 96:152–156.

355. Otero, T.F., S. Villanueva, M.T. Cortes, S.A. Cheng, A. Vazquez, I. Boyano, D. Alonso, and R. Camargo. 2001. *Synth Met* 119:419–420.

356. Otero, T.F., and J.M. Sansinena. 1997. *Bioelectrochem Bioenerget* 42:117–122.

357. Immerstrand, C., K. Holmgren-Peterson, K.E. Magnusson, E. Jager, M. Krogh, M. Skoglund, A. Selbing, and O. Inganas. 2002. *Mrs Bull* 27:461–464.

358. Wang, Y.J., X.H. Wang, X.J. Zhao, J. Li, Z.S. Mo, X.B. Jing, and F.S. Wang. 2002. *Macromol Rapid Commun* 23:118–121.

359. Randriamahazaka, H., F. Vidal, P. Dassonville, C. Chevrot, and D. Teyssie. 2002. *Synth Met* 128:197–204.

360. Vidal, F., J.F. Popp, C. Plesse, C. Chevrot, and D. Teyssie. 2003. *J Appl Polym Sci* 90:3569–3577.

361. Malhotra, B.D., R. Singhal, A. Chaubey, S.K. Sharma, and A. Kumar. 2005. *Curr Appl Phys* 5:92–97.

362. Cosnier, S. 1998. *Can J Chem Eng* 76:1000–1007.

363. Cosnier, S., A. Le Pellec, R.S. Marks, K. Perie, and J.P. Lellouche. 2003. *Electrochem Commun* 5:973–977.

364. Cosnier, S., R.S. Marks, J.P. Lellouche, K. Perie, D. Fologea, and S. Szunerits. 2000. *Electroanalysis* 12:1107–1112.

365. Kojima, K., T. Yamauchi, M. Shimomura, and S. Miyauchi. 1998. *Polymer* 39:2079–2082.

366. Ram, M.K., M. Adami, S. Paddeu, and C. Nicolini. 2000. *Nanotechnology* 11:112–119.

367. Reiter, S., K. Habermuller, and W. Schuhmann. 2001. *Sens Actuat B Chem* 79:150–156.

368. Sharma, A.L., R. Singhal, A. Kumar, K.K. Rajesh Pande, and B.D. Malhotra. 2004. *J Appl Polym Sci* 91:3999–4006.

369. Sung, W.J., and Y.H. Bae. 2000. *Anal Chem* 72:2177–2181.

370. Sung, W.J., and Y.H. Bae. 2003. *Biosens Bioelectron* 18:1231–1239.

371. Shimomura, M., N. Kojima, K. Oshima, T. Yamauchi, and S. Miyauchi. 2001. *Polym J* 33:629–631.

372. Piro, B., L.A. Dang, M.C. Pham, S. Fabiano, and C. Tran-Minh. 2001. *J Electroanal Chem* 512:101–109.

373. Piro, B., V.A. Do, L.A. Le, M. Hedayatullah, and M.C. Pham. 2000. *J Electroanal Chem* 486:133–140.

374. Raffa, D., K.T. Leung, and F. Battaglini. 2003. *Anal Chem* 75:4983–4987.

375. Kong, Y.T., M. Boopathi, and Y.B. Shim. 2003. *Biosens Bioelectron* 19:227–232.

376. Gambhir, A., A. Kumar, B.D. Malhotra, B. Miksa, and S. Slomkowski. 2002. *E-Polymers*.

377. Lakard, B., G. Herlem, S. Lakard, A. Antoniou, and B. Fahys. 2004. *Biosens Bioelectron* 19:1641–1647.

378. Nakaminami, T., S. Ito, S. Kuwabata, and H. Yoneyama. 1999. *Anal Chem* 71:1928–1934.

379. Rajesh Kaneto, K. 2005. *Curr Appl Phys* 5:178–183.

380. Ramanavicius, A., K. Habermuller, E. Csoregi, V. Laurinavicius, and W. Schuhmann. 1999. *Anal Chem* 71:3581–3586.

381. Pande, R., G.C. Ruben, J.O. Lim, S. Tripathy, and K.A. Marx. 1998. *Biomaterials* 19:1657–1667.

382. Cha, J.H., J.I. Han, Y. Choi, D.S. Yoon, K.W. Oh, and G. Lim. 2003. *Biosens Bioelectron* 18:1241–1247.

383. Cai, H., Y. Xu, P.G. He, and Y.Z. Fang. 2003. *Electroanalysis* 15:1864–1870.

384. Ramanaviciene, A., and A. Ramanavicius. 2004. *Anal Bioanal Chem* 379:287–293.

385. Piro, B., M.C. Pham, and T. Ledoan. 1999. *J Biomed Mater Res* 46:566–572.

386. Piro, B., J. Haccoun, M.C. Pham, L.D. Tran, A. Rubin, H. Perrot, and C. Gabrielli. 2005. *J Electroanal Chem* 577:155–165.

387. Korri-Youssoufi, H., and A. Yassar. 2001. *Biomacromolecules* 2:58–64.

388. Korri-Youssoufi, H., and B. Makrouf. 2002. *Analyt Chim Acta* 469:85–92.

389. Jiang, X.H., and X.Q. Lin. 2005. *Analyst* 130:391–396.

390. Livache, T., B. Fouque, A. Roget, J. Marchand, G. Bidan, R. Teoule, and G. Mathis. 1998. *Anal Biochem* 255:188–194.
391. Pernaut, J.M., and J.R. Reynolds. 2000. *J Phys Chem B* 104:4080–4090.
392. Piro, B., T.A. Nguyen, J. Tanguy, and M.C. Pham. 2001. *J Electroanal Chem* 499:103–111.
393. Darain, F., S.U. Park, and Y.B. Shim. 2003. *Biosens Bioelectron* 18:773–780.
394. Korri-Youssoufi, H., C. Richard, and A. Yassar. 2001. *Mater Sci Eng C Biomimet Supramol Systems* 15:307–310.
395. Makhloufi, L., H. Hammache, and B. Saidani. 2000. *Electrochem Commun* 2:552–556.
396. Zhang, L., and S.J. Dong. 2004. *J Electroanal Chem* 568:189–194.
397. Yang, X., C.O. Too, L. Sparrow, J. Ramshaw, and G.G. Wallace. 2002. *Reactive Funct Polym* 53:53–62.

19

Internal Fields and Electrode Interfaces in Organic Semiconductor Devices: Noninvasive Investigations via Electroabsorption

Thomas M. Brown
and Franco Cacialli

The operation and physics of optoelectronic devices based on conjugated molecules such as light-emitting diodes (LEDs) and photovoltaic diodes, where organic thin films are sandwiched between different electrodes, is critically influenced by the barriers at electrode/semiconductor interfaces and by the electric fields present within the different functional layers of the devices.

In this chapter we will review a powerful type of modulation spectroscopy: electroabsorption (EA) spectroscopy. Such technique is particularly interesting, since it enables probing of completed and operational devices for the determination of: (a) the internal fields that arise either from application of an external bias, from equilibration of the Fermi energy, or from fixed or injected charges, and their accumulation at trap sites or at heterointerfaces, and (b) the evolution of the injection barrier heights when different materials are used to optimize charge injection into the device. The wealth of experimental information provided by EA measurements has not only enabled the testing of theoretical models for the energy level lineup at (organic) semiconductors/electrodes interfaces but also the drawing of significant conclusions on contact optimization and the device physics underpinning the operation of efficient electrodes such as poly(ethylene-dioxythiophene): poly(styrene-sulphonic acid) (PEDOT:PSS) anodes and LiF(CsF)/Al or LiF(CsF)/Ca/Al cathodes.

19.1 Introduction

LEDs based on conjugated molecules are technologically attractive for their great application potential. Since their discovery [1,2], which has created huge interest in the field of organic optoelectronics, the aim of much research has been channeled into understanding the physical mechanisms underlying their operation, in order to spur the development of efficient devices. Both the size of the electrode/semiconductor barrier heights of LEDs, whose simplest structure consists of an emitting organic film sandwiched by two electrodes, and the magnitude and distribution of the electric fields across the diode structure are recognized to be crucial quantities that govern device physics and operation of LEDs [3], as is also the case for photovoltaic cells and other optoelectronic devices. In this chapter we will review EA spectroscopy in its use as a powerful experimental technique with which it has become possible to gauge the internal electric fields, the energy level lineup, and the barrier heights on completed and operational devices.

The chapter is organized in the following sections: in Section 19.2 we will briefly outline injection mechanisms in polymer diodes, present model theories on how the energy levels of metals and semiconductors line up at the interface between the two and how built-in internal fields arise in polymer diodes; Section 19.3 introduces EA spectroscopy and its use as a tool to gauge internal electric fields in diodes and the energy level lineups at polymer/electrode interfaces; in Section 19.4 and Section 19.5 we will review examples on how EA has been used to study and research the internal fields in organic diodes and the energy level lineup of its contacts, becoming an important tool for the understanding of the physics of organic diodes.

19.2 Injection Barriers, Interfaces, and Built-In Electric Fields in Organic Diodes

19.2.1 Injection

Charge injection is a process crucial to the operation of semiconductor devices, and depends greatly on the size of the metal/semiconducting-polymer Schottky barriers. Compared to inorganic systems, Schottky barriers in undoped conjugated polymers are typically found to depend much more strongly on the electronic characteristics of the injecting contact [4–8], making injection in polymer light-emitting diodes (PLEDs) highly dependent on the electrode material.

Mechanisms of charge injection into organic materials are less established than those for conventional inorganic semiconductors. Two injection mechanisms [9], borrowed from inorganic semiconductor device physics, have been applied to describe the functional dependence of the current on barrier height, electric field, and temperature: (a) Fowler–Nordheim (FN) quantum mechanical tunneling and (b) thermionic emission. Although at high fields (F) and barriers (Φ_B) the FN model gives a reasonable fit for the functional dependence of the current [8,10] [$I \propto F^2 \exp(-\kappa\Phi_B^{3/2}/F)$], it is not able to quantitatively account for the experimental I–V characteristics. It has been shown that the qualitative dependence similar to that of FN tunneling may be accidental [11–13].

Thermionic emission over the metal/semiconductor interface yields a current of the form [9,14,15], $I = A^* T^2 \exp(-\Phi_B^{\text{eff}}/kT)$ where A^* is known as the effective Richardson constant. The field dependence is included in the effective barrier height, Φ_B^{eff}, which is reduced compared to the triangular barrier height, by the image force potential by a factor proportional to the square root of the field over the dielectric constant of the material. For low barriers and fields, polymer diodes reveal thermally activated injection typical of thermionic emission [9,14,15], although the standard thermionic emission equation is unable to give an accurate quantitative prediction of the field dependence and the actual magnitude of the currents involved.

The difficulty of extending mechanisms valid for crystalline semiconductors to highly amorphous ones arises because charge transport in the latter no longer consists of free propagation in extended states, but rather occurs through hopping between molecular sites [16]. Based on this consideration, conduction in PLEDs starts with a carrier either tunneling or being thermally injected from the Fermi level of the metal electrode into a polaron level close to the contact. Transport then occurs by hopping among an energetic distribution of localized states [11,12,17–20].

Because of the localization and low mobilities, most carriers are injected into molecules close to the contact. For conduction to occur, the carriers need to overcome the random barriers due to disorder [18] and, crucially, the barrier/coulombic well of their image potential in which they are trapped. Only few carriers are able to do this at low fields and end up recombining at the metallic interface, resulting in currents which are considerably lower than those predicted by inorganic semiconductor models [11,18,21–24]. Higher electric fields will strongly increase the probability of carriers surmounting the barriers and of those subsequently drifting toward the opposite electrode under the effect of the field in the bulk of the polymer layer.

Although carrier injection in conjugated polymers cannot be described accurately by a single equation, the probability of injection depends strongly (i.e., exponentially) on the barrier height between the Fermi level of the metal and some average state in the disorder-broadened distribution of the polymer's conduction levels [17]. This strong phenomenological dependence on Φ_B has been extensively reported in the literature [6–8,10,23,25].

However, in some cases a reduction in the barrier height does not always translate into a strong increase in the current passing across the device. Two limiting regimes for the current are in fact found to exist in single carrier diodes (i.e., for devices in which the current is dominated by either holes or electrons). For large Schottky energy barriers or large mobilities, the current is injection limited, i.e., it is limited by the rate of carrier injection from the contact into the organic solid. In this instance the current is strongly dependent on the height of the injection barrier. For low energy barriers or mobilities, the current is instead space charge limited (SCL). In this regime, the drift of the charge carriers away from the contact occurs at a rate slower than injection. The resulting accumulation of space charge disturbs the local electric field and ultimately limits the current. Davids et al. [22,23] have shown that, in polymers with a mobility of $\sim 10^{-6}$ cm^2 V^{-1} s^{-1}, lowering the barrier below 0.3–0.4 eV does not lead to any significant enhancement in the current flow. Thus, in those conditions, the current was found to be SCL [26]. It is also possible to probe different regimes as a function of applied bias: in a diode with nonohmic contacts, the current can be electrode-limited at low fields and SCL at high fields [27]. Note that the classic expression for the trap-free SCL current ($I \propto \mu V^2$, where μ is the constant carrier mobility and V is the applied voltage [28]) is not readily applicable to organic semiconductors, since conduction in these amorphous materials does not occur through delocalized conduction and valence bands.

Transport of carriers through conjugated polymers has been extensively studied and established device models for organic diodes [17,19,22,25,29–33] consist of drift diffusion equations for carrier transport, field-dependent carrier mobility [16], combined with Poisson's equation for the electrostatic potential, which is able to take into account perturbations to the field due to the injected space charge. Ioannidis et al. [19] have shown that their model, which falls in the above category, can account for a current of the form $I \propto V^\beta$ ($\beta > 2$), under significant injection conditions. This same V^β functional dependence was observed in previous studies [26,34] and was attributed to trapped charge-limited current, in correspondence with inorganic semiconductors. Extensive confirmation of the importance of SCL current has finally been provided by the careful work of Blom and coworkers that focused on PLEDs based on derivatives of poly(p-phenylene vinylene), PPV [34–36].

Fabricating an efficient LED involves achieving a truly bipolar current, where the electron and hole currents are balanced. The presence of both carriers increases the complexity of the physics of transport and injection [37], since the introduction of carriers of opposite sign can reduce the net space charge considerably, thus affecting the electric field. A fraction of the carriers will also recombine. Doping of the semiconductor by both anode and cathode material affects not only the emissive and conduction properties of semiconducting layer but also the injection efficiencies from the electrodes into the conjugated polymers [38–41]. Furthermore, the presence of interfaces and trapped states throughout the heterostructure diodes [3,42–46], where a significant concentration of carriers can build up, will create internal fields that can dominate the physics of the device and have an influence as great as the magnitude of barrier heights.

19.2.2 Metal/Semiconductor Energy Lineup and Barrier Heights

We have shown in Section 19.2.1 that in many cases injection can depend strongly on the Schottky barrier height (i.e., the energy difference at the interface between the Fermi level of the metal and the conduction levels in the conjugated polymer). It is thus crucial to understand how the energy levels of the metal and the organic semiconductor align in order to predict the magnitude of these barriers.

In general, both Schottky barriers (for the metal/semiconductor interface) and band discontinuities (for semiconductor/semiconductor heterojunctions) play a major role in determining injection, transport, and optical properties of a semiconductor device and have thus received much attention from both the experimental and theoretical standpoint [47–49]. Accurate theoretical predictions and experimental determinations are necessary in order to design state of the art devices. It is crucial to keep injection barriers to a minimum, to reduce operating voltages (and powers) and therefore increase power efficiency and operational longevity. Ideally, the accuracy required for the energy barriers and band offsets, to predict the optoelectronic and transport properties of a device, is of the order of kT ($\sim$26 meV at room temperature). Although progress has been made in the general understanding of the problem over the years, such a limit generally remains unattained by both experiments and theory. This shows how complex the study and the interpretation of the alignment mechanisms is, and also how difficult it is to fabricate an "ideal" and reproducible interface. It is important to bear in mind that the metal/semiconductor interface can be a complex quasi–two-dimentional (2D) system with unknown atomic geometry, interdiffusion, transfer of charge, defect states, and new chemical compound formation [50], that can all greatly affect the electronic alignment at the interface. Nevertheless, extensive semiconductor research has yielded some well-controlled experiments and general theoretical models, which can, in various cases, reproduce or predict such an alignment, with a precision in the tenth of an eV range.

The aim of "model" theories [51], which predict band offsets through the macroscopic parameters of the materials, is to find or specify a reference energy level (E_r) for the metals and semiconductors. Upon formation of the heterojunction, one merely lines up the reference levels in the two materials. In this case, the valence band offset between two materials is simply obtained as the difference between the two valence band maxima, measured relatively to such a reference level. E_r enables one to define an absolute scale for all semiconductors or metals. Once E_r is calculated, we need to know only the semiconductor's valence band maximum E_v (or the highest occupied molecular orbital (HOMO) levels in molecules) relative to E_r for each material, which following Tersoff's approach [51] we denote as:

$$E_V^r \equiv E_V - E_r \qquad (19.1)$$

As shown in Figure 19.1, upon heterojunction formation between materials (1) and (2), the reference level approach consists of assuming that E_r is the same on both sides of the interface:

$$E_r(1) \equiv E_r(2) \qquad (19.2)$$

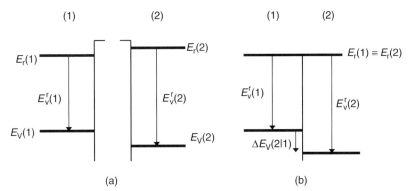

FIGURE 19.1 Formation of a heterojunction between two materials (1) and (2): (a) before and (b) after contact. Indicated are the reference energy levels, E_r, and the valence band maxima, E_V, relative to E_r (E_V^r). Upon contact, the reference levels are brought to alignment and the offset between the valence band maxima is simply calculated as: $\Delta E_V(2|1) = \Delta E_V^r(2|1) = E_V^r(2) - E_V^r(1)$. (From Tersoff, J. *Heterojunction band discontinuities—Physics and device applications*, eds. G. Margaritondo, and F. Capasso, Amsterdam: North-Holland. 1987. With permission.)

The alignment of the respective reference levels yields a valence band discontinuity equal to:

$$\Delta E_V(2|1) = \Delta E_V^r(2|1) = E_V^r(2) - E_V^r(1) \tag{19.3}$$

Similarly, the Schottky barrier height for holes is the difference between the valence band maximum of the semiconductor and the metal Fermi level, measured with respect to E_r.

It is important to note that in these model theories, the lineup of the energy levels can in principle be derived from experimentally measurable or bulk (in the charge neutrality model [51]) properties of the material, rather than relying on ab initio numerical calculations of the complete electronic structure of a specified interface. Although this introduces a degree of approximation, it enables us to focus more narrowly on the underlying physics behind the interfacial band alignment.

The simplest example of a reference level consists in the Fermi level of a metal: in a metal/metal junction it is the Fermi energies that align. Thus, in order to determine the discontinuities between the bands of the two metals, one merely needs to know the energy levels relative to E_F, which is taken as the reference level [$E_F \equiv E_r$(metal)]: no complex numerical calculations are necessary at the interface.

In this section, three major theories explaining how the electronic bands of semiconductors and metals align after contact will be outlined. These are the Schottky, Bardeen, and Tersoff theories. These models are employed to predict the Schottky barrier height at the metal/semiconductor junction. Comprehensive reviews on the general issue of heterojunctions, which also include other lineup theories, can be found in [47–50,52,53].

19.2.2.1   Schottky's Model ($E_r \equiv E_{VAC}$)

One of the earliest models for the metal/semiconductor contact is due to Schottky. Here the reference level, E_r, is the vacuum level E_{VAC} at the solid's surface. This is a valid approximation if one neglects dipole layers and charged interface states. In this case, the Schottky barrier height for electrons is given by (Figure 19.2a):

$$\Phi_B(e^-) = \phi_M - X_S \tag{19.4}$$

where X_S is the electron affinity of the semiconductor and ϕ_M is the work function of the metal. Similarly, from aligning the vacuum levels of the metal and semiconductor at the interface, the barrier height for holes is determined by the difference between the ionization potential of the semiconductor ξ_S and ϕ_M:

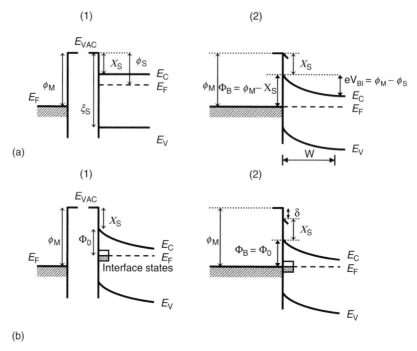

FIGURE 19.2 Energy band diagrams of metal-semiconductor (n-type) contacts, before (1) and after formation (2): (a) according to the Schottky model and (b) according to the Bardeen model. In (a) the vacuum levels align. In (b) Φ_B and the built-in potential, V_{BI}, are pinned by interface states. Φ_0 is the pinned value of the barrier height, δ is the resulting interfacial dipole, W is the depletion region width and V_{BI} is the contact or built-in potential.

$$\Phi_B(h^+) = \xi_S - \phi_M \tag{19.5}$$

Equation 19.4 and Equation 19.5 are based on the assumption that the metal/semiconductor interface is a simple superposition of the surfaces of the two semiinfinite solids. However, in general this is not the case, as it is necessary to consider charge transfer or rearrangement (e.g., due to surface states) occurring at the interface, which can introduce an interfacial dipole layer. The latter will alter the interfacial energy level lineup, introducing a discontinuity in the electrostatic potential at the interface.

Denoting the interface charge density as $\rho(I)$ and that of the separate surfaces forming the heterojunction $\rho(j)_{j\ =\ 1,2}$, the Schottky barriers can be expressed through Equation 19.4 and Equation 19.5 only if $\rho(I) = \rho(1) + \rho(2)$. In general there will be a variation in the charge density, induced by transfer/rearrangement of charge, after the heterojunction is formed, and given by:

$$\delta\rho = \rho(1) + \rho(2) - \rho(I) \tag{19.6}$$

$\delta\rho$ can introduce a finite dipole moment. The Schottky model for predicting the barrier height will be violated by the potential step δ, associated to $\delta\rho$. Consequently, the metal/semiconductor barriers will have shifted (Figure 19.2b):

$$\Phi_B(h^+) = \xi_S - \phi_M + \delta \tag{19.7}$$

and

$$\Phi_B(e^-) = \phi_M - X_S - \delta \tag{19.8}$$

The shortcoming of the Schottky model in many real heterojunctions between metals and semiconductors lies in the fact that it ignores the role of dipoles and of interface states at these junctions. The latter are, instead, central to the Bardeen model.

19.2.2.2 Bardeen Model: The Pinned Limit

Although strong dependence of the barrier height, Φ_B, with the metal work function, ϕ_M, is observed in some ionic semiconductors, the majority of metal/inorganic semiconductor contacts do not appear to obey Schottky's rule. In many semiconductors, the barrier is a less sensitive function of ϕ_M than given by Equation 19.4 and Equation 19.5 and, in some cases, it is independent of ϕ_M.

The insensitivity of the barrier height to the metal work function in covalently bonded semiconductors was first explained by Bardeen. He proposed a model where E_F is pinned by a high density of localized states present on the semiconductor surface itself. If the density is high enough, upon contact with the metal, these states can accommodate all the charge transferred from the metal without noticeably moving E_F. Thus, the charge transfer creates an interface dipole, δ, that tends to align E_F in the metal to that of the surface states in the semiconductor. Therefore, the barrier is pinned to a value:

$$\Phi_B = \Phi_0 \tag{19.9}$$

dependent on the number and energetic distribution of surface states, but almost independent of what metal (or ϕ_M) is employed in making the contact (Figure 19.2b).

19.2.2.3 Tersoff's Model ($E_r \equiv$ Charge Neutrality Level)

An interesting step toward an understanding of the nature of the Fermi level pinning is the concept of charge neutrality level (CNL) outlined in depth by Tersoff [51,54,55]. To introduce such a model we consider the case of a metal first. When two metals are brought together, electrons will flow from the metal with lower electron affinity to the one with the higher electron affinity. Transfer of charge creates oppositely charged (Debye) layers within each metal, and thus an interfacial dipole, which tends to align the two Fermi levels. When full alignment occurs, the dipole is the difference between the two work functions and the interface is in equilibrium. Conversely, in the case of a junction between two highly insulating, nonpolarizable solids, rearrangement of charge is expected to be negligible and $\delta\rho \approx 0$. The case of a semiconductor falls between these two limits.

According to Tersoff's approach, which has often been able to predict the energy level lineup in (inorganic) semiconductor junctions with a good degree of success [48], one can define an energy, E_n, in the solid, denoted the CNL. The CNL plays a role analogous to the Fermi level of the metal ($E_r \equiv E_n$, in Equation 19.4), in the sense that, upon formation of a heterojunction between two materials, the two CNLs tend to align, i.e., $E_n(1) \equiv E_n(2)$. In a metal E_n coincides with E_F. In a semiconductor E_n is in general located somewhere close to midgap.

The main idea of this theory is the presence of a dipole layer at the interface, which can be formed as a consequence of interface states present in the forbidden energy gap of the semiconductor and which can bring the reference levels to alignment. As mentioned earlier, the simple case is that of a metal/metal junction, in which, upon contact, electrons flow from one metal (with lower ϕ) to the other (with higher ϕ). This charge transfer results in a dipole, which brings the two Fermi levels to equilibrium. At equilibrium, the dipole is exactly equal to the mismatch between the work function of the metals.

Similarly, in a metal/semiconductor junction, in which the metal work function lies between the valence band maximum and the conduction band minimum, the tails of the metallic wave function decrease exponentially in the adjacent semiconductor gap states, inducing transfer of charge, which in turn creates the dipole. The mismatch between the metal Fermi level and E_n is reduced by this dipole (by a factor inversely proportional to the semiconductor's optical dielectric constant, ε_∞, in first approximation [51]). Only in the case of semiconductors with a large optical dielectric constant ε_∞ or high density of interface states is the Fermi level almost completely pinned at E_n. In this case, since E_n is an intrinsic property of the semiconductor, the Schottky barrier height of a particular semiconductor is independent of the metal (and its work function) utilized as contact. In order to define the CNL, one must remember that gap states come from both the valence and conduction bands. States closer to the valence (conduction) band have a predominantly valence (conduction) character. To be the equivalent of the Fermi level, E_n in the semiconductor should represent the natural energy separating the filled states from the empty states. Tersoff defines E_n as the point where the gap states cross over from valence band type (with bonding character) to conduction band type (with antibonding character). This crossing occurs somewhere near midgap. The physical meaning of the CNL is still under discussion. In the case of tetrahedral semiconductors such as silicon and germanium, some researchers [56] have identified E_n with the energy of the hybrid orbital sp^3, i.e., with the unsaturated "dangling" bond, E_{db}, which, having nonbonding character, lies near the middle of the forbidden energy gap.

The theoretical approach to band lineups outlined above, yields a metal/semiconductor barrier height of the form [53]:

$$\Phi_B = S(\phi_M - X_S) + (1 - S)\Phi_0 \tag{19.10}$$

where S is a constant. Extensive experimental determinations on inorganic semiconductor/metal contacts, confirm that, even though the ideal Schottky model generally fails, there is still a linear relation between Φ_B and ϕ_M as expressed by Equation 19.10. The proportionality factor, $S = |d\Phi_B/d\phi_M|$, is known as the interface parameter and is a characteristic of each semiconductor [49,55,57,58]. Φ_0 represents the pinned value of the barrier for a specific semiconductor surface. Note that Equation 19.10 represents the weighted average of the Schottky ($S = 1$) and the Bardeen ($S = 0$) limits. It is important to remark that S practically vanishes for common semiconductors such as Si, Ge, and GaAs, which means that the barrier height is almost independent of ϕ_M and of which metal is employed as contact. Most cases fall in an intermediate situation between the $S = 1$ and the $S = 0$ limits. Common values of S are 0.1 for covalent and III–V semiconductors and 0.3 for ionic II–VI semiconductors [53].

$1/\varepsilon_\infty$ can be used as a rough estimate for the interface parameter [51,55], S, since it delivers the proper limits: for $\varepsilon \to \infty$ (as in a metal), $S = 0$ and complete pinning occurs. When $\varepsilon \to 1$ (no screening, as in an ideal insulator) the Schottky limit of $S \to 1$ is approached. For semiconductors such as Si and Ge, since $\varepsilon \sim 10$, the actual lineup corresponds to 10% of the Schottky rule lineup plus 90% of the neutrality rule lineup. In this case, the neutrality rule represents a good approximation: the barriers are pinned close to the neutrality level as observed in experiments. The interface parameter S is also linked to the density of surface states, D_S, of the semiconductor. In the model described by Sze [9], the pinned limit ($S = 0$) is achieved for $D_S \to \infty$. The Schottky limit corresponds to no interface states, i.e., $D_S \to 0$.

19.2.2.4 The Conjugated Polymer/Metal Junction

The Schottky model of Section 19.2.2.1 has been used extensively to predict the barrier height in conjugated polymer diodes and has been shown in many cases to be at least qualitatively acceptable in determining good injecting contacts. Although interfacial dipoles also need to be considered [59–61] for a correct quantative prediction of barrier heights, generally, in PLEDs low work function electrodes make good electron injectors and high work function electrodes make good hole injectors. Both current–voltage [8] and EA [4] experiments have shown that, for metals whose work function lies inside the polymer π–π* gap, the Schottky model can be often a good first approximation in predicting the dependence of the barrier height upon the work function of the metal as expressed in Equation 19.4 and Equation 19.5.

Two factors corroborate the validity of the Schottky model:

(a) The semiinsulating properties and low dielectric constant (ε) of conjugated polymers ($\varepsilon \in [2,3]$) that result in a small $\delta\rho$ in Equation 19.6.

(b) A negligible concentration of surface states, compared to inorganic semiconductors. The majority of surface states in silicon are due to orbitals which are not able to bond because of the lack of neighboring silicon atoms at the silicon–vacuum interface. The covalent chemical bonds are necessarily broken at nonepitaxial interfaces. This results in a relatively high concentration of dangling bonds. In contrast, chemical bonds are not necessarily broken at interfaces involving organic materials [5] and restrictions due to crystalline order do not arise. Thus, in conjugated polymers, unsaturated bonds are comparatively rare. Also, interfacial chemical reactions that do occur do not always introduce a large concentration of states in the energy gap that would otherwise pin the barrier [5].

However, there are many factors which affect the interfacial electronic structure and lineup at the organic semiconductor/metal contact, that can cause the alignment not to follow the Schottky model [59,62]. Significant dipoles have been detected in the literature [59–61]. These factors include charge transfer but also chemical reactions, diffusion at the interface and the presence of contaminants which also depend on the atmosphere in which the structures are fabricated and experiments carried out. In general, organic LEDs are found to be very sensitive to fabrication conditions, which makes it of paramount importance to use polymers with low levels of impurities/traps (unless these are introduced intentionally) and to keep evaporation and test conditions under the highest control in order to have reproducible and more interpretable results. Even in highly controlled experiments, organic materials have been shown to react with the evaporated metal contact, especially with those reactive metals with a low work function [63–68]. New states are created in the gap and a shift in the interfacial energy level alignment can occur. Furthermore, adsorbates are known to shift the work function by rearranging the electronic charge and by adding elementary dipoles [50]. These can be inserted intentionally to shift the work function and the barrier heights in devices [69–71]. However, adsorption of molecules can also occur unintentionally as most electrode evaporations for organic devices are carried out, not in ultra high vacuum, but in chambers where the pressures are $\sim 10^{-6}$ mbar and contaminants may be present [67,72]. In-depth reviews on metal/organic interfaces and the lineup conditions, with many examples and references therein, can be found in [59,60]. The lack or occurrence of pinning and the Schottky model for polymer–electrode interfaces will be investigated in this chapter not only in the case of metal contacts, but also when incorporating injecting interlayers, both polymeric and inorganic, between the electrode and the organic layer.

19.2.3 The Built-In Potential

Another characteristic of metal/semiconductor contacts at equilibrium is the presence of an internal electrostatic field, illustrated in Figure 19.2 by bands which are not flat. When a metal with work function ϕ_M is brought into contact with a semiconductor with work function ϕ_S, the Fermi levels align at thermal equilibrium. Figure 19.2a illustrates the case of an n-type semiconductor and a metal with $\phi_M > \phi_S$. For the E_F of the two materials to be coincident upon contact, electrons are transferred from the semiconductor to the metal. In the n-type semiconductor a charged depletion region of ionized donors, W, is formed (Figure 19.2a). To maintain charge neutrality, the positively ionized donors of the depletion layer are balanced by a negative sheet charge in the metal. This results in the generation of the contact or built-in potential, which prevents further net electron transfer between the two solids. In an ideal Schottky model, $eV_{BI} = \phi_M - \phi_S$ and the barrier height to electron injection is given by Equation 19.4: $\Phi_B = \phi_M - X_S$. This barrier is responsible for the rectifying properties of the junction (Figure 19.2). As discussed in Section 19.2.2, the presence of interfacial states and dipoles can have a significant effect on the alignment of the energy levels and thus on the value of V_{BI}. V_{BI} in the Bardeen diagram of Figure 19.2b is effectively pinned by the high concentration of surface states and is almost independent of ϕ_M.

The depletion width at equilibrium can be written as:

$$W \cong \sqrt{\frac{2\varepsilon_S V_{BI}}{eN_D}} \tag{19.11}$$

where N_D is the concentration of donor states. Under the "abrupt" approximation (donors are all ionized up until a distance W from the interface and not ionized from a distance $> W$) an electric field is only present in the depletion region and is zero thereafter so that the electronic bands and parameters remain unaffected. Thus for sufficiently long metal/semiconductor/metal structures ($d \gg W$, where d is the thickness of the semiconductor), a field is present only in regions close to the interfaces. If the distance between the two electrodes is shortened, there exists a semiconductor layer thickness, of the order of the sum of the two depletion regions, for which a built-in field is present across the entire structure. This thickness can be calculated from the value of ε_S, N_D and V_{BI}. It is important to remark that the vast majority of organic diodes are fabricated with pure undoped polymers, and that, although a small unwanted doping concentration is leftover from the synthesis of the polymers and from impurities, W is typically large compared to commonly doped inorganic semiconductors. Indeed, for some conjugated polymers the concentration can be extremely low. In photoemission experiments on polyfluorene polymers, on which most of the devices of this chapter are based, Greczynski et al. [73] detected no band bending across the polymer or surface charging effects for distances as long as 110–160 nm, which implied that the depletion region was in the micrometer range [73,74]. To quantify the level of intrinsic doping, the insertion of typical values for the PLED built-in potential ($V_{BI} \sim 1$ V) and the dielectric constant of a polymer ($\varepsilon_S \sim 3$) in Equation 19.11, yields a depletion region width of $W \cong 1.8$ μm for a doping concentration of $N_D = 10^{14}/cm^3$. This is around ten times the length of the emitting layer in a PLED, which means that devices made with a polymer of this purity are fully depleted at equilibrium and that space charge due to doping/impurities is negligible. The diode sandwich structures also remain fully depleted up to concentrations in the $10^{16}/cm^3$ range ($W \sim 180$ nm), resulting in an internal field being present throughout the entire polymer layer (typical thickness for the organic layer in PLEDs is $\sim$100 nm).

Since the polymer can be synthesized to such a high purity, a more appropriate description for the space charge is that given by trapped charges, rather than ionized impurities. Traps may arise, e.g., from conformational defects of the organic semiconductors. It is then the density of the traps in the polymer that determines band bending [14]. In their detailed model for poly(2-methoxy, 5-(2′-ethyl-hexyloxy)-1,4 phenylene vinylene) (MEH-PPV), Davids et al. [75] included a density of trap states of the order of 10^{16} cm^{-3} positioned at ± 0.6 eV from midgap. In that study [75], it was shown that, at this trap concentration (and obviously at lower values), the diode structures are fully depleted in reverse and weak forward bias (i.e., before significant injection occurs) and that the electric field is essentially uniform in the bulk of the polymer layer. The theoretical work was confirmed by impedance and capacitance experiments on MEH-PPV diodes [76].

Therefore, for LEDs fabricated with high purity polymers, the space charge is small enough so that it does not significantly perturb the electric field in the organic layer (note that it is only for biases lower than those necessary for significant injection to occur). In that case, the energy diagram of a PLED can be well approximated by drawing rigid bands. For these metal/polymer/metal structures, it is important to highlight two characteristics.

In the reverse to weak forward bias regime, the electric field is spatially uniform across the semi-insulating polymer layer and the rigid band model is appropriate, until significant charge injection occurs. In this voltage range, the field is uniform and simply given by $F = V/d$, where d is the thickness of the organic layer and V is the sum of the applied bias and the built-in potential, which is due to the mismatch between the work functions of the two metal electrodes.

The Schottky barrier can scale directly with the metal work function for those metals whose Fermi level lies within the polymer bipolaron gap. At zero bias and at thermal equilibrium, a built-in potential (V_{BI}) is present across the polymer layer of the LED. V_{BI} is generated upon equilibration of the Fermi level throughout the heterostructure, which occurs by transfer of charge from the electrode with the lower electron affinity/work function (i.e., cathode) to the metal with the higher work function (anode). It has been shown that in some cases, the built-in potential can, with good approximation, be equal to the difference in work function of the two metal contacts [4]:

$$V_{BI} = \Delta\phi/e = [\phi(anode) - \phi(cathode)]/e \qquad (19.12a)$$

Metal/polymer interfaces which do follow the Schottky model theory, and PLEDs, for which Equation 19.12a is valid, have been often observed in experiments where deposition and test conditions were carefully controlled [4,6,8,73,74,77,78].

Note however, that Equation 19.12a holds only in the absence of charging of surface states, dopants, bipolarons or polarons, which can alter the interfacial energy level alignment and ultimately the value of V_{BI}. In the more general case, dipoles need to be included and 19.12a becomes:

$$V_{BI} = \Delta\phi/e = [\phi(anode) + \delta(anode) - \phi(cathode) - \delta(cathode)]/e \qquad (19.12b)$$

Furthermore, the work function is a parameter which is highly sensitive to the state of the surface of a solid [79,80]. Experimental determinations of the ϕ of a surface can therefore vary by several tenths of an eV, even in similar experimental conditions, and can also depend on which measurement technique is employed. For example, although gold work function is usually reported to be 5.1 eV in UHV, it has been demonstrated that it can be as low as 4.2 eV and as high as 5.4 eV and a range of values in between depending on the surface cleanliness and ambient conditions [79,81–83]. Similar variations in the barrier height, Φ_B, are thus expected.

It becomes apparent from Equation 19.12a and Equation 19.12b that accurately measuring the internal fields and built-in potential of diodes would not only assist in the understanding of the physics of the device but also permit the gauging of the energy level lineup of the diodes' electrodes. EA spectroscopy is a powerful tool as it enables the investigator to do this, not on model structures, but on real finished devices so that results are directly comparable to the current and luminance or photovoltaic characteristics of the device under study.

The next section introduces EA spectroscopy. In a macroscopic description, the EA signal can be viewed as a nonlinear field-induced optical effect. In molecular solids, the microscopic origin of EA can often be ascribed to the Stark effect. The quadratic dependence of EA on the electric field will be derived. Subsequently we will focus on EA as an optical probe of the internal electrostatic fields in organic diodes, outlining some examples in the following sections.

19.3   Modulation Spectroscopy

Modulation spectroscopy [84] is an effective optical technique to study the physics of both bulk semiconductors and also of heterostructures. In particular, it can be employed to study the properties of an actual working device, yielding data which can be directly compared to the characteristics of the device itself.

Modulation spectroscopy deals with the measurement and interpretation of changes in the optical spectra when modifying the measurement conditions. This can be achieved by applying an oscillatory perturbation such as an electric (EA) or magnetic field, heat pulse (thermomodulation) or stress (piezomodulation). Since electroluminescent diodes are operated by the application of an external electric field, EA is a particularly pertinent and powerful type of spectroscopy for the study of these devices.

19.3.1   EA Spectroscopy

EA spectroscopy measures the normalized variation of the transmission of light through a sample (conjugated polymer films in this study) upon application of a periodic electric field [85,86].

For normal incidence, the transmitted intensity of the light through a sample, ignoring multiple reflections and interference, may be written as [87,88]:

$$I_T = I_0(1 - R)^2 e^{-\alpha d} \qquad (19.13)$$

where I_0 is the incident intensity, R is the reflection coefficient (assumed to be the same at the front and back surfaces), α is the absorption coefficient, and d is the sample thickness.

The application of a field, F, changes both the absorption and reflection coefficients of the sample. The field-induced change of the transmitted intensity is given by:

$$\frac{\partial I_T}{\partial F} = -I_0 e^{-\alpha d} \cdot \left[d(1-R)^2 \frac{\partial \alpha}{\partial F} + 2(1-R)\frac{\partial R}{\partial F} \right] \tag{19.14}$$

Dividing Equation 19.14 by the unperturbed intensity (Equation 19.13) yields:

$$\frac{\Delta T}{T} = \frac{\Delta I_T}{I_T} = -d\Delta\alpha + \frac{2}{1-R}\Delta R \tag{19.15}$$

For the typical operating conditions of EA experiments, the second term of Equation 19.15 can, to good approximation, be neglected [87,88]:

$$\frac{\Delta T}{T} \cong -d\Delta\alpha \tag{19.16}$$

The field-induced normalized variation in transmission, measured with an EA spectrometer, is thus directly proportional to the field-induced variation in the absorption coefficient.

19.3.2 EA as a Field-Induced Nonlinear Optical Effect

The variation in absorption due to the electric field modulation (Equation 19.16) is a nonlinear optical effect. We now consider the origin of nonlinear behavior in materials. In a classical description [89–91], the electric field interacts with the charges (q) in an atom through the force (qF), which displaces the centre of the electron density away from the nucleus. This results in charge separation and thus in a field-induced dipole μ. For an assembly of atoms, the average summation over all atoms ultimately gives rise to the bulk polarization **P** vector of the material. **P** opposes the externally applied field and is given by:

$$\mathbf{P} = \varepsilon_0 \chi \mathbf{F} \tag{19.17}$$

χ is the electric susceptibility tensor, which relates the dielectric displacement **D** to the electric field **F**:

$$\mathbf{D} = \varepsilon_0 \varepsilon_r \mathbf{F} = \varepsilon_0 \mathbf{F} + \mathbf{P} = \varepsilon_0 (1+\chi)\mathbf{F} \tag{19.18}$$

ε_r is the dielectric constant that can be expressed in terms of χ and the index of refraction N:

$$\varepsilon_r = (1+\chi) = N^2 \tag{19.19}$$

Because of resonances in molecules and solids associated with electronic and nuclear motions, the quantities in Equation 19.19 are complex. The complex index of refraction is given by:

$$N = n + ik, \tag{19.20}$$

where n is the refractive index and k is an attenuation factor or extinction coefficient. k is a measure of the loss in power of a wave propagating through the material and is proportional to the absorption coefficient α. The plane wave, used for describing the propagation of the electromagnetic field in a material in the x direction with a velocity ($v = c/N$), can be written as [92]:

$$E = E_0 \exp\left[i\omega\left(\frac{Nx}{c} - t\right)\right] = E_0 \exp\left[i\omega\left(\frac{nx}{c} - t\right)\right] \exp\left(\frac{-\omega kx}{c}\right) \tag{19.21}$$

which leads to the proportionality relation between α and k:

$$\alpha = \frac{4\pi}{\lambda} k = \frac{2\pi}{n\lambda} \text{Im}(\varepsilon_r) \tag{19.22}$$

where λ is the wavelength of the light.

For an ideal linear optical material, the quantities in Equation 19.19 (χ, ε_r, N) are independent of the electric field strength. However, the linear dependence of one physical quantity on another, as in Equation 19.17, is often an approximation valid only over a limited range of values. More generally, one can express the polarization in a power series of the field strength:

$$\mathbf{P} = \varepsilon_0(\chi^{(1)}\mathbf{F} + \chi^{(2)}\mathbf{FF} + \chi^{(3)}\mathbf{FFF} + \cdots) \tag{19.23}$$

where $\chi^{(n)}$ is the nth order susceptibility tensor. For most materials, orders higher than $\chi^{(3)}$ are extremely difficult to observe [91]. $\chi^{(2)}$ vanishes in media with inversion symmetry since the polarization must change sign when the field is reversed. Therefore, for centrosymmetric conjugated polymers, the polarization can be expressed as:

$$\mathbf{P} = \varepsilon_0\chi_{\text{eff}}\mathbf{F} = \varepsilon_0(\chi^{(1)} + \chi^{(3)}\mathbf{FF})\mathbf{F} \tag{19.24}$$

Combining Equation 19.24 with Equation 19.19, the variation of χ_{eff} with the electric field can be expressed as:

$$\Delta\chi_{\text{eff}} = \chi_{\text{eff}}(\text{F}) - \chi_{\text{eff}}(0) = \chi^{(3)}\text{FF} = 2N\Delta N = \Delta\varepsilon_r \tag{19.25}$$

Note that the field-induced changes in the physical quantities of the material found in the above equations (including $\Delta\alpha$) have a quadratic dependence with the electric field (Equation 19.25). A quadratic dependence on applied field is indeed generally found for the EA signal in experiments.

Equation 19.25 also allows the calculation of the third order nonlinear susceptibility, which, for isotropic materials, is given by [88,93]:

$$\chi^{(3)} = \frac{2N\Delta N}{F^2} \tag{19.26}$$

The variation in the imaginary part of the complex refractive index can be derived from Equation 19.20, i.e., $\text{Im}(\Delta N) = \Delta k$. Importantly then, Δk can be extracted from the EA response (Equation 19.16):

$$-(1/d)\Delta T/T \cong \Delta\alpha = (4\pi/\lambda)\Delta k \tag{19.27}$$

On the other hand, the variation in the real part $\text{Re}(\Delta N) = \Delta n$ can be calculated from the EA spectrum by the Kramers–Kroenig equation:

$$\Delta n(\omega) = \frac{2}{\pi}\int_0^\infty \frac{\omega'\Delta k(\omega')}{\omega'^2 - \omega^2}d\omega' = \frac{c}{\pi}\int_0^\infty \frac{\Delta\alpha(\omega')}{\omega'^2 - \omega^2}d\omega' \tag{19.28}$$

Similarly the real part of the refractive index can be derived from the absorption spectrum:

$$n(\omega) = \frac{c}{\pi} \int\limits_{0}^{\infty} \frac{\alpha(\omega')d\omega'}{\omega'^2 - \omega^2} \qquad (19.29)$$

Thus, the third order nonlinear susceptibility $\chi^{(3)}$ in Equation 19.26 can be calculated from a Kramers–Kroenig analysis of the absorption and EA spectra of the conjugated polymer film (Equation 19.27 through Equation 19.29).

Typical values of $\chi^{(3)}$ for conjugated polymers are in the 10^{-12}–10^{-10} esu range [91]. Large nonlinear susceptibilities are desirable in the field of nonlinear optics/photonics. Photonics has been the subject of much attention, since it is the analog of electronics in that it describes how photons can be used to acquire, transmit, and process information. Organic molecules are particularly interesting because they exhibit large nonresonant optical nonlinearities as a result of their unique chemical structure (π bonding). For this reason, organic materials can exhibit fast response times compared to inorganic systems, crucial in switching applications. Furthermore, organic materials have many benefits associated with the flexibility and ease of processing and fabrication.

The relation between the EA response and the macroscopic nonlinear properties of the material has been drawn by a classical approach. In Section 19.3.3, the mechanism behind the field-induced variation will be explored by a quantum mechanical approach. In molecular solids and in conjugated polymers, where the states are localized to a high degree, the change in α with the electric field is in general ascribed to a Stark shift of the molecular energy levels.

19.3.3   EA Response and the Stark Shift

Under the presence of an electric field, the Hamiltonian of the unperturbed molecular electronic state $|\psi_j\rangle$ is modified by the addition of the following perturbation [94]:

$$H' = e\mathbf{F} \cdot \mathbf{r} = eF_z \qquad (19.30)$$

where $-e$ is the charge of the electron and the electric field is directed along the z direction. From perturbation theory, the first order correction to the energy of state $|\psi_j\rangle$ is simply the $\langle \text{bra}||\text{ket}\rangle$ of the state over the operator H':

$$\Delta E_j^{(1)} = e\mathrm{F}\left\langle \psi_j | z | \psi_j \right\rangle \qquad (19.31)$$

This term (Equation 19.31), linear in F, is zero for a nondegenerate system with no permanent electric dipole moment, whose Hamiltonian is unaffected by the parity operation [94]. In centrosymmetric nondegenerate polymers with no permanent dipole moment, the linear Stark effect ensues from disorder [95]. In a semiclassical approach, the shift in energy caused by a permanent dipole moment $\mathbf{m}_j$ can be expressed as:

$$\Delta E_j^{(1)} = -\mathbf{m}_j \cdot \mathbf{F} \qquad (19.32)$$

In conjugated polymers, permanent dipoles can arise from charge transfer states, defects, or from disorder that introduces asymmetry in the charge density along a conjugated segment [95–97]. Equation 19.31 is also known as the linear Stark effect, named after the proportionality dependence on the electric field.

For states with no intrinsic dipole moment, a second order term of the perturbation series needs to be considered [94]:

$$\Delta E_j^{(2)} = e^2 F^2 \sum_{l \neq j} \frac{\left| \langle \psi_l | z | \psi_j \rangle \right|^2}{E_j - E_l} \tag{19.33}$$

which can also be expressed as [97–99]:

$$\Delta E_j^{(2)} = -\frac{1}{2} p_j F^2 \tag{19.34}$$

where p is called the dipole polarizability. This term, known as the quadratic Stark effect, originates from the presence of a dipole moment induced by the field (hence the quadratic dependence on F) and is responsible for causing a redshift in the absorption.

Ultimately, the absorption spectrum is affected by both the linear and quadratic field-induced Stark effects. The electric field induces a change in the transition energy $\Delta E(F)$, expressed as a combination of both terms from Equation 19.32 and Equation 19.34 [87,100]:

$$\Delta E(F) = E(F) - E(0) = -(\mathbf{m}_f - \mathbf{m}_i) \cdot \mathbf{F} - 1/2 \Delta p F^2 \tag{19.35}$$

$\mathbf{m}_f - \mathbf{m}_i$ are the dipole moments in the final excited state and the initial ground state typically induced by disorder [95]. Similarly, Δp is the variation of polarizability upon excitation.

The field-induced change in absorption, measured in an EA experiment (Equation 19.16), can be expressed in terms of ΔE as a McLaurin series, truncated at the second term since the spectral shifts are small (typically tens of μeV for fields used in EA [97,99]):

$$\Delta \alpha = \left\langle \frac{\partial \alpha}{\partial E} \Delta E \right\rangle + \frac{1}{2} \left\langle \frac{\partial^2 \alpha}{\partial^2 E} \Delta E^2 \right\rangle \tag{19.36}$$

Inserting Equation 19.35 in Equation 19.36 yields the change in α in terms of the modulating electric field. Isotropic averaging over randomly oriented molecules leads to $\langle \mathbf{m} \cdot \mathbf{F} \rangle = 0$. Thus, the only contribution to the first term of Equation 19.36 (linear in ΔE and proportional to the first derivative of α) comes from the quadratic Stark effect ($-1/2 \Delta p F^2$ in Equation 19.35) which is always present. Isotropic averaging of $\mathbf{m} \cdot \mathbf{F}$ does however give a nonzero contribution to the second term of Equation 19.36 quadratic in ΔE (and proportional to the second derivative of α). For randomly oriented dipoles, isotropic averaging of the linear Stark effect term yields [87]:

$$\left\langle (\Delta \mathbf{m} \cdot \mathbf{F})^2 \right\rangle = 1/3 (\Delta \mathbf{m} \cdot \mathbf{F})^2 \tag{19.37}$$

Thus, inserting Equation 19.35 in Equation 19.36 yields the field-induced variation of the absorption in terms of the (linear and quadratic) Stark effect:

$$\Delta \alpha = \frac{1}{2} \left[-\Delta p \frac{\partial \alpha}{\partial E} + \frac{1}{3} (\Delta m)^2 \frac{\partial^2 \alpha}{\partial^2 E} \right] F^2 \tag{19.38}$$

A third contribution which needs to be added to Equation 19.38 arises from transfer of oscillator strength to higher lying excited states, also predicted by the Stark effect [97,98]:

$$\frac{\Delta f_{ij}}{f_{ij}} = e^2 F^2 \frac{\left| \langle \psi_l | z | \psi_j \rangle \right|^2}{(E_i - E_j)^2} \tag{19.39}$$

This term, which is in general smaller than those of Equation 19.38 is also quadratic in F, and in isotropic conjugated polymers can be approximated with a contribution following the absorption line shape.

Combining Equation 19.38 and Equation 19.39, the EA spectrum, which is proportional to $\Delta\alpha$, can be modelled as a linear combination of the absorption spectrum and its first and second derivatives:

$$\Delta\alpha = \left[a\alpha + b\frac{\partial\alpha}{\partial E} + c\frac{\partial^2\alpha}{\partial^2 E} \right] F^2 \tag{19.40}$$

From this fit, important parameters such as polarizabilities upon excitation can be extracted. In general, the EA spectra of conjugated polymers comprise a combination of all three contributions ($a,b,c \neq 0$), with the relative weight of the second derivative lineshape depending greatly on the degree of disorder present in the polymer film.

The above analysis, based on the Stark effect on localized molecular states, is found to work well for conjugated polymers, especially close to the absorption edge region. However, other effects may also become important depending on the material and the experimental conditions. For example, in solids where the electron is truly delocalized and a continuum of states is present, the lineshape of the spectrum follows the third derivative of the absorption in the low-field region [101]. For higher fields the quantum mechanical approach of Franz and Keldysh is necessary [86]. The latter can be dominant in crystalline inorganic semiconductors but is not readily visible for molecular or polymeric solids where the coherence length is not sufficiently long [97]. The Franz–Keldysh oscillations have, however, been observed in high quality crystals of polydiacetylenes [99]. In media with a high piezoelectric coefficient, changes in the index of refraction due to piezoelectricity may also become important [102].

In this chapter we will not dwell on the spectroscopic side of the EA technique which is used to study the susceptibility and excited states of conjugated polymers [95,97–99,102–108], but will focus solely on its field dependence. It is important to point out from the preceding analysis that the EA signal is expected to have a quadratic dependence on the magnitude of the electric field, F. As will be described in Section 19.3.4, this dependence is crucial in order to extract the magnitude of internal fields and the built-in potential across the polymer film in PLEDs, and is also the the key to monitor variations in barrier heights.

19.3.4   Measurement of the EA Signal in Organic Semiconductors

As described in Section 19.3.3, EA measures the normalized variation of the transmission of light through a material upon application of an electric field. In typical EA experiments on conjugated polymers (and equivalently on organic semiconductors in general—although in the following we will refer explicit to polymers the same considerations and results also hold true for small molecular weight organic semiconductors), this can be accomplished by two sample structures. The first involves depositing metal electrodes on a transparent substrate in an interlocking finger geometry (Figure 19.3a). The separation of the fingers is generally in the 10–100 μm range depending on the electrode deposition processes. Polymer or organic films are then cast over the electrodes. Light is incident

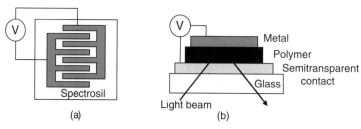

FIGURE 19.3 Sample structures used in electroabsorption experiments. (a) Top view of the interdigitated electrode configuration and (b) side view of the vertical diode structure.

perpendicularly to the sample surface. It travels through the polymer between the electrode fingers, and is measured by a photodetector at the other side of the substrate. A voltage is applied across the electrodes to create an electric field in the organic, which gives rise to an EA signal (Section 19.3.2 and Section 19.3.3) of the form:

$$\Delta T / T \cong \chi(h\nu) F^2 \qquad (19.41)$$

where $\chi(h\nu)$ contains the spectral dependence. The EA signal has a quadratic F^2 dependence on the electric field. Because $\Delta T / T$ is small (10^{-7}–10^{-4}), even for high fields, a lock-in technique is necessary. In order to achieve electric fields of $\sim 10^5$ V/cm, often required to obtain measurable signals, high sinusoidal voltage amplitudes (V_{AC}) of around 10^2–10^3 V are applied to the contacts. The interdigitated electrode array is the preferred geometry for spectroscopic measurements of polymer films and is best suited for investigations of the electronic excitations and nonlinear properties of conjugated polymers.

The second way of creating an electric field in a polymer film, and of studying its EA signal, is to use a vertical electrode/polymer/electrode structure. Semitransparent electrodes are required in order for light to be transmitted through this sandwich structure. Only one semitransparent contact is actually necessary since the light can be shone onto the PLED at a $\sim 45°$ angle, with the back electrode acting as a mirror (Figure 19.3b). This also increases the distance the photons have to travel through the polymer layer. Note that the metal/polymer/metal sample geometry coincides with the basic PLED structure. The fact that EA measurements can be performed directly on conjugated polymer diodes, makes EA a powerful technique since it enables the investigator to compare the EA results directly with the electronic and light-emitting or photovoltaic characteristics of the device. Typical diode thicknesses of ~ 100 nm mean that sinusoidal voltages are low ($\sim 10^{-1}$–10^0 V) in contrast to those required in the interdigitated geometry. EA experiments on diodes are particularly helpful for investigating the electric field inside the polymer devices, an aspect of EA spectroscopy we will now focus on. A schematic diagram of a typical EA apparatus is depicted in Figure 19.4.

In EA experiments, the electric field in the polymer layer, arising from the application of a sinusoidal voltage of the form $V = V_{DC} + V_{AC} \cos(\omega t)$, consists of the superposition of a direct current (DC) component (F_0) and an alternating current (AC) component (F_{AC}):

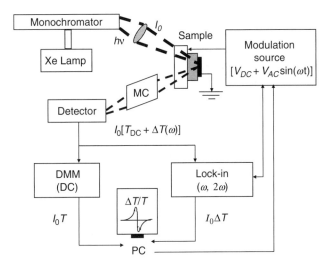

FIGURE 19.4 Electroabsorption spectrometer including the digital multimeter (DMM) for detecting the DC component ($I_0 T$) and the lock-in amplifier for detecting the AC component ($I_0 \Delta T$) generated by the sinusoidal voltage of the modulation source. The two components of the signal are fed into the computer (PC), which calculates the ratio to give the EA signal ($\Delta T / T$). The latter is independent of the intensity of the monochromatized light (I_0). MC is the second monochromator placed between the sample and the detector.

$$F = F_0 + F_{AC} \cos(\omega t) \tag{19.42}$$

The EA response to such a field can be determined by inserting Equation 19.42 in Equation 19.41, which after some simple trigonometry yields the following expression:

$$\Delta T/T(h\nu) \propto \chi(h\nu)[F_{AC}^2(1 + \cos(2\omega t))/2 + 2F_{AC}F_0 \cos(\omega t) + F_0^2] \tag{19.43}$$

$\Delta T/T$ contains a component at the fundamental harmonic frequency and a component at the second harmonic frequency. Employing a phase-sensitive lock-in technique one can measure the 2ω component of the EA signal:

$$\Delta T/T(h\nu, 2\omega) \propto \chi(h\nu)(F_{AC}^2/2) \cos(2\omega t) \tag{19.44}$$

and, in the presence of a DC field, also the 1ω component:

$$\Delta T/T(h\nu, \omega) \propto 2\chi(h\nu)F_0 F_{AC} \cos(\omega t) \tag{19.45}$$

It is clear from Equation 19.44 and Equation 19.45 that the EA signal provides information on the size of the field in the polymer film. Note that the DC component of the electric field, F_0, is not due only to the applied DC voltage across the electrodes (F_{DC}), but comprises also the internal or "built-in" electric field (F_{BI}) present in the organic semiconductor: $F_0 = F_{DC} + F_{BI}$. As discussed in Section 19.2, the internal field is generated in semiconductor/metal structures by the equilibration of the Fermi level throughout the different components of the metal/polymer/metal heterostructure and is also strongly influenced by any space charge present in the semiconducting organic layer or at the metal/polymer interfaces.

19.3.5   EA in Organic Semiconductor Diodes as a Probe for Internal Fields and Energy Level Lineups

Section 19.2 described that, in many cases, the Schottky model for the metal/polymer contact has been shown to be at least qualitatively valid, i.e., that barrier heights scale quite strongly with the work functions of the electrodes. This is particularly interesting, as for typical PLEDs fabricated with a calcium cathode ($\phi \sim 2.8$ eV) and a gold anode ($\phi \sim 5.1$ eV), the built-in potential can be over 2 V. Considering that the thickness of the PLED's active layer is $\sim$100 nm, a large built-in field (as large as $\sim 2 \times 10^5$ V/cm) can be present across the polymer. The actual magnitude of the electric field in PLEDs can indeed be measured accurately by EA spectroscopy using the method first reported by Campbell et al. in 1996 [4].

In Section 19.2 we also saw that in high purity films, when carrier injection and trapping is negligible, the charge density in the organic films can be low enough that the electric field across the bulk of the film is uniform [5,75,76]. This is visualized by rigidly tilted bands in the energy diagram of the PLED (Figure 19.5). For diodes in which injection is an efficient process, this condition is met only for reverse biases and small forward biases. For less efficient diodes, the range can be broader. In this uniform field case, the expression for the size of the electric field, constant across the bulk of the active layer, is straightforward:

$$F = V/d \tag{19.46}$$

Combining Equation 19.44 with Equation 19.47, the EA signal generated from the application of a bias of the form:

$$V = V_{DC} + V_{AC} \cos(\omega t) \tag{19.47}$$

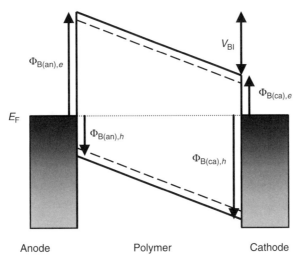

FIGURE 19.5 Schematic energy level diagram of an asymmetric metal/polymer/metal structure at equilibrium ($e = 1$) with negligible space charge. The solid lines represent the formation energy of polarons and the dashed lines that of bipolarons. The built-in electrostatic potential (V_{BI}) and the barriers (Φ_B) to holes and electrons are indicated. In the absence of pinning due to charged gap states, the Schottky model predicts $V_{BI} = \Delta\phi$, where ϕ are the work functions of the electrodes. (From Campbell, I.H., and D.L. Smith. *Solid state physics*. San Diego: Academic Press, pp. 1–117, 2001. With permission.)

is:

$$\Delta T/T(h\nu, 2\omega) \propto \chi(h\nu)(V_{AC}^2/2)\cos(2\omega t) \tag{19.48}$$

at 2ω. The signal at 1ω is given by:

$$\Delta T/T(h\nu,\omega) \propto 2\chi(h\nu)V_0 V_{AC}\cos(\omega t) \tag{19.49}$$

Note that V_0 is the sum of the applied voltage (V_{DC}) and the built-in electrostatic potential (V_{BI}):

$$V_0 = V_{DC} - V_{BI} \tag{19.50}$$

with the convention of $V_{DC} > 0$ for a positive bias applied to the anode, and V_{BI} positive for $\phi(\text{anode}) > \phi(\text{cathode})$. An internal electric field, $F_{BI} = V_{BI}/d$, generated by the mismatch between the work function of the diode's contacts, is present at equilibrium.

It is possible to determine the value of V_{BI} by monitoring the $\Delta T/T(1\omega)$ signal (Equation 19.49) as a function of V_{DC}. V_{BI} is easily calculated as the bias at which the EA(1ω) signal vanishes (i.e., $V_0 = V_{DC} - V_{BI} = 0$). V_{BI} can also be measured by calculating the ratio of the $\Delta T/T(1\omega)$ and $\Delta T/T(2\omega)$ signals.

Analysis of the energy level diagram of a conventional sandwich structure diode (Figure 19.6) yields the following relation between the energy gap (E_G) of the polymer, the built-in potential, and the barrier heights:

$$\Phi_B(\text{cathode}) + eV_{BI} + \Phi_B(\text{anode}) = E_G \tag{19.51}$$

where e is the elementary electronic charge.

The strategy that can be used to gauge the relative energy level positions is to build a series of devices differing only in the contact material of one of the electrodes, keeping the polymer layer and the other

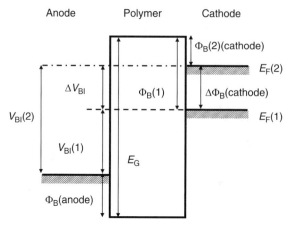

FIGURE 19.6 Schematic diagram of the energy levels of a sandwich-structure PLED at flat band ($e = 1$). Φ_B(cathode) [Φ_B(anode)] is the barrier height defined as the difference between the polymer LUMO [HOMO] and the Fermi energy of the electrode. At thermodynamic equilibrium the Fermi levels align, generating a built-in potential across the polymer layer. In this example, by replacing cathode (1) with cathode (2), the built-in potential (V_{BI}) increases by the difference in the Fermi energies. The variation (ΔV_{BI}) can be measured by electroabsorption spectroscopy and reflects a shift, of equal magnitude, in the barrier height ($\Delta\Phi_B$) at the electrode/polymer interface.

electrode unaltered [109]. For example in Figure 19.6: [E_G = constant, Φ_B(anode) = constant]. Thus, a linear dependence between V_{BI} and the barrier height at the modified electrode/polymer interface [i.e., Φ_B(cathode)] can be established. Accordingly, any variation measured in the V_{BI} across the set of devices is evidence of a variation in the barrier height equal in magnitude and opposite in sign, or explicitly:

$$e\Delta V_{BI} = -\Delta\Phi_B(\text{cathode}) \qquad (19.52)$$

Variations in V_{BI} will therefore have a profound effect on the current density vs. voltage (J–V) and EL (electroluminance) characteristics of the devices since Φ_B controls injection of charged carriers.

Since the energy levels of the amorphous conjugated polymers under study consist of a manifold of electronic states broadened by disorder, the definition of the Schottky energy barrier is inherently ambiguous. In internal photoemission experiments, this was defined as the difference between the metals Fermi energy (E_F) and the polymers lowest polaron level, and E_G represents the single particle energy gap [4]. When a distribution of molecular states is considered, the relevant energy levels in the polymer may be approximated to some average state in the polaron distribution [17]. In polymers with long conjugation lengths, these energy levels, relevant for conduction, can lie close [110] to the highest occupied and lowest unoccupied molecular levels (HOMO and LUMO), which are also widely employed as the polymer reference levels. We emphasize that the exact definition of the position of these states does not affect our discussion since we determine shifts in the cathode Fermi level and in the barrier height ($\Delta\Phi_B$ from ΔV_{BI}), rather than the absolute values of Φ_B. $\Delta\Phi_B$ is thus independent of the polymer reference level used.

To use Equation 19.52, it is important to employ high purity polymers with low trap concentrations, since the EA technique and Equation 19.49 through Equation 19.51 are based on the electric field (F) in the bulk of the polymer layer being uniform [4,111]. This can be confirmed by capacitance measurements [75] in the reverse to small forward bias regime. It is also essential to protect the samples during measurement procedures from contact with the atmosphere, either by encapsulation or by placing them in a vacuum chamber, in order to avoid degradation/contamination [112].

Since the probing beam of the EA technique passes through the whole thickness (and also the different layers) of the diode, the EA signal is the integrated contribution over all the thickness. It follows that in the presence of space charge and a nonuniform electric field, the expression for $\Delta T/T$ will also require the inclusion of the spatially dependent plane wave, representing the propagation of the light in the solid, and a spatial integration [113]. The presence of significant space charge could cause unwanted nonlinearity/hysteresis in the $\Delta T/T$ signal [112]. The electric field F will not be uniform also for biases where strong injection occurs (this may occur shortly after flat band because of the high injection efficiency of some of the devices reported here). The $\Delta T/T$ signal, which in general will not be linear anymore, will also include contributions from EL, bleaching of the linear absorption, excited state absorption, screening, and charge modulation [114–117]. In samples where significant injection only occurs at voltages considerably higher than V_{BI}, the $\Delta T/T$ signal continues to depend linearly on the applied bias even for voltages greater than V_{BI}. In diodes where barriers are minimized and injection occurs at biases close to V_{BI}, it is necessary to operate in a broad voltage range below V_{BI} to avoid these effects: in this case, if no significant space charge is present in the device (e.g., trapped or intrinsic charge), the EA signal is linear with V_{DC} at least in the reverse to weak forward bias regime, i.e., $V_{DC} \leq V_{BI}$.

We have seen that the energy levels of the materials that form the heterostructure may line up in different ways, with vacuum level alignment occuring in particular experimental conditions. As noted, interfacial interactions, such as charge transfer or dipole layer formation, can also alter the energy level alignment of the Fermi level at the electrode/semiconductor interface [59,60,118]. Dipoles, in particular, can be formalized by a discontinuity in the vacuum level, which will induce a shift in the energy level lineup, modifying the interfacial barrier heights (Φ_B) and V_{BI} (see Equation 19.12b). Thus, relying on the energy levels (including work functions) of the isolated components, when describing a heterostructure may lead to inaccuracies. Since Equation 19.51 is based on the energy diagram of a completed heterostructure (Figure 19.6), the above interfacial effects are already included, and therefore the relation $\Delta\Phi_B \cong -e\Delta V_{BI}$ is always maintained. This is providing that no strong chemical reactions, doping or filling of trapped states occur which modify the energy band of the polymer throughout an extended region. The fact that EA can measure these shifts on the actual completed diode makes this technique a uniquely powerful tool for the investigation of device physics.

It is also important to note that, inherently, EA measures a spatial average value of V_{BI}, since the area of the probe beam that is focused onto the LED is relatively large (a few mm^2). Similarly, Φ_B is best described in terms of an effective barrier height. Any local nonhomogeneity, present in the contact and in the emitting layer (i.e., uneven/varying electrode surface, polymer blends), will contribute to the average value of V_{BI} (and ΔV_{BI}, $\Delta\Phi_B$) and may be important in determining the "limiting" parameters of the operating characteristics (e.g., turn-on voltage, maximum efficiency).

To measure the Fermi level lineup of the diode electrodes according to the concept illustrated in Figure 19.6 and Equation 19.51 and Equation 19.52, it is fundamental that the field created by intrinsic and extrinsic space charge such as injected or trapped charge and impurities must be negligible with respect to that caused by the simple equilibration of the Fermi level of the electrode/semiconductor/electrode structure. For the energy levels of the semiconductor to simply tilt rigidly (constant field) rather than bend (nonuniform field), the voltage range has to be limited to biases below the injection limit. On the other hand, in some of the examples reported in the following sections, we will show that EA can also be employed effectively in applied voltage ranges where injection and charge accumulation does occur. In this second regime, if effects of EL, charge modulation, and the bleaching of absorption are screened out, EA spectroscopy can be a useful tool to gauge the effect of injected or trapped charges on the physics of devices where the presence of space charge is inevitable or even wanted (e.g., injected carriers, intentional, or unintentional doping, accumulation of charge at heterointerfaces in multilayer structures, trapped charge). In these cases, any deviation from the linear dependence of the EA signal on applied bias is a strong indication of nonuniform electric fields arising in the device and EA can be exploited to measure the fields and concentration of accumulated space charge at the various interfaces or trap sites across the device.

In what follows, we will now further review how EA spectroscopy has been employed, in some of the available literature, to gauge the built-in fields, the energy level lineup at organic semiconductor/electrode interfaces, and the physics of device operation [4,5,40,42,43,46,69,70,77,109,112,114,116,117,119–143].

19.4   The EA Method as a Probe for the Energy Level Line-up of Metallic Electrodes and Internal Field Distribution in Organic Devices

Campbell et al. [4] employed the EA method to measure the internal electric fields set up by the contact potentials in a variety of metal/MEH-PPV/metal diodes. The EA spectrum of the diodes measured at a fixed AC voltage exhibit a peak at ~2.1 eV. In order to extract the built-in potential, the energy of the probe beam was kept fixed (at 2.1 eV) and the EA signal monitored as a function of applied bias (Figure 19.7). V_{BI} coincides with the bias at which the EA signal (at the fundamental frequency) is null (Equation 19.49 and Equation 19.50).

The measured values of V_{BI} were compared against those calculated from the difference between the work functions to which V_{BI} would be equal to in the Schottky model according to the following equation:

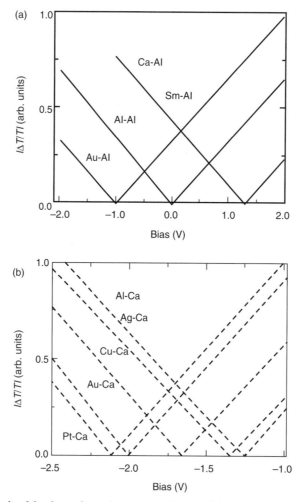

FIGURE 19.7   Magnitude of the electroabsorption response (energy of the probe beam = 2.1 eV) as a function of bias for (a) metal/MEH-PPV/Al structures and (b) metal/MEH-PPV/Ca structures. V_{BI} is extracted by interpolating at which bias the EA signal meets the zero axis. (From Campbell, I.H., T.W. Hagler, D.L. Smith, and J.P. Ferraris. *Phys. Rev. Lett.*, 76, 1900–1903, 1996. With permission.)

$$V_{BI}(\text{Schottky model}) = [\phi(\text{anode})/e - \phi(\text{cathode})/e]/d \qquad (19.12a)$$

and plotted in Figure 19.8. These results in combination with internal photoemission measurements of the Schottky barrier heights (Φ_B) [4,5], demonstrated that the metal contacts to MEH-PPV are accurately described by the ideal Schottky picture (Equation 19.12a) for most metals, and that the Schottky barriers indeed scale with the work functions of the metal contacts, as inferred from current–voltage investigations [8]. Only for metals whose work function lies outside the bipolaron (or charge transfer) energy gap, does the built-in potential saturate and not follow Schottky's predictions, i.e., $V_{BI} \neq \Delta\phi$. Theoretical calculations [75,77] (solid line of Figure 19.8) showed that the Fermi level of these low (high) work function metals, such as Ca and Sm (Pt), are in fact pinned close to the bipolaron levels (Figure 19.5), which explains the V_{BI} saturation for large work function differences between electrodes. The comparison of the V_{BI} and Φ_B results also demonstrated that the charge transfer bipolaron gap (determined from the V_{BI} saturation) is nearly as large as the single particle polaron gap (determined from Φ_B measurements); in fact, the polaron and bipolaron levels are indistinguishable within the experimental error of ~0.1 eV [4]. This means that the bipolarons were not found to be strongly bound and significantly influence device behavior [4,5].

Another valuable application of the EA method was to determine the acceptor level introduced by C_{60} when blended with MEH-PPV [128]. These MEH-PPV/C_{60} systems are particularly interesting for their photoconductive and photovoltaic properties [144,145]. Figure 19.9 shows the energy level diagrams for the pristine and C_{60}-doped MEH-PPV diodes. In the pristine case, the metal Fermi levels are pinned close to MEH-PPV's polaron levels and the value of V_{BI} is close to that of the polymer single particle polaron energy gap (Figure 19.9a). C_{60} introduces an electron acceptor energy level below the electron polaron level of MEH-PPV, which pins the E_F of calcium (Figure 19.9b). Since the C_{60} level lies within the gap of MEH-PPV, the built-in potential is visibly reduced. The extent of the V_{BI} reduction (~0.6 V) [128] indicates how deep the C_{60} state is from the electron polaron level of MEH-PPV.

Another application of the EA method was to determine the shift in the work function of a metal (and the consequent Schottky barrier shift when used as a contact in diodes) that can be induced by self-assembled monolayers (SAMs) [69,70,146–148]. The metal work function can be controlled by attaching a monolayer of polar molecules to the surface of the metal. Because of the inherent ordering of the monolayers (especially for thiol-based adsorbates on copper gold or silver electrodes), the molecular dipoles are oriented relative to

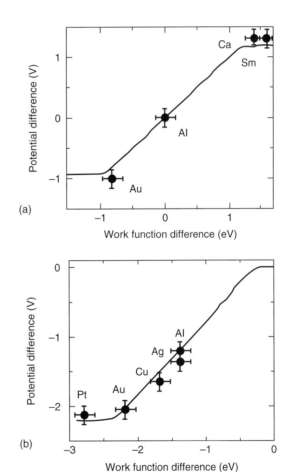

FIGURE 19.8 Experimental (points) and calculated (solid line) potential difference across (a) metal/MEH-PPV/Al and (b) metal/MEH-PPV/Ca structures as a function of the work function difference of the contacts. (From Campbell, I.H., T.W. Hagler, D.L. Smith, and J.P. Ferraris. *Phys. Rev. Lett.*, 76, 1900–1903, 1996. With permission.)

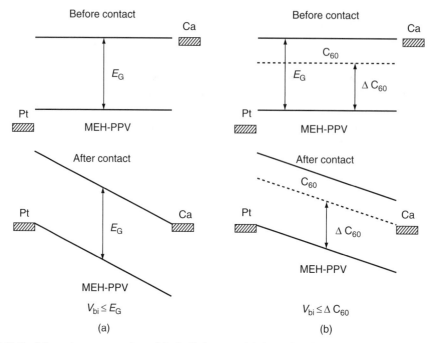

FIGURE 19.9   Schematic representation of the built-in potentials in undoped (a) and C_{60}-doped (b) MEH-PPV diodes. The upper panels show the relative alignment of the energy levels before contact. The lower panels show the built-in potential for the structures after contact. (From Heller, C.M., I.H. Campbell, D.L. Smith, N.N. Barashkov, and J.P. Ferraris. *J. Appl. Phys.*, 81, 3227–3231, 1997. With permission.)

the metal surface [5] and produce a surface potential shift of the form:

$$\Delta \theta = N_{mol}\left(\frac{\mu_{mol}}{\varepsilon} + \mu_{M^+S^-}\right) \tag{19.53}$$

where N_{mol} is the areal density of molecules, μ_{mol} is the dipole moment of the molecule perpendicular to the surface and $\mu_{M^+S^-}$ is the screened dipole moment of the metal$^+$$-S^-$ bond perpendicular to the surface (S being the sulphur atom of the SAM molecule which is expected to bond to the electrode). The measured changes in the built-in potential for copper electrodes were -0.3 V and 0.2 V [69], depending on which SAM was used in the fabrication of the LED electrode. These changes were found to increase to up to ~ 0.8 V when employing SAMs on silver contacts [70]. The EA and work function results [5,69,70] demonstrated that these thiol-based SAMs can be used to either increase or decrease the injection barriers. In the latter case, this does not always translate into improved charge injection because of the carrier blocking properties of the thiol molecules [5].

  Investigations of the internal field have also been used to study the field distribution and shed light on the physics of multilayer structures [43,46,131,135,137]. EA was employed to measure the density of accumulated charges at the interfaces between the different organic layers composing the multilayer diodes under both reverse and forward bias conditions [43]. To extract the magnitude of the electric field in each layer of a multilayer device from EA data, reference EA spectra were found for each individual layer under a known bias or electric field. In order to distinguish the signals from one layer to the other, it is important that the spectral position of the EA peaks and features of each layer are well separated and preferably the EA contribution of the other layers is small on the prominent features of each layer.

  In all these studies, the electric field distribution along all the organic semiconducting layers that make up the device was found to be approximately equal in reverse bias regime. Forward bias represents the more interesting regime, as the injected charge creates local fields that can differ strongly from layer to layer.

In general, in the forward bias regime experimental difficulties arise since the measured $\Delta T/T$ signal can become dominated by modulation of the EL signal, bleaching of the linear absorption, charge modulation, and screening due to injected and trapped charge (see previous sections). Furthermore, as also occurs in the case of devices made with organic blends, the wavelength at which one of the layers (or components in the case of a blend) shows a prominent peak, may coincide with an energy of the probe beam that is absorbed significantly by another layer (component). The resulting excitation, migration, separation, and accumulation of charges, if significant, may create local fields that perturb the fields under normal operation. In the studies of multilayer diodes reported here, it is assumed that the contribution of these effects was negligible compared to the EA signal arising from the electric fields present during normal operation of the diodes. Campbell et al. [115] have shown that in MEH-PPV devices, for injected carrier densities below about 5×10^{17} cm^{-3}, the modulation of the optical properties is still dominated by EA effects.

Accumulation of injected charges at interfaces between the layers of the multilayer device is definitely significant enough to create a highly nonuniform electric field distribution among the different layers of the multilayer diodes in all the reported studies. Campbell and collaborators [43] showed that in a multilayer organic LED composed of two electrodes, a hole transport layer, a light-emitting layer, and an electron transport layer, the magnitude of the DC electric field along the three layers was nonuniform as a result of accumulation of electrons (holes) at the interface between the light-emitting layer and the hole (electron) transport layer.

Rohlfing et al. [46] carried out a study on bilayer diodes composed of a 4,4′-*bis*[*N*-(1-naphthyl)-*N*-phenylamin]-biphenyl (α-NPD) hole transport layer and a *tris*-(8-hydroxyquinoline) aluminum (Alq) emitting layer, sandwiched between an ITO (indium tin oxide) anode and a Mg:Al cathode, extracting the average fields across the two organic layers from EA spectra as a function of applied bias. In reverse bias the fields are approximately equal. In forward bias the field is considerably stronger (up to roughly one order of magnitude) in the Alq layer compared to the NPD layer (Figure 19.10). Results also show that the field in the NPD remains lower than that of Alq but increases for forward biases above the turn-on voltage. Interestingly, Martin et al. [137] carried out a similar study, but

accompanying the experimental results with a one-dimensional (1D) time-independent drift–diffusion device model. The simulation results show that the electric field is uniform not only in reverse bias, as seen by Rohlfing et al. [46], but also in weak forward bias regime (below the turn-on voltage) since injection of carriers is not significant below turn-on. The field in the hole transport layer starts to decrease significantly above the turn-on voltage. The combination of simulations and experiments show that the magnitude and distribution of the electric field within the layers of the diode are critically affected by the two injection barriers at the opposite electrodes and by the the offsets between the HOMO and LUMO levels at the heterointerface between the two organic layers. It is these parameters that have a sizable impact on the accumulation of net charge (electrons and holes) at the organic/organic heterointerface, and hence the electric field distribution in the device.

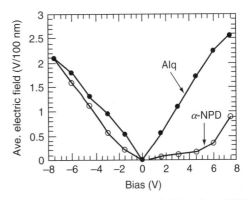

FIGURE 19.10 Average electric field in the α-NPD layer (open circles) and in the Alq layer (closed circles) of an α-NPD/Alq double-layer device due to DC bias only (excluding the built-in potential). (From Rohlfing, F., T. Yamada, and T. Tsutsui. *J. Appl. Phys.*, 86, 4978–4984, 1999. With permission.)

Rohlfing and coworkers [46] also performed EA measurements that allowed the determination of not just the average electric field in one layer, but of the electric field as a function of the position z within each organic layer, $F(z)$. This was accomplished by doping the organic layer with small concentrations of dopant that show a high intrinsic EA signal. Thus, the measurement of the EA signal of a series of Alq devices with the doped layer located at different positions z within the Alq layer provides information on the electric field distribution. The dopant material must have a very strong EA response compared to the host material in order to compensate for the small dopant concentration. Their results on single layer Alq devices suggest that the electric field in the Alq single layer is inhomogeneous in forward bias, and that the electric field near the ITO electrode (anode) is larger than that near the Mg:Ag electrode. Instead of introducing a thin doping layer to gauge the electric field distribution, $F(z)$, across one of the layers of the device by EA spectroscopy, Hiramoto et al. [129] introduced a "floating" Au electrode in the middle of the Alq layer (in a subsection of the device). The potential at this Au electrode was monitored during device operation. A series of devices was built with the Au electrode positioned at different (100 nm) intervals along the thickness of the Alq layer in order to map the electric field distribution. Qualitatively, the results of Hiramoto et al. [129] confirm that, during operation, most of the electric field falls across the Alq layer, rather than across the hole transport layer.

Other EA experiments confirm that under a DC bias, the electric field even within a single semiconductor layer is not uniform, not only in LEDs but also in solar cells [131,135]. The EA response from the bulk of the layer and at its interfaces, both metal/organic and organic/organic were shown to vary significantly, so that the detected EA signal is actually a linear combination of both components. Separating the two components by comparison of the EA spectra at different DC biases shows that strong electric fields develop at these interfaces.

Further insight into the transient response of PLEDs following the application of an electric field was also provided by EA spectroscopy. In particular, Giebeler et al. [112,127] showed that applying an electric bias to PLEDs induces a counter field with both a static and dynamic component. This field builds up within a few seconds of turning on the device, increases in magnitude with the operating voltage, and decays exponentially with a time constant of several seconds. The turn-on voltage of PLEDs increases as the device irreversibly degrades. The origin of the counter field and the decay dynamics could be explained by bulk carrier traps (which dominate in this case any variation in the work function of the electrode that may have occurred) which were formed when the device came into contact with air, either during fabrication and/or during measurement causing degradation (both reversible and irreversible) in the device characteristics.

The EA technique was also applied by Gao et al. [142] and deMello et al. [126] on light-emitting electrochemical cells (LECs) [125,126,142,149], a particular type of device where the active layers are intentionally heavily doped, to demonstrate the basic physics of these structures. The two studies show contrasting results. Gao et al. [142] presented a sharp increase in the electric fields above a threshold voltage as the DC bias applied to the LEC is increased. These EA results were compared to capacitance voltage and current voltage characteristics of the LEC, and were attributed to the formation of an electrochemical junction between an n-type (near the cathode) and p-type (near the anode) ion doped regions. In the work by deMello et al. [126], instead, EA clearly revealed that the electric field is null in the bulk of the active layer at room temperature and under steady state driving conditions for a range of applied biases. This was attributed to the migration of mobile ions toward the two electrodes under the influence of the applied electric field. Since the concentration of ions was intentionally high, the high density of charges at the interface was able to screen the field in the bulk of the electrochemical cell. Because of the redistribution of mobile charges, the local field was almost completely cancelled in the bulk of the polymer film [125,126]. Nearly complete screening of the electric field in the bulk of the active layer has also been detected in PLEDs [116,117] that incorporate a PEDOT:PSS anode, and will be discussed in the subsequent section.

Thus EA can be utilized to accurately measure the built-in potential of organic diodes and to gauge the electric fields not only across the active layers of (multilayer) devices but also their spatial distribution within a layer as a function of applied voltage (in both forward and reverse bias). In the next chapter we will see how EA has been used to probe the energy level lineup and the effect on device operation of nonmetallic (multilayer) electrodes, such as PEDOT:PSS anodes, LiF(CsF)/Al and LiF(CsF)/Ca/Al cathodes, that are used in the fabrication of PLEDs to deliver efficient carrier injection into the active light-emitting layer.

19.5 EA as a Probe of Internal Fields in Diodes with PEDOT:PSS Anodes and Multilayer Cathodes

19.5.1 PEDOT:PSS

Since the discovery of PLEDs [1], ITO has been the predominant choice as the anode electrode because of its good transparency in the visible spectrum and low resistivity. Such desirable properties, however, are counterbalanced by the difficulty of achieving uniformity amongst ITO suppliers, by a work function which is lower than desired, and by the inability to fulfil the industry's aim of all-polymer device fabrication. Furthermore, ITO is not a well-defined system (e.g., because of stoichiometric and structural inhomogeneities) and its physical properties are highly sensitive to the fabrication and preparation processes [150]. Polymeric anodes, such as PANI and PEDOT doped with PSS have proved to be good hole injectors in spite of possessing a relatively low conductivity [39,151–155]. PEDOT:PSS was first incorporated in polymer diodes by Scott et al. [156] and Carter et al. [153] in 1997. The ITO/PEDOT bilayer anode (also used as a contact in the production of efficient photodiodes and solar cells [157,158]) has given the most promising results as it combines the conductivity and transparency of ITO with the desirable properties of PEDOT yielding an increase in device efficiency and lifetime, and a reduction in the operating voltage [159,160]. This is ascribed to: (a) the ability of the conducting polymer to make a repeatable clean interface with good adhesion and hole injection into the EL polymer, (b) PEDOT's ability to suppress the oxidation process caused by the migration of oxygen out of the ITO into the emissive film [152,153,156], and (c) in great part to the beneficial hole injecting properties that the PEDOT film brings about as a result of the Schottky barrier lowering, interfacial doping, and electron accumulation at the anodic interface [39,42,44,78,116,119,123]. We will also review examples of how EA spectroscopy has been used to investigate the role and function of PEDOT:PSS in polymer diodes and will see that the insertion of this layer has a variety of effects.

We will start with the combined EA–Current–Luminance investigations of Brown et al. [119,123], in which the effects of a PEDOT:PSS covered ITO anode were compared to that of a bare ITO anode (cleaned in acetone and IPA) on the device characteristics and on the built-in potential of poly(4–4′-diphenylene diphenylvinylene) (PDPV) based LED.

Figure 19.11 shows the current density (top panel) and the luminance (bottom panel) dependence on applied bias of (a) ITO/PDPV/Ca/Al and (b) ITO/PEDOT:PSS/PDPV/Ca/Al. Clearly, the incorporation of the 60 nm thick PEDOT:PSS layer has a dramatic influence on the device's characteristics. Both the current and luminance increase considerably at a fixed voltage. A substantial reduction in operating voltages (V_{OP}) is also observed. At 0.1 cd/m^2 (and also at 0.1 mA/cm^2), V_{OP} decreases from over 30 V for the structure without PEDOT, to less than 20 V for the structure with PEDOT.

The EA(1ω) response of the two devices was measured as a function of applied DC bias (Figure 19.12). Since significant injection only occurs at several volts above V_{BI} in both devices, the EA signal is a linear function of V_{DC} over the whole range (Figure 19.12). In the ITO/PDPV/Ca case (a), the bias necessary to neutralize the built-in field across the PDPV layer is 2.0 V. The magnitude of the built-in potential increases by 0.5 to 2.5 V, when switching from an ITO-only anode to the ITO/PEDOT:PSS bilayer (b). Since the two LEDs differ only in the anode material and V_{BI} is expected to be equal to the difference in the work functions of the electrodes (if the Schottky model is found to be valid), an increase in V_{BI} would signal that the work function of PEDOT:PSS is larger than that of ITO.

To ascertain the possible presence of interface dipole effects in Figure 19.12, work function measurements of both ITO and PEDOT:PSS surfaces were carried out through a Kelvin Probe (KP) [119]. The work functions determined were: 4.8 ± 0.1 eV for ITO, and 5.2 ± 0.1 eV for PEDOT:PSS. Note that the difference between the two determinations (0.4 eV) is independent of the reference used. A similar

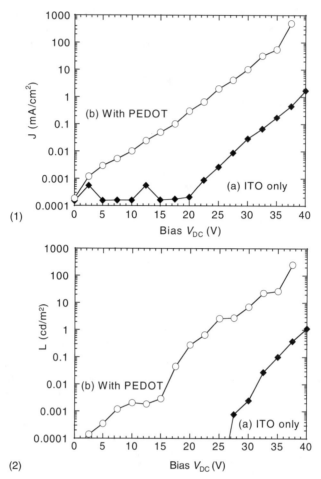

FIGURE. 19.11 Current density (1) and luminance (2) as a function of applied voltage for the (a) ITO/PDPV/Ca/Al and (b) ITO/PEDOT:PSS/PDPV/Ca/Al structures. (From Brown, T.M., and F. Cacialli. *J. Polym. Sci. B. Polym. Phys.*, 41, 2649–2664, 2003. With permission.)

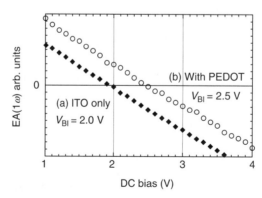

FIGURE 19.12 Electroabsorption response at 2.95 eV (V_{AC} = 1 V) as a function of applied DC voltage for the (a) ITO/PDPV/Ca/Al and (b) ITO/PEDOT:PSS/PDPV/Ca/Al diodes. The bias at which the response is null corresponds to the built-in potential, V_{BI}. (From Brown, T.M., et al. *Appl. Phys. Lett.*, 75, 1679–1681, 1999. With permission.)

increase in the work function was measured with ultraviolet photoelectron spectroscopy (UPS) in a parallel study [78], although, the absolute values were ~0.2/0.3 eV lower. UPS and KP work functions are not expected to agree, in view of the different nature of the techniques and especially because of the very different state of the surfaces during the measurement [161,162]. Other recent work function determinations [134,163,164] show that the work function of PEDOT:PSS does fall in the 5.2 – 5.3 eV range although it can be varied by subjecting it to different treatment and doping procedures [133,163,165]. Two important pieces of information have come out from the combination of the above EA and KP measurements: (1) the increase in work function is equal, within the experimental error, to the increase in the built-in potential observed when using PEDOT:PSS as a hole injector in the LED in lieu of the bare ITO electrode; (2) the KP values are also fully consistent with the cathode/anode work function mismatch estimated by subtracting the work function of calcium (2.8 eV) [63] from the KP readings of the ITO and PEDOT:PSS work functions, i.e., 2.0 ± 0.2 eV and 2.4 ± 0.2 eV respectively, indicating also that little band bending or dipole-induced vacuum level shifts are to be expected at either interface. In simple terms, the Schottky model is applicable at both polymer/contact interfaces and that $V_{BI} \approx \phi(\text{anode}) - \phi(\text{Ca})$, where the anode is either (a) ITO or (b) PEDOT:PSS as also confirmed by photoemission experiments on similar sandwich structures [74]. An important consequence of the increase in V_{BI} with PEDOT:PSS insertion is the barrier height reduction at the anode/emitter interface. In the first instance, the injection of carriers from the contacts into the emitting layer will be controlled by the offset between the Fermi level of each electrode and the relevant conduction levels in the polymer (here approximated to the HOMO or LUMO). Since the barrier height is lowered by ~0.5 eV by incorporating PEDOT:PSS, a significant enhancement in hole injection ensues. Such an effect is conspicuous in the EL characteristics of the LEDs, the operating voltages of which are reduced by a factor of two (Figure 19.11). Furthermore, the efficiency was typically ~0.11 cd/A for the devices without PEDOT, and increased by ~30%, on average, after the insertion of the PEDOT:PSS hole injecting layer. The observation of a large increase in EL in concomitance with a large decrease of the hole injection barriers is a strong indication that, in the diodes with the ITO anode and the low work function Ca cathode, the current is not dominated by holes [166] but that the electron contribution to the current is important.

Murata et al. [44] have suggested that the PEDOT:PSS layer also acts as a blocking layer [45,133] for electrons due to surface phase separation of the PSS component of PEDOT:PSS films, as observed in x-ray photoelectron spectroscopy experiments [167]. The buildup of electrons at the anode would then increase the field at the interface considerably. The result is a very significant improvement in hole injection and thus of overall current (consequently of light emission in PLEDs).

A similar effect was detected recently by Brewer et al. [42,117] and Lane et al. [116] through its manifestation on the internal electric field within the active emitting layer of PLEDs fabricated with a PEDOT:PSS anode and a Ca or Ba cathode. In these studies, EA spectroscopy was extended to biases above turn-on using a differential mode lock-in technique [168] to eliminate, from the $\Delta T/T$ signal, the modulated EL which becomes dominant once the diode emits light. Identifying spectral regions in which the measured EM signal was entirely due to the Stark effect, with negligible contamination from charge-induced modulation, permitted them to show that the field-induced modulation at both 1ω and 2ω fell to zero above turn-on (Figure 19.13).

By a combination of EA spectroscopy and modelling studies Lane et al. and Brewer et al. [42,116,117] propose that the bulk internal field is zero under ordinary operating conditions, with the injected electrons at the cathode becoming trapped close to the anode causing the potential to drop preferentially at the PEDOT:PSS/organic interface and fully screening the bulk semiconductor from the external applied field. The screening effect is only seen in polyfluorene-based LEDs with a PEDOT:PSS anode (therefore responsible for the electron traps) and that also have a cathode that injects electrons efficiently so that the traps get populated. Screening of the field in the bulk of the organic semiconductor under forward bias injecting conditions may not be exclusive to PEDOT:PSS, as recent studies by Hiromitsu et al. [130,131] have detected a similar effect in ZnPC Schottky barrier cells when Au is used as electrode. We note that although it is in theory possible that a zero overall $\Delta T/T$ signal can originate from the

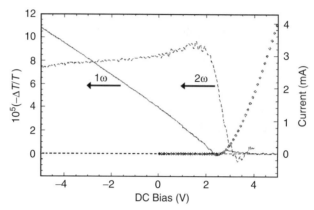

FIGURE 19.13 Left axis: DC bias dependence of the 1ω (solid line) and 2ω (dashed line) EA signals of a ITO/PEDOT:PSS/polyfluorene/Ca/Al diode. The 2ω EA signal was multiplied by a factor of 4 for ease of comparison. Right axis: current–voltage characteristics of the device (symbols). The EL intensity is proportional to the current. (From Lane, P.A., J.C. deMello, R.B. Fletcher, and M. Bernius. *Appl. Phys. Lett.*, 83, 3611–3613, 2003. With permission.)

nulling of opposite contributions at different depths in the active layer, it is very unlikely that this is the case for an extended range of voltages as reported by the authors. Similarly, it could be argued that, taken on its own, the vanishing of both the first and second harmonic of $\Delta T/T$ only indicates that the AC field is zero, and that the analytical expression of Equation 19.49 is unusable to gain information on the DC field. However, we note that the interpretation of de-Mello and collaborators has also been guided by separate experimental evidence for electron trapping at the hole injecting contact (e.g., van Woudenbergh and collaborators [169]) together with device simulations that demonstrate how with reasonable device parameters field-screening arises as a natural consequence of the electron trapping.

Overall, the strong influence PEDOT:PSS has in improving hole injection is due to the combined effect of the reduction of the barrier height [119], p-type interfacial doping of the semiconducting layer [39], and the high field strength at the anode interface due to accumumulated electronic charges [42,116].

In a work published in 2000, improved anodic injection and high LED efficiencies were achieved by Ho et al. [133] by using the polyelectrolyte route to deposit self-assembled anodes [133,170]. High injection efficiencies were obtained by progressively de-doping a series of self-assembled PEDOT:PSS interfacial layers creating a stepped and graded electronic anodic profile. Holes are first injected from the ITO anode into the gap states of the highly doped PEDOT, then transported down less doped PEDOT states that possess larger ionization potential values and injected into the light-emitting polymer over a lower effective barrier. EA experiments were utilized to show that there was an increase in the built-in potential of 0.18 ± 0.05 eV when the uniform interlayer was replaced by the graded interlayer confirming that a polymeric gradient doped profile could be sustained. By employing these graded anodes, and an electron-blocking interlayer at the anodic interface, it was possible to obtain remarkably efficient PLEDs [133].

Khan et al. [134] have recently studied degradation mechanisms in LEDs composed of a polyfluorene-based blue polymer emitter sandwiched between an ITO/PEDOT:PSS anode and an Al cathode. In this device, hole injection is initially ohmic. However, after electrical stress ($J = 12.5$ mA/cm^2, fields $\sim 10^6$ V/cm, $t = 16.75$ h) the degradation/decrease in current density was attributed to loss of ohmic injection as mobility measurements did not reveal significant change. Furthermore, the V_{BI} extracted from EA measurements (Figure 19.14) decreased drastically from 1.4 to 0.6 V after stress, evidence of a lowering of the work function of PEDOT:PSS. Corroborating measurements revealed that the work function of the PEDOT:PSS surface (measured with a KP after delaminating the blue emitter and cathode from the device after degradation) had indeed decreased significantly from its initial value. Additionally, the

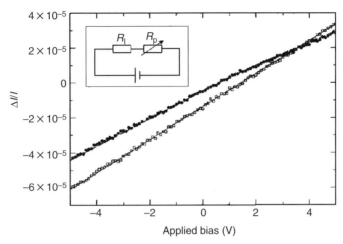

FIGURE 19.14 Variation of $\Delta T/T$ as a function of applied DC voltage. The open squares are for the unstressed device and the solid circles are for the stressed device. The inset shows a simple electrical model, where the resistance of the polymer is R_P and the interface resistance is R_I. (From Khan, R.U.A., D.D.C. Bradley, M.A. Webster, J.L. Auld, and A.B. Walker. *Appl. Phys. Lett.*, 84, 921–923, 2004. With permission.)

EA(1ω) vs. V_{DC} plots had a constant slope on prestressed devices and a double slope after stress. This indicated that an interfacial resistance R_I at the PEDOT/polymer interface was generated in series with the resistance of the bulk of the polymer R_P in the stressed device, perhaps arising from dedoping of PEDOT:PSS [171] and/or a chemical reaction between the two layers during electrical stress.

19.5.2 Cathodes Incorporating LiF and CsF Interlayers

An ideal cathode electrode for OLED applications is highly conductive, stable, provides a low barrier to electron injection, and does not degrade the light-emitting organic layer upon contact. In practice materials possessing all the properties have not yet been discovered and, for this reason, more elaborate structures have been devised. Low work function metals, such as Ca, Mg, Li, Cs, Sr, and Sm, are highly reactive [64,65,68,172–174] and have been shown to interact with the organic layer onto which they are evaporated and quench luminescence. Furthermore, these metals can also interact with the environment leading to quick degradation. Nevertheless, calcium capped with a protective layer of aluminum has been a common choice for the cathode as it has been shown to lead to devices with satisfactory luminescent characteristics if encapsulated properly. Barium has also been recently employed. An alternative consists of inserting a thin insulating layer (organic [69,175,176] or inorganic [177,178]) between a stable metal (Al) and the emitting organic film. Devices with fluoride/Al bilayer cathodes have shown substantial electron injection capability, high efficiency, and low turn-on voltages, with lithium-fluoride (LiF) and caesium-fluoride (CsF) yielding particularly successful results [40,71,109,120, 122,177,179–192]. However, for low electron affinity polymers [193], both calcium and fluoride/Al bilayers may not be efficient enough electron injectors, making the drive toward producing even more effective and nonhighly reactive electron-injecting contacts for wide gap conjugated polymers an ongoing one. Brown et al. [109] showed that by evaporating a 10 nm Ca film between the LiF(CsF) and the Al electrode (i.e., LiF(CsF)/Ca/Al cathodes), electron injection can be further enhanced.

Lack of consensus over the mechanism behind the enhancement of electron injection brought about by these thin layers points to the likely presence of several contributions, and different interpretations comprise enhanced electron tunneling from the Al due to a large voltage drop across the insulating interlayer [175,179,180], longer electron attenuation length [194], reduction of the image force effect [195], reduction of interfacial reactions of the polymer with the metal electrode [183,196], interfacial doping by the fluoride metal [65,187,190,196,197] or ions, dissociation of the fluoride to form a low work function contact [183], and reduction of the barrier height to electron injection [72,120,183,198–201]. In Section 19.5.2.1, we present EA investigations in which the changes in the barrier height at the cathode/polymer interface were monitored as a result of the cathode modifications with fluoride interlayers [40,109,120]. A strong correlation between V_{BI} and the devices EL characteristics was found, providing clear and direct evidence that the incorporation of the injecting interlayers leads to a marked reduction of the electron injection barrier.

Here we will discuss the results relative to two different sets of cathodic systems: (1) LEDs with LiF/Al cathodes for different LiF thicknesses; (2) LEDs with multilayer cathodes (including CsF/Al, LiF/Al, CsF/Ca/Al and LiF/Ca/Al). Further details on these investigations can be found in [109,120].

19.5.2.1   LEDs with LiF/Al Cathodes

These LEDs consisted of a 70 nm thick blue polyfluorene-based light-emitting polymer spun over an ITO/PEDOT:PSS(70 nm) anode. A thin layer of LiF of varying thickness ($d = 0$–11 nm) was thermally evaporated over the polymer, followed by the evaporation of an Al electrode (200 nm) at pressures of $\sim 10^{-6}$ mbar. All the fabrication processes were performed in a nitrogen glove-box and the completed LEDs encapsulated using an epoxy-resin for protection against degradation processes and agents (e.g., oxygen and water) during test procedures.

EA(1ω) vs. V_{DC} plots of the type shown, e.g., in Figure 19.7 and Figure 19.12 were utilized to extract the built-in potential for each device of the series of LEDs with varying LiF thickness. The field was uniform across the bulk of the active layer, for the ranges of V_{DC} (reverse to small forward bias) employed for the study as confirmed by capacitance/voltage measurements [114] comparable to those of Davids et al. [75]. This is not surprising since polyfluorenes can be synthesized to a very high purity ($\sim 10^{14}$ impurity sites/cm^3) [73], making them excellent candidates for EA experiments [114]. For the plot of the EA(1ω) vs. V_{DC} results the reader is referred to [120] (an analysis of the EA spectra of polyfluorene can be found in [202]). In Figure 19.15 we present the measured V_{BI} value for each diode of the series. In Figure 19.15 we have also incorporated the V_{BI} determination (increasing its error bar by 0.2 eV) of the second series of devices [109] [d(LiF) = 3.6 nm], whose investigations we will describe in Section 19.5.2.2, since the diodes' fabrication details were exactly the same except for the fact that the blue polyfluorene emitter was from a slightly different and more efficient batch (however with approximately the same energy levels). Although the EL characteristics between the two batches are not directly comparable, it is expected that the energy level alignment at the electrode/polymer interface in the same

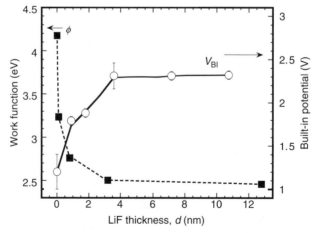

FIGURE 19.15   The built-in potential (circles), extracted from EA(1ω) vs. V_{DC} plots (from Brown, T.M., et al. *J. Appl. Phys.*, 93, 6159–6172, 2003; Brown, T.M., et al. *Appl. Phys. Lett.*, 77, 3096–3098, 2000), for PLEDs with LiF/Al cathodes and PEDOT:PSS anodes and the work function (squares) of LiF/Al bilayers (from Shaheen, S.E., et al. *J. Appl. Phys.*, 84, 2324–2327, 1998. With permission.) as a function of LiF thickness.

deposition conditions should be approximately analogous since the two blue polyfluorenes are very similar in composition especially and importantly in their energy levels.

The built-in potential of the bare Al electrode is $V_{BI} = 1.2$ V. For the device with the 0.9 nm-thick LiF layer, the bias required to null the built-in field is 1.8 V. This corresponds to a sizable increase. V_{BI} increases further with LiF thickness and at $d \sim 3.6$ nm has roughly reached the saturation value of 2.3 V belonging also to the devices with $d \cong 7$ and 11 nm. We have seen in Section 19.3 (Figure 19.6) that this large increase in V_{BI}, observed when evaporating LiF, is evidence of the LiF layer lowering the barrier height at the cathode by an equivalent amount: $\Delta\Phi_B = -e\Delta V_{BI}$. It is important to recognize that the saturation of V_{BI} may either be due to saturation of the LiF/Al bilayer work function or to the Fermi level being strongly pinned close to the polymer polaron levels. The occurrence of the latter can be excluded as a higher V_{BI} (and thus a lower barrier) has been measured in the same type of structures, but with different cathode materials (see Section 19.5.2.2). Work function saturation is consistent with photoemission experiments by Shaheen et al. [183] (and also by Heil et al. [38]), who have studied the LiF/Al system as a function of LiF thickness in the same (0–13 nm) range. Their work shows that the work function of Al decreases sharply after evaporation of a thin LiF overlayer, and reaches saturation at a LiF thickness of about 3 nm. A remarkably similar trend is manifested in the V_{BI} results of Figure 19.15, although the overall built-in potential variation (ΔV_{BI}) with LiF coverage is smaller than the variation in the work function observed by Shaheen et al. The degree to which the variation of the work function contributes to the alteration of the barrier height can be described in terms of the interface parameter $S = |d\Phi_B/d\phi|$ [55,57]. In the case of the LiF/Al diodes presented here, we calculate the ratio between the maximum variation in V_{BI} [i.e., $\Delta V_{BI} = V_{BI}(7 \text{ nm}) - V_{BI}(0 \text{ nm}) = 1.1 \pm 0.25$ V] upon the incorporation of LiF, and the maximum variation of the work function from [183] for the same LiF thickness (i.e., $\Delta\phi_{(cathode)} = 1.7 \pm 0.1$ eV). This yields $S = 0.65 \pm 0.2$. Thus, although the dipole contribution is significant (i.e., ~0.6 eV) the fact that $\Delta V_{BI} \cong 0.65*\Delta\phi_{(cathode)}$ is clear evidence of strong correlation between barrier lowering and the reduction of the LiF/Al bilayer work function. In the case of PEDOT:PSS (Section 19.5.1), an increase of the anode work function brought about an increase of V_{BI} of the same magnitude (and thus an equivalent reduction in the height of the barrier). This indicated that the vacuum levels of the emissive polymer and the contiguous electrodes (Ca and ITO or PEDOT:PSS) were continuous and aligned, and that $S \cong 1$. This corresponds to the Schottky–Mott model, where the variation in barrier height ($\Delta\Phi_B \sim e\Delta V_{BI}$) mirrors the change in the work function of the electrodes. It should be noted that this feature is common to a variety of polymeric semiconductors/metal junctions [4,73]. This is in striking contrast with the case of common 3D bonded elemental inorganic semiconductors such as silicon and germanium. The completely pinned limit (Bardeen-limit, $S = 0$) is closely approached by metal/silicon diodes for which the barrier height is almost independent of the metal work function ($S \leq 0.15$) [49,55,57]. However, for other semiconductor/metal systems the value of S is found to fall in a broad 0–1 range [55,57,58]. We also note that Tersoff and Mönch find a strong correlation between S and the inverse of the optical dielectric constant (ε_∞) of inorganic semiconductors [49,55]. For example, the optical dielectric constant is relatively large for Si and Ge ($\varepsilon_\infty > 10$) and the interface parameter S is low. In these semiconductors, the presence of a high density of surface gap states results in a strong degree of pinning of the metal Fermi energy and consequently of the injection barrier. Instead, EL polymers are characterized by essentially semiinsulating properties (large gap and low ε_∞) with a low concentration of gap states which suggest that S is closer to 1, as indeed found for many metal/polymer systems, a further indication of the purity and very low density of surface states in these materials [73,76].

It is also important to note that the evaporation conditions in the photoemission experiment are different (including the order in which the layers are evaporated) and carried out in a distinctive environment (ultra high vacuum), to which surface properties, such as the work function ϕ, are very sensitive. In fact, the energy level lineup of the LiF/Al heterojunctions is found to vary in the literature [38,40,72,120,183,198,199,203].

The fact that $S = 0.65 \pm 0.2 < 1$ may be a result of a moderate degree of E_F pinning and the formation of a dipole, which may be due to transfer of charge at the polymer/LiF interface, to the introduction of surface states (e.g., broadening of the EL spectrum does occur as shown in ref. [109]) or to a superior interfacial bonding between LiF and the polymer [55]. However, this may also be the result of the different evaporation conditions (including the order in which the layers are evaporated) used in the experiments, which can affect the magnitude of the work function reduction and the energy level lineup of the cathodic heterojunction [38,72,183,198,199,203]. Note that the magnitude of the built-in potential across the bulk of the polymer layer will reflect the consequences of interfacial reactions and dipoles that alter the energy level lineup. Since EA measurements are performed on the PLED itself, EA results are directly comparable to the LED's EL characteristics, eliminating problematic issues of comparability concerning deposition conditions.

The two main mechanisms [38,179,190,196,198,203] expected to be behind the reduction of ϕ (and Φ_B) in the presence of the fluoride interlayer are: (a) interaction of the fluoride with the metal and organic layers and its dissociation. The liberated Li would not only dope the polymer but, importantly, either create a low work function contact (Li has a low work function [204–206]) or, in the form of Li^+ ions, build a doped region of space charge at the cathode/polymer interface [40,41]; (b) a dipole-induced work function change due to either the large dipole moment of the oriented fluoride molecules [183,207,208] or the interfacial transfer of charge from the adsorbed fluoride layer [50] (in particular from the alkali metal atoms [71]) to the Al cathode.

Arguments both in favor [38,190,196] and against [179,198,203] the dissociation of LiF have been brought forward and depend crucially on evaporation and fabrication conditions [209]. It has been shown that LiF dissociation with Al is thermodynamically favorable only in the presence of water molecules [38] or in the presence of Alq$_3$ in small molecule organic LEDs [197]. In addition, a possible dipole has also been linked to chemisorbed H_2O [72]. The hygroscopic nature of LiF may promote the involvement of water molecules whose partial pressure is not negligible in the vacuum conditions ($\sim 10^{-6}$ mbar) typically used for evaporation. Jin et al. [40] have shown that inserting LiF between an Ag cathode and a light-emitting polymer actually reduces electron injection (and has no effect on the V_{BI} of the diode). This provided evidence that in the case of Ag the LiF remains an insulator, and thus impedes electron injection, while it does interact with Al.

Supporting the "dissociation" description are observations of barrier height lowering in devices with Al cathodes alloyed with metallic Li (and other low work function metals) [6]. In addition, enhanced electron injection has not only been observed in cathodes made with Al:Li alloys (difficult to evaporate reproducibly [178]) and Li/Al bilayers [210], but also with Al coevaporated with LiF or CsF (as opposed to their use as interlayers [184]). The reaction of the fluoride layer with Al, extending to a thickness of the order of nanometers [190], will also affect the transport properties of the interlayer itself and increase its conductivity significantly [120,187]. Note that in the experiments reported here [120], significant injection is measured with fluoride layers up to ~ 10 nm thick for which direct tunneling mechanisms would be improbable indicating that the LiF does not remain intact as an insulator. One possibility that may also contribute partially to improved electron injection in devices where holes are majority carriers is that the LiF layer may actually be beneficial in stopping holes (that remain majority carriers) from exiting the device at the cathode creating a build-up of positive charge and a high field at the interface.

Jin et al. [40] have recently conducted EA experiments on PLEDs fabricated with Al and LiF/Al cathodes that show similar results to those of Brown et al. [120] reported here (but quantitatively different due to different emitting polymer and fabrication conditions). They also correlate the large increase in the built-in potential, observed when inserting a thin LiF layer, to a corresponding decrease of the electron injection barrier. In their paper, Jin and coworkers propose a model, also recently put forward by Zhao and White. [41], in which a Li^+ *n*-doped region is created in the organic layer close to the Al electrode. This space charge layer is not as thin as an interfacial dipole that shifts the work function at a heterojunction between a metal and a semiconductor or another metal, but being just a few nm wide its effect is almost equivalent: the result is a lowering of the effective barrier height (although the charge has an interfacial region of a few nm to surmount). Although the presence of Li^+ ions may contribute to render electron injection more efficient it is not mutually exclusive with the reduction of the work function since a strong lowering of ϕ has also been reported on isolated LiF/Al bilayers [38,72,183].

The effects of LiF thickness on the EL characteristics of the LEDs are reported in Figure 19.16 alongside V_{BI} for comparison. Figure 19.16 confirms that LED performance is enhanced dramatically (lower turn-on voltage, higher luminance L, and maximum efficiency η) upon evaporation of a thin layer of LiF over the Al electrode. We wish to emphasize that this is in clear correspondence with the increase in V_{BI} and the related reduction of the cathodic barrier height. L and η do not, however, continue to increase with d, but have already reached a maximum at ~ 7 nm and decrease for the thicker 11 nm layer. LED performance worsens for thicker layers.

In order to interpret the data, we observe that, although a detailed analysis is complicated since other effects may also contribute to injection, including interfacial doping of the cathode/polymer interface, injection will depend strongly on LiF thickness (d), the barrier height (Φ_B) and electric field (F). Φ_B is strongly lowered by the LiF interlayer. Clearly, the barrier height is reduced progressively with d (from the Al case), until it reaches a plateau: the overall reduction achieved is large [$V_{BI}(7\text{ nm}) - V_{BI}(0\text{ nm}) = 1.1 \pm 0.25$ V]. This leads to a strong enhancement of the rate of electron injection from the cathode, which promotes a better balance of electron and hole currents, and, therefore, an improvement in device performance. Dissociation and reactions with Al, which greatly increase the conductivity of the LiF layer, are expected to occur only for the thinner layers, since for thicker layers the LiF should retain more of its insulating properties. Any tunneling mechanisms would also depend greatly on thickness. Increasing the thickness of LiF should thus have detrimental effects on the EL characteristics of the LED (higher turn-on voltage, lower luminance and efficiency).

Evidently, electron injection is a balance between the reduction of the cathodic barrier height (which initially increases with LiF thickness) and the conductivity of the LiF layer itself (which decreases with

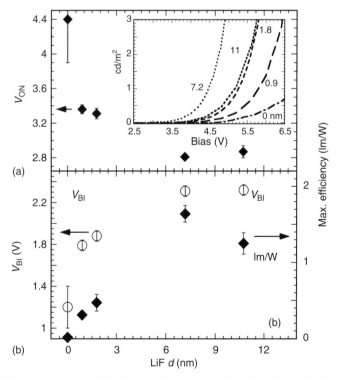

FIGURE 19.16 LED turn-on voltage and luminance characteristics (inset) are shown in the upper panel (a) as a function of the LiF thickness d (nm) of the cathodic injecting interlayer. The lower panel (b) shows the built-in potential (open circles) and the maximum efficiency (lm/W) as a function of d (nm). (From Brown, T.M., et al. *Appl. Phys. Lett.*, 77, 3096–3098, 2000. With permission.)

thickness). Figure 19.16, which only includes results from the series of devices of [120], would indicate that the optimum LiF thickness for efficient electron injection in the devices reported in [109,120] should occur somewhere between 1.8 and 7 nm, a thickness for which V_{BI} has reached a plateau. However, by also integrating the results of Section 19.5.2.2 ([109], $d = 3.6$ nm) into the results of this subsection ([120], Figure 19.14), as explained in the text , it is likely that this optimum thickness range is narrowed with an upper limit of 3.6 nm, since Figure 19.15 indicates that by increasing the LiF thickness beyond this value does not bring about further significant reductions in barrier height.

19.5.2.2 LEDs with Al, Ca, LiF/Al, CsF/Al, LiF/Ca/Al and CsF/Ca/Al Cathodes

In Section 19.5 EA spectroscopy was employed to give direct and quantitative evidence of a large lowering ($\sim$1.1 eV) of the cathodic barrier height upon the insertion of a thin LiF layer between Al and the emitting polymer. CsF/Al has also been used as an efficient electron injector in lieu of LiF [109,187]. Brown and coworkers [109,122] have subsequently shown that by evaporating calcium (lower work function and higher reactivity compared to Al) over the fluoride interlayers as opposed to aluminum (i.e., LiF/Ca/Al and CsF/Ca/Al cathodes) a further reduction in the barrier heights to electron injection can be achieved. This is especially important for injection in devices based on low electron affinity polymers [193]. In Section 19.5.2.2 we review these investigations [109] on a series of LEDs fabricated with ITO/PEDOT:PSS anodes, a blue polyfluorene light-emitting polymer and the following multilayer cathodes: (1) Al, (2) Ca(50 nm)/Al, (3) LiF/Al, (4) LiF/Ca(10 nm)/Al, (5) CsF/Al, (6) Ca(1.5 nm)/Al, and (7) CsF/Ca(10 nm)/Al. The thickness of the LiF and CsF layers was 3.6 nm and that of Al was 200 nm.

First of all, the V_{BI} values of the above PLEDs were extracted from the plots of the EA response vs. DC bias [109]. The results are reported in Table 19.1 and are utilized to draw the relative energy level positions of the different cathodes in Figure 19.17 (shown at flat band) according to Equation 19.51 and Equation 19.52.

In the Al cathode case V_{BI} is (1.2 $\pm$ 0.1) V. Inserting LiF, Ca, or CsF interlayers increases the V_{BI} sizably. The largest value for V_{BI} which we were able to determine, however, belongs to the LEDs fabricated with LiF/Ca/Al cathodes [$V_{BI} = (2.71 \pm 0.05)$ V]. The EA response of the CsF/Ca/Al, which deviates from the linear V_{DC} dependence and crosses the zero axis at biases lower than the other multilayer cathode devices, is reported in [109].

The increase of V_{BI} is connected directly to a reduction of the work function of the metallic-only cathodes when replacing Al with Ca/Al: the magnitude of V_{BI} for the two LEDs ($V_{BI} = 1.2$ V for Al and 2.4 V for Ca) is consistent with the built-in potential being equal to the difference between the work function of the cathode (Al or Ca) and that of the PEDOT:PSS anode ($\phi \cong 5.2$ eV) as in the ideal Schottky case: $V_{BI} \cong [\phi(\text{anode}) - \phi(\text{cathode})]/e$.

The strong correlation between V_{BI} and the work function of LiF/Al bilayers, for which the interface parameter $S = |d\Phi_B/d\phi| = 0.65 \pm 0.2$, is documented in [120] and Section 19.5.2.1. Although the exact dominant mechanism that leads to the reduction of the barrier heights upon inserting LiF/Al is still not fully pinpointed, not least because it depends strongly on fabrication materials and conditions, there is convincing evidence that CsF dissociates [187] more easily than LiF to liberate low work function Cs (and/or ions), even under UHV conditions [207,211]. Evaporation of a fluoride/Ca cathode could give rise to a lower barrier compared to fluoride/Al as Ca, being more reactive than Al would cause more dissociation and liberation of low work function Li or Cs (and/or ions creating an interfacial space charge area [40]), resulting in a lower effective barrier height averaged over the pixels' area and a more highly doped region at the interface. Note that since $\phi(\text{Cs}) < \phi(\text{Li}) < \phi(\text{Ca})$ [79,205,206], CsF can make a better electron injecting interlayer than LiF as indeed observed in Table 19.1 (LiF being less reactive yields devices that have a higher efficiency). In the case of a dipole contributing to the shift of the energy levels of Al or Ca closer toward the LUMO of the polymer upon insertion of the fluoride interlayer, the magnitude of such dipole (and thus the injecting capabilities of the interlayer) would also follow the experimental observations of Brown and coworkers [109] as it has been linked to the Pauling electronegativity of the fluoride metal ion [71].

TABLE 19.1 The Built-in Potential (V_{BI}), the Variation of the Built-in Potential from the Al-Only Device (ΔV_{BI}), the Maximum Efficiency (η), the Operating Voltage (V_{OP}) Defined as the Bias at which the Luminance (L) is 100 cd/m², the Operating Voltage (V_{OP}) Defined as the Bias at which $L = 0.01$ cd/m², and the Current Density (I) at a Bias of 5 V for the Diodes with Different Cathodes (All the Cathodes were Capped with a 200 nm Al Layer). The Values Represent Averages Taken on ~8 Pixels. For the CsF/Ca Device See Text.

Device	Al	Ca(50 nm)/Al	LiF/Al	LiF/Ca/Al	CsF/Al	Ca(1.5 nm)/Al	CsF/Ca/Al
V_{BI} (V)	1.22 ± 0.1	2.43 ± 0.05	2.31 ± 0.05	2.71 ± 0.05	2.55 ± 0.1	2.23 ± 0.05	0.8 ± 0.1
ΔV_{BI} (V)	0	1.21	1.09	1.49	1.33	1.01	
η(cd/A)	(4.4 ± 1.8)10^{-2}	3.5 ± 0.4	5.42 ± 0.27	3.44 ± 0.50	3.06 ± 0.80	2.0 ± 0.2	
V_{OP} (V) [$L=100$ cd/m²]	12 ± 2	4.1 ± 0.1	4.19 ± 0.15	3.57 ± 0.15	3.64 ± 0.07	6.45 ± 0.10	3.3 ± 0.1
V_{ON} (V)	4.2 ± 0.3	2.64 ± 0.1	2.74 ± 0.1	2.66 ± 0.1	2.5 ± 0.1	2.94 ± 0.16	2.6 ± 0.1
L (cd/m²) at 5 V	0.21 ± 0.18	496 ± 53	457 ± 122	1596 ± 280	1055 ± 185	10 ± 2	1147 ± 346
I (mA/cm²) at 5 V	0.99 ± 0.3	15 ± 2	8.5 ± 2.6	45 ± 7	37 ± 7	0.82 ± 0.10	142 ± 25

Source: From Brown, T.M., et al. *J. Appl. Phys.*, 93, 6159–6172, 2003.

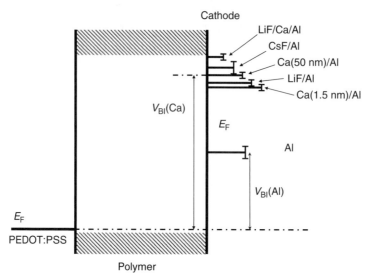

FIGURE 19.17   Schematic diagram ($e = 1$) of the average relative energy positions of the Fermi levels at flat band for the series of diodes with different multilayer cathode materials, drawn according to the built-in potential results reported in [109] and Table 19.1.

For the LED with the Cs/Ca/Al cathode, the EL efficiency is low (Table 19.1) and a new peak (shifted toward the red with respect to the main blue peak) appeared in the EL spectrum as reported [109]. This suggests that evaporation of the reactive Ca on the CsF leads to extensive dissociation of the CsF, to a much greater extent than when only Al [187,211] is evaporated on CsF. Cs can then diffuse and extensively dope the polymer layer and chemically react with it, creating space charge which gives rise to a nonlinear EA response [109] which makes definition of a V_{BI} nontrivial or not meaningful for the device as a whole and is the reason why CsF/Ca/Al is not included when discussing the correlation between V_{BI} and the EL characteristics. The new peak in the EL spectra is also an indication of the presence of a large concentration of new states in the energy gap [65,67,203,212] which can quench luminescence [64,65,173] and alter the energy level lineup, perhaps also pinning the Fermi of the CsF/Ca/Al level toward midgap [109].

As is apparent from Figure 19.17, the CsF/Al cathodes produce a smaller average barrier height than both LiF/Al and Ca. Devices with LiF/Ca/Al cathodes are expected to have the smallest average Φ_B consistently with the considerations outlined above. The lower effective electron barriers brought about by these multilayer cathodes lead to enhanced electron injection. This promotes larger electron (and hole currents as described later), leading to an increase in exciton formation and ultimately in the emission of light. Note that, especially for devices with LiF/Ca/Al and CsF/Al cathodes, the Φ_B reduction may not equal that of the work function ϕ, as the latter may be offset by strong pinning of the Fermi level occurring near the polymer polaron levels since the turn-on voltage for the LiF/Ca/Al and CsF/Al devices essentially coincides with V_{BI}.

The experimental EL characteristics of these LEDs are indeed extremely sensitive to the magnitude of the barrier heights [213], as is manifest in Figure 19.18. This figure displays the average V_{BI} values in conjunction with the operating voltages (V_{OP}) necessary to maintain a luminance of 100 cd/m^2, typical for display applications. Both LiF/Ca/Al and CsF/Ca/Al [whose luminance, however, degrades rapidly at higher voltages (Table 19.1)] show a particularly low V_{OP}, followed by CsF/Al, Ca, and then LiF/Al. As expected, Figure 19.18 shows that those LEDs with a larger built-in potential, hence with a smaller barrier to electron injection, require lower driving voltages (V_{OP} is plotted on a logarithmic scale while V_{BI} is on a linear one), consistent with carrier injection depending strongly on the barrier height ($\Phi_B = $ constant $- V_{BI}$).

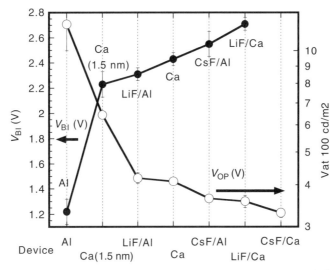

FIGURE 19.18 V_{BI} and the operating voltage (V_{OP}) at which the luminance is 100 cd/m^2 for PLEDs fabricated with different multilayer cathodes. (From Brown, T.M., et al. *J. Appl. Phys.*, 93, 6159–6172, 2003. With permission.)

Since both the luminance and the current increase with V_{BI}, the increase in luminance is a direct consequence of a larger current. This is important, because in LEDs where holes remain majority carriers, the overall current can remain roughly unvaried even when employing cathodes that greatly enhance both electron injection and luminance. In the experiments reported both luminance and overall current are greatly affected by the barrier-reducing cathodes, indicating that a truly bipolar transport should be taken into account. A concomitant increase in light emission and efficiency, together with a large increase in the current with enhanced electron injection, suggests that a strong supply of holes to form excitons for radiative recombination is present in all devices. Furthermore, we observe that the current in the diodes incorporating an interlayer is much larger than in the hole-dominated Al device, although the anodic barrier is common to all LEDs. Based on the above evidence it was concluded that the enhanced electron current acts as a positive feedback mechanism that also enhances the hole contribution to the overall current. This effect can be ascribed to the presence of the PEDOT:PSS causing accumulation of electrons at the anode interface and giving rise to a strong field that promotes hole injection as described [42,44,117,214] and outlined in the section on PEDOT.

Table 19.1 reveals that, compared to the Al-only device, the maximum efficiency increases by two orders of magnitude upon incorporation of LiF/Al, Ca, CsF/Al, and LiF/Ca cathodic interlayers. The highest maximum efficiency belongs to the LiF/Al LED {(5.42 ± 0.27) cd/A}. Conversely CsF/Ca/Al LEDs have a low efficiency (~0.8 cd/A) due to a very large current. The luminance of CsF/Ca/Al LEDs is large at low biases but degrades at higher voltages. The blue PLEDs with LiF/Ca/Al cathodes exhibit the highest luminance at 5 V {(1596 ± 143) cd/m^2} and also retain a high maximum efficiency {(3.44 ± 0.50)cd/A}. The LiF/Al devices exhibit the highest efficiency, although not the highest V_{BI}. This may be due to the LiF/Al diode presenting better quality interfaces with Ca, CsF/Al, CsF/Ca, LiF/Ca interacting more with the organic layer [196,199,203,211], thus reducing the PL efficiency [173]. LiF/Al devices may also maximize out-coupling and/or strike the most favorable balance of electron and hole currents.

In summary it is important to emphasize that cathode optimization involves, to a large extent, developing relatively stable contacts that provide a low electron injection barrier, while inducing minimum degradation to the luminescent efficiency of the emitting polymer layer. For the series of devices investigated by Brown and coworkers, devices with LiF/Ca/Al cathodes and PEDOT:PSS anodes yielded excellent electron-injecting properties while maintaining a high efficiency and also possessed a turn-on voltage that essentially coincided with the built-in potential within the experimental error, indicating an effective minimization of both barrier heights. The EA determinations [109,119,120] can also be used as a qualitative test for the energy level lineup theories presented in Section 19.2. For ITO, PEDOT, Al, and Ca electrodes the Schottky model for energy level lineup at the electrode/semiconductor interface was found to be valid within ~0.2 eV. For LiF/Al cathodes, although the estimated dipole contribution is significant (~0.6 eV), there remains a strong correlation between barrier heights and work function as the estimated interface parameter is $S = 0.65 \pm 0.2$, a high number compared to metal-silicon junctions for which the barrier height is almost independent of the metal work function, i.e., $S \leq 0.15$. Hence, although a quantitative description needs to take into account dipoles which can be sizable, the Schottky model for many of the devices investigated [109,119,120] was found to be a better first order approximation than the completely pinned limit (e.g., Bardeen model) except for those electrodes that may be either pinned at the polymer (bi)polaron levels [4] or which react extensively with the polymer layer introducing a large concentration of deep gap states (e.g., CsF/Ca/Al).

19.6 Conclusions

We have reviewed EA spectroscopy and its use as a powerful experimental tool to monitor, on completed and operational organic diodes: (a) the internal fields that arise from application of an external bias, equilibration of the Fermi energy, injection of charge carriers, and their accumulation at trap sites or at the heterointerfaces, and (b) the evolution of the injection barrier heights when different materials are used in order to optimize charge injection into the device. The results yielded by EA and their interpretation by researchers, also with the aid of device modelling, have not only enabled to gauge the electric fields in multilayer devices but also to determine the electric field distribution within an active layer of a device, to test theoretical models for the energy level lineup at the interface between conjugated polymers and their electrodes and also to draw significant conclusions on contact optimization and the device physics that underlies the operation of efficient electrodes for PLEDs such as PEDOT:PSS, LiF/Al, LiF(CsF)/Al, and LiF(CsF)/Ca/Al.

References

1. Burroughes, J.H., et al. 1990. Light-emitting diodes based on conjugated polymers. *Nature* 347:539–541.
2. Tang, C.W., and S.A. Vanslyke. 1987. Organic electroluminescent diodes. *Appl Phys Lett* 51:913–915.
3. Friend, R.H., et al. 1999. Electroluminescence in conjugated polymers. *Nature* 397:121–128.
4. Campbell, I.H., T.W. Hagler, D.L. Smith, and J.P. Ferraris. 1996. Direct measurement of conjugated polymer electronic excitation energies using metal/polymer/metal structures. *Phys Rev Lett* 76:1900–1903.
5. Campbell, I.H., and D.L. Smith. 2001. *Solid state physics*. San Diego: Academic Press, pp. 1–117.
6. Naka, S., et al. 1997. Electrical properties of organic electroluminescent devices with aluminium alloy cathode. *Synth Met* 91:129–130.
7. Campbell, I.H., and D.L. Smith. 1999. Schottky energy barriers and charge injection in metal/Alq/metal structures. *Appl Phys Lett* 74:561–563.
8. Parker, I.D. 1994. Carrier tunneling and device characteristics in polymer light-emitting diodes. *J Appl Phys* 75:1656–1666.
9. Sze, S.M. 1969. *The physics of semiconductor devices*. New York: Wiley.
10. Marks, R.N., D.D.C. Bradley, R.W. Jackson, P.L. Burn, and A.B. Holmes. 1993. Charge injection and transport in poly(*p*-phenylene vinylene) light-emitting diodes. *Synth Met* 55–57:4128–4133.
11. Arkhipov, V.I., E.V. Emelianova, Y.H. Tak, and H. Bassler. 1998. Charge injection into light-emitting diodes: Theory and experiment. *J Appl Phys* 84:848–856.
12. Wolf, U., V.I. Arkhipov, and H. Bässler. 1999. Current injection from a metal to a disordered hopping system. I. Monte Carlo simulation. *Phys Rev B* 59:7507–7513.

13. Barth, S., et al. 1999. Current injection from a metal to a disordered hopping system. III. Comparison between experiment and Monte Carlo simulation. *Phys Rev B* 60:8791–8797.

14. Greenham, N.C., and R.H. Friend. 1995. *Solid state physics.* San Diego: Academic Press, pp. 1–149.

15. Rikken, G.L.J.A., D. Braun, E.G.J. Staring, and R. Demandt. 1994. Schottky effect at a metal–polymer interface. *Appl Phys Lett* 65:219–221.

16. Bässler, H. 1993. Charge transport in disordered organic photoconductors. *Phys Status Solidi B* 175:15.

17. Conwell, E.M., and M.W. Wu. 1997. Contact injection into polymer light-emitting diodes. *Appl Phys Lett* 70:1867–1869.

18. Garstein, Y.N., and E.M. Conwell. 1996. Field-dependent thermal injection into a disordered molecular insulator. *Chem Phys Lett* 255:93–98.

19. Ioannidis, A., E. Forsythe, Y. Gao, M.W. Wu, and E.M. Conwell. 1998. Current–voltage characteristic of organic light-emitting diodes. *Appl Phys Lett* 72:3038–3040.

20. Vestweber, H., et al. 1995. Majority carrier injection from ITO anodes into organic light-emitting diodes based upon polymer blends. *Synth Met* 68:263–268.

21. Davids, P.S., S.M. Kogan, I.D. Parker, and D.L. Smith. 1996. Charge injection in organic light-emitting diodes: Tunneling into low mobility materials. *Appl Phys Lett* 69:2270–2272.

22. Davids, P.S., I.H. Campbell, and D.L. Smith. 1997. Device model for single carrier organic diodes. *J Appl Phys* 81:3227–6325.

23. Campbell, I.H., P.S. Davids, D.L. Smith, N.N. Barashkov, and J.P. Ferraris. 1998. The Schottky energy barrier dependence of charge injection in organic light-emitting diodes. *Appl Phys Lett* 72:1863–1865.

24. Scott, J.C., and G.G. Malliaras. 1999. Charge injection and recombination at the metal–organic interface. *Chem Phys Lett* 299:115–119.

25. Crone, B.K., I.H. Campbell, P.S. Davids, and D.L. Smith. 1998. Charge injection and transport in single-layer organic light-emitting diodes. *Appl Phys Lett* 73:3162–3164.

26. Campbell, A.J., D.D.C. Bradley, and D.G. Lidzey. 1997. Space-charge limited conduction with traps in poly(phenylene vinylene) light-emitting diodes. *J Appl Phys* 82:6326–6342.

27. Wolf, U., S. Barth, and H. Bässler. 1999. Electrode versus space-charge-limited conduction in organic light-emitting diodes. *Appl Phys Lett* 75:2035–2037.

28. Lampert, M.A., and P. Mark. 1970. *Current injection in solids.* New York: Wiley.

29. Blom, P.W.M., M.J.M. de Jong, and M.G. van Munster. 1997. Electric-field and temperature dependence of the hole mobility in poly(*p*-phenylene vinylene). *Phys Rev B* 55:R656–R659.

30. Blom, P.W.M., and M.J.M. de Jong. 1998. Electrical characterization of polymer light-emitting diodes. *IEEE J Sel Top Quant Elect* 4:105–112.

31. Pinner, D.J., R.H. Friend, and N. Tessler. 1999. Transient electroluminescence of polymer light-emitting diodes using electrical pulses. *J Appl Phys* 86:5116–5130.

32. Malliaras, G.G., and J.C. Scott. 1999. Numerical simulations of the electrical characteristics and the efficiencies of single-layer organic light-emitting diodes. *J Appl Phys* 85:7426–7432.

33. Tessler, N. 2000. Transport and optical modeling of organic light-emitting diodes. *Appl Phys Lett* 77:1897–1899.

34. Blom, P.W.M., M.J.M. de Jong, and J.J.M. Vleggaar. 1996. Electron and hole transport in poly(*p*-phenylene vinylene) devices. *Appl Phys Lett* 68:3308–3310.

35. Blom, P.W.M., M.J.M. de Jong, and S. Breedijk. 1997. Temperature dependent electron-hole recombination in polymer light-emitting diodes. *Appl Phys Lett* 71:930–932.

36. Blom, P.W.M., M.J.M. de Jong, and M.G. van Munster. 1997. Electric-field and temperature dependence of the hole mobility in poly(*p*-phenylene vinylene). *Phys Rev B Cond Matt* 55:R656–R659.

37. Pinner, D.J. 2000. Ph.D. thesis, Cambridge.

38. Heil, H., et al. 2001. Mechanisms of injection enhancement in organic LEDs through an Al/LiF electrode. *J Appl Phys* 89:420–424.

39. Wang, J.Z., J.F. Chang, and H. Sirringhaus. 2005. Contact effects of solution-processed polymer electrodes: Limited conductivity and interfacial doping. *Appl Phys Lett* 87:083503-1–3.

40. Jin, Y.D., et al. 2004. Role of LiF in polymer light-emitting diodes with LiF-modified cathodes. *Org Electron* 5:271–281.

41. Zhao, W., and J.M. White. 2005. Dramatically improving PLED performance by doping with inorganic salt. *Appl Phys Lett* 87:103503.

42. Brewer, P.J., et al. 2005. Role of electron injection in polyfluorene-based light-emitting diodes containing PEDOT:PSS. *Phys Rev B* 71:205209.

43. Campbell, I.H., M.D. Joswick, and I.D. Parker. 1995. Direct measurement of the internal electric field distribution in a multilayer organic light-emitting diode. *Appl Phys Lett* 67:3171–3173.

44. Murata, K., S. Cinà, and N.C. Greenham. 2001. Barriers to electron extraction in polymer light-emitting diodes. *Appl Phys Lett* 79:1193–1195.

45. Ho, P.K.H., M. Granström, R.H. Friend, and N.C. Greenham. 1998. Ultrathin self-assembled layers at the ITO interface to control charge injection and electroluminescence efficiency in polymer light-emitting diodes. *Adv Mater* 10:769–774.

46. Rohlfing, F., T. Yamada, and T. Tsutsui. 1999. Electroabsorption spectroscopy on *tris*-(8-hydroxy-quinoline) aluminum-based light-emitting diodes. *J Appl Phys* 86:4978–4984.

47. Flores, F., and C. Tejedor. 1987. On the formation of semiconductor interfaces. *J Phys C Solid State Phys* 20:145–175.

48. Margaritondo, G., and F. Capasso. 1987. *Heterojunction band discontinuities—Physics and device applications.* Amsterdam: North-Holland.

49. Mönch, W. 1986. On the present understanding of Schottky contacts. *Festkörperprobleme* XXVI:67–88.

50. Lüth, H. 1993. *Springer series in surface sciences*, Vol. 15, ed. M. Cardona, 372, 429. Berlin/Heidelberg: Springer.

51. Tersoff, J. 1987. *Heterojunction band discontinuities—Physics and device applications*, eds. G. Margaritondo, and F. Capasso, Amsterdam: North-Holland.

52. Margaritondo, G. 1988. *Electronic structure of semiconductor heterojunctions.* Milana: Jaca Book.

53. Magaud, L., and F. Cyrot-Lackmann. 1996. *Encyclopedia of applied physics*, ed. G.L. Trigg, pp. 573–591. New York: VCH Publishers.

54. Tersoff, 1984. Theory of semiconductor heterojunctions: The role of quantum dipoles *J Phys Rev B* 30:4874.

55. Tersoff, J. 1985. Schottky barriers and semiconductor band structures. *Phys Rev B* 32:6968–6971.

56. Lannoo, M., and P. Friedel, eds. 1991. *Atomic and electronic structure of surfaces.* Berlin/Heidelberg: Springer.

57. Kurtin, S., T.C. McGill, and C.A. Mead. 1969. Fundamental transition in the electronic nature of solids. *Phys Rev Lett* 22:1433–1436.

58. Schlüter, M. 1978. Chemical trends in metal-semiconductor barrier heights. *Phys Rev B* 17:5044–5047.

59. Ishii, H., K. Sugiyama, E. Ito, and K. Seki. 1999. Energy level alignment and interfacial electronic structures at organic/metal and organic/organic interfaces. *Adv Mater* 11:605–625.

60. Salaneck, W.R., S. Stafström, and J.-L. Brédas. 1996. *Conjugated polymer surfaces and interfaces: Electronic and chemical structure of interfaces for polymer light emitting devices.* Cambridge: Cambridge University Press.

61. Kahn, A., N. Koch, and W. Gao. 2003. Electronic structure and electrical properties of interfaces between metals and pi-conjugated molecular films. *J Polym Sci B* 41:2529–2548.

62. Hill, I.G., A. Rajagopal, A. Kahn, and Y. Hu. 1998. Molecular level alignment at organic semiconductor–metal interfaces. *Appl Phys Lett* 73:662–664.

63. Park, Y., et al. 1996. Energy level bending and alignment at the interface between Ca and phenylene vinylene oligomer. *Appl Phys Lett* 69:1080–1082.

64. Park, Y., V.E. Choong, B.R. Hsieh, C.W. Tang, and Y. Gao. 1997. Gap-state induced photoluminescence quenching of PPV oligomer and its recovery by oxidation. *Phys Rev Lett* 78:3955–3958.
65. Kido, J., and T. Matsumoto. 1998. Bright organic electroluminescent devices having a metal-doped electron-injecting layer. *Appl Phys Lett* 73:2866–2868.
66. Bharathan, J.M., and Y. Yang. 1998. Polymer/metal interfaces and the performance of polymer light-emitting diodes. *J Appl Phys* 84:3207–3211.
67. Birgerson, J., M. Fahlman, P. Broms, and W.R. Salaneck. 1996. Conjugated polymer surfaces and interfaces: A mini review and some new results. *Synth Met* 80:125–130.
68. Stoessel, M., et al. 2000. Cathode-induced luminescence quenching in polyfluorenes. *J Appl Phys* 87:4467–4475.
69. Campbell, I.H., et al. 1997. Controlling charge injection in organic electronic devices using self-assembled monolayers. *Appl Phys Lett* 71:3528–3530.
70. Campbell, I.H., et al. 1996. Controlling Schottky energy barriers in organic electronic devices using self-assembled monolayers. *Phys Rev B* 54:14321–14324.
71. Fujikawa, H., et al. 2000. Organic electroluminescent devices using alkaline-earth fluorides as an electron injection layer. *J Lumin* 87–9:1177–1179.
72. Schlaf, R., et al. 1998. Photoemission spectroscopy of LiF coated Al and Pt electrodes. *J Appl Phys* 84:6729–6736.
73. Greczynski, G., M. Fahlman, and W.R. Salaneck. 2000. Electronic structure of hybrid interfaces of poly(9,9-dioctylfluorene). *Chem Phys Lett* 321:379–384.
74. Greczynski, G., T. Kugler, and W.R. Salaneck. 2000. Energy level alignment in organic-based three-layer structures studied by photoelectron spectroscopy. *J Appl Phys* 88:7187–7191.
75. Davids, P.S., A. Saxena, and D.L. Smith. 1995. Nondegenerate continuum model for polymer light-emitting diodes. *J Appl Phys* 78:4244–4252.
76. Campbell, I.H., D.L. Smith, and J.P. Ferraris. 1995. Electrical impedance measurements of polymer light-emitting diodes. *Appl Phys Lett* 66:3030–3032.
77. Campbell, I.H., et al. 1996. Probing electronic state charging in organic electronic devices using electroabsorption spectroscopy. *Synth Met* 80:105–110.
78. Kugler, T., W.R. Salaneck, H. Rost, and A.B. Holmes. 1999. Polymer band alignment at the interface with indium tin oxide: Consequences for light emitting devices. *Chem Phys Lett* 310:391–396.
79. Michaelson, H.B. 1977. The work function of the elements and its periodicity. *J Appl Phys* 48:4729–4733.
80. Zangwill, A. 1988. *Physics at surfaces*, pp. 317–324. Cambridge: Cambridge University Press.
81. Hansen, W.N., and K.B. Johnson. 1994. Work function measurement in gas ambient. *Surf Sci* 316:373–382.
82. Hansen, W.N., and G.J. Hansen. 2001. Standard reference surfaces for work function measurements in air. *Surf Sci* 481:172–184.
83. Saville, G.F., P.M. Platzman, G. Brandes, R. Ruel, and R.L. Willet. 1995. Feasibility study of photocathode electron projection lithography. *J Vac Sci Technol B* 13:2184–2188.
84. Cardona, M. 1969. *Modulation spectroscopy.* New York: Academic Press.
85. Aspnes, D.E. 1980. *Handbook on semiconductors*, eds. T.S. Moss, and M. Balkanski, pp. 109–154. Amsterdam: North-Holland.
86. Pollak, F.H. 1994. *Handbook on semiconductors*, eds. T.S. Moss, and M. Balkanski, pp. 527–635. Amsterdam: North-Holland.
87. Weiser, G. 1973. Absorption and electroabsorption on amorphous films of polyvinylcarbazole and trinitrofluorene. *Phys Status Solidi A* 18:347–359.
88. Phillips, S.D., et al. 1989. Electroabsorption of polyacetylene. *Phys Rev B* 40:9751–9759.
89. Butcher, P., and D. Cotter. 1990. *The elements of nonlinear optics.* Cambridge: Cambridge University Press.
90. Bruce, W.D., and D. O'Hare. 1992. *Inorganic materials.* Chichester: Wiley.

91. Prasad, P.N., and D.J. Williams. 1991. *Introduction to nonlinear optical effects in molecules and polymers.* New York: John Wiley & Sons.

92. Bhattacharya, P. 1994. *Semiconductor optoelectronic devices.* Englewood Cliffs, NJ: Prentice-Hall.

93. Uchiki, H., and T. Kobayashi. 1988. New determination of electro-optic constants and relevant nonlinear susceptibilities and its application to doped polymer. *J Appl Phys* 64:2625–2629.

94. Bransden, B.H., and C.J. Joachain. 1983. *Physics of atoms and molecules.* New York: Longman Scientific and Technical.

95. Horvath, A., G. Weiser, G.L. Baker, and S. Etemad. 1995. Influence of disorder on the field-modulated spectra of polydiacetylene films. *Phys Rev B* 51:2751–2758.

96. Weiser, G., and A. Horvath. 1998. *Chem Phys* 227:153.

97. Harrison, M.G., et al. 1999. Electro-optical studies of a soluble conjugated polymer with particularly low intrachain disorder. *Phys Rev B* 60:8650–8658.

98. Horvath, A., H. Bässler, and G. Weiser. 1992. Electroabsorption in conjugated polymers. *Phys Status Solidi* 173:755–764.

99. Horvath, A., G. Weiser, C. Lapersonne-Meyer, M. Schott, and S. Spagnoli. 1996. Wannier excitons and Franz-Keldysh effect of polydiacetylene chains diluted in their single crystal monomer matrix. *Phys Rev B* 53:13507–13514.

100. Sebastian, L., G. Weiser, and H. Bässler. 1981. Charge transfer transitions in solid tetracene and pentacene studied by electroabsorption. *Chem Phys* 61:125–135.

101. Aspnes, D.E., and J.E. Rowe. 1972. Resonant nonlinear optical susceptibility: Electroreflectance in the low-field limit. *Phys Rev B* 5:4022–4030.

102. Heldmann, C., L. Brombacher, D. Neher, and M. Graf. 1995. Dispersion of the electro-optical response in poled polymer films determined by Stark spectroscopy. *Thin Solid Films* 261:241–247.

103. Liess, M., et al. 1997. Electroabsorption spectroscopy of luminescent and nonluminescent pi-conjugated polymers. *Phys Rev B* 56:15712–15724.

104. Liess, M., Z.V. Vardeny, and P.A. Lane. 1999. Electromodulation of charge-transfer photoexcitations in pristine and C60-doped conjugated polymers. *Phys Rev B* 59:11053–11061.

105. Premvardhan, L., L.A. Peteanu, P.-C. Wang, and A.G. MacDiarmid. 2001. Electronic properties of the conducting form of polyaniline from electroabsorption measurements. *Synth Met* 116:157–161.

106. Rohlfing, F., and D.D.C. Bradley. 1997. Optical non-linearity in beta-carotene: New insight from electroabsorption spectroscopy. *Chem Phys Lett* 277:406–416.

107. Rohlfing, F., and D.D.C. Bradley. 1998. Non linear Stark effect in polyazomethine and PPV. *Chem Phys* 227:133–151.

108. Weiser, G. 2004. Comparative electroabsorption studies of organic and inorganic solids. *J Lumin* 110:189–199.

109. Brown, T.M., et al. 2003. Electronic line-up in light-emitting diodes with alkali-halide/metal cathodes. *J Appl Phys* 93:6159–6172.

110. Shuai, Z., J.-L. Brédas, and W.P. Su. 1994. Nature of photoexcitations in poly(paraphenylene vinylene) and its oligomers. *Chem Phys Lett* 228:301–306.

111. Michelotti, F., S. Bussi, L. Dominici, M. Bertolotti, and Z. Bao. 2002. Space charge effects in polymer-based light-emitting diodes studied by means of a polarization sensitive electroreflectance technique. *J Appl Phys* 91:5521–5532.

112. Giebeler, C., et al. 1999. Optical studies of electric fields in MEH-PPV LEDs. *Appl Phys Lett* 74:3714–3716.

113. Aspnes, D.E., and A. Frova. 1969. Influence of spatially dependent perturbations on modulated reflectance and absorption of solids. *Solid State Commun* 7:155.

114. Brown, T.M. 2001. Ph.D. thesis, Cambridge University.

115. Campbell, I.H., D.L. Smith, C.J. Neef, and J.P. Ferraris. 2001. Optical properties of single carrier polymer diodes under high electrical injection. *Appl Phys Lett* 78:270–272.

116. Lane, P.A., J.C. deMello, R.B. Fletcher, and M. Bernius. 2003. Electric field screening in polymer light-emitting diodes. *Appl Phys Lett* 83:3611–3613.

117. Brewer, P.J., P.A. Lane, A.J. deMello, D.D.C. Bradley, and J.C. deMello. 2004. Internal field screening in polymer light-emitting diodes. *Adv Func Mat* 14:562–570.

118. Hill, I.G., D. Milliron, J. Schwartz, and A. Kahn. 2000. Organic semiconductor interfaces: Electronic structure and transport properties. *Appl Surf Sci* 166:354–362.

119. Brown, T.M., et al. 1999. Built-in field electroabsorption spectroscopy of polymer light-emitting diodes incorporating a doped poly(3,4-ethylene dioxythiophene) hole injection layer. *Appl Phys Lett* 75:1679–1681.

120. Brown, T.M., et al. 2000. LiF/Al cathodes and the effect of LiF thickness on the device characteristics and built-in potential of polymer light-emitting diodes. *Appl Phys Lett* 77:3096–3098.

121. Brown, T.M., and F. Cacialli. 2001. Energy level line-up in polymer light-emitting diodes via electroabsorption spectroscopy. *IEE Proc Optoelectr* 148:74.

122. Brown, T.M., et al. 2001. Efficient electron injection in blue-emitting polymer light-emitting diodes with LiF/Ca/Al cathodes. *Appl Phys Lett* 79:174–176.

123. Brown, T.M., and F. Cacialli. 2003. Contact optimization in polymer light-emitting diodes. *J Polym Sci B Polym Phys* 41:2649–2664.

124. Cheng, X.-M., S.-W. Yao, C.-Q. Li, T. Manaka, and M. Iwamoto. 2004. Measurement of surface potential at metal/organic-material interfaces by electro-absorption method. *Chin Phys Lett* 21:2026.

125. deMello, J.C., J.J.M. Halls, S.C. Graham, N. Tessler, and R.H. Friend. 1998. Ionic space-charge effects in polymer light-emitting diodes. *Phys Rev B* 57:12951–12963.

126. deMello, J.C., J.J.M. Halls, S.C. Graham, N. Tessler, and R.H. Friend. 2000. Electric field distribution in polymer light-emitting electrochemical cells. *Phys Rev Lett* 85:421–424.

127. Giebeler, C., S.A. Whitelegg, D.G. Lidzey, P.A. Lane, and D.D.C. Bradley. 1999. Device degradation of PLEDs studied by electroabsorption measurements. *Appl Phys Lett* 75:2144–2146.

128. Heller, C.M., I.H. Campbell, D.L. Smith, N.N. Barashkov, and J.P. Ferraris. 1997. Chemical potential pinning due to equilibrium electron transfer at metal/C60-doped polymer interfaces. *J Appl Phys* 81:3227–3231.

129. Hiramoto, M., K. Koyama, K. Nakayama, and M. Yokoyama. 2000. Direct measurement of internal potential distribution in organic electroluminescent diodes during operation. *Appl Phys Lett* 76:1336–1338.

130. Hiromitsu, I., Y. Murakami, and T. Ito. 2003. Electroabsorption study of the inner electric field in phthalocyanine/perylene solar cells. *Synth Met* 137:1385–1386.

131. Hiromitsu, I., Y. Murakami, and T. Ito. 2003. Electric field in phthalocyanine/perylene heterojunction solar cells studied by electroabsorption and photocurrent measurements. *J Appl Phys* 94:2434–2439.

132. Hiromitsu, I., and G. Kinugawa. 2005. Correlation between inner electric field and photocurrent of Zn-phthalocyanine Schottky-barrier cells. *Jpn J Appl Phys Part 1* 44:60–66.

133. Ho, P.K.H., et al. 2000. Molecular-scale interface engineering for polymer light-emitting diodes. *Nature* 404:481–484.

134. Khan, R.U.A., D.D.C. Bradley, M.A. Webster, J.L. Auld, and A.B. Walker. 2004. Degradation in blue-emitting conjugated polymer diodes due to loss of ohmic hole injection. *Appl Phys Lett* 84:921–923.

135. Lane, P.A., et al. 2000. Electroabsorption studies of phthalocyanine/perylene solar cells. *Solar Energy Mater Solar Cells* 63:3–13.

136. Liess, M., et al. 1999. Charge injection into OLED's during operation studied by electroabsorption screening. *Synth Met* 102:1075–1076.

137. Martin, S.J., G.L.B. Verschoor, M.A. Webster, and A.B. Walker. 2002. The internal electric field distribution in bilayer organic light-emitting diodes. *Org Electron* 3:129–141.

138. Yamada, T., F. Rohlfing, D. Zou, and T. Tsutsui. 2000. Internal electric field in *tris*-(8-hydroxyquinoline) aluminium (Alq) based light-emitting diode. *Synth Met* 111–112:281–284.

139. Yoon, J., J.J. Kim, T.W. Lee, and O.O. Park. 2000. Evidence of band bending observed by EA studies on PLED with ionomer/Al or LiF/Al cathode. *Appl Phys Lett* 76:2152–2154.

140. Whitelegg, S.A., et al. 2000. Optical studies of polymer light-emitting diodes using electroabsorption measurements. *Synth Met* 111–112:241–244.

141. Nuesch, F., M. Carrara, M. Schaer, D.B. Romero, and L. Zuppiroli. 2001. The role of copper phthalocyanine for charge injection into organic light emitting devices. *Chem Phys Lett* 347:311–317.

142. Gao, J., A.J. Heeger, I.H. Campbell, and D.L. Smith. 1999. Direct observation of junction formation in polymer light-emitting electrochemical cells. *Phys Rev B* 59:R2482–R2485.

143. Zimmermann, B., M. Glatthaara, M. Niggemanna, M. Riedea, and A. Hinscha. 2005. Electroabsorption studies of organic bulk-heterojunction solar cells. *Thin Solid Films* 493:170–174.

144. Sariciftci, N.S., and A.J. Heeger. 1995. Photophysics of semiconducting polymer C60 composites—A comparative study. *Synth Met* 70:1349–1352.

145. Halls, J.J.M., K. Pichler, R.H. Friend, S.C. Moratti, and A.B. Holmes. 1996. Exciton diffusion and dissociation in a poly(*p*-phenylenevinylene)/C-60 heterojunction photovoltaic cell. *Appl Phys Lett* 68:3120.

146. de Boer, B., A. Hadipour, M.M. Mandoc, T. van Woudenbergh, and P.W.M. Blom. 2005. Tuning of metal work functions with self-assembled monolayers. *Adv Mater* 17:621–625.

147. Morgado, J., et al. 2003. Self-assembly surface modified indium-tin oxide anodes for single-layer light-emitting diodes. *J Phys D-Appl Phys* 36:434–438.

148. Appleyard, S.F.J., S.R. Day, R.D. Pickford, and M.R. Willis. 2000. Organic electroluminescent devices: Enhanced carrier injection using SAM derivatized ITO electrodes. *J Mater Chem* 10: 169–173.

149. Pei, Q., G. Yu, C. Zhang, Y. Yang, and A.J. Heeger. 1995. Polymer light-emitting electrochemical-cells. *Science* 269:1086–1088.

150. Kim, J.S., et al. 1998. Indium-tin oxide treatments for single- and double-layer polymeric light-emitting diodes: The relation between the anode physical, chemical, and morphological properties and the device performance. *J Appl Phys* 84:6859–6870.

151. Gustafsson, G., et al. 1992. Flexible light-emitting diodes made from soluble conducting polymers. *Nature* 357:477–479.

152. Karg, S., J.C. Scott, J.R. Salem, and M. Angelopoulos. 1996. Increased brightness and lifetime of polymer light-emitting diodes with polyaniline anodes. *Synth Met* 80:111–117.

153. Carter, S.A., M. Angelopoulos, S. Karg, P.J. Brock, and J.C. Scott. 1997. Polymeric anodes for improved polymer light-emitting diode performance. *Appl Phys Lett* 70:2067–2069.

154. Sirringhaus, H., et al. 2000. High-resolution inkjet printing of all-polymer transistor circuits. *Science* 290:2123–2126.

155. Campbell, A.J., et al. 2000. Transient and steady-state space-charge-limited currents in polyfluorene copolymer diode structures with ohmic hole injecting contacts. *Appl Phys Lett* 76:1734–1736.

156. Scott, J.C., S.A. Carter, S. Karg, and M. Angelopoulos. 1997. Polymeric anodes for organic light-emitting diodes. *Synth Met* 85:1197–1200.

157. Granström, M., et al. 1998. Laminated fabrication of polymeric photovoltaic diodes. *Nature* 395:257–260.

158. Arias, A.C., M. Granström, D.S. Thomas, K. Petritsch, and R.H. Friend. 1999. Doped conducting-polymer–semiconducting-polymer interfaces: Their use in organic photovoltaic devices. *Phys Rev B* 60:1854–1860.

159. Carter, J.C., et al. 1997. Operating stability of light-emitting polymer diodes based on poly (*p*-phenylene vinylene). *Appl Phys Lett* 71:34–36.

160. Kim, J.S., R.H. Friend, and F. Cacialli. 1999. Improved operational stability of polyfluorene-based organic light-emitting diodes with plasma-treated indium-tin oxide anodes. *Appl Phys Lett* 74:3084.

161. Henrich, V.E., and P.A. Cox. 1994. *The surface science of metal oxides*, 7–10. Cambridge: Cambridge University Press.

162. Kim, J.S., et al. 2000. Kelvin probe and ultraviolet photoelectron spectroscopy measurements of indium tin oxide workfunction: A comparison. *Synth Met* 111–112:311–314.

163. Huang, J., P.F. Miller, J.S. Wilson, A.J. deMello, and J.C. deMello. 2005. Investigation of the effects of doping and post-deposition treatments on the conductivity, morphology, and work function of poly(3,4-ethylenedioxythiophene)/poly(styrene sulfonate) films. *Adv Funct Mater* 15:290.

164. Tengstedt, C., et al. 2005. Study and comparison of conducting polymer hole injection layers in light emitting devices. *Org Electron* 6:21–33.

165. Snaith, H.J., H. Kenrick, M. Chiesa, and R.H. Friend. 2005. Morphological and electronic consequences of modifications to the polymer anode PEDOT:PSS. *Polymer* 46:2573–2578.

166. Scott, J.C., et al. 1999. Hole limited recombination in polymer light-emitting diodes. *Appl Phys Lett* 74:1510–1512.

167. Greczynski, G., T. Kugler, and W.R. Salaneck. 1999. *Thin Solid Films* 354:129–135.

168. Pires, M.P., P.L. Souza, and J.P. von der Weid. 1996. New improved technique to measure photoreflectance. *Bras J Phys* 26:252–255.

169. van Woudenbergh, T., P.W.M. Blom, and J.N. Huiberts. 2003. Electro-optical properties of a polymer light-emitting diode with an injection-limited hole contact. *Appl Phys Lett* 82:985–987.

170. Ho, P.K.H. 2000. Ph.D. thesis, Cambridge University.

171. Kim, J.S., P.K.H. Ho, C.E. Murphy, N. Baynes, and R.H. Friend. 2002. Nature of non emissive black spots in PLEDs by in situ micro Raman spectroscopy. *Adv Mater* 14:206–209.

172. Choong, V., et al. 1996. Dramatic photoluminescence quenching of phenylene vinylene oligomer thin-films upon submonolayer Ca deposition. *Appl Phys Lett* 69:1492–1494.

173. Choong, V.E., et al. 1997. Effects of Al, Ag, and Ca on luminescence of organic materials. *J Vac Sci Technol A* 15:1745–1749.

174. Parthasarathy, G., C. Shen, A. Kahn, and S.R. Forrest. 2001. Lithium doping of semiconducting organic charge transport materials. *J Appl Phys* 89:4986–4992.

175. Kim, Y.E., H. Park, and J.J. Kim. 1996. Enhanced quantum efficiency in polymer electroluminescence devices by inserting a tunneling barrier formed by Langmuir–Blodgett films. *Appl Phys Lett* 69:599–601.

176. Jung, G.Y., et al. 2000. The effect of insulating spacer layers on the electrical properties of polymeric Langmuir–Blodgett film light emitting devices. *J Phys D Appl Phys* 33:1029–1035.

177. Li, F., H. Tang, J. Anderegg, and J. Shinar. 1997. Fabrication and electroluminescence of double-layered organic light-emitting diodes with the Al_2O_3/Al cathode. *Appl Phys Lett* 70:1233–1235.

178. Wakimoto, T., et al. 1997. Organic EL cells using alkaline metal compounds as electron injection materials. *IEEE Trans Electron Dev* 44:1245.

179. Hung, L.S., C.W. Tang, and M.G. Mason. 1997. Enhanced electron injection in organic electroluminescence devices using an Al/LiF electrode. *Appl Phys Lett* 70:152–154.

180. Jabbour, G.E., et al. 1997. Highly efficient and bright organic electroluminescent devices with an aluminum cathode. *Appl Phys Lett* 71:1762–1764.

181. Kido, J., and Y. Iizumi. 1998. Fabrication of highly efficient organic electroluminescent diodes. *Appl Phys Lett* 73:2721–2723.

182. Matsumura, M., and Y. Jinde. 1998. Analysis of current–voltage characteristics of organic light-emitting diodes having a LiF/Al cathode and an Al-hydroxyquinoline/diamine junction. *Appl Phys Lett* 73:2872–2874.

183. Shaheen, S.E., et al. 1998. Bright blue organic light-emitting diode with improved color purity using a LiF/Al cathode. *J Appl Phys* 84:2324–2327.

184. Jabbour, G.E., B. Kippelen, N.R. Armstrong, and N. Peyghambarian. 1998. Aluminum based cathode structure for enhanced electron injection in electroluminescent organic devices. *Appl Phys Lett* 73:1185–1187.

185. Ishii, M., T. Mori, H. Fujikawa, S. Tokito, and Y. Taga. 2000. Improvement of organic electroluminescent device performance by in situ plasma treatment of indium-tin-oxide surface. *J Lumin* 87–9:1165–1167.

186. Kobayashi, H., et al. 2000. A novel RGB multicolor light-emitting polymer display. *Synth Met* 111:125–128.

187. Piromreun, P., et al. 2000. Role of CsF on electron injection into a conjugated polymer. *Appl Phys Lett* 77:2403–2405.

188. Wang, J.F., et al. 1999. Oxadiazole metal complex for organic light-emitting diodes. *Adv Mater* 11:1266–1269.

189. Donat-Bouillud, A., et al. 2000. Light-emitting diodes from fluorene-based pi-conjugated polymers. *Chem Mater* 12:1931–1936.

190. Hung, L.S., C.W. Tang, M.G. Mason, P. Raychaudhuri, and J. Madathil. 2001. Application of an ultra-thin LiF/Al bilayer in organic surface-emitting diodes. *Appl Phys Lett* 78:544–546.

191. Chan, M.Y., et al. 2003. Efficient CsF/Yb/Ag cathodes for OLEDs. *Appl Phys Lett* 82:1784–1786.

192. Chin, D.C., L. Duan, M.-H. Kim, S.T. Lee, and H.K. Chung. 2004. Effects of cathode thickness and thermal treatment on the design of balanced blue LEP devices. *Appl Phys Lett* 85:4496–4498.

193. Janietz, S., et al. 1998. Electrochemical determination of the ionization potential and electron affinity of poly(9,9-dioctylfluorene). *Appl Phys Lett* 73:2453–2455.

194. Wolf, U., and H. Bässler. 1999. Enhanced electron injection into light-emitting diodes via interfacial tunneling. *Appl Phys Lett* 74:3848–3850.

195. Tutis, E., M.N. Bussac, and L. Zuppiroli. 1999. Image force effects at contacts in organic light-emitting diodes. *Appl Phys Lett* 75:3880–3882.

196. Le, Q.T., et al. 2000. Photoemission study of aluminum/tris-(8-hydroxyquinoline) aluminum and aluminum/LiF/tris-(8-hydroxyquinoline) aluminum interfaces. *J Appl Phys* 87:375–379.

197. Mason, M.G., et al. 2001. Interfacial chemistry of Alq(3) and LiF with reactive metals. *J Appl Phys* 89:2756–2765.

198. Mori, T., H. Fujikawa, S. Tokito, and Y. Taga. 1998. Electronic structure of 8-hydroxyquinoline aluminum/LiF/Al interface for organic electroluminescent device studied by ultraviolet photoelectron spectroscopy. *Appl Phys Lett* 73:2763–2765.

199. Yoshimura, D., et al. 1999. Electronic structure of Alq(3)/LiF/Al interfaces studied by UV photoemission. *Synth Met* 102:1145–1146.

200. Yang, X.H., Y.Q. Mo, W. Yang, G. Yu, and Y. Cao. 2001. Efficient polymer light-emitting diodes with metal fluoride/Al cathodes. *Appl Phys Lett* 79:563–565.

201. Masenelli, B., D. Berner, M.N. Bussac, F. Nuesch, and L. Zuppiroli. 2001. Simulation of charge injection enhancements in organic light-emitting diodes. *Appl Phys Lett* 79:4438–4440.

202. Cadby, A.J., et al. 2000. Film morphology and photophysics of polyfluorene. *Phys Rev B* 62:15604–15609.

203. Greczynski, G., M. Fahlman, and W.R. Salaneck. 2000. An experimental study of poly(9,9-dioctylfluorene) and its interfaces with Li, Al, and LiF. *J Chem Phys* 113:2407–2412.

204. Lang, N.D., and W. Kohn. 1971. Theory of metal surfaces: Work function. *Phys Rev B* 3:1215–1222.

205. Greczynski, G. Work function of Li = 2.2–2.3 eV. Personal communication.

206. Friedlein, R. Work function of Li = 2.6–2.7 eV. Personal communication.

207. Greczynski, G., M. Fahlman, and W.R. Salaneck. 2001. Hybrid interfaces of poly(9,9-dioctylfluorene) employing thin insulating layers of CsF: A photoelectron study. *J Chem Phys* 114:8628–8636.

208. Brabec, C.J., S.E. Shaheen, C. Winder, N.S. Sariciftci, and P. Denk. 2002. Effect of LiF/metal electrodes on the performance of plastic solar cells. *Appl Phys Lett* 80:1288–1290.

209. Jonsson, S.K.M., W.R. Salaneck, and M. Fahlman. 2005. Photoemission of Alq3 and C60 films on Al and LiF/Al substrates. *J Appl Phys* 98:014901.

210. Kido, J., K. Nagai, and Y. Okamoto. 1993. Bright organic electroluminescent devices with double-layer cathode. *IEE Trans Electron Dev* 40:1342–1344.

211. Greczynski, G., W.R. Salaneck, and M. Fahlman. 2001. Hybrid interfaces in polymer-based electronics. *Synth Met* 121:1625–1628.

212. Ettedgui, E., H. Razafitrimo, Y. Gao, B.R. Hsieh, and M.W. Ruckman. 1996. Schottky barrier formation at the Ca/PPV interface and its role in tunneling at the interface. *Synth Met* 78:247–252.

213. Malliaras, G.G., J.R. Salem, P.J. Brock, and C. Scott. 1998. Electrical characteristics and efficiency of single-layer organic light-emitting diodes. *Phys Rev B-Condens Matt* 58:R13411–R13414.

214. Morgado, J., R.H. Friend, and F. Cacialli. 2002. Improved efficiency of light-emitting diodes based on polyfluorene blends upon insertion of a poly(*p*-phenylene vinylene) electron-confinement layer. *Appl Phys Lett* 80:2436–2438.

20

Electrochromism of Conjugated Conducting Polymers

Aubrey L. Dyer and
John R. Reynolds

20.1 Introduction

Electrochromic materials exhibit a change in absorbance, reflection, or transmission of electromagnetic radiation induced by an electrochemical oxidation–reduction reaction. The term electrochromism is analogous to thermochromism (change in color produced by heat) and photochromism (change in color produced by light) [1]. Most often, the color change is between a transmissive state and a colored state, but many materials switch between multiple colored states and are referred to as polyelectrochromic materials. Some of the most commonly studied electrochromic materials include transition metal oxides (e.g., tungsten oxide, WO_3) organic molecules (e.g., viologens), and inorganic complexes (e.g., Prussian Blue [PB]) [2–5]. With many of these materials, along with conducting polymers, the electrochromic effect is reversible and can occur not only in the visible region but also in the ultraviolet (UV), infrared (IR), and even microwave regions of the spectrum [6–8]. Conjugated polymers are another class of materials that are increasingly studied for their electrochromic properties. These polymers offer better processability, faster switching, and color tunability than their inorganic and molecular counterparts. Conjugated polymers can be either anodically coloring (color upon oxidation) or cathodically coloring (color upon reduction), and can exhibit multiple colored states in the same material. Through minimal structural modifications, these polymers can be made soluble in common organic solvents, which allow straightforward preparation of polymer films for large-scale applications.

20.1.1 Metal Oxide Electrochromism

The phenomenon of electrochromism was first noted in the high bandgap semiconductor WO_3 in 1969 [9]. Since then, it has been the most widely studied material in the field of electrochromics. Films of WO_3 can be prepared by various techniques such as thermal evaporation in vacuo, electrochemical oxidation of tungsten metal, chemical vapor deposition, sol–gel methods, and RF sputtering. Thin films of tungsten trioxide, with all tungsten sites as oxidation state W^{VI}, are transparent. Upon electrochemical reduction, W^V sites are generated to yield an intense blue color. The electrochromic effect occurs by intervalence charge transfer between adjacent W^V and W^{VI} sites. The generalized equation can be written as:

$$\underset{\text{(transparent)}}{WO_3} + xM^+ + xe^- \rightarrow \underset{\text{(blue)}}{M_xWO_3} \tag{20.1}$$

where M represents the metal cation (e.g., Li^+ and H^+) [2,10–12]. Many other transition metal oxides have been studied for their electrochromic properties, e.g., oxides of V, Mo, Nb, and Ti that are cathodically coloring and Ni, Co, and Ir that are anodically coloring [2].

20.1.2 Viologen Electrochromism

The viologens are 1,1′-disubstituted-4,4′-bipyridinium salts with the general structure as shown in Scheme 20.1. The most extensively studied material from the viologen family is methyl viologen (MV) [2,13]. Of the three redox states, the colorless dication is the most stable. One electron reduction forms the colored radical cation. The stability of the radical cation state is attributed to the delocalization of the radical electron through the π-framework. The electrochromism occurs due to optical charge transfer between the (formally) +1- and 0-valent nitrogens. Color control of the radical cation can be obtained by varying the nitrogen substituent to obtain the appropriate molecular orbital energy levels.

MV switches from a colorless state to a purple-blue state, whereas 4-cyanophenyl viologen switches from a colorless state to green upon reduction to the radical cation. When fully reduced, the viologens show color of low intensity because no optical charge transfer or internal transition corresponding to wavelengths in the visible region occurs.

20.1.3 Prussian Blue Electrochromism

Polynuclear transition metal hexacyanometallates, e.g., PB (iron(III) hexacyanoferrate(II)), are a class of insoluble mixed-valence compounds that have the general formula $M'_k[M''(CN)_6]_l$ where M′ and M″ are transition metals with different formal oxidation numbers and k and l are integrals [2,3,5,14]. PB thin films are formed by electrochemical reduction of solutions containing iron(III) and hexacyanoferrate(III) ions. Partial electrochemical oxidation of PB films yields Prussian green (PG). When fully oxidized, Prussian brown (PX) results, which is golden yellow. Reduction of PB yields transparent Prussian white (PW). Electrochemical redox reactions for the three different color states are shown in Scheme 20.2.

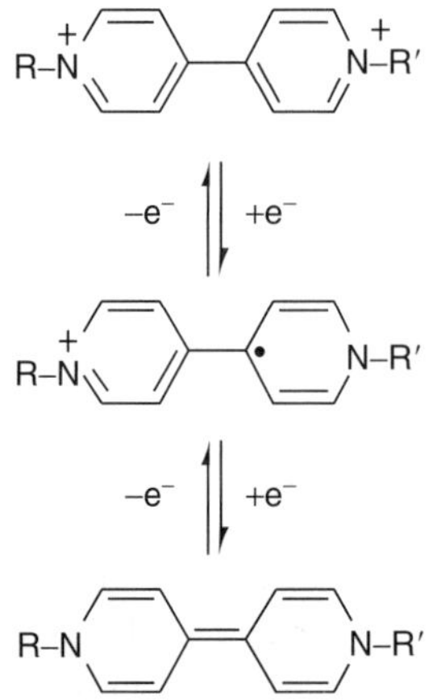

20.1.4 Conducting Polymer Electrochromism

Even though not as developed as the other materials discussed, conjugated conducting polymers have increasingly gained attention because of their

SCHEME 20.1 Redox switching for viologens.

electrochromic properties. This is due to the attainability of higher contrast ratios and faster response times, which are properties sought after in electrochromic displays [2,3,5,13,15]. Conjugated polymers also offer improved processability through electrochemical deposition, spin coating, drop casting, and even spray coating

$$3[Fe^{III}Fe^{II}(CN)_6]^- \longrightarrow [Fe^{III}_3\{Fe^{III}(CN)_6\}_2 \{Fe^{II}(CN)_6\}]^- + 2e^- \quad (20.2)$$
$$\text{PB} \qquad\qquad\qquad\qquad \text{PG}$$

$$[Fe^{III}Fe^{II}(CN)_6]^- + e^- \longrightarrow [Fe^{III}Fe^{III}(CN)_6] + e^- \quad (20.3)$$
$$\text{PB} \qquad\qquad\qquad \text{PX}$$

$$[Fe^{III}Fe^{II}(CN)_6]^- + e^- \longrightarrow [Fe^{II}Fe^{II}(CN)_6]^{2-} \quad (20.4)$$
$$\text{PB} \qquad\qquad\qquad \text{PW}$$

SCHEME 20.2 Redox process for Prussian blue.

and printing. Another advantage of these polymers over other electrochromic materials is the higher degree of color tunability through structural modification.

This chapter provides an overview of the electrochromic properties of conjugated conducting polymers and their utilization in devices. Section 20.2 illustrates how electrochromism arises in conjugated polymers and how color tuning is possible through structural modification. Section 20.3 covers the families of electrochromic polymers from thiophene, pyrrole, aniline, carbazole (Cz), alkylenedioxythiophene, and alkylenedioxypyrrole. Section 20.4 discusses multicomponent electrochromic polymers including copolymers, polymer blends, laminates, and composites. Section 20.5 ends with a short introduction to the types of electrochromic devices (ECDs) constructed utilizing these conducting polymers.

Unfortunately, due to the large amount of work carried out in this field, a review of this type cannot be fully comprehensive. Hence, we have selected examples that are state-of-the-art representatives and recently developed in the field. Wherever possible, we direct the reader to the primary literature where significantly more examples and details await the interested scientist.

20.2 Electrochromism in Conjugated Polymers

All conjugated polymers are potentially electrochromic with the intrinsic optical properties determined by the polymer's π–π* transition, which is characterized by the polymer's bandgap and λ_{max} [5,16,17]. The value of the optical bandgap is determined by the onset of the π–π* transition for the polymer in the undoped state.

As a demonstration of the breadth of these systems, we compare a series of polymers ranging from poly(*p*-phenylene) (PPP), poly(*p*-phenylene vinylene) (PPV), polythiophene (PT), and poly(isothianaphthene) (PITN). PPP is an anodically coloring polymer with a bandgap of 3.0 eV [18,19]. This polymer and its derivatives are widely studied for their light emission properties and are utilized in light-emitting diodes (LEDs). Due to high oxidation potential and sensitivity of the redox-doped state to oxygen and water, PPP and its vinyl analog, PPV with a bandgap of 2.3–2.4 eV [18,20], are not utilized in the field of electrochromics.

PT has a lower bandgap of 2.1–2.2 eV, along with a higher HOMO level for more facile oxidation, allowing it to redox dope at practically accessible potentials and it exhibits two colored states [18,21]. Due to its good environmental stability and ease of functionalization, PT and its derivatives are some of the most extensively studied electrochromic conjugated polymers. While thiophene itself has a relatively high oxidation potential, substitution on the thiophene ring with electron-donating units produces a polymer with a low oxidation potential that can exhibit various colors in the neutral state or upon oxidation. For example, the addition of an ethylenedioxy bridge to PT reduces the bandgap by 0.5 eV, due to the two electron-donating oxygen atoms adjacent to the thiophene unit for poly(3,4-ethylenedioxythiophene) (PEDOT) [22,23]. In addition to possessing a lower oxidation potential, PEDOT is also transmissive to visible light, highly conducting, and quite stable in the oxidized state [22,24]. PITN was the first reported example of an especially low bandgap cathodically coloring polymer with a value of 1.0 eV [25]. Unfortunately, due to a number of synthetic and electrochemical reasons, PITN and its derivatives have not found a high level applicability in electrochromic polymer research.

With this select set, it is evident that conjugated polymers have bandgap values that cover a wide range and this can lead to a broad color palette through structural modification of the parent polymer. This versatility in color tuning is a unique property of the organic electrochromic systems and illustrates the growing interest and value of conjugated polymers in electrochromic displays and devices.

20.2.1 Inducing Charged States

For a neutral conjugated polymer, as shown in Figure 20.1a with PT as an example, the electronic transition that occurs is from the valence band to the conduction band with the minimum energy difference being equal to the bandgap, E_g [5,16,26]. At low oxidation levels, electrons are removed from the valence band and a radical cation, which behaves as a polaronic carrier of charge, is formed that is delocalized over a polymer segment, as shown in Figure 20.1b.

The oxidation is accompanied by a relaxation of the aromatic structural geometry of the polymer chain toward a quinoid-like structure. Charge balance by electrolyte anions yields a p-doped material, where p-doping and n-doping (reduction) are adopted from classical semiconductor theory [5,16,26]. The removal of π-electrons from the valence band creates half-filled polaron levels that are symmetrical about the bandgap center, causing the emergence of new electronic transitions at lower energies and a shift of the absorption maximum to higher wavelengths.

As the polymer is oxidized further, a dication is formed, a bipolaronic charge carrier with the cations coupled to one another, that is delocalized over the same polymer segment. These bipolaron levels are unoccupied, only electronic transitions from the top of the valence band can occur, which is shown in Figure 20.1c [5,16,26].

Figure 20.2 shows the absorption spectroelectrochemical series for the PT derivative, poly(2,2'-dimethyl-3,4-propylenedioxythiophene) (PProDOT-Me$_2$). The neutral polymer is blue and characterized by an absorption maximum centered at 550 nm with a bandgap of 1.7 eV. As the potential applied to the polymer film is increased to more positive values, an absorption begins to emerge between 700 and 1100 nm while the π–π* absorption begins to decrease in intensity. At slightly higher oxidation levels, an even lower energy absorption band begins to emerge above 1250 nm. As illustrated in Figure 20.1, when the polymer approaches full oxidation, the bipolaron band absorption begins to dominate at the expense of the polaron and π–π* transitions. This is seen in the absorption spectra as the formation of a continuous absorption band at lower energies as the polaron and π–π* bands disappear from the spectra giving the polymer a very transmissive light blue color.

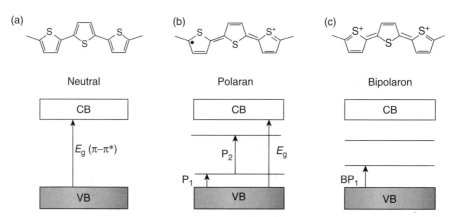

FIGURE 20.1 Geometric structure of a thiophene trimer showing the allowed electronic transitions for the (a) neutral, (b) polaron, and (c) bipolaron states. *Note:* The forbidden transitions are not shown for clarity.

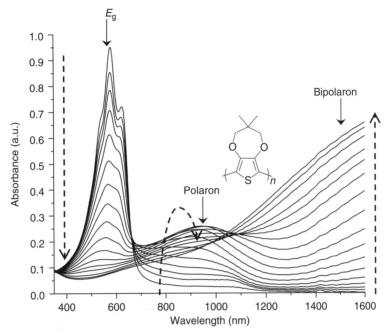

FIGURE 20.2 Spectroelectrochemistry of PProDOT-(Me)$_2$ showing electronic transitions for the bandgap (E_g), polaron, and bipolaron. Dashed arrows show direction of spectral growth or recession. (Dyer, A., 2005, unpublished data.)

20.2.2 Structural Control with Respect to Electrochromism

In electrochromics, conducting polymers offer an advantage over other materials due to their high degree of color tailorability. Through structural modification, the positions of the HOMO and LUMO levels, and thus the energy gap can be adjusted, yielding color changes of both the neutral and doped forms [15,22,27]. One method in which this can be accomplished is by varying the degree of π-overlap along the polymer backbone through steric interactions. The interactions cause a break in the effective conjugation length and increase the polymer bandgap.

Figure 20.3 shows the 3D geometries of the dimeric units representative of the polymeric repeat for three poly(3,4-propylenedioxypyrrole) (PProDOP) derivatives and the twisting induced between the pyrrole rings upon *N*-substitution. When comparing poly(3,4-ethylenedioxypyrrole) (PEDOP) (Figure 20.3a) and PProDOP (Figure 20.3b), the increase in the size of the ring from six- to seven-membered causes a steric interaction that leads to a small decrease in the π-overlap. Even though there is no evident torsional angle between the pyrrole rings for both polymers, in structures from simple calculations, the energy difference between the twisted conformation on the alkylenedioxy ring for PEDOP and the boat conformation for PProDOP causes an increase in the bandgap [28–30]. The result is an increase in the bandgap of 0.2 eV for PProDOP (2.2 eV) compared with PEDOP (2.0 eV). The bandgap can be further increased by substitution of the ring nitrogen inducing a twist in the polymer backbone that decreases the effective π-conjugation. For example, *N*-methyl PProDOP (Figure 20.3c) has a calculated degree of twisting between the pyrrole rings increased from $\theta = 0°$ to 59.3° and bandgap at 3.0 eV; 0.8 eV higher than that for the unsubstituted polymer. By increasing the size of the *N*-substituent to a bulky ethoxyethoxyethanol (Figure 20.3d), the bandgap is further increased to 3.4 eV with a twist angle $\theta = 71.6°$.

Another approach to bandgap modification is through control of the electronic character by electron-donating or electron-accepting substituents or corepeat units [27]. The HOMO and LUMO levels can be adjusted, yielding polymers with higher or lower bandgaps relative to the parent homopolymer, e.g.,

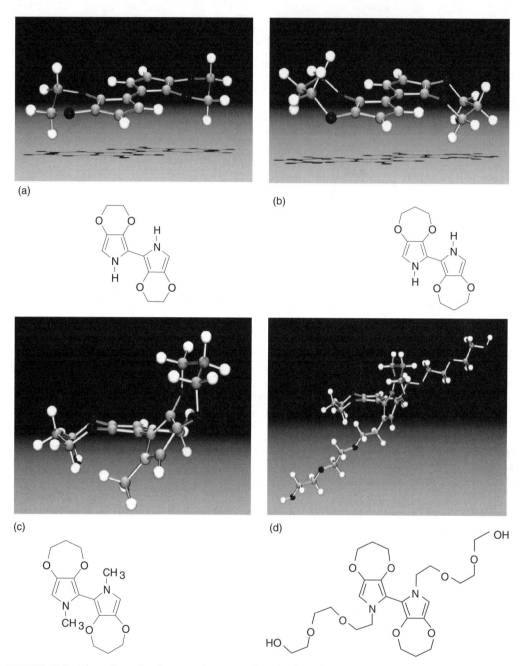

FIGURE 20.3 Three-dimensional geometric perspectives for four dioxypyrrole derivatives. The value of the torsional angle θ indicates the degree of twisting between the two rings. Geometries and values obtained after optimizations at the B3LYP/6-31G (d) level of theory using the program Gaussian 98[30]. (a) PEDOP, $\theta = 0°$, (b) ProDOP, $\theta = 0°$, (c) *N*-Me ProDOP, $\theta = 59.3°$, (d) *N*-Gly ProDOP, $\theta = 71.6°$. (Perdomo, A., 2005, unpublished data.)

poly(bis-3,4-ethylenedioxythiophene-cyano vinylene) (PBEDOT-CNV) [31]. The cyano vinylene acts as an electron-accepting conjugated spacer between two EDOT electron-donating units that greatly reduces the LUMO level. The effect is a significant lowering of the polymer bandgap to approximately 1.1–1.2 eV. Numerous examples demonstrating these concepts are discussed further in later sections.

Many of the polymers that will be discussed throughout this chapter have been utilized in a variety of applications beyond electrochromism that include supercapacitors [32–34], mechanical actuators [34–36], polymer light-emitting diodes (PLEDs) [34,37–39], polymer photovoltaic devices [34,37–39], and sensors [40–42]. Section 20.3 focuses on the electrochromic properties of some of the most common poly(heterocycles) and their derivatives. The specific families to be discussed are PTs, polypyrroles, poly(3,4-alkylenedioxyheterocycles), polyanilines, and polycarbazoles.

20.3 Families of Electrochromic Polymers

Electrochromic properties of polymers, which include their hue, color saturation, contrast or luminance change, cycle life, and electrochromic efficiency, can be controlled by structural modification and/or electrochemical switching conditions. Influence over many material properties that include solubility, thermal and environmental stability, oxidation potential, and conductivity can also be achieved through structural modification.

This section focuses on the influence of structure of the repeat unit on the colors available in the neutral, oxidized, and in some instances the reduced forms for the various families of electrochromic polymers. The section also covers the electrochromism of parent polymers and the result that the addition of substituents with a variety of functionalities has on the electrochromism.

20.3.1 Polythiophenes

Conjugated PTs have been intensively investigated because of their environmental stability in the neutral form and the structural versatility of the thiophene ring making synthetic modification straightforward and facile. The first reported electrochemical polymerization and electrochemical switching of both a substituted and unsubstituted thiophene was in 1983 by Garnier and coauthors [43,44] with the films being red in the neutral state, switching to blue upon oxidation and black-green upon reduction [45].

The absorption spectroelectrochemistry for PT (1) during p-doping and n-doping is shown in Figure 20.4 and Figure 20.5, respectively, with both showing the $\pi-\pi^*$ absorption decrease as lower energy absorption bands increase. Table 20.1 presents the polymer repeat unit structures along with colors of all redox states for all thiophene-based polymers discussed.

1
(E_g = 2.0–2.2 eV)

The difficulty of obtaining uniform polymer films, due to polymer instability (overoxidation) under the conditions of electrochemical polymerization and/or side reactions at the β-position, has sparked research toward the development of alternative electrochemical polymerization methods or thiophene derivatives with lower monomer oxidation potentials and improved physical properties. In an effort to lower the thiophene monomer oxidation potential, electrochemical polymerization has been carried out in the strong Lewis acid, boron trifluoride ethylether (BFEE) [46]. The resulting polymer films were of good quality, homogeneous, and showed improved electroactivity over thiophene films prepared in common organic solvents. Another method explored is the production of PT through the polymerization of oligothiophenes (dimers, trimers, and tetramers) [43,44,47]. The result is monomer units that have lower electrochemical oxidation potentials for polymerization, and polymer films that contain fewer structural defects due to less α–β couplings, yet exhibit similar colors in the neutral and oxidized states.

One thiophene-containing polymer family developed combines the attractive morphological properties of insulating polymers with the electrochromic properties of conjugated polymers by creating

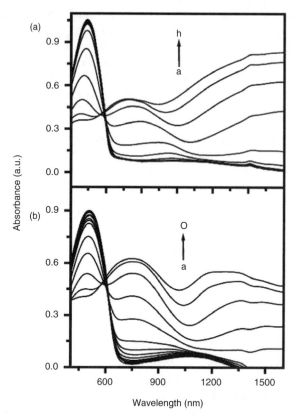

FIGURE 20.4 Spectroelectrochemistry for (a) polythiophene between the potentials of −0.2 and +1.1 V vs. Ag/Ag$^+$ and (b) poly(3-methylthiophene) between the potentials of −0.1 and +0.9 V vs. Ag/Ag$^+$. (From Alkan, S., *Adv. Funct. Mater.*, 13, 331, 2003. With permission.)

nonconjugated polymers with pendant oligothiophenes. To this effort, vinyl (**2a,b**) [48–51] and methacrylate (**3a,b**) [52,53] polymers with oligothiophene pendant groups have been created that contain the ease of processability of the nonconjugated polymers and the unique optical properties of the oligothiophene units. It was found that the polymers switched from colors such as yellow or orange when neutral to green and purple upon oxidation with the colors achieved being dependent on the length of the oligothiophene unit and the extent of cross-linking with the absorption maxima shifting to longer wavelengths as the pendant thiophene chain length increased [49–53]. Figure 20.6 shows the absorption spectroelectrochemistry of the methacrylate polymer with pendant sexithiophene units, which switches from orange to blue.

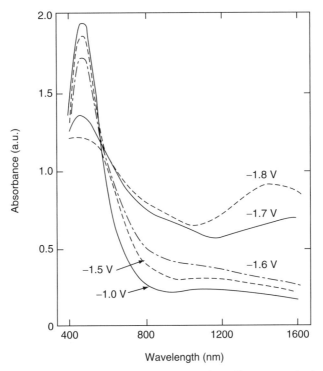

FIGURE 20.5 Absorption spectroelectrochemistry for a polythiophene film upon n-doping. (From Aizawa, M., *J. Chem. Soc. Chem. Commun.*, 25, 264, 1985. With permission.)

Much attention has been devoted to the addition of a variety of substituents to either the 3-position or the 3- and 4-positions of the thiophene ring [54–59]. By increasing the length of the alkyl chain in poly(3-alkylthiophenes) from methyl (**4a**) to octyl (**4b**), the polymer maintains a red to blue color when switched from neutral to p-doped while the solubility increases dramatically yielding an easily processable polymer. The conjugation length of poly(3-alkylthiophenes) can be tailored by modifying steric effects through regiochemistry in electropolymerizable multiring monomers [60,61]. It was shown that comparing the head-to-tail (**5a**), head-to-head (**5c**), and tail-to-tail (**5e**) conformations in bi(methylthiophene), the color of the neutral polymer was red, yellow, or orange, respectively, and upon oxidation, all are blue-violet. When the alkyl chain length was increased to hexyl, the colors of the neutral polymers were orange-red for the head-to-tail polymer (**5b**) or pale green for both the head-to-head (**5d**) and tail-to-tail (**5f**) polymers with all three switching to blue upon oxidation.

	R_1	R_2	R_3	R_4
a.	H	CH_3	H	CH_3
b.	H	C_6H_{13}	H	C_6H_{13}
c.	H	CH_3	CH_3	H
d.	H	C_6H_{13}	C_6H_{13}	H
e.	CH_3	H	H	CH_3
f.	C_6H_{13}	H	H	C_6H_{13}

4
a. R = H
b. R = C_8H_{17}

TABLE 20.1   Polythiophenes

Polymer Number	Neutral Color	Oxidized Color	Refs.
1	Red	Blue (black-green)[a]	43–47
2a	Yellow-orange	Blue-purple	48–51
2b	Yellow	Green	48–51
3a	Yellow	Purple	52,53
3b	Orange	Blue	52,53
4a,b	Red	Blue	54–59
5a	Red	Blue	60,61
5b,e	Orange-red	Pale blue	60,61
5c	Yellow	Blue-violet	60,61
5d,f	Pale green	Pale blue	60,61
6a	Red-orange	Slate green	62
6b,c	Red	Blue	62
7a	Red	Blue	63
7b	Yellow	Blue	64
8	Red	Black	65
9	Red-purple	Blue-purple	66
10a	Red	Gray-blue	67–69
10b	Red	Light green (colorless)[b]	67–69
10c,d	Red	Blue	67–69
11	Red	Pale blue (violet)[c] and (orange)[d]	70
12	Red	Colorless (black)[e]	71
13a-f	Blue	Transmissive yellow	29,72–75
14a	Deep blue	Transmissive tan	76
14b	Blue-green	Pale blue-gray	79,80
15b	Red	Purple (purple)[a]	82
16a,b	Red	Blue	83
17	Yellow	Blue	83

[a]n-doping.
[b]Partial oxidation.
[c]First reduction.
[d]Second reduction.
[e]Reduction.

In an effort to produce electroactive PTs with improved chemical and thermal stabilities, a series of (3-fluoromethyl)thiophenes were developed [62]. The three thiophene derivatives generated were poly(3-fluoromethyl)thiophene (**6a**), poly(3-difluoromethylthiophene) (**6b**), and poly(3-trifluoromethylthiophene) (**6c**). The polymers required a higher oxidation potential than their methyl analogs with poly(3-fluoromethylthiophene) switching from red-orange when neutral to slate green when p-doped and both poly(3-difluoromethylthiophene) and poly(3-trifluoromethylthiophene) switching from red when neutral to blue on oxidation.

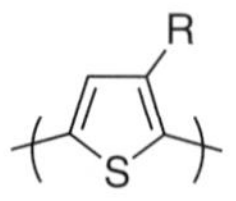

6
a. R = CH$_2$F
b. R = CHF$_2$
c. R = CF$_3$

Through the introduction of ester groups at the 3-position of the thiophene ring, electrochromic polymers were formed that were more hydrophilic than poly(3-alkylthiophene)s due to the polar ester group substitution. Two examples, poly(3-methyl-butyric acid 2-thiophen-3-yl-ethyl ester) (**7a**) [63]

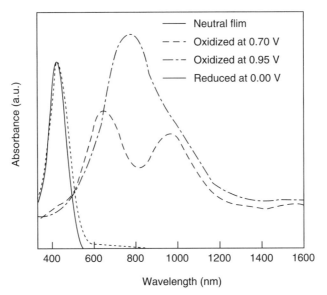

FIGURE 20.6 Absorption spectroelectrochemistry of a film of poly[(dioctyl-sexithiophen-5-yl)methyl methacrylate] (PMADOc6T) in the neutral, partially oxidized, fully oxidized, and reneutralized states. All potentials are referenced to Ag/Ag$^+$. (From Ohsedo, Y., *J. Polym. Sci. Part B: Polym. Phys.*, 41, 2471, 2003. With permission.)

and poly(octanoic acid 2-thiophen-3-yl-ethyl ester) (**7b**) [64] showed a large change in the color of the neutral polymer with the methyl-butyric acid polymer as red and the octanoic acid polymer as yellow when changing from a branched to a linear substituent. Both polymers switched to blue in their oxidized states. In contrast, poly(thiophene-3-acetic acid) (**8**) changes from red to black upon oxidation [65].

By incorporating phenyl groups as the substituent in the 3-position in PTs, controlling the steric hindrance can result in polymers with an array of bandgaps yielding a variety of colors. For example, poly[3-(4-octyl-phenyl)-thiophene] (**9**) changes in color from reddish purple in the neutral state to a bluish purple in the p-doped state [66]. Films of poly(3-[ω-(p-methoxyphenoxy)alkyl]) thiophenes switch from deep red in the neutral form to gray-blue, light green, and blue in the oxidized form when increasing alkyl chain lengths from butyl (**10a**), to hexyl (**10b**), and both decyl (**10c**) and dodecyl (**10d**), respectively [67–69].

Other functionalities that have been added to thiophenes include redox-active pendant groups such as anthraquinone and viologen [70,71]. These polymers offer the ability to reach multiple colored states due to the electrochromism of both the thiophene polymer and the pendant group combined.

Poly(*N*-methyl-*N'*-(6-thiophene-3-ylhexyl)-4,4'-bipyridinium) (**11**) films are red when neutral and switch to a transmissive pale blue upon oxidation. The polymer films then switch to violet at the first viologen reduction and orange at the second viologen reduction. Similarly, a thiophene with a carboxyanthraquinone substituent (**12**) produces a red polymer film in the neutral state that switches to nearly colorless upon oxidation and switches to black upon the production of the anthraquinone radical anion.

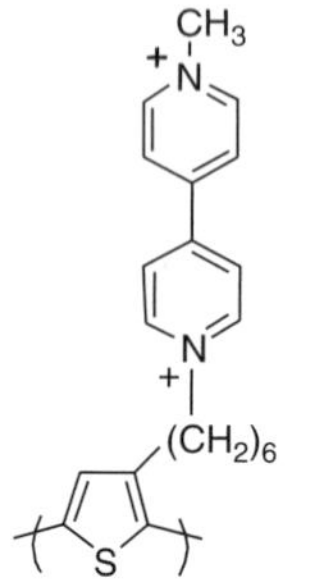

Some of the most successful efforts to date to produce low bandgap conjugated polymers have been in the area of fused ring heterocycles of thiophenes. The first low gap polymer was PITN (**13a**), with a bandgap of 1 eV [25,72,73]. The success of this polymer is due to the reduction in the degree of bond length alternation, which in turn leads to a reduction in the bandgap. In PT, the energy difference between the aromatic and quinoid forms produces a polymer with a large degree of bond length alternation. However, the quinoid form of PITN is stabilized while the aromatic form is destabilized, reducing the energy difference between the two, yielding a lower degree of bond length alternation.

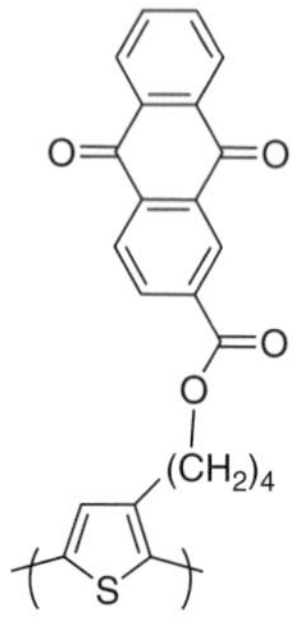

PITN was the first reported example of a polymer that is highly transmissive in its conducting state with a dark blue in the neutral state and transmissive greenish-yellow in the oxidized state. While the low bandgap and conducting transmissive state is promising for the field of conducting polymer electrochromics, practical applications of this polymer have been prevented by the polymer insolubility, poor environmental stability, and the difficulty in reproducibly forming homogeneous, even films.

Efforts have been directed toward stabilizing both the p-doped and n-doped forms of PITN through the introduction of electron-donating and electron-withdrawing substituents, respectively. By the addition of hexyloxy groups at the 5- and 6-positions (**13b**), the combination of the electron-donating ability of the substituents and the long alkyl chains induces both an improved stability in the doped state over PITN, and improved solubility in organic solvents with little change in the polymer bandgap [74]. The introduction of methyl groups (**13c**), fluorines (**13d, e**), and chlorine (**13f**) to the PITN structure produced polymer films that, while insoluble, have their redox potentials for n-doping shifted to more positive, accessible values with a modest affect on the polymer bandgap (~0.1 eV increase) [75].

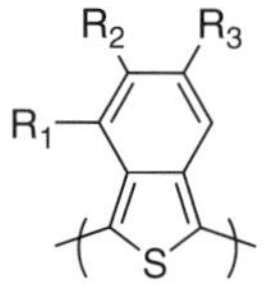

13
(E_g = 1 eV)

	R_1	R_2	R_3
a.	H	H	H
b.	H	OC_6H_{13}	OC_6H_{13}
c.	H	CH_3	H
d.	F	H	H
e.	H	F	H
f.	H	Cl	Cl

Other low bandgap fused ring systems include those based on thieno[3,4-*b*]thiophenes. The reported bandgap of the unsubstituted polymer (**14a**) is 0.85 eV with the polymer film switching between a deep blue in the neutral form and a transmissive tan in the oxidized state [76]. Water-dispersible forms of the polymer have been formed when thieno[3,4-*b*]thiophene is chemically polymerized in the presence of poly(styrene sulfonic acid) [77,78]. Through the addition of solubilizing alkyl groups, e.g., decyl (**14b**), polymer films that switch from a blue-green in the neutral state to a pale blue-gray oxidized state with a slightly higher bandgap value of 1.2 eV can be produced [79,80].

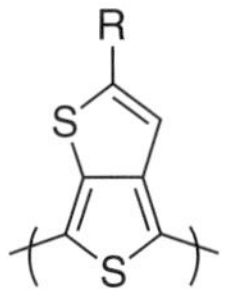

14
(E_g = 0.85–1.2 eV)
a. R = H
b. R = $C_{10}H_{21}$

The bandgap of PTs has also been shown to be lowered by forcing planarity to occur with adjacent rings as with polymers based on cyclopentadithiophenes, and naphthodithiophenes. Polymers based on cyclopentadithiophene have been made with substituents on the cyclopentane ring that range from disubstitution of cyano groups (**15a**) to substitution with both cyano and nonafluorobutylsulfonyl groups (**15b**) [81,82]. These polymers have been suggested to show a bandgap below 1 eV with the films of the cyano, nonafluorobutylsulfonyl-substituted polymer switching between red in the neutral state to purple in both the p-doped and n-doped states [82]. Another polymer in which planarity is forced through coupling bithiophene repeat units is one based on naphthodithiophene [83]. Films of this polymer switch from either red or yellow to dark blue depending on the arrangement of the thiophene units relative to the benzene bridging segment. Poly(naphtho[2,1-*b*;3,4-*b*']dithiophene (**16a,b**) is red in the neutral state, while poly(naphtho[1,2-*b*;4,3-*b*']dithiophene (**17**) is yellow due to the low degree of electronic delocalization caused by the disruption of aromaticity in the polymer.

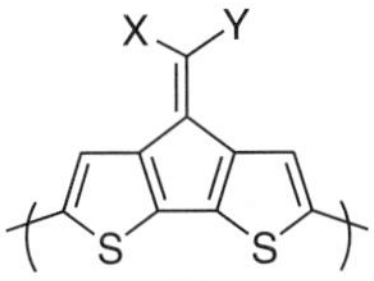

15
(E_g = <1 eV)
a. X = CN, Y = CN
b. X = CN, Y = $SO_2C_4F_9$

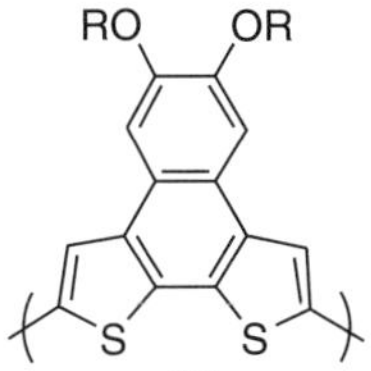

16
a. R = C_8H_{17}
b. R = 2-ethylhexyl

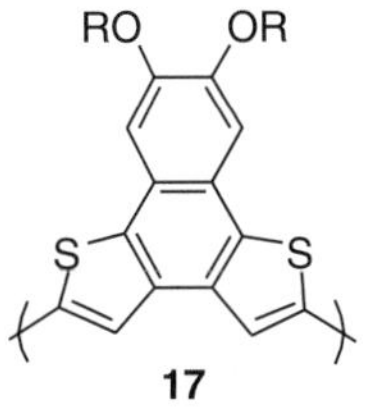

17
R = $C_{10}H_{21}$

20.3.2 Polypyrroles

Polypyrroles are a class of conjugated polymers that have been extensively studied due to their low oxidation potential leading to stable conductors, along with the compatibility of their electroactivity in aqueous media. Since the first report on the electrochemical preparation of conductive polypyrrole films [84–87], interest in the electronic properties of this polymer and its derivatives has increasingly grown.

In early work, it was shown that flexible polymer films could be prepared with conductivities as high as 105 S cm^{-1} in the doped state [88]. It was also shown that copolymer films of pyrrole and *N*-methylpyrrole could be prepared by electrolysis of a mixture of the two monomers [84,85]. Given that the homopolymer of *N*-methylpyrrole has a lower conductivity than the unsubstituted polypyrrole, the resulting film conductivity can be tuned by varying the copolymer composition. In the thin film form, polypyrole (**18a**) exhibits a yellow to brown-black electrochromism upon oxidation [86].

Table 20.2 lists the pyrrole-based polymers for this section along with their colors in the neutral and oxidized states. While *N*-substitution on the pyrrole ring causes a marked decrease in film conductivity and a higher monomer oxidation potential, varying the substituent from methyl (**18b**) to ethyl (**18c**), propyl (**18d**), butyl (**18e**), or phenyl (**18f**) does not change the colors of the resulting films in both oxidized and neutralized states [89].

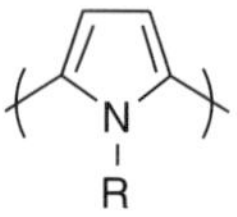

18

a. R = H (E_g = 2.7 eV)
b. R = CH$_3$
c. R = C$_2$H$_5$
d. R = C$_3$H$_7$
e. R = C$_4$H$_9$
f. R = phenyl

Not only do oxidatively doped polypyrrole films show significant environmental stability, but a number of these films can also be electrochemically prepared and switched in both organic and aqueous solvents [90]. While poly(*N*-methylpyrrole) was the only polymer of the series that exhibited a comparable electrochromic contrast in both solvents, *N*-benzyl (**19a**), *N*-tolyl (**19b**), and the parent polypyrrole showed a decrease in electrochromic contrast when comparing results obtained using organic and aqueous electrolytes. Polymers prepared using *N*-phenyl and *N*-benzoylpyrrole (**20**) showed no electrochromic switching in aqueous solutions.

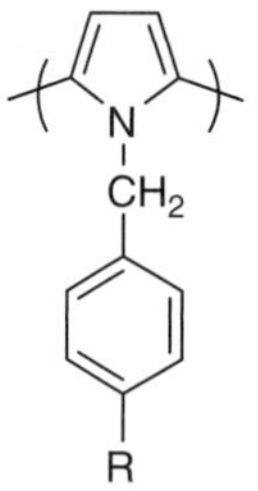

19

a. R = H
b. R = CH$_3$

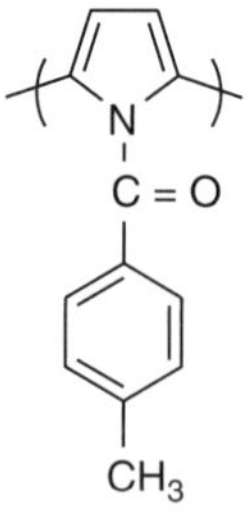

20

TABLE 20.2   Polypyrroles

Polymer Number	Neutral Color	Oxidized Color	Refs.
18a	Yellow (yellow)[a,b]	Black (violet, dark blue, or brown)[a,b]	84–88 (97,98)[a,b]
18b-f	Yellow	Black	89
21a-e	Red	Blue	91
24	Transmissive yellow	Gray-blue	94–96
26	Transmissive green	Green	99

[a]Dodecylsulfate anion-doped film.
[b]Dodecylbenzenesulfonate-doped film.

Substitution at the *N*-position of the pyrrole ring allows the production of increasingly complex repeat unit structures. This is evident in pyrrole polymers substituted with various benzylideneamino groups as illustrated by structures **21a–e** [91]. Of these monomers, those with highly electron-withdrawing NO$_2$ (**21e**) and CN (**21d**) groups at the *para* position of the phenyl ring yielded polymers soluble in polar aprotic solvents such as dimethyl sulfoxide and *N,N*-dimethylformamide. Alternatively, monomers in which the phenyl ring was unsubstituted (**21a**) or substituted with a chlorine (**21c**) or methoxy (**21b**) group yielded insoluble polymer films. The insolubility of these polymers might be due to cross-linking at the electron-donating aromatic unit during oxidative electropolymerization. All of these polymers exhibited red to blue electrochromism upon oxidation.

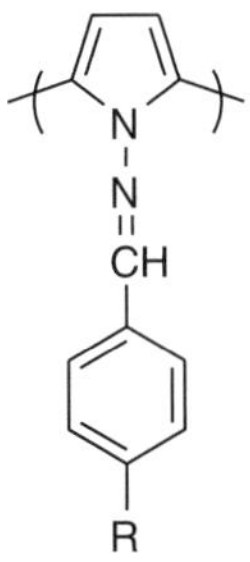

R
21
a. R = H
b. R = OCH$_3$
c. R = Cl
d. R = CN
e. R = NO$_2$

For the purpose of producing polymer films that can act catalytically, electroactive groups such as MV (**22**) and anthraquinone (**23**) have also been added at the *N*-position of the monomer unit with polymers subsequently produced by electropolymerization of the pyrrole unit [92,93].

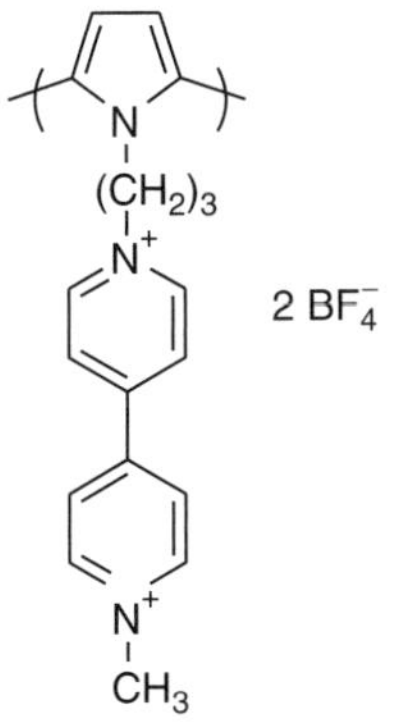

2 BF$_4^-$

22 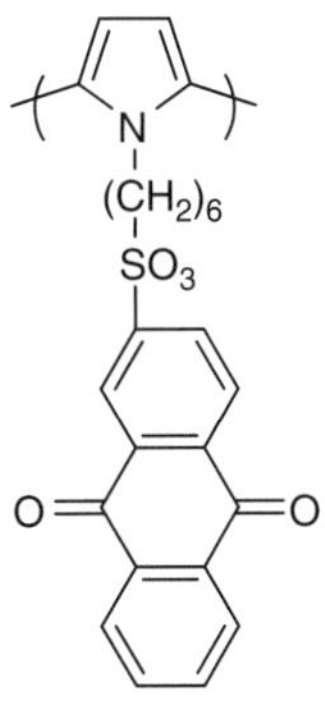 **23**

Analogous to PTs, pyrrole polymers have been polymerized from the dimer with similar results. For example, poly(*N,N'*-dimethyl-2,2'-bipyrrole), poly(NNDMBP) (**24**) exhibits a blue-shifted onset of absorbance of the neutral polymers [94–96]. This results in an increased visible region transparency of the film in the neutral state. The polymer is pale, transmissive yellow in the neutral state and switches to gray-blue when oxidized. In addition, the dimer oxidizes at a lower oxidation potential and the resulting polymer film has an increased conductivity over that produced from the monomer.

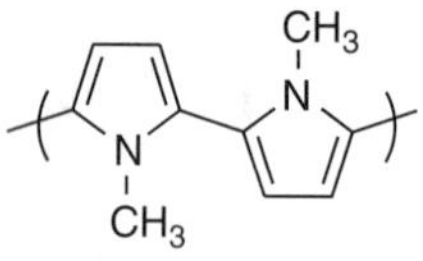

24

In an effort to produce polymer films with improved mechanical properties in aqueous solutions, pyrrole was electrochemically polymerized in the presence of surfactants such as sodium dodecylsulfate and sodium dodecylbenzenesulfonate. In the case of dodecylsulfate anion, the polymers switched between a transmissive yellow in the neutral state to violet when partially oxidized and brown when fully oxidized [97]. When electropolymerized in the presence of dodecylbenzenesulfonate anion, the polymer films switched between a transmissive yellow when neutral to dark blue when oxidized [98]. In both cases, the polymers showed an improved electrochemical stability over polymer films produced with other inorganic counterions. This has been attributed to the binding of the surfactant dopant within the film during charge compensation since there is less mechanical distortion of the polymer film.

In contrast to the *N*-substituted polypyrroles, substitution at the 3-position results in polymers with a similar oxidation potential as the unsubstituted pyrrole and a higher conductivity. Some examples include methyl substituted bipyrroles (**25a–c**) [96], poly(3-hydroquinonylpyrrole) (**26**) [99], and poly(3-alkylsulfonate pyrrole)s (**27a–c**) [100]. Unfortunately, because 3-substitution of the pyrrole ring is synthetically demanding and the resulting polymer shows a higher degree of sensitivity to air in the neutral state, more focus has been placed on *N*-substitution as opposed to 3-substitution of the pyrrole monomer unit.

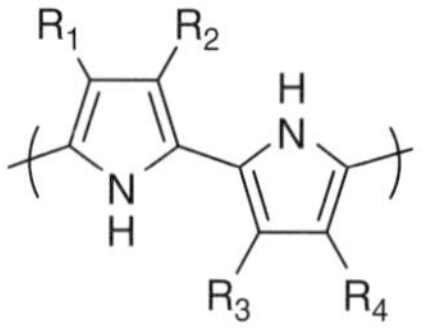

25

	R$_1$	R$_2$	R$_3$	R$_4$
a.	CH$_3$	H	H	CH$_3$
b.	H	CH$_3$	CH$_3$	H
c.	H	CH$_3$	H	CH$_3$

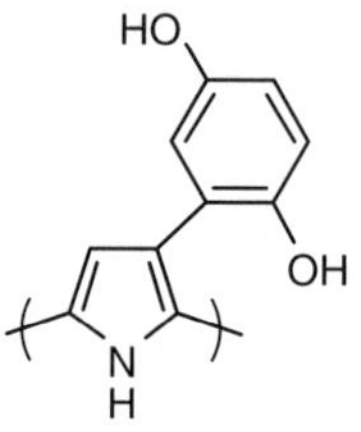

26

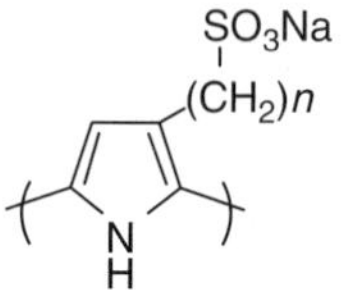

27

a. $n = 3$
b. $n = 4$
c. $n = 6$

Polypyrroles continue to be some of the most extensively studied conjugated polymers due to a relatively high conductivity and enhanced stability under ambient conditions. They also show promise for utilization in applications ranging from capacitors, electromagnetic shielding, secondary batteries, and even biological applications given that pyrroles retain their electronic activity in aqueous electrolyte solutions. Even though these polymers have received much attention for their redox and electronic properties, there is little focus on their electrochromic properties. The exception to this is the alkylenedioxypyrroles that will be discussed in Section 20.3.6.

20.3.3 Polycarbazoles

While less intensively studied than other conducting polymers covered in this chapter, polycarbazoles have found the most use as electroluminescent polymers. In the area of electrochromics, carbazoles have most commonly been utilized as a unit in electrochromic hybrid heterocycle systems, which will be covered later in this chapter. Most commonly, substitution occurs at the *N*-position. While this can offer increased solubility in a variety of solvents, there is not much variation in the colors achieved.

Unsubstituted films of polycarbazole (**28a**) show a yellow to green electrochromism upon oxidation at relatively low potentials [101,102]. Upon *N*-substitution, the green color remains at either partial or full oxidation. Organic-soluble poly(*N*-butyl-3,6-carbazolediyl) (**28b**) has been prepared and films cast onto transparent electrodes for electrochromic studies [103]. This polymer shows three colored states switching from colorless when neutral to green when partially oxidized and finally blue when fully oxidized. When the alkyl chain length is extended resulting in poly(*N*-dodecylcarbazole) (**28c**), the polymer films also give a colorless to green to blue electrochromism [104].

In an attempt to produce carbazole polymers soluble in aqueous solutions, oligoether groups have been attached to the carbazole unit at the *N*-position (**28d**) and the polymer prepared by chemical polymerization and electrochemical polymerization [105]. Due to the oligoether substituents, electro-chemical polymerization can occur in aqueous solutions without the need for a cosolvent. Polymer films switch between a highly transmissive state to deep green upon oxidation. The self-doped polymer, poly[3,6-carbaz-9-yl)propanesulfonate] (**28e**), has also been produced, which is water-soluble and switches from a transmissive neutral state to a dark green oxidized state [106].

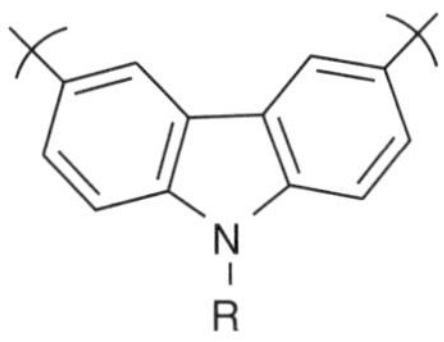

28

a. R = H
b. R = C_4H_9
c. R = $C_{12}H_{25}$
d. R = $(C_2H_4O)_3CH_3$
e. R = $C_3H_6SO_3^-$ TBA$^+$

Electrochromic polymer films have been made where the carbazole unit is not part of the main chain of the polymer, e.g., poly-*N*-vinylcarbazole (**29**) [107–109]. When electrochemically polymerized, it has been shown that the cation radical of the carbazole unit easily dimerizes to form a 3,3′-bicarbazolyl that is more easily oxidized than the parent monomer. These films also show a colorless to green transition upon oxidation. Similarly, films of poly[3-(3-bromocarbazol-9-yl)propyl]methylsiloxane (**30**) have been prepared where the carbazole dimerizes creating a cross-linked film that switches between colorless and green as with the other polymers [110].

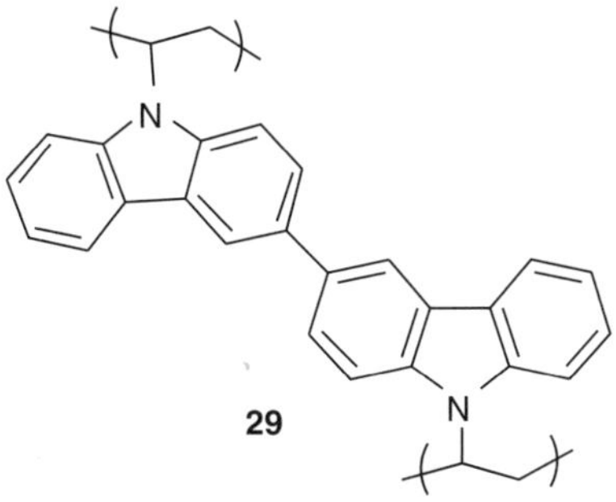

29 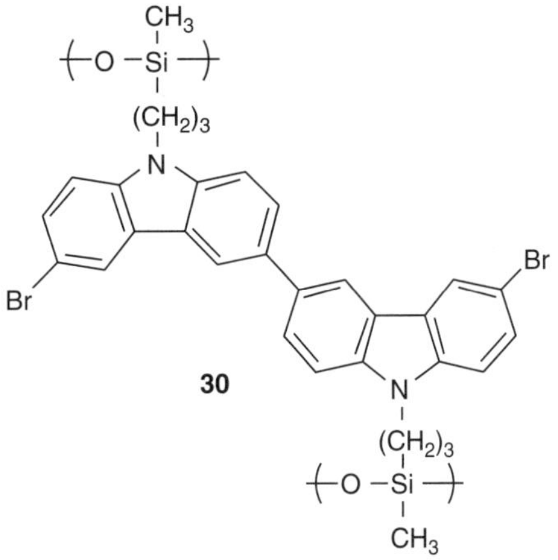   **30**

TABLE 20.3 Polyanilines

Polymer Number	Neutral Color	Oxidized Color	Refs.
31a,b,d,e.	Transmissive yellow	Violet (green and blue)[a]	111–116, 121–128, 129–131
31c	Transmissive yellow	Blue (green)[a]	123–125
32a-c	Transmissive yellow	Violet (green and blue)[a]	129–131
33a	Transmissive yellow	Blue	122
33b	Transmissive yellow	Blue	132
34a	Transmissive yellow	Blue (green)[a]	124,133
34b	Yellow	Dark green	134
35	Yellow-green	Brown (dark green)[a]	124,133
36	Yellow-green	Dark blue	135

[a]Partially oxidized.

20.3.4 Polyaniline

Polyaniline stands out as one of the oldest and most studied conducting and electrochromic polymer. The first report of the electrochemical preparation of polyaniline (**31a**, Table 20.3 where all structures are shown as fully neutralized poly(phenyleneamines) for simplicity) dates back to 1862 when Letheby described the production of a dark green film on an electrode upon oxidation of aniline under acidic conditions [111]. Polyaniline can be prepared either chemically or electrochemically in aqueous acidic solutions [112–114]. This system is unique in that thin polymer films are shown to exhibit multiple colors depending on redox state as shown in Table 20.3 for polyaniline and substituted derivatives. Figure 20.7 shows the cyclic voltammogram for polyaniline indicating the colors seen at each potential and Scheme 20.3 shows the structural changes occurring upon oxidation and proton loss [115].

When fully neutralized, thin films of polyaniline are transmissive yellow and insulating [116–118]. On oxidation, the film turns green and becomes conducting. Further oxidation produces an even deeper green followed by proton loss and production of a blue color in the film and eventually violet. The violet pernigraniline form is also insulating. When immersed in solutions of a pH greater than 4, the films turn

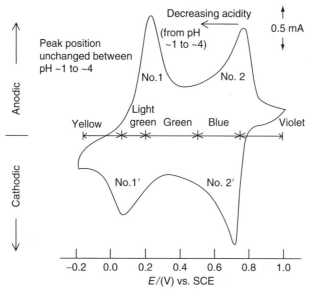

FIGURE 20.7 Cyclic voltammogram of polyaniline (emeraldine base). Colors given represent those seen at potential regions indicated. (From MacDiarmid, A., *Synth. Met.*, 18, 393, 1987. With permission.)

black and no longer switch color [112,118]. If the polyaniline films are prepared under neutral or basic conditions, the films are black and show no electrochromism [117]. The full color range of yellow to violet is not stable to repeated cycling and the films begin to degrade. If kept within the yellow to green region, the films are very stable and can be repeatedly switched more than 10^6 times [116].

Polyaniline films have not only been shown to exhibit electrochromism in the visible region, but also in the microwave and far-IR regions of the electromagnetic spectrum. A polyaniline film doped with camphorsulfonic acid and incorporated into a solid state microwave shutter demonstrated that the transmittance and reflectance of X-band microwave energy could be modulated [6]. At a wavelength of 10 GHz, the shutter could be switched between 4.8% transmission when the polymer is oxidized and 42% transmission when the polymer is neutral. When utilized in a reflective device configuration in combination with poly(diphenylamine), polyaniline yields a high reflective modulation in the far-IR [119,120]. This device shows a reflectance contrast of 53% at 10.5 μm, 28% at 16.5 μm, and 46% at 620 nm.

In an attempt to produce polymer films soluble in organic and aqueous solvents, there has been much work focused on producing substituted polyanilines. By polymerizing *ortho-* and *meta-*toluidines, polyaniline ring substituted with a methyl group (**31b**) can be produced [121–128]. Since polymerization preferentially proceeds by a head-to-tail mechanism, *ortho*-substituted polymers are produced at a higher yield than the *meta*-substituted polymers [124–126,128]. The polymer exhibits the same colors when switched electrochemically as unsubstituted polyaniline, yellow-green-blue-violet. The substituted polymers also show better stability of the full color range to multiple cycles and the full range occurs over a smaller potential window than with the unsubstituted polymer, and with faster switching speeds [121,125,128].

31

	R_1	R_2
a.	H	H
b.	CH_3	CH_3
c.	C_2H_5	C_2H_5
d.	OCH_3	OCH_3
e.	OC_2H_5	OC_2H_5

Ethyl-substituted aniline (**31c**) produces polymers with similar properties as the methyl-substituted polymers except that the films switch from green to blue on oxidation rather than blue to violet [123–125]. Substitution with bulky alkyl and aryl groups such as 2,5-di-*tert*-butylaniline, 2,6-diethylaniline, 2-benzylaniline, and 2-propylaniline does not produce polymer films [121,124]. This can be attributed to the increased steric hindrance by the bulky alkyl groups which prevent formation of head-to-tail couplings during polymerization.

It has been shown that substitution with alkoxy groups (poly(methoxyaniline) (**31d**), poly(ethoxyaniline) (**31e**)) produces polymer films more easily and yields films of slightly higher conductivity, depending on the substituents, than those monosubstituted with methyl groups [129–131]. This is due to the reduced steric hindrance induced by the methoxy group relative to the methyl groups reducing the torsion angle between repeat units. The polymers synthesized give similar colors to the unsubstituted polyaniline. A similar effect is seen in polyanilines substituted at the 2,5 positions (poly(2,5-dimethoxyaniline) (**32a**), poly(2,5-dimethylaniline) (**32b**), and poly(2-methoxy-5-methylaniline) (**32c**)) with the dimethylaniline exhibiting a lower conductivity than the methoxy-substituted polymers. The spectroelectrochemical series for poly(2,5-dimethoxyaniline) is shown in Figure 20.8. While substitution directly on the ring of aniline generally produces polymers of lower conductivity than polyaniline,

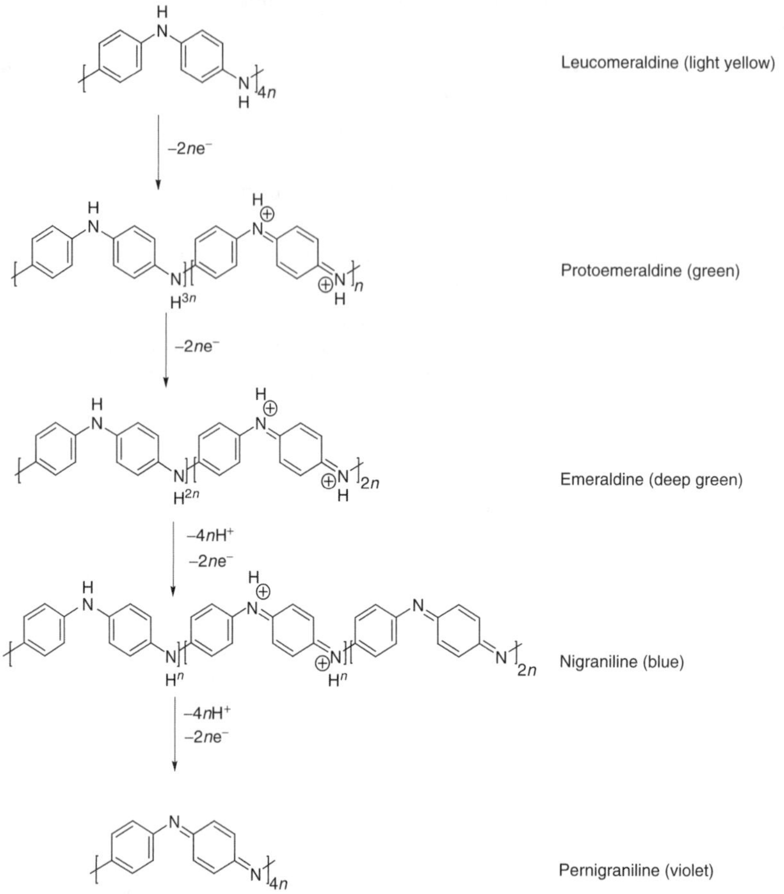
Leucomeraldine (light yellow)
−2ne⁻
Protoemeraldine (green)
−2ne⁻
Emeraldine (deep green)
−4nH⁺
−2ne⁻
Nigraniline (blue)
−4nH⁺
−2ne⁻
Pernigraniline (violet)

SCHEME 20.3 Oxidation of polyaniline with associated colors for each redox state.

poly(2,5-dimethoxyaniline) has a conductivity close to that of the parent polymer with improved solubility in organic solvents [129–131].

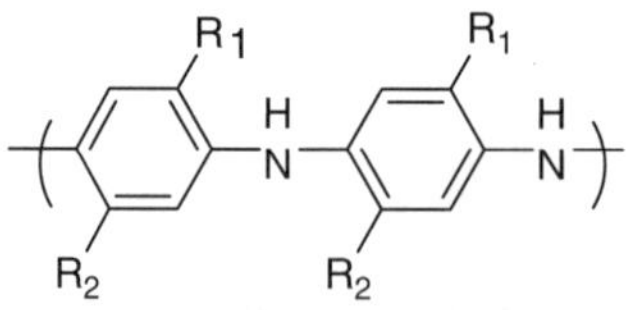

32

	R_1	R_2
a.	OCH_3	OCH_3
b.	CH_3	CH_3
c.	OCH_3	CH_3

Substitution with an electron acceptor, such as chlorine for poly(2-chloroaniline) (**33a**), yields polymers that switch between yellow in the neutral form to blue in the oxidized form [122]. Sulfonic acid ring-substituted polyaniline (**33b**) can be prepared by direct chemical reaction of polyaniline with concentrated sulfuric acid [132]. This polymer film switches from transmissive yellow to blue even in solutions of elevated pH up to 7.

33

a. R = Cl
b. R = SO₃H

Substitution at the nitrogen in aniline can also produce electrochromic polymers upon polymerization. Poly(*N*-methyl aniline) (**34a**) polymerizes at the *para* position in a C–N coupling with the methyl group replacing the acidic hydrogen [124,133]. The polymer switches reversibly from yellow to green and finally blue and has a conductivity comparable with the ring-substituted polyanilines. By placement of a phenyl ring on the nitrogen, polymerization proceeds by a C–C coupling with production of a polyaniline-*p*-phenylene repeat unit (**35**) being the most likely structure [124,133]. Another example of an *N*-substituted polyaniline is poly(aniline *N*-butylsulfonate) (**34b**) that switches reversibly between yellow and dark green with high redox stability in organic solvents and is soluble in water [134].

34

a. R = CH₃
b. R = C₄H₈SO₃⁻

35

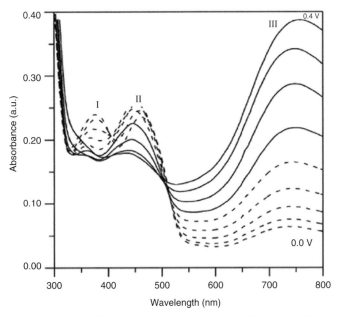

FIGURE 20.8 Absorption spectroelectrochemistry for a thin film of poly(2,5-dimethoxyaniline) between the potentials of 0 and 0.4 V vs. Ag/AgCl. The labels I, II, and III refer to the reduced state, intermediate oxidized state, and fully oxidized state absorption bands, respectively. (From Huang, L., *Synth. Met.*, 130, 155, 2002. With permission.)

By reacting the leucomeraldine base polyaniline with NaH, deprotonation of the NH occurs giving an anion of the polyaniline [135]. The polymer can then be reacted with glycidyl phenyl ether to cause a ring-opening polymerization yielding a graft copolymer of polyaniline with oligoether side chains at the *N*-position (**36**). The polymer is soluble in common organic solvents such as chloroform, THF, and benzene. Due to the increased steric hindrance of the grafted side chain, the color of the neutral polymer film is yellowish-green and switches to dark blue. Alternatively, side chains of polyaniline can be grafted onto an insulating polymer chain. An example includes the grafting of branches of polyaniline on nitrilic rubber, a copolymer of butadiene and acrylonitrile [136]. Nitrilic rubber was chosen as the main chain polymer due to its resistance to attack by strong acids. The graft copolymer exhibits the mechanical properties of a cross-linked elastomer while retaining the electrochemical and electrochromic properties of polyaniline with the polymer switching from yellow to green and finally blue-green.

36

Polyaniline is one of the earliest studied conjugated polymers and remains as one of the most researched, especially in the field of electrochromics. While potentially quite useful as a stable and cost-effective conductor, it is a bit unfortunate that substituted polyanilines exhibit little change in the electrochromic properties as a function of substituent identity and position. Substituted and unsubstituted polyanilines have the potentially useful property of being polyelectrochromic in thin film form, but structural modification has not been shown to increase the range of colors available.

20.3.5   Poly(3,4-Alkylenedioxythiophene)s

As illustrated previously, PT derivatives have been the focus of much research due to their unique combination of electronic properties that include electrical conductivity, electrochromism, and structural versatility. These PTs typically have a moderate bandgap, which is in the middle of the visible spectrum while others such as polyanilines and polypyrroles have higher bandgaps. This gives the PTs a colored appearance (most normally red) in the neutral state and they tend to become more absorbing and opaque upon oxidation (Table 20.4).

Although this color change can be useful, another desirable color change for electrochromic windows or displays occurs when the polymer switches from a highly colored state to a highly transmissive state upon oxidation, making it cathodically coloring [2,4,66]. This type of polymer has a low bandgap with its interband transition in the visible region of the spectrum, with an onset close to the near-IR (NIR). Upon oxidation, the absorption maximum will shift to the NIR region and the material becomes transmissive in the visible region. Historically, there have been few electrochromic conjugated polymers that exhibit this property [25,72].

One method that has been utilized to achieve low bandgap polymers is to synthesize polymerizable fused ring heteroaromatic monomers. This eliminates known competing polymerization routes in thiophene through the 3- and 4-carbon positions, which are now blocked. One of the first and most highly studied polymers synthesized to this end is poly(3,4 ethylene dioxythiophene) (PEDOT) (**37**) [22–24,137]. This alkoxy-substituted PT has a bandgap of ~1.6 eV, which is lower than the unsubstituted and alkyl-substituted PTs that have higher oxidation potentials causing their oxidized and conducting forms to be less stable under ambient conditions. The presence of the two electron-donating oxygen atoms adjacent to the thiophene unit decreases the polymer oxidation potential by raising the

TABLE 20.4 Poly(3,4-alkylenedioxythiophene)s

Polymer Number	Neutral Color	Oxidized Color	Refs.
37,38	Dark blue	Transmissive sky blue	22–24,137,139
39	Dark blue	Transmissive sky blue[a]	140
40a	Dark blue	Transmissive sky blue	38,141–143
40b,c	Dark purple	Transmissive light gray	138,141–143
40d,e	Dark blue	Transmissive sky blue	142–144
41	Brown-yellow	Dark green	145
42a,b	Dark blue	Transmissive sky blue	142,143,146
43a,b	Dark blue	Transmissive sky blue	146,149
43c	Dark red-purple	Transmissive sky blue	147
43d-f	Dark blue-purple	Transmissive sky blue	150–152
43g	Dark blue	Transmissive sky blue	150,153
44a,b	Dark blue-purple	Transmissive sky blue	150–152
45	Dark red-purple	Dark blue (transmissive blue-gray)[b]	154

[a] As-cast film color.
[b] Intermediate oxidation state color.

energy of the valence band electrons while at the same time causing little perturbation on the energy of the conduction band, which leads to the decrease in the polymer electronic bandgap [138].

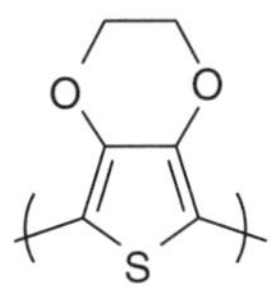

37
(E_g = 1.6 eV)

PEDOT was first developed by Bayer in an attempt to produce a conducting polymer with a high degree of order (due to the lack of α–β and β–β couplings), which could be easily oxidized and stable. Even though the polymer was initially insoluble and not processable, the films did possess impressive electronic and electrochromic properties. This polymer exhibits a low monomer oxidation potential, is quite stable under ambient and elevated temperature conditions, and has a high conductivity in the oxidatively doped, electrically conducting state. Another desirable property of this polymer is that it is cathodically coloring. Thin films of the polymer are opaque dark blue in the neutral state and switch to a very transmissive sky blue in the oxidized state. As shown in Figure 20.9, the neutral film shows a bandgap edge at around 1.5 eV and the light sky blue appearance in the doped film is caused by the presence of a tailing absorption in the range of 1.6–2.0 eV. The ability of PEDOT to easily switch between an absorptive state and a transmissive state quite rapidly and at low oxidation potentials makes it a useful material for electrochromic applications. In fact, PEDOT has dominated much of the literature in the last two decades both as an electronic conductor and as an active electrochromic material [22,23,137].

Similar to the previously mentioned polymers, attempts have been made toward polymerizable EDOT-based oligomers that have lower oxidation potentials than their respective monomers and longer conjugation lengths. BiEDOT (**38**) has an oxidation potential of 1.4 V lower than that of thiophene and 0.6 V lower than that of EDOT [139]. The dimer easily polymerizes, is air stable, and when polymerized, the resultant polymer has identical electrochromic properties to PEDOT. TerEDOT has also been synthesized with a slightly lower oxidation potential than BiEDOT. Although it is somewhat unstable and can be difficult to handle and store, it electropolymerizes to afford an electrochromic PEDOT.

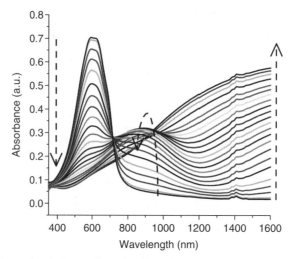

FIGURE 20.9 Spectroelectrochemical series for a thin film of PEDOT. Dashed arrows show direction of spectral growth or recession. (Dyer, A., 2005, unpublished data.)

38

While an invaluable tool in producing conjugated polymers on conducting substrates, electropolymerization has limitations that include a lack of primary structure verification and characterization along with the inability to synthesize large quantities of processable polymer. To overcome the insolubility of PEDOT, a water-soluble polyelectrolyte, poly(styrenesulfonate) (PSS) was incorporated as the counterion in the doped PEDOT to yield the commercially available PEDOT/PSS (Baytron P) (**39**), which forms a dispersion in aqueous solutions [140]. While this polymer finds most of its application as a conductor for antistatic films, solid state capacitors, and organic electronic devices, its electrochromism is distinct and should not be ignored.

39

Another route to yield soluble forms of PEDOT has been the introduction of solubilizing alkyl groups on the ethylenedioxy bridge. It has been found that the nature of the substitution has little effect on the extent of conjugation, which can be seen by the electronic bandgaps for all polymers occurring at approximately 1.7 eV [138,141–144]. Polymers have been produced with alkyl chains ranging from one carbon (PEDOT-Me) (**40a**) to 14 carbons (PEDOT-C14) (**40c**). While PEDOT-Me is insoluble and synthesized electrochemically, longer alkyl chain-substituted polymers can be synthesized both electrochemically and chemically [142,143]. PEDOT-C8 (**40b**) and -C14 switch from an opaque deep purple in the neutral form to transmissive light gray in the oxidized form as both films and in solution, as shown in Figure 20.10 for PEDOT-C14 [138,141].

The substituted PEDOTs provide excellent cathodically coloring electrochromic properties that include a high optical contrast in the visible region and high redox stability. The higher optical contrast over unsubstituted PEDOT comes from significantly less optical absorption in the visible region for the substituted PEDOTs when oxidized due to repression of the NIR tail in the conducting form. The longer alkyl chain-substituted polymers are synthesized by oxidative polymerization yielding solution processable polymers that have similar optical properties as the electrochemically polymerized materials.

The higher contrasts and faster switching speeds seen in these polymer films have been explained by the long alkyl chains acting to separate the conjugated backbones from one another, effectively leading to a more expanded morphology [138,141–143]. It is thought that this facilitates better counterion transport through the polymer film, allowing higher doping levels to be attained. Similarly, phenyl substitution on the ethylene bridge in PEDOT-Ph (**40d**) provides a substituent that is nearly orthogonal to the thiophene ring, suggesting that the phenyl group will separate the polymer chains even further [142,143]. This polymer exhibits switching speeds comparable with that of PEDOT-C14.

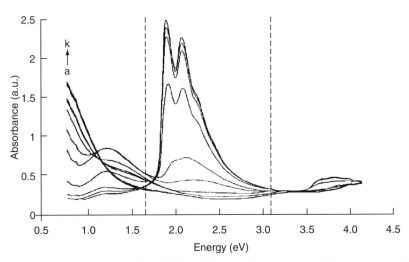

FIGURE 20.10 Spectroelectrochemistry for PEDOT-C14 between the potentials of (a) −0.8 V vs. Ag/Ag$^+$ and (k) +0.8 V vs. Ag/Ag$^+$. (From Sankaran, B., *Macromolecules*, 30, 2582, 1997. With permission.)

The incorporation of fluorine into insulating polymers has been previously studied because of their low surface energies, chemical inertness, high thermal stability, and enhanced rigidity. Perfluorinated materials have been shown to be useful in commercial applications ranging from nonadhesive surfaces, convective coolants, insulating materials, and membranes. Most EDOT polymers and their similar analogs are oxygen and moisture sensitive in their neutral form. In an effort to enhance the environmental stability of these polymers, highly hydrophobic perfluoroalkyl substituents have been added to PEDOT, exemplified by the pentadecafluoro-octanoate derivative of PEDOT (PEDOT-F) (**40e**) [144]. This polymer can be polymerized by electrochemical polymerization with a slightly more anodic oxidation potential compared with unsubstituted EDOT. This is due to the electronegative nature of the perfluoroalkyl substituent in combination with steric factors. The polymer produced has a bandgap of 1.65 eV with subsecond switching speeds attainable. The polymer switches between a dark blue neutral state and a transmissive sky blue oxidized state with over a 60% contrast in the visible region. The oxidized form of the polymer film is highly hydrophobic with a contact angle of 110° while the wettable PEDOT has a contact angle of less than 30°.

Compared with PEDOT, poly(3,4-ethylenedithiathiophene) (EDTT) (**41**) has been polymerized where the two oxygens are replaced by two sulfurs in the β-position of the thiophene ring [145]. The all-sulfur analogs can be synthesized chemically and electrochemically, yielding a polymer completely soluble in 1-methyl-2-pyrrolidinone and partially soluble in tetrahydrofuran and chloroform. The bandgap of the chemically polymerized polymer is 2.19 eV with the bandgap of the electrochemically polymerized polymer slightly lower at 2.14 eV. In contrast to PEDOT, PEDTT is not optically transparent in the doped state, switching between brown-yellow in the neutral state and dark green in the oxidized state.

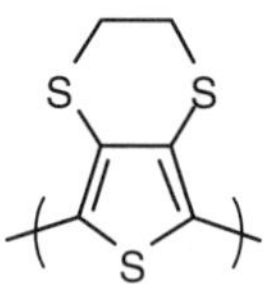

41

(E_g = 2.15–2.19 eV)

It has been shown that not only the nature of the substituent has little effect on the extent of conjugation for alkylenedioxythiophenes, but the size of the alkylenedioxy ring also has little effect on the polymer conjugation length as all polymers have similar bandgap values. On the other hand, when increasing the ring size on the alkylenedioxy bridge, switching speeds can be increased for poly(3,4-propylenedioxythiophene) (PProDOT) (**42a**) and poly(3,4-butylenedioxythiophene) (PBuDOT) (**42b**) [142,143,146]. When increasing the bridge size from ethylene to propylene and butylene, it is seen that the oxidation potentials for the polymers are increased over that of PEDOT. This occurs because of a decrease in the electron-donating ability of the oxygens due to the larger alkylenedioxy ring size, which allows for larger degrees of freedom for movement of the oxygen lone pair electrons. Regardless of the higher oxidation potential, PProDOT and PBuDOT both show increased switching speeds, higher optical contrast (54% for PProDOT and 63% for PBuDOT), and a relatively high in situ conductivity for PProDOT.

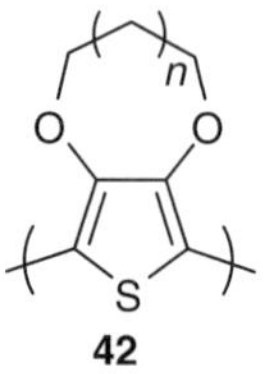

42

a. *n* = 1
b. *n* = 2

Symmetrical disubstitution can be performed on the central carbon atom of the propylene bridge in PProDOT. Additionally, dialkylation at this central carbon atom allows for the production of regiosymmetric polymers and induction of solubility when the alkyl chain is of sufficient length [146,147]. When disubstituted with methyl groups (PProDOT-Me$_2$) (**43a**), the polymer exhibits a particularly high transmittance contrast ratio of 78% at λ_{max}. This higher contrast can be attributed to the attenuation of the peak in the NIR region upon oxidation, yielding higher transparency in the visible region as can be seen when comparing the absorbance around 800 nm for both PEDOT (Figure 20.9), and PProDOT-Et$_2$ (Figure 20.11), when in their fully oxidized states. As shown in Figure 20.12, PProDOT-Me$_2$ also shows IR electrochromism with the IR absorption strong throughout the NIR and mid-IR region [148]. At 1.8 μm, the reflectance contrast approaches values greater than 90%, and values near 60% for longer wavelengths between 4 and 5 μm.

Diethyl-substitution (PProDOT-Et$_2$) (**43b**) also yields a polymer with a high contrast in the visible region with values up to 75% as evident by the spectroelectrochemical series in Figure 20.11 [149]. PProDOT-Et$_2$, like PProDOT-Me$_2$, switches between a highly absorbing dark blue neutral state and a highly transmissive sky-blue oxidized state with subsecond switching speeds. The high optical contrasts and fast switching speeds for both PProDOT-Me$_2$ and PProDOT-Et$_2$ are the direct result of the tetrahedral substitution pattern, which causes the alkyl groups to be positioned above and below the plane of the polymer chain. This separates the polymer chains allowing high doping levels due to facile counterion movement. This also leads to inhibition of π-stacking, causing a decrease in the polymer in situ conductivity.

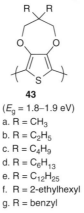

43
(E_g = 1.8–1.9 eV)
a. R = CH$_3$
b. R = C$_2$H$_5$
c. R = C$_4$H$_9$
d. R = C$_6$H$_{13}$
e. R = C$_{12}$H$_{25}$
f. R = 2-ethylhexyl
g. R = benzyl

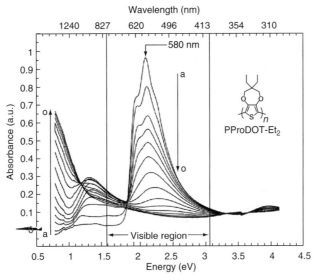

FIGURE 20.11 Absorbance spectroelectrochemistry for a thin film of PProDOT-Et$_2$ between the potentials of (a) −0.1 V vs. Ag/Ag$^+$ and (o) 0.9 V vs. Ag/Ag$^+$. (From Gaupp, C., *Macromol. Rapid Commun.*, 23, 885, 2002. With permission.)

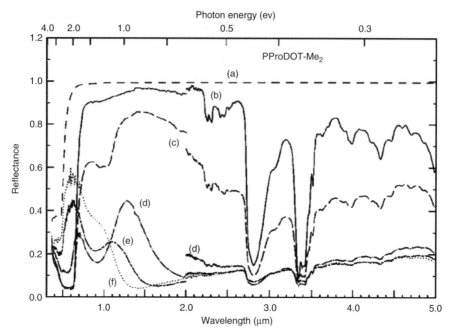

FIGURE 20.12   Reflectance spectroelectrochemistry for a reflective electrochromic device with PProDOT-Me$_2$ as the active electrochromic material. The applied voltages are between (b) -1.5 V and (f) 1.0 V with spectra shown relative to control gold surface (a). (From Schwendeman, I., *Adv. Mater.*, 13, 634, 2001. With permission.)

Upon increasing the dialkyl chain length to butyl, polymers are produced that are soluble in organic solvents that include tetrahydrofuran, toluene, chloroform, and methylene chloride [147]. Not only does solubility allow for complete structural characterization, but also allows for the development of processable polymers that opens up the possibilities for applications. The dibutyl-substituted polymer, PProDOT-Bu$_2$ (**43c**), can be prepared chemically, with films being solution-cast, or electrochemically polymerized from the monomer. It was seen that while films prepared with both methods have the same bandgap of 1.8 eV and switch from a deep red-purple in the neutral state to a highly transmissive sky blue when oxidized, the electrochemically prepared films exhibit faster switching speeds and higher visible region contrast values. The solution-cast films have a change of transmittance of 45% while the electrochemically prepared films show a value of 75% in the visible region. Polymers substituted with longer alkyl chain lengths, including the dihexyl (PProDOT-Hx$_2$ (**43d**)), didodecyl (PProDOT-(C$_{12}$H$_{25}$)$_2$ (**43e**)), and dioctadecyl (PProDOT-(C18)$_2$ (**44a**)) derivatives are soluble in common organic solvents and switch between a dark blue-purple and a transmissive sky blue with subsecond switching times, similar to PProDOT-(C$_4$H$_9$)$_2$ [150–152]. Figure 20.13 shows the evolution of the absorption spectra as the potential applied to a thin film of PProDOT-Hx$_2$ spray-cast onto indium-doped tin oxide (ITO) is changed. These polymers also have high optical contrast with the dihexyl polymer exhibiting a 74% contrast and the didodecyl polymer showing a 77% contrast.

OR   OR

44

($E_g \sim 1.9$ eV)

a. R = C$_{18}$H$_{37}$

b. R = 2-ethylhexyl

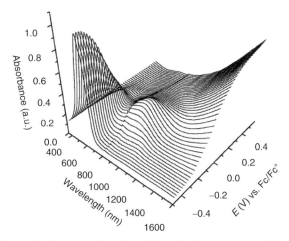

FIGURE 20.13 Three-dimensional spectroelectrochemical series for a spray-cast film of PProDOT-Hx$_2$. (From Reeves, B., *Macromolecules*, 37, 7559, 2004. With permission.)

When the alkyl substituent is branched, as in PProDOT-(EtHx)$_2$ (**43f**) and PProDOT-(CH$_2$OEtHx)$_2$ (**44b**), the polymers have higher bandgaps ($\sim$1.9 eV) and the electrochromic color change occurs over a smaller voltage range when compared with the polymers with linear substituents [151,152]. The higher bandgaps are due to a sharper onset on the lower energy side of the $\pi-\pi^*$ absorbance in the visible region. Both the sharper absorbance onset and narrow switching range have been proposed to be due to a limited effective conjugation length in the polymer solid state when the branched substituents are present. The branched polymers also switch between a dark blue-purple color in the neutral state and a transmissive sky blue in the oxidized state.

To further increase the interchain separation, ProDOT polymers were synthesized with dibenzyl substitution on the propylenedioxy ring [150,153]. The benzyl substituents have been shown to orient themselves orthogonally to the thiophene ring with the groups on opposite sides of the thiophene plane allowing further interchain separation and higher electrochromic contrasts. Unlike the dialkyl-substituted polymers, PProDOT-Bz$_2$ (**43g**) is completely insoluble, but does exhibit a very high contrast of 89% at λ_{max} with the polymer film switching between a dark blue neutral state and a transmissive state in the oxidized form.

The myriad of substitution patterns available for PProDOT allow not only the synthesis of disubstituted polymers, but also the linking of two ProDOT molecules at the 2-position on the propylenedioxy bridge [154]. This ProDOT derivative, poly(spiroBiProDOT) (**45**), exhibits increased functionality, having the ability to cross-link during oxidative electropolymerization. In the monomer, the x-ray crystal structure shows that the thiophene rings, oxygen atoms, and the spiro carbon exist within the same plane. This is due to the alkylenedioxy rings being large enough to allow a small distortion in the tetrahedral angle of the spiro carbon, twisting the C8 and C6 carbons out of plane. The bandgap of this polymer is similar to the other ProDOTs and exhibits three colored states with the emergence of a second absorption in the visible region that shifts to lower wavelengths as the polymer is being oxidized. The neutral polymer film is a dark red-purple, switching to a transmissive blue-gray at intermediate oxidation potentials, finally switching to a dark blue when fully oxidized.

45

The alkylenedioxythiophene classes of electrochromic polymers are some of the most extensively studied polymers to date. These polymers have a relatively low oxidation potential, are stable in their oxidized state, and exhibit the desirable electrochromic property of switching between a colored and a transmissive state. Not only do they possess impressive electrochromic properties, but also electrochemical properties with high conductivity values. While substitution increases electrochromic performance with higher contrast ratios and faster switching speeds, there is a decrease in the maximum conductivity available. Alkyl and alkoxy substitution also increases the solubility of these polymers in common organic solvents making the polymers more processable. This allows for more options in film deposition methods, with films being produced using drop casting, spin casting, and even spray casting.

20.3.6   Poly(3,4-Alkylenedioxypyrrole)s

As with thiophenes, an alkylenedioxy bridge can be attached to the 3- and 4-positions of pyrroles. This allows polymerization to occur exclusively at the 2- and 5-positions of the resulting 3,4-alkylenedioxy-pyrrole, yielding a polymer with fewer structural defects from undesired α–β couplings that can typically occur with polypyrroles that are highly sensitive to overoxidation [87]. Additionally, the alkylenedioxy bridge adds electron density, resulting in reduced monomer and polymer oxidation potentials. The combination of the electron-rich character of the pyrrole with the alkylenedioxy substituent causes an increase in the HOMO level and provides a polymer with what is likely the lowest oxidation potential for p-type polymers published to date. The low oxidation potential affords an advantage in that polymerization occurs under milder conditions, and the resulting polymer exhibits increased ambient stability in the doped conducting state [28–30,155,156].

46

($E_g = 2.0$ eV)

While the addition of the alkylenedioxy moiety produces higher HOMO levels, pyrroles have higher LUMO levels than thiophenes, yielding a polymer with a higher bandgap than the analogous poly(alkyl-enedioxythiophene) yet a lower bandgap than the unsubstituted pyrrole. This is evident in PEDOP (**46**),

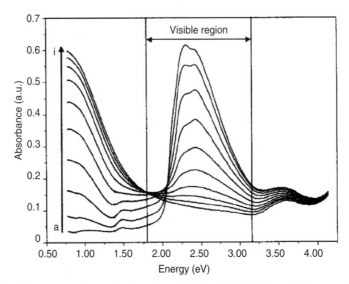

FIGURE 20.14   Absorbance spectroelectrochemistry for a thin film of PEDOP stepped between potentials of (a) −0.87 V vs. Ag/Ag$^+$ and (j) −0.37 V vs. Ag/Ag$^+$. (From Gaupp, C., *Macromolecules*, 33, 1132, 2000. With permission.)

which has a bandgap value of ~2.0 eV, whereas PEDOT has a bandgap of ~1.6 and pyrrole with a value of 2.7 eV [28,29,155,157,158]. As shown in Figure 20.14, the $\pi-\pi^*$ transition for PEDOP occurs in the visible region, and upon oxidation, is depleted with a strong lower energy transition induced in the NIR. The polymer switches between a bright red neutral state to a highly transmissive blue-gray in the oxidized state.

Due to tailing of the broad NIR absorption in the oxidized state, PEDOP exhibits a lower optical contrast of 59% compared with PProDOP (**47a**), which has an optical contrast of 70% [29,155,159]. The higher optical contrast for PProDOP is due to the redshifting of the NIR band, allowing a higher degree of light transmission along with a higher bandgap (2.2 eV) resulting from an increased steric distortion caused by the seven-membered ring. Films of PProDOP switch between orange in the neutral form to a light gray-blue in the doped form with an intermediate brown state. Both PEDOP and PProDOP exhibit quite high long-term redox switching stability with PEDOP retaining more than 96% and PProDOP retaining more than 90% of their electroactivity after 3000 potential switches, while polypyrrole retains only 80% under identical conditions. Both polymers switch relatively rapidly with films attaining about 90% of their full optical contrast within 0.25 s in comparison with polypyrrole, which reaches the same contrast value in over 1 s.

As with the poly(alkylenedioxythiophene) series, substitution on the central carbon atom of the propylene bridge offers the ability to induce increased solubility in various solvents. An increase in the optical contrast can also be achieved through increasing the ring size and addition of alkyl groups on the alkyl bridge [29,155,159–161]. This is evident with dimethyl-substituted PProDOP-$(CH_3)_2$ (**48b**), which has a $\Delta\%$ T of 76% at 534 nm and retains 90% of its electroactivity after undergoing 40,000 potential switches. PProDOP-$(CH_3)_2$, along with the monomethyl-substituted PProDOP (PProDOP-CH_3) (**48a**) and poly(3,4-butylenedioxypyrrole) (PBuDOP) (**47b**) exhibit a bandgap value of 2.2 eV and switch from orange, when fully neutralized, to a red-brown or orange-brown when partially oxidized, and finally to a light blue when fully oxidized as shown in Table 20.5.

47
$(E_g = 2.2$ eV$)$
a. $n = 1$
b. $n = 2$

48
$(E_g = 2.2$ eV$)$
R_1 R_2
a. CH_3 H
b. CH_3 CH_3

Similarly with PEDOT, alkylsulfanyl derivatives of PXDOP, poly(3,4-methylenedithiopyrrole) (PMDSP) (**49a**), and poly(3,4-ethylenedithiopyrrole) (PEDTP) (**49b**) have also been synthesized in which the alkylenedioxy bridge is replaced with a alkylenedisulfanyl bridge [162,163]. The spectro-electrochemical series for PEDTP is shown in Figure 20.15. The polymers exhibit low oxidation

TABLE 20.5   Poly(3,4-alkylenedioxypyrrole)s

Polymer Number	Neutral Color	Oxidized Color	Refs.
46	Red	Light sky blue	29,155–158
47a	Orange	Light gray-blue (brown)[a]	29,155,156,159
47b	Orange	Light blue-gray (orange-brown)[a]	29,155,156,159–161
48a	Orange	Light blue (brown-red)[a]	29,155,156,159–161
48b	Orange	Light blue (red-brown)[a]	29,155,156,159–161
50a	Purple	Blue (dark green)[a]	136,155,159,164,165
50d	Colorless	Blue-gray (pale pink, pink-gray)[a]	136,155,159,164,165
50e	Colorless	Gray-blue (pale pink, pink, tan, and gray-green)[a]	136,155,159,164,165

[a]Partially oxidized

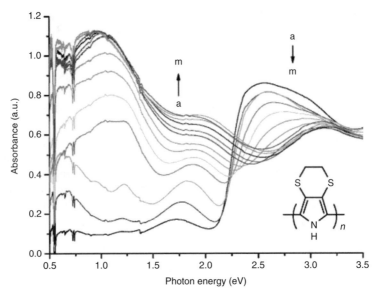

FIGURE 20.15   Absorbance spectroelectrochemistry of PEDTP at applied potentials between (a) −0.9 V and (m) 0.6 V vs. Fc/Fc$^+$. (From Li, H., *Macromolecules*, 39, 2049, 2006. With permission.)

potentials and lower bandgap values than polypyrrole with PMDSP having a bandgap of 1.8 eV and PEDTP having a bandgap value of 2.1 eV.

49

(E_g = 1.8–2.1 eV)

a. n = 1

b. n = 2

In an effort to produce polymers that have a high bandgap and color upon oxidation while maintaining the low oxidation potentials seen with the PXDOPs, a series of PProDOPs that are N-substituted have been produced [155,156,159,164,165]. The increase in the steric effect of N-substitution causes a shift of the π–π* transition to higher energies, which places the λ_{max} of the neutral polymer in the UV region of the spectrum resulting in a higher bandgap. By moving the π–π* transition into the UV region, the transmissivity in the visible region is increased for the neutral polymer. The polymers have bandgap values that are considerably higher than PProDOP and range from 3.0 eV for the N-propanesulfonated PProDOP (PProDOP-N-PrS) (**50e**) and the methyl-substituted polymer (PProDOP-N-Me) (**50a**), to 3.4 eV for the propyl- (PProDOP-N-Pr) (**50b**), octyl- (PProDOP-N-Oct) (**50c**), and ethoxyethoxyethanol-substituted (PProDOP-N-Gly) (**50d**) polymers [30,155,159,165]. By shifting the absorbance of the neutral polymer into the UV region, they become anodically coloring, being transmissive in the neutral state and becoming colored when oxidized.

50

(E_g = 3.0–3.4 eV)

a. R = CH_3

b. R = C_3H_7

c. R = C_8H_{17}

d. R = $(C_2H_4O)_2CH_2OH$

e. R = $C_3H_6SO_3Na$

Both PProDOP-*N*-Gly and PProDOP-*N*-PrS are colorless when neutral, switching through colored intermediate states with PProDOP-*N*-Gly becoming a gray-blue and PProDOP-*N*-PrS becoming a blue-gray when fully oxidized as can be seen for the high transmittance throughout the entire visible region for PProDOP-*N*-Gly in Figure 20.16. The *N*-Me and *N*-Pr polymers retain a residual coloration in the neutral state with PProDOP-*N*-Me switching from a light purple to a dark green, and finally blue when fully oxidized. The spectroelectrochemical series for the polymers is shown in Figure 20.17. In addition to being highly transmissive and nearly colorless in the neutral state, PProDOP-*N*-PrS has the possibility of being self-doping and water-soluble due to the presence of a sulfonate group at the terminus of the propyl chain. This polymer has also been shown to increase the transmissivity and optical contrast of an electrochromic window device when utilized as the anodically coloring material along with PProDOT-Me$_2$ as the cathodically coloring material [165]. Not only do the *N*-alkyl PXDOPs show complimentary optical properties with PXDOTs, but they also have the required electrochemical compatability and reversibility necessary for practical application in an ECD.

In summary, through the introduction of an alkylenedioxy bridge on pyrrole, the polymers produced combine the desirable properties of the PXDOTs such as stability, high optical contrast, and lower bandgap while capitalizing on the intrinsic properties of the pyrrole heterocycle, such as low oxidation potential and aqueous compatability, which are ideal for biomaterials applications. At the same time, *N*-derivatization allows the introduction of substituents that range from nonpolar to polar, and even ionic functionalities. The ability to derivatize at two different positions allows for a wider range of possibilities, including substitution at both the *C*- and *N*-positions. Even though the PXDOPs exhibit attractive properties ranging from having a high bandgap while maintaining a low oxidation potential, along with electrochemical stability, good conductivity, multicolor electrochromism, and aqueous compatability, there remain some drawbacks that hinder extensive interest in these polymers.

Where synthesis of thiophene and PXDOT monomers are relatively simple and the polymerizations uncomplicated, synthesis of XDOP polymers is somewhat complex. Additional synthetic steps must be undertaken to protect the unfunctionalized nitrogen atom, and the desired low oxidation potential of the monomer presents a disadvantage during synthesis in that the monomers must be handled carefully to avoid decomposition during various synthetic steps [28,156–158,166]. The production of the *N*-alkylated pyrroles, on the other hand, provides properties not accessible with other conducting polymers where the *N*-substituent provides both protection during synthesis and is an active portion of the final polymer structure for property control.

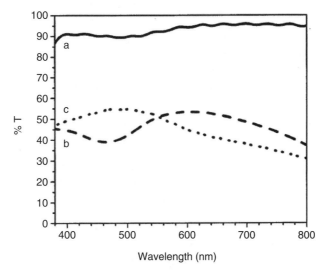

FIGURE 20.16 Percentage transmittance for a thin film of PProDOP-*N*-Gly in the (a) neutral state, (b) partially oxidized state, and (c) fully oxidized state. (From Sonmez, G., *Macromolecules*, 36, 639, 2003. With permission.)

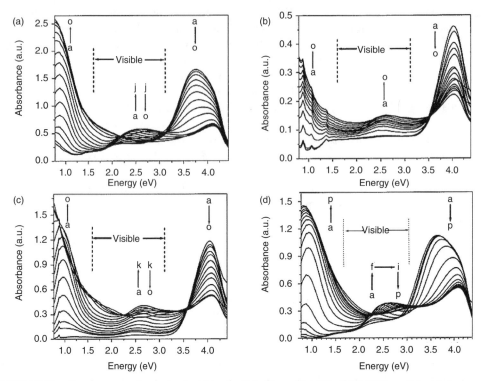

FIGURE 20.17   Absorbance spectroelectrochemistry for thin films of (a) PProDOP-*N*-Me switched between −0.5 and 0.7 V, (b) PProDOP-*N*-Pr switched between −0.4 and 0.6 V, (c) PProDOP-*N*-Gly switched between −0.2 and 0.7 V, and (d) PProDOP-*N*-PrS switched between −0.4 and 0.5 V. All potentials reported are vs. Fc/Fc$^+$. (From Sonmez, G., *Macromolecules*, 36, 639, 2003. With permission.)

20.4   Hybrid Electrochromic Systems

While the previously discussed conjugated polymers exhibit promising electrochromic properties, such as fast switching speeds, high optical contrasts, and multicolor capabilities, there remain limitations in the properties ultimately accessible for practical display and device applications. Through the incorporation of conjugated conducting polymers with other electroactive or electrochromic materials, the electrochromic properties available can be expanded to yield materials that have even higher optical contrasts, faster switching speeds, a larger color palette, lower oxidation potentials, and variable bandgaps. Various methods include the utilization of hybrid repeat units, copolymerizations, multipolymer blends, laminates, and composites, all of which will be discussed in this section.

20.4.1   Multiring Hybrid Repeat Unit Polymers and Copolymers

The combination of the repeat unit species typically used in electrochromic homopolymers such as the thiophene, furan, pyrrole, dioxythiophene and dioxypyrrole rings with one another and with other conjugated linkages (e.g., vinylenes, phenylenes, fluorenes, and carbazoles) provides color control through both the electronic character of the backbone and by steric interactions between repeat units (Table 20.6).

Thienylenevinylene hybrid polymers provide a simple example of how the introduction of a conjugated spacer influences the electronic properties of the resulting polymer compared with the parent homopolymer. Single substitution of a polyheterocycle with electron-donating units, e.g., alkoxy groups, raises the HOMO energy level causing a narrowing of the bandgap. On the other hand, 3,4-dialkoxy

TABLE 20.6   Multi–Ring Hybrid Repeat Unit Polymers and Copolymers

Polymer Number	Neutral Color	Oxidized Color	Refs.
51a-e	Deep blue	Transmissive light gray	167–169
52	Deep purple	Transmissive sky blue	171–173
53a-f	Opaque blue	Transmissive light blue	31,174
55a-d	Blue	Red	180–181
57a-d	Pale red	Deep blue (green)[a]	186,187
57e	Deep red-purple	Pale blue (purple-blue, blue and gray-blue)[a]	30,188
57f	Red	Blue-black	189
58	Crimson	Light blue (purple, magenta, and pale-tan/cream)[a]	190
59a-f	Transmissive yellow	Blue-green	175,191
60	Red	Blue-purple (pale blue)[b]	192–194
61a-e	Orange	Black (dark gray)[b]	195,196
63	Lavender	Pale blue	192,197
64	Lime green	Light gray (burgundy red and dark gray)[b]	192,193
66	Blue	Transmissive near-colorless	201,202
67a-e	Dark blue-green	N/A	203,204
68a-f	Dark blue-green	N/A	204,205
69a,b	Green	Light brown	206,207
71a	Olive green	Dark green	210,211
71c	Red-tan	Dark blue	210,211
73	Yellow	Pink (green)[a]	212
74	Yellow	Gray (light orange)[a]	213
75	Orange	Green-gray (slate gray)[a]	213
76a	Blue	Blue (green)[a]	214
76b	Red	Blue (green)[a]	212,214
76c	Orange	Blue (green)[a]	212,214
76d	Yellow	Blue (green)[a]	212,214

[a]Intermediate oxidation states.
[b]n-doping.

substitutions on thiophenes yield polymers with increased bandgaps due to the steric interactions forcing the thiophene rings out of coplanarity. To overcome these steric issues, introduction of a vinylene unit separates the rings and allows the polymer to attain a planar conformation. On comparison of poly(3,4-dibutoxythiophene) and poly(3,4-dibutoxythienylene vinylene) (**51a**) a large increase in absorption maximum of 150 nm and concurrent reduction in bandgap is observed [167]. Poly(thienylene vinylenes) of mono- (**51b–d**) and dialkoxy-substituted (**51a,e**) thiophenes exhibit bandgaps ranging between 1.5 and 1.7 eV with polymer films switching from a deep blue neutral state to transmissive light gray on oxidation [167–169].

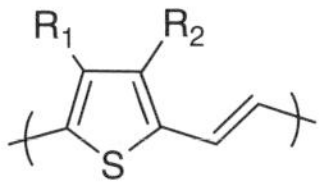

51

(E_g = 1.5–1.7 eV)

	R_1	R_2
a.	OC_4H_9	OC_4H_9
b.	H	CH_3
c.	H	OC_2H_5
d.	H	$OC_{12}H_{25}$
e.	$OC_{12}H_{25}$	$OC_{12}H_{25}$

By assembly of a three-ring monomer in which an electron-rich and electropolymerizable moiety is used as an external heterocycle with other conjugated internal units, the oxidation potential of the D-Π-D monomer can be lowered relative to the single ring monomer. Early work showed poly[bis(2-thienyl)-1,2-vinylene] to reduce the bandgap of PT from 2.0 to 1.8 eV [170]. A further decrease in the polymer bandgap is induced through the introduction of the alkylenedioxy bridge to produce 1,2-bis(2-(3,4-ethylenedioxy)thienyl)vinylene (**52**) [171–173]. This polymer switches between deep purple in the neutral state and transmissive sky blue in the oxidized state with a bandgap of 1.4 eV.

52

(E_g = 1.4 eV)

As a means of lowering the bandgap even further, cyano groups have been incorporated into the poly(bis(heterocycle)vinylene) system at the vinylene unit as illustrated by structures **53**. Through the alternation of heterocycle electron donors and various electron acceptors, LUMO level reductions narrow the gaps. On comparison of a series of donor–acceptor–donor (DAD) monomers in which the strength of the donor heterocycle is increased, bandgaps ranging from 1.6 to 1.1 eV were attained. As illustrated in Figure 20.18, the HOMO of the donor heterocycle mainly contributes to the polymer valence band while the polymer conduction band is determined by the LUMO of the acceptor unit. While keeping the acceptor moiety unchanged, the external donor units can be varied by increasing the electron-rich character following the trend of EDOP > EDOT > Th, as discussed in previous sections. By increasing the electron donor strength, the bandgap is lowered from 1.6 eV for PBTh-CNV (**53a**), 1.4 eV for PEDOT-CNV-Th (**53b**), 1.2 eV for PTh-CNV-EDOT (**53c**), and 1.1 eV for PBEDOT-CNV (**53d**), PTh-CNV-EDOP (**53e**), and PEDOT-CNV-EDOP (**53f**) [31,174]. Due to the low bandgap of all of these polymers, they are all found to switch between an opaque blue in the neutral state to a transmissive light

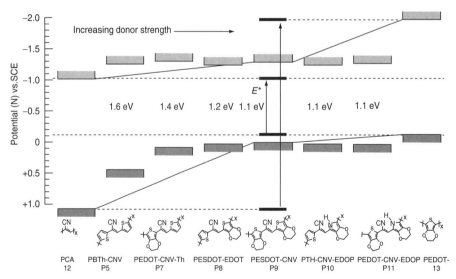

FIGURE 20.18   Approximate levels of the valence band and conduction band in a series of poly(bis(heterocycle)-cyanovinylene)s. (From Thomas, C., *J. Am. Chem. Soc.*, 126, 16440, 2004. With permission.)

blue in the oxidized state. While many other examples of homologous series could be culled from the extensive literature in this area, the above examples illustrate how a relatively closely matched series of polymers can have broadly varied gaps (>2 eV to near 1 eV) and neutral film colors ranging through orange, red, purple, and blue.

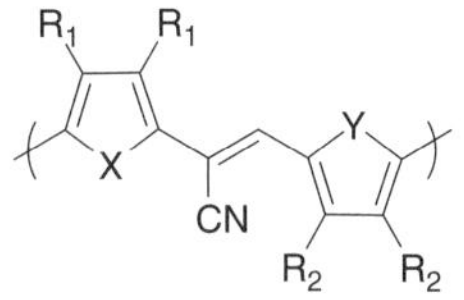

53

(E_g = 1.1–1.6 eV)

	X	Y	R_1	R_2
a.	S	S	H	H
b.	S	S	–OCH$_2$CH$_2$O–	H
c.	S	S	H	–OCH$_2$CH$_2$O–
d.	S	S	–OCH$_2$CH$_2$O–	–OCH$_2$CH$_2$O–
e.	S	NH	H	–OCH$_2$CH$_2$O–
f.	S	NH	–OCH$_2$CH$_2$O–	–OCH$_2$CH$_2$O–

Much interest has been focused on the bis(2-heterocycle)phenylene polymers due to the possibility of mono- and disubstitution (symmetrically and asymmetrically) on the phenylene ring, with the variety of heterocycle units available. A relatively high bandgap of 2.3 eV is seen for the unsubstituted bis(thienyl)phenylene polymer (**54a**) [175–177]. On substitution of the phenyl ring with methyl groups (**54b**), the bandgap is increased to 2.6 eV due to introduction of torsional strain in the polymer repeat unit. As seen before, substitution with alkoxy groups (**54e**) lowers the bandgap to 2.1 eV because of the electron-donating nature of the substituents and low energy barrier to attain a nearly planar polymer. These polymers can also be made soluble in common organic solvents through an increase in the alkyl (**54b–d**) and alkoxy (**54e–j**) chain length [176,178,179].

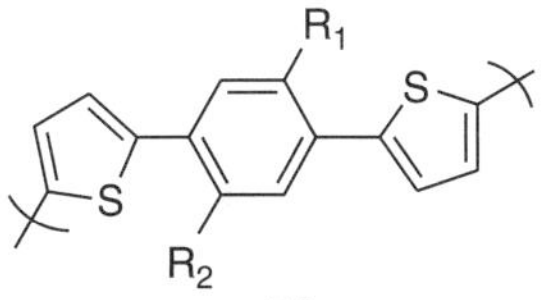

54

(E_g = 2.1–2.6 eV)

	R_1	R_2
a.	H	H
b.	CH$_3$	CH$_3$
c.	C$_6$H$_{13}$	C$_6$H$_{13}$
d.	C$_{12}$H$_{25}$	OCH$_3$
e.	OCH$_3$	OCH$_3$
f.	OCH$_3$	OC$_7$H$_{15}$
g.	OC$_7$H$_{15}$	OC$_7$H$_{15}$
h.	OC$_{12}$H$_{25}$	OC$_{12}$H$_{25}$
i.	OC$_{16}$H$_{33}$	OC$_{16}$H$_{33}$
j.	OC$_{20}$H$_{41}$	OC$_{20}$H$_{41}$

Through the attachment of oligoethylene oxide chains on the phenylene ring (**55a–d**), a polymer is produced that has solubility in polar solvents and shows improved film formation and adhesion to glass substrates [180,181]. The polymers switch at relatively low oxidation potentials showing a blue to red electrochromism. Even after 10,000 switching cycles, the polymers showed the same electroactivity in water as in acetonitrile. Similar trends can be seen in bandgap variation for polymers of pyrrole (**56a–c**) or furan (**56d–h**) combined with disubstituted *p*-phenylene units [182–185]. The pyrrole-phenylene polymers exhibit bandgaps ranging between 2.3 and 2.4 eV, lower than that for either the pyrrole or *p*-phenylene homopolymers.

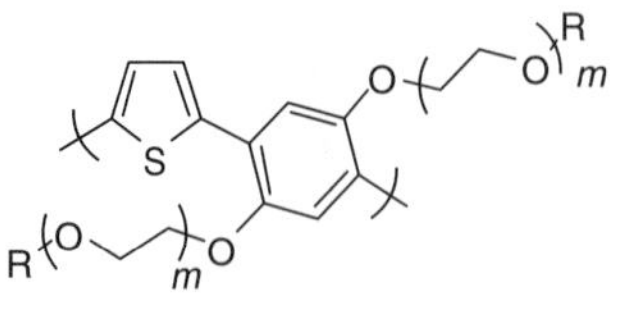

	X	R_1	R_2	$(E_g = 2.3–2.4 \text{ eV})$
a.	NH	H	H	
b.	NH	OCH_3	OCH_3	
c.	NH	$OC_{12}H_{25}$	$OC_{12}H_{25}$	
d.	O	H	H	
e.	O	CH_3	CH_3	
f.	O	OCH_3	OCH_3	
g.	O	OC_7H_{15}	OC_7H_{15}	
h.	O	OCH_3	OC_7H_{15}	

55

	m	R
a.	2	CH_3
b.	2	C_4H_9
c.	3	CH_3
d.	1	C_4H_9

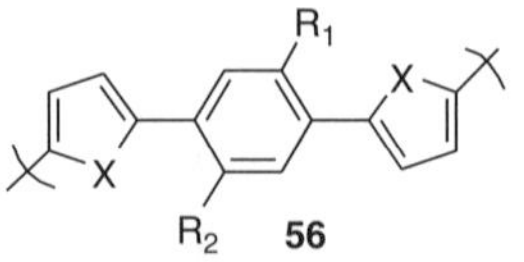

Through the combination of EDOT and substituted phenylene units, electron-rich polymers with low oxidation potentials have been created. The bandgap for hybrid polymers of EDOT and dialkoxybenzene is reduced in comparison with the thienyl dialkoxybenzene polymers due to the four electron-donating substituents on the polymer repeat unit [186,187]. These EDOT-dialkoxybenzene (**57a–d**) polymers exhibit bandgaps ranging from 1.8 to 2.2 eV, depending on the alkoxy chain length, and show three colored states being pale red in the neutral state, green at intermediate oxidation states, and deep blue when fully oxidized. Unsymmetrical substitution of a branched alkoxy group on the phenylene ring yields a polymer that exhibits a similar bandgap (1.95 eV) and continuously changing colored states [30,188]. Poly(1,4-bis(2-[3,4,-ethylenedioxy]thienyl)-2-methoxy-5-2″-ethylhexyloxybenzene) (**57e**) is deep red-purple when fully neutralized and pale blue when fully oxidized. The polymer exhibits a purple-blue to blue to gray-blue transition at intermediate potentials.

The electrochromic properties can be further tuned in an EDOT-phenylene type polymer through the 2,5-substitution of electron-withdrawing groups, e.g., fluorines (**57f**) [189]. Alteration of the electronic properties of this polymer is expected to be largely due to electronic effects rather than steric interactions (as with alkyl and alkoxy groups) due to the small size of the fluoro substituents. The effect is an increase in the monomer oxidation potential by 200–300 mV and a resulting bandgap of 1.9 eV, 0.1 eV higher than the nonfluorinated polymer. The polymer switched from red when fully neutralized to blue-black when fully oxidized.

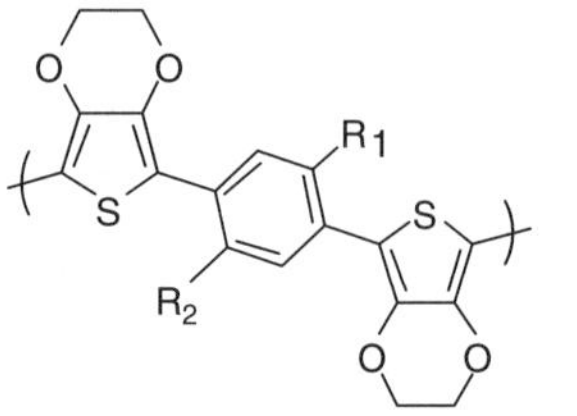

57
$(E_g = 1.8–2.2 \text{ eV})$

	R_1	R_2
a.	OCH_3	OCH_3
b.	$OC_{16}H_{33}$	$OC_{16}H_{33}$
c.	OC_7H_{15}	OC_7H_{15}
d.	$OC_{12}H_{25}$	$OC_{12}H_{25}$
e.	OCH_3	OEtHx
f.	F	F

By substitution on the phenylene ring with a pendant group that is also electrochromic, the available electrochromic states are increased for the resulting polymer. Given that poly(bis(EDOT)benzene) (PBEDOT-B) and viologen have distinctly different colored states with their redox potentials distinctly separate, substitution of a viologen moiety at the 2-5-positions of the phenylene ring (**58**) results in a polymer in which the multiple colored states arise from the polymer and the viologen substituent separately [190]. When the polymer is fully reduced (viologen and PBEDOT-B neutral) the film is crimson color. At a low oxidation potential, the film is purple due to the viologen unit being in its radical cation state. On further oxidation, the film turns magenta due to the viologen being in its dication state while the conjugated polymer backbone remains fully neutralized. When an even higher oxidation potential is applied, the film turns a pale tan/cream with the PBEDOT-B partially oxidized and the viologen remaining in its dication state. At a fully oxidized state, both polymer and pendant group fully oxidized, the film is transmissive light blue.

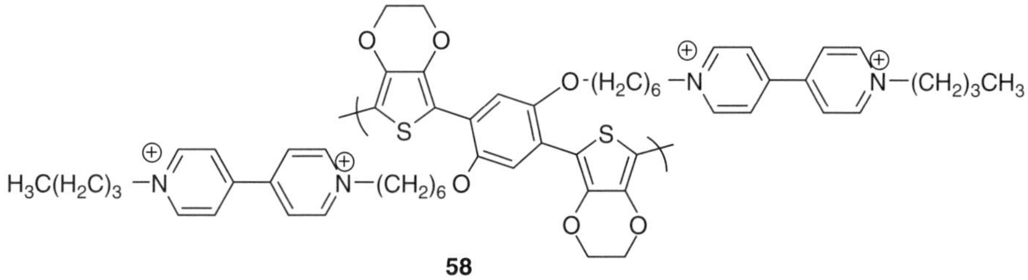

58

Increasing the number of phenyl rings in a bis(2-thienyl)phenylene polymer is shown to have little effect on the resulting colors or bandgap of the polymer film. For example, poly-4,4′-bis(2-thienyl) biphenyl (**59a**) is transmissive yellow when neutral and green when oxidized [175,191]. Even when the phenylene rings are fused, as in the poly(bis(2-thienyl)naphthalene) (**59b–f**) polymers, there is little change in the electrochromic properties. Polymers with unsubstituted (**59b,c**) and methoxy-substituted (**59d–f**) naphthalene units exhibit a bandgap between 2.4 and 2.9 eV and are transmissive yellow in the neutral state and bluish-green when oxidized. It was shown that as the number of electron-donating methoxy groups substituted on the naphthalene unit increased, there was a slight decrease in the oxidation potential of the monomer.

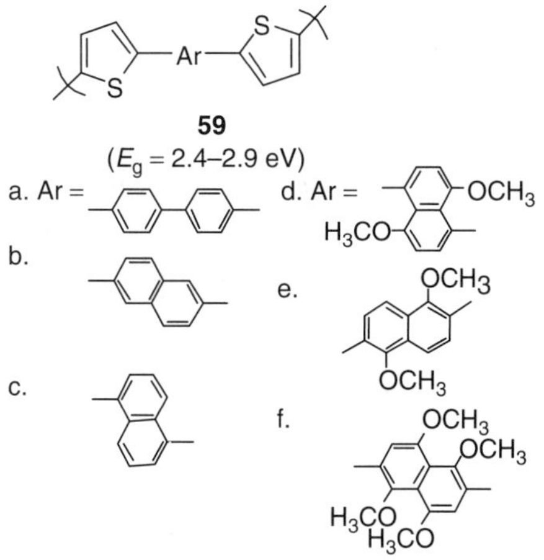

59
(E_g = 2.4–2.9 eV)

The electronic properties of the DAD hybrids discussed earlier can be further tuned by the introduction of stronger acceptor units, such as electron-deficient nitrogen-containing heterocycles. Combinations of EDOT and pyridine (**60**) have a bandgap of 1.9 eV, exhibiting multiple colored states through both p-doping and n-doping [192–194]. The polymers can be switched between a pale blue reduced state (n-doped), a red neutral state, and a blue-purple oxidized state (p-doped). The electron-accepting power of the azoheterocycle unit can be further increased by increasing the number of nitrogens in the phenylene from one to either two (for pyridazine) (**61a–e**) or four (for tetrazine)(**62**) [195,196]. The DAD-type hybrid polymer with alkylthiophenes as the D-unit and pyridazine as the A-unit are orange when neutral, dark gray when reduced, and black when oxidized with the polymers containing long alkyl substituents on the thiophene being soluble in organic solvents.

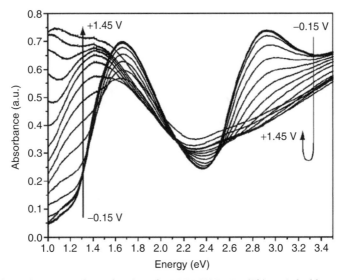

By the introduction of pyridopyrazine units, the polymer bandgap can be further decreased. Poly(5,8-bis[2-thienyl]-2,3-diphenylpyrido[3,4-*b*]pyrazine) (**63**) has a bandgap of 1.6 eV and switches between a lavender neutral state and a pale blue oxidized state [192,197]. The polymer bandgap can be further decreased by increasing the donor strength with EDOT-pyridopyrazine hybrid polymers. Poly(5,8-bis[EDOT]-2,3-diphenylpyrido[3,4-*b*]pyrazine) (**64**) has a bandgap of 1.2 eV with a more positive reduction potential than the pyrazine polymers previously mentioned. The absorbance spectroelectrochemistry for the p-doping of this polymer is shown in Figure 20.19 [192,193]. This polymer switches between a neutral lime green, an oxidized light gray, a reduced burgundy red, and a dark gray on further reduction.

FIGURE 20.19   Absorption spectroelectrochemistry for PBEDOT-PyrPyr(Ph)$_2$ switched between the potentials of -0.15 and 1.45 V vs. SCE. (From DuBois, C., *Adv. Mater.*, 14, 1844, 2002. With permission.)

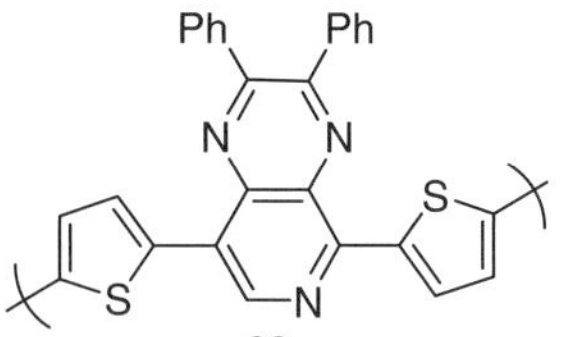

63
(E_g = 1.6 eV)

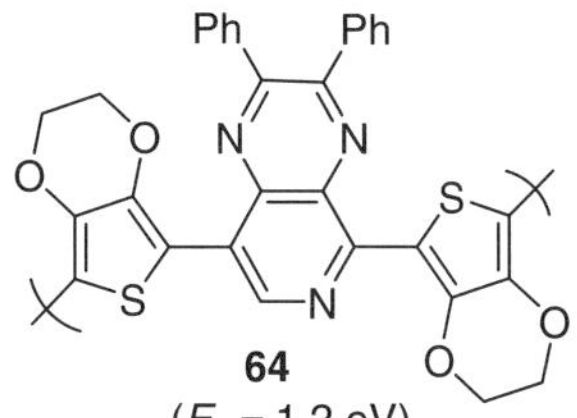

64
(E_g = 1.2 eV)

In further efforts to lower the bandgap of conjugated polymers, repeat units that stabilize quinoidal resonance structures have been developed leading to relaxation of bond length alternation [198]. One such polymer was produced by combining thiophene (unsubstituted (**65a,b,d,e**) and alkyl-substituted (**65c**)) with isothianaphthene (unsubstituted (**65a**), chlorine- (**65b,c**), and alkyl-substituted (**65d,e**)), which exhibit bandgaps between 1.7 and 1.4 eV [199,200]. Acceptor character of the isothianaphthene unit has been induced by the attachment of an alkyl-dicarboxylic imide functionality onto the benzene ring [201,202]. The material incorporating benzo[*c*]thiophene-*N*-2'-ethylhexyl-4,5-dicarboxylic imide and EDOT (**66**) produces a polymer with a bandgap of 1.1 eV. The film is blue in its neutral state and highly transmissive in the oxidized state with large contrasts occurring in the NIR region.

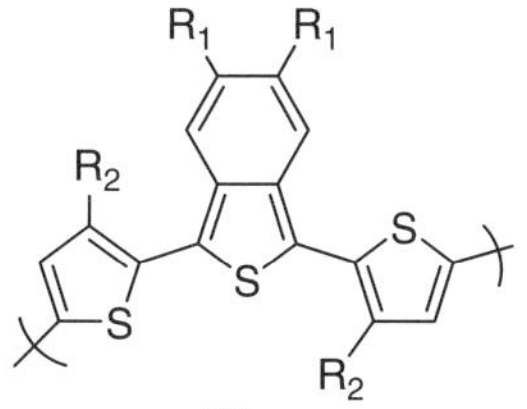

65
(E_g = 1.4–1.7 eV)

	R_1	R_2
a.	H	H
b.	Cl	H
c.	Cl	C_8H_{17}
d.	SC_8H_{17}	H
e.	C_8H_{17}	H

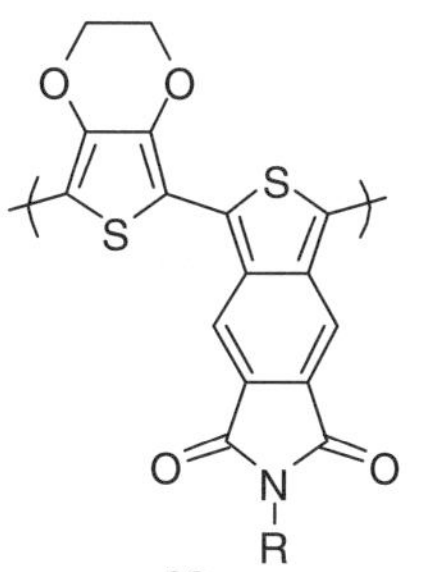

66
(E_g = 1.1 eV)
R = 2-ethylhexyl

In addition to the polymerization of materials incorporating both aromatic donor and quinoid acceptor units to reduce the bandgap, an effort to decrease steric repulsions between donor and acceptor units has also been utilized to reduce the polymer bandgap. When compared with the thiophene–isothianaphthene–thiophene polymer, the use of multiring heterocycles as the acceptor unit is expected to result in an efficient π-electron delocalization along the polymer main chain due to the lack of steric repulsions between the hydrogen atom of the fused benzene ring and the adjacent thiophene. It has been shown through x-ray structural analysis that 5,7-di(2-thienyl)thieno[3,4-*b*]pyrazine (**67a–e**) has an almost coplanar conformation [203]. Polymers incorporating unsubstituted (**67a–d**) or alkyl-substituted (**67e**) thiophene with thienopyrazine exhibit a bandgap of 1.0 eV and are a dark blue-green in the neutral state [203,204]. Similarly, polymers of thiophene with quinoid acceptors such as quinoxaline (**68a**), 2,1,3-benzothiadiazole (**68b**), [1,2,5]thiadiazolo[3,4-*g*]quinoxaline (**68c**), benzo[1,2-*c*;3,4-*c'*]bis[1,2,5]-thiadiazole (**68d**), pyrazino[2,3-*g*]quinoxaline (**68e**), and [1,2,5]thiadiazolo[3,4-*e*]thieno[3,4-*b*]pyrazine (**68f**) have been produced [204,205]. These materials exhibited bandgaps that ranged from 1.4 eV to values possibly as low as 0.3 eV and were dark blue-green in the neutral state.

67
(E_g = 1.0 eV)

R_1	R_2
a. H	H
b. CH_3	H
c. C_6H_{15}	H
d. $C_{13}H_{27}$	H
e. $C_{16}H_{13}$	$C_{16}H_{13}$

68

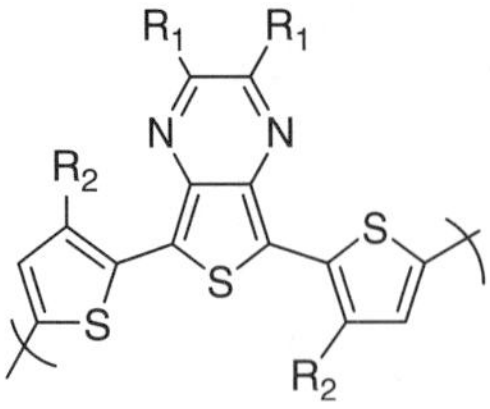

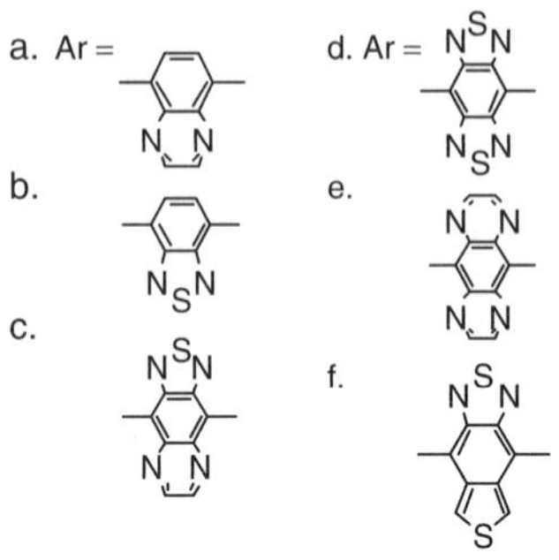

By the disubstitution on the poly(thieno[3,4-*b*]pyrazine) unit with thiophenes, a polymer backbone that contained two isolated conjugated chains, a polymer that is green in the neutral state has been achieved [206,207]. One chain has an absorption at long wavelengths in the red region of the spectrum, while the other chain absorbs in the blue region below 500 nm, thus transmitting green light. This polymer, 2,3-di(thien-3-yl)-5,7-di(thien-2-yl)thieno[3,4-b]pyrazine (**69a**), and its alkyl-substituted analog (**69b**), are soluble in organic solvents and exhibit a bandgap of 1.15 eV. The polymer switches between green in the neutral state and a light brown in the oxidized state. Through the introduction of a stronger donor unit with EDOT compared to thiophene, polymers combining EDOT and thieno[3,4-*b*] pyrazine (**70**) have been synthesized that exhibit low bandgaps, reportedly as low as 0.36 eV, that have been shown to be stable to repeated cycling for n-doping [208,209].

69
(E_g = 1.15 eV)
a. R = H
b. R = C_8H_{17}

70
(E_g = 0.36 eV)

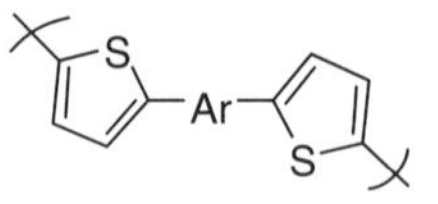

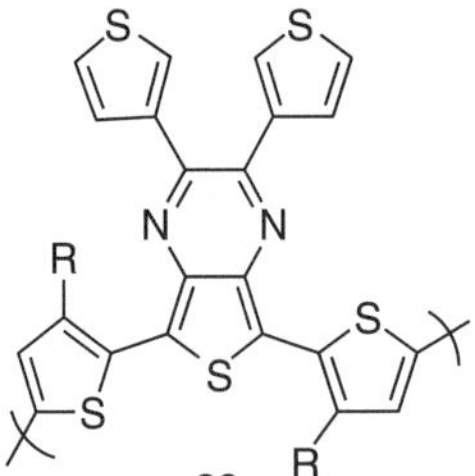

Moderate bandgap polymers have been synthesized by the incorporation of bis-heterocycles (thiophene, EDOT, and pyrrole) with fluorene and carbazole units [210,211]. The polymer poly (bis(2-thienyl)-9,9-didecylfluorene) (**71a**) exhibits a bandgap of 2.3–2.4 eV with the films in the fully neutral state being olive green and switching to a dark green on oxidation. Similar bandgap values have been seen for didecylfluorene-pyrrole (**71b**) polymers. By increasing the electron-richness of the donor unit with an EDOT (**71c**), the bandgap is reduced to 2.15–2.2 eV with the polymer film switching between reddish-tan in the neutral state and dark blue when oxidized.

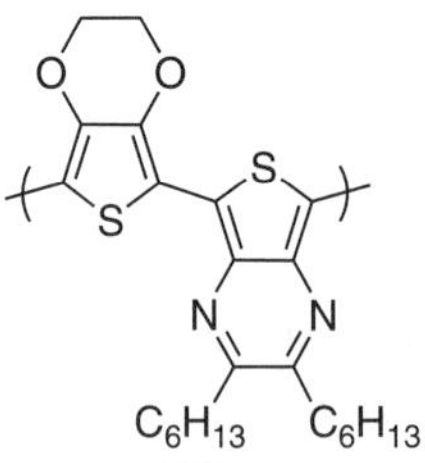

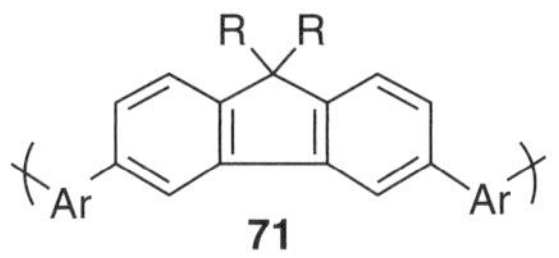

71
($E_g = 2.15\text{--}2.4$ eV)

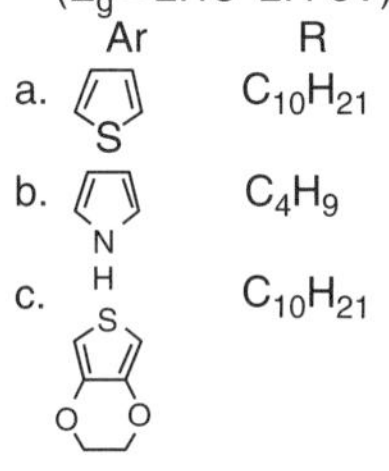

	Ar	R
a.	(thiophene)	$C_{10}H_{21}$
b.	(pyrrole)	C_4H_9
c.	(EDOT)	$C_{10}H_{21}$

Through the polymerization of the external heterocycles thiophene, EDOT, pyrrole, and benzothiadiazole, with 3,6-linked carbazole, three distinct color states can be seen in the resulting polymer. This is due to the conjugation breaks at the nitrogen in the carbazole, isolating the radical cations from one another and not allowing them to combine. On further oxidation, a dication is formed, yielding a second oxidized state and third color. Another feature of the carbazole unit is that *N*-substitution has little effect on the monomer oxidation potential. Since the *N*-substituted site is removed from the fused external heterocycle rings, there is no torsional distortion of the monomer structure to cause a change in the oxidation potential.

The carbazole-containing polymers exhibit bandgaps in the range of 2.2–2.5 eV, with the poly(3,6-bis(pyrrole-2-yl)-9-ethylcarbazole) (**72**) on the higher end of the range [212]. The combination of *N*-octylcarbazole and thiophene (**73**) switches between a yellow neutral state, a faint green radical cation state, and a pink dication state. By the addition of another thiophene to the repeat unit in the *N*-octylcarbazole-bithiophene polymer (**74**), the absorbance of the polymer is shifted to slightly longer wavelengths [213]. This polymer is also yellow in its neutral state, and changes to a light orange in the radical cation state, and gray in the dication state. A donor–acceptor polymer of the electron-poor benzothiadiazole and the relatively electron-rich carbazole (**75**) shows an even further shift in absorbance to longer wavelengths. This polymer film is orange in its neutral state, green-gray in the radical cation state, and slate gray in the dication state.

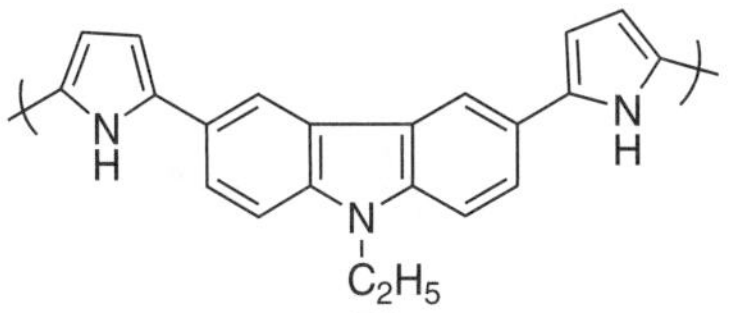

72
($E_g = 2.5$ eV)

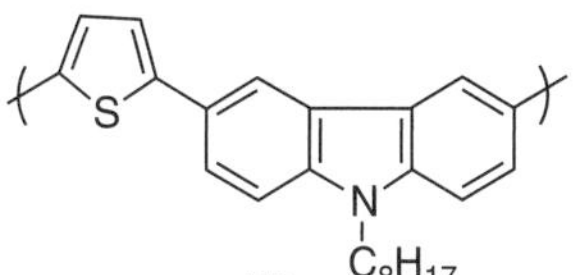

73

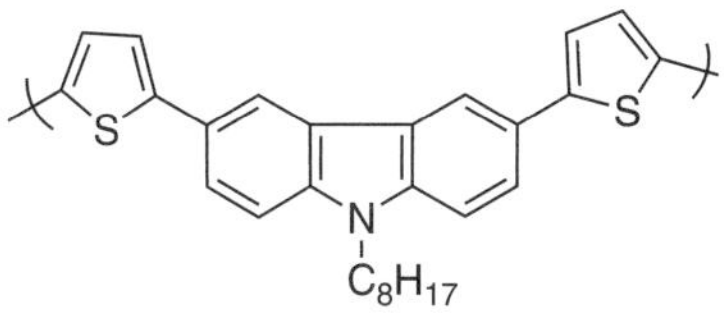

74

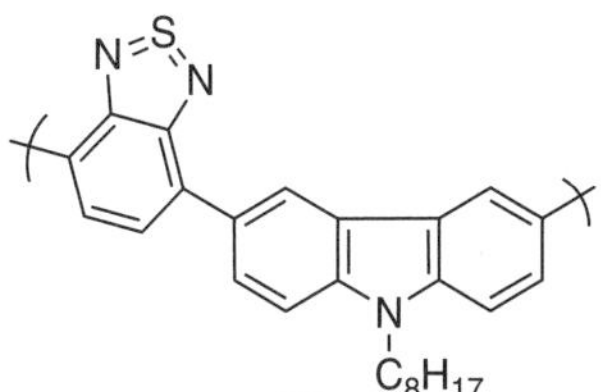

75

In addressing color control in true copolymers, it has also been shown that the colors of the neutral state for copolymers of 2,2′-bis(3,4-ethylenedioxythiophene) and 3,6-bis-(2-(3,4-ethylenedioxy) thienyl)-*N*-methylcarbazole can be tuned by varying the comonomer compositions (**76**) [214]. As seen in Figure 20.20, the λ_{max} of the neutral copolymer is observed to shift to lower wavelengths as the carbazole to BiEDOT ratio is increased compared with 100% BiEDOT. This gives colors to neutral polymers ranging from blue (100% BiEDOT), red (70:30 BiEDOT:Cz), orange (20:80 BiEDOT:Cz), and yellow (100% BEDOT-*N*MeCz). These copolymers, regardless of comonomer ratios, switch to a green cation radical state and finally a blue dication state.

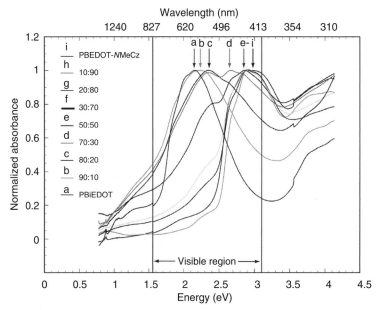

76

	m	*n*
a.	1.0	0
b.	0.7	0.3
c.	0.2	0.8
d.	0	1.0

Through the hybrid incorporation and copolymerization of various conjugated units, the electronic and electrochromic properties of the resulting polymers can be varied to cover colors across the entire visible spectrum. In this manner, the advantages of different conjugated units (i.e., low oxidation potential or solubilizing substituents) can be utilized while playing on the interaction of the comonomers (i.e., with the donor–acceptor systems).

FIGURE 20.20   Neutral absorbance spectra for a series of copolymers of various ratios of BiEDOT copolymerized with BEDOT-*N*-Methyl carbazole. (From Gaupp, C., *Macromolecules*, 36, 6305, 2003. With permission.)

20.4.2 Electrochromic Blends, Laminates, and Composites

As outlined in Section 20.4.1, polymerization of conjugated multiring units allows the production of electrochromic polymers with bandgaps, oxidation potentials, and colored states that vary from one extreme to the next. While this method leads to modification of the electrochromic and electronic properties of the polymer at the molecular level, other methods look to modify not only the electronic and electrochromic properties, but also the physical structures of the electrochromic materials. These techniques include the utilization of blends, laminates, and composites [215–218]. This section discusses the alterations of physical properties produced when conjugated polymers are combined with inorganic electrochromics (e.g., PB and Wo$_3$), organic molecules (e.g., dyes), insulating polymers (e.g., Hydrin C), and other conjugated polymers. Due to the diversity of approaches and materials available for combination, we select a few examples in order to delineate the preparation methods without attempting to exhaustively review all examples in the literature. As soluble and processable electrochromic polymers become more available to researchers in the field, it is expected that this research area will develop even more rapidly.

Through the preparation of either blends or composites of conjugated electrochromic polymers with insulating polymers, the electrochromic and electronic properties of the electrochromic polymer can be typically preserved while introducing the favorable mechanical properties and processability associated with the insulating polymer. For example, when Hydrin C (a vulcanizable elastomeric copolymer of epichlorohydrin and ethylene oxide) is coated onto an electrode surface followed by electrochemical polymerization of 4,4′-dipentoxy-2,2′-bithiophene, a film is produced that switches between dark blue in the neutral state to colorless in the oxidized state [219].

Films prepared in a similar manner with Hydrin C are those utilizing poly(N,N'-dimethyl-2,2′-bipyrrole) [220], and polypyrrole [221]. Electrochromic films of Hydrin C and poly(o-methoxyaniline) have also been produced in which the aniline polymer is chemically polymerized in the presence of p-toluene sulfonic acid and blended with Hydrin C with the blend cast from solution [219]. Another example in which an electrochromic polymer was electrochemically polymerized in the presence of an insulating polymer is that of polypyrrole-poly(ether urethane) or polypyrrole-poly(ethylene-co-vinyl alcohol) composite films [222]. Both films switched between a yellow reduced state to a bluish brown oxidized state, similar to polypyrrole.

A composite film of polyaniline and chlorosulfonated polyethylene was produced by the polymerization of aniline in an emulsion of water and xylene containing the chlorosulfonated polyethylene with dodecylbenzene sulfonic acid acting as a surfactant and dopant [223]. The polymer film switches between a light yellow on reduction and a dark green on oxidation with the composite behaving like a thermoplastic elastomer when the polyaniline content is between 12 and 18 wt%. Polymer blends in which chemically polymerized poly(o-alkoxyaniline)s, e.g., poly(o-methoxyaniline) and poly(o-ethoxy-aniline), were mechanically blended into an insulating polymer matrix of poly(vinylidene fluoride) or poly(acrylonitrile) [219,224]. Both the blends produced are electrochromic with the polymers exhibiting the same electrochromic behavior as the pure conducting polymer with free-standing films that are flexible and elastomeric.

A method that has been used to achieve a wide range of colors, or broaden the absorption in the visible region, has been the combination of two or more electrochromic polymers into a single film. Through blending of poly(N-phenyl-2-(2′-thienyl)-5-(5″-vinyl-2″-thienyl)pyrrole (PSNPhS) and poly(N-vinylcarbazole) (PVK) at various mass ratios, polymer films were produced that have colors that are a combination of the pure polymers [225]. The neutral state of PVK is colorless and PSNPhS is yellow and in the oxidized state, PVK is green and PSNPhS is red. The intensity of the yellow color of the polymer blend in the neutral state is dependent on the amount of PSNPhS in the blend. When the blends are oxidized, the colors range from shades of green, brown, and tan, and also dependent on the amount of PSNPhS present.

Polymer films that combine the optical properties of PEDOT and polypyrrole have been produced by either casting a solution of the PEDOT/PSS suspension followed by electrochemical polymerization of pyrrole [226] or *N*-methylpyrrole [217], or electrochemical polymerization of *N*-methylpyrrole onto an electrode followed by casting the PEDOT/PSS [227]. Other electrochromic films comprised of two conducting polymers that are laminates of polyaniline with PEDOT, and polyaniline with poly (*N*-methylpyrrole) [227].

Composites of conducting polymers, e.g., polyaniline and PEDOT, with polyacids, e.g., poly (2-acrylamido-2-methyl-1-methyl-1-propanosulfonic acid) (PAMPS), have been shown to be electrochromic. The polyacid acts as a dopant for the polymer film with the optical properties of the composite being contributed by the conducting polymer. The composites are formed by either chemical or electrochemical polymerization of the electrochromic component monomer in the presence of the polyacid. Films of polyaniline-PAMPS switch from yellow to green and finally to blue on oxidation [228,229]. Composite films of PEDOT and PAMPS show similar electrochromic properties to PEDOT with the films switching from dark blue in the neutral state to light sky blue in the oxidized state [140,230,231].

The technique of layer-by-layer deposition has been utilized in the production of electrochromic films of polyanions and polycations in which one or both of the polymers used are electrochromic [232,233]. Figure 20.21 shows the spectroelectrochemistry for a layer-by-layer polymer film of PEDOT-PSS as the polyanion and poly(hexyl viologen) as the polycation [234]. Both PEDOT-PSS and the polyviologen are cathodically coloring, switching from a highly transmissive oxidized state to a dark blue reduced state. The composite polymer film then switches between a highly absorptive purple-blue when reduced to a transmissive sky blue when oxidized with a transmittance change of 82% at 525 nm.

Composite electrochromic films have been produced through the incorporation of small organic molecules into a conjugated polymer film. This has been achieved by the electropolymerization of the conducting polymer in the presence of an organic molecule. Two molecular electrochromes that have been used are 2,2′-azinobis(3-ethylbenzothiazoline-6-sulfonate) (ABTS) [235] and indigo carmine [236]. Both of these molecules are dianionic and therefore act as dopant ions in the polymer film. Composites of polypyrrole and ABTS exhibit multiple colored states, switching from brown to greenish blue on oxidation and yellow on reduction. These colors are a result of the electrochromism exhibited by polypyrrole and the colorless to greenish blue electrochromism shown on oxidation of the ABTS [235].

Some of the most extensively studied composites are the conducting polymers with inorganic electrochromic materials. One such inorganic electrochrome is PB. Composites have been formed

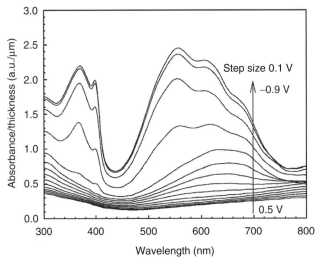

FIGURE 20.21 Absorption spectroelectrochemistry for a 40-bilayer film of a composite of poly(hexyl viologen) and PEDOT-PSS at potentials between −0.9 and 0.5 V vs. SCE. Absorbance values are normalized to the polymer film thickness. (From DeLongchamp, D., *Chem. Mater.*, 15, 1575, 2003. With permission.)

between PB and, polymers such as polyaniline, polypyrrole, and poly(3-methylthiophene). The composite films can be formed either by deposition (casting or electrochemical polymerization) of the polymer film, followed by electrochemical deposition of the PB film, by electrochemical polymerization of the polymer film and PB deposition simultaneously from the same solution, or the layer-by-layer method.

When the first method is used, the PB film is deposited directly on the conducting polymer film, with the polymer acting as an extended electrode. Composites of polyaniline and PB switch from pale yellow when reduced to blue when oxidized, the combination of the colors exhibited by the host materials [237,238]. When PB is combined with polypyrrole in a composite film, the absorbance of the film is extended to a wider region of the visible spectrum compared with the pure polypyrrole. These films switch from pale brown in the reduced state to bluish green in the oxidized state [238–240].

Composite films of PB and poly(3-methylthiophene) capitalize on the transparent state of the reduced PB and the similar colored states of both PB and poly(3-methylthiophene) in their respective oxidized states. This composite switches from deep red, a contribution of poly(3-methylthiophene) in the neutral state, to dark blue in the oxidized state [241]. In addition to the additive electrochromic properties displayed by the materials in the composites, the improved adhesion of PB to the polymer films in comparison with electrode surfaces, along with efficient control of the amount of the composite materials deposited, are the advantages seen with this method.

Other methods produce composite films that are alternating, interpenetrating layers of conducting polymer and PB. One of the methods utilizes galvanostatic deposition of both polyaniline and PB by switching the applied current back and forth between a positive value for the polymerization of aniline and a negative value for the deposition of PB for discrete intervals in solutions containing the precursor materials [242]. Similar results were seen for a composite of polyaniline and the PB-like nickel hexacyanoferrate produced in the same manner [243]. Another method, layer-by-layer deposition, utilizes the electrostatic attraction between the polycation, polyaniline, and a negatively ionized PB nanoparticle dispersion to produce alternating layers of the materials to form a composite film [244]. The composite switches between a highly transmissive pale yellow in the reduced state, deep green intermediate state, and dark navy blue in the oxidized state.

Similar to the PB-conducting polymer composites, composite films can be prepared with conducting polymers and transition metal oxides. Composite films have been prepared by the electrostatic layer-by-layer technique with polyaniline and vanadium oxide with the electrochemical properties dominated by the vanadium oxide layers and the optical properties dominated by the polyaniline layers [245,246]. Composites have also been made between the water-soluble poly(2-(3-thienyloxy)ethanesulfonic acid) and vanadium oxide by mixing the two materials at different mole ratios in aqueous solutions. These films switched between an orange color when fully reduced, to yellow-green at intermediate potentials, and to dark blue when fully oxidized [247].

Electrochromic composite films using polypyrrole and phosphotungstate are prepared by the electropolymerization of pyrrole in aqueous solutions containing the sodium salt of phosphotungstate. Since these two materials switch in potential ranges that do not overlap, there is an increase in the number of colors produced in the composite film. Alone, phosphotungstate switches from a blue-purple color when reduced to colorless when oxidized, while polypyrrole switches from yellow when neutral to blue when oxidized. The composite film switched from yellow when fully reduced to red at intermediate potentials, to blue when fully oxidized. This can be seen in Figure 20.22, with the blue to yellow transition being attributed to the polypyrrole component and the red color being attributed to the phosphotungstate component within the film [248].

Polyaniline–tungsten oxide composite films, utilizing two of the most widely studied electrochromic materials, have been produced [218]. The films are prepared by alternating the potential applied to the electrode between positive potentials and negative potentials, the aniline polymerization potential and tungsten oxide deposition potential, respectively. This has been shown to produce an alternating layered structure within the film, which switches between dark blue at negative potentials and violet-green at positive potentials.

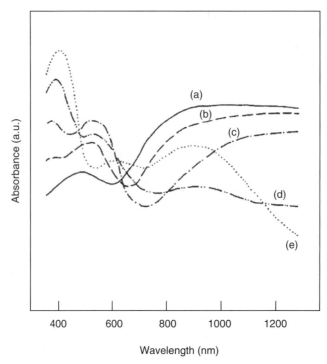

FIGURE 20.22 Absorption spectroelectrochemistry of a polypyrrole/phosphotungstate composite at (a) 0.2, (b) 0.0, (c) −0.2, (d) −0.4, and (e) −0.6 V vs. SCE. (From Shimidzu, T., *J. Chem. Soc. Faraday Trans.*, 841, 3941, 1988. With permission.)

20.5 Electrochromic Devices

ECDs are designed to modulate absorbed, transmitted, or reflected incident electromagnetic radiation. This is accomplished through the application of an electric field across the electrochromic materials within the device. The device acts as an electrochemical cell where electrochemical reactions occur between two redox-active materials that are separated by an electrolyte. Often, an ECD includes two electrochromic materials that have complementary electronic and optical properties allowing both electrochromes to contribute to the optical response of the device.

In a dual polymer ECD, the redox-active material at both electrodes is an electrochromic polymer, while in a hybrid device the material at the working electrode is an electrochromic polymer and the material at the counter electrode is an inorganic (e.g., WO_3 or PB) or molecular organic (e.g., polyviologen) electrochrome [4,5,16,239,249,250].

Electrolytes utilized in ECDs are required to have a high conductivity, be have transparency in the wavelength range used, have relatively large electrochemical windows, and have low volatility. Typical electrolytes that have been used are gel electrolytes, solid electrolytes, and ionic liquids [35,165,251–254]. Recent interest has grown in the use of room temperature ionic liquids in polymer ECDs due to lack of a solvent, enhancement of device lifetime, and enhancement of switching speeds [255–257].

The electrodes used in the ECDs are dependent on the type of device: absorptive/transmissive (window type) or absorptive/reflective. Typical electrodes are the transparent conductors such as ITO, fluorine-doped tin oxide (SnO_2:F), and PEDOT/PSS for the absorptive/transmissive window-type ECDs and reflective metals such as gold in reflective display-type devices. Single-walled carbon nanotubes (SWNTs) have emerged as an alternative to transparent electrodes such as ITO, with comparable transparency in the visible region and far superior transparency in the wavelength range of 2–5 μm [258].

In the field of electrochromics, especially electrochromic displays and devices, there are various measurements undertaken to characterize the electrochromic properties of the material or device. Such measurements include in situ colorimetry, composite coloration efficiency, and electrochromic switching speed studies, along with spectroelectrochemistry [15].

20.5.1 Absorptive/Transmissive ECDs

The absorptive/transmissive-type ECD operates with a reversible switching of the electrochromic materials between a colored state and a bleached state. Both working electrode and counter electrode are transparent so that light can pass through the device [4,5,15,250]. For flexible devices, ITO, SWNT, or PEDOT/PSS deposited onto a plastic such as poly(ethylene terepthalate) (PET) have been used [258,259]. When deposited in the doped form and dried, PEDOT/PSS films, to a thickness of 300 nm, are relatively transmissive in the visible region ($\sim$75% T), have a relatively low resistivity (500 $\Omega/\square$), and adhere to the plastic substrate in most common electrolyte solutions. The polymer films were demonstrated to be useable over the operating range of the device with no loss in conductivity or transmissivity.

To achieve high contrast values when switching a window device between the bleached and colored states, two complementary electrochromic materials are utilized, one cathodically coloring and the other anodically coloring. Figure 20.23 shows the schematic of the device, which is representative of the general construction of most transmissive/absorptive devices. The anodically coloring polymer has a high bandgap and is transmissive in the neutral state. Upon oxidation, the polymer becomes colored and absorbs light in the visible region. The cathodically coloring polymer has a low bandgap and is colored in the neutral state, becoming transmissive upon oxidation. When both polymers are sandwiched together in a device and an external voltage applied, the device switches between a colored state and a transmissive state with the colored state being the combined colors of both polymers.

For example, a device containing PProDOT-Me$_2$ (**43a**) and PProDOP-NPrS (**50e**) was constructed [15,148,151,165]. PProDOT-Me$_2$ is a cathodically coloring polymer with a high contrast in the visible region (78% at 580 nm), switching between a dark purple-blue in the neutral state and a highly transmissive light blue in the doped state. PProDOP-NPrS is an anodically coloring polymer with a high bandgap ($E_g = 3.0$ eV) giving it an almost clear, transmissive neutral state, switching to a gray-green upon doping. The dual polymer device has a contrast value in the visible region of 68% and switches in under 0.5 s between the voltages of $\pm$1.5 V.

20.5.2 Reflective ECDs

As mentioned, electrochromism is not confined to the visible region of the spectrum but can extend to the NIR, mid-IR, and microwave regions. In order to take advantage of this electrochromism at longer

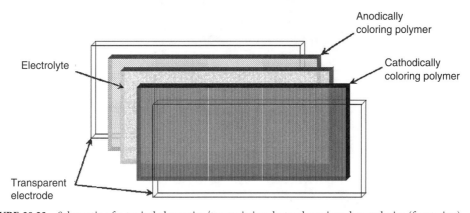

FIGURE 20.23 Schematic of a typical absorptive/transmissive electrochromic polymer device (front view).

wavelengths, ECDs have been constructed that can operate beyond the visible region. The typical device design is an outward facing reflective electrode onto which the active electrochromic polymer is deposited. The most commonly used reflective electrode material is gold deposited onto a flexible, ion permeable substrate.

One example is gold coated onto Mylar in which slits have been cut into the Mylar film to allow ions from the electrolyte to diffuse between the slits and contact the active polymer layer [250]. Another example uses a porous plastic membrane, e.g., nylon, polycarbonate, polyester, or polysulfone, which is also coated with a layer of gold. The gold-coated porous substrate allows faster ion diffusion and thus faster and more uniform switching of the active polymer layer than with the slitted-type device [120,250,260,261].

As with the absorptive/transmissive device, a complementary electrochromic polymer layer is also deposited onto the counter electrode. The device is a sandwich-type design with the counter electrode behind the working electrode, and an electrolyte layer between both. The electrochromic polymer on the counter electrode does not contribute any of its optical properties to the device, but only contributes to the balancing of the electroactive sites, allowing high Faradaic reversibility.

The top layer is in contact with an electrolyte and window transparent to the wavelengths measured. Typical windows include ZnSe for NIR to mid-IR, glass for NIR to visible, polyethylene for visible through mid-IR, and polyolefin for visible through far-IR. A device schematic illustrating a typical device design is shown in Figure 20.24. For example, using the same polymer described previously for the absorptive/transmissive device, a reflective ECD has been constructed with PProDOT-Me$_2$ as the active polymer and PBEDOT-NMeCz (**76d**) as the counter electrode polymer [148,250]. When the PProDOT-Me$_2$ layer is fully neutralized, it is strongly absorbing in the visible region, giving a low reflectance for the device in this region, and is transparent at wavelengths longer than 0.9 μm so that the gold electrode is reflecting in this region.

When the active polymer is fully oxidized, it is transmissive in the visible region, showing the reflective gold underneath, and is highly absorbing in the IR, as the gold layer is not seen at these longer wavelengths. This device shows reflectance contrast ratios, ΔR, of 55% in the visible region at 0.6 μm, greater than 80% in the NIR between 1.3 and 2.2 μm, and greater than 50% in the mid-IR between 3.5 and 5.0 μm.

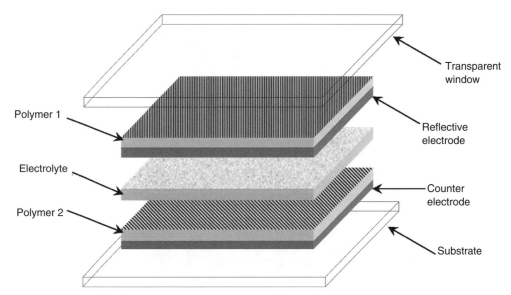

FIGURE 20.24   Schematic of a typical absorptive/reflective electrochromic polymer device (top–down view).

Reflective ECD construction that has recently gained attention is the patterning of devices for display applications. By depositing the reflective metal electrode in a defined pattern, higher resolution pixel devices can be constructed that take advantage of the various colors available in the neutral polymer by depositing different polymers onto each pixel. Patterned devices have been constructed by methods that include metal vapor deposition, line patterning, screen printing, inkjet printing, and microcontact printing. Not only can the metal electrodes be patterned with these methods, but soluble electrochromic polymers can also be patterned directly onto the electrode surface using inkjet printing and micro-contact printing [217,262–264].

The combination of various device platforms, along with electrochromic polymers available in colors that span the entire range of the visible spectrum and beyond, allows the construction of displays that can be tailored to fit most needs in real applications. While these devices have shown enhanced performance in colors available, optical memory, and low power consumption as compared to devices constructed of other electrochromic materials, there are still drawbacks to be addressed.

Much of the current research in the field of electrochromic polymer focus on increasing the lifetimes of these devices, along with the switching speeds and environmental stability. Devices incorporating these electrochromic polymers have demonstrated fast switching, with the devices fully switching at subsecond rates [15,151]. These devices have also been shown to exhibit long lifetimes with values that reach into millions of cycles reported [35,254,255].

20.6 Conclusions

Conjugated conducting polymers, not new to the field of electrochromics, have experienced a surge of interest for their applications in electrochromic displays and windows in the past decade. Yet, much recent work has focused on utilization of these materials in areas outside the typical display/window device, such as the previously mentioned mechanical actuators, LEDs, photovoltaics, capacitors, and antistatic coatings. Meanwhile, the field of electrochromics of conducting polymers sees the introduction of new polymers with ever-improving physical properties, such as processability and environmental stability, in a wide range of colors and spectral properties.

Acknowledgments

Funding at the University of Florida electrochromic polymers program by the AFOSR, DARPA (HIDE Program) and ARO MURI program is gratefully acknowledged. JRR recognizes the contributions and thanks the many students, postdocs, and collaborators who have been involved with our electrochromic polymers research.

References

1. Platt, J.R. 1961. Electrochromism, a possible change of color producible in dyes by an electric field. *J Chem Phys* 34:862–863.
2. Monk, P.M.S., R.J. Mortimer, and D.R. Rosseinsky. 1995. *Electrochromism: Fundamentals and applications.* Weinheim, Germany: Wiley-VCH.
3. Mortimer, R.J. 1997. Electrochromic materials. *Chem Soc Rev* 26:147–156.
4. Rosseinsky, D.R. and R.J. Mortimer. 2001. Electrochromic systems and the prospects for devices. *Adv Mater* 13:783–793.
5. Somani, P.R. and S. Radhakrishnan. 2002. Electrochromic materials and devices: Present and future. *Mater Chem Phys* 77:117–133.
6. Rose, T.L., S. D'Antonio, M.H. Jillson, A.B. Kon, R. Suresh, and F. Wang. 1997. A. Microwave shutter using conductive polymers. *Synth Met* 85:1439–1440.
7. Franke, E.B., C.L. Trimble, J.S. Hale, M. Schubert, and J.A. Woollam. 2000. Infrared switching electrochromic devices based on tungsten oxide. *J Appl Phys* 88:5777–5784.

8. Topart, P. and P. Hourquebie. 1999. Infrared switching electroemissive devices based on highly conducting polymers. *Thin Solid Films* 352:243–248.

9. Deb, S.K. 1969. Novel electrophotographic system. *Appl Opt Suppl* 3:192–195.

10. Deb, S.K. 1973. Optical and photoelectric properties and color centers in thin films of tungsten (VI) oxide. *Philos Mag* 27:801–822.

11. Granqvist, C.G. 1999. Progress in electrochromics: Tungsten oxide revisisted. 1999. *Electrochim Acta* 44:3005–3015.

12. Granqvist, C.G. 2000. Electrochromic tungsten oxide films: Review of progress 1993–1998. *Solar Energy Mater Solar Cells* 60:201–262.

13. Mortimer, R.J. 1999. Organic electrochromic materials. *Electrochim Acta* 44:2971–2981.

14. Mortimer, R.J. and J.R. Reynolds. 2005. In situ colorimetric and composite coloration efficiency measurements for electrochromic Prussian blue. *J Mater Chem* 15:2226–2233.

15. Argun, A.A., P.-H. Aubert, B.C. Thompson, I. Schwendeman, C.L. Gaupp, J. Hwang, N.J. Pinto, D.B. Tanner, A.G. MacDiarmid, and J.R. Reynolds. 2004. Multi-colored electrochromism in polymers: Structures and devices. *Chem Mater* 16:4401–4412.

16. Panero, S., S. Passerini, and B. Scrosati. 1993. Conducting polymers: New electrochromic materials for advanced optical devices. *Mol Cryst Liq Cryst* 229:97–109.

17. Patil, A.O., A.J. Heeger, and F. Wudl. 1988. Optical properties of conducting polymers. *Chem Rev* 88:183–200.

18. Yang, S., P. Olishevski, and M. Kertesz. 2004. Bandgap calculations for conjugated polymers. *Synth Met* 141:171–177.

19. Lee, C.H., G.W. Kang, J.W. Jeon, W.J. Song, S.Y. Kim, and C. Seoul. 2001. Photoluminescence and electroluminescence of vacuum-deposited poly(*p*-phenylene) thin film. *Synth Met* 117:75–79.

20. Obrzut, J. and F.E. Karasz. 1987. Ultraviolet and visible spectroscopy of poly(paraphenylenevinylene). *J Chem Phys* 87 (4):2349–2358.

21. Chung, T.-C., J.H. Kaufman, A.J. Heeger, and F. Wudl. 1984. Charge storage in doped poly(thiophene): Optical and electrochemical studies. *Phys Rev B* 30:702–710.

22. Groenendaal, L.B., F. Jonas, D. Freitag, H. Pielartzik, and J.R. Reynolds. 2000. Poly(3,4-ethylenedioxythiophene) and its derivatives: Past, present, and future. *Adv Mater* 12:481–494.

23. Groenendaal, L.B., G. Zotti, P.-H. Aubert, S.M. Waybright, and J.R. Reynolds. 2003. Electrochemistry of poly(3,4-alkylenedioxythiophene) derivatives. *Adv Mater* 15:855–879.

24. Heywang, G. and F. Jonas. 1992. Poly(alkylenedioxythiophene)s new, very stable conducting polymers. *Adv Mater* 4:116–118.

25. Wudl, F., M. Kobayashi, and A.J. Heeger. 1984. Poly(isothianaphthene). *J Org Chem* 49:3382–3384.

26. Brédas, J.L. and G.B. Street. 1985. Polarons, bipolarons, and solitons in conducting polymers. *Acc Chem Res* 18:309–315.

27. Roncali, J. 1997. Synthetic principles for bandgap control in linear π-conjugated systems. *Chem Rev* 97:173–205.

28. Gaupp, C.L., K. Zong, P. Schottland, B.C. Thompson, C.A. Thomas, and J.R. Reynolds. 2000. Poly(3,4-ethylenedioxypyrrole): Organic electrochemistry of a highly stable electrochromic polymer. *Macromolecules* 33:1132–1133.

29. Schottland, P., K. Zong, C.L. Gaupp, B.C. Thompson, C.A. Thomas, I. Giurgiu, R. Hickman, K.A. Abboud, and J.R. Reynolds. 2000. Poly(3,4-alkylenedioxypyrrole)s: Highly stable electronically conducting and electrochromic polymers. *Macromolecules* 33:7051–7061.

30. Sönmez, G., H. Meng, Q. Zhang, and F. Wudl. 2003. A highly stable, new electrochromic polymer: Poly(1,4-bis(2-(3′,4′-ethylenedioxy)thienyl)-2-methoxy-5-2″-ethylhexyloxybenzene). *Adv Funct Mater* 13:726–731.

31. Thomas, C.A., K. Zong, K.A. Abboud, P.J. Steel, and J.R. Reynolds. 2004. Donor-mediated band gap reduction in a homologous series of conjugated polymers. *J Am Chem Soc* 126:16440–16450.

32. Arbizzani, C., M. Catellani, M. Mastragostino, and C. Mingazzini. 1995. N- and p-doped poly-dithieno[3,4-B:3′,4′-D] thiophene: A narrow band gap polymer for redox supercapacitors. *Electrochim Acta* 40:1871–1876.

33. Carlberg, J.C. and O. Inganas. 1997. Poly(3,4-ethylenedioxythiophene) as electrode material in electrochemical capacitors. *J Electrochem Soc* 144:L61–L64.

34. Gurunathan, K., A.V. Murugan, R. Marimuthu, U.P. Mulik, and D.P. Amalnerkar. 1999. Electrochemically synthesised conducting polymeric materials for applications towards technology in electronics, optoelectronics and energy storage devices. *Mater Chem Phys* 61:173–191.

35. Lu, W., A.G. Fadeev, B. Qi, E. Smela, B.R. Mattes, J. Ding, G.M. Spinks, J. Mazurkiewicz, D. Zhou, G.G. Wallace, D.R. MacFarlane, S.A. Forsyth, and M. Forsyth. 2002. Use of ionic liquids for π-conjugated polymer electrochemical devices. *Science* 297:983–987.

36. Smela, E. 2003. Conjugated polymer actuators for biomedical applications. *Adv Mater* 15: 481–494.

37. Kim, Y.-G., B.C. Thompson, N. Ananthakrishnan, G. Padmanaban, S. Ramakrishnan, and J.R. Reynolds. 2005. Variable band gap conjugated polymers for optoelectronic and redox applications. *J Mater Res* 20:3188–3198.

38. Huang, F., H.L. Wang, M. Feldstein, A.G. MacDiarmid, B.R. Hsieh, and A.J. Epstein. 1997. Application of thin films of conjugated polymers in electrochemical and conventional light-emitting devices and in released photovoltaic devices. *Synth Met* 85:1283–1284.

39. Yu, G. and A.J. Heeger. 1997. High efficiency photonic devices made with semiconducting polymers. *Synth Met* 85:1183–1186.

40. McQuade, D.T., A.E. Pullen, and T.M. Swager. 2000. Conjugated polymer-based chemical sensors. *Chem Rev* 100:2537–2574.

41. Huang, J., S. Virji, B.H. Weiller, and R.B. Kaner. 2004. Nanostructured polyaniline sensors. *Chem Eur J* 10:1314–1319.

42. Li, Y. and M. Yang. 2002. Humidity sensitive properties of substituted polyacetylenes. *Synth Met* 129:285–290.

43. Garnier, F., G. Tourillon, M. Gazard, and J.C. DuBois. 1983. Organic conducting polymers derived from substituted thiophenes as electrochromic material. *J Electroanal Chem* 148:299–303.

44. Gazard, M., J.C. DuBois, M. Champagne, F. Garnier, and G. Tourillon. 1983. Electrooptical properties of thin films of polyheterocycles. *J Phys* C3:537–542.

45. Aizawa, M., S. Watanabe, H. Shinohara, and H. Shirakawa. 1985. Electrochemical cation doping of a polythienylene film. *J Chem Soc Chem Commun* 25:264–265.

46. Alkan, S., C.A. Cutler, and J.R. Reynolds. 2003. High quality electrochromic polythiophenes via BF$_3$–Et$_2$O electropolymerization. *Adv Funct Mater* 13:331–336.

47. Druy, M.A. and R.J. Seymour. 1983. Poly(2,2′-bithiophene): An electrochromic conducting polymer. *J Phys* C3:595–598.

48. Imae, I., K. Moriwaki, K. Nawa, N. Noma, and Y. Shirota. 1995. Synthesis and electrical properties of novel electrochemically-doped vinyl polymers containing 'end-capped' quaterthiophene and quinquethiophene as pendant groups. *Synth Met* 69:285–286.

49. Imae, I., K. Nawa, Y. Ohsedo, N. Noma, and Y. Shirota. 1997. Synthesis of a novel family of electrochemically-doped vinyl polymers containing pendant oligothiophenes and their electrical and electrochromic properties. *Macromolecules* 30:380–386.

50. Nawa, K., I. Imae, N. Noma, and Y. Shirota. 1995. Synthesis of a novel type of electrochemically doped vinyl polymer containing pendant terthiophene and its electrical and electrochromic properties. *Macromolecules* 28:723–729.

51. Nawa, K., K. Miyawaki, I. Imae, N. Noma, and Y. Shirota. 1993. Polymers containing pendant oligothiophenes as a novel class of electrochromic materials. *J Mater Chem* 3:113–114.

52. Ohsedo, Y., I. Imae, N. Noma, and Y. Shirota. 1996. Synthesis and electrochromic properties of a methacrylate polymer containing pendant terthiophene. *Synth Met* 81:157–162.

53. Ohsedo, Y., I. Imae, and Y. Shirota. 2003. Synthesis and electrochromic properties of a new family of methacrylate polymers containing pendant oligothiophenes. *J Polym Sci Part B: Polym Phys* 41:2471–2484.

54. Elsenbaumer, R.L., K.Y. Jen, G.G. Miller, and L.W. Shacklette. 1987. Processible, environmentally stable, highly conducting forms of polythiophene. *Synth Met* 18:277–282.

55. Collomb-Dunand-Sauthier, M.-N., S. Langlois, and E. Genies. 1994. Spectroelectrochemical behaviour of poly(3-octylthiophene): Application to electrochromic windows with polyaniline and iridium oxide. *J Appl Electrochem* 24:72–77.

56. Gustafsson, J.C., O. Inganäs, and A.M. Andersson. 1994. Conductive polyheterocycles as electrode materials in solid state electrochromic devices. *Synth Met* 62:17–21.

57. Lère-Porte, J.-P., J.J.E. Moreau, and C. Torreilles. 2001. Highly conjugated poly(thiophene)s—Synthesis of regioregular 3-alkylthiophene polymers and 3-alkylthiophene/thiophene copolymers. *Eur J Org Chem* 7:1249–1258.

58. Elsenbaumer, R.L., K.Y. Jen, and R. Oboodi. 1986. Processible and environmentally stable conducting polymers. *Synth Met* 15:169–174.

59. McCullough, R.D. and P.C. Ewbank. 1998. *Handbook of conducting polymers*, 2nd ed., eds. T.A. Skotheim, R.L. Elsembaumer, and J.R. Reynolds. New York: Marcel Dekker, pp. 229–258.

60. Arbizzani, C., A. Bongini, M. Mastragostino, A. Zanelli, G. Barbarella, and M. Zambianchi. 1995. Polyalkylthiophenes as electrochromic materials: A comparative study of poly(3-methylthiophenes) and poly(3-hexylthiophenes). *Adv Mater* 7:571–574.

61. Mastragostino, M., C. Arbizzani, A. Bongini, G. Barbarella, and M. Zambianchi. 1993. Polymer-based electrochromic devices—I. Poly(3-methylthiophenes). *Electrochim Acta* 38:135–140.

62. Ritter, S.K., R.E. Noftle, and A.E. Ward. 1993. Synthesis, characterization, and oxidative polymerization of 3-(fluoromethyl)thiophenes. *Chem Mater* 5:752–754.

63. Lee, C., K.J. Kim, and S.B. Rhee. 1995. The effects of ester substitution and alkyl chain length on the properties of poly(thiophene)s. *Synth Met* 69:295–296.

64. Camurlu, P., A. Cirpan, and L. Toppare. 2005. Conducting polymers of octanoic acid 2-thiophen-3-yl-ethyl ester and their electrochromic properties. *Mater Chem Phys* 92:413–418.

65. Giglioti, M., F. Trivinho-Strixino, J.T. Matsushima, L.O.S. Bulhões, and E.C. Pereira. 2004. Electrochemical and electrochromic response of poly(thiophene-3-acetic acid) films. *Sol Energy Mater Sol Cells* 82:413–420.

66. Gustafsson-Carlberg, J.C., O. Inganäs, M.R. Andersson, C. Booth, A. Azens, and C.G. Granqvist. 1995. Tuning the bandgap for polymeric smart windows and displays. *Electrochim Acta* 40:2233–2235.

67. Iraqi, A., J.A. Crayston, and J.C. Walton. 1995. Synthesis, spectroelectrochemistry, and thermochromism of regioregular head-to-tail oligo- and poly(3-aryloxyhexylthiophenes). *J Mater Chem* 5:1831–1836.

68. Ribeiro, A.S., W.A. Gazotti, Jr., P.F. dos Santos Filho, and M.-A. De Paoli. 2004. New functionalized 3-(alkyl)thiophene derivatives and spectroelectrochemical characterization of its polymers. *Synth Met* 145:43–49.

69. Ribeiro, A.S., V.C. Nogueira, P.F. dos Santos Filho, and M.-A. De Paoli. 2004. Electrochromic properties of poly{3-[12-(p-methoxyphenoxy)dodecyl]thiophene}. *Electrochim Acta* 49:2237–2242.

70. Ko, H.C., S.-A. Park, W.-K. Paik, and H. Lee. 2002. Electrochemistry and electrochromism of the polythiophene derivative with viologen pendant. *Synth Met* 132:15–20.

71. Iraqi, A., J.A. Crayston, and J.C. Walton. 1998. Covalent binding of redox active centres to preformed regioregular polythiophenes. *J Mater Chem* 8:31–36.

72. Kobayashi, M., N. Colaneri, M. Boysel, F. Wudl, and A.J. Heeger. 1985. The electronic and electrochemical properties of poly(isothianaphthene). *J Chem Phys* 82:5717–5723.

73. Yashima, H., M. Kobayashi, K.-B. Lee, D. Chung, A.J. Heeger, and F. Wudl. 1987. Electrochromic switching of the optical properties of polyisothianaphthene. *J Electrochem Soc* 134:46–52.

74. Hung, T.-T. and S.-A. Chen. 1999. The synthesis and characterization of soluble poly(isothia-naphthene) derivative: Poly(5,6-dihexoxoyisothianaphthene). *Polymer* 40:3881–3884.

75. King, G. and S.J. Higgins. 1995. Synthesis and characterisation of novel substituted benzo[*c*]thio-phenes and polybenzo[*c*]thiophenes: Tuning the potentials for n- and p-doping in transparent conducting polymers. *J Mater Chem* 5:447–455.

76. Lee, K. and G.A. Sotzing. 2001. Poly(thieno[3,4-*b*]thiophene). A new stable low band gap con-ducting polymer. *Macromolecules* 34:5746–5747.

77. Lee, B., V. Seshadri, and G.A. Sotzing. 2005. Water dispersible low band gap conductive polymer based on thieno[3,4-*b*]thiophene. *Synth Met* 152:177–180.

78. Lee, B., V. Seshadri, and G.A. Sotzing. 2005. Poly(thieno[3,4-*b*]thiophene)-poly(styrene sulfonate): A low band gap, water dispersible conjugated polymer. *Langmuir* 21:10797–10802.

79. Pomerantz, M., X. Gu, and S.X. Zhang. 2001. Poly(2-decylthieno[3,4-*b*]thiophene-4,6-diyl). A new low band gap conducting polymer. *Macromolecules* 34:1817–1822.

80. Pomerantz, M. and X. Gu. 1997. Poly(2-decylthieno[3,4-*b*]thiophene). A new soluble low-bandgap conducting polymer. *Synth Met* 84:243–244.

81. Ferraris, J.P. and T.L. Lambert. 1991. Narrow bandgap polymers: Poly-4-dicyanomethylene-4*H*-cyclopenta[2,1-*b*;3,4-*b'*]dithiophene (PCDM). *J Chem Soc Chem Commun* 1268–1270.

82. Ferraris, J.P., C. Henderson, D. Torres, and D. Meeker. 1995. Synthesis, spectroelectrochemistry and application in electrochromic devices of an n- and p-dopable conducting polymer. *Synth Met* 72:147–152.

83. Tovar, J.D. and T.M. Swager. 2001. Poly(naphthodithiophene)s: Robust, conductive electrochro-mics via tandem cyclization-polymerizations. *Adv Mater* 13:1775–1780.

84. Kanazawa, K.K., A.F. Diaz, R.H. Geiss, W.D. Gill, J.F. Kwak, J.A. Logan, J.F. Rabolt, and G.B. Street. 1979. 'Organic Metals': Polypyrrole, a stable synthetic 'metallic' polymer. *J Chem Soc Chem Commun* 19:854–855.

85. Diaz, A.F., K.K. Kanazawa, and G.P. Gardini. 1979. Electrochemical polymerization of pyrrole. *J Chem Soc Chem Commun* 635–636.

86. Diaz, A.F., J.I. Castillo, J.A. Logan, and W.-Y. Lee. 1981. Electrochemistry of conducting polypyrole films. *J Electroanal Chem* 129:115–132.

87. Pfluger, P., G. Weiser, J.C. Scott, and G.B. Street. 1986. *Handbook of conducting polymers*, 1st ed., ed. T.A. Skotheim. New York: Marcel Dekker, pp. 1369–1381.

88. Wynne, K.J. and G.B. Street. 1985. Poly(pyrrol-2-ylium tosylate): Electrochemical synthesis and physical and mechanical properties. *Macromolecules* 18:2361–2368.

89. Diaz, A.F., J. Castillo, K.K. Kanazawa, and J.A. Logan. 1982. Conducting poly-*N*-alkylpyrrole polymer films. *J Electroanal Chem* 133:233–239.

90. Bjorklund, R., S. Andersson, S. Allenmark, and I. Lundström. 1985. Electrochromic effects of conducting polymers in water and acetonitrile. *Mol Cryst Liq Cryst* 121:263–270.

91. Murakami, Y. and T. Yamamoto. 1999. Synthesis of new polypyrroles by oxidative polymerization of *N*-(benzylideneamino)pyrroles and properties of the polymers. *Polym J* 31:476–478.

92. Bidan, G., A. Deronzier, and J.-C. Moutet. 1984. Electrochemical coating of an electrode by a poly(pyrrole) film containing the viologen (4,4'-Bipyridinium) system. *J Chem Soc Chem Commun* 1185–1186.

93. Audebert, P., G. Bidan, and M. Lapkowski. 1986. Reduction by two successive one-electron transfers of anthraquinone units bonded to electrodeposited poly(pyrrole) films. *J Chem Soc Chem Commun* 887–889.

94. Gazotti, Jr., W.A., M.-A. De Paoli, G. Casalbore-Miceli, A. Geri, and G. Zotti. 1999. A solid-state electrochromic device based on complementary polypyrrole/polythiophene derivatives and an elastomeric electrolyte. *J Appl Electrochem* 29:753–757.

95. De Paoli, M.-A., G. Casalbore-Miceli, E.M. Girotto, and W.A. Gazotti. 1999. All polymeric solid state electrochromic devices. *Electrochim Acta* 44:2983–2991.

96. Benincori, T., E. Brenna, and F. Sannicolò. 2000. Steric and electronic effects in methyl-substituted 2,2′-bipyrroles and poly(2,2′-bipyrroles)s: Part I. Synthesis and characterization of monomers and polymers. *Chem Mater* 12:1480–1489.

97. De Paoli, M.A., S. Panero, P. Prosperi, and B. Scrosati. 1990. Study of the electrochromism of polypyrrole/dodecylsulfate in aqueous solutions. *Electrochim Acta* 35:1145–1148.

98. Peres, R.C.D., V.F. Juliano, M.-A. De Paoli, S. Panero, and B. Scrosati. 1993. Electrochromic properties of dodecyclbenzenesulfonate doped poly(pyrrole). *Electrochim Acta* 38:869–876.

99. Kon, A.B., J.S. Foos, and T.L. Rose. 1992. Synthesis and properties of poly(3-hydroquinonylpyrrole). *Chem Mater* 4:416–424.

100. Havinga, E.E., W. ten Hoeve, E.W. Meijer, and H. Wynberg. 1989. Water-soluble self-doped 3-substituted polypyrroles. *Chem Mater* 1:650–659.

101. Verghese, M.M., M.K. Ram, H. Vardhan, S.M. Ashraf, and B.D. Malhotra. 1996. Polycarbazole-film-coated electrodes as electrochromic devices. *Adv Mater Opt Electron* 6:399–402.

102. Verghese, M.M., M.K. Ram, H. Vardhan, B.D. Malhotra, and S.M. Ashraf. 1997. Electrochromic properties of polycarbazole films. *Polymer* 38:1625–1629.

103. Pelous, Y., G. Froyer, D. Adès, C. Chevrot, and A. Siove. 1990. Spectroelectrochemical studies of electrochromic poly(N-butyl-3,6-carbazolediyl) films. *Polym Commun* 31:341–342.

104. Chevrot, C., E. Ngbilo, K. Kham, and S. Sadki. 1996. Optical and electronic properties of undoped and doped poly(N-alkylcarbazole) thin layers. *Synth Met* 81:201–204.

105. Tran-Van, F., T. Henri, and C. Chevrot. 2002. Synthesis and electrochemical properties of mixed ionic and electronic modified polycarbazole. *Electrochim Acta* 47:2927–2936.

106. Qiu, Y.-J. and J.R. Reynolds. 1990. A self-doped polymer with both cation and anion exchange properties. *J Electrochem Soc* 137:900–904.

107. Lacaze, P.C., J.E. DuBois, A. Desbene-Monvernay, P.L. Desbene, J.J. Basselier, and D. Richard. 1983. Polymer-modified electrodes as electrochromic material. Part III. Formation of poly-N-vinylcarbazole films on transparent semiconductor ITO surfaces by electropolymerization of NVK in acetonitrile. *J Electroanal Chem* 147:107–121.

108. Desbene-Monvernay, A., P.C. Lacaze, J.E. DuBois, and P.L. Desbene. 1983. Polymer-modified electrodes as electrochromic material. Part IV. Spectroelectrochemical properties of poly-N-vinylcarbazole films. *J Electroanal Chem* 152:87–96.

109. Davis, F.J., H. Block, and R.G. Compton. 1984. The electrochemical oxidation of poly(N-vinylcarbazole) films. *J Chem Soc Chem Commun* 890–892.

110. Booth, T.W., S. Evans, and J.M. Maud. 1989. Novel electrochromic films via anodic oxidation of poly[3-(3-bromocarbazol-9-yl)propyl]methylsiloxane. *J Chem Soc Chem Commun* 196–198.

111. Letheby, H. 1862. XXIX—On the production of a blue substance by the electrolysis of sulphate of aniline. *J Chem Soc* 15:161–163.

112. Prakash, R. 2002. Electrochemistry of polyaniline: Study of the pH effect and electrochromism. *J Appl Polym Sci* 83:378–385.

113. Diaz, A.F. and J.A. Logan. 1980. Electroactive polyaniline films. *J Electroanal Chem* 111:111–114.

114. Kobayashi, T., H. Yoneyama, and H. Tamura. 1984. Electrochemical reactions concerned with electrochromism of polyaniline film-coated electrodes. *J Electroanal Chem* 177:281–291.

115. MacDiarmid, A.G., L.S. Yang, W.S. Huang, and B.D. Humphrey. 1987. Polyaniline: Electrochemistry and application to rechargeable batteries. *Synth Met* 18:393–398.

116. Kobayashi, T., H. Yoneyama, and H. Tamura. 1984. Polyaniline film-coated electrodes as electrochromic display devices. *J Electroanal Chem* 161:419–423.

117. Kaneko, M., H. Nakamura, and T. Shimomura. 1987. Multicolor electrochromism of polyaniline film. *Makromol Chem Rapid Commun* 8:179–180.

118. Watanabe, A., K. Mori, Y. Iwasaki, Y. Nakamura, and S. Niizuma. 1987. Electrochromism of polyaniline film prepared by electrochemical polymerization. *Macromolecules* 20:1793–1796.

119. Chandrasekhar, P., G.C. Birur, P. Stevens, S. Rawal, E.A. Pierson, and K.L. Miller. 2001. Far infrared electrochromism in unique conducting polymer systems. *Synth Met* 119:293–294.

120. Chandrasekhar, P., B.J. Zay, G.C. Birur, S. Rawal, E.A. Pierson, L. Kauder, and T. Swanson. 2002. Large, switchable electrochromism in the visible through far-infrared in conducting polymer devices. *Adv Funct Mater* 12:95–103.

121. Leclerc, M., J. Guay, and L.H. Dao. 1988. Synthesis and properties of electrochromic polymers from toluidines. *J Electroanal Chem* 251:21–29.

122. Cattarin, S., L. Doubova, G. Mengoli, and G. Zotti. 1988. Electrosynthesis and properties of ring-substituted polyanilines. *Electrochim Acta* 33:1077–1084.

123. Wei, Y., W.W. Focke, G.E. Wnek, A. Ray, and A.G. MacDiarmid. 1989. Synthesis and electrochemistry of alkyl ring-substituted polyanilines. *J Phys Chem* 93:495–499.

124. Dao, L.H., M. Leclerc, J. Guay, and J.W. Chevalier. 1989. Synthesis and characterization of substituted poly(anilines). *Synth Met* 29:E377–E382.

125. Leclerc, M., J. Guay, and L.H. Dao. 1989. Synthesis and characterization of poly(alkylanilines). *Macromolecules* 22:649–653.

126. Foot, P.J.S. and R. Simon. 1989. Electrochromic properties of conducting polyanilines. *J Phys D: Appl Phys* 22:1598–1603.

127. D'Aprano, G., M. Leclerc, and G. Zotti. 1993. Steric and electronic effects in methyl and methoxy substituted polyanilines. *J Electroanal Chem* 351:145–158.

128. Mortimer, R.J. 1995. Spectroelectrochemistry of electrochromic poly(*o*-toluidine) and poly(*m*-toluidine) films. *J Mater Chem* 5:969–973.

129. Huang, L.-M., T.-C. Wen, and A. Gopalan. 2002. In situ UV–visible spectroelectrochemical studies on electrochromic behavior of poly(2,5-dimethoxy aniline). *Synth Met* 130:155–163.

130. Zotti, G., N. Comisso, G. D'Aprano, and M. Leclerc. 1992. Electrochemical deposition and characterization of poly(2,5-dimethoxyaniline): A new highly conducting polyaniline with enhanced solubility, stability and electrochromic properties. *Adv Mater* 4:749–752.

131. Gazotti, Jr., W.A., M.J.D.M. Jannini, S.I.C. de Torresi, and M.-A. De Paoli. 1997. Influence of dopant, pH and potential on the spectral changes of poly(*o*-methoxyaniline): Relationship with the redox processes. *J Electroanal Chem* 440:193–199.

132. Li, C. and S. Mu. 2004. Electrochromic properties of sulfonic acid ring-substituted polyaniline in aqueous and non-aqueous media. *Synth Met* 144:143–149.

133. Comisso, N., S. Daolio, G. Mengoli, R. Salmaso, S. Zecchin, and G. Zotti. 1988. Chemical and electrochemical synthesis and characterization of polydiphenylamine and poly-*N*-methylaniline. *J Electroanal Chem* 255:97–110.

134. Kim, E., K.-Y. Lee, M.-H. Lee, J.-S. Shin, and S.B. Rhee. 1997. All solid-state electrochromic window based on poly(aniline *N*-butylsulfonate)s. *Synth Met* 85:1367–1368.

135. Yasuda, T., I. Yamaguchi, and T. Yamamoto. 2003. Preparation of *N*-grafted polyanilines with oligoether side chains by using ring-opening graft copolymerization of epoxide, and their optical, electrochemical and thermal properties and ionic conductivity. *J Mater Chem* 13:2138–2144.

136. Tassi, E.L., M.-A. De Paoli, S. Panero, and B. Scrosati. 1994. Electrochemical, electrochromic and mechanical properties of the graft copolymer of poly(aniline) and nitrilic rubber. *Polymer* 35:565–572.

137. Pei, Q., G. Zuccarello, M. Ahlskog, and O. Inganäs. 1994. Electrochromic and highly stable poly(3,4-ethylenedioxythiophene) switches between opaque blue–black and transparent sky blue. *Polymer* 35:1347–1351.

138. Sankaran, B. and J.R. Reynolds. 1997. High-contrast electrochromic polymers from alkyl-derivatized poly(3,4-ethylenedioxythiophenes). *Macromolecules* 30:2582–2588.

139. Sotzing, G.A., J.R. Reynolds, and P.J. Steel. 1997. Poly(3,4-ethylenedioxythiophene) (PEDOT) prepared via electrochemical polymerization of EDOT, 2,2′-bis(3,4-ethylenedioxythiophene) (BiEDOT), and their TMS derivatives. *Adv Mater* 9:795–798.

140. Heuer, H.W., R. Wehrmann, and S. Kirchmeyer. 2002. Electrochromic window based on conducting poly(3,4-ethylenedioxythiophene)-poly(styrene sulfonate). *Adv Funct Mater* 12:89–94.

141. Kumar, A. and J.R. Reynolds. 1996. Soluble alkyl-substituted poly(ethylenedioxythiophenes) as electrochromic materials. *Macromolecules* 29:7629–7630.

142. Kumar, A., D.M. Welsh, M.C. Morvant, F. Piroux, K.A. Abboud, and J.R. Reynolds. 1998. Conducting poly(3,4-alkylenedioxythiophene) derivatives as fast electrochromics with high-contrast ratios. *Chem Mater* 10:896–902.

143. Welsh, D.M., A. Kumar, M.C. Morvant, and J.R. Reynolds. 1999. Fast electrochromic polymers based on new poly(3,4-alkylenedioxythiophene) derivatives. *Synth Met* 102:967–968.

144. Schwendeman, I., C.L. Gaupp, J.M. Hancock, L.B. Groenendaal, and J.R. Reynolds. 2003. Perfluoroalkanoate-substituted PEDOT for electrochromic device applications. *Adv Funct Mater* 13:541–547.

145. Wang, C., J.L. Schindler, C.R. Kannewurf, and M.G. Kanatzidis. 1995. Poly(3,4-ethylenedithiathiophene). A new soluble conducting polythiophene derivative. *Chem Mater* 7:58–68.

146. Welsh, D.M., A. Kumar, E.W. Meijer, and J.R. Reynolds. 1999. Enhanced contrast ratios and rapid switching in electrochromics based on poly(3,4-propylenedioxythiophene) derivatives. *Adv Mater* 11:1379–1382.

147. Welsh, D.M., L.J. Kloeppner, L. Madrigal, M.R. Pinto, B.C. Thompson, K.S. Schanze, K.A. Abboud, D. Powell, and J.R. Reynolds. 2002. Regiosymmetric dibutyl-substituted poly(3,4-propylenedioxythiophene)s as highly electron-rich electroactive and luminescent polymers. *Macromolecules* 35:6517–6525.

148. Schwendeman, I., J. Hwang, D.M. Welsh, D.B. Tanner, and J.R. Reynolds. 2001. Combined visible and infrared electrochromism using dual polymer devices. *Adv Mater* 13:634–637.

149. Gaupp, C.L., D.M. Welsh, and J.R. Reynolds. 2002. Poly(ProDOT-Et$_2$): A high-contrast, high-coloration efficiency electrochromic polymer. *Macromol Rapid Commun* 23:885–889.

150. Mishra, S.P., K. Krishnamoorthy, R. Sahoo, and A. Kumar. 2004. Synthesis and characterization of monosubstituted and disubstituted poly(3,4-propylenedioxythiophene) derivatives with high electrochromic contrast in the visible region. *J Polym Sci Part A: Polym Chem* 43:419–428.

151. Cirpan, A., A.A. Argun, C.R.G. Grenier, B.D. Reeves, and J.R. Reynolds. 2003. Electrochromic devices based on soluble and processable dioxythiophene polymers. *J Mater Chem* 13:2422–2428.

152. Reeves, B.D., C.R.G. Grenier, A.A. Argun, A. Cirpan, T.D. McCarley, and J.R. Reynolds. 2004. Spray coatable electrochromic dioxythiophene polymers with high coloration efficiencies. *Macromolecules* 37:7559–7569.

153. Krishnamoorthy, K., A.V. Ambade, M. Kanugo, A.Q. Contractor, and A. Kumar. 2001. Rational design of an electrochromic polymer with high contrast in the visible region: Dibenzyl substituted poly(3,4-propylenedioxythiophene). *J Mater Chem* 11:2909–2911.

154. Reeves, B.D., B.C. Thompson, K.A. Abboud, B.E. Smart, and J.R. Reynolds. 2002. Dual cathodically and anodically coloring electrochromic polymer based on a spiro bipropylenedioxythiophene [poly(spiroBiProDOT)]. *Adv Mater* 14:717–719.

155. Thomas, C.A., K. Zong, P. Schottland, and J.R. Reynolds. 2000. Poly(3,4-alkylenedioxypyrrole)s as highly stable aqueous-compatible conducting polymers with biomedical implications. *Adv Mater* 12:222–225.

156. Walczak, R.M. and J.R. Reynolds. 2006. Poly(3,4-alkylenedioxypyrroles): The PXDOPs as versatile yet underutilized electroactive and conducting polymers. *Adv Mater* 18:1121–1131.

157. Zotti, G., S. Zecchin, and G. Schiavon. 2000. Conducting and magnetic properties of 3,4-dimethoxy- and 3,4-ethylenedioxy-capped polypyrrole and polythiophene. *Chem Mater* 12:2996–3005.

158. Merz, A., R. Schropp, and E. Dötterl. 1995. 3,4-Dialkoxypyrroles and 2,3,7,8,12,13,17,18-octaalkoxyporphyrins. *Synthesis* 7:795–800.

159. Zong, K. and J.R. Reynolds. 2001. 3,4-Alkylenedioxypyrroles: Functionalized derivatives as monomers for new electron-rich conducting and electroactive polymers. *J Org Chem* 66:6873–6882.

160. Zong, K., K.A. Abboud, and J.R. Reynolds. 2004. A palladium-catalyzed synthesis of 2-alkylidenepyrrolo[*c*]-1,4-dioxanes: Synthesis of 3,4-(*cis*-1,2-dimethyl)ethylenedioxypyrrole. *Tetrahedron Lett* 45:4973–4975.

161. Xu, C., L. Liu, S.E. Legenski, D. Ning, and M. Taya. 2004. Switchable window based on electrochromic polymers. *J Mater Res* 19:2072–2080.

162. Li, H. and C. Lambert. 2005. 3,4-Methylenedithiopyrrole: Convenient synthesis and application as a novel monomer for electroactive polymers. *J Mater Chem* 15:1235–1237.

163. Li, H., C. Lambert, and R. Stahl. 2006. Conducting polymers based on alkylthiopyrroles. *Macromolecules* 39:2049–2055.

164. Sönmez, G., I. Schwendeman, P. Schottland, K. Zong, and J.R. Reynolds. 2003. *N*-substituted poly(3,4-propylenedioxypyrrole)s: High gap and low redox potential switching electroactive and electrochromic polymers. *Macromolecules* 36:639–647.

165. Schwendeman, I., R. Hickman, G. Sönmez, P. Schottland, K. Zong, D.M. Welsh, and J.R. Reynolds. 2002. Enhanced contrast dual polymer electrochromic devices. *Chem Mater* 14:3118–3122.

166. Sönmez, G., P. Schottland, K. Zong, and J.R. Reynolds. 2001. Highly transmissive and conducting poly[(3,4-alkylenedioxy)pyrrole-2,5-diyl] (PXDOP) films prepared by air or transition metal catalyzed chemical oxidation. *J Mater Chem* 11:289–294.

167. Blohm, M.L., J.E. Pickett, and P.C. Van Dort. 1993. Synthesis, characterization, and stability of poly(3,4-dibutoxythiophenevinylene) copolymers. *Macromolecules* 26:2704–2710.

168. Havinga, E.E., C.M.J. Mutsaers, and L.W. Jenneskens. 1996. Absorption properties of alkoxy-substituted thienylene–vinylene oligomers as a function of the doping level. *Chem Mater* 8:769–776.

169. Jen, K.-Y., H. Eckhardt, T.R. Jow, L.W. Shacklette, and R.L. Elsenbaumer. 1988. Optical, electrochemical, and conductive properties of poly(3-alkoxy-2,5-thienylenevinylenes). *J Chem Soc Chem Commun* 215–217.

170. Martinez, M., J.R. Reynolds, S. Basak, D.A. Black, D.S. Marynick, and M. Pomerantz. 1988. Electrochemical synthesis and optical analysis of poly[(2,2′-dithienyl)-5,5′-diylvinylene]. *J Polym Sci Part B: Polym Phys* 9:911–920.

171. Fu, Y., H. Cheng, and R.L. Elsenbaumer. 1997. Electron-rich thienylene–vinylene low bandgap polymers. *Chem Mater* 9:1720–1724.

172. Sotzing, G.A., J.R. Reynolds, and P.J. Steel. 1996. Electrochromic conducting polymers via electrochemical polymerization of bis(2-(3,4-ethylenedioxy)thienyl) monomers. *Chem Mater* 8:882–889.

173. Sotzing, G.A. and J.R. Reynolds. 1995. Poly[*trans*-bis(3,4-ethylenedioxythiophene)vinylene]: A low band-gap polymer with rapid redox switching capabilities between conducting transmissive and insulating absorptive states. *J Chem Soc Chem Commun* 703–704.

174. Sotzing, G.A., C.A. Thomas, J.R. Reynolds, and P.J. Steel. 1998. Low band gap cyanovinylene polymers based on ethylenedioxythiophene. *Macromolecules* 31:3750–3752.

175. Mitsuhara, T., K. Kaeriyama, and S. Tanaka. 1987. New electrochromic polymers. *J Chem Soc Chem Commun* 764–765.

176. Child, A.D., B. Sankaran, F. Larmat, and J.R. Reynolds. 1995. Charge-carrier evolution in electrically conducting substituted polymers containing biheterocycle/*p*-phenylene repeat units. *Macromolecules* 28:6571–6578.

177. Danieli, R., P. Ostoja, M. Tiecco, R. Zamboni, and C. Taliani. 1986. Poly[1,4-di-(2-thienyl)benzene]: A new conducting polymer. *J Chem Soc Chem Commun* 1473–1474.

178. Ruiz, J.P., J.R. Dharia, J.R. Reynolds, and L.J. Buckley. 1992. Repeat unit symmetry effects on the physical and electronic properties of processable, electrically conducting, substituted poly(1,4-bis(2-thienyl)phenylenes). *Macromolecules* 25:849–860.

179. Reynolds, J.R., J.P. Ruiz, A.D. Child, K. Nayak, and D.S. Marynick. 1991. Electrically conducting polymers containing alternating substituted phenylene and bithiophene repeat units. *Macromolecules* 24:678–687.

180. Lère-Porte, J.-P., J.J.E. Moreau, F. Serein-Spirau, C. Torreilles, A. Righi, J.-L. Sauvajol, and M. Brunet. 2000. Synthesis, orientation and optical properties of thiophene–dialkoxyphenylene copolymers. *J Mater Chem* 10:927–932.

181. Silva, R.A., F. Serein-Spirau, M. Bouachrine, J.-P. Lère-Porte, and J.J.E. Moreau. 2004. Synthesis and characterization of thienylene–phenylene copolymers with oligo(ethylene oxide) side chains. *J Mater Chem* 14:3043–3050.

182. Larmat, F., J. Soloducho, A.R. Katritzky, and J.R. Reynolds. 2001. Electronic structure and charge transport mechanism in poly[1,4-bis(pyrrol-2-yl)phenylene]. *Synth Met* 124:329–336.
183. Reynolds, J.R., A.D. Child, J.P. Ruiz, S.Y. Hong, and D.S. Marynick. 1993. Substituent effects on the electrical conductivity and electrochemical properties of conjugated furanyl phenylene polymers. *Macromolecules* 26:2095–2103.
184. Reynolds, J.R., A.R. Katritzky, J. Soloducho, S. Belyakov, G.A. Sotzing, and M. Pyo. 1994. Poly[1,4-bis(pyrrol-2-yl)phenylene]: A new electrically conducting and electroactive polymer containing the bipyrrole-phenylene repeat unit. *Macromolecules* 27:7225–7227.
185. Sotzing, G.A., J.R. Reynolds, A.R. Katritzky, J. Soloducho, S. Belyakov, and R. Musgrave. 1996. Poly[bis(pyrrol-2-yl)arylenes]: Conducting polymers from low oxidation potential monomers based on pyrrole via electropolymerization. *Macromolecules* 29:1679–1684.
186. Irvin, J.A. and J.R. Reynolds. 1998. Low-oxidation-potential conducting polymers: Alternating substituted *para*-phenylene and 3,4-ethylenedioxythiophene repeat units. *Polymer* 39:2339–2347.
187. Rauh, R.D., F. Wang, J.R. Reynolds, and D.L. Meeker. 2001. High coloration efficiency electrochromics and their application to multi-color devices. *Electrochim Acta* 46:2023–2029.
188. Sönmez, G., H. Meng, and F. Wudl. 2004. Organic polymeric electrochromic devices: Polychromism with very high coloration efficiency. *Chem Mater* 16:574–580.
189. Irvin, D.J. and J.R. Reynolds. 1998. Tuning the band gap of easily oxidized bis(2-thienyl)- and bis(2-(3,4-ethylenedioxythiophene))-phenylene polymers. *Polym Adv Technol* 9:260–265.
190. Ko, H.C., S. Kim, H. Lee, and B. Moon. 2005. Multicolored electrochromism of a poly{1,4-bis[2-(3,4-ethylenedioxy)thienyl]benzene} derivative bearing viologen functional groups. *Adv Funct Mater* 15:905–909.
191. Sankaran, B., M.D. Alexander, Jr., and L.-S. Tan. 2001. Synthesis, emission and spectroelectrochemical studies of bithienylnaphthalene systems. *Synth Met* 123:425–433.
192. DuBois, Jr., C.J., F. Larmat, D.J. Irvin, and J.R. Reynolds. 2001. Multi-colored electrochromic polymers based on BEDOT-pyridines. *Synth Met* 123:425–433.
193. DuBois, Jr., C.J. and J.R. Reynolds. 2002. 3,4-Ethylenedioxythiophene–pyridine-based polymers: Redox or n-type electronic conductivity? *Adv Mater* 14:1844–1846.
194. Irvin, D.J., C.J. DuBois, Jr., and J.R. Reynolds. 1999. Dual p- and n-type doping in an acid sensitive alternating bi(ethylenedioxythiophene) and pyridine polymer. *Chem Commun* 2121–2122.
195. Yasuda, T., Y. Sakai, S. Aramaki, and T. Yamamoto. 2005. New coplanar (ABA)$_n$-type donor–acceptor π-conjugated copolymers constituted of alkylthiophene (unit A) and pyridazine (unit B): Synthesis using hexamethylditin, self-organized solid structure, and optical and electrochemical properties of the copolymers. *Chem Mater* 17:6060–6068.
196. Audebert, P., S. Sadki, F. Moinmandre, and G. Clavier. 2004. First example of an electroactive polymer issued from an oligothiophene substituted tetrazine. *Electrochem Commun* 6:144–147.
197. Lee, B.-L. and T. Yamamoto. 1999. Syntheses of new alternating CT-type copolymers of thiophene and pyrido[3,4-*b*]pyrazine units: Their optical and electrochemical properties in comparison with similar CT copolymers of thiophene with pyridine and quinoxaline. *Macromolecules* 32:1375–1382.
198. Brédas, J.L., A.J. Heeger, and F. Wudl. 1986. Towards organic polymers with very small intrinsic band gaps. I. Electronic structure of polyisothianaphthene and derivatives. *J Chem Phys* 85:4673–4678.
199. Musmanni, S. and J.P. Ferraris. 1993. Preparation and characterization of conducting polymers based on 1,3-di(2-thienyl)benzo[*c*]thiophene. *J Chem Soc Chem Commun* 172–173.
200. Vangeneugden, D.L., R.H.L. Kiebooms, D.J.M. Vanderzande, and J.M.J.V. Gelan. 1999. A general synthetic route towards soluble poly(1,3-dithienylisothianaphthene) derivatives. *Synth Met* 101:120–121.
201. Cravino, A., M.A. Loi, M.C. Scharber, C. Winder, H. Neugebauer, P. Denk, H. Meng, Y. Chen, F. Wudl, and N.S. Sariciftci. 2003. Spectroscopic properties of PEDOTEHIITN, a novel soluble low band-gap conjugated polymer. *Synth Met* 137:1435–1436.

202. Meng, H., D. Tucker, S.Y.C. Chaffins, R. Helgeson, B. Dunn, and F. Wudl. 2003. An unusual electrochromic device based on a new low-bandgap conjugated polymer. *Adv Mater* 15:146–149.

203. Kitamura, C., S. Tanaka, and Y. Yamashita. 1994. Synthesis of new narrow bandgap polymers based on 5,7-di(2-thienyl)thieno[3,4-*b*]pyrazine and its derivatives. *J Chem Soc Chem Commun* 1585–1586.

204. Kitamura, C., S. Tanaka, and Y. Yamashita. 1996. Design of narrow-bandgap polymers. Syntheses and properties of monomers and polymers containing aromatic-donor and *o*-quinoid-acceptor units. *Chem Mater* 8:570–578.

205. Tanaka, S. and Y. Yamashita. 1995. Syntheses of narrow band gap heterocyclic copolymers of aromatic-donor and quinonoid-acceptor units. *Synth Met* 69:599–600.

206. Sönmez, G., H.B. Sonmez, C.K.F. Shen, R.W. Jost, Y. Rubin, and F. Wudl. 2005. A processable green polymeric electrochromic. *Macromolecules* 38:669–675.

207. Sönmez, G., C.K.F. Shen, Y. Rubin, and F. Wudl. 2004. A red, green, and blue (RGB) polymeric electrochromic device (PECD): The dawning of the PECD era. *Angew Chem Int Ed* 43:1498–1502.

208. Akoudad, S. and J. Roncali. 1999. Electrogenerated poly(thiophenes) with extremely low bandgap. *Synth Met* 101:149.

209. Akoudad, S. and J. Roncali. 1998. Electrogenerated poly(thiophenes) with extremely narrow bandgap and high stability under n-doping cycling. *Chem Commun* 2081–2082.

210. Tsui, B., J.L. Reddinger, G.A. Sotzing, J. Soloducho, A.R. Katritzky, and J.R. Reynolds. 1999. Electroactive and luminescent polymers: New fluorene-heterocycle-based hybrids. *J Mater Chem* 9:2189–2200.

211. Larmat, F., J.R. Reynolds, B.A. Reinhardt, L.L. Brott, and S.J. Clarson. 1997. Comparative reactivity of thiophene and 3,4-(ethylenedioxy)thiophene as terminal electropolymerizable units in bis-heterocycle arylenes. *J Polym Sci Part A: Polym Chem* 35:3627–3636.

212. Sotzing, G.A., J.L. Reddinger, A.R. Katritzky, J. Soloducho, R. Musgrave, J.R. Reynolds, and P.J. Steel. 1997. Multiply colored electrochromic carbazole-based polymers. *Chem Mater* 9:1578–1587.

213. Witker, D. and J.R. Reynolds. 2005. Soluble variable color carbazole-containing electrochromic polymers. *Macromolecules* 38:7636–7644.

214. Gaupp, C.L. and J.R. Reynolds. 2003. Multichromic copolymers based on 3,6-bis(2-(3,4-ethylenedioxythiophene))-*N*-alkylcarbazole derivatives. *Macromolecules* 36:6305–6315.

215. Anand, J., S. Palaniappan, and D.N. Sathyanarayana. 1998. Conducting polyaniline blends and composites. *Prog Polym Sci* 23:993–1018.

216. Gomez-Romero, P. 2001. Hybrid organic–inorganic materials—In search of synergic activity. *Adv Mater* 13:163–174.

217. Brotherston, I.D., D.S.K. Mudigonda, J.M. Osborn, J. Belk, J. Chen, D.C. Loveday, J.L. Boehme, J.P. Ferraris, and D.L. Meeker. 1999. Tailoring the electrochromic properties of devices via polymer blends, copolymers, laminates and patterns. *Electrochim Acta* 44:2993–3004.

218. Gangopadhyay, R. and A. De. 2000. Conducting polymer nanocomposites: A brief overview. *Chem Mater* 12:608–622.

219. Gazotti, Jr., W.A., G. Casalbore-Miceli, S. Mitzakoff, A. Geri, M.C. Gallazzi, and M.-A. De Paoli. 1999. Conductive polymer blends as electrochromic materials. *Electrochim Acta* 44:1965–1971.

220. Gazotti, W.A., G. Casalbore-Miceli, A. Geri, A. Berlin, and M.A. De Paoli. 1998. An all-plastic and flexible electrochromic device based on elastomeric blends. *Adv Mater* 10:1522–1525.

221. De Paoli, M.-A. and D.J. Maia. 1994. Polypyrrole–poly(epichlorohydrin-*co*-ethylene oxide) blend: An electroactive, electrochromic and elastomeric material. *J Mater Chem* 4:1799–1803.

222. Shibata, M., K.-I. Kawashita, R. Yosomiya, and Z. Gongzheng. 2001. Electrochromic properties of polypyrrole composite films in solid polymer electrolyte. *Eur Polym J* 37:915–919.

223. Xie, H.-Q. and Y.-M. Ma. 2000. Preparation of conductive polyaniline/chlorosulfonated polyethylene composites via in situ emulsion polymerization and study of their properties. *J Appl Polym Sci* 76:845–850.

224. Malmonge, L.F. and L.H.C. Mattoso. 1995. Electroactive blends of poly(vinylidene fluoride) and polyaniline derivatives. *Polymer* 36:245–249.

225. Meeker, D.L., D.S.K. Mudigonda, J.M. Osborn, D.C. Loveday, and J.P. Ferraris. 1998. Tailoring electrochromic properties using poly(*N*-vinylcarbazole) and poly(*N*-phenyl-2-(2′-thienyl)-5-(5″-vinyl-2″-thienyl)pyrrole) blends. *Macromolecules* 31:2943–2946.

226. Inganäs, O., T. Johansson, and S. Ghosh. 2001. Phase engineering for enhanced electrochromism in conjugated polymers. *Electrochim Acta* 46:2031–2034.

227. Boehme, J.L., D.S.K. Mudigonda, and J.P. Ferraris. 2001. Electrochromic properties of laminate devices fabricated from polyaniline, poly(ethylenedioxythiophene), and poly(*N*-methylpyrrole). *Chem Mater* 13:4469–4472.

228. Hechavarría, L., H. Hu, and M.E. Rincón. 2003. Polyaniline-poly(2-acrylamido-2-methyl-1-propanosulfonic acid) composite thin films: Structure and properties. *Thin Solid Films* 441:56–62.

229. Hu, H., L. Hechavarría, and J. Campos. 2003. Optical and electrical responses of polymeric electrochromic devices: Effect of polyacid incorporation in polyaniline film. *Solid State Ionics* 161:165–172.

230. Sönmez, G., P. Schottland, and J.R. Reynolds. 2005. PEDOT/PAMPS: An electrically conductive polymer composite with electrochromic and cation exchange properties. *Synth Met* 155:130–137.

231. Kirchmeyer, S. and K. Reuter. 2005. Scientific importance, properties and growing applications of poly(3,4-ethylenedioxythiophene). *J Mater Chem* 15:2077–2088.

232. Laurent, D. and J.B. Schlenoff. 1997. Multilayer assemblies of redox polyelectrolytes. *Langmuir* 13:1552–1557.

233. Schlenoff, J.B., H. Ly, and M. Li. 1998. Charge and mass balance in polyelectrolyte multilayers. *J Am Chem Soc* 120:7626–7634.

234. DeLongchamp, D.M., M. Kastantin, and P.T. Hammond. 2003. High-contrast electrochromism from layer-by-layer polymer films. *Chem Mater* 15:1575–1586.

235. Song, H.-K., E.J. Lee, and S.M. Oh. 2005. Electrochromism of 2,2′-azinobis(3-ethylbenzothiazoline-6-sulfonate) incorporated into conducting polymer as a dopant. *Chem Mater* 17:2232–2233.

236. Girotto, E.M. and M.-A. De Paoli. 1998. Polypyrrole color modulation and electrochromic contrast enhancement by doping with a dye. *Adv Mater* 10:790–793.

237. Leventis, N. and Y.C. Chung. 1990. Polyaniline–Prussian Blue novel composite material for electrochromic applications. *J Electrochem Soc* 137:3321–3322.

238. Somani, P., A.B. Mandale, and S. Radhakrishnan. 2000. Study and development of conducting polymer-based electrochromic display devices. *Acta Materialia* 48:2859–2871.

239. Leventis, N. and Y.C. Chung. 1992. New complementary electrochromic system based on polypyrrole-Prussian Blue composite, a benzylviologen polymer, and poly(vinylpyrrolidone)/potassium sulfate aqueous electrolyte. *Chem Mater* 4:1415–1422.

240. Somani, P. and S. Radhakrishnan. 1998. Electrochromic response in polypyrrole sensitized by Prussian Blue. *Chem Phys Lett* 292:218–222.

241. Leventis, N. and Y.C. Chung. 1992. Poly(3-methylthiophene)-Prussian Blue: A new composite electrochromic material. *J Mater Chem* 2:289–293.

242. Jelle, B.P. and G. Hagen. 1998. Electrochemical multilayer deposition of polyaniline and Prussian Blue and their application in solid state electrochromic windows. *J Appl Electrochem* 28:1061–1065.

243. Kulesza, P.J., K. Miecznikowski, M. Chojack, M.A. Malik, S. Zamponi, and R. Marassi. 2001. Electrochromic features of hybrid films composed of polyaniline and metal hexacyanoferrate. *Electrochim Acta* 46:4371–4378.

244. DeLongchamp, D.M. and P.T. Hammond. 2004. Multiple-color electrochromism from layer-by-layer-assembled polyaniline/Prussian Blue nanocomposite thin films. *Chem Mater* 16:4799–4805.

245. Ferreira, M., F. Huguenin, V. Zucolotto, J.E. Pereira da Silva, S.I. Córdoba de Torresi, M.L.A. Temperini, R.M. Torresi, and O.N. Oliveira, Jr. 2003. Electroactive multilayer films of polyaniline and vanadium pentoxide. *J Phys Chem B* 107:8351–8354.

246. Huguenin, F., M. Ferreira, V. Zucolotto, F.C. Nart, R.M. Torresi, and O.N. Oliveira, Jr. 2004. Molecular-level manipulation of V_2O_5/polyaniline layer-by-layer films to control electrochromogenic and electrochemical properties. *Chem Mater* 16:2293–2299.

247. Wu, C.-G. and M.-H. Chung. 2004. Water-soluble poly(2-(3-thienyloxy)ethanesulfonic acid)/V_2O_5 nanocomposites: Synthesis and electrochromic properties. *J Solid State Chem* 177:2285–2294.

248. Shimidzu, T., A. Ohtani, M. Aiba, and K. Honda. 1988. Electrochromism of a conducting polypyrrole–phosphotungstate composite electrode. *J Chem Soc Faraday Trans 1* 84:3941–3949.

249. Granqvist, C.G., E. Avendaño, and A. Azens. 2003. Electrochromic coatings and devices: Survey of some recent advances. *Thin Solid Films* 442:201–211.

250. Reynolds, J.R., K. Zong, I. Schwendeman, G. Sonmez, P. Schottland, A.A. Argun, and P.-H. Aubert. 2004. Electrochromic polymers and polymer electrochromic devices. United States Patent 6,791,738, 2004.

251. Vondrák, J., M. Sedlaříková, J. Reiter, and T. Hodal. 1999. Polymer gel electrolytes for electrochromic devices. *Electrochim Acta* 44:3067–3073.

252. Bernard, M.-C., A.H.-L. Goff, and W. Zeng. 1998. Elaboration and study of a PANI/PAMPS/WO_3 all solid-state electrochromic device. *Electrochim Acta* 44:781–796.

253. Pawlicka, A., D.C. Dragunski, K.V. Guimarães, and C.O. Avellaneda. 2004. Electrochromic devices with solid electrolytes based on natural polymers. *Mol Cryst Liq Cryst* 416:105–112.

254. Lu, W., A.G. Fadeev, B. Qi, and B.R. Mattes. 2004. Fabricating conducting polymer electrochromic devices using ionic liquids. *J Electrochem Soc* 151:H33–H39.

255. Lu, W., A.G. Fadeev, B. Qi, and B.R. Mattes. 2003. Stable conducting polymer electrochemical devices incorporating ionic liquids. *Synth Met* 135–136:139–140.

256. Lu, W., B.R. Mattes, A.G. Fadeev, and B. Qi. 2003. Stable conjugated polymer electrochromic devices incorporating ionic liquids. United States Patent 6, 667, 825.

257. Lu, W., B.R. Mattes, and A.G. Fadeev. 2004. Long-lived conjugated polymer electrochemical devices incorporating ionic liquids. United States Patent 6, 828, 062.

258. Wu, Z., Z. Chen, X. Du, J.M. Logan, J. Sippel, M. Nikolou, K. Kamaras, J.R. Reynolds, D.B. Tanner, A.F. Hebard, and A.G. Rinzler. 2004. Transparent, conductive carbon nanotube films. *Science* 305:1273–1276.

259. Argun, A.A., A. Cirpan, and J.R. Reynolds. 2003. The first truly all-polymer electrochromic devices. *Adv Mater* 15:1338–1341.

260. Bennett, R.N., W.E. Kokonaski, M.J. Hannan, and L.G. Boxall. 2004. Improved electrode for display devices. United States Patent W09416356.

261. Chandrasekhar, P. 1999. Electrochromic display device. United States Patent W09944093.

262. Argun, A.A., M. Berard, P.-H. Aubert, and J.R. Reynolds. 2005. Back-side e lectrical contacts for patterned electrochromic devices on porous substrates. *Adv Mater* 17:422–426.

263. Argun, A.A. and J.R. Reynolds. 2005. Line patterning for flexible and laterally configured electrochromic devices. *J Mater Chem* 15:1793–1800.

264. Aubert, P.-H., A.A. Argun, A. Cirpan, D.B. Tanner, and J.R. Reynolds. 2004. Microporous patterned electrodes for color-matched electrochromic polymer displays. *Chem Mater* 16:2386–2393.

21

Photoelectron Spectroscopy of Conjugated Polymers

M.P. de Jong, G. Greczynski,
W. Osikowicz, R. Friedlein,
X. Crispin, M. Fahlman,
and W.R. Salaneck

21.1   Introduction

Photoelectron spectroscopy (PES) is a technique based on the photoelectric effect, which was first documented in 1887 by Hertz and explained in 1905 by Einstein. The use of soft x-ray sources led to the development of x-ray photoelectron spectroscopy (XPS), originally known as electron spectroscopy for chemical analysis (ESCA) [1], indicating the applicability of the method to studies of chemical properties. In parallel with the development of XPS, ultraviolet photoelectron spectroscopy (UPS) [2], i.e., PES based on ultraviolet photon sources, emerged as a tool for studying the valence electronic structure of gaseous and solid samples. However, the increasing use of the continuous spectral distribution of synchrotron radiation [3,4] as a photon source has made the historical terminology less meaningful. At present, PES has become a widely used technique for studying the chemical and electronic structure of solid matter and gases. Particularly, the method is very useful for studies on conjugated polymers and their interfaces because (1) it provides maximum amount of chemical and electronic information within a single technique, (2) it is essentially nondestructive to most organic systems, and (3) the method is extremely surface sensitive. Recently, the range of photoelectron spectroscopies has been extended to include resonant processes occurring near the x-ray absorption threshold. Resonant photoemission (RPE) associated with these processes provides additional information on local electronic structure and charge vibrational coupling.

The purpose of this chapter is to illustrate the usefulness of PES for studying the electronic and chemical structure of conjugated polymer surfaces and interfaces. As this handbook deals with all aspects of conjugated polymers, the background information on conjugated polymers is kept to a minimum here. It is further assumed that the reader is familiar with some of the unique electronic structural issues in the physics of conjugated polymers, i.e., the concepts of solitons, polarons, and bipolarons.

This chapter is organized as follows: First, the basic principles of PES, applied to molecular systems, are outlined including brief discussions of the reference energy level problem, RPE, and the interpretation of PES spectra with the help of quantum chemical calculations. Next, selected PES studies on the valence electronic structure of π-conjugated polymers that incorporate features chosen to be important and representative are reviewed, considering both the pristine and the n-doped state. In the case of poly(p-phenylenevinylene) (PPV), we discuss RPE, based on core-excitation and nonradiative decay, as a probe of charge vibrational coupling and localization of states. Subsequently, we focus on the issue of energy level alignment at interfaces between π-conjugated polymers and metallic substrates, which is highly relevant for tuning charge injection barriers in polymer-based devices. Finally, particular attention is given to the (surface) chemical and electronic properties of the commercially successful polymer blend poly(3,4-ethylenedioxythiophene)/poly-styrenesulfonate (PEDOT–PSS).

21.2 Photoelectron Spectroscopy

21.2.1 X-Ray and Ultraviolet Photoelectron Spectroscopy

The basic principles of PES are presented and some specific features in a photoelectron spectrum which are helpful in understanding the examples in this chapter are discussed below. Since this chapter deals with thin films of π-conjugated polymers, the emphasis will be on molecular-solids aspects. For a more in-depth discussion of the technique, relative to organic polymeric systems, readers can refer [5–7].

The fundamental process in molecular photoionization is represented by

$$M_0 + h\nu \rightarrow M_+^* + e^-, \tag{21.1}$$

where M_0 is the isolated neutral molecule in the ground state, $h\nu$ is the ionizing photon, M_+^* is the positive molecular ion in the excited state, and e^- is the photoelectron that carries kinetic energy E_k. If E_k is sufficiently large, then the escaping electron and the molecular ion are not strongly coupled. The energy balance according to Equation 21.1 is then given by

$$E_0 + h\nu = E_+^* + E_k, \tag{21.2}$$

where E_0 is the total energy of the neutral molecule in the ground state and E_+^* is the total energy of the positive molecular ion in an excited state. The basic equation used to derive binding energies E_B^V from photoelectron spectra can be written as

$$E_B^V = E_+^* - E_0 = h\nu - E_k. \tag{21.3}$$

The photon energy $h\nu$ is known and the photoelectron kinetic energy distribution is measured to deduce the binding energy E_B^V. Although intuitively one associates the binding energy with the energy of the

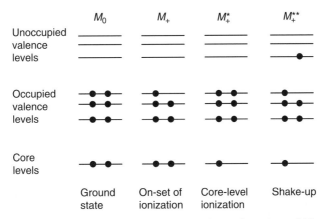

FIGURE 21.1 Schematic illustration of some single-electron molecular configurations, which are discussed in the text.

single electron in the neutral ground state of the molecule, it is important to note that in reality the binding energy corresponds to the total energy difference between the initial ground state and various final ionized states. The electronic structures of M_0 and M_+^* are usually modeled in terms of single-electron states. Several schematic single-electron molecular configurations are shown in Figure 21.1. The different panels illustrate: M_0, the neutral molecule in the ground state, where all of the electrons in the molecule occupy only the lowest allowed energy levels; M_+, ionization from the HOMO level, corresponding to the ground state of the (photoionized) molecular cation; M_+^*, a core-ionized molecule; and M_+^{**}, a "shake-up" phenomenon, where an electron is excited across the gap simultaneously with a core-level ionization event [1,8,9].

All of the various possible final states corresponding to a given initial state contribute in principle to a photoelectron spectrum. In practice, however, most of the intensity usually resides in one main line. The main line corresponds to a particular simple final excited state of the molecular ion formed by the direct removal of a single electron from a specific initial state of the molecule. Other types of final states, e.g., shake-up events, usually have low intensities and can only be seen as weak satellites on the main line. An idealized photoelectron spectrum, or energy distribution curve (EDC), presented as the number of electrons emitted per time unit vs. the binding energy, is shown in Figure 21.2, with the corresponding

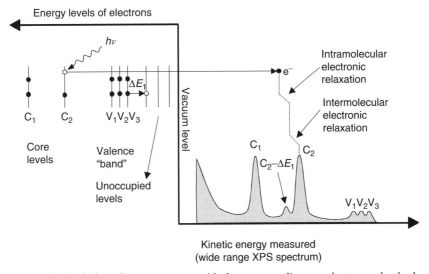

FIGURE 21.2 An idealized photoelectron spectrum with the corresponding one-electron molecular levels.

molecular energy levels. Within a one-electron picture, there is a one-to-one correspondence between the peaks in a photoelectron spectrum and the one-electron molecular levels in the neutral molecule. An additional small feature appearing on the low kinetic energy side of the main peak, labeled C_2, is a shake-up satellite: the escaping electron loses kinetic energy, ΔE_1, as a result of an excitation from an occupied level (V_3) to an unoccupied level simultaneously with the electron emission from a core level, C_2. There are relaxation effects that occur during the photoelectron emission event, which can be rationalized, in a *classical picture*, as follows. An electron in a photoelectron emission event leaves a molecule typically within $\sim 10^{-15}$ s. The nuclear geometric relaxation time is $\sim 10^{-13}$ s (which corresponds to an optical phonon frequency), while the corresponding electronic relaxation time is $\sim 10^{-16}$ s [10]. Thus, the main line corresponds to the binding energy of an electron in a ionized molecule where the electrons have had time to relax, i.e., the hole is fully screened, but the nuclei are frozen during the process; this is referred to as the adiabatic peak.

The electronic relaxation effects are usually divided into *intra-*, and *inter*molecular relaxation effects. The intramolecular electronic-relaxation effects occur in response to the creation of a photohole on an isolated molecule, i.e., in the gas phase. In the solid state, there are also intermolecular relaxation effects occurring due to electronic and atomic polarization of the molecules surrounding the particular molecule on which the photohole is created. The intermolecular relaxation energy is given to the escaping electron and thus the kinetic energy is increased, i.e., the corresponding peak in the photoelectron spectra appears at lower binding energy compared to photoelectron emission from the same level in the gas phase. This energy difference, i.e., the polarization energy, is typically 1–3 eV [11] throughout the valence region, and depends usually on the local aggregation state and type of crystalline phase [12]. The relaxation energy for core levels, however, can be much larger due to a higher degree of localization of the core-holes [9,10].

With XPS, it is possible to perform qualitative and even (semi)quantitative analysis of chemical composition in the near surface region of a solid sample. Although the core electrons are not involved in the chemical bonds, core-level XPS provides a wealth of information about chemical environment [1]. Changes in the valence electron density will be reflected as small, but significant shifts in the core-level binding energies. Of course, XPS can be used (and has been extensively used) for the studies on the valence band region. However, the photoionization cross section for the valence levels in XPS is approximately one order of magnitude lower than that for the core levels, which leads to more time-consuming experiments.

It is possible to obtain useful information about the valence π-electronic structure of a molecular ion through "shake-up structures" in XPS spectra, appearing as weak satellites on the low binding-energy side of the main line. The shake-up structure reflects the spectrum of the 1-electron-2-hole states generated in connection with photoionization processes, and can sometimes be used as "finger prints" of certain molecular constituents by comparison with shake-up spectra of known systems.

For studies of the valence band region, UPS has two advantages over XPS. The first is that high-energy resolution can be obtained from standard laboratory photon sources, i.e., He resonance lamps: the full width at half maximum (FWHM) of the He lines is ~ 30 meV. However, in most cases, this advantage is not crucial, as the natural line widths in condensed molecular solids, arising from both homogeneous and inhomogeneous broadening effects, can be as high as 1 eV at room temperature [6,13,14]. The second advantage of UPS is the higher photoionization cross section for electrons in the valence region. Generally, the cross section increases when the photon energy approaches the threshold ionization energy of a particular electronic state. Within the same atomic subshell, this dependence is steeper for p-type as compared to s-type orbitals, so that the probability for photoionization from p-type orbitals increases when moving from soft x-rays to ultraviolet radiation. Particularly, the atomic carbon C(2p) and C(2s) cross sections are equal at ~ 60 eV photon energy, whereas the C(2p)/C(2s) cross-section ratio is ~ 5 times higher at ~ 20 eV. This situation is reversed already at ~ 90 eV photon energy. Therefore, the frontier valence states of π-conjugated materials, which originate mostly from C(2p) atomic orbitals, are strongly enhanced in UPS spectra as compared to XPS spectra. This simple argumentation, however, should be treated only as a "rule of thumb," because for molecular systems the particular symmetries of the molecular orbitals might also influence the relative intensities of 2p vs. 2s derived spectral features.

Valence band spectra provide information about the electronic and chemical structure of the system, as many of the valence electrons participate directly in chemical bonding. The experimental UPS spectra are sometimes evaluated by the "fingerprint" method, i.e., in comparison with known standards. A much better approach is to utilize comparison with the results of appropriate model quantum chemical calculations [7]. In combination with quantum chemical calculations it is possible to specify the electronic structure in terms of atomic or molecular orbitals or in terms of band structure. Some basic considerations and procedures are outlined in Section 21.2.4.

21.2.2 Reference Energy Level Problem

When performing PES on solid samples, the problem of establishing an appropriate reference level often complicates the interpretation of the spectra, as the binding energy scale must be defined with respect to an unambiguous reference. In contrast to gas phase measurements, where the vacuum level serves as a natural reference, measurements for solid state samples are performed with reference to the Fermi level, E_F, of the photoelectron spectrometer by the way the spectroscopy works. To obtain the vacuum level E_{VAC}^{∞}, it is a common practice to add the work function (Φ) of the sample, measured directly by PES [10] to its Fermi Level E_F (see Figure 21.3). However, in the most general case, the local character Φ introduces uncertainties for obtaining E_{VAC}^{∞}. An important point to note here is that the work function, Φ, is measured *just outside the surface* ($\sim 10^{-4}$ cm) [15,16]. Therefore, the work function may vary from one position on the surface to another, depending on the changes in the electrostatic potential $e\phi_{outer}$ of the various exposed surfaces [17,18]. Note that $e\phi_{outer}$ stems from the termination of the periodic (in crystalline samples) ionic/electronic charge density, leading to a distortion of the lattice near the surface accompanied by appreciable electric fields. In such a case, the spectrum is referred to the "surface vacuum level" [19,20], which is not a meaningful reference. Due to these difficulties, in solid state PES the binding energies cannot be referred exactly to E_{VAC}^{∞} *in general*, as in the case of gaseous samples. However, if at least an approximate reference to E_{VAC}^{∞} is required the best option is to average the values of the work functions corresponding to the different surfaces, e.g., crystal faces [21].

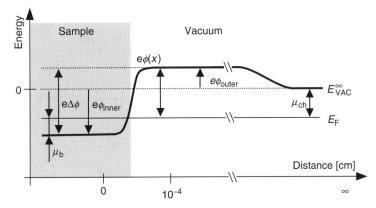

FIGURE 21.3 Schematic representation of quantities related to the reference level problem. (Adapted from Brazovskii, S.A., and Kirova, N.N. 1981. *JEPT Lett.*, 33, 4, 1981.) The relevant quantities are the Fermi level E_F, the vacuum level E_{VAC}^{∞}, the work function Φ, the internal (bulk) chemical potential μ_b, the chemical potential μ_{ch}, the electrostatic potential as a function of the distance to the surface $\phi(x)$, the volume average of $\phi(x)$ inside the surface ϕ_{inner}, the volume average of $\phi(x)$ outside the surface ϕ_{outer}, and the surface electrostatic potential $\Delta\phi = \phi_{outer} - \phi_{inner}$.

The situation is much easier if the samples are good conductors and their reference to E_F is satisfactory. If the sample is in good electrical contact with the spectrometer, the sample Fermi level is in equilibrium with E_F of the spectrometer, as shown in Figure 21.4. The photoelectron that has crossed the surface of the sample, still being close enough to the surface (e.g., 10^{-4} cm), has a kinetic energy E'_k. While traveling to the entrance slit of the analyzer the photoelectron is either accelerated or retarded by an amount that equals the difference between the work function of the sample, Φ, and the work function of the spectrometer Φ_{sp} (the so-called contact potential difference). Thus, when the photoelectron reaches the analyzer, its kinetic energy E_k is given by

$$E_k = E'_k - e(\Phi_{sp} - \Phi). \qquad (21.4)$$

On the other hand, the photoelectron emitted from a core level with binding energy E_B^F, referred to the Fermi level, carries a kinetic energy equal to

$$E'_k = h\nu - e\Phi - E_B^F. \qquad (21.5)$$

Substituting Equation 21.5 in Equation 21.4 gives the relation between the measured kinetic energy and the corresponding binding energy, referred to E_F:

$$E_k = h\nu - E_B^F - e\Phi_{sp}. \qquad (21.6)$$

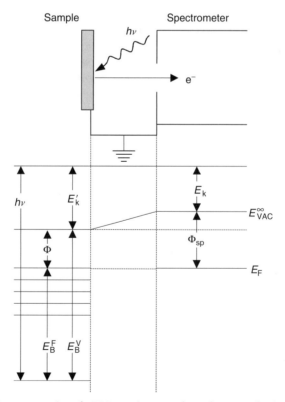

FIGURE 21.4 Schematic representation of a PES experiment performed on a conducting sample. The sample is in good electrical contact with the spectrometer and the sample Fermi level can be easily aligned to the Fermi level of the spectrometer by means of charge transfer. Therefore, the Fermi level of the spectrometer forms an unambiguous reference level.

An important point to note is that for *conducting samples* the photoelectrons originating from a given core level will always appear in the spectrometer with the same kinetic energy, no matter what the work function of the sample is. This has two crucial implications: (1) any change to the sample work function related to a modification of its surface dipole part (e.g., by adsorbing atoms or molecules on the surface) does not shift the positions of the binding energy lines with respect to E_F of spectrometer (this holds at least for bulk atoms), (2) although the discussed work function modifications do not shift bulk core-levels as observed by the spectrometer, they will shift the core levels with respect to E_{VAC}^{∞} [15].

The Fermi level cannot automatically be used as a reference level if thermodynamic equilibrium between spectrometer and sample is uncertain. As a matter of fact, this is often the case for π-conjugated polymers and molecules, as shown by many PES investigations [22–27]. A discussion of some of these cases is given in Section 21.4.

21.2.3 Resonant Photoemission

Development of the "third generation" synchrotron sources in the 1990s, driven by the progress in the design of electron-guiding magnets and insertion devices such as wigglers and undulators, has significantly improved the capabilities to study the electronic structure of matter. Here, we briefly consider the spectroscopy of electrons emitted in the nonradiative decay of core-excited states, an example of a developing synchrotron-based technique that is promising for detailed studies on the ground and excited state electronic structure of organic molecular materials. Core-excited states created on x-ray absorption (Figure 21.5 case A) decay either through radiative processes, i.e., through emission of x-rays, or nonradiatively by emitting electrons. The latter dominates for the lighter elements that form the main constituents of the organic materials considered here. In the nonradiative case, the so-called participator (case B) and spectator decay (case C) processes are distinguished. As participator decay results in the same one-hole final state as reached in direct photoionization, the related spectroscopy is often called resonant photoemission (RPE). The fundamentally different pathways to the same final states in RPE vs. PES give access to entirely different information.

The transition matrix elements in the two processes are different: dipole (PES) vs. Coulomb-type (for the decay process in RPE), and each has its own molecular dynamics. These different matrix elements and the involvement of an intermediate, core-excited state in RPE often result in altered intensity ratios of the spectral features [28,29] and a different relative population of vibrational states [28]. Since core excitations are site-specific for different elements or chemically shifted atoms [30], with well-defined and separated absorption edges, RPE may supply information on the local electronic structure [31–33].

Considering π-conjugated polymers, the degree of electronic localization in both the occupied and the unoccupied bands impacts the nonradiative decay processes, through an interplay among vibrational broadening, dispersion of the bands, and the dependence of the Auger matrix elements on the electron wave number [29]. This allows one, in principle, to discriminate between localized and delocalized bands, although the interpretation of the decay spectra is by no means straightforward. In RPE, as the ionization process is not vertical in nature as it is in PES, the occupation of *different vibrational* sublevels of a given electronic final state generally depends on the excitation energy. Moreover, one has to consider broadening due to the spectral function of the radiation [34–36], specific vibrational structures in the core-excited and final

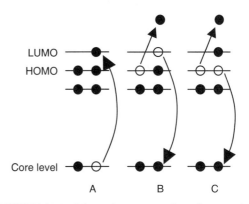

FIGURE 21.5 Schematic representation of core-excitation (A) and decay (B,C). The core-excited state (A) can decay nonradiatively through Auger-like transitions: participator decay (B) and spectator decay (C).

states [37], quenching [38] and dephasing [39] of transitions, as well as charge transfer to systems in contact [40,41].

The physics related to the dispersion of RPE final states as a function of excitation energy is well studied for atoms and relatively small molecules (with less than ~10 atoms) [42]. For solids and large molecules, the situation remains largely unexplored, probably due to complexity issues. RPE has rarely been used to study relatively large organic molecules or carbon-based solid materials; among the few exceptions are graphite [43], C_{60} [44,45], polystyrene [46], and pure as well as nitrogen-substituted benzene [46,47]. In Section 21.3.4, we review RPE processes in the conjugated polymer poly-(*para*-phenylenevinylene).

21.2.4 Interpretation of Valence Band Spectra through Theoretical Modeling

As discussed above, the interpretation of valence band UPS spectra relies either on a comparison with photoelectron spectra on reference materials or on theoretical model calculations. During the last decades, the use of quantum chemical calculations in the interpretation of experimental UPS valence band spectra has become almost standard [48–50], due to progress in theoretical methods and, in particular, a dramatic increase in computational power. A one-electron approximation of the neutral molecules is commonly applied, which gives a one-to-one correspondence between the peaks in the photoelectron spectrum and the molecular energy levels (see Figure 21.2). According to Koopman's theorem, the numerical values of the calculated binding energies are commonly set to the eigenvalues of the molecular orbitals in the neutral ground state of the molecule.

Important effects to consider when relying on quantum chemical calculations are the various relaxation effects that occur during a photoelectron emission event. It is usually necessary to put in "by hand" (or in some more sophisticated theoretical fashion) corrections for the relaxation phenomena that account for differences between the molecular orbitals of the neutral molecule and the molecular ion. Theoretical model calculations for the interpretation of UPS valence band spectra of molecular solids are usually performed on isolated molecules, i.e., not taking into account any intermolecular interaction. This is justified by the fact that, for organic molecules, the intermolecular van der Waals interaction between the molecules in the solid state is rather weak. There is a rigid shift throughout the valence region, due to polarization of the surrounding molecules caused by the positive charge created on the photoionized molecule, which stabilizes the final ion state and leads to a decrease in binding energy [9]. In addition, the energy scale of the calculated spectrum is routinely compressed to compensate for the relaxation and electron correlation effects that are neglected in Koopman's approximation [51].

Among the theoretical methods used for the interpretation of, primarily, UPS valence band data are semiempirical Hartree–Fock methods—such as the modified neglect of diatomic overlap (MNDO) [52,53] and Austin Model 1 (AM1); [54] the nonempirical Valence Effective Hamiltonian (VEH) pseudopotential method; [55,56] ab initio Hartree–Fock techniques; the local spin density (LSD) approximation [57,58]; and Density Functional Theory (DFT). A common strategy for calculating the density of valence states for large conjugated systems, for both pristine and charged (doped) systems, is to use a "two-step procedure"; the ground state geometry is first optimized by applying a semiempirical method (e.g., AM1 or MNDO); the optimized geometries then serve as input for electronic structure calculations performed with, e.g., the VEH method. The AM1 and MNDO methods are known to yield reliable geometries for large organic molecules. The applicability of the VEH model in interpreting photoelectron valence band spectra of conjugated polymers is well established [7,59–61]. VEH parameters for these systems have been determined by fitting to double-zeta-quality ab initio results on model molecules [55,56]. Calculations on conjugated molecules using ab initio Hartree–Fock, LSD, or first principles DFT methods have, of course, to be restricted to relatively small systems due to the dramatically increasing computational efforts. However, in some cases one can gain reliable detailed information for large systems (polymers) by performing calculations on adequate model systems (e.g., oligomers).

21.3 Valence Band PES of Pristine and n-Doped π-Conjugated Polymers

21.3.1 Polyacetylene

Trans-polyacetylene (*trans*-PA) has so far been one of the most studied conjugated polymers, because it is the polymer that actually opened the field of "conducting polymers" and can be considered as the prototype of conjugated polymers. *Trans*-PA has a particularly simple geometrical structure (see the inset in Figure 21.6); a planar zigzag configuration with alternating single and double bonds of 1.44 and 1.36 Å [62]. The band structure is very simple and well suited for demonstrating the relationship between band structure, density of valence states (DOVS), and experimental valence band spectra.

Figure 21.6 shows, from top to bottom: (1) the experimental valence band spectra of *trans*-PA recorded with photon energies of 27 and 50 eV; (2) the DOVS convoluted from the VEH calculation; and (3) the energy band diagram, from which the DOVS has been calculated [48]. The energy scale is given with respect to the Fermi level in the UPS spectra. The input geometry for the VEH calculations has been taken from AM1 geometry optimizations on an oligomer of *trans*-PA. Differences when using experimental values instead are small and are not found to affect any of the results (the AM1-optimized bond lengths are 1.44 and 1.35 Å).

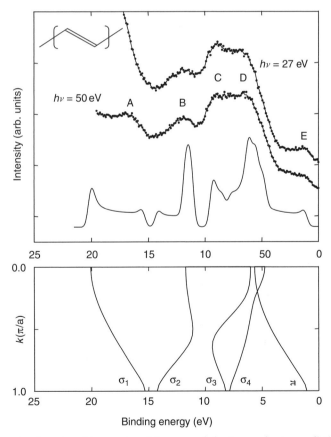

FIGURE 21.6 Valence band spectra of *trans*-polyacetylene, recorded using synchrotron radiation at 27 and 50 eV of photon energy, and the corresponding DOVS derived from VEH calculations. The VEH band structure is shown in the lower part of the figure. (From Lögdlund, M., Salaneck, W.R., Meyers, F., Brédas, J.L., Arbuckle, G.A., Friend, R.H., Holmes, A.B., and Froyer, G., *Macromolecules*, 26, 3815–3820, 1993.)

The electronic structure of *trans*-PA can be rationalized in the following manner [56]. In the absence of dimerization, i.e., with equal C—C and C=C bond lengths, polyacetylene would be a "regular polyene", and the unit cell would consist of a single (—CH—) group. At least within a one-electron theory, there would exist three occupied valence bands. In localized bond orbital terminology, these would correspond to a C—C σ-band, a C—H σ-band, and a π-band, the latter being derived from the remaining p_z atomic orbital on each C-atom. Since there would be one electron in each p_z atomic orbital on each C-atom, the occupied π-band would be only half filled and the electronic structure of the polymer would have a metallic character. Because of a Peierls transition (or, equivalently in molecular terminology, a Jahn-Teller distortion) acting to lower the energy of the π-electrons by distorting the σ-bonds, and additional electron correlation effects, the system is dimerized with alternating single and double C—C bonds. The unit cell in *trans*-PA contains two carbon atoms, corresponding to (—CH=CH—) units. Therefore, the width of the one-dimensional Brillouin zone is only half of that for the case of the (hypothetical) regular polyene, and each band of the regular polyene is split into two bands. In the dimerized unit cell, there are thus four occupied σ-bands and one fully occupied π-band. Energy gaps open up at the Brillouin zone edge, near 8 and 15 eV in Figure 21.6, where each pair of bands from the regular polyene structure is folded back to form the dimerized-chain band structure. The Peierls gap in the π-band corresponds to the energy gap of ~1.5 eV in *trans*-PA. In addition, at the points where two σ-bands of similar symmetry intend to cross, there is an avoided band crossing.

From the calculations, the different peaks in the UPS spectra can be assigned to different bands: Peak A corresponds to electrons from a band derived almost exclusively from C(2s) atomic orbitals, i.e., the C—C backbone. Peak B is derived from the flat portion of the σ_2-band near the zone center. Peaks C and D correspond to electrons from the σ_3- and σ_4-bands derived from combinations of C—C and C—H bands at points in the Brilluoin zone where there is a high density of states, i.e., at the zone center near 9 eV and at the zone edge near 8 eV. The broad Peak E has its origin in the strongly dispersed π-band, which also contributes to Peak D. The valence band edge corresponds to the highest portion of the occupied π-band, at the zone edge (at ~1 eV in the calculated band structure of Figure 21.6).

21.3.2   Poly(*p*-Phenylenevinylene)

PPVs are the widely studied conjugated polymers (see Figure 21.7) that are used as active layers in light emitting devices. Here, we consider unsubstituted PPV as a representative example. In Figure 21.7, the experimental UPS HeI and HeII valence band spectra of PPV, prepared from the tetrahydrothiophenium precursor route, are compared with the DOVS as calculated from the VEH band structure [48]. The corresponding band structure is shown in the bottom part of the Figure. The energy scale is again fixed relative to the experimental Fermi level. The PPV is taken to be planar in the VEH calculations used here, since neutron-diffraction measurements on oriented PPV at room temperature have shown that the ring torsion angles, i.e., the twist of the phenyl rings out of the vinylene plane, are in the order of 7° ± 6° [63]. Such small torsion angles result in negligible effects on the calculated electronic band structure compared to that of the fully coplanar conformation.

In the UPS spectra of PPV, the various peaks are assigned from the details in the VEH calculations: Peaks A, B, and C originate from electrons in σ-bands; Peak D is built-up from contributions from the four highest σ-bands, the lowest π-band, and a small portion from the relatively flat part of the second π-band; Peak E is derived from the next highest π-band that is extremely flat, since it corresponds to electronic levels fully localized on the bonds between *ortho* carbons within the phenyl rings. In general, a flat band results in a high intensity peak in the DOVS, since there are many states per unit energy just at the flat band. There also are small contributions to Peak E from the second and fourth π-bands. Finally, Peak F is derived from the top part of the highest π-band. The larger dispersion of the top π-band results in lower intensity in the UPS spectra.

The unit cell of PPV consists of a styrene-like unit. On attaching a vinylene group to a benzene molecule, the doubly degenerate outermost π-orbitals split into two orbitals separated in energy [64–66].

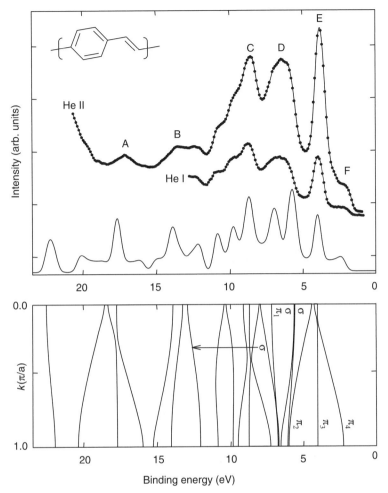

FIGURE 21.7 He I and He II UPS valence band spectra of PPV compared with DOVS derived from VEH calculations. The VEH band structure is shown in the lower part of the figure. (From Lögdlund, M., Salaneck, W.R., Meyers, F., Brédas, J.L., Arbuckle, G.A., Friend, R.H., Holmes, A.B., and Froyer, G., *Macromolecules*, 26, 3815–3820, 1993.)

The π_3-band in PPV is localized on the *ortho* carbons, i.e., it consists of localized states. The wave function that has a nonzero intensity at the *para* carbon atoms, which are adjacent to the vinylene groups, interacts with the π-state of the vinylene group leading to the generation of the dispersed π-bands π_2 and π_4. The top of the π_4-band has almost equal contributions from the vinylene and the phenylene groups. The difference between the top of band π_4 and the flat band π_3 is ~2 eV. The bandwidth of π_2 is ~2.0 eV, i.e., almost the same degree of delocalization as π_4. Finally, the bandwidth of π_1 is ~0.6 eV. This band is related to the $1a_{2u}$ orbital in benzene.

21.3.3 n-Type Doping of Poly(*p*-Phenylenevinylene)s with Sodium

The first direct measure of multiple, resolved gap states in a doped conjugated *polymer* was reported by Fahlman et al., in the case of sodium doped PPV [67]. The UPS spectra indicated a slight decrease in the work function as the first monolayers (equivalent) of sodium atoms were deposited on the surface. XPS data showed that the sodium atoms diffuse uniformly throughout the thin film (a few tens of nm in thickness). At ~40% doping, defined as the Na/monomer ratio, a large change in the work function of

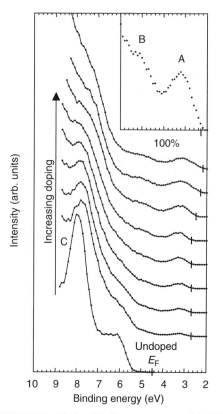

FIGURE 21.8 The low binding-energy part of the He I UPS spectra of PPV recorded during successive exposure to sodium (From Fahlman, M., Beljonne, D., Lögdlund, M., Friend, R.H., Holmes, A.B., Brédas, J.L., and Salaneck, W.R., *Chem. Phys. Lett.*, 214, 327–332, 1993), plotted relative to the vacuum level.

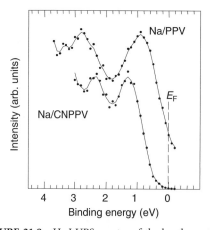

FIGURE 21.9 He I UPS spectra of the bandgap states of CNPPV and PPV at 100% doping level. (From Fahlman, M., Bröms, P., dos Santos, D.A., Moratti, S.C., Johansson, N., Xing, K., Friend, R.H., Holmes, A.B., Brédas, J.L., and Salaneck, W.R., *J. Chem. Phys.*, 102, 8167–8174, 1995.)

~1.2 eV occurred, followed by a slight decrease as the doping level approaches 100%, i.e., one sodium atom per "monomer" repeat unit.

Simultaneously with the 1.2 eV change in the work function, at intermediate (40%) doping levels, two new states appear at the low binding-energy side of the valence band edge. The intensities of the new states increase uniformly with the doping as shown in Figure 21.8 (note that the spectra are plotted relative to the vacuum level). The separation between the two new peaks is ~2.0 eV at maximum doping level, with the lower binding energy peak positioned at ~3.2 eV. At this doping level, the charges in this nondegenerate ground state polymer can be accommodated in either two polaron bands or two bipolaron bands. From model calculations performed using the VEH technique for the fully doped system, the new states are assigned to two doping induced bipolaron bands. The lack of a significant density of states at the Fermi level, as would be expected for a polaron situation, further supports this assignment. For chemically doped systems, such bipolaron states can be strongly stabilized by the presence of the counterions, due to screening of the Coulomb repulsion between the two negative charges forming the bipolaron [68].

In the following study, the interaction between sodium and a cyano-substituted poly(dihexyloxy-*p*-phenylenevinylene) (CNPPV) has been investigated [69]. As expected, the evolution is very reminiscent of that for the unsubstituted PPV. Again, two new states appear upon sodium doping, and no density of states is detected at the Fermi level, consistent with a bipolaron band model. However, there are some differences between the results of Na-doped CNPPV and PPV. The experimental peak-to-peak splitting of the two bipolaron peaks is ~1.05 eV in CNPPV, compared to ~2.0 eV for the sodium-doped PPV. The region where the doping-induced states appear in the UPS spectra, recorded at saturation doping, for the two systems is shown in Figure 21.9. The difference in splitting between the gap states is caused by a stronger confinement of the bipolaron wave functions in CNPPV. AM1 calculations show that the bipolarons in CNPPV are confined on cyano-vinylene-ring-vinylene-cyano segments along the polymer backbone; the phenyl rings included in those segments can accept nearly twice as much charge as the phenyl rings outside the sequence, which are almost unperturbed [69]. The bipolaron levels appear deeper in the gap as a result of the confinement of the charge carriers.

21.3.4 Resonant Photoemission of Poly(*p*-Phenylenevinylene)

RPE spectra of PPV, recorded throughout the C(1s)→π* resonance ranging from $h\nu = 284.2$ eV to $h\nu = 286$ eV, are shown in Figure 21.10 [29]. Participator decay events are easily identified as sharp features with strongly varying intensities as a function of the photon energy [44]. The broad, unstructured signal sloping upward with increasing binding energy is related to spectator processes. In addition to these resonant features, a small contribution is present due to the usual valence band photoelectron emission that occurs at any photon energy. Resonant enhancement of all structures between ~2 and 13 eV can be seen, as the photon energy is tuned to

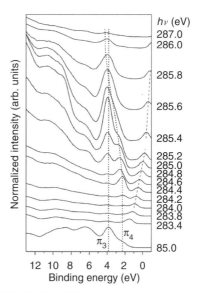

FIGURE 21.10 RPE spectra for photon energies throughout the C(1s)→π* resonance, and off-resonance PES spectrum recorded with a photon energy of 85 eV. The position of the C(1s) line excited by second-order light is indicated by a dashed line.

C(1s)→π* resonances [29]. The resonant behavior vanishes for photon energies above ~286 eV. The resonant enhancement is the strongest for the two structures at lowest binding energy. A comparison with the nonresonant valence-band spectrum ($h\nu = 85$ eV) shows that these two strongly resonating features correspond to the π₃- and π₄-bands of PPV. The nondispersing peak at ~3.8 eV binding energy, related to the localized π₃-band, appears at a photon energy of ~284.8 eV, i.e., in the middle of the C(1s)→π* resonance. A very similar feature has been observed for polystyrene as the single-dominating π feature [70]. This peak arises from states localized on the phenylene group. For PPV, a second, strongly dispersing feature starts to appear at a threshold energy of $h\nu = 284.2$ eV, on the low binding-energy side of the spectrum. This structure is related to the π₄-band. As the photon energy is increased, the peak disperses to higher binding energies, and merges with the nondispersing π₃-derived peak. At the onset of the resonance at $h\nu = 284.2$ eV, RPE emission seems to occur only from a small part of the delocalized band, corresponding to a distinct part of the Brillouin zone. The question arises whether by changing the photon energy different parts of the 1D Brillouin zone are accessed.

To interpret the RPE spectra, the participator decay process in PPV has been investigated theoretically [29]. Although a thorough description of the theory and numerical methods lies well beyond the scope of this chapter, a brief outline is given. PPV is modeled as an ideal infinite chain, motivated by the large conjugation length of the material studied (20–25 repeat units). The infinite chain model has two strong advantages: there are no end effects, and the Bloch theorem is valid. The cross sections for participator decay events can be described by the Kramers–Heisenberg formula [33]. This formula contains the electronic matrix elements of photoabsorption and decay, and a sum over the many-mode Franck–Condon amplitudes referring to the vibrational part of the RPE process. Based on this formalism, two different mechanisms can be identified that could cause the experimentally observed differences in dispersion of the resonating features corresponding to the π₃- and π₄-bands: (1) conservation of electron momentum in the decay process, and (2) excitation processes involving energy transfer to vibrational degrees of freedom. The strong dispersion of the RPE peak related to the π₄-band could intuitively be explained in terms of conservation of momentum, since increasing the excitation energy places the core-excited electron at lower *k*-values in the (core-hole distorted) unoccupied π₅*-band.

However, theoretical analysis shows that there is no special correlation between the momenta of the electrons excited into the unoccupied band and the electrons emitted from the occupied band, meaning that the intuitive picture has to be dismissed. Instead, it turns out that the overlap between the vibrational density in the ground, intermediate, and final states of the RPE process strongly influences the spectral shapes: the vibrational broadening is significantly larger for excitation into delocalized as compared to localized unoccupied bands. The observed dispersion behavior thus depends strongly on the degree of electronic localization of the states involved in the RPE process, by the coupling between electronic and vibrational states.

To summarize, using RPE experiments in combination with thorough theoretical analysis, it was possible to correlate the differences in the dispersion behavior of individual RPE peaks of *nonordered* thin films of PPV with the localized or delocalized nature of the corresponding π-bands and the charge-vibrational coupling in the polymer.

21.4   Energy Level Alignment at Metal–Polymer Interfaces

In any polymer-based charge-injection or charge-extraction devices, tuning of the energy barriers at the interfaces with the electrodes is essential. Consequently, the energy level alignment at metal–organic interfaces has been subject to a large number of studies [20,22,23,25,26,71–73]. In this section, we focus on the alignment of the energy levels at polymer-on-metal interfaces prepared under ambient conditions (in air), by spin-coating techniques. For spin-coated polymer films on metallic substrates, no interfacial chemistry is expected to occur. The metal surfaces, which may in principle be reactive or contain dangling bonds, will be passivated due to the presence of, e.g., residual hydrocarbons or solvents. Therefore, energy level alignment at interfaces prepared by spin-coating in air is expected to be completely different as compared to "ideal" interfaces formed by vapor-deposition of organic molecules on ultraclean metal surfaces in UHV. Although preparation under ambient conditions complicates the situation from a fundamental point of view, the information obtained on such "real-life" interfaces is directly applicable to the analysis of real device performance. For "ideal" interfaces, prepared in UHV, the vacuum levels typically do not align. Recent PES studies [20] emphasize the importance of the formation of interfacial dipoles, due to charge transfer across the interface, interfacial electronic states, chemical interactions, and image forces, among others. For passivated surfaces, the intrinsic interfacial dipole is, in most cases, determined by the modification of the metal electron density tail that is induced by the presence of any adsorbates [74].

21.4.1   Nearly Insulating vs. Conducting Polymers on Metallic Substrates

We first describe how the parameters relevant to polymer-on-metal interfaces are derived from a UPS spectrum. In Figure 21.11, the full UPS spectra of both a gold substrate and a polymer poly(9,9-dioctylfluorene) (PFO) overlayer on a gold substrate are shown. The nomenclature and structure of the diagrams are now standard, and follow that of Seki and coworkers, and Kahn and coworkers [72,75–77]. Note that the UPS spectrum of a sputter-cleaned gold sample is shown, for which the work function is considerably higher compared to gold that is kept in the air, with adsorbed hydrocarbon contamination.

The essential parameters are the following: (1) E_F^{VAC}, the energy difference between the vacuum level of the polymer on top of the gold substrate and the Fermi level of the clean gold substrate; (2) E_{VB}^F, the offset between the Fermi level of the substrate and the valence band edge of the polymer; (3) Δ, the energy offset between the vacuum level of the substrate and the vacuum level of the substrate covered by the polymer film; (4) Φ_{Au}, the work function of the clean substrate (in this example gold). Three related issues should be emphasized. First, the sum of E_F^{VAC} and E_{VB}^F is equal to the ionization potential (IP) of the polymer, which is a specific property of the material, independent of the energy level alignment at the interface. Second, the position of the Fermi level in the polymer film cannot be measured directly in UPS since, with no electrons "at" the Fermi energy in the bandgap of the polymer, it is difficult to "see"

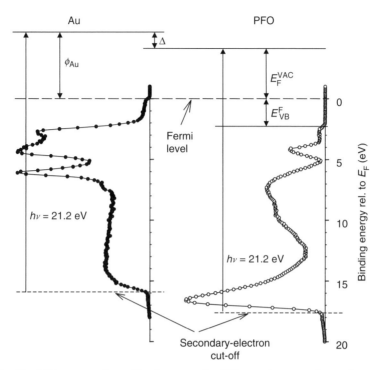

FIGURE 21.11 The UPS spectra of a gold substrate and a PFO over-layer are shown along with the energy differences discussed in the text.

the Fermi level. Its position (energy) in the polymer bandgap must be deduced from the properties of the system [73]. Finally, for nearly insulating polymer samples with a low density of free charge carriers, the work function Φ of the polymer film on top of the gold substrate is *not an intrinsic property of the polymer film itself* (in contrast to IP) and can be easily modified by the surface potential of the underlying substrate.

We now illustrate this last point by measurements of the energy level alignment of two nearly insulating polymers, PFO [78] and poly(bis-(2-dimethyloctylsilyl)-1,4-phenylenevinylene), or bis-DMOS-PPV [22], on various substrates. It turns out that the position of the energy levels of the polymer layer, measured with respect to the Fermi level, can be changed by as much as 1.2 eV [78], by changing the work function of the metallic substrate. Two different approaches have been used for varying the substrate work function: (1) by using various substrate materials with substantially different work functions [78] and (2) by in-situ modification of the work function of an indium-tin-oxide (ITO) substrate [22].

Figure 21.12 shows changes in E_F^{VAC} of PFO films of different thicknesses, in the range of 50–1100 Å, as a function of the work function of the three different substrates: Al with native oxide (denoted Al_xO_y), Si with native oxide (denoted SiO_2), and UV-ozone treated gold (denoted Au), with Φ equal to 3.9, 4.4, and 5.4 eV, respectively. In each case, E_F^{VAC} follows the Schottky–Mott limit [75,79] (long-dashed line), although a small deviation in slope can be seen. As the Fermi energy changes from 3.9 eV, for aluminum with native oxide, to 5.4 eV, for UV-ozone treated gold, E_F^{VAC} increases from 3.9 to 5.2 eV. This implies a negligible interfacial dipole Δ and therefore alignment of the vacuum levels at the interface between PFO and these conducting substrates. It is important to point out that the substrates considered here all have passivated surfaces. Apparently, spin-coating a polymer on top of such surfaces induces essentially no further modification of the work function. The case of sputter-cleaned gold, shown in Figure 21.11, is different in this respect: a significant reduction of the work function, from 5.2 to 4.6 eV (see Figure 21.11), is observed as compared to the gold substrate covered with PFO. This is mainly due to the reduction of the electron density tail protruding from the gold surface by coating the surface by any

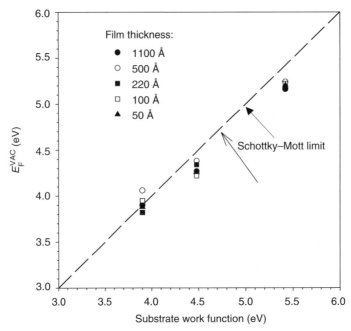

FIGURE 21.12   Dependence of $E_{\mathrm{F}}^{\mathrm{VAC}}$ on the work function of the substrate for the PFO films of different thickness: 50, 100, 220, 500, and 1100 Å.

overlayer [74]. This is not the result of any specific interaction between gold and PFO, in fact solvent-cleaned gold with residual hydrocarbon contamination shows a similar work function of ~4.6 eV. The surface of UV-ozone treated gold is passivated by means of oxidation [80], which additionally increases the work function to 5.4 eV.

The Schottky–Mott-like alignment of the energy levels is recognized from a rigid shift in binding energy of the entire UPS spectrum, up to 1.3 eV, when the work function (substrate) is varied. In particular, $E_{\mathrm{VB}}^{\mathrm{F}}$ changes by the same amount as $E_{\mathrm{F}}^{\mathrm{VAC}}$ since the difference between these parameters is given by the IP of the polymer. This finding is very important for charge-injection devices like polymer light emitting devices, since such significant shifts cause large changes in the barrier-heights for carrier-injection at the interfaces between the polymer and the electrode contacts.

The PFO films follow the Schottky–Mott limit rather closely *independently of the film thickness*. Even for the thickest film, 1100 Å, the dependence of $E_{\mathrm{F}}^{\mathrm{VAC}}$ on the substrate work function is evident. This means that the depletion region extends at least to this distance from the interface. Such large depletion lengths are due to the very low concentrations of free charge carriers in the PFO polymer layer, thanks to the high purity of PFO that is confirmed by the exceptionally good quality of the UPS spectrum [81], and the absence of any O(1s) signal in the XPS spectra. However, at this low concentration of free charge carriers, the position of the Fermi level within the organic layer is completely determined by the properties of the substrate.

Analogous behavior of $E_{\mathrm{F}}^{\mathrm{VAC}}$ for bis-DMOS-PPV [22] on ITO is observed when, instead of changing the substrate, the work function is modified in situ by exposure to x-rays (see Figure 21.13). Upon exposure to the x-ray radiation, the work function of ITO decreases from 4.9 eV down to 4.4 eV. At the same time, $E_{\mathrm{F}}^{\mathrm{VAC}}$ of bis-DMOS-PPV decreases from 4.7 to 4.2 eV, i.e., linearly with the substrate work function, although a small misalignment of E_{VAC} occurs at this interface ($\Delta = -0.2$ eV, close to the experimental error).

Figure 21.13 contains data for poly(4-styrenesulfonic acid) (PSSH). For this polymer, there is a significant (0.4 eV) offset Δ between the E_{VAC}s, both before and after exposure to the x-rays. This is a

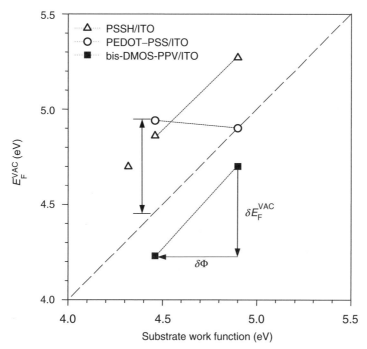

FIGURE 21.13 Changes in E_F^{VAC} induced by in-situ modification of the substrate work function. In analogy to the previous case PEDOT–PSS shows essentially different behavior from the rest of polymers shown. E_F^{VAC} of PEDOT–PSS does not change upon decrease of Φ from 4.9 to 4.4 eV, in contrast to PSSH and bis-DMOS-PPV.

clear evidence for the formation of an interfacial dipole that is likely the result of the acidic character of PSSH that can protonate the ITO surface [22]. The dipole shifts E_{VAC} of PSSH upwards with respect to E_{VAC} of ITO. Still, the work function of PSSH on top of ITO changes linearly with the substrate work function (Schottky–Mott limit), as is expected for this insulating polymer.

Now we briefly consider an *electrically conducting* polymer blend, poly(3,4-ethylene dioxythiophene)-poly(4-styrenesulfonate) (PEDOT–PSS), on a metallic substrate. Regarding energy level alignment, PEDOT–PSS shows a typically metallic behavior [22], in that the Fermi level of the conducting polymer blend is aligned with that of metallic substrates by means of charge transfer across the interface, as is expected for two metals in contact. Decreasing the ITO substrate work function in situ by 0.5 eV does not affect the PEDOT–PSS work function at all (see Figure 21.13). The apparent E_{VAC} alignment for PEDOT–PSS on the ITO substrate prior to the x-ray exposure (ITO work function 4.9 eV) is coincidental. Clearly, PEDOT–PSS does not follow the change in the work function of the ITO substrate. In contrast to PFO or bis-DMOPS-PPV, there are enough free charge carriers that, in response to the modification of the surface potential of ITO, can be transferred to the substrate such that the common Fermi level can be maintained. This charge transfer results finally in the formation of a contact potential difference or, in other words, an interfacial dipole ($\Delta = 0.5$ eV).

21.4.2 Energy Level Alignment as a Function of Na-Doping at Poly(9,9-Dioctylfluorene)/Metal Interfaces

The energy level alignment at interfaces between PFO and metallic substrates changes drastically on doping with alkali metal atoms. The two substrates considered here are Al with native oxide (denoted Al_xO_y) and UV-ozone treated Au, with a large difference in work function (1.3 eV), such that the changes in the parameters of interest, E_F^{VAC} and E_{VB}^F, are expected to be large. The PFO film thickness was ~100 Å. Following vapor deposition of sodium, the Na-atoms form ions by donating charge to the polymer chains, and are uniformly distributed within the polymer film, as determined by angle-dependent XPS data.

Figure 21.14 shows the changes in the energy level alignment as a function of Na-doping. Starting from the bottom, the successive panels show the variation of E_{VB}^F, E_F^{VAC} and their sum, plotted against the Na concentration (determined by XPS) for PFO/Al$_x$O$_y$ (white dots) and PFO/Au (black dots). The most important observations are: (1) as the doping level increases, E_{VB}^F increases as well while E_F^{VAC} decreases, such that their sum remains constant; (2) both parameters, E_{VB}^F and E_F^{VAC}, tend to reach the same limits (dashed lines) *independent of the substrate*; and (3) for PFO on Al$_x$O$_y$ the saturation values (corresponding to horizontal lines) for E_{VB}^F and E_F^{VAC} are reached at considerably lower Na-concentration than for PFO on Au, as is discussed below.

There are two main reasons for the observed changes in E_{VB}^F and E_F^{VAC}. First, Na-doping introduces new electronic states in the π-system of the polymer chains [82]. The added electrons lead to the

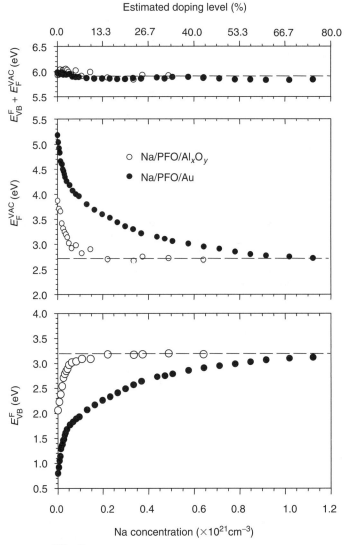

FIGURE 21.14   Changes in E_F^{VAC}, E_{VB}^F, and their sum upon increasing the doping concentration in a PFO film. The horizontal axis, which is common for all three panels, is given in terms of the average concentration of sodium atoms (bottom) as well as the percentage doping level (top).

generation of self-localized states, bipolaron states in PFO, which appear in the bandgap such that the position of the Fermi level shifts. The Fermi level moves gradually to a position half way between the lower binding energy bipolaron state and the conduction band edge [83], i.e., halfway between a filled band (the upper bipolaron band) and an empty band (the conduction band). Second, the generation of charge carriers in the polymer, in principle, allows band bending to occur. As significant amounts of charge carriers become available on the polymer side of the interface, the equilibrium between the conducting substrate and the organic layer becomes determined by charge transfer across the interface. For low dopant concentrations, the depletion length might still be larger than the film thickness. Hence, shrinking of the depletion region caused by an increase density of charge carriers will result in a continuous shift of the energy levels of the PFO towards higher binding energy. This process is observed until the depletion length becomes shorter than the film thickness, beyond which it is not observable by surface-sensitive UPS, even if it should continue with further doping.

E_F^{VAC} tends to the same saturation value (2.7 eV) upon doping, *independent of the substrate type used*. E_{VB}^F tends to 3.2 eV in the same way. Once the amount of charge carriers in the doped PFO film is high enough, the energy band parameters become *completely independent of the work function of the substrate*, in agreement with the above presented results for electrically conducting films of PEDOT–PSS on ITO. The saturation values of E_F^{VAC} and E_{VB}^F are reached at quite different doping levels for the different substrates: 0.2×10^{21} atoms/cm^3 for Al$_x$O$_y$ vs. 10^{21} atoms/cm^3 for Au. This difference is a consequence of the fact that in the latter case more charge is necessary to bring the energy levels in equilibrium.

21.4.3 Consequences for Charge Injection Barriers

To conclude this section, we consider an energy band diagram (see Figure 21.15) for PFO on various substrates [78]. The E_{VAC} of the substrate and of the spin-coated polymer film is aligned ($\Delta \approx 0$). For metal substrates with work functions varying from 3.9 eV (for Al) up to 5.4 eV (for Au), the Fermi level of the metal substrate always falls within the bandgap of PFO. This has important consequences for

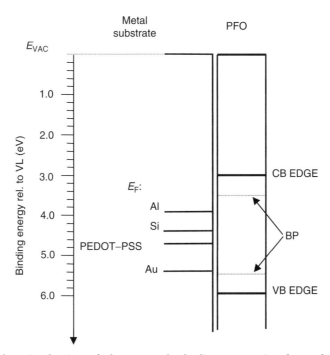

FIGURE 21.15 Schematic drawing of the energy level alignment at interfaces of PFO and different metallic substrates. In all cases, the Fermi level of the substrate (E_F) is situated between the negative and positive bipolaron levels.

charge transfer across the metal–polymer interface. First, there is a very low density of the extrinsic charge carriers in PFO, typically on the order of 10^{16} cm^{-3} [78,84,85]. Thus, charge transfer from the polymer to metal, if any, is very small, even if the Fermi energy of the metal should be situated below the impurity levels of PFO. On the other hand, within the model for metal–polymer interfaces based on the nondegenerate continuum model of Brazovskii and Kirova [86], the necessary condition for the charge transfer from the metal to polymer to occur requires the Fermi level of the metal to be above the negative bipolaron formation energy per particle (or below the positive bipolaron formation energy per particle). Furthermore, electronic charges are stored in PFO in the form of bipolarons [87]. The negative bipolaron formation energy per particle estimated from the alkali metal doping of PFO is ~2.4 eV, relative to the valence band edge [88]. This is, however, only a lower limit in the present case, since the presence of counterions (to supply electrons in doping) shifts the bipolaron energy levels deeper into the bandgap. For both PFO/Au and PFO/Al, the Fermi level of the metal is situated below the negative bipolaron formation energy per particle (or above the positive bipolaron formation energy per particle), thus charge transfer from the metal to polymer cannot occur [89]. Consequently, there is no electrostatic potential drop in a contact region due to this mechanism, and in the absence of other factors that could possibly influence the electrostatic potential at this interface (chemical bonding, etc.), alignment of E_{VAC} results. In the absence of significant charge transfer, the electrochemical potential in the polymer layer is determined by the work function of the substrate. It can be tuned within the limits given by the positive and negative bipolaron formation energy per particle. Beyond this limit, charge transfer from metal to polymer is allowed, and the Fermi energy is pinned at the value corresponding to the bipolaron formation energy [89].

21.5 Poly(3,4-Ethylenedioxythiophene)-Polystyrenesulfonate

Prepared using standard oxidative chemical or electrochemical polymerization methods, poly(3,4-ethylenedioxythiophene), or PEDOT (also denoted as PEDT), was initially found to be an insoluble polymer, yet exhibited some very interesting properties. In addition to a high electrical conductivity (~550 S cm^{-1}) [90,91] and metallic behavior [90,92–96], PEDOT was found to be almost transparent and highly stable in thin oxidized films [91,97–100]. The solubility problem was circumvented by resorting to a water-dispersible polyelectrolyte, poly(styrenesulfonic acid) (PSS), as the charge balancing counterion during polymerization, yielding PEDOT–PSS (Figure 21.16). This polymer blend displays good film-forming properties, high electrical conductivity (~10 S cm^{-1}), high transmission of visible light, and excellent chemical (environmental) stability [91]. Both PEDOT and PEDOT–PSS have found their way into commercial applications, not the least as antistatic coatings in photographic films [101,102]. The electronic and ionic (cation) conductivities [103] make PEDOT–PSS attractive for capacitors [101,102,104–106] and high-power Li batteries [107–109]. New applications for PEDOT–PSS are emerging in the field of organic-based (opto)electronics; in organic-based light emitting devices (OLEDs) [110–117], photovoltaic cells [118–121], and electrochromic displays [100,122–127].

In this section, several PES studies on selected properties of PEDOT–PSS that are relevant to certain applications in organic electronics are reviewed. The unique initial-state binding energies of the sulfur atoms in the PEDOT and the PSS portions of the polymer blend make it possible to distinguish PEDOT from PSS within PEDOT–PSS films. We focus on the analysis of the elemental composition, the PEDOT-to-PSS ratio, some aspects of the granular morphology, and some additional chemical information (for example, an estimate of the doping level).

21.5.1 XPS of S(2p) and O(1s) Core Levels of a Commercial PEDOT–PSS Blend

We first consider PEDOT–PSS, Baytron P obtained from Bayer AG (Germany), in the form of an aqueous dispersion, with an overall PEDOT-to-PSS molar ratio equal to 0.8 [128]. For sample preparation,

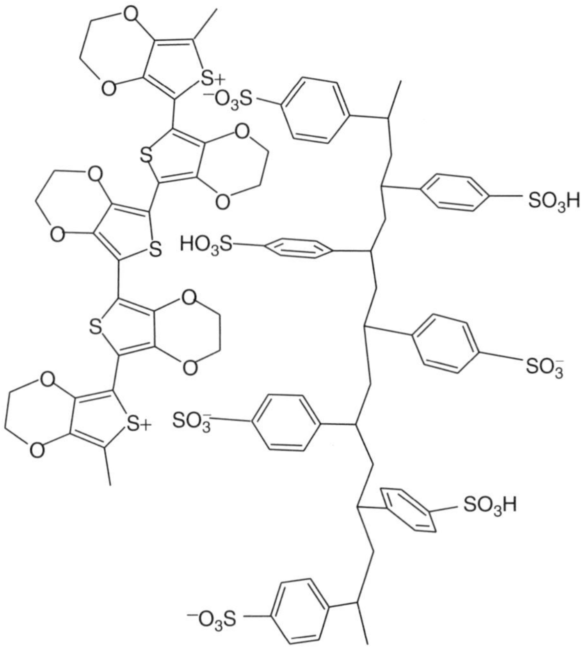

FIGURE 21.16 Sketch of the chemical structure of PEDOT–PSS with a doping level of ~0.33. The PEDOT-chain is on the left-hand side and the polyanion PSS on the right-hand side.

the PEDOT–PSS suspension was first filtered using a 1 μm glass-fiber filter in series with a 1 μm polysulfone-based filter, in order to remove possible lumps. The ultrathin films were spin-coated on optically flat Si(100) substrates. The resulting films had a thickness, estimated (DEKTAK instrument), in the range of 350–450 Å. Samples were heated in UHV at 120°C for ca. 2 h before the measurements in order to remove possible water contamination.

The S(2p) and O(1s) core-level spectra of a PEDOT–PSS film are shown in Figure 21.17a and Figure 21.17b, respectively [129,130]. S(2p) core-level features are composed of spin-split doublets, $S(2p_{1/2})$ and $S(2p_{3/2})$, of well known energy splitting (1.2 eV), intrinsic line width, and relative intensity ratio (1:2). Considering the composition of the material [128], the S(2p) spectrum of PEDOT–PSS is expected to have several different contributions from S-atoms in different chemical environments, with a dominant signal corresponding to undoped PEDOT sites and the PSSH part of PSS. The contribution due to the latter, containing strongly positively charged S-atoms since they are bonded to three oxygen atoms, should appear at considerably higher binding energy than the PEDOT S(2p)-signal. In fact, the complexity of the S(2p) spectra displayed in Figure 21.17a, implies that several sulfur-containing chemical species are present in the film. Although the deconvolution is not straightforward, the lower binding energy peaks, at 164.5 and 165.6 eV can be assigned to the sulfur atoms in PEDOT, on the basis of earlier PES work [131], and results of quantum chemical calculations, which suggest a stabilization of the S(2p) core level of PSSH by 5.8 eV with respect to that of PEDOT [129]. Although in the present case the energy splitting between the spin-split components is close to the ideal value, the intensity ratio is not correct. This is due to the presence of p-doped regions in the PEDOT polymer chain resulting in S(2p) contributions that are shifted towards higher binding energy with respect to the neutral sulfur sites. The positive charges are not localized on one monomer unit, but delocalized over several adjacent rings [132], resulting in a spread in the S(2p) binding energy values: the positive charge on a particular PEDOT ring depends on the distance from the PSS^- counterion. Hence, the charge distribution along a partly oxidized PEDOT-chain results in a broad spectrum of S(2p) binding energies, giving rise to an asymmetric tail on the higher binding-energy side of the (neutral) PEDOT sulfur signal.

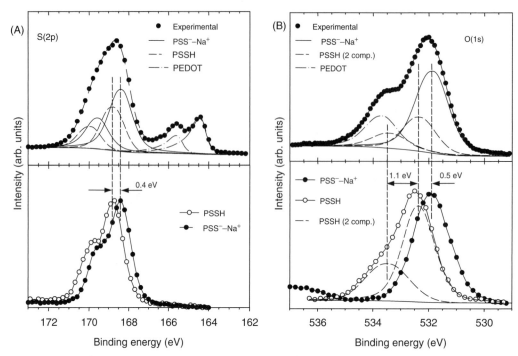

FIGURE 21.17 The S(2p) (a) and O(1s) (b) core-level spectra from (1) as-prepared films of PEDOT–PSS (*upper panels*), and (2) thin films of PSSH and PSS⁻–Na⁺ used as the reference (*bottom panels*).

The higher binding energy feature in the S(2p) spectrum, near 169 eV, can be assigned to the sulfur atoms in PSS. Since no spin-splitting is apparent, it is evident that also in this case sulfur occurs in a number of different chemical states, closely bunched in binding energy. According to the chemical structure, three chemically different sulfur atoms are expected, namely those in PSSH, PSS⁻–Na⁺, and PSS⁻, the latter of which are the counterions to PEDOT⁺. We assume for now that the PSSH and PSS⁻–Na⁺ species dominate the intensity of the spectral feature at 169 eV binding energy for the following reasons: (1) the expected PSSH-to-PSS⁻ ratio is 4:1; (2) the intensity of the PSS–Na⁺ contribution to the S(2p) signal, as estimated from the amount of Na-atoms detected by XPS, is comparable to that of PSSH and thus is considerably higher than that from PSS⁻; and finally (3) the intensity of the sulfur signal from PSS⁻ is expected to be significantly lower than that from PEDOT⁺ because just one PSS⁻ counterion can affect several sulfur sites in a polymer backbone of PEDOT. Since the signal from PEDOT⁺ sites is already quite weak, the contribution from PSS⁻ should consequently be negligible. Therefore, in the deconvolution procedure, the PSS⁻ part is omitted. This assumption is further verified below.

The bottom panel of Figure 21.17a shows the S(2p) spectra of reference films of PSSH and PSS⁻–Na⁺. Since the carbon atoms in both compounds are in almost identical chemical states, the C(1s) core-level signal serves as a good reference level for the alignment of S(2p) spectra from PSSH and PSS⁻–Na⁺. Such aligned spectra show a relative energy shift of 0.4 eV, with the PSSH part being stabilized with respect to PSS⁻–Na⁺. These data enable deconvolution of the PEDOT–PSS S(2p) feature near 169 eV into two spin-split doublets. The higher binding energy component with the S(2p$_{3/2}$) peak at 168.8 eV (dashed-line) is associated with PSSH, whereas the lower binding energy component with the S(2p$_{3/2}$) peak at 168.4 eV (solid line) corresponds to the PSS⁻–Na⁺ salt. The PSS⁻–Na⁺-to-PSSH molar ratio derived from the fit equals 1.4.

The O(1s) spectrum of a PEDOT–PSS film is shown in Figure 21.17b. Again, a number of different chemical species are expected to contribute to the core-level signal. In analogy to the S(2p) spectrum, a combination of the a priori knowledge of the chemical composition and the measurements performed on the reference compounds has been used in order to deconvolute the O(1s) spectrum. The bottom panel of Figure 21.17b shows the O(1s) spectra from the same set of reference films as used for the S(2p) spectrum (pure PSSH and PSS⁻–Na⁺). First of all, the O(1s) spectrum of PSSH contains two peak components, separated by about 1.1 eV, which are displayed with dashed lines in Figure 21.17b. The stronger, lower binding energy peak corresponding to PSSH, at 532.4 eV, originates from the oxygen atoms bound to the sulfur atoms in the sulfonic acid groups of PSSH. The small peak at 533.5 eV corresponds to the hydroxyl-oxygen atoms. The intensity ratio of 2:1 is in agreement with the chemical composition of PSSH. The higher binding energy component is broadened, corresponding to the formation of hydrogen bonds involving the hydroxyl groups. On the other hand, the oxygen peak in the O(1s) spectrum for PSS⁻–Na⁺ films is completely symmetric, implying that the sodium cation resides at a bridging position with respect to all three oxygen atoms in the sulfonate group.

When properly aligned using C(1s) data, the O(1s) peak from PSS⁻–Na⁺ is ~0.5 eV lower in binding energy than the stronger component of the PSSH signal (cf. bottom panel of Figure 21.17b). The energy splitting measured in the reference samples has been used for the deconvolution of the PEDOT–PSS spectrum in Figure 21.2b (the peak intensity ratio of 2:1 and the energy splitting of 1.1 eV between the two PSSH oxygen components were kept constant throughout the fitting procedure). Finally, the highest binding energy component in the O(1s) spectrum of PEDOT–PSS (dash-dotted line at 533.7 eV), necessary in order to obtain a good fit, is assigned to the oxygen atoms in the dioxyethylene bridge of PEDOT [129]. The relative intensities of the deconvoluted peaks indicate a PSS⁻–Na⁺-to-PSSH ratio of 1.6, in agreement with the values obtained from the deconvolution of the sulfur S(2p) and S(2s) spectra (1.4 and 1.3, respectively). No contributions from PEDOT⁺ and PSS⁻ sites are explicitly included in the deconvolution of the O(1s) spectrum. This is justified because the oxygen atoms reside in the non-conjugated dioxyethylene segments, and are therefore not influenced significantly by the positive charge delocalized along the PEDOT backbone. Additionally, the PSS⁻ contribution has not been considered here for the same reasons as in the case of S(2p) spectrum.

An important point is that the experimental PSS-to-PEDOT ratio in the surface region probed by XPS (including both the PSSH and the PSS⁻–Na⁺ components) is equal to 3.5, as determined by a quantitative analysis of the S(2p) core-level spectra. This is in contradiction to the expected ratio of 1.2, which is the PSS-to-PEDOT molar ratio in the blend [128]. Therefore, we conclude that phase segregation takes place, with an excess of PSS (both PSSH and PSS⁻–Na⁺) in the surface region. The issue will be discussed in more detail below.

21.5.2 Valence Band UPS

The UPS spectra of the valence band (VB) region of PEDOT–PSS again reflect the phase segregation in the blend. Figure 21.18 shows the He II (40.8 eV photons) UPS spectra of a HCl-treated PEDOT–PSS film, i.e., with the PSS⁻–Na⁺ component removed [130], and a reference PSSH film. Since the probing depth in UPS is much less than in XPS, the spectra reflect only the very outermost surface region (most of the signal comes from the first 10–15 Å [133]). Except for minor features at binding energies higher than 15 eV, the spectra are essentially identical. This implies that the surface of the PEDOT–PSS film is dominated by PSSH species, as was expected from the XPS measurements discussed above. Also shown in the figure is the PSSH DOVS generated from the output of quantum chemical calculations performed at the ab initio level [129,130].

Taking advantage of the high photoionization cross section for C(2p)-derived states in UPS (He I), one can in fact observe the density of states in the region close to the Fermi level (see Figure 21.19). The observation of a finite density of states at the Fermi level is a direct confirmation of the metallic character of this material. It must be emphasized here that the intensity of this signal is extremely low with respect to the rest of the spectral features in the VB region (by ~1/1000), due to phase segregation resulting in a low concentration of PEDOT within the near-surface region probed by UPS. The fact that the Fermi edge can be detected at all is entirely due to the lack of any PSSH-signal intensity in this binding energy range (2 eV binding energy).

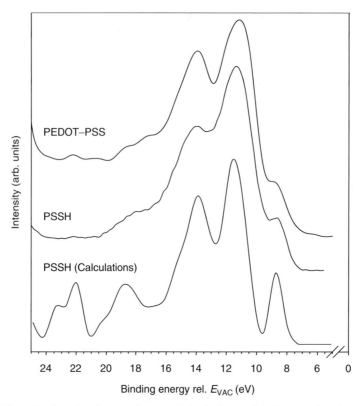

FIGURE 21.18 The UPS valence band spectra (He II radiation) of PSSH (*black dots*) and acid-treated PEDOT–PSS (*gray dots*). The *solid line* shows the calculated DOVS.

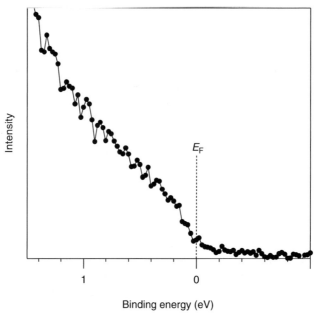

FIGURE 21.19 The Fermi edge of PEDOT–PSS observed by UPS (He I).

21.5.3 Electropolymerized vs. Chemically Prepared Systems

One of the problems associated with PEDOT–PSS is that the exact composition is generally difficult to determine. The blend that is obtained using poly(sodium-*p*-styrenesulfonate) (PSSNa), either chemically via Fe(III) salts [103] or electrochemically [134], has been investigated by means of energy dispersive x-ray (EDX) analyses, but the results were rather uncertain. The sulfonate/thiophene molar ratio $R_{S/T}$ was found to be in the range of 0.2–3 for samples made chemically [103] and ~0.2 for those made electrochemically [134]. The electrical conductivity appears to increase with a decrease of $R_{S/T}$, with typical values around 1 and 10 S cm^{-1} for $R_{S/T} = 2.5$ and 0.25, respectively [103].

Here, we consider electropolymerized 3,4-ethylenedioxythiophene (EDT), prepared with different supporting electrolytes (see [135]): polystyrenesulfonic acid (PSS), *p*-toluenesulfonic acid (Tos), and tetrabutylammonium perchlorate (Bu$_4$NClO$_4$). The anion produced from the dissociation of toluene-sulfonic acid is also called tosylate (ϕ-SO$_3^-$). Additionally, we address chemically prepared PEDOT–PSS, in a water emulsion, sodium free (<0.5 ppm), provided by Agfa Gevaert N.V. None of these blends contains PSS$^-$–Na$^+$, as was the case for Baytron P discussed above. The conductivity values σ obtained for the polymers are summarized in Table 21.1. PEDOT/Tos is the most conductive (450 S cm^{-1}). The polyanion-based materials give lower conductivities; 80 S cm^{-1} for electropolymerized PEDOT–PSS and 0.03 S cm^{-1} for chemically polymerized PEDOT–PSS.

Electrochemical- and UV-Vis spectroscopic analysis [135] of PEDOT/Tos and electropolymerized or chemically prepared PEDOT–PSS do not show any significant differences. As the electrochemical and spectroscopic properties, i.e., the electronic characteristics, are related to the conjugated backbone of PEDOT, differences may be expected to occur in the nonelectroactive anionic component of the polymers. This aspect of the polymer structure, i.e., the anion composition, is expected to have significant consequences for the conductivity, as reported below.

Figure 21.20a shows S(2p) spectra of pure PSS (S(2p$_{3/2}$) at 167.8 eV) and PEDOT/ClO$_4^-$ (S(2p$_{3/2}$) at 163.3 eV) for comparison with the spectra of the various blends. Note again the asymmetric tail on the higher binding-energy side of the PEDOT/ClO$_4^-$ S(2p)-feature, due to the positive charge stored along the doped PEDOT-chain, partially localized on the sulfur atoms [132]. The intensity ratio between the S(2p) XPS signals from PSS and PEDOT gives directly the $R_{S/T}$ ratio. Figure 21.20b shows the S(2p) spectrum of the chemically polymerized PEDOT–PSS dispersion, which clearly has the highest $R_{S/T}$ ratio (3.8). The fact that this blend also has the lowest conductivity (0.03 S cm^{-1}) suggests that the concentration of the excess insulating polymer PSS in the film is important. Figure 21.20c shows the S(2p) XPS spectrum of the electrochemically prepared PEDOT-PSS for which a much lower $R_{S/T}$ ratio (0.68) is found. Hence, the electrochemical PEDOT–PSS sample contains more PEDOT than PSS. Additionally, absorption spectroscopy measurements indicate that a single PSS chain interacts with many shorter PEDOT-chains [135]. It can thus be concluded that the average distance between adjacent PEDOT-chains is small, which favors hopping of charge between PEDOT-chains. Hence the electro-polymerized PEDOT–PSS can be seen as an organic salt, very compact and conductive. This explains the higher conductivity found for the electrochemical PEDOT–PSS (80 S cm^{-1}) as compared to the chemically prepared PEDOT–PSS (0.03 S cm^{-1}). The even higher conductivity (450 S cm^{-1}) that results when PSS is replaced by Tos is once again related to the $R_{S/T}$ ratio (0.28 for PEDOT/Tos). In this case, the organic layer is mainly composed of PEDOT, while the tosylate groups in the film are solely involved

TABLE 21.1 Amount (Molar Fraction) of the Three Different Sulfonate Groups and Their Sum ($R_{S/T}$) Relative to EDT Content in Polymers. Comparison Is Made with Conductivity (σ)

Polymer	SO$_3^-$PEDOT$^+$	SO$_3^-$Na$^+$	SO$_3^-$H$^+$	$R_{S/T}$	σ (S cm^{-1})
PEDOT–PSS[a]	0.30–0.35	—	3.5	3.8	0.03
PEDOT–PSS	0.35	—	0.33	0.68	80
PEDOT/Tos	0.22	—	0.06	0.28	450

[a]Chemically prepared.

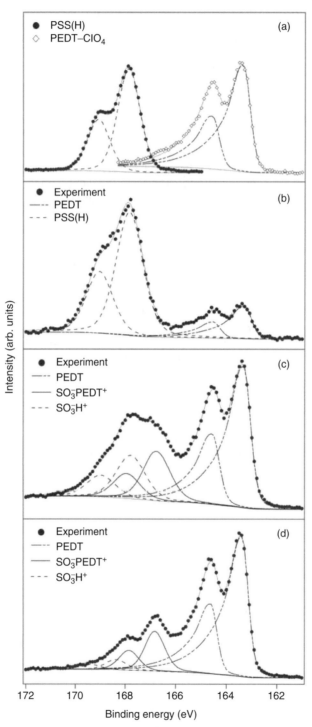

FIGURE 21.20 S(2p) signals from (a) PSS and PEDOT/ClO$_4$, (b) chemically prepared PEDOT–PSS, (c) electropolymerized PEDOT–PSS, and (d) electropolymerized PEDOT/Tos. To compare the binding energies of the core levels from the measurements performed on the conductive (PEDOT–PSS) and insulating PSS samples, the hydrocarbon C(1s) line is used as a common reference. For the conductive samples PEDOT–PSS, PEDOT/Tos, and PEDOT/ClO$_4$, the Fermi level is the reference level.

TABLE 21.2 Values of Binding Energy (in eV) of Various Components of the S(2p) Signal in Polymers (Shifts from PEDOT Peak in Round Brackets)

Polymer	PEDOT	SO_3^-PEDOT$^+$	SO_3^-H$^+$
PEDOT–PSS	163.3	166.8 (3.5)	167.8 (4.5)
PEDOT/Tos	163.4	166.8 (3.4)	168.3 (4.9)

in the charge neutralization of the positive charges carried by PEDOT. This is in agreement with the study by Aasmundtveit et al. [136] who demonstrated that PEDOT/Tos is an organic crystal composed of one tosylate anion for every four EDT units in the unit cell.

We now address the S(2p) spectrum of the electropolymerized PEDOT–PSS film (Figure 21.20c) in more detail. As was pointed out previously, the signal at higher binding energy (168 eV) arises from the presence of PSS in the film. This signal is composed of two types of sulfur atoms (only one S(2p) doublet with a 2:1 ratio cannot fit this spectrum), corresponding to the sulfonic acid group (SO_3^-H$^+$, doublet at 167.8 eV, dashed) and the sulfonate group (SO_3^-PEDOT$^+$, doublet at 166.8 eV, solid line), respectively. The S(2p) signal of SO_3^-PEDOT$^+$ appears at a lower binding energy than SO_3^-H$^+$ because of the different chemical environments and polarization energies of the hole produced on photoemission. Since peak deconvolution shows that there is an equal amount of SO_3^-PEDOT$^+$ and SO_3^-H$^+$ moieties, the ratio between the SO_3^-PEDOT$^+$ and the PEDOT signals (0.35) gives a direct estimation of the doping level. The results of the curve fitting procedure are summarized in Table 21.1 and Table 21.2. The S(2p) spectrum of PEDOT/Tos (Figure 21.20d) shows that the tosylate-related S(2p) feature (S(2p) doublet at 166.8 eV) appears at a binding energy similar to the SO_3^-PEDOT$^+$ component observed in the electropolymerized PEDOT–PSS. Thus for PEDOT/Tos, the tosylate contribution arises essentially from tosylate anions SO_3^-PEDOT$^+$ involved in the doping process. The doping level of PEDOT/Tos is estimated to be 0.22, equivalent to one charge per 4–5 EDT monomeric units.

On electropolymerization, the fundamental difference between a polymeric (PSS) counteranion and a small (Tos) counteranion lies in the presence of sulfonate groups in PSS that are not involved in the doping process, which is apparent from the high density of sulfonate groups in PSS compared to the density of positive charges carried by PEDOT. Therefore, while half of the sulfonate groups are used to neutralize the doping charge, the other half cannot be excluded from the film as they are attached to the PSS chain. In contrast, for PEDOT/Tos, only the amount necessary to neutralize the doping charge in PEDOT is incorporated into the film during the electropolymerization. Another difference between the blends containing monomeric vs. polymeric counterions is the doping level, which is 0.35 for PEDOT–PSS and 0.22 for PEDOT/Tos. Although the doping level is higher for PEDOT–PSS, this polymer is much less conductive than PEDOT/Tos. Thus, the reduced conductivity of PEDOT–PSS has to be explained by the higher average hopping distance, which appears to overwhelm the doping-level effect.

21.5.4 Surface Composition Revisited, Photon Energy Dependent XPS

As discussed, the electropolymerized PEDOT–PSS ($\sigma = 80$ S cm^{-1}, $R_{S/T}$ ratio $= 0.68$) has a completely different composition than the chemically polymerized PEDOT–PSS ($\sigma = 0.03$ S cm^{-1}, $R_{S/T}$ ratio $= 3.8$). Electrochemical quartz crystal microbalance (EQCM) analyses have shown that the composition of the electropolymerized PEDOT–PSS is not dependent on the anion concentration [135]. This indicates that the mechanism of synthesis of the polymer strongly influences its composition. During electropolymerization, the first formed (and doped) oligomers are very close to the metal electrode as charge transfer occurs at a tunnel distance. Consequently, those doped oligomers interact electrostatically with the closest sulfonate anions of a PSS chain at the vicinity of the electrode. This mechanism leads to a high concentration of PEDOT in the PEDOT–PSS film, independent on the anion concentration.

The chemically prepared PEDOT–PSS contains an excess of PSS, which is likely due to the micellar nature of the emulsion. During polymerization, EDT monomers are in a bath with an excess of PSS. An oxidizing agent is added and the solution is stirred. Consequently, the first doped oligomers that are formed are homogeneously distributed throughout the solution. A stable emulsion with particles or micelles is obtained upon vigorous stirring. An excess of PSS possibly surrounds those PEDOT–PSS particles. When the dispersion is coated onto a substrate, a polymer film is formed with a granular surface [129,137]. The origin of the grains may be related to the particles in the emulsion and/or a partial segregation of the PSS chains. The insulating PSS domains are responsible for the low conductivity as compared to the electropolymerized PEDOT–PSS.

A nondetrimental depth profile analysis of the PEDOT–PSS films obtained from the emulsion [130] was carried out by photon energy dependent XPS using synchrotron radiation (Max-Lab, Lund, Sweden). Changing the energy of incident photons alters the kinetic energy of a photoelectron from a given core level which, in turn, determines the particular value of the inelastic mean free path and thus allows one to chose the probing depth. The self-explanatory idea behind the experiment is sketched in Figure 21.21.

Variations in the $R_{S/T}$ ratio as a function of depth are clearly visible: as the intensity of the signal from the PEDOT increases (due to the increase in probing depth), the PSS part of the S(2p) spectrum decreases. This is a consequence of the fact that the surface region of PEDOT–PSS is dominated by PSS species. The details of changes in $R_{S/T}$ are summarized in Figure 21.22, where the latter quantity is plotted as a function of electron kinetic energy. From these data, the thickness of the surface-segregated PSS layer was estimated to be between 30 and 40 Å.

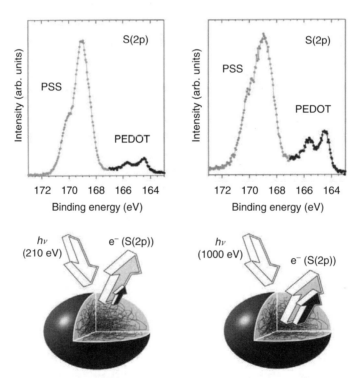

FIGURE 21.21 Tuning the photon energy allows for depth profiling in the PEDOT–PSS polymer blend. In the cross section of the PEDOT–PSS grain shown at the top of the figure, the dark areas correspond to PEDOT-rich regions and the light areas correspond to PSS-rich regions. The size and the color (light vs. dark) of the arrows representing photoelectrons ejected from the sample are correlated with the intensity of the signal (size) from a particular region within the grain (color), determined by the effective probing depth for a given photon energy.

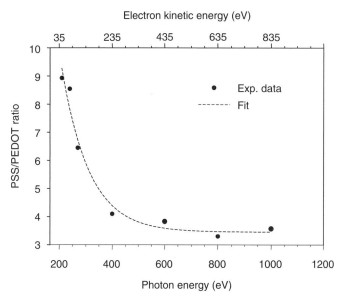

FIGURE 21.22 Evolution of $R_{S/T}$ vs. the energy of incident photons (*scale at the bottom*) and electron kinetic energy (*scale on top*).

21.6 Synopsis

The usefulness of PES for the characterization of the chemical and electronic structure of π-conjugated polymers and their interfaces with metallic substrates has been outlined. By employing a combined experimental–theoretical approach, the valence electronic structure of representative pristine and n-doped polymers has been discussed. The combination of experiment with theory supplies more information than either method can provide alone. By combining RPE again with theoretical modeling, additional information on the interplay between electronic and vibrational degrees of freedom in the formation of charge carriers and excitations in conjugated polymers is obtained.

The alignment of energy levels at interfaces between π-conjugated polymers and metallic substrates has been discussed. In case of pristine, nearly insulating PFO and poly(bis-(2-dimethyloctylsilyl)-1,4-phenylenevinylene), Schottky–Mott-like alignment occurs. This means that the position of the polymer energy levels changes strongly (by 1.2 eV for the cases considered here) with respect to the Fermi level as the substrate work function is changed. In contrast, for the *electrically conducting* polymer blend, PEDOT–PSS as well as for n-doped PFO, transport of free charge carriers across the interface leads to a common Fermi level and the formation of a contact potential difference or, in other words, an interfacial dipole.

A detailed PES study of the commercially highly successful polymer blend PEDOT–PSS has been presented. Following the assignment of the numerous features in the core-level XPS spectra of a commercial PEDOT–PSS blend, the differences in terms of sulfonate to thiophene ratio ($R_{S/T}$), doping level and conductivity of various blends prepared either electrochemically or chemically by Fe(III) salts were addressed. It turns out that $R_{S/T}$ determines the conductivity rather than the doping level, through an increase in the average hopping distance due to the incorporation of excess PSS. By means of photon energy dependent XPS using synchrotron radiation of chemically prepared (commercial) PEDOT–PSS films, the surface region was found to be dominated by PSS species. Such films have a granular surface, possibly related to the particles in the emulsion from which the films were prepared and/or a partial segregation of the PSS chains.

Acknowledgments

Research on conjugated polymers and molecules in Linköping is supported in general by grants from the Swedish Research Council and the Swedish Foundation for Strategic Research through the Center for Advanced Molecular Materials and the Center of Organic Electronics. The authors also like to acknowledge the important input from all of our other collaborators, whose names are omitted due to lack of enough space. However, we do acknowledge specifically the long-time theory and experimental collaborations with S. Stafström and coworkers in Linköping, J.L. Brédas et al. at the Georgia Institute of Technology, Atlanta, USA, and R. Lazzaroni and coworkers at the University of Mons-Hainaut, Mons, Belgium.

References

1. Siegbahn, K., C. Nordling, G. Johansson, J. Hedman, P.F. Heden, V. Hamrin, U. Gelius, T. Bergmark, L.O. Werme, R. Manne, and Y. Baer. 1971. *ESCA applied to free molecules*. Amsterdam: North-Holland.
2. Turner, D.W., C. Baker, A.D. Baker, and C.R. Brundle. 1970. *Molecular photoelectron spectroscopy*. London: Interscience.
3. Winick, H., and S. Doniach. 1980. *Synchrotron radiation research*. New York: Plenum.
4. Bachrach, R.Z. 1992. *Synchrotron radiation research: Advances in surface and interface science*, Vol. 1. New York: Plenum.
5. Briggs, D., and M.P. Seah. 1983. *Practical surface analysis*. Chichester: John Wiley & Sons.
6. Salaneck, W.R. 1985. *CRC Crit Rev Solid State Mater Sci* 12:267–296.
7. Salaneck, W.R., S. Stafström, and J.L. Brédas. 1996. *Conjugated polymer surfaces and interfaces*. Cambridge: Cambridge University Press.
8. Freund, H.-J., E.W. Plummer, W.R. Salaneck, and R.W. Bigelow. 1981. *J Chem Phys* 75:4275–4284.
9. Freund, H.-J., and R.W. Bigelow. 1987. *Phys Scripta* T17:50–63.
10. Fadley, C.S. 1978. *Basic concepts of x-ray photoelectron spectroscopy*. London: Academic Press.
11. Sato, N., K. Seki, and H. Inokuchi. 1981. *J Chem Soc Faraday Trans II* 77:1621.
12. Friedlein, R., X. Crispin, M. Pickholz, M. Keil, S. Stafstrom, and W.R. Salaneck. 2002. *Chem Phys Lett* 354:389–394.
13. Salaneck, W.R., C.B. Duke, W. Eberhardt, E.W. Plummer, and H.J. Freund. 1979. *Phys Rev Lett* 45:280–288.
14. Duke, C.B., W.R. Salaneck, T.J. Fabish, J.J. Ritsko, H.R. Thomas, and A. Paton. 1978. *Phys Rev B* 18:5717–5739.
15. Broughton, J.Q., and D.L. Perry. 1978. *Surf Sci Rep* 74:307–317.
16. Ashcroft, N.W., and N.D. Mermin. 1976. *Solid state physics*. Philadelphia: Saunders College.
17. Smoluchowski, R. 1941. *Phys Rev* 60:661.
18. Cahen D., and A. Kahn. 2003. *Adv Mater* 15:271–277.
19. Hagstrum, H.D. 1976. *Surf Sci* 54:197.
20. Ishii, H., E. Sugiyama, E. Ito, and K. Seki. 1999. *Adv Mater* 11:605–625.
21. Egelhoff, W.F., Jr. 1986. *Surf Sci Rep* 6:253–415.
22. Kugler, T., W.R. Salaneck, H. Rost, and A.B. Holmes. 1999. *Chem Phys Lett* 310:391–396.
23. Greczynski, G., T. Kugler, and W.R. Salaneck. 2001. *J Appl Phys* 88:7187–7191.
24. Greczynski, G., and W.R. Salaneck. 2001. *Appl Phys Lett* 79:3185.
25. Hill, I.G., A. Rajagopal, A. Kahn, and Y. Hu. 1998. *Appl Phys Lett* 73:662–664.
26. Seki, K., E. Ito, and H. Ishii. 1997. *Synth Met* 91:137–142.
27. Koch, N., C. Chan, A. Kahn, and J. Schwartz. 2003. *Phys Rev B Condens Mat Mater Phys* 67:195330–195331.
28. Sundin, S., F.K. Gel'mukhanov, H. Agren, S.J. Osborne, A. Kikas, O. Bjorneholm, A. Ausmees, and S. Svensson. 1997. *Phys Rev Lett* 79:1451–1454.

29. Friedlein, R., S.L. Sorensen, A. Baev, F. Gel'mukhanov, J. Birgerson, A. Crispin, M.P. de Jong, W. Osikowicz, C. Murphy, H. Ågren, and W.R. Salaneck. 2004. *Phys Rev B* 69:125204.

30. Stöhr, J. 1992. *NEXAFS spectroscopy*, Vol. 25. Berlin: Springer.

31. Sorensen, S.L., S.J. Osborne, A. Ausmees, A. Kikas, N. Correia, S. Svensson, A. Naves de Brito, P. Persson, and S. Lunell. 1996. *J Chem Phys* 105:10719–10724.

32. Luo, Y., H. Ågren, J. Guo, P. Skytt, N. Wassdahl, and J. Nordgren. 1995. *Phys Rev A* 52:3730–3736.

33. Gel'mukhanov, F., and H. Ågren. 1999. *Phys Rep* 312:87–330.

34. Gel'mukhanov, F., and H. Ågren. 1994. *Phys Rev A Atom Mol Opt Phys* 49:4378.

35. Gel'mukhanov, F., and H. Ågren. 1994. *Phys Lett A* 193:375.

36. Aksela, S., E. Kukk, H. Aksela, and S. Svensson. 1995. *Phys Rev Lett* 74:2917.

37. Gel'mukhanov, F., and H. Ågren. 1996. *Phys Rev A* 54:3960.

38. Braicovich, L., M. Taguchi, F. Borgatti, G. Ghiringhelli, A. Tagliaferri, N.B. Brookes, T. Uozumi, and A. Kotani. 2001. *Phys Rev B (Cond Matter Mater Phys)* 63:245115.

39. Gel'mukhanov, F., A. Baev, Y. Luo, and H. Ågren. 2001. *Chem Phys Lett* 346:437.

40. Gortel, Z.W., and D. Menzel. 2001. *Phys Rev B* 64:115416.

41. Brühwiler, P.A., O. Karis, and N. Martensson. 2002. *Rev Mod Phys* 74:703–740.

42. Svensson, S., and A. Ausmees. 1997. *Appl Phys A Mater Sci Process* 65:107.

43. Brühwiler, P.A., A.J. Maxwell, C. Puglia, A. Nilsson, S. Andersson, and N. Mårtensson. 1995. *Phys Rev Lett* 74:614.

44. Brühwiler, P.A., A.J. Maxwell, A. Nilsson, R.L. Whetten, and N. Martensson. 1992. *Chem Phys Lett* 193:311–316.

45. Brühwiler, P.A., A.J. Maxwell, P. Rudolf, C.D. Gutleben, B. Wastberg, and N. Martensson. 1993. *Phys Rev Lett* 71:3721–3724.

46. Menzel, D., G. Rocker, H.-P. Steinruck, D. Coulman, P.A. Heimann, W. Huber, P. Zebisch, and D.R. Lloyd. 1992. *J Chem Phys* 96:1724–1734.

47. Eberhardt, W., R. Dudde, M.L.M. Rocco, E.E. Koch, and S. Bernstorff. 1990. *J Electron Spectrosc Relat Phenom* 51:373–382.

48. Lögdlund, M., W.R. Salaneck, F. Meyers, J.L. Brédas, G.A. Arbuckle, R.H. Friend, A.B. Holmes, and G. Froyer. 1993. *Macromolecules* 26:3815–3820.

49. Brédas, J.L. 1997. *Synth Met* 84:3–10.

50. Salaneck, W.R., and J.L. Brédas. 1994. *Organic materials for electronics*. Amsterdam: North-Holland, p. 15.

51. Cornil, J., S. Vanderdonckt, R. Lazzaroni, D. dos Santos, G. Thys, H. Geise, L.-M. Yu, M. Szablewski, D. Bloor, M. Lögdlund, W. Salaneck, N. Gruhn, D. Lichtenberger, P. Lee, N. Armstrong, and J. Brédas. 1999. *Chem Mater* 11:2436–2443.

52. Dewar, M.J.S., and W. Thiel. 1977. *J Am Chem Soc* 99:4899.

53. Dewar, M.J.S., and W. Thiel. 1977. *J Am Chem Soc* 99:4907.

54. Dewar, M.J.S., E.G. Zoebisch, E.F. Healy, and J.J.P. Stewart. 1985. *J Am Chem Soc* 107:3902.

55. Brédas, J.L., R.R. Chance, R. Silbey, G. Nicolas, and P. Durand. 1981. *J Chem Phys* 75:255–267.

56. André, J.M., J. Delhalle, and J.L. Brédas. 1991. *World scientific lecture and course notes in chemistry*, Vol. 2. Singapore: World Scientific.

57. Hohenberg, P., and W. Kohn. 1964. *Phys Rev* 136B:864.

58. Kohn, W., and L.J. Sham. 1965. *Phys Rev* 140A:1133.

59. Brédas, J.L., and W.R. Salaneck. 1986. *J Chem Phys* 85:2219.

60. Orti, E., and J.-L. Bredas. 1992. *J Am Chem Soc* 114:8669.

61. Brédas, J.L., and W.R. Salaneck. 1995. *Organic electroluminescence*, eds. D.D.C. Bradley, and T. Tsutsui. Cambridge: Cambridge University Press.

62. Yannoni, C.S., and T.C. Clarke. 1983. *Phys Rev Lett* 51:1191–1193.

63. Mao, G., J.E. Fischer, F.E. Karasz, and M.J. Winokur. 1993. *J Chem Phys* 98:712.

64. Johnstone, R.A.W., and F.A. Mellon. 1973. *J Chem Soc Faraday Trans II* 69:1155.

65. Maier, J.P., and D.W. Turner. 1973. *J Chem Soc Faraday Trans II* 69:196.

66. Rabalais, J.W., and R.J. Colton. 1972/1973. *J Electron Spectr Rel Phenom* 1:83.
67. Fahlman, M., D. Beljonne, M. Lögdlund, R.H. Friend, A.B. Holmes, J.L. Brédas, and W.R. Salaneck. 1993. *Chem Phys Lett* 214:327–332.
68. Crispin, A., X. Crispin, M. Fahlman, D.A. Dos Santos, J. Cornil, N. Johansson, J. Bauer, F. Weissortel, J. Salbeck, J.L. Bredas, and W.R. Salaneck. 2002. *J Chem Phys* 116:8159–8167.
69. Fahlman, M., P. Bröms, D.A. dos Santos, S.C. Moratti, N. Johansson, K. Xing, R.H. Friend, A.B. Holmes, J.L. Brédas, and W.R. Salaneck. 1995. *J Chem Phys* 102:8167–8174.
70. Kikuma, J., and B.P. Tonner. 1996. *J Electr Spec Rel Phenom* 82:41.
71. Ishii, H., K. Sugiyama, and K. Seki. 1997. Interfacial electronic structures of organic/metal interfaces studied by UV photoemission. *SPIE97 Proc.* (San Diego), pp. 228–237.
72. Rajagopal, A., and A. Kahn. 1998. *J Appl Phys* 84:355–358.
73. Salaneck, W.R., M. Logdlund, M. Fahlman, G. Greczynski, and T. Kugler. 2001. *Mater Sci Eng R Rep* 34:26.
74. Crispin, X., V. Geskin, A. Crispin, J. Cornil, R. Lazzaroni, W.R. Salaneck, and J.-L. Brédas. 2002. *J Am Chem Soc* 124:8131–8141.
75. Seki, K., and H. Ishii. 1998. *J Electron Spectrosc Relat Phenom* 88–91:821–830.
76. Sugiyama, K., D. Yoshimura, T. Miyamae, T. Miyazaki, H. Ishii, Y. Ouchi, and K. Seki. 1998. *J Appl Phys* 83:4928–4238.
77. Wu, C.I., Y. Hirose, H. Sirringhaus, and A. Kahn. 1997. *Chem Phys Lett* 272:43–47.
78. Greczynski, G., M. Fahlman, and W.R. Salaneck. 2000. *Chem Phys Lett* 321:379–384.
79. Rhoderick, E.H., and R.H. Williams. 1988. *Metal-semiconductor contacts.* Oxford: Clarendon Press.
80. Krozer, A., and M. Rodahl. 1997. *J Vac Sci Technol A Vacuum Surf Films* 15:1704–1709.
81. Heeger, A.J., I.D. Parker, and Y. Yang. 1994. *Synth Met* 67:23–29.
82. Lögdlund, M., P. Dannetun, S. Stafström, W.R. Salaneck, M.G. Ramsey, C.W. Spangler, C. Fredriksson, and J.L. Brédas. 1993. *Phys Rev Lett* 70:970–973.
83. Andre, J.M., J. Delhalle, and J.L. Brédas. 1991. *Quantum chemistry aided design of organic polymers.* Singapore: World Scientific.
84. Greenham, N.C., and R.H. Friend. 1994. *Solid State Phys* 49:1–149.
85. Schott, M. 1994. *Organic conductors, fundamentals and applications*, ed. J.-P. Farges. New York: Marcel Dekker, pp. 539–646.
86. Brazovskii, S.A., and N.N. Kirova. 1981. *JEPT Lett* 33:4.
87. Salaneck, W.R., R.H. Friend, and J.L. Brédas. 1999. *Phys Rep* 319:231–251.
88. Greczynski, G., M. Fahlman, and W.R. Salaneck. 2000. *J Chem Phys* 113:2407–2412.
89. Davids, P.S., A. Saxena, and D.L. Smith. 1995. *J Appl Phys* 78:4244.
90. Pettersson, L.A.A., F. Carlsson, O. Inganäs, and H. Arwin. 1998. *Thin Solid Films* 313–314:356–361.
91. Groenendaal, L., F. Jonas, D. Freitag, H. Pielartzik, and J.R. Reynolds. 2000. *Adv Mater* 12:481–494.
92. Kiebooms, R., A. Aleshin, K. Hutchison, and F. Wudl. 1997. *J Phys Chem B* 101:11037–11039.
93. Aleshin, A.N., R. Kiebooms, H. Yu, M. Levin, and I. Shlimak. 1998. *Synth Met* 94:157–159.
94. Aleshin, A.N., R. Kiebooms, and A.J. Heeger. 1999. *Synth Met* 101:369–370.
95. Aleshin, A., R. Kiebooms, R. Menon, F. Wudl, and A.J. Heeger. 1997. *Phys Rev B* 56:3659–3663.
96. Chang, Y., K. Lee, R. Kiebooms, A. Aleshin, and A.J. Heeger. 1999. *Synth Met* 105:203–206.
97. Heywang, G., and F. Jonas. 1992. *Adv Mater* 4:116.
98. Dietrich, M., J. Heinze, G. Heywang, and F.J. Jonas. 1994. *J Electroanal Chem* 369:87.
99. Winter, I., C. Reese, J. Hormes, G. Heywang, and F. Jonas. 1995. *Chem Phys* 194:207–213.
100. Pei, Q., G. Zuccarello, M. Ahlskog, and O. Inganäs. 1994. *Polymer* 35:1347–1351.
101. Jonas, F., and G. Heywang. 1994. *Electrochim Acta* 39:1345.
102. Jonas, F., and J.T. Morrison. 1997. *Synth Met* 85:1397–1398.
103. Lefebvre, M., Z. Qi, D. Rana, and P.G. Pickup. 1999. *Chem Mater* 11:262.
104. Ghosh, S., and O. Inganäs. 1999. *Adv Mater* 11:1214–1218.
105. Ghosh, S., and O. Inganäs. 2000. *J Electrochem Soc* 147:1872–1877.
106. Kudoh, Y., K. Akami, and Y. Matsuya. 1999. *Synth Met* 102:973–974.

107. Noel, V., H. Randriamahazaka, and C. Chevrot. 2000. *J Electroanal Chem* 489:46–54.
108. Arbizzani, C., M. Mastragostino, and M. Rossi. 2002. *Electrochem Commun* 4:545–549.
109. Murugan, A.V., C.-W. Kwon, G. Campet, T. Maddanimath, B.B. Kale, and K. Vijayamohanan. 2002. *J Power Sources* 105:1–5.
110. Karg, S., J.C. Scott, J.R. Salem, and M. Angelopoulos. 1996. *Synth Met* 80:111–117.
111. Carter, J.C., I. Grizzi, S.K. Heeks, D.J. Lacey, S.G. Latham, P.G. May, O.R. de los Panos, K. Pichler, C.R. Towns, and H.F. Wittmann. 1997. *Appl Phys Lett* 71:34–36.
112. Carter, S.A., M. Angelopoulos, S. Karg, P.J. Brock, and J.C. Scott. 1997. *Appl Phys Lett* 70:2067–2069.
113. Berntsen, A.J.M., P. Van de Weijer, Y. Croonen, C.T.H.F. Liedenbaum, and J.J.M. Vleggaar. 1998. *Philips J Res* 51:511–525.
114. Kim, J.S., P.K.H. Ho, D.S. Thomas, R.H. Friend, F. Cacialli, G.-W. Bao, and S.F.Y. Li. 1999. *Chem Phys Lett* 315:307–312.
115. Brown, T.M., J.S. Kim, R.H. Friend, F. Cacialli, R. Daik, and W.J. Feast. 1999. *Appl Phys Lett* 75:1679–1681.
116. Arias, A.C., M. Granstrom, K. Petritsch, and R.H. Friend. 1999. *Synth Met* 102:953–954.
117. Ouyang, J., T.-F. Guo, Y. Yang, H. Higuchi, M. Yoshioka, and T. Nagatsuka. 2002. *Adv Mater* 14:915–918.
118. Roman, L.S., W. Mammo, L.A.A. Pettersson, M.R. Andersson, and O. Inganäs. 1998. *Adv Mater* 10:774–777.
119. Ding, L., M. Jonforsen, L.S. Roman, M.R. Andersson, and O. Inganäs. 2000. *Synth Met* 110:133–140.
120. Dhanabalan, A., J.K.J. van Duren, P.A. van Hal, L.J. van Dongen, and R.A.J. Janssen. 2001. *Adv Funct Mater* 11:255–262.
121. Huynh, W.U., J.J. Dittmer, and A.P. Alivisatos. *Science* 295:2425–2427.
122. Gustafsson, J.C., B. Liedberg, and O. Inganäs. 1994. *Solid State Ionics* 69:145–152.
123. Sankaran, B., and J.R. Reynolds. 1997. *Macromolecules* 30:2582–2588.
124. Sapp, S.A., G.A. Sotzing, and J.R. Reynolds. 1998. *Chem Mater* 10:2101.
125. DeLongchamp, D., and P.T. Hammond. 2001. *Adv Mater* 13:1455–1459.
126. Heuer, H.W., R. Wehrmann, and S. Kirchmeyer. 2002. *Adv Funct Mater* 12:89–94.
127. Cutler, C.A., M. Bouguettaya, and J.R. Reynolds. 2002. *Adv Mater* 14:684–688.
128. Bayer 1997. *Baytron P, product information.* Bayer, Germany.
129. Greczynski, G., T. Kugler, and W.R. Salaneck. 1999. *Thin Solid Films* 354:129–135.
130. Greczynski, G., T. Kugler, M. Keil, W. Osikowicz, M. Fahlman, and W.R. Salaneck. 2001. *J Electron Spectrosc Relat Phenom* 121:1–17.
131. Xing, K.Z., M. Fahlman, X.W. Chen, O. Inganäs, and W.R. Salaneck. 1997. *Synth Met* 89:161–165.
132. Dkhissi, A., D. Beljonne, R. Lazzaroni, F. Louwet, L. Groenendaal, and J.L. Brédas. 2003. *Int J Quant Chem* 91:517–523.
133. Nielsen, P., D.J. Sandman, and A.J. Epstein. 1975. *Solid State Commun* 17:1067.
134. Li, G., and P.G. Pickup. 2000. *PCCP* 2:1255.
135. Zotti, G., S. Zecchin, G. Schiavon, F. Louwet, L. Groenendaal, X. Crispin, W. Osikowicz, W. Salaneck, and M. Fahlman. 2003. *Macromolecules* 36:3337–3344.
136. Aasmundtveit, K.E., E.J. Samuelsen, L.A.A. Pettersson, O. Inganäs, T. Johansson, and R. Feidenhans'l. 1999. *Synth Met* 101:561–564.
137. Ghosh, S., J. Rasmusson, and O. Inganäs. 1998. *Adv Mater* 10:1097–1099.

22

Ultrafast Exciton Dynamics and Laser Action in π-Conjugated Semiconductors

Z. Valy Vardeny and
O. Korovyanko

22.1 Introduction

22.1.1 Basic Properties of π-Conjugated Polymers

The π-conjugated polymer systems have been intensively studied during the last 25 years. They form a new class of semiconductor electronic materials with potential applications such as organic light-emitting diodes (OLEDs) [1–3], thin-film transistors (TFTs) [4], photovoltaic cells [5], spin-valve devices [6], and optical switches and modulators [7]. As polymers, these organic semiconductors have a highly anisotropic quasi-one-dimensional (1D) electronic structure that is fundamentally different from the structures of conventional inorganic semiconductors. This has two consequences: (1) their chainlike structure leads to strong coupling of the electronic states to conformational excitations peculiar to the 1D system [8] and (2) the relatively weak interchain binding allows diffusion of dopant molecules into the structure (between chains), whereas the strong intrachain carbon–carbon bond maintains the integrity of the polymer [8]. In their neutral form, these polymers are semiconductors with an energy gap of ≈ 2 eV. However, most of the polymers can be easily doped with various p- and n-type dopants, increasing their conductivity by many orders of magnitude; conductivities in the range of $\approx 10^3$–10^4 S/cm are not unusual [9]. The ability to dope these organic semiconductors to metallic conductivities has attracted the Nobel Prize in Chemistry to Alan Heeger, Alan McDiarmid, and Hideki Shirakawa in 2000.

The simplest example of the class of conducting polymers is polyacetylene $(CH)_x$, which is depicted in Figure 22.1. It consists of weakly coupled chains of CH units forming a pseudo-1D lattice. The stable isomer is *trans*-$(CH)_x$, in which the chain has a zigzag geometry; the *cis*-$(CH)_x$ isomer, in which the chain has a backbone geometry, is unstable at room temperature or under high illumination. Simple conducting polymers like polyacetylene are planar, with three of the four carbon valence electrons forming sp^2 hybrid orbitals (σ bonds), while the fourth valence electron is in a π orbital perpendicular to the plain of the chain. The σ bonds are the building blocks of the chain skeleton and are thus responsible for the strong elastic force constant of the chain. The π orbitals form the highest occupied molecular orbitals (HOMO) and the lowest unoccupied molecular orbitals (LUMO), which together span an energy range of ≈ 10 eV [8]. *trans*-$(CH)_x$ is a semiconductor with a gap $E_g \approx 1.5$ eV, which has two equivalent lowest energy states having two distinct conjugated structures [8]. Other polymers shown in Figure 22.1, e.g., *cis*-$(CH)_x$, polythiophene (PT), poly(p-phenylene vinylene) (PPV), and polyfluorene

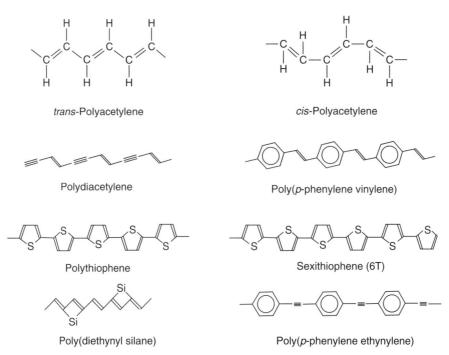

FIGURE 22.1 Backbone structure of some π-conjugated polymers and oligomers. (From Vardeny, Z.V. and Wei, X., *Handbook of conducting polymers*, 2nd ed., eds. T.A. Skotheim, R.L. Elsenbaumer, and J. Reynolds, Marcel Dekker, New York, 1998. With permission.)

(PFO), have a nondegenerate ground state structure, and, in contrast to $t\text{-}(CH)_x$ they are highly luminescent.

The properties and dynamics of optical excitations in π-conjugated polymers are of fundamental interest because they play an important role in the potential applications. However, in spite of intense studies of the linear and nonlinear optical properties, the basic model for the proper description of the electronic excitations in conducting polymers is still somewhat controversial. 1D semiconductor models [10], in which electron–electron (e–e) interaction has been ignored, have been successfully applied to interpret a variety of optical experiments in π-conjugated polymers [8]. In these models, the strong electron–phonon (e–p) interaction leads to rapid self-localization of the charged excitations, "polaronic effect." Then optical excitations across E_g, which is now the Peierls gap, are entirely different from the electron–hole pairs of conventional semiconductors. Instead, the proper description of the quasiparticles in $trans\text{-}(CH)_x$ is that 1D domain walls or solitons (S) separate the two degenerate ground-state structures [10]. As a result of their translational invariance, solitons in $trans\text{-}(CH)_x$ are thought to play the role of energy- and charge-carrying excitations. Since the soliton is a topological defect, it can be either created in soliton–antisoliton (S–S) pairs or be created in polyacetylene chains with odd numbers of CH monomers upon isomerization from $cis\text{-}(CH)_x$. The same Hamiltonian that has predicted soliton excitations in $trans\text{-}(CH)_x$ predicts polarons as a distinct solution when a single electron is added to the *trans* chain [8]. For the nondegenerate ground-state (NDGS) polymers mentioned above, adding an extrinsic gap component to the electron–phonon Hamiltonian results in polarons and bipolarons as the proper descriptions of their primary charge excitations. Singlet and triplet excitons have been also shown to play a crucial role in the photophysics of π-conjugated polymers [11–16]. However, their existence in theoretical studies can be justified only when electron–hole interaction and electron correlation effects are added to the Hamiltonian [17,18]. Recently, two-dimensional (2D)-type charge excitations, or 2D polarons have been suggested to explain the properties of charge carriers in planar polymers and polymer that form lamellae such as regioregular poly-hexyl thiophene (P3HT) [19,20]. An exceptional success is the discovery of photoinduced charge transfer from the π-conjugated polymer chain onto C_{60} molecules that are blended together [21,22]. In this case an ultrafast charge transfer dissociates the photogenerated exciton in the π-conjugated chain forming polaron pairs (PPs) where one polaron resides on the chain whereas the other polaron charges the C_{60} molecule. Section 22.1.2.2 and Section 22.1.2.3 deal with the variety of short-lived charged and neutral photoexcitations in doped and undoped π-conjugated polymers and thus a brief summary of their physical properties is in order.

For the description of the intramolecular excitations (and vibrations) the symmetry of the single polymer molecule usually dictates the wavefunction nomenclatures. Most π-conjugated polymers belong to the C_{2h} point group symmetry, which has a center of symmetry [8]. The properties of odd and even wavefunctions are then determined by the properties of the C_{2h} group, irreducible representations of which is given in Table 22.1. Notations such as $1B_u$ and mA_g are reserved for the intrachain excitons, where the index before the nomenclature denotes the order within the exciton band, and the subscripts g and u stand for even (gerade) and odd (ungerade) parity. Singlet excitons are distinguished from triplet exciton by the number on the left; e.g., 1^1B_u stand for singlet states, whereas 1^3B_u denotes triplet states. The dipole moment component in the direction of the chain is the strongest; its irreducible representation is B_u. Therefore, optical transitions involving B_u and A_g pair of states are symmetry allowed. On the contrary dipole transitions involving two A_g states are forbidden. However, a two-photon process allows the optical transition between two A_g states, since its irreducible representation is also A_g.

TABLE 22.1 Character Table of the Group C_{2h}

C_{2h}	E	C_2	i	σ_x
A_g	1	1	1	1
B_g	1	−1	1	−1
A_u	1	1	−1	−1
B_u	1	−1	−1	1

The e–e interaction is important even in the simplest example of *trans*-$(CH)_x$, and therefore it is more comfortable to use the "exciton (or correlated) notation" for the various excited states of π-conjugated polymers. In this notation the ground state is $1A_g$; the excited states are either even symmetry excitons, mA_g, or odd symmetry excitons, kB_u [23,24]. The importance of the e–e interaction in *trans*-$(CH)_x$ can be concluded from the fact that the $2A_g$ excited state (determined experimentally [25,26] by two-photon absorption [TPA]) is located below the first optically allowed excited state exciton, the $1B_u$. The e–e interaction cannot be ignored in any π-conjugated polymer, and it has significant effects on various optical properties such as photoluminescence (PL), electroabsorption (EA), and third-order optical susceptibilities. In cases where the bond alternation is relatively small, the ordering of the odd and even symmetry lowest excited states is $E(2A_g) < E(1B_u)$ [17,26]. When the "effective" bond alternation is relatively large, the ordering of these states is reversed, resulting in strong PL emission band. An excellent example is the poly(diphenyl acetylene) (PDPA), where the polymer chain has a *t*-$(CH)_x$ backbone structure, and also contains bulky side groups. It has been shown [27,28] that even this polymer has a degenerate ground state structure that supports soliton excitations; nevertheless, it is highly luminescent since, unlike *t*-$(CH)_x$ the large bond alternation in PDPA pushes the $2A_g$ exciton to be above the $1B_u$ exciton and this allows PL emission to appear. Also in PPV-type polymers, as another example, the benzene ring in the backbone structure gives rise to a large effective bond alternation for the extended π electrons [17] and therefore leads to high PL efficiency and improved OLED devices. It is thus noteworthy to realize that the Coulomb interaction among the π electrons, even when it is not dominant, leads to behavior qualitatively different from the prediction of single-particle Hückel model [23,24].

The soliton excitation in *t*-$(CH)_x$ is an amphoteric defect that can accommodate zero, one, or two electrons [8,29,30]. The neutral soliton (S^0, spin $\frac{1}{2}$) has one electron; positively and negatively charged solitons ($S^\pm$, spin 0) have zero or two electrons, respectively. Within the framework of the Su–Schrieffer–Heeger (SSH) model [10] Hamiltonian, which contains electron–phonon (e–p) inter-actions, but does not contain e–e or 3D interactions, it has been shown that a photoexcited electron–hole (e–h) pair is unstable toward the formation of a soliton–antisoliton (S–S) pair [31]. Subsequently, it was demonstrated [32] that as a consequence of the Pauli principle and charge conjugation symmetry in *t*-$(CH)_x$, the photogenerated soliton and antisoliton excitations are oppositely charged. The study of photoexcited *t*-$(CH)_x$, however, has revealed several unexpected phenomena, which were not predicted by the SSH model of the soliton (for a review see [18,25]). Most important is that an overall neutral state as well as charged excitation has been observed; this neutral state has been correlated with S^0 transitions [33]. This finding together with the absence (in undoped *trans*-$(CH)_x$) of optical transitions at the midgap level, where transitions of neutral and charged solitons should have appeared according to the SSH picture, has shown that e–e interaction in *trans*-$(CH)_x$ cannot be ignored. In this case the electronic gap in *trans*-$(CH)_x$ is partially due to the electron correlations rather than entirely due to electron–phonon interaction as in the SSH model. Under these circumstances, the nature of the photoexcitations in *trans*-$(CH)_x$ may be very different from that predicted by the SSH Hamiltonian. These findings have stimulated photophysical research in all π-conjugated polymers, mostly the NDGS polymers. The previous picture of photogeneration of bound soliton–antisoliton pairs in NDGS polymers has been modified. On the contrary, photogeneration of singlet excitons [15] and consequently triplet excitons [34] and/or PPs [35–38] has been demonstrated in many such π-conjugated polymers.

The origin of the branching process that determines the relative photoproduction of neutral vs. charged photoexcitations in the class of π-conjugated polymers is not very well understood. One possible explanation of the branching process is offered by the Onsager theory, which has successfully explained charge photoproduction in disordered materials and in molecular crystals [39]. The difficulty with this approach is that π-conjugated polymers are quasi-1D semiconductors for which the Onsager theory based on the e–h Coulomb attraction may not be applicable. In addition, the application of this theory for 1D semiconductors results in negligible quantum efficiency for charge photoproduction under weak electric fields, contrary to some experimental results. To solve this problem, it was suggested that the 1D–3D interplay is important in the photophysics of π-conjugated polymers [40]. In the proposed model, *intrachain* excitation results in a neutral state, which is an exciton with relatively large binding energy, whereas *interchain* excitation may produce separate charges on the neighboring chains, or PP if the interchain geminate recombination is overcome. A demonstration [41] of the important role of interchain excitation was that long-lived charged excitations in oriented films are more efficiently photogenerated with light polarized perpendicular to the polymer chain direction than paralled to it. On the other hand, the demonstration that charged polarons can be photogenerated in isolated PT and PPV chains [42] in both solutions and solid forms may challenge the common view of the unique importance of 3D interaction for charge photoproduction in π-conjugated polymers. In any case the instantaneous photogeneration of charged polarons in π-conjugated polymer chains with absorption bands in resonance with the PL band may preclude laser action to occur [35–39], which requires that the stimulated emission (SE) cross section is larger than the excited state absorption cross section.

In recent years, it has been recognized that many characteristics of the excited states of π-conjugated polymers are in essence very similar to those of the corresponding finite length oligomers. *trans*-β-carotene, e.g., is a π-conjugated molecule that consists of a backbone of 11 double bonds, similar to *trans*-$(CH)_x$, which has fixed length of about 22 carbons. Upon doping, it was observed that the charge is stored in a spinless stable configuration accompanied by structural relaxation [43]. Another example is the thiophene oligomers. Here, several oligomers ranging from bis-thiophene through sexithiophene [44,45] to octa- and deca-thiophene were studied [46]. At low doping levels, the radical ions of all the oligomers show the characteristics of polarons (i.e., charged defect, with spin $\frac{1}{2}$ accompanied by bond relaxation). In another work [47], the long-lived charge photoexcitations in sexithiophene (Figure 22.1) were identified as polarons because of the similarity of their absorption spectrum to that of the radical cations. It thus appears natural to use oligomers of various lengths to characterize the excited states of the longer polymers. In this chapter, we explore laser action in distyryl-benzene (DSB); a PPV-type oligomer with three repeat units [48].

In this chapter, we review the studies of ultrafast photoexcitations in π-conjugated polymers; the backbone structure of several polymers and an oligomer is schematically depicted in Figure 22.1. In general, we have studied photoexcitations in such polymers in a broad time interval from femtoseconds to milliseconds and spectral range from 0.1 to 3.5 eV. However, in this chapter we review only the transient studies in the femtosecond to picosecond time domains. The main experimental technique described herein is transient photomodulation (PM), which gives information complementary to that obtained by transient PL, which is limited to radiative processes, or transient photoconductivity (PC), which is sensitive to high mobility photocarriers. The PM method, in contrast, is sensitive to nonequilibrium excitations in all states. The interested reader may find studies of long-lived photo-excitations in π-conjugated polymers in other review articles and chapters.

22.1.2   Optical Transitions of Photoexcitations in π-Conjugated Polymers

The best way to detect and characterize short- and long-lived photoexcitations in the class of π-conjugated polymers is to study their optical absorption using the PM technique. As a consequence of their localization they give rise to gap states in the electron and phonon level spectra, respectively. The scheme of the PM experiments is the following. The polymer is photoexcited with above-gap light pulses and then the changes in the optical absorption of the sample are probed in a broad spectral range from the IR to visible using several light pulse sources. The PM spectrum essentially obtains difference spectra, i.e., the difference in the optical absorption ($\Delta\alpha$) of the polymer when it contains a nonequilibrium carrier concentration and that in the equilibrium ground state. Therefore, the optical transitions of the various photoexcitations are of fundamental importance. In this section, we discuss and summarize the states in the gap and the associated electronic transitions of various photoexcitations in π-conjugated polymers; the Infrared-active vibrations (IRAVs) related to the charged excitations are also briefly summarized.

Rather than discussing the various electronic states in π-conjugated polymers in terms of bands (e.g., valence and conducting bands), which might be the proper description of the infinite chains, or discrete levels with proper symmetries, which should be used for oligomers or other finite chains, we prefer the use of HOMO, LUMO, and singly occupied molecular orbital (SOMO). In the semiconductor description of the infinite chain, HOMO is the top of the valence band, LUMO is the bottom of the conduction band, and SOMO is a singly occupied state in the forbidden gap. In this case $E_g = $ LUMO $-$ HOMO, where E_g is the charge gap opposed to the optical gap that is related to the first allowed exciton. On the other hand, in the excitonic description of the correlated infinite chain, HOMO is the $1A_g$ state and LUMO is the $1B_u$ exciton; in this case $E_g = E(1B_u)$. In the case of $E(1B_u) > E(2A_g)$, E_g is still the optical gap and thus again $E_g = E(1B_u)$. In finite chains, HOMO and LUMO are discrete (isolated) levels with definite symmetries. These symmetries are extremely important for possible optical transitions, since one-photon absorption (OPA) can take place between states of *opposite* parity representations, e.g., g $\to$ u or u $\to$ g. This is true in the singlet manifold as well as in the triplet manifold. On the other hand TPA can take place between states of the same parity, which from the ground state A_g is g $\to$ g. Excitations in degenerate and NDGS polymers are discussed separately.

22.1.2.1   Optical Transitions of Solitons in Polymers with Degenerate Ground State

The semiconductor model for the PM spectrum associated with the soliton (S) transitions is shown in Figure 22.2 [29]. The amphoteric S defect has ground state S^0, negatively charged state S^-, and positively charged state S^+; both S^+ and S^- are spinless. Charge-conjugated symmetry is also assumed. The charge states are unrelaxed (S^-, S^+) during a time shorter than the relaxation time of the lattice around the defects; at longer times they are relaxed $\left(S_r^-, S_r^+\right)$. The energy of the unrelaxed S^- state differs from the energy of the S^0 by the (bare) electron correlation energy $U = E^- - E^0$; E^0 is the energy of S^0, and E^- and E^+ are the energies of the unrelaxed states S^- and S^+, respectively (Figure 22.2). The relaxed states differ from the unrelaxed states by the relaxation energy $\Delta E_r^- = E^- - E_r^-$, $\Delta E_r^+ = E_r^+ - E^+$. The relaxed state energy E_r and the ground state energy E^0 differ by the effective correlation energy $U_{eff} = E_r^- - E^0 = U - \Delta E_r^-$. The optical transitions of the soliton defects are therefore $\delta S^\mp$ from S_r^- (at E_r^-) to the LUMO level and from the HOMO level into S^+, and δS^0 from S^0 at E^0 into the LUMO and from the HOMO into S^- at E^-. If E_g of *trans*-$(CH)_x$ is known ($E_g \approx 1.5$ eV), then

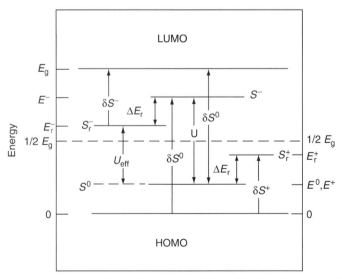

FIGURE 22.2   Energy levels and associated optical transitions of charged ($S^\pm$) and neutral (S°) soliton excitations in *trans*-$(CH)_x$. The parameters U, U_{eff}, and ΔE_r are defined in the text. (From Vardeny, Z.V. and Tauc, J., *Philos. Mag. B*, 52, 313, 1985. With permission.)

all the energy levels in Figure 22.2 may be determined from the optical transitions δS^0 and $\delta S^{\mp}$, respectively. In particular, U_{eff} (Figure 22.2) can be directly determined from the relation:

$$U_{\text{eff}} = U - \Delta E_r = \delta S^0 - \delta S^{\pm} \tag{22.1}$$

$$2\delta S^0 = E_g + U \tag{22.2}$$

Equation 22.2 shows that we can determine U from a single transition $(\underline{\delta S}^0)$ if E_g is known and charge conjugation symmetry exists. In more general case in which S^0 is replaced by a polaron level, this transition should be associated with the SOMO level in the gap. We used this relation to determine U in NDGS polymers from the optical transitions associated with the polaron and bipolaron levels in the gap [49].

22.1.2.2 Optical Transitions of Charged Excitations in NDGS Polymers

The proper description of charged excitations in NDGS polymers has been the polaron ($P^{\pm}$) [8,50], which carries spin $\frac{1}{2}$, and the spinless bipolaron ($BP^{\pm}$) [49,51].

22.1.2.2.1 Polaron Excitation

The states in the gap and the associated optical transitions for P^+ are shown in Figure 22.3a. The polaron energy states in the gap are SOMO and LUMO, respectively, separated by $2\omega_0(P)$ [52]. Then three optical transitions P_1, P_2, and P_3 are possible [52–54]. In oligomers, the parity of the HOMO, SOMO, LUMO, and LUMO + 1 levels alternates; they are g, u, g, and u, respectively. Therefore, the transition vanishes in the dipole approximation, and the polaron excitation is then characterized by the appearance of two correlated optical transitions below E_g. Even for long chains in the Hückel approximation, transition P_3 is extremely weak, and therefore the existence of two optical transitions upon doping or photogeneration indicates that polarons were created [51,54]. Unfortunately, polaron transitions have not been calculated for an *infinite* correlated chain. This is a possible disorder-induced relaxation of the optical selection rules that may cause ambiguity as to the number of optical transitions associated with polarons in "real" polymer films.

22.1.2.2.2 Bipolaron Excitation

The states in the gap and the possible optical transitions for BP^{2+} are given in Figure 22.3b. There are two unoccupied energy states separated by $2\omega_0(BP)$: the LUMO and LUMO + 1, which are deeper in the

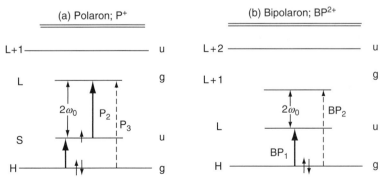

FIGURE 22.3 Energy levels and associated optical transitions of positive polaron and bipolaron excitations. The full and dashed arrows represent allowed and forbidden optical transitions, respectively. H, S, and L are HOMO, SUMO, and LUMO levels, respectively; and u and g are odd and even parity representations, respectively. $2\omega_0(P)$ and $2\omega_0(BP)$ are assigned. (From Vardeny, Z.V. and Wei, X., *Handbook of conducting polymers*, 2nd ed., eds. T.A. Skotheim, R.L. Elsenbaumer, and J. Reynolds, Marcel Dekker, New York, 1998. With permission.)

gap than corresponding states for P^+. Two optical transitions are then possible: BP_1 and BP_2. In short oligomers, again the parity of HOMO, LUMO, and LUMO + 1 alternates (g, u, and g, respectively), and therefore the BP_2 transition vanishes [49]. In this case, the BP is characterized by a single transition below E_g. Even in the approximation for an infinite chain the BP_2 transition is very weak. Electron correlation and disorder-induced relaxation of the optical selection rule, however, may cause the BP_2 transition to gain intensity, and therefore BPs with one strong transition at low energy and a second, weaker transition at higher energy should not be unexpected in "real" films.

22.1.2.3 Optical Transitions of Neutral Excitations in NDGS Polymers

Upon excitation, a bound e–h pair or an exciton (X) is immediately generated. By definition the exciton is a neutral, spinless excitation of the polymer. Following photogeneration the exciton may undergo several processes. It may recombine radiatively by emitting light in the form of fluorescence (FL), which is the light source of OLED devices. It may also recombine nonradiatively through recombination centers by emitting phonons. Excitons may also be trapped (X_t), either by a self-trapping process undergoing energy relaxation (this can be envisioned as a local ring rotation) or by a trapping process at defect centers. Excitons may also undergo an intersystem crossing into the triplet manifold, creating a long-lived triplet (T) state. Finally, a singlet exciton may disassociate into a PP either on the same chain (but on two different segments) or onto two different chains. Since we are interested in short-lived excitations, we deal only with trapped singlet excitons (X_t), and PPs (PP), where the T exciton is mentioned in passing. The energy levels and possible optical transitions of all three excitations are shown in Figure 22.4. For X_t and T we adopt the correlated picture in which the notations for different many-body exciton levels, which follow the group theory representations, are A_g and B_u, respectively.

22.1.2.3.1 *Singlet Excitons*

Two important exciton levels ($1B_u$ and mA_g) and a double excitation-type level (BX or kA_g) are shown in Figure 22.4a [15]; their electron configurations are also shown for clarity. We consider mA_g to be an excited state above the $1B_u$ level, whereas the BX or kA_g excited state may be due to a biexciton state, i.e., a bound state of two $1B_u$ excitons. The mA_g level is known to have strong dipole moment coupling to $1B_u$, as deduced form the various optical nonlinear spectra of π-conjugated polymers analyzed in terms of the "four essential states" model [23,24]. We therefore expect two strong optical transitions to form the following $1B_u$ photogeneration: X_1 and X_2, as shown in Figure 22.4a. Due to exciton self-trapping, however, we do not know whether X_1 would maintain its strength, since the relaxed $1B_u$ state may no longer overlap well with the mA_g state. X_2, on the other hand, will be always strong regardless of the $1B_u$ relaxation, since there is always room for a second exciton photogeneration on a chain following the photoproduction of the first exciton.

From TPA and EA spectra in soluble derivatives of PT and PPV polymers, we know that mA_g is about 0.8 eV higher than $1B_u$ [34,55]. We therefore expect the X_1 transition to be in the mid-IR spectral range, at about 0.8 eV or higher due to the exciton relaxation energy. The BX level, on the other hand, has not as yet been directly identified in π-conjugated polymers, although a weak two-photon state, termed as kA_g was identified in recent nonlinear optical spectroscopy [56,57].

22.1.2.3.2 *Triplet Excitons*

The most important electronic states in the triplet manifold are shown in Figure 22.4b. The lowest triplet level is 1^3B_u, which is lower that of $1B_u$ by the singlet–triplet energy splitting Δ_{ST}. In principle, 1^3B_u can directly recombine to the ground state by emitting photons (leading to phosphorescence [PH]) or phonons. But the transition is spin-forbidden and therefore extremely weak, leading to the well-known long triplet lifetime. The other two levels shown in Figure 22.4b are the m^3A_g level, which is equivalent to mA_g in the single manifold and the TX level, which is a complex composed of a triplet exciton and a singlet exciton bound together; their electronic configuration is also shown in Figure 22.4b for clarity.

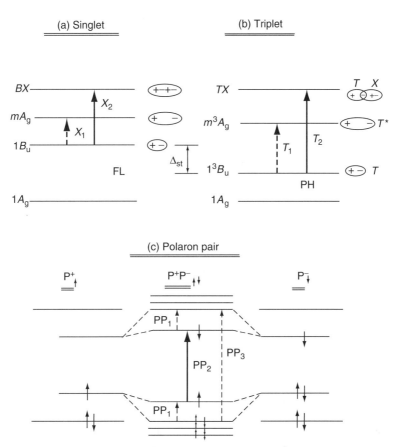

FIGURE 22.4 Energy levels, optical transitions, and emission bands associated with singlet and triplet excitons and PP excitations, respectively. The symbols $1A_g$, $1B_u$, mA_g, and BX are the ground state, lowest allowed exciton, most strongly coupled even-parity exciton, and biexciton level, respectively; T is the lowest triplet level (1^3B_u). Full and dashed arrows are for allowed and forbidden transitions, respectively, and wiggly arrows are for emission bands; FL is fluorescence and PH is phosphorescence. (From Vardeny, Z.V. and Wei, X., *Handbook of conducting polymers*, 2nd ed., eds. T.A. Skotheim, R.L. Elsenbaumer, and J. Reynolds, Marcel Dekker, New York, 1998. With permission.)

As in the case of single excitons, we expect for triplets two strong transitions T_1 and T_2 (Figure 22.4b). T_1 is into the m^3A_g level, and from Figure 22.4a and Figure 22.4b it is clear that is possible to estimate Δ_{ST} from the relation

$$\Delta_{ST} = T_1 - X_1 \qquad (22.3)$$

since m^3A_g and mA_g levels should not be far from each other. Δ_{ST} has recently been directly measured in π-conjugated polymers from phosphorescence emission involving heavy atoms and other techniques [58–62], and Equation 22.3 has been confirmed. Unfortunately, we do not know whether T_1 is indeed strong, since, as for singlet excitons, the relaxed triplet may not well overlap with the m^3A_g state, leading to a decrease in T_1 intensity. In contrast, it is quite certain that transition T_2 into the TX level is strong, because it is always possible to photogenerate a second (singlet) exciton close to a previously formed triplet exciton. However this transition has not as yet identified in π-conjugated polymers.

22.1.2.3.3 Polaron Pairs

A PP [35,36] is a bound pair of two oppositely charged polarons, P^+ and P^-, formed on two adjacent chains. The strong wavefunction overlap leads to large splitting of the P^+ and P^- levels as shown in

Figure 22.4c. Following the same arguments given previously for polaron transitions, we expect three strong transitions, PP_1–PP_3. For a loosely bound PP, these transitions are not far from transitions P_1–P_3 of polarons. However, for a tightly bound PP excitations, we expect a single transition, PP_2 to dominate the spectrum, as PP_1 is considered to be intraband with traditional low intensity and PP_3 is close to the fundamental transition and therefore difficult to observe. In this case, there are mainly two states in the gap, and the excitation is also known as a neutral BP (BP^0) or a polaronic exciton. However, the PP_2 transition is close in spirit to transition X_2 discussed above for excitons, as a second electron is also promoted to the excited level in the case of PP. Then from the experimental point of view, it is not easy to identify and separate the transitions of a trapped exciton (X_t) from those of a tightly bound PP of (BP^0) in the PM spectra. However, they may differ in their PADMR spectra [63].

22.1.2.4   Photoinduced Infrared-Active Vibrational Modes

The neutral π-conjugated polymer chain is free of excess charges. The neutral chain has a set of Raman active A_g vibrations that are strongly couple to the electronic bands by the e–p coupling [64]. These vibrations have been termed as amplitude modes (AM) since they modulate the electronic gap, 2Δ in the notation of Peierls gap [65]. The AM vibrations have been the subject of numerous studies and reviews since they play a crucial role in resonant Raman scattering dispersion with the laser excitation, and therefore can show important properties of the coupled electronic levels [66,67]. The most successful description of the AM-type vibrations was advanced by Horowitz [65], and its application to RRS by Vardeny et al. [64] and Ehrenfreund et al. [67].

When charges are added onto the chain some of the Raman active modes become IRAVs, since there is excess charge on the chain that may easily couple to EM radiation. The IRAVs that are manifested as peaks in the absorption spectrum accompany all the charge excitations mentioned above. In various studies, the appearance of photoinduced IRAVs is taken in the literature as evidence of photogenerated charge excitation onto the chain [68]. Usually, the IRAVs have immense oscillator strength, which is actually comparable with the strength of the electronic transitions. The reason for this excess strength is the small kinetic mass of the correlated charge excitation that translates into a large dipole moment [68]. Sometimes the IRAVs do not appear as peaks in the absorption but rather as dips or antiresonances (ARs) [69]. This happens when the electronic transitions overlap in energy with the IRAVs. In this case Fano resonance occurs between the two types of transitions, which results in the appearance of ARs. Despite the Fano-type AR, still the AM model can accurately describe the absorption spectrum, as was recently demonstrated for the ARs of DP excitations in P3HT [69]. Here, we want to briefly give several equations to describe both the doping-induced and photoinduced IRAVs in some polymers, as well as the ARs in the PM spectrum in other polymers.

An important ingredient of the AM model is that all IRAVs are interconnected and contribute to the same phonon propagator [65]. It has been therefore defined as the pinned, many-phonon-propagator, $D_\alpha(\omega) = D_0(\omega)/[1 - \alpha_p D_0(\omega)]$, where α_p is the polaron-vibrational pinning parameter and $D_0(\omega)$ is the bare phonon propagator. The latter is given [65] by: $D_0(\omega) = \Sigma_n d_{0,n}(\omega)$, and $d_{0,n}(\omega) = \lambda_n/\lambda \left\{ \left(\omega_n^0\right)^2 / \left[\omega^2 - \left(\omega_n^0\right)^2 - i\delta_n\right] \right\}$, where ω_n^0, δ_n, and λ_n are the bare phonon frequencies, their natural linewidth (inverse lifetime) and electron–phonon (e–p) coupling constant, respectively, and $\Sigma\lambda_n = \lambda$, which is the total e–p coupling. In the nonadiabatic limit, for a polaron current coupling to phonons, $f(\omega)$ that influences the conductivity, $\sigma(\omega)$ (and hence also shows in the absorption spectrum, since imaginary $(\sigma) \approx \alpha$), there is a correlated contribution, $g(\omega)$ from the most strongly coupled phonons by the e–p coupling. The function $g(\omega)$ was given in the random phase approximation (RPA) by [66,69]:

$$g(\omega) = \omega^2/E_r^2 [-\lambda D_\alpha(\omega)f^2(\omega)]/[1 + 2\lambda D_\alpha(\omega)\Pi_\phi(\omega)] \tag{22.4}$$

where $\Pi_\phi(\omega)$ is the phonon self-mass correction due to the electrons, which can be approximated for charge density wave by the relation: $2\lambda\Pi_\phi(\omega) = 1 + c(\omega)$ [69]. In the charge density wave approximation $c(\omega)$ is given by: $c(\omega) = \lambda\omega^2 f(\omega)/E_r^2$ [66], but since $\omega > E_r'$ $f(\omega)$ changes slowly over the phonon linewidth, it was approximated by a constant, C. The RPA terms in Equation 22.4 contain the sharp structure of $D_\alpha(\omega)$, whereas additional non-RPA terms contribute a smooth background term.

In general, the sharp features in the conductivity spectrum, $\sigma(\omega)$ are given by [69]:

$$\sigma(\omega) \approx [1 + D_0(\omega)(1 - \alpha')]/[1 + D_0(\omega)(1 + C - \alpha)] \qquad (22.5)$$

where α' is a constant that replaces a smooth electronic response; in the CDW approximation [66] $\alpha' = \alpha_p$, which was defined above for the trapped polaron excitation.

The poles of Equation 22.5, which can be found from the relation: $D(\omega) = -(1 - \alpha_p + C)^{-1}$, give peaks (or IRAVs) in the conductivity (absorption) spectrum. These absorption bands are very strong and can be taken as signature of charges added to the chain; actually we use the IRAVs to identify the charge state of photoexcitations in the ps PM spectra of π-conjugated polymers [70,71]. On the other hand, the zeros in Equation 22.5, which can be found by the relation: $D_0(\omega) = -(1 - \alpha_p)^{-1}$, give the indentations (or ARs) in the conductivity (absorption) spectrum. These will be also used to identify the charge states of photoexcitations in polymers with quasi-2D (lamellae) structure [72].

22.1.2.5 Nonlinear Optical Spectroscopy Related to Exciton Transient Response

In this section, we briefly review nonlinear optical spectroscopy that are related to exciton response in π-conjugated polymer systems that will be used in this chapter. This includes EA and TPA spectroscopies.

22.1.2.5.1 Electroabsorption

EA has provided a sensitive tool for studying the band structure of inorganic semiconductors [73] and their organic counterparts [74–77]. Transitions at singularities of the joint density of states respond particularly sensitively to an external field, and are therefore lifted from the broad background of the absorption continuum. The EA sensitivity decreases, however in more confined electronic materials, where electric fields of the order of 100 kV/cm are too small of a perturbation to cause sizable changes in the optical spectra. As states become more extended by intermolecular coupling, they respond more sensitively to an intermediately strong electric field, F since the potential variation across such states cannot be ignored compared with the separation of energy levels. EA thus may selectively probe extended states and is particularly effective for organic semiconductors, which traditionally are dominated by excitonic absorption.

One of the most notable examples of the application of EA spectroscopy to organic semiconductors is polydiacetylene, in which EA spectroscopy was able to separate absorption bands of quasi-1D excitons from that of the continuum band [78]. The confined excitons were shown to exhibit a quadratic Stark effect, where the EA signal scales with F^2 and the EA spectrum is proportional to the derivative of the absorption with respect to the photon energy. In contrast, the EA of the continuum band scales with $F^{1/3}$ and showed Frank–Keldish (FK) type oscillation in energy. The separation of the EA contribution of excitons and continuum band was then used to obtain the exciton binding energy in polydiacetylene, which was found to be ~0.5 eV [78].

Following the success of EA spectroscopy in polydiacetylene, this spectroscopy has been applied to a variety of π-conjugated polymer films including those given in Figure 22.1 [77]. A typical EA spectrum of MEH-PPV is given in Figure 22.5 and is analyzed according to the four essential states introduced by Mazumdar and collaborators [23,76], e.g., $1A_g$, $1B_u$, mA_g, and nB_u, respectively. The EA spectrum contains the following features: (a) at energies close to the optical gap a first derivative of the absorption is revealed; this is the Stark shift of the $1B_u$ exciton and can be used to determine the average $1B_u$ energy of the individual π-conjugated polymer and (b) an induced absorption feature at about 0.7 eV from $E(1B_u)$ due to the transfer of oscillator strength from the allowed exciton ($1B_u$) to the most coupled forbidden exciton (mA_g). This feature clearly unravels the energy of the mA_g exciton $E(mA_g)$, which may be taken as the lowest value for the exciton binding energy in the π-conjugated polymer. Sometimes a derivative absorption feature is also observed; this is due to a strongly coupled B_u exciton, the so-called

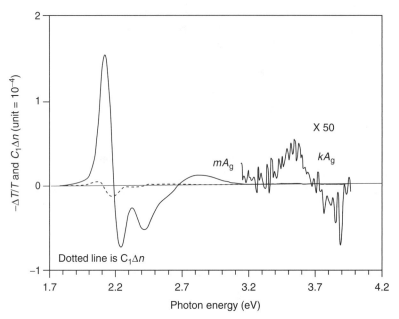

FIGURE 22.5   Typical EA spectrum for MEH-PPV film. The states $1B_u$, mA_g, and kA_g are assigned. (From Frolov, S.V., et al., *Phys. Rev. Lett.*, 78, 729, 1997. With permission.)

nB_u, which marks the continuum band threshold. The determination of $E(mA_g)$ is important in ps spectroscopy because photogenerated excitons may have a strong optical transition into this state that determines the photon energy of the PA_1 band in their transient PM spectrum (Figure 22.4).

22.1.2.5.2   Two-Photon Absorption Spectroscopy

In π-conjugated polymers the optical transitions between the ground state $1A_g$ and the B_u excitonic states are allowed; in particular, the transition between $1A_g$ and $1B_u$ dominates the absorption spectrum [79,80]. However, the optical transitions between $1A_g$ and excitonic A_g states are allowed by TPA. Therefore, TPA spectroscopy has been used for π-conjugated polymer to get information about the photon energies of A_g excitonic states in these materials [81]. Again this information is important since transitions between photogenerated $1B_u$ excitons into the A_g states are dipole allowed and thus dominate the PA spectrum of excitons in π-conjugated polymers [34] (Figure 22.7).

A typical TPA spectrum in comparison with the linear absorption is shown in Section 22.3.1 for the DOO-PPV polymer (Figure 22.7). Two TPA bands are typically observed, e.g., mA_g and the kA_g, where $E(mA_g) > E(kA_g)$ [57]. Thus, the optical transitions of photogenerated excitons in this polymer into the two A_g states may explain the typical excitonic PA that contains two bands: PA_1, i.e., $1B_u$–mA_g and PA_2, i.e., $1B_u$–kA_g. However, if $E(2A_g) < E(1B_u)$, then the photogenerated $1B_u$ excitons may quickly undergo internal conversion into the lowest energy exciton, i.e., $2A_g$. In this case the transitions PA_1 and PA_2 from the $1B_u$ excitons would be quickly replaced by the transition $2A_g$–nB_u from the lower exciton $2A_g$ [82]. This situation may occur in t-$(CH)_x$ where $E(2A_g) < E(1B_u)$ [18].

22.1.3   Properties of Excitons in π-Conjugated Polymers

Due to the nanometer-scale confinement of the electron wavefunction in two dimensions and its extension over many repeat units in the third dimension, π-conjugated polymers possess quasi-1D electronic states and, consequently 1D-type density of states. The lateral quantization may result in structured conduction and valence bands with subbands and van Hove singularities. The properties of

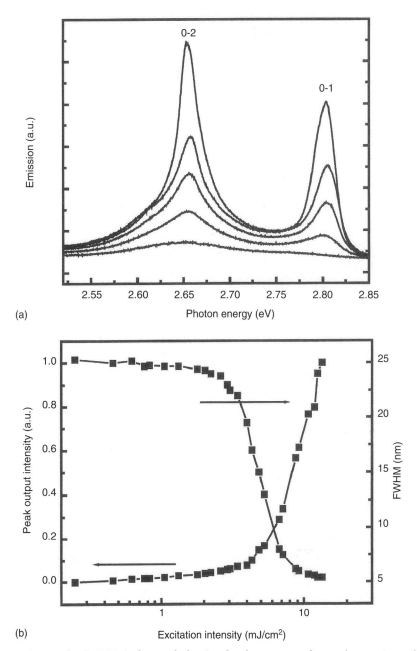

FIGURE 22.6 Laser action in DSB single crystal, showing the phenomenon of spectral narrowing and nonlinear input–output relation close to the threshold excitation for lasing.

such electronic systems are determined by the ratio of two characteristic energies, i.e., the electron–electron Coulomb interaction, E_{Coul} and the confinement energy, ΔE of lateral quantization (energy separation, ΔE between the van Hove singularities). In all known π-conjugated systems, $E_{Coul} > \Delta E$. Under these conditions it is rather impossible to consider electrons and holes as independent, non-interacting particles as in semiconductor quantum dots or nanorods [83]. Therefore, π-conjugated

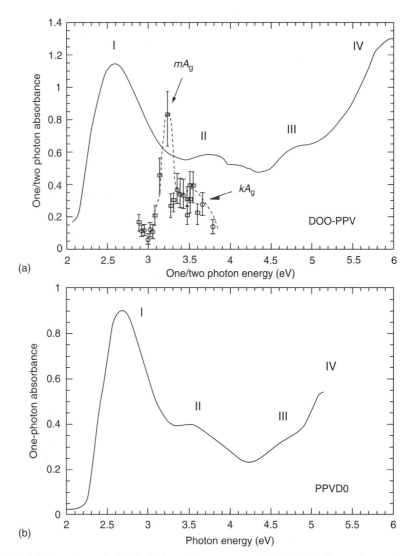

FIGURE 22.7 (a) One-photon (solid line) absorption spectrum of DOO-PPV film, and two-photon (circles) absorption spectrum of DOO-PPV solution. (b) One-photon absorption spectrum of PPVD0 film. Bands I, II, III, and IV, as well as mA_g and kA_g states are assigned. (From Frolov, S.V., et al., *Phys. Rev. B*, 65, 205209, 2002. With permission.)

polymers constitute a unique class of electronic systems in which $\Delta E \sim E_{Coul}$, and thus their electronic energy spectrum is determined by a complex interplay between quantum confinement and Coulomb e–h interaction. This regime is largely unexplored in theoretical studies; it is the subject of various computational models [84].

Typical lifetime of the lowest exciton in luminescent π-conjugated polymers is of the order of 100 ps. There are two processes that contribute to exciton lifetime; radiative and nonradiative recombination. Exciton radiative lifetime in π-conjugated polymers was determined to be ~1 ns; it actually fits well with the exciton size in the dipole approximation [86]. The binding energy of excitons in π-conjugated polymers is still debated, but a consensus seems to put it ~0.5 eV [84,85]. This value is in agreement with the exciton binding energy in polydiacetylene PTS [78] of which value is universally accepted. With such large binding energy, the solution of the Coulomb problem in 1D obtained by Loudon in 50th shows that the exciton state grabs most of the oscillator strength from the interband transition [87]. As a result the interband transition strength is very small [85], and because of inhomogeneity is in fact invisible in the absorption process, unless sophisticated modulation techniques are employed.

In a "band model" that describes excitations in polymers, absorption of a photon by a photogenerated carrier in a continuum band is not allowed by two particle energy and quasimomentum conservation rules [8]. Furthermore, photogenerated carriers in more common semiconductors do not show structured PA bands; instead their PA is in the form of a structureless Drude free carrier absorption that peaks at low energies [88]. On the contrary, in the "exciton model" for describing the photoexcitations in polymers, photon absorption by a low-lying exciton that results in the creation of higher energy exciton is, in fact an allowed two-particle process.

Excitons in π-conjugated polymers strongly interact with optical phonons. A typical interaction energy is ~0.2 eV [64], and optical phonons are coupled to dipole-allowed transitions, frequently increase their oscillator strength or even enable dipole-forbidden transitions. If one or two optical phonons are involved, then the transitions are termed as 0–1 and 0–2, respectively [89]. Lattice distortions due to optical phonons are extended, and frequently exceed the spatial extent of exciton wavefunction (~5–6 repeat units). Release of excess energy to phonons, referred to as internal conversion, is a primary step of ultrafast exciton relaxation, and is dramatically dependent on chain environment [90]. In films of nonaggregated π-conjugated polymers, phonons follow the symmetry of the polymer chain. As a result, 1D electron–phonon interaction is effective. However, for polymer chains in aggregates, solid matrices, and solutions, the electron–phonon interaction is weaker [91].

The time constant for exciton excess energy relaxation within 1D density of homogeneously broadened states is ~1 ps [90]. In the next step of the energy relaxation process excitons migrate to the bottom of the inhomogeneously broadened density of states, i.e., to the energetically favorable states with larger conjugation length by the incoherent intermolecular energy transfer [55], termed Förster energy transfer. This effect is absent in solution, since the average interchain distance is relatively large. Förster energy transfer within the excitons density of states in films gives rise to transient PL redshift on a 10-ps timescale [90]. For studying this process in detail, polarization memory dynamics is usually monitored. The time-dependent degree of polarization, $P = [\Delta T(\text{par}) - \Delta T(\text{per})]/[\Delta T(\text{par}) + \Delta T(\text{per})]$, reveals transient change of an average transition dipole moment orientation. From the $P(t)$ decay it is possible to obtain the transient exciton dynamics within the polymer chains and its dipole moment reorientation [55].

22.1.4 Experimental Setup for Measuring Transient and cw Responses

For measuring the transient photoexcitation response in the fs to ns time domain we have used the fs two-color pump–probe correlation technique with linearly polarized light beams. We have used two laser systems; a high repetition rate, low power laser for the mid-IR spectral range [92] and a relatively lower repetition rate, high power laser system for the near-IR and visible spectral range [93].

22.1.4.1 Low Intensity fs Laser System

For the transient PM spectroscopy in the spectral range between 0.55 and 1.05 eV our ultrafast laser system was a 100-fs titanium:sapphire oscillator operating at a repetition rate of about 80 MHz, which pumped an optical parametric oscillator (OPO) (Tsunami, Opal, Spectra-Physics) where both signal and idler beams were respectively used as probe beams [92]. The pump beam was extracted either from the fundamental at ≈1.6 eV or from its second harmonics at 3.2 eV. To increase the signal/noise ratio, an acousto-optical modulator operating at 85 KHz was used to modulate the pump beam intensity. For measuring the transient response at time t with ≈150-fs time resolution, the probe pulses were mechanically delayed with respect to the pump pulses using a translation stage; the time $t = 0$ was obtained by a cross-correlation between the pump and probe pulses in a nonlinear optical crystal. Typically, the laser pump intensity was kept lower than 5 μJ/cm^2 per pulse, which corresponds to ≈10^{16}cm^{-3} initial photoexcitation density per pulse in the polymer films. This low density avoids the complications usually encountered with high power lasers such as bimolecular recombination and SE; both processes tend to increase the recombination rate of the photoexcited excitons.

The transient PM signal, $\Delta T/T(t)$ is the fractional changes ΔT in transmission T, which is negative for photoinduced absorption, PA and positive for photoinduced bleaching, PB at photon energy above the absorption onset, and SE for photon energies below the optical gap when overlapping the PL spectral range. The pump beam was directed to pass though a polarization rotator that changed its polarization to be $45°$ to that of the probe beam. An optical polarizer was used on the transmitted probe beam to analyze the changes of transmission ΔT for both parallel, $\Delta T_{\parallel}$, and perpendicular, $\Delta T_{\perp}$, polarizations of the pump and probe beams. The transient polarization memory, $P(t) = (\Delta T_{\parallel} - \Delta T_{\perp})/(\Delta T_{\parallel} + \Delta T_{\perp})$, was then calculated to study its decay. The pump and probe beams were carefully adjusted to get complete spatial overlap on the film, which was kept under dynamic vacuum. In addition, the pump–probe beam walk with the translation stage was carefully monitored and the transient response was adjusted by the beam walk-measured response.

22.1.4.2   High Intensity fs Laser System

For the near-IR and visible range we used the high intensity laser system [93]. This laser system was based on a homemade Ti:sapphire regenerative amplifier that provides pulses of 100 fs duration at photon energies of 1.55 eV, with 400 μJ energy per pulse at a repetition rate of 1 kHz. The second harmonic of the fundamental pulses at 3.1 eV was used as the pump beam. The probe beam was either a white light supercontinuum within the spectral range from 1.1 to 2.8 eV, which was generated using a portion of the Ti:sapphire amplifier output in a 1 mm-thick sapphire plate, or a signal output of an optical parametric amplifier (TOPAS, Light Conversion Ltd.) in the spectral range from 0.7 to 1.02 eV. To improve the signal-to-noise ratio in our measurements, the pump beam was synchronously modulated by a mechanical chopper at exactly half the repetition rate of the Ti:sapphire laser system ($\cong$500 Hz). The probe beam was mechanically delayed with respect to the pump beam using a computerized translation stage in the time interval, t up to 200 ps. The beam spot size on the sample was about 1 mm in diameter for the pump beam and about 0.4 mm diameter for the probe beam. The pump beam intensity was set below 300 μJ per pulse, which is below the signal saturation limit. The wavelength resolution of this system was about 8 nm using a $\frac{1}{8}$-m monochromator having a 1.2 mm exit slit, which was placed in the probe beam after it had passed through the sample. The transient spectrum $\Delta T/T(t)$ was obtained using a phase-sensitive technique with a resolution in $\Delta T/T \approx 10^{-4}$ that corresponds to a photoexcitation density of about 10^{17} cm^{-3}; this is below the threshold exciton density for exciton–exciton annihilation by bimolecular recombination kinetics.

22.1.4.3   cw Optical Measurements

The ps spectroscopy and laser action phenomena described very often are complemented by measuring the equivalent optical spectra under steady state conditions, using cw spectroscopic techniques [94] described inSection 22.1.4.3.1.

22.1.4.3.1   PL Spectrum

For the cw-polarized PL emission study, we used the fundamental of the Ti:sapphire laser system at 1.6 eV operating at full power (1.5 W) to excite polymers with low optical gap such as t-(CH)$_x$ and an Ar$^+$ laser to excite polymer samples with optical gap throughout the visible spectral range. The PL emission was collected by a lens with large F-number, and spectrally and spatially filtered to eliminate the relatively strong excitation laser intensity. A polarizer was used to select the PL emission either parallel or perpendicular to the polarization of the pump beam, and a polarization scrambler in front of the monochromator was used to detect the two PL components through the spectrometer. The collected PL emission was then directed onto the exit slit of a $\frac{1}{4}$-m monochromator with 1 nm resolution; a nitrogen-cooled germanium photodiode was used for the light detection in the near- and mid-IR spectral ranges, and a silicon photodiode was used for PL in the visible/near-IR spectral range.

22.1.4.3.2 PM Spectrum

For the cw spectroscopy we used a standard PM setup at low temperatures [94 and references therein]. The excitation beam was an Ar^+ laser with several lines in the visible spectral range and UV at 353 nm, which was modulated with a mechanical chopper at a frequency of ~300 Hz. The probe beam was extracted from a tungsten lamp in the spectral range of 0.25 to 3 eV. A combination of various diffraction gratings, optical filters, and solid state detectors (e.g., silicon, germanium, and indium antimonite) was used to record the PM spectra. The spectral resolution was about 2 nm in the visible spectral range and 4–10 nm in the near-IR range, with $\Delta T/T$ resolution of $\approx 10^{-6}$.

22.1.4.3.3 Optically Detected Magnetic Resonance Measurements

For the optically detected magnetic resonance (ODMR) measurements the sample was mounted in a high-Q microwave cavity at 3 GHz equipped with a superconducting magnet [94]. Microwave resonant absorption leads to small changes, ΔPL in the PL intensity. By scanning the external magnetic field, H the relative change in PL is measured to unravel the magnetic resonance signal for the various photoexcitations. Spin $\frac{1}{2}$ species give magnetic resonance at $H_0 \approx 1008$ gauss, whereas triplet excitons with spin 1 show a relatively sharp magnetic resonance signal at "half-field" below $H_{1/2} = 504$ gauss. Any deviation from $H_{1/2}$ is proportional to the zero-field splitting parameters, which can be used to calculate the triplet wavefunction extent; the "full-field" resonance of the triplet excitons [94] is beyond the scope of this chapter.

22.2   Classification of Laser Action Phenomena

At high excitation intensity a substantial exciton density is generated in the sample. Each exciton has an optical cross section, σ for absorbing and emitting photons. The cross-section, σ is dependent on the photon energy, ω so that it may be either positive or negative depending on the ability of the exciton to absorb or emit light efficiently. In general:

$$\sigma(\omega) = \sigma_e(\omega) - \sigma_a(\omega), \tag{22.6}$$

where $\sigma_e(\omega)$ is the emitting cross section to the ground state, and $\sigma_a(\omega)$ is the absorption cross section to upper states. It turns out that $\sigma_e(\omega)$ spectrum is similar to the PL spectrum, but modified by ω^{-3} so that the emitting cross section is stronger in the red part of the spectrum [95]. On the other hand, $\sigma_a(\omega)$ follows the photoinduced absorption bands from the photogenerated exciton quite exactly. When $\sigma(\omega) > 0$, then SE may exist in the sample. In this case, the optical gain, $g(\omega)$ is defined similar to the absorption, i.e., $g(\omega) = N\sigma(\omega)$ is the gain per unit length in the medium, where N is the exciton number density. In the case of SE, the intensity increase, ΔI at the end of a slab of length dx is given by $\Delta I = g(\omega)dx$. Equivalently [95], at the end of a length L in the gain medium, the intensity $I(L)$ is given by the formula

$$I(L) = \beta[\exp(gL) - 1] \tag{22.7}$$

where β is a constant that depends on the excitation geometry and wavelength. For laser action or SE condition, the optical gain must be larger than the absorption loss in the medium; or equivalently, for getting an increase in the light intensity, the inequality [96]

$$\exp[(g - \alpha)L] > \kappa\tau_0, \tag{22.8}$$

must apply, where α is the absorption at the SE wavelength, κ is the loss rate of light due to effects such as scattering, imperfections, and light leakage in the cavity, and τ_0 is the time needed for the SE pulse to transverse the cavity length L. From the above discussion it is clear that optical gain that leads to laser action may occur in neat media within the spectrum such as $g > \alpha_{\text{eff}}$, where α_{eff} is the depleted (or bleached) absorption due to the excited state density of chromophores. For a two-level system, this inequality would never occur unless there is *population inversion* in the medium so that the density of excited chromophores is larger than the density of chromophores left in the ground state [97]. However, dye molecules and π-conjugated systems show a broad PL spectrum due to relatively strong electron–phonon coupling that leads to ample "phonon side bands" or "phonon replica" described by the Huang–Reiss formalism [89]. In this case the inequality $g(\omega) > \alpha_{\text{eff}}(\omega)$ may easily occur for the nth phonon replica at $h\omega = E_{\text{g}} - n h\nu$ below the absorption edge, E_{g}. In addition, because of the ω^{-3} factor in the gain spectrum coefficient, the nth phonon replica would take over to show SE and lasing even without population inversion that is needed for a two-level system [98]; all that is really needed is that gain overcomes loss in the system (Equation 22.8).

The laser threshold intensity, I_{th} is defined by Equation 22.8 (that is an approximation for low intensity) as the excitation intensity at which the photogenerated exciton density is sufficiently large to create an optical gain in the cavity. Usually I_{th} is large and thus can be achieved mainly using pulsed excitation; thus, relatively strong pulsed excitation lasers are used for studying laser action phenomena. For $I > I_{\text{th}}$, optical gain takes over and the relation of the output intensity vs. the excitation intensity, $I_{\text{out}}(I_{\text{ex}})$ abruptly increases its slope at I_{th}; $I_{\text{out}}(I_{\text{ex}})$ keeps the same increased slope until a saturation intensity, I_{sa} is reached in I_{ex} where saturation occurs, and thus the slope decreases again. This forms a famous S-shape curve for $I_{\text{out}}(I_{\text{ex}})$, which defines laser action (Figure 22.6) [98]. In addition, the emission spectrum dramatically changes at I_{th}. For $I > I_{\text{th}}$ the SE process takes over the spontaneous emission process and thus light at photon frequency within the gain spectrum, where $g(\omega) > \alpha(\omega)$ prevails. Under these conditions a dramatic spectral narrowing (SN) is usually measured, where the emission bandwidth substantially narrows to follow the spectrum at which gain occurs [99–109]. In addition, the emission time dramatically decreases due to the process of SE [103]. In fact the emission time of an ASE pulse cannot be longer than τ_0, the transverse time in the cavity.

In conclusion, the three factors that define laser action in a gain medium are therefore SN, sudden change in $I_{\text{out}}(I_{\text{ex}})$ dependence, and sudden decrease in the emission time. An example of laser action in DSB single crystal is given in Figure 22.6 [110]; it is seen that the increase in slope of $I_{\text{out}}(I_{\text{ex}})$ is accompanied by a strong SN.

Five different laser action processes discussed in this chapter are as follows:

1. Amplified spontaneous emission (ASE)
2. Superradiance (Dicke type)
3. Superfluorescence (SF)
4. Cavity lasing (with optical feedback)
5. Random lasing (RL)

Processes 1–3 and 5 occur without the necessity of optical feedback (or mirrors), and thus have been termed as "mirrorless lasing" [111]. Processes 2–5 are also coherent, whereas ASE is not. In addition, processes 2 and 3 are examples of *cooperative* emission, whereas ASE is a *collective* emission process. The superradiance process is very similar to the SF laser action process, except that in superradiance the system is prepared coherently from $t = 0$, whereas in SF the system evolves in time to be coherent at $t > 0$ [111].

22.2.1 Amplified Spontaneous Emission

ASE abundantly occurs whenever the condition of SE satisfied by Equation 22.8 is met [96,105]. In this type of mirrorless lasing, the optical pulse transverse only once through the gain medium, where the intensity at the end of the gain medium exponentially depends on the optical length and gain as given by Equation 22.7. Thus, at large excitation intensities $\ln(I_{\text{out}})$ depends linearly on the excitation intensity, I_{ex} and cavity length, L. ASE is usually measured in a cuvette where the laser dye (or polymer) is in a solution. The laser excitation is usually in the form of a stripe having length L that can be varied [96].

In this way the exponential dependence of I_{out} on L can be easily verified. Several commercially available pulsed lasers operate in the ASE mode. Examples are copper-vapor laser and the gas lasers based on HF molecules.

22.2.2 Superradiance and Superfluorescence

Superradiance and SF are examples of *cooperative* emission processes. At low densities following optical excitation, excitons or any other optical emitters decay by spontaneous emission. The individual emitters then act independently of the radiation field, their respective phases are completely random, and their emission is characterized by a radiative decay time τ. At high exciton densities in the case of SE, however this simple behavior may change due to the strong interaction between the excitons through their own radiation field [112–114]. The decay of the excited state then occurs by SF in a much shorter characteristic time. SF is a cooperative spontaneous emission from an ensemble of electrical dipole emitters [115]. Depending on the emitter dynamics, this emission may have a high degree of coherence and therefore resemble laser emission with feedback. Following the original work of Dicke on superradiance [116] the effect has been theoretically studied in details [112–115]. However, its experimental observations in solid state systems are rare [117–119]. In SF, the common radiation field that overlaps different emitters induces the initial ordering process among the emitters. This leads to the buildup of correlation among the dipole moments belonging to different emitters, and the ensemble of phase-locked emitters, each having a dipole moment μ, acquires a macroscopic dipole moment $N_c\mu$, where N_c is the number of correlated emitters [112,116]. This macroscopic dipole moment then radiates spontaneously in a well-defined direction depending on the geometry of the sample with a higher rate and much stronger peak intensity than the total emission of the independent dipoles. The *coherent* radiation power is then proportional to $(N_c\mu)^2$ and the emission emerges in a short pulse with time duration of order τ/N_c [116]. In order to observe SF, the gain coefficient of the emitters must be large and the sample length must be small compared to the distance that the radiation can travel in the medium within the dipole dephasing time T_2^*. In this case radiation coming from any dipole can be strongly amplified and transmitted to another dipole and vice versa before any of these dipoles spontaneously emits or loses its phase coherence.

22.2.3 Cavity-Based Lasers

Cavity-based laser, a "true laser," where the active medium is bound within a cavity of length L, has optical feedback such as mirrors for a linear Fabry–Perot case [97], a ring for whispering gallery mode of operation [120], or distributed feedback-type cavities [121]. In this type of laser the SE beam travels many times inside the resonator within the time duration T_Q determined by the laser Q-factor [97], which is given by the formula [122]:

$$Q = 2\pi nL/\lambda(\sqrt{r}/1 - r) \tag{22.9}$$

where n is the refraction index, and $r = R\exp(gL)$ where R is the mirror reflectivity. The cavity Q can be directly estimated from the laser emission spectrum by measuring the linewidth of the emission modes, $Q = \lambda/d\lambda$ where $d\lambda$ is the mode full width at half maximum [97]. The longitudinal laser modes are in essence the Fourier transform (FT) of the transient pulse that travels within the cavity. The mode spacing, $\Delta\lambda$ in the laser emission spectrum is then given by the formula [122]:

$$\Delta\lambda = \lambda^2/2nL \tag{22.10}$$

The longitudinal mode spacing may be very small for gas lasers such as Ar^+ laser with L of the order of 1 m, so that it is difficult to detect them. However, in this chapter we discuss relatively small laser length of a few tens of microns and thus it is relatively easy to detect the laser longitudinal modes using a small spectrometer with 0.2 nm resolution. The FT of the emission spectrum gives a more accurate value of nL [122], which may then be used to explain the emission lines more precisely [123]. The expected intensity of the FT of a Fabry–Perot cavity was derived before [122]. It consists of a series of equally spaced diminishing lines with $\Delta d = nL/\pi$ (or $nD/2$ for cylindrical cavities) where d is the FT pathlength variable.

22.2.4   Random Lasers

RL is a relatively newly discovered process. In most familiar lasers great care is required to configure the system for obtaining lasing. Many of the difficulties are encountered in obtaining proper alignment of mirrors, which produce feedback throughout the various gain media by creating resonators with high quality factor, Q. However, several disordered systems with optical gain have shown laser action that is not due to simple ASE with no easily identifiable cavities. These systems include neodymium powders [124], dyes and polymers mixed with scatterers in films and solutions [125,126], films of π-conjugated polymers [127], semiconductor powders [128,129], and synthetic opals infiltrated with π-conjugated polymers and dyes [130,131]. The laser action process in these systems does not rely on carefully imposed resonant cavities for the necessary feedback, but arises from multiple scattering in the disordered medium in spite of the general belief that scattering is detrimental to lasing. Therefore, this type of laser action is termed as "random laser" [132]. The potential resonant loops formed from multiple scattering are not engineered at all in these systems. Coherent backscattering (CBS) measurements of light in such disordered systems have shown the possibility of recurrent scattering events [133] that indicates the existence of multiple scattering loops in the medium. The oxymoron term random laser may therefore be a good description for this type of laser action.

The fascinating subject of RL has recently attracted many experimental and theoretical groups [133,134]. One reason for this is the relation of random cavities in disordered media to light prelocalization, i.e., the analogy of RL modes with "nearly trapped" electron states in disordered media [135,136]. Although the RL experiments with semiconductor powders were indeed done in strong scattering gain media, the light mean free path, l^* was of the order of the laser wavelength, λ and therefore close to localization [128,129]. On the contrary RL measurements in DOO-PPV thin films described in this chapter were performed in optical gain systems where $l^* \cong 15\lambda$ [136]. We thus conclude that the role of l^* in the process that leads to RL is not completely understood at the present time.

22.2.5   Experimental Setup for Studying Laser Action

For studies of laser action in organic semiconductors polymer films with uniform thickness, d, ranging from 0.5 to 4 μm were slowly spin-coated from fresh chloroform solutions onto quartz substrates. The variation in d was typically less than 5% over the film length of 1 mm [96]. For polymers in solutions we thoroughly mixed the polymer powder in good solvents such as THF or chlorophorm typically with concentration of few milligram per milliliter. Then the polymer solution was placed in a transparent cuvette with flat windows, in which the side windows were tiled to avoid optical feedback.

The excitation source for these measurements was typically a frequency-doubled regenerative laser amplifier producing 100 ps pulses at 500 Hz and $\lambda = 530$ nm for PPV-based polymers, whereas we used a frequency-tripled laser at $\lambda = 353$ nm for excitation of polymer films emitting in the blue. For time-resolved measurements, we sometimes used a doubled Ti:sapphire amplifier laser system with 100-fs time resolution. The pump laser beam was typically focused on the polymer film or cuvette using a striped geometry. The length of the stripe-like excitation area could be varied from 100 μm to 6 mm using a variable-width slit in front of the sample, which could block parts of the pump beam. The polymer emission was collected from either the front or the side of the substrate and spectrally analyzed using either a 0.25-m spectrometer or a 0.6-m triple spectrometer. All experiments were performed in a dynamic vacuum at room temperature.

For microring lasers we used cylindrically shaped thin polymer films that were prepared by dipping commercially available optical fibers into saturated chloroform solutions [120]. Thin polymer rings were consequently formed around the glass cylindrical core following fast drying in air. The estimated thickness of the deposited polymer rings was about 2–3 μm. For these measurements the light emitted from the excited polymer ring was collected in the plane of the ring with a round lens and spectrally analyzed using a 0.6-m spectrometer and a CCD array with spectral resolution of about 1 Å.

Microdisks are small photolithographically defined circular resonance structures with an inherently high optical quality factor, Q. One use of such structures is to fabricate microlasers based on high luminescent materials [137–140]. Other uses are to investigate material optical properties such as laser threshold [141], spontaneous emission efficiency [142], and its time dynamics [143]. Photolithographic fabrication and reactive ion etching techniques borrowed from semiconductor manufacturing allow the fabrication of microdisks from films of DOO-PPV [123]. A polymer film of ~1-μm thickness is spin cast onto a glass substrate. Photoresist is then spun on top of the polymer and circles are patterned into the resist. The polymer is then etched in plasma and the remaining resist is removed. The microdisk devices were photoexcited in vacuum to avoid photodegradation. The laser emission was collected with a 1-mm diameter fiber optic placed several millimeters from the device. The emission was sent through a $\frac{1}{2}$-m spectrometer, detected with a charge-coupled device, CCD, and recorded on a personal computer. The overall spectral resolution of the collection setup was 0.02 nm. The polymer microdisk reviewed here had a typical diameter of 8–70 μm and a thickness of 1 μm.

For lasing from opals, microcrystalline opals with crystal sizes of 20–100 μm were prepared from crystallizing colloidal suspensions of nearly monodispersed SiO_2 balls with diameters D varying between 190 and 300 nm, as described elsewhere [144]. A typical opal slab size was 1 mm × 1 cm × 1 cm. After the complete penetration of the solution, the opals became semitransparent due to close matching between the refractive indices of solvents and silica ($\Delta n \approx 0.01$). As a result, light scattering from the silica nanoballs was relatively weak: from transmission measurements in the spectral range between 550 and 650 nm for opals infiltrated with ethylene glycol, we estimated the mean photon diffusion length $\ell^* \geq 0.5$ mm [145]. We also found [136] that Bragg scattering stop bands, which are known to exist in opals, did not influence the ASE spectra. The opal slabs soaked in solutions that contained the gain media were placed inside 1 × 1 cm quartz cuvette and photoexcited using the pulsed laser system described above. The stripe-like excitation at intensities above the ASE threshold resulted in the emission from the side of the slab of an ~5° divergent beam directed along the stripe axis.

22.3 Ultrafast Dynamics of π-Conjugated Polymer Films and Solutions

Polymer photophysics is determined by a series of alternating odd (B_u) and even (A_g) parity excited states that correspond to one-photon and two-photon allowed transitions, respectively [23]. Optical excitation into either of these states is followed by subpicosecond nonradiative relaxation to the lowest excited state [90]. This relaxation is due to either vibrational cooling within vibronic sidebands of the same electronic state, or phonon-assisted transitions between two different electronic states. In molecular spectroscopy [146], the latter process is termed internal conversion. Internal conversion is usually the fastest relaxation channel that provides efficient nonradiative transfer from a higher excited state into the lowest excited state of the same spin multiplicity. As a result, the vast majority of molecular systems follow Vavilov–Kasha's rule, stating that FL typically occurs from the lowest excited electronic state and its quantum yield is independent of the excitation wavelength [91].

Luminescent π-conjugated polymers, like many other complex molecular systems, are expected to follow Vavilov–Kasha's rule. The independence of exciton generation yield on the excitation wavelength has indeed been demonstrated for PPV-type polymers [86]. This observation indicated that internal conversion is likely to be the dominant relaxation channel between different excited states in π-conjugated polymers. However, there are other processes, which may interfere and successfully compete with the internal conversion. Among the known competing processes are singlet exciton fission and exciton dissociation [91]. The first process creates two triplet excitons with opposite spins, whereas the second process generates charge carriers.

22.3.1 Exciton Dynamics in DOO-PPV and PPVD0 Polymers

22.3.1.1 Ground and Excited States

Figure 22.7 shows the OPA and TPA spectra, $\alpha(\omega)$ of DOO-PPV and PPVD0 films [55]. Absorption bands marked I, II, III, and IV in the OPA spectra are observed in virtually all PPV derivatives [79]. These bands have been identified as optical transitions between π (occupied) and π* (unoccupied) molecular orbitals (MO). MOs in PPV polymers have been traditionally classified as delocalized (*d* and *d**) and localized (*l* and *l**); the former are delocalized across all carbon atoms, whereas the latter have nodes at *para* positions of the benzene rings, which result in charge confinement at the rings. Calculations show that bands I and II originate from $d \rightarrow d^*$ transitions, band IV is due to $l \rightarrow l^*$, and band III involves degenerate transitions $d \rightarrow l^*$ and $l \rightarrow d^*$ [79]. A PPV chain is a molecular system with a center of inversion, thus its excited states have been classified as A_g (*gerade*) and B_u (*ungerade*). Since the polymer ground state is described by a totally symmetric wavefunction (i.e., an A_g state), optical transitions are allowed only to the B_u states (states of the opposite parity). Thus, band I in Figure 22.7 describes the lowest allowed state, i.e., $1B_u$ exciton at $E(1B_u) = 2.2$ eV. Excitation to bands II–IV produces higher B_u states with very different electron–hole distributions [147]. Each of these excited states has different potential energy surfaces (the energy dependence on the nuclear coordinates), and therefore, their relaxation pathways may differ.

The A_g states, which are not observable by OPA, may be found from the TPA spectrum, as shown in Figure 22.7 inset for the DOO-PPV solution. This spectrum was obtained in DOO-PPV solutions using a Z-scan technique [57]. From these measurements, as well as EA measurements [56] there were found two prominent two-photon allowed states, mA_g and kA_g, at $E(mA_g) \approx 3.1$ eV and $E(kA_g) \approx 3.5$ eV, respectively. The theory of A_g states in PPV polymers is less developed as compared to that of B_u states [84], partly due to the scarcity of the relevant experimental data. Several recent theoretical studies, however have attempted to elucidate the complex nature of the A_g states in PPV [81]. These calculations show a broad manifold of A_g states, only two of which appear prominently in the nonlinear optical spectra. Since these states are two-photon allowed with strong coupling to the $1B_u$ exciton, they also appear in the PA spectrum from $1B_u$ excitons, as explained in Section 22.1.1.

Figure 22.8 shows the transient PM spectra of DOO-PPV and PPVD0 films measured at $t = 2$ ps, where the prominent features are assigned due to optical transitions of the relaxed $1B_u$ excitons [148]. These features include a vibronically broadened SE band at $\hbar\omega > 1.7$ eV, and two PA bands at $\hbar\omega < 1.7$ eV marked as PA_1 and PA_2. The SE band closely resembles the respective PL spectrum, and thus it describes the radiative transitions of the $1B_u$ exciton to the ground state ($1A_g \leftarrow 1B_u$). PA_1 and PA_2 have been attributed to transitions from $1B_u$ to mA_g and kA_g, respectively. Indeed, the peak positions of PA_1 at 1.0 eV and PA_2 at 1.4 eV match with the energy differences between the two A_g states and $1B_u$ (see Figure 22.7a). Furthermore, the decay dynamics of the SE, PA_1, and PA_2 bands are identical up to ~300 ps, as shown in Figure 22.9a for the PPVD0 film. Similar results were obtained for the DOO-PPV films [34].

In general, the PA_2 dynamics differ from the SE or PA_1 dynamics due to contributions from other species, such as triplets, polarons, and PPs [34,55]. Specifically, in the case of pristine PPVD0 films PA_2 appears to have a long-lived component, which is also observed in DOO-PPV [34]. The long-lived component was attributed to triplet excitons produced from the singlet excitons by intersystem crossing [62]. There are also discrepancies between the PA_1 and PA_2 dynamics on the subpicosecond timescale. The match between the PA_1 and SE dynamics, on the other hand is almost perfect. Their ps decays are virtually identical, and the rise-time dynamics are also very close to each other, as shown in Figure 22.9b.

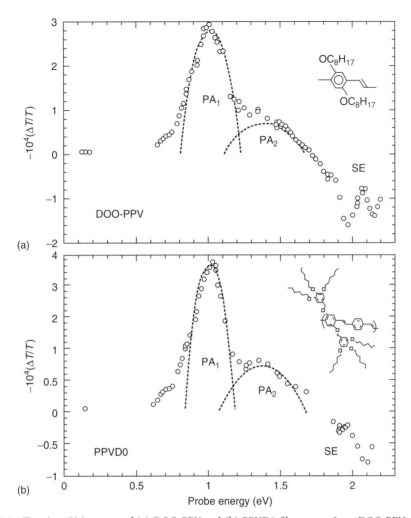

FIGURE 22.8 Transient PM spectra of (a) DOO-PPV and (b) PPVD0 films at $t = 2$ ps. DOO-PPV and PPVD0 repeat units are shown in the upper and lower *insets*, respectively. The dashed lines schematically mark the PA$_1$ and PA$_2$ bands; SE is the stimulated emission band. (From Frolov, S.V., et al., *Phys. Rev. B*, 65, 205209, 2002. With permission.)

From the ps decay we infer the average exciton lifetime, τ to be about 200 ps for DOO-PPV and 300 ps for the PPVD0 films, respectively.

A mid-IR spectral range between 0.11 and 0.21 eV (Figure 22.8) deserves special attention. This region corresponds to the absorption of IRAVs, which have been used for the identification of photogeneration of charge excitations [70,71]. A small transient PM signal was found in this range with a flat, featureless spectrum. This transient mid-IR PM response cannot be attributed to the IRAV modes, which are characterized by several pronounced narrow lines [71]. Figure 22.10a shows a comparison [55] between the PA dynamics at 8.2 μm (0.15 eV) and at 1.0 eV: both PA decays (Figure 22.10a) and onsets (Figure 22.10b) are identical. We therefore attribute the mid-IR PM response to $1B_u$ absorption, possibly the low-energy tail of the excitonic PA$_1$ band.

22.3.1.2 Excited States Relaxation Dynamics

We adopt the nomenclature of molecular spectroscopy to describe the excitation and relaxation processes in PPV derivatives, since the polymer photophysics is similar to the photophysics of large organic molecules [146]. Figure 22.11 shows schematically the configuration coordinate diagram of all

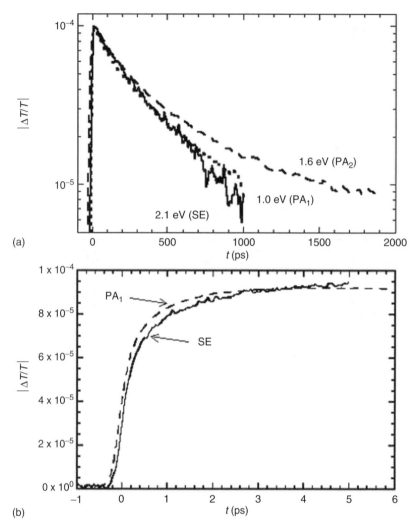

FIGURE 22.9   (a) $\Delta T/T$ decays at $E = 2.1$ eV (solid line), 1.6 eV (broken line), and 1.0 eV (dashed line) in PPVD0 films. (b) SE (solid line) and PA$_1$ (broken line) rise dynamics in PPVD0 films. (From Frolov, S.V., et al., *Phys. Rev. B*, 65, 205209, 2002. With permission.)

identifiable low-energy states in the family of PPV derivatives [55]. Due to the coupling to the nuclear coordinates (described by an effective configuration coordinate), each electronic state is characterized by a potential energy surface as shown by the parabolas in Figure 22.11. In this diagram the most probable optical transition corresponds to the vertical line connecting two different energy surfaces of opposite parities. Several different excitation manifolds are distinguished, marked by dashed boxes: the inhomogeneously broadened lowest singlet exciton ($1B_u$), the manifold of A_g states (mA_g through kA_g), the lowest triplet exciton (1^3B_u), and the manifold of charge excitations and charge transfer states (polarons and PPs). Other B_u states are not shown, since their dipole couplings are small. In this description it has been assumed that the essential photophysics is given by intrachain interactions and ignore possible complications due to interchain interactions. This assumption is supported by the similarity between the optical properties of solid films and their respective dilute solutions [34].

π-Conjugated polymers are generally characterized by significant inhomogeneous broadening [149]. This is particularly evident from the featureless band I in the linear absorption spectra (Figure 22.7), which shows a well-defined vibronic progression based on the $1B_u$ exciton. Instead, the distribution in

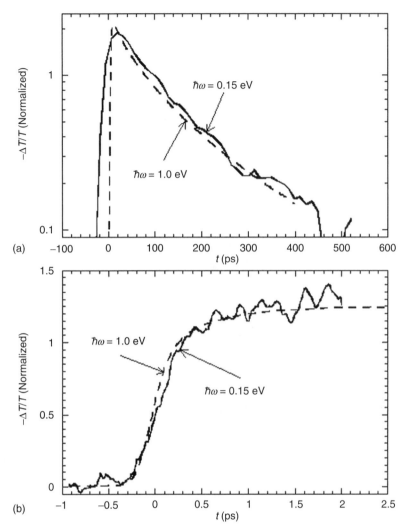

FIGURE 22.10   (a) $\Delta T/T$ decays at $E = 0.15$ eV (solid line) and 1.0 eV (broken line) in DOO-PPV films. (b) $\Delta T/T$ rise dynamics at $E = 0.15$ eV (solid line) and 1.0 eV (broken line) in DOO-PPV films. (From Frolov, S.V., et al., *Phys. Rev. B*, 65, 205209, 2002. With permission.)

the chain conjugation length leads to a broad distribution of $1B_u$ energies, which in turn smears out the vibronic structure of band I. This disorder also hinders the accurate identification of other excited states above the $1B_u$. The disorder is due to both polymer chain length fluctuations and conjugation length distribution within a single chain (determined by the amount of kinks and other defects on the chain). The inhomogeneous broadening thus opens an additional relaxation pathway for the excited states, which consequently results in exciton energy and spatial diffusion in the polymer film [150]. The diffusion occurs in the direction of longer conjugation chain segments, in which the exciton energy is lower.

Relaxation within the $1B_u$ manifold can be studied by monitoring the fs dynamics of SE, PA_1, or PA_2, which approximately follow the $1B_u$ population dynamics. Additional information about relaxation can be obtained from the polarization anisotropy decay, which is defined as a ratio between PM components with the probe beam polarizations parallel (XX) and perpendicular (XY) to the pump beam polarization. Figure 22.12 shows the PA_1 dynamics in DOO-PPV films in two different timescales, for two probe polarizations, and the ratios between them. As shown in Figure 22.9b, the rise-time response of PA_1 is

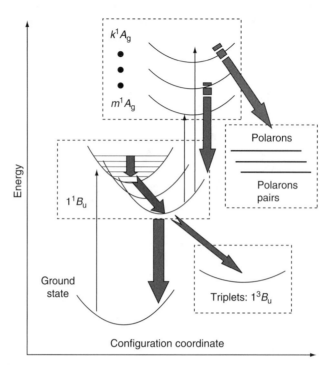

FIGURE 22.11   Configuration coordinate diagram of low-energy-excited states in π-conjugated polymers. Various excitation manifolds are marked by dashed line boxes. Narrow vertical arrows show optical transitions, whereas broad arrows indicate nonradiative relaxation pathways. (From Frolov, S.V., et al., *Phys. Rev. B*, 65, 205209, 2002. With permission.)

the same as that of SE. The rise-time kinetics is easily resolved with the 100-fs time resolution of the experiments. The excitation energy in these measurements was 3.2 eV; this means that photon absorption occurs into the highest vibronic sidebands of the $1B_u$ exciton and possibly some other allowed states (nB_u) below band II. Since the shorter chain segments have higher $1B_u$ energies, this excitation may also preferentially excite short conjugation chain segments. Thus, subsequent relaxation of the inhomogeneously broadened, vibrationally hot $1B_u$ excitons includes both phonon emission and exciton diffusion among different chain segments (Figure 22.11). The first process may affect the PA_1 magnitude, whereas the second process manifests itself in the decay of polarization anisotropy (or polarization memory).

Accordingly, the PA_1 onset dynamics can be attributed to cooling toward the new configuration equilibrium within the $1B_u$ manifold. This vibrational relaxation initially occurs with a time constant of 300 fs, which is followed by a slower component with a time constant of 830 fs. The onset dynamics may also be influenced by the exciton diffusion. However, the polarization decay on the sub-ps timescale shows a slower dynamics characterized by a time constant of 1.6 ps, suggesting that the exciton spatial diffusion occurs on a longer timescale compared to that of the intrachain exciton cooling. As shown in Figure 22.12b, the polarization ratio decay, and thus spatial diffusion, continues until PA_1 becomes isotropic at about 150 ps.

In an attempt to separate the intrachain (vibrational) and interchain (diffusion) relaxation channels within the $1B_u$ excitation manifold, the PA_1 decay and polarization anisotropy were measured in a very dilute solution of DOO-PPV in chloroform (few milligram perliter). Figure 22.13 [55] shows the polarized PA_1 dynamics in DOO-PPV dilute solution in two timescales (the poor sensitivity is due to the low DOO-PPV concentration in the solution). A major difference between the dynamics in films and dilute solutions was found in the decay of polarization memory. The PA_1 polarization memory in solution is completely preserved, at least up to 1 ns. This indicates that DOO-PPV chains in chloroform, which is considered to be a good solvent, are *uncoiled and straight*, and that exciton diffusion is limited

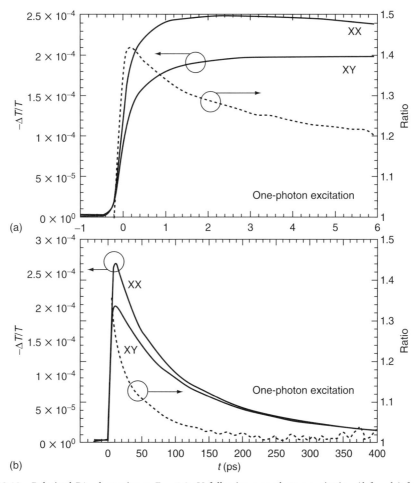

FIGURE 22.12 Polarized PA_1 dynamics at $E = 1.0$ eV following one-photon excitation (*left* scale) for the probe polarization parallel (XX) and perpendicular (XY) to the pump polarization, and the polarization anisotropy ratio (*right* scale) in DOO-PPV films in (a) short and (b) long timescales. (From Frolov, S.V., et al., *Phys. Rev. B*, 65, 205209, 2002. With permission.)

to the diffusion within single chains, suspended and isolated from each other by the solvent. On the contrary, the PA_1 rise-time dynamics in solutions is very similar to those in films, which indicates that the intrachain processes primarily determine the PA_1 sub-ps dynamics in films. We therefore conclude that spectral relaxation resulting from interchain diffusion is negligible in DOO-PPV films during at least few ps.

22.3.1.3 "Three-Beam" Spectroscopy

In order to clarify the relaxation of the A_g states, three-beam transient PM technique was used to monitor the exciton dynamics following optical reexcitation from $1B_u$ to mA_g or kA_g [148]. In this technique, as shown in Figure 22.14a, $1B_u$ excitons are initially populated by excitation from the ground state by the first pump pulse (1) at $\hbar\omega_1 = E_1 > E(1B_u)$. Following this the photogenerated $1B_u$ is reexcited after a delay time t_2 by a second pump pulse (2) at $\hbar\omega_2 = E_2$, tuned to a specific exciton transition (within PA_1 or PA_2 bands). The resulting exciton PM dynamics is monitored by a probe pulse (3) at $\hbar\omega_3 = E_3$ at a delay time t_3 using either an absolute or relative measuring mode. In the absolute mode ΔT due to both pump pulses is measured, whereas in the relative mode only the change, δT, due to the second pump pulse is detected using a double-frequency modulation (DM) technique [151]. For the

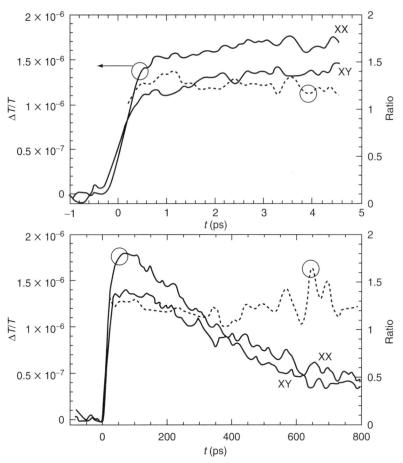

FIGURE 22.13   Same as in Figure 22.12 but for a dilute DOO-PPV solution. (From Frolov, S.V., et al., *Phys. Rev. B*, 65, 205209, 2002. With permission.)

DM technique, the first pump beam is modulated at 1 MHz, whereas the second pump beam is modulated at 1 kHz; the signal (δT) is first electronically mixed with the 1 MHz square wave, and then measured with a lock-in amplifier referenced at 1 kHz. The second pump switching efficiency, η, can be defined as $|\delta T/\Delta T|$. Figure 22.14b shows the experimental setup, in which two Ti:sapphire mode-locked lasers and the synchronously pumped opal produce the three pulsed beams, e.g., the lower Ti:sapphire laser produces the frequency-doubled first pump pulse; the unused portion of the upper Ti:sapphire laser produces the second pump pulse; and the signal port of the opal generates the probe pulse. In this arrangement, the jitter between the first and second pump pulses does not limit the temporal resolution of the recovery dynamics after reexcitation, since the second pump and probe pulses are perfectly synchronized. The two pump and one-probe beam polarizations were set to be parallel to each other. As discussed previously, among the A_g states there are only two prominent states with strong dipole coupling to $1B_u$; mA_g and kA_g. The mA_g relaxation is studied first.

22.3.1.3.1 mA_g Relaxation Dynamics

The mA_g relaxation dynamics were measured by the three-beam technique following $1B_u$ reexcitation at $E_2 = 1.0$ eV (within the PA$_1$ band) and four different t_2 [55]. The resulting decay of $1B_u$ population is observed by monitoring the SE dynamics (δSE) at $E_3 = 2.0$ eV, as shown in Figure 22.15a for the PPVD0 film. It is seen that η is independent of t_2, which indicates that only $1B_u$ excitons are involved in the

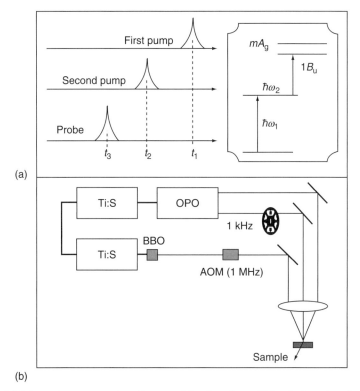

(a)

(b)

FIGURE 22.14   (a) Schematic representation diagram of the three-beam PM technique that shows two excitation pulses, and one probe pulse; and the associated optical transitions induced in the polymer film. (b) Schematic representation diagram of the three-beam experimental setup. Ti:S are Ti:sapphire lasers; OPO is an optical parametric oscillator; BBO is barium borate crystal doubler; and AOM is acousto-optic modulator. (From Frolov, S.V., et al., *Phys. Rev. B*, 65, 205209, 2002. With permission.)

reexcitation process [148]. The mA_g quickly relaxes back to $1B_u$ by internal conversion with a time constant of about 200 fs (Figure 22.15a inset) [55]. The origin of a slower δT decay component with a time constant of ~3 ps is unclear at the present time. For $E_2 = 0.8$–1.1 eV over 99% of mA_g excitons recover back to $1B_u$ within 10 ps following reexcitation. Similar mA_g relaxation dynamics are measured in the dilute polymer solutions.

In addition to $1B_u$ depletion, as measured by transient δSE (Figure 22.15a inset), it was observed [55] a concomitant transient PA from the reexcited mA_g excitons at $E_3 = 1.35$–1.6 eV with identical δPA dynamics to that of δSE. Figure 22.15b shows δPA_2 dynamics at $E_3 = 1.53$ eV following reexcitation into mA_g; the transient increase in PA (δPA) has the same decay as δSE and thus is attributed to the same process. This δPA can be tentatively assigned to a transition from mA_g into B_u states that corresponds to band III in Figure 22.7, since $E(mA_g) + E_3 = E(\text{band III}) = 4.7$ eV.

22.3.1.3.2 kA_g Relaxation Dynamics

A dramatically different δT dynamics was found when E_2 was increased up to 1.6 eV (now within the PA_2 band), when the kA_g state is reached [55]. In these measurements the $1B_u$ recovery dynamics at $E_3 = 0.6$–1.0 eV (within PA_1) was probed, as shown in Figure 22.16a inset. Figure 22.16a shows the PA_1 dynamics with and without the second pump pulse at $E_2 = 1.6$ eV and $t_2 = 6$ ps. Reexcitation of the $1B_u$ exciton into kA_g does not result in an ultrafast recovery back to the $1B_u$. Figure 22.16b shows the $1B_u$ population decay probed at $E_3 = 0.6$ eV (within PA_1) with and without the second pump pulse at $t_2 = 25$ ps. As can be seen from the normalized δT decay (Figure 22.16b inset) the $1B_u$ recovery, if any is very

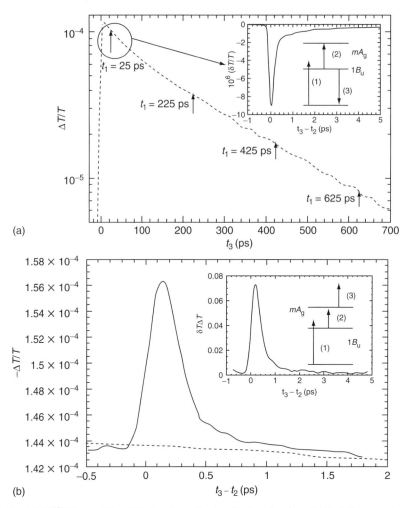

FIGURE 22.15 (a) $\Delta T/T$ decay at $E_3 = 2.0$ eV without (broken line) and with (solid line) the second pump pulse at $E_2 = 1.0$ eV, for four different t_2 values in PPV-D0 films. The *inset* shows the corresponding $\delta T/T$ decay and the three-beam energy diagram. (b) $\Delta T/T$ decay at $E_3 = 1.6$ eV without (broken line) and with (solid lines) the second pump pulse at $E_2 = 1.0$ eV in PPVD0 films. The *inset* shows the corresponding $\delta T/T$ decay, and the three-beam energy diagram. (From Frolov, S.V., et al., *Phys. Rev. B*, 65, 205209, 2002. With permission.)

slow compared to the exciton recombination process. It was also found that kA_g relaxation could be probed directly by monitoring a transient PA that originates from kA_g at $E_3 = 1.6$–1.8 eV. It was concluded that kA_g decays with high quantum yield into a relatively long-lived state, other than $1B_u$. Similar kA_g relaxation dynamics were observed in polymer solutions showing that the long decay from kA_g is an intrinsic process of the isolated polymer chain.

Few different relaxation routes (other than internal conversion back to $1B_u$) can be envisioned for kA_g [148]. (i) One route is the internal conversion directly to the ground state. The Franck-Condon factors for this nonradiative transition, however, are smaller than those for the $1B_u$ state itself and thus makes this process highly improbable. (ii) The second route is singlet fission, where one kA_g singlet exciton decomposes into two triplets with opposite spins. (iii) The third route is exciton dissociation into free charges (or polarons). Both (ii) and (iii) routes are energetically possible and in fact, they may occur concurrently. However, it is known [34] that triplets are characterized by a lifetime of few microseconds at room temperature and have a strong PA band that peaks at 1.45 eV. This long-lived PA band was

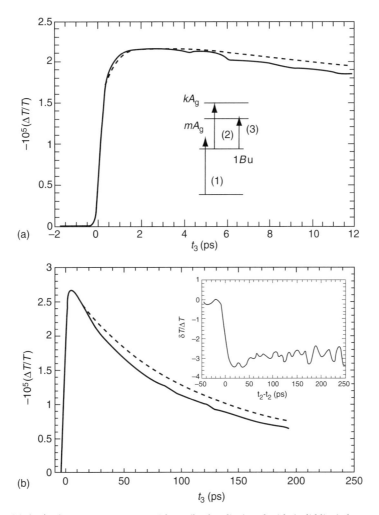

FIGURE 22.16   (a) $\Delta T/T$ decay at $E_3 = 1.0$ eV without (broken line) and with (solid line) the second pump pulse at $E_2 = 1.6$ eV and $t_2 = 6$ ps in DOO-PPV films. The *inset* shows the three-beam energy diagram. (b) $\Delta T/T$ decay at $E_3 = 0.6$ eV without (broken line) and with (solid line) the second pump pulse at $E_2 = 1.6$ eV and $t_2 = 25$ ps in DOO-PPV films. The *inset* shows the normalized δT decay. (From Frolov, S.V., et al., *Phys. Rev. B*, 65, 205209, 2002. With permission.)

measured in DOO-PPV films at the modulation frequency of 10 kHz, using a steady state PM setup described in [155], and it does not fit the δPA spectrum from the kA_g. Thus, singlet fission does not appear to be supported by the transient spectroscopy measurements. Further measurements using changes in PL induced by the three-beam technique were needed to elucidate the kA_g relaxation route. Using this technique it was concluded that route (iii), the exciton dissociation is the most probable cause for the slow kinetics of the kA_g relaxation rate.

The three-beam data show a clear difference between the mA_g and kA_g relaxation pathways in PPV derivatives, and illustrate the different characters of these states. The classification of A_g states in PPV consisting of two distinct classes is therefore justified. In the earlier studies mA_g was described as an excited $1B_u$ exciton with energy close to the continuum edge; whereas kA_g was assigned to a biexciton (Figure 22.4). The latter assignment, however has been questioned [81]. The diagrammatic exciton basis valence band approach was used [17] to describe the low energy excited states in PPV. The low energy even parity states are produced as combinations of charge-transfer configurations (one-excitation, in which an electron is moved to the neighboring unit on the same chain) and triplet–triplet configurations (two-excitations, which are composed of two coupled triplets on different units with overall zero spin), all of which involve the delocalized MOs (d and d^*) much the same way as B_u states corresponding to bands I and II. The high energy even parity states contain configurations of the same charge transfer one-excitations, however the triplet–triplet configurations here involve both delocalized and localized MOs (d and d^*, l and l^*). This may render the kA_g state to have a very different relaxation routes compared to the mA_g state [81,152].

22.3.2 Excitation Dynamics in Pristine and C_{60}-Doped MEH-PPV

22.3.2.1 Pristine MEH-PPV Films and Solutions

MEH-PPV polymer apparently shows very different properties than those of DOO-PPV and PPVD0 polymers discussed in Section 22.3.1. Due to its side group the MEH-PPV polymer is able to form an ordered phase when films are cast from "bad" solvents such as toluene [153,154]. In such films two or more polymer chains are coupled together due to increased interchain interaction, which renders the benzene rings of the polymer backbone to face each other. Under these conditions the primary excitations and polaron wavefunctions may acquire enhanced 2D delocalization that is in fact spread over two or more chains. Such delocalized photoexcitations were identified in other polymers with improved order such as PFO, mLPPP [54], and regioregular P3HT [72], which will be discussed later.

The improved order in films cast from toluene may also be induced by prolonged illumination at low temperatures, as recently discovered at the Technion [155]. Delocalized photoexcitations among adjacent chains may lead to several characteristic properties that are not common in isolated chains. These are [71]: (i) reduced PL quantum efficiency, (ii) PL redshift, (iii) relatively large generation of PP excitations, rather than intrachain excitons, (iv) more substantial delayed PL due to PP recombination, (v) reduced formation efficiency of triplet excitons since the intersystem process is hampered by the interchain delocalization [156], and (vi) increased exciton dissociation efficiency in C_{60}-doped polymer films. It is thus important to review the ultrafast excitation dynamics in MEH-PPV in comparison with those of DOO-PPV.

Figure 22.17 shows the absorption and PL spectra of MEH-PPV in toluene solution and film cast from toluene solution [157]. It is seen that although the peak absorption is the same, occurring at ~2.5 eV, there is enhanced absorption tail in the film. In addition, the PL spectrum in the film is redshifted and broadened compared to that in the solution. The absorption tail cannot be formed in the film due to a different conjugation length distribution in the chains, since the film was cast from the same solution and thus the conjugation length distribution is the same for the polymer chains in

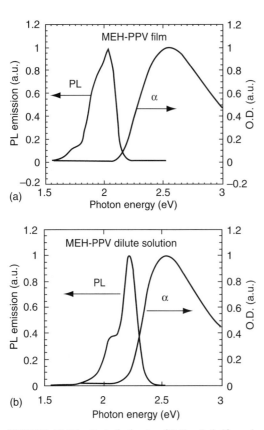

FIGURE 22.17 Optical density [O.D. $\sim (\alpha(\omega))$] and photoluminescence (PL) spectra of (a) MEH-PPV film and (b) MEH-PPV dilute solution. (From Sheng, C.X., Ph.D. thesis, University of Utah, 2005, unpublished.)

the film and solution. We therefore conjecture that the film contains substantial amount of "coupled" chains where two or more chains are coupled together by an enhanced inter-chain transfer integral caused by the chain arrangement in the film. This coupling causes a redshift in the absorption and PL spectra as seen in Figure 22.17a [158]. The featureless PL spectrum in the film compared to that in the solution point out to a more delocalized excitons in the film having a reduced Huang–Riess coupling parameter.

The transient spectrum of the primary excitations in MEH-PPV solution at $t \approx 0$ is shown in Figure 22.18a. Figure 22.18b shows the transient dynamics of the various bands in Figure 22.18a spectrum [157]. There are two PA bands, PA_1 at 1 eV and PA_2 at 1.6 eV, and an SE band at 2.2 eV, similar to the transient PM spectrum of DOO-PPV. But in the case of MEH-PPV the PA spectrum was extended to the mid-IR range in regions that the toluene solution does not show significant IRAV bands; still a third PA band was not observed in the transient PM spectrum. This is significant since the transient PM spectrum in DOO-PPV shown before could, in principle, contain a third PA band in the mid-IR spectral range. It is also seen that all bands decay together up to 1 ns and thus belong to the same excitation, which we identify as singlet intrachain exciton based on the appearance of the SE band.

The transient PM spectrum of MEH-PPV film at $t \approx 0$ ps is shown in Figure 22.19 [157]. There are two PA bands and one SE band similar to those in MEH-PPV solution. However, the most striking difference between the PM spectra in MEH-PPV film and solution is the appearance of a third PA band in the film that peaks at ~0.35 eV. The relative weaker PA band in the near IR range in the film's PM spectrum and the splitting of the SE band into two components are also seen. We identify the two components as SE and PB of the absorption, since the PL and absorption in the film are redshifted compared to those in the solution. This also may be the reason that the PA band in the near-IR range is weaker in the film.

We also attempted to measure possible photoinduced IRAVs in the film. None could be identified at room temperature, but significantly, photoinduced IRAVs were observed at 80 K (Figure 22.20a). The ps photoinduced IRAV spectrum is compared with that in the steady state PA measurements of MEH-PPV/C_{60} mixture, where ample photogenerated polarons have been previously observed [159]. The IRAV spectrum in the ps time domain of pristine MEH-PPV at 80 K, and that of steady state MEH-PPV/C_{60} mixture are identical. In addition, the IRAV's photogeneration is instantaneous as seen in Figure 22.20b. The reason why photoinduced IRAVs appear at 80 K, but not at room temperature is not clear at the present time. However, from the observation of IRAV in the PM spectrum of pristine MEH-PPV we conclude that *charged excitations are instantaneously photogenerated in the film*; this is consistent with several ps transient measurements completed recently at different laboratories [70,71]. Since the two PA bands in the film at 1.0 and 1.6 eV, respectively, appear also in solution, where charge

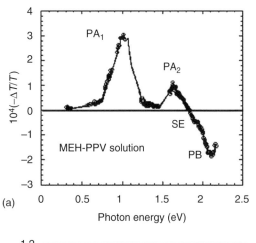

(a)

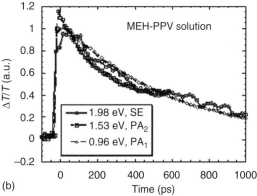

(b)

FIGURE 22.18 (a) The transient PM spectrum of MEH-PPV dilute solution at $t \approx 0$ ps. Various PA, PB, and SE bands are assigned. (b) Transient ps decay dynamics up to 1 ns at various probe energies. (From Sheng, C.X., Ph.D. thesis, University of Utah, 2005, unpublished.)

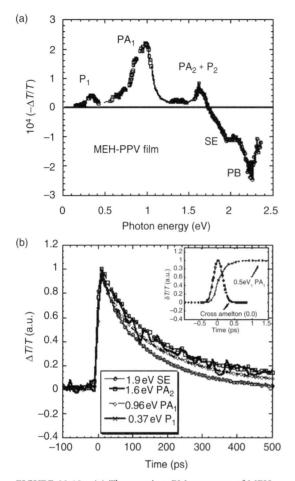

FIGURE 22.19 (a) The transient PM spectrum of MEH-PPV film at $t \approx 0$ ps. Various PA, PB, and SE bands are assigned. (b) Transient ps dynamics at various probe energies up to 500 ps. The *inset* shows the PA$_1$ onset dynamics in more detail; the dashed square line shows the pump–probe cross-correlation trace. (From Sheng, C.X., Ph.D. thesis, University of Utah, 2005, unpublished.)

excitations are not usually photogenerated, the PA band at 0.35 eV is also identified due to charge excitations. The most probable charge excitations in polymers are polarons [8], and thus we identify the 0.35 eV PA as due to P$_1$ band of photogenerated polarons in MEH-PPV chains. Since there are no accompanying IRAV at room temperature we conclude that the PPs generated at room temperature may be highly correlated and thus the IRAV pinning parameter may be large, and consequently the IRAVs may be weak [65]. At low temperature the photogenerated PPs may dissociate reducing the pinning parameter and increasing the IRAV intensities, so that the polaron-related IRAVs may be better observed. The PM spectrum was also extended in DOO-PPV films toward the mid-IR range, thus completing the ps spectrum in Figure 22.8; no extra PA band was identified in the energy range from 0.2 to 0.6 eV [157]. Therefore, the appearance of P$_1$ PA band in MEH-PPV film is unique to this polymer. This is in agreement with previous measurements of ps photogeneration of charge excitations in MEH-PPV using transient PC and IRAVs [70,71].

22.3.2.2 C$_{60}$-Doped MEH-PPV Films

It is known that C$_{60}$ doping of many π-conjugated polymers promotes a strong photoinduced charge transfer process (by weight), where the photogenerated excitons dissociate in record time into positive polaron on the polymer chain and negative polaron on the C$_{60}$ molecule [21,22]. In 50% C$_{60}$ doping of MEH-PPV it was measured that the photoinduced dissociation occurs with time constant of ~50 fs [160]. It was not clear, however whether this fast process occurs with the same time constant in MEH-PPV films with less C$_{60}$ dopants. Since the exciton and polaron PA in the mid-IR range are separated so nicely in the transient PM spectrum of MEH-PPV in the mid-IR spectral range, then the mid-IR range is ideal to measure the dynamics of the photoinduced charge transfer in MEH-PPV with less dopant density.

Figure 22.21a shows the mid-IR PM spectra of 10% C$_{60}$-doped (by weight) MEH-PPV film at various delay times; $t \approx 0$, 2, and 100 ps following the pulse excitation [157]. It is seen that the PA$_1$ exciton band at 0.9 eV gradually disappears from the PM spectrum as the exciton dissociation occurs. In parallel with the decrease of PA$_1$, photoinduced IRAV at ~0.18 eV gradually appears in the PM spectrum (Figure 22.21b). This shows that photogenerated excitons decay into charge polarons indicating that exciton dissociation indeed occurs between the polymer chains and C$_{60}$ molecules [21,22]. However, the charge transfer reaction is not as fast as that measured in the 50% C$_{60}$ doping [160]. From Figure 22.21b we conclude that the IRAV dynamics is delayed with respect to the exciton PA$_1$ instantaneous response.

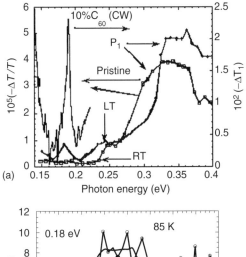

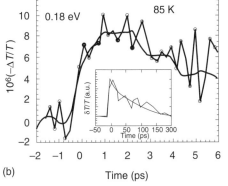

FIGURE 22.20 (a) Transient PM spectra of pristine MEH-PPV film at $t = 0$ ps at 85 K and room temperature, respectively. (b) $\Delta T/T$ onset dynamics at $\hbar\omega = 0.18$ eV and 85 K with a line "to guide the eye"; the *inset* is $\Delta T/T$ decay dynamics at $\hbar\omega = 0.18$ eV at 85 K. (From Sheng, C.X., Ph.D. thesis, University of Utah, 2005, unpublished.)

We estimate that the IRAVs are photogenerated within ~2 ps in the 10% C_{60}-doped film, in contrast to the time constant of about 50 fs in 50% C_{60}-doped films [160]. This indicates that the photoinduced charge transfer reaction in C_{60}-doped films is actually limited by the exciton diffusion toward C_{60} molecules close to the polymer chains. Consequently, the exciton wavefunction in MEH-PPV films is not as extended as previously thought. The same conclusion was drawn for C_{60}-doped DOO-PPV films, where it was estimated [161] that the exciton diffusion constant to reach the C_{60} molecules is of the order of 10^{-4} cm^2/s.

22.3.3 Photoexcitations in Regiorandom and Regioregular Poly-3 Alkyl Thiophenes

For many of the optoelectronic applications, the degree of structural order of the π-conjugated polymer's active layer has been recognized as one of the key parameters governing the photophysics and consequently its specific application for the optimal device performance. Whereas interchain interactions are usually detrimental for the π-conjugated polymers use in LEDs due to weak optical coupling to the ground state [162], charge transport in other applications such as thin-film field-effect transistors (TET) and solar cells requires good contacts between neighboring conjugated chains in the film. In addition, the quasi-1D electronic structure of π-conjugated polymers results in strong self-localization of the electronic excitations, which may also limit the carrier mobility to values of the order of 10^{-4} cm^2/V s [163].

Much higher mobilities, of the order of 0.1 cm^2/V s, have been recently obtained, however with regioregular-substituted poly-3-hexyl thiophene (RR-P3HT) (see Figure 22.22b inset) in FET [164]. Such films were also successfully used in an FET device to drive an OLED that demonstrated an all-organic display pixel [165]. The reason for the dramatic increase in carrier mobilities is that self-organization of RR-P3HT chains results in a lamellae structure perpendicular to the film substrate [164], where 2D sheets are formed by strongly interchain interaction due to the short interchain distance of the order of 3.8 Å. Delocalization of the charge carrier has been invoked to be the reason for the high interlayer mobility. Recent optical studies of RR-P3HT films, where delocalized polaron excitations on adjacent chains have been measured using charge-induced optical techniques in FET [164,166], and direct photogeneration in thin films [172] have confirmed this assumption. On the contrary, P3HT films casted from polymer chains having regiorandom (RRa) order (see Figure 22.22a inset) do not form ordered lamellae and the obtained field-effect carrier mobilities in FET are consequently small [164]. The main reason for the reduced carrier mobility is the lack of sufficiently strong interchain interaction that is caused by the chain-like film morphology. Thus, the effect of the strong interchain interaction on excitons in P3HT films should be quite interesting [162].

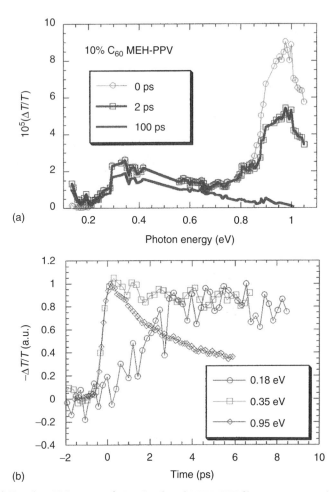

FIGURE 22.21   (a) Transient PM spectra of 10% C_{60}-doped MEH-PPV film at $t \approx 0$, 2, and 100 ps following the pulse excitation. (b) Transient PM responses at various probe energies. (From Sheng, C.X., Ph.D. thesis, University of Utah, 2005, unpublished.)

The samples used for these studies were thin films of RR-P3HT (92% RR order) that were grown in Jensen's laboratories in the Netherlands, as well as RR-P3HT and RRa-P3HT that were purchased from Aldrich. The films were cast from chloroform or xylene solutions (usually at a concentration of a few milligram per milliliter) onto sapphire or quartz substrates. Special care was taken to minimize contamination of the powders and films by oxygen and water at ambient conditions.

Figure 22.22 shows the absorption and PL spectra of RRa-P3HT and RR-P3HT films at room temperature. The strong absorption band over the gap is due to $\pi-\pi^*$ transitions, and according to Kasha's rule, the PL emission comes from the lowest exciton in the system. We note the redshift of the RR-P3HT absorption and PL bands with respect to those in RRa-P3HT, which is caused by the superior order in the lamellae [164]. The planar order leads to polymer chains with longer conjugations caused by fewer defects, such as twists and radicals on the chains. In addition, the absorption and the PL bands of the RR-P3HT film show pronounced structures due to phonon replica, indicating that the polymer chains in this film are more homogeneous than those in RR-P3HT films. In spite of the superior order, the PL quantum efficiency, η in RR-P3HT is measured to be 0.5%, which is an order of magnitude lower than in RRa-P3HT ($\eta \approx 8\%$). The PL η decrease in RR-P3HT cannot be explained by an increase in the nonradiative decay rate, because this film contains fewer defects and intersystem crossing to the triplet manifold is absent. Therefore, the PL η decrease in RR-P3HT is due to a weaker *radiative* transition of

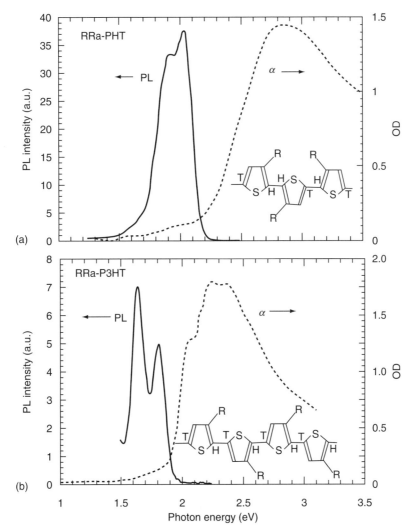

FIGURE 22.22 The room temperature absorption and photoluminescence spectra of (a) RRa-P3HT and (b) RR-P3HT films. The respective polymer regio-order structures are given in the *insets*. (From Korovyanko, O.J., et al., *Phys. Rev. B*, 64, 235122, 2001.)

the lowest lying excitons in this film. The weaker optical transition may be due to a larger interchain contribution for the lowest excitons in the lamellae, compared with the usual intrachain excitons in RRa-P3HT. Indeed an interlayer separation of 3.8 Å in RR-P3HT lamellae causes a stronger interchain–interlayer interaction, as recently calculated using numerical quantum chemical methods [158,162]. For an H-aggregate chain configuration this interaction leads to splitting of the HOMO and LUMO levels, so that the lower, redshifted LUMO level becomes optically forbidden [162]. The higher splitted LUMO level is optically allowed in this model [162], and may be directly seen in absorption, which should be blueshifted with respect to the PL.

The ps dynamics studies in P3HT further elucidated the excitons characteristic properties [156]. Figure 22.23 shows the transient PA spectra of P3HT films with both regio-orders measured at $t = 200$ fs and $t \approx 5$ ps following the pulse excitation. The transient PA spectra of both films show three PA bands (PA_1, PA_2, and PA_3), but only the fs spectrum of RRa-P3HT contains an SE band that is due to the photogenerated excitons. The lack of SE in RR-P3HT was explained in Ref. [158] as due to the much

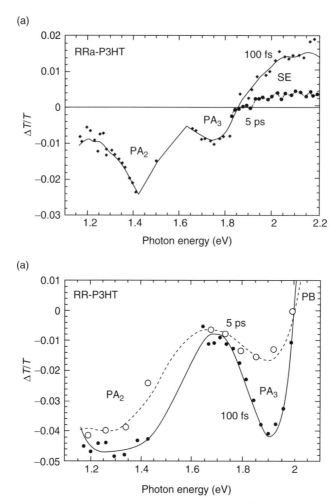

FIGURE 22.23   The transient PA spectra of RRa-P3HT (a) and RR-P3HT (b) measured at $t \approx 100$ fs (full diamonds and circles) and $t = 5$ ps (full and empty circles), respectively following the pulse excitation. Various PA, SE, and PB bands are assigned. (From Korovyanko, O.J., et al., *Phys. Rev. B*, 64, 235122, 2001.)

smaller oscillator strength for the photogenerated excitons in this material, confirming the above conclusion that the lowest lying excitons in RR-P3HT films are, in fact optically forbidden [158].

To better characterize the primary excitations in RRa-P3HT that are revealed in the ultrafast PA spectra, the transient dynamics of each PA band was studied and compared with that of the SE band [156]. Figure 22.24a shows that the dynamics of PA_1 and PA_2 are similar to that of SE; the bands initially decay together, but the SE band disappears from the spectrum at ~15 ps (Figure 22.23a) that is caused by the competing effect of PA_3. Due to their correlated dynamics it was concluded that PA_1, PA_2, and SE belong to the same species, i.e., intrachain excitons. Similar conclusion was drawn before for the correlated excitonic PA and SE in other π-conjugated polymer films [34,167–171].

On the contrary, PA_3 decay in RRa-P3HT is much slower than that of SE and PA_2 (Figure 22.24a) and therefore it does not involve intrachain excitons. In fact it decays only by a factor of about two from its initial, $t \approx 0$ value within 200 ps. Therefore, it was concluded that the PA_3 band is due to interchain, trapped PPs. Similar conclusion was drawn previously in ps studies of other π-conjugated polymer films [167–170]. Such excitations may also be instantaneously generated onto two adjacent chains at a place on the chains close to their "contact-point," where the interchain distance is the smallest. In view of these

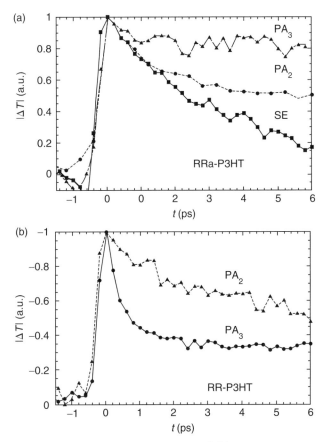

FIGURE 22.24 Transient PA and SE decays in (a) RRa-P3HT and (b) RR-P3HT up to 6 ps. (From Korovyanko, O.J., et al., *Phys. Rev. B*, 64, 235122, 2001.)

results it is clear that in π-conjugated polymer films with superior order simultaneously with intrachain excitons, PPs on adjacent chains may also be photogenerated. If the polarons reside on different chains then the PP species can also give rise to photoinduced IRAV. However, as a result of the Coulomb attraction between the oppositely charged polarons, the PP species are trapped near the contact point of the adjacent chains at which they had been originally generated. The slow dynamics of the PP species may also be obtained by the decay of the polarization memory, $P(t)$. Indeed we measured a long $P(t)$ transient decay for the PA_3 band, which is much longer than that of PA_1 and PA_2.

The PA dynamics of two of the PA bands in RR-P3HT films are shown in Figure 22.24b. As in RRa-P3HT it was found [156] that the ps transient decays of PA_2 and PA_3 in this film are not correlated. Moreover the decay dynamics of PA_2 depends on the excitation intensity, I showing faster decay at higher I that is perhaps due to exciton–exciton annihilation, whereas the decay dynamics of PA_3 is independent from I. From this and the similarity with the PA transient spectrum in RRa-P3HT discussed above, it is concluded [156] that PA_2 and PA_3 in RR-P3HT are likewise due to excitons and geminate PPs, respectively. The lack of ps SE band in RR-P3HT shows that the excitons in the lamellae are not regular intrachain excitons. This may explain the reason why the excitonic PA bands in this film, i.e., PA_1 and PA_2 (Figure 22.23b) have a much broader spectrum than that in RRa-P3HT (Figure 22.23a). The observed broadening may be due to the splitting of the even parity states (A_g) that are optically coupled to the lower lying excitons (or "$1B_u$"). Similar to the LUMO level [162] these A_g states may also split due to the increased interlayer interchain coupling in the lamellae causing excess broadening to their optical transitions.

The band PA_3 in RR-P3HT (Figure 22.23b) is substantially stronger than that in RRa-P3HT (Figure 22.23a). This shows that PP generation is more efficient in RR-P3HT films. This is not surprising since the polymer chains in the lamellae are longer and closer to each other, so that the "contact points" between any two adjacent chains are therefore more extended. Moreover the extended contact points for the polymer chains in the lamellae lead to a more mobile PP species in RR-P3HT compared with that in RRa-P3HT. This results in a faster PA_3 decay in RR-P3HT compared with that in RRa-P3HT (Figure 22.24a and Figure 22.24b). PA_3 transient decay in RR-P3HT was fit with a two-component decay function; a fast component using a $t^{-1/2}$ decay with a 0.5 ps lifetime, and a slow component. The $t^{-1/2}$ decay is in agreement with the diffusion-limited geminate recombination considered for the photogenerated geminate soliton–antisoliton pair recombination in t-$(CH)_x$ [172,173]. Similar to RRa-P3HT, the slow PA_3 component in RR-P3HT decays only slightly during the 200-ps time interval of measurements. The two PA_3 decay components were therefore interpreted as due to PPs that are generated in the lamellae, which are more mobile and consequently have faster dynamics, and PPs that are generated in the disordered portions of the polymer film. The latter, similarly as in RRa-P3HT are less mobile and consequently have slower dynamics. This two-component dynamics is also apparent in $P(t)$ decay obtained at probe energies within the PA_3 band. $P(t)$ was found to quickly decay to a plateaus level within about 1 ps, similar to the PA_3 fast component decay [156]. This showed that the fast PA_3 component belongs to the more mobile species, such as PPs in the lamellae.

22.3.4 Photoexcitations in Poly(diphenyl-acetylene)

The disubstituted polyacetylene, i.e., polydiphenyl-acetylene) (PDPA), of which backbone structure is shown in Figure 22.25 inset, is unique among the class of π-conjugated polymers [27]. On the one hand, PDPA has a strong PL band that has been used in optoelectronic applications such as OLED [174], and solid state lasers [27,175]. On the other hand, this polymer has been shown to have a degenerate ground state [27], which supports topological soliton excitations [8]. It has been shown that the PL quantum efficiency in π-conjugated polymers is associated with the order of the lowest lying excited states with odd ($1B_u$) and even ($2A_g$) parities [176,177]. If the energies $E(1B_u) < E(2A_g)$, then the polymer is strongly luminescent; conversely, if $E(2A_g) < E(1B_u)$, then the polymer is only weakly luminescent. It has been shown by Mazumdar et al. that in PDPA $E(1B_u) < E(2A_g)$ in spite of its polyene backbone [178,179]. Surprisingly, it was discovered [27] that the steady state PM spectrum of PDPA films contains long-lived soliton excitations (neutral solitons, $S°$, as well as charged solitons, $S^{\pm}$), for which the

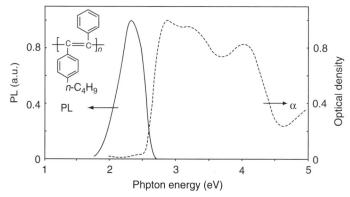

FIGURE 22.25 Room temperature optical spectra of PDPA-nB_u film absorption, α (in terms of optical density) and photoluminescence (PL). The polymer repeat unit is shown in the *inset*. (From Korovyanko, O.J., et al., *Phys. Rev. B*, 67, 035114, 2003. With permission.)

photogeneration mechanism has remained a mystery. A persistent debate exists whether soliton excitations are by-products of photogenerated intrachain excitons; or conversely, excitons are unstable toward the formation of soliton–antisoliton pairs. Other types of photoexcitations, such as polarons were also discovered in PDPA in solution [180], where again, their photogeneration mechanism has remained largely unclear.

The primary photoexcitation dynamics in PDPA solutions and films in the fs to ps time domain using transient PM spectroscopy were extensively studied [182]. The PDPA polymer used was a disubstituted biphenyl derivative of *trans*-polyacetylene, where one of the hydrogen-substituted phenyl groups was attached to a butyl group, which is referred to as PDPA-nB_u (Figure 22.25 inset) [181]. The polymer films were cast on sapphire substrates from a toluene solution; the same solution was used for measuring the photoexcitation dynamics in a PDPA-nB_u solution.

The optical absorption and PL spectra of a PDPA-nB_u film at room temperature (Figure 22.25) have been studied in detail [28]; the respective spectra in PDPA-nB_u solution are very similar. The relatively broad absorption band with an onset at 2.65 eV, and peaks at about 2.85 eV and 3.05 eV, respectively are due to delocalized π–π* transitions involving optical transitions from the ground state ($1A_g$) to the first odd parity exciton band ($1B_u$), and phonon replicas. This absorption band is broadened by the inhomogeneity in the sample caused by a distribution of the polymer chain conjugation lengths. The band at 4 eV is due to delocalized to localized transitions [178]. The featureless PL band with an onset at 2.65 eV and peak at 2.4 eV somewhat resembles the first absorption band, with an apparent Stokes shift of about 0.45 eV between the peaks of the respective bands. It was found that the PL emission has a quantum efficiency of about 50% both in solid films and solutions, which is considered to be relatively large in the class of π-conjugated polymers, and is thus in agreement with the assumed order $E(1B_u) < E(2A_g)$ [179].

The ultrafast excitation dynamics in dilute PDPA-nB_u solution were studied by the transient PM spectra as shown in Figure 22.26a [182]. Upon photoexcitation (or at "$t = 0$") an SE band at 2.4 eV and two PA bands with peaks at 1.05 eV (PA_1) and 2.0 eV (PA_2), respectively are formed. SE band is polarized preferentially parallel to the pump polarization, i.e., mainly along the polymer chains, and therefore it is assigned as due to intrachain excitons. Since the SE and the cw PL bands (Figure. 22.25 and Figure 22.26a) are essentially the same, the PL in PDPA-nB_u is attributed to intrachain excitons with dipole moment lying along the polymer chains. Figure 22.26 inset shows that the SE lifetime is about 120 ps, similar to those of the two PA bands [182]. This demonstrates that intrachain excitons in PDPA-nB_u have a strong SE band in the visible spectral range and two correlated PA bands in the visible/near-IR spectral range, similar to intrachain exciton spectra in other luminescent π-conjugated polymers [167–171].

Figure 22.26b shows the room temperature steady state PM spectrum of PDPA-nB_u in solution. The PM spectrum consists of two correlated PA bands at 0.25 and 2.35 eV, respectively, followed by PB of the π*–π transition (not shown). The lower energy PA band is correlated with photoinduced IRAV seen at energies below ~0.2 eV. It was therefore concluded that the underlying long-lived species are charged, and, in accordance with previous studies using PA-detected magnetic resonance [1], have spin $\frac{1}{2}$. Therefore, the two PA bands are due to long-lived charged polarons (Figure 22.26b inset). Compared to the ps transient results we conclude that the long-lived polarons are generated in PDPA-nB_u solution by exciton dissociation, and are therefore not the primary photoexcitations. This was also recently seen in another ps transient dynamics study of PDPA in solution, where the time in which excitons dissociate into polarons was measured to be about 200 ps [180].

Figure 22.27a shows the ps transient PM spectra in a PDPA-nB_u film [182]. In addition to the SE band and two PA bands at 1.03 (PA_1) and 2.0 eV (PA_2), respectively, which, as in Figure 22.26a for PDPA-nB_u solution are due to photogenerated $1B_u$ excitons, the transient PM spectrum in the polymer film also shows a PA band at 1.7 eV (PA_g). At 200 ps PA_g significantly narrows (Figure 22.27a inset); it becomes the dominant PA feature in the PM spectrum at longer delay times. The transient dynamics of the various PM bands (Figure 22.28) show that SE shares the same dynamics as PA_1 and PA_2 having a lifetime of about 50 ps, but it does not show the same dynamics as PA_g. In fact, whereas SE has an

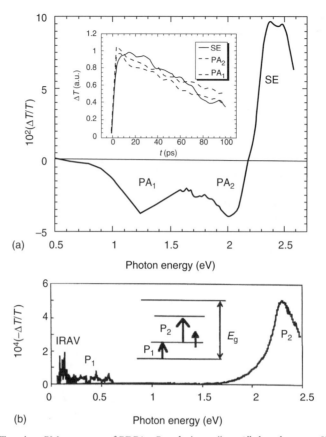

FIGURE 22.26 (a) Transient PM spectrum of PDPA-nB_u *solution* at "$t = 0$" that shows an SE and two correlated PA bands of excitons; the onset of a third PA band is assigned. The *inset* shows the decay kinetics of the SE (full line) and two PA bands (dashed lines). (b) The steady state PM spectrum at 80 K that shows two other correlated PA bands and photoinduced IRAVs of polaron excitations. The *inset* shows the energy levels and optical transitions associated with a positively charged polaron excitation. (From Korovyanko, O.J., et al., *Phys. Rev. B*, 67, 035114, 2003. With permission.)

exponential decay, the decay kinetics of PA_g is complicated. This band decays much faster in the first few ps, but basically stops decaying after about 30 ps. Two possible scenarios can explain this latter dynamics: either PA_g changes from one type of photoexcitation to another during the time period of about 10 ps, or the photoexcitations associated with PA_g experience "random walk-type" geminate recombination with recombination time of about 10 ps [172].

Figure 22.27b demonstrates the steady state PM spectrum in the PDPA-nB_u film at 80 K. In contrast to the PM spectrum in PDPA-nB_u solution that shows two PA bands (Figure 22.26b) and associated IRAVs, in PDPA films only a single PA band (δS) at 1.65 eV dominates the steady state PM spectrum, and no IRAVs are observed. Using PA-detected magnetic resonance it was shown [27] that δS is associated with spin $\frac{1}{2}$ excitations; also the lack of IRAVs shows that δS is due to neutral excitations. This reversed spin–charge relationship, which contrasts other known spin $\frac{1}{2}$ excitations in condensed matter physics, is unique to soliton excitations in degenerate ground state π-conjugated polymers [8]. Therefore, δS is identified as due to neutral solitons ($S°$), of which optical transitions were shown in Figure 22.2. From the similarity of the transient PA_g band at 200 ps and the slightly redshifted ($\approx$0.1 eV) steady state δS band, then PA_g at $t > 10$ ps was concluded to be also due to neutral solitons. Solitons are topological excitations that cannot participate in interchain hopping [8]. This means that the transient PA_g decay up

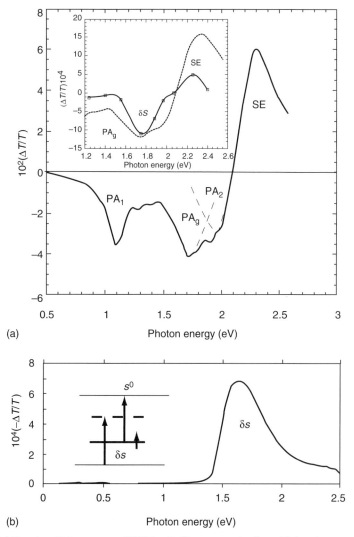

FIGURE 22.27 (a) Transient PM spectrum of PDPA-nB_u *film* measured at "$t = 0$" that shows an SE and several PA bands. The higher energy PA band is decomposed into two separate PA bands, PA_g and PA_2, respectively. The *inset* compares the normalized transient PM spectra at $t = 0$ (full line) and $t = 200$ ps (dotted line), where PA_g transforms into δS. (b) The steady state PM spectrum of the PDPA-nB_u *film* at 80 K shows a single, neutral PA band, δS. The *inset* shows the two degenerate optical transitions δS associated with a neutral soliton excitation, $S°$. (From Korovyanko, O.J., et al., *Phys. Rev. B*, 67, 035114, 2003. With permission.)

to about 10 ps (Figure 22.28a) is due to ultrafast $\overline{S}S$ recombination rather than formation of a new transient species.

Figure 22.29 schematically shows the model originally proposed [182] for the ultrafast energy relaxation processes in PDPA films. It contains two relaxation channels [183,184]; ionic, by $1B_u$ and covalent, by $2A_g$, which is populated following an ultrafast phonon-assisted internal conversion from the photogenerated $1B_u$ excitons. PA_g at short time is thus due to transitions from $2A_g$ (dark) excitons. As in long-chain polyenes [185] and t-$(CH)_x$ [186] these excitons are subject to ultrafast recombination dynamics and this explains the ultrafast decay dynamics seen in Figure 22.28a. In degenerate ground state polymers $2A_g$ is unstable with respect to the formation of soliton excitations and therefore undergoes fission into two neutral $\overline{S}S$ pairs, $2A_g \Rightarrow 2(S° + \overline{S}°)$ [18,184], followed by further separation

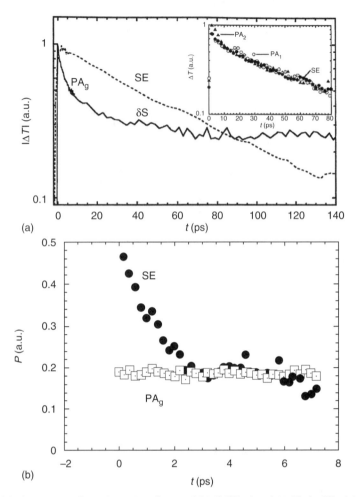

FIGURE 22.28 (a) The transient decay dynamics of PA_g and δS (full line) and SE (dashed line) in PDPA-nB_u film. The *inset* compares the transient decay dynamics of SE, PA_1, and PA_2. (b) The transient decay of the polarization memory, $P(t)$ measured at the SE band (2.3 eV, open circles) and PA_g (1.7 eV, full triangles). (From Korovyanko, O.J., et al., *Phys. Rev. B*, 67, 035114, 2003. With permission.)

into individual neutral solitons; the latter state has a slightly narrower PA band (δS) compared to PA_g. A similar separation between ionic and covalent relaxation channels happens in other, nondegenerate ground state π-conjugated polymers [170], except that the $2A_g$ in the covalent channel separates into two triplets rather than into two soliton pairs. The triplets are stable in nondegenerate ground state π-conjugated polymers [184,187,188] and thus soliton pairs are not formed. Comparing the data in PDPA-nB_u and the prototype degenerate ground state polymer, t-$(CH)_x$ it was suggested that in t-$(CH)_x$ the $2A_g$ fission process occurs in the sub-ps time domain [189]. This may happen in t-$(CH)_x$ since $E(2A_g) \ll E(1B_u)$.

The contrast between the steady state PM spectra in PDPA-nB_u solution, that shows long-lived polarons, and films, that show long-lived neutral solitons (Figure 22.26b and Figure 22.27b, respectively) is quite astonishing and points to a radical change in the photoexcitation dynamics in the two polymer chain environments. The underlying mechanism for this apparent difference may be the suppression of the covalent channel in PDPA-nB_u solution. The solvent thermal bath having many low-energy vibrations may affect the hot exciton thermalization rate in the ionic channel in polymer solution. Since in PDPA-nB_u, $E(1B_u) < E(2A_g)$, then in solution the covalent relaxation channel cannot be reached

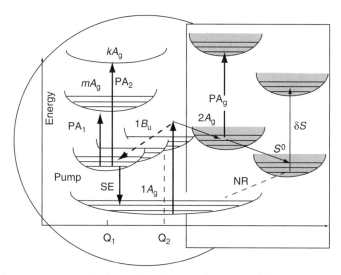

FIGURE 22.29 Schematic representation in configuration coordinates, Q of the energy levels, relaxation processes and optical transitions in the two relaxation channels of PDPA-nB_u, namely the ionic (*left*) and covalent (*right*). The full vertical lines are optical transitions, whereas the dashed lines represent relaxation processes. The different parabolas associated with $1B_u$ stand for intrachain exciton levels of polymers with various conjugation lengths, where the dashed line represents interchain hopping. mA_g and kA_g are upper energy states with even parity, which can be reached from the $1B_u$ exciton by optical transitions (PA$_1$ and PA$_2$, respectively). (From Korovyanko, O.J., et al., *Phys. Rev. B*, 67, 035114, 2003.)

following the ultrafast hot exciton thermalization. Since solitons are by-products of $2A_g$ fission in the covalent relaxation channel [18] then the affected symmetry of low-energy vibrations may explain the lack of soliton photoexcitations in PDPA-nB_u in solution [180,182].

22.4 Laser Action in π-Conjugated Polymers

The first report on SN and laser action in π-conjugated polymers was published by Moses, who reported SE and lasing from MEH-PPV in solution [190]. Yan et al. [191,192], first reported pump–probe type SE in MEH-PPV in both films and solutions. Another breakthrough was reported in 1996 when three groups independently reported laser action in π-conjugated polymer films [99,100,103]. The demonstration of dramatic SN in PPV-type polymer films [99–103] and also in other π-conjugated polymers in the form of solutions and thin solid films [104–107,193–201] has stirred widespread interest to the phenomenon of laser action in organic semiconductors. SN in π-conjugated polymers occurs at relatively low excitation intensities and is typically accompanied by substantial excitonic lifetime shortening [103,202]. High optical gain, as explained in Section 22.2 is required for all laser action phenomena. An additional requirement for "true" lasing is the presence of optical feedback, which typically results in well-defined cavity-dependent laser modes. Such laser action has indeed been demonstrated in Fabry–Perot type resonators [99,190] and also in planar and cylindrical [120] microcavities. However, SN has been observed also in thin π-conjugated polymer films, where the existence of an optical feedback mechanism necessary for lasing is not obvious. In this case explanations involving "mirrorless lasing" phenomena, such as ASE [105,200,203] and SF [103,204], have been invoked. Whereas both lasing and ASE are the direct result of SE processes, SF is a cooperative spontaneous process, which is due to the buildup of a macroscopic optical dipole moment ensuing from coherent interactions between the photogenerated excitons by their radiation electromagnetic field [96].

Interest in the phenomenon of laser action in π-conjugated polymers continues to grow. This is largely due to the possible applications of these polymers as active laser media in future plastic laser diodes. Hence, better understanding of the various mechanisms leading to SN and the criteria determining their respective contributions is beneficial. In this section we review the phenomenon of cavity lasers, and "mirrorless" nonlinear emission in thin films and solutions of a soluble derivative of PPV, i.e., 2,5-dioctyloxy PPV (DOO-PPV). We also describe the characteristic properties of RL and also discuss SF in DSB single crystals. We show how to separate the contributions of SF and ASE to the SN, and conclude that SF is dominant in thin films with poor optical confinement in small illumination area, whereas ASE prevails in dilute solutions and thin films with superior optical confinement [96].

22.4.1 Amplified Spontaneous Emission in Solutions and Thin Films of DOO-PPV

The ASE process occurs in a gain medium whenever the optical confinement is superior, or the decoherence time is short (Section 22.2). The confinement can be described by the radiation leakage rate κ, whereas the cooperation among chromophores may be quantified by the Arrechi-Courtens time [112] τ_c, which is inversely proportional to the chromophore density in the medium. The value of $(\kappa\tau_c)^{-1}$ accounts for the relative number of photons emitted by the ASE process [96]. This happens in both polymer solutions (in a cuvette), or in neat polymer films of thickness ~100 nm deposited on glass substrates due to optical confinement formed by the film waveguiding properties.

22.4.1.1 Spectral Narrowing in Dilute DOO-PPV Solutions

When diluting the DOO-PPV chromophore concentration in solution the ability of photoexcited excitons to communicate among themselves by their EM radiation field diminishes. Under these condition, the communication time is longer than the decoherence time ($\tau_c > T_2^*$) and therefore ASE prevails over SF laser action [96]. Figure 22.30a shows the emission spectra of a dilute DOO-PPV solution in chloroform (concentration of ~2 g/L), which were measured at various excitation intensities using transverse photoexcitation with an excitation area in the shape of a stripe, as shown in the inset. The cuvette with solution was tilted in order to avoid cavity-related lasing to occur due to the reflections of its sides. The obtained SN is significantly different from that found in DOO-PPV films. The peak wavelength here is at 590 nm rather than 630 nm in thin films, and the final linewidth is 20 nm instead of 7 nm found in films. The dependence of the emission peak intensity I_{ASE} at 590 nm on the excitation intensity I is shown in Figure 22.30b. It is seen that I_{ASE} grows exponentially with I, which is consistent with a simple ASE process [97] described by

$$I_{se} = \beta(e^{(\gamma-\alpha)L} - 1) \tag{22.11}$$

where β is a constant that depends on the excitation geometry, and γ and α are the optical gain and loss coefficients at the peak intensity wavelength λ (~590 nm). Since γ is linear with I in the first approximation, then $\ln(I_{se})$ ~I_p at large I_{se} is in agreement with the data and fit shown in Figure 22.30b. These results show that indeed SN in dilute DOO-PPV solutions is due to a single-pass ASE.

22.4.1.2 ASE in DOO-PPV Films with Superior Optical Confinement

In superior quality films prepared using spinning speeds of 100–300 rpm, low scattering planar DOO-PPV waveguides are formed where a large portion of the polymer emission is optically confined inside the polymer film. In this case optical losses are small and this increases the value of $(\kappa\tau_c)^{-1}$. The improved DOO-PPV films were characterized by more uniform thickness with about 3% variation per 1 mm length, whereas the regularly spin casted films have thickness variations of 15%–25% per 1 mm length [96]. The superior DOO-PPV films also showed much better optical confinement, which in part may be explained by their better surface quality. Using the refraction index of the glass substrate, $n_s = 1.46$, and that of the DOO-PPV film, $n_f = 1.7$, [105] it was estimated that the maximum fraction f of

emission waveguided inside the film [205] is $f = \sqrt{1 - (n_s/n_f)^2} = 0.51$.

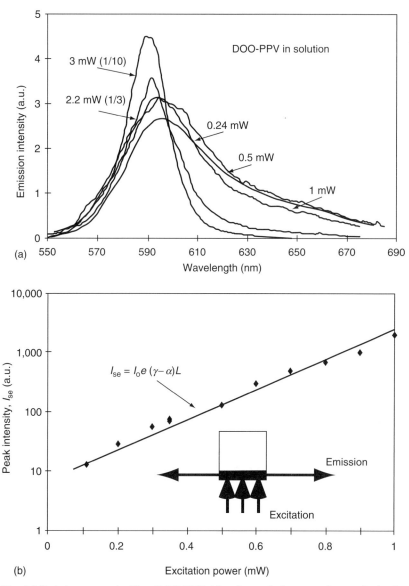

FIGURE 22.30 (a) Emission spectra in dilute DOO-PPV chloroform solutions at various excitation intensities. (b) The dependence of the emission peak intensity at 590 nm on the excitation power; the line through the data points is a fit using the ASE model (Equation 22.11). The *inset* shows the experimental setup for measuring the nonlinear emission from the polymer solutions. (From Frolov, S.V., Vardeny, Z.V., and Yoshino, K., *Phys. Rev. B*, 57, 9141, 1998. With permission.)

The spectrally narrow SE (bandwidth of ~8 nm) from such films was observed only in the direction parallel to the film surface, whereas the emission perpendicular to the film surface remained spectrally broad (~80 nm) even for $I > I_0$ [96]. This means that SE is enhanced due to waveguiding along the film, where the emission experiences the largest gain. To prove the existence of ASE in such films the directional emission was measured along the film, where an excitation area in the shape of a narrow stripe (~100 μm wide) was used. As a result, ASE was predominantly emitted along the axis of the stripe, parallel to the film surface. The directional ASE appeared in the form of a thin narrow beam propagating outside the excitation area, where it was scattered on the edge of the film [96]. A part of this scattered light was trapped inside the quartz substrate; it was collected by a round lens in front of the mono-chromator and used for the spectral analysis of the DOO-PPV emission.

Figure 22.31 shows the directional ASE spectra obtained by increasing either the excitation intensity I (a) or the excitation stripe length L (b) [96]. The results are virtually identical: in both cases SN of the polymer emission was observed above certain threshold values for both I and L. This directional SE can be successfully modeled using the ASE approximation and Equation 22.11. Since γ has a maximum at $\lambda \sim 630$ nm, then I_{ASE} (630 nm) experiences the maximum gain; whereas amplification at other λs is relatively smaller. Consequently, this nonlinear amplification process leads to SN when either I or L increases.

I_{ASE} dependence on L at different I was also measured (Figure 22.32a) [96]. In accordance with Equation 22.11, I_{ASE} grows exponentially at small L. This allowed to estimate the effective gain coefficients: $(\gamma - \alpha)$ ~ 70 cm^{-1} for $I = 0.6$ MW/cm^2 ($N \sim 4 \times 10^{17}$ cm^{-3}) and ~ 40 cm^{-1} for $I = 0.4$ MW/cm^2 ($N \sim 2.5 \times 10^{17}$ cm^{-3}). From these measurements it was obtained the relation $\gamma \propto 170 \times I$ (MW/cm^2) (or $\gamma \propto N\sigma$, where $\sigma \sim 2.5 \times 10^{-16}$ cm^2), and $\alpha \sim 30$ cm^{-1} at 630 nm. In the ASE model, there exists a saturation emission intensity I_{sat} at which the optical gain saturates as I_{se} approaches I_{sat}. As a result, the emission rate at $\lambda = 630$ nm is approximately equal to the pump excitation rate at $\lambda = 532$ nm for I_{ASE} close to I_{sat}; in the case of loss-limited gain saturation, I_{ASE} completely stops growing [97]. From the onset of gain saturation

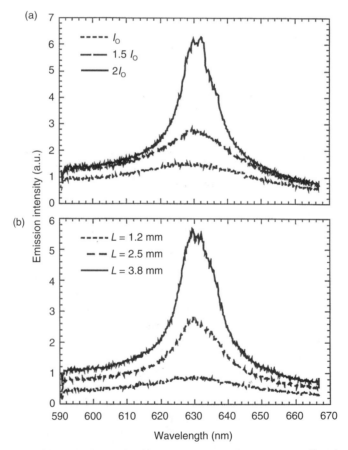

FIGURE 22.31 Spectra of directional stimulated emission in a superior DOO-PPV film obtained by increasing either (a) the excitation intensity, I or (b) the excitation length, L. The respective intensities and lengths are given, where $I_0 \sim 0.2$ MW/cm^2. (From Frolov, S.V., Vardeny, Z.V., and Yoshino, K., *Phys. Rev. B*, 57, 9141, 1998. With permission.)

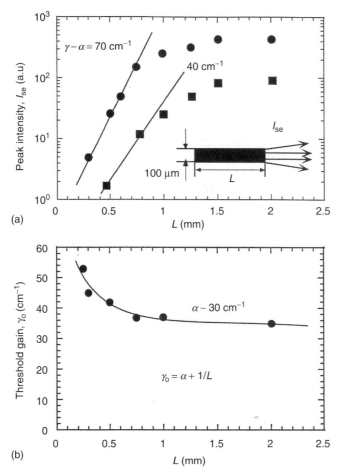

FIGURE 22.32   (a) The peak emission intensity dependence on L for excitation intensity of 0.6 MW/cm^2 (circles) and 0.4 MW/cm^2 (squares). The schematic illumination and collection methods are shown in the *inset*. (b) The stimulated emission threshold gain coefficient γ_0 obtained for various L. The lines through the data points are fit using Equation 22.12 as shown. (From Frolov, S.V., Vardeny, Z.V., and Yoshino, K., *Phys. Rev. B*, 57, 9141, 1998. With permission.)

in Figure 22.32a it was estimated [96] that I_{sat} is of the order of 5×10^7 W/cm^2. Moreover using $I_{sat} = hc/\lambda\sigma\tau$ [97] (τ is the exciton lifetime), $\sigma \sim 2.5 \times 10^{-16}$ cm^2 and $\tau \sim 300$ ps [4] it was calculated that I_{sat} to be of the order of 4×10^6 W/cm^2, which is consistent with these measurements.

I_0 for ASE can be defined by the onset of nonlinear amplification at 625 nm, which occurs when $(\gamma_0 - \alpha)L = 1$; this condition may be rewritten as follows [96]:

$$\gamma \geq \alpha + \frac{1}{L}. \tag{22.12}$$

Figure 22.32b shows the threshold gain γ_0 measured at various L; γ_0 was calculated from the measured threshold intensity and the previously determined relation between I and γ. Using Equation 22.12 the functional dependence in Figure 22.32b was successfully modeled, and from the fit α at $\lambda = 630$ nm was obtained; $\alpha \sim 30$ cm^{-1}, in agreement with the above estimate from the relation between I and γ. The estimated value of α also agrees with values of subgap absorption coefficient previously measured in thin films of PPV and its derivatives [206], which range from 30 to 70 cm^{-1} at $\lambda = 625$ nm.

It can be also seen from Figure 22.32b that the ASE threshold for $L > 1$ mm is mainly determined by α. This is an important conclusion for the quest of plastic lasers. A similar threshold condition may occur in a laser cavity where the cavity finess Q, which determines the threshold for lasing, is limited by self-absorption; $Q \propto Q_{abs} = 2\pi/\alpha\lambda$ [97]. From α value measured above it was estimated Q_{abs} to be about 3000; this Q value also determines the lowest attainable laser threshold and linewidth for a DOO-PPV polymer laser [120]. α in DOO-PPV films is likely to be determined by both self-absorption and scattering, and thus will vary from film to film. We conjecture that it may be possible to further decrease α and consequently lower the threshold for both ASE and lasing by improving the polymer and film qualities. In addition, other luminescent conducting polymers may have even lower optical losses, and thus may be more suitable for laser applications. Subgap absorption α values of less than 1 cm^{-1} have been obtained in polydiacetylene films [207]. If such high optical transparency is achieved in highly luminescent conducting polymers, then Q_{abs} would be of the order of 10^5, which would, in turn, lower the ASE threshold in such films by almost two orders of magnitude compared with the best DOO-PPV films at the present time.

22.4.1.3   Transient ASE Dynamics in DOO-PPV Films

The transient emission response of a DOO-PPV film under the conditions of laser action was measured using the gated frequency upconversion technique with 300-fs time resolution [202]. Figure 22.33a and Figure 22.33b show the emission dynamics using front emission geometry for $I < I_0$ and $I > I_0$, respectively below and above the threshold intensity for lasing. A narrow SE band is formed at the 0–1 transition at ~630 nm for $I > I_0$. Furthermore, the SE dynamics is much faster than that of the spontaneous emission or PL obtained at $I < I_0$ even at wavelengths other than at 630 nm. This indicates that in addition to the fast SE process, the exciton energy relaxation within the density of exciton states (DOS) distribution in the film was much faster at higher I. The faster relaxation rate is probably caused by the SE process, which rapidly depopulates the most strongly coupled excitons. The energy migration dynamics is illustrated by fitting the emission decay at few selected wavelengths (λ) with a double exponential function $A\exp(-t/\tau_1) + B\exp(-t/\tau_2)$ with time constants τ_1 that describes the spectral diffusion process and τ_2 describing the exciton recombination time. The obtained fitted τ_2 values [202] were in the range of 200–300 ps. For $I < I_0$, $\tau_1 \approx 25$ ps and remained relatively unchanged across the 0–1 PL band. For $I > I_0$, however τ_1 was measured to be substantially shorter (≈ 2.5 ps at 630 nm) than those for $I < I_0$. This indicates that the energy relaxation is much faster for $I > I_0$. It was estimated from Figure 22.33b that the exciton energy relaxation time τ_{Re}, is of the order of 5 ps. This relatively long energy relaxation process indicates that the most strongly coupled excitons are generated in the film for a much longer time than the actual excitation pulse duration itself. One immediate consequence of this prolonged generation process for the most strongly coupled excitons was postulated by a delay emission with respect to the spontaneous emission, or PL, as shown in Figure 22.33c for 630 nm. It is seen that whereas the PL onset occurs at $t = 0$ (for both $\lambda = 580$ and 700 nm), the ASE at $\lambda = 630$ nm reaches its maximum at a time delay $\tau_D = 2$ ps.

More detailed insight into the SE dynamics in the DOO-PPV film could be obtained for various excitation intensities I and excitation stripe length L using edge emission geometry, as shown in Figure 22.34 [202]. Figure 22.34a shows the SE decay at 630 nm for a fixed stripe length $L = 600$ μm and various values of I ($>I_0$). It is seen that the SE delay time, τ_D decreases with I; and at the same time the SE decay acquires a second "bump" at about 5 ps. The clearer indication of a series of "relaxation bumps" in the SE decay can be seen in Figure 22.34b measured at various L for constant excitation intensity. The first SE maximum at this intensity occurs for $L = 300$ μm at $\tau_D \approx 4$ ps; whereas, the second maximum occurs now at 7.5 ps. At longer L, both the "relaxation" period T' in the SE decay and τ_D dramatically decreases. Both the time delay and the second SE process may be the manifestations of the prolonged exciton generation coupled with SE propagation in the film. However they can also be due to SF oscillation as described in Section 22.4.5. For describing the exciton–photon dynamics in the illuminated stripe at high I the following coupled rate equations with a time-dependent exciton generation rate $R(t)$ were developed [202].

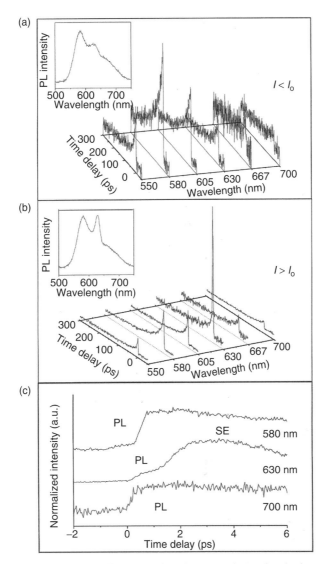

FIGURE 22.33 Transient emissions at different wavelengths across the PL band of DOO-PPV film at 300 K measured with a round spot excitation geometry. (a) $I = 20\ \mu J/cm^2$ per pulse ($I < I_0$) and (b) $I = 50\ \mu J/cm^2$ per pulse ($I > I_0$), where $I_0 = 40\ \mu J/cm^2$ per pulse, is the threshold intensity for stimulated emission, SE (or laser action). The *insets* to (a) and (b) show the time-integrated PL spectra for each case. (c) Normalized emission decay of (b) up to 6 ps for three different wavelengths (λ), where $\lambda = 580$ nm and 700 nm represent regular PL emission, and $\lambda = 630$ nm is the SE peak. (From Lee, C.W., et al., *Chem. Phys. Lett.*, 314, 564, 1999. With permission.)

$$\partial N(x)/\partial t = R(t) - N(x,t)/\tau - N(x,t)\int \sigma_e(\lambda)\left[I^+(x,t,\lambda) + I^-(x,t,\lambda)\right]\,d\lambda, \tag{22.13}$$

$$\pm dI^\pm(x,t,\lambda)/dx = N(x,t)\sigma_e(\lambda)I^\pm(x,t,\lambda) + N(x,t)E(\lambda,t)g^\pm(x)/\tau_{Ra} - \kappa(\lambda)I^\pm(x,t,\lambda) \tag{22.14}$$

The generation term was chosen to be: $R(t) = R_0\left[1 - \exp(-t/\tau_h)\right]\exp(-t/\tau_m)$, where R_0 is a constant, τ_h is the hot exciton thermalization time ($\tau_h \approx 0.5$ ps) [90] and τ_h is the energy migration time at high I; where $d/dx = \partial/\partial x + (n_r/c)\partial/\partial t$. In the above equations, N is the most strongly coupled exciton density, and $I^\pm$ is the SE propagation wave along the illuminated stripe to the right (+) and left (−)

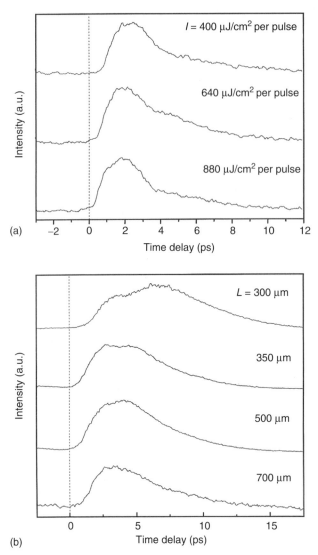

FIGURE 22.34 Picosecond SE dynamics of DOO-PPV film at 630 nm at various excitation intensities and excitation lengths. (a) Constant $L = 600$ μm for various I and (b) constant $I = 300$ μJ/cm² per pulse for various L. (From Lee, C.W., et al., *Chem. Phys. Lett.*, 314, 564, 1999. With permission.)

direction, respectively. In these equations, $\sigma_e(\lambda)$ is the SE optical cross-section spectrum, with $\sigma_e(\lambda) \approx 10^{-16}$ cm² and $\lambda_0 = 630$ nm [103], τ is the exciton lifetime ($\tau \approx 250$ ps in DOO-PPV [103]), τ_{Ra} is the radiative lifetime ($\tau_{Ra} \approx 1$ ns in PPV-based polymers [86]) and $\kappa(\lambda)$ is the absorption loss rate in the absorption tail [10], where $\kappa(\lambda) = \alpha(\lambda)c/n_r$; here $\alpha(\lambda)$ is the absorption coefficient, c is the speed of light in vacuum, and n_r is the polymer refractive index ($n_r = 1.7$). In Equation 22.13 and Equation 22.14, $E(\lambda)$ is the 0–1 PL band (Γ) at low intensity of which spectrum is taken to be Lorentzian with 30 nm width, that is normalized by the PL quantum yield of 25% [6]; i.e., $\int_{\Gamma} E(\lambda)d\lambda = 0.25$; $g^{\pm}(x)$ in Equation 22.14 are geometrical factors describing the fraction of PL emission that is emitted along the stripe into a solid angle at which the edge is seen from position x. The 0–1 PL band $\Gamma(\lambda)$ was divided into 21 different λs and, therefore, there were overall 43 coupled differential equations including both left (I^-) and right (I^+) light propagations.

Equation 22.13 and Equation 22.14 were solved numerically at different I and L, to simulate the experimental results (Figure 22.34). Figure 22.35 shows [202] the simulated transient SE intensity at the right edge of the illuminated stripe $I^+(x = L, t, \lambda = \lambda_0)$ for various generation rates R_0, and fixed $L = 300$ μm. In agreement with the data in Figure 22.34, it is seen that the SE time delay, τ_D decreases with R_0, reaching the limit $\tau_D = \tau_h$ at very high intensities. Since L was kept constant, these simulations then clearly demonstrate that τ_D is not simply due to a propagation effect, but also follows the coupled exciton density buildup, similar to the coupled rate equations for SF emission discussed in Section 22.5. The simulated result also shows that a clear second, more delayed SE bump is formed at sufficiently high I at about 5 ps (Figure 22.35b), and this is also in agreement with the experimental result in Figure 22.34a for $L = 300$ μm. Therefore, the second SE buildup is not caused by reflection at the film edge, or any other feedback effects that were not included in the model. To explain this second SE band it was noted that at high I the SE decays faster (~2 ps) than the energy migration time τ_M (~5 ps). Then, following the ultrafast decay caused by the SE in the film the most strongly coupled exciton density *grows*

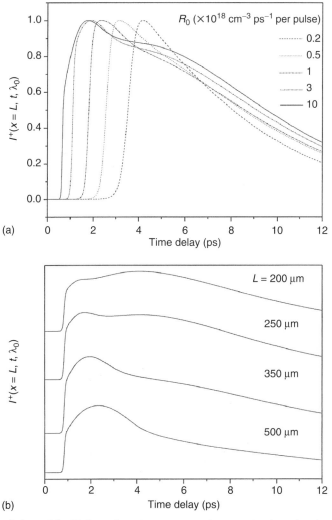

FIGURE 22.35 Simulations of the SE dynamics at $\lambda_0 = 630$ nm using Equation (22.13) and Equation (22.14) with parameters given in the text for (a) normalized I^+ $(x = L, t, \lambda_0)$ at constant $L = 300$ μm at various R ranging from 0.2 to $10R_0$, where $R_0 = 10^{18} \text{cm}^{-3} \text{ ps}^{-1}$; and (b) normalized I^+ $(x = L, t, \lambda_0)$ at constant $R = 5R_0$ for several L. (From Lee, C.W., et al., *Chem. Phys. Lett.*, 314, 564, 1999. With permission.)

again due to the prolonged generation term, $R(t)$. If R_0 is sufficiently large, then a second, more delayed SE may be formed, in agreement with the data. However this explanation cannot describe an oscillatory emission in time where several SE bumps are seen, because the generation term cannot be extended for much longer times. Alternatively, if the generation term $R(t)$ cannot be extended in time, then the experimental results that include the bump at about 5 ps cannot be described by an ASE model, and thus another, more exotic laser action process such as SF may be involved. This happens in the case of laser action from DSB single crystal described in Section 22.5.

Figure 22.35b shows $I^+(x = L, t, \lambda_0)$ for a fixed R_0 at various stripe lengths, L [202]. In agreement with the data in Figure 22.34a, it is seen that the SE bump gradually diminishes with L, whereas τ_D increases with L due to the increased propagation time in the film. To understand the dependence of the second SE peak on L, the SE depopulation of the exciton density is more efficient at large L, and this may prevent an effective second SE buildup to occur in the transient emission [206]. This model, however cannot simulate all the details of the experimental results. For example, a more rapid SE rise time is seen in the simulation compared to the experiment. However the simulations are in reasonable agreement with the essential features of the experimental data, showing that the model (Equation 22.13 and Equation 22.14) is correct. This validates the interpretation for both the SE delay and the appearance of the second SE bump, and their dependence on L. We therefore conclude that the relatively long energy migration time within the exciton DOS distribution, which consequently leads to the prolonged generation of excitons with superior radiative coupling, dominates the SE dynamics in DOO-PPV film.

22.4.2   Cylindrical Microlasers of DOO-PPV

"True" feedback-related lasing in the class of π-conjugated polymers was first demonstrated in a Fabry–Perot resonator using a dilute solution of MEH-PPV [190]. Similar results were obtained later with solutions of other π-conjugated polymers [193]. Early time-resolved studies showed that unlike laser dyes, conducting polymers do not experience concentration quenching and, therefore, may exhibit optical amplification, or gain, even when they are prepared as thin films [191]. However, because the absorption length in neat, undiluted polymer films are much shorter than that in solutions, it is much more difficult to use films as gain media in open laser cavities formed with external mirrors [107,207]. Thus, work in this area has mainly concentrated on microstructures, such as planar [99] and cylindrical microcavities [208], distributed feedback lasers [121,209] and other configurations using waveguiding films.

22.4.2.1   Microring Lasers

The microring and microdisk cavity structures are schematically shown in Figure 22.36. In both cases a thin, uniform polymer film forms the entire cavity of the laser. The main advantage of such microlasers is the ease with which they can be produced, particularly the microring lasers. Typically, an optical fiber is dipped into a saturated chloroform solution of a polymer with high optical gain, which after quick evaporation uniformly coats the fiber and produce a complete cylinder of ~1 μm in thickness, 100 μm or more in length and a diameter, D that is predetermined by the size of the fiber. Alternatively, any cylindrical substrate could be used with equal success, e.g., lasing was demonstrated using metal wires and also polyaniline fibers [208]. The fabrication of microdisks is slightly more difficult; thin spin-coated films are photolithographically etched to produce microdisk arrays of various diameters. The substrate is usually either quartz or indium tin oxide (ITO)-coated glass.

An important advantage of a cylindrical microcavity is its relatively high finesse, or quality factor Q [120]. Light in such cavities is confined inside the gain medium by total internal, practically lossless reflections; the radiation leakage is due to the cavity surface curvature and light scattering from imperfections. In comparison, a planar microcavity always experiences losses due to imperfect reflections from the two highly reflective mirrors that form the microcavity [99,195]. Optical modes inside a cylinder are given by the solution of the 2D Helmholtz equation [97], which leads to longitudinal modes that satisfy the equation:

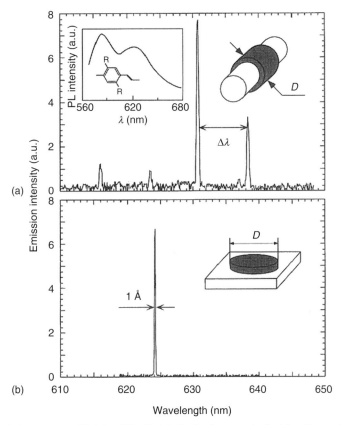

FIGURE 22.36 Emission spectra of DOO-PPV cylindrical microlasers excited with 100 ps pulses at 532 nm, with intensities I above the laser threshold excitation intensity $I_0 \sim 100$ pJ/pulse; the intermode separation $\Delta\lambda$ is assigned. The *insets* show schematically the microlaser structures, where D is the outer diameter. (a) Microring laser with $D = 11$ μm; the polymer repeat unit ($R = OC_8H_{17}$) and the PL band for $I < I_0$ are shown in the *inset*. (b) Microdisk with $D = 8$ μm that shows a single longitudinal laser mode having spectrometer resolution-limited linewidth. (From Vardeny, Z.V., et al., *SPIE*, 3797, 1, 2000. With permission.)

$$M\lambda_M = 2\pi R n_{eff} \tag{22.15}$$

where R is the μ-cavity radius, n_{eff} is the effective mode refractive index, and M is the mode index. These longitudinal laser modes are also classified by another index, K, according to their radial intensity distribution inside the disk [210,211].

Equation 22.15 in fact describes the longitudinal modes of a ring resonator formed by the thin polymer waveguide with the total length of $2\pi R$. TE modes (polarization in the plane of waveguide and parallel to the fiber axis) with $K = 1$ have the highest Q and thus the lowest threshold intensity, I_0. These modes may dominate the spectrum of the microring laser for very thin polymer films, which can be seen in Figure 22.37a. From Equation 22.15 an expression for the intermodal separation, $\Delta\lambda = (\lambda_{M-1} - \lambda_M)$ is obtained:

$$\Delta\lambda = \frac{\lambda^2}{2\pi R n_{eff}} \left(1 - \lambda \frac{\partial n_{eff}}{\partial\lambda}\right)^{-1} \tag{22.16}$$

Assuming negligible dispersion in Equation 22.16 it was found that for PDPA-nB_u films, for example $n_{eff} \approx 1.75$ (this value was not measured previously to the lasing experiment).

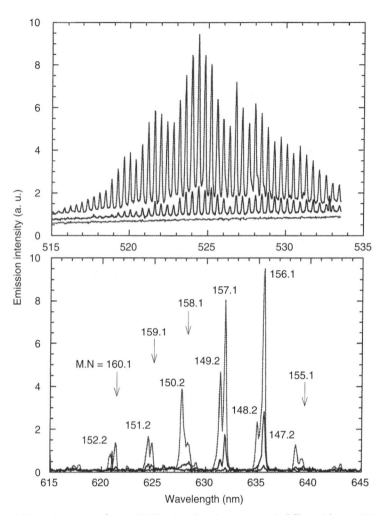

FIGURE 22.37 (a) Emission spectra from a PDPA-nB_u microring on an optical fiber with $D = 125$ μm at different excitation intensities below and above laser action threshold. The excitation intensities from top to bottom are: 1, 0.7, and 0.6 μJ, respectively. (b) Emission spectra from a DOO-PPV microring on a 20-μm diameter fiber. The excitation intensities from top to bottom are: 165, 90, and 65 μJ, respectively. M and N indices are assigned to each laser mode (see text). (From Vardeny, Z.V., et al., *SPIE*, 3797, 1, 2000. With permission.)

The cavity Q-factor can be generally defined as $Q = \omega t_c$, where t_c is the decay lifetime of a cavity mode [97]; thus the longer is the photon lifetime inside the cavity, the higher is Q. The value Q may be influenced by various contributions and near laser threshold is give by [121]:

$$Q^{-1} = Q_{cav}^{-1} + Q_{scat}^{-1} + Q_{abs}^{-1} \tag{22.17}$$

where Q_{cav}^{-1} describes radiation losses determined by the cavity geometry for a given mode, Q_{scat}^{-1} is due to scattering from imperfections inside and on the cavity surface, and Q_{abs}^{-1} is determined by self-absorption of the unexcited gain medium according to $Q_{abs} = 2\pi n/\alpha\lambda$. Q_{cav} is known to strongly depend on M and K [211]. It is maximum for $K = 1$ for which the corresponding modes were termed as "whispering gallery" modes; for $M > 20$ it was found in DOO-PPV microrings that $Q_{cav} > 10^4$ [210]. In measurements of both microring and microdisk lasers made from pristine polymer films, it was found that typical Q values were on the order of 3×10^3 [120,209]. It was concluded that despite rather high Q_{cav} values, Q of a polymer microcavity is usually limited by scattering losses and material absorption, Q_{scat}^{-1} and Q_{abs}^{-1}, respectively. Q_{scat}^{-1} can be somewhat minimized by making smoother microactivity surfaces and purifying the polymer solution. Q_{abs}, on the other hand is determined by α, and thus is difficult to change. However, it is possible to dilute the polymer with various blends of organic dyes, oligomers [99,109], and polymers [107,207].

Figure 22.37a shows the emission spectra obtained from a PDPA-nB_u microring deposited on a fiber with diameter $D = 125$ μm [98]. The broad, featureless PL band at low excitation intensities, I, is transformed into a multimode ring laser spectrum at higher I; this transition into the lasing regime is accompanied by a kink in the emission intensity vs. I at the threshold excitation intensity, I_0 [120,208]. The wavelength of each laser mode, $\lambda(M,K)$, is given by Equation 22.16.

Figure 22.37b shows [98] the emission spectra of DOO-PPV microring laser with $D = 20$ μm. These spectra cannot be adequately described by Equation 22.16, since more than one set of longitudinal modes are observed. However, it is possible to model them using two lowest order TE modes with $K = 1$ and 2 (TM modes were not observed in thin microrings). As a result of such modeling, the M and K index numbers may be assigned to each laser line, as shown in Figure 22.37b. The only fitting parameter is n_{eff}, which was found to be 1.680 and 1.677 for $K = 1$ and 2, respectively. Higher order waveguided modes (with $K = 2$) are expected to have lower n_{eff} due to their deeper penetration inside the glass fiber.

In order to avoid light propagation inside the optical fiber, thin metal wires were used as a cylindrical core for the microring polymer laser [212]. Although an absorptive metal surface may quench SE and thus prevents lasing, a thicker (>5 μm) polymer film helps to isolate the modes from the metal core and thus minimize the optical losses. Figure 22.38a shows the emission spectra obtained from a DOO-PPV microring ($D = 35$ μm) deposited on a 25-μm diameter aluminum wire [98]. At low excitation intensities the spectrum is dominated by a single set of equidistant longitudinal modes. However, Figure 22.38a shows that at higher intensities additional modes with a higher threshold appear in the emission spectrum. Assuming negligible dispersion, from $\Delta\lambda$ in Figure 22.38a $n_{eff} = 2.23$ was calculated using Equation 22.16, which is significantly higher than the value estimated from Figure 22.37b, where $n_{eff} = 1.7$. Lower n_{eff} for the thinner microrings on glass fibers indicates that the laser modes in such cavities are not confined to the polymer film, but also partly propagate inside the glass core, where the refractive index is low (~1.4). The modes in thick microrings, however, are fully contained inside the polymer and presumably have higher n_{eff}.

22.4.2.2 Microdisk Lasers

22.4.2.2.1 Simple Microdisk Lasers

It was found that polymer microdisk lasers behave similarly to thick microring lasers. Typically, a single microdisk with a diameter ranging from 8 to 128 μm is photoexcited by a focused laser beam. Unlike microrings, however microdisks provide good lateral confinement for the laser modes. In addition, it is easy to achieve a complete and uniform excitation of the whole microdisk area. The mode structure of the disk microcavity is also described by Equation 4.6. In fact, the spectra of the microdisk lasers are virtually indistinguishable from those of microrings, as shown in Figure 22.36b. Using Equation 4.6 and again assuming zero dispersion for the DOO-PPV microdisk laser shown in Figure 22.36b with $D = 16$ μm, it was calculated that $n_{eff} = 2.22$, which is close to n_{eff} obtained for thick DOO-PPV microring deposited on metal wires.

22.4.2.2.2 Multimode Microdisk Lasers

Many times it has been found that more than a single series of longitudinal modes survives lasing in microdisks. In such cases it was found that a careful FT helps to separate the contributions of the different modes series. An example of this complication is given below. Figure 22.38a [123] shows the emission spectrum for the 55-μm diameter microdisk laser measured above the threshold intensity. There are many well-spaced and narrow emission lines. A closer examination of the spectrum reveals two series of modes, one series, a, has larger amplitude than the other, b. For the larger intensity peaks,

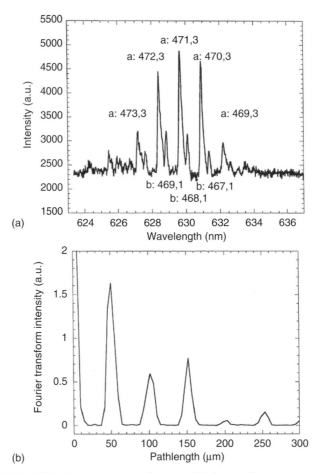

FIGURE 22.38 (a) DOO-PPV emission spectrum of a microdisk having diameter $D = 55$ μm above the laser threshold intensity. Various laser modes with indices *a* and *b* are assigned (see text). (b) Fourier transform of the emission spectrum shown in (a). (From Polson, R.C., Vardeny, Z.V., and Chinn, D.A., *Appl. Phys. Lett.*, 81, 1561, 2002. With permission.)

the spacing $\Delta\lambda_a$ for the modes averages 1.27 nm, whereas for the smaller peaks the spacing $\Delta\lambda_b$ averages to 1.31 nm. Equation 22.16 for mode spacing of a Fabry–Perot cavity can be used where the roundtrip distance $2L$ is replaced by microdisk circumference πD. The mode spacing values would suggest that the product of index of refraction and diameter, nD is different for the two modes series present in the microdisk laser.

The very narrow emission lines seen in Figure 22.38 indicate that the cavity quality factor, Q is relatively high, on the order of 3000, and only little emission escapes during each cycle. The cylindrical geometry of the disk allows the wave equation to be separated into different analytic functions in each of the r and θ directions, with the radial direction consisting of Bessel functions. Since the Q value is high, then an approximation can be made that the fields go to zero at the polymer–air interface. For the field to be zero everywhere along the interface, the argument of the integer Bessel function, $J_s(kr)$ must be zero at the boundary [123],

$$J_s(2\pi nR/\lambda) = 0 \qquad (22.18)$$

where R is the disk radius. Bessel functions have many zeros, so this condition can be written as

$$X_{st} = 2\pi nR/\lambda \qquad (22.19)$$

where X_{st} indexes the tth zero of Bessel function of order s [213].

In order to describe the microdisk modes, the product of nR needs be known to several decimal places. An accurate value of nR may come directly from the FT of the emission spectrum [122]. Figure 22.38b is the FT of the emission spectrum in Figure 22.39a. If the units of the emission spectrum are measured in wavevector ($k = 2\pi/\lambda$) then the unit of the FT is length [214]. In Figure 22.38b a single series of well-spaced peaks can be observed. The FT gives peaks at nR of 50.7 μm. The physical disk diameter, $D = 2R$ is 55 μm and from the measured value of nR it gave 1.84 as the effective index of refraction. This value indicates that the fields of these modes are entirely contained within the polymer disk since the index of refraction of the DOO-PPV polymer is 1.8, which was measured by elipsometry on an unetched polymer film.

In Figure 22.38a the strongest peak in the emission spectrum is at $\lambda_{0,a} = 629.65$ nm. Investigation of Bessel functions with the first zero at the polymer–air interface near s_0 revealed that $s = 491$ fitted reasonably well with an expected wavelength $\lambda_{0,a} = 629.67$ nm for the main emission peak. Neighboring emission peaks correspond to successive integer values of Bessel functions. The entire series of main peaks could thus be fit with a series of Bessel functions as seen in Figure 22.38a. The greatest discrepancy is just 0.08 nm for the series of seven peaks described by a series of Bessel functions $468 < s < 474$ and the product of $nR = 50.7$ μm. In this way of fitting laser spectra there are no adjustable parameters in the fit [123].

The minor peaks of the emission spectrum seen in Figure 22.38a, namely the b-series deserves a different treatment. Neither the first nor the third zeros of any Bessel functions accurately describes these peaks. The spacing $\Delta\lambda_b$ of these peaks is larger than $\Delta\lambda_a$ and this indicates that the product nD is smaller for these longitudinal laser modes. The FT does not seem to show a second cavity, Figure 22.38b, where only singular harmonics were present. The spacing of the minor peaks, $\Delta\lambda_b$, is about 3% larger than the major peaks; the spacing of points in the FT is 4.6 μm, which is roughly 3% of the product of nD, or 101.4 μm. The next point below 101.4 μm in the FT of Figure 22.38a occurs at 96.8 μm, and this was then used to determine nD for fitting the minor peaks of the spectrum. The same fitting procedure used to fit the major peaks was then used to fit the most intense minor peak occurring at $\lambda_{0,b} = 630.09$ nm. The first zero of Bessel function $s = 468$ gives a wavelength $\lambda_{0,b} = 630.06$ nm. The neighboring minor peaks fit nicely in a series with $nD = 96.8$ μm. Again, the series of six minor peaks was described with no adjustable parameters with $467 < s < 472$.

The different effective refractive indices were argued to be due to the different electric field distribution of the two mode series. Whereas the field distribution of series (a) showed only a single node at the microdisk interface, the field distribution of series (b) showed three nodes ($K = 3$), and maxima where the field modes (a) are small (Figure 22.39). In this way several series modes can survive together the limited optical gain in the microdisk [123].

22.4.2.2.3 *Laser Action Obtained with Longer Pulse Excitations*

The emission spectra of polymer microlasers reviewed here were obtained using 100 ps pulse excitation. Although the duration of such excitation is much longer than the photon lifetime in the microcavity ($t_c = Q/\omega \sim 1$ ps), it is of the order of the exciton lifetime in the polymer film and therefore may not be considered quasi-continuous [196]. The effects of longer pulse excitation were then studied, and it was found that the 10 ns pulse excitation also resulted in efficient lasing. Figure 22.40 compares [98] the emission spectra obtained from a single DOO-PPV microdisk ($D = 32$ μm) using 100 ps pulses (a) and 10 ns pulses (b). The mode structures in both cases are essentially identical. The main difference is the pronounced broadening and blueshift of the laser lines in the case of 10 ns excitation. The blueshift may still be observed with 100 ps pulses, but at higher intensities. Both of these effects are highly detrimental to the performance of the laser. It was speculated [98] that this might be due to excessive heating of the polymer film, since most of the excitation energy is spent on heating of the polymer film and the longer pulse excitation does provide more energy. The spectral blueshift (manifested in the decrease observed in λ_M),

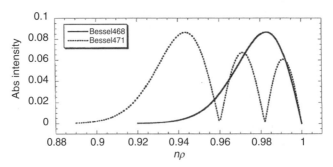

FIGURE 22.39   Field distribution for Bessel functions $s_{0,a} = 471$ and $s_{0,b} = 468$ that describe the two laser mode series (a) and (b) of the polymer microdisk shown in Figure 22.38. The two Bessel functions are normalized so that the first zero of $s_{0,b} = 468$ and the third zero of $s_{0,a} = 471$ are at the boundary. (From Polson, R.C., Vardeny, Z.V., and Chinn, D.A., *Appl. Phys. Lett.*, 81, 1561, 2002. With permission.)

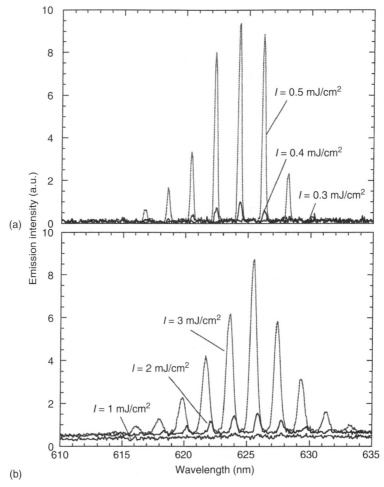

FIGURE 22.40   Emission spectra of a DOO-PPV microdisk with $D = 32\ \mu m$ at different excitation intensities using (a) 100 ps pulse duration and (b) 10 ns pulse duration. (From Vardeny, Z.V., et al., *SPIE*, 3797, 1, 2000. With permission.)

however, indicated that either D or n_{eff} decreases as I increases, and this could not be simply explained by an increase in temperature. It was then speculated [98] that the blueshift is caused by an optically induced lowering of the polymer refractive index at high excitation densities, which is caused by the nonlinear refractive index of the polymer. Also, the substantial line broadening in the case of long pulse excitation could be attributed to the reduction of the Q-factor, which may be due to either microcavity deformations caused by overheating, or additional absorption losses from triplet exciton population buildup [34].

22.4.3 Random Lasers in Films and Photonic Crystals

In 1968, V. Letokhov calculated the optical properties of a random medium that both amplifies and scatters light [215]. He concluded his studies by advancing the idea that laser action is possible in these media due to the process of "diffusive feedback" [216]. A propagating light wave in such systems makes a long random walk before it leaves the medium and is amplified in between the scattering events, giving rise to light "trapping." This random walk can be orders of magnitude longer than a straight line along which waves would have left the medium if no scatterers were present. However the "light diffusion" approach does not take into account two important physical effects, namely interference that can be regarded as precursor of light localization [217] and spatial correlations of light on scales much larger than the light mean free path, i.e., "mesoscopic" phenomena [218]. In spite of these omissions the ability to perform multiple scattering with gain has opened up a whole new field of research.

For a long time Letokhov's pioneering work [215] was not followed up by experiments until N. Lawandy and coworkers suspended TiO_2 particles into solutions of laser dyes [125]. In such systems SN of the emission was observed when sufficiently large density of scatterers was introduced into the gain medium. This phenomenon was debated in the literature [132]; it was proposed to be due to ASE in the gain medium, since the SN was always followed by a nonlinear input vs. output intensity dependence. More recently extrinsic scatterers were introduced into π-conjugated polymer films [126], and the analysis of the obtained SN was made using the same approach as that of the earlier work [125]. Another stage in random laser research was reached when two groups, one working with polymer films [127], and the other working with ZnO powders [128] discovered that the SN phenomenon is, in fact followed at higher excitation intensities by a finer spectral structure that contains much narrower laser-like spectral lines (of order 0.1 nm, see Figure 22.41). This new laser action phenomenon shows the dominance of resonant random cavities in the disordered gain media at high excitation intensity [129]. It was also shown [127,219] that: (i) the narrow lines vary with the illumination spot on the sample, (ii) they are only weakly polarized along the polarization of the excitation beam, and (iii) the number of narrow lines increases as the excitation intensity increases [127,204]. Recently, similar fine spectral lines were also discovered at high excitation intensities of dyes and π-conjugated polymers infiltrated into opal photonic crystals [130,131] (Figure 22.42) showing that this new type of laser action at high excitation intensities is *generic*.

As discussed above the underlying cavity length of laser emission lines can be obtained from the FT [122] of the emission spectrum. In the FT spectrum the length of the resonator loop, L is given by the relation $L = 2d_1\pi/n$, where d_1 is the shortest FT length at which a peak is apparent in the transformed spectrum. The average of many FT spectra is needed to convincingly show the dominance of a specific random cavity in the film [136], but the FT spectrum of Figure 22.41 may already contain the right peaks to demonstrate an important property of RL that it is dominated by specific random cavities in the gain medium. The specific cavity length that dominates the RL in the DOO-PPV film shown in Figure 22.41 was obtained from Equation 22.16, the fundamental d_1 and $n = 1.7$ to be $L = 185$ μm. The light mean free path, l^* in disordered media can be measured by the coherence backscattering (CBS) technique [218]. Applying the CBS technique to thin polymer films of DOO-PPV it was found [220] that $l^* \approx 9$ μm. Taking the naive approach that the random cavity is created from few scatterers [128], then from the obtained cavity length and l^* that was found above, it may be calculated that about 20 scatterers are involved in such a resonator (see a possible closed loop with scatterers in Figure 22.41).

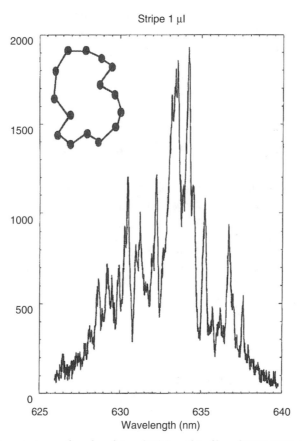

FIGURE 22.41   Emission spectrum of random lasers (RL) in a thin film of DOO-PPV at excitation intensity of 1 μJ/pulse, which is above the laser threshold excitation I' for RL action. Many uncorrelated laser modes can be observed. The *inset* shows schematically the way light forms a close loop from 15 randomly distributed scatterers. (From Polson, R.C., Chipouline, A., and Vardeny, Z.V., *Adv. Mater.*, 13, 760, 2001. With permission.)

    To clearly unravel hidden features in the RL emission spectra that are otherwise not easily accessible, ensemble averaging of many PFT spectra collected from *different* individual excitation pulses was performed. Figure 22.43 shows the ensemble average of PFT performed on spectra such as the DOO-PPV [221]. The most striking phenomenon is that the averaged PFT spectrum *does not smooth out with increasing the averaging number j, but instead develops rather sharp features* at $d_1 \approx 22$ μm with about five harmonics up to $d_6 \approx 132$ μm. This surprising result not only confirms that the random cavity scenario is valid [136], but also shows that a *dominant laser cavity* exists in the DOO-PPV film. It appears in the film under many different illuminated pulses, and therefore is not averaged out with j; in fact it is universal [222]. Using $n = 1.7$ for DOO-PPV films we obtain from $d_1 = nL_0/\pi$ an ensemble-averaged random cavity pathlength, $L_0 \approx 9.3$ μm. Also from the CBS albedo cone of the film, we obtained $l^* \approx 9$ μm. We thus have for the RL resonator $L_0 \gg \lambda$, and $l^* \gg \lambda$ that falls in the category of weak light-scattering regime. We do not really know what are the random laser cavities that are formed [223] in the film; but we know that they should be "clean," i.e., relatively free of scatterers, which otherwise would easily scatter the light out of the resonator. Apparently the formation of the random resonators is due to some kind of *long-range disorder*, which does not affect l^*.

    For explaining the existence of a dominant random cavity in RL of organic disordered gain media, it was noted that L_0 sharp determination in the ensemble-averaged PFT spectra may come from the competition of two opposing effects, both with steep dependence on L; one diminishes with L, whereas

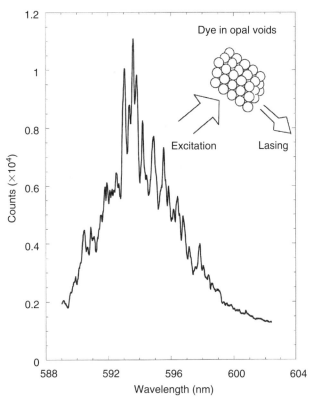

FIGURE 22.42 Random laser emission spectrum of a DOO-PPV in toluene solution that is infiltrated into an opal photonic crystal. The *inset* shows the opal, which is composed of silica spheres in an FCC lattice and the laser excitation and collection geometries. (From Polson, R.C., Chipouline, A., and Vardeny, Z.V., *Adv. Mater.*, 13, 760, 2001. With permission.)

the other increases with L. In contrast with the other RL regime of *strong scattering* [124], light scatterers in the *weak scattering* regime act to destroy coherent lasing. We therefore conclude that to overcome loss, a lasing random resonator must not contain a scatterer in its perimeter, where whispering gallery-type modes are formed. However, such random resonators are scarce, since the probability that light is not scattered after a distance L is $\exp(-L/l^*)$; this is the steeply *decreasing* function of L mentioned above [222]. For a given optical gain, γ in the film there is a minimum clean random cavity, L_γ, which allows lasing. Formally L_γ may be obtained from the condition that at $L = L_\gamma$ gain overcomes loss, or $\gamma L_\gamma \approx 1$. Lasing may occur for all $L > L_\gamma$, however clean cavities with large L are scarce, and thus the dominant RL cavity occurs at $L = L_\gamma$, of which value is also determined by the light scattering by l^*. A computer simulation was recently performed [136] to reproduce the average PFT procedure and the dominant cavity found for RL of polymer films.

One way of studying temporal coherence in laser systems is by measuring photon statistics [224]. In this technique the transient laser emission properties are measured using pulsed excitation and a time-resolved setup [225]. The transient emission curve generated by each pulse above the laser threshold intensity is divided into time intervals that are smaller than the emission coherence time. The number of photons is then measured in each time interval and for each pulse, and a photon number histogram is calculated to obtain the probability distribution function (PDF) of the photons for each time interval. Photon statistics is achieved separately for each time interval, and correlation between different time intervals or between different wavelengths of the emission spectrum can be also studied. It is expected that for coherent radiation the Poisson distribution determines the PDF, whereas for noncoherent light

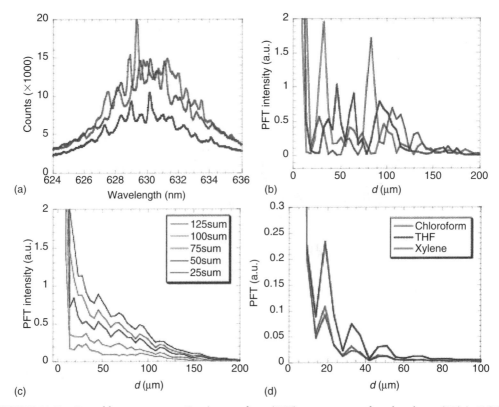

FIGURE 22.43 Ensemble average power Fourier transform (PFT) spectroscopy of random lasers (RL) in DOO-PPV polymer film. (a) Three RL emission spectra of a DOO-PPV film spin casted from toluene solution, which were collected from three different illuminated stripes on the film. (b) The PFT spectra of the emission spectra shown in (a). (c) Ensemble-averaged PFT spectra of RL emission spectra as in (a) and (b), which were averaged from different illuminated stripes over the film area. The numbers in the upper right corner denote the number of RL spectra collected in the averaging process. (d) Ensemble-averaged PFT spectra of 125 different RL emission spectra of DOO-PPV films spin casted from solutions of different solvents; green stands for xylene; red for chloroform; and blue for THF. (From Polson, R.C., *Ph.D. thesis*, University of Utah, 2002, unpublished.)

the PDF is expected to follow a Bose–Einstein distribution around zero number of photons [225,226]. Figure 22.44 shows the PDF obtained from a DOO-PPV film above the threshold intensity for RL [145]. The photon histogram, $P(N)$, was measured at 630 nm at one of the random laser modes (Figure 22.44). It is seen that $P(N)$ does not peak at zero number of photons, $N = 0$; on the contrary, it gets a maximum at $N = 5$. The theoretical curve through the data points is a fit using a Poisson distribution, where $P(N) = (N')^N e^{-N'}/N!$ and N' is the mean photon number ($N' = 5$). Figure 22.44a also shows that Bose-type statistics does not fit the data at all. It was therefore concluded [145,226] that the emission seen in the random laser regime is indeed coherent and hence the word "laser" to describe this phenomenon is justified.

22.4.4 Superfluorescence in Organic Gain Media

The main features of SF, such as excitation intensity dependence, emission pulse shortening, and time delay, can be described within a simplified semiclassical approach, which uses Maxwell–Bloch equations while neglecting the dipole–dipole interaction [113,114]. It was shown by Bonifacio and Lugatio [113] that in a mean field approximation the system of noninteracting emitters is described by the "damped pendulum" equations with two driving terms, as given below:

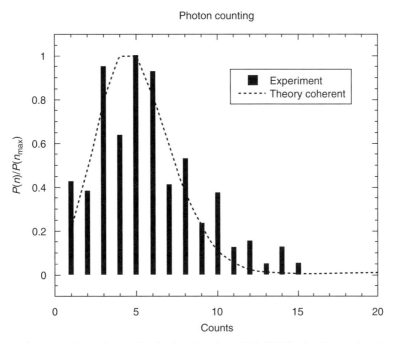

FIGURE 22.44 The normalized photon distribution function, $P(N)$ (full line), of a random laser mode. The dashed line through the data points is a fit using Poisson distribution around $N = 5$, proving that the random laser emission is indeed a coherent process. (From Polson, R.C., Chipouline, A., and Vardeny, Z.V., *Adv. Mater.*, 13, 760, 2001. With permission.)

$$d[(S_T)^2 + (S_z)^2]/dt = 0 \qquad (22.20)$$

$$d[(A_T)^2 + S_z]/dt = -2\kappa(A_T)^2 \qquad (22.21)$$

$$d^2(S_z)/dt^2 + (\kappa + 1/2T_2^*)dS_z/dt = -G(t)[2(S_T)^2 + 4(A_T)^2 S_z] \qquad (22.22)$$

where S_z is the exciton population, S_T describes the cooperative, macroscopic dipole moment of the system, A_T is the photon number operator of the emitted electromagnetic field, κ is the radiation leakage rate out of the active volume, and T_2^* is the inhomogeneous dephasing time. The right-hand side in Equation 22.22 has a time-dependent generation factor $G(t)$ given by: $G(t) = [(b_0)^2/V]\exp(-t/T_2^*)$, where V is the active medium volume, and b_0 is the coupling constant, which is proportional to the exciton oscillator strength [113]. Equation 22.20 and Equation 22.21 are the consequence of the momentum and energy conservation, respectively, where the emission intensity is given by:

$$I(t) = 2\kappa(A_T)^2 \qquad (22.23)$$

It is apparent that the right-hand side of Equation 22.22 contains two driving terms, which correspond to the macroscopic dipole moment of the excitons $(S_T)^2$ and their radiated electromagnetic field $(A_T)^2$, respectively. The first term gives rise to cooperative radiation or SF, whereas the second term is the source of ASE. It thus becomes clear that, in general the resultant laser action emission contains contributions from both SF and ASE. Whether the laser action emission is dominated by SF or ASE is determined within this model by the values of $\kappa\tau_c$ and T_2^*. Here, τ_c is the Arrechi-Courtens cooperation time [112] given by $\tau_c = (b_0\sigma N/V)^{-1}$. The value of $(\kappa\tau_c)^{-1}$ accounts for the number of photons emitted by the ASE process [96]. Therefore, in the case of strong optical confinement, where $\kappa\tau_c < 1$, ASE is the dominant radiation process. On the other hand, for weak optical confinement where $\kappa\tau_c > 1$, cooperative emission is the primary laser action process [113]. This also requires that $T_2^* > 1/\kappa$. Thus, the case of pure SF can be identified by the following conditions:

$$1/k < \tau_c < T_2^* \qquad (22.24)$$

22.4.4.1 Spectral Narrowing in DOO-PPV Films with Poor Optical Confinement

Measurements showed [227] that the laser action emission from poorly prepared DOO-PPV thin films remained isotropic at all excitation intensities, and no well-defined waveguiding was observed. Similar results were obtained using other luminescent π-conjugated polymer films [105]. This isotropic emission pattern could be attributed to poor film quality that consequently leads to strong light scattering inside the polymer film and on its surface. Optical confinement in such films is poor and the characteristic length for the nonlinear emission process L_s is on the order of few microns. On the other hand, an ASE process requires a substantially longer characteristic length for appreciable amplification [227], thus SF appears to be a more appropriate explanation for the isotropic narrow-band emission pattern found in poor quality polymer films.

Instead of intensity I_0, the onset of SN may be characterized by a threshold excitation power P_0. Using the same DOO-PPV films that were used before [227], the dependence of P_0 was measured [96] as a function of the diameter, D of the round excitation area as shown in Figure 22.45. It can be seen that P_0 is practically constant in the range of 50 μm $< D <$ 300 μm; this behavior cannot be explained by a simple ASE process. Ignoring losses and assuming that light amplification occurs in the direction of maximum excitation length (i.e., D) parallel to the film surface, a simple ASE process at threshold satisfies the relation $g_0 D \sim 1$, where g_0 is the threshold gain coefficient given by $N_0\sigma$, where N is the photogenerated exciton density and σ is the optical emission cross section. For intermediate excitation intensities, where the bimolecular recombination is negligibly small, $N_0 \sim P_0/V$ and $V = \pi D^2 d/4$ (where d is the film thickness), leading to $P_0 \sim \pi D d/4\sigma$ at threshold. This functional dependence on D cannot explain the P_0 independence on D for D values up to 300 μm as seen in Figure 22.45. On the contrary, an

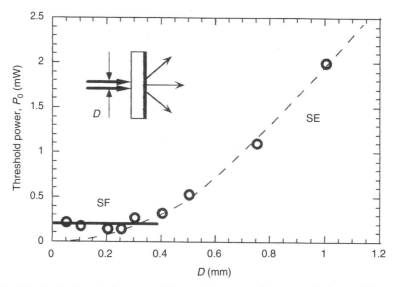

FIGURE 22.45   The threshold excitation power P_0 for obtaining spectral narrowing in isotropically emitting DOO-PPV films at various diameters D. The full and dashed lines through the data points are the expected dependencies for SF and ASE, respectively (see text). The *inset* illustrates the excitation setup, where D is the diameter of the round illuminated area on the polymer film. (From Frolov, S.V., Vardeny, Z.V., and Yoshino, K., *Phys. Rev. B*, 57, 9141, 1998. With permission.)

SF process is governed by the total *number* of photogenerated excitons and thus is determined by the excitation *power* [115,116], which implies that P_0 remains constant and independent on D for excitation areas within the SF cooperation length [113]. SF is therefore a more plausible explanation of the P_0 independence on D at small values of D. From the length of the plateau in Figure 22.45 it was estimated that the maximum cooperation length, l_c, of excitons in DOO-PPV films is ~300 μm. In the semi-classical approximation l_c is given by $l_c = c\tau_c/2n$ [113]. Moreover, the SF conditions (Equation 22.24) require that $T_2^* > \tau_c = 2nl_c/c$ [113], from which we estimate a relaxed exciton dephasing time T_2^* to be ~3 ps. The right-hand side of condition (Equation 22.24) can thus be also satisfied for films with poor optical confinement, where it was estimated $\kappa \sim c/nL_s$ $(10 - 100 \text{ fs})^{-1}$. It was therefore suggested that SF is dominant in such poor quality DOO-PPV films for $D < 300$ μm. However, for $D > l_c$, the SF conditions (Equation 22.24) are no longer satisfied, indicating that in this case ASE is the primary emission process [96]. Accordingly, P_0 dependence on D in Figure 22.45 for $D > 300$ μm may be approximated by $P_0 \sim D^2$ as indicated in the figure.

22.4.4.2 Superfluorescence in Distyryl-Benzene Single Crystals

The optical and electronic properties of π-conjugated organic semiconductors are dependent on their morphology [228]. It is not surprising therefore that single crystal of π-conjugated oligomers have very different properties than those of polymer thin films discussed so far in this chapter. The optical properties of DSB single crystal, for example, have been the focus of attention in a number of excellent research projects [48,89,229,230]. In particular, it was found that singlet excitons in low defect DSB single crystals are highly cooperative at low temperature, and in fact their dipole moment conspire to show SF radiative transition even at low excitation intensity [229,230]. It is therefore instructive to investigate laser action in DSB single crystal at high excitation intensities to decisively show the existence of SF in organic semiconductors.

High quality DSB single crystals were successfully prepared in our laboratories [231]; their crystal structure was determined to be orthorhombic, where the DSB molecules are arranged in layers of Herringbone-type structure. Unique among the group of molecular crystals found in DSB was a high PL quantum yield of ~65% at room temperature [48]. This indicates that DSB single crystals are excellent candidate for laser action. The room temperature absorption and PL spectra of DSB single crystal are shown in Figure 22.46 [232]. The optical gap is at ~3.0 eV, and an apparent Stokes shift of about 0.2 eV is evident. This large Stokes shift is not due to exciton relaxation; rather it is caused by the disappearance of the 0–0 transition at room temperature. Figure 22.6 in Section 22.2 shows laser action in the form of SN of the PL band when excited at high intensities [232]. Simultaneously, nonlinear input–output emission intensity is also seen; the SN and the change in the emission slope are evidence for laser action, as discussed in Section 22.2.

The transient emission in DSB under the conditions of laser action was measured by the technique of gated upconversion with 150-fs time resolution [233]. The transient laser emission is shown in Figure 22.47 together with the pulse autocorrelation function that determines the time $t = 0$ as well as the time resolution in our measurements. It can be seen that the laser emission in DSB has a delayed peak formed at about 5 ps after the pulsed excitation, followed by several oscillation ringing that last for ~100 ps; this is typical for SF emission process [113–115]. The transient oscillatory emission response was studied at several stripe illumination lengths, temperature, and polarization and was found to be in agreement with the model of transient SF dynamics.

The transient emission signal of Figure 22.47 could be well fit with the SF "damped pendulum" model (Equation 22.20 through Equation 22.22), in which the exciton density and the emission photons are coupled together by a strong, coherent interaction identified as a retarded dipole interaction [113]. The solution of these equations under the conditions $\kappa\tau_c \approx 1$ gives an oscillatory behavior where the "pendulum" crosses several times through the "south pole." In this model [113], the peaks in the transient emission are described by a $\text{sech}^2(t/\tau_R)$, where $\tau_R = \kappa\tau_c^2$, and the time separation, τ_D between the peaks is the same as the delay time, which is given by the relation: $\tau_D = \frac{1}{2}\tau_R \ln(N_c)$, where N_c is the correlated number of excitons in the system. For the pendulum ringing solution described here, where

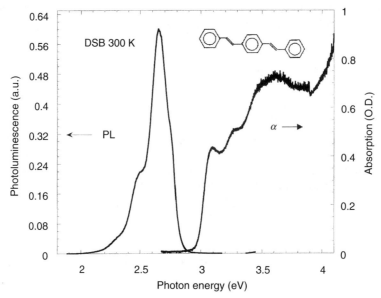

FIGURE 22.46 The absorption and PL spectra of a DSB single crystal at 300 K. The *inset* shows the DSB oligomer. (From Wu, C.C., et al., *Synth. Met.*, 137, 939, 2003. With permission.)

$\kappa\tau_c \approx 1$, it follows that $\tau_R \approx \tau_c$. Then from the width of the SF peaks in the transient emission response we get $\tau_R \sim 5$ ps, and thus N_c is of the order of 10^5 excitons. In addition, from the SF relaxation oscillation we conjecture that the SF in DSB single crystal maintains its coherence properties up to $T_2^* \approx$ 100 ps, and thus indeed $\tau_c < T_2^*$, which is needed for SF (Equation 22.24).

We therefore conclude that the emission oscillation seen in Figure 22.47 is an unambiguous proof that the laser emission process in DSB single crystal is coherent radiation [110]. Also the good fit with

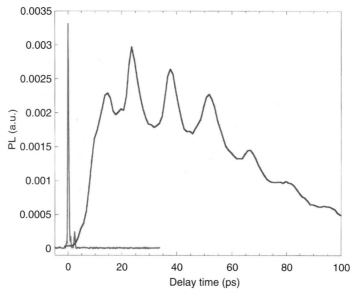

FIGURE 22.47 The transient emission spectrum of DSB single crystal at the 0–1 band (Figure 22.46) measured using an upconversion setup. The transient oscillations are due to superfluorescence (laser action coherent process) in the film.

the pendulum model of Equation 22.22 reinforces this conclusion. This shows that SF radiation may be formed in organic semiconductors, and its influence in laser action in these materials cannot be ignored [202].

22.5 Summary

In this chapter, we reviewed the ultrafast dynamics of photoexcitations in π-conjugated polymers in the time domain from 100 fs to 200 ps in a broad spectral range, and its relation to laser action where these materials serve as optical gain media. The different excitations that play a role in this short time domain were introduced in Section 22.1. They were separated into neutral and charged excitations and include solitons in degenerate ground state polymers; singlet and triplet excitons, and PPs in nondegenerate ground state polymers. Photoinduced IR-active vibrations that accompany the photogenerated charged excitations in π-conjugated polymers were also briefly reviewed. The saga of "band vs. exciton" models to explain the primary photoexcitations in π-conjugated semiconductors was also reviewed in relation with nonlinear optical spectroscopy of these materials. The five different laser action phenomena that are known to exist in organic optical gain media were introduced in Section 22.2. These are ASE, SF, superradiance, lasing in optical cavities, and RL with coherent and noncoherent feedback.

In Section 22.3 we reviewed the ultrafast photoexcitation dynamics in various π-conjugated polymers using the polarized pump and probe correlation technique in PM. Firstly, we dealt with soluble derivatives of the PPV polymer, i.e., DOO-PPV and MEH-PPV. We showed that the primary excitations in these polymers are singlet excitons with two strong photoinduced absorption bands in the near-and mid-IR spectral range, and SE band in the visible spectral range that may lead to laser action at high excitation conditions. DOO-PPV was also investigated using the sophisticated "three-beam technique." We showed that reexcitation of the photoexcited excitons in $1B_u$ into the mA_g state at about 0.8 eV above results in ultrafast relaxation of the order of 200 fs. However reexcitation to the kA_g state at energies 1 eV and more the $1B_u$ exciton results in charge separation. MEH-PPV polymer is different from all other PPV derivatives in that it can form aggregates, or chains with strong interchain interaction. This destabilizes the photogenerated excitons resulting in instantaneous charge separation. Photoinduced charge separation in C_{60}-doped polymers is also reviewed and contrasted to the properties of photo-induced excitons in pristine samples. Other polymers that are reviewed are P3HT and PDPA; the first forms two polymers with different regio-orders, whereas the latter is a degenerate ground state polymer. In these two polymer excitons are still the primary excitations. Except that in RR-P3HT, similar to MEH-PPV charges may easily separate into PPs, and in PDPA there is another relaxation route, i.e., the covalent route that results in generation of neutral solitons.

In Section 22.4 we reviewed different laser action phenomena found in π-conjugated semiconductors in various forms such as solutions, this films, microcavities, and single crystals. Each laser action phenomenon that was introduced in Section 22.2 is shown to exist in these materials. ASE is dominant in solutions and thin films at intermediate excitation intensities, whereas RL with coherent and incoherent feedback is dominant at high excitation intensities. In microcavities such as microrings and microdisks we showed the physics and applications of lasing with coherent feedback including longitudinal laser modes and their classification. Finally in DSB single crystals at high excitation intensities we prove the existence of the elusive phenomenon of SF in organic semiconductors, by studying the lasing time dependent that shows coherence manifested in collective oscillation.

Acknowledgments

We would like to mention a list of collaborators, postdocs, and graduate students at the University of Utah Physics Department over the years 1995–2005, without which this work would never have been completed. The list includes: A. Chipouline, M.C. DeLong, S.V. Frolov, W. Gellermann, I.I. Gontia, J.D. Huang, S.A. Jeglinski, X.M. Jiang, J.M. Leng, P.A. Lane, G. Levina, M. Liess, R. Meyer, R. Österbacka, R.C. Polson, M.E. Raikh, C.X. Sheng, M.N. Shkunov, M. Tong, M. Wohlgenannt, and C.G. Yang; they all worked a certain amount of time in Utah. We also thank our collaborators S. Mazumdar from Arizona; E. Ehrenfreund, the Technion; G. Lanzani, Milan; M. Ozaki, A. Fujii, and K. Yoshino, Osaka; A.A. Zakhidov, R.A. Baughman, and J.P. Ferraris, U.T. Dallas; R.V. Gregory, Clemson; K.S. Wong, Hong Kong; D. Chinn, Sandia; and R.A.J. Janssen, Eindhoven; and T. Masuda from Kyoto.

This work was supported over the years 1989–2005 by the U.S. Department of Energy, grants No. FG-03-89ER45490 and FG-04-ER46109 and by the National Science Foundation grants Nos. DMR 97-32820 and DMR 02-02790.

References

1. Burroughes, J.H., et al. 1990. *Nature* 347:539.
2. Forrest, S.R. 2004. *Nature* 428:911.
3. Malliaras, G. and R.H. Friend. 2005. *Phys Today* 58:53.
4. Garnier, F., et al. 1994. *Science* 265:1684.
5. Granström, M., et al. 1998. *Nature* 395:257.
6. Xiong, Z.H., et al. 2004. *Nature* 427:821.
7. Etemad, S., G.L. Baker, and Z.G. Soos. 1994. *Molecular nonlinear optics: materials, physics, and devices*, ed. J. Zyss. San Diego: Academic Press, p. 433.
8. Heeger, A.J., et al. 1988. *Rev Mod Phys* 60:781.
9. Skotheim, T.A., R.L. Elsenbaumer, and J. Reynolds. 1998. *Handbook of conducting polymers*, 2nd ed. New York: Marcel Dekker.
10. Su, W.P, J.R. Schrieffer, and A.J. Heeger. 1979. *Phys Rev Lett* 42:1698.
11. Robins, L., J. Orenstein, and R. Superfine. 1986. *Phys Rev Lett* 56:1850.
12. Swanson, L.S., J. Shinar, and K. Yoshino. 1990. *Phys Rev Lett* 65:1140.
13. Wei, X., et al. 1992. *Phys Rev Lett* 68:666.
14. Kersting, R., et al. 1993. *Phys Rev Lett* 70:3820.
15. Leng, J.M., et al. 1994. *Phys Rev Lett* 72:156.
16. Yan, M., et al. 1994. *Phys Rev Lett* 72:1104.
17. Soos, Z.G. and S. Ramasesha. 1984. *Phys Rev B* 29:5410.
18. Tavan, P. and K. Schulten. 1987. *Phys Rev B* 36:4337.
19. Sirringhaus, H., et al. 1999. *Nature* 401:685.
20. Österbacka, R., et al. 2000. *Science* 287:839.
21. Sariciftci, N.S., et al. 1992. *Science* 258:474.
22. Yoshino, K., et al. 1993. *Jpn J Appl Phys Part 2* 32:L140.
23. Dixit, S.N., D. Guo, and S. Mazumdar. 1991. *Phys Rev B* 43:6781.
24. Guo, D., et al. 1993. *Phys Rev B* 48:1433.
25. Kohler, B.E., C. Spangler, and C. Westerfield. 1988. *J Chem Phys* 89:5422.
26. Soos, Z.G., S. Ramasesha, and D.S. Galvao. 1993. *Phys Rev Lett* 71:1609.
27. Gontia, I.I., et al. 1999. *Phys Rev Lett* 82:4058.
28. Gontia, I.I., et al. 2002. *Phys Rev B* 66:075215.
29. Vardeny, Z.V. and J. Tauc. 1985. *Phys Rev Lett* 54:1844.
30. Vardeny, Z.V. and J. Tauc. 1985. *Philos Mag B* 52:313.
31. Su, W.P. and J.R. Schrieffer. 1980. *Proc Natl Acad Sci USA* 77:5626.
32. Ball, R., W.P. Su, and J.R. Schrieffer. 1983. *J Phys C* 44:429.
33. Orenstein, J. 1998. *Handbook of conducting polymers*, 2nd ed., eds. T.A. Skotheim, R.L. Elsenbaumer, and J. Reynolds. New York: Marcel Dekker, p. 1297.
34. Frolov, S.V., et al. 1997. *Phys Rev Lett* 78:4285.
35. Yan, M., et al. 1994. *Phys Rev Lett* 72:1104.
36. Mizes, H.A. and E.M. Conwell. 1994. *Phys Rev B* 50:11243.
37. Korovyanko, O.J. and Z.V. Vardeny. 2002. *Chem Phys Lett* 365:361.
38. Korovyanko, O.J., et al. 2001. *Phys Rev B* 64:235122.

39. Pai, D.M. and R.C. Enck. 1975. *Phys Rev B* 11:5163.

40. Siddiqui, A.S. 1984. *J Phys C* 17:683.

41. Rothberg, L., et al. 1987. *Phys Rev B* 36:7529.

42. Botta, C., et al. 1993. *Phys Rev B* 48:14809.

43. Ehrenfreund, E., et al. 1992. *Chem Phys Lett* 196:84.

44. Fichou, D., et al. 1990. *Synth Met* 39:243.

45. Bauerle, P., et al. 1993. *Angew Chem Int Ed Engl* 32:76.

46. Zotti, G., et al. 1993. *Chem Mater* 5:620.

47. Poplawski, J., et al. 1994. *Mol Cryst Liq Cryst* 256:407.

48. Wu, C.C., et al. 2003. *Synth Met* 139:735.

49. Lane, P.A., X. Wei, and Z.V. Vardeny. 1996. *Phys Rev Lett* 77:1544.

50. Ziemelis, K.E., et al. 1991. *Phys Rev Lett* 66:2231.

51. Vardeny, Z.V., et al. 1986. *Phys Rev Lett* 56:671.

52. Campbell, D.K., et al. 1982. *Phys Rev B* 26:6862.

53. Fesser, K., et al. 1983. *Phys Rev B* 27:4804.

54. Wohlgenannt, M., X.M. Jiang, and Z.V. Vardeny. 2004. *Phys Rev B* 69:241204.

55. Frolov, S.V., et al. 2002. *Phys Rev B* 65:205209.

56. Liess, M., et al. 1997. *Phys Rev B* 56:15712.

57. Meyer, R.K., et al. 1997. *SPIE* 3145:219.

58. Monkman, A.P., et al. 2001. *J Chem Phys* 115:9046.

59. Österbacka, R., et al. 1999. *Phys Rev B* 60:R11253.

60. Monkman, A.P., et al. 1999. *Chem Phys Lett* 307:303.

61. Romanoskii, Y.V., et al. 2000. *Phys Rev Lett* 84:1027.

62. Monkman, A.P., et al. 2001. *Phys Rev Lett* 86:1358.

63. Lane, P.A., X. Wei, and Z.V. Vardeny. 1997. *Phys Rev B* 56:4626.

64. Vardeny, Z.V., et al. 1983. *Phys Rev Lett* 51:2326.

65. Horovitz, B. 1982. *Solid State Commun* 41:729.

66. Horovitz, B., H. Gutfreund, and M. Weger. 1978. *Phys Rev B* 17:2796.

67. Ehrenfreund, E., et al. 1987. *Phys Rev B* 36:1535.

68. Vardeny, Z.V., J. Orenstein, and G.L. Baker. 1983. *Phys Rev Lett* 50:2032.

69. Österbacka, R., et al. 2002. *Phys Rev Lett* 88:226401.

70. Miranda, P.B., D. Moses, and A.J. Heeger. 2001. *Phys Rev B* 64:081201.

71. Mizrahi, V., et al. 1999. *Synth Met* 102:1182; Mizrahi, V., et al. 2001. *Synth Met* 119:507.

72. Österbacka, R., et al. 2001. *Synth Met* 116:317.

73. Willardson, R.K. and A.C. Beers, eds. 1972. *Semiconductors and semimetals*, vol.9. New York: Academic Press.

74. Sebastian, L., G. Weiser, and H. Bassler. 1989. *Chem Phys* 61:125.

75. Weiser, G. 1994. *Phys Rev B* 45:14076.

76. Jeglinski, S., et al. 1992. *Synth Met* 50:557.

77. Jeglinski, S., et al. 1992. *Synth Met* 50:504.

78. Sebastian, L. and G. Weiser. 1981. *Phys Rev Lett* 46:1156.

79. Chandross, M., et al. 1997. *Phys Rev B* 55:1486.

80. Rice, M.J. and Yu. N. Garstein. 1994. *Phys Rev Lett* 73:2504.

81. Chakraverti, A. and S. Mazumdar. 1999. *Phys Rev B* 59:4839.

82. Sony, P. and A. Shukla. 2005. *Phys Rev B* 71:165204.

83. Klimov, V.J. 2000. *J Phys Chem B* 104:6112.

84. Barford, W. 2004. *Electronic properties of conjugated polymers*. Oxford: Clarendon Press.

85. Rohlfing, M. and S.G. Louie. 1999. *Phys Rev Lett* 82:1959.

86. Harison, N.T., et al. 1996. *Phys Rev Lett* 77:188.

87. Eliott, R.J. and R. Loudon. 1960. *J Phys Chem Solids* 15:196.

88. Pankove, J.J. 1971. *Optical processes in semiconductors*. New York: Dover Publication.

89. Wu, C.C., et al. 2005. *Phys Rev B* 71:R-081201.
90. Kersting, R., et al. 1994. *Phys Rev Lett* 73:1440.
91. Pope, M. and C.E. Swenberg. 1999. *Electronic processes in organic crystals and polymers*, 2nd ed. Oxford: Oxford University Press.
92. Sheng, C.X., et al. 2005. *Phys Rev B* 71:125427.
93. Korovyanko, O., et al. 2004. *Phys Rev Lett* 97:017403.
94. Vardeny, Z.V. and X. Wei. 1998. *Handbook of conducting polymers*, 2nd ed., eds. T.A. Skotheim, R.L. Elsenbaumer, and J. Reynolds. New York: Marcel Dekker, p. 639.
95. Österbacka, R., et al. 2003. *J Chem Phys* 118:8905.
96. Frolov, S.V., Z.V. Vardeny, and K. Yoshino. 1998. *Phys Rev B* 57:9141.
97. Yariv, A. 1989. *Quantum electronics.* New York: John Wiley & Sons.
98. Vardeny, Z.V., et al. 2000. *SPIE* 3797:1.
99. Tessler, N., G.J. Denton, and R. Friend. 1996. *Nature* 382:395.
100. Hide, F., et al. 1996. *Science* 273:1833.
101. Brower, H.J., et al. 1996. *Adv Mater* 8:935.
102. Gelink, G., et al. 1997. *Chem Phys Lett* 265:320.
103. Frolov, S.V., et al. 1997. *Phys Rev Lett* 78:729.
104. Long, X., et al. 1997. *Chem Phys Lett* 78:729.
105. Denton, G.J., et al. 1997. *Adv Mater* 9:547.
106. Zenz, C., et al. 1997. *Appl Phys Lett* 71:2566.
107. Wegmann, G., et al. 1998. *Phys Rev B* 57:R-4218.
108. Berggren, M., et al. 1997. *Nature* 389:466; *Appl Phys Lett* 71:2230; *Adv Mater* 9:968.
109. Kozlov, V.G. 1997. *Nature* 389:362.
110. Wu, C.C., et al., to be published.
111. Siegman, A.E. 1986. *Lasers.* California: University Science Book.
112. Arecchi, F.T. and E. Courtens. 1970. *Phys Rev A* 2:1730.
113. Bonifacio, R. and L.A. Lugiato. 1975. *Phys Rev A* 11:1507.
114. MacGillivray, J.C. and M.S. Feld. 1976. *Phys Rev A* 14:1169.
115. Vrehen, Q.H.F. and H.M. Gibbs. 1982. Dissipative systems in quantum optics. *Topics in current physics*, vol. 27, ed. R. Bonifacio. Berlin: Springer.
116. Dicke, R.H. 1954. *Phys Rev* 93:99.
117. Florian, R., L.O. Schwan, and D. Schmid. 1984. *Phys Rev A* 29:2709.
118. Auzel, F., S. Hubert, and D. Meichenin. 1988. *Europhys Lett* 7:459.
119. Wang, H.Z., et al. 1995. *Phys Rev Lett* 74:4079.
120. Frolov, S.V., et al. 1997. *Phys Rev B* 56:R-4363.
121. Shkunov, M.N., et al. 2002. *Adv Funct Mater* 12:21.
122. Polson, R.C., G. Levina, and Z.V. Vardeny. 2000. *Appl Phys Lett* 76:3858.
123. Polson, R.C., Z.V. Vardeny, and D.A. Chinn. 2002. *Appl Phys Lett* 81:1561.
124. Gouedard, C., et al. 1993. *J Opt Soc Am B*, 10:2358.
125. Lawandy, N.M., et al. 1994. *Nature* 368:436.
126. Hide, F., et al. 1996. *Chem Phys Lett* 256:424.
127. Frolov, S.V., et al. 1999. *Phys Rev* 59:R-5284.
128. Cao, H., et al. 1999. *Phys Rev Lett* 82:2281.
129. Cao, H., et al. 2000. *Phys Rev Lett* 84:5584.
130. Frolov, S.V., et al. 1999. *Opt Commun* 162:241.
131. Yoshino, K., et al. 1999. *Appl Phys Lett* 74:2590.
132. Wiersma, D.S., M.P. van Albada, and A. Lagendijk. 1995. *Nature* 373:203.
133. Wiersma, D.S., et al. 1994. *Phys Rev Lett* 74:4193.
134. Siddique, M., et al. 1996. *Opt Lett* 21:450.
135. Apalkov, V.M., et al. 2002. *Phys Rev Lett* 89:016802.
136. Polson, R.C. and Z.V. Vardeny. 2005. *Phys Rev B* 71:045205.

137. McCall, S.L., et al. 1992. *Appl Phys Lett* 60:289.

138. Chu, D.Y., et al. 1993. *IEEE Photon Technol Lett* 5:1353.

139. Kuwata-Gonokami, M., et al. 1995. *Opt Lett* 20:2093.

140. Mair, R.A. 1998. *Appl Phys Lett* 72:1530.

141. Slusher, R.E., A.F.J. Levi, and U. Mohideen. 1993. *Appl Phys Lett* 63:1310.

142. Chin, M.K., D.Y. Chu, and S.T. Ho. 1994. *J Appl Phys* 75:3302.

143. Luo, K.J., J.Y. Xu, and H. Cao. 2001. *Appl Phys Lett* 78:3397.

144. Yoshino, K., et al. 1997. *Jpn J Appl Phys* 36:L714.

145. Polson, R.C., A. Chipouline, and Z.V. Vardeny. 2001. *Adv Mater* 13:760.

146. Birks, J.B. 1970. *Photophysics of aromatic molecules.* London: Wiley-Interscience.

147. Kohler, A., et al. 1998. *Nature (London)* 392:903.

148. Frolov, S.V., et al. 2000. *Phys Rev Lett* 85:2196.

149. Heun, S., et al. 1993. *J Phys Condens Matter* 5:247.

150. Nguyen, T.Q., et al. 2000. *Science* 288:652.

151. Frolov, S.V. and Z.V. Vardeny. 1998. *Rev Sci Instrum* 69:1257.

152. Bradford, W., R.J. Bursill, and Yu. M. Lauzentiev. 1998. *J Phys Cond Matt* 10:6429; 1999. *Phys Rev B* 59:9987.

153. Rothberg, L. 2002. *Proceedings of the International School of Physics "Enrico Fermi" Course CXLIX,* eds. V.M. Agranocich and G.C. La Rocca. Amsterdam: IOS Press, p. 299.

154. Wang, P., C.M. Cuppoletti, and L. Rothberg. 2003. *Synth Met* 137:1461.

155. Drori, T., E. Ehrenfreund, and Z.V. Vardeny, to be published.

156. Jiang, X.M., et al. 2002. *Adv Funct Mater* 12:587.

157. Sheng, C.X. 2005. *Ph.D. thesis,* University of Utah, unpublished.

158. Belgonne, D., et al. 2001. *Adv Funct Mater* 11:1.

159. Wei, X., et al. 1996. *Phys Rev B* 53:2187.

160. Brabec, C., et al. 2001. *Chem Phys Lett* 340:232.

161. Frolov, S.V., et al. 1998. *Chem Phys Lett* 286:2.

162. Cornil, J., et al. 2001. *Adv Mater* 13:1053.

163. Horowitz, G. 1998. *Adv Mater* 10:365.

164. Sirringhaus, H., et al. 1999. *Nature* 401:685.

165. Sirringhaus, H., N. Tessler, and R.H. Friend. 1998. *Science* 280:1741.

166. Brown, P.J., et al. 2001. *Phys Rev B* 63:125204.

167. Klimov, V.I., et al. 1998. *Phys Rev B* 58:7654.

168. Silva, C., et al. 1998. *Chem Phys Lett* 23:277.

169. Stevens, M.A., et al. 2001. *Phys Rev B* 63:165213.

170. Kraabel, B., et al. 2000. *Phys Rev B* 61:8501.

171. Dogariu, A., A.J. Heeger, and H. Wang. 2000. *Phys Rev B* 61:16183.

172. Vardeny, Z.V., et al. 1982. *Phys Rev Lett* 49:1657.

173. Shank, C., et al. 1982. *Phys Rev Lett* 49:1661.

174. Tada, K., et al. 1995. *Jpn J Appl Phys* 34: L1087.

175. Frolov, S.V., et al. 1997. *Jpn J Appl Phys* 36: L1268.

176. Soos, Z.G., et al. 1992. *Chem Phys Lett* 194:341.

177. Mazumdar, S. and D. Guo. 1994. *J Chem Phys* 100:1665.

178. Ghosh, H., A. Shukla, and S. Mazumdar. 2000. *Phys Rev B* 62:12763.

179. Shukla, A. and S. Mazumdar. 1999. *Phys Rev Lett* 83:3944.

180. Gustafson, T.L., et al. 2001. *Synth Met* 116:31.

181. Tada, K., et al. 1998. *Proc SPIE* 3145:171.

182. Korovyanko, O.J., et al. 2003. *Phys Rev B* 67:035114.

183. Hayden, G.W. and E.J. Mele. 1986. *Phys Rev B* 34:5484.

184. Lanzani, G., et al. 2001. *Phys Rev Lett* 87:187402.

185. Lucullier, R., et al. 1998. *Phys Rev Lett* 80:4068.

186. Shank, C.V., et al. 1983. *Phys Rev B* 28:6095.

187. Su, W.P. 1995. *Phys Rev Lett* 74:1167.

188. Bursil, R.J. and W. Bradford. 1999. *Phys Rev Lett* 82:1514.

189. Adachi, S., et al. 2002. *Phys Rev Lett* 89:027401.

190. Moses, D. 1992. *Appl Phys Lett* 60:3215.

191. Yan, M., et al. 1994. *Phys Rev B* 49:9419.

192. Yan, M., et al. 1995. *Phys Rev Lett* 75:1992.

193. Brouwer, H.J., et al. 1995. *Appl Phys Lett* 66:3404.

194. Holzer, W., et al. 1996. *Adv Mater* 8:974.

195. Schulzgen, A., et al. 1998. *Appl Phys Lett* 72:269.

196. Bulovic, V., et al. 1998. *Science* 279:553.

197. Frolov, S.V., et al. 1998. *Adv Mater* 10:871.

198. Nisoli, M., et al. 1999. *Phys Rev B* 59:11328.

199. Jeung, S.C., et al. 1999. *Appl Phys Lett* 74:212.

200. Duan, V., V. Tran, and B.J. Schwartz. 1999. *Chem Phys Lett* 74:2767.

201. Virgilo, T., et al. 1999. *Appl Phys Lett* 74:2767.

202. Lee, C.W., et al. 1999. *Chem Phys Lett* 314:564.

203. McGehee, M.D., et al. 1999. *Synth Met* 102:1030.

204. Shahbazyan, T.V., M.E. Raikh, and Z.V. Vardeny. 2000. *Phys Rev B* 61:13266.

205. Hogelnik, H. 1988. *Guided-wave optoelectronics*, ed. T. Tamir. Berlin: Springer.

206. Wong, K.S., et al. 2000. *Synth Met* 111–112:497.

207. Kumar, N.D., et al. 1997. *Appl Phys Lett* 71:999.

208. Frolov, S.V., et al. 1998. *Appl Phys Lett* 72:2811.

209. Kawabe, Y., et al. 1998. *Appl Phys Lett* 72:141.

210. Slusher, R.E., et al. 1993. *Appl Phys Lett* 63:1310.

211. McCall, S.L., et al. 1992. *Appl Phys Lett* 60:289.

212. Frolov, S.V., et al. 1998. *Appl Phys Lett* 72:1802.

213. Jackson, J.D. 1975. *Classical electrodynamics*, 2nd ed. New York: John Wiley & Sons.

214. Hofsttetter, D. and R.L. Thorton. 1998. *Appl Phys Lett* 72:404.

215. Ambartsumyan, R.V., et al. 1966. *Zh Eksp Teor Fiz* 51:724; 1967. *Sov Phys JETP* 24:481.

216. Letokhov, V.S. 1967. *Zh Eksp Teor Fiz* 53:1442; 1968. *Sov Phys JETP* 26:835.

217. John, S. 1991. *Phys Today* 44:32.

218. Berkovits, R. and S. Fang. 1994. *Phys Rep* 238:135.

219. Frolov, S.V., et al. 2000. *IEEE J Quant Electron* 36:2.

220. Polson, R.C., J.D. Huang, and Z.V. Vardeny. 2001. *Synth Met* 119:7.

221. Polson, R.C. 2002. *Ph.D. thesis*, University of Utah, unpublished.

222. Polson, R.C., E.M. Raikh, and Z.V. Vardeny. 2002. *CR Phys* 3:509.

223. Muzumdar, S., et al. 2004. *Phys Rev Lett* 93:053903.

224. Loudon, R. 1983. *The quantum theory of light*, 2nd ed. Oxford: Oxford University Press.

225. Zacharakis, G., et al. 2000. *Opt Lett* 25:923.

226. Loa, H., et al. 2001. *Phys Rev Lett* 86:4524.

227. Frolov, S.V., et al. 1996. *Jpn J Appl Phys* 35:L1371.

228. Schwartz, B.J. 2003. *Annu Rev Phys Chem* 54:141.

229. Meinardi, S.F., et al. 2002. *Phys Rev Lett* 64:157403, and references therein.

230. Spano, F. 2003. *J Chem Phys* 118:981, and references therein.

231. Wu, C.C., et al. 2003. *Synth Met* 137:939.

232. Wu, C.C. 2005. *Ph.D. thesis*, University of Utah, unpublished.

Index